2016 21st OptoElectronics and Communications Conference (OECC 2016) held jointly with 2016 International Conference on Photonics in Switching (PS 2016)

Niigata, Japan
3-7 July 2016

Pages 1-602

IEEE Catalog Number: CFP1699A-POD
ISBN: 978-1-5090-2147-5

Copyright © 2016, The Institute of Electronics, Information and Communication Engineers
All Rights Reserved

***This publication is a representation of what appears in the IEEE Digital Libraries. Some format issues inherent in the e-media version may also appear in this print version.*

IEEE Catalog Number: CFP1699A-POD
ISBN (Print-On-Demand):   978-1-5090-2147-5
ISBN (Online):   978-4-88552-305-2
ISSN: 2155-8507

Additional Copies of This Publication Are Available From:

Curran Associates, Inc
57 Morehouse Lane
Red Hook, NY 12571 USA
Phone: (845) 758-0400
Fax: (845) 758-2633
E-mail: curran@proceedings.com
Web: www.proceedings.com

TABLE OF CONTENTS

OPTICAL TRANSPORT TO OPEN TRANSPORT IN CARRIER NETWORKS 1
Vishnu Shukla

SILICON PHOTONIC INTEGRATED CIRCUITS FOR COHERENT COMMUNICATIONS 3
Chris Doerr

ENERGY LIMITATIONS IN DATA TRANSMISSION AND SWITCHING 5
Rod Tucker

OPTICAL PACKET SWITCHING: MYTH, FACT, AND PROMISE 7
Ken-ichi Kitayama

ULTRAHIGH SPEED AND HIGH SPECTRAL EFFICIENCY TRANSMISSION USING OPTICAL NYQUIST
PULSES 9
Masataka Nakazawa

ROLL-OFF FACTOR DEPENDENCE OF SYSTEM PERFORMANCE IN 1.28 TBIT/S/CH-525 KM NYQUIST
PULSE TRANSMISSION 12
Koudai Harako ; Daiki Suzuki ; Toshihiko Hirooka ; Masataka Nakazawa

A NOVEL DEMULTIPLEXING SCHEME FOR NYQUIST OTDM SIGNAL USING A SINGLE IQ
MODULATOR 15
Lei Yue ; Deming Kong ; Yan Li ; Jiangchuan Pang ; Miao Yu ; Jian Wu

1.76-TBIT/S SUPERCHANNEL GENERATION BASED ON INP IQ MODULATOR AND OPTICAL
FREQUENCY COMB 18
K. Solis-Trapala ; T. Inoue ; H. Nguyen Tan ; S. Namiki

C-BAND SPANNING FREQUENCY COMB BASED ON AN OUTPUT PHASE STABILIZED MODE-LOCKED
LASER 21
Mark Pelusi ; Hung Nguyen Tan ; Karen Solis-Trapala ; Takashi Inoue ; Shu Namiki

10-GBAUD QPSK SIGNAL SIMULTANEOUS TRANSMISSION IN THE T AND C BANDS FOR ULTRA-
BROADBAND PHOTONIC TRANSPORT SYSTEM 24
Akihiro Murano ; Shoko Yamada ; Atsushi Kanno ; Naokatsu Yamamoto ; Hideyuki Sotobayashi

SUBCARRIER POLARIZATION MANIPULATION USING ORTHOGONAL DUAL PHASE-MODULATION
IN FIBER... 27
Tomoyuki Kato ; Takahito Tanimura ; Takeshi Hoshida ; Thomas Richter ; Robert Elschner ; Carsten Schmidt-Langhorst ; Colja Schubert ; Shigeki Watanabe

CASCADED ALL-OPTICAL SUB-CHANNEL ADD/DROP MULTIPLEXING FROM A 1-TB/S SUPER-
CHANNEL HAVING 2-GHZ GUARD-BANDS 30
M. Song ; E. Pincemin ; B. Baeuerle ; A. Josten ; D. Hillerkuss ; J. Leuthold ; R. Rudnick ; D. M. Marom ; S. Ben-Ezra ; J. F. Ferran ; S. Sygletos ; A. Ellis ; J. Zhao ; G. Thouenon ; C. Betoule ; J. M. Rivas ; D. Klonidis ; I. Tomkos

1024 QAM, 7-CORE FIBER/MULTI-CORE EDFA TRANSMISSION OVER 100 KM WITH AN
AGGREGATED SPECTRAL EFFICIENCY OF 109 BIT/S/HZ.. 33
Masato Yoshida ; Keisuke Kasai ; Toshihiko Hirooka ; Masataka Nakazawa

ACCURATE ANALYSIS OF CROSSTALK BETWEEN LP11 DEGENERATE MODES DUE TO OFFSET
CONNECTION USING EXACT EIGENMODES 36
Yasuo Kokubun ; Seiya Miura ; Tatsuhiko Watanabe

NEW METHOD FOR MEASURING INTER-CORE CROSSTALK IN MULTI-CORE FIBERS USING NEAR-
INFRARED CAMERA 39
Shota Saitoh ; Yoshimichi Amma ; Yusuke Sasaki ; Katsuhiro Takenaga ; Kazuhiko Aikawa

MULTICORE FIBER FOR BI-DIRECTIONAL TRANSMISSION 42
Tomohiro Gonda ; Katsunori Imamura ; Ryuichi Sugizaki

SINGLE-MODE MULTICORE FIBER FOR DENSE SPACE DIVISION MULTIPLEXING 45
Yusuke Sasaki ; Yoshimichi Amma ; Katsuhiro Takenaga ; Shoichiro Matsuo ; Kazuhiko Aikawa ; Kunimasa Saitoh ; Toshio Morioka ; Yutaka Miyamoto

DYNAMIC SKEW MEASUREMENTS IN 7, 19 AND 22-CORE MULTI CORE FIBERS 48
G. M. Saridis ; B. J. Puttnam ; R. S. Luís ; W. Klaus ; Y. Awaji ; G. Zervas ; D. Simeonidou ; N. Wada

IMPULSE RESPONSE ANALYSIS OF AIR-HOLE ADDED COUPLED SIX-CORE FIBERS..................... 51
Yasuyuki Chida ; Takeshi Fujisawa ; Kunimasa Saitoh

HIGH SPATIAL DENSITY FEW-MODE MULTI-CORE FIBER WITH LOW DIFFERENTIAL MODE DELAY
CHARACTERISTICS 54
Taiji Sakamoto ; Takashi Matsui ; Kunimasa Saitoh ; Shota Saitoh ; Katsuhiro Takenaga ; Shoichiro Matsuo ; Yuki Tobita ; Nobutomo Hanzawa ; Kazuhide Nakajima ; Fumihiko Yamamoto

ULTRAFAST COMPLEX IMPULSE RESPONSE MEASUREMENT OF 2-MODE FIBERS BY USING LINEAR
OPTICAL SAMPLING 57
Naoto Kono ; Fumihiko Ito ; Ryo Maruyama ; Nobuo Kuwaki

PMD MEASUREMENTS OF TWO-MODE FIBERS USING MODE COUPLER BASED ON FIXED
ANALYZER TECHNIQUE 60
Shota Igarashi ; Masaharu Ohashi ; Yuji Miyoshi ; Hirokazu Kubota ; Taiji Sakamoto ; Takashi Matsui ; Kazuhide Nakajima

HIGH SPEED VCSEL-BASED LINKS FOR USE IN DATA CENTERS AND HPC 63
Daniel M. Kuchta

HIGH SPEED MODULATION OF TRANSVERSE COUPLED CAVITY VCSELS 64
Fumio Koyama

OSA MODULE WITH ×4 MINI MT AND 45° FIBER MIRRORS FOR 48 GB/S OPTICAL LINK 67
Feng-Cheng Hsu ; Tsu-Shiuh Wu ; Fang-Zeng Lin ; Dung-Chin Yang ; Chun-Yen Peng ; Ann-Kuo Chu

FABRICATION OF HCG MEMS VCSELS FOR ATHERMAL OPERATIONS 70
Shunya Inoue ; Masanori Nakahama ; Akihiro Matsutani ; Takahiro Sakaguchi ; Fumio Koyama

**VCSEL-INTEGRATED BRAGG REFLECTOR WAVEGUIDE AMPLIFIER WITH SINGLE-MODE OUTPUT
POWER OVER 10 MW** 73
Xiaodong Gu ; Masanori Nakahama ; Akihiro Matsutani ; Fumio Koyama

HIGH POWER NON-MECHANICAL BEAM SCANNER BASED ON VCSEL AMPLIFIER 76
Masanori Nakahama ; Xiaodong Gu ; Akihiro Matsutani ; Takahiro Sakaguchi ; Fumio Koyama

**RECONFIGURABLE OPTICAL ROUTERS AND BUFFERS BUILT ON THE SILICON-ON-INSULATOR
PLATFORM** 79
Linjie Zhou ; Liangjun Lu ; Shuoyi Zhao ; Xinyi Wang ; Jianping Chen

**POLARIZATION-INDEPENDENT C-BAND TUNABLE FILTER BASED ON CASCADED SI-WIRE
ASYMMETRIC MACH-ZEHNDER INTERFEROMETER** 82
*Keijiro Suzuki ; Ken Tanizawa ; Satoshi Suda ; Hiroyuki Matsuura ; Kazuhiro Ikeda ; Yojiro Mori ; Ken-ichi Sato ; Shu Namiki ;
Hitoshi Kawashima*

**FUNDAMENTAL OPERATION OF A PHASED ARRAY SWITCH USING FERROELECTRIC LIQUID
CRYSTAL CLADDINGS** 85
W. Kanakubo ; K. Nakatsuhara ; M. Takeda

CONTROL, CONFIGURATION AND STABILIZATION OF PHOTONIC INTEGRATED CIRCUITS 88
Andrea Melloni ; Andrea Annoni ; Stefano Grillanda ; Daniele Melati ; Francesco Morichetti

LARGE-SCALE SILICON PHOTONIC SWITCHES 91
Ming C. Wu ; Tae Joon Seok ; Sangyoon Han ; Niels Quack

WAVEFRONT TRANSMISSION FOR REMOTE OPTICAL BEAM FORMING 94
Mitsumasa Nakajima ; Junji Sakamoto ; Keita Yamaguchi ; Kenya Suzuki ; Toshikazu Hashimoto

STACKED WAVELENGTH SELECTIVE SWITCH DESIGN FOR LOW-COST CDC ROADMS 97
Haining Yang ; Brian Robertson ; Peter Wilkinson ; Daping Chu

FIRST DEMONSTRATION OF DYNAMIC MODE-SWITCHING BY USING OPTICAL MODE SWITCH 100
Ryan Imansyah ; Luke Himbele ; Shota Oe ; Haisong Jiang ; Kiichi Hamamoto

ULTRA-FAST SUPER-RESOLUTION IMAGING BY TIME-STRETCH STRUCTURED ILLUMINATION 103
Yuxi Wang ; Qiang Guo ; Hongwei Chen ; Minghua Chen ; Sigang Yang

TIME-FREQUENCY PHOTONIC SIGNAL PROCESSING 106
*G. Cincotti ; T. Konishi ; T. Murakawa ; T. Nagashima ; S. Shimizu ; M. Hasegawa ; K. Hattori ; M. Okuno ; S. Mino ; A. Himeno
; N. Wada ; H. Uenohara*

**ALL-OPTICAL NYQUIST-WDM TO NYQUIST-OTDM CONVERSION FOR FLEXIBLE OPTICAL
NETWORKS** 109
Satoshi Shimizu ; Gabriella Cincotti ; Naoya Wada

OPTICAL SERIAL-TO-PARALLEL CONVERSION BASED ON FRACTIONAL OFDM SCHEME 112
*M. Hiraoka ; T. Nagashima ; B. Karanov ; G. Cincotti ; S. Shimizu ; T. Murakawa ; M. Hasegawa ; K. Hattori ; M. Okuno ; S.
Mino ; A. Himeno ; N. Wada ; H. Uenohara ; T. Konishi*

**MODULATION FORMAT CONVERSION FROM QPSK TO 16QAM USING DELAY LINE
INTERFEROMETER AND SPECTRAL SHAPING FILTER** 115
Kazuya Mori ; Hiroki Kishikawa ; Nobuo Goto

TIME-FREQUENCY PACKED VCSEL-BASED IM/DD TRANSMISSION FOR WDM ACCESS NETWORKS 118
Antonio Malacarne ; Francesco Fresi ; Gianluca Meloni ; Tommaso Foggi ; Luca Potì

ELASTIC WDM SWITCHING FOR SCALABLE DATA CENTER AND HPC INTERCONNECT NETWORKS 121
Adel A. M. Saleh ; Akhilesh S. P. Khope ; John E. Bowers ; Rod C. Alferness

PROGRAMMABLE OPTICAL PROCESSORS FOR SIGNAL PROCESSING AND MONITORING 124
Simon B. Poole ; Patrick Blown ; Qing Li ; Ralf Stolte ; Ian Clarke

PATTERN-INDEPENDENT WAVELENGTH CONVERSION OF PAM SIGNALS IN SOAS 127
Ryosuke Matsumoto ; Giampiero Contestabile ; Yuki Yoshida ; Akihiro Maruta ; Ken-ichi Kitayama

**EXPERIMENTAL DEMONSTRATION OF ALL-OPTICAL FEC CODING SCHEME WITH
CONVOLUTIONAL CODE** 130
Yohei Aikawa ; Hiroyuki Uenohara

**FIRST DEMONSTRATION OF GEOGRAPHICALLY UNCONSTRAINED CONTROL OF AN INDUSTRIAL
ROBOT BY JOINTLY EMPLOYING SDN-BASED OPTICAL TRANSPORT NETWORKS AND EDGE
COMPUTE** 133
*N. Yoshikane ; T. Sato ; Y. Isaji ; C. Shao ; T. Marco ; S. Okamoto ; T. Miyazawa ; T. Ohshima ; C. Yokoyama ; Y. Sumida ; H.
Sugiyama ; M. Miyabe ; T. Katagiri ; N. Kakegawa ; S. Matsumoto ; Y. Ohara ; I. Satou ; A. Nakamura ; S. Yoshida ; K. Ishii ; S.
Kametani ; J. Nicho ; J. Meyer ; S. Edwards ; P. Evans ; T. Tsuritani ; H. Harai ; M. Razo ; D. Hicks ; A. Fumagalli ; N.
Yamanaka*

THZ PHOTONICS-WIRELESS TRANSMISSION OF 160 GBIT/S BITRATE 136
Xianbin Yu ; Shi Jia ; Hao Hu ; Pengyu Guan ; Michael Galili ; Toshio Morioka ; Peter U. Jepsen ; Leif K. Oxenløwe

**LOW NOISE, REGENERATION OF OPTICAL FREQUENCY COMB-LINES FOR 64QAM ENABLED BY
SBS GAIN** 139
*Mark Pelusi ; Amol Choudhary ; Takashi Inoue ; David Marpaung ; Benjamin Eggleton ; Hung Nguyen Tan ; Karen Solis-Trapala
; Shu Namiki*

20.7-TB/S REPEATER-LESS TRANSMISSION OVER 401.1-KM USING QSM FIBER AND XPM COMPENSATION VIA TRANSMITTER-SIDE DBP 142

Yue-Kai Huang ; Ezra Ip ; Shaoliang Zhang ; Fatih Yaman ; John D. Downie ; William A. Wood ; Aramais Zakharian ; Jason Hurley ; Snigdharaj Mishra ; Yoshiaki Aono ; Eduardo Mateo ; Yoshihisa Inada

1.2-TB/S MIMO-LESS TRANSMISSION OVER 1 KM OF FOUR-CORE ELLIPTICAL-CORE FEW-MODE FIBER WITH 125-µM DIAMETER CLADDING 145

Giovanni Milione ; Ezra Ip ; Yue-Kai Huang ; Philip Ji ; Ting Wang ; Ming-Jung Li ; Jeffery Stone ; Gaozhu Peng

DEMONSTRATION OF A HYBRID SILICON EVANESCENT QUANTUM DOT LASER 148

Bongyong Jang ; Katsuaki Tanabe ; Satoshi Kako ; Satoshi Iwamoto ; Tai Tsuchizawa ; Hidetaka Nishi ; Nobuaki Hatori ; Masataka Noguchi ; Takahiro Nakamura ; Keizo Takemasa ; Mitsuru Sugawara ; Yasuhiko Arakawa

SILICON PHOTONIC 32 × 32 STRICTLY-NON-BLOCKING BLADE SWITCH AND ITS FULL PATH CHARACTERIZATION 151

Ken Tanizawa ; Keijiro Suzuki ; Satoshi Suda ; Hiroyuki Matsuura ; Kazuhiro Ikeda ; Shu Namiki ; Hitoshi Kawashima

WIRELESS NETWORK EVOLUTION TOWARD 5G NETWORK 154

Fumiyuki Adachi

5G WIRELESS NETWORK RESEARCH IN CHINA - NON-UNIFORM WIRELESS DENSE NETWORKS 157

Xiaofeng Tao

ACTIVITIES OF THE FIFTH GENERATION MOBILE COMMUNICATIONS PROMOTION FORUM (5GMF) IN JAPAN 160

Akira Matsunaga

HIGH CAPACITY MOBILE FRONTHAUL AND BACKHAUL NETWORK RESEARCH IN KOREA 163

Seungjoo Hong ; Sungmin Cho

CONVERGENCE OF PHOTONICS AND ELECTRONICS FOR TERAHERTZ WIRELESS COMMUNICATIONS - THE ITN CELTA PROJECT 166

Idelfonso Tafur Monroy

FIBER-WIRELESS FRONTHAUL: THE LAST FRONTIER 169

Gee-Kung Chang ; Lin Cheng

OPTICAL ACCESS NETWORK TECHNOLOGY FOR 5G WIRELESS FRONT/BACKHAUL NETWORK 172

Jun Terada ; Tatsuya Shimada ; Tatsuya Shimizu ; Akihiro Otaka

COST-EFFECTIVE OPTICAL TRANSMITTERS FOR NEXT-GENERATION MOBILE FRONTHAUL NETWORKS 175

B. G. Kim ; Hoon Kim ; Y. Chung

ULTRA-DENSE SPACE-DIVISION-MULTIPLEXING OPTICAL FIBERS 178

Tetsuya Hayashi

RECENT PROGRESS IN MODE MULTIPLEXING DEVICES USING MULTI-PLANE LIGHT CONVERSION 181

Guillaume Labroille ; Pu Jian ; Nicolas Barré ; Bertrand Denolle ; Jean-François Morizur

SWITCHING SOLUTIONS FOR SPATIAL AND SPECTRAL MULTIPLEXED NETWORKS 184

Dan M. Marom

MODE-MULTIPLEXED TRANSMISSION OVER MULTIMODE FIBERS 187

R. Ryf ; H. Chen ; N. K. Fontaine

ULTRA-HIGH-CAPACITY TRANSMISSION OVER FEW-MODE MULTI-CORE FIBERS 190

Koji Igarashi ; Takehiro Tsuritani ; Itsuro Morita

ADVANCED MIMO SIGNAL PROCESSING FOR DENSE SDM TRANSMISSION USING MULTI-CORE FEW-MODE FIBERS 193

K. Shibahara ; T. Mizuno ; D. Lee ; Y. Miyamoto

OPTICAL SWITCHING PERFORMANCE METRICS FOR SCALABLE DATA CENTERS 196

Keren Bergman ; Sébastien Rumley

ADVANCED OPTICAL INTERCONNECTION TECHNOLOGIES BASED ON SILICON PHOTONICS FOR FUTURE DCS 199

Takahiro Nakamura ; Junichi Tsuchida ; Kazuhiko Kurata

THE ROLE OF PHOTONICS IN FUTURE EXASCALE DATA SYSTEMS 202

S. J. Ben Yoo

BURST-MODE OPTICAL PACKET PROCESSING TECHNOLOGIES 205

Salah Ibrahim ; Ryo Takahashi

HIGH PERFORMANCE DCN ARCHITECTURE BASED ON FLOW-CONTROLLED OPTICAL SWITCHING SYSTEM 208

Nicola Calabretta ; Wang Miao ; Fulong Yan ; Oded Raz

EARLY DAYS OF VAD METHOD 211

Tatsuo Izawa

FIBER AND FIBER BASED TECHNOLOGY AFTER VAD DEVELOPMENT - CONTRIBUTION TO TRANSOCEANIC SUBMARINE NETWORKS - 213

Hiroo Kanamori

DEVELOPMENT OF OPTICAL FIBERS FOR LONG-HAUL TRANSMISSION 216

Haruki Ogoshi

SPECIALTY FIBERS AND OPTICAL FIBER DEVICES FOLLOWING DEVELOPMENT OF VAPOR-PHASE AXIAL DEPOSITION 219

Kenji Nishide

NEW OPTICAL FIBER DEVELOPMENT BASED ON OVD TECHNOLOGY 222

Ming-Jun Li

HOLLOW CORE FIBRES FOR DATA TRANSMISSION..225
D. J. Richardson ; Y. Chen ; N. V. Wheeler ; J. R. Hayes ; T. Bradley ; S. R. Sandoghchi ; G. T. Jasion ; E. Numkam Fokoua ; Z. Liu ; R. Slavik ; M. N. Petrovich ; F. Poletti

OPTICAL FIBERS FOR FUTURE COMMUNICATION SYSTEMS..227
Kazuhide Nakajima

OPTICAL NETWORKING AND NODE TECHNOLOGIES FOR CREATING COST EFFECTIVE BANDWIDTH ABUNDANT NETWORKS..230
Ken-ichi Sato

RECONFIGURABLE WDM MULTICAST SUPPORTING CONTENT DELIVERY FOR CONTENT DELIVERY NETWORK BASED ON SOA AND TB-WSS..233
Ze Li ; Min Zhang ; Danshi Wang ; Yue Cui

METRO-EMBEDDED CLOUD PLATFORM WITH ALL-OPTICAL INTERCONNECTIONS FOR VIRTUAL DATACENTER PROVISIONING..236
Hongxiang Guo ; Gang Chen ; Dongxu Zhang ; Xiaoyuan Cao ; Jian Wu ; Takehiro Tsuritani

FLEXIBLE AND COST-EFFECTIVE OPTICAL METRO NETWORK WITH PHOTONIC-SUB-LAMBDA AGGREGATION CAPABILITY..239
Masahiro Nakagawa ; Kana Masumoto ; Kyota Hattori ; Toshiya Matsuda ; Masaru Katayama ; Katsutoshi Koda

A SCALABLE OPTICAL NETWORK ARCHITECTURE BASED ON WSS FOR INTRA DATA CENTER................242
Aijun Liu ; Yongmei Sun ; Hongxiang Wang ; Yuefeng Ji

FRAME LENGTH AVERAGING METHOD FOR MULTI-CARRIER AGGREGATION TRANSPORT................245
Toru Homemoto ; Toshiya Matsuda ; Masaru Katayama ; Katsutoshi Koda

METHOD TO SHARE A BURST-MODE RECEIVER BETWEEN CONTINUOUS- AND BURST-MODE TRANSMITTER FOR VM MIGRATION OF CLOUD EDGE..248
Kyota Hattori ; Masahiro Nakagawa ; Toshiya Matsuda ; Masaru Katayama ; Katsutoshi Koda

REMOTELY CONTROLLABLE WDM-PON TECHNOLOGY FOR WIRELESS FRONTHAUL/BACKHAUL APPLICATION..251
Michael H. Eiselt ; Christoph Wagner ; Mirko Lawin

DYNAMIC RECONFIGURATION OF COORDINATED MULTI-POINT TRANSMISSION IN HIGH DENSITY SMALL CELL NETWORK USING CENTRALIZED WAVELENGTH SWITCHING IN DWDM PON................254
Goji Nakagawa ; Kyosuke Sone ; Setsuo Yoshida ; Shoichiro Oda ; Motoyuki Takizawa ; Hirokazu Shimada ; Yasuhiko Aoki ; Jens C. Rasmussen

25-GB/S ASE-SEEDED WDM PON..257
Qikai Hu ; Changyuan Yu ; Pooi-Yuen Kam ; Hoon Kim

MITIGATION OF OPTICAL BEAT INTERFERENCE IN N-DIMENSIONAL CAP-PON UPLINK TRANSMISSION..260
Jiale He ; Lei Deng ; Lu Shi ; Mengfan Cheng ; Ming Tang ; Songnian Fu ; Minming Zhang ; Deming Liu ; Perry Ping Shum

PERFORMANCE INVESTIGATION FOR LDPC-CODED UPSTREAM TRANSMISSION SYSTEMS IN IM/DD OFDM-PONS..263
Xiaoxue Gong ; Lei Guo ; Jingjing Wu

MOBILE BACKHAUL AND FRONTHAUL SYSTEMS..266
Frank J. Effenberger

HIGHLY LINEAR W-BAND TRANSMITTER BASED ON OFC AND HETERODYNE UP-CONVERSION................269
Jinhui Luo ; Shangyuan Li ; Xiaoping Zheng ; Hanyi Zhang ; Bingkun Zhou

SELF-PHASE MODULATION BASED SIGNAL DISTORTIONS OF OPTICAL SSB-SC SIGNAL WITH PILOT CARRIER..272
Pilot Carrier ; K. I. Amila Sampath ; Katsumi Takano ; Manabu Sato

TRANSMISSION OF 28-GB/S DUOBINARY SIGNALS OVER 45-KM SSMF USING 1.55-μM DIRECTLY MODULATED LASER..275
Chuanbowen Sun ; S. H. Bae ; Hoon Kim ; Y. C. Chung

A NOVEL AMPLITUDE DECISION BASED SYMBOL SYNCHRONIZATION METHOD FOR REAL-TIME IMDD-OOFDM SYSTEMS..278
Weiliang Wu ; Junjie Zhang ; Zhen Zhang ; Han Dun ; Qianwu Zhang

STUDY ON OPTICAL MODULATION AMPLITUDE OF 4-LEVEL PULSE AMPLITUDE MODULATION WITH ELECTROABSORPTION MODULATOR..281
Riu Hirai ; Nobuhiko Kikuchi ; Takayoshi Fukui

NOVEL ELECTRICAL DISPERSION COMPENSATION TECHNIQUE FOR IMDD-BASED SYSTEMS................284
Kazuki Tanaka ; Takashi Kobayashi ; Akira Agata ; Kosuke Nishimura

TRANSMITTER DIGITAL PRE-DISTORTION TECHNIQUES FOR BAND-LIMITED OPTICAL NETWORKS..287
Fotini Karinou ; Zhao Yu ; Nebojsa Stojanovic ; Changsong Xie

IMPROVEMENT OF BANDWIDTH LIMITATION AND CHROMATIC- DISPERSION TOLERANCES USING 4-LEVEL/7-LEVEL CODING PAM FOR DIRECT-DETECTION SYSTEM................................290
Akira Masuda ; Shuto Yamamoto ; Hiroki Kawahara ; Shingo Kawai ; Mitsunori Fukutoku

POWER BUDGET IMPROVEMENT OF 25-GB/S OPTICAL LINK BASED ON 1.5-μM 10G-CLASS VCSEL................293
Jingjing Zhou ; Changyuan Yu ; Gurusamy Mohan ; Hoon Kim

RZ-PAM4 TRANSMISSION WITH QUASI-FOURIER-TRANSFORM-LIMITED GAIN-SWITCHED PULSE SOURCE..296
Songyuan Dai ; Masanori Hanawa

OVER 400 GBIT/S DIGITAL COHERENT CHANNELS FOR OPTICAL TRANSPORT NETWORK................299
Yutaka Miyamoto

EXPERIMENTAL DEMONSTRATION OF A PROGRAMMABLE 400-GBPS DMT TRANSCEIVER WITH POLICY-BASED CONTROL 302

Yutaka Kai ; Ryo Okabe ; Masato Nishihara ; Toshiki Tanaka ; Tomoo Takahara ; Jens C. Rasmussen ; F. Javier Vílchez ; Laia Nadal ; Josep M. Fàbrega ; Michela Svaluto Moreolo

320 GBIT/S, 256 QAM LD-BASED COHERENT TRANSMISSION OVER 160 KM WITH AN INJECTION-LOCKED HOMODYNE DETECTION TECHNIQUE 305

Yixin Wang ; Keisuke Kasai ; Masato Yoshida ; Masataka Nakazawa

COHERENT INTERFERENCE REDUCTION IN SINGLE-FIBER BIDIRECTIONAL SYSTEM FOR 100 GB/S SHORT DISTANCE APPLICATIONS 308

Xu Zhou ; Ning Deng

200GBIT/S NYQUIST 16-QAM HALF-CYCLE SUBCARRIER MODULATION TRANSMISSION WITH DUAL-POLARIZATION DIRECT DETECTION 311

Kaiheng Zou ; Yixiao Zhu ; Fan Zhang ; Zhangyuan Chen

CONNECTIVITY TECHNIQUES OF MCF, FOR DEPLOYMENT TO PRACTICAL USE 314

Tsunetoshi Saito ; Kengo Watanabe

INVESTIGATION OF CONNECTOR DAMAGE CAUSED BY HIGH POWER TRANSMISSION IN OPTICAL CONNECTOR WITH DUST 317

Chisato Fukai ; Kotaro Saito ; Ryo Koyama ; Mitsuru Kihara ; Toshio Kurashima

INPUT/OUTPUT CHANNEL COUPLING DEVICES FOR SDM AND MDM 320

Yasuo Kokubun

SPINNING EFFECT IN FEW-MODE FIBER WITH DISTRIBUTED LONG PERIOD GRATINGS 323

Jian Fang ; Byoung Yoon Kim ; William Shieh

RECENT DEVELOPMENTS OF MULTICORE MULTIMODE FIBER AMPLIFIERS FOR SDM SYSTEMS 326

Kazi S. Abedin

4-LP MODE DISTRIBUTED RAMAN AMPLIFICATION TECHNIQUE WITH GRADED-INDEX MULTI-MODE FIBER TRANSMISSION LINE 329

Masaki Wada ; Taiji Sakamoto ; Takayoshi Mori ; Takashi Yamamoto ; Kazuhide Nakajima

MFD MEASUREMENT OF A SIX-MODE FIBER WITH LOW-COHERENCE DIGITAL HOLOGRAPHY 332

Yuta Wakayama ; Hidenori Taga ; Takehiro Tsuritani

FABRICATION OF HELICAL FIBER GRATING AND ITS APPLICATION TO FLAT-TOP BAND-REJECTION FILTER 335

Peng Wang ; Gen Inoue ; Ramanathan Subramanian ; Hongpu Li

TRI-SECTION SIDE-POLISHED POLARIZATION MAINTAINING FIBRE POLARIZER 338

Xinyue Wang ; Kaige Chen ; Zhongwei Tan ; Ziyu Wang

ULTRASONIC WELDING OF PLASTIC OPTICAL FIBERS ONTO COMPOSITE MATERIALS 341

Shumpei Shimada ; Hiroki Tanaka ; Kazuhiko Hasebe ; Neisei Hayashi ; Yutaka Ochi ; Takahiro Matsui ; Itaru Nishizaki ; Yukihiro Matsumoto ; Yosuke Tanaka ; Hitoshi Nakamura ; Yosuke Mizuno ; Kentaro Nakamura

ULTRA-LOW LOSS, ULTRA-LARGE AEFF OPTICAL FIBERS FOR UNDERSEA NETWORKS 344

Sergey Ten

HIGH-SPEED 1.55-µM VCSELS FOR DATACOM AND TELECOM APPLICATIONS 347

Silvia Spiga ; Alexander Andrejew ; Gerhard Boehm ; Markus-Christian Amann

EQUALIZER-FREE 2-KM SMF TRANSMISSION OF 106-GBIT/S 4-PAM SIGNAL USING OPTICAL TRANSMITTER/RECEIVER WITH 50 GHZ BANDWIDTH 350

Shigeru Kanazawa ; Satoshi Tsunashima ; Yasuhiko Nakanishi ; Yoshifumi Muramoto ; Hiroshi Yamazaki ; Yuta Ueda ; Wataru Kobayashi ; Hiroyuki Ishii ; Hiroaki Sanjoh

LOW THERMAL RESISTANCE VCSEL ARRAY ADOPTED TUNNEL JUNCTION DESTRUCTION USING PROTON IMPLANTATION 353

Shohei Oshida ; Masashi Suhara ; Tomoyuki Miyamoto

INFLUENCE OF WAVELENGTH DEVIATION ON PHASE LOCKED VCSEL ARRAY USING TALBOT EFFECT 356

Yuki Komori ; Tomoyuki Miyamoto

DEMONSTRATION OF AN ULTRA-LOW THRESHOLD PHONON LASER WITH COUPLED MICROTOROID CAVITIES IN VACUUM 359

Mingming Zhao ; Guanzhong Wang ; Zhiqiang Jin ; Yingchun Qin ; Zhang-qi Yin ; Xiaoshun Jiang ; Min Xiao

FLEX RATE TRANSMISSION AND CHALLENGES FOR LONG HAUL TRANSMISSION 362

Fred Buchali ; Wilfried Idler

95.2TB/S TRANSOCEANIC SEVEN-CORE FIBER TRANSMISSION USING OPTICAL PRE-FILTERED 200GBIT/S-BASED PDM-QPSK WDM SIGNALS 365

Yu Kawaguchi ; Koki Takeshima ; Daiki Soma ; Takehiro Tsuritani

96GBAUD NYQUIST-PDM-QPSK SIGNAL TRANSMISSION OVER 12,120KM USING DP-AM-DAC AND DECISION-FEEDBACK EQUALIZER 368

Kengo Horikoshi ; Fukutaro Hamaoka ; Asuka Matsushita ; Munehiko Nagatani ; Hiroshi Yamazaki ; Akihide Sano ; Toshikazu Hashimoto ; Hideyuki Nosaka ; Kazushige Yonenaga ; Akira Hirano ; Yutaka Miyamoto

ENABLING TECHNOLOGIES FOR HIGH SPECTRAL EFFICIENCY AND HIGH CAPACITY TRANSMISSION OVER TRANSOCEANIC DISTANCES 371

Matt Mazurczyk

HIGH-SPEED AVALANCHE PHOTODIODE FOR DATA CENTER NETWORKS 374

Masahiro Nada ; Takuya Hoshi ; Hideaki Matsuzaki

LOW VOLTAGE HIGH SPEED SI-GE AVALANCHE PHOTODIODES 377

Zhihong Huang ; Cheng Li ; Di Liang ; Kunzhi Yu ; Marco Fiorentino ; Raymond G. Beausoleil

11-GBPS 16-QAM OFDM RADIO OVER FIBER DEMONSTRATION USING 100 GHZ HIGH-EFFICIENCY PHOTORECEIVER BASED ON PHOTONIC POWER SUPPLY 380
T. Umezawa ; K. Kashima ; A. Kanno ; K. Akahane ; A. Matsumoto ; N. Yamamoto ; T. Kawanishi

DESIGN FOR HIGH SPEED OPERATION OF DOUBLE MICRORING RESONATOR-LOADED MACH-ZEHNDER 2×2 QUANTUM WELL OPTICAL SWITCH 383
Naoki Kawaguchi ; Kento Hori ; Taro Arakawa ; Yasuo Kokubun

MODE-DIVISION MULTIPLEXING LINBO₃ MODULATOR USING DIRECTIONAL COUPLER 386
Yutaro Kodama ; Yuya Yamaguchi ; Atsushi Kanno ; Tetsuya Kawanishi ; Masayuki Izutsu ; Hirochika Nakajima

MULTI-LEVEL HIGH SPEED BURST-MODE RECEIVERS 389
Xin Yin ; J. Van Kerrebrouck ; G. Coudyzer ; Johan Bauwelinck

MULTI-RATE COHERENT BURST-MODE PDM-QPSK OPTICAL RECEIVER FOR FLEXIBLE OPTICAL NETWORKS 392
José Manuel Delgado Mendinueta ; Hideaki Furukawa ; Satoshi Shinada ; Naoya Wada

AN INP MONOLITHICALLY INTEGRATED MULTIWAVELENGTH TRANSMITTER WITH DIRECT MODULATION 395
N. Andriolli ; P. Velha ; P. Tommasino ; M. Chiesa ; G. B. Preve ; A. Trifiletti ; M. Romagnoli ; G. Contestabile

SELF-CLOCKING SYNCHRONIZED OPTICAL DEMULTIPLEXING USING FOUR-WAVE MIXING IN A QUANTUM-DOT SOA 398
Liang Yang ; Tomoya Yatsu ; Motoharu Matsuura

DEMONSTRATION OF BI- AND MULTI-STABILITY IN A HIGH ORDER RING RESONATOR 401
Li Jin ; Alessia Pasquazi ; Luigi Di Lauro ; Marco Peccianti ; Edwin Y. B. Pun ; David J. Moss ; Roberto Morandotti ; Brent E. Little ; Sai T. Chu

OPTICAL PACKET AND CIRCUIT INTEGRATED RING NETWORK 404
Hideaki Furukawa

4-CHANNEL SILICON PHOTONIC MODE UNSCRAMBLER 407
Andrea Melloni ; Andrea Annoni ; Emanuele Guglielmi ; Marco Carminati ; Giorgio Ferrari ; Nicola Peserico ; Stefano Grillanda ; Marc Sorel ; Francesco Morichetti

MODE DIVISION MULTIPLEXING SWITCH FOR ON-CHIP OPTICAL INTERCONNECTS 410
Xinru Wu ; Ke Xu ; Daoxin Dai ; Hon Ki Tsang

PLASMONIC-ORGANIC HYBRID (POH) MODULATORS 413
A. Melikyan ; K. Koehnle ; M. Lauermann ; R. Palmer ; S. Koeber ; S. Muehlbrandt ; P. C. Schindler ; D. L. Elder ; S. Wolf ; M. Sommer ; L. R. Dalton ; D. Van Thourhout ; W. Freude ; M. Kohl ; J. Leuthold ; C. Koos

IN-PLANE SATURABLE ABSORPTION OF GRAPHENE ON A SILICON SLOT WAVEGUIDE 416
Jiaqi Wang ; Zhenzhou Cheng ; Hon Ki Tsang ; Chester Shu

GRAPHENE-INDUCED ON-DEMAND NANOCAVITY BASED ON SI PHOTONIC CRYSTAL 419
Hisashi Chiba ; Masaya Notomi

SILICON SLOT WAVEGUIDE RING RESONATOR: FROM SINGLE RESONANCE TO ENVELOPE INDEX SENSING 422
Weiwei Zhang ; Samuel Serna ; Xavier Le Roux ; Laurent Vivien ; Eric Cassan

SIMULTANEOUS MEASUREMENT OF STRAIN AND TEMPERATURE USING PI-PHASE-SHIFTED FIBER BRAGG GRATING ON POLARIZATION MAINTAINING FIBER 425
Jiageng Chen ; Qingwen Liu ; Xinyu Fan ; Zuyuan He

SPECTRAL EFFICIENT GROUPED ROUTING NETWORK THAT APPLIES DYNAMIC OPTICAL PATH GROOMING 428
Yuki Terada ; Yojiro Mori ; Hiroshi Hasegawa ; Ken-ichi Sato

FIELD TRIAL OF VIRTUAL TRANSPORT NETWORK SERVICES BASED ON HIERARCHICAL CONTROL OVER MULTI-DOMAIN OTN NETWORKS 431
Ruiquan Jing ; Chengliang Zhang ; Junjie Li ; Yiran Ma ; Qian Hu ; Xiaoli Huo ; Yongli Zhao ; Jiayu Wang ; Baoquan Rao ; Chen Qiu

A STATIC TRAFFIC GROOMING ALGORITHM FOR ELASTIC OPTICAL NETWORKS WITH ADAPTIVE MODULATION 434
Takafumi Tanaka ; Tetsuro Inui ; Wataru Imajuku

HARDWARE SCALE ANALYSIS AND PROTOTYPE DEVELOPMENT OF FLEXIBLE WAVEBAND ROUTING OXCS 437
Tomohiro Ishikawa ; Masaki Niwa ; Koh Ueda ; Yojiro Mori ; Hiroshi Hasegawa ; Suresh Subramaniam ; Ken-ichi Sato ; Osamu Moriwaki

SCALABLE AND DYNAMIC OPTICAL NETWORK ARCHITECTURE 440
Vincent W. S. Chan

3.5-GBIT/S QPSK SIGNAL TRANSMISSION IN BEAM FORMING OF 60-GHZ INTEGRATED PHOTONIC ARRAY-ANTENNA 443
Kotoko Furuya ; Takayoshi Hirasawa ; Masayuki Oishi ; Shigeyuki Akiba ; Jiro Hirokawa ; Makoto Ando

REAL-TIME TRANSMISSION OF 5-GBIT/S PI/4-SHIFT DQPSK SIGNAL OVER MILLIMETER-WAVE RADIO-OVER-FIBER LINKS 446
Abdelmoula Bekkali ; Takashi Kobayashi ; Kosuke Nishimura ; Nobuhiko Shibagaki ; Kenichi Kashima ; Yosuke Sato

MOBILE FRONTHAUL OPTICAL LINK FOR LTE-A SYSTEM USING DIRECTLY-MODULATED 1.5-μM VCSEL 449
Byung Gon Kim ; S. H. Bae ; Hoon Kim ; Y. C. Chung

POWER-OVER-FIBER TRANSMISSION USING 1.3-μM DUAL-CHANNEL RADIO-OVER-FIBER SIGNALS IN A DOUBLE-CLAD FIBER 452
Akira Yoneyama ; Yamato Minamoto ; Motoharu Matsuura

INTERNET OF THINGS WITH OPTICAL CONNECTIVITY, NETWORKING, AND BEYOND 455
  Philip N. Ji ; Ting Wang
STANDARDIZATION OF OPTICAL ACCESS TECHNOLOGIES: PROGRESS AND FUTURE PROSPECTS 458
  Jun-ichi Kani
TACTILE INTERNET CAPABLE PASSIVE OPTICAL LAN FOR HEALTHCARE 459
  Elaine Wong ; Maluge Pubuduni Imali Dias ; Lihua Ruan
ANALYSIS OF POWER CONSUMPTION IN MOBILE BACKHAUL NETWORK WITH DENSELY DEPLOYED SMALL CELLS UNDER DYNAMIC TRAFFIC BEHAVIOR 462
  Kyosuke Sone ; Inwoong Kim ; Xi Wang ; Yasuhiko Aoki ; Hiroyuki Seki ; Jens C. Rasmussen
OPTIMAL DESIGN AND BACKHAULING OF SMALL-CELL NETWORK: IMPLICATION OF ENERGY COST 465
  Chathurika Ranaweera ; Elaine Wong ; Christina Lim ; Chamil Jayasundara ; Ampalavanapillai Nirmalathas
A MOBILE FRONTHAUL SYSTEM ARCHITECTURE FOR DYNAMIC PROVISIONING AND PROTECTION 468
  Qianmei Yang ; Ning Deng ; Xu Zhou ; Chun-Kit Chan
EFFECT OF CHROMATIC DISPERSION ON CABLE RE-ROUTING OPERATION SUPPORT SYSTEM WITH NO SERVICE INTERRUPTION 471
  Kazutaka Noto ; Masaaki Inoue ; Hiroshi Watanabe ; Yusuke Koshikiya ; Keiji Okamoto ; Tetsuya Manabe
MULTIDIMENSIONAL MODULATION FORMATS FOR OPTICAL COMMUNICATION 474
  Magnus Karlsson
OPTIMIZATION OF NETWORKS CARRYING SUPER-CHANNELS WITH DIFFERENT MODULATION FORMATS FOR MAXIMUM SPECTRAL EFFICIENCY AND REACH 477
  Olga Vassilieva ; Tomohiro Yamauchi ; Shoichiro Oda ; Inwoong Kim ; Motoyoshi Sekiya ; Takeshi Hoshida ; Yasuhiko Aoki ; Jens C. Rasmussen ; Tadashi Ikeuchi
OPTIMUM CAPACITY UTILIZATION IN SPACE-DIVISION MULTIPLEXED TRANSMISSION SYSTEMS WITH FEW-MODE FIBERS 480
  Georg Rademacher ; Klaus Petermann
COMPARISON OF FOUR DIMENSIONALLY CODED VERSUS HYBRID MODULATION 483
  Han Sun
JOINT LINEAR AND NON-LINEAR ADAPTIVE PRE-DISTORTION OF HIGH BAUD RATE TRANSMITTERS FOR HIGH-ORDER MODULATION FORMATS 486
  Bernhard Spinnler ; Ginni Khanna ; Stefano Calabrò ; Erik De Man ; Uwe Feiste ; Thomas Bex ; Heinrich von Kirchbauer
FIELD DEMONSTRATION OF MODULATION FORMAT ADAPTATION BASED ON PILOT-AIDED OSNR ESTIMATION USING 400GBPS/CH REAL-TIME DSP 489
  Kazushige Yonenaga ; Kengo Horikoshi ; Seiji Okamoto ; Mitsuteru Yoshida ; Yutaka Miyamoto ; Masahito Tomizawa ; Takeshi Okamoto ; Hidemi Noguchi ; Jun-ichi Abe ; Junichiro Matsui ; Hisao Nakashima ; Yuichi Akiyama ; Takeshi Hoshida ; Hiroshi Onaka ; Kenya Sugihara ; Soichiro Kametani ; Kazuo Kubo ; Takashi Sugihara
QPSK ASSISTED CARRIER PHASE RECOVERY FOR HIGH ORDER QAM 492
  Xiaofei Su ; Liang Dou ; Zhenning Tao ; Takeshi Hoshida ; Jens C. Rasmussen
BLOCK CARRIER-PHASE RECOVERY WITH RECURSIVE NOISE ADAPTIVE KALMAN FILTERING FOR 16-QAM SIGNALS 495
  Zheng Bofang ; Chester Shu
HIGHLY ASE TOLERANT PASS-BAND SHAPE MONITOR FOR CASCADED ROADMS 498
  Guoxiu Huang ; Shoichiro Oda ; Tomohiro Yamauchi ; Setsuo Yoshida ; Goji Nakagawa ; Yasuhiko Aoki ; Zhenning Tao ; Jens C. Rasmussen
DIGITAL NONLINEAR DISTORTION COMPENSATION 501
  Zhenning Tao ; Liang Dou ; Ying Zhao ; Bo Liu ; Lei Li ; Tomofumi Oyama ; Takeshi Hoshida ; Jens C. Rasmussen
IMPACT OF LINK SYMMETRY ON NONLINEAR NOISE MITIGATION USING SPECTRAL INVERSION IN SUPERCHANNEL TRANSMISSION 504
  Inwoong Kim ; Olga Vassilieva ; Paparao Palacharla ; Motoyoshi Sekiya ; Tadashi Ikeuchi
ACHIEVABLE RATES COMPARISON FOR PHASE-CONJUGATED TWIN-WAVES AND PM-QPSK 507
  Tobias A. Eriksson ; Abel Lorences-Riesgo ; Pontus Johannisson ; Tobias Fehenberger ; Peter A. Andrekson ; Magnus Karlsson
KNN-BASED DETECTOR FOR COHERENT OPTICAL SYSTEMS IN PRESENCE OF NONLINEAR PHASE NOISE 510
  Danshi Wang ; Min Zhang ; Meixia Fu ; Zhongle Cai ; Ze Li ; Yue Cui ; Bin Luo
OSNR MONITORING BY DEEP NEURAL NETWORKS TRAINED WITH ASYNCHRONOUSLY SAMPLED DATA 513
  Takahito Tanimura ; Takeshi Hoshida ; Jens C. Rasmussen ; Makoto Suzuki ; Hiroyuki Morikawa
TRELLIS CODED OPTICAL MODULATION USING QAM CONSTELLATIONS 516
  Emmanuel Le Taillandier de Gabory ; Tatsuya Nakamura ; Hidemi Noguchi ; Wakako Maeda ; Sadao Fujita ; Jun'ichi Abe ; Kiyoshi Fukuchi
DIFFERENTIAL CODING IN POLARIZATIONS WITH HALF-SYMBOL-PERIOD TIMING OFFSET FOR COHERENT PM-QPSK TO IMPROVE PHASE-NOISE TOLERANCE 519
  Guo-Wei Lu ; Takahide Sakamoto ; Yukiyoshi Kamio
CROSSOVER BLOCK MODULATION WITH COMPLEMENTARY CODES SUPERPOSITION 522
  Tsuyoshi Yoshida ; Keisuke Kojima ; Toshiaki Koike-Akino ; David Millar ; Keisuke Dohi ; Keisuke Matsuda ; Kieran Parsons ; Kazuo Kubo ; Kenichi Uto ; Takashi Sugihara
TURBO EQUALIZATION FOR DUOBINARY-SHAPED SIGNALS IN SUPER-NYQUIST WDM SYSTEMS 525
  Shuai Yuan ; Koji Igarashi

ENHANCED BLIND MODULATION FORMATS RECOGNITION USING CONNECTED COMPONENT ANALYSIS WITH QUADRUPLE ROTATION 528
Tianwai Bo ; Jin Tang ; Calvin Chun-Kit Chan

BROADBAND AMPLIFICATION CHARACTERISTICS ON THE CASCADE BI-DOPED AND ER-DOPED OPTICAL FIBER AMPLIFIERS 531
Mikoto Takahashi ; Daiki Higuchi ; Mizuki Ohara ; Yusuke Fujii ; Naoto Yoshimoto ; Soichi Kobayashi

ALL-OPTICAL DYNAMIC GAIN CONTROL OF REMOTELY PUMPED ERBIUM-DOPED FIBER AMPLIFIER 534
Kokoro Kitamura ; Kenta Udagawa ; Hiroji Masuda

PHASE ADJUSTMENT BY WAVELENGTH TUNING OF PARAMETRIC PUMP ON RAMAN ASSISTED PHASE SENSITIVE AMPLIFIER 537
Y. Cao ; F. Alishahi ; Y. Akasaka ; M. Ziyadi ; A. Mohajerin-Ariaei ; A. Almaiman ; T. Ikeuchi ; S. Takasaka ; R. Sugizaki ; A. E. Willner

POLARIZATION PULLING IN FIBER OPTICAL PARAMETRIC AMPLIFIERS 540
S. H. Wang ; P. K. A. Wai

NONLINEAR IMPAIRMENTS MITIGATION AIDED BY LOW-NOISE OPTICAL FREQUENCY COMBS 543
Ping Piu Kuo

1.6 ~ 1.8 μM BAND HYBRID BROADBAND LIGHT SOURCE CONSISTING OF CASCADED SLD AND TDFA 544
Kazuya Ota ; Du Xiaoen ; Sho Tujita ; Kousuke Senda ; Fumiki Hanafuji ; Jun Ono ; Kazuaki Mise ; Yoshiharu Shimose ; Hiroshi Mori ; Osanori Koyama ; Hirotaka Ono ; Tatsuro Endo ; Makoto Yamada

1.8 μM HIGH-ORDER MICRORING RESONATOR MODE-LOCKED LASER USING A CARBON NANOTUBE 547
K. S. Tsang ; Jie Wang ; Li Jin ; Victor Ho ; Jack Cheung ; Yanny Tsang ; Alessia Pasquazi ; Ray Man ; Sai T. Chu ; A. Ping Zhang ; Hwa-yaw Tam ; P. K. A. Wai

WIDELY TUNABLE, SINGLE-LONGITUDINAL MODE BRILLOUIN/ERBIUM-DOPED FIBER LASER 550
Huan Wu ; Bofang Zheng ; Chester Shu

CONTINUOUSLY TUNABLE MICROWAVE PHOTONIC FILTER BASED ON A WAVELENGTH-SPACING-TUNABLE MULTIWAVELENGTH LASER 553
Seungmin Lee ; Young Bo Shim ; Young-Geun Han

REAL-TIME INTERROGATION TECHNIQUE USING FOURIER DOMAIN MODE-LOCKED FIBER LASER 556
Ik Su Jo ; Seungmin Lee ; Sanggwon Song ; Kwang Wook Yoo ; Young-Geun Han

QUANTUM DOT LASERS FOR SILICON PHOTONICS 559
Yasuhiko Arakawa

LOW TEMPERATURE LASING CHARACTERISTICS OF GAINASP DOUBLE-HETERO LASER INTEGRATED ON INP/SI SUBSTRATE USING DIRECT WAFER BONDING 561
Tetsuo Nishiyama ; Keiichi Matsumoto ; Junya Kishikawa ; Toshiki Sukigara ; Yuya Onuki ; Naoki Kamada ; Tomokazu Kanke ; Kazuhiko Shimomura

FIRST DEMONSTRATION OF MODE SELECTIVE LIGHT SOURCE BY USING ACTIVE MULTIMODE INTERFEROMETER LASER DIODE 564
Bingzhou Hong ; Takuya Kitano ; Akio Tajima ; Haisong Jiang ; Kiichi Hamamoto

QUANTUM-CASCADE LASERS: LINE-NARROWING AND SUPPRESSION OF FLICKER-NOISE 567
Masamichi Yamanishi ; Toru Hirohata ; Saverio Bartalini ; Paolo De Natale

IMPROVED EQUALIZING CHARACTERISTICS IN PRE-EQUALIZING ELECTRO-OPTIC MODULATOR WITH POLARIZATION-REVERSED STRUCTURES 570
Tomohiro Ohno ; Hiroshi Murata ; Yasuyuki Okamura

BROADBAND DUAL-POLARIZATION DUAL-PARALLEL MACH ZEHNDER MODULATOR BASED PHOTONIC MICROWAVE PHASE SHIFTER 573
Tong Niu ; Erwin H. W. Chan ; Xudong Wang ; Xinhuan Feng ; Bai-Ou Guan

PHOTONIC RADIO FREQUENCY DOWN-CONVERTER BASED ON PARALLEL ELECTRO-ABSORPTION MODULATORS IN KU/KU BAND FOR SPACE APPLICATIONS 576
Jordane Thouras ; Benoit Benazet ; Herve Leblond ; Christelle Aupetit-Berthelemot

ELECTRO-OPTIC MODULATOR USING MILLIMETER-WAVE GAP-EMBEDDED PATCH ANTENNA WITH STACKED STRUCTURE 579
Hironori Aya ; Yusuf Nur Wijayanto ; Atsushi Kanno ; Tetsuya Kawanishi ; Hiroshi Murata ; Yasuyuki Okamura

ROBUSTNESS FOR TRACKING ERROR IN FMF BASED FSO RECEIVER 582
T. Ishikawa ; K. Hosokawa ; S. Takahashi ; M. Arikawa ; Y. Ono ; T. Ito ; K. Fukuchi

IMPEDANCE MATCHED GAN LD PACKAGE FOR DIRECT OFDM COMMUNICATION AT 14 GBPS 585
Yu-Fang Huang ; Tsai-Chen Wu ; Yu-Chieh Chi ; Cheng-Ting Tsai ; Wei Wang ; Tien-Tsorng Shih ; Hao-Chung Kuo ; Gong-Ru Lin

OCT PRECODING FOR OFDM-BASED INDOOR VISIBLE LIGHT COMMUNICATIONS 588
Yang Hong ; Lian-Kuan Chen

NONLINEAR COMPENSATION OF 850 NM VCSEL WITH DISCRETE MULTI-TONE MODULATION EMPLOYING A VOLTERRA-WIENER FILTER 591
Ta-Ching Tzu ; Chia-Chien Wei ; Jun-Jie Liu ; Chun-Yen Chang ; Kai-Lun Chi ; Xin-Nan Chen ; Jin-Wei Shi ; Jyehong Chen

PRE-LEVELED 16-QAM OFDM MODULATION OF AN 850-NM VCSEL FOR 56-GBIT/S TRANSMISSION 594
Cheng-Ting Tsai ; Chun-Yen Pong ; Yun-Chen Wu ; Shan-Fong Leong ; Yu-Chieh Chi ; Chao-Hsin Wu ; Tien-Tsorng Shih ; Jian Jang Huang ; Hao-Chung Kuo ; Wood-Hi Cheng ; Gong-Ru Lin

MULTIMODE FIBERS FOR TELECOMMUNICATIONS AND IMAGING 597
Joel Carpenter ; Benjamin J. Eggleton ; Jochen Schröder

NEXT-GENERATION MULTIMODE FIBERS 600
Pierre Sillard ; Marianne Bigot-Astruc ; Denis Molin ; Adrian Amezcua-Correa

STATISTICAL TREATMENT OF IEEE SPREADSHEET MODEL FOR VCSEL-MULTIMODE FIBER TRANSMISSIONS 603
Xin Chen ; John Abbott ; Dale Powers ; Doug Coleman ; Ming-Jun Li

SINGLE-CARRIER 1-TB/S SIGNAL GENERATION USING HIGH-SPEED INP MUX-DACS AND AN INTEGRATED CSRZ-OTDM MODULATOR 606
Hiroshi Yamazaki ; Akihide Sano ; Munehiko Nagatani ; Yutaka Miyamoto

COST EFFECTIVE FRACTIONAL OFDM RECEIVER USING ARRAYED WAVEGUIDE GRATING 609
T. Nagashima ; G. Cincotti ; T. Murakawa ; S. Shimizu ; M. Hasegawa ; K. Hattori ; M. Okuno ; S. Mino ; A. Himeno ; N. Wada ; H. Uenohara ; T. Konishi

SUB-FF-CAPACITANCE PHOTONIC-CRYSTAL PHOTODETECTOR TOWARDS FJ/BIT ON-CHIP RECEIVER 612
Kengo Nozaki ; Shinji Matsuo ; Takuro Fujii ; Koji Takeda ; Masaaki Ono ; Abdul Shakoor ; Eiichi Kuramochi ; Masaya Notomi

4-CHANNEL SYNCHRONOUS THZ-WAVE GENERATORCOMPOSED OF ARRAYED UTC-PDS AND ANTENNAS 615
Goki Sakano ; Jun Haruki ; Kazuki Sakuma ; Kazutoshi Kato

DEMONSTRATION OF MICRO-PROJECTION ENABLED SHORT-RANGE COMMUNICATIONS FOR 5G 618
Hsi-Hsir Chou ; C. -Y. Tsai

THERMAL EMISSION CONTROL BY PHOTONIC CRYSTALS 621
Susumu Noda

MID IR APPLICATIONS OF SI PHOTONICS 622
G. Z. Mashanovich ; J. Soler Penades ; V. Mittal ; G. S. Murugan ; A. Z. Khokhar ; C. J. Littlejohns ; S. Stankovic ; A. Ortega-Monux ; G. Wanguemert-Perez ; R. Halir ; I. Molina-Fernandez ; C. Alonso-Ramos ; D. Benedikovic ; A. Villafranca ; P. Cheben ; J. J. Ackert ; A. P. Knights ; J. S. Wilkinson ; M. Nedeljkovic

SILICON PHOTONIC INTEGRATED CIRCUITS FOR HIGH-CAPACITY OPTICAL COMMUNICATIONS 624
Po Dong

LENS-INTEGRATED FAN-IN/FAN-OUT DEVICE FOR MULTI-CORE FIBER 627
Osamu Shimakawa ; Hajime Arao ; Tomomi Sano

FOUR-MODE-SELECTIVE PHOTONIC LANTERN BASED ON TWO-LAYER POLYMER WAVEGUIDE BRANCHES 630
Yunfei Wu ; Kin Seng Chiang

EXCITATION OF LP$_{21B}$ AND LP$_{02}$ MODES WITH PLC-BASED TAPERED WAVEGUIDE FOR MODE-DIVISION MULTIPLEXING 633
Yoko Yamashita ; Yuhei Ishizaka ; Nobutomo Hanzawa ; Takeshi Fujisawa ; Taiji Sakamoto ; Takashi Matsui ; Kyozo Tsujikawa ; Fumihiko Yamamoto ; Kazuhide Nakajima ; Kunimasa Saitoh

WIDELY SWITCHING THE ORBITAL ANGULAR MOMENTUM MODES WITH INTEGRATED COBWEB EMITTER 636
Yu Wang ; Peng Zhao ; Xue Feng ; Yidong Huang

COMPLIANT POLYMER INTERFACE FOR FIBER CONNECTION OF NANOPHOTONIC WAVEGUIDES 639
Hidetoshi Numata

HIGHLY EFFICIENT CIRCULAR HOLES ADDED A-SI:H GRATING COUPLER WITH METAL MIRROR FOR 3D OPTICAL INTERCONNECTS 642
JoonHyun Kang ; Yuki Kuno ; Kazuto Ito ; Yusuke Hayashi ; Junichi Suzuki ; Il-Ki Han ; Tomohiro Amemiya ; Nobuhiko Nishiyama ; Shigehisa Arai

300-MM ARF-IMMERSION LITHOGRAPHYTECHNOLOGY BASED SI-WIRE GRATING COUPLERS WITH HIGH COUPLING EFFICIENCY AND LOW CROSSTALK 645
Yohei Sobu ; Seok-Hwan Jeong ; Yu Tanaka ; Ken Morito

GRATING-BASED MODE CONVERTERS FABRICATED BY ONE-STEP PHOTOLITHOGRAPHY 648
Wei Jin ; Kin Seng Chiang

SIGNIFICANT LOSS REDUCTION OF 3.0 DB/CM ON CORE-TOP ETCHED WAVEGUIDE FOR VERTICAL MULTI-MODE INTERFERENCE 651
Ryosuke Sakata ; Kazuhiro Tanabe ; Haisong Jiang ; Kiichi Hamamoto

OPTICAL SYSTEM ARCHITECTURES AND CONTROL FOR DATA CENTER NETWORKS 654
Dimitra Simeonidou

EXPLICIT WAVELENGTH RESOURCE REALLOCATION TO ENSURE CRITICAL COMMUNICATION LINES IN EMERGENCY SITUATIONS 655
Masaki Shiraiwa ; Takaya Miyazawa ; Hideaki Furukawa ; Yoshinari Awaji ; Naoya Wada

REMOTE EXPERIMENTS WITH TEST-PLANE ON A DISTRIBUTED OPTICAL SWITCHED NETWORK TESTBED 658
Sugang Xu ; Masaki Shiraiwa ; Hiroaki Harai ; Yoshinari Awaji ; Naoya Wada

FLEXIBLE-BANDWIDTH OPTICAL INTERCONNECTS FOR DATACOM NETWORKS 661
Roberto Proietti ; Paolo Grani ; Zheng Cao ; S. J. Ben Yoo

LOAD BALANCING IN SWITCH-FABRIC TYPE OF TORUS OPS DATA CENTER NETWORKS WITH HYBRID OPTOELECTRONIC ROUTERS 664
Yue-Cai Huang ; Yuki Yoshida ; Salah Ibrahim ; Ryo Takahashi ; Ken-ichi Kitayama

SPECTRUM ASSIGNMENT AND UPDATE METHOD BASED ON ADAPTIVE SOFT RESERVATION IN ELASTIC OPTICAL NETWORKS 667
Hideki Tode ; Akira Fukushima ; Yosuke Tanigawa ; Yusuke Hirota

A STUDY ON SPECTRUM ASSIGNMENT AND RESOURCE SHARING METHOD IN ELASTIC OPTICAL PACKET AND CIRCUIT INTEGRATED NETWORKS 670
Ken Nagatomi ; Seitaro Sugihara ; Shohei Fujii ; Yusuke Hirota ; Hideki Tode ; Takashi Watanabe

OPTIMIZED MULTICAST SCHEDULING IN DATACENTER OPTICAL BURST RING NETWORK 673
Gang Chen ; Dongxu Zhang ; Hongxiang Guo ; Jian Wu ; Xiaoyuan Cao ; Noboru Yoshikane ; Takehiro Tsuritani ; Itsuro Morita

TUNABLE OPTICAL TECHNOLOGIES FOR NEXT GENERATION OPTICAL ACCESS SYSTEM 676
Kota Asaka

SYMMETRIC 4×25-GBIT/S TWDM-PON TRANSMISSION BY USING SPECTRUM RESHAPING 679
Yuan-Yuan Sung ; Yu-Chang Liu ; Puspa Devi Pukhrambam ; Hung-Wen Hung ; San-Liang Lee

AUTO-CONFIGURATION OPTICAL AMPLIFIER FOR FLEXIBLE ACCESS NETWORK DESIGN 682
Takuya Tsutsumi ; Yu Nakayama ; Shunsuke Kanai ; Manabu Kubota ; Akihiro Otaka

INTEGRATED SOI-BASED RECEIVER MODULE FOR TWDM-PON 685
Y. Hsu ; X. R. Wu ; L. Y. Wei ; K. Xu ; C. W. Hsu ; J. Y. Sung ; C. H. Yeh ; H. K. Tsang ; C. W. Chow

1GBPS FULL-DUPLEX 5GHZ FREQUENCY SLOTS UDWDM FLEXIBLE METRO/ACCESS NETWORKS BASED ON VCSEL-RSOA TRANSCEIVER 688
Jose A. Altabas ; David Izquierdo ; Jose A. Lazaro ; Ignacio Garces

SURVIVABLE MULTIPATH PROVISIONING WITH CONTENT CONNECTIVITY IN ELASTIC OPTICAL DATACENTER NETWORKS 691
Tao Gao ; Shanguo Huang ; Bingli Guo ; Xin Li ; Qian Kong ; Yu Zhou ; Wenzhe Li ; Wanyi Gu

DISTRIBUTED 4K-VIDEO CAMERA MONITORING SERVICE ON ETHERNET PON SYSTEM USING HYBRID BANDWIDTH ALLOCATION FOR SECURE COMMUNITY 694
Akihisa Shoji ; Naoto Yoshimoto

DEMONSTRATION OF CONTROLLING CHROMATIC DISPERSION AT SOA WITH MACH-ZEHNDER INTERFEROMETRIC MEASUREMENT SYSTEM 697
Yusuke Yamanaka ; Yuki Fujimura ; Kazutoshi Kato

HYBRID METAL INSULATOR METAL PLASMONIC WAVEGUIDE AND RING RESONATOR 700
Prateeksha Sharma ; Dinesh Kumar V.

TUNABLE SINGLE BANDPASS MICROWAVE PHOTONIC FILTER BASED ON PHASE COMPENSATED SILICON-ON-INSULATOR MICRORING RESONATOR 703
Wenjian Yang ; Xiaoke Yi ; Shijie Song ; Suen Xin Chew ; Liwei Li ; Linh Nguyen

FULL-BAND DIRECT-CONVERSION RECEIVER USING MICROWAVE PHOTONIC I/Q MIXER 706
Yue Zheng ; Jianqiang Li ; Qiang Lv ; Jianwei Zhou ; Yuting Fan ; Feifei Yin ; Yitang Dai ; Kun Xu

OUTPUT TIMING ADJUSTMENT MECHANISM OF OPTICAL AND ELECTRONIC COMBINED BUFFER FOR OPTICAL PACKET SWITCHING 709
Takahiro Hirayama ; Hiroaki Harai

DESIGN OF A 1×2 WAVELENGTH SELECTIVE SWITCH USING AN ARRAYED-WAVEGUIDE GRATING WITH FOLD-BACK PATHS ON A SILICON PLATFORM 712
Hideaki Asakura ; Koki Sugiyama ; Hiroyuki Tsuda

SUPPRESSION OF PHASE NOISE INDUCED BY OPTICAL INTERFERENCE IN OPTOELECTRONIC OSCILLATORS 715
Huanfa Peng ; Xiaofeng Peng ; Yongchi Xu ; Cheng Zhang ; Lixin Zhu ; Weiwei Hu ; Zhangyuan Chen

SDM NETWORKING SCHEMES COMPARED FOR COMPUTATION COMPLEXITY AND EFFICIENCY 718
Miri Blau ; Dan M. Marom

TOPOLOGY RECONSTRUCTION STRATEGY WITH THE OPTICAL SWITCHING BASED SMALL WORLD DATA CENTER NETWORK 721
Tingting Yang ; Dongxu Zhang ; Hongxiang Guo ; Jian Wu

TOLERANCE TO LASER FREQUENCY DEVIATION OF NYQUIST-WDM AND TIME-FREQUENCY-PACKING MODULATION FORMATS 724
Francesco Fresi ; Gianluca Meloni ; Fabio Cavaliere ; Luca Potì

BRANCHING RATIO OF OPTICAL COUPLER FOR CABLE RE-ROUTING OPERATION SUPPORT SYSTEM WITH NO SERVICE INTERRUPTION 727
Hiroshi Watanabe ; Kazutaka Noto ; Yusuke Koshikiya ; Tetsuya Manabe

INVESTIGATION OF SWITCH REDUCTION OF A MULTI-DIMENSIONAL OPTICAL NODE FOR SPATIAL DIVISION MULTIPLEXING NETWORKS 730
Hiroki Sasago ; Hiroyuki Uenohara

BIT-ERROR-RATE PERFORMANCE IN OPTICAL 16QAM RECOGNITION BY MAXIMUM OUTPUT WITH OPTICAL WAVEGUIDE CIRCUIT 733
Kensuke Inoshita ; Hiroki Kishikawa ; Nobuo Goto

ANALYSIS OF A SI-NANOCRYSTAL STRIP-LOADED WAVEGUIDE FOR NONLINEAR APPLICATIONS 736
Pengjiang Wei ; S. H. Wang ; Brent E. Little ; Sai Tak Chu

INTEGRATION OF MICRO DATA CENTER WITH OPTICAL LINE TERMINAL IN PASSIVE OPTICAL NETWORK 739
Bingliang Yang ; Zitian Zhang ; Kuo Zhang ; Weisheng Hu

AMPLIFIED SPONTANEOUS EMISSION NOISE INFLUENCE ANALYSIS ON OPTICAL QUANTIZATION USING SOLITON SELF-FREQUENCY SHIFT 742
Y. Yamasaki ; T. Nagashima ; M. Hiraoka ; T. Konishi

LARGE DATA TRANSFERS IN WDM NETWORKS WITH NODE STORAGE 745
Da Feng ; Weiqiang Sun ; Weisheng Hu

20 GB/S OPTICAL SWITCHED DQPSK TRANSMISSION OVER 50 KM SSMF AND 7 KM DCF 748
Pacharapon Chulok ; Suvit Nakpeerayuth ; Duang-rudee Worasucheep ; Naoya Wada

VARIABLE GAIN PI-BASED CYCLIC SLEEP CONTROL WITH ANTI-WINDUP TECHNIQUE FOR QOS-AWARE AND ENERGY-EFFICIENT ETHERNET PONS 751
Takahiro Kikuchi ; Ryogo Kubo

EXPERIMENTAL DEMONSTRATION OF A COST EFFECTIVE BIDIRECTIONAL COHERENT OFDM-PON 754

Rongrong Chen ; Caixia Kuang ; Yingxiong Song ; Min Wang ; Rujian Lin ; Qianwu Zhang

SUB-CARRIER SHARING IN OFDM-PON FOR 5G MOBILE NETWORKS SUPPORTING RADIO-OVER-FIBRE 757

Sheng Xu ; Sugang Xu ; Yoshiaki Tanaka

CHANNEL-REUSE IMDD-BASED 40 GB/S/λ 16-QAM NYQUIST-SCM DOWNSTREAM AND 20 GB/S/λ NYQUIST 4 PAM UPSTREAM WDM-PON 760

Zhiguo Zhang ; Bingbing Zhang ; Yanxu Chen ; Bingchang Hua ; Xue Chen

EFFICIENT MOBILE FRONTHAUL FOR SIMULTANEOUS TRANSMISSION OF 4G AND FUTURE MOBILE SIGNALS 763

Pham Tien Dat ; Atsushi Kanno ; Naokatsu Yamamoto ; Tetsuya Kawanishi

FREQUENCY-STABILIZED MULTI-TONE SIGNAL GENERATION BY OPTICAL FREQUENCY LOCKED-LOOP AND OPTICAL FREQUENCY COMB 766

Jinsei Oba ; Kosuke Kamikawa ; Atsushi Kanno ; Naokatsu Yamamoto ; Hideyuki Sotobayashi

THIRD-ORDER HARMONICS SUPPRESSION IN TWO-TONE SIGNAL GENERATION USING A DUAL-PARALLEL MACH-ZEHNDER MODULATOR 769

Kazunori Osato ; Moriya Nakamura

SELF-INTERFERENCE CANCELLATION FOR 2×2 MIMO IN-BAND FULL-DUPLEX RADIO-OVER-FIBER SYSTEMS 772

Yunhao Zhang ; Shilin Xiao ; Yinghong Yu ; Shaojie Zhang ; Lu Zhang ; Ling Liu ; Haiyun Xin

DYNAMIC VIRTUAL OPTICAL NETWORK MAPPING BASED ON SWITCHING CAPABILITY AND SPECTRUM FRAGMENTATION IN ELASTIC OPTICAL NETWORKS 775

Hongxiang Wang ; Xin Xin ; Jiawei Zhang ; Yongmei Sun ; Yuefeng Ji

FIBER-WIRELESS AND FIBER-IVLLC CONVERGENCES 778

Bo-Rui Chen ; Hung-Hsien Lin ; Chang-Jen Wu ; Chun-Yu Lin ; Chung-Yi Li ; Hai-Han Lu

VISIBLE LIGHT ENCRYPTION SYSTEM USING CAMERA IMAGE SENSOR 781

Chin-Wei Hsu ; Kevin Liang ; Hung-Yu Chen ; Liang-Yu Wei ; Chien-Hung Yeh ; Yang Liu ; Chi-Wai Chow

NON-ORTHOGONAL MULTIPLE ACCESS WITH MULTICARRIER PRECODING IN VISIBLE LIGHT COMMUNICATIONS 784

Xun Guan ; Yang Hong ; Calvin Chun-Kit Chan

6.6-GBIT/S PHASE MODULATED IMPULSE-RADIO GENERATION FROM RZ-OOK OPTICAL SIGNALS FOR OPTICAL-WIRELESS SYSTEMS 787

Yuri Ohara ; Hdeaki Kawahara ; Tomoki Kishida ; Kengo Nabika ; Saeko Oshiba

SIMULTANEOUS ALL-CHANNEL OTDM DEMULTIPLEXING BASED ON COMPLETE OPTICAL FOURIER TRANSFORMATION 790

Pengyu Guan ; Mads Lillieholm ; Kasper Meldgaard Røge ; Toshio Morioka ; Leif Katsuo Oxenløwe

EXPERIMENTAL DEMONSTRATION OF 7.09GB/S DML BASED REAL-TIME OPTICAL OFDM TRANSMISSION WITH SPECTRAL EFFICIENCY UP TO 6.93BITS/S/HZ OVER 50KM SSMF 793

Qianwu Zhang ; Han Dun ; Chen Qian ; Bingyao Cao ; Mingzhi Mao ; Junjie Zhang ; Min Wang

DISPERSION COMPENSATION SCHEME FOR TIME-FREQUENCY-DOMAIN MULTIPLEXED 4X/8X ULTRA-WIDEBAND MULTI-CARRIER SIGNALS 796

Takahide Sakamoto

REAL-TIME VISIBLE LIGHT COMMUNICATION SYSTEM BASED ON 2ASK-OFDM CODING 799

Yuankai Xue ; Yinan Hou ; Shilin Xiao ; Yunhao Zhang ; Lu Zhang

SPECTRAL OVERLAP OF TWO BANDWIDTH VARIABLE NYQUIST-WDM SIGNALS TO RESOLVE WAVELENGTH CONFLICT IN ELASTIC OPTICAL NETWORKS 802

Shuang Gao ; Lingchen Huang ; Chun-Kit Chan

HIGHLY LINEAR OPTICAL W-BAND RECEIVER FRONT-END BASED ON OPTICAL PROCESSING 805

Ziang Li ; Shangyuan Li ; Xiaoping Zheng ; Jinhui Luo ; Hanyi Zhang ; Bingkun Zhou

DISTANCE-ADAPTIVE ROUTING, MODULATION LEVEL AND SPECTRUM ALLOCATION (RMLSA) IN K-NODE (EDGE) CONTENT CONNECTED ELASTIC OPTICAL DATACENTER NETWORKS 808

Xin Li ; Shanguo Huang ; Shan Yin ; Bingli Guo ; Yongli Zhao ; Jie Zhang ; Min Zhang ; Wanyi Gu

EXPERIMENTAL DEMONSTRATION OF 10-GB/S DIRECT DETECTION OPTICAL OFDM TRANSMISSION WITH TRELLIS-CODED 8PSK SUBCARRIER MODULATION 811

Yiming Lou ; Zhenming Yu ; Minghua Chen ; Hongwei Chen ; Sigang Yang ; Shizhong Xie

POWER EFFICIENT OPTICAL OFDM TRANSMISSION WITH PHASE MODULATION AND DIRECT DETECTION 814

Zhenhua Feng ; Qiong Wu ; Ming Tang ; Rui Lin ; Ruoxu Wang ; Jiale He ; Songnian Fu ; Lei Deng ; Deming Liu ; Perry Ping Shum

A 3-D ADAPTIVE LOADING ALGORITHM FOR DIRECT DETECTION OPTICAL OFDM SYSTEM 817

Xi Chen ; Zhenhua Feng ; Ming Tang ; Huibin Zhou ; Songnian Fu ; Deming Liu

A ROBUST TIMING ESTIMATION METHOD FOR DDO-OFDM SYSTEMS 820

Zhen Zhang ; Yingxiong Song ; Junjie Zhang ; Yufeng Cai ; Weiliang Wu ; Qianwu Zhang

PERFORMANCE COMPARISON OF NYQUIST-4PPM-QPSK AND QPSK IN UNREPEATERED TRANSMISSION SYSTEMS 823

Jiangchuan Pang ; Yan Li ; Miao Yu ; Lei Yue ; Sujie Fan ; Deming Kong ; Jian Wu

SIGNAL LIGHT CARRIER AUTOMATIC PHASE-LOCK OPERATION TO OPTICAL FREQUENCY GRID COMB 826

Yudai Hisata ; Akira Mizutori ; Masafumi Koga

MAXIMUM RATIO COMBINING CHARACTERISTICS AFFECTED BY LASER PHASE NOISE FOR WAVELENGTH DIVERSITY DIGITAL COHERENT SYSTEM...829

Akira Naka ; Kohei Saito ; Kazuki Tomita ; Hideki Maeda

MULTI-LEVEL PRE-EQUALIZATION USING ANALOG FIR FILTERS BASED ON 28-NM FD-SOI FOR 20-GB/S 4-PAM MULTI-MODE FIBER TRANSMISSION...832

Ryoichiro Nakamura ; Kenta Amino ; Kawori Sekine ; Kazuyuki Wada ; Moriya Nakamura

BLIND POLARIZATION DE-MULTIPLEXING OF TIME DOMAIN HYBRID QAM SIGNALS BASED ON RADIUS-DIRECTED LINEAR KALMAN FILTER...835

Qun Zhang ; Wen Jiang ; Yanfu Yang ; Xian Zhou ; Kangping Zhong ; Yong Yao

RELATIONSHIP BETWEEN ROLL-OFF FACTOR AND TRANSMISSION DISTANCE IN NYQUIST OTDM SCHEME BASED ON CORRELATION DETECTION WITH EDFA REPEATERS...838

Takahiro Oguro ; Yuji Miyoshi ; Hirokazu Kubota ; Masaharu Ohashi

ELASTIC-BANDWIDTH ACCESS WITH SPECTRUM-SLICED INCOHERENT LIGHT SOURCE IN WDM-PON...841

Jing Zhang ; Qikai Hu ; Changyuan Yu ; Xingwen Yi ; Kun Qiu

EQUALIZATION OF OPTICAL NONLINEAR WAVEFORM DISTORTION USING NEURAL-NETWORK BASED DIGITAL SIGNAL PROCESSING...844

Shotaro Owaki ; Moriya Nakamura

PERFORMANCE ANALYSIS OF EKF-BASED POLARIZATION AND PHASE TRACKING FOR HIGH ORDER QAM SIGNALS...847

Wen Jiang ; Yanfu Yang ; Qun Zhang ; Kangping Zhong ; Xian Zhou ; Yong Yao

POST-FEC PERFORMANCE COMPARISON OF PILOTAIDED PHASE UNWRAP AND TURBO DIFFERENTIAL DECODING FOR CYCLE SLIP MITIGATION...850

Tangqing Liu ; Yan Li ; Huizhong Chen ; Haiquan Cheng ; Deming Kong ; Jian Wu

CHARACTERIZATION OF SOLITON FUSION PHENOMENON BASED ON SOLITONS' EIGENVALUE...853

Gihan Weerasekara ; Akihiro Maruta

FEASIBILITY STUDY OF 100G ETHERNET WITH CARRIERLESS AMPLITUDE AND PHASE MODULATION...856

Hirotaka Ochi ; Shinya Sasaki

BANDWIDTH ENHANCEMENT EQUALIZATION ENABLING A 40-GBPS PAM4 TRANSMISSION VIA A 9.5-GHZ PHOTORECEIVER...859

Jun-Jie Liu ; Chia-Chien Wei ; Ta-Ching Tzu ; Chun-Yen Chuang ; Kai-Lun Chi ; Xin-Nan Chen ; Jin-Wei Shi ; Jyehong Chen

INFLUENCE OF POINTING ERRORS ON ERROR PROBABILITY OF INTER-SATELLITE LASER COMMUNICATIONS...862

Qian Wang ; Tianyu Song ; Ming-Wei Wu ; Tomoaki Ohtsuki ; Mohan Gurusamy ; Pooi-Yuen Kam

TEST AND VERIFICATION OF VEHICLE POSITIONING BY INFRARED SIGNAL-DIRECTION DISCRIMINATION...865

Wern-Yarng Shieh ; Ti-Ho Wang ; Kun-Lung Chung ; Zong-Yu Tsai ; Wei-Jhih Huang

NOISE TOLERANT OPTICAL DETECTION IN CMOS IMAGE SENSOR BASED VISIBLE LIGHT COMMUNICATION...868

Joon-Woo Lee ; Se-Hoon Yang ; Yong-Hwan Son ; Sang-Kook Han

A CW, ALL-FIBER 2100 NM HO DOPED FIBER LASER PUMPED AT 1950 NM...871

Jiachen Wang ; Sang Bae Lee ; Kwanil Lee

SECURITY STRATEGY AGAINST MULTIPOINT EAVESDROPPING IN ELASTIC OPTICAL NETWORKS...874

Wei Bai ; Hui Yang ; Yongli Zhao ; Jie Zhang ; Yuanlong Tan ; Xiaoxu Zhu ; Huixia Ding

SWITCHABLE PASSIVELY MODE-LOCKED THULIUM-DOPED FIBER LASER BASED ON SINGLE-WALL CARBON NANOTUBES...877

Zhichao Wu ; Songnian Fu ; Minming Zhang ; Kai Jiang ; Jue Song ; Ping Shum ; Deming Liu

EVALUATION OF ALCOHOL CONCENTRATION IN SAKE USING 1.7 μM BAND TM^{3+}-TB^{3+}-DOPED TUNABLE FIBER RING LASER...880

Fumiki Hanafuji ; Kousuke Senda ; Xiaoen Du ; Sho Tujita ; Kazuya Ota ; Jun Ono ; Osanori Koyama ; Hirotaka Ono ; Tatsuro Endo ; Makoto Yamada

GENERATION OF 100 GHZ PULSE TRAIN FROM A PHASE MODULATED HYBRID MODE-LOCKED ER-FIBER LASER WITH AN INTRA-CAVITY ETALON...883

Sheng-Min Wang ; Yinchieh Lai

INFRARED LUMINESCENCE INVESTIGATION OF BISMUTH AND ERBIUM CO-DOPED FIBER...886

Chunsheng Li ; Binbin Yan ; Xiaojuan Wu ; Xinzhu Sang

ENERGY-EFFICIENT DUAL-PUMP DEGENERATED PHASE SENSITIVE AMPLIFIERS...889

Mingyi Gao ; Takayuki Kurosu ; Shu Namiki ; Gangxiang Shen

GAIN SATURATION AND RECOVERY OF AN ERBIUMDOPED FIBER AMPLIFIER MEASURED BY TEMPORALLY RESOLVED PROBE TECHNIQUES...892

Keiji Kuroda ; Yuzo Yoshikuni

HIGH GROSS GAIN OF SINGLE-MODE CR-DOPED FIBERS EMPLOYING ON-LINE GROWTH SYSTEM...895

Chun-Nien Liu ; Gia-Ling Cheng ; Nan-Kuang Chen ; Yi-Chung Huang ; Pi-Ling Huang ; Sheng-Lung Huang ; Wood-Hi Cheng

DOUBLE INSCRIBING METHOD WITH CO_2 LASER FOR SUITABLE SPECTRUM OF LPFG USED IN MULTIPOINT TEMPERATURE SENSOR...898

Akihiro Kusama ; Mamoru Iida ; Osanori Koyama ; Syo Takasuka ; Toshinori Murakami ; Makoto Yamada

SIMULTANEOUS INTERROGATION OF MULTIPLE CORRELATION PEAKS IN A BOCDA SYSTEM BY TIME-DOMAIN DATA PROCESSING...901

Gukbeen Ryu ; Kwang Yong Song ; Gyu-Tae Kim ; Sang Bae Lee ; Kwanil Lee

RISK-AWARE VIRTUAL NETWORK EMBEDDING IN OPTICAL DATA CENTER NETWORKS 904
Weigang Hou ; Lei Guo ; Cunqian Yu ; Yue Zong

SIMULTANEOUS MEASUREMENT OF CURVATURE AND TEMPERATURE BASED ON THIN CORE ULTRA-LONG PERIOD FIBER GRATING 907
Wenjun Ni ; Ping Lu ; Chao Luo ; Xin Fu ; Deming Liu ; Jiangshan Zhang

METAL-ASSISTED HIGH SENSITIVITY MICRO MACH-ZEHNDER FIBER TEMPERATURE SENSORS 910
Kuan-Hao Lin ; Nan-Kuang Chen ; Shien-Kuei Liaw ; Wood-Hi Cheng

THE LIQUID LENGTH EFFECTS ON THE MACH-ZEHNDER INTERFEROMETER INDUCED BY TWO LIQUID SECTIONS IN A PHOTONIC CRYSTAL FIBER 913
Jia-Hong Liou ; Chin-Ping Yu

HIGH-SENSITIVITY HUMIDITY SENSOR COMPOSED OF OPTICAL FIBER COATED WITH SOL-EL DERIVED POROUS SILICA 916
Nobuaki Tsuda ; Hideki Fukano ; Shuji Taue

A HIGH-SENSITIVITY TWO-DIMENSIONAL INCLINOMETER BASED ON TWO ETCHED-CHIRPED FIBER GRATINGS 919
Hung-Ying Chang ; Kuan-Ting Li ; Po-Chia Huang ; Jung-Sheng Chiang ; Nai-Hsiang Sun ; Wen-Fung Liu

A POLYMER-COATED HOLLOW CORE FIBER FABRY-PÉROT INTERFEROMETER FOR SENSING LIQUID LEVEL 922
Teng-Wei Fu ; Yuan-Jie Yang ; Jun-Han Lin ; Tung-Yuan Yeh ; Pin Han ; Cheng-Ling Lee

MULTI-CHANNEL LASING CHARACTERISTICS FOR LINEAR-CAVITY FIBER SENSOR SYSTEM USING SOA AND FIBER BRAGG GRATING ELEMENTS 925
Kazuto Takahashi ; Mao Okada ; Hiroki Kishikawa ; Nobuo Goto ; Yi-Lin Yu ; Shien-Kuei Liaw

NOVEL SOFT-CLADDING OPTICAL FIBER FOR DISTRIBUTED PRESSURE SENSING 928
Bin Zhou ; Lin Htein ; Zhengyong Liu ; A. Ping Zhang ; Chao Lu ; Hwa-yaw Tam

NOVEL BIDIRECTIONAL REFLECTIVE SEMICONDUCTOR OPTICAL AMPLIFIER 931
G. de Valicourt ; A. Maho ; A. Le liepvre ; R. Brenot ; A. Velázquez ; Y. K. Chen

SEMICONDUCTOR OPTICAL AMPLIFIER IN AWG-STAR NETWORK WITH WAVELENGTH PATH RELOCATION FUNCTION 934
Takumi Niihara ; Minoru Yamaguchi ; Osanori Koyama ; Hiroaki Maruyama ; Kazuya Ota ; Makoto Yamada

PROPOSAL OF AN ORCHESTRATOR-TO-ORCHESTRATOR INTERFACE USING THE CONTROL ORCHESTRATION PROTOCOL 937
Yoshiaki Inoue ; Jun Matsumoto ; Satoru Okamoto ; Naoaki Yamanaka

40GB/S OPTICAL RECEIVER USING HIGH-GAIN MULTI-LEVEL ACTIVE FEEDBACK WITH SERIAL INDUCTOR PEAKING 940
Cheng-Ta Chan ; Oscal T. -C. Chen

FREQUENCY CHIRP PROPERTIES WITH DATA PATTERN DEPENDENCE IN QUANTUM-DOT SOAS 943
Hiroki Hoshino ; Norihiko Ninomiya ; Motoharu Matsuura

OPTICAL SENSOR BASED ON MACH-ZEHNDER INTERFEROMETER USING ORBITAL ANGULAR MOMENTUM 946
Haozhe Yan ; Shangyuan Li ; Bian FengKai ; Xiaoping Zheng ; Hanyi Zhang ; Bingkun Zhou

PRECISE MEASUREMENT OF MICROWAVE EVANESCENT FIELDS ALONG FIBERGLASS-REINFORCED PLASTIC MORTAR PIPE USING ELECTRO-OPTIC SENSOR FOR NONDESTRUCTIVE INSPECTION 949
Yoshiyuki Azuma ; Fumiaki Ueno ; Hiroshi Murata ; Yasuyuki Okamura ; Tadahiro Okuda ; Masaya Hazama

PLASMON-INDUCED TRANSPARENCY BASED ON SIDECOUPLED STUB AND HEXAGONAL RESONATORS AND ITS SENSING CHARACTERISTICS 952
Tianye Huang ; Minming Zhang ; Songnian Fu

OPTICAL CHARACTERISTICS OF INP/GALNAS CORE-MULTISHELL NWS GROWN BY SELF-CATALYTIC VLS MODE 955
Kohei Takano ; Takehiro Ogino ; Keita Asakura ; Takao Waho ; Kuzuhiko Shimomura

S-K GROWTH OF INAS QUANTUM DOTS ON DIRECTLY-BONDED INP/SI SUBSTRATE USING MOVPE 958
Naoki Kamada ; Toshiki Sukigara ; Keiichi Matsumoto ; Junya Kishikawa ; Tetsuo Nishiyama ; Yuya Onuki ; Kazuhiko Shimomura

HYBRID ELECTRO-OPTIC POLYMER MODULATORS 961
Feng Qiu ; Shiyoshi Yokoyama

PHOTODETECTION FREQUENCY RESPONSE CHARACTERIZATION FOR HIGH-SPEED GE-PD ON SI WITH AN EQUIVALENT CIRCUIT 964
Jeong-Min Lee ; Minkyu Kim ; Stefan Lischke ; Lars Zimmernman ; Seong-Ho Cho ; Woo-Young Choi

SUB-μM ELECTRODE SPACING SOI-PIN PHOTODIODE FABRICATED BY CMOS COMPATIBLE PROCESS 967
Hiroya Mitsuno ; Takeo Maruyama ; Koichi Iiyama

ARBITRARY OUTPUT PORT SELECTION IN MULTI-MODE FIBER NETWORKS USING MODE DIVISION MULTIPLEXING 970
Yuki Morizumi ; Hirokazu Kobayashi ; Katsushi Iwashita

COMPARISON OF TWO PHOTODETECTOR LINEARITY CHARACTERIZING SYSTEMS 973
Youxin Liu ; Yongqing Huang ; JiaRui Fei ; Yangan Zhang ; Xiaomin Ren ; Kai Liu ; Xiaofeng Duan

OPTIMIZATION OF TEMPERATURE-DEPENDENT 850 NM VCSELS WITH DIFFERENT OXIDE-CONFINED APERTURE SIZES 976
Chun-Yen Pong ; Cheng-Ting Tsai ; Yun-Chen Wu ; Shan-Fong Leong ; Yu-Chieh Chi ; Gong-Ru Lin ; Chao-Hsin Wu

ENHANCEMENT OF LIGHT EXTRACTION EFFICIENCY OF INGAN LIGHT-EMITTING DIODES WITH MICROHOLES ARRAY AND NANO-ROUGHENED ZNO STRUCTURE 979
Ming Wang ; Zhigang Zang ; Xiaosheng Tang

CONCENTRIC CIRCULAR HIGH INDEX CONTRAST GRATINGS REFLECTOR WITH FOCUSING ABILITY 982
Wenjing Fang ; Yongqing Huang ; Xiaofeng Duan ; Kai Liu ; Ren Xiaomin ; Jun Wang

FABRICATION OF WAVEGUIDE-TYPE MIRRORS FOR RED-GREEN-BLUE LASER BEAM MULTIPLEXERS 985
S. Tanaka ; A. Nakao ; S. Yokokawa ; S. Hayashiguchi ; T. Katsuyama ; K. Nakajima ; N. Ikeda ; Y. Sugimoto

MULTIPLE-ACCESS AND TWO-WAY VISIBLE LIGHT COMMUNICATION WITH IMAGE SENSOR AND LED ARRAY 988
Tomoki Kondo ; Ryotaro Kitaoka ; Wataru Chujo

BROADBAND POLYMER 3-DB COUPLER USING SHORTCUT TO ADIABATICITY BASED OPTIMIZATION 991
Hung-Ching Chung ; Shuo-Yen Tseng

MULTIMODE THREE BRANCH POLYMER SPLITTER 994
Václav Prajzler ; Radek Maštera

2-STAGE CASCADED SILICON PHOTONIC PBS BASED ON MACH ZEHNDER DELAY INTERFEROMETERS 997
Kohei Morita ; Hiroyuki Uenohara

AN ULTRA-COMPACT AND LOW-LOSS WAVELENGTH DEMULTIPLEXER EMPLOYING THE PHOTONIC-CRYSTAL-LIKE METAMATERIAL STRUCTURE 1000
Feiya Zhou ; Luluzi Lu ; Minming Zhang ; Songnian Fu ; Deming Liu

ENERGY ANALYSIS FOR DYNAMIC CACHE STORAGE IN VIDEO ON DEMAND SERVICES 1003
Zhongwei Tan ; Chuanchuan Yang ; Yu Yang ; Zhaopeng Xu ; Xinyue Wang ; Ziyu Wang

POLARIZING BEAM SPLITTER WITH FOCUSING ABILITY BASED ON SUB-WAVELENGTH GRATINGS 1006
Wang Ying ; Huang Yongqing ; Guo Yanan ; Fang Wenjing ; Ren Xiaomin

SILICON PHOTONIC TE POLARIZER USING ADIABATIC WAVEGUIDE BENDS 1009
Bruna Paredes ; Humaira Zafar ; Marcus S. Dahlem ; Anatol Khilo

DESIGN OF A SI ARRAYED-WAVEGUIDE GRATING USING DISTRIBUTED BRAGG REFLECTORS 1012
Takahiro Inaba ; Hiroyuki Tsuda

FREQUENCY/ENERGY-TIME HYPER-ENTANGLED PHOTON PAIR GENERATION BASED ON A SILICON MICRO-RING RESONATOR 1015
Jing Suo ; Wei Zhang ; Shuai Dong ; Yidong Huang ; Jiangde Peng

ON THE CONTROL OF THE MICRORESONATOR OPTICAL FREQUENCY COMB IN TURING PATTERN REGIME VIA PARAMETRIC SEEDING 1018
Jinghao Wang ; Minming Zhang ; Meifeng Li ; Yuanwu Wang ; Deming Liu

GENERALIZED POLYNOMIAL CHAOS EXPANSION FOR PHOTONIC CIRCUITS OPTIMIZATION 1021
Daniele Melati ; Tsui-Wei Weng ; Luca Daniel ; Andrea Melloni

A GAUSSIAN BEAM WRITTEN SAMPLED FIBER GRATING FOR SUB-PS TIME DELAY LINES 1024
Weiqian Zhao ; Haifeng Qi ; Yunchuan Zhang ; Mingya Shen

PERIODICALLY POLED LINBO3 RIDGE WAVEGUIDE FOR HIGH-GAIN PHASE-SENSITIVE AMPLIFIER 1027
Tadashi Kishimoto ; Koji Inafune ; Yoh Ogawa ; Hitoshi Murai

PRELIMINARY RESEARCH OF MCF COUPLING TO OPTICAL MODE SWITCH CONFIGURATION 1030
Xiaoyang Cheng ; Luke Himbele ; Ryan Imansya ; Haisong Jiang ; Kiichi Hamamoto

NUMERICAL STUDY OF POWER COUPLING BETWEEN INDEX-ANTIGUIDED SLAB WAVEGUIDES 1033
Chang-I Hsieh ; Chih-Hsien Lai

SDN/NFV ORCHESTRATION OF MULTI-TECHNOLOGY AND MULTI-DOMAIN NETWORKS IN CLOUD/FOG ARCHITECTURES FOR 5G SERVICES 1036
Ricard Vilalta ; Arturo Mayoral ; Ramon Casellas ; Ricardo Martínez ; Raul Muñoz

SOFTWARE-DEFINED OPTICAL TRANSMISSION AND NETWORKING WITH FUNCTIONAL SERVICE DESIGN 1039
Xiaoyuan Cao ; Noboru Yoshikane ; Koki Takeshima ; Ion Popescu ; Takehiro Tsuritani ; Itsuro Morita

MULTI-STRATUM OPTIMIZATION WITH ROUTING RADIO AND SPECTRUM ALLOCATION FOR CLOUD RADIO OVER FIBER NETWORKS 1042
Hui Yang ; Wei Bai ; Ao Yu ; Jie Zhang

HIGHLY EFFICIENT ADAPTIVE BANDWIDTH ALLOCATION ALGORITHM FOR ELASTIC LAMBDA AGGREGATION NETWORK 1045
Hiroyuki Saito ; Naoki Minato ; Shuko Kobayashi ; Hideaki Tamai

RESILIENT VIRTUAL OPTICAL NETWORK PROVISIONING OVER SOFTWARE-DEFINED OPTICAL NETWORKS 1048
Xi Wang ; Inwoong Kim ; Qiong Zhang ; Paparao Palacharla ; Tadashi Ikeuchi

NFV AND SDN FOR NEXT TELECOM CLOUD AND CORE NETWORKING 1051
Michiaki Hayashi

MULTILAYER VIRTUAL INFRASTRUCTURE MAPPING IN IP OVER WDM NETWORKS 1054
Zilong Ye ; Philip N. Ji

WEIGHTED ATTACK-EVASION ROUTING AND WAVELENGTH ASSIGNMENT ALGORITHM AGAINST HIGH POWER JAMMING IN OPTICAL NETWORKS 1057
Liangkai Huang ; Yongli Zhao ; Wei Wang ; Jie Zhang

COHERENT OPTICAL COMMUNICATION TECHNOLOGY 1060
Kazuro Kikuchi

ON THE ORBITAL ANGULAR MOMENTUM (OAM) OF LIGHT IN FIBER 1061
Siddharth Ramachandran

GAIN CONTROL IN MULTI-CORE ERBIUM/YTTERBIUM-DOPED FIBER AMPLIFIER WITH HYBRID
PUMPING 1062
Makoto Yamada ; Hirotaka Ono ; Tsukasa Hosokawa ; Kentaro Ichii

EFFECTIVE AREA MEASUREMENT OF TWO-MODE FIBER USING BIDIRECTIONAL OTDR
TECHNIQUE 1065
Masaharu Ohashi ; Shun Asuka ; Yuji Miyoshi ; Hirokazu Kubota

FIBER-OPTIC GUIDED-ACOUSTIC-WAVE BRILLOUIN SCATTERING OBSERVED WITH PUMP-PROBE
TECHNIQUE 1068
Neisei Hayashi ; Heeyoung Lee ; Yosuke Mizuno ; Kentaro Nakamura

DISTRIBUTED HYDROSTATIC PRESSURE MEASUREMENT BASED ON BRILLOUIN DYNAMIC
GRATING IN POLARIZATION MAINTAINING FIBERS 1071
Yong Hyun Kim ; Hong Kwon ; Jeongjun Kim ; Kwang Yong Song

A MULTICAVITY FIBER FABRY-PÉROT INTERFEROMETER FOR SENSING MULTIPLE PARAMETERS 1074
Wei-Kang Chang ; Meng-Shan Wu ; Chung-Hao Tseng ; Pin Han ; Cheng-Ling Lee

HIGH SENSITIVITY FIBER OPTIC CURRENT SENSOR BASED ON RECIRCULATING FIBER LOOP 1077
Yemeng Tao ; Jiangbing Du ; Lin Ma ; Shuai ; Yinping Liu ; Wenjia Zhang ; Zuyuan He

FABRICATION AND CHARACTERIZATION OF LANTHANUM BOROALUMINOSILICATE GLASS FIBER
FOR MAGNETO-OPTICAL DEVICE APPLICATIONS... 1080
Kadathala Linganna ; Seongmin Ju ; Bok Hyeon Kim ; Won-Taek Han

HIGH SENSITIVITY CURVATURE SENSOR BASED ON MODAL INTERFEROMETER FOR VIBRATION
DETECTION 1083
Li Liu ; Ping Lu ; Shun Wang ; Hao Liao ; Wenjun Ni ; Xin Fu ; Xinyue Jiang ; Deming Liu

ULTRA-HIGH-RESOLUTION OTDR BASED ON LINEAR OPTICAL SAMPLING 1086
Shuai Wang ; Xinyu Fan ; Bin Wang ; Guangyao Yang ; Qingwen Liu ; Zuyuan He

POLARIZATION FADING ELIMINATION IN PHASE-EXTRACTED OTDR FOR DISTRIBUTED FIBER-
OPTIC VIBRATION SENSING... 1089
Guangyao Yang ; Xinyu Fan ; Bin Wang ; Qingwen Liu ; Zuyuan He

ORTHOGONALLY-POLARIZED PULSE PAIR BOTDA SENSOR WITH THREE-TONE PROBE.................... 1092
Yiming Tao ; Xiaobin Hong ; Zhisheng Yang ; Wenqiao Lin ; Jian Wu

PROOF OF CONCEPT FOR BRILLOUIN OPTICAL CORRELATION-DOMAIN REFLECTOMETRY
ASSISTED BY SPECTRAL SLOPE 1095
Heeyoung Lee ; Neisei Hayashi ; Yosuke Mizuno ; Kentaro Nakamura

SIMPLIFIED OPTICAL CORRELATION-DOMAIN REFLECTOMETRY WITHOUT REFERENCE PATH 1098
Makoto Shizuka ; Neisei Hayashi ; Yosuke Mizuno ; Kentaro Nakamura

PHASE NOISE MITIGATION FOR LONG-RANGE OFDR USING ULTRAFAST FREQUENCY SWEEP 1101
Bin Wang ; Xinyu Fan ; Shuai Wang ; Guangyao Yang ; Qingwen Liu ; Zuyuan He

HIGH-SPEED GERMANIUM-BASED WAVEGUIDE ELECTRO-ABSORPTION MODULATOR.................... 1104
*P. De Heyn ; S. A. Srinivasan ; P. Verheyen ; R. Loo ; I. De Wolf ; S. Balakrishnan ; G. Lepage ; D. Van Thourhout ; M.
Pantouvaki ; P. Absil ; J. Van Campenhout*

LOW-VOLTAGE CARRIER-DEPLETION SILICON MACH-ZEHNDER MODULATOR AT HIGH
TEMPERATURES WITHOUT THERMO-ELECTRIC COOLING 1107
*Norihiro Ishikura ; Kazuhiro Goi ; Hiroki Ishihara ; Shinichi Sakamoto ; Kensuke Ogawa ; Tsung-Yang Liow ; Xiaoguang Tu ;
Guo-Qiang Lo ; Dim-Lee Kwong*

HIGH-EFFICIENCY SILICON OPTICAL MODULATOR USING A SIN-STRIP LOADED WAVEGUIDE ON
THE PHOTONIC SOI PLATFORM 1110
Guangwei Cong ; Yuriko Maegami ; Morifumi Ohno ; Makoto Okano ; Koji Yamada

1.3-µM DFB LASER µ-PLATFORM; LIGHT SOURCE SUITABLE FOR SILICON PHOTONICS PLATFORM.................... 1113
*Takanori Suzuki ; K. R. Tamura ; Koichiro Adachi ; Aki Takei ; Akira Nakanishi ; Kazuhiko Naoe ; Kouji Nakahara ; Shigehisa
Tanaka*

HYBRID SILICON-BASED TUNABLE LASER WITH INTEGRATED REFLECTIVITY-TUNABLE MIRROR.................... 1116
G. de Valicourt ; C. Gui ; A. Melikyan ; P. Dong ; C-M. Chang ; A. Maho ; R. Brenot ; Y. K. Chen

MEMBRANE DISTRIBUTED-REFLECTOR LASERS 1119
Shigehisa Arai ; Nobuhiko Nishiyama ; Tomohiro Amemiya ; Takuo Hiratani ; Daisuke Inoue

ULTRA LOW POWER CONSUMPTION OPERATION OF SOA ASSISTED EXTENDED REACH EADFB
LASER (AXEL) 1122
Wataru Kobayashi ; Naoki Fujiwara ; Takahiko Shindo ; Shigeru Kanazawa ; Koichi Hasebe ; Hiroyuki Ishii ; Mikitaka Itoh

UNCOOLED 25.78 GB/S TRANSMISSION OVER 10 KM USING A 1.3 µM DIRECTLY MODULATED DFB
LASER IN A TO-CAN PACKAGE 1125
Seiki Nakamura ; Mizuki Shirao ; Masamichi Nogami

STABILIZATION OF 14XX-NM HIGH POWER SEMICONDUCTOR LASER BY SINGLE FIBER BRAGG
GRATING CONFIGURATION 1128
Hideaki Hasegawa ; Taketsugu Sawamura ; Junji Yoshida ; Noriyuki Yokouchi

FAST WAVELENGTH SWITCHING OF DFB LD.. 1131
Yuto Ueno ; Keita Mochizuki ; Kiyotomo Hasegawa ; Masamichi Nogami ; Hiroshi Aruga

OPTICAL CHIRP AND AMPLITUDE PROCESSING USING EAM INTEGRATION 1134
Yi-Jen Chiu ; Shin-Wei Shen ; Jui-Pin Wu ; Kuo-Chun Chang ; Chia-Chien Wei

CONCURRENT DWDM TRANSMISSION WITH RING MODULATORS DRIVEN BY A COMB LASER WITH 50GHZ CHANNEL SPACING 1137
M. Ashkan Seyedi ; Chin-Hui Chen ; Marco Fiorentino ; Daniil Livshits ; Alexey Gubenko ; Sergey Mikhrin ; Vladimir Mikhrin ; Raymond G. Beausoleil

MEASUREMENT OF VECTORIAL RESPONSE OF IQ MODULATOR USING OPTICAL INTERFERENCE 1140
Yuya Yamaguchi ; Kazuki Seki ; Naoki Takahashi ; Atsushi Kanno ; Tetsuya Kawanishi ; Masayuki Izutsu ; Hirochika Nakajima

INTEGRATED STOKES VECTOR ANALYZER ON INP 1143
Samir Ghosh ; Yuto Kawabata ; Takuo Tanemura ; Yoshiaki Nakano

PROPOSAL OF COMPACT TE/TM POLARIZATION SWITCH BASED ON MICRORING RESONATOR 1146
Keita Suzuki ; Tomoki Hirayama ; Taro Arakawa

ADVANCES IN SECOND ORDER NONLINEAR EFFECT IN SILICON 1149
P. Damas ; X. Le Roux ; M. Berciano ; G. Marcaud ; C. Alonso-Ramos ; D. Benedikovic ; E. Cassan ; D. Marris-Morini ; L. Vivien

MULTIPLE OPTICAL CARRIER GENERATION USING MULTIPLE QPM DEVICE 1152
Kazuki Nakamura ; Hin Channa ; Masaki Asobe ; Takeshi Umeki ; Hirokazu Takenouchi

FABRICATION OF HIGH OPTICAL QUALITY FACTOR FREE-STANDING AS2S3 MICRODISK RESONATORS ON A SILICON CHIP 1155
Mingxiao Zhao ; Mingming Zhao ; Xiaoshun Jiang ; Yuan Chen ; Jiyang Ma ; Min Xiao

WAVELENGTH MODULATION SPECTROSCOPY OF FORMALDEHYDE USING 3µM DFG LASER 1158
Ryohei Fujisawa ; Masaki Asobe ; Akira Katoh ; Shigeru Yamaguchi ; Akio Tokura ; Hirokazu Takenouchi

CHIRP-FREE SPECTRAL COMPRESSION OF PARABOLIC PULSES IN SILICON NITRIDE CHANNEL WAVEGUIDES 1161
Chao Mei ; Jinhui Yuan ; Kuiru Wang ; Xinzhu Sang ; Chongxiu Yu

SLOW LIGHT DEVICES IN SILICON PHOTONICS 1164
Toshihiko Baba

DESIGN OF DOUBLE-SLOTTED HIGH-Q PHOTONIC CRYSTAL NANOCAVITY FILLED WITH ELECTRO-OPTIC POLYMER 1167
Masahiro Nakadai ; Ryotaro Konoike ; Yoshinori Tanaka ; Takashi Asano ; Susumu Noda

EXPERIMENTAL REPORT FOR DISPERSION ENGINEERING OF SILICON SLOT PHOTONIC CRYSTAL WAVEGUIDES 1170
Samuel Serna ; Weiwei Zhang ; Xavier Le Roux ; Laurent Vivien ; Eric Cassan

SILICON POLARIZING BEAM SPLITTER BASED ON ASYMMETRIC SLOT WAVEGUIDE 1173
Jijun Feng ; Ryoichi Akimoto ; Heping Zeng

PARITY-TIME SYMMETRIC COUPLED RESONATOR WAVEGUIDE WITH PHOTONIC CRYSTAL NANOCAVITIES 1176
Kenta Takata ; Masaya Notomi

NXN WAVELENGTH SELECTIVE SWITCHES 1179
Hisato Uetsuka ; Shu Namiki ; Keiichi Sasaki

WAVELENGTH SELECTIVE SWITCH FOR MULTI-CORE FIBER BASED SPACE DIVISION MULTIPLEXED NETWORK WITH CORE-BY-CORE SWITCHING CAPABILITY 1182
Kenya Suzuki ; Mitsumasa Nakajima ; Keita Yamaguchi ; Goh Takashi ; Yuichiro Ikuma ; Kota Shikama ; Yuzo Ishii ; Mikitaka Itoh ; Mitsunori Fukutoku ; Toshikazu Hashimoto ; Yutaka Miyamoto

DEMONSTRATION OF 1,440×1,440 FAST OPTICAL CIRCUIT SWITCH FOR DATACENTER NETWORKING 1185
Koh Ueda ; Yojiro Mori ; Hiroshi Hasegawa ; Hiroyuki Matsuura ; Kiyo Ishii ; Haruhiko Kuwatsuka ; Shu Namiki ; Toshio Watanabe ; Ken-ichi Sato

32×32 SILICON PHOTONIC SWITCH 1188
Dritan Celo ; Dominic J. Goodwill ; Jia Jiang ; Patrick Dumais ; Chunshu Zhang ; Fei Zhao ; Xin Tu ; Chunhui Zhang ; Shengyong Yan ; Jifang He ; Ming Li ; Wanyuan Liu ; Yuming Wei ; Dongyu Geng ; Hamid Mehrvar ; Eric Bernier

SILICON PHOTONICS OPTICAL SWITCH BASED ON RING RESONATOR 1191
Antoine Descos ; M. Ashkan Seyedi ; Chin-Hui Chen ; Marco Fiorentino ; François Vincent ; David Penkler ; Bertrand Szelag ; Raymond G. Beausoleil

OPTICAL SWITCHING FUNCTIONS USING INTEGRATED SILICON-BASED DEVICES 1194
G. de Valicourt

INTEGRATED FAT-TREE OPTICAL SWITCH WITH CASCADED MZIS AND EAM-GATE ARRAY 1197
Yusuke Muranaka ; Toru Segawa ; Ryo Takahashi

INTEGRATED INP OPTICAL SWITCH MATRICES PERFORMANCE FOR PACKET DATA NETWORKS 1200
Ripalta Stabile

1310NM HIGH-CAPACITY WAVEBAND SWITCH NODE FOR FLAT OPTICAL DATA CENTER NETWORKS 1203
Wang Miao ; Huug de Waardt ; Nicola Calabretta

Author Index

Plenary Talk 1

Optical Transport to Open Transport in Carrier Networks

Vishnu Shukla
Verizon Labs, USA

Network traffic, dominated by data and video flows, is growing unabated. This is exacerbated by dynamic and shifting traffic patterns of mobile and cloud-based services. Today's static, manual optical transport networks are not optimized to meet customer needs for flexible, on-demand services. This situation challenges transport network operators and the supplier ecosystem to improve the network's ability to dynamically adapt to the needs of applications and services (residing mostly in data centers); i.e. make the network more programmable and increase network efficiency and agility.

SDN and virtualization promise to simplify transport network control by adding management flexibility and programmatic network element control to enable the rapid services development and provisioning. The centralized network-wide management and control of SDN can drive improvements in network efficiency and speed in terms of service acceleration. SDN may also reduce the total cost of ownership for optical switches by moving control and management planes from embedded processors to general-purpose common off-the-shelf (COTS) hardware and virtualized software.

Regular patterns of time of day and day of week usage are seen for specific classes of users and access networks. For example; enterprise users generate most traffic during weekdays and normal business hours. On the other hand, a consumer typically generates most traffic during nights and weekends. This means that time of day sharing between enterprise and consumer usage patterns is possible. Rearranging transport network topologies to interconnect networking and computing more economically based upon time of day and day of week to better serve these predictable phases of behaviors could significantly reduce overall networking costs. Furthermore resiliency of the network can be improved through coordinated multi-layer restoration techniques implemented through a centralized network control function that includes a view of both packet and optical layer topologies and the ability to steer the re-allocation of resources in response to network failures. This talk will present what is needed to make open transport SDN more widely deployable in service provider network.

Biography

Vishnu Shukla

Vishnu Shukla (M'86-SM'10) Received the M.Sc. and Ph.D. degrees from the University of Wales, U.K.

Vishnu Shukla is a Principal Technologist at Verizon. He has more than 30 years experience in optical network design and development, technology transfer and overall technology planning. His current responsibilities include development of the next generation network architecture, analyzing the current state-of-the-art technology trends and developments and evaluation of its strategic impact on Verizon. He founded and leads Verizon Interoperability Forum which facilitates early introduction of strategic technologies and services in the network. Vishnu contributes to Verizon's participation in multi-disciplined standards concerning leading edge technologies. He is the President and on the Board of Directors at the Optical Networking Forum and active member of the ONF.

Dr. Shukla is a Senior Member of IEEE and member of Optical Society of America (OSA). He has been the OFC Technical Program Co-Chair (2011-2012) and General Chair (2013-2014).

Plenary Talk 2

Silicon Photonic Integrated Circuits for Coherent Communications

Chris Doerr
Acacia Communications, 1301 Route 36, Hazlet, NJ 07730, USA
chris.doerr@acacia-inc.com

Coherent communications is transmitting an optical signal with an advanced format down an optical fiber and receiving it with an optical coherent receiver. An optical coherent receiver interferes the incoming signal with a local continuous-wave laser in a dual-polarization, dual-quadrature 90° hybrid with balanced photodiode pairs. The primary advantages of coherent communications are high spectral efficiency, high sensitivity, and digital compensation of transmission impairments.

Traditionally, coherent communications was limited to undersea and long-haul applications because it had a high price, large footprint, and high power consumption. Indeed, old 100-Gb/s coherent transceivers occupied an entire line card and consumed more than 100 W.

However, today, coherent transceivers are in pluggable modules the size of a smart phone and consume less than 30W. They are low cost enough to use in metro and inter-data-center applications. The digital signal processor (DSP) power has been reduced mainly by going to smaller transistors. The cost and footprint has been reduced mainly by integrating multiple optical and electronic functions into chips.

The most popular material platforms for opto-electronic integration are indium phosphide (InP) and silicon. InP has a direct bandgap and consequently produces efficient lasers and modulators. However, InP waveguides are formed from a quaternary epitaxy and thus have a relatively low component yield. Silicon, on the other hand, has an indirect bandgap, but the waveguides are single element and come from the original boule and thus have a very high yield. Finally, silicon photonics is ideal for co-packaging with silicon electronics, resulting in high-fidelity high-speed RF connections between the optics and electronics. This hot, non-hermetic environment precludes an integrated laser anyways. Thus silicon photonics is nearly ideal for coherent transceivers.

We will discuss the current state and future directions of silicon photonics in coherent transceivers. We will show that there is a still a great deal of size and cost reduction that can be done, ultimately resulting in short-reach coherent transceivers based on silicon photonics.

Biography

Chris Doerr

Christopher R. Doerr earned a B.S. in aeronautical engineering and a B.S., M.S., and Ph.D. in electrical engineering from the Massachusetts Institute of Technology. He was a pilot in the U.S. Air Force. Since joining Bell Labs in 1995, Doerr's research has focused on integrated devices for optical communication. He received the OSA Engineering Excellence Award in 2002. He is a Fellow of IEEE and OSA. He was Editor-in-Chief of IEEE Photonics Technology Letters from 2006-2008. He was an Associate Editor for the Journal of Lightwave Technology from 2008-2011. He was awarded the IEEE William Streifer Scientific Achievement Award in 2009 and Microoptics Conference Award in 2013. He became a Bell Labs Fellow in 2011. He joined Acacia Communications in 2011 as Associate Vice President of Integrated Photonics.

Plenary Talk 3

Energy limitations in data transmission and switching

Rod Tucker
University of Melbourne

The expanding volume of data on the global Internet is causing continuous growth in energy consumption, both in the network and in the attached data centres. Data centre operators have been fighting rising energy consumption for some time. And more recently, in response to rising electrical power bills, network operators are also beginning to think seriously about how they can improve the energy efficiency of their networks. Fortunately, there have been continuous improvements in the energy efficiency of commercial communications and switching equipment. But there is no end in sight to the growth of the so-called data tsunami, and improvements in equipment energy efficiency are barely keeping up with what is needed to constrain growth of the total energy consumption of the global network.

Where is this leading us? And what are the future prospects for continued improvements in energy efficiency? In this talk, I will identify some of the key energy bottlenecks in the network. I will consider opportunities for improved performance and explore some of the technological and theoretical limitations on the lower bound of energy consumption in transmission and switching. I will compare energy consumption of state-of-the-art commercial devices and systems with these lower-bound limitations and address the question of how the gap between theory and practice can be closed.

Biography

Rod Tucker

Rod Tucker is an Emeritus Professor at the University of Melbourne. He is a Fellow of the IEEE, the Optical Society of America, the Australian Academy of Science, and the Australian Academy of Technological Sciences and Engineering. He has previously held positions at the Plessey Company, AT&T Bell Laboratories, Hewlett Packard Laboratories, and Agilent Technologies. He has served on a government-appointed panel of experts that advised on the establishment of a national broadband fiber-based access network in Australia. Rod was Founding Director of the University of Melbourne's Centres for Ultra-Broadband Information Networks (CUBIN) and Energy-Efficient Telecommunications (CEET).

Plenary Talk 4

Optical Packet Switching : Myth, Fact, and Promise

Ken-ichi Kitayama
The Graduate School for the Creation of New Photonics Industries, Hamamatsu Japan
kitayama@gpi.ac.jp

To cope with the surge of data traffic due to cloud computing and other emerging web applications, warehouse-scale data centers (DCs) have to be sustainable in the energy-efficiency and throughput scalability. Current DC network uses commodity electronic switches, and there remains challenges how to adopt optical switching technologies. In this talk, for the starter preceding works relevant to optical switching technologies in DC networks are briefly reviewed. Next, we will focus on the optical packet switching (OPS) technology in DC networks and discuss whether the OPS can be a solution to the good sustainability and enhance the throughput. We introduce our work on co-deployment of OPS and optical circuit switching (OCS), conducted within a five-year-long NICT-funded R&D program until March 2016. It features the OPS along with an instantaneous optical path on-demand, so-called *Express Path* in the torus network topology, which enables the flow management in the presence of mice and elephant type of traffic. A key to the co-deployment of OPS and OCS is an intelligent contention resolution strategy, including prioritized buffering scheme and back-last deflection routing. The performance evaluation for the 4^6 (6-dimension and 4-radix) torus network topology having 4,096-node shows that the packet loss probability below 10^{-4} against traffic load of practical interest, up to ~0.7 can be achieved. Finally, we present a prototype of hybrid optoelectronic packet router (HOPR) having 100 Gbps (25 Gbps × 4-wavelength) links and its enabling device technologies based upon silicon photonics, including label processor, optical switch, and optoelectronic shared buffer. The HOPR prototype aims at the throughput of 1.28 Tbps with the total power consumption of 120 W, resulting in an excellent energy-efficiency of 90 mW/Gbps and low-latency of 100 ns. We hope that lingering myths on OPS will be resolved by the above-mentioned facts and promising use case.

Biography

Ken-ichi Kitayama

Dr. Ken-ichi Kitayama received the B.E., M.E., and Dr. Eng. degrees from Osaka University, in 1974, 1976, and 1981, respectively. In 1976 he joined the NTT Laboratories. In 1982-1983, he spent a year as a Research Fellow at the University of California, Berkeley. In 1995, he joined the Communications Research Laboratory (Presently, NICT), Tokyo. He retired from Osaka University at the end of March 2016, and became the Professor Emeritus. Since April 2016 he is with the Graduate School for the Creation of New Photonics Industries (GPI), Hamamatsu, Japan.

His research interests are optical packet switching, optical signal processing, radio-over-fiber (RoF) system, and the next-generation optical access systems such optical code division multiple access (OCDMA). He has published more than 310 journal papers, written a book, "Optical Code Division Multiple Access -A Practical Perspective-" (Cambridge University Press, 2014). He holds about 40 patents. He is currently serving as the Associate Editor of IEEE/OSA J. Optical Communications and Networking (JOCN). He is the *Life Fellow* of IEEE and the *Fellow* of IEICE, Japan.

Ultrahigh Speed and High Spectral Efficiency Transmission Using Optical Nyquist Pulses

Masataka Nakazawa

Research Institute of Electrical Communication, Tohoku University, 2-1-1 Katahira, Aoba-ku, Sendai, 980-8577 Japan
nakazawa@riec.tohoku.ac.jp

Abstract: *Recent progress is presented on ultrahigh-speed and high spectral-efficiency (SE) TDM transmission using non-coherent and coherent Nyquist pulses, including a single-channel 5.12-Tbit/s, 300-km DQPSK transmission and 3.84-Tbit/s, 150-km 64-QAM transmission with 6.4-bit/s/Hz SE.*
Keywords: *Nyquist pulse, digital coherent transmission, OTDM, QAM, spectral efficiency*

I. INTRODUCTION

In recent years, a single-channel transmission capacity exceeding 1 Tbit/s has become an important research target because of the need to cope with the rapid growth in Internet traffic [1]. Optical time division multiplexing (OTDM) is a simple and direct approach for increasing the baud rate beyond the speed and bandwidth limitation of electronics. In general, however, OTDM requires an ultrashort pulse width, and the inherently broad bandwidth of RZ optical pulses is disadvantageous in terms of spectral efficiency. To overcome this bottleneck, we recently proposed a novel OTDM transmission scheme using an optical Nyquist pulse [2]. The waveform of an optical Nyquist pulse is defined as the sinc-like impulse response of a Nyquist filter. In contrast to ordinary RZ pulses, the tail of a Nyquist pulse does not undergo exponential decay but approaches zero slowly accompanied by a periodic oscillation. Therefore we can employ the interleaving of the Nyquist pulse at every zero-crossing period of the sinc-like pulse without causing intersymbol interference (ISI) in spite of the pulse overlap. This results in a significant bandwidth reduction in OTDM transmission. Nyquist OTDM demultiplexing can be realized by using an optical sampling technique [3] or the orthogonal property of a sinc function [2, 4]. Because of these outstanding properties, optical Nyquist pulses have attracted a lot of attention as ideal pulses for high-speed and highly spectrally efficient OTDM transmission.

This paper presents the generation of an optical Nyquist pulse and ultrafast and spectrally efficient transmission using non-coherent and coherent Nyquist pulses.

II. PRINCIPLE OF NYQUIST PULSE AND ITS TDM TRANSMISSION

Figure 1 shows the basic configuration for Nyquist OTDM transmission. In the transmitter, we first generate an optical Nyquist pulse whose amplitude waveform and frequency spectrum are given by

$$r(t) = \frac{\sin(\pi t/T)}{\pi t/T} \frac{\cos(\alpha \pi t/T)}{1-(2\alpha t/T)^2}, \quad R(f) = \begin{cases} 1, & 0 \le |f| \le \dfrac{1-\alpha}{2T} \\ \dfrac{1}{2}\left\{1-\sin\left[\dfrac{\pi}{2\alpha}(2T|f|-1)\right]\right\}, & \dfrac{1-\alpha}{2T} \le |f| \le \dfrac{1+\alpha}{2T} \\ 0, & |f| \ge \dfrac{1+\alpha}{2T} \end{cases} \tag{1}$$

where T is the symbol period, and α ($0 \le \alpha \le 1$) is known as a roll-off factor. The pulse power has an oscillating tail but it becomes zero at $t = nT$ (n: integer), and therefore there is no ISI at any symbol interval after the OTDM despite a strong overlap with neighboring pulses. Optical Nyquist pulses can be generated from a mode-locked laser [3] or a coherent CW light source followed by sideband generation using a phase modulator [5], by manipulating the generated spectrum into a Nyquist profile with a programmable optical filter such as a liquid crystal spatial modulator [6]. We have also proposed a Nyquist laser that can directly emit an optical Nyquist pulse train with a high OSNR, where the programmable optical filter is installed inside the cavity [7].

In the receiver, unlike conventional OTDM demultiplexing schemes, data only at the ISI-free point has to be extracted from the overlapped data sequence with an ultranarrow sampling gate ①. Alternatively, coherent Nyquist pulses can be demultiplexed using a homodyne detection with a Nyquist local oscillator (LO) pulse due to the time-domain orthogonality of Nyquist pulses ② [11]. Photo-mixing of the data signal with a Nyquist LO pulse on a PD can easily extract a tributary from the overlapping data sequence with a high SNR, and the leakage from other channels can be completely suppressed.

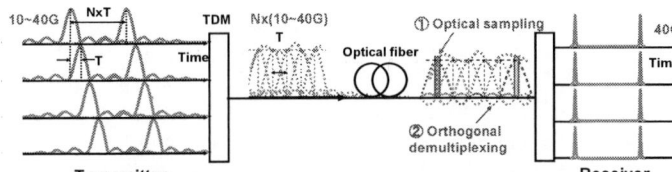

Fig. 1. Basic configuration for Nyquist OTDM transmission.

III. Single-Channel 5.12 Tbit/s (1.28 Tbaud), DQPSK Non-Coherent Nyquist TDM Transmission

We first carried out a single-channel 5.12-Tbit/s non-coherent Nyquist pulse transmission over 300 km using polarization-multiplexed DQPSK at 1.28 Tbaud. Figure 2(a) shows our experimental setup. We used a 40 GHz mode-locked fiber laser (MLFL) as a transmitter pulse source. The MLFL emits a 1.5 ps Gaussian pulse at a center wavelength of 1541 nm. The Nyquist pulse was generated from the Gaussian pulse by spectral broadening with a highly nonlinear dispersion-flattened fiber (HNL-DFF) and pulse shaping with a programmable optical filter. Figure 2(b) shows the Nyquist pulse generated for a 1.28 Tbaud transmission. The roll-off factor α was set at 0.5, and the zero-crossing period was 780 fs, which corresponds to 1.28 Tbaud after OTDM. The Nyquist pulses were modulated with DQPSK at 40 Gbaud and multiplexed to 1.28 Tbaud with an OTDM bit interleaver. After polarization multiplexing, 5.12 Tbit/s data were finally obtained. The 5.12 Tbit/s Nyquist OTDM signal was launched into a 300 km transmission link. The transmission link consisted of 4×75 km spans, where each span was composed of a 50 km SMF and a 25 km inverse dispersion fiber (IDF) in which the CD and dispersion slope were both compensated. The loss of each span was compensated for with an EDFA and a Raman amplifier. The launched power was optimally set at 6 dBm. On the receiver side, after polarization demultiplexing, the 1.28 Tbaud OTDM signal was demultiplexed to 40 Gbaud using an all-optical nonlinear optical loop mirror (NOLM) switch. Here we set the switching gate width at 500 fs, which was optimized by the trade-off between the larger distortion from adjacent channels due to a broader gate width and the OSNR decrease after demultiplexing due to a narrower gate width. After OTDM demultiplexing, a 40 Gbaud data signal was separated from the sampling pulse spectrum with a 20 nm optical filter, demodulated with a one-bit delay interferometer, and received with a 40 GHz balanced photo diode (PD).

Figure 2(c) shows the BER characteristics for a 5.12 Tbit/s-300 km transmission. Error-free operation was achieved for a back-to-back transmission at a received power of −15 dBm, but the BER curve had an error floor at 2.5×10^{-5} after a 300 km transmission in a single polarization. This impairment was caused by OSNR degradation, where the OSNR was reduced from 40 dB to 31.5 dB after the 300 km transmission. In the polarization-multiplexed transmission, the BER was further degraded by about two orders of magnitude from the single polarization, which was attributed to the second-order PMD-induced crosstalk. The BER was 1.5×10^{-3} at a received power of −15 dBm, which was below the forward-error correction (FEC) threshold of 2×10^{-3} (7 % overhead). As a result, a single-channel 5.12 Tbit/s (4.8 Tbit/s net data rate) signal was successfully transmitted over 300 km using non-coherent Nyquist pulses in a 1.92 THz bandwidth, which corresponds to a spectral efficiency of 2.5 bit/s/Hz. This is the longest transmission distance yet achieved at 5.12 Tbit/s, in which the high PMD tolerance of Nyquist pulses contributes greatly to the large increase in the transmission distance.

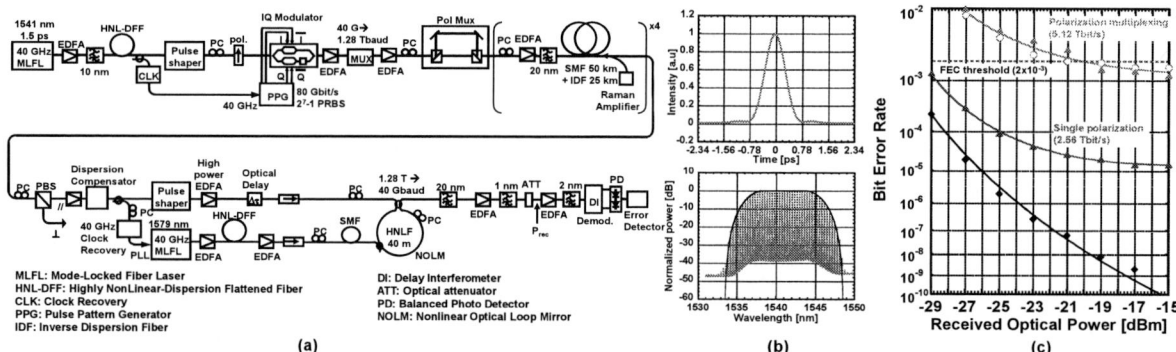

Fig. 2. 5.12 Tbit/s/ch-300 km non-coherent Nyquist pulse transmission. (a) Experimental setup, (b) waveform and spectrum, (c) BER characteristics.

IV. Single-Channel 3.84 Tbit/s (320 Gbaud), 64 QAM Coherent Nyquist TDM Transmission

Next we describe a 3.84 Tbit/s/ch, PDM-64 QAM orthogonal TDM transmission experiment using coherent Nyquist pulses. Figure 3(a) shows the experimental setup. As a coherent pulse source, we used a 1.55 μm, 0.95 ps, 10 GHz MLFL, in which the optical frequency was stabilized to an HCN linear absorption line. The laser output power was 50 mW and the OSNR was 48 dB. Since a coherent pulse was directly generated by the laser, the OSNR of the obtained Nyquist pulse (roll-off factor $\alpha = 0.5$) could be kept as high as 48 dB, which is 10 dB higher than that obtained with the combination of a CW laser and an optical comb generator [5]. The Nyquist pulse was then data modulated with a 10 Gbaud, 64 QAM signal using an IQ modulator, where we pre-compensated for the nonlinear-phase rotation induced by self-phase modulation (SPM) during transmission. The Nyquist pulse was then time-domain multiplexed to 320 Gbaud using a delay-line bit interleaver, and after polarization multiplexing and combining the pilot tone, the 3.84 Tbit/s PDM-64 QAM signal was launched into a 150 km transmission link. The optical bandwidth including the pilot tone was as narrow as 500 GHz. After the transmission, the 320 Gbaud Nyquist OTDM signal was coupled to a 90° optical hybrid circuit with a 10 GHz Nyquist LO pulse train at the receiver. The Nyquist LO pulse was generated from a CW-LO

10

(frequency-tunable CW fiber laser) and a comb generator. Here, the CW-LO was phase-locked to the data signal by using the pilot tone via an optical phase-locked loop (OPLL), and the optical comb generator was driven by a 10 GHz clock that was extracted from the transmitted OTDM data. The 320 Gbaud Nyquist OTDM data were then synchronously homodyne-detected with the Nyquist LO pulse. The demultiplexed 10 Gbaud I and Q data can be obtained after balanced detection based on the time-domain orthogonality of the Nyquist pulse [2, 4]. After A/D conversion with an 80 Gsample/s (66 GHz) digital oscilloscope, the 10 Gbaud data were demodulated offline in a digital signal processor (DSP) with a 10 GHz demodulation bandwidth.

Figure 3(b) show the constellations of a demultiplexed 10 Gbaud, 64 QAM signal at a received power of 0 dBm for back-to-back and 150 km transmissions, and the BER characteristics are shown in Fig. 3(c). A BER below the forward error correction (FEC) threshold 2×10^{-2} was achieved after 150 km with a power penalty of 3.8 dB from the back-to-back transmission, reaching the FEC threshold with a 7% overhead. The present result potentially yields a net spectral efficiency of as high as 6.4 bit/s/Hz even when taking account of the 20% FEC overhead.

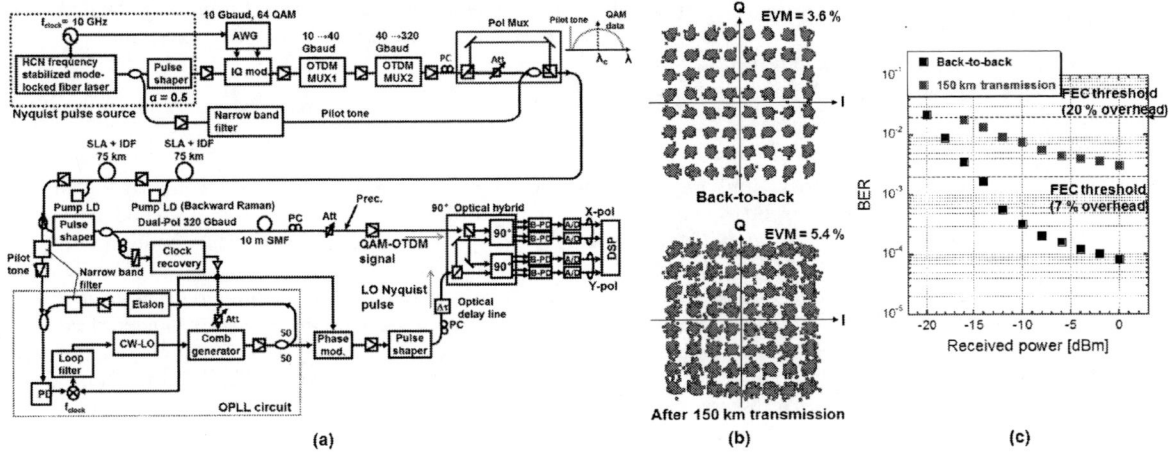

Fig. 3. Single-channel 3.84 Tbit/s, 64 QAM coherent Nyquist pulse transmission. (a) Experimental setup, (b) constellations before and after transmission, (c) BER characteristics.

V. CONCLUSIONS

We described non-coherent and coherent Nyquist pulse transmissions, which easily exceeded 1 Tbit/s/ch. The simultaneous achievement of an ultrahigh channel capacity and a high spectral efficiency was realized with coherent Nyquist pulses by taking advantage of their time-domain orthogonality. This scheme is potentially scalable to a higher symbol rate such as 1 Tbaud and a higher multiplicity of, for example, 256 QAM, and thus it is expected to offer the possibility of realizing ultrahigh-speed and high spectral efficiency transmission in, for example, a single-channel 10 Tbit/s and 10 bit/s/Hz regime.

REFERENCES

[1] G. Raybon, J. Cho, A. Adamiecki, P. Winzer, A. Konczykowska, F. Jorge, J. Dupuy, M. Riet, B. Duval, K. Kim, S. Randel, D. Pilori, B. Guan, N. K. Fontaine, and E. Burrows, "Single carrier high symbol rate transmitterfor data rates up to 1.0 Tb/s," OFC 2016 Th3A.2, March 2016.
[2] M. Nakazawa, T. Hirooka, P. Ruan, and P. Guan, "Ultrahigh-speed "orthogonal" TDM transmission with an optical Nyquist pulse train," Opt. Express vol. 20, pp. 1129-1140, January, 2012.
[3] K. Harako, D. Suzuki, T. Hirooka, and M. Nakazawa, "2.56 Tbit/s/ch (640 Gbaud) polarization-multiplexed DQPSK non-coherent Nyquist pulse transmission over 525 km," Opt. Express, vol. 23, pp. 30801-30806, November, 2015.
[4] K. Harako, D. O. Otuya, K. Kasai, T. Hirooka, and M. Nakazawa, "High-performance TDM demultiplexing of coherent Nyquist pulses using time-domain orthogonality," Opt. Express, vol. 22, pp. 29456-29464, December, 2014.
[5] D. O. Otuya, K. Kasai, T. Hirooka, and M. Nakazawa, "Single-channel 1.92 Tbit/s, 64 QAM coherent Nyquist orthogonal TDM transmission with a spectral efficiency of 10.6 bit/s/Hz," J. Lightwave Technol. vol. 34, pp. 768-775, January, 2016.
[6] G. Baxter, S. Frisken, D. Abakoumov, H. Zhou, I. Clarke, A. Bartos, and S. Poole, "Highly programmable wavelength selective switch based on liquid crystal on silicon switching elements" OFC/NFOEC 2006, OTuF2, March 2006.
[7] M. Nakazawa, M. Yoshida, and T. Hirooka, "The Nyquist laser," Optica vol. 1, pp. 15-22, July, 2014.
[8] K. Yoshida, K. Kasai, M. Yoshida, and M. Nakazawa, "HCN frequency-stabilized 10 GHz mode-locked fiber laser," ISUPT/EXAT 2015, P-34, July 2015.

Roll-off Factor Dependence of System Performance in 1.28 Tbit/s/ch-525 km Nyquist Pulse Transmission

Koudai Harako, Daiki Suzuki, Toshihiko Hirooka, and Masataka Nakazawa
Research Institute of Electrical Communication, Tohoku University, 2-1-1 Katahira, Aoba-ku, Sendai-shi, Miyagi-ken, 980-8577
Japan
harako@riec.tohoku.ac.jp

Abstract: We evaluate the roll-off factor dependence of a Nyquist pulse in a 1.28 Tbit/s-525 km transmission. The optimum value is 0.5, and for lower values an overlap with neighboring symbols results in larger nonlinear impairments.
Keywords: Pulse shaping, Multiplexing

I. INTRODUCTION

Increasing the serial line rate toward 1 Tbit/s and beyond has been the subject of intensive research following the development of 100 and 400 Gbit/s transport systems. The single-carrier bit rate has been approaching 1 Tbit/s through the use of ETDM and advanced multi-level modulation, such as 856 Gbit/s with polarization-division multiplexed (PDM) QPSK at 107 Gbaud [1] and 864 Gbit/s with PDM-64 QAM at 72 Gbaud [2]. However, achieving an ETDM bit rate beyond 1 Tbit/s remains a difficult task due to fundamental limitations as regards the sampling rate and bandwidth of digital-to-analog (DA) and analog-to-digital (AD) converters and their trade-off with the effective number of bits (ENOB). This makes it difficult to increase the baud rate and multiplicity (and thus the spectral efficiency). On the other hand, OTDM can easily overcome these limitations, and has already achieved a serial bit rate in the Tbit/s regime [3-5]. Conventionally, OTDM had difficulties as regards tolerance to chromatic dispersion (CD) and polarization-mode dispersion (PMD) and increasing the spectral efficiency because of the large bandwidth occupation associated with short RZ pulses. To overcome this bottleneck, we have developed an OTDM transmission technique using an optical Nyquist pulse (Nyquist OTDM) and demonstrated highly spectral-efficient Tbit/s transmission [6]. In this scheme, optical sinc (roll-off factor $\alpha = 0$) or quasi-sinc ($0 < \alpha \leq 1$) pulses are bit-interleaved to a higher symbol rate with an interval equal to the period of zero crossing in the tail, where intersymbol interference (ISI) is not induced even if the adjacent tributaries strongly overlap. This enables the signal bandwidth to be greatly reduced. Based on this approach, we have demonstrated a 1.92 Tbit/s/ch (160 Gbaud) transmission of a PDM-64 QAM signal within a bandwidth of only 170 GHz, which potentially yields a spectral efficiency of 10.6 bit/s/Hz [7].

The transmission of such an ultrahigh-speed signal over long distances has also been a major challenge. RZ pulses are generally vulnerable to CD and PMD, but their tolerance can be greatly improved by using Nyquist pulses thanks to their narrow bandwidth. A Nyquist pulse with lower α values has a narrower spectral width and is particularly advantageous in terms of spectral efficiency and CD/PMD tolerance, but other impairments such as nonlinearity and dependence on the roll-off factor remains an issue for long-distance transmission. In this paper, we compare the transmission performance for different roll-off factors in a 1.28 Tbit/s-525 km DQPSK transmission, and find the optimum α value while taking account of linear and nonlinear impairments both experimentally and analytically.

II. EXPERIMENTAL SETUP

Figure 1 shows the experimental setup for a 1.28 Tbit/s (640 Gbaud) DQPSK transmission over 525 km using non-coherent Nyquist pulses. In the transmitter, a 40 GHz Nyquist pulse train was first generated from a mode-locked fiber laser (MLFL) emitting a 1.5 ps Gaussian pulse at 1541 nm, followed by spectral broadening in a highly nonlinear fiber (HNLF) and an LCoS programmable optical filter. A Nyquist pulse with a different α value was generated with a properly designed filter profile. The pulse waveforms and optical spectra of the generated Nyquist pulses with $\alpha = 0$, 0.5, and 1 are shown in Fig. 2. It can be seen that they accurately fit an ideal profile. To deal with the imperfect spectral flatness in the EDFA gain profile and other components, which is critical for maintaining the flat-top spectral shape of the Nyquist pulse, we applied pre-emphasis to the intensity profile at the filter so that an ideal Nyquist spectrum could be obtained during the 525 km transmission. The Nyquist pulse was then data modulated with a 40 Gbaud DQPSK signal using an IQ modulator, and then time-domain multiplexed to 640 Gbaud using a delay-line bit interleaver. The 1.28 Tbit/s DQPSK signal was then launched into a 525 km transmission link, in which the optimal launch power of 10 dBm was chosen for all α values. Each span of the link consists of a 50 km single-mode fiber (SMF) with an anomalous dispersion of +20 ps/nm/km and a 25 km inverse dispersion fiber (IDF) with a normal dispersion of -40 ps/nm/km, whose loss was compensated by an erbium-doped fiber amplifier (EDFA). First-order PMD was mitigated with

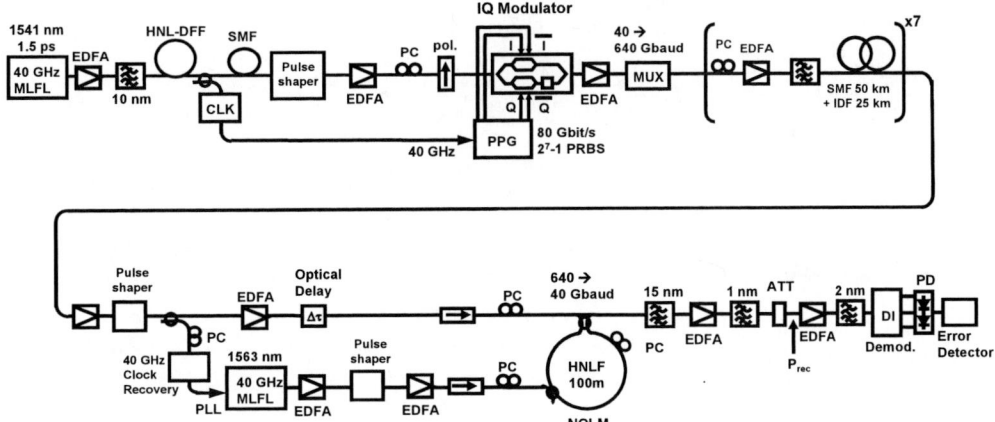

Fig. 1. Experimental setup for 1.28 Tbit/s (640 Gbaud) DQPSK transmission over 525 km using non-coherent Nyquist pulses. Abbreviations are defined in the text.

polarization controllers (PC) so that the signal was coupled to the principal state of polarization of the fiber link. We chose an in-line optical filter bandwidth of 10 nm for $\alpha = 0$ and 0.2 and 15 nm for $\alpha = 0.5$, 0.7, and 1

After the transmission, the Nyquist OTDM signal was passed through another programmable optical filter, whose amplitude and phase profiles compensated for the non-flat EDFA gain in the receiver and the residual CD and dispersion slope in the transmission link, respectively. The Nyquist OTDM signal was then demultiplexed to 40 Gbaud using a nonlinear optical loop mirror (NOLM) as an optical sampler, which was driven by a 40 GHz Nyquist control pulse at 1563 nm. The Nyquist control pulse was obtained from another 40 GHz mode-locked fiber laser operating at 1563 nm, which was PLL-operated with a 40 GHz clock extracted from the transmitted data, and shaped into a Nyquist pulse in the same way as the transmitter. The sampling gate width was optimized at 1.0 ps taking account of the trade-off with the SNR of the demultiplexed signal. Finally, the 40 Gbaud DQPSK signal was demodulated with a one-bit delay interferometer (DI) and received with a balanced photo-detector (PD).

III. EXPERIMENTAL AND ANALYTICAL RESULTS

We measured the bit error rate (BER) after a 525 km transmission for different α values. Figure 3 shows the relationship between α and BER. As can be seen, the BER has its minimum value for $\alpha = 0.5$ and increases for both lower and higher α values. To explain this dependence, we evaluated analytically the influence of nonlinearity in a Nyquist pulse transmission. We applied an analytical model for nonlinear interactions developed in [8] to a Nyquist TDM transmission. Following [8], the nonlinear impairments arising from an interaction between three pulses, $u_l u_m u_n^{*}$ (u_k: a pulse centered at $t = kT$, T: symbol interval), can be described as a noise with a power of

$$\sigma_{NL}^2 = (\eta_{SPM} + \eta_{XPM} + \eta_{FWM})P_{av}^3 \qquad (1)$$

which is proportional to the cube of the transmission power P_{av}^3 with the coefficient given by

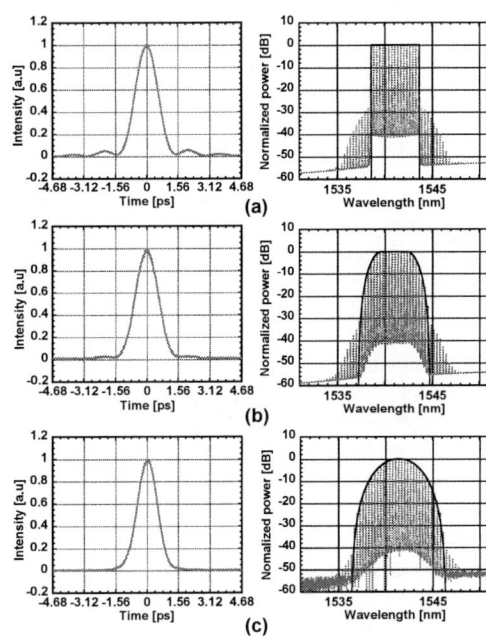

Fig. 2. The waveform (left) and spectrum (right) of a Nyquist pulse for a 640 Gbaud transmission with $\alpha = 0$ (a), 0.5 (b), and 1 (c). The waveform was measured with a cross correlator.

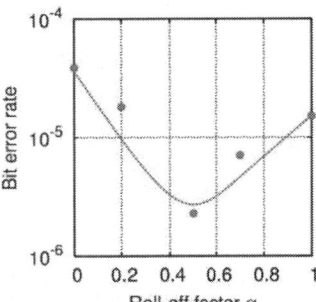

Fig. 3. BER vs. α in a 1.28 Tbit/s-525 km transmission.

$$\eta = \frac{\gamma^2 (P_p / P_{av})^3}{T} \int_{-\infty}^{\infty} \sum_{l,m=-N/2}^{N/2} \sum_{l',m'=-N/2}^{N/2} \left\langle a_l a_m a_{l+m}^* a_{l'}^* a_{m'}^* a_{l'+m'} \right\rangle Y_{l,m}(f) Y_{l',m'}^*(f) H(f) df \qquad (2a)$$

$$Y_{l,m}(f) = \int_0^z \frac{a^2(z')}{2\pi |\beta_2 z'|} \hat{s}\left(f - \frac{lT}{2\pi\beta_2 z'}\right) \hat{s}\left(f - \frac{mT}{2\pi\beta_2 z'}\right) \hat{s}\left(-f + \frac{(l+m)T}{2\pi\beta_2 z'}\right) \exp\left(i \frac{lmT^2}{\beta_2 z'}\right) dz' \qquad (2b)$$

Here, β_2 and γ are the dispersion and nonlinear coefficients, $a^2(z) = \exp(-\alpha z)$ takes account of the loss and gain profile, $\hat{s}(f)$ is the spectrum of a Nyquist pulse, P_p is the peak power of each pulse, N is the number of pulses under consideration, a_k is a random data pattern at $t = kT$, and $H(f)$ is the transfer function of optical filters. The nonlinear impairments that occurs at $t = 0$ due to SPM, XPM, and FWM can be obtained from the combination (l, m) of $l = m = 0$ ($|u_0|^2 u_0$), $l = 0$, $m \neq 0$ ($|u_l|^2 u_0$) or $l \neq 0$, $m = 0$ ($|u_m|^2 u_0$), and $l = m \neq 0$ ($u_l^2 u_{2l}^*$) or $l \neq m \neq 0$ ($u_l u_m u_{l+m}^*$), respectively.

We computed the XPM and FWM coefficients, η_{XPM} and η_{FWM} respectively, for various α values using Eqs. (1) and (2). The result is shown in Fig. 4(a). Both η_{XPM} and η_{FWM} decrease as α increases from 0, but they begin to increase for $\alpha > \sim 0.6$. This trend is in good agreement with Fig. 3. This behavior can be explained from the fact that the

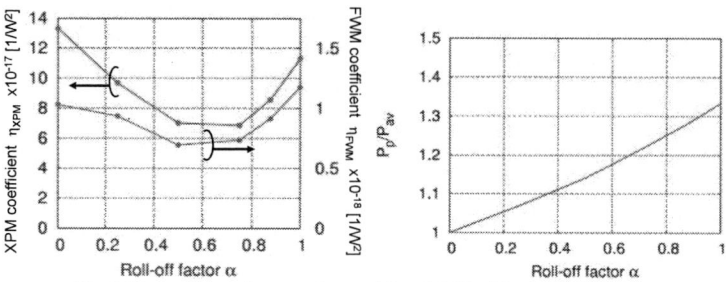

Fig. 4. Dependence of η_{XPM} and η_{FWM} (a) and P_p/P_{av} (b) on roll-off factor α.

overlap between neighboring pulses decreases as α increases, but at the same time, the peak power of a Nyquist pulse also increases under a fixed average power (i.e., the increase in P_p/P_{av} in Eq. (2a)). The P_p/P_{av} value as a function of α is plotted in Fig. 4(b). It can be seen that, when α increases to 1, P_p/P_{av} increases to 1.33 times that of $\alpha = 0$. Thus the optimum roll-off factor for the Nyquist pulse transmission is approximately 0.5.

IV. Conclusions

We compared the transmission performance for different roll-off factors of a Nyquist pulse in 1.28 Tbit/s-525 km transmission. The optimum roll-off factor was $\alpha \sim 0.5$. Smaller α values are preferable in terms of greater CD and PMD tolerance while a strong overlap between neighboring symbols results in large impairments due to XPM and FWM. On the other hand, larger α values lead to a higher peak power and again increase the nonlinearity.

Acknowledgment

This work was supported by Grant-in-Aid for Specially Promoted Research (26000009) by Japan Society for the Promotion of Science (JSPS).

References

[1] G. Raybon, A. Adamiecki, P. J. Winzer, M. Montoliu, S. Randel, A. Umbach, M. Margraf, J. Stephan, S. Draving, M. Grove, and K. Rush, "All-ETDM 107-Gbaud PDM-16QAM (856-Gb/s) transmitter and coherent receiver," ECOC 2013, PD2.D.3.

[2] S. Randel, D. Pilori, S. Corteselli, G. Raybon, A. Adamiecki, A. Gnauck, S. Chandrasekhar, P. J. Winzer, L. Altenhain, A. Bielik, and R. Schmid, "All-Electronic Flexibly Programmable 864-Gb/s Single-Carrier PDM-64-QAM," OFC 2014, Th5C.8.

[3] M. Nakazawa, T. Yamamoto, K. R. Tamura, "1.28 Tbit/s-70 km OTDM transmission using third- and fourth-order simultaneous dispersion compensation with a phase modulator," Electron. Lett. 36, 2027-2029 (2000).

[4] H. C. M. Mulvad, M. Galili, L. K. Oxenløwe, H. Hu, A. T. Clausen, J. B. Jensen, C. Peucheret, and P. Jeppesen, "Demonstration of 5.1 Tbit/s data capacity on a single-wavelength channel" Opt. Express 18, 1438-1443 (2010).

[5] T. Richter, E. Palushani, C. Schmidt-Langhorst, M. Nölle, R. Ludwig, and C. Schubert, "Single wavelength channel 10.2 Tb/s TDM-data capacity using 16-QAM and coherent detection," OFC 2011, PDPA9.

[6] M. Nakazawa, T. Hirooka, P. Ruan, and P. Guan, "Ultrahigh-speed "orthogonal" TDM transmission with an optical Nyquist pulse train," Opt. Express 20, 1129-1140 (2012).

[7] D. O. Otuya, K. Harako, K. Kasai, T. Hirooka, and M. Nakazawa, "Single-channel 1.92 Tbit/s, 64 QAM coherent orthogonal TDM transmission of 160 Gbaud optical Nyquist pulses with 10.6 bit/s/Hz spectral efficiency," OFC 2015, M3G.2.

[8] S. Kumar, S. N. Shahi, and D. Yang, "Analytical modeling of a single channel nonlinear fiber optic system based on QPSK," Opt. Express 20, 27740-27755 (2012).

A Novel Demultiplexing Scheme for Nyquist OTDM Signal Using a single IQ Modulator

Lei Yue, Deming Kong, Yan Li, Jiangchuan Pang, Miao Yu, and Jian Wu*

State Key Laboratory of Information Photonics and Optical Communications,
Beijing University of Posts and Telecommunications, Beijing 100876, China
jianwu@bupt.edu.cn

Abstract: Demultiplexing of a 160 Gbaud N-OTDM signal to 40 Gbaud is experimentally demonstrated using a commercially available IQ modulator. 6.8 dB required-OSNR improvement (@BER=1E-4) is observed compared with Gaussian sampling with the same pulse-width.

Keywords: Nyquist OTDM; demultiplexing; coherent sampling; IQ modulator.

I. INTRODUCTION

With the continuing growth of the Internet traffic, various effective techniques have been proposed to meet the high transmission capacity demand in optical backbone networks. Nyquist optical time-division multiplexing (N-OTDM) is attracting ever-growing research attention [1]. In addition to the substantial advantage of realizing inter-symbol-inference (ISI) free transmission with high spectral efficiency (SE), N-OTDM signal also shows strong insusceptibility to chromatic dispersion and polarization-mode dispersion even at an ultrahigh symbol rate [2]. However, an essential challenge is the demultiplexing process at the receiver side in which the ISI-free point must be extracted from the overlapped pulses in the time-domain. Frequency domain solution based on optical Fourier transformation [3] is proposed while the received N-OTDM signal need to be time-to-frequency converted. Time domain solutions utilizing time domain orthogonality of Nyquist pulses [4-6] are made to achieve a relative better detection performance. But usually these methods require complex setups for the generation of Nyquist pulses. Recently full-band coherent detection of N-OTDM signal with a continuous wave (CW) as local oscillation (LO) has been demonstrated [7]. However, such method requires a high-speed digital oscilloscope and limited by the electronic bandwidth, especially for high baudrate N-OTDM signals.

In this paper, a novel demultiplexing scheme for N-OTDM signal using a single commercially available IQ modulator (i.e. DPMZM, Dual Parallel Mach-Zehnder Modulator) has been demonstrated through a 160 Gbaud to 40 Gbaud demultiplexing experiment. And 6.8 dB required-OSNR improvement (@BER=1E-4) is observed compared with Gaussian sampling with the same pulse-width.

II. PRINCIPLE

The principle of generating Nyquist-shaped pulses using a single IQ modulator can be [8] illustrated through Fig. 1 (a). A CW light is directly launched into the modulator via a polarization-maintaining fiber. Only one RF input is used to drive one child MZM. By properly adjusting the amplitude of the RF signal and two DC bias voltages, an optical comb with a rectangular shape can be obtained at the output of IQ modulator. Insets (a) and (b) show the simulated output spectra from MZM1 and MZM2, respectively. The light wave from the two child MZMs are further interfered at the output of the modulator. The flatness of the comb can be adjusted by precisely turning the DC3 bias voltage. Insets (c) and (d) depict the optical spectrum and temporal waveform of generated pulses. The high-order sideband is by ~30 dB suppressed and the temporal profile of generated pulse is of Nyquist shape.

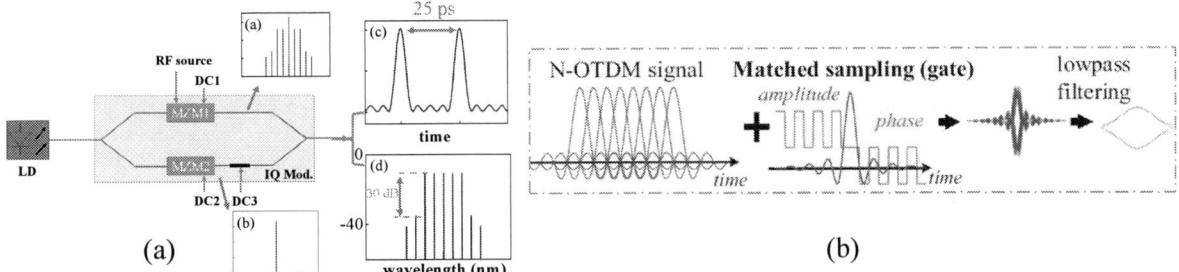

Fig.1 Principle of (a) Nyquist pulse generation using an IQ modulator and simulation results, (b) Coherent sampling of N-OTDM signal using Nyquist pulse (coherent matched sampling).

Fig. 1 (b) shows the principle of coherent matched sampling of N-OTDM signal using a Nyquist pulse. The N-OTDM signal is sampled by baserate Nyquist pulses. Large chirp will be imposed on the signal because of the occurrence of phase jumps between each lobe within the pulse duration of the Nyquist pulse, while the center of the sampled tributary remains un-chirped. When a low-pass filter is applied to the sampled signal, the target tributary is demultiplexed while other tributaries are mostly filtered out. In addition, the amplitude profile of the sampling pulses maximizes the power of the target tributary while minimizes the power of other tributaries. In another way, ISI from adjacent tributaries will be greatly suppressed. Therefore, this sampling method exhibits better demutiplexing performance for N-OTDM signal [4]. It should be noted that the low-pass filtering required for matched sampling is realized automatically by finite receiver bandwidth.

III. EXPERIMENT SETUP

Fig. 2 shows the experiment setup where the IQ modulator based pulse source is applied for the demultiplexing of a 160 Gbaud N-OTDM signal. A 1.8 ps Gaussian-shaped pulse source with a repetition rate of 40 GHz based on pulse carving and linear pulse compression [9] is used to support the generation of the N-OTDM signal. The generated optical pulse train is modulated by a 40 GHz NRZ 2^7-1 pseudo random bit sequence (PRBS). After being amplified and multiplexed to 160 Gbaud in time domain employing a passive delay-line based multiplexer (MUX×4), the signal is spectral manipulated into a Nyquist profile with a roll factor of approximately zero through the pulse shaper (WSS, Finisar Waveshaper 4000s). Then the generated N-OTDM signal is fed into an optical modulation analyzer (OMA, Agilent N4391A) for sampling and coherent detection.

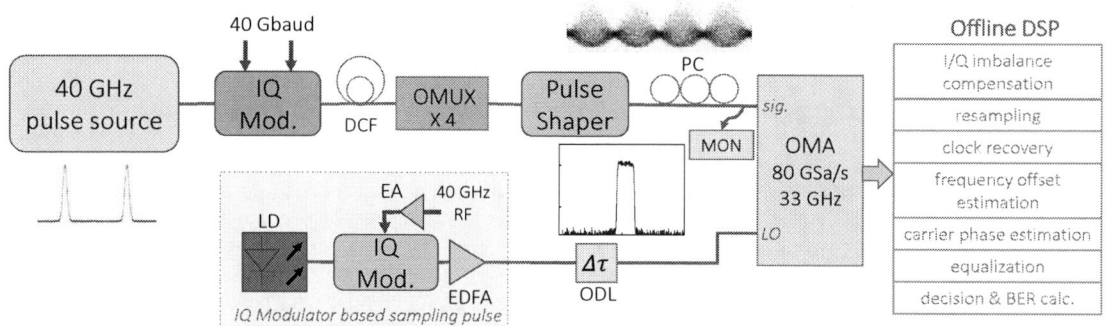

Fig. 2 Experimental setup. (LD: laser diode; EDFA: Erbium-doped fiber amplifier; DCF: dispersion compensating fiber; OMUX: optical multiplexer; PC: polarization controller; EA: electrical amplifier; ODL: optical delay line; OMA: optical modulation analyzer; MON: monitor.). Insets show the eye diagrams of 40 GHz pulse source and generated 160 Gbaud N-OTDM signal.

For the sampling pulses, the continuous wave (CW) light at 1550.276 nm with optical power of 15.5 dBm is directly launched into a commercially available IQ modulator (Fujitsu FTM7961EX) for the generation of the local Nyquist pulse train. The insertion loss of the IQ modulator is compensated by a low noise optical amplifier (EDFA). The generated sampling pulse train is then mixed with the N-OTDM signal as a LO source for coherent detection, where demultiplexing and homodyne detection are carried out simultaneously. After coherent detection and analog-to-digital conversion, the signal is offline processed with most common digital signal processing algorithms, including I/Q imbalance compensation, resampling, Gardner clock recovery, frequency offset estimation, carrier phase recovery and finally QPSK decision and BER calculation. 1 million samples (2 million bits) are collected for calculating each BER value.

IV. RESULTS AND DISCUSSIONS

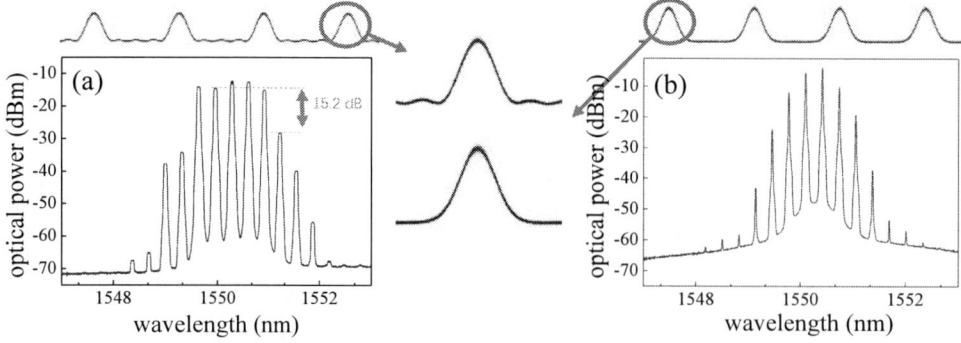

Fig. 3 Optical spectrums and temporal waveforms of (a) the sampling pulses, (b) the Gaussian pulses

Fig. 3 (a) shows optical spectrum and temporal waveform of the IQ modulator based sampling pulses. The high-order sideband of the generated frequency comb is suppressed by 15.2 dB. In order to further evaluate the detection performance of the proposed scheme, coherent sampling of N-OTDM signal with standard Gaussian pulses is also demonstrated for comparison. Since the detection performance is greatly influenced by the sampling pulse-width [10], here both trains with the same pulse-width is used for comparison. Fig. 3 (b) shows optical spectrum and temporal waveform of the generated Gaussian sampling pulses. The 40 GHz pulse source is spectral manipulated into Gaussian shape with full-width at half-maximum (FWHM) pulse-width of 4.50 ps (with a spectral bandwidth of 100 GHz).

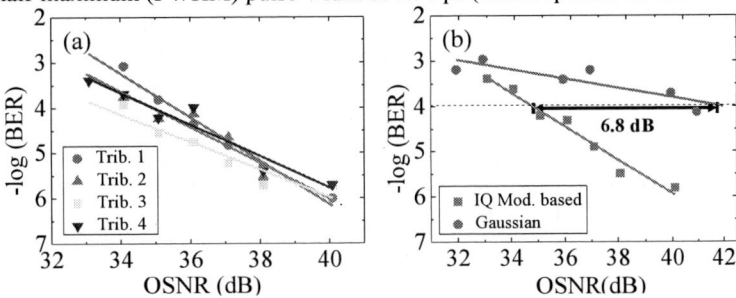

Fig.4 (a) Experimental BER results of demultiplexed 4 tributaries using IQ modulator based sampling pulse; (b) Average BER results of 4 tributaries using IQ modulator generated sampling pulses and Gaussian pulses, respectively.

Fig. 4 (a) depicts the measured bit-error ratio (BER) results of all 4 tributaries using the IQ modulator based Nyquist sampling pulse as a LO source for coherent detection. It can be seen that all tributaries are able to be sampled out and correctly detected. For comparison, coherent sampling using Gaussian pulses is also demonstrated. Fig. 4 (b) gives the average BER results of 4 tributaries with Gaussian sampling pulses and IQ modulator based Nyquist sampling pulses. It should be note that the pulse-width for both scenarios is the same (i.e. 4.50 ps). The performance difference indicates that the proposed scheme outperforms Gaussian sampling. And about 6.8 dB required-OSNR improvement (@BER=1E-4) is observed.

V. CONCLUSIONS

We have demonstrated a simple demultiplexing scheme for N-OTDM signal using only a single commercially available IQ modulator. Demultiplexing of 160 Gbaud N-OTDM signal down to 4×40 Gbaud tributaries is experimentally demonstrated. This scheme features a simple structure and enables high performance detection of N-OTDM signal. A 6.8 dB required-OSNR improvement (@BER=1E-4) is observed compared with Gaussian sampling with the same pulse-width.

ACKNOWLEDGMENT

This work was partly supported by 863 program 2013AA014202, 973 program 2014CB340100, NSFC program 61505011, 61475022, 61331008, 61205031, 61307055, 61335009, fund of state key laboratory of IPOC (BUPT), the fundamental research funds for the central universities, and the China Postdoctoral Science Foundation.

REFERENCES

[1] M. Nakazawa, et al., "Ultrahigh-speed orthogonal TDM transmission with an optical Nyquist pulse train," Opt. Express 20, 1129-1140 (2012).
[2] T. Hirooka, et al., "Linear and nonlinear propagation of optical Nyquist pulses in fibers," Opt. Express 20, 19836-19849 (2012).
[3] H. Hu, et al., "320 Gb/s Nyquist OTDM received by polarization-insensitive time-domain OFT," Opt. Express 22, 110-118 (2014).
[4] D. Kong, et al., "A Novel Detection Scheme for Nyquist Optical Time-division Multiplexed Signal with Coherent Matched Sampling," Proceedings of the Optical Fiber Communication Conference (OFC) 2015, W3C.4.
[5] T. Hirooka, et al., "Ultrafast Nyquist OTDM demultiplexing using optical Nyquist pulse sampling in an all-optical nonlinear switch," Opt. Express 23, 20858-20866 (2015).
[6] J. R. Stroud, et al., "All-optical demultiplexing of Nyquist OTDM using a Nyquist gate," Proceedings of the Conference on Lasers and Electro-Optics (CLEO) 2014, SW1J.4.
[7] J. Zhang, et al., "High Speed All Optical Nyquist Signal Generation and Full-band Coherent Detection," Sci. Reports 4, 6156 (2014).
[8] J. Wu, et al., "Investigation on Nyquist pulse generation using a single dual-parallel Mach-Zehnder modulator," Opt. Express 22, 20463-20472 (2014).
[9] Y. Ji, et al., "A Phase Stable Short Pulses Generator Using an EAM and Phase Modulators for Application in 160-GBaud DQPSK Systems," IEEE Photonics Technology Letters 24, 64-66 (2012).
[10] L. Yue, et al., "Investigation on Pulse-width and Roll-off Factor of Sampling Pulses in Coherent Matched Sampling of a Nyquist Optical Time-division Multiplexed Signal," Proceedings of the Asia Communications and Photonics Conference (ACP) 2015, AS3F.3.

1.76-Tbit/s Superchannel Generation Based on InP IQ Modulator and Optical Frequency Comb

K. Solis-Trapala, T. Inoue, H. Nguyen Tan, and S. Namiki

National Institute of Advanced Industrial Science and Technology (AIST)
1-1-1 Umezono, Tsukuba, Ibaraki 305-8568, Japan
k.solis-trapala@aist.go.jp

Abstract: We study five-subcarrier 44-Gbaud DP-Nyquist-16QAM superchannel generation using a commercial high-speed InP IQ modulator and optical frequency comb for the first time. The 1.76-Tbit/s superchannel with 7.04-bit/s/Hz spectral efficiency is successfully operated over 160-km SSMF.

Keywords: (photonic integrated technology, high speed modulator, coherent transmitter)

I. INTRODUCTION

The tremendous growth of data traffic is naturally extending the demand for high capacity links in the metro area to support the efficient delivery of the bandwidth-intensive services and applications. To facilitate such capacity increase, the widespread use of digital coherent technology is attractive as long as the key building blocks, i.e. the coherent transceivers, are cost-effective, compact, pluggable, and power efficient. Undoubtedly, photonic integration is an enabler technology [1] and a rapid progress has been made in the development of terabit/s class transmitter and receiver photonic integrated circuits [2]. A good example is the recently demonstrated wavelength multiplexed InP dual-polarization quadrature phase-shift keying (DP-QPSK) transmitter chip integrating as many as 1700 functions to deliver a more than 2 Tbit/s capacity in a 1 THz superchannel (40x57 Gbit/s DP-QPSK on 25 GHz grid) [3]. In parallel, the generation of higher-order quadrature amplitude modulation (QAM) employing InP IQ modulators (IQM) has also improved dramatically [4-6] and QAM as high as 64 has been generated employing digital signal pre-compensation [6]. This adds up to the compelling advantages, i.e. smaller footprint and lower driving voltage, offered by the semiconductor platform over its Lithium-Niobate (LN) counterpart. Moreover for additional cost, space and energy savings, optical comb (OC) technology may be considered as the WDM light source in coherent transmission; a comb based transmitter spanning over a broad wavelength range has recently been demonstrated [7]. Importantly, integrated solutions have also been proposed [8,9], including those based on InP platform [9,10]. In general, a transmitter architecture based on OC and IQMs is able to provide high-quality, equally- and tightly-spaced carriers and it has been proved its operability with high order modulation formats such as dual-polarization 16-ary QAM (DP-16QAM) generated however from the more traditional LN modulators [7]. Additionally, as it is well known, Nyquist pulse shaping can help to maximize the bandwidth utilization efficiency.

Till now, terabit/s class superchannel transmitters based on photonic integrated technology rely on multiple wavelengths from a laser array in combination with polarization multiplexing and phase modulation formats such as QPSK [2,3]. In view of the escalation in capacity and integration, the use of an optical comb instead is promising for a hardware efficient solution; however, its operation with InP IQM has not been reported yet. Also to reach efficiently terabit/s capacities, with the continuous progress in electronics and digital signal processing (DSP), an increased baud-rate and high-order QAM are expected to be exploited in a trade of capacity for reach. While an increased symbol rate generally increments the energy consumption of electronic devices, the number of subcarriers can be reduced for a fixed total capacity, which could save the energy as a whole. Hence, in this paper, we demonstrate for the first time the use of a commercial high-speed integrated InP IQ modulator and an optical frequency comb source based on cascaded intensity and phase modulators [11], to realize a high capacity transmission implemented through high baud-rate and high order QAM. Specifically, we study a superchannel comprising five digitally shaped Nyquist subcarriers at 44-Gbaud, modulated with DP-16QAM format. The 1.76-Tbit/s gross line-rate signal occupying only a 250 GHz total bandwidth, is compatible with the 50-GHz ITU grid for WDM systems and achieves a high spectral efficiency of 7.04 bit/s/Hz, which is promising towards the development of photonically integrated multi-Tbit/s coherent transmitters. The superchannel was successfully transmitted over 160 km of standard single-mode fiber (SSMF) with ample margin.

II. EXPERIMENT AND RESULTS

Figure 1(a) depicts the experimental setup to generate and transmit the 44-Gbaud DP-Nyquist-16QAM superchannel. It consists of four parts, namely 1) the comb generation, 2) the superchannel generation, 3) the transmission link and 4) the receiver. At first, a 25 GHz spaced optical frequency comb was generated based on electro-optic modulators [8, 11]. A ~100 kHz linewidth, tunable external-cavity laser (TLS) at 193.1 THz was modulated in a cascade of intensity

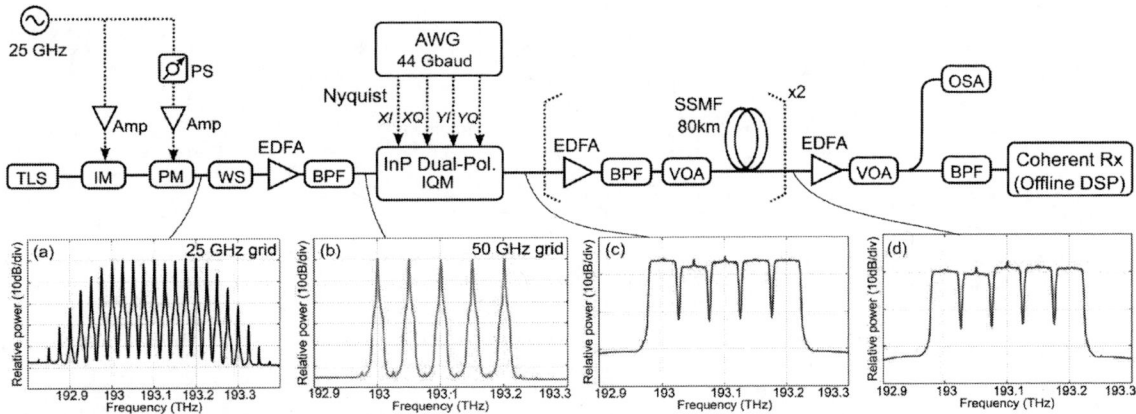

Fig. 1. Experimental demonstration of the generation and 160 km transmission of the 1.76 Tbit/s superchannel employing optical frequency comb and integrated InP IQM. A practical implementation would include some means of distribution (e.g. arrayed waveguide grating) of the several carriers into multiple IQMs and a coupling stage afterwards. Insets: (a) Comb spectrum (25 GHz spacing), (b) five carriers spanning from 193 THz to 193.2 THz (50 GHz spacing) to form the superchannel in which 44-Gbaud DP-16QAM modulation format and Nyquist spectral shaping (roll-off factor of 0.1) were used, as shown in (c) before transmission and (d) after 160 km transmission. Resolution bandwidth is 0.02nm. OSA: optical spectrum analyzer. All other abbreviations are defined in the text.

modulator (IM) and phase modulator (PM) driven by a common 25 GHz RF sinuosidal clock. A phase shifter (PS) in the RF path of the PM was carefully adjusted such that the input optical pulses experienced a linear phase shift induced by the leading edge of the sinusoidal RF clock. Afterwards, an LCOS-based wave-shaper (WS) selected five of the comb lines with a spacing of 50 GHz and introduced minimal channel equalization that reduced the comb flatness from originally ~1.5 dB to ~0.3 dB, as can be seen from the insets of Fig. 1(a-b) where the OC and five carriers' spectra are shown. Next, to realize the superchannel generation, the five carriers were amplified and modulated using a commercial integrated dual-polarization IQM based on InP (Teraxion P/N IQM-CBG-E21) with a 3-dB bandwidth of 35 GHz and Vπ of only 1.5 V. Note that thanks to the low Vπ value, external driver amplifiers where not necessary to generate a high quality signal, as it will be shown later. The four-channel modulating signal was derived from an arbitrary waveform generator (AWG, Keysight M8195A) having a 3-dB analog bandwidth of 20 GHz and a sampling rate of 64 GS/s. Offline digital signal processing was performed in advance, where a 16-QAM signal encoded from PRBS with a length of 2^{15} was generated at 44 Gbaud for each polarization, and Nyquist filtering with a roll-off factor of 0.1 was applied after resampling to 64GS/s, also frequency-domain equalization was performed to compensate for the waveform distortion by the analog response of the AWG. We note that both polarizations have the same data pattern but with a delay corresponding to 13107 bits (74.5 ns). The five-subcarrier superchannel was then transmitted over a link which consisted of two spans of 80 km SSMF that employed EDFA for span loss compensation. A 4.3-nm optical band pass filter (BPF) suppressed the amplified spontaneous emission noise from the EDFA and a variable optical attenuator (VOA) was used to set the launched power into the fiber. At the receiver, the signal was amplified and the subcarrier under consideration selected by a 100-GHz BPF, with no strict requirements owing to the built-in selectivity characteristic enabled by the coherent receiver detection. The signal was then demodulated through offline digital signal processing where dispersion compensation and bit-error-counting were performed.

In our laboratory demonstration a single IQM was used, in contrast to a practical implementation where some means of carrier distribution (e.g. arrayed waveguide grating) into multiple IQMs, followed by a coupling stage would be employed. Instead here, four of the carriers acted as interferers during the modulation process, also limiting the optical signal-to-noise ratio (OSNR); yet we will show that thanks to the Nyquist spectral shaping, the impact on the signal under test is minimal and the OSNR adequate for the realization of the superchannel transmission. We note that the subcarrier spacing of 50 GHz corresponds to only ~14% above the bandwidth required for an ideal 44-Gbaud Nyquist signal, enabling a high spectral efficiency of 7.04 bit/s/Hz. The 1.76 Tbit/s superchannel spectra occupying only 250 GHz bandwidth as can be seen from Fig. 1(c-d), is compatible with the 50-GHz ITU grid for WDM systems. Likewise, the well-known DC bias wavelength-dependence of the InP IQM can be appreciated; the spectra was measured when the DC bias was adjusted for the subcarrier at 193 THz. However, as mentioned, in a practical implementation this would be eliminated, since multiple IQMs would be employed, each one adjusted to operate at the relevant wavelength. Before the superchannel transmission, we measured the signal back-to-back (BtB) characteristics. In particular, Fig. 2(a) plots the BtB signal bit-error-rate (BER) as a function of OSNR for the five 50-GHz spaced subcarriers. In each case, the DC bias of the IQM was carefully adjusted favoring the subcarrier under test. For reference, the BtB curves of a single free running laser (seed laser) and a single comb line after being selected using the WS, are also included. It is observed that the comb source shows negligible excess penalty with respect to the free running laser. Furthermore, the five subcarriers when tested, exhibit similar BtB performance, and the benefits of the Nyquist shaping are evident with only ~1 dB penalty with respect to the 'single comb line' curve at 10^{-3} BER. With this, we evaluated the transmission performance after the two 80-km SSMF spans, each with an average ~16-dB loss. We

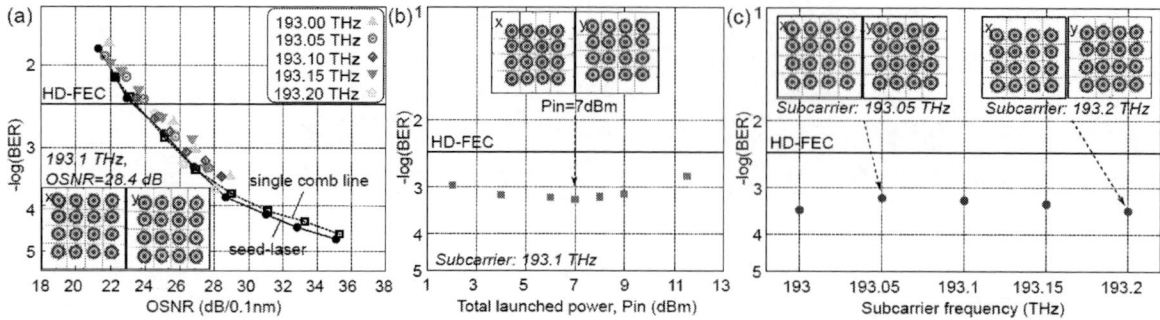

Fig. 2. Generation and transmission performance evaluation of the 1.76 Tbit/s superchannel employing 44-Gbaud DP-16QAM modulation format and Nyquist spectral shaping. a) Back-to-back characteristics, b) transmission performance versus total launched power for the subcarrier at 193.1 THz, and (c) transmission performance of each subcarrier at the optimum launch power of 7 dBm. Sample constellations for each polarization tributary are also shown.

characterized the BER as a function of the total launched power as shown in Fig. 2(b). At the optimum value, which was found to be 7 dBm, we evaluated the performance of each subcarrier as shown in Fig. 2(c), once more, adjusting in each case the IQM DC bias condition. As it is observed, the subcarriers are very similar and exhibit clear constellation diagrams after the 160-km transmission, as those shown as samples in Fig. 2(b-c). In average, the BER after the 160-km transmission was of 4.8×10^{-4}, being well below 3.8×10^{-3} which is the limit for hard-decision (HD) forward-error correction (FEC) with 7% overhead.

III. CONCLUSIONS

We studied the generation and transmission of a 1.76-Tbit/s superchannel exploiting high baud rate and high order QAM. The superchannel comprised five, 44-Gbaud, DP-16QAM subcarriers occupying only 250-GHz total bandwidth. The high capacity superchannel was achieved employing Nyquist spectral shaping and technology with integration potential, including a commercial integrated high-speed InP IQ modulator with $V\pi$ as low as 1.5 V, and an optical frequency comb, for a hardware efficient solution. The gross 1.76-Tbit/s line-rate signal with a high spectral efficiency of 7.04 bit/s/Hz was successfully operated over 160-km SSMF with still transmission margins with respect the threshold limit for HD-FEC; a total reach of ~320 km is roughly estimated. Contrary to this laboratory demonstration, we note that in a practical implementation, where the number of employed comb lines for carrier generation will be likely equivalent to the number of IQMs, a high-quality 44-Gbaud Nyquist DP-16QAM superchannel with higher optical signal-to-noise ratio could be potentially delivered in a photonic integrated platform. This paves the way for next generation multi-Tbit/s coherent transmitters, with the expected eventual development of high speed electronics to enable >44 Gbaud rates.

ACKNOWLEDGMENT

The authors would like to thank A. Albores-Mejia and H. Kuwatsuka of AIST for fruitful discussions. This work was supported in part by Project for Developing Innovation Systems of MEXT, Japan.

REFERENCES

[1] W. Forysiac and D. S. Govan, "Progress toward 100-G digital coherent pluggables using InP-based photonics," Journal of Lightwave Technology, vol. 32, pp. 2925-2934, August 2014.

[2] R. Nagarajan et al., "Terabit/s class InP photonic integrated circuits," Semicond. Sci. Technol., vol. 27, 094003, August 2012.

[3] J. Summers et al., "Monolithic InP-based coherent transmitter photonic integrated circuit with 2.25 Tbit/s capacity," Electron. Lett., vol. 50, pp. 1150-1152, July 2014.

[4] E. Yamada et al., "86 Gbit/s PDM 16-QAM signal transmission using InP optical IQ modulator." Electron. Lett., vol. 48, pp. 1486-1487, November 2012.

[5] S. Chandrasekhar et al., "Small-form-factor all-InP integrated laser vector modulator enables the generation and transmission of 256-Gb/s PDM-16QAM modulation format," Proc. OFC2013, PDP5B.6.

[6] N. Kikuchi et al., "High-speed optical 64QAM signal generation using InP-based semiconductor IQ modulator," Proc. OFC2014, M2A.2.

[7] A. H. Gnauck et al., "Comb-based 16-QAM transmitter spanning the C and L bands," IEEE Phot. Technol. Lett., vol. 26, pp. 821-824, April 2014.

[8] M. Doi et al., "Advanced LiNbO₃ optical modulators for broadband optical communications," IEEE J. Sel. Top. Quantum Electron., vol. 12, pp. 745-750, August 2006.

[9] N. Dupuis et al., "InP-based comb generator for optical OFDM," J. Lightw. Technol., vol. 30, pp. 466-472, February 2012.

[10] N. Yokota et al., "Nonlinearity of semiconductor Mach–Zehnder modulator for flat optical frequency," Photon. Technol. Lett., vol. 27, pp. 2219-2221, November 2015.

[11] H. Nguyen Tan et al., "First demonstration of wavelength translation for 1.376-Tbit/s DP-QPSK Nyquist OTDM signal," Proc. OFC2015, M3F.4.

C-band Spanning Frequency Comb Based on an Output Phase Stabilized Mode-locked Laser

Mark Pelusi[1], Hung Nguyen Tan[2], Karen Solis-Trapala[2], Takashi Inoue[2], Shu Namiki[2]

[1]Centre for Ultrahigh bandwidth Devices for Optical Systems (CUDOS), IPOS,
School of Physics, University of Sydney, NSW 2006, Australia
[2]National Institute of Advanced Industrial Science and Technology (AIST), Tsukuba, Ibaraki 305-8568, Japan

Abstract: A C-band spanning optical frequency comb is generated from an actively modelocked fiber laser using cross-phase modulation in highly nonlinear fiber before broadening in a 2nd stage. Comblines modulated with 96 Gb/s DP-64QAM achieve bit-error rates $<4.5 \times 10^{-3}$.

Keywords: Coherent communications, Fiber lasers, Nonlinear optical devices

I. INTRODUCTION

Optical frequency combs are of growing interest for future high data capacity communication systems for enabling tens or hundreds of WDM channels to be obtained from a single source with lower cost, power consumption and footprint than for conventional single frequency laser modules. The advantages are even stronger as WDM systems evolve to use more advanced data modulation formats, which increasingly rely on sophisticated narrow linewidth lasers [1], as well as improved suppression of the Nonlinear Shannon limit in long distance transmission, which has been shown more effective for a fixed frequency spacing between WDM channels [2]. Various techniques have been demonstrated for generating broadband frequency combs based primarily on using a CW laser seed with either electro-optic modulation and nonlinear propagation in highly nonlinear fiber (HNLF) [3], [4], or nonlinear ring resonator waveguides [5]. Nonlinear propagation of short pulses obtained directly from an ERGO mode-locked laser have also been produced [6]. The challenges, in general, are obtaining high power, low noise comb lines with flexible frequency spacing, in an easily reproducible design.

In this paper, an alternative frequency comb suitable for 64QAM modulation is demonstrated using a 40 GHz harmonically mode-locked fiber laser (MLFL) producing short pulses at a flexible, high repetition rate. A low, noise broadband comb suitable for 64QAM modulation is achieved by first using cross phase cross phase modulation (XPM) of the MLFL with a CW laser probe in a HNLF to overcome the phase instability of the MLFL pulses due to the long cavity length [7], and then broadening the frequency comb obtained by propagation in a second HNLF. Modulating extracted comb-lines with a 96 Gb/s DP-64 QAM signal demonstrates a bit error rate (BER) below the FEC limit of 4.5×10^{-3} (7% overhead) for channels within a bandwidth containing over 40 comb lines. An OSNR penalty as low as 1.5 dB is also measured relative to a reference laser, confirming the low noise frequency comb obtained.

II. EXPERIMENT

Figure 1 shows the schematic of the frequency comb source. A MLFL (Pritel Inc., UOC-40) was driven by a 40 GHz synthesizer to produce ≈2 ps pulses at a 40 GHz repetition rate, with ultra-low timing jitter, and 1559.5 nm center wavelength. To overcome the phase instability of the MLFL output and obtain a low noise frequency comb, the pulses were co-propagated with a CW probe at different wavelength so that the low jitter pulse waveform induced through the Kerr effect XPM sidebands around the probe, free in principle, from the phase and wavelength drift of the MLFL pulses

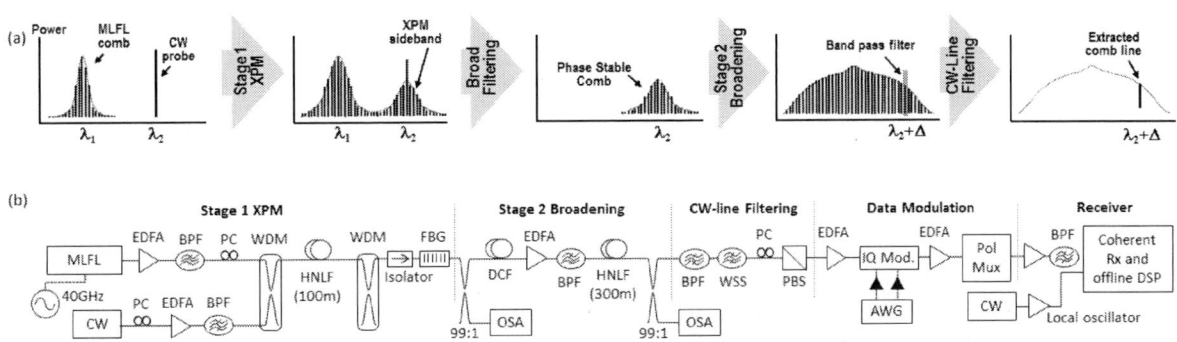

Fig. 1. (a) Schematic, and (b) experimental set-up of 40 GHz comb source using 1st stage XPM of a 40 GHz MLFL with a CW probe, followed 2nd stage comb broadening, before comb line extraction, and data modulation with a 96 Gb/s DP-64 QAM signal, followed by BER evaluation.

21

[7], [8]. The low timing jitter of the pulses also plays a crucial role in this work for ensuring the linewidth of the generated comb lines are minimally broadened, so that individual comb-lines can then be modulated with phase noise sensitive, advanced data format signals, without degradation. The XPM stage was implemented by using a WDM coupler to combine the MLFL output pulses with a CW laser (Teraxion PS-TNL with <100 kHz linewidth) at 1546 nm, before launching into a 100m long HNLF. Both the MLFL pulses and CW probe were optimally co-polarized in the HNLF using polarization controllers (PC) to maximize the XPM sideband generation, and their respective launch powers were adjusted by erbium-doped fiber amplifiers (EDFA). For an average pulse launch power of $\approx$100 mW, the induced peak phase shift of the CW probe was expected to be $\approx$1.7π for the HNLF parameter of $\gamma = 21$ W^{-1}km^{-1}, and considering both the pulse dispersion and walk-off between the CW probe and pulses to be negligible for the HNLF having a zero dispersion wavelength of 1551 nm, and a dispersion slope of 0.02 ps/nm^2.km. The optical spectrum before and after the HNLF in Fig. 2(a) shows the XPM sidebands generated around the CW probe wavelength, along with four wave mixing (FWM) products at other wavelengths.

The XPM frequency comb was extracted by another WDM coupler with passband $\approx$<1548 nm. To improve the power flatness of the frequency comb spectrum, the strong CW probe was suppressed by a fiber Bragg grating (FBG) with a narrow reflection bandwidth of $\approx$0.3 nm centered just slightly off the CW probe wavelength. The frequency comb was then amplified, and filtered to remove the FWM products at the shorter wavelengths, and launched into a 2nd HNLF with 300 m length, $\gamma = 30$ W^{-1}km^{-1} and dispersion of -0.7 ps/nm.km at 1550 nm. The frequency comb bandwidth was also maximized by optimizing the frequency chirp of the input pulses with a dispersion of ≈-0.44 ps/nm from a short length of dispersion-compensating fiber (DCF) inserted before the EDFA. Figure 2(b) compares the input and output optical spectrum from the 2nd HNLF for a launch power of $\approx$350 mW, and shows the broadened frequency comb obtained, spanning beyond the C-band from 1530 to 1565 nm.

A single comb line was then extracted by a wavelength-selective switch (WSS) with $\approx$10 GHz bandwidth, before 64 QAM data modulation. A tunable bandpass filter (BPF) with 2 nm bandwidth was also used before the WSS to slice out a portion of the comb spectrum for a safer, lower input power to the WSS. The selected comb line was then passed through a polarization beam splitter (PBS) to ensure linearly polarized input to a LiNbO$_3$ IQ modulator (IQ-mod.) that was driven by a two channel arbitrary waveform generator (AWG) outputting uncorrelated 64QAM I and Q data channels of 2^{11}-1 PRBS at 8 Gbaud. A polarization multiplexing (Pol-Mux) emulator then combined two replicas of the signal on dual orthogonal polarization (DP) states to give 96 Gb/s DP-64QAM output. At the receiver, an EDFA and 0.5 nm BPF preceded a polarization diversity coherent receiver using another Teraxion PS-TNL as the local oscillator that was tuned to the signal. The detector outputs were captured by a 4 channel realtime oscilloscope for offline DSP and signal analysis, including BER counting.

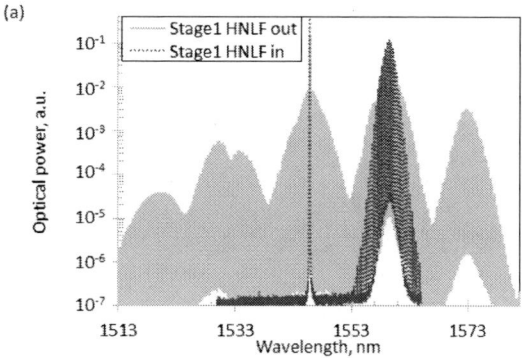 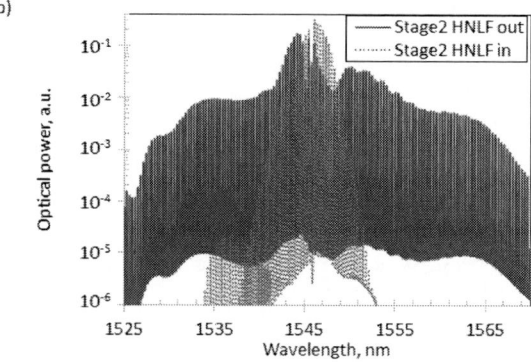

Fig. 2. 40 GHz MLFL based frequency comb source generation: Optical spectra before and after (a) Stage 1 XPM of a MLFL at 1559.5 nm wavelength, with a CW probe at 1546 nm, in a 100m HNLF to produce a phase stable 40 GHz frequency comb, followed by (b) stage 2 comb broadening in a 300 m HNLF. RBW = 0.03 nm for all traces.

III. 64 QAM MODULATION RESULTS

The performance of the C-band spanning comb was evaluated for individual comb lines that were extracted and modulated with the 96 Gb/s DP-64QAM signal. The BER count for all measurements was taken as the average of both polarization channels. For the comb line at 1544.9 nm wavelength near the CW probe, the OSNR penalty was measured to be 1.5 dB at a BER of 4.5×10^{-3} (FEC threshold with 7% overhead), relative to the reference case of the CW probe laser connected directly to the IQ-mod., and tuned to 1545 nm wavelength. The output signal constellation was also similar to the reference case as shown in Fig. 3(a), with only $\approx$0.8 dB degradation of the calculated Q^2-factor. On the other hand, the signal constellation was unmeasurable if modulating any comb line directly from the MLFL, confirming the effective phase stabilization by XPM. For modulating other comb lines, the BER was measured to be below 4.5×10^{-3} over the spectral range containing over 40 comb lines from 1537.5 to 1552.5 nm, as shown in Fig. 3(b), confirming the broadband, low noise comb obtained. The BER was also only moderately higher than for the reference laser case over the same range as shown. The observed degradation of the BER outside the 15 nm range was due to the decreasing comb-line OSNR, and phase noise. This could be improved by more sophisticated design [3], [4], including optimization of the HNLF parameters and input pulse seed, which should also help flatten the comb power spectrum.

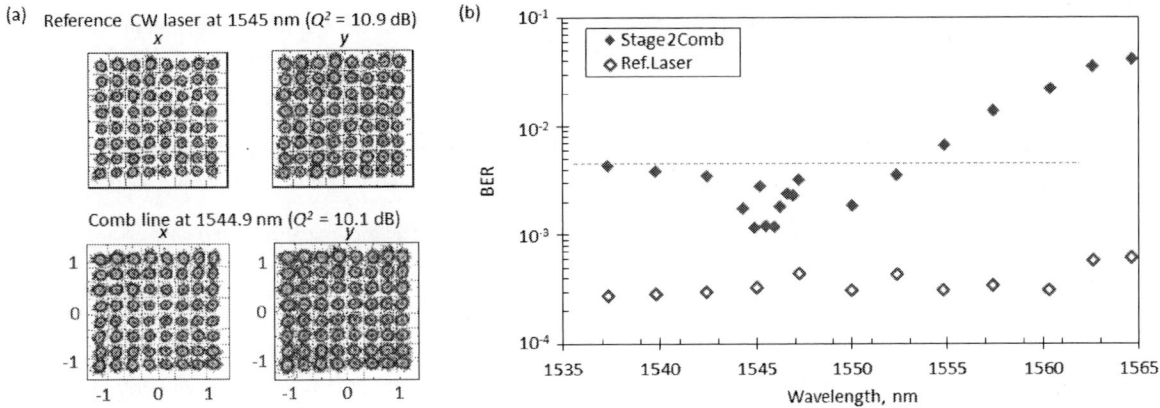

Fig. 3. 96 Gb/s DP-64QAM signal generation from 40 GHz MLFL based comb source: (a) constellations of data modulated comb lines, and (b) BER.

IV. CONCLUSIONS

A broadband, low noise optical frequency comb was demonstrated using an actively modelocked fiber laser by phase stabilizing the output using cross phase modulation with a CW probe in HNLF before further broadening in another HNLF. Measurements for individual comb-lines modulated with a 96 Gb/s DP-64QAM signal confirmed both low OSNR penalty relative to a reference laser, and a BER below 4.5×10^{-3} over a wide spectral range containing over 40 comb lines from 1537.5 to 1552.5nm. The measurements demonstrate the potential for low noise, broadband frequency generation applicable to phase noise sensitive, higher order QAM format, using harmonically mode-locked fiber lasers.

ACKNOWLEDGMENT

This research was supported by the Project for Developing Innovation Systems of MEXT, Japan, and the Australian Research Council (ARC) Future Fellowship and CUDOS programs (FT110101037, CE110001018). Dr. Hung Nguyen Tan also acknowledges the support of the University of Sydney International Research Collaboration Award.

REFERENCES

[1] T. Pfau, S. Hoffmann, and R. Noé, "Hardware-efficient coherent digital receiver concept with feedforward carrier recovery for M-QAM constellations," J. Lightwave Technol., vol. 27, pp. 989-999, 2009.

[2] E. Temprana, E. Myslivets, B.P.-P. Kuo, L. Liu, V. Ataie, N. Alic, and S. Radic, "Overcoming Kerr-induced capacity limit in optical fiber transmission", Science, vol. 348,pp. 1445-1448, 2015.

[3] V. Ataie, E. Temprana, L. Liu, E. Myslivets, B.P.-P. Kuo, N. Alic, and S. Radic, "Ultrahigh count coherent WDM channels transmission using optical parametric comb-based frequency synthesizer," J. Lightwave Technol. vol. 33, pp. 694-699, 2015.

[4] E. Temprana, V. Ataie, B.P.-P. Kuo, E. Myslivets, N. Alic, and S. Radic, "Low-noise parametric frequency comb for continuous C-plus-L-band 16-QAM channels generation," Opt. Express vol. 22, pp. 6822-6828, 2014.

[5] J. Pfeifle et al., "Coherent terabit communications with microresonator Kerr frequency combs", Nature Photon. Vol. 8, pp. 375–380, 2014.

[6] D. Hillerkuss, et al., "Single-laser 32.5 Tbit/s Nyquist WDM transmission," J. Opt. Commun. Netw. vol. 4, pp. 715-723, 2012.

[7] H. Chen, X. Gu, M.a Chen, and S. Xie, "Ultrashort optical pulse phase stabilization using cross-polarization modulation method", Opt. Eng. 51(4), 040508, 2012.

[8] D.J. Jones, S.A. Diddams, M.S. Taubman, S.T. Cundiff, L.-S. Ma, and J.L. Hall, "Frequency comb generation using femtosecond pulses and cross-phase modulation in optical fiber at arbitrary center frequencies" Opt. Lett. vol. 25, pp. 308-310 2000.

10-Gbaud QPSK Signal Simultaneous Transmission in the T and C bands for Ultra-Broadband Photonic Transport System

Akihiro Murano[1], Shoko Yamada[1], Atsushi Kanno[2], Naokatsu Yamamoto[2], Hideyuki Sotobayashi[1]

(1) Aoyama Gakuin University, 5-10-1 Huchinobe, Chuo-ku, Sagamihara-shi, Kanagawa 252-5258, Japan
(2) National Institute of Information and Communications Technology (NICT), 4-2-1 Nukui-kitamachi, Koganei, Toyko 184-8795, Japan
E-mail: a5412139@aoyama.jp, sotobayashi@ee.aoyama.ac.jp

Abstract: *We successfully demonstrate 10-Gbaud QPSK simultaneous transmission in the T and C bands. Observed bit error rates are within a forward error correction limit of 2×10^{-3} under homodyne coherent detection configuration.*

Keywords: *holey fiber, optical communication, optical transmission system, quadrature phase-shift keying*

I. INTRODUCTION

Sharp increase in the transmission capacities of photonic transport networks is in high demand and is being developed by using alternative wavebands. Conventional wavebands such as the C and L band have already been utilized in deployed photonic transport networks [1]. However, frequent switching path in photonic networks is required for optical paths owing to the limited available bandwidth, even wavelength-division-multiplexing (WDM) techniques using these bands. Therefore, pioneering and developing advanced transport technology in ultra-broadband such as the T-band (Thousand-band: 1.000–1.260 nm, 61.9-THz) is required not only high capacity transport networks but also load reduction of switching path. The T band have potential broad-bandwidth feature owing to an ytterbium (Yb^{3+})-doped fiber amplifier (YDFA) utilized as an optical amplifier [2]. Furthermore, photonic crystal fibers (PCFs) have been already developed for single-mode transmission for broad bandwidth signals in the form of endlessly single-mode (ESM) optical fibers [3]. Therefore, the ultra-coarse WDM techniques using the T band allows drastic increase of transmission capacity with a reduction of the energy consumption related on optical switching paths in photonic networks. This is because sparse wavelength channels can exclude a kind of thermoelectric controller for stabilization of the optical devices. Previously, we had successfully demonstrated WDM transmission in the T-band with on-off keying signals [4]. Apart from above techniques, multilevel modulation and demodulation using a coherent detection technique are important for spectral efficiency, even in the broad-bandwidth wavebands.

In this study, we demonstrate 10-Gbaud quadrature phase-shift keying (QPSK) signal WDM transmission in the T and C bands. The transmission system is based on a homodyne coherent detection optimized for the T band and C band with an offline digital signal processing (DSP).

II. CONFIGURATION OF ULTRA-BROADBAND PHOTONIC TRANSPORT SYSTEM FOR THE T AND C BANDS

Figure 1 shows the experimental setup over the ESM-holey-fiber (HF) under homodyne coherent detection configuration used for the demonstration. This 4-km-long ESM-HF based on the photonic crystal fibers has the total dispersion characteristics of –43.8 ps/nm in the T band and 191.2 ps/nm in the C band, that would be zero at the wavelength of approximately 1200 nm, by optimization of dispersion characteristics and its mode field diameter [5].

On the transmitter side, a wavelength-tunable GaAs-based quantum dot (QD) semiconductor laser was used as the narrow-linewidth light source (approximately 200 kHz) for the T band. The carrier wavelength was tuned to 1069.65 nm. A wavelength-tunable external cavity laser (ESL) with a 100-kHz linewidth at a wavelength of 1550.07 nm was used as the C-band light source. A pseudo-random binary sequence (PRBS, the length of $2^{15} - 1$ bits) data signal of a non-return-to-zero at 10 Gb/s was output from each channel of a pulse pattern generator (PPG). In the T band, the QPSK signal was generated by cascaded optical modulators, that are serially connected by a dual-drive Mach–Zehnder (DD-MZM) optical modulator and an optical phase modulator (PM), with a two-channel PPG. During the QPSK signal generation, the binary phase shift keying (BPSK) signal is generated by the DD-MZM under differential operation condition. Subsequently, the QPSK signal is generated by adding phase rotation of 90-degree by the PM with one independent port of the PPG [6]. On the other hand, the QPSK signal for the C band was generated using an integrated optical IQ modulator with a two-channel PPG. Thereafter, the respective optical signals were amplified using YDFA and an Erbium-doped fiber amplifier (EDFA). To combine the optical signals in the T and C bands, WDM coupler were used at both ends of the transmission path, and finally, the signals was transmitted over the 4-km-long ESM-HF.

On the receiver side, transmitted signals were separated by the WDM coupler, and then, the QPSK signals were launched into single-polarization optical homodyne coherent receivers optimized for T and C bands, respectively. The signals were mixed with optical local oscillator (LO) signals in 90-degree optical hybrid couplers through polarization controllers (PC). A light source for the T and C bands were split by optical couplers, whose branch ratio are 50:50, set before the modulators generated for the QPSK signals; the split signals were utilized as an optical LO signal for optical coherent detection. It should be noted that the other 4-km-long ESM-HFs, whose lengths are same as the signal transmission line, are used in the LO paths for the homodyne configuration. Both LO signals were amplified to approximately 7.0 dBm by a YDFA and EDFA.

The optical outputs for in-phase (I) and quadrature-phase (Q) components from the optical hybrids were detected by two balanced photodetectors (BPD), whose 3-dB waveband was approximately 1060-1565 nm. The electrical outputs from BPDs were acquired by a 20-GS/s real-time oscilloscope. The carrier phase and IQ data recovery of an electrical signal demodulated in each optical waveband were implemented using an offline DSP. Digitized signals were adaptively equalized by a finite impulse response filter with a tap length of 20 by the constant modulus algorithm, and it was equalized using phase-noise estimation and compensation in the same manner as a conventional optical coherent detection system for the C band. In addition, bit error rates (BERs) were measured using the compensated signal data. The crosstalk in WDM systems is also evaluated.

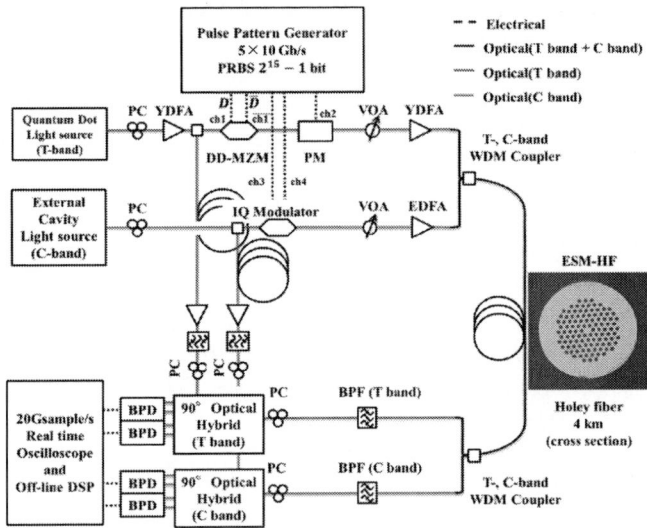

Fig. 1. Ultra-broadband photonic transport system over a holey fiber transmission line for simultaneous transmissions in the T and C bands.

III. EVALUATION OF WDM TRANSPORT SYSTEM WITH THE T AND C BANDS

Figure 2 (a) and (b) show the optical spectra observed with and without the 4 km-long ESM-HF transmission line at the 1069.65 nm and 1550.07 nm. The insertion loss of the transmission line including the losses of the couplers and connectors was confirmed to be 11.4 dB in the T band. The insertion loss in the C band was also evaluated 10.2 dB. The resultant transmission loss of the ESM-HF is estimated approximately 0.7–0.8 dB/km in the T band and 0.4–0.5 dB/km in the C band, respectively.

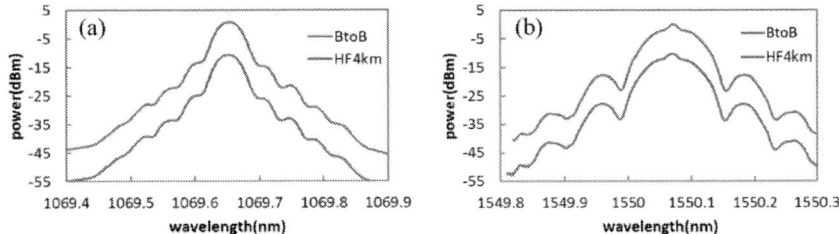

Fig. 2. Optical spectra of QPSK signals of over the ESM-HF in (a) the T band and (b) the C band

Figure 3 (a) and (b) show the estimation of crosstalk by observed optical spectra in the T band and the C band, respectively, by connecting each input port selectively. Crosstalk of 37.1 dB in the T band and 46.6 dB in the C band were measured. The difference may be caused by the transmittance with each port of WDM couplers. Figure 4 shows constellation maps of demodulated QPSK signals under back-to-back and 4-km-ESM-HF transmission conditions in (a)-(b) for the T band and (c)-(d) for the C band. Clear symbol separation was observed with an error vector magnitude (EVM) of 22.3%-RMS and 21.6%-RMS for the T-band back-to-back and ESM-HF transmission conditions, respectively. On the other hand, in the C band, the obtained EVMs were 16.1%-RMS and 16.7-RMS under back-to-back and ESM-HF transmission conditions, respectively. A small degradation of the EVM could be caused by the error of sensitive bias control and optical polarization control for the cascaded optical modulators, which are serially connected by a DD-MZM optical modulator and an optical PM.

Figure5 (a) and (b) show BERs dependences on optical signal-to-noise ratio (OSNR). Observed BERs were within a forward error correction (FEC) limit of the BER of 2×10^{-3} when the OSNR was higher than approximately 17 dB in both T and C bands. These results indicate that simultaneous WDM transmission with 10-Gbaud QPSK signals were realized over 4-km-long ESM-HF in the T and C bands.

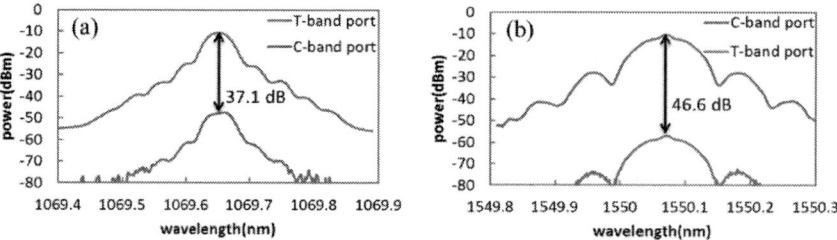

Fig. 3. Estimation of crosstalk with optical spectra after HF transmission through the BPF in (a) the T band and (b) the C band

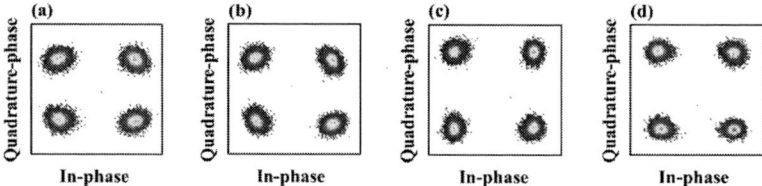

Fig. 4. Constellation maps under B-to-B and ESM-HF transmission conditions in (a)-(b) the T band and (c)-(d) the C band

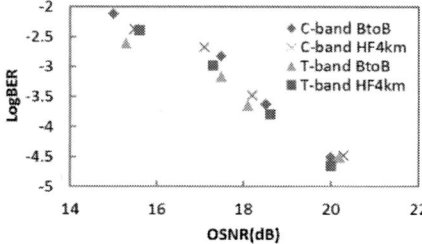

Fig. 5. BERs vs. OSNR in the case of B-to-B and ESM-HF transmission condition in the T and C band

IV. CONCLUSIONS

We demonstrated 10-Gbaud QPSK WDM transmission over 4-km-long ESM-HF in the T and C bands. The measured BERs were within the FEC limit under the homodyne coherent detection by a commonly developed offline DSP. Ultra-broadband coherent photonic transport systems with a large number of wavelength channels in T band promise more high-capacity transmission systems.

ACKNOWLEDGMENT

The authors thank the staff of Furukawa Electric Co., Japan Koshin Kogaku Co., Japan, Sevensix, Inc., Japan, and the Photonic Device Laboratory of NICT.

REFERENCES

[1] T. Morioka, "New Generation Optical Infrastructure Technologies: 'EXAT Initiative' Towards 2020 and Beyond," Proc. of OECC/ACOFT, FT4, Hong Kong SAR (China), 2009.

[2] R. Paschotta, J. Nilsson, A. C. Tropper, and D. C. Hanna, "Ytterbium-doped fiber amplifiers," IEEE J. Quantum Electron., vol. 33. No. 7, pp. 1049-1056, 1997.

[3] T. A. Birks, J. C. Night, and P. St. J. Russell, "Endlessly single-mode photonic crystal fiber," Opt. Lett., vol. 22. No. 13, pp. 961-963, 1997.

[4] N. Yamamoto, Y. Omigawa, K. Akahane, T. Kawanishi, and H. Sotobayashi, "Simultaneous 3 × 10 Gbps optical data transmission in 1-μm, C-, and L-wavebands over a single holey fiber using an ultra-broadband photonic transport system," Optics Express., vol. 18. No. 5, pp. 4695-4700, 2010.

[5] K. Imamura, K. Mukasa, Y. Mimura, and T. Yagi "Multi-core holey fibers for the long-distance (> 100 km) ultra large capacity transmission," Proc. of Euro. Conf. Optical Fiber Communication (OFC 2009), p. OTuC3, 2009.

[6] H. Y. Choi, T. Tsuritani, and I. Morita, "Effects of LN Modulator Chirp on Performance of Digital Coherent Optical Transmission System," The 10th international Conference on Optical Internet (COIN 2012), pp.50-51, 2012.

Subcarrier Polarization Manipulation Using Orthogonal Dual Phase-Modulation in Fiber

Tomoyuki Kato[1], Takahito Tanimura[1], Takeshi Hoshida[1], Thomas Richter[2], Robert Elschner[2],
Carsten Schmidt-Langhorst[2], Colja Schubert[2], and Shigeki Watanabe[1]

[1]Fujitsu Laboratories Ltd., 4-1-1 Kamikodanaka, Nakahara-ku, Kawasaki 211-8588, Japan
[2]Fraunhofer Institute for Telecommunications, Heinrich Hertz Institute, Einsteinufer 37, 10587 Berlin, Germany
kato.tom@jp.fujitsu.com

Abstract: *We propose and demonstrate a subcarrier polarization manipulation using two orthogonally polarized local subcarrier signals. Polarization angle of a 25-Gb/s QPSK subcarrier is arbitrary controlled between orthogonal states without affecting 25-GHz spaced adjacent subcarriers.*

Keywords: *Nonlinear optics, subcarrier multiplexing, all-optical networks*

I. INTRODUCTION

For improvement of transmission capacity, optical signals should be densely packed in a given bandwidth. We have investigated an optical frequency-division multiplexing scheme by using frequency conversion in nonlinear fibers [1]. The scheme achieves multiplexing with precise frequency allocation since it is based on frequency control using stable electrical oscillators. The technique has been utilized for subcarrier (SC) multiplex of coherent optical orthogonal frequency-division multiplexed (CO-OFDM) and Nyquist wavelength-division multiplexed (WDM) superchannels at spatially distributed nodes using free-running lasers [2]. By using the master carrier of the SC multiplexing as a frequency and polarization reference, a state of polarization (SOP) of the multiplexed SC signals can be controlled.

Polarization manipulation on multiplexed signals in such optical networks is attractive from the viewpoints of transmission characteristics improvement [3] and optical switching function [4]; scrambling or control of the relative SOP of individual multiplexed subcarrier may mitigate the nonlinear distortions occurring in the transmission line. Moreover, optical switching based on polarization in single polarization systems offers a very high spectral selectivity which cannot be achieved using conventional optical wavelength selective switches (WSSs).

In this paper, we propose a novel scheme for the polarization manipulation. Two new functions, i.e. distributed optical subcarrier multiplexing with variable SOP and in-line optical subcarrier polarization rotation, are experimentally demonstrated for the first time, to our knowledge. By using frequency conversion in a nonlinear fiber with two orthogonally polarized local SCs [5], a single-polarization 25-Gb/s QPSK subcarrier is multiplexed with variable polarization angle and is rotated to arbitrary polarization angle without affecting the adjacent SCs.

II. PRINCIPLE OF SUBCARRIER POLARIZATION MANIPULATION

Here we consider two basic operations for polarization manipulation: addition of a SC with variable polarization angle and polarization rotation of a selected SC out of multiplexed subcarriers (Fig. 1). In each function, the SC signals are transferred to the master carrier by optical frequency conversion in all-optical modulator (AOM), which consists of a combiner (WDM coupler), a nonlinear fiber (NLF) and two local SC generators with identical local data modulation. A free-running laser diode is used as a light source in each local SC generator. Each local SC consists of a continuous wave (CW) signal and a modulated data signal which is orthogonally polarized at the NLF input and has a frequency separation of $\Delta \nu$. In the NLF, the master carrier at the optical frequency ν_0 is phase-modulated with the local SC-A and the local SC-B with the same subcarrier frequency $\Delta \nu$ by cross-phase modulation (XPM).

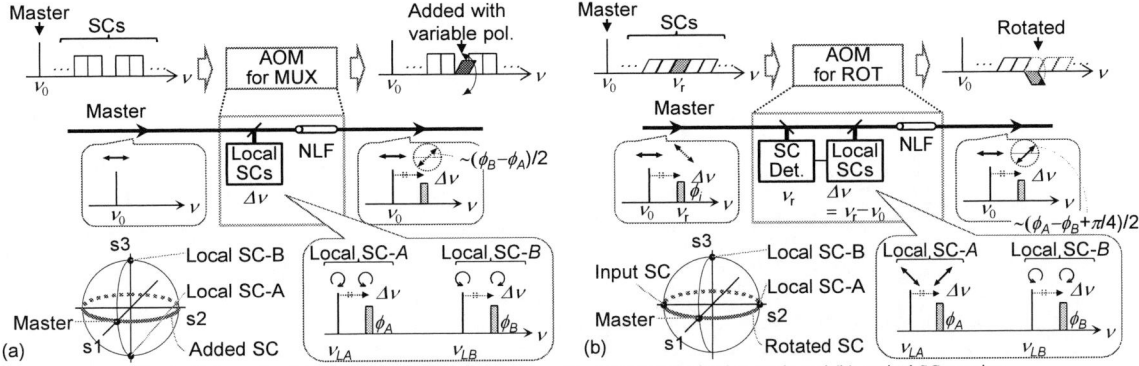

Fig. 1. Schematic of (a) the optical SC multiplexing with variable polarization angle and (b) optical SC rotation with illustrated optical spectra and SOP on the Poincaré sphere.

For the SC multiplexing (MUX), the two local SCs (local SC-A and local SC-B) should be mutually orthogonal in SOP, while they should be circularly polarized with respect to the linearly polarized master carrier, as depicted in the inset of Fig. 1(a). The resultant new SC is the sum of the two XPM components,

$$\mathbf{E} = [(E_A e^{j\phi_A} + E_B e^{j\phi_B})\mathbf{h} + (jE_A e^{j\phi_A} - jE_B e^{j\phi_B})\mathbf{v}]e^{2\pi j(\nu_0 + \Delta\nu)t}, \quad (1)$$

where $\mathbf{h}$ and $\mathbf{v}$ are the unity field vectors along horizontal and vertical polarization, E_A and E_B are converted fields with the local SC-A and the local SC-B, and ϕ_A and ϕ_B are subcarrier phase of the local SC-A and local SC-B, respectively. By tuning the subcarrier phase difference, $\Delta\phi_M \equiv \phi_B - \phi_A$, and adjusting the efficiencies to be $E_A = E_B$, the SOP of the multiplexed subcarrier is changed on the equator of the Poincaré sphere as shown in the inset of Fig. 1(a).

To realize the SC rotation (ROT), an SC contained in the input of the node is polarization-rotated without affecting the polarization of the master carrier and surrounding SCs. It is almost the same configuration as the SC multiplexing, but the modulated data signal in the local SCs is generated based on the detected signal of the incoming SC. The other difference is the polarization setting of the local SCs. The local SC-A is orthogonally polarized with respect to the incoming SC (E_i, ϕ_i) and the local SC-B is circularly polarized as depicted in inset of Fig. 1(b). The resulting SC is the sum of the three components,

$$\mathbf{E} = [(E_i e^{j\phi_i} + E_A e^{j\phi_A} + E_B e^{j\phi_B})\mathbf{h} + (-E_i e^{j\phi_i} + E_A e^{j\phi_A} + jE_B e^{j\phi_B})\mathbf{v}]e^{2\pi j(\nu_0 + \Delta\nu)t}. \quad (2)$$

By tuning the subcarrier phase difference, $\Delta\phi_R \equiv \phi_A - \phi_B + \pi/4$ while $\phi_i - \phi_A = \pi/2$, and adjusting the efficiencies to be $E_i = E_A = \sqrt{2}\,E_B/2$, the SOP of the SC can be changed on the equator of the Poincaré sphere as shown in inset of Fig. 1(b). In particular, for the case of $\phi_A - \phi_B = 3\pi/4$, the SOP of the multiplexed SC is switched from parallel polarization to orthogonal polarization and vice versa. Thus a subsequent PBS can switch the individual SCs to different receivers.

III. EXPERIMENTAL DEMONSTRATION

Figure 2 shows the experimental setup. The master carrier with the optical frequency $\nu_0 = 193.500$ THz and the optical power of 10 dBm was input to the AOM-1, where three SCs are generated. Three 12.5-GBd QPSK SCs were multiplexed with 25-GHz spacing at AOM-1. Polarization manipulation, MUX and ROT of the center SC with the subcarrier frequency of 100 GHz are realized by AOM-2. Then the signal was finally detected and evaluated using a digital coherent receiver (Coh. Rx). The relative polarization angle of the SCs was measured by adjusting a polarization controller (PC-1) with respect to a polarizer before optical spectrum analyzer (OSA).

To generate the local SCs for AOM-1 and AOM-2, an optical frequency comb was prepared by modulating the output of a CW laser (191.850 THz, ~100 kHz linewidth) at $f_0 = 12.5$ GHz. The WSS extracted components of the optical frequency comb for the local SCs. The components for SCs were modulated with an IQ-modulator (IQM) driven with a pulse pattern generator (PPG) and generated single-polarization 25-Gb/s QPSK signals. The CW lights and the QPSK signals were coupled in orthogonal polarization with a PBS. After the PBS, the CW light and the QPSK signals for AOM-1 and those for AOM-2 were divided and extracted with optical bandpass filters.

At AOM-1, the master carrier and local SC-1 were combined and launched into the highly nonlinear fiber (HNLF-1; 500 m, 15 W^{-1}km^{-1}). The component extracted with the WSS at $\nu_{L1} = 191.800$ THz was applied for the CW light of local SC-1 and those at $\nu_{L2} = 191.875$ THz, $\nu_{L3} = 191.900$ THz, and $\nu_{L4} = 191.925$ THz were applied for SCs to be multiplexed at AOM-1. The optical power was set to 15 dBm for the CW light and −5 dBm for the SC signals. The master was adjusted to horizontally linear polarization with PC-2 and the local SC-1 was adjusted to −45-degree linear polarization for SC signal and 45-degree linear polarization for the CW light with PC-3. After the HNLF-1, QPSK signals with subcarrier frequencies of 75, 100, 125 GHz were multiplexed on the master carrier as shown in Fig. 3(a).

At AOM-2, the master carrier with multiplexed SCs, local SC-A and local SC-B were combined and launched into HNLF-2 (500 m, 15 W^{-1}km^{-1}). The master was adjusted to horizontally linear polarization with PC-4. The components extracted by the WSS at $\nu_{L5} = 191.600$ THz and $\nu_{L6} = 191.700$ THz were applied for the CW light and the SC signal of local SC-A and those at $\nu_{L7} = 192.000$ THz and $\nu_{L8} = 192.100$ THz were applied for the CW light and the SC signal of local SC-B, respectively. The local SC-A and the local SC-B were coupled by adjusting the symbol-delay with optical delay lines (τ_1, τ_2). For the MUX operation, the optical power of each component was set to 15 dBm for the CW light and −5 dBm for the QPSK signals and the local SCs were adjusted to circular polarization with PC-5 and PC-6. For the

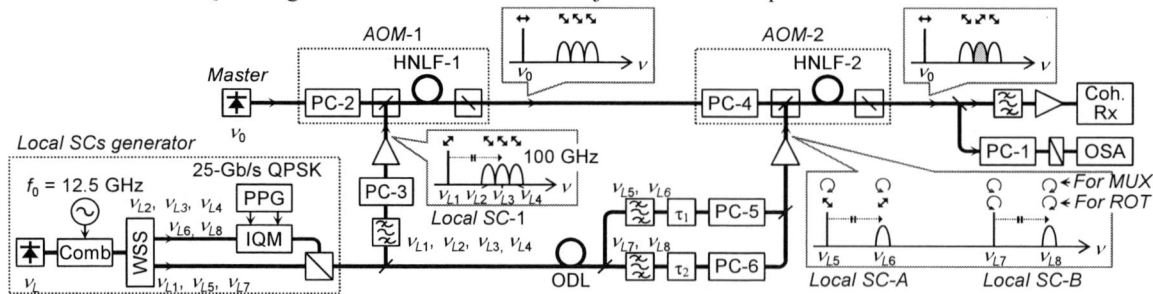

Fig. 2. Experimental setup.

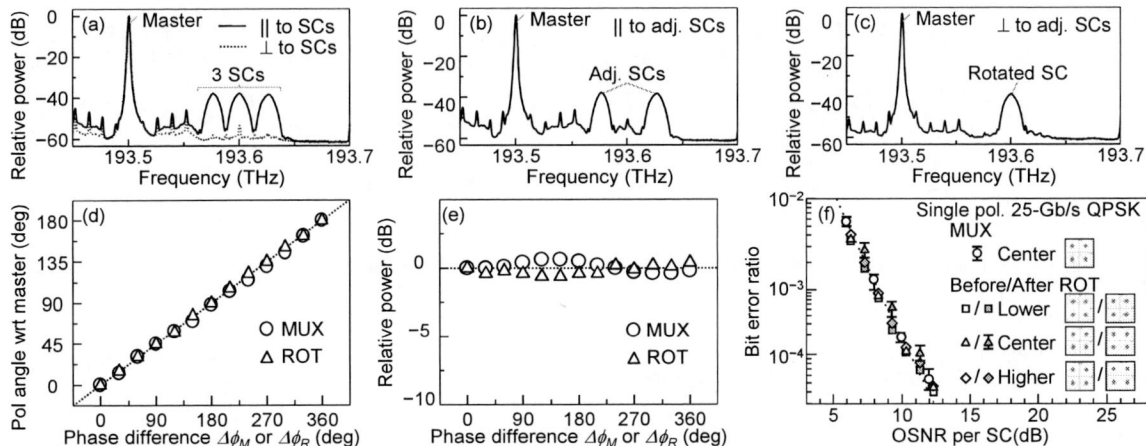

Fig. 3. Optical spectra (Res. 0.01 nm) of (a) AOM-2 input, (b) AOM-2 output with polarizer parallel aligned to adjacent SCs, and (c) AOM-2 output with polarizer orthogonally aligned to adjacent SCs, (d) polarization angle change of the SC and (e) optical power variation of the SC by tuning the phase difference of the local SCs, (f) BER characteristics and the typical constellation diagram (OSNR = 20 dB/SC) of the received SCs.

ROT operation, the configuration is the same as the MUX case, while the SC signal of the local SC-*A* was re-adjusted to 45-degree linear polarization with PC-5 and the optical power of the local SC-*B* was increased by 3 dB. The QPSK signals for AOM-1 and AOM-2 were the same optical modulator output and symbol-delay of them was adjusted by an optical delay line (ODL) before launching the local SCs to HNLF-2 for the proof-of-concept demonstration. After the HNLF-2, the polarization of the center SC (100 GHz) was rotated with keeping the adjacent SCs. Figures 3 (b) and 3 (c) show the corresponding optical spectra for the case of $\phi_A-\phi_B=3\pi/4$ in which the center SC is rotated by 90-degree from being copolarized with the adjacent subcarriers to being orthogonally polarized to the adjacent subcarriers.

First, we measured the polarization angle change at the AOM-2 by detuning the phase difference between the local SC-*A* and the local SC-*B*. The phase difference was changed by using the WSS. The MUX case and the ROT case were evaluated by turning off and on the local SC-1 in AOM-1, respectively. The polarization angle was confirmed to linearly change with the phase difference in the both cases as shown in Fig. 3(d). Uniform optical power over the SOP change was achieved in each condition as shown in Fig. 3(e). The deviation of <1 dB was mainly attributed to the optical power imbalance among the local SCs.

Second, bit error ratio (BER) characteristics were measured for multiplexing with variable polarization and the three SCs before and after the center SC was rotated, as shown in Fig. 3 (f). The BER performances were measured with the OSNR defined by the OSNR per SC and 0.1-nm reference noise bandwidth. SC selection at the receiver was achieved by aligning the emission frequency of the local oscillator laser with the center of the particular SC. The BER of the center SC was measured for five different phase differences of 0, 90, 180, 270, and 360 degrees. The BER variation is indicated by error bars in Fig. 3(f). As shown in Fig. 3(f), we achieved a very good in-line performance with an OSNR penalty of less than 1 dB at a BER of 10^{-3}. We observed variation of the measured BER due to the optical power deviation of the SC. No significant distortion in the constellation diagram after the processing was observed as shown in the inset of Fig. 3(f).

IV. CONCLUSIONS

We proposed and demonstrated variable polarization subcarrier multiplexing and subcarrier polarization rotation of a multiplexed subcarrier by using frequency conversion with two orthogonally polarized local SCs. In the subcarrier multiplexing scheme, a single-polarization 25-Gb/s QPSK subcarrier was multiplexed with arbitrary polarization angle by tuning the phase difference between the local SCs. In the subcarrier polarization rotation scheme, one incoming single-polarization 25-Gb/s QPSK subcarrier within three 25-GHz spaced subcarriers was arbitrary controlled to be rotated between two orthogonal polarization states without affecting the adjacent SCs.

REFERENCES

[1] S. Watanabe, T. Kato, R. Okabe, R. Elschner, R. Ludwig, and C. Schubert, "All-optical data frequency multiplexing on a single-wavelength carrier light by sequentially provided cross-phase modulation in fiber," *IEEE JSTQE* **18**, 577-584, 2012.

[2] T. Richter, C. Schmidt-Langhorst, R. Elschner, L. Molle, S. Alreesh, T. Kato, T. Tanimura, S. Watanabe, J. K. Fischer, C. Schubert, "Distributed 1-Tb/s all-optical aggregation capacity in 125-GHz optical bandwidth by frequency conversion in fiber," *ECOC2015*, PDP.2.5.

[3] T. Ito, T. Ono, Y. Yano, K. Fukuchi, H. Yamazaki, M. Yamaguchi, K. and Emura, "Feasibility study on over 1 bit/s/Hz high spectral efficiency WDM with optical duobinary coding and polarization interleave multiplexing," *OFC'97*, TuJ1.

[4] R.H. Stolen and A. Ashkin, "Optical Kerr effect in glass waveguide," *APL* **22**, 294-296, 1973.

[5] T. Kato, T. Tanimura, T. Hoshida, T. Richter, C. Schmidt-Langhorst, R. Elschner, C. Schubert, and S. Watanabe, "In-line optical signal multiplexing by polarization-insensitive fiber frequency conversion," *OFC2016*, Th2A.5.

Cascaded All-Optical Sub-Channel Add/Drop Multiplexing from a 1-Tb/s Super-Channel Having 2-GHz Guard-Bands

M. Song[1], E. Pincemin[1], B. Baeuerle[2], A. Josten[2], D. Hillerkuss[2], J. Leuthold[2], R. Rudnick[3], D. M. Marom[3], S. Ben-Ezra[4], J. F. Ferran[5], S. Sygletos[6], A. Ellis[6], J. Zhao[7], G. Thouenon[1], C. Betoule[1], J.M. Rivas[8], D. Klonidis[8] and I. Tomkos[8]

[1]Orange Labs Networks, Lannion, France. [2]ETH Zurich, Switzerland. [3]Hebrew University, Jerusalem, Israel.
[4]Finisar Ltd., Israel, [5]W-Onesys, Barcelona, Spain. [6]Aston University, Birmingham, UK.
[7]University College Cork, Cork, Ireland. [8]Athens Information Technology, Marousi, Greece.
Email: erwan.pincemin@orange.com

Abstract: We demonstrate cascaded 100-Gb/s sub-channel add/drop from a 1-Tb/s multi-band OFDM super-channel having 2-GHz inter-sub-channel guard-bands within a recirculating loop via a hierarchical ROADM using high-resolution filters, showcasing 1000-km transmission reach and five ROADM node passages for the add/drop sub-channel when hybrid Raman-EDFA is implemented.

Keywords: All-optical networks, Switching, Circuit, Optical Communications

I. Introduction

Elastic optical networking (EON) [1] is one of the key enablers for increasing fiber capacity utilization by adapting the network resources to the dynamically varying traffic demands, thus avoiding the requirement for over-provisioning [2]. This adaptability is determined by the characteristics of the defined elastic super-channel (Sp-Ch) and corresponding switching technology capabilities. Network traffic routing can be performed transparently at the Sp-Ch level in a similar way to today's WDM networks, but with spectrally adaptive channels and switching elements [3]. Although EON addresses the capacity-on-demand issue, it depends on costly and power consuming electronic aggregation/grooming functions (i.e. OTN) at network nodes. Any processing of the Sp-Ch contents requires the reception of the whole super-channel at the node and, subsequently, the electronic processing and switching/grooming of its sub-channel (Sb-Ch) contents. Recently, a new generation of elastic switching solutions based on high spectral resolution (HSR) optical filters and wavelength selective switches (WSSs) [4-6] has emerged. These filtering and switching elements offer dynamic all-optical traffic aggregation/grooming (AOTG) at the Sb-Ch level [7]. Such solutions can achieve ultra-fine switching granularity, resulting in significantly enhanced spectral utilization and reduced global network cost compared to electronic-based alternatives [7].

In this paper we demonstrate, for the first time, elastic networking with cascaded all-optical grooming-capable ROADMs able to perform sub-channel add/drop operations inside a 1-Tb/s (10×100 Gb/s) multi-band dual-polarization 16QAM-OFDM super-channel spanning 200 GHz, with 2-GHz inter-sub-channel guard-bands. We show that five (respectively, four) of these novel elastic optical cross-connects (or network ROADMs) can be cascaded in a 2×100-km G.652 fibre-based recirculating loop equipped with hybrid Raman-EDFA (respectively, pure EDFA), enabling 1000-km (respectively, 800-km) maximum transmission distance for the add/drop Sb-Ch. Compared to previous sub-system experiments, the HSR optical filters are steeper, the guard-band is halved, the modulation used for OFDM is 16-QAM (leading to 100 Gb/s per Sb-Ch and 5 bit/s/Hz spectral efficiency), and the Sp-Ch carries 1 Tb/s [8,9]. The obtained performance shows possibility of AOTG for the next generation of flexible optical transport networks.

II. Experimental Set-Up

The experimental set-up is depicted in Fig. 1. In the transmitter, the 1-Tb/s OFDM Sp-Ch is generated by combining ten 100-kHz linewidth external cavity lasers (ECL) spaced by 20 GHz, which feed two complex MZ modulators (CMZM). Two independent pairs of DAC (15 GHz, 64 GSa/s) generate the odd and even Sb-Chs, respectively. This ensures that the data carried by neighboring Sb-Chs are de-correlated. Using dual-polarization (DP) 16QAM-OFDM over 18 GHz of Sb-Ch bandwidth (BW) yields a raw bit rate of 144 Gb/s per Sb-Ch, and provides 2-GHz guard-band between Sb-Ch. After removing various overheads, such as FEC, cyclic prefix, training sequence and pilots, the nominal bit rate is 100 Gb/s. Six additional laser diodes (LD) are used to produce six other OFDM Sb-Chs (three on each side of the 1 Tb/s Sp-Ch), which are introduced to emulate fibre nonlinearities. They are located at 10 GHz from the last right and left Sb-Ch of the Sp-Ch. The 16 Sb-Chs are combined by two stages of 8:1 and 2:1 polarization-maintaining (PM) couplers. Polarization-maintaining Erbium-doped fibre amplifiers (PM-EDFA) are inserted to balance coupler and CMZM losses. We emulated a dual-polarization signal by combining the signal with a delayed replica on the orthogonal polarization. Finally, the 16 Sb-Chs are loaded with 53 additional WDM channels, each carrying 128-Gb/s DP-QPSK modulation, which fill the EDFA bandwidth and emulate fibre nonlinearities.

The transmitter is connected to the recirculating loop through an acousto-optic (AO) switch. The loop constitutes two 100-km spans of G.652 fibre equipped with either pure EDFA or hybrid Raman-EDFA. In the first configuration, EDFAs have 20-dB gain and 4.5-dB noise figure. In the second, two Raman pumps at 1435 nm and 1455 nm offer 12 dB of backward Raman gain, with an additional EDFA compensating for the remaining loss. The ROADM node is inserted after the fibre spans. A dynamic gain equalizer and a synchronous polarization scrambler achieve power equalization and polarization rotation of the WDM multiplex after each loop round-trip.

At the receiver side, the Sb-Ch under measurement and its immediate neighbors are selected by a square flat-top optical pass-band filter (OPBF) of 0.5 nm bandwidth, and detected by a polarization diverse coherent receiver using a 100-kHz linewidth ECL as local oscillator (LO). The signals are converted back to the digital domain by ADC embedded into a real-time storage oscilloscope (20 GHz, 50 GSa/s). The LO wavelength is tuned to the center of the Sb-Ch under measurement. The off-line OFDM digital signal processing is detailed in [9].

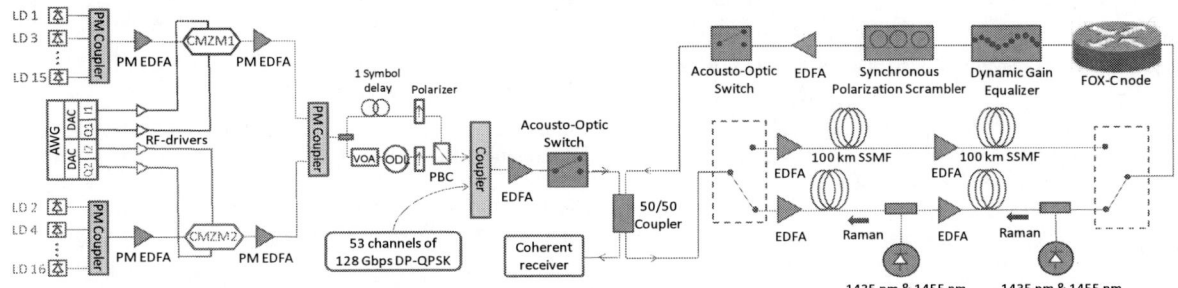

Fig. 1: Set-up of the OFDM transmitter and the 2×100-km G.652 fiber-based recirculating loop equipped with pure EDFA or Raman-EDFA.

The hierarchical structure of the ROADM node is shown in Fig. 2. It comprises two levels of switching: one for Sp-Ch routing at the fibre/network level, and a second for processing (i.e. Sb-Ch add/drop) at the Sp-Ch level. At the first level, a flexgrid LCoS-based WSS extracts the 1-Tb/s Sp-Ch and sends it to its lower output for handling, with the remaining WDM multiplex on its top output. At the second stage, the selected Sp-Ch is replicated with a 1:2 coupler. The first replica is used for the "drop" operation. The HSR optical pass-band filter (HSR-OPBF) extracts the sixth Sb-Ch from the Sp-Ch. The second copy on the upper arm of the 1:2 coupler is sent into a HSR optical stop-band filter (HSR-OSBF), which deletes this sixth Sb-Ch from the Sp-Ch. A second 1:2 coupler duplicates the extracted sixth Sb-Ch, with one copy used for measurements and optimization of the HSR-OPBF bandwidth and the other recombined ("add" signal) with the remaining Sp-Ch (from which the sixth Sb-Ch has been removed) by a 2:1 coupler. Decorrelating fibres are inserted for temporally shifting the Sp-Ch and the sixth Sb-Ch during recombination, to emulate crosstalk over the sixth Sb-Ch and its neighbors. The node terminates with a second flexgrid LCoS-WSS, which reinserts the reconstituted Sp-Ch into the WDM multiplex. Signal spectra before and after the node, and at the output of the HSR filters are shown in the right-hand part of Fig. 2. The principle and architecture of HSR filters used here and developed during the project are presented in [4,5].

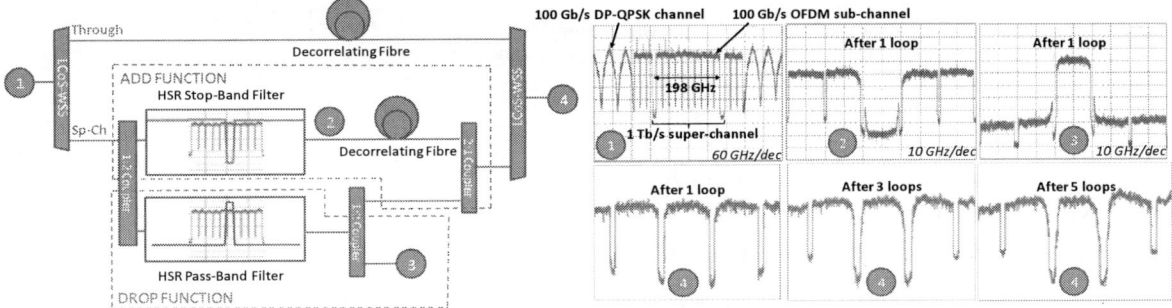

Fig. 2: On the left hand side, set-up of the ROADM node with its two levels of optical switching at the super-channel and sub-channel levels. On the right hand side, signal spectra at various positions of the ROADM node set-up: (1) before the node, (2-3) at the output of the HSR stop-band and pass-band filter after 1 loop round-trip, (4) at the output of the node after 1, 3 and 5 loop round-trips.

III. RESULTS AND DISCUSSIONS

The impact of fine resolution Sb-Ch add/drop is first evaluated in back-to-back. The ROADM node and noise loading elements are placed between the transmitter and receiver. Fig. 3 presents the OSNR sensitivity curves of the central Sb-Ch (n°6) and left-right Sb-Chs (n°5 and 7) with and without the node. An error floor ~1-2×10^{-3} appears on the sensitivity curves of the central and adjacent Sb-Ch of 18-GHz BW, due to crosstalk and attenuation on the subcarriers of the Sb-Ch edges. To limit the filtering impact, two other configurations with edge subcarriers eliminated from the BER calculation are investigated, corresponding to a BW support of 16.5 GHz and 15 GHz. Reducing the BW does not modify the OSNR sensitivity without the ROADM node, but significantly improves the performance with it. For a 15-GHz BW, the OSNR penalty of the add/drop operation is limited to ~1 dB at 1×10^{-3} BER.

OECC/PS2016

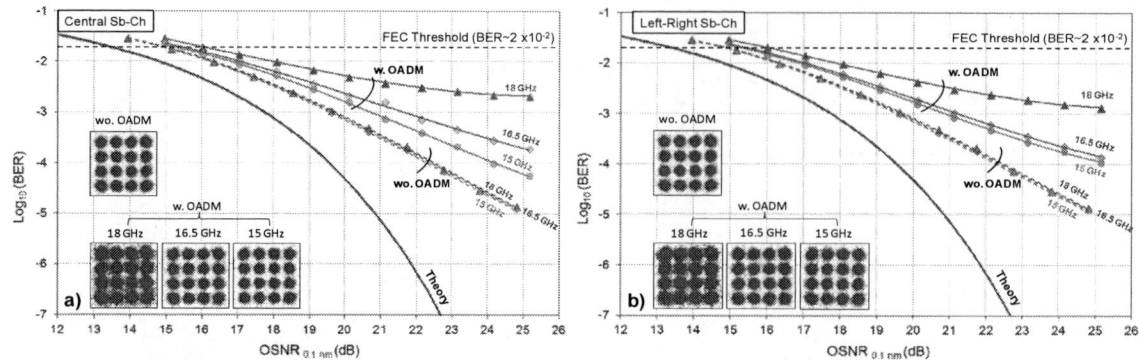

Fig. 3: BER versus OSNR (in 0.1 nm) in BtB with (w.) and without (wo.) the ROADM node for various BER calculation BW (i.e. 18 GHz, 16.5 GHz and 15 GHz) for (a) central Sb-Ch and (b) left-right Sb-Ch. Insets show constellations w. and wo. the ROADM node at OSNR= 22.5 dB.

Cascadability of the high resolution Sb-Ch add/drop is studied in a recirculating loop. The transmission results with and without the ROADM node for BER calculation BW of 18 GHz and 15 GHz are shown in Fig. 4. Without the node, the BER is below 2×10^{-2} after 1000 km and 1400 km for pure EDFA and hybrid Raman-EDFA, respectively. With the node and at 18-GHz BW (Fig. 4a), error free transmission (considering a FEC threshold of 2×10^{-2}) is achieved over 400 km and 800 km for the central Sb-Ch using pure EDFA and hybrid Raman-EDFA, respectively. Increasing the FEC threshold to 3.8×10^{-2} [10] allows transmission over 800 km and 1000 km, respectively. These distances correspond to 4 and 5 loop round-trips with one add/drop operation in each loop round-trip. The left-right sub-channel reach in the same configuration is 600 km and 800 km respectively, slightly worse than that of the central Sb-Ch. This is due to lower extinction ratio of HSR-OPBF with respect to HSR-OSBF (see spectra n°2 and n°3 of Fig. 2) which results in higher crosstalk. Decreasing the BW from 18 GHz to 15 GHz gives significant performance improvements for the central Sb-Ch, however, the left-right Sb-Chs only exhibit a slight performance improvement due to this crosstalk (Fig. 4b).

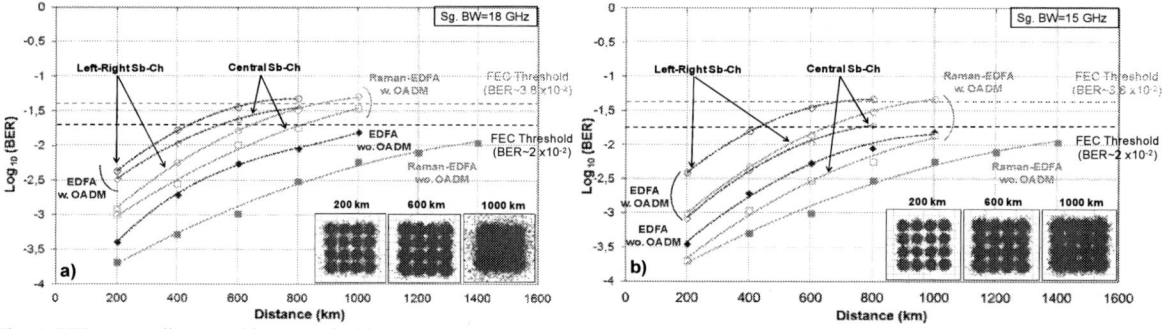

Fig. 4: BER versus distance with (w.) and without (wo.) the ROADM node for BER calculation BW of (a) 18 GHz and (b) 15 GHz. Insets show constellations of the central Sb-Ch with the ROADM node and hybrid Raman-EDFA after 200, 600 and 1000 km (1,3 and 5 loops).

IV. CONCLUSIONS

We demonstrated cascadability of high spectral resolution add/drop of 100 Gb/s Sb-Ch from a 1 Tb/s 16QAM-DP-OFDM Sp-Ch with 2-GHz guard-band between adjacent Sb-Chs. All-optical add/drop of Sp-Ch and Sb-Ch opens the way to flexible spectrum and cost-efficient meshed EON, addressing the looming capacity crunch.

ACKNOWLEDGMENT: We acknowledge funding by the EC under the FP7 program, project FOX-C (grant no. 318415).

[1] M. Jinno et al., "Spectrum-efficient and scalable elastic optical path network: architecture, benefits, and enabling technologies," IEEE Comm. Mag., vol. 47, pp. 66–73, 2009.

[2] O. Gerstel et al., "Elastic optical networking: a new dawn for the optical layer?" IEEE Comm. Mag., vol. 50, pp. s12–s20, 2012.

[3] S. Poole et al., "Bandwidth-flexible ROADMs as network elements," Opt. Fiber Commun. Conf., Los Angeles, USA, 2011, paper OTuE1.

[4] R. Rudnick et al., "Sub-banded / single-sub-carrier drop-demux and flexible spectral shaping with a fine resolution photonic processor," European Conf. Opt. Commun., Cannes, France, 2014, Postdeadline paper 4.1.

[5] R. Rudnick et al., "One GHz Resolution Arrayed Waveguide Grating Filter with LCoS Phase Compensation", Opt. Fiber Commun. Conf., San Francisco, USA, 2014, paper Th3F.7.

[6] S. Sygletos et al., "A Novel Architecture for All-Optical Add-Drop Multiplexing of OFDM Signals", European Conf. Opt. Commun., Cannes, France, 2014, paper We.1.5.4.

[7] G. Thouenon et al., "Electrical v/s Optical Aggregation in Multi-layer Optical Transport Networks," Photonics in Switching, Florence, Italy, 2015, pp. 28-30.

[8] A. Klekamp et al., "Transmission Reach of Optical-OFDM Super-channels with 10-600 Gb/s for Transparent Bit-Rate Adaptive Networks," European Conf. Opt. Commun., Geneva, Switzerland, 2011, paper Tu.3.K.2.

[9] E. Pincemin et al., "Multi-Band OFDM Transmission at 100 Gbps With Sub-Band Optical Switching," IEEE/OSA J. Lightwave Techn., vol. 32, pp. 2202-2219, June 2014.

[10] M. Yang et al., "Design of efficiently encodable moderate-length high-rate irregular LDPC codes," IEEE Trans. on Communications, vol. 52, pp. 564-571, April 2004.

MB2-5

OECC/PS2016

1024 QAM, 7-core Fiber/Multi-core EDFA Transmission over 100 km with an Aggregated Spectral Efficiency of 109 bit/s/Hz

Masato Yoshida, Keisuke Kasai, Toshihiko Hirooka, and Masataka Nakazawa
Research Institute of Electrical Communication, Tohoku University, 2-1-1 Katahira, Aoba-ku, Sendai-shi,
Miyagi-ken, 980-8577 Japan
masato@riec.tohoku.ac.jp

Abstract: We report the first 1024-QAM, 7-core fiber (60 Gbit/s x 7) transmission over 100 km. The spectral efficiency per core reached 15.6 bit/s/Hz, which corresponds to an aggregate spectral efficiency as high as 109 bit/s/Hz.

Keywords: Multi-level QAM coherent optical transmission, multi-core fiber

I. INTRODUCTION

To expand the transmission capacity per fiber toward the > 1 Pbit/s region, the combination of three "multi" technologies, i.e., multi-level modulation, multi-core fiber (MCF), and multi-mode control, is expected to be the ultimate approach [1]. Specifically, 1 Exabit/s·km transmission experiments have been reported that used a combination of multi-level modulation and MCF [2],[3], where a duobinary QPSK or 16 QAM signal was transmitted over more than 1000 km with multi-core erbium-doped fiber amplifiers (MC-EDFAs). However, multiple relay MCF transmission with a QAM multiplicity of more than 32 has not yet been reported.

In this paper, we demonstrate the first 1024 QAM, 7-core fiber transmission over a 100 km with our ultra-multilevel coherent QAM optical transmission technology [4]. An MC-EDFA was installed after a 55 km transmission. Thus, a 420 Gbit/s (60 Gbit/s x 7 cores) data signal was successfully transmitted with an optical bandwidth of 3.6 GHz, resulting in an aggregate spectral efficiency of 109 bit/s/Hz per fiber.

II. EXPERIMENTAL SETUP FOR 1024 QAM, 7-CORE FIBER TRANSMISSION OVER 100 KM

The experimental setup for 1024 QAM, 7-core fiber transmission is shown in Fig. 1. As a coherent light source, we used a 1.5 μm acetylene frequency-stabilized fiber ring laser (1538.8 nm) with a linewidth of 4 kHz. The output of the laser was modulated by an IQ modulator with a 3 Gsymbol/s 1024 QAM baseband signal produced by an arbitrary waveform generator (AWG) operating at 12 Gsample/s with a 12-bit resolution. We employed a raised-cosine Nyquist filter with a roll-off factor α = 0.2 at the AWG so that the bandwidth of the QAM signal was reduced to 3.6 GHz. In addition, we adopted a pre-equalization process based on frequency domain equalization (FDE) with an FFT size of 16384 to provide high-resolution compensation for distortions caused by individual components such as the AWG and the IQ modulator. Furthermore, the pilot tone signal used for the optical phase-locked loop (OPLL) in the receiver was also generated at the AWG and embedded in the signal at a frequency down-shifted by 1.8 GHz from the center. The power of the pilot tone signal was set - 20 dB lower than the 1024 QAM signal, which was optimally chosen to

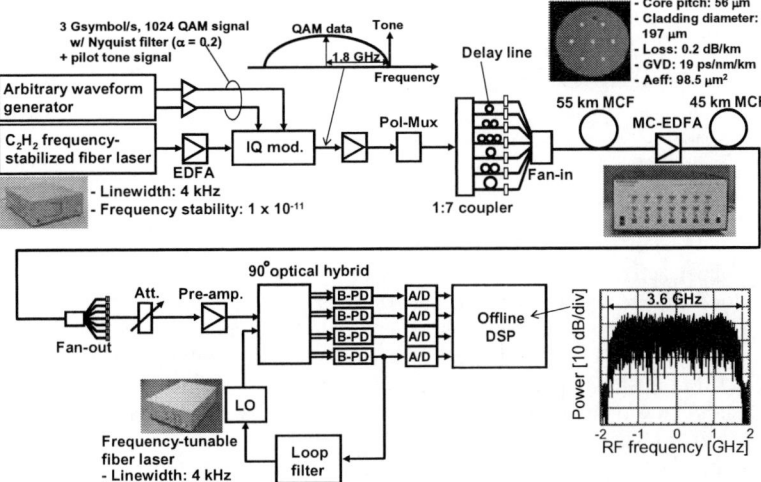

Fig. 1. Experimental setup for 1024 QAM, 7-core (60 Gbit/s x 7) coherent transmission over 100 km with an MC-EDFA as an optical repeater.

maximize the OSNR of the QAM signal and minimize the phase error in the OPLL circuit. The optical 1024 QAM signal was then orthogonally polarization-multiplexed, and 60 Gbit/s data were obtained.

In order to employ space division multiplexing, the obtained QAM signal was amplified to + 8 dBm and then split into 7 paths with a relative delay between them for decorrelation. The 7 channels were then sent to 7 input SMF ports of a fan-in device to couple them to 7 cores of a 100 km MCF transmission line. An MC-EDFA was installed as an optical repeater. The MC-EDFA is based on a core-pumping scheme in which separate 980 nm single-mode laser diodes (LDs) are used to pump each of the cores [5]. The gain and noise figure for an input signal power of - 10 dBm were more than 20 dB and 5.0 dB, respectively. The MCF fan-in device was composed of thinly-clad fibers bundled in a glass capillary [6], where the insertion loss and crosstalk were < 0.3 dB and < - 60 dB, respectively. The 100 km transmission line was composed of 55 km and 45 km MCFs with seven homogeneous cores. The core pitch of the MCF was 56.1 μm and the cladding diameter was 197 μm. The transmission loss of the MCF was 0.2 dB/km at 1550 nm, and the dispersion and effective area were 19 ps/nm/km and 98.5 μm², respectively. The QAM signals were coupled into each core with a launch power of - 2 dBm, which was chosen as the optimum value with which to maximize the OSNR and minimize the nonlinear impairments.

After a 100 km MCF transmission, the 7 core channels were spatially demultiplexed with the fiber bundle-type fan-out device, and fed into a QAM receiver. At the receiver, the QAM signal was homodyne-detected at a polarization-diverse 90° optical hybrid. As a local oscillator (LO), we used a frequency-tunable fiber laser whose phase was locked to the pilot tone transmitted with the data signal via the OPLL. The phase noise of the IF signal was 0.37 degrees. After detection with balanced photodiodes, the QAM data were A/D-converted (40 Gsample/s) and processed with a DSP in an off-line condition. In the DSP, we compensated for fiber nonlinearities and dispersion simultaneously by using a digital back-propagation (DBP) method. We employed a split-step Fourier analysis of the Manakov equation, which describes the pulse propagation in a fiber with dispersion, SPM, and XPM between two orthogonal polarizations. After that, we adopted FDE with an FFT size of 16384 to compensate for residual distortions caused by hardware imperfections in the receiver. Finally, the compensated QAM signal was demodulated into binary data, and the BER was evaluated.

III. EXPERIMENTAL RESULTS

We first evaluated the crosstalk of the 100 km MCF transmission link between the center core and the outer cores with a conventional transmission method. The crosstalk from the six outer cores to the center core was - 43 dB. In considering the theoretical BER values for a 1024 signal, an error-free condition with an FEC (assuming BER=2 x 10⁻³) can be obtained with an SNR of 33.5 dB. Therefore, the crosstalk of - 43 dB is negligible as regards the demodulation of a 1024 QAM signal, which enables us to realize QAM transmissions with high multiplicities of 1024.

Figure 2 shows the optical spectra of 1024 QAM signals before and after transmission, which were measured at the center core. The OSNR after a 100 km transmission was 38.6 dB as shown in Fig. 2(b), which was determined by the ASE noise at the MC-EDFA (NF = 5 dB). As shown in Fig. 2(c), the OSNR after pre-amplification with an EDFA installed in front of the receiver was reduced to 36.3 dB due to the ASE noise from the pre-amplifier (NF = 4 dB).

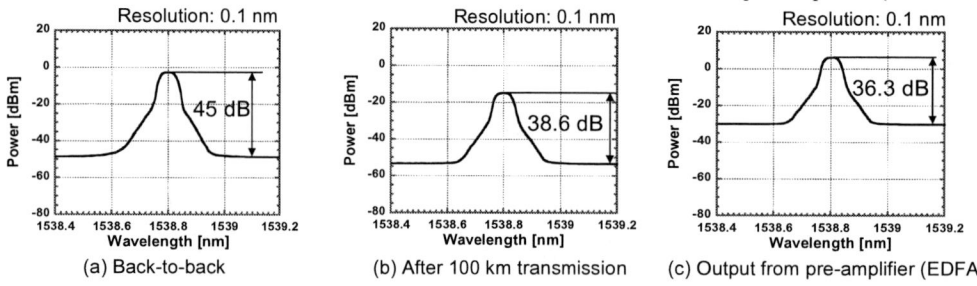

Fig. 2. Optical spectra of 1024 QAM signals before and after transmission, measured at the center core.

Figure 3(a) and 3(b) show the constellations of the 1024 QAM signals at the center core with a received power of - 15 dBm, measured for back-to-back and 100 km transmissions, respectively. The error vector magnitude (EVM) increased slightly from 0.85 % to 1.05 %. The EVM degradation after the transmission was caused by the imperfect linear and nonlinear compensation of the total system. Here, a constellation after a 100 km transmission obtained without compensation for the XPM between the two polarizations is also shown in Fig. 3(c). Among various linear and nonlinear transmission impairments, we can easily pre-compensate for chromatic dispersion and SPM without the DBP method. However, the distortions caused by the XPM between the two polarizations still remain and degrade the performance of a 1024 QAM transmission as shown in Fig. 3(c). This indicates that compensation for XPM is important for a 1024 QAM transmission although the transmission distance is as short as 100 km. Figure 4 shows the BER characteristics for each spatial channel after a 100 km transmission as a function of the received power defined by the input power to the pre-amplifier. After 100 km transmission, all the channels achieved a BER below the FEC limit with a power penalty of 7.5 dB from back-to-back. It can also be seen that the BERs for core 1 and cores 2~7 are almost

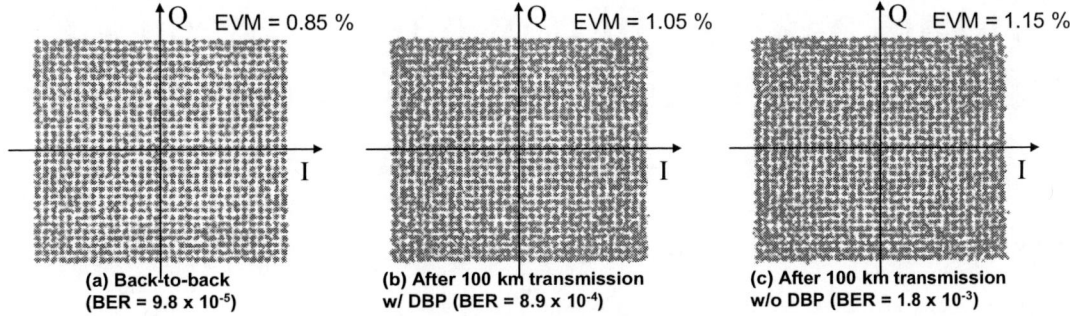

Fig. 3. Constellations of 1024 QAM signals before and after transmission, measured at the center core with a received power of - 15 dBm.

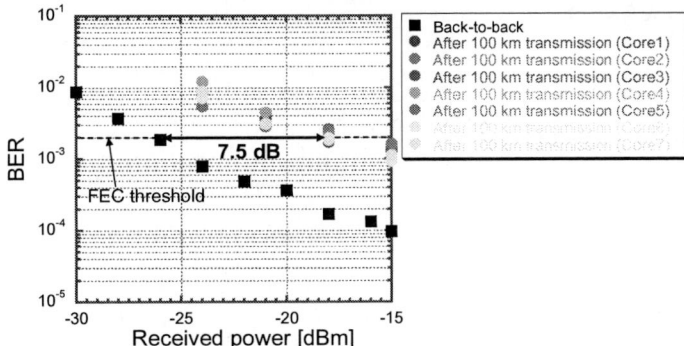

Fig. 4. BER characteristics for 420 Gbit/s (60 Gbit/s x 7 core) polarization-multiplexed 1024 QAM transmission over 100 km.

identical, although the center core is generally most susceptible to crosstalk due to the surrounding outer cores. This indicates that the influence of crosstalk on the transmission performance remains negligible over 100 km.

In the present experiment, a capacity of 60 Gbit/s was transmitted in each core by the polarization-multiplexed 1024 QAM signal within a bandwidth of 3.6 GHz, and thus the potential spectral efficiency reached 60 Gbit/s / 3.6 GHz / 1.07 = 15.6 bit/s/Hz when the 7% FEC overhead was taken into account. Therefore, the present single-mode MCF transmission experiment provided an aggregate spectral efficiency of 109 bit/s/Hz per fiber.

IV. CONCLUSIONS

We described a 1024 QAM polarization-multiplexed transmission at 3 Gsymbol/s over a 100 km 7-core fiber, MC-EDFA, with a total bit rate of 420 Gbit/s (60 Gbit/s x 7 cores) and a potential spectral efficiency as high as 15.6 bit/s/Hz per core. This corresponds to an aggregate spectral efficiency of 109 bit/s/Hz, which is the highest value ever achieved in a single-mode MCF. The obtained results indicate that the influence of MCF crosstalk is still negligible over this distance.

ACKNOWLEDGMENT

This work is supported by the National Institute of Information and Communications Technology (NICT), Japan, as part of the "R&D of Innovative Optical Communication Infrastructure."

REFERENCES

[1] M. Nakazawa, "Giant leaps in optical communication technologies towards 2030 and beyond," European Conference on Optical Communication (ECOC 2010), Plenary Talk, September (2010)

[2] K. Igarashi, T. Tsuritani, I. Morita, Y. Tsuchida, K. Maeda, M. Tadakuma, T. Saito, K. Watanabe, R. Sugizaki, and M. Suzuki, "1.03-Exabit/s·km super-Nyquist-WDM transmission over 7,326-km seven-core fiber," ECOC 2013, PDP3.E.3.

[3] T. Kobayashi, H. Takara, A. Sano, T. Mizuno, H. Kawakami, Y.Miyamoto, K. Hiraga, Y. Abe, H. Ono, M. Wada, Y. Sasaki, I. Ishida, K. Takenaga, S. Matsuo, K. Saitoh, M. Yamada, H. Masuda, and T. Morioka, "2 × 344 Tb/s propagation-direction interleaved transmission over 1500-km MCF enhanced by multicarrier full electric-field digital back-propagation," ECOC 2013, PDP3.E.4.

[4] S. Beppu, M. Yoshida, K. Kasai, and M. Nakazawa, "2048 QAM (66 Gbit/s) single-carrier coherent optical transmission over 150 km with a potential SE of 15.3 bit/s/Hz," Opt. Express 23, 4960-4969 (2015).

[5] Y. Tsuchida, K. Maeda, K. Watanabe, T. Ito, K. Fukuchi, M. Yoshida, Y. Mimura, R. Sugizaki, and M. Nakazawa, Multicore EDFA for DWDM transmission in full C-band," Optical Fiber Communication Conference (OFC 2013) JW2A.16.

[6] K. Watanabe, T. Saito, K. Imamura, and M. Shiino, "Development of fiber bundle type fan-out for multicore fiber," Opto Electronics and Communications Conference (OECC 2012) 5C1–2.

MC1-1

Accurate analysis of crosstalk between LP₁₁ degenerate modes due to offset connection using exact eigenmodes

Yasuo Kokubun, Seiya Miura, Tatsuhiko Watanabe

Yokohama National University, Graduate School of Eng., 79-5 Tokiwadai, Hodogaya-ku, Yokohama, Japan 240-8501
kokubun-yasuo-sd@ynu.ac.jp

Abstract: The crosstalk between LP_{11} degenerate modes due to offset connection is accurately analyzed using exact eigenmodes and it is shown the propagation characteristics of few-mode fibers must be analyzed by eigenmode not by LP mode.

Keywords: mode division multiplexing, few mode fiber, LP mode, exact eigenmode, connection crosstalk

I. INTRODUCTION

LP mode[1] is widely used to describe the light propagation in a few mode fiber. It has been known, however, that an LP mode is no more than an approximated mode and is expressed by the linear combination of exact eigenmodes, such as HE, EH, TE, and TM modes[2]. Since the light emitted from LD is linearly polarized and the input signal can be regenerated at the output end using MIMO-DSP, even though the transmission channels are mixed during the transmission, the mode division multiplexing has been demonstrated based on LP modes[3]-[8]. In addition, since the eigenmodes constituting an LP mode are quasi-degenerated, the propagation characteristics of few mode fibers have been expressed using the LP mode. Therefore, there has been no report on the fact that some kind of analysis based on the LP mode leads false conclusions. One of such false conclusions is that the orthogonally polarized LP modes can't be coupled to each other by the offset connection or by the perturbation like core-cladding boundary.

In this study, the authors analyze the evolution of field profile of LP₁₁ degenerate modes in terms of exact eigenmodes and show how the coupling between LP₁₁ degenerate modes at the connection point with axial offset can be expressed in terms of exact eigenmodes. It is shown from this accurate analysis that the field distribution of LP₁₁ degenerate modes varies along with the propagation and is expressed by the elliptical polarization, of which ellipticity and the direction of polarization rotation depend on the local position. As a result, it is revealed that the propagation characteristics such as the crosstalk due to the offset connection should be analyzed by the eigenmodes not by the LP modes.

II. RELATION BETWEEN LP MODES AND EXACT EIGENMODES

According to the definition of exact eigen-modes of multi-mode round optical fiber[1], the electric field profile can be expressed in terms of cylindrical coordinate. On the other hand, since LP modes[2] have linearly polarized electric field as summarized in Table I, the electric field should be transformed into Cartesian coordinate. For example, the x and y components of electric field of LP_{11x}^{even} mode are expressed by the following equations.

$$E_x^{LP11-e} = j\frac{\beta_{aM}}{\kappa} A_{LP11}^e e^{j(\omega t - \beta_{aM} \cdot z)}$$
$$\times J_1(\kappa r)\cos(\theta)\left[2\cos(\delta\beta_M \cdot z)\right], \quad (1)$$

$$E_y^{LP11-e} = j\frac{\beta_{aM}}{\kappa} A_{LP11}^e e^{j(\omega t - \beta_{aM} \cdot z)}$$
$$\times J_1(\kappa r)\sin(\theta)\left[2j\sin(\delta\beta_M \cdot z)\right]. \quad (2)$$

TABLE I DEFINITION OF DEGENERATE LP MODES

where β_{aM} and $\delta\beta_M$ are the average and difference of propagation constants of HE₂₁ and TM₀₁ modes defined by

$$\beta_{aM} = \frac{\beta_{HE21} + \beta_{TM01}}{2}, \quad \delta\beta_M = \frac{\beta_{HE21} - \beta_{TM01}}{2}, \quad (3)$$

respectively. It is seen from Eqs. (1) and (2) that the electric field of LP_{11x}^{even} mode is no longer linearly polarized even if the x- polarized light is incident on the input end, and is elliptically polarized because the phases of x and y components differ by $\pi/2$. In addition, the LP_{11x}^{even} mode is evolved into LP_{11y}^{odd} mode at the propagation distance of

---

*This work was supported by the National Institute of Information and Communication Technology (NICT), Japan, under "R&D of Innovative Optical Fiber and Communication Technology".

$$z = \frac{L_b^{\mathrm{TMH}}}{2} = \frac{\pi}{2\delta\beta_M} \quad . \tag{4}$$

L_b^{TMH} is in the order of several meters for V=3.3-3.8[9]. Thus the LP_{11x}^{even} and LP_{11y}^{odd} modes are the same mode group, called TMH mode group[9], because these modes consist of HE_{21}^{even} and TM_{01} modes.

On the other hand, LP_{11x}^{odd} mode evolves into LP_{11y}^{even} mode in the same way, but the period L_b^{TEH} is in the order of 10-20cm for V=2.4-3.8[9], which is much shorter than L_b^{TMH}. LP_{11x}^{odd} and LP_{11y}^{even} modes consist of HE_{21}^{odd} and TE_{01} modes.

The above relation between quasi-degenerate LP modes and exact eigenmodes is expressed by the following transform matrix.

$$\begin{bmatrix} LP_{11-x}^{\mathrm{even}} \\ LP_{11-y}^{\mathrm{odd}} \\ LP_{11-x}^{\mathrm{odd}} \\ LP_{11-y}^{\mathrm{even}} \end{bmatrix} = \frac{1}{\sqrt{2}} \cdot \begin{bmatrix} 1 & -1 & 0 & 0 \\ 1 & 1 & 0 & 0 \\ 0 & 0 & 1 & -1 \\ 0 & 0 & 1 & 1 \end{bmatrix} \begin{bmatrix} TM_{01} \\ HE_{21}^{\mathrm{even}} \\ HE_{21}^{\mathrm{odd}} \\ TE_{01} \end{bmatrix} \tag{5}$$

Now the matrix in Eq.(5) is defined as M1. The evolution of field profile of TMH group and TEH group modes along with the propagation are illustrated in Fig.1 (a) and (b). This evolution of field profile was observed by changing the wavelength and the result is shown in Fig.2 for the case of LP_{11x}^{even} mode input (TMH group). Fig.2 (a) – (d) show the evolution of intensity profile corresponding to the profile at z=0 to that at z=0.25L_b shown in Fig.1 (a).

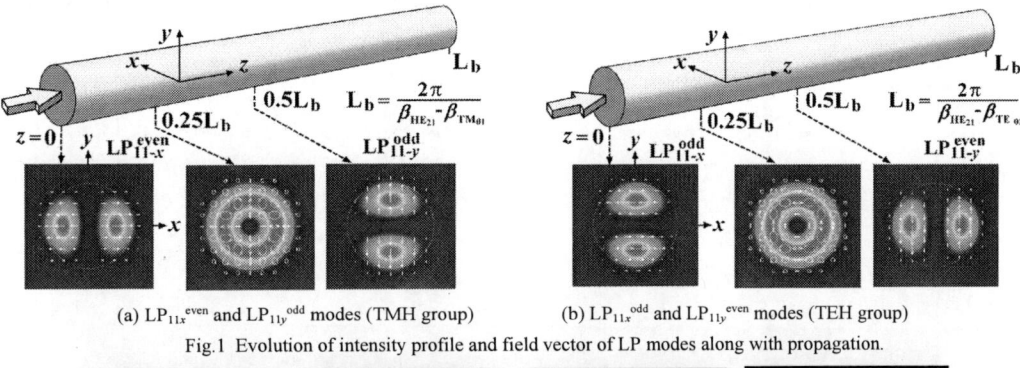

(a) LP_{11x}^{even} and LP_{11y}^{odd} modes (TMH group)   (b) LP_{11x}^{odd} and LP_{11y}^{even} modes (TEH group)

Fig.1 Evolution of intensity profile and field vector of LP modes along with propagation.

(a) λ=1550.5 nm   (b) λ=1551.5 nm   (c) λ=1552.5 nm   (d) λ=1553.5 nm

Fig. 2 Measured NFPs of SI-FMF for LP_{11x}^{even} mode input. (Fiber length is 100m, Δ=0.35%, 2a=13.3μm)

III. ACURATE ANALYSIS OF COUPLING CROSSTALK USING EXACT EIGENMODES

At the connection point of few mode fibers, mode coupling occurs if axial offset is induced between two cores. The field coupling coefficient C^E between exact eigenmodes can be calculated by the overlap integral given by

$$C^E = \frac{\int_0^{2\pi}\int_0^\infty \boldsymbol{E}_t^{\mathrm{in}} \cdot \boldsymbol{E}_t^{\mathrm{out}*} \, r\,dr\,d\theta}{\sqrt{\left[\int_0^{2\pi}\int_0^\infty |\boldsymbol{E}_t^{\mathrm{in}}|^2 \, r\,dr\,d\theta\right] \cdot \left[\int_0^{2\pi}\int_0^\infty |\boldsymbol{E}_t^{\mathrm{out}}|^2 \, r\,dr\,d\theta\right]}} \tag{6}$$

where $\boldsymbol{E}_t^{\mathrm{in}}$ and $\boldsymbol{E}_t^{\mathrm{out}}$ are the transverse components of electric field at the input and output ends. Using this field coupling coefficient, the coupling between eigenmodes at the connection point with offset d is expressed by

$$\begin{bmatrix} TM_{01} \\ HE_{21}^{\mathrm{even}} \\ HE_{21}^{\mathrm{odd}} \\ TE_{01} \end{bmatrix}_{\mathrm{out}} = \begin{bmatrix} P(d) & Q(d,\Theta) & R(d,\theta) & 0 \\ Q(d,\theta) & P(d) & 0 & -R(d,\Theta) \\ R(d,\Theta) & 0 & P(d) & Q(d,\Theta) \\ 0 & -R(d,\Theta) & Q(d,\Theta) & P(d) \end{bmatrix} \cdot \begin{bmatrix} TM_{01} \\ HE_{21}^{\mathrm{even}} \\ HE_{21}^{\mathrm{odd}} \\ TE_{01} \end{bmatrix}_{\mathrm{in}} \tag{7}$$

where Θ is the azimuth angle and the components $P(d)$, $Q(d, \Theta)$, and $R(d, \Theta)$ are illustrated in Fig.3 for d=0-11μm. The matrix in Eq.(7) is defined as M3. The Θ dependences of $Q(d, \Theta)$ and $R(d, \Theta)$ are periodic with the period of $\pi/2$ and the phase of periodicity differs by $\pi/4$. Since the polarization of input light is usually linearly polarized, either of LP_{11x}^{even}, LP_{11y}^{odd}, LP_{11x}^{odd}, and LP_{11y}^{even} modes is excited at the input end. However, the field profile varies as illustrated in Fig.1 along with the propagation, the coupling between LP modes at the connection point can be expressed by the product of matrices as follows.

$$
\begin{bmatrix} LP_{11-x}^{even} \\ LP_{11-y}^{odd} \\ LP_{11-x}^{odd} \\ LP_{11-y}^{even} \end{bmatrix}_{out} = [M1]\cdot[M3]\cdot[M2]\cdot[M1]^{-1}\cdot \begin{bmatrix} LP_{11-x}^{even} \\ LP_{11-y}^{odd} \\ LP_{11-x}^{odd} \\ LP_{11-y}^{even} \end{bmatrix}_{in}
\tag{8}
$$

where matrix M2 is the diagonal matrix expressing the propagation of eigenmodes, of which diagonal elements are $\exp(-j\beta_\nu z)$ (ν: either of eigenmodes).

Using this analytical method, the power coupling efficiency between LP modes can be calculated. For example, the power coupling efficiency from LP_{11x}^{even} mode input to LP_{11x}^{even} and LP_{11y}^{odd} modes was calculated as shown in Fig.4. When the phase difference $\delta\beta_M z=0$, the coupling occurs from LP_{11x}^{even} to LP_{11x}^{even} mode and there is no coupling between LP_{11x}^{even} and LP_{11y}^{odd} modes because the polarization directions of these two modes are orthogonal to each other. The coupling also occurs from LP_{11x}^{even} to LP_{11x}^{odd} mode, but the power coupling efficiency is quite small (<20dB) because the axes of odd symmetry are orthogonal to each other for these two modes.

On the other hand, however, when the phase difference $\delta\beta_M z=\pi/4$, the coupling between LP_{11x}^{even} and LP_{11y}^{odd} modes occurs opposite to the analysis using LP modes. Therefore, the propagation characteristics of few mode fibers should be analyzed by exact eigenmodes not by LP modes.

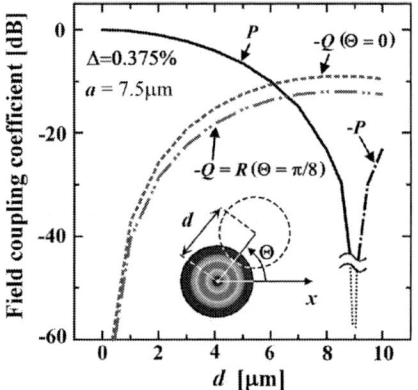

Fig.3 Field coupling coefficient of eigen-modes against offset at connection point.

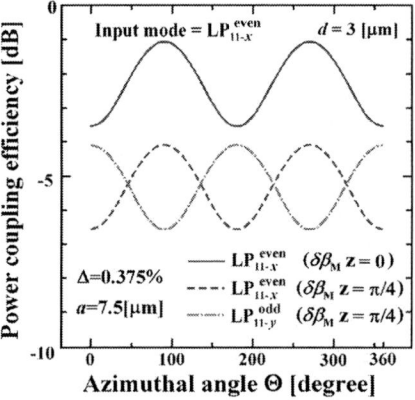

Fig.4 Power coupling efficiency between LP_{11x}^{even} and LP_{11y}^{odd} modes for different propagation distances.

IV. CONCLUSIONS

Since the electro-magnetic field distribution of few mode fiber is no longer linearly polarized during the propagation, the propagation characteristics such as the crosstalk due to the offset connection and the mode coupling resulting from the perturbation should be analyzed using the exact eigenmode.

REFERENCES

[1] D. Gloge, "Weakly guiding fibers," Appl. Opt., vol.10, no.10, pp.2252-2258, Oct. 1971.
[2] E. Snitzer, "Cylindrical dielectric waveguide modes," J. Opt. Soc. of Am., vol.51, no.5, pp.491-498, May 1961.
[3] Ip, N. Bai, et al., "88×3×112-Gb/s WDM Transmission over 50 km of Three-Mode Fiber with Inline Few-Mode Fiber Amplifier," ECOC2011, Geneva, Th.13.C.2, 2011.
[4] V.A.J.M. Sleiffer, et al., "73.7 Tb/s (96X3x256-Gb/s) mode-division-multiplexed DP-16QAM transmission with inline MM-EDFA," ECOC2012, Amsterdam, Th.3.C.4, Sept. 2012.
[5] K. Shibahara, et al., "Dense SDM (12-core x 3-mode) Transmission over 527 km with 33.2-ns Mode-Dispersion Employing Low-Complexity Parallel MIMO Frequency-Domain Equalization," OFC2015, Los Angeles, Th5C.3, March 2015.
[6] J. Sakaguchi, et al., "Realizing a 36-core, 3-mode Fiber with 108 Spatial Channels," OFC2015, Los Angeles, Th5C.2, March 2015.
[7] D. Soma, et al., "2.05 Peta-bit/s Super-Nyquist-WDM SDM Transmission Using 9.8-km 6-mode 19-core Fiber in Full C band," 29th European Conference on Optical Communication (ECOC2015), Valencia, PDP.3.2, Oct. 1, 2015.
[8] R. Ryf, et al., "10-Mode Mode-Multiplexed Transmission over 125-km Single-Span Multimode Fiber," ECOC2015, Valencia, PDP.3.3, Oct. 1. 2015.
[9] H. Kogelnik and P. J. Winzer, "Modal Birefringence in Weakly Guiding Fibers," J. Lightwave Technol., vol.30, no.14, pp.2240-2245, July 2015.

OECC/PS2016

New Method for Measuring Inter-Core Crosstalk in Multi-Core Fibers Using Near-Infrared Camera

Shota Saitoh, Yoshimichi Amma, Yusuke Sasaki, Katsuhiro Takenaga, and Kazuhiko Aikawa

Advanced Technology Laboratory, Fujikura Ltd., 1440, Mutsuzaki, Sakura, Chiba, 285-8550, Japan.
shota.saito@jp.fujikura.com

Abstract: A new method using a near-infrared camera for measuring inter-core crosstalk in multi-core fibers is presented. We show that an appropriate evaluation of output power enables good estimation of crosstalk in dual-core and 7-core fibers.
Keywords: Multi-core fibers, Inter-core crosstalk

I. INTRODUCTION

In order to exceed the capacity limits of optical-transmission systems based on conventional single-mode fibers (SMFs), considerable attention has been paid to space-division multiplexing, including multi-core fibers (MCFs), few-mode fibers, and few-mode MCFs (FM-MCFs). High-density MCFs with more than 20 cores have been recently reported [1–4]. In such MCFs, the measurement of the inter-core crosstalk (XT) by the conventional method using a power meter (PM) [5–6] is time-consuming because these MCFs have many cores and the number of alignments with each core for receiving output power increases.

In this paper, we propose a new method for measuring XT using a near-infrared (NIR) camera that does not need alignment for receiving power. The measurement results obtained by the proposed method employing dual-core fibers (DCFs) and 7-core fibers (7CFs) agree well with those obtained by the conventional method. We also show that the proposed method enables the simultaneous measurement of multiple XTs.

II. BRIEF EXPLANATION OF NEW METHOD FOR MEASURING XT

Figure 1 shows the conventional method using a PM [5]. A single-core fiber (SCF) is spliced to excite one core of a MCF, and another SCF is aligned to the opposite end of the MCF to receive the output power. This method is simple and allows the easy measurement of MCFs that do not have many cores. In addition, the wavelength-sweeping method [6] is available for the statistical characterization of the XT behavior. However, for high-density MCFs, the number of alignments increases; consequently, the measurement requires time and effort.

Figure 2 shows the new method employing an NIR camera. We use the camera and lenses to take a near-field picture from the MCF instead of the PM and the SCF. As shown in Fig. 2, there is no need to align the SCF with each core. We also employ filters for masking or absorbing the output power from the excited core. Typically, the inter-core XT in uncoupled MCFs should be approximately -20 dB or lower, in the case of using multi-level modulation [7]. This means that the output power from the excited core is far stronger than the XT power and can destroy the photodetectors of the camera. Therefore, we insert a spatial filter to partly mask the output power from the excited core in the lens barrel for taking a picture of the XT. When we take a picture of the excited core itself, a neutral-density (ND) filter (30 dB absorption) is inserted. A variable optical attenuator (VOA) is used to control the output power, which appears in the picture and is evaluated as described in Section III.

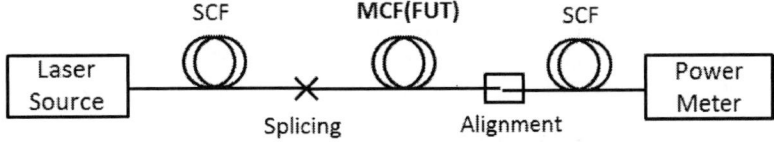

Fig. 1. Conventional method for measuring XT by a PM.

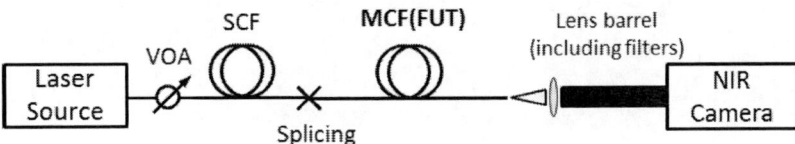

Fig. 2. New method for measuring XT by an NIR camera.

39

III. PREPARATION FOR EVALUATION OF OUTPUT POWER

Before measuring the XT, we must know the relationship between the output power from the fiber end and the pictures taken by the camera. Therefore, we took pictures with various output powers from the fiber and calculated the sum of the luminosity. The experimental setup was the same as that shown in Fig. 2, except that the MCF was not spliced to the SCF, and the output power from the SCF was observed. The wavelength of the laser source was 1.55 μm, and a standard SMF was used as the SCF. The output power was controlled by the VOA from -30 to -60 dBm. Figure 3 shows the relationship between the sum of the luminosity and the product of the output power and shutter speed. The relationship was almost linear. Using this result, the output power was evaluated appropriately.

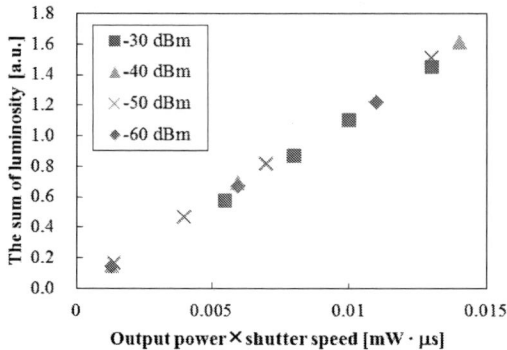

Fig. 3. Relationship between the sum of the luminosity and the output power × shutter speed.

IV. XT MEASUREMENT BY THE NIR CAMERA

We measured two MCFs using the NIR camera: a DCF and a 7CF. Figures 4(a) and (b) show the cross sections of the MCFs. Table I shows the characteristics of the MCFs.

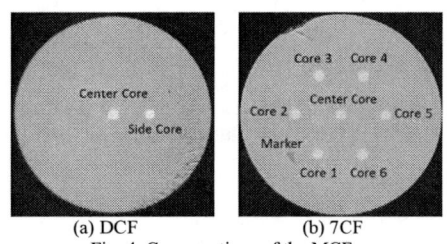

(a) DCF   (b) 7CF
Fig. 4. Cross sections of the MCFs.

TABLE I
CHARACTERISTICS OF THE DCF AND 7CF

Item	DCF	7CF
Length [km]	1.3	5.8
Cladding diameter [μm]	163.1	179.8
Average core-to-core pitch [μm]	30.3	40.5
Attenuation loss @ 1.55 μm [dB/km]	≤ 0.43	≤ 0.23
Cable-cutoff wavelength [μm]	≤ 1.46	≤ 1.22

The measurement setup was the same as that shown in Fig. 2. The SCF was spliced to the center core of each MCF. Then, the XT from the center core to the side core in the DCF was measured. Similarly, in the 7CF, we measured the XTs from the center core to the six outer cores. First, we diminished the output power from the center core of the MCF to approximately -35 dBm so that it was not too bright for the photodetectors. After we confirmed the location of the center core in a picture, the center core was masked by the spatial filter. Next, we increased the output power until XTs were visible in the picture. Finally, we recorded a video while sweeping the wavelength and averaged the frames of the video. The wavelength range was 1.545 to 1.555 μm.

Figures 5(a)–(c) show examples of the XT images for the DCF, which were averaged and extracted from the video. Figures 6(a)–(c) show examples of the XT images for the 7CF. Clearly, the XTs in both MCFs blinked randomly during the sweeping time. As shown in Fig. 6, the proposed method enabled the simultaneous measurement of multiple XTs.

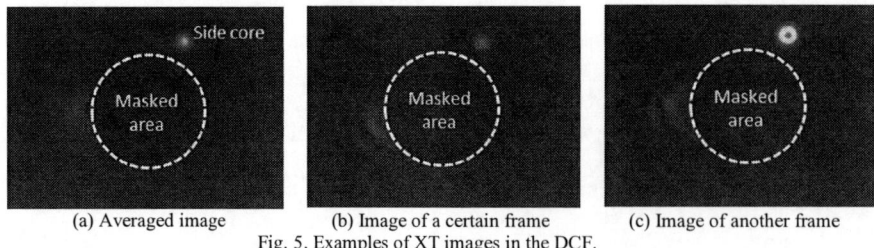

(a) Averaged image   (b) Image of a certain frame   (c) Image of another frame
Fig. 5. Examples of XT images in the DCF.

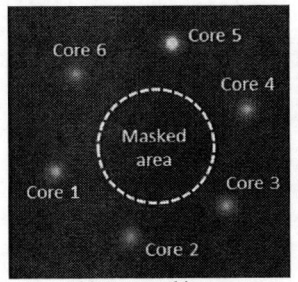 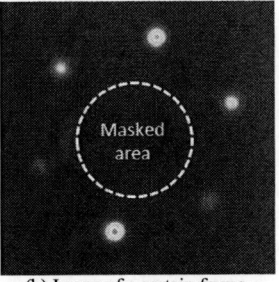 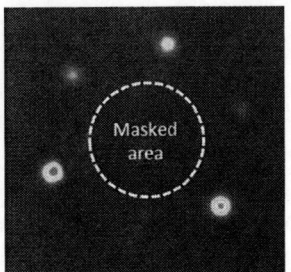

| (a) Averaged image | (b) Image of a certain frame | (c) Image of another frame |

Fig. 6. Examples of images in the 7CF.

The XT values were estimated using Figs. 3, 5(a), and 6(a), along with the pictures of the excited cores obtained using the ND filter. Table II shows the XT values obtained by the proposed method (N = 3) and the conventional method using a PM. These values agree with each other, and the differences between the results are less than 1.5 dB. We consider that the proposed method is suitable for measuring the XT of high-density MCFs. However, for measuring XTs far smaller than -30 dB, the output field of an excited core may not be sufficiently masked by the spatial filter that was used in this study. Consequently, observing the XT in a picture is difficult. In this case, an improved spatial filter for measuring small XTs should be considered.

TABLE II

XT VALUES MEASURED BY THE PROPOSED METHOD AND THE CONVENTIONAL METHOD

Fiber	Combination of cores	XT by the proposed method [dB]			XT by the conventional method [dB]
		1st time	2nd time	3rd time	
DCF	Center-Side	-30.1	-30.3	-29.8	-30.8
7CF	Center-Core 1	-25.1	-25.0	-25.2	-25.3
	Center-Core 2	-25.3	-25.6	-25.6	-26.2
	Center-Core 3	-25.1	-25.2	-25.6	-25.1
	Center-Core 4	-25.8	-25.6	-25.6	-26.6
	Center-Core 5	-24.9	-24.2	-24.1	-25.5
	Center-Core 6	-25.4	-25.6	-25.7	-26.4

V. CONCLUSIONS

We proposed a new method for measuring the XT in MCFs by using an NIR camera. Compared with the conventional method employing a PM, the proposed method does not need alignment with each core for receiving output power. Furthermore, the proposed method enables the simultaneous measurement of multiple XTs. We believe that this method has the potential to reduce the time and effort needed for measuring the XT of high-density MCFs.

ACKNOWLEDGEMENT

This work was partly supported by the EU-Japan coordinated R&D project on "Scalable and Flexible optical Architecture for Reconfigurable Infrastructure (SAFARI)" of the Ministry of Internal Affairs and Communications (MIC) of Japan and EC Horizon 2020.

REFERENCES

[1] B.J. Puttnam et al., "2.15 Pb/s transmission using a 22 core homogeneous single-mode multi-core fiber and wideband optical comb," Proc. of 41st European Conf. on Optical Communication (ECOC2015), Valencia (Spain), Oct. 2015, PDP.3.1.

[2] Y. Amma et al., "High-density multicore fiber with heterogeneous core arrangement," Proc. of 2015 Optical Fiber Communications Conf. and Exhibition (OFC2015), Los Angeles (USA), Mar. 2015, Th4C.4.

[3] Y. Sasaki et al., "Quasi-single-mode homogeneous 31-core fibre," Proc. of 41st European Conf. on Optical Communication (ECOC2015), Valencia (Spain), Sep. 2015, We.1.4.4.

[4] J. Sakaguchi et al., "Realizing a 36-core, 3-mode fiber with 108 spatial channels," Proc. of 2015 Optical Fiber Communications Conf. and Exhibition (OFC2015), Los Angeles (USA), Mar. 2015, Th5C.2.

[5] K. Imamura et al., "Multi-core holey fibers for the long-distance (>100 km) ultra large capacity transmission," Proc. of 2009 Optical Fiber Communication Conf. and Exposition and National Fiber Optic Engineers Conf. (OFC/NFOEC2009), San Diego (USA), Mar. 2009, OTuC3.

[6] T. Hayashi et al., "Characterization of crosstalk in ultra-low-crosstalk multi-core fiber," J. Lightw. Technol., vol. 30, no. 4, pp. 583-589, Feb. 2012.

[7] T. Hayashi et al., "Behavior of inter-core crosstalk as a noise and its effect on Q-factor in multi-core fiber," IEICE Trans. Commun., vol. E97-B, no. 5, pp. 936-944, May 2014.

Multicore fiber for bi-directional transmission

Tomohiro Gonda, Katsunori Imamura, and Ryuichi Sugizaki
Furukawa Electric co., Ltd
gonda.tomohiro@furukawa.co.jp

Abstract: *The optimum structure of MCF for bi-directional transmission was studied. Based on the investigation of core pitch and cladding diameter, 16-core fiber design was found to be the most feasible for the bi-directional transmission system.*
Keywords: *MCF, Crosstalk, Square Lattice Structure*

I. INTRODUCTION

There are continuous demands for transmission capacity increase. High density transmission like WDM is effective to enlarge the transmission capacity however a total input power into the core is limited because of a fiber fuse phenomenon. The effectiveness of space division multiplexing to overcome the limitation of transmission capacity is revealed by the various experimental results [1-6]. Among these investigations, multicore fiber is the most simple and flexible technology to suit for the conventional optical communication system. There are two types of multicore fibers, namely few-mode and single-mode multicore fibers. The few-mode multicore fiber is widely investigated which can drastically enhance the transmission capacity by utilizing mode division multiplexing technology however it has difficulty in preparing the input/output devices. On the other hand, single mode multicore fiber is the most feasible technology which is relatively easy to make the input/output devices. Increase of multiplexing degree of single mode multicore fiber is simply realized by the increase of the number of cores in the multicore fiber. So far, single-mode multicore fibers with 19-cores, 22-cores, and 31-cores were reported however increase of cladding diameter can't be prevented to realize higher core numbers. Increase of cladding diameter can be a cause of fiber break due to reliability degradation because of fiber bending [7]. Realizing suppression of cladding diameter increase and enlargement of core numbers simultaneously are key of improving the characteristics of single-mode multicore fiber.

The size of cladding diameter is determined by the core pitch between cores. Because the core pitch is limited by the crosstalk, core pitch decrease with suppressing the crosstalk is a key point to avoid the cladding diameter increase. It is reported that core pitch can be decreased with suppressing the crosstalk by utilizing the bi-directional transmission, namely Bi-directional Signal Assignment (BSA) [8, 9]. In this method, by transmitting signals in the adjacent cores for the opposite direction, the crosstalk in the system can be drastically decreased. In this report, we discuss about an optimum structure and core numbers to realize feasible single-mode multicore fibers.

II. DESIGN OF MCF WITH SQUARE LATTICE STRUCTURE

A. Study on comparison of various structure of multicore fibers

The most popular structure of multicore fiber is 7-core structure which is based on the triangle lattice structure (TLS). This structure is efficient because all cores are arranged in the same core pitch. The hexagonal closely packed structure, namely 7-, 19-, 37-cores, has been widely investigated so far. Regarding the crosstalk increase, we need to consider the relationship between crosstalk and core pitch. In case of TLS, there are 6 adjacent cores in maximum, so 7.8 dB crosstalk increase will be occurred and it can be a big challenge to realize a low XT transmission line. On the other hand, we can adopt the square lattice structure (SLS) which is suitable for the BSA transmission. By assigning the signal through cores opposite to that of adjacent core, the crosstalk can be suppressed drastically. This method is effective in terms of practical transmission system where 2-direction transmission in one optical cable is used. To suppress the influence of crosstalk from adjacent cores, some technique has been reported. From the reference [9], the effect of suppression of crosstalk occurred from back scattering light in the adjacent cores are about -20 dB. Other structures which can utilize the BSA has been reported, namely, One Ring Structure (ORS) and Dual Ring Structure (DRS) [10]. It was reported that the 12-core fiber with SLS has the merit to realize the lowest XT between the structures for 12-core fiber with BSA transmission. So in this study, we focused on the SLS multicore fiber for the long distance BSA transmission line.

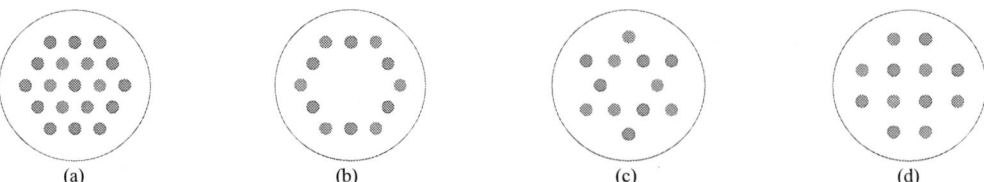

(a)   (b)   (c)   (d)
Fig. 1. Bi-directional core assignments in (a) 19-core HCPS, (b) 12-core ORS, (c) 12-core DRS, (d) 12-core SLS [10]

B. Design of Multicore Fiber with Square Lattice Structure

We investigated the optimum design of SLS multicore fiber with a trench assisted type refractive index core profile shown in Fig. 2. Fig. 3 and Table I. show the cross sectional view and optical properties of 16-core multicore fiber which was reported in ref. [9]. Here, the core pitch was 37.5 μm and cladding thickness, namely the distance between the cladding edge and the center of the most peripheral core, was 38.0 μm. Fig. 4 shows the relationship between cladding thickness and excess loss at the wavelength of 1625 nm. The excess loss at the most peripheral cores should satisfy the requirement that it is less than 0.001 dB/km at 1625 nm to certainly assure the good signal transmission characteristics over C+L band. On the other hand, Fig. 5 shows the relationship between core pitch and crosstalk. The requirement for crosstalk depends on the system where the multicore fiber will be used, so we set the target value as less than -40dB after 100 km for example. In case of the reported 16-core fiber, both of the cladding thickness and core pitch were larger than the value which satisfy the target properties, so good optical properties were obtained by the fabricated 16-core fiber.

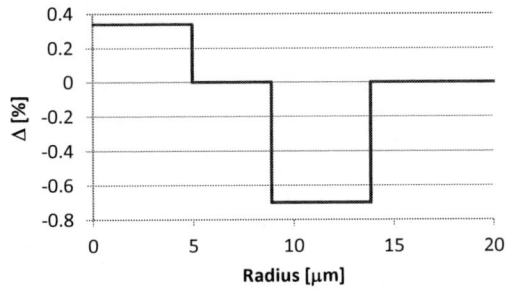

Fig. 2. Designed Refractive index profile

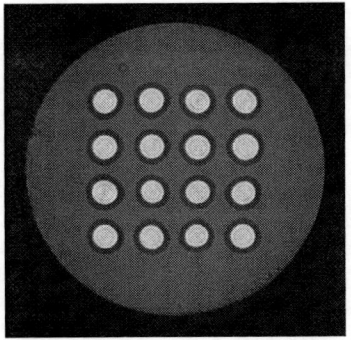

Fig. 3. Cross section of 16-core fiber [9]

TABLE I
STRUCTURAL AND OPTICAL PROPERTIES OF FABRICATED 16-CORE FIBER [9]

Item	Value
Length	55 km
Cladding diameter	235 μm
Core pitch	37.5 μm
Aeff @1550 nm	87 μm^2
XT @1550 nm *	-40.5 dB

**) After 55 km transmission, R=140 mm

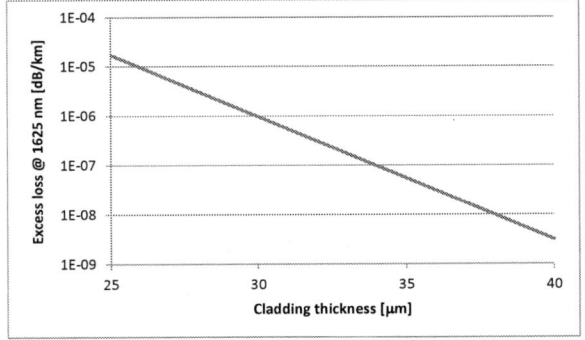

Fig. 4. Relationship between cladding thickness and excess loss

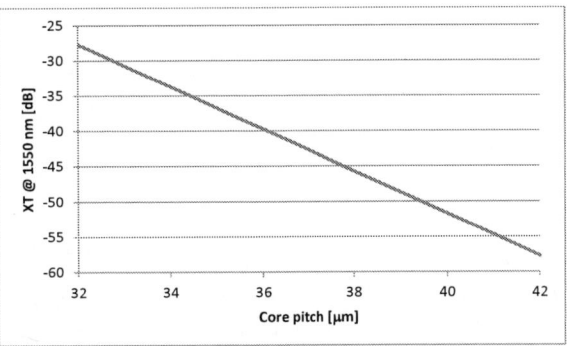

Fig. 5. Relashionship between core pitch and crosstalk

C. Optimum core numbers of multicore fiber with Square Lattice Structure

Regarding the SLS, we can consider some candidate structures which have different core numbers, namely 4, 5, 9, 12, 16, 21, 24, 25-core fibers. The cross section images of these SLS multicore fibers are shown in TABLE II. In case of each core number, cladding diameters which have same core pitch and cladding thickness as fiber in the former subsection are shown in this table. Additionally, core density of each structure is also shown in this table. The core density is defined as the core number assuming the same cladding diameter with conventional fibers, namely 125 μm. In comparison of the core density of each structure, 12 and 21-cores have much merit of increasing the core density efficiently. On the other hand, considering easiness to use in the real deployment situation, unit of 2, 4, 8, 12- pairs of fibers are widely used. From this point of view, 4, 12, 16, 24-core multicore fibers are the most suitable structure to fit the real system. Additionally, because the largest cladding diameter should be less than 250 μm from the reliability point of view, core numbers are limited less than 21. As a result, it can be said that 16-core fiber is a most feasible structure for BSA transmission.

TABLE II
MULTICORE FIBERS WITH SQUARE LATTICE STRUCTURE (SLS)
(CLADDING DIAMETER CORRESPOND TO THE CASE OF CORE PITCH = 37.5 μM AND CLADDING THICKNESS = 38.0 μM)

Core numbers	4	5	9	12	16	21	24	25
Cross section								
Cladding diameter	129	151	182	195	235	244	267	288
Core density	3.8	3.4	4.2	5.0	4.5	5.5	5.3	4.7

III. CONCLUSIONS

The optimum structure of multicore fiber to realize bi-directional signal assignment transmission was investigated. Base on the investigation of core pitch and cladding diameter, 16-core fiber design is the most feasible for the bi-directional transmission system.

ACKNOWLEDGMENT

We acknowledge NEC Corporation for the fruitful discussions. This research is supported by the National Institute of Information and Communications Technology (NICT), Japan under "R&D of Innovative Optical Fiber and Communication Technology".

REFERENCES

[1] J. Sakaguchi, et al., "109-Tb/s (7x97x172-Gb/s SDM/WDM/PDM) QPSK transmission through 16.8-km homogeneous multi-core fiber," in OFC/NFOEC 2011, PDPB6 (2011).

[2] J. Sakaguchi, et al., "19-core fiber transmission of 19x100x172-Gb/s SDM-WDM-PDM-QPSK signals at 305Tb/s," in OFC/NFOEC 2012, PDP5C.1 (2012).

[3] H. Takahashi, et al., "First Demonstration of MC-EDFA-Repeatered SDM Transmission of 40 x 128-Gbit/s PDM-QPSK Signals per Core over 6,160-km 7-core MCF," in ECOC 2012, Th.3.C.3 (2012).

[4] H. Takara, et al., "1.01-Pb/s (12 SDM/222 WDM/456 Gb/s) Crosstalk-managed Transmission with 91.4-b/s/Hz Aggregate Spectral Efficiency," in ECOC 2012, Th.3.C.1 (2012).

[5] B. J. Puttnam, et al., "2.15 Pb/s Transmission Using a 22 Core Homogeneous Single-Mode Multi-Core Fiber and Wideband Optical Comb," in ECOC 2015, PDP3.1 (2015).

[6] D. Soma, et al., "2.05 Peta-bit/s Super-Nyquist-WDM SDM Transmission Using 9.8-km 6-mode 19-core Fiber in Full C band," in ECOC 2015, PDP3.2 (2015).

[7] K. Imamura, et al., "A Study on Reliability for Large Diameter Multi-core Fibers," in IWCS 2011, P-2 (2011).

[8] T. Ito, et al., "Experimental Evaluation of the Reduction of the Received Signal Penalty due to Inter-core Crosstalk in a Lossy 7-core Multicore Fiber by Using Bidirectional Signal Assignment," in IEEE Summer Topical Meetings 2013, WC3.2 (2013).

[9] M. Arikawa, et al., "Crosstalk Reduction Using Bidirectional Signal Assignment over Square Lattice Structure 16-Core Fiber for Gradual Upgrade of SSMF-Based Lines," in ECOC 2015, Th.1.2.3 (2015).

[10] F. Ye, et al., "High-count Multi-Core Fibers for Space-Division Multiplexing with Propagation-Direction Interleaving," in OFC/NFOEC 2015, Th4C.3 (2015).

MC1-4 (Invited)

OECC/PS2016

Single-mode Multicore Fiber for Dense Space Division Multiplexing

Yusuke Sasaki[1], Yoshimichi Amma[1], Katsuhiro Takenaga[1], Shoichiro Matsuo[1], Kazuhiko Aikawa[1], Kunimasa Saitoh[2], Toshio Morioka[3], and Yutaka Miyamoto[4]

[1] Fujikura Ltd., 1440, Mutsuzaki, Sakura, Chiba, 285-8550, Japan
[2] Graduate School of Information Science and Technology, Hokkaido University, Sapporo, 060-0814, Japan
[3] Department of Photonics Engineering, Technical University of Denmark, Kgs. Lyngby, 2800, Denmark
[4] NTT Network Innovation Laboratories, NTT Corporation, Yokosuka, Kanagawa, 239-0847, Japan
yusuke.sasaki@jp.fujikura.com

Abstract: Single-mode multicore fiber (SM-MCF) is attractive for high-capacity transmission. Our fabricated SM-MCFs achieve high core count and low crosstalk with a cladding diameter of 230 μm. Characteristics of fan-in/fan-out for the SM-MCFs are also investigated.
Keywords: Multicore fiber, crosstalk, heterogeneous, quasi-single mode, fan-in/fan-out, splice

I. INTRODUCTION

Dense space division multiplexing (DSDM) over multicore fibers (MCFs) is expected to overcome the capacity limit of current optical communication systems that use single-mode fibers (SMFs). A main approach for realizing DSDM is using few-mode multicore fibers (FM-MCFs), which are MCFs whose cores support few-mode transmission. However, FM-MCF transmission involves massive signal processing, such as MIMO, in order to separate the coupled modes, and also requires a technology for equalizing mode dependent loss. Therefore, long-haul transmission over 500-km is achieved only by an FM-MCF with spatial multiplicity of 36 [1], and spatial multiplicity over 100 is realized by FM-MCFs shorter than 10-km [2, 3]. In contrast, SM-MCFs are well compatible with the present digital coherent technology, and are attractive to popularize high-capacity transmission over MCFs. The first transmission capacity over 1 Pbit/s and a capacity-distant product over 1 Ebit/s·km have been achieved by 12-core SM-MCFs, respectively [4, 5]. In addition, a 22-core SM-MCF has achieved the highest transmission capacity of 2.15 Pbit/s [6]. This paper describes design issues of crosstalk managed SM-MCFs suitable for DSDM transmission with spatial multiplicity over 30, and reviews our recent work.

II. CROSSTALK REDUCTION METHODS OF SM-MCF

One of the main issues in designing SM-MCF is inter-core crosstalk (XT). Low XT is preferable for realizing long distance transmission with high-order multilevel modulation. For example, QPSK and 32QAM transmission with a Q-penalty of 0.5 dB requires XT values of less than −19 dB and −29 dB, respectively [4]. Approaches for realizing low XT are as follows: (A) enlarging core pitch, (B) reducing A_{eff}, (C) introducing index trench, (D) employing heterogeneous core arrangement, and (E) using quasi-single-mode (QSM) transmission.

Enlarging core pitch leads to a large cladding diameter, however mechanical reliability limits cladding diameter dimensions [7]. Reduced A_{eff} is also inadequate for long-haul transmission due to non-linear effects. Introducing index trench, shown in Fig. 1, is an effective technique to improve XT characteristics. However, one issue with the trench assisted structure is that the cutoff wavelength of inner cores rapidly shifts to a longer wavelength, if the core pitch drops below a certain threshold [8]. This is because the confinement of the LP_{11} mode in the inner cores is increased by the index trench of the surrounding cores. The minimum core pitch of a trench assisted MCF is thus also limited by the cutoff wavelength, not just by the XT. Another approach involves employing a heterogeneous core arrangement with cores that have different effective indices (n_{eff}). Figure 2 illustrates the XT behavior as a function of the bending radius. In the range of bending radius of lower than the threshold bending radius (R_{pk}), XT degrades due to phase matching. On the contrary, in the range of bending radius of over R_{pk}, XT is drastically reduced because phases do not match. By setting R_{pk} below the effective bending radii in the cables, we can reduce the XT of a SM-MCF. In general, the varied-n_{eff} cores have different A_{eff}, i.e. heterogeneous A_{eff}, which is not preferable due to difficulties in equalizing splice loss and optical signal-to-noise ratio (OSNR) across all cores. Heterogeneous SM-MCF with homogeneous A_{eff} is preferable [9]. QSM transmission is a technique, which uses only the fundamental mode of a two-LP mode MCF. The high

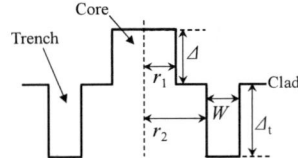

Fig. 1. Schematic of trench-assisted structure.

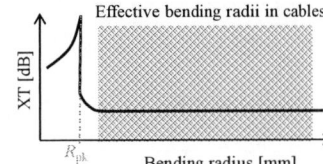

Fig. 2. Schematic of the XT behavior of a heterogeneous MCF.

confinement of the fundamental mode in a two-LP core, which has far longer cutoff wavelength compared to conventional SMFs, enables XT reduction. However, the remaining undesired LP$_{11}$ mode, which is generated by inter-mode crosstalk in fibers and at splice points, deteriorates the quality of the fundamental-mode signals. Therefore, to reduce signal deterioration, the intentional elimination of the LP$_{11}$ mode is required.

III. FABRICATED SM-MCFs

We fabricated a heterogeneous 30-core fiber with four types of core (Fiber A) based on the approach (C) and (D) [10], and a homogeneous 31-core fiber (Fiber B) based on the approach (C) and (E) [11], respectively. Figure 3 and Fig. 4 show cross-sections and core assignments of both fibers. Table I summarizes the core parameters. Designed 22-m cutoff wavelength of fiber A is lower than C-band, while that of Fiber B is longer than L-band. A_{eff} of both fibers was set to 80 μm^2 at 1550 nm, which is the same as conventional SMFs. For Fiber B, we adopted two methods to eliminate the LP$_{11}$ mode across all cores. The first is spreading of the LP$_{11}$ mode in inner cores to the outer cores by designing XT between the LP$_{11}$ modes (XT$_{11-11}$) to be as large as possible. The second is a selective absorption of the LP$_{11}$ mode in outer cores into the coating by adequate controlling of outer cladding thickness. They can be achieved due to the large field size difference between the LP$_{11}$ mode and the LP$_{01}$ mode. Table II summarizes measured dimensions of the two fibers, which are almost the same as the fiber designs. Figure 5 shows measured one-span XT of Fiber A at the bending diameter of 310 mm and at 1550 nm. XTs of all combinations of Fiber A were less than −50 dB at 9.6 km. Figure 6 shows measured one-span XT of Fiber B. Measuring conditions were same as Fiber A. As calculations, XT$_{11-11}$ was bigger than XT between the LP$_{01}$ mode (XT$_{01-01}$) by 30 dB.

Figure 7 shows the relationship between core count and the worst case XT for Fiber A, Fiber B, and reported SM-MCFs. The worst case XT is calculated by assuming that all neighboring cores are excited against each core. Fiber A exhibits both a lower XT and higher core count compared to reported SM-MCFs. Fiber B achieved the highest core count in SM-MCFs despite its relatively high XT. It should be noted that Fiber A involves carefully designed four types of cores with different effective indices, which is difficult to manufacture in large scale. It will take further research to realize an MCF that combines easy fabrication, higher core count, and lower XT characteristics.

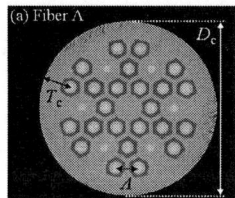
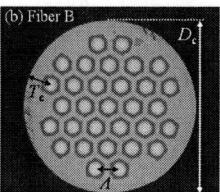

Fig. 3. Cross-sections of fabricated fibers.

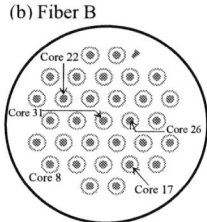

Fig. 4. Core assignments of fabricated fibers.

TABLE I CORE PARAMETERS OF FIBER A AND FIBER B

		r_1 [μm]	Δ [%]	r_2/r_1	W/r_1	Δ_t [%]
Fiber A	Core 1	4.76	0.338	1.7	1.0	−0.7
	Core 2	4.62	0.305	1.7	1.0	−0.7
	Core 3	4.47	0.273	1.7	1.2	−0.7
	Core 4	4.68	0.388	-	-	-
Fiber B		5.17	0.450	1.7	0.9	−0.7

TABLE II MEASURED DIMENSIONS

Item	Fiber A	Fiber B
Core pitch (Λ) [μm]	29.6	31.6
Outer cladding thickness (T_c) [μm]	35.6	31.5
Cladding diameter (D_c) [μm]	228	231
Length [km]	9.6	11

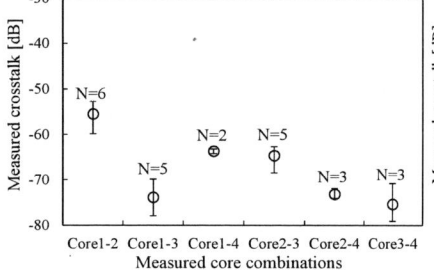

Fig. 5. Measured core-to-core XT of Fiber A.

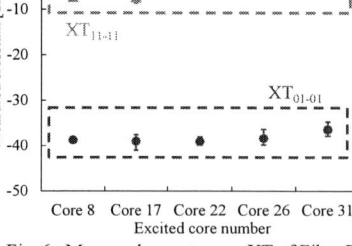

Fig. 6. Measured core-to-core XT of Fiber B.

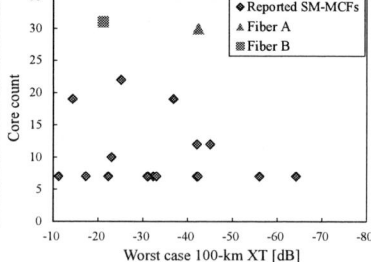

Fig. 7. Relationship between the worst case XT and core count of SM-MCFs.

IV. CONNECTION TECHNIQUE FOR SM-MCFs

Fan-in/fan-out (FI/FO) devices are necessary for using SM-MCFs in optical transmission lines. The main characteristics required for FI/FO are low XT and low insertion loss as well as transmission fibers. Many types of FI/FO have been proposed so far: waveguide type [12], free-space optics type [13], and fused fiber type [14]. Our work prepared a waveguide type FI/FO (W-FI/FO) and a free-space optics type FI/FO (F-FI/FO) for both Fiber A and Fiber B. Figure 8 shows photographs of the W-FI/FO and F-FI/FO. Figure 9 shows measured insertion loss per pair of FI/FOs at

1550 nm. The F-FI/FOs realized an insertion loss of lower than 1.5 dB. Figure 10 shows the measured total XT of FI/FOs at 1550 nm. Total XT was calculated by aggregating output power from all ports other than the excited port. The F-FI/FOs realized XT of around −40 dB. Those low insertion loss and low XT characteristics are helpful for high capacity SM-MCF transmission.

Splice technique is also required to use FI/FO with transmission fibers. Figure 11 shows splice loss between Fiber A. In order to realize low splice loss across all cores, we used an automatic core alignment system with end view image, and a swing electrode system to heat a wide region uniformly [15]. Although splice loss of outer core is tend to be high, due to the misalignment caused by the rotation error, splice loss is realizable under 0.3 dB.

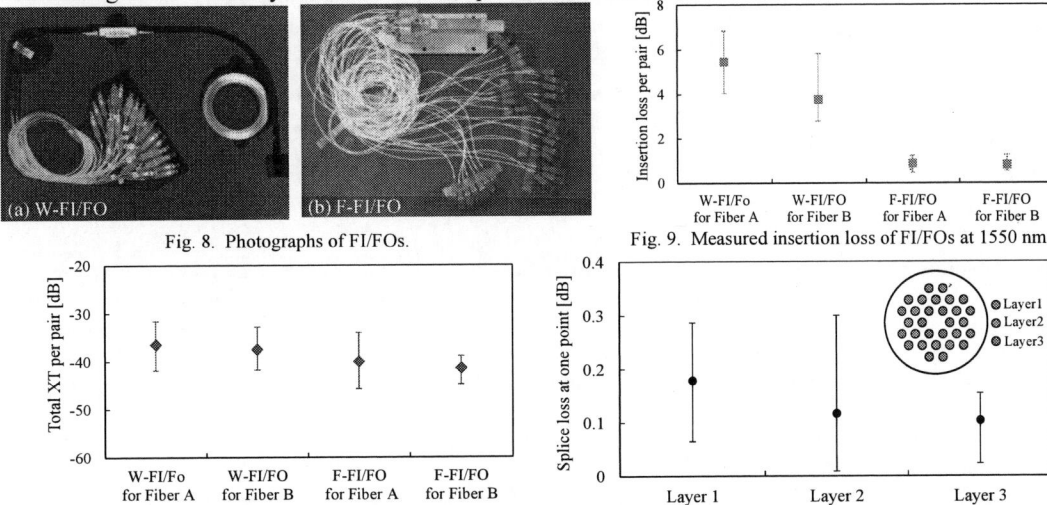

Fig. 8. Photographs of FI/FOs.

Fig. 9. Measured insertion loss of FI/FOs at 1550 nm.

Fig. 10. Measured total XT of FI/FOs at 1550 nm.

Fig. 11. Measured splice loss at 1550 nm.

V. CONCLUSIONS

Our recent work on SM-MCF was reviewed. We achieved SM-MCFs of high core count of 30 and 31 with a cladding diameter of 230 μm by utilizing heterogeneous core arrangement or quasi-single mode technology. We also evaluated characteristics of FI/FO for high core count SM-MCF, and confirmed that free-space optics type FI/FO achieved insertion loss of lower than 1.5 dB and XT of around −40 dB.

ACKNOWLEDGMENT

Part of this work is supported by the EU-Japan coordinated R&D project on "Scalable And Flexible optical Architecture for Reconfigurable Infrastructure (SAFARI)" by the Ministry of Internal Affairs and Communications (MIC) of Japan and EC Horizon 2020.

REFERENCES

[1] K. Shibahara *et al.*, "Dense SDM (12-core x 3-mode) transmission over 527 km with 33.2-ns mode-dispersion employing low complexity parallel MIMO frequency-domain equalization," Proc. of OFC2015, Th5C.3.

[2] J. Sakaguchi *et al.*, "Realizing a 36-core, 3-mode fiber with 108 spatial channels," Proc. of OFC2015, Th5C.2.

[3] D. Soma *et al.*, "2.05 peta-bit/s super-nyquist-WDM SDM transmission using 9.8-km 6-mode 19-core fiber in full C band," Proc. of ECOC2015, PDP3.2.

[4] H. Takara *et al.*, "1.01-Pb/s (12 SDM/ 222WDM/ 456 Gb/s) crosstalk-managed transmission with 91.4-b/s/Hz aggregate spectral efficiency," Proc. of ECOC2012, Th.3.C.1.

[5] T. Kobayashi *et al.*, "2 x 344 Tb/s propagation-direction interleaved transmission over 1500-km MCF enhanced by multicarrier full electric-field digital back-propagation," Proc. of ECOC2013, PD3.E.4.

[6] B. J. Puttnam *et al.*, "2.15 Pb/s transmission using a 22 core homogeneous single-mode multi-core fiber and wideband optical comb," Proc. of ECOC2015, PDP3.1.

[7] S. Matsuo *et al.*, "Large-effective-area ten-core fiber with cladding diameter of about 200 μm," Opt. Lett., vol. 36, no. 23, pp. 4626-4628, Dec. 2011.

[8] K. Takenaga *et al.*, "Reduction of crosstalk by trench-assisted multi-core fiber," Proc. of OFC/NFOEC2011, OWJ.4.

[9] K. Saitoh, *et al.*, "Low-crosstalk multi-core fibers for long-haul transmission", Proc. of SPIE, vol. 8284, no. 82840I, Jan. 2012.

[10] Y. Amma *et al.*, "High-density multicore fiber with heterogeneous core arrangement," Proc. of OFC2015, Th4C.4.

[11] Y. Sasaki *et al.*, "Quasi-single-mode homogeneous 31-core fibre," Proc. of ECOC2015, We.1.4.4.

[12] P. Mitchell *et al.*, "57 channel (19x3) spatial multiplexer fabricated using direct laser inscription," Proc. of OFC2014, M3K.5.

[13] W. Klaus *et al.*, "Free-space coupling optics for multicore fibers," Proc. of 2012 IEEE Summer Topicals Meeting, paper WC3.3.

[14] H. Uemura *et al.*, "Fused taper type fan-in/fan-out device for multicore EDF," Proc. of OECC/PS2013, TuS1-4.

[15] Y. Amma *et al.*, "Low-loss fusion splice technique for multicore fiber with a large cladding diameter," Proc. of 2013 IEEE Summer Topicals Meeting, MC2.1.

Dynamic Skew Measurements in 7, 19 and 22-core Multi Core Fibers

G. M. Saridis[1,2], B. J. Puttnam[1], R. S. Luís[1], W. Klaus[1], Y. Awaji[1], G. Zervas[2],
D. Simeonidou[2] and N. Wada[1]

[1] National Institute of Information and Communications Technology (NICT), Tokyo, Japan
[2] High-performance Networks Group, University of Bristol, United Kingdom
E-mail: george.saridis@bristol.ac.uk

Abstract: We report simultaneous dynamic inter-core skew measurements between 7 cores of several homogeneous MCFs. The largest variation was 4.33 picoseconds for 31km span with diminishing influence of mechanical vibrations, temperature, core-layout and wavelength observed.
Keywords: multi-core fibers; dynamic skew; transmission; space division multiplexing; optical communications

I. INTRODUCTION

Optical networks and communications systems are a key part of the global communications infrastructure, underpinning the digital economy, supporting intensive data communication needs of industry, commerce, academic institutions, governments and individuals worldwide. However, the current optical communication infrastructure is constantly challenged by the soaring traffic demands dictated by the modern Internet of Things (IoT) trends, "Big Data" establishment, intense social networking and high definition (HD) content transferring. Space-division-multiplexing (SDM) technologies [1] and networking approaches [2] have widely been proposed to enable cost-effective large scale capacity increase in transmission capacity with multiple fiber cores [3] or spatial modes [4] being used to increase the number of transmission channels in a single fiber. In particular, homogeneous single-mode (SM) multi-core fibers (MCFs) offer a good opportunity for adoption of high-capacity SDM technology in the near term having been shown to support high spectral efficiency (SE) modulation formats and wide band operation without the complexity of high-order multiple input-multiple output (MIMO) based receivers [3]. They can also be fabricated with relatively low SDM-crosstalk (XT) and a smaller core diameter than equivalent FM-MCFs, making splicing and handling easier [5]. Furthermore, the relative uniformity of cores in SM-MCFs with homogeneous cores supports spatial super channels (SSCs) for shared, transmitter hardware, DSP resources and simplified switching [6]. SSCs are also compatible with other system features relying on correlated propagation delay such as self-homodyne detection with pilot-tone transmission [7] or multi-dimensional modulation or coding [8] across cores.

However, the success of such systems depends to some extent on the magnitude of variation in propagation delay between signals travelling in different cores. As has been an issue for parallel datalinks since the introduction of electronic parallel transfer buses, this dynamic inter-core skew has implications for achievable baud-rates, transmission distances, receiver design and complexity of digital signal processing (DSP). Hence, characterizing it over extended time periods is crucial to fully understand the usefulness of such systems. Previously, the inter-core skew in MCFs was investigated by considering the skew between core pairs in a single 7-core MCF [10]. Here, we use a novel experimental set-up to extend that study and simultaneously measure dynamic inter-core skew of all cores of a 7 core fiber as well and apply the same set-up to investigate the skew in multiple cores of high core count MCFs with 3 ring structures for the first time. With fibers up to 22 cores, we observe that dynamic skew fluctuations can be caused by vibrations and temperature changes with lower dependence on core layout, fiber design and even smaller impact of wavelength observed.

II. EXPERIMENTAL DESCRIPTION

The experimental set-up for simultaneously measuring skew fluctuation between multiple cores is shown in Fig. 1.

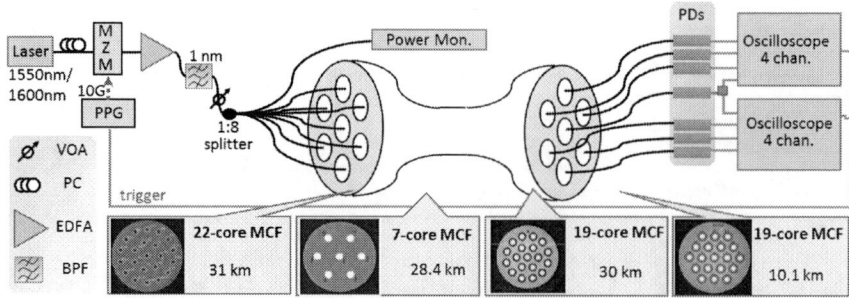

Fig. 1. Experimental set-up for dynamic skew measurements. Insets: the studied MCF

On the transmitter side, the light source was a C/L-band tunable external cavity laser (ECL), modulated by a Mach-Zehnder modulator (MZM) driven by a pulse pattern generator (PPG) at 10Gb/s. The modulation format on-off keying (OOK) using a 2^7-1 pseudo random binary sequence (PRBS). The generated signal was amplified by an erbium-doped fiber amplifier (EDFA) and then filtered by a 1 nm band-pass filter (BPF) to limit the amplified spontaneous emission (ASE) noise bandwidth. Next, a variable optical attenuator (VOA) was used to control input power before dividing in a 1 x 8 power splitter. Seven copies of the same signal were injected in seven different cores of the MCF while the eighth output of the splitter was used to monitor the injected power to the fiber. The injected power was maintained between 4 and 6 dBm depending on the fiber and their losses with 3 fibers using 3D waveguides to couple light in MCF cores with the exception being the 30 km 19-core fiber which used free-space couplers.

After fiber transmission, the signals were received by seven individual 10G photodetectors (PDs). The resulting electrical signals were sent to two identical high sampling rate oscilloscopes, each equipped with four 33 GHz channels. In order for the oscilloscopes to have a common reference, the electrical signal from the center core's PIN, was electrically split and fed to both scopes which were both triggered by a pattern synchronisation signal from the PPG. The relative dynamic skew between each core and centre core was estimated by cross-correlating the corresponding cores, as in [9]. The homogeneous MCFs that were studied here, comprised various designs and characteristics, including core arrangement design, core pitch, core number, etc. As a result, the cores under test were carefully picked for each fiber, in order to explore any possible effects from the above-mentioned diversities. The fibers' detailed specs along with references are shown in Table I with facet photos shown as inset in Fig. 1.

TABLE I
MULTI CORE FIBER SPECS

MCF type	Length (km)	Average Insertion Loss (dB/km)	Core pitch (um)	Cladding diameter (um)	Core arrangement	Ref.
7-core	28.4 km	0.184	44.2	160	hexagonal	[9]
19-core	10.1 km	0.227	35	200	hexagonal	[10]
19-core	30 km	0.285	39 & 37,6	220	circular	[11]
22-core	31 km	0.2	41 & 48	260	circular	[3]

III. RESULTS AND DISCUSSION

Skew fluctuations were observed for a period of 18 hours along with temperature monitoring. Fig. 2 depicts the resulting dynamic skew plots along with the rate of change (dSkew/dt) of those skew variations obtained for the fibers under analysis. The various core layouts with core numbering and colors matching the plot lines are also included as insets. The center core (no. 1) was used as a reference core to which the relative dynamic skew of the rest of the selected cores was measured. The skew between core *n* and *m* is symbolized as *n-m* and is represented by the color of each core. Fig. 2-a and Fig. 2-b show the evolution of dynamic skew for the 7-core MCF firstly with the laser tuned in 1550 nm and then in 1600 nm. In both cases, the maximum skew fluctuation was observed in the *1-5* core pair with values of 2.3 and 3.7 ps respectively. Although greater skew fluctuations was observed for the longer wavelength case, considering the statistical fluctuations and influence of external factors discussed below, there is not strong evidence for a wavelength dependence of dynamic skew variation. Indeed, as discussed below, the abrupt change in measured dynamic skews for all cores after 8 hours in Fig. 2-a was most likely caused by mechanical vibrations.

Over all the measurements, both vibrations and temperature appear to influence the measured skew values. Measurements of the ambient lab temperature show that it was varied by only two degrees over all measurement period but the frequency of the temperature fluctuations appear to have some correlation with the measured fiber skew. This is most evident in Fig 2-c and Fig. 2-d, showing the data for the 19-core fibers; although we note that the reduced length of the 10.1 km fiber could also be a factor for reduced skew fluctuations, since longer spans are more likely to be affected by temperature, vibrations or fiber imperfections. This fiber exhibited a maximum skew fluctuation of 0.79 ps for the *1-17* core pair with a reduced dynamic skew fluctuation rate which is just 0.002 ps/s for the same cores.

The impact of vibrations on inter-core skew is best observed in the results from the 22-core MCF, in Fig. 2-e, which show that measurements taken in periods with the least activity and resulting sources of mechanical vibrations, such as the early hours of morning (between 9 to 15 elapsed hours), result in less dynamic skew fluctuations compared with daytime measurements. In contrast to the other fibers mounted on isolated optical benches, this fiber was mounted on a trolley with the least mechanical isolation and positioned close to the main lab door. The above fact and the resulting movement of people and equipment in close proximity to the fiber and mechanical vibrations from door closing were observed to cause large step changes in the measured skew. Indeed, the largest of all measured skew values was with this fiber, being 4.33 ps with 0.0178 ps/s dSkew/dt for the *1-2* core pair.

Over all plots, although the cores with the largest dynamic skew variations tend to be outer cores, suggesting that higher skew is more likely for cores with larger physical separation, the effect is not constantly apparent to allow firm conclusions of this point.

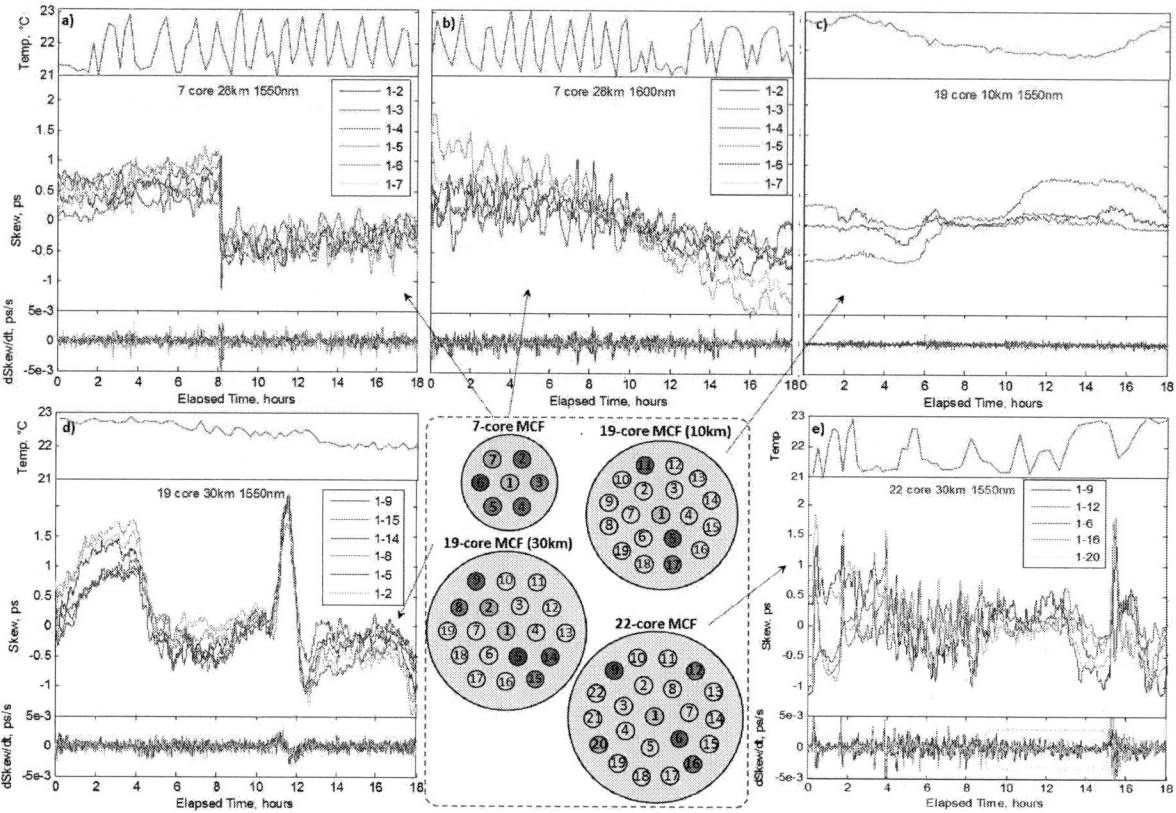

Fig. 2. Dynamic skew (ps) and its evolution rate (psec/sec) measurements for the MCFs under test with the associated temperature over a period of 18 hours. Insets: the selected cores for each skew measurement.

IV. CONCLUSION

We have investigated the dynamic inter-core skew in high-core count MCFs with three different core layouts and simultaneously in all cores of a 7-core fiber. We observed that, although skew is often larger in the outer ring cores than in the inner ones, it mostly appears to become more prone to mechanical vibrations and temperature changes.

ACKNOWLEDGMENTS

This work was supported by the NICT Internship Research Fellowship and by the EPSRC grant EP/I01196X: The Photonics HyperHighway.

REFERENCES

[1] D. J. Richardson, et al., "Space-division multiplexing in optical fibres," Nature Photonics, Vol. 7, pp. 354–362, 2013.

[2] G. M. Saridis et al., "Survey and Evaluation of Space Division Multiplexing: From Technologies to Optical Networks," IEEE Communications Surveys & Tutorials, vol. 17, no. 4, pp. 2136–2156, 2015.

[3] B. J. Puttnam et al. "2.15 Pb/s Transmission using a 22 core homogeneous single-mode multi-core fiber and wideband optical comb," ECOC, PDP 3-1, 2015.

[4] N. K. Fontaine. "Characterization of multi-mode fibers and devices for MIMO communications ", Proc. SPIE 9009, Next-Generation Optical Communication: Components, Sub-Systems, and Systems III, 90090A, 2013.

[5] T. Hayashi, et al., "Design and fabrication of ultra-low crosstalk and low-loss MCF," OPEX, 19 (17), pp 16576–16592, 2011.

[6] M .D. Feuer et al., "Joint digital signal processing receivers for spatial superchannels" IEEE PTL. Let. 24, 1957-60 2012.

[7] B. J. Puttnam et al., "High-capacity self-homodyne PDM-WDM-SDM transmission in a 19-core fiber," Optics Express, 22 (18), pp. 21185-91, 2014.

[8] B. J. Puttnam et al., "Modulation formats for multi-core fiber transmission," Optics Express 22 (26), pp. 32457-32469, 2014.

[9] R. S. Luís, et al., "Time and Modulation Frequency Dependence of Crosstalk in Homogeneous Multi-Core Fibers," IEEE J. of Lightwave Technology, 34 (2) pp. 441-447, 2016.

[10] J. Sakaguchi et al., "19-core fiber transmission of 19×100×172-Gb/s SDM-WDM-PDM-QPSK signals at 305Tb/s," OFC, pp. 1-3, 2012.

[11] J. Sakaguchi et al., "19-core MCF transmission system using EDFA with shared core pumping coupled in free-space optics," ECOC, pp. 1-3, 2013.

MC2-2

Impulse Response Analysis of Air-hole Added Coupled Six-core Fibers

Yasuyuki Chida, Takeshi Fujisawa, and Kunimasa Saitoh

Graduate School of Information Science and Technology, Hokkaido University, Sapporo 060-0814, Japan.
chida@icp.ist.hokudai.ac.jp

Abstract: *Impulse response of coupled six-core fibers is analyzed based on nonlinear Schrödinger equation, showing good agreement between theory and reported measured results. It is shown that air-hole added structure reduced the group delay spread further.*
Keywords: *Mode division multiplexing, Group delay spread, Coupled core fibers, Impulse response*

I. Introduction

Mode division multiplexing (MDM) technologies have been actively investigated for increasing the capacity of optical communication system and various types of few-mode fibers (FMFs) for MDM have been proposed [1]. In FMFs, the group delay spread (GDS) is one of the major problems since the GDS increases the burden of the MIMO receiver [2]. Therefore, reducing the GDS is crucial for constructing MDM link. A coupled multicore fiber (CMCF) is a kind of FMFs and promising candidate for small GDS fibers [3,4]. Recently, it was reported that by investigating impulse response (IR) of coupled six-core fibers (6CF) the measured GDS is considerably smaller than the theoretical GDS calculated by the differential modal group delay (DMGD) of the supermodes of 6CF [4]. The reduced GDS is attributed to strong modal coupling due to various perturbations in the fiber (micro-, macro-bending, and twisting). Since in 6CF, we can reduce the effective refractive index difference (Δn_{eff}) between supermodes by proper fiber design, we can use the perturbations in the fiber effectively. Investigating IRs of FMF is a simple way to measure GDS and IRs are very important information for MIMO receivers. It was shown theoretically that the GDS of the multimode fibers are proportional to the length of the fiber, L, in the weak coupling regime (the perturbation is not strong) and the square root of L in strong coupling regime [5]. Also, even if the fiber is in weak coupling regime, the GDS is proportional to the square root of L for long distance transmission. To engineer the performance of 6CFs, general theoretical methods, which can treat all the coupling regimes, as well as new fiber structure reducing GDS are highly desired.

In this work, we analyzed IRs of 6CF based on multimode generalized nonlinear Schrödinger equation (GNLSE) for the first time. For the perturbations inducing modal coupling in the fiber, we employ microbending effect [6,7] and it is shown that weak and strong coupling regime as well as the transmission between two regimes can be naturally treated. Reported measured [4] and calculated waveform of the IRs of 6CF are in very good agreement, showing the validity of the theory. Furthermore, it is shown that by placing air-holes between cores, we can reduce Δn_{eff} and DMGD, using the symmetry of supermodes. The IR analysis of the air-hole added fiber shows the GDS is considerably reduced compared with 6CF without air-holes.

II. Impulse response analysis of coupled six-core fiber

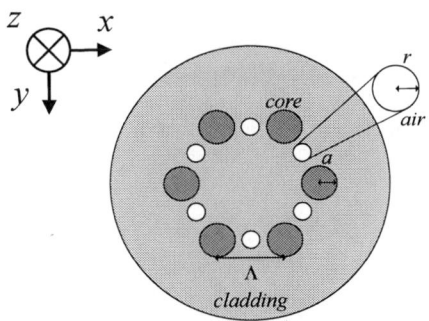

Fig. 1 Cross section of 6CF with air-holes.

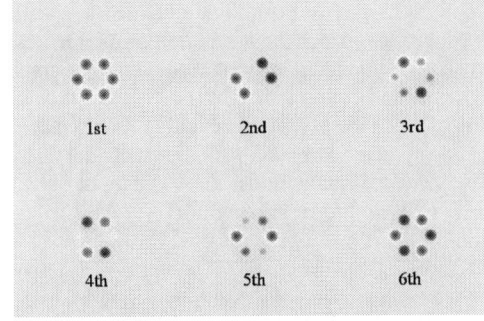

Fig. 2 The electric field distributions of supermode in 6CF.

Figure 1 shows a cross-section of 6CF with air-holes. The cores are arranged in regular hexagon with the core pitch of $\Lambda = 28.0$ μm, core radius $a = 5.55$ μm, relative refractive index difference $\Delta = 0.32\%$ [4] and r is the air-hole radius. In this section, we consider 6CF without air-holes. Figure 2 shows six supermodes of 6CF. The second and third, fourth and fifth modes are degenerate and the calculated Δn_{eff} and DMGD between 1st and 6th mode are 2.0×10^{-6} and 67.6 ps/km.

51

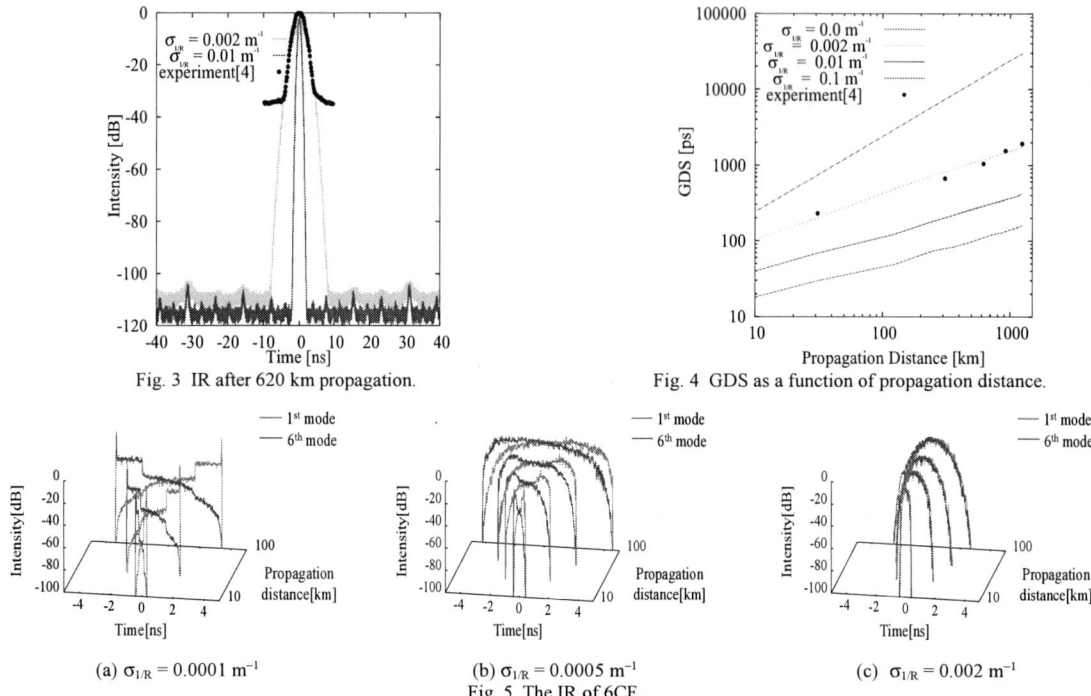

Fig. 3 IR after 620 km propagation.

Fig. 4 GDS as a function of propagation distance.

(a) $\sigma_{1/R} = 0.0001$ m^{-1}

(b) $\sigma_{1/R} = 0.0005$ m^{-1}

(c) $\sigma_{1/R} = 0.002$ m^{-1}

Fig. 5 The IR of 6CF.

The second order chromatic dispersion, third order dispersion, fiber loss are 20.6 ps/km/nm, 0.059 ps/km/nm^2, 0.236 dB/km [4]. The carrier wavelength is 1550 nm. The IRs of 6CF can be calculated by solving multimode GNLSE given by

$$\frac{\partial A_m}{\partial z} = -j\beta_{0m}A_m - \beta_{1m}\frac{\partial A_m}{\partial t} + j\frac{\beta_{2m}}{2}\frac{\partial^2 A_m}{\partial z^2} + \frac{\beta_{3m}}{6}\frac{\partial^3 A_m}{\partial z^3} - \frac{\alpha_m}{2}A_m - j\sum_{npq}\gamma_{mnpq}A_n^*A_pA_q e^{-j\Delta\beta_{mnpq}} - j\sum_{m\neq n}\kappa_{mn}A_m \qquad (1)$$

where A_m is the complex electric field of mth mode, t is the temporal coordinate normalized to 1st mode. β_{0m}, β_{1m} β_{2m} β_{3m} and α_m are the propagation constant, group delay, second order chromatic dispersion, third order dispersion and fiber loss of the mth mode. γ_{mnpq}, $\Delta\beta$, and κ_{mn} are the nonlinear constant, inter-modal phase mismatch, and inter-modal coupling coefficient. Here, the nonlinearity of the fiber is neglected with the assumption that the input power is small. It should be noted that β_{0m} term is usually omitted in NLSE and treat only the envelope in z-direction. However, to treat group delay change due to random linear coupling, the inclusion of β_{0m} term is indispensable [8]. To solve (1), we divide the equation into two steps [9]. First, β_{1m}, β_{2m}, β_{3m}, and α_m terms are solved by well-known split-step fourier method (SSFM) and obtain temporal field distributions including the effect of GD, dispersion and loss. Next, the equation is solved for β_{0m} and κ_{mn} terms with the temporal field obtained in the first step. In this step, analytical solutions can be used [8]. This treatment is necessary since the oscillation of β_{0m} term is so fast that it is difficult to solve this term with SSFM.

For the perturbation inducing modal coupling, we consider microbending effect [6,7]. The total fiber length L is divided into N segments and each segment of the fiber has the random bending R_x and R_y given by Gaussian distribution with the mean value of $\mu = 0$ m^{-1}, standard deviation (STD) $\sigma_{1/Rx,y} = \sigma_{1/R} = 1/R$. $\sigma_{1/R} = 0$ m^{-1} indicates that the fiber is straight and the magnitude of the perturbation increases with increasing $\sigma_{1/R}$. ΔL is 10 m, which is the correlation length of polarization mode dispersion in a single mode fiber. The input optical pulse is 10 ps Gaussian pulses for all modes. Figure 3 shows the IRs after 620 km propagation with $\sigma_{1/R} = 0.002$ and 0.01 m^{-1}. Output pulses of six modes are the similar Gaussian distribution for these $\sigma_{1/R}$ and show one of the pulses of all modes. Experimental result taken from [4] is also plotted and calculated and measured results are in very good agreement for $\sigma_{1/R} = 0.002$ m^{-1}. The pulse width is narrowed if $\sigma_{1/R}$ increases. We evaluate the pulse width by fitting the waveform with Gaussian distribution and used the STD as the GDS [3]. Therefore, this evaluation is only valid for the strong coupling regime. Figure 4 shows the GDS as a function of propagation distance for $\sigma_{1/R} = 0$, 0.002, 0.01, 0.1 m^{-1}. For $\sigma_{1/R} = 0$ m^{-1}, we plot the value obtained by the DMGD of the fiber [2]. While GDS is proportional to L for $\sigma_{1/R} = 0$ m^{-1}, it is proportional to the square root of L for other $\sigma_{1/R}$, showing the fiber is in strong coupling regime. If the $\sigma_{1/R}$ increases, the magnitude of the perturbation increases in fibers, and therefore, the spread of the pulse is suppressed. Figure 5 (a), (b), and (c) show the IR for $\sigma_{1/R} = 0.0001$, 0.0005, and 0.002 m^{-1}, respectively. Only the waveforms of 1st and 6th modes are shown. For $\sigma_{1/R} = 0.0001$ m^{-1}, each modes have their own peak and the plateau is observed between the modes. However, for $\sigma_{1/R} = 0.0005$ m^{-1}, the level of plateau becomes higher as it propagates and all the waveforms become Gaussian distribution, indicating the transition from weak to strong coupling regime. Finally, for $\sigma_{1/R} = 0.002$ m^{-1}, all the pulses are strongly mixed and the GDS is significantly reduced.

III. IR analysis of 6CF with air-holes

From Fig. 2, we can see that the field distributions of 1st and 6th order modes have even and odd symmetries between two cores. Therefore, if we place the air-hole between cores, the n_{eff} of 1st mode is much reduced than that of 6th mode, leading to smaller Δn_{eff} and stronger modal coupling. Figure 6 shows calculated Δn_{eff} and DMGD between 1st and higher-order mode as a function of r. Here, DMGD is calculated between 1st and 6th mode. As expected, Δn_{eff} is reduced by placing the air-holes and the reduction is large for large r. Figure 7 shows the IRs of 6CF with $r = 0$, 1, and 3 μm after 1240 km propagation for $\sigma_{1/R} = 0.002$ m^{-1}. For the same $\sigma_{1/R}$, the pulse width is reduced for larger air-hole. Calculated GDSs are 1790, 562, and 92 ps for $r = 0$, 1, and 3 μm fibers. GDS can be reduced almost 95% by placing the air-holes with $r = 3$ μm. These results indicate that the 6CF with air-holes is useful for reducing GDS under microbending perturbation.

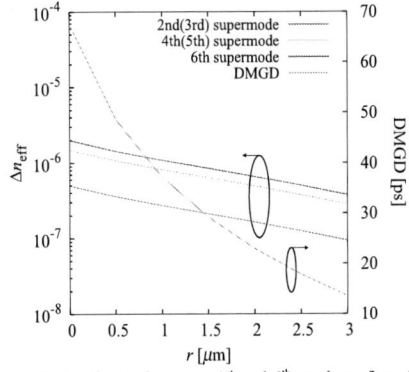

Fig. 6 Calculated Δn_{eff} between 1st and 6th mode as function of r.

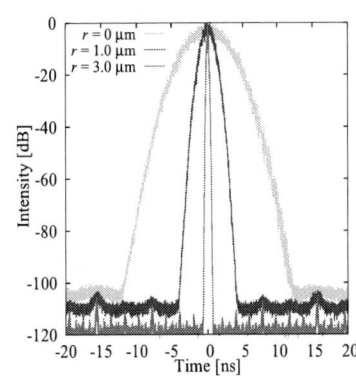

Fig. 7 IR after 1240 km for $\sigma_{1/R} = 0.002$ m^{-1}.

IV. Conclusion

We simulated optical pulse propagation and obtained the IRs of 6CFs by using multimode GNLSE. Calculated waveform and GDS are in very good agreement with the reported measured results. Furthermore, the IRs of 6CF with air-holes are analyzed and numerical results show that the GDS can be reduced further under microbending perturbation.

Acknowledgement

A part of this work was supported by the National Institute of Information and Communication Technology (NICT). Japan under "R&D of innovative Optical Fiber and Communication Technology".

References

[1] D.J. Richardson et al., "Space division multiplexing in optical fibers," Nat. Photonics, vol. 7, no. 5, pp. 354-362, 2013.
[2] S.O. Arik et al., "MIMO signal processing for mode-division multiplexing," Signal Processing Magazine, vol. 31, no. 2, pp. 25-34, 2014.
[3] R. Ryf et al., "Space-division multiplexed transmission over 4200-km 3-core microstructured fiber," OFC, PDP5C2, 2011.
[4] R. Ryf et al., "1705-km transmission over coupled-core fibre supporting 6 spatial modes," ECOC2014, PD.3.2, 2014.
[5] K.-P. Ho et al., "Statics of group delays in multimode fiber with strong mode coupling," J. Lightw. Technol., vol. 29, no. 21, pp. 3119-3128, 2011.
[6] T. Fujisawa et al., "A principal mode analysis of strongly-coupled 3-core fibers," ECOC, We.1.4.6, 2015.
[7] T. Fujisawa et al., "Impulse response analysis of strongly-coupled three core fibers," FiO2015, FM1E.3, 2015.
[8] M.B. Shemirani et al., "Principal modes in graded-index multimode fiber in presence of spatial- and polarization-mode coupling," J. Lightw. Technol., vol. 27, no. 10, pp. 1248-1261, 2009.
[9] Y. Xiao et al., "Theory of intermodal four-wave mixing with random linear mode coupling in few mode fibers," Optics Express, vol. 22, no. 26, pp. 32039-32059, 2014.

High Spatial Density Few-mode Multi-core Fiber with Low Differential Mode Delay Characteristics

Taiji Sakamoto[1], Takashi Matsui[1], Kunimasa Saitoh[2], Shota Saitoh[3], Katsuhiro Takenaga[3], Shoichiro Matsuo[3], Yuki Tobita[2], Nobutomo Hanzawa[1], Kazuhide Nakajima[1], Fumihiko Yamamoto[1]

[1]Nippon Telegraph and Telephone Corporation, 1-7-1 Hanabatake, Tsukuba, 305-0805, Japan,
[2]Graduate School of Information Science and Technology, Hokkaido University, Sapporo, 060-0814, Japan,
[3]Fujikura Ltd. 1440, Mutsuzaki, Sakura, 285-8550, Japan

Abstract: Inter-core crosstalk suppressed 4-LP mode 12-core fiber with the highest relative core multiplicity factor is realized using a feasible cladding diameter. We obtain low differential mode delay characteristics by optimizing the trench-assisted graded-index core profile.
Keywords: (Multi-core fiber, differential mode delay, inter-core crosstalk)

I. INTRODUCTION

Space division multiplexing technologies have been investigated intensively with the aim of achieving much greater transmission capacity. A transmission capacity exceeding 1 Ebit/s·km [1] or 2 Pbit/s [2] was demonstrated by using a single-mode multi-core fiber (SM-MCF). To further increase capacity, a few-mode MCF (FM-MCF) is a promising candidate since it can increase the number of spatial channels by increasing the number of modes per core [3]. Recently, various FM-MCFs have been proposed and MCFs with more than 100 spatial channels have been achieved [4, 5]. There are various parameters to be considered as regards telecommunication network use. Spatial density is an important parameter because the cladding diameter is limited to a certain value to take account of the mechanical reliability issue [6], and thus the number of cores to be deployed in the MCF is also limited while maintaining low inter-core crosstalk characteristics. A low differential mode delay (DMD) characteristic is also important for FM-MCF in terms of the complexity of the MIMO processing at the receiver.

In this paper, we investigate the optimum design for a high spatial density FM-MCF with low crosstalk and low DMD characteristics. We show that a trench-assisted graded-index core profile can reduce the inter-core crosstalk and realize a low DMD value, and 4-LP mode MCF can improve the spatial density compared with 2-LP mode MCF. Finally, we fabricate 4-LP mode 12-core fiber and realize the highest relative core multiplicity factor of 49 while maintaining low inter-core crosstalk of less than -27 dB/100 km and a low DMD value of less than 0.43 ns/km within the C-band [7].

II. DESIGN OF FM-MCF WITH HIGH SPATIAL DENSITY

We first discuss the maximum target cladding diameter of the FM-MCF taking account of the mechanical reliability of the fiber [8]. Figure 1 shows the relative failure probability compared with that of conventional 125 μm-cladding fiber as a function of the fiber cladding diameter D. In our calculation we assumed a bending radius R of 30 mm, a stress corrosion parameter n of 16, and a lifetime of 20 years. We found that the mechanical reliability can be improved by increasing the proof level during fiber manufacture, and the same failure probability can be obtained for MCF with a D of 250 μm and 2% proof testing as for conventional fiber with 1% proof testing. As 1~2 % is considered a typical proof testing range we targeted a D of less than 250 μm in our design.

We then investigated numerically the impact of the increased core/mode number in the FM-MCF on the spatial

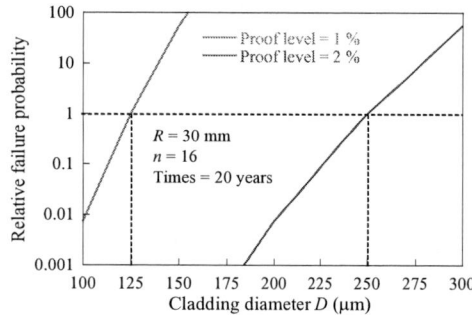

Fig. 1 Relative failure probability vs. cladding diameter of MCF with 1 or 2 % proof testing

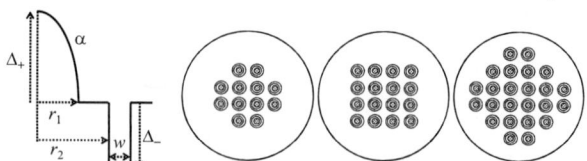

Fig. 2 Graded-index trench-assisted core profile and proposed core arrangements

Table I Example of core profile with 2-LP or 4-LP mode

LP mode	r_1 (μm)	r_2/r_1	w/r_1	D_+ (%)	D_- (%)	Max. |DMD| within C-band	A_{eff} at 1550 nm (μm²)			
							LP_{01}	LP_{11}	LP_{21}	LP_{02}
2	9.0	1.2	0.6	0.57	-0.7	<100 ps/km	90.1	180.3	-	-
4	10.1	1.2	0.7	0.72	-0.7		89.8	179.4	240.0	181.4

density. Figure 2 shows the refractive index profile and a cross-sectional image of the MCF. A trench-assisted graded-index (TA-GI) profile is employed because of its DMD controllability. We assumed a square lattice-based core arrangement since it can reduce the number of neighboring cores and can relax the tendency of the cutoff wavelength to become longer, and we considered 12, 16 and 24 cores in this study. We designed two types of core profile supporting the 2- and 4-LP modes in the C-band to investigate the impact of mode number on spatial density. The effective area A_{eff} of the LP_{01} mode at 1550 nm was designed to be 90 μm^2. The cutoff wavelength λ_c was set at 1530 nm to support the 2- or 4-LP mode in the C-band. The bending loss α_b was designed to be less than 0.001 dB/km at 1625 nm at a bending radius R of 140 mm to suppress the excess loss in the operation window. Table I summarizes example parameters for the TA-GI profile. Here, the α parameter of the core was about 2.0 for both the 2- and 4-LP mode cores, and we obtained a maximum absolute value for the differential modal delay (DMD) of less than 100 ps/km at 1550 nm for both core profiles.

Figure 3 (a) shows the core pitch Λ dependence of the crosstalk (solid lines) and the 22-m cutoff wavelength λ_c (dashed lines). Here, the crosstalk represents the core-to-core value between the highest order propagation modes in neighboring cores. The black and red lines show the results obtained for 2- and 4-LP cores, respectively. The dash-dotted line shows the target crosstalk value and λ_c level. We found that the minimum Λ value was restricted by λ_c rather than by the crosstalk, and the allowable Λ becomes large as the mode number increases. Minimum Λ values of 39.0 and 43.4 μm are needed for 2- and 4-LP cores, respectively, to ensure the desired few-mode operation in the C-band. Figure 3 (b) shows the relationship between the α_b value of the highest propagation mode and cladding thickness t calculated at 1625 nm and $R = 140$ mm. The black and red lines show the 2- and 4-LP cores, respectively. It can also be seen that the allowable cladding thickness becomes large as the mode number increases, and t values of 37.9 and 43.9 μm or more are needed for 2- and 4-LP cores, respectively, to suppress α_b to below 0.001 dB/km.

Next we compare the spatial density of 2- or 4-LP mode MCF with 12, 16 and 24 cores as shown in Fig. 4. Here, the core multiplicity factor (CMF) [9] was used, which is calculated as

$$CMF = \sum_i A_i /(\pi r^2),\qquad(1)$$

where A_i is the effective area of the ith propagation mode and r is the cladding radius of the MCF. The relative CMF (RCMF) is derived by normalizing the CMF of the MCF with that of conventional single-mode fiber. We confirmed that the RCMF of 2-LP mode MCF cannot be improved even by increasing the core number from 12 to 16, and a further increase in the mode number results in a larger cladding diameter than our target value of 250 μm, which may degrade the mechanical reliability of the fiber. By contrast, an RCMF of more than 50 can be expected if we introduce a 4-LP core into a 12-core structure while keeping D at less than 230 μm.

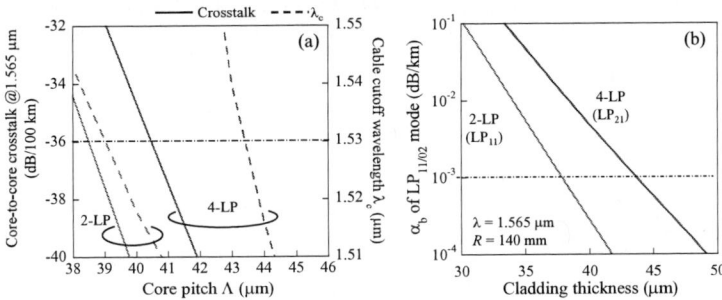
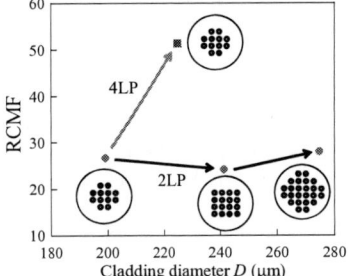

Fig. 3 (a) Core pitch dependence of crosstalk at 1565 nm and λ_c, (b) cladding thickness dependence of α_b

Fig. 4 Relationship between RCMF and cladding diameter in various MCF structure

III. PROPERTIES OF 4-LP 12-CORE FIBER

Based on the previous design consideration, we fabricated a 4.2 km-long 4-LP 12-core to achieve a high RCMF. Figure 5 shows a cross-sectional photograph. The cladding diameter D was 227 μm. Λ and t were well controlled at 43.0-44.5 μm and 43.4-44.8 μm, respectively, against the target values (43.4 μm and 43.9 μm). Table II summarizes the optical properties of the fabricated 4-LP 12-core fiber. The cutoff wavelength of the LP_{31} mode λ_c was less than 1530 nm for all cores. The attenuation coefficient at 1550 nm measured with the cutback method was 0.23±0.01 dB/km for the LP_{01} mode. We also obtained an attenuation coefficient of 0.29±0.03 dB/km when we launched all the modes, LP_{01}, LP_{11}, LP_{21}, and LP_{02}, simultaneously. The A_{eff} of the LP_{01} mode was 87±5 μm at 1550 nm as we designed. The α_b values were extremely low at less than 0.1 dB/turn at $R = 5$ mm and under a full-mode launching condition.

We measured the core-to-core crosstalk between neighboring cores and evaluated the total crosstalk value from all neighboring cores (max. 4 cores). We used a planar lightwave circuit (PLC) type mode multiplexer [10] to excite predominantly the LP_{21} mode. We confirmed that the crosstalk of all neighboring cores was successfully suppressed below -31 dB even after a 100 km transmission. On the other hand, the crosstalk value was slightly degraded by a few decibels when we launched all modes simultaneously. We consider that the slight degradation in the crosstalk was mainly due to the LP_{02} mode propagation in the MCF since the crosstalk of the LP_{02} mode is larger than that of the other modes. We believe that the crosstalk characteristics can be further improved by slightly increasing the Λ value above the target value of 43.4 µm. We also measured the DMD characteristics of four modes in the MCF. The maximum DMD in the C-band was well controlled below 0.43 ns/km in all the cores. Thus, we realized FM-MCF with a high spatial density while maintaining low crosstalk and low DMD characteristics. Finally, we compare our MCF with various MCFs in terms of spatial density. Figure 6 shows the RCMF values of various MCFs as a function of the number of spatial channels. We classified the MCF into three types: SM-MCF, non-DMD controlled FM-MCF, and DMD controlled FM-MCF. We found that the fabricated FM-MCF successfully realized the highest RCMF of about 49, which is an almost twofold improvement over recently reported non-DMD controlled FM-MCF. Moreover, a significant RCMF improvement of more than 30 was achieved based on the reported DMD controlled FM-MCF. As a result, our inter-core crosstalk suppressed low DMD FM-MCF has the potential to be employed in a long-distance and large-capacity transmission system using advanced multi-level modulation.

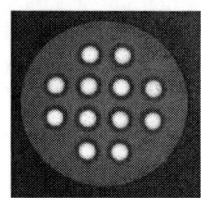

Figure. 5 Cross-section
D = 227 µm
Λ = 43.7±0.7 µm
t = 44.0±0.5 µm

Table II Properties of fabricated fiber

22 m Cutoff λ_c		(µm)	<1.53
Loss [1]	LP_{01}	(dB/km)	0.23±0.01
	All	(dB/km)	0.29±0.03
A_{eff} of LP_{01} [1]		(µm²)	87±5
Bending loss α_b [2]		(dB/turn)	<0.1
Crosstalk [3] (Four core)	LP_{21}	(dB/100km)	<-31
	All	(dB/100km)	<-27
DMD [3]		(ps/km)	<430

(1: 1550 nm, (2: 1565 nm, R = 5 mm, (3: C-band

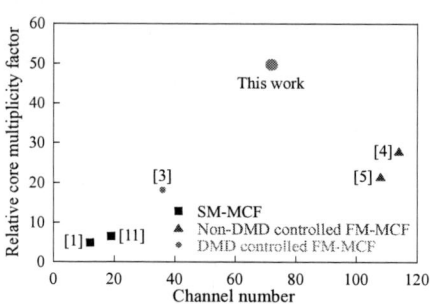

Fig. 6 RCMF value of various MCFs as a function of the number of spatial channels

IV. CONCLUSIONS

We investigated the optimum design for realizing a high spatial density FM-MCF with low crosstalk and low DMD characteristics. We showed numerically that increasing the mode number from 2-LP to 4-LP was beneficial for improving the RCMF value. We successfully fabricated 4-LP 12-core fiber and realized experimentally the highest reported RCMF value of 49 with a 227 µm cladding diameter. Our FM-MCF has the potential to be used for a large-capacity and long-distance transmission since its crosstalk and DMD characteristics were well controlled at less than -27 dB/100 km and 0.43 ns/km, respectively.

ACKNOWLEDGMENT

Part of this research is supported by NICT, Japan as part of the "R&D of Innovative Optical Fiber and Communication Technology" program.

REFERENCES

[1] T. Kobayashi et al., "2 x 344 Tb/s propagation-direction interleaved transmission over 1500-km MCF enhanced by multicarrier full electric-field digital back-propagation," Proc. of 39th ECOC2013, PDP3.4.

[2] P. J. Puttnam et al. "2.15 Pb/s transmission using a 22 core homogeneous single-mode multi-core fiber and wideband optical comb, " Proc. of 41st ECOC2015, paper 1056.

[3] K. Shibahara et al., "Dense SDM (12-core × 3-mode) transmission over 527 km with 33.2-ns mode-dispersion employing low-complexity parallel MIMO frequency-domain equalization," Proc. of 38th OFC2015, paper Th5C3.

[4] K. Igarashi et al, "114 space-division-multiplexed transmission over 9.8-km weakly-coupled-6-mode uncoupled-19-core fibers," Proc. of 38th OFC2015, paper Th5C4.

[5] J. Sakaguchi et al., "Realizing a 36-core, 3-mode fiber with 108 spatial channels, " Proc. of 38th OFC2015, paper Th5C2.

[6] S. Matsuo et al., "Large-effective-area ten-core fiber with cladding diameter of about 200 µm," Opt. Lett., Vol.23, p. 4626 (2011).

[7] T. Sakamoto et al, "Few-mode multi-core fibre with highest core multiplicity factor, " Proc. of 41st ECOC2015, paper We.1.4.3.

[8] Y. Mitsunaga et al, "Failure prediction for long length optical fiber based on proof testing, " J. Appl. Phys., Vol. 53, p. 4847 (1982).

[9] K. Takenaga et al., "A large effective area multi-core fiber with an optimized cladding thickness," Opt. Express, Vol.19, p. B543, (2011).

[10] N. Hanzawa et al., "Four-mode PLC-based multi/demultiplexer with LP11 mode rotator on one chip for MDM transmission," Proc. of 40th ECOC2014, We.1.1.1.

[11] K. Imamura et al., "19-core fiber with new core arrangement to realize low crosstalk," Proc. 14th OECC2014, paper Tu5C-3.

Ultrafast Complex Impulse Response Measurement of 2-Mode Fibers by Using Linear Optical Sampling

Naoto Kono[*], Fumihiko Ito[*], Ryo Maruyama[**], and Nobuo Kuwaki[***]
[*]Interdisciplinary Faculty of Science and Engineering, Shimane University
[**] Optical Fiber Division, Fujikura Ltd.
[***] Corporate R&D Planning Division, Fujikura Ltd.
ito@ecs.shimane-u.ac.jp

Abstract: Complex impulse responses of 2-mode optical fibers are observed by linear optical sampling oscilloscope with 12 ps resolution, where two independent and highly coherent passive mode-locked lasers are used for probe and local beams.
Keywords: 2-mode optical fiber, impulse response, optical sampling, coherence

I. INTRODUCTION

As the fundamental capacity limit of single mode fibers is perceived commonly, space-division multiplexing technologies which use few-mode fibers (FMFs), multi-core fibers (MCFs), and their combination, have attracted increased attention [1]. When characterizing FMFs, the complex impulse response is very essential [2], not only in manufacturers' test, but also in examining cables installed by operators, since it provides us numerous parameters, which deeply concern transmission system design, such like the differential mode delay (DMD), chromatic and polarization mode dispersions (CD and PMD), and mode coupling strength in the fiber. For diagnosing the impulse response, both time resolution and high sensitivity (wide dynamic range) are required for the measurement.

In order to realize impulse response measurement with fine time resolution, an FMCW-based technology has been developed [2], where ps-level resolution has been achieved. However, there is needed a reference single-mode path, whose length is similar to that of the measured FMF, so that its length should be adjusted to match that of the measured FMF. Such a reference path is unavailable, in particular, for installed cables.

In this paper, we demonstrate a novel impulse response measurement of FMFs, by using the linear optical sampling [3, 4], especially, based on the periodic pulse characterization [5, 6]. FMFs of several lengths are characterized without any severe change of the experimental setup, thanks to the unnecessity of the reference path. The complex impulse responses are measured with ~10 ps time resolution, and the results successfully characterize the chromatic dispersion for each spatial mode and DMD. In addition, mode coupling distribution in fiber is observed with over 45 dB.

II. EXPERIMENTS AND DISCUSSIONS

Figure 1 shows the experimental setup. Two passive mode-locked lasers which oscillate at around 10 MHz repetition rate were used for yielding probe and local pulses. The probe pulse is introduced into the 2-mode fiber under test (FUT), by a single-mode fiber which is spliced to FUT with a minimal offset, hence, LP_{01} is dominantly exited at the input end. After propagating through FUT, the probe pulse was coupled to another single-mode fiber spliced with moderate offset, in order to receive LP_{01} and LP_{11} evenly. The probe pulse (complex impulse response) is detected by polarization diversity I-Q detector, where it is combined with the local pulse. The detected I-Q signals were sampled with the sampling clock synchronized to the local pulse and A/D converted. The repetition rate of the probe pulse is

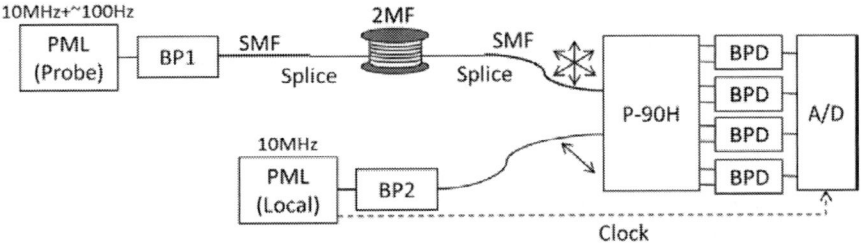

Fig. 1 Experimental setup. 2MF: 2-mode fiber under test (FUT). PML: Passive mode-locked fiber laser oscillating at 1550 nm. BP1: Optical bandpass filter (0.3 nm). BP2: Optical bandpass filter (1.0 nm). P-90H: Polarization diversity optical 90-degree hybrid. BPD: Balanced photodetector. A/D: A/D converter.

slightly (~100 Hz) larger than that of the local pulse, as a result, their collision points move about 1 ps for every acquisition which is repeated at 100 MHz. The pulse spectra were shaped by optical bandpass filters with 0.3 nm and 1.0 nm bandwidths for probe and local pulses, respectively. Consequently, suppose that the pulses are transform-limited Gaussian, the pulse widths are estimated to be ~12 ps and ~3 ps, respectively.

Figure 2 (top row) shows an example of the observed impulse responses in 23.5 km-long two-mode fiber. Here, the black and red lines show In-phase (I) and Quadrature (Q) signals, respectively. We see that a small pulse (LP_{11}) first arrived and a large one (LP_{01}) accompanies it, thus, the DMD of the fiber is negative. It is also seen that the pulse width is broadened considerably from the initial value, and the frequencies of the interference fringes are larger at the front edges of the pulses than those at the trailing edges. That is understood as the effect of the chromatic dispersion, but it should be noted that the observed phasor is rotating in the complex space in a anti-clockwise direction, in other words, the beat note oscillates at a positive frequency. Therefore, the movement of the fringes means that the chirp direction is of blue shift (anomalous dispersion).

The issue on the coherence time of the lasers is essential to understand the observed dispersion effect. Prior to the above experiment, we conducted a subsidiary experiment to estimate the coherence time, where the two pulse lasers interfere with the divided portions of a single continuous laser beam, hence, the phase difference bet ween the two obtained interference signals corresponds to the phase difference between the two pulse lasers (Because the phase of the CW laser is canceled out.). The result shows that the coherence of the two pulse lasers is maintained at least for 100 µs, namely, while the system collects 1000 samples, in other words, about 1000 ps of the time range of the oscilloscope, even with the free running lasers. That means the interference fringe reconstructed in Fig. 2 correctly reflects the phase of the probe pulse.

The center row of Fig. 2 shows the spectral phases of the interference signals, either of which is obtained by the sections corresponding to LP_{11} and LP_{01} pulses, respectively (shown by the dashed squares.). The spectral phases fit very well to parabolic curves shown by dashed lines, thus, the chirping of the pulses is purely parabolic. The same measurement was accomplished for the initial probe pulse before propagating in FUT, and comparing the spectral phases, we can estimate the 2-nd order dispersion coefficients of each mode. The results are β_2=-25.9 ps^2/km (20.3 ps/nm/km) for LP_{01} and β_2=-26.4 ps^2/km (20.7 ps/nm/km). The standard deviations of eight-time measurements of β_2 were 0.17 ps^2/km for LP_{01} and 0.32 ps^2/km for LP_{11}, respectively. These results show that the chromatic dispersion is slightly larger in LP_{11}. The bottom row of Fig. 2 shows the complex impulse response where the effect of the

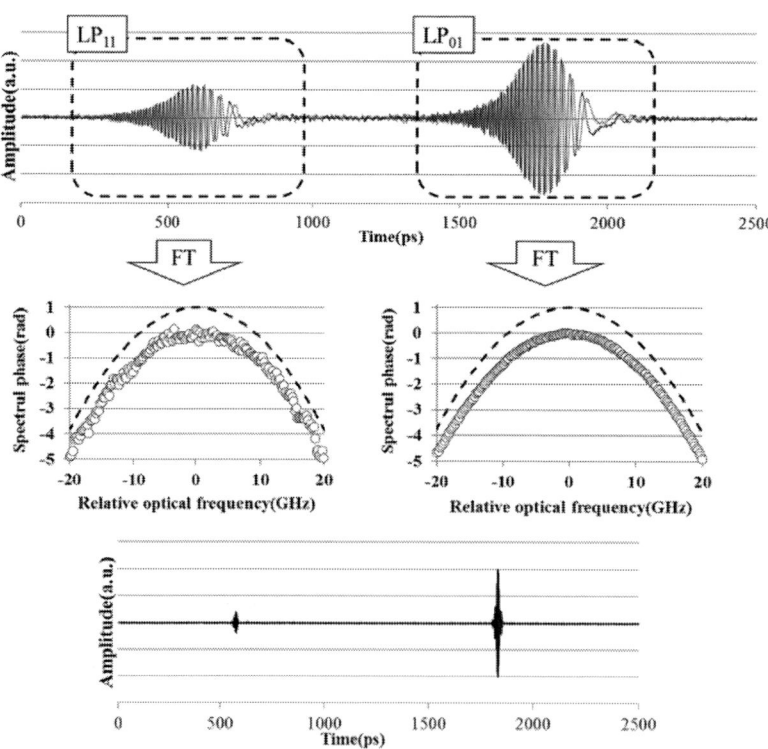

Fig. 2 Example of the observed complex impulse response of 2-mode fiber (top), and its spectral phase (center), and the complex impulse response with numerical dispersion compensation.

chromatic dispersion is numerically compensated for. It is seen that the time resolution determined by the initial pulse width is realized and the DMD is accurately estimated even when the pulse broadening due to the second-order dispersion is considerable, particularly, in ultra-low DMD fibers [7].

Figure 3 shows the impulse response observed in another 2-mode fiber of 2.1 km length. Here, the power is plotted in dB scale. The mode coupling which occurred in FUT during the propagation is clearly seen. It is also seen that the noise level of the measurement is about 45 dB smaller than the signal peak power, while the time resolution is maintained to be ~ 10 ps. Such ultrafast and high sensitivity nature would be attractive for characterizing mode coupling, particularly for localizing the mode coupling point in the fiber with high spatial resolution.

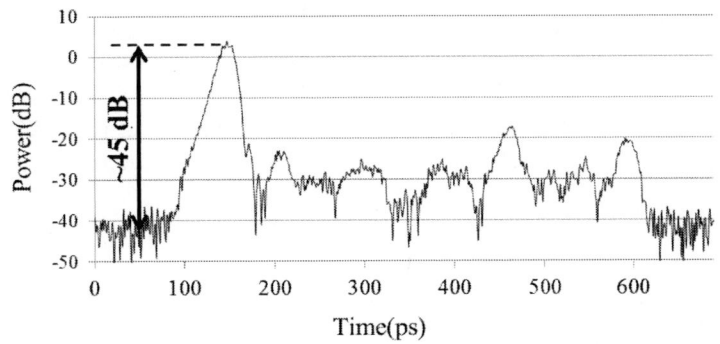

Fig. 3 Mode coupling distribution observed in 2.1 km long 2-mode fiber.

III. CONCLUSIONS

The complex impulse responses of 2-mode fibers have been observed based on linear optical sampling, with about 10 ps time resolution and 45 dB dynamic range. The measurements precisely characterize the chromatic dispersions of different spatial modes and the numerical dispersion compensation can provide accurate DMD estimation even when the pulse broadening due to second-order dispersion is considerable compared to the intermodal group delay difference (DMD). Unlike other type interferometer based measurements, the demonstrated technique needs no reference path of compatible length.

REFERENCES

[1] R-J. Essiambre, R. W. Tkach, and R. Ryf, "Fiber nonlinearlity and capacity: Single-mode and multi-mode fibers," in Optical Fiber Telecommunications VIB, I. P. Kaminov, T. Li, and A. E. Willner, Eds. Oxford: Academic Press, 2013, pp. 1-43.

[2] N. K. Fontaine, R. Ryf, M. A. Mestre, B. Guan, X. Palou, S. Randel, Y. Sun, L. Grüner-Nielsen, R. V. Jensen,. and R. Lingle, Jr., "Characterization of Space-Division Multiplexing Systems using a Swept-Wavelength Interferometer," in Proc. The Optical Fiber Communication Conference and Exposition/National Fiber Optic Engineers Conference, 2013, paper OW1K.2.

[3] C. Dorrer, C. R. Doerr, I. Kang, R. Ryf, J. Leuthold, and P. J. Winzer, "Measurement of eye diagrams and constellation diagrams of optical sources using linear optics and waveguide technology," J. Lightwave Technol., Vol. 23, no. 1, pp. 178-186, Jan. 2005.

[4] K. Okamoto and F. Ito, "Dual-channel linear optical sampling for simultaneously monitoring ultrafast intensity and phase modulation," J. Lightwave Technol., Vol. 27, pp. 2169-2175, June 2009.

[5] C. Dorrer, "Complete characterization of periodic optical sources by use of sampled test-plus-reference interferometry," Opt. Lett., Vol. 30, pp. 2022-2024, Aug. 2005.

[6] I. Coddington, W. C. Swann, and N. R. Newbury, "Coherent linear optical sampling at 15 bits of resolution," Opt. Lett., Vol. 34, pp. 2153-2157, July 2009.

[7] R. Maruyama, N. Kuwaki, S. Matsuo, and M. Ohashi, "Two mode optical fibers with low and flattened differential modal delay suitable for WDM-MIMO combined system," Opt. Exp., Vol. 22, pp. 14311-14321, June 2014.

PMD Measurements of Two-Mode Fibers Using Mode Coupler Based on Fixed Analyzer Technique

Shota Igarashi[1], Masaharu Ohashi[1,*], Yuji Miyoshi[1], Hirokazu Kubota[1], Taiji Sakamoto[2], Takashi Matsui[2] and Kazuhide Nakajima[2]

[1]Osaka Prefecture University, 1-1, Gakuen-cho, Naka, Sakai, Osaka, 599-8531 Japan
[2]NTT Access Network Service Systems Laboratories, 1-7-1, Hanabatake, Tsukuba, Ibaraki, 305-0805 Japan
*ohashi@eis.osakafu-u.ac.jp

Abstract: *We propose PMD measurements of the LP_{11} mode in two-mode fibers using mode coupler based on the fixed analyzer technique. For the first time, we have successfully measured the PMD values of the LP_{11} mode.*
Keywords: *polarization mode dispersion (PMD), two-mode fiber, mode coupler, fixed analyzer technique*

I. INTRODUCTION

In recent years, the transmission capacity of a single-mode fiber (SMF) has been approaching the transmission limit [1]. Mode-division multiplexing (MDM) systems using few-mode fibers (FMF) have been studied and extensively developed as space-division multiplexing (SDM) systems to achieve an optical communication system beyond the transmission limit [2-4]. Therefore, FMFs' transmission characteristics such as mode-dependent loss, differential modal group delay, chromatic dispersion, and polarization mode dispersion (PMD) are very important issues that should be considered when designing MDM systems. These parameters and their measurement techniques have been studied. Notably, there have been few reports of PMD measurements of FMFs.

In this paper, we propose a technique for measuring the PMD of the LP_{11} mode in a long two-mode fiber (TMF) using a mode coupler based on the fixed analyzer technique [5]. We achieve PMD measurements by launching only the LP_{01} or LP_{11} mode into the test TMF using a mode coupler. To the best of our knowledge, this is the first study wherein the PMDs of the LP_{01} and LP_{11} modes in a 30-km-long TMF have been successfully estimated. We confirm that our technique can be applied to the PMD measurements of the LP_{01} and LP_{11} modes in the TMF.

II. THEORETICAL BACKGROUND OF PMD

PMD for an SMF is defined as the linear average, $<\Delta\tau>$, of DGD values, $\Delta\tau(v)$, over a given optical frequency range from v_1 to v_2 [6]:

$$PMD = \langle \Delta\tau \rangle = \frac{\int_{v_1}^{v_2} \Delta\tau(v)dv}{v_2 - v_1}. \tag{1}$$

There exist multiple test methods for the PMD measurements of SMFs: the Stokes parameter evaluation technique [7], the state of polarization method [8], the interferometric method [9], and the fixed analyzer technique [5].

We applied the fixed analyzer technique to the PMD measurement of the LP_{11} mode in a TMF because its measurement setup is the simplest among the abovementioned techniques. If the light of the LP_{11} mode is launched into a TMF, its PMD can be measured using the fixed analyzer technique as well as a conventional SMF. Figure 1 shows the proposed schematic of the fixed analyzer technique for PMD measurements of the LP_{11} mode in a test TMF.

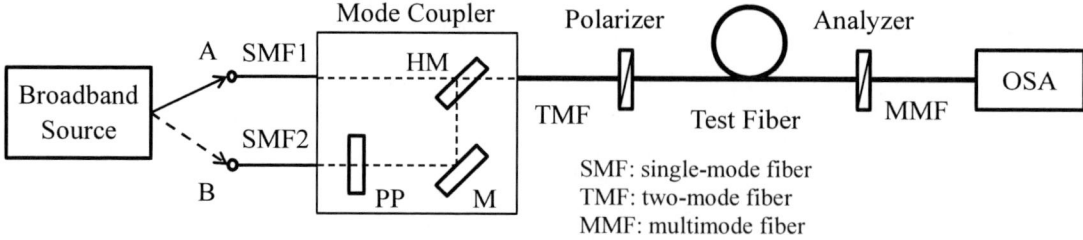

Fig. 1. Proposed schematic of the fixed analyzer technique for the PMD measurement of the LP_{11} mode.

The light of the LP_{01} mode from the broadband source can be converted to that of the LP_{11} mode using the phase plate in the mode coupler, as shown in Fig. 1. The mode coupler is inserted before a polarizer to excite the LP_{11} mode into the TMF. The output power from the test fiber is received by the optical spectrum analyzer (OSA) via the analyzer.

The measurement procedure is as follows. The light from the broadband source is launched into the test TMF through the mode coupler and polarizer. Next, the output power $P_A(\lambda)$ from the test fiber is received by the OSA via the analyzer. Then, the analyzer is rotated 90° with respect to the orientation used above, and the output power $P_B(\lambda)$ is received again by the OSA. The power ratio R (λ) is obtained as in [6]:

$$R(\lambda) = \frac{P_A(\lambda)}{P_A(\lambda) + P_B(\lambda)}. \tag{2}$$

The PMD can be measured by the power ratio $R(\lambda)$. The function $R(\lambda)$ should be obtained at equally spaced wavelength intervals from a minimum wavelength of λ_1 to a maximum wavelength of λ_2. N_e is the number of extrema (both maximums and minimums) minus one within the window. The formula for the PMD, $<\Delta\tau>$, is defined as in [6]:

$$\langle \Delta \tau \rangle = \frac{k N_e \lambda_1 \lambda_2}{2c(\lambda_2 - \lambda_1)}, \tag{3}$$

where c is the speed of light in vacuum and k is a mode-coupling factor, which equals 1.0 in the absence of strong mode coupling and 0.82 in the limit of strong mode coupling [6].

III. Experimental Results

To confirm the practicality of our technique, we measured the PMD of LP_{01} and LP_{11} modes in a 30-km-long TMF. The measurement setup is shown in Fig. 1. We used a superluminescent diode whose operating wavelength was 1.45–1.625 μm as the broadband source. A phase-plate-based mode coupler was used, as shown in Fig. 1.

First, we observed the field pattern before the polarizer and at the end of the test fiber using an infrared vidicon camera. Figures 2 and 3 show the field patterns (a) before the polarizer and (b) at the end of the test fiber, respectively, when launching the LP_{01} or LP_{11} mode into the test fiber. At the bottom of each figure, the intensity of the field pattern along a line is shown as a function of the radius.

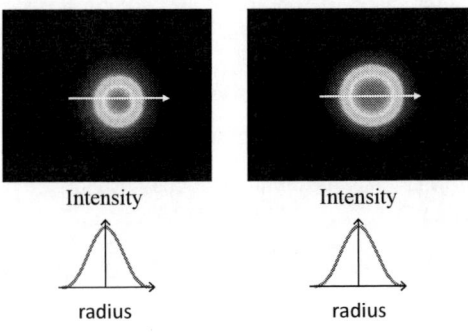
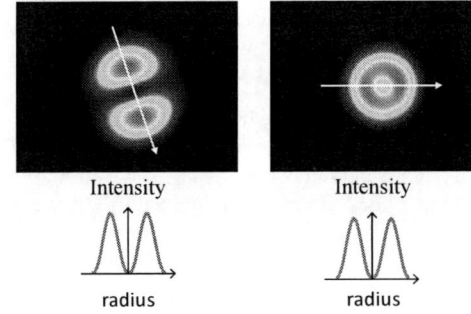

Fig. 2. Field pattern when launching the LP_{01} mode into the test fiber: (a) before the polarizer and (b) at the end of the test TMF.

Fig. 3 Field pattern when launching the LP_{11} mode into the test fiber: (a) before the polarizer and (b) at the end of the test TMF.

When launching the LP_{01} mode into the test fiber, we observed the LP_{01} mode pattern as a Gaussian intensity before the polarizer and at the end of the test fiber. Conversely, when launching the LP_{11} mode into the test fiber, we observed the LP_{11} mode pattern as the intensity shown in Fig. 3 at both positions. Therefore, we confirmed that the PMD measurements of the LP_{01} and LP_{11} modes can be measured using the mode coupler with the fixed analyzer technique.

Figures 4 and 5 show the respective experimental power ratios in Eq. (2) of LP_{01} and LP_{11} modes. Using Eq. (3), the PMD values of the LP_{01} and LP_{11} modes were calculated to be 0.37 ps and 0.48 ps, respectively. The mode-coupling factor k was set to 0.82 for PMD estimation. Next, we estimated the PMD coefficient for the PMD normalized to the measurement length.

For random mode coupling, the PMD coefficient is the PMD divided by the square root of the length and is usually expressed in units of $ps/km^{1/2}$ [6]. According to the definition of the PMD coefficient, PMD coefficients for the LP_{01} and LP_{11} modes were determined to be $0.067\ ps/km^{1/2}$ and $0.088\ ps/km^{1/2}$, respectively. It was found that the PMD of the LP_{11} mode was larger than that of the LP_{01} mode for the test TMF.

As a result, for the first time, we successfully measured the PMD of LP_{01} and LP_{11} modes in the TMF using the mode coupler with the fixed analyzer technique.

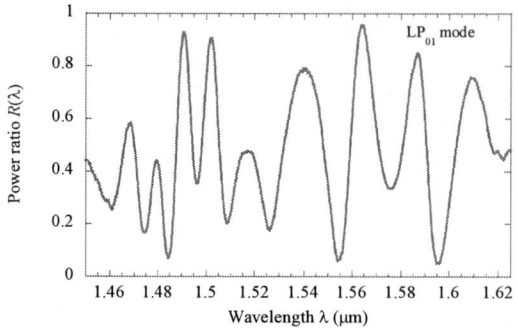

Fig. 4. Power ratio, $R(\lambda)$, as a function of the wavelength when launching the LP_{01} mode into the test fiber.

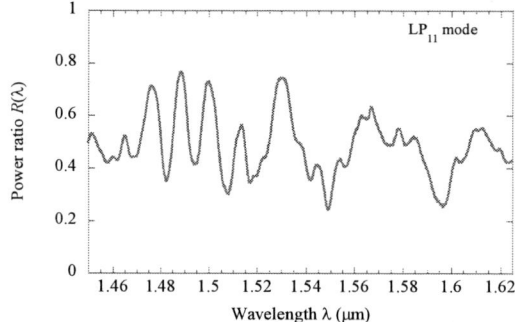

Fig. 5. Power ratio, $R(\lambda)$, as a function of the wavelength when launching the LP_{11} mode into the test fiber.

IV. CONCLUSIONS

We proposed a technique for measuring the PMD of the LP_{11} or LP_{01} mode in a TMF using a phase-plate-based mode coupler based on the fixed analyzer technique. Field patterns before the polarizer and at the end of test fiber were observed to confirm whether each mode can be excited in the test fiber. As a result, this was confirmed using the mode coupler. To the best of our knowledge, this is the first study wherein the PMD values of LP_{01} and LP_{11} modes have been successfully estimated for the 30-km-long TMF.

ACKNOWLEDGMENT

This research is partially supported by the National Institute of Information and Communications Technology (NICT), Japan, under "Research on Innovative Optical Fiber Technology."

REFERENCES

[1] T. Morioka, "New generation optical infrastructure technologies: "EXAT initiative: Towards 2020 and beyond," OECC 2009, FT4, 2009.

[2] D. Soma et al., "2.05 Peta-bit/s Super-Nyquist-WDM SDM Transmission Using 9.8-km 6-mode 19-core Fiber in Full C band ECOC2015, 1047, 2015.

[3] B. J. Puttnam et al., "2.15 Pb/s Transmission using a 22 core homogeneous single-mode multi-core fiber and wideband optical comb," ECOC2015, 1056, 2015.

[4] K. Igarashi et al., "114 space-division-multiplexed transmission over 9.8-km weakly-coupled-6-mode uncoupled-19-core fibers," OFC2015, Th5C.4, 2015.

[5] C. D. Poole and D. L. Favin, "Polarization-mode dispersion measurements based on transmission spectra through an analyzer," J. Lightwave Technol., vol. 12, pp. 917-929, 1994.

[6] ITU-T Recommendation G.650.2, "Definitions and test methods for statistical and non-linear related attributes of single-mode fiber and cable," 2005.

[7] B. L. Heffner, "Automated measurement of polarization mode dispersion using Jones matrix eigen analysis," IEEE Photon. Technol. Lett., vol. 4, pp. 1066-1069, 1992.

[8] Y. Namihira and J. Maeda, "Polarization mode dispersion measurements in optical fibers," SOFMC'92, Boulder, pp. 145-150, 1992.

[9] N. Gisin, J. P. Von der Weid, and J. P. Pellaux, "Polarization mode dispersion of short and long single-mode fibers," J. Lightwave Technol., vol. 9, pp. 821-827, 1991.

MD1-1

High Speed VCSEL-based links for use in data centers and HPC

Daniel M. Kuchta
IBM Research Staff Member, IBM

This tutorial will cover the use and application of directly modulated Vertical Cavity Surface Emitting Lasers (VCSELs), high speed direct detection receivers and multi-mode fiber for data centers and High Performance Computing (HPC) applications. Topics will include packaging, advances in modulation performance to 71 Gb/s and 50 Gb/s high temperature operation to 90C, higher fiber bandwidth density using multi-core multimode fiber and Short Wavelength Division Multiplexing (SWDM) using extended window single fiber. Other topics will include the use of equalization to extend the bit rate and the use of CMOS to lower the link power consumption.

Biography

Daniel M. Kuchta

Daniel M. Kuchta (IEEE SM · 7) is a Research Staff Member in the Communication and Computation Subsystems Department at the
IBM Thomas J. Watson Research Center. He received B.S., M.S., and Ph.D. degrees in Electrical Engineering and Computer Science from the University of California at Berkeley in 1986, 1988, and 1992, respectively. He subsequently joined IBM at the Thomas J. Watson Research Center, where he has worked on high-speed VCSEL characterization, multimode fiber links, and parallel fiber optic link research. Dr. Kuchta is an author/coauthor of more than 125 technical papers and inventor/co-inventor of at least 20 patents.

MD2-1 (Invited)

OECC/PS2016

High Speed Modulation of
Transverse Coupled Cavity VCSELs

Fumio Koyama

Laboratory for Future Interdisciplinary Research of Science and Technology, Tokyo Institute of Technology, Japan
koyama@pi.titech.ac.jp

Abstract: We review the bandwidth-enhancement of transverse-coupled-cavity VCSELs with an engineered oxide aperture. The noticeable modulation-bandwidth enhancement can be exhibited owing to "photon-photon resonance effect". Also, the same structure can be used for modulator-VCSEL integrations.

Keywords: VCSEL, coupled cavity, high-speed modulation

I. INTRODUCTION

Vertical cavity surface-emitting lasers (VCSELs) are the most popular choice as the ultra-fast and energy-efficient transmitters for use in optical interconnection of datacenters and supercomputers [1-4]. Multi-mode VCSELs are commonly used for their high output power and easy alignment with a multi-mode fiber [1,3]. Improvements on the wafer level have made much progress especially on the heat control and reliability [5]. High speed operations at data rates beyond 40 Gbps were demonstrated through the careful optimization of epi-wafers and device structures [6, 7], and recently reached at to 71 Gbps with electrical equalization employed [8]. But there still remain difficulties in increasing the frequency response to go over 20 GHz, which is mostly limited by the relaxation oscillation frequency and paracitics. Innovations on the device design and modulation schemes have shown a possibility in dramatical increase in the modulation bandwidth for example by using injection locking [9, 10] and modulator integrations [11, 12]. As an alternative, an exciting approach was proposed and demonstrated using a coupled cavity structure functioning like optical equalization [13]. A 980 nm-band transverse coupled-cavity VCSEL has shown a 3dB bandwidth over 29 GHz [10], and a single-mode coherent twin VCSELs exhibits a 37 GHz bandwidth after resonance tuning [14].

In this paper, we review the progress of transverse-coupled-cavity VCSELs for increasing the modulation bandwidth and the monolithic lateral modulator integration.

II. TRANSVERSE-COUPLED CAVITY VCSELS

This transverse coupled cavity VCSEL is monolithically fabricated on an 850 nm-band GaAs-based epitaxial wafer grown by a Metal Organic Chemical Vapor Deposition (MOCVD). A 3D schematic view of a typical design is shown in Figs. 1(a) and (b) [15]. The VCSEL is on the left side, with an feedback cavity attached to the right. Proton implantation penetrating the whole top distributed Bragg reflectors (DBR)-mirror was carried out between the two sections for current confinement, in addition to the forming of an oxide aperture by wet-oxidization. In order to enhance the lateral coupling, a tapered aperture width change is introduced, making battle-shaped oxide aperture. It is because the cutoff-wavelength of the VCSEL side will be red-shifted due to heat, which prevents the light to couple to the feedback cavity. The length of the feedback cavity is critical for selecting a suitable mode for bandwidth enhancement. The mesa sizes for VCSEL and feedback cavity are 10×14 μm^2 and 20×24 μm^2, respectively, with an oxidization distance of 5 μm from the side in design. The proton-implanted electrical isolation width is 3 μm. A photo of the fabricated device is shown in Fig. 1(c). The basic physics for the bandwidth enhancement is similar to the optical feedback scheme in edge emitting lasers, which has been successfully employed for DFB/DBR lasers with an extended cavity [16-18]. A several μm long feedback cavity is long enough for a few tens GHz spacing of photon-photon resonance thanks to slowing light in lateral guiding in the VCSEL structure [12]. We pumped only a VCSEL cavity while the coupled feedback cavity is un-pumped. Proton implantation was carried out for avoiding current leakage into the feedback cavity.

III. BANDWIDTH ENHANCEMENT OF TCC VCSELS

Small-signal responses of the directly modulated laser at different biases were measured. As shown in Fig. 2(a), an even larger f_{3dB} of 29 GHz is obtained with a bias of only 4.3 mA. We did not include the photo-detector (PD, Newport 1414-50, $f_{3dB} > 25GHz$) calibration in this result thus a calibration is needed. The calibrated response is also plotted in Fig. 2 in purple. We can see that the intrinsic device f_{3dB} exceeds 30GHz, which is almost three times than that of the device without feedback and the higher bandwidth ever reported so far. Large-signal modulation was also carried out as shown in Fig. 3. We use non-return-to-zero (NRZ) signals consisting of pseudo-random binary sequences (PRBS) and a peak-to-peak voltage swing of 500 mV_{pp}. The VCSEL bias current here is 6.2 mA without pumping the feedback cavity. Eye-patterns for bitrates from 30 Gbps to 40 Gbps are shown in Fig. 2(b).

OECC/PS2016

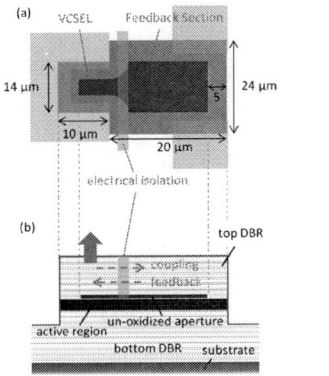

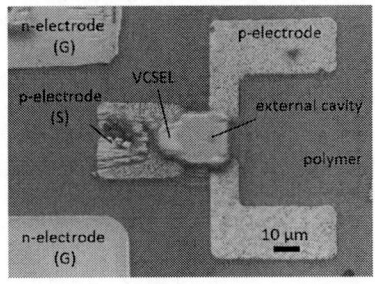

(c)

Fig. 1. (a), (b) Schematic view of the 850 nm-band transverse cavity surface-emitting laser and (c) top-view of one fabricated device. [15]

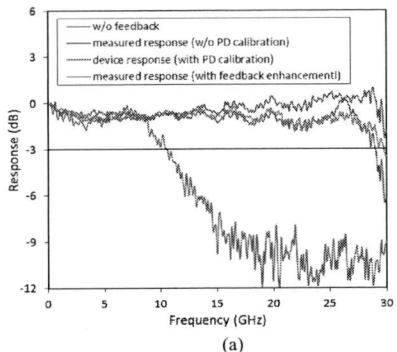

(a)

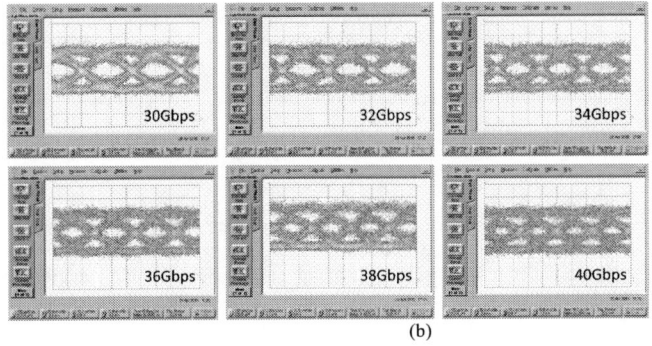

(b)

Fig. 2. Measured small signal response and large signal response of 850nm TCC VCSEL. [15]

IV. MODULATOR-INTEGRATED VCSEL

The structure of the electro-absorption modulator is similar to our previously demonstrated slow-light modulator [19]. Highly efficient modulation can be realized in the slow-light modulator because the group index in the modulator is estimated over 150. By applying a negative bias on the modulator section, strong electro-absorption occurs in the quantum wells thus the light intensity can be modulated at the output. Proton implantation was carried out for electrical isolation between the VCSEL and the modulator. The electrical isolation is over 1MΩ, which is large enough so that the leakage current is negligible for reverse-bias modulator voltages. The output from the VCSEL and the modulator is separately measured by collecting each power through a multi-mode fiber. We demonstrated an ultra-compact (13μm long) electro-absorption slow-light modulator laterally-integrated with a 980-nm VCSEL as shown in Fig. 3(a) [20]. The total device length is as small as 25 μm. A 3-dB small signal modulation bandwidth of over 30 GHz is obtained thanks to the resonance effect in the coupled cavities as shown in Fig. 3(c). The bandwidth is far beyond the relaxation oscillation frequency of directly modulated VCSELs. Static extinction ratios of 4 and 8 dB are obtained at reverse bias voltages of 0.5 and 1.3 V for a 13 μm long modulator, respectively.

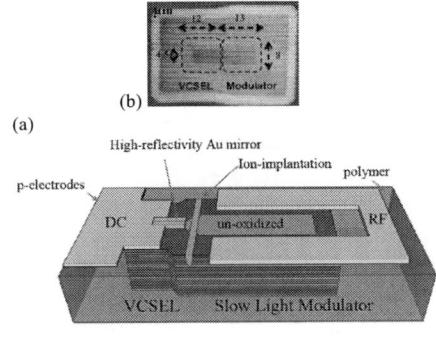

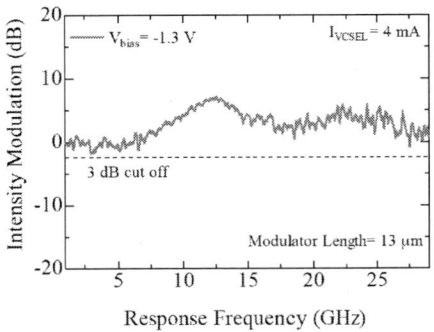

(c)

Fig. 3. (a) Schematic structure, (b) oxidation formation of the modulator integrated VCSEL and (c) measured small signal response. [20]

V. CONCLUSIONS

We reviewed our study on the high-speed modulation of transverse-coupled-cavity VCSELs. The noticeable modulation-bandwidth enhancement can be exhibited owing to "photon-photon resonance effect". We realized a quasi-single mode transverse-coupled cavity VCSEL with a smaller bow-tie shape aperture, which makes current injection unnecessary in a feedback cavity. We demonstrated the modulation-bandwidth enhancement of a quasi-single mode VCSEL with a passive optical-feedback-cavity. The 3-dB modulation bandwidth can reach at 30 GHz, which is 3 times larger than a conventional VCSEL without optical feedback. Eye opening of large signal modulations at 40 Gbps was obtained. The modeling and future prospect toward the bandwidth enhancement beyond 40GHz will also be presented. In addition, we demonstrated the lateral integration of an ultra-compact electro-absorption modulator with VCSEL. We obtained a sub-volt low driving voltage, and the bandwidth beyond 30 GHz. Our ultra-compact modulator integrated VCSEL can boost the modulation speed far beyond the direct modulation bandwidth for use in next-generation computing and data-center networks.

ACKNOWLEDGMENT

This work was supported by JSPS KAKENHI (#26630152).

REFERENCES

[1] K. Iga, Jpn. IEEE J. Sel. Top. Quantum Electron. 6, 1204 (2000).
[2] F. Koyama, Opt. Rev. 21, 893 (2014).
[3] A. Larsson, IEEE J. Sel. Top. Quantum Electron. 17, 1552 (2011).
[4] D. A. B. Miller, Proc. IEEE 97, 1166 (2009).
[5] S. Kamiya, T. Kise, M. Funabashi, M. Suzuki, A. Imamura, K. Hiraiwa, T. Nakamura, H. Shimizu, T. Ishikawa, and A. Kasukawa, presented at OECC2013, 18th OptoElectronics and Communications Conference, 2013.
[6] P. Moser, J. A. Lott, P. Wolf, G. Larisch, H. Li, and D. Bimberg, Electron. Lett., 50 (2014) 1369–1371.
[7] P. Westbergh, E. P. Haglund, E. Haglund, R. Safaisini, J. S. Gustavsson and A. Larsson, Electron. Lett., 49, (2013) 1021.
[8] D. M. Kuchta, A. V. Rylyakov, F. E. Doany, C. L. Schow, J. E. Proesel, C. W. Baks, P. Westbergh, J. S. Gustavsson and A. Larsson, IEEE Photon. Technol. Lett., 27 (2015) 577.
[9] X. Zhao, Y. Zhou, C. J. Chang-Hasnain, W. Hofmann, and M. C. Amann, Opt. Express 14, 10500 (2006).
[10] L. Chrostowski, X. Zhao, and C. J. Chang-Hasnain, IEEE Trans. Microwave Theory Tech. 54, 788 (2006).
[11] T. D. Germann, W. Hofmann, A. M. Nadtochiy, J. H. Schulze, A. Mutig, A. Strittmatter, and D. Bimberg, Opt. Express 20, 5099 (2012).
[12] H. Dalir, and F. Koyama, Appl. Phys. Lett. 7, 112101 (2014).
[13] H. Dalir, and F. Koyama, Appl. Phys. Lett. 103, 091109 (2013).
[14] S. T. M. Fryslie, M. P. T. Siriani, D. F. Siriani, M. T. Johnson, and K. D. Choquette, IEEE Photonics Technol. Lett., 27, 415 (2015).
[15] X. Gu, M. Nakahama, A. Matsutani, M. Ahmed, A. Bakry and F. Koyama, Applied Physics Express, vol. 8, no. 8, pp. 82702-1-4, (2015).
[16] M.Vallone, P.Bardella and I.Montrosset, IEEE J. of Quant. Electron., 47, (2011) 1269-1276.
[17] U. Troppenz, J. Kreissl, W. Rehbein, C. Bornholdt, T. Gaertner, M. Radziunas, A. Glitzky, U. Bandelow, and M. Wolfrum, 32nd European Conference on Optical Communication, Th4.5.5 (2006)
[18] G. Morthier, R. Schard, and O. Kjebon, IEEE J. Quantum Electron. 36, (2000) 1468–1475.
[19] G. Hirano, F. Koyama, K. Hasebe, T. Sakaguchi, N. Nishiyama, C. Caneau, C. E. Zah, Optical Fiber Communications Conference, PDP34, (2007).
[20] H. Dalir and F. Koyama., Applied Physics Express, vol. 7, no. 11, pp. 112101-1-3 (2014).

OSA module with ×4 Mini MT and 45⁰ Fiber Mirrors for 48 Gb/s Optical Link

Feng-Cheng Hsu, Tsu-Shiu Wu, and Fang-Zeng Lin
LinkWell Opto-electronics Corporation, 6F, 1-27 & 1-28 Kuo-Jian Rd., Cianjhin Dist, Kaohsiung, Taiwan
Dung-Chin Yang, Chun-Yen Peng, and Ann-Kuo Chu*
Department of Photonics, National Sun Yat-sen University, Kaohsiung, Taiwan
Chu5066@faculty.nsysu.edu.tw

Abstract: *OSA with ×4 mini MT and 45⁰ fiber mirrors was proposed. The total optical loss from VCSEL to PD is less than 4dB. The eye diagrams of the OSA modules at 48Gb/s were demonstrated.*
Keywords: *fiber optical link, optical subassembly module, ×4 mini MT, 45⁰ fiber mirror*

I. INTRODUCTION

High-speed optical links are utilized to meet increasing demands of internet traffic and to overcome the limitations of electrical interconnects due to skin effect. The speed and quality of data transmissions in short-reach applications, such as data processing and storage in a data center, were significantly improved using optical interconnecting technology. Recently, high performance optical links with optical subassembly (OSA) modules based on lens, Si optical bench, and 45° fiber mirror technologies [1-3] were successfully demonstrated. In this paper, OSA with ×4 mini MT and 45° fiber mirrors was proposed for high-speed optical links. The proposed OSA is unique because the lateral spacing among the fiber mirrors was precisely determined simply by plugging and curing the ×4 MMF to a commercially-available ×4 mini MT. In this way, active devices of OSA can be passively aligned to 45° fiber mirror by matching the centers of the mirrors to the active areas of the devices. The optical coupling losses from VCSEL array to 45° fiber mirror and from 45° fiber mirror to PD array were measured. In addition, the fabricated OSA modules demonstrated successful 48 Gb/s optical signal transmission for high-speed optical links.

II. STRUCTURE AND COUPLING LOSS OF ×4 MINI MT WITH 45⁰ FIBER MIRRORS

The photo of the ×4 mini MT with 45° fiber mirrors and its coupling to VCSEL array are shown in Fig. 1. The ×4 mini MT with 45° fiber mirrors was fabricated by plugging and curing a ×4 MMF to a modified ×4 mini MT ferrule. The 45° inclined facets of the MMFs were formed by polishing technique using a special designed fixture. The 45° fiber mirrors were obtained by coating the facets of the MMF with Cr/Al thin film. The passive alignment between the ×4 45° fiber mirrors and the VCSEL array were obtained by matching the centers of the fiber mirrors to the active areas of the VCSEL array.

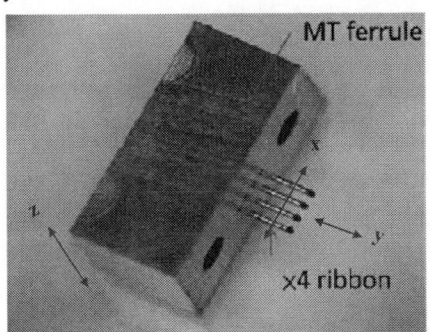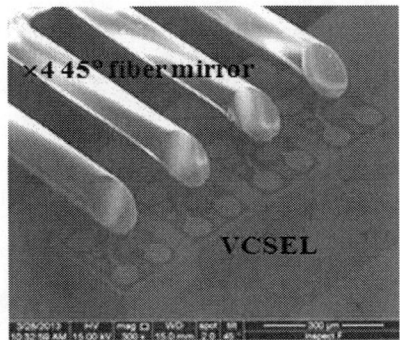

Fig. 1. The photos of the ×4 mini MT with 45° fiber mirrors and its coupling to VCSEL array.

Fig. 2 compares the measured and the calculated coupling loss between the VCSEL array and the ×4 mini MT with 45° fiber mirrors. The 45° fiber mirrors were positioned on the top of the VCSEL, and then moved the fiber mirrors in both x and y directions to measure the optical losses and the -3 dB alignment tolerance. The spacing between the bottom of the fiber mirrors to the VCSEL was 30 μm. The measured lowest optical loss from the VCSEL to the fiber mirrors was -1.9 dB, as compared with -1.6 dB of the calculation result at the position with no shifts along both x and y directions. The coupling loss of the 45° fiber mirrors to the PD was less than that of the VCSEL, which was around -1.0 dB, due to large aperture diameter of the PD active area. As shown in the figure, the -3 dB alignment tolerance along x direction was ±13 μm. However, the -3 dB alignment tolerance along y direction was not symmetry because of the asymmetry structure of the inclined fiber facets. The alignment tolerances on positive x and negative x directions were 17 μm and −11 μm, respectively.

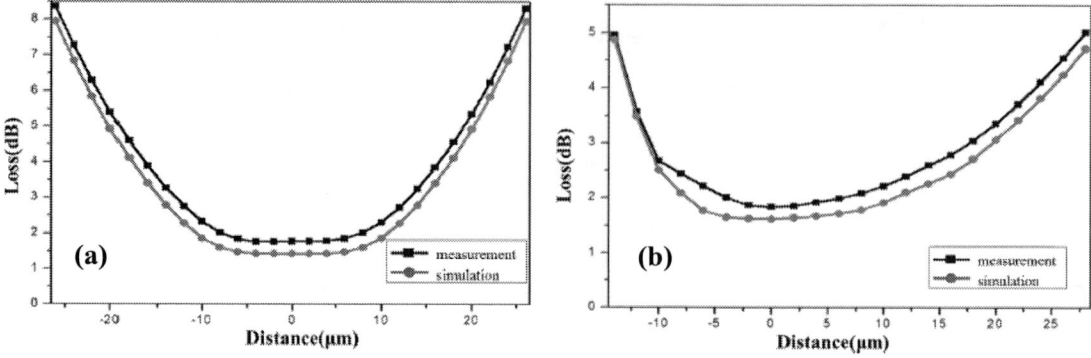

Fig. 2. The measured and the calculated coupling losses between the VCSEL array and the ×4 mini MT with 45° fiber mirrors along (a) x direction and (b) y direction.

III. ×4 OPTICAL LINK IMPLEMENTATION

The schematics of the setup for eye diagram measurement together with the photos of the fabricated ×4 OSA TX and RX modules based on QSFP standard are shown in Fig. 3. The ×4 mini MT with 45° fiber mirrors and VCSEL/PD array were passively aligned and were bounded on the submount to form the optical engine of TX/RX. The two QSFP modules were connected using an OM3 MMF fibers array terminated with ×4 mini MT on both ends. The length of the MMF was 50 m. The eye diagrams at the speeds of 6.0 Gb/s and 12 Gb/s per channel were measured.

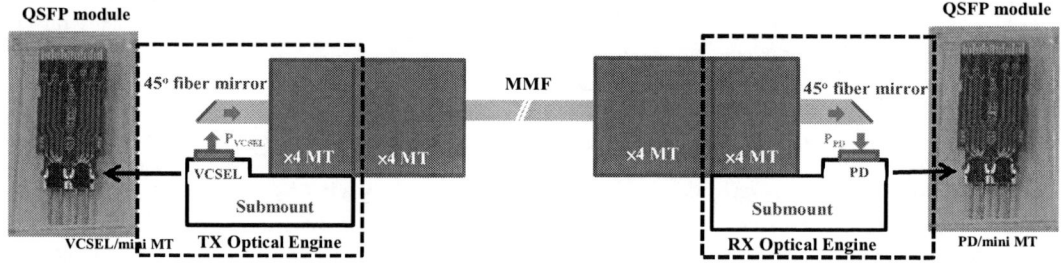

Fig. 3. The schematics of the setup for eye diagram measurement together with the photos of the fabricated ×4 OSA TX and RX modules

Fig. 4 shows the eye diagrams of the ×4 high-speed optical link measured at 6 Gb/s/ch and 10Gb/s/ch. To measure the eye diagrams of the optical link, a pulse pattern generator generated an electrical voltage signal which was converted into an optical signal by the TX IC and ×4 VCSEL array, and using the OSA module on the TX, the optical signal was transmitted through the MMF. The optical signal was then passed to the ×4 PD array by the OSA module located on the RX and finally the photocurrent generated by the PD array was converted to an electrical voltage signal with the RX IC to feed to the oscilloscope. As illustrated in Fig. 4, the measured clear and uniform eyes suggested that good optical link was obtained. The peak-to-peak jitters were 15.6 ps and 15.1 ps at 6Gs/s and 12 Gb/s data rate, respectively. The measured rise time and fall time were 36.7 and 31.1 ps at 6 Gb/s sata rate and 32.0 and 29.3 ps at 12 Gb/s data rate.

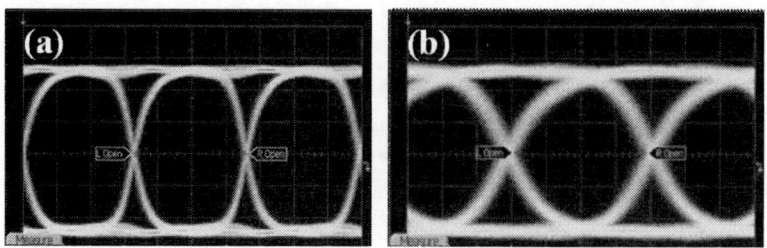

Fig. 4. The eye diagrams of the ×4 optical link measured at (a) 6 Gb/s/ch and (b) 12Gb/s/ch.

IV. CONCLUSIONS

The 48 Gb/s optical link with QSFP standard was demonstrated using the ×4 OSA module based on the architecture of the mini MT with the 45° fiber mirrors. The optical coupling losses of the module from VCSEL array via two fiber mirrors to PD array were less than 4 dB. The alignment tolerances of the ×4 45° fiber mirrors to the VCSEL array were larger than 10 μm suggesting the proposed architecture can be utilized to fabricate OSA modules with low optical losses. In addition, the fabricated OSA modules demonstrated successful 48 Gb/s signal transmission for high-speed optical link. The proposed architecture offers a unique alternative to fabricate OSA module, and may well lead to novel applications in a high-speed optical link.

REFERENCES

[1] Hak-Soon Lee, Jun-Young Park, Sang-Mo Cha, Sang-Shin Lee, Gyo-Sun Hwang and Yung-Sung Son, "Ribbon plastic optical fiber linked optical transmitter and receiver modules featuring a high alignment tolerance," Optics Express, Vol. 19, Issue 5, pp. 4301-4309 (2011).

[2] Chin-Ta Chen, Hsu-Liang Hsiao, Po-Kuan Shen, Guan-Fu Lu, Hsiao-Chin Lan, Yun-Chih Lee, Chia-Chi Chang, Jen-Yu Li, Ajay Nedle, Shuo-Fu Chang, Yo-Shen Lin and Mount-Learn Wu, "Miniaturized bidirectional optical subassembly using silicon optical bench with 45-deg micro-reflectors in short-reach 40-Gbit/s optical interconnects," Opt. Eng. 51(11), 115005 (2012).

[3] Jamshid Sangirov, Gwan-Chong Joo, Jae-Shik Choi, Do-Hoon Kim, Byueng-Su Yoo, Ikechi Augustine Ukaegbu, Nguyen T. H. Nga, Jong-Hun Kim, Tae-Woo Lee, Mu Hee Cho, and Hyo-Hoon Park, "40 Gb/s optical subassembly module for a multichannel bidirectional optical link", Optics Express, Vol. 22, Issue 2, pp. 1768-1783 (2014).

Fabrication of HCG MEMS VCSELs for athermal operations

Shunya Inoue[1], Masanori Nakahama[1], Akihiro Matsutani[2], Takahiro Sakaguchi[1], Fumio Koyama[1]

[1]Precision and Intelligence Laboratory, Tokyo Institute of Technology,
4259 Nagatsuta, Midoriku, Yokohama 226-8503, JAPAN
[2]Semiconductor and MEMS Processing Center, Tokyo Institute of Technology,
4259 Nagatsuta, Midoriku, Yokohama 226-8503, JAPAN
inoue.s.ak@m.titech.ac.jp

Abstract: *We fabricated HCG-MEMS-VCSELs with a thermally actuated membrane mirror. The HCG mirror is actuated by a bimorph beam thanks to different thermal expansion coefficients. The wavelength shows thermal blue-shift, exhibiting a possibility of athermal operations.*
Keywords: *HCG, Tunable VCSEL, atermal laser, nanoimprint lithography*

I. INTRODUCTION

Recently, data traffic is increasing explosively by spreading of internet services such as video on-demand and music streaming services. To deal with the demand, wavelength division multiplex (WDM) in backbone networks have been widely used for its large capacity. On the other hand, toward the next generation access networks such as NG-PON2 as well as optical interconnects, WDM has been considered to increase the traffic capacity [1]. To realize this new network, low cost tunable light sources are required. There are several lasers that are widely used. A DFB laser array is the integration of DFB lasers with different lasing wavelengths [2]. External cavity lasers offer wide tuning range, but the assembling cost is high [3]. SG-DBR lasers show fast wavelength tuning but complex current control is needed for tuning [4].

Vertical cavity surface emitting lasers with micro electromechanical systems (MEMS VCSELs) have lots of merits as a candidate of the low cost WDM light source. It offers wide and continuous wavelength tuning, low power consumption and possibility of athermal operation. Widely tunable MEMS VCSELs have been reported from several groups [5,6]. They use a top mirror actuated either by thermal bimorph effect or electrostatic force. Wide tuning range and high speed wavelength tuning have been demonstrated. For the light sources of WDM, thermal insensitivity of wavelengths is very important. Because wavelength separation is several nm and lasing wavelength would shift by temperature change, a semiconductor laser for use in WDM networks needs temperature control. In our group, athermal operation of a tunable MEMS VCSEL has been demonstrated using a bimorph cantilever [7]. An Al 85% composition AlGaAs strain control layer is loaded beneath the top DBR mirror. Athermal wavelength tuning was realized in 30 nm wavelength range. In conventional VCSELs, a distributed Bragg reflector is used to form a high-reflective mirror. High-index contrast subwavelength grating (HCG) was proposed by Prof. Chang-Hasnain et al[8]. HCG offers a lot of unique features, for example, broadband and high reflectivity, high Q resonances and polarization selectivity. These features have been used for VCSELs, filters, hollow waveguides, ultra-thin lens, beam scanners and so on [9,10].

To realize athermal and widely tunable VCSELs, we introduce HCG as an thermally actuated MEMS mirror. HCG consists of a very thin single layer and thus we expect high speed and low voltage wavelength tuning simultaneously. In this paper, we demonstrate HCG-MEMS-VCSELs with a thermally actuated membrane mirror. The lasing wavelength shows thermal blue-shift, exhibiting a possibility of athermal operations by adjusting the membrane structure.

II. DESIGN AND FABRICATION OF HCG MEMS VCSEL

We fabricated HCG VCSELs with MEMS structure to realize tunable and athermal lasers. Figure 1 shows the device structure we designed and fabricated. The top HCG mirror consists of a membrane cantilever structure, which includes a strain control layer.

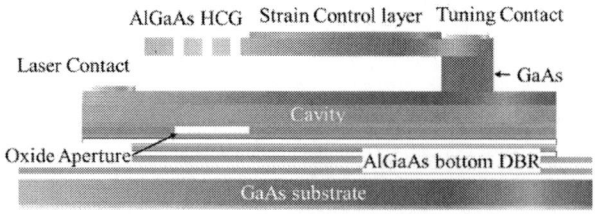

Fig.1 Device structure of HCG MEMS VCSELs

First we explain the principle of athermal operation. While the lasing wavelength of a MEMS VCSEL changes by the deflection of the membrane mirror, it is also changed by temperature. When a laser is heated, refractive index increases and the effective cavity length becomes longer. We found that the two phenomena are independent and can be compensated with each other. If the displacement of the membrane due to temperature changes is properly controlled, the thermal wavelength shift is canceled. A cantilever consists of two thin layers with different thermal expansion coefficients, which induces the downward displacement of the mirror as shown in Fig. 2. Thus, we expect the athermal operation by controlling the thermal expansion coefficients of the two layers,.

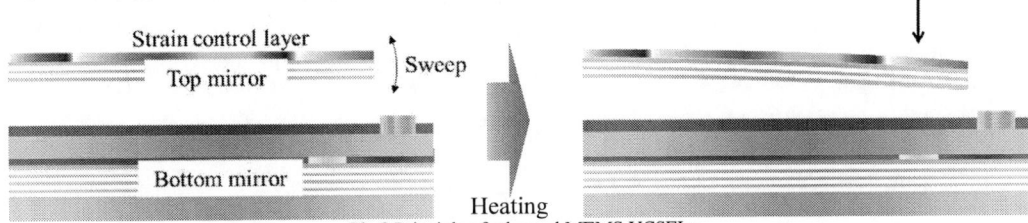

Fig.2 Principle of athermal MEMS VCSELs

A. Athermal operation with bimorph cantilever

We considered several materials as the strain control layer. Because a strain control layer locates on the MEMS cantilever, its thermal expansion coefficient should be larger than that of the HCG material ($Al_{0.65}Ga_{0.35}As$ with thermal expansion coefficient of 5.6 ppm/K). We choose Au, Cr, Ti and Al lower composition $Al_{0.3}Ga_{0.7}As$. Their thermal expansion coefficinets are 14.2, 6.2, 8.5 and 6.0 ppm/K, respectively.

The deflection of the cantilever is expressed by the equation (1) and Figure 3 shows the calculated deflection of each structure. By increasing the thickness of the strain control layer or the difference in thermal expansion coefficients, the deflection becomes larger. Given that the thermal wavelength drift is typically 0.07 nm/K and the effective cavity length of the VCSEL is 7 times wavelength, the required athermal MEMS deflection is 0.5 nm/K. We chosen 20 nm thick $Al_{0.3}Ga_{0.7}As$ for a strain control layer.

$$\Delta x = \frac{3L^2 t_c t_d E_c E_d (t_c + t_d)(t_c E_c + t_d E_d)}{t_c^4 E_c^2 + 4t_c^3 t_d E_c E_d + 6t_c^2 t_d^2 E_c E_d + t_d^4 E_d^2} \cdot \frac{(\alpha_c - \alpha_d)\Delta T}{t_c E_c + t_d E_d + (t_c E_c \alpha_d + t_d E_d \alpha_c)\Delta T} \tag{1}$$

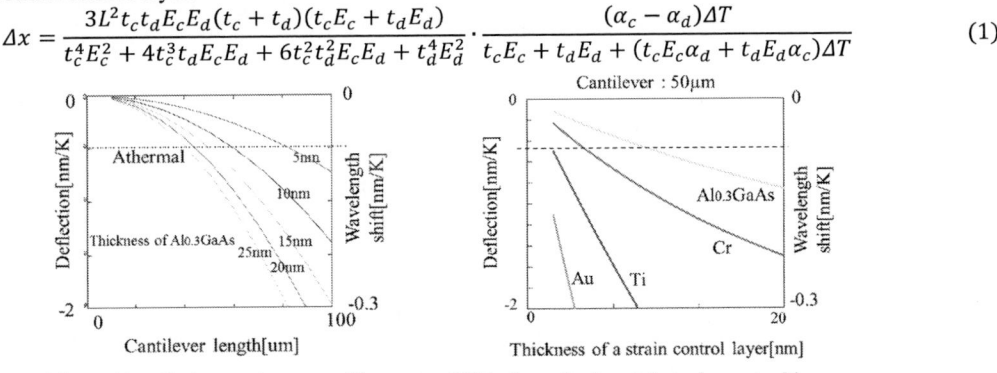

With an Al0.3GaAs strain control layer   With deposited metal strain control layers

Fig.3 Athermal condition for each strain control layer

B. HCG and VCSEL fabrication

Because HCG is grating structure with subwavelength period ($< \mu$m), EB lithography is typically used for HCG fabrication. On the other hand, we utilized a nanoimprint lithography for high-throughput fabrication and low cost manufacturing. Nanoimprint lithography is the method to transfer fine patterns on photoresist using a mold. Figure 4 shows the top view of the fabricated HCG. The HCG was precisely formed according to the mold design (period 464 nm, grating bar width 290nm). Fine HCG patterns could be formed as shown in the figure.

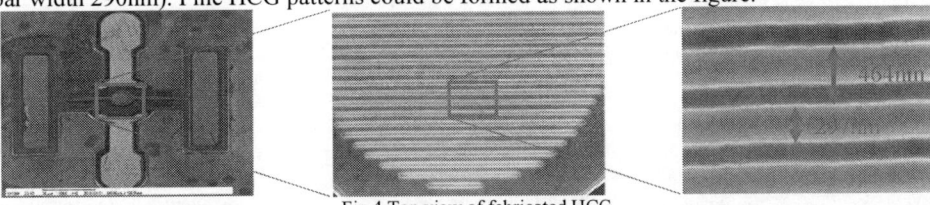

Fig.4 Top view of fabricated HCG

In HCG MEMS VCSEL fabrication process, we carried out the HCG fabrication first, followed by cantilever and mesa formation, oxidation for making current confinement aperture and electrode deposition. Finally we carried out the selective etching of a GaAs sacrificial layer underneath the HCG mirror and critical point drying for making a membrane structure.

III. DEVICE CHARACTERIZATION

The fabricated HCG VCSEL with an oxide aperture of 5 μm in diameter shows threshold currents of 1.8mA~3.4mA and the maximal output power of 0.7mW. The initial deflection was different according to cantilever length, so lasing wavelength was different. The initial deflection of VCSELs with 80 μm and 200μm long beam are 1.8 μm and 4.2 μm then lasing wavelength are 987 nm and 978 nm, respectively.

Figure 5 shows the temperature dependence of lasing wavelengths. TEC temperature increase caused the blue shift in wavelength, which is opposite for conventional semiconductor lasers. In this result, although wavelength shift was not linear, average thermal coefficient is -0.12nm/K and bridge deflection rate is estimated to be -1.3nm/ K. This negative thermal wavelength drift shows the menbrane mirror's downward deflection. By optimizing the bridge length, we can expect athermal wavelength VCSELs.

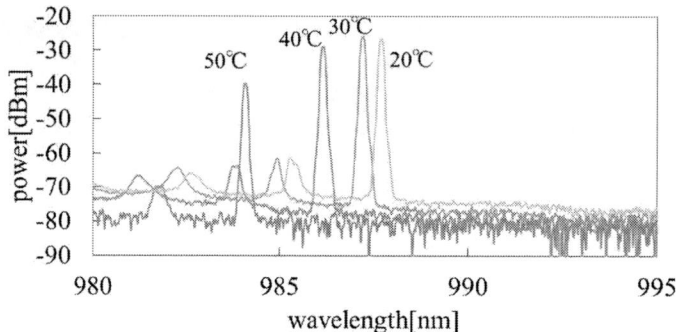

Fig.5 Lasing wavelength shift by temperature

IV. CONCLUSIONS

We designed and fabricated athermal HCG MEMS VCSELs with a 20 nm thick $Al_{0.3}Ga_{0.7}As$ strain control layer. We fabricated HCG MEMS VCSELs using nanoimprint lithography followed by selective etching of a GaAs sacrificial layer. The fine HCG was successfully fabricated with a Si mold. The fabricated VCSEL showed the negative temperature dependence of lasing wavelengths. The optimization of the structure enables us to realize athermal and tunable HCG VCSELs.

ACKNOWLEDGMENT

This work was supported by JSPS KAKENHI Grant Number 15J11948.

REFERENCES

[1] http://www.itu.int/rec/T-REC-G.989.1-201303-I, https://www.itu.int/rec/T-REC-G.989.2/e, http://www.itu.int/itu-t/workprog/wp_item.aspx?isn=9119

[2] H. Ishii, K. Kasaya, H. Oohashi, Y. Shibata, H. Yasaka and K. Okamoto, "Widely wavelength-tunable DFB laser array integrated with funnel combiner," IEEE Selected Topics in Quantum Electronics, Vol. 13, no.5, pp. 1089-1094, Oct. 2007.

[3] Harvey, K. C., and C. J. Myatt. "External-cavity diode laser using a grazing-incidence diffraction grating," Optics Letters Vol. 16, no.12, no. 910-912, June 1991.

[4] Silva, C. F. C., A. J. Seeds, and P. J. Williams. "Terahertz span> 60-channel exact frequency dense WDM source using comb generation and SG-DBR injection-locked laser filtering," IEEE Photonics Technology Letters, Vol.13, no. 4, pp. 370-372, April 2001.

[5] C. Gierl, T. Gruendl, P. Debernardi, K. Zogal, C. Grasse, H. A. Davani, G. Böhm, S. Jatta, F. Küppers, P. Meißner, and M.-C. Amann, "Surface micromachined tunable 1.55 μm-VCSEL with 102 nm continuous single-mode tuning," Optics Express, Vol. 19, no.18, pp. 17336-17343, Aug. 2011

[6] V. Jayaraman, G.D. Cole, M. Robertson, A. Uddin and A. Cable, "High-sweep-rate 1310 nm MEMS-VCSEL with 150 nm continuous tuning range," Electronics letters, Vol. 48, no. 14, pp. 867-869, July 2012

[7] M. Nakahama, T. Sakaguchi, A. Matsutani, and F. Koyama, "Precise Wavelength Tuning of MEMS VCSELs enabling 110-ch Operations," Conference on Lasers and Electro-Optics/Pacific Rim. Optical Society of America, 2015.

[8] Y. Zhou, C. Y. Huang, C. Chase, V. Karagodsky, M. Moewe, B. Pesala, F. G. Sedgwick and C. J. Chang-Hasnain, "High-index-contrast grating (HCG) and its applications in optoelectronic devices," IEEE Selected Topics in Quantum Electronics, Vol. 15, no. 5, pp. 1485-1499, Oct. 2009.

[9] F. Lu, F. G. Sedgwick, V. Karagodsky, C. Chase, and C. J. Chang-Hasnain, "Planar high-numerical-aperture low-loss focusing reflectors and lenses using subwavelength high contrast gratings," Optics express Vol. 18, no.12, pp. 12606-12614, June 2010.

[10] H. Weiwei, and C. J. Chang-Hasnain, "Optical phased array for far field beam steering with varied HCG." SPIE OPTO. International Society for Optics and Photonics, 2012.

VCSEL-Integrated Bragg Reflector Waveguide Amplifier with Single-mode Output Power over 10 mW

Xiaodong Gu[1], Masanori Nakahama[1], Akihiro Matsutani[2], Fumio Koyama[1]

[1]Photonics Integration System Research Center, P&I Lab., Tokyo Institute of Technology, Japan
[2]Semiconductor and MEMS Processing Center, Technical Department, Tokyo Institute of Technology, Japan
gu.xiaodong@ms.pi.titech.ac.jp

Abstract: *We present the designing, fabrication and device characteristics of a vertical-cavity surface-emitting laser-integrated Bragg reflector waveguide amplifier. A single-mode output over 10 mW is obtained by only 500 µm-long slow-light propagation and radiation.*

Keywords: *VCSEL, Semiconductor optical amplifier, slow-light*

I. INTRODUCTION

Enhancing the single-mode output power of a semiconductor laser has always been an ultimate task for both academic and industry. Especially in recent years, the fast development of drones, robots and autonomous cars require high-power, high-quality and low-cost sensing solutions. Vertical-cavity surface-emitting laser (VCSEL) is a strong candidate for its compact size and low production cost. However, its output-power for single-mode operation is still limited to a few milliwatts. Various approaches have been made such as surface trimming for suppressing higher-order mode [1], but the output power still cannot go over 10 mW. Two-dimensional arraying of VCSELs can be used to obtain 10 Watt-class output power but the coherency of the wavelength and phase are a severe problem [2]. In our group, we proposed and demonstrated the lateral integration of a VCSEL and slow-light waveguide amplifier [3,4], but the output power was not large due to weak coupling efficiency and low amplification gain. In this work, we improved the whole device structure and tried to pump the amplifier over its threshold condition. Single-mode power over 10 mW was obtained with high beam quality. The proposed device can be a high-quality laser source for use in long-haul optical communication systems, Optical Coherence Tomography applications and various laser sensing systems.

II. DEVICE STRUCTURE AND WORKING PRINCIPLES

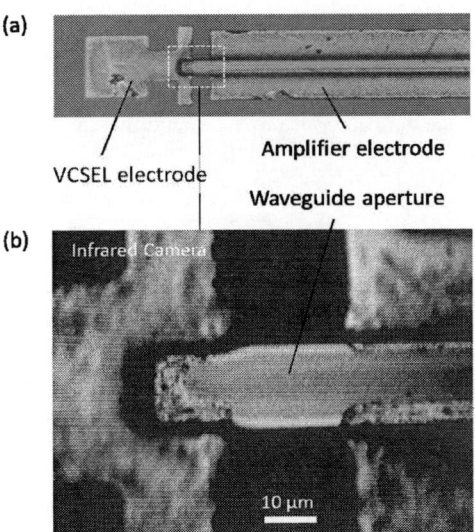

Fig. 1. Top-view of one fabricated device: (a) optical microscope, (b) infrared carema.

The device we designed has mainly two sections integrated laterally: one VCSEL on the left and a long waveguide amplifier attached on the right side. The two sections were electrically isolated via an ion implantation process. We show the top view of a fabricated device from the microscope in Fig. 1(a) and an infrared camera-captured image in Fig.

1(b). The aperture width of the waveguide is designed to be slightly wider than that of the VCSEL, in order to get a unidirectional and high-efficiency lateral coupling. The amplifier is designed to be a few millimeters for getting higher optical gain and at the same time avoiding the reflection at the waveguide end. When some current is injected to the VCSEL section, slow-light is laterally coupled to the waveguide. Because the cutoff wavelength at the laser side is smaller than the amplifier side, radiation beam emitting from the waveguide surface will not be vertical but lean down to the propagation direction. It is interesting that if we inject current through the amplifier electrode, the coupled slow-light intensity will increase intensively. Especially when the amplifier is biased above its threshold condition, its vertical emission will be suppressed while a large output at the tilted angle can be obtained.

III. MEASUREMENT RESULTS AND DISCUSSIONS

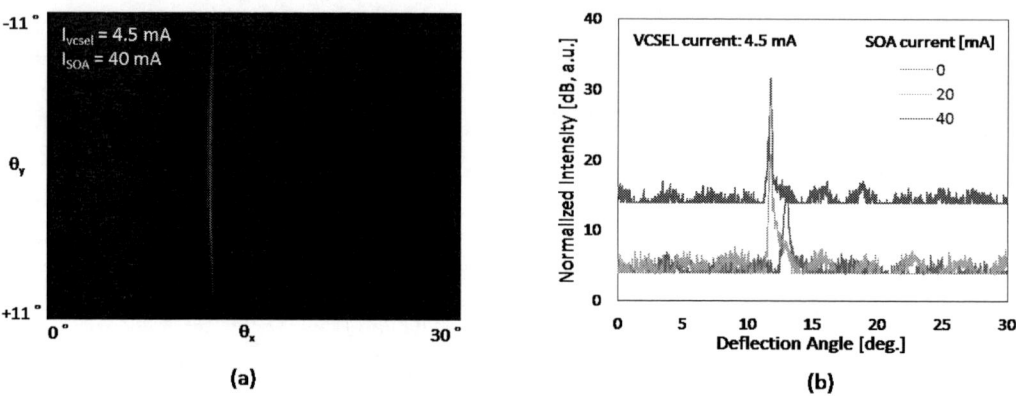

Fig. 2. (a) Far-field pattern and (b) profiles of the measured device for typical VCSEL and amplifier currents.

Firstly we observed the far-field patterns of the device at typical currents for VCSEL and amplifier sections. With a VCSEL current of 4.5 mA, especially clear single mode coupling and radiation are observed. The FFP image and profiles are illustrated in Fig. 2 (a) and (b). Extremely sharp beam can be seen at a deflection angle of 12 degrees. The different background noise level for different SOA currents are due to the change in optical attenuator during measurement. Actual background noise for I_{SOA} = 40 mA is the same as that of 0 mA and 20 mA. Comparing with the radiation without amplifier, optical gain of 4.6dB and 15.4dB are obtained for amplifier current of 20 mA and 40 mA. We can see that a much larger amplification is obtained after 20 mA, by only doubling the injection currents. In order to see the amplification current dependence, a photo-detector is put above the device with a tilted angle of 12 degrees. The captured power with and without VCSEL input are shown in Figure 3. The deducted power between the two can be used to estimate the output power of the deflection beam. We found that the threshold current of the amplifier itself is about 20~30 mA. Therefore, it can be seen that the output amplification is largely enhanced after the threshold condition.

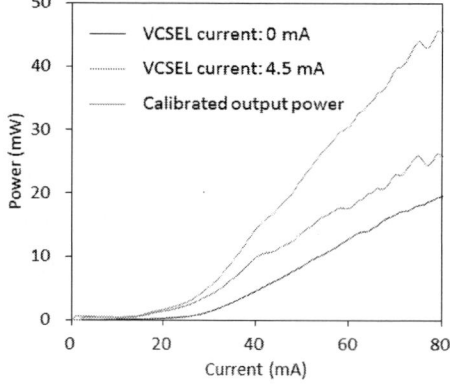

Fig. 3. Output power captured by a photo-detector above the device for different VCSEL currents.

Although very sharp deflection beam has been confirmed, it is needed to see in which mode the device works. A multi-mode fiber is used to measure the device spectrum. The fiber is also tilted at 12 degrees. Two spectrum results are shown in Fig. 4 with the same VCSEL current and different SOA currents. It is found that the device works as single-mode until 44 mA current injection into the amplifier. A side-mode suppression ratio (SMSR) of 15dB is obtained. Based on the L/I and spectra results, it is shown that a single mode output of 10.8 mW is obtained in this compact device. Considering the capturing loss to the photo-detector, the actual output power is even larger.

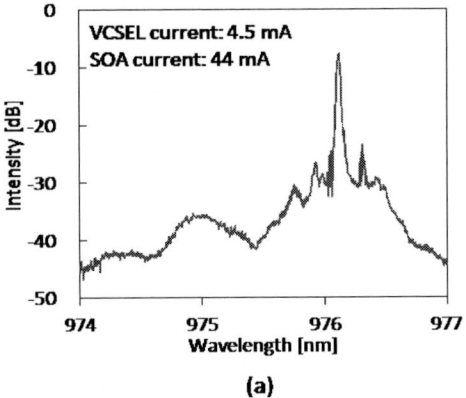 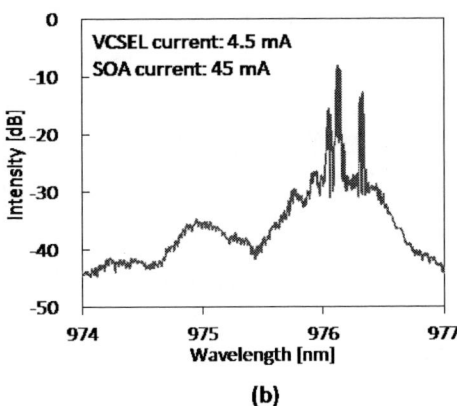

(a)   (b)

Fig. 4. Spectra captured by a multi-mode fiber for different amplifier current, with 4.5 mA VCSEL current injection.

In this device, the current injection at the amplifier section is not uniform yet, therefore the actual propagation distance of the slow-light is only few hundreds of micrometers, although the device is 2 mm in design. If the electrode pattern and thickness are optimized, or a cascaded amplification is carried out, the actual output can be further enhanced proportionally to the propagation distance. In addition, if the waveguide top-mirror reflectivity and photo-luminance peak are optimized, output power can be efficiently enhanced while keeping the device size compact.

IV. Conclusions

We present a 10 mW-class single-mode semiconductor amplifier. The device is an integration of a vertical-cavity surface-emitting laser and a slow-light Bragg reflector waveguide amplifier. Slow-light is coupled from the VCSEL to the waveguide and amplified along the propagation direction. The device output is strongly enhanced when the two sections are both at lasing conditions. Single-mode output power of 10.8 mW is captured by a photo-detector tilted to the waveguide deflection angle, with an optical gain over 15dB. By optimizing the wafer and electrodes designs, further enhancement of the single mode power to 100 mW or even Watt-class is promising.

ACKNOWLEDGMENT

This work was supported by JSPS KAKENHI (#26630152).

REFERENCES

[1] A. Haglund, J. S. Gustavsson, J. Vukusic, P. Modh, and A. Larsson "Single Fundamental-Mode Output Power Exceeding 6 mW From VCSELs With a Shallow Surface Relief, " IEEE Photon. Tech. Lett., vol. 16, no. 2, pp. 368-370, Feb 2004.

[2] J-F. Seurin., G. Xu., A., Miglo, Q. Wang, R. V. Leeuwen, Y. Xiong, W-X. Zou, D. Li, J. D. Wynn, V. Khalfin, and C. Ghosh., "High-power vertical-cavity surface-emitting lasers for solid-state laser pumping," Vertical-Cavity Surface-Emitting Lasers XVI, Proc. of SPIE, vol. 8276, 827609-1~10, 2012.

[3] T. Shimada, A. Matsutani, and F. Koyama, "Lateral integration of vertical-cavity surface-emitting laser and slow light Bragg reflector waveguide devices," App. Optics, vol. 53, no. 9, pp.1766, March 2014.

[4] M. Nakahama, T. Shimada, and F. Koyama, "Lateral integration of MEMS VCSEL and slow light amplifier boosting single mode power," IEICE ELEX, vol. 9, no.6, pp.544-551, 2012.

MD2-5

High Power Non-mechanical Beam Scanner based on VCSEL Amplifier

Masanori Nakahama[1], Xiaodong Gu[1], Akihiro Matsutani[2], Takahiro Sakaguchi[1] and Fumio Koyama[1]

[1] Photonics Integration System Research Center, [2] Semiconductor and MEMS Processing Center, Tokyo Institute of Technology
4259-R2-22Nagatsuta, Midori-ku, Yokohama 226-8503, Japan
e-mail: nakahama.m.aa@m.titech.ac.jp

Abstract: We demonstrate beam-steering and amplification operation of a novel VCSEL amplifier, exhibiting continuous beam steering of 25° and beam divergence below 1.1°. The output power reaches at 260 mW under pulsed operation.
Keywords: Semiconductor lasers, semiconductor optical amplifiers, Scanners.

I. INTRODUCTION

An optical beam scanner is a key element for use in various applications such as laser radars, laser displays, 3D scanners and free-space optical communications. A mechanical beam scanner has been widely used, but they are bulky and their steering speed is limited by mechanical parts. Therefore, non-mechanical solid state scanner is attracting much attention for compact LiDAR applications in recent days. Phased array beam steering devices based on silicon photonics were reported [1, 2], but there still remain critical issues to be solved toward high-resolution beam steering. We proposed and demonstrated a beam steering device based on a VCSEL structure as a passive dispersive element, showing the record high-resolution beam steering [3, 4]. But it is a challenge to obtain high pulsed output power for time of flight sensing applications, which typically needs over 10 W for LiDAR application. In this paper, we demonstrate the high-resolution beam-steering and high-power operation of a novel VCSEL amplifier.

II. NON-MECHANICAL BEAM SCANNER BASED ON VCSEL STRUCTURE

A schematic of the VCSEL amplifier is shown in Fig. 1(a). It is similar to our reported beam scanner based on a Bragg reflector waveguide structure [3, 4]. Its epitaxial structure is the same as oxide-confinement type VCSELs. An input light is coupled from a coupling region, where the top DBR is partly removed, through a tilted lensed fiber. The light couples to a lateral propagation mode having a small group velocity so called "slow light", which is conceptually illustrated in the figure as zig-zag rays. Since the propagation constant of the slow light mode is highly dispersive around the cut-off condition, large change in output beam angle can be obtained by sweeping the input wavelength.

Due to the lower reflectivity of the top DBR, a portion of the propagating light is radiated from the amplifier surface which forms a uniform wave fronts. The light decays exponentially due to radiation and absorption loss, but its intensity becomes uniform along the VCSEL amplifier by injecting a current above the lasing threshold of the VCSEL. In this case, power of the amplified light is proportional to the amplifier length, and the beam divergence is getting smaller and smaller with increasing the amplifier length. Therefore, we can obtain sharp beam steering and high power simultaneously. The proposed device is easy to integrate with a VCSEL laterally, and a slow light mode can be excited efficiently. Thus, we can expect an ultra-compact, high-resolution and high power beam scanner. A microscope image of the fabricated VCSEL amplifier is shown in Fig. 1(b). Pumping current is injected from the surface and backside electrodes. Output light is emitted from the emitting window.

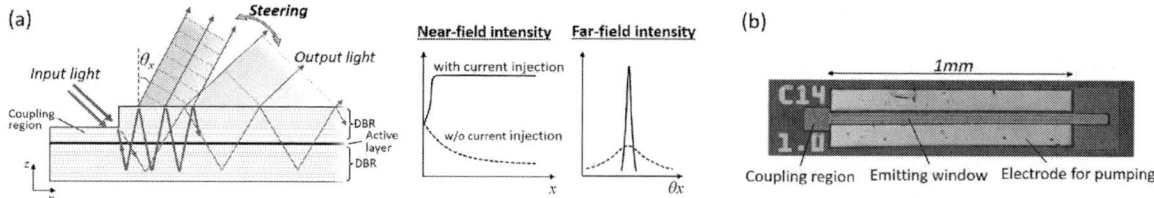

Fig. 1. (a) Schematic structure of a VCSEL amplifier we propose. (b) Microscope image of a fabricated device.

III. EXPERIMENT

We measured far field patterns (FFPs) of the output beams from a 1 mm-long device for different input wavelengths. Figure 2 shows the peak angle of the captured beams as a function of input wavelength. Continuous steering angle of 24.5° was obtained by sweeping the input wavelength for 18 nm. The angular dispersion is as large as 1.4 degree/nm, which is 2 times larger than that of typical virtually-imaged phased array (VIPA) scanners [5] thanks to the larger refractive index of III-V semiconductor materials.

Smaller beam divergence is one of the critical characteristics for beam scanning devices as well as large beam steering angle. Figure 3 shows the beam divergence for different input wavelength. The amplifier was driven by a pulse current. Pulse width and cycle were 1 μs and 0.1 ms, respectively. We defined the beam divergence as a full-width-at-half-maximum of beam intensity in *x*-direction. A sharp beam with divergence of 0.26° was observed at input wavelength of 970 nm. For higher current, the divergence is almost unchanged at 970 nm, but it is increased to 1.1° at 960 nm. It might be partly due to a non-uniformity of injection current along the amplifier, which can be improved by using a thicker electrode and multiple wire-bonding. The details are under study.

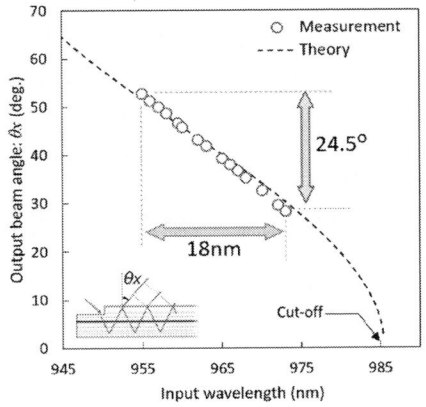

Fig. 2. Peak angle of the captured beams as a function of input wavelength

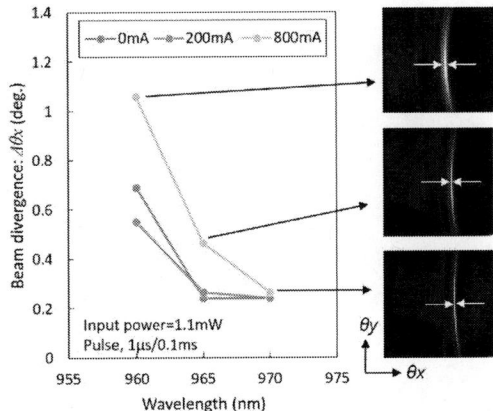

Fig. 3. Beam divergence for different input wavelength

Next we measured the amplification characteristics of the VCSEL amplifier under pulse operation. Output beam was captured with a large area (5 mm in diameter) InGaAs photodiode which was set few centimeters away from the device under measurement. In addition, the photodiode was tilted from vertical by 32-52 degrees so that the vertical lasing of the amplifier was not captured. Figure 4 shows the captured power as a function of input power for different wavelengths and injection currents. The input power was calculated by subtracting the coupling loss of 11 dB from the power of external tunable laser. The captured powers were below 1mW for every input wavelength when no current was injected. The power corresponds to the spontaneous emission and lasing of the amplifier. On the other hand, as increasing the injection current, captured power dramatically increased. A small input power of 1.1 mW was amplified to over 100 mW for every input wavelength at 800 mA. The maximum power of 260 mW was obtained at the input wavelength of 965 nm which is close to the gain peak. The chip gain corresponds to as high as 23 dB. We expect much larger power by reducing the reflectivity of the top DBR and introducing an efficient heat sinking.

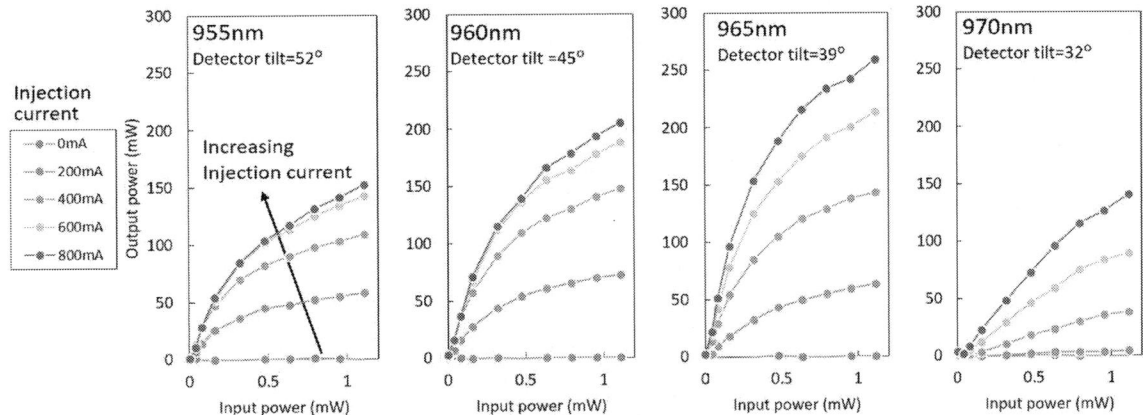

Fig. 4. Captured power as a function of input power (coupling loss of 11 dB was subtracted from the power of external tunable laser) for different wavelength. The amplifier was driven under pulsed operation. The photodiode was tilted from vertical so that the vertical lasing of the amplifier was not captured.

In order to make sure that the captured power was not from the lasing of the amplifier, we measured the spectra. Output beam was directly coupled to a multi-mode fiber which is tilted from the vertical. Figure 5 shows the measured spectra for each input wavelength under pulsed operation. In order to distinguish the reflected light from amplified light, spectra measured without current injection are superimposed.

Strong single peak was observed at every input wavelength. No other peaks were seen. In addition, intensity of peaks at 800 mA were at least 30 dB higher than peaks observed when no current was injected. Consequently, it was confirmed that the measured power in Fig. 4 was the amplified output.

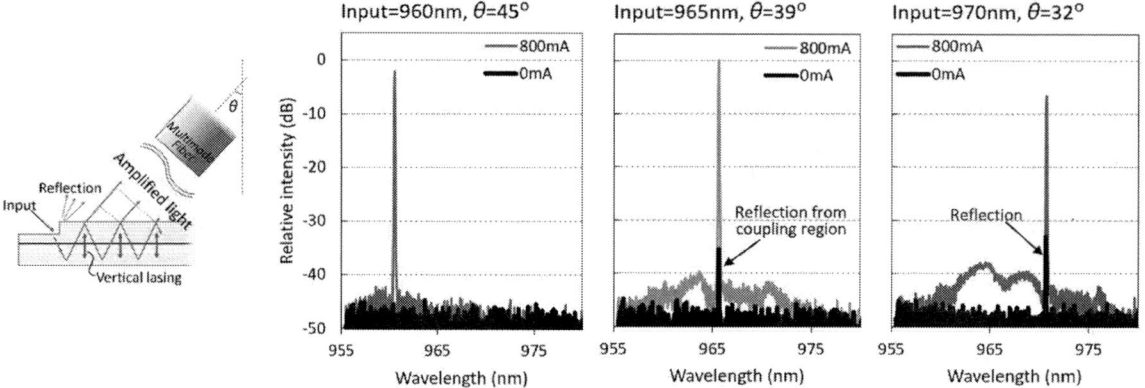

Fig. 5. Spectra of amplified light for each input wavelength. In order to distinguish the reflected light from amplified light, spectra measured without current injection are superimposed.

IV. CONCLUSIONS

We proposed and demonstrated the high-resolution beam-steering and amplification of novel VCSEL slow light amplifier. A beam steering of 25° and sharp beam divergence of 0.26° were obtained. The output power is as high as 260 mW with a chip gain of 23 dB. The output power at the moment is limited by poor heat sinking, so we expect much higher power by using die-bonding process. The result shows a possibility of beam steering function, boosting high single-mode power toward Watt-class operations with maintaining a good spectral purity and beam quality. The proposed amplifier could be laterally integrated with a VCSEL. The high-power and high-resolution beam VCSEL scanner will open up new VCSEL applications including LiDAR for automobile and industry robots.

ACKNOWLEDGMENT

This work was supported by JSPS KAKENHI (#26630152).

REFERENCES

[1] K. Van Acoleyen, W. Bogaerts, R. Baets, "Two-dimensional Dispersive Off-chip Beam Scanner Fabricated on Silicon-On-Insulator," IEEE Photon. Technol. Lett., 23(17), pp.1270-1272, (2011).

[2] J. Doylend, M. Heck, J. Bovington, J. Peters, L. Coldren, and J. Bowers, "Two-dimensional free-space beam steering with an optical phased array on silicon-on-insulator," Opt. Express, vol. 19, pp. 21595-21604 (2011).

[3] X. Gu, T. Shimada, F. Koyama, "Giant and high-resolution beam steering using slow-light waveguide amplifier," Opt. Express 19, 22675-22683 (2011)

[4] X. Gu, T. Shimada, A. Matsutani and F. Koyama, "Miniature Nonmechanical Beam Deflector Based on Bragg Reflector Waveguide With a Number of Resolution Points Larger Than 1000," IEEE Photonics Journal, vol. 4, pp. 1712-1719, Oct. 2012.

[5] M. Shirasaki, "Large angular dispersion by a virtually imaged phased array and its application to a wavelength demultiplexer," Opt. Lett. 21, 366-368 (1996).

ME1-1 (Invited)

Reconfigurable optical routers and buffers built on the silicon-on-insulator platform

Linjie Zhou*, Liangjun Lu, Shuoyi Zhao, Xinyi Wang, and Jianping Chen

State Key Laboratory of Advanced Optical Communication Systems and Networks, Department of Electronic Engineering
Shanghai Jiao Tong University, Shanghai 200240, P. R. China
*ljzhou@sjtu.edu.cn

Abstract: We report our recent progress on large-scale integrated silicon optical routers and buffers. The optical router is made of a 16×16 switch fabric with low loss and loss crosstalk. The optical buffer can provide nanosecond-range continuous delay tuning.

Keywords—silicon photonics, optical router, optical switch, integrated photonics

I. INTRODUCTION

The communication capacity in data centers is fast approaching Petabyte/s [1], which demands the scale-up of the current network architecture to accommodate more network nodes and bandwidth while retaining low lost and low power consumption. The conventional electrical switching can no longer satisfy the ever-growing complexity in data center networks due to its power hungry performance. Optical switching at the packet level is one of the candidate technologies that can be adopted to deal with the volatile traffic in data center networks. To establish an efficient optical network, reconfigurable optical routers and buffers are indispensable. Optical routers and buffers can be implemented based on various material platforms, such as silica, silicon-on-insulator (SOI), InP *etc*. The SOI platform is quite attractive due to its high electrical and optical integration capability and can be made low cost for potential mass production.

Optical routers based on SOI switch fabrics have seen considerable progress in recent years [2]. Previously, we used multimode interferometers (MMIs) and Mach-Zehnder interferometers (MZIs) to build 4×4 optical switches with broadband operation [3-5]. To further reduce the size and power consumption of switch elements, we also developed double-ring assisted MZI (DR-MZI) switches [6, 7]. The DR-MZI can have a higher switching extinction ratio but at the cost of a narrower optical bandwidth.

As light cannot be stopped and photons cannot be directly stored in any media, optical buffers are usually realized by converting optical signals to other recordable symbols or using delay lines to temporally trap the optical signals. A delay line can be made based on two approaches by either slowing down the light velocity in the waveguides or increasing the optical path length. We ever used reflective-type microring resonators to achieve continuous time delay up to 100 ps [8]. To enlarge the tuning range, we also implemented a switchable delay line

based on cascaded MZI switches and delay waveguides [9]. The delay tuning range is 1.27 ns with a minimum step of 10 ps. It can be regarded as a digital delay line, in contrast to the microring continuous delay line.

In this paper, we report our recent process on reconfigurable optical routers and buffers on the SOI platform. In particular, we demonstrate a 16×16 non-blocking silicon optical switch based on a Benes network of thermally tunable MZIs. It exhibits low loss and low crosstalk for both cross and bar states. As for the optical buffers, we combine the digital and continuous delay lines to achieve nanosecond-range continuous delay.

II. OPTICAL ROUTER

A. Structure

Figure 1(a) shows the optical router structure based on a 16×16 switch fabric in a Benes architecture. It incorporates 56 basic switch elements composed of thermally tunable MZIs. The switch elements are arranged in an 8×7 matrix. Each optical path passes 7 switch elements, resulting in balanced output. The Benes architecture is reconfigurable non-blocking. The 16×16 switch has 16! unduplicated states, achievable by setting each switch element to either "cross" or "bar" state.

There are two commonly used refractive-index tuning approaches in silicon: the electro-optic (EO) and the thermo-optic approaches. Although the EO approach based on free-carrier dispersion effect has a fast response speed, it inevitably incurs free-carrier absorption loss. The inset shows the structure of the switch element made of a 2×2 MZI with silicon resistive microheaters integrated in both arms for TO tuning [4]. The active arm length is 400 μm, isolated by air trenches to prevent thermal crosstalk. The resistive microheater is formed by an *n-i-n* type structure in which the n^+ doping is positioned in the slab near the waveguide intrinsic region. Heat is generated in the waveguide when current flows through the high-resistivity waveguide cross section, leading to high thermal efficiency.

Figure 1(b) shows the fabricated chip using CMOS-compatible processes. Grating couplers were used to couple light in and out of the chip. The electrical pads along the chip edges were wire-bonded to a printed circuit board. The chip size is 7 mm×3.6 mm.

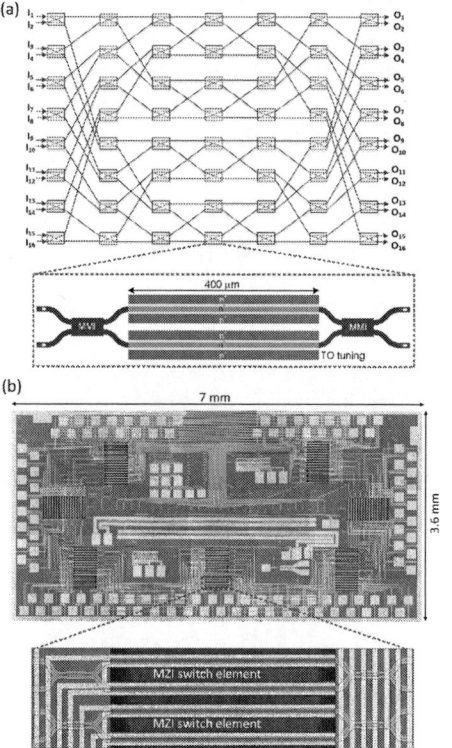

Fig. 1 (a) Schematic of the 16×16 optical switch with a Benes architecture. Inset shows the structure of the MZI switch element. (b) Optical microscope image of the switch chip. Inset shows the zoom-in view.

B. Experimental results

As the MZI arm length is relatively long, a small variation in width can result in a significant phase difference between the two arms. Therefore, each switch element is originally not at the exact "cross" or "bar" state. TO tuning, therefore, is needed to first correct the phase errors so that each switch element is at the "cross" state, which is the starting point for the following switching operation. The average power consumption for phase correction is around 9 mW. The top panel in Fig. 2 shows the measured spectra for output ports O1 to O3. The spectra are normalized to a test waveguide. It can be seen that only one spectrum in each plot shows high transmission, corresponding to the optical routing path. The others all have low transmission, causing small crosstalk to the main path. As the phase correction is performed at 1560 nm, the extinction ratio reaches the maximum (>30 dB) around that wavelength.

To perform switching, the TO power is further increased to induce a π phase change in one arm so that the state is flipped to the "bar" state. The average power consumption for π phase shift is less than 20 mW. It depends on the required input-to-output mapping to set the states of all switch elements. In the worst case, all switch elements are tuned to the "bar" state. The bottom panel in Fig. 2 shows the transmission spectra for O1 to O3 to compare with the previous case. It is clearly seen that the optical path is switched from I9-O1, I10-

O2, I11-O3 to I1-O1, I2-O2, I3-O3, respectively. The on-chip loss and crosstalk are close, suggesting the switch fabric has balanced performances for any state.

To further reveal the transmission variation among all optical paths, we group all 256 spectra together in one plot as shown in Fig. 3. The on-chip loss varies from 4 to 8 dB for the all-cross state and 4 to 7 dB for the all-bar state at 1560 nm. The optical bandwidth for extinction ratio >20 dB is more than 20 nm.

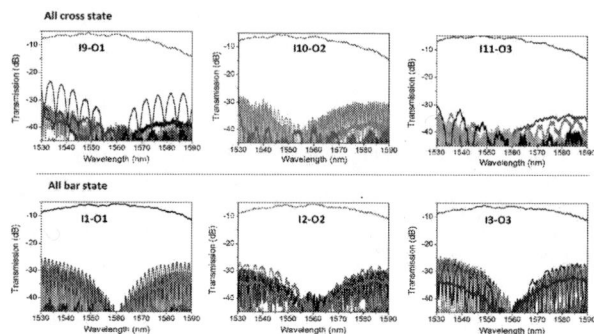

Fig. 2 Measured transmission spectra for output O1 to O3 of the 16×16 switch when the switch elements are all at the cross state (top pannel) and bar state (bottom pannel).

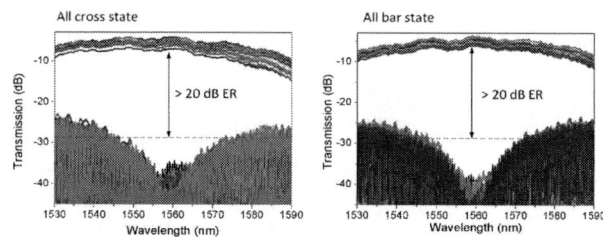

Fig. 3 Grouped spectra for all 256 input-to-output transmissions at the all-cross and all-bar states.

III. OPTICAL BUFFER

A. Structure

The optical buffer is composed of two parts, an array of micoring resonators and a 7-bit switchable delay line, as shown in Fig. 4(a). The microring resonators generate continuous time delay by shifting the resonance wavelengths. As the performance of a single microring resonator is limited by the delay-bandwidth product, multiple microrings are used to increase the buffering capacity. Here we use differential drive to tune the resonators in which half rings are blue-shifted and half red-shifted with respect to the operation wavelength. In this way, the group delay dispersion can be greatly reduced. In the switchable delay line part, the delay is changed by choosing different optical paths upon reconfiguring the switch states. In our design, we set the delay increment step of the switchable delay line to be smaller than the maximum delay provided by the microring array. Thus, the entire structure can generate continuous delay with both coarse and fine tuning features.

The optical buffer is based on 60-nm-thick silicon waveguides to reduce the waveguide propagation loss. The insets in Fig. 4(a) show the structures of the microring resonators, the MZI switch, and the variable optical attenuator (VOA). All of them are thermally tunable by integrating TiN heaters on top of silicon waveguides. The purpose of VOA is to eliminate the leakage optical power from the main optical path to other paths due to the limited extinction ratio of the MZI switches, so that the signal-to-noise ratio of the delayed optical signals can remain high. Figure 4(b) shows the mask layout of the entire chip. Figure 4(c) shows the microscope image of the fabricated chip. The chip has a footprint of 5 mm×4.7 mm.

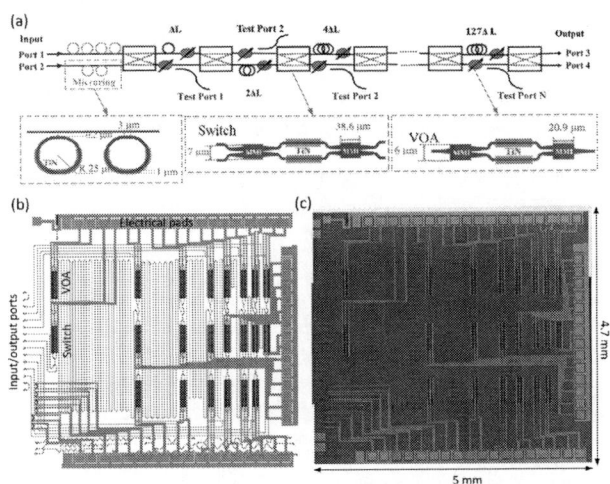

Fig. 4 (a) Architecture of the optical buffer composed of a microring continuous delay line and a switchable digital delay line. (b) Mask layout of the optical buffer chip. (c) Optical microscope image of the fabricated chip.

We characterized the optical buffer by measuring the relative delay of an optical pulse passing through the chip. Figure 5(a) illustrates the optical pulses after the four-microring delay line. The delay can be continuously tuned from 0 to 10 ps by applying different voltages onto the resonators. Figures 5(b)-5(e) illustrate the optical pulses after various digital delays (with a resolution of 10 ps) given by the switchable delay line. The maximum delay is 1.27 ns when the longest optical path is selected. Figure 2(f) illustrates the optical pulses when the switchable delay is fixed at the maximum while the microring delay is continuously tuned. It reveals that the optical buffer can provide continuous delay tuning up to 1.28 ns. The on-chip insertion loss for the maximum delay is <13 dB.

CONCLUSIONS

We have demonstrated an optical router using a 16×16 non-blocking MZI switch and an optical buffer using microring resonators cascaded with a switchable delay line on the SOI platform. These two chips can be applied to enable all-optical switching in further data center networks to satisfy the high-capacity, low-latency and low-power demand in big data era.

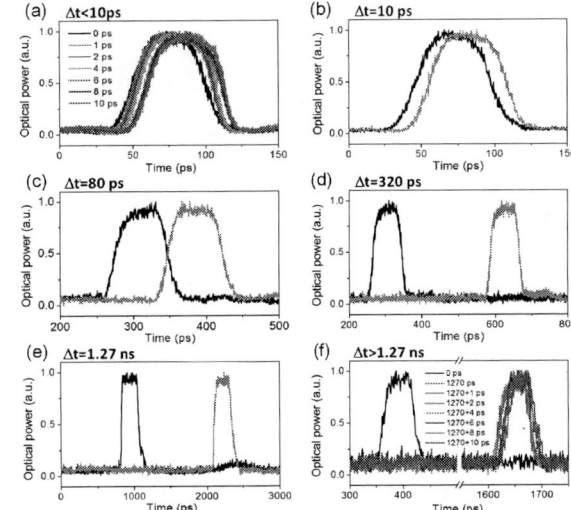

Fig. 5 (a) Optical pulses delayed the microring delay line. (b)-(e) Optical pulses delayed by the switchable delay line. Black curves: reference pulses; red curves: delayed pulses. (f) Optical pulses delayed by the entire strucutre.

ACKNOWLEDGEMENTS

This work was supported in part by the 863 program (2013AA014402), the National Natural Science Foundation of China (NSFC) (61422508), the Shanghai Rising-Star Program (14QA1402600).

REFERENCES

[1] Cisco Networks white paper, "Cisco global cloud index: forecast and methodology, 2013-2018." (Cisco Systems, 2013), http://www.cisco.com/c/en/us/solutions/collateral/service-provider/global-cloud-indexgci/Cloud_Index_White_Paper.html.

[2] Y. Li, Y. Zhang, L. Zhang, and A. W. Poon, "Silicon and hybrid silicon photonic devices for intra-datacenter applications: state of the art and perspectives [Invited]," *Photonics Research,* vol. 3, pp. B10-B27, 2015.

[3] Z. Li, L. Zhou, L. Lu, S. Zhao, D. Li, and J. Chen, "4×4 nonblocking optical switch fabric based on cascaded multimode interferometers," *Photon. Res.,* vol. 4, pp. 21-26, 2016/02/01 2016.

[4] L. Lu, L. Zhou, Z. Li, X. Li, and J. Chen, "Broadband 4×4 non-blocking silicon electro-optic switches based on Mach-Zehnder interferometers," *IEEE Photon. J.,* vol. 7, p. 7800108, 2015.

[5] L. Lu, L. Zhou, S. Li, Z. Li, X. Li, and J. Chen, "4×4 non-blocking silicon thermo-optic switches based on multimode interferometers," *J. Lightwave Technol.,* vol. 33, pp. 857-864, 2015.

[6] L. Lu, L. Zhou, Z. Li, D. Li, S. Zhao, X. Li, *et al.,* "4×4 silicon optical switch based on double-ring assisted Mach-Zehnder interferometers," *IEEE Photon. Technol. Lett.,* vol. 27, pp. 2457-2460, 2015.

[7] L. Lu, L. Zhou, X. Li, and J. Chen, "Low-power 2×2 silicon electro-optic switches based on double-ring assisted Mach–Zehnder interferometers," *Opt. Lett.,* vol. 39, pp. 1633-1636, 2014.

[8] J. Xie, L. Zhou, Z. Zou, J. Wang, X. Li, and J. Chen, "Continuously tunable reflective-type optical delay lines using microring resonators," *Optics Express,* vol. 22, pp. 817-823, 2014.

[9] J. Xie, L. Zhou, Z. Li, J. Wang, and J. Chen, "Seven-bit reconfigurable optical true time delay line based on silicon integration," *Optics Express,* vol. 22, pp. 22707-22715, 2014.

Polarization-Independent C-Band Tunable Filter Based on Cascaded Si-Wire Asymmetric Mach-Zehnder Interferometer

Keijiro Suzuki[1,*], Ken Tanizawa[1], Satoshi Suda[1], Hiroyuki Matsuura[1], Kazuhiro Ikeda[1], Yojiro Mori[2], Ken-ichi Sato[1,2], Shu Namiki[1], and Hitoshi Kawashima[1]

[1]National Institute of Advanced Industrial Science and Technology (AIST), 16-1 Onogawa, Tsukuba, Ibaraki, 305-8569, Japan
[2]Department of Electrical Engineering and Computer Science, Nagoya University, Furo-cho, Chikusa-ku, Nagoya, Aichi, 464-8603, Japan
*k.suzuki@aist.go.jp

Abstract: We fabricated and characterized a polarization-independent tunable band-pass filter, composed of a four-stage Si-wire asymmetric Mach-Zehnder interferometer. The filter exhibited full tunability in the C-band, 2.2 nm bandwidth, and 2 dB on-chip loss.

Keywords: Silicon photonics, optical integrated circuits, optical filter

I. INTRODUCTION

Silicon photonics, which is based on complementary metal-oxide-semiconductor (CMOS) process, is a promising platform to realize highly integrated and ultra-compact optical circuits. Additionally, silicon photonics can provide mass-production leading to low-cost. Using the platform, large-scale photonic integrated circuits such as multi-port switches [1, 2] and transceivers [3] have been demonstrated. Among numerous applications, tunable filters are important devices for telecom and datacom signal transmission systems to deal with tens of wavelength division multiplexed signals [4], which are therefore expected to be integrated on the silicon photonics platform for compact-size, low-cost, and low-power consumption. Such silicon based tunable filters have been reported in several device configurations such as a micro-ring [5, 6], micro-ring with Mach-Zehnder interferometer (MZI) [7, 8]. However, passband wavelength tunability is limited in approximately 10 nm due to the limited free-spectral-range (FSR) of the ring resonator. To expand tunable range, device configuration other than micro-ring is necessary. In addition, polarization independence is crucial for practical applications. One option to obtain the polarization independence is to use on-chip integrated polarization beam splitters (PBSs) and polarization rotators (PRs) for polarization diversity [5]. However, an on-chip PR requires vertical asymmetry, and thus significant process modification for fabrication. Another option is possible by using the off-chip circulator and PBS. This scheme provides simple fabrication with a standard CMOS process.

In this study, we fabricated and characterized a four-stage cascaded asymmetric-MZI filter which is polarization independent and tunable in the C-band (λ = 1.530 – 1.565 µm). The filter configuration is simple, and the characteristics of the fabricated device agree well with calculations.

II. DEVICE DESIGN

The layout of the tunable filter is illustrated in Fig. 1. The tunable filter consists of cascaded four-stage asymmetric MZIs, and their 3-dB couplers are directional coupler. The FSR of the first-stage asymmetric MZI is designed to be the FSR of the whole tunable filter, and the FSR of the Nth-stage asymmetric MZI is set to be a half of the one of the N-1-stage asymmetric MZI, as shown in Figs. 2(a)-(d). Output of the fourth-stage MZI is connected to a shutter which is used for blocking optical leakage during passband wavelength tuning. All MZIs have phase shifters on their two arms to adjust their center wavelength. Figure 2(e) shows an expected transmission spectrum of the filter, in which wavelength components outside the passband are suppressed with the cascaded asymmetric MZIs.

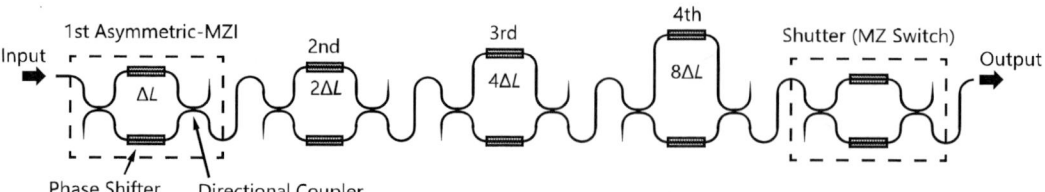

Fig. 1. Schematic of tunable filter based on cascaded Mach-Zehnder interferometer (MZI). ΔL is path-length difference between two arms of the first MZI.

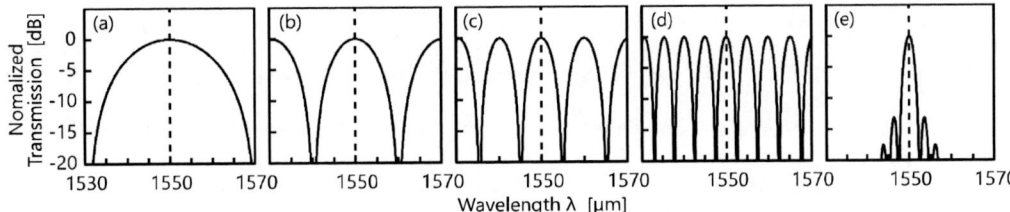

Fig. 2. Calculated transmission spectra. Free spectral range of the filter is 40 nm. (a) The first stage asymmetric Mach-Zehnder interferometer. (b) The second stage. (c) The third stage. (d) The fourth stage. (e) Cascaded four-stage filter.

We calculated the tunability and the passband characteristics of the tunable filter by using the finite element method and the transfer matrix analysis. In the calculation, the polarization was transverse-magnetic (TM) -like mode, and wavelength dependence of the effective indices was taken into account. We designed the FSR of the tunable filter to be 40 nm to cover the C-band, in which the FSR of each asymmetric MZI should be 40 nm, 20 nm, 10 nm, and 5 nm, respectively. The phase shifters of each asymmetric MZI were adjusted to tune the passband wavelength. Figure 3(a) shows calculated transmission spectra of the tunable filter, demonstrating that the passband center wavelength can be set to desired value in the C-band. Figure 3(b) compares the passband at the different center wavelength. The passband shape are slightly different depending on the center wavelength. The 3 dB bandwidth and the crosstalk is estimated to be 2.3 nm (287.5 GHz), and -13 dB, respectively.

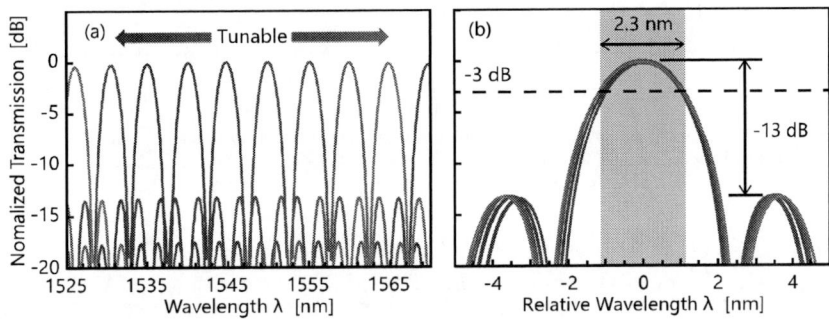

Fig. 3. Calculated transmission characteristics of tunable filter. (a) Passband tunability in C-band. (b) Comparison of passband characteristics at different center wavelength.

III. RESULTS AND DISCUSSION

The cascaded asymmetric MZI filter was fabricated on a silicon-on-insulator wafer (top: 220 nm, buried oxide: 3 μm) by using e-beam lithography and reactive ion etching. The microscopic image of the fabricated device is shown in Fig. 4. The phase shifters are thermooptic, and it consists of platinum heaters and are controlled through gold electric wires. These metal components were fabricated by photolithography and e-beam evaporation. The gold electric wires were connected to an external fan-out circuit by wire-bonding, and the heaters were operated with an external heater driver. The optical input/output was the edge-coupling, and a circulator and a PBS were used for a polarization independent operation.

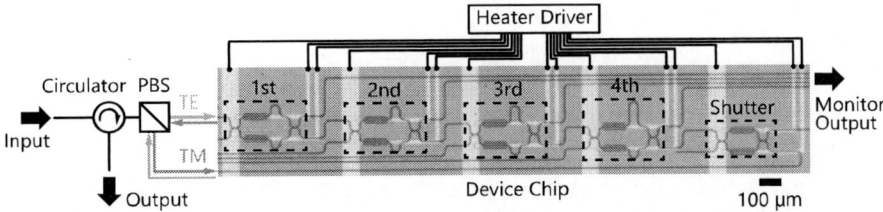

Fig. 4. Fabricated tunable filter. PBS: polarization beam splitter. The circulator and the PBS are fiber components, not integrated on the device chip.

The fabricated tunable filter was evaluated by using an amplified spontaneous emission (ASE) light source and an optical spectrum analyzer (OSA). The light from the ASE light source propagates the circulator and the PBS. The transverse electric (TE) –like and the TM-like components of the light were separated, then both components were coupled to the tunable filter as the TM-like mode. Both components propagated the tunable filter in the clockwise or the counter clockwise way, then combined at the PBS. The combined light was led to another output port of the circulator, then analyzed with the OSA. The heaters on the asymmetric MZIs were optimized while observing the transmission spectrum.

Figure 5(a) shows the measured transmission spectra of the fabricated tunable filter. We found that the center wavelength of the fabricated filter is tunable in the C-band, and the spectra agree well with calculated one shown in Fig. 3(a). The measured passband characteristics are summarized in Fig. 5(b). The measured 3 dB bandwidth and the crosstalk are 2.2 nm and -13 dB, respectively. These values are good agreement with the analytical results. The on-chip loss, which was defined as a fiber-to-fiber insertion loss without losses of the circulator, the PBS, and coupling, was estimated to be 2 dB. The fiber-to-fiber insertion loss was ~15 dB, including 2 dB loss of the circulator and the PBS, 5.5 dB × 2 coupling loss, and 2 dB on-chip loss. Additionally, we evaluated polarization dependent loss (PDL) and differential group delay (DGD) with an optical component analyzer (N7788B, Agilent). The PDL and the DGD were ~0.5 dB and ~1 ps, respectively.

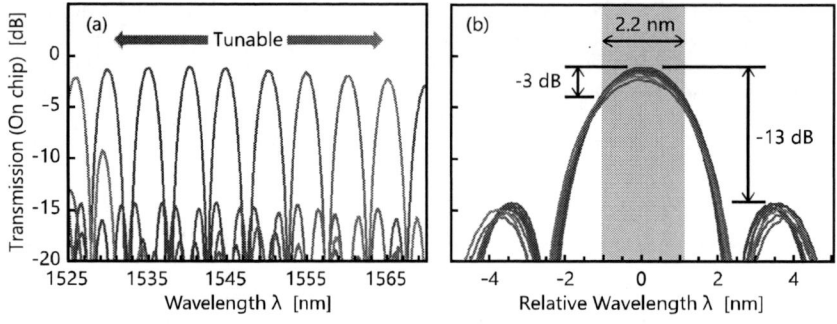

Fig. 5. Measured transmission characteristics of fabricated tunable filter. (a) Passband tunability in C-band. (b) Comparison of passband characteristics at different center wavelength.

IV. CONCLUSIONS

We have demonstrated the filtering characteristics of the cascaded four-stage asymmetric MZI filter. The filter shows full tunability in the C-band, 2 dB on-chip loss, ~0.5 dB PDL, and ~1 ps DGD. The band width can be modified with asymmetric MZI stage number. We believe that up to eight-stage filter (50 GHz bandwidth) is enough possible by taking into account a room temperature fluctuation.

ACKNOWLEDGMENT

This work was partly supported by the Project for Developing Innovation Systems of MEXT, Japan. The authors are grateful to Mr. K. Tashiro for his technical assistance in device assembly.

REFERENCES

[1] K. Tanizawa, K. Suzuki, M. Toyama, M. Ohtsuka, N. Yokoyama, K. Matsumaro, M. Seki, K. Koshino, T. Sugaya, S. Suda, G. Cong, T. Kimura, K. Ikeda, S. Namiki, and H. Kawashima, "Ultra-compact 32 × 32 strictly-non-blocking Si-wire optical switch with fan-out LGA interposer," *Opt. Express,* vol. 23, pp. 17599-17606, June 2015.

[2] K. Suzuki, G. Cong, K. Tanizawa, S.-H. Kim, K. Ikeda, S. Namiki, and H. Kawashima, "Multiport optical switches integrated on Si photonics platform," *IEICE Electronics Express,* vol. 11, pp. 20142011, December 2014.

[3] T. Nakamura, Y. Urino, and Y. Arakawa, "High-density silicon optical interposer for inter-chip interconnects," *Proc Spie,* vol. 9133, pp. 91330R, April 2014.

[4] R. Younce, J. Larikova, and Y. Wang, "Engineering 400G for colorless-directionless-contentionless architecture in metro/regional networks," *J. Opt. Commun. Netw.,* vol. 5, pp. A267-A273, October 2013.

[5] C. Li, J. H. Song, J. Zhang, H. Zhang, S. Chen, M. Yu, and G. Q. Lo, "Silicon polarization independent microring resonator-based optical tunable filter circuit with fiber assembly," *Opt. Express,* vol. 19, pp. 15429-15437, August 2011.

[6] A. Leliepvre, R. Brenot, G.-H. Duan, and A. Maho, "Fast tunable silicon ring resonator filter for access networks," in *Optical Fiber Communication Conference,* Los Angeles, California, 2015, pp. Tu3E.5.

[7] Y. Ding, M. Pu, L. Liu, J. Xu, C. Peucheret, X. Zhang, D. Huang, and H. Ou, "Bandwidth and wavelength-tunable optical bandpass filter based on silicon microring-MZI structure," *Opt. Express,* vol. 19, pp. 6462-6470, March 2011.

[8] M. S. Rasras, D. M. Gill, S. S. Patel, K.-Y. Tu, Y.-K. Chen, A. E. White, A. T. S. Pomerene, D. N. Carothers, M. J. Grove, D. K. Sparacin, J. Michel, M. A. Beals, and L. C. Kimerling, "Demonstration of a fourth-order pole-zero optical filter integrated using CMOS processes," *J. Lightwave Technol.,* vol. 25, pp. 87-92, January 2007.

ME1-3

Fundamental operation of a phased array switch using ferroelectric liquid crystal claddings

W. Kanakubo, K. Nakatsuhara, M. Takeda

Kanagawa Institute of Technology, 1030, Shimo-ogino, Atsugi-shi, Kanagawa, 243-0292, Japan
knakatsu@ele.kanagawa-it.ac.jp

Abstract: *A Si-phased array waveguide with ferro-electric liquid crystal claddings was proposed and fabricated. Switching operation of the fabricated device was demonstrated for the first time.*
Keywords: *Optical switching devices, Liquid-crystal device*

I. INTRODUCTION

Large-scale $1 \times N$ and $N \times N$ optical switches are key devices in constructing optical-path-control circuits in photonic networks [1]. Phased array structures allow $1 \times N$ optical switches preserving the small size regardless of the switch scale increases [2]. Silicon waveguides formed on silicon-on-insulator (SOI) wafers are suitable for high-density photonic node circuits [3,4]. Generally, the thermo-optical effect is utilized for Si photonics devices in order to achieve the optical function. However, these devices consume power to sustain the state and increase the power consumption of the overall photonic network. As an alternative, we have been studying Si waveguide devices that have ferroelectric liquid crystal (FLC) cladding [5,6]. FLC materials have relatively high speed response of sub-microseconds and bistable characteristics that eliminate the state-sustaining voltage required for the devices [7,8]. Latching operation due to the bistability in FLC materials that had possibilities of achieving ultralow power consumption and free thermal interference in high-density photonic integrated circuits was reported [9]. In the present paper, we report the first demonstration in the fabricated Si-phased array waveguide having FLC cladding.

II. OPERATION PRINCIPLE AND DEVICE STRUCTURE

Figure 1 shows a schematic diagram of our proposed 4×4 Si-phased array waveguide that has the FLC cladding. It consists of four input waveguides, two slab waveguides, twenty arrayed waveguides and four output waveguides. The optical waveguides are rib waveguides using the Si guide layer in an SOI wafer. The switching characteristic is achieved by the phase change in the array waveguides with the FLC molecules, which are inclined $\pm\theta_{tilt}$ around the orientation direction by applied electric fields.

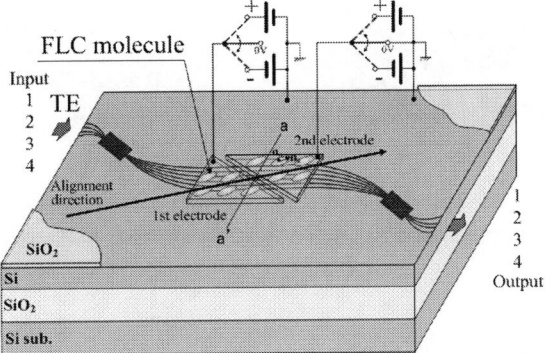

Fig. 1. Schematic diagram of the phased array waveguide with FLC cladding.

The shape of the phase shifter regions having FLC cladding is a triangle. We employed 20 phase shifters. The spacing between two neighboring arrays is 30 μm. A unit FLC length in phase shifter is 60 μm. The width and the length of the regions are 600 μm and 1200 μm, respectively.

Figure 2 shows a cross section of the fabricated phased array waveguide with FLC cladding. The DC voltage was applied between the upper ITO electrode and Si guide layer for the switching operation. The molecular orientation of the FLC inclines at $\pm\theta_{tilt}$ angles tilt with respect to the initial rubbing direction depending on the polarity of the applied voltage.

OECC/PS2016

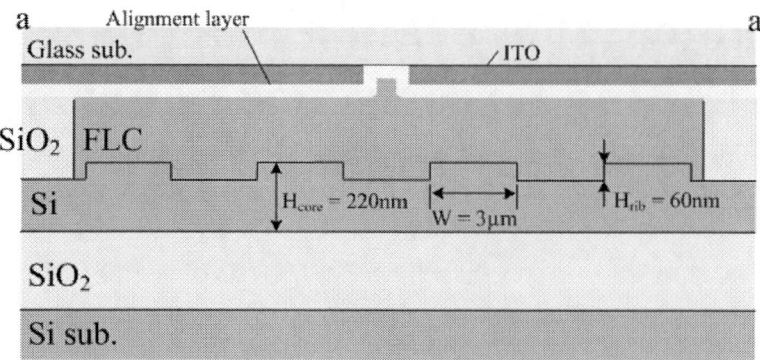

Fig. 2. Cross section of the fabricated device.

III. EXPERIMENTS

We fabricated the optical switch using the structure described above. The phased array waveguide was formed on the SOI wafer by CF_4/Ar-RIE. The measured width of waveguides was 3.0 μm as designed. The core guide layer was 220 nm high. The rib waveguide was 60 nm high. The size of the optical switch was 2.5 mm × 6.2 mm. The FLC with the refractive indices of the ordinary index no=1.51 and the extra-ordinary index n_e=1.653 and θ_{tilt}=26.7 degrees were used in our experiment. The optical axes of the FLC molecules were initially orientated by the alignment layer that was coated on the upper indium tin oxide (ITO) electrode. The 700 nm SiO_2 layer was sputtered on a Si guide layer to provide the spacing for the FLC layer as shown in Fig. 2. The SiO_2 layer over the straight waveguide region was selectively eliminated so that the FLC cladding covered only the phased array waveguide. The evanescent field in the FLC layer exponentially decays as the distance from the Si guide layer increases, so the attenuation due to the ITO electrode was negligible. The FLC material and the alignment polymer used in the experiment were AZ Electronic Materials FELIX-016/100 and Nissan Chemical Industries RN-1744, respectively. We used a vacuum injection process [10] to form the FLC layer.

We used the amplified spontaneous emission as a light source. A TE-polarized light was launched into the input port through a polarization maintaining fiber that was butt-coupled to the waveguide core.

The switching characteristics of the fabricated device were measured by applying at +15V, -15V. Figure 3 shows the near field image of output light when the incident light was launched from the input port 3. Figure 4 shows the wavelength characteristics of the fabricated device in the switching operation. The device has four output ports, We have applied to the positive and negative voltage to the left and right of the electrode.

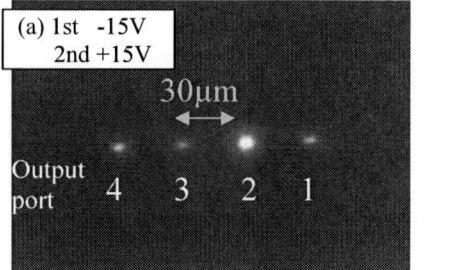 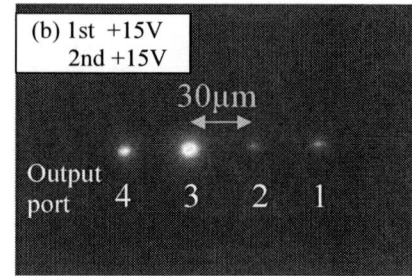

Fig. 3. Near field images.

86

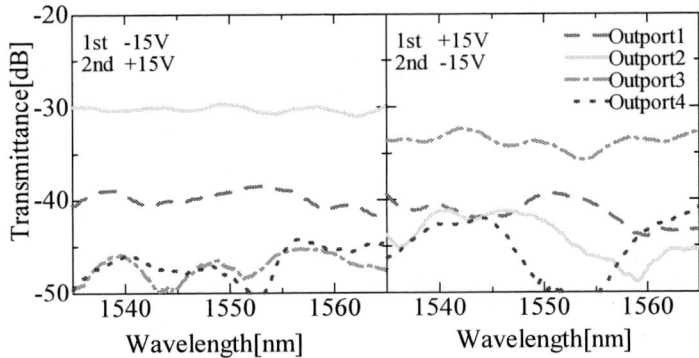

Fig. 4. Wavelength characteristics in the fabricated device.

The output light was switched by changing the polarities of the applied voltage on the 1st and 2nd electrodes. The extinction ratio of 9.5dB was obtained at 1550nm wavelength by applying voltages at -15V on 1st and at +15V on 2nd electrodes. While, the extinction ratio of 5.6dB was obtained by applying voltages at +15V on 1st and at -15V on 2nd electrodes. Although the switching operation has been demonstrated, the extinction ratios are insufficient for practical applications. Improving the fabrication process and structure of the Si-phased array waveguide would increase the extinction ratio between output waveguides. The alignment condition, and the injection condition of the FLC are also being investigated to improve the switching characteristics.

IV. CONCLUSIONS

Silicon phased array switch featuring FLC cladding was designed and fabricated. The switching operation of the fabricated device was demonstrated for the first time. Further improvement in the switching characteristics is expected by revising the design and fabrication process.

ACKNOWLEDGMENT

This study was partially supported by JST A-STEP FS stage search type (AS262Z01757H) and JSPS KAKENHI (B) Grant Number 15H04018. Also we would like to thank Mr. Toshiaki Nonaka of AZ Electronic Materials for providing a ferroelectric liquid crystal FELIX-016/100.

REFERENCES

[1] T. Tanemura, Y. Nakano, "Design and scalability analysis of optical phased-array 1×N switch on planar lightwave circuit," IEICE Electronics Express, Vol.5, No.16, pp603-609, 2008.

[2] S. Katayose, Y. Hashizume, A. Mori and M. Itoh, "Design and fabrication of low loss 1×8 silica-based phased array switch with low power consumption," 18th Microoptics Conference (MOC'13), Tokyo, Japan, Oct. 27 - 30, 2013.

[3] B. Jalali, and S. Fathpour, "Silicon Photonics," J. Light. Technol., vol. 24, no. 12, pp. 4600-4615, Dec. 2006.

[4] W. Bogaerts, S. K. Selvaraja, P. Dumon, J. Brouckaert, K. D. Vos, D. V. Thourhout, and R. Beats, "Silicon-on-Insulator Spectral Filters Fabricated With CMOS Technology," IEEE J. Sel. Topics Quantum Electron., vol. 16, no. 1, pp. 33-44, Jan./Feb. 2010.

[5] R. Hoshi, K. Nakatsuhara, and T. Nakagami, "Optical switching characteristics in Si-waveguide asymmetric Mach-Zehnder interferometer having a ferro-electric liquid crystal cladding," IEE Electronics Letters, vol. 42, no. 11, pp. 635–636, 2006.

[6] A. Kato, K. Nakatsuhara, and T. Nakagami, "Tunable Wavelength Selective Operation in Grating Silicon Waveguide having Ferroelectric Liquid Crystal Cladding," IEEE Photonics Conference 2011, MJ6, pp. 81-82, 2011.

[7] R. Asquini and A. d'Alessandro, "A bistable optical waveguided switch using a ferroelectric liquid crystal layer", LEOS 2000, pp. 119-120, 2000.

[8] E. P. Pozhidaev, V. G. Chigrinov, T. Du, "Fast switching bistable ferroelectric liquid crystal switches as a new optical elements for photonics applications", OECC2009, DOI : 10.1109/OECC.2009.5218309, 2009.

[9] K. Nakatsuhara, A. Kato, Y Hayama, "Latching operation in a tunable wavelength filter using Si sampled grating waveguide with ferroelectric liquid crystal cladding," Optics Express, DOI : 10.1364/OE.22.009597, vol.22, no.8, pp.9597-9603, 2014.

[10] A. Kato, K. Nakatsuhara, Y. Hayama, "Switching Operation in Tunable Add-Drop Multiplexer with Si-Grating Waveguides Featuring Ferroelectric Liquid Crystal Cladding," J. Light. Technol., vol. 32, no. 22, pp. 4464-4470, 2014.

ME1-4

OECC/PS2016

Control, configuration and stabilization of photonic integrated circuits

Andrea Melloni[1], Andrea Annoni[1], Stefano Grillanda[1], Daniele Melati[1], and Francesco Morichetti[1]

[1]Dipartimento di Elettronica, Informazione e Bioingegneria, Politecnico di Milano, via Ponzio 34/5, 20133 Milano, Italy
Author e-mail address: andrea.melloni@polimi.it

Abstract: Recent results on tools and strategies to manage complex PICs to reliably set, hold, control and steer the desired working point through feedback-controlled algorithms are presented. Examples of PICs controlled employing different strategies are shown.
Keywords: *Photonic Integrated Circuits, Control, Silicon Photonics*

I. INTRODUCTION

The explosive growth of silicon photonics driven by datacenters, sensing, automotive and telecommunication applications demands the development of integrated photonic circuits (PICs) with unprecedented complex architectures. The current technologies already enable device miniaturization and aggregation of many photonic components and functionalities onto the same chip, yet the lack of suitable control technologies is today perceived as a "grand challenge" and major obstacle to the advent of large-scale photonic integrated systems. The photonic chip is becoming an increasingly complex system operating in a multiphysical dynamic environment together with electronics, radiofrequency, thermal management and microfluidics and sensing elements. Advanced tools are essential to reliably set, hold, control and steer the desired working point of the system through feedback-controlled algorithms.

The possibility to control the PIC is a key requirement for different reasons and in many applications: i) reconfigurability of the circuit to provide the required functionality such as in routers, cross connects, tunable bandwidth filters, reconfigurable add-drop multiplexers, etc.; ii) adaptively circuits that modify their behavior depending on the state of the system such as signal polarization state, signal to noise ratio, crosstalk, eye aperture, BER, etc.; iii) locking or stabilization of the circuit in a well defined state independently of the temperature, electrical fluctuations, drifts, stress, aging, etc; iv) compensation of fabrication tolerances and technological non-uniformities. All these requirements become even more critical when dealing with wavelength selective devices, such as microrings resonators, high bit rate signals operating in coherent domain, dense WDM systems and densely integrated PICs realized on semiconductor photonic platforms, such as silicon on insulator (SOI) or indium phosphide (InP). Local and global feedback control tools and strategies will soon become the ordinary way to operate even for simple circuits.

Examples of complex circuits with feedback control discussed here are an 8x8 router and a wavelength locking platform for 4×10 Gbit/s L-Band Si-Photonic multiplexer and carver.

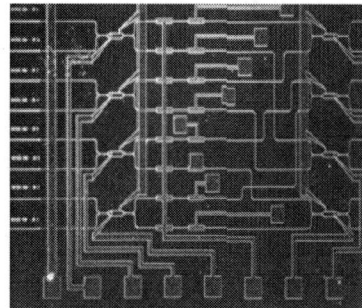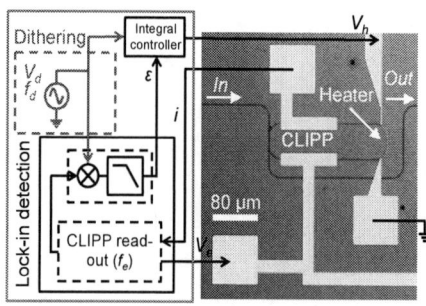

Figure 1: Example of integrated photonic circuits that can be controlled by a dedicated CMOS ASIC: (from left to right) an 8x8 Silicon Photonic Router with heathers and CLIPPs, photonic chip mounted on a PCB with two CMOS ASICs for the control and readout of CLIPPs and a detail of a ring resonator with its feedback control loop.

In order to realize a feedback control loop, light monitors and actuators are required, in addition to an electronic control unit and control strategies. Monitoring approaches based on conventional photodetectors are not effectively scalable to large-scale PICs due to the need for multi-point light tapping. In this work, we report on our recent achievements on the development of an in-line transparent integrated detector, named ContacLess Integrated Photonic Probe (CLIPP), that can monitor the light intensity in semiconductor waveguides [1, 5] without introducing any photon absorption in excess to the waveguide propagation [1] by exploiting the sensing of changes in the waveguide properties cause by the passage of optical signals [2]. The CLIPP can be used to monitor signals on different polarizations [3],

88

different modes and wavelength [3, 7] and its signal can be exploited for feedback loop to perform automatic control, locking and tuning of photonic integrated devices [4].

II. 8x8 ROUTER

The first considered example of complex circuit control is shown in Fig. 2. The circuit is an 8x8 Silicon Photonic Router [6] composed by 12 MZI arranged in 3 stages; each MZI is monitored with two CLIPPs integrated at both the output ports (for a total of 24 monitoring point throughout the circuit) and controlled using a NiCr thermo-optical heater. The pervasive presence of CLIPPs allows to monitor the circuit in every desired point and control the working points of the various switching elements individually. Fig 2 (a) and (b) show the automatic reconfiguration of the stages needed to rout signals from input port 8 (I_8) to output port 8 (O_8); Fig 2(a) is a schematic representation of the routing to identify the elements involved in the path, note that all the CLIPPs involved are labelled accordingly to their position. Fig. 2(b) shows the monitoring signals acquired throughout the circuit and used to tune, individually and sequentially, the stages involved in the routing path.

As a first step (Tuning A), stage A is tuned to maximize the optical power in the lower output waveguide; this condition is reached when the difference between the signals provided by CLIPP A8 (black dashed line) and CLIPP A7 (blue dotted line) of Fig. 2(b1) is maximum. Once stage A is tuned, its state is continuously monitored and feedback locked, while the following stages of the circuit are sequentially tuned. In the second step (Tuning B), all the power is routed in the lower output waveguide of stage B, by maximizing the difference between the signals from CLIPPs B8 (purple dash-dotted line) and B8 (orange solid line). Likewise, the third stage C is configured by looking at the signal provided by CLIPPs C7 (grey dashed line) and C8 (brown solid line) of Fig. 2(b2).

PD signals of Fig. 2(b3) confirm the effectiveness of the CLIPP-assisted reconfiguration procedure. It is worthwhile to note that, even though some information on the current state of the circuit could be inferred from the output PD signals, CLIPP-assisted on-chip local monitoring provides direct information on the status of every single element of the entire architecture. For instance, during the tuning of the stage A, the three PD output O6, O7 and O8 all increase since the MZI is redirecting the light to the lowest four ports of the switch fabric. Therefore, to optimize the working point of this MZI, all the eight output ports (O1-O8) should be simultaneously monitored; in contrast, by exploiting on-chip monitoring only two CLIPPs (A7 and A8) need to be used. Similar considerations apply to any switching element whose output ports are not directly connected to external PDs.

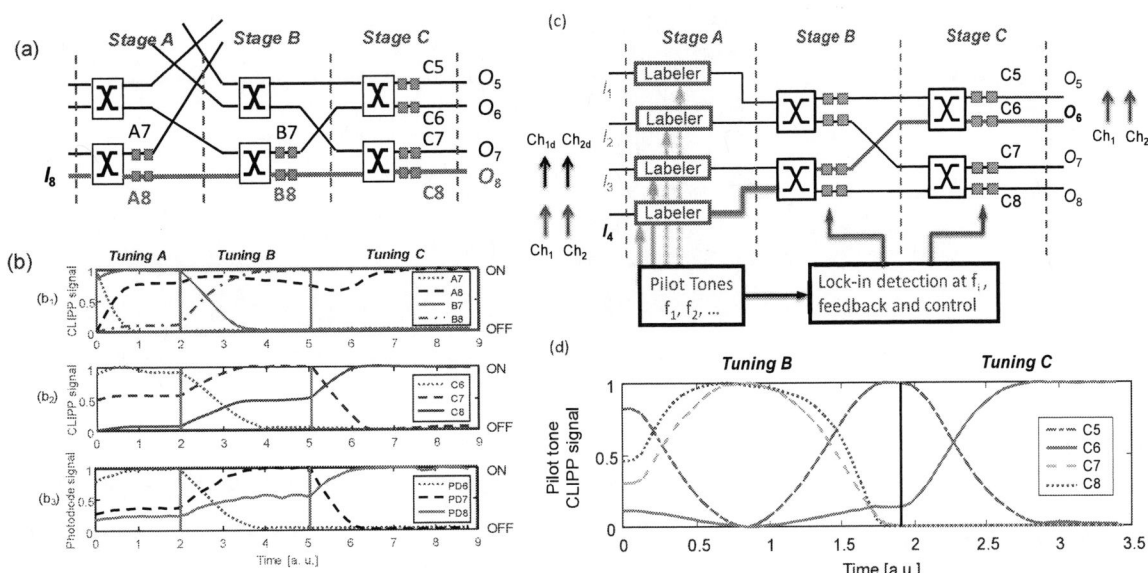

Figure 2: Example of an 8x8 silicon switch fabric controlled by feedback loop enabled by CLIPP sensors and pilot tones. Left: CLIPP assisted automated routing of data signals. CLIPPs throughout the circuit allow to monitor and control, individually and sequentially, the switching elements needed for the routing. (a) shows the path and the elements involved, (b) shows the acquired signals (b1 and b2) compared with external convention photodiodes (b3). Right: lightpath tracking using on-chip labelling: (a) scheme of the routing realized, note that the first stage this time is used to generate on-chip label for the different input channels. (b) CLIPPs signal after the stage B and stage C; on-chip labels allow to monitor individually each channels without interference from the other channels routed by the switching matrix.

When multiple optical signals are simultaneously injected at different input ports of the switch fabric, CLIPP detectors are used to identify channels coming from specific input ports regardless of the presence of other concurrent channels injected at the other input ports if these are labelled on-chip, using the MZI from the first stage, with a tone with an intensity modulation amplitude of a few percent and a frequency in the order of some kilohertz. Fig. 2 (c) shows the CLIPP assisted lightpath tracking of concurrent signals distinguished by using the on-chip labelling through pilot tones; locking of each element can be performed by both minimizing or maximizing the contribute from a channel observed by a CLIPP. Fig. 2(d) shows the signals acquired by the CLIPPs in stage C when several channels are present simultaneously but the CLIPPs are demodulated to distinguish a particular channel by its on-chip generated label tone.

III. 4 WAVELENGTHS TRANSMITTER

The second example of feedback-enabled complex Si photonic circuit we show is a 4-channel 4x10Gbit/s multiplexer based on micro ring resonators (MRRs), in conjunction with CLIPPs, to both multiplex and carve four 10 Gbit/s wavelength division multiplexing (WDM) channels in the L-band from four directly modulated lasers (DMLs). A sketch of the control system for each MRR is reported in Fig. 3(a). The MRR heater voltage is set according to an integral control law, employing an error signal ε obtained from the ratio between the input CLIPP (CLIPP$_{IN}$) and the CLIPP inside the MRR (CLIPP$_{MRR}$). Figures 3(b) and (c) show the eye diagrams of the 10 Gbit/s DML signal respectively at the output of the DML and after the Si MRR. The signal of each DML (Q-factor = 3.6, ER = 2.6 dB) is transmitted through a MRR, that improves its quality by enlarging the eye aperture, thus achieving ER = 8.8 dB and a Q-factor of 4.2. Each MRR is constantly locked to the DML wavelength to compensate drift of the working point of the ring or of the laser wavelength; Fig. 3(d) show the eye diagram when the DML is carved using the MRR but the locking is OFF. Figure 3(e) and (f) report respectively the time-dependent ER of the signal after the MRR and the voltage applied to the heater, when the locking system is ON (time < 5 min) and OFF (time > 5min). The corresponding eye diagrams of the signal after the MRR are shown in Figs. 3(c) and (d). Although the TEC of the Si chip is OFF, when the feedback control is active the eye aperture remains well open and the ER of the signal is stable around 8.8 dB [Fig. 3(c)] (Q-factor is about 4.2 dB). The beneficial effect of the feedback is confirmed when the locking system is switched OFF: the MRR wavelength drifts under the effect of the ambient temperature variations. As a consequence, the DML signal and the MRR drift apart, degrading the quality of the signal (ER down to 6.5 dB and Q-factor down to 3.5 as shown in Figs. 3(d-e)).

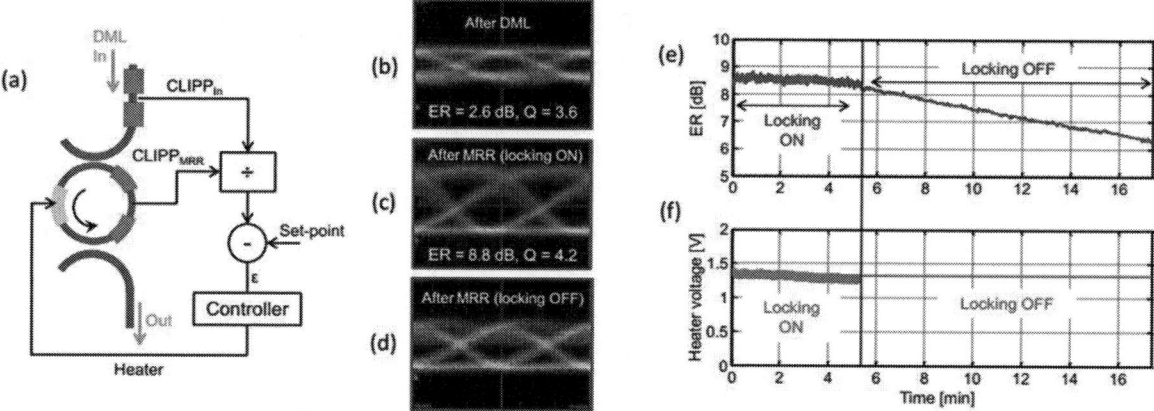

Fig. 3: (a) Wavelength locking scheme of each MRR of the Si photonic transmitter; 10 Gbit/s eye diagram (b) after the DML and after the MRR carver with the feedback control (c) ON and (d) OFF; (e) Extinction-ratio and (f) MRR heater voltage as a function of time when the locking is ON and OFF.

REFERENCES

[1] F. Morichetti et al., "Non-invasive on-chip light observation by contactless waveguide conductivity monitoring," IEEE J. Sel. Top. Quantum Electron. 20, 1–10, 2014

[2] S. Grillanda et al., "Light-induced metal-like surface of silicon photonic waveguides," Nat. Commun. 6:8182, 2015

[3] S. Grillanda et al., "Non-invasive monitoring of mode-division multiplexed channels on a silicon photonic chip," J. Ligthwave Technol., vol. 33, no. 6, pp. 1197-1201, Mar. 2015.

[4] S. Grillanda et al., "Non-invasive monitoring and control in silicon photonics using CMOS integrated electronics," Optica 1, 129-136, 2014

[5] D. Melati et al., "Contactless integrated photonic probe for light monitoring in InP-based devices. IET Opt. 9, 2015

[6] A. Annoni; E. Guglielmi; M. Carminati; S. Grillanda; P. Ciccarella; G. Ferrari; M. Sorel; M. Strain; M. Sampietro; A. Melloni; F. Morichetti, "Automated Routing and Control of Silicon Photonic Switch Fabrics," in IEEE Journal of Selected Topics in Quantum Electronics, vol. PP, no. 99, 2016

[7] F. Morichetti, A. Annoni, S. Grillanda, N. Peserico, M. Carminati, P. Ciccarella, G. Ferrari, E. Guglielmi, M. Sorel, A. Melloni, "All-optical mode unscrambling on a silicon photonic chip", arXiv preprint arXiv:1512.06762

ME2-1 (Invited)

OECC/PS2016

Large-Scale Silicon Photonic Switches

Ming C. Wu[1*], Tae Joon Seok[1], Sangyoon Han[1], and Niels Quack[2]

[1]Electrical Engineering and Computer Sciences Department and Berkeley Sensor and Actuator Center
University of California, Berkeley, CA 94720, USA
[2]Institute of Microengineering, Ecole Polytechnique Fédérale de Lausanne (EPFL), Lausanne, Switzerland
*mingwu@berkeley.edu

Abstract: Large-scale (50x50 and 64x64) integrated photonic switches have been realized by combining silicon photonics with efficient micro-electro-mechanical-system (MEMS) switching mechanisms. Low on-chip loss (0.058dB/port) and sub-microsecond switching time have been achieved.

Keywords: Silicon photonics, optical switching devices, integrated photonic switches

I. INTRODUCTION

Cloud computing and data-intensive applications (big data) have dramatically increased the size and number of datacenters [1]. On-demand, high-bandwidth and low-latency communications between the servers within datacenters as well as between datacenters (virtual machine migration) are essential to achieve rapid responses and reliable operation. Currently, datacenter networks employ electronic packet switching. However, it is increasingly challenging for the electronic switch to keep up with the ever-increasing data rates. Optical switching has been proposed to augment, or enhance, the electronic switching in various ways [2]–[10].

The focus of this paper is on integrated optical space switches (OSS), also called optical circuit switches (OCS), that can be potentially used in datacenter networks. Large-scale OCS's are commercially available in 3D MEMS (micro-electro-mechanical systems) (320x320 [11]) and piezoelectric (384x384 [12]) technologies. However, these switches employ bulky free-space optics and have long switching time (10s of milliseconds). Integrating the entire switch monolithically offers many advantages such as smaller size/weight, lower cost, faster switching speed and lower power consumption. 32x32 switches have been demonstrated in silica-based planar Lightwave circuits (PLC) [13]. This might be the limit of PLC as the footprint of the switch is already larger than 100 cm^2. Silicon photonics is an attractive platform for large-scale OCS as the integration density is more than 100x higher than in PLC. In the past few years, several NxN silicon photonic switches have been reported [7], [14]–[17]. However, most of these switches have limited port counts, limited by the cumulative losses of the multi-stages switching architecture.

Recently, we have reported a new silicon photonic switch that is fundamentally more scalable [18]–[20]. The switch employs a passive crossbar architecture and MEMS switching mechanisms with very high ON-OFF ratio (> 60 dB). Light goes through the switching element only once in the switch fabric, independent of the number of ports. In this paper, we will review the state of the art of silicon photonic switches, and focus on the principle, design, construction, and most recent results of our MEMS silicon photonic switches. Finally, we will discuss the possible scaling limit of the switch.

II. SWITCH DESIGN AND OPERATION PRINCIPLE

The schematic of the switch is shown in Fig. 1(a). It consists of N horizontal and N vertical waveguides forming passive crossbar architecture. Advances in silicon photonics have made it possible to design waveguide crossings with very low loss (< 0.01 dB) and low crosstalk (< -40 dB). Light stays in the horizontal waveguide until a switch is turned on at the selected cross point. We employ a pair of adiabatic couplers suspended on top of the waveguides as the switching element. In the OFF state, the couplers are far above the waveguide (> 1 μm) and there is literally no loss or interference on light propagation (Fig. 1(b)). In the ON state, the couplers are physically pulled down to ~ 100 nm from the waveguides, and nearly 100% of light is coupled through the two couplers to the orthogonal waveguide (Drop port).

The MEMS switching mechanisms offer distinctive advantages that are essential for this scalable architecture. The free-carrier plasma or thermo-optic effect produces a small perturbation of the refractive index (a few %). They are often used in conjunction with Mach-Zehnder interferometers (MZI) or ring resonators to achieve complete switching. Unfortunately, the MZIs have non-negligible loss for both ON and OFF states, and the ring resonator works only for critically matched wavelength. MEMS, on the other hand, can physically move the waveguide and achieve 100% switching in a very small footprint. More importantly, the OFF state of the switch has extremely low loss (< -60dB), making the passive crossbar structure possible.

Prototype switches have been fabricated on 6-inch silicon-on-insulator (SOI) substrates at Marvell Nanofabrication Lab at UC Berkeley. The thickness of the bus waveguides is 220 nm, which is compatible with the process used by many other silicon photonic foundries.

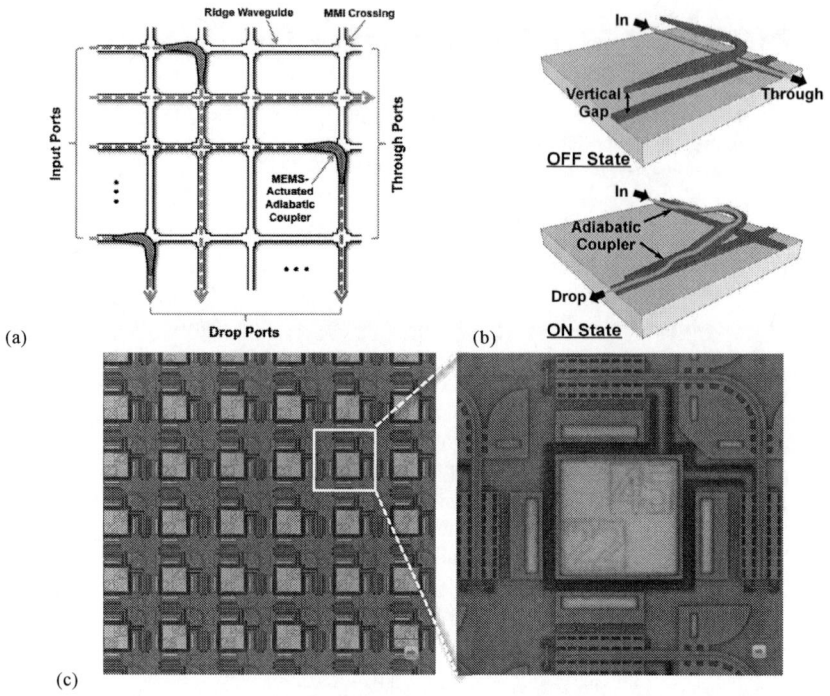

Fig. 1. (a) Schematic illustrating passive crossbar architecture and MEMS switching mechanisms. (b) Schematic illustrating the principle of the switching unit. In the OFF state (top), the adiabatic couplers are far above the bus waveguides. In the ON state (bottom), light is coupled to the couplers and transferred to the orthogonal waveguide. (c) Microscope photograph of the switch chip (partial) and close-up view of a single switching unit.

III. MEASUREMENT RESULTS

We have made several generations of MEMS silicon photonic switches. The 50x50 switch we reported earlier has an on-chip loss of 8.5 dB [19]. More recently, we have reported a 64x64 switch with an even lower on-chip loss of 3.7 dB [20]. The loss-to-port-count ratio (0.058 dB/port) is significantly lower than other silicon photonic switches (0.62 dB/port in [21]). Because the MEMS actuator only needs to move over a micrometer, it can be designed to have high resonance frequency (> 100 kHz) and fast switching time. Both switches we reported exhibit sub-microsecond switching time. This is 10,000x faster than the 3D MEMS switches. Moreover, with monolithic construction, the entire switch fits comfortably on a 1x1 cm^2 die. The switch has been shown to operate for more than 10 billion cycles [22].

IV. CONCLUSIONS

We have shown that the MEMS silicon photonic switches with passive crossbar architecture are fundamentally more scalable than other silicon photonic switches. We have experimentally demonstrated high performance, low-loss switches with 50x50 and 64x64 ports. The on-chip loss is only 0.058 dB/port. We expect even lower loss with improved fabrication process, and the port count should be scalable to several hundred ports.

ACKNOWLEDGMENT

This work is supported in part by NSF Center for Integrated Access Network (CIAN) #EEC-0812072 and DARPA E-PHI program.

REFERENCES

[1] C. Kachris, K. Bergman, and I. Tomkos, *Optical Interconnects for Future Data Center Networks*. Springer Science & Business Media, 2012.
[2] N. Farrington, G. Porter, S. Radhakrishnan, H. H. Bazzaz, V. Subramanya, Y. Fainman, G. Papen, and A. Vahdat, "Helios: a hybrid electrical/optical switch architecture for modular data centers," *ACM SIGCOMM Comput. Commun. Rev.*, vol. 41, no. 4, pp. 339–350, 2011.
[3] G. Wang, D. G. Andersen, M. Kaminsky, K. Papagiannaki, T. S. E. Ng, M. Kozuch, and M. Ryan, "c-Through: Part-time Optics in Data Centers," in *Proceedings of the ACM SIGCOMM 2010 Conference*, New York, NY, USA, 2010, pp. 327–338.

[4]   A. Vahdat, H. Liu, X. Zhao, and C. Johnson, "The emerging optical data center," in *Optical Fiber Communication Conference*, 2011, p. OTuH2.

[5]   H. Liu, F. Lu, A. Forencich, R. Kapoor, M. Tewari, G. M. Voelker, G. Papen, A. C. Snoeren, and G. Porter, "Circuit Switching Under the Radar with REACToR," in *Proceedings of the 11th USENIX Conference on Networked Systems Design and Implementation*, Berkeley, CA, USA, 2014, pp. 1–15.

[6]   L. Schares, B. G. Lee, F. Checconi, R. Budd, A. Rylyakov, N. Dupuis, F. Petrini, C. L. Schow, P. Fuentes, O. Mattes, and C. Minkenberg, "A Throughput-Optimized Optical Network for Data-Intensive Computing," *IEEE Micro*, vol. 34, no. 5, pp. 52–63, Sep. 2014.

[7]   B. G. Lee, N. Dupuis, P. Pepeljugoski, L. Schares, R. Budd, J. R. Bickford, and C. L. Schow, "Silicon Photonic Switch Fabrics in Computer Communications Systems," *J. Light. Technol.*, vol. 33, no. 4, pp. 768–777, Feb. 2015.

[8]   Y. Xia, M. Schlansker, T. E. Ng, and J. Tourrilhes, "Enabling topological flexibility for data centers using omniswitch," in *Proceedings of the 7th USENIX Conference on Hot Topics in Cloud Computing*, 2015, pp. 4–4.

[9]   Z. Zhu, S. Zhong, L. Chen, and K. Chen, "Fully programmable and scalable optical switching fabric for petabyte data center," *Opt. Express*, vol. 23, no. 3, p. 3563, Feb. 2015.

[10]   P. Samadi, K. Wen, J. Xu, and K. Bergman, "Software-defined optical network for metro-scale geographically distributed data centers," *Opt. Express*, vol. 24, no. 11, p. 12310, May 2016.

[11]   http://www.calient.net/products/s-series-photonic-switch/

[12]   http://www.polatis.com/series-7000-384x384-port-software-controlled-optical-circuit-switch-sdn-enabled.asp

[13]   S. Sohma, T. Watanabe, N. Ooba, M. Itoh, T. Shibata, and H. Takahashi, "Silica-based PLC Type 32 x 32 Optical Matrix Switch," in *European Conference on Optical Communications, 2006. ECOC 2006*, 2006, pp. 1–2.

[14]   L. Chen and Y. Chen, "Compact, low-loss and low-power 8×8 broadband silicon optical switch," *Opt. Express*, vol. 20, no. 17, pp. 18977–18985, Aug. 2012.

[15]   K. Suzuki, K. Tanizawa, T. Matsukawa, G. Cong, S.-H. Kim, S. Suda, M. Ohno, T. Chiba, H. Tadokoro, M. Yanagihara, Y. Igarashi, M. Masahara, S. Namiki, and H. Kawashima, "Ultra-compact 8 × 8 strictly-non-blocking Si-wire PILOSS switch," *Opt. Express*, vol. 22, no. 4, p. 3887, Feb. 2014.

[16]   K. Tanizawa, K. Suzuki, M. Toyama, M. Ohtsuka, N. Yokoyama, K. Matsumaro, M. Seki, K. Koshino, T. Sugaya, S. Suda, G. Cong, T. Kimura, K. Ikeda, S. Namiki, and H. Kawashima, "Ultra-compact 32 × 32 strictly-non-blocking Si-wire optical switch with fan-out LGA interposer," *Opt. Express*, vol. 23, no. 13, p. 17599, Jun. 2015.

[17]   L. Lu, S. Zhao, L. Zhou, D. Li, Z. Li, M. Wang, X. Li, and J. Chen, "16 × 16 non-blocking silicon optical switch based on electro-optic Mach-Zehnder interferometers," *Opt. Express*, vol. 24, no. 9, p. 9295, May 2016.

[18]   S. Han, T. J. Seok, N. Quack, B.-W. Yoo, and M. C. Wu, "Large-scale silicon photonic switches with movable directional couplers," *Optica*, vol. 2, no. 4, pp. 370–375, Apr. 2015.

[19]   T. J. Seok, N. Quack, S. Han, R. S. Muller, and M. C. Wu, "Highly Scalable Digital Silicon Photonic MEMS Switches," *J. Light. Technol.*, vol. 34, no. 2, pp. 365–371, Jan. 2016.

[20]   T. J. Seok, N. Quack, S. Han, R. S. Muller, and M. C. Wu, "Large-scale broadband digital silicon photonic switches with vertical adiabatic couplers," *Optica*, vol. 3, no. 1, p. 64, Jan. 2016.

[21]   K. Tanizawa, K. Suzuki, M. Toyama, M. Ohtsuka, N. Yokoyama, K. Matsumaro, M. Seki, K. Koshino, T. Sugaya, S. Suda, and others, "32x32 Strictly Non-Blocking Si-Wire Optical Switch on Ultra-Small Die of 11x25 mm^2," in *Optical Fiber Communication Conference*, 2015, p. M2B–5.

[22]   T. J. Seok, N. Quack, S. Han, W. Zhang, R. S. Muller, and M. C. Wu, "Reliability study of digital silicon photonic MEMS switches," in *Group IV Photonics (GFP), 2015 IEEE 12th International Conference on*, 2015, pp. 205–206.

Wavefront Transmission for Remote Optical Beam Forming

Mitsumasa Nakajima, Junji Sakamoto, Keita Yamaguchi, Kenya Suzuki, and Toshikazu Hashimoto

NTT Device Technology Labs, NTT Corporation, 3-1 Morinosato Wakamiya, Atsugi, Kanagawa, Japan
nakajima.mitsumasa@lab.ntt.co.jp

Abstract:
A wavefront transmission technique is proposed for remote optical beam forming. As a demonstration, a spatially encoded optical wavefront was transmitted using a seven-core fiber to form the output beam from a waveguide-based phased array.

Keywords: Optical phased array, Wavefront modulation, Multicore fiber, Spatial light modulator

I. INTRODUCTION

Electromagnetic phased array devices at radio frequency have been widely studied because of their attractive applications such as wireless communications and ladders [1]. Recently, the concepts of this microwave phased-array beam forming technology have been extended to the optical domain because the much shorter optical wavelength holds promise for large-scale integration of phased array devices. Thanks to recent advances in nanophotonic technology, various types of optical phased array (OPA) have been proposed and demonstrated [2-4]. However, they usually require electrical drivers in order to individually control the optical phases. The footprint of the drivers is generally much larger than that of the OPA devices, which limits their potential applications.

In this study, we propose wavefront transmission using a multicore fiber in order to steer the optical beams remotely. In the proposed architecture, the size of an OPA device is only determined by that of the OPA device because the phase controller is separated from it. As a demonstration of the concept, we encoded the optical wavefront by a liquid-crystal-on-silicon (LCOS) based spatial light modulator (SLM). The encoded wavefront was transmitted by a seven-core fiber in order to form the output beam from a remotely established optical frontend.

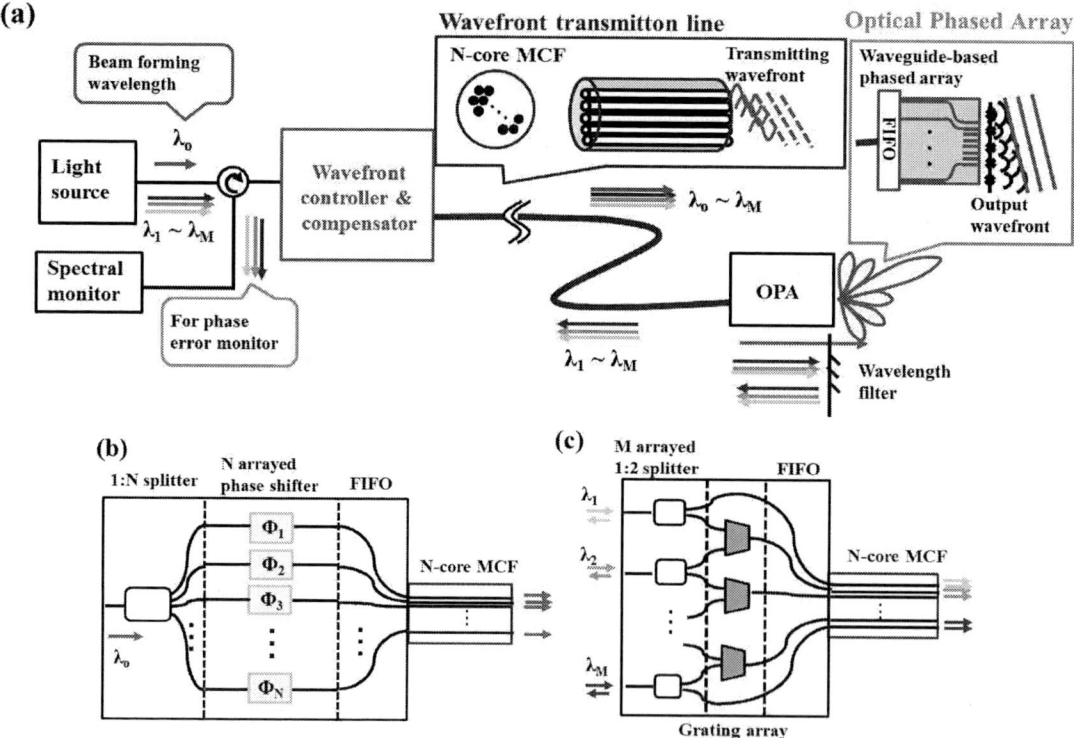

Fig. 1 Schematics of (a) proposed remotely controlled optical phased-array system and wavefront controller for (b) target wavelength and for (c) monitoring wavelength.

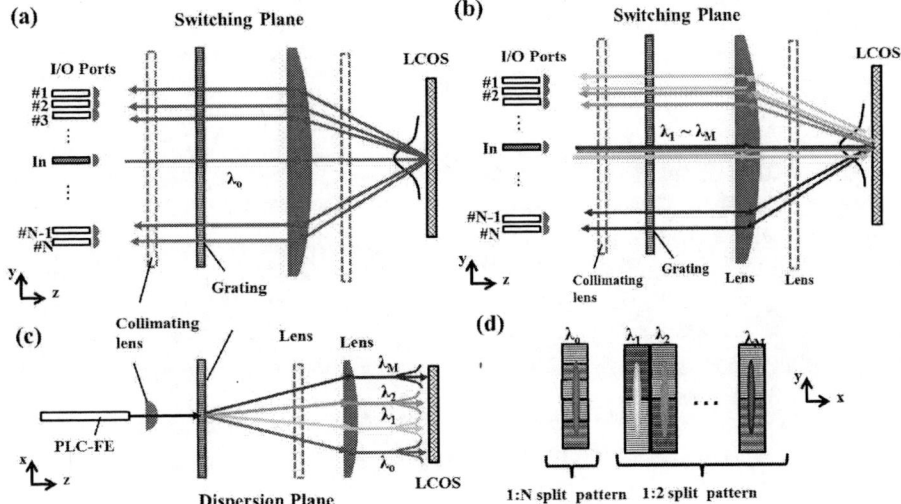

Fig.2 Detailed optics of wavefront controllers in switching plane for (a) target and (b) monitoring wavelength, and (c) dispersion plane. (d) Typical phase pattern displayed on SLM.

II. CONCEPT OF REMOTE OPTICAL PHASED ARRAY SYSTEM

Figure 1(a) shows the architecture of the proposed OPA system. This system consists of three components: a wavefront controller, a wavefront transmitting line, and an OPA. In this system, an optical signal with wavelength of λ_o inputs to the optical wavefront controller so that the phases of N-split signals are controlled individually as shown in Fig. 1(b). The output signals from the wavefront controller propagate through respective cores of an N-core multicore fiber, in which the relative phase of transmitting lights are almost kept. Finally, the transmitted signals are entered to the OPA and are radiated to free-space. As the beam shapes from the OPA are determined by the optical phase profiles of the transmitted signals, we can vary the output beam shape remotely using the wavefront controller. In a real environment, however, optical phases in multicore fibers fluctuated due to disturbances such as mechanical vibration and/or temperature variation of the fiber. It is needed to monitor and compensate those fluctuations to form the output beam shape accurately. Therefore, we also implemented the monitor and compensation technique. In order to monitor such phase fluctuations, optical signals with other wavelengths from λ_1 to λ_M are also transmitted through the same system. These signals are split and input into two cores in the multicore fiber and take the same path as the target wavelength of λ_o, as shown in Fig.1(c). These reference signals are reflected by a wavelength filter at the edge of the OPA. The returned signals re-enter the wavefront controller, which forms multi-lane delay line interferometers. By monitoring every interfered signal, we can determine and compensate for the phase fluctuations in the multicore fiber.

To simplify the optics of the proposed wavefront controller, we also propose an integrated setup. Figures 2(a)-(c) are explanatory schematics of the optics system of the proposed wavefront controller. It consists of I/O ports, a dispersion grating, lenses, and an LCOS-based SLM. Importantly, this optics is the same as that of typical wavelength selective switches (WSS) which are common devices in photonic networks [5]. An input signal coming from the input port is focused by a Fourier lens onto the SLM plane for each wavelength. The optical function of each separated areas can be defined by changing a phase pattern on each areas of the SLM. For example, if we divided the corresponding area of the LCOS into M regions, we can realize M-splitting function. By using this programmable feature of SLM, the wavefront control function and 1:2 splitting function for the monitors can be integrated as shown in Fig. 2(d). For the signal wavelength of λ_o, the area is separated into N region in order to realize a 1:N splitting function and the phase shift function. On the other hand, the two different phase patterns are displayed on the monitor wavelength of λ_1 toλ_M so that a simple delay line interferometric function is realized. Detailed methods for beam splitting and phase control are described in elsewhere [6].

III. EXPERIMENTAL DEMONSTRATION

As a demonstration of our concept, we constructed a remote OPA system. 1x20 WSS, A 10-m-long seven-core fiber with 10m long and a waveguide-based phased array were employed as the wavefront controller, the wavefront transmitting line, and the OPA, respectively. Figure 3 shows the designed and fabricated 20 arrayed waveguide-based OPA. The separations of output waveguides are set to 9 μm near the edge of the OPA, because the separation determines the maximum steering angle. In order to joint the OSA and the seven-core fiber, a FI/FO fiber is employed. We placed an IR camera near the OPA and measure the beam profiles of the output beams. The results are shown in Figs. 4(a)-(c). As shown in Fig. 4(a), the beam profiles without any wavefront control show multiple beam spots, which suggest sidelobes caused by phase inconsistency. On the other hand, the levels of the sidelobes were drastically suppressed when the wavefront was controlled [see Figs. 4(b) and (c)]. In addition, the beam spot can be controlled by the slope of the defined wavefront. These results mean that remote optical beam forming is successfully achieved with our proposed architecture.

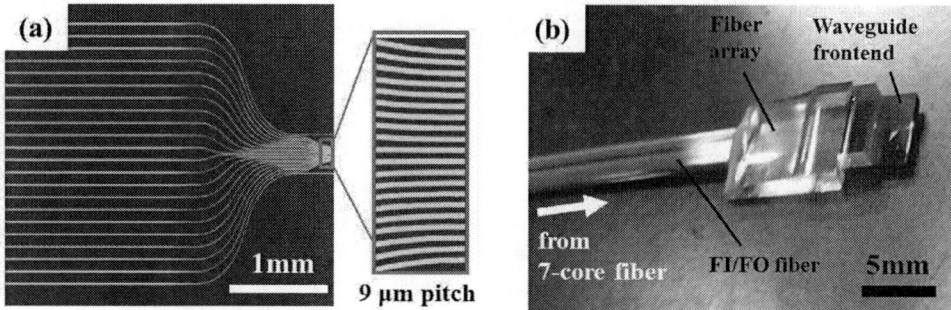

Fig. 3 (a) Designed and (b) fabricated waveguide-based optical phased array

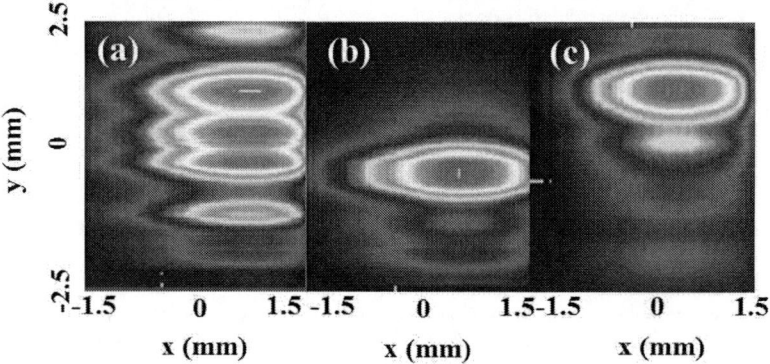

Fig. 4 Far field profiles from OPA devices (a) without wavefront control, (b) with flat phase profile, and (c) with sloped phase profile

IV. Conclusions

We proposed a wavefront transmission using multicore fiber in order to steer the optical beams remotely and phase error monitoring and compensating technique. We demonstrated the proposed concept using a conventional WSS, seven-core fiber, and waveguide-based OPA. The output beam was successfully controlled remotely by using our proposed OPA system.

References

[1] R. C. Hansen, "Phased Array Antennas", Wiley, 1998.

[2] J. Sun, E. Timurdogan, A. Yaacobi, E. S. Hosseini, and M. R.Watts, "Large-scale nanophotonic phased array", Nature, vol. 493 pp.195-199 (2003).

[3] D. Kwong, A. Hosseini, J. Covey, Y. Zhang, X. Xu,H. Subbaraman, and R. T. Chen, "On-chip silicon optical phased array for two-dimensional beam steering". Opt. Lett., vol. 39, pp.941–944 (2014).

[4] D. P. Resler, D. S. Hobbs, R. C. Sharp, L. J. Friedman, and T. A. Dorschner, "High-efficiency liquid-crystal optical phased-array beam steering", Opt. Lett., vol. 21, pp. 689-691 (1996).

[5] Y. Ikuma, K. Suzuki, N. Nemoto, E. Hashimoto, O. Moriwaki, and T. Takahashi, "Low-loss Transponder Aggregator Using Spatial and Planar Circuit", J. Lightwave Technol., vol. 34. pp.67-72 (2015).

[6] M. Nakajima, N. Nemoto, K. Yamaguchi, J. Yamaguchi, K. Suzuki, and T. Hashimoto, "In-band OSNR Monitors Comprising Programmable Delay Line interferometer Integrated with Wavelength Selective Switch by Spatial and Planar Optical Circuit", OFC 2016, Th2A.-11, Anaheim (2015).

ME2-3

Stacked Wavelength Selective Switch Design for Low-cost CDC ROADMs

Haining Yang[1], Brian Robertson[1], Peter Wilkinson[1,2], and Daping Chu[1,2] *

[1] Roadmap Systems Ltd, St John's Innovation Centre, Cowley Road, Cambridge CB4 0WS, UK
[2] Department of Engineering, University of Cambridge, 9 JJ Thomson Avenue, Cambridge CB3 0FA, UK
* daping.chu@roadmapsystems.co.uk, dpc31@cam.ac.uk

Abstract: A highly flexible stacked switch module is proposed, where 48 independent flex-spectrum 1×12 WSSs can be realised on a single 4k LCOS device. This design can handle all the switching operations within a CDC ROADM.

Keywords: Wavelength Selective Switch (WSS), Reconfigurable optical add/drop multiplexer (ROADM)

I. INTRODUCTION

A wavelength selective switch (WSS) is the key building block for a modern reconfigurable optical add/drop multiplexer (ROADM) [1], which enables reconfigurable optical networks. A typical WSS is able to selectively route individual wavelength division multiplexed (WDM) channels entering its input fibre port to any of the output fibre ports according to the software configuration that is remotely controlled by the service providers. In recent years, phase-only liquid crystal on silicon (LCOS) spatial light modulators (SLMs) have become the technology of choice for WSSs, due to their software upgradable nature and support for flexible spectrum switching [2]. Efforts [3-6] have also been made to improve the port count, crosstalk levels and passband shapes in LCOS WSSs. WSSs are usually based on the 'disperse-and-select' optical design, where the WDM channels from the input port are diffracted along the dispersion axis at the LCOS plane, before being switched to the target output ports according to the sub-holograms displayed on the corresponding areas of the LCOS device. Due to the limited number of pixels available on the current generation LCOS devices, anamorphic optics are invariably used in these designs to convert the input signals to elongated beams at the LCOS plane. Correspondingly, output ports are arranged along a switching axis that is orthogonal to the dispersion axis. Although such an approach is able to increase the port count in one axis, it fails to fully exploit the two dimensional (2D) nature of the LCOS pixel array. Moreover, for such a configuration, all the undesirable diffraction orders due to the LCOS quantization effects will also appear along this switching axis, which makes it fundamentally difficult to suppress crosstalk, especially in WSSs with high port counts.

In this paper, we proposed a stacked WSS module, which does not use anamorphic optics. It utilises 2D beam steering and can incorporate 48 independent 1×12 flex-spectrum WSSs on a single 4k LCOS device within one module. This module can also be configured to realise 12×12 contentionless wavelength cross-connect (WXC), for a low-cost, small-footprint add/drop solution for colourless, directionless and contentionless (CDC) ROADMs.

II. OPTICAL DESIGN

The principle of the proposed stacked module is shown in Fig. 1, which depicts M 1×N WSSs (M=3 and N=8 in this example) plus an array of objective lenses (L_A), a relay system (L_1 and L_2) and DEMUX optics (P_g). Each of the 1×N WSSs has a fibre array cluster, which consists of 1 input, N output fibre ports and the corresponding micro-lens array. These clusters are arranged along the y-axis, each acting as an independent 1×N WSS, with each input light beam illuminating a spatially distinct row of sub-holograms (e.g., S_1, S_2, and S_3). The WDM input is launched into each WSS via the central fibre in the corresponding cluster. The objective lens generates a beam waist of radius ω_o at plane P_o. The

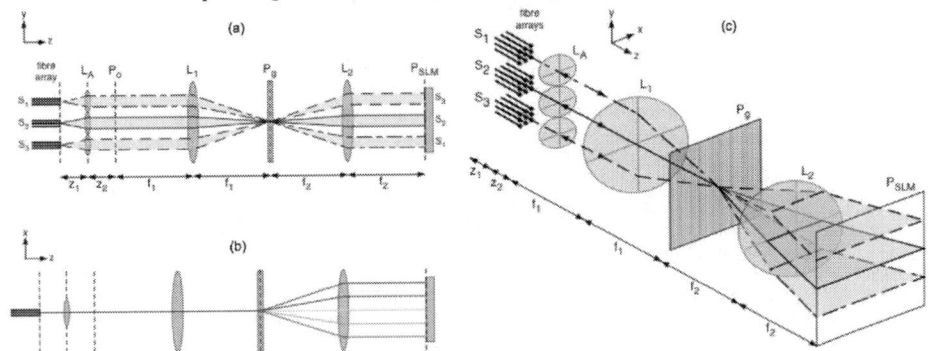

Fig. 1. Design principle for the stacked WSSs based on a single LCOS device, (a) side view, (b) top view and (c) system view.

4f relay system images this beam waist at the SLM plane (P_{SLM}). The static grating (P_g) imparts an angular displacement of $\beta_G(\lambda)$ to each wavelength channel in the x-z plane. Signal beams associated with these wavelengths illuminate separate sub-holograms displayed on the LCOS device. The sub-hologram width can be adjusted to enable flex-spectrum switching. The sub-hologram for a wavelength channel could be a grating of period T, orientated at an angle of φ with respect to the local xy-coordinate system, that diffracts the light beam such that it leaves the LCOS SLM with a propagation vector of $k(\rho, \varphi, \lambda)$, where ρ is the angle of the vector with respect to the local z-axis. The diffracted beam is subsequently imaged at P_o by the relay system. The L_A in the output optics shown in Fig. 2 converts the propagation vector of a wavelength channel, $k(\rho, \varphi, \lambda)$, to a beam position that is offset from the optical axis. The angle is controlled such that the beam is concentric with respect to the intended output fibre, thereby maximising coupling efficiency as shown in Fig. 2(a). A secondary lenslet array, L_F, focuses the wavelength channels into the output fibre array.

Our development work has shown that it is possible to use 50×50 pixels for each 50GHz channel within the C-band. By using a standard 4k LCOS device (such as the Jasper JD2704 with 4096x2400 pixels), 48 independent WSSs can be stacked along the y-axis of a single chip. In the absence of anamorphic optics, the un-modulated input signal to each sub-hologram has a circular beam shape on P_{SLM}, which is designed to cover 31×31 pixels, and hence achieve the 4th order super Gaussian passband shape. The small number of pixels covered by the beam limits the maximum grating period to 10 pixels, which is larger than the minimum period of 7 pixels required to realise sufficient switching efficiency and reasonably low crosstalk level for two switchable positions along a given axis. The circular beam on P_{SLM} allows 2D steering, giving 8 switchable output ports arranged on a Cartesian grid, as shown in Fig. 2(b). The number of the switchable output ports can be further increased, given the same beam steering range, to 12 if the fibre ports are arranged in a hexagonal pattern, as shown in Fig. 2(c).

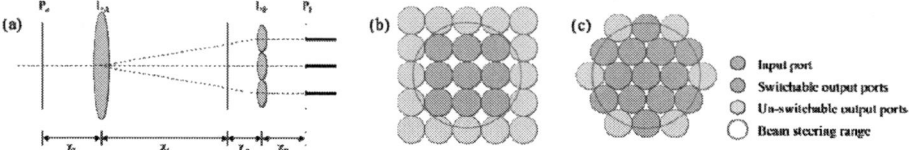

Fig. 2. (a) Output optics design; 2D beam steering over the fibre ports arranged on (b) a Cartesian grid and (c) a hexagonal grid.

III. CROSSTALK SUPPRESSION

Crosstalk from a blazed grating on an LCOS occurs due to phase level quantization and pixel fringing fields whereby the electric field due to the voltage applied to a pixel leaks towards neighbouring pixels. The optical replay field incorporating the edge effect is shown for a blazed grating with a period of 7 pixels in Fig. 3(b), where a series of discrete diffraction orders can be seen. These may coincide with a number of fibre ports in the conventional LCOS WSS design, with a one-dimensional linear fibre port array and anamorphic optics, leading to high crosstalk levels. In the design illustrated in Fig. 1, however, only the -1st diffraction order of the blazed grating can coincide with an un-targeted port, in either the Cartesian or hexagonal arrangement. In this specific case, the -1st diffraction order will cause a crosstalk of about -20 dB, which can be suppressed by using advanced computer generated hologram algorithms for a specific order [5]. More effectively, we propose a general hardware approach, with the same principle as the wavefront encoding technique [4], to suppress crosstalk by building an optical asymmetry into the system via a matched array of axicon spatial filters positioned at plane P_o. Specifically, each of the lenslets L_A shown in Fig. 2(a), has an associated axicon element of wedge angle β, as shown in Fig. 3(a). In operation a Gaussian beam enters the switch via the central fibre of each fibre cluster, and is relayed to plane P_o, where the axicon phase element imparts a radially linear phase delay. The wavefront leaving the axicon is imaged by lenses L_1 and L_2 at the LCOS plane, with each wavelength channel being separated by the DEMUX grating such that it illuminates a separate sub-hologram. If no phase pattern were displayed on a sub-hologram, the light would simply reflect off the LCOS device, and be re-imaged at the axicon element. Due to the phase reversal of light on reflection, the overall action of the axicon is to double the radially linear phase shift, causing the light to be focused by L_A into a ring pattern centred about the input fibre. As a result, for a specific propagation vector, $k(p,\varphi,\lambda)$ the sub-hologram must also display a phase pattern that compensates for the axicon filter in a lock-and-key approach that maximizes coupling efficiency for the +1st order. Consequently, all other mth diffraction orders are focused into ring patterns with a radius dependent on (m-1), where m is the diffraction order. Fig. 3(c) illustrates the corresponding sub-hologram and an optimized axicon filter (wedge angle of $\beta = 0.38°$), which reduces the -1st order crosstalk power by 11.8 dB, with only a penalty of 0.25 dB on insertion loss.

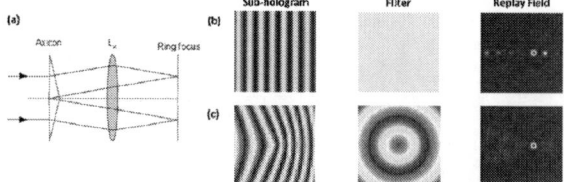

Fig. 3. (a) Illustration of an axicon operation; (b) replay field using a blazed grating with a period of 7 pixels with no filter; (c) equivalent replay field using an axicon matched spatial filter at plane P_o with a wedge angle of $\beta = 0.38°$.

IV. APPLICATIONS

The large number of independent WSSs in this module allows various potential applications. In particular, this module offers two possible low-cost and small-footprint CDC add/drop solutions for ROADMs. Fig. 4(a) shows the current CDC add/drop solution based on multicasting switches including splitters and space switches. Although such a multicasting switch has the advantages of small footprint and low cost, it is not scalable due to the insertion loss in the splitters. Consequently, amplifier arrays are required to maintain a sufficient signal level, which increases the cost and power consumption. The second solution shown in Fig. 4(b) replaces the splitters with an array of WSSs so that the insertion loss no longer scales with the add/drop port count. However, the large number of WSS modules required here significantly increases the cost and footprint. This conundrum can be solved by using our stacked WSSs module in an M 1×N WSS configuration, as shown in Fig. 4(c). Our stacked WSSs module is able to efficiently route WDM channels in this configuration, with minimum increase in the cost and footprint due to sharing of common optics and the LCOS device between its 48 independent WSSs. As shown in Fig. 4(d), our stacked WSSs module can be further used as a WXC, which is able to realise contentionless wavelength switching between multiple input and output ports. In this case, space switches are no longer required to achieve CDC add/drop, further increasing the system integration. The operational principle of the WXC is given in Fig. 4(e) based on an example of a 4×4 WXC, in which the output ports are paired between WSSs. In this specific example, each WSS still has 8 spare ports for adding or dropping wavelengths to or from corresponding directions. Our module is able to realise 12×12 WXC using 24 of the independent WSSs. The rest 24 1×12 WSSs within our module can be either used to construct a second 12×12 WXC or deployed as standard 1×12 WSSs at the transit part of ROADM, according to the needs of the network operator. In the latter case, all the switching operations within a ROADM could be handled by our stacked WSS module.

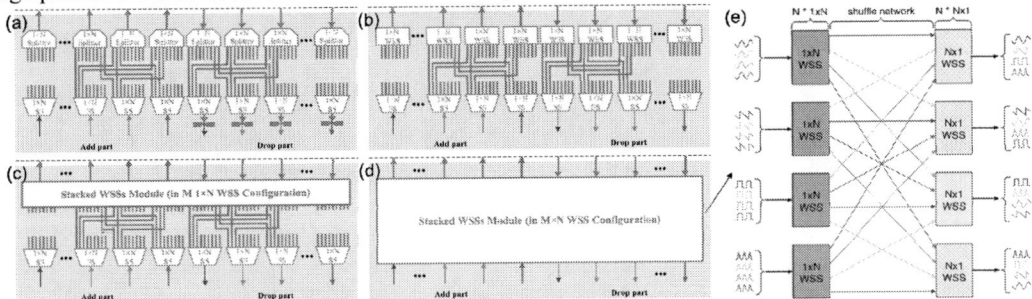

Fig. 4. CDC add/drop solutions for ROADM: (a) multicasting switch; (b) WSS array; (c) stacked WSSs module in M 1×N configuration; (d) stacked WSSs module in M×N configuration; (e) 4×4 WXC based on 8 WSSs.

V. CONCLUSIONS

We have proposed a stacked WSS module, in which 48 independent 1×12 flex-spectrum WSSs can be realised on a single 4k LCOS device. The proposed module can maximise the switching capability in terms of port count per pixel, and minimise the cost and energy consumption per port. The crosstalk in such WSSs is primarily due to -1st diffraction order of the sub-holograms, which can be reduced by building an optical asymmetry into the optical system, via an array of axicon matched spatial filters. Our results show an 11.8 dB reduction in the crosstalk with only a 0.25 dB insertion loss penalty. This proposed module is highly flexible and it can be configured as an array of 1×N WSSs, an N×N WXC with N up to 12 or any combination thereof. This offers a low-cost and small-footprint switching and CDC add/drop solution for ROADMs.

REFERENCES

[1] T. A. Strasser, and J. L. Wagener, "Wavelength-selective switches for ROADM applications," IEEE J. Sel. Top. Quantum Electron. 16(5), 1150-1157 (2010).

[2] S. Frisken, G. Baxter, D. Abakoumov, H. Zhou, I. Clarke, and S. Poole, "Flexible and grid-less wavelength selective switch using LCOS technology," in Optical Fiber Communication Conference/National Fiber Optic Engineers Conference, paper OTuM3 (2011).

[3] K. Suzuki, Y. Ikuma, E. Hashimoto, K. Yamaguchi, M. Itoh, and T. Takahashi, "Ultra-High Port Count Wavelength Selective Switch Employing Waveguide-Based I/O Frontend," in Optical Fiber Communication Conference, paper Tu3A.7 (2015).

[4] B. Robertson, Z. Zhang, H. Yang, M. M. Redmond, N. Collings, J. Liu, R. Lin, A. M. Jeziorska-Chapman, J. R. Moore, W. A. Crossland, and D. P. Chu, "Reduction of crosstalk in a colourless multicasting LCOS-based wavelength selective switch by the application of wavefront encoding," Proc. SPIE 8284, 82840S (2012).

[5] H. Yang, B. Robertson, and D. Chu, "Crosstalk reduction in holographic wavelength selective switches based on phase-only LCOS devices," Optical Fiber Communication Conference, paper Th2A.23 (2014).

[6] C. Pulikkaseril, L. A. Stewart, M. A. F. Roelens, G. W. Baxter, S. Poole, and S. Frisken, "Spectral modeling of channel band shapes in wavelength selective switches," Opt. Express 19(9), 8458-8469 (2011).

First Demonstration of Dynamic Mode-Switching by Using Optical Mode Switch

Ryan Imansyah, Luke Himbele, Shota Oe, Haisong Jiang and Kiichi Hamamoto

Interdisciplinary Graduate School of Engineering Science, Kyushu University, Kasuga, Fukuoka 816-8580, Japan
E-mail : 3ES15027K@s.kyushu-u.ac.jp

Abstract: *Dynamic mode switching is demonstrated by using optical mode switch. The switch utilizes phase shift based-on current injection. Less than 60 ns switching-time for the both 0th-to-1st and 1st-to-0th was confirmed for the first time.*

Keywords: *dynamic mode-switching, optical mode switch*

I. INTRODUCTION

Optical switch is attractive because it decreases the power consumption of the router by eliminating the optical-electrical-optical (OEO) signal exchange [1-4]. One of the optical switch is based on Si-wire waveguide, known as the optical spatial switch. One of the advantage by using Si-wire waveguide is the possibility to realize revolution of on-chip communication [5]. Moreover, silicon photonics offers a merit of dense footprint and also nanosecond scale reconfiguration time using electro-optic phase shifter [6,7]. Scalability for high integration has been one critical issue for optical spatial switch [8]. Thus, to realize the optical switching we have proposed optical mode switch to overcome this scalability issue [9-11].

As mode is switched by using phase shift based-on current injection, it is possible to switch mode dynamically, however, there has been no report of dynamic mode-switching so far. In this work, dynamic mode-switching is demonstrated for the first time. At least less than 64 ns switching time, for the both 0th-to-1st and 1st-to-0th, was confirmed for the first time. More than 1200 ns status-holding was also confirmed successfully.

II. CONCEPT AND DESIGN

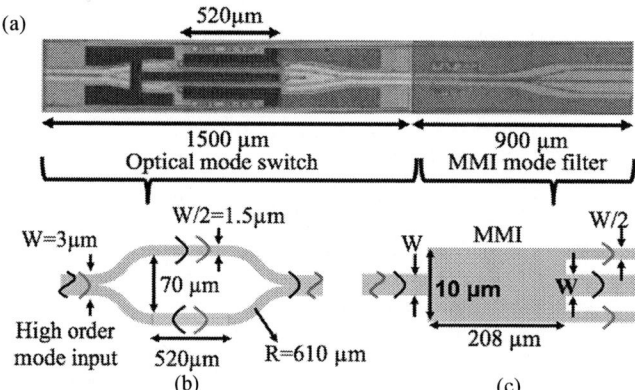

Fig. 1. Device configuration. (a) Top view of implemented device, (b) configuration detail of optical mode switch and (c) MMI mode filter.

Figure 1 (b) shows the schematic view of this optical mode switch. The configuration of the optical mode switch is similar to Mach-Zehnder interferometer, however, the difference is its width in input and output waveguide to realize the optical mode switch than the interference modulation. The width should permit the mode combining of the fundamental mode and the first order mode, therefore, the both Y-junctions work mode-separator and mode combiner, apart from power divider and power coupler. The input and output waveguide is designed to be 3 µm while ones of each arm to be 1.5 ☐m. S-bend waveguides are used in the both Y-junctions to realize the proper mode-splitter and mode-combiner with a high fabrication tolerance. We designed the radius of S-bend waveguide as 610 µm with the opening space between arms was 70 µm for ion implantation. While the arms are designed to have the width of 520 µm with one of the arms is employed as refractive index change region.

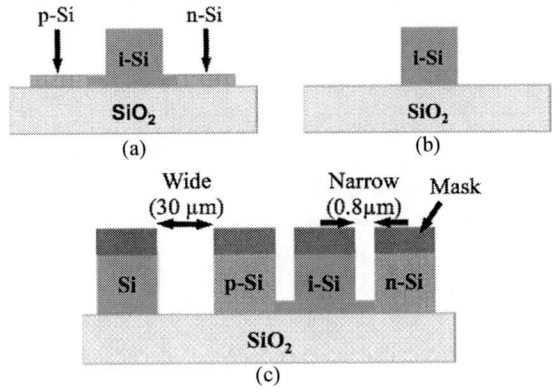

Fig. 1. Waveguide structure. (a) common pin structure, (b) other waveguide structure, and (c) pin trench structure.

Commonly, the refractive index change region is fabricated as a pin structure in Fig. 2 (a). Meanwhile, the other region of the waveguide is fully etched and has a structure like in Fig. 2 (b). These structures make the fabrication process is complicated with at least two step of dry etching. To simplify this fabrication process into one single-step dry etching, pin trench structure is used [10]. For the ion implantation, the p-doped and n-doped regions are implanted with a dosage of 2×10^{15} ions/cm2, and the ion beam energies were 120 keV and 45 keV, respectively. By using this structure and ion implantation dosage, the refractive index change region is realized to make a a π phase difference between arms.

In order to evaluate the optical mode switching, MMI mode filter is integrated so that the power of each mode and the crosstalk between them is evaluated by using filter output [11] (see Fig. 1(c)). MMI mode filter is used for the purpose to distinguish the fundamental and first-order mode to distinguish mode-status in this work (usually this filter is not necessary). Through this filter, injected fundamental mode propagates toward the center port while first order mode propagates toward two side output ports of the MMI filter. Figure 1(a) shows the top view of the optical mode switch that was integrated with MMI mode filter.

III. EXPERIMENTAL RESULTS

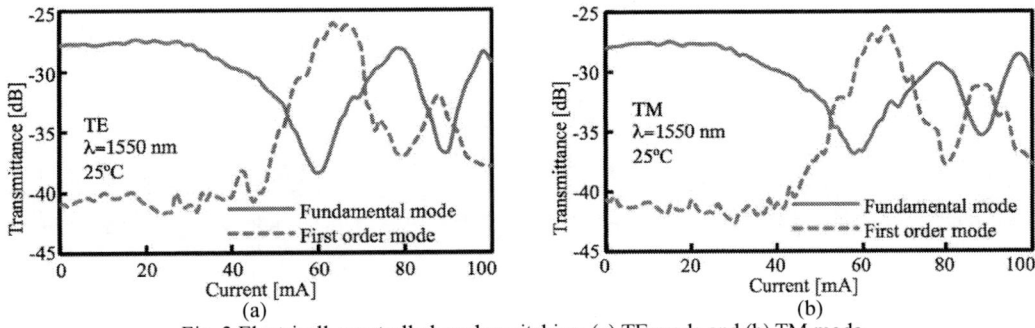

Fig. 3 Electrically controlled mode switching. (a) TE mode and (b) TM mode.

Figure 3 shows the electrically controlled mode switching result for the both of TE and TM mode. Blue (solid) and red (dash) lines show the transmittance of the fundamental mode and first order mode, respectively. Here, λ=1550 nm with fundamental mode was injected into the device. When the current is increased gradually, fundamental mode power decreases slightly while first order mode power increases up to 42 mA (3.6 V). From 42 mA current injection, both of the power changed dramatically with the dip of fundamental mode was at 58 mA while the peak of the first-order mode was at 62 mA, showing that the cross-state occurred at approximately 60 mA (5.7V). The light output power shows the highest at -28 dB with the crosstalk of approximately -10 dB. This crosstalk is relatively high, however, it can be decreased by improving the fabricated dimension of the device. There is a mismatch of 0.3 μm MMI mode filter width. However, this output power is too small to be detected by the oscilloscope. To compensate this insertion loss, an erbium-doped fiber amplifier (EDFA) was used after the output light was coupled by using lensed fiber.

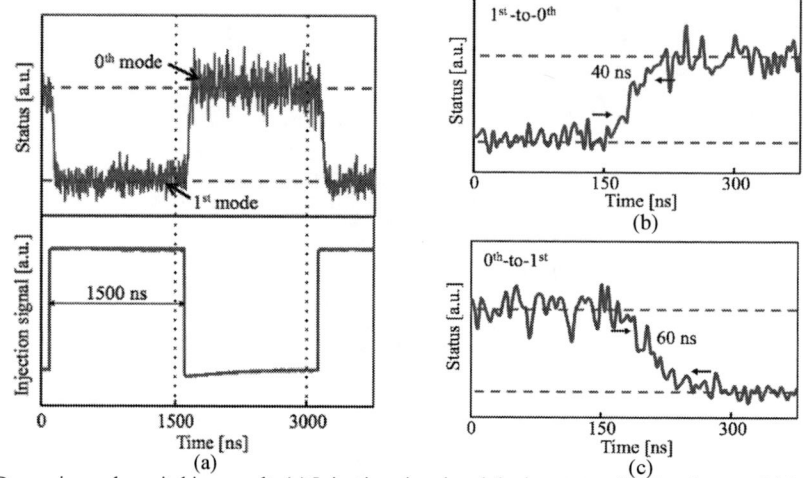

Fig. 4. Dynamic mode-switching result. (a) Injection signal and device status, (b) rise time, and (c) fall time.

Figure 5 (a) shows device status from the injection signal of a 1500 ns with 2 V peak-to-peak square pulse. The status was evaluated by monitoring the output power at the fundamental mode port of MMI filter. When the current is set to cross-state (approx.60 mA), the device status shows the lower optical power level, which corresponds to the first order mode, while the device status shows higher optical power level, which corresponds to fundamental order mode, in the case of current is set to bar-state. Figure 5 (b) shows the switching time of 40 ns for 1st-to-0th switching, while the switching time of 60 ns for 0th-to-1st switching was observed as shown in Fig. 5 (c). These switching time can be reduced by the reduction of current path resistance through the pin trench structure.

IV. CONCLUSIONS

We have confirmed dynamic optical mode switching for the first time. At least less than 60ns mode switching time, for the both 0th-to-1st and 1st-to-0th, was confirmed for the first time.

ACKNOWLEDGMENT

This work has been supported by NICT, Japan. The authors thank to Dr. Y. Hinokuma for his technical advice and discussion.

REFERENCES

[1] D. J. Blumenthal, J. E. Bowers, L. Rau, H. F. Chou, S. Rangarajan, W. Wang and K. N. Poulsen, "Optical signal processing for optical packet switching networks," IEEE Opt. Comm., vol. 41, No. 2, pp. S23-S29, 2003.

[2] H. J. S. Dorren, M. T. Hill, Y. Liu, N. Calabretta, A. Srivatsa, F. M. Huijskens, H. de Waardt and G. D. Kohe, "Optical packet switching and buffering by using all-optical signal processing methods," IEEE J. Light. Technol., vol. 21, no.1, pp. 2-12, 2003.

[3] P. Dong, S. Liao, H. Liang, R. Shafiiha, D. Feng, G. Li, X. Zheng, A. V. Krishnamoorthy and M. Asghari, "High-speed and broadband electro-optic silicon switch with submilliwatt switching power,," Tech. Dig. OFC 2011, OWZ4, 2011.

[4] J. V. Campenhout, W. M. J. Green, S. Assefa and Y. A. Vlasov, "Low-power, 2×2 silicon electro-optic switch with 110-nm bandwidth for broadband reconfigurable optical networks" Opt. Express, vol. 17, no. 26, pp. 24020-24029, 2009.

[5] A. V. Krishnamoorthy, R. Ho, X. Zheng, H. Schwetman, J. Lexau, P. Koka, G. Li, I. Shubin, and J. E. Cunningham, "computer systems based on silicon photonic interconnects," Proc. IEEE 97(7), 1337–1361 (2009).

[6] J. Xing, Z. Li, P. Zhou, X. Xiao, J. Yu, and Y. Yu, "Nonblocking 4x4 silicon electro-optic switch matrix with pushpull drive," Optics Letters, vol. 38, no. 19, p. 3926, Oct. 2013.

[7] B. G. Lee, A. V. Rylyakov, W. M. J. Green, S. Assefa, C. W. Baks, R. Rimolo-Donadio, D. M. Kuchta, M. H. Khater, T. Barwicz,C. Reinholm, E. Kiewra, S. M. Shank, C. L. Schow, and Y. A. Vlasov, "Monolithic Silicon Integration of Scaled Photonic Switch Fabrics, CMOS Logic, and Device Driver Circuits," J. Lightwave Technol., vol. 32, no. 4, pp. 743–751, Feb. 2014.

[8] N. Dupuis, "Technologies for Fast, Scalable Silicon Photonics Switches," Proc. Photonics in Switching, 2015, p.100.

[9] R. Imansyah, T. Tanaka, L. Himbele, H. Jiang and K. Hamamoto, "Mode Crosstalk Evaluation on Optical Mode Switch," Proc. IBP, P3.2, p.82, 2015.

[10] R. Imansyah, L. Himbele, H. Jiang and K. Hamamoto, "Single-Step Dry-Etched Lateral PIN by Using Trench Structure for Optical Mode Switch," Proc. PS, ThI2-7, pp.81-84, 2015.

[11] R. Imansyah, L. Himbele, H. Jiang and K. Hamamoto, "First Demonstration of Electrically Controlled Mode Switching," Proc. MOC, D3, pp.50-51, 2015.

Ultra-fast super-resolution imaging by time-stretch structured illumination

Yuxi Wang, Qiang Guo, Hongwei Chen, Minghua Chen, Sigang Yang

National Laboratory for Information Science and Technology,
Department of Electronic Engineering, Tsinghua University, Beijing, China, 100084
chenhw@mail.tsinghua.edu.cn

Abstract: A new strategy to generate illuminating pattern for structured illumination microscopy based on the method of optical pulse laser time-stretch open up the possibilities to conduct ultra-fast (50MHz pattern generation rate) super-resolution imaging.

Keywords: super-resolution, structured illumination microscopy, time-stretch

INTRODUCTION

According to the theory of Ernst Abbe, the lateral resolution of the optical microscope is fundamentally limited by optical diffraction. It is impossible for traditional optical microscopy to obtain a better resolution than half the wavelength of light. Several recent methods can go well beyond this limit. Among them, a method known as structured-illumination microscopy (SIM) find its advantages lies in its high frame rate, reported to be more than 10Hz over fields of view of 8x8 um[1].

And now there shows a great potential for the imaging speed of SIM to be further improved. We here invite a technique known as time-stretch imaging to generate the structured illumination pattern, and we name this method as TS-SIM. The technique of time-stretch has proven to work successfully in ultra-fast microscopy [2]. In a typical SIM system, a spatial light modulator (SLM) or a digital micro-mirror device (DMD) often works as the component to generate a modulated spatial pattern. The pattern generation rate is largely limited by the mechanical mirror scanning speed or the response time of DMD/SLMs, both of which are typically on the order of kHz. While on the other hand, the time-stretch method can easily improve the pattern generation rate to dozens of MHz.

EXPERIMENTS

A. Resolution improvement by structured illumination

SIM is often used to improve lateral two-dimensional (2D) resolution and need several 2D patterns [3]. The time-stretch method could either generate a one-dimensional (1D) modulated scanning line or a 2D pattern [4]. But in order to get pattern with better accuracy, we here generate 1D patterns for a line-scan mode SIM to enhance the imaging resolution on axial direction as an exhibition. The working principle can be explained as this (Fig.1): As for conventional illumination, the optical transfer function (OTF) of the imaging system filters away the high frequency component in the spatial frequency domain (Fig.1.(a)), thus limiting the resolution of detected image (Fig.1.(b)). While as for structured illumination, the sinusoidal modulated incident intensity introduces a frequency shift and moves the higher frequency component into the passband of the OTF (Fig.1.(c)). In this way, when reconstructed with proper algorithm, a super-resolution imaging result could be gotten (Fig.1.(d)). The ratio of modulation frequency to the cut-off frequency decides how much is the resolution improved, which is up to 100% in linear SIM. To calculate the component with higher frequency, the target should be detected at least three times by different pattern of different initial phase ϕ (e.g. 0°,90° and 180°). The new method to generate these pattern will be mentioned below.

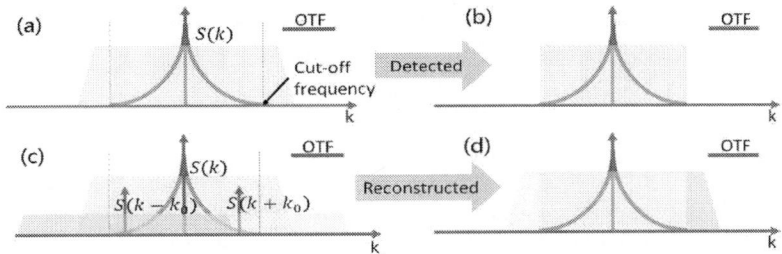

Fig. 1. Principle of achieving super-resolution by SIM.

This work is supported by NSFC under Contracts 61120106001, 61322113, 61271134; by Tsinghua University Initiative Scientific Research Program; by the young top-notch talent program sponsored by Ministry of Organization, China;

B. Ultra-fast pattern generation by time-stretch

The main principle of time-stretch is to build up a mapping between the spatial distributions of the illuminating light source onto its serial time-domain waveform. Specifically, a chirped laser pulse is employed as the optical source. When propagating through a dispersive component with group-velocity dispersion (GVD), a process known as amplified dispersive Fourier transform (ADFT) is conducted and build up the mapping from the spectrum of the laser pulse onto its time domain waveform. Afterwards, with the help of a spatial disperser, the spectrum of laser pulse is further mapped onto different spatial positions. In this way, when modulating a designed pattern onto the time domain waveform of the laser pulse, a spatial-distributed pattern could be generated through the two steps mapping from time to wavelength then to space. High-speed modulating instruments, for example an arbitrary waveform generator (AWG), could help to act this modulation on time domain.

To go in details, we consider the experimental set-up depicted in Fig.2. A broadband pulse laser is employed as the optical source (center wavelength 1560nm, 3dB broadband 10nm and pulse period 50.0MHz). This laser source is then divided into two path. One path, after propagating through a distance of dispersive compensating fiber to conduct the ADFT process, is dispersed on time domain, thus build up the mapping from time domain to wavelength domain. On the other hand, a 5GHz high-order harmonic of the original source is filtered out to serve as the external clock for a high-speed pulse pattern generator (PPG). The PPG outputs a rectangular wave signal, with the period of T1 (several times the period of the original source). For the sake of a better synchronization, this rectangular wave is needed as a trigger signal for an arbitrary waveform generator (AWG, 40GSa/s), which generates a predefined pattern A. Then on, this pattern is modulated on to the time domain waveform of the first path mentioned above. Till this step, the time domain as well as wavelength domain is modulated with pattern. Next, when collimate this pulse laser into free space and disperse this laser beam by a diffraction grating (grating period d =1/600mm), a serial of spatial-distributed patterns will be gotten.

To give an example, we may assume the rectangular wave period to be 80.0ns, 4 times the period of the original laser pulse, and the predefined pattern A consists of 4 parts of equal length (P1, P2, P3 and P4). Each quarter is encoded with a section of sinusoidal signal of 4 different initial phase. When modulating pattern A onto the time-stretched laser pulses, every 4 neighboring pulses are of 4 different waveform, which eventually ends in 4 different pattern modes periodically illuminating onto the imaging target.

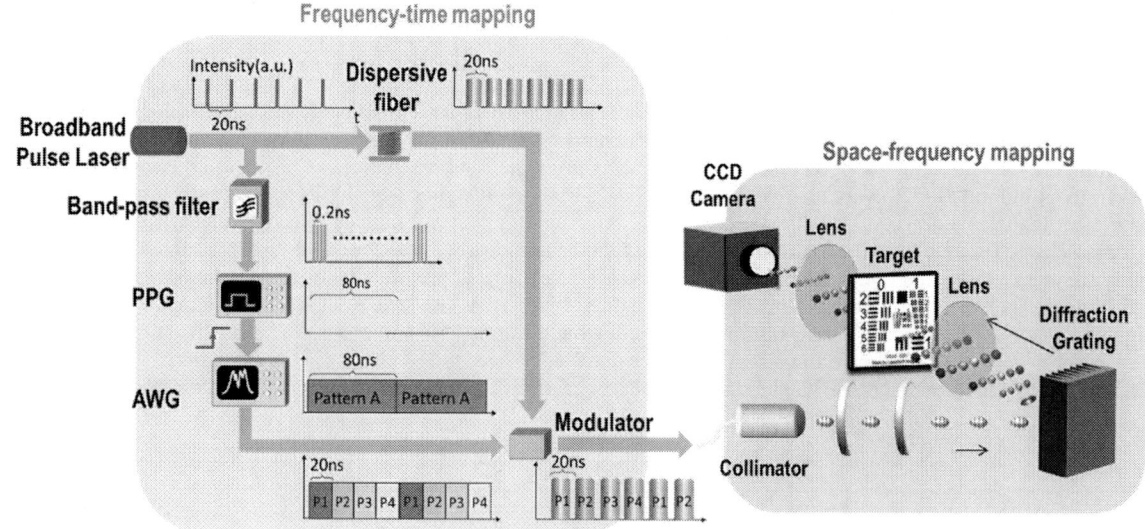

Fig. 2. Schematic of the TS-SIM system.

Actually, according to the principle of this method, the pattern generation rate find its limitation to be equal to the repetition rate of the laser pulse, 50MHz here in our set-up, more than 3 orders of magnitudes faster than other contemporary methods. But in an entire system, the pattern generation rate should match the frame rate of the imaging detector, for example a CCDs camera with the frame rate up to 20 Hz in our design.

C. Results

Here we experimentally demonstrate a TS-SIM system. A USAF1951 standard resolution test plate is chosen as the imaging target. The Abbe resolution limit is estimated to be δx ~12.5um calculated by,

$$\delta x \approx \frac{f}{w} \cdot \lambda \qquad (1)$$

where f = 100mm is the focal length of the objective lens, λ = 1560nm represents the center wavelength, and W =

12.5mm the input beam waist.

The results are shown below in Fig.3. As a comparison, image Fig.3.(a) was captured in the conventional illumination mode, and Fig.3.(b) by TS-SIM. As the results show, the resolution limit is tested to be around 13.9um (element 5-2 on test plate) for the conventional method, and around 9.84um (element 5-5) for TS-SIM, improving 40% of the resolution. Fig.3.(c) and (d) give out the line-scan waveform from the selected position in Fig.3.(a) and (b).

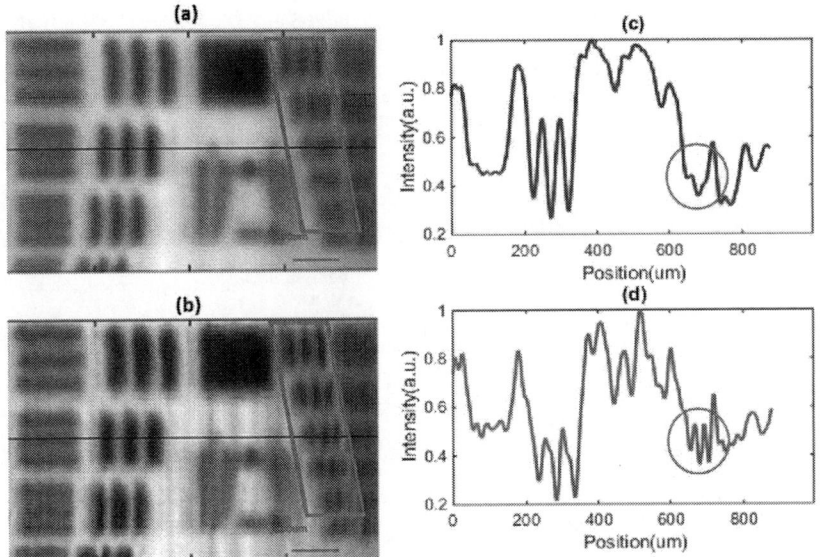

Fig. 3. Images obtained by TS-SIM and conventional line-scan microscopy for USAF1951 resolution test target.

CONCLUSIONS

In conclusion, we present a new method of generating illuminating pattern for SIM based on the time-stretch strategy. This method is capable to alter illuminating pattern at an ultra-high speed of 50MHz, 3 orders of magnitudes faster than contemporary methods. And we experimentally proved the TS-SIM system can help to improve axial resolution by a factor of 1.4. Moreover, there is a great potential to further improve the pattern generation rate by directly take use of pulse laser source with higher repetition rate. And besides super-resolution, this time-stretch pattern-generating strategy may also find its application for other purpose, such as optical sectioning, surface profiling or phase imaging, etc.

REFERENCES

[1] P. Kner, B.B.Chhun, E.R. Griffis, L. Winoto and M.G.L. Gustafsson, "Super-resolution video microscopy of live cells by structured illumination," Nature Methods, VOL.6 NO.5, MAY 2009.

[2] K.Goda, K.K.Tsia, and B.Jalali, "Serial time-encoded amplified imaging for real-time observation of fast dynamic phenomena", Nature, 458(7242), 1145–1149 (2009).

[3] M. G. L. GUSTAFSSON, "Surpassing the lateral resolution limit by a factor of two using structured illumination microscopy", Journal of Microscopy, Vol. 198, Pt 2, May 2000, pp. 82-87.

[4] A.C.S. CHAN, A.K.S. LAU, K.K.Y. WONG, E.Y. LAM, and K.K. TSIA, "Arbitrary two-dimensional spectrally encoded pattern generation—a new strategy for high-speed patterned illumination imaging", Optica, Vol. 2, No. 12, December 2015.

Time-Frequency Photonic Signal Processing

G. Cincotti[1], T. Konishi[2], T. Murakawa[2], T. Nagashima[2], S. Shimizu[3], M. Hasegawa[2],
K. Hattori[4], M. Okuno[4], S. Mino[4], A. Himeno[4], N. Wada[3], H. Uenohara[5]

(1) Engineering Department University Roma Tre, via V. Volterra 62, I-00143 Rome, Italy
(2) Graduate School of Engineering, Osaka University, Japan
(3) NICT 4-2-1, Nukui-Kitamachi, Koganei, Tokyo 184-8795, Japan
(4) NTT Electronics Co. Ltd. 6700-2 to,Naka-shi, Ibaraki,311-0122, Japan
(5)Tokyo Institute of Technology 4259 Nagatsuta, Midoriku, Yokohama 226-8503, Japan
gabriella.cincotti@uniroma3.it

Abstract: We demonstrate an ultimate flexible approach for simultaneous OFDM and N-OTDM transmission, using intermediate grids between time and frequency axes. We achieve open eye diagrams, and performance below the FEC limit in 89.2-km field-trial.
Keywords: Multiplexing; Optical communications; Coherent communications.

I. INTRODUCTION

Adaptive, efficient and flexible spectrum allocation, as well as agile bit-rate granularities are the key technologies to fully exploit the available bandwidth in a single wavelength, single core, single-mode fiber [1]. Orthogonal frequency division multiplexing (OFDM) and Nyquist-optical time division multiplexing (N-OTDM) are the competing approaches to generate Tb/s superchannels, multiplexing subcarriers in frequency, or interleaving short *sinc*-shaped pulses in time [2, 3]. Both methodologies have pros and cons, mainly related to physical impairments and implementation issues. For instance, OFDM is more resistant to linear effects, such as chromatic dispersion, since the transmitted symbol is stretched in time, but, with an increased peak-to-average power ratio (PAPR), it becomes more sensitive to nonlinear effects. In general when selecting a time or frequency multiplexing approach, a trade-off is achieved, sacrificing spectral efficiency to a higher capacity, or system performances to a reduced complexity. The ultimate efficiency in physical resource exploitation can be achieved only in truly flexible systems, where it is possible to switch from time to frequency multiplexing and vice versa, through intermediate grids.

The use of fractional Fourier transform (FrFT), instead of conventional Fourier transform (FT) subcarriers, allows us to introduce the largest degree of flexibility in optical networking, moving from time to frequency multiplexing and viceversa. OFDM and N-OTDM systems are implemented by the same TXs, and have different RXs. In the TX, arrayed-waveguide gratings (AWG) or a wavelength selective switches (WSS) are used as waveshapers. The proposed approach introduces ultimate flexibility allowing us to receive either OFDM subcarriers in parallel or a serial N-OTDM signal. We have already demonstrated the record generation of 4.88 ps sinc-shaped N-OTDM pulses, using the same WSS-based TX, and a dispersion compensating fiber (DCF) [4]; the WSS induces a quadratic phase modulation (QPM) on an input pulse, and the time-lens effect in the DCF compresses, properly shapes and delays the sinc pulses to generate a N-OTDM signal. An OFDM/OTDM approach has been recently proposed, based on time-lens effect and QPM [5], and we have demonstrated hybrid 4-ch multiplexing in a back-to-back setup, using an intermediate grid between time and frequency [6].

In the present paper, we report on the field trial of a 4-ch 40Gb/s hybrid OFDM/N-OTDM, over the 89.2-km JGN X test bed. We use WSSs both at the transmitter (TX) and receiver (RX), for system reconfigurability; compared to conventional OFDM experiments with the same architecture [7], we achieve a large PAPR reduction, without affecting the spectral efficiency and energy consumption per bit [8, 9].

II. HYBRID OFDM/N-OTDM APPROACH

In a conventional OFDM scheme, complex data s_n with high-order modulation are transmitted in parallel using N FT subcarriers, within a symbol of duration T

$$\phi_n^1(t) = s_n \cdot rect\left(\frac{t}{T}\right) \cdot e^{j2\pi\frac{n}{T}t} \quad n = 1,2,..,N. \tag{1}$$

If all the subcarriers are phase modulated, the theoretical PAPR upperbound is N [8]. This value can be reduced using FrFT subcarriers [9]

$$\phi_n^p(t) = s_n \cdot rect\left(\frac{t}{T}\right) \cdot e^{-j\pi\left\{\left[n^2\sin^2\left(p\frac{\pi}{2}\right)+\frac{t^2}{T^2}\right]\cot\left(p\frac{\pi}{2}\right)-2\frac{n}{T}t\right\}} \quad n = 1,2,..,N \tag{2}$$

where p is the fractional parameter ($p=1$ for conventional FT). The FrFT introduces a $p\pi/2$ rotation in the time-frequency plane that is compensated at the RX by applying a complementary $-p\pi/2$ rotation in the OFDM RX, as it is shown in Fig. 1 (a) [6]. The FrFT subcarriers are orthogonal over a symbol duration and their chirped behavior largely

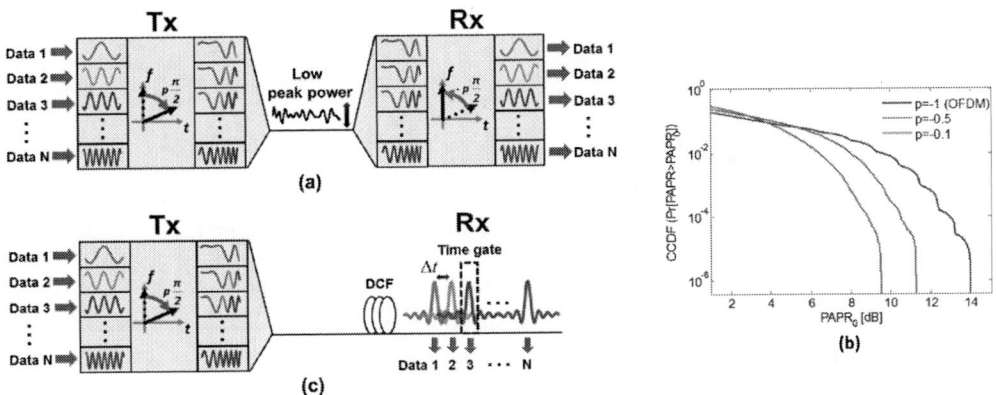

Fig. 1. Schematic diagram of (a) OFDM system, (b) PAPR performance (c) N-OTDM system.

reduces the PAPR, as it is shown in Fig. 1(b) [9]. The same WSS-based TX can be used also to generate a *sinc*-shaped N-OTDM signal, if it is followed by a DCF, with matched dispersion parameter $D=cT^2\tan(p\pi/2)/\lambda^2$, where c is the speed of light and λ the central wavelength (Fig. 1(c)) [4]. We observe that the WSS not only generates the *sinc*-shaped pulses, but it also multiplexes and properly delays the streams, without using any delay synchronization, splitter and coupler, as in conventional approaches [3]. In the N-OTDM RX, the WSS is replaced by time-gating modules (Fig. 1(c)).

III. FIELD TRIAL THROUGH 89.2-KM JGN-X TEST BED

Figure 2 shows the field trial for a 40 Gb/s hybrid OFDM/N-OTDM transmission through the 89.2-km JGN-X test bed. We take $p=1.5$ ($3\pi/4$ rotation in the time-frequency plane) at the TX, and $p=0.5$ ($\pi/4$ rotation) at the RX, to reduce the PAPR. Since the available WSS has only 4 input ports, we could transmit only 4 adjacent independent subcarriers (ch 1,2,3,4), with the maximum spectral efficiency; a large number of subcarriers can be generated using an AWG) [10]. The laser source is a 10-GHz mode locked laser diode (MLLD), with 1.5 ps pulse width (full width at half maximum) and 193.245 THz center frequency. The pulse train is differentially binary phase shift keying (DBPSK) modulated by 10 Gb/s pseudo random bit sequence (PRBS) with a pattern length of $2^{31}-1$. The 10 Gb/s DBPSK modulated signal is split into 4, and different delays are applied using optical delay lines (ODL), for synchronization and pattern decorrelation. The signals are fed to the input ports of the 4×1 WSS (Finisar 4000S) at the TX, to generate a 40 Gb/s hybrid OFDM/N-OTDM signal that is transmitted through the 89.2 km JGN-X test bed; dispersion is compensated by a dispersion compensation module (DCM). At the RX, a 1×1 WSS (Finisar 1000S) is used, and we suitably change the filtering function to separately receive the four subcarriers. Finally, the output signal is time-gated by a LN-IM. The eye diagrams of the four received subcarriers are shown in Fig. 3(a), and we observe a clear eye opening. Figure 3(b) shows the bit error rate (BER) measured as a function of the received power, which is largely below the FEC-limit (2×10^{-3}). The intermediate subchannels 2 and 3 present the worse performance, because they suffer for larger inter-carrier interference (ICI), due to the large time gating interval (~50 ps). System performances can be enhanced reducing the gating interval.

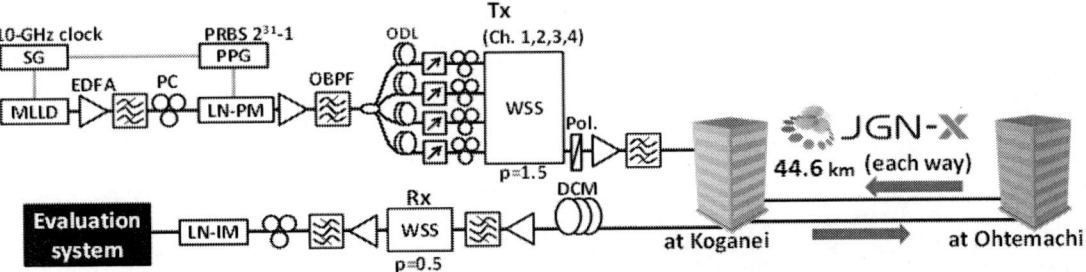

Figure 2 Experimental setup for DQPSK modulated OFDM transmission using 4 Fr-FT subcarriers.

107

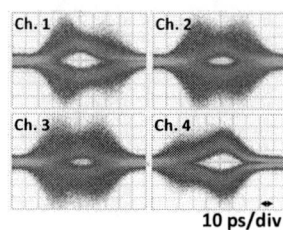 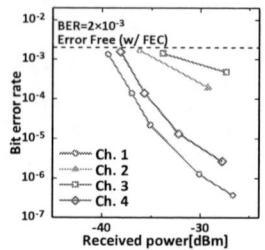

(a)   (b)

Figure 3 (a) Eye diagram of the received FrFT subcarriers after time gating (b) BER performance of the 4 FrFT subcarriers.

IV. HYBRID OFDM/N-OTDM EXPERIMENT

To demonstrate the hybrid OFDM/N-OTDM capability, we have used the same experimental setup of Fig. 2, without the test bed. In this case, we have transmitted 6 (over 12) Fr-FT subcarriers (ch 1,2,3,6,7,8), considering p=1.7884 at the TX to generated a 60 Gb/s hybrid OFDM/N-OTDM signal, whose spectrum is shown in Fig. 4(a). For the OFDM RX (Fig. 1(a)), we use a WSS with p=0.2116, and Fig. 4(b) shows the eye diagrams of 3 received subcarriers, after time-gating. For the N-OTDM RX, we replace the WSS with a 25-km single mode fiber (instead of a DCF) with dispersion parameter D=435 ps/nm. Fig. 4(e) shows the eye diagrams of the N-OTDM signal; the width of the *sinc*-shaped pulse is $T\tan(p\pi/2)$=34.5 ps.

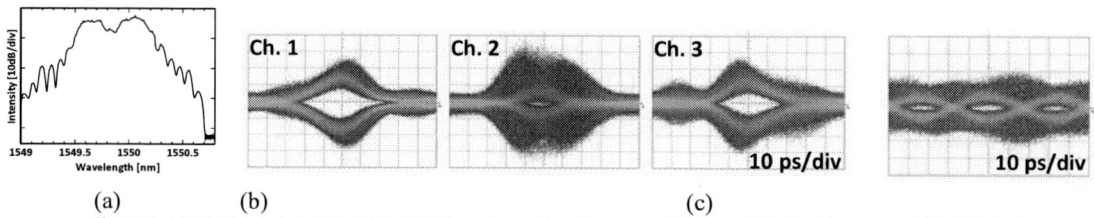

(a)   (b)   (c)

Figure 4. (a) 60 Gb/s hybrid OFDM/N-OTDM spectrum, Eye diagrams of (b) three OFDM subcarriers, (c) N-OTDM signal.

V. CONCLUSIONS

We have demonstrated a field trial over the 89.2 km JGN X test bed of a hybrid OFDM/N-OTDM system using an intermediate time/frequency grid. The new approach is less sensitive to nonlinearities, compared to conventional OFDM, without affecting the spectral efficiency and the BER performances, that are below the FEC limit. Ultimate multiplexing flexibility is also demonstrated receiving either OFDM subchannels in parallel, or N-OTDM pulses.

ACKNOWLEDGMENTS

This work was supported by STARBOARD project of MIC Strategic Harmonized International R&D Promotion Programme (SHIP), and the Italian Ministry of University and Research, ROAD-NGN project (PRIN2010).

REFERENCES

[1] M. Jinno, et al., "Spectrum-efficient and scalable elastic optical path network: architecture, benefits, and enabling technologies," IEEE Commun Mag., vol. 47, no. 11, pp. 66-73 (2009).

[2] D. Hillerkuss, et al., "26 Tbit s−1 line-rate super-channel transmission utilizing all-optical fast Fourier transform processing," Nat. Photon., vol. 5, pp. 364–371 (2011).

[3] M. Nakazawa et al., "Ultrahigh-speed "orthogonal" TDM transmission with an optical Nyquist pulse train," Opt. Express, vol. 20, pp.1129-1140 (2012).

[4] G. Cincotti et al. "Flexible power-efficient Nyquist-OTDM transmitter, using a WSS and time-lens effect," OFC 2015, paper W3C.5.

[5] A. Clausen et al., "All-optical signal processing of OTDM and OFDM signals based on time-domain optical Fourier transformation., " in ICTON 2014..

[6] T. Murakawa, et al., "Fractional OFDM transmitter and receiver for time/frequency multiplexing in gridless, elastic networks," in OFC 2015, paper W3C.

[7] L. B. Du et al., "Flexible all-optical OFDM using WSSs," in OFC 2013, paper pdp 5B.9.

[8] C. W. Korevaar, et al., "Peak-to-average power reduction by rotation of the time-frequency representation," in GROBECOM 2013, p. 3722.

[9] T. Nagashima, et al, "PAPR management of all-optical OFDM signal using fractional Fourier transform for fibre nonlinearity mitigation," 41th European Conference on Optical Communication (ECOC), Valencia, Spain 2015.

[10] G. Cincotti, "Optical signal processing using AWGs," in ECOC 2014, paper OTu2I.3.

MF1-2

All-Optical Nyquist-WDM to Nyquist-OTDM Conversion for Flexible Optical Networks

Satoshi Shimizu[1], Gabriella Cincotti[2], and Naoya Wada[1]

1: National Institute of Information and Communications Technology, 4-2-1, Nukui-Kitamachi, Koganei, Tokyo
2: Engineering Department, University Roma Tre, via V. Voltera 62, I-00143 Rome, Italy
sshimizu@nict.go.jp

Abstract: *2×6.25 and 2×12.5 Gbaud Nyquist-WDM signals are successfully converted to 12.5 and 25 Gbaud Nyquist-OTDM signals, respectively. The converted signals showed clear symbol separation in constellation and the OSNR penalty was around 3dB.*
Keywords: *Nyquist-WDM, Nyquist-OTDM, Flexible Optical Networks*

I. INTRODUCTION

For the full utilization of spectral resources in the optical fiber links, not only spectral efficient modulation techniques but also time and frequency division multiplexing techniques are intensively investigated. On the other hand, flexible and dynamic optical channel control is also an important technology for future optical network, where versatile communications are co-exist and handled on a same platform [1-3]. The optical Nyquist pulse coding is one of the most spectral efficient modulation techniques and both optical time division multiplex (Nyquist-OTDM) and wavelength division multiplex (Nyquist-WDM) are the promising technologies for spectral efficient large capacity transmission [4-5], although they have different aspects in the network applications. The Nyquist-OTDM is the better solution for an ultra-high capacity single channel transmission such as over 1 Tbit/s transport system, whereas the Nyquist-WDM is better for the high granular flexible channel control, such as add-drop operation at the network node [3]. For future highly flexible optical network, they are expected to co-exist in the same networks, and seamless conversion between time and frequency domain is needed, to achieve both large capacity transport and high granular channel control. In our previous work, we have demonstrated an all-optical multiplexing(MUX)-format conversion from Nyquist-OTDM to Nyquist-WDM, by using a simple optical phase modulator and an optical Nyquist filter [6].

In this paper, we demonstrate the inverse conversion of MUX-format, where Nyquist-WDM is converted to Nyquist-OTDM, by using the same configuration as in the previous work; an optical phase modulator and an optical Nyquist filter. A Nyquist-WDM signal is phase modulated by a sinusoidal wave with the same frequency of the signal baud-rate. The phase modulation broadens the signal spectrum generating the multi-copies of Nyquist signal spectrum. The optical Nyquist filter, which has the same bandwidth as the original Nyquist-WDM signal, reshapes the spectrum extending the bandwidth of each channel to the double. Aligning the timing between channels carefully to keep the time domain orthogonality, the signal becomes a Nyquist-OTDM signal. In the experiments, 2 x 6.25 Gbaud and 2 x 12.5 Gbaud Nyquist-WDM signals are converted to 12.5 Gbaud and 25 Gbaud Nyquist-OTDM signals, respectively. The error-vector-magnitude (EVM) was measured for both WDM and OTDM signals and the optical signal-to-noise ratio (OSNR) penalties were 2 and 3 dB.

II. EXPERIMENTAL SETUP AND OPERATION PRINCIPLE

Figure 1 shows the experimental setup of Nyquist-WDM signal transmitter and MUX-format converter. A 2 x 6.25 Gbaud Nyquist-WDM signal is generated by using arbitrary waveform generator (AWG) and an optical IQ-modulator, whereas carrier-suppressed return-to-zero (CS-RZ) modulation and optical Nyquist filtering are used to generate 2 x 12.5 Gbaud Nyquist-WDM signal [7], as shown in Fig. 1(a) and (b), respectively. The sampling-rate and analogue bandwidth of the AWG are 12 GSa/s and 5 GHz, respectively, and the analogue frequency response is digitally compensated. The modulation format is quadrature phase-shift-keying (QPSK) and the data sequences for I- and Q-channels are 2^{23}-1 and 2^{15}-1 pseudo random binary sequences (PRBSs), respectively. In both two cases, the frequency spacing of two laser light sources is set to exactly the same as the signal baud-rate frequencies; there is no guard-interval between channels. The generated Nyquist signals are multiplexed by a 3 dB coupler, after aligning their polarizations by individual polarization controllers. The multiplexed signal is sent to the MUX-format converter. The input signal is phase-modulated by a sinusoidal wave to broaden the signal spectrum, generating the multi-copies of Nyquist spectrum, as shown in Fig. 1(c). The following optical Nyquist filter, which has the same bandwidth and center frequency as the input Nyquist-WDM signal, filters the overlapping-band of two channels. The applied modulation voltage for the phase-modulator is 0.913 V_π where center three tones show the same intensity level [6]. In this way, the bandwidth of each WDM channel is extended to the double and therefore the symbol Nyquist pulse is temporally shrunk to the half of the original one, keeping the time domain orthogonality, as illustrated in the bottom of Fig. 1(c). According to the phase modulation theory, the lower-frequency side-band has inverted phase with respect to the higher-frequency side-band. Therefore, Nyquist pulses of two channels output from the optical Nyquist filter have half-symbol

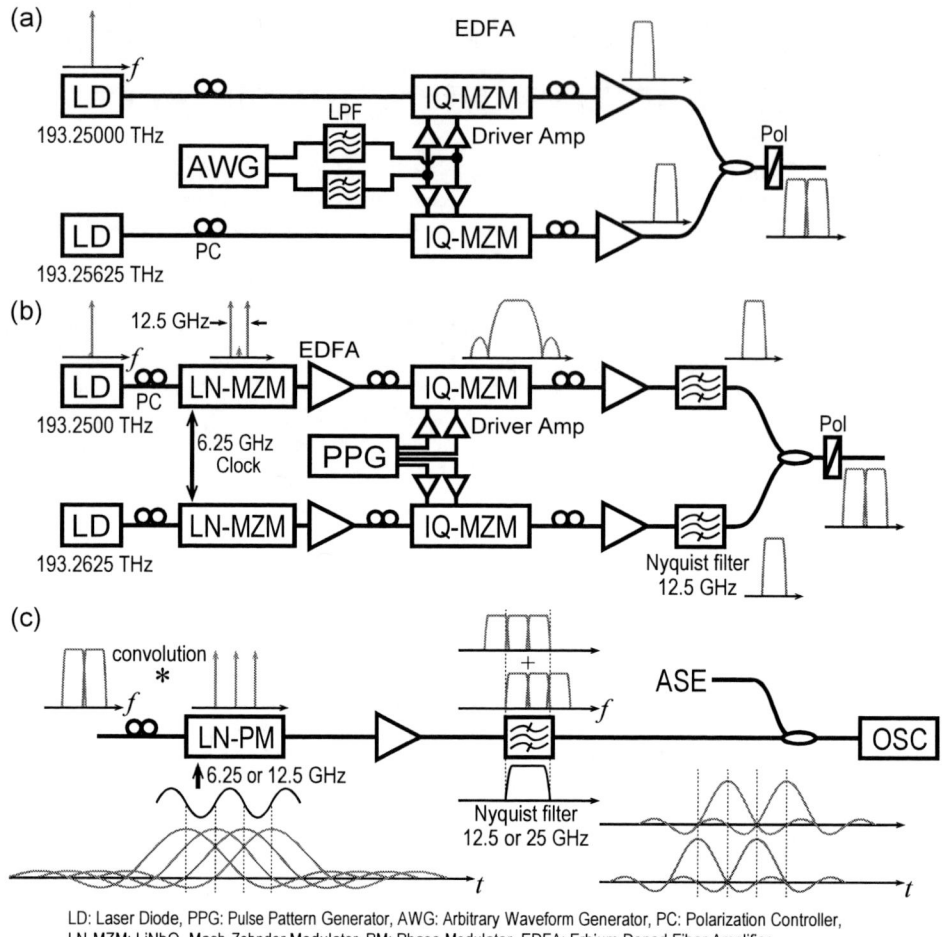

LD: Laser Diode, PPG: Pulse Pattern Generator, AWG: Arbitrary Waveform Generator, PC: Polarization Controller,
LN-MZM: LiNbO₃ Mach-Zehnder Modulator, PM: Phase Modulator, EDFA: Erbium Doped Fiber Amplifier,
POL: Polerizer, VOA: Variable Optical Attenuator, FE: Front End, ASE: Amplified Spontaneous Emission

Fig. 1. Experimental setup for (a) 2 x 6.25 Gbaud Nyquist-WDM transmitter, (b) 2 x 12.5 Gbaud Nyquist-WDM transmitter, and (c) Nyquist-WDM to Nyquist-OTDM converter.

offset with each other, and they are in orthogonal relation in time domain. In this way, the output signal becomes a Nyquist OTDM signal. To evaluate the system operation, we have measured the EVM by using optical modulation analyzer (OMA), which has 13 GHz electrical bandwidth and 40 GSa/s sampling rate.

III. RESULTS

Figure 2 shows the spectra and eye-diagrams of the transmitted Nyquist-WDM signals and the received Nyquist-OTDM signals. The eye-diagrams of Nyquist-WDM signals are individually measured in single channel condition. The Nyquist-WDM signals are successfully converted to a single Nyquist signal with double the baud-rete of original signal.

Figure 3 shows the measured EVMs as functions of OSNR and the insets are constellation plots for Nyquist-WDM and -OTDM signals. The channels of Nyquist-WDM signal are digitally de-multiplexed after receiving the signal by OMA. In both conversion cases, the constellation shows clear separations of the symbol points. In the EVM measurements, the Nyquist-OTDM signals show 2 and 3 dB OSNR penalties for 12.5 Gbaud and 25 Gbaud signal, respectively. We think these penalties stem from the timing jitter between the Nyquist-WDM signal and the clock signal for the phase-modulator, and the effect of timing jitter is more serious for the high baud-rate signal.

IV. SUMMARY

We have experimentally demonstrated all-optical multiplexing format conversion from Nyquist-WDM to Nyquist-OTDM by using a simple phase modulator and an optical Nyquist filter, which is the same as the one used in the inverse conversion [6]. The phase-modulator broadens the signal spectrum generating the multi-copies of the Nyquist spectrum, and the optical Nyquist filter extract bandwidth-doubled Nyquist signal for each WDM channel. Keeping the time domain orthogonality between channels, the output from the Nyquist filter becomes the Nyquist-OTDM signal. The spectra and eye-diagrams show that 2 x 6.25 Gbaud and 2 x 12.5 Gbaud Nyquist-WDM signals are successfully converted to 12.5 Gbaud and 25 Gbaud Nyquist-OTDM signals, respectively. The EVM measurement revealed that the converted OTDM signals showed clear symbol separation with 2 and 3 dB OSNR penalties for each case.

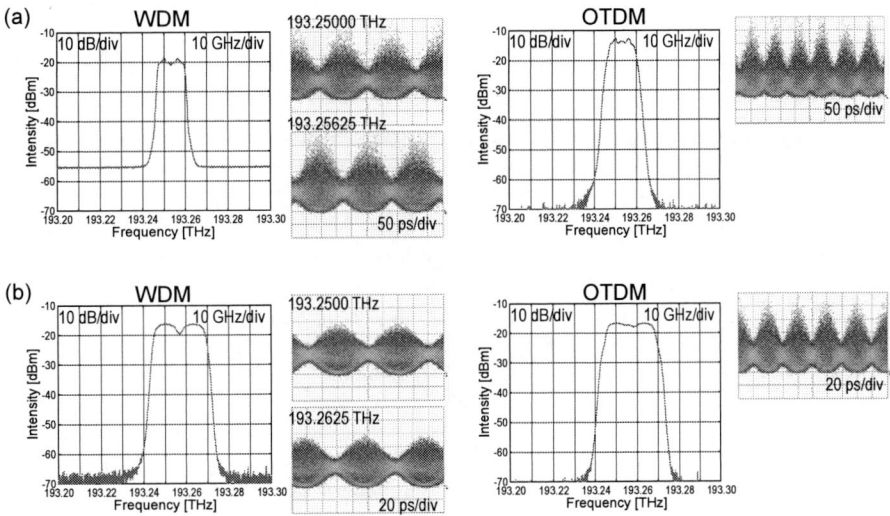

Fig. 2. The measured spectra and eye-diagrams of (a) 2 x 6.25 Gbaud to 12.5 Gbaud and (b) 2 x 12.5 Gbaud to 25 Gbaud conversion.

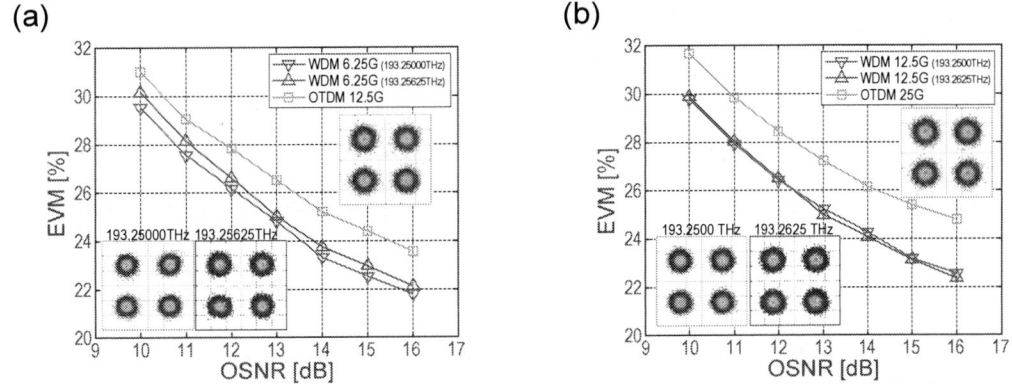

Fig. 3. The measured EVMs as functions of OSNR. (a) 2 x 6.25 Gbaud to 12.5 Gbaud and (b) 2 x 12.5 Gbaud to 25 Gbaud conversion.

REFERENCES

[1] O. Gerstel, M. Jinno, A. Lord, and S. J. Ben Yoo, "Elastic Optical Networking: A New Dawn for the Optical Layer?," IEEE Communications Magazine, S12-S20, February 2012.

[2] N. Amaya, M. Irfan, G. Zervas, K. Banias, M. Garrich, I. Henning, D. Simeonidou, Y. R. Zhou, A. Lord, K. Smith, V. J. F. Rancano, S. Liu, P. Petropoulos, D. J. Richardson, "Gridless Optical Networking Field Trial: Flexible Spectrum Switching, Defragmentation and Transport of 10G/40G/100G/555G over 620-km Field Fiber," European Conference on Optical Communication 2011 (ECOC2011), Th.13.K.1, 2011.

[3] S. Satoshi, G. Cincotti, and N. Wada, "Demonstration of Multi-hop Optical Add-Drop Network with High Frequency Granular Optical Channel Defragmentation Nodes," Optical Fiber Communication Conference 2015 (OFC2015), M2I.4, 2015.

[4] M. Nakazawa, T. Hirooka, P. Ruan, and P. Guan, "Ultrahigh-speed "orthogonal" TDM transmission with an optical Nyquist pulse train," Opt. Express 20, 1129-1140, 2012.

[5] J. Yu, Z. Dong, H. Chien, Z. Jia, D. Huo, H. Yi, M. Li, Z. Ren, N. Lu, L. Xie, K. Liu, X. Zhang, Y. Xia, Y. Cai, M. Gunkel, P. Wagnel, H. Mayer, and A. Schippel, "Field Trial Nyquist-WDM Transmission of 8×216.4Gb/s PDM-CSRZ-QPSK Exceeding 4b/s/Hz Spectral Efficiency," Optical Fiber Communication Conference 2012 (OFC2012), PDP5D.3, 2012.

[6] S. Shimizu, G. Cincotti, and N. Wada, "All-Optical Nyquist-OTDM to Nyquist-WDM conversion for Highly Flexible Optical Networks," Optical Fiber Communication Conference 2016 (OFC2016), W3D.1, 2016.

[7] S. Shimizu, G. Cincotti, and N. Wada, "Demonstration of No-Guard-Interval 6 x 25 Gbit/s All-Optical Nyquist WDM System for Flexible Optical Networks by using CS-RZ signal and Optical Nyquist Filtering," OptoElectronics and Communications Conference 2014 (OECC2014), MO1B-1, 2014.

Optical Serial-to-Parallel Conversion based on Fractional OFDM scheme

M. Hiraoka[1], T. Nagashima[1], B. Karanov[1], G. Cincotti[2], S. Shimizu[3], T. Murakawa[1], M. Hasegawa[1], K. Hattori[4], M. Okuno[4], S. Mino[4], A. Himeno[4], N. Wada[3], H. Uenohara[5], and T. Konishi[1]

(1) Graduate school of Engineering, Osaka University, Suita, Osaka 565-0871, Japan
(2) Engineering Department, University Roma Tre, via V. Volterra 62, I-00143 Rome, Italy
(3) NICT 4-2-1, Nukui-Kitamachi, Koganei, Tokyo 184-8795, Japan
(4) NTT Electronics Co. Ltd. 6700-2 to, Naka-shi, Ibaraki, 311-0122, Japan
(5) Tokyo Institute of Technology 4259 Nagatsuta, Midoriku, Yokohama 226-8503, Japan
hiraoka@photonics.mls.eng.osaka-u.ac.jp

Abstract: We have proposed a Fr-OFDM based serial-to-parallel conversion scheme and experimentally verified the preliminary operation using a 10 GHz serial signal. The experimental results show that a serial data is successfully received in parallel.
Keywords: Serial-to-Parallel Conversion, Fractional Fourier Transform, OFDM, N-OTDM

I. INTRODUCTION

Serial-to-parallel conversion is an indispensable interface technique for bandwidth matching between broad and narrow bandwidth systems [1-5]. Time-stretch technique using a chirped optical pulse [1,2] or a colored pulse train [3-5] has been proposed for serial-to-parallel conversion in front of an ultrawide-band photonic time-stretch analog-to-digital (A/D) converter. They can be regarded as a kind of wavelength division multiplexing (WDM) techniques and such WDM-based approaches have a potential to upgrade their conversion capacity by using a much denser orthogonal frequency division multiplexing (OFDM) scheme. A Nyquist pulse train is the densest serial signal [6] and a time lens technique based on four wave mixing (FWM) has been proposed for format conversion between high-speed serial Nyquist optical time division multiplexing (N-OTDM) signals and several lower rate WDM ones [7]. Fractional Fourier transform (FrFT) allows us to introduce the largest degree of flexibility in an intermediate domain between time and frequency [8-12] and it can exploit new multiplexing scheme, Fractional OFDM (Fr-OFDM). Our proposed Fr-OFDM scheme can realize Nyquist OTDM-to-OFDM conversion and vice versa without any nonlinear signal processing and it is equivalent to serial-to-parallel conversion. We have already demonstrated the generation of 4.88 ps *sinc*-shaped N-OTDM pulses, using a wavelength selective switch (WSS), and a dispersion compensating fiber (DCF) [10]; the WSS induces a quadratic phase modulation (QPM) on an input pulse, and the time-lens effect in the DCF compresses, properly shapes and delays the *sinc*-shaped pulses to generate a N-OTDM signal.

In this paper, we propose Fr-OFDM based serial-to-parallel conversion scheme and verify the preliminary operation using a 10 GHz serial signal. Our proposed Fr-OFDM based serial-to-parallel conversion is expected to incorporate both benefits of previous serial-to-parallel conversion techniques.

II. PROPOSED FR-OFDM BASED SERIAL-TO-PARALLEL CONVERSION

Figure 1 shows a schematic diagram of a system of Fr-OFDM based serial-to-parallel conversion which can convert a *sinc*-shaped N-OTDM serial signals into Fr-OFDM parallel signals using Fr-OFDM based transmitter (Tx) and receiver (Rx). The impulse response h_n^p along with time axis and the corresponding transfer function H_n^p along with frequency axis of the nth Fr-OFDM subcarrier are given by,

$$h_n^p(t) = \left| \sin\left(p\frac{\pi}{2} \right) \right|^{-\frac{1}{2}} \cdot e^{j\frac{\pi}{4}\left\{ p - \text{sign}\left[\sin\left(p\frac{\pi}{2} \right) \right] \right\}} \cdot e^{j\pi\left[\left\{ \frac{t^2}{N^2} + n^2 \sin^2\left(p\frac{\pi}{2} \right) \right\} \cot\left(p\frac{\pi}{2} \right) - \frac{2nt}{T} \right]}, \quad (1)$$

$$H_n^p(f) = \left| \cos\left(p\frac{\pi}{2} \right) \right|^{-\frac{1}{2}} \cdot e^{j\frac{\pi}{4}\left\{ p + \text{sign}\left[\cot\left(p\frac{\pi}{2} \right) \right] \right\}} \cdot e^{-j\pi\left[\left\{ n^2 \sin^2\left(p\frac{\pi}{2} \right) \right\} \tan\left(p\frac{\pi}{2} \right) + 2nfT \tan\left(p\frac{\pi}{2} \right) \right]} \cdot e^{-j\pi\left\{ f^2 T^2 \tan\left(p\frac{\pi}{2} \right) \right\}}, \quad (2)$$

where t is time, f is frequency, T is the symbol duration, and p is the fractional parameter ($p=1$ for standard FT), N is the total number of subcarriers, respectively [8-10]. While a conventional Fourier Transform (FT) performs a $\pi/2$ rotation in the time-frequency plane, the FrFT makes a rotation of an intermediate angle θ ($=p\pi/2$), as shown in Fig. 1. In particular, the rotation angles corresponding to $p=2m$ and $p=2m+1$ ($m=0,1,2,...$) are the time and frequency axes, respectively. The QPM factor in H_n^p induces a chirp effect and the effect is reversibly convertible with dispersion through fiber transmission. According to Eq. (2), OFDM subcarriers can be converted into a *sinc*-shaped N-OTDM pulse train via Fr-OFDM after passing through a dispersive fiber, with matched dispersion parameter $D=f^2T^2\tan(p\pi/2)$

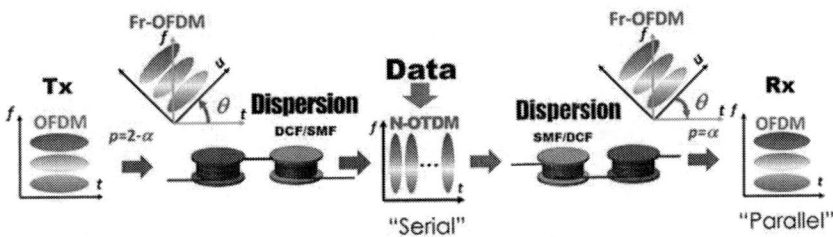

Fig. 1 Schematic diagram of FrFT based serial-to-parallel conversion: A *sinc*-shaped N-OTDM pulse train is prepared for optical sampling. Data is superimposed on the prepared optical sampling pulse and parallel signals are provided after FrFT based serial-to-parallel conversion.

(left hand in Fig. 1) [11, 12]. This N-OTDM pulse train is used as an optical sampling pulse train. A data signal is superimposed by modulating the optical sampling pulse train. The modulated optically sampled serial signal can be reversely converted into the modulated OFDM subcarriers via Fr-OFDM after passing through another dispersive fiber, with matched opposite dispersion parameter $D'=-f^2T^2\tan(p\pi/2)$ (right hand in Fig. 1). A WSS can work as Fr-OFDM based Tx and Rx as well as arrayed waveguide grating (AWG) based optical filters [13]. It doesn't only generate the *sinc*-shaped pulses, but it also multiplexes and properly delays the streams, without using any delay synchronization, splitter and coupler, as in conventional approaches [9].

III. EXPERIMENTAL VERIFICATION USING 10GHz SERIAL DATA SIGNAL

Figure 2 (a) illustrates the experimental setup for the proposed Fr-OFDM based serial-to-parallel conversion. A mode-locked laser diode (MLLD) was used as the seed pulse light source; the repetition frequency and the center wavelength are 10 GHz and 1540 nm, respectively. For preparation of an optical sampling pulse train, Fr-FT parameter p and fiber dispersion D corresponding p was to be -0.156 and 319 ps/nm, respectively. The number of subcarriers was set as three and three subcarriers are converted into triple *sinc*-shaped pulses with a 25 ps interval for optical sampling. An oscilloscope was used for measuring a seed optical pulse and temporal waveform of the triple *sinc*-shaped pulses, as shown in Fig. 2 (b) and (c), respectively. We noted that distortion of the waveform in Fig. 2 (b) and (c) are attributed to responsibility of a photo detector. Optical delay lines (ODLs) were used in order to adjust the delay of the three seed pulses. These seed optical pulses were filtered by WSS (Finisar 4000S) based on Fr-OFDM scheme; the filter combines Fr-OFDM filters and fiber dispersion. A 10 GHz sine wave as a data signal drives a LiNbO$_3$ intensity modulator (LN-IM) and it is temporally adjusted so as to reduce signal intensity of the 2nd pulse corresponding to the 2nd OFDM subcarriers. The modulated optically sampled triple *sinc*-shaped pulses can be reversely converted into the modulated OFDM subcarriers via Fr-OFDM. We used WSS (Finisar 1000S), and changed the filter function to separately receive three modulated OFDM subcarriers. For serial-to-parallel conversion, a modulated signal is filtered by a corresponding FrFT function. Figure 3 (a), (b), and (c) illustrate temporal waveforms of 3 channels before 10 GHz modulation, respectively. The arrows show the position of the time gate at the Rx and three signals have a certain level within the time gate before 10GHz modulation. Figure 3 (d), (e), and (f) illustrate temporal waveforms of 3 channels after 10 GHz modulation, respectively. After 10 GHz modulation, only the 2nd channel's signal intensity clearly reduces in comparison with those of the other channels, while cross-talk due to an inter channel interference degrades their signal separation quality a little bit. From these preliminary experimental results, it can be seen that a serial data was successfully converted into a corresponding parallel data and they are received at each channel of the Fr-OFDM based Rx.

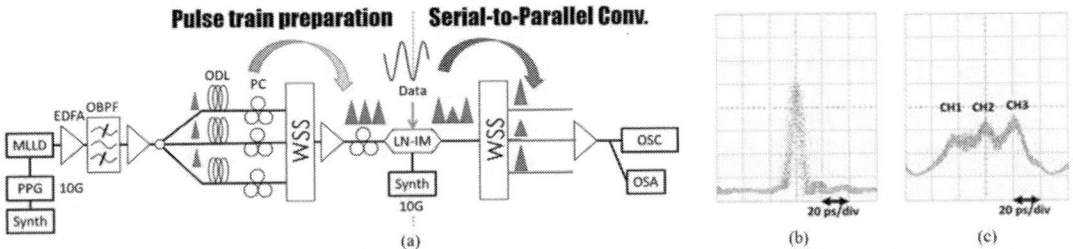

Fig. 2 Experimental verification; (a) experimental setup for the proposed Fr-OFDM based serial-to-parallel conversion. MLLD: Mode-Locked Laser Diode, PPG: Pulse Pattern Generator, Synth: Synthesizer, EDFA: Erbium-Doped Fiber Amplifier, OBPF: Optical Band-Pass Filter, ODL: Optical Delay Line, PC: Polarization Controller, WSS: Wavelength Selective Switch, LN-IM: LiNbO$_3$ Intensity Modulator, OSC: Oscilloscope, OSA; Optical Spectrum Analyzer, (b) measured waveform of seed optical pulse, and (c) measured waveform of triple *sinc*-shaped optical pulses.

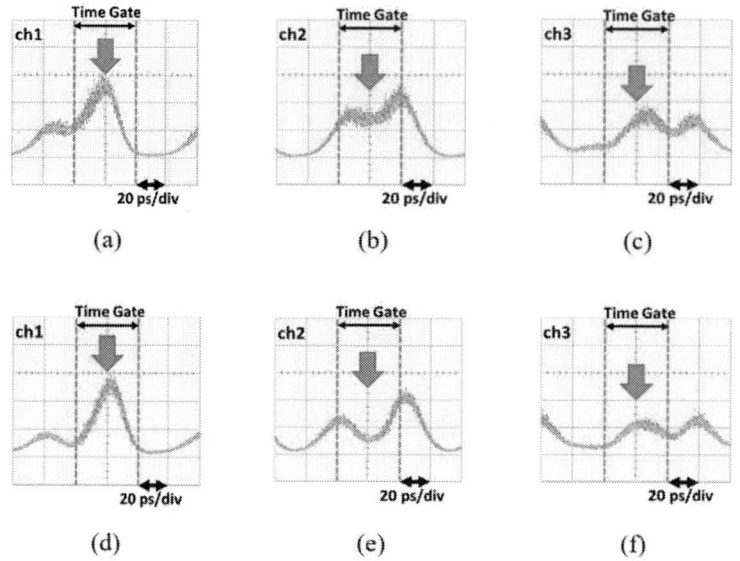

(a) (b) (c)

(d) (e) (f)

Fig. 3 Experimental results of the proposed serial-to-parallel conversion scheme using a 10 GHz serial signal; unmodulated waveforms at (a) 1st channel, (b) 2nd channel, and (c) 3rd channel, respectively. Modulated waveforms at (d) 1st channel, (e) 2nd channel, and (f) 3rd channel, respectively. The arrows show the position of a time gate at each channel.

IV. CONCLUSION

We have proposed a Fr-OFDM based serial-to-parallel conversion scheme and verified the preliminary operation using a 10 GHz serial signal. To upgrade its performance including sampling number furthermore, a cross-talk due to an inter channel interference should be carefully removed, which degrades the signal separation quality. Since our proposed scheme can theoretically achieve the densest optical sampling because it utilizes a Nyquist optical pulse as an optical sampling pulse in time domain. In addition, our proposed Fr-OFDM based serial-to-parallel conversion scheme is very power efficient because it does not require any nonlinear signal processing. Such an advantage would introduce the largest degree of flexibility in bandwidth matching interface techniques for optical signal processing.

ACKNOWLEDGMENT

This work was supported by STARBOARD project of MIC Strategic Harmonized International R&D Promotion Programme (SHIP), and the Italian Ministry of University and Research, ROAD-NGN project (PRIN2010).

REFERENCES

[1] Y. Han, et al. "Photonic time-stretched analog-to-digital converter: Fundamental concepts and practical considerations," Journal of Lightwave Technology, vol. 21, no. 12, pp. 3085-3103 (2003).

[2] Ali M. Fard, et al., "Photonic time‐stretch digitizer and its extension to real‐time spectroscopy and imaging," Laser & Photonics Reviews, vol. 7, no. 2, pp. 207-263 (2013).

[3] A. Khilo, et al., "Photonic ADC: overcoming the bottleneck of electronic jitter," Opt. Exp., vol. 20, no. 4, pp. 4454-4469 (2012).

[4] Franz X. Kärtner, et al., "Progress in Photonic Analog-to-Digital Conversion," OFC 213, paper OTh3D.5.

[5] Amir H. Nejadmalayeri, et al., "Attosecond Photonics for Optical Communications," OFC 2012, paper OM2C.1.

[6] M. Nakazawa, et al., "Ultrahigh-speed "orthogonal" TDM transmission with an optical Nyquist pulse train," Opt. Exp., vol. 20, pp.1129-1140 (2012).

[7] E. Palushani, et al. "OTDM-to-WDM conversion based on time-to-frequency mapping by time-domain optical Fourier transformation," IEEE Journal of Selected Topics in Quantum Electronics, vol. 18, no. 2, pp. 681-688 (2012).

[8] G. Cincotti et al. "Optical OFDM based on the fractional Fourier transform," PS 2014, paper PTIB.3.

[9] T. Murakawa, et al., "Fractional OFDM transmitter and receiver for time frequency multiplexing in gridless, elastic networks," OFC 2015, paper W3C.1.

[10] G. Cincotti et al. "Flexible power-efficient Nyquist-OTDM transmitter, using a WSS and time-lens effect," OFC 2015, paper W3C.5.

[11] T. Konishi, et al., "Fractional-OFDM Transmission for Time/Frequency Multiplexing in Elastic Networks," ACP 2015, paper ASu4H.1.

[12] T. Nagashima, et al., "PAPR management of all-optical OFDM signal using fractional Fourier transform for fibre nonlinearity mitigation," ECOC 2015, P. 5. 11 (Valencia, Spain, 27 Sept. - 1 Oct. 2015).

[13] G. Cincotti et al. "OFDM to OTDM conversion by optical fractional Fourier transform," APC 2014, paper PT1B.3.

MF1-4

Modulation Format Conversion from QPSK to 16QAM Using Delay Line Interferometer and Spectral Shaping Filter

Kazuya Mori, Hiroki Kishikawa and Nobuo Goto
Department of Optical Science and Technology, Tokushima University
2-1 Minamijosanjima-cho, Tokushima 770-8506, Japan
kishikawa.hiroki@tokushima-u.ac.jp

Abstract: We propose an all-optical modulation format conversion from QPSK to 16QAM based on the coherent superposition using a delay line interferometer. BER performance is assessed numerically by signal power, OSNR, and signal linewidth as parameters.
Keywords: optical processing, modulation format, QPSK, 16QAM,

I. INTRODUCTION

Conversion between different levels of multi-level modulation format is one of the promising functions to realize adaptive modulation and demodulation for elastic and flexible networking with efficient use of the fiber resources. Various all-optical techniques have been studied for the conversion from lower-order to higher-order modulation formats to increase spectral efficiency (SE) [1,2]. Among different m-ary phase-shift keying (PSK) formats, the authors proposed a passive interference method to convert from binary PSK (BPSK) to quadrature PSK (QPSK) [3] and the same principle was further applied to convert to quadrature amplitude modulation (QAM) in [4]. In order to increase SE on this conversion principle, occupied bandwidth should be reduced accordingly to the converted format. We have studied bit error rate (BER) performances using spectral shaping filters [5] after the conversion to reduce the bandwidth.

In this paper, we apply the format conversion method with spectral shaping filter for QPSK to 16QAM conversion. BER performances are numerically verified with signal power, optical signal to noise ratio (OSNR) and laser linewidth as parameters.

II. CONVERSION PRINCIPLE

Fig. 1 illustrates the schematic conversion principle of the modulation format convert from QPSK to 16QAM, in which two QPSK signals having a certain level of amplitude and a twice amplitude are coherently superposed and converted to a 16QPSK signal. In order to realize the scheme, we use a delay line interferometer (DLI) which is usually employed for receiving differential PSK signals.

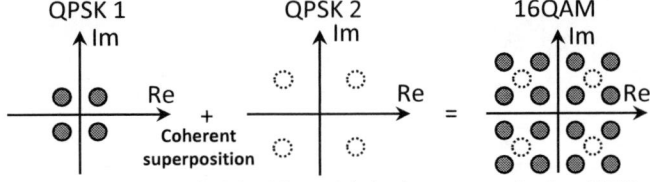

Fig. 1. Schematic conversion principle of the modulation format conversion from QPSK to 16QAM

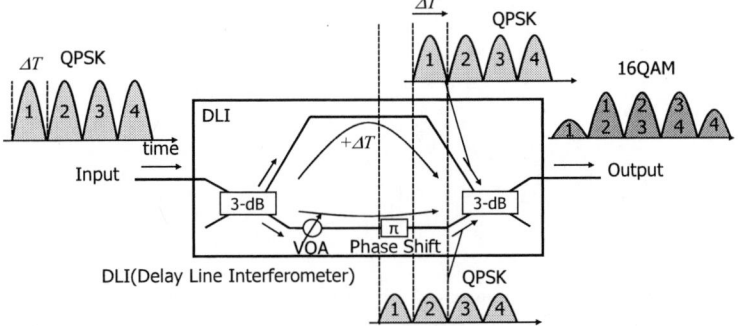

Fig. 2. Schematic diagram of the conversion using delay line interferometer.

This research was supported in part by JSPS KAKENHI (15H06443) and The Nakajima Foundation.

OECC/PS2016

Fig. 2 shows how to achieve the conversion using DLI. For instance, when temporally arranged four QPSK pulses with symbol rate of $1/\Delta t$ are incident into the DLI with free spectral range of $1/\Delta t$, pulses going through the upper side arm experience time delay of Δt, and pulses going through the lower side arm are attenuated and phase adjusted by π. Then, these pulses are coherently superposed by the latter 3-dB coupler, thereby being converted to 16QAM stream. The 16QAM stream includes both essential pulses for demodulation as #1&2 and #3&4, and informationally redundant unwanted pulses as #1, #2&3, and #4. Therefore, we eliminate the unwanted pulses by using additional intensity modulator. Then, we use a Gaussian optical filter as a spectral shaping filter to reduce the occupied bandwidth of the converted 16QAM signal to a half of the QPSK signal. Detailed filter design is described in [5].

III. NUMERICAL SIMULATION

We evaluate the conversion performance numerically using Optisystem (Optiwave Systems Inc.). The conversion system shown in Fig. 3 consists of three stages as QPSK signal generation, QPSK to 16QAM format conversion and 16QAM signal receiver. A QPSK signal at 28 Gbaud is generated using a 0 dBm laser source at f=193.2 THz with a pseudorandom binary sequence (PRBS) of 2^{15} -1 at bit rate of R_0=56 Gb/s. Then, the signal is amplified and the ASE

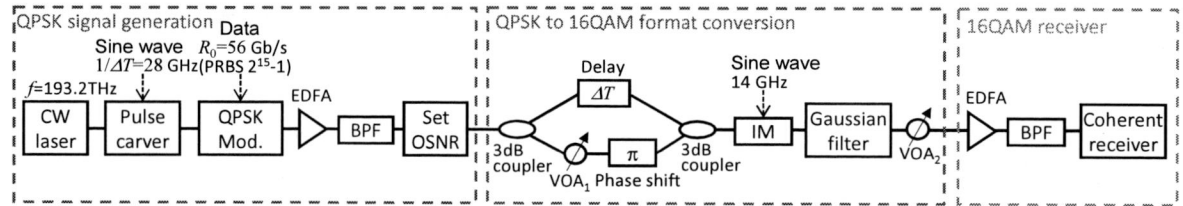

Fig. 3. Simulation setup

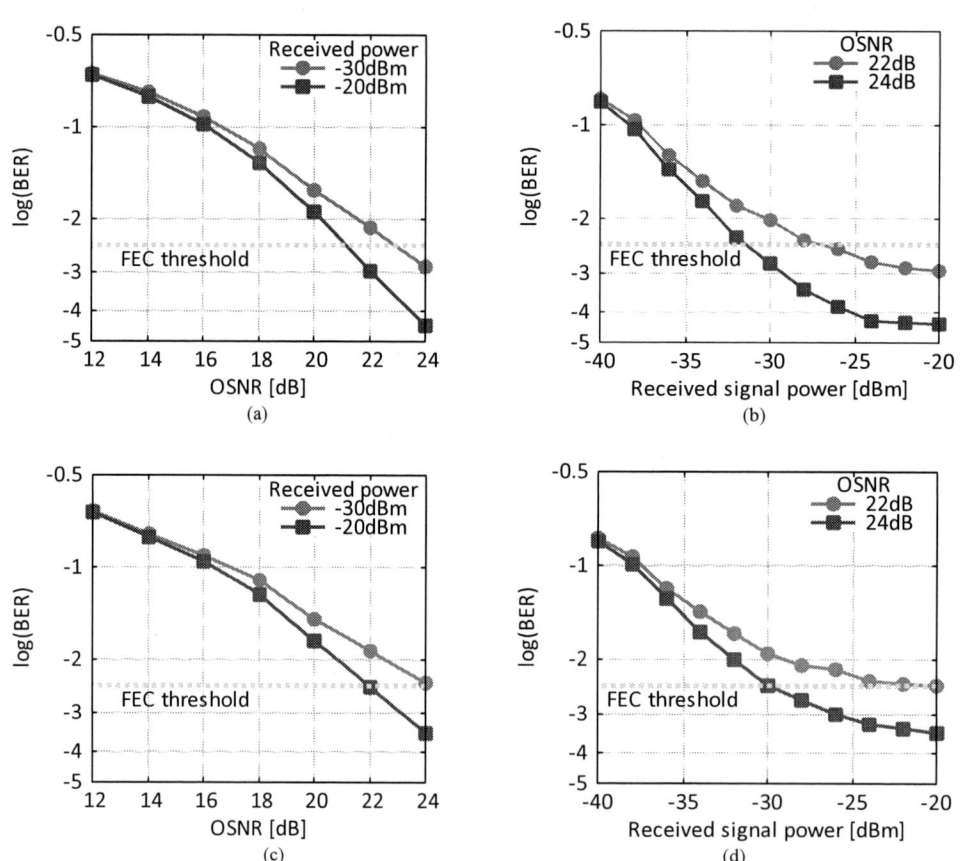

Fig.4 Numerical simulation results; the BER performances as a function of (a) OSNR and (b) received signal power with linewidth of 0 Hz, and (c) OSNR and (d) received signal power with linewidth of 10 MHz.

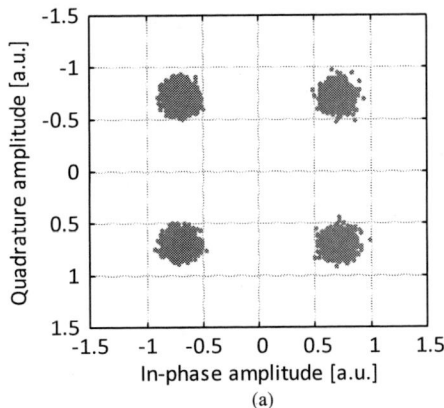 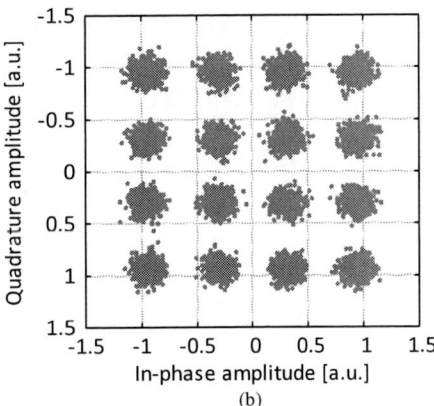

Fig.5 Constellation diagrams of (a) the original QPSK signal and (b) the converted 16QAM signal. The OSNR of original QPSK signal is 24dB.

noise is added for BER calculation. The converted 16QAM signal is generated at the DLI and unwanted pulses are suppressed by the intensity modulator (IM). The bandwidth of the 16QAM signal is reduced to a half by the following Gaussian filter. The received power is adjusted by the variable optical attenuator (VOA$_2$). Then the 16QAM signal is received coherently with a self-homodyne local oscillator of 5 dBm. In this simulation, we use a simple hard decision method for demodulating the 16QAM signal because we assume only the ASE noise, shot noise and thermal noise are sources of signal degradation. Therefore, we do not employ digital demodulation techniques such as polarization demultiplexing, frequency offset error correction, carrier phase recovery, and so on. The BER is calculated by directly counting bit errors over 2^{15} bits.

Fig. 4(a) shows BER as a function of OSNR of the original QPSK signal with the received signal power as a parameter. The linewidth of the CW laser is set to 0 Hz. It is found that higher OSNR shows better BER and also higher received signal power shows lower BER. Both BER curves are below the reference threshold corresponding to 7% overhead hard decision forward error correction (FEC) threshold of $\log_{10}(3.8\text{x}10^{-3}) = -2.42$, thus representing error-free conversion. Fig. 4(b) shows BER as a function of the received signal power with OSNR of the original QPSK signal as a parameter. It is also found that higher received power and higher OSNR show better BER values. In Figs. 4(c) and 4(d), similar but slightly higher BER are achieved when the laser linewidth is set to 10 MHz. Sample constellation diagrams of the original QPSK signal and the converted 16QAM signal at OSNR=24dB are plotted in Figs. 5(a) and 5(b) in which each average amplitude is normalized to 1. It is found that 16QAM constellation is successfully achieved.

IV. CONCLUSION

We have proposed an all-optical modulation format conversion from QPSK to 16QAM based on coherent superposition by using DLI. We numerically demonstrate the system performance by BER as functions of OSNR and received power and by constellation diagrams. We will consider experimental verification as a future work.

REFERENCES

[1] K. Mishina, A. Maruta, S. Mitani, T. Miyahara, K. Ishida, K. Shimizu, T. Hatta, K. Motoshima, and K. Kitayama, "NRZ-OOK-to-RZ-BPSK, "modulation-format conversion using SOAMZI wavelength converter," J. Lightw. Technol., vol. 24, no. 10, pp. 3751-3758, Oct. 2006.

[2] G. W. Lu, E. Tipsuwannakul, T. Miyazaki, C. Lundström, M. Karlsson, and P. A. Andrekson, "Format conversion of optical multilevel signals using FWM-based optical phase erasure," J. Lightw. Technol., vol. 29, no. 16, pp. 2460-2466, Aug. 2011.

[3] H. Kishikawa, P. Seddighian, N. Goto, S. Yanagiya, and L. R. Chen, "All-optical modulation format conversion from binary to quadrature phase-shift keying using delay line interferometer," IEEE Photonics Conference, WO2, Oct. 2011.

[4] F. Parmigiani, L. Jones, J. Kakande, P. Petropoulos, and D. J. Richardson, "Modulation format conversion employing coherent optical superposition," Opt. Exr., vol.20, no.26, pp.B322–B330, Dec. 2012.

[5] M. Mihara, Y. Shinohara, H. Kishikawa, N. Goto, and S. Yanagiya, "Modulation format conversion from BPSK to QPSK using delayed interferometer and pulse shaping filter," IEEE Photonics Conference, MD2.5, Oct. 2014.

Time-frequency packed VCSEL-based IM/DD transmission for WDM access networks

Antonio Malacarne[1], Francesco Fresi[2], Gianluca Meloni[2], Tommaso Foggi[2] and Luca Poti[2]

1: Scuola Superiore Sant'Anna, Via G. Moruzzi 1, 56124 Pisa, Italy
2: CNIT, Via G. Moruzzi 1, 56124 Pisa, Italy
antonio.malacarne@sssup.it

Abstract: A commercial 4G 1550nm-VCSEL is employed for wavelength- and polarization-independent 14Gb/s optical transmission up to 25km of SMF, thanks to the time-frequency packing technique that is, for the first time, employed in an IM-DD system.
Keywords: *Vertical cavity surface emitting lasers; WDM access networks; Faster-than-Nyquist techniques.*

I. INTRODUCTION

Long-wavelength vertical-cavity surface-emitting lasers (LW-VCSELs) are considered attractive light sources for high-speed short- and medium-reach optical communications thanks to the remarkable energy and cost effectiveness and the possibility to operate at high temperature so as to relax the operational cost related to thermal stabilization [1]. For these reasons, their use has been recently investigated for all applications where massive low-cost production of energy efficient transceivers is targeted, such as optical interconnects, access and metro networks.

Direct intensity modulation (IM) of VCSEL allows low driving current (and low power consumption) as well as low-cost direct detection at the receiver. The data rate is limited by the modulation bandwidth that typically reaches up to 10 GHz for commercial devices [2] and up to 18 GHz for research prototypes [3]. To increase the data rate, 4-level pulse amplitude modulation (4-PAM) is often preferred, thus doubling the data rate with a reasonable hardware cost and complexity [4]. However, the large frequency chirp induced by direct modulation of VCSELs in interaction with chromatic dispersion heavily limits both the data rate and optical reach in transmission systems based on standard single-mode fiber (SMF) in the 1550 nm window. Different approaches have been proposed to mitigate this issue, such as injection locking of VCSEL to another laser [5] or the use of electronic dispersion compensation [6]. Concerning access networks applications, recent efforts focus on techniques to enhance both the data rate and the fiber reach in case of bandwidth-limited directly-modulated (DM) lasers, still keeping on-off-keying (OOK) modulation format [7]. In this case the adopted simplified analog coherent detection is wavelength-dependent and could hardly be exploited in case of DM-VCSELs, due to their intrinsic phase and frequency noise.

In this work, we propose and experimentally demonstrate a novel solution for increasing both the data rate and the optical reach thanks to the so called time-frequency packing (TFP) technique [8], while preserving both simple OOK IM obtained through DM-VCSEL and simple wavelength-independent direct detection (DD) with an avalanche photodiode (APD). Primarily, TFP was developed for coherent optical systems and then applied by the same Authors in [9] to improve the spectral efficiency of 125 Gb/s dual-polarization quadrature-phase shift keying (DP-QPSK) WDM signal obtained through external modulation of VCSEL in a regional network scenario (~800km). The same concept has been successfully applied here for the first time to an IM/DD system, through an extremely simplified digital signal processing (DSP) unit (e.g. neither frequency synchronization nor phase error recovery) which only takes into account the photo-detected signal intensity (regardless of the phase information). In the presented experiment a 14 Gb/s line rate (12.5 Gb/s net bit rate) is achieved with a 4G commercial VCSEL (by RayCan™) and transmission over 25 km of standard SMF is demonstrated at 1550 nm. Indeed, thanks to the TFP technique, a bit rate three times higher than the employed VCSEL operation speed is achieved, dispersion tolerance is improved and a 19-dB power budget without optical amplification is experimentally demonstrated. Additional measures over 45 km of SMF demonstrate error free operation at 9.2 Gb/s and a 16-dB power budget. The system is intrinsically polarization- and wavelength-independent and does not require either coherent detection or DSP at the transmitter.

II. OPERATION PRINCIPLE

The proposed system is based on the TFP principle [8], whose method lies on increasing the signaling rate for a fixed pulse bandwidth (or, equivalently, reducing the pulse bandwidth for a fixed signaling rate), thus saving some bandwidth resources at the expense of introducing intersymbol interference (ISI). The intentionally induced ISI is then coped with by a maximum a-posteriori probability (MAP) detector. This way, it was shown that the spectral efficiency can be greatly increased, by trading-off between MAP detector complexity and performance. We here present an IM-DD system, which therefore entails substantial simplifications with respect to the already demonstrated optical TFP for coherent detection [9]; although the receiver side processing is remarkably simplified as, for instance, frequency synchronization and phase error recovery are not necessary and clock recovery can be simplified as well, on the other hand the presence of the square-law detector implies a different channel model. The schematic of the proposed solution is depicted in Fig. 1(a). The input bit stream is randomly generated and encoded in a sequence {c} by using 64800-bit low-density parity check (LDPC) codes in order to assemble a sequence of codewords, each preceded by a known 448-bit pattern (or header) for synchronization purposes. At the receiver the signal is then sampled at one sample per symbol. The received samples read

$$r_k = \sum_{m=1}^{L} \sum_{n \leq m}^{L} c_m c_n h(n-k, m-k) + w_k \qquad (1)$$

where w_k is filtered noise, $h(n,m)$ is the discrete time system response [10] and L is the channel memory. Then frame synchronization is performed and afterwards a block of 10k bits is used as a pilot for obtaining an estimate of the system impulse response. The impulse response coefficients are employed by the MAP detector, which accounts for a fixed amount of channel memory L_r that can be lower than the actual memory L (in this case entailing a loss of performance). MAP detector iteratively exchanges soft information with the LDPC decoder providing reliable decisions on the transmitted codewords. Results are hence presented both as bit-error rate (BER) computations by employing practical LDPC codes and as achievable bit rate lower bounds (ABR) [8].

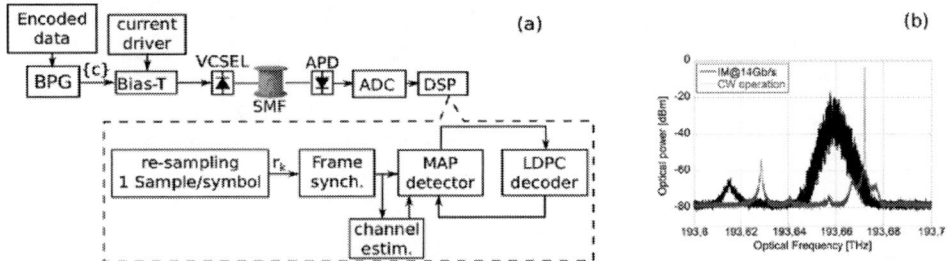

Fig. 1. Experimental setup and DSP block diagram (a); optical spectrum in CW operation and for OOK IM at 14 Gb/s (b).

III. VCSEL CHARACTERIZATION AND TRANSMISSION EXPERIMENTS

First, VCSEL characterization is conducted to investigate the VCSEL behavior under different operation condition so as to find the optimum working point. The employed device is a commercial VCSEL supplied by RayCan™ for 4 Gb/s OOK single mode transmission. Optical, electrical and RF characterization are shown in Fig. 2. The optical output power is measured as a function of the bias DC current for different temperatures (Fig. 2(a)). A threshold current of 1 mA is measured and a roll-over occurs between 7 and 8 mA where the maximum output power is reached. Optical power decreases as the device temperature increases, ranging from 0.6 mW (T = 20°C) down to 0.33 mW (T = 50°C). Fig. 2(b) shows the corresponding diode voltage leading to typical electrical power consumption below 35 mW. S_{21} parameter is measured through a vector network analyzer for different bias current values. Results are reported in Fig. 2(c) showing, as expected, an increasing small-signal modulation bandwidth as the bias current increases. The frequency chirp induced by direct modulation is experimentally measured as in [11] for a peak-to-peak RF voltage of 0.6 V, resulting in a transient peak-to-peak chirp of 8.5 GHz and an adiabatic chirp equal to 1.5 GHz.

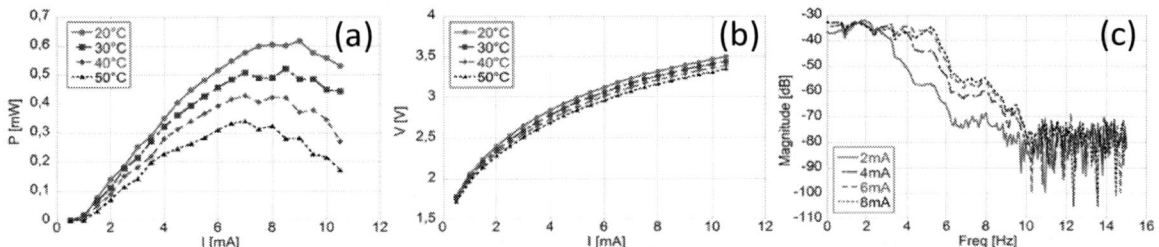

Fig. 2. Optical (a), electrical (b) and RF (c) VCSEL characterization for different operating conditions.

TFP technique for IM-DD is validated through transmission experiments in access-metro scenario for demonstrating the advantages in terms of bit rate for a given VCSEL analog bandwidth. In our experiment, the employed VCSEL operates at 20°C with 8mA-bias current, providing an output optical power of -2.2 dBm with an overall power consumption of 43 mW (including 17 mW for thermal stabilization). The experimental setup is depicted in Fig. 1(a): the VCSEL gets directly modulated by a bit pattern generator (BPG) providing LDPC encoded data with 0.6 V peak-to-peak. A bias-T is employed to combine the RF signal with the bias DC current. After propagation, the received signal is detected through an APD and digitized by a real-time (RT) scope acting as analog-to-digital converter (ADC). In our experiment the RT scope operates at 25 GSample/s with an analog bandwidth of 5 GHz. Digital data are then re-sampled by the DSP in order to operate at 1 Sample/bit. Targeting a 25-km optical reach, the ABR is evaluated for different line rates and received optical powers. A channel memory $L_r = 2$ is chosen to keep the MAP detector complexity low. Fig. 3(a) shows that the maximum ABR = 13 Gb/s is reached for a line rate of 14 Gb/s and a received power of -18 dBm. Error free operation is demonstrated by employing a LDPC code with a rate $R_c = 8/9$ corresponding to a net bit rate of 12.5 Gb/s with a corresponding energy consumption per bit of 3.4 pJ/bit including 1.3pJ/bit for thermal stabilization. Comparative measures showing the required turbo decoder iterations versus received optical power are performed. Results for transmission up to 25 km and 12.5 km of SMF and in the back-to-back (BtB) case are depicted in Fig. 3(b), showing a power budget of about 19 dB and 21 dB for 25 and 12.5 km respectively. Optical spectrum for continuous wave (CW) operation and IM at 14 Gb/s are reported in Fig. 1(b). A side mode suppression ratio higher than 40 dB can be observed. As additional measure, the ABR for a longer distance of 45 km is measured. The maximum ABR = 10.2 Gb/s is reached for a line rate of 11 Gb/s and a received power of -18 dBm. In this case error free operation is obtained with a code rate $R_c = 5/6$ corresponding to a net bit rate of 9.2 Gb/s (Fig. 3(c)).

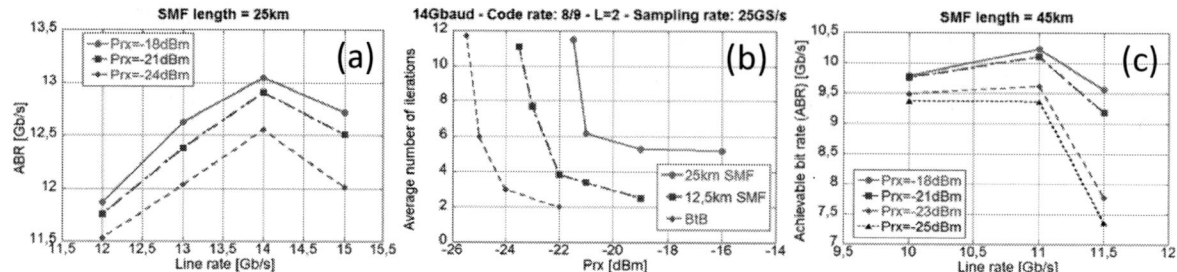

Fig. 3. Achievable bit rate lower bound (ABR) vs line rate for different received powers and 25km of SMF (a); average turbo decoder iterations vs received power for a line rate of 14 Gb/s, code rate 8/9 and different propagation lengths (b); ABR vs line rate for 45 km of SMF (c).

IV. CONCLUSIONS

A simple polarization- and wavelength-independent IM/DD transmission system has been proposed and experimentally validated by using a low-cost commercial 4G DM-VCSEL at the transmitter and an APD at the receiver. Thanks to the TFP technique, adapted and employed here in an IM/DD system for the first time, a net bit rate (12.5 Gb/s) three times higher than the employed VCSEL nominal speed is achieved and successfully transmitted over 25 km of SMF with a 19-dB power budget and energy consumption of 3.4pJ/bit including thermal management. A net bit rate of 9.2 Gb/s is achieved for 45 km of SMF with a 16-dB power budget. TFP is performed at the receiver by a low-complexity DSP ($L_r = 2$) unit, whereas the narrow-filtering of the 14Gb/s (gross rate) electrical signal is implemented at the transmitter by the limited VCSEL bandwidth. The system does not require chromatic dispersion compensation, optical amplification, coherent detection and DSP at the transmitter. The low complexity and energy consumption of the system, together with the potential low cost, make the proposed solution a good candidate for next-generation WDM access network applications.

ACKNOWLEDGMENT

This work is partly supported by the European project "RAPIDO" (619806) and the H2020 project "ROAM" (645361).

REFERENCES

[1] Ortsiefer M. et al., "Long wavelength high speed VCSELs for long haul and data centers", in Proc. OFC2014, paper W4C.2, 9-13 March 2014.

[2] http://www.vertilas.com/sites/default/files/Downloads/vertilas_communications_v5_0_0.pdf

[3] Malacarne A. et al., "High-speed long-wavelength VCSELs for energy-efficient 40 Gbps links up to 1 km without error correction," in Optical Fiber Communications Conference and Exhibition (OFC), Tu2H.1, 22-26 March 2015

[4] F. Karinou et al., "IM/DD vs. 4-PAM using a 1550-nm VCSEL over short-range SMF/MMF links for optical interconnects," in Proc. OFC 2013, paper OW4A.2, 2013.

[5] A. Gatto, A. Boletti, P. Boffi, and M. Martinelli, "Adjustable-chirp VCSEL-to-VCSEL injection locking for 10-Gb/s transmission at 1.55 μm," Opt. Express 17, 21748-21753 (2009).

[6] J. Zhou et al., "1.5-μm, 21.4-Gbps 4-PAM VCSEL Link for Optical Access Applications" in Proc. OFC2015, paper Th2A.54, 2015.

[7] R. Corsini, M. Presi, M. Artiglia and E. Ciaramella, "10-Gb/s Long-Reach PON System With Low-Complexity Dispersion-Managed Coherent Receiver," in IEEE Photonics Journal, vol. 7, no. 5, pp. 1-8, Oct. 2015.

[8] G. Colavolpe and T. Foggi, "Time-frequency packing for high capacity coherent optical links," IEEE Trans. Commun., vol. 62, pp. 2986–2995, Aug. 2014

[9] Meloni G., Malacarne A., Fresi F., Potì L., "6.27 bit/s/Hz spectral efficiency VCSEL-based coherent communication over 800km of SMF," in Optical Fiber Communications Conference and Exhibition (OFC), Th2A.30, 22-26 March 2015

[10] W. Chung, "Channel estimation methods based on volterra kernels for MLSD in optical communication systems," Photonics Technology Letters, IEEE, vol. 22, no. 4, pp. 224–226, 2010

[11] C. Laverdiere, A. Fekecs and M. Tetu, "A new method for measuring time-resolved frequency chirp of high bit rate sources," in IEEE Photonics Technology Letters, vol. 15, no. 3, pp. 446-448, March 2003.

MF2-1 (Invited)

OECC/PS2016

Elastic WDM Switching for Scalable Data Center and HPC Interconnect Networks

Adel A M Saleh A hilesh S P   hope   ohn E Bowers Rod C Alferness

Electrical and Computer Engineering Department, University of California Santa Barbara, California, USA

AdelSaleh@ece.ucsb.edu

Abstract: *An elastic W M switch, suitable for silicon photonic integration, was recently proposed by the authors. ere, we summari e its characteristics, and show how it can enhance the scalability and performance of data centers and PCs.*

Keywords: (ptical nterconnects, ata Centers, PC, Silicon Photonics, ynamic etwor s)

I INTRODUCTION

The use of fast optical circuit switching in the interconnect networks of data centers and high performance computers (HPCs) has been advocated by various researchers [1]-[5]. Recently, we proposed an elastic, microring-based, WDM, crossbar switch suitable for such applications [6]. The switch is compatible with silicon photonic integration, and thus, given the steady progress in that field [7], promises to have small size, cost and power requirements. The purpose of this paper is to show how one can exploit these properties to cost-effectively achieve enhanced performance and scalability of future data centers and HPCs, by incorporating a large number of such switches in their interconnect fabrics. The architecture, operational principles and unique characteristics of the WDM switch are summarized in Sec. II. Examples of its application in folded Clos and HyperX interconnect networks are presented in Secs. III and IV, respectively.

II ELASTIC DM CROSSBAR S ITC

Figure 1a shows the $\times N$, M-wavelength, WDM switch proposed in [6], which has ^{2} cross-points (L-blocks), each having microrings, as shown in Fig. 1b. The microrings are individually tuned to drop up to wavelengths from any input to any output. This elastic connectivity is a key feature of the switch, which leads to a greatly reduced latency [6]. Generally, $1 \leq \ \leq M$ and $M \leq \ $; and, typically, $\ll M$. An input-buffered transmitter and a receiver are shown in Fig. 1c. The optical spectrum is depicted in Fig. 1d, showing open slots, to which the microrings would be tuned when they are not required to drop any wavelength. The M operating wavelengths and the open slots are assumed to fall within one free spectral range of the microrings to enable individual control of each wavelength.

The switch operation requires central control, and, as shown in Fig. 1e, assumes equal time slots that are much longer than the switching time (which includes both the time needed by a centralized algorithm to compute the input-output wavelength assignments for each time slot, and the time needed to tune and stabilize the microrings). Our wavelength assignment algorithm is illustrated in Fig. 2, which generalizes known algorithms for determining the port connectivity in input-buffered packet switches [8]. For each time slot, the algorithm receives an admissible $\times N$ connection matrix A, which indicates the input-output wavelength traffic demands. The matrix A is expanded, using a *graph matching algorithm*, into the sum of M *permutation matrices*, each representing the input-output connectivity for one of the M wavelengths. The algorithm runs in polynomial time, which is estimated to take less than 1 μs. In addition, since each microring needs to be tunable across the entire wavelength spectrum, *thermal tuning* is required, which is estimated to need on the order of 1 μs. Thus, the transmission time slot shown in Fig. 1(e) can be on the order of 10 μs (or longer).

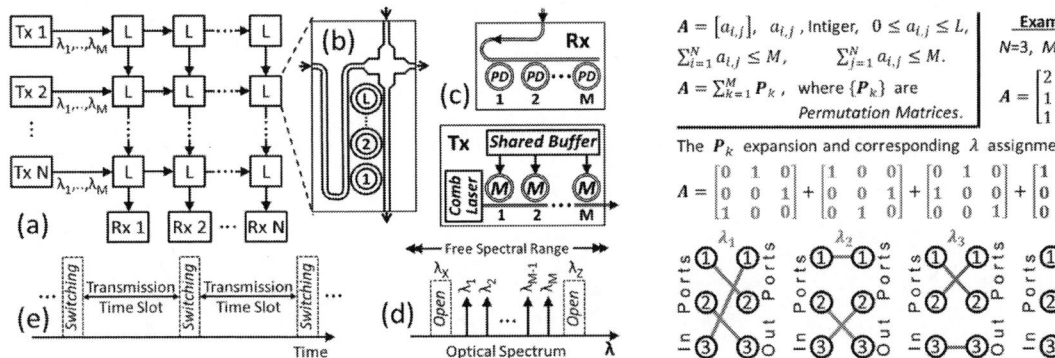

Fig a The $\times N$ WDM switch.   Microring-based, -wavelength switching point, with a low-crosstalk multimode crossing. **c** M-wavelength transmitter (Tx) & receiver (Rx) - **PD**: Photodetector, **M**: Modulator. **d** Optical spectrum. **e** Timing diagram.

Fig Properties of the connection matrix, A, which is expandable into the sum of *permutation matrices*; and an example showing the expansion and the λ assignments.

This research was supported by the American Institute for Manufacturing of Integrated Photonics (AIM Photonics).

The switching mode described above is actually one of three possible switching modes, which are listed below:

1) <u>Quasi-static circuit switching</u>: In this mode, the switching time is longer than the packet time, which applies to HPC interconnect applications, where the processors interchange very short ($\ll 1$ μs) packets. In this case, the connectivity pattern of the switch is set to best match the workload, and is left fixed for the duration of each computational task. This mode also applies to data centers to respond to relatively slow workload variations.

2) <u>Fast time-slot switching</u>: This is the main switching mode discussed above, which is applicable when the time slot duration (e.g., 10 μs) is much larger than the switching time. The input-output connectivity for each time slot is configured in real time in response to varying input-output packet traffic demands (i.e., matrix A above).

3) <u>Ultrafast time-division-multiplexed (TDM) switching</u>: This mode augments the above modes by using an end wavelength, e.g., λ_1 in Fib. 2d, to add periodic, ultrafast-switched, TDM, all-to-all connectivity within each time slot, to enable the interconnected devices to exchange very short ($\ll 1$ μs) messages. To realize this, one microring is reserved within each L-block to be rapidly tuned, via *current-in ection*, between λ_1 and the adjacent open slot, i.e., λ_X. This way, the ultra-fast-switched channel would not affect the other connections.

III APPLICATION IN FOLDED CLOS INTERCONNECT NET OR S

A folded Clos, leaf/spine data center is shown in Fig. 3a, which has $2P$ leaf switches and P spine switches, all having the same capacity of $2PR$ (b/s), where R is the bit rate of a fiber link. For R $\geq$ 100G, WDM fiber links with integrated transceivers [9],[10] are typically employed. For example, for M-wavelength links with r (b/s) per wavelength, R = 100G; 400G and 1,000G can be implemented, respectively, with $M = 4$, r = 25G; $M = 16$, r = 25G and $M = 20$, r = 50G.

Note from Fig. 3a that the total number of servers is limited to $2MP^2$ because one cannot add any more leaf or spine switches to the topology. However, if the different wavelengths from a switch port are fanned-out to different switches, then more leaf and spine switches, can be added, and hence, also servers. This is done in Fig. 3b by adding a layer of $\times N$, M-wavelength WDM switches, leading to a factor of   increase in the number of servers. (A similar technique was proposed in [3], using arrayed waveguide grating routers (AWGRs) instead WDM switches.) Of course, adding a layer of packet switches instead of the WDM switches would also lead to an increased number of servers, which is the idea behind the multi-level folded Clos architecture. But, there are two problems with that approach: First, latency will be increased due to the added layer of packet switches, while the WDM switches will add no latency when operating in the circuit switching mode discussed above. Second, because the WDM switches are compatible with silicon photonic integration, as mentioned in Sec. I, they will have lower cost and power consumption than the packet switches.

In addition to the above scalability, the WDM switching layer also enables reconfigurability of the data center, e.g., to respond to slow or fast workload variations, using, respectively, the circuit switching mode or the time-slot switching mode (possibly augmented by the TDM switching mode). The goal is to enable elastic bandwidth connectivity between the various port pairs of the packet switches to better match workload variations. This leads to a better utilization of the available resources (e.g., servers and transponders), and thus, improved performance (e.g., throughput and latency).

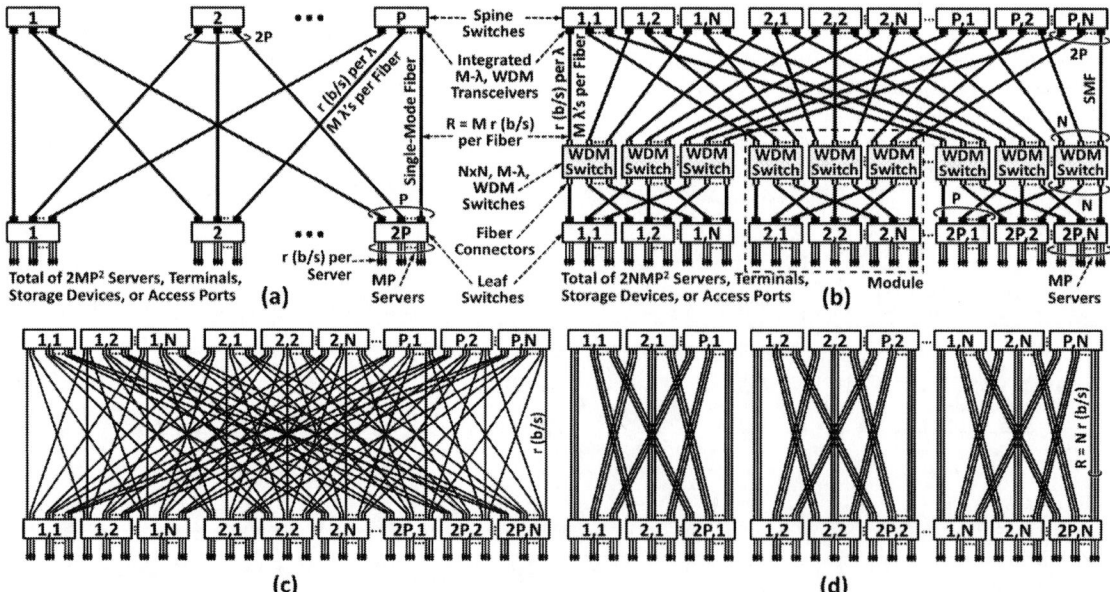

Fig   a A folded Clos leaf/spine data center architecture with WDM fiber links.   An   -fold increase in the number of servers is obtained by adding a layer of   × , M-wavelength WDM switches (each representing either two   × switches, one for each direction, or one bidirectional 2 ×2 switch). For simplicity, we assume that $M = $, in Figs. c and d. **c** With all the WDM switches in Fig. b set to an AWGR routing pattern, a dense, folded Clos topology results. **d** With all the WDM switches in Fig. b set to a straight-through pattern,   , uncoupled, folded Clos, topologies result.

Two extreme configurations of the interconnection network of Fig. 3b are shown in Figs. 3c and 3d. The dense folded Clos topology of Fig. 3c results when all the WDM switches are set to an AWGR pattern (as was done in [3]). The same topology would result if the packet switches had -times as many ports, interconnected with -times as many single-wavelength fibers, which would not be a scalable solution. The topology of Fig. 3d results when all the WDM switches are configured to pass the input wavelengths straight through the switch. In this case, the data center is divided into uncoupled, folded Clos partitions with -wavelength links. One can get a rich set of interconnection topologies, to respond to varying workloads, by configuring the various WDM switches differently from the above two extremes.

As a design example: a leaf/spine system with $P = 16$, r $= 25$ Gb/s per λ, $M = 16$ λ's per fiber (i.e., packet switches with 32×400G ports), requires $4P^2 = 1{,}024$ WDM switches (with $= 16$-ports) to support $2\,MP^2 = 131{,}072$ servers.

IV APPLICATION IN perX INTERCONNECT NET OR S

HyperX topology [11] is an important class of multi-dimensional, interconnect networks that encompasses other well know types of data center and HPC interconnect networks, such as the Flattened Butterfly and the Hypercube. In general, it requires a very dense fiber interconnection fabric, as depicted in Fig. 4a, which shows a small, *regular* HyperX network, having $= 3$ dimensions and $S = 4$ packet switches or routers per dimension, yielding a total of $S = 64$ switching nodes. Large data centers and HPCs would require a much larger value for S, which could result in very complex fiber interconnect fabric. Some packaging techniques are proposed in [11] to simplify the system wiring. The wiring can be further simplified and the number of fibers can be greatly reduced by incorporating WDM switching as shown in Fig. 4c. The switches also enable the reconfigurability of the interconnect topology, thus making it possible to adjust the bandwidth connectivity among the various packet switches to better match workload variations.

As a design example, consider a HyperX system with $= 3$, $S = 16$, $= 1$, $T = 16$, which supports $S = 4{,}096$ routers of a radix of $(S-1)$ $T = 61$, and $TS = 65{,}536$ servers. This requires $(-1)S^{(-1)} = 512$ WDM switches, with $= 2\,S = 32$ ports and (with one λ per channel) $M = (S-1) = 45$ λ's per port, which are challenging but achievable numbers.

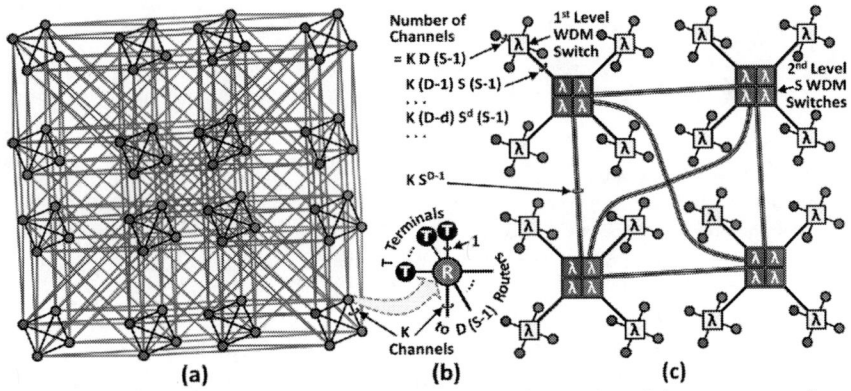

Fig 4 a A small, *regular*, HyperX interconnect network with $= 3$ dimensions and $S = 4$ packet switches or routers per dimension. Details of a switching node, showing a connectivity of channels between corresponding routers, and T terminals, processors or servers connected to each router. **c** Inserting WDM switches (labeled 'λ') results in greatly simplifying the wiring, reducing the number of fibers, and enabling reconfigurability.

V CONCLUSIONS

A flexible WDM switch of potentially small size, cost and power was presented. It was argued that incorporating a large number of such switches in the interconnect networks of data centers and HPCs can enhance their performance and scalability in a cost-effective way. For example, adding WDM switching was shown to simplify the system wiring, greatly reduce the number of required fibers, scale-out the system, and enable reconfigurability of the system. One can also use spatial switching with fiber ribbons or multi-core fibers instead of, or in addition to, WDM switching [3],[12].

REFERENCES

[1] G. Porter, et al, "Integrating microsecond circuit switching into the data center," ACM SIGCOMM '13, Aug 2013.
[2] H. Liu, et al, "Circuit switching under the radar with REACToR," 11th USENIX Symposium on NSDI, 2014.
[3] A.A.M. Saleh, "Scaling-out data centers using photonics technologies," Photonics in Switching Conference, July 2014.
[4] R. Yu, et al, "A scalable silicon photonic chip-scale optical switch for HPC systems." Optics Express, Dec 2013.
[5] D. Nikolova, et al, "Scaling silicon photonic switch fabrics for data center interconnection networks," Optics Express, Jan 2015
[6] A.S.P. Khope, et al, "Elastic WDM crossbar switch for data centers," Optical Interconnects Conference, May 2016.
[7] T. Komljenovic, et al, "Heterogeneous silicon photonic integrated circuits," Journal of Lightwave Technology, Jan 2016.
[8] P. Giaccone, et al, "Randomized scheduling algorithms for high-aggregate bandwidth switches," IEEE JSAC, May 2003.
[9] C. Cole, "Future datacenter interfaces based on existing and emerging optics technologies," IEEE Summer Topicals, 2013.
[10] B. R. Koch, et al, "Integrated silicon photonic laser sources for telecom and datacom," OFC/NFOEC, 2013.
[11] J.H. Ahn, "HyperX: topology, routing, and packaging of efficient large-scale networks," ACM SC '09, Nov 2009.
[12] A.A.M. Saleh, "Evolution of the architecture and technology of data centers towards exascale and beyond," OFC 2013.

Programmable Optical Processors for Signal Processing and Monitoring

Simon B. Poole, Patrick Blown, Qing Li, Ralf Stolte, Ian Clarke

Finisar Australia, 21 Rosebery Ave, Rosebery, NSW 2018, Australia

simon.poole@finisar.com

Abstract: *Recent developments in programmable optical signal processing and monitoring are reviewed with emphasis on non-telecom applications including laser pulse shaping and intra-data-center switching.*

Keywords: *optical switching, wavelength processing; WSS; optical filtering*

I. INTRODUCTION

LCoS-based Programmable Optical Processors based on WSS technology have, over the past few years, established themselves as a core equipment capability for many optical R&D laboratories[1]. Indeed, a quick search in the IEEE Xplore database provides nearly 350 papers over the past 7 years which refer to these devices as part of their experimental equipment.

In this paper, we will briefly review the principles of operation of an LCoS-based programmable optical processor and then look at some of the main applications of such devices, with a particular emphasis on applications outside of the main DWDM/telecom applications space, and including areas such as software-definable interferometers (Section III), femtosecond lasers (Section IV), RF and microwave signal processing (Section V) and data-centre switching networks (Section VI).

II. PRINCIPLE OF OPERATION

Figure 1 shows a simplified schematic of the operating principle of a programmable optical processor. Light from an input fiber (within a fiber array) is spectrally dispersed and imaged onto the two-dimensional LCoS array containing a large number of individual pixels (around 1000 x 500). The voltage applied to each pixel is individually controlled, which, then controls the phase retardance of the liquid crystal layer immediately above the electrode. The phase front of the reflected light can therefore be varied on a pixel-by-pixel basis, allowing for very precise control. For example, in the vertical axis of the array (the "switching" dimension), generating a phase ramp results in steering the reflected beam before retracing back through to the fiber array. The slope of this phase ramp allows controllable coupling into a desired output port. This vertical dimension can thus also be classified as the 'switching' dimension. Subtle deviations from the ideal

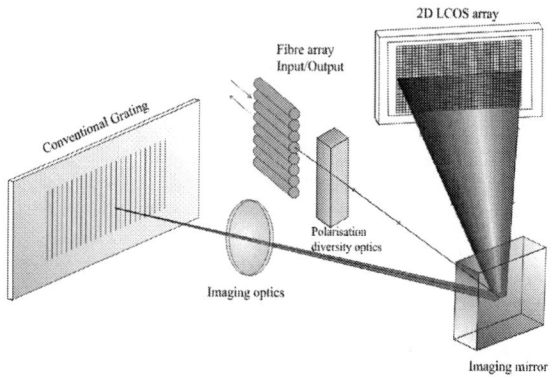

Figure 1 Simplified schematic of a multiport programmable optical processor

phase ramp introduce controllable decoupling into the output fiber, which gives precise amplitude control, better than 0.01 dB. The horizontal axis in Fig. 1 is the "spectral" dimension and each of the pixel columns can be freely programmed with a desired attenuation and relative phase level be appropriate choice of the phase 'image' on the LCoS.

III. PROGRAMMABLE POWER-SPLITTING AND INTERFEROMETERS (THE "OPTICAL FPGA")

Early work by Frisken [2] showed that using an LCoS-based WSS, light of a given wavelength could be split to multiple output locations to implement a drop-and-continue function for telecommunications networks. Subsequent work by Schröder [3] showed that it was possible to create a delay line interferometer within a WSS, which was then used to measure the OSNR in a polarization-multiplexed 40 GHz signal.

More recently, Schröder [4] also combined the ability of the programmable optical processor to create arbitrary power splitting with the ability to simultaneously create software-defined interferometers within the processor to demonstrate devices with variable, wavelength-dependent splitting of light to multiple outputs as well as complex wavelength-dependent splitters and interferometers, including nested interferometer structures. They termed this capability the "Optical FPGA". A typical example of what can be achieved is the all-optical DFT filter, created using 3 output ports of a 1 x 4 Programmable Optical Processor, shown in Figure 2.

This capability has recently been extended to allow the creation of arbitrary optical transfer functions with up to 1 x 16 splitting capability [5].

IV. ULTRA-SHORT PULSE LASERS

High-power ultrashort laser pulses are used in many applications including high precision material ablation, laser induced breakdown spectroscopy, x-ray emission generation etc where the ability to control the duration and shape of the pulse provides additional parameters to optimize the application. A number of techniques to reshape short optical pulses exist, with varying degrees of accuracy, reconfigurability and flexibility. However, in most approaches, spectral manipulation of the phase and/or amplitude of an optical pulse is used to provide the desired change in temporal characteristics through a process known as Fourier-domain pulse shaping. This was first demonstrated using a spatial light modulator (SLM) by Wiener [6] in the mid 1980s.

Fig. 2. Insertion loss of the three drop ports (solid black, dashed red, dash-dot green) of an all-optical DFT filter The theoretical response of the second (dashed red) drop port is shown (dotted blue) for comparison.

This is no longer just an area of research, and pulse optimization techniques based on programmable optical processors are increasingly being used in commercial systems, including systems with multiple stages of post-optimisation amplification [7].

V. RADIO-FREQUENCY ARBITRARY WAVEFORM GENERATION (RF-AWG)

An extension of the capabilities of ultra-short pulse-shaping is to photonic-assisted RF-AWG techniques where spectral-domain Fourier Transform pulse shaping, combined with frequency-to-time mapping, provides advantages in terms of programmability and attainable signal complexity. In these techniques the tailored spectrum of an ultrashort optical pulse is mapped to the time domain by stretching the optical pulses through a dispersive medium (eg a length of single-mode fibre) to give an optical intensity profile that is a scaled version of the tailored spectrum. The optical intensity waveform is then mapped to the RF domain using a high-speed photodetector. Here, the programmable optical processor provides independent manipulation of the spectral amplitude and phase of the input optical waveform

Work by Rashidinejad, *et al* [8] showed that the time-bandwidth product (TBWP) of this type of process is set by the number of independent pulse shaper control elements and provides a TBWP twice as large as conventional frequency-to-time mapping techniques. Fig. 3 shows the experimental arrangement and generated down-chirp waveforms with a 16 ns time aperture and frequency content from 7 to 45 GHz, for a TBWP of ~589. Pulse shaping was carried out over the full optical C-band (1525-1565 nm, ~5 THz optical bandwidth)

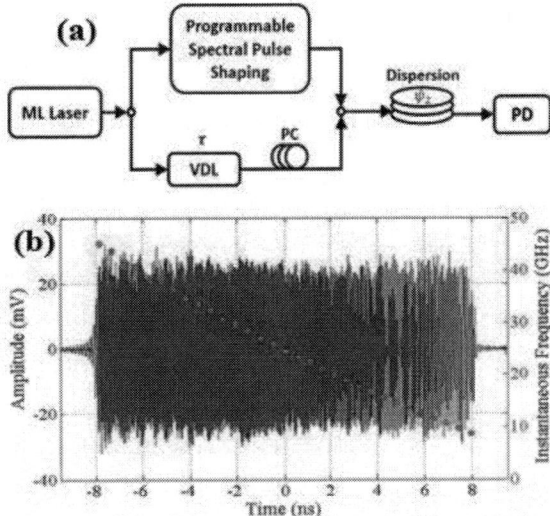

Fig. 3. Interferometric (passband) RF-AW G (a) schematic. (b) detected waveform and instantaneous frequency for millimeter wave down-chirp generation spanning 7-45 GHz with TBWP of ~589 from [8]

with ~10 GHz shaping resolution at 3 dB. The ratio of optical bandwidth to spectral resolution is ~500, consistent with the maximum TBWP achieved.

VI. Reconfigurable port capability

Reconfigurable DWDM networks are now well established for both long-haul and metro telecommunications but the application of WDM to intra-data-center networks has, to date, been limited. High-port-count programmable optical processors with up to 20 ports are now available with software reconfigurable port allocation to create different multi-dimensional switching and signal processing units with port configurations from 1x19 to 10 x 10.. This capability was used by Yan *et al* [9] who constructed a reconfigurable, multidimensional all-optical Top-of-Rack switch as part of a low-latency optically-switched network for data-center applications. They used three 20-port programmable optical processors per switch and an OpenDaylight agent to communicate between the network controller and the programmable optical processors. The optical processors were software configured as either 4x16 or 8x12 wavelength-programmable switching matrices as part of a switching architecture which supported both intra- and inter-rack wavelength-switched networking (Fig. 4).

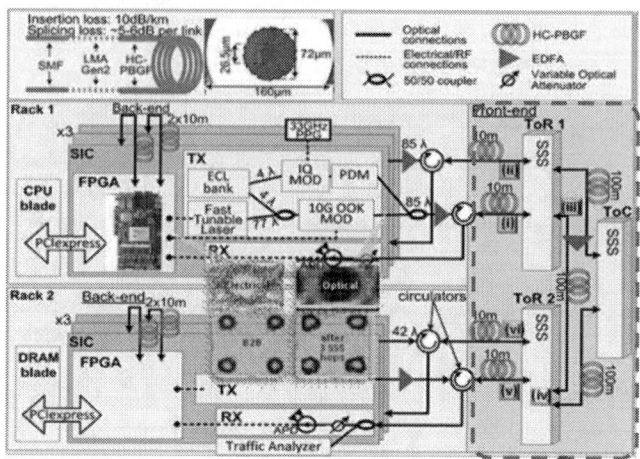

Fig. 4. Multidimensional reconfigurable ToR/ToC data centre switch [9] The programmable optical processors are marked "SSS" (Spectrum Selective Switch) in lower RH corner

VII. Conclusions

We have summarized some of the advanced applications of LCoS-based programmable optical processors, focusing on non-telecoms applications. The flexibility and capabilities of the core LCoS platform will continue to provide many opportunities for application of the technology to further novel applications.

References

[1] M. Roelens, *et al*, "Applications of LCoS-based programmable optical processors," Optical Fiber Communications Conference and Exhibition (OFC), 2014, San Francisco, CA, 2014.

[2] S. J. Frisken, et al, "High performance 'Drop and Continue' Functionality in a Wavelength Selective Switch," in Proc OFC 2006, paper PDP14.

[3] J. Schröder, *et al*, "Simultaneous multi-channel OSNR monitoring with a wavelength selective switch," Opt. Express **18**, 22299-22304 (2010)

[4] J. Schröder et al, "An optical FPGA: Reconfigurable simultaneous multi-output spectral pulse-shaping for linear optical processing," Opt. Express 21, 690-697 (2013)

[5] www.finisar.com/optical-instrumentation/waveshaper-16000s-multiport-optical-processor

[6] A. M. Wiener, *et al*, "High resolution femtosecond pulse shaping," in *JOSA- B*, 1988. 8:1563-1572

[7] M. Mielke et al., "100 µJ, 20 W Femtosecond Fiber Laser for Precision Industrial Micro-Machining", in Proc. CLEO 2013, paper ATh3K (2013)

[8] A. Rashidinejad, *et al*, "Recent Advances in Programmable Photonic-Assisted Ultrabroadband Radio-Frequency Arbitrary Waveform Generation," in *IEEE Journal of Quantum Electronics*, vol. 52, no. 1, pp. 1-17, Jan. 2016. doi: 10.1109/JQE.2015.2506987

[9] Y. Yan *et al.*, "All-Optical Programmable Disaggregated Data Centre Network Realized by FPGA-Based Switch and Interface Card," in *Journal of Lightwave Technology*, vol. 34, no. 8, pp. 1925-1932, April15, 15 2016.

Pattern-Independent Wavelength Conversion of PAM signals in SOAs

Ryosuke Matsumoto[1*], Giampiero Contestabile[2], Yuki Yoshida[1], Akihiro Maruta[1], and Ken-ichi Kitayama[1]

(1) Graduate School of Engineering, Osaka University, 2-1 Yamadaoka, Suita, 565-0871 Osaka, Japan
(2) TeCIP Institute, Scuola Superiore Sant'Anna, Via Moruzzi 1, 56124 Pisa, Italy
E-mail[*] : matsumoto@pn.comm.eng.osaka-u.ac.jp

Abstract: We propose a scheme for pattern-free wavelength-conversion of PAM-signals in SOAs. By exploiting co-propagation of two complementary PAM signals in a saturated-SOA, the proposed scheme, in simulations, achieves 5-dB OSNR improvement with respect of XGM.
Keywords: Pulse amplitude modulation (PAM), Semiconductor optical amplifier (SOA), Wavelength conversion

I. INTRODUCTION

Wavelength conversion enhances not only the flexibility and the capacity of network but also the network transparency and interoperability. For last few decades, a lot of theoretical and experimental studies on wavelength conversion have been made and experimentally done using various types of optical devices such as semiconductor optical amplifiers (SOAs) [1], highly nonlinear fibers (HNLFs) [2], and periodically poled lithium niobates (PPLNs) [3]. Among these conversion schemes, SOA-based wavelength converters have advantageous properties of small size, low energy consumption, and possibility to be integrated with other components. Up to the present, several SOA-based wavelength converters have been proposed to convert the wavelengths of on-off-keying (OOK), phase-shift-keying (PSK), and quadrature-amplitude-modulation (QAM) signals [4, 5]. However, there has been no report on wavelength conversion for pulse amplitude modulation (PAM) format.

The limiting factor when converting PAM signals in SOAs are the effects related to gain saturation. When a PAM signal propagates through a saturated-SOA, gain saturation leads to two serious detrimental effects which degrade the signal quality. The first one is the unequal extinction ratios (ERs) of the intensity levels. The other one is the gain fluctuation dependent on input data patterns, which is generally called pattern effect. To solve the former ER distortion, we have recently proposed a wavelength converter cascading two XGM in SOAs [6]. The reported wavelength converter can remove the unequal intensity intervals by exploiting cross gain modulation (XGM) in the cascaded SOAs. Thanks to this, we have demonstrated wavelength conversion of PAM-4 signals without ER degradation. However, there is no significant mitigation of pattern effects in the converting. The detrimental waveform distortion caused by pattern effects severely restricts system performance, especially in case of cascaded conversions. A pattern compensation scheme is mandatory for realizing wavelength conversion free from pattern limitations.

In this paper, we propose an improved SOA-based wavelength converter to overcome pattern limitations for PAM format. The proposed wavelength converter uses two SOAs and compensates pattern effects by co-propagating two PAM signals at different wavelength with inverted logic in an SOA. Numerical simulations demonstrate the effectiveness of proposed compensation scheme showing an improvement of the optical signal-to-noise ratio (OSNR) penalty for the conversion of a 50-Gbit/s PAM-4 signal from 6.5 dB in case of XGM to 1.5 dB only.

II. SCHEMATIC OF THE PAM WAVELENGTH CONVERTER WITH PATTERN EFFECTS COMPENSATION

The operation principle of the proposed wavelength converter for pattern compensation of PAM-4 signal is illustrated in Fig 1. The proposed system consists of a particular double-stage XGM in SOAs for the wavelength conversion and compensation of pattern effects [7]. An input PAM-4 signal is split into two lines, the upper line for a second-stage SOA and the lower line for a first-stage SOA. At the first-stage SOA, the PAM-4 signal is transferred into a probe light by using XGM, which is selected by an optical filter with a passband wavelength of λ_2. At the same time,

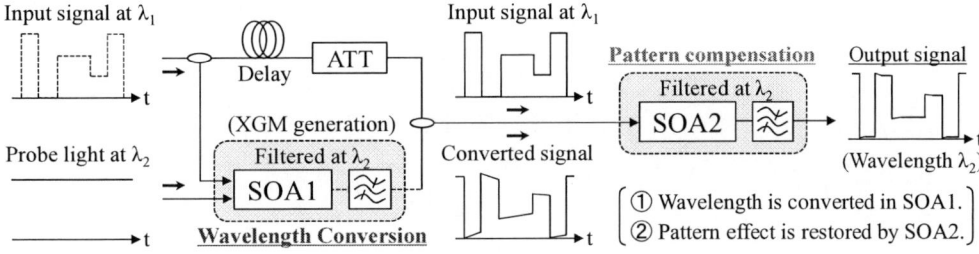

Fig. 1. Schematic diagram of proposed wavelength converter compensating pattern effects.

127

the wavelength-converted signal is distorted by pattern effects in the SOA. The distorted signal is coupled with the PAM-4 signal sent in the upper line. The two signals are synchronized power equalized and injected into the second SOA. In the second SOA, the two signals with opposite logic propagate while sharing optical gain. During the propagation, lower intensity levels experience compressed gain by the other signal with higher intensity, whereas higher intensity levels obtain more gain without gain depletion as co-propagating lower intensity levels. Therefore, the complementary signal pair can suppress gain fluctuation through the cross-gain compression, resulting in pattern compensation of the SOA-based wavelength converter for PAM format [7].

III. NUMERICAL ANALYSIS

We performed numerical simulations to prove the feasibility of the proposed wavelength converter. Four different systems are compared in the simulation. Figure 2(a) shows a block diagram of back-to-back system. At the transmitter, 25-Gbaud PAM-4 signal is generated from 50-Gbit/s data pattern with 2^{11}-1 pseudo random binary sequence (PRBS). The PAM-4 symbol sequence at 25-Gbaud is shaped by a fourth-order Bessel filter with a 36-GHz bandwidth and translated to the electrical field by a square root operation. The generated PAM-4 signal is sent out from the transmitter, where the optical power and center wavelength are 0 dBm and 1550 nm, respectively. The transmitted signal is detected through the square-low operation with an additive white Gaussian noise (AWGN) to determine the OSNR. The detected signal is passed through a matched filter and demodulated by a PAM-4 decoder. Finally, the system performance is evaluated by means of the BER calculated from the demodulated signal. The same transmitter and receiver are used in the other systems.

A simulation setup of the conventional XGM wavelength converter is shown in Fig 2(b). In the wavelength converter, a PAM-4 signal with 1542 nm and a probe light with 1557 nm are launched into an SOA. In the SOA, the modulated information is transferred onto the probe light by the XGM effect. By selecting the modulated probe light at an optical filter, this scheme operates wavelength conversion from 1550 nm to 1557 nm. Figure 2(c) illustrates a previously reported wavelength converter [6]. The system consists of two SOAs in cascade connection, and utilizes the XGM effect in each SOA. In the same way as Fig 2(b), a PAM-4 signal with 1542 nm is converted to a probe light 1 with 1550 nm in SOA1. At the same time, the gain saturation characteristic also becomes inverted due to intrinsic nature of the XGM effect. The inverted signal is launched into SOA 2 together with probe light 2 with a wavelength of 1557 nm. The SOA 2 inverts the polarity of the signal wavelength-converted in SOA 1, and restores the unequal ERs of the inverted signal. In other words, SOA 2 has a role to compensate for the ER degradations generated in SOA 1. Although the cascaded XGM can largely reduce the ER distortion, an additional probe light is required and the conversion efficiency is low for twice XGM wavelength conversion. Figure 2(d) depicts the simulation model of the proposed wavelength converter compensating pattern effects. As explained in the above section, the proposed system coverts the signal wavelength in SOA1 and compensates pattern effects in SOA2. The major advantage over the cascaded XGM is the pattern compensation with simple structure. In these simulations, the SOA parameters corresponding to 40 Gbit/s are adopted for wavelength conversion of a 50-Gbit/s PAM-4 signal [8]. The small signal gain of the SOA is 17 dB, and the gain recovery time calculated in the SOA is about 23 ps. Moreover, probe lights 1 and 2 are fixed at optical power of 5 dBm and 10 dBm, respectively. These power levels are optimized for operating the cascaded XGM and the proposed wavelength converter.

Figure 3 reports the eye-diagram of detected signal in each case. A 2^{11}-1 PRBS of 5,000 PAM-4 symbols was tested in the simulations to capture the eye-diagrams. In the back-to-back case, an ideal waveform without ER distortions and

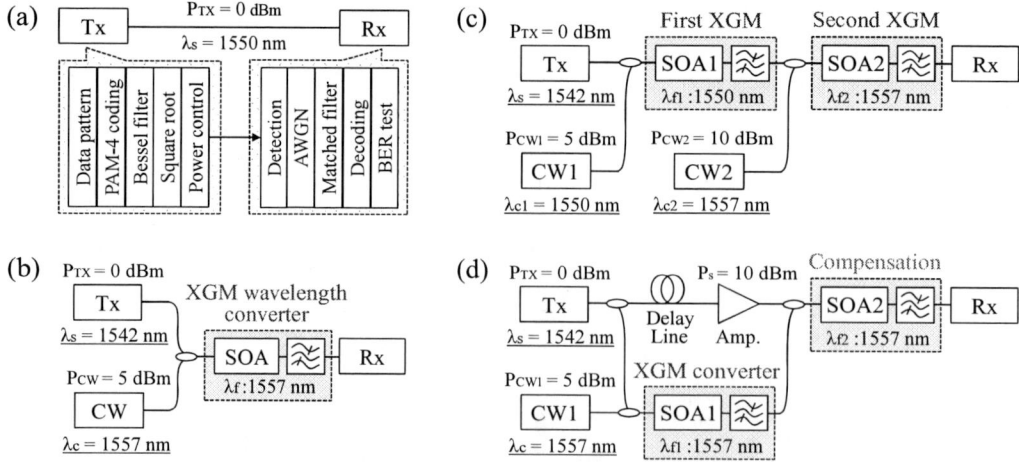

Fig. 2. Simulation models. (a) back-to-back, (b) conventional XGM wavelength conversion, (c) cascaded XGM, (d) proposed scheme.

128

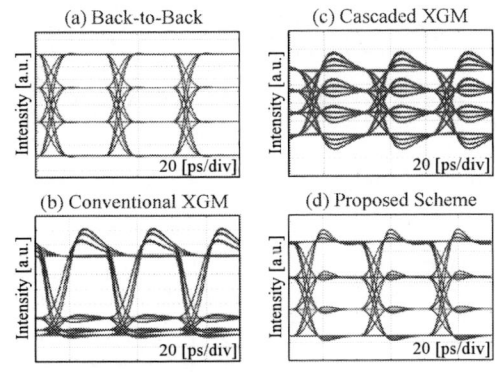

Fig. 3. Detected eye-diagram at the receiver.

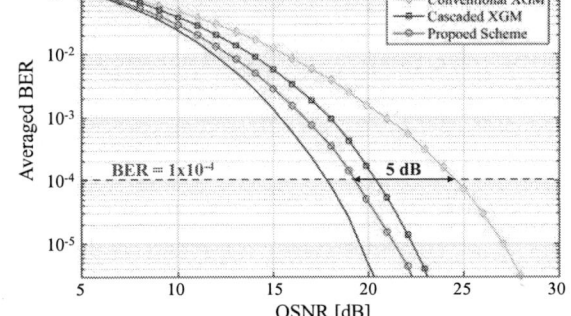

Fig. 4. BER performance as a function of OSNR.

pattern effects is obtained as shown in Fig. 3(a). After XGM wavelength conversion, the waveform of PAM-4 signal is readily distorted due to the ER degradation and pattern effect [see Fig. 3(b)]. The distorted signal is reshaped in SOA2, and thus unequal ERs between each intensity level are coped with as indicated in Fig 3(c). However, the reshaped signal is still suffering from pattern effects. On the other hand, a clear eye-opening is observed in the pattern compensation case as shown in Fig 3(d). This result shows that the proposed wavelength converter is effective for reducing pattern effects.

We have also evaluated the BER performances of four types of the system. Figure 4 depicts the averaged BER performances over 100 iterations as a function of OSNR at the receiver. The BER performances should be compared at the error-free region (BER < 10^{-9}), but 500,000 PAM-4 symbols are transmitted per trial in terms of proceeding times. By reducing an OSNR penalty from 6.5 dB to 1.5 dB at BER = 1×10^{-4}, the proposed system has improved the 5-dB OSNR from the conventional scheme. The 1-dB OSNR improvement is also confirmed in comparison with the cascaded XGM case. From these results, we can conclude that a proof-of-concept demonstration of the wavelength converter free from pattern effects for PAM format is successfully achieved. In this demonstration, there is a limited OSNR difference between the cascaded XGM and the proposed wavelength converters. This is because pattern effects were not dominant over system performances at BER = 1×10^{-4}. A significantly larger OSNR improvement is expected at much lower BER values.

IV. CONCLUSIONS

We have proposed a wavelength converter for PAM signals compensating pattern effects and showing 5-dB improved OSNR penalty in respect of the conventional XGM wavelength conversion. The scheme based on a double-stage XGM using logically inverted signals in the second stage can potentially be very useful for the optical processing of PAM signals in various applications especially when cascaded processing is required.

ACKNOWLEDGMENT

The authors would like to thank A. Agata of KDDI laboratory and C. Stamatiadis of Technical University of Berlin for their continuous supports throughout the work. This work has also been supported by JSPS KAKENHI Grant Numbers 15J12474 (2015-2017).

REFERENCES

[1] T. Durhuus, B. Mikkelsen, C. Joergensen, S. Danielsen, and K. Stubkjaer, "All-Optical Wavelength Conversion by Semiconductor Optical Amplifiers," J. Lightwave Technol., vol. 14, no. 6, pp. 942-954, June 1996.
[2] M. Ragab, Wavelength Conversion using Nonlinear Effects in Optical Fibers, 1st ed., LAP LAMBERT Academic, 2013.
[3] R. Roussev, C. Langrock, J. Kurz, and M. Fejer, "Periodically Poled Lithium Niobate Waveguide Sum-Frequency Generator for Efficient Single-Photon Detection at Communication Wavelengths," Opt. Lett., vol. 29, no. 13, pp. 495-497, July 2004.
[4] S. Nakamura, Y. Ueno, and K. Tajima, "168-Gb/s All-Optical Wavelength Conversion with a Symmetric-Mach-Zehnder-Type Switch," IEEE Photon. Technol. Lett., vol. 13, no. 10, pp. 1091-1093, October 2001.
[5] G. Contestabile, Y. Yoshida, A. Maruta, and K. Kitayama, "Ultra-Broad Band, Low Power, Highly Efficient Coherent Wavelength Conversion in Quantum-dot SOA," Opt. Exp., vol. 20, no. 25, pp. 27902-27907, December 2012.
[6] R. Matsumoto, Y. Yoshida, C. Stamatiadis, A. Maruta, and K. Kitayama, "A Novel Wavelength Converter for PAM Signal based on XGM in Cascaded SOAs," Proc. Asia Comm. and Photon. Conf. (ACP2015), Hong Kong (China), Nov. 2015, AS3J.3.
[7] G. Contestabile, R. Proietti, N. Calabretta, and E. Ciaramella, "Cross-Gain Compression in Semiconductor Optical Amplifiers," J. Lightwave Technol., vol. 25, no. 3, pp. 915-921, March 2007.
[8] T. Hatta, T. Miyahara, Y. Miyazaki, K. Takagi, K. Motoshima, K. Mishina, A. Maruta, and K. Kitayama, "Polarization-insensitive monolithic 40 Gbps SOA-MZI wavelength converter with narrow active waveguides," J. Sel. Topics Quantum Electron., vol. 13, no. 1, pp 32-39, 2007.

MF2-4

Experimental Demonstration of All-Optical FEC Coding Scheme with Convolutional Code

Yohei AIKAWA* and Hiroyuki UENOHARA*

*Precision and Intelligence Laboratory, Tokyo Institute of Technology
4259 Nagatsuta, Midori-ku, Yokohama, Kanagawa 226-8503, Japan
E-mail address: aikawa.y.aa@m.titech.ac.jp

Abstract: *We experimentally demonstrate all-optical FEC coding scheme consisting of an all-optical HNLF-based XOR gate with convolutional code. The proposed scheme offers 3.4 dB net coding gain at BER=10^{-9} with DPSK-modulated RZ-format signals at 10 Gbps.*
Keywords: *Optical signal processing, Optical FEC, Optical XOR*

I. INTRODUCTION

With the rapid increase of Internet traffic owing to the startling advancement in mobile-network and video-distribution services, the demand for large capacity optical networks has been growing. The technology of wavelength-defragmentation is one of the key functions to efficiently utilize the spectrum resources, however, it induces a dynamic noise change owing to the variation in their bandwidths [1]. To solve such problems, we have proposed an all-optical FEC coding scheme in the middle of the optical networks with no need for optical-to-electrical-to-optical converters or demodulating-decoding-coding-modulating processing in contrast to the electrical FEC coding scheme (Fig.1) [2].

Two-types of all-optical FEC, syndrome decoding of convolutional code and cyclic code have been reported because they can be applied to optical signal processing without buffer components [3][4]. However, these techniques are not well suited for advanced photonic networks, because of their relatively low coding gain. On the other hand, maximum-likelihood estimation (MLSE) of convolutional code, which can yield a high coding gain, is of considerable interest. Although the feasibility of the optical FEC technologies was analytically presented in the previous paper, there have been few experimental investigations so far.

In this paper, we experimentally demonstrated an all-optical FEC coding scheme with MLSE of convolutional code for realizing dynamic noise compensation. Experimental results of the proposed scheme for the performance and effectiveness of the all-optical processing will be presented.

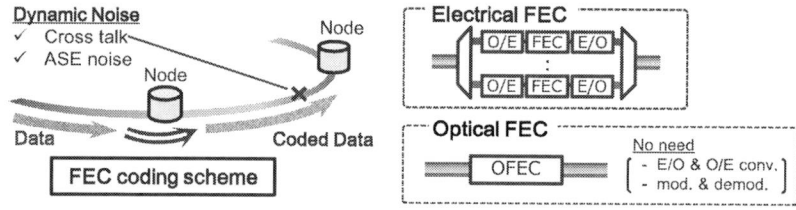

Fig. 1. Conceptual image of all-optical FEC coding scheme for dynamic noise-compensation technologies.

II. OPERATING PRINCIPLES

Fig.2 shows a diagram of the optical FEC coding circuit used to implement the MLSE of convolutional code. In this circuit, the injected signals to the coding circuit are encoded by a $(7,5)_8$ coding scheme whose generating polynomial is equal to $G(D)= (1+D+D^2, 1+D^2)$. The proposed circuit mainly comprises two optical exclusive-OR (XOR) gates and delay line, which correspond to "+" and "D" in the equation of generating polynomial.

When a DPSK-modulated signal at the wavelength λ_A is injected into this circuit, it is divided equally into three signals. Two of these undergo wavelength conversions to λ_B and λ_C, respectively, by use of the degenerate four-wave mixing (FWM) in the highly non-linear fiber (HNLF) in each wavelength converter (WC). Subsequently, they are delayed in 1- and 2-bit time duration, respectively, by propagating in single-mode fibers with comparable lengths corresponding to the delay times. In the upper XOR gate, the input signal (λ_A), 1-bit delayed signal (λ_C) and 2-bit delayed signal (λ_B) are injected into the HNLF, and the idler at a wavelength of λ_{idler} ($\lambda_{idler}^{-1}=\lambda_A^{-1}-\lambda_B^{-1}+\lambda_C^{-1}$) is obtained. Similarly, in the lower XOR gate, three input signals at λ_A, λ_B, and CW probe at λ_C are injected into the HNLF, and the idler at λ_{idler} is obtained. Therefore, these output signals correspond to the results of the $(7,5)_8$ coding scheme.

Fig 2 (a) shows the operating principles of a FWM-based three-input optical XOR gate. An idler at a wavelength of λ_{idler}, with an optical frequency of $f_{idler}=f_A-f_B+f_C$ is generated by FWM in the HNLF, where f_A, f_B, and f_C represent the optical frequencies of λ_A, λ_B, and λ_C. Under this condition, the phase of the idler is $\varphi_{idler}=\varphi_A -\varphi_B+\varphi_C$, where φ_A, φ_B, and φ_C represent the optical phases of λ_A, λ_B, and λ_C, according to the phase matching condition. With the input signals having a phase of either 0 or π owing to the DPSK format, the generated phase is 0, π. Then, $\varphi_{idler}=\varphi_A -\varphi_B+\varphi_C$ is simplified as $\varphi_{idler}=\varphi_A \oplus \varphi_B \oplus \varphi_C$. If we change the signal C into a CW probe, the phase of the idler is similarly expressed as $\varphi_{idle} =\varphi_A -\varphi_B=\varphi_A \oplus \varphi_B$ by canceling the effect of φ_C during demodulating processing because of its continuous phase change (Fig 2 (b)).

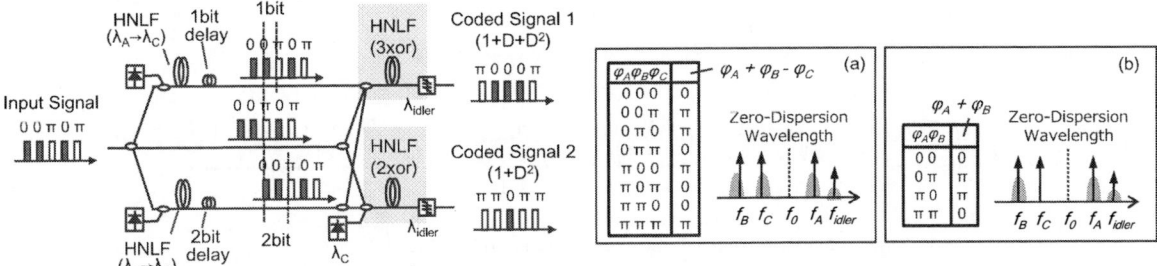

Fig. 2. Diagram of all-optical coding circuit and operating principles of (a) three-input XOR gate, (b) two-input XOR gate.

III. EXPERIMENTAL SETUP

The experimental setup for all-optical FEC coding scheme consisting of three signal sources is shown in Fig.3. In this experiment, three signal sources were utilized to confirm the feasibility of the proposed scheme. As depicted in Fig.3, 10.72 Gbps DPSK-modulated NRZ signals were generated using three tunable laser diodes (TLDs) emitting at λ_A=1547.5 nm, λ_B=1554.5 nm, and λ_C=1550.5 nm, respectively, and a Lithium Niobate phase modulator (LN-PM) which was driven by a pulse pattern generator with pseudo-random bit sequence (2^{23}-1 PRBS). Then, NRZ signals was converted into RZ-format by a LN intensity modulator (LN-IM). In the case of coded signal 1, three DPSK signals A, B, and C were separated by an optical couplers, then relatively delayed in 1- and 2-bit duration corresponding to the $(7,5)_8$ coding scheme using optical delay lines. In the case of coded signal 2, the signal C was changed into CW probe.

The signals were coupled, and they were amplified by an erbium-doped fiber amplifier (EDFA) to be around 13 dBm, respectively. The amplified signals were injected into an HNLF, with a length of 600 m, a dispersion slope of 0.006 ps nm^{-2} km^{-1}, a non-linear coefficient of 10.8 W^{-1} km^{-1}, and zero-dispersion wavelength of 1529 nm.

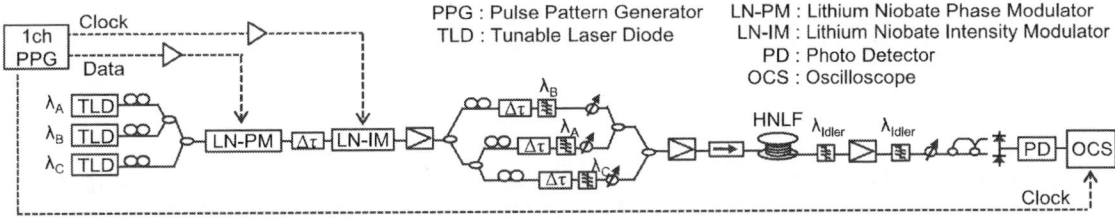

Fig. 3. Experimental setup for all-optical FEC coding scheme consisting of three signal sources

IV. EXPERIMENTAL RESULTS

Fig. 4 shows the observed temporal waveforms of the signal A, B, C, and the idler, respectively. The signals were demodulated using a balanced photo detector following a one-bit delayed interferometer with free spectral range of 10.72 GHz. The lower and upper peaks of the waveforms in Fig. 4 are the bit pattern 0 and 1, respectively, which correspond to phase differences of 0 and π, respectively. As can be seen in Fig. 4, the coded signals with a precise bit pattern corresponding to the XOR operation between any input signals are obtained. The Q-factors of 7.8 and 8.8 are obtained in the case of coded signal 1 and coded signal 2, respectively.

The performance of all-optical FEC coding was experimentally investigated in various signal powers, and the results are indicated in Fig.5. Firstly, in the coded signal 2, the average power of CW probe (λ_C) was fixed at 13 dBm, and relative power was normalized based on 13 dBm, the power of signal A and B was changed from 7 dBm to 19 dBm, respectively. In Fig. 5, the black circles and triangles indicate the dependences of Q-factor on signal power in various signal power of A and B. In both cases, high-quality coded signals are achieved under the condition that both signals have equal power. Then, in the case of coded signal 1, the average power of signal A and B were fixed at 13 dBm, the power of signal C are changed. The power dependence of Q-factor is indicated by black squares in Fig. 5. The maximum Q-factor is also achieved under the condition where all inputs are set at the same powers of 13 dBm.

Finally, we investigated bit error rate (BER) for both coded signal 1 and 2 with 2^7-1 PRBS. Fig. 6 shows the BER results of back-to-back (B2B), coded signal 1, and coded signal 2, respectively. In this experiment, the error-free operation can be achieved for all evaluated signals, and the power penalties of 0.7 and 0.5 dB for coded signal 1 and 2

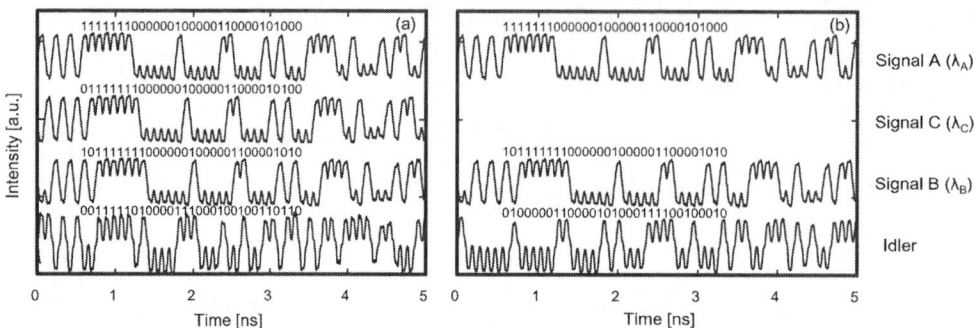

Fig. 4. Waveforms for all-optical FEC coding scheme: (a) coded signal 1, (b) coded signal 2.

at BER=10^{-9} are achieved. The decoded signal of the $(7,5)_8$ coding scheme is theoretically derived by using complementary error function with additive white Gaussian noise (AWGN) channel. Therefore, the proposed scheme offers 3.4 dB net coding gain at BER=10^{-9} owing to its negligible small power penalties.

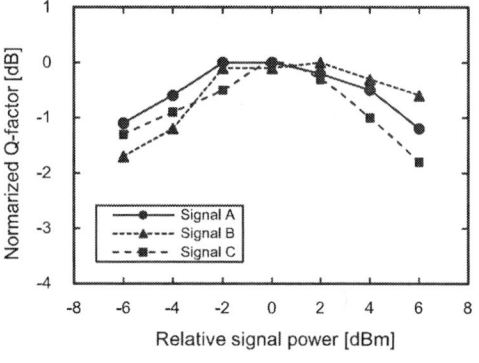

Fig. 5. Dependence of normalized Q-factor on relative signal power

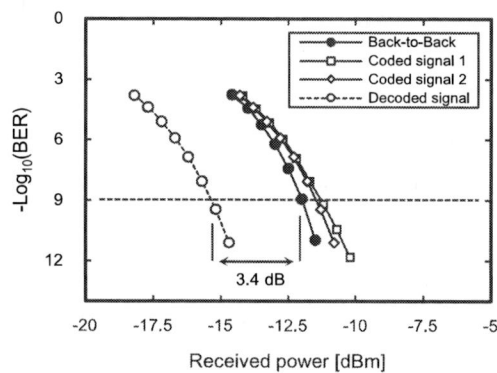

Fig. 6. BER investigation of coded signal 1 and coded signal 2

V. CONCLUSIONS

In this paper, we experimentally demonstrated an all-optical FEC coding scheme with convolutional code. The proposed scheme consists of three-signal sources and was operated by DPSK-modulated RZ format signal at 10 Gbps. The experimental results accurately correspond to the $(7,5)_8$ coding scheme between any input signals for both coded signal 1 and 2. For the optimized signal power condition, maximized Q-factor for both coded signals is achieved under the condition where all inputs into the HNLF are set to have equal power. The proposed scheme theoretically offers 3.4 dB net coding gain at BER=10^{-9} owing to its negligible small power penalties.

ACKNOWLEDGMENT

This work has supported by the Ministry of Education , Culture, Sports, Science and Technology of Japan #26-1211, and by the Ministry of Internal Affairs and Communications #0159-0093 (SCOPE). The authors would like to thank Prof. Emeritus K. Iga, Prof. Emeritus K. Kobayashi, Prof. F. Koyama, and Assoc. Prof. T. Miyamoto of Tokyo Institute of Technology for helpful discussions.

REFERENCES

[1] M. Jinno, H. Takara, B. Kozicki, Y. Tsukishima, Y. Sone, and S. Matsuoka, "Spectrum-Efficient and Scalable Elastic Optical Path Network: Architecture, Benefits, and Enabling Technologies," IEEE Commun. Mag., vol. 47, no. 11, pp. 66-73, 2009.
[2] Y. Aikawa and H. Uenohara, "Analytical investigation of all-optical FEC coding scheme with convolutional code," Int. Conf. on Photonics in Switching (PS 2015), Florence (Italy), pp.259-261, 2015.
[3] M. Suzuki and H. Uenohara, "Investigation of all-optical error detection circuit using SOA-MZI-based XOR gates at 10 Gbit/s," Electron. Lett., vol. 45, no. 4, pp. 224-225, 2009.
[4] Y. Aikawa, S. Shimizu, and H. Uenohara, "Demonstration of All-Optical Divider Circuit Using SOA-MZI-Type XOR Gate and Feedback Loop for Forward Error Detection," IEEE J. Lightwave Technol., vol. 29, no. 15, pp. 2259-2266, 2011.

PD1-1

OECC/PS2016

First Demonstration of Geographically Unconstrained Control of an Industrial Robot by Jointly Employing SDN-based Optical Transport Networks and Edge Compute

N.Yoshikane[1], T. Sato[2], Y. Isaji[2], C. Shao[3], T. Marco[3], S. Okamoto[2], T. Miyazawa[4], T. Ohshima[5], C. Yokoyama[5], Y. Sumida[5], H. Sugiyama[6], M. Miyabe[7], T. Katagiri[7], N. Kakegawa[8], S. Matsumoto[8], Y. Ohara[9], I. Satou[9], A. Nakamura[10], S. Yoshida[11], K. Ishii[11], S. Kametani[11], J. Nicho[12], J. Meyer[12], S. Edwards[12], P. Evans[12], T. Tsuritani[1], H. Harai[4], M. Razo[3], D. Hicks[3], A. Fumagalli[3], N. Yamanaka[2]

[1]KDDI R&D Labs., [2]Keio University, [3]The University of Texas at Dallas, [4]NICT, [5]NTT Communications, [6]Redhat, [7]Fujitsu, [8]IXIA communications, [9]OA Laboratory, [10]Toyo Corporation, [11]Mitsubishi Electric, [12]SwRI, yoshikane@kddilabs.jp

Abstract: Geographically unconstrained remote control of an industrial robot for surface blending over multi-domain SDN-based optical transport networks is demonstrated, showing that cloud/edge computing technology improves controllability of the industrial robot.
Keywords: Industrial robot, Software-Defined Networking, Optical transport network, Cloud/Edge computing, Cloud robotics.

INTRODUCTION

Industrial robots have been widely deployed to achieve efficient mass-production and cost-saving in the manufacturing sector. Since current industrial robot applications have been designed for repetitive and high-volume tasks, manufacturers do not have viable solutions for handling a variety of compute-intensive unstructured tasks. Cloud robotics [1] − where robots are connected to the cloud and can leverage the virtually unlimited compute and communication resources in the cloud to carry out the unstructured tasks − is a possible solution to overcome this drawback. For supporting time-constrained unstructured tasks which require data sets of considerable size to be exchanged between the robot production floor and the cloud, the development and adoption of new technologies such as high-speed flexible networks employing software-defined networking (SDN) technology are of the essence to flexibly interconnect the production floor to the cloud/edge sites. The introduction of an SDN-enabled edge node in support of IoT services by employing end-to-end SDN orchestration of integrated cloud and network resources has been reported recently [2]. However, no studies have been reported on the applicability of SDN-enabled optical transport network and cloud/edge computing technologies to supporting cloud robotics.

This paper introduces for the first time the remote control of an industrial robot performing *surface blending* by employing SDN-based optical transport networks. The SDN orchestrator enables the industrial robot located on the production floor and the robot control system located in the cloud/edge network to jointly and successfully carry our this unstructured task while minimizing both data transfer latency and volume across the transport network. Finally, the paper shows experimental validation of the proposed system.

PROPOSED SYSTEM ARCHITECTURE FOR REMOTE CONTROL OF INDUSTRIAL ROBOT

Figure 1 (a) and (b) show, respectively, the proposed system architecture and concept for remote control of an industrial robot with SDN-enabled optical transport networks and cloud/edge compute nodes. For the network configuration, a data center and an edge compute node located at the network edge are interconnected through a number of network domains.

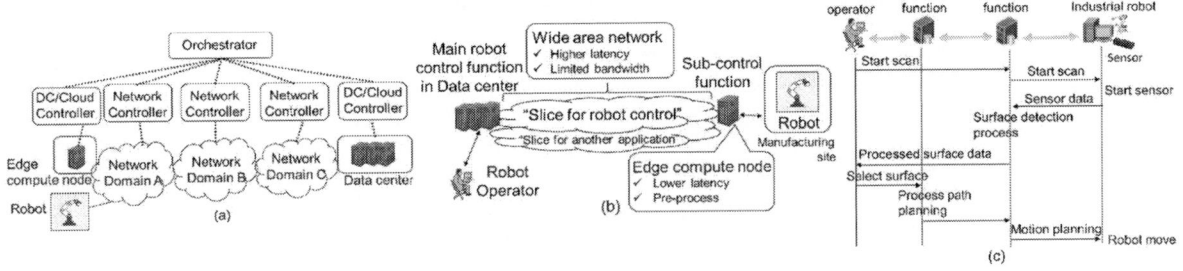

Fig. 1. (a) Proposed system architecture, (b) Concept of remote control of robot, (c) Robot control sequence.

133

OECC/PS2016

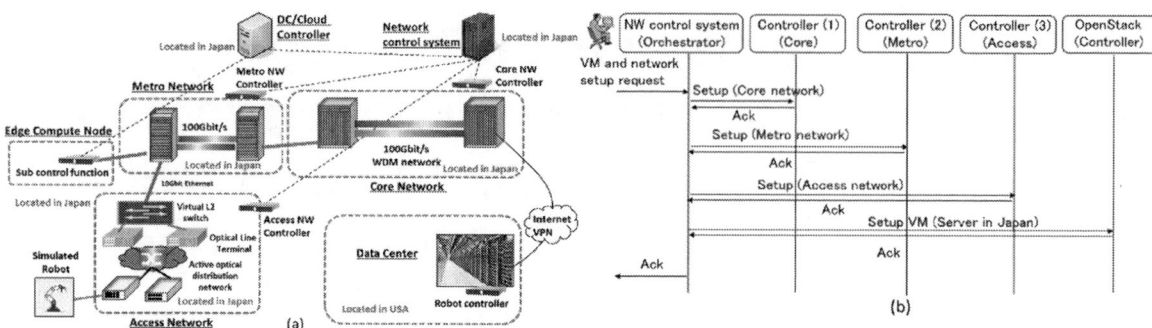

Fig. 2. (a) Testbed configuration, (b) Message workflow for path provisioning and virtual machine setup.

Each network domain, the data center and the edge compute node are controlled by a domain specific network controller and a cloud/edge controller, respectively. A holistic network orchestration (performed by the Network control system in Fig. 1 (a)) is introduced for end-to-end flow provisioning as well as setting up virtual machine (VM) on the cloud/edge resources at both the datacenter and edge compute node. The industrial robot system consists of a robot arm and controller based on the robot operating system (ROS) framework [3]. Fig. 1 (c) shows the robot control sequence for the surface blending of a metal piece. The ROS control functions are separated into two parts; 1) one is the main control function, which is implemented in the datacenter in order to assist and interact with a robot operator. 2) the other is a sub-control function, which is implemented at the edge compute node to assist and interact with the robot in order to reduce latency and minimize the data traffic volume that is going back and forth across the networks. The task of the robot is twofold. It first scans the piece with a 3D sensor, sending the collected 3D raw data to a server at the edge compute node (i.e., sub-control function). The edge compute node processes the raw data to reconstruct a digital image of the surface that needs to be blended. The digital image of the surface is then transmitted to a server in the datacenter, which makes use of an appropriate set of algorithms to compute the blending path for the robot, defined as the sequence of movements that the robot has to execute in order to blend the surface with a grinding tool attached to the robot. Finally, the blending path is sent to the robot to be executed.

TESTBED IMPLEMENTATION

Figure 2 (a) shows the testbed configuration. The data planes of the testbed are implemented at the NICT premises in Tokyo, Japan and at the University of Texas at Dallas premises in Dallas, TX, USA. The testbed in Japan consists of three different network domains: a WDM core network domain, a metro network domain, and elastic lambda aggregation network (EλAN) [4] domain (i.e., access network domain). The core network domain comprises two WDM nodes and an SDN controller. Each node is equipped with 100-Gb/s optical transponders to provide a 100-Gb/s wavelength transmission pipe for the attached metro network. The metro network comprises two packet optical transport systems (POTS) with an SDN controller. The POTS metro network domain provides a 100-Gb/s transmission pipe for the attached edge compute node, controlled by an OpenStack [5] controller, and the access network. The sub-control function to control the industrial robot for surface detection and motion planner is implemented at the edge compute node. The access network is constructed from EλAN and its associated SDN controller. The robot arm is simulated by a computer [3] connected to the access network. In the control plane, the network orchestrator is connected to the WDM SDN controller, the POTS SDN controller, the EλAN SDN controller in the access network, and the OpenStack controller. As a southbound interface (SBI) of the network orchestrator, a RESTful interface is utilized. The network orchestrator is based on the OdenOS architecture [6] and each domain is orchestrated by the orchestrator located at the KDDI Labs premises in Saitama, Japan, and connected through a virtual private network (VPN) connection.

The testbed in the USA consists of a datacenter emulated by a server located at the University of Texas at Dallas premises in Dallas, USA. The main robot control function (i.e., blending path computation) is implemented on the server and is controlled by the robot operator. The network connection between Japan and USA is established using a VPN. Fig. 2 (b) shows the message workflow between the controllers and the orchestrator, which requests VM setup and path provisioning over the three network domains. The VM setup request is submitted to the OpenStack controller, while the interconnection setup requests among the datacenter, the VM at the edge compute node and the simulated robot are submitted to each of the domain specific controller for execution.

USE CASE AND RESULTS: ROBOT OPERATION FOR SURFACE BLENDING

The proposed concept is validated with a surface blending use case, where the combination of the SDN-enabled optical transport networks and the edge compute node assist an industrial robot arm in carrying out the unstructured task. In the demonstration, an end-to-end data flow between the server for the main robot control function in the USA and the simulated robot via the edge compute node is provisioned. A virtual machine is setup at the edge compute node to assist

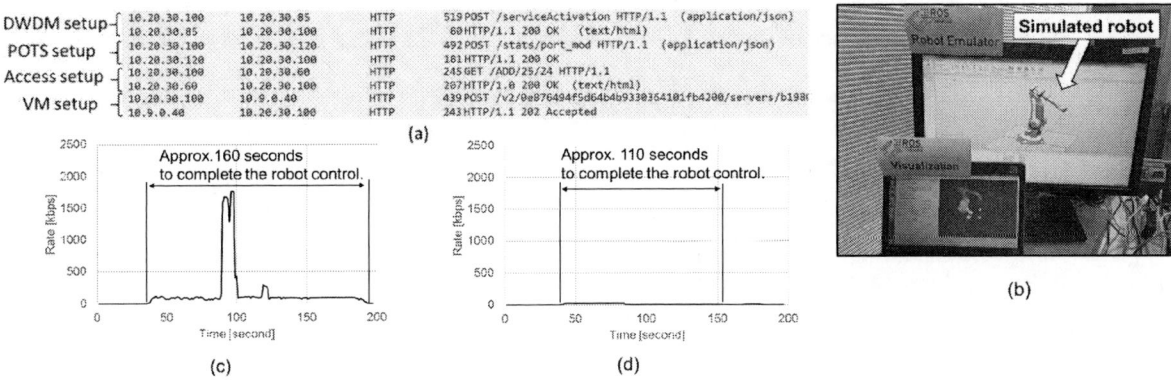

Fig. 3. (a) Captured message for network provisioning and VM setup. (b) Photo of the simulated robot, (c) Data receive rate at the robot controller without the sub-controller at the edge compute node, (d) Data receive rate at the main controller with the sub-controller at the edge compute node.

the industrial robot with processing the collected 3D raw data, which is pre-defined in the computer that simulates the robot arm. Next, the robot control software at the edge compute node is turned up and the robot control is started by the robot operator. Fig. 3 (a) shows the message exchange between the network control system and each domain SDN controller to provide the resource slice required for controlling the robot and the running the VM. The setup delay for the end-to-end flow is about 3 s, and the VM setup by the DC/cloud controller is in the order of 10 s. The average network latency between the edge compute node and the simulated robot is approximately 0.5 milliseconds, while the average latency between the server in the USA and the edge compute node is approximately 70.5 milliseconds (one way). Fig. 3 (b) shows the photograph of the simulated robot arm utilized in the demonstration. By utilizing this simulated robot, it is possible to compare the system controllability with and without the sub-controller running at the edge compute node. Fig. 3 (c) and (d) show the obtained data receive rate measured at the main controller with and without the sub-controller running at the edge compute node, respectively. When the sub-controller is not utilized for the robot control, the peak data rate is over 1600 kbps and it takes approximately 160 seconds to complete the robot control procedure of the main controller as shown in Fig.3 (c). When the sub-controller is running at the edge compute node, since the 3D sensor data is transferred only to the sub-controller over the access network, only the image of the detected surface is sent to the main controller in the datacenter. Therefore, the data transmission requires significantly less bandwidth (less than 30 kbps) compared to the previous case without the sub-controller as shown in Fig.3 (d) and it takes about 110 seconds to complete the robot control procedure of the main controller in this case. Overall, when the surface blending operation is supported by the combination of SDN-enabled networks and edge compute node, the completion time for the surface blending operation is reduced by about 30 %, thanks to the resource slice specifically provisioned to the robot control and sub-controller function running at the edge compute node.

CONCLUSION

Geographically unconstrained remote control of an industrial robot performing surface blending was, for the first time, achieved and demonstrated by employing SDN-enabled optical transport networks and cloud/edge computing technology.

ACKNOWLEDGMENTS

This work was partially supported by "Research and Development of Elastic Optical Aggregation Network," the commissioned research of NICT, and both NSF (grants # CNS-1111329, CNS-1405405, CNS-1409849, ACI-1541461, and CNS-1531039), USA, and NICT, Japan, under the JUNO (Joint Japan-US Network Opportunity) program, NICT "ACTION project". This work was promoted by Keihanna Interoperability Working Group. The authors would like to thank Furukawa electric for providing a wavelength-selective switch. The authors also would like to thank Mr. K. Miwa, Mr. R. Mikami and Mr. T. Uemura of NICT for their support in the experiment.

REFERENCES

[1] B. Kehoe, S. Patil, P. Abbeel, and K. Goldberg, "A Survey of Research on Cloud Robotics," IEEE Transactions on automation science and engineering, vol. 12, No.12, April 2015, pp.398-409.

[2] R. Vilalta, A. Mayoral, D. Pubill, R. Cassellas, R. Martinez, J. Serra, C. Verikoukis, and R. Munos, "End-to-End WDN Orchestration of IoT Services Using an SDN/NFV-enabled Edge Node," OFC2016, W2A.42.

[3] Robot Operating System (ROS) Industrial, http://rosindustrial.org/

[4] T. Sato, K. Tokuhashi, H. Takeshita, S. Okamoto, and N. Yamanaka, "A Study on Network Control Method in Elastic Lambda Aggregation Network (EλAN)," IEEE Conference on High Performance Switching and Routing (HPSR 2013), S2.1, July 2013.

[5] OpenStack Project, http://www.openstack.org/

[6] Y. Iizawa, and K. Ishida, "Multi-layer and Multi-domain Network Orchestration by ODENOS," OFC2016, Th1A.3.

THz Photonics-Wireless Transmission of 160 Gbit/s Bitrate

Xianbin Yu[1,2*], Shi Jia[2,3], Hao Hu[2], Pengyu Guan[2], Michael Galili[2], Toshio Morioka[2], Peter U. Jepsen[2], Leif K. Oxenløwe[2]

[1]College of Information Science and Electronic Engineering, Zhejiang University, Hangzhou 310027, China
[2]DTU Fotonik, Department of Photonics Engineering, Technical University of Denmark, DK-2800, Lyngby, Denmark
[3]School of Electronic Information Engineering, Tianjin University, Tianjin, China
E-mail: xyu@zju.edu.cn

Abstract: We present a record bitrate wireless transmission in the THz band above 300 GHz by successfully demonstrating a 160 Gbit/s photonics wireless link operating in the 300-500 GHz band based on a single THz emitter.
Keywords: THz photonics, RF photonics, THz communications.

I. INTRODUCTION

The demand of wireless data rates have grown explosively in the past decades, mobile data traffic is forecasted to increase 10-fold between 2014 and 2019 by Cisco, reaching 24.2 exabytes per month by 2019 [1]. To be capable of accommodating the ever increasing traffic in the access networks, wireless links in the future are highly desired to operate at a data rate of well beyond 100 Gbit/s, eventually up to Tbit/s. 100 Gbit/s Ethernet (100GbE) was already established in 2009, and indeed, optical fiber communication systems have been technically demonstrated with a capability of delivering several tens of Tbit/s. Therefore wireless links which are transparently, seamlessly compatible with the existing high capacity optical network infrastructure, will be beneficial and appreciated for last-mile wireless access. In addition to backhauling mobile data streams, ultrafast wireless communication technologies are also expected to significantly benefit many other bandwidth-hungry applications, such as wireless transmission of ultrahigh definition video, wireless transfer of large volume data, ultrafast intra/inter-chip data exchange, fast restoration of wired connections in disaster areas, and so on.

Recently, a lot of efforts have been contributed to develop photonics-aided high-speed wireless communication systems. 100 Gbit/s and beyond wireless connections have been achieved in the sub-THz band (200-300 GHz) [2][3] and in the W-band (75-110 GHz) incorporating optical polarization division multiplexing (PDM) [4] and spatial multiple-input-multiple-output (MIMO) techniques [5]. To move further, the THz band (300 GHz-10 THz) between the millimeter-wave and infrared radiation has been considered as the 'Next Frontier' to meet the eventual bitrate target of Tbit/s [6][7], due to its much larger frequency bandwidth comparing to the millimeter-wave band or sub-THz band, as well as less atmospheric disturbances compared to optical wireless links.

Up to date, significant progress on the development of THz photonics-wireless communication systems have been also achieved, for instance, tens of Gbit/s transmission have been enabled [9]-[13] by using ultra-broadband uni-travelling carrier photodiodes (UTC-PDs) [8] as photo-mixing emitters and Schottky diodes as electronic receivers. We have recently contributed to this development by demonstrating 60 Gbit/s Nyquist-QPSK wireless transmission at 400 GHz using a UTC-PD and a Schottky receiver [14], representing the highest reported bitrate in the frequency band above 300 GHz. However, the capacity potential of THz wireless systems has not been achieved yet. The use of free running lasers for heterodyne photo-mixing in the UTC-PD in the previous work [14] intrinsically leads to a THz beat-note with high phase noise, which seriously degrades the performance of spectrally efficient modulation formats requiring highly pure THz carriers.

In this paper, we significantly reduce the THz phase noise by creating a coherent optical frequency comb and compensating phase decorrelation of photo-mixing tones induced by the optical path difference. This scheme not only enables generation of high quality THz carriers, but also allows exploitation of large THz bandwidth in the 300-500 GHz band for communication, and up to 160 Gbit/s bitrate wireless transmission is successfully demonstrated. To the best of our knowledge, this is a new record in the THz band above 300 GHz, which pushes THz communication bitrate beyond 100 Gbit/s.

II. EXPERIMENTAL SETUP

The experimental configuration is shown in Fig. 1. We first optically create a frequency comb based on two concatenated phase modulators (PMs), both of which are driven by an amplified 25 GHz sinusoidal signal. An optical tunable delay line in-between is used to match the phase of the two-stage modulation, in order to improve the signal-to-noise ratio (SNR) of the optical tones in the comb used for the 300-500 GHz carrier generation. Subsequently, a programmable wavelength selective switch (WSS, Finisar 4000S) is employed to separate one optical comb line, to be used as an optical local oscillator (LO), from 8 other comb lines, to be used for optical baseband modulation. The LO line and the 8 data lines are 300-500 GHz apart, and the WSS additionally performs the equalization of the comb lines. The digital baseband data signal is generated from an arbitrary waveform generator (AWG) and modulated onto the lightwave at an in-phase (I) and quadrature (Q) modulator. In the experiment, we modulate a Nyquist quadrature phase shift keying (QPSK) PRBS 2^7-1 signal and perform Nyquist pulse shaping in the AWG by applying a square root raised cosine filter with 0.1 roll-off factor. The even-order and the odd-order channels after the IQ modulator are de-correlated by using a second WSS and a fiber delay line. With respect to the optical LO, we compensate the path by using a piece of matched fiber, since the path difference between the optical LO and the 8 data lines de-correlates their phases.

The LO and the 8 data lines are eventually directed on to the UTC-PD based THz photo-mixing emitter, generating 8 carrier frequencies centered around 400 GHz with 25 GHz spacing. The UTC-PD has an extremely fast photo-response and high responsivity (DC responsivity of 0.15 A/W here). Before launching into the UTC-PD, the optical signals are polarized to minimize the polarization dependency of the UTC-PD. In the free space path, a pair of THz lenses is used to collimate the THz beam, in order to reduce the free space propagation loss in a 50 cm line-of-sight (LOS) communication link. At the receiver side, a 12-order harmonic THz Schottky mixer operating in the frequency range of 300-500 GHz is used to down-convert the received THz signal into an intermediate frequency (IF) signal. The mixer is driven by a 31-36 GHz tunable electrical LO signal. The IF output is amplified by a chain of electrical amplifiers with 42 dB gain, and is finally demodulated and analyzed in a broadband real time sampling oscilloscope (63 GHz Keysight DSOZ634A Infiniium).

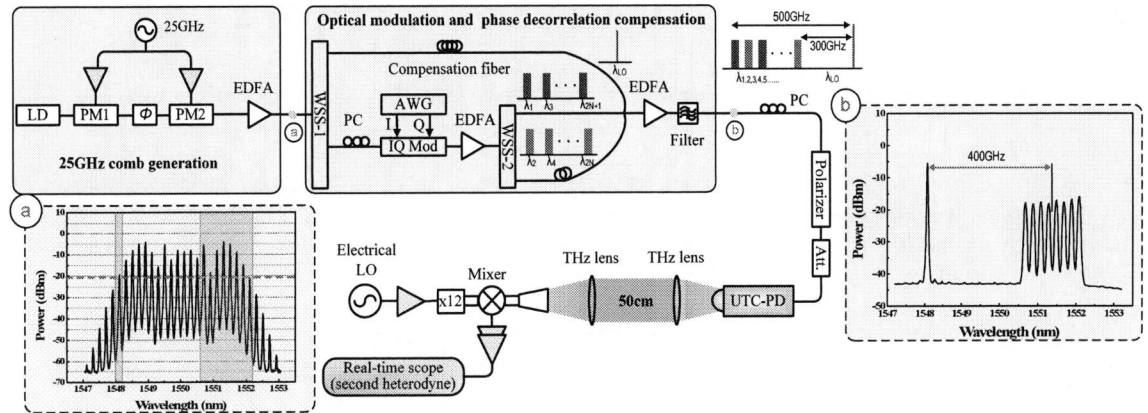

Fig. 1. Experimental configuration of the 300-500 GHz photonics-wireless communication link. PM: phase modulator, WSS: wavelength selectable switch, AWG: arbitrary waveform generator, LO: local oscillator, Att.: attenuator, UTC-PD: uni-travelling carrier photodiode. a) Generated optical frequency comb spectrum. b) WSS-prepared 8-channel centered 400 GHz with 25 GHz spacing before the UTC-PD.

III. EXPERIMENTAL RESULTS

The 25 GHz spaced optical frequency comb is shown in the inserted spectrum in Fig. 1(a). The comb extends beyond 5 nm and the desired optical tones (in shaded regions) for heterodyne generation of 300-500 GHz carriers present more than 40 dB carrier-to-noise ratio, which is mainly limited by the 25 GHz modulation index. The data modulated 8 channels are used with the LO to generate the THz signal around 400 GHz, as seen in Fig. 2(c). We measure the phase noise performance of the THz carriers after wireless propagation, as shown in Fig. 2(a). Compared to an electrical sinusoidal signal from a synthesizer, photo-mixing of two coherent tones directly from the comb generates a THz carrier (350 GHz in Fig.2(a)) with 10 dB higher phase noise at 10 kHz offset frequency. This is because the photo-mixing of the two phase-correlated harmonic tones, originating from the same laser, adds their individual phase noise. When the LO path after WSS1 is not compensated, the phase noise of the generated THz carrier is much worse due to phase de-correlation. When compensating the path difference with a piece of 49.5 m fiber, the same phase noise performance as for coherent beating can be achieved.

The combined optical spectrum before the UTC-PD is illustrated in Fig. 1(b). 8 tones from the 25 GHz comb occupy 200 GHz overall operation bandwidth. Photo-mixing of these 8 tones with the LO wavelength in the UTC-PD correspondingly generates the 8-channel 25 GHz spaced THz signals with a center frequency of around 400 GHz. The data modulation is 10 Gbaud Nyquist QPSK per channel, resulting in an aggregated bit rate of 160 Gbit/s. The baud rate is less than half of the channel spacing, for the sake of reducing interference from the neighbor channel in the down-conversion in the demonstration.

We measure bit error rate (BER) performance for all the 8 channels after 50 cm wireless transmission, as shown in Fig. 2(b). It can be seen that all channels have achieved a BER performance below the hard decision forward error correction limit (FEC, BER 3.8e-3 with 7% overhead). The BER performance is evaluated from the error-vector magnitude (EVM) of the processed constellations. For illustration purposes, two constellation diagrams corresponding to BER of 4e-4 and 7e-3 are also displayed. We observe that the BER performance of the 8 channels is approximately classified into 3 clusters: The 400 GHz channel is the best, the 375 GHz and 500 GHz channels are the worst, the 325-, 350-, 425- and 450 GHz channels are in-between, and the penalty between clusters is about 1 dB. This penalty can be explained by the unflat conversion loss of the Schottky mixer based receiver, as shown in Fig. 2(c). The receiver conversion loss fluctuates over the 300-500 GHz frequency range, 375 GHz and 500 GHz bands exhibit the largest conversion loss and 400 GHz lest, which comply well with the BER performance observation. Here the conversion loss is investigated by measuring the down-converted analogue IF power at 6 GHz without modulation and it is also reflected in the 8-channel electrical spectrum in Fig. 2(c).

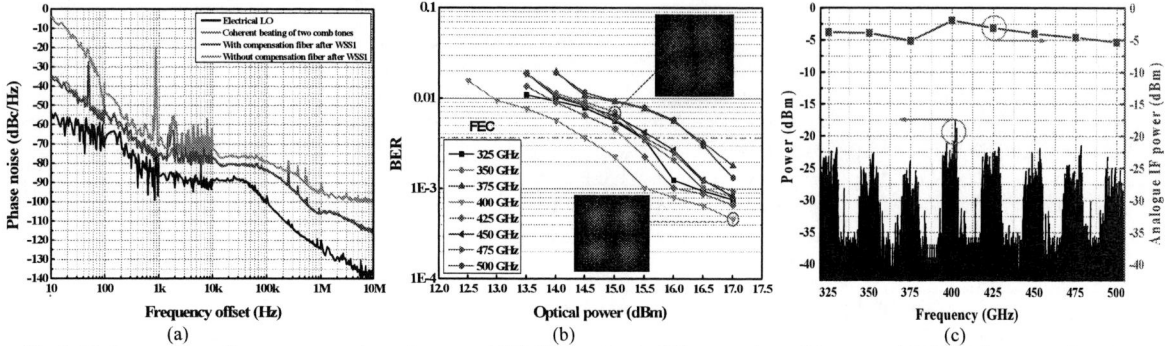

(a) (b) (c)

Fig. 2. (a) phase noise performance comparison of generated 350 GHz carriers in different regimes, (b) measured BER performance after 50 cm wireless transmission for 8 channels in 300-500 GHz band, (c) 8-channel electrical spectrum and frequency dependent conversion loss of the receiver.

IV. CONCLUSIONS

We have experimentally demonstrated an ultrafast THz wireless link in the 300-500 GHz band by using a single UTC-PD. By significantly improving the purity of generated THz carriers, the record 160 Gbit/s bitrate above 300 GHz band has been successfully achieved. This achievement reveals the potential of THz communication links accommodating beyond 100 Gbit/s bitrates, for bringing high capacity in optical networks to ultrafast wireless access applications.

ACKNOWLEDGMENT

We would like to thank the support by the ERC-PoC project TWIST, the Danish center of excellence CoE SPOC, and the Chinese Scholarship Council (CSC).

REFERENCES

[1] http://www.cisco.com/c/en/us/solutions/collateral/service-provider/ip-ngn-ip-next-generation-network/white_paper_c11-481360.pdf.
[2] S. Koenig, et al, "Wireless sub-THz communication system with high data rate," *Nature Photonics,* vol. 7, pp. 977-981, 2013.
[3] H. Shams, et al, "100 Gb/s multicarrier THz wireless transmission system with high frequency stability based on a gain-switched laser comb source," *IEEE Photonics J.,* vol. 7, no. 3, pp. 1-11, 2015.
[4] X. Pang, et al, "100Gbps hybrid optical fiber-wireless link in the W-band (75-110 GHz)", *Optics Express,* vol. 19, pp. 24944-24949, 2011.
[5] X. Li, et al, "A 400G optical wireless integration delivery system," *Optics Express,* vol. 21, no. 16, pp. 18812-18819, 2013.
[6] I. F. Akyildiz, et al, "Terahertz band: next frontier for wireless communications," *J. Physical Communication,* vol.12, pp. 16-32, 2014.
[7] X. Yu, et al, "The prospects of ultra-broadband THz wireless communications," *2014 International Conference on Transparent Optical Networks (ICTON) 2014,* Paper Th.A3.3.
[8] T. Ishibashi, et al., "Uni-traveling-carrier photodiodes for terahertz applications," *IEEE J. Select. Topics in Quantum Electron.,* vol. 20, no. 6, pp. 79-88, 2014.
[9] H.-J. Song, et al, "24Gbps data transmission in 300GHz band for future thz communications," *Electron. Lett.,* vol. 48, pp. 953-954, 2012.
[10] T. Nagatsuma, et al, "Terahertz wireless communications based on photonics technologies," *Optics Express,* vol. 21, no. 20, pp. 23736-23747, 2013.
[11] T. Nagatsuma, et al, "Enabling technologies for real-time 50Gbit/s wireless transmission at 300GHz," *2015 ACM NANOCOM,* Boston.
[12] G. Ducournau, et al, "Ultrawide-bandwidth single-channel 0.4-THz wireless link combining broadband quasi-optic photomixer and coherent detection," *IEEE Transactions on THz Science and Technology,* vol. 4, pp. 328-337, 2014.
[13] G. Ducournau, et al, "32Gbps QPSK transmission at 385GHz using coherent fibre-optic technologies and THz double heterodyne detection," *Electronics Letters,* vol. 51, no. 12, pp. 915-917, 2015.
[14] X. Yu, R. Asif, M. Piels, D. Zibar, M. Galili, T. Morioka, P. U. Jepsen, L. K. Oxenløwe, "60Gbit/s 400GHz Wireless Transmission," *Photonics in Switching 2015,* Postdeadline paper PDP1, 2015.

Low Noise, Regeneration of Optical Frequency Comb-Lines for 64QAM Enabled by SBS Gain

Mark Pelusi[1], Amol Choudhary[1], Takashi Inoue[2], David Marpaung[1], Benjamin Eggleton[1], Hung Nguyen Tan[2], Karen Solis-Trapala[2], Shu Namiki[2]

[1]Centre for Ultrahigh bandwidth Devices for Optical Systems (CUDOS), School of Physics, University of Sydney, NSW, Australia
[2]National Institute of Advanced Industrial Science and Technology (AIST), Tsukuba, Ibaraki, Japan
m.pelusi@physics.usyd.edu.au

Abstract: We demonstrate narrow-band spectral-line amplification by Stimulated Brillouin Scattering (SBS) of a parametrically generated frequency comb, enabling lower noise carriers for 96-Gb/s-DP-64-QAM with bit-error rate <4.5×10⁻³ over the C-band. Scalability to SBS gain-combs is also shown.
Keywords: Stimulated Brillouin Scattering, Laser sources, Coherent communications.

I. INTRODUCTION

Optical frequency combs can be a transformative solution for next generation high bit rate WDM communications to provide tens or hundreds of individual wavelength carriers for data modulation, at lower cost, energy consumption and footprint than multiple single frequency laser modules. The potential is even stronger as systems evolve to employ more spectrally efficient formats such as 64-QAM, requiring more sophisticated narrower linewidth lasers (<100 kHz) [1], than the common DFB lasers with >1 MHz linewidth. The challenges are the more stringent requirements on narrow carrier linewidth (<100 kHz), large peak carrier to noise power ratio (CNPR), flexible frequency separation in 10-100 GHz range, and high power (>10 mW), over the wide wavelength window around 1550 nm used for long distance transmission. So far, the most viable, broadest frequency combs have been obtained by the parametric Kerr effect for either a CW laser in ring resonator waveguides [2], or optical pulses in highly nonlinear fiber (HNLF), from CW lasers with electro-optic modulation [3], or mode-locked lasers [4]. Generating the broadest frequency combs, or post-amplification, however, usually comes at the cost of degraded CNPR, which limits performance, so regeneration schemes that can recover a high CNPR, without compromising other properties are needed. This has been limited so far to preceding the comb generation with a nonlinear amplifying loop mirror [3], or laser injection locking [5].

In this paper, we demonstrate for the first time, a new approach using narrow-band spectral-line amplification by Stimulated Brillouin Scattering (SBS), to boost the comb-line CNPR and enable its modulation with advanced data modulation formats. By applying ≈14 dB gain to select spectral lines from the C-band spanning 40 GHz frequency comb obtained by parametric comb generation in HNLF, we show improved bit error rate (BER) at the receiver (Rx) after modulation with 96 Gb/s DP-64-QAM signals, to < 4.5×10⁻³ (the hard decision FEC limit), for spectral lines across the C-band range, achieving double the usable bandwidth. The Q^2-factor is also improved by up to 3 dB within the bandwidth where the Q^2-factor is measurable without SBS. The SBS gain is obtained without the need for extra CW lasers and frequency locking, by feedback propagating the comb-line itself for pumping SBS, after its frequency upshift by electro-optic modulation (EOM). We also show the scalability of the scheme to amplify multiple comb lines simultaneously by pumping SBS with a frequency comb, to achieve a BER < 4.5×10⁻³ within a bandwidth containing 10 comb lines. The results highlight the capability of SBS with limited narrow gain bandwidth on the order of 10 MHz, to significantly suppress noise power around the carrier frequency to boost the CNPR for enabling 64-QAM modulation.

II. CONCEPT

Figure 1(a) shows the schematic for regenerating optical frequency combs using SBS gain. Unlike for previous applications of SBS gain to frequency combs [6-8], this work focusses on signal regeneration, for demonstrating the

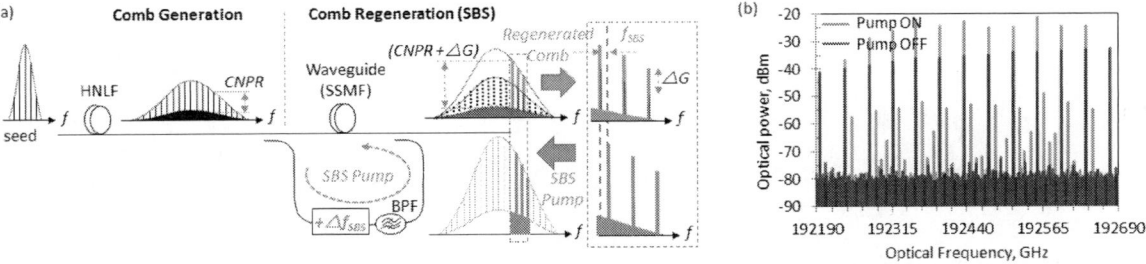

Fig. 1. Regeneration of parametrically generated optical frequency comb by SBS gain. (a) Schematic of SBS pumped by frequency-shifted comb lines. (b) Measured high resolution optical spectrum of SBS gain stage output set for parallel 10-line SBS comb pump, comparing SBS pump on vs. off.

performance gain from improved CNPR on signals relying on a carrier linewidth <100 kHz. This was implemented by propagating the frequency comb through an optical waveguide together with a backward propagating CW wave(s) to pump SBS. Instead of the usual SBS pumping by CW lasers that would require frequency locking to the comb for stable output power, the pump was derived from the comb itself by frequency upshifting a tapped off portion after EOM at frequency f_{SBS} [9], determined by the Doppler effect of the induced refractive index grating moving at the acoustic wave velocity, in order to align the SBS gain peak with the target comb-line frequency. A bandpass optical filter (BPF) in the pump path selected the comb-lines for SBS gain, by passing only the corresponding CW pump line(s) at f_{SBS} offset. The power of each pump line was set above the SBS threshold, to deliver a gain, $\triangle G$, such that spectral noise outside the SBS gain bandwidth is suppressed relative to the carrier for improved CNPR. Importantly, the power spectral density of the noise surrounding the dominant CW pump typically remains below threshold, so it doesn't corrupt the SBS gain.

III. EXPERIMENT

Figure 2(a) shows the experimental set-up. A 40 GHz optical frequency comb spanning the C-band (Fig. 2(b)) was generated from a 40 GHz harmonically mode-locked fiber laser (MLFL) by first phase stabilizing its output using cross phase modulation (XPM) in HNLF, before parametric comb broadening in a 300 m HNLF, as described in [10]. A 5 nm portion was then extracted by a BPF for select line amplification by SBS in 4.46km standard single mode fiber (SSMF). To derive the pump line(s), half of the input power was tapped off by a coupler and passed through a phase modulator driven by 2V p-p sinewave at f_{SBS} to form EOM sidebands that were then extracted by a programmable LCOS filter with ≈10 GHz spectral resolution. The upper frequency single sideband filter (SSBF) output at f_{SBS} was then amplified, and launched via a circulator in the reverse direction of the SSMF, whose optimum f_{SBS} for centering the gain peak with the forward propagating comb line reduced linearly with wavelength from 11.0 GHz at 1530 nm to 10.76 GHz at 1565 nm.

The CNPR of the frequency comb spectrum in Fig. 2(b) ranged from ≈26 dB at 1530 nm to 41 dB at 1546 nm (for measurement RBW = 0.03 nm), while the target output CNPR for minimum BER of the 64-QAM signal was >36 dB. This was satisfied by SBS with CW pump power ≈10mW per line giving $\triangle G$ of 13~14dB. Following SBS, a comb line was extracted by a BPF, and modulated with a 96 Gb/s DP-64 QAM signal obtained by driving an IQ-modulator with 8Gbaud 2^{11}-1 PRBS signals from an arbitrary waveform generator (AWG), before polarization multiplexing (Pol.-mux) and performance evaluation following coherent detection with a CW laser (<100 kHz linewidth), as described in [10].

IV. RESULTS AND DISCUSSION

Figure 3(a) shows the BER versus wavelength for modulating the 96 Gb/s DP-64QAM signal on isolated lines from the C-band comb in Fig. 2(b), for both with and without the SBS stage included in the set-up. This was compared to the reference case of connecting a wavelength tunable CW laser (linewidth <100kHz) directly to the IQ-modulator input. In case of no SBS stage, a BER below the 4.5×10^{-3} FEC limit was limited to only those comb lines between 1537.5-1552.5 nm. While with the SBS stage included for single line CW line pumping tuned across the C-band with optimized f_{SBS} for $\triangle G$ of generally 13~14dB, the BER remained $<4.5\times10^{-3}$ from 1530-1565nm, more than doubling the usable bandwidth. The constellations figures (inset) at 1562.6 nm, showed a 3 dB improvement in Q^2-factor from 6.3 to 9.3 dB. SBS also enabled a $\triangle Q^2$ penalty ≤1 dB from the reference laser for comb lines between 1530-1562.3nm. All BER and Q^2-factors were calculated from the PRBS and constellations, respectively, and averaged over both polarization channels, x and y.

The scalability of the SBS amplifier to regenerate multiple comb lines simultaneously was also demonstrated. The change from the previous single line SBS was to program the SSBF to extract multiple frequency upshifted sidebands after EOM, and scale the total pump power with the number of lines. The experiment tested a 10 line SSBF design, with the 5 nm BPF before the SBS stage tuned to 1557.3 nm where a BER below the FEC limit was not achievable without SBS. The EOM used f_{SBS} = 10.81 GHz, giving $\triangle G$ of mostly 11~15 dB for the 40 GHz spaced, 10 comb-lines between 1555.7 to 1558.7 nm. Figure 1(b) shows the high resolution optical spectrum trace of the SBS gain stage output, with

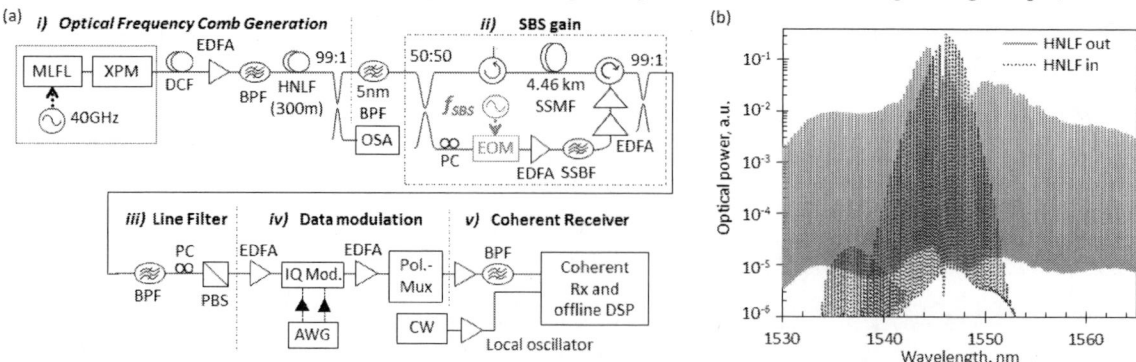

Fig. 2. (a) Experimental set-up of i) 40 GHz frequency comb generation in HNLF from output phase stabilized MLFL, ii) SBS gain, (iii) comb-line isolation, iv) data modulation and v) coherent receiver. (b) Optical spectra before and after spectral comb broadening in HNLF. RBW = 0.03 nm.

the SBS pump power on and off, showing the improved CNPR. The observed higher frequency sideband accompanying each SBS amplified comb line was residual pump feedthrough, which did not impact performance. Fig. 3(b) shows the BER vs. wavelength performance comparing the 4 cases of the reference laser, and frequency comb, for both without and with the SBS gain stage included, pumped by either a single CW spectral line, or parallel 10 line comb. It shows a BER below the FEC limit of 4.5×10^{-3} was achieved for all 10 comb-lines, and was even slightly better than for 1-pump SBS at 1557.3nm, explained by the higher $\triangle G$ used of 15dB vs. 13dB, benefiting both CNPR and SBS gain bandwidth. The observed parabolic dependence with wavelength is from the expected detuning of the comb line frequency from the peak SBS gain for a fixed f_{SBS}, which was the main limit for processing a larger number of comb lines with this set-up.

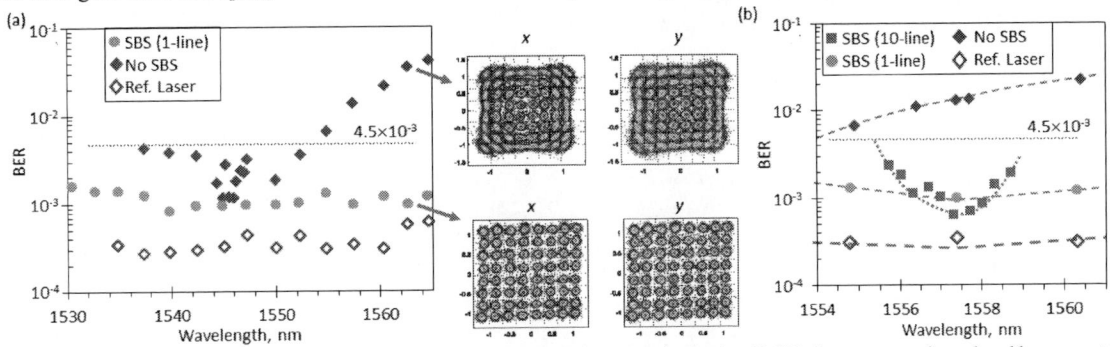

Fig. 3. BER of 96 Gb/s DP-64-QAM signal modulated on spectral line from C-band spanning 40 GHz frequency comb produced by parametric spectral broadening in HNLF of 40 GHz MLFL output pulses after output phase stabilization by XPM, for with and without SBS gain stage included, using SBS pumped by either (a) single spectral line, and (b) parallel 10-line 40 GHz frequency comb, compared to reference case of CW laser connected directly to IQ modulator. (Inset) Signal constellations of data modulated comb lines at 1562.4 nm, both with and without SBS gain stage.

V. CONCLUSIONS

In conclusion, improved carrier to noise power ratio for a C-band spanning optical frequency comb was demonstrated for enabling 64-QAM modulation. Applying line by line SBS gain of ≈13-14 dB, improved the BER to below the FEC limit for comb lines modulated with a 96 Gb/s DP-64-QAM signal, achieving more than double the usable bandwidth. The Q^2-factor penalty from the ideal reference laser case was improved by up to 3 dB, to be <1 dB over most of the C-band. The scheme was also shown scalable to simultaneous, multi-comb-line regeneration by using a frequency comb pump, enabling a BER below the FEC limit over a wavelength range containing 10 lines. These results demonstrate the regeneration capability of SBS for enabling broadband frequency combs to be modulated with higher order QAM.

ACKNOWLEDGMENT

This research was supported by the Project for Developing Innovation Systems of MEXT, Japan, and the Australian Research Council (ARC) Future Fellowship, Laureate Fellowship, DECRA and CUDOS programs (FT110101037, DE150101535, FL120100029, CE110001018). Dr. Hung Nguyen Tan also acknowledges the support of the University of Sydney International Research Collaboration Award.

REFERENCES

[1] T. Pfau, S. Hoffmann, and R. Noé, "Hardware-efficient coherent digital receiver concept with feedforward carrier recovery for M-QAM constellations", J. Lightwave Technol., vol. 27, pp. 989-999, 2009.

[2] J. Pfeifle et al., "Coherent terabit communications with microresonator Kerr frequency combs", Nature Photon. Vol. 8, pp. 375–380, 2014.

[3] V. Ataie, E. Temprana, L. Liu, E. Myslivets, B.P.-P. Kuo, N. Alic, and S. Radic, "Ultrahigh count coherent WDM channels transmission using optical parametric comb-based frequency synthesizer", J. Lightwave Technol. vol. 33, pp. 694-699, 2015.

[4] D. Hillerkuss, et al., "Single-laser 32.5 Tbit/s Nyquist WDM transmission", J. Opt. Commun. Netw. vol. 4, pp. 715-723, 2012.

[5] A. Lorences-Riesgo, T.A. Eriksson, A. Fülöp, P.A. Andrekson, and M. Karlsson, "Frequency-comb regeneration for self-homodyne superchannels", J. Lightwave Technol. vol. 34, pp. 1800-1806, 2016.

[6] J. Subías, C. Heras, J. Pelayo, and F. Villuendas, "All in fiber optical frequency metrology by selective Brillouin amplification of single peak in an optical comb", Opt. Express vol. 17, pp. 6753-6758, 2009.

[7] H. Al-Taiy, N. Wenzel, S. Preußler, J. Klinger, and T. Schneider, "Ultra-narrow linewidth, stable and tunable laser source for optical communication systems and spectroscopy," Opt. Lett., vol. 39, pp. 5826-5829, 2014.

[8] W. Li, W. T. Wang, W. H. Sun, J. G. Liu and N. H. Zhu, "Generation of flat optical frequency comb using a single polarization modulator and a brillouin-assisted power equalizer", IEEE Photonics Journal, vol. 6, pp. 1-8, 2014.

[9] T. Tanemura, Y. Takushima, and K. Kikuchi, "Narrowband optical filter, with a variable transmission spectrum, using stimulated Brillouin scattering in optical fiber", Opt. Lett., vol. 27, pp. 1552-1554, 2002.

[10] M. Pelusi, H. Nguyen Tan, K. S.-Trapala, T. Inoue, and S. Namiki, "C-band spanning frequency comb based on an output phase stabilized mode-locked laser", OptoElectronics and Communications Conference (OECC), July 2016, Nigata, Japan.

PD1-4 OECC/PS2016

20.7-Tb/s Repeater-less Transmission Over 401.1-km Using QSM Fiber and XPM Compensation via Transmitter-side DBP

Yue-Kai Huang[1], Ezra Ip[1], Shaoliang Zhang[1], Fatih Yaman[1], John D. Downie[2], William A. Wood[2], Aramais Zakharian[2], Jason Hurley[2], Snigdharaj Mishra[3], Yoshiaki Aono[4], Eduardo Mateo[5], Yoshihisa Inada[5]

[1] NEC Laboratories America, 4 Independence Way, Princeton, NJ 08550, USA
[2] Corning Inc., Corning, NY 14831, USA. [3] Corning Optical Communications, Corning Incorporated, Wilmington, NC 28405, USA
[4] Converged Network Division, NEC Corporation, Abiko, Japan [5] Submarine Network Division, NEC Corporation, Tokyo, Japan
Author e-mail address: kai@nec-labs.com

Abstract: *We achieved record repeaterless transmission of 20.7-Tb/s over a 401.1-km link comprising hybrid spans of QSM and single-mode fibers by applying multi-channel transmitter-side digital back-propagation. We also demonstrate error-free DWDM transmission of real-time 200G channels.*
Keywords: *Repeater-less transmission, digital back propagation, cross-phase modulation compensation.*

I. INTRODUCTION

Repeater-less transmission, assisted by Raman amplification and remote optically pumped amplifiers (ROPA), is a practical solution for high data rate transmission over long distances without inline amplifiers. Recently, digital coherent transmission has enabled advanced modulation formats which have higher spectral efficiency (SE), allowing increased system capacity in repeater-less links. Dual-polarization (DP) 16-ary quadrature amplitude modulation (QAM) was demonstrated for a 355-km link, achieving a capacity of 15.4-Tb/s [1], while DP-32QAM has been demonstrated in a repeater-less multi-core fiber experiment, achieving 17.2-Tb/s per core at a distance of 204 km [2].

It is possible to increase system capacity and reach by using low-loss, large-effective area (A_{eff}) quasi-single mode (QSM) fibers. In conventional single-mode fibers, it is difficult to obtain ever larger A_{eff} at satisfactory bending loss. By relaxing the requirement that the fiber be purely single-moded, QSM fiber can achieve large A_{eff} and good mode confinement, enabling higher signal powers without excessive nonlinear penalties. QSM fiber transmission was recently demonstrated for ultra-long haul submarine links using digital signal processing (DSP) designed to compensate inter-symbol interference arising from multipath interference (MPI) [3.4]. For transmission over a single repeater-less span, the deleterious effects of MPI caused by destructive interference will be smaller because the system length is reduced.

In this paper, we report the transmission of 108×32-Gbaud DP-16QAM channels over a 401.1 km link comprising hybrid spans of QSM fiber and Vascade® EX2000 fiber. A record single-core capacity of 20.7 Tb/s is achieved. The use of transmitter-side multi-channel digital backpropagation (DBP) to combat both self-phase modulation (SPM) and cross-phase modulation (XPM) enables the use of wider optical spectrum, as the shorter wavelength channels were launched at higher powers to equalize optical signal-to-noise ratio (OSNR) across the C-band.. We also show the compatibility of QSM fiber with commercially available transceiver by demonstrating errorless transmission of 12×200-Gb/s DP-16QAM channels over the same link.

II. EXPERIMENTAL SETUP

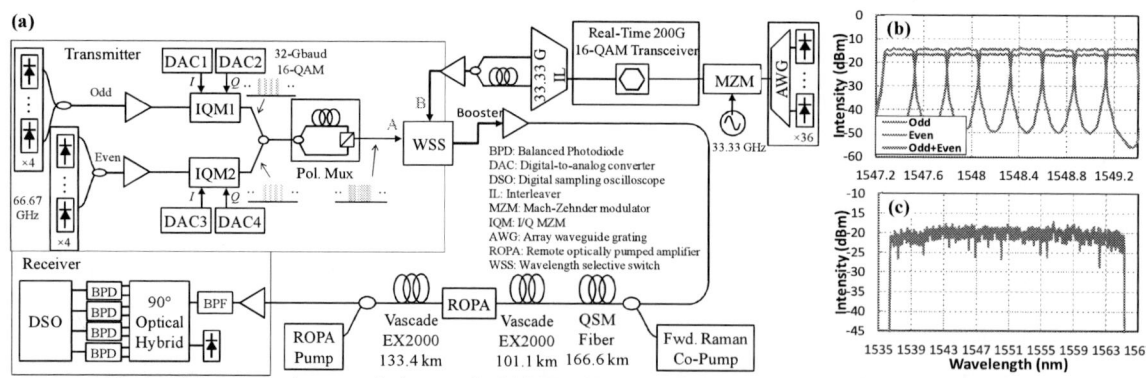

Fig. 1. (a) Experimental setup; optical spectra for (b) 8 DP-16QAM test channels and (c) 108 combined C-band channels.

The experimental setup is shown in Fig. 1a. Eight external cavity lasers (ECLs) at a channel spacing of 33.33 GHz were used for the channels under test. These were divided into odd and even groups; each group is modulated by a different Mach-Zehnder in-phase/quadrature modulators (IQMs) driven by the output of 8-bit 64-GSa/s digital-to-analog converters (DACs). To generate the I/Q drive signals, we encode raw data from a pseudorandom binary sequence (PRBS) of length of $2^{18}-1$ with a binary regular LDPC (32500, 24375, 0.75) encoder with girth 8 and column weight of 3. The encoded bits are interleaved and mapped to 16-QAM (or 32-QAM) symbols. The sequence was then truncated to the memory length of the DAC which was 16384 symbols (32768 samples). Raised-cosine pulse shaping with 1% roll-off was applied, followed by optional transmitter-side DBP, and pre-compensation of the DAC response. The modulated odd- and even-carriers are combined using a polarization maintaining (PM) 3-dB coupler. The resulting signal was then split again; with one signal rotated to the orthogonal polarization followed by a decorrelation delay of 256 ns (8192 symbols) equal to half the DAC memory length. We multiplex the 8×32-Gbaud DP-16QAM channels under test (Fig. 1b) with dummy channels using a flexible-band wavelength selective switch (WSS). To create the dummy channels, 36 distributed feedback (DFB) lasers at 100 GHz spacing were first modulated by a Mach-Zehnder modulator (MZM) driven by a 33.33 GHz clock to produce three optical carriers per laser. The resulting 108 carriers are simultaneously modulated by an IQM inside the real-time 200G transceiver. To decorrelate neighboring channels, a 33.3-GHz interleaver was used to split odd- and even-channels, followed by delay decorrelation and passive coupling. The spectrum after combining the channels under test (A in Fig. 1a) with the dummy channels (B) is shown in Fig. 1c.

The 401.1 km link in this experiment comprises of hybrid spans of QSM fiber and single-mode Vascade EX2000 fiber whose A_{eff} are ~200 µm^2 and 112 µm^2, respectively. The average attenuations for the two fiber types are 0.157 and 0.158 dB/km. Including connector and splice losses, the total link loss is 69.1 dB. Transmission is assisted by an MPB forward-Raman co-pump with four laser diodes at 1400, 1410, 1426 and 1454 nm delivering a maximum total nominal pump power of 2.8 W. A remote optically pumped amplifier (ROPA) is inserted at 133.4 km from the receiver. The ROPA is pumped using MPB's third-order Raman fiber laser (RFL) [5,6], which has a 5.4 W pump laser at 1276 nm and a 1485 nm seed laser whose power was set to 62.75 mW for optimal system performance. The Raman/ROPA pump wavelengths are all below the QSM fiber cutoffs with no observable detrimental effects.

At the receiver, a channel of interest is coherently detected by mixing the signal with a local oscillator (LO) laser in a dual-polarization optical 90° hybrid followed by balanced photodetection. The I/Q waveforms of the two polarizations are sampled and digitized using a quad-channel 50-GSa/s digital sampling oscilloscope (DSO) with 20 GHz bandwidth. Offline DSP was used to recover the transmitted data: the captured waveforms are first re-sampled to two samples per symbol. When CD and/or nonlinearity of the fiber channel are compensated by the receiver, we apply either chromatic dispersion compensation (CDC) using a frequency-domain equalizer (FDE), or digital backpropagation (Rx-DBP) which will compensate SPM nonlinearity. An adaptive butterfly-structure time-domain equalizer (TDE) is used to recover the symbols. The equalized symbols are sent to the offline LDPC decoder. For bit-error rate (BER) counting, five data sets of 20 µs duration (3.2 million symbols) were processed to estimate each BER measurement.

The use of DAC at the transmitter of our setup optionally enabled transmitter-side DBP (Tx-DBP). In order to perform Tx-DBP correctly on both polarizations, the DAC data pattern needs to be twice the decorrelation delay (i.e., 16384 symbols) since the two polarizations are delayed copies of each other. Furthermore, as the transmitter has knowledge of the data patterns modulating all (odd- and even-) carriers, it is possible to perform multi-channel DBP to compensate both SPM and XPM [7,8].

III. RESULTS AND DISCUSSIONS

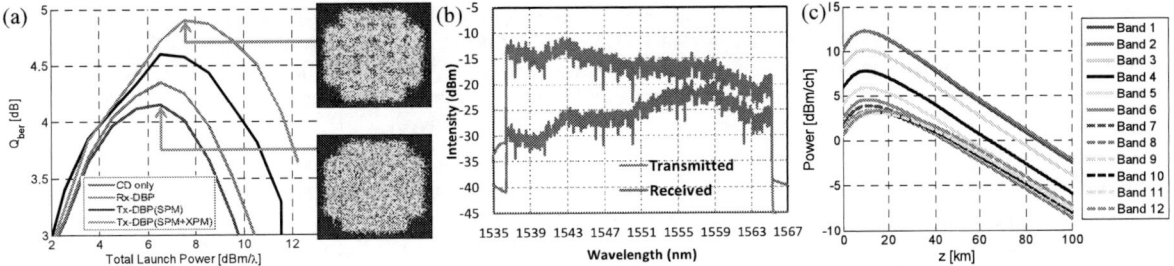

Fig. 2. (a) Q. vs launch power for DP-32QAM; (b) optical spectra for 20.7-Tb/s transmission; (c) power profiles vs. distance for different bands.

We first assessed the effectiveness of the interchannel nonlinear compensation via DBP by transmitting only the 8 channels under test, each modulated with 32-Gbaud DP-32QAM at a net SE of 7.2-b/s/Hz. We define the Q-factor from measured BER as $Q_{ber} = 20log_{10}\left(\sqrt{2}erfc^{-1}(2\,BER)\right)$. Fig. 2a shows Q_{ber} vs launch power sweep when different DSP algorithms are used. With CDC, the optimum Q is only 4.1 dB at 6.5 dBm/λ. Using 8-channel Tx-DBP to compensate SPM+XPM, the optimum Q is increased to 4.9 dB at 7.5 dBm/λ – a net Q improvement of 0.8 dB. The insets in Fig. 2a show the recovered signal constellations using CDC and using 8-channel Tx-DBP at their optimal launch powers. We observe that when single-channel DBP is performed (SPM compensation only), Tx-DBP outperforms Rx-DBP because noise corrupts the received signal, and due to the nonlinear nature of the Manakov equation, perturbations in the initial waveform grow. Tx-DBP benefits from being performed on a "clean" waveform distorted only by quantization noise and any uncompensated DAC nonlinearity & impulse response.

To demonstrate high capacity across the C-band, we transmit DP-16QAM on all the carriers. The reduction in constellation size is necessary because of decreased optimal channel launch power and insufficient Raman pump power. The optical spectra at launch and after transmission are shown in Fig. 2b. An 8 dB tilt in launch power is intentionally created to realize flat OSNR across the C-band after transmission. Differential link loss across the C-band is due to Raman power transfer and ROPA gain tilt. Since nonlinearity scales with signal power, the large power difference between different parts of the spectrum meant we needed to generate different pre-distorted waveforms via DBP for different parts of the C-band. The power profiles vs distance for different bands (Band 1 being the shortest λ's) are shown in Fig. 2c. The BER of all 108 channels was measured by sweeping the 8 test channels across the C-band, and using the WSS to insert them correctly with the dummy channels (Fig. 1a). The result in Fig. 3a shows that with CDC only, most channels in the short-wavelength region (< 1545 nm) are below the FEC threshold due to nonlinearity. As these were launched at the highest power, Tx-DBP is also the most effective in this region. The use of Tx-DBP yielded Q improvement of 0.2 to 0.6 dB in Bands 1 to 4, enabling all channels to have Q-factors above the FEC limit of 4.33 dB (BER = 5.0×10^{-2}) for 33.33%-overhead LDPC. All channels are recovered error-free after FEC decoding. We estimated FEC threshold offline simulation shown in Fig. 3b, which assumes gray mapping for 16-QAM and an additional 0.15 dB to reduce post-FEC BER from 1×10^{-9} to 1×10^{-15}, as shown in LDPC design with similar code girth and codeword length [9]. The total capacity realized by this experiment is $108 \times 32 \times 8/1.3333 = 20.74$ Tb/s, at net SE of 5.76 b/s/Hz.

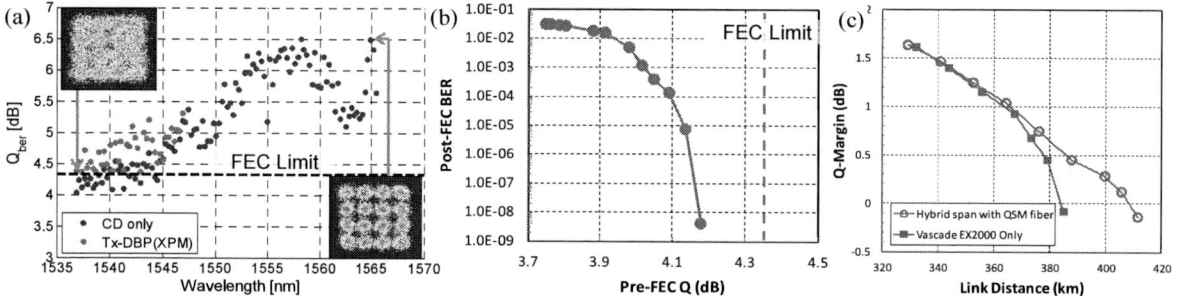

Fig. 3. (a) Q vs. wavelength after 401.1-km transmission; (b) offline FEC simulation; (c) Q-margin vs. emulated distance for real-time 200G.

We also conducted real-time 200G DP-16QAM transmission over the same link to evaluate the effectiveness of QSM fiber for reach enhancement in repeater-less systems. We replaced all the channels in the previous experiment by a DWDM 12×200-Gb/s DP-16QAM signal. Twelve ECLs at 37.5-GHz spacing were multiplexed and simultaneously modulated by a single IQM inside the real-time 200G transceiver. Odd- and even-channels were separated using a 37.5-GHz interleaver, followed by delay decorrelation and passive coupling. Error free transmission was achieved after 401.1 km for all 12 channels. To emulate different transmission distances for the cases of hybrid span or single-mode-fiber-only span, we removed either the 101.1-km Vascade EX2000 or the QSM fiber span and inserted a variable optical attenuator (VOA) before the ROPA. By adjusting its attenuation value, we found that the real-time 12×200G system can tolerate higher total link loss in case of hybrid QSM+Vascade EX2000 configuration compared with Vascade EX2000 only, as shown in Fig. 3c. The ~4 dB improvement in loss-budget (~24 km increase in system reach) comes from the larger A_{eff} of QSM fiber, which allows higher signal and Raman pump powers. Whereas fiber nonlinearity limited the optimal Raman pump power to 1.6 W when using only Vascade EX2000, the forward Raman module could operate at its maximum output of 2.8 W when using QSM fiber as the first span. Our experiment also confirmed that ISI arising from MPI is sufficiently small that it has no impact on our real-time transceiver, as the BER results using hybrid spans and Vascade EX2000 only are the same at short distances.

IV. CONCLUSIONS

We successfully transmitted 20.7 Tb/s over a 401.1-km hybrid QSM and single-mode fiber link assisted by SPM+XPM compensation using Tx-DBP. To our best knowledge, the achieved capacity and capacity-distance product are records for single-core repeater-less transmission. The use of interchannel nonlinear compensation via Tx-DBP can yield Q improvement of 0.2 to 0.6 dB for short wavelength channels when a tilted launch power profile is used to equalize OSNR. We also demonstrated real-time 200-Gb/s DP-16QAM transmission for the same link, showing that suppressed MPI makes QSM fiber compatible with commercially available transceivers.

ACKNOWLEDGMENT

The authors acknowledge Dr. Wallace Clements, Dr. Serguei Papernyi, Bernard Shum-tim, and Kris Sanapi from MPB Communications for their invaluable knowledge and assistance on Raman/ROPA equipment operation.

REFERENCES

[1] D. Mongardien, et. al., ACP 2015, paper AM3D.2, (2015).

[2] H. Takara, et. al., J. Lightwav. Technol., 33(7), pp.1473 (2015)

[3] F. Yaman, et al., OFC 2015, Th5C.7, Los Angeles, (2015).

[4] Q. Sui, et al., Proc. OFC'14, Paper M3C.5 (2014).

[5] V.Karpov, et. al., Proc. Suboptic 2004, Paper We.8.8, (2004).

[6] S.Papernyi, et. al., Proc. OFC 2002, PDP FB4, (2002)

[7] E. Mateo, et al., Opt. Exp., 16(20), pp. 16124 (2008).

[8] E. Temprana, et al., Science Mag, 348, pp. 1445 (2015).

[9] K. Sugihara, et. al., Proc. OFC 2013, OM2B.4 (2013).

1.2-Tb/s MIMO-less Transmission Over 1 km of Four-Core Elliptical-Core Few-Mode Fiber with 125-μm Diameter Cladding

Giovanni Milione[1,*], Ezra Ip[1], Yue-Kai Huang[1], Philip Ji[1], Ting Wang[1],
Ming-Jung Li[2,*], Jeffery Stone[2], Gaozhu Peng[2]

[1]NEC Laboratories America, Inc., 4 Independence Way, Princeton, NJ 08540 (USA)
[2]Corning, Inc., Sullivan Park, Corning, NY 14831 (USA)
[*]gmilione@nec-labs.com; *lim@corning.com

Abstract: *We demonstrate 1.2-Tb/s (300-Gb/s per λ) WDM+SDM transmission over 1 km of four-core elliptical-core few-mode fiber with 125-μm diameter cladding, using only 4-PAM and OOK with direct detection and no MIMO.*

Keywords: (060.2330) Fiber optics communications; (060.4230) Multiplexing

INTRODUCTION

Due to the unabated growth in data center traffic, the line rates and distances required for intra-data center optical fiber connections have continually increased. It is possible to realize a line rate of 100 Gb/s using parallel single-mode fibers, e.g., 4 lanes of single-mode fibers can scale 25-Gb/s transmission rates to 100 Gb/s [1]. However, to realize future generation line rates of 400-Gb/s and beyond will require other solutions [2].

It is possible to use multi-level signaling such as 4-PAM, to double the data rate per transceiver to 50-Gb/s. Wavelength-division multiplexing (WDM) can also be used to provision parallel channels. Coarse WDM (CWDM) is attractive compared with dense WDM (DWDM) due to the lower cost of CWDM components. But since CWDM reduces the number of λ channels, a high channel count per fiber can be realized using CWDM in conjunction with space-division-multiplexing (SDM). An SDM fiber uses multiple cores, and possibly multiple spatial modes per core to realize parallel spatial channels. SDM fibers can help lower cost by reducing number of fiber connections and fiber patch cords needed in a data center. To date, SDM fibers with many as 114 spatial channels have been demonstrated using 19 cores [4-6]. To mitigate modal crosstalk in such fibers, however, coherent-detection and multiple-input-multiple-output (MIMO) digital signal processing is required; while to achieve high core count, >125 μm cladding is required. High transceiver cost, and mechanical reliability issues make such fibers unsuitable for data centers [7,8]. For SDM fibers to be a viable data center solution, it is desirable to: (a) increase the number of spatial channels within the confines of a 125 μm diameter cladding, and (b) allow independent detection of the spatial channels without MIMO. The highest spatial channel count reported to date for 125-μm diameter cladding fiber is eight single-mode cores [9,10].

Recently, elliptical-core few-mode fibers (EC-FMF) were introduced as an SDM solution [11-13]. Due to birefringence arising from the core's ellipticity, the normally degenerate modes of the LP_{11} mode group become non-degenerate, and propagate with negligible mode coupling in the fiber. When used in conjunction with spatial multiplexers/demultiplexers (S-MUX/DEMUX) with low crosstalk, MIMO-less SDM with direct detection is possible. The use of multiple elliptical cores is an attraction candidate for realizing large number of spatial channels within a 125 μm diameter cladding.

In this work, we demonstrate 1.2-Tb/s (300-Gb/s per λ) MIMO-less SDM+CWDM transmission over 1 km of 125-μm diameter fiber with four elliptical cores, each core supporting LP_{01} and one LP_{11} modes (i.e., 8 spatial channels per core). 300-Gb/s per λ is achieved by transmitting 56-Gb/s 4-level pulse amplitude modulation (PAM-4) signals over the four spatial modes supported by two of the cores, while transmitting 28-Gb/s on-off-keying (OOK) signals in the four spatial-modes in the other two cores.

125-MICRON-DIAMETER CLADDING FOUR-CORE ELLIPTICAL CORE FEW-MODE FIBER

A cross section of the four-core elliptical-core few-mode fiber (4C-EC-FMF) is shown in Fig. 1(a). The 1 km of 4C-EC-FMF has a 125 μm diameter cladding, and was wound on a 15-cm diameter spool. Each core had major and minor diameters of ~17 μm and ~19 μm, respectively. The cores were arranged in a circular array, and the spacing between neighboring cores is ~55 μm. A parabolic refractive index profile was used for the cores, and a low index trench surrounded each core in order to reduce inter-core crosstalk. At ~1550 nm, each core supports two spatial-modes – the fundamental LP_{01} mode and one LP_{11} mode (the other LP_{11} mode is cutoff). Intensity profiles of the spatial modes of each core at 1550 nm after 1 km are shown in Fig. 1(b)-(i). In total, our 4C-EC-FMF supports 8 spatial channels over four cores.

OECC/PS2016

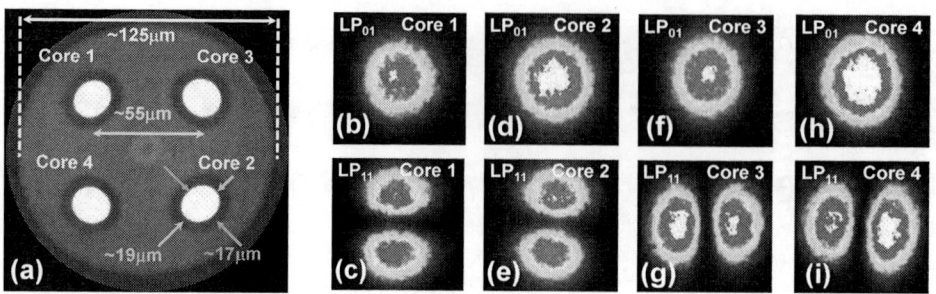

Fig. 1: (a) Cross section of the 4-core elliptical-core few-mode fiber, (b)-(i) Intensity profiles of the spatial modes of each core.

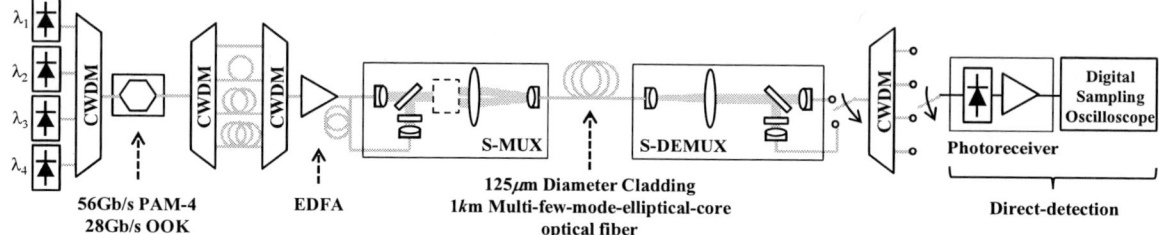

Fig. 2: Experimental setup.

EXPERIMENTAL SETUP

The experimental setup is shown in Fig. 2. Four external cavity lasers (λ_1=1537nm, λ_2=1553nm, λ_3= 1569nm, λ_4= 1585nm) were multiplexed using a CWDM multiplexer (MUX). These were simultaneously modulated by a Mach-Zehnder modulator (MZM) driven by a 4-PAM signal generated using an 8-bit 64-GSa/s digital-to-analog converter (DAC). The four-level waveform generated by the DAC comes from interleaving a pseudorandom binary sequences (PRBS) of length of $2^{31}-1$, then truncating to the memory length of the DAC which is 16384 symbols (32768 samples). Pre-compensation of the DAC's impulse response and the nonlinear characteristic of the MZM is then applied. We decorrelate the signals modulating the four wavelengths by passing the MZM output through a CWDM demultiplexer (DEMUX) followed by decorrelation fibers followed by a CWDM MUX. The resulting signal was the amplified by a C+L-band EDFA with output power of 22 dBm.

To load light into the 4C-EC-FMF, we constructed a spatial multiplexer using collimators, passive splitters and phase plates [11,12]. Splitters and mirrors were used to create four decorrelated copies of the mode-division multiplexed light beam. These were then coupled into the 4C-EC-FMF via telescopic launch using two lenses. At the output of the 4C-EC-FMF, a S-DEMUX was aligned to the core being detected, and the LP_{01} and LP_{11} modes were demultiplexed using collimators, passive splitters and phase plates. The mode being measured was passed through a CWDM DEMUX, and the wavelength of interest was detected by a high-speed photoreceiver, whose output is sampled and digitized using a 50-GSa/s digital sampling oscilloscope (DSO) with 20 GHz bandwidth. Offline processing was used to recover the transmitted bits. For every bit-error rate (BER) counting, we captured 20 data sets of 20 μs duration (1.12×10^7 symbols).

EXPERIMENTAL RESULTS

Fig. 3(a) show the signal spectra measured for the CWDM signal at the S-MUX input, and at the LP_{01} and LP_{11} outputs of the S-DEMUX for one of the cores. We then measured the crosstalk of our 4C-EC-FMF, and the results are shown in Fig. 3(b). It is observed that inter-core crosstalk is always below −30 dB, with the worst inter-core crosstalk being between LP_{01} of core 1 to LP_{01} of core 2. For modal crosstalk between the modes of the same core, it is observed that there exists two "good" cores (1 and 2) with modal crosstalk < −20 dB, and two "bad" cores (3 and 4) whose modal crosstalk are as high as −12 dB. We speculate that the relatively high crosstalk values even for the "good" cores is due to spooling, as previously, the crosstalk we have measured for a single-core EC-FMF was less than −25 dB [12].

For the pair of good cores, Fig. 3(c) shows sweeps of BER vs received power at 1553 nm, with the BER averaged over the two good cores. The black curve is the baseline back-to-back performance of our 4-PAM system. We transmitted 4-PAM on one mode individually at a time using one of the good cores (No SDM, 1 core). It is observed that the LP_{01} and LP_{11} modes have power penalties of ~0.6 and ~1 dB, respectively, compared to back-to-back at the threshold BER of 3.8×10^{-3} for 7% hard-decision (HD-FEC). This power penalty arises from polarization-mode dispersion (PMD) caused by birefringence, which is an unavoidable property of propagation through an elliptical core [12]. We then transmitted 4-PAM on both modes simultaneously on one of the good cores (SDM, 1 core). Intermodal crosstalk further increased the power penalties of the LP_{01} and LP_{11} modes to 0.8 and 2.5 dB, respectively. Finally, we

146

transmitted 4-PAM on all modes and cores simultaneously (SDM, All 4 cores). No further power penalty is observed due to the very low level of inter-core crosstalk compared with inter-modal crosstalk.

For the "good" cores, the results of Fig. 3(c) show it is possible to detect 4-PAM without MIMO. For the "bad" cores, however, inter-modal crosstalk is too high. Consequently, there are two ways of realizing 300-Gb/s transmission per λ: (A) transmit 4-PAM on only *one* of the LP_{01} or LP_{11} modes on the bad cores (the LP_{01} mode is better as per Fig. 3(c)); or (B) transmit mode-division multiplexed OOK on the bad cores. We tested both configurations. Fig. 3(d) shows the results for configuration (A), and Fig. 3(e) for configuration (B). These figures show the BER measured at all four wavelengths for all the modes and cores. Note that the BERs of LP_{01} and LP_{11} of the cores 1 and 2 are the same in each plot. The BERs of cores 3 and 4 in Fig. 3(d) were obtained by transmitting LP_{01} and LP_{11} individually, showing that either of these modes can be used to realize BER below the FEC limit. Fig. 3(e) shows that the BER obtainable for the "bad" cores is substantially lower using mode-division multiplexed OOK, as OOK is inherently much more tolerant to noise, crosstalk and PMD. In fact, no bit errors were measured for most of the modes and wavelengths of the "bad" cores, and these are shown as having BER of "10^{-7}". Nevertheless, PMD resulted in some bit errors for three of the wavelengths on the LP_{11} mode. Overall, all of the BER results shown in Fig. 3(d) and (e) are below the FEC threshold of 3.8×10^{-3}, confirming that 1.2-Tb/s transmission is realized.

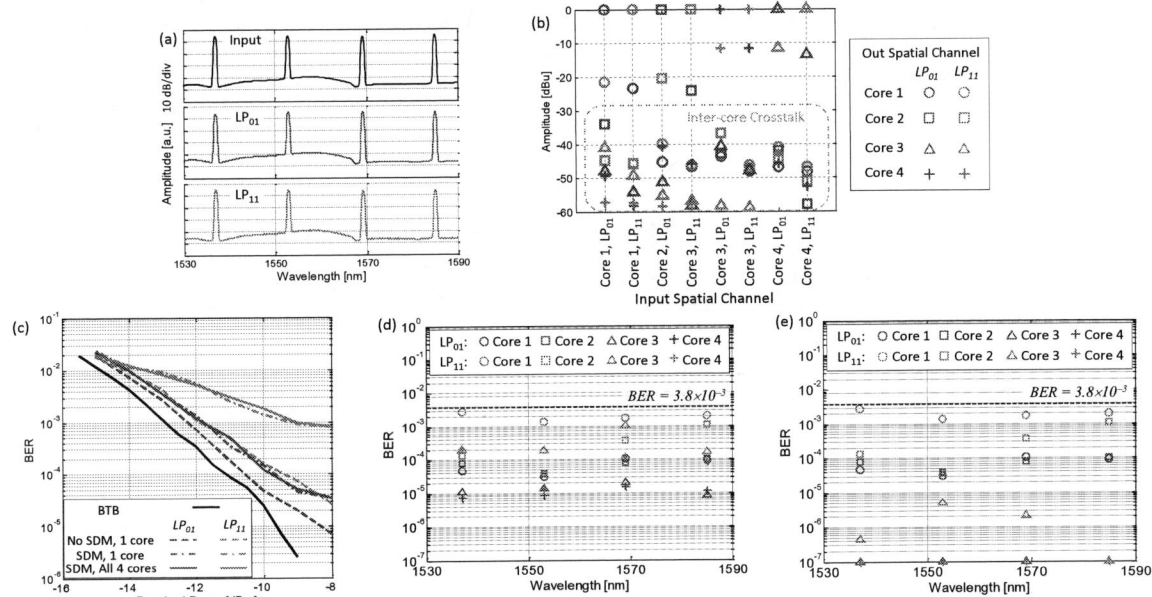

Fig. 3. (a) Signal spectra, (b) Crosstalk measurement, (c) BER vs received power, and BER measurements for 1.2-Tb/s transmission realized using (d) Configuration A, (e) Configuration B.

CONCLUSION

We demonstrated 1.2 Tb/s (300 Gb/s per λ) transmission over 1 km of 4-core elliptical-core few-mode fiber with 125 μm cladding diameter, using only direct detection. Our results are realized by transmitting 56-Gb/s 4-PAM WDM+SDM on cores 1 and 2 having low inter-modal crosstalk < −20 dB and 28-Gb/s OOK WDM+SDM on cores 3 and 4 having high inter-modal crosstalk of −12 dB. It is also possible to transmit 56-Gb/s 4-PAM WDM using only one of the modes of cores 3 and 4. SDM provides an attractive way of increasing the capacity per fiber while keeping the number of wavelength channels low to enable the use of low-cost coarse WDM (CWDM) components.

REFERENCES

[1] IEEE P802.3ba 40Gb/s and 100Gb/s Ethernet Task Force.
[2] IEEE P802.3bs 200Gb/s and 400Gb/s Ethernet Task Force.
[3] D. Richardson et al., Nat. Photonics vol. 7, pp. 354-362, 2013.
[4] N. Fontaine et al. in OFC 2016, Post-Deadline paper Th5C.1
[5] T. Mizuno et al. in OFC 2016, Post-Deadline paper Th5C.3
[6] S. Matsuo et al., Opt. Lett. 36(23), 4626-4629 (2011).
[7] T. Sakamoto et al. in OFC 2016, Post-Deadline paper Th5A.
[8] M. Chagnon et al., in OFC 2015, Post-deadline paper ThFB.2.

[9] T. Hayashi et al., in OFC 2015, Post-deadline paper Th5C.6.
[10] D. L. Butler, in OFC 2015, paper Tu3I.1.
[11] J. Sakaguchi et al., in OFC 2015, Post-deadline paper Th5C.2.
[12] E. Ip et al., Opt. Express 23(13), 17120–17126 (2015).
[13] N. Riesen et al., Opt. Express 22, 29855-29861 (2014)
[14] S. G. Leon-Saval, Opt. Express 22, 1036-1044 (2014)
[15] C. Montero-Orille et al. Appl. Opt. 52, 2332-2339 (2013).

PD2-2

OECC/PS2016

Demonstration of a hybrid silicon evanescent quantum dot laser

Bongyong Jang[1,2], Katsuaki Tanabe[2,3], Satoshi Kako[2], Satoshi Iwamoto[1,2], Tai Tsuchizawa[4], Hidetaka Nishi[4],
Nobuaki Hatori[5], Masataka Noguchi[5], Takahiro Nakamura[5], Keizo Takemasa[6], Mitsuru Sugawara[6], and
Yasuhiko Arakawa[1,2]

[1]Institute of Industrial Science, The University of Tokyo, Tokyo 153-8505, Japan
[2]Institute for Nano Quantum Information Electronics, The University of Tokyo, Tokyo 153-8505, Japan
[3]Department of Chemical Engineering, Kyoto University, Kyoto 615-8510, Japan
[4]NTT Device Technology Laboratories, Atsugi, Kanagawa 243-0198, Japan
[5]Photonics Electronics Technology Research Association, Tsukuba, Ibaraki 305-8569, Japan
[6]QD Laser Incorporation, Kawasaki, Kanagawa 210-0855, Japan
byjang@iis.u-tokyo.ac.jp, arakawa@iis.u-tokyo.ac.jp

Abstract: *A hybrid silicon quantum dot laser evanescently coupled to a silicon waveguide has been demonstrated for the first time. We have observed laser emission from the silicon waveguide and achieved high-temperature operation over 100 °C.*
Keywords: *Quantum dot laser, Hybrid silicon laser, Silicon photonics, Photonic integrated circuit, Wafer bonding*

I. INTRODUCTION

III-V compound semiconductor light sources wafer-bonded onto silicon waveguides are promising components for the realization of photonic integrated circuits (PICs) for next-generation ultra-low-power-computation and telecommunications [1,2]. In the last decade, several varieties of III-V quantum well-based lasers on silicon waveguides have been developed using wafer bonding technology [2-4]. Quantum dots (QDs) are considered more suitable as an optical gain media for lasers in PIC applications, due to their superior lasing characteristics over quantum well-based structures [5]. To date, high temperature operation of QD lasers on Si substrates [6], and also silicon optical interposers utilizing QD lasers [7], have been successfully demonstrated.

In this presentation, we report the fabrication of a hybrid silicon InAs/GaAs QD laser on a silicon waveguide, which utilizes evanescent optical coupling between the QD gain region and the silicon waveguide with adiabatic taper structures. The laser cavity is defined by the waveguide and DBRs that were fabricated into the silicon waveguide. The device lases at a wavelength of 1255 nm, with emission emerging from the silicon waveguide facet. In addition, high temperature operation is achieved at temperatures above 100 °C. This demonstration is a significant step towards the realization and high-density integration of low-power-consuming laser light sources for PICs.

II. A HYBRID SILICON EVANESCENT QUANTUM DOT LASER

A. Device design

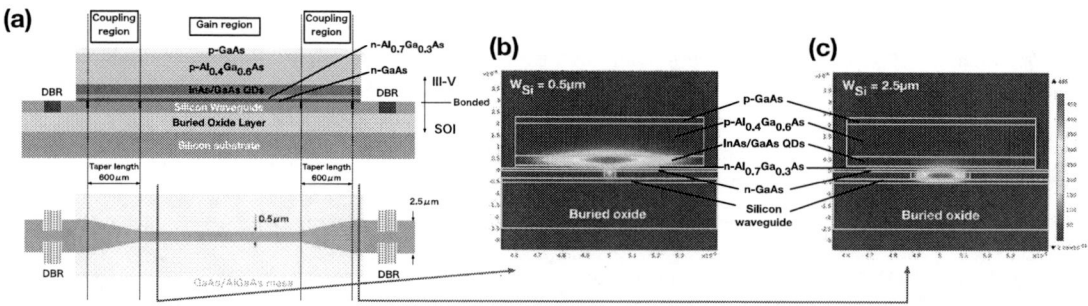

Fig. 1. (a) Side view of the hybrid silicon evanescent quantum dot laser and top view of the silicon waveguide with mesa area. (b, c) Calculated cross-sectional optical mode distributions for silicon rib waveguide regions of width 0.5 μm (b) and 2.5 μm (c), respectively.

Figure 1 (a) shows illustrations of the hybrid silicon evanescent QD laser. The laser cavity was defined by two DBRs spaced 5.4 mm apart. The DBRs consisted of 100-pairs of side-wall gratings, which were formed during the silicon-waveguide fabrication process. The pitch of the gratings was designed to be 193.3 nm, with a 50% duty cycle, so that the center wavelength of the DBR's stop band was at 1255 nm. The bandwidth of the stop band was measured to be 20 nm, and the reflectivity was measured to be higher than 90%. An InAs/GaAs QD laser structure with a 150 nm thin lower cladding layer was layer-transferred onto an array of silicon rib waveguides by means of direct wafer bonding [8] and the subsequent removal of the GaAs substrate. The GaAs/AlGaAs mesa structure (length: 5 mm, width: 20 μm) was

148

formed between the DBRs directly on top of the waveguide, which itself consisted of a 3-mm-long gain region and two 600-μm-long tapered mode-transformers. The tapered mode-transformers act to provide a mode transition between the GaAs/AlGaAs mesa and silicon waveguide by varying the width of the silicon rib waveguide from 0.5 μm to 2.5 μm. The width and taper length of the silicon waveguide were designed to maximize the modal gain and optical coupling. The design/optimization process was performed using confinement-factor and beam-propagation calculations using finite element and beam propagation methods, respectively. The calculated mode distributions for Si waveguides of width 0.5 μm and 2.5 μm are shown in Fig. 1 (b) and (c), respectively. The calculated confinement factors are 75% (Γ_{QDs} in Fig. 1 (b)) and 85% ($\Gamma_{Si\,waveguide}$ in Fig. 1 (c)), respectively, and the coupling efficiency is 70%.

B. Device fabrication

The silicon rib waveguide and DBRs were fabricated on a 4-inch silicon-on-insulator (SOI) wafer using electron-beam lithography and dry etching with an etch depth of 300 nm (the SOI wafer consisted of a 500 nm thick silicon layer and 2 μm thick buried oxide layer). The laser wafer die bonded to the SOI wafer die at 300 °C in ambient air for 3 h under a uniaxial pressure of 0.1 MPa after native-oxide removal of the both bonding surfaces. After the bonding process, the GaAs substrate was removed by selective chemical wet etching. The laser structure was formed through a four-step lithography process: definition of the GaAs/AlGaAs mesa, patterning of the SiO2 passivation layer, patterning of the metal electrodes, and finally, removal of the DBR protection area.

C. Experimental results

Figure 2 shows measured light-current curves of the hybrid silicon evanescent QD laser under 2 kHz, 200 ns pulsed current injection for various operating temperatures. The hybrid silicon laser operates at over 100 °C, and a maximum lasing temperature of 110 °C was achieved. The threshold current at room temperature is 334 mA, and analysis of the threshold at varying operating temperatures allows us to calculate a characteristic temperature T_0 of 303 K near room temperature, and 60 K at around 100 °C, respectively.

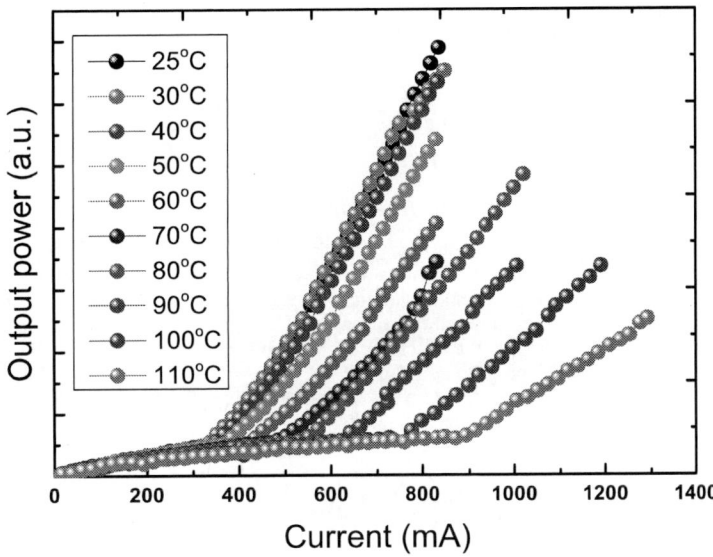

Fig. 2. Light-current curves of the hybrid silicon evanescent QD laser at varied temperatures.

In Fig. 3 (a) we present electroluminescence spectra measured from the device under drive currents of 320 and 650 mA, corresponding to spontaneous emission and lasing emission, respectively. A calculated transmission spectrum of the DBR is also shown in order to highlight the overlap between the lasing wavelength (1255 nm) and the DBR stop band (center: 1255 nm, width: 20 nm). A dip centered at 1252 nm is also observed in the middle of the spontaneous emission spectrum, corresponding to the DBR stop band. The small difference in the center wavelengths between the calculated DBR stop-band and the observed dip may be attributed to fabrication errors in the DBR gratings. This result provides clear evidence that the DBRs define the laser cavity of the device, and that the emission is collected from the silicon waveguide, which is efficiently coupled to the InAs/GaAs QDs. In addition, we have observed a lasing emission from silicon waveguide facet. Fig. 3 (b) shows the silicon waveguide facet with lasing emission patterns. This image further confirms the operation of the laser with evanescent coupling between the InAs/GaAs QD gain medium and the silicon waveguide.

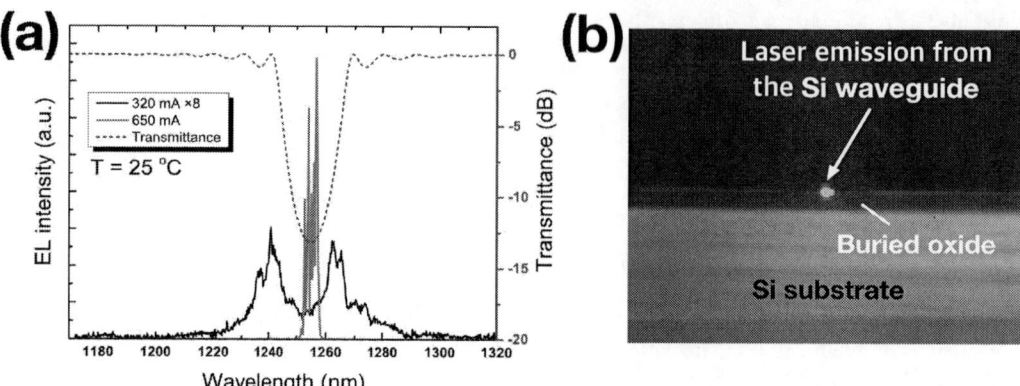

Fig. 3. (a) Electroluminescence spectra from the device under a current injection of 320 mA (spontaneous emission, black) and 650 mA (lasing emission, red) measured at 25 °C. The calculated DBR transmission spectrum is also shown (blue). (b) Near-field image of the silicon waveguide facet with a laser emission under 650 mA pulsed current injection.

III. CONCLUSIONS

We have demonstrated a hybrid silicon laser with a QD gain for the first time. The laser structure was integrated by direct wafer bonding onto an SOI-waveguide wafer with tapered couplers and integrated DBRs. Our device exhibited room-temperature lasing, and a threshold current of 334 mA. Lasing emission was clearly observed from the silicon waveguide facet, and high temperature operation was achieved at temperatures up to 110 °C with a T_0 value of 303 K near room temperature and 60 K around 100 °C. This demonstration opens up a new pathway towards the realization of future silicon optical interposers with ultra-low-power consumption and high-density light-source integration.

ACKNOWLEDGMENT

The authors would like to thank Yuan-Hsuan Jhang, Mark Holmes, and Yasutomo Ota of University of Tokyo for their fruitful discussions. This work was supported by the New Energy and Industrial Technology Development Organization.

REFERENCES

[1] Y. Urino, T. Shimizu, M. Okano, N. Hatori, M. Ishizaka, T. Yamamoto, T. Baba, T. Akagawa, S. Akiyama, T. Usuki, D. Okamoto, M. Miura, M. Noguchi, J. Fujikata, D. Shimura, H. Okayama, T. Tsuchizawa, T. Watanabe, K. Yamada, S. Itabashi, E. Saito, T. Nakamura, and Y. Arakawa, "First demonstration of high density optical interconnects integrated with lasers, optical modulators, and photodetectors on single silicon substrate," Opt. Express, vol. 19, no. 26, pp. B159–B165, 2011.

[2] A. W. Fang, H. Park, O. Cohen, R. Jones, M. J. Paniccia, and J. E. Bowers, "Electrically pumped hybrid AlGaInAs-silicon evanescent laser," Opt. Express, vol. 14, no. 20, p. 9203, 2006.

[3] B. Ben Bakir, A. Descos, N. Olivier, D. Bordel, P. Grosse, E. Augendre, L. Fulbert, and J. M. Fedeli, "Electrically driven hybrid Si/III-V Fabry-Pérot lasers based on adiabatic mode transformers," Opt. Express, vol. 19, no. 11, pp. 10317–10325, May 2011.

[4] D. Inoue, J. Lee, T. Hiratani, Y. Atsuji, T. Amemiya, N. Nishiyama, and S. Arai, "Sub-milliampere threshold operation of butt-jointed built-in membrane DFB laser bonded on Si substrate," Opt. Express, vol. 23, no. 6, pp. 7771–7778, Mar. 2015.

[5] Y. Arakawa and H. Sakaki, "Multidimensional quantum well laser and temperature dependence of its threshold current," Appl. Phys. Lett., vol. 40, no. 11, pp. 939–941, 1982.

[6] K. Tanabe, T. Rae, K. Watanabe, and Y. Arakawa, "High-Temperature 1.3 μm InAs/GaAs Quantum Dot Lasers on Si Substrates Fabricated by Wafer Bonding," Appl. Phys. Express, vol. 6, no. 8, p. 082703, Dec. 2013.

[7] Y. Urino, N. Hatori, K. Mizutani, T. Usuki, J. Fujitaka, K. Yamada, T. Horikawa, T. Nakamura, and Y. Arakawa, "First demonstration of athermal silicon optical interposers with quantum dot lasers operating up to 125 C," J. Lightwave Technol., vol. 33, no. 6, pp. 1223–1229, 2015.

[8] K. Tanabe, K. Watanabe, and Y. Arakawa, "III-V/Si hybrid photonic devices by direct fusion bonding," Sci. Rep., vol. 2, no. 349, pp. 1–6, Apr. 2012.

OECC/PS2016

Silicon Photonic 32 × 32 Strictly-Non-Blocking Blade Switch and Its Full Path Characterization

Ken Tanizawa*, Keijiro Suzuki, Satoshi Suda, Hiroyuki Matsuura,
Kazuhiro Ikeda, Shu Namiki, and Hitoshi Kawashima
National Institute of Advanced Industrial Science and Technology (AIST)
Author e-mail address: ken.tanizawa@aist.go.jp

Abstract: We demonstrate all 1024 path connections of a compact 32×32 strictly-non-blocking thermooptic Si-wire switch. Advanced 300-mm wafer fabrication, dense electronic flip-chip packaging, and pulse-width-modulation heater control are successfully integrated.
Keywords: Optical switch, Silicon photonics, Optical fiber communications.

I. INTRODUCTION

Optical path switching of large-granular data such as video streaming data is effective for reducing total energy consumption in optical networks. Recently, dynamic optical switching is introduced in conjunction with smart network management, and dynamic networking is demonstrated for telecom and datacom applications [1],[2]. A key component is low-cost and densely-integrated optical switch (SW) fabrics with a few dozen port counts and strictly-non-blocking operation. The emerging Si photonics that exploits economic supplier chain for CMOS devices is a particularly promising platform. Various Si-photonic strictly-non-blocking SW fabrics were demonstrated [3]-[6], where the port count was limited up to 8 × 8. Recently, a 32 × 32 thermooptic SW fabricated by advanced CMOS-compatible process [7] and a 64 × 64 SW with waveguide-MEMS actuation [8] have been developed. However, their experimental demonstrations were limited: only sampled path connections were evaluated. The technical challenges are packaging and control of a large number of electronic I/Os, i.e. more than 2000 for heater control in [7].

This paper reports full path characterization of a 32 × 32 strictly-non-blocking optical blade SW using a Si-photonic SW chip. To the best of our knowledge, this is the first demonstration of full path connections of a Si photonic strictly-non-blocking SW with a port count higher than 8 × 8. In the SW, 1024 thermooptic Mach-Zehnder (MZ) element SWs are integrated on a small chip size of 11×25 mm^2 [7]. For dense and reliable electronic contacts, the chip is flip-chip bonded to a 0.5-mm pitch land grid array (LGA) interposer, and all 2112 electronic I/Os are successfully fanned out. The packaging density on a circuit board, which is practically limited by electronic packaging as well as chip footprint, is greatly improved compared with a Silica-based 32 × 32 optical SW [9]. We develop a control FPGA circuit based on pulse width modulation and demonstrate full 1024 path connections. On-chip loss and crosstalk in the worst scenario are 14.5 dB and less than -20 dB around center wavelength, respectively.

II. DEVICE

A. SW chip fabrication

A standard path-independent-insertion-loss (PILOSS) topology [10] is employed for the 32 × 32 optical SW. Figure 1 shows an SEM image of the SW chip. The 2 × 2 element SW is based on a thermooptic MZ interferometer, and the waveguide intersection is based on a directional coupler. 1024 MZ element SWs and 961 intersections are integrated on the chip with a size of 11×25 mm^2. The chip is picked up from the wafer we fabricated in [7]. Since the number of components on the chip is large, highly uniform process is essential. We employed advanced CMOS fabrication using 193-nm immersion lithography on 300-mm diameter silicon-on-insulator wafer [11].

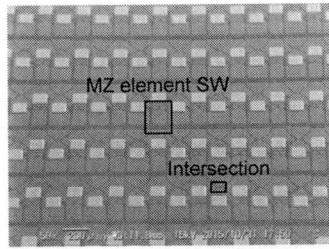

Fig. 1. SEM image of the SW chip.

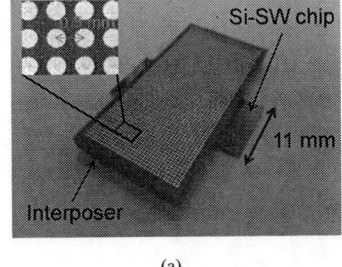

(a)

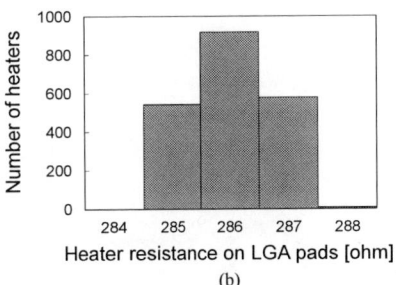

(b)

Fig. 2. Flip-chip package of the SW: (a) Picture and (b) Histogram of heater resistances.

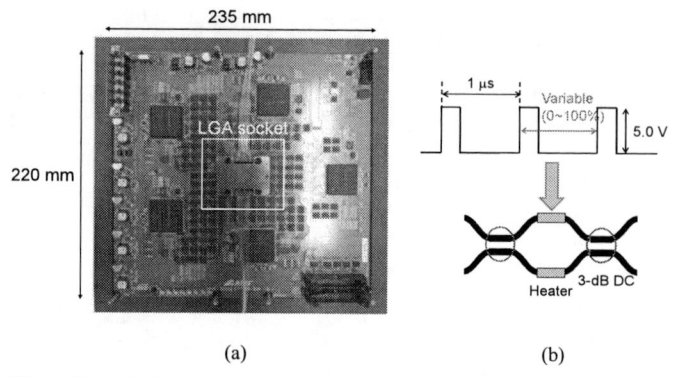

(a)   (b)

Fig. 3. Control circuit board of the 32 × 32 SW: (a) Picture and (b) PWM control.

Fig. 4. 32 × 32 1-RU blade SW.

B. Flip-chip packaging

The 32 × 32 optical SW has 2112 electronic I/Os in total, including 1024 × 2 (two heaters in an MZ element SW) = 2048 I/Os for controlling heaters and 64 I/Os for connecting to ground. If these pads are simply arranged in an outer perimeter for wire-bonding connection, the chip size becomes undesirably large. We have proposed flip-chip packaging using an LGA interposer to achieve the small chip size [7]. Figure 2 (a) shows an image of the packaged SW chip. Pad arrangement on the chip is converted to 0.5-mm pitch LGA by flip-chip bonding the multilayer ceramic interposer. The connection is checked through the measurement of heater resistance on the LGA pad. Figure 2 (b) shows the histogram of all heater resistances. 100 % connectivity and small resistance variation of 1 % have been achieved. This flip-chip packaging derives benefits from the small Si-wire SW chip and greatly increases packaging density (0.5-mm pitch grid array) on a printed circuit board (PCB).

C. Pulse-width-modulation heater control

For fully controlling the packaged 32 × 32 SW, we newly develop a control circuit. Figure 3 (a) shows the picture of the PCB. The control circuit consists of FPGAs (Altera Cyclone IV) and off-the-shelf buffer ICs. Pulse width modulation (PWM) is employed for controlling the heaters [12]. As shown in Fig. 3 (b), rectangular waveform with a modulation frequency of 1 MHz and 5.0-V amplitude is generated. The modulation frequency is enough faster than the heater response of approximately 30 kHz. The duty cycle can be changed individually from 0 to 100 % with less than 0.02 % resolution. The packaged 32 × 32 SW is mounted on the PCB by using an LGA socket. Optical fiber arrays are attached for optical I/Os. The footprint of the PCB is 220 mm × 235 mm. Since the FPGAs and ICs occupy a large area of the PCB, further drastic footprint reduction is expected if an application specific integrated circuit (ASIC) is developed. Even in the present form, it can be mounted in a 1-RU blade together with DC power supplies, as shown in Fig. 4. The blade is designed based on a concept of open and disaggregated rack.

III. RESULTS

We calibrated the SW and measured transmission characteristics of all 1024 (= 32 × 32) paths. A CW laser at 1545 nm was used. The measurement wavelength is a center of the MZ element SW. Figure 5 shows the results. We measured transmission of all 32 output ports, including dark ports, for each path setting: there is 1024 × 32 = 32768

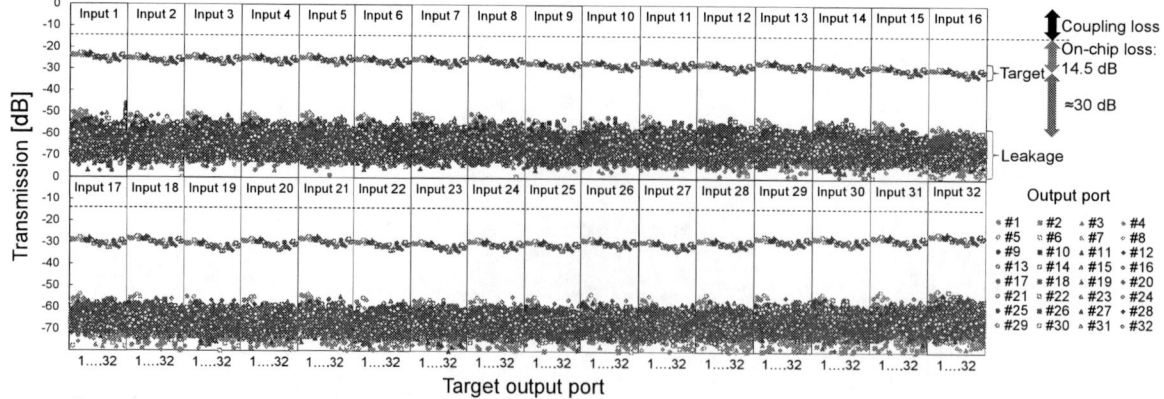

Fig. 5. Transmission characteristics of all 1024 paths on the 32 × 32 SW.

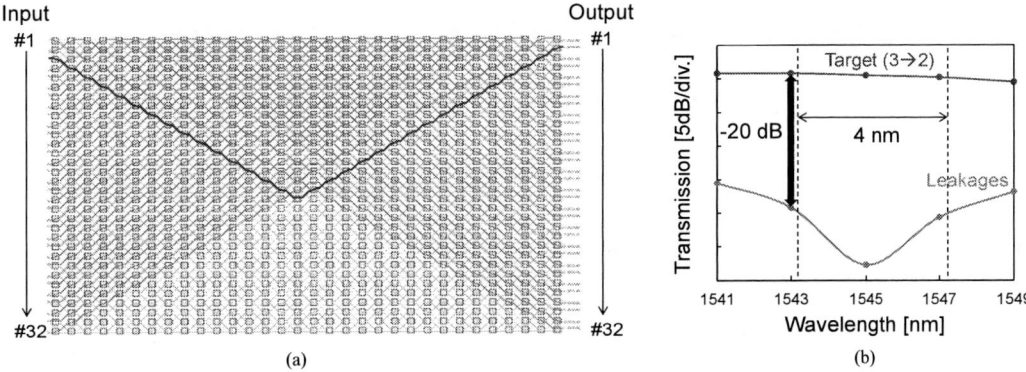

Fig. 6. Crosstalk measurement in the worst scenario: (a) A switch state where the blue path (from input #3 to output #2) has the worst crosstalk, and (b) Transmission characteristics at the output #2 in the switch state.

plots on the figure. One of 32 plots for each path is a target output, and the others are leakages from the path to a non-target port. The leakage is approximately 30 dB lower than the target output. Thus, full 1024 path connections are successfully demonstrated. The insertion loss is 28.4 ± 5.7 dB. We estimate that the loss variation is caused by variation in the fiber coupling efficiency. The chip is equipped with simple spot size converters based on inverse taper; however, the mode field diameter is not large enough for low-loss butt coupling with the fiber array. The coupling loss for the input and output, estimated from the loss measurement of straight waveguides on the chip, is 13.9 dB on average, and alignment tolerance is very small. Further reductions of the loss variation as well as the coupling loss are expected with advanced spot size converters. The on-chip loss, excluding the coupling loss, is approximately 14.5 dB. The on-chip loss consists of 7.6-dB propagation loss (2.0 dB/cm) and 6.9-dB excess loss. The excess loss is accumulated by passing through 31 intersections and 32 MZ element SWs. Further optimization of both the fabrication process and design are necessary for lower loss. Average power consumptions of the element SW are 46.8 and 6.6 mW for the bar and cross states, respectively. The total power consumption of the SW is estimated to be $0.0468 \times 32 + 0.0066 \times 992 = 8.0$ W.

We discuss crosstalk of the SW when lights at the same power level are simultaneously launched into all 32 inputs. Since crosstalk dominantly occurs when two paths cross at the element SW or the intersection, the worst scenario is as follows. Figure 6 (a) shows one of switch states in the worst scenario. The path from input #3 to output #2, highlighted in blue color, and the other 31 red paths cross 29 times at the element SWs and 30 times at the intersections. A path consists of 31 cross-state SWs, 1 bar-state SW, and 31 intersections in the PILOSS topology. Two paths on the SW never cross each other either at the element SWs in the first and last columns or at the bar-state element SW. Hence, the blue path has path crossings at almost all possible element SWs and intersections, and this switch state is considered to be the worst scenario. The worst scenario also exists in the line-symmetric switch state. We measured leakage power from the red paths at the output #2 when the element SWs on the blue path were calibrated precisely. Figure 6 (b) shows the transmission characteristics of the target output (input #3 to output #2) and the leakage. The crosstalk, which is defined as the difference between them, is -20 dB in a bandwidth of 4 nm around 1545 nm. Wider operating bandwidth is required for WDM transmission. A prospective approach is to employ double-gate switch configuration [13].

IV. CONCLUSIONS

We demonstrated full path connections of a 32×32 strictly-non-blocking Si-photonic SW. Highly uniform wafer fabrication using CMOS process facilities for integrating 1024 MZ element SWs, reliable flip-chip packaging of a large number of electronic I/Os, and PWM control of all the heaters were solidly integrated. Lower loss, wider operating bandwidth, and input-polarization insensitivity are future challenges.

ACKNOWLEDGMENT

This work was supported in part by Project for Developing Innovation Systems of MEXT, Japan. The device fabrication was supported by TIA SCR in AIST.

REFERENCES

[1] J. Kurumida, et al., *ECOC 2014*, PD.1.3.
[2] N. Farrington et al., *OFC 2013*, OW3H.3.
[3] S. Nakamura, et al., *OFC 2011*, OTuM2.
[4] L. Chen and Y.-k. Chen, *Opt. Express*, 20, 18977, 2012.
[5] K. Suzuki, et al., *Opt. Express*, 22, 3887, 2014.
[6] T. Tanemura, et al., *ECOC 2015*, P.2.9.
[7] K. Tanizawa, et al., *Opt. Express*, 23, 17599, 2015.

[8] T. J. Seok, et al., *Optica*, 3, 64, 2016.
[9] S. Sohma, et al., *ECOC 2006*, OThV4.
[10] T. Shimoe, et al., *OFC 1987*, WB2.
[11] T. Mogami, et al., *OECC2015*, JWeE.43.
[12] G. W. Cong et al., *OFC 2015*, M2B.7.
[13] T. Goh, et al., *J. Lightwave Technol.*, 17, 1192, 1999.

S1-1 (Invited)

OECC/PS2016

Wireless Network Evolution
Toward 5G Network

Fumiyuki Adachi

Research Organization of Electrical Communication, Tohoku University
2-1-1 Katahira, Aoba-ku, Sendai, Miyagi, 980-8577 Japan
adachi@ecei.tohoku.ac.jp

Abstract: In this paper, we discuss about the technical issues toward 5G and introduce the recent advances in distributed antenna cooperative signal transmission techniques which provides high quality broadband data services over an entire macro-cell area.

Keywords: 5G, LTE/LTE-A, distributed antenna, cooperative signal transmission

I. INTRODUCTION

After 35 years from its birth in Dec. 1979, mobile communications networks have evolved into the 4th generation (4G) in 2015 and have become an infrastructure of our modern society. We witnessed the new generation approximately every 10 years. In 1G and 2G, the coverage expansion was the most important target. From 2G to 3G, there was a big leap in the radio transmission data rate. The major communications service was the voice in 1G and 2G networks. In 3G, high speed data communication of up to 2Mbps was targeted. Since the start of 3G services, video communications have been getting increasingly popular. In 3.9G long-term evolution (LTE) and 4G LTE-Advanced (LTE-A), much higher quality video communications and close-to-1Gbps broadband data services will become more and more popular. LTE-A services started in March 2015 in Japan. In 5G, much broader data services (>1Gbps/user) are expected.

In this paper, we discuss about the technical issues toward 5G and then, introduce the recent advances in distributed antenna cooperative signal transmission techniques which provides high quality broadband data services over an entire macro-cell area.

II. TECHNICAL ISSUES TOWARD 5G

The mobile data traffic volume in 2020 will reach about 1,000 times of 2010 owing to the increasing popularity of smartphones. Therefore, the energy-efficiency (bits/J) will become an important technical issue as well as the spectrum-efficiency (bps/Hz/km^2). Furthermore, in addition to traditional trend to enrich the broadband data services, new services are expected, e.g., IoT related massive device connections and mobile machine control by wireless. Therefore, the radio resource management will become another important technical issue in 5G.

5G networks will be a small-cell network. By reducing the cell size (i.e., small-cell network), the same frequency can be reused at near locations and the transmit power of user terminal (UE) can be significantly lowered, thereby, the spectrum- and energy-efficiencies can simultaneously be improved. Beside straightforwardly reducing the cell size, there are two promising approaches. One approach is to use a large number of collocated antennas at the macro-cell BS (MBS), i.e., massive MIMO beamforming. Another approach is to spatially deploy a large number of antennas over a macro-cell area [1]. This is called distributed MIMO or distributed antenna small-cell network. Distributed antennas are connected with a virtual MBS by low latency coherent optical link to forms a virtual macro-cell as illustrated in Fig. 1. Radio signal processing and resource management (e.g., scheduling) are performed at the virtual MBS. One or more distributed antennas close to a user are selected for signal transmission; this can be considered an advanced version of coordinated multipoint (CoMP) transmission [2] of LTE-A.

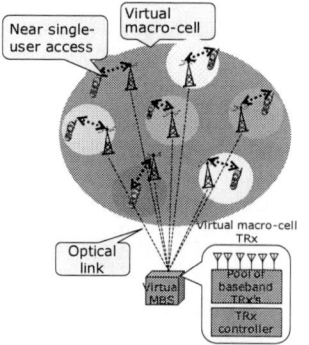

Fig. 1 Virtual macro-cell using distributed antennas.

154

The optical link between a virtual MBS and each distributed antenna can be implemented by using either radio over fiber (RoF), common public radio interface (CPRI) for optical transmission of digitized baseband I/Q signals, or fully coherent optical transmission. The use of RoF link allows all radio signal processing to be implemented at a virtual MBS. However, nonlinearity of RoF link causes a serious problem. Using CPRI, distributed antenna side needs simple RF modulator/amplifier only, but very high speed digital transmission of multi-Gbps is required over fiber. On the other hand, the use of fully coherent optical communication allows to treat the optical transmission and radio transmission similarly; the difference between two is only the carrier frequency (Fig. 2). The concatenation of optical link and radio link can be treated as an equivalent radio link if the same data modulation is used over both the optical link and radio link.

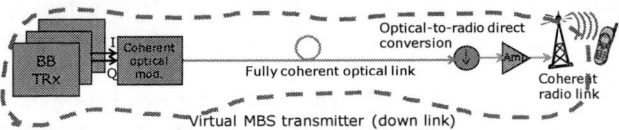

Fig. 2 Fully coherent optical link (downlink).

III. DISTRIBUTED ANTENNA COOPERATIVE SIGNAL TRANSMISSION

To provide high quality broadband data services over an entire macro-cell area, cooperative signal transmission using one or more distributed antennas can be employed. Since antennas are spatially distributed unlike massive MIMO, the shadowing loss problem can be alleviated as well as multipath fading problem. Below, transmission performance of distributed antenna cooperative signal transmission techniques (space-time block coded (STBC) diversity, multiuser MIMO (MU-MIMO), and blind selected mapping (SLM)) (Fig. 3) is presented. 7 virtual macro-cells consisting of 7 distributed antennas each are considered. Frequency-domain block signal transmission (OFDM downlink and SC-FDE uplink) of 128 subcarriers/100MHz band is assumed. FDMA is assumed as the multi-access technique for STBC diversity. 2 UEs having 2 antennas each is randomly located in each virtual macro-cell area. 4 distributed antennas are selected out of 7 distributed antennas based on the instantaneous received signal strength criterion. The transmission performance is measured at UE of the center virtual macro-cell. The OFDM downlink transmission performances with STBC diversity and MU-MIMO, obtained by computer simulation, are plotted in Fig. 4. Figure 5 plots the peak-to-average power ratio (PAPR) of uplink SC signal with blind SLM.

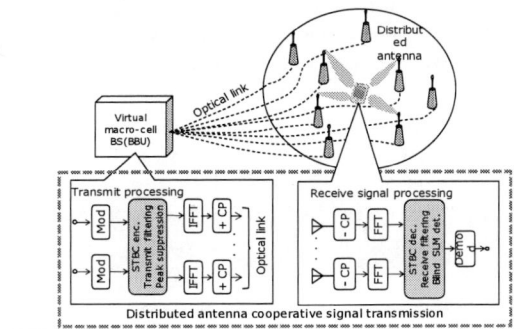

Fig. 3 Distributed antenna cooperative signal transmission (OFDM downlink).

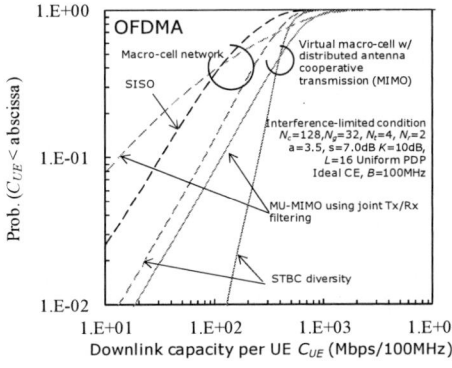

Fig. 4 Transmission performance (OFDM downlink).

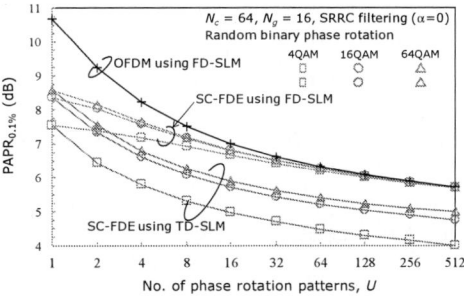

Fig. 5 PAPR reduction by blind SLM.

A. STBC diversity and MU-MIMO

Simple Alamouti code [3] can be used for STBC diversity. By combining transmit frequency-domain equalization (FDE) with downlink STBC diversity (by combining receive FDE with uplink STBC diversity), an arbitrary number of distributed antennas can be used to obtain the full spatial diversity gain although limiting the number of UE antennas to 6 [4]. It can be seen from Fig. 5 that STBC diversity can significantly improves the downlink capacity compared to the case of collocated antennas at the MBS.

In MU-MIMO, multiple users simultaneously transmit their signals using the same frequency unlike FDMA. Inter-user interference (IUI) and inter-antenna interference (IAI) as well as inter-symbol interference (ISI) due to the channel frequency-selectivity limit the achievable transmission performance. Hence, suppression of IUI, IAI, and ISI becomes an important issue for MU-MIMO. Recently, the author's group proposed two distributed antenna MU-MIMO schemes: BD-SVD and MMSE-SVD [5]. Similar to STBC diversity, it can be seen from Fig. 4 that MU-MIMO significantly improves the downlink capacity compared to the case of collocated antennas at the MBS. MU-MIMO provides lower outage probability than STBC diversity in a high downlink capacity region. However, in a low downlink capacity region, the outage probability is higher with MU-MIMO than with STBC diversity. This is due to strong co-channel interference (CCI) from adjacent virtual macro-cell. Suppression of CCI is our future study.

B. Blind SLM

When the distributed antenna cooperative transmission is used, the transmit signal PAPR increases. Therefore, some PAPR reduction technique is still necessary for uplink transmission if the mm wave band is used. SLM is an efficient PAPR reduction technique which multiplies an appropriate phase rotation sequence to the transmit signal either in the frequency-domain or in the time-domain. SLM requires the side information (phase-rotation sequence information) for signal detection at a receiving side. Recently, the author's group proposed a blind SLM which requires no side information and hence, causes no spectrum efficiency degradation [6]. The blind SLM can reduce the transmit signal PAPR by about 3 dB with a slight bit error rate performance degradation (ideal transmit filtering of zero roll-off factor is assumed) (see Fig. 5).

IV. CONCLUSIONS

In this paper, after overviewing the wireless evolution of mobile communications technology over the past 35 years, we discussed about the technical issues toward 5G. Then, we introduced the recent advances in distributed antenna cooperative signal transmission techniques. The target of mobile communications technology is to achieve as high speed data transmission as possible under the limited available bandwidth and power while suppressing the CCI to a certain degree (no perfect cancellation of co-channel interference is intended). A key of realizing 5G networks is the advanced utilization of spatial domain (massive MIMO and distributed antennas). The antenna will become one of radio resource in addition to frequency, time, and power. The scheduling algorithm which efficiently allocates the limited resource among a huge number of mobile communications devices is a paramount technical issue in 5G and beyond.

ACKNOWLEDGMENT

This paper includes a part of results of "The research and development project for realization of the fifth-generation mobile communications system," commissioned to Tohoku University by The Ministry of Internal Affairs and Communications (MIC), Japan.

REFERENCES

[1] F. Adachi, et al., "Distributed antenna network for gigabit wireless access," Int. J. of Electronics and Commun. (AEUE), Elsevier, Vol. 66, Issue 6, pp. 605-612, 2012. doi: 10.1016/j.aeue.2012.03.010.

[2] M. Sawahashi, et al., "Coordinated multipoint transmission/reception techniques for LTE-advanced [coordinated and distributed MIMO]," IEEE Wireless Commun., Vol. 17, Issue 3, pp.26-34, June 2010.

[3] S. M. Alamouti, "A simple transmit diversity technique for wireless communications," IEEE J. Sel. Areas Commun., Vol. 16, No. 8, pp. 1451–1458, Oct. 1998.

[4] R. Matsukawa, T. Obara, and F. Adachi, "Frequency-Domain Space-Time Block Coded Transmit/Receive Diversity For Single-Carrier Distributed Antenna Network," IEICE Communications Express, Vol. 2, No. 4, pp. 141-147, 15 April, 2013.

[5] S. Kumagai and F. Adachi, "Joint Tx/Rx MMSE filtering for single-carrier MU-MIMO uplink," Proc. 2015 IEEE APWCS 2015, Singapore, 19-21 Aug. 2015.

[6] A. Boonkajay and F. Adachi, "A blind selected mapping technique for low-PAPR single-carrier signal transmission," Proc. ICICS 2015, Singapore, 2-4 Dec. 2015.

S1-2 (Invited)   OECC/PS2016

5G Wireless Network Research in China
– Non-Uniform Wireless Dense Networks

Xiaofeng Tao

Beijing University of Posts and Telecommunications, China.
e-mail address: taoxf@bupt.edu.cn

Abstract: *Non-uniform wireless dense networks will be a deployment scenario in the future 5G mobile networks. This paper will introduce the challenges and methods in terms of the network modeling and architecture design in such networks.*

Keywords: *aggregation of dense heterogeneous networks, stochastic geometry*

I. INTRODUCTION

The worldwide rollout of the fourth-generation (4G) mobile networks began in 2010. In the meantime, people started to seek the possibilities of the fifth-generation (5G) mobile networks, a new type of networks that aims to cope with future challenges, such as the increasing data traffic and new mobile applications [1]. After considering the requirements of various services and applications, the International Telecommunication Union (ITU) proposed some key capabilities of the IMT-2020 compared to the IMT-Advanced, such as the 10 Gbit/s peak data rate, 100 Mbit/s user experienced data rate, 100x network energy efficiency and 3x spectrum efficiency [2]. To achieve those IMT-2020 key capabilities, it is expected that the 5G technology would interconnect base stations (BSs) and hot spots, creating non-uniform wireless dense networks, as shown in Fig. 1. On the other hand, the co-existence of BSs and hot spots would introduce severer inference and thus requires more complicated radio resource allocation schemes. Moreover, the 5G mobile networks may be redesigned to satisfy users' high privacy and security requirements. In summary, how to utilize the limited radio-frequency spectrum resources in non-uniform wireless dense networks for different over-the-top (OTT) applications will be a big challenge in the 5G networks.

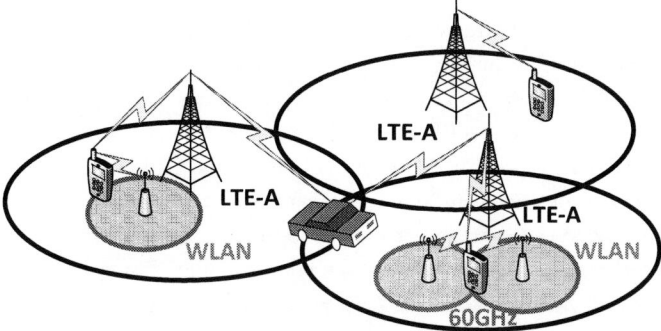

Fig.1: 5G network example that consists of cellular networks and hot spot networks.

II. WIRELESS DENSE NETWORK ANALYSIS BASED ON STOCHASTIC GEOMETRY

The locations of BSs and hot spots tend to be randomly distributed in any wireless dense network, which makes the cellular hexagon model invalid. Therefore, it is necessary to find a new modeling or analytic method for wireless dense networks. Recently, stochastic geometry has become a common method for the random network modeling and it has been widely applied to the modeling and analysis of random networks.

Point processes are one of the most important concepts in stochastic geometry. A point process is a set of points in a certain region and the properties of the point distribution can be inferred by the space averaging. Any wireless network can be modeled as a set of points in a certain region because any point represents a transmitter (e.g. BS, AP) or a receiver (mobile terminal). The Poisson point process (PPP) and its variants (e.g. the Cox process, the cluster process) are widely accepted in the network modelling and analysis. Now let us take the PPP as an example. For a point in a random network, its aggregated interference (AI) from some other points may produce a predominant effect on the network performance. The AI is the sum of all the single-link interference around that point and the single-link interference is a function of distance and channel fading between two points. The complex nature of AI will make the probability distribution function (PDF) of AI difficult to obtain. Stochastic geometry provides an approach to analyze the statistical properties of AI by using the Laplace transform and the probability generating function. Furthermore, the statistical properties of SINR can be obtained for calculating the network capacity and coverage.

In the literature, Dhillon et al. [3] considered a *K*-tier heterogeneous network (HetNet) with *K* independent PPPs and analyzed the network coverage performance with several one-tier access strategies. Wu, Tao and Li [4] proposed a *K*-tier cooperative access strategy, in which the selection of the cooperative tiers is made by channel state and inter-cell offloading requirement. The analytical and simulation results indicate that the access threshold can be used to perform the offloading and mobile management effectively, as well as improve the network coverage by 18% as is shown in Fig.2. Furthermore, consider the location relevance of macro BSs and hot spots. The general *K*-tier HetNet is modeled when the distance between an arbitrary macro BS and a hot spot is greater than a positive number [5]. The analytical and simulation results indicate that the coverage probability of edge users in macro BS is improved by 14% as is shown in Fig.3. In addition, to improve the performance of secrecy transmission in HetNets, Wu et al. [6] proposed a threshold-based user association strategy, in which users could adjust the threshold according to their own computational capabilities and security requirements. It is shown that the secrecy capacity is four times better when the artificial noise technology is introduced [7].

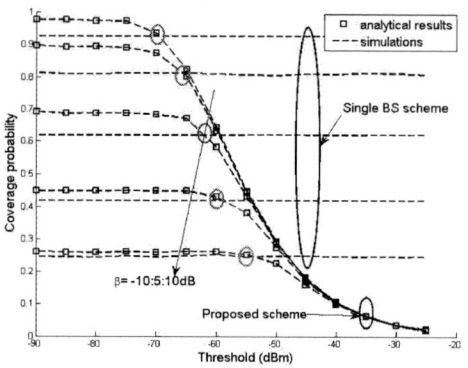
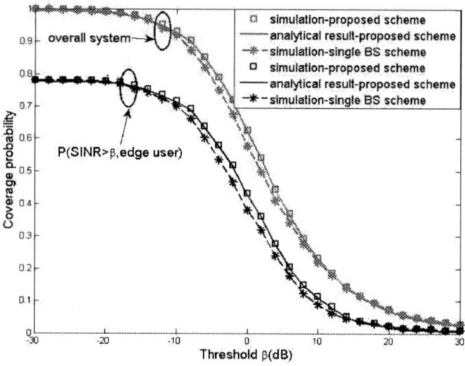

Fig. 2: Coverage probability [4].   Fig.3: Coverage probability [5].

III. WIRELESS DENSE NETWORK DESIGN

A system that supports dense and aggregated networks can be designed based on the aggregation and cooperation techniques with some modifications to the physical layer, the data link layer and the network layer. All these techniques remain the implementations of their lower or higher layers intact but have their distinct pros and cons:

- The physical cooperation technique is difficult to implement. However, it can not only extend the cell coverage but also perform the interference coordination to further improve the network performance.
- The data link cooperation is moderately difficult to implement but has medium computational complexity.
- The network layer cooperation is easy to implement but has high computational complexity.

Fig.4 shows a system model to verify the data link cooperation technique for dense and aggregated networks. This network aggregates two types of networks: a cellular network and a WiFi network with three hot spots. Fig. 5 shows a testbed of a dense aggregate network based on this system model. Note that this testbed is capable of aggregating many networks, such as LTE/Gbps/WiFi. In our testbed, the cellular network is responsible for the control signal and the maintenance signal while the WiFi network is responsible for user data transmission. In other words, the control plane and the user plane are separated. Such network architecture is particularly useful in future scenarios because only the handover function is operated in the cellular network.

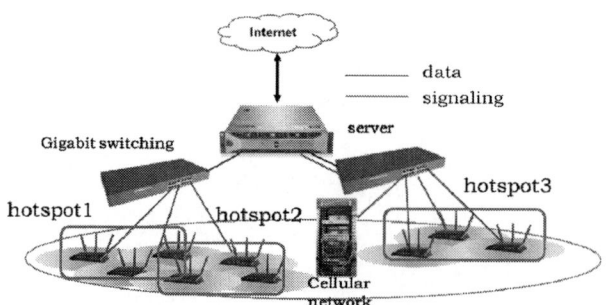

Fig. 4: System model of a dense and aggregated network.

158

Fig. 5: Prototype system of a dense and aggregated network.

The data rate in the testbed of Fig. 5 can achieve up to 7.5 Gbps based on the current network setup. Table I compares the measured rates with the theoretical rates. It is seen from the table that the measured spectrum efficiency and the peak data rate are close to the 5G requirements. The measured energy efficiency and the area traffic capacity are significantly higher than the 5G requirements. However, there is still a big gap between the measured and theoretical results in all aspects.

TABLE I

COMPARISONS BETWEEN THE MEASURED AND THEORETICAL RESULTS IN 5G

Metric	KPI	Measured result (3WiFi bands+1 cellular band) 220MHz, 100m^2	Theoretical value (3WiFi bands+1 cellular band) 260MHz, 100m^2
Peak data rate	10Gbps	7.5Gbps	13Gbps
Spectrum efficiency	10x	34bps/Hz (6.8x)	51.2bps/Hz (10.24x)
Power efficiency	10x	10^{-9}J/s (40x)	$5.8*10^{-9}$J/s (232x)
Traffic	25x	7.5Tbps/km^2 (75x)	13Tbps/km^2 (130x)

IV. CONCLUSIONS

Future 5G mobile networks will make non-uniform dense wireless network possible because it will integrate an increasing number of cellular networks and hot spot networks. In this paper, we first introduce a theoretical method based on stochastic geometry to model such non-uniform dense networks. We then show a real testbed that was designed and built in China for the 5G dense and aggregated network. The measured results indicate that the 5G requirements can be achieved by the aggregation and cooperation techniques.

REFERENCES

[1] X. Xu, D. Wang, X. F. Tao, and S. Tommy, "Resource pooling for frameless network architecture with adaptive resource allocation," Sciece China Inf. Sci., vol. 56, no. 12, pp. 1–12, 2013.
[2] ITU, "IMT Vision -- Framework and overall objectives of the future development of IMT for 2020 and beyond," 2014.
[3] H. S. Dhillon, R. K. Ganti, F. Baccelli, and J. G. Andrews, "Modeling and analysis of k-tier downlink heterogeneous cellular networks," IEEE J. Sel. Areas Commun., vol. 30, no. 3, pp. 550–560, 2012.
[4] H. Wu, X. Tao, and N. Li, "Coverage Analysis for K-Tier Heterogeneous Networks with Multi-Cell Cooperation," in 2015 IEEE Globecom Workshops (GC Wkshps), 2015, pp. 1–5.
[5] H. Wu, X. Tao, and N. Li, "Coverage Analysis for CoMP in Two-Tier HetNets With Nonuniformly Deployed Femtocells," IEEE Commun. Lett., vol. 19, no. 9, pp. 1600 – 1603, 2015.
[6] H. Wu, X. Tao, J. Xu, and N. Li, "Secrecy outage probability in multi-RAT heterogeneous networks," IEEE Commun. Lett., vol. 20, no. 1, pp. 53 – 56, 2016.
[7] N. Li, X. Tao, H. Wu, J. Xu, and Q. Cui, "Large System Analysis of Artificial Noise Assisted Communication in the Multiuser Downlink: Ergodic Secrecy Sum-rate and Optimal Power Allocation," IEEE Trans. Veh. Technol., vol. PP, no. 99, pp. 1–1, 2015.

OECC/PS2016

Activities of the Fifth Generation Mobile Communications Promotion Forum (5GMF) in Japan

Akira Matsunaga
Acting Chair, Technical Committee of 5GMF
KDDI Corporation
ak-matsunaga@kddi.com

Abstract: *This document addresses activities of 5GMF (The Fifth Generation Mobile Communications Promotion Forum) in Japan, including roadmap, potential applications, spectrum implications, and 5G integrated verification trial.*
Keywords: *5GMF, 5G,2020, Japan*

I. INTRODUCTION

5GMF (The Fifth Generation Mobile Communications Promotion Forum) was established in September 2014 in Japan wiith the following objectives:

To promote R&D in 5G mobile technology and research and study of 5G mobile standardization

To collect information relating to 5G mobile and exchange thereof with other organizations

To correspond and coordinate with related organizations on issues related to 5G mobile

To promote awareness and disseminate information related to 5G mobile

This document addresses activities of 5GMF, including roadmap, potential applications, spectrum implications, and 5G integrated verification trial.

II. ROADMAP TOWARDS IMPLEMENTATION OF 5G MOBILE IN JAPAN

Study on Spectrum for 5G as a preparation of WRC-19 with attention to international harmonization.

Translated from "Final Report from the Radio Vision Council" Ministry of Internal Affairs and Communications, Japan, December 2014

Fig.1 Roadmap towards Implementation of 5G Mobile in Japan

160

III. POTENTIAL APPLICATION ENABLED BY 5G MOBILE

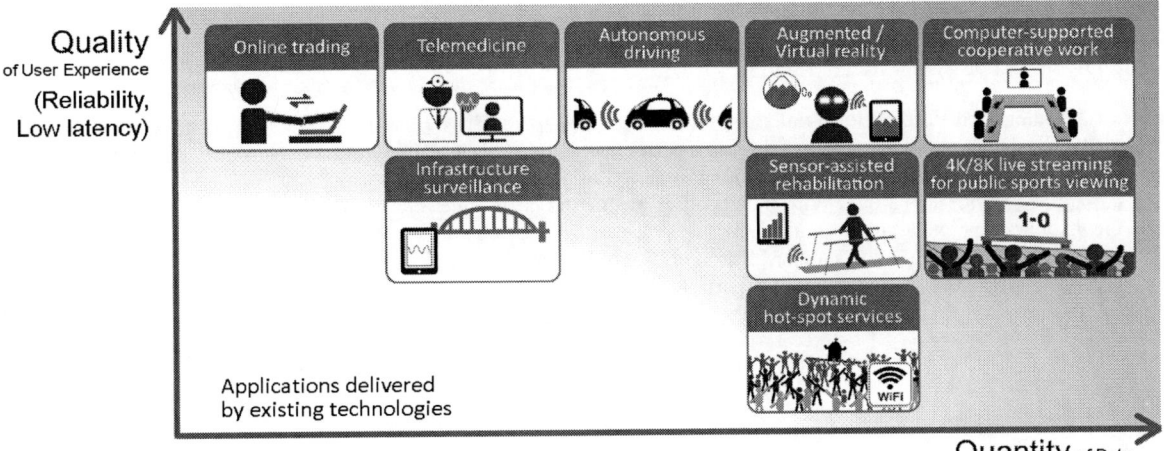

Fig.2 Potential Application enabled by 5G mobile

IV. SPECTRUM RANGES ABOVE 6GHZ AND CHARACTERISTICS

Table 1. Spectrum Ranges above 6GHz and Characteristics

Spectrum range	Low (6 – 30GHz)	Middle (30 – 60GHz)	High (60 – 100GHz)
Practical range of a contiguous spectrum bandwidth (Note 1)	Approx. 300MHz - 1.5GHz	Approx. 1.5GHz - 3GHz	Approx. 3 – 5GHz
Coverage example (Note 2)	Several 100m– Approx. 1km	⟷	Several 10m– Approx. 100m
Deployment scenario	Different scenarios for mobile communication are possible (Outdoor, Indoor, Outdoor to indoor, Hotspot and so on)	⟷	Scenarios for wider bandwidth and dense deployment (Indoor, Hotspot and so on)

(Note 1) These values are contiguous spectrum bandwidth considering fractional bandwidth of 5% with respect to the carrier frequency. The fractional bandwidth of 5% is derived by reviewing the existing 3GPP frequency bands. The values do neither represent required spectrum bandwidth (spectrum demand), nor imply any actual spectrum assignment which is subject to administrative authorities. This bandwidth is desired to be contiguous, in terms of the efficient use of spectrum and implementation (on the other hand, considering 5G applications (Mobile Broadband, M2M and so on), bandwidth of several 100MHz to several GHz is desired, however it is necessary to consider actually available bandwidth for 5G in each range).

(Note 2) The coverage values can vary depending on radio propagation condition, deployment scenario, applicable radio technologies and so on .

V. 5G INTEGRATED VERIFICATION TRIAL

Objectives :
- Estimation of research and development result in 5GMF activities
- Estimation of the operation of an overall 5G mobile
- Demonstration of outputs on 5GMF to domestic and international organizations
- Promotion for practical use of 5G mobile

Aims of 5G Integrated Verification Trial
- Launch an integrated verification trial in FY2017 connecting radio access, network, and applications to ensure smooth commercialization of 5G systems by 2020.
- Evaluate 5G services in actual environments, after evaluating basic functions in the initial phase, in order that the views of application developers and users are reflected as research and development efforts progress.
- Contribute to global R&D and standardization efforts of 5G systems by having an open environment for the verification trial in Japan, allowing for participation of relevant parties, enterprises and universities involved in 5G systems research worldwide, thereby accelerating the widespread adoption of 5G systems.

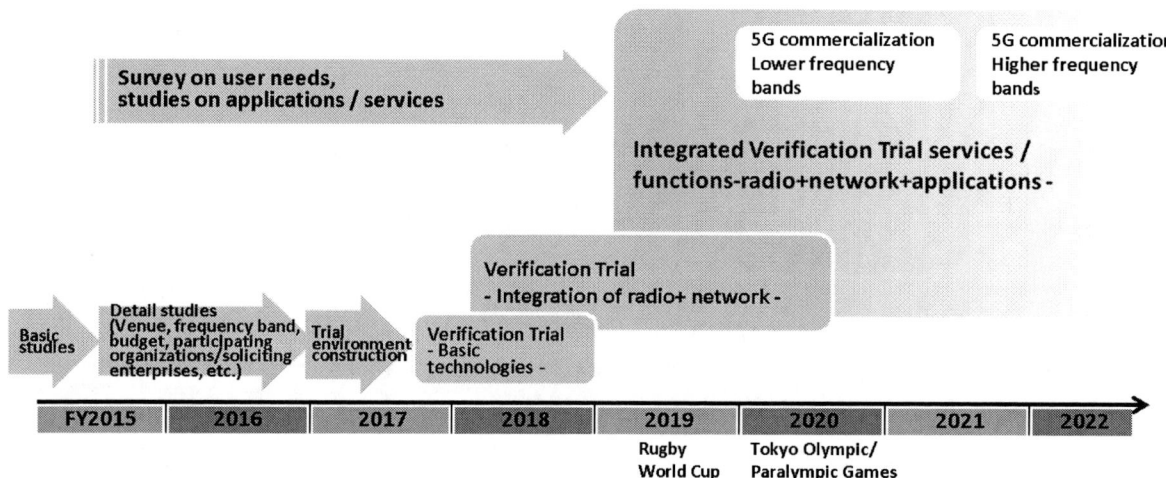

Fig.3   5G Integrated Verification Trial Schedule Overview

S1-4 (Invited)

OECC/PS2016

High capacity mobile fronthaul and backhaul network research in Korea

Seungjoo Hong, Sungmin Cho

SK telecom, 6 Hwangsaeul-ro 258beon-gil, Bungdang-gu, Seongnam-si, Gyeonggi-do, 13595, Republic of Korea
Seungjoo.hong@sk.com, Sungmin@sk.com

Abstract: *This paper describes SK telecom's mobile fronthaul solution for 4G LTE network. It also describes next generation mobile fronthaul and backhaul solution based on colorless multi sub-channel CWDM technology for beyond 4G and 5G mobile network.*

Keywords: *Fronthaul, Backhaul, Multi sub-channel CWDM, C-RAN, LTE, 5G*

I. INTRODUCTION

SK telecom has developed its own mobile fronthaul system which can support both common public radio interface (CPRI) and open base station architecture initiative (OBSAI) that requires low latency and strict clock synchronization. [1][2] Through it, SK telecom has been providing high quality mobile data service to our customers by not only the first 4G long term evolution (LTE) commercialization of cloud/centralized radio access network (C-RAN) base in South Korea July, 2011 but also the first to build a nationwide 4G LTE network.

1st generation mobile fronthaul system which was developed by SK telecom can support 76 links in total including 2.5 Gbps of CPRI and 3.1 Gbps of OBSAI and the evolved 2nd generation mobile fronthaul system can be applicable for up to 4.9 Gbps of CPRI and 6.1 Gbps of OBSAI in the same system platform. All equipment is operating in real field of 4G LTE network in SK telecom. Currently, SK telecom are developing 3rd generation mobile fronthaul system considering beyond 4G and 5G mobile network.

II. CURRENT MOBILE FRONTHAUL OF SK TELECOM

Both 1st and 2nd generation mobile fronthaul system of SK telecom provides wavelength division multiplexing (WDM) based O(optical)-E(electrical)-O(optical) wavelength conversion function and sub-rate multiplexing function together. Sub-rate multiplexing function which is similar to the way that maps the CPRI and OBSAI signal into the 10G optical transport network (OTN) frame, it uses specific frame structure to satisfy a required low latency from mobile fronthaul and also uses wavelength division multiplexing by using dense wavelength division multiplexing (DWDM) wavelength band from 1530 nm to 1560 nm. WDM based O-E-O wavelength conversion function uses all wavelength band except water peak band (1371 nm and 1391 nm) of G.652 fiber and DWDM wavelength band (1531nm and 1551nm) within coarse wavelength division multiplexing (CWDM) bands, to secure more wavelength resource from conventional CWDM, SK telecom has developed the dual sub-channel and multi sub-channel CWDM technology and they have been commercializing since 2011.

A. Dual sub-channel CWDM technology

Dual sub-channel CWDM can provide two sub-wavelengths within CWDM passband of +/- 6.5 nm and it can increase two times capacity of wavelength resource and optical link compared to conventional CWDM. While still using the existing infrastructure of CWDM filters installed, dual sub-channel CWDM can be increase optical link from 18 to 36 for unidirectional application or from 9 to 18 for bidirectional applications. [3][4][5][7]

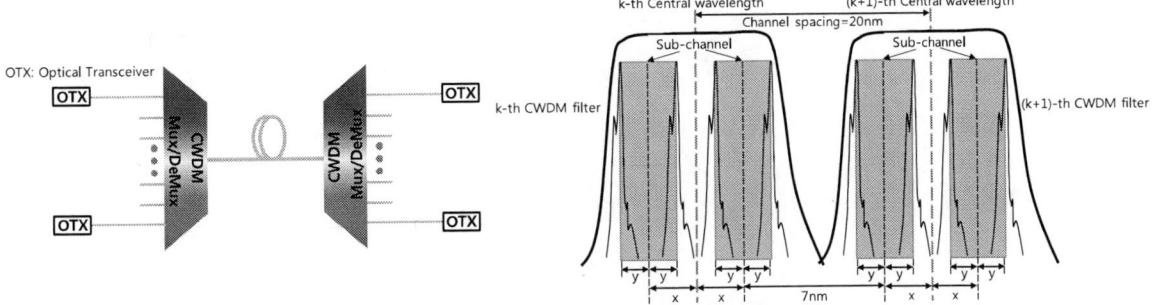

Fig. 1. Concept of dual sub-channel CWDM

This research was funded by the MSIP (Ministry of Science, ICT & Future Planning), Korea in the ICT R&D Program 2016

B. Fixed multi sub-channel CWDM technology

Multi sub-channel CWDM can provide multiple sub-wavelengths within CWDM passband of +/- 6.5 nm and it can increase up to 16 times capacity of wavelength resource and optical link compared to conventional CWDM with reference of 0.8 nm channel spacing. Currently, SK telecom has developed 6 sub-channel CWDM providing 1.6 nm sub-channel spacing by using fixed optical transceiver and it have been commercializing since 2014. [6]

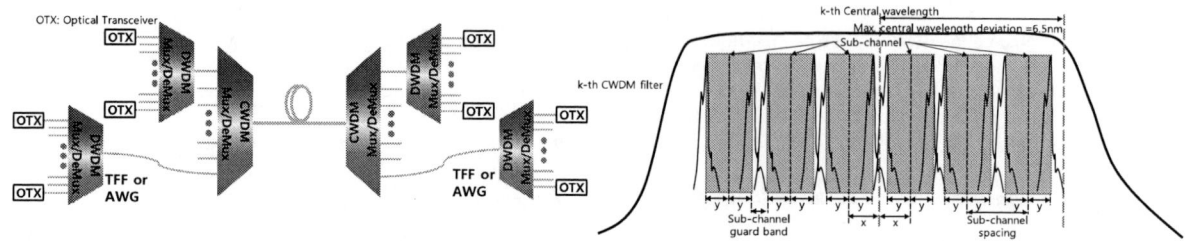

Fig. 2. Concept of multi sub-channel CWDM

III. NEXT GENERATION MOBILE FRONTHAUL AND BACKHAUL OF SK TELECOM

A. Colorless multi sub-channel CWDM Technology

Because multi sub-channel CWDM can freely be located DWDM filters of thin film filter (TFF) or athermal waveguide grating (AWG) between conventional CWDM filters and optical transceivers, it is possible to constitute flexible node topology as well as to transmit more high speed signal than now, when considering the next generation mobile network that is required denser small cell compared to 4G LTE. In addition, because it is advantages to reduce capital expenditures (CapEx) and operating expenditures (OpEx) by using passive node and tunable optical transceiver, SK telecom is developing colorless multi sub-channel CWDM of tunable optical transceiver base for the next generation mobile fronthaul and backhaul solution. [8][9]

B. Architecture of next generation mobile fronthaul and backhaul

The following figure shows architecture of next generation fronthaul and backhaul which is developing in SK telecom. By using all CWDM wavelength bands except water peak band (1371 nm and 1391 nm) of G.652 fiber as shown in Fig. 3, SK telecom is developing colorless multi sub-channel CWDM technology providing 256 wavelengths in total.

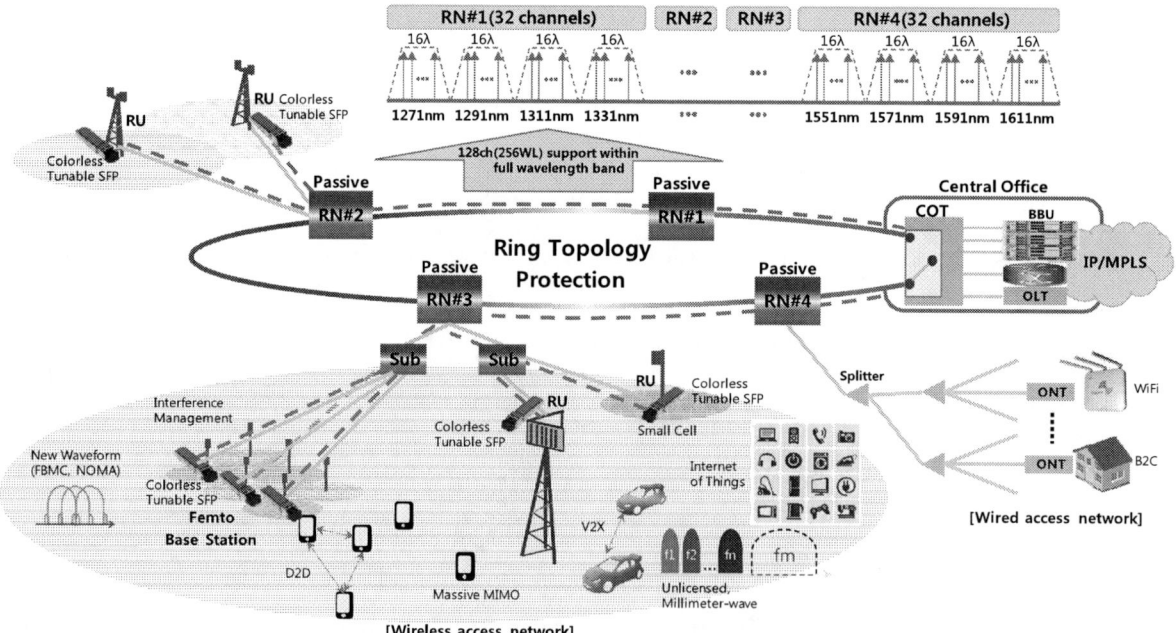

Fig. 3. Architecture of SK telecom's next generation mobile fronthaul and backhaul

As SK telecom in developing colorless multi sub-channel CWDM technology consists of 4 main nodes and from 2 to 4 sub node which is connected to each main node, it is possible to constitute flexible node according to environment of operators as well as to support ring topology in order to provide reliable services for network operation, SK telecom is also considering time division multiplexing passive optical network (TDM-PON) integration with colorless multi sub-channel technology in order to support future wired and wireless access integration and backhaul.

IV. CONCLUSIONS

Because low latency and broadband bandwidth of wired and wireless access network are key factors for next generation mobile network, it is necessary to develop transparent transport technology of L0 base providing rich wavelength resource.

According to this trend, SK telecom is developing the various ways to maximize mobile network performance and to dramatically reduce CapEx and OpEx of mobile network than 4G LTE, and it also believe that colorless multi sub-channel CWDM technology which can solve these problems is one of the strong candidates.

REFERENCES

[1] CPRI Specification Version 7.0, 2015.
[2] OBSAI Reference Point 3 Specification Version4.2, 2010.
[3] ITU-T G.695, "Optical interface for coarse wavelength division multiplexing applications", 2015.
[4] ITU-T G 694.2, "Spectral grids for WDM applications: CWDM wavelength grid", 2003.
[5] Y.K. Kwon, et al, "Optical transceiver for CWDM networks with multi sub-channel interface", OFC 2014, Th3J2.
[6] Seungjoo. Hong, "Requirement of next generation mobile fronthaul", FSAN December 2015.
[7] Seungjoo. Hong and Y.K. Kwon, "Technical feasibility of bidirectional sub-channel CWDM technology", FSAN April 2016.
[8] Seungjoo. Hong and B.W. Kim, "Technical feasibility of colorless sub-channel CWDM technology", FSAN April 2016.
[9] Seungjoo. Hong and Y.S. Kim, "Wavelength routed ODN for multi sub-channel CWDM", FSAN April 2016.

S1-5 (Invited)

OECC/PS2016

Convergence of photonics and electronics for Terahertz wireless communications – the ITN CELTA project

Idelfonso Tafur Monroy

Technical University of Denmark and ITMO University
idtm@fotonik.dtu.dk

Abstract: Terahertz wireless communications is expected to offer the required high capacity and low latency performance required from short-range wireless access and control applications. We present an overview of some the activities in this area in the newly started H2020 ITN project CELTA: convergence of electronics and photonics technologies enabling Terahertz applications.
Keywords: Terahertz, wireless communications, beam forming, photonics; electronics.

I. INTRODUCTION

CELTA is the acronym for Convergence of Electronics and Photonics Technologies for Enabling Terahertz Applications a H2020 ITN project comprising 15 PhD projects. CELTA aims to produce the next generation of researchers who will enable Europe to take a leading role in the multidisciplinary area of utilising Terahertz technology for applications involving components and complete systems for sensing, instrumentation, imaging, spectroscopy, and communications. All these technologies are keys to tackling challenges and creating solutions in a large number of focus areas relevant for the societal challenges identified in the Horizon 2020 programme. To achieve this objective, CELTA [1] integrates multidisciplinary scientific expertise, complementary skills, and experience working in academia and industry to empower ESRs to work in interdisciplinary teams, integrate their activities, share expertise, and promote a vision of a converged co-design and common engineering language between electronics and photonics for Terahertz technologies.

CELTA will introduce the strategy of converged electronics and photonics co-design in its research programme and makes a special effort on establishing a common engineering language in its training programme across the electronics, photonics and applications disciplines. We believe this common engineering language and converged co-design is mandatory to make the next logical step towards efficient and innovative solutions that can reach the market. Within CELTA a series of summer schools will be organized whose detailed compendium of lectures on state-of-the-art technology, soft skills and entrepreneurship is accompanied by a research programme that focuses on THz key technologies. CELTA ESRs will develop three demonstrators: beam steering technology for communication applications, a photonic vector analyser for spectroscopy and materials characterisation, and a THz imager for sensing applications. This paper will further focus on the aspects of Terahertz wireless communications that are investigated at the Metro-Access group [2] of the Technical University of Denmark.

II. THZ WIRELESS COMMUNICATIONS

THz-radiation (100 GHz-10 THz) is a frequency range at the junction between electronics and photonics that has a large potential for new applications such as in terabit wireless communications, sensing, imaging, spectroscopy, astronomy, security and industrial inspection among others.

A. Beam steering

Beam steering is referred to the ability to adjust the beam pointing, both in position and angle, of radiation of emitted signals. Beam steering is a crucial functionality required for several applications, however in CELTA we will focus on beam steering of Terahertz signals for imaging, sensing and communications. For wireless communications, the desirable features of sought solutions include flexibility, integration compactness and high resolution. Complying with such requirements poses an interesting challenge and at the same time opens an exciting opportunity for contributing with novel approaches combining both photonics and electronics techniques. One particular challenge is related to the broad bandwidth nature of such wireless signals for data transmission and the required dynamicity for end-user communication purposes. In this project, we will explore a number of approaches based on the combination of technologies such as multicarrier optical sources, orbital angular momentum, multicore optical fibre and hybrid photonic-electronic and digital signal processing. The result of this project will be part of one of the three prototypes that all ESRs in CELTA will jointly develop and demonstrate by the end of the ITN action.

B. Orbital Angular momentum and beam shaping

A radio signal carrying orbital angular momentum (OAM) is characterized by featuring a helical phase front, with a vortex at the beam axis [3]. An integer number characterizes the helical modes, depending on the number of full twists the phase front undergoes within one wavelength; these helical modes are also called topological charges. As we can see in Fig. 1 (right), OAM beams are intrinsically shaped and therefore we would like to explore techniques for both beam forming and multiple-input multiple output architectures. Due to these properties, OAM MIMO signaling may significantly increase the available transmission capacity without increasing the required bandwidth, by using different topological charges [3, 4]. Two radio signals sharing the same frequency may be modulated using different vortex modes – i.e. with different topological charges – and although being transmitted using the same frequency and path, they are orthogonal due to the characteristics of the OAM. This leads to an increase in capacity that cannot be achieved by any other radio frequency technique [4].

There are several approaches to generate radio beam with OAM [5]. Among those are circular antenna array, spiral phase plates, and periodic frequency selective surfaces, among others. Due to its simplicity and the possibility of using 3D printing with a wide range of materials, the spiral phase plate (SPP) may be regarded as the easiest to implement approach. In Figure 1 (right) an example is given o such an SPP.

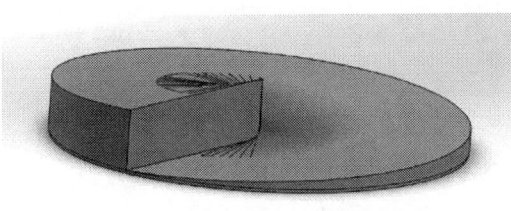 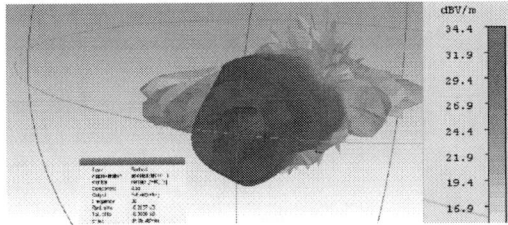

Fig. 1. (Left) Dielectric cylindrical spiral phase plate design to induce a helical phase with topological charge 1. (Right) Simulation results for far field of an example of vortex RF radiation at 80 GHz.

C. 3D printed spiral phase plates (SPP)

Currently, at our group we are investigating the design and manufacturing of spiral plates by 3D printing. A number of different SPPs were designed and manufactured using the BQ Prusa i3 Hephestos 3D printer and corresponding heated print bed. The details regarding the design process, materials and 3D printing flow are presented in [4].

Fig. 3.Example of designed and printed SPP.

III. SUMMARY

Terahertz wireless communications will open new possibility for high capacity and low latency data transfer in a new range of wireless application such as Tactile Internet and cyber control. In the presentation we will present further details both on the CELTA project as whole as well as on the developments in key technologies for Terahertz wireless communications. We will discuss the challenges and opportunities in converging electronics and photonics for Terahertz applications.

ACKNOWLEDGMENT

The author would like to thank the Terahertz wireless team at the Metro-Access group of the Technical University of Denmark: Adrián Ruiz Salazar, Simon Rommel, Eldar Anufriyev, Juan José Vegas Olmos, and O. Rybalko of the Department of Electrical Engineering, Technical University of Denmark for many fruitful discussions and his valuable input. The author further thank the Danish Council for Independent research for partly funding this research through the mmW-SPRAWL project, the H2020 ITN CELTA under grant number 675683 of Call: H2020-MSCA-ITN-2015 and Fujitsu Labs America.

REFERENCES

[1] G. Eason, B. Noble, and I.N. Sneddon, "On certain integrals of Lipschitz-Hankel type involving products of Bessel functions," Phil. Trans. Roy. Soc. London, vol. A247, pp. 529-551, April 1955.

[2] [CELTA INT web page: www.celta-itn.eu

[3] [Metro-Access group web page: metroaccess.dk

[4] [Q. Zhu, T. Jiang, D. Qu, D. Chen, and N. Zhou, "Radio Vortex; Multiple-Input Multiple-Output Communication Systems With High Capacity," IEEE Access, 2015; 3: 2456–2464.

[5] Q. Zhu, T. Jiang, Y. Cao, K. Luo, and N. Zhou, "Radio vortex for future wireless broadband communications with high capacity," IEEE Wireless Commun., 2015; 22: 98–104.

[6] Adrián Ruiz Salazar, Simon Rommel, Eldar Anufriyev, Idelfonso Tafur Monroy, Juan José Vegas Olmos, "Rapid Prototyping by 3D Printing for Advanced Radio Communications at 80 GHz and Above", submitted to 4M/IWMF conference 2016.

S1-6 (Invited)

OECC/PS2016

Fiber-Wireless Fronthaul: the Last Frontier

Gee-Kung Chang and Lin Cheng

School of Electrical and Computer Engineering, Georgia Institute of Technology, Atlanta, GA 30332, USA

geekung.chang@ece.gatech.edu

Abstract: We study and evaluate a centralized mobile fronthaul architecture supporting multiple wireless access structures based on fiber-wireless technologies to improve access efficiency and resource sharing for ubiquitous heterogeneous coverage in future radio access networks.

Keywords: Fiber-wireless, heterogeneous network, millimeter-wave, mobile fronthaul, radio access network, radio-over-fiber.

I. INTRODUCTION

The approach of 5th generation communications introduces new type of use cases such as enhanced mobile broadband and carrier aggregation (CA), massive machine type communications (MTC)/Internet of things (IoT), and ultra-reliable/low-latency communications [1]. These cases directly drive the expansion of heterogeneous networks (HetNets) in terms of inclusive standardization, broad spectrum, diverse radio interface, dynamic aggregation, and ubiquitous coverage [2]. Mobile fronthaul is the last frontier that provides proximity facing mobile users and endows greatest challenges. A solid fronthaul design is the key approach to build efficient pipes from HetNets to users in terms of coverage, throughput, bandwidths, multi-service, quality, cost, and the compatibility to a long-term development. In 5G RANs, as cell sizes become smaller, a huge amount of fronthaul links are needed to distribute high-density, high-frequency, and high-performance small cells. Multi-tier cells will coexist in a HetNet to fulfill different demands and provide ubiquitous coverage, inheriting legacy cells and as well construct new-style small cells [3]. These require the fronthaul design be high-capacity, light-weight, flexible, and transparent to services. Furthermore, advanced technologies such as 3D connectivity and multi-user superposition transmission (MUST) highly rely on fronthaul capabilities and bring stringent synchronization and system stability requirements on fronthaul designs [4].

In current RANs, digital fronthaul solutions such as CPRI and OBSAI are straight-forward and robust approaches to fulfill basic LTE bandwidth needs [5],[6]. However, their low efficiency will require unaffordable high-speed transceivers and limit any further bandwidth improvement when they are applied in 5G RANs. In CPRI, as an example, a 20-MHz LTE signal takes up to 10-Gb/s fronthaul rate. When CA and MUST are applied, the speed will eventually consume all transceiver capacity. Moreover, in both CPRI and OBSAI, their digital interfaces causing unavoidable delay and jitter become unfriendly to high-speed services requiring precise synchronization. To solve all the problems that a distributed digital system may cause, radio-over-fiber (RoF) technologies are developed as an analog solution that avoids digital components or extra processing overhead over fronthaul. On top of a centralized RAN (C-RAN) [7], RoF provides high-level centralization and minimizes complexity distribution. All baseband functions and most RF signaling are realized in a central office (CO) where resources are shared among all fronthaul links even if the links have various structures and serve different types of cells. In addition, RoF facilitates the generation and distribution of millimeter-wave (MMW) signals, the new exploitation for 5G spectrum, in optical approaches as MMW signaling burdens fronthaul systems much more than low-frequency microwave (MW) signals.

In this paper, we study a fiber-wireless fronthaul architecture based on RoF technologies to provide centralization for multiple fronthaul structures in a C-RAN. By realizing centralization and resource sharing, all fronthaul structures with simplified fronthaul links and light-weight remote access units (RAUs) share baseband and RF signaling. Their low-cost and flexible distribution supports large-amount high-density small cells while their all-analog and transparent transmissions support dynamic multi-band services.

II. CENTRALIZED MULTI-STRUCTURE FIBER-WIRELESS FRONTHAUL

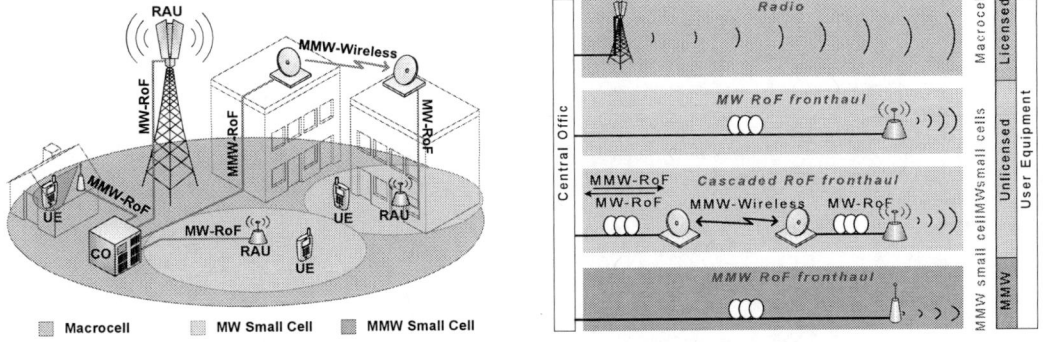

Fig. 1. Application and architecture of centralized multi-structure fiber-wireless fronthaul.

A multi-structure mobile fronthaul architecture based on RoF technologies and C-RAN hierarchy is depicted in Fig. 1. In this architecture, a CO centralizes functions for both macrocells and small cells, including baseband processing as well as radio signaling. Three small-cell fronthaul structures are illustrated here, including two for MW bands and one for MMW bands. On top of MW small cells in unlicensed bands, the MMW small cell (e.g. 60 GHz) provides very-high-throughput coverage for minor-mobility indoor users. As a result of centralization and all-analog transmission, no baseband or digital processing exists between the CO and UE in any fronthaul structure. Besides cost and management savings, this simplification leads to an important property that the propagation delay is only determined by the fronthaul length. This is a promising improvement from traditional CPRI systems where digital processing causes jitters. Furthermore, in the RoF based fronthaul structures shown in Fig. 1, all latencies are predictable, stable, and easily compensable at the CO so that multi-point technologies such as licensed-assisted CA can get precise synchronization.

MW RoF fronthaul is the most straight-forward form of a RoF system with the highest level of fronthaul simplicity. In MW RoF fronthaul, RF signals are carried by lightwave and directly detected at the RAU or CO to recover RF signals. The whole optical system including modulation, demodulation, and fiber propagation can be seen as transparent in terms of the delivery of RF signals with various modulation formats such as OFDM, FBMC, and UFMC.

The cascaded RoF fronthaul structure meets environmental requirement in urban areas as continuous wiring between the CO and cell sites sometimes can be impossible or cost-prohibit. At the same time, it maintains a centralized structure and all-analog transmission consisting of MMW-RoF, MMW-wireless, and MW-RoF links. From users' view, cascaded RoF fronthaul has no difference than MW RoF fronthaul and is also transparent to RF signals though it involves multiple links and MMW frequencies for more complex deployment scenarios.

In MMW RoF fronthaul, ultra-high throughput services or multiple low-frequency services can be directly carried over MMW bands. In the downlink of the fronthaul, the generation, propagation, and detection of MMW signals are the same as the case in the MMW-RoF link in cascaded RoF fronthaul, while in uplink the collected MMW signals are down-converted to an intermediate frequency (IF) before it is carried onto lightwave at the RAU.

III. EXPERIMENT

In this section we evaluate the end-to-end downlink performances of the centralized architecture consisting of three types of fiber-wireless fronthaul structures serving different frequency bands in different environmental scenarios. With most complexity centralized in a CO, the three fronthaul structures deliver LTE-A signals to the UE. The setup consists of a CO, MW RoF fronthaul connecting RAU1, cascaded RoF fronthaul connecting RAU2, MMW RoF fronthaul connecting RAU3, and a set of UE, as shown in Fig. 2.

The CO centralizes all baseband processing and shared RF signaling functions. LTE-A signals at unlicensed spectrum (LTE-U) are generated by a Vertex FPGA chip, DAC, and up-converters. The FPGA used for the generation of base-band LTE-A signals establishes real-time 16-QAM mapping and OFDM modulation. The OFDM signals follow the format of LTE-A -- a 20-MHz bandwidth, 1,200 subcarriers with spacing of 15 kHz, an IFFT/FFT size of 2,048, extended CP mode, and 60 OFDM symbols as a half frame with the 6th used as a PSS. After DACs and shared frequency up-conversion, the LTE-U signals have a central frequency of 2.462 GHz (IEEE 802.11 channel 11). The signals are modulated onto lightwave and transmitted through different fronthaul structures.

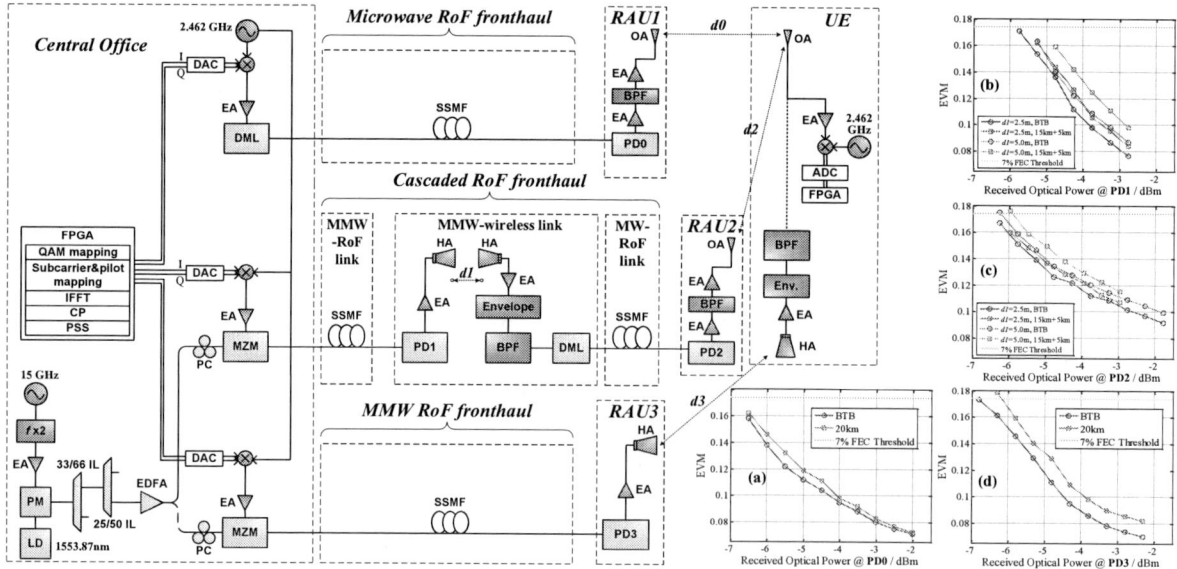

Fig. 2. Experimental setup and received optical power test results at different positions.

Another function of the CO is to generate shared optical MMW carriers. As shown in Fig. 2, an optical signal from a laser diode (LD) at a wavelength of 1,553.87 nm is applied to a 40-GHz phase modulator (PM). The PM is driven by a 30-GHz tone, which is double the 15-GHz microwave source, creating MMW combs. After passing two inter-leavers (ILs) of 33/66 GHz and 25/50 GHz, a 60-GHz optical MMW carrier is formed. After amplification, the optical MMW carrier is fed into MZMs to be intensity-modulated by the LTE-U signals for different fronthaul structures.

The first fronthaul structure we setup is the MW RoF fronthaul. The LTE-U signal is carried onto lightwave by using a directly modulated laser (DML) after an electrical amplifier (EA). The DML has fixed output power of 6 dBm. The lightwave signal is transmitted for 20 km over a standard single mode fiber (SSMF). At RAU1, the LTE-U signal is detected by a 2.5-GHz photo detector (PD), filtered by a band-pass filter (BPF), and amplified by a low-noise amplifier and a power amplifier. A pair of 3-dBi omni-directional antennas (OAs) is used for wireless transmissions.

The cascaded RoF fronthaul setup shown in Fig. 2 consists of a MMW-RoF link, a MMW-wireless link, and a MW-RoF link over the fronthaul. After intensity modulation at the CO, the optical MMW carrier carrying the LTE-U signal is launched into the cascaded RoF fronthaul for a first MMW-RoF transmission over a curl of 15-km SSMF. For MMW-wireless transmission, the signal is detected by a 60-GHz PD, amplified by an EA, and then radiated by a V-band horn antenna (HA). The received MMW signal is amplified and down-converted to the MW band by using a 60-GHz envelope detector. After a BPF the MW signal drives a DML, which has fixed output power of 6 dBm and a wavelength of 1,547.44 nm. The succeeding link including RAU2 has the same setup as in MW RoF fronthaul.

The third fronthaul structure in the setup is a MMW-RoF link that provides a MMW small-cell coverage instead of in MW bands. The fronthaul setup along with RAU3 has the same configuration as the MMW-RoF link and the transmitter side of the MMW-wireless link in cascaded RoF fronthaul.

At the UE side, for MW small-cell signals, the UE down-converts the amplified RF signal from 2.462 GHz to baseband. After ADC, the baseband signal is processed in an FPGA chip, functionally including synchronization, OFDM demodulation, and equalization. Real-time error vector magnitude (EVM) values of equalized symbols are shown in a real-time manner. For MMW small cells, a V-band down-converter is used to recover the LTE-U signal from the 60-GHz carrier so that succeeding processing can be carried on in the same way as in MW small cells. The V-band down-converter has the same setup as the receiver side of the MMW-wireless link in cascaded RoF fronthaul.

We measure received-optical-power performance by adding an attenuator and changing optical power before PD0, PD1, PD2, and PD3 separately. For MW RoF fronthaul, Fig. 2(a) shows EVM curves under the change of received optical power at PD0. The MW wireless distance is fixed to $d0 = 2$ m. 20-km fiber transmission induces an approx. 0.3-dB power penalty with respect to back-to-back (BTB) transmission. For cascaded RoF fronthaul, Fig. 2(b) shows EVM curves under the change of received optical power at PD1. The MW-wireless distance is fixed to $d2 = 2$ m, and the optical power at PD2 is at 2 dBm during the test. As shown in Fig. 2(b), 15-km + 5-km fiber transmission induces an approx. 0.5-dB power penalty with respect to BTB transmission. As the MMW-wireless link is part of the fronthaul, we give results at $d1 = 2.5$ m as well as 5 m. In both BTB and 15-km + 5-km fiber transmission, the power penalty caused by doubling $d1$ is lower than 0.7 dB, which reveals the feasibility of centralized MMW generation. Similarly, Fig. 2(c) shows EVM curves under the change of received optical power at PD2. The MW-wireless distance is fixed to $d2 = 2$ m, and the optical power at PD1 is fixed to 0 dBm during the test. According to the curves, 15-km + 5-km fiber transmission also has an approx. 0.5-dB power penalty with respect to BTB transmission, which is constant with the case in Fig. 2(b). The power penalty caused by doubling MMW-wireless distance is approx. 1 dB, which is higher than the case in (b), because of the succeeding MW-RoF transmission. Comparing (a) and (c), the MMW parts in cascaded RoF fronthaul induce impairment including approx. 1.5-dB and 2-dB power penalties with respect to MW RoF fronthaul when $d1$ is 2.5 m and 5 m, respectively. These penalties decrease when received optical power at PD0 and PD2 is reduced. For MMW RoF fronthaul, Fig. 2(d) shows EVM curves under the change of received optical power at PD3. The MMW-wireless distance is fixed to $d3 = 2.5$ m. 20-km fiber transmission also induces an approx. 0.5-dB power penalty with respect to BTB transmission. By comparing the BTB curves in b) and (d), we can find that the succeeding link after MMW wireless in cascaded RoF fronthaul induces a 0.5-dB power penalty on average with respect to MMW RoF fronthaul.

IV. CONCLUSIONS

We have studied and experimentally evaluated multi-tier fiber-wireless fronthaul architectures. In this report, we design and optimize resource sharing and service provisioning through efficient centralization and function partition in mobile fronthaul. Using this approach, we can realize high capacity and high performance in 5G environment to meet the challenges in the last frontier of heterogeneous mobile data networks.

REFERENCES

[1] NGMN, "NGMN 5G White Paper", 2015
[2] R. Hu et al, IEEE Comm. Mag. 52(5), 2014
[3] D. Lopez-Perez et al, IEEE Wireless Com. 18(3), 2011
[4] 4G America, "Mobile Broadband Evolution Towards 5G", 2015

[5] Common public radio interface specification, 6.1, 2014
[6] Open base station architecture initiative V2.0, 2006
[7] China Mobile, C-RAN white paper, 2011

Optical Access Network Technology for 5G Wireless Front/Backhaul Network

Jun Terada, Tatsuya Shimada, Tatsuya Shimizu, and Akihiro Otaka

NTT Access Network Service Systems Laboratories, NTT Corporation, 1-1, Hikarinooka, Yokosuka-shi, Kanagawa, JAPAN

{terada.jun, shimada.tatsuya, shimizu.tatsuya, ootaka.akihiro}@lab.ntt.co.jp

Abstract: This paper describes trends in 5G mobile networks and RAN technologies. Cloud RAN, which uses switches to enable flexible MFH networking, and two promising candidates of MFH in Cloud RAN are discussed.

Keywords: Mobile Communication, 5G, RAN, Cloud-RAN, MFH, Switching network

I. INTRODUCTION

Mobile devices such as smart phones and tablet are becoming our favorite communication tools. Accordingly, mobile communication services have shown significant growth. Currently, many operators around the world are deploying Long Term Evolution (LTE) and LTE-Advanced to offer faster access with lower latency and higher frequency-usage efficiency than their predecessors, the third generation (3G) and 3.5G. Given these trends, there is inexorable demand for new mobile communications systems with even greater capabilities, namely 5G system [1]. A Japanese national project started in 2015 to develop 5G mobile network technologies including optical technologies. The 5G mobile network is expected to support multiple types of radio access technologies (RATs) in several frequency bands (multi-band). This will support various network requirements such as the specific latency, reliability, and throughput targets for each service. These services include IoT, high-resolution video streaming, and augmented reality. In this paper, we describe trends in future mobile networks and introduce attractive optical network technologies.

II. RAN EVOLUTION

The mobile base station's architecture has evolved with the progress of mobile networks. Fig. 1 shows 3 such architectures: conventional Distributed radio access network (Distributed-RAN), Centralized-RAN and Cloud-RAN. In the Distributed-RAN architecture, all functions of the base station are at each antenna site, while the baseband functions are concentrated in baseband unit (BBU) and radio functions such as an amplifier and antenna are set at the remote radio head (RRH). The link between BBU and RRH, called the mobile fronthaul (MFH), is based on digital radio over fiber (DRoF). Common Public Radio Interface (CPRI) and Open Base Station Architecture Initiative (OBSAI) are the de facto standards for MFH transmission [2, 3]. Transmission at constant bit rate is assumed. Cloud RAN is an enhanced Centralized-RAN with virtualization. Each BBU can communicate with any RRH via a switching network as shown in Fig. 1 (c). As regards MFH in Cloud-RAN, there are two options, one is the current DRoF-based MFH and the other is a new MFH with a redefined functional split between BBU and RRH. DRoF-based MFH is easy to upgrade because the BBU and RRH have basically the same architecture as the current base station. The architecture of the redefined BBU/RRH variant reduces MFH optical bandwidth. The exact splitting point is still undetermined and is being discussed all over the world. This kind of MFH will be packet-based and used when the base station communicates with user equipment.

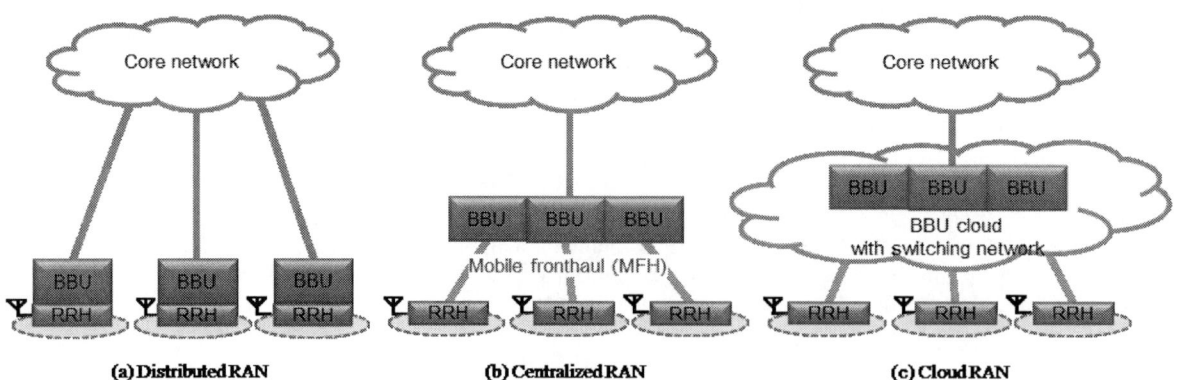

Fig. 1. Architecture of base stations.

172

III. FUTURE MOBILE FRONTHAUL NETWORK

As described above, the future MFH is assumed to have a switching network. BBU will be much more centralized in a ring network as shown in Fig. 2. This kind of MFH has an issue where the switching function is located in the network. Solutions roughly fall into two categories as shown in Fig. 3.

One locates a switch at each central office hosting a BBU (white box) and a ring network is set between central offices (Fig 3(a)). This enables network flexibility but high costs and high latency are big concerns. Low latency transmission is very important for MFH because of the round trip time (RTT) requirements between BBU and user equipment (UE). IEEE 802.1CM is currently discussing the transport of time-sensitive fronthaul streams in Ethernet bridge networks. MFH between the last switch and RRH is assumed to use PON technology as well as point-to-point optical fiber. A 10 Gb/s class TDM-PON system can achieve downstream latency of the order of several tens of microseconds to network equipment. Due to the dynamic bandwidth allocation (DBA) algorithm based on the status reporting method, the upstream latency is more than 1 ms, which is too large to support MFH transmission. One approach to reducing DBA latency is using fixed bandwidth allocation (FBA). By using FBA, no control signal between OLT and ONU for bandwidth allocation is transmitted, which reduces the latency to several tens of microseconds. However, with FBA operation, the bandwidth of each connection is fixed, and efficient bandwidth usage based on statistical multiplexing cannot be achieved. One solution is to reuse the scheduling information of the mobile system [4, 5] by setting a new coordination I/F between BBU and OLT. In the LTE system, the base station allocates uplink bandwidth for each UE and also passes scheduling information to UE with 4 ms timing advance. Therefore, the required upstream bandwidth of the PON can be calculated using the information of the mobile system, and grant data reaches each ONU before uplink signal reception. This scheme is currently being discussed in FSAN/ITU-T. It minimizes the latency of TDM-PON and total latency, excluding fiber delay, is theoretically of order of several tens of microseconds. Measurements on a prototype with the proposed scheme show latency values under 50 µs. To implement this method, the base station and OLT need a cooperative interface to transmit mobile scheduling information.

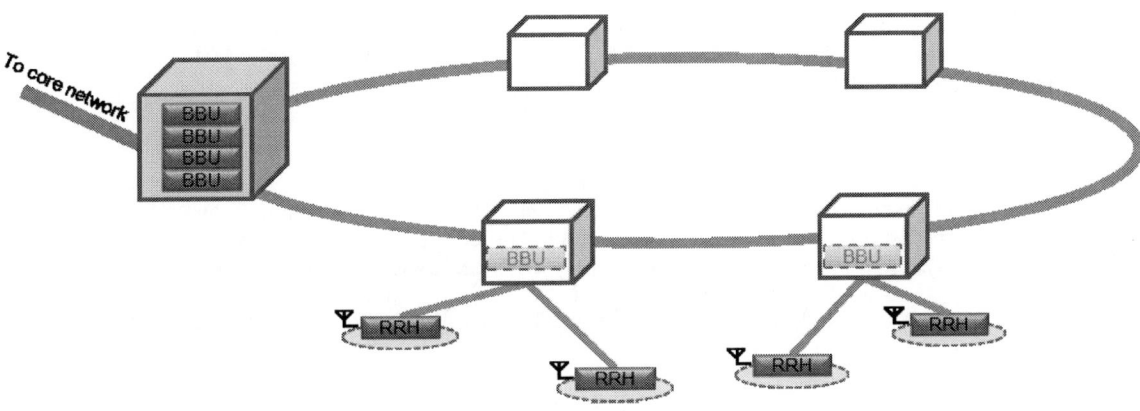

Fig. 2. MFH with a ring network

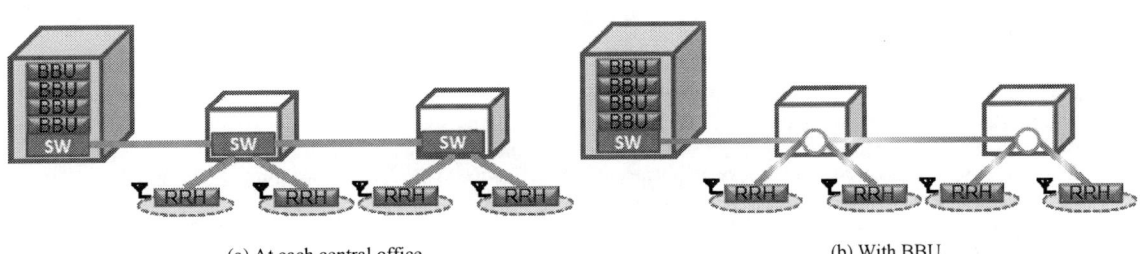

(a) At each central office.　　　　　　　　　　　　　　(b) With BBU.

Fig. 3. Location of switches

OECC/PS2016

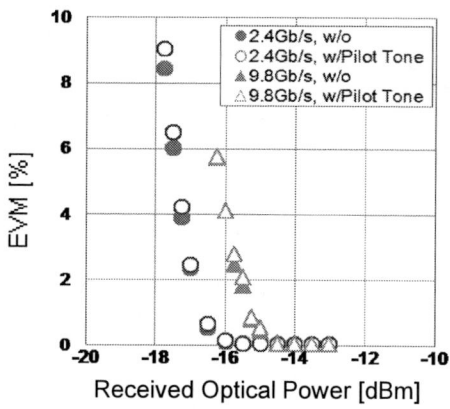 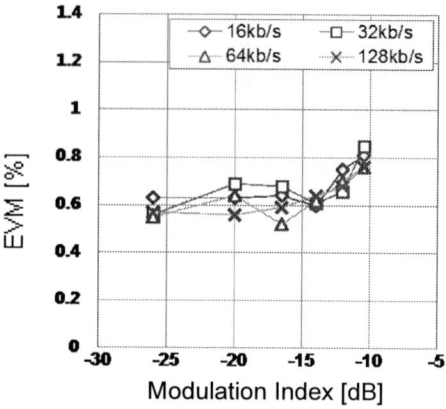

Fig. 4. EVM performance with AMCC pilot tone.

The architecture type locates a switch at only the central office with centralized BBUs (blue box) as shown in Fig. 3(b). WDM and PON technologies as well as WDM/TDM-PON can be used to combine several fibers into one fiber. While this enables flexibility, a management and control channel is required that is independent if the main signal. It is being discussed in ITU-T G.989.3 as the auxiliary management and control channel (AMCC). G.989 series covers the 40 Gb/s class PON system (NG-PON2: Next Generation PON2). AMCC carries Physical Layer Operation, Administration and Maintenance (PLOAM) messages for wavelength allocation, wavelength calibration etc. A WDM-PON system with electrically embedded AMCC channel has been proposed [6]. A 128 kb/s AMCC channel with -20 dB modulation index at 500 kHz carrier frequency does not affect error vector magnitude (EVM) performance of LTE signals employing CPRI signals (option 3: 2.4576 Gb/s, option 7: 9.8304Gb/s) as shown in Fig. 4. This kind of DRoF-based MFH requires more than 10 times the optical bandwidth of the wireless system. Bandwidth more than 10 Gb/s will be needed in the future given the vision of 10 Gb/s class user bandwidth with high frequencies and beam forming using a lot of antennas. A line rate of 24 Gb/s has been added to the CPRI interface specification. The transceiver is assumed to be that of the 25 Gb/s Ethernet discussed in IEEE P802.3. Moreover, combining digital coherent detection with high level modulation is promising for enhancing the capacity per wavelength [7].

IV. CONCLUSIONS

Trends in 5G mobile networks and RAN technologies were described. Cloud RAN, which uses switches to enable flexible MFH networks, and two promising candidates of MFH in Cloud RAN were mentioned. One sets a switch at each central office and establishes a ring network between central offices. IEEE 802.1CM is being discussed for time sensitive networking. The other candidate sets a switch at only the central office with centralized BBUs. WDM and PON technologies as well as WDM/TDM-PON can combine several fibers to form a single fiber. WDM-PON with AMCC has been proposed and shows good EVM performance. Future MFH requires much more optical bandwidth and some enabling techniques have been proposed.

ACKNOWLEDGMENT

This technical paper includes a part of results of "The research and development project for realization of the fifth-generation mobile communications system" commissioned by The Ministry of Internal Affairs and Communications, Japan.

REFERENCES

[1] NTT DOCOMO, "DOCOMO 5G White Paper," Jul. 2014.
[2] Common Public Radio Interface (CPRI); Interface Specification, V.7.0,(2015).
[3] Open Base Station Architecture Initiative (OBSAI); Interface Specification, (2006).
[4] T. Tashiro, et al., "A Novel DBA Scheme for TDM-PON based Mobile Fronthaul," Proc. OFC, Tu3F–3, Mar. 2014.
[5] H. Ou et al., "Passive Optical Network Range Applicable to Cost-effective Mobile Fronthaul," Proc. IEEE ICC, May 2016. (accepted)
[6] K. Honda et al., "WDM Passive Optical Network Managed with Embedded Pilot Tone for Mobile Fronthaul," Proc.ECOC, We.3.4.4, Sep. 2015.
[7] T. Hirooka et al., "Optical and wireless-integrated next-generation access network based on coherent technologies," Proc. Photonics West 2016

S1-8 (Invited)

OECC/PS2016

Cost-Effective Optical Transmitters for Next-Generation Mobile Fronthaul Networks

B. G. Kim, Hoon Kim, and Y. Chung

School of Electrical Engineering, KAIST, 291 Daehak-ro Yuseong-gu, Daejeon, Korea 34141

Email: hoonkim@kaist.ac.kr

Abstract: We report on the use of cost-effective optical transmitters such as vertical-cavity surface emitting lasers and directly modulated lasers for radio-over-fiber-based mobile fronthaul networks. Technical challenges associated with those optical transmitters are presented.

Keywords: mobile fronthaul, radio-over-fiber technology, directly modulated laser, VCSEL.

CLOUD-RAN AND MOBILE FRONTHAUL NETWORKS

The cloud (or centralized) radio access network (C-RAN), thanks to its capability of providing centralized operation, administration, and maintenance (OA&M) of the RAN, has recently gained a great deal of interest among the mobile service. In this architecture, the baseband processing of digitized wireless signals is all performed at centralized locations by co-locating baseband units (BBUs), while remote radio heads (RRHs) simply transmit and receive the wireless signals at their corresponding cell sites, acting as remote antennas. The major benefits of the C-RAN architecture can be categorized into three areas: site simplification, load balancing, and facilitation of technologies related to inter-cell interferences such as coordinated multipoint (CoMP) [1]. Centrally co-located BBUs not only considerably curtail the OA&M costs of cell sites (including site leases and repair works) but also allow new cell sites to be installed in fast and cost-effective manners. A set of centralized BBUs can also be pooled together to cope with dynamic shifts of traffic loads for flexible load balancing among cell sites. Lastly, the CoMP techniques could be readily implemented to improve the capacity and fairness of throughput across the cellular networks when the BBUs for neighboring cells are co-located.

The mobile fronthaul networks, which interconnect BBUs and RRHs in the C-RAN architecture, are usually implemented by using digital fiber-optic interfaces such as the Common Radio Public Radio Interface (CPRI), Open Base Station Architecture Initiative (OBSAI), or Open Radio equipment Interface (ORI). Major technical problems related to these fronthaul networks are (1) the enormous transmission capacity required for future wireless communication systems and (2) the latency incurred by the fronthaul equipment [5]. It is envisaged that the current digital fiber-optic interfaces should provide several hundred Gb/s of capacity for a single cell cite supporting three sectors and multiple-input multiple-output (MIMO) antennas for the 5th generation (5G) mobile communication systems. The fronthaul latency including the round-trip propagation delay as well as the processing delay by the fronthaul equipment should be less than a couple of microseconds even though the overall latency requirements are specific to the air interface and media access control of the 5G networks to come [6].

MOBILE FRONTHAUL NETWORK USING THE RADIO-OVER-FIBER TECHNOLOGY

The radio-over-fiber (RoF) technology is attracting a great deal of attention as a fiber-optic transmission technology for mobile fronthaul networks [7]-[11]. Instead of sampling and digitizing the wireless signals, and then multiplexing them using the time-division multiplexing technique, we can multiplex numerous wireless signals carried over different intermediate frequencies (i.e., different from the frequencies adopted for wireless transmission) of in the frequency domain without format conversion, and deliver them over optical fiber by using the RoF technology. Multiplexing the wireless signals could be performed either by utilizing digital signal processing or analog frequency converters [7]-[11]. Therefore, we can minimize the bandwidth of devices required for the transmission of wireless signals and the additional latency occurring during the format conversions from/to the wireless signal to/from the digital fiber-optic interface signal and vice versa.

The technical challenges associated with the RoF technology for mobile fronthaul networks include the stringent requirements of device linearity and high carrier-to-noise ratio (CNR). Nonlinearity of optical transmitter typically limits the maximum modulation index of the signal, and thus the maximum signal-to-noise ratio. The CNR requirement for 64-QAM-modulated LTE-A signals to satisfy the error-vector magnitude performance specified in the 3GPP specifications is 22.0 (assuming that the system performance is limited by additive white Gaussian noise) [12]. Since the shot or thermal noises of the receiver are usually the major sources of noise in RoF systems, this CNR requirement could not be easily satisfied unless the link loss is low and/or the modulation index is high. Due to this tight CNR requirement, the RoF systems typically adopt the point-to-point architecture.

OECC/PS2016

COST-EFFECTIVE RADIO-OVER-FIBER LINK FOR MOBILE FRONTHAUL NETWORK

In this paper, we report on the use of cost-effective optical transmitters such as vertical-cavity surface emitting lasers (VCSELs) and directly modulated lasers (DMLs) for RoF-based mobile fronthaul networks. VCSELs, thanks to their low-cost manufacturability, on-wafer testability, high energy efficiency, and direct modulation capability, are quickly becoming the light source of choice for various short-reach transmission systems. Recent technical advancement in the performance of long-wavelength (e.g., beyond 1.3 μm) VCSELs provides us with a possibility of utilizing these highly cost-effective devices for short- and intermediate-reach applications [13]-[18]. The linear relationship between the driving current and output power of VCSELs also makes them attractive for RoF-based systems. Nonetheless, the nonlinearity of the laser diode still exists and eventually limits the maximum modulation index of the RoF signals. One octave allocation of the signals driving the laser diode can alleviate the linearity requirements of the device at the expense of its available bandwidth since the even-order harmonics fall outside the signal band. The performance of VCSEL-based RoF system is also limited by the receiver noise due to the intrinsically low output power of VCSELs [19]. In the symposium, we present the maximum number of 20-MHz-bandwidth component carriers (CCs) for LTE-A signals when commercially available VCSELs are utilized for RoF-based mobile fronthaul networks.

Edge-emitting directly modulated lasers produce much higher light power than VCSELs and can support a larger capacity in RoF-based mobile fronthaul networks. For example, compared to the case of using a 1.5-μm VCSEL with an output power of -5 dBm, a considerable improvement in the number of CCs for mobile fronthaul network (e.g., a factor of ~40) can be achieved by using a DML having an output power of 6 dBm. In this system, one octave allocation of the subcarrier-multiplexed CCs also helps to increase the modulation index of the signals. It also serves to avoid other nonlinearities from falling on the signal band, including signal distortions induced by the interplay between laser chirp and fiber chromatic dispersion [20].

Since the RoF-based mobile fronthaul transmission systems require a high CNR, they are vulnerable to multi-path interferences possibly caused by Rayleigh back-scattering and optical reflections in the link. We also present in this symposium the effects of multi-path interference on the performance of RoF-based mobile fronthaul transmission system based on DMLs [21].

ACKNOWLEDGMENT

This work was supported by the IT R&D programs of MSIP (Ministry of Science, ICT and Future Planning), Korea [15ZI1300, Development of compact radio & dense digital base station technologies based on RoF technology].

REFERENCES

[1] U. Dotsch, M. Doll, H.-P. Mayer, F. Schaich, J. Segal, and P. Sehier, "Quantitative analysis of split base station processing and determination of advantageous architectures for LTE," *Bell Labs Tech. J.*, vol. 18, pp. 105-128, 2013.

[2] A. Checko, H. Christiansen, Y. Yan, L. Scolari, G. Kardaras, M. Berger, and L. Dittmann, "Cloud RAN for mobile networks-a technology overview," *IEEE Comm. Surv. Tut.*, vol. 17, pp. 405-426, 2015.

[3] K. Tanaka and A. Agata, "Next-generation optical access networks for C-RAN", in *Proc. OFC/NFOEC*, 2015, paper Tu2E.1.

[4] A. Pizzinat, P. Chanclou, F. Saliou, and T. Diallo, "Things you should know about fronthaul," *J. Lightwave Technol.*, vol. 33, pp. 1077-1083, 2015.

[5] S. H. Bae, H. K. Shim, U. H. Hong, H. Kim, A. Agata, K. Tanaka, M. Suzuki, and Y. Chung, "25-Gb/s TDM optical link using EMLs for mobile fronthaul network of LTE-A system," *IEEE Photon. Technol. Lett.*, vol. 27, pp. 1825-1828, 2015.

[6] N. Gomes, P. Chanclou, P. Turnbull, A. Magee, and V. Jungnickel, "Fronthaul evolution: from CPRI to Ethernet," *Opt. Fiber Technol.*, vol. 26, pp. 50-58, 2015.

[7] X. Liu, F. Effenberger, N. Chand, L. Zhou, and H. Lin, "Efficient mobile fronthaul transmission of multiple LTE-A signals with 36.86-Gb/s CPRI-equivalent data rate using a directly-modulated laser and fiber dispersion mitigation," in *Proc. ACP*, 2014, pp. 1-3, Paper AF4B.5.

[8] S. H. Cho, H. S. Chung, C. Han, S. Lee, and J. H. Lee, "Experimental demonstration of next generation cost-effective mobile fronthaul with IFoF technique," in *Proc. OFC*, 2015, paper M2J.5.

[9] M. Zhu, X. Liu, N. Chand, F. Effenberger, and G.-K. Chang, "High-capacity mobile fronthaul supporting LTE-Advanced carrier aggregation 8×8 MIMO," in *Proc. OFC*, 2015, paper M2J.3.

[10] F. Effenberger and X. Liu, "Power-efficient method for IM-DD optical transmission of multiple OFDM signals," *Opt. Express*, vol. 23, pp. 13571-13579, 2015.

[11] C. Ye, X. Hu, X. Huang, Q. Chang, X. Sun, Z. Gao, S. Xiao, and K. Zhang, "A first demonstration of a PON-based analog fronthaul solution supporting 120 20 MHz LTE-A signals for future Het-Net radio access," in *Proc. ACP*, 2015, paper ASu3E.2.

[12] 3GPP TS 36.104 version 11.9.0 Release 11, 2014.

[13] N. Nishiyama, C. Caneau, J. Downie, M. Sauer, and C. Zah, "10-Gbps 1.3 and 1.55-μm InP-based VCSELs: 85 °C 10-km error-free transmission and room temperature 40-km transmission at 1.55-μm with EDC," in *Proc. OFC/NFOEC* 2006, paper PDP23.

[14] F. Fidler, C. Hambeck, P. Winzer, and W. Leeb, "4×10-Gb/s CWDM transmission using VCSELs from 1531 nm to 1591 nm," in *Proc. ECOC* 2006, paper WE4.5.1.

[15] R. Rodes, J. Estaran, B. Li, M. Muller, J. Jensen, T. Gruendl, M. Ortsiefer, C. Neumeyr, J. Rosskopf, K. Larsen, M. Amann, and I. Monroy, "100 Gb/s signal VCSEL data transmission link," in *Proc. OFC/NFOEC* 2012, paper PDP5D.10.

[16] Z. Al-Qazwini, J. Zhou, and H. Kim, "1.5-μm 10-Gb/s VCSEL link for optical access applications," *IEEE Photon. Technol. Lett.*, vol. 25, no. 22, pp. 2160-2163, Nov. 2013.

[17] J. Zhou, C. Yu, and H. Kim, "Transmission performance of OOK and 4-PAM signals using directly modulated 1.5-μm VCSEL for optical access network," *J. Lightw. Technol.*, vol. 33, no. 15, pp. 3243-3249, Aug. 2015.

[18] F. Karinou, C. Prodaniuc, N. Stojanovic, M. Ortsiefer, A. Daly, R. Hohenleitner, B. Kogel, and C. Neumeyr, "Directly PAM-4 modulated 1530-nm VCSEL enabling 56 Gb/s/λ data-center interconnects," *IEEE Photon. Technol. Lett.*, vol. 27, no. 17, pp. 1872-1875, Sep. 2015.

[19] B. G. Kim, H. Kim, and Y. Chung, "Mobile fronthaul optical link for LTE-A system using directly-modulated 1.5-μm VCSEL," in Proc. OECC 2016.

[20] C. Han, M. Sung, S.-H. Cho, H. S. Chung, S. M. Kim, and J. H. Lee, "Impact of dispersion-induced second-order distortions in multi-IFoF-based mobile fronthaul link for C-RAN," in *Proc. OFC/NFOEC* 2016, paper Tu2B.4.

[21] B. G. Kim, H. Kim, K. Tanaka, T. Kobayashi, K. Nishimura, M. Suzuki, and Y. Chung, "Analysis of performance degradations induced by multipath interferences in RoF-based mobile fronthaul network implemented by using directly modulated lasers," Submitted to ECOC 2016.

S2-1 (Invited)

OECC/PS2016

Ultra-Dense Space-Division-Multiplexing Optical Fibers

Tetsuya Hayashi

Optical Communications Laboratory, Sumitomo Electric Industries, Ltd., 1, Taya-cho, Sakae-ku, Yokohama, 244-8588 Japan
Author e-mail address: t-hayashi@sei.co.jp

Abstract: *This talk will review ultra-dense SDM fibers for optical communications, such as the multi-core fiber with the improved spatial capacity, few-mode multi-core fibers supporting 100+ spatial modes, and the 125-µm-cladding 8-core fiber optimized for O-band short-reach interconnects.*
Keywords: *Space-division multiplexing, SDM, multi-core fiber, MCF*

I. INTRODUCTION

Spatial division multiplexing (SDM) is a strong candidate technology to overcome the capacity limit of single-mode fiber transmission systems [1,2]. Various kinds of SDM fibers has been actively studied [2,3], and fiber transmission capacity records have been bettered exclusively by the experiments using SDM fibers recently [4–6].

This paper reviews the research and development on the optical fibers for space-division multiplexing having ultra-dense spatial mode density from recent our studies [7–13].[1]

II. HIGH-SPATIAL-SPECTRAL-EFFICIENCY MULTI-CORE FIBER

In the case of the MCF, the spatial capacity, i.e., the aggregate capacity per unit cross-sectional area, does not necessarily increase with the increase of spatial channel capacity. By assuming dispersion-uncompensated Nyquist WDM transmission—for approximating nonlinear interference as AWGN [14]— and high symbol rate—for approximating XT as AWGN [15]—, the XT-affected SNR in the MCF core can be analytically expressed [16]. Since the Shannon limit of the SE for a polarization mode can be calculated from the SNR, the maximal SE (SE_{lim}: SE limit) at the optimum input power in the MCF can be calculated from the XT-affected SNR. Similarly, we can calculate the upper limit (SSE_{lim}) of the spatial SE (SSE) defined as the aggregate SE per cross-sectional area (A_{cs}) and expressed as: $SSE \equiv \sum SE/A_{cs}$. Figure 1 shows an example of the SE_{lim} and SSE_{lim} of an MCF (for details of the calculation conditions, c.f. [7]). A steep change of the SE_{lim} in the horizontal direction is caused by the large dependence of the XT on the core pitch Λ. We can understand that A_{eff} enlargement improves the SE_{lim} but requires larger Λ for avoiding an XT increase[1].). On the other hand, SSE_{lim} improves at smaller A_{eff} & Λ. The smaller A_{eff} induces larger nonlinear noises but can improve the light confinement and the core density. The effect of the core density increase is much larger than the nonlinear noise increase. As an example, we can realize a 31-core fiber with the A_{eff} of ~57 µm² and the cladding diameter of 225 µm, which can be calculated to have more than 21 times better capacity per fiber [8].

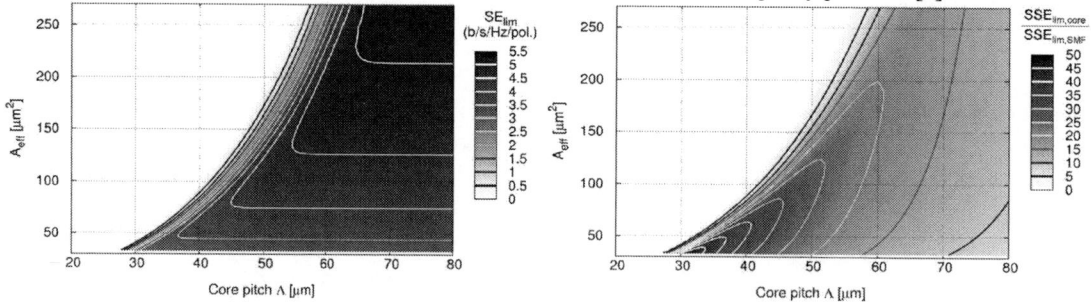

Fig. 1. Calculated maximal SE (left) and SSE (right) of an MCF [7].

III. FEW-MODE MULTI-CORE FIBERS SUPPORTING MORE THAN 100 SPATIAL MODES

To achieve a factor of 100 of improvement of spatial channel count per fiber [1], we designed and fabricated two kinds of few-mode multi-core fibers.

Three-mode 36-core Fiber with 108 spatial modes

One is the three-mode 36-core fiber (3M36CF, 108 spatial modes in total) [9,10]. The 3M36CF has the

---

[1] This research is supported in part by the National Institute of Information and Communications Technology (NICT), Japan.
[1] Note that the A_{eff} increase over 80 µm² needs the depressed/trench-assisted cladding structure for suppressing macro/microbend losses, but these results were calculated with the simple step-index core with the matched cladding for showing the design trend.

heterogeneous double cladding structure [17] with three types of cores, as shown in Fig. 2(a). All the core types have identical core diameter and Δ ($\Delta_+ + \Delta_-$), with 0.1% absolute Δ difference. All the cores support LP_{01} and LP_{11} modes and the effective index difference between core types was > 0.001 for all combinations of LP_{01} and LP_{11} modes. Hence, the inter-core XT is suppressed when the fiber bending radius is larger than the critical bending radius, $R_{pk} \sim 5$ cm, at a core pitch of 34 µm [18]. The 36 cores were arranged on a 4-layer hexagonal lattice, as shown in Fig. 2. One core in the outmost layer was omitted to serve as a marker. The cross section and the optical properties of the fabricated 5.5-km 3M36CF are shown in Fig. 2(b) and Table 1, respectively. The cladding diameter was 306 µm. Though the Λ was fabricated to be 34 µm, the Λ of 40 µm can be realizable with the same cladding diameter of 306 µm without any degradation in the coating-leakage loss of the outmost core, according to the simulation using the full-vectorial finite-element imaginary distance beam propagation method [19]. In such Λ-expanded configuration, the inter-core XT can be lower than −30 dB after 100 km propagation.

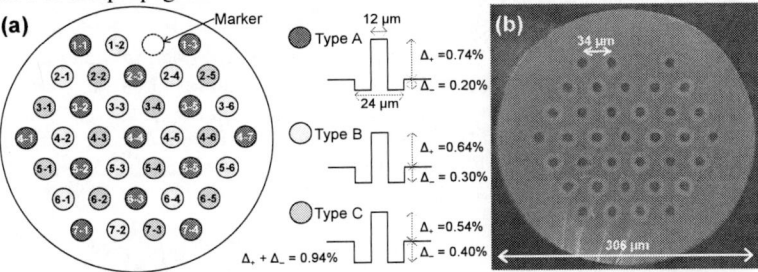

Fig. 2. (a) The design of the 3M36CF, and (b) a cross section of the fabricated fiber.

TABLE I OPTICAL PROPERTIES OF THE FABRICATED 3M36CF AT 1550 NM.

	Mode	Core Type A	Core Type B	Core Type C
Transmission loss [dB/km][†]	$LP_{01}+LP_{11}$	0.242–0.266	0.281–0.292	0.262–0.308
Cable cutoff wavelength [nm][†]	LP_{21}	1458–1526	1346–1357	1199–1245
Effective area [µm²][‡]	LP_{01} / LP_{11}	77 / 105	74 / 102	76 / 110
DMD [ps/m][‡]	Between LP_{11} and LP_{01}	7.4	7.1	6.3
Inter-core crosstalk[‡] [dB] after 5.5 km (5 cm correlation length assumed)		A-B	B-C	C-A
	LP_{01}-LP_{01} / LP_{01}-LP_{11}	−98 / −61	−86 / −32	−88 / −48
	LP_{11}-LP_{01} / LP_{11}-LP_{11}	−77 / −54	−60 / −31	−33 / −32

† Measured values. ‡Calculation from measured refractive index profiles.

A. Six-mode 19-core Fiber with 114 spatial modes

Another FM-MCF with 100+ spatial modes is the six-mode 19-core fiber (6M16CF, 114 spatial modes in total) [11]. The core index profile and the cross section of a fabricated 6M19CF are shown in Fig. 3(a,b). All the cores of the 6M19CF have the homogeneous graded-index profile. The optical characteristics are also summarized in Table 2. According to the fabricated fiber profile, the IC-XT was calculated to be < −80 dB/100 km—the IC-XT < −30 dB/100km can be realized when $\Lambda \geq 48.7$ µm; thus, the cladding diameter can be reduced to ~266 µm can be realized if we set Λ as 50 µm—. Δn_{eff}s between the mode groups were more than 3.9×10^{-3}. Therefore, according to [20], IM-XTs are expected to be very low during the fiber propagation. Δn_{eff} between the LP21 and LP02 modes was suppressed as designed. The differential mode group delays (DMDs) within each mode group was not suppressed well for the LP21 and LP02 mode group. This is because the higher-order mode fields spread over the non-GI cladding and not highly confined in the GI profile—the DMD can be suppressed when the mode fields are highly confined in the GI profile—. It has been reported that the index trench can mitigate such effect on the higher-order modes [21,22].

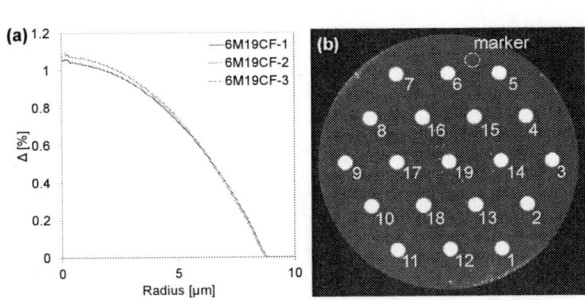

Fig. 3. (a) The index profiles of the fabricated cores of the 6M19CF. (b) The cross sections of the fabricated 6M19CF-2 [11].

TABLE II OPTICAL PROPERTIES OF A FABRICATED 6M19CF [11].

Items	Values			
Core Δ [%]	~1.08			
Core index gradient α	~2.0			
Core diameter [µm]	17.4			
Core pitch [µm]	63.4 (62.5–64.0)			
Cladding diameter [µm]	318			
Transmission loss [dB/km] (LP_{01})	0.388 (0.314–0.467)			
λ_{cc} [nm]	1443 (1431–1458)			
IC-XT [dB/100km][*]	−80			
Modal characteristics	LP_{01}	LP_{11}	LP_{02}	LP_{21}
Δn_{eff}[*] from LP_{01} [$\times 10^{-3}$]	0	−4.09	−8.14	−8.17
A_{eff} [µm²][*]	64.3	128.7	140.2	179.4
DMD [ns/km][*] against LP_{01}	0	0.30	−2.0	−0.65

*: Calculated.

IV. EIGHT-CORE FIBER WITH THE CLADDING DIAMETER OF 125 MICROMETER FOR OPTICAL INTERCONNECTS

We also designed and fabricated a homogeneous trench-assisted 8-core MCF (8CF) with the cladding diameter of 125 μm for short-reach optical interconnects. The standard 125-μm cladding is compatible with the conventional fiber fabrication/cabling/connecting technologies, and its mechanical reliability has been proven in the field use [23]. The eight cores are good granularity to realize full-duplex 100-Gb/s transmission using the conventional 25-Gb/s transceiver technologies. Figure 4(a) shows the cross-section of the fabricated MCF. A trench-assisted core index profile was employed for simultaneously realizing strong light confinement, single-mode operation, low chromatic dispersion, and low-splice-loss MFD, in O-band. All the cores were equally-spaced with 31-μm pitch on a common circle. The pitch deviation from the design was well controlled to be less than ±0.3 μm owing to the simple one-ring core layout. To pack 8 cores into the 125-μm cladding, we designed the thin minimum core-to-coating distance of 22 μm by ensuring the leakage loss suppression in O-band but accepting higher leakage loss in longer wavelength bands. The attenuation spectra averaged over all the cores, shown in Fig. 4(b), reflect the impact of the high leakage loss in the longer wavelength bands. However, we can see that the transmission window is still open in the O-band as we designed. The average of the XT between neighboring cores of the MCF was −53.7 dB after 13.14 km (−64.9 dB after 1 km). The optical characteristics in O-band and the cable cutoff wavelength (λ_{cc}) of each core are shown in Table 1, which are compatible with ITU-T G.652 as far as in O-band. We also fabricated an indoor MCF cable including 12 MCFs (96 cores) in the 3-mm outer diameter as shown in Fig. 4(c–e), which achieved the core density of 13.6 cores/mm², the highest value ever reported.

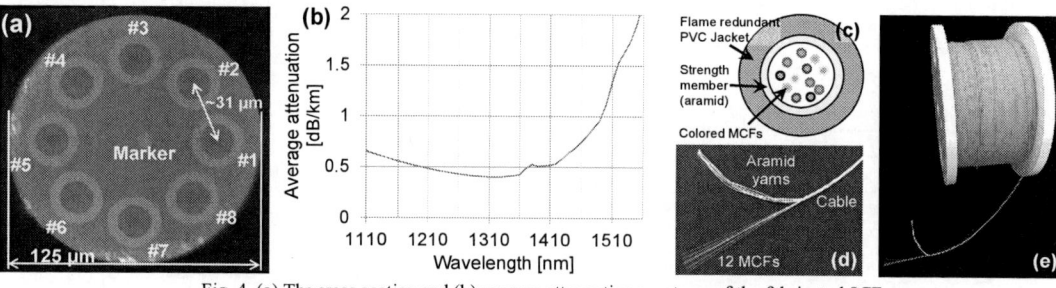

Fig. 4. (a) The cross section and (b) average attenuation spectrum of the fabricated 8CF, and (c) a schematic cross section and (d,e) photos of the fabricated MCF cable.

TABLE III OPTICAL PROPERTIES OF THE FABRICATED 8CF.

	Transmission loss [dB/km]	λ_{cc} [nm]	MFD [μm]	λ_0 [nm]	Macrobend loss [dB/turn] at R=3mm
λ [nm]	1310	n/a	1310	n/a	1310
Avg.	0.388	1229	8.4	1319	0.029
Min.–Max.	0.346–0.412	1217–1238	8.3–8.5	1318–1320	0.023–0.034
ITU-T G.652	≤0.5	≤1260	8.6–9.5 ± 0.6	1300–1324	n/a

V. CONCLUSIONS

In the research field, diverse kinds of SDM fibers have been developed and proposed for realizing the SDM high-capacity/-density transmission, and the record transmission capacity has been updated with them in recent years. To make the SDM fibers commercially deployable options in the near future, the SDM fiber has to be designed based on the specific requirements and limits by assuming specific applications (e.g., short-reach, ultra-long-haul, core, access, etc.). Of course, we need to take account of fiber evaluation/characterization, connecting, cabling, amplification, and so on.

REFERENCES

[1] T. Morioka, in *OECC2009*, p. FT4. [2] P. J. Winzer, *Opt. Photonics News*, vol. 29, no. 3, pp. 28–35, Mar. 2015. [3] P. Sillard, presented at *ISUPT/EXAT 2015*, Kyoto, 2015, p. T2.2. [4] B. J. Puttnam et al., in *ECOC2015*, p. PDP.3.1. [5] D. Soma et al., in *ECOC2015*, p. PDP.3.2. [6] H. Takara et al., in *ECOC2012*, p. Th.3.C.1. [7] T. Hayashi and T. Sasaki, in *IEEE Photon. Soc. Sum. Top.*, 2013, p. MC2.4. [8] T. Nakanishi et al., in *OFC2015*, p. Th3C.3. [9] J. Sakaguchi et al., in *OFC2015*, p. Th5C.2. [10] J. Sakaguchi et al., *J. Lightw. Technol.*, vol. 34, no. 1, pp. 93–103, Jan. 2016. [11] T. Hayashi et al., in *OFC2016*, to be presented as paper W1F.4. [12] T. Hayashi et al., in *OFC2015*, p. Th5C.6. [13] T. Hayashi et al., *J. Lightw. Technol.*, vol. 34, no. 1, pp. 85–92, Jan. 2016. [14] P. Poggiolini, *J. Lightw. Technol.*, vol. 30, no. 24, pp. 3857–3879, Dec. 2012. [15] T. Hayashi et al., *IEICE Trans. Commun.*, vol. E97.B, no. 5, pp. 936–944, May 2014. [16] T. Hayashi et al., *Opt. Express*, vol. 20, no. 26, pp. B94–B103, Nov. 2012. [17] Y. Kokubun and T. Watanabe, in *Microoptics Conference (MOC), 17th*, Sendai, Japan, 2011, p. K-5. [18] T. Hayashi et al., in *ECOC2010*, p. We.8.F.6. [19] K. Saitoh and M. Koshiba, *J. Quantum Electron.*, vol. 38, no. 7, pp. 927–933, 2002. [20] R. Maruyama et al., in *OFC2015*, p. M2C.1. [21] T. Mori et al., in *OFC2013*, p. OTh3K.1. [22] P. Sillard et al., in *ECOC2015*, p. Mo.4.1.2. [23] T. Volotinen et al., in *IWCS2014*, pp. 47–54.

Recent Progress in Mode Multiplexing Devices using Multi-Plane Light Conversion

Guillaume Labroille, Pu Jian, Nicolas Barré, Bertrand Denolle, Jean-François Morizur

CAILabs SAS, 8 rue du 7e d'Artillerie, 35000 Rennes, France
guillaume@cailabs.com

Abstract: We report a 10 spatial mode multiplexer based on the technique of Multi-Plane Light Conversion. The device shows insertion loss below 4 dB and 26 dB mode selectivity across the full C+L-band.
Keywords: Space division multiplexing; Optical communication; Mode multiplexer; Mode selectivity; Passive optical component

I. INTRODUCTION

Due to the ever increasing data capacity requirement in optical fibers and to the approaching non-linear Shannon limit for single-mode fibers (SMFs), Space Division Multiplexing (SDM) [1] has been proposed as a means to increase fiber capacity. Parallel spatial paths can be exploited through multiple modes of a multi-mode fiber (MMF) [2], multiple cores of a multi-core fiber (MCF) [3], or a combination of both [4]. In the case of transmission in a MMF, one of the key elements in the transmission system is the spatial multiplexer (MUX), capable of converting N SMFs into N spatial channels of the MMF, and the spatial demultiplexer (DEMUX) performing the inverse operation.

Two strategies exist for handling the spatial channels of the MMF. On the one hand, one can launch the input channels into a random set of orthogonal combination of the eigenmodes of the MMF [2]. The channels are thus strongly coupled, and demultiplexing the channels requires full modal diversity at the receiver and complex digital signal processing (DSP). On the other hand, one can use fibers and components in which spatial modes are weakly- or un-coupled, therefore limiting the DSP to the joint detection of degenerate mode groups [5].

In both strategies, there is an interest in mode-selective MUX and DEMUX: indeed, weakly coupled technique requires that the modal cross-talks generated by the MUX/DEMUX stay small enough, while mode selectivity insures low mode-dependent loss (MDL) and possibility to compensate differential mode group delay [6], which relieve the complexity of DSP in strongly coupled transmissions. Other important features for mode MUX and DEMUX are low insertion loss (IL) and large bandwidth of operation for compatibility with wavelength division multiplexing.

Several techniques have been proposed for mode selective MUX. Primarily, binary phase plate converters [7] have been widely used due to their simplicity, but they suffer from large intrinsic IL. Another technique largely investigated is mode-selective photonic lantern [8] using dissimilar input SMFs, which presents the possibility of very low IL; however experimentally achieved mode selectivity remains low due to the complexity of fiber engineering. Other techniques based on fused couplers [9, 10] have also been reported.

Here we demonstrate a highly mode-selective, 10-mode MUX and DEMUX based on Multi-Plane Light Conversion (MPLC). This technique has been previously reported for 6-mode multiplexers and used in various transmission experiments [4, 5]; we present here the extension of these results to 10 modes, with good IL and high selectivity over the full C+L band.

II. MULTI-PLANE LIGHT CONVERSION

Multi-Plane Light Conversion (MPLC) is a technique that allows to perform any unitary spatial transform. Theoretically, any unitary spatial transform can be implemented by a certain succession of transverse phase profiles separated by optical Fourier transforms (OFT). In particular, the conversion of N separate input Gaussian beams into N orthogonal propagation modes of a fiber, i.e. spatial multiplexing, can be considered as a unitary spatial transform and therefore can be achieved with MPLC [11]. An example of spatial multiplexing for 3 modes is shown in Fig. 1a. The unitarity of the transform insures that there is no intrinsic loss in the mode conversion. Losses in a MPLC only occur due to imperfect optical elements (e.g. coating). The inverse unitary transform, given by using the MPLC in the reverse direction, implements the demultiplexing operation of the same modes.

In order to reduce the footprint of the MPLC as well as decreasing the complexity of aligning free-space optical elements, the MPLC is experimentally implemented using a multi-pass cavity, in which the successive phase profiles are all printed on a single reflective phase plate. The cavity is formed by a mirror and the reflective phase plate, and performs the successive phase profiles and optical transforms.

Two multiplexers supporting 10 spatial modes are fabricated using an implementation with 14 reflections on the phase plate. The multi-pass cavity for a 10-mode multiplexer is shown in Fig. 1b. The system converts light from 10 input SMFs into the first 10 modes of a conventional graded-index MMF with 50 mm core diameter. Fig. 1c shows the output modes in free-space (before coupling into a MMF) when using a super luminescent diode (SLD) centered at 1550 nm as light source. The optical losses, from input SMF to free-space output, are below 2 dB.

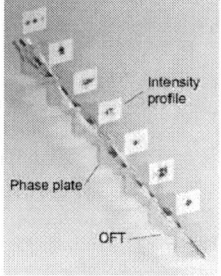 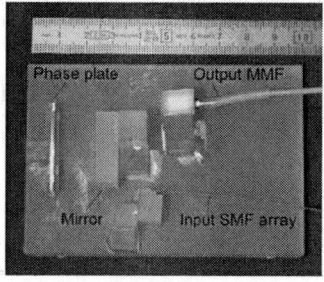 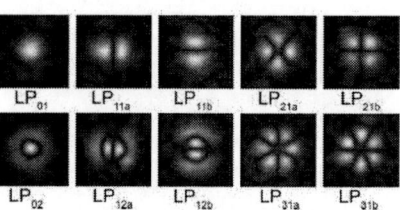

Fig. 1. Multi-Plane Light Conversion. Left: Schematics of a 3-mode MUX. Center: Photo of a 10-mode MUX. Right: Free-space output of the 10-mode MUX with a SLD source.

III. CHARACTERIZATION

We characterize the 10-mode multiplexers using the setup shown in Fig. 2. The transmission matrix of a back-to-back system comprising a MUX, 10 meters of MMF and a DEMUX is measured using an optical switch and a multichannel optical power meter. Two types of input light sources are used: a SLD, centered at 1550 nm with FWHM bandwidth of 50 nm, and a tunable distributed feedback laser (DFB) able to cover full C+L band (1530 nm to 1630 nm). Measurements with low coherence SLD source are not affected by multi-path interferences; therefore they are more stable and they do not fluctuate with fiber position. These measurements give similar results as measurements with DFB averaged across the C-band.

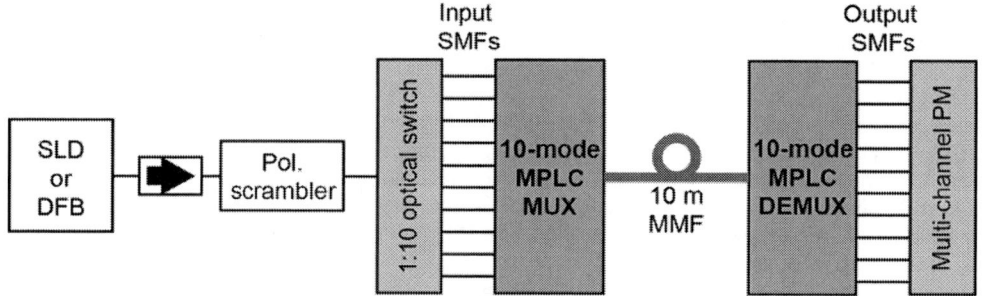

Fig. 2. Characterization setup: Light source is either a SLD or a tunable DFB. After an optical isolator and a polarization scrambler, an optical switch and a multi-channel optical power meter (PM) are used to characterize the system

We measure the matrix of output powers $P_{i,j}$ for power in output j when light is on input i, allowing to retrieve the coupling efficiency and the modal cross-talk for all modes. In order to remove the effect of intra-mode group mixing inside the fiber in the calculation of MUX performance, for one input of a mode group we consider its output power as the sum of all the outputs of the same mode group. For example, the coupling efficiency of LP_{11a} input is given by $(P_{LP11a,LP11a} + P_{LP11a,LP11b})/P_{in}$, where P_{in} is the input power, equal for all input fibers. In the same way, the cross-talk is averaged between outputs of the same mode group. For example, the cross-talk from input LP_{01} to mode group LP_{11} is given by $XT(LP_{01} \rightarrow LP_{11a}) = XT(LP_{01} \rightarrow LP_{11b}) = (P_{LP01,LP11a} + P_{LP01,LP11b})/(2P_{LP01,LP01})$.

IV. RESULTS

Table 1 shows the measured cross-talk and coupling efficiency matrix for a back-to-back system using SLD source. Assuming that the MUX and DEMUX are identical, the average insertion loss for a single MUX is 3.4 dB and the average cross-talk is -26 dB.

TABLE I
CROSS-TALK AND COUPLING EFFICIENCY FOR A BACK-TO-BACK SYSTEM USING A SLD.

Unit: dB	Output LP_{01}	Output LP_{11a}	Output LP_{11b}	Output LP_{21a}	Output LP_{21b}	Output LP_{02}	Output LP_{31a}	Output LP_{31b}	Output LP_{12a}	Output LP_{12b}	Coupling efficiency
Input LP_{01}	N/A	-26.16	-26.16	-20.62	-20.62	-20.62	-24.01	-24.01	-24.01	-24.01	-6.81
Input LP_{11a}	-25.02	N/A	N/A	-23.72	-23.72	-23.72	-22.60	-22.60	-22.60	-22.60	-6.16
Input LP_{11b}	-25.29	N/A	N/A	-22.47	-22.47	-22.47	-22.13	-22.13	-22.13	-22.13	-7.24
Input LP_{21a}	-25.26	-24.86	-24.86	N/A	N/A	N/A	-24.63	-24.63	-24.63	-24.63	-6.26
Input LP_{21b}	-17.27	-23.60	-23.60	N/A	N/A	N/A	-27.16	-27.16	-27.16	-27.16	-6.63

Input LP$_{02}$	-22.07	-23.56	-23.56	N/A	N/A	N/A	-26.86	-26.86	-26.86	-26.86	-7.07
Input LP$_{31a}$	-22.55	-22.57	-22.57	-23.99	-23.99	-23.99	N/A	N/A	N/A	N/A	-6.85
Input LP$_{31b}$	-21.29	-22.99	-22.99	-24.18	-24.18	-24.18	N/A	N/A	N/A	N/A	-7.35
Input LP$_{12a}$	-23.74	-21.06	-21.06	-24.31	-24.31	-24.31	N/A	N/A	N/A	N/A	-6.97
Input LP$_{12b}$	-21.54	-18.67	-18.67	-23.89	-23.89	-23.89	N/A	N/A	N/A	N/A	-6.99

The same measurements are made across C+L band using the tunable DFB source and shown in Fig. 3 for one multiplexer. The average insertion loss over full C+L band is 3.4 dB with a maximum of 5.8 dB, the average cross-talk is -25 dB with a maximum of -17 dB, and the average mode dependent loss is 1.2 dB. Finally, polarization dependent loss is measured to be < 0.2 dB for all modes.

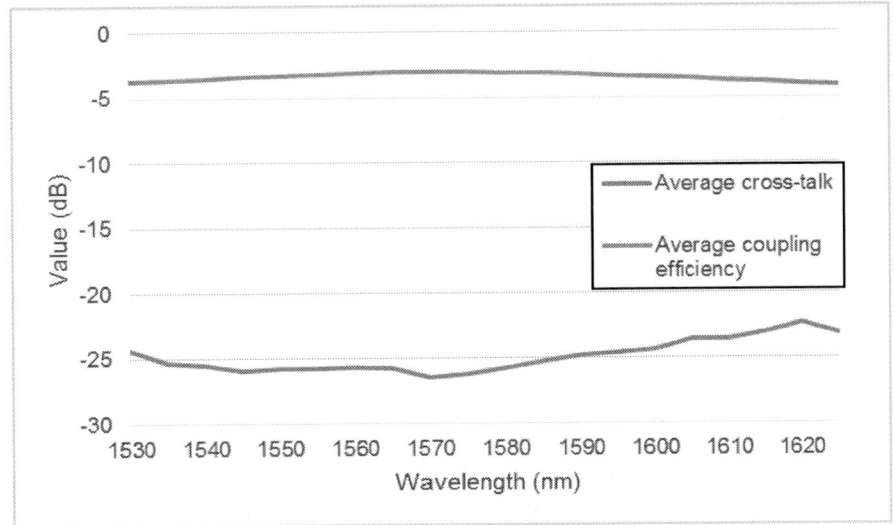

Fig. 3. Cross-talk and coupling efficiency for a single MUX across C+L band.

V. CONCLUSION

We have demonstrated mode-selective, 10-mode multiplexer and demultiplexer based on Multi-Plane Light Conversion. The multiplexers exhibit the highest reported mode selectivity in a 10 mode multiplexer (average cross-talk -26 dB), as well as a good coupling efficiency (insertion loss below 4 dB after coupling in the MMF) and very low mode dependent loss (average MDL 1.2 dB), over the full C+L band.

REFERENCES

[1] D.J. Richardson et al., Nature Photonics 7, 354 (2013).
[2] R. Ryf et al., ECOC 2015, PDP 3.3.
[3] B.J. Puttnam et al., ECOC 2015, PDP 3.1.
[4] D. Soma et al., ECOC 2015, PDP 3.2.
[5] P. Genevaux et al., OFC 2015, W1A.5..
[6] R. Ryf et al., ECOC 2014, PD.3.5.
[7] R. Ryf et al., J. Lightwave Technol. 30, 521 (2012).
[8] A.M. Vel´asquez-Benitez et al., ECOC 2015, Tu.3.3.2.
[9] S. Chang et al., Opt. Express 22, 14229 (2014).
[10] S. Gross et al., OFC 2015, W3B.2.
[11] G. Labroille et al., Opt. Express 22, 15599 (2015).

Switching solutions for spatial and spectral multiplexed networks

Dan M. Marom
Applied Physics department Hebrew University of Jerusalem
danmarom@mail.huji.ac.il

Abstract: Space division multiplexing offers a transmission capacity multiplier in fiber links, in advance of the looming 'capacity crunch' projected for SMF-based systems. We present efficient and cost effective switching solutions for WDM-SDM channel routing.
Keywords: Optical communications, space-division multiplexing, wavelength-division multiplexing, ROADM, WSS.

I. INTRODUCTION

Optical networks have evolved considerably over the past 20 years, since the introduction of WDM and optical amplifiers [1]. Fixed channel add and drop have been replaced with reconfigurable optical add/drop at ROADM nodes, enabled by wavelength-selective switch (WSS) technology [2]. We have witnessed sustained exponential network capacity growth during this time, and have reaped its benefits in ubiquitous Internet access. But at present we are see warning signs on the horizon which challenge how we can continue to evolve the infrastructure and offer continued exponential growth opportunities on an economically viable basis. Space division multiplexing (SDM) systems based on new types of fibers, e.g. few mode fiber (FMF), multicore fiber (MCF) and their hybridization (FM-MCF), all offer additional spatial conduits for information transmission with significant additional opportunities for capacity enhancements [3]. Expanding fiber link channel counts to the spatial domain requires further innovation and flexibility to route these channels at network nodes. SDM-WDM ROADM functionality and additional photonic switching innovations in support of the spatial domain are discussed in this paper.

II. SDM-WDM SWITCHING STRATEGIES

Single-mode based ROADM solutions provided flexibility across the wavelength domain, enabling the switching of wavelength channels. Employing both WDM and SDM within fiber links expands the channel count M-fold, with M being the number of spatial modes or cores. However, whereas wavelength channels do not intermix in a linear medium, spatial channels might. When spatial channels mix, the information can still be unraveled by employing multiple input, multiple output (MIMO) signal processing across the mixed channels. For MIMO processing to function, all the mixed channels have to be present and jointly processed. Hence when mixing occurs, the mixed channels must be kept as a group and cannot be independently switched to distinct network destination. We distinguish between three mixing situations influencing the switching strategies: (i) *no spatial mixing*, when cores are isolated or several single mode fibers are used to create a spatial group, (ii) *mixing between all spatial modes*, which may occur in multimode fibers or in fibers with coupled cores, and (iii) *mixing within subgroups*, which may occur in few-mode, multicore fibers or irregularly spaced multicore fibers. With these mixing situations under consideration, we identify four main SDM-WDM switching strategies, categorized according to the switching granularity (i.e. the smallest switched element) [4]:

I. Space granularity:
    Switching is performed in the spatial domain across the entire optical communication band (all WDM channels), forfeiting any wavelength flexibility. The switching hardware is devoid of any dispersing means, thus relying on relatively simple space switches. This strategy is only applicable to the no spatial mixing transmission case.

II. Space-wavelength granularity:
    Switching can be prescribed down to the single spatial and wavelength channel granularities in the fiber, exploiting the full switching potential of an SDM-WDM transmission system. Due to its offered finest granularity (resulting in largest flexibility), its potential implementations are the most cumbersome. As in the space granularity case, space-wavelength granularity requires that the spatial modes/cores do not mix.

III. Wavelength granularity:
    Contrary to space-granularity, which forfeited access to the spectral domain, wavelength granularity sacrifices the spatial domain in order to simplify the switching hardware. Implementations rely on modified WSS in support of joint switching of all M spatial channels on a wavelength basis. This scenario is the only one applicable when all the modes/cores intermix, as they must remain together as a group throughout the network.

IV. Fractional space, full wavelength granularity:
    For fibers where mixing is limited to within spatial subgroups, this strategy adopts the wavelength granularity approach within the subgroups, and each subgroup can be independently switched on a wavelength basis.

Attractive features of this strategy are the intermediate routing flexibility and moderate implementation complexity.

Possible implementations for the four switching strategies are shown in Figure 1 and detailed in the next sub-section. While the above switching strategies are geared for different mixing situations experienced in fiber transmission, the more restrictive solutions can always be applied to the less restrictive ones. For example, the wavelength granularity and fractional space-wavelength granularity approaches can be adopted for transmission systems having no spatial mode/core mixing. In addition, for fibers having no spatial mixing the switching may offer the ability to switch a wavelength channel from one spatial mode/core on an ingress fiber to a different one at the egress fiber. This SDM 'lane change' can be used to circumvent some blocking situations, and can be recognized as the equivalent of 'wavelength conversion' in WDM networking. However, whereas lane changes can be performed with simple optical switches and is an attractive networking feature, wavelength conversion requires more complicated solutions that have not been adopted in commercial systems.

III. SDM-WDM ROADM ARCHITECTURES

Implementing the different SDM-WDM ROADM nodes can be accomplished through many of the technologies that have been developed for SMF, as well as some complementary ones. Foremost of these is a spatial multiplexer (mux), which is used for combining M input single mode channels into an SDM fiber, ideally in a one-to-one mapping between mode/core and individual SMF. Employing such SDM demultiplexers (demux) at the ingress ROADM ports separates out the spatial channels to a discrete SMF set for subsequent switching. Considering the space granularity approach, fiber switches are then required to route the entire communication band on each separated spatial channel to its destination. Two switching scenarios can be considered: (1) routing from an ingress spatial channel is performed to the same spatial channel on the egress fiber, i.e. without spatial lane change, and (2) unrestricted routing is performed permitting spatial lane changes. Switching without lane changes can be realized by a bank of small optical cross connects (OXC), one per spatial channel (Figure 1-A). The input/output port count of an individual OXC (devoted to a spatial channel) is at minimum the number of directions, D, i.e. D×D OXC, and M such OXC are required. In support of lane changes, one large OXC should handle all the node traffic (Figure 1-B). In this case the OXC port count is (D·M)×(D·M). A significant risk associated with an architecture based on a single large OXC is its failure impact. This can be circumvented by a second OXC placed in parallel for protection, but this adversely impacts the cost, space requirement, and fiber routing complexity. Without lane changes using a bank of smaller OXC, a single switch failure impacts only 1/M of the node traffic.

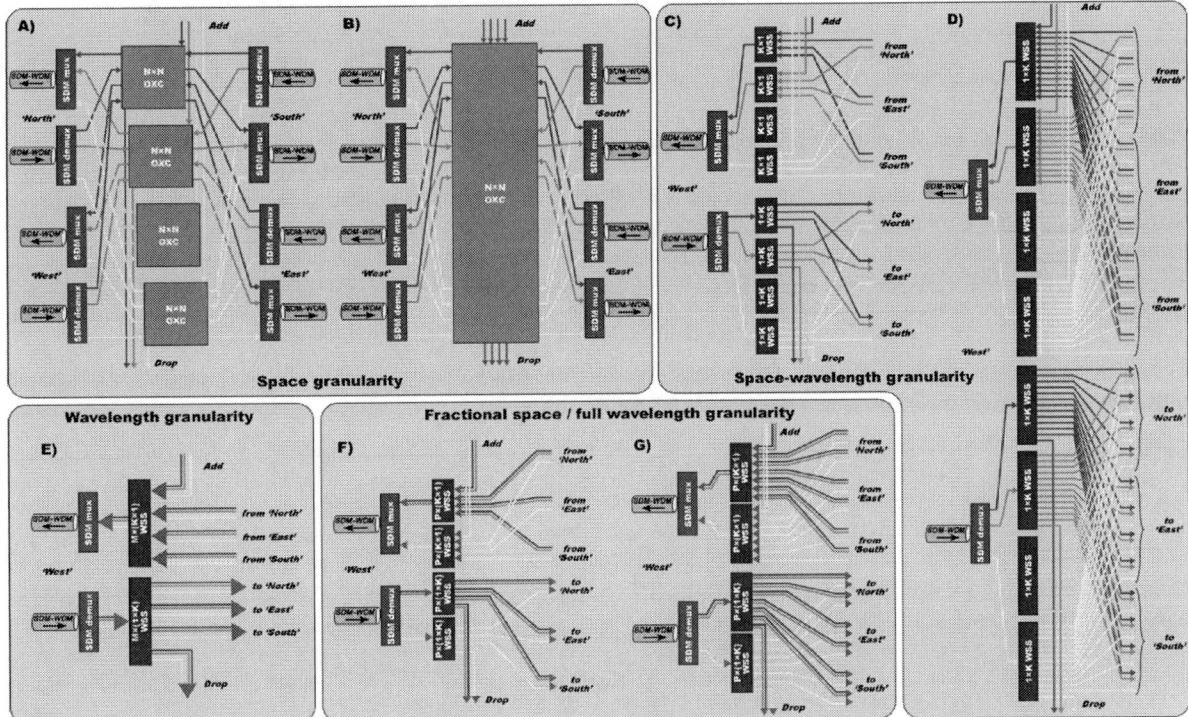

Figure 1 – Proposed implementations of SDM-WDM switching strategies for different switching granularities, drawn for four spatial modes and degree-four node. (A) Space granularity without SDM lane changes. (B) Space granularity with SDM lane changes. (C) Space-wavelength granularity without SDM lane changes. (D) Space-wavelength granularity with SDM lane changes. (E) Wavelength granularity. (F) Fractional space and full wavelength granularity without lane changes. (G) Fractional space and full wavelength granularity with lane changes. Cases (C-G) show the switching hardware for one direction only (West).

Space-wavelength granularity offers the finest switching granularity at the expense of implementation complexity. Here too lane changes may be supported at the cost of additional complexity. Access to the space-wavelength granularity is achieved by first spatially demultiplexing the signal, followed by placing a WSS on each spatially separated channel for routing to other egress ports. If routing to the same spatial channel on the other ports, then the number of WSS output ports needs to be D, with routing to D-1 other destinations and at least one drop port (Figure 1-C, with only one direction shown for simplicity). If spatial lane change is desired, then each ingress WSS should reach all other destinations and spatial channels, requiring the WSS to support M·(D-1)+drop output ports (Figure 1-D). The fiber wiring can become unwieldly in this case. What is evident is that the number of WSS increases M-fold, with the fiber capacity. Hence, barring any technology improvements such as multi-packing of WSS into a single module [5], the cost of this solution increases linearly with the capacity, and hence fails to deliver a cost advantage.

Implementation simplicity can be obtained by sacrificing flexibility, as has been the case for space granularity. In wavelength granularity, all spatial channels are routed as a whole per wavelength channel. This functionality can be provided by a single WSS modified to support joint switching over all SDM channels on a wavelength basis by employing spatial diversity (Figure 1-E) [6]. In the case of FMF, joint switching within the WSS can be implemented without requiring spatial diversity and its associated space mux/demux, providing further implementation simplicity [7].

As a compromise between switching granularity size and implementation complexity and cost, the joint switching concept of spatial channels can be applied to subgroups of the spatial channel count. Hence if the M spatial channels are divided into subgroups of size P, then the number of WSS required to route subgroups is M/P. This fractional space, full wavelength granularity solution can be implemented without lane changes (or rather subgroup change), see Figure 1-F, or with, Figure 1-G. The latter, while more cumbersome, is tolerable if the subgroup count, M/P, is a small number.

To better gauge the scaling of an SDM-WDM ROADM solution, we can place reasonable values for the node degree and spatial channel count. Let us assume that D=4 to cover the most prevalent network node degree size, and that the SDM channel count is M=12 (for an order of magnitude capacity increase). Space granularity can then be implemented with twelve 4×4 OXC, or a single 48×48 OXC in support of spatial lane changes (actual values will be slightly larger in support of add/drop ports). Space-wavelength granularity will require 2M=24 WSS per direction, which adds up to 96 WSS for the ROADM node of degree four. If no lane changes are required, then each WSS needs to support 1×4 and 4×1 routing capability. If lane changes are desired, then the port count of each WSS expands to 1×37. Should wavelength granularity be implemented, the ROADM node would require 8 WSS (one per ingress and egress port of the node, as occurs in today's SMF-based optical networks). However, the WSS should jointly switch twelve spatial channels at once to one of three directions plus one drop port (at least), which is denoted as a 12×(1×4) joint switching WSS. Let us now assume that the 12 spatial channels can be separated to four groups of size 3 (perhaps a four core fiber, where each core supports three modes), though other options can be considered. Hence each direction (ingress and egress) requires eight joint switching WSS, and 32 are required for the entire node. For the case without subgroup lane change, each WSS should support 3×(1×4), while the WSS capability grows to 3×(1×13) for lane changes.

Based on the discussion above, it should be clear that the choice of switching granularity directly affects the specific SDM-WDM ROADM architecture, the routing algorithms and related network performance. For instance, for wavelength granularity, a spatial superchannel will occupy all the M spatial modes for a given wavelength slot. While this constraint on the creation of spatial superchannels will limit the flexibility to switch, add, and drop single SDM channels, it simplifies the network control and provisioning since fragmentation will be limited to the wavelength domain only. In the case of an SDM system not exhibiting mixing among its spatial channels, fractional space-full wavelength granularity or wavelength granularity can still be used, to take advantage of its implementation simplicity and cost benefits. However, fragmentation in both the spatial and wavelength domains might occur. It becomes then necessary to study new routing and defragmentation algorithms that take into account this new spatial dimension.

The authors gratefully acknowledge the funding by the European Community's Seventh Framework Program (FP7/2007-2013) under grant agreement n° 619732 (INSPACE).

REFERENCES

1. T. E. Stern, G. Ellinas, and K. Bala. *Multiwavelength optical networks: architectures, design, and control*. Cambridge University Press, 2009.
2. D. T. Nelson, et al., "Wavelength selective switching for optical bandwidth management," Bell Labs Technol. J. 11(2), 105–128, 2006
3. K. Nakajima, et al., "Transmission Media for an SDM-Based Optical Communication System", IEEE Commun. Mag. 53(2), 44-51, 2015.
4. D. M. Marom and M. Blau, "Switching Solutions for WDM-SDM Optical Networks," IEEE Commun. Mag. 53(2), 60-68, 2015.
5. Y. Ikuma, et al., "Low-Loss Transponder Aggregator Using Spatial and Planar Optical Circuit," J. Lightwave Technol. 34(1), pp. 67-72, 2016.
6. L. E. Nelson, et al., "Spatial Superchannel Routing in a Two-Span ROADM System for Space Division Multiplexing," J. Lightwave Technol. 32(4), 783-789, 2014.
7. D. M. Marom, et al., "Wavelength-selective switch with direct few mode fiber integration," Opt. Exp. 23(5), 5723-5737, 2015.

S2-4 (Invited)

Mode-Multiplexed Transmission over Multimode Fibers

R. Ryf[1], H. Chen[1], N. K. Fontaine[1],

[1]Nokia Bell Labs, 791 Holmdel-Keyport Rd, Holmdel, NJ, 07733, USA

Roland.Ryf@nokia.com

Abstract: We report about mode-multiplexed transmission over multimode fiber with 10 spatial modes. In particular we describe the latest progress in transmission experiments, mode-multiplexers, optical amplifiers and wavelength selective switches.

OCIS codes: 060.4510, 060.1660, 060.2280, 040.1880, 060.4230.

1. Introduction

In mode-division multiplexing (MDM) multiple modes are used as independent transmission channels in order to increase either the capacity or the performance of an optical link. The concept of using modes as independent transmission paths, has been obvious since the dawn of fiber optic communication but technically difficult to realize even for short length of fibers, mainly because of lack of mode multiplexers with high mode selectivity and the coupling between modes present in real fiber, in particular the twist and bend induced coupling typically present within degenerate or nearly degenerate modes.

The situation dramatically changed over the last ten years thanks to the advent of digital coherent receiver technologies and the development of new multimode optical components. All key optical components are now available for mode counts up to 15 spatial modes, resulting in a potential capacity that is 15 times larger than the capacity of the standard single mode fiber (SSMF).

In this contribution we will review the latest developments in key optical components and the most promising fibers and devices for mode-multiplexed communication over optical fibers.

2. Optical Fibers for Mode-Multiplexing

The main fiber requirements for mode-multiplexed transmission are simple: The fiber should support a precise number of modes, and the loss should be low (comparable to standard SMF) and similar for all supported modes. This can practically be achieved if the modes are well confined and the leakage into cladding modes minimized by design in order to keep bending losses under control. Additionally the resulting modes should either have low intermodal coupling between non degenerate modes or alternatively, if coupling is unavoidable, coherent detection and digital signal processing (DSP) techniques can be applied to digitally undo the coupling after detection.

Fiber with low intermodal coupling between non degenerate modes are best achieved by keeping large phase-velocity differences between the modes. In currently existing fibers, the crosstalk will grow linearly with the length of the fiber, and can typically be kept low enough to not require any form of crosstalk suppression for propagation distances of up to about 50 km [1].

To reach transmission distances of 500 km and above, DSP based crosstalk suppression is required. In order to keep the complexity of the DSP low, the relative pulse propagation delay between signals traveling in different modes should be kept small (< 10 ns). This requires fibers with small differential group delays (DGD) between modes, which can be achieved by using a graded-index doping profile like commonly used in conventional multimode fiber (OM3 for example) well established for short reach applications.

Optimized graded-index (GI) fiber profile for fiber supporting 10 spatial modes [2,3] and 15 spatial modes [4,5] have been fabricated, with optical properties that are close to SSMFs and show DGDs that are < 200 ps/km. The DGD can be further reduced by combining fibers with DGDs of opposite sign, combining fibers with the correct length. The method is referred as DGD-compensation and can be applied at different places within a transmission system either by using compensating fibers or as lumped compensation using a compensating device where multiple external single mode fiber delays are used to delay signals from demultiplexed modes by the correct amount.

An alternative promising fiber that supports multiple modes is the coupled-core multicore fiber (CC-MCF), which is a multicore fiber where the cores are placed close enough to introduce significant coupling between cores for distances as short as a few meters. The resulting fiber offers a significantly larger core density than the conventional uncoupled-core multicore fiber, but still can make use of the high performance cores used in large effective area low loss single mode fibers. The modes of the CC-MCF consist of the super-modes of the multicore structure guided by the core ensemble, therefore the CC-MCFs have typically larger effective areas compared to GI multimode fibers, and also SSMFs. Note however that it is not necessary to couple transmitted signals into the super-modes of the fiber, but a simple core-launch will have the same effect as the strong coupling will immediately distribute the signal launched core signals into all super-modes.

The performance of CC-MCFs strongly depends on the core-to-core spacing. If the core spacing is large, the fiber will behave like an uncoupled-core MCF and each signal will propagate with the group velocity of the corresponding cores, which for CC-MCF are nominally identical, but in practice small variation will cause arrival delays between different core signals. If the core spacing is too small, the super-modes will show considerable amount of DGD, that is much larger than the typical DGD values of a GI-multimode fiber. There is however and optimum core spacing, where the interplay between the core-to-core coupling and the super-mode DGD produce impulse response width that are short compared to the impulse response width of the GI-MMFs and grows with the square root of the fiber length instead of growing linearly as would happen in the absence of the core-to-core coupling.

Recently a 4-core CC-MCF with less than 6.1 ps/$\sqrt{\text{km}}$ impulse response width was demonstrated [6], the fiber also shows a low loss of < 0.158dB/km which combined with large core effective are of 120μm, is expected to perform well at transoceanic distances. A major advantage of the CC-MCFs is also their nonlinear performance which is expected to be better than for a single mode fiber with the same core parameters [7].

3. Mode Multiplexers for Mode-Multiplexed Transmission

Numerous mode-multiplexer technologies have been proposed in the recent years, the most promising in terms of optical performance and scalability to larger number of modes are currently the photonic lanterns and the multiplane phase-mask mode converter. The first consists of an adiabatic transition starting from N single-mode fibers into a step-index core multimode fiber. The lantern can be made mode selective, therefore showing a one-to-one correspondence between the single-mode fibers and specific modes groups at the multimode end of the lantern. The fabrication of mode selective 10 mode lantern is reported in [8]. Lanterns supporting 10 modes have a typical insertion loss range of 0.5 to 2.5 dB, where the loss was smallest for the LP01 mode, whereas the highest loss was observed for the 4th mode group (LP12, LP31). Detailed characterization of a photonic lantern pair connected using a short piece of 10-mode fiber revealed a total mode dependent loss (DL) of 8.7 dB, which is considerable, but acceptable for single span experiments. Further improvements can be expected by improving the lantern-fabrication techniques and by optimizing the transition between the step-index profile of the lantern and the trench assisted graded-index profile of the transmission fiber. The second device is based on multiple phase-masks serially arranged in multiple planes between incoming and outgoing light patterns [9]. The incoming light pattern consists of Gaussian beams generated by a single-mode fiber array and the outgoing light pattern consists of the desired modes of the multimode fibers. Low loss and high modal accuracy can be obtained when enough phase masks are introduced, typically 2 phase masks for each spatial mode are required. The multiple phase masks are typically manufactured on a common substrate and the phase masks are traversed by using a single high reflectivity mirror to form a cavity-like arrangement. Devices supporting 10 modes are commercially available and have a typical insertion loss < 5 dB for each device whereas a MDL of 4.25 dB was measured for a multiplexer pair connected by 20-m 10-mode fiber.

4. Wavelength Selective Switches Supporting Multimde Fibers

Wavelength selective switches (WSSs) are the key components to build wavelength routed fiber-optic networks. A conventional $1 \times M$ WSS has one input port carrying a wavelength-division multiplex (WDM) signal, and is capable to steer any input wavelength channel to any of the desired M output ports.

The concept can be extended to mode multiplexing by slightly modifying the optical design of the WSS such that multimode ports instead of single mode ports are supported. This simple approach can scale to large number of modes, up to 36 modes are supported in the device reported in [10]. The drawback of this approach is that by using multiple modes as input the passband edges become mode dependent, and therefore slightly larger guard bands between WDM channels are required. This is however a justifiable compromise, if we consider that the total switching capacity is still considerably increased and also future WDM channels are expected to occupy wider bandwidth to accommodate the higher line interface of transponders. Uncompromised performance can be obtained by fully demultiplexing the multimode signal into single mode domain and use the concept of "joint switching" [11], where a single steering element is used to steer multiple single-mode channels at the same time.

5. Optical Amplification for Multimode Fibers

Optical amplification is a key requirement for cost effective transmission over optical fibers, and amplifiers that can amplify multiple modes at the same time are therefore essential. The major difficulty for multimode amplifiers has been the excessive mode dependent gain (MDG). The issue has been solved for both distributed and lumped amplification.

In distributed Raman amplification, where the transmission fiber itself is used as amplifying medium, low MDG is obtained in GI-fiber by simply injecting the same amount of power in all modes (including degenerate modes) of the highest mode group supported by the fiber [12].

Low MDG in lumped erbium doped amplifiers can be obtained by using cladding pumped step-index fibers with a constant erbium doping profile.The principle is expected to scale gracefully to larger number of modes. Cladding pumping uses an efficient uncooled high power multimode pump laser that is coupled to a small cladding diameter active fiber by using side pumping. Recently an active fiber supporting 10 spatial modes, a core diameter of 24 μm and a cladding diameter of 74 μm has been demonstrated [13]. The fiber has a nominal constant doping profile in the core with an erbium concentration of 800 ppm. When pumped with 18 W pump power from a 980-nm mulitmode laser, the amplifier provides at total output power of 21 dBm and has a small signal gain of 18 dB and a noise figure < 6 dB. The MDL of the MM-EDFA spliced between two PL-SMUXs was measured only 2.3 dB added MDL compared to the the PL-SMUXs back-to-back, when measured across the entire C-band.

6. Recent Transmission experiment

Mode-division multiplexing (MDM) in fibers supporting 10 and 15 spatial modes has recently been demonstrated for single fiber spans for distances up to 125 km [14]. With the availability of 10-mode components even high performance and more advanced demonstrations are possible. For example, in [15] we show a full C-band transmission distance of 104 km in a recirculating loop, and show up to 121-km transmission with a capacity of 111.4 Tb/s using a cladding pump multimode amplifier as inline amplifier.

7. Conclusions

Our results show that mode-multiplexing can be scaled to 10 modes using multimode components and amplifiers. In particular cladding pumped multimode amplifiers can provide low mode dependent gain and scales to a large number of modes.

Acknowledgements

This work was supported by the ICT R&D program of MSIP/IITP, Republic of Korea. (R0101-15-0071, Research of mode-division-multiplexing optical transmission technology over 10 km multi-mode fiber).

References

1. P. Genevaux, M. Salsi, A. Boutin, F. Verluise, P. Sillard, and G. Charlet, "Comparison of QPSK and 8-QAM in a three spatial modes transmission," *IEEE Photonics Technology Letters*, vol. 26, no. 4, pp. 414–417, 2014.
2. P. Sillard, D. Molin, M. Bigot-Astruc, K. de Jongh, and F. Achten, "Micro-bend-resistant low-DMGD 6-LP-mode fiber," in *Optical Fiber Communications Conference and Exhibition (OFC), 2016*, 2016, p. Th1J.5.
3. P. Sillard, D. Molin, M. Bigot-Astruc, H. Maerten, D. V. Ras, and F. Achten, "Low-dmgd 6-lp-mode fiber," in *Optical Fiber Communications Conference and Exhibition (OFC), 2014*, March 2014, p. M3F.2.
4. P. Sillard, et.al., "Low-differential-mode-group-delay 9-lp-mode fiber," *Journal of Lightwave Technology*, vol. 34, no. 2, pp. 425–430, Jan 2016.
5. R. V. Jensen, ET.AL., "Demonstration of a 9 lp-mode transmission fiber with low dmd and loss," in *Optical Fiber Communication Conference.* Optical Society of America, 2015, p. W2A.34.
6. T. Hayashi, et.al., "125-μm-cladding coupled multi-core fiber with ultra-low loss of 0.158 db/km and record-low spatial mode disp. of 6.1 ps/$\sqrt{km}$," in *Optical Fiber Communication Conference 2016*, Th5A.1.
7. S. Mumtaz, R. Essiambre, and G. Agrawal, "Reduction of nonlinear impairments in coupled-core multicore optical fibers," in *Photonics Society Summer Topical Meeting Series, 2012 IEEE*, 2012, pp. 175–176.
8. A. M. Velazquez-Benitez, et.al., "Scaling the fabrication of higher order photonic lanterns using microstructured preforms," in *Optical Communication (ECOC) 2015* , p. Tu.3.3.2.
9. G. Labroille, P. Jian, N. Barré, B. Denolle, and J.-F. Morizur, "Mode selective 10-mode multiplexer based on multi-plane light conversion," in *Optical Fiber Communication Conference 2016*, Th3E.5.
10. H. Chen, et.al., "Wavelength selective switch for commercial multimode fiber supporting 576 spatial channels," in *submitted to ECOC 2016*
11. L. Nelson, et.al. "Spatial superchannel routing in a two-span ROADM system for space division multiplexing," *IEEE/OSA Journal of Lightwave Technology*, vol. 32, no. 4, pp. 783–789, 2014.
12. M. Esmaeelpour, et.al., "Transmission over 1050-km few-mode fiber based on bidirectional distributed raman amplification," *Journal of Lightwave Technology*, vol. 34, no. 8, pp. 1864–1871, April 2016.
13. N. K. Fontaine, et.al., "Multi-mode optical fiber amplifier supporting over 10 spatial modes," in *Optical Fiber Communication Conference Postdeadline Papers.* Optical Society of America, 2016, p. Th5A.4.
14. R. Ryf, et.al., "10-mode mode-multiplexed transmission over 125-km single-span multimode fiber," in *Optical Communication (ECOC), 2015 European Conference on*, Sept 2015, pp. 1–3.
15. R. Ryf, et.al., "10-mode mode-multiplexed transmission with inline amplification,"*submitted to ECOC 2016*

Ultra-high-capacity Transmission over Few-mode Multi-core Fibers

Koji IGARASHI[1,2], Takehiro TSURITANI[1], Itsuro MORITA[1]

1: KDDI R&D Laboratories, Inc. 2-1-15 Ohara, Fujimino-shi, Saitama, 356-8502, Japan
2: Osaka University, 2-1 Yamadaoka, Suita, Osaka, 565-0871 Japan
E-mail: iga@ comm.eng.osaka-u.ac.jp

Abstract: *We present ultra-dense-S M Super- y uist-W M transmission over . - m six-mode nineteen-core fibers with a spatial multiplicity of 114, achieving a transmission capacity of 2.0 Pbit/s with a record aggregate spectral efficiency of 4 bit/s/ in C band.*
Keywords: *Space division multiplexing, Wavelength division multiplexing*

I. INTRODUCTION

Space-division multiplexing (SDM) using multi-core fibers (MCFs) and/or few-mode fibers (FMFs) is promising technique [1] in order to overcome the capacity crunch in single-core single-mode fibers (SMFs) due to the fiber nonlinearity [2] and the fiber fused phenomena [3]. Fig. 1 shows the relationship between the fiber capacity and the aggregate spectral efficiency in the recently-reported transmission experiments. Red open triangles and blue open triangles indicate the results with single-mode MCFs and single-core FMFs. For comparison, the results with conventional SMF transmission are shown with black open triangles. With SDM techniques, the spectral efficiency has been improved. In particular, the use of MCFs drastically improves the fiber capacity, and then, the 2.15 Pbit/s transmission experiment was achieved using 52-km twenty-two core fibers [4], whereas the capacity in the SMF transmission has been restricted to be around 100 Tbit/s [5].

For achieving the higher spectral efficiency, the hybrid techniques of MCFs and FMFs, namely few-mode multi-core fibers, have been proposed. The capacity and the spectral efficiency are plotted by red closed circles in Fig. 1. A 1.05 Pbit/s transmission was demonstrated by using MCF with twelve single-mode cores and two three-mode cores [6], and then, dense SDM transmission over 527-km three-mode twelve-core fibers was reported [7]. The ultra-dense SDM transmission experiments with a spatial multiplicity over 100 have been eventually achieved [8-10].

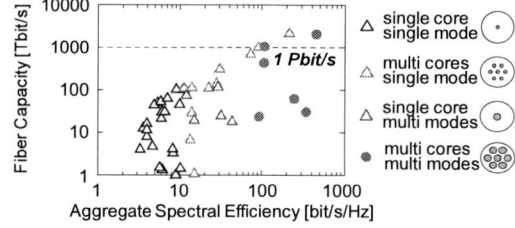

Fig. 1. Fiber capacity and aggregate spectral efficiency in the transmission experiments reported so far.

In this paper, we review the ultra-dense SDM transmission using six-mode nineteen-core fibers (6M-19CF), in which the record spatial multiplicity of 114 is achieved. In order to confirm applicability of 6M-19CF to the ultra-high-capacity transmission, we demonstrate the Super-Nyquist wavelength-division-multiplexed (WDM) transmission using 9.8-km 6M-19CF, achieving the capacity of 2.05 Pbit/s and the record spectral efficiency of 456 bit/s/Hz [10].

II. C ARACTERISTICS OF M- CF

It is desirable to suppress the crosstalk between SDM channels because it reduces the MIMO complexity, while the crosstalk can be compensated for by using multiple-input multiple-output (MIMO) processing in the receiver. The crosstalk between cores is suppressed by increasing the core pitch and enlarging the cladding diameter. The crosstalk between modes due to the mode coupling can be suppressed by enlarging the effective refractive index difference between modes. For suppression the mode coupling, the high refractive index ratio of the core and the clad, Δ, is required. We fabricated 6M-19CF which was designed so that the crosstalk between the cores and the modes were suppressed, resulting in the relatively large cladding diameter over 300 μm and high Δ over 1%.

We conducted experiments with a transmission span which was composed of 9.8-km 6M-19CF for the transmission line, and fan-in and fan-out devices, as shown in Fig. 2(a). A photograph of the 6M-19CF cross-section is shown in Fig. 2(b). The core diameter and the core pitch were 17 μm and 62 μm, respectively. The cladding diameter was 318 μm. The refractive index profile is shown in Fig. 2(c). The graded-index profile with a parameter $\alpha \sim 2$ was adopted, and Δ was 1.1%. In this case, four types of LP modes with degenerated modes ($LP_{01}/LP_{11a}/LP_{11b}/LP_{21a}/LP_{21b}/LP_{02}$) can be sustained. We measured the core-to-core crosstalk when each mode was excited. Figure 2(d) shows the core-to-core crosstalk when each mode was excited to the typical outer core (core 1), the typical inner core (core 13), and the center core (core 19). The horizontal axis indicates the distance between cores normalized by the core pitch. The crosstalk from adjacent core was suppressed to be < -50 dB, and the total core-to-core crosstalk from all other cores was maintained to be < -40 dB even for the center core. In the center core, the differential mode delay (DMD) between LP_{01} and LP_{11} was measured to be 0.65 ns/km at the wavelength of 1550 nm, and that between LP_{01} and LP_{21}/LP_{02} was 2.03 ns/km.

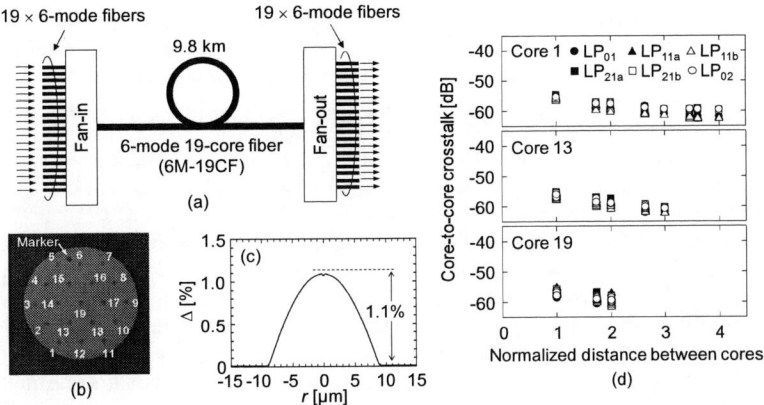

Fig. 2. (a) Configuration of 6M-19CF transmission span. (b) Cross-section of 6M-19CF. (c) Refractive index profile. (d) Measured core-to-core crosstalk of the typical outer core (core1), the typical inner core (core 13), and the center core (core19).

III. EXPERIMENTAL SETUP

In order to confirm applicability of 6M-19CF to the ultra-high capacity transmission, we demonstrated ultra-dense SDM/WDM transmission experiments in the C band. Here, we introduced Super-Nyquist WDM technique based on duobinary shaping, achieving higher spectral efficiency than that of conventional Nyquist-shaped signals [11-13]. The duobinary shaping reduces the signal bandwidth to half of the signal baudrate, suppressing the linear crosstalk from adjacent WDM channels even when WDM spacing is set to smaller than the signal baudrate. Although the duobinary-shaped signals inevitably suffer from inter-symbol interference (ISI), ISI can be compensated for by maximum likelihood sequence estimation (MLSE) in the receiver. In this experiment, 15-Gbaud duobinary-shaped signals were wavelength-multiplexed with WDM spacing of 12.5 GHz. The spectral efficiency of 4 bit/s/Hz was achieved for the single SDM channel even when LDPC-based FEC with 20% overhead [14] was used.

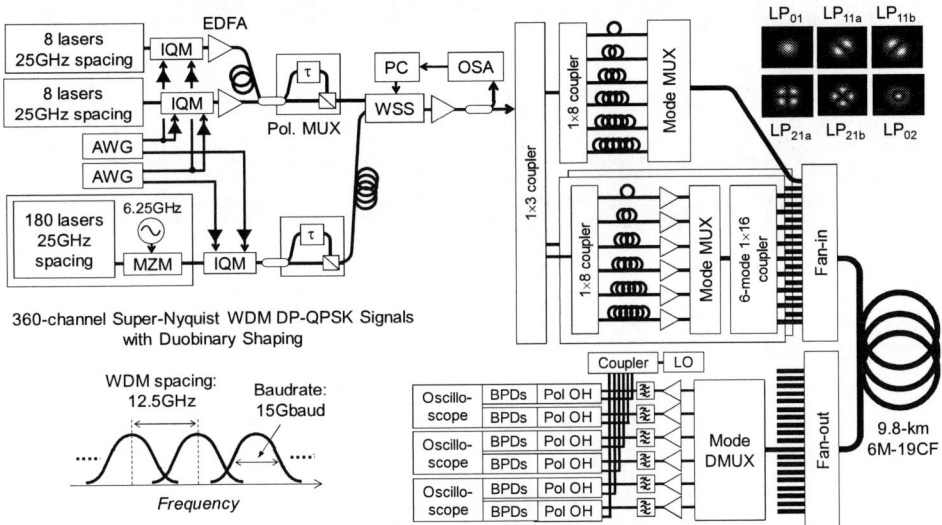

Fig. 3. Experimental setup for 2.05 Pbit/s ultra-dense SDM/WDM transmission.

Figure 3 shows the experimental setup for ultra-dense SDM/WDM transmission. In the transmitter based on three rail configuration, two rails were used for measured even and odd WDM channels, and other rail was for the dummy WDM channels to maintain OSNR and nonlinear effects. The measured even and odd channels were independently modulated by optical modulators (IQMs), which were driven by two streams of electrical duobinary-shaped signals. Using two arbitrary waveform generators (AWGs) with a sampling rate of 50 GSample/sec, the duobinary-shaped signals for and components were independently generated based on offline processing as follows: Pseudo-random bit sequences (PRBSs) with a period of $2^{15} - 1$ were up-sampled to 2 sample/symbol, and then, square-root duobinary shaping was

191

performed. The duobinary-shaped samples were rate-converted and sent to digital-analog converters embedded in AWGs. The even and odd channels were combined with polarization multiplexing. In the third rail for dummy WDM channels, 360 tone with 12.5 GHz spacing were generated and modulated and polarization-multiplexed in the same manner as those of the measured SDM channels. Combining the measured channels and the dummy channels by a wavelength selective switch (WSS), we obtained 360-channel Super-Nyquist WDM DP-QPSK signals with line rate of 60 Gbit/s. The net bit rate was calculated to be 50 Gbit/s, assuming the use of 20%-overhead FEC.

The WDM signals were split into three branches. One was used for measured SDM channels, and others were for dummy SDM channels. For the measured SDM channels, the WDM signals were divided into six copies with fiber delay lines. After the six delayed signals were mode-multiplexed by using multi-plane light conversion (MPLC) [15], the six-mode-multiplexed signal was launched into the measured core of 6M-19CF through the fan-in device. For the dummy SDM channels, two branches were split into two sets of six paths with decorrelation, and then, they were mode-multiplexed. After six-mode multiplexed signals were divided by two sets of 1:16 splitters, 18 SDM channels output from the splitters were launched into the remaining 18 cores.

After the transmission of 9.8-km 6M-19CF, the mode-multiplexed signals output from the measured core were spatial-demultiplexed by the fan-out device and the mode demultiplexer. The six demultiplexed signals were simultaneously detected based on heterodyne reception. The detected electrical signals were stored by six synchronized oscilloscopes with a sampling rate of 50 GSample/sec. The stored samples were off-line processed as follows: After down-conversion to baseband and Nyquist shaping, 12×12 MIMO equalization was performed. The tap size was set to 1,000, and the tap coefficients were updated based on LMS algorithm so that the equalized samples became duobinary-shaped samples. After MLSE, bit errors were counted.

IV. EXPERIMENTAL RESULTS

We measured bit error rates (BERs) of all 41,040 SDM/WDM channels after the transmission. Figure 4 shows the measured BERs of the typical outer core (core 1), the typical inner core (core 13), and the center core (core 19). Closed circles, closed triangles, open triangles, closed squares, open squares, and open circles indicate the results of LP_{01}, LP_{11a}, LP_{11b}, LP_{21a}, LP_{21b}, and LP_{02}, respectively. Although BERs of the remaining 16 cores are not shown in this paper, all measured BERs were maintained to be smaller than 2.3×10^{-2}, which did not exceed the threshold BER of 2.7×10^{-2} of 20%-overhead FEC [14]. In this experiment, the transmission capacity of 2.05 Pbit/s with the record aggregate spectral efficiency of 456 bit/s/Hz was achieved in the C band.

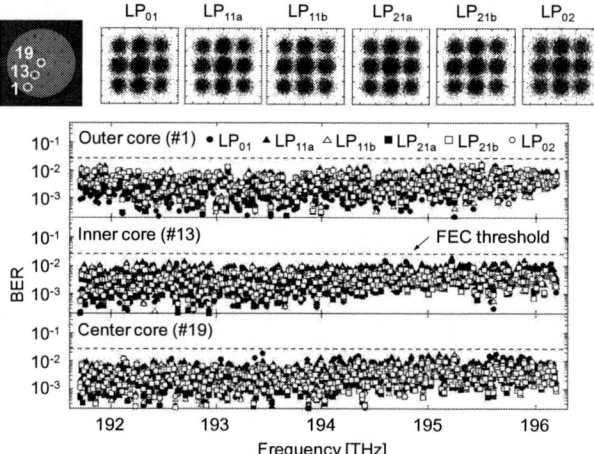

Fig. 4. Measured BERs of six-mode-multiplexed WDM signals in the core 1, 13 and 19.

V. CONCLUSIONS

We presented ultra-dense SDM Super-Nyquist WDM transmission experiments using 9.8-km 6M-19CF. We achieved the transmission capacity of 2.05 Pbit/s and the record aggregate spectral efficiency of 456 bit/s/Hz, and confirmed the feasibility of ultra-high-capacity transmission with 6M-19CF.

REFERENCES

[1] D. J. Richardson *et al.*, Nature Photonics, 7, 354, 2013.
[2] R.-J. Essiambre *et al.*, J. Lightwave Technol., 28, 662, 2010.
[3] R. Kashyap and K. J. Blow, Electron. Lett., **4**, 47, 1988
[4] B. J. Puttnam *et al.*, ECOC2015, **PDP** , 2015.
[5] A. Sano *et al.*, OFC2012, **PDP C** , 2012.
[6] D. Qian *et al.*, FIO2012, **F C**, 2012.
[7] T. Mizuno *et al.*, OFC2014, **Th B** , 2014.
[8] J. Sakaguchi *et al.*, OFC2015, **Th C** , 2015.

[9] K. Igarashi *et al.*, OFC2015, **Th C 4**, 2015.
[10] D. Soma *et al.*, ECOC2015, **PDP** , 2015.
[11] J. Li *et al.*, J. Lightwave Technol., , 1664, 2012.
[12] K. Igarashi *et al.*, ECOC2013, **PD E** , 2013.
[13] J. Zhang *et al.*, OFC2014, **Th B** , 2014.
[14] D. Chang *et al.*, OFC2012, **O** 4, 2012.
[15] G. Labroille *et al.*, Optics Express, , 15599, 2014.

OECC/PS2016

Advanced MIMO Signal Processing for Dense SDM Transmission Using Multi-core Few-mode Fibers

K. Shibahara[1], T. Mizuno[1], D. Lee[1], and Y. Miyamoto[1]

[1]NTT Network Innovation Laboratories, NTT Corporation, 1-1 Hikari-no-oka, Yokosuka, Kanagawa, 239-0847 Japan

shibahara.kouki@lab.ntt.co.jp

Abstract: We review the recent progress of long-haul SDM transmission experiments and the issues to be addressed. Also described are advanced MIMO processing techniques we have developed to achieve low-complexity DMD compensation and MDL-tolerant transmission.

Keywords: SDM transmission; differential mode delay; mode dependent loss; frequency-domain equalization; space-time coding

I. INTRODUCTION

An astonishing breakthrough achieved in the last few years' fiber-optic communication research to overcome capacity limit per optical fiber is space-division-multiplexing (SDM) transmission that employs multi-core (MC) and/or multi-mode fibers (MMFs). To date spatial multiplicity has risen to more than 100 by employing MC-MMFs with high-count cores, each of which supports 3 [1] or 6 spatial modes [2-3]. To increase the scalability toward transmission systems with petabit/s capacity, we have developed dense SDM (DSDM) transmission technologies with a spatial multiplicity over 30 [4-5]. Recent experiments demonstrated multi-petabit/s transmission by using a 22-core homogeneous MCF [6] and a 19-core × 6-mode MC-MMF [7]. Extending transmission reach is another important issue to be addressed for future practical applications of DSDM transmission systems. To achieve this, it is essential to develop technologies to transmit DSDM signals over long distance with high reliability for both MCF transmission and MMF transmission. In terms of optical signal propagation, MCF transmission is almost equivalent to single-mode fiber (SMF) transmission except for inter-core crosstalk. In fact, a few studies have already succeeded in performing thousands-of-km MCF transmission experiments in the past few years. A demonstration of 12-core MCF transmission over 14,350 km with capacity-distance product of 1.51 Pb/s*km was reported in [8]. We achieved the long-haul DSDM transmission over 1,644.8 km by using low-crosstalk 32-core heterogeneous MCF [9]. On the other hand, extending MMF transmission reach is restricted by some inherent issues in mode-division-multiplexed (MDM) signal propagation. Figure 1 shows recent progress of MMF transmission over 100 km. As of today, while transmission experiments over distances of several thousand km have been reported for coupled-core MCF (CC-MCF) [10-11], transmission reach for MMF is still limited to around 1,000 km [12-14]. The major challenge to achieve long-haul MMF transmission is to combat with differential mode delay (DMD) and mode dependent loss (MDL).

This paper briefly reviews methodologies currently reported to mitigate or compensate for these issues. We also describe advanced MIMO signal processing schemes we have developed and previously reported in [4-5, 15]. Our parallel MIMO frequency-domain equalization (FDE) method, which does not require optical DMD management, successfully compensates for DMD with significantly low complexity. Furthermore, our space-time coding (STC) method, which does not impose additional insertion loss or MDL, achieves MDL-tolerant transmission.

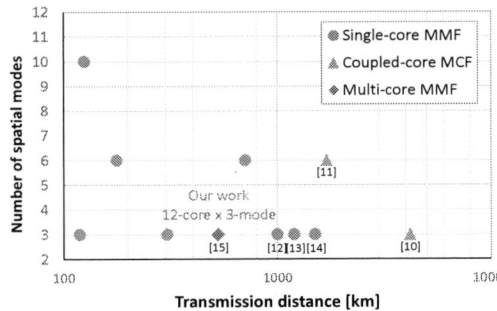

Fig. 1. Recent progress of MMF transmission over 100 km..

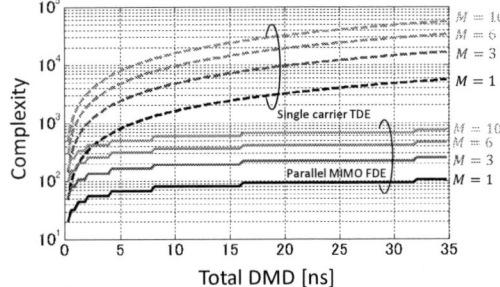

Fig. 2. Required complexity for DMD compensation. M denotes the number of spatial channels.

II. ISSUES IN LONG-HAUL MMF TRANSMISSION

A. Differential mode delay (DMD)

In the same way that signals are distorted by polarization mode dispersion (PMD) occurring in SMF, optical signals in certain MMF modes suffer from a combined effect of multiple modal crosstalk and DMD in linear propagation regimes. When a Gaussian pulse is launched into MMF, it is observed as pulse spreading in the time domain. Methodologies to combat this effect currently reported include employing graded-index (GI) MMF [5, 12-14], CC-MCF [10-11], optical DMD compensation [13-14], and digital DMD compensation approaches [4-5]. The GI-MMF approach successfully suppresses DMD due to its characteristic refractive index profile, which decreases as the distance from the center of the core increases. In [5], it was reported that DMD could be decreased by a factor of nine by switching from step-index (SI) MC-FMF to GI MC-FMF. Another promising fiber-based approach to decrease DMD is CC-MCF. This approach is designed to utilize strong inter-core crosstalk by reducing core spacing so that signals propagating in CC-MCF behave as "super-mode" signals [10-11]. The most significant advantage it features is that accumulated DMD scales with the square root of transmission distance. As for optical DMD compensation, it is a straightforward approach that employs concatenated multiple opposite-sign-DMD fiber segments to cancel accumulated DMD [13-14]. Although these techniques are promising ways to decrease or compensate for DMD, it does not mean that they fully compensate for total DMD and mode coupling effects. Consequently, it is preferable to apply receiver-side digital DMD compensation in conjunction with the above-cited optical approaches to cancel DMD effects, especially for long-haul MMF transmission. One of the most popular DSP techniques is adaptive MIMO equalization. Its advantages are that it can be easily implemented and that it works adaptively for temporal channel variations. However, as transmission distance increases, a larger number of equalizer taps is required to compensate for overall DMD, which increases MIMO equalizer complexity.

To mitigate the complexity, we proposed to employ parallel MIMO FDE in [4-5] rather than conventional single-carrier time-domain equalization (TDE). In the proposed parallel MIMO FDE, two distinguishing schemes are employed. The first is low-symbol-rate multicarrier transmission. A single carrier is divided into individual subcarriers to reduce the required number of equalizer taps [4]. With this approach, MIMO processing can be individually performed by each subcarrier in parallel. The second is employing an FDE algorithm. The use of FDE helps to significantly reduce the number of multiplications in convolution steps for output/update calculations; thus it directly suppresses the complexity. On the basis of the complexity estimation provided in [5], we can compare the required complexity for single-carrier TDE and parallel MIMO FDE. In this work the complexity is defined as the number of complex multiplications per symbol per mode for 10-Gbaud signals (in the TDE case) or for 10-FDM 1-Gbaud signals (in the parallel MIMO FDE case). Figure 2 compares the complexity for various total DMD values in spatial mode number cases of 1, 3, 6, and 10. It turns out that, while the complexity for single-carrier TDE scales linearly with the total DMD, employing parallel MIMO FDE offers significant complexity reduction. The complexity of parallel MIMO FDE is only about 1.5 % relative to that of an equivalent single-carrier TDE for a total DMD of 30 ns.

B.   Mode dependent loss (MDL)

In MDM systems, signals experience different gain or loss depending on their excited modes. The gain and loss are respectively known as differential mode gain (DMG) and MDL, although we here regard them identically as "MDL" for simplicity. The MDL phenomenon arises from inline components, including multi-mode amplifiers, couplers, and MMFs due to different attenuation coefficients among modes. Since MDL induces non-unitary distortions, it severely impairs MDM signals even after MIMO equalization.

A number of studies have focused on reducing MDL itself or MDL-induced impairments. In [16], it was reported that a 2-LP-mode ring-core (RC) FM erbium-doped fiber amplifier (EDFA) had been fabricated to reduce DMG. The authors demonstrated that DMG was suppressed to less than 1.8 dB between two LP modes. Another optical approach is introducing inline mode scramblers (MSs) in MDM systems [17-19]. Although they are designed to offer an ideal mode coupling among MDM signals, it was pointed out that introducing MSs themselves may induce additional MDL in a realistic application [19]. The development of a free-space optics MDL equalizer was reported in [20]. It directly compensated for MDL by 3 dB by utilizing a spatial filter that imposes larger attenuation to the LP_{01} light relative to the LP_{11} light. The impact of MDL can also be partially mitigated by using advanced DSP techniques. In [17], the authors numerically demonstrated the effectiveness of applying receiver-side maximum-likelihood detection (MLD) to MDL-impaired signals. Gathering channel state information (CSI) at a transmitter would be effective for MDL impact mitigation. However, it is likely to be unsuitable since overall MDL varies statistically according to a random matrix model analysis [21]. When no knowledge of CSI at a transmitter is available, one of the attractive digital approaches is STC, which utilizes diversity of both space and time. A few studies can be found that applied STC to MDM transmission. In [22], the authors numerically demonstrated that STC techniques based on linear threaded algebraic STC in conjunction with MLD mitigated MDL-induced penalty in a 6-mode SDM system. In [23], it was reported that a round-robin coding scheme was employed in both a transmitter and a receiver to equalize spatial inter-channel performance by interleaving symbols over all modes and polarizations. In [15], we proposed an STC method that spreads the spatial channels at the transmitter in the digital domain by using Hadamard transform (HT). It improved MDL-impaired signal performance because it intersperses distortions of particular mode signals over all mode signals at the receiver-side MIMO equalization stage.

We here numerically investigate the effects brought by our STC scheme and the MS method based on the matrix channel propagation model. Figure 3 depicts the concept of the model with three spatial modes used in this work. An

MDM transmission system is divided into K spans, each of which contains an MS (if used), an MMF, and an MM amplifier. When STC is used, an ST encoder is inserted before a mode multiplexing device. In the figure, S, M_k, and T_k respectively denote a 3 × 3 HT matrix, a random unitary matrix of the k-th span, and an MDL matrix of the k-th span. At the end of each span, a unitary matrix U_k is introduced that regulates the degree of mode coupling between adjacent spans. The unitary matrix is constructed as $U_k = R_z(\alpha_k)R_y(\beta_k)R_x(\gamma_k)$ with Euler angles α_k, β_k, and γ_k, where R_i ($i = x,\ y,\ z$) denotes a rotation matrix about the i-axis in three dimensions. Euler angle parameters α_k, β_k, and γ_k respectively obey von Mises distribution $f(\kappa, \mu)$ with shape parameter κ and location parameter μ. In this work we set $(\kappa, \mu) = (100, 0)$, which corresponds to a weakly coupling regime. Then an entire channel matrix J can be obtained by matrix multiplications as $J = (\prod_{k=1}^{K} U_k T_k M_k)S$. Note that DMD, polarization effects, and fiber nonlinearities are not considered here to concentrate on MDL effects. In the BER measurements, MDM 10-Gbaud QPSK signals are launched into a considered MMF transmission system. The BER is calculated from J and the minimum mean square error (MMSE) equalization matrix $W = (J^*J^T + 1/\gamma_0 I)^{-1}J^*$, where γ_0, I, $*$, and H respectively denote SNR, an identity matrix, a complex conjugate operation, and a transpose operation. We varied the MDL value in the range from 0.1-0.7 dB/span and created 500 channel realizations for each MDL configuration. Signals are loaded with ASE noise before entering the coherent receiver to set received OSNR to 18 dB in the zero-MDL case. Figure 4 depicts the results obtained for mean Q-factors after propagation over 13 spans, which are calculated from averaged BERs over all spatial modes. When no MS is used, the STC methods obviously improved Q-factor performance relative to the case without STC, since signal performance is averaged over spatial channels [15]. If we introduce MS, a further Q-factor increase was obtained even without STC. This is because the use of MS brought strong coupling effects to the MMF link, thus letting the overall MDL decrease. The combined use of MS and STC slightly outperformed the use of MS alone. At the MDL value of 0.7 dB/span, the STC, MS, and MS-STC methods respectively provided Q-factor increases of 1.2, 2.1, and 2.3 dB. These results demonstrate that our STC method enables us to enhance MDL-tolerance, especially in weakly coupling regimes. However, we also emphasize that the STC method works for ideal mode coupling at a transmitter without introducing additional insertion loss or MDL since it can be applied in the digital domain.

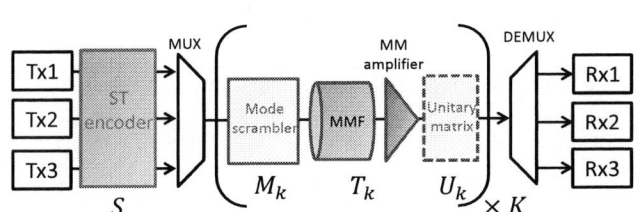

Fig. 3. MDM system model.

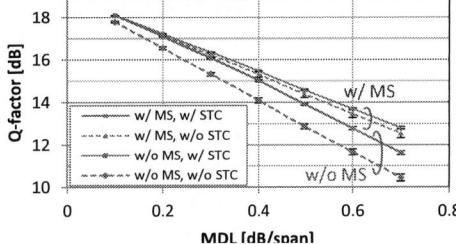

Fig. 4. Mean Q-factors as a function of MDL per span. Error bars are represented as a standard deviation of 500 channel realizations.

III. CONCLUSIONS

In this paper we have reviewed recent SDM/DSDM transmission progress and the technologies to combat DMD and MDL effects. We also described two schemes we have recently developed for long-haul MMF transmission. The parallel MIMO FDE effectively suppresses the complexity required in digital DMD compensation. The STC method enhances MDL tolerance by equalizing signal performance among spatial channels. For the purpose of achieving thousands-of-km DSDM transmission, further research is needed that will contribute to innovation through collaboration between advanced SDM devices and MIMO signal processing techniques.

REFERENCES

[1] J. Sakaguchi et al., Proc. OFC, Th5C.2, 2015.
[2] K. Igarashi et al., Proc. OFC, Th5C.4, 2015.
[3] T. Sakamoto et al., Proc. OFC, Th5A.2, 2016.
[4] T. Mizuno et al., Proc. OFC, Th5B.2, 2014.
[5] K. Shibahara et al., JLT, **34**(4), 2015.
[6] B. J. Puttnam et al., Proc. ECOC, PDP.3.1, 2015.
[7] D. Soma et al., Proc. ECOC, PDP.3.2, 2015.
[8] A. Turukhin et al., Proc. OFC, Th4C.1, 2016.
[9] T. Mizuno et al., Proc of OFC, Th5C.3, 2016.
[10] R. Ryf et al., Proc. OFC, PDP5C.2, 2012.
[11] R. Ryf et al., Proc. ECOC, PD.3.2, 2014.
[12] E. Ip et al., Proc. OFC, PDP5A.2, 2013.
[13] S. Randel et al., Proc. OFC, PDP5C.5, 2012.
[14] R. Ryf, et al., Proc. OFC, W4J.2, 2014.

[15] K. Shibahara et al., Proc. OFC, Th4C.4, 2016.
[16] H. Ono et al., Electronics Letters, **51**(2), 172-173, 2015.
[17] A. Lobato et al., Proc. ECOC Th.2.C.3, 2013.
[18] T. Mori et al., Tu2D.2, 2015.
[19] E. Ip et al., Proc. OFC, W4I.4, 2016.
[20] T. Mizuno et al., Proc. ECOC, P5.9, 2015.
[21] K. P. Ho, and M. K. Joseph, JLT, **29**(24), 3719-3726, 2011.
[22] E. Awwad et al., Proc. ICC2015, 5228-5234 (2015).
[23] J. V. Weerdenburg et al., Optics Express, **23**(19), 24759-24769, 2015.

S3-1 (Invited)

OECC/PS2016

Optical Switching Performance Metrics for Scalable Data Centers

Keren Bergman and Sébastien Rumley

Department of Electrical Engineering, Columbia University, 530 West 120th Street, New York, NY 10027
rumley@ee.columbia.edu

Abstract: *Optical switching can address some key challenges associated with scaling the communications infrastructure in data centers. We review the critical metrics that would enable significant performance gains and thereby wide adoption in data centers.*
Keywords: *Optical interconnects, Data Centers, Optical Switching*

I. INTRODUCTION

The performance of Data Center interconnects (DCIs) is gaining intense interest as cloud based applications span over a growing number of servers, trigger rising traffic volumes, and are increasingly sensitive to network congestion. Until recently, DCIs were primarily designed to ensure the connectivity of single servers with the Internet, and to support "North-South", generally latency tolerant traffic. Nowadays, DCIs are becoming the main backbone of higher-order computing architectures, supporting massive amounts of "East-West" traffic at high throughputs. Data centers are also increasingly operated with virtualization software for sharing servers among multiple users, further amplifying bandwidth demands across DCIs.

These transformations do not only require servers' network interfaces to support increased bandwidths (10Gbps today, 40 and 100Gbps in the near term), but also require the interconnect itself to cope with intensely growing traffic volumes. This means drastically limited over-subscription levels, as illustrated in Figures 1a and 1b. To interconnect 20,000 servers with 32 ports switches (following a generally adopted 3-levels fat-tree typed topology), a minimum of 668 switches are required (646 at the first level, 21 at the second level, and 1 at the last level), as well as 667 internal (i.e. switch-to-switch) links. This infrastructure is sufficient as long as servers communicate very sporadically, i.e. average network utilization of a single server remains below 0.1%. However, as requirements for bandwidth grow, potentially to the point where full-bisectional bandwidth is provisioned to support full utilization of all servers simultaneously, the number of switches and internal links required balloon – in this particular example, by factors up to ~7 for the switches and ~92 for the internal links. For a large data center involving 50,000 servers, the increase in terms of internal links is even higher (~125x). When network utilizations is very low, large data center scaling permits a better amortization of the interconnect cost. By contrast, when the interconnect utilization levels are high, as is the overwhelming applications traffic trend, cost grows super-linearly with data-center scaling [1].

The additionally required internal links, moreover, are predominantly higher-cost long reach since they span between racks (as opposed to short reach, intra-rack links - Fig 1c). Under minimal traffic requirements, these long reach links represent a relatively small portion of all links (less than 4% for the 20,000 servers, 32 port example), but under the highest traffic requirements, the ratio is reversed (75% in the example). This further amplifies the growing cost.

Altogether, the shift to high network throughput cloud applications can raise by orders of magnitude the cost of DCIs. This motivates the development of new technologies that can counter-balance this cost explosion.

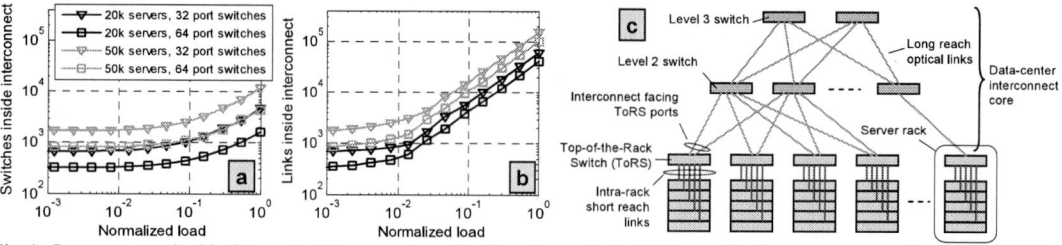

Fig. 1. Resources required in data-center interconnect as a function of expected traffic (normalized to server communication capabilities).

II. MOTIVATION FOR OPTICAL SWITCHING

Optical cables account for a substantial part of the long-distance links present in current data centers, and in the future this proportion is destined to further rise. Copper based links can span over relatively long distances (10GBase-T: 100m, 40GBase-T: 30m, 40GBase-CR4: 7m), but are subject to higher link latency as they rely on digital signal processing and advanced coding formats [2]. Moreover, as line-rates rise to 40Gbps or beyond, it becomes simply more economical to rely on fiber based cables for distances larger than 5-10m. This break-even distance is expected to decrease in the next years.

Provided that the majority of links involved in the DCI core (Fig. 1c) are (or will be) fiber optics based, the opportunity to use transparent optical switches instead of conventional, electrical packet based ones has been investigated. Such an all-optical architecture can drastically limit the number of hops taken by messages in the interconnect core. In the ideal case, each message can be optically emitted once at the Top-of-the-rack Switch (ToRS), kept in the optical domain at all times inside the DCI core, and received at the ToRS it is destined to. Transparent switching does not provide benefits in terms of number of installed fibers, but can help to drastically limit the number of optical emitters and receivers (transceivers). Assuming negligible optical switching time and ideal arbitration, this number can be reduced to the number of interconnect facing ToRS ports (Fig. 1b), plus several units attached to the data-center internet gateway. Under the highest traffic requirements, these interconnect facing ToRS are at most as numerous as servers. For the 20,000 servers, 32 port example, the transparent switching can represent a factor 3 decrease in the number of required transceivers.

III. OPTICAL SWITCHING CAVEATS

The alluring picture of transparent switching is incomplete, however, as several hurdles must be taken into account before considering a transparent DCI core to be a viable option.

A. Switching time. Optical switches replacing electrical ones in the interconnect core must provide negligible switching times. To be considered as negligible, a switching time must be a) significantly shorter than the median data exchange duration and b) smaller than the latency tolerated by cloud applications. By data exchange, we mean an uninterruptible sequence of bits sent by one client to another over an optical, circuit switched, link.

If condition a) is not fulfilled, the DCI throughput will be impacted: no data can be transmitted while an optical switch is adapting its state. Furthermore, after each switch state adaptation, the end receiver must also recover its synchronization. If, after each data exchange of duration d_{data} the link is silent for a time $d_{switching+sync} = d_{data}$, the link throughput is limited to 50%. A second link operated in parallel becomes thus necessary. If optical switches configurations are based on standard Ethernet frames, a data exchange is at most ~1500 bytes long. At 10Gbps and 40Gbps, the *maximum* duration is thus of 1.2 us and 0.3 us respectively. This requires switching times <120ns and <30ns respectively to keep throughput above 90%. Furthermore, as revealed by Benson et al. [3], median *flow* sizes in data-center fall generally in the 100 to 1000 byte range. Each flow being composed of one Ethernet frame at least, this further pushes switching time requirements to 80ns (10Gbps) and 20ns (40Gbps) if median size is 1000 bytes, and to 8ns and 2ns if most data exchanges are smaller than 100bytes. Electrical packet switches are not subject to this throughput issue as long as the bandwidth of the internal crossbar connecting input to output queues is high enough, which is generally the case. Also note that even shorter switching times are required if link rates rise to 100Gbps or more.

As for condition b), one can take the latency displayed by current interconnects as a reference for "tolerated" latency, although applications may tolerate higher latencies. An Ethernet 10G switch adds an average latency of 0.5 to 1us at each hop [4]. This translates in a typical server-to-server latency of the order of several microseconds. As a point of comparison, node-to-node latency in supercomputing environments falls below or close to the microsecond [5].

Nanosecond scale switching times (1ns – 100ns) have often been demonstrated with electro-optically tuned switches [6], yet these devices generally inflict significant power penalties to optical signals [7]. This threatens scalability as detailed hereafter. If median data exchange durations can be made larger, microsecond scale switching times are acceptable. Microsecond switching times can also be tolerated if cost of optical switches is kept low. Hence, in this case, the cost of the extra transceivers required to keep throughput unchanged can be counter-balanced. Microsecond switching times can be achieved with thermal effects or with MEMS-actuated planar switches [8]. Switching times above 10us, in contrast, may affect application performance too much to be considered.

B. Arbitration. In Section 2, an ideal arbitration has been assumed: schedules of all data exchange, as well as their routing, are ideally picked up to minimize average latency, maximize throughput, or a combination thereof, *while ensuring that no network resource is allocated more than once.* Realizing such an ideal arbitration is an extremely complex optimization problem. Optical switches are unable to buffer data exchange for arbitration durations: consequently, end-to-end reservations are necessary to ensure that all the resources required for a data exchange to happen are free at the same time. Operating with end-to-end reservations is not only more demanding in terms of optimization efforts, but also lead to a fragmentation of the occupancies over time, synonym of crippled throughput. Furthermore, collecting resource availabilities, and dispatching reservation decision from and across the DCI is challenging at scale.

As a result, arbitration cannot be ideal and necessarily translates into diminished throughput and increased latency. The former can be compensated by provisioning links and switches (in which case the necessity of having low cost switches applies again). The latter can be improved by maintaining circuit on after use [9], and/or by mean of circuit prefetching [10]. If circuit maintenance strategy is applied, provisioning more links also helps to reduce latency.

C. Cost. By transitioning to a transparent DCI *while keeping the same bandwidth*, the number of transceiver required can be reduced by a potentially significant factor. However, to compensate for switching times and arbitration latency, bandwidth must be boosted which comes at the expense of additional transceiver. Furthermore, the insertion of one or more switches modifies the optical budget, as each switch might induce a power penalty ranging from 1 to 10dB or more. More expensive transceivers providing extra optical budget, e.g. longer-reach ones, might thus be required.

From a transceiver cost point of view, if regular transceivers support optical switching and if arbitration and switching time effects require twice more bandwidth to be provisioned, the transition to optical switching, in the 20,000 server case, can provide a 33% cost reduction. However, if twice more expensive "wide budget" transceivers are required, the total transceiver cost rises by 33% in that example.

Transitioning to transparent DCI, however, permits additionally to replace electrical packet switches by optical switches. An inspection of 10Gb Ethernet switch prices [11] shows that every for 10Gbps interconnect every switch port cost around 100$ (350$ for 40Gbps). The cost of an optical port should remain below those prices for 32 ported switches. If much larger port counts are available, the number of levels in the interconnection network that replaces the fat-tree can be reduced, which also limits the number of switches required. For instance, for the 20,000 servers example, having 320 ports switches available instead of 32 reduces the number of ports required by 30%.

D. Power consumption. One can expect the transceivers number reduction and the replacement of electrical routers by transparent switches to diminish the power consumption. Some optical switches, however, do consume a non negligible amount of power to maintain their configuration. The consumption of a future ring resonator based switching fabric with 128 ports, for instance, has been estimated to ~47mW per port. With 10Gbps per port, this translates into 4.7 pJ/bit. For SOA/MZI based switches, hundreds of mW per port must be accounted [12], translating into tens of pJ/bit. As a point of comparison, best-in-class electrical packet routers consume 20-30 pJ/bit. Consumption of optical switches should typically be limited to such values.

On the transceiver side, power consumption is affected by the higher launch optical powers required to overcome extra power penalties introduced by switches. If these power penalties are of 10dB, laser consumption is multiplied by a factor of ten. Lasers are not the dominant energy consumers in transceivers *today*, but this may no longer be the case in the future. If laser is the only consumption of a transceiver, a transceiver facing an additional 10dB power penalty will consume as much as 10 point-to-point transceivers, negating power savings obtained by diminishing the transceiver number.

E. Scalability. Optical power penalties of switches, eventually, limit the scalability of transparent DCI architectures. If end-to-end power penalties exceed the amplest transceiver power budgets, optical amplifiers become necessary in between the switches. This rises both interconnect cost and power consumption.

In general, the power penalty associated to the interconnect core should be limited to 20dB. If three switch stages are sufficient to provide full connectivity, this fixes the fiber-to-fiber power penalty for individual switches to around 6.5 dB. This is typically the case if 320 ports switches are used in a 20,000 server data-center. However, if fives or even seven stages are necessary, constraint on individual switch penalty becomes severe, especially for electro-optical effect based switches, capable of sub-microsecond switching time.

IV. CONCLUSION

Optical switching in data centers offers potential advantages in terms of cost and power consumption. However to realize these gain, critical design metrics cannot be neglected. Mitigating these photonic technology challenges requires some device level advances, but above all subtle trade-offs to be identified and addressed at the architecture level.

REFERENCES

[1] S. Rumley, S. D. Hammond, A. Rodrigues, K. Bergman, "Design Methodology for Optimizing Optical Interconnection Networks in High Performance Systems", ISC-HPC conference, 2015.

[2] Z. Zhang, et al. "A 47 Gb/s LDPC Decoder with Improved Low Error Rate Performance", Symp. on VLSI Circuits, 2009.

[3] T. Benson, et al., "Network Traffic Characteristics of Data Centers in the Wild", SIGCOMM, New Dehli, India, 2010.

[4] H. Subramoni, P. Lai, M. Luo and D. K. Panda, "RDMA over Ethernet – A Preliminary Study", IEEE Cluster, 2009.

[5] D. Chen, et al., "The IBM Blue Gene/Q Interconnection Network and Message Unit", Supercomputing, Seattle, WA, 2011.

[6] T. Shiraishi, et al., "Scalability of Silicon Photonic Enabled Opticaly Connected Memory", IEEE Optical Interconnect, 2014.

[7] D. Nikolova, S. Rumley, D. M. Calhoun, Q. Li, R. Hendry, P. Samadi, K. Bergman., "Scaling silicon photonic switch fabrics for data center interconnection networks," Optics Express 23(2), 2015.

[8] T. Joon Seok, N. Quack, S. Han, R. S. Muller, M. C. Wu, "Large-scale broadband digital silicon photonic switches with vertical adiabatic couplers", Optica (3) 1, 2016.

[9] K. Wen, et al., "Reuse Distance Based Circuit Replacement in Silicon Photonic Interconnection Networks for HPC", IEEE Symposium on High Performance Interconnects, 2014.

[10] K. Wen, S. Rumley, J. Wilke, K. Bergman, "Latency-avoiding Dynamic Optical Circuit Prefetching Using Application-specific Predictors," ISC-HCP ExaComm Workshop, 2015.

[11] http://www.colfaxdirect.com

[12] S. Liu, et al., "Low Latency Optical Switch for High. Performance Computing With. Minimized Processor Energy. Load", JOCN 3(7), 2015.

Advanced Optical Interconnection Technologies Based on Silicon Photonics for Future DCs

Takahiro Nakamura, Junichi Tsuchida, and Kazuhiko Kurata

Photonics Electronics Technology Research Association

Author e-mail address: t-nakamura@petra-jp.org

Abstract: We developed a chip-scale optical transceiver, "optical I/O core", which is $\times$ mm², has a maximum capacity of 300 Gbps, and consumes a low amount of power, mW/Gbps. A 2 -Gbps/ch error-free transmission using our optical / cores with a wide-bandwidth S was observed.

Keywords: o t ca core s co oto cs w de ba dw dt

I INTRODUCTION

High-speed large-data processing and flexible change in both data amount and analysis scheme have been required for applying large-scale computers, such as those for data centers (DCs) or high performance computers (HPCs), to big data analysis or artificial intelligence. However, in current DCs, server utilization is less than 50% without optimization processing. Furthermore, it takes a few weeks to provision new services. One solution to these problems is disaggregated server for high server utilization and flexibility proposed by the Open Computer Project [1]. In the disaggregated server, all the components in conventional server racks, such as CPUs, memories, networks, storages are divided into each component. After that, the same components are aggregated and shared over server racks. This disaggregated method requires high-speed optical interconnection of over 100 Gbps due to the need to connect each component at a distance. The demand for bandwidth expansion in DCs and for HPCs has been increasing, including the introduction of this new server. For example, Infiniband, one of the standardizations for interconnection among server nodes, has a twofold increase in bandwidth every two years, and optical interconnection of 100 Gbps has already been introduced. On the other hand, PCIe4.0, one of the standardizations for interconnection in a server's node, has used an electrical interconnection of 16 Gbps. Furthermore, PCIe5.0 will require an interconnection of 32 Gbps around 2020, which must be optical. As indicated above, the bandwidth expansion not only among servers but also among chips in DCs have been increasing and the penetration of optical interconnection has been progressing.

However, active optical cables (AOCs), one of the optical transceivers among servers, are large, about 9×2 cm², and consume a large amount of power, about 30 mW/Gbps. Therefore, it is difficult to mount AOCs around LSI chips. Then, we developed a compact chip-scale optical transceiver, "optical I/O core", which takes the function of only OE/EO out from the conventional optical transceiver.

II CONCEPT AND SPECIFICATIONS OF OPTICAL I/O CORE

Figure 1 shows a structural comparison of our developed optical I/O core and a traditional optical transceiver. Traditional optical transceivers generally involve packaging optical chips and their driver ICs on printed circuit boards with optical and electrical interfaces. These optical transceivers require customized parts due to fixed optical and electrical interfaces, size, and so on. On the other hand, our optical I/O core is a chip-scale optical transceiver integrating only common optical chips and their driver ICs. Therefore, it can be used with various components because of being able to change optical and electrical interfaces according to requests from customers of component companies. Our optical I/O core will contribute to expanding optical interconnects not only to AOCs in DCs but also to consumer electronics such as TVs and cars.

Figure 2 shows a photograph and cross-sectional view of our developed optical I/O core. The optical I/O core is 5×5 mm², enables transmission at the maximum capacity of 300 Gbps, and consumes a small amount of power, 5 mW/Gbps [2]. The main technologies of the optical I/O core for miniaturization, high transmission capacity, and low consumption power are silicon photonics, CMOS-LSIs, and their assembly technologies. Silicon photonics technology enables the downsizing of an optical integrated circuit to one-hundredth of the conventional planar lightwave circuit by high-index silicon optical waveguides instead of glass optical waveguides. The silicon photonics integrated circuit consists of optical waveguides, MOS-type Mach-zehnder optical modulators, Ge photodiodes [3], and grating couplers [4]. Furthermore, we downsized our optical I/O core by assembling three-dimensionally CMOS-LSIs on the silicon photonics integrated circuits. We used a three-dimensional polymer waveguide, "optical pin", instead of an optical lens [5], as the optical interconnect. Through glass via (TGV) is used as the electrical interconnect. Both interconnects contribute to downsizing the optical I/O core due to its three-dimensional structure. We also achieved transmission capacity of 300 Gbps by 12 parallel channels at 25 Gbps per channel. For low power consumption, we also developed an optical modulator driver and a photo diode trans-impedance amplifier by 28-nm CMOS process.

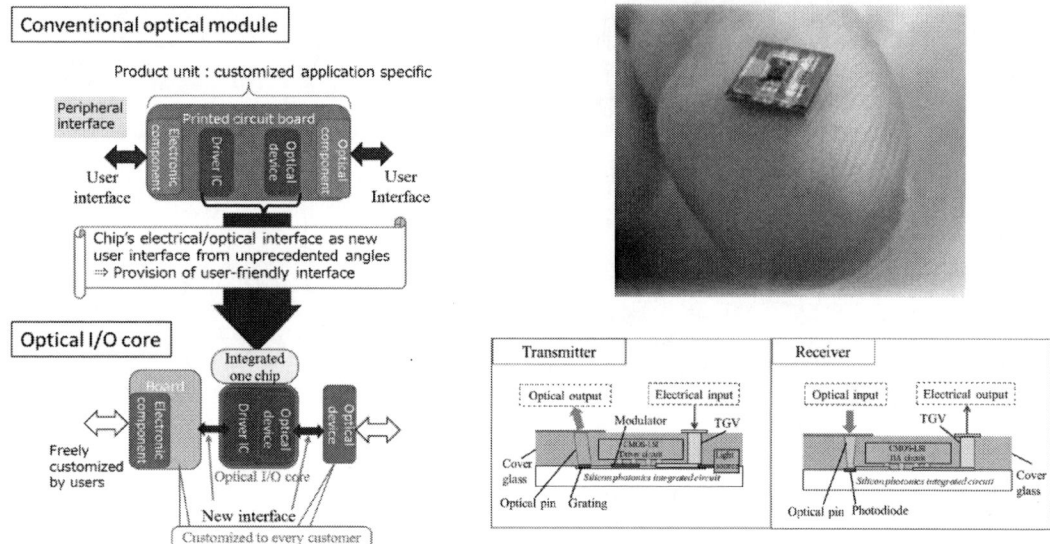

Figure 1. Structure in our developed optical I/O core and traditional optical transceiver.

Figure 2. The photograph and cross-sectional view of developed optical I/O core.

III APPLICATION TO IDE-BAND IDT LSIs

Because our optical I/O core is much smaller than conventional optical modules, as mentioned in Section II, several optical I/O cores can be mounted around wide-bandwidth LSIs such as field-programmable gate arrays (FPGAs). This means that optical I/O cores enable one LSI to connect the other LSIs on boards and in chassis with low power consumption by directly converting high-speed electrical signals to optical signals without degradation. This may be a low-cost solution without designing high-speed electrical wiring on boards and using high-cost low-loss printed circuit boards. Furthermore, since optical fibers are compact and lightweight compared to electrical wires, the volume of wires and connectors can be reduced and cooling space can be secured using our optical I/O core. Connecting wide-bandwidth LSIs, such as FPGAs or CPUs, by optical I/O cores will be effective for servers' scale-up with associated CPUs and for disaggregation to achieve free connection among CPUs or memories.

Figure 3 shows a wide-bandwidth LSI board with, 16 optical I/O cores mounted around it. We demonstrated that the electrical signal of 25 Gbps from the wide-bandwidth LSI was converted to an optical signal by an optical I/O core, and the optical signal was looped back to the other optical I/O core. Then the electrical signal converted from the optical one returned to the wide-bandwidth LSI. Therefore, a 25-Gbps/ch error-free transmission using our optical I/O cores with a wide-bandwidth LSI was observed. When transmitting by only the electrical signal, CDR or FEC LSIs are required due to the attenuation and deterioration caused by the high-speed signal. Therefore, if the electrical signal can be converted to an optical signal before deterioration, the optical signal will enable transmission over several hundred of meters without CDR or FEC. This application is expected as a low-power consumption application.

IV CONCLUSIONS

We developed a chip-scale optical transceiver, "optical I/O core", which is 5×5 mm^2, and has a maximum capacity of 300 Gbps, and consumes a low amount of power, 5 mW/Gbps. A 25-Gbps/ch error-free transmission using our optical I/O cores with a wide-bandwidth LSI was demonstrated. This technology will enable wide-bandwidth LSIs, such as CPUs and FPGAs, to transmit high-speed and high-capacity signals over several hundred meters with low-power consumption.

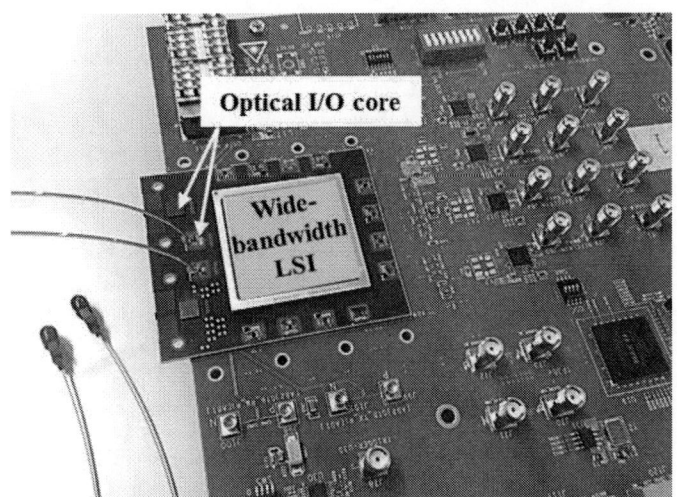

Figure 3. Optical I/O core with the wide-band width LSI.

AC NO LEDGMENT

This research is partly supported by New Energy and Industrial Technology Development Organization (NEDO).

REFERENCES

[1] http://www.opencompute.org/
[2] K. Yashiki, K. Mizutani, J. Ushida, Y. Suzuki, M. Kurihara, M. Tokushima, J. Fujikata, Y. Hagihara, and K. Kurata, "25-Gbps error-free operation of chip-scale Si-photonics optical transmitter over 70℃ with integrated quantum dot laser," Th1F.7, OFC2016.
[3] J. Fujikata, S. Takahashi, M. Takahashi, M. Noguchi, T. Nakamura, and Y. Arakawa, "High-performance MOS-capacitor-type Si optical modulator and surface-illumination-type Ge photodetector for optical interconnection," J. J. A. P. 55, pp.04EC01-1-04EC01-5, 2016.
[4] M. Tokushima, J. Ushida, T. Uemura and K. Kurata, "Shallow-grating coupler with optimized anti-reflection coating for high-efficiency optical output into multimode fiber," Appl. Phys. Express 8(9), 092501, 2015.
[5] T. Uemura, A. Ukita, K. Takemura, M. Kurihara, D. Okamoto, J. Ushida, K. Yashiki, and K. Kurata, "125-μm-pitch × 12-channel "optical pin" array as I/O structure for novel miniaturized optical transceiver chips," Electronic Components and Technology Conference (ECTC), pp.1305-1309, 2015.

S3-4 (Invited)

OECC/PS2016

The Role of Photonics in Future Exascale Data Systems

S. J. Ben Yoo

Department of Electrical and Computer Engineering, University of California, Davis, CA 95616, USA

sbyoo@ucdavis.edu

Abstract: *We discuss the role of photonics in our pursuit for future exascale data systems. High-throughput, low-latency, low-contention, and high-radix switching together with energy-efficient silicon photonics intimately integrated with electronics will be vitally important.*

Keywords: *exascale computing, data systems, optical interconnects, high radix switch.*

I. INTRODUCTION

We rely on computing and data systems for everything from healthcare and climate predictions to entertainment and financial transactions. In the healthcare sector alone, we are seeing rapid transitions in data processing from two-dimensional images to three-dimensional or hyper-spectral real-time 3D images. However, today's data centers and computing systems have reached limitations to scale further. As Figure 1. indicates, the exponential growth of the computing performance has slowed and flatlined since 2012, and the energy efficiency is also projected to flatline at ~0.5 nJ/FLOP. Typical data centers are consuming megawatts of power, and the desire to realize exascale computing is seriously challenged by the power consumption projected at 0.5 GW for 1 Exaflop/s.

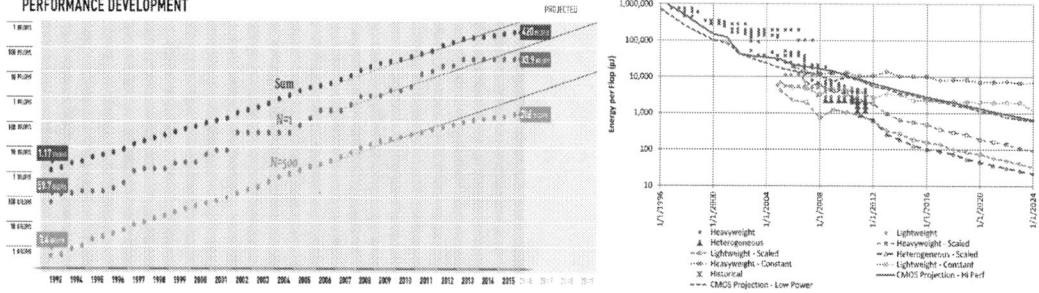

Figure 1. [Left] Rapid growth of supercomputers performance. The logarithmic y-axis shows performance in GFLOPS. The red line denotes the fastest supercomputer in the world at the time. The yellow line denotes supercomputer no. 500 on TOP500 list. The dark blue line denotes the total combined performance of supercomputers on TOP500 list.[1] [Right] Energy per Flop - Historical and Projected [2]

II. FLAT ALL-TO-ALL INTERCONNECTION INDEPENDENT OF DISTANCE AND BANDWIDTH

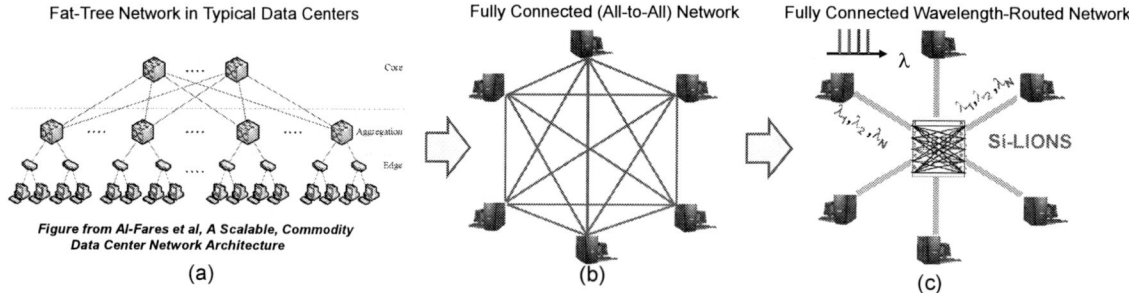

Figure 2. Proposed evolution of data center networking topology from the traditional fat-tree topology to flat fully-connected topology made possible by wavelength routed optical interconnection.

Amdahl's law suggests that a system with balanced computation, memory, and communications performs best across most applications, but today's computing systems are typically unbalanced by more than two orders of magnitude. As Figure 2 (a) shows, typical fat-tree network topologies deployed in data centers can save capital equipment costs by deploying inexpensive electrical switches with low radix and low bandwidth, but incur high energy consumption and large latency for low throughput. On the other hand, all-to-all optical interconnection topology in Figure 2 (b), which can be simplified by wavelength routing in Figure 2 (c) can offer far greater throughput without the need for arbitration or contention resolution. What makes a practical implementation of Figure 2 (c) possible is wavelength routing capability of arrayed waveguide router (AWGR) shown in Figure 3.

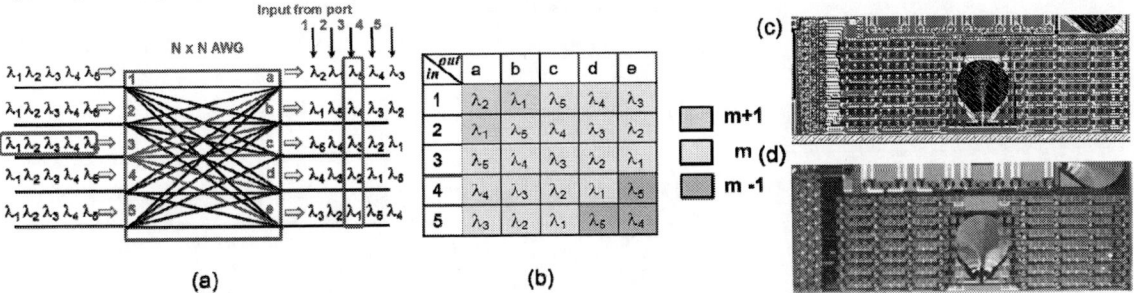

Figure 3. $N \times N$ cyclic frequency arrayed waveguide grating router (AWGR)'s (a) wavelength routing property ($N = 5$ example), (b) a wavelength assignment table, (c) a mask layout of a Low-latency Interconnect Optical Network Switch (LIONS) utilizing 8 x 8 arrayed waveguide grating, (d) fabricated silicon photonic LIONS[3].

Using AWGRs, we can realize Low-latency Interconnect Optical Network Switch (LIONS) as shown in **Error! Reference source not found.** with (a) all passive components at the core a $N \times N$ cyclic AWGR at the core with each linecard containing k_t transmitters, k_r receivers, and electrical switches containing k_t transmitters, $1 \le k\_r$, $k\_t \le N$, (b) active LIONS [4] consisting of an AWGR at the core with each linecard containing a tunable wavelength transmitter, and k_r receivers. When $k\_t = N$, no arbitration or contention resolution is necessary.

III. ACTIVE AND PASSIVE LIONS IN SCALABLE HIERARCHICAL DATA SYSTEM ARCHITECTURE

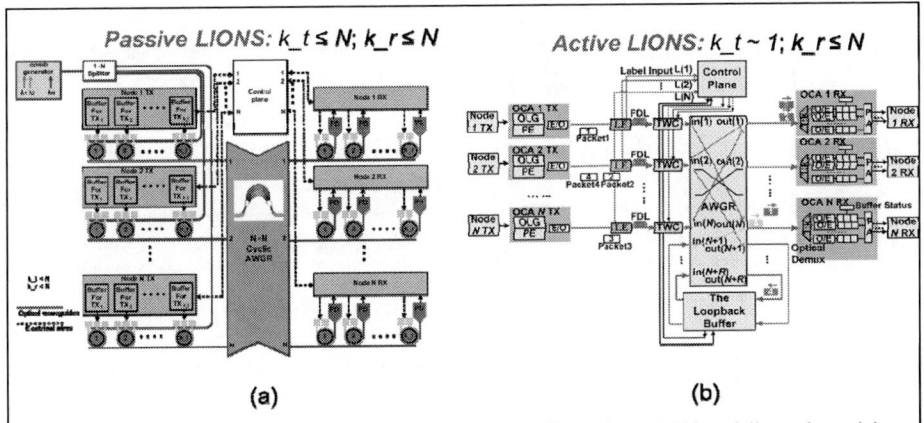

The active and passive LIONS can be combined to create hierarchically interconnected data systems as shown in Figure 5 and Figure 6. As shown, passive LIONS are more effective at the lower hierarchy providing completely distributed control planes and all-to-all connectivity, while active LIONS with tunable linecards offering high-throughput packet switching at aggregated data rates. Figure 6 also

Figure 4. (a) passive LIONS [3] consisting of a $N \times N$ cyclic AWGR at the core with each linecard containing k_t transmitters, k_r receivers, and electrical switches containing k_t transmitters, $1 \le k\_r$, $k\_t \le N$, (b) active LIONS [4] consisting of an AWGR at the core with each linecard containing a tunable wavelength transmitter, k_r receivers, and optional loopback buffer[5].

shows recursive all-to-all topology of hierarchical LIONS architecture (H-LIONS) [6].

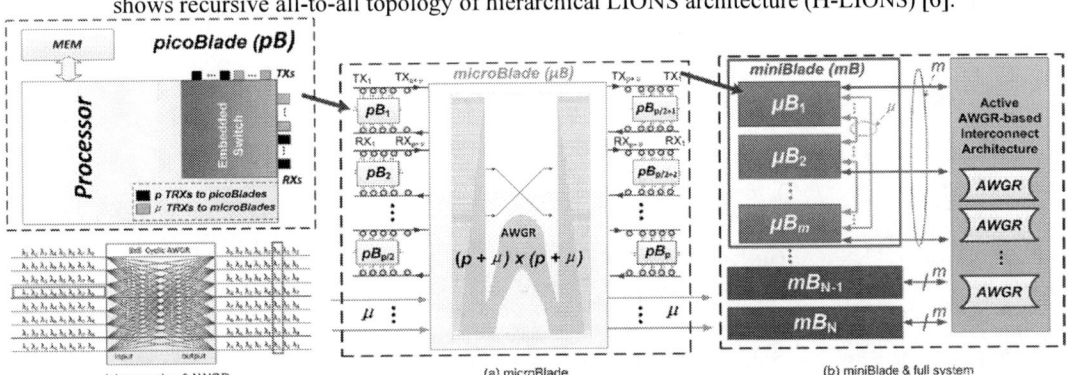

Figure 5. Hierarchical scaling of AWGR interconnected data centers with picoBlades at the lowest hierarchy, microBlades, miniBlades, and macroBlades at the higher hierarchies.

OECC/PS2016

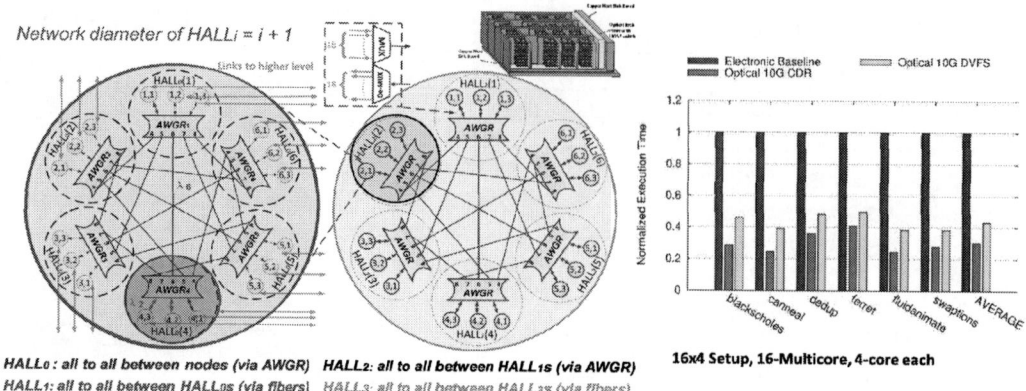

HALLᵢ: all to all between nodes (via AWGR)   HALL₂: all to all between HALL₁s (via AWGR)
HALL₁: all to all between HALL₀s (via fibers)   HALL₃: all to all between HALL₂s (via fibers)

16x4 Setup, 16-Multicore, 4-core each

Figure 6. Hierarchical LIONS (H-LIONS) [6], where all-to-all interconnection topology is repeated at each hierarchy.

IV. TOWARDS EXASCALE DATA SYSTEM

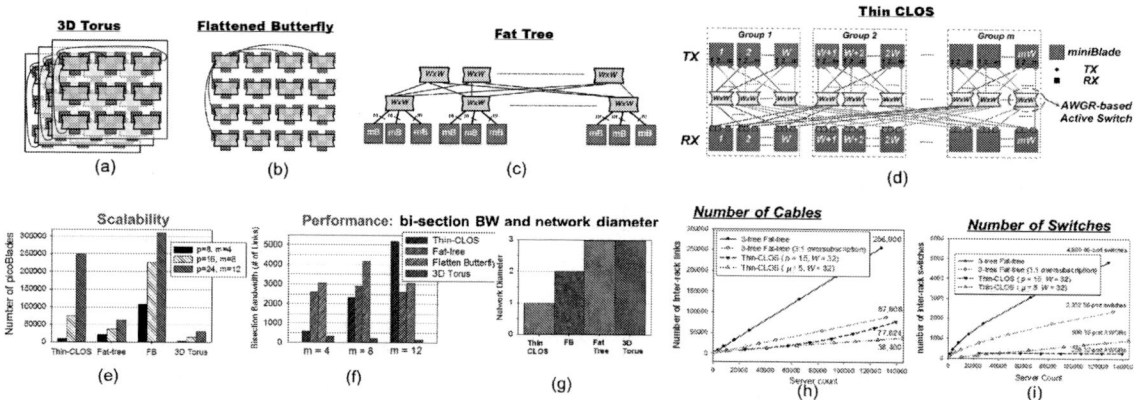

Figure 7. Top-hierarchy topology choices for interconnecting H-LIONS [6]: (a) 3D Torus, (b) Flattened Butterfly, (c) Fat Tree, (d) Thin CLOS, (e) total number of picoblades supported in example topologies, (f) bisection bandwidth, (g) network diameter, (h) required scalability of the number of cables, (i) required scalability of the number of switches.

At the top of the hierarchy (e.g. MacroBlades), one can utilize the same Active LIONS architecture to interconnect the lower hierarchy networks (e.g. MiniBlades) or utilize simpler ThinCLOS with reduced number of wavelengths as shown in Figure 7 (d). Comparison of this architecture to (a) 3D Torus, (b) Flattened Butterfly, (c) Fat Tree indicate (e, h, i) superior scalability as shown in terms of support of a large number of compute nodes (>140,000) with minimal number of cables and switches while providing (f) high bi-section bandwidths and (g) lowest network diameters.

V. SUMMARY

Photonics will play essential roles in realizing exascale computing systems in the future. High-throughput, low-latency, low-contention, and high-radix wavelength-routing switching together with energy-efficient silicon photonics supporting all-to-all interconnection are expected to be key enabling attributes for exascale data systems.

ACKNOWLEDGMENT

This work was supported in part under DoD Agreement Number: W911NF-13-1-0090.

REFERENCES

[1] Top500. (2016). *TOP 500 Supercomputer Sites.* Available: http://www.top500.org
[2] P. M. Kogge and T. J. Dysart, "Using the TOP500 to trace and project technology and architecture trends," presented at the Proceedings of 2011 International Conference for High Performance Computing, Networking, Storage and Analysis, 2011.
[3] R. Yu, S. Cheung, Y. Li, K. Okamoto, R. Proietti, Y. Yin, *et al.*, "A scalable silicon photonic chip-scale optical switch for high performance computing systems," *Optics Express,* vol. 21, pp. 32655-32667, 2013/12/30 2013.
[4] X. Ye, Y. Yin, S. J. B. Yoo, P. Mejia, R. Proietti, and V. Akella, "DOS: a scalable optical switch for datacenters," presented at the Proceedings of the 6th ACM/IEEE Symposium on Architectures for Networking and Communications Systems, 2010.
[5] X. H. Ye, R. Proietti, Y. W. Yin, S. J. B. Yoo, and V. Akella, "Buffering and Flow Control in Optical Switches for High Performance Computing," *Journal of Optical Communications and Networking,* vol. 3, pp. A59-A72, Aug 2011.
[6] Z. Cao, R. Proietti, and S. J. B. Yoo, "Hi-LION: Hierarchical Large-Scale Interconnection Optical Network With AWGRs [Invited]," *Journal of Optical Communications and Networking,* vol. 7, pp. A97-A105, 2015/01/01 2015.

Burst-Mode Optical Packet Processing Technologies

Salah Ibrahim and Ryo Takahashi

NTT Device Technology Laboratories, NTT Corporation

3-1 Morinosato-Wakamiya, Atsugi-shi, Kanagawa Pref., 243-0198, Japan

ibrahim.salah@lab.ntt.co.jp

Abstract: *We present a review for the burst-mode optical packet processing technologies demanded in optical packet switched (OPS) networks with focus on recent research trends.*

Keywords: *Optical processing devices, Data Center networks, Optical packet switching*

I. INTRODUCTION

The increase of bandwidth-hungry applications and convergence of telecommunication services are steadily pushing for higher data-rates at all the transport network domains namely the core, metro and access parts [1-3], whereas data center (DC) networks have been facing unprecedented demands and new requirements [4]. In order to cope with these conditions, moving forward towards a more dynamic photonic network is a powerful approach that enhances the utilization granularity of the network available bandwidth, and enables sharing the physically-limited resources among a wider range of users. In such dynamic networks, the transmission of optical data takes place in smaller blocks e.g., optical bursts or packets [5], without having any signal transmitted in-between these blocks i.e., burst-mode (BM) transmission; besides switching data in the optical domain without performing optical-electrical-optical conversion (OEO). BM technologies are indispensable for realizing such networks as they are the set of technologies that enables the transmission, manipulation and reception of optical packets. Currently the deployment of BM transmission in the conventional network domains is limited to the access part, whereas a more immediate opportunity for BM technologies lies in the field of DC networks [6-10] or data-processing networks in general, as an optical packet-based approach can provide a radical solution for the critical scalability issue currently confronted by these networks. Moreover reducing the latency of such networking nodes by using advanced BM technologies can enable a new range of low-latency applications on a wider network scale; paving the way for a more powerful processing capability than ever attained before.

Standards are already present specifying the requirements for the upstream packet-part of the access network [3], where the nature of this network implies the need for a highly sensitive BM receiver capable of supporting a wide dynamic range. The speed of clock recovery is another important spec of the BM receiver and as the considered data rate gets higher, the packet length decreases and the need for faster clock recovery becomes more critical. Unlike the access network, an optical packet-based DC network is still in the research phase where the conducted research work takes very different forms. The proposed networks can be classified in different ways; as on one hand there is the fully optical packet switching (OPS)-based networks and the hybrid networks in which OPS and optical circuit switching (OCS) are simultaneously deployed either with different sets of hardware [11] or with a unified platform [12]. There is also the all-optical approach where an AWG-based network is considered [13], compared to a hybrid network where electrical processing partially takes place [12]. The duration of the considered packets is another important feature, and it can vary from 10's of nanoseconds up to 10's of microseconds as for example in the case of time-slotted ring networks [14]. By considering the key elements required for realizing such networks in a more general sense, the following key processing elements are required; a means to recognize the packet destinations i.e., label processor, a switching unit, and a packet-buffering unit, in addition to burst-mode amplification.

II. KEY PROCESSING ELEMENTS

A. Label Processing

An overview for different research directions on labeling schemes can be found in [15], and the key points can be summarized as follow; a) Owing to the technical difficulty in realizing an ultrafast-bit label processor that can operate in a burst-mode fashion, the great majority of the research efforts are directed towards avoiding the ultrafast processing difficulty by using a substituent labeling scheme. One example for that is identifying packets with different wavelengths. However such solutions suffer a strict scalability issue that directly limits their practical deployment, b) Performing clock recovery can enable efficient label processing but the currently available solutions for clock extraction takes very long time compared to the short packet length, as for example if we consider a 100-Gbps packet which is around 120 ns, the packet label is only a few nanoseconds and the clock recovery should even be a fraction of that number. c) A promising approach is using a self-triggered device that can operate without clock recovery. We have recently reported an enhanced implementation for such principle with a self-triggered label processor for 25-Gbps packets that operates with a power consumption of 120 mW [16].

B. Switching

A long set of requirements needs to be satisfied by the optical switch; where in terms of transparency, the bit-rate (10~400 Gbps), packet format (coherent, WDM), wavelength and polarization transparencies are all critical. Moreover fast switching characteristics, low power consumption, high extinction ratio, low crosstalk, ease of controllability and compactness are all demanded as well. To meet these requirements, different types of optical switches have already been researched [17]. Examples include the matrix switch that consists of cascading smaller switches, phased-array switch, wavelength routing switch [18], and broadcast-and-select (B&S) switch [19]. Among them the B&S switch is an attractive choice because it can meet most of these requirements to a good extent, and despite its limited port-count scalability, it can still allow a highly scalable network when for example a Torus topology is considered [12]. For higher port counts, an AWG-based switch can be a proper solution, whereas the combination of AWGs and B&S switches can be used to further increase the port count [20, 21]. To compensate for the inherently large insertion loss caused by optical power splitting at the B&S switch, semiconductor optical amplifiers (SOAs) are usually employed as optical gates. However, the SOA severely degrades the optical signal quality due to its large ASE noise, pattern dependence i.e., gain saturation, and nonlinear effects (four-wave mixing and cross-gain modulation) especially for WDM packets. The SOA also requires a high-speed, high-current driver, that results in a power-hungry switch. To solve these issues, we have proposed an EAM-based B&S optical switch as a replacement for the SOA deployment [22], and demonstrated good switching operation with low power consumption.

C. Buffering

Buffering optical packets is essential for performing contention resolution among colliding packets, and demanded as well for enabling higher network functionalities such as packets regeneration, QoS, format conversion, etc. Realizing an optical memory has attracted a wide research interest [23, 24] but a reliable solution is still missing. The currently available solutions can be classified into all-optical passive buffering where a fiber delay line (FDL) is employed [25], or electrical buffering where a BM interface is utilized to interface the optical packets into and out of an electrical memory. Recovering the packet clock is indispensable for performing the optical-electrical interfacing and vice versa. The conventional way to do that is to use a phase locked loop (PLL) where preamble bits are required, and a relatively long locking time takes place [26]. Using a phase picking method [27] instead is currently used for lower data rates (10 Gbps). The oversampling architecture [28] is another circuit-based alternative, where recently a clock recovery time of 31 ns has been reported [29] based on the combination of oversampling mechanism and a successive approximation algorithm implemented with CMOS circuitry. A completely different approach for BM clock generation involves the formation of a single pulse upon packet arrival and then the generation of a train of pulses by circulating this pulse with a precisely controlled time in a unity-gain loop throughout the whole packet presence time [30].

D. Amplification and Equalization

Having packets with variable durations and duty cycles directly affects the dynamics of the EDFA and thus results in unavoidable gain fluctuation. Several approaches have been proposed to mitigate such fluctuations [31, 32]. This includes a transient-suppressed EDFA which has a wider fiber core that reduces the gain fluctuation but without completely eliminating it, active compensation methods where the power of the incoming packet is monitored by a PD and the EDFA pump power is controlled accordingly, in addition to compensation by using a dummy optical signal to keep a constant power level at the EFDA all over the time.

III. CONCLUSIONS

Burst-mode technologies are the set of technologies that enables the transmission, manipulation and reception of discontinuous optical packets and thus they are the basis for future optical packet-based networks that will play a key role in data processing networks and beyond.

REFERENCES

[1] Cisco visual networking index: Forecast and Methodology, 2014–2019, Cisco Systems, Inc., San Jose, CA, USA, 2015.
[2] A. Pizzinat, P. Chanclou, T. Diallo, et al.: 'Things you should know about front haul', J. Lightwave Technol., 2015, 33, (5), pp. 1077–1083.
[3] ITU-T G.989 series, 40-gigabit-capable passive optical networks (NG-PON2).
[4] T. Benson et al., "Network traffic characteristics of data centers in the wild," in Proc. ACM, New York, NY, USA, 2010, pp. 267–280.
[5] S. J. Ben Yoo, "Optical packet and burst switching technologies for the future photonic internet," J. Lightwave Technol. 24, 2006, 4468-4492.
[6] C. Kachris, K. Kanonakis, and I. Tomkos, "Optical interconnection networks in data centers: Recent trends and future challenges," IEEE Commun. Mag., vol. 51, no. 9, pp. 39–45, Sept. 2013.

[7] R.G. Beausoleil, M. McLaren, and N.P. Jouppi, "Photonic architectures for high-performance Data Centers," IEE J. Selected Topics in Quantum Electronics, vol. 19, no.2, 2013.

[8] W. Zhang, H. Wang, and K. Bergman, "Next-generation optically-interconnected high-performance data centers," J. Lightwave Technol., vol. 30, no. 24, p. 3836–3844, 2012.

[9] L. Schares, X. J. Zhang, R. Wagle, D. Rajan, P. Selo, S. P. Chang, J. Giles, K. Hildrum, D. Kuchta, J. Wolf, and E. Schenfeld, "A reconfigurable interconnect fabric with optical circuit switch and software optimizer for stream computing systems," in Optical Fiber Communication Conf. (OFC), 2009, paper OTUA1.

[10] M. Al-Fares, A. Loukissas, and A. Vahdat, "A scalable, commodity data center network architecture," in Proc. ACM SIGCOMM, 2008, pp. 1–12.

[11] N. Farrington, G. Porter, S. Radhakrishnan, H. Bazzaz, V. Subramanya, Y. Fainman, G. Papen, and A. Vahdat, "Helios: A hybrid electrical/optical switch architecture for modular data centers," in Proc. ACM SIGCOMM, 2010, pp. 339–350.

[12] R. Takahashi, T. Segawa, S. Ibraihm, T. Nakahara, H. Ishikawa, A. Hiramatsu, Y-C. Huang, K-I. Kitayama, "Torus Data Center network with smart flow control enabled by hybrid optoelectronic routers, " J. Opt. Comm., vol.7, no. 12, 2015.

[13] Y. Yin, R. Proietti, X. Ye, C. J. Nitta, V. Akella, and S. J. B. Yoo, "LIONS: An AWGR-based low latency optical switch for high-performance computing and data centers," IEEE J. Sel. Top. Quantum Electron., vol. 19, no. 2, 3600409, 2013.

[14] Y. Pointurier, G. de Valicourt, J.E. Simsarian, J. Gripp, and F. Vacondio, "High Data Rate Coherent Optical Slot Switched Networks: A Practical and Technological Perspective", IEEE Communications Magazine, August 2015.

[15] S. Ibrahim, H. Ishikawa, T. Nakahara, Y. Suzakai, and R. Takahashi, "A novel optoelectronic serial-to-parallel converter for 25-Gbps 32-bit optical label processing," IEICE Trans. Electron. E97-C(7), 773–780 (2014).

[16] S. Ibrahim, T. Nakahara, H. Ishikawa, and R. Takahashi, "Burst-mode optical label processor with ultralow power consumption," Optics Express, vol.24, no.7, 2016.

[17] G. I. Papadimitriou, C. Papazoglou, and A. S. Pomportsis, "Optical switching," Wiley-Interscience, 2006.

[18] J. Gripp, M. Duelk, J. E. Simsarian, A. Bhardwaj, P. Bernasconi, O. Laznicka, and M. Zirngibl, "Optical switch fabrics for ultra-high-capacity IP routers," J. Lightwave Technol. Vol.21, No.11, p. 2839, Nov. 2003.

[19] K. Sone, S. Yoshida, Y. Kai, G. Nakagawa, G. Ishikawa, and S. Kinoshita, "High-speed 4×4 SOA switch subsystem for DWDM systems," Proc. 16th Opt-Electronics and Communication Conference (OECC), 776-777, 2011.

[20] Yong-Kee Yeo, Zhaowen Xu, Dawei Wang, Jianguo Liu, Yixin Wang, and Tee-Hiang Cheng, "High-speed optical switch fabrics with large port count, " Opt. Express 17, 10990-10997 (2009).

[21] K. Sato, H. Hasegawa, and T. Niwa, "A large-scale wavelength routing optical switch for data center networks," IEEE. Commun. Mag., vol. 51, no. 9, pp. 46–52, 2013.

[22] T. Segawa, M. Nada, M. Nakamura, Y. Suzaki, and R. Takahashi, "An 8×8 broadcast-and-select optical switch based on monolithically integrated EAM-gate array," Proc. ECOC2013, We.4.B.1.

[23] R. S. Tucker, P-C. Ku, and C. J. Chang-Hasnian, "Slow-light optical buffers: capabilities and fundamental limitations," vol.23, no.12, 2005.

[24] K. Nosaki, A. Shinya, S. Matsuo, Y. Suzaki, T. Segawa, T. Sato, Y. Kawaguchi, R. Takahashi, and M. Notomi, "Ultralow-power all-optical RAM based on nanocavities," Nature photonics 6, 248-252, 2012.

[25] T. Zhang, K. Lu and J. R. Jue, "Shared fiber delay line buffers in asynchronous optical packet switches, " in *IEEE Journal on Selected Areas in Communications*, vol. 24, no. 4, pp. 118-127, 2006.

[26] S. Porto, C. Antony, A. Jain, D. Kelly, D. Carey, G. Talli, P. Ossieur, and P. D. Townsend, "Demonstration of 10 Gbit/s burst-mode transmission using a linear burst-mode receiver and burst-mode electronic equalization," J. Opt. Commun. Netw. 7(1), A118–A125 (2015).

[27] R. Yu, R. Proietti, S. Yin, J. Kurumida, and S. J. Ben Yoo, "10-Gb/s BM-CDR circuit with synchronous data output for optical networks," IEEE Photon. Technol. Lett., vol. 25, no. 5, 2013.

[28] N. Suzuki, K. Nakura, T. Suehiro, M. Nogami, S. Kosaki and J. Nakagawa, "Over-sampling based burst-mode CDR technology for high-speed TDM-PON systems," *Optical Fiber Communication Conference and Exposition (OFC/NFOEC), 2011 and the National Fiber Optic Engineers Conference*, Los Angeles, CA, 2011, pp. 1-3.

[29] A. Rylyakov, et al., "A 25 Gb/s burst-mode receiver for low latency photonic switch network," IEEE J. Solid-state circuits, vol.50, no.12, 2015.

[30] Tatsushi Nakahara and Ryo Takahashi, "Self-stabilizing optical clock pulse-train generator using SOA and saturable absorber for asynchronous optical packet processing," Opt. Express 21, 10712-10719 (2013).

[31] Y. Awaji, H. Furukawa, B. J. Puttnam, and N. Wada, "Burst-mode optical amplifier," OFC, OThI4, 2010.

[32] Masaki Shiraiwa, Yoshinari Awaji, Hideaki Furukawa, Satoshi Shinada, Benjamin J. Puttnam, and Naoya Wada, "Performance evaluation of a burst-mode EDFA in an optical packet and circuit integrated network, " Opt. Express 21, 32589-32598 (2013).

High Performance DCN Architecture Based on Flow-controlled Optical Switching System

Nicola Calabretta, Wang Miao, Fulong Yan, and Oded Raz

COBRA Research Institute, Eindhoven University of Technology, Eindhoven, the Netherlands

n.calabretta@tue.nl

Abstract: *We present a novel high performance flat DCN employing bufferless and distributed sub-microsecond optical switches with wavelength, space, and time switching operation. Numerical and experimental investigations indicate potential scalability of the DCN to Petabit/s capacity.*

Keywords: *Optical switch, data center network, wavelength division multiplexing*

I. INTRODUCTION

Data center applications are creating new communication patterns with more than 75% of the total data center traffic processed within data centers (server to server and rack to rack) [1, 2]. To handle this traffic, innovative architectural and technological innovations to the underlying interconnect networks are required in order to enable the scalable growth both in communicating endpoints and traffic volume, while decreasing the costs and the energy consumption. Nowadays, building scalable data center network (DCN) architectures that interconnect a large amount of servers with electrical switches will either require multiple-layer solution or utilize customized core commercial switches, which introduce extra latency, high cost, and high power consumption. This is becoming a more stringent issue as the data rate increases. 100 Gb/s per port Ethernet switches have already been deployed in data centers and 400 Gb/s and beyond is under development [3]. However these seemingly high-speed Ethernet switches are using multiple electrical lanes (4-8) to switch such high capacity links which effectively dramatically reduces the radix of the switches [http://www.broadcom.com/products/Switching/Data-Center/BCM56960-Series].

In contrast, optical switching technologies can transparently switch high data rate/format signals, reducing the required number of switch ports and avoiding costly O/E/O in hierarchical multiple-stage architecture. Optical circuit switches (OCS) with large port-count are commercially available. However, the milliseconds configuration time introduces large latency and prevents efficient statistical multiplexing of data packets. On the other hand, fast optical switches with sub-microseconds reconfiguration time has been so far demonstrated only for a moderate port count [4]. Recently, a scalable optical packet switch (OPS) with modular structure has been investigated and experimentally verified in a simplified star topology DCN architecture [5]. In this work, we report on a novel flat DCN architecture that employs distributed buffer-less optical switches with optical flow-control to implement scalable DCN architecture. The novel all-optical DCN architecture enables large-scale interconnectivity by using feasible optical switches with moderate number of ports.

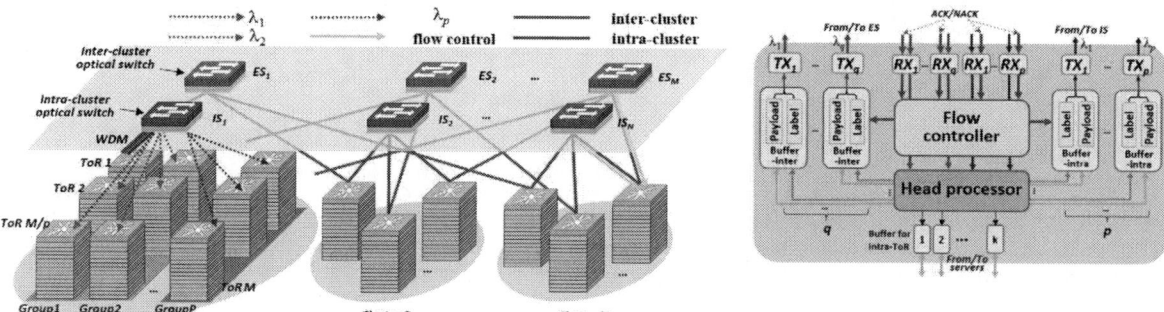

Figure 1. Novel flat DCN architecture based on optical switch with fast flow control.

Figure 2. ToR schematic diagram

II. DCN ARCHITECTURE AND OPTICAL CROSS-CONNECT SWITCH

The proposed flat DCN architecture based on two parallel inter- and intra-cluster networks is shown in Fig. 1. It consists of N clusters, and each cluster groups M racks by using an intra-cluster optical switch (IS). Each rack contains k servers interconnected by an electronic ToR switch. The ToR switch is equipped with two wavelength division multiplexing (WDM) bi-directional optical links. One optical link is used to interconnect the ToR to the IS for intra-cluster communication. The second optical link interconnects the ToR to the inter-cluster optical switch (ES). The i-th ES interconnects the i-th ToR of each cluster, with i = 1,...,N. It can be seen from Fig.1 that a single-hop communication is needed between racks of the same cluster, while at most two hops are needed to interconnect racks of different clusters. It is worth to notice that the proposed DCN architecture allows ToRs to be interconnected via different path connections, increasing network fault-tolerance. As shown in Fig. 1, the number of interconnected ToRs (and servers) scales as N ×M. By using a feasible 64 × 64 port IS and ES, up to 4,096 ToRs and then 163,840 servers (in case each ToR groups 40 servers) can be interconnected. The functional blocks of the ToR are illustrated in Fig. 2. The ToR aggregates the traffic coming from the k servers. Assume that each server produces a traffic at data rate b_serv. Therefore, ToR aggregates a total traffic of k × b_serv. Part of this traffic is exchanged between intra-rack servers (intra-rack traffic), and the rest of traffic is directed to servers in the same cluster (intra-cluster traffic) and servers located in different clusters (inter-cluster traffic). Packets which should travel between servers in the rack are directly processed and forwarded by the TOR. Packets which travel between racks in a cluster or between clusters are served using k intra/inter-rack buffer queues in the TOR. Both intra and inter cluster interfaces operate in a similar manner. Take the case of the intra-cluster interface. This interface consists of p WDM transceivers (TX in the figure) with dedicated electronic buffers that interconnects the ToR to the IS optical switch with optical flow controlled links. According to the packet header destination, each buffer stores the traffic destined to a group of intra-cluster ToRs. The WDM transceiver forwards the stored traffic destined to a group of M/p ToRs, and in general the buffer of the transceiver at λ_i stores and forwards the traffic destined to the intra-cluster ToR (i-1)×M/p+1, ..., ToR i×M/p (Group i), with i=1, ..., p. As an example shown in Fig. 1, with regards to ToR 1, the buffer of the transceiver at λ_1 stores and forwards the traffic destined to the intra-cluster ToR 2, ToR 3, ..., ToR M/p (Group 1). Moreover, the allocation of the wavelengths should meet the requirement that for the ToRs in the same group, the wavelengths used to address the same ToR destination should not overlap with each other. The copy of the stored packet, combined with an optical label generated by the ToR, is forwarded to the IS. The optical label, processed on-the-fly by the IS optical switch, determines the forwarding of the packet to the ToR destination. Due to the buffer-less operation of the optical switch, if a contention occurs at the IS optical switch, the optical flow control provides a NACK to retransmit the stored packet. When no contention occurs an ACK signal is generated to release the packet from the buffer. for the sake of flexibility we allow the the inter-cluster interface to consists of q parallel WDM transceivers instead of p channels. The number of p and q WDM transceivers depends on the required capacity to guarantee a target over-subscription. Multiple 25 Gb/s or 50 Gb/s p and q WDM transceivers can be tailored according to the expected intra- and inter cluster data traffic. The IS optical switch is based on broadcast and select architecture as shown in Fig. 3 (for ES, just substitute q for p and N for M). It processes in parallel the multiple WDM input packets by using autonomously controlled 1 × F optical switches. Note that F is equal to M/p for IS, and F equals N/q for ES since each of the p (or q) WDM transceivers groups the traffic destined to M/p intra-cluster ToRs (or N/q inter-clusters ToRs). This allows to scale the port count of the switch to M × M (or N × N) by using p × M (or q × N) parallel smaller 1 × F switches with moderate broadcast splitting losses and therefore lower OSNR degradation. At the optical switch node, the optical label is extracted and processed by the switch controller, while the optical payload is transparently switched by the 1 × F switch. According to the optical label, the switch controller enables the fast (nanosecond) SOA-based gates of the 1 × F switch. Multicast operation is also supported by enabling multiple SOA-based gates. The use of SOA gates has two advantages: fast nanosecond switching time and optical amplification of the signal to compensate the splitting losses of the broadcast and select architecture. Moreover, the proposed technologies allow photonic integration of the switch in a single chip [6]. The control complexity and the configuration time are largely determined by the label processing time.

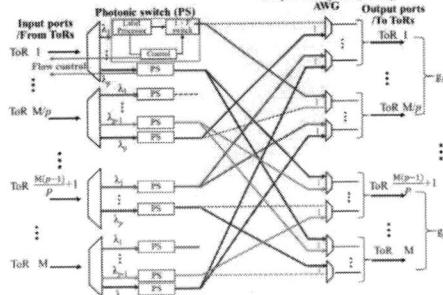

Using our in-band RF tone labeling technique and the distributed control have resulted in port-count independent nanosecond reconfiguration time allowing operation on large as well as small flows exploiting statistical multiplexing. thanks to the modular structure and parallel processing of channels, the possible contentions among the F input ports can be resolved in a distributed manner which significantly minimizes the processing latency. The packet with higher priority is forwarded to one or several of the F possible output ports and the others are blocked and the corresponding flow control signals are generated and sent back to the ToRs. According to the received ACK (or NACK), the flow controller at the ToR releases (or retransmits) the packets stored in the buffers.

Fig. 3. Schematic of the optical switch architecture.

III. PERFORMANCE ANALYSES OF THE DCN ARCHITECTURE

To investigate the scalability, DCNs consisting of 5,760 (medium size) and 100,000 (large size) servers with 10 Gb/s link have been considered. The packet size is set to 1000 Bytes at the ToR output that each slot is 800 ns. The total data rate of the optical links to the IS and ES is 200 Gb/s (4 × 50 Gb/s WDM channels). Therefore, the designed oversubscription of the system is 1:1. The round trip time (RTT) for ToRs with direct connection is 560 ns, including the delay to process the labels and control the switches (60 ns), and the optical link (2 × 50 = 100 m) latency (500 ns). OMNeT++ [7] has been employed as the simulation tool to fully investigate the performances of the proposed DCN under packet transmission. The operation of the system is synchronous and the simulations run for 1 × 108 time slots. At each time slot, the server generates a packet according to a binomial distribution with a fixed load, thus there is a certain probability to have a new packet at each server. All servers are programmed to operate independently with the same load.

Traffic is classified into 3 categories (inter-cluster, intra-cluster and intra-ToR). In each category, packets destinations are chosen randomly between all possible destinations according to a uniform distribution. Considering the facts that most of the traffic is exchanged within the ToR and cluster, two traffic distributions with different ratios of 3 categories are studied. Case A is the large inter-cluster traffic while case B is light inter-cluster traffic. Different buffer size for 200 Gb/s link ports are considered (40 KB and 60 KB). Since the inter-cluster traffic will pass both the intra-cluster link and the inter-cluster link, the real oversubscription of case A and case B is 1:1 and 3:4 respectively. Figure 4 illustrates the average server-to-server latency including the latency to switch the packet to the server inside the rack as a function of the load. 8.1 μs server-to-server latency has been obtained at load of 0.4 for both medium and large size DC and different buffer sizes. It is clearly shown that the average latency of large size DC is greater than medium size DC, and the latency with buffer size 60 KB is greater than buffer size 40 KB. For larger load, the effect of the buffer on the system performance of latency is visible. The average latency differences between buffer size 40 KB and 60KB become large as the load increases. For load above 0.4, latency increases rapidly as it may take increasingly more attempts for the packet to be successfully transmitted, before the systems saturates at high loads. When the load is close to 1, the average latency of both medium size and large size DC under case B increases rapidly and exceeds case A for the reason that the packet loss of case A is higher than case B, and the packet loss ratio cannot be neglected, thus there are more packets going through the inter- cluster in case B, which contributes to the larger latency. Fig. 5 shows the packet loss of the system. A packet loss less than 10^{-5} for load of 0.5 has been achieved for medium size DC under case A, while the packet loss is less than10^{-7} for large size DC at load of 0.4 under case A. In agreement with the curve of the average latency, the packet loss increases with the load. For load exceeding 0.5 (case A, large size DC) and 0.6 (case A, medium size DC), the packets loss is unavoidable as the links and the buffers are fully occupied, and larger buffers do not help to decrease packet loss but create extra latency. Despite the influence of the DC size, the packet loss of case A is greater than case B simply because case A has more inter-cluster traffic and the buffer will be filled quicker compared with case B.

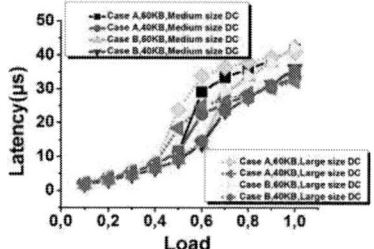

Fig. 4 Server to server average latency

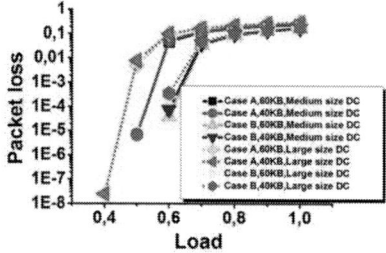

Fig. 5. Packet loss versus load

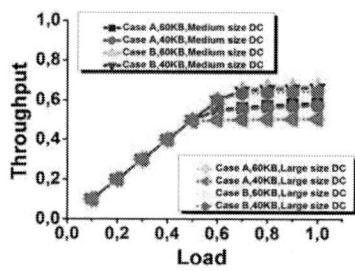

Fig. 6. Normalized throughput

Fig. 6 reports the normalized network throughput of both DC sizes under different conditions. As shown in Fig. 6, the normalized throughput increases as the load increase. However, for case A, large size DC starts to saturate at load of 0.5 and medium size DC saturates at load of 0.6. And both size DCs saturate at load of 0.7 under case B. These slight differences can be attributed to the fact that under case A, there is more inter-cluster traffic which needs to pass in two hops, therefore it takes up more link bandwidth compared with intra-cluster traffic.

IV. CONCLUSIONS

We propose a novel flat DCN architecture based on distributed buffer-less optical switches with fast flow control. The performance of the proposed architecture was numerically assessed with OMNeT++. Preliminary experimental validation using a 4 × 4 optical switch prototypes are reported in [8]. The numerical results show that the proposed DCN architecture allows for 8.1 μs end- to-end average latency and less than 10^{-7} packet loss for traffic load of 0.4 when interconnecting 100,000 servers. Those results indicate that a scalable DCN can be built by using feasible optical switches with moderate switch radix, which provides a promising solution for the DCN architecture.

ACKNOWLEDGMENT

The authors would like to thank FP7 COSIGN project (NO. 619572) for supporting this work.

REFERENCES

[1] Cisco global cloud index: Forcast and Methodology, CISCO, 2013.

[2] T. Benson et al., "Network Traffic Characteristics of Data centers in the Wild," Proc. ACM, New York, 2010.

[3] M. Alizadeh, et al. "Less is more: trading a little bandwidth for ultra-low latency in the data center" 9th USENIX conference on Networked Systems Design and Implementation, 2012.

[4] C. Kachris et al., Optical Interconnects for Future Data Center Networks, Springer, 2013.

[5] W. Miao, et al. "Novel flat datacenter network architecture based on scalable and flow-controlled optical switch system." Optics Express 22.3, 2465-2472, 2014.

[6] N. Calabretta, K. Williams, and H. Dorren, "Monolithically Integrated WDM Cross-Connect Switch for Nanoseconds Wavelength, Space, and Time Switching," in Proc. ECOC 2015.

[7] OMNeT++ Network Simulation, http://www.omnetpp.org.

[8] W. Miao, et al. " Petabit/s Data Center Network Architecture with Sub-microseconds Latency Based on fast optical packet switches" Proceeding Ecoc 2015, 2015.

Early Days of VAD Method

Tatsuo Izawa

Chitose Institute of Science and Technology
t-izawa@photon.chitose.ac.jp

Abstract: *Early history of the development of VAD method for mass-production of optical fibers for telecommunications is presented. Today, more than 300milion km fibers are made in the world every year, and 60% of the fibers are made by VAD method. There were many difficulties in the development of the method, the biggest hurdle was the consolidation process, and it was solved by the introduction of helium gas in zone-heating furnace.*
Keywords: *(optical fiber, VAD, MCVD, silica glass)*

I. Prologue

Soon after I joined NTT research Laboratories in 1970, I started research on the optical fibers and planar light-wave circuits for telecommunications. However, Corning group reported the success of the fabrication of low-loss fibers in September of the same year. The materials and the processes were not reported, I estimated the materials from their publications and made an optical fiber with TiO_2 doped silica glass [1] for the core material. The loss of the fiber was almost infinity, however I reminded that the origin of absorption loss of TiO_2 doped silica glass might be the defects of oxygen atoms. After heat treatment of the fiber in oxygen atmosphere, the loss was decreased down to about 20 dB/km as reported by Corning group. Although the loss was acceptable for telecommunication systems, many small cracks were introduced on the surface of the fiber by the heat treatment and the fiber became fragile.

I noticed that there were many issues to be solved to realize an optical fiber for practical telecommunication systems. The most important issue was dopant for the core glass. From my investigation, I concluded that the best dopant for the core glass is GeO_2. However, it was not easy to make GeO_2 doped silica glass because of the difference of vapor pressure at high temperature. It seemed impossible to make GeO_2 doped silica glass with the similar method of making TiO_2 doped silica glass.

II. Development of New Method [2,3]

MacChesney reported very smart method called MCVD [4] in1974. GeO_2 doped silica core fibers were made very easily by using the method, however, it seemed uneasy to make a large size preforms. So the target of our research was focused to develop a mass production method of optical fibers.

The idea of new method was very simple, fine particles of GeO_2 doped silica glass were formed at low temperature by hydrolysis of raw materials and the fine glass particles were deposited on a starting material to make a porous preform of core part. Then, silica glass particles for cladding were deposited on the core part as shown in Fig. 1. The porous preform was consolidated to a transparent preform by zone-heating in a furnace at high temperature. It was estimated that the evaporation of GeO_2 was prevented by the pure silica glass cladding layer in the consolidation process.

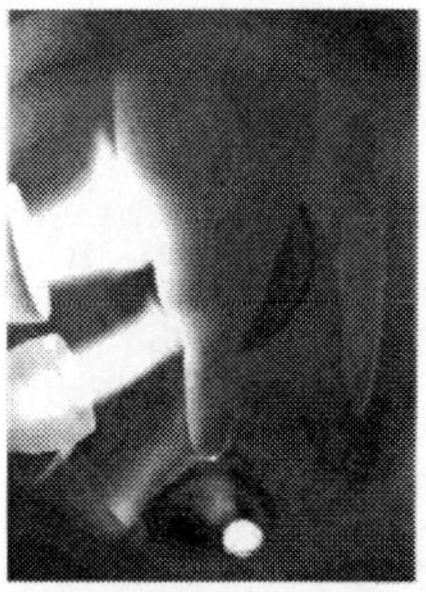

Fig. 1. Porous preform fabrication; lower burner is for core and the upper two burners are for cladding.

There were many difficulties in making a porous preforms with well controlled shape, we could solve the issues by improving the facilities and apparatus. The insurmountable difficulties ware in the consolidation process, it seemed impossible to make a transparent preform without leaving fine bubbles as shown in Fig. 2. Just before giving up the new process, we introduced helium gas into the consolidation furnace and all the issues are solved as shown in Fig. 3. The thermal conductivity of helium gas is much higher than other inert gas such as nitrogen or argon gas. So the temperature profile in radial direction of the heating zone could be uniform. I had been hesitating the use of helium gas because of the high cost.

This process was named Vapor-phase Axial Deposition (VAD) method and has good features such as easy to make big preforms, easy to remove OH and no dip in the center of core. After the success of the fundament process of VAD method, we constructed mini-plants in our laboratory for mass production of optical fibers. Many preforms and fibers were made and tested for more the one year. With these data, the process was refined.

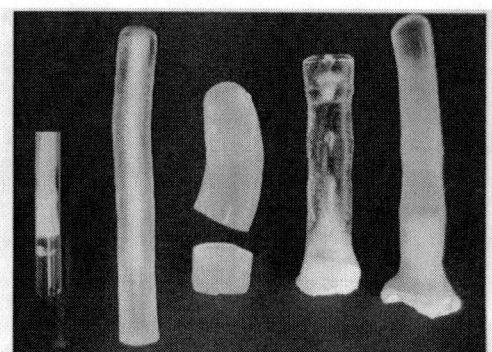

Fig. 2. Failed samples in the consolidation process.

Fig. 3. Consolidated preform (upper part) in helium gas atmosphere.

III. Introduction Barrier of VAD fibers to Field Tests

Graded index fibers were adopted for early field test of communication system in Japan, and VAD method was not good for making graded index fibers with well controlled index profile. In principle, it is possible to fabricate graded index fibers by VAD method with multiple torches for making core porous preform. However, it was not easy to develop the facility for GI fiber in short time. Therefor VAD method was considered a BAD method for practical production by system engineers and engineers in cable makers at that time.

After a while, single mode fibers were adopted for optical fiber communication systems because of the success of low loss connection technologies. VAD method came back again to a positon of the practical fabrication method with the adoption of single mode fibers for the major telecommunication network systems not only for long distance but also for FTTH. Nowadays, more than 300 million km fibers for telecommunication have been fabricated every year in the world, and 60% of them are made by VAD method.

IV. Epilogue

Many researchers and engineers were joined to the development group of VAD after the success of the fundamental process. They refined the process very rapidly and built factories for mass-production lines of optical fibers. The period I worked as a member of optical fiber research group was very short. Soon after my presentation of VAD method at IOOC77, I was forced to move to a basic research group, where I started the development of planar light-wave circuits (PLC) using similar method with VAD. The period of my research on PLC was also short time, however, young researchers succeeded my work and developed many useful devices such as splitters and AWGs.

[1] Jeff Hecht "City of Light, the Story of Fiber Optics", Oxford Univ. Press (1999)
[2] T. Izawa and S. Sudo, "Optical Fibers: Materials and Fabrication", KTK Scientific Publishers (1987)
[3] Tatsuo Izawa, IEEE J. Selected Topics in Quantum Electronics, "Early days of VAD process", vol. 6, pp.1220-1227, Nov/Dec 2000
[4] J. B. MacChesney, "Preparation of low loss optical fibers using simultaneous vapor phase deposition and fusion", Proc. 10th Int. Congr. Glass, vol. 6, pp. 40-44, 1974

S4-2 (Invited)

OECC/PS2016

Fiber and Fiber Based Technology after VAD Development
- Contribution to Transoceanic Submarine Networks -

Hiroo Kanamori

Sumitomo Electric Industries, Ltd. 1, Taya-cho, Sakae-ku, Yokohama 244-8588, Japan
kanamri-hiroo@sei.co.jp

Abstract: The development of VAD optical fibers for transoceanic submarine cable networks, from standard single-mode fibers for TPC-3 in 1989 to ultra-low loss fibers with large mode area for the latest digital coherent systems, is reviewed.

Keywords: optical fiber, VAD, single-mode, transmission capacity, transmission loss, chromatic dispersion

I. INTRODUCTION

In 1985, the development of the VAD (vapor-phase axial deposition) method led to successful completion of Japanese terrestrial optical trunk lines from Asahikawa, Hokkaido to Kagoshima, Kyushu [1]. In parallel, submarine optical communication systems had been developed, and TPC-3, the first transpacific optical cable network, was laid in 1989 [2]. Currently, a total length of submarine optical cables installed in the world has reached a million kilometers, the equivalent of twenty five times around the Earth. The transmission capacity for transoceanic systems has been exponentially increasing as shown in Fig.1 and TABLE I. In this history, optical fibers have been drastically advanced with the progress of epoch-making technologies such as DFB-LDs, erbium-doped fiber amplifiers (EDFAs), dense wavelength division multiplexing (DWDM) systems, and digital coherent transmission systems. This paper describes VAD-based optical fibers developed for each generation of transoceanic submarine optical cable networks around Japan.

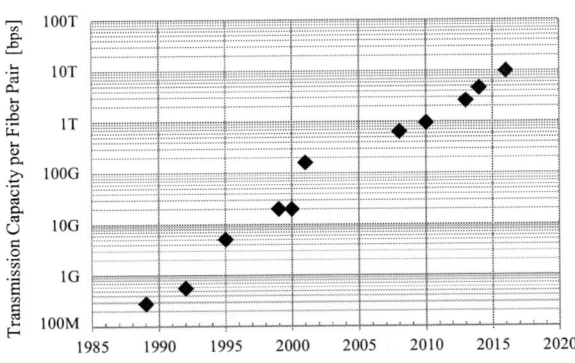

Fig. 1. Evolution of Transmission Capacity for Transoceanic Submarine Systems around Japan

TABLE I
REPRESENTATIVE TRANSOCEANIC SUBMARINE SYSTEMS AROUND JAPAN

	Ready for Service	Capacity per Channel [Gbps]	Number of Channel	Capacity per Fiber Pair [Gbps]	Number of Fiber Pair	Capacity of Cable [Gbps]	Ref.
TPC-3	1989	0.28	1	0.28	2	0.56	[2]
TPC-4	1992	0.56	1	0.56	2	1.12	[3]
TPC-5CN	1995	5.0	1	5.0	4	20	[4]
SEA-ME-WE 3	1999	2.5	8	20	4	80	[3]
China-US	2000	2.5	8	20	4	80	[3]
Japan-US CN	2001	10	16	160	4	640	[3]
TPE	2008	10	64	640	4	2,560	[5]
UNITY	2010	10	96	960	8	7,680	[6]
TPE	2011	40	32	1,280	4	5,120	[5]
SJC	2013	40	64	2,560	6	15,360	[7]
SJC	2014	40	116	4,640	6	27,840	[7]
FASTER	2016	100	100	10,000	6	60,000	[8]

II. EVOLUTION OF OPTICAL FIBERS FOR TRANSOCEANIC SUBMARINE OPTICAL CABLE NETWORKS

A. Single-mode fibers for 1.3 μm systems

In the early phase of the commercial deployment of VAD fibers, only an inner portion of an optical fiber was synthesized by the VAD method and a silica tube made from natural quartz was used for an outer portion of a fiber. However, it was difficult to draw long fibers through a proof test level as high as 2.0 % required for submarine application. It was because defects in natural quartz made a fiber weak. In addition, loss increase due to hydrogen was found to be enhanced by the impurities contained in natural quartz. In order to overcome the problems, single-mode fibers (SMFs) wholly synthesized by the VAD method was developed, where an outer portion of the fiber is synthesized by the VAD method as well instead of using a silica tube, resulting in optical fibers sufficiently applicable to submarine optical networks [9]. In 1989, TPC-3 was eventually installed, employing wholly synthesized VAD fibers. In the system for TPC-3, named OS-280M, the operating wavelength was 1.3 μm, the transmission capacity per a fiber pair was 280 Mbps, and the repeater span was 53 km.

213

B. Pure silica core fibers for 1.55 μm systems

Although the wavelength window at around 1.55 μm is the band showing minimum transmission loss for silica-based optical fibers, 1.3 μm was used in TPC-3 because of the high chromatic dispersion of 17-18 ps/nm/km at 1.55 μm for a standard SMF. The situation was changed by the commercialization of DFB-LDs, single longitudinal mode light sources, which enabled 1.55 μm operation with a standard SMF having zero dispersion wavelength at 1.3 μm.

The transmission loss of standard SMFs had been already reduced to as low as 0.20 dB/km, which was considered to be an intrinsic limit dominated by scattering of silica glass. In order to overcome the limit, scattering increment due to GeO_2, which is used as an additive to increase refractive index of a core, should be removed. This challenge was surmounted by a pure silica core fiber with a cladding doped with fluorine which decreases refractive index of silica glass, as shown in Fig.2 (b), and a minimum transmission loss of 0.154 dB/km was attained [10]. Moreover, it was confirmed pure silica core fibers show no OH loss increment due to hydrogen and less radiation induced loss owing to chemical stability of pure silica glass.

Additionally, in order to improve sensitivity against bending, a new fiber design with a cut-off wavelength shifted to around 1.45 μm was proposed [11], and has been standardized as ITU-T G.654 [12]. In 1992, TPC-4 [3] was installed adopting pure silica core / cut-off shifted fibers. The capacity per a fiber pair was 560 Mb/s, and the repeater span was as large as100-150 km.

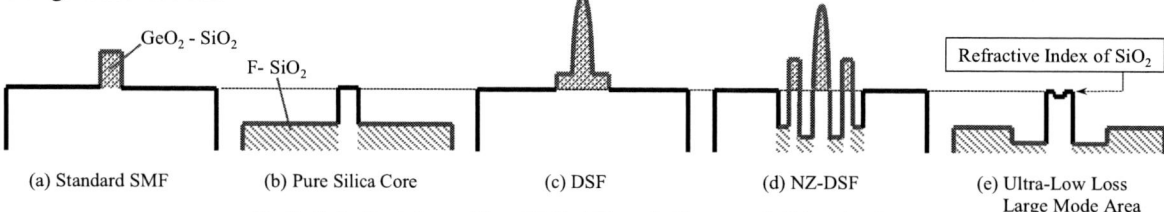

Fig. 2. Refractive index profiles of Optical Fibers for Transoceanic Submarine Systems

C. Dispersion shifted fibers for 1.55 μm systems with EDFAs

In the early 1990's, an erbium doped fiber amplifier (EDFA) was dramatically improved, and promptly applied to transoceanic submarine systems. In TPC-5CN [4] installed in 1996, the transmission capacity was increased up to 5 Gbps thanks to EDFAs, where dispersion shifted fibers (DSFs) were employed. DSFs have a zero-dispersion wavelength at around 1.5 μm. While a couple of refractive index profiles had been proposed for DSFs, "dual shape core" profile [12], as shown in Fig.2 (c), became the most popular in Japan because of the suitability for the VAD method. DSFs had been already installed for NTT's repeaterless submarine links [3] as well as terrestrial trunk lines in the late 1980s.

D. NZ-DSF, and DCF for DWDM systems

In the mid 1990s, DWDM technologies achieved rapid development. One of the spearhead of transoceanic DWDM systems was SEA-ME-WE3 [3], which started as 8 wavelengths x 2.5 Gbps and eventually upgraded up to 40 wavelengths x 10 Gbps. In the time frame of 2000-2001, vast amount of DWDM systems were laid [3]. As increase of wavelength number and transmission speed [5-7], signal quality degradation due to nonlinear optical effect such as four wave mixing became serious in case of a zero dispersion transmission media. Hence, non-zero dispersion shifted fibers (NZ-DSFs), in which some amount of chromatic dispersion occurs at the operating wavelength, were actively developed [14-15], whereas the residual dispersion was compensated by a dispersion compensating fiber (DCF) [16-19].

Dispersion managed fibers such as DSF, NZ-DSF, and DCF usually require a large waveguide dispersion obtained by a large refractive index difference between a core and a cladding, leading to small mode area. On the other hand, large mode area is preferred to suppress nonlinearity, although a fiber with large mode area is tend to be sensitive to bending. In order to provide a solution to address the trade-off requests for mode areas of a dispersion managed fibers, complicated refractive index profiles with a couple of ring core layers, as shown in Fig.2 (d), had been proposed [15-19].

Polarization mode dispersion (PMD) was another important issue in the stage to step over 10 Gbps. Since PMD attributes to non-circularity of a fiber, much efforts in manufacturing processes had been taken to keep circularity of preforms and fibers.

E. Ultra-low loss pure silica core fibers with large mode area for digital coherent systems

The emergence of digital coherent technology in the mid 2000s completely changed the requirements for optical fibers. A pioneering transoceanic system employing digital coherent technology is FASTER [8], where a transmission capacity per a fiber pair reaches 10 Tbps (100 wavelengths x 100 Gbps).

In digital coherent systems, since phase information of optical signals is kept even in detected electrical signals, linear distortion of signals due to dispersion of optical fibers can be compensated by digital signal processors. As a result, requirements for optical fibers in long haul transmission are focused on a lower attenuation and larger mode area for less nonlinearity. Accordingly, ultra-low loss pure silica core fibers with effective mode areas (A_{eff}) larger than 130 μm^2 have been developed [20] [21]. Figure 3 is a trajectory of the transmission loss for optical fibers. For mass produced fibers, an average attenuation of 0.154 dB/km has been realized [20] as shown in Fig. 4.

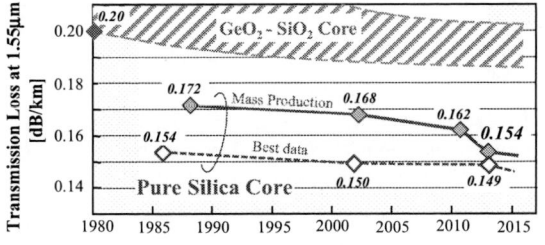

Fig. 3. Improvement of Transmission Loss

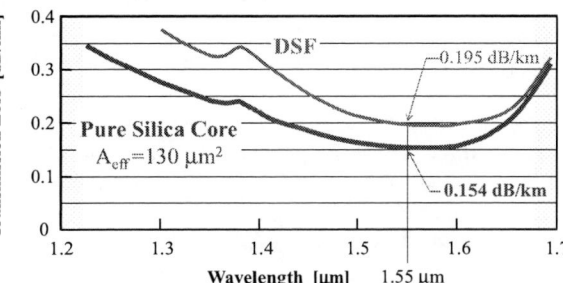

Fig. 4. Typical Loss Spectrum of Pure Silica Core Fiber
with Large Mode Area in Mass production

III. CONCLUSIONS

The first generation of optical fibers for submarine application was standard SMFs for 280 Mbps systems operated at 1.3 µm. Then, pure silica core fibers combined with DFB-LDs operated at 1.55 µm, DSFs with EDFAs, NZ-DSFs and DCFs for DWDM systems had followed. The latest is ultra-low loss pure silica core fibers with large mode area for digital coherent systems, supporting to ever expanding global information networks.

REFERENCES

[1] S. Shimada, "Status and Future Trends in Terrestrial Optical Fiber Systems in Japan," IEEE Comm. Mag., Vol. 25, Issue 10, pp. 18-21, October 1987.

[2] K. Amano, and Y. Iwamoto, "Optical Fiber Submarine Cable Systems," IEEE J. Lightwave Tech. Vol. 8, No. 4, pp. 595-609, April 1990.

[3] S. Akiba, and S. Nishi, "Submarine Cable Network Systems," NTT Quality Printing Service Co., 2001, pp. 405-407.

[4] W. C. Barnett, H. Takahira, J. C. Baroni, and Y. Ogi, "The TPC-5 Cable Network," IEEE Comm. Mag., Vol. 34, Issue 2, pp. 36-40, February 1996.

[5] http://www.submarinenetworks.com/systems/trans-pacific/tpe

[6] http://submarinenetworks.com/systems/trans-pacific/unity.

[7] http://www.submarinenetworks.com/systems/intra-asia/sjc.

[8] http://www.nec.com/en/press/201408/global_20140811_01.html

[9] S. Ito, H. Sato, F. Mizutani, K. Tsuneishi, and H. Kanamori, "High-Strength Long Length Single-Mode Fiber Synthesized by the VAD Method," IEEE J. Lightwave Tech., Vol. LT-4, No. 8, pp. 1067-1070, August 1986.

[10] H. Kanamori, H. Yokota, G. Tanaka, M., Watanabe, Y. Ishiguro, I. Yoshida, T. Kakii, S. Ito, Y. Asano, and S. Tanaka, "Transmission Characteristics and Reliability of Pure-Silica-Core Single-Mode Fibers," IEEE J. Lightwave Tech., Vol. LT-4, No. 8, pp. 1144-1150, August 1986.

[11] Y. Namihira, Y. Horiguchi, Y. Kuwazuru, N.Nunokawa, and Y. Iwamoto, "Optimum Fiber Parameters of Low-Loss Single-Mode Optical Fibres for Use in 1.55µm-Wavelength Region," Electron. Lett. Vol. 23, No. 18, pp. 963-964, August 1987.

[12] https://www.itu.int/rec/T-REC-G.654/en

[13] N. Kuwaki, M. Ohashi, C. Tanaka, N. Uesugi, S. Seikai, and Y. Negishi, "Characteristics of Dispersion-Shifted Dual Shape Core Single-Mode Fibers," IEEE J. Lightwave Tech., Vol. LT-5, No. 6, pp. 792-797, June 1987.

[14] T. Kato, S, Ishikawa, Y. Suetsugu, and M. Nishimura, "Low non-linearity dispersion-shifted fibers employing dual-shape core and depressed cladding," OFC TuN2, pp.66, February 1997.

[15] E. Sasaoka, and T. Kato, United States Patent 6,157,754, December 2000.

[16] M. Hirano, T. Kato, K. Fukuda, K. Tamamno, M. Onishi, Y. Makio. and M. Nishimura, "Novel dispersion flattened link consisting of new NZ-DSF and dispersion compensating fiber module," ECOC2000, vol.1, 2.4.4, pp.99-100, September 2000

[17] T. Kato, M. Hirano, K. Fukuda, A. Tada, M. Onishi, and M. Nishimura, "Design optimization of dispersion compensating fibers for NZ-DSF considering nonlinearity and packaging performance," OFC2001, TuS6, March 2001.

[18] M. Tsukitani, T. Kato, E. Yanada, M. Hirano, M. Nakamura, Y. Ohga, M. Onishi, E. Sasaoka, Y. Makio, and M. Nishimura, "Low-Loss Dispersion-Flattened Hybrid Transmission Lines Consisting of Low Nonlinearity Pure Silica Core Fibres and Dispersion Compensating Fibres," Electron. Lett. Vol. 36, No. 1, pp. 64-65, January 2000.

[19] M. Hirano, A. Tada, T. Kato, M. Onishi, Y. Makio, and M. Nishimura, "Dispersion compensating fibers over 140 nm-bandwidth," ECOC 2001,Th.M.1.4, pp.494-495, September 2001.

[20] M. Hirano, T. Haruna, Y. Tamura, T. Kawano, S. Ohnuki, Y. Yamamoto, Y. Koyano, and T. Sasaki, "Record Low Loss, Record High FOM Optical Fiber with Manufacturable Process", OFC/NFOEC Postdeadline Papers PDP5A.7, March 2013.

[21] S. Makovejs, C. C. Roberts, F. Palacios, H. B. Matthews, D. A. Lewis, D. T. Smith, P. G. Diehl, J. J. Johnson, J. D. Patterson, C. R. Towery, S. Y. Ten, "Record-Low (0.1460 dB/km) Attenuation Ultra-Large A_{eff} Optical Fiber for Submarine Applications," OFC Postdeadline Papers Th5A.2, March 2015.

Development of optical fibers for long-haul transmission

Haruki Ogoshi
Furukawa Electric co., Ltd
Ogoshi@ch.furukawa.cp.jp

Abstract: *After advent of VAD, many technologies for manufacturing optical fibers are progressed. I review the historical topics and state-of-the art low non-linear fibers including improvement of fiber reliability.*
Keywords: *VAD, Optical fibers, reliability, non-linearity*

I. INTRODUCTION

There is no doubt that one of the key issues for spreading optical communication systems is an improvement of processes for producing optical fibers. After the first experiment utilizing installed optical fiber cable in 1974, many projects of optical communication systems using optical fiber cables are started. Firstly, the replacement of using metallic cables to optical fiber cables in the telecommunication systems are started from long-haul systems. At the present, optical fibers are reached to every premise by fiber-to-the-home (FTTH) systems. Continual effort by engineers for optical fibers and cables are highly contributed for high quality communication systems.

At the first stage of the optical communication systems, multimode fibers with graded refractive index profile (GI-MMF) are used. Since GI-MMF has larger core, reducing splicing loss by connecters and fusion splicing is easy. It was advantageous for legacy networks. The key points to achieve the high-quality link with GI-MMF are loss budget and spectral bandwidth. The loss budget is secured by high quality fibers with low attenuation loss and low splice loss by their larger core sizes. On the other hand, spectral bandwidth is almost determined by quality of fibers. Spectral bandwidth of GI-MMF is strongly depending on refractive index profile of the core. Especially, a subtle fluctuation at the center of the core region causes larger degradation for its quality. In 1980's, GI-MMF by MCVD and OVD method was disadvantageous because they have index distortion in the center of the core by collapsing process. It causes their lower spectral bandwidth. On the other hand, refractive index profile at the center of GI-fibers by VAD method is very smooth since its entire core region is controlled by a single torch.

Processes making Single-mode fibers (SMF) have many difficulties due to its small core size. Since small offset of the core location from the center of cladding makes larger impacts on splice loss, core concentricity error should be strictly minimized. This is also advantageous for VAD method because there are no asymmetrical processes. Additionally, elimination of glass interfaces in the core region is effective to avoid OH loss peak appeared around 1390nm. This technology enables production of the fibers for entire O-L band transmission such as Allwave[TM] fiber. Since cladding area of SMF is over 150 times larger than core area, process making cladding area is also important for both cost reduction and quality improvement, respectively. Former preform by VAD had problems on cladding made by natural quartz glass which could buy in cheaper price. In order to solve quality issue, VAD technique is adopted for all synthesized fiber.

Though VAD technologies are invented in Japan, it is spread to all over the world. Thanks to its high purity and cost effectiveness, standard SMF which comply with ITU-T G.652 is mostly produced by VAD method. Because of its low loss less than 0.20dB/km at 1550nm region and low nonlinearity, standard SMF is generally used for terrestrial long-haul systems. However, further improvement to reduce nonlinearity is demanded for ultra-high-density transmission systems with digital coherent detection.

In this report, mass productivity and quality improvement by progress of VAD technologies and trend for reducing nonlinearity of optical fibers are described.

II. IMPROVEMENT OF FIBER RELIABILITY AND QUALITY USING VAD

Since contamination of the materials for core region of fiber preform should be strictly avoided for reducing attenuation loss, all researchers for optical fibers are invent several method to obtain high purity core material. MCVD method is the particular example to obtain high purity cores. In MCVD process, oxidation of halide gas materials, which is inputted from the one end of substrate tube with oxygen, and deposition of high purity glasses at the inside of the substrate tube are achieved by heating from outside of the tube. The collapsing process is done with vacuuming in the center hole and highly heating from outside of substrate tube after core deposition process. MCVD process is free from other materials, so contamination including OH group does not exist in the core region. However, purity of substrate tube which is used for cladding region did not discussed because purity of cladding is not mainly contributed to attenuation loss.

In 1980's, the same quartz glass as substrate tube for MCVD are used to jacketing materials for VAD core rod. For making SMF preform, 10 times larger cladding diameter comparing with core diameter is needed. Since the field of signal light propagating in the SMF is larger than core area, attenuation loss would be increased if the surface between core and cladding is contaminated. In order to avoid the contamination in the cladding area around core region, purity of the cladding area facing core region should be higher. In order to remove the interface of the each synthesized glass around core area, multi torch VAD is developed. In the present core rods with around four times larger cladding diameter against center core which is enough for reducing attenuation loss of optical fibers, can be produced. At the stage, extremely high purity glass for outer most cladding region is not required for reducing attenuation loss.

At the next stage, quality of tubes for outer most cladding is focused. First point is break rate of the fiber. At that period purity of tube is not high enough and some contamination would be included in the tube. In order to improve the break rates of the fibers, contamination including in the cladding materials should be avoided. We tried to adopt Jacketing VAD (JVD) method which is modified and improved VAD technique for jacketing process to improve the break rate of the fibers. This trial was effective to reduce the break rate as 10 times lower than classical approached one. The further merit was brought to entire synthesized preform on geometrical quality. Since ovality of substrate tube was not perfect, some problems by geometrical errors are observed. Our approach adopting JVD method into making preform for SMF is effective for improving core concentricity error and non-circularity of cladding.

Thanks to improvement of geometrical aspects, Polarization Mode Dispersion (PMD) of the fiber is also improved. PMD is almost depending on circularity of the core and imbalance of lateral stress to the core. PMD was not discussed when the fibers with entire synthesized preform were deployed. At the moment, electrical repeaters are used for optical transmission systems. After advent of Erbium doped Fiber Amplifier (EDFA), transmission length between 3R repeaters had been extended. By this change, new problems for optical fibers were occurred such as chromatic dispersion (CD) and PMD. For solving problems on CD limit, new fibers such as Dispersion Shifted Fibers (DSF), Non-Zero Dispersion Shifted Fibers (NZ-DSF) and Dispersion Compensating Fibers (DCF) were studied and deployed. Problems on CD were solved by designing new fibers. On the other hand, there are no solutions for improving PMD by fiber design approach. However, this problem had already been solved by making fibers by entirely synthesized technique. It was good news for researchers for this method. It was a one of the good proof for adopting entirely synthesized technique for making optical fibers.

III. New Fibers for Long Haul applications

Commercial deployment of 100Gbps long-haul transmission systems using technologies such as polarization multiplexed (PolMux) QPSK, coherent receivers, digital signal processors, and EDF amplified spans of standard single mode fiber (SSMF) is happening today. However, the growing internet traffic will require continuing maximization of the fiber capacity through the use of more spectrally efficient modulation formats [1]. These denser signal constellations are more susceptible to amplifier and nonlinear noise thereby requiring higher OSNR (optical signal to noise ratio). For example, 400Gbps with the 16-QAM format requires almost 10dB higher OSNR compared to QPSK [2]. In addition, ultra wideband transmission over the C+L bands using complimentary Raman/EDF amplification will further support increased fiber capacity [3]. The deployment of next generation optical fibers along with the use of Raman amplification is expected to be needed to achieve these higher OSNR targets.

The development of next generation fiber designs has emphasized reducing attenuation, α, and increasing the effective area, Aeff. Although decreasing fiber attenuation can reduce ASE noise it also implicitly increases the nonlinear effective length thereby increasing the nonlinear impairments, so the overall benefit to system performance of reducing α is not clear. To address this, simulations using the well know Gaussian Noise model [4], [5] have been used to develop fiber OSNR Margin Figures of Merit (FOM) comparing the system performance benefit of next generation fibers [6]. The performance benefit of fiber with reduced α and/or increased Aeff were studied by Balemarthy in four different amplifier configurations: (a) EDFA only (b) hybrid Raman/EDFA with counter-pumping and 11.5dB Raman gain (c) All Raman with counter-pumping, (d) All Raman with co- and counter-pumping with 50% co-pumping. The OSNR Margin FOM is defined as

$$ FOM \approx \left| a * \left(\alpha - \alpha_{ref} \right) L_s - b * 10 \log_{10} \left(\frac{A_{eff}}{A_{eff,ref}} \right) \text{ dB} \right| \qquad (1) $$

where α_{ref} and $A_{eff,ref}$ are the loss and effective area of the reference fiber, respectively, Ls is the amplifier span length and a and b are the constants whose values for each amplification case are listed in table 1.

Amplification technology	a	b
(a) EDFA only	1.0	0.9
(b) Hybrid Raman/EDFA with counter-pumping and 11.5 dB Raman gain	1.0	0.8
(c) All Raman with counter pumping	1.05	0.65
(d) All Raman with co- and counter-pumping with 50% co-pumping	1.0	0.45

Clearly from table 1, a 1dB improvement in effective area results in a 1dB (or better) improvement in OSNR margin. In contrast, a 1dB reduction in span loss always results in less than 1dB of OSNR margin improvement. With Raman amplification, the relative advantage of lower attenuation decreases further. This aspect of next generation optical fiber design is particularly important, given that the additional OSNR provided by Raman amplification will be essential to increasing the fiber capacity and reach.

TeraWave™ fiber is a next generation Ge-doped, depressed-cladding, cutoff-shifted fiber meeting ITU-T G.654.B recommendation, and fabricated using state of the art VAD processing. The typical Aeff at 1550 nm is 125 um2 and the typical 1550 nm loss when installed in high-density terrestrial cable designs is less than 0.184 dB/km. The micro-bending and macro-bending performance is consistent with deployment using standard splicing and installation practices. Typical arc fusion splice loss between TeraWave™ fibers is 0.03 dB using standard splicing technology. TeraWave™ fiber has been used in numerous state-of-the-art hero experiments published by a wide range of research teams [for example see 7-10].

IV. CONCLUSIONS

I summarized progress and future prospective for fibers for long-haul transmission systems. VAD is superior method for mass production on all productivity, reliability and quality. I expect that reducing non-linearity is the most important issue on fibers for long-haul transmission systems. It is no doubt that VAD method is the most effective production method for future fibers.

REFERENCES

[1] X. Zhou et al, OFC2012, PDP5C.6
[2] J. Proakis, *Digital Communications,* 3rd Ed. McGraw Hill
[3] B. Zhu et. al. OFC2015, paper W3G.4, March 22-26, 2015
[4] P. Poggiolini et al, IEEE PTL 23(11), 2011
[5] P. Poggiolini, IEEE JLT 30(24), 2012
[6] K. Balemarthy, et. al., ECOC2013, Sept 22-26 2013
[7] X. Liu et al, OFC 2012, postdeadline paper PDP5B
[8] Nolle, et al, Optics Express, Nov 21, 2011, Vol 19, No 24
[9] Raybon et al, OFC 2012, paper OTu2A.1
[10] Zhu et al, OFC 2016 March 20-24, 2016

S4-4 (Invited)

OECC/PS2016

Specialty Fibers and Optical Fiber Devices Following Development of Vapor-phase Axial Deposition

Kenji Nishide

Advanced Technology Laboratory, Fujikura Ltd., 1440, Mutsuzaki, Sakura, Chiba, 285-8550, Japan
kenji.nishide@jp.fujikura.com

Abstract: Optical fibers fabricated using vapor-phase axial deposition exhibit excellent performance and have been used globally for a long time. We present specialty fibers, optical fiber devices, and fiber lasers fabricated using these optical fiber technologies.

Keywords: Optical fiber, Vapor-phase axial deposition, PANDA fiber, Fiber coupler, Bend-insensitive fiber, Fiber Bragg grating, Fiber laser

I. Introduction

In 1966, Dr. Kao presented a feasible technique for attaining low-loss optical fiber transmissions by eliminating impurities in glass [1]. After that, following reports from Corning Inc., Bell Laboratories, Nippon Telegraph and Telephone Corporation (NTT), and Japanese manufacturers started research and development (R&D) of optical fibers. In 1976, NTT and Fujikura Ltd. developed fibers with an attenuation of 0.47 dB/km at a wavelength of 1.2 μm [2]. In 1975, NTT, Fujikura Ltd., Furukawa Electric, and Sumitomo Electric established a joint R&D team that aimed to develop practical silica–glass optical fibers. Then, in 1977, during the course of this collaborative work, the vapor-phase axial deposition (VAD) method was invented by Dr. Izawa and the other members of the joint R&D team. The VAD method was highly suitable for the mass production of optical fibers. In 1980, we reported ultimately low-hydroxyl-content (low-OH) optical fibers fabricated using the VAD method [3]. Figure 1(a) shows the loss spectrum of the fibers.

Fujikura has globally supplied large volumes of optical fibers fabricated using the VAD method. We have developed several kinds of specialty fibers and optical fiber devices using both the VAD method and other technologies. In this paper, as examples of specialty fibers and optical fiber devices, we introduce polarization-maintaining and attenuation-reducing (PANDA) fibers, optical fiber couplers using PANDA fibers, fiber Bragg gratings (FBGs), bend-insensitive fibers, and fiber lasers based on the technologies of specialty fibers and optical fiber devices.

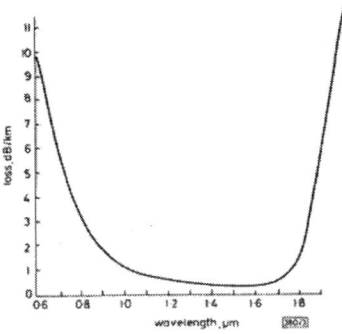

Fig. 1(a). Loss spectrum of ultimately low-OH-content optical fibers [3].

Fig. 1(b). Cross section of the PANDA fiber.

II. PANDA Fiber

In a normal single-mode fiber, the axial asymmetry of the core and/or disturbances cause mode coupling between orthogonally polarized light sources. Emitted light propagates along the fiber by changing its polarization state randomly. The PANDA fibers have excellent mode-coupling stability performance [4,5]. Figure 1(b) shows the cross section and refractive index profile of the PANDA fibers. The core and cladding of the preforms are fabricated using the VAD method, and the stress-applying parts are fabricated using the modified chemical vapor deposition method. The PANDA fibers are used for fabricating pigtail fibers of optical sources and optical transmission devices that require stable optical signal transmission while maintaining polarization.

219

III. OPTICAL FIBER COUPLER

Optical fiber couplers are indispensable for optical fiber networks. There are several types of optical fiber couplers. Our product line-up includes a polarization beam combiner (PBC), which can deliver a high output power from the pump sources with low loss. Figure 2 shows the structure and function of all-fiber PBCs [6]. The PBCs have been used in optical submarine repeaters because submarine systems require higher reliability than conventional special-structure PBCs.

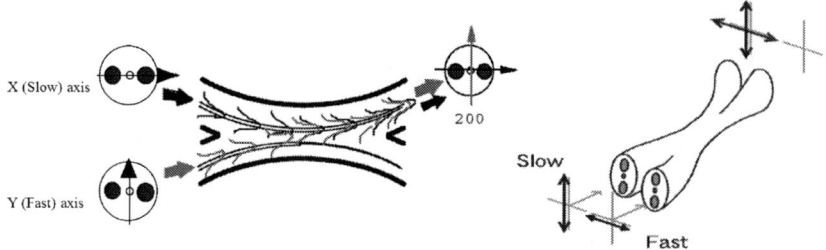

Fig. 2. Structure and function of all-fiber PBCs [6].

IV. BEND-INSENSITIVE FIBER

Figure 3(a) shows the refractive index profile of a standard single-mode fiber (SSMF) used in telecommunications. As the refractive index profile design for the SSMF is considered to influence its macro- and micro-bending characteristics, the fibers are not suitable for situations with extremely small bending, such as in limited spaces in houses. In 2004, Dr. Matsuo of Fujikura Ltd. reported a trench index profile in Fig.3(b). The bend-insensitive fibers with the index profile have excellent bending performance and low splice loss. The bending loss of the fiber is 0.1 dB/turn at 1650 nm for a bending diameter of 20 mm and the mode field diameter (MFD) is 8.1 μm at 1310 nm [7]. The VAD method is also used for fabricating fibers. The demand for bend-insensitive fibers has steadily increased in recent years; therefore, to satisfy this demand, we fabricated fibers with improved characteristics in addition to a large MFD to reduce the splice losses.

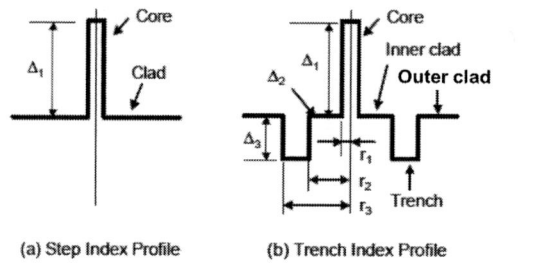

Fig. 3. Refractive index profiles of (a) SSMF and (b) bend-insensitive fiber [7].

V. FIBER BRAGG GRATING

When optical fibers containing Ge-doped SiO_2 are irradiated by ultraviolet rays, the defects in the glass network and the refractive index of the glass increase [8]. Figure 4 shows the fabrication method and basic design of a FBG. Using this design, we fabricated several types of FBG products, which can be assembled to form optical components. For example, FBG-based products are used in the monitoring systems of fiber to the home (FTTH) to reflect particular wavelengths of light in the FBG devices.

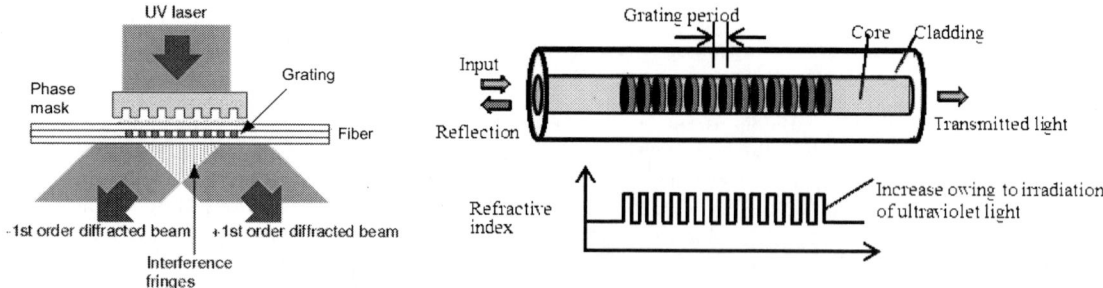

Fig. 4. Fabrication method and basic design of FBG.

VI. FIBER LASER

A fiber laser employs an optical fiber with a core doped with rare earth elements, which is used as the gain medium. In particular, a fiber laser that employs an active fiber doped with Yb can be excited in the infrared region at a wavelength of approximately 1 μm. Such fiber lasers are superior in every aspect, such as output power, beam quality, power and space efficiencies, and reliability, compared to conventional lasers using either a solid crystal or a gas as the gain medium. Figure 5 shows the basic configuration of a fiber laser [9]. High-efficiency Yb-doped fibers have been developed using rare-earth-doped fiber production technologies [10]. Photodarkening-induced excess loss value of high Al-concentration type is reduced to one-fourth that of the low Al-concentration type. The Yb-doped fibers, FBGs, and optical fiber devices are employed in various products.

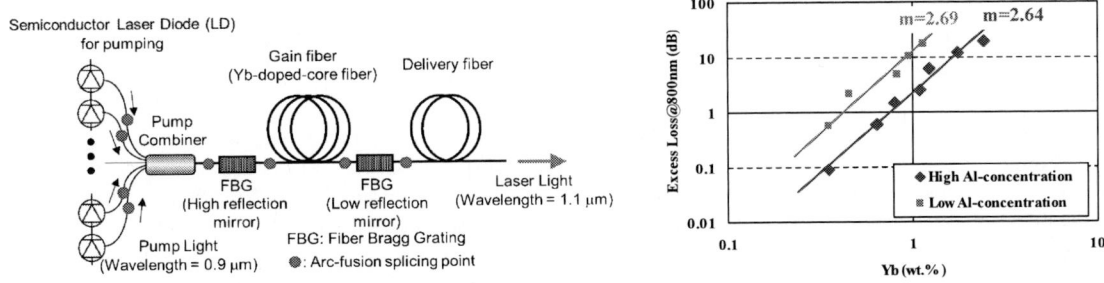

Fig. 5. Basic configuration of a fiber laser and reported examples of improved Yb-doped fibers [9,10].

VII. CONCLUSIONS

Specialty fibers and optical fiber devices fabricated using the VAD method were presented. In collaboration with the advanced VAD method, technologies for specialty fibers and optical fiber devices have undergone significant developments. Both technologies contribute to the development of optical communications network.

REFERENCES

[1] C. K. Kao and G. A. Hockham, "Dielectric fiber surface waveguides for optical frequencies," Proc. of IEE, vol. 113, July 1966, pp. 1151-1158.

[2] M. Horiguchi and H. Osanai, "Spectral losses of low-OH-content optical fibers," Electronics Lett., vol. 12, pp. 310-312, June 1976.

[3] T. Moriyama, O. Fukuda, K. Sanada, K. Inada, T. Edahiro and K. Chida, "Ultimately low OH content V.A.D optical fibers," Electron. Lett., vol. 16, pp. 698-699, August 1980.

[4] T. Hosaka, K. Okamoto, T. Miya, Y. Sasaki and T. Edahiro, "Low-loss single polarisation fibers with asymmetrical strain birefringence," Electron. Lett., vol. 17, pp. 530-531, July 1981.

[5] K. Himeno, M. Sawada, A. Wada, Y. Kikuchi, F. Suzuki and R. Yamauchi, "Polarization-maintaining optical fibers," Fujikura Giho (in Japanese), pp. 1-9, October 1993.

[6] D. Tanaka, H. Sasaki, R. Matsumoto, K. Nishide and R. Yamauchi, "980 nm and 1480 nm pump power doublers for EDFAs by using a novel PANDA fiber type polarization beam combiner," 16th Annual National Fiber Optic Engineers Conference (NFOEC) session A7, Denver (USA), August 2000.

[7] S. Matsuo, M. Ikeda and K. Himeno, "Bend-insensitive and low-splice-loss optical fiber for indoor wiring in FTTH," Proc. of 2004 Optical Fiber Communications Conf. and Exhibition (OFC2004), Los Angeles (USA), March 2004, ThI3.

[8] K. O. Hill, Y. Fujii, D. C. Johnson, and B. S. Kawasaki, "Photosensitivity in optical fiber waveguides: Application to reflection filter fabrication," Appl. Phys. Lett., vol. 32, pp. 647-649, May 1978.

[9] K. Himeno, "Fiber laser and their advanced optical technologies of Fujikura" Fujikura Technical Review, pp. 33-37, March 2013.

[10] T. Kitabayashi, M. Ikeda, M. Nakai, T. Sakai, K. Himeno and K. Ohashi, "Population inversion factor dependence of photodarkening of Yb-doped fibers and its suppression by highly aluminum doping," Proc. of 2006 Optical Fiber Communications Conf. and Exhibition (OFC2006), Anaheim (USA), March 2006, OThC5.

New Optical Fiber Development Based on OVD Technology

Ming-Jun Li

Corning Incorporated, Science & Technology, SP-AR-02-5, Corning NY 14831, USA
lim@corning.com

Abstract: *This paper reviews recent development of optical fibers using the OVD technology. New optical fibers for long haul and data center applications as well as few mode fibers and multicore fibers for SDM are presented.*

Keywords: *OVD, single- mode fiber, multimode fiber, universal fiber, few-mode fiber, multicore fiber*

I. INTRODUCTION

The concept of optical communications can be traced back to the 19 century. In1880, Alexander Graham Bell did the first experiment of using optical wave for a phone conversation [1]. But no progress was made in telecommunications using light for 80 years due to lack of powerful light sources and low loss transmission media. In 1960, laser technology was developed [2] that solved the light source problem, but the problem of lack of a good transmission medium remained, which attracted researchers' attention. In 1966, Charles Kao predicted that loss of silica fiber could be below 20 dB/km [3], which would be practical for optical communications. His prediction encouraged scientists to discover clear glass for low loss fibers. In 1970, Corning's scientists, Keck, Maurer and Shultz made the first low loss fiber of 16 dB/km [4-5], which was a milestone in fiber optic communications. After the demonstration of the first low loss fiber, the fiber loss decreased very quickly. By 1973, the loss of less than 5 dB/km was reported at 850 nm [6], approaching the intrinsic scattering limit at that wavelength. Researchers started to explore longer wavelengths where the intrinsic scattering loss is lower. In 1976, the first fiber with loss of 0.47 dB/km at 1200 nm was reported [7]. Within three years, the fiber attenuation reached 0.2 dB/km at 1550 nm [8]. Today, the state of art of low loss has reached 0.146 dB/km [9]. Corning's first low loss fiber has revolutionized the telecommunications industry, which received the IEEE Milestones Award in 2012 [10].

The loss optical fiber is enabled by the chemical vapor deposition (CVD) process that was invented by Frank Hyde, a Corning scientist in 1942 [11]. This process allows making high purity fused silica glass with metallic impurities of less than a few parts per billion. In the CVD process, vapors of chemicals such as $SiCl_4$ and $GeCl_4$ are oxidized to form glass soot. The soot is collected inside a glass tube in an inside vapor deposition (IVD) process [12] or on a ceramic rod in an outside vapor deposition (OVD) process [13]. The soot is consolidated to form a preform that is drawn to an optical fiber. Both the OVD and IVD processes were originally proposed by Keck and Schultz [12-13]. They fully developed the OVD process [15] while two variations of IVD process were pursued by other companies. One variation of IVD is the modified chemical vapor deposition (MCVD) process [16]. The other variation of IVD is the plasma chemical vapor deposition (PCVD) that uses plasma-activated chemical reactions to form soot particles [17]. Another variation of the OVD process is the vapor axial deposition process (VAD) [18]. In this process, a perform fabrication is accomplished by depositing soot at the end of a silica rod and growing the soot preform along the axial direction. The VAD process received the IEEE Milestones Award last year, which is being celebrated at the 2016 OECC conference. Now the processes of OVD, MCVD, PCVD and VAD are four major process platforms for making optical fibers today.

Corning selected the OVD process due to its process consistency, superior glass quality, ability to scale to large volume, and independence from requiring a third party supply of key glass components. Corning has been continuously improving the OVD process and fiber designs to deliver new fibers for various applications. In this paper, we present recent development in optical fibers based on the OVD technology.

II. OPTICAL FIBERS FOR LONG HAUL TRANSMISSION

With the new development in coherent detection and digital signal processing technologies, transmission impairments from fiber dispersion effects can be compensated digitally. The two most important fiber parameters for high capacity long-haul transmission are now fiber attenuation and effective area.

The total attenuation of an optical fiber is the addition of intrinsic loss factors such as Rayleigh scattering, infrared absorption and ultraviolet absorption, and extrinsic loss factors such as absorption due to transition metals and OH ions, scattering due to waveguide imperfections and loss due to fiber bending effects. The intrinsic loss factors can be eliminated or controlled to minimum levels in the preform and fiber manufacturing processes. For the intrinsic factors, the most important one is the Rayleigh scattering loss that is related to density and concentration fluctuations. By optimizing the glass composition and manufacturing processes, ultra-low-loss fiber with typical attenuation around 0.16 dB/km can be made using the OVD process. The lowest loss that has been reported is 0.146 dB/km [9].

The fiber macro- and micro-bend losses are important factors for fiber attenuation, especially for fibers with large effective areas. To increase the effective area while keeping good bending performance, the core refractive index profile needs to be carefully designed. For micro-bend loss, fiber coating materials are also important. With optimization of both fiber designs and coating materials, a fiber with an effective area of 150 μm^2 has been demonstrated and is now commercially available.

The benefits of ultra-low loss and large effective area in high data rate transmission has been demonstrated experimentally. For 256 Gbit/s PM-16QAM transmission, the increases in transmission reach of the ultra-low-loss fibers with attenuation of 0.16 dB/km and with effective areas of 82, 112, 150 μm^2, were about 32%, 80%, and 115%, respectively, compared to the standard single-mode fiber with an effective area of 82 μm^2 and attenuation of 0.192 dB/km [20].

The effective area can be increased further by increasing the fiber cutoff wavelength beyond the operating wavelength. In this case, the fiber can be used as a quasi-single-mode fiber by launching the light into the fundamental mode. It has been shown that the effective area of LP_{01} mode can be increased to over 200 μm^2 using this approach [21].

III. OPTICAL FIBERS FOR DATACENTERS

Multimode fiber (MMF) has been evolving to meet higher data rate and longer distance demands. High bandwidth MMF requires very accurate profile controls. With process improvements in the OVD technology, OM4 MMF can be produced with effective bandwidth greater than 4700 MHz.km. Also, Corning has been leading in bend-insensitive MMF (BIMMF) that offers the benefits of reduced sizes of cable, hardware and equipment to improve data center operations. The bending loss of the BIMMF is more than 10 times lower than the conventional MMF [22]. For a 15 mm bend diameter, the bending loss is less than 0.2 dB at 850 nm wavelength.

For VCSEL based 850 nm systems, chromatic dispersion becomes a limit beyond OM4 bandwidth. To reduce the chromatic dispersion effects, one solution is to move to longer operating wavelengths. Recently, MMFs optimized at 1060 nm and 1310 nm were developed and used in 25 Gb/s system experiments. System reaches of 300 m and 820 m were demonstrated at 1060 and 1310 nm, respectively [23-24].

Although MMF offers low cost system solutions, the data rate and link length are limited by the modal dispersion. For longer links required in emerging mega-scale data centers, single-mode fiber is used. Recently, we have proposed a new type of fiber, called the universal fiber that can accommodate both multimode transmission at 850 nm and single-mode transmission at 1310 nm [25]. The fiber is designed as MMF with smaller core than conventional MMF but with its mode field diameter of the fundamental mode similar to that of standard single mode fiber at 1310 nm and 1550 nm. A universal fiber was produced and demonstrated 100 m and 50 m transmission using 10Gb/s and 25 Gb/s VCSEL based transceivers at 850 nm, respectively, and 2 km single-mode 25 Gb/s transmission at 1310 nm.

IV. OPTICAL FIBERS FOR SPACE DIVISION MULTIPLEXING

Space division multiplexing (SDM) is a promising technology for future capacity growth. Two approaches for SDM have been proposed: one is to use MMF, and the other one is to use multicore fiber (MCF).

Mode-division multiplexing in MMF requires multiple-input-multiple-output (MIMO) digital signal processing (DSP) to separate the modes due to mode coupling. Conventional MMF with 1% core delta and 50 μm core diameter has more than 100 modes, which requires very high complex DSP. So far, research efforts have been focused on few-mode fiber (FMF) transmission systems. For FMF transmission, it is desirable to minimize differential mode group delay (DMGD) to reduce decoding complexity. FMFs have been designed and produced with graded index profiles. DMGDs as low as 20 ps/km were achieved [26]. A DMGD managed approach has been proposed and has demonstrated DMGD of less than 6 ps/km across the C-band [27].

Long distance FMF transmission has been demonstrated using a Corning developed FMF. With a low DMGD FMF and optimized FM-EDFA, wavelength-and-mode-division multiplexed transmission over a fiber re-circulating loop with a 50-km span was demonstrated over 500 km with 146λ-channels, and over 1,000 km with 16λ-channels [28].

FMF is attractive for datacenter applications if both coherent detection and MIMO DSP are eliminated. An elliptical core FMF (EC-FMF) is proposed to reduce random mode mixing effects. An EC-FMF with ovality 0.4% was developed and achieved crosstalk of less than -26 dB between the LP_{11e} and LP_{11o} modes at 1.3 μm and 1,55 μm. Further demonstrated was the first few-mode SDM transmission with no coherent detection and MIMO DSP [29].

MCF is another potential fiber for SDM. Various MCFs were produced using hexagonal structures with 7 and 14 cores [30], and linear structures with 1x2, 1x4, 2x4 and 2x8 [31-33], and a 1x8 ribbon fiber [34].

High capacity transmission over a Corning MCF has been demonstrated. Using a 14-core hybrid MCF with 12 single mode cores and 2 few mode cores, 1 Pb/s total transmission capacity was achieved [30].

MCF is also attractive for short reach applications to increase the bandwidth-distance capability and to meet the power and density needs in data centers and high performance computer interconnects. Short reach transmission was demonstrated with 8-core and 2-core fibers. For the 8-core fiber transmission, a 1490 nm SiP transceiver array operated at 25 Gb/s 1490 nm was coupled into the MCF [32]. Error free transmission was achieved over 200 m of MCF. Another experiment was done using a 2-core fiber together with fan-outs and connectors. The transmitter was directly modulated at 25 Gb/s with an operating wavelength of 1490 nm [33]. Error free transmission was demonstrated using bi-directional traffic with a small power penalty of <0.5 dB.

REFERENCES

[1] A. G. Bell, "On the Production and Reproduction of Sound by Light", Am. J. Sci., Third Series XX (118), pp. 305–324, 1880.

[2] T. H. Maiman, "Stimulated optical radiation in ruby", Nature 187 (4736), pp 493–494, 1960.

[3] C. K. Kao and G. A. Hockham, "Dielectric-fiber surface waveguides for optical frequencies", Proc. IEE, vol. 133, pp. 1151-1158, 1966.

[4] F. P. Kapron, D. B. Keck, and R. D. Maurer, "Radiation losses in glass optical waveguides", Trunk Telecom. Guided Waves, IEE, pp.148-153, 1970.

[5] F. P. Kapron, D. B. Keck, and R. D. Maurer, "Radiation losses in glass optical waveguides", Appl. Phys. Lett., vol. 17, pp. 423-425, 1970.

[6] D.B. Keck, R.D. Maurer and P.C. Schultz, "On the ultimate lower limit of attenuation in glass optical waveguides", Appl. Phys. Lett., vol. 22, no. 7, pp.307-309, 1973.

[7] M. Horiguchi and H. Osanai, "Spectral losses of low-OH-content optical fibres", Electron. Lett., vol. 12, pp. 310-312, 1976.

[8] T. Miya et al., "Ultimate low-loss single-mode fibers at 1.55 mm", Electron. Lett., vol. 15, pp.106-108, February 15 1979.

[9] S. Makovejs et al., "Record-low (0.1460 dB/km) attenuation ultra-large Aeff optical fiber for submarine applications," OFC 2015, Th5A.2.

[10] http://ethw.org/Milestones:World's_First_Low-Loss_Optical_Fiber_for_Telecommunications,_1970.

[11] J. F. Hyde, "Method of making a transparent article of silica", U.S. Patent 2272342, Feb. 10, 1942.

[12] D. B. Keck and P. C. Schultz, "Method of producing optical waveguide fibers", U.S. Patent 3711262, Jan. 16, 1973.

[13] D. B. Keck, P. C. Schultz P, and F. Zimar, "Method of forming optical waveguide fibers", U.S. Patent 3737292, June 5, 1973.

[14] P. C. Schultz, "Fabrication of optical waveguides by the outside vapor deposition process," Proc. IEEE, vol. 68, no. 10, pp.1187-1190, 1980.

[15] P. C. Schultz, "Fabrication of optical waveguides by the outside vapor deposition process," Proc. IEEE, vol. 68, no. 10, pp.1187-1190, 1980.

[16] J. B. MacChesney, P. B. O'Connor, F. V. DiMarcello, J. R. Simpson, and P. D. Lazay, "Preparation of low loss optical fibers using simultaneous vapor deposition and fusion," in Proc. 10th Int. Congr. Glass, Kyoto, Japan, 1974.

[17] P. Geittner, D. Kuppers, and H. Lydtin, "Low-loss optical fibers prepared by plasma-activated chemical vapor deposition (CVD)", Appl. Phys. Lett., vol. 28, no. 11, pp. 645-646, June 1976.

[18] T. Izawa, S. Sudo, and H. Hanawa, "Continuous fabrication process for high-silica fiber preforms," Trans. IEICE, vol. E62, pp. 779–785, Nov. 1979.

[19] L. Maksimov, A. Anan'ev, V. Bogdanov, T. Markova, V. Rusan1, O. Yanush, "Inhomogeneous structure of inorganic glasses studied by Rayleigh, Mandel'shtam-Brillouin, Raman scattering spectroscopy, and acoustic methods", IOP Conf. Series: Materials Science and Engineering 25 (2011) 012010, doi:10.1088/1757899X/25/1/012010.

[20] J. D. Downie, M.-J. Li, and S. Makovejs, "Optical Fibers for Flexible Networks and Systems", J. Opt. Commun. Netw. 8(7), A1-A11, 2016.

[21] F. Yaman, et al., "First Quasi-Single-Mode Transmission over Transoceanic Distance using Few-mode Fibers", OFC2015, paper Th5C.7, 2015.

[22] M.-J. Li, P. Tandon, D. C. Bookbinder, S. R. Bickham, K. A. Wilbert, J. S. Abbott, and D. A. Nolan, "Designs of Bend-Insensitive Multimode fibers", OFCNFOEC2011, paper JThA3, 2011.

[23] X. Chen et al, "25 Gb/s Transmission over 820m of MMF using a Multimode Launch from an Integrated Silicon Photonics Transceiver", ECOC2013, postdeadline paper pd-4-f-5, 2013.

[24] T. Kise et al., "Development of 1060nm 25 -Gb/s VCSEL and Demonstration of 300m and 500m System Reach using MMFs and Link optimized for 1060nm", OFC2014, paper Th4G.3, 2014.

[25] X. Chen, J. Hurley, J. Stone, J. Downie, I. Roudas, D. Coleman, and M.-J. Li "Universal Fiber for Both Short-reach VCSEL Transmission at 850 nm and Single-mode Transmission at 1310 nm", OFC2016, paper Th4E.4, 2016.

[26] M.-J. Li, E. Ip, and Y. Huang, "Large effective area FMF with low DMGD", Invited talk, IEEE Summer Topicals 2013, paper MC3.4, Waikoloa HI, July 8-10, 2013.

[27] M.-J. Li et al., "Low Delay and Large Effective Area Few-mode Fibers for Mode-division Multiplexing", OECC2012, paper 5C3-2, 2012.

[28] E. Ip, et al., "146λ×6×19-Gbaud Wavelength- and Mode-Division Multiplexed Transmission over 10×50-km Spans of Few-Mode Fiber with a Gain-Equalized Few-Mode EDFA", OFC2013, paper PDP5A.2, 2013.

[29] E. Ip, G. Milione, M.-J. Li, N. Cvijetic, K. Kanonakis, J. Stone, G. Peng, X. Prieto, C. Montero, V. Moreno, and J. Liñares, "SDM transmission of real-time 10GbE traffic using commercial SFP + transceivers over 0.5km elliptical-core few-mode fiber", Optics Express vol. 23, no. 13, pp. 17120-17126, 2015.

[30] D. Qian, et al., "1.05Pb/s Transmission with 109b/s/Hz Spectral Efficiency using Hybrid Single- and Few-Mode Cores", Frontiers in Optics 2012, paper FW6C.3, 2012.

[31] M.-J. Li, B. Hoover, V. N. Nazarov, and D. L. Butler, "Multicore Fiber for Optical Interconnect Applications", OECC2012, paper 5E4-2, 2012.

[32] D. L. Butler, M.-J. Li, S. Li, K. I. Matthews, V. N. Nazarov, A. Koklyushkin, R. L. McCollum, Y. Geng and J. P. Luther, "Multicore optical fiber and connectors for high bandwidth density, short reach optical links", IEEE Photonics Conference 2012, paper MB2, 2012.

[33] Y. Geng, S. Li, R. McCollum, M.-J. Li, R. L. McClure, A. V. Koklyushkin, K. I. Matthews, D. L. Butler, "High-speed bidirectional dual-core fiber transmission system for high density short-reach optical interconnects" (Invited Paper), Photonics West 2015, paper 9390-8, San Francisco, California, February 7-12, 2015.

[34] O. N. Egorova,1, S. L. Semjonov, A. K. Senatorov, M. Y. Salganskii, A. V. Koklyushkin,V. N. Nazarov, A. E. Korolev, D. V. Kuksenkov, Ming-Jun Li, and E. M. Dianov, "Multicore fiber with rectangular cross-section", Opt. Lett., vol. 39, no. 7, pp.2168-2170, 2014.

S4-6 (Invited)

OECC/PS2016

Hollow Core Fibres for Data Transmission

D.J. Richardson, Y. Chen, N.V. Wheeler, J.R. Hayes, T. Bradley, S.R. Sandoghchi
G.T. Jasion, E. Numkam Fokoua, Z. Liu, R. Slavik, M. N. Petrovich, and F. Poletti

Optoelectronics Research Centre, University of Southampton, Highfield, Southampton, SO17 1BJ, UK
djr@orc.soton.ac.uk

Abstract: *We review recent progress in the development of hollow core optical fibres for telecoms/datacoms, highlighting in particular their suitability for low-latency/time-sensitive applications, and their ultimate potential for broadband, ultralow loss (and nonlinearity) transmission.*

Keywords: *Optical fibres; optical fibre communications*

Optical fibre technology plays an ever increasing role in modern society, enabling for example the global internet and communication systems that most of the world now takes for granted, and more recently, providing a new generation of high power laser devices that are finding ever more uses across an increasingly wide range of manufacturing, scientific and medical applications. After more than 50 years of continuous development, and the enormous commercial uptake of fibre technology, it is perhaps surprising that there is much left to discover and improve. However, to the contrary, many would contend that we are currently in a golden period for optical fibre research, with several new forms of fibre under development and many intriguing device and application opportunities emerging.

Arguably one of the most exciting prospects over the past two decades has been the development of hollow core fibre (HCF) technology, of which there are two primary categories: Hollow-Core Photonic Band Gap Fibre (HC-PBGF) and Antiresonant fibre (ARF) which guide light in a fundamentally different way to the majority of fibres used around the world today. Both of these fibre types allow light to be confined over a range of resonant frequencies within a low-index, hollow core surrounded by a glass microstructure. This consequently provides for a number of intriguing optical properties including ultralow optical nonlinearity, excellent power handling capabilities, low latency, and also hold the prospect of ultralow losses, both at conventional wavelengths (e.g. around 1550 nm) and at longer wavelengths (i.e., into the mid-IR) at which conventional solid silica fibres effectively cease to transmit light.

The exact guidance properties of both HC-PBGFs and ARFs are determined by the complex nature of the microstructured cladding – which typically comprises a geometrical arrangement of thin struts and nodes (which form where the struts meet). Until recent times it was largely considered that PBGFs and ARFs themselves relied on fundamentally different guidance mechanisms, however there is now a greater appreciation that the underlying physics is the same and determined by the anti-resonant properties of the individual glass struts and nodes in the fibre cross-section. The guidance properties of HC-PBGFs (which tend to be lower loss and narrower bandwidth) are primarily defined by the properties of the nodes, whilst those of ARFs (which tend to have higher losses and broader bandwidths) are primarily determined by the strut properties. As we shall discuss this improved understanding is now leading to new ARF designs with the potential both for low loss and broad bandwidth. The properties of HCFs are unique relative to conventional solid optical fibres and point to a host of exciting application opportunities, both in terms of existing uses of fibre optics (e.g., in telecommunications and the delivery of high power laser light), and new application areas besides (e.g., gas-based linear/ nonlinear optics and lasers, and particle guidance).

In this talk we will mostly review the current status of the application of HCFs in telecommunications with the primary focus on HC-PBGFs, as to date these have yielded the lowest losses and been studied most for data transmission. We will begin with a brief discussion of how such fibres are fabricated, describing some of the new modelling tools and characterization techniques that have recently been developed to allow the reliable production of fibres with excellent structural characteristics extending over lengths in excess of 10km. We will review our experimental studies that confirm and quantify many of the novel optical properties of these fibres, which include ultralow latency, ultralow nonlinearity, low temperature sensitivity of propagation delay, high resistance to exposure to ionizing radiation and low loss transmission at wavelengths well beyond the transparency window of silica. We will then discuss the best results to date in terms of usable bandwidth and loss, and review the state-of-the- art in terms of transmission experiments - both at conventional C-band wavelengths and at wavelengths around 2000nm (the minimum predicted loss window for this fibre type). Finally, we will present our very latest results on the fabrication of ARFs capable of supporting single-mode transmission over more than an octave of bandwidth and hence allowing data transmission at key wavelengths around 1000nm, 1550nm and 2000nm through the same fibre. We conclude the talk with a discussion of the relative merits and prospects for both HCPBGFs and ARFs in various telecoms/datacoms applications.

Optical Fibers
for Future Communication Systems

Kazuhide NAKAJIMA

Access Network Service Systems Laboratories, NTT Corporation

nakajima.kazuhide@lab.ntt.co.jp

Abstract: This paper reviews the progress made in optical fiber technology mainly since 2000. Two key terms, namely "low loss" and "space division multiplexing" are picked out and their relationship with fabrication techniques is considered.

Keywords: Optical fiber, Attenuation coefficient, Space division multiplexing, Fabrication technique.

I. INTRODUCTION

The IEEE milestone-award-winning vapor-phase axial deposition (VAD) technique [1] has enabled the mass-production of low-loss and high-quality single-mode preforms, and has strongly supported the worldwide spread of optical fiber communication. Conventional single-mode fiber (SMF), with a germanium-doped core and a zero dispersion wavelength in the 1310 nm region, has been widely adopted in access, metro and long-haul optical transmission systems. SMF appears to be a universal optical fiber for today and for the future, however, its available transmission bandwidth is limited to around 100 Tbit/s. Therefore, it is important to enhance the optical signal-to-noise ratio (OSNR) characteristic by improving the quality of optical fiber and cable. It would also be advantageous to obtain an additional transmission window by introducing a new design concept.

In this paper, I review recent progress on optical fiber technology for future optical transmission systems. This paper mainly focuses on work undertaken since 2000, and considers two key terms *"Low Loss Optical Fibers"* and *"Optical Fibers for Space Division Multiplexing (SDM)"*. Finally, I will note that key technical innovations in harmony with fabrication technology is essential if we are to realize future optical communication links that can be used for tens of years into the future.

II. OPTICAL FIBER TECHNOLOGY AFTER 2000

A. Progress on Low-Loss Optical Fibers

Figure 1 (a) shows the historical improvements made in the attenuation coefficients of various optical fibers. The black circles show the optical loss reduction in conventional SMFs. After Kapron achieved 20 dB/km in 1970 [2], the attenuation coefficient of SMF was greatly reduced thanks to intensive worldwide research using multiple fabrication techniques such as chemical vapor deposition (CVD) [2], modified-CVD (MCVD), and VAD [1]. In 1979, Miya realized an attenuation coefficient of 0.2 dB/km [3] around 1550 nm, which is comparable to the attenuation coefficient of current SMF. This historical progress shows that research on various fabrication techniques has played an important role with respect to the more than one hundredfold optical loss reduction achieved over a period of 10 years. The blue circles show the optical losses of photonic crystal fibers (PCFs). PCF is expected to be a low-loss transmission medium since it can be composed of a single material by using a number of air holes periodically arranged in the fiber cross-section. Moreover, PCF enables single-mode operation in an arbitrary wavelength region. Thus, it is a candidate as a

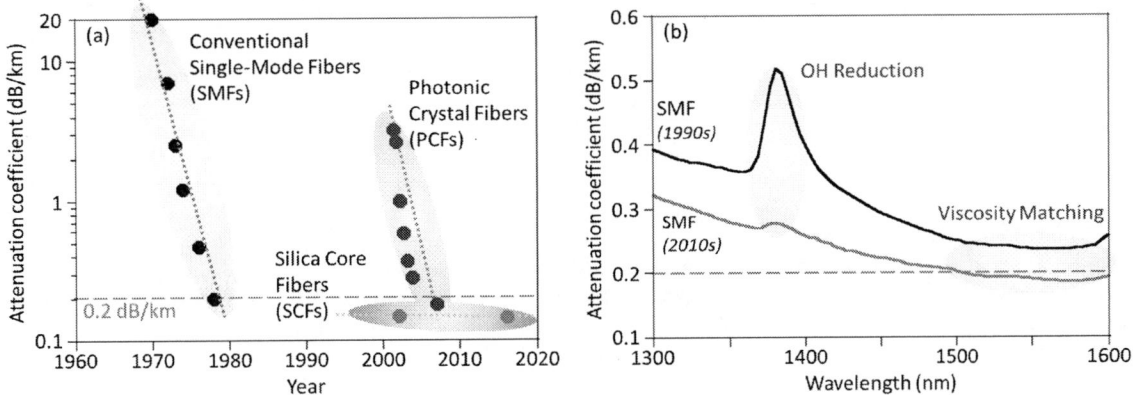

Fig. 1. Optical loss properties of various optical fibers.
(a) History of attenuation coefficient reduction, (b) Example wavelength dependence of attenuation coefficient.

low-loss and wideband transmission medium. After Russell succeeded in fabricating a PCF [4] in the late 1990s, the attenuation coefficient of silica based PCF improved rapidly, and Tajima realized the lowest reported attenuation coefficient of less than 0.2 dB/km in 2007 [5]. It is interesting to note that intensive research successfully reduced the optical loss more than one hundredfold within one decade as achieved with conventional SMFs in the 1970s. Here, it should be noted that PCF research has involved two innovative fabrication technologies, namely the stack & draw and rod in tube techniques. Although the technical principles behind these fabrication techniques were established before 2000, the huge loss reduction in PCF could not have been achieved without the steady progress made on these fabrication technologies. The red circles show the attenuation coefficients of silica core fibers (SCFs). Recent progress on digital coherent technology has required optical fiber with a lower attenuation coefficient and lower nonlinearity rather than a tailored dispersion property. With this as the background, Corning broke the lowest loss record with a value of 0.1460 dB/km [6] after an interval of fourteen years. Moreover, Sumitomo revealed the mass-producibility of low-loss SCF [7]. These achievements have promoted new standardization activities in relation to optical fiber technology. Traditionally, SCF has been used specifically in submarine optical links to realize ultra-long-haul transmission systems. ITU-T is now planning to establish SCF for terrestrial links as a potential way of realizing a future digital coherent system with a bit rate of 100 Gbit/s or beyond.

Here, we cannot disregard the steady historical progress in fabrication technologies when we look at the state of the art of low-loss optical fiber. Figure 1 (b) shows example optical loss spectra measured with two conventional SMFs with a germanium-doped core. The black and red lines show SMFs fabricated in the 1990s and 2010s, respectively. This figure clearly shows the impact of technological progress on the attenuation coefficient over the last twenty years. I believe that two key technologies, namely OH reduction [8] and viscosity matching [9], have played important roles in terms of realizing the historical optical loss improvement, despite the principle of these two technologies being already established in the 1980s. Although the loss reduction in SMF including SCF over the last twenty years is not as large as that achieved in the 1970s, a 0.01 dB/km improvement in the attenuation coefficient provides a 1 dB OSNR increase in a 100 km link. When we look at the possibility of realizing optical fiber with lower loss, we find that Tsujikawa predicted that the intrinsic loss of silica fiber can be reduced to 0.13 dB/km or less by optimizing the fictive temperature [10]. He also pointed out that an intrinsic loss of less than 0.10 dB/km might be feasible by using phosphorus-doped silica glass. Therefore, I believe that these historical researches and discussion show the potential for realizing ultra-transparent optical fiber for future optical links if we continue steady work on both optical fiber and fabrication technologies.

B. Optical Fibers for SDM

Recently, the future "*capacity crunch*" faced by SMF has been remarked upon, and this has encouraged us to look for a new class of optical fiber. SDM transmission is one of the solutions being considered as a way of overcoming the "*capacity crunch*", and intensive research has been conducted over the past ten years. Generally speaking, optical fiber can use two spatial channels, "*core*" and "*mode*". Figure 2 (a) shows the historical progress made on the number of spatial channels obtained with various types of SDM fibers. The black, red and blue circles show data for multi-mode fiber (MMF), multi-core fiber (MCF) and MM-MCF, respectively. It can be seen from Fig. 2 (a) that the number of spatial channels obtained with MMF and MCF has been gradually increasing in the 2010s. Sillard realized a 9LP-mode (15-spatial channel) MMF, and revealed that a 50-μm core MMF offers 36 spatial channels [11]. Mizuno revealed the applicability of a 32-core fiber to a 1600 km transmission [12]. However, it appears to be difficult to obtain 50 or more spatial channels using "*mode*" or "*core*". Figure 2 (a) also shows that the number of spatial channels can be effectively increased by using the "*mode*" and "*core*" multiplicity with MM-MCF. Recently, three research groups achieved MM-MCFs with more than 100 spatial channels. Sakamoto succeeded in packing 114 spatial channels into a cladding diameter of less than 250 μm [13].

Figure 2 (b) shows the numerical relationship between the relative core multiplicity factor (RCMF) and cladding diameter [13]. Here, the RCMF corresponds to the effective core density including all modes and cores, and it is normalized with the CMF value of conventional SMF. The red and blue symbols represent 3- and 6-mode cores, respectively. We optimized each refractive index profile so that the effective area of the LP01 mode becomes 80 μm^2 at 1550 nm with a satisfactory low macrobending loss in the highest transmission mode. The schematic cross-sections in Fig. 2 (b) show the core arrangements, and the number in brackets shows the number of spatial channels as "*mode*" × "*core*". The core pitch was determined so that the crosstalk with all neighboring cores was -30 dB/100km or less. We revealed that the RCMF can be effectively increased by introducing a 6-mode core in place of the 3-mode core. Moreover, it is found that a 114-spatial channel can be obtained by using 19 hexagonally arranged cores while keeping the cladding diameter at less than 250 μm. In general, a thicker cladding diameter greatly degrades the long-term reliability of optical fiber. However, we can expect to realize a satisfactory reliability level comparable to that of conventional SMF by setting the maximum cladding diameter at 250 μm and introducing a 2% proof test during the fabrication process. Thus, it can be said that a 114-channel MM-MCF with a cladding diameter of less than 250 μm shows that these SDM fibers have the potential to be used in actual optical terrestrial networks.

The green circles in Fig. 2 (a) show the number of spatial channels reported using a coupled-MCF. If we want to realize a long-haul transmission using mode-multiplexing technology, it is very important to manage the differential

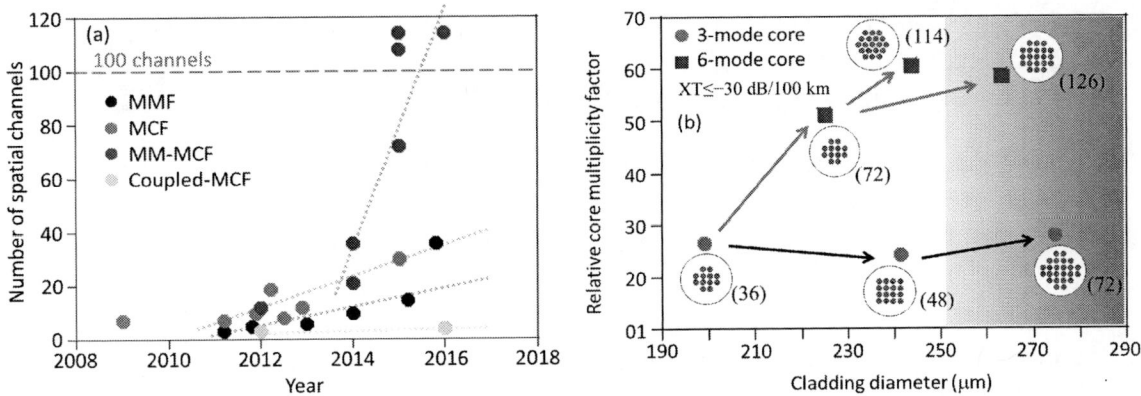

Fig. 2. Recent research on optical fibers for SDM transmission.
(a) Historical progress on number of spatial channels, (b) Example relationship between relative core multiplicity factor and cladding diameter of few-mode multicore fibers.

mode group delay (DMGD) so that it is as small as possible since the DMGD directly increases multi-input multi-output (MIMO) complexity. Here, coupled-MCF can be considered a medium for mode multiplexing long-haul transmission because the coupling between cores potentially reduces the DMGD. Moreover, the coupled MCF intrinsically increases the core multiplicity in a unit area by reducing the core pitch. Sakamoto revealed that spatial mode dispersion (SMD) can be reduced effectively by optimizing the relationship between core pitch and twisting rate along the fiber length [14]. Hayashi realized a four-coupled-core fiber with a low loss of less than 0.158 dB/km and with an ultra-low SDM of less than 6.1 ps/√km by using silica core technology [15]. These results imply the possibility of realizing mode-multiplexing long-haul transmission with lower MIMO complexity using coupled-MCF with low DMGD although the current number of spatial channels is fewer than ten.

Here, it should be noted that all this notable research on SDM fibers has been undertaken based on the steady background provided by the fabrication technologies. In fact, SDM fiber research is not so novel, and various MMFs and MCFs have been investigated since the 1970s. The latest refractive index profile control technology directly improves the modal transmission property in MMFs. Stack & draw and rod in tube technologies, which have been greatly improved as a result of the PCF research performed in the 2000s, support and accelerate MCF research. However, these fabrication technologies may also have limitations in terms of preform size and cross-sectional complexity. Intrinsically, the maximum fiber length drawn from a constant size preform decreases as the cladding diameter increases with MCF. Moreover, we have to prepare multiple rods with different diameters if we want to use the stack & draw technique for MCF with a non-hexagonal core arrangement. Recently, Nagashima proposed a modified rod in tube technique for handling larger volume MCF preforms [16]. Arai and Fukumoto investigated a new fabrication technique that enables one-step cladding construction with an arbitrary core arrangement [17], [18]. Cook also tried to fabricate a holey fiber preform using a 3D-printer [19]. With this challenging research providing the background, I hope we will realize an SDM based optical communication link as a result of continuous efforts to develop both fiber and fabrication technologies.

III. CONCLUSIONS

This paper reviewed optical fiber technology for future communication systems. I focused on two key features, namely "*Low Loss*" and "*SDM*", and provided an overview of example milestone research conducted after 2000. This technical review revealed that a new class of fiber compatible with key fabrication technologies will enable us to realize a future optical communication link that will be used for tens of years.

REFERENCES

[1] T. Izawa et al., IEICE Trans., **E62**, p. 779, 1979.
[2] F. P. Kapron et al., Appl. Phys., Lett., **17**, p. 423, 1970.
[3] T. Miya et al., Electron. Lett., **15**, p. 106, 1979.
[4] T. A. Birks et al., Opt. Lett., **22**, p. 961, 1997.
[5] K. Tajima, in Proc. ECOC'07, PD2.1, 2007.
[6] S. Makovejs et al., in Proc. OFC'15, Th5A.2, 2016.
[7] M. Hirano et al., in Proc. OFC'13, PDP5A.7, 2013.
[8] F. Hanawa et al., Electron. Lett., **16**, p. 699, 1980.
[9] M. Ohashi et al., Photon. Technol. Lett., **5**, p 812, 1993.
[10] K. Tsujikawa et al., in Proc. OFC'04, WI5, 2004.
[11] P. Sillard et al., in Proc. ECOC'15, Mo.4.1.2, 2015.
[12] T. Mizuno et al., in Proc. OFC'16, Th5C.3, 2016.
[13] T. Sakamoto et al., in Proc. OFC'16, Th5A.2, 2016.
[14] T. Sakamoto et al., in Proc. OFC'16, W1F.5, 2016.
[15] T. Hayashi et al., in Proc. OFC'16, Th5A.1, 2016.
[16] T. Nagashima et al., IEICE Soc. Conf., B10-14, 2016.
[17] S. Arai et al., IEICE Tech. Rep., **OFT2015-73**, 2016.
[18] R. Fukumoto et al., IEICE Tech. Rep., **OCS2014-37**, 2014.
[19] K. Cook et al., Opt. Lett., **40**, p. 3966, 2015.

ThA1-2 (Invited)

OECC/PS2016

Optical Networking And Node Technologies For Creating Cost Effective Bandwidth Abundant Networks

Ken-ichi Sato

Department of Electrical Engineering and Computer Science, Nagoya University
Furo-cho, Chikusa-ku, Nagoya, Aichi, 464-8603, Japan
sato@nuee.nagoya-u.ac.jp

Abstract: We present a novel networking concept and a control algorithm that attains link and node cost reduction simultaneously. The proposal can be implemented using existing hardware technologies, OXCs/ROADMs that utilize LCOS WSSs, and an enhanced control algorithm, Grouped Routing scheme, that is aware of the subsystem modular OXC configuration. Numerical experiments demonstrate that the spectral efficiency is improved by more than 20% while the number of WSSs needed can be reduced by about 80%.

Keywords: OXC, ROADM, grouped-routing, waveband, subsystem-modular OXC

I. INTRODUCTION

Internet traffic is continuously increasing in the world, and the recent yearly increase rate was about 50% in Japan [1]. Expanding the optical link bandwidth and node throughput are perpetual goals to be sought. Here, one notable point in network development is that metro traffic, which remains within the metro area, surpassed core network traffic in 2014, and is expected to grow nearly twice as fast as core network traffic [2]. This is spurred by datacenter development in metro areas and the advancement of contents delivery networks. Optical networking technologies are now mostly applied to metro-core networks, but in the future they will penetrate deeper into metro networks to support metro access. The requirements of core and metro networking are different. First, we briefly discuss the different requirements and highlight some important challenges to be resolved. Next we present a viable solution that makes use of grouped routing technologies and a scalable OXC/ROADM architecture, which makes the network highly expandable in a very cost effective manner.

II. REQUIREMENTS OF METRO NETWORK

The requirements of metro networks are different from core networks in many aspects, some of which are summarized in Table 1. First of all, the larger traffic requires multiple fibers between adjacent nodes which demands much larger port count OXCs/ROADMs. Transmission distances between nodes are smaller than those in core networks, and hence the OXCs must be much more cost-effective than the core network equivalents. As a result, OXCs to be applied to metro networks must be scalable and very cost-effective. The second point is the increased number of nodes traversed (see Fig. 1). For example, when considering a ring-based configuration and two fiber ring networks (at present 16 nodes per ring is the typical maximum) are connected optically, which is done electrically in present networks, the number of optically traversed nodes can be more than 30 considering optical path protection. This increased number of nodes imposes severe requirements on OXC performance, since the spectrum narrowing caused by traversing many WSSs incurs severe OSNR penalties [5,6]. These requirements, high-scalability (pay-as-you-grow capability), cost-effectiveness, and high-performance must be met simultaneously, which has been hard and so far hindered wide OXC deployment in metro networks.

TABLE I COMPARISONS OF OXCs FOR METRO AND CORE NETWORKS

	Core Network	Metro Network
Transmission Distance	Large	Small
Acceptable Node Cost	Larger	Smaller
Number of Nodes Traversed	Small	Large
Add/drop Ratio	Large	Small
Fiber Degree	Small	Small to Large (Large Deviation)

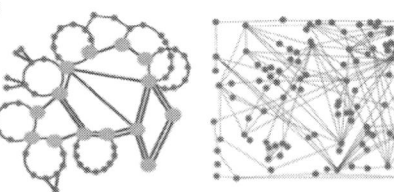

Figure 1. Examples of metro networks [3,4].

III. SUBSYSTEM MODULAR OXC ARCHITECTURE

The number of fibers connected to a node (fiber degree) increases as traffic grows, while the number of adjacent nodes (node degree) does not change if the physical network topology does not change. Exploiting the difference between the node degree and the fiber degree, we have already clarified that setting an appropriate restriction on the routing performance of the OXC express switch can well maintain the available network performance (blocking ratios or fiber utilization) when we adopt intra-node blocking aware Routing and Wavelength Assignment (RWA), while the necessary OXC hardware scale can be substantially reduced [7,8]. Among the various approaches available, the subsystem modular OXC architecture (Fig. 2) [7] that utilizes multiple small-scale OXCs has been proved to be very effective. The architecture retains the broadcast-&-select approach and yields large fiber degree nodes with significantly reduced hardware scale. Detailed assessments [9] have confirmed the effectiveness of newly developed add/drop architectures [10] and a network design algorithm with protection [11]. Although the above architectures introduce a marginal offset on routing capability, they can be used in a similar way to conventional contentionless OXCs without concern for the differences between node degree and fiber degree; only the network design/control algorithm need to be modified. Figure 2 compares conventional and subsystem modular OXCs. With the subsystem modular architecture, the hardware cost can be substantially reduced since it requires much fewer and smaller degree WSSs. The number of interconnection fibers used to connect WSSs/optical couplers is significantly reduced as well, which greatly simplifies OXC hardware complexity. The subsystem modular architecture also allows graceful modular growth, since adding subsystems is easy. With the conventional architecture, the number of WSSs traversed per node ranges from 2 to 4 in this example (Fig. 3), while that for the proposed architecture ranges from 1 to s (# of subsystems, in this example s=5) since an optical path may traverse multiple subsystems in a node. However, it is shown that the traverse of multiple subsystems in a node can be effectively suppressed (more than 80% of optical paths traverse only one WSS in the node) and hence the proposed subsystem architecture greatly reduces the maximum number of traversed WSSs (see Fig. 3) and the average total optical loss at nodes compared to the conventional architecture. All the benefits shown in Fig. 3 are achieved at the cost of a marginal path accommodation efficiency offset (~1%, Fig. 3).

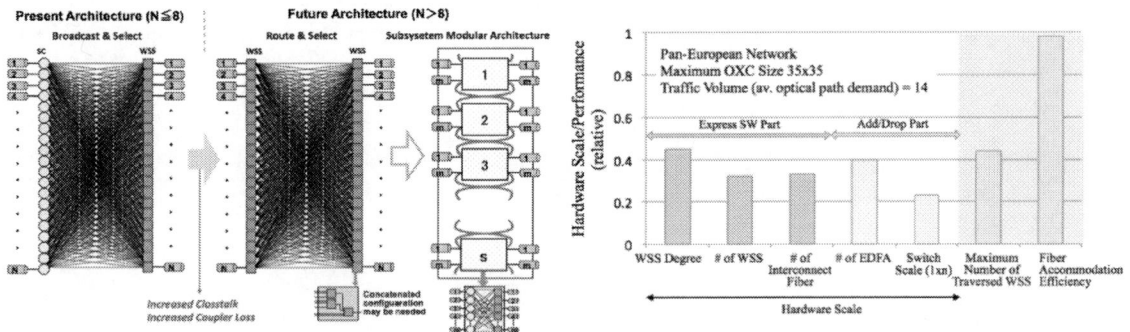

Fig. 2. Comparisons of present and subsystem modular OXC architectures.

Fig. 3. Effectiveness of the subsystem modular OXC.

IV. GROUPED ROUTING NETWORK

In Grouped Routing networks [12,13], the wavelengths in a fiber are divided into several wavelength groups called "Grouped Routing Entities" (GREs). Routing is done at the GRE granularity level while add/drop operations at a node are processed at the wavelength granularity level as shown in Fig. 4. GRE granularity routing defines virtual pipes (GRE pipes); wavelength paths can be added/dropped at any intermediate node on the route of a GRE pipe (See Fig. 4). GRE pipes differ from waveband paths in conventional hierarchical optical path networks since GRE pipes do not offer path functions such as termination as defined by ITU-T [14]. This is a key difference from the conventional waveband "path". It may then form a closed loop as seen in Fig. 4, and a GRE pipe can be regarded as a virtual fiber. Furthermore, GRE pipes differ from "super-channels" in that sub-carrier signals can not be dropped(/added) from each super-channel along the route from source to destination node.

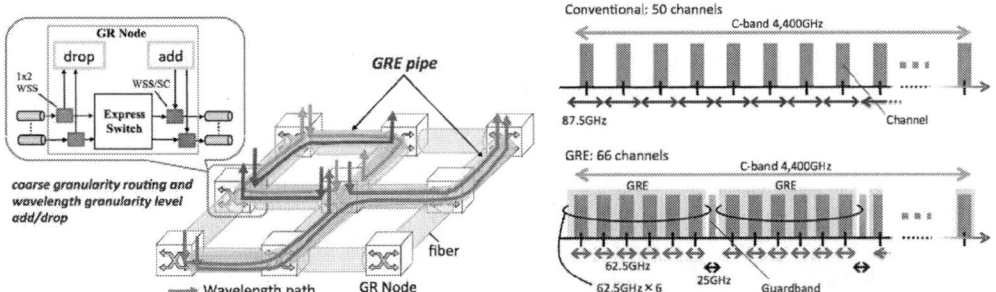

Fig. 4. Grouped Routing network.

Fig. 5. Frequency allocation example (400 Gbps DP 16QAM signals)

This grouped routing substantially mitigates spectrum narrowing at nodes, since channel by channel optical filtering at WSS is removed and hence channels in a GRE can be as densely packed as possible and an appropriate guard band is inserted only between GREs (see Fig. 5). Drop operation at the OXC imposes a filtering impairment on adjacent paths in a GRE, and hence a new network design and control algorithm that can limit the frequency of adjacent path drop operations needs to be developed [15]. Figure 5 depicts a possible frequency allocation when transmitting 400 Gbps polarization-multiplexed 16QAM signals and limiting the worst number of adjacent paths dropped to 3 [15]. Fig. 6 shows the performance of the proposed strategy for four different network topologies [15]. In this graph, label T represents forecasted traffic volumes (average optical paths between a node pair). Accommodated traffic volumes increase compared to the conventional network by more than 20% in all topologies. The results for larger topologies are better since there are more fibers when T is the same. One of the beauties of the GRE scheme is that it can be implemented with present LCOS based WSSs; no hardware changes are required, only the control is changed.

V. NETWORK CONTROL THAT SUBSTANTIALLY REDUCES LINK AND NODE COSTS SIMULTANEOUSLY

Combining the subsystem modular OXC and Grouped Routing scheme enhances the reduction in network cost. A network control algorithm that simultaneously considers both schemes while maximally exploiting their benefits in combination has been developed [16]. Numerical experiments demonstrate that the proposed network with subsystem modular OXCs and the GR scheme achieves 22% improvement in fiber capacity and 80% reduction in the necessary number of WSSs, simultaneously, compared to conventional networks. The normalized number of necessary fibers are shown in Fig. 7 for the GR network and GR plus subsystem modular OXC network. The results are normalized by those of the conventional network. The effectiveness of the Grouped Routing network is enhanced as traffic volume increases because paths are more effectively routed as groups. The subsystem modular OXC marginally increases fibers needed (almost +1%). Our proposal (GR+Subsystem) achieves good performance for the tested network topologies, especially when traffic volume is large. When the average number of path demands is set at 20, the necessary fibers can be reduced by 21.9% for the 5x5 poly-grid network and 20.5% for the USnet network. In Fig. 7 the total number of 1x9 WSSs needed is also shown. The number increases with the traffic, however, the increase is much stronger in conventional networks. The proposed network reduced the number of necessary WSSs by 81.7% for the 5x5 poly-grid network and 79.9% for the USnet when average path demand was 20.

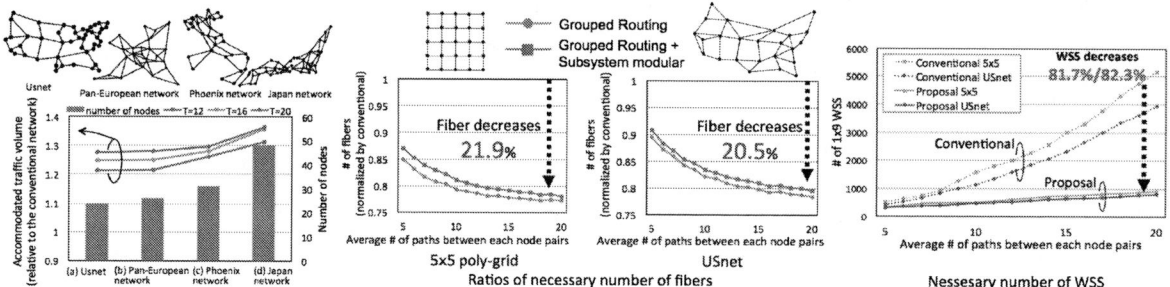

Fig. 6. Performance of the Grouped Routing networks   Fig. 7. Performance of the proposed networks compared to present configuration.

VI. CONCLUSIONS

The technologies presented here yield substantial link and node cost reductions simultaneously, and will greatly expand optical networking capabilities: a lot of fibers can be accommodated within an OXC and the number of OXCs that can be traversed is increased. Thus the applicable extent of optical networking can be greatly expanded and the traffic volume supported by the network is also magnified. Please note that the networks can be created using present hardware technologies; OXCs that utilize current LCOS WSSs. Only an enhanced control algorithm, the Grouped Routing scheme that is aware of subsystem modular OXC configuration, is needed.

ACKNOWLEDGMENT: This work was partly supported by NICT and KAKENHI (26220905).

REFERENCES

[1] Ministry of Internet Affairs and Communications, http://www.soumu.go.jp/main_content/000402062.pdf
[2] Cisco Visual Networking Index: Forecast and Methodology, 2014-2019 White Paper.
[3] G. Wellbrock, OFC/NFOEC 2013 Market Watch Panel Sessions, Panel IV: Metro 100G Applications, March 21, 2013.
[4] R. Younce, et al., IEEE/OSA JOCN, Vol. 5, Issue 10, pp. A267-A273, 2013.
[5] A. Morea et al., OFC 2014, Th1E.4.
[6] M. Filer et al., OFC 2014, Th1I.2.
[7] Y. Iwai et al., Proc. ACP 2012, ATh2D.2, China, 2012.
[8] K. Sato, Proc. ONDM 2014, S7-1, Stockholm, May 2014.
[9] Y. Tanaka et al., Optics Express Vol. 23, Issue 5, pp. 5994-6006, 2015.
[10] H. Ishida et al., IEEE/OSA JOCN, Vol.7, Issue 6, pp. 586–595, 2015.
[11] K. Sato et al., Proc. OECC, M01A-3, Melbourne, July 2014.
[12] Y. Taniguchi et al., OFC/NFOEC 2012, JW2A.2, 2012.
[13] Y. Taniguti et al., J. Opt. Commun. Netw. 5(7), 774-783, 2013.
[14] ITU-T Recommendation G.783, Digital transmission systems.
[15] Y. Terada et al., OFC 2015, Th1I.6, 2015.
[16] Y. Terada et al., OFC 2016, W2A.48, 2016.

OECC/PS2016

Reconfigurable WDM Multicast Supporting Content Delivery for Content Delivery Network Based on SOA and TB-WSS

Ze Li, Min Zhang*, Danshi Wang, Yue Cui

State Key Laboratory of Information Photonics and Optical Communications,
Beijing University of Posts and Telecommunications, Beijing 100876, China
E-mail: mzhang@bupt.edu.cn

Abstract: *We propose a reconfigurable WDM multicast scheme supporting content delivery from origin servers to cache servers in CDN based on SOA and TB-WSS. One-to-six/seven/eight 10 Gb/s QPSK of reconfigurable WDM multicasts have been successfully demonstrated.*

Keywords: *Multicast; wavelength division multiplexed; Content Delivery Network; four-wave mixing; semiconductor optical amplifier.*

I. INTRODUCTION

The growing number of IPTV and multimedia content over the Internet has become an important motivation of improving Internet performance in recent years, as a solution to which, content delivery networks (CDNs) are systems used to improve accessibility and reliability by replicating data objects and caching them at multiple locations in the network as shown in Fig. 1 (a) [1]. The emergence of CDNs allows the content be replicated to the servers close to the clients, which results in fast and reliable Internet services [2]. With the success of commercial CDN applications, such as Akamai [3] and Limelight [3], CDNs have attracted more and more attention from both the industry and academic communities. Multicast function is crucial to content delivery. How to effectively realize content multicasting from the origin servers to cache servers is a key technique of CDN. Currently, content multicast from the origin servers to cache servers is supported only in the IP layer which suffers the power consumption and cost growth [4]. From physical realization, the capability of multicast directly in the optical domain is a target for energy saving, fast service provisioning, and high network resource efficiency. Due to the greater multicast bandwidth for big data sharing, the natural evolution is toward optical WDM multicast, which can efficiently deliver a stream of information from one input wavelength to multiple wavelengths, to avoid the wavelength conflict and spectrum resource overlap.

In this paper, we propose a reconfigurable WDM multicast scheme with large bandwidth supporting content delivery from origin servers to cache servers in CDN based on semiconductor optical amplifier (SOA) and tunable bandwidth wavelength selective switch (TB-WSS). With 15 GHz minimum bandwidth and 6.25 GHz bandwidth setting resolution, the TB-WSS can respond to the multicast services with different bandwidth granularity and different modulation wavelength. Through a multicast module (MM) based on four-wave mixing (FWM) in SOA, it is flexible to select the downlink wavelength for multicasting and multicast to multiple wavelengths. It also plays a role of physical backup in case of the big data migration or the network disaster which providing reliability and emergency protection for Internet. To conceptually prove the scheme, one-to-six/seven/eight 10 Gb/s QPSK WDM multicasts have been successfully demonstrated and one-to-more WDM multicasts are also possible with different pump mechanism.

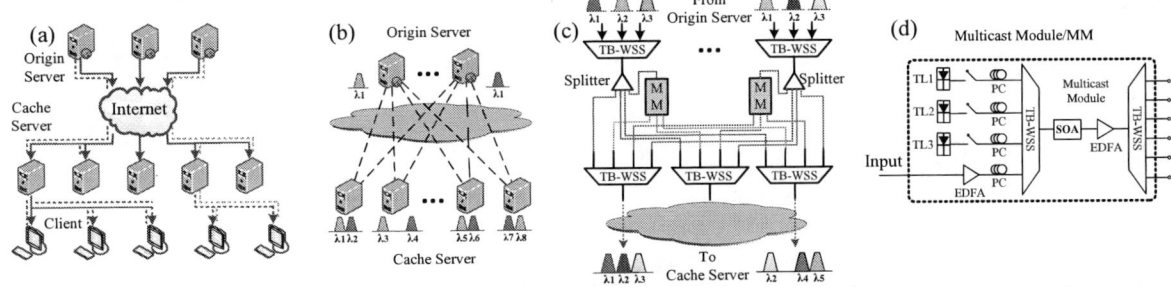

Fig. 1. (a) conceptual structure of CDN; (b) content delivery of multi service from multi source servers (c) principle and structure of proposed reconfigurable WDM multicast supporting content delivery for CDN; (d)the structure of multicast module.

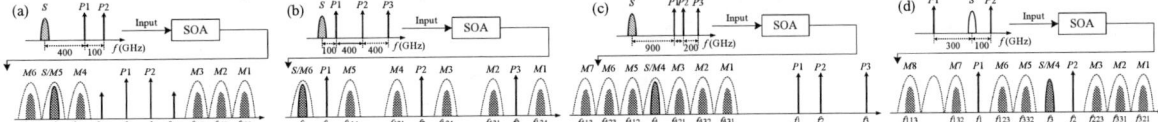

Fig. 2. Schematic diagram of reconfigurable WDM multicast operation based on FWM in SOA.

233

II. PROPOSED SCHEME AND PRINCIPLE

The network scenario of the optical multicast of content delivery in CDN are depicted in Fig. 1 (b), where different color blocks represent different multicast services delivered from origin servers to cache servers. A service from an origin server at wavelength $\lambda 1$ needs to be multicast simultaneously to cache servers and some service connections may be failed due to that wavelength $\lambda 1$ is unavailable in some links between origin server and cache servers. What's more, if another origin server multicast another service at wavelength $\lambda 1$, such two multicast services will also be blocked because of the wavelength conflict. It need WDM multicast which deliver a stream of information from wavelength $\lambda 1$ to other wavelengths in this case. Therefore, the optical multicast of content delivery in CDN should not only have the power splitter when the spectrum resource is sufficient, but also be equipped with WDM multicast module (MM) when the wavelength conflicts along the route.

The structure and principle of proposed reconfigurable WDM multicast supporting content delivery for CDN and the structure of MM are depicted in Fig. 1 (c) and (d). All the downlink services from an origin server multiplexed together through a TB-WSS on the origin server side, while the corresponding services transmitted to the same cache server also multiplex through a TB-WSS on the cache server side. The output signal of TB-WSS from origin server is split into all the input of TB-WSS which is located on the origin server side and the input of MM. With this way, the cache server can share the services from origin server but to be selected the services which need to be transmitted by the TB-WSS located on the cache server side. Through MM, we can achieve reconfigurable WDM multicast and solve the problem of wavelength conflict. In the MM, a switch is used to initiate the multicast function depending on demand and an SOA is utilized to generate the multicast signals via the FWM effect. Flexibly selected by a tunable optical bandpass filter (OBPF), the downstream signal which needs multicast is firstly amplified by an erbium-doped fiber amplifier (EDFA) and then coupled into the SOA with one or more CW pump lights. The polarization states of the signal and pumps are adjusted through three polarization controllers (PCs). The number of newly generated idle light via FWM is related to the power and polarized states of the signal light and pumps [5]. Meanwhile, due to the phase matching conditions and the rule of energy conservation [5], the wavelengths of pumps should be properly adjusted to make sure that the FWM-yielded signals contain the information of original signal and locate in the desired multicast wavelength channels. The viable options are shown in Fig. 2. The input signal S, pump lights $P1$ and $P2$ interact in SOA and after FWM, five, six or seven idlers which contain the information of original signal are generated. Therefore, together with the original signal, 1-to-6, 1-to-7, 1-to-8 WDM multicast is achieved. Amplified by an EDFA, the output of the SOA is selected by a TB-WSS to get the desired multicast signals. The output of MM is connected to the input of TB-WSS located on the cache server side to transmit the corresponding multicast signal.

The proposed reconfigurable WDM multicast scheme with large bandwidth, fast service provisioning and high resource efficiency is effectively support content delivery for CDN and avoid the wavelength conflict and spectrum resource overlap. It also plays a role of physical backup in case of the big data migration or the network disaster which providing reliability and emergency protection for Internet.

III. EXPERIMENT AND RESULTS

To prove of our scheme, we design the experiment setup depicted in Fig.3. An amplified spontaneous emission (ASE) laser and an optical spectrum analyzer (OSA) are used to test the characteristic of TB-WSS. The testing characteristics of TB-WSS are shown in Fig.3. With 15 GHz minimum bandwidth and 6.25 GHz bandwidth setting resolution, the TB-WSS can respond to the multicast services with different bandwidth granularity and adapt to flex-grid networks.

As shown in Fig. 3 (c), the wavelength from tunable laser (TL3) is fed into an IQ modulator driven by a repeated pseudo-random binary sequence with the length of 2^{15}-1 at 10 Gb/s to generate the QPSK signal for multicasting. Pumps are coupled with the amplified downstream signal. All the wavelengths are assigned with a 100 GHz bandwidth

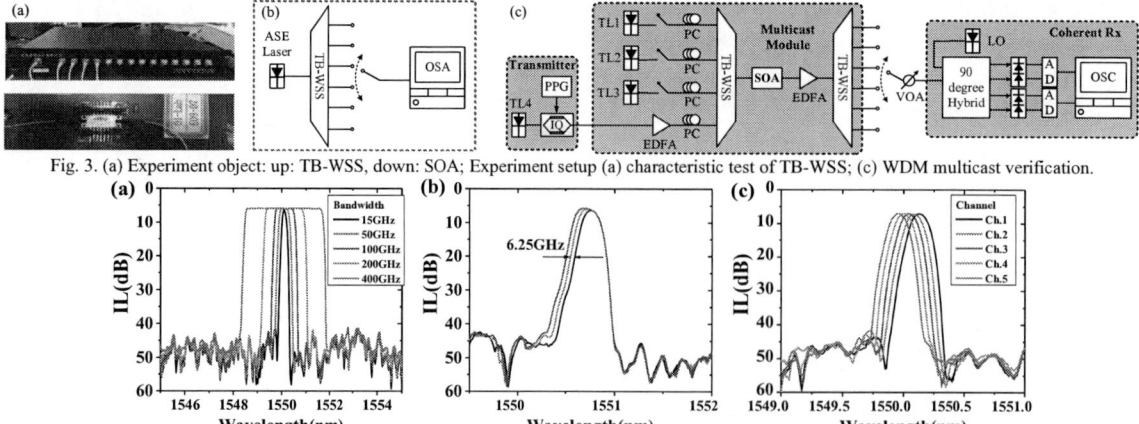

Fig. 3. (a) Experiment object: up: TB-WSS, down: SOA; Experiment setup (a) characteristic test of TB-WSS; (c) WDM multicast verification.

Fig. 4. TB-WSS: (a) Function of tunable bandwidth; (b) Measured bandwidth setting resolution; (C) variant center wavelength with minimum bandwidth.

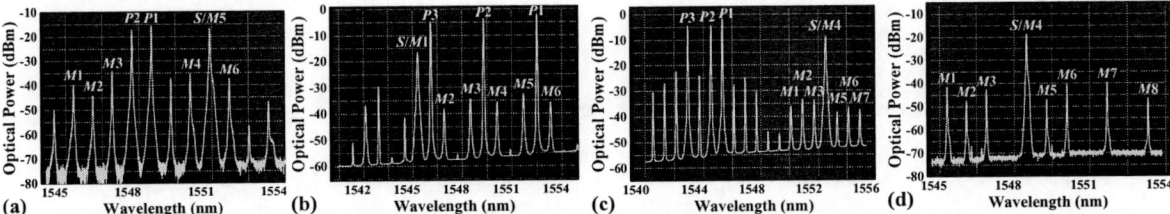

Fig. 5. Optical spectrums (a) one to six of QPSK; (b) another one to six of QPSK; (c) one to seven of QPSK; (d) one to eight of QPSK.

which observed the ITU-T standards. The combined signals are injected into an SOA with a bias current of 300 mA. The selected signal is detected by a coherent receiver and sent to the digital oscilloscope (OSC) with the sample rate of 80 GSa/s. The bit-error rate (BER) is got by off-line signal processing.

For instance, the optical spectrums of 1-6/1-7/1-8 multicast measured at the output of SOA are shown in Fig. 5. In these cases, all idlers containing the information of original signal are generated via FWM, which is achieved as predicted in the principle analysis above. In the case of 1-6/1-8, the characteristics of the converted signals, such as wavelength, conversion efficiency (CE), optical signal-to-noise rate (OSNR) and power penalty (PP), are summarized in Table 1. The BER performance of 1-6 and 1-8 multicasts under BTB condition are shown in Fig. 6. All the multicasts exhibit better BER performance than FEC threshold of 3.8×10^{-3}. Therefore, it proves the feasibility of proposed scheme.

TABLE I

TABLE 1. SUMMARY OF WAVELENGTH, CONVERSION EFFICIENCY, OSNR AND POWER PENALTY OF THE MULTICAST SIGNAL.

One to Six Multicasts					One to Eight Multicasts				
Signal	Wavelength (nm)	CE (dB)	OSNR (dB)	PP (dBm)	Signal	Wavelength (nm)	CE (dB)	OSNR (dB)	PP (dBm)
M1	1544.52	-32.73	24.53	2.32	M1	1545.32	-23.22	31.13	0.28
M2	1545.32	-22.33	34.66	0.71	M2	1546.12	-29.56	24.24	1.45
M3	1546.12	-26.74	30.25	1.89	M3	1546.92	-24.6	28.92	0.93
M4	1550.12	-18.49	35.93	0.44	M4	1548.52	/	52.42	/
M5	1550.92	/	53.61	/	M5	1549.32	-29.01	23.03	1.59
M6	1551.72	-20.04	33.57	1.32	M6	1550.12	-21.84	29.83	0.69
/	/	/	/	/	M7	1551.72	-21.56	28.76	1.08
/	/	/	/	/	M8	1553.32	-28.52	20.97	2.08

Fig. 6. (a) the constellation and I/Q eye-diagrams of modulated QPSK; BER curves: (b) one to six multicasts; (c)one to eight multicasts.

IV. CONCLUSIONS

A reconfigurable WDM multicast scheme with large bandwidth supporting content delivery from origin servers to cache servers in CDN has been proposed and demonstrated. With 15 GHz minimum bandwidth and 6.25 GHz bandwidth setting resolution, the TB-WSS can respond to the multicast services with different bandwidth granularity. Through a multicast module based on FWM in SOA, it is flexible to select the downlink wavelength for multicasting and multicast to multiple wavelengths. It also plays a role of physical backup in case of the big data migration or the network disaster which providing reliability and emergency protection for Internet. One-to-six/seven/eight 10 Gb/s QPSK WDM multicasts have been successfully demonstrated. All of the signals exhibit better BER performance than FEC threshold of 3.8×10^{-3} and one-to-more WDM multicasts are also possible with different pump mechanism.

ACKNOWLEDGMENT

This study is supported by NSFC Project No.61372119 and Doctoral Scientific Fund of MOE China No.20120005110010.

REFERENCES

[1] A. Vakali et al. "Content delivery networks: Status and trends," IEEE Computer Magazine, 7(6), Pages 68-74, 2003.
[2] J. Dilley et al. "Globally distributed content delivery," IEEE Internet Computing, 13(7), 50-58, 2002.
[3] Yang, Chenkai, et al. "Replica placement in content delivery networks with stochastic demands and M/M/1 servers," Performance Computing and Communications Conference 2014, 1-8, 2014.
[4] D. Wang et al. "Multifunctional switching unit for add/drop, wavelength conversion, format conversion, and WDM multicast based on bidirectional LCoS and SOA-loop architecture," Optics Express, 22(18), 21847-21858, 2014.
[5] A. E. Willner et al., "All-optical signal processing," Journal of Lightwave Technology, 32(4), 660-680-755, 2014.

ThA2-1 (Invited)

OECC/PS2016

Metro-embedded Cloud Platform with All-optical Interconnections for Virtual Datacenter Provisioning

Hongxiang Guo[1], Gang Chen[1], Dongxu Zhang[1], Xiaoyuan Cao[2], Jian Wu[1], Takehiro Tsuritani[2]

[1]Beijing University of Posts and Telecommunications, No.10 Xitucheng Road, Haidian District, Beijing 100876, China
[2]KDDI R&D Laboratories Inc, 2-1-15 Ohara, Fujimino-shi, Saitama 356-8502, Japan
Email: hxguo@bupt.edu.cn

Abstract: *We propose a novel cloud platform based on multiple metro-distributed micro datacenters (mDCs) with intra-mDC optical burst switching ring and inter-mDC burst over wavelength switched optical network interconnections. Dynamic virtual datacenter provisioning was experimentally demonstrated.*

Keywords: *metro-embedded datacenter, VDC, OBS, SDN.*

I. INTRODUCTION

To match the rapid development of cloud applications, more and more servers are consolidated into a few large-scale datacenters (DCs) as a unified cloud platform. This centralized hosting model is expected to achieve better economies of scale, but DCs may also suffer from some limitations such as over-provisioned resources, less flexibility, poor accessibility and high energy consumption. In this context, there is a realistic trend [1, 2] that DC resources are preferred to be more geographically distributed and closer to users in order to deliver reliable and elastic services with lower access latency. Meanwhile, virtual datacenter (VDC) is intensively addressed as one of the most important service models on future cloud platforms, which facilitates efficient, elastic and fast infrastructure provisioning in a "pay-as-you-go" manner that both benefits user experience and reduces service provider's cost [3].

In this paper, we introduce a metro-embedded cloud platform, in which conventional centralized resource clusters are broken down to small pieces, referred to as micro DCs (mDCs), spreading across metro areas. To enable data communications with high bandwidth, low latency and high energy efficiency among all hosted resources, advanced all-optical interconnection technologies including wavelength switched optical network (WSON) as well as fine-grained optical burst switching (OBS) are employed. Moreover, OpenFlow-based SDN [4] & OpenStack [5] technologies are used to enable elastic VDC provisioning and intelligent resource optimization.

II. PROPOSED METRO-EMBEDDED CLOUD PLATFORM

As shown in Fig. 1, the principle of metro-embedded cloud platform is to disperse smaller-scaled resource pools (i.e. mDCs) within the metropolitan areas which can be optically interconnected and easily accessed by end users. Specifically, each mDC employs a broadcast-and-select based OBS ring [6], with a dedicated ring controller (RC) to synchronize all burst switching nodes and distribute dynamic bandwidth allocation instructions to them. So a logical

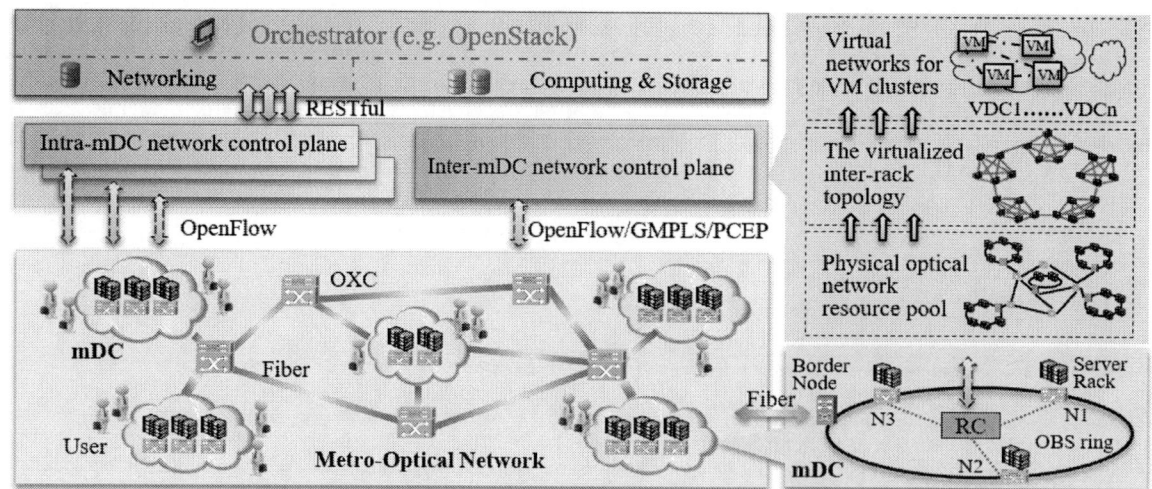

Fig. 1 Architecture of metro-embedded cloud platform with all-optical interconnections

full-mesh network capable of adaptive bandwidth adjustment among server racks can be achieved. As for the inter-mDC interconnections, a virtualized network slice is provided by the metro optical network with transparent wavelength switching capability, and then the OBS-over-WSON paradigm [7] can be applied. To holistically organize the distributed mDCs and network resources, a hierarchical software-defined networking (SDN) control plane is adopted. Note that a centralized solution (e.g. OpenFlow) is suitable for the intra-mDC network as well as the virtualized inter-mDC slice, but the physical metro optical network can still employ distributed approaches (e.g. GMPLS) if required. All network controllers have so-called northbound APIs (e.g. RESTful [8]), through which an orchestrator (e.g. OpenStack) is at the top to integrate distributed computing/storage/networking resources together and provide differentiated VDC services to various clients.

The merits of such a platform design are manifold. First, all-optical sub-wavelength switching guarantees flexible data communications with high efficiency, low latency and low energy consumption both within and between mDCs. Second, the resources are geographically distributed, so not only the reliability is improved significantly, but also miscellaneous (e.g. energy-minimized or location-based) resource optimization strategies can be applied to maximize the resource utilization and improve user experience. At last, the overall architecture is highly modularized, enabling easy incremental expansion.

III. EXPERIMENTAL DEMONSTRATION

A prototype platform with three mDCs (mDC_a~mDC_c) interconnected by a metro WSON network is implemented as showed in Fig. 2. In each mDC, several server racks with different numbers of vCPUs are deployed, and accordingly an OBS ring consisting of one RC and several switching nodes is used to interconnect these computing resources. Specifically, mDC_a has three switching nodes which are developed based on Xilinx FPGAs with extended 10 Gbps burst mode transceivers and crystal optical switches with 300 ns switching speed, while mDC_b and mDC_c each has one switching node capable of inter-mDC optical burst communication. The metro WSON network is constructed by using four commercial optical cross connects (OXCs). In the control plane, RYU [9] and POX [10] based controllers are used for OBS ring and metro networks respectively. Through RESTful APIs, an orchestrator based on OpenStack is able to obtain network layer resource information and program the whole cloud platform with JSON messages.

In the demonstration, we supposed that a large amount of data mainly located in mDC_a needs to be analyzed, so a VDC request (phase I) for three virtual machines (VM1~3) with 2.5 Gb/s interconnection links (corresponding to burst length of 40 us) was generated. Upon this request, the orchestrator immediately sent out JSON messages (GET) through RESTful APIs to obtain the latest status of networking as well as computing resources and then attempted to embed VM1 and VM2 into mDC_a and VM3 into mDC_b with consideration of data locality as well as robustness. After that, the interconnection among VMs were accordingly established by configuring (POST) two intra_mDC rings and an inter-mDC lightpath via OXC_1~3. The control plane signaling is shown in the upper part of Fig. 3 (Procedure ①~⑤).

Then, in order to improve data processing capability, we initiated a VDC modification request (phase II) attempting to add a new VM4 into mDC_a. Note that we also changed the instance type of VM3 from "small" to "large" in order to process the extra data stream from VM4 in mDC_a. Similarly, the orchestrator would check current resource pools, and obviously it would find out that mDC_a had enough computing resource to host VM4 while the enlarged VM3 needed to be migrated from mDC_b to mDC_c. Thus, as shown in the middle part of Fig. 3, the orchestrator informed both intra- and inter- mDC networks (POST) to firstly set up a connection between mDC_b and mDC_c for VM3 migration. After that, in order to support the newly added VM4 in N_a1, the connection bandwidth between N_a1 and N_a2 was doubled (the burst length was changed from 40 us to 80 us) and also a new connection between mDC_a and mDC_c was established (the burst length for N_a1 to N_c1 and N_a2 to N_c1 were 80 us and 40 us respectively).

Finally, we evaluated the service recovery operation on the initially generated VDC in phase I. A disruption of VDC service was emulated by manually introducing a fiber cut between OXC_1 and OXC_3. After the metro-network

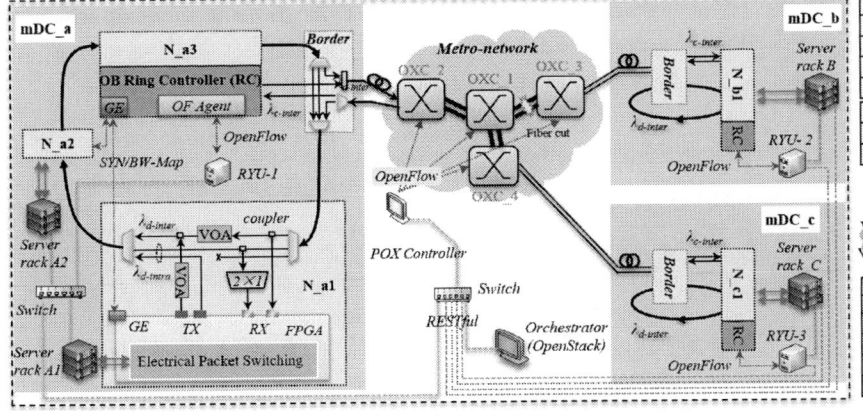

Fig. 2 Experiment testbed

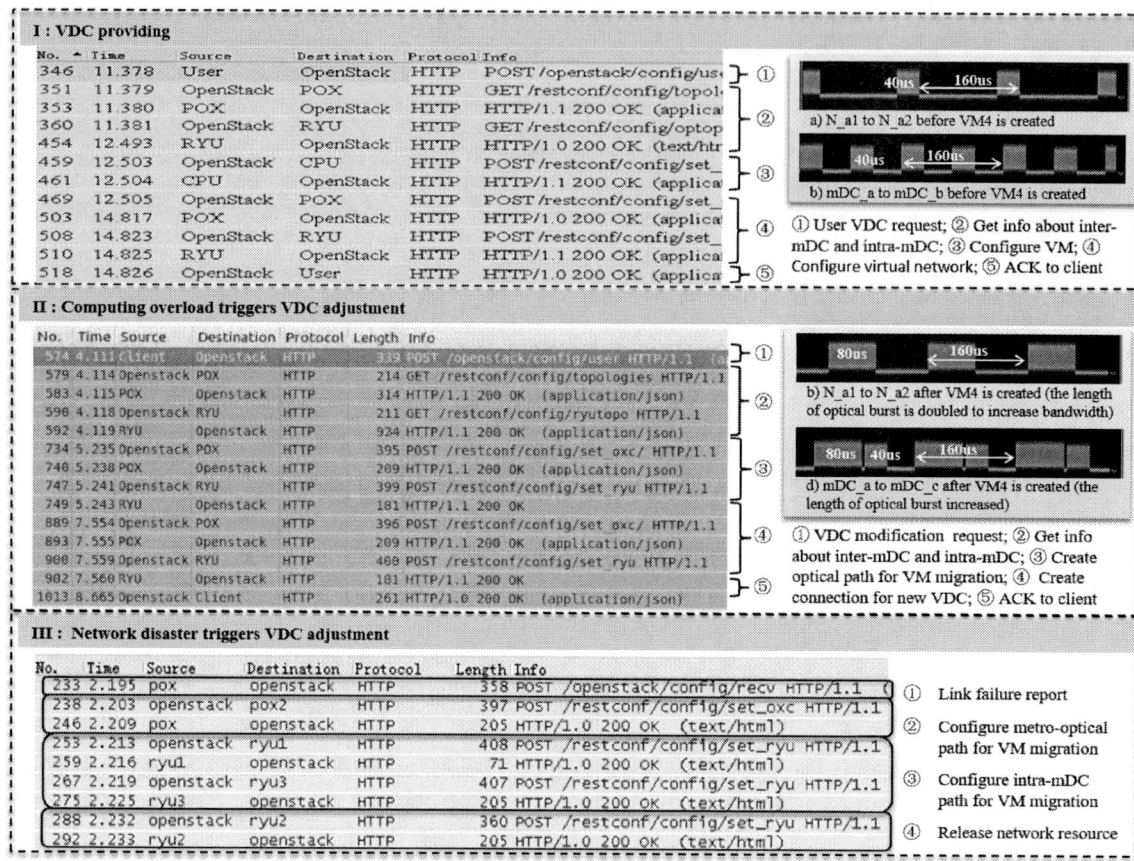

Fig. 3 Experiment results

controller (POX) detected this link failure, it would firstly try a network layer recovery. However, as no detour path against the failure point was available, the POX controller would choose to send a message (containing the fault type and location) to the orchestrator in expectation of upper layer recovery. Then the orchestrator would search for idle computing resources, namely mDC_c in this experimental demonstration, to take over the disrupted VDC service. As a result, a new connection between mDC_a and mDC_c was configured by using JSON messages (POST) through RESTful APIs, and then the backup image of VM3 was migrated from mDC_a to mDC_c to restore the VDC service, instead if there was no backup mechanism, a new VM with the same instance type of VM3 could be created. At last, the orchestrator also sent a message to release the previously occupied networking and computing resources in mDC_b.

IV. CONCLUSIONS

We have presented a novel cloud platform where resource clusters are embedded into metropolitan areas by using all-optical interconnection networks including OBS ring and WSON, and a centralized SDN based control plane is used to holistically organize computing, storage and networking resources. Dynamic VDC provisioning with high reliability on this platform was successfully demonstrated.

ACKNOWLEDGMENT

This work was partly supported by NSFC program (Grant No. 61331008, 61471054), SRFDP program (No. 20130005110013) and the opening fund of State Key Laboratory of Computer Architecture.

REFERENCES

[1] "Data Center 2025: Exploring the Possibilities", available online: http://www.emersonnetworkpower.com/en-US/Latest-Thinking/Data-Center-2025/Pages/default.aspx

[2] G. Chen, et al. Proc. OECC, PDP2C.3 (2015).

[3] E. S. Correa, et al. IEEE Latin America Transactions, vol. 13, no. 5, pp. 1661-1670 May 2015

[4] SDN, https://www.opennetworking.org/index.php

[5] Openstack, http://www.openstack.org/

[6] N. Farrington, et al. Proc. OFC, OW3H.3 (2013).

[7] D. Zhang, et al. Proc. ONDM, Wed2.4 (2012).

[8] RESTful API, http://www.restapitutorial.com/

[9] RYU, http://osrg.github.io/ryu/

[10] POX, https://github.com/noxrepo/pox.

ThA2-2

OECC/PS2016

Flexible and Cost-Effective Optical Metro Network with Photonic-Sub-Lambda Aggregation Capability

Masahiro Nakagawa, Kana Masumoto, Kyota Hattori, Toshiya Matsuda, Masaru Katayama, and Katsutoshi Koda

NTT Network Service Systems Laboratories, NTT Corporation, 3-9-11, Midori-cho, Musashino-shi, Tokyo, 180-8585 Japan
nakagawa.masahiro@lab.ntt.co.jp

Abstract: This paper presents a cost-effective optical metro network called "Photonic Sub-Lambda transport network" suitable for low-loaded rural areas, and presents the results of network cost evaluation and experimental demonstration of photonic-sub-lambda aggregation.

Keywords: Metro networks, Optical burst transport, Photonic-sub-lambda aggregation

I. INTRODUCTION

New network services including 4K/8K video and various cloud services are increasing traffic demands and driving the need for more flexible network infrastructure. Indeed, metro networks are currently facing dramatic changes in traffic characteristics such as the peak-to-average ratio [1, 2]. Metro networks connect access and core networks, in which connectivity patterns and/or bandwidth demand dynamically change [3]. Moreover, metro networks must cost-effectively meet the requirements for various residential and business services. In addition, another important attribute of metro networks to be considered is that traffic characteristics strongly depend on the country, region or area; there are large differences in the amount of traffic in densely populated urban areas and in sparsely populated rural areas [4, 5]. Therefore, for network architecture to be suitable for these different areas, it must enable overall network cost reductions. Although several flexible optical metro network architectures have been proposed [6-9], the cost of high-speed optical switches and/or coherent transceivers (TRXs) is considerable in rural areas. This paper proposes transport network suitable for rural areas called a "Photonic Sub-Lambda transport network" (PSL network). This network enables flexible resource utilization without the need to resort to traditional electronic switching or high-end optical devices. Moreover, we present the results of fundamental evaluation of network cost and discuss the effectiveness of the PSL network. We also present a prototype demonstration that verifies the feasibility of the PSL network.

II. PHOTONIC SUB-LAMBDA TRANSPORT NETWORK (PSL NETWORK)

The outline of the PSL network is shown in Fig. 1. Here, a single fiber ring is depicted for simplicity, although a two fiber ring is usually deployed in practical solutions. In Fig. 1, a particular node (core node) is connected to the core network, while the other nodes (access nodes) are connected to access networks. Every node has optical burst adaptors (OBAs) and optical passive devices such as couplers. Wavelength filters and AWGs can also be used. In addition, optical repeaters (REPs) are installed in some of the access nodes for metro-scale transmission. A schematic of an OBA is shown in Fig. 2, where the traffic is encapsulated in an optical burst and transmitted with particular timing at particular wavelength(s). The timing and wavelengths are controlled adaptively by a controller at the core node. More specifically, the controller periodically gathers network information from each OBA, and allocates sub-lambda granularity bandwidth (burst length and/or the number of bursts) to each path between OBA pairs according to network conditions. Then, each OBA operates in accordance with the allocation, which is periodically notified from the

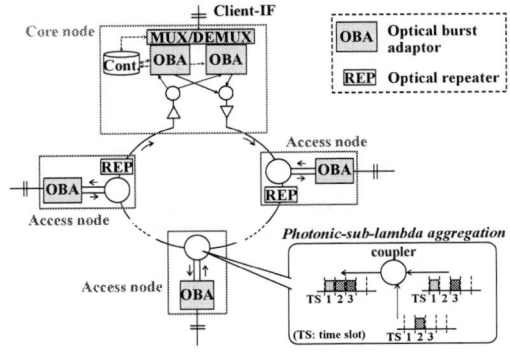

Fig. 1. Outline of a single-ring PSL network.

Fig. 2. Schematic of the optical burst adaptor.

controller through the OBA at the core node. Furthermore, optical burst transmission is managed so that burst collisions in the optical domain can be avoided, which achieves the *photonic-sub-lambda aggregation*. Photonic-sub-lambda aggregation is a promising technology that enables multiple bursts/paths to be multiplexed in the TDM manner in the optical domain (as shown in Fig. 1). This approach differs from the traditional electronic-domain aggregation and grooming approaches that rely on header processing and electronic switching. Therefore, network resources such as wavelength channels and TRXs can be shared by many paths while drastically reducing O/E/O conversions and electronic functions. Indeed, each OBA in each node can be shared by many paths, and the number of OBAs required at each node can be flexibly determined to suit network conditions such as traffic volume. Thus, compared to conventional wavelength-routed networks, a large numbers of TRXs and wavelength channels can be saved especially in low-loaded rural areas, which leads to a reduction of network costs.

Optical burst transmission is already a mature technology and has been widely deployed in several PON systems in access networks. In addition, NG-PON2 technology [10] makes optical burst transmissions with a WDM feasible for practical use. Therefore, some commodity PON devices such as TRX and MAC/PHY LSI can be utilized in our network, hence enabling much lower network costs. In such cases, to meet metro network requirements, REPs are key for economical design for rural metro networks (described in next section).

III. NETWORK COST CASE STUDY

We evaluated network costs to clarify the benefits of the PSL network. In this paper, we apply a discrete component approach proposed in [11], which is suitable for evaluation of rural areas since traffic volume in such areas may be quite small compared to the overall transmission capacity, and the "cost per bit" metric is not suitable. Therefore, just as in [11], the total network cost can be calculated by the summation of each component cost multiplied by the required number of components under the given traffic conditions. Following, we present a model to evaluate network costs and our assumptions, and then we discuss the advantages of our PSL network.

We assumed a 9-node bidirectional ring network in which one node (the core node) is connected to the core network and the other nodes (access nodes) are connected to access networks. Specifically, the nodes are interconnected by two counter-rotating fiber rings, where one of these fiber rings is assumed to be used for protection. Hence, in this evaluation, we consider cost per single fiber ring for transporting data. Each fiber link can support W wavelength, while the bandwidth of each wavelength is divided into time intervals known as time slots (TSs) on the PSL network. Each wavelength has a 10-Gbps capacity. Accommodated traffic flows between each core and access node pair, and the volume of each flow is uniformly distributed as A [Gbps]. In addition, we compared ROADM-based wavelength-routed network and packet transport network (PTN) employing MPLS-TP, both of which are now commercially used in metro networks. Simplified ROADM and PTN node architectures are shown in Fig. 3. Table I summarizes the relative costs of components in each node/network, which is based on [12, 13]. Please note that OBAs and REPs in the PSL network are assumed to be implemented by using PON devices/components.

In this study, the required number of transponders/TRXs is a key metric for evaluating the total network cost. Indeed, in the ROADM network, the required number of transponders only depends on the number of nodes when $0 \leq A \leq 10$. On the other hand, in the PTN, the required numbers of 10G line card and 10G TRXs depend not only on the number of nodes but also on the traffic volume to be accommodated because traffic grooming in the electronic domain can be used. In the PSL network, same as PTN, the required number of OBAs depends on the traffic volume because of the photonic-sub-lambda aggregation capability. Please note that the effect of FEC (e.g., RS (255, 223)) must be considered when using PON devices. Another important parameter for the PSL network is the number of REPs, although each access node is assumed to be equipped with a REP for simplicity. The calculated total network costs of the architectures are shown in Fig. 4. The results reveal that the PSL network can reduce network costs by 45-60% compared to the ROADM network. This comparison verifies that the photonic-sub-lambda aggregation, i.e., flexible optical resource utilization and resource sharing, and the application of PON devices are effective in rural-area metro networks. In addition, compared to PTN, the PSL network can attain lower costs except when $A \leq 1.25$, because relatively expensive REPs (in this paper, REPs are assumed to support optical burst with WDM technology, same as [13]) installed in each access node lead to higher network costs in very-low-loaded areas. Therefore, low-cost and low-complexity REPs and optimal placement algorithms are required to increase the benefits of the PSL network.

IV. EXPERIMENTAL DEMONSTRATION OF PHOTONIC-SUB-LAMBDA AGGREGATION IN A RING NETWORK

A primitive experiment was conducted to demonstrate the feasibility of the photonic-sub-lambda aggregation using the PSL network prototype. The network configuration is depicted in Fig. 5 and a photograph of the prototype system is shown in Fig. 6. Please note that tunable wavelength filter was equipped at each access node OBA while fixed-tuned wavelength filter was equipped at each core node OBA. In nodes #2 and #3, the Ethernet frame received from tester #2 was encapsulated and transmitted to node #1 as a 1532.68 nm-wavelength optical burst. In node #2, optical bursts from OBAs in nodes #2 and #3 were aggregated in the optical domain without any electrical processing or buffering. At node #1, optical burst signals from two nodes were received by one burst-mode TRX. An example of the received waveform is shown in Fig. 7. The received optical bursts were decapsulated and sent to tester #1. We verified that no optical burst collision was measured, and predictable performance was achieved in terms of latency and jitter. Therefore, the feasibility of the photonic-sub-lambda aggregation was successfully demonstrated. The PSL network can efficiently accommodate 10GE interfaces and various services in rural-metro areas where the actual traffic volume may be relatively small compared to wavelength bandwidth by providing optical paths of sub-lambda granularity bandwidth.

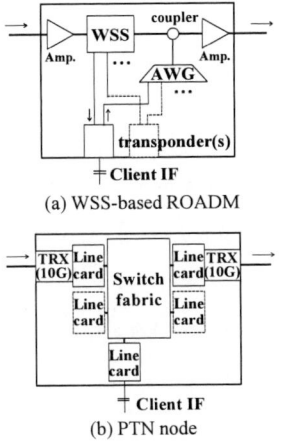

(a) WSS-based ROADM

(b) PTN node

Fig. 3. Comparative architectures.

TABLE I
RELATIVE COMPONENT COSTS

Devices / Components		Relative cost
Common	Optical coupler	0.6
	Amplifier	15
	AWG (1 : N)	$0.3 \times N$
ROADM	10G transponder	18.75
	WSS (1×4)	37.5
	WSS (1×9)	75
PTN	Switch fabric	1.45 /10G
	10G line card w/ 10G TRX	9.84
	1G×10 line card	1.87
	1G TRX	0.37
PSL network	core node	OBA (10G colored)	7.6
		MUX/DEMUX	1 /10G
	access node	OBA (10G tunable)	3.1
		REP	40

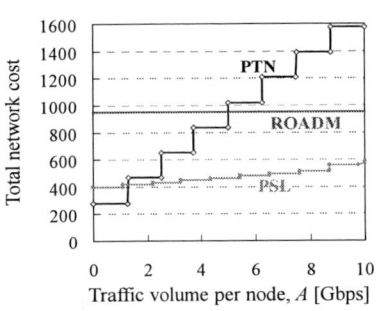

Fig. 4. Cost comparison of the three architectures.

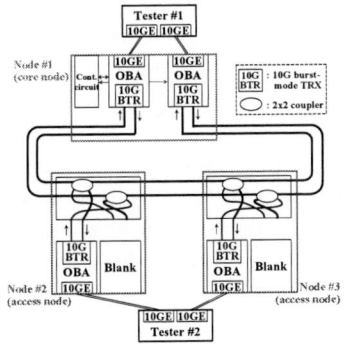

Fig. 5. Experimental setup.

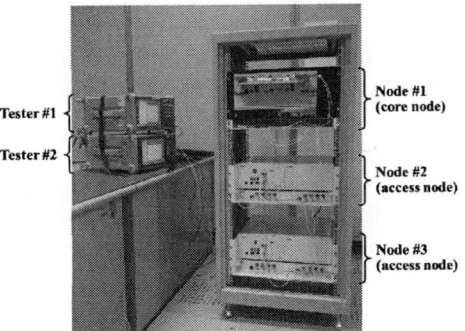

Fig. 6. Photograph of the prototype.

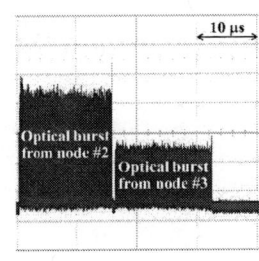

Fig. 7. A captured waveform.

V. CONCLUSIONS

We proposed a cost-effective optical metro network called the "Photonic Sub-Lambda transport network" (PSL network). The PSL network enables photonic-sub-lambda aggregation, which can efficiently accommodate traffic with limited network resources. We also discussed the advantages of the PSL network in terms of network costs and clarified that a 45%+ reduction in total network costs can be achieved compared with traditional ROADM-based wavelength-routed metro networks. Moreover, we successfully demonstrated photonic-sub-lambda aggregation in a prototype ring network.

REFERENCES

[1] Cisco Visual Networking Index: Forecast and Methodology, *Cisco White Paper*, May 2015.
[2] Bell Labs Metro Network Traffic Growth: An Architecture Impact Study, *Alcatel-Lucent Strategic White Paper*, Dec. 2013.
[3] Q. Liu, *Proc. OFC 2014*, M3B.3, Mar. 2014.
[4] Ericsson whitepaper: Full Service Broadband Metro Architecture, Nov. 2007.
[5] Ericsson whitepaper: Microwave Towards 2020, Sept. 2014.
[6] D. Chiaroni et al., *Proc. OFC 2010*, OThN3, Mar. 2010.
[7] G. S. Zervas et al., *Optics Express*, vol. 19, no. 26, pp. B509–B514, Dec. 2011.
[8] K. Hattori et al., *Proc. ECOC 2012*, We.3.D, Sept. 2012.
[9] P. Gavignet et al., *Proc. PS 2015*, pp. 43–45, Sept. 2015.
[10] D. Nesset, *JLT*, vol. 33, no. 5, pp. 1136–1143, Mar. 2015.
[11] A. Bianco et al., *JOCN*, vol. 5, no. 1, pp. 81–91, Jan. 2013.
[12] F. Rambach et al., *JOCN*, vol. 5, no. 3, pp. 210–225, Mar. 2013.
[13] FP7 OASE project deliverable D4.2.2, June 2013.

A Scalable Optical Network Architecture Based on WSS for Intra Data Center

Aijun Liu, Yongmei Sun, Hongxiang Wang and Yuefeng Ji

State Key Laboratory of Information Photonics and Optical Communications, Beijing University of Posts and Telecommunications,
Beijing, 100876, China
Email: {laj, ymsun, wanghx, jyf}@bupt.edu.cn

Abstract: A scalable optical network architecture with low diameter based on WSS for intra data center is proposed. Simulations show that it can outperform Fat Tree architecture and is suitable to cluster traffic.

Keywords: data center; wavelength selective switch; optical network

I. INTRODUCTION

The exponential growth of traffic due to data-intensive cloud applications has driven a surge of redesigning the network architecture for data center. Particularly, the network architecture of intra data center is a fundamental challenge, since the most traffic (above 74%) remains within intra data center nowadays [1]. Currently, data center employ the traditional electrical switching, yet which can not satisfy the need of bandwidth of traffic expansion due to limited speed.

To overcome the above mentioned limitation, optical switching which avoids optical-electrical-optical (O/E/O) conversion and holds huge transmission capacity can benefit intra data center network. Various innovative all-optical or optical/electrical hybrid architectures were proposed to replace traditional tier electrical switching. Many of those employ micro-electro-mechanical system (MEMS), arrayed waveguide grating router (AWGR) or wavelength selective switch (WSS), with the target of improving network capacity and scalability while reducing end-to-end latency and energy consumption [2-6].

To enable the advent of cloud applications boosting the demand of bandwidth, many papers focused only on the scalability of the network architecture, but did not consider the diameter (i.e., the longest "shortest path") of the network architecture [3-6]. However, the network performance depends not only on the capacity of the network, but is also inversely proportional to the average path length [7]. Low diameter can lead to low average path length, since the major of the path length is equal to the network diameter. In addition, as for routing and wavelength assignment (RWA) in WDM optical network without wavelength conversion, the wavelength along the routing should comply with wavelength continuity constraint, and the shorter path with less constraint on consecutive links obtains more available wavelengths than longer path. Therefore, the path length plays a significant role within the design of the network architecture, especially for optical network.

In this paper, we propose a scalable WSS (SWSS) optical network architecture for intra data center with low diameter. The performance of the proposed network architecture are evaluated numerically in terms of blocking ratio and bandwidth occupation ratio, which are compared with an all optical Fat-tree data center architecture.

II. HIGHLY SCALABLE ARCHITECTURE BASED ON WSS

A WSS is typical asymmetric 1 x N optical component with one common port and N wavelength ports, and can be configured to carry any subset of the wavelengths from common port to N wavelength ports (all ports refer to the wavelength ports in the following paper, if not any special indication). WSS possesses flexible wavelength selective feature while has limitation in terms of constructing network due to asymmetric feature. Thus, how to interconnect WSS to a symmetric building block which is used to construct a larger architecture is a primary problem.

To overcome the above mentioned problem, complete graph, which enables any two nods communication and has only one hop between any adjacent nodes, is a good candidate architecture. What's more, as for the architecture between different building blocks, multi tier tree architecture is wildly employed to construct large intra data center. However, we still employ complete graph instead of multi tier tree architecture, since each tier of tree architectures can increase the network diameter by two. In contrast, the complete graph can only increase the network diameter by one. Therefore, we propose a SWSS optical network architecture enlightened by complete graph for intra data center based on WSS. The SWSS is a two tier architecture which has low network diameter and shorter average shortest path length (ASPL), and can be constructed in two steps. The first is to construct basic construction unit (BCU) as a symmetric building block (in Fig. 1), and the second is to connect different BCU to construct SWSS (in Fig. 2). The specific constructing way of the SWSS is followed:

First, in the example of BCU shown in Fig. 1 in the case of N=4 which presents the number of WSS ports, the BCU is consisted of four WSS and corresponding network components. The N ports of each WSS are divided into two groups, and the one with N-1 ports are internal ports used for intra BCU connection, while the other with one port is external port used for inter BCU connection. The internal ports of each WSS are fully connected with the internal ports of every other WSS to form a BCU, i.e., a complete graph with N WSS (the external port of each WSS is shown by blue arrow in Fig. 1). The BCU is a symmetric $N/2$ x $N/2$ building block, which is suitable to the cluster traffic yielded by the adjacent network nodes within BCU in cloud applications [8]. Furthermore, each WSS is also connected by the corresponding network components, including MUX/DEMUX, Top of Rack (ToR) switch, and servers. The MUX/DEMUX can multiplex/demultiplex different wavelengths ejected by transceivers into single fiber connecting to the common port of the corresponding WSS, the ToR switch is equipped with W WDM transceivers which convert electrical signal into optical signal, and servers are connected to its ToR switch by electrical links. Electrical links are omitted and the other 3 network components are expressed by rectangle dashed in the Fig. 1. In BCU, the ID of each WSS is made up of a 1-tuple [a_0] and takes value from [0, N) which represents the address space.

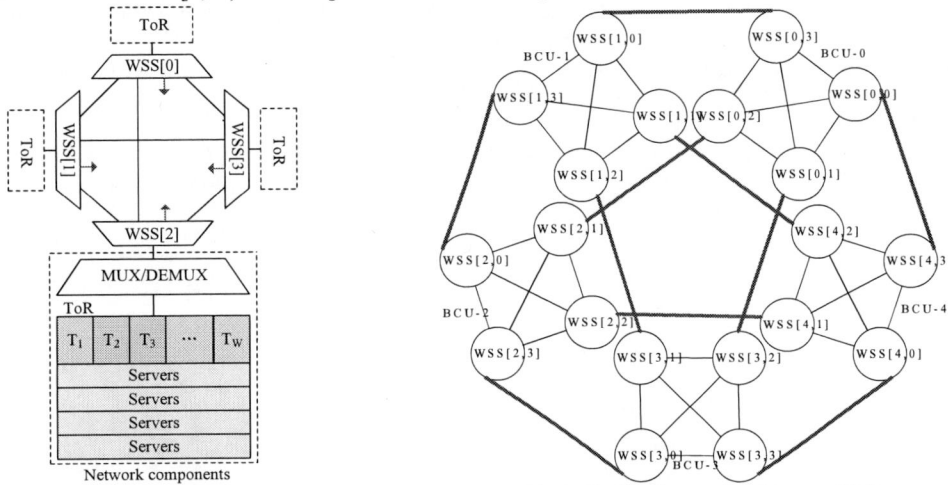

Fig. 1. The basic construction unit (N=4) Fig. 2. The SWSS network architecture (N=4)

Second, the SWSS network architecture is constructed by N+1 BCU shown in Fig. 2 in the case of N=4. Each BCU, if treated as a virtual node with N external ports, is connected fully to every other virtual node to form a virtual complete graph connected by blue bolded lines shown in Fig. 2. Since virtual node has N external ports, the SWSS network architecture can only have N+1 BCU. As a result, the external port of each WSS within the same BCU is connected to different BCU to form a symmetric architecture with N(N+1) nodes and low diameter (source node is ToR switch and the diameter is 5). N determines the network architecture size and node degree with N. In contrast, if the SWSS employ two tier tree architecture, its diameter is 6. In the SWSS architecture, the ID of each BCU is also made up of a 1-tuple [a_1] and takes value from [0, N+1), thus the ID of each WSS in SWSS is made up of a 2-tuple [a_1, a_0], as illustrated in Fig. 2.

III. NUMERICAL EVALUATION

First, a numerical comparison between Fat Tree and SWSS is performed under the same node degree N. From TABLE I, it can be seen that Fat Tree has better scalability and more bisection bandwidth (BiW) than SWSS, which leverages more network nodes, more links, thus obtains longer diameter and ASPL. However, many data center applications may not need full BiW possessed by Fat Tree, and longer ASPL can lower the performance of the network. In contrast, SWSS employ two tier complete graph architecture, which with low network diameter and shorter ASPL can benefit the RWA of the optical network.

TABLE I
COMPARISON OF FAT TREE AND SWSS

Architecture	Degree	Diameter	BiW	# Nodes	# Links	# Servers	ASPL
Fat Tree	N	6	$\dfrac{N^3}{8}$	$\dfrac{5N^2}{4}$	$\dfrac{3N^3}{4}$	$\dfrac{N^3}{4}$	$\dfrac{6N^3 - 2N^2 - 4N - 8}{N^3 - 4}$
SWSS	N	5	$\dfrac{(N+1)^2}{4}$	$N(N+1)$	$\dfrac{N}{2}\left(N^2 + 3N + 2\right)$	$N(N+1)$	$\dfrac{5N^2 + N - 3}{N^2 + N - 1}$

Second, we evaluate the SWSS network architecture compared with Fat Tree through Visual Studio 2010. For the sake of fairness, both network architectures have the same network capacity. Since there are more links in Fat Tree, this implies that the SWSS has more wavelengths per fiber. Connection request are launched and terminated by different ToR switch. The arrival of the traffic follows Poisson distribution and the holding time follows a negative exponential distribution with unit mean. The bandwidth by each connection request is one wavelength and we employ first fit wavelength assignment algorithm [9].

Fig. 3 shows the blocking ratio variation of the both network architectures as random traffic load grows under different network size. It's obvious that the blocking ratio of the both network architectures increases as traffic load grows, and the bigger network size experiences the low blocking ratio. Moreover, the SWSS outperforms the Fat Tree up to 25% performance improvement under high load. One reason is that the network capacity increases along with N increases, thus the blocking ratio decreases with the network capacity increasing. Another is the SWSS obtains shorter ASPL which not only utilities less bandwidth to support more flows, but also has more chance to be allocated wavelength due to less constraint on RWA. Fig. 4 describes the bandwidth occupation ratio variation of the both network architectures as random traffic load grows. As traffic load increases, the bandwidth occupation ratio also increases and the bigger network size experiences the lower bandwidth occupation ratio. Moreover, the SWSS achieves up to 47% performance improvement compared with the Fat Tree under high load. One reason is that the more network architecture capacity is, the lower the bandwidth occupation ratio is. Another is SWSS has lower blocking ratio which can support more flows to make use of the network bandwidth.

Fig. 5 presents the blocking ratio variation of the both network architectures as traffic load grows under three different traffic distribution with $N=4$. Here, we define traffic locality factor (TLF), which means a ratio of within cluster traffic and between cluster traffic, i.e., the ratio of within BCU traffic and between BCU traffic in SWSS, and within pod traffic and between pod traffic in Fat Tree, respectively. The random traffic distribution exhibits the highest blocking ratio in both networks, because more traffic travelling multiple hops routing consume excessive network resources and have less chance to find available wavelengths. When TLF=4/1, the blocking ratio is lowest, since more traffic have shorter routing with more chance to find available wavelengths. It is worth nothing that the blocking ratio of the SWSS decreases up to 42% compared with its random traffic distribution under high load. The reason is the SWSS has worse BiW, while the major of the network bandwidth is located in BCU which is more suitable to cluster traffic.

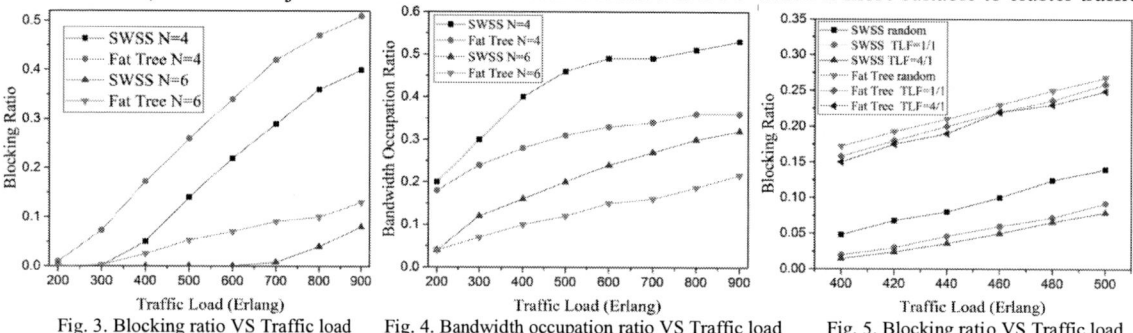

Fig. 3. Blocking ratio VS Traffic load   Fig. 4. Bandwidth occupation ratio VS Traffic load   Fig. 5. Blocking ratio VS Traffic load

IV. CONCLUSIONS

In this paper, we propose a scalable optical network architecture based on WSS. The proposed SWSS leveraging two tier complete graph architecture offers huge network capacity and shorter network diameter. Numerical simulations show that the SWSS architecture outperforms the Fat Tree up to 25% and 47% performance improvement in terms of blocking ratio and bandwidth occupation ratio under random traffic distribution with high load, respectively. Furthermore, the SWSS is suitable to cloud applications owing to the major of its bandwidth located in BCU.

ACKNOWLEDGMENT

This work has been supported by NSFC project (No.61331008) and 863 program (No.2013AA014501).

REFERENCES

[1] http://www.cisco.com/c/en/us/solutions/collateral/service-provider/global-cloud-index-gci/Cloud_Index_White_Paper.html.
[2] G. Wang, et al., "c-Through: Part-time Optics in Data Center," SIGCOMM 2010, pp. 327-338.
[3] N. Farrington, et al., "A 10 us Hybrid Optical-Circuit/Electrical-Packet Network for Datacenter," OFC 2013, OW3H.3.
[4] R. Proietti, et al., "Scalable and Distributed Optical Interconnect Architecture based on AWGR for HPC and Data Centers," OFC 2014, Th2A.59.
[5] Z. Zhu, et al., "Scalable and Topology Adaptive Intra-data Center Networking Enabled by Wavelength Selective Switching," OFC 2014, Th2A.60.
[6] D. Zhang, et al., "A Deterministic Small-World Topology based Optical Switching Network Architecture for Data Centers," ECOC 2014, p.6.11.
[7] A. Singla, et al., "Jellyfish: Networking Data Center Randomly," NSDI 2012.
[8] T. Benson, et al., "Network Traffic Characteristics of Data Center in the Wild," IMC 2010, pp. 267-280.
[9] S. Du, et al., "Power-efficient RWA in dynamic WDM optical networks considering different connection holding times," Sci China Inf Sci, 2013, 56(4): 1-9.

Frame Length Averaging Method for Multi-carrier Aggregation Transport

Toru Homemoto, Toshiya Matsuda, Masaru Katayama, and Katsutoshi Koda

NTT Network Service Systems Laboratories, NTT Corporation
9-11, Midori-Cho 3-Chome Musashino-shi, Tokyo 180-8585 Japan
{homemoto.toru, matsuda.toshiya, katayama.masaru, koda.katsutoshi}@lab.ntt.co.jp

Abstract: Frame Length Averaging (FLA) method bundles multiple pluggable DWDM interfaces to behave as a high-rate transport link. We demonstrate FLA has 95% throughput of an aggregated link with low frame jitter and no frame-order reversal.
Keywords: IP-Optical Integration; Link Aggregation; Frame Length Averaging; Circuit Design

I. INTRODUCTION

The rapid increase in internet traffic has led to greater demand for high-capacity optical transport systems at low cost. IP-Optical integration focuses on integrating router/switch function and optical transport equipment. Its objective is to reduce the system-wide count of optical-electrical signal conversion for saving energy and simplify the system. To achieve IP-Optical integration, "Black Link" approach [1] has been investigated, and its specifications are already part of the ITU-T Recommendation G.698.1 and G.698.2. This method implements Dense Wavelength Division Multiplexing (DWDM) transponder functions on pluggable interfaces of the router/switch such as Small Form-factor Pluggable (SFP), leaving only wavelength multiplexing and de-multiplexing functions on optical transport equipment. The pluggable 11.1-Gbps DWDM interface with integrated G.709 encapsulation (OTN Framer) and Forward Error Correction (FEC) has also been developed and tested [2].

However, latest DWDM transponder functions usually include advanced modulation techniques such as Dual Polarization-Quadrature Phase Shift Keying (DP-QPSK) in addition to high-powered lasers, OTN Framers and FEC calculations. They require high energy supplements, making it hard to integrate into small form at the same transmission rate as that of short-range Ethernet links. To deal with this problem, we have focused on bundling a multiple low-rate pluggable DWDM interface to behave as a high-rate one; we call it "Multi-carrier Aggregation (MCA)" in this paper. Fig. 1 illustrates the model case of MCA transport, where t lanes of N-Gbps transport links behave as M-Gbps Transport links and carry M-Gbps Ethernet ingress traffic to egress transparently. In the proceeding section, we discuss the detail approaches to construct MCA transport.

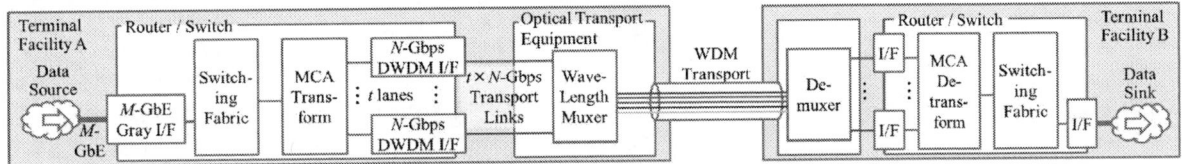

Fig. 1: Model case of multi-carrier aggregation transport

II. CONSTRUCTING MULTI-CARRIER AGGREGATION TRANSPORT

An Ethernet Link Aggregation (LAG) [3] is a simple way to construct MCA transport. We call these interfaces to be bundled "elementary links". They work fine in use to accommodate a lot of flows such as the Internet's backbone link, where hash-based algorithms distribute Ethernet frames into each elementary links equally. But its single-flow throughput is limited to the speed of one elementary link because frames in a flow usually contain the same header and generates the same hash keys, resulting in all of them being redirected into a specific elementary link.

Therefore it might not be suitable for enterprise-wide area network (WAN) or data center interconnect (DCI), where a single flow would use the whole bandwidth of aggregated links. Moreover, these services usually require fixed data rate, no frame-order reversal and low frame jitter on customer demand because frame-order reversal and frame jitter causes severe degradation in Transmission Control Protocol (TCP) performance.

From the discussion, we summarized the conditions to bundling methods below and then compare three approaches.
1. Bundling on Layer 2 (Ethernet) or higher layer to use pluggable DWDM interface without any customizing.
2. Elementary link is itself a transport link such as OTN, guaranteeing frame order between ingress and egress.
3. Bundled link can be used for a whole bandwidth by a single flow; it keeps frame order and has low frame jitter.

A. Pure Round Robin (Pure RR) -Based MCA

We can use Round Robin (RR) load balancing algorithm to distribute frames: frame 1 to elementary link 1's transmission queue, frame 2 to queue 2, and so on. If the next candidate queue is full, the dispatcher will wait until the queue is ready, so we call it "Pure RR". This is because the receiver can easily guarantee the frame order by just picking up frames from queue 1, queue 2, and so on consecutively. Its throughput of aggregated link will be lower than the sum of elementary links because the frame length variation results unbalanced length of the queues. If one of the queues becomes full, the dispatcher will stall in spite of the other queues have space left. Both the implementation cost and frame jitter seems to be low when the ingress traffic load is low, because the frame is dispatched into a target queue immediately after its arrival.

B. Fair Load Balancing(FLB)-Based MCA with Embedded Sequence Number

While pure RR-based MCA suffers from lower throughput, we can use a kind of fair load balancing schemes as an alternative. There are preceding studies of them, discussed in [4]. A typical fair load balancing scheme dispatches the frame into queues at several probabilities, achieving balanced length of the queue and maximum throughput of aggregated elementary links. On the other hand, the sequence number field must be installed on frames at the transmitter as the receiver side can distinguish which queue to pick up the next frame, guaranteeing the frame order. This matter causes slight throughput degradation depending on the length of sequence number field and average frame length. The implementation cost seems to be higher than that of pure RR because of the frame dispatching and picking up algorithm. The frame jitter will vary depending on which queue the frame is dispatched.

C. Frame Length Averaging(FLA)-MCA

Formerly in [5], we introduced frame length averaging (FLA) technique as another approach to balance length of queues, achieving high throughput. Although FLA could be applied for any pair of N-Gbps elementary link and t-lanes bundling, we assume 10 Gbps as N and 4 lanes as t for explanation and performance evaluation in this work.

On the transmitter side, input frame sequence is handled every four frames (FRx), with x meaning variables 1 to 4. The FLA-MCA transform unit extracts 4 payloads (Dx) and divides each one into 4 fragments with the same length (D1-1 to D1-4, D2-1 to D2-4, and so on). Each converted frame (FR'x) consists of the original x-th frame's header (OHx), original frame size in 2 byte field (Sx), concatenated payload fragment (D1-1 to D1-4, D2-1 to D2-4, and so on), and footer (F'x) regenerated on the data as before, shown in Fig. 2. Because converted frame grows 2 bytes in size by Sx field, the maximum transfer unit (MTU) on the input sequence must be less than that of transport link by 2 bytes.

On the receiver side, the FLA-MCA de-transform unit waits until the 4 converted frames arrival and restores the 4 original frames using the frame size written in the Sx field. After de-transforming, the frame order is guaranteed as that of the original sequence. FLA-MCA seemed to exploit maximum throughput of aggregated elementary links excepting the portion of additional size field usage. Ideally, maximum throughput is 97% in the worst case of all ingress frames consisted of 64 bytes minimal frames. The implementation cost expected to be higher because of payload scattering and gathering procedure. Its frame length averaging results in low frame jitter.

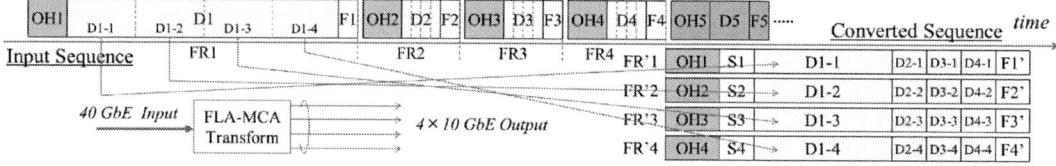

Fig. 2: FLA-MCA transformation

We summarized the characteristics of three approaches to construct MCA in Table I. FLA seems to have improvement in latency stability compared to FLB method with sequence number. On the other hand, we can find out the trade-off relationship between pure RR and FLA in terms of throughput and implementation cost. Therefore we have performed throughput and implementation cost evaluation for pure RR-based MCA and FLA-MCA in next section.

TABLE I
CHARACTERISTICS OF METHODS TO CONSTRUCT MCA TRANSPORT LINK

	THROUGHPUT	IMPLEMENTATION COST	LOWER FRAME JITTER
PURE RR	DEGRADED	GOOD	GOOD
FLB WITH SEQ. NUMBER	GOOD	DEGRADED	DEGRADED
FLA	GOOD	DEGRADED	GOOD

III. PERFORMANCE EVALUATION

A. Throughput and Jitter Estimation

To estimate the performance of MCA algorithm, an accurate processing model of electronic circuit is required. Then we have implemented pure RR-based MCA and FLA-MCA on the register transfer level (RTL) circuit model and simulated on a circuit simulator. In the RTL simulation, the internal behavior of the electronic circuit was simulated clock-by-clock. The experimental setup is shown in Fig. 3.

A 40-Gbps frame generator creates traffic according to the Paleto distribution [6] on the length of a data sequence, then divides the data sequence into frames with 1500bytes payload and places the remainder into the last frame. In the rate limiter, the token bucket algorithm is implemented to regulate the traffic load of the transmitter. The traffic policer unit has a buffer memory in the ingress side. If there is no space left to receive a frame, the traffic policer unit will drop the frame that was determined as frame loss in the receiver. We also evaluated the pure RR-based frame distributor and frame aggregator for performance comparison. We obtained throughput (Gbps in Fig. 4) when the traffic load varied from 10- to 40-Gbps and average latency and standard deviation (nanoseconds in Fig. 5 and Fig. 6).

Using FLA load balancing, achieved throughput increased constantly along with the traffic load, whereas the pure RR-based load balancing had a peak value around 30-Gbps and no growth at heavier traffic loads. At 40-Gbps of ingress traffic, FLA load balancing achieved throughput of around 95% of the traffic load without frame-order reversal. The standard deviation of frame jitter in MCA is suppressed under 0.1μs at 32.5-Gbps or higher, while pure RR yields 0.16μs at 25-Gbps. Higher FLA jitter at lower load is caused by the behavior of FLA transform waits until every 4 frames arrival. The timeout algorithm might improve the result that allows FLA conversion of 3 or less frames.

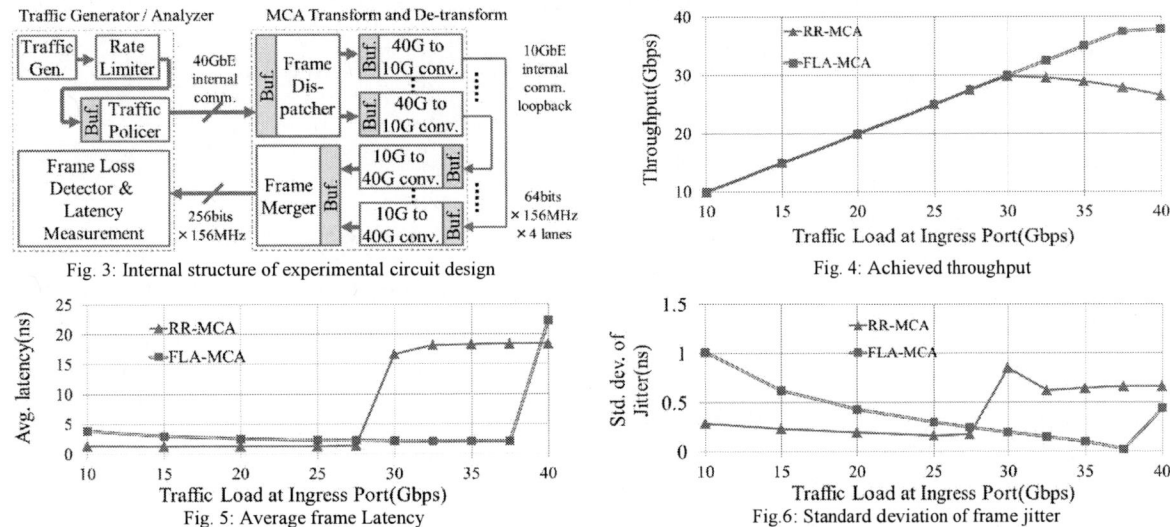

Fig. 3: Internal structure of experimental circuit design

Fig. 4: Achieved throughput

Fig. 5: Average frame Latency

Fig.6: Standard deviation of frame jitter

B. Implementation Cost Estimation

The RTL circuit designs for throughput evaluation is also able to be synthesized, placed, and routed on Field Programmable Gate Array (FPGA) using Electronic Design Automation (EDA) tools, so we have evaluated implementation costs based on logic utilization (Arithmetic Logic Units, ALMs) and memory usage of FPGA. We used Altera Quartus II 13.1 Software as EDA tool and Altera Stratix V GX as target FPGA device and used clock timing constraints of 6.4ns period (156MHz). We also show estimated equivalent gates when the design is implemented on Application Specific IC (ASIC) by using rough estimation of 1ALM = 30 gates. The results are shown in the table below, shows FLA-MCA requires reasonable logic usage although 4 times larger than pure RR.

TABLE II
CIRCUIT INSTALLATION RESULT ON FPGA (ALTERA STRATIX V GX EA7N2F45C2)

	LOGIC UTILIZATION	BLOCK SRAM USAGE
PURE RR	3584 ALMs (1.5% OF THE FPGA, ABOUT 100 KGATES)	944KBITS
FLA-MCA	13984 ALMs (6.0% OF THE FPGA, ABOUT 400 KGATES)	1198KBITS

IV. CONCLUSION

We discussed several load balancing methods to construct MCA transport, which enables bundling low-rate pluggable DWDM interfaces to behave as a high-rate one. We designed and evaluated the circuit model of a pure RR-based MCA and FLA-MCA, revealing that FLA method can exploit 95% throughput of aggregated links while a pure RR-based one shows 75%. We also showed FLA-MCA with 4 lanes of 10-Gbps can be implemented about 400 kgates in ASIC. It will be extensible to a low-cost substitution of latest DWDM transponder such as 100G, 400G, and so on.

REFERENCES

[1] M. Gunkel et al., "Elastic Black Link for Future Vendor Independent Optical Networks," OFC, Th1l.3, 2015.

[2] P. R. Morkel et al., "Integrated IP-Optical Networks. Demonstration of DWDM Router-to-Router IP Transport Over 574km SMF Fiber Link Using 11.1Gbit/s OTN Pluggable Interface with Integrated G.709 and FEC," OFC/NFOEC, NME4, 2008.

[3] IEEE Std. 802.3ad-2000.

[4] S. Kobayashi et al., "Packet-based lane bundling for terabit-LAN," APCC, pp. 1-5, 2008.

[5] T. Matsuda et al., "Zero-dispersion DWDM Transmission and Multi-carrier Aggregation for Beyond 10G Metro Network," ECOC, pp. 1-3, 2015.

[6] G. Kramer et al., "IPACT: A Dynamic Protocol for an Ethernet PON (EPON)," IEEE Comm. Magazine 40.2 (2002): pp. 74-80.

ThA2-5

Method to share a burst-mode receiver between continuous- and burst- mode transmitter for VM migration of cloud edge

Kyota Hattori, Masahiro Nakagawa, Toshiya Matsuda, Masaru Katayama, and Katsutoshi Koda

NTT Network Service Systems Laboratories, NTT Corporation, Tokyo, Japan
hattori.kyota@lab.ntt.co.jp

Abstract: *We study a future metro network architecture based on NG-PON2 systems. We propose a method to reduce the optical receivers and electrical switches by diverting NG-PON2's circuits and transmitters to support the cloud edge cost-effectively.*
Keywords: *Future metro network, NG-PON2, supporting cloud edge*

I. INTRODUCTION

Metro networks (NWs) will experience significant traffic growth to accommodate both traditional "vertical" traffic that involves carrying consumer traffic to the IP edge and "horizontal" traffic transmitted between data centers [1]. Both kinds of traffic will become more dynamic through application of cloud computing such as cloud edges (CEs) [2] and cloud data centers. A key technology facilitating this will be virtual machine (VM) migration according to resource usage of CEs. Therefore, future metro NWs have to maintain bandwidth efficiency using VM migration in order to achieve cost-effectiveness. To achieve this, we previously proposed the Optical Layer-2 Switch NW (OL2SW-NW) [3-5], which allows the bandwidth in a metro NW to be shared with optical TDM paths according to traffic volume in order to construct the metro NW cost-effectively. The links of OL2SW-NW consist of optical SWs. Therefore, in case of the application of the OL2SW-NW to the low demand area which accommodates the low volume traffic, OL2SW-NW will become a little cost-expensive caused by the deployment of optical SWs. Here, we propose a future metro NW architecture based on the Next-Generation Passive Optical Network stage 2 (NG-PON2) [6] system, with the link consisting of optical passive devices. Proposed NW can connect the Access NW to CEs distributed to the different locations with supporting VM migrations. Specifically, the proposed NW can share a burst-mode (BM) receiver between a continuous-mode (CM) transmitter and BM transmitters. This makes it possible to reduce the optical receiver for the accommodation of the user traffic cost-effectively with supporting CEs. We show the experimental results and feasibility of receiving both BM and CM signals at a BM receiver.

II. ARCHITECTURE OF PROPOSED NW

The proposed NW consists of the technology of burst transmission and optical passive devices on the transmission link without the application of the electrical L2SW. Fig. 1 shows the proposed NW based on NG-PON2 system, which is a WDM/TDM ring NW located between the access and core NWs. One of the advantages of applying access NW technology such as NG-PON2 to a metro NW is that it is possible to reduce the Capital Expenditure. Because access NW devices are mass-produced, they must be low NW cost. Here, we label the nodes that constitute the proposed NW; that is, the one connected to access NW is called an A-node and the one connected to core NW is called a C-node. Also, either C-node plays a role of master node, which has the functions of dynamic bandwidth allocation and measurement of the delay. Each node has ether-burst converters (EBC) that convert between Ethernet frames and BM signals. To apply the technology of an NG-PON2 system to the proposed NW, C-node's EBC is based on the function of optical subscriber unit (OSU), which converts BM signals from A-nodes to Ethernet frames because C-nodes have to aggregate the traffic from access NW to Core NW. On the other hand, A-node's EBC is based on the function of optical network unit (ONU), which converts variable Ethernet frames from L2SW in the access NW into BM signals. These make it possible to apply the NG-PON2 transceivers to each node's EBC. That is to say, the C-node's transmitter is CM while

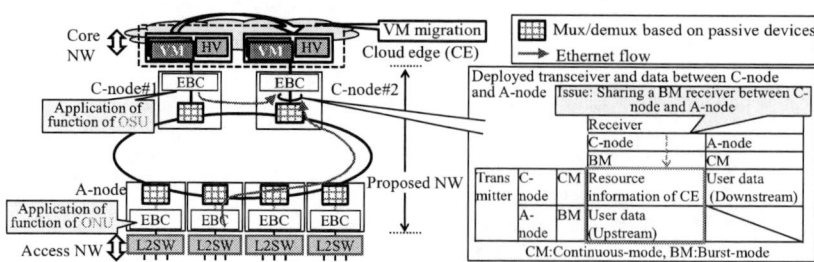

Fig. 1. Architecture of proposed NW.

OECC/PS2016

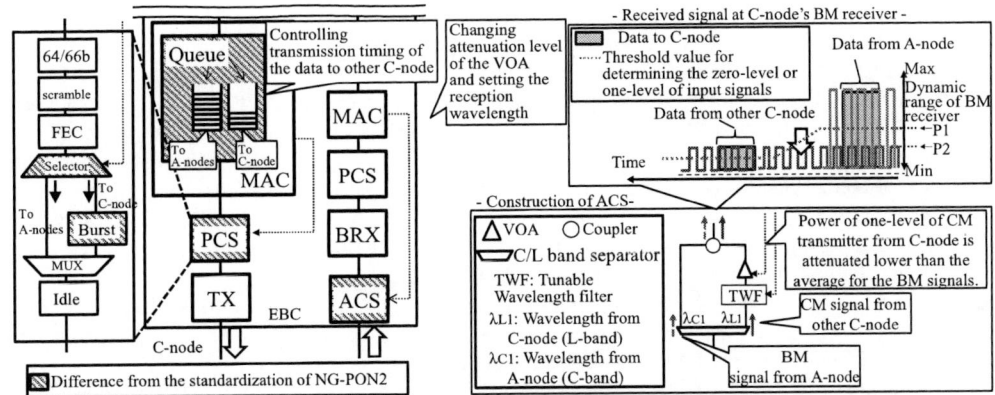

Fig. 2. Summary of the proposed method to share a BM receiver between a CM transmitter and BM transmitters.

the C-node's receiver is BM and the A-node's transmitter is tunable BM while the A-node's receiver is tunable CM. Here, we assume that an A-node belongs to one C-node simultaneously, and the reachability between CEs is only via the proposed NW. Then, switching CEs according to VM migration is achieved by learning Media Access Control (MAC) address of the VMs and changing the wavelength of transmission and the reception wavelength dynamically at the EBC of A-node. Besides, to increase the extent of the proposed NW and allow for a high user count, reach extenders comprising optical amplifiers are required, suitably placed at each node to compensate for the split loss and propagation loss in both uplink and downlink bi-directions [7].

III. ISSUE WITH COMMUNICATION BETWEEN CEs VIA PROPOSED NW

In the proposed NW, it is necessary to transmit and receive the information of hardware resources between the CEs to support VM migrations. This means that C-nodes have to send and receive the data from CEs between each other. To achieve this, when applying the NG-PON2 system in unadornment, each C-node has to prepare the additional receiver to receive the CM signals from other C-node and deploy the L2SW to avoid the collision of data between an A-node and other C-node. This leads to an increase in the NW cost. Therefore, for solving this issue, we propose the method to share a BM receiver of C-node between a CM transmitter of C-node and BM transmitter of A-node using the transceiver of NG-PON2 systems.

IV. PROPOSED METHOD TO SHARE A BM RECEIVER BETWEEN CM AND BM TRANSMITTER

The proposed method can share a C-node's BM receiver between other C-node's CM transmitter and A-node's BM transmitter. Fig. 2 shows the summary of the proposed method. The proposed method consists of (a) attachment of burst header and controlling of transmission timing at sender side of C-node and (b) Attenuator for CM Signal (ACS) deployed at receiver side of C-node. (a) is treated at the Physical Coding Sublayer (PCS) and the MAC at sender side of C-node. Improvements of these PCS and MAC are achieved to add the same circuits of PCS and MAC of ONU to that of OSU. The signals destined for other C-node are switched to the "Burst" element of PCS, attaching the burst header by the installment of the selector, which controls every data signal by MAC. Also, the timing of transmission of the data to other C-node is controlled by the deployment of queues for C-node and A-nodes taking into account the delay between C-nodes measured by the MAC of C-node. This is to avoid the collision between data from C-node via CM transmitter and data from A-node via BM transmitter. Furthermore, the "Burst" element in the PCS of C-node attaches the physical synchronization blocks (PSBu), such as the preamble and burst delimiter, to the data destined for other C-node in order to perform the reception of CM signals with overlapping BM signals and burst synchronization processing at the C-node. If there is no signal destined for both A-nodes and C-node, idle signals are inserted. This is because the CM transmitter is deployed in the A-nodes. Meanwhile, (b) attenuates the CM signal from the C-node. Specifically, the power of one-level of CM transmitter from C-node (P1) is attenuated lower than the average for the BM signals (P2) from the A-node (Fig. 2). By this mechanism, a BM receiver at C-node can receive the data both from a CM transmitter of C-node and BM transmitter of A-nodes by reusing the transceiver, PCS, and MAC of NG-PON2 systems.

V. EVALUATION

To evaluate the proposed method, we conducted an experiment on receiving signals both from a CM and BM transmitter at a BM receiver using a prototype system. The experimental setup is shown in Fig. 3. We set two C-nodes and an A-node. C-node #2 receives the signals from both A-node and C-node #1. C-node #2 allocates the bandwidth and controls the transmission timing for both A-node and C-node #2 by measuring the delay between each node. In front of BM receiver at the C-node #2, we set an ASC, which consists of wavelength filter (WF) and a variable optical attenuator (VOA). WF separates the wavelengths into the wavelength of CM transmitter (; λ_{L1}, L-band) from C-node #1

249

and the wavelength of BM transmitter (; λ_{C1}, C-band) from A-node. Then, the power of the signal of CM transmitter from C-node#1 was attenuated at the VOA. To confirm that the signals transmitted from the A-node and C-node #1 can be received at C-node #2, we measured the receiving data at C-node#2 using a real-time oscilloscope and traffic generator. The evaluation indices are the response speeds in the BM receiver to ensure the required guard band with the proposed method and Bit Error Rate (BER). The dynamic range of BM receiver used in this experiment was from -6 to -28 dBm. We set the differences of the power for the one level between BM signals and CM signals to 4 dB at the ACS. Figs. 4 show the experimental results: (a) both BM signals and CM signals received before the BM receiver, (b) these signals received after the BM receiver, and (c) the timing of receiving data at C-node #2, respectively. As you can see in Figs. 4, the power of CM transmitter from C-node #1 was attenuated to -24 dBm with 4 dB gap compared to that of BM signals. We found that the BM receiver was able to receive both the data from C-node #1 and A-node. This is because the proposed method controls both the data from C-node#1 in the CM signal included in the PSBu and the transmission timing from C-node#1. Then, the response time was less than 50 ns. Even for taking into account the measurement error of the delay between each node, required guard band becomes below 100 ns, which is a feasible length. Also, we then measured the BER without FEC (Fig. 4 (d)). The received power of CM signals taken at the FEC limit (BER of 10^{-3}) was -24 dBm. Then, the power budget was able to acquire 18 dB with application of FEC. According to these results, we find that the proposed method enables sharing of a C-node's BM receiver between both other C-node's CM transmitter and an A-node's BM transmitter in a proposed NW.

VI. CONCLUSIONS

We proposed a method for sharing a BM receiver between a CM transmitter and BM transmitters. We evaluated the proposed method by using a prototype system based on a NG-PON2 system. The results showed that the proposed method could be feasible by including the burst header, controlling the transmission timing for the data on the CM signals, and attenuating the power of the CM signals.

REFERENCES

[1] P. Momtahan, "Evolving Metro Networks to support 400G and Beyond," ECOC, Market Focus, 2013.

[2] A. Misawa et al., "Proposal on Virtual Edge Architecture Using Virtual Network Function Live Migration with Wavelength ADM," Proc. APCC2015, 15-PM1-E.1, 2015.

[3] K. Hattori et al., "Optical Layer-2 switch network based on WDM/TDM nano-sec wavelength switching," Proc. ECOC, We.3.D.5, 2012.

[4] K. Hattori et al., "Optical TDM Fast Reroute Method for Bufferless Metro Ring Networks with Arbitrary Fiber Length," Proc. ECOC, P.6.1, 2015.

[5] K. Hattori et al., "Optical Layer 2 Switch Network with Bufferless Optical TDM and Dynamic Bandwidth Allocation," IEICE TRANSACTIONS on Electronics, Vol. E99-C, No.2, pp.189-202, Feb. 2016.

[6] ITU-T Rec. G.989.2, "40-Gigabit-capable passive optical networks 2 (NG-PON2): Physical media dependent (PMD) layer specification," 2014.

[7] M. Fujiwara et al., "Field trial of 100-km reach simmetric-rate 10G-EPON system using automatic level controlled burst-mode SOAs," J. Lightw. Technol., vol. 31, no. 4, pp. 634 - 640, Feb. 2013.

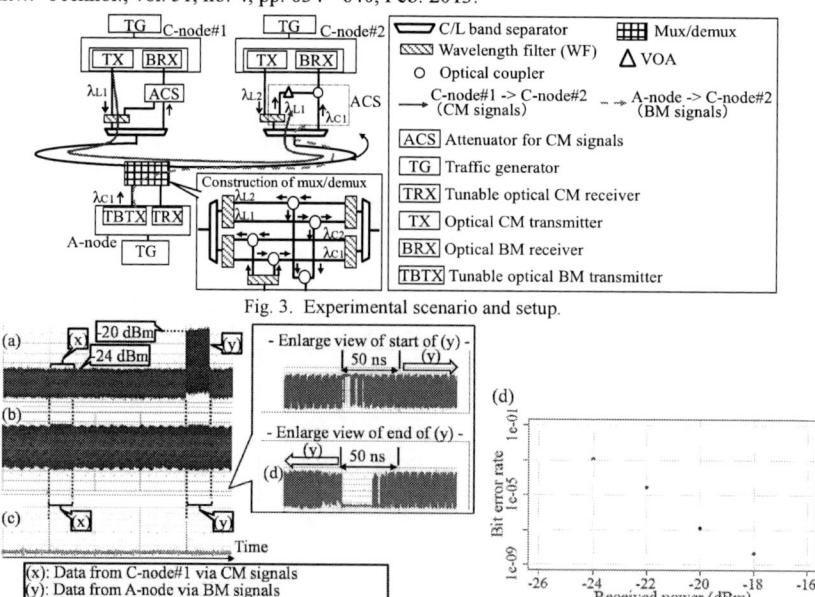

Fig. 3. Experimental scenario and setup.

Figs. 4. Receiving signals (a) before BM receiver, (b) after BM receiver, (c) the timing of receiving data at C-node #2, and (d) BER for the received power of CM signals.

ThA3-1 (Invited)

OECC/PS2016

Remotely Controllable WDM-PON Technology for Wireless Fronthaul/Backhaul Application

Michael H. Eiselt[1], Christoph Wagner[1,2], Mirko Lawin[1]

[1]ADVA Optical Networking, Maerzenquelle 1-3, 98617 Meiningen, Germany
[2]Technical University of Denmark (DTU), Dept. of Photonics Engineering, Ørsteds Plads, Build. 343, 2800 Kgs. Lyngby, Denmark
meiselt@advaoptical.com

Abstract: *Low-cost WDM-PON solutions for fronthaul and backhaul applications will include remotely controlled tail-end transceivers. We report on control aspects of these transceivers and how standardization is evolving to enable these applications.*
Keywords: *tunable laser, WDM-PON, pilot tone*

I. INTRODUCTION

A strong increase in traffic in mobile networks has led to increased capacity demand on the optical backhaul network between the air interface and the core network. In addition, the demand by network operators to consolidate and concentrate network functions has led to a separation of RF signal generation (and processing) in centralized base band units (BBU's) and RF-band / baseband conversion in remote radio heads (RRH's). This further increased the capacity demand on the link between BBU and RRH. This connection type is commonly called "front haul".

To meet these two increased capacity demands, optical connections between core network and BBU (backhaul) or between BBU and RRH (front haul) are necessary. As numerous connections are concerned, a novel type of low-cost passive WDM network is currently being developed for this purpose.

A unified network concept, integrating backhaul and front haul network sections, is shown in Figure 1 as an output of the ongoing 5G-Xhaul project [1]. For both, front haul and backhaul sections, a central location (BBU hotel or router) is connected on point-to-point links to distributed locations (RRH or BBU). To meet the lowest cost demands and to simplify fiber and connections handling at the central location, wavelength division multiplexing (WDM) technology is inevitable for this type of systems.

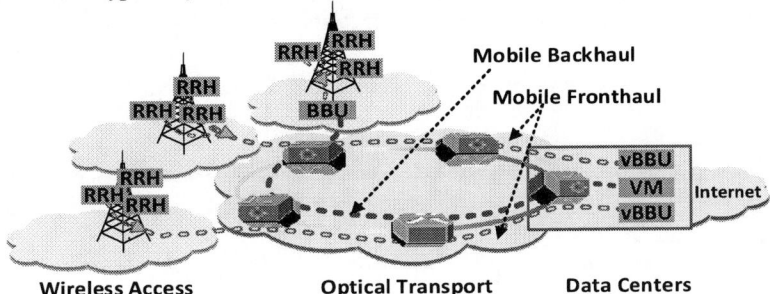

Fig. 1. Integrated backhaul and front haul network [1]. BBU: base band unit, RRH: remote radio head, VM: virtual machine, vBBU: virtual BBU

Current standardization efforts in ITU-T SG15 are also directed towards this application area. Q.6/15 is defining the physical layer interfaces in the framework of the project G.metro. To meet lowest cost requirements, the distributed lasers (called tail-end in Q.6 terms) must be wavelength agnostic, i.e. they need to automatically set their wavelengths to match the connected port on the central side (called head-end). This will also require the definition of a management channel between head-end and tail-end and a labelling of the tail-end to head-end signal.

In this paper, we will present the principle setup of a self-tunable system and discuss the issues resulting from the setup. Especially, we will consider the cross-talk induced to other channels, when a laser is searching for its correct wavelength, investigate the use of pilot tones as label for the automatically tuning wavelengths, and look into the communication between head-end and tail-end of the system.

II. ARCHITECTURE OF THE WDM-PON SYSTEM

To meet the demands of both, fronthaul (FH) and backhaul (BH) applications, multiple point-to-point connections between a central location (e.g. router for BH, or BBU hotel for FH) and multiple distributed locations (eg. BBU for

---

This work has been supported by the EU Horizon 2020 5G-PPP project 5G-XHaul and by the Marie Curie ABACUS project.

BH, or RRH for FH) are required. To reduce the number of fiber cables at the central location, wavelength division multiplexing (WDM) must be employed. Filters in the network then route the channels to the respective distributed locations. This distribution can be based on a star architecture, as shown in Fig. 2a, where the wavelength split is done in an arrayed waveguide grating (AWG), or in a dropline architecture, as shown in Fig. 2b, where wavelengths are added and dropped at a few OADM locations in a fiber ring. The dropline architecture could also enable a fiber protection by terminating both ends of the line at the central location into a "horseshoe" configuration.

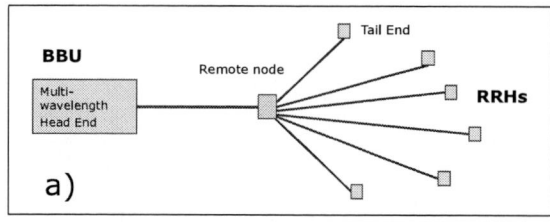 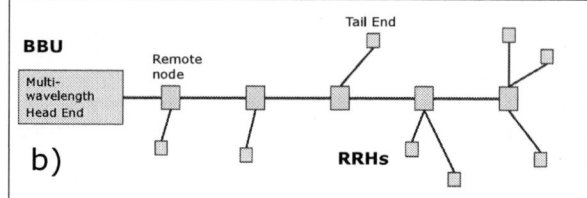

Fig. 2. WDM-PON architectures. a) star, b) add-drop line

In any architecture, the distributed transceivers in the tail-ends need to adapt their transmission wavelengths to the filter port they are connected to. However, to reduce cost, the wavelength control itself needs be removed from the tail-end transceiver and be placed at the head-end. This way, control components (wavelength locker etc.) can be shared between all tail-ends, but some communication and routines need to be provided to enable the finding of the correct wavelength.

III. CROSS TALK EFFECTS IN THE WDM-PON SYSTEM

To maintain a low cost of the tail-end transmitter, wavelength calibration of the laser needs to be minimized. Instead, the laser will sweep across all available wavelengths, until it fits the filter port it is connected to and it can be detected by the head-end receiver. While optical signals lasing outside the target channel are in general suppressed by the multiplexing filter, reduced isolation at the border between two channels can lead to non-negligible cross talk even after two filters. For instance, the concatenation of two flat-top AWG filters provides an isolation of less than 15 dB for signals injected into a neighboring multiplexer port and separated by half of the channel spacing from the central wavelength, as shown in Figure 3. As the channels might see different insertion losses between tail-end and head-end, they are received at the head-end with some power range, potentially leading to an interference of a (strong) tuning channel onto a (weak) working channel. This leads to the requirement that tuning channels need to reduce their transmit power until it is secured that they do not transmit outside their target channel.

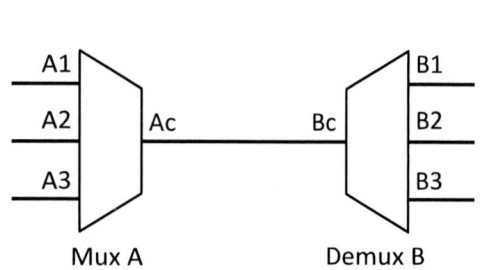 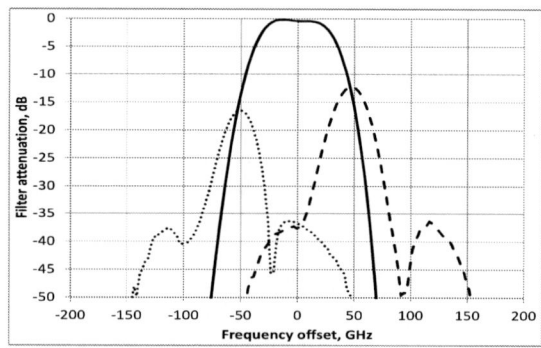

Fig. 3. Filter function of two cascaded 100-GHz AWGs. Solid line: Target channel (A2 -> B2), dotted and dashed lines: cross-talk, when laser is connected to upper and lower neighboring port of multiplexer filter (A1->B2 or A3->B2).

A second manifestation of low filter isolation and large loss range is a mis-detection of a tuning tail-end in the neighbor channel. To avoid this operational problem, each tail-end is assigned a pilot tone label, which can be detected in the respective head-end receiver. Only if this label is detected with sufficient power, the presence of the channel is assured.

IV. WAVELENGTH LABELLING USING PILOT TONES

Wavelength labelling using pilot tones has been used in long-haul and submarine systems for quite some time. In a WDM-PON system, these tones can be used to distinguish the wavelengths at the head-end and to monitor their presence in the correct channel, even if cross-talk power is observed in the channels. These tones can also be used to distinguish the wavelengths in a periodic wavelength control element, like a wavelength locker. Here, all wavelengths are monitored simultaneously, and the tone amplitudes serve as a proxy for the respective optical power values [2].

One issue with low frequency pilot tones can occur in optically amplified systems. Slow power variations of the optical signal can lead to a gain variation in an EDFA, even with state-of-the-art gain control, and cross-modulation of the pilot tone can occur on other channels. This issue impacts the choice of the pilot tone frequencies, as the cross-gain modulation can be overcome by selecting sufficiently high frequencies, e.g. 20 kHz.

V. MANAGEMENT COMMUNICATION CHANNEL

Precise wavelength monitoring can be performed at the head-end for all upstream wavelength channels. To control the wavelength in the tail-end, the monitoring information needs to be transmitted to the tail-end. Several communication protocols provide space for a low bit-rate management communication channel. However, this would require terminating the protocol directly at the tail-end receiver, or communicating the management information from the terminal equipment at the tail-end location into the transceiver. It would also limit the application of the WDM-PON system, as no transparent protocol transmission would be allowed. Therefore, it is preferable to use an overhead communication channel, which can be easily added at the head-end transmitter and can be extracted at the tail-end receiver. Direct envelop modulation of the optical signal, like for the pilot tones, appears to be a good option requiring little extra hardware effort. While there are several options for the modulation format, e.g. pilot tones with ASK, PSK, or FSK data modulation, direct modulation of the digital data onto the optical channel appears to be the method with lowest effort and cost.

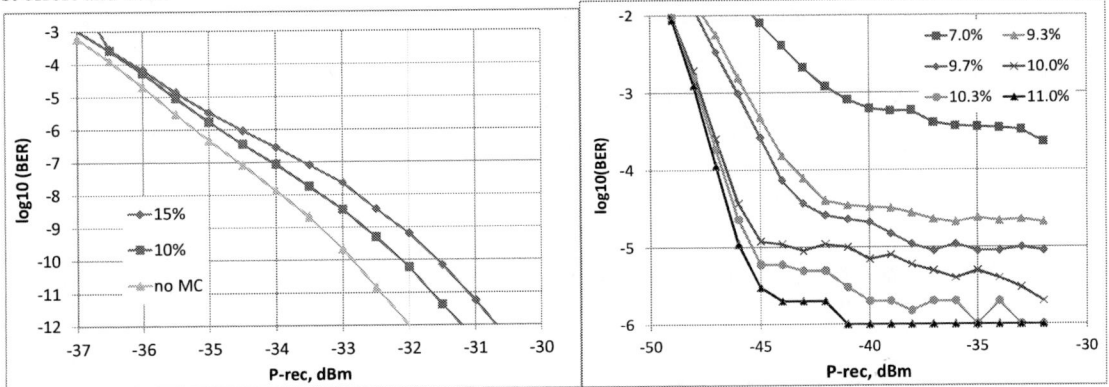

Fig. 4. Left: Receiver sensitivity of 2.5 Gb/s PRBS-31 signal in the presence of 100 kb/s envelop modulation with different modulation depths. Right: Receiver sensitivity of 100 kb/s overhead channel with different modulation depths, modulated onto a 2.5 Gb/s PRBS-31 optical signal.

The envelope modulation leads to a power penalty in the optical channel. This sets an upper limit on the modulation depth of the envelope modulation. On the other hand, any modulation of the optical channel in the spectral range of the overhead communication channel will deteriorate the communication channel. This can be counteracted by choosing a sufficiently large modulation index of the envelop modulation.

Fig. 4 shows, on the left hand side, the receiver sensitivity for a 2.5 Gb/s PRBS-31 in the presence of a100-kb/s overhead channel with different modulation depths. It can be seen that, for a BER of 10^{-12}, a penalty of less than 1.5 dB is obtained with a 15% modulation depth. The right hand side of Fig. 4 shows the receiver sensitivity of the 100-kb/s overhead channel for various modulation depths. Different error floors can be observed for different modulation depths. A BER of 10^{-6} is achieved with a modulation depth of 11%. This BER can easily be corrected using few checksum bits. The modulation depth could therefore be chosen between 11% and 15%.

VI. CONCLUSIONS

WDM-PON systems can be used for front-haul and back-haul of mobile communication radio signals in an integrated transmission system. To maintain lowest equipment and operational cost, automated tuning and centralized control of the tail-end equipment is required. We have discussed several issues arising from this operation. Limited isolation in filters can lead to cross-talk during the tuning procedure, such that a tuning tail-end needs to reduce its transmit power. To still reliably detect the channel at the head-end, each wavelength can be labelled by a distinct pilot tone. This pilot tone also aids in performing a wavelength control at a central wavelength locker. Finally, we discussed the performance of a management channel to transport control information between head-end and tail ends, demonstrating a sufficient performance of direct envelope modulation at lowest cost and hardware effort

REFERENCES

[1] 5G-PPP project "5G-Xhaul", Deliverable D2.1 "Requirements, Specifications and KPIs Document", available at www.5g-xhaul-project.eu/.

[2] S. Pachnicke et al., "Tunable WDM-PON system with centralized wavelength control," J. Lightw. Technol., vol. 34 (2106) 2, pp. 812-818, February 2016.

Dynamic Reconfiguration of Coordinated Multi-Point Transmission in High Density Small Cell Network Using Centralized Wavelength Switching in DWDM PON

Goji Nakagawa(1), Kyosuke Sone(1), Setsuo Yoshida(2), Shoichiro Oda(2), Motoyuki Takizawa(1), Hirokazu Shimada(1), Yasuhiko Aoki(1), and Jens C. Rasmussen (2)

(1) Fujitsu Limited, (2) Fujitsu Laboratories Ltd., 4-1-1 Kamikodanaka, Nakahara-ku, Kawasaki 211-8588, Japan
Author e-mail address: gnakagawa@jp.fujitsu.com

Abstract: *We proposed DWDM-PON architecture, which utilize wavelength selective switch with tunable OLT for dynamic reconfiguration of coordinated multi-point transmission in high density small cell network. We successfully demonstrated the dynamic routing and monitoring behavior employing pilot tone scheme.*
Keywords: *DWDM-PON, wavelength selective switch, superimposition*

I. INTRODUCTION

Mobile data traffic is drastically increasing year by year because of the large deployment of high speed mobile terminals, such as smart phones and tablet PCs, as well as huge number of sensor devices for Internet of Things (IoT) applications. Cloud radio access network (C-RAN) architecture with centralized base-band units (C-BBUs) and remote radio heads (RRHs) is one of the promising architecture to support high throughput for mobile terminals by using coordinated multi-point (CoMP) transmission and common public radio interface (CPRI) to satisfy severe synchronization requirement to transport the digitized wireless signal waveform. In high density small cell networks, the connection between BBUs and RRHs to support CoMP transmission can be changed due to the dynamic behavior of mobile traffic demand and dynamic reconfiguration of the fronthaul network is one of the challenges of future mobile access network. DWDM PON is suitable for such high speed mobile fronthaul network because of large bandwidth and low latency with dedicated wavelength channels [1-3] and one possible implementation of DWDM-PON architecture over splitter based optical distribution network to support dynamic reconfiguration of CoMP transmission is utilize tunable wavelength ONU, which consists of tunable lasers and tunable filters [4-5]. In the tunable ONU implementation, however, wide tuning range of tunable laser and filter is necessary and it makes the monitoring and operation of ONU complex. Another implementation is to use matrix switch between OLTs and BBUs, but it needs additional switching element.

In this paper, we propose new DWDM-PON architecture employing tunable transceivers with a wavelength selective switch (WSS) for OLT and fixed wavelength lasers and filters for ONU to realize dynamic reconfiguration of CoMP transmission by centralized switching of wavelength path. We demonstrate the routing and monitoring of wavelength path between C-BBU-OLT site and ONU-RRH site employing ASK (amplitude shift keying) supervisory signal superimposition and simple and robust ASK message monitors at OLT site.

II. PROPOSED ARCHITECTURE

Figure 1 shows the implementations of DWDM PON architecture, which previously demonstrated (Fig.1 (a)), and our proposal (Fig.1 (b)). In the former architecture, fixed wavelength is assigned to each transceiver which is connected to each port of wavelength mux/demux component at OLT, and tunable transceiver is used at ONU, where each wavelength is remotely controlled from OLT site by means of auxiliary management and control channel (AMCC) scheme [6] for example. In our proposed architecture, we use tunable transceivers and WSS, which have been matured and used in metro/core ROADM network, in OLT site, and fixed wavelength is assigned to each transceiver for each ONU. Compared with the previous architecture, our proposed architecture has the benefit of simple wavelength management by using centralized wavelength switching function at OLT site rather than the distributed wavelength switching at remote ONU site.

Fig.2 shows the coordinated behavior between BBU and OLT in the central office by assigning fixed wavelength to each ONU, result in the reconfiguration of BBU-RRH relationship caused by changes in heavy traffic spots of densely deployed small cells. On the day 1 (Fig.2 (a)), both wavelength of tunable Tx and WSS ports are assigned to $\lambda 6$, $\lambda 7$, $\lambda 9$, $\lambda 10$, which correspond to heavy traffic area in the ONU (RRH) site. This setting can be easily reconfigured by changing the wavelength assignment of Tx and WSS ports ($\lambda 11$, $\lambda 12$, $\lambda 15$, $\lambda 16$) to accommodate traffic shift and CoMP transmission on the day 2 (Fig.2 (b)). In this manner, we make it easy to realize the centralized controlling system at the central office site in an existing passive optical distribution network (ODN) for C-RAN application.

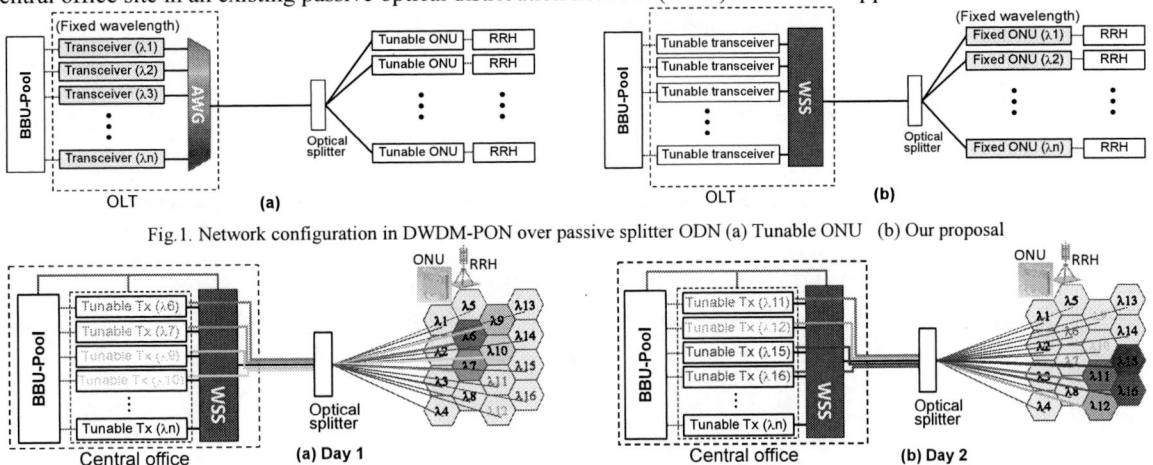

Fig.1. Network configuration in DWDM-PON over passive splitter ODN (a) Tunable ONU (b) Our proposal

Fig.2. The coordinated behavior between BBU and OLT

III. EXPERIMENTAL SETUP

In order to confirm the feasibility of our proposed architecture, we prepared the experimental setup as shown in Fig.3. One of the key functions of the proposed architecture is the management of active optical components and wavelength paths, and we equipped the ASK message monitor at OLT site, which is connected to WSS and ODN to isolate the cause of failure among ONUs, transceiver at OLT site, and WSS and to monitor the status of wavelength path. 10 Gbit/s NRZ main signal with PRBS 31 was generated by DWDM SFP+ transceiver at both ONU and OLT site. The ASK message at 1 kbit/s with Manchester encoding was modulated onto a 100 kHz carrier, and then superimposed onto the main signal by variable optical attenuator (VOA) with 10% modulation index, where we did not observe any main signal degradation by superimposition of ASK message. The modulation index is defined as the ratio of the voltage amplitude between the message channel and the average of main signal. The ASK message pattern and its modulation index were programmed by the function generator. We used 1x9 WSS at OLT site for WDM channel assignment at each transceiver port. For simultaneous monitoring of both OLT and ONU signals to realize the simple and robust monitoring, we arranged the optical splitter (SPL) in a triangle layout with a pair of 9:1 and 1:1 splitting ratio and the output is connected to a ASK message monitor. The ASK message monitor consists of low speed PD, bandpass filter, and decoder to extract the bit patterns from the data stream. In this experiment, we used digital storage oscilloscope (DSO: sampling ratio; 400 kS/s) for offline signal processing.

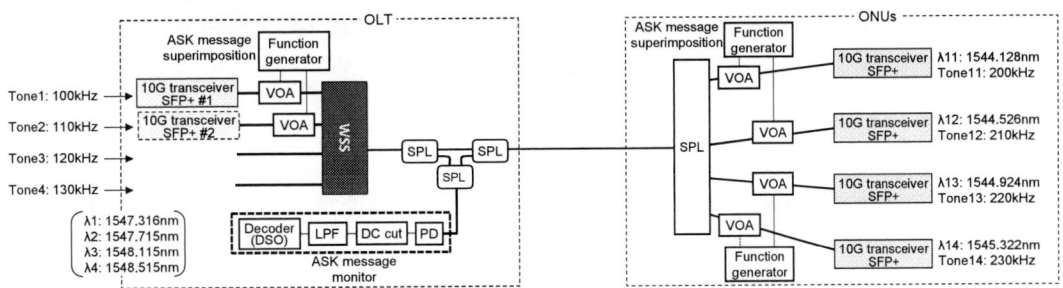

Fig.3. Experimental setup and wavelength / pilot tone assignments. VOA : variable optical attenuator, SPL : optical splitter, PD : photodetector, LPF : low pass filter, DSO : digital storage oscilloscope.

IV. RESULTS AND DISCUSSION

At first, we measured the pilot tone through the ASK message monitor to check the detection of transmission line failure at the ONU site. In this experiment, pilot tone frequency at ONU site corresponds to wavelength, which is assigned to each ONU, while pilot tone frequency at OLT site corresponds to WSS port number for the detection of port failure of WSS. Note that the tone number was detected in accordance with the port number, not the wavelength number to isolate the failure of WSS and transceiver at OLT site. We measured the pilot tone from transceiver at OLT site through WSS to check the detection of reconfiguration operation process as shown in Fig.4. At the initial setting (Fig.4 (a)), SFP+ #1 was connected to the port 1 of WSS with λ1. Reconfiguration operation was to set the SFP+ #2 at port 2

255

with λ1 and set the SFP+ #1 at port 1 with unused λ2. When wavelength of WSS port 1 was changed from λ1 to λ2, no pilot tone was detected from transceivers at OLT site (Fig.4 (b)) because the un-provisioned wavelength was set to WSS port 1. After wavelength of SFP+ #1 was changed from λ1 to λ2, the proper provisioning was confirmed by detecting pilot tone #1 with λ2 (Fig.4 (c)). Next, we set the WSS port 2 at λ1, but the monitor state stayed the same. Finally, after provisioning the new SFP+ #2 at port 2 with λ1, the path set up was confirmed by detecting the pilot tone #2 with λ1 (Fig.4 (d)). Throughout this experiment, pilot tone from ONU site (λ12-tone12 and λ13-tone13) was also detected thanks to triangle layout of SPLs for ASK message monitor.

Next, we extracted the bit patterns of an ASK message from the multiplexed supervisory ASK messages detected by ASK massage monitor. In this experiment, we simultaneously monitored four different massages modulated on each different frequency carrier from different transceivers at OLT site (Fig.5 (a)), and stored the multiplexed signal by DSO through PD detection. The stored signal was FFT-operated, and filtered to extract the desired signal as shown in Fig.5 (a). After inverse FFT operation, 32 bit message sequence was obtained as shown in Fig.5 (b), which confirms the error free operation. Similarly, supervisory ASK message from ONU site can be monitored simultaneously in the frequency axis (Fig.5 (a)) and the bit patterns can be extracted by using the same scheme.

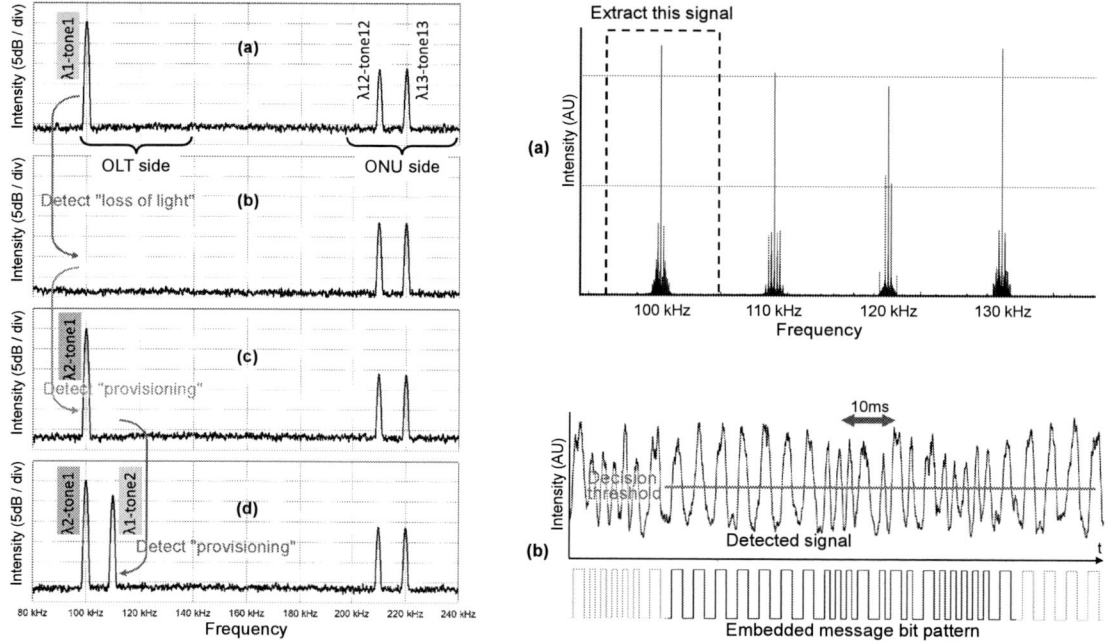

Fig.4. Detection of reconfiguring operation process

Fig.5. (a)ASK message multiplexed pilot tone and (b) its pattern extraction.

V. SUMMARY

We have proposed new DWDM-PON architecture employing WSS and tunable transceiver for OLT, and fixed wavelength for ONU to realize the centralized controlling system in mobile fronthaul network. We have successfully demonstrated the routing and monitoring behavior and confirmed that this configuration is applicable to dynamic reconfiguration of CoMP transmission in high density small cell network.

ACKNOWLEDGMENT

This paper includes a part of results of "The research and development project for realization of the fifth-generation mobile communications system" commissioned by The Ministry of Internal Affairs and Communications, Japan.

REFERENCES

[1] Derek Nesset, "NG-PON2 Technology and Standards," IEEE J. Lightwave Technol., vol. 33, pp. 1136-1143 (2015)
[2] Vincent O'Byrne, "PON Evolution for Residential and Business Applications," in OFC 2014, W1D.1, (2014)
[3] S. Kaneko et al., "In-Service Wavelength Tuning Technology in WDM/TDM-PONs for Multiple-Service Convergence," in ECOC 2015, Tu.3.1.1 (2015)
[4] Recommendation ITU-T G.989.2 (2014), 40-Gigabit-capable passive optical networks (NG-PON2): Physical Media Dependent Layer Specification,
[5] Ning Cheng et al., "World's First Demonstration of Pluggable Optical Transceiver Modules for Flexible TWDM PONs", in ECOC 2013, PD4.F.4
[6] Kazuaki Honda et al., "WDM Passive Optical Network Managed with Embedded Pilot Tone for Mobile Fronthaul," in ECOC 2015, We.3.4.4 (2015)

ThA3-3

OECC/PS2016

25-Gb/s ASE-Seeded WDM PON

Qikai Hu[1,2], Changyuan Yu[3], Pooi-Yuen Kam[1], and Hoon Kim[4]

[1]Dept. of Electrical & Computer Engineering, National University of Singapore, Singapore 117576
[2]National University of Singapore (Suzhou) Research Institute, China 215123
[3]Dept. of Electronic and Information Engineering, The Hong Kong Polytechnic University, Hong Kong
[4]Dept. of Electrical Engineering, KAIST, 291 Daehak-ro, Yuseong-gu, Daejeon, Korea 34141
E-mail: hoonkim@kaist.ac.kr

Abstract: We experimentally demonstrate, for the first time, a 25-Gb/s/channel amplified spontaneous emission seeded WDM passive optical network by exploiting the ultra-narrow spectrum pre-slicing, offset optical filtering, and electronic dispersion compensation techniques.
Keywords: Spectrum slicing, incoherent light, optical access network, WDM.

I. Introduction

Wavelength-division-multiplexed (WDM) passive optical networks (PONs), thanks to their high capacity, future-proof upgradability, and secure privacy, have been regarded as a promising solution for broadband optical access networks. However, the widespread deployment of such networks have been hindered mainly by relatively expensive WDM components including wavelength-specific light sources [1]. In order to lower the implementation, operation, and maintenance costs of WDM-PON systems, extensive attention has been paid to cost-effective colorless transmitters such as amplified spontaneous emission (ASE) seeded optical transmitters [2-6]. In WDM-PON systems based on these optical transmitters, a broadband ASE light is spectrum-sliced by an arrayed waveguide grating (AWG) and fed as seed lights to colorless optical modulators including reflective semiconductor optical amplifiers (SOAs) and electro-absorption modulators (EAMs). Thus, these WDM-PON systems could be implemented cost-effectively without requiring costly wavelength-specific optical transmitters. The WDM-PON systems based on ASE-seeded optical transmitters have been approved by ITU-T as an international standard for PONs providing 1.25-Gb/s/channel data services [7]. However, the potential of the ASE-seeded optical transmitters to offer greater than 10-Gb/s/channel services is very unclear. This is because the performance of these optical transmitter is limited by the spontaneous-spontaneous beat noise of the incoherent ASE light. The signal-to-noise ratio (SNR) of the ASE-seeded optical transmitter is governed by B_o/B_e, where B_o is the optical linewidth and B_e is the receiver electrical bandwidth [2]. The use of a gain-saturated SOA has been proposed to suppress the beat noise [8], but a large B_o should be still exploited to accommodate high-speed signals [4, 5]. For example, it is necessary to have B_o larger than 0.5 nm to accommodate 10-Gb/s data rate. However, a large B_o makes the signal susceptible to fiber chromatic dispersion (CD) and optical filtering (by AWGs in the network), and more importantly it limits the total capacity of WDM-PON system [3]. For these reasons, all the previously demonstrated ASE-seeded WDM-PON systems were limited to 10 Gb/s/channel.

In this paper, we experimentally demonstrate, for the first time to our knowledge, a 25-Gb/s/channel ASE-seeded WDM-PON system. In order to make the 25-Gb/s signal robust against the CD and optical filtering, we employ the ultra-narrow-linewidth ASE seed light [3]. However, the ASE light having an ultra-narrow linewidth suffers from large spontaneous-spontaneous beat noise. To suppress this noise, we utilize a gain-saturated SOA and offset optical filtering [8, 9]. The waveform distortions caused by CD are mitigated by using the electronic dispersion compensation (EDC). With the aid of these techniques, we are able to transmit the 25-Gb/s on-off keying (OOK) signal over 20-km standard single-mode fiber (SSMF).

II. Experiment and Results

Fig. 1 (a) shows the experimental setup. A wideband ASE generated from an erbium-doped fiber amplifier (EDFA) with no input signal is first spectrum-sliced by a fiber Fabry-Perot (FFP) filter. The FFP filter used in this experiment has a 3-dB bandwidth of 0.005 nm (=700 MHz) and a free-spectral range of 0.82 nm. This ultra-narrow spectrum pre-slicing localizes the optical power of the ASE seed light at around the center of the passbands of AWGs in the network. Thus, it

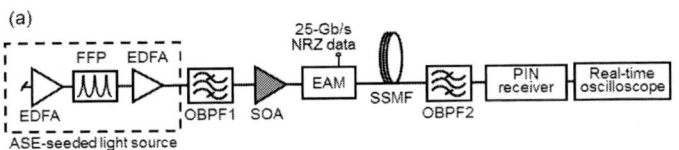 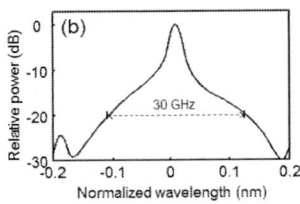

Fig. 1 (a) Experimental setup. (b) Optical spectrum of the 25-Gb/s OOK signal measured at the output of the modulator.

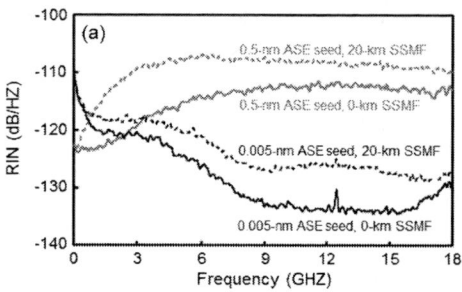 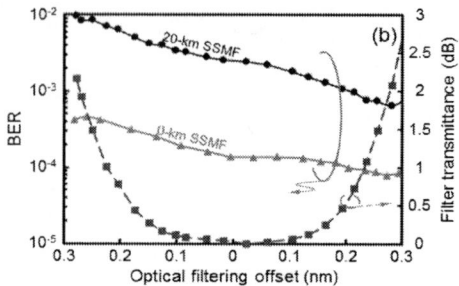

Fig. 2 (a) RIN of ASE seed lights for two different linewidths measured at the output of the SOA. (b) Measured BER performance of the 25-Gb/s OOK signal as a function of the optical filtering offset. Also plotted is the power loss incurred by offset filtering.

not only minimizes the power loss of the signal incurred by the AWGs in the network, but also greatly improves the energy consumption of the ASE seed light source [6]. Due to the periodicity of the passbands of the FFP, we need only one FFP filter for the generation of multiple WDM lights. We select one of the periodic passbands of the spectrum pre-sliced ASE by using an optical bandpass filter (OBPF), which emulates an AWG at the central office. The ultra-narrow ASE seed light at 1542.37 nm is then sent to a gain-saturated SOA for the suppression of the beat noise inherent in the incoherent ASE light. The SOA has a small signal gain of 22 dB and a saturation input power of -15 dBm when biased at 200 mA. The optical power of the ASE seed light into the SOA is set to be 1.5 dBm. Thus, the SOA operates in the saturation regime. For data modulation, the ASE seed light is fed to an EAM driven by a 25-Gb/s non-return-to-zero signal. The modulation bandwidth of the EAM is measured to be 15 GHz. Fig. 1(b) shows the optical spectrum of the signal measured at the output of the EAM. Due to the ultra-narrow linewidth of the ASE seed light, the spectral width of the signal is determined by data modulation and its 20-dB spectral width is measured to be 30 GHz. This signal can be fit into a dense WDM grid with a 50-GHz channel spacing. The 25-Gb/s OOK signal is launched into SSMF and then fed to OBPF2 (passband center=1542.14nm, 3-dB bandwidth=0.64 nm) before direct detection. This optical filter emulates an AWG located at the remote node. The detected signal is sampled by using a 100-Gsample/s oscilloscope and processed off-line. In order to compensate for the CD and band-limitation of the modulator, we employ a half-symbol-spaced 4-tap feed-forward equalizer followed by a 2-tap decision-feedback equalizer. The tap coefficients of the equalizer is determined by the minimum mean-square-error criterion.

We first show the efficacy of the use of ultra-narrow-linewidth ASE seed light. Fig. 2(a) shows the relative intensity noise (RIN) of the two ASE seed lights having different linewidths measured at the output of the SOA. No data modulation is applied to observe the RIN only. When the linewidth of the ASE seed light is 0.5 nm, the RIN is lower than -120 dB/Hz below 3 GHz. The RIN increases to >-113 dB/Hz beyond 9 GHz since the high-frequency RIN is not suppressed by the SOA. After transmission over 20-km SSMF, the RIN rises considerably since the spontaneous-spontaneous beat noise suppressed by the SOA is restored by CD [10]. The SOA suppresses the beat noise by creating an intensity correlation between different wavelength components of the ASE seed light. However, this correlation is vulnerable to wavelength-dependent phenomena and thus can be broken by CD [10]. On the other hand, the ASE seed light having a 0.005-nm linewidth has much lower RIN than the wide-linewidth ASE seed light. Except at the frequencies lower than 1 GHz, the RIN is measured to be lower than -120 dB/Hz. This RIN is also kept relatively low after 20-km transmission. This is because of the ultra-narrow linewidth of the ASE seed light. Thus, the ASE seed light with an ultra-narrow linewidth outperforms the wide-linewidth ASE seed light in terms of RIN and tolerance to CD, and consequently is more suitable for 25-Gb/s transmission.

Next we investigate the BER improvement produced by the optical offset filtering. Fig. 2(b) shows the BER performance of the signal as a function of the wavelength offset between the ASE seed light and OBPF2. The offset optical filtering is achieved by tuning the passbands of the FFP filter without touching OBPF2. Thus, this offset filtering technique does not increase the complexity of the WDM-PON system. The results clearly show that the BER is improved when the ASE seed light is detuned to longer wavelengths. For example, the BER measured after 20-km transmission is reduced from 2.4×10^{-3} to 7.5×10^{-4} when the wavelength of the ASE seed light shifts from 1542.14 to 1542.37 nm (i.e., 0.23-nm offset). This performance improvement should be attributed to the fact that a gain-saturated SOA creates a correlation between amplitude and frequency (or chirp) of the ASE seed light and then the offset optical filtering converts the chirp into amplitude to produce destructive interference with the existing beat noise of the ASE seed light [9]. Although 0.23-nm offset filtering increases the link loss by 1.1 dB, it could be readily compensated by slightly increasing fiber launch power. The optimum amount of filtering offset remains unchanged regardless of the transmission distance.

Fig. 3(a) shows the measured receiver sensitivity (at a BER=10^{-3}) of the 25-Gb/s OOK signal as a function of the transmission distance. Without using the EDC and offset optical filtering, we achieve a receiver sensitivity of -15.7 dBm, but after 6-km transmission we observe an error floor at around 1.8×10^{-3}. In the presence of EDC at the receiver, the

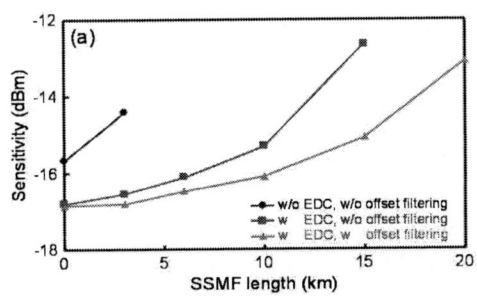 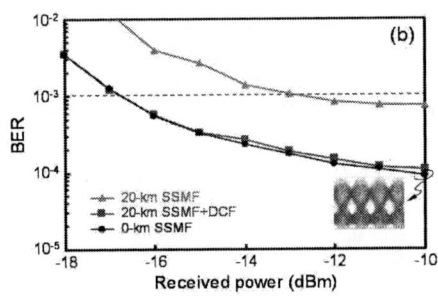

Fig. 3 (a) Measured sensitivity (BER=10^{-3}) versus the transmission distance. (b) Measured BER as a function of received power. The inset is the eye diagram of signal at the output of the EDC.

receiver sensitivity is improved by 1.1 dB in the back-to-back condition since the EDC compensates for the band-limitation of the modulator. The EDC also compensates for the waveform distortions caused by CD and enables us to transmit the 25-Gb/s signal up to 15-km-long SSMF. Fig. 3(a) shows that the efficacy of offset optical filtering is improved as the transmission distance increases. This is because the correlation between amplitude and frequency (or chirp) of the seed light is enhanced with the transmission distance [9]. With the aid of the EDC and offset filtering, we are able to transmit the 25-Gb/s signal over 20-km-long SSMF.

Fig. 3(b) shows the measured BER curves of the 25-Gb/s OOK signal. The 0.23-nm offset filtering is employed and the EDC is used at the receiver. In the back-to-back condition, the receiver sensitivity is measured to be -16.8 dBm. After transmission over 20-km-long SSMF, the receiver sensitivity is degraded to -12.8 dBm, but we are able to achieve BERs lower than the forward error correction threshold of 10^{-3}. The penalty after the transmission is caused by the enhanced spontaneous-spontaneous beat noise, as explained in the previous page. Thus, the BER performance after transmission can be considerably improved by using optical dispersion compensation. Fig. 3(b) also shows the measured BER curve when a dispersion compensating fiber (DCF) is inserted in the network. The amount of dispersion and insertion loss of the DCF are -400 ps/nm and 3.0 dB, respectively. The result shows that when the fiber CD is compensated, the performance after 20-km transmission is almost the same as the back-to-back performance. Thus, in WDM-PON systems having a large number of channels, a DCF module could be used in the systems to compensate for the fiber dispersion all over the WDM channels.

III. CONCLUSIONS

We have experimentally demonstrated a 25-Gb/s/channel WDM-PON system based on ASE-seeded light sources. Thanks to the spectrum pre-slicing by using an ultra-narrow FFP filter, offset optical filtering, and electronic dispersion compensation, we successfully transmit a 25-Gb/s OOK signal over 20-km SSMF at BER<10^{-3}.

ACKNOWLEDGMENT

This work is supported by the AcRF Tier 2 Grant MOE2013-T2-2-135 from MOE Singapore and Grants (61501313, 61571316 and 61302112) from National Natural Science Foundation of China.

REFERENCES

[1] F. Saliou et al., "WDM PONs based on colorless technology," Opt. Fiber Technol., vol. 26, pp.126-134, 2015.
[2] J. Lee et al., "Spectrum-sliced fiber amplifier light source for multichannel WDM applications," IEEE Photon. Technol. Lett., vol. 5, pp. 1458-1461, 1993.
[3] Z. Al-Qazwini et al., "Ultra-narrow spectrum-sliced incoherent light source for 10-Gb/s WDM PON," J. Lightw. Technol., vol. 30, pp. 3157-3163, 2012.
[4] J.-Y. Kim et al., "400 Gb/s (40×10 Gb/s) ASE injection seeded WDM-PON based on SOA-REAM," OFC, 2013, OW4D.4.
[5] S.-H. Yoo et al., "Pulsed-ASE-seeded DWDM optical system with interferometric noise suppression," Opt. Express, vol. 22, pp. 8790-8797, 2014.
[6] H. H. Lee et al., "Demonstration of performance enhanced pre-spectrum sliced seed light for low-cost seeded WDM-PON architecture," Opt. Fiber Technol., vol. 19, pp. 126-131, 2015.
[7] Multichannel seeded DWDM applications with single-channel optical interfaces, ITU-T G.698.3, 2012.
[8] M. Munroe et al., "Spectral broadening of stochastic light intensity-smoothed by a saturated semiconductor optical amplifier," J. Quantum Electron., vol. 34, pp. 548-551, 1998.
[9] Q. Hu et al., "Performance improvement of ultranarrow spectrum sliced incoherent light using offset filtering," IEEE Photon. Technol. Lett., vol. 26, pp. 870-873, 2014.
[10] H. Kim et al., "Impact of dispersion, PMD, and PDL on the performance of spectrum-sliced incoherent light sources using gain-saturated semiconductor optical amplifiers," J. Lightw. Technol., vol. 24, pp. 775-784, 2006.

Mitigation of optical beat interference in N-dimensional CAP-PON uplink transmission

Jiale He[1], Lei Deng[1*], Lu Shi[1], Mengfan Cheng[1], Ming Tang[1], Songnian Fu[1], Minming Zhang[1], Deming Liu[1] and Perry Ping Shum[2]

[1]Next generation Internet Access National Engineering lab (NGIA), School of Optoelectronic Science and Engineering, Huazhong University of Sci.&Tech. (HUST), Wuhan, China
[2]School of Electrical and Electronic Engineering, Nanyang Technological University, Singapore
*denglei_hust@mail.hust.edu.cn

Abstract: To reduce OBI noises in N-dimensional-CAP-PON uplink transmission over a single wavelength, a novel technique based on OLT-side coherent detection and DSP algorithm is proposed. 10Gb/s CAP-PON uplink transmission over 50-km-SSMF is also experimentally demonstrated.
Keywords: optical beat noise; carrierless amplitude and phase modulation; digital signal process; blind phase search; cascaded mulit-modulus algorithm

I. INTRODUCTION

As ever increasing bandwidth demands for multimedia applications in multi-user access networks, the next generation passive optical network (PON) becomes important to provide high capacity and flexible access service [1]. To accommodate multiple users, a 10 Gb/s optical frequency division multiplexing access (OFDMA) PON systems has been proposed and demonstrated [2]. Besides, carrierless amplitude and phase modulation (CAP) PON is also regarded as a potential technique due to its lost cost and energy-consuming [3]. Moreover, it can be extended to higher dimensions, which could be used to support flexible optical service and more users in PON system [4].

For the practical implementation of a low-cost PON system, colorless optical network units (ONUs) without wavelength-specific laser source is of the most important issue. To address this issue, wavelength reuse concept has been widely investigated [2]. However, for the upstream transmission with wavelength reuse concept, both OFDMA and N-dimensional CAP signal will suffer from the optical beat interference (OBI) noises which are generated at the optical line terminal (OLT) due to the beating of multiple ONUs with the same wavelength. For OFDMA-PON system, optical carrier suppression at the ONUs combined with coherent detection at the OLT could effectively eliminate the OBI between each ONUs, but IQ mixtures and RF sources are used to carry the complex-OFDM signals [2]. An alternative way is suppressing the OBI peak [5], but RF source is also required to generate a RF clipping tone signal for OBI reduction. For CAP-PON system, an 11×5×9.3 Gb/s WDM-CAP-PON has been achieved recently [3], however, the upstream transmission and the influence of OBI in CAP-PON systems with colorless ONUs are not considered yet.

In this paper, we propose a method to achieve high-speed upstream N-dimensional CAP-PON transmission over a single wavelength. By using the ONU-side carrier suppression and OLT-side coherent detection, the OBI noises between ONUs could be eliminated. To further recover signals from frequency offset and phase noise between ONU and LO, a modified BPS algorithm and a CMMA-based equalizer are optimally designed. Both experimental and simulated setups are analyzed for the demonstration of this novel technique in 10 Gb/s CAP-PON uplink transmission over a 50-km standard single mode fiber (SSMF).

II. PRINCIPLE

Assuming each ONU is allocated with one CAP filter. In the transmitter of each ONU, the Mach-Zehnder modulator (MZM) is biased at its power null point and driven with one CAP signal. The optical filed of ith ONU can be expressed as $E_i = A_i \, exp(j(\omega_i t + \emptyset_i))$, $A_i = \sqrt{P_i} \, cos(\pi v_i / v_\pi + \pi/2)$, where v_i is the ith CAP electronic signal, P_i, ω_i and $\emptyset_i$ are the power, frequency and phase of optical carrier respectively. After all the ONUs being coupled in one fiber over the same wavelength, the optical filed can be written as $E_s = \sum_{i=1}^{n} A_i exp(j(\omega_i t + \emptyset_i))$. Similarly, LO can be expressed as $E_0 = \sqrt{P_0} exp(j(\omega_0 t + \emptyset_0))$. After 90-degree optical hybrid, the obtained signal could be expressed as:

$$
\begin{aligned}
I_i &= R[(E_s + E_o)(E_s + E_o)^* - (E_s - E_o)(E_s - E_o)^*]/2 \\
&= R[\sum_{i=1}^{N} A_i^2 + A_o^2 + 2\sum_{i=1}^{N}\sum_{j=i}^{N} A_i A_j \cos((\omega_i - \omega_j)t + (\phi_i - \phi_j)) + 2\sum_{i=1}^{N} A_i A_o \cos((\omega_i - \omega_o)t + (\phi_i - \phi_o)) \\
&\quad - (\sum_{i=1}^{N} A_i^2 + A_o^2 + 2\sum_{i=1}^{N}\sum_{j=i}^{N} A_i A_j \cos((\omega_i - \omega_j)t + (\phi_i - \phi_j)) + 2\sum_{i=1}^{N} A_i A_o \cos((\omega_i - \omega_o)t + (\phi_i - \phi_o)))]/2 \\
&= 2R\sum_{i=1}^{N} A_i A_o \cos(\varphi_{io}),
\end{aligned}
\tag{1}
$$

where $\varphi_i = (\omega_i - \omega_0)t + (\emptyset_i - \emptyset_0)$, it represents the frequency offset and phase noise between the ith ONU and the LO. Similarly, the quadrature electronic signal can be written as $I_q = 2R\sum_{i=1}^{n} A_i A_0 sin(\varphi_i)$. The MZM is biased at its power null point, the IQ components can be also expressed as:

$$I_i = 2R \sum_{i=1}^{n} A_i A_j \cos(\varphi_i) = 2R \sum_{i=1}^{n} \sqrt{P_i P_0} \cos\left(\frac{\pi v_i}{v_\pi} + \frac{\pi}{2}\right) \cos(\varphi_i) \approx -2R \frac{\pi}{v_\pi} \sum_{i=1}^{n} \sqrt{P_i P_0} v_i \cos(\varphi_i), \quad (2)$$

$$I_q \approx -2R \frac{\pi}{v_\pi} \sum_{i=1}^{n} \sqrt{P_i P_0} v_i \sin(\varphi_i), \quad (3)$$

where v_i is the ith CAP signal, P_i is the optical power of ith ONU. After CAP demodulation and normalization, we can get the n streams of all ONUs, and the ith stream can be written as $E_i = a_i \cos(\varphi_i)$, $E_q = a_i \sin(\varphi_i)$, and a_i is the normalized mapped signal. To remove the frequency offset and phase noise, we combine IQ component of each ONU into a complex format.

$$E = E_i + jE_q = a_i\big(\cos(\varphi_i) + j\sin(\varphi_i)\big) = a_i \exp(j\varphi_i) \quad (4)$$

The phase component φ_i would cause the rotation of a_i in the complex filed. It has a similar feature in coherent optical communication, so the phase noise estimator is also suitable here. The original phase noise estimator is aimed at IQ modulation scheme such as QAM, so the estimator used here should be modified to match our PAM scheme. It can be also noticed that each ONU has its own frequency offset and phase noise, so they can be estimated respectively.

III. EXPERIMENTAL AND SIMULATION SETUP AND RESULTS

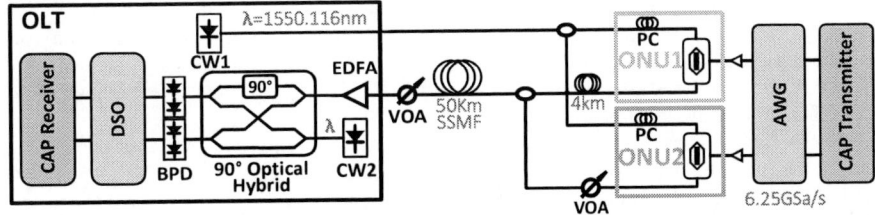

Fig. 1. Experimental setup for CAP-PON upstream transmission with carrier suppression and coherent detection

Fig. 1 shows the experimental setup of CAP-PON upstream transmission with carrier suppression and coherent detection. A group of 4D CAP filters with an upsampling factor of 5 is allocated to two ONUs. A data stream is mapped into PAM4 scheme and filtered by the 4D CAP filters to generate the CAP signals of two ONUs in Matlab offline. The net data rate of 5Gb/s per ONU (6.25Gsa/s × 2 × 2/5) baseband CAP signal is produced by an arbitrary waveform generator (AWG, TekAWG7122B). After electrical amplifier, the two CAP signals are used to modulate two 100 kHz-linewidth continuous waves (CW) at two MZMs biased at their power null point. These two CWs are split by a coupler from an external cavity laser (ECL, λ = 1550.116nm). Subsequently, these two optical carriers are coupled together and launched into a 50 km SSMF. It should be noticed that the 4 km transmission length difference between ONU_1 and ONU_2 is used to not only break the correlation between the upstream signals but also emulate the different locations of different ONUs. In OLT, the optical CAP signals are received by a coherent receiver. After balanced detection, the IQ components are captured by a 100GSa/s digital sampling oscilloscope (DSO) with 25GHz analog bandwidth (Tektronix, DPO72504DX). A digital signal process (DSP) based CAP receiver is used for offline process.

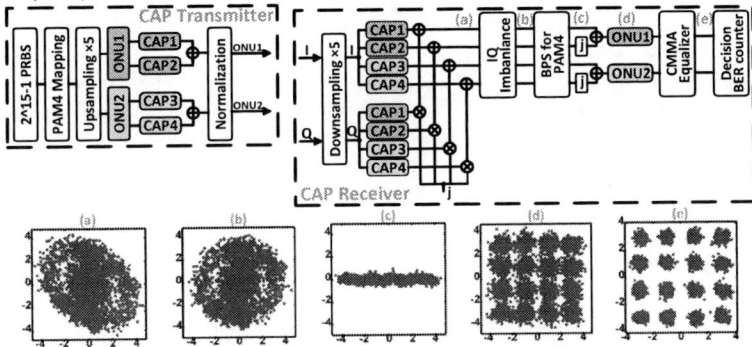

Fig. 2. The algorithm diagram of CAP transmitter and receiver in experimental setup.

Fig. 2 illustrates the algorithm diagram of the CAP transmitter and receiver in experimental setup, and all the constellations shown in the insets of Fig. 2 are driven from the experimental data. As mentioned above, the frequency offset and phase noise between each ONU and LO should be removed by the DSP. A total of eight streams are demodulated from the IQ components after the matched CAP filters. Fig. 2 (a) shows the combination of IQ components of the first stream. However, IQ imbalance would cause an elliptical constellation of the digitized received signals when considering the laser linewidth [6]. We use the ellipse correction method (EC method) to correct the ellipse to an ideal circle [6] as shown in Fig. 2 (b). The modified blind phase search (BPS) algorithm for PAM4 is used to estimate the frequency offset and phase noise to get a PAM4 scheme from Fig. 2 (b). The BPS algorithm also has a good tolerance to frequency offset and Fig. 2 (c) shows the constellation of the received signal after the BPS algorithm. Due to the fact that the two data stream of each ONU are propagated in the same optical link, so they can be combined as I/Q components for channel estimation, and the cascaded multi-modulus algorithm (CMMA) based equalizer is designed for channel

261

estimation in our experiment [7]. Fig. 2 (d) and (e) show the constellations of the received CAP signals before and after equalization respectively. In our experiment, a total number of 2×10^5 bits are used for BER counting.

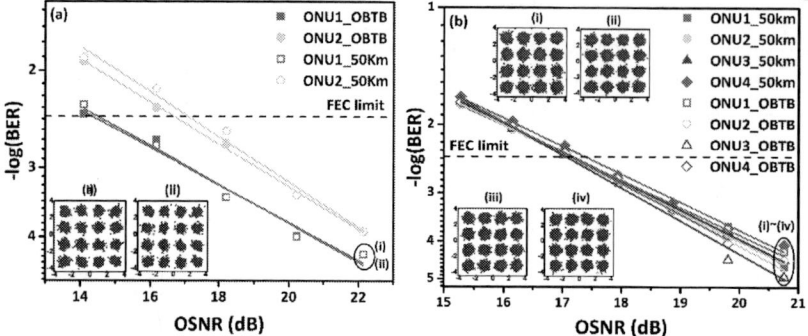

Fig. 3. The experimental (a) and simulated (b) BER performance by using the proposed technique.

Fig. 3 (1) shows the experimental BER performance of two ONUs by using the proposed technique in both optical back-to-back (OBTB) and 50 km SSMF cases. The optical signal to noise ratio (OSNR) is measured by an optical spectrum analyzer (OSA) before the 90-degree optical hybrid. By using the proposed technique, all the ONUs could achieve BER performance below the forward error correction limit (FEC) at 3.8×10^{-3}. The received constellation of two CAP signal (5 Gb/s per ONU) by using the proposed method are shown in the insets (a) and (b) of Fig. 3. It can be clearly noticed that 2.2 OSNR penalty between ONU1 and ONU2 are observed, and this can be attributed to the different performances of optical and electrical components, and the observed OSNR value of the optical CAP signal at the output of each ONU are not the same actually. Only two ONUs are tested in our experiment due to the limited number of MZMs in our laboratory. To verify the advantages of the proposed method in more users' scenario, the same architecture setup is simulated for four ONUs by using VPI Transmission Maker. In the simulation, 8-D CAP transversal filters are designed and each ONU is allocated two filters. The CAP modulation and demodulation are kept same as described above (the used upsampling factor is 12, the used laser linewidth is 100 kHz and each ONU has a net data rate of 6.67 Gbit/s (20GSa/s×2×2/12)). Fig. 3 (2) illustrates the BER performance of four ONUs with our proposed technique in both OBTB and 50km SSMF cases. The same as shown in Fig. 3 (1), all the OBI noises could be removed completely by using the proposed method. In this time, all the ONUs have almost the same BER performance due to the fact that the same optical and electrical components are used. The received constellations of four CAP signal (6.67 Gb/s per ONU) by using the proposed method in 50km SSMF case are shown in the insets (a)~(d) respectively.

IV. CONCLUSION

We have experimentally demonstrated a novel N-D CAP-PON system to achieve high-speed upstream transmission over a single wavelength. The OBI noises among ONUs in the upstream are suppressed after coherent receiver. The BPS algorithm is used to remove frequency offset and phase noise between ONU and LO. Theoretical analysis, experimental and simulated results show that all the OBI noises in the upstream transmission can be removed perfectly, and 10 Gb/s 4-D CAP-PON uplink transmission could be successfully achieved over a 50-km SSMF. This technique has potential applications in low-cost and high speed CAP-PON system for upstream transmission.

ACKNOWLEDGMENT

This work was supported by the National "863" Program of China (No. 2015AA016904) and the National Nature Science Foundation of China (NSFC) under Grant No. 61307091, 61331010 and 61107087.

REFERENCES

[1] L. G. Kazovsky, W. T. Shaw, D. Gutierrez, N. Cheng, and S.-W. Wong, "Next-generation optical access networks," J. Lightwave Technol., vol. 25, no. 11, pp. 3428-3442, November 2007.

[2] N. Cvijetic, D. Qian, J. Hu, and T. Wang, "Orthogonal frequency division multiple access PON (OFDMA-PON) for colorless upstream transmission beyond 10 Gb/s," IEEE J. Sel. Areas Commun., vol. 28, no. 6, pp. 781–790, August 2010.

[3] L. Zhang, B. Liu, X. Xiang, and Y. Wang, "10× 70.4-Gb/s dynamic FBMB/CAP PON based on remote energy supply," Opt Express, vol. 22, no. 22, pp. 26985-26990, October 2014.

[4] J. He, L. Shi, L. Deng, M. Cheng, M. Tang, S. Fu, M. Zhang P. Shum, and D. Liu, "Novel design of N-dimensional CAP filters for 10 Gb/s CAP-PON system," Opt. Lett., Vol. 40, no. 10, pp. 2409-2412, May 2015.

[5] S. M. Jung, S. M. Yang, K. H. Mun, and S. K. Han, "Optical beat interference noise reduction by using out-of-band RF clipping tone signal in remotely fed OFDMA-PON link," Opt. Express, vol. 22, no. 15, pp. 18246-18253, July 2014.

[6] S. H. Chang, H. S. Chung, and K. Kim, "Impact of quadrature imbalance in optical coherent QPSK receiver," IEEE Photon. Technol. Lett., vol. 21, no. 11, pp. 709-711, June 2009.

[7] S. J. Savory. "Digital coherent optical receivers: algorithms and subsystems," IEEE J. Sel. Top. Quantum Electron., vol. 16, no. 5, pp. 1164-1179, October 2010.

Performance investigation for LDPC-coded upstream transmission systems in IM/DD OFDM-PONs

Xiaoxue Gong, Lei Guo*, Jingjing Wu

School of computer science and engineering, Northeastern University, Shenyang, 110819, China
*guolei@cse.neu.edu.cn

Abstract: To mitigate interferences of subcarrier-to-subcarrier intermixing and ONU-to-ONU beating, we design a LDPC-coded and spectrum-efficient upstream transmission system. Simulation results demonstrate the receiver sensitivity is improved 3.6 dB under QPSK after 100 km transmission.

Keywords: *IM/DD OFDM-PONs; upstream transmission system; interference mitigation; LDPC codes*

I. INTRODUCTION

There have been some works focusing on the downstream transmission in intensity-modulation direct-detection orthogonal frequency division multiplexing passive optical networks (IM/DD OFDM-PONs), but it will be more challenging for us to investigate upstream transmission since interferences will occur when multiple optical network units (ONUs) simultaneously send their data to an optical line terminal (OLT). Aside from subcarrier-to-subcarrier intermixing interferences (SSII) induced by square-law detection, the same laser frequency for sending data from different ONUs results in ONU-to-ONU beating interferences (OOBI) at the OLT receiving side. SSII and OOBI produce unwanted frequency components that may fall within the range of the desired OFDM signal spectrum, thus leading to the degradation of transmission performance [1, 2].

Correspondingly, we introduce low density parity check (LDPC) codes previously used in optical communications to our designed multipoint-to-point upstream transmission system in IM/DD OFDM-PONs, with the objective of combating SSII and OOBI. In addition, a novel radio frequency (RF) upconversion scheme is proposed for achieving the improvement of spectrum efficiency. This scheme converts the baseband OFDM single to the frequency of the signal bandwidth in positive frequency components, leaving no guard band between the optical carrier and signal subcarriers. Our contributions can be summarized as follows. 1) A novel RF upconversion scheme is proposed by us, which doubles the spectrum efficiency compared to the traditional approach. 2) LDPC codes are introduced within the multipoint-to-point upstream transmission system of IM/DD OFDM-PONs for the first time, in order to mitigate SSII and OOBI. Numerical simulation results indicate that the proposed system can effectively combat SSII and OOBI, and then the receiver sensitivity is improved 3.6 dB under QPSK after 100 km standard single-mode fiber (SSMF) transmission.

II. DESIGN OF OUR LDPC-CODED UPSTREAM TRANSMISSION SYSTEM

Figure 1(a) describes the schematic of our LDPC-coded upstream transmission system in IM/DD OFDM-PONs. This system adopts a tree-based topology where two ONUs are connected to an OLT via distribution fibers, a combiner and a feeder fiber. As shown in Fig. 1(a), an electrical OFDM transmitter, a digital-to-analog converter (DAC), a RF upconvertor and an optical modulation module are deployed in each ONU. Correspondingly, a photon detector (PD), a low-pass filter (LPF), a RF downconvertor, an analog-to-digital converter (ADC) and an electrical OFDM receiver are deployed in the OLT. The structures of LDPC-coded OFDM transmitter and receiver are depicted in Fig. 1(b) and (c).

For the LDPC-coded OFDM transmitter shown in Fig. 1(b), data streams $\{b_1, b_2, ..., b_G\}$ are encoded into corresponding code words $\{C_1, C_2, ..., C_G\}$ by using the identical (J, K) LDPC with code rate K/J, where K and J are the number of information bits and code words, respectively. These code words are multiplexed into a stream of bits that will be mapped into M-ary phase-shift keying (MPSK) symbols before serial to parallel (S/P) conversion. Note that every transmitted symbol S_i takes its value from the element of a MPSK constellation set $D=\{d_n \mid n=0,1,2,...,M\text{-}1\}$. To achieve channel equalization, training symbols (TSs) are inserted at the front of the symbols owned by each ONU. Next, these symbols are modulated into partial overlapping and orthogonal subcarriers through N-point inverse fast Fourier transform (IFFT). These subcarriers can be used to carry useful data except direct current (DC) and Nyquist subcarriers. Subsequently, a cycle prefix (CP) is added, and the parallel-to-serial (P/S) conversion is then executed to get discrete OFDM signals which will be uploaded into DACs. It is worthy addressing two points: 1) every ONU modulates its data symbols on pre-assigned several of totally N subcarriers and leaves the remaining subcarriers zero-padded. In addition, according to the bandwidth requirement of ONUs, the OLT can allocate any number of data-carrying subcarriers; 2) we insert a certain amount of TSs at the beginning of data-carrying symbols for the individual ONU so that a separate channel estimation and equalization can be conducted.

OECC/PS2016

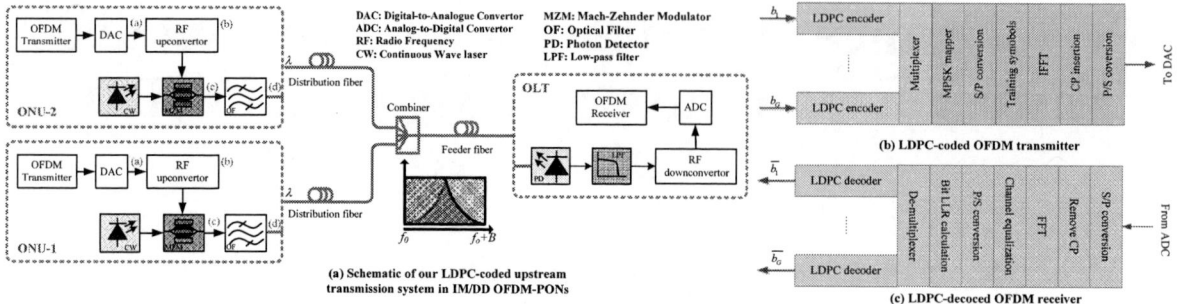

Fig. 1. Our designed transmission system.

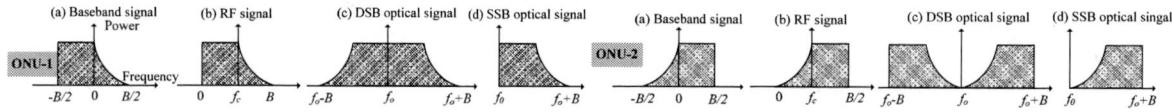

Fig. 2. Signal spectrum of ONU-1 and ONU-2.

Next, as shown in Fig. 1(a), we upconvert baseband complex signals resulted by DACs to f_{RF} through I-Q modulation before modulating a continuous wave (CW) laser using a Mach-Zehnder modulator (MZM). Here, f_{RF} is equal to the OFDM signal bandwidth in positive frequency components, so that there is no guard band between the optical carrier and the signal band, which doubles the spectrum efficiency obtained by using the traditional RF upconversion. Figure 2 illustrates the signal spectrum of ONU-1 and ONU-2 in different points relative to Fig. 1(a). Note that the RF bandwidth B of each ONU is given, and f_c is set to $B/2$ in our upconversion method because the corresponding spectrum efficiency can arrive to 100% in theory (50% for the traditional setting of $1.5B$). We also consider these two ONUs have the same optical carrier frequency w_0 and occupy the same number of subcarriers, e.g., $N/2$. In ONU-1, data symbols are carried by the subcarriers indexed from $N/2+1$ to N, while zeroes are assigned to the remaining subcarriers. Inversely, in ONU-2, the subcarriers indexed from 1 to $N/2$ carry data symbols.

At the combiner shown in Fig. 1(a), the optical OFDM signals coming from different ONUs via distribution fibers are merged into the whole OFDM signal that finally arrives at the OLT receiver along a feeder fiber. In particular, only one receiver is configured to process the received optical OFDM signal, which greatly reduces the cost and complexity of the OLT. In the OLT receiver, the square-law PD based on direct detection is firstly adopted to acquire the electrical OFDM. Subsequently, a LPF is employed to filter the out-of-band noise, and the received electrical OFDM signal is then processed with the inverse procedure adopted at the transmitter side. Note that the channel estimation and equalization is performed based on the known TSs, in order to get received data symbols R. The noise power spectrum density σ^2 of our proposed upstream transmission system is also estimated for bit log likelihood ratio (LLR) calculation. More specifically, we first compute the conditional probability $P(d_n|R)$ based on σ^2 according to Eq. (1), and then the LLR value $\lambda(P_i^m)$ of the mth bit corresponding to the ith received symbol R_i is determined by Eq. (2). Here, Φ_+ is the set of n satisfying the condition that the mth bit corresponding to d_n takes the value of 0; Φ_- is the set of n satisfy the condition that the mth bit corresponding to d_n takes the value of 1; S_i^m denotes the value of the mth bit corresponding to the ith transmitted symbol S_i.

$$P(d_n|R) = P(R|d_n) = \exp\left(-\frac{(x-x_n)^2 + (y-y_n)^2}{2\sigma^2}\right), R = x + j \cdot y, d_n = x_n + j \cdot y_n \quad (1)$$

$$\lambda(P_i^m) = \ln\left(\frac{P((S_i^m = 0)|R_i)}{P((S_i^m = 1)|R_i)}\right) = \ln\left(\frac{\sum_{n \in \Phi_+} P((S_i = d_n)|R_i)}{\sum_{n \in \Phi_-} P((S_i = d_n)|R_i)}\right) \quad (2)$$

As mentioned above, based on received symbols R, we can obtain the initial LLR value of all bits which will be used as the initialized information required by binary LDPC decoders. These initial LLR values are then de-multiplexed into G groups, each of which is decoded by using the log-domain belief propagation (BP) algorithm [3], so that the received bit sequence $(\bar{b}_1, \bar{b}_2, ..., \bar{b}_G)$ is determined. Finally, the bit error ratio (BER) of ONUs will be computed.

III. SIMULATIONS AND ANALYSIS

Through the co-simulation of VPI transmission Maker 9.1 and MATLAB software, we emulate our LDPC-coded upstream transmission system consisting of an OLT and two ONUs in Fig. 1(a). The LDPC-coded OFDM signals from ONUs are generated off-line by using MATLAB and then uploaded into VPI transmission Maker 9.1. The finally received electrical signal is downloaded to MATLAB for off-line process. In each ONU, (J=8192, K=6144) LDPC codes with the code rate of 0.75 are employed to encode information bits into code words. The number of iterations for log-domain BP decoding is 50. The total number N of subcarriers is 256, i.e., the size of an IFFT or FFT module is 256. In ONU-1, data symbols are carried by the subcarriers indexed from 129 to 256, while zeroes are assigned to the remaining subcarriers. Inversely, in ONU-2, the subcarriers indexed from 1 to 128 carry data symbols. It should be

264

pointed out that no data is carried by two subcarriers including DC component (the first subcarrier) and Nyquist component (the 129th subcarrier). The CP length is 32. The channel estimation and equalization is accomplished by inserting 5 TSs at the front of data symbols owned by each ONU. The symbol rate of two ONUs is 5 GS/s, which corresponds to the bandwidth of 2.5 GHz in positive frequency components. As a result, the bit rate of two ONUs is 5 Gb/s under QPSK modulation format. The baseband OFDM signal is upconverted to 2.5 GHz. The central frequency of CW laser is 193.1 THz. Figure 3 first depicts the spectrum of baseband signals, RF signals and optical signals for ONU-1 and ONU-2, respectively. Figure 3(a) illustrates the imaged electrical spectrum of ONU-1 baseband signal since negative spectrum components cannot be displayed in spectrum analyzers. The electrical spectrum of ONU-2 baseband signal is shown in Fig. 3(b). The electrical spectrums of ONU-1 and ONU-2 RF signals are demonstrated in Fig. 3(c) and (d). We can see that the power of those zero-padding subcarriers is smaller than that of data-carrying subcarriers. Figure 3(e) and (f) show the optical spectrum of ONU-1 and ONU-2. In theory, the proposed RF upconversion method can achieve 100% spectrum utilization, while the conventional approach only has 50% spectrum utilization. However, the final spectrum efficiency of our system will be degraded by introducing LDPC codes. Even so, the final spectrum efficiency of our system still yields an improvement of 25% since the code rate is 75%.

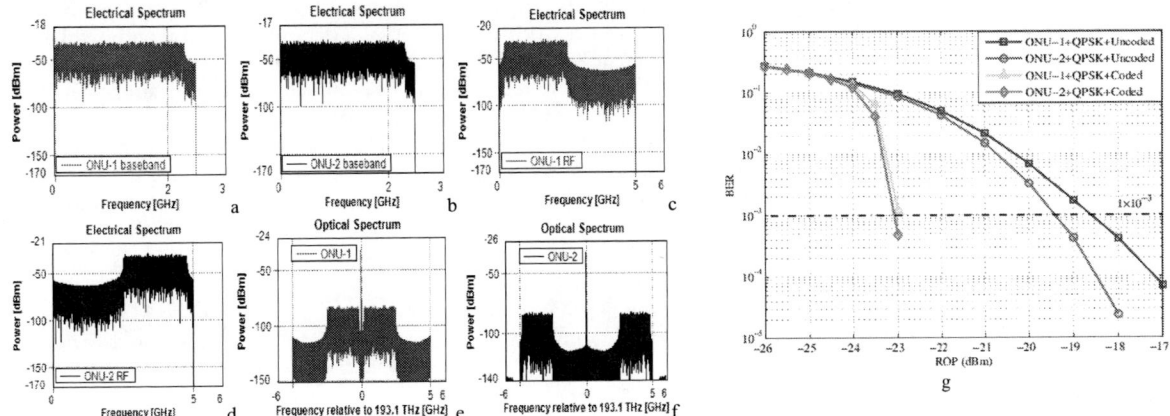

Fig. 3. Simulation results: (a) (b) baseband signals, (c) (d) RF signals, (e) (f) optical signals, (g) BER vs. ROP

Under QPSK modulation format, the measured BER as a function of received optical power (ROP) with and without LDPC codes after 100 km SSMF transmission is illustrated in Fig. 3(g) for ONU-1 and ONU-2, respectively. As expected, when the ROP increases, the corresponding BER becomes low. The ROP becomes the receiver sensitivity with the BER of 1×10^{-3} (FEC limitation). Without using LDPC codes, the receiver sensitivity of ONU-1 and ONU-2 is -19.3 dBm and -19.6 dBm, respectively. In addition, when the ROP becomes smaller, both two ONUs have almost the same BER since the key BER-limiting factor is the system noise associated with the ROP. However, the BER gap between ONUs gradually emerges as the ROP becomes larger. Moreover, the BER performance of ONU-2 is better than that of ONU-1. This is because the extra noise interferences suffered by two ONUs become the main BER-limiting factors. In particular, the extra noise interferences induced by SSMI and OOBI are more serious in low frequency components occupied by ONU-1, thus leading to a relatively high BER value of ONU-1. While under the case of using LDPC codes, the BER of both ONUs drops sharply as the ROP increases, and exhibits an error-free transmission when the ROP is larger than -23 dBm. The receiver sensitivity of ONU-1 and ONU-2 is -23 dBm and -23.1 dBm, which has been improved by 3.7dB and 3.5dB, respectively, when compared to the case without using LDPC codes.

IV. CONCLUSIONS

In IM/DD OFDM-PONs, aside from SSII induced by square-law detection, the same laser frequency for data sending from ONUs results in OOBI at the OLT receiving side. We designed a novel LDPC-coded upstream transmission system. The extensive simulation results demonstrated that the proposed system effectively alleviated SSMI and OOBI, and then improved the receiver sensitivity of ONUs.

REFERENCES

[1] C. Ju, X. Chen, N. Liu, and L. Wang, "SSII cancellation in 40Gbps VSB-IMDD OFDM system based on symbol pre-distortion," in *Proc*. ECOC, paper. P.7.9 (2014).

[2] C. Ju, X. Chen, N. Liu, Q. Zhang, and H. Wang, "OFDM PON downstream scheme with symbol pre-distortion and scalable receiver frontend," in *Proc*. OFC, paper. Th1H.4 (2015).

[3] Y. Chang, A. Casado, M. Chang, and R. Wesel, "Lower-complexity layered belief-propagation decoding of LDPC codes," in *Proc*. ICC, 1155-1160 (2008).

ThB1-1 (Invited)

OECC/PS2016

Mobile Backhaul and Fronthaul Systems

Frank J. Effenberger

Futurewei Technologies, 400 Crossing Blvd., Bridgewater, NJ 08807
frank.effenberger@huawei.com

Abstract: *The radio access network is growing fast, and the coming 5th generation systems promise to accelerate this trend. This paper reviews the potential solutions that address the significant challenges to the fiber backhaul and fronthaul.*
Keywords: *IMT2020; fronthaul; backhaul; optical access.*

I. INTRODUCTION

The rapid growth of wireless access systems is quite remarkable. This was mainly triggered through the introduction of smart phones and tablet computers into the market, and the deployment of the 4G LTE systems that can support them more efficiently. This transformed the cellular network from a voice and messaging system to a general purpose data network. Once that happened, the explosion of software applications drove data rates as high as the network can bear.

This growth has fuelled the deployment of more and more 4G systems, and this motivates the optimization of how 4G systems are constructed. A central theme of this optimization is the separation of the purely radio functions into the remote radio unit (RRU) and purely DSP functions in the baseband unit (BBU). The interconnection of these devices is the common public radio interface (CPRI) [1]. Originally designed to be a simple data link between BBUs at the base and the RRU at the top of the tower, this link has become stretched longer and longer, until significant (10~20km) distances are covered. This naturally led researchers to contemplate centralized radio access networks (C-RAN), with BBU-hotel concepts, or a data-center-based pool of BBU resources. However, as 4G is already in advanced deployment, these ideas and concepts are not so commonly deployed yet.

The growth has also raised interest in 5G systems, which promise to address not only the core data networking demand but new applications. Fifth generation wireless systems aim to increase total capacity by 1000×, to decrease latency to ~1ms, and to support 100× more devices in a given area. These 'headline' capabilities are envisioned to serve the needs of wireline access replacement / demand hot-spot (e.g., stadium events), telematics (e.g., driverless cars), and the Internet of Things, respectively.

An interesting point here is that these diverse requirements are not meant to be delivered at the same time. The vision is that the network should be application-driven, and the network will be adapted to suit the application(s) at hand. Since multiple physical networks would be too expensive and difficult to operate, these networks will have to be virtualized. This has led to the concept of network slicing, where applications can request any peculiar combination of network resources (spectrum, modulation, media access control, coding, etc). However, some of the aspects being virtualized are quite low in the networking stack: different modulation schemes are being considered for each big application. This is big reason why C-RAN architecture is attractive, since if all the raw data can be brought to the data center, then the system's capability is most flexible.

Another design concept being considered in 5G networks is the elimination of the cell-centric concept, in preference for the user-centric design. In cellular networks up through 4G, the basic unit of the network is the cell: a territory where one base station is the controller. Users move through cells, and the system handles the hand-off when a change of cell is needed. This inevitably results in less performance at the cell edges, and indeed cell edge performance is a typical key performance indicator. In the user-centric design, the user is in communication with all the RRU's that are his nearest neighbors, using coordinated multi-point (CoMP) techniques. In the general case of multiple diverse RRU's, the signal for a user can be maximized at his physical location, reducing the cell edge issue. The key enabler of such techniques is the dynamic and fine-grained control of the RRUs over a geographically relevant area. This is yet another reason why C-RAN is popular.

The remaining sections of this paper will address the situation of wireless backhaul and fronthaul, considering both the requirements of the wireless systems and the suitable transport solutions.

II. WIRELESS BACKHAUL

The backhaul network links the BBU to the core network. For the most part, backhaul uses Ethernet in modern wireless networks as its handoff interface. Interface rates can be 100 Mb/s or 1 Gb/s, and in future we might see 10 Gb/s interfaces. These signals can be carried either natively or in any of the many systems that can carry Ethernet, such as optical transport network (OTN), passive optical network (PON), and digital subscriber line (DSL).

In the current 4G core network, the BBU's need to be connected to the mobility management entity and the serving gateway, and these are connected to the packet data gateway and home subscriber server. Additionally, adjacent BBUs can communicate via the X2 interface, as can the adjacent servers and gateways. While the logical connectivity is somewhat complex, the use of Ethernet / Internet Protocol (IP) networks allow for a single common infrastructure to connect all these core functions. The logical connectivity can be enforced at the network layer, through the use of VLANs or other transport constructs. In the past, the core functions tended to be implemented on purpose built platforms; however, many of these functions are quite suitable for virtualization (lower computational complexity, generally packet-oriented processing). As a result, the core network is likely to be the first part of the wireless network to be moved to the data center. What this means for the backhaul network is that the BBU (wherever they happen to be) will be connected to a centralized data center using packet transport networks with a hub-and-spoke traffic pattern.

In terms of aggregated demand, current BBUs consume some fraction of a gigabit Ethernet interface, and this will be growing incrementally over the foreseeable future. For a typical metro area of perhaps 100 sectors, total capacity of 100 Gb/s is required, and this does not stress the currently deployed metro or access networks. Indeed, there are several 4G networks being served using G-PON, and when that system exhausts 10 Gb/s PONs are ready to assume the load.

The one requirement of wireless that goes beyond ordinary packet transport is the delivery of accurate time. Some wireless standards require time of day accurate to microsecond levels. Conventionally, this is obtained from global positioning system (GPS) receivers; however, in some deployments GPS is not so available. In particular, many future deployments will be indoors, and so the GPS signal will be blocked. Thus, precision timing needs to be carried over the backhaul network. On the conventional segments of the packet network, IEEE1588 precision time protocol (PTP) can be used, because the propagation delays are relatively symmetric. On some access networks like PON or DSL, the propagation delay is asymmetric, and this must be compensated by those networks. These methods have been described in ITU recommendations, and they leverage the knowledge of delay that the access systems already possess. Such methods have been implemented, and they deliver sub-100 ns accuracy timing.

III. WIRELESS FRONTHAUL

As defined in the Metro Ethernet Forum, fronthaul is the network that connects the RRU location to the BBU location. This definition was contemplated in the sense of a "classic" C-RAN or 4G network, where the RRU contains no baseband processing. These issues will be considered in section A below. However, there are current investigations on changing the functional split between the remote unit and the centralized unit. These will be addressed in section B.

A. Classic fronthaul

The classic fronthaul is the system wherein the RRU contains only RF and time domain digital functions. The components directly adjacent to the fronthaul interfaces are the digital to analog converter (DAC), and the analog to digital converter (ADC). In this case, all of the over-the-air information is transported to/from the BBU, and the RRU exhibits minimal control-plane functions. There is management plane interaction, and the fronthaul system must care for this management information. CPRI defines a framing structure that has both wireless sample data and management data segments, with a fixed ratio of 15 bits of samples for every bit of management.

The first thing to notice about the CPRI is the extreme requirements on its signal fidelity. Its frequency must not deviate by more than 2 parts per billion, and the bit error rate must be better than 10^{-12}. The origin of the former is that the CPRI system is line-timed, meaning that the DAC and ADC derive their frequency from the CPRI interface. Any deviation of frequency will produce incorrect RF channel placement, which is forbidden by regulation. These specifications raise significant transport issues [1-2].

A rough order of magnitude of fronthaul capacity can be estimated as follows. The current CPRI data rate resulting from a 20 MHz wide 4G LTE signal is approximately 1.25 Gb/s. The bandwidth driver for 5G will be the goal of 1000 times increase in capacity. This will be achieved by increasing spectrum, cell density, and frequency re-use by 10 times each. Increases of spectrum width and cell density produce direct increases in front-haul capacity. The 10 time increase in multiple input multiple output (MIMO) gain requires more than linear increases in fronthaul bandwidth. A commonly discussed MIMO degree is 64. Hence, the total bandwidth required for one sector of 5G would be 800 Gb/s.

Clearly, this C-RAN bandwidth problem is a shocking result. The costs of this transport would outweigh the presumed benefits of the 5G wireless system. There are already some remediation methods that have been proposed. For example, CPRI has a fairly high overhead bandwidth ratio. This is being addressed in some of the newer interface standards being developed now. The number of bits per sample can be adjusted, or the samples can be compressed. These compression methods can achieve perhaps a 2x reduction in total capacity, or 400 Gb/s.

Given this daunting problem, one can consider what optical systems could be used to address it. There are four solutions that could be considered: multi-fiber, WDM, OFDM, and analog, which are discussed below.

The simplest solution is to use simple point to point fibers to connect inexpensive data-com transceivers between the remote and central sites. The cheapest data rate at the current time is 10 Gb/s. The difficulty here is that to serve each 400 Gb/s sector would require 40 fiber pairs. This is obviously too fiber-hungry for a realistic network deployment. Currently, 25 Gb/s interfaces are starting to rise in popularity, and so this could reduce the fiber requirement here to ~16. The relevant standards in this case would likely be those developed in the IEEE P802.3.

To reduce the fiber count further, WDM can be employed. For instance, 100 Gb/s interfaces that employ 4 wavelengths of 25 Gb/s each are commonplace now. This can get the system to 4 fiber pairs per sector, which starts to be reasonable. Using even more wavelengths can get us to a single fiber per sector, which is indeed an attractive design point from an equipment perspective. However, all of the WDM approaches suffer from the fact that while fiber is conserved the equipment still has a large number of opto-electronic components. The relevant standards here would be the ITU-T G.989 series (NG-PON2) and G.9802 (Multi-wavelength access systems).

If we set our goal at larger capacity per wavelength, then we must consider more spectrally efficient use of the channel. This leads us to the use of OFDM over the optical link. If we begin with optics with bandwidth typical for a 25 Gb/s, then the RF bandwidth is about 20 GHz. Employing OFDM could perhaps get us to 5 b/s/Hz, producing a 100 Gb/s link on each wavelength. This approach reduces the requirement for WDM but it doesn't entirely eliminate it. Also, carrying an OFDM signal (5G wireless) over another OFDM signal (optical) is inefficient.

Taking this thought to the next step, directly carrying the wireless signal over the optical link seems more efficient and natural [3-6]. In this case, the different MIMO channels would be multiplexed using frequency, time, or other multiplexing method. Assuming we begin with a 20 GHz bandwidth optical link, frequency division multiplex can easily fit 64 channels of 200 MHz in this space using reasonable guard bands. So this system can achieve the goal of carrying an entire sector on a single channel. Unlike earlier RoF systems, this application is unique in that all these channels are being multiplexed in a single location. Hence, the necessary RF processing can be accomplished digitally, using DSP techniques. The viability of this new kind of digital frequency multiplexing based RoF has recently been verified experimentally. Of course, nothing comes without a drawback, and in this case, the analog optical transmission will induce some loss of signal fidelity. The back-to-back fidelity limited by the resolution of DAC's and ADC's can be engineered to <1% error vector magnitude (EVM). This technology has been described in ITU-T G.sup.55. This technology can be combined with WDM-PON to build a very useful high capacity system.

B. Modified fronthaul

This section tends toward speculation of what might be developed, because nobody knows what the modified fronthaul system will be. Nevertheless, we can make some general comments about it. The fundamental tradeoff in play here is that when more processing is transferred to the RRU, less transport bandwidth is required. Starting from the CPRI, there is the PHY (possibly broken into sub-layers), then the MAC (again broken into sub-layers), and then higher layers. There is little bandwidth saving motivation to go beyond the MAC.

Moving the PHY to the remote results in considerable bandwidth savings, because this layer terminates the MIMO processing as well as the fundamental detection process. The MIMO data reduction could reduce the transport demand by the difference between the MIMO degree (64) and the actual MIMO gain (10), or 6~7 times. This processing can be accomplished in the time domain, and it preserves the analog information needed to do coordinated transmission and reception techniques. Furthermore, it can be done largely independently of signal format.

Remoting the PHY detection process would reduce bandwidth by the difference between the CRPI sampling bit rate (30 bits/sample) and the user's actual peak bit rate (4~5 bits/symbol tone), or another 6~7 times. The detected signal is also less phase sensitive, and so packet transport becomes a real possibility. Unfortunately, detection requires the knowledge of the signal format, and so the 5G flexibility would have to move to the remote, making it more complex. Also, detection fundamentally discards information, and thus makes advanced CoMP processes impossible. Still, coordinated scheduling and beam forming could be possible.

Remoting the MAC would reduce bandwidths largely by suppressing the silent intervals on the link. The fronthaul bandwidth would be dependent on actual network activity, and not the "always-on" characteristic of classic fronthaul. Such a scheme could certainly use packet transport schemes, as there is little time sensitivity. However, a remote MAC moves most of the BBU out to the remote, and the difference between this and backhaul is hardly worth it.

IV. CONCLUSIONS

The relationship between wireless systems and the wireline transport systems that serve them is in flux. Backhaul systems of many flavors are available, and are seeing large deployment. C-RAN and fronthaul faces the gap between huge demand and limited bandwidth supply. It is likely that both the transport system and the fronthaul interface will be modified to meet each other at a common point. On supply, spectrally efficient or compressed optical formats such as analog transmission are likely to be used. On demand, some functions such as MIMO processing are likely to be moved to the remote site. These can close the gap, and achieve an efficient yet feature-rich 5G wireless system.

REFERENCES

[1] A. Pizzinat et al., "Things you should know about fronthaul," Proc. ECOC, Tu.4.2.1, Cannes (2014).
[2] J. Kani, "Solutions for future mobile fronthaul and access-network convergence," Proc. OFC, W1H.1, Anaheim (2016).
[3] N. Shibata et al., "256-QAM 8 wireless signal transmission with DSP-assisted analog RoF for mobile front-haul in LTE-B," Proc. OECC, 129-131, Melbourne (2014).
[4] X. Liu, et al., "Efficient mobile fronthaul via DSP-based channel aggregation," J. Lightwave Tech., **34**, no. 6, p. 1556 (2016).
[5] CPRI Specification V7.0, "Common Public Radio Interface (CPRI); Interface Specification," October 2015.
[6] X. Liu et al., "CPRI-compatible efficient mobile fronthaul transmission via equalized TDMA achieving 256 Gb/s CPRI-equivalent data rate in a single 10-GHz-bandwidth IM-DD Channel," Proc. OFC, W1H.3, Anaheim (2016).

ThB1-2

OECC/PS2016

Highly Linear W-band Transmitter Based on OFC and Heterodyne Up-conversion

Jinhui Luo, Shangyuan Li, Xiaoping Zheng, * Hanyi Zhang and Bingkun Zhou

Tsinghua National Laboratory for Information Science and Technology, Department of Electronic Engineering, Tsinghua University, Beijing 100084, China

* xpzheng@tsinghua.edu.cn

Abstract: *A highly linear W-band millimeter-wave over fiber transmitter incorporating optical frequency comb is proposed. By linearized modulation of two coherent comb teeth using a dual-parallel Mach-Zehnder modulator, an experimental SFDR of 122.7dB·Hz$^{2/3}$ is demonstrated.*

Keywords: *W-band; Fiber optics links and subsystems; Modulation*

I. INTRODUCTION

With the development of multimedia service driving the demand for increasing wireless communication capacity, a higher carrier frequency at millimeter wave (MMW) bands such as the V-band (57-64 GHz), the W-band (75-110GHz) and the terahertz has drawn intense research because of the broad license-free bandwidth as much as 7GHz [1]. MMW over fiber (MMoF) is a promising technology to combine the merits of both fiber optics and MMW communication, such as low loss, immune to electromagnetic interference and large bandwidth for MMW signal [2]. Nevertheless, limited by the device bandwidth and the state of art manufacturing techniques, it's hard to directly modulate W band radio signals on an optical carrier. A widely used method [3] is heterodyne mixing of two free-running lasers or two coherent optical carriers at a photodetector (PD) where broadband signals are up converted to W band or higher frequency. An optical frequency comb (OFC), which has a series of highly coherent optical carrier, can also be used as a heterodyne local oscillator (LO) which up-converts signals to W-band or even higher frequency by selecting two comb teeth of the OFC [4]. Such a transmitter provides a simple way to generate MMW at W-band or terahertz band and reduces the requirements for expensive MMW components. However, due to the nonlinearity of an electro-optic modulator, a heterodyne up-conversion transmitter suffers from the intermodulation distortions (IMDs) which lay inner the signal band and the spurious free dynamic range (SFDR) is limited. To improve the SFDR performance, numerous methods have been proposed. The main ideas behind can be divided into two categories, i.e., introducing additional nonlinear distortions which is opposite to the existing ones, or analyzing the optical spectrum contributors of the IMD3s and suppress them directly in the optical domain [5]. However, none of the methods aforementioned is designed for a W-band MMoF system based on OFC and heterodyne frequency up-conversion.

We have proposed a highly linear radio-over-fiber transmitter incorporating a single-drive dual-parallel Mach-Zehnder modulator (SD-DPMZM) [6]. An improved SFDR can be reached by optimizing the three bias points of the modulator and the two origins of the IMD3 cancel each other. In this article, we introduce this method to a W-band transmitter based on heterodyne frequency up-conversion using an OFC source. A SFDR of 122.7 dB·Hz$^{2/3}$ is experimentally obtained for a heterodyne up-conversion and optical down-conversion W-band system working at the frequency of 95GHz. The SFDR is improved by 22.8 dB compared to a system utilizing conventional quadrature biased MZM.

II. SYSTEM DESIGN

The proposed system design is shown in Fig. 1. A flat OFC is generated by cascading intensity and phase modulators driven directly by a sinusoidal waveform. By adjusting the direct current (DC) bias of the intensity modulator and the input of the sinusoidal source, more than 15 highly coherent comb lines within 1 dB spectral power variation can be obtained [7]. Two comb teeth spaced at $f_{RF} = n * \Delta f$ are selected by an arrayed waveguide grating (AWG) as optical carriers for heterodyne frequency up-conversion, where f_{RF} is the central frequency of the W-band signal and Δf is the frequency interval of the comb teeth. A full band utilization of W-band ranging from 75 to 110GHz can be achieved by adjusting the frequency interval of the comb teeth and the number of intervals. Then the two comb teeth, illustrated as Fig. 1.B, are modulated in the SD-DPMZM. One of the sub-modulators, denoted as MZM-1, is driven by an intermediate frequency (IF) signal $V_m(t)$ and biased at V_1. The other sub-modulator (MZM-2) and the parent modulator (MZM-3) are kept unmodulated and biased at V_2 and V_3, respectively.

After modulation in the SD-DPMZM, the two comb teeth are feed into a photodetector (PD), where the IF signal is up-converted to W-band. However, limited by the test conditions, it's difficult to directly measure the W-band spectrum. To evaluate the SFDR of the transmitter, optical heterodyne down-conversion is adopted. That is, a coherent comb tooth is reserved as optical LO (the green arrow in Fig. 1.B) and the W-band signal is down-converted to IF, which is in the measurement range of an electrical signal analyzer (ESA-1). As a comparison, the SFDR of the system with electrical frequency down-conversion is also measured by ESA-2.

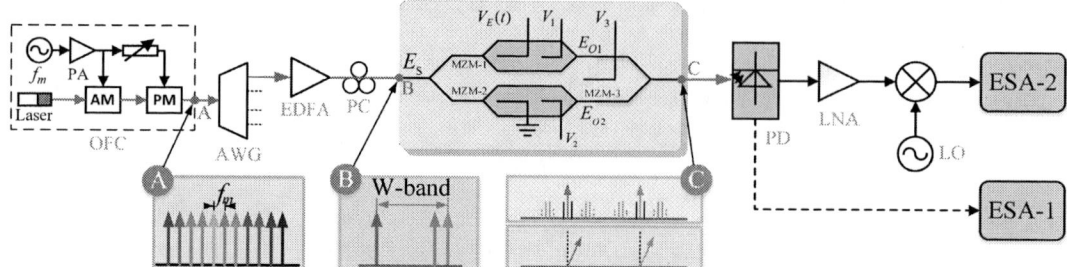

Fig. 1. Block diagram and optical spectra of the proposal, with an inside look at the structure of the DPMZM

To investigate the IMDs of the W-band transmitter, a two-tone signal is applied to the SD-DPMZM. The optical spectra at the output of MZM-1 and MZM-2, denoted as E_{O1} and E_{O2} , are shown as Fig. 2(a) and Fig. 2(b), respectively. By adjusting the bias point of the three MZMs to an optimized position, the phase of the comb teeth is opposite to that of their sidebands at the output of the SD-DPMZM, shown as Fig. 2(c). There are two optical contributors to the IMD3 in electrical domain after heterodyne beating at the PD. One is the beating results between the optical carrier and the IMD3 components in optical domain, the other is the beating results between the 1st-order IMD components and the 2nd-order IMD components. When biased at the optimized position, the IMD3s from the two origins have equivalent intensity and opposite phase and the IMD3 at the output of the SD-DPMZM is eliminated, as shown in Fig. 2(d). As a result, the SFDR of the transmitter is greatly improved.

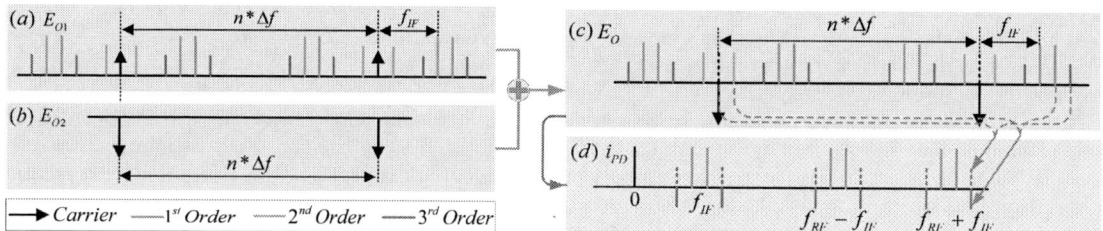

Fig. 2. Spectrum evolution of the W-band transmitter. The optical spectra of (a) MZM-1, (b) MZM-2, (c) MZM-3. (d) The electrical spectra after PD.

III. EXPERIMENTS

In the experiment, the RF source used to drive the OFC source is 18GHz. More than 15 highly coherent comb teeth are generated and then feed into an optical spectrum processor (OSP, Finisar Waveshaper 4000s). Two comb teeth spaced at 90GHz (5*18GHz) are selected by the OSP as optical carrier and then modulated in the DPMZM (Covega, Mach-10) by two RF tones centered at 5.0GHz, with frequency separation of 40MHz. The optical spectra after modulation compose of two optical carriers and one optical LO, as shown in Fig. 3. Then the two-tone signal is up-converted to W-band with the frequencies of 94.98GHz and 95.02GHz. Limited by the bandwidth of the signal analyzer (SA, Agilent N9030A) used, the spectra of the W-band two-tone signal can't be directly measured. So a coherent optical LO is reserved and the two-tone signal is down-converted to 22.98GHz and 23.02GHz.

The main tone (MT) and IMD3 power after PD vary along with the three bias voltage applied to the DPMZM. By adjusting the bias voltage of the DPMZM, the IMD3 are eliminated and the carrier to interference ratio (CIR) is improved from 36 dB to 65 dB, as shown in Fig. 4.

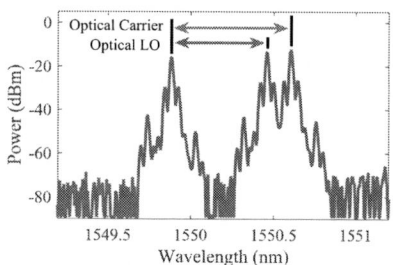

Fig. 3 Optical spectra at the output of the DPMZM.

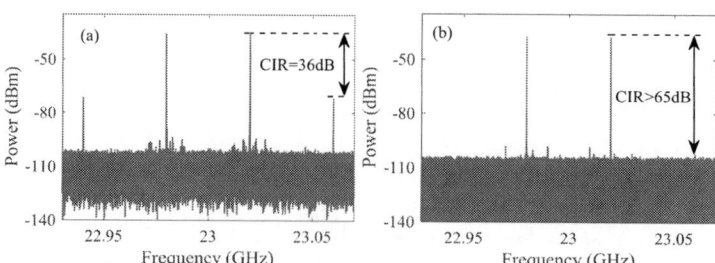

Fig. 4 Measured CIR after down-conversion (a) conventional MZM, (b) optimized.

OECC/PS2016

The three MZMs of the DPMZM are biased at 0.548V, 3.4V and -1.3V. The noise floor, which is dominated by shot noise and thermal noise, is -158.3 dBm/Hz and the measured SFDR is 122.7 dB·Hz$^{2/3}$, as shown in Fig. 5(a). An improvement of 22.8 dB can be achieved compared to a conventional quadrature biased MZM system.

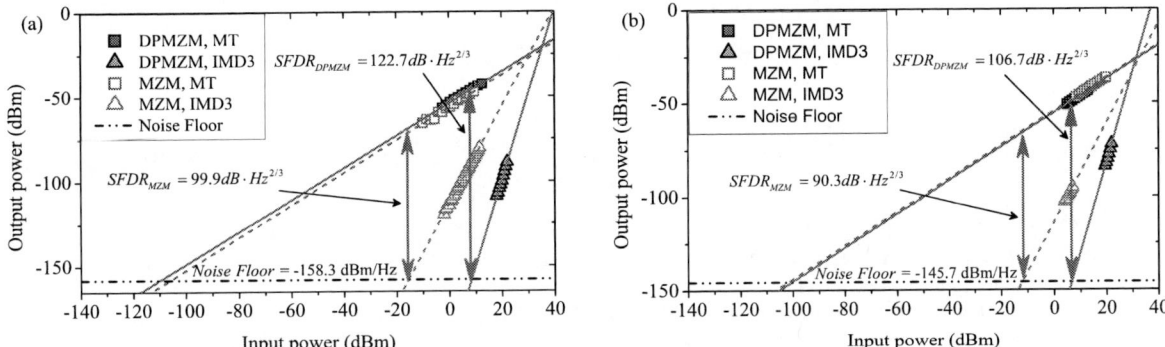

Fig. 5 SFDR of the optimized DPMZM system and conventional MZM system, (a) with heterodyne optical down-conversion, (b) with electrical frequency down-conversion.

Furthermore, to evaluate the linearity of the proposed W-band transmitter, an W-band electrical balanced mixer (Farran BMC-10) is adopted to down-convert the W-band two tone signal to IF. An improvement of 16.4 dB is achieved compared to a conventional quadrature biased MZM system with the same electrical mixer, as shown in Fig. 5(b).

IV. CONCLUSIONS

In conclusion, a highly linear W-band transmitter based on OFC and heterodyne up-conversion is proposed. By selecting two coherent comb teeth as optical carrier, IF signals are up-converted to W-band and a SFDR of 122.7 dB·Hz$^{2/3}$ are experimentally reached when the DPMZM are biased at optimized points.

ACKNOWLEDGMENT

This work was partly supported by National Basic Research Program of China (973 project) under grant No. 2012CB315603-04; National Nature Science Foundation of China (NSFC) under grant No. 61420106003，61435006, 61321004.

REFERENCES

[1] T. Nagatsuma, T. Takada, H.-J. Song, K. Ajito, N. Kukutsu, and Y. Kado, "Millimeter- and THz-wave photonics towards 100-Gbit/s wireless transmission," Proc. 23rd Annu. Meeting IEEE Photon. Soc., Denver, CO (2010), pp. 385–386.
[2] J. Beas, G. Castanon, I. Aldaya, A. Aragón-Zavala, & G. Campuzano, (2013) "Millimeter-wave frequency radio over fiber systems: a survey." IEEE Communications Surveys & Tutorials, vol. 15, 2013, pp. 1593-1619.
[3] J. Yu, Z. Jia, L. Yi, Y. Su, and G. Chang, "Optical millimeter-wave generation or up-conversion using external modulators," IEEE Photon. Technol. Lett. vol. 18, 2006, pp. 256–267.
[4] S. A. Diddams, "The evolving optical frequency comb [Invited]," J. Opt. Soc. Am. B 27, 2010, B51–B62.
[5] K. Xu, R. Wang, Y. Dai, F. Yin, J. Li, Y. Ji, & J. Lin (2014), "Microwave photonics: radio-over-fiber links, systems, and applications [Invited]." Photonics Research, vol. 2, 2010, B54-B63.
[6] S. Li, X. Zheng, H. Zhang, & B. Zhou. "Highly linear millimeter-wave over fiber transmitter with subcarrier upconversion." In CLEO: Science and Innovations (p. JWA3). Optical Society of America.
[7] Y. Dou, H. Zhang, & M. Yao, "Improvement of flatness of optical frequency comb based on nonlinear effect of intensity modulator" Optics letters, vol. 18, 2011, pp. 2749-2751.

ThB1-3

OECC/PS2016

Self-phase Modulation Based Signal Distortions of Optical SSB-SC Signal with Pilot Carrier

K. I. Amila Sampath, Katsumi Takano, and Manabu Sato

Graduate School of Science and Engineering, Yamagata University, 4-3-16 Jonan, Yonezawa, Yamagata, 992-8510, Japan.
kiamila.sampath.jp@ieee.org

Abstract: *SPM based signal distortions of pilot carrier added OSSB-SC signal are numerically analyzed. With increasing pilot carrier power, SPM threshold of pilot carrier added OSSB-SC signal reaches that of OSSB-EC signal and saturates.*
Keywords: *Optical communications, PAPR, Hilbert transform, Fiber transmission.*

I. INTRODUCTION

Compared with intensity modulation, optical single sideband (OSSB) modulation utilizes only a half of the baseband spectrum, presenting a two-fold increase in spectral efficiency. Because of narrower bandwidth OSSB signal is dispersion tolerant. Moreover, dispersion compensation can be done in electrical domain after direct detection using simple methods such as microstrip lines [1]. OSSB signal can be generated by filtering intensity modulated signal with an optical bandpass filter [2] or by suppressing one side band using phase-shift method [3]. Along with the spectral efficiency, carrier suppressed OSSB (OSSB-SC) signal is beneficial in terms of energy efficiency because the entire energy is being transmitted is used for data transmission. Higuma et al. first demonstrated OSSB-SC signal generation by phase shift method using LiNbO₃ (LN) vector modulator and reported high sideband suppression [4]. However, high peak-to-average power ratio (PAPR) of OSSB-SC signal increases self-phase modulation (SPM) based fiber nonlinear impairments limiting the maximum reach [5]. Since peaky Hilbert-transformed signal component which is used for spectral suppression is the origin of high PAPR [6], emitted carrier OSSB (OSSB-EC) signal can reduce PAPR and signal distortions accordingly. However, this PAPR reduction comes only at the expense of sideband suppression degradation [7].

In this paper, we propose adding a pilot carrier to OSSB-SC signal which avoids sideband suppression degradation while reducing PAPR. We numerically analyze SPM based signal distortions of pilot carrier added OSSB-SC signal. Signal distortions caused by SPM are evaluated and compared with those of OSSB-EC signal using fiber transmitted signal eye diagrams.

II. OPTICAL SSB MODULATION BY PHASE-SHIFT METHOD

A schematic of phase-shift method OSSB transmitter is shown in Fig. 1(a). Two sub-Mach-Zhender interferometers (sub-MZIs) of LN vector modulator are driven by baseband signal $V_B(t)$ and its Hilbert transform $V_H(t)$. The Hilbert transform is defined in Eq. (1) where sgn($\cdot$) is signum function.

$$H(\omega) = \begin{cases} -j\,\text{sgn}(\omega) & (\omega \neq 0) \\ 0 & (\omega = 0) \end{cases} \quad (1)$$

Because of π phase difference between the spectral sidebands, a sideband suppressed signal can be generated by combining the light of upper and lower arms of LN vector modulator. Carrier suppression is implemented by biasing two sub-MZIs to their transmission null points. As can be seen in Fig. 1(a), Hilbert transformed signal becomes peaky at the transition points of Non-return-to-zero (NRZ) coded baseband signal. These peaks of Hilbert transformed signal cause peaks in modulator output signal. Peak-to-average power ratio (PAPR) of OSSB-SC signal becomes high causing signal degradations at the receiver due to SPM induced phase shifts during fiber transmission.

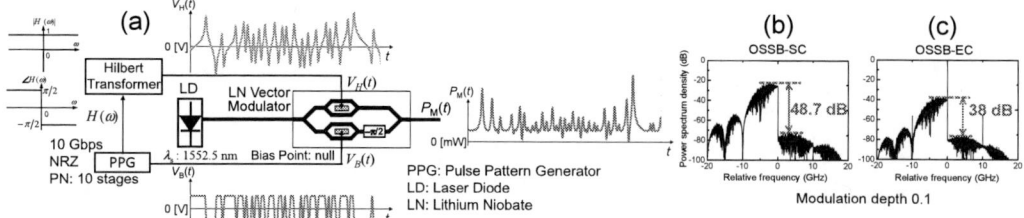

Fig. 1. (a) OSSB-SC transmitter and modulator input, output waveforms, (b) and (c) power spectra of OSSB-SC and OSSB-EC signals at modulation depth of 0.1.

272

Since the presence of optical carrier hides the peaks form Hilbert transformed signal component, PAPR of OSSB-EC signal decreases. OSSB-EC signal can be generated by simply changing the bias point of the lower sub-MZI from null point to quadrature point where the baseband signal is modulated. However, changing the bias point brings on the high order harmonics in the spectrum by increasing the power of suppressed sideband due to the nonlinearity of the sub-MZI. Fig. 1(b) and (c) portray power spectra of OSSB-SC and OSSB-EC signals respectively at modulation depth of 0.1. Modulation depth is defined as the ratio of baseband signal peak voltage to V_π where V_π is half-wave voltage of LN vector modulator. We define sideband suppression ratio (SSR) as the difference of maximum power spectral densities of unsuppressed and suppressed sidebands. As depicted in Fig. 1(b), (c) 48.7 dB SSR of OSSB-SC signal drops by more than 10 dB when bias point changes from null to quadrature point to emit the carrier. To avoid the SSR degradation due to emission of carrier, pilot carrier can be added to OSSB-SC signal.

III. SYSTEM DESCRIPTION

SPM based signal distortions of pilot carrier added OSSB-SC signal were analyzed using the fiber transmission model illustrated in Fig. 2. The light of 1552.5 nm in wavelength from continuous wave (CW) laser was split by fiber coupler A and its output ports were connected to LN vector modulator and Erbium-doped fiber amplifier (EDFA). The input light to the modulator was amplitude modulated by 10 Gb/s NRZ coded baseband (PN stage 10) signal and its Hilbert transform at modulation depth of 0.1. Both sub-MZIs of LN vector modulator were biased to null points of their transmission curves. The optical power of the light entering to the EDFA was amplified and adjusted by an optical attenuator after the amplification. Phase of the amplified light was matched with the phase of LD output before combining with the modulator output signal at fiber coupler B. The output of coupler B was launched into 100-km long single mode fiber (SMF). SMF has an attenuation of 0.2 dB/km, dispersion coefficient 17 ps/nm/km, and effective core area of 80 μm^2. Dispersion compensating fiber of 21.5 km with an attenuation of 0.45 dB/km, dispersion coefficient −80 ps/nm/km, and effective core area of 14 μm^2 was used to compensate fiber chromatic dispersion. Transmitted signal was detected by a phase-diversity homodyne detector consisting of a balanced receiver and a local oscillator (LO).

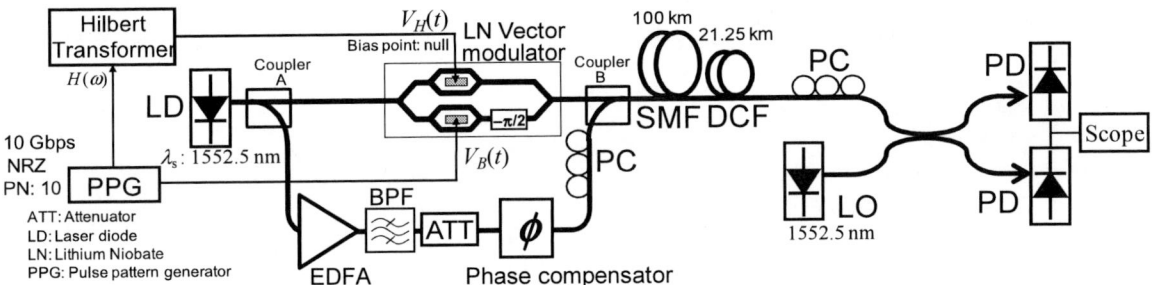

Fig. 2. Simulated fiber transmission setup of pilot carrier added OSSB-SC signal.

Since this study is focused on SPM based signal distortions, amplitude spontaneous (ASE) noise form EDFA was ignored for simplicity. For the same reason, LD and LO were assumed to have 0 Hz linewidths and identical phase. Under the assumption that two photodiodes of the balanced receiver have ideal 20 GHz rectangular frequency response, transmitted signal was detected when LO power-to-received signal power ratio was 20 dB. Fiber transmission was calculated by solving non-linear Schrödinger equation using split-step Fourier method [8].

IV. RESULTS AND DISCUSSION

Fig. 3 compares the fiber input intensity waveforms, their power spectra and received signal eye diagrams of OSSB-SC, pilot carrier added OSSB-SC and OSSB-EC signals at average fiber input power of 9 dBm. Received signal eye diagram of OSSB-SC signal was distorted considerably due to the peaks of intensity waveform. Peaks of fiber input intensity waveform disappeared with introduction of optical carrier. As a result, PAPR was decreased and eye openings of received signal were improved. Carrier-to-LSB (lower sideband) main lobe power ratio of OSSB-EC signal becomes 14.3 dB at modulation depth of 0.1. It is also noticeable that almost similar eye opening to OSSB-EC can be achieved for pilot carrier added OSSB-SC signal at pilot carrier-to-LSB main lobe power ratio of 10 dB.

SPM based signal distortions were evaluated using eye diagrams of the received signal. Eye opening (*EO*) is defined as the ratio of inside and outside eye openings as depicted in the inset of Fig. 4(a). Eye opening penalty (*EOP*) is defined as follows:

$$EOP = \frac{E_R}{\left(\alpha_{SMF}L_{SMF} \times \alpha_{DCF}L_{DCF}\right)E_T}, \qquad (2)$$

where $E_{R,T}$ are *EO*s of received and transmitted signals respectively. α and L are loss coefficients and fiber lengths respectively. *EOP* is plotted against average fiber input power in Fig. 4(a) when the pilot carrier power was varied. *EOP* increases with average fiber input power because of signal distortions due to nonlinear phase shift caused by SPM. Increasing pilot carrier power decreases *EOP* for a given average fiber input power. We define SPM threshold as the average fiber input power which *EOP* becomes 1 dB. Fig. 4(b) presents the relation of SPM threshold and PAPR of pilot carrier added OSSB-SC signal. With increasing pilot carrier power, SPM threshold power increases due to PAPR reduction. And SPM threshold power starts to saturate around 5.4 dBm which is the SPM threshold of studied system for OSSB-EC signal. Further increasing of pilot carrier power does not improve SPM threshold power in spite of PAPR reduction. This can be attributed to the phase mismatch between the pilot carrier and LO.

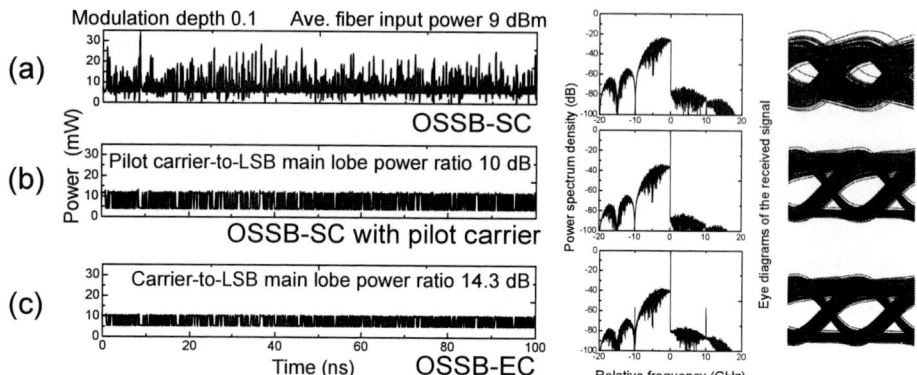

Fig. 3. Fiber input intensity waveforms, their power spectra and signal eye diagrams after fiber transmission (a) OSSB-SC, (b) OSSB-SC with pilot carrier, (c) OSSB-EC.

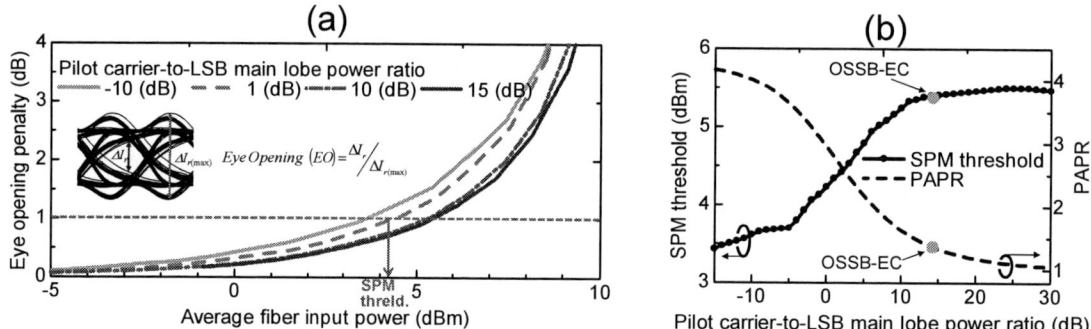

Fig. 4. (a) Eye opening penalty (EOP) of pilot carrier added OSSB-SC signal, inset: definition of eye opening (EO), (b) comparison of SPM threshold power and PAPR of pilot carrier added OSSB-SC and OSSB-EC signals.

V. CONCLUSIONS

SPM based signal distortions of pilot carrier added OSSB-SC signal were numerically analyzed after phase-diversity homodyne detection. Introduction of pilot carrier reduces PAPR of OSSB-SC signal resulting an increase of received signal eye opening and SPM threshold power accordingly. However, it was found that the maximum SPM threshold power of pilot carrier added OSSB-SC signal cannot exceed the SPM threshold of OSSB-EC signal at phase-diversity homodyne detection.

ACKNOWLEDGMENT

This work was partly supported by MEXT (KAKENHI No. 15K13985), Marubun Research Promotion Foundation, and Sato Yo International Scholarship Foundation (No. 644).

REFERENCES

[1] M. Sieben, et al., IEEE J. Lightwave Technol., vol. 17, no. 10, pp. 1742-1749, Oct., 1999.
[2] S. Bigo, IEEE J. Sel. Topic Quan. Electron., vol. 10, no.2, pp. 329- 340, March 2004.
[3] B. P. Lathi and Z. Ding, Modern Digital and Analog Communication Systems, 4th ed., Oxford, 2009.
[4] K. Higuma, et al., Electron. Lett., vol. 37, no. 8, pp. 515-516, April 2001.
[5] K. Takano, et al., Electron. Lett., vol. 40, no. 18, pp. 1150-1151, Sept., 2004.
[6] K. Takano, et al., Opt. Express vol. 19, no. 10, pp. 9699-9707 May 2011.
[7] K. Takano, et al., IEICE Trans., on Commun., vol. E88-B, no. 5, pp. 1994-2003, May 2005.
[8] G.P Agrawal, Nonlinear Fibe Optics, 4th ed., Academic, 2010.

ThB1-4

Transmission of 28-Gb/s Duobinary Signals over 45-km SSMF Using 1.55-μm Directly Modulated Laser

Chuanbowen Sun[1], S. H. Bae[2], Hoon Kim[1,2], and Y. C. Chung[2]

[1]Graduate School for Green Transportation, KAIST, 291 Daehak-ro Yuseong-gu, Daejeon, Korea 34141
[2]School of Electrical Engineering, KAIST, 291 Daehak-ro Yuseong-gu, Daejeon, Korea 34141
Email: hoonkim@kaist.ac.kr

Abstract: We demonstrate the transmission of 28-Gb/s 3-level electrical duobinary signals generated by a directly modulated laser over 45-km SSMF. The duobinary signals driving the laser are obtained by either a low-pass filter or delay-and-add circuit.
Keywords: Duobinary, direct modulation, modulation format

INTRODUCTION

Recently, there have been growing interests in 25-Gb/s/channel systems operating in the 1.55-μm window for short- and intermediate-haul fiber-optic links. By using the 1.55-μm window instead of the conventional 1.3-μm window, we can readily increase the system capacity by using the dense wavelength-division-multiplexing technology and take the advantage of erbium-doped fiber amplifiers for boosting the signal powers. Previously, the 25-Gb/s/channel systems have been implemented by using expensive and lossy external modulators such as Mach-Zehnder modulators and electro-absorption modulators. However, it is highly desirable to utilize low-cost optical transmitters such as directly modulated lasers (DMLs) for the cost-effective implementation of short- and intermediate-haul 25-Gb/s/channel transmission systems.

Major technical challenges associated with the use of 1.55-μm DMLs for high-speed (e.g., >25 Gb/s) transmission are the limited modulation bandwidth of the devices and the chirp-induced signal distortions. The bandwidth of commercially available DMLs is typically limited to <15 GHz. Thus, multi-level modulation formats such as 4-level pulse amplitude modulation (PAM4) could be used to be accommodated within the modulation bandwidth of DMLs [1-2]. A large chirp accompanied by the direct current modulation of laser diode gives rise to signal waveform distortions when combined with fiber chromatic dispersion. The dispersion-limited transmission distance of the DML-based system can be extended by using optical dispersion compensation such as a chirped fiber Bragg grating and a delay interferometer [3-4].

In this paper, we experimentally demonstrate the transmission of 28-Gb/s 3-level electrical duobinary signals over 45-km standard single-mode fiber (SSMF) by using a 1.55-μm DML. In order to generate the 28-Gb/s electrical duobinary signals, we employ two different duobinary encoders, a low-pass filter (LPF) and a delay-and-add (DAA) circuit. The extinction ratio (ER) of the optical signal is optimized to maximize the transmission distance without using bulky and lossy optical dispersion compensation modules. The electrical equalization technique is utilized at the receiver to compensate for the fiber chromatic dispersion.

EXPERIMENTS AND RESULTS

Fig. 1(a) shows the experiment setup. We first send a pseudo-random binary sequence (length=2^{15}-1) either to an LPF or DAA circuit to generate 28-Gb/s electrical duobinary signals. The LPF used as an encoder is a Bessel filter

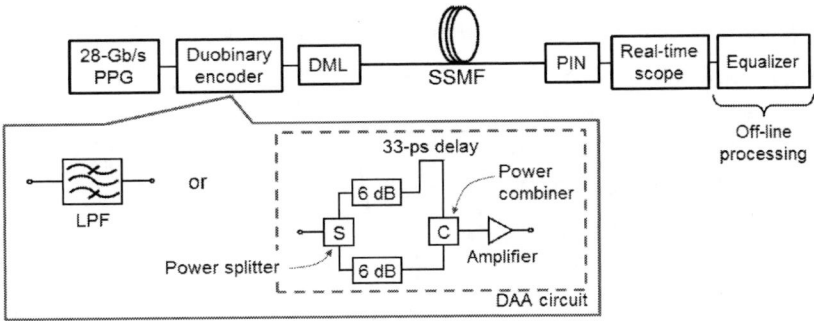

Fig. 1. Experiment setup.

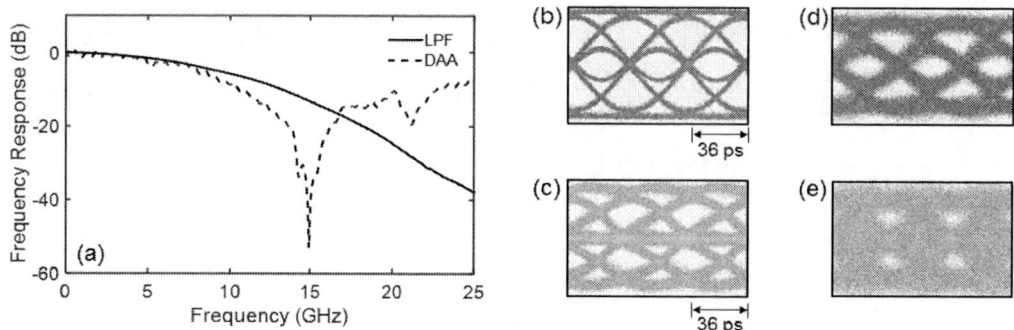

Fig. 2 (a) Measured frequency responses of the LPF and DAA circuit. Eye diagrams of 28-Gb/s electrical duobinary signals generated by using (b) LPF and (c) DAA circuit. Eye diagrams of the optical signals when the duobinary signals are generated by using (d) LPF and (e) DAA circuit.

having a 3-dB bandwidth of 7.0 GHz, as shown in Fig. 2(a). The DAA circuit is implemented by using a power splitter, an electrical delay line, and a power combiner. The electrical delay is set to be ~33 ps, which approximately corresponds to the symbol duration of the signal. Two 6-dB attenuator pads are inserted between the power splitter and combiner to mitigate the adverse effects of electrical reflections occurring between these devices. Thus, an electrical amplifier is used at the output of the power combiner. The frequency response of the DAA circuit is depicted in Fig. 2(a). The 3-dB bandwidth of the circuit is measured to be 6.2 GHz, which is slightly narrower than that of the LPF. However, the DAA circuit has a periodic frequency response and thus exhibits higher transmittance at frequencies higher than 17 GHz. Fig. 2(b) and (c) show the eye diagrams of the electrical duobinary signals measured at the outputs of the LPF and DAA circuit, respectively. Both of them are three-level signals and have two eyes between the levels. The eye diagram in Fig. 3(c) forms with thicker lines due to the electrical amplifier used for the DAA circuit. The electrical duobinary signals are fed to a DML operating at 1549.9 nm. It emits an optical power of 7.4 dBm when biased at 75 mA. Fig. 2(d) and (e) show the eye diagrams of the optical signals at the output of the DML when the LPF and DAA circuit were utilized, respectively. The 28-Gb/s duobinary signals are launched into SSMF and detected by using a PIN receiver. The signals are then digitized by using a sampling scope running at 80 Gsample/s for off-line processing. In order to compensate for the signal distortions caused by the fiber dispersion, we employ a half-symbol-spaced 17-tap feed forward equalizer followed by a 7-tap decision-feedback equalizer. Finally, bit-error rate (BER) is calculated by direct error counting.

We first optimize the ER of the signal. A higher ER is beneficial for improving the receiver sensitivity but it broadens the spectral width of the signal, leading to poorer transmission performance. Fig. 3(a) shows the measured receiver sensitivities (@ BER=10^{-3}) as a function of ER. The ER is defined as the intensity ratio between the highest and lowest levels. In the back-to-back condition, the receiver sensitivities are improved as the ER increases. After 40-km transmission, however, the optimum ER is found to be 1.5 dB, regardless of the type of duobinary encoders. We attribute this to the fact that the optical spectral width at the output of the DML is mainly governed by the peak-to-peak driving current of the electrical signal, which in turn, determines the ER of the signal. Comparing the performances of the two duobinary encoders, we observe that the signal generated by using the LPF slightly outperforms the one by the DAA circuit. For example, we achieve a receiver sensitivity of -10.3 dBm after 40-km transmission when the LPF is used whereas the sensitivity is measured to be -8.7 dBm in the case of using the DAA circuit. The slightly superior

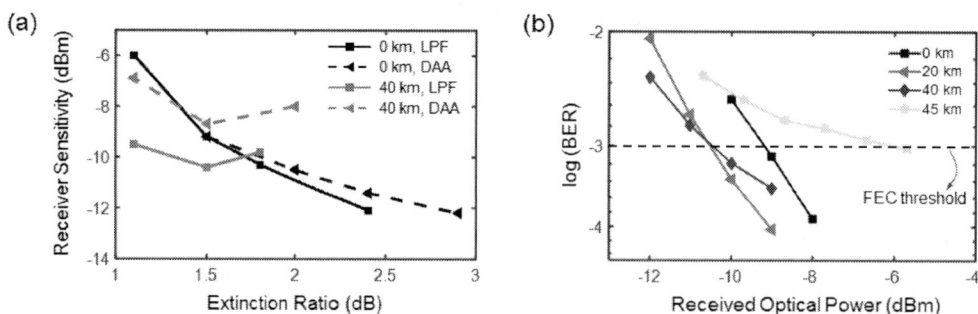

Fig. 3. (a) Measured receiver sensitivity as a function of ER for two different transmission distances.
(b) Measured BER curves of the duobinary signals generated by using the LPF. The ER is set to be 1.5 dB.

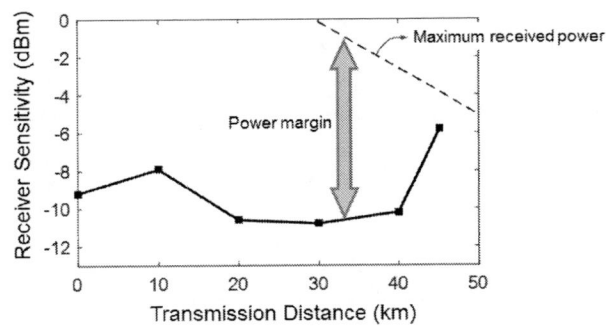

Fig. 4. Measured receiver sensitivity as a function of transmission distance over SSMF. The duobinary signal is generated by using the LPF and the ER is set to be 1.5 dB.

performance of the LPF duobinary signal should be ascribed to the better waveform quality from the encoder [as observed in Fig. 2(b) and (c)].

Based on the performance comparison of the duobinary encoders and ER optimization shown in Fig. 3(a), we choose the LPF as a duobinary encoder and set the ER to be 1.5 dB. Fig. 3(b) shows the measured BER curves for various transmission distances. The back-to-back receiver sensitivity is measured to be -9.2 dBm. After 20- and 40-km transmissions, we have slight improvement in the receiver sensitivity. However, the slope of the BER curve starts to be gentle after 30-km transmission and thus we have a sensitivity penalty of 3.4 dB after 45-km transmission. An error floor was observed at around 4×10^{-3} after 50-km transmission.

Fig. 4 shows the measured receiver sensitivity as a function of the transmission distance over SSMF. As the transmission distance increases, the receiver sensitivity is at first degraded up to 10 km but it is improved to ~40 km. This is because of the self-steepening effect [5-6] and duobinary-like phase characteristics of the signals [7]. Also shown with the dashed line in this figure is the maximum received power of the signal. It has a slope of -0.25 dB/km due to the fiber attenuation at 1.55 µm. Thus, the difference between the maximum received signal power and the receiver sensitivity represents the power margin of the system. The results show that we can achieve a power margin larger than 13 dB after 20-km transmission. This implies that we could exploit this DML-based 28-Gb/s system for power-splitting optical access networks supporting 16 subscribers. Alternatively, the margin could be used to accommodate the losses of two waveguide grating routers in wavelength-division-multiplexed passive optical networks.

CONCLUSIONS

We have successfully transmitted the 28-Gb/s duobinary signals generated by using a 1.55-µm DML over 45-km long SSMF. The electrical duobinary signals are generated by using either a low-pass filter or a delay-and-add circuit. The results show that the low-pass filtered 28-Gb/s duobinary signal slightly outperforms the delay-and-added signal due to the absence of electrical amplifier and electrical reflections occurring in the encoder. The results also show that we can achieve power margins larger than 13 and 7 dB after 20- and 40-km transmissions, respectively. Therefore, the DML-based 28-Gb/s system could be used for the cost-effective implementation of optical access networks.

ACKNOWLEDGEMENT

This work was supported by the National Research Foundation of Korea (NRF) grant funded by the Korea government (MSIP) (2015R1A2A1A05001868).

REFERENCES

[1] Y. Matsui, T. Pham, T. Sudo, G. Carey and B. Young, "112-Gb/s WDM link using two directly modulated Al-MQW BH DFB lasers at 56 Gb/s," Proc. of Optical Fiber Communications Conference (OFC 2015), California (US), Mar. 2015, Th5B.6.

[2] L. Frejstrup Suhr, J. J. Vegas Olmos, B. Mao, X. Xu, G. N. Liu and I. T. Monroy, "Direct modulation of 56 Gbps duobinary-4-PAM, " Proc. of Optical Fiber Communication Conference (OFC 2015), California (US), Mar. 2015, Th1E.7.

[3] Z. Li, L. Yi, X. Wang and W. Hu, "28 Gb/s duobinary signal transmission over 40 km based on 10 GHz DML and PIN for 100 Gb/s PON," Optics express, vol. 23, pp. 20249-20256, Jul. 2015.

[4] Z. Al-Qazwini and H. Kim, "Directly modulated laser driven by low-bandwidth duobinary signals," IEEE Photonics Technology Letters, vol. 22, pp. 1306-1308, Sept. 2010.

[5] P. J. Corvini and T. L. Koch, "Computer simulation of high-bit-rate optical fiber transmission using single-frequency lasers," J. Lightwave Technology, vol. 5, pp. 1591-1595, Nov. 1987.

[6] P. Winzer et al., "10-Gb/s upgrade of bidirectional CWDM systems using electronic equalization and FEC," J. Lightwave Technology, vol. 23, pp. 203-210, Jan. 2005.

[7] J. Zhou, C. Yu, and H. Kim, "Transmission performance of OOK and 4-PAM signals using directly modulated 1.5-µm VCSEL for optical access network," J. Lightwave Technology, vol. 33, pp. 3243-3249, Aug. 2015.

ThB1-5

A Novel Amplitude Decision based Symbol Synchronization Method for Real-time IMDD-OOFDM Systems

Weiliang Wu, Junjie Zhang, Zhen Zhang, Han Dun, and Qianwu Zhang
The Key Laboratory of Specialty Fiber Optics and Optical Access Network, Shanghai University,
149 Yanchang Road, Shanghai 200072, China
*zhangqianwu@shu.edu.cn

Abstract: *A novel symbol synchronization method based on amplitude decision for real-time IMDD-OOFDM systems is experimental demonstrated. Synchronization without correlation and multiplication can be achieved within 5 samples error at the received optical power of -20dBm.*
Keywords: *symbol synchronization, amplitude decision, real-time IMDD-OOFDM*

I. INTRODUCTION

Recently, intensity modulation and direct detection OOFDM (IMDD-OOFDM) has been considered as one of the candidate technologies for next generation PONs (NG-PONs) [3] due to its simple structure and low DSP complexity [1,2]. Robust and precise synchronization techniques are extremely important for the high-speed OOFDM systems. Although much research attention has been drawn on symbol synchronization [4,5], some problems as the inherent plateau in timing metric and huge consumption of computing resource are still annoying [6,7].

In this paper, a novel symbol synchronization method based on amplitude decision for real-time IMDD-OOFDM systems is experimental demonstrated. Synchronization without correlation and multiplication can be achieved within 5 samples error at the received optical power of -20dBm over 25-km standard single-mode fiber (SSMF).

II. FPGA-BASED REAL-TIME OFDM-PON TRANSMISSION SYSTEM DESCRIPTION

An OFDM frame is constructed as shown in Fig. 1(a). In this experiment, the leading-zeros, used as symbol synchronization, is at the beginning of one OFDM frame, which number is set to 80 samples. The number of OFDM subcarriers is set to 64, but due to the Hermitian symmetry only 32 are defined: 30 carry data, the other two are filled with zeros which are DC and high-frequency subcarrier, respectively. The length of cyclic prefix (CP) is 16 samples, 30 sub-carries in the positive frequency bins are modulated with QPSK symbols, and the clipping ratio is set to 11.5dB. The OFDM samples are generated offline using MATLAB, and they are sent to the 12-bit DAC with the sampling rate of 4 GSa/s. So the raw signal bit rate is $(30×2)/64×4 = 3.75$ Gb/s. The electric signal is filtered by a 2 GHz bandwidth low pass filter (LPF). Then the electric amplifier (EA) which adjusts the amplitude of signal to the driving voltage of the 1550 nm directly modulated liner optical isolated distributed feedback (DFB) laser. After 25 km SSMF transmission with variable optical amplifier (VOA), the optical signal is detected by a PIN detector. Another LPF is used as an anti-alias filter. After that an EA is connected which ensures the amplitude of signal to reach the full span of 4 GSa/s 10 bit ADC. The block diagram of the real-time IMDD-OOFDM system is shown in Fig. 1(b). In this transmission system, both of the transmitter and receiver are realized by Virtex-6 XC6VLX240T from Xilinx. The internal signals of FPGA, including OFDM frame loss number, OFDM frame error number and bit errors in the real-time receiver are sampled by Xilinx integrated logic analyzer ChipScope.

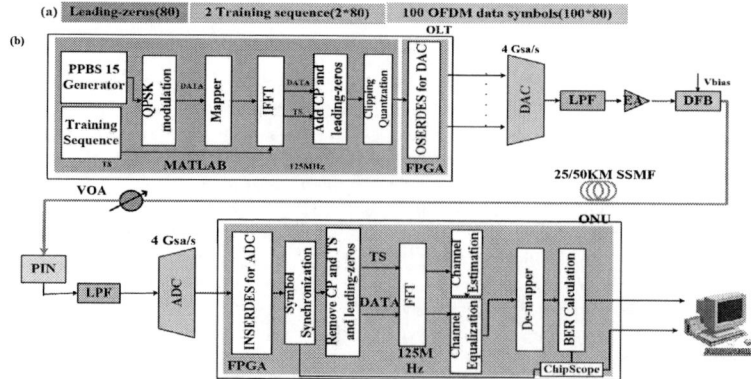

Fig. 1. (a) OFDM time-domain frame structure. (b) Experimental setup of the real-time IMDD-OOFDM system

III. LOW-COMPLEXITY AND HIGH-ACCURACY SYMBOL SYNCHRONIZATION

A. symbol synchronization technique based on amplitude decision

The symbol timing recovery procedure is divided into two stages: tri-level conversion and leading-zeros detection. In the first stage, The received OFDM signal is converted into tri-level signal by two thresholds through comparator, the threshold of comparator is set to ± 50. Moreover, to avoid OFDM frame loss, the last sample of leading-zeros is design to be a high peak, so no matter what the training sequence is, the tri-level signal must be non-zero at the end of leading-zeros. In the second stage, the synchronization position is acquired where the end of leading-zeros is. The received signal and its tri-level signal are presented in Fig. 2 (a), (b). Meanwhile, the received signal whose amplitude is higher or lower than the threshold is decided as ± 1 while the others are decided as 0. Since the optical fiber channel is stable, the leading-zeros can be extracted from tri-level signal accurately.

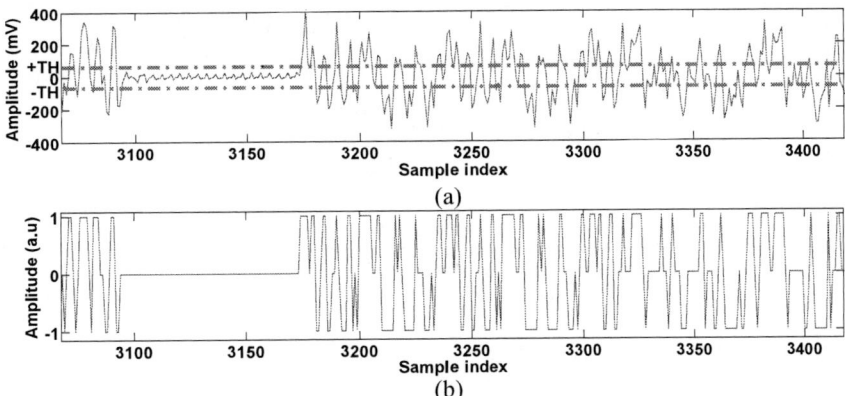

Fig. 2. (a) Received OFDM signal in time-domain. (b) The tri-level signal after conversion

B. Complexity comparisons

In the proposed symbol synchronization technique, the leading-zeros instead of training sequence and cycle prefix are used as symbol synchronization. And no correlation or multiplication is applied in the proposed technique, the resource consumption is much lower than other symbol synchronization techniques proposed in OOFDM systems.

The complexity comparisons between the proposed symbol synchronization technique, CP-based technique [4], TS-based technique using cross-correlation [6] and preamble-based on auto-correlation [7] is presented in Table I.

These techniques are used for high-speed OOFDM receiver systems, and they all process Np samples in parallel. For a fair comparison of these four synchronization algorithm, they are used in the IMDD-OOFDM system with real-valued OFDM signal generation and detection, in which only symbol synchronization is considered. The technique proposed in [6] makes the cross-correlation computation work with 1 bit quantization for input and reference signals, so that the XNORs could take place of multipliers. The technique proposed in [4] make the auto-correlation computation with full bit-width, and it also needs to find a maximum (MAX) to determine the correct timing position of the start of the data symbol, which takes more computing consumption than using threshold detection (TH). In [7] the received signals are cross-correlated with the training sequence which is 7 bits quantized. This correlation operation is implemented using additions and subtractions, so multiplication is removed and a large area is saved. Our technique has much lower complexity than others since it only needs comparator.

TABLE I
COMPLEXITY COMPARISON

ALGORITHM	Multipliers	Adders	XNORs	MAX/TH
CP-BASED ALGORITHM [4]	$2Np + Ncp + 1$	$2(Np + Ncp + 1)$	0	MAX,TH
TS-BASED ALGORITHM [6]	0	$Np(N_{TS} - 1)$	$Np(Nts)$	2TH
PREAMBLE-BASED ALGORITHM [7]	0	$Np(N_{TS} - 1)$	0	TH
PROPOSED ALGORITHM	0	0	0	2TH

Proposed symbol synchronization has been modeled using the Verilog hardware description language. The number of slice registers and slice LUTs used in the symbol synchronization is 557 and 735, respectively. The synchronization resource consumption of the proposed technique is only 6%. To the best of our knowledge it is the lowest computing complexity in OOFDM systems.

IV. EXPERIMENTAL RESULTS

According to the proposed symbol synchronization, the real-time measured curves of BER versus received optical power with 500000 OFDM frames for optical back-to-back and 25-km SSMF transmission are depicted in Fig. 3(a). We can see that the minimum received optical power required for achieving a forward error correction (FEC) BER limit of 3.8×10^{-3} is -17.2 dBm, compared with the back-to-back (OBTB) case, at a BER of 3.8×10^{-3}, the power penalty is about 0.5dB. The system performance verifies that the proposed symbol synchronization technique is highly accurate in IMDD-OOFDM systems.

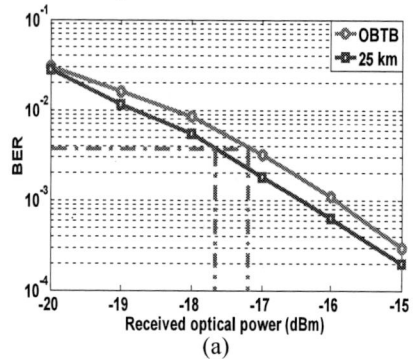 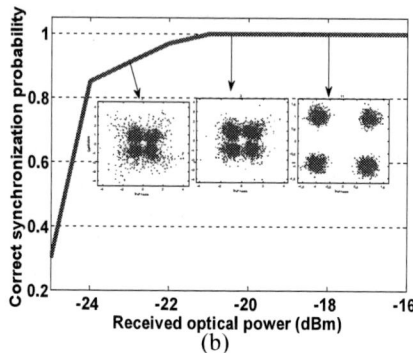

(a)   (b)

Fig. 3. (a) BER versus versus received optical power with modulation format of QPSK for back-to-back and after 25 km SSMF (b) correct synchronization probability as a function of received optical power.

The correct synchronization probability as a function of received optical power was obtained as shown in Fig. 3(b). We can see that the correct synchronization maintain a higher level until the received optical power reaches -20dBm. The results indicate that the proposed technique can get stable correct synchronization even in low received optical power. However, when the received optical power meets -25dBm, the correct synchronization probability decrease dramatically, it is because that received signal's amplitude is almost the same as the thresholds in such a condition. In addition, the QPSK constellations of the OFDM signal with received optical power of -18.1dBm, -20.2dBm, -22.9dBm are also shown in Fig. 3(b). As can be seen from the experimental result, the proposed technique can achieve synchronization with high accuracy and sensitivity.

V. CONCLUSIONS

In this work, a novel symbol synchronization method based on amplitude decision for real-time IMDD-OOFDM systems is experimental demonstrated. Synchronization without correlation and multiplication can be achieved within 5 samples error at the received optical power of -20dBm over 25-km SSMF.

ACKNOWLEDGMENT

This work was supported in part by the Nature Science Foundation of China (Project No. 61132004, 61275073, 61420106011) and the Shanghai Science and Technology Development Funds (Project NO. 13J1402600, 14511100100, 15511105400).

REFERENCES

[1] Q. W. Zhang, E. Hugues-Salas, Y. Ling, H. B. Zhang, R. P. Giddings, J. J. Zhang, M. Wang, and J. M. Tang, "Record-high and robust 17.125 Gb/s gross-rate over 25 km SSMF transmissions of real-time dual-bandoptical OFDM signals directly modulated by 1 GHz RSOAs," Opt. Exp., vol. 22, no. 6, pp. 6339-6348, 2014.

[2] Cvijetic, N, "OFDM for next-generation optical access network," J. Lightw. Technol., vol. 30, no. 4, pp. 384-398, Feb. 2012.

[3] C. Hao, X. Yang, M. Bi, H. He ,W. Hu, "A simple and accurate timing synchronization algorithm for IMDD optical OFDM," Opto-Electronics and Communications Conference (OECC 2012). July 2, 2012, pp. 166-167.

[4] X. Q. Jin, R. P. Giddings, E. Hugues-salas, and J. M. Tang, "Real-time experimental demonstration of optical OFDM symbol synchronization in directly modulated DFB laser-based 25 km SMF IMDD systems," Opt. Express, vol. 18, no. 20, pp. 21100-21110, 2010.

[5] X. Jin and J. Tang, "Optical OFDM synchronization with symbol timing offset and sampling clock offset compensation in real-time IMDD systems," IEEE Photon. J., vol. 3, no. 2, pp. 187-196, Apr. 2011.

[6] M. Chen, J. He, Z. Cao, J. Tang, L. Chen, and X. Wu, "Symbol synchronization and sampling frequency synchronization techniques in real-time DDO-OFDM systems," Opt. Commun., vol. 326, pp. 80-87, 2014.

[7] S. Chen, Q. Yang and W. Shieh, "Demonstration of 12.1-Gb/s single band real-time coherent optical OFDM reception," Opto-Electronics and Communications Conference (OECC 2010). July 15, 2010, pp. 472-473.

ThB2-1

OECC/PS2016

Study on Optical Modulation Amplitude of 4-level Pulse Amplitude Modulation with Electroabsorption Modulator

Riu Hirai*, Nobuhiko Kikuchi* and Takayoshi Fukui**

* Center for Technology Innovation, Hitachi Ltd., Yoshida-cho 292, Totsuka, Yokohama, Kanagawa, 244-0817 Japan.
**Oclaro Japan, Inc., 4-1-55 Oyama Chuo-ku, Sagamihara, Kanagawa, 252-5250 Japan
Tel: +81-50-3154-9482, E-mail Address: riu.hirai.tq@hitachi.com

Abstract: For power budget improvement of 28-Gbaud Nyquist-PAM4 signaling with an EA modulator, we increase optical modulation amplitude by asymmetric nonlinearity compensation and DC bias optimization. We achieve 1.53-dB power budget improvement and demonstrate a 50-km SSMF transmission.

Keywords: Optical fiber transmission, Pulse amplitude modulation, Digital signal processing

I. INTRODUCTION

Recently, since the capacity of intra/inter data center network explosively expands due to the spread of video contents service or social network service, high-capacity transmission scheme has been strongly required. On this background, 400-Gb/s Ethernet Task Force, namely IEEE P802.3bs, has been promoting the standardization of 400-Gigabit Ethernet (400GbE) in order to meet the demand for higher speed of optical/electrical interfaces. So far, 4-level pulse amplitude modulation (PAM4) as an optical modulation scheme has been adopted for the first time in the standard of Ethernet; therefore twice higher-speed optical interfaces can be realized with the same channel bandwidth. Nevertheless, there are a few obstacles to realize optical multilevel intensity modulation with sufficient power budget. An EAM is suitable for high-speed PAM signal generation and its use for 2- and 10-km transmission is discussed in IEEE P802.3bs, however its modulation characteristic shows strong nonlinearity and also asymmetry to applied voltage and its linear modulation region is limited around 50-% extinction point. Instead, even if the amplitude is increased beyond the linear modulation region, its receiver sensitivity is degraded due to distorted eye opening. Therefore, in order to increase the power budget of EAM-based PAM signaling, its optical modulation amplitude (OMA) should be increased by using careful nonlinearity compensation of an EAM's modulation characteristic combined with bias point optimization. So far, a PAM4 driver IC has been reported, which is able to adjust asymmetrically each level of PAM4 [1], but the optimization of power budget of PAM4 signals with OMA and bias-point optimization is not shown.

In this paper, in order to improve power budget of 28-Gbaud IM/DD Nyquist-PAM4 signaling [2, 3], we increase optical modulation amplitude with using nonlinear region by digital asymmetric nonlinear compensation (NLC) and also with DC bias optimization. As a result, we improve power budget by 1.53 dB comparing with linear modulation region case. In this study we increase power budget up to 17.6 dB at most, which greatly exceeds the required power budget for 10-km SMF transmission of 400GbE by 9.1 dB [4], and we successfully demonstrate a 50-km unamplified SSMF transmission.

II. EXPERIMENTAL SETUP

A. Introduction of nonlinear compensation for increase of OMA

In this paper we compare the effect of both symmetric and asymmetric NLCs. Since the modulation characteristic of EAMs at 50-% extinction is nearly linear and can be approximated by sinusoidal function, we implemented symmetric NLC with arcsine function. The arcsine function well-known for a NLC scheme of an optical mach-zehnder modulator (MZM). In contrast, for larger OMA, we study inverse function of an EAM's modulation characteristic as another NLC (we call asymmetric NLC which can be used at arbitrary bias point.). We accurately measure an asymmetric modulation characteristic of an EAM and introduce the inverse function to a NLC block of transmitter-side digital signal processing as shown in figure 1(a). Also, figure 1(b) shows histograms comparison of digital signals without and with asymmetric NLC. After the asymmetric NLC is applied, the amplitude of digital signals diverges exponentially and then amplitude clipping is required appropriately in order to suppress it within the effective range of digital-to-analogue converters. In this study, we determine to clip amplitude beyond 2.76 times of standard deviation σ of the signal distribution.

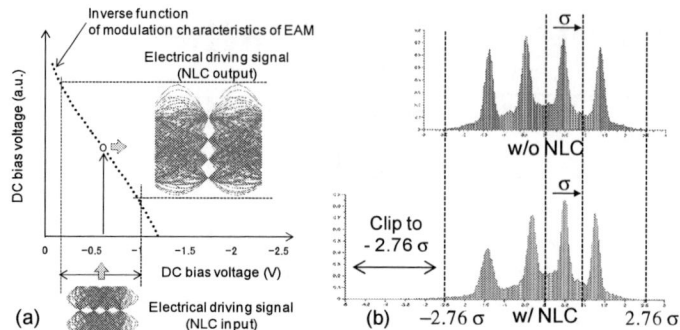

Figure 1 (a) Asymmetric NLC function of an EA modulator and Nyquit-PAM4 waveforms.(b) Histograms of digital signals output from NLC

B. Configuration of a 28-Gbaud Nyquist-PAM4 transceiver

Figure 2 shows the DSP block diagrams and experimental setup. At the transmitter side, first, PAM4 symbol sequences are encoded at 28 Gbaud with length 32768 symbols. The PAM4 signal is shaped into bandwidth-reduced spectrum (28 x (1+α) GHz) with raised cosine filter (101 taps, roll off factor α=0.1) at 2-sps (sample per symbol) sampling. At the NLC block, either symmetric NLC (shortly arcsine function) or asymmetric NLC, described above, is applied to the Nyquit-PAM4 signal to compensate a modulation characteristic of an EAM. And also pre-equalizing process is performed in order to compensate frequency responses of mainly the following DAC and other electrical/optical devices with 41-tap FIR filter. After amplitude clipping, an electrical driving signal is outputted from 56-GSa/s 8-bit DAC (3dB-bandwidth: 15 GHz). At the end of the transmitter, a 28-Gbaud optical Nyquist-PAM4 signal is generated by using a 1.3-μm band Electroabsorption Modulation Laser (100G CFP2-LR4 TOSA produced by Oclaro, DC bias voltage at extinction ratio of 20 dB is about -2.5 V). In this study, we control OMA with adjusting amplitude of electrical driving signals by three RF fixed attenuators, and the respective amplitudes are adjusted to 0.60, 0.95 and 1.35 Vpp (peak-to-peak), which are defined as actual eye opening from L00 to L10 of PAM4 at the EML input without device losses. On the other hand, at the receiver side, the front-end, which consists of a PIN-PD and a TIA, detects the intensity of optical signals. The received signal is AD converted by a real-time digital oscilloscope (3dB-bandwidth: 20GHz) with the sampling rate of 50 GSa/s, and stored to its memory for off-line DSP. In the Rx-side DSP, first, the data is up-sampled to 2-sps. Next, after applying root-raised cosine filter matching to the Tx-side filter, a digital adaptive equalizer (41-tap FIR filter) is performed in order to reduce the ISI (Inter-symbol interference) using the Least Mean Square (LMS) algorithm. Finally, the BER is evaluated from the data after clock recovery and symbol decision. In addition, 50-km SSMF (insertion loss 17.6 dB) is prepared for the optical transmission line.

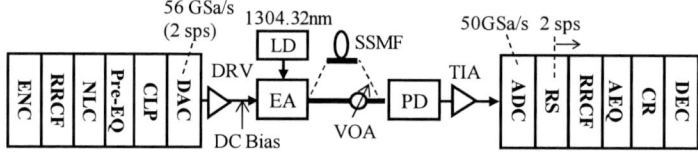

Figure 2 Experimental setup of 28-Gbaud IM/DD Nyquist-PAM4 transmitter, receiver and optical fiber transmission line. ENC: PAM4 encoder, RRCF: root-raised-Cosine filter, NLC: modulator non-linear compensator, pre-EQ: linear equalizer, CLP: Clipping, DAC: DA converter, LD: laser diode, DRV: Linear driver amplifier, VOA: Variable optical attenuator, PD: Photo-diode, TIA: Trans-impedance amplifier, ADC: AD converter, RS: re-sampling, AEQ: adaptive EQ, CR: clock recovery, DEC: PAM4 decoder , sps: Sample per symbol

III. RESULTS

A. Comparison between symmetric and asymmetric nonlinear compensation

First, we compare compensation capability of asymmetric NLC with that of symmetric NLC. In this section, we adjust DC bias voltage to -0.74 V which is a bit shifted from the center of linear region but corresponds to the optimum sensitivity. Figure 2(a) shows the BER characteristics without NLC when the amplitudes of electrical driving signals are adjusted to 0.60, 0.95 and 1.35 Vpp. Considering with the BER curves, beyond 0.95 Vpp, the slopes of BER curves become gentler and the level of BER floor rises up due to nonlinearity of the EAM. Also the asymmetric reduction of eye openings and the degradation of EVM (error vector magnitude) are observed as shown in figure 2(b). Assuming 28-Gbaud Nyquist-PAM4 modulation within just linear modulation region, we can obtain power budget less than or equal to 14.68 dB at BER=2E-3. Next, the BER curves after symmetric and asymmetric NLC are shown in figure 2(c) at 0.95 Vpp and (d) at 1.35 Vpp, respectively. Figure 2(c) presents slight difference between both NLCs. On the other hand, with asymmetric NLC, the BER floor is mitigated to less than BER=1E-4 and also 1.21-dB power budget improvement is obtained comparing with linear region case. Additionally, 1.28-dB OMA increase is measured with a digital communication analyzer.

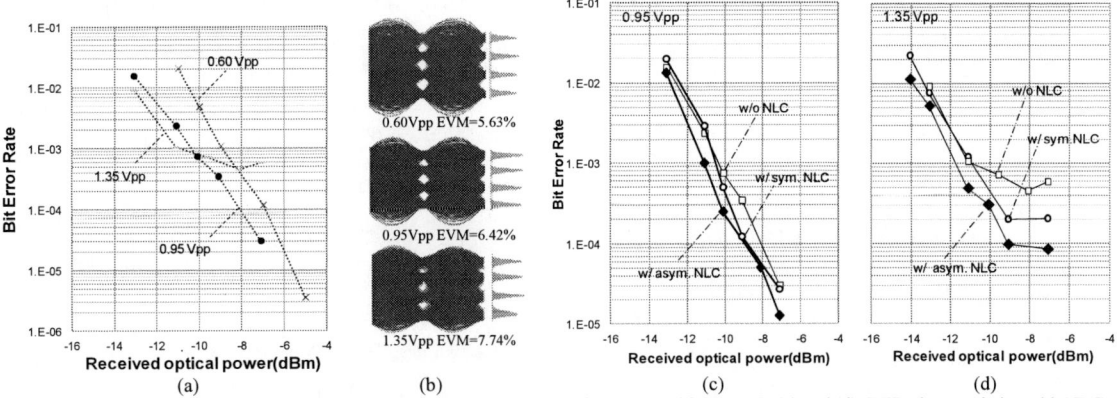

Figure 3(a) and (b): BER characteristics, received waveform and histograms without NLC, (c) and (d): BER characteristics with NLC.

B. Improvement of power budget with asymmetric nonlinear compensation and DC bias optimization

Since compensation capability of asymmetric NLC has been confirmed in the above, we can modulate multilevel signals beyond linear region of an EAM. In addition, we can expect more power budget improvement by optimization of DC bias voltage when modulating with nonlinear region, because deeper DC bias results in shrink OMA despite the same amplitude of driving signal as shown in figure 4(a). Therefore, we evaluate power budget at BER=2E-3 depending on DC bias voltage. Figure 4(b) and (c) show power budget variation at 0.95 Vpp and at 1.35 Vpp, respectively. In both cases with asymmetric NLC, the DC bias voltage, where its power budget becomes the maximum, varies 0.04~0.08 V shallower from the optimum voltage without NLC (-0.74 V). In case of 1.35 Vpp, 0.32-dB power budget improvement (or 0.46-dB OMA increase) is achieved with optimizing DC bias -0.66 V. Conclusively, it is clear that optimization of DC bias voltage shows advantages for power budget improvement thanks to introduction of asymmetric NLC.

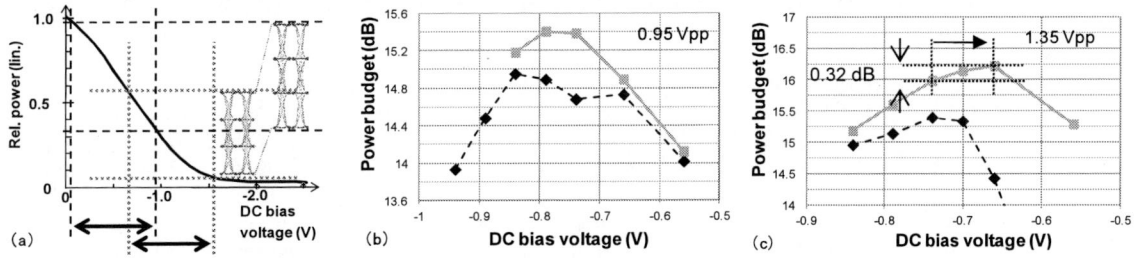

Figure 4 (a) shows OMA depends on DC bias voltage. (b) and (c) show power budget variation versus DC bias voltage at 0.95 and 1.35 Vpp, respectively. (Diamond symbols show without NLC and square symbols show with NLC.)

C. Investigation of the maximum power budget

At the end, we investigate the maximum power budget with the same EML whose LD bias current is increased from 80 mA to 100 mA and DC bias voltage is adjusted the optimum value (-0.66 V) at 1.35 Vpp. As a result, we demonstrate an up-to 50-km SSMF transmission (insertion loss 17.6 dB) at less than BER=2E-3. Therefore we confirm 28-Gbaud Nyquist-PAM4 signaling has 17.6-dB power budget at most if using this commercially available EML for CFP2-LR4 TOSA. Comparing with the required power budget (8.5 dB) for 10-km SSMF transmission of 400GbE which includes channel insertion loss and allocated penalties, we achieve more than twice power budget.

IV. CONCLUSIONS

In this paper, we confirm that asymmetric nonlinear compensation and also DC bias optimization lead to increase of optical modulation amplitude and power budget improvement of 28-Gbaud Nyquist-PAM4 signaling with an EAM. As a result, we achieve 9.1-dB larger power budget than the requirement for 10-km SSMF transmission of 400GbE.

REFERENCES

[1] T. Kishi, et al., "A 56-Gb/s PAM4 InP HBT Driver IC Compensating for Nonlinearity of Extinction Curve of EAM" in CSICS 2015, New Orleans, LA, USA, October 2015, O.2.

[2] N. Kikuchi, e. al., "Intensity-Modulated / Direct-Detection (IM/DD) Nyquist Pulse-Amplitude Modulation (PAM) Signaling for 100-Gbit/s/l Optical Short-reach Transmission," in ECOC 2014, Cannes, France, September 2014, P.4.12..

[3] R. Hirai, et al., "Feasibility study of 100G/lambda Nyquist-PAM4 with commercially available 1.3um/1.5um EML," in IEEE P802.3bs 400GbE Task Force meeting, Ottawa, Canada, September 2014

[4] J. D'Ambrosia, "IEEE P802.3bs Baseline Summary" in http://www.ieee802.org/3/bs/baseline_3bs_0715.pdf.

Novel Electrical Dispersion Compensation Technique for IMDD-based Systems

Kazuki Tanaka, Takashi Kobayashi, Akira Agata and Kosuke Nishimura

KDDI R&D Laboratories Inc., 2-1-15 Ohara, Fujimino-shi, Saitama, 356-8502, Japan

au-tanaka@kddilabs.jp

Abstract: *We propose a novel electrical dispersion compensation (EDC) technique in frequency domain for IMDD-based systems and demonstrate the effectiveness in a transmission experiment of 40 Gbit/s 4-PAM signals over 20 km SMF in C-band.*

Keywords: *EDC, FDE, IMDD, Chromatic dispersion, Optical access systems*

I. INTRODUCTION

With rapid growth of mobile data traffic caused by high speed mobile devices such as smartphones and tablet computers, the mobile base stations are expected to make their cell coverage smaller to increase their capacity in the future. In general, if many base stations are allocated closer in an area, the interference among them might be an issue in mobile network systems. In order to increase the capacity and reduce the interference among base stations, Centralized Radio Access Network (C-RAN) architecture has been proposed, and its deployment has been launched [1,2]. In C-RAN, baseband units (BBUs), which have been allocated in each antenna site in conventional distributed RAN (D-RAN) architecture, are centralized to one central office (CO) to enable coordinated control of multiple remote radio heads (RRHs). The optical link between BBUs in a CO and RRHs in an antenna site is referred to as "fronthaul" in C-RAN, while one is called "backhaul" in D-RAN.

However, C-RAN has a drawback in exchange of the advantage mentioned above. The fronthaul link needs around sixteen times as much capacity as a backhaul one. In ref. [3], optical access systems will be required to have the transmission capacity of 40 Gbit/s by C-RAN. What we should think about in such systems is to mitigate and/or compensate the waveform distortion due to the chromatic dispersion caused by optical fiber transmission. To overcome the issue, various types of techniques have been already proposed and demonstrated; One is based on electrical dispersion compensation (EDC) in time domain (TDE: Time Domain Equalizer) [4,5] for intensity-modulation and direct-detection (IMDD)-based systems which requires to optimize parameters depending on the modulation format and line rate causing the complex system operation. Also, EDC techniques in frequency domain (FDE: Frequency Domain Equalizer) have been proposed [6,7], which only require to decide the value of accumulated chromatic dispersion. Regarding FDE techniques, digital coherent optical transmission systems are considered in ref. [6], while, for IMDD-based systems, only single side-band (SSB) modulation format is studied in Ref. [7] because only SSB modulation with direct detection can remain information on the optical phase in the detected base-band electrical signals which is conventionally necessary for FDE methods [8]. To realize simple and cost-effective optical access systems for C-RAN, IMDD-based systems in a C-band wavelength without SSB modulation format are strongly preferable.

In this paper, we propose a novel FDE technique for IMDD-based systems. Using the technique, we successfully demonstrate 40 Gbit/s 4-PAM signals transmission over 20 km SMF at 1550 nm wavelength.

II. PROPOSED DISPERSION COMPENSATION TECHNIQUE

In this section, we present the principal of our novel FDE scheme for IMDD-based systems applicable to any intensity-modulation scheme. The electric field $E(z,t)$ at the propagation distance of z can be expressed using the nonlinear Schrödinger equation as follows:

$$i\frac{\partial E}{\partial z} = -i\frac{\alpha}{2}E + \frac{\beta_2}{2}\frac{\partial^2 E}{\partial t^2} - \gamma|E|^2 E \qquad \left(\because \beta_2 = -\frac{\lambda^2 D}{2\pi c}\times 10^{-6}\right) \tag{1}$$

Where β_2 is the second order mode-propagation constant proportional to the chromatic dispersion of the optical fiber, α is the attenuation constant to quantify the fiber losses, γ is the nonlinear coefficient, λ is the wavelength of the optical signal, D is the chromatic dispersion of the optical fiber and c is the speed of light. If the Fourier transform is defined by $\widetilde{E}(z,\omega)$, $\widetilde{E}(z,\omega)$ is given by (2) without consideration for the term related to the nonlinear coefficient in the equation of (1),

$$\widetilde{E}(z,\omega) = \widetilde{E}(0,\omega)\exp\left[i\frac{\beta_2}{2}\omega^2 z\right]\exp\left[-\frac{\alpha}{2}z\right] \tag{2}$$

$$= \widetilde{E}(0,\omega)\cos\theta\cdot\exp\left(-\frac{\alpha}{2}z\right) + i\widetilde{E}(0,\omega)\sin\theta\cdot\exp\left(-\frac{\alpha}{2}z\right) \qquad \left(\because \theta = \frac{\beta_2}{2}\omega^2 z\right) \tag{3}$$

Where ω is the angular frequency. If we assume $E(0,t)$ is a real and even function in time domain, $\widetilde{E}(0,\omega)$ is also a real and even function in frequency domain. Therefore, the real part and imaginary parts of $\widetilde{E}(z,\omega)$ are equivalent to the first and second terms of the equation (3), respectively. If the absolute value of the ratio between the first and the second terms, $|sin\theta/cos\theta| = |tan\theta|$, is small, $\widetilde{E}(z,\omega)$ can be approximated as follows:

$$\widetilde{E}(z,\omega) \approx \widetilde{E}(0,\omega)\cos\theta \cdot \exp\left(-\frac{\alpha}{2}z\right) \tag{4}$$

The equation (4) shows $\widetilde{E}(z,\omega)$ is a real function which can be obtained by a receiver side in conventional IMDD scheme. Though we assumed the electric field $E(0,t)$ was even function above, $E(0,t)$ and $E(z,t)$ are not an even function in real systems. Therefore, the process to make $E(z,t)$ an even function is necessary to obtain only the real part in frequency domain and just to take the consequent processes of a real function. Without the process, $\widetilde{E}(z,\omega)$ includes the factor of the incorrect imaginary part. Fig. 1 shows the process of a proposed FDE-based EDC technique to compensate the influence of chromatic dispersion without use of the information on the optical phase. At first, we separate a received intensity waveform and make a waveform block with a length of N samples, and then make an even function block by connecting a copy of the waveform block inversely as illustrated in Fig. 1 (a). Second, we apply Fourier transform to the signal obtained by the first process as Fig. 1 (b). Third, we equalize the signal after the process in Fig. 1 (b) by $H(\omega) = \widetilde{E}(0,\omega)/\widetilde{E}(z,\omega)$, the inverse function of the transform one in the equation (4), as shown in Fig. 1 (c). Finally, we apply inverse Fourier transform to the signal in Fig. 1 (c) and extract the additional data to get the original waveform. Actually, the performance of the proposed technique may be degraded by the losses of the phase factor of received signals because the second part in the equation of (3) is not complete zero.

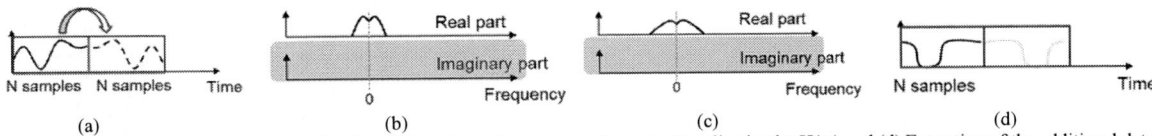

(a)   (b)   (c)   (d)

Fig. 1. FDE process of (a) Making even function, (b) Applying Fourier transform, (c) Equalization by H(ω) and (d) Extraction of the additional data

Now, though we assumed the $|sin\theta/cos\theta| = |tan\theta|$ was small in the equation (3) and (4), we evaluate the value in a transmission condition. When we think about transmission of 40 Gbps 4-PAM signals on 20 km SMF at 1550 nm wavelength and $D = 17.5$ ps/nm/km, the value of the $|tan\,\theta|$ as a function of $f = \omega/2\pi$ is shown in Fig. 2. If the frequencies are 0 to 9.4 GHz, 16.4 to 21.1 GHz and so on, the $|tan\,\theta|$ is less than 1.0. The reason why we set the SMF transmission distance to 20 km is most of common fronthaul links have a distance limitation of around 20 km. If the transmission distance is shorter, the frequency range with the first $|tan\,\theta| < 1$ is broader and other frequency ranges with $|tan\,\theta| < 1$ are in higher frequencies. Fig. 3 shows the power spectrum of 4-PAM signals with frequency resolution of 250 MHz. We found that the frequencies with the value of $|tan\,\theta| < 1$ and the frequency range up to 5.2 GHz accounts for around 96 % and 90 %, respectively, of the total power of the 4-PAM signals. In general, digital signals have most of the power components in the lower frequencies. Therefore, the proposed technique for digital signals transmission will be more effective than for analog one. Though it may be difficult to compensate the waveform distortion for some frequencies, we can expect to partially compensate the influence of chromatic dispersion by the proposed technique.

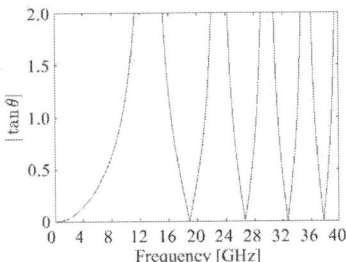

Fig. 2. |tanθ| as a function of the frequency after 20 km SMF transmission

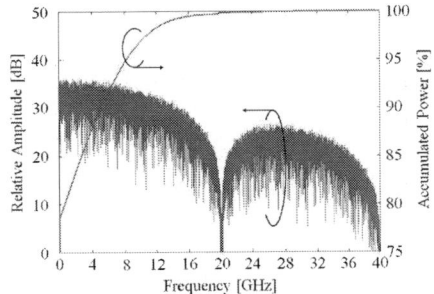

Fig. 3. Power spectrum of 4-PAM signals

III. TRANSMISSION EXPERIMENT

To confirm the feasibility of the proposed EDC technique, we have conducted a 40 Gbit/s 4-PAM transmission experiment over 20 km SMF using C-band wavelength. The experimental setup is shown in Fig. 4. At the transmitter side, the output of a tunable laser diode operating at 1550 nm was modulated by 4-PAM signals by a Mach-Zehnder modulator. The transmitted data pattern was set to a $2^{15}-1$ pseudorandom binary sequence and encoded to 4-PAM signals by mapping every two data bits into one 4-PAM symbol. The bit rate was set to 41.95 Gbit/s, assuming four

Common Public Radio Interface (CPRI) [9] Option 7 channels (4×9.83 Gbit/s) and the use of the conventional RS(255, 239) code or the Super-FEC, both having a redundancy of 6.69 % [10]. The extinction ratio was set to 8.1 dB to attain the linearity of the modulation characteristics. The modulated optical signal was fed into the 20 km SMF. The fiber input power was measured to be 3.4 dBm. The output of the SMF was detected by a PIN photodiode with bandwidth of 50 GHz, and was sampled at 100 GS/s by a digital sampling oscilloscope. The length of the sampled data was 2^{14} symbols. In the off-line DSP process at the receiver side, we implemented the proposed EDC process. To mitigate the influence of inter-block interference (IBI) on each waveform block with a length of N samples described in fig. 1 (a), the overlap FDE [11] was applied.

To evaluate the EDC performance, the receiver sensitivity was measured for back-to-back and after 20 km SMF transmission. The measured results are shown in Fig. 5 including the received 4-PAM eye diagrams before and after the EDC, after 20 km SMF transmission. The BER after 20 km transmission with and without EDC were measured to be 7.9×10^{-4} and 4.6×10^{-5}, respectively, at the received optical power of 0.7 dBm, which was the maximum power using the experimental setup. The results indicate that if the well-known RS(255, 239) code (FEC limit: 1.8×10^{-4}) is applied, error-free transmission cannot be achieved without the proposed EDC. If a more sophisticated FEC code, such as the Super-FEC (FEC limit: 3.3×10^{-3}), is applied, the margin in receiver sensitivity is expanded from 1.9 dB to 3.0 dB thanks to the proposed EDC. In this experiment, we did not completely compensate the influence of the chromatic dispersion. However, the effectiveness of the proposed technique was clearly demonstrated.

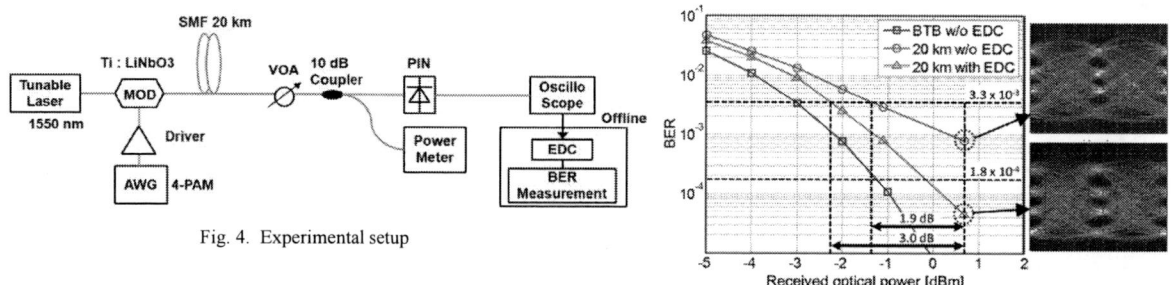

Fig. 4. Experimental setup

Fig. 5. BER curve for 41.95 Gbps 4-PAM signals

IV. CONCLUSIONS

We proposed a novel FDE-based EDC technique without the use of information on the optical phase for IMDD-based systems. In addition, we have successfully demonstrated the effectiveness of the proposed technique in a transmission experiment of 40 Gbit/s 4-PAM signal over 20 km SMF at 1550 nm wavelength. Thanks to the proposed technique, we can expect to deploy simple and cost-effective optical access systems for C-RAN.

REFERENCES

[1]   " C-RAN The Road Towards Green RAN", White Paper, http://labs.chinamobile.com/cran/wp-content/uploads/2014/06/20140613-C-RAN-WP-3.0.pdf, Dec., 2013.

[2]   S. Namba, T. Matsunaka, T. Warabino, S. Kaneko and Y. Kishi, "Colony-RAN Architecture for Future Cellular Network," Future Network & Mobile Summit, pp. 1-8, Berlin, 2012.

[3]   A. Agata, M. Oishi and K. Tanaka, "Performance Enhancement of Optical Access Network in C-RAN using Nonlinear Quantization-based Compression," Proc. OECC, TU3A-2, Melbourne, 2014

[4]   B.Y. Cao, M.L. Deng, R.P. Giddings, X. Duan, Q.W. Zhang, M. Wang and J.M. Tang, "RSOA Intensity Modulator Frequency Chirp-Enabled 40Gb/s over 25km IMDD PON Systems," Proc. OFC, W1J.3, Los Angeles, 2015.

[5]   C. Chen, X. Tang and Z. Zhang, "Transmission of 56-Gb/s PAM-4 over 26-km Single Mode Fiber Using Maximum Likelihood Sequence Estimation," Proc. OFC, Th4A.5, Los Angeles, 2015.

[6]   K. Ishihara, T. Kobayashi, R. Kudo, Y. Takatori, A. Sano, E. Yamada, H. Masuda and Y. Miyamoto, "Frequency-domain Equalization for Optical Transmission Systems," Electron. Lett., Vol. 44, no. 14, pp. 870-871, 2008.

[7]   J. Zhang, W. Fang, C. Hou, X. Liu, X. Zheng and N. Chi, "Single Carrier Frequency Domain Equalization based on SSB Modulation," Proc. ACP, WL59, Shanghai, 2009.

[8]   K. Yonenaga and N. Takachio, "A Fiber Chromatic Dispersion Compensation Technique with an Optical SSB Transmission in Optical Homodyne Detection Systems," Photon. Technol. Lett., Vol. 5, no. 8, pp. 949-951, 1993.

[9]   CPRI Specification V6.0, [Online] http://www.cpri.info/downloads/CPRI_v_6_0_2013-08-30.pdf, 2013.

[10]   ITU-T Rec. G.975.1: "Forward error correction for high bit-rate DWDM submarine systems", 2004.

[11]   R. Kudo, T. Kobayashi, K. Ishihara, Y. Takatori, A. Sano and Y. Miyamoto, "Coherent Optical Single Carrier Transmission Using Overlap Frequency Domain Equalization for Long-Haul Optical Systems," J. Lightwave Technol., Vol. 27, no. 16, pp. 3721-3728, 2009.

Transmitter Digital Pre-distortion Techniques for Band-Limited Optical Networks

Fotini Karinou[1], Zhao Yu[1,2], Nebojsa Stojanovic[1] and Changsong Xie[1]

[1]Huawei Technologies Duesseldorf GmbH, European Research Center, 80992, Munich, Germany

[2]Now with the Technical University of Hamburg-Harburg, Eißendorfer Straße 40 D, 21073, Hamburg.

Fotini.Karinou@huawei.com

Abstract: *We experimentally investigate the impact of different transmitter's pre-distortion techniques for 28 Gb/s/λ IM/DD transmissions over 80 km in DCF-free band-limited systems. OSNR requirements of the proposed scheme are presented as well.*

Keywords: *(060.0060) Fiber optics and optical communications; (120.4820) Optical systems.*

I. INTRODUCTION

Annual global IP traffic, including services such as video on demand and cloud computing, is foreseen to surpass the zettabyte limit in the next years [1]. This leads to an exploding high-capacity demand in the interconnection between the core and the access networks which in turn calls for the deployment of 100-G metro networks in near future [2]. These networks should be able to serve several km of reach and at the same time reduce the required cost-per-bit by employing low-cost transmitter and receiver sub-systems. In order to enable for such cost-efficient technologies, inexpensive schemes such as intensity modulation/direct detection (IM/DD) together with cost-efficient, low-bandwidth components are preferable among other more complex and expensive candidate solutions such as optical duo-binary (ODB), or 4-pulse amplitude modulation (4-PAM) [3]. Towards this direction, in a previous work we have reported 28 Gb/s IM/DD transmission over dispersion compensation fiber (DCF)-free links by employing inexpensive <10-GHz commercially available optical components both at the transmitter and the receiver side and by using a high-performance digital signal processing (DSP)-based scheme at the receiver to compensate for signal distortion [4].

In this paper, first, we extend our investigation and we report the optical signal to noise ratio (OSNR) requirements of our proposed solution for point-to-point links up 80-km and we thoroughly investigate by experiment various transmitter pre-distortion techniques to further optimize the performance. This is realized by the investigation of different filter shape profiles, i.e., flat, Gaussian or Bessel, used for pre-compensation of the limited transmitter's bandwidth. It is shown that after optimum filter shaping employed at the transmitter, the system performance can be optimized by 3 dB at BER=1×10^{-3} enabling for more than 2.5 order of magnitude improvement in BER performance for the same receiver requirements after transmission over 80 km. At the same time, the error floors caused by the transmitter's bandwidth limitations are significantly suppressed. Our results indicate that the DSP techniques employed at the transmitter can be a useful tool in order to enable 28 Gb/s per wavelength transmission in future 112 G (4×28 Gb/s) metropolitan optical networks.

II. EXPERIMENTAL SETUP

The experimental setup used to evaluate the performance of the system is shown in Fig. 1. At the transmitter side, a pseudo-random bit sequence of length 2^{15}-1 is used to generate the offline data using *Matlab* and pre-distortion of the signal is performed in order to compensate for the bandwidth limitations. Then, the generated data are up-sampled to 2 Samples/symbol and are loaded to a 11-GHz analogue bandwidth DAC. The DAC operates at 56 GSamples/s leading to the generation of 28 Gb/s IM/DD electrical signal at its differential outputs. These electrical signals are fed into a 25 GHz 3-dB bandwidth RF driver and then to a bias-T in order to modulate a 10-GHz transmitter optical sub-assembly (TOSA). The TOSA consists of DFB laser and EAM modulator integrated in the same module. A voltage equal to V_{EAM}=-0.512 V is applied through the second port of the bias-T, in order to set the bias voltage of the EAM so as to obtain the optimum performance at the receiver side. TOSA's output optical eye diagram is shown as inset in Fig. 1. Next, the 28 Gb/s IM/DD optical signal is launched over 80 km of single-mode fiber (SMF), is amplified by an EDFA, and filtered through a 0.4-nm optical band-pass filter (OBF). At the receiver side, the received optical power is varied using a variable optical attenuator (VOA) and the signal is detected on a 9-GHz 3-dB bandwidth receiver optical sub-assembly (ROSA). The ROSA consists of a p-i-n photodiode integrated together with a trans-impedance amplifier (TIA). The two differential electrical outputs of the ROSA are fed into a real-time oscilloscope operating at 50 GSamples/s, and the samples are stored and processed offline using DSP. In the offline DSP, the signal is first re-sampled to two samples per symbol, then an automatic gain control (AGC) unit is used to adjust the signal amplitude to the suitable level and an MLSE algorithm is used to equalize the channel distortion and decode the received waveform [5]. The BER is calculated at the output of the MLSE. Note here that a 16-states and a 64-states MLSE is used for the offline processing of the experimental data for the back-to-back and the 80 km transmission case, respectively. First, in order to investigate the OSNR tolerance in the presence of residual CD, we carry out BER vs. OSNR measurements

OECC/PS2016

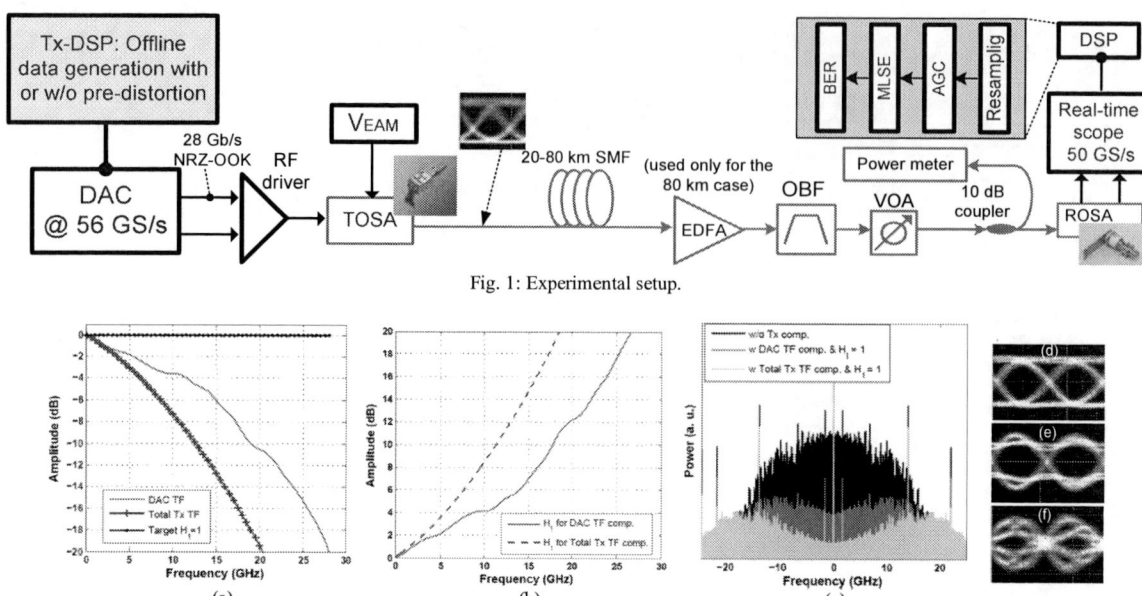

Fig. 1: Experimental setup.

Fig. 2 (a) DAC TF characteristic (solid line); Total Tx TF characteristic (crosses); H_t to be obtained after Tx pre-compensation (dotes); (b) H_I (inverse TF) used to compensate only for the DAC's TF, and for the Total TF assuming $H_t = 1$; (c) DAC output spectra in three cases: without employing any compensation, in black, with DAC's TF compensation assuming a flat H_t, in dark grey, and with Total TF compensation assuming a flat H_t, in ligth grey color; (d), (e), (f): 28 Gb/s NRZ-OOK electrical eye-diagram for the aforementioned cases, respectively.

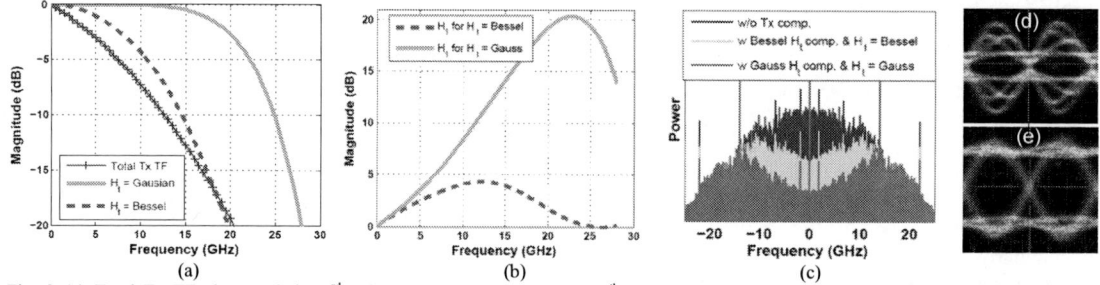

Fig. 3 (a) Total Tx TF characteristic, 3rd order 20-GHz Gaussian and 5th order 10-GHz Bessel H_t target TFs; (b) H_I (inverse TF) used to compensate for the Total TF characteristic assuming a 3rd order 20-GHz Gaussian (in green) and a 5th order 10-GHz Bessel (in purple) H_t; (c) DAC output spectra for the three aforementioned cases; (d), (e): eye-diagrams at the output of the DAC for the aforementioned cases.

using an additional EDFA for ASE loading not shown in the schematic due to space limitations. For these measurements the ROSA input power is kept to Pin=-6 dBm. Next, the transmission performance in terms of sensitivity requirements is evaluated in three cases. First when no pre-distortion is employed in the Tx-DSP, and secondly, when pre-distortion techniques are applied in order to compensate for the limited components' bandwidth. For the latter case, two transfer functions (TFs) are measured: i) the DAC's TF (solid line w/o markers in Fig. 2(a)); and (ii) the system's end-to-end TF, mentioned as "Total TF" in the remainder of the paper (crosses in Fig. 2(a)). Comparing these two curves we observe that the 3-dB bandwidth of the DAC's TF is 11 GHz while the Total TF 3-dB bandwidth is 7 GHz. In this context, we investigate two sub-cases: First, we compensate for the limited bandwidth of the DAC TF and secondly, of the Total TF. To do that, we set as a desired final target function (H_t) a flat-profile one ($H_t=1$), depicted as a dotted line in Fig. 2(a), and according to that assumption the inverse H_1 is calculated. H_1 is shown in solid and dashed lines in Fig. 2(b) for the DAC and the Total TF compensation cases, respectively. Then, the generated data are filtered using H_1 and are loaded to the DAC. The DAC's output spectra, for three cases i.e., i) without employing any transmitter pre-distortion, ii) with DAC TF compensation, and iii) with Total TF compensation, are shown in Fig. 2(c) in black, dark grey and light grey color, respectively. The eye-diagrams at the output of the DAC for the three aforementioned cases are also shown in Fig. 2(d)-(f). Finally we investigate the impact on the transmission performance of different pre-distortion techniques by approximating the Ht by a Bessel- and a Gaussian- profile to further improve the performance. The corresponding Figures i.e., the targeted Gaussian and Bessel TFs, the inverse H_1, the DAC's output spectra and the electrical eye-diagrams after pre-distortion, for these two cases, are shown in Fig. 3(a)-(e) as well.

288

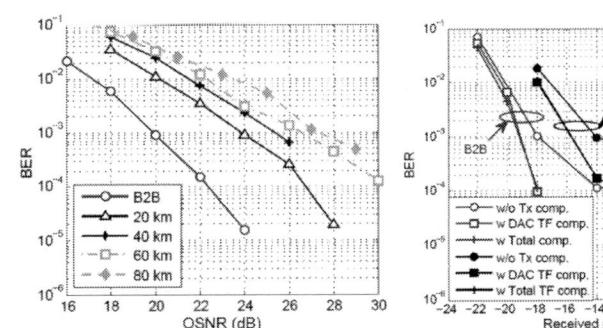

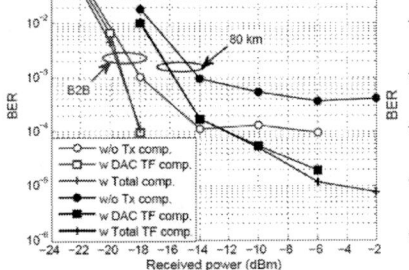

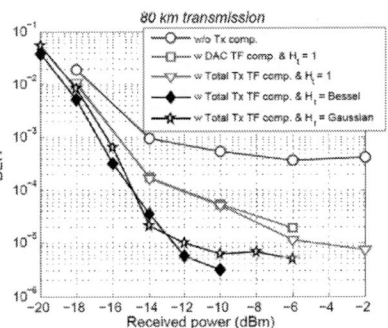

Fig. 4: BER vs OSNR for back-to-back, 20 km, 40 km, 60 km, and 80 km.

Fig. 5: BER vs received optical power for BTB and 80-km transmission.

Fig. 6: BER vs received optical power for 80-km transmission implementing different H_t filter profile approximations.

III. RESULTS AND DISCUSSION

The BER vs OSNR requirements for the back-to-back case, as well as after transmission over 20 km, 40 km, 60 km and 80 km are shown in Fig. 4. As we can observe, a 4 dB, 5 dB, 6 dB, and 7 dB OSNR penalty is observed at BER=1×10^{-3} for 20 km, 40 km, 60 km, and 80 km transmission compared to the back-to-back case, respectively. In all cases transmission below the hard-decision forward error correction (HD-FEC) limit, i.e, BER=1×10^{-3} is achieved. The BER vs the received optical power for the back-to-back case when no Tx pre-distortion is employed is shown in Fig. 5 in open circles. In that case we can see that an error floor attributed to the limited transmitter's bandwidth appears at BER$\sim10^{-4}$ for received optical power greater than -14 dBm when no pre-distortion is employed in the transmitter. The performance improves significantly and the error floor is eliminated when we compensate either for only the DAC's TF or the Total TF in the Tx-DSP. Results for the 80-km transmission case are shown in Fig. 5 as well, with filled markers. The filled circles in Fig. 5 correspond to the 80-km transmission without employing any DSP at the transmitter and reveals that an error floor at BER=4×10^{-4} imposed by the limited system's bandwidth governs the system's performance for input power levels greater that $\sim$-12 dBm. BER performance gets improved by 2 dB at BER=1×10^{-3} and the error floor is eliminated when we compensate either for only the DAC's TF or for the Total TF. Moreover, we can see that there is a 3 dB degradation for the 80-km transmission case compared to the back-to-back case at BER=1×10^{-3} HD-FEC limit when transmitter pre-distortion is employed, attributed to CD-induced penalty in the link. It is worth noting here that in both back-to-back and 80-km transmission cases the optimum performance is achieved when pre-distortion is used to compensate only for the DAC TF while further compensation of the end-to-end system's TF does not lead to a further improvement of the performance as it is not possible to compensate further for bandwidth limitations. Results are further optimized when approximating the H_t by a Bessel and a Gaussian profile one as shown in Fig. 6, where the BER vs. received optical power performance is compared after the investigation of different transmitter pre-distortion techniques for the 80-km transmission case. As we can see, further optimization of the filter shape profile of the H_t, i.e., when we approximate it by a 3rd order/20 GHz Gaussian or a 5th order/10 GHz Bessel leads to performance improvement by 2.6 dB and 3 dB, respectively, at BER=1×10^{-3}, compared to the case when no pre-compensation is employed. In this way, BER$<1\times10^{-5}$ values are achieved for both Gaussian- and Bessel- shape profile H_t, for the received optical power levels greater than -12 dBm. Moreover, the Bessel approximation results in the elimination of the BER error floor compared to the Gaussian one. Note that the order and the bandwidth of either Bessel or Gaussian TF has been scanned and optimized to achieve the optimum BER performance shown in Fig. 6.

IV. CONCLUSIONS

We experimentally evaluated the OSNR and sensitivity requirements for 28 Gb/s IM/DD signal transmission based on low-bandwidth commercially available components. We showed that the transmission performance over 80 km can be significantly optimized by employing data pre-distortion at the transmitter DSP, thus eliminating the need for DCF deployment in WDM metropolitan networks. Bandwidth limitations are imposed mainly by the DAC, and can be optimally compensated using a 5th order, 10 GHz Bessel-profile total transfer function system approximation.

REFERENCES

[1] http://www.cisco.com/c/en/us/solutions/collateral/service-provider/ip-ngn-ip-next-generation-network/white_paper_c11-481360.html.

[2] ITU-T Recommendation G.709, Interfaces for the Optical Transport Network (OTN), Dec. 2009.

[3] K. Szczerba, P. Westbergh, E. Agrell, M. Karlsson, P. A. Andrekson, and A. Larsson, "Comparison of intersymbol interference power penalties for OOK and 4-PAM in short-range optical links," J. Lightw. Technol., vol. 31, no. 22, p. 3525 (2013).

[4] F. Karinou, N. Stojanovic, and Z. Yu, "Towards cost-efficient 100G metro networks using IM/DD, 10-GHz components and MLSE receiver," IEEE/OSA J. Lightw. Technol., vol. 33, no. 99, pp. 4109 - 4117, July 2015.

[5] G. Forney, "Maximum-likelihood sequence estimation of digital sequences in the presence of intersymbol interference," Trans. Inf. Theory, vol. 18, no. 3, p. 363 (1972).

ThB2-4

OECC/PS2016

Improvement of Bandwidth Limitation and Chromatic-Dispersion Tolerances Using 4-Level/7-Level Coding PAM for Direct-Detection System

Akira Masuda, Shuto Yamamoto, Hiroki Kawahara, Shingo Kawai, and Mitsunori Fukutoku

NTT Network Innovation Laboratories, NTT Corporation, 1-1 Hikarinooka, Yokosuka, Japan

masuda.akira@lab.ntt.co.jp

Abstract: *We have achieved improved bandwidth limitation and chromatic-dispersion tolerances in a 4-level/7-level coding PAM scheme for a direct-detection system. Simulation results showed the scheme enabled improved 112-Gb/s transmission. The scheme's feasibility was also experimentally confirmed.*
Keywords: *PAM, direct detection, coding, bandwidth limitation, chromatic dispersion*

I. INTRODUCTION

With mobile 5G networks coming into increasingly widespread use, many assume that the majority of future data traffic will comprise mobile traffic. To meet the expected rapid growth in mobile traffic, metro-access networks with higher capacity and lower costs will be required more than ever. In this regard direct-detection systems have attracted much attention; in particular, much study has been devoted to discrete multitone modulation (DMT) [1, 2] and 4-level pulse amplitude modulation (PAM-4) [3, 4]. In general, DMT requires complex digital signal processing (DSP) such as fast Fourier transformation (FFT) at the transmitter and receiver sides and subcarrier generation where an adequate modulation format is chosen for each subcarrier. On the other hand, signal processing for PAM-4 is easier than that for DMT. However, since PAM-4 has low tolerance to bandwidth limitation, it requires the use of broad-bandwidth electrical devices in a transceiver. Furthermore, its low tolerance to chromatic dispersion (CD) limits transmission distance in C-band transmission. To mitigate the bandwidth limitation penalty in PAM-4 transmission, methods using duobinary decoding at the receiver have been reported [5, 6]. While these methods improve signal quality very much when narrow-band devices are used, it is necessary to apply maximum likelihood sequence estimation (MLSE) to receiver DSP and this means that the amount of signal processing is enormous.

In this paper, we describe a simple scheme we propose to improve bandwidth limitation and CD tolerance of PAM-4 where 4-level/7-level coding is applied at the transmitter side. We show that the 4-level/7-level coding PAM (CPAM) scheme reduces signal performance degradation through numerical simulations where 112-Gb/s CPAM signals are transmitted. We also show that the scheme's feasibility was experimentally confirmed in a back-to-back situation.

II. 4-LEVEL/7-LEVEL CODING PAM

In this section, we first introduce the coding procedure of CPAM. Next, we show the principle for CPAM improving the bandwidth limitation and CD tolerances.

The coding procedure of CPAM is defined as

$$C(t) = \begin{cases} S(t) & when \quad P(t) = 1 \\ -S(t) & when \quad P(t) = -1 \end{cases} \quad (1),$$

where $S(t)$ is 4-level data ($S = 0, 1, 2, 3$) before encoding and $C(t)$ is 7-level data ($C = -3, -2, -1, 0, 1, 2, 3$) after encoding. Parameter $P(t)$ is a management sign ($P = -1, 1$), which is updated according to the following rules:

$$P(t) = \begin{cases} P(t-1) & when \quad S(t) \neq 0 \\ -P(t-1) & when \quad S(t) = 0 \end{cases} \quad (2).$$

As shown in Fig. 1, 7-level signals generated in this manner are assigned to 7-level optical amplitude. Note that the n-th level and $-n$-th level are assigned so that they have the same optical intensity with the optical-phase difference of π. As a result, 7-level optical amplitude modulation signals equivalent to 4-level optical intensity modulation signals are generated.

Figure 2 depicts examples of transmission waveforms of the conventional PAM-4 and CPAM. As it shows, there is no occurrence of high-frequency level change in CPAM because CPAM utilizes symbols with optical phase π. This means that CPAM has high tolerance to bandwidth limitation. It is also expected to mitigate inter-symbol interference (ISI) induced by CD because optical phase inversion is always maintained between the two sides of the 0-th level symbol where the ISI from neighboring symbols are canceled in CPAM. In the proposed CPAM scheme, received signals are demodulated to original data through conventional direct detection without any decoders at the receiver side because 7-level optical amplitude corresponds to 4-level optical intensity.

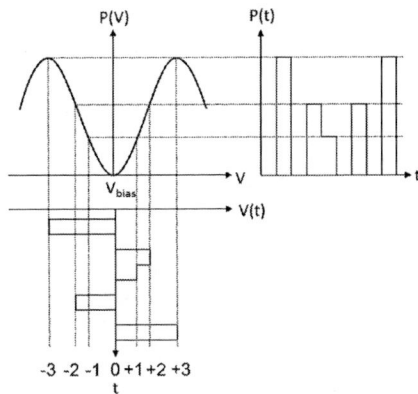

Fig. 1. Relationship between modulation voltage and optical intensity in 7-level amplitude modulation.

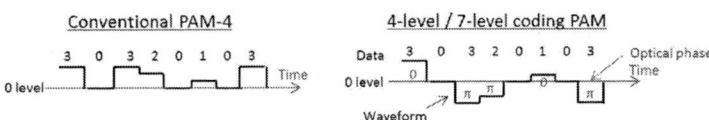

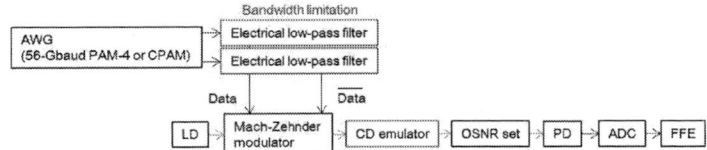

Fig. 2. Optical waveforms of conventional PAM-4 and CPAM.

Fig. 3. Simulation setup.

III. NUMERICAL SIMULATION

We performed 112-Gb/s optical transmission simulation to examine the CPAM scheme's performance at a baud rate of 56 Gbaud. Figure 3 shows the simulation setup. To evaluate the bandwidth limitation tolerance of CPAM, we limited the signal bandwidth by using an electrical low-pass filter at the transmitter side after DSP comprising a 4-level/7-level encoder and a raised-cosine filter with a 0.1 roll-off factor. The laser source linewidth was 1 MHz. In this simulation, we added CD to the link in order to evaluate CD tolerance. Optical signals were directly detected by a photo detector (PD) and resampled to two samples per symbol. Then they were demodulated by using a feed-forward equalizer (FFE) based on decision-directed least mean square (DD-LMS).

Figure 4 shows the relationships between received optical signal -to -noise ratio (OSNR) and bit -error -rate (BER) for CPAM and conventional PAM-4. As shown, for 26-GHz bandwidth limitation the OSNR penalty in PAM-4 is about 2 dB at the BER of 3.8E-3, which corresponds to the hard-decision forwarding error correction (HD-FEC) limit. On the other hand, the OSNR penalty in CPAM is less than 0.5 dB. In CPAM, bandwidth limitation narrower than signal baud rate can be carried out without signal quality degradation. This means that it is possible to use narrower-band electrical devices than those used with conventional PAM-4.

Figure 5 shows the relationship between received OSNR and BER for dispersive links. The performances in conventional PAM-4 and CPAM are almost equal at the accumulated CD of 0 ps/nm. By contrast, the performance of CPAM for accumulated CD is better than that of PAM-4. For accumulated CD it is 32 ps/nm and CPAM can achieve BER of 3.8E-3 at the OSNR of 32 dB, while conventional PAM-4 cannot achieve BER of 3.8E-3. The CD tolerance can be increased by applying CPAM.

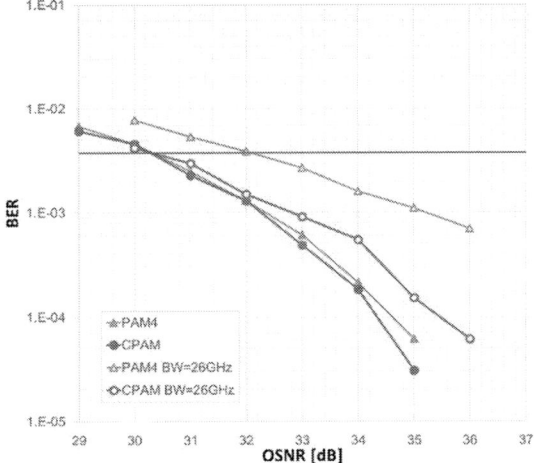

Fig. 4. Bandwidth limitation tolerances of PAM-4 and CPAM.

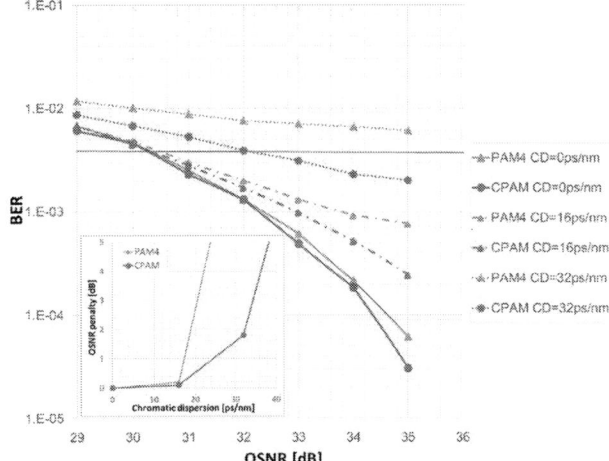

Fig. 5. CD tolerances of PAM-4 and CPAM.

IV. EXPERIMENT

We experimentally evaluated the performances of 16-Gb/s PAM-4 and 16-Gb/s CPAM in order to confirm the feasibility of CPAM. Figure 6 shows the experimental setup. The PAM-4 and CPAM signals were generated by using an arbitrary waveform generator (AWG). The electrical bandwidth of the AWG is 20 GHz which is sufficiently wider than the signal baud rate, so the bandwidth limitation is negligible.

Figure 7 shows the eye diagrams of PAM-4 and CPAM for 4-GHz bandwidth limitation. Figure 8 shows the relationship between received OSNR and BER for CPAM and conventional PAM-4 with or without bandwidth limitation. The limitation is performed by a digital filter in which 3-dB bandwidth is 4 GHz. Where there is no bandwidth limitation, the difference in required OSNR at the BER of 3.8E-3 between PAM-4 and CPAM is less than 1 dB. This means that CPAM can be generated with the OSNR penalty of less than 1 dB. Where there is bandwidth limitation, the OSNR penalty in CPAM is 1 dB while that in PAM-4 is 3 dB. This means that CPAM has higher bandwidth limitation tolerance than PAM-4 and that applying CPAM mitigates the requirement for bandwidth of electrical devices.

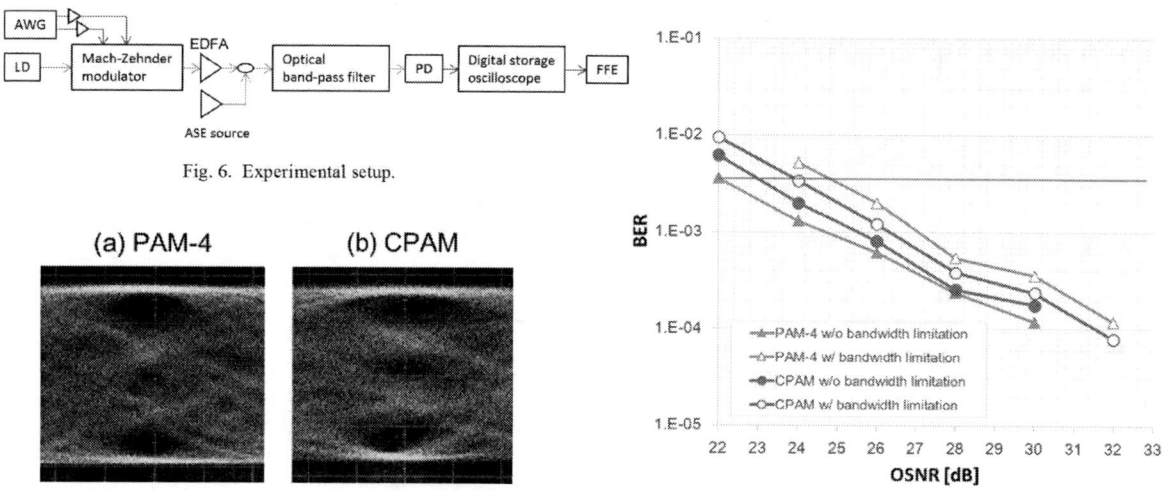

Fig. 6. Experimental setup.

Fig. 7. Eye diagrams of (a) PAM-4 and (b) CPAM with 4-GHz bandwidth limitation.

Fig. 8. Relationship between received OSNR and BER for PAM-4 and CPAM with or without bandwidth limitation.

V. CONCLUSION

This paper described a coding PAM (CPAM) scheme we propose that effectively improves bandwidth limitation and chromatic dispersion (CD) tolerances in a direct-detection system. We performed numerical simulation in order to evaluate the performances of 112-Gb/s PAM-4 and CPAM. The simulation results showed that the CPAM scheme reduced bandwidth limitation penalty by 1.5 dB for 26-GHz bandwidth limitation and achieved BER of 3.8E-3 where the accumulated CD was 32 ps/nm. We also experimentally confirmed that CPAM is feasible, that CPAM signals can be generated with an OSNR penalty of less than 1 dB, and that the CPAM scheme reduces bandwidth limitation penalty from 3 to 1 dB where there is a 4-GHz bandwidth limitation for 16-Gb/s signals.

REFERENCES

[1] S. Randel, et al., "100-Gb/s discrete-multitone transmission over 80-km SSMF using single-sideband modulation with novel interference-cancellation scheme," Proc. of ECOC, Mo. 4. 5. 2, 2015

[2] F. Li, et al., "Real-time direct-detection of quad-carrier 200Gbps 16QAM-DMT with directly modulated laser," Proc. of ECOC, Mo. 4. 5. 5, 2015

[3] H. Yamazaki, et al., "160-Gbps Nyquist PAM4 transmitter using a digital-preprocessed analog-multiplexed DAC," Proc. of ECOC, PDP 2.2, 2015

[4] M. A. Mestre, et al., "Direct detection transceiver at 150-Gbit/s net data rate using PAM 8 for optical interconnects," Proc. of ECOC, PDP 2.4, 2015

[5] N. Stojanovic, et al., "Performance and DSP complexity evaluation of a 112-Gbit/s PAM-4 transceiver employing a 25-GHz TOSA and ROSA," Proc. of ECOC, Mo. 4. 5. 5, 2015

[6] T. Zuo, et al., "112-Gb/s duobinary 4-PAM transmission over 200-m multi-mode fiber," Proc. of ECOC, P. 5. 19, 2015

ThB2-5

Power Budget Improvement of 25-Gb/s Optical Link Based on 1.5-μm 10G-Class VCSEL

Jingjing Zhou[1,2], Changyuan Yu[3], Gurusamy Mohan[1], and Hoon Kim[4*]

[1]Dept. of Electrical & Computer Engineering, National University of Singapore, Singapore, 117583
[2]National University of Singapore (Suzhou) Research Institute, China 215123
[3]Dept. of Electronic and Information Engineering, The Hong Kong Polytechnic University, Hong Kong
[4]School of Electrical Engineering, Korea Advanced Institute of Science and Technology (KAIST), Korea 34141
E-mail: hoonkim@kaist.ac.kr

Abstract: *We present a significant performance improvement of a 25-Gb/s OOK signal generated from a 1.5-μm 10G-class VCSEL (bandwidth: ~7 GHz) by using a delay interferometer.*
Keywords: *VCSEL, optical access network, equalization*

I. INTRODUCTION

To satisfy the increasing demands of transmission capacity in optical access networks, short- and intermediate-haul fiber-optic links are constantly required to scale up their capacity by increasing the number of channels and/or data rate per channel. Current 10-Gb/s optical access networks are envisaged to be upgraded to 25 Gb/s or higher bit rate in the near future [1]. However, cost-effective implementation of 25-Gb/s transmission systems operating at 1.5 μm is still challenging due to expensive and lossy high-speed external modulators and/or dispersion compensation modules (DCMs) typically required for such systems. In order to bring down the cost per bit of 25-Gb/s systems lower than 10-Gb/s counterpart, it is necessary to utilize highly cost-effective optical transmitters such as vertical-cavity surface-emitting lasers (VCSELs), but not to employ lossy and bulky DCMs.

VCSELs, thanks to their high energy efficiency, direct modulation capability, and excellent manufacturability and testability, have been widely adopted as transmitters for data-center and short-haul applications. Recent advancement in reliability and performance of long-wavelength VCSELs also opens up the possibility of using these attractive devices for optical access networks [2]. However, one of the biggest concerns about the use of 1.5-μm VCSELs for optical access applications would be a tight power budget, which comes from poor tolerance to fiber chromatic dispersion. Direct current modulation of laser diodes always accompanies a large frequency chirp originating from the dependence of the refractive index on the carrier density in lasers. A short cavity length of VCSELs, compared to edge-emitting laser diodes, makes this effect prominent and thus the high-speed signals generated by using 1.5-μm VCSEL signals become very sensitive to fiber dispersion. As a result, the power budget of the 1.5-μm VCSEL based optical access networks is significantly limited due to the low output power of the VCSEL and poor dispersion tolerance.

In this paper, we report on the improvement in power budget of a 25-Gb/s optical access link based on 1.5-μm 10G-class VCSELs by using a delay interferometer (DI). The simple passive optical device not only significantly increases the extinction ratio (ER) of the 25-Gb/s on-off keying (OOK) signal generated by using the directly-modulated VCSEL but also improves the eye opening of the signal by partly compensating the band-limitation of the VCSEL. Our experimental results show that the use of DI improves the power budget of the 25-Gb/s optical link by up to 5.9 dB despite the insertion loss of the DI.

II. EXPERIMENT AND RESULTS

Fig. 1(a) shows the experimental setup. A 25-Gb/s non-return-to-zero signal is generated by multiplexing two 12.5-Gb/s pseudo-random bit sequences (length $2^{15}-1$) from a pulse pattern generator (PPG), and then directly fed to a 1.54-μm VCSEL. No electrical amplifier is used to drive the laser diode. The VCSEL used in the experiment is a 10G-rated device packaged in a transistor-outline can, as shown in the inset of Fig. 1(b). The VCSEL emits an optical power of -0.8 dBm when biased at 14 mA. The 3-dB modulation bandwidth of this device is measured to be ~7 GHz, as shown in Figure 1(b). Due to the limited bandwidth of the VCSEL, the eye diagram at the output of the VCSEL is severely closed as shown in Fig. 1(d) even though the eye diagram of the driving electrical signal is wide open [shown in Fig. 1(c)]. The output signal from the VCSEL is sent to a DI before transmission over standard single-mode fiber (SSMF). Two free-spectral range (FSR) values of DI are investigated in this experiment: 16.1 and 25 GHz. The phase of the DI is set to be at the quadrature point of the DI's transmittance curve. As a result, the ER of the signal is dramatically enhanced after the DI, which is clearly shown in Fig. 1(e). However, due to non-uniform frequency modulation (FM) response of the VCSEL especially at low frequencies, the OOK signal suffers from severe performance degradation when the frequency-modulated VCSEL signal is converted to the OOK signal by using the DI [3]. This problem can be alleviated by utilizing a DC-balanced line coding with low-frequency depletion capability, whilst, it is not employed in this experiment. This is

OECC/PS2016

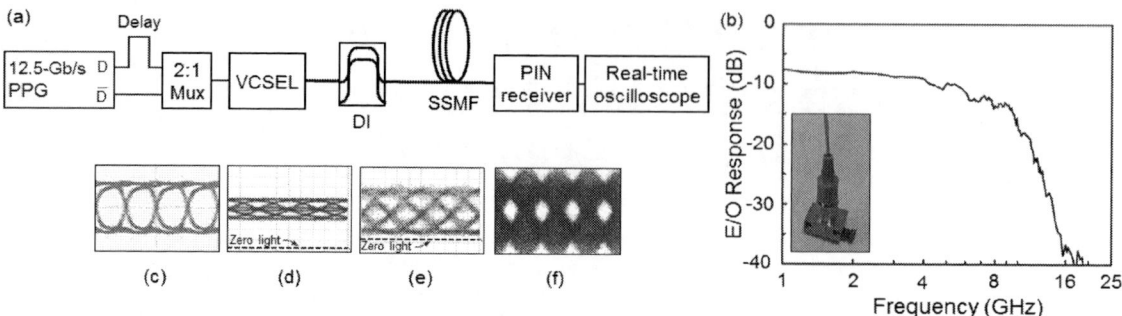

Fig. 1. (a) Experimental setup. (b) Measured E/O response of the VCSEL used in the experiment. The inset shows the photograph of the VCSEL. (c) Eye diagram of the 25-Gb/s OOK electrical signal driving the VCSEL. Eye diagrams of the optical signal measured (d) at the output of the VCSEL and (e) after DI. (f) Eye diagram of the signal measured at the receiver when the electrical equalization is applied.

because the signal generated by multiplexing two complementary signals in this experiment already exhibits low-frequency depletion. However, in a real system, line coding is needed to combat the signal degradations caused by the non-uniform FM response of the VCSEL [3]. After transmission, the signal is detected by a PIN receiver and then sampled by using a digital sampling scope (sampling rate: 100 Gsample/s) for offline processing. In order to compensate for the signal distortions caused by the fiber dispersion and limited bandwidth of VCSEL, we employ a half-symbol-spaced 11-tap feed-forward equalizer followed by a 4-tap decision-feedback equalizer. Finally, the number of bit errors is counted directly for bit-error rate (BER) measurement.

We first optimized the ER of the 25-Gb/s OOK signal. The ER is defined as the intensity ratio between long marks and long spaces at the output of the VCSEL and thus the effects of band-limitation are excluded. Fig. 2(a) and (b) show the measured receiver sensitivities (at BER=10^{-3}) of the signal as a function of ER for various transmission distances when 25- and 16.1-GHz DIs are applied, respectively. When the 25-GHz DI is applied [see Fig. 2(a)], the receiver sensitivity is improved in the back-to-back condition as the ER increases. However, due to large frequency chirp of high-ER signals, the optimum ER for the best receiver sensitivity shifts towards a lower value as the transmission distance increases. For example, the optimal ER is found to be 3.0 dB after 10-km transmission, and it is further reduced to 2.7 dB at 20 km. For the 16.1-GHz DI, the receiver sensitivity is improved as the ER increases, as shown in Fig. 2(b). The optimal ER is found to be 3.4 dB from 0- to 20-km transmissions. Higher ER values would not improve the receiver sensitivity after transmission because it considerably broadens the spectral width of the signal. Also an electrical amplifier needs to be used to drive the VCSEL. Fig. 2(c) shows the ER measured at the output of the DI versus the ER measured at the input of the DI. First, it is observed that the ER of signal is significantly increased after the application of DI. This is because the marks and spaces have different wavelengths due to the adiabatic chirp of the VCSEL and they experience different transmittance by the DI. Second, a bigger improvement in ER after DI is achieved as the ER of VCSEL's output signal (i.e., signals before DI) increases. In addition, a DI with a smaller FSR exhibits a steeper transmittance slope with respect to wavelength and thus shows a bigger improvement in ER. For instance, from Fig. 2(c), the ER is increased to 11.0 dB when a DI with an FSR of 25 GHz is employed for the 25-Gb/s OOK signal having a 3.0-dB ER (before the DI). It is further improved by additional 3.3 dB when the FSR of DI is changed to 16.1 GHz.

Next, we investigate the transmission performance of the 25-Gb/s OOK signal. Based on the results found in Fig. 2, we set the ERs of signals to be 3.4 and 2.7 dB at the output of the VCSEL when 16.1- and 25-GHz DIs are applied, respectively. Fig. 3(a) shows the BER curves after 0- and 20-km transmissions. With the 25-GHz DI, the receiver

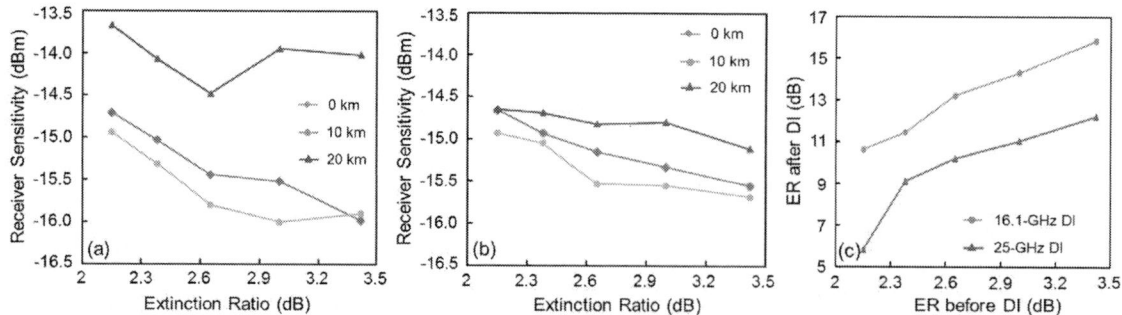

Fig. 2. Receiver sensitivity as a function of the ER of the signal for various transmission distances when the FSRs of DI are (a) 25 and (b) 16.1 GHz, respectively. (c) ER measured after the DI versus ER before the DI.

294

OECC/PS2016

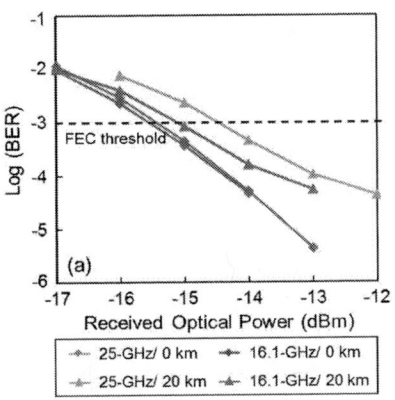 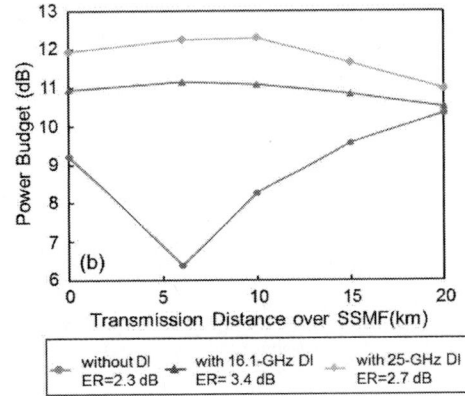

| 25-GHz/ 0 km | 16.1-GHz/ 0 km |
| 25-GHz/ 20 km | 16.1-GHz/ 20 km |

| without DI
ER=2.3 dB | with 16.1-GHz DI
ER= 3.4 dB | with 25-GHz DI
ER=2.7 dB |

Fig. 3 (a) BER curves of the 25-Gb/s OOK signal after 0- and 20-km transmissions. (b) Measured power budget of the 25-Gb/s OOK signals as a function of transmission distance over SSMF when no DI, 16.1-, and 25-GHz DIs are applied, respectively. The corresponding ERs measured at the output of the VCSEL are 2.3, 3.4 and 2.7 dB, respectively.

sensitivities of 25-Gb/s OOK signal are measured to be -15.4 and -14.5 dBm after 0- and 20-km transmission, respectively. Thus, the transmission penalty after 20-km transmission is measured to be merely 0.9 dB. When the FSR of the DI is reduced to 16.1 GHz, the receiver sensitivities of the OOK signal are improved by 0.1 and 0.6 dB after 0- and 20-km transmissions, respectively. On the other hand, the use of DI inevitably introduces additional insertion loss in the transmission link. The use of DI for optical access networks can be justified as long as the sensitivity improvement brought by the DI is larger than its insertion loss. Therefore, we plot the power budget (i.e., the fiber launch power minus the receiver sensitivity) of the transmission link as a function of transmission distance, as displayed in Fig. 3(b). The fiber launch power should be decreased by a few decibels when the DI is employed. Also plotted in Fig. 3(b) for comparison is the power budget of the OOK signal generated without using the DI. In this case, the ER is set to be 2.3 dB, which is found to be optimum for the maximum transmission distance [4]. The results show that the power budget improvement is clearly achieved by using the DI. For instance, we achieve 2.7- and 1.7-dB power budget improvements compared with the case without DI at the back-to-back condition when the FSRs of DIs are 25 and 16.1 GHz, respectively. It is worth mentioning that we achieve better receiver sensitivities in Fig. 3(a) when the FSR of the DI is 16.1 GHz than the case with 25-GHz FSR. However, a 25-GHz DI shows larger power budget in Fig. 3(b) because it has lower insertion loss than the 16.1-GHz DI. The results also show that when the DI is employed, the power budget is gradually degraded as the transmission distance increases. On the other hand, in the absence of DI, the power budget of the OOK signal degrades up to 6 km and then improves until 20 km. Thus, we achieve the largest improvement in power budget of 5.9 dB after 6-km transmission by using the DI having an FSR of 25 GHz. The results also show that the effectiveness of using the DI is maintained up to 20 km.

III. Conclusions

We have reported on the improved power budget of the 25-Gb/s optical access link based on 10G-class VCSELs by using a delay interferometer. This simple passive optical device not only significantly enhances the extinction ratio of the signal but it also serves to compensate partly for the band-limitation of the VCSEL. The results show that despite the insertion loss of the DI, we achieve up to 5.9-dB improvement in power budget in a 20-km long optical link. Due to the periodicity of the DI's transmittance, a single DI can also be used for multiple channels in a WDM system. Thus, we believe that 1.5-μm, 10G-class VCSELs together with a delay interferometer could be used for 25-Gb/s passive optical networks having a low splitting ratio (e.g., <10).

Acknowledgements

The authors would like to thank the supports of AcRF Tier 2 Grant MOE2013-T2-2-135 from MOE Singapore and Grant 61501313 from National Natural Science Foundation of China.

References

[1] J. Man *et al.*, "25-Gb/s and 40-Gb/s faster-than-Nyquist PON based on low-cost 10G-class optics," in *Proc. ACP*, 2015, paper AM1F.1.

[2] F. Fidler *et al.*, "4×10-Gb/s CWDM transmission using VCSELs from 1531 nm to 1591 nm," in *Proc. ECOC*, 2006, paper WE4.5.1.

[3] Z. Al-Qazwini *et al.*, "1.5-μm 10-Gb/s VCSEL link for optical access applications," *IEEE Photon. Technol. Lett.*, vol. 25, no. 22, pp. 2160-2163, 2013.

[4] J. Zhou *et al.*, "25-Gb/s OOK and 4-PAM transmission over >35-km SSMF using directly modulated 1.5-μm VCSEL," in *Proc. OFC*, 2016, paper Th1G.7.

RZ-PAM4 transmission with quasi-Fourier-transform-limited gain-switched pulse source

Songyuan Dai   Masanori Hanawa

University of Yamanashi, 4-3-11 Takeda, Kofu, Yamanashi 400-8511, Japan

E-mail: hanawa@yamanashi.ac.jp

Abstract: RZ-PAM4 signal transmission over 30 km of SMF based on quasi-Fourier-transform-limited gain-switched short optical pulse source is demonstrated thanks to both timing jitter suppression by optical feed-back and wavelength chirp reduction by narrowband optical filtering.

Keywords: Gain-switched short pulse, Timing jitter, Wavelength chirp, RZ-PAM4

I. INTRODUCTION

Gain switched-laser diode (GS-LD) is a very convenient optical device for generating pico-second width optical short pulses easily. It has potentials in optical measurements [1], multi-photon imaging [2], time-resolved spectroscopy [3], and so on. However due to its oscillating principle, output pulses from GS-LDs hold large timing jitter (over 7 ps) and wavelength chirp, thus a GS-LD is hardly considered as an optical source for fiber-optic communication systems. To solve the timing jitter problem, Nonaka et al. came up with the self-seeding scheme and suppressed the timing jitter to 1.0 ps range [4]. On the other hand Chen et al. reported nearly chirp free pulse generation by inserting an optical band pass filter (OBPF) after a GS-LD [5] without discussing on jitter reduction.

The authors have reported on 10-Gbit/s QPSK×2ch. Fourier-encoded synchronous optical code division multiplexing (FE-SOCDM) system [6], which is considered as an optically-encoded and digitally-decoded orthogonal frequency division multiplexing (OFDM) system. Although the system realized a high spectral efficiency of 2 bit/s/Hz by means of QPSK format, it required a stable short optical pulse source and a digital coherent receiver having a narrow line width local laser, wideband photo-detectors, high-speed analog-to-digital converters, and a digital signal processor enabling massive signal processing in real time. When the number of bits per symbol is limited small, i.e. 2 bit/symbol, the use of the QPSK format is inefficient from the view point of system cost. PAM4 format also realizes 2 bit/symbol, same as the QPSK format, and it results in significantly simpler receiver configuration.

In this paper, we firstly report RZ-PAM4 transmission based on GS-LD by combining timing jitter suppression by self-seeding and wavelength chirp reduction by narrowband optical filtering simultaneously. The experimental results for 5 Gbit/s RZ-PAM4 transmission over a 30 km of single mode fiber (SMF) are reported below.

II. QUASI-FOURIER-TRANSFORM-LIMITED PULSE GENERATION FROM GS-LD

In a GS-LD, accumulation of carrier density by injection current fluctuates due to noise, resulting in wobble of oscillation timing called timing jitter. The self-seeding scheme [4] uses optical feed-back of some part of generated optical pulses to regularize pulse oscillation timing in the gain-medium. Because of this optical feed-back, the carrier density in the gain-medium is forced to exceed the oscillation threshold for the avalanche effect. By setting the amount and timing of the feed-back adequately, the pulse oscillation timing is regulated and the timing jitter is suppressed. What are required for this scheme are a gain-medium such as a distributed feed-back laser diode (DFB-LD) without an isolator in its casing, an optical delay line with a proper length, and a partial reflector with a proper reflectivity. This scheme requires small injection current to guarantee single peak pulses without tails, resulting in small pulse amplitude and rather wide pulse width more than 30ps.

Chen et al. proposed the use of optical narrowband filter to reduce wavelength chirp. On the contrary to the GS-LDs with self-seeding, it requires rather large injection current to have large pulse amplitude and narrow pulse width. By such large injection current, the output has a narrow width and high peak pulse followed by a long tail. The leading high peak pulses and the long tails consist of different wavelength components and the long tails have longer wavelength components than the leading high peak pulses; thus by cutting-off the long tails with sharp optical filtering, only leading pulses with high-peak and narrow width are obtained. The larger injection current is, the smaller timing jitter of the generated pulses comparing with a basic GS-LD scheme without any additional equipment. It, however, is not enough for communication applications in the range of our preliminary experiments.

Figure 1 shows the proposed quasi-Fourier-transform-limited optical short pulse source combining above two schemes. To make tuning of optical delay line length between the gain-medium and the partial reflector easy, a narrowband optical band pass filter (Alnair Laboratory BVF-200C) was placed after the optical isolator. The repetition rates of optical pulses were chosen at 5 patterns: 0.5, 0.8, 1.0, 1.25, and 2.5 GHz. The relationship between temporal pulse width and spectral bandwidth, full-width measured at -30 dB from the spectral peak, of the generated pulses are plotted in Figure 2. The open marks are for the proposed scheme and the filled marks are for the GS-LD with self-seeding only. Clearly, the generated pulses by the proposed scheme are closer to the line showing the Fourier-transform limit and it is confirmed that the generated pulses are quasi-Fourier-transform-limited. The shortest optical pulse width of 10.28 ps was obtained at 0.8 GHz of the repetition rate, while the lowest chirp of 0.13 nm was observed at 2.5 GHz. Figure 3 shows optical spectra and temporal waveforms for both the proposed scheme and the self-seeding only. At both repetition rates of 0.8 GHz and 2.5 GHz, shorter optical pulses preserving small timing jitter could be obtained by spectral slicing of the longer wavelength region.

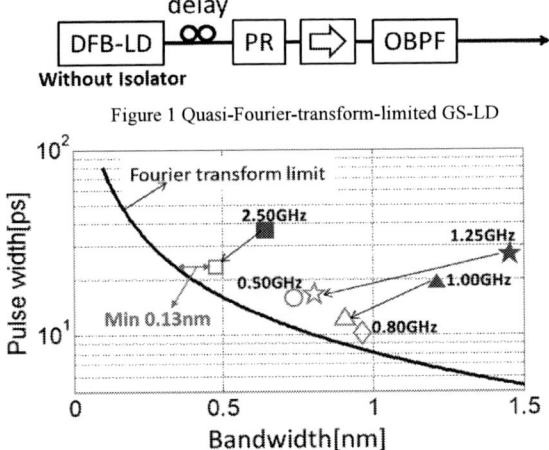

Figure 1 Quasi-Fourier-transform-limited GS-LD

Figure 2 Temporal pulse width vs. -30dB spectral bandwidth

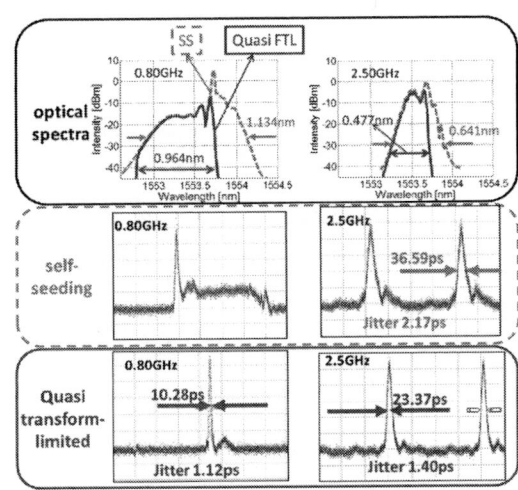

Figure 3 Comparison of optical spectra and temporal waveforms for both the self-seeding only and the proposed schemes

III. 5 GBIT/S PAM4 TRANSMISSION WITH QUASI-FOURIER-TRANSFORM-LIMITED GAIN-SWITCHED PULSES

RZ-PAM4 signals were generated from the quasi-Fourier-transform-limited optical pulses at the repetition rate of 2.5 GHz. The pulses are modulated by an IQ modulator by setting the two Mach-Zehnder modulator (MZM) outputs to have amplitudes $(1, \sqrt{0.5})$ or $(1, 0)$ as well as choosing the phase difference between two MZMs zero. PAM4 signals obtained by this modulation scheme are shown in Figure 5. Clear eye-openings are observed by using the proposed quasi-Fourier-transform-limited optical pulses.

The experimental setup is shown in Figure 6. The transmitted RZ-PAM4 signals were power-adjusted using both an erbium doped fiber amplifier and a variable optical attenuator, and transmitted over SMFs of 10, 20, and 30 km. Only bit error rates (BER) for the lower arm of the IQ modulator were measured with an error detector and total BERs for the whole PAM4 signals were estimated using the relationship $BER_{2\,bit/sym} = \frac{3}{2} BER_{1\,bit/sym}$ by assuming all the error rates at three eye-openings were the same.

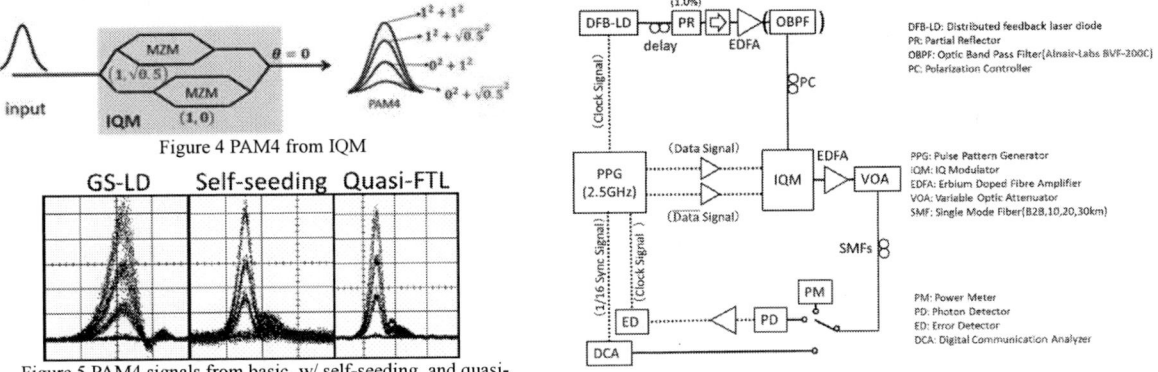

Figure 4 PAM4 from IQM

Figure 5 PAM4 signals from basic, w/ self-seeding, and quasi-Fourier-transform-limited GS-LDs

Figure 6 Experimental setup

The estimated BER of both the self-seeding and the proposed scheme are plotted in Figure 7. The PAM4 signals generated from the quasi-Fourier-transform-limited GS-LD were successfully transmitted over the 30 km of SMF with BER under 10^{-9} with an average received power of -3.5dBm while the self-seeding scheme got an error floor above 10^{-9}. Figure 8 shows temporal waveforms for both schemes at around 10^{-7} of BER. Due to no spectral slicing in the self-seeding only scheme, the strong wavelength chirp are remained in the source pulses; thus when the distance gets longer, the pulses in the self-seeding only get wider more than in the proposed scheme, resulting in more BER. On the other hand, the quasi-Fourier-transform-limited gain-switched PAM4 signals suffered less chromatic dispersion in transmission and had better BER. This experimental result demonstrates that the proposed quasi-Fourier-transform-limited GS-LD can be an optical source for 5 Gbit/s PAM4 communication systems.

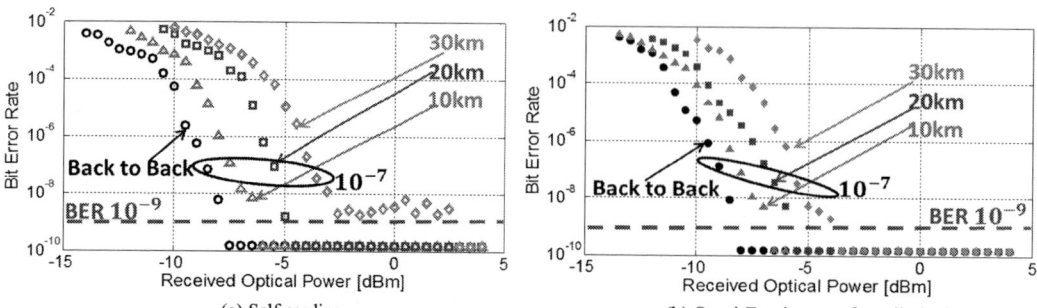

(a) Self-seeding   (b) Quasi-Fourier-transform-limited

Figure 7 BER characteristics for RZ-PAM4 transmission with both the proposed and the self-seeding only GS-LDs

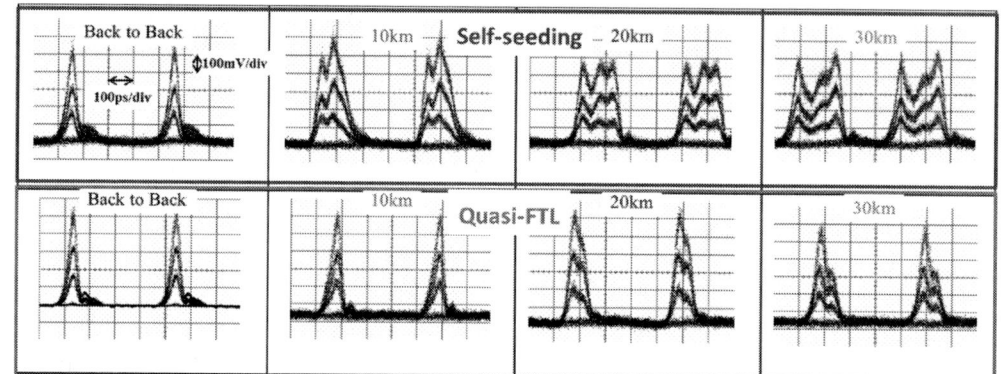

Figure 8 Temporal waveforms of RZ-PAM4 signals (up: self-seeding only scheme, down: proposed scheme)

IV. CONCLUSIONS

For simpler implementation of spectral efficient OCDM systems, RZ-PAM4 transmission with simple gain-switched optical pulse source was experimentally investigated. With optical feed-back to suppress timing jitter and optical spectral slicing to reduce wavelength chirp, a quasi-Fourier-transform-limited short optical pulse source was implemented with a GS-LD and 5 Gbit/s RZ-PAM4 signals generated from the GS-LD were successfully transmitted over standard SMF with BER under 10^{-9}. Also due to suffering less chromatic dispersion in long distance transmission, we can say the GS-LD with the proposed scheme has a potential in multi-channel communication systems such as optical time domain multiplexing and OCDM, as a cheaper and simpler optical short pulse source.

REFERENCES

[1] Kenji Wada, Satoru Matsukura, Amaka Tanaka, Tetsuya Matsuyama, and Hiromichi Horinak, "Precise Measurement of Optical Fiber Length Using a Gain-Switched Distributed Feedback Laser with Delayed Optical Feedback", Opt. Express 23, 18, pp. 23013-23020, 2015

[2] H. Yokoyama, H. Guo, T. Yoda, K. Takashima, K. Sato, H. Taniguchi, and H. Ito, "Two-photon bioimaging with picosecond optical pulses from a semiconductor laser," Opt. Express 14, 8, pp3467–3471, 2006

[3] A. Sato, S. Kono, K. Saito, K. Sato, and H. Yokoyama, "A high-peak-power UV picosecond-pulse light source based on a gain-switched 1.55 microm laser diode and its application to time-resolved spectroscopy of blueviolet materials," Opt. Express 18, 3, pp2522–2527, 2010.

[4] Koji Nonaka, Hiroaki Mizuno, Hongbin Song, Nobuyasu Kitaoka and Akihito Otani, "Low-Time-Jitter Short-Pulse Generator Using Compact Gain-Switching Laser Diode Module With Optical Feedback Fiber Line", JJAP, 47, 8, pp6754-6756, 2008

[5] Shaoqiang Chen, Masahiro Yoshita, Aya Sato, Takashi Ito, Hidefumi Akiyama, and Hiroyuki Yokoyama "Dynamics of short-pulse generation via spectral filtering from intensely excited gain-switched 1.55-μm distributed-feedback laser diodes" , Opt. Express, 21, 9,pp10597-10605, 2013

[6] Yasuhiro Okamura, Osamu Iijima, Satoshi Shimizu, Naoya Wada, and Masanori Hanawa, "Simultaneous detection of 10-Gbit/s QPSK × 2-ch. Fourier-encoded synchronous OCDM signals with digital coherent receiver" , Opt. Express, 21, 3, pp3298-3307, 2013

Over 400 Gbit/s Digital Coherent Channels for Optical Transport Network

Yutaka Miyamoto

NTT Network Innovation Laboratories   e-mail address: miyamoto.yutaka@lab.ntt.co.jp

Abstract: *This paper reviews the recent R&D challenges of digital coherent optical channel transmission with channel rates over 400 Gbit/s. The promising DSP and preprocessing technologies for 1-Tbit/s channel long-haul transmission are addressed.*
Keywords: *Digital coherent, 1 Tbit/s, OTN*

I. INTRODUCTION

Internet traffic continues to increase because of the evolution and rapid adoption of various mobile and cloud services in Japan. The digital coherent transmission system [1] is very promising to accommodate such high capacity demands sustainably. 100 Gbit/s digital coherent optical transport systems[2] have recently been installed in Japan and all over the world. Further research and development of digital coherent transport technologies with channel rates over 400 Gbit/s are being conducted [3-5] for real commercial deployment as shown in Fig.1..

This paper reviews recent R&D progress in 400-Gbit/s channel based digital coherent transport for OTN with novel features such as super channel, adaptive modulation format and flexible grid. The novel high-speed and nonlinear-tolerant preprocessing technologies for 1 Tbit/s digital coherent channel are also addressed.

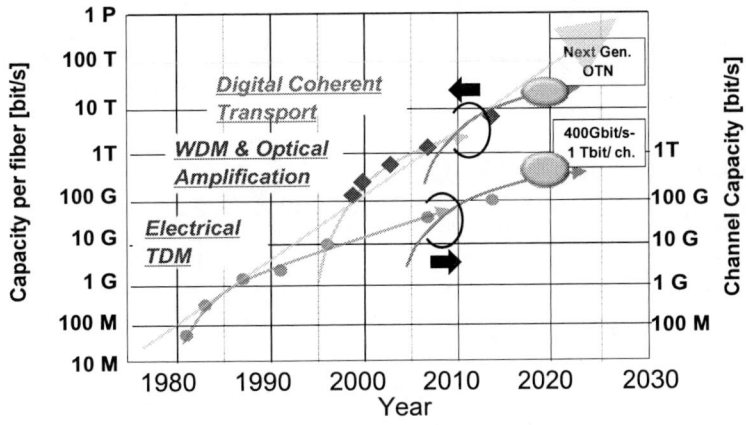

Fig. 1 Capacity evolution of NTT Commercial System

II. OVER 400 GBIT/S DIGITAL COHERENT CHANNELS

A. Super channel with adaptive modulation format in flexible grid

The choice of modulation format is the key in enhancing OTN spectral efficiency and reach. Figure 2 shows the relationship between spectral efficiency (SE) and required OSNR for various modulation formats normalized against 100 Gbit/s polarization division multiplexed quadrature phase shift keying (PDM QPSK). 400 Gbit/s PDM 16 QAM channel format doubles the SE, however, a 9.7-dB OSNR enhancement is required to match the regenerative repeater spacing of 100 Gbit/s PDM QPSK channels. To relax the severe tradeoff between SE and OSNR, various kinds of flexibilities have been introduced for 400 Gbit/s channel transmission. One of the main features is multi carrier transmission (super channel) with an adaptive modulation format scheme and the flexible grid. In the case of 400 Gbit/s long-reach optical paths, 4 pairs of 100-Gbit/s subcarriers with PDM QPSK format are used, while short-reach optical paths use 2 pairs of 200-Gbit/s subcarriers with 16 QAM. To determine the modulation formats suitable for a given optical path, inline OSNR monitoring in the receiver has been proposed [6]. Since the required signal bandwidth depends on the modulation format, ITU-T has standardized flexible signal bandwidth with 6.25-GHz granularity. Digital to analog converters offering more than 64 Gsymbol/s have been introduced to control the transmitted signal bandwidth by Nyquist filtering within the assigned signal bandwidth (ex. 37.5 GHz =6 x 6.25 GHz slots). A novel automatic bias control circuit was also introduced to stabilize the adaptive operation of the I/Q modulator[7].

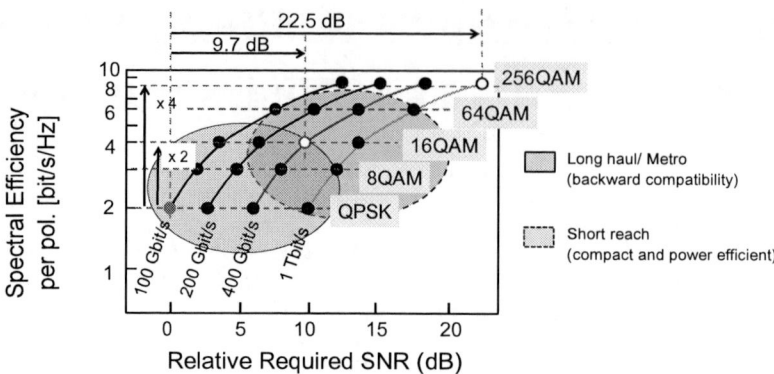

Fig. 2 Trade of between spectral efficiency and required OSNR in various modulation formats.

Nonlinearity compensation[8], adaptive equalization[9] and adaptive advance FEC[10] have been studied to enhance the OSNR tolerance and implemented using DSP ASICs.

B. Field experiment using 400 Gbit/s super channels

Several field experiments have been conducted to verify the above-mentioned DSP functions. Adaptive modulation format with inline OSNR monitoring were tested; QPSK was switched to 16 QAM according to the monitored OSNR of the optical path in dispersion shifted fiber (DSF) field fiber under offline processing [11]. The WDM transmission performance was achieved with both 100 Gbit/s PDM-QPSK channels and 400 Gbit/s PDM-16QAM super channels in DSF field-installed fiber with high polarization mode dispersion (PMD) under offline processing [5]. Recently, a WDM field experiment using standard single mode fiber (SSMF) field fiber was conducted in JGN-X using a newly developed real-time DSP ASIC, and more than 22 Tbit/s capacity transmission using 400 Gbit/s super channel with adaptive modulation functions was confirmed[12].

III. TECHNOLOGIES FOR HIGH-SYMBOL-RATE SYSTEM

A. Advanced DSP for Novel architecture of high symbol rate DAC and ADC

Advanced DSP are extensively studied to enhance the OSNR tolerance and nonlinearity tolerance using higher order QAM, such as novel coded modulation[13] and probabilistically shaped constellations[14]. For compact and cost effective realization of the next generation digital coherent channels over 400 Gbit/s, the technologies realizing high symbol rate system are attractive. Figure 3 shows the impact of high symbol rate system in 400 Gbit/s and 1 Tbit/s digital coherent channels

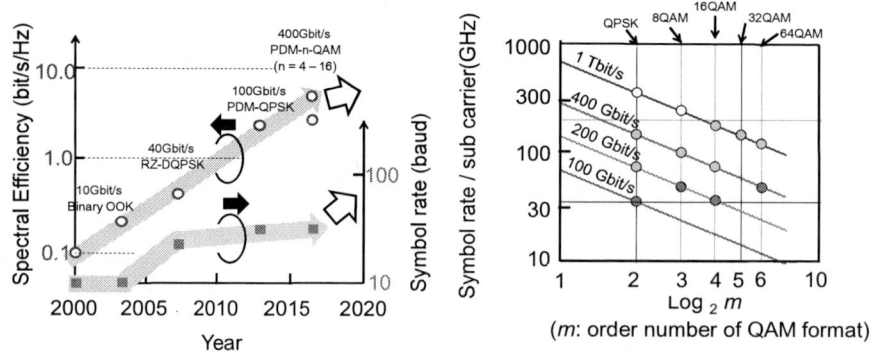

(a) Evolution of symbol rate in commercial systems.　　(b) Combination of symbol rate and modulation format.
Fig. 3 Impact of 400 Gbit/s and 1 Tbit/s digital coherent channels on high symbol rate systems.

As shown in Fig.3(a), trying to increase the symbol rate is no long effective for enhancing the SE since the Nyquist-filtered signal spectra already fill the available path band of the optical reconfigurable add drop multiplexer node in OTN. Figure (b) shows the possible combinations of modulation format and symbol rate for 400-Gbit/s and 1 Tbit/s digital coherent channels. Today's CMOS-ASIC can realize the combinations shown by the red plots with symbol rates of around 40 Gsymbol/s. The evolution of the fine CMOS-ASIC process has the possibility of realizing the single-carrier 400-Gbit/s and 1 Tbit/s digital coherent channels, shown by the yellow plots, if the DAC and ADC bandwidth can be enlarged with advanced DSP functions.

B. Advanced Electrical and Optical Pre Processing

Figure 4 shows the electrical and optical preprocessing that enhance the performance of high-symbol-rate digital coherent channel transmission. Figure 4(a) shows the example of electrical preprocessing for enlarging the DAC bandwidth. The digitally preprocessed analog multiplexing DAC (DP-AM DAC) configuration effectively utilizes today's DSP resources and the DAC bandwidth has been doubled by using the InP HBT analog multiplexer and CMOS-ASICs[16]. Since the circuit configuration of the DP AM DAC is symmetric, this is very attractive for balancing the analog performance of each lane, and 160 Gbit/s single PAM4 transmission [15]and 300 Gbit/s single-carrier PDM QPSK long-haul transmission[16] have been experimentally confirmed. Figure 4(b) shows the example of optical preprocessing that can reduce DSP complexity. The complementary spectrum inversion optical parametric converter (CSI-OPC) greatly reduces DSP complexity with regard to chromatic dispersion compensation and nonlinearity compensation based on PPLN waveguides without sacrificing the SE [17]. These preprocessing technologies are very promising for enhancing the next generation high-symbol-rate digital coherent channel transport in OTN

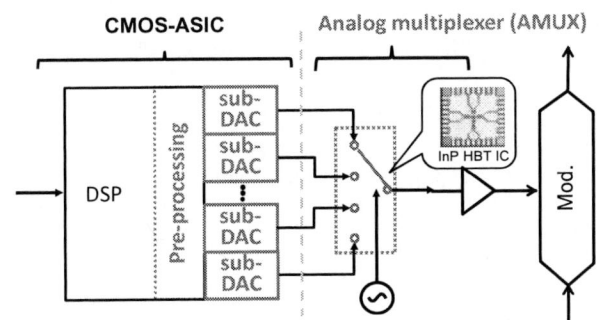 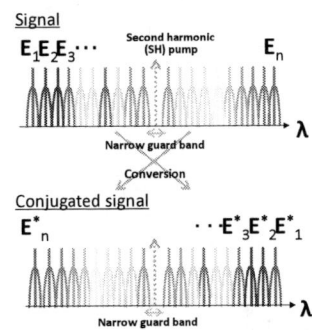

(a)Digitally-preprocessed Analog Multiplexing DAC   (b) Complementary Spectrum Inversion OPC

Fig. 4. Electrical and optical preprocessing for high symbol rate systems.

IV. CONCLUSIONS

Super channels with adaptive modulation are introduced for OTN with a flexible frequency grid and channel rates of over 400 Gbit/s. The transmission performance of 400 Gbit/s super channels with higher QAM orders, up to 16 QAM, have been successfully tested in the field experiments. Advanced preprocessing operations such as DP-AM DAC and CSI-OPC were introduced to enhance the transmission performance of the next generation digital coherent channels with over 400 Gbit/s.

ACKNOWLEDGMENT

This work is partly supported by the R&D project commissioned by the Ministry of Internal Affairs and Communications (MIC) of Japan and the R&D projects commissioned by the National Institute of Information and Communications Technology (NICT).

REFERENCES

[1] K. Kikuchi, *J. Lightwave Technol.,* Tutorial Review, vol.34, no.1, pp157-179, 2016.

[2] E. Yamazaki et al., *Opt. Exp.,* vol. 19, no. 14, pp. 13139–13184, 2011.

[3] Y. Miyamoto et al., *Proc. OFC/NFOEC2012,* Invited paper NTu2E.3, 2012.

[4] http://www.ntt.co.jp/news2014/1409e/140904a.html.

[5] H. Maeda et al., *Proc. OFC2016,* Invited paper W1K.4, 2016.

[6] S. Okamoto *et al., Proc OECC/PS2013,* paper TuR2-4, 2013.

[7] Hiroto Kawakami et al., *Optics Express,* vol. 22, no. 23, p. 28163, 2014.

[8] T. Hoshida et al., *Proc. ECOC2014,* Paper We.3.3.1, 2014.

[9] K. Fukuchi et al., *Proc. OFC2015,* Paper Tu3B.3, 2015.

[10] K. Sugihara, et al, *Proc. OFC2013* OM2B.4, 2013.

[11] S. Okamoto et al., *Proc. of OFC 2016,* Th2A.2. 2016.

[12] K. Yonenaga et al., *Proc OECC/PS2016,* to be appeared.

[13] K. Kojima et al., *Proc. ECOC2014,* paper 3.25, 2014

[14] F. Buchali et al., *J. Lightwave Technol.,* vol.34, no.7, pp1599, 2016.

[15] H. Yamazaki et al., *Proc. ECOC2015,* postdeadline paper PDP2.2, 2015.

[16] K. Horikoshi et al., *Proc OECC/PS2016,* to be appeared.

[17] T. Umeki et al., *Proc. ECOC2016,* paper We.2.6.2, 2015.

Experimental Demonstration of a Programmable 400-Gbps DMT Transceiver with Policy-based Control

Yutaka Kai(1), Ryo Okabe(1), Masato Nishihara(1), Toshiki Tanaka(1), Tomoo Takahara(1), Jens C. Rasmussen(1), F. Javier Vílchez(2), Laia Nadal(2), Josep M. Fàbrega(2), Michela Svaluto Moreolo(2)

(1) Fujitsu Ltd., 1-1 Kamikodanaka 4-chome, Nakahara-ku, Kawasaki 211-8588, Japan
(2) Centre Tecnològic de Telecomunicacions de Catalunya (CTTC), Av. C. F. Gauss 7, 08860, Castelldefels, Spain
E-mail: kai@jp.fujitsu.com

Abstract: We propose a programmable 400-G discrete multi-tone (DMT) transceiver with policy-based control. We experimentally demonstrate the 400-G DMT transceiver enabling either capacity, distance, or energy efficiency priority in a four-node photonic mesh network.
Keywords: Discrete Multi-Tone, programmable optical transceiver, policy-based control

I. INTRODUCTION

Network traffic for wireless back/front-haul and data centers has been increasing every year, and recent research efforts have focused on the efficient use of network resources to support this traffic exponential growth. In particular, for a metro-access/regional network, simple low-cost solutions enabling high capacity are needed. Therefore, in this context, a beyond-100-Gbps-class system is strongly required for cost-sensitive applications. A suitable candidate is the discrete multi-tone (DMT) technology [1, 2]. A DMT transceiver can provide high capacity transmission and rate/distance adaptability by adjusting the number of bits/subcarrier thanks to the DMT features, as it is an OFDM-based multi-carrier modulation with simple configuration/implementation.

In addition, multiple advanced functionalities can be enabled by the DMT transceiver by suitably programming and dynamically (re)configuring specific parameters/characteristics for an efficient and flexible usage and management of the network resources [3]. Particularly, an adaptive DMT transceiver can be remotely programmed according to the network priority (e.g. high-capacity, long-distance and low power consumption) to meet the traffic demand, channel bandwidth/path/state and energy efficiency requirements. Therefore, we propose to enable a policy-based control in the optical transceivers in view of its integration in a software defined network (SDN) based control plane. Figure 1 shows the configurations of the 400-G DMT transceiver according to the proposed policy-based control. The optical transceiver, consisting of four lanes with variable rate up to 100-G each, enables rate/distance adaptability, in order to fulfil the priority target. 75-G, 50-G or either 25-G can be enabled yielding the 3/4, 1/2 or 1/4 of the maximum capacity per lane, to ensure the best effort. In case of capacity priority, the target is maximizing the achievable rate on each one of the four lanes. Whereas, adopting the distance priority policy, the achievable reach is extended at the expense of the capacity per lane, by adaptively loading the transceiver subcarriers with the number of bits suitable for meeting the target performance. Finally, the number of active transceiver lanes is varied according to the energy efficiency policy: the corresponding transmitters and receivers are enabled/disabled to target the power consumption requirement. We demonstrate the functionalities of the proposed DMT transceiver with policy-based control in a four-node photonic mesh network (CTTC ADRENALINE testbed) [4].

For the assessment of the capacity priority policy, we evaluate the transmission of 400-G (100-Gbps x 4-lane) on a 35-km single mode fiber (SMF) link of the ADRENALINE network (Fig. 2). Then, we validate the distance priority policy on 150-km non-zero dispersion-shifted fiber (NZDSF) link. Last, we analyze the priority policy of the energy efficiency, targeting the transmission of 100-G/1-lane on the 35-km SMF link.

II. 400G DMT TRANSCEIVER WITH POLICY-BASED PRIORITY CONTROL

Figure 2 shows the configuration of the DMT transceiver and the experimental setup of the ADRENALINE network testbed. The laser diode (LD) wavelengths set for the four lanes (LD1, LD2, LD3, and LD4) are 1553.33nm, 1552.52nm, 1551.72nm and 1550.92nm, respectively; they are modulated by external Mach-Zehnder modulators (MZMs) and multiplexed. The DMT signal is generated by off-line processing and is converted to an analog signal by a 64 GS/s, 8 bits digital-to-analog converter (DAC). The transmitted DMT signal is received by a photodetector (PD) through the demultiplexer (DEMUX). The received optical power is maintained to +0 dBm by means of a variable optical attenuator (VOA). The received signal is converted to digital by a 64 GS/s, 8 bits analog-to-digital converter (ADC) and subsequently demodulated. The subcarrier number and the cyclic prefix of the DMT signal are 1024 and 16, respectively. The input power to the EDFA1 and the OXC-2 is +4.7 dBm, and +18 dBm, respectively. The dispersion of the transmitted signals is compensated by the tunable dispersion compensator (TDC) at the receiver side.

The priority policy of the transceiver can be set to capacity, distance, and energy efficiency. When the capacity priority policy is adopted, the total capacity is maximized and the target bitrate of each lane is determined by the OSNR reference value based on Fig. 3. If an OSNR is over 43.8 dB, the target is set to 110.3-Gbps. The target bitrate is decreased due to the low OSNR. For this evaluation, we consider a target bit error ratio (BER) of 4.5×10^{-3}, assuming CI-BCH-FEC with 7% overhead (OH) of Ethernet [5].

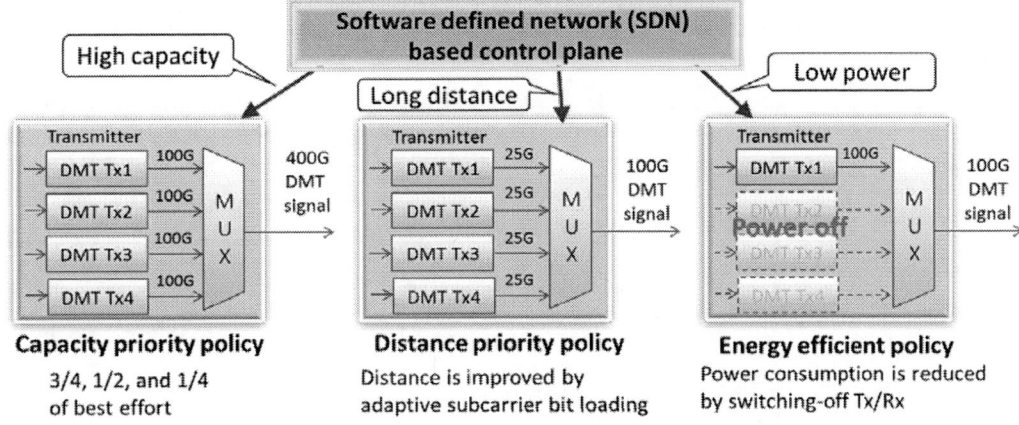

Figure 1. The configuration of the transceiver with policy-based control

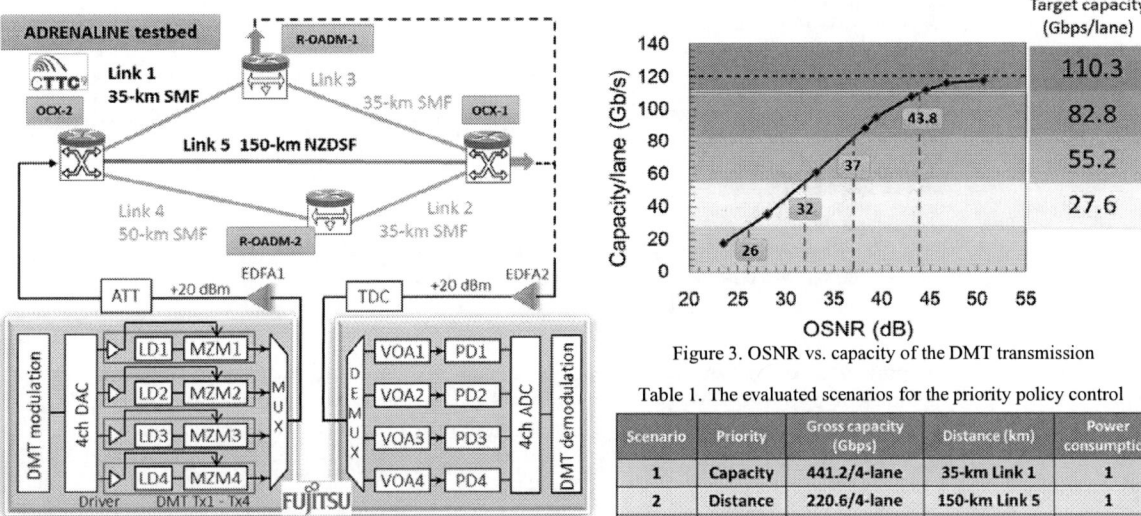

Figure 2. Experimental setup of the ADRENALINE testbed and the configuration of the proposed DMT transceiver

Figure 3. OSNR vs. capacity of the DMT transmission

Table 1. The evaluated scenarios for the priority policy control

Scenario	Priority	Gross capacity (Gbps)	Distance (km)	Power consumption
1	Capacity	441.2/4-lane	35-km Link 1	1
2	Distance	220.6/4-lane	150-km Link 5	1
3	Energy efficiency	110.3/1-lane	35-km Link 1	0.25

III. RESULTS AND DISCUSSION

We evaluate the feasibility of the DMT transceiver with policy-based control when operated in the ADRENALINE testbed. Table 1 shows the evaluated scenarios to demonstrate the priority control capability/policy. Figure 4 shows the received optical spectra before demultiplexing, the bit allocation, SNR profile of lane-1, and the BER characteristic for the (1) capacity, (2) distance, and (3) energy efficiency priority scenarios.

In Scenario 1, the transceiver is set to operate according to the high capacity priority over the 35-km SMF of ADRENALINE link 1. We obtain a high OSNR value of more than 44.8 dB, as shown in Fig 4(a)-(1). Therefore, the net bitrate of the transceiver is set to 400-Gbps (110.3-Gbps gross per lane). We achieve the successful DMT transmission of 400-Gbps/4-lane over 35-km SMF at the target BER, as shown in Fig. 4(d)-(1).

In Scenario 2, the DMT transceiver is configured to operate with the distance priority policy over the 150-km NZDSF of ADRENALINE link 5. Since the obtained OSNR is lower than 37.0 dB, the target bitrate is set to a gross value of 55.2-Gbps/lane, according to Fig. 3. As a result, we achieve a net data rate of 200-Gbps/4-lane over 150-km NZDSF by adaptively assigning the number of bits per symbol per subcarrier.

Last, in Scenario 3, we evaluate the effectiveness of adopting the energy efficient priority policy. When the network traffic is sparse, the power consumption of all 4-lane transmissions is reduced by switching-off the transmitters and receivers of some lanes. According to the traffic, the priority policy can be changed from capacity (Scenario 1) to energy efficiency (Scenario 3). Therefore, in order to assess this case for 100-Gbps transmission, the 3-lanes of the transceiver are disabled (instead of enabling 4 lanes at 27.6-Gbps gross). We achieve successful transmission (below FEC limit) of 100-Gbps with only 1-lane over link 1 (35-km path), as shown in Fig. 4. In this case, the power consumption of the ADRENALINE network elements, such as OXC, ROADM, and EDFAs, is not taken into account. The energy saving achieved with the DMT transceiver operating with scenario 3 policy, where 3 over 4 lanes are switched-off, is 75% compared to the scenarios where all the 4 lanes are enabled, as indicated in table 1.

All scenarios analyzed within the ADRENALINE network are successfully demonstrated, as shown in Fig. 4. Therefore, we confirm the effectiveness of introducing a policy-based control for operating the DMT transceiver with high-capacity, long-distance, and energy efficiency priority, by achieving a transmission of more than 400-G/4-lane, 200-G/4-lane, and 100-G/1-lane, respectively. Thus, a policy-based control of the transceivers enables to flexibly and efficiently manage the available network resources.

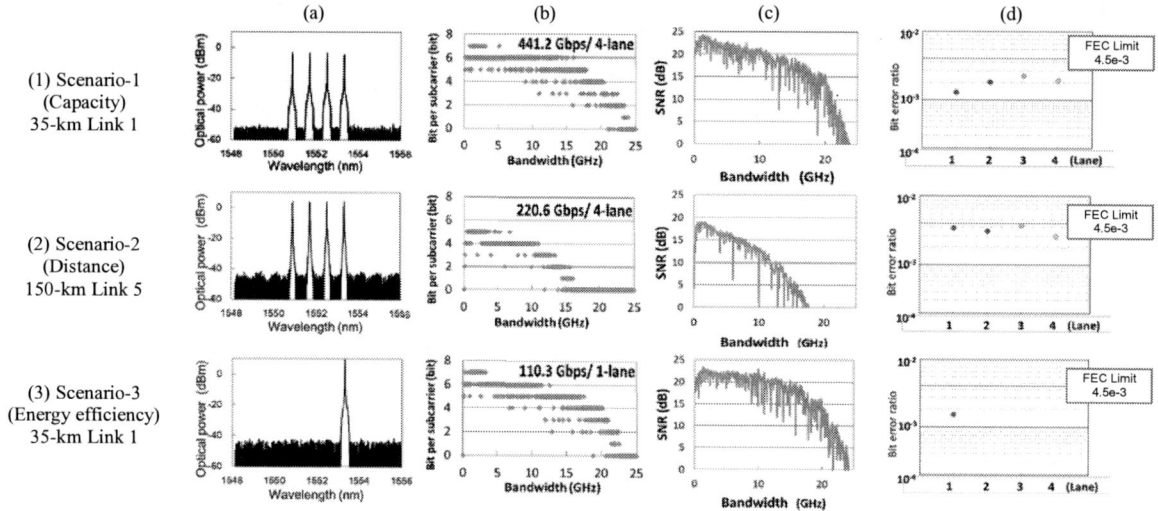

Figure 4. (a) Optical spectra, (b) bit allocation, (c) SNR profile and (d) BER per lane of
(1) capacity, (2) distance, and (3) energy efficiency priority scenarios

IV. CONCLUSION

We propose to use a policy-based control in programmable/adaptive optical transceivers. We experimentally demonstrate the functionalities of a 400-G DMT transceiver enabling operation policy within the ADRENALINE network. For the high capacity priority, we achieve a DMT transmission of more than 400-Gbps/4-lane (110.3-Gbps /lane) over 35-km SMF. Next, for the long distance priority, we achieve 200-Gbps/4-lane (55.2-Gbps/lane) over 150-km NZDSF by adaptively decreasing the bits/subcarrier. Last, for the energy efficient priority, we demonstrate that energy saving is achieved proportionally to the number of disabled lanes, by achieving 100-G/1-lane transmission over the 35-km path. Thus, the proposed policy-based DMT transceiver is an attractive candidate for future metro-access networks and data center applications, featuring rate/distance adaptability, energy efficiency and cost-effective architecture.

ACKNOWLEDGMENT

This work is partly supported by the EU-Japan coordinated R&D project on "Scalable and efficienT oRchestrAstion of ethernet services Using Software-defined and flexible optical networkS (STRAUSS)"by the Ministry of Internal Affairs and Communications (MIC) of Japan and EC FP7.

REFERENCES

[1] W. Yan, et al., "100Gb/s Optical IM-DD Transmission with 10G-Class Devices Enabled by 65Gsamples/s CMOS DAC Core", Proc. OFC, OM3H.1, Anaheim(2013).
[2] T. Tanaka, et al., "Experimental Demonstration of 448-Gbps+ DMT Transmission over 30-km SMF", Proc. OFC, M2I.5, San Francisco (2014).
[3] M. Svaluto Moreolo, et. al., "SDN-enabled Sliceable BVT Based on Multicarrier Technology for Multi-Flow Rate/Distance and Grid Adaptation," J. Lightwave Technol., vol. 34, no. 6, March 2016.
[4] R. Vilalta, et al., "The SDN/NFV Cloud Computing Platform and Transport Network of the ADRENALINE Testbed," in Proc. IEEE Conf. Network Softwarization, 13-17 April 2015, London (UK).
[5] M. Scholten, et. al., "Continuously-Interleaved BCH (CI-BCH) FEC delivers bestin class NECG for 40G and 100G metro applications", Proc. OFC2010, NTuB3, (2010).

320 Gbit/s, 256 QAM LD-based Coherent Transmission over 160 km with an Injection-locked Homodyne Detection Technique

Yixin Wang, Keisuke Kasai, Masato Yoshida, and Masataka Nakazawa

Research Institute of Electrical Communication, Tohoku University, 2-1-1 Katahira, Aobaku, Sendai, 980-8577, Japan

yixin@riec.tohoku.ac.jp

Abstract: *We transmitted a 320 Gbit/s, 20 Gsymbol/s 256 QAM signal over 160 km by using an LD-based injection-locked homodyne detection circuit. This is the highest bit rate yet achieved in a 256 QAM coherent transmission.*

Keywords: *Coherent communications, injection locking, semiconductor lasers.*

I. INTRODUCTION

Low phase noise carrier-phase synchronization between transmitted data and a local oscillator (LO) is indispensable for the demodulation of a high multi-level modulation signal in a coherent optical transmission. Several multi-level coherent transmissions have been achieved by employing carrier-phase estimation with digital signal processing (DSP) [1] or an optical phase-locked loop (OPLL) [2].

An injection-locking scheme is very attractive for realizing low phase noise carrier-phase synchronization with a simple receiver configuration [3]. This scheme has already been utilized for 8 and 16 orthogonal frequency-division multiplexing (OFDM)-quadrature amplitude modulation (QAM) transmissions [4] [5]. However, in these experiments, a residual carrier with a high optical signal-to-noise ratio (OSNR) placed at the center of the data was used as an injection seed to realize low noise carrier-phase synchronization. Since the maximum amplitude is determined by the residual carrier, the OSNR of the data signal inevitably degrades. Therefore, it has been difficult to apply the scheme to higher-order QAM transmission. To overcome this bottleneck, we have developed an injection-locked homodyne detection technique, where a pilot tone is added near the data signal and used as a high OSNR seed. With this setup, we have successfully performed a transmission experiment with a 140 Gbit/s, 10 Gsymbol/s 128 QAM [6] and an 80 Gbit/s, 5 Gsymbol/s 256 QAM signal [7].

In this paper, we further increase the symbol rate of 256 QAM transmission up to 20 Gsymbol/s, and realize a 320 Gbit/s injection-locked coherent transmission over 160 km by using high resolution frequency domain equalization (FDE) [8] and digital back-propagation (DBP) [9] to compensate for waveform distortions. 320 Gbit/s 256 QAM data were transmitted within an optical bandwidth of 24.5 GHz, where the potential spectral efficiency (SE) reached as high as 10.9 bit/s/Hz. This is the highest bit rate and SE yet achieved in a 256 QAM transmission.

II. EXPERIMENTAL SETUP FOR 320 GBIT/S, 256 QAM INJECTION-LOCKED COHERENT OPTICAL TRANSMISSION

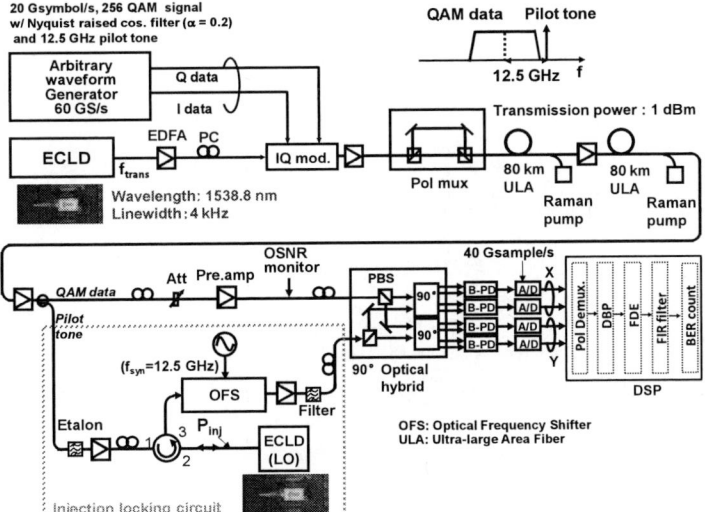

Fig. 1. Experimental setup for 320 Gbit/s pol-mux, 256 QAM-160 km coherent transmission.

Figure 1 shows our experimental setup for a 320 Gbit/s, polarization-multiplexed (pol-mux), 256 QAM-160 km coherent transmission with an injection-locked homodyne detection circuit. The optical source for the transmitter was a 4 kHz linewidth, InP-based external cavity LD (ECLD) emitting at 1538.8 nm with an external Bragg grating on a silica planar lightwave circuit [10]. The output of the transmitter was coupled to an IQ modulator, where the coherent light was modulated with a 20 Gsymbol/s, 256 QAM signal and a pilot tone signal generated from an arbitrary waveform generator (AWG) running at 60 Gsample/s. A Nyquist filter with a roll-off factor of 0.2 was adopted at the AWG, which enabled us to reduce the bandwidth of the QAM signal to 24 GHz. The frequency of the pilot tone signal was up-shifted by 12.5 GHz against the center of the data. At the AWG, pre-equalization was employed to compensate for waveform distortions caused by individual components such as the IQ modulator and AWG by using a 99-tap finite impulse response (FIR) digital filter. The pol-mux was carried out by using a polarization division multiplexing emulator. These signals were transmitted over two 80 km spans of ultra large area (ULA) fiber with a launch power of 1 dBm. The loss of passing the transmission fiber was compensated for by using EDFAs and Raman amplifiers. In each span, the Raman amplifiers provided a 9.5 dB gain that contributed to a total gain of 16.4 dB.

At the receiver, the transmitted signal was split into two arms. On one arm, the pilot tone signal was injected into an ECLD as an LO. The output signal of the ECLD was frequency downshifted by 12.5 GHz for homodyne detection. The 256 QAM signal was homodyne-detected with the LO by using a polarization-diverse 90-degree optical hybrid and four balanced photo-detectors (B-PDs). The detected data signals were then A/D-converted using a digital oscilloscope (40 Gsample/s, 16 GHz bandwidth) and demodulated with DSP in an offline condition. For DSP, we used a 12 GHz low-pass digital filter to eliminate the pilot tone signal. We compensated for fiber nonlinearities and chromatic dispersion simultaneously by using a DBP method. Then we adopted FDE with a frequency resolution of 4.88 MHz to compensate for residual distortions caused by hardware imperfections. Finally, the bit error rate (BER) was evaluated from 123 kbit data.

III. EXPERIMENTAL RESULTS

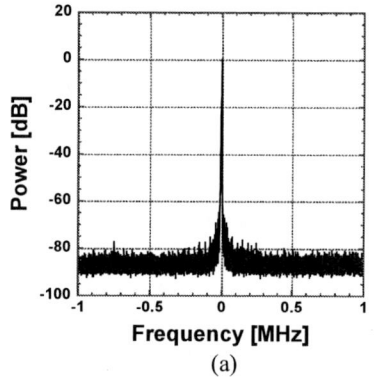 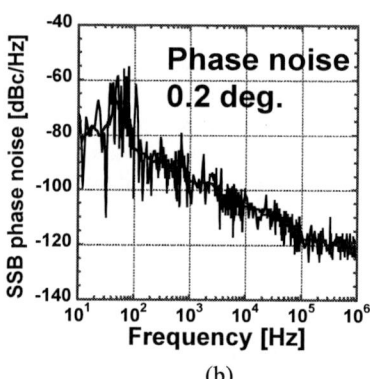

Fig. 2. IF spectrum of beat between pilot and injection-locked LO in (a) 2 MHz span and, (b) its SSB phase noise spectrum (10 Hz~1 MHz).

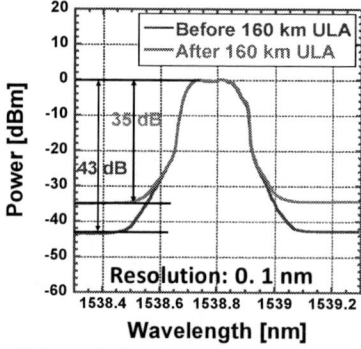

Fig. 3. Optical spectra of 20 Gsymbol/s 256 QAM data signal before and after 160 km transmission.

Figure 2 (a) shows the intermediate frequency (IF) spectrum of the beat signal between the injection-locked LO and the pilot tone, measured after a 160 km transmission within a 2 MHz span. Figure 2 (b) shows its single-side band (SSB) phase noise spectrum. The phase noise variance (RMS) of the IF signal, estimated by integrating the SSB noise power spectrum from 10 Hz to 1 MHz, was as small as 0.2 degrees.

Figure 3 shows the optical spectra of the 20 Gsymbol/s 256 QAM data signal before and after a 160 km transmission measured with a 0.1 nm resolution. After the 160 km transmission, the OSNR of the data signal was degraded from 43 to 35 dB.

Figure 4 (a) shows the BER characteristics of a 20 Gsymbol/s, 256 QAM transmission. The blue line shows the results under a back-to-back condition. The red line shows the transmission results. The error floor is observed under a back-to-back condition even with a high OSNR, which is mainly due to the insufficient signal-to-noise ratio of the data signal generated by the AWG. After transmission, a BER below 2×10^{-2} (20 % forward error correction (FEC) threshold) was realized. Figure 4 (b) and (c) show the constellations for back-to-back and 150 km transmissions, respectively. The error vector magnitudes were 1.87 and 2.74 %, respectively. In this experiment, 320 Gbit/s data were transmitted within an optical bandwidth of 24.5 GHz, which corresponds to a potential SE of 10.9 bit/s/Hz taking account of the 20% FEC overhead.

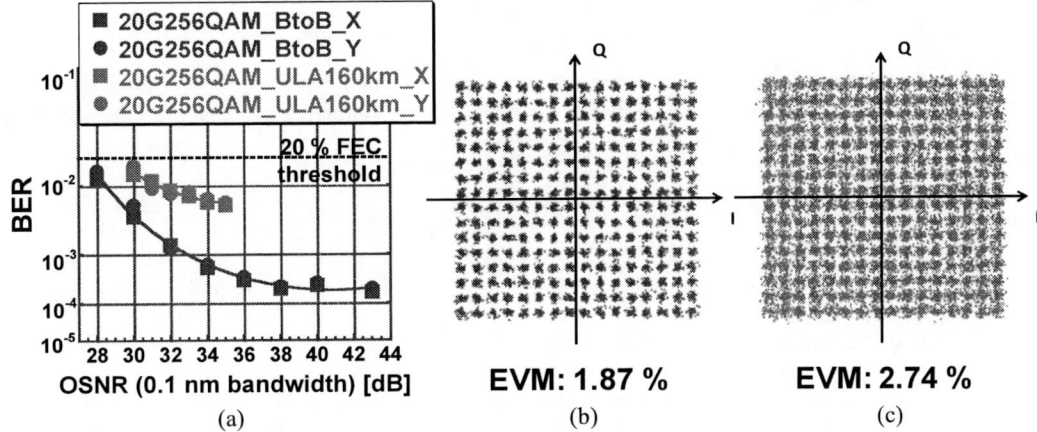

Fig.4. (a) BER characteristics for pol-mux, 20 Gsymbol/s, 256 QAM (320 Gbit/s)-160 km transmission, and constellations for (b) back-to-back and (c) 160 km transmission.

IV. CONCLUSIONS

We succeeded in transmitting a pol-mux, 20 Gsymbol/s 256 QAM signal (320 Gbit/s) over 160 km by using an LD-based injection-locked homodyne detection circuit. The potential SE reached as high as 10.9 bit/s/Hz. This is the highest bit rate and SE yet achieved in a pol-mux 256 QAM coherent transmission.

ACKNOWLEDGMENT

This work is supported by "The research and development project for the expansion of radio spectrum resources" of the Ministry Internal affairs and Communications, Japan.

REFERENCES

[1] R. Noe, "PLL-free synchronous QPSK polarization multiplex/diversity receiver concept with digital I&Q baseband processing," IEEE Photon. Technol. Lett., vol. 17, pp. 887-889, April 2005.

[2] J. Hongo, K. Kasai, M. Yoshida, and M. Nakazawa, "1-Gsymbol/s 64-QAM coherent optical transmission over 150 km," IEEE Photonics Technol. Lett., vol. 19, pp. 638–640, May 2007.

[3] S. K. Ibrahim, S. Sygletos, R. Weerasuriya, and A. D. Ellis, "Novel real-time homodyne coherent receiver using a feed-forward based carrier extraction scheme for phase modulated signals," Opt. Exp., vol. 19, pp. 8320–8326, April 2011.

[4] S. Adhikari, S. Sygletos, A. D. Ellis, B. Inan, S. L. Jansen, and W. Rosenkranz, "Enhanced self-coherent OFDM by the use of injection locked laser," Proc. of Optical Fiber Communication Conf. (OFC 2012), Los Angeles (USA), Mar. 2012, JW2A.

[5] Z. Liu, D. S. Wu, D. J. Richardson, and R. Slavik, "Homodyne OFDM using simple optical carrier recovery," Proc. of Optical Fiber Communication Conf. (OFC 2014), San Francisco (USA), Mar. 2014, W4K.3.

[6] Y. Wang, S. Beppu, K. Kasai, M. Yoshida, and M. Nakazawa, "140 Gbit/s, 128 QAM LD-based coherent transmission over 150 km with an injection-locked homodyne detection technique," Proc. of Asia Communications and Photonics Conf. (ACP 2014), Shanghai (China), Nov. 2014, ATh1E.3.

[7] K. Kasai, Y. Wang, S. Beppu, M. Yoshida, and M. Nakazawa, "80 Gbit/s, 256 QAM coherent transmission over 150 km with an injection-locked homodyne receiver," Opt. Exp., vol. 23, pp. 29174-29183, Oct. 2015.

[8] K. Ishihara, T. Kobayashi, R. Kudo, Y. Takatori, A. Sano, E. Yamada, H. Masuda, and Y. Miyamoto, "Coherent optical transmission with frequency-domain equalization," Proc. Of European Conf. on Optical Communications (ECOC 2008), Brussels (Belgium), Sep. 2008, We2E3.

[9] C. Paré, A. Villeneuve, P. -A. Bélanger, and N. J. Doran, "Compensating for dispersion and the nonlinear Kerr effect without phase conjugation," Opt. Lett., vol.21, pp. 459-461, April 1996.

[10] L. Stolpner, S. Lee, S. Li, A. Mehnert, P. Mols, S. Siala, and J. Bush, "Low noise planar external cavity laser for interferometric fiber optic sensors," Proc. of International Society for Optical Engineering (SPIE 2008), San Diego (USA), vol. 7004, pp. 700457-1-700457-4, April 2008.

Coherent Interference Reduction in Single-Fiber Bidirectional System for 100 Gb/s Short Distance Applications

Xu Zhou, Ning Deng

Fixed Networks Research, Huawei Technologies Co., Ltd., Shenzhen, China
sean.zhouxu@huawei.com, ning.deng@huawei.com

Abstract: *We propose bidirectional transmission on same wavelengths in a single fiber for 100-Gb/s short distance applications. By utilizing alternate-mark inversion coding, coherent interference induced by reflection and Rayleigh backscattering is suppressed significantly.*

Keywords: bidirectional transmission, data center, alternate-mark inversion coding

I. INTRODUCTION

To handle ever increasing data traffic in recent years, warehouse-scale data center are being deployed world widely. For <300-meter connection 850-nm vertical cavity surface emitting lasers (VCSELs) and multimode fiber (MMF) are predominant. However, for further expansion and connections of large data centers, short distances in the range of 2 km and 10 km (e.g. 100GE-LR) with single mode fiber (SMF) are becoming popular [1,2].

An existing scheme for such short-distance connection at 100 Gb/s is using so-called LAN-WDM, which consists of four 1.3-µm-window wavelengths with each operated at about 25 Gb/s. Uncooled lasers and limited number of wavelengths are essential for low cost. Two fibers are used in this scheme, with one for each direction.

As optical fiber is becoming among the top factors of material and operational cost in data centers, single-fiber bidirectional operation is highly expected. With the use of the four wavelengths in both directions, however, the coherent interference induced by reflections of fiber connectors and Rayleigh backscattering (RB) along the fiber will severely limit the transmission performance [3]. In literature coherent interference reduction schemes have been proposed, which used differential phase-shift keying (DPSK) and intensity-modulation in opposite direction for the suppression of coherent interference [4,5]. However, such a system requires an optical demodulator for DPSK signal, which may arouse cost issues, and also makes the optical modules at each end of the fiber be different.

In this paper, we proposed a novel single-fiber bidirectional scheme for four-channel 100-Gb/s applications. We inherit to use existing optics in current commercial short-distance 100-Gb/s (e.g. 100GE) optical modules. To reduce the coherent interference, we utilize the alternate-mark inversion (AMI) line coding in one direction, which can still be received by existing photodetector. The proposed scheme maintains the simple structure in the transceivers, and achieves the unified optical transceiver at each end of the fiber. We perform an experiment of 10-km single fiber bidirectional transmission at a lane of 25 Gb/s to verify the proposed scheme.

II. PROPOSED SCHEME AND EXPERIMENTAL SETUP

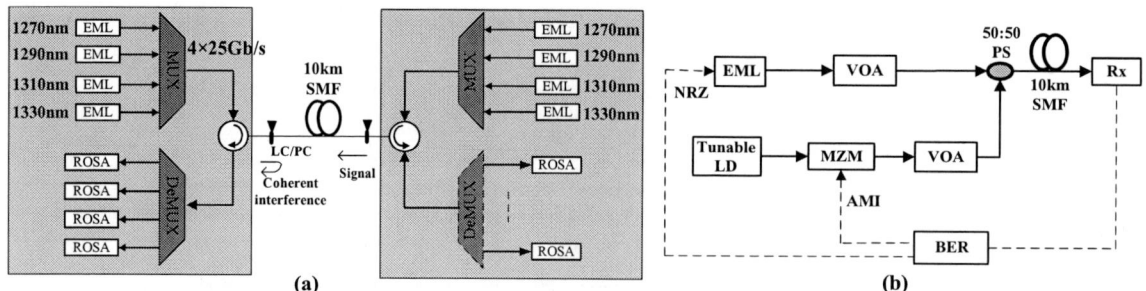

Fig.1. (a) The proposed 100-Gb/s single-fiber bidirectional system architecture and (b) the experimental setup

Fig.1(a) shows our proposed system architecture for the bidirectional transmission on same wavelengths in single fiber for 100 Gb/s short-distance applications. Four LAN-WDM wavelengths are used and each wavelength is operated at about 25 Gb/s. It is noted that unified, i.e. exactly the same optical transceivers are placed at each end of the fiber link. Unified transceiver is an important feature for easy and low-cost field deployment, and this would not be possible by using different set of wavelengths for the two directions. At each end a circulator is used for separating the signals of the two directions. To alleviate the coherent interference of the two directions due to connector reflections and Rayleigh backscattering (RB), we propose to use NRZ coding in one direction and AMI coding in the other direction, in which the optical spectra have a lot reduced overlap. Fig. 1(b) shows the experimental setup. To precisely emulate the amount of coherent interference and clearly study the impact, we transmit the two signals (one in NRZ, the other in AMI) in the same direction. The 25-Gb/s NRZ signal and AMI signal are generated and passed through 10km SMF before being received. Variable optical attenuator (VOA) is used to control the amount of coherent interference of one line coding while the other serves as the signal. The tunable laser is used to emulate the wavelength offset between the two signals.

The generation and detection for the AMI signal is shown in Fig. 2. After correlative coding, the original binary signal converts to a three-level signal in electrical domain [6]. When electrical signal is modulated onto the optical carrier, the Mach-Zehnder modulator (MZM) is biased at the null point. Hence the two level-optical signals are generated finally as shown at bottom of Fig. 2. The detection for AMI signal is the same as the NRZ, by using a single detector. A low-cost version of the transmitter is an externally modulated laser (EML) with MZM integrated [7-9].

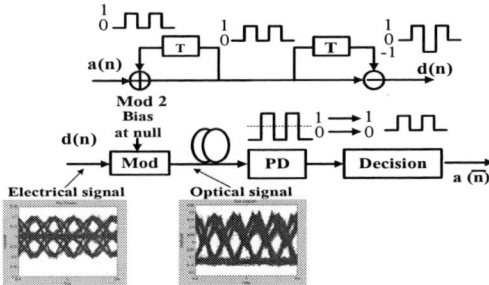

Fig. 2. Generation and detection for the AMI signal.

Fig. 3. Optical spectrum in (a) wavelength alignment and (b) wavelength offset.

III. RESULTS AND DISCUSSION

Fig. 3 depicts the optical spectra of the AMI and NRZ signal. The spectral energy of NRZ signal is concentrated around optical carrier. On the contrary, the optical carrier of the AMI signal is suppressed and much spectral content is shifted to the two sides. As the spectral energy of the AMI signal is shifted with respect to the NRZ signal, the impact of coherent interference will be decreased in the case of alignment of central wavelengths, as shown in Fig. 3(a) in 0.1nm/D and RES=0.02nm. Then we study the case of central wavelength offset by sweeping wavelength of the tunable laser to the worst impairment point, as shown in Fig. 3(b). Here we discuss the worst case of the coherent interference power. Assume that the maximum reflections of the LC/PC connectors (in TIA 568-B.3) is 26 dB, the RB noise induced

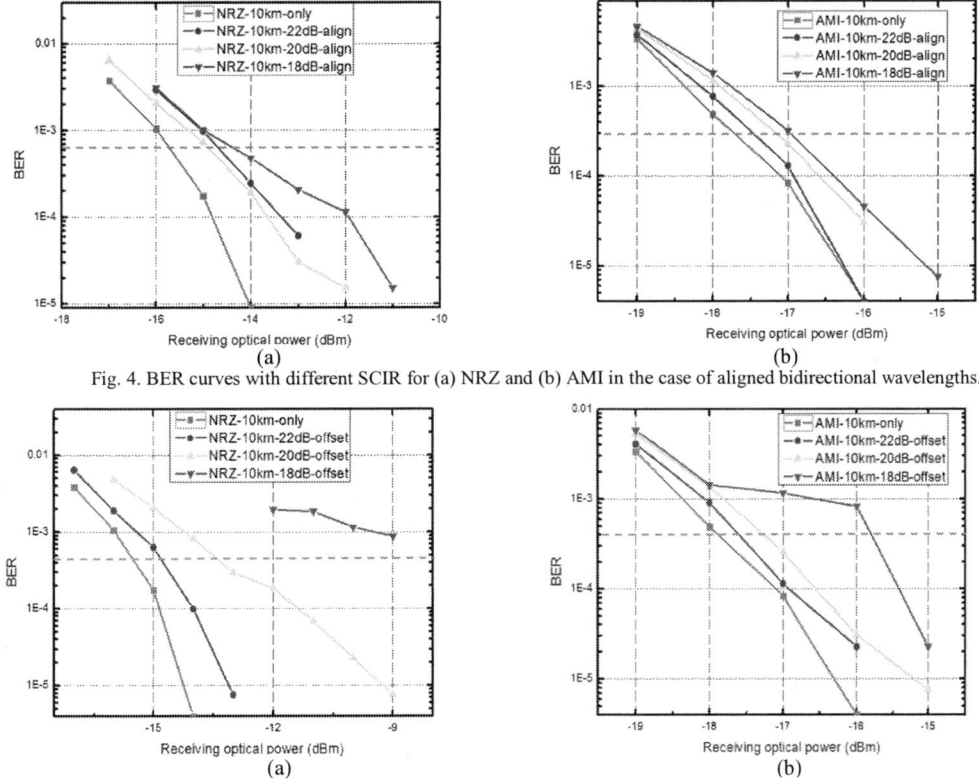

Fig. 4. BER curves with different SCIR for (a) NRZ and (b) AMI in the case of aligned bidirectional wavelengths.

Fig. 5. BER curves with different SCIR for (a) NRZ and (b) AMI in the case of offset bidirectional wavelengths.

by the fiber is 30 dB, and maximum channel loss (in 100GE-LR of IEEE802.3) is 6 dB. Hence, when both two transmitter power are equal to P_{tx}, the worst signal to coherent interference ratio (SCIR) is $(P_{tx} - 6) - (P_{tx} - 26) = 20$ dB.

Fig.4 shows the bit error ratio (BER) curves at different SCIR for (a) NRZ and (b) AMI in the case of aligned central wavelengths. Because the overlapping portion between the NRZ and AMI signal is small, the BER performance for the AMI signal is slightly deteriorated when the SCIR increased. The NRZ signal is more vulnerable than the AMI signal due to the concentrated energy around the optical carrier.

The BER curves with different SCIR in the case of offset central wavelengths are shown in Fig. 5. We can see that the BER performance for both NRZ signal and AMI signal degraded significantly when the SCIR increased. The receiver sensitivity for the NRZ signal is -13 dBm at BER of 3×10^{-4} (at FEC threshold) at the worst interference case SCIR=20 dB. And the BER floor for NRZ signal is observed at SCIR=18 dB. This indicates that the channel loss is limited to 8 dB, resulting in limited transmission distance. The AMI signal performs better than the NRZ signal against the coherent interference.

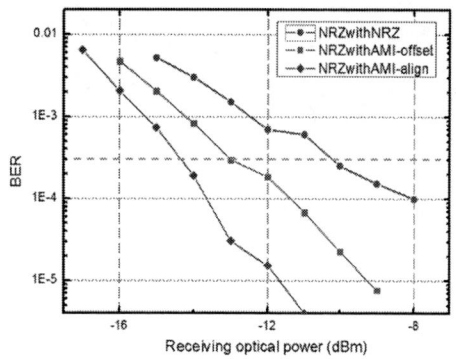
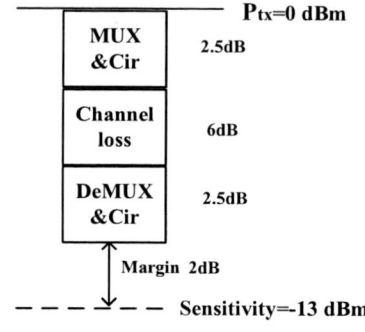

Fig. 6. BER curves compared with current scheme   Fig. 7. Power budget in our proposed scheme.

Fig.6 shows the BER curves compared with the traditional NRZ/NRZ bidirectional transmission, i.e. with and without using our scheme. It shows that our scheme outperforms the NRZ/NRZ transmission by 3 dB in terms of receiver sensitivity (at BER of 3×10^{-4}). Such improvement in receiver sensitivity is very important as can be seen from the power budget analysis in Fig. 7. We adopt the values in 100GE-LR (10 km) specification (IEEE 802.3) as the loss parameters for main optical components and optical link.

IV. CONCLUSIONS

We propose to replace the two fibers with a single fiber for 100 Gb/s (4×25 Gb/s) short-distance applications. The coherent interference from the same wavelength of the opposite direction induces severe performance degradation. We propose to use AMI line coding for one direction and NRZ coding for the other direction to suppress the coherent interference. Experiment with a 25-Gb/s lane shows our scheme achieves over 3-dB better receiver sensitivity compared with using NRZ in both directions at the same wavelength in a single fiber. With our scheme, power budget requirement is met for the 100GE-LR or comparable applications.

REFERENCES

[1] Rabinovich R., et al., "40Gb/s & 100Gb/s ethernet long-reach host board channel design," IEEE Communications Magazine, vol.51, no.11, pp. 152-158, 2013.

[2] Fumito N., et al., "High-Speed Avalanche Photodiode for 100-Gbit/s Ethernet," Proc. OFC, M3B.5, 2015.

[3] Mats G., et al., "Statistical Analysis of Interferometric Crosstalk:Theory and Optical Network Examples," J. Lightw. Technol., vol.15, no.11, pp. 2006-2019, 1997.

[4] Chow C. W., et al., "Rayleigh Noise Reduction in 10-Gb/s DWDM-PONs by Wavelength Detuning and Phase-Modulation-Induced Spectral Broadening," Photon. Technol. Lett., vol.19, no.6, pp. 423–425, 2007.

[5] Xu J., et al., "Rayleigh Noise Reduction in 10-Gb/s Carrier-Distributed WDM-PONs Using In-Band Optical Filtering," J. Lightw. Technol., vol.29, no.24, pp. 3632–3639, 2011.

[6] Winzer P. J., et al., "40-Gb/s alternate-mark-inversion return-to-zero (RZAMI) transmission over 2,000 km," IEEE Photon.Technol. Lett., vol.15, no.5, pp. 766–768, 2003.

[7] Lange, S., et al., "Low switching voltage InP-based travelling wave electrode Mach-Zehnder modulator monolithically integrated with DFB-laser for 60 Gb/s NRZ," Proc. OFC,Th4E.1, 2015.

[8] Griffin, R.A., et al., "InP Mach–Zehnder Modulator Platform for 10/40/100/200-Gb/s Operation," IEEE Journal of Selected Topics in Quantum Electronics, vol.19, no.6, 3401209, 2013.

[9] Valicourt, G. de., et al., "Ultra-Compact Monolithic Integrated InP Transmitter at 224 Gb/s with PDM-2ASK-2PSK modulation," Proc. OFC,Th5C.3, 2014.

ThB3-5

200Gbit/s Nyquist 16-QAM Half-cycle Subcarrier Modulation Transmission with Dual-Polarization Direct Detection

Kaiheng Zou, Yixiao Zhu, Fan Zhang, and Zhangyuan Chen

State Key Laboratory of Advanced Optical Communication System and Networks
Peking University, Beijing 100871, P. R. China
fzhang@pku.edu.cn

Abstract: *We demonstrated single channel dual-polarization direct detection transmission over 80km SSMF of 200Gb/s Nyquist 16-QAM half-cycle single side-band subcarrier modulation. The net bit rate is 182.5Gb/s considering frame redundancy and 7% overhead for FEC.*

Keywords: *direct detection; polarization division multiplexing; short reach*

I. INTRODUCTION

With the growing demand for broadband services, coherent detection systems attract much attentions as it can achieve high performance in long haul transmission. However, coherent receivers require complex hardware including local oscillators, hybrids, balanced photodiodes and analog-to-digital converters (ADCs). In contrast, direct-detection (DD) systems only need one single-ended photodiode and one ADC for each polarization in the receiver, which have advantages of lower cost and easier integration. Therefore it is more attractive in short reach transmission and some metro applications.

Polarization division multiplexing (PDM) has been applied to DD systems recently. A dual-polarization half-cycle 14Gbaud 16-quadrature amplitude modulation (QAM) 4km transmission with a single channel 112Gb/s capacity is reported [1]. In [1], a polarization controller (PC) and one polarization beam splitter (PBS) is used at the receiver. A single channel 224Gb/s PDM-PAM-4 10km transmission at the wavelength of 1.3μm based on Stokes vector detection is reported [2]. In [2], complex Stokes vector receiver including PBS, 90° hybrid and balanced photodiodes is applied. The polarization state could be tracked with the detected stokes vector, thus no PC is required in [2]. Recently, a quaternary polarization-multiplexed subsystem for intensity modulation direct detection optical links is proposed in [3]. Transmission over 2km with the net bit rate of 100Gb/s (4×27GBaud) and 120Gb/s (4×32GBaud) is demonstrated [3].

In this paper, we demonstrated a single channel dual-polarization direct detection transmission over 80km standard single mode fiber (SSMF) at the wavelength around 1550nm based on 25GBaud Nyquist 16-QAM half-cycle single side-band subcarrier modulation. In our receiver, a finely adjusted PC is required. Compared to Stokes vector receiver, the structure is simple and the cost is much lower. The signal-signal beat interference (SSBI) is compensated with an iterative way at receiver side [4]. A post equalization method is applied to compensate the enhanced in-band noise in bandwidth limited systems [5]. No in-line chromatic dispersion compensation is used. The net bit rate is 182.5Gb/s with the consideration of frame redundancy and the 7% overhead for hard-decision forward error correction (HD-FEC).

II. EXPERIMENTAL SETUP

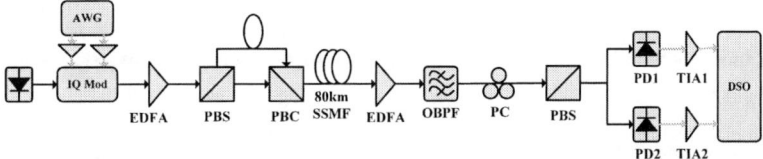

Fig.1. Experimental setup. AWG: arbitrary waveform generator, IQ Mod: IQ modulator, EDFA: erbium-doped fiber amplifier, PBS: polarization beam splitter, PBC: polarization beam combiner, SSMF: standard single-mode fiber, OBPF: optical band-pass filter, PC: polarization controller, PD: photodiode, TIA: transimpedance amplifier, DSO: digital storage oscilloscope.

The experimental setup of the 16-QAM dual-polarization (DP) Nyquist single side-band subcarrier modulation (Nyquist-SSB-SCM) transmission over 80km SSMF with direct detection is shown in Fig.1. The arbitrary waveform generator (AWG) (Keysight M8195A) operating at 64GSa/s generates the 25GBaud half-cycle Nyquist-SSB-SCM 16QAM signal. It drives the IQ modulator to generate the optical signal. The IQ modulator is biased above the null point and the bias is finely adjusted to achieve a desired carrier power for the SSB signal. The signal is amplified by an erbium-doped fiber amplifier (EDFA) after the modulator output. Then a PDM emulator is used to emulate a PDM signal. The PDM emulator consists of a PBS, an optical delay line and a polarization beam combiner (PBC). The delay of the optical delay line is 497 symbols.

311

The optical signal is launched into 80km SSMF. At the receiver, the signal is first amplified by an EDFA to compensate the fiber loss and then filtered by an optical band-pass filter (Yenista Optics XTM-50) to remove out-of-band noise. Then a PC is used before a PBS. The two signal outputs of the PBS are detected by two photodiodes with 50GHz bandwidth and subsequently amplified by two TIAs with 20GHz bandwidth, respectively. The electrical signals is sampled by a real-time digital storage oscilloscope (Keysight DSA-X 96204Q) operating at 80GSa/s to perform off-line digital signal processing (DSP) later. The PC is finely adjusted to confirm the signals of the two polarizations are separated into the two outputs of the PBS, respectively. The transmitted and the received optical spectrum of the signal with the resolution of 0.02nm is shown in Fig.2(b).

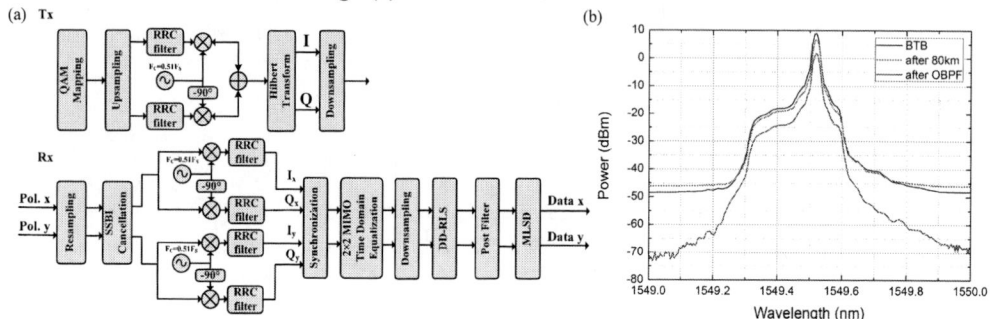

Fig.2. (a) Diagram of digital signal processing. RRC: root raise cosine, SSBI: signal-signal beat interference, MIMO: multi-input multi-output, DD-RLS: decision direct-recursive least square, MLSD: maximum-likelihood sequence decision. (b) Transmitted and received spectrum of the signal.

The diagram of the DSP is shown in Fig.2(a). At the transmitter, the bit stream is mapped to 16-QAM first. After 64 times up-sampling, the signal is digitally shaped using root raise cosine filter with a roll-off factor of 0.01. The preamble includes two 64-symbol synchronization sequences and four 497-symbol training sequences with the same length of zero sequences followed. 87040 data symbols are transmitted after the preamble. The structure of the transmitted DP signal frame is shown in Fig.3. The frame is then up-converted by a subcarrier with the frequency 0.51 times to the symbol rate, which results in a nearly half-cycle SCM. A digital Hilbert transform filter is applied to the signal. Finally the signal is 25 times down-sampled before driving the optical IQ modulator.

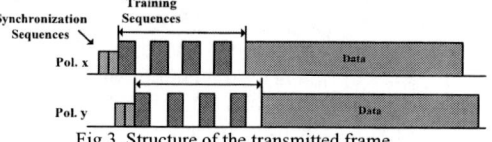

Fig.3. Structure of the transmitted frame.

At the receiver, the SSBI is first compensated in each branch of the two polarizations, respectively. The SSBI and its compensation technique is explained in our previous work [4]. Then the signal is down-converted and matched filtered. After synchronization, the signals of two polarizations are equalized with a 2×2 multi-input multi-output training sequence based time domain equalization. A $T_s/4$ (T_s means the symbol time period) spaced finite impulse response (FIR) filter is extracted from the training sequence with the taps updated by the recursive least square (RLS) algorithm. The FIR filter is subsequently used to equalize the data. Then a T_s spaced decision-direct RLS filter is respectively used for two polarizations to improve the signal quality. As the AWG has an analog bandwidth of approximately 20GHz and the TIAs have the bandwidth of 20GHz, which is smaller than the signal bandwidth of 25GHz. In such a bandwidth-limited system, linear equalization would enhance the in-band high frequency noise. Therefore, a post equalization method is applied, in which the signal is digitally post filtered and maximum likelihood sequence decision (MLSD) is used for the partial-response signal [5]. Finally, the bit error rate (BER) is measured by error counting.

III. RESULTS

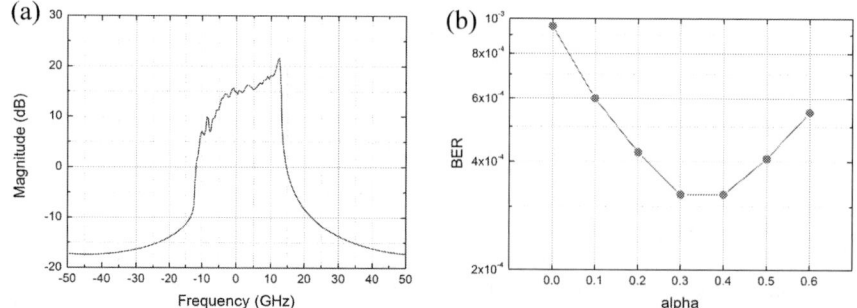

Fig.4. (a) Received electrical spectrum of the baseband signal. (b) BER as a function of the digital post filter tap α

We first design the digital post filter. The received spectrum of the baseband signal after down-converting is shown in Fig.4(a). The best post filter has the similar shape to the channel response [5]. However, the received spectrum is asymmetry because of the subcarrier modulation. It is hard to find a simple digital filter similar to the asymmetry channel response. Therefore, we still use a two tap low-pass post filter to achieve a low complexity of the MLSD. The impulse response of the post filter is $h(t) = 1 + \alpha\delta(t - T_s)$. The tap α is optimized in Fig.4(b) with the back-to-back signal. When α is 0, no post filter is applied. The optimal α comes to 0.3. With the post filter and MLSD, the back-to-back BER is reduced from 9.5×10^{-4} to 3.2×10^{-4}. The MLSD is performed for the real and the imaginary components of the two signals independently.

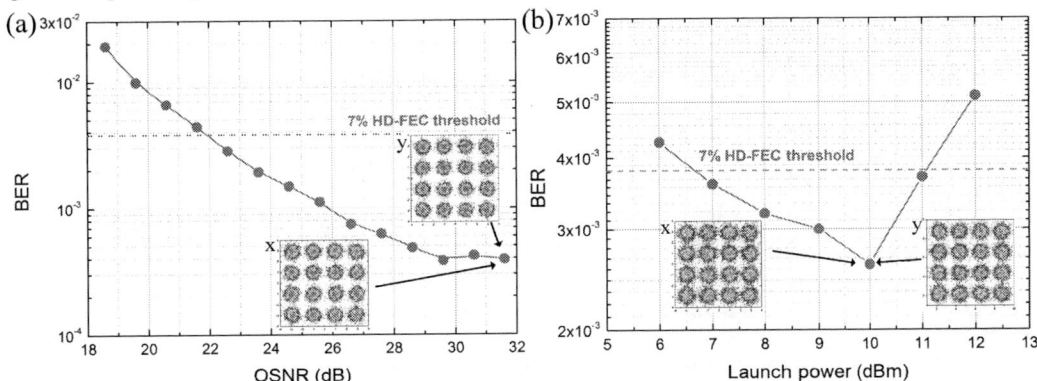

Fig.5 (a) BER as a function of the OSNR at back-to-back scenario. The insets are the constellations of the two polarizations before post filter. (b) BER as a function of the launch power over 80km SSMF. The insets are the constellations of the two polarizations before post filter.

Fig.5(a) shows the back-to-back BER as a function of the optical signal to noise ratio (OSNR) with the resolution of 0.1nm. In our OSNR measurement, the optical carrier power is not included. When the OSNR is larger than 29.5dB, the BER curve becomes flat. The required OSNR for the HD-FEC threshold with 7% overhead is about 22dB. The insets of the Fig.5(a) are the constellations of the two polarizations before the digital post filter when the OSNR is 31.6dB.

Fig.5(b) shows the BER as a function of the launch power over 80km SSMF transmission. The optimal launch power is 10dBm and the BER is 2.6×10^{-3}, which is below the HD-FEC threshold with 7% overhead (3.8×10^{-3}). The insets of the Fig.5(b) are the constellations of the two polarizations before the digital post filter when the launch power is 10dBm. The measured OSNR after 80km SSMF transmission is 27.2dB when the launch power is 10dBm. Compared to the BER curve in Fig.5(a), the OSNR penalty is about 4dB with the similar BER at back-to-back scenario.

IV. CONCLUSIONS

In conclusion, we demonstrated a single channel dual-polarization direct detection transmission over 80km SSMF based on 25GBaud Nyquist 16-QAM half-cycle single side-band subcarrier modulation. The gross bit rate is 200Gb/s (25GBaud×4×2) and the net bit rate is 182.5Gb/s. The BER after transmission is 2.6×10^{-3} after both SSBI compensation and post equalization.

ACKNOWLEDGMENT

This work was supported by National Natural Science Foundation of China (No. 61535002, No. 61475004).

REFERENCES

[1] A. S. Karar, and J. C. Cartledge, Proc. ECOC'2012, Postdeadline paper Th.3.A.4, 2012.
[2] M. Morsy-Osman, M. Chagnon, M.Poulin, S. Lessard, and D. V. Plant, Proc. ECOC'2014, Postdeadline paper Pd.4.4, 2014.
[3] J. Estarán, M. A. Usuga, E. P. da Silva, M. Piels, M. I. Olmedo, D. Zibar, and I. T. Monroy, J. Lightwave Technol., **33**, 1408, 2015.
[4] K. Zou, Y. Zhu, F. Zhang, and Z. Chen, Proc. ACP'2015, Postdeadline paper AM4A.4, 2015.
[5] J. Li, E. Tipsuwannakul, T. Eriksson, M. Karlsson, and P. A. Andrekson, J. Lightwave Technol., **11**, 1664, 2012.

Connectivity Techniques of MCF, for Deployment to Practical Use

Tsunetoshi Saito, Kengo Watanabe

Furukawa Electric Co., Ltd. 6, Yawata-kaigandori, Ichihara, Chiba, 290-8555, Japan
tsune@ch.furukawa.co.jp

Abstract: *Multicore fiber (MCF) that is the hope of next generation optical fiber has been paid a much attention and its researches and experiments keep showing fruitful results. One of the key for practical use of MCF is connectivity. In this paper, MCF connectivity techniques ready for practical use are introduced.*

Keywords: *(Multicore Fiber (MCF), Space Division Multiplexing (SDM), Fan-in, Fan-out, Optical Connector)*

I. INTRODUCTION

MCF and the systems using it must be promising candidates for next generation networks that will break barriers of the limitation of conventional optical fibers. There are many experimental results of large capacity transmission using MCF [1, 2]. Furthermore, studies for applying MCF to optical interconnections are also getting active [3]. For practical use of MCF, connectivity such as fan-in/out or MCF connectors are essential. We have been studying and developing them especially from the point of view of productivity and installability. In this Paper, we introduce digest of our techniques and works of fan-in/out and MCF connectors with focusing practical use.

II. FAN-IN/OUT

Keys for practical realization of fan-in/out are optical performances, compactness and productivity. Figure 1 shows the classification of fan-in/out technics.

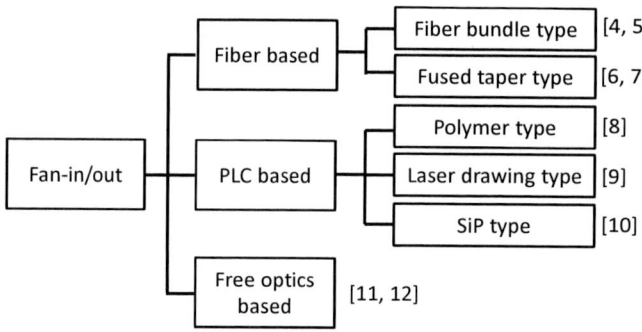

Fig. 1. Classification of fan-in/out technics.

Among several types of fan-in/out has been reported, we have been focusing on fiber bundle type which is the simplest because it's easy principle. We developed self-packing method to realize precise core pitch without restriction by capillary which needs accurate control of both hole diameter and cladding diameter of thin cladding fibers. The fabrication method of a fiber bundle is shown in Fig. 2.

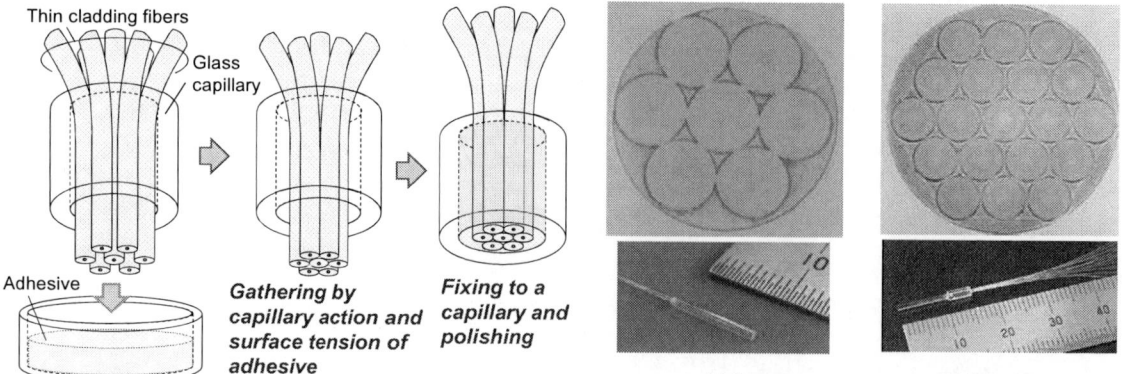

Fig. 2. Fabrication method of the fiber bundle

Fig. 3. Appearances and end faces of fiber bundles

(a) 7-core   (b) 19-core

At first, 7 thin-cladding fibers are coarsely aligned in a capillary whose hole diameter is sufficiently larger than stacked cladding diameter. Then the end of fibers is immersed into adhesive with low viscosity. Adhesive is sucked up due to capillary action and the 7 thin-cladding fibers are brought together to closest packed structure by surface tension of adhesive. After curing adhesive, the bundled fibers are fixed in a glass capillary with adhesive. Finally, end face is polished and the fiber bundle is completed. We also developed fiber bundle with 19 thin fibers. Figure 3 shows appearances and micrographs of end faces of developed 7 and 19-core fiber bundles. Thin cladding fibers are arrayed into closest packed structure despite large clearance of capillary hole. Precise core pitch less than ±0.25 μm and ±0.40 μm was confirmed for 7 and 19-core fiber bundles respectively [5]. Fan-in/out is realized by aligning and bonding the fiber bundle and MCF. Low insertion loss less than 0.2 dB and extremely small cross talk less than -65 dB was confirmed for 7-core type (figs. 4 & 5). Good optical Performances are also confirmed to 19-core Type. [13]

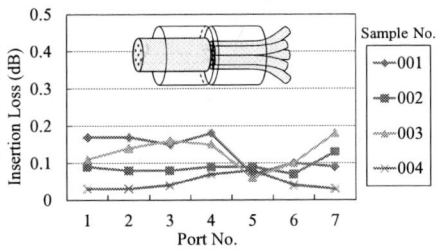

Fig. 4. Insertion loss of fan-in/out with 7-core MCF

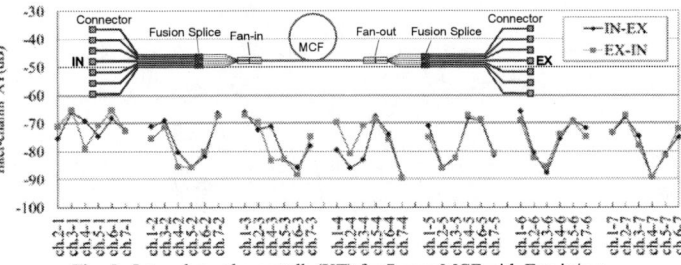

Fig. 5. Inter-channel crosstalk (XT) for 7-core MCF with Fan-in/out

Stability for environments was also confirmed for packaged fan-in/out of 7-core MCF (Fig. 6). Figure 7 shows the result of temperature cycling test and the table I is the summary of temperature cycling, vibration, and impact tests.

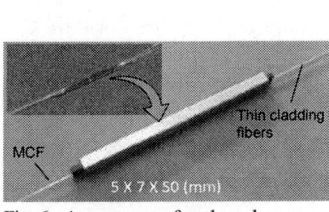

Fig. 6. Appearance of packaged fan-in/out

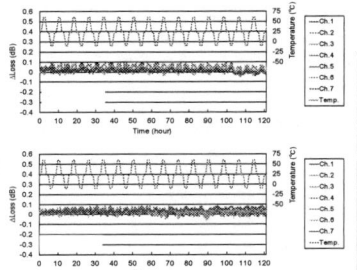

Fig. 7. Result of temperature cycling test

TABLE I : RESULTS OF ENVIRONMENT TESTS FOR 2 SAMPLES

Test Items	Test conditions	Standards	Loss variation
Temperature cycling	-10 to 60°C, 15 Cycles	IEC61753-1 Category C	<0.1dB
Vibration	1.5mm P-P amplitude, 10 to 55Hz /min, 2 hours /X,Y, Z axis	IEC61753-1 Category O	<0.1dB
Impact	100G, 6ms, 5 Times /X, Y, Z axis, 2 directions	JIS C5983	<0.1dB

III. CONNECTOR TYPE FAN-IN/OUT

For single MCF connector, K. Sakaime and R. Nagase developed MU type MCF connector with us [14-16]. By incorporate Oldham's coupling to MU type MCF connector, ferrule floating and suppression of MCF rotation are simultaneously satisfied. We developed connector type fan-in/out utilizing this connector. Figure 8 shows the appearance and the schematic drawing. We confirmed insertion loss less than 0.35 dB and low reflection less than -45 dB by PC in all cores for 3 samples. Stability for environmental conditions was also confirmed for 2 samples (table II).

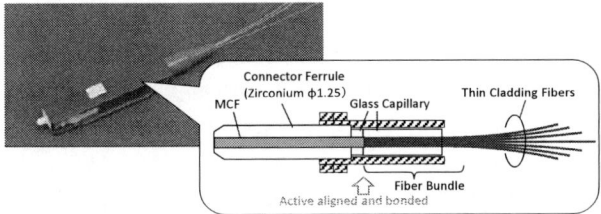

Fig. 8. Appearance and schematic drawing of MU type Fan-in/out

TABLE II : RESULTS OF ENVIRONMENT TESTS FOR 2 SAMPLES

Test Items	Test conditions	Standards	Loss variation
Temperature cycling	-10 to 60°C, 5 Cycles	IEC61753-1 Category C	<0.1dB
Damp Heat	40°C , 95%, 96Hours	IEC61753-1 Category C	<0.1dB
Vibration	1.5mm P-P amplitude, 10 to 55Hz /min, 2 hours /X,Y, Z axis	IEC61753-1 Category O	<0.1dB
Impact	100G, 6ms, 5 Times /X, Y, Z axis 2 directions	JIS C5983	<0.1dB

IV. MULTIPLE FIBER CONNECTOR

Multiple MCF connectors are also key items for practical use especially applying MCF to ultra-high density optical wiring such as in data centers or central offices. We developed 8-MCF connector [17, 18]. Newly developed structure using alignment member dramatically improves complexity and yield of rotational alignment of all MCF (Fig. 9). We also developed the method of precise mechanical polish for protruded fibers and realized physical contact over all MCFs and cores by pressing force less than 18 N. Figure 10 shows the appearance and end face of the 8-MCF connector. Figure 11 shows the result of connection loss and return loss. Although connection loss of MCF 5 is little bit high, it is still less than 0.9 dB and it can be improved by improving dimensions of the parts.

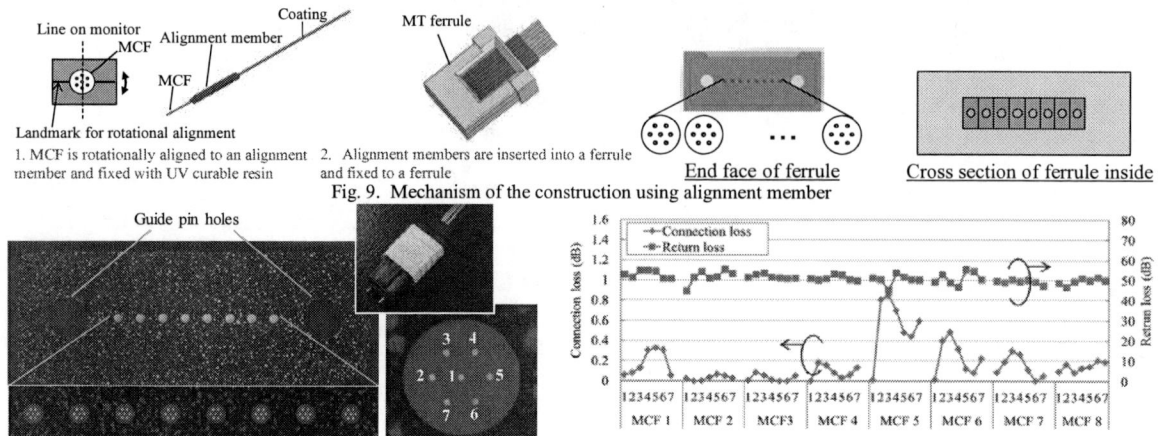

Fig. 9. Mechanism of the construction using alignment member

Fig. 10. Appearance and end face of 8-MCF connector

Fig. 11. Connection loss and return loss of 8-MCF connector

V. CONCLUSIONS

MCF connectivity techniques ready for practical use are introduced. Compact fiber bundle type fan-in/out has good optical properties and stability for environment conditions. It is also suitable for mass production because precise core arrangement is realized by easy self-packing method by surface tension of adhesive. MU type fan-in/out is suitable for high density wiring systems. MPO type 8-MCF connector is remarkable breakthrough for ultra-high density wiring. PC connection for all 56 cores with pressing force less than 18 N was realized.

ACKNOWLEDGMENT

This work was supported by the National Institute of Information and Communications Technology (NICT), Japan under "Research on Innovative Optical Communication Infrastructure".

REFERENCES

[1] H. Takara et al., "1.01-Pb/s (12SDM/222WDM/456Gb/s) crosstalk-managed transmission with 91.4-b/s/Hz aggregate spectral efficiency," Proc. ECOC 2012, Th.3.C (2012).

[2] K. Igarashi et al., "Super-Nyquist-WDM transmission over 7,326-km seven-core fiber with capacity-distance product of 1.03 Exabit/s•km," Opt. Exp., vol. 22, no. 2, pp. 1220-1228, (2014).

[3] T. Hayashi et al., "125-µm-Cladding 8-Core multi-core fiber realizing ultra-high-density cable suitable for O-Band short-Reach optical Interconnects," Proc. OFC 2015, Th5C.6 (2015).

[4] K. Watanabe et al., "Development of fiber bundle type fan-out for 19-core multicore fiber," Proc. OECC 2014, Mo1E2 (2014)

[5] T. Saito et al., "Confirmation of core pitch accuracy of fiber bundle type fan-out for MCF," Proc. IEEE Summer Topicals 2014, TuE2.3 (2014)

[6] B. Zhu et al., "Seven-core multicore fiber transmissions for passive optical network," Opt. Express, 18, 11, 11117-11122 (2010)

[7] H. Uemura et al., "Fused Taper Type Fan-in/Fan-out Device for 12 Core Multi-Core Fiber ," Proc. OECC 2014, Mo1E4 (2014)

[8] Watanabe et al., "Laminated polymer waveguide fan-out device for uncoupled multi-core fibers," Opt. Express, Vol. 20, No. 24, 26317-26325 (2012)

[9] R. R. Thomson et al., "Ultrafast-laser inscription of a three dimensional fan-out device for multicore fiber coupling applications," Opt Express Vol. 15 No. 18, 11691-11697 (2007).

[10] Ding et al., "On-chip Grating Coupler Array on the SOI Platform for Fan-in/Fan-out of Multi-core Fibers with Low Insertion Loss and Crosstalk," Proc. ECOC 2014, We.1.1.3 (2014)

[11] Arao et al., "Compact Multi-core Fiber Fan-in/out Using GRIN Lens and Microlens Array," Proc. OECC 2014, Mo1E1 (2014)

[12] Y. Tottori et al., "Improved return loss of fan-in/fan-out device for circular core array multi-core fiber using free space optics ," Proc. IEEE Summer Topicals 2014, TuE2.2 (2014)

[13] K. Watanabe et al., "Compact Fan-out for 19-core Multicore Fiber, with High Manufacturability and Good Optical Properties," Proc. OECC 2015, PWe.31 (2015).

[14] R. Nagase et al., "MU-type multicore fiber connector," Proc. IWCS 2012, pp. 823-827 (2012).

[15] K. Sakaime et al., "Mechanical characteristics of MCF connector," Proc. IEEE Summer Topicals 2014, pp. 172-173, (2014).

[16] K. Sakaime et al., "Mechanical characteristics of MU-Type MCF connector," Proc. MOC 2015, H54 (2015).

[17] K. Watanabe et al., "Development of MPO type 8-multicore fiber connector," Proc. OFC 2015, W4B.3 (2015).

[18] K. Watanabe et al., "MPO type 8-multicore fiber connector with physical contact connection," J. Lightw. Technol., vol. 33, no. 24, pp. 1-7 (2015)

ThC1-2

OECC/PS2016

Investigation of connector damage caused by high power transmission in optical connector with dust

Chisato Fukai, Kotaro Saito, Ryo Koyama, Mitsuru Kihara, and Toshio Kurashima

NTT Access Network Service Systems Laboratories, NTT Corporation, 1-7-1 Hanabatake, Tsukuba, Ibaraki, 305-0805, Japan

fukai.chisato@lab.ntt.co.jp

Abstract: We have investigated connector damage with various dust attachment patterns and with various optical power losses. A high power transmission with a power loss exceeding 50 mW might damage connectors with a dust attachment pattern.

Keywords: optical connector, connector damage, dust attachment pattern, optical power loss

I. INTRODUCTION

Recently, various optical fiber communication technologies have been studied with the aim of achieving large capacity transmission. Distributed Raman amplification, wavelength division multiplexing, and mode division multiplexing, which are examples of those technologies, require high power optical signal transmission in the optical fiber [1], [2]. It is important to investigate the influence of a high power transmission on the optical medium. Several reports have investigated the damage to optical fibers and optical components caused by high power transmission [3]-[6]. For example, an optical connector is reportedly damaged by high power transmission when contamination such as carbon black is attached to a connector endface [3]. However, certain points remain unclear. The dust condition on a connector endface might depend on the connector damage caused by a high power transmission.

In this paper, we investigate experimentally a condition where a connector containing dust is damaged by a high power transmission. We focus on the pattern with which the dust is attached to the connector endface and the optical power loss calculated from the input power and the splice loss.

II. INFLUENCE OF DUST ATTACHMENT PATTERN ON CONNECTOR DAMAGE

Figure 1 shows our experimental setup. A Raman fiber laser operating at a wavelength of 1450 nm was used as a high power source. The optical power was injected into an SC connector via an isolator. The maximum output power from the SC connector was 5.5 W. Moreover, we prepared a conventional SC connector as a test connector whose splice loss and return loss were less than 0.5 dB and more than 40 dB, respectively. The power level at which connector damage is observed in a connector with carbon black is increased by increasing the mode field diameter (MFD) [3]. To confirm the MFD dependence of the connector damage in a connector containing dust, three kinds of fibers with different MFDs were inserted in the test connectors. We used a single-mode fiber (SMF), a dispersion shifted fiber (DSF), and a pure silica core fiber (PSCF) and their MFDs at 1450 nm were 9.8, 6.9, and 11.1 μm, respectively.

In our experiment, the connector containing dust was exposed to a high optical power for one minute. We used dust with a size of several micrometers to several tens of micrometers. To confirm the connector damage, we observed the connector endfaces with a microscope. In addition, we measured the optical power before and after fiber insertion into the test connector by using an optical power meter, and calculated the splice loss.

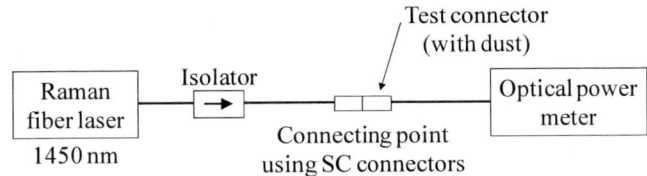

Fig. 1. Experimental setup.

Table I shows the splice losses and connector endfaces of one test sample. Table I (a), (b), and (c), respectively, show a good endface without dust, an endface with dust before a high power transmission, and an endface damaged by a high power transmission of 1 W. In Table I (a), we can see the fiber core when no dust was attached to the connector endface. In Table I (b), dust with a size of 31.5 μm was attached to the entire core, and the splice loss was 0.68 dB. As shown in Table I (c), a discolored area nearly the same size as the dust was confirmed at the center of the fiber after the high power was transmitted, and the splice loss was 4.85 dB. In the discolored area, it appears that the dust absorbed part of the input power and burned. As a result, the fiber core melted and the connector endface was not smooth.

317

OECC/PS2016

TABLE I
SPLICE LOSSES AND CONNECTOR ENDFACES OF ONE TEST SAMPLE

	(a)	(b)	(c)
Splice loss	less than 0.5 dB	0.68 dB	4.85 dB
connector endface			

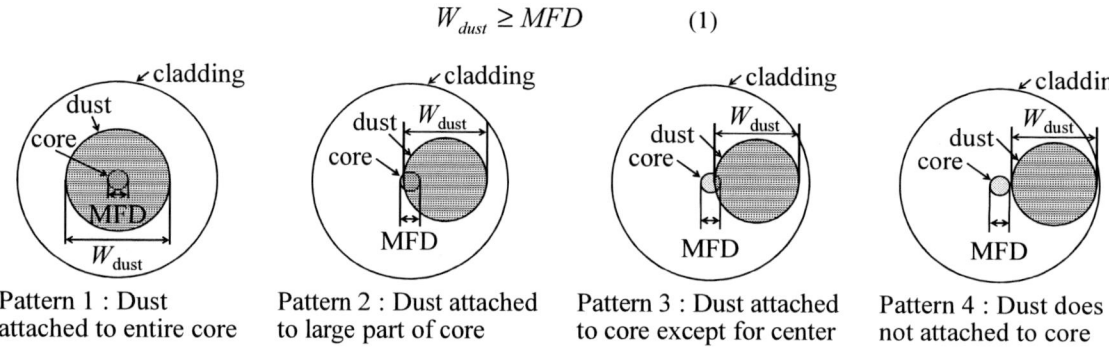

Figure 2 shows the pattern of dust attachment on the connector endface. We classify the attachment between a fiber core and dust into 4 patterns. Pattern 1 is where dust is attached to the entire core. Pattern 2 is where dust is attached to a large part of the core. Namely, in pattern 2, the dust is attached to the center of the core. Pattern 3 is where the dust is attached to the core except for its center. Pattern 4 is where no dust is attached to the core. As the dust size is expressed by equation (1), we used dust with a width W_{dust}, which is larger than MFD of the fiber.

$$W_{dust} \geq MFD \qquad (1)$$

Pattern 1 : Dust attached to entire core

Pattern 2 : Dust attached to large part of core

Pattern 3 : Dust attached to core except for center

Pattern 4 : Dust does not attached to core

Fig. 2. Pattern of dust attachment on the connector endface.

Figure 3 shows the number of damaged/undamaged connectors with each pattern. The black and white parts show the results obtained for the damaged and undamaged connectors, respectively. Figure 3 (a), (b), and (c) show the results obtained for SMF, DSF, and PSCF, respectively. For each fiber, we prepared ten samples of each pattern. Figure 3 shows that damaged connectors were obtained with patterns 1 and 2, and the connectors with pattern 3 and 4 were undamaged. It has been reported that the optical field pattern of a conventional SMF can resemble a Gaussian distribution [7]. In patterns 1 and 2, the dust overlaps the center of the core, which is the peak of the field pattern. Even if the dust does not overlap the entire core, it appears that thermal absorption occurs at the center of the core to which dust was attached and the connector was damaged. In this experiment, we used DSF and PSCF whose field patterns differ from a Gaussian distribution. However, the peaks also appear near the center of the core for those field patterns. Therefore, we considered that the thermal absorption caused by the dust occurred in DSF and PSCF as in SMF, and the connectors were damaged only with patterns 1 and 2.

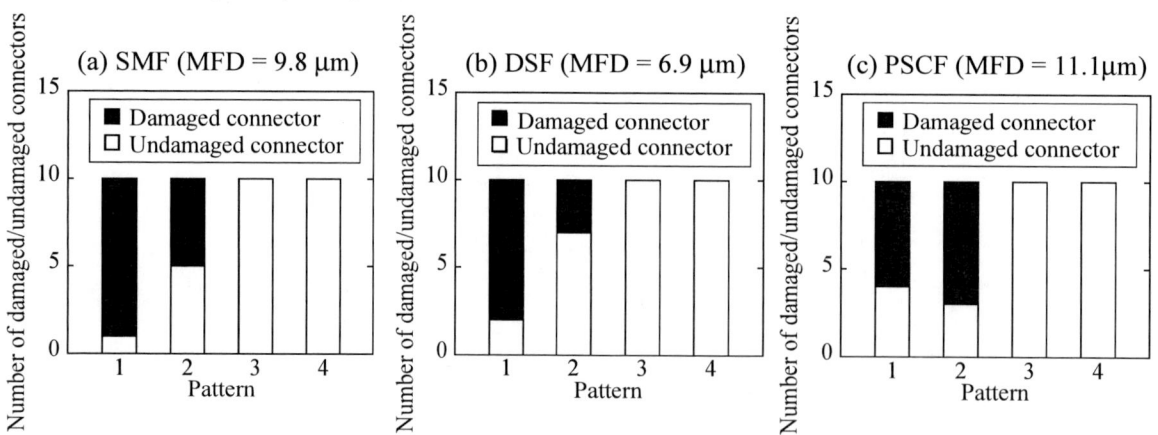

Fig. 3. Number of damaged/undamaged connectors with each pattern.

III. INFLUENCE OF OPTICAL POWER LOSS ON CONNECTOR DAMAGE

We investigated the optical characteristics of patterns 1 and 2 because we wanted to clarify the input power condition of the connector damaged by a high power transmission. Figure 4 shows the optical power loss P_{LOSS} as a function of the input power P_{IN} and the splice loss α. The filled and unfilled symbols show the results obtained for the damaged and undamaged connectors, respectively. The circles, squares, and triangles show the results for SMF, DSF, and PSCF, respectively. We prepared more than forty samples of each fiber. Moreover, the gray area in this figure shows the P_{LOSS} (W) expressed by equation (2).

$$P_{LOSS}(W) = P_{IN}(W) - 10^{[\{P_{IN}(dB) - \alpha(dB)\}/10]} \qquad (2)$$

From Fig. 4, α values in the 0.08 to about 20 dB range were obtained when a high power was transmitted to the connector in the presence of dust. In general, α in an SC connector without dust was less than 0.5 dB and was usually about 0.1 to 0.2 dB. Some of α values for a connector with dust were the same as those for a connector without dust, but most splice losses exceeded 0.2 dB because the input power transmission was impeded by the dust. In Fig. 4, we also found that the connector damage gathers in the gray area. Specifically, we confirmed that connector damage occurs when the P_{LOSS} is more than 50 mW. Moreover, the minimum P_{LOSS} value at which the connector is damaged 50 to 100 mW in the SMF, DSF, and PSCF. It is reported that the power level at which connector damage is observed is increased by expanding the MFD from 10 to 80 μm [3]. In our study, the difference between the MFDs of the fibers was several micrometers. The dependence of the P_{LOSS} on the MFD was not clearly observed because the difference between the MFDs of the fibers was very small compared with the values reported in Ref. 3.

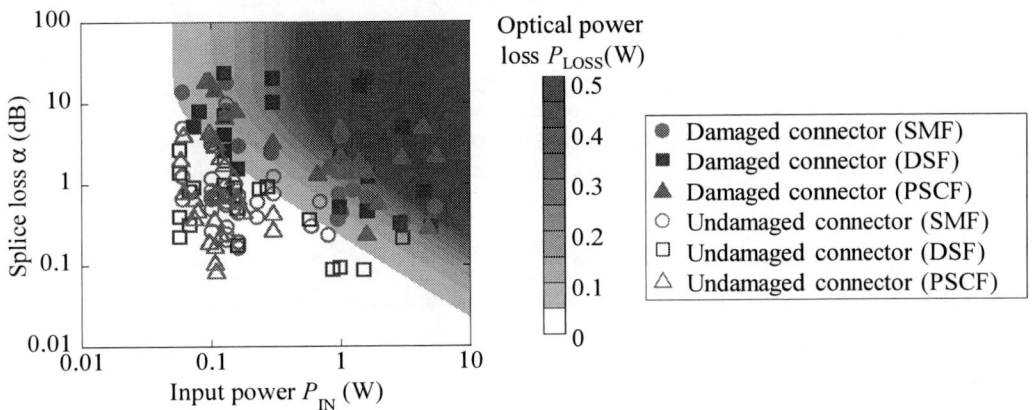

Fig. 4. Optical power loss as a function of input power and splice loss.

IV. CONCLUSIONS

We investigated experimentally a condition where a connector containing dust is damaged by a high power transmission. We showed that connector damage occurs when the dust overlaps the center of a fiber core. We also showed that optical power loss is very important as regards connector damage. We confirmed that there is a high possibility that dust attached to a connector endface causes connector damage when conventional optical fibers such as SMF, DSF, and PSCF are used as a transmission line with an optical power loss exceeding 50 mW.

REFERENCES

[1] H. Masuda, M. Tomizawa, Y. Miyamoto, and K. Hagimoto, "High-performance distributed Raman amplification systems with limited pump power," IEICE Trans. Commun., vol. E89-B, pp. 715-723, March 2006.

[2] R. Ryf, M. Esmaeelpour, N. K. Fontaine, H. Chen, A. H. Gnauck, R. -J. Essiambre, J. Toulouse, Y. Sun, and R. Lingle, Jr., "Distributed Raman amplification based transmission over 1050-km few-mode fiber," Proc. of 41st European Conference on Optical Communication (ECOC 2015), Valencia (Spain), Sep. 2015, ID: 0760.

[3] M. De Rosa, J. Carberry, V Bhagavatula, K. Wagner, and C. Saravanos, "High-power performance of single-mode fiber-optic connectors," J. Lightwave Technol., vol. 20, pp. 879-885, May 2002.

[4] S. Yanagi, S. Asakawa, and R. Nagase, "Characteristics of fibre-optic connector at high-power optical incidence," Electron. Lett., vol. 38, pp. 977-978, Aug. 2002.

[5] K. Hogari, K. Kurokawa, and I. Sankawa, "Influence of high-optical power light launched into optical fibers in MT connector," J. Lightwave Technol., vol. 21, pp. 3344-3348, Dec. 2003.

[6] Y. Shuto, S. Yanagi, S. Asakawa, M. Kobayashi, and R. Nagase, "Fiber fuse phenomenon in step-index single-mode optical fibers," IEEE J. Quantum. Electron., vol. 40, pp. 1113-1121, Aug. 2004.

[7] D. Marcuse, "Loss analysis of single-mode fiber splices," Bell Sys. Tech. J., vol. 56, pp. 703-718, May-June 1977.

ThC1-3 (Invited)

OECC/PS2016

Input/Output Channel Coupling Devices for SDM and MDM

Yasuo Kokubun

Yokohama National University, 79-5 Tokiwadai, Hodogaya-ku, Yokohama, Japan 240-8501
kokubun-yasuo-sd@ynu.ac.jp

Abstract: *Input/output devices such as fan-in/fan-out and mode MUX/DEMUX devices are inevitable for SDM and MDM transmissions. In this review, the recent advances of these devices as well as their principles are introduced.*
Keywords: *Space division multiplexing, Mode division multiplexing, Fan-in/fan-out device, Mode MUX/DEMUX*

I. INTRODUCTION

Since the birth of low loss optical fiber and semiconductor laser in early 1970's, the optical fiber communication technology has brought dramatic improvement in transmission capacity. Owing to many inventions such as optical fiber amplifier, wavelength division multiplexing, high order multilevel modulation, digital coherent detection, and recent 3m technologies[1] involving space division multiplexing (SDM) and mode division multiplexing (MDM), the world record of transmission capacity has reached over 2 Pb/s[2][3], and is now transfigured to a new generation photonic network.

In the SDM and MDM transmissions, special input/output devices such as fan-in/fan-out devices for SDM and mode multi/demultiplexers for MDM are inevitable to discriminate transmission channels. In addition, the mode analysis and measurement technologies are needed to evaluate the transmission characteristics of each mode channel and also the mode excitation ratio and/or inter-channel crosstalk in the MDM transmission.

In this invited talk, I will review the recent advances of input/output devices for multi-core fibers (MCFs) and few-mode fibers (FMFs) as well as their operating principle.

II. FAN-IN/FAN-OUT DEVICES FOR MULTI-CORE FIBERS

Fan-in/fan-out devices are inevitable to couple input and output space channels to corresponding cores of MCF. To save the space, the categories are introduced. There have been reported the tapered-fiber bundle type[4], thin fiber bundle type[5], bulk optics type using lens coupling[6], laser inscription waveguide type[7], and stacked waveguide type using laminated polymer waveguide (LPW)[8].

III. MODE MULTI/DEMULTIPLEXERS FOR FEW-MODE FIBERS

The mode multi/demultiplexers are briefly categorized as summarized in Table I. These are the space-beam type using phase plate[9]-[11], spot-based mixer-type[12],[13], fiber-type including photonic lantern[14]-[17] and fiber directional coupler[18]-[20], waveguide-type including Si waveguide[21]-[23], laser inscribed waveguide photonic lantern[24], waveguide directional coupler[25],[26], MMI waveguide[27],[28], and mode-evolutional serial branching type[29],[30].

In the space-beam type using phase plate, the operating principle is similar to the pattern recognition of transverse field profile of modes, and so translucent mirror is needed to split or combine beams which results in the branching loss. The Mach-Zehnder interferometer with mirror inversion in one arm can MUX/DEMUX degenerate modes without the beam splitting/combining loss[11]. In the photonics lantern with dissimilar cores including tapered fiber-type and laser inscribed waveguide-type, since the position of input port corresponds to the mode pattern, this type can be used as the mode multiplexer. It is needed, however, to control the phase difference between two input ports corresponding to degenerate modes, and so this type is difficult to be used as mode demultiplexer. Silicon waveguide type can connect an FMF perpendicularly to the surface of Si waveguide using grating coupler, and can couple both orthogonal polarizations using the cross-type grating. Therefore this type is applicable to FM-MCFs. There still remains, however, the problem of high loss. Fiber directional coupler type is low loss and applicable to degenerate modes. Waveguide directional coupler type and mode-evolutional serial branching type are also applicable to degenerate modes. The scalability to FM-MCFs remains the issue to be solved in these type of mode multiplexers.

IV. MODE ANALYSIS TECHNOLOGIES

Mode analysis technologies are briefly categorized into two methods, one is based on the measurement of propagation characteristics of each mode such as the propagation constant, group delay, dispersion, and the distribution of coupling coefficient in the propagation direction, and the other is based on the measurement of the excitation ratio and

TABLE I
CATEGORIZATION AND FEATURES OF MODE MULTI/DEMULTIPLEXERS

	Need of beam splitter* (loss)	Wavelength dependence	Applicability to degenerate modes	Integration
Phase plate	Yes	Weak	Yes	Not suited
Spot-based mixer	Yes	None	Yes	Not suited
Photonic lantern	Partially yes	Weak	Yes, additional device needed	Yes (waveguide)
Si photonics	No, but high loss	Large	Yes	Suited
Fiber directional coupler	None	Weak	Yes	Not suited
Waveguide directional coupler	None	Weak	Yes	Suited
MMI waveguide	None	Weak	Yes	Suited
Mode evolutional serial branching waveguide	None	Very small	Yes	Suited

crosstalk among modes guided in an FMF, which are obtained from the observation of transverse electromagnetic field profile or intensity profile. The categorization is summarized in Table II.

In the mode analysis aiming at measuring the mode excitation ratio, it is very important to discriminate accurately the mode including not only the amplitude coefficient but also the polarization state and the phase, because the degenerate LP mode consists of different polarization as well as field pattern as shown in Table III. To realize the measurement of complex amplitude of modes at the output end of FMF, we developed a novel mode analysis method named mode analysis from intensity profiles through angled polarizer (IPAP method)[38].

TABLE II
FONT: TIMES NEW ROMAN / (SIZE 8)
TYPE SIZES AND STYLES FOR OECC/PS 2016 PAPER

	Name of methods	Features
Method based on propagation characteristics (β, delay, back scattering)	Mode analyzer using prism coupler, and FBG.	m-line measurement by difference of propagation constant.
	ToF(time of flight), OTDR	Difference of propagation time between modes
Method based on field profile (NFP, FFP etc.)	OLCI (Optical Low Coherence Interferometry) technique	ASE source and Fourier transformation, GVD evaluation by time-wavelength mapping.
	Mode expansion coefficient method	Overlap integral with theoretical field
	S^2 (Spatial and Spectral) imaging method	Group velocity and intensity of modes, need of broadband source and LP_{01} dominant.

V. EIGENMODE IN FMFs

In the mode division multiplexing, the most important issue which should not be forgotten is that MIMO digital processing is inevitable even if the fiber length is very short. This is caused by not only the mode conversion due to the macro-/micro-bending and perturbations but also the modal evolution of LP modes, which is named modal birefringence[41]. This evolution of modal field profile is caused by the difference of propagation constants of true eigenmodes[39] which constitute an LP mode[40]. As is well known, since the concept of LP mode is derived by approximating $\Delta \ll 1$ in the eigenvalue equation[40], the propagation constant of LP mode does not satisfy the eigenvalue equation. Since $HE_{\ell+1,m}$ and $EH_{\ell-1,m}$ modes (TE_{0m} and TM_{0m} modes when $\ell=1$) are quasi-degenerate, i.e. the propagation constants are very close to each other, the notation and concept of $LP_{\ell m}$ mode have been widely used in many analysis and experiments of mode division multiplexing. However, since the propagation constants of quasi-degenerate eigenmodes are slightly different[41], the intensity profile varies along with the propagation distance[42]-[44]. In addition, the field vector also evolves as shown in Fig.1[45]. Therefore, even when a linearly polarized mode (LP mode) is launched at the input port of FMF, the field profile is no more the same as that of the LP mode. The mode analysis and measurement should take into account this effect[45].

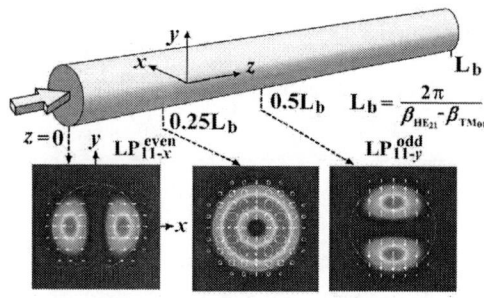

Fig.1 Evolution of intensity profile and field vector of LP_{11} mode along with propagation.

TABLE III DEFINITION OF DEGENERATE LP MODES

	LP_{01}	LP_{11}^{even}	LP_{11}^{odd}
x - polarization			
y - polarization			

For the same reason, the mode demultiplexer can not discriminate accurately the LP mode without crosstalk. MIMO tells us what mode was launched at the input port but the problem of complexity of DSP process remains.

REFERENCES

[1] T. Morioka, Optoelectronics and Communication Conference (OECC), FT4, 2009.
[2] B. J. Puttnam, et al., ECOC2015, Valencia, PDP.3.1, 2015.
[3] D. Soma, et al., (ECOC2015, Valencia, PDP.3.2, 2015.
[4] B. Zhu, et al., Optics Express, vol.18, no.11, pp.11117-11122, 2010.
[5] K. Watanabe, T. Saito, Y. Tsuchida, K. Maeda, M. Shiino, OECC/PS2013, Kyoto, TuS1-5, 2013.
[6] Y. Tottori, T. Kobayashi, and M. Watanabe, Photon. Technol. Lett., vol.24, no.21, pp.1926-1928, 2012.
[7] R. R. Thompson, et al., Opt. Express, vol. 15, no. 18, pp.11691-11697, 2007.
[8] T. Watanabe, Y. Kokubun, Optics Express, vol. 20, no.24, pp.26317-26325, 2012.
[9] E. Ip, et al., ECOC2011, Geneva, Th.13.C.2, 2011.
[10] H. Chen and T. Koonen, OFC/NFOEC2013, Anaheim, OTh1B.4, 2014.
[11] D. Soma, K. Takeshima, K. Igarashi, and T. Tsuritani, ECOC2014, Cannes, We.1.1.6, 2016.
[12] R. Ryf, N. K. Fontaine, and R-J. Essiambre, Photon. Technol. Lett., vol.24, no.21, pp.1973-1976, 2012.
[13] H. Chen, et al., Photon. Technol. Lett., vol.25, no.24, pp.2474-2477, Dec. 2013.
[14] S. G. Leon-Saval, T. A. Birks, J. Bland-Hawthorn, M. Englund, Opt. Lett. vol.30, no.19, pp.2545?2547, 2005.
[15] N. K. Fontaine, R. Ryf, J. Bland-Hawthorn, S. G. Leon-Saval, Optics Express, vol.20, no.24, pp.27123-27132, 2012.
[16] N. K. Fontaine, S. G. Leon-Saval, R. Ryf, Joel R. S. Gil, B. Ercan, J. Bland-Hawthorn, ECOC2013, London, PD1.C.3, 2013.
[17] S. G. Leon-Saval, et al., Opt. Express, vol.22, no.1, pp.1036-1044, 2014.
[18] F. Saitoh, K. Saitoh, M. Koshiba, Optics Express, vol.18, no.5, pp.4709-4716, 2010.
[19] A. Li, Xi Chen, A. Al Amin, W. Shieh, Photon. Technol. Lett., vol.24, no.21, pp.1953-1956, 2012.
[20] K. Takenaga, et al., ECOC2014, Cannes, Tu.4.1.4, 2014.
[21] A. M. J. Koonen, H. Chen, H. P. A. van den Boom, O. Raz, Photon. Technol. Lett., vol.24, no.21, pp.1961-1964, 2012.
[22] Y. Ding, H. Ou, J. Xu, C. Peucheret, Photon. Technol. Lett., vol.25, no.7, pp.648-651, 2013.
[23] H. Chen, R. van Uden, C. Okonkwo, T. Koonen, Optics Express, vol.22, no.26, pp.31582-31594, 2014.
[24] N. K. Fontaine, R. Ryf, OECC/PS2013, Kyoto, MR202, July 2013.
[25] N. Hanzawa, et al., Opt. Exp. vol. 22, no. 24, pp. 29321-29330, 2014.
[26] S. Gross, N. Riesen, J. D. Love, and M. J. Withford, OFC2015, Los Angeles, W3B.2, 2015.
[27] R. Takakura, M. Jizodo, A. Fujino, T. Tanaka, K. Hamamoto, Jpn. J. Appl. Phys., vol. 53, 08MB10, 2014.
[28] R. Imansyah, L. Himbele, H. Jiang, K. Hamamoto, 20th Microoptics Conference (MOC'15), Fukuoka, D3, Sep. 2015
[29] T. Watanabe, Y. Kokubun, IEEE Photincs Journal, vol.7, no.6, 7103311, 2015.
[30] T. Watanabe, K. Kojima, Y. Kokubun, IEICE Electron. Express, vol.13, no.1, pp.1-12, 2016.
[31] K. Iga, Y. Kokubun, Trans. IECE of Japan, vol.E60, no.1, pp.1-7, 1977.
[32] A. Obeysekara, F. Poletti, D. J. Richardson, ECOC2012, Amsterdam, Tu.1.F.5, 2012.
[33] R. A. Barankov, P. Steinvurzel, S. Ramachandran, OFC/NFOEC2012, Los Angeles, OM3C.2, 2012.
[34] M. Nakazawa, M. Yoshida, T. Hirooka, Optics Express, vol.22, no.25, pp.31299-31309, 2014.
[35] R. Gabet, E. Le Cren, C. Jin, M. Gadonna, B. Ung, Y. Jaouen, M. Thual, S. LaRochelle, ECOC2014, Canne, Th.1.4.2, 2014.
[36] S. Berdague, P. Facq, Appl. Opt., vol.21, no.11, pp.1950-1955, 1982.
[37] D. R. Gray, et al., OFC2015, Los Angeles, W4I.6, 2015.
[38] Y. Kokubun, T. Watanabe, K. Morita, R. Kawata, 20th Microoptics Conference (MOC'15), Fukuoka, D2, 2015.
[39] E. Snitzer, "Cylindrical dielectric waveguide modes," J. Opt. Soc. of Am., vol.51, no.5, pp.491-498, 1961.
[40] D. Gloge, "Weakly guiding fibers," Appl. Opt., vol.10, no.10, pp.2252-2258, 1971.
[41] H. Kogelnik, P. J. Winzer, J. Lightwave Technol., vol.30, no.14, pp.2240-2245, July 2015.
[42] J. von Hoyningen-Huene, R. Ryf, P. J. Winzer, Optics Express, vol.21, no.15, pp.18097-18110, 2013.
[43] E-L. Lim, Q. Kang, M. Gecevicius, F. Poletti, S. Alam, and D. J. Richardson, OFC/NFOEC2013, OTu3G.2, Anaheim, 2013.
[44] D. J. Richardson, "Fiber Amplifiers for SDM Systems," OFC/NFOEC2013, Anaheim, OTu3G.1, 2013.
[45] Y. Kokubun, S. Miura, T. Watanabe, submitted to OECC/PS2016, Niigata, 2016.

Spinning effect in few-mode fiber with distributed long period gratings

Jian Fang,[1,*] Byoung Yoon Kim,[2] and William Shieh[1]

[1]Department of Electrical and Electronic Engineering, the University of Melbourne, VIC 3010, Australia
[2]Department of Physics, Korea Advanced Institute of Science and Technology, Daejeon 305-701, South Korea
*jianf1@student.unimelb.edu.au

Abstract: The spinning in few-mode fiber with distributed long period gratings can dramatically reduce the DMD at 1550nm. The effective bandwidth and DMD can be further improved with frequency modulation in the spin profile.
Keywords: Few-mode fiber, differential mode delay, long period gratings.

I. INTRODUCTION

Mode-division multiplexing (MDM) via few-mode fibers (FMFs) has recently attracted great attention due to its capability to overcome the capacity limit of single-mode fibers [1]. In order to recover the transmitted signals through digital signal processing (DSP) and multiple-input multiple-output (MIMO), few-mode fibers with low differential mode delay (DMD) are desirable. Apart from the optimization of fiber profile [2] and the group delay compensation method [3], introducing strong mode coupling has become another promising approach to reduce DMD in FMF [4]. Strong intra-group coupling, which is defined as the coupling within the same spatial mode groups, can be easily generated by distributed random perturbation such as fiber spinning [5]. However, strong coupling among different mode groups requires sufficiently large perturbation of refractive index, which is more difficult than intra-group coupling. To meet this requirement, long period gratings (LPGs) based mode scramblers have been proposed to use at regular intervals along the fiber link [6]. Few-mode fibers with distributed random long period gratings are also proposed to induce inter-group coupling [7]. However, in contrast with relying on random modal coupling, it is also important to find a deterministic spin method to guarantee the strong inter-group coupling which is easier to implement at fiber drawing. In this paper, we use long period gratings to generate mode coupling, and explore the spin effect on DMD in few mode fibers with distributed long period gratings.

II. SINUSOIDALLY SPUN LPGS IN FEW-MODE FIBERS

For simplicity, a step-index two-mode fiber, which supports LP_{01}, LP_{11a} and LP_{11b} modes, is used to analyze the spin effect with distributed LPGs in FMF. The fiber profile is given in Fig.1 (a), with parameters $n_1 = 1.4505$, $n_2 = n_4 = 1.4440$, $n_3 = 1.435$, $r_1 = 7$ μm, $r_2 = 10$ μm, $r_3 = 14$ μm and $r_4 = 62.5$ μm. A trench ($r_2 \sim r_3$) is added outside the core to reduce the overlap between core modes and cladding modes. LPGs are then written in the FMF by single-side UV illumination, resulting in an asymmetric cross-section refractive index change, which will generate strong mode coupling between LP_{01} mode and two LP_{11} modes, as shown in Fig.1 (b). The angle between y axis and the exposed direction of UV light and is assigned as ψ. If the UV exposure direction rotates sinusoidally during the grating writing process, then a sinusoidal rotary LPG is obtained. This kind of rotary LPG can be manufactured either by writing gratings in the twisted fiber [8] or writing the gratings in a spun fiber during the pulling process [9]. The distributedly-spun LPG then can be modeled as a concatenation of many rotary LPGs, as shown in Fig.1 (c). The field coupling among LP modes in LPG can be described by the coupled mode equations (CMEs) [10] given by $d\mathbf{A}(z)/dz =$

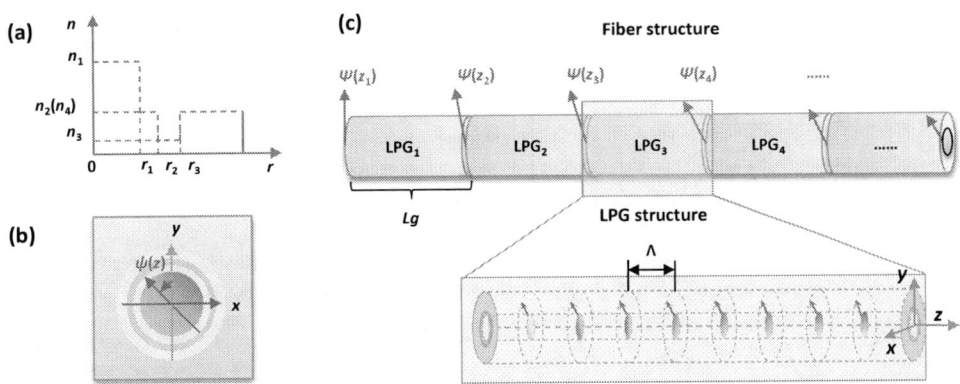

Fig. 1. (a) Fiber profile. (b) Cross-section index change of LPG with exposure angle $\psi(z)$. (c) Structure of FMF with distributed LPGs.

$-[\alpha/2+j\mathbf{K}(z)]\mathbf{A}(z)$, where vector $\mathbf{A}(z)$ is the complex amplitude of the LP modes and $\boldsymbol{\alpha}$ is a diagonal matrix denoting the power attenuation. $\mathbf{K}(z) = [K_{v-\mu}]$ is the coupling matrix with the elements derived from the integration of the transverse normalized electric fields of mode v and μ, as well as the transverse refractive index distribution $\Delta n_g\,(x,\,y) = \Delta n_g\exp[-\zeta\,(x,\,y)]$, where Δn_g is the grating strength and $\zeta\,(x,y)$ is the shape function defined in [7]. Both the refractive indices and the electric fields for core modes and cladding modes are calculated by the matrix method with a precise multi-layer model [11]. Transmission loss of core modes due to cladding mode coupling is evaluated by numerically solving the coupled mode matrix and calculating the total power of cladding modes when only the core modes are injected. 3 core modes and 27 cladding modes which have the largest coupling coefficients are selected for loss calculation. To obtain more accurate results, the fourth-order Runge-Kutta (RK4) method is used. When the grating strength $\Delta n_g = 2\times10^{-5}$, the loss of the LP_{01}, LP_{11a} and LP_{11b} modes are -5.58×10^{-7} dB/m, -5.84×10^{-7} dB/m and -1.16×10^{-6} dB/m, respectively, which are acceptable for long distance transmission.

Since the transmission loss is insignificant, the coupled mode matrix can be reduced to a 3×3 matrix containing only core modes, and the analytical expressions of CMEs are obtained as $\mathbf{A}(z) = \mathbf{T}(z)\mathbf{A}(z_0)$ [10], which $\mathbf{T}(z)$ is assumed to be the transfer matrix with $\psi(z_0) = 0$. Transfer matrix with a rotation angle ψ is $\mathbf{H}(z) = \mathbf{R}(\psi)\mathbf{T}(z)\mathbf{R}^+(\psi)$, where $\mathbf{R}(\psi)$ is the axis rotation matrix. For sinusoidal spin profile, the rotation can be written as $\psi(z) = \psi_0\sin(2\pi z/p)$, where ψ_0 is the spin amplitude and p is the spin period.

III. SPINNING EFFECT ON DMD REDUCTION

The total transfer matrix $\mathbf{M}$ can be obtained by multiplying the transfer matrix of $\mathbf{H}(z)$ of a regular interval sequentially, each of which has a sinusoidal rotation angle $\psi(z)$. The group delays $\boldsymbol{\tau} = (\tau_1, \tau_2, \tau_3)$ is evaluated via the eigenvalues of the group delay operator $\mathbf{G} = j(d\mathbf{M}/d\omega)\mathbf{M}^+$. Then DMD can be calculated through the standard deviation of group delays. Simulations have been carried out with grating strength of 2×10^{-5} and fiber length of 10km. The LPG period Λ is 673.82 μm. Results are shown in Fig. 2. We use the spin induced reduction factor (SIRF), which is defined as the ratio between the DMD of fiber with spun LPGs and the DMD of the fiber without LPG, to evaluate the effect of spinning. Interestingly, we find that DMD is remarkably reduced with some certain spin amplitudes and spin period. As shown in Figs. 2 (a)-(b), when $\psi_0 = 7.1$ rad and $p = 2$ m (marked as #1 in Fig. 2 (a)), SIRF reaches its minimum value. Figs. 2 (d)-(e) show the DMD of various spin profile versus the fiber length. When there are no LPGs, DMD in two-mode fiber increases linearly to 18.5 ns at 10 km. However, when the fiber is written by sinusoidal spin LPGs with $\psi_0 = 7.1$ rad and $p = 2$ m, DMD is dramatically reduced to 10.98 ps at 10 km. DMDs with the (#1) spin profile from 1549 nm to 1551 nm are calculated, as shown in Fig. 2 (c). Although the DMD is remarkably low at 1550 nm, the reduction does not perform well for other wavelengths. As an example, DMD is up to 2.11 ns at 1551nm, which is only 9 times smaller than the DMD of fiber without spin LPGs at the length of 10 km. This phenomenon is due to the fact the optimal spin profile (#1) is specially chosen for 1550 nm, while for other wavelength it does not possess the same improvement. That

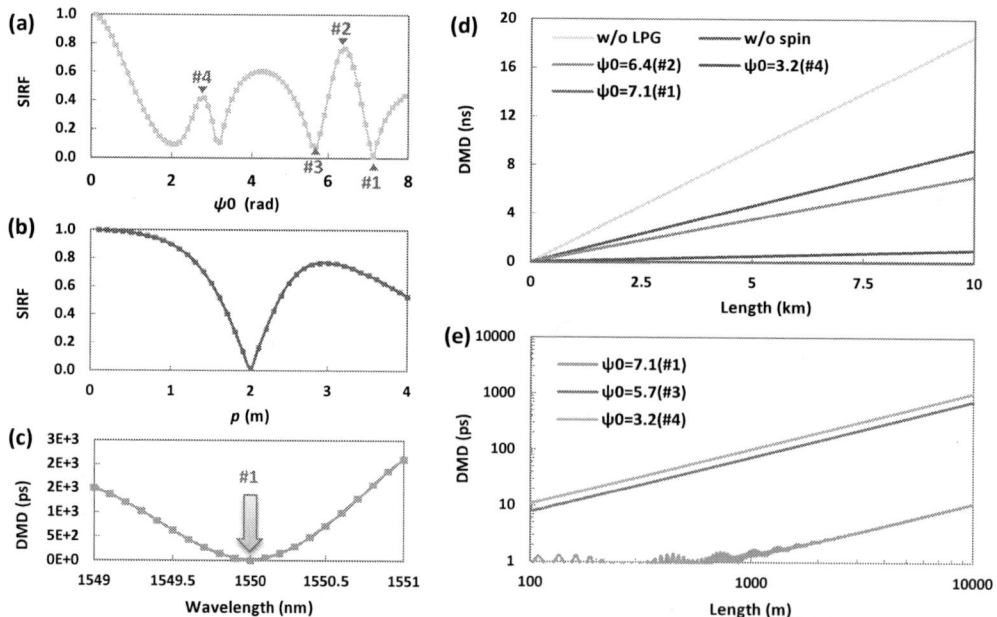

Fig. 2. Effect of sinusoidal spin at 1550 nm. (a) SIRF versus spin amplitude ψ_0 when $p = 2$ m (b) SIRF versus spin period p when $\psi_0 = 7.1$ rad for (c) Spin profile when $\psi_0 = 7.1$ rad and $p = 2$ m. (d) DMDs versus fiber length when $p = 2$ m for fiber without LPG, DG-FMF without sinusoidal spin, $\psi_0 = 6.4$, 3.2 and 7.1 rad, respectively. (e) Log-plot of DMDs for $\psi_0 = 7.1$, 5.7 and 3.2 rad when $p = 2$ m. (f) DMDs versus wavelength at 10 km fiber length with $\psi_0 = 7.1$ rad and $p = 2$ m.

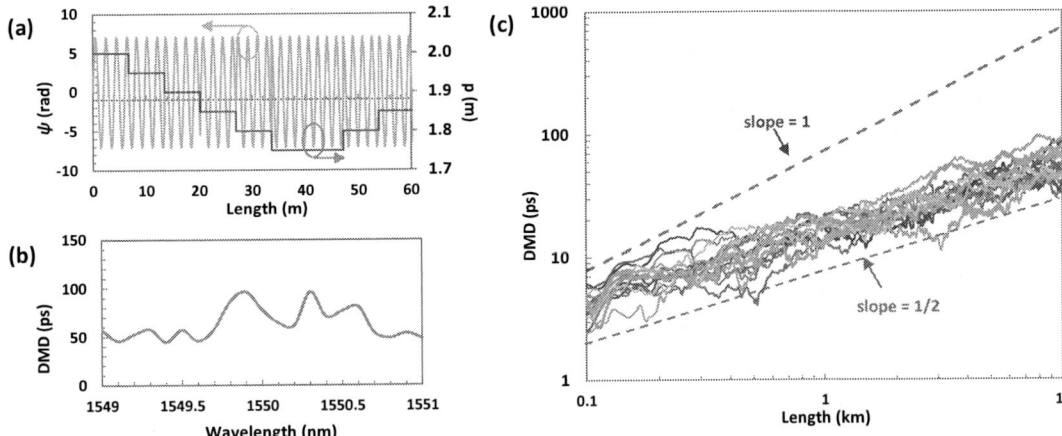

Fig. 3. (a) Frequency modulated (FM) spin profile. (b) DMD versus wavelength at 10km. (c) Comparison between ordinary sinusoidal spin profile and FM spin profile. Purple dashed line: log-plot of DMD with sinusoidal spin profile #3 ($\psi_0 = 5.7$ rad, $p = 2$ m); yellow dashed line: an indicative curve with slope of ½; other lines: DMDs of different wavelengths (1549~1551 nm) with FM spin profile.

means the sinusoidal spin profile, though it can reduce the modal delay at a certain wavelength, may not be feasible for wideband application.

To overcome this problem, frequency-modulation (FM) is utilized in the sinusoidal spin profile. We firstly calculate the SIRF versus spin period p for 1549~1551 nm. When ψ_0 is fixed to 7.1 rad, the optimal spin period p is 1.75 m and 1.8 m for 1549 nm and 1551 nm, respectively. For other wavelengths between 1549~1551 nm, optimal p occurs in the interval of [1.75, 2]. Then we change the value of p periodically, from 1.75 to 2.0, as shown in Fig. 3(a), and calculate the group delays at the length of 10 km. Fig. 3 (b) indicates that although the DMDs for various wavelength have a fluctuation between 50 ps and 100 ps, they are still small enough for MIMO and DSP. In addition, Fig. 3 (c) gives the log-plot of DMDs versus the fiber length. When ordinary sinusoidal spin profile (#3) is used, the DMD has a slope of 1, meaning that it grows linearly with the fiber length. However, when FM spin profile is utilized, DMDs approximately scale with the square-root of the fiber length. This result suggests that frequency modulation in spin profile can not only increase the bandwidth, but it can also further decrease the DMD for long-haul transmission.

IV. CONCLUSIONS

In this paper, we have investigated the spinning effect on DMD reduction in few-mode fibers with distributed long period gratings. Results show that by optimizing the sinusoidal spin profile, differential mode delay can be dramatically reduced to 10.98 ps at 10 km for 1550 nm. The effective bandwidth and DMD can be further improved by applying frequency modulation in the spin profile.

REFERENCES

[1] D. J. Richardson, J. M. Fini and L. E. Nelson, "Space-division multiplexing in optical fibres," Nat. Photonics, vol. 7, pp. 354-362, 2013.

[2] P. Sillard, M. Bigot-Astruc, D. Molin, "Few-Mode Fibers for Mode-Division-multiplexed Systems," J. Lightwave Technol. Vol 16, pp. 2824-2829, 2014

[3] T. Sakamoto, T. Mori, T. Yamamoto and S. Tomita, "Differential mode delay managed transmission line for WDM-MIMO system using multi-step index fiber," J. Lightwave Technol. vol. 30, pp. 2783-2787, 2012.

[4] K.-P. Ho and J. M. Kahn, "Statistics of group delays in multimode fiber with strong mode coupling," J. Lightwave Technol. vol. 29, pp. 3119-3128, 2011.

[5] L. Palmieri, "Modal Dispersion Properties of Few-Mode Spun Fibers," OFC 2015, paper Tu2D.4.

[6] S. Ö. Arik, D. Askarov and J. M. Kahn, "MIMO DSP Complexity in Mode-Division Multiplexing," OFC 2015, paper Th1D.1.

[7] J. Fang, A. Li and W. Shieh, "Low-DMD few-mode fiber with distributed long-period fiber grating," Opt. Lett. vol. 40, pp. 3937-3940, 2015.

[8] T. Zhu, K. S. Chiang, Y. J. Rao, C. H. Shi, M. Liu, "Characterization of Long-Period Fiber Gratings Written by CO_2 Laser in Twisted Single-Mode Fibers," J. Lightwave Technol. vol. 27, pp. 4863-4869, 2009.

[9] H. Guo, F. Liu Y. Yuan, H. Yu and M. Yang, "Ultra-weak FBG and its refractive index distribution in the drawing optical fiber," Opt. Express, vol. 23, pp. 4829-4838, 2015.

[10] W.-P. Huang, "Coupling-mode theory for optical waveguides: an overview," J. Opt. Soc. Am. A vol. 11, pp. 963, 1994.

[11] Q. Chen, J. Lee, M. Lin, Y. Wang, S. Yin, Q. Zhang and K. M. Reichard, "Investigation of tuning characteristics of electrically tunable long-period gratings with a precise four-layer model," J. Lightwave Technol. vol. 24, pp. 2954-2962, 2006.

ThC2-1 (Invited)

OECC/PS2016

Recent Developments of Multicore Multimode Fiber Amplifiers for SDM Systems

(Invited Paper)

Kazi S. Abedin

OFS Laboratories, 19 Schoolhouse Road, Somerset, New Jersey 08873.
kabedin@ofsoptics.com

Abstract: We will present experimental demonstration of core and cladding pumped MCF amplifiers, and designing of a few-mode MCF amplifier with an extended rare-earth doped region that provides uniform modal gain in C-band, with cladding pumping.

Keywords: Multicore, multimode, erbium doped fiber amplifier, space division multiplexing

I. INTRODUCTION

Space division multiplexing (SDM) which relies on transmitting optical signals using multiple cores or multiple modes in a fiber has drawn immense interests lately as a potential means for enhancing the capacity of optical transmission systems. To compensate for losses incurred in SDM transmission links, various forms of amplifiers have already been developed, which includes few-mode [1-4], bundled, multi-element and multicore [5-12]. In few-mode EDFA the core is designed to support a few modes that could be simultaneously amplified by using suitable pump source(s). In multicore fiber amplifier, cores are embedded in a common glass cladding. EDFAs based on multicore are particularly attractive due to its simplicity, compactness, lower cost, and reduced power consumption. Depending on how amplifiers are being pumped, MC-EDFAs can be mainly divided into two different categories, which are core- [5-7] and cladding-pumped [8-12].

In this paper, we report on the recent development of MC-EDFAs. We will present the amplification, noise properties of MC-EDFAs that can be pumped through the cores or through the cladding. We further show using numerical simulation a novel design of a few-mode multicore fiber that enables uniformly amplifying various modes supported by each cores of the fiber.

II. CORE-PUMPED 7-CORE EDFA

The architecture of the core-pumped multicore fiber amplifier that uses a tapered fiber bundled (TFB) couplers as fan-in and fan-out devices is shown in Fig. 1(a) [5]. Here SDM signal from passive MCF is split into individual channels using a TFB coupler, and combined with pump radiation using WDM couplers, and then launched into MC-EDF using another TFB coupler. TFB couplers were created by tapering adiabatically a bundle of specially designed fibers by a predetermined ratio so that the core-to-core pitch at the tapered end matches with that of the MC-EDF.

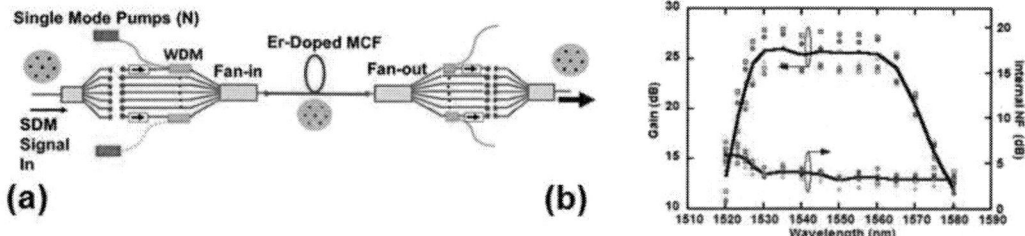

Fig. 1. (a) Structure of a core-pumped-MC- EDFA. (b) Net gain and internal NF measured for the 7 different cores of the MC-EDFA. The length of the gain fiber was 15 m. Solid line represents average.

Figure 1(b) shows net gain and internal NF measured in the C-band for the seven cores. For an input signal power of -15 dBm and a pump power of 146 mW (at 980nm), an average net gain of 25 dB was obtained from the amplifier. From the passive losses (maximum of 5 dB) in the TFBs and splices, the internal gain was estimated to be 30 dB and the internal NF over the whole C-band was found to be ~4 dB.

III. Cladding Pumped 7-core EDFA

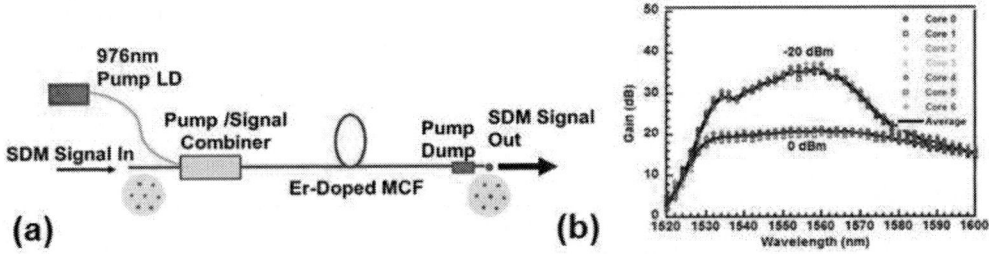

Fig. 2. (a) Schematic diagram of a 7-core EDFA with side-coupled cladding pumping. The length of MC-EDFA was 34 m. (b) Gain measured for the 7 different cores of the MC-EDFA, for input signal powers of 0 and -20dBm. The launched pump power was 4.7 W. The central core is represented by core 0.

The schematic of a cladding pump 7-core fiber amplifier where multimode pump is coupled from the side of the multicore gain fiber is shown in Fig 2(a) [9]. Here, signal can be launched through the MC input end of the pump signal combiner, and multimode pump is incident through the multimode pump port. To extend the gain spectrum to C-band, a relatively short length (34m) of erbium-doped fiber was used. Figure 2(b) shows the gross gain versus wavelength measured in the seven cores for input signal powers of -20 and 0 dBm. The maximum gross gain was about 36 dB near 1560 nm, and gain over 25 dB was obtained over a bandwidth of ~40 nm. For input signal powers of 0 dBm, the NF was ~8 dB and 5 dB for the signal wavelength of 1530 nm and 1560nm, respectively.

IV. Cladding-Pumped Multicore Amplifiers Supporting Few Modes: Design

Rare-earth doped fibers that are used in single mode optical amplifiers have a core with step-like refractive-index profile and a rare-earth doped region that has a radius the same as the core or slightly smaller. When the core size is made larger in order to support and amplify higher order modes, the overlap integral of modal intensity profile over the rare-earth doped region becomes different for different modes. For the higher order modes, a greater fraction of light remains outside the doped region which results in a smaller overlap integral and thus a smaller gain. To reduce the differential modal gain, a number or solutions have been proposed recently, which include, i) using reconfigurable few mode pump field [1,2], ii) inclusion of a ring-like doped region [3].

Here, we have investigated few mode multicore fiber amplifier with extended doped region pumped through a highly multimoded cladding that can inherently provide equal gain to all the modes supported by the core [11]. The proposed few mode fiber has a core supporting few higher-order modes and a rare-earth doped region, which consists of the whole core region and a region that sufficiently extends beyond the core in order to encompass the optical field of the few-order modes supported by the core. When such a fiber is pumped through the cladding, a uniform population inversion can be achieved across the doped region, achieving equal gain for different signal modes. The fiber design has the advantage that the differential modal gain can be minimized among a larger number of modes, and over a broader range of input pump power.

A schematic diagram of the structure of the core, rare-earth doped region, and the pump region of the proposed few-mode rare earth fiber is shown in Fig. 3. The core has a radius a_1 which is large enough to support four fiber modes, e.g. LP_{01}, LP_{11}, LP_{21}, and LP_{02}. The region doped with rare-earth ions has a radius a_2, sufficiently large to fully encompass the field of each of the modes LP_{mn}. The rare earth doped region is uniformly pumped through an inner cladding which is larger than the size of the doped region, i.e. $a_3 \geq a_2$.

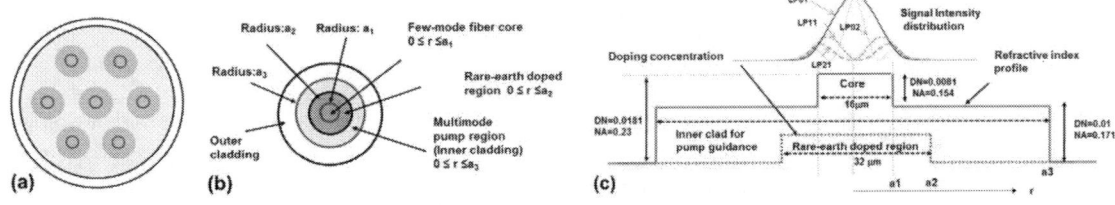

Fig. 3. Schematic diagram of the few-moded doped fiber to equalize gain. (a) with multiple cores, (b) single core. (c) the refractive index profile and doping distribution of the few-moded gain fiber

Figure 4(a) shows the gain and NF of four modes LP_{01}, LP_{11}, LP_{21}, and LP_{02} numerically calculated [11] for different radii of the doped region varied between 8 to 16 μm. The length of the gain fiber is 6 m. The input signal power is -20dBm (at 1550 nm) and the pump power is held constant at 1.24 mW/μm². When the core rare earth doped region has a size of 8μm, i.e. same as the core, the differential small-signal modal gain between LP_{01} and LP_{02} is around

3 dB and decreases with an increase in the size of the rare earth doped region. The differential gain can be kept below 1dB when radius of rare earth doped region is increased, anywhere in the range from 10 to 16 μm, i.e. 25% to 100% larger than the core size. Differential modal gain reaches a minimum value of 0.2 dB when a_2 is ~11.5 μm. The NFs for the signal modes are around 3.5 dB. Fig. 4(b) shows the gain and noise figure plotted as a function of input signal power for different modes for signal wavelength of 1530nm. The small signal gain was 33 dB and the NF was less than 3.8 dB for input signal with powers up to 1 mW.

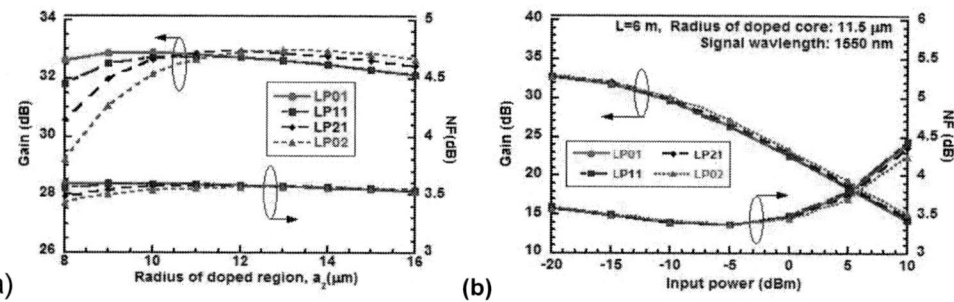

Fig. 4. (a) Gain for different signal modes calculated as a function of radius of rare-earth doped region. Core radius is 8 μm (V=5.0). (b) Gain and NF of versus input signal power. Signal wavelength is 1550 in. The fiber is assumed to have uniform erbium doping with a concentration of $6.89 \times 10^{24}/m^3$.

A pump intensity of 1.24 mw/μm² corresponds to a pump power of 1 W when the inner cladding radius is chosen to be 16 μm(a_3) for a single core fiber amplifier. While making multicore fiber using this core structures, the cladding size needs to be increased accordingly so as to include multiple cores. Also, the rare earth-doped region which is outside of the respective few-moded core needs to be index-matched with the cladding. If we can incorporate 7 such cores in a cladding diameter of 110μm, a pump power of ~12W will be required to maintain the same amount of pump intensity.

V. CONCLUSION

We have reviewed recent development of core- and cladding-pumped multicore EDFAs for amplifying SDM signals. Core-pumping, while involving higher cost, has the advantage of allowing independent control of gain in each core by adjustment of pump power. Cladding pumping, on the other hand, requires fewer optical components, and has the potential to use low-cost, energy efficient multimode diodes. By employing side-coupled pumping technique, small-signal gain >20 dB could be achieved throughout the C-band. We have also shown that by extending the doped region beyond the core of few-moded doped fiber, and pumping with a multi-moded pump, we can realize few-mode MC-EDFA with minimal modal differential gain.

REFERENCES

[1] N. Bai, E. Ip, T. Wang, and G. Li, "Multimode fiber amplifier with tunable modal gain using a reconfigurable multimode pump," Opt. Express, vol. 19, no. 17, pp. 16601-16611, 2011.

[2] Y. Jung, S. Alam, Z. Li, A. Dhar, D. Giles, I. P. Giles, J. K. Sahu, F. Poletti, L. Grüner-Nielsen, and D. J. Richardson, "First demonstration and detailed characterization of a multimode amplifier for space division multiplexed transmission systems," Opt. Express, vol. 19, no. 26, pp. B952-B957, 2011.

[3] M. Salsi, D. Peyrot, G. Charlet, S. Bigo, R. Ryf, N. Fontaine, M. A. Mestre, S. Randel, X. Palou, C. Bolle, B. Guan, G. Le Cocq, L. Bigot, and Y. Quiquempois, "A Six-Mode Erbium-Doped Fiber Amplifier," in Proc. ECOC 2012, Amsterdam, The Netherlands, pp. 1-3, paper Th.3.A.6.

[4] E. L. Lim, Y. Jung, Q. Kang, T. C. May-Smith, N. H. L. Wong, R. Standish, F. Poletti, J. K. Sahu, S. Alam, and D. J. Richardson, "First Demonstration of Cladding Pumped Few-moded EDFA for Mode Division Multiplexed Transmission," in Optical Fiber Communication Conference, OSA Technical Digest (online) (Optical Society of America, 2014), paper M2J.2.

[5] K. S. Abedin, T. F. Taunay, M. Fishteyn, M. F. Yan, B. Zhu, J. M. Fini, E. M. Monberg, F.V. Dimarcello, and P. W. Wisk, "Amplification and noise properties of an erbium-doped multicore fiber amplifier," Opt. Exp. 19, 16715-16721 (2011).

[6] Y. Tsuchida, K. Maeda, K. Watanabe, T. Saito, S. Matsumoto, K. Aiso, Y. Mimura, and R. Sugizaki, "Simultaneous 7-Core Pumped Amplification in Multicore EDF through Fibre Based Fan-In/Out," in European Conference and Exhibition on Optical Communication, OSA Technical Digest (online) (Optical Society of America, 2012), paper Tu.4.F.2.

[7] J. Sakaguchi, W. Klaus, B, J. Puttnam, J. M. D. Mendinueta, Y. Awaji, N. Wada, Y. Tsuchida, K. Maeda, M. Tadakuma, K. Imamura, R. Sugizaki, T. Kobayashi, Y. Tottori, M. Watanabe, and R. V. Jensen, "19-core MCF transmission system using EDFA with shared core pumping coupled via free-space optics," Opt. Express 22, 90-95 (2014).

[8] K. S. Abedin, T. F. Taunay, M. Fishteyn, D. J. DiGiovanni, V.R. Supradeepa, J. M. Fini, M. F. Yan, B. Zhu, E. M. Monberg, and F.V. Dimarcello, "Cladding-pumped erbium-doped multicore fiber amplifier," Opt. Express 20, 20191-20200 (2012).

[9] Kazi Abedin, John Fini, Taunay Thierry, V. Supradeepa, Benyuan Zhu, Man Yan, Lalit Bansal, Eric Monberg, and David DiGiovanni, "Multicore Erbium Doped Fiber Amplifiers for Space Division Multiplexing Systems," J. Lightwave Technol. 32, 2800-2808 (2014).

[10] K. Maeda, Y. Tsuchida, S. Takasaka T. Saito, K. Watanabe, T. Sasa, R Sugizaki, K. Takeshima, T. Tsuritani, "Cladding Pumped Multicore EDFA with Output Power Over 20dBm Using a Fiber Based Pump Combiner," OECC2015, paper PWe.30.

[11] K. S. Abedin, M. F. Yan, J. M. Fini, T. F. Thierry, L. K. Bansal, B. Zhu, E. M. Monberg, and D. J. DiGiovanni, "Space division multiplexed multicore erbium-doped fiber amplifiers", Journal of Optics, 1-9 (2015).

[12] C. Jin, B. Huang, K. Shang, H. Chen, R. Ryf, R. Essiambre, N. Fontaine, G. Li, L. Wang, Y. Messaddeq, and S. LaRochelle, "Efficient annular cladding amplifier with six, three-mode cores," European Conf. Exhibition Optical Communication, Valencia, Spain, 2015, Paper PDP.2.1.

4-LP Mode Distributed Raman Amplification Technique with Graded-index Multi-mode Fiber Transmission Line

Masaki Wada, Taiji Sakamoto, Takayoshi Mori, Takashi Yamamoto, Kazuhide Nakajima

NTT Access Network Service Systems Laboratories, NTT Corporation 1-7-1, Hanabatake, Tsukuba, Ibaraki, 305-0805, Japan
wada.masaki@lab.ntt.co.jp

Abstract: *We demonstrate the first distributed-Raman amplifier supporting the 4-LP mode. A low-differential modal-gain of less than 1-dB is numerically and experimentally realized with a 5-dB on-off gain by using graded-index fiber and a single-LP_{21}-mode pump.*

Keywords: *Distributed Raman amplifier, Space division multiplexing, Graded-index fiber*

INTRODUCTION

The expected transmission capacity with conventional fiber is restricted to around 100 Tbit/s because of the incident power limitation caused by the fiber fuse phenomenon or fiber non-linear effect [1]. To deal with this capacity crunch, mode-division multiplexing (MDM) using few-mode fiber (FMF) has been intensively investigated [2, 3]. An MDM transmission system requires few-mode components such as a MUX/DEMUX [4] and an amplifier [5-8]. Specifically, a few-mode amplifier is essential if we are to realize a long-haul MDM transmission system. An important requirement for a few-mode amplifier is a reduction in the differential modal gain (DMG) to avoid power deviation among the signal modes. Studies of few-mode EDFAs have reported reductions in differential mode gain (DMG) [5, 6], and distributed Raman amplification (DRA) in FMF [7, 8]. In a recent DRA study, long-haul transmission with 2-LP modes was demonstrated experimentally [7], and low-DMG 4-LP mode Raman amplification using step index (SI) fiber has been investigated numerically using a simultaneous LP_{21} and LP_{02} mode pumping technique [8].

In this paper, we propose a low DMG 4-LP mode DRA technique using graded index (GI) FMF with a single LP-mode pump configuration. First, we calculate the modal gain characteristics needed to reduce DMG using a GI-FMF transmission line and confirm that single LP_{21} mode pumping can obtain a low DMG due to LP_{21} and LP_{02} mode degeneration for both the signal and pump lights. Then, we experimentally investigate the modal gain characteristics using a 71 km GI-FMF and realize a low DMG with a deviation of less than 1 dB using backward pumped light with a single LP_{21} mode.

4-LP MODE RAMAN AMPLIFICATION WITH GRADED-INDEX FEW-MODE FIBER TRANSMISSION LINE

The evolution of a signal S_m and backward pump P_n^- power along the longitudinal axis of the FMF in a DRA system can be described by

$$\frac{dS_m}{dz} = -\alpha_s S_m + \gamma_R \left(\sum_n f_{n,m} P_n^- \right) S_m \qquad (1)$$

$$\frac{dP_n^-}{dz} = \alpha_p P_n^- + \frac{\lambda_s}{\lambda_p} \gamma_R \left(\sum_n f_{n,m} S_m \right) P_n^- \qquad (2)$$

where α_s and α_p are the loss coefficients at signal and pump wavelengths, respectively, γ_R is related to Raman gain coefficient. The intensity overlap integral $f_{n,m}$ between signal and pump mode is defined as

$$f_{n,m} = \frac{\iint_{-\infty}^{+\infty} S_m(x,y) P_n(x,y) dx dy}{\iint_{-\infty}^{+\infty} S_m(x,y) dx dy \iint_{-\infty}^{+\infty} P_n(x,y) dx dy} \qquad (3)$$

The DMG value is controlled by the intensity overlap integral $f_{n,m}$, which can be adjusted by changing the pump launch mode or the core profile of the fiber. In a previous study [7], 4-LP mode Raman amplification using step index (SI) fiber was numerically investigated and a low DMG of 0.13 dB was realized by using LP_{21} and LP_{02} mode pump lights simultaneously, where the power ratio of the LP_{21} and LP_{02} modes was set at 7:3. Thus, Raman amplification with SI-FMF requires multiple pump modes to realize a low DMG, and precise control of the pump mode ratio. We propose 4-LP mode Raman amplification using GI-FMF to realize a low DMG value with a single LP-mode pump configuration. We assume that the LP_{21} and LP_{02} modes of the signal and pump lights in GI-FMF are one mode group because they have similar propagation constants and are strongly coupled along the longitudinal axis. We regard the LP_{21} and LP_{02} modes of the signal and pump lights as one mode group with the same DRA characteristics. Figure 1 shows the relative refractive index profile of a GI core with trench fiber. In our calculation we assumed the same core profile as described in [9]. The effective areas for propagated modes at a wavelength of 1550 nm are also shown Fig. 1 (a). We assume the pump and signal wavelengths to be 1450 and 1550 nm, respectively. We calculated the wavelength dependence of the differential propagation constant of the $LP_{21p,s}$ and $LP_{02p,s}$ modes ($\Delta\beta_{21-02}$) of SI-FMF and GI-FMF as shown in Fig. 1(b). Here the subscripts "s" and "p" represent signal and pump modes, respectively. We confirmed that GI-FMF realized a low $\Delta\beta_{21-02}$ of 50 rad/m over the pump and signal band compared with the value of 2500 rad/m obtained with SI-FMF. It is known that a low $\Delta\beta$ of less than 1000 rad/m leads to strong coupling between propagated modes [10]. We then calculated the modal gains in a Raman amplifier using a GI-FMF transmission line. First, we evaluated the gain characteristics obtained with the two kinds of calculations where we assumed negligible or strong coupling between the $LP_{21s,p}$-$LP_{02s,p}$ modes. Figure 2 (a) shows the modal gains as a function of the pump mode power ratio of the LP_{21p} and LP_{02p} modes using backward pumping without mode coupling. The DMG can be reduced to 0.4 dB when the mode ratio between the LP_{21p} and LP_{02p} modes for the pump light becomes 64:36. Figure 2 (b) shows the modal gains as a function of the pump mode power ratio of the LP_{11p} and LP_{21+02p} modes when we assume strong mode coupling between the $LP_{21,p}(LP_{21,s})$ and $LP_{02,p}(LP_{02,s})$ modes. The modal gains of the LP_{21s} and LP_{02s} modes have the same value. We believe that the field pattern of superposed modes between the LP_{21} and LP_{02} modes changes along the propagation direction when we launch the LP_{21} or LP_{02} mode into GI-FMF. However, in this calculation, we easily assumed that the modal profile of the $LP_{21s,p}$ +$LP_{02s,p}$ mode group consists of the average power profile of the two modes in the fiber. We revealed numerically that a low DMG of less than 0.3 dB is realized for an on-off gain of 5 dB with a single LP_{21p} mode as shown in Fig. 2 (b).

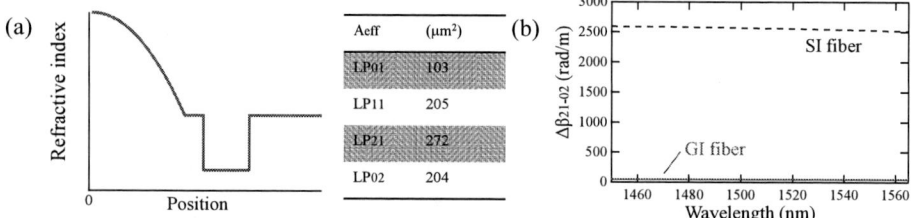

Fig.1 (a) Refractive index profile of the GI core with trench fiber used for the calculations and the effective areas of the propagated modes at a wavelength of 1550 nm, (b) the wavelength dependence of the $\Delta\beta_{21-02}$ of the GI-FMF.

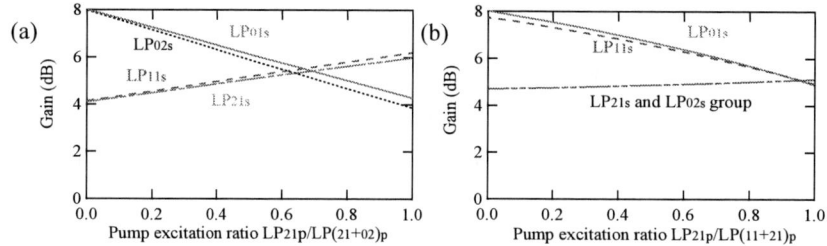

Fig.2 Modal gains as a function of the pump mode power ratio of the LP_{21p} and LP_{02p} modes using backward pumping (a) without and (b) with mode coupling.

EXPERIMENTAL SETUP AND RESULTS

We experimentally investigated the Raman amplification characteristic using a 71 km GI FMF transmission line (6-LP mode GI-FMF; 53 km + 4-LP mode GI-FMF; 18 km). Figure 3 shows the configuration of our experimental setup. We used a super-luminescent diode (SLD) source as a signal light operating at wavelengths over the C-band. The signal broadband light was connected to a multi-plane light type mode converter [4] and the mode-converted signal was injected into the GI FMF transmission line. Figure 1 also shows near field patterns (NFPs) after the mode converter obtained with our experimental setup, and we observed clear NFPs for the LP_{01s}, LP_{11s}, LP_{21s} and LP_{02s} modes. Figure 4 shows the gain spectrum obtained with the various pump mode configurations. We confirmed that the LP_{21s} and LP_{02s} modes had almost the same on-off gain whatever the pump conditions, so we believe that the LP_{21s} and LP_{02s} modes can be treated as the same mode group in this GI transmission fiber due to the strong coupling realized with a small $\Delta\beta_{02-21}$. The on-off gain spectrum with LP_{01p} mode pumping is shown in Fig. 4(a). The highest gain was obtained for the LP_{01s} mode, and the gains for the LP_{11s} mode and LP_{21s},-LP_{02s} mode group exhibited a lower value.

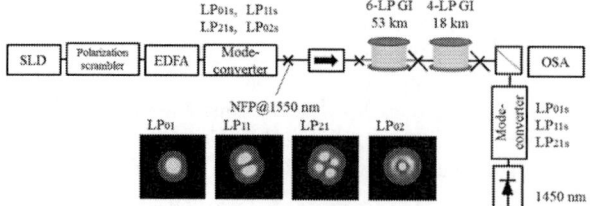

Fig.3 Configuration of our experimental setup and NFPs for the LP_{01s}, LP_{11s}, LP_{21s} and LP_{02s} modes after a mode converter

The measured DMG value at 1550 nm was 2.0 dB with an on-off gain of around 5 dB for the LP_{01s} mode when the launched pump power into the GI FMF was 750 mW. The DMG decreased to 1.3 dB when the LP_{11p} mode was used as shown in Fig. 4 (b). With LP_{21p} mode pumping, a lower DMG value of 0.8 dB was achieved at a wavelength of 1550 nm as we expected. Figure 4(d) shows the effect of pump mode variations in the DMG at 1550 nm between the LP_{01s} mode and higher order modes. The DMG values were reduced by using a higher order pump mode and we realized a DMG reduction in the 1.9 to 0.8 dB range with an on-off gain of 5 dB.

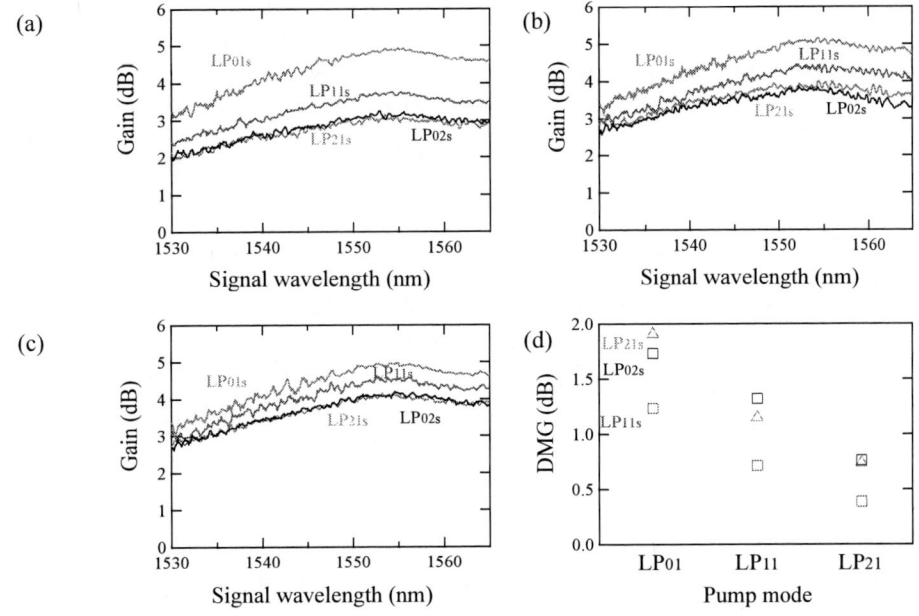

Fig. 4 Gain spectra of the various modes for the (a) LP_{01p}, (b) LP_{11p}, and (c) LP_{21p} modes and (d) the effect of pump mode variations

CONCLUSIONS

We first experimentally demonstrated 4-LP mode Raman amplification using a GI fiber transmission line. We evaluated the difference in the gain characteristics of 4 LP modes numerically and confirmed that a low DMG of less than 0.5 dB was realized with an on-off gain of 5 dB with a single LP_{21p} mode on the assumption of strong mode coupling between the LP_{21} and LP_{02} modes. Finally, we investigated modal gain characteristics experimentally using a 71 km GI-FMF and realized a low DMG with a deviation of 0.8 dB when the on-off gain was 5 dB using backward pumping with a single LP_{21p} mode.

REFERENCES

[1] R-J. Essiambre, G. Foschini, and P. Winzer and G. Kramer "Capacity Limits of Fiber-Optic Communication Systems," *OFC/NFOEC*, OThL1, 2009.

[2] R. Ryf, S. Randel, A. H. Gnauck, C. Bolle, A. Sierra, S. Mumtaz, M. Esmaeelpour, E. C. Burrows, R. Essiambre, P. J. Winzer, D. W. Peckham, A. H. McCurdy, and R. Lingle, "Mode-division multiplexing over 96 km of few-mode fiber using coherent 6 × 6 MIMO processing," *J. Lightw. Technol.*, vol. 30, no. 4, pp. 521–531, 2012.

[3] E. Ip, M. Li, K. Bennett, Y. Huang, A. Tanaka, A. Korolev, K. Koreshkov, W. Wood, E. Mateo, J. Hu, and Y. Yano, "146λ × 6 × 19-Gbaud Wavelength-and Mode-Division Multiplexed Transmission Over 10 × 50-km Spans of Few-Mode Fiber With a Gain-Equalized Few-Mode EDFA," *J. Lightwave Technol.* vol. 32, no. 4, pp. 790-797, 2014.

[4] G. Labroille, B. Denolle, P. Jian, P. Genevaux, N. Treps, and J-F. Morizur, "Efficient and mode selective spatial mode multiplexer based on multi-plane light conversion," *Opt. Express.*, 22, 15599, 2014.

[5] Y. Jung, Q. Kang, J. K. Sahu, B. Corbett, J. O'Callagham, F. Poletti, S. U. Alam, and D. Richardson, "Reconfigurable modal gain control of a few-mode EDFA supporting six spatial modes, " *Photon. Technol. Lett.*, vol. 26, pp. 1100-1103 2014.

[6] M. Wada, T. Sakamoto, T. Mori, T. Yamamoto, N. Hanzawa, and F. Yamamoto, "Modal gain controllable 2-LP-mode fiber amplifier using PLC type coupler and long-period grating," *J. Lightw. Technol.*, vol. 32, pp. 4092-4098 2014.

[7] R. Ryf, M. Esmaeelpour, N.K. Fontaine, H. Chen, A. H. Gnauck, R.-J. Essiambre, J. Toulouse, Y. Sun, and R. Lingle, Jr. "Distributed Raman Amplification based Transmission over 1050-km Few-Mode Fiber," *ECOC*, Tu.3.2.3, 2015.

[8] R. Ryf, R. Essiambre, J. Hoyningen-Huene, and P. Winzer, "Analysis of Mode-Dependent Gain in Raman Amplified Few-Mode Fiber," in Optical Fiber Communication Conference, OSA Technical Digest, paper OW1D.2. 2012.

[9] T. Mori, T. Sakamoto, M. Wada, T. Yamamoto, and F. Yamamoto, "Few-mode fibers supporting more than two LP modes for mode-division-multiplexed transmission with MIMO DSP," *J. Lightw. Technol.*, vol. 32, pp. 2468-2479 2014.

[10] T. Mori, T. Sakamoto, M. Wada, T. Yamamoto, and K. Nakajima, "Strongly-coupled Two-LP-mode Ring-core Fiber with Optimized Index Profile Considering S-bend Model," *OFC*, W1F. 6 2016.

MFD Measurement of a Six-Mode Fiber with Low-Coherence Digital Holography

Yuta Wakayama, Hidenori Taga, and Takehiro Tsuritani

KDDI R&D Laboratories Inc., 2-1-15 Ohara, Fujimino-shi, Saitama, 356-8502 Japan
yu-wakayama@kddilabs.jp

Abstract: This paper presents an application of low-coherence interferometry for measurement of mode field diameters (MFDs) of a few-mode fiber and shows its performance compared with another method using a mode multiplexer.

Keywords: Few-mode fiber, Mode field diameter, Low-coherence digital holography

I. INTRODUCTION

Mode-division multiplexed (MDM) transmission using few-mode fibers (FMFs) are considerably affected by modal crosstalk (MXT) and mode-dependent loss (MDL) [1]. The MXT and MDL are induced by mismatching of mode fields at splicing/contacting points of few-mode fibers/devices; therefore, the estimation of mode field mismatching is necessary to design and develop few-mode components and MDM systems [2,3]. Mismatching of mode fields can be evaluated based on mode-field diameters (MFDs) [4]. In general, the MFDs of existing modes in a FMF can be measured by individually exciting each mode into the FMF; however, the MFD cannot be measured correctly in case that MXT at a mode exciter is considerably large. In this paper, we show an application of low-coherence digital holography [4] as a technique for measurement of MFDs in a 6-mode fiber without mode exciters.

II. EXPERIMENTS

A. Setup with a mode multiplexer

Figure 1 shows an experimental setup for measurement of MFDs in a 6-mode fiber with a mode multiplexer. The mode multiplexer used in this evaluation is based on a multi-plane light conversion [5] and its MXT is lower than −15dB. The measured fiber was designed with the graded-index profile with the relative index difference of 1.1% and the core diameter of 17.6 μm so as to propagate four LP modes (six spatial modes). In this case, the design value of the MFDs for LP_{01}, LP_{11}, LP_{21}, and LP_{02} were 8.9 μm, 12.8 μm, 16.2 μm, and 16.5 μm, respectively. As a probe light, a broadband ASE source was filtered into Gaussian-shape spectrum at the center wavelength of 1550 nm with the full width at half maximum (FWHM) of 10 nm. The probe light was sequentially connected to the input ports of the mode multiplexer as exciting four LP modes individually. After the sample fiber, the probe light was collimated by an objective lens and its near-field pattern was captured by the near infrared camera constructed with the detector array of InGaAs. This camera has the dynamic range of 14bits, i.e., the pixel values were ranged from 0 to $2^{14}-1$. The pixel size and the number of pixels were 30×30 μm² and 320×256, respectively. The MFDs were evaluated from the near-field patterns of each mode with the determination of Petermann I [6]

$$d = 2\sqrt{\frac{2 \iint I(x,y)(x^2+y^2)dxdy}{\iint I(x,y)dxdy}}, \tag{1}$$

where $I(x,y)$ is the image of the near-field pattern emitted from the fiber under test, and x and y are a vertical direction and a horizontal direction on the captured images, respectively. In general, the image $I(x,y)$ includes background noise of the camera caused by dark current, bias noise, background light, and so on. In the evaluation of Eq. (1), an image data of the background noise was captured by shutting out the probe light, and it was subtracted from $I(x,y)$ for maximizing the peak signal-to-noise ratio (PSNR). A variable optical attenuator (VOA) was used for changing power of the probe light when evaluating the MFDs for different PSNRs.

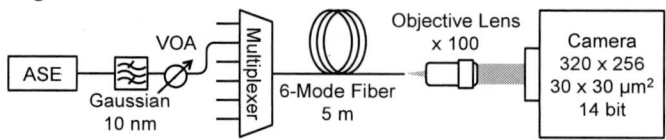

Fig. 1. Experimental setup for MFD measurement with a mode multiplexer.

B. Setup with low-coherence digital holography

In the MFD measurement with low-coherence digital holography, the probe light was generated in the same way with the setup using a mode multiplexer, as shown in Fig. 2. The filtered light was divided into a probe arm and a reference arm. The probe light and the reference light were entered into the 6-mode fiber and a standard single-mode fiber, respectively. On the probe arm, the emitted light from the fiber under test was collimated with the objective lens and separated into two polarizations by a calcite. On the reference arm, the length of the light path was shifted by a translation stage for scanning the temporal delay of each mode in the fiber under test. The step size of the translation stage and the bandwidth of the light source determine the resolution in the time region. Consequently, the near-field pattern can be evaluated by

$$|f(x,y;\tau)|^2 = \left| IFFT\left\{ FFT\left[\frac{I(x,y;\tau) - I_{\mathrm{p}}(x,y) - I_{\mathrm{r}}(x,y)}{\sqrt{I_{\mathrm{r}}(x,y)}\exp(-i\theta)} \right] W \right\} \right|^2, \qquad (2)$$

where τ is temporal delay corresponding to the relative position of the translation stage, θ is the angle between the probe light and the reference light on the camera, W is a circular window function for extraction of a mode field, $I(x,y;\tau)$ is a interference fringe, $I_{\mathrm{p}}(x,y)$ and $I_{\mathrm{r}}(x,y)$ are the intensity distribution of the probe light and the reference light, respectively. In this experiment, the MFD of each mode was also calculated using Eq. (1).

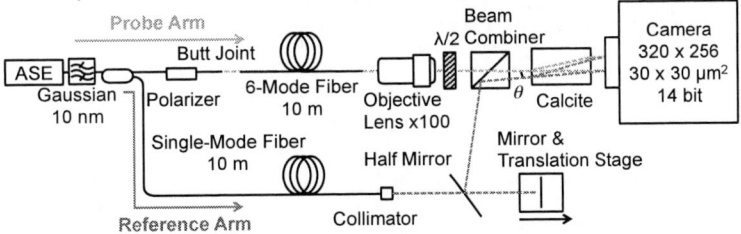

Fig. 2. Experimental setup for MFD measurement with low-coherence digital holography.

III. RESULTS AND DISCUSSION

A. *MFD Measurement with a mode multiplexer*

Firstly, we evaluate the MFDs with a mode multiplexer. Figure 3 shows near-field patterns and a background image, which were averaged over ten measurements with the setup in Fig. 1. The intensity distribution of each mode was mingled with their degenerate modes and shaped as a spot and/or a ring by using a low-coherence light source. Figure 4 presents the projected intensity (solid line) onto the horizontal axis and the intensity level of $1/e^2$ (dashed line). The average value of the background noise was ~200. This noise level deteriorates PSNR and causes measurement error because Eq. (1) includes integrals over whole images. Indeed, without subtraction of the noise image, the MFDs of LP_{01}, LP_{11}, LP_{21}, and LP_{02} were estimated from Eq. (1) as 30.8 μm, 26.0 μm, 23.6 μm, and 30.2 μm, respectively. These MFD values were obviously overestimated comparing to Fig. 4. In contrast, with subtraction of the noise image, the average noise level could be reduced from ~200 to ~3; thus, the MFDs were calculated as 10.5 μm, 12.7 μm, 15.5 μm, and 16.3 μm, respectively.

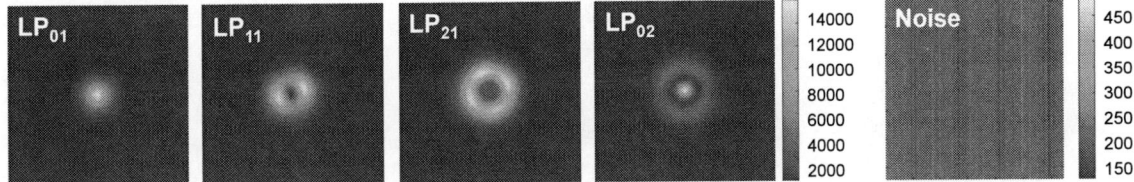

Fig. 3. Near-field patterns with a mode multiplexer and noise image when blocking out the probe light.

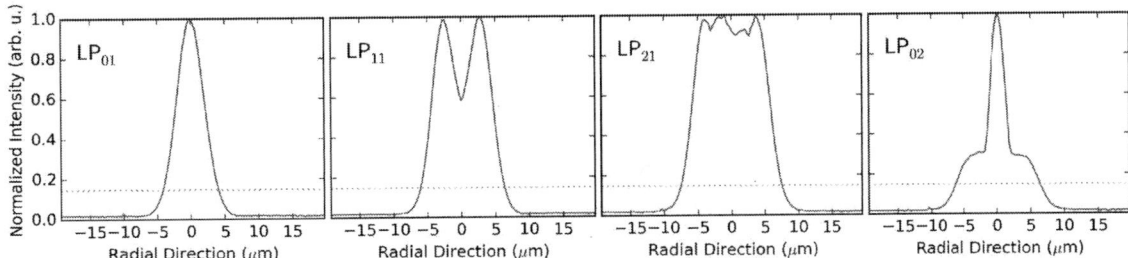

Fig. 4. Projections of intensity distributions on horizontal axis.

B. *MFD Measurement with low-coherence digital holography*

Next, we demonstrate the performance of the low-coherence digital holography for MFD measurement. Figure 5 shows captured images and an impulse response. The fringe image $I(x,y;\tau=0)$ was observed at the relative time τ of ±0.0 ps in the impulse response. The reference image $I_{\mathrm{r}}(x,y)$ was captured while blocking out the probe light. The image without fringe patterns $I_{\mathrm{p}}(x,y) + I_{\mathrm{r}}(x,y)$ was captured when the translation stage was significantly shifted from each peak in the impulse response. From these images, complex amplitude $f(x,y;\tau=0)$ of LP_{01} was clearly extracted using Eq. (2). In the time region, the FWHM of LP_{01} was 0.5 ps; thus, every mode of the 6-mode fiber could be recognized even without a mode multiplexer. In this experiment, the height of peaks on each of modes strongly depended on the condition on the input face of the fiber under test; therefore, the incident condition was adjusted while fixing the position of the translation stage on each mode. Figure 6 shows images obtained from Eq. (2). Comparing to Fig. 3, the noise level was dramatically suppressed thanks to the window function and intensity compensation by the probe and reference images. As the result, the MFDs of LP_{01}, LP_{11}, LP_{21}, and LP_{02} were obtained as 8.5 µm, 12.2 µm, 14.8 µm, and 15.0 µm, respectively.

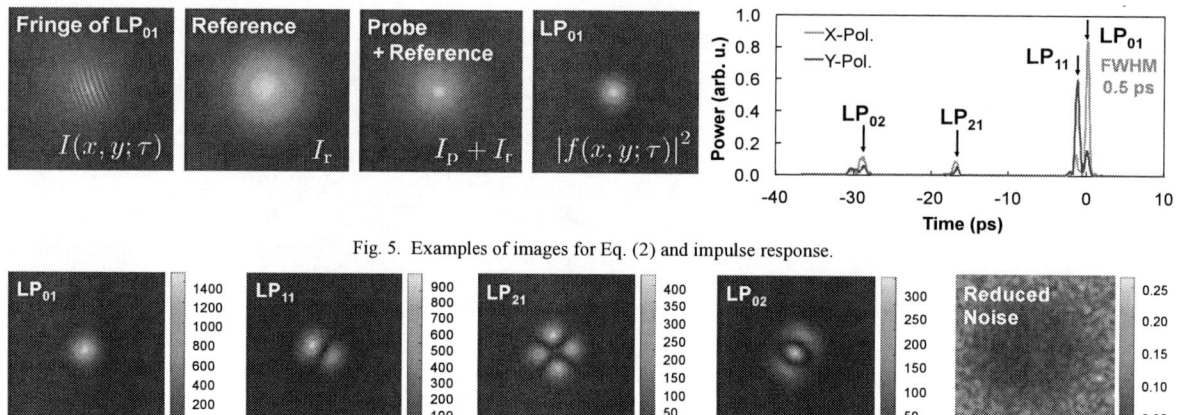

Fig. 5. Examples of images for Eq. (2) and impulse response.

Fig. 6. Near-field patterns for evaluation of MFD with low-coherence digital holography.

C.   Comparison of MFD measurements with a mode multiplexer and low-coherence digital holography

Finally, the aforementioned two methods for MFD measurements were compared based on PSNR, as shown in Fig. 7. The open symbols show the measured MFDs with the mode multiplexer while adjusting the incident power of probe light by VOA shown in Fig. 1. The MFDs measured with low-coherence digital holography are plotted as closed symbols. The maximum of PSNR was limited due to the dynamic range of the camera and the residual noise level after subtraction of the noise image. For the open symbols, the maximum of PSNR was 36.6dB, however, we found that it was improved up to 43.6dB for the closed symbols because the noise level was successfully reduced less than 1.0 as shown in Fig. 6. The four curves were numerically obtained from the designed values of MFDs based on Hermite-Gaussian functions [7]. These curves were in good agreement with both the open and closed symbols. This result indicates that MFDs in a FMF can be measured without a mode multiplexer using low-coherence digital holography.

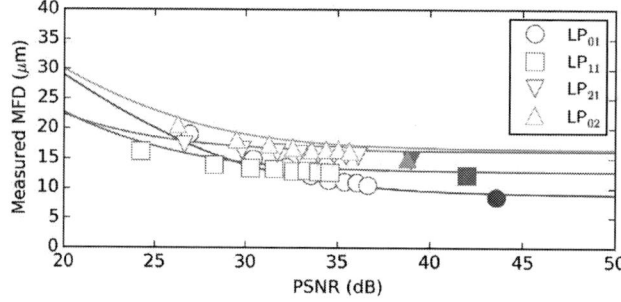

Fig. 7. Measured MFD vs. PSNR with a mode multiplexer (open symbols) and low-coherence digital holography (closed symbols).

IV. CONCLUSIONS

We measured the MFDs of 6-mode fibers with the proposed low-coherence interferometer and presented a performance comparison between the proposed method and a general method using a mode multiplexer. We found that the MFDs of the FMF could be measured by low-coherence digital holography even without a mode multiplexer.

A part of the research results have been achieved by the Commissioned Research of the National Institution of Communications Technology (NICT), Japan.

REFERENCES

[1] T. Sakamoto *et. al.*, JLT, **31**(13), 2192, 2013.
[2] R. Ryf *et. al.*, JLT, **30**(4), 521, 2012.
[3] D. M. Maron *et. al.*, OE, **23**(5), 5723, 2015.
[4] Y. Wakayama *et. al.*, ECOC, P.1.19, 2015.

[5] G. Labroille *et. al.*, OE, **22**(13), 15599, 2014.
[6] K. Petermann and R. Kühne, JLT **4**(1), 2, 1986.
[7] M. B. Shemirani *et. al.*, JLT, **27**(10), 1248, 2009.

Fabrication of Helical Fiber Grating and Its Application to Flat-top Band-Rejection Filter

Peng Wang, Gen Inoue, Ramanathan Subramanian, and Hongpu Li[*]

Faculty of Engineering, Shizuoka University, 3-5-1, Johoku, Hamamatsu, 432-8561, Japan

*ri.kofu@shizuoka.ac.jp

Abstract: *We propose and demonstrate a novel method to produce a flat-top band-rejection filter, which is realized by consecutively cascading two helical long-period fiber gratings (HLPG) with opposite helicities. Unlike most of the other LPG-based flat-top filters obtained to date, the proposed HLPGs have no complex apodization in the grating's amplitude and only a short length (less than 4.6 cm) is needed, which thus makes this kind of HLPGs easily and particularly suitable to be fabricated by using the CO_2 laser technique. As typical example, a flat-top filter with a bandwidth of ~13 nm@0.5dB and ~15 nm@1dB has been successfully obtained.*

Keywords: Filters; Gratings; Phase shift; Fiber optics components.

I. INTRODUCTION

Helical fiber grating refers to a fiber where there exists a periodical screw-type index-modulation along the fiber axis. In the past few decades, helical long-period fiber gratings (HLPG) have attracted a great research interest and found versatile applications, such as the lateral stress, temperature, and torque sensors, all-fiber band-rejection filter, mode-converter for micromanipulation, and conversion of orbital angular momentum beams, etc. [1-5]. On the other hand, it is well known that long-period fiber grating (LPG) can generally be used as a broad band-rejection filter with a bandwidth of several tens of nanometers. However, bandwidth and spectral profile of the resulted notch are extremely difficult to be controlled during its fabrication. Especially for a LPG with a broad flat-top rejection-band, which is strongly desirable in both the fiber communication and fiber laser system, it has rarely been realized due to some realistic limitations in fabrication. In this study, we propose and demonstrate a novel method for the fabrication of a HLPG with a flat-top band-rejection spectrum, which is realized by continuously cascading two helical LPGs but with opposite helicities. Unlike most of the previous approaches, the proposed LPGs does not require a complex apodization structure and the required length is shorter than several centimeters. It can be easily and repeatedly fabricated, and thus available for practical application.

II. EXPERIMENTAL RESULTS FOR THE SPECTRUM OF THE CASCADED HLPGS

The setup for fabrication of the HLPGs is shown in Fig. 1, in which there consists of a CO_2 laser, translation stages, a fiber rotation motor, and a testing system for measuring the transmission spectrum the fabricated HLPG. To fabricate HLPG, firstly the selected fiber (fixed at the clamp and the center of rotator) is homogenously heated to softly fused status through a sapphire tube, meanwhile the heated fiber is homogenously twisted through the rotation motor. In addition, a small weight (8g) is added on the left side of the fiber as shown in Fig. 1, which can provide a little longitudinal stress to the fiber and thus makes the fiber straight all the time during the fabrication process. Period of the HLPG is precisely controlled by the speeds of the fiber-moving stage and the rotator. Noted that in our experiment, the sapphire tube is fixed at a spatial site but the fiber is moveable, which can continuously move through the tube by driving the motored stage 3. Since the sapphire tube rather than the fiber is directly heated by the CO_2 laser, the passed fiber within the tube region can be homogeneously heated and twisted, which facilitates the fabrication of HLPG with almost 100% yielding-rate [7].

In order to realize a flat-top band-rejection filter, two HLPGs which are successively cascaded but with opposite helicities were utilized here as shown in the inset of Fig. 1. The detailed procedures are described below. At beginning, the first section of the SMF was heated and continuously twisted to the clockwise direction (observe in the direction from right to the left side) until the cHLPG was completely fabricated, which has a grating period of 648 μm and a length of 22.68 mm (35 periods). Transmission spectrum of this grating is shown in Fig. 2(a) as labeled by black solid line. It is seen that like the general LPGs, there exists a deep notch (loss band) in the transmission spectrum, and the loss depth and its peak-wavelength are ~26 dB and 1574.50 nm, respectively. Sooner after the cHLPG was produced, the neighboring part of the utilized SMF was then successively twisted to the counterclockwise direction until the fabrication of ccHLPG was completed, which has the same period and grating length as those of the first grating (i.e., cHLPG). Then we measured the transmission spectrum of these two cascaded HLPGs, which was shown in Fig. 2(a) labeled by the red solid line. Frank speaking, the result obtained here is totally beyond our expectations, a fine flat-top filter with a rejection-band of ~15 nm@0.5dB and ~20 nm@1dB has been successfully obtained. Fig. 2(b) shows a magnification part of the Fig. 2(a) at the central wavelength of 1574.50 nm region. It is obviously seen that there exists two small peaks (labeled with an arrow of LCP and RCP) in the flat-top band and central wavelength of the band almost lies in the same position as that of the cHLPG (i. e, black line shown in Fig. 2(a)). To make further clear the mechanism of the resulted band-rejection filter, we also investigated the polarization dependence of the transmission spectrum by adding a fiber polarizer (Thorlabs: IPP1550SM-FC) and a fiber polarization controller (PC) right after the ASE source, the measuring results are shown in Fig. 3, where two particularly polarization status for the input light: a left circular polarization (LCP) and a right circular polarization (RCP) are particularly selected. However, due to the polarization scrambling effect most probably existed in the connecting fiber (SMF) between the PC and the two HLPGs, 100% of LCP and RCP light cannot be obtained in our case, where a small portions of the light with a random-polarization (RP) status are inevitable. To compare Fig. 2(b) with the Fig. 3, it can be seen that the two peaks existed in the flat-top band (Fig. 2(b)) are strongly dependence on the polarization status of the incident light, which may be attributed to the two kinds of circular polarization light, respectively. In our case, due to the core-cladding eccentricity in the pristine fiber [5-7], it is believed that the newly circular birefringence may also be generated, which would split the degenerated HE_{11} mode (with effective index of n_{eff}) into two de-degenerate modes: one is the right circular polarization (RCP) mode with effective index n_{eff}^{RCP} and the other is the left circular polarization (LCP) mode with effective index n_{eff}^{LCP}.

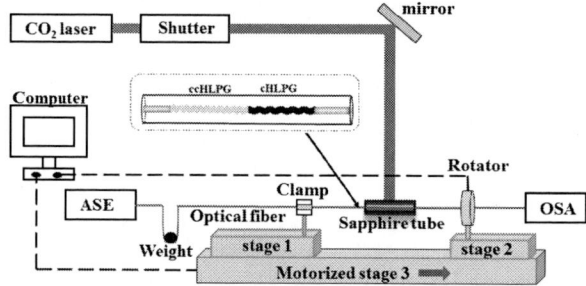

Fig. 1. Experimental setup for fabrication of the HLPG based on CO_2 laser.

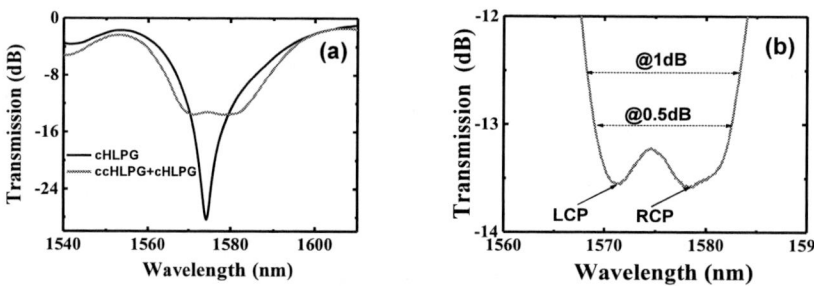

Fig. 2 (a) Transmission spectrum of the cascaded HLPGs, where the solid black line shows the spectrum of the first grating cHLPG and Fig. 2 (b) a magnification part of the Fig. 2(a) near the center wavelength region.

Then we measured the transmission spectrum of these two cascaded HLPGs, which was shown in Fig. 2(a) labeled by the red solid line. Frank speaking, the result obtained here is totally beyond our expectations, a fine flat-top filter with a rejection-band of ~15 nm@0.5dB and ~20 nm@1dB has been successfully obtained. Fig. 2(b) shows a magnification part of the Fig. 2(a) at the central wavelength of 1574.50 nm region. It is obviously seen that there exists two small peaks (labeled with an arrow of LCP and RCP) in the flat-top band and central wavelength of the band almost lies in the same position as that of the cHLPG (i. e, black line shown in Fig. 2(a)). To make further clear the mechanism of the resulted band-rejection filter, we also investigated the polarization dependence of the transmission spectrum by adding a fiber polarizer (Thorlabs: IPP1550SM-FC) and a fiber polarization controller (PC) right after the ASE source, the measuring results are shown in Fig. 3, where two particularly polarization status for the input light: a left circular polarization (LCP) and a right circular polarization (RCP) are particularly selected. However, due to the polarization scrambling effect most probably existed in the connecting fiber (SMF) between the PC and the two HLPGs, 100% of LCP and RCP light cannot be obtained in our case, where a small portions of the light with a random-polarization (RP) status are inevitable. To compare Fig. 2(b) with the Fig. 3, it can be seen that the two peaks existed in the flat-top band (Fig. 2(b)) are strongly dependence on the polarization status of the incident light, which may be attributed to the two kinds of circular polarization light, respectively. In our case, due to the core-cladding eccentricity in the pristine fiber [5-7], it is believed that the newly circular birefringence may also be generated, which would split the degenerated HE_{11}

mode (with effective index of n_{eff}) into two de-degenerate modes: one is the right circular polarization (RCP) mode with effective index n_{eff}^{RCP} and the other is the left circular polarization (LCP) mode with effective index n_{eff}^{LCP}.

To verify the above assumption, the circular birefringence (CB: $n_{eff}^{RCP} - n_{eff}^{RCP}$) of 5 HLPGs with different lengths were measured using a cut-back method. A tunable continuous DFB laser, two linear polarizer and a power detector were utilized to measure the change of the polarization direction after the grating. The measuring results at wavelength of 1560 nm are shown in Fig. 4, which shows the relationship between the grating length and the rotation angle of the polarization direction. As we expected, the output polarization state keeps linearly polarized but rotated by an angle θ due the HLPG-induced CB. Moreover it is easily seen that there exists a linear relationship between the grating length and the rotation angle, and the slope is ~1.11°/mm, which corresponds to a magnitude of ~$1.0*10^{-5}$ for the induced CB. We believe that the result we obtained above will pave many new applications of the HLPG in the near future.

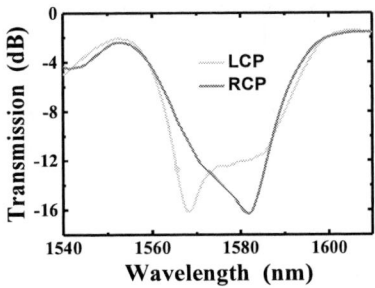

Fig. 3. Polarization properties for the transmission spectrum of the cascaded HLPGs.

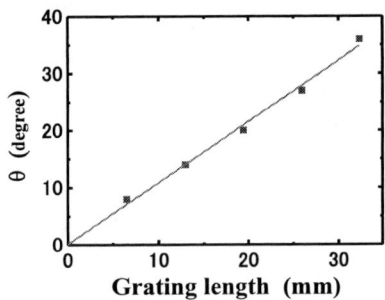

Fig. 4. The relationship between the grating length and the rotation angle.

III. CONCLUSIONS

A simple and robust method enabling to fabricate Helical LPG with a flat-top rejection-band has been proposed and demonstrated. Unlike most of the previous approaches, the proposed LPGs have a relative short length (less than 4.6 cm) and do not require a complex apodization in grating's amplitude, which makes this kind of HLPGs particularly suitable to be fabricated by using the robust CO_2 laser writing technique. As an example, a flat-top band-rejection filter with a bandwidth of ~13 nm@0.5dB and ~15 nm@1dB has been obtained, which is the broadest one reported to date, based on our knowledge, and may find further applications to the fields of fiber communication, fiber sensing, and all-optical information processing.

REFERENCES

[1] C. D. Pool, C. D. Townsend, and K. T. Nelson, "Helical-grating two-mode fiber spatial-mode coupler," J. Lightwave Technol. 9, pp. 598–604 (1991).
[2] V. I. Kopp, V. M. Churikov, J. Singer, N. Chao, D. Neugroschl, and A. Z. GenacK, "Chiral fiber gratings," Science 305, pp. 74-75 (2004).
[3] S. Oh, K. Lee, U. Paek, and Y. Chung, "Fabricatin of helical long-period fiber gratings by use of a CO_2 laser," Opt. Lett. 29, pp. 1464-1466 (2004).
[4] W. Shin, B. Yu, Y. Noh, Z. Lee, and D. Ko, "Bandwidth-tunable band-rejection filter based on helicoidal fiber grating pair grating of opposite helicitied," Opt. Lett. 32, pp. 1214-1216 (2007).
[5] O.V. Ivanov, "Fabrication of long-period fiber gratings by twisting a standard single-mode fiber," Opt. Lett. 30, pp. 3290-3292 (2005).
[6] G. Shvets, S. Trendafilov, V. I. Kopp, D. Neugroschl, and A. Z. Genack, "Polarization properties of chiral fiber gratings," J. Opt. A: Pure Appl. Opt. 11, pp. 074007-1-10 (2009).
[7] P. Wang and H. Li, "Helical long-period grating formed in a thinned fiber and its application to refractometric sensor," Appl. Optics 55, pp. 1430-1434 (2016).

ThC3-2

OECC/PS2016

Tri-section Side-polished Polarization Maintaining Fibre Polarizer

Xinyue Wang[1], Kaige Chen[1], Zhongwei Tan[2] & Ziyu Wang[2]*

[1]College of Information and Electrical Engineering, China Agricultural University, Beijing, China
[2]State Key Laboratory of Advanced Optical Communication Systems and Networks,
Department of Electronics, Peking University, Beijing, China
E-mail: *wangziyu@pku.edu.cn

Abstract: we propose a tri-section side-polished polarization maintaining fibre polarizer, and analyse the polarization mechanism theoretically. Experimental results show that the expectation of the output polarization extinction ratio of the polarizers is up to 32.45dB.

Keywords: fibre polarizer, polarization extinction ratio, polarization mechanism

I. INTRODUCTION

Optical polarizers have generated great interest in high-speed coherent fibre optic communications, fibre optic gyroscopes, interferometric sensors and quantum communications. Several types of optical polarizers with different polarization mechanism have been reported [1-4]. The in-line polarization maintaining fibre (PMF) polarizer, which is fabricated directly onto PMFs, with no interruptions to the optical path, is compatible with most fibre optic systems. Here we propose a tri-section side-polished PMF polarizer, which is based on metal-coated fibre polarizer and analyse the polarization mechanism theoretically. This tri-section polarizers offer a higher polarization extinction ratio (PER), lower insertion loss, and reduce production cost of the optical path. Experimental results show that the expectation of the output PER of tri-section side-polished PMF polarizers is up to 32.45 dB, and the success rate for making the polarizers with PER > 24 dB is 100 %. This tri-section side-polished PMF polarizer therefore has wide application in optical communication systems, optical sensor systems, etc.

II. THEORETICAL BASIS

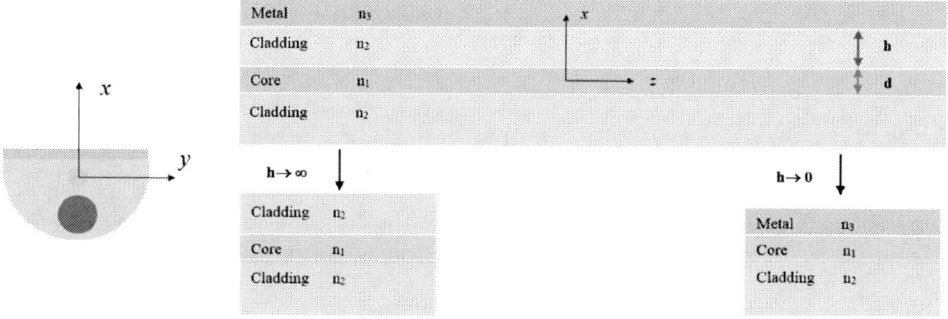

Fig. 1. Theoretical models of tri-section side-polished PMF polarizers

Fig. 1 shows theoretical models of side polished PMF polarizers, where d is the diameter of the fibre core, h is the distance between the core and metal, n_1 and n_2 are the refractive index of fibre cores and fibre cladding respectively, and n_3 is the complex refractive index of the metal. In this model, the guided wave in the multilayer metal-clad waveguide is assumed to propagate in the z direction, x is perpendicular to the metal plane, and y to be invariant. Suppose the electromagnetic field has the form $\varphi(x)\exp[j(\omega t - \beta z)]$, where φ is the major electromagnetic component, and the form of φ is determined by the solutions of wave equation, $\beta=\beta'+i\beta''$ is the complex propagation constant, β' is the phase constant, and β'' is the attenuation constant. The loss of optical intensity along the z direction can be obtained from β''. Electromagnetic field consists of transverse electric (TE) and transverse magnetic (TM) fields. For the TE mode, it consists of E_y component of the electric field and H_x, H_z component of the magnetic field. For the TM mode, only the components of E_x and E_z and H_y need to be considered. The mode characteristics are determined by applying the boundary conditions, and β is expressed as solutions of the eigenvalue equation[5-7].

This work was funded by State Key Laboratory of Advanced Optical Communication Systems and Networks, China, and the Fundamental Research Funds for the Central Universities.

$$\kappa_1 d = m\pi + \arctan\left(\gamma_{12} \cdot \frac{\alpha_2}{\kappa_1}\right) + \arctan\left(\gamma_{13} \cdot \frac{\alpha_3}{\kappa_1}\right) \tag{1}$$

$$\kappa_1 = (k_0^2 \varepsilon_1 - \beta^2)^{1/2} \tag{2}$$

$$\alpha_2 = (\beta^2 - k_0^2 \varepsilon_2)^{1/2} \tag{3}$$

$$\alpha_3 = (\beta^2 - k_0^2 \varepsilon_3)^{1/2} \tag{4}$$

Where $k_0 = 2\pi/\lambda$, and ε_1, ε_2, ε_3 are the dielectric constants of the three layers, m is the order of mode $m=0,1,2,3\ldots$. For the TE mode, $\gamma_{12} = \gamma_{13} = 1$. For the TM mode, $\gamma_{12} = \frac{\varepsilon_1}{\varepsilon_2}$ $\gamma_{13} = \frac{\varepsilon_1}{\varepsilon_3}$.

If the side polished surface is far from the fibre core (h→∞), as shown in Fig. 1, the theoretical models is simplify to a 3-layer symmetric planar optical waveguide consisting of a dielectric film sandwiched between two dielectric layers of lower refractive index. The propagation constant $\beta = \beta' + i\beta''$ is real ($\beta''=0$) for lossless dielectric, and satisfies $k_0 n_2 < \beta < k_0 n_1$. The parameters of the PMF we used are as follows: wavelength $\lambda=850$ nm, core diameter $d=5\mu m$, core refractive index $n_1=1.456$, cladding refractive index $n_2=1.446$. Therefore, for the PM fibre we used, the values of propagation constant for TE_0 and TM_0 mode are $\beta_{TE}=10752286$ and $\beta_{TM}=10752223$, respectively.

If the side polished surface reaches the fibre core (h=0), as shown in Fig. 1, the theoretical models is a 3-layer asymmetric metal/dielectric/dielectric layer structures. The propagation constant $\beta = \beta' + i\beta''$ is complex for metals, and the phase constant β' and the attenuation constant β'' of each mode are obtained by the numerical analysis of the eigenvalue equation concerning the complex propagation constant. The metal film of the PMF polarizer we used is aluminium (Al) with the refractive index of $n_3=2.08-7.15j@850nm[8]$. Therefore, the values of propagation constant of TE_0 mode is $\beta_{TE}=10749274-22j$, and the propagation constant of TM_0 mode is $\beta_{TM}=11015119-74046j$, respectively. The attenuation constant β''_{TM} of TM_0 mode is much higher than that of TE_0 mode. This means that TM modes suffer greater loss than TE modes with the same propagation distance. That is the polarization mechanism of metal-coated side-polished PMF polarizers.

III. EXPERIMENTS AND DISCUSSIONS

It is worth to point out that the theoretical analysis is based on multi-layer planar optical waveguide. In other words, the polished face is parallel to the core of the fibre in z direction, without any curve. Therefore conventional method to improve the output polarization extinction ratio of side-polished PMF polarizer is to increase the length of polished face, that means you have to increase the radius of curvature of the polish face or increase the polish depth of the fibre. In this paper, we propose a new method to increase the output PER of PMF polarizer. This method is illustrated by following the production steps as shown in Fig2. A PANDA fibre of 125 μm cladding diameter is stretched over the curved face of an aluminium substrate. The length, radius of curvature, and the width of the substrate are about 25 mm, 250 mm, and 2 mm, respectively. We use a power meter to monitor the optical intensity pass through the side-polished fibre during polishing. We first polish centre of the fibre and then polish the front end and back end of the fibre. The fibre's output power should be monitored carefully to ensure that the evanescent field has been reached and the core of the PMF is not penetrated. We finally deposited an aluminium film of 40 nm in thickness on a portion of the polished face [2, 3].

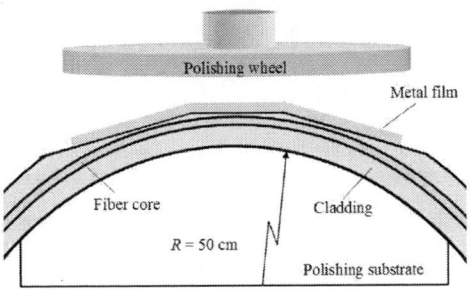

Fig. 2. Experimental setup used to fabricate tri-section side-polished PMF polarizers

Fig. 3. Shows the experimental results for the tri-section PMF polarizers. The low PER light of an SLED at a wavelength of 850 nm is launched into the fibre polarizers to be tested. A Glan-Taylor (G-T) prism, is placed at the output end of the fibre polarizer to evaluate the polarization dependence of the output power. The output light intensity from the polarizer is measured using an optical detector attached to an optical power meter. The G-T prism is rotated in steps of 10 degrees, and the output power recorded for each rotation of the prism. The polar plot of the normalized output light intensity of the tri-section polarizer of No. cau139, as a function of G-T prism angles in steps of 10° is shown in Fig. 3a. Fig. 3b. shows the output PER of 21 samples. Experimental results show that 76% tri-section polarizers offer output PER over 30 dB. The expectation of the output PER of tri-section polarizers is 32.45 dB and the success rate for making the polarizers with PER > 24 dB is 100 %.

OECC/PS2016

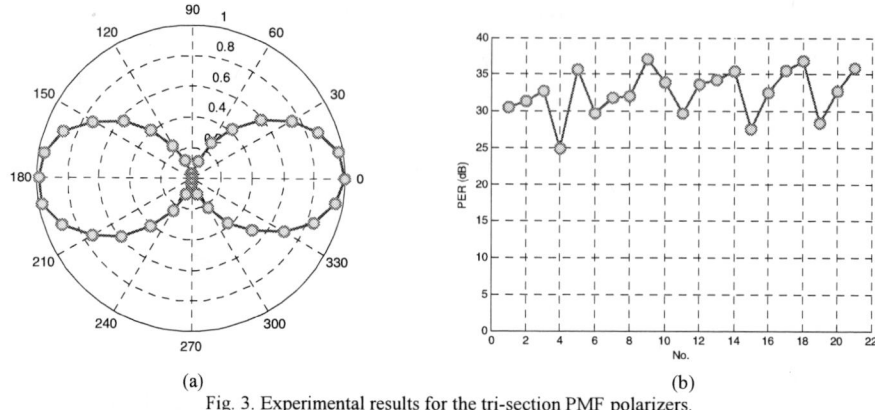

(a)　　　　　　　　　　(b)

Fig. 3. Experimental results for the tri-section PMF polarizers.

Fig. 4 shows the optical image of side-polished PMF polarizer viewed by Fujikura fibre Fusion Splicers FSM-45PM and LZM-100. Fig. 4a shows a cross-sectional image of a PANDA fibre, where the diameter of the fibre cladding and fibre core are 125μm and 5μm, respectively. Fig. 4b is the sideview of side-polished PMF polarizer. Fig. 4c and Fig. 4d are cross-sectional images of side-polished PMF polarizers, one with the side-polished face parallel to the fast axis as shown in Fig. 4c, another with the side-polished face parallel to the slow axis as shown Fig. 4d.

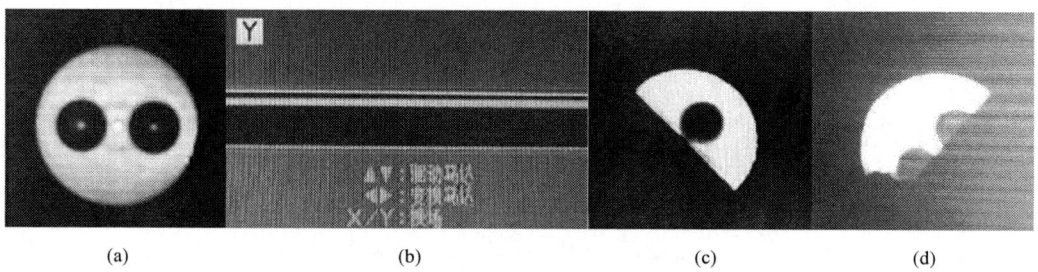

(a)　　　　　　　(b)　　　　　　　(c)　　　　　　　(d)

Fig. 4. Optical image of the tri-section side-polished PMF polarizer

IV. CONCLUSION

In summary, we report theoretical and experimental investigations into a tri-section side-polished PMF polarizer. Numerical calculations reveal that the polarization mechanism originates from the different attenuation of the TE and TM modes. Experimental results show that this tri-section polarizers offer a high output PER and a simple fabrication step. The expectation of the output PER of tri-section side-polished PMF polarizers is 32.45 dB and the success rate for making the PMF polarizers with PER > 24 dB is 100 %. Therefore, this tri-section side-polished PMF polarizers has wide application in optical sensor systems, optical communication systems, etc.

REFERENCES

[1] Crespi, A. et al., "Integrated photonic quantum gates for polarization qubits," Nature Communication, Vol.2, 2011, 566
[2] X. Wang and Z. Wang, "Self-aligning polarization strategy for making side polished polarization maintaining fiber devices" Opt. Express Vol.18, 2009, pp. 49-55
[3] X. Wang, C. Wu, & Z. Wang, "In-line fiber-optical polarizer with high extinction ratios and low insertion loss," Microwave Opt. Technol. Lett. Vol.51, 2009, pp. 1763-1765
[4] Q. Bao, et al. "Broadband graphene polarizer" Nature Photonics, Vol. 5, 2011 pp.411-415
[5] I. P. Kaminow, W. L. Mammel, and H. P. Weber, "Metal-Clad Optical Waveguides: Analytical and Experimental Study" Applied Optics, Vol. 13, No. 2, 1974, pp. 396-405
[6] G. Wu and Z. Wang, "Propagation characteristics of multi-coating D-shaped optical fibers," J Optic Pure Appl Optic, Vol. 8, 2006, pp. 450–453
[7] Li, G. Y. & Xu, A. S. "Analysis of the TE-pass or TM-pass metal-clad polarizer with a resonant buffer layer," J. Lightwave Technol. Vol.26, 2008, pp. 1234–1241
[8] M. A. Ordal, et al. "Optical properties of the metals Al, Co, Cu, Au, Fe, Pb, Ni, Pd, Pt, Ag, Ti, and W in the infrared and far infrared," Applied Optics, Vol. 22, No. 7, 1983, pp. 1099-1119

340

Ultrasonic Welding of Plastic Optical Fibers onto Composite Materials

Shumpei Shimada[1], Hiroki Tanaka[1], Kazuhiko Hasebe[1], Neisei Hayashi[2], Yutaka Ochi[3], Takahiro Matsui[3], Itaru Nishizaki[4], Yukihiro Matsumoto[5], Yosuke Tanaka[6], Hitoshi Nakamura[7], Yosuke Mizuno[1], and Kentaro Nakamura[1]

[1]Tokyo Institute of Technology, 4259-R2-26, Nagatsuta-cho, Midori-ku, Yokohama 226-8503, Japan
[2]The University of Tokyo, 4-6-1, Komaba, Meguro-ku, Tokyo 153-8904, Japan
[3]Toray Industries, Inc., 9-1, Oe-cho, Minato-ku, Nagoya 455-0024, Japan
[4]Public Works Research Institute, 1-6, Minamihara, Tsukuba, Ibaraki 305-8516, Japan
[5]Toyohashi University of Technology, 1-1, Hibarigaoka, Tempaku, Toyohashi, Aichi 441-8580, Japan
[6]Tokyo University of Agriculture and Technology, 2-24-16, Naka-cho, Koganei, Tokyo 184-8588, Japan
[7]Tokyo Metropolitan University, 1-1, Minami-osawa Hachioji-shi, Tokyo 192-0397, Japan
{shimada, ymizuno, knakamur}@sonic.pi.titech.ac.jp

Abstract: We demonstrate the ultrasonic welding of plastic optical fibers onto carbon-fiber-reinforced plastics for advanced sensing. The relationships among the welding time, preload, optical loss, and adhesive force are fully evaluated. High-speed monitoring is also performed.
Keywords: Plastic optical fibers, composite materials, ultrasonic welding, optical fiber sensing

I. INTRODUCTION

The aging degradation and seismic damage of steel structures have recently turned out to be a serious social problem. One of the methods for their efficient maintenance is to utilize carbon-fiber-reinforced plastics (CFRPs)—composite materials suitable for uses where a high strength-to-weight ratio and rigidity are required—as the reinforcing materials [1]–[6]. CFRPs can be fabricated in various ways; one low-cost technique is called vacuum-assisted resin transfer molding (VaRTM) [4],[5], where the resin is infused into dry fabrics formed on a mold under vacuum pressure. In the meantime, there is a considerable demand for health-monitoring techniques of the CFRP-reinforced parts [5],[6].

One of the most promising monitoring techniques is fiber-optic sensing with distributed strain and temperature measurement capability [7]–[10]. As glass optical fibers commonly used in such sensors are relatively easily damaged, we have been paying attention to the use of plastic optical fibers (POFs) [11]–[13] with much larger core diameters (up to ~1 mm) and extremely high flexibility. By using POFs, the practicality in the fields will be greatly enhanced with less risk of fiber breakage. Conventionally, adhesive materials (such as epoxy glue) were often used to fix the POFs on the surface of the CFRPs, but this method poses two problems: (1) the waiting time required for the hardening of the adhesive is not short (up to several minutes or longer), and (2) the deformation of the CFRP, which should be detected with the fiber sensor, is absorbed by the adhesive layer.

In this work, to mitigate these problems, we develop a new technique for fixing the POFs on the surface of CFRPs, which is based on so-called ultrasonic welding [14]–[17]. In experiment, the POFs are successfully welded onto the CFRP surfaces within a short time (in the order of seconds). The optical propagation loss of the POFs is measured as a function of welding time. The adhesive force is also investigated.

II. PRINCIPLE AND EXPERIMENTAL SETUP

Ultrasonic welding is one of the most popular methods for joining thermoplastics because of its high-speed low-cost operation and ease of automation [14]–[16]. To give a typical example, ultrasonic oscillation at several tens of kilohertz is applied to stacked sheets of thermoplastics under some preload. The boundaries of the plastics then generate frictional heat, leading to melting and welding [16]. This technique is suitable for POFs with relatively low glass-transition temperature (~100°C) [11] and not applicable to glass optical fibers.

In the experiment, we used POF samples with a core diameter of 0.98 mm, a cladding diameter of 1.0 mm, and an optical propagation loss of 0.25 dB/m at 650 nm wavelength, which were mainly composed of polymethyl methacrylate—a thermoplastic material. In contrast, the base material of the CFRP sheet samples (fabricated by VaRTM), 0.85 mm in

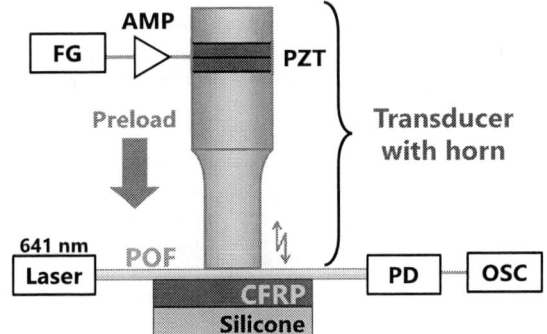

Fig. 1. Schematic of experimental setup. AMP: amplifier, CFRP: carbon-fiber-reinforced plastic, FG: function generator, OSC: oscilloscope, PD: photodiode, POF: plastic optical fiber, PZT: lead zirconate titanate.

thickness, was epoxy resin (known as a thermoset plastic) [18], and thus the CFRP samples themselves did not melt. However, melted POFs were so firmly welded to the rough surface of the CFRP samples that the POFs followed the deformation (strain and/or bending) of the CFRPs.

During ultrasonic welding of the POF, according to the change in the cross-sectional shape of its fiber core, additional optical loss is inevitably induced. Therefore, the transmitted light power was plotted as a function of the welding time (defined as the time after the ultrasonic oscillation started to be applied). The experimental setup is depicted in Fig. 1. A semiconductor laser at a wavelength of 641 nm and an output power of 1.5 mW was used as a light source. We employed a Langevin-type transducer [19] possessing a cylindrical horn with a flat tip (tip diameter: 30 mm; oscillation frequency: 29 kHz; oscillation amplitude: 31 μm), the position of which was fixed after the initial preload was set to a desired value (verified using a weight scale). A silicone rubber sheet was placed at the bottom of the CFRP sheet to absorb the ultrasonic oscillation. The transmitted light was guided to a photo detector and its electrical output was monitored using an oscilloscope; this configuration enabled the transmitted light power measurement on a real-time basis during the welding process. The change in the cross-sectional view of the POF during the welding process was simultaneously monitored using a high-speed camera at 1000 fps.

The POF-to-CFRP adhesion force was also evaluated. First, 100-mm-long POF samples were welded onto the CFRP sheets under different preloads (welded length: ~30 mm; welding time: 1 s). Then, as shown in Fig. 2, the welded POFs were nipped with tweezers (tip width: ~5 mm) connected to a force gauge, which was gradually lifted upward. In this evaluation, the POF-to-CFRP adhesion force was defined as the force at which part of the welded bottom of the POFs started to be peeled.

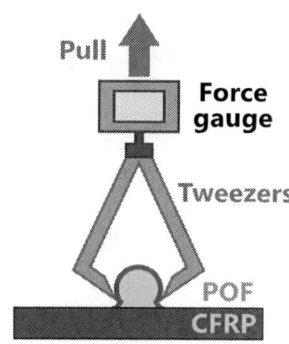

Fig. 2. Schematic setup for measuring the POF-to-CFRP adhesion force. CFRP: carbon-fiber-reinforced plastic, POF: plastic optical fiber.

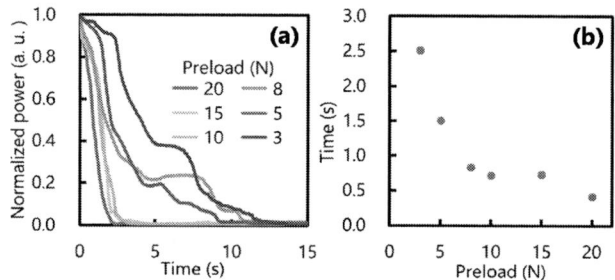

Fig. 3. (a) Temporal variations of normalized transmitted light power under different initial preloads. (b) Preload dependence of the time required until the normalized power decreases to 0.8.

III. EXPERIMENTAL RESULTS

A. Transmitted light power vs welding time

The transmitted light power dependence on the welding time is shown in Fig. 3(a). The initial preloads were varied from 3 up to 20 N. The vertical axis was normalized so that the optical power under preload before welding was 1. With increasing preload (resulting in larger deformation of the POF), the optical power tended to decrease rapidly. To quantitatively evaluate this behavior, we plotted the preload dependence of the welding time required for the normalized transmitted power to decrease to 0.8 (Fig. 3(b)). In this result, the welding time became drastically long when the preload was lower than ~8 N. Note that, in Fig. 3(a), the transmitted power ultimately decreased to 0 for all the preloads. By repetitive experiments, we found that this was either because the cross-section of the POF was completely crushed (for relatively high preloads) or because the POF was cut (for relatively low preloads). The latter can be explained as follows. As the melting of the POF advanced, a gap was generated between the horn and the POF; then, owing to the deflection of the POF, the ultrasonic oscillation was locally applied to a certain point of the POF. This mechanism also serves as the reason for the unstable

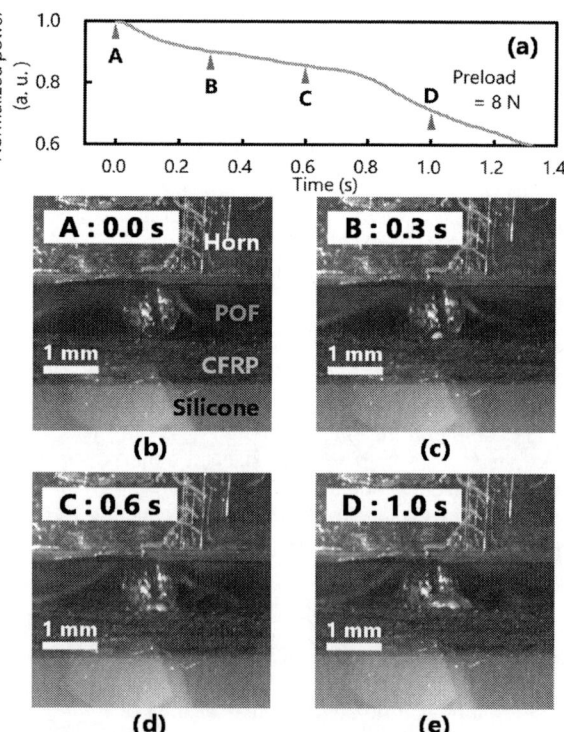

Fig. 4. (a) Temporal variation of normalized transmitted light power under an 8-N preload (magnified view of Fig. 3(a)). (b-e) High-speed micrographs of the ultrasonically welded POF at 0.0, 0.3, 0.6, and 1.0 s, respectively.

behavior observed at later than 5 s for relatively low preloads (see Fig. 3(a)).

B. High-speed monitoring of POF welding

Subsequently, we observed the change in the cross section of the POF during the ultrasonic welding onto the CFRP surface using a high-speed camera. The initial preload was set to 8 N. Figure 4(a) shows the normalized transmitted light power plotted as a function of the welding time t, and the photographs taken at the corresponding welding times (t = 0, 0.3, 0.6, and 1.0 s) are shown in Fig. 4(b)–(e), respectively. The melting of the POF was found to occur only on the surface that touched the CFRP surface. This is because the friction coefficient at the boundary between the POF and the rough surface of the CFRP was larger than that at the boundary between the POF and the smooth surface of the transducer horn, resulting in the generation of relatively significant frictional heat at the POF-to-CFRP boundary. This behavior is preferable from the standpoint of suppressing the undesired increase in optical loss.

C. Adhesive force of welded POF

Finally, the adhesive force of the welded POF was measured with respect to the preload (Fig. 5). For each preload, the same measurement was performed 5 times, and their average values were plotted as points with error bars of their standard deviations. For low preloads of < 5 N, the POF was not welded onto the CFRP surface. With increasing preload, the adhesive force was enhanced; this behavior is valid, considering that the higher preload induces larger deformation of the POF (at t = 1 s in this measurement), leading to a larger melted area and its more profound infiltration into the rough surface of the CFRP. Thus, a trade-off relationship between the optical loss and the adhesive force was experimentally verified. Note that the adhesive force can be significantly augmented by welding multiple points of a single POF. Continuous welding of a single POF in a longitudinal direction could be another method for obtaining a high adhesive force.

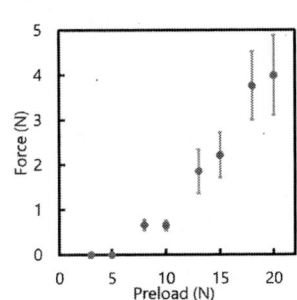

Fig. 5. Adhesive force plotted with respect to the preload (welding time t = 1 s).

IV. CONCLUSIONS

Exploiting the ultrasonic welding technique, we demonstrated a new technique for rapidly fixing POFs on CFRP surfaces. The time required for the POF welding was in the order of seconds, which was by far shorter than those of conventional methods. The optical propagation loss of the POFs and the POF-to-CFRP adhesive force were then quantitatively evaluated, and we found that there is a trade-off relationship between the two parameters. The adhesive force will be drastically enhanced either by the multipoint/continuous welding of the POF or by the use of special CFRPs, the base material of which comprises thermo-plastics. We anticipate that this technique can also be used to fix the geometry of POFs to be embedded in CFRPs, especially during the VaRTM process, and that this result will be a useful guideline in applying POF-based sensing technology to the health monitoring of composite materials.

ACKNOWLEDGMENTS

This work was supported by MLIT Construction Technology Research and Development Subsidy Program, by JSPS KAKENHI Grant Numbers 25709032, 26630180, and 25007652, and by research grants from the Iwatani Naoji Foundation, the SCAT Foundation, and the Konica Minolta Science and Technology Foundation.

REFERENCES

[1] N. Uddin, *Developments in Fiber-Reinforced Polymer (FRP) Composites for Civil Engineering* (Woodhead Publishing, Oxford, 2013).
[2] M. A. Masuelli, *Fiber Reinforced Polymers – The Technology Applied for Concrete Repair* (InTech, Croatia, 2013).
[3] C. R. Dandekar and Y. C. Shin, Int. J. Mach. Tools Manuf. **57**, 102 (2012).
[4] L.-Y. Lin, J.-H. Lee, C.-E. Hong, G.-H. Yoo, S. G. Advani, Comp. Sci. Technol. **66**, 2116 (2006).
[5] N. Takeda, Int. J. Fatigue **24**, 281 (2002).
[6] T. Hamouda, A-F. M. Seyam, and K. Peters, Comp. Part B: Eng. **78**, 79 (2015).
[7] T. Horiguchi and M. Tateda, J. Lightwave Technol. **7**, 1170 (1989).
[8] T. Kurashima, T. Horiguchi, H. Izumita, S. Furukawa, and Y. Koyamada, IEICE Trans. Commun. **E76-B**, 382 (1993).
[9] Y. Mizuno, W. Zou, Z. He, and K. Hotate, Opt. Express **16**, 12148 (2008).
[10] K. Hotate and T. Hasegawa, IEICE Trans. Electron. **E83-C**, 405 (2000).
[11] M. G. Kuzyk, *Polymer Fiber Optics: Materials, Physics, and Applications* (CRC Press, Boca Raton, 2006).
[12] Y. Mizuno and K. Nakamura, Appl. Phys. Lett. **97**, 021103 (2010).
[13] N. Hayashi, Y. Mizuno, and K. Nakamura, J. Lightwave Technol. **32**, 3397 (2014).
[14] J. Tsujino, M Hongoh, R. Tanaka, R. Onoguchi, and T. Ueoka, Ultrasonics **40**, 375 (2002).
[15] M. R. Rani and R. Rudramoorthy, Ultrasonics **53**, 763 (2013).
[16] A. Benatar, *Power Ultrasonics*, eds. J. A. Gallego-Juárez and K. F. Graff (Woodhead Publishing, Cambridge, 2015).
[17] Y. Mizuno, S. Ohara, N. Hayashi, and K. Nakamura, Electron. Lett. **50**, 1384 (2014).
[18] M. C. Lafarie-Frenot and F. Touchard, Comp. Sci. Technol. **52**, 417 (1994).
[19] C.-H. Yun, T. Ishii, K. Nakamura, S. Ueha, and K. Akashi, Jpn. J. Appl. Phys. **40**, 3773 (2001).

ThC3-4 (Invited)

OECC/PS2016

Ultra-low loss, ultra-large Aeff optical fibers for undersea networks

Sergey Ten

Corning Incorporated, One Riverfront Plaza, SP-DV-01-05, Corning, NY 14831

tens@corning.com

Abstract: Silica core optical fiber technology enabled a distinct class of ultra-low attenuation optical fibers (record is 0.14 0 d / m). arge effective area (≤1 0 μm²) silica core fibers became the solution of choice in the high capacity undersea communication systems.

Keywords: ptical communications, optical fiber design and fabrication, undersea communications

I INTRODUCTION

This year scientific community celebrates 50 years since Charles Kao published a seminal paper predicting that optical waveguides made with silica could achieve attenuation better than 20 dB/km [1]. In 1970, Donald Keck and his colleagues experimentally showed the first silica fiber with the attenuation below 20 dB/km [2]. Only 15 years later (See Fig.2), commercial single mode optical fibers achieved attenuation of <0.26 dB/km at 1550 nm [3] and in 1988 the first undersea optical fiber cable was deployed across Atlantic (TAT-8), enabling a spectacular growth of high capacity undersea systems (see Fig. 1) from single channel regenerated 1310 nm systems to 1550 nm dense wavelength division multiplexed (DWDM) amplified optical systems. Since the first deployment of optical fiber in the undersea cable the progress in transmission systems and improvement in optical fiber attributes have been mutually linked. Transmission system benefited from lower loss shifting from 1310 nm window to 1550 nm, but in order to minimized the impact of higher chromatic dispersion (CD) at 1550 nm dispersion shifted fibers were developed. The use of DWDM enabled by Erbium doped optical amplifier (ED   development of non-zero dispersion shifted fibers (NZ   ber solutions [4].

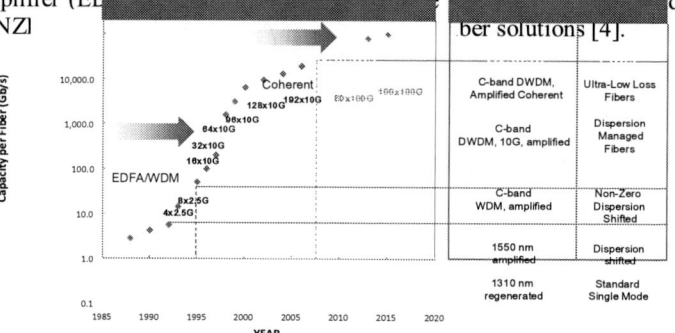

Fig. 1. Capacity (per optical fiber) of the installed undersea cables and the corresponding optical fiber solutions.

Recently, the adoption of coherent transponders set the roadmap for increasing spectral efficiency by using higher level M-QAM formats [5]. In addition, coherent receiver enabled CD and PMD compensation through digital signal processing, which led to the elimination of CD managements in the cable, and undersea links switched to cable with one fiber type (with positive CD). Fundamentally, increase in spectral efficiency requires higher optical signal to noise ratio (OSNR) that can be achieved by lower attenuation and lower nonlinearity that is inversely proportional to the fiber effective area (Aeff). This trend has led to the adoption of large Aeff ultra-low attenuation silica core fibers that, together with coherent transmission technology, become the solution of choice for transoceanic undersea systems [5].

II ULTRA-LO LOSS SILICA CORE OPTICAL FIBERS

Ge-doped standard single-mode fiber, i.e., optical fiber compliant with ITU-T G.652 is the most widespread fiber type with total deployed length exceeding 2 billion kilometers. Attenuation of typical Ge-doped silica G.652 fiber is shown in Fig.2b together with the main transmission bands used in modern optical networks. The total attenuation of an optical fiber is a sum of different attenuation components that have unique spectral dependence:

$$\alpha = \alpha_S + \alpha_T = \alpha_S + (\alpha + \alpha + \alpha_{TM} + \alpha + \alpha_M) \qquad (1)$$

The largest component is Rayleigh scattering α_{RS} that asymptotically scales with wavelength as $\sim 1/\lambda^4$ and becomes dominant at shorter wavelengths, e.g. at 1310 nm. The attenuation at longer wavelengths (>1650 nm) is dominated by infrared absorption $\alpha_{IR} \sim \exp(-a/\lambda)$. Additional (typically smaller) components are ultraviolet absorption α_{UV} in silica and extrinsic loss factors, such as absorption due to transition metals α_{TM}, absorption due to OH ions α_{OH}, and scattering due to waveguide imperfections α_{IM}.

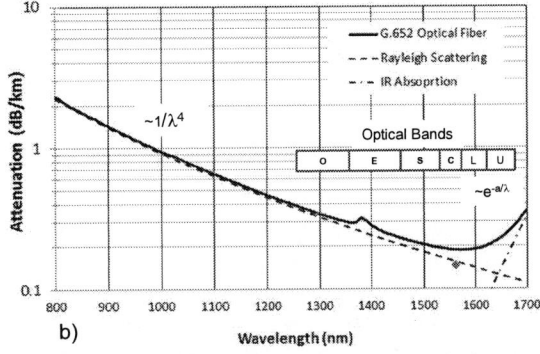

Year	Record Attenuation* (dB/km)	Attenuation of G.652 at 1550 nm (dB/km)
1970	20*	
1973	5**	
1976	0.47***	
1979	0.20	
1986	0.154	0.26
2001	0.152	0.20
2002	0.1495	0.20
2004		0.19
2007		0.18
2013	0.1480	0.17
2015	0.1460	

*In 1550nm–1600nm window unless wavelength is designated
* 632.8 nm, ** 850 nm *** 1200nm

a) b)

Fig. 2 a) Table showing notable attenuation results and attenuation of commercially available ITU-T G.652 optical fibers. b) Attenuation of typical commercially available G.652 optical fiber together with Rayleigh and IR Absorption components of attenuation. Red diamond shows the best achieved attenuation of 0.1460 dB/km [8]

For the intrinsic factors, the most important is the Rayleigh scattering loss α_S that can be expressed by the sum $\alpha_S = \alpha_\rho + \alpha_c$ of two contributions from density fluctuations α_ρ and concentration fluctuations α_c:

$$\alpha_\rho = \frac{8\pi^3}{3\lambda^4} n^8 p^2 \beta_T \quad T_f \qquad \text{and} \qquad \alpha_c \quad \frac{1}{\lambda^4}\left(\frac{\partial n}{\partial C}\right)^2 \langle \Delta C^2 \rangle T_f \qquad (2)$$

The density fluctuations in pure silica that cause Rayleigh scattering depend on the fictive temperature, T_f, which is determined as the temperature where the glass structure is the same as that of the super cooled liquid: where λ is the wavelength of incident light, p the photoelastic coefficient, n is the refractive index, k_B is the Boltzmann constant and β_T is the isothermal compressibility. In the presence of dopant, e.g. Germania of certain concentration ΔC, there is an additional attenuation component due to concentration fluctuations. Both components of Rayleigh scattering are caused by frozen-in density fluctuation and are proportional to the fictive temperature T_f. Concentration fluctuations are present in the Ge-doped fibers that are the most widespread fiber designs. The associated attenuation component is reduced by transitioning to Ge-free core fiber designs where the refractive index profile is defined by a silica core and a Fluorine-doped cladding (see Fig. 3a and 3b). In Ge-doped fibers most of the optical power is concentrated in the Ge-doped region with higher additional scattering due to concentration fluctuations (shaded area in Fig 3a) whereas in silica core optical fiber only the "tail" of an optical mode will experience the scattering due to concentration fluctuations in the F-doped silica (Fig. 3b), hence lower total scattering is achieved.

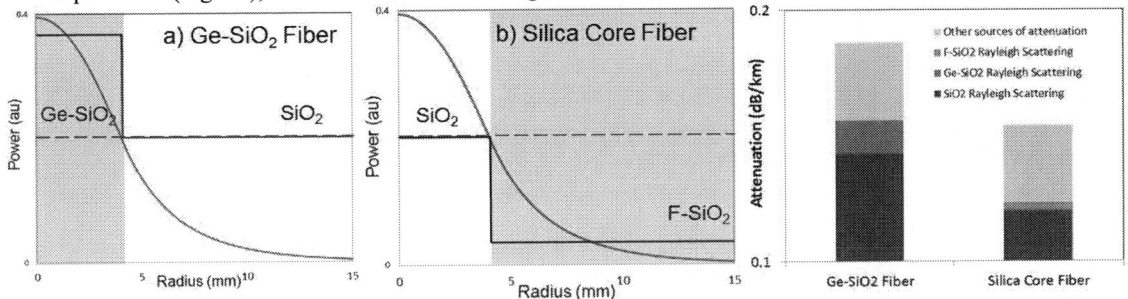

Fig 3. a) and b) Refractive index and power profiles of the LP_{01} mode in Ge-doped and pure silica core standard single mode fibers c) Rayleigh scattering components of attenuation in the respective fibers. Contribution of scattering in F-doped silica is lower due to the lower fraction of optical power localized in the cladding

The other path to lower attenuation consists of reducing fictive temperature, T_f. The most known method for this is viscosity modification to lower the glass viscosity that allows density fluctuations to dissipate more efficiently together with process improvement during the drawing process [6]. The contaminants due to transition metals can be practically eliminated in the fiber preform manufacturing processes by chemical vapor deposition techniques using raw materials with high chemical purity. Waveguide imperfection loss is caused by the geometry fluctuation at the core and cladding boundary. The boundary fluctuations are mainly due to the residual stress which is induced during the manufacturing process. The residual stress depends on the magnitude of the viscosity difference between the core and cladding and the fiber drawing tension. The stress can be reduced by matching the viscosity of the core and cladding [7].

As a result of all aforementioned improvements, silica core fiber have dominated the ultra-low attenuation record results since 1986 (See Fig. 2a), culminating in the current record attenuation of 0.1460 dB/km at 1560 nm [8].

Moreover, suppliers of undersea optical fibers routinely manufacture silica core fibers with the attenuation of <0.160 dB/km at 1550 nm, thus creating the distinct class of ultra-low loss silica core optical fibers that is also gaining adoption in the terrestrial networks [9].

III ULTRA LARGE Aeff SILICA CORE FIBERS

Further improvement in optical fiber performance can be achieved by lowering nonlinearity of optical fiber that is proportional to $\gamma \sim n_2/Aeff$. Compared to Ge-doped fibers, silica core fibers have lower nonlinear coefficient n_2 [10] that provides ≤10% improvement in γ for practical fiber designs. Larger improvements in γ can be achieved by increasing the Aeff. The first generation silica core fibers that were used in the unrepeatered undersea systems had an Aeff similar to that of G.652 fiber i.e. ~80 μm^2 at 1550 nm. However, the second generation silica core fiber increased the Aeff to ~110 μm^2. The two key challenges that optical fiber designers have to overcome to develop larger Aeff fibers are to insure that macrobend specifications (loss at the specified bend radius) are met and to minimize attenuation increase due to microbending. The latter is achieved by using optical fiber coating systems (primary and secondary) with "softer" primary coating i.e. lower modulus [11]. The former is done by careful profile optimization with respect to cable cutoff λ_{cc} and optical mode field diameter (MFD). It is known that increase in λ_{cc} improves macrobend losses, however, λ_{cc} cannot increase above the shortest wavelength of signal band (e.g. 1530 nm) due to the requirement to maintain single mode operation. This requirement is also captured in the ITU-T G.654 Recommendation with which most advanced undersea fibers comply. Today the latest generation of large Aeff silica core fibers achieves Aeff of 130-150 μm^2 [8,11].

In order to take the full advantage of large Aeff fiber, the splice loss between transmission fiber and repeater pigtails that typically have smaller Aeff (~80 um²) must be minimized. There has been a view that large Aeff fibers suffer from higher splice loss with pigtail fiber due to the abrupt transition from optical mode with small to large MFD [12]. Splice optimization enables adiabatic optical mode transition from one fiber to another (so called tapering technique) and the reduction of splice loss to 0.15-0.17 dB range has been demonstrated for the fibers with Aeffs of 150 μm^2 and 80 μm^2 [12]. Simulations show the feasibility of further splice loss reduction to 0.043 dB, by optimizing the tapers to deeper and asymmetric shapes [12].

In order to continue increasing Aeff, some of the existing requirements must be relaxed (e.g. macrobend specifications) or use advances in transmission technology to mitigate the negative impact from the induced transmission penalties. For example, a λ_{cc} increase above 1530 nm may result in multi path interference (MPI) that appears as an additional noise at the receiver, degrading the BER performance. The optical fibers with very large Aeff (180-220 μm^2) and λ_{cc} above 1530 nm are sometimes called quasi-single mode fibers (QSMFs) [13]. MPI in QSMFs is caused by the simultaneous propagation of the fundamental LP_{01} mode and higher order modes, e.g. LP_{11} modes, and coupling between them at the splices and during propagation in the optical fiber itself. MPI mitigation in receiver DSP was demonstrated using decision-directed least mean square algorithm [13] and by QSMF profile and span optimization [14]. Recently very high spectral efficiency of 8.3 bits/s/Hz was demonstrated at the cross-Atlantic distances of 6375 km without additional DSP mitigation [14], suggesting that QSMF could be a viable candidate for another step forward in Aeff.

IV CONCLUSIONS

The relatively short (45-year history) of silica-based optical fibers demonstrated unprecedented reduction in optical attenuation from 20 dB/km to 0.1460 dB/km. Rayleigh scattering is the biggest component of optical attenuation in the C-band used in the undersea communications systems and silica core optical fiber designs enable ultra-low attenuation levels by removing GeO₂ from the fiber core. In addition the nonlinearity reduction due increase in Aeff (up to 150 μm^2) created a distinct class of ultra-low loss, ultra-large Aeff silica core optical fibers that became the optical fiber solution of choice for the very high capacity undersea communication systems.

REFERENCES

[1] K. C. Kao and G. A. Hockham, "Dielectric-fiber surface waveguides for optical frequencies," Proc. Inst. Elect. Eng., 113, p.1151, 1966
[2] F. P. Kapron, D. B. Keck, R. D. Maurer, "Radiation losses in glass optical waveguides", Trunk Telecom. Guided Waves, IEE, p. 148, 1970
[3] Corning® Corguide® SMF™ specifications, 1984 and Corning® Corguide® SMF-28™ specifications
[4] S. Ten, "Advanced fibers for Submarine networks" Proc. Suboptic 2007
[5] Gabriel Charlet, "Utra high ca[acity transoceanic transmission" paper TH2A.4, Suboptic 2016
[6] K. Saito and A. J. Ikushima "Structural relaxation enhanced by Cl ions in silica glass" Applied Physics Letters v73, p.1209 (1998)
[7] M. Tateda, M. Ohashi, K. Tajima, and K. Shiraki, "Design of Viscosity-Matched Optical Fibers", IEEE Phot. Tech. Lett., v4, p. 1023, 1992
[8] S. Makovejs et al., "Record-low (0.1460 dB/km) attenuation ultra-large Aeff optical fiber for submarine applications," OFC 2015, Th5A.2.
[9] Huang Junhua et. al. "Small Size Ultra Low Attenuation 48-fiber OPGW for Long Repeater-Span Application" IWCS Conf, 2011, p.229
[10] K. Nakajima and M. Ohash "Dopant Dependence of Effective Nonlinear Refractive Index in GeO2- and F-Doped Core Single-Mode Fibers"
[11] S. Ohnuki et. al " Manufacturing of Aeff enlarge pure silica core fiber oultra low attenuation of 0.154 dB/km" SubOptic 2013, poster EC11
[12] S. Makovejs "Reduction in splice loss between fibers with dissimilar effective areas" SubOptic 2016, paper TU1A – 1
[13] Qi Sui, "256 Gb/s PM-16-QAM Quasi-Single-Mode Transmission over 2600 km using FMF with MPI Compensation" OFC 2014, M3C.5
[14] S. Zhang, et. al. "Capacity-Approaching Transmission over 6375 km at Spectral Efficiency of 8.3 bit/s/Hz" OFC 2016, paper Th5C.2.

High-Speed 1.55-µm VCSELs
for Datacom and Telecom Applications

Silvia Spiga, Alexander Andrejew,
Gerhard Boehm, and Markus-Christian Amann
Walter Schottky Institut, TU München, Am Coulombwall 4, D-85748 Garching, Germany
Silvia.Spiga@wsi.tum.de

Abstract: We present stationary and modulation characteristics of a short-cavity vertical-cavity surface-emitting laser (VCSEL) with 17 GHz bandwidth and emitting at 1.55 µm. The results of several bit-error rate experiments performed with comparable VCSELs and different modulation formats are compared. Net bit rate up to 87.5 Gb/s have been demonstrated with a single VCSEL, as well as transmission distances up to 960 km.
Keywords: (140.7260) Vertical cavity surface emitting lasers; (060.4080) Modulation.

I. INTRODUCTION

Directly-modulated VCSELs are particularly attractive light sources for high-speed applications due to their cost effectiveness, energy efficiency, and small footprint [1]. InP-based VCSELs emitting at long wavelengths such as 1.55 µm and 1.3 µm have gained large interest due to the low losses in silicon waveguides and silica optical fibers.

In this work, we review the bit-error rate experiments performed with 17-GHz bandwidth VCSELs over different transmission distances, and by using different modulation formats and receivers.

II. DESIGN AND STATIONARY CHARACTERISTICS

The VCSELs presented in this work implement a short-cavity design as presented in [2]. The optical gain is provided by eight 1.2-% compressively strained AlGaInAs quantum wells embedded between a n-doped InP layer and a highly p-doped AlInAs cladding. Current confinement is achieved by a p^+-AlGaInAs/n^+-GaInAs buried tunnel junction (BTJ), which is highly conducting in a circularly-shaped area with diameter (d_{BTJ}) of 5 µm, while current blocking outside this area is obtained by a reverse-biased p^+n-junction. The presented VCSEL employ a dielectric outcoupling distributed Bragg reflector (DBR) and a hybrid bottom DBR with reflectivity of 99.3% and 99.9%, respectively. An electroplated gold substrate provides electrical contact and works as an efficient heat sink. The contact pads are insulated by means of benzocyclobutene (BCB).

The power-current and voltage-current characteristics for temperature ranging from 20°C up to 90°C are shown in Fig. 1 (a) and (b), respectively. A maximum output power of 4.1 mW is reached at 20°C for 15.9-mA bias current. At a heat sink temperature of 90°C, the maximum output power exceeds 1.4 mW for a bias current of 10.5 mA. The threshold current ranges from 1.0 mA at 20°C to 2.3 mA at 90°C. The series resistance of the laser is 55 Ω and the turn-on voltage slightly decreases by increasing the heat-sink temperature.

The dissipated power and the wall-plug efficiency versus bias current are plotted in Fig. 1 (c). The dissipated power is slightly depending on the heat sink temperature and increases with the bias current. The highest wall-plug efficiency achieved is 25% for a bias current of 4.7 mA and 4.3-mW dissipated power at 20-°C heat sink temperature. For operation at 90°C, the maximum wall-plug efficiency achieved is 12% for 6-mA bias current.

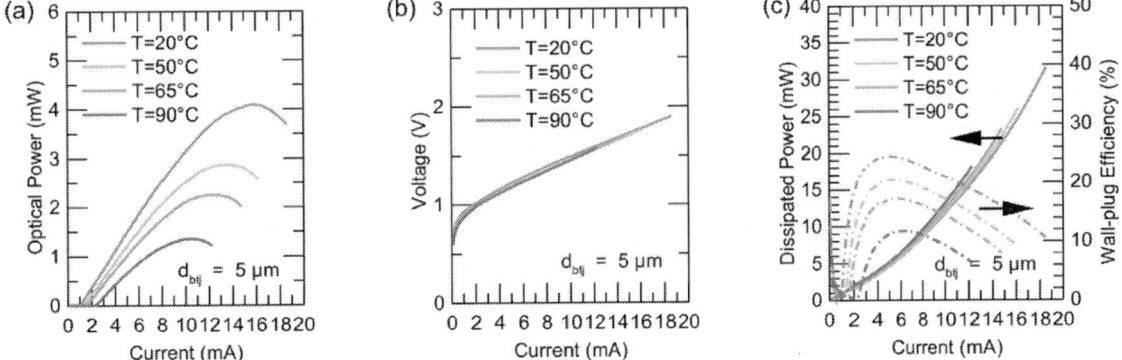

Fig. 1. For temperatures ranging from 20°C up to 90°C, (a) power-current characteristics, (b) voltage-current characteristics, (c) dissipated power and wall-plug efficiency of a VCSEL with tunnel junction diameter of 5 µm and 17 GHz bandwidth.

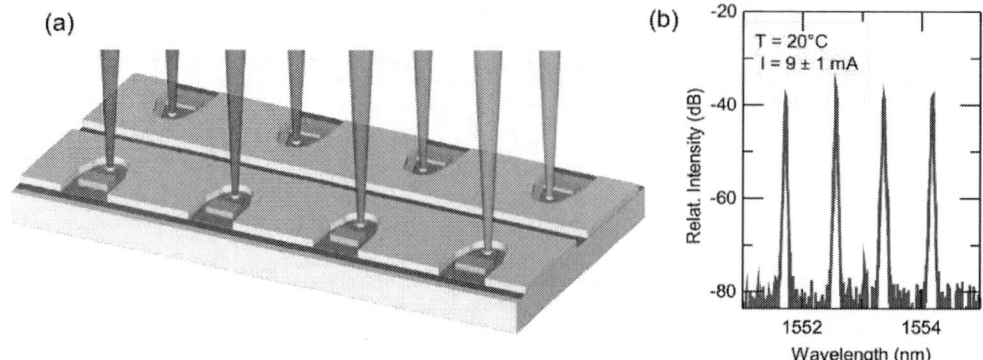

Fig. 2. (a) Schematic image of a 2x4 VCSEL array emitting at 4 different wavelengths. (b) Room-temperature optical spectrum of a 2x4 VCSEL array biased at around 9 mA.

In Fig. 2 (a), a schematic image of a 2x4 VCSEL array emitting at 4 different wavelengths is presented. VCSEL arrays provides an energy efficient and compact solution from wavelength division multiplexing (WDM) and polarization division multiplexing (PDM). The measured optical spectrum of the eight VCSELs is presented in Fig. 2 (b) where of each wavelength a couple of VCSELs emitting at the same wavelength have perpendicular polarization.

III. BIT-ERROR RATE EXPERIMENTS

The 3-dB small-signal bandwidth of the devices presented in Section II is around 17 GHz [2]. In this section, the result of bit-error rate experiments performed with comparable VCSELs and different modulation formats are compared in Table 1.

The need for low-latency and energy-effective optical links motivates the use of non-return-to-zero (NRZ) modulation. The highest bandwidth achieved with this modulation format in back-to-back (BTB) configuration is 56 Gb/s without forward-error-correction (FEC) or digital-signal-processing (DSP) [3]. In this experiment, a 0.13 µm BiCMOS driver with 2-tap feed-forward equalization (FFE) has been used.

Using direct detection and advanced modulation formats such as 4-level pulse-amplitude modulation (4-PAM), net bit rates in excess of 40 Gb/s are transmitted over 1 km of single-mode fiber (SMF). These results were achieved by means of a 0.13 µm SiGe BiCMOS driver with 2-tap FFE architecture [4].

TABLE 1
COMPARISON OF THE BIT-ERROR RATE EXPERIMENTS WITH A 17 GHZ VCSEL

Single VCSEL net bit rate (Gb/s)	SMF Length (km)	Modulation format	Transmitter	Detection	Reference
10	17.3	NRZ	DAC	Direct	[5]
25	4.2				[2]
35	BTB				[2]
40	1	4-PAM	0.13 µm SiGe BiCMOS 2-tap FFE driver	Direct	[4]
42	400	4-PAM	DAC	Choerent 20% SD-FEC, DSP	[6]
44	960*	3-PAM	DAC	Choerent 20% HD-FEC, DSP	[7]
50	2	NRZ	0.13 µm SiGe BiCMOS 2-tap FFE driver	Direct	[3]
56	BTB				[3]
79	4	DMT	72 GS/s DAC	Direct 20% HD-FEC, DSP	[8]
88	0.5				[8]
96	BTB				[8]

* EDFA+DGEF

Digital coherent detection at the receiver allows the recovery of the amplitude and phase of an optical carrier and the compensation by using DSP of chromatic dispersion (CD) and polarization-mode dispersion (PMD). The combination of digital coherent detection and 4-PAM and 3-PAM results in net bit rates in excess of 40 Gb/s and transmission distances typical of metro networks (several hundreds of kilometers). A digital to analog converter (DAC) was used to generate these multi-level PAM signals. The longest distance achieved with this system was 960 km where a 4x80 km Erbium-doped-fiber amplifier (EDFA) recirculating loops were alternated with a dynamic gain equalization filter (DGEF) to block amplified spontaneous emission, and with an EDFA [7].

In order to make full use of the bandwidth of the transmission link, discrete multi-tone (DMT) and direct detection is a suitable candidate for short-reach optical links. With this modulation format, 96-Gb/s net bit rate have been demonstrated in BTB configuration [8].

IV. WHAT'S NEXT?

In the past years, remarkable advances have been made in the development of high-speed 1.5-μm VCSELs, which have shown maximum modulation bandwidths in excess of 20 GHz by means of different approaches. A maximum small-signal bandwidth of 22 GHz was achieved by means of an ultra-short cavity design [9]. Furthermore, a combination of a highly-strained active region with a double-mesa and ultra-short cavity VCSEL design has been proposed to further enhance the bandwidth of 1.5-μm VCSELs beyond 25 GHz [10, 11].

V. CONCLUSIONS

The results of bit-error rate experiments performed with 1.5-μm 17-GHz short-cavity VCSELs are compared. The highest net bit rate transmission of 96 Gb/s was achieved by DMT modulation and direct detection, while the longest distance of 960 km was achieved by using 3-PAM modulation and coherent detection.

ACKNOWLEDGMENT

This work was supported by the European Commission through the FP7 project MIRAGE (ref.318228). The authors would like to thank Vertilas GmbH, Munich, Germany for discussions and technical support.

REFERENCES

[1] E. Kapon and A. Sirbu, "Long-wavelength VCSELs: Power-efficient answer," *Nature Photonics,* vol. 3, pp. 27-29, 2009.

[2] M. Müller, W. Hofmann, T. Grundl, M. Horn, P. Wolf, R. D. Nagel, *et al.*, "1550-nm High-Speed Short-Cavity VCSELs," *Selected Topics in Quantum Electronics, IEEE Journal of,* vol. 17, pp. 1158-1166, 2011.

[3] D. M. Kuchta, F. E. Doany, L. Schares, C. Neumeyr, A. Daly, K. B, *et al.*, "Error-free 56 Gb/s NRZ modulation of a 1530 nm VCSEL link," in *Optical Communication (ECOC), 2015 European Conference on,* 2015, pp. 1-3.

[4] W. Soenen, R. Vaernewyck, Y. Xin, S. Spiga, M. C. Amann, K. S. Kaur, *et al.*, "40 Gb/s PAM-4 Transmitter IC for Long-Wavelength VCSEL Links," *Photonics Technology Letters, IEEE,* vol. 27, pp. 344-347, 2015.

[5] M. C. Amann, E. Wong, and M. Müller, "Energy-efficient high-speed short-cavity VCSELs," presented at the Optical Fiber Communication Conference and Exposition (OFC/NFOEC), 2012 and the National Fiber Optic Engineers Conference, 2012.

[6] C. Xie, S. Spiga, P. Dong, P. Winzer, M. Bergmann, B. Kögel, *et al.*, "400-Gb/s PDM-4PAM WDM System Using a Monolithic 2x4 VCSEL Array and Coherent Detection," *Lightwave Technology, Journal of,* vol. 33, pp. 670-677, 2015.

[7] C. Xie, P. Dong, P. Winzer, C. Gréus, M. Ortsiefer, C. Neumeyr, *et al.*, "960-km SSMF transmission of 105.7-Gb/s PDM 3-PAM using directly modulated VCSELs and coherent detection," *Optics Express,* vol. 21, pp. 11585-11589, 2013/05/06 2013.

[8] C. Xie, P. Dong, S. Randel, D. Pilori, P. J. Winzer, S. Spiga, *et al.*, "Single-VCSEL 100-Gb/s Short-Reach System Using Discrete Multi-Tone Modulation and Direct Detection," in *Optical Fiber Communication Conference,* Los Angeles, California, 2015, p. Tu2H.2.

[9] S. Spiga, D. Schoke, A. Andrejew, M. Müller, G. Boehm, and M.-C. Amann, "Single-Mode 1.5-μm VCSELs with 22-GHz Small-Signal Bandwidth," in *Optical Fiber Communication Conference,* Anaheim, California, 2016, p. Tu3D.4.

[10] S. Spiga, D. Schoke, A. Andrejew, G. Boehm, and M.-C. Amann, "Enhancing the small-signal bandwidth of single-mode 1.5-μm VCSELs " in *Optical Interconnects Conference,* 2016.

[11] S. Spiga, A. Andrejew, G. Boehm, and M.-C. Amann, "Single-Mode 1.5-μm VCSELs with Small-Signal Bandwidth beyond 20 GHz " in *Transparent Optical Networks (ICTON), 2016 17th International Conference on,* 2016, pp. 1-4.

ThD1-2

OECC/PS2016

Equalizer-free 2-km SMF transmission of 106-Gbit/s 4-PAM signal using optical transmitter/receiver with 50 GHz bandwidth

Shigeru Kanazawa[1], Satoshi Tsunashima[1], Yasuhiko Nakanishi[1], Yoshifumi Muramoto[1], Hiroshi Yamazaki[2], Yuta Ueda[1], Wataru Kobayashi[2], Hiroyuki Ishii[2], and Hiroaki Sanjoh[1]

1. NTT Device Innovation Center, NTT Corporation, 3-1 Morinosato Wakamiya, Atsugi City, Kanagawa, Japan
2. NTT Device Technology Laboratories, NTT Corporation, 3-1 Morinosato Wakamiya, Atsugi City, Kanagawa, Japan
Kanazawa.shigeru@lab.ntt.co.jp

Abstract: *Using a fabricated optical transmitter and receiver with a 50-GHz bandwidth, we obtained a bit-error rate of less than 2 x 10^{-4} under error-free conditions using KP4 FEC without an equalizer.*
Keywords: *EADFB laser, electroabsorption modulator (EAM), flip-chip interconnection, 100 Gbit/s, Ethernet.*

I. INTRODUCTION

Recently, data-center traffic has been increasing explosively because of the rapid growth of cloud services. To cope with this trend, 400 gigabit Ethernet (400GbE) is currently being standardized by an IEEE task force [1]. 8 x 26.6-Gbaud 4-level pulse amplitude modulation (4-PAM) signals are sure to be used for 400GbE long-reach (such as 2- and 10-km) applications and optical transmitters operating at 50 Gbit/s have been reported [2-3]. However, there is a strong demand for a compact transceiver with low power consumption. And a smaller number of lanes is desirable if we are to reduce the power consumption and downsize the transceiver. Therefore, 100-Gbit/s/λ operation is needed.

With a conventional 4-PAM signal optical network, an equalizer is often used to compensate for limited optical transmitter and receiver bandwidth. If an equalizer-free network is realized, the power consumption and transceiver size can both be reduced. However, the conventional wire interconnection electroabsorption modulator integrated with a DFB laser (EADFB laser) module does not have a sufficiently high modulation bandwidth with a 3-dB bandwidth of more than 56 GHz and a flatter frequency response for equalizer-free transmission [4-5]. Therefore, in our previous work, using a high-frequency and integrated design based on a flip-chip interconnection technique (Hi-FIT), we fabricated a high-speed EADFB laser module that provides a high modulation bandwidth with a 3-dB bandwidth of over 56 GHz and a flat frequency response. And we observed clear eye opening without an equalizer for 56-Gbaud 4-PAM operation [6-7].

In this work, we modified the module design and fabricated a Hi-FIT EADFB laser module with a 3-dB bandwidth of more than 59 GHz. We also fabricated a high-speed optical receiver, which includes a maximized-induced-current photodiode (MIC-PD) [8] and a wide-bandwidth electrical amplifier. This receiver has a 3-dB bandwidth of more than 50 GHz. Using the fabricated optical modules, we observed a bit-error rate (BER) of less than 2 x 10^{-4} under error-free conditions using KP4 forward error correction (FEC) without an equalizer after a 2-km transmission through single-mode fiber (SMF).

II. OPTICAL TRANSMITTER DESIGN

Figure 1 shows the schematic structure of the Hi-FIT EADFB laser module. When the conventional wire interconnection technique is used, the E/O response degrades due to the parasitic inductance of the wire between the RF circuit board and the EA modulator (EAM). In contrast, with Hi-FIT, the flip-chip interconnection circuit board is connected to the EAM and the RF circuit board through Au bumps with a height of 30 μm, as shown in the figure. And the termination resistance is integrated with the flip-chip interconnection circuit board. Accordingly, the Hi-FIT requires no wire at all and provides a high modulation bandwidth and flat frequency response. In addition, in the fabricated module, the wire between the RF circuit board and the package was shorter than that in the previous one [7-8] because the assembly height was optimized to correspond with the top of the RF circuit board and the package circuit. Therefore, the length of these wires was suppressed to less than 100 μm, and the E/O response of this module was likely to be improved compared to that of the previous one. For this module, a V connector was used as an RF connector.

Figure 2 shows the E/O response of the fabricated module. The length of the EAM was 100 μm. The chip temperature was 25 °C, and the LD bias current was 100 mA. The EA bias voltage was -2.0 V. The 3-dB bandwidth exceeded 59 GHz, and the E/O response was sufficiently flat at less than 45 GHz. The measured peak wavelength of this module was 1304.7 nm at an LD current of 50 mA and a temperature of 25°C.

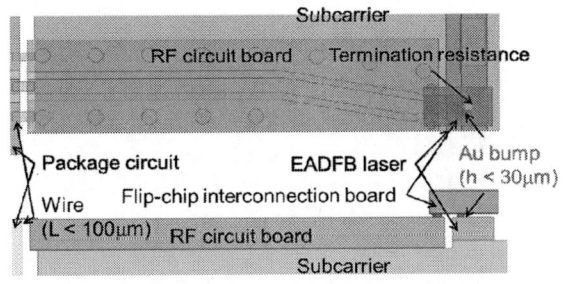

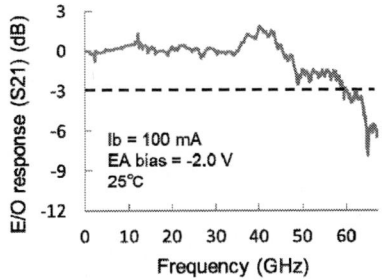

Fig. 1 Schematic structure of Hi-FIT EADFB laser module

Fig. 2 E/O response of Hi-FIT EADFB laser module

III. OPTICAL RECEIVER DESIGN

We fabricated an optical receiver that included the MIC-PD and a wide-bandwidth electrical amplifier as shown in Fig. 3. The MIC-PD had a higher modulation bandwidth and efficiency than a conventional pin photodiode (pin-PD) and the 3-dB bandwidth of the MIC-PD exceeded 50 GHz [8]. The responsivity of the MIC-PD was about 0.4 A/W. The electrical amplifier is commercially available and has a measured 3-dB bandwidth of more than 55 GHz and a measured gain of about 13 dB. Because these devices have a 3-dB bandwidth of more than 50 GHz, this fabricated module is promising receiver for equalizer-free transmission under 106-Gbit/s 4-PAM operation. A V connector was used as an RF connector.

Figure 3 shows the O/E response of this receiver. The optical input power was set at -10 dBm. The 3-dB bandwidth exceeded 50 GHz and the frequency response was flat.

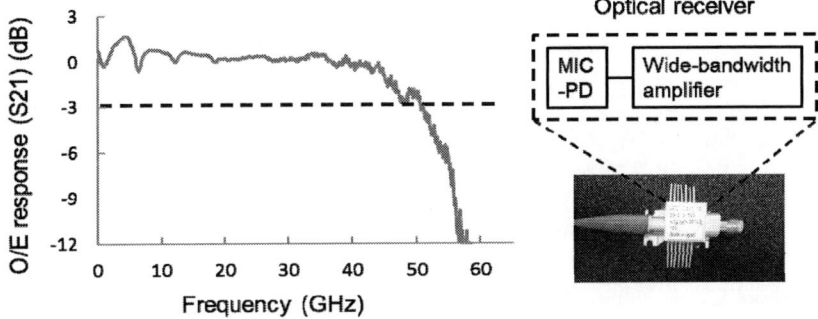

Fig. 3 O/E response and photograph of optical receiver

IV. TRANSMISSION EXPERIMENTS

We performed a 106-Gbit/s 4-PAM transmission experiment over a 2-km SMF using the fabricated high-speed optical transmitter and receiver. Figure 4 shows the measurement setup for the 106-Gbit/s transmission experiment. A 53.2-Gbaud 4-PAM, PRBS 2^7-1 signal was generated with a 3-bit digital-to-analog converter (DAC). And the modulation amplitudes of the levels were set at 0.7, 0.65, and 0.7 Vpp to compensate for the EAM's non-linearity. The LD current was 80 mA, and the EA bias voltage was -2.53 V. The chip temperature was set at 25 °C. The received electrical signal was sampled and digitized using a real-time storage oscilloscope with a sampling rate of 160 GSamples/s and a bandwidth of 62 GHz. And the digitized signal was demodulated by offline digital signal processing, in which only symbol-timing recovery, decision, and error counting were performed and no equalizer was used.

Figure 5 shows the eye diagrams of the optical and received electrical signal for a back-to-back (BtoB) configuration and after a 2-km SMF transmission. The modulation output power was +3.8 dBm and the outer extinction ratio was 7.3 dB at point A in Fig. 4. Clear eye openings were obtained without an equalizer even after a 2-km SMF transmission. With 106-Gbit/s 4-PAM operation, the BER characteristics were measured for a back-to-back configuration and a 2-km SMF transmission as shown in Fig. 6. For the 2-km SMF transmission, the BER was less than 2×10^{-4}, which is an error-free condition using KP4 FEC [1], at an average received power exceeding -1 dBm. And there is almost no power penalty compared with that for the back-to-back configuration.

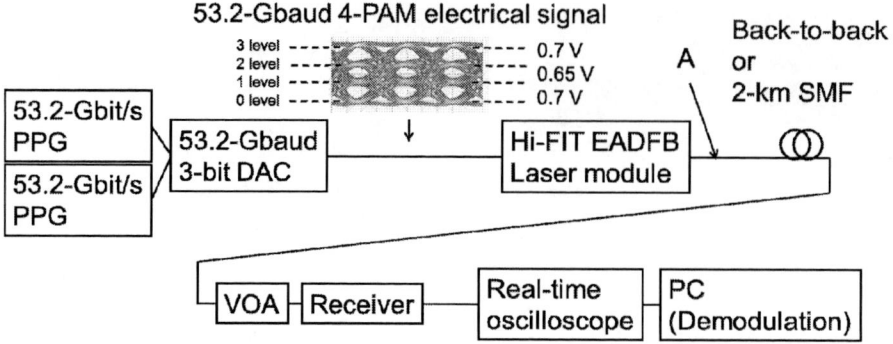

Fig. 4 Measurement setup for 106-Gbit/s 4-PAM transmission experiment

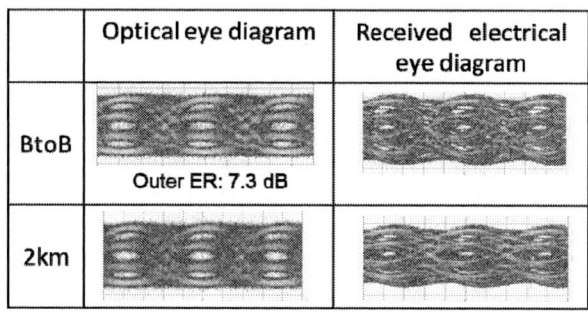

Fig. 5 Eye diagrams under 106-Gbit/s operation

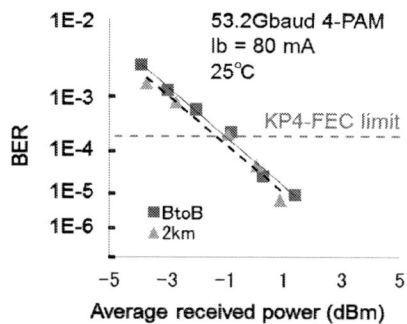

Fig. 6 BER characteristics under 106-Gbit/s operation

V. CONCLUSIONS

We fabricated a Hi-FIT EADFB laser transmitter and a high-speed optical receiver. The 3-dB bandwidths of the optical transmitter and receiver exceeded 59 and 50 GHz, respectively. Using these high-speed optical modules, we obtained a BER of less than 2×10^{-4} under error-free conditions using KP4 FEC without an equalizer even after a 2-km SMF transmission. These results indicate that high-speed (> 50 GHz) optical modules are required for a low-power consumption and compact transceiver operating at 100 Gbit/s/λ.

REFERENCES

[1] http://www.ieee802.org/3/bs/
[2] Y. Morita et al., "1.3 μm 28 Gb/s EMLs with hybrid waveguide structure for low-power-consumption CFP2 transceivers," in Proc OFC2013, OTh4H.5.
[3] W. Kobayashi et al., "Advantages of EADFB laser for 25 Gbaud/s 4-PAM (50 Gbit/s) modulation and 10 km single-mode fibre transmission," Electron. Lett., vol. 50, no. 9, pp 683-685, 2014.
[4] C. Xu et al., "Performance improvement of 40-Gb/s electroabsorption modulator integrated laser module with two open-circuit stubs," IEEE Photon. Technol. Lett., vol. 24, no. 22, pp. 2046-2048, 2012.
[5] T. Fujisawa et al., "50 Gbit/s uncooled operation (5–85°C) of 1.3 μ m electroabsorption modulator integrated with DFB laser," Electron. Lett., vol. 49, no. 3, pp. 204-205, 2013.
[6] S. Kanazawa et al., "Flip-chip interconnection lumped-electrode EADFB laser for 100-Gb/s/λ transmitter," IEEE Photon. Technol. Lett., vol.27, no. 16, pp. 1699-1701, 2015.
[7] S. Kanazawa et al., "56-Gbaud 4-PAM (112-Gbit/s) operation of flip-chip interconnection lumped-electrode EADFB laser module for equalizer-free transmission," in Proc. OFC 2016, W4J.1.
[8] Y. Muramoto et al., "InP/InGaAs pin photodiode structure maximising bandwidth and efficiency," Electron. Lett. Vol. 39, no. 24, pp. 1749-1750, 2003.

ThD1-3

Low Thermal Resistance VCSEL Array Adopted Tunnel Junction Destruction using Proton Implantation

Shohei Oshida, Masashi Suhara, and Tomoyuki Miyamoto

P & I Lab., Tokyo Institute of Technology, 4259 Nagatsuta, Midori-ku, Yokohama 226-8503, Japan

shohei.oshida@ms.pi.titech.ac.jp

Abstract: A tunnel junction destruction vertical cavity surface emitting lasers (TJD-VCSEL) is proposed towards high output and high efficiency light source. Fabrication process of the proposed structure was established and low thermal resistance property was revealed.

Keywords: VCSEL, proton implantation, tunnel junction, thermal resistance, optical wireless power transmission

I. INTRODUCTION

Following the progress of wireless communication system, wireless power transmission (WPT) system is expected. However, conventional methods of electromagnetic induction and magnetic field resonance have issues of large leakage of electromagnetic wave. Therefore, these methods have limits of transmission power and distance. Optical wireless power transmission (OWPT) has attracted attention as another WPT technology [1]. The OWPT is expected to have advantages of high transmission power based on a sharp directional beam, scalability of transmission power, and long distance transmission capability.

As the light source of the OWPT, a surface emitting laser (VCSEL) [2] is promising. Advantages of the VCSEL is high power COD-free operation and ease of 2D integration. These characteristics are beneficial in this new application. The high power conversion efficiency of the VCSEL array is a key towards practical realization of the OWPT. In addition, in a power supply application of the VCSEL array, reduction of the light source module (chip) size will become another requirement. The chip size of the VCSEL array becomes small by decreasing the array pitch. However, there is a trade-off relation between reduction of the array pitch and increase of the power conversion efficiency through the thermal crosstalk between each element of VCSELs. Therefore, reduction of the thermal resistance is indispensable for high-output power density and high power conversion efficiency of the VCSEL array.

In this study, we investigated a low thermal resistance VCSEL that adopted tunnel junction destruction by proton implantation technique for formation of a mesa-free structure. The target device is a 980 nm wavelength range GaAs-based VCSEL. Since the mesa formation for the AlAs selective oxidation process is not needed, AlAs or high Al composition AlGaAs which have low thermal resistance was introduced in a distributed Bragg reflector (DBR) of the VCSEL. In addition, the mesa-free structure is advantageous because there is no thermal separation structure in the in-plane direction.

II. DEVICE STRUCTURE

An AlAs selective oxide confinement structure shown in Fig. 1(a) which enables confinement of the light and the current is used in the most conventional VCSEL. This structure is fabricated by the mesa formation and the oxidation process. Low thermal resistance AlAs/GaAs DBR can't be used in this structure because AlAs is introduced as the selective oxidation layer. In addition, the mesa structure restricts the heat spread to the in-plane direction. High thermal resistance of the oxidized AlAs layer is also a problem. Therefore, in the case of the AlAs selective oxide confinement structure, there are difficult problems in order to further reduce the thermal resistance.

As a mesa-free current confinement structure, a proton implantation structure has been commercialized. In this structure, an AlAs/GaAs DBR can be used essentially. From the numerical analysis, the thermal resistance of a proton implanted VCSEL using an AlAs/GaAs DBR is expected to become 1/3 in comparison with that of a AlAs selective oxide confinement structure. However, abrupt hetero-interface of the AlAs/GaAs DBR will be required instead of the usual gradient composition interface to achieve low thermal resistance. Since an abrupt interface of the DBR especially in p-type shows excessively high electrical resistance, the combination of the conventional proton implantation structure and an AlAs/GaAs DBR is not realistic.

In this work, a tunnel junction destruction VCSEL (TJD-VCSEL) shown in Fig. 1(b) is proposed. In order to obtain a tunnel junction destruction structure for forming a current confinement structure, proton implantation into the tunnel junction was performed. By applying of the tunnel junction, low resistance n-type AlAs/GaAs DBR with abrupt hetero-interface can be used as a top DBR. By the numerical analysis, the electrical resistance is expected to become less than half of the conventional VCSELs when the resistance of tunnel junction is ignored. Furthermore, low absorption loss in the top DBR and low contact resistance in an n-type top electrode are expected.

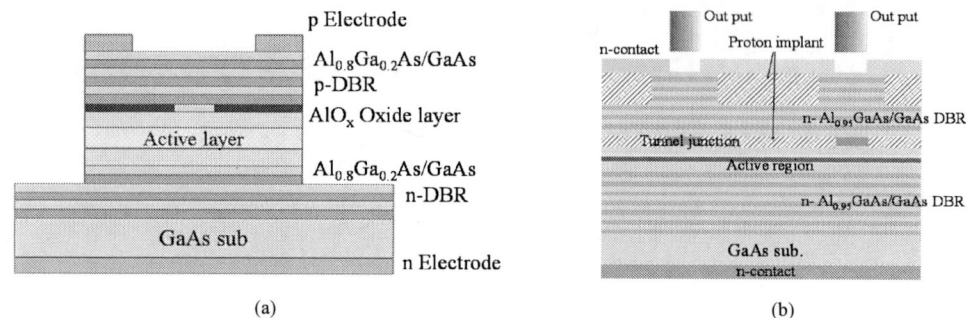

(a)   (b)

Fig. 1. Schematic structures of (a) conventional and (b) TJD-VCSEL.

III. FABRICATION AND CHARACTERIZATION OF DEVICES

Prior to the fabrication of the proposed VCSEL, implantation condition of the tunnel junction destruction was evaluated. The proton implantation energy and the special distribution of proton was simulated numerically using TRIM software. Based on the simulation results, three test samples with different proton dose density of 3×10^{12}, 3×10^{13}, and 3×10^{14} cm^{-2} were prepared. Figure 2 shows I-V characteristics of the test samples. In the cases of 3×10^{12} and 3×10^{13} cm^{-2}, the tunnel junctions were conductive. On the other hand, insulating state was confirmed in the case of 3×10^{14} cm^{-2}. This is the first observation of destruction condition of the GaAs-based tunnel junction using proton implantation.

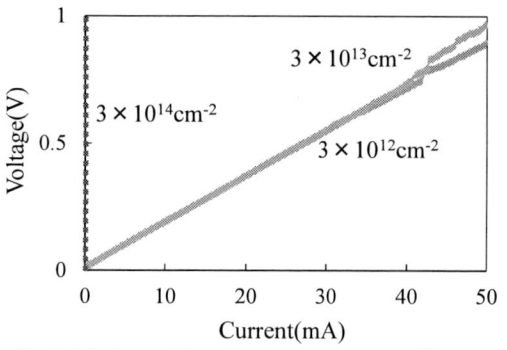

Fig. 2. I-V characteristics of proton implanted tunnel junction.

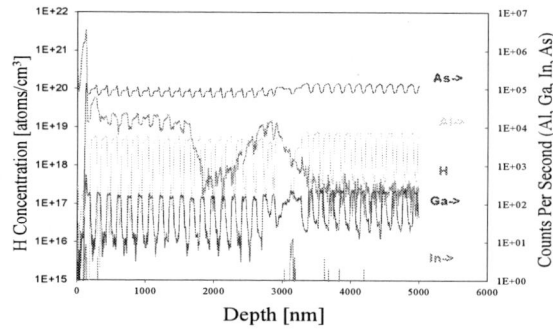

Fig. 3. SIMS profile of proton implanted wafer

Two types of the TJD-VCSEL were fabricated using the DBR composition of $Al_{0.95}Ga_{0.05}As$/GaAs and $Al_{0.8}Ga_{0.2}As$/GaAs. A current confinement structure using tunnel junction destruction was formed by proton implantation with a dose density of 3×10^{14} cm^{-2}. From the SIMS profile shown in Fig. 3, high density Hydrogen atom was confirmed at the tunnel junction which was inserted into the second DBR pair from the active layer. The mask size of the TJD aperture diameter was from 5 μm to 40 μm. Figure 4(a) shows a top-view of the fabricated VCSELs. A flat structure was confirmed because there is no need of the mesa structure. I-L-V characteristics of the TJD-VCSELs are shown in Fig. 4(b). CW lasing operation was observed. Because the layer structure and doping amount of the tunnel junction was not optimized, excess voltage was appeared. However, the applicability of the TJD structure using proton implantation to the VCSEL fabrication process was clarified.

Next, the thermal characteristics of the TJD-VCSEL are discussed. Figure 5 shows measured and numerically analyzed thermal resistance of the VCSELs. In this graph, the aperture diameter was corrected by considering the shape of the resist mask. Compared with the oxide confinement structure, low thermal resistance of the TJD-VCSEL was observed. The improvement was mainly caused by the mesa-free structure because clear improvement of the $Al_{0.95}Ga_{0.05}As$/GaAs DBR was not confirmed. One of the reasons is correction error of the shape of the current confinement aperture size. Figure 6 shows measurement results of the thermal crosstalk of the TJD and oxide confinement VCSEL (conventional). Thermal crosstalk of the TJD-VCSEL was increased. This result means fast heat spread of the TJD-VCSEL in the in-plane direction.

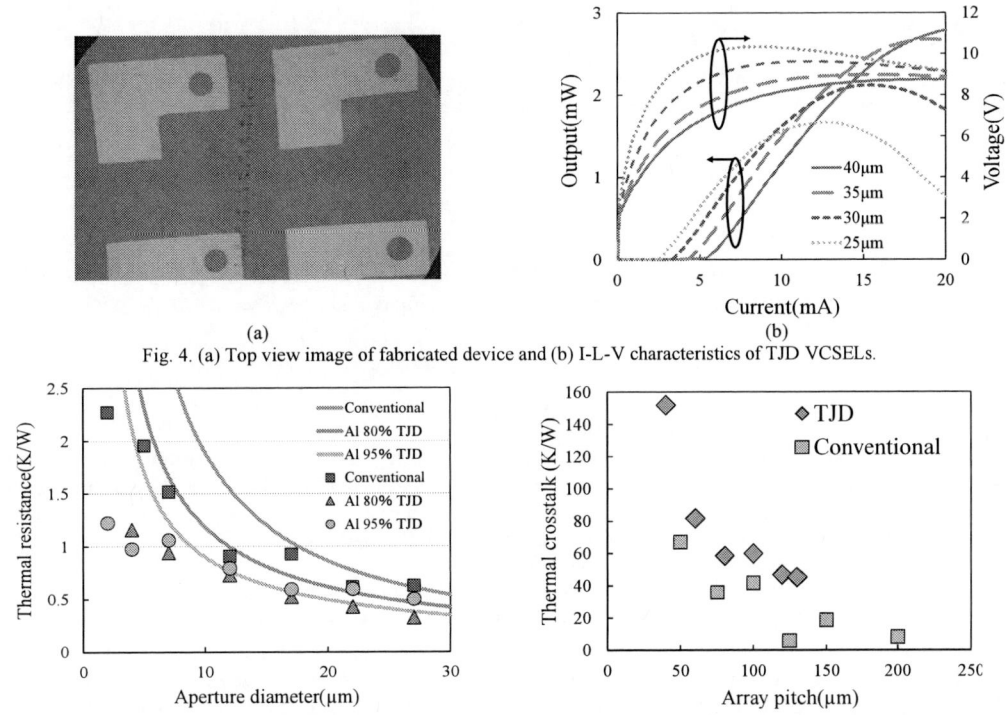

(a) (b)

Fig. 4. (a) Top view image of fabricated device and (b) I-L-V characteristics of TJD VCSELs.

Fig. 5. Measured and calculated thermal resistance of TJD and conventional VCSEL.

Fig. 6. Measured thermal crosstalk of TJD and conventional VCSEL

IV. CONCLUSIONS

In this work, we proposed a TJD-VCSEL towards the high output and high efficiency VCSEL array. Proton implantation condition of the tunnel junction destruction was revealed and continuous wave lasing operation was observed. Furthermore, through the thermal property investigation, the effectiveness of the TJD-VCSEL was confirmed in terms of the thermal resistance and the thermal crosstalk. We believe a high power density and high efficiency VCSEL array can be realized using an optimized tunnel junction.

REFERENCES

[1] K. Iga, "Vertical-cavity surface-emitting laser: Its conception and evolution," Jpn. J. Appl. Phys., vol. 47 pp. 1-10, 2008.
[2] M. Hirota, S. Iio, Y. Ohta, and T. Miyamoto, 20th Microoptics Conference (MOC'15), Japan (Fukuoka), Oct. 2015, H86.

ThD1-4   OECC/PS2016

Influence of Wavelength Deviation on Phase Locked VCSEL Array using Talbot Effect

Yuki Komori and Tomoyuki Miyamoto
P & I Lab., Tokyo Institute of Technology, 4259 Nagatsuta, Midori-ku, Yokohama 226-8503, Japan
komori.y.ab@m.titech.ac.jp

Abstract: *Numerical analysis of the phase-locked VCSEL array using the Talbot effect is investigated considering the wavelength deviation. A narrow pitch and a large number of array are significant. A monolithically integrated Talbot-VCSEL structure is proposed.*
Keywords: *VCSEL array, Talbot effect, phase locking, injection locking, high output power*

I. INTRODUCTION

In recent years, a vertical cavity surface emitting laser (VCSEL) array has attracted attention as a high-power light source. The VCSEL has excellent characteristics, such as low threshold, circular beam, and two-dimensional (2D) integration [1]. Though output power of a standard single element of the VCSEL is small, it is possible to obtain a high output power by 2D array integration. In such a VCSEL array, individual VCSEL is lasing independently. If the wavelength and the phase of laser elements can be synchronized, the beam quality becomes improved and the beam shape can be easily controlled. Such a phase locked VCSEL array is advantageous in high power laser applications of the optical wireless power transmission [2], laser machining, laser heat treatment, laser display, and laser illumination.

A few technologies of phase locking of a VCSEL array have been investigated. One uses a strong lightwave coupling between laser elements for the phase locking. In this configuration, it is difficult to widen the array pitch. Therefore, it causes difficulty of precise control of device characteristics due to fabrication difficulty of a densely packed array. In addition, it causes difficulty of the heat dissipation due to narrow array pitch, and thus, it deteriorates the output power.

A phase locked VCSEL array using the Talbot effect is another technology [3]. Since the Talbot-VCSEL can vary the array pitch, it is possible to fabricate an array which has excellent heat dissipation. Numerical simulation [3] and device fabrication using relatively complicated module [4] of the Talbot-VCSEL have been reported. However, the phase locked operation with a simple configuration has not been reported. This might be due to insufficient device design of the Talbot-VCSEL without considering the influence of the wavelength deviation in the array.

In this study, we investigated the detail of the required conditions for phase locking of the Talbot-VCSEL and propose a monolithic device configuration.

II. FUNDAMENTAL CONDITION OF THE TALBOT-VCSEL

The Talbot effect is the phenomenon of interference of a periodic array light source. When the periodic array pattern of a coherent light aligned infinitely, the reconstruction image is exactly the same distribution as the light source pattern at a Talbot length. The Talbot length $2Z_0$ is expressed by the following equation using the array pitch p, wavelength λ and natural number m.

$$2z_0 = \frac{2mp^2}{\lambda} \qquad (1)$$

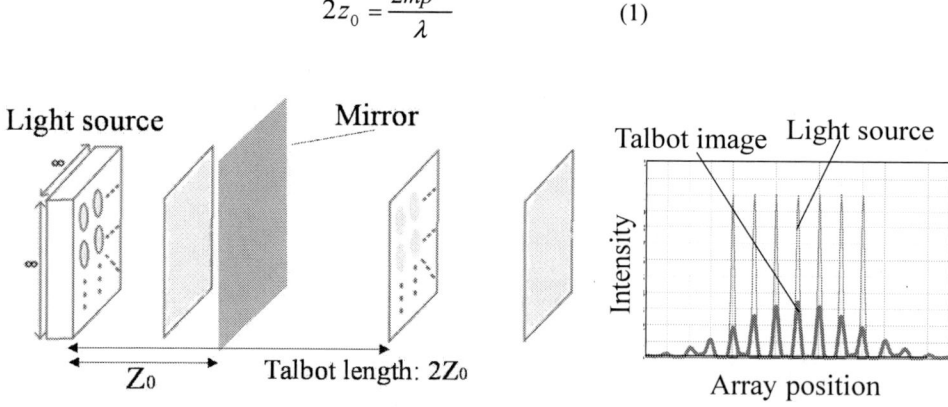

Fig. 1. Conceptual diagram of Talbot effect   Fig. 2 Calculated Talbot image of finite array at Talbot length

A concept diagram of the Talbot-VCSEL is shown in Fig. 1. In this configuration, reconstructed light by the Talbot effect is used as the light of the mutual injection locking of a VCSEL array by placing a mirror at half of the Talbot length.

There are two problems for construction of the Talbot-VCSEL. One is that the actual array has finite period. Although reconstruction image of light at the Talbot length in the case of finite period array shows the same array pitch as the light source, the light intensity becomes weaker than that of the light source. In addition, the reconstruction light intensity gradually decreases at the edge of the array as calculated in Fig. 2. Another critical problem is that the elements of the actual array operate independently and the lasing wavelength is not the same. Though it is necessary to synchronize the lasing wavelength and the phase completely in order to achieve the Talbot effect, the condition is satisfied after the phase locking. In the case of independent lasing condition, similar effect to the Talbot effect known as the Lau effect can be observed, however the reconstructed image is deteriorated. These problems decrease the intensity of the reconstructed image, and therefore mutual injection locking should be achieved by the careful design of the array configuration.

III. TALBOT EFFECT IN VIEW OF WAVELENGTH DEVIATION

In this study, numerical simulation of the Talbot and Lau effect for a phase-locked VCSEL array are performed under conditions of wavelength deviation and finite period number of the array.

A. Influence analysis of wavelength deviation

Firstly, we show influence of the Talbot (Lau) effect under the wavelength deviation using theoretical formula. The relationship between the Talbot length shown in eq. (1) and the coherent length $L_c = \lambda^2/\Delta\lambda$ is considered, where $\Delta\lambda$ is standard deviation of the wavelength in the array. If the Talbot length $2Z_0$ is longer than the coherent length L_c, the phase shift of the lightwave becomes large, and the interference, which is fundamental phenomena of the Talbot effect, will not occur. Therefore, the Talbot length as well as the array pitch is limited by the wavelength deviation. Actually, the length is required to be $L_c/2$ or less for avoiding the interference at opposite phase. In order to improve the interference condition, sufficient shorter length than L_c are necessary. The required condition for array pitch is shown in the next equation,

$$p \leq \frac{\lambda}{\sqrt{2a \cdot \Delta\lambda/\lambda}} \qquad (2)$$

where a is parameter for appropriate interference. As a practical condition of λ=980 nm, $\Delta\lambda$=0.03 nm, and a=6, the array pitch is about 50 μm or less.

Under the required array pitch, peak intensity dependence on the wavelength deviation was analyzed numerically using the Fresnel-Kirchhoff integration. An example of calculated light intensity distribution of the light source and the Talbot image are shown in Fig. 3(a) for a 15 × 15 array with each beam diameter of 9 μm. Due to the wavelength deviation of the $\Delta\lambda$=0.1 nm, the Talbot effect becomes incomplete and the result corresponds to the Lau effect. In this study, the Talbot effect was evaluated by the peak intensity ratio of the reconstruction image to the light source at the array center. The peak intensity dependence on the wavelength deviation for different array pitch is shown in Fig. 3(b). Black line shows the required condition of the injection locking [6]. The peak intensity of the narrow pitch is higher than that of wide pitch. In the case of the array pitch of 50 μm or less, the sufficient Talbot effect for the mutual injection locking is expected when the wavelength deviation is less than about 0.04 nm. The deviation corresponds to 0.004% at wavelength of 980 nm. Though the wafer thickness deviation is directly related with the wavelength deviation of the VCSEL, 1 mm size of a part of a 3 inch wafer satisfy the thickness deviation of 0.004%.

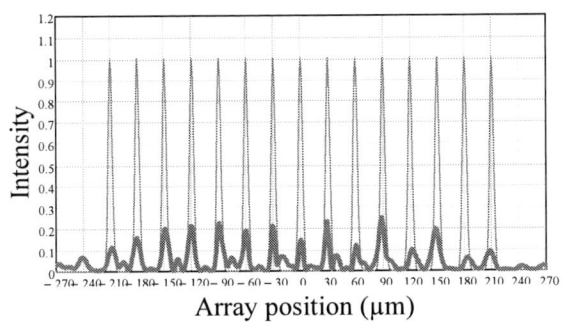

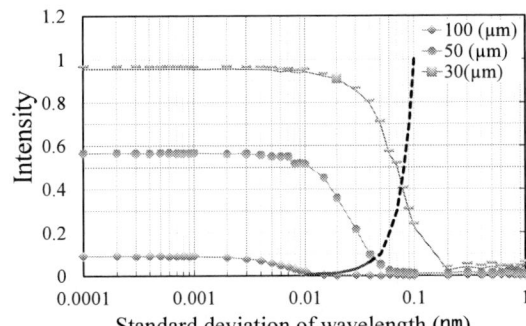

Fig. 3. (a) Example of Talbot image with wavelength deviation ($\Delta\lambda$ =0.1 nm) (b) Peak intensity dependence on the wavelength deviation and
(15×15 array, Aperture is 9 μmφ.)   required condition of injection locking (black line).

B.   Influence analysis of finite array

Next, the influence of the array size ($N \times N$) is discussed. Figure 4 shows calculated results of the peak intensity dependence on the number of array N. Though the peak intensity becomes large by increasing N, it becomes saturated below 0dB. In the case of finite array, the Talbot effect is achieved more easily by increasing N. This is because a large array size increases the number of elements for interference. On the other hand, spot size of each element is finite, and therefore, the intensity of increased number of array becomes saturated.

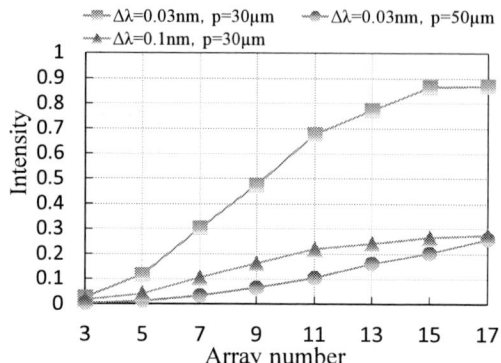

Fig. 4. Peak intensity dependence on the number of array

C.   Concept of monolithic integrated Talbot-VCSEL

The Talbot length is 1.837 mm for an array pitch of 30 µm, and 5.102 mm for 50 µm pitch at λ=980 nm. The external mirror configuration is required because mm scale distance from the VCSEL to the Talbot mirror is required, and therefore complex module is indispensable. However, considering the refractive index of GaAs of 3.527, a half Talbot length of 30 µm pitch becomes 260 µm which is within the thickness of a standard substrate. Therefore, it is possible to form a monolithically integrated structure that uses a substrate back surface as a Talbot mirror. The schematic proposed structure is shown in Fig. 5. We are now fabricating this type of the Talbot-VCSEL.

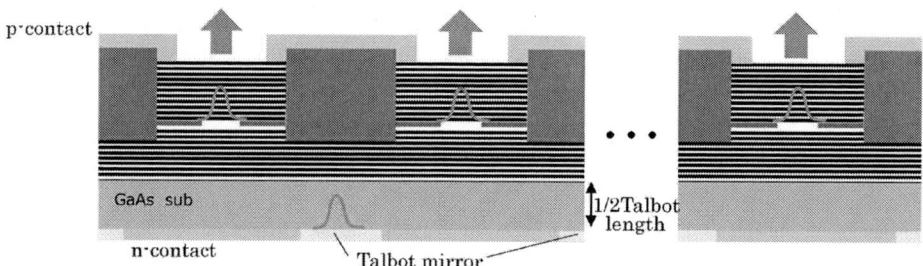

Fig. 5. Concept structure of monolithic Talbot-VCSEL

IV. CONCLUSIONS

In this study, injection locking condition of the Talbot-VCSEL was analyzed considering influence of the wavelength deviation in the array. The detailed effect of the array pitch, the number of array, and the light beam size was analyzed. From the consideration of the effect of structural parameters, a monolithically integrated Talbot-VCSEL structure was proposed with the array pitch of 30 µm. The operation of the phase-locked VCSEL array is expected by using the analyzed results and a designed structure.

REFERENCES

[1] K. Iga, "Vertical-cavity surface-emitting laser: Its conception and evolution," Jpn. J. Appl. Phys., vol. 47 pp. 1-10, 2008.
[2] M. Hirota, S. Iio, Y. Ohta, and T. Miyamoto, 20th Microoptics Conference (MOC'15), Japan (Fukuoka), Oct. 2015, H86.
[3] E. Ho, F. Koyama and K. Iga, "Effective reflectivity from self-imaging in a Talbot cavity and its effect on the threshold of a finite 2-D surface emitting laser array," Appl. Opt., vol. 29, pp. 5080-5085, 1990.
[4] S. Sanders, R. Waarts, D. Nam, D. Welch, D. Scifres, J. C. Ehlert, W. J. Cassarly, J. M. Finlan, and K. M. Flood, "S. Sanders, R. Waarts, D. Nam, D. Welch, D. Scifres, J. C. Ehlert, W. J. Cassarly, J. M. Finlan, and K. M. Flood," Appl. Phys. Lett., vol. 64, pp. 1478-1480, 1994.
[5] J. Y. Lawy, G. H. M. van Tartwijk, and G. P. Agrawal, "Effects of transverse-mode competition on the injection dynamics of vertical-cavity surface-emitting lasers," Quantum Semiclass. Opt., vol. 9, pp. 737-747, 1997.

ThD1-5

Demonstration of an Ultra-Low Threshold Phonon Laser with Coupled Microtoroid Cavities in Vacuum

Mingming Zhao[1], Guanzhong Wang[1], Zhiqiang Jin[1], Yingchun Qin[1], Zhang-qi Yin[2], Xiaoshun Jiang[1,*], and Min Xiao[1,3]

[1]National Laboratory of Solid State Microstructures and College of Engineering and Applied Sciences, Nanjing University, Nanjing, 210093, China
[2]Center for Quantum Information, Institute for Interdisciplinary Information Sciences, Tsinghua University, Beijing 100084, China
[3]Department of Physics, University of Arkansas, Fayetteville, Arkansas 72701, USA
*jxs@nju.edu.cn

Abstract: *We report an ultra-low threshold phonon laser using coupled microtoroid cavities in vacuum with a novel coupling method. The measured lasing threshold is as low as 1.3 μW.*
Keywords: *microcavity, optomechanics, phonon laser*

I. INTRODUCTION

Over decade efforts, cavity optomechanics has now become a very active research field with focus on the interaction between electromagnetic radiation and mechanical motions in optical cavities [1]. In this research activity, silica microtoroid cavities have been established as a powerful platform partly due to their high quality optical [2] and mechanical [3] properties. To name a few, silica microtoroids have been largely exploited in the studies of mechanical oscillation [4], optomechanically induced transparency [5] and optomechanical cooling [6] with single microcavity.

Only a few years ago, with the use of a compound microcavity system in air, the Vahala's group reported the realization of a phonon laser [7], the phonon analog of an optical laser in a two-level system. However, in their work two microresonators were all located at the chip edges which lead to the asymmetry of the silicon pillars. As a result, such asymmetry deleteriously influenced the mechanical mode and prevented the possibility of fabricating ultra-thin pillars. Consequently, a low mechanical quality factor is typically associated with this type of coupling geometry.

To achieve a higher mechanical quality factor as well as to stabilize the system, here we demonstrate a two-level phonon-laser action with two coupled microtoroid resonators in vacuum by employing a different but new coupling method. In comparison with the previous work, our system not only exhibits a number of new features but also yields the lasing threshold of 1.3 μW, which is 5 times lower than that reported in Ref. [7].

II. SAMPLE AND MEASUREMENT

A. Sample Preparation

As schematically shown in figure 1(a), the coupled microtoroid system consists of a microtoroid A and an inverted microtoroid B. In the experiment, the microtoroid A, fabricated by two times XeF_2 dry etching, has an ultra-thin silicon pillar (Fig. 1(b)) which ensures a high mechanical Q-factor [3]. The microtoroid B with a normal silicon pillar is fabricated at the corner of a silicon chip (Fig. 1(c)). The position and temperature of the two microcavities are further controlled by attocube nanopositioners and thermoelectric cooler (TEC) elements for precision control and stabilization.

OECC/PS2016

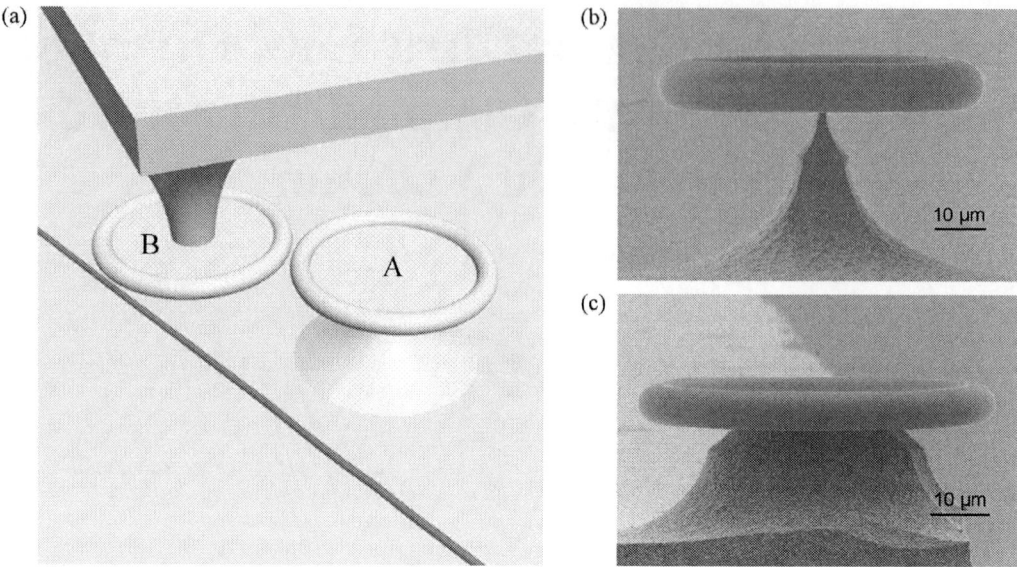

Fig.1. (a) Schematic diagram of the coupling geometry between two microtoroid resonators. (b) and (c) Scanning electron microscope images of the microtoroid with tiny silicon pillar and the microtoroid at the corner of a silicon chip, respectively.

B. Phonon Laser Measurement

In the experiment, two microtoroid cavities with intrinsic optical Q-factors of 9.7×10^7 and 9.3×10^7 and scattering-induced mode splittings of 11.2 MHz, 7.6 MHz, were used to form the coupled system. By carefully tuning the positions and temperature of the two microtoroids, we can obtain a supermode with a mode splitting equal to the mechanical frequency of the microtoroid A. The mechanical frequency for the fundamental radial breathing mode of microtoroid A is measured to be 57.5 MHz, and the mechanical Q-factor is 9000 in vacuum. To excite the phonon laser, the pump laser is locked at the blue supermode using a wavelength meter. When the pumped optical power is operated above the threshold, an oscillatory transmission can be clearly observed in the time domain (Fig. 2(a)). The high performance of our system allows us to have a much lower threshold. As shown in figure 2(c), the measured phonon lasing threshold is about 1.3 µW, which is much lower than 7 µW reported in the work of Ref. [7].

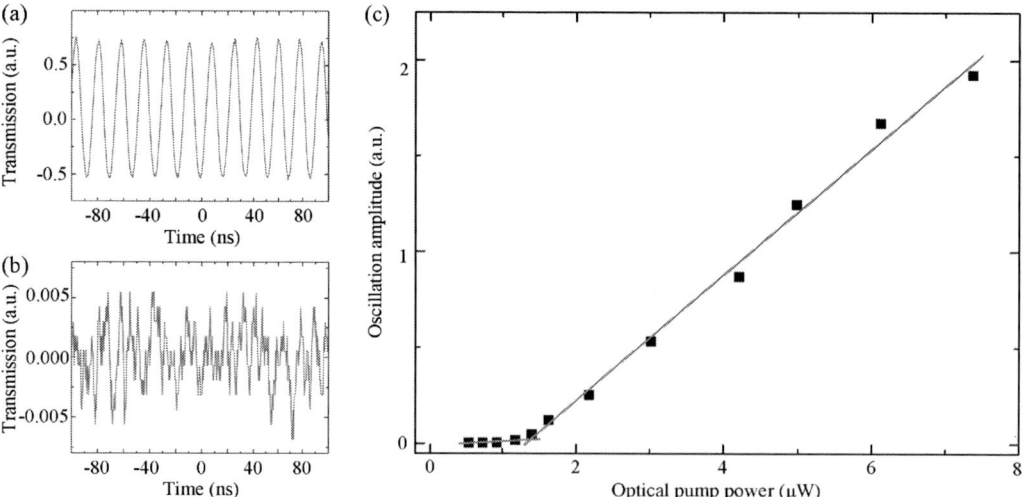

Fig. 2. (a) and (b) Optical transmitted powers above and below the threshold, respectively. (c) Measured oscillation amplitude versus optical pump power.

III. CONCLUSIONS

In conclusion, we demonstrated a compound structure of two microtoroids through a new coupling method for multi-mode cavity optomechanics. In particular, we have successfully realized an ultra-low threshold phonon laser operating in vacuum. It is expected that our system will find other applications in multi-mode optomechanical cooling.

360

ACKNOWLEDGMENT

This work was supported by the National Basic Research Program of China (2012CB921804), the National Natural Science Foundation of China (nos. 61435007, 11574144 and 11321063) and the Natural Science Foundation of Jiangsu Province (BK20150015).

REFERENCES

[1] M. Aspelmeyer, T J Kippenberg, and F Marquardt, "Cavity optomechanics," Rev. Mod. Phys, vol. 86, pp. 1391-1452, December 2014.

[2] D. K. Armani, T. J. Kippenberg, S. M. Spillane, and K. J. Vahala, "Ultra-high-Q toroid microcavity on a chip," Nature, vol. 421, pp. 925-928, February 2003.

[3] G. Anetsberger, R. Rivière, A. Schliesser, O Arcizet, and T. J. Kippenberg, "Ultralow-dissipation optomechanical resonators on a chip," Nat. Photonics, vol. 2, pp. 627-633, September 2008.

[4] T. Carmon, H. Rokhsari, L. Yang, T. J. Kippenberg, and K. J. Vahala, "Temporal behavior of radiation-pressure-induced vibrations of an optical microcavity phonon mode," Phys. Rev. Lett, vol. 94, 223902, June 2005.

[5] S. Weis, R. Rivière, S. Deléglise, E. Gavartin, O. Arcizet, A. Schliesser, and T. J. Kippenberg, "Optomechanically induced transparency," Science, vol. 330, pp. 1520-1523, December 2010.

[6] A. Schliesser, R. Rivière, G. Anetsberger, O. Arcizet, and T. J. Kippenberg, "Resolved-sideband cooling of a micromechanical oscillator," Nat. Physics, vol. 4, pp. 415-419, May 2008.

[7] I. S. Grudinin, H. Lee, O. Painter, and K. J. Vahala, "Phonon laser action in a tunable two-level system," Phys. Rev. Lett, vol. 104, 083901, February 2010.

Flex Rate Transmission and Challenges for Long Haul Transmission

Fred Buchali, Wilfried Idler

Nokia, Bell Labs, Lorenzstr. 10, D-70435 Stuttgart, Germany
Fred.buchali@nokia.com

Abstract: *We review flexible bitrate transponder options combining variable QAM formats and a variable baudrate. We show that high spectral efficiency and high baudrate systems suffer from limited converter ENOB, furthermore high baudrate systems suffer from nonlinear distortions.*
Keywords: *coherent transmission, DSP, flexible transponder, high order modulation formats*

I. INTRODUCTION

Flexible transponders have been introduced particularly with the beginning of the coherent age since massive digital signal processing (DSP) is available on optical transponders chipsets. First the variability targeted especially diverging requirements in transmission reach. Today flexible transponders are strongly supported by high speed and high resolution digital to analog converters (DAC) as well as analog to digital converters (ADC) [1]. Intelligent digital signal processors (DSPs) built the second important block which offers a suitable adaptation to the respective transmission formats. Therewith easy modifications of the transmission modes are enabled by switching between different operational modes of the transponders.

Today's transponders apply mainly QAM modulation formats in a dual polarization (dp) implementation at a variable baudrate. By selecting a suitable combination of baudrate and order of modulation format the data rate and the spectral efficiency can be adapted to the requirements of the channel or to the requirements of the application. Off coarse the granularity of regular QAM formats coarse leading to larger reach variations if varying the QAM order step by step [2]. A more fine granular variation can be achieved if time domain multiplexed hybrid modulation (TD-HM) formats are applied [3-5]. These formats combine 2 or more modulation formats; the usage of first and second format is variable and enables numerous steps to make the granularity in spectral efficiency and reach more fine granular.

In this paper we report on the opportunities and limitations of flexible transponders. Starting from the capacity boundaries different options for flexibility will be discussed. Next these options will be assessed taking the characteristics of real transponders especially the characteristics of real DACs into account. Several schemes will be compared by experimental investigations. Finally future opportunities especially to close the gap to the Shannon capacity will be discussed.

II. FLEX TRANSPONDERS USING REGULAR QAM FORMATS

First we assess the options and characteristics of regular QAM formats regarding its capacity considering an additional white Gaussian noise (AWGN) transmission channel (cf. Fig. 1.). Typically we are considering a 28% overhead for FEC and protocol which enables the operation of the systems at capacities shown in Tab. 1. The bitrates for the considered modulation formats QPSK up to 64QAM ranges from 100 to 300 Gb/s and the required SNR values vary strongly between 3.8 and 15.0 dB. It is well known that in a first approximation the transmission reach scales linar with the required SNR leading to huge reach differences between the formats. Whereas for QPSK distances far beyond 10.000 km have been demonstrated [5] for 64QAM a reach of 300 km has been found [6]. These numbers clearly indicate that optical transmission systems with a high spectral efficiency can be implemented, but a drastic reduction in maximum transmission distance has to be taken into account. Moreover all these formats suffer from a remaining gap to the Shannon capacity in the range of 0.9 to 1.1 dB in SNR. If applying TD-HM formats intermediate steps between adjacent QAM formats can be implemented [3] and the spectral efficiency or maximum transmission distance can be adapted to address the channel requirements

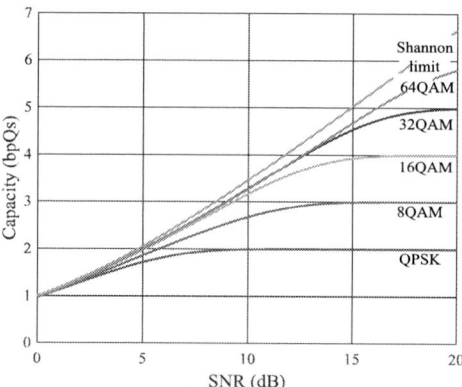

Fig. 1. Capacity versus SNR for few QAM formats

Format	Capacity [bit/s/Hz]	Bitrate [Gb/s]	Req. SNR [dB]
QPSK	1.56	100	3.8
8QAM	2.34	150	7.75
16QAM	3.125	200	9.85
32QAM	3.91	250	12.33
64QAM	4.69	300	15.03

Tab. 1. Main characteristics of investigated modulation formats operated at 32 Gbaud.

in an improved way. Recently shaped QAM formats have been proposed to close the gap to the Shannon capacity [7,8]. The extensive exploration of these formats is still ongoing.

III. TRANSPONDER CHARACTERISTICS

The implementation of higher order QAM formats strongly depends on the capabilities of the converters (DACs and ADCs) used in the transmitters and receivers. The converters are typically integrated together with the DSP in a single chip using CMOS technology, which is the only technology offering the integration option. In this paper we investigate converters with 88 GSa/s sampling rate, 8 bit physical resolution, which enable conversion of highly oversampled data carrying higher order modulation. In the DSP the QAM data are modulated. In addition to that a pulseshaping is applied to generate spectra with e.g. raised cosine shape. Finally we are applying a digital pre-emphasis, which compensates for the complete transmitter chain consisting of the converter itself, of the driver amplifier and of the optical modulator. Therefore a DAC with an effective resolution much higher than the number of modulation levels is required for the in-phase or quadrature components. As a rule of thumb for these converters one bit in resolution is required for the digital pre-emphasis and one bit for the pulseshaping. Finally 4 bit and 5 bit resolution are required in case of 16QAM modulation and 64QAM modulation, respectively. The effective resolution of converters is benchmarked by the effective number of bit (ENOB). This number is typically derived by signal to noise and distortion (SINAD) measurements. Single carrier measurements are often applied, but they do not consider the conditions of broad band data signals. Previously we proposed a new preemphased prime frequency multicarrier (PPFM) measurement to assess the DAC performance, which takes the broadband characteristics and the digital preemphasis into account. The measured ENOB is plotted vs. the baudrate in Fig. 2 and a 1 bit slope has been observed if doubling the baudrate from 32 Gbaud to 64 Gbaud. Up to ~55 Gbaud the application of a 64QAM modulation is feasible to implement, whereas up to 64 Gbaud a 32QAM modulation is feasible.

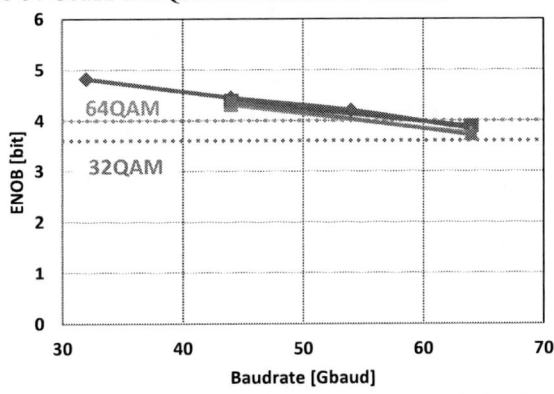

Fig. 2. ENOB derived by PPFM measurements vs. the baudrate which corresponds to the multicarrier bandwidth.

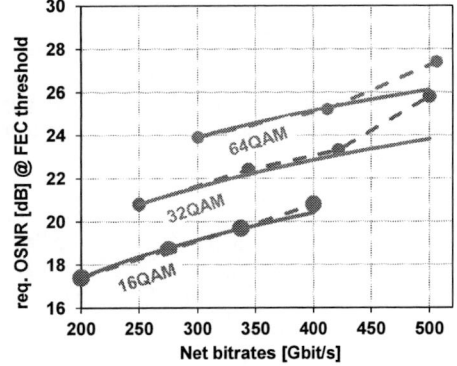

Fig. 3. Required OSNR vs. net bitrate of the investigated QAM formats.

Optical back to back (B2B) measurements at variable OSNR have been performed for 16QAM, 32QAM and 64QAM at baudrates ranging from 32 Gbaud up to 64 Gbaud. The bitrates range from 200 Gb/s up to ~500 Gb/s (cf. Tab. 2.). At the target BER of $4 \cdot 10^{-2}$ we measured the required optical signal to noise ratio (OSNR, cf. Fig. 3.). The solid lines indicate the expected proportionality of the required OSNR with the baudrates. For all modulation formats we found an excellent scaling except both 500 Gb/s modes basing on 32QAM and 64QAM, respectively. As concluded from the ENOB measurements the margin in resolution is exhausted for both 500 Gb/s modes leading to an significant implementation penalty.

Baudrate [Gbaud]	16QAM	32QAM	64QAM
32	200	250	300
44	275	344	412
54	337	422	506
64	400	500	-

Tab. 2. Bitrates in Gb/s for the investigated combinations of baudrate and QAM format.

IV. TRANSMISSION EXPERIMENTS

Optical transmission systems are operated typically at an optimal fiber launch power, where the noise and the nonlinear distortions limit the performance similarly and the sum of both is minimal. We investigated this optimized launch power for the variable baudrate signals carrying different modulation formats. Additional results from [3] and [4] have been added, too. As seen in Fig. 4. the launch power scales less strong with the baudrate: for a doubling of baudrate we have 2 dB increased launch power, only. The reduction of sensitivity for doubling the baudrate is 3 dB from which 2 dB can be compensated by an increased launch power. Nevertheless the nonlinear distortions have to be taken into account. Finally we measured the reach for variable baudrate signals and 3 modulation formats (cf. Fig. 5.). All reach values have been scaled to the highest reach achieved for 16QAM and 32 Gbaud and evaluated in dB. For the lowest order 16QAM the increase of baudrate leads to a ~2 dB reach penalty, from which 1 dB is due to the launch power and 0.5 dB is due to the sensitivity penalty. Towards higher order formats the reach penalty is increasing. Off course higher order formats suffer from larger losses in sensitivity (cf. Tab. 1.), the measured reach penalties are slightly beyond these estimates at lowest baudrate. Towards 64 Gbaud 32QAM signals suffer from an increased reach penalty (+ ~1 dB). 64QAM signals are feasible to transit up to 54 Gbaud, where a +2.5 dB excess penalty has been observed.

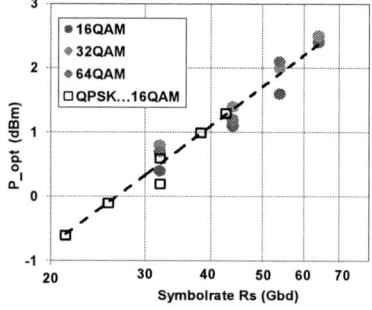

Fig. 4. Optimum fiber launch power vs. variable symbol rate.

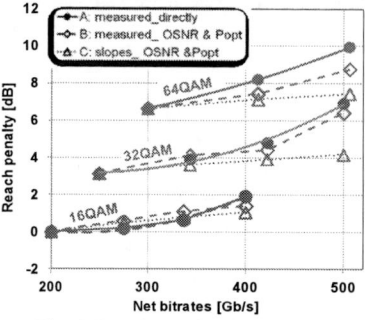

Fig. 5. Reach penalty for variable modulation format and baudrate.

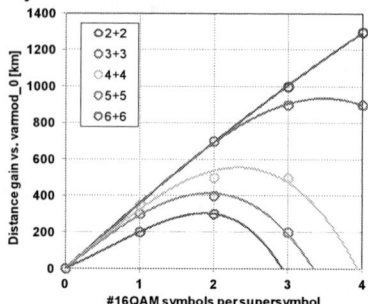

Fig. 6. Reach gain of TD-HM in presence of WSS in the link.

Finally we investigated the advantage of finer granular variation of the spectral efficiency by applying a combination of 8QAM and 16QAM signals (TD-HM formats) at 200 Gb/s bitrate. We defined supersymbols consisting of 4 symbols. Mode 0 (cf. Fig. 6.) applies 4 symbols carrying 16QAM operated at 32 Gbaud, mode 4 consists of 4 symbols carrying 8QAM operated at 43 Gbaud, and the intermediate modes contain a mixture of 8QAM and 16QAM symbols at intermediate baudrates. In that experiment we placed a variable number of WSS on transmit side and same number on receive side, the bandwidth of the transmission channel is narrowed step by step. Without bandwidth limitation clearly the mode 4 using 8QAM leads to highest transmission distances. But, if bandwidth limitation is used the maximum transmission reach is achieved at a reduced baudrate even if applying a less sensitive modulation format.

V. CONCLUSIONS

In conclusion we reviewed the challenges of variable transponders. The upper limits of flexible transponders regarding spectral efficiency and baudrate strongly depend on the effective resolution of the converters. We showed that standard QAM formats are the first option for the variability and in combination with a variable baudrate a multitude of applications with different spectral efficiency, variable spectral efficiency and variable reach requirements can be addressed. Furthermore using time domain hybrid modulation formats a finer granularity in spectral efficiency can be achieved which support the adaptation of transponders to the transmission channel in an improved way. The increase of bitrate at constant spectral efficiency requires the increase of transponder baudrates, which is limited by the converters ENOB. For each order of modulation format an individual maximum baudrates has to be taken into account. Finally the nonlinear distortions have to be taken into account, which lead to additional reach limitations for high baudrate data.

REFERENCES

[1] F. Buchali et al.," Preemphased Prime Frequency Multicarrier Bases ENOB Assessment and its Application for Optimizing a Dual-Carrier 1-Tb/s QAM Transmitter", in Proc. of OFC 2016, paper Th3A4.

[2] W. Idler et al., "Experimental Study of Symbol-Rates and MQAM Formats for Single Carrier 400 Gb/s and Few Carrier 1 Tb/s", in Proc. of OFC 2016, paper Tu3A.7

[3] W. Idler et al., "Hybrid Modulation Formats Outperforming 16QAM and 8QAM in Transmission Distance and Filtering with Cascaded WSS", Proc. OFC 2015, M3G4

[4] F. Buchali et al: „Optimization of Time-Division Hybrid-Modulation and its Application to Rate Adaptive 200Gb Transmission, ECOC 2014, paper Tu.4.3.1.

[5] F. Buchali et al., "Performance and Advantages of 100Gb/s QPSK/8QAM Hybrid Modulation Formats", Proc. OFC 2015, Th2A16

[6] F. Buchali et al., "Implementation of 64QAM at 42.66 GBaud using 1.5 Sample per Symbol DAC and demonstration of up to 300 km fiber transmission", Proc. OFC 2014, M2A1, San Francisco (2014).

[7] T. Lotz, et al., "Coded PDM-OFDM transmission with shaped 256-iterative-polar-modulation achieving 11.15-b/s/Hz intrachannel spectral efficiency and 800-km reach," J. Lightw. Technol., vol. 31, no. 4, pp. 538–545, Feb. 2014.

[8] F. Buchali, G. Böcherer, W. Idler, L. Schmalen, P. Schulte, and F. Steiner, "Experimental demonstration of capacity increase and rate-adaptation by probabilistically shaped 64-QAM," in Optical Communication (ECOC), 2015 European Conference on. Valencia, Spain, IEEE, 2015, pp. 1–3.

ThD2-2

OECC/PS2016

95.2Tb/s Transoceanic Seven-core Fiber Transmission Using Optical Pre-filtered 200Gbit/s-based PDM-QPSK WDM Signals

Yu Kawaguchi, Koki Takeshima, Daiki Soma, and Takehiro Tsuritani

KDDI R&D Laboratories Inc., 2-1-15 Ohara, Fujimino-shi, Saitama, 356-8502 Japan
uu-kawaguchi@kddilabs.jp

Abstract: 200Gb/s-based 95.2-Tb/s transmission using 64-Gbaud PDM-QPSK signals over 6,160-km seven-core fibers has been successfully demonstrated by optimization of optical pre-filtering without high-speed digital analog converter (DAC).
Keywords: Coherent communications, Multicore fiber, Ultra-long haul transmission

I. INTRODUCTION

It is expected that a rapid increase of internet traffic in the next decade due to the huge increase in the amount of traffic per person and the number of internet-connected devices worldwide [1]. In light of this, optical fiber transmission technology would be continuously evolving with higher channel speed. Recently, beyond 100-Gb/s techniques for 200G/400G/1Tb/s transmission system have been actively studying toward commercial deployment into metro/datacenter optical networks [2, 3]. In addition, in ultra-long-haul transmission over transoceanic distance, 54-Tb/s transmission was reported with 180 and 202.5-Gb/s channels over single-core single mode fibers (SMFs) [4]. However, it is difficult to achieve over 100-Tb/s transmission capacity over single-core SMFs because of a physical limit. Space division multiplexing transmission with multi-core fibers (MCFs) is a promising candidate to overcome such limit and transoceanic distance transmission over 100 Tb/s has been achieved [5, 6]. In such demonstrations, however, the channel bit rate is limited to lower than 100 Gb/s. Therefore, the feasibility of ultra-long haul MCF transmission with beyond 100-Gb/s channels should be investigated. To obtain beyond 100-Gb/s signals, there are two approaches. One is the use of high-order quadrature amplitude modulation (QAM) such as code modulation technique [7, 8] and another is the use of higher symbol-rate signals [9-11]. Considering core-to-core crosstalk and large transmission loss in MCF, the use of higher symbol-rate signals is more suitable in MCF transmission.

In this paper, ultra-long haul MCF transmission using PDM-QPSK signals with 64-Gbaud symbol-late is experimentally demonstrated. 64-Gbaud PDM-QPSK signals are generated in simple way using optical pre-filtering [9-11] without high-speed DAC for equalizing limited bandwidth of the optoelectronic components in the transmitter and the receiver. 70-GHz-spaced sixty-eight 200-Gb/s signals with 2.85 bit/s/Hz spectral efficiency are transmitted over 6,160-km MCF using seven-fold recirculating loop.

II. EXPERIMENTAL SETUP

Figure 1 shows the experimental setup. The even and odd channels were generated by multiplexing thirty-four external cavity lasers (ECLs), respectively, and then they were independently modulated by two IQ modulators (IQMs) driven by 64-Gb/s electrical binary signals, which were produced by multiplexing two 32-Gb/s patterns with pseudo-random bit sequence (PRBS, 2^{15}-1) from a pulse pattern generator (PPG). As a result, 140-GHz-spaced 64-Gbaud QPSK WDM signals without any spectral shaping are generated. After that, the QPSK WDM signals were simultaneously pre-filtered using a WaveShaper® [12] to enhance higher frequency components of the signals. Even and odd channels were combined onto a 70-GHz frequency grid using a polarization-maintaining fiber coupler and then polarization-multiplexed by polarization-multiplexing emulator (PME) with over 889-symbol delay between x and y polarizations for decorrelation. After that, these signals were fed into a wavelength selective switch (WSS) to automatically equalize signal powers of 68 channels by monitoring each signal power using an optical spectrum analyzer. Consequently, we obtained 68-channels 70-GHz-spaced 64-Gbaud PDM-QPSK signals optimized by optical pre-filtering in C-band. The nominal data-rate is 200 Gb/s assuming a LDPC-based soft-decision forward-error-correction (SD-FEC) with 25.5% overhead [13].

For multi-core fiber (MCF) transmission, the 68-channel WDM signals were launched into a specially configured seven-fold recirculating loop consisting of a span of 40-km seven-core fiber, a seven-core erbium-doped fiber amplifiers (EDFA), external gain tilt filters (GTFs), and optical switches (SWs). The seven loops were synchronously operated as reported in [14]. The seven re-circulating loops shared a common load switch and then identical copies of the WDM signals through a power splitter with different delays (at least 160-symbol delay) between cores for the signal decorrelation were launched into each optical SW [5, 6]. After amplification by the individual-core-pumped seven-core EDFA with fiber-based fan-in /fan-out (FI/FO) devices followed by external GTFs, each WDM signal was launched into each core of the 40-km MCF with FI/FO. The output signals from one core were sent to the re-circulating loop input of the next core, in a cyclic fashion. The core-to-core configuration can average out variations in span loss, dispersion, and other component imperfections [14]. A WSS and a single-core EDFA were inserted at input of the seventh core for gain equalization, and the gain flatness was automatically controlled by the WSS at input of the seventh core. A polarization scrambler was inserted at the input of the fifth core.

The transmitted signals were received by a digital coherent receiver after pre-amplification and channel selection with an optical band-pass filter (OBPF) with a bandwidth of 1 nm. Electrical signals from the receiver were stored in sets of 800k samples by using two synchronized digital oscilloscopes with 36-GHz analog bandwidth operating at 80 GSample/s. The stored data were processed offline by digital signal processing (DSP) as follows: The received signals were re-sampled to two sample/symbol. After chromatic dispersion compensation in the frequency domain, polarization de-multiplexing and signal equalization were performed by half-symbol-spaced FIR filters with 41 taps, which were adapted by the constant modulus algorithm (CMA). Finally, the signals were decoded and bit error rate (BER) was calculated.

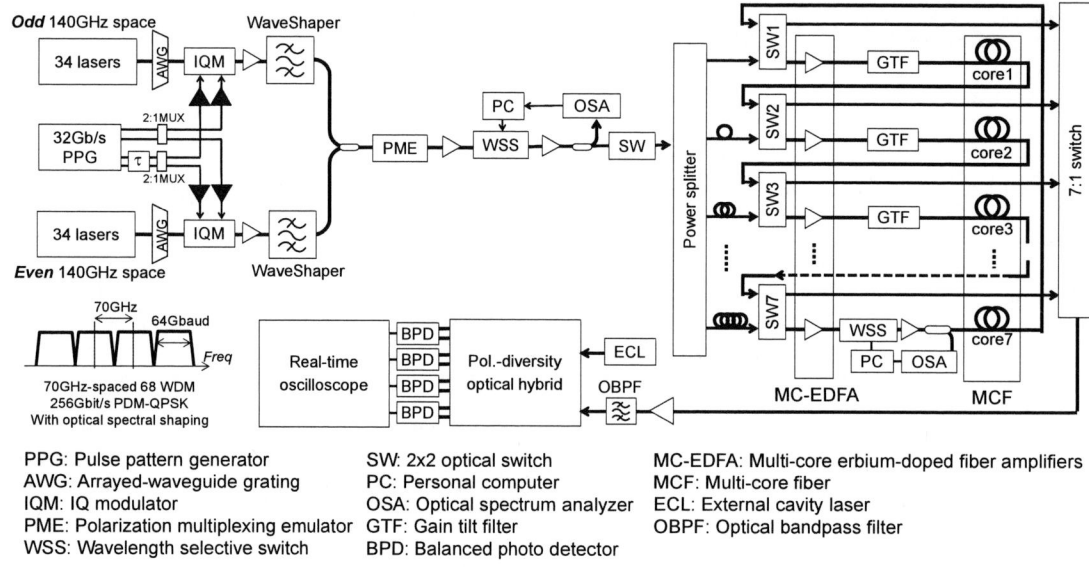

PPG: Pulse pattern generator
AWG: Arrayed-waveguide grating
IQM: IQ modulator
PME: Polarization multiplexing emulator
WSS: Wavelength selective switch

SW: 2x2 optical switch
PC: Personal computer
OSA: Optical spectrum analyzer
GTF: Gain tilt filter
BPD: Balanced photo detector

MC-EDFA: Multi-core erbium-doped fiber amplifiers
MCF: Multi-core fiber
ECL: External cavity laser
OBPF: Optical bandpass filter

Fig. 1. Schematic of experimental setup.

III. EXPERIMENTAL RESULT

First, we investigated the optimal pre-filtering depth (Δd) which was the difference between the maximum power and the center wavelength power on optical filter response characteristics. In this experiment, the spectrum of pre-filter was controlled with 2.5-GHz resolution and its shape was set to the inversely proportional to the signal spectrum in 70-GHz range at the center. Figure 2 shows Q-factor as a function of pre-filtering depth. In this measurement, we set an OSNR to 21 dB/0.1nm. Note that optical filter frequency response with pre-filtering depth of 0 dB corresponds to flat-top filter response. As shown in figure, it is found that the highest Q-factor is obtained when pre-filtering depth is set to 6~9 dB, and Q-factor is improved 1 dB compare to the case of flat-top filter response (Δd = 0 dB).

Next, we evaluated OSNR characteristics of 64-Gbaud PDM-QPSK signals. Figure 3 shows OSNR characteristics for single channel with/without pre-filtering and pre-filtered neighboring channels. From aforementioned results, pre-filtering depth was set in 6 dB. We found that pre-filtering technique resulted in improvement of Q-factor of about 1.0 dB for single channel. Additionally, even in WDM channels, Q-factors of pre-filtered channels were 0.5 dB better than that of single channel without pre-filtering. Next, we prepared WDM channels (68 channels, Frequency: 191.284 ~ 195.974 THz) and investigated the dependence of transmission distance. Figure 4 shows measured Q-factors and OSNRs as a function of transmission distance of 64-Gbaud PDM-QPSK signal. In this case, we measured the performance of only a center channel (ch.35). The launched power to MCF was set to 0.8 dBm/ch. The Q-factor decreased according to OSNR degradation as the transmission distance increased, but sufficiently high Q-factor was obtained even after 6,160-km transmission with the center channel. Additionally, Q-factor degraded in proportion to OSNR, which indicates that the significant performance degradation due to core-to-core crosstalk and fiber nonlinearity was not occurred even in ultra-long-haul transmission. Note that since the crosstalk of the optical switch in each loop input was relatively larger, the penalty caused by the crosstalk of the input optical switch was included in this Q factor penalty in particular for the shorter transmission distance.

Finally, we investigated the performance of all WDM channels (68 channels x 7 cores = 476 channels) after 6,160-km transmission. Figure 5 shows optical spectra before and after transmission and Figure 6 shows the measured Q-factors of all WDM channels after 6,160-km transmission. We confirmed that all WDM channels had Q-factors larger than 4.95 dB which is FEC limit [13].

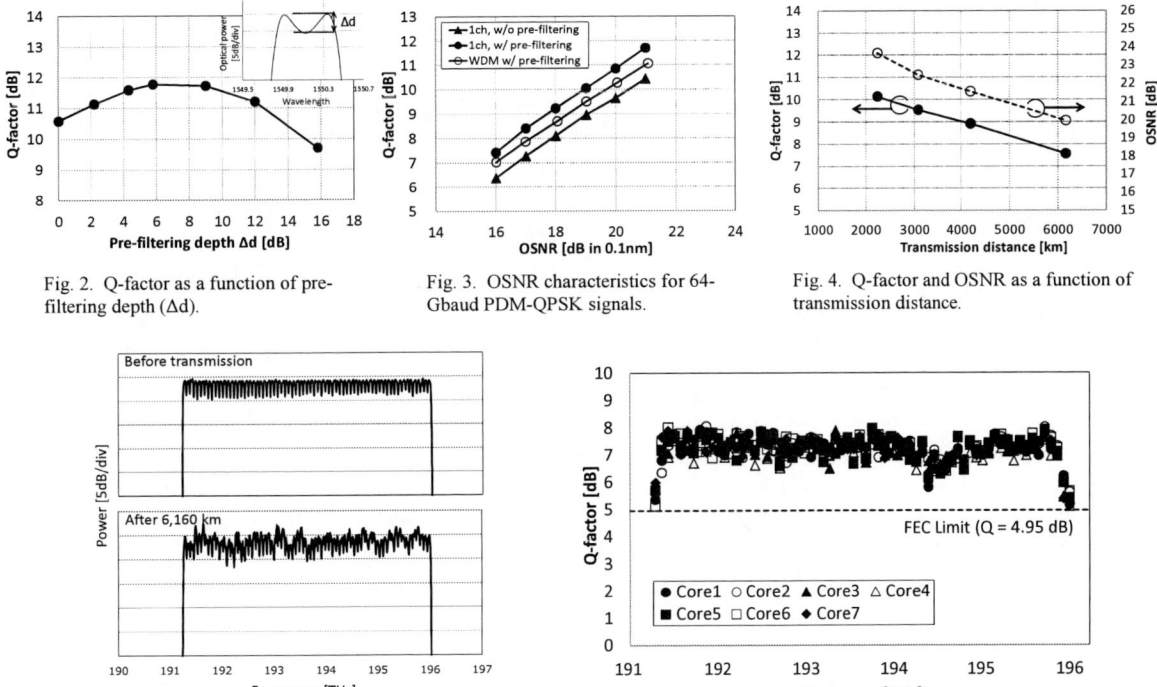

Fig. 2. Q-factor as a function of pre-filtering depth (Δd).

Fig. 3. OSNR characteristics for 64-Gbaud PDM-QPSK signals.

Fig. 4. Q-factor and OSNR as a function of transmission distance.

Fig. 5. Optical spectrum of before and after 6,160-km transmission.

Fig. 6. Q-factor of all WDM channels after 6,160-km transmission.

IV. CONCLUSIONS

We have successfully demonstrated 95.2-Tb/s seven-core fibers transmission using 70-GHz-spaced 64-Gbaud-based pre-filtered PDM-QPSK signals over 6,160km, which indicates the possibility of transoceanic transmission with the capacity of close to 100 Tb/s using 200-Gb/s signal and multicore fibers.

ACKNOWLEDGMENT

We would like to acknowledge Teledyne LeCroy for providing measurement equipment. This work was partly supported by "R&D of Innovative Optical Communication Infrastructure," commissioned research of the National Institute of Information and Communications Technology (NICT), Japan.

REFERENCES

[1] Cisco White Paper, Cisco Visual Networking Index: Forecast and Methodology, 2013–2018.
[2] http://www.oiforum.com/wp-content/uploads/OIF-Tech-Options-400G-01.0.pdf
[3] T. J. Xia, et al., OFC 2014, Tu2B. 1 (2014).
[4] J.-X. Cai, et al., ECOC 2014, PD.3.3 (2014).
[5] K. Igarashi, et al., Optics Express, Vol. 21, Issue 15, pp. 18053-18060 (2013).
[6] K. Igarashi, et al., ECOC 2013, PD3.E.3 (2013).
[7] J.-X. Cai, et al., OFC 2015, PDP Th5C.8 (2015).
[8] F. Yaman, et al., OFC 2014, PDP Th5B.5 (2014).
[9] G. Raybon, et al., Photonics Technology Letters, vol. 23, no. 22, pp.1667-1669 (2011).
[10] R. Rios-Müller, et al., ECOC2014, PD.4.2 (2014).
[11] A. Sano, et al., OFC2015, M3G.3 (2015).
[12] https://www.finisar.com/sites/default/files/downloads/waveshaper_1000s_product_brief_11_14.pdf
[13] K. Sugihara, et al., OFC/NFOEC 2013, OM2B.4 (2013).
[14] S. Chandrasekhar, et al., ECOC2011, Th.13.C.4 (2011).

OECC/PS2016

96Gbaud Nyquist-PDM-QPSK Signal Transmission over 12,120km using DP-AM-DAC and Decision-feedback Equalizer

Kengo Horikoshi[1], Fukutaro Hamaoka[1], Asuka Matsushita[1], Munehiko Nagatani[2], Hiroshi Yamazaki[2], Akihide Sano[1], Toshikazu Hashimoto[2], Hideyuki Nosaka[2], Kazushige Yonenaga[1], Akira Hirano[1], and Yutaka Miyamoto[1]

[1] NTT Network Innovation Labs., NTT Corporation, 1-1 Hikari-no-oka, Yokosuka, Kanagawa, Japan
[2] NTT Device Technology Labs., NTT Corporation, 3-1 Morinosato-Wakamiya, Atsugi, Kanagawa, Japan
Email:horikoshi.kengo@lab.ntt.co.jp

Abstract: *We demonstrate a high baud-rate data transmission using digital-preprocessed-analog-multiplexed DAC and spectral-narrowing mitigation algorithm. Over 12,120 km transmission of 96Gbaud PDM-QPSK WDM signal with 100-GHz of spacing has been achieved.*
Keywords: *Optical transmitters, Adaptive filters, Digital signal processing, Decision feedback equalizers*

I. INTRODUCTION

In order to accommodate with increasing demand for data transmission, beyond 100Gb/s optical data transport technologies such as 400Gbit/s class per-channel capacity are being studied intensively[1-5]. Raising the Baud-rate is inevitable to achieve higher per-channel capacity and longer transmission distances, simultaneously. The key problem for high Baud-rate signaling is the bandwidth limitation of electric devices such as analog-to-digital converters (ADCs) and digital-to-analog converters (DACs). DAC bandwidth is especially important since bandwidth narrowing at the transmitter degrade eventual signal quality more severely than that at the receiver.

One approach that greatly eases the bandwidth-limitation is to synthesize signals in the optical domain [4,5]. A more direct way is to enhance bandwidth of electric devices. Using InP hetero-junction-bipolar-transistor (HBT) based technologies, electric device bandwidth can be enhanced dramatically [6]. Recently, high baud-rate signal generation scheme using digital-preprocessed-analog-multiplexed DAC (DP-AM-DAC) is proposed in [7], which uses high-speed analog-multiplexer (A-MUX) to obtain high-baud-rate signal.

In this paper, we demonstrate a 96Gbaud data transmission scheme using DP-AM-DAC and spectral-narrowing tolerant digital signal processing (DSP) algorithm. InP-HBT based A-MUX provides both wide-bandwidth and good connectivity with DSP with relatively low complexity. A penalty from bandwidth limitation still seen at 96Gbaud was mitigated by spectrum-narrowing tolerant receiver DSP algorithm. We achieved over 10,000-km transmission with 307Gbps per lambda channel capacity and 3.07 bit/s/Hz spectral efficiency (SE).

II. COHERENT TRANSMISSION SYSTEM USING DP-AM- DAC AND DECISION FEEDBACK EQUALIZER

The DP-AM-DAC scheme uses two sub-DACs combined with an A-MUX. In case of coherent transmission using QAM modulation format, each in- and quadrature-phase signals are generated by a DP-AM-DAC. So we employed two DP-AM-DACs to obtain a QPSK signal with four sub-DACs.

The simplified schematic for DP-AM-DAC and its digital processing procedure are illustrated in Figure 1. Two sub-DACs generates partial analog signal in electric domain, and the A-MUX integrates the partial signals to synthesize a high-speed main signal. By applying digital preprocessing in frequency domain at the before sub-DACs, it is possible to generate arbitrary signal at the output of A-MUX..

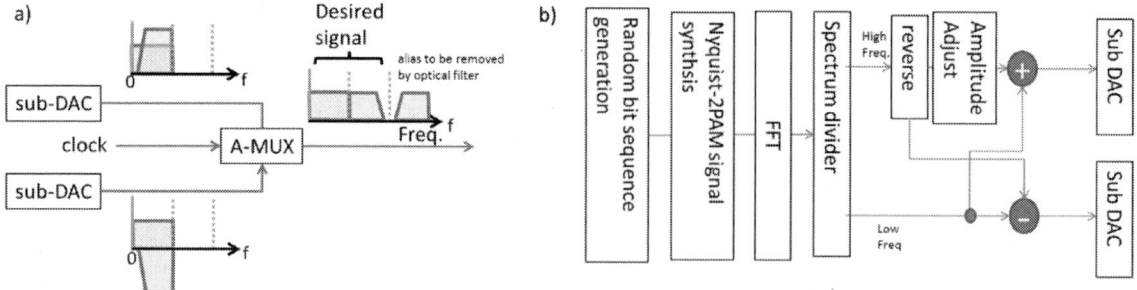

Figure 1 a): simplified schematic of A-MUX operation, b): Transmitter DSP

Decision feedback equalizer is employed to compensate bandwidth narrowing, which still occur even in the case with DP-AM-DAC. We previously reported mitigation of bandwidth narrowing penalty using maximum-likelihood sequence estimation [8], which was so effective but computation cost was high. DFE is not so powerful in bandwidth narrowing compensation as MLSE, but computation cost was low and practical option in real systems.

III. EXPERIMENTAL SETUP

Figure 2 illustrates simplified schematic of optical transmitter and receiver. The digital preprocessing procedure is as follows. Nyquist 2-PAM signal is numerically generated. In order to generate partial signals, the Nyquist 2-PAM signal is converted to frequency domain and divided into high and low frequency part. Amplitude of the high frequency signal is boosted to compensate principle frequency-response of A-Mux on the high frequency signal. The dividing position is 24GHz, which corresponds half of A-MUX clock frequency of 48-GHz. The high frequency part of signal is reversed about frequency. Low-frequency part and high frequency part are added to form first partial signal, and subtracted to form second partial signal. Two 2-PAM signals are preprocessed to form two set of partial signal pairs, which were eventually used to compose a QPSK signal.

The two pairs of partial signals are loaded onto a 4-channel arbitrary waveform generator (AWG). The two AWG channels generate partial signals with a sampling rate of 64 GSa/s; the outputs are fed to two A-MUXs to synthesize I and Q lane signals. Since the sub-DACs have to generate the 24-GHz bandwidth of the partial signals, half the Nyquist frequency of the 96Gbaud main signal, AWG sampling rate must be >48GSa/s. A clock signal for A-MUX was generated using a synthesizer and a frequency doubler. Since the clock signal needs to be synchronized to the partial signals from AWG, the synthesizer also provides a clock signal to the AWG via a frequency divider. I- and Q-lane main signals from the two A-MUXs are fed to wideband linear amplifiers and put into an optical IQ-modulator to form a Nyquist-QPSK optical signal.

The optical signal is passed through a programmable optical filter to remove aliasing spectrum. A polarization division multiplexing emulator using delay-line and polarization beam coupler was used to generate a PDM-QPSK signal. Figure 3 shows the transmission line setup. In the WDM transmission experiment, we used two DP-AM-DAC based optical transmitters. Each transmitter was provided with four wavelengths of CW laser light with 200-GHz frequency spacing and generated even and odd WDM channel signals. Signals from the two transmitters are coupled to make an 8-channel 100-GHz spacing WDM signal. The transmission line was a 404-km recirculation loop composed of four spans of 101-km pure-silica-core fiber (PSCF). Hybrid optical amplifiers using EDFAs and backward Raman pumping were employed. Distance dependence of received OSNR is shown in Figure 7.

The optical coherent receiver was composed of four balanced photo-diodes (BPDs) with a bandwidth of 70GHz and an optical hybrid. Signals from the BPDs were digitized by a 160-GS/s digital storage oscilloscope (DSO).

Receiver side digital signal processing (DSP) is done offline. A simplified block diagram of the offline DSP is given in Figure 5. The offline DSP was composed of CMA-based adaptive equalizer, FFT-4th power based frequency offset compensator [9], and Viterbi-Viterbi carrier phase recovery block. A symbol point adjuster algorithm was employed to compensate constellation alignment distortion. A decision feedback equalizer was employed to compensate the bandwidth narrowing yielded by the electrical devices.

IV. RESULTS

OSNR dependence of Q-factor of the received signal in back-to-back configuration is shown in Figure 4. OSNR, the difference between measured Q-factor and theory, is about 1.8 dB at the received Q-factor of 6.4-dB. The difference from theory may due to the bandwidth limitation of A-MUX and ENoB of AWG and DSO.

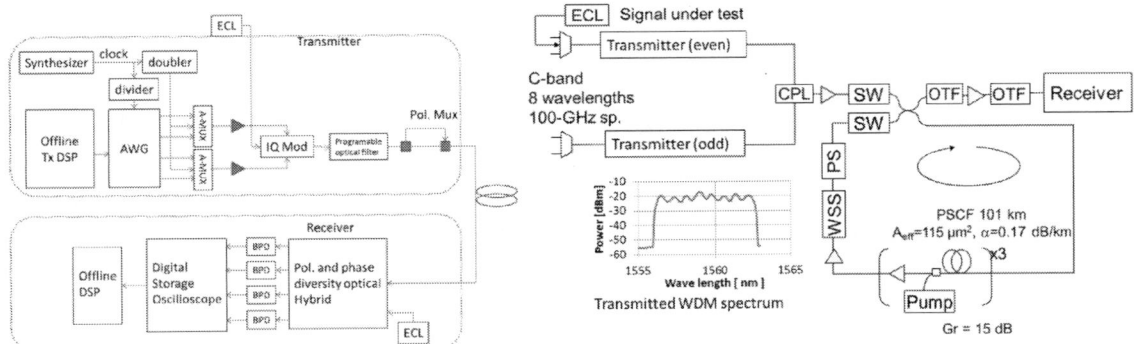

Figure 2: Transmitter and receiver   Figure 3: Transmission line

Figure 6 shows transmission distance dependence of the received Q-factor. By using symbol point adjuster and DFE, received Q-factor is enhanced by 0.4dB. This corresponds to approximately 500-km of reach extension. The transmission reach is more than 10,100 km assuming 20% overhead FEC with 6.4-dB of Q-limit [10], in which case the payload capacity is 307Gbit/s per channel and spectral efficiency (SE) is 3.07 bit/s/Hz. Assuming 25% overhead FEC with 5.0-dB of Q-limit [11], transmission distance is extended to >12,120km, with 288Gbit/s of per channel capacity and 2.88 bit/s/Hz of SE. Because of the limitations of the experimental system, we did not measure received Q-factor for distances over 12,120-km. The maximum reach for 25% overhead FEC is estimated to be 12,800-km.

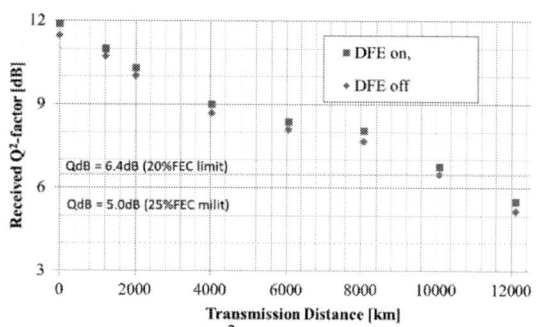

Figure 6: Received Q^2-factor after transmission

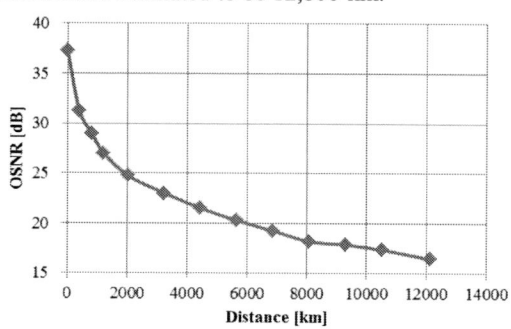

Figure 7: Measured OSNR at each distance

V. CONCLUSION

We demonstrated a 96Gbaud Nyquist-PDM-QPSK signal transmission by using DP-AM-DAC. The DFE contributed in expanding the transmission reach by ~500 km. The transmission reach of 10,100 km was achieved assuming 20% overhead FEC, with 307Gbit/s channel capacity and 3.07 bit/s/Hz of SE. When 25% overhead FEC was assumed, a transmission distance of 12,120-km was achieved, with 288-Gbit/s channel capacity and 2.88-bit/s/Hz of SE.

ACKNOWLEDGMENT

The transmission experiments were based on joint research with Japan's National Institute of Information and Communications Technology (NICT), and were performed using NICT's testing equipment in part

This work was partly supported by "The research and development project for the Tera-bit optical network technologies towards big data era" of the Ministry of Internal Affairs and Communications, Japan.

REFERENCES

[1] R. Rios-Müller et al, "1-Terabit/s Net Data-Rate Transceiver Based on Single-Carrier Nyquist-Shaped 124 GBaud PDM-32QAM", OFC 2015 Th5B

[2] G. Raybon et al, "160-Gbaud coherent receiver based on 100-GHz bandwidth 240-GS/s analog-to-digital conversion", OFC2015 M2G.1

[3] S. Randel et al, "All-Electronic Flexibly Programmable 864-Gb/s Single-Carrier PDM-64-QAM", OFC 2014 Th5C.8

[4] David Odeke Otuya et al, "Single-Channel 1.92 Tbit/s, 64 QAM Coherent Orthogonal TDM Transmission of 160 Gbaud Optical Nyquist Pulses with 10.6 bit/s/Hz Spectral Efficiency", OFC2015 M3G.2

[5] H. Mardoyan *et al.*, "Transmission of Single-Carrier Nyquist-Shaped 1-Tb/s Line-Rate Signal over 3,000 km", OFC2015 W3G.2

[6] Akihide Sano *et al.*, "5 x 1-Tb/s PDM-16QAM Transmission over 1,920 km Using High-Speed InP MUX-DAC Integrated Module", OFC2015 M3G.3

[7] H. Yamazaki *et al.*, "160-Gbps Nyquist PAM4 Transmitter Using a Digital-Preprocessed Analog-Multiplexed DAC", ECOC2015 PDP 2.2

[8] K. Horikoshi *et al.*, "Spectrum-narrowing tolerantsignal-processing algorithm usingvmaximum-likelihood sequence estimation for coherent optical detection", Electronics Letters vol.47(10)(2011)

[9] T. Nakagawa *et al.*, "Non-Data-Aided Wide-Range Frequency Offset Estimator for QAM Optical Coherent Receivers",OFC.2011.OMJ1

[10] K. Onohara *et al*, "Soft-Decision-Based Forward Error Correction for 100 Gb/s Transport Systems", Journal of selected topics in quantum electronics, vol.16(5) (2010)

[11] K. Sugihara, *et al*, "A Spatially-coupled Type LDPC Code with an NCGof 12 dB for Optical Transmission beyond 100 Gb/s", OFC2013 OM2B.4(2013)

ThD2-4 (Invited)

OECC/PS2016

Enabling Technologies for High Spectral Efficiency and High Capacity Transmission over Transoceanic Distances

Matt Mazurczyk

TE SubCom, 250 Industrial Way West, Eatontown, NJ USA
mmazurczyk@subcom.com

Abstract: We review advances in the technologies used to enable high capacity transoceanic distance transmission including transmitter DSP, advanced modulation formats, and C+L transmission. Demonstrations including transmission of 49.3Tb/s over 9,000km are presented.
Keywords: Coherent communications; Fiber optics communications

I. INTRODUCTION

The demonstrated capacity of transoceanic systems, typically defined to have lengths from 6,000 km to more than 10,000 km continues to grow. Figure 1 reviews the results of recent single core coherent capacity demonstrations at these lengths beginning with the 2009 coherent PDM-QPSK result of 7.2 Tb/s at 7,200 km [1-17]. A significant increase in capacity results from the use of advanced modulation formats and transmitter digital signal processing in the form of near Nyquist PDM-16QAM based coded modulation (CM) [9] to provide higher spectral efficiency (SE), enabling a C-band only capacity of 30 Tb/s at 6,630 km. A second significant increase in capacity is provided by increasing the optical transmission bandwidth to include the L-Band to provide ~9 THz of transmission bandwidth. Results include 49.3 Tb/s at 9,100km using PDM-16QAM based CM formats with C+L band EDFAs [13] and 54 Tb/s over 9150 km using similar formats and hybrid C-Raman EDFAs [14]. A further increase in capacity results from the use of 64QAM based CM to achieve 61.9Tb/s at 5,920km using hybrid Raman-EDFA amplification [14]. This paper reviews these results as well as a number of technology enablers including transmitter digital signal processing, advanced modulation formats, and the use of C+L band transmission.

II. TRANSMITTER DIGITAL SIGNAL PROCESSING AND DIGITAL-TO-ANALOG CONVERTERS

The introduction of ultra-high speed digital-to-analog converters and transmitter digital signal processing has enabled precise control and flexibility over the transmitted electrical field. This allows direct synthesis of arbitrary electrical fields, enabling higher modulation formats beyond PDM-QPSK to PDM-nQAM formats such as PDM-16QAM or PDM-64QAM as well as more generalized PDM-APSK formats. Along with flexible control over the constellation and modulation format, the spectrum of the transmitted signals can be shaped to avoid linear crosstalk and allow for near Nyquist spacing without introducing ISI [18]. See Figure 2. Additionally, the use of super Nyquist transmission in which the channel spacing is less than the symbol rate has been investigated [1, 16]. However, the receiver DSP techniques required can be computationally complex and may be difficult to generalize for modulation formats beyond PDM-QPSK.

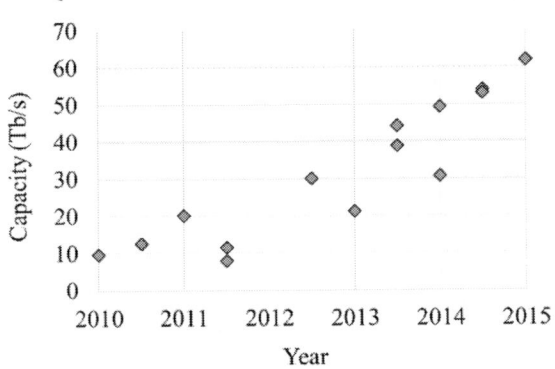

Figure 1: Transoceanic coherent capacity demonstrations

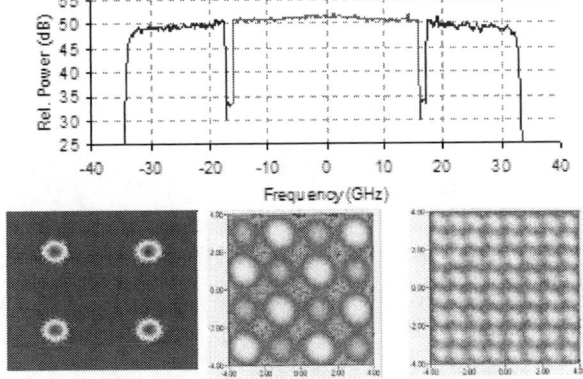

Figure 2: Spectrum of near Nyquist WDM channels (top) QPSK, 16QAM CM, and 64QAM constellations (bottom)

371

III. Advanced Modulation Formats, Coded Modulation and Variable Spectral Efficiency

Despite their long distance, transoceanic links can be designed to produce receive OSNR well in excess of what is required for reliable PM-QPSK transmission with FEC coding. Advanced modulation formats including PDM-16QAM, PDM-64QAM and others use this margin to provide significantly greater SE and capacity. These higher order modulation formats can also be used to form the basis for CM formats [19]. These formats design the error correction codes together with the modulation format to increase the Euclidean distance between valid sequences of two dimensional symbols. CM formats can also be used in conjunction with an outer FEC code to provide the benefits of iterative decoding. Table 1 shows details of several CM formats used in ultra-long haul high capacity demonstrations that range in SEs from 2.4 to 8.0 b/s/Hz. All of the formats use the same outer 20% LDPC code. The table illustrates an additional benefit of CM which is that a range of variable spectral efficiencies and receiver sensitivities can be obtained by changing the CM scheme while keeping the symbol rate and LDPC code fixed.

In long transoceanic lengths with large bandwidths that can exceed 9 THz, some variation in OSNR with wavelength can be expected due to variations in noise figure, fiber loss, as well as fiber nonlinear effects arising from differences in path-averaged power as well as Raman gain. In this case, a variety of CM formats can be used across the optical bandwidth to tailor the spectral efficiency of each channel to the available effective SNR [13]. Figure 3 shows a case in which two 16QAM CM formats were used over 9100 km to deliver 49.3Tb/s [13]. Calculations show that by using a larger number of CM formats capacity could be increased further to 51.2 Tb/s.

Other techniques have also been demonstrated to provide variable spectral efficiency and receiver sensitivity. These include the approach of hybrid QAM, where each channel sends a pre-determined ratio of two different SE constellations, for example QPSK and 16QAM [21]. The SE can then be varied by changing the ratio of the number of symbols sent with each constellation. An additional technique is the use of multi-rate FEC [13] in which the modulation format is fixed but different FEC codes are used to provide variable spectral efficiency.

TABLE 1: Details of CM used in high capacity transoceanic demonstrations							
SPC Ratio	Constel-lation	SE (b/s/Hz)	Overall OH	Error Correction Threshold (dB)	Capacity [Tb/s]	Distance [km]	Ref.
5/6 SPC	64QAM	8.0	44%	3.6	61.9	5920	[14]
15/16 SPC	16QAM	6.0	28%	4.9	30.4	10,290	[17]
9/12 SPC	16QAM	4.9	60%	3.35	44.1	9100	[11]
10/12 SPC & 9/12 SPC	16QAM	5.4 4.86	44% 60%	4.1 3.35	49.3	9,100	[13]
9/12 SPC	16QAM	4.9	60%	3.35	44.1	9,100	[22]
3/4 SPC	QPSK	2.4	60%	3.4		>60,000	[21]

IV. Expanded Optical Bandwidth Using C+L Transmission

Ultimately, the capacity available in C-band at transoceanic distances is limited by the available BW. In practice, penalties from the nonlinear Kerr effect limit the effective SNR and spectral efficiency that can be achieved [24]. To grow capacity further, recent demonstrations have extended the transmission bandwidth beyond the 40nm C-band into the L-band to provide bandwidths in excess of 70 nm. Amplification can be provided for this extended bandwidth either with the use of separate C and L band EDFAs with band-splitters and combiners or with the use of hybrid Raman-EDFAs.

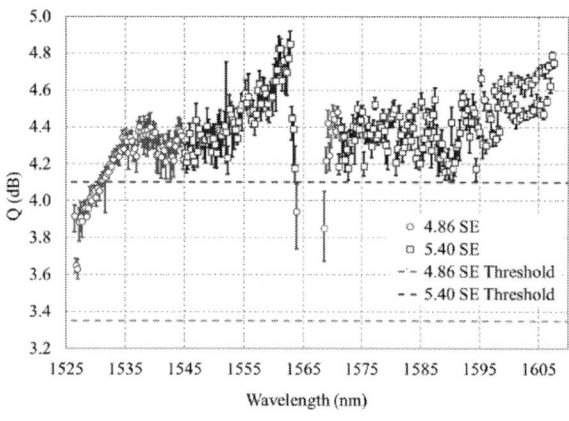

Figure 3: C+L EDFA transmission results after 9,100km

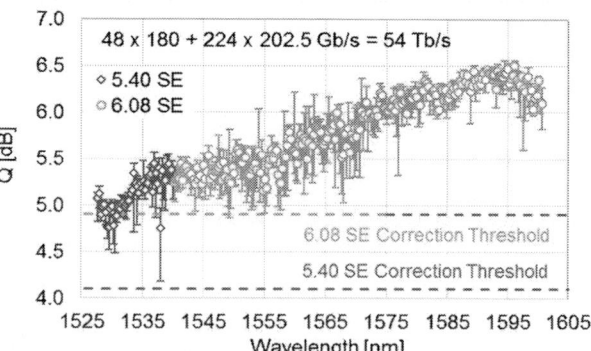

Figure 4: Hybrid Raman-EDFA transmission results after 9,150 km

The effectiveness of increasing the transmission bandwidth as well as the relative merits of these two amplification schemes can be investigated by comparing a demonstration of 49.3 Tb/s over 9,100 km using C+L EDFA amplification [13] and a hybrid Raman-EDFA amplification demonstration of 54 Tb/s over 9,150 km [15]. The use of hybrid

Raman-EDFAs allowed for only a 10% increase in capacity at similar distances, despite having a significantly lower theoretical noise figure. This is due in part to lower path averaged power (see Figure 6) required to achieve the optimal balance of ASE and non-linear penalty [15, 25]. Analysis further shows that this modest increase in capacity required approximately double the electrical power to supply the Raman-EDFAs, a significant concern for undersea systems in which power needs to be distributed from shore over distances of at least half the system length.

An additional design feature of these experiments is that the amplification is designed to allow all channels across the band to operate at the peak of their performance curves. In conjunction with variable SE formats this allows for maximization of capacity rather than only maximizing the performance of the worst channels. The variations of ASE, noise figure and non-linear penalties across the bandwidth are accounted for in this optimization. Confirmation of this approach can be seen in pre-emphasis data of Figure 6 that shows that channels across the whole transmission bandwidth have optimal performance at the same level of pre-emphasis [15].

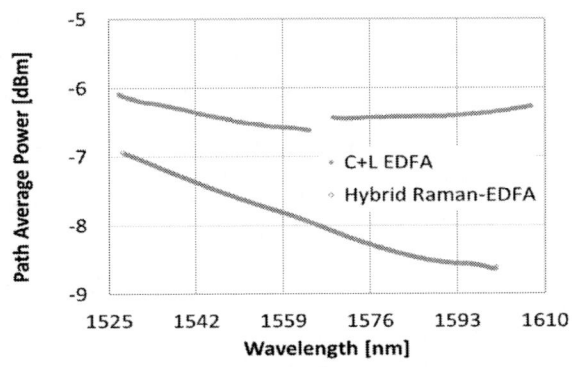

Figure 5: Path average channel power for C+L systems with amplification designed for maximum performance across the band

Figure 6: Pre-emphasis for channels across the transmission band with amplification designed for maximum performance across the band

V. CONCLUSIONS

A number of technologies have been reviewed that have contributed to the ~5x increase in demonstrated undersea capacity over the last several years. The use of advanced modulation formats and CM have enabled higher spectral efficiencies and the ability tailor each channels SE to its effective receive SNR. The use of extended amplification bandwidth also enables a significant capacity increase.

ACKNOWLEDGMENT

The author is pleased to acknowledge his colleagues at TE SubCom including J.-X. Cai, H. G. Batshon, A. Turukhin, Y. Sun, O. Sinkin, C.R. Davidson, D. Foursa, M. A. Bolshtyansky, A. Pilipetskii, and Neal S. Bergano

REFERENCES

[1] H. Masuda et al., Proc. OFC2009, PDPB5
[2] G. Charlet et al., Proc. OFC2009, PDPB6
[3] M. Salsi et al., Proc. ECOC 2009, PD2.5
[4] J.-X. Cai et al., Proc. OFC2010/NFOEC, PDPB.10
[5] J.-X. Cai et al., Proc. ECOC2010, PDPB.1
[6] J.-X. Cai et al., Proc. OFC2011/NFOEC, PDPB.4
[7] M. Salsi et al., Proc. ECOC2011 Th.13.C.5
[8] D. Qian et al., Proc. ECOC 2011 Th.13.K.3
[9] M. Mazurczyk et al., Proc. ECOC2012 Th.3.C.2.pdf
[10] H. Zhang et al., Proc. OFC/NFOEC 2013, PDP5A.6
[11] D. Foursa et al., Proc. ECOC2013, PD3E.1
[12] M. Salsi et al., Proc. ECOC2013, PD3E.2
[13] J.-X. Cai et al., Proc. OFC2014, Th5B.4
[14] F. Yaman et al., Proc. OFC2014, Th5B.5.
[15] J.-X. Cai et al.," Proc. ECOC2014, PD.3.3
[16] A. Ghazisaeidi et al., Proc. ECOC2014, PD.3.4
[17] J.-X. Cai et al., Proc. OFC2015, Th5B.3
[18] M. Mazurczyk, JLT, Vol. 32, No. 16 2915-2924
[19] H.G. Batshon et al., Proc. OECC 2015,ThA.92
[20] K. Igarashi, Proc. OECC2015 , JMoA.42

[21] J. –X. Cai, et al., Opt. Exp., vol. 22, no. 8,
[22] J.-X. Cai, et al., Proc. OFC 2014, OTu2B.3
[23] W. R. Peng, et al. OECC C 2011, 8D2-4
[24] R.J. Essiambre et al, JLT, Vol. 28, No.4, 662-669
[25] P. Poggiolini, JLT, vol. 30, no. 24, pp. 3857-3879

OECC/PS2016

High-speed avalanche photodiode for data center networks

Masahiro Nada, Takuya Hoshi, and Hideaki Matsuzaki

NTT Device Technology Laboratories, NTT Corporation, 3-1 Morinosato Wakamiya, Atsugi, Kanagawa Japan

nada.masahiro@lab.ntt.co.jp

Abstract: This paper presents high-speed avalanche photodiodes (APDs) featuring a unique hybrid absorber and their application for future large capacity datacenters. The APDs exhibit high-sensitivity operation for 50-Gbit/s NRZ optical signal as well as for 28-Gbaud PAM4.

Keywords: Avalanche photodiodes; Ethernet; Optical-fiber communications systems; Datacenter.

I. INTRODUCTION

The huge explosion of computing and storage capacity in datacenters has led to need for higher bit rates and more effective use of bandwidth. One major way to increase the capacity in datacenter networks is to utilize a higher-order modulation (HOM) format such as 400 Gbit/s Ethernet (400GbE), which is mainly supported by a baud rate of 25 Gbaud with 4-level pulse-amplitude modulation (PAM4) [1]. Although utilizing the 25-Gbaud PAM4 format helps an effective increase of the bit rate per lane, datacenters still strongly require an increase in the baud rate; namely, a baud rate of over 50 Gbaud in order to reduce number of lanes, minimize transceiver size, and increase port density.

Regardless of the type of signal format, transmission distance is important for datacenter networks. Transmission distances of up to 10 km, which has been discussed in 400GbE standardization, will cover only 60% of inter-datacenter networks, while 40 km can cover up to 95% [2]. Although improving the receiver sensitivity so that the transceiver covers 40-km transmission reach is compelling, which should be achieved while maintaining low power consumption and small transceiver size, it is a technical challenge due to the lack of sufficient responsivity of conventional a pin-photodiode (pin-PD). One way to boost the receiver sensitivity is to use a semiconductor optical amplifier (SOA) with a pin-PD, but this sacrifices power consumption and transceiver size.

The avalanche photodiode (APD) is an attractive alternative to the SOA to ensure low power consumption and small transceiver size owing to its internal electrical gain. Different from APDs designed for 100GbE, APDs for beyond 100GbE have to have good linearity against optical input power along with large gain, responsivity, and bandwidth to meet the requirements of future datacenter networks utilizing HOM or a baud rate of over 50 Gbaud. In this paper we describe the optimized design of an APD for such an application. The APD features a unique absorber consisting of undoped and p-doped layers, which can relax trade-offs among responsivity, bandwidth, and linearity. Using an APD-based receiver, we achieved 40-km transmission with both the 25G-baud PAM4 and 50-Gbaud NRZ signal formats.

II. DESIGN AND CHARACTERISTICS OF APDs

Figure 1(a) shows a schematic view of a fabricated APD. In order to apply the APD to the HOM format, its linearity should be improved. A hybrid absorber containing p-doped and undoped InGaAs is a key design for optimizing the linearity while ensuring large bandwidth and responsivity. The reasons the hybrid absorber can provide this good performance are as follows [3, 4]. The linearity of an APD is degraded mainly due to the space charge effect in the undoped absorber as shown in Fig. 1 (b). When optical input power increases at the operating-bias condition, photo-generated holes accumulate around the p-doped field control layer. These accumulated holes cancel out the charges of acceptors in the depleted p-doped field control layer. This results in lowering the electric field of the avalanche layer and then reducing the gain of the APD. Linearity in responsivity is thus degraded. Thinning the undoped absorber can suppress the space charge effect and improve the linearity of the APD, but it will degrade the responsivity. If we intentionally replace a part of the undoped absorber on the p-type contact layer side with a p-doped absorber as shown in Fig. 1(c), the distance that the holes in the undoped part have to travel can be reduced, and then accumulation of the holes is effectively suppressed while keeping absorber thickness, that is, without sacrificing responsivity.

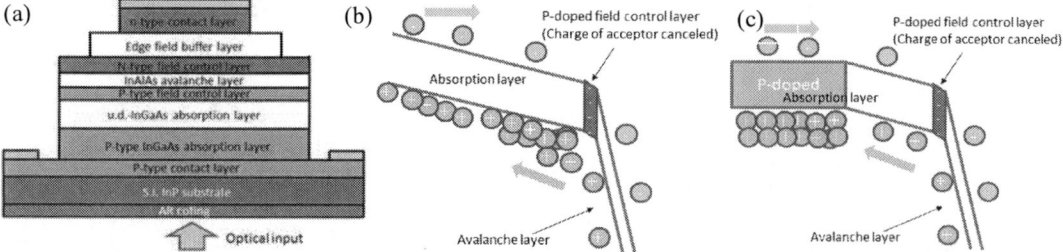

Fig. 1. (a)Schematic cross-sectional view of APD. Band diagrams of the APD for the (b) depleted absorber and (c) hybrid absorber.

374

In what follows, the linearity of two APDs with undoped absorbers of different thicknesses is compared. One has a 300-nm undoped absorber and the other has a 600-nm one. Fig. 2(a) shows the responsivity compression with increasing optical input power for both APDs. In order to eliminate an effect of the difference in responsivity on unit gain, output photocurrent instead of the optical input power is selected as the horizontal axis [5]. The thinner one maintains the compression of less than 1 dB up to the output photocurrent of 0.9 mA, while the thicker one does so only up to the output photocurrent of 0.46 mA. Significant improvement in the linearity can be obtained by reducing the portion of the undoped absorber thickness against the whole thickness of the hybrid absorber.

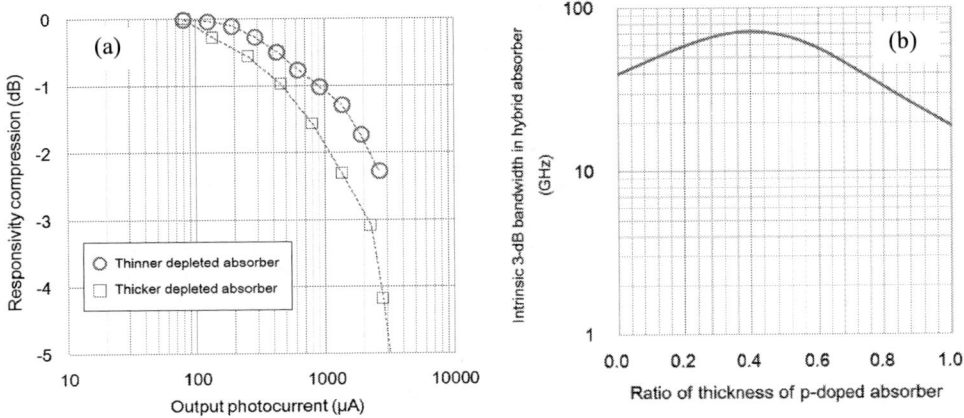

Fig. 2. (a)Output photocurrent dependence of compression of the linearity of two APDs with different depleted-absorber thicknesses. (b) Change in intrinsic f_{3dB} in hybrid absorber against ratio of thickness of p-doped absorber.

The hybrid absorber also contributes to extending bandwidth owing to the difference in carrier transport mechanisms between the p-doped and undoped absorbers. In the hybrid absorber, the electron and hole transit times are respectively dominated by diffusion time in the p-doped absorber and drift time in the undoped one. The carrier transit time can be minimized by optimizing the ratio of p-doped and undoped absorber thickness, once the total thickness of the absorber is determined in terms of obtaining the required responsivity. Fig. 2(b) shows the intrinsic 3-dB bandwidth (f_{3dB}) estimated from carrier transit time in the hybrid absorber against the ratio of thickness of the p-doped absorber when the total absorber thickness is 600 nm. As the thickness ratio increases, the intrinsic f_{3dB} also increases, and it reaches a maximum value of 70 GHz at the ratio of 0.40. Thus, the hybrid absorber can improve both linearity and bandwidth without reducing the total absorber thickness, and therefore, avoid the degradation of responsivity.

Fig. 3 shows the gain-bandwidth characteristics of the fabricated APD. The APD contains a 600-nm-thick hybrid absorber with the optimized thickness ratio of the p-doped and undoped absorbers, a 90-nm-thick InAlAs avalanche layer, and a 300-nm-thick edge-field buffer layer. The responsivity at unity gain was 0.69 A/W at 1300-nm wavelength. The fabricated APD exhibits maximum f_{3dB} of 35 GHz and large f_{3dB} of 20 GHz even at a gain of 10. The obtained characteristics are good enough for the 28-Gbaud PAM4 and 50-Gbaud NRZ operation.

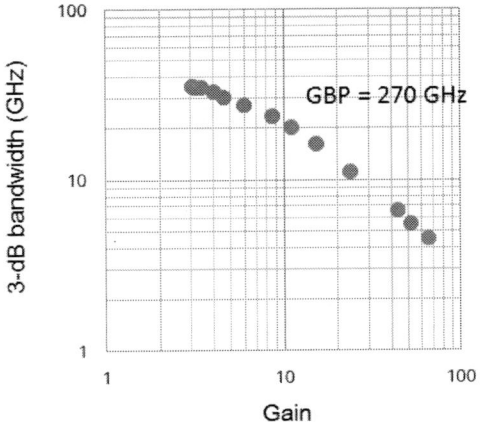

Fig. 3. Gain-bandwidth characteristics of fabricated APD.

III. PERFORMANCE OF APD-BASED RECEIVER

We investigated the basic characteristics of the APD-based receiver for 50-Gbit/s transmission with 28-Gbaud PAM4 and 56-Gbaud NRZ modulation formats. Here, considering the use of FEC in practical systems, the overhead of the bit rate was taken into account. In the 28-Gbaud PAM4 and 56-Gbaud NRZ transmission tests, we respectively used a commercially available linear trans-impedance amplifier (TIA) with an f_{3dB} of 20 GHz and a limiting TIA with an f_{3dB} of 28 GHz. Both experiments were done with a 1.3-μm electro-absorption modulator integrated with a DFB laser (EML) featuring an f_{3dB} of over 30 GHz [6]. The optical input signal with pseudorandom bit sequences (PRBSs) lengths of 2^{15}-1 was injected into the APD-based receiver through a 40-km single-mode fiber with a loss of 0.47 dB/km. The electrical output signal from the APD-based receiver was demodulated by offline digital signal processing, in which we used an adaptive feed-forward equalizer (FFE). Fig. 4(a) shows bit-error-rate (BER) characteristics of the APD-based receiver in the 28-Gbaud PAM4 test for back-to-back and 40-km transmission. For both transmission conditions, the TAP number of FFE was 17. The APD-based receiver exhibited a received power of -17.0 dBm at a BER of less than 2.3 × 10^{-4} for each transmission condition, which corresponds to an error-free operation with KP4 FEC [7]. The BER rapidly degraded for the received power over -13.0 dBm because of the input saturation for the TIA [3]. Fig. 4(b) shows the BER characteristics in the 56-Gbaud NRZ test with the same TAP number. As in the case of the 28-Gbaud PAM4 format, we succeeded in 40-km transmission. Obtained received power at a BER of less than 2.3 × 10^{-4} was -20.8 dBm for 40-km transmission. The BER reached the lower limit of measurement for received power of over -17.0 dBm for both conditions. In addition, we investigated dependence of the BER on the TAP number of FFE for 40-km transmission with 56-Gbit/s NRZ format. We set the received power to -20 dBm, where the BER is sufficiently small for the KP4 FEC limit. As shown in Fig 4(c), the BER stayed almost constant from 17 to 11 TAPs, and then it gradually degraded. The BER of 2.1 ×10^{-3} obtained at 3 TAPs is low enough for error-free operation if we assume stronger FEC, which would allow a BER of 4.4 x 10^{-3} [8].

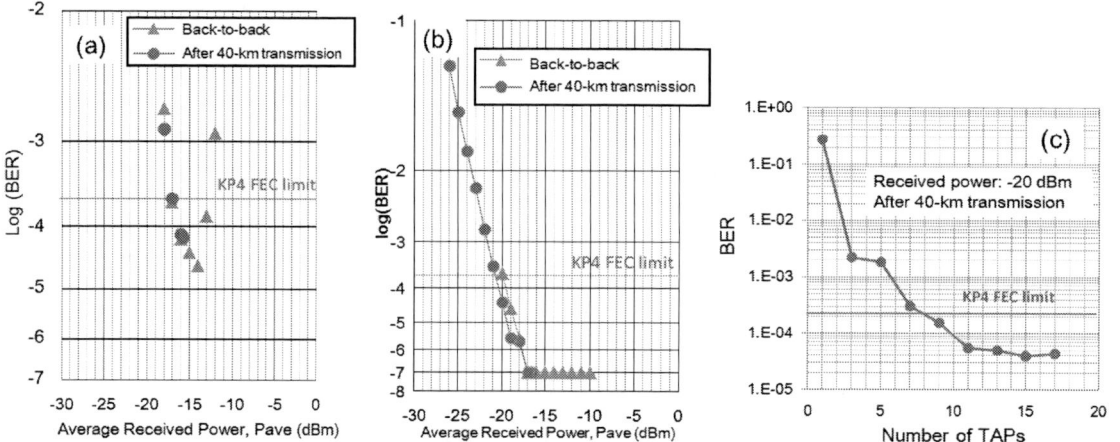

Fig. 4. BER characteristics of the APD receiver for (a)28-Gbaud PAM4, and (b) 56-Gbaud NRZ signals. (c) TAP number dependence of BER for 56-Gbit/s NRZ signals. The received power was set to -20 dBm at the measurement.

IV. CONCLUSIONS

We presented the design and performance of an APD applicable to 28-Gbaud PAM4 and 56-Gbaud NRZ transmission formats. The APD features a unique hybrid absorber that provides large bandwidth and good linearity with large responsivity. Using an optical receiver made with the APD, we achieved 40-km transmission with both 28-Gbaud PAM4 and 56-Gbaud NRZ formats. Even when the TAP number is reduced to 3, the BER is maintained at less than 2.1 × 10^{-3} for 56-Gbaud NRZ format. The presented APD has sufficient performance for realizing high-speed optical receivers for long-range transmission with low power consumption.

ACKNOWLEDGMENT

The authors thank T. Ishibashi, Y. Muramoto, F. Nakajima, T. Yoshimatsu T. Ohno, S. Kanazawa, H. Yamazaki for valuable discussions and suggestions, and S. Kodama for continuous encouragement.

REFERENCES

[1] http://www.ieee802.org/3/400GSG/.
[2] http://www.ieee802.org/3/400GSG/public/14_01/sone_400_01_0114.pdf
[3] M. Nada et al., Optics Express, **23**, 247801 (2015).
[4] M. Nada et al., Optics Express, **22**, 209867 (2014).
[5] Q. Zhou et al., IEEE Photon. Technol. Lett., **25**, 907 (2013).
[6] W. Kobayashi et al., Electron. Lett., **50**, 683 (2014).
[7] J. Man et al., in proc. OFC2014, M2E.7 (2014).
[8] M. Olmendo et al., in proc. OFC2014, M2E.5 (2014).

Low Voltage High Speed Si-Ge Avalanche Photodiodes

Zhihong Huang, Cheng Li, Di Liang, Kunzhi Yu, Marco Fiorentino, Raymond G. Beausoleil

Hewlett Packard Laboratories, Palo Alto, CA 94304

Zhihong.huang@hpe.com

Abstract: We present a Si-Ge Avalanche Photodiode (APD) design and achieved a breakdown voltage of -10V, a speed of 25GHz, a gain-bandwidth product (GBP) of 276GHz and operation up to 30Gbps data rate.

Keywords: Photodetectors, Avalanche Photodiodes

I. INTRODUCTION

The increasing demand for bandwidth and power efficiency in data centers has accelerated the development of photonic interconnects and optical switches. A key component of a switch with optical input/output ports is a CMOS-compatible photodiode at telecommunication wavelengths. Silicon-germanium p-i-n photodiodes have been widely developed in the past 10 year and have achieved good performances compared to their III-V counterparts [1]. Si-Ge photodiodes have the advantage of low cost and large scale manufacturability. In order to further reduce the power consumption and cost avalanche photodiodes with low voltage supply and high speed have been studied. Previously, Si-Ge based APD have achieved a sensitivity of -30dBm at 10Gbps, but such APDs operated above 25V and at a data rate of 10Gbps [2,3,4]. Although low voltage Ge-only APDs have been studied recently and have achieved operation below 10V, the use of germanium as both the absorption and multiplication region increases the dark current, restricts the APD gain-bandwidth product preventing the device from operating at both high gain and high speed. In this paper, we first discuss the design considerations for achieving low voltage and high speed APDs. Then we demonstrate an APD with a breakdown voltage of 10V, and a speed of 25GHz. We experimentally demonstrate a gain-bandwidth product of 276GHz. This design enables the use of APD in data centers. Its use can greatly lower the laser power and total energy consumption in an optical link. Alternatively, a single laser can provide power for multiple WDM channels, hence reducing the system cost.

II. DESIGN AND MEASUREMENT

Our waveguide APD is designed with a separate absorption-charge-multiplication (SACM) structure. With proper charge-layer doping, a high electric field can be confined in the silicon multiplication region and a low field in germanium. Fig. 1(a) shows a schematic of the device and the inset shows the layer doping and thicknesses for each layer. In order to achieve low voltage operation, the thickness of the intrinsic silicon multiplication region was reduced to ~100nm. This enables less than 10V operating voltage while still producing an electric field in silicon sufficient for carrier multiplication. Optically, the waveguide design allows light travel through silicon and evanescently couple to the germanium absorption region, making the responsivity independent of the germanium thickness. This gives us flexibility in design and makes the device capable of working at both high speed and high responsivity. During operation, photo-generated carries in germanium drift into silicon. Once it gets into the silicon, the high electric field causes impact ionization and generate more electron-hole pairs. Since silicon has a much lower impact ionization ratio than most III-V materials, many more electrons are generated than holes in the multiplication process, effectively reducing noise figure and increasing the gain-bandwidth product.

Fig. 1 (b) shows the current versus voltage (IV) curve of a 4μm×10μm device (design S1). This device has a DC punch-through voltage of -2V and a breakdown voltage of -10V. The DC punch through voltage is indicated by the small hump in the photo current. At this bias the conduction band energy barriers between silicon and germanium is lowered and more photo-generated electrons can overcome the barrier and move into silicon. The gain is determined by comparing the responsivity of the APD with a p-i-n photodiode having the same structure but without charge layer doping (design S3). Fig. 1 (c) shows another device (design S2) with a higher charge layer doping where the electric field is confined completely in silicon. Without electric field in germanium, photo-generated carriers cannot drift into silicon, hence no photocurrent is collected. The IV curve also shows that this device has a breakdown voltage of -8V. This also indicates that in the S1 device there is a 8V drop across the silicon intrinsic region and a 2V drop in germanium. Fig. 1 (d) shows the IV curve for the S3 device which has the same structure but no charge layer doping. Since it lacks a p-type charge layer doping, the electric field is applied on germanium even at a small bias and saturation current occurs near 0V. However since the electric field in silicon is low, no multiplication process occurs in silicon and this device doesn't show gain during normal operation. A number of devices for designs S1 and S3 were measured. The best responsivity of 1.05A/W, corresponding to an internal quantum efficiency of 84% at 1550nm, was measured in a 4μm×50μm device.

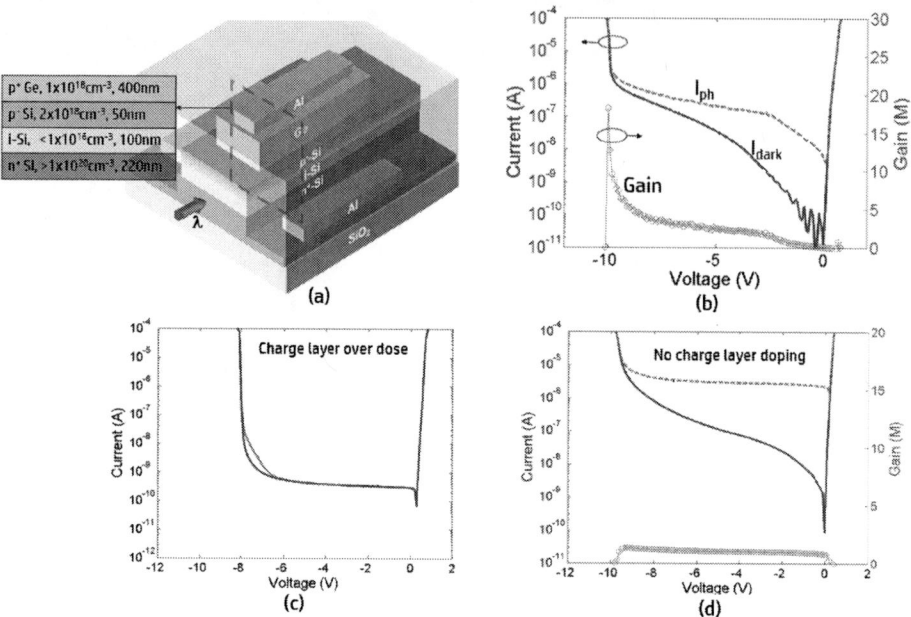

Fig. 1: (a) a schematic of the waveguide APD structure. The inset shows the layer thickness and doping concentrations for each layer for design S1. (b) IV and gain curves for a 4μm×10μm APD device with moderate charge-layer doping (S1 design). (c) IV curve for APD with high charge-layer doping (S2 design), no photocurrent is observed. (d) IV and gain curves for a device without charge-layer doping, unitary gain is observed.

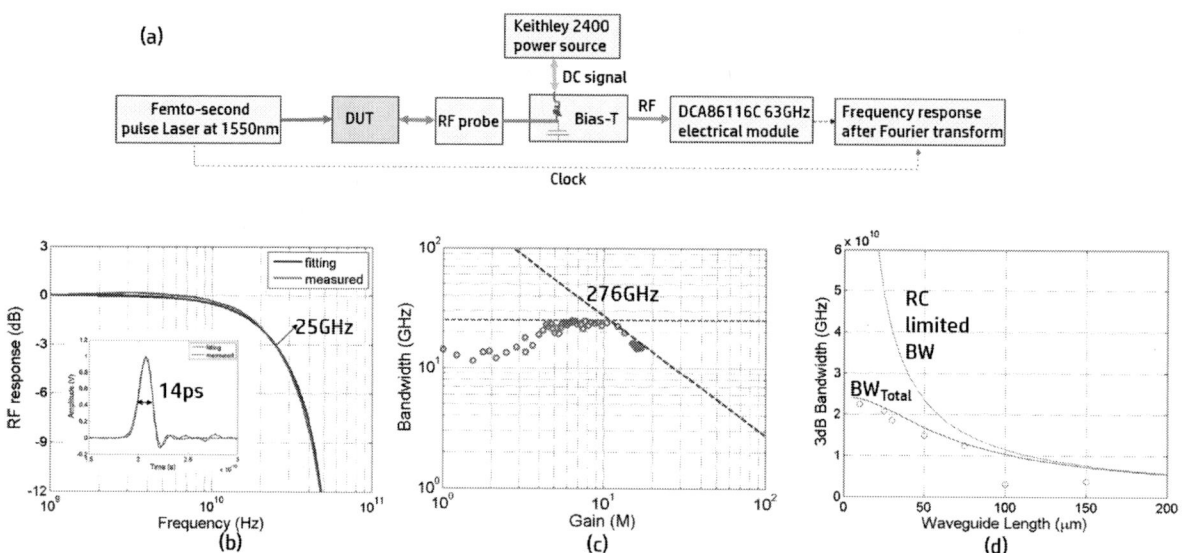

Fig. 2. (a) Impulse response measurement setup to measure bandwidth. (b) Bandwidth for the same 4x10um device, and the inset shows the measured impulse response. (c) Measured gain-bandwidth-product for the same device. (d) Extrapolated bandwidth for devices with 4um width and various lengths.

The APD bandwidth is characterized by an impulse-response method and the measurement setup is shown in Fig. 2 (a). A femtosecond pulsed laser is coupled to the device and generates electron hole pairs within tens of femtoseconds [5]. Since the germanium layer is p-type doped, holes as the majority carrier recombine quickly, leaving electrons as the unipolar carrier and traveling from germanium to silicon at the saturation velocity. The APD time response combines the carrier transit time and the RC delay which broadens the electrical pulse. We use a DCA sampling scope to measure the electrical impulse and retrieve the frequency-domain response by using Fourier transform. Fig. 2 (a) shows the RF response for a 4μm×10μm APD at gain of 5 and the inset shows the impulse response from the sampling scope. The

378

APD pulse has a full-width-half-maximum (FWHM) of 14ps, corresponding to a unity gain bandwidth of 25GHz. Fig. 2(b) plots the bandwidth vs. gain for the same device. The maximum gain-bandwidth product of 276GHz is obtained when gain is 12 and bandwidth of 23. Fig. 2(c) shows the measured bandwidth for a series of APDs with 4μm width and various lengths. The blue solid line indicates the calculated total bandwidth considering the effects from carrier transit time and RC delay, where the RC delay is plotted in the green solid line.

We also measured the large signal response by directly probing the APD on wafer without external amplifiers. A CW laser source was coupled to a 40Gb/s electro-optic modulator and modulated by a 2^7-1 PRBS source. The modulated optical signal is coupled to the APD and the eye diagram was measured with a DCA sampling scope. Fig. 3 (a)-(c) show the eye diagrams of the 4μm×10μm APD at various biases and at data rate of 12.5, 25 and 30Gbps, respectively. At unity gain, the signal to noise ratio (S/N) is too low to observe eye opening. At a gain of 2, the device shows the eye opening, and at gain of 5.9, the eye is wide open with S/N of 9 at 12.5Gbps. At data rate of 30Gbps an open eye is still observed at a gain of 4.

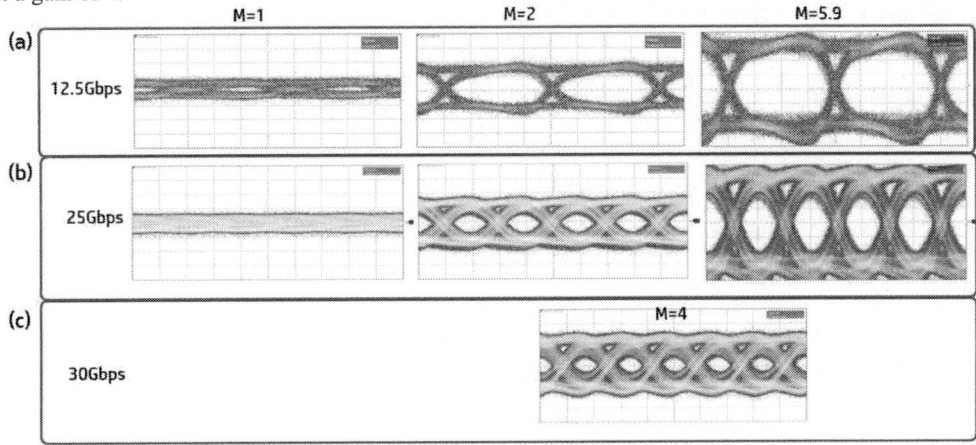

Fig. 3: (a) and (b) are the measured APD eye diagrams directly taken from a DCA86100C sampling scope with 12.5Gbps and 25Gbps data rate with gain of 1, 2 and 5.9, respectively. (c) is the eye diagram at 30Gbps at gain of 4.

III. CONCLUSIONS

We have demonstrated a low voltage Si-Ge APD that operates at -10V and 25Gbps. Error free detection up to 30Gbps is obtained. The design consideration for high speed low voltage APD is also discussed.

ACKNOWLEDGMENT

The Si-Ge APDs are fabricated at Institute of Microelectronics (IME), Singapore. The authors would like to thank Ning Duan, Lianxi Jia (Larry) and Chee-Wei Tok for device fabrication. The authors would also like to thank Jason Pelc, Xingyu Zhang for their helpful discussions about device measurements.

REFERENCES

[1] Jurgen, M. et. al.. "High-performance Ge-on-Si photodetectors." Nature Photonics 4.8 (2010): 527-534.
[2] Kang, Y. et al. "Monolithic germanium/silicon avalanche photodiodes with 340 Ghz gain–bandwidth product", Nature Photonics, volume 3, pages 59–63, 2009.
[3] Assefa, S. et al. "Reinventing germanium avalanche photodetector for nanophotonic on-chip optical interconnects," Nature, volume 464, pages 80–84, 2010.
[4] Virot, L. et al. "Germanium avalanche receiver for low power interconnects," Nature communications, volume 5, 2014.
[5] Sze, S. et. al. "Physics of semiconductor devices," John wiley & sons, 2006.

ThD3-3

OECC/PS2016

11-Gbps 16-QAM OFDM Radio over Fiber Demonstration using 100 GHz High-Efficiency Photoreceiver based on Photonic Power Supply

T. Umezawa[1], K. Kashima[2], A. Kanno[1], K. Akahane[1], A. Matsumoto[1], N. Yamamoto[1] and T. Kawanishi[1, 3]

1: National Institute of Information and Communications Technology (NICT)
4-2-1, Nukui-Kitamachi, Koganei, Tokyo 184-8795, Japan
2: Hitachi-kokusai Electric Inc., 3: Waseda University

Abstract: A 100 GHz photoreceiver was designed, fabricated, and integrated with a pHEMT amplifier, based on a photonic power supply. In this paper, a high data rate photonic wireless communication over 11 Gbps is demonstrated.

Keywords: Radio over Fiber, Power over Fiber, High speed photoreceiver

I. INTRODUCTION

In wireless communications, the data traffic increases rapidly every year, as well as in fixed optical fiber communications, where the trend for data traffic is not saturated yet. In the next generation of advanced wireless communication services, a high data rate over 10 Gbps, high capacity, low latency, and low cost are expected. The 2 km radio cover area for present macro-cell should be divided into plenty of small-cells in order to mitigate heavy data traffic in the one macro-cell. On the other hand, radio over fiber is a promising technology for access networks such as mobile back-haul and front-haul in wireless communication. Additionally, high data rates over 100 Gbps per fiber are already achieved in fixed optical fiber communications. From these points of views, high data rate transmission through optical fibers should be introduced near end-user points and be converted to radio signals from simple antennas in small-cells. A high speed photoreceiver is one of key components in RoF communications. Many studies investigating high-power photodetectors (PDs) with high-optical input were reported previously [1]. However, high-efficiency, high conversion gain, and high-speed PDs with high output should be considered to take low optical power level, which is in plenty of small-cells, into account. The 60 GHz band is well known and commonly used for license-free wireless communication; however, long range wireless transmission cannot be expected due to large atmospheric attenuation. According to the white paper in 5G, a higher carrier frequency over 30 GHz can be applied for single-carrier waveforms, and advanced OFDM with MIMO has to be developed at low-carrier frequencies [2]. The 90 to 110 GHz band is a very attractive frequency band, which involves not only low atmospheric attenuation [3] but also a wider bandwidth compared to that in the microwave region. Moreover, power over fiber (PoF) technology, which provides electric power supply through optical fibers, is a compatible technology with a good affinity to RoF technology. It is already commercially available for the high-voltage power transmission monitoring system instead of for the metal cable usage because of the intrinsic high-electric isolation [4]. In this study, we report high conversion gain high-speed photoreceiver based on photonic power supply through PoF. Then, a high data rate RoF communication was demonstrated at the 100 GHz region. The detailed design of the 100 GHz photoreceiver module and the experimental results on photonic wireless communication would be discussed.

II. DESIGN FOR HIGH CONVERSION GAIN 100-GHZ PHOTORECEIVER BASED ON POF

Fig. 1(b) shows a schematic diagram of a 100 GHz photoreceiver, which consists of hybrid integration with a high-speed PD and an amplifier in a small metal package. The 100 GHz PD can work under zero-bias condition, which plays the role of reducing a number of power fiber cables from phonic power supply. Assuming a low-electric power supply from PoF technology, the low-power consumption 100 GHz amplifier should be designed and fabricated. The hybrid integration with 100 GHz PD and amplifier would be extracted in the package. Note that the power fiber is directly connected to the radio frequency (RF) amplifier only, and not necessarily to the PD. The photoreceiver is designed to work well in the 100 GHz band through a data fiber and a power fiber. RoF technology with PoF technology is a good candidate for small-cell operation in advanced wireless access network. We developed zero-bias operational high speed PD beyond 100 GHz, which consisted of high p-doped photo-absorption layer and low carrier concentration carrier collector layer in a UTC structure. The detail is reported in our previous work [5]. The measured optical frequency response had a wide 3 dB bandwidth over 110 GHz. In the measured linearity, we confirmed good one and no output saturation up to -1.5 dBm at 100 GHz. A high-gain, low-power consumption 100 GHz RF amplifier was designed using an AlGaAs based 0.1 μm pHEMT process. The maximum oscillation frequency of 240 GHz was recognized for a transistor. The small chip (2.8 mm × 4 mm) includes a 4-gain stage with an input/output stage. Driving the bias voltage to 3 V, a high gain of 20 to 22 dB in S21 can be measured in 85 to 110 GHz, where the total power consumption became 219 mW. For a high-speed package attached with a 1 mm-co-axial connector, a microstrip line (MSL) was

380

designed to coplanar waveguide (CPW) conversion interface with low insertion loss and return loss in a frequency range up to 110 GHz. Then, the insertion loss as low as 1 dB at 100 GHz, can be measured. [6] For the hybrid integrated photoreceiver module, which was implemented with three above key components (Fig. 2), the frequency response of S21 in frequency range of 85 to 110 GHz was measured and compared with the simulation results. It was found that there was a good agreement between those two data, and the relative gain was peaked at 96 GHz. The 3 dB bandwidth from 96 GHz was recognized as 7 GHz. At 1 mA photo-current, +7.8 dBm RF output power could be successfully achieved at 96 GHz, which implied that there was a 24 dB higher gain than that in single PD without the amplifier. The maximum RF output power was saturated with +9.5 dBm at 96 GHz.

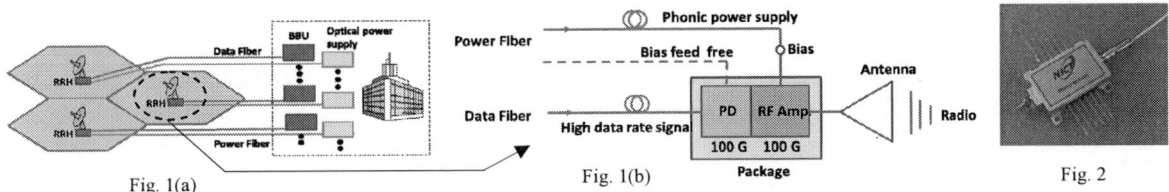

Fig. 1(a) Fig. 1(b) Fig. 2

Fig. 1 (a) Schematic viewgraph for advanced access network using RoF and PoF, (b) Schematic diagram for a 100 GHz photoreceiver integrated with a RF amplifier based on photonic power supply, Fig. 2 Picture of 100 GHz photoreceiver integrated with an RF amplifier

III. RESULTS AND DISCUSSION FOR 96 GHZ RoF DEMONSTRATION USING PHOTONIC POWER SUPPLY

At first, we investigated the 96 GHz carrier spectrum for the 100 GHz photoreceiver integrated with an amplifier using photonic power supply. The experimental setup for the photoreceiver is shown in Fig. 3. A data fiber and two power fibers were directly coupled to the zero-bias operational UTC-PD and photonic power devices for the RF amplifier, respectively. The 1 mm coaxial connector attached to the photoreceiver module was converted to WR-10 waveguide, and connected with a horn antenna. No post amplifier was used between the photoreceiver and horn antenna. In the receiver side, a horn antenna attached with a spectrum analyzer was set near the radio launching antenna. By using a 96 GHz high SNR two-tone signal generator, the peak power from the hybrid integrated photoreceiver was adjusted to approximately +0 dBm. After measuring the RF spectrum at 10 kHz span, the photoreceiver was switched to a single-chip photoreceiver in order to reveal the noise affection from the amplifier. Fig. 4 shows the measured spectrum (a) with / (b) without the RF amplifier for the 100 GHz photoreceiver. With the RF amplifier, the single-side band (SSB) phase noise was roughly estimated to be -103 dBc/Hz at 5 kHz offset. The SSB phase noise of -105 dBc/Hz at 5 kHz was also estimated without the RF amplifier. The insignificant difference between these two spectrums was confirmed. As a result, the hybrid integrated photoreceiver operated by the photonic power supply could be fully applied for a photonic wireless communication demonstration.

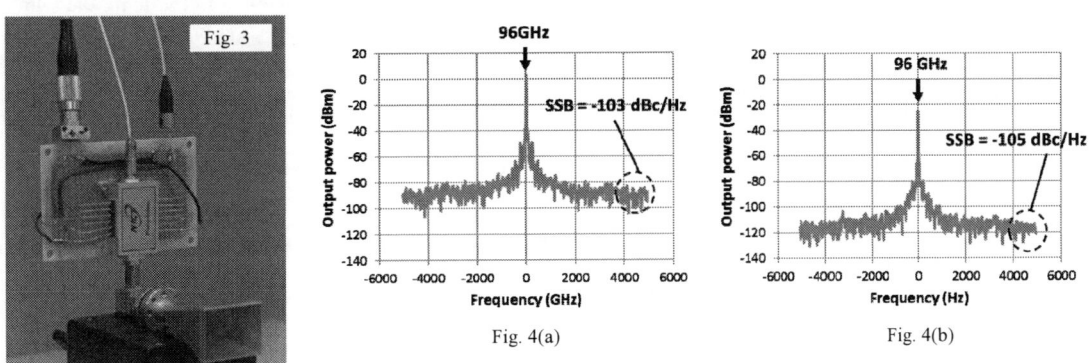

Fig. 4(a) Fig. 4(b)

Fig. 3 Experimental setup of a photoreceiver for a photonic wireless communication based on photonic power supply, Fig. 4 Measured RF spectrum (a) with (a) / (b) without the RF amplifier for the 100 GHz photoreceiver

Next, the high-data rate RoF was demonstrated using the above PD with PoF. The experimental setup is shown in Fig. 5. The intermediate frequency (IF) carrier of 85 GHz was created by four times multiplication in an optical two-tone signal generator. The 85 GHz two-tone signal was introduced to an array waveguide apparatus, and was separated to an upper-side band and a lower-side band. By applying intensity modulation from an arbitrary waveform generator (IF = 7G Hz, 16-QAM, OFDM, data rate = 11 Gbps), and the additional sub-upper and lower side band were also formed near the main lower side band. After filtering the main lower band with the sub-upper side band, the sub-lower side band and the main upper side band were combined by an optical coupler, where the difference of the two-tone peak corresponded to 92 GHz. Thus, the 92 GHz carrier with 11 Gbps with optical signal generation could be achieved. The signal was introduced to a 10 km data fiber and a wireless transmission distance of 1 m was fixed. The received 92 GHz carrier frequency was down-converted to 15 GHz, and the 16-QAM, OFDM signal was analyzed by an oscilloscope. The 16-QAM constellation and EVM were measured with different optical power conditions. As a result, the clear constellation map was successfully obtained (Fig. 6) above +1.8 dBm. The EVMs of 11.7% and 13.8% were estimated at optical input power level of +4.8 dBm and +2.8 dBm, respectively. By changing the optical input power to the photoreceiver, the bit error rate (BER) was measured as shown in Fig. 7. The BER of 1×10^{-3}, which is the threshold level for the forward error correction (FEC) with 7% overhead, could be achieved at a small optical input power level of +0.5 dBm. To take the 7% FEC into account, the data rate with error-free condition would be as high as 10.3 Gbps.

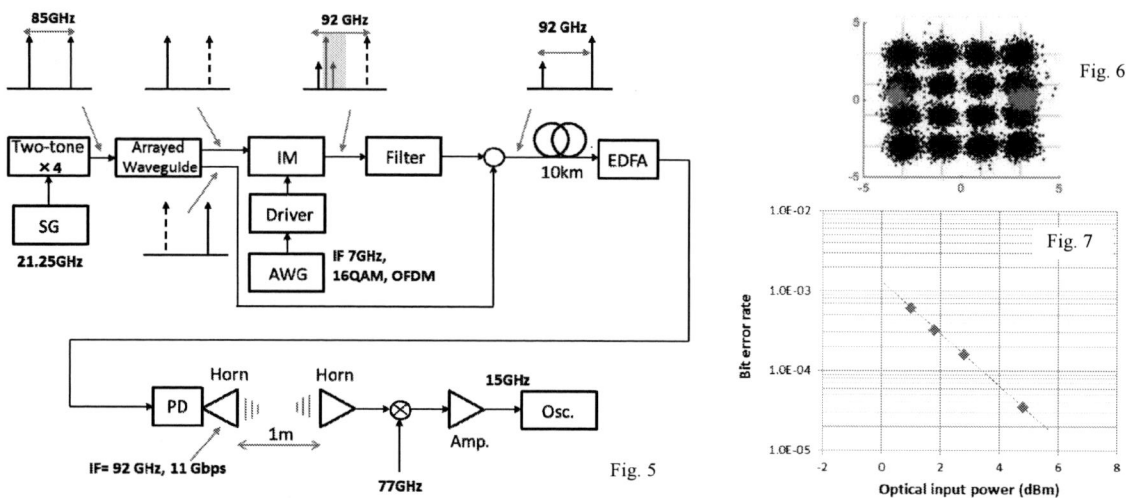

Fig. 5 Experimental setup for 11 Gbps RoF demonstration, Fig. 6 Measured 16-QAM constellation map with 11 Gbps OFDM signal, by using photonic power supply. The IF carrier frequency of 92 GHz, 10 km fiber, and 1 m wireless distance were fixed, Fig. 7 BER curve as a function of optical input power in the photoreceiver

IV. CONCLUSIONS

We designed and fabricated a 100 GHz photoreceiver integrated with a pHEMT amplifier, which could be operated by a small photonic power supply. The developed zero bias operational UTC-PD contributed to reducing the number of power fibers, and lowering the power consumption of the AlGaAs pHEMT amplifier. Hybrid integration in the 100 GHz region was designed with a high-speed metal package. A 92 GHz carrier photonic wireless transmission demonstration was carried out by applying a photonic power supply through an optical fiber to the amplifier. Under the data transmission conditions (16-QAM, OFDM, 11 Gbps), the BER of 1×10^{-3} could be successfully achieved at a low optical input level of +0.5 dBm. The high data rate of 10.3 Gbps could be expected for the error-free condition, to take the 7% FEC into account. We believe that the developed high speed photoreceiver will be a good candidate for small-cell access networks based on RoF technology.

ACKNOWLEDGMENT

This study was conducted as a part of a research project supported by the Japanese Government funding for "R&D to Expand Radio Frequency Resources" by the Ministry of Internal Affairs and Communications.

REFERENCES

[1] S. Kodama and H. Ito, "UTC-PD Based Optoelectronic Components for High-Frequency and High-Speed Applications,",IEICE Trans. Electron., Vol. E90-C, No. 2, February (2007)
[2] NTT DOCOMO, Inc. "DOCOMO 5G white paper," July 2014
[3] Federal Communications Commission Office of Engineering and Technology, "Millimeter Wave Propagation: Spectrum Management Implications,"https://transition.fcc.gov/Bureaus/Engineering_Technolog /Documents/bulletins/oet70/oet70a.pdf
[4] Laser focus world, http://www.laserfocusworld.com/articles/print/volume-42/issue-1/features/photonic-frontiers-photonic-power-delivery-photonic-power-conversion-delivers-power-via-laser-beams.html
[5] T. Umezawa, K. Akahane, N. Yamamoto, A. Kanno, K. Inagaki, and T. Kawanishi, "Zero-Bias Operational Ultra-Broadband UTC-PD above 110 GHz for High Symbol Rate PD-Array in High-Density Photonic Integration," M3C.7, Proc. of OFC 2015
[6] T. Umezawa, K. Katshima, A. Kanno, K. Akahane, A. Matsumoto, N. Yamamoto and T. Kawanishi, "High-efficiency W-band hybrid integrated photoreceiver module using UTC-PD and pHEMT amplifier," 9747-12, Proc. of Photonics West 2016

ThD3-4

Design for High Speed Operation of Double Microring Resonator-Loaded Mach-Zehnder 2×2 Quantum Well Optical Switch

Naoki Kawaguchi, Kento Hori, Taro Arakawa, and Yasuo Kokubun

Graduate School of Engineering, Yokohama National University, 79-5, Tokiwadai, Hodogaya-ku, Yokohama, 240-8501, Japan
kawaguchi-naoki-sn@ynu.jp, arakawa-taro-vj@ynu.ac.jp

Abstract: *We discuss theoretically the design of a quantum-well double microring resonator-loaded Mach-Zehnder 2×2 path switch for high-speed operation. Calculated results show the switching time can be reduced to 54 ps using 0.4 V push-pull drive.*

Keywords: *Microring Resonator, Mach-Zehnder Interferometer, Optical Path Switch, Quantum Well*

I. INTRODUCTION

A Mach-Zehnder interferometer (MZI) optical path switch is one of the key devices for high speed optical networks, and various MZI switches has been proposed and developed such as Si MZI switches and compound semiconductor MZ switches [1-3]. In addition, path switches based on microring resonator-loaded MZI (MRR MZI) are also promising [4] because high speed and low voltage operations are expected due to a non-linear phase shift effect in an MRR [5]. Although we have investigated the MRR MZI modulators and path switches, the design for high speed operation has not been fully discussed yet.

In this paper, we discuss theoretically the design of InGaAs/InAlAs multiple quantum well double-MRR MZI 2×2 optical path switches for high speed operation.

II. DEVICE STRUCTURE AND OPERATION PRINCIPLE

Fig. 1(a) shows the schematic top view of the proposed 2×2 optical path switch based on a double-MRR MZI. Two racetrack-shaped MRRs (Rings 1 and 2) are loaded to both arms of the MZI Directional couplers are used for coupling between the MRR and the arm waveguide. The schematic cross sectional view of a high-mesa InP waveguide at the directional coupler is shown in Fig. 1(b). An InGaAs/InAlAs multiple five-layer asymmetric quantum well (FACQW) [6] is assumed in the 300 nm-thick core layer. Reverse voltage is applied to the core layer through the pin junction and the refractive index of the core layer is changed due to the quantum confined Stark effect (QCSE) in the FACQW. To reduce propagation loss caused by p-doped cladding layer, a non-doped InP upper layer is located on the core layer. A shallow trench is used to control a coupling coefficient easily.

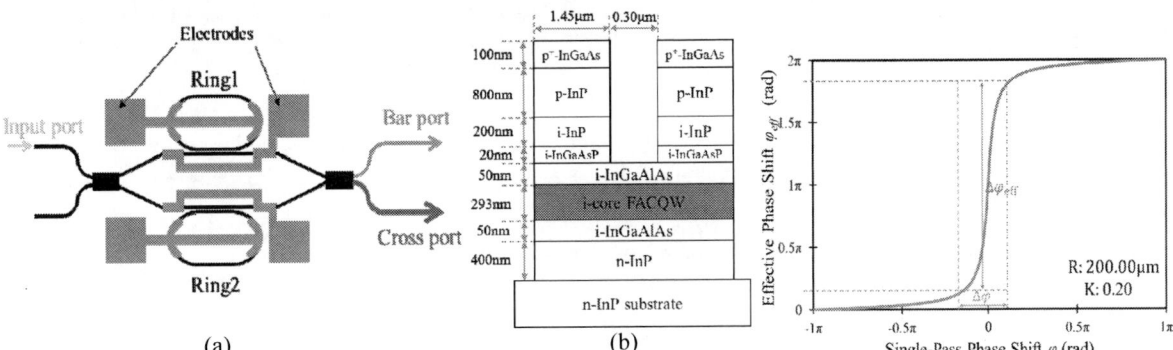

Fig. 1. (a) Schematic top view of proposed 2×2 double MRR-MZI optical path switch.
(b) Schematic cross-sectional view of waveguide at directional coupler.

Fig. 2. Effective phase shift φ_{eff} as functions of single-pass phase shift φ in ring resonator.

Fig. 2 shows the effective phase shift φ_{eff} as a function of a single-pass phase shift φ for coupling efficiency K=0.20, round-trip length of the MRR L_{ring} =200 μm, and propagation loss of 5.0 cm^{-1}. In the vicinity of φ =0, that is, in the on-resonance state, the marked nonlinearity and the single-pass phase shift are strongly enhanced. Here, we define the phase shift enhancement factor F_{pe} as [7]

$$F_{pe} = \frac{3\pi/2 - \pi/2}{\left.\varphi\right|_{\varphi_{eff}=3\pi/2} - \left.\varphi\right|_{\varphi_{eff}=\pi/2}} = \frac{\pi}{\left.\varphi\right|_{\phi_{eff}=3\pi/2} - \left.\varphi\right|_{\phi_{eff}=\pi/2}}. \qquad (1)$$

Using this enhancement of the phase shift in one arm of an MZI, the driving voltage of the MZM can be significantly reduced. If a microring resonator is coupled with one arm of the MZM and a change in the effective phase shift $\Delta\varphi_{eff}$ of π rad from $\varphi_{eff}=\pi/2$ to $3\pi/2$ rad is used, the output light can be modulated with a small change in the single-pass phase shift $\Delta\varphi$.

In the MZI, the length of one upper arm is larger than the other by 0.5π rad, and the round-trip length of Ring 1 is larger than that of Ring 2 by 1.5π rad. By the push-pull drive of the effective phase φ_{eff} in Rings 1 and 2, that is, by changing φ_{eff} in Ring 1 by $0.5\,\pi$ rad (from 0.75π to $1.25\,\pi$ rad) and that in Ring 2 by $-0.5\,\pi$ rad (from 1.25π to 0.75π), the optical path changes from Cross to Bar states.

III. DEVICE DESIGN FOR HIGH SPEED OPERATION

We theoretically discuss the switching time of the MRR MZI switch. In our simulation, propagation loss in the waveguide is assumed to 5.0 cm^{-1}, which is measured value in our experiment, and a coupling loss at the directional coupler are neglected. The switching time of the device is determined by the Q factor of the MRRs, and the Q factor depends on the device parameters such as a coupling efficiency at the directional couplers and a round-trip length of the MRRs. Fig. 3(a) shows the dependence of the switching time and a phase-shift enhancement factor on the round-trip length when the coupling efficiency K is 0.2. The phase shift enhancement factor defined as $\Delta\phi_{eff}/\Delta\phi$ in the vicinity of $\phi=0$ rad. Both switching time and phase shift enhancement factor decrease as the decrease in the round-trip length. There is a tradeoff between decreasing the switching time and reducing the operation voltage. Fig. 3(b) shows the dependence of the switching time and phase-shift enhancement factor on the coupling efficiency K when the round-trip length of the MRR is fixed to 200 μm. Both switching time and phase shift enhancement factor increase as the decrease in the coupling efficiency K due to the increase in the Q factor of the MRR.

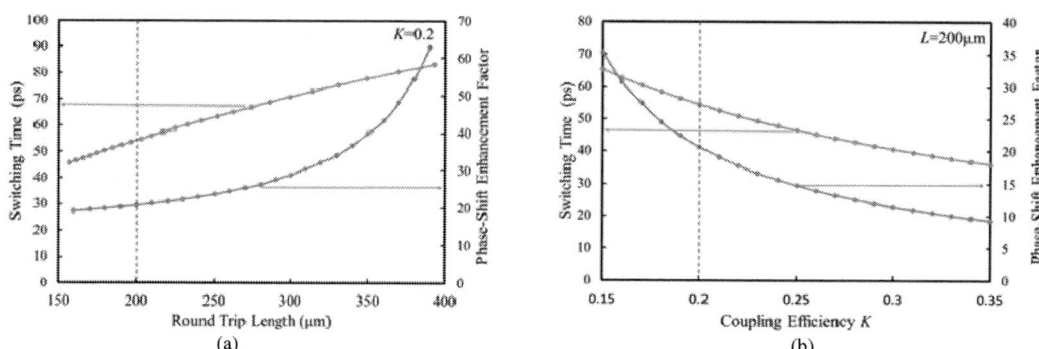

Fig. 3. (a) Dependence of switching time and phase-shift enhancement factor on the round-trip length when coupling efficiency K is 0.2. (b) Dependence of switching time and phase-shift enhancement factor on coupling efficiency K when round-trip length of MRR is fixed to 200 μm.

Considering the balance between an insertion loss of the device and the operation voltage, we choose the round-trip length of 200 μm and the coupling efficiency of 0.20 as optimized values. In that case, the phase shift enhancement factor F_{pe} of the MRR is 20.7, which corresponds to the switching time of 54 ps. The device parameters for the simulation are listed in Table 1.

Table 1. Device parameters of double MRR MZI optical path switch.

Parameters	Values
Round-trip length of Ring 1	200.00 μm
Round-trip length of Ring 2	199.64 μm
Coupling efficiency K	0.20
Coupling length	55.86 μm
Free spectral range of FSR	2.0 nm
Quality factor	1.1×10^4

Switching characteristics of the device were calculated using the device parameters in Table 1. Fig. 5 shows the spectra of the transmittance at Cross port. The reverse bias voltage V_{bias} is assumed to be 8.0 V, and the total applied voltage is $V_{bias}+\Delta V$ where is a driving voltage. When ΔV is 0 V, Cross and Bar ports are OFF and ON states, respectively. When ΔV is changed to 0.4 V, the states of Cross and Bar ports are changed to ON and OFF, respectively, and the optical path is exchanged, as shown in Fig. 5. Although the insertion loss is relatively high (7.4 dB), the calculated results show that the low voltage operation at as low as 0.4 V is possible in the compact MRR MZI switch. The insertion loss can be reduced by reducing the propagation loss of the waveguide.

On the basis of the above device design, we are now fabricating the MMR MZI switch using electron beam lithography and inductively coupled plasma reactive ion etching (ICP-RIE) etching based on Cl_2 gas, as shown in Fig. 6.

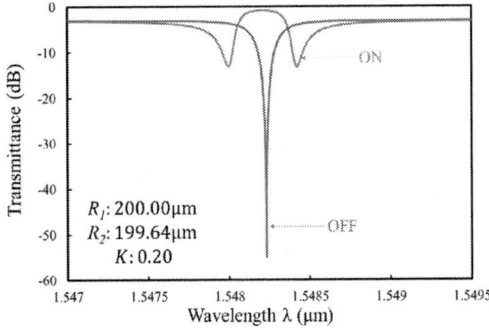

Fig. 4. Calculated spectral response at Cross port of double MRR-MZI 2×2 optical pass switch.

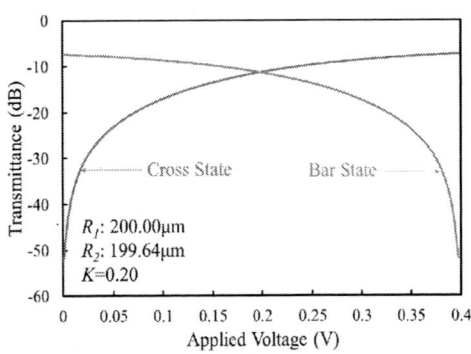

Fig. 5. Calculated static path switching characteristics of double MRR-MZI 2×2 optical pass switch.

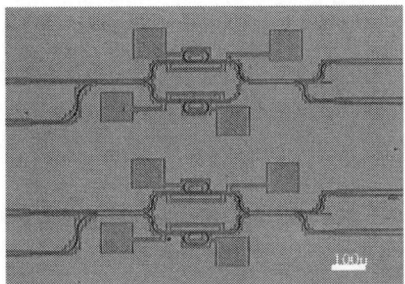

Fig. 6. Microscopic image of fabricated MRR MZI optical pass switch.

IV. CONCLUSIONS

We have discussed theoretically the design of the InGaAs/InAlAs FACQW double-MRR MZI 2×2 optical path switche for high speed operation. The optimized design has double MRRs with the round-trip length of approximately 200 μm and the coupling efficiency of 0.20. The calculated driving voltage and the switching time were 0.4 V and 54 ps, respectively. The calculated results show that the MRR MZI optical path switch is promising for high speed and low-voltage path switching.

ACKNOWLEDGMENT

This work was partly supported by a Grant-in-Aid for Scientific Research B from the Ministry of Education, Culture, Sports, Science and Technology, Japan (No. 15H03577) and the Fujikura foundation, Japan.

REFERENCES

[1] K. Suzuki, K. Tanizawa, T. Matsukawa, G. Cong, S.-H. Kim, S. Suda, M. Ohno, T. Chiba, H. Tadokoro, M. Yanagihara, Y. Igarashi, M. Masahara, S. Namiki, and H. Kawashima, "Ultra-compact 8 × 8 strictly-non-blocking Si wire PILOSS switch," Opt. Express, vol. 22, pp.3887-3894, 2014.

[2] Y. Ueda, N. Koyama, K. Kambayashi, S. Fujimoto, K. Utaka, T. Shiota, and T. Kitatani, "4 × 4 InAlGaAs/InAlAs Optical-Switch Fabric by Cascading Mach–Zehnder Interferometer-Type Optical Switches With Low-Power and Low-Polarization-Dependent Operation," IEEE Photon. Technol. Lett., vol. 24, pp.757-759, 2012.

[3] R. Gautam, H. Kaneshige, H. Yamada, R. Katouf, T. Arakawa, and Y. Kokubun, "Therm-Optically-Driven Silicon Microring-Resonator-Loaded Mach-Zehnder Modulator for Low-Power Consumption and Multiple-Wavelength Modulation," Jpn. J. Appl. Phys., vol. 53, 022201, 2014.

[4] M. Nishimura, T. Arakawa and Y. Kokubun, Design of Low-Voltage 2×2 Optical Switch Based on Quantum Well Microring-Enhanced Mach-Zehnder Interferometer, MSST 2013

[5] J. E. Heebner and R. W. Boyd, "Enhanced all-optical switching by use of a nonlinear fiber ring resonator," Optics Lett., vol.24, no.12, pp.847-849, 1999.

[6] T. Arakawa, T. Toya, M. Ushigome, K. Yamaguchi, T. Ide, and K. Tada: "InGaAs/InAlAs Five-Layer Asymmetric Coupled Quantum Well Exhibiting Giant Electro refractive Index Change", Jpn. J. Appl. Phys., vol.50, 032204, 2011.

[7] H. Kaneshige, R. Gautam, Y. Ueyama, R. Katouf, T. Arakawa, and Y. Kokubun, "Low-voltage quantum well Microring-enhanced Mach-Zehnder modulator", Opt. Express, vol. 21, pp. 16888-16900, 2013.

ThD3-5

Mode-Division Multiplexing LiNbO₃ Modulator Using Directional Coupler

Yutaro KODAMA[1], Yuya YAMAGUCHI[1], Atsushi KANNO[2],
Tetsuya KAWANISHI[1,2], Masayuki IZUTSU[1,3] and Hirochika NAKAJIMA[1]

[1]Waseda University
3-4-1 Okubo, Shinjuku, Tokyo 169-8555, Japan
[2]National Institute of Information and Communications Technology (NICT)
4-2-1 Nukui-kitamachi, Koganei, Tokyo 184-8795, Japan
[3]Japan Society for the Promotion of Science, San Francisco Center
2001 Addison Street Suite 260 Berkeley, CA 94704, USA
E-mail: y-kodama@moegi.waseda.jp

Abstract: *We designed and fabricated a LiNbO₃ integrated optical modulator for mode-division multiplexing, where optical signals in two modes can be independently controlled and combined together. Mode crosstalk was smaller than -20dB in C-band.*
Keywords: *LiNbO₃, Optical modulator, Mode-division multiplexing, Directional coupler*

I. INTRODUCTION

With increase of data traffic in optical fiber communications, modulation techniques have evolved in commercial systems. Initially, intensity modulation (IM) was commonly used. Recently, coherent detection scheme has made it possible to detect optical phase shift and multilevel amplitude modulation techniques, such as phase-shift-keying (PSK) and quadrature amplitude modulation (QAM) [1] [2]. Quadrature PSK (QPSK, or 4-ary QAM) is realized by a dual-parallel Mach-Zehnder modulator (DPMZM) comprised of two sub MZMs, where real and imaginary components are modulated independently. 100 Gb/s polarization-division-multiplexed QPSK (PDM-QPSK) has been realized by using an integrated modulator consisting of two DPMZMs [3].

Transmission capacity over a standard single-mode fiber (SMF) is limited to about 100 Tb/s, due to nonlinear effect in the fiber. In order to exceed this limit, space-division multiplexing (SDM) and mode-division multiplexing (MDM) have attracted attention because optical power density can be reduced in fibers. Dense space-division multiplexing (DSDM) transmission using 12-core × 3-mode fiber was reported [4].

MDM transmission requires mode multi/demultiplexer (MUX/DEMUX). The existing mode MUXs multiplex optical signals outside the modulator, causing an additional coupling loss between the modulator and mode MUX and also require an additional polarization controller at the input side of MUX [5].

In this paper, we propose two-mode multiplexing optical integrated modulator which can modulate and multiplex the fundamental mode and the first-order mode on a single LiNbO₃ (LN) chip using Ti diffusion waveguides as shown in Fig. 1. By integrating the mode MUX with modulator, the connection loss of optical fiber to mode MUX decreases, and MDM transmission is realized with fewer devices and equipment. In addition, combination with the existing PDM-IQ modulator, and transmission exceeding 400 Gb/s on a single chip can be realized with MDM-PDM-IQ modulator. Figure 2 is a concept of this modulator using x-cut LN.

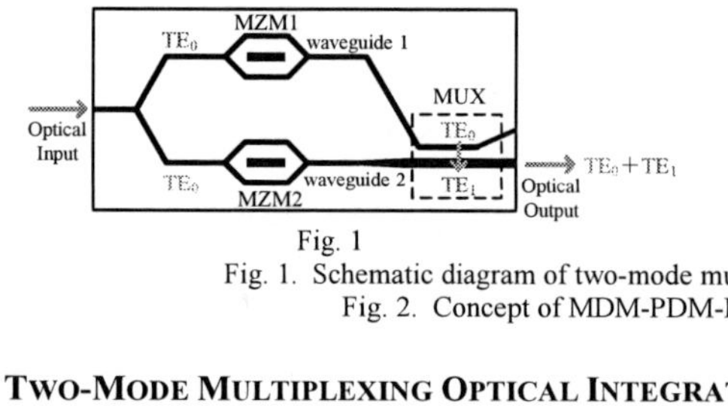

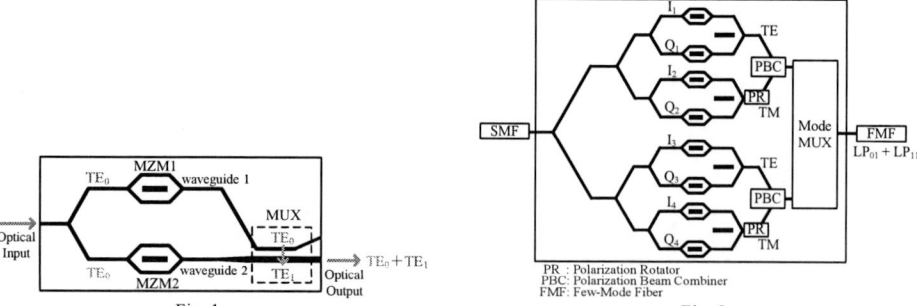

Fig. 1

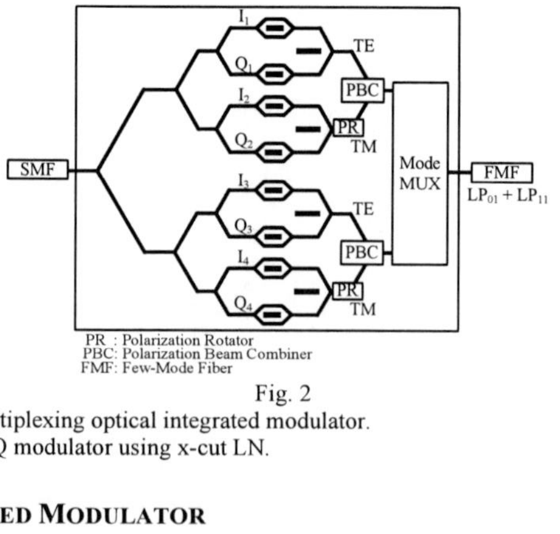

Fig. 2

Fig. 1. Schematic diagram of two-mode multiplexing optical integrated modulator.
Fig. 2. Concept of MDM-PDM-IQ modulator using x-cut LN.

II. DESIGN OF TWO-MODE MULTIPLEXING OPTICAL INTEGRATED MODULATOR

Two-mode multiplexing optical integrated modulator consists of two paralleled MZMs and mode MUX based on directional coupler. We use x-cut LN substrate so it modulates the TE mode light. In the mode MUX, the fundamental mode (TE_0) of waveguide 1 is converted to the first-order mode (TE_1) of waveguide 2. The TE_0 and TE_1 modes are multiplexed and outputted from waveguide 2 (port 2). In order to induce this mode conversion, the effective refractive index (n_e) of the TE_0 mode of waveguide 1 should be equal to that of the TE_1 of waveguide 2. Figure 3(a) is simulation result of the dependence of n_e of the TE_0 and TE_1 modes as functions of the waveguide width (w) using BPM (beam propagation method) simulation. We set the width of waveguide 1 (w_1) equal to the MZMs, 6 μm. To match n_e of the TE_0 and TE_1 modes, n_e of the TE_1 of waveguide 2 (w_2) was required to be equal to n_e of the TE_0 mode at w = 6 μm. From the simulation result, w_2 was determined to 10.3 μm.

We calculated interaction length (L) and coupling ratio of the TE_1 mode as functions of waveguide gap (G). Coupling ratio was defined to the maximum value of overlap integral with the TE_1 mode. L was defined to the half of the length of straight waveguide which coupling ratio of the TE_1 mode becomes peak. Figure 3(b) is the simulation result of the dependence of L and coupling ratio as functions of G. From the simulation results, L became longer when G became larger. We selected G = 8 μm where relatively high coupling ratio and short L was obtained. L became 740 μm when G was 8 μm. Figure 4 is the structure of the mode MUX.

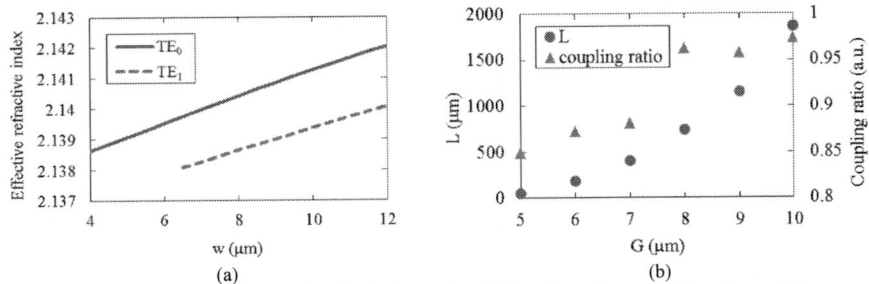

(a)   (b)

Fig. 3. (a) Effective refractive index dependence of w. (b) L and coupling ratio dependence of G.

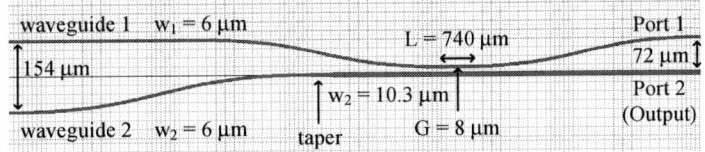

Fig. 4. Structure of Mode MUX.

III. Property of Fabricated Two-Mode Multiplexing Optical Integrated Modulator

We fabricated two-mode multiplexing optical integrated modulator. First, we investigated ON/OFF switching of the both modes of the fabricated modulator. We input CW light into the modulator with TE mode and applied DC voltage on two MZMs. Figure 5 shows near field patterns (NFPs) at a wavelength of 1550 nm when switching two MZMs ON/OFF. The TE_1 mode pattern was observed when MZM1 was switched on. The TE_0 mode pattern was clearly observed when MZM2 was switched on. From Fig. 5, the intensity of the TE_1 mode is weaker than that of the TE_0 mode. We considered that L was not the optimal length.

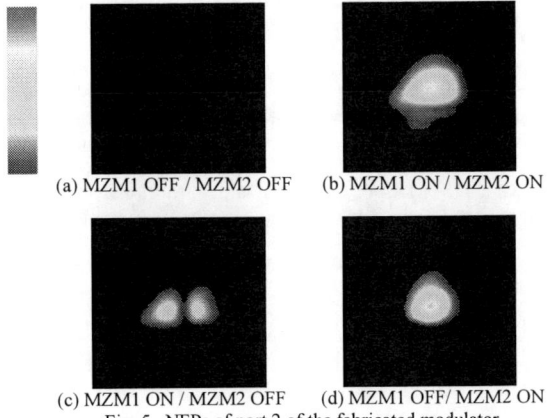

Fig. 5. NFPs of port 2 of the fabricated modulator.

IV. MEASUREMENT OF MODE CROSSTALK

We fabricated a chip to measure mode crosstalk as shown in Fig. 6. The mode MUX and DEMUX are integrated on an LN chip. In the mode MUX, waveguide gap at input ports is about 1 mm, and w_2 is 6 μm at input port in order to input CW light by using SMF. The waveguide 2 includes the tapers to convert from single-mode waveguide to multi-mode waveguide. The mode DEMUX has the reversed structure of the mode MUX. Mode crosstalk means the light propagating port A to D and port B to C. We defined the mode extinction ratio as the optical power ratio at each output port as shown in table I. We measured wavelength dependence of mode extinction ratio. Figure 7 shows the experimental results on wavelength dependence of the mode extinction ratio. The obtained mode extinction ratio was higher than 20 dB in C-band.

TABLE I. Mode extinction ratio

Port	Mode extinction ratio
C	A to C / B to C
D	B to D / A to D

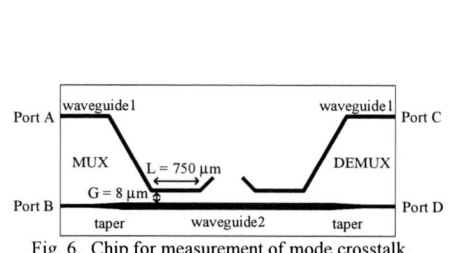

Fig. 6. Chip for measurement of mode crosstalk.

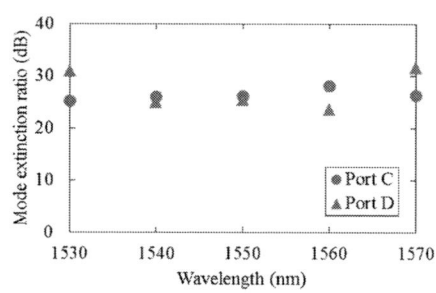

Fig. 7. Mode extinction ratio dependence of wavelength.

V. CONCLUSION

We proposed two-mode multiplexing modulator which has two MZMs and mode MUX on a single LN chip. The fabricated modulator worked as ON/OFF switching device of the fundamental mode and the first-order mode successfully. We also measured the mode extinction ratio on another device. The mode extinction ratio was higher than 20 dB in C-band. Our proposed technique can be applied to MDM-PDM-IQ modulation, where expected bitrate would exceed 400 Gb/s.

ACKNOWLEDGMENT

A part of this research is the result of the research commissioned by the National Institute of Information and Communications Technology (NICT) entitled "Agile Deployment Capability of Highly Resilient Optical and Radio Seamless Communication Systems."

REFERENCES

[1] K. Kikuchi, "Digital coherent optical communication systems: fundamentals and future prospect", IEICE Electronics Express, Vol. 8, No. 20, pp. 1642-1662, Oct 2011.

[2] M. Nakazawa, T. Hirooka, M. Yoshida, and K. Kasai, "Ultrafast coherent optical transmission," IEEE J. Sel. Topics Quantum. Electron, Vol. 18, No. 1, pp. 363–376, Jan/Feb 2012.

[3] H. Yamazaki, T. Yamada, T. Goh, and A. Kaneko, "PDM-QPSK modulator with a hybrid configuration of silica PLCs and LiNbO₃ phase modulators," J. Lightwave Technol, Vol. 29, No. 5, pp.721–727, March 2011.

[4] T. Mizuno, T. Kobayashi, H. Takara, A. Sano, H. Kawakami, T. Nakagawa, Y. Miyamoto, Y. Abe, T. Goh, M. Oguma, T. Sakamoto, Y. Sasaki, I. Ishida, K. Takenaga, S. Matsuo, K. Saitoh, and T. Morioka, "12-core x 3-mode Dense Space Division Multiplexed Transmission over 40 km Employing Multi-carrier Signals with Parallel MIMO Equalization," Proc. OFC, PDP5B.2, 2014.

[5] N. Hanzawa, K. Saitoh, T. Sakamoto, T. Matsui, K. Tsujikawa, M. Koshiba, and F. Yamamoto, "Two-mode PLC-based mode multi/demultiplexer for mode and wavelength division multiplexed transmission," Opt. Express, Vol. 21, No. 22, pp. 25752-25760, Nov 2013.

Multi-Level High Speed Burst-Mode Receivers

Xin Yin, J. Van Kerrebrouck, G. Coudyzer, and Johan Bauwelinck

INTEC Department, Ghent University – imec – iMinds, Technologiepark 15, 9052 Gent-Zwijnaarde, Belgium.
Author e-mail address: xin.yin@intec.ugent.be

Abstract: We review recent burst-mode receiver technologies beyond 10Gb/s for broadband access and fast optical-switching networ s, focusing on multi-level modulations. The performance of newly developed 2 Gb/s 3-level burst-mode receiver using low-cost 10Gb/s AP s is also presented.
Keywords: burst-mode receiver, pulse amplitude modulation, 3-level duobinary, passive optical networ s, optical pac et switching.

I INTRODUCTION

The Internet has become the ubiquitous tool that is transforming the lives of all of us. New broadband applications in the field of entertainment, commerce, and social interactions demand increasingly higher data rate. In order to support such bandwidth growth in optical access networks, both ITU-T and IEEE standard bodies, during past years, endorsed their efforts in developing various passive optical networks (PON) standards up to 10 Gb/s line rate. In parallel, upstream 10 Gb/s burst-mode receivers (BM-RX), the most critical PMD components of PON systems, have been experimentally demonstrated and reviewed in [1]. Meanwhile, the rapid growth of the Internet traffic is boosting the requirement of higher capacity data center networks (DCN). In such networks, optical packet switch (OPS) with nanosecond time-scale operation could be a preferred solution to efficiently realize the flat DCN. Besides a necessity in broadband TDM-PON access networks, the BM-RX is also considered as a key enabling technology for fast packet-based optical switching systems [2].

Recently, a number of researches have been working on higher serial rate TDM-PONs beyond 10 Gb/s [3][4][7]. Based on those efforts, the investigation of a practical low-cost solution for high serial rate PONs is gaining a lot of interest in IEEE NG-EPON initiative. IEEE NG-EPON is currently focusing its efforts on 25-Gb/s single-wavelength and 50-Gb/s two-wavelength pair solutions [5]. This is in line well with the recent rapid shift in datacenter from 10-Gb/s technologies and processes to 25 Gb/s, including both VCSEL and single-mode optics. However, increasing burst-mode line rate beyond 10Gb/s remains very challenging, due to limitations in the cost, availability and performance of optical components. For instance, while APDs have been used extensively in 10G-class PONs to improve the receiver sensitivity, high-speed 25Gb/s APD devices are still not widely available as low-cost 10Gb/s APDs. Therefore, multi-level modulation needs to be taken into consideration as a new dimension to effectively support a high network capacity without increasing usable bandwidth of those critical components. In this paper, we review the most applicable multi-level modulation schemes for burst-mode operation, including 3-level duobinary modulation and pulse amplitude modulation (PAM).

II MULTI-LEVEL BURST-MODE MODULATION

So far, only non-return to zero (NRZ) on-off keying (OOK) has been specified in the ITU and IEEE PON standards. As the NRZ is a simple form of pulse amplitude modulation, i.e. PAM-2, a natural upgrade path is to use 4 or 8 levels in the amplitude. However increasing number of amplitude levels will reduce the eye height accordingly, resulting in a worse signal to noise ratio (SNR). Therefore, PAM-4 appears to be a ready candidate of multi-level burst-mode modulation [6]. Recently, increasing line rate with 3-level duobinary has received a lot of attention [3][4]. Duobinary data encoding is a form of partial response signaling, which allows reducing bandwidth with a mount of intentionally-controlled inter-symbol interferences (ISI). Different from optical duobinary (ODB), the 3-level BM detection de-stresses optical components requirement and therefore lower the cost of the burst-mode transceivers [7].

Fig. 1 illustrates the difference in eye diagrams for NRZ, PAM-4, and 3-level duobinary. Since PAM-4 transmits 2 bits per symbol, the unit interval (UI) of PAM-4 is twice that of NRZ, which significantly relaxes the required minimum transmission bandwidth. For instance, as shown in Fig. 1(a) and 1(b), when normalized receiver bandwidth (with respect to data rates) is less than 0.5, the eye height of the NRZ signal significantly drops while little impact is shown in the PAM-4 case. On the other hand, PAM-4 eyes have 16 different transitions against NRZ with 4 transitions. The skewed transitions result in extra horizontal eye closure and cause PAM-4 more vulnerable to offset and ISI. The simulated eye width versus different receiver bandwidth is also shown in Fig. 1, indicating smaller horizontal eye open in UI for PAM-4 modulation. The 3rd option, duobinary modulation, has a better SNR than PAM-4 because of larger vertical eye opening. Unlike PAM-4, duobianry has no skewed crossing edges that makes its BM clock-and-data recovery (BM-CDR) synchronization faster and more reliable. The duobinary encoder can usually be implemented in two different approaches, either a low-pass filter (LPF) with ¼ data rate bandwidth [7] or using a digital delay-and-add filter. In Fig. 1(c) the duobinary signal is generated with a delay-and-add filter in the transmitter. In this case, both vertical eye height and horizontal eye width decrease gently with receiver bandwidth, showing a nice compromise between NRZ and PAM-4.

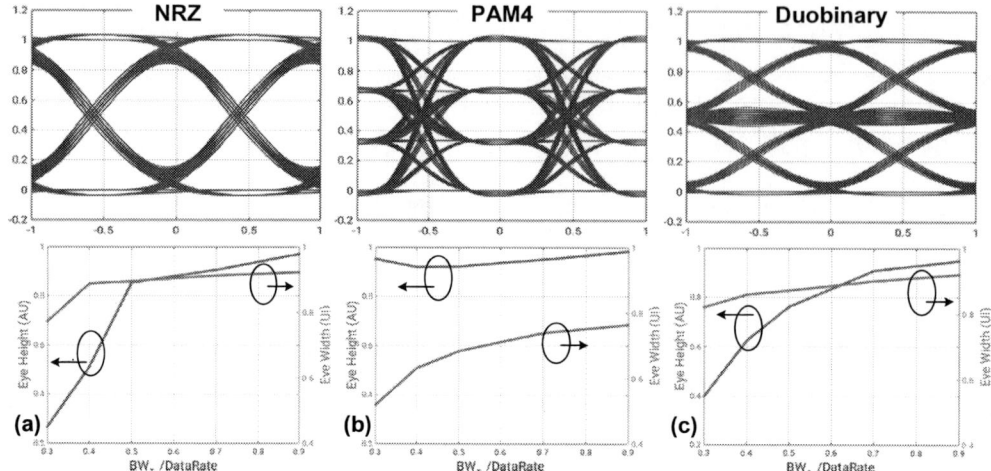

Fig. 1. Eye diagram, normalized eye height and eye width of (a) NRZ (b) PAM-4 (c) 3-level duobinary using delay-and-add filter.

III PAM-4 BURST-MODE EXPERIMENTS

In [6] we presented burst-mode PAM-4 transmission at 20 Gb/s in a TDM-PON, exploiting a low-drive chirped EAM transmitter and a 10Gbit/s linear BM-RX. The received upstream 4-PAM data was stored with a 50 GS/s real-time scope for the purpose of symbol-by-symbol signal decoding with a multi-level slicer. No post-distortion or additional post-processing technique was applied to the received signal. Since the transmission function of the EAM transmitter is highly non-linear (Fig. 2(a)), biasing in its low and high bias regime will result in asymmetric eyes in large-signal operation. In order to compensate for the non-linear transfer function of the EAM, the electrical PAM-4 signal is pre-distorted 67% by means of simply varying the amplitude ratio of the constituent binary PRBS signals before the electrical power combiner.

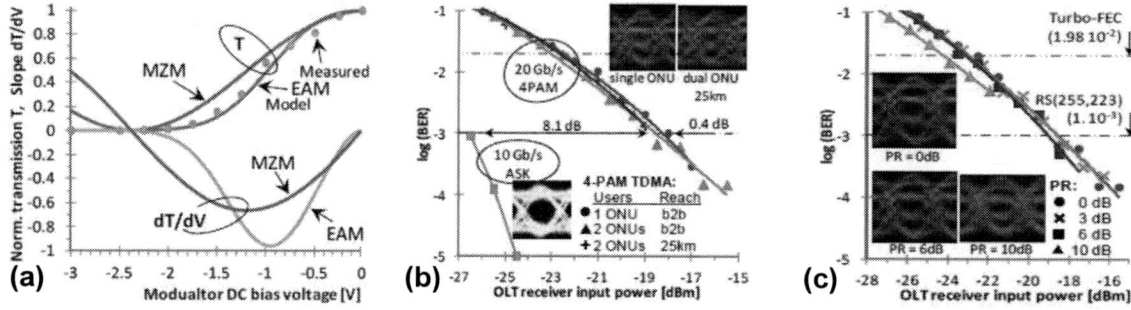

Fig. 2. (a) Transmitted PAM-4 optical signal (b) BM BER bursts with equal packet power (c) BM BER for PAM-4 at different loud/soft ratio.

The BER measurements for a configuration with a single ONU and two ONUs with equal packet power are presented in Fig. 2(b). With respect to 10 Gb/s NRZ transmission, the 20 Gb/s TDM 4-PAM experiences a power penalty of 8.1 dB. The excessive penalty is primarily explained by the eye closure caused by the ringing/ISI in the transmitter eyes. For the worst TDM scenario, there is no penalty for a loud/soft ratio up to 6 dB compared with equal-power packets. In case of 10 dB power difference there was a limitation in the power budget; however, the BER did not worsen. In another experiment, the BM-RX is used with MZM based burst-mode transmitters (BM-TXs). As can be seen in Fig. 2(a), compared to EAM the linearity of the MZM is clearly better and leads to improved transmitted signal quality without pre-distortion. The burst-mode PAM-4 experiments with MZM at ≥20 Gb/s are currently being performed with a further optimized linear BM-RX and will be discussed during the presentation.

IV -LEVEL DUOBINAR BURST-MODE EXPERIMENT

A newly developed 3-level APD BM-RX [7] was evaluated using the 25Gb/s experimental set-up as shown in Fig. 3. Two 1.3 μm BM-TXs named TX #1 and TX #2, are alternately sending burst packets. At the OLT, the linear BM APD-TIA was integrated with the 3-level decoder IC. After 3-level duobinary decoding, one of the 4 de-multiplexed outputs at 6.25 Gb/s was fed into a BER analyzer for real-time BER measurement. A gated semiconductor optical amplifier (SOA) was used to increase the optical output power of the TX #2 to +5dBm, in order to generate a sufficiently large loud/soft ratio for this experiment.

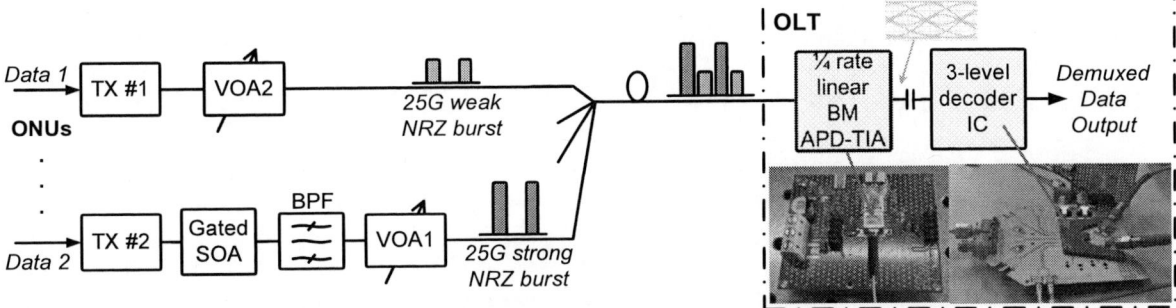

Fig. 3. The burst-mode experimental setup with the 25 Gb/s 3-level BM-RX prototype.

Fig. 4(a) shows the measured received bursts and the output signal of the linear BM APD-TIA. The applied 25 Gb/s burst packets consist of a 245 ns preamble and a 1800 ns payload of 2^7-1 pseudo random bit sequence (PRBS) patterns. The guard time between bursts is set to 15 ns. The measured B2B and BM BER curves are presented in Fig. 4(b). The measured 25Gb/s BM sensitivity at a pre-FEC BER of 1E-3 is -22.4dBm. The BM-RX was also assessed in different loud/soft ratio. For a typical OPS system, assuming a loud/soft ratio less than 10 dB, the BM penalty was negligible ($\leq$ 0.6 dB). Note that the burst-mode 3-level reception can reach BER below 1E-10 (BM sensitivity -16.3 dBm for 0 dB loud/soft): this allows FEC-free operation for OPS systems, which could be a critical factor for ultra-low latency optical links in DCNs.

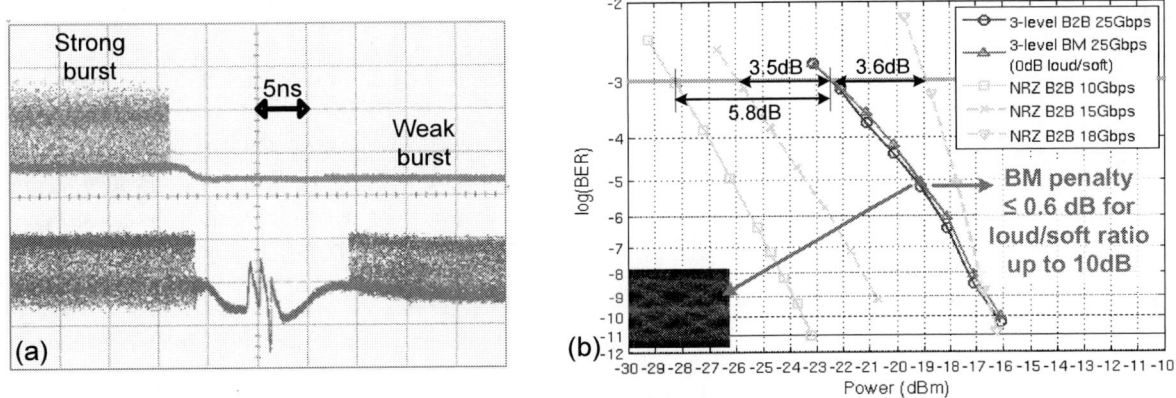

Fig. 4. (a) Loud-soft bursts waveform (b) BER measurement results for 25 Gb/s 3-level burst-mode transmission.

V CONCLUSIONS

We have compared and demonstrated experimentally 20-25 Gb/s PAM-4 and 3-level duobinary in burst-mode transmissions. It is shown that the BM-RX employing simpler multi-level modulation schemes can pave the way for a further upgrade of next-generation broadband access and fast optical-switching networks.

AC NO LEDGMENT

The authors would like to thank the support from the European Union FP7 under grant agreement n. 318137 (Collaborative project "DISCUS"), IWT, Special Research Fund of Ghent University, and Sumitomo Electric Devices Innovations, Inc for their professional linear BM APD-TIA Assembly.

REFERENCES

[1] X. Z. Qiu et al., "Fast synchronization 3R burst-mode receivers for passive optical networks [invited tutorial]," JLT, vol. 32, pp. 644-659, 2014.

[2] M. Nada et al., "25-Gbit/s burst-mode optical receiver using high-speed avalanche photodiode for 100-Gbit/s optical packet switching," Opt. Express, vol. 22, pp. 443-449, 2014.

[3] V. Houtsma et al., "Demonstration of symmetrical 25 Gbps TDM-PON with 31.5 dB optical power budget using only 10 Gbps optical components," ECOC, PDP4.3, 2015.

[4] X. Yin et al., "40-Gb/s TDM-PON Downstream with Low-Cost EML Transmitter and 3-Level Detection APD Receiver," OFC, Tu3C.1, 2016.

[5] C. Knittle, "IEEE 100G-EPON," OFC, Th1I.6, 2016.

[6] J. Verbrugghe et al., "Quaternary TDM-PAM as upgrade path of access PON beyond 10Gb/s," Optics Express, Vol. 20, no. 26, pp. B15-B20, 2012.

[7] X. Yin et al., "An Asymmetric High Serial Rate TDM-PON With Single Carrier 25 Gb/s Upstream and 50 Gb/s Downstream," J. Lightwave Technology, vol. 34, pp. 819-825, 2016.

OECC/PS2016

Multi-rate coherent burst-mode PDM-QPSK optical receiver for flexible optical networks

José Manuel Delgado Mendinueta, Hideaki Furukawa, Satoshi Shinada and Naoya Wada

Photonic Network System Laboratory, National Institute of Information and Communications Technology (NICT), 4-2-1 Nukui-Kitamachi, Koganei, Tokyo 184-8759, Japan.
Author e-mail address: mendi@nict.go.jp

Abstract: We numerically investigate a PDM-QPSK multi-rate coherent burst-mode optical receiver capable of receiving 3 different line-rates. The line-rate detection algorithm has low implementation complexity and is insensitive to polarization rotations and frequency offset.
Keywords: Coherent burst-mode receivers; Flexible optical networks; Optical packet switching.

I. INTRODUCTION

Flexible hybrid optical networks, simultaneously supporting optical circuit switching (OCS) and optical packet switching (OPS), have been recently proposed to achieve diversification of services, adequate resource allocation, and efficient energy consumption in datacenter or metro/core WDM networks [1]. A fundamental component of such flexible hybrid optical networks is the adaptive optical receiver, able to dynamically adapt to the line-rate and/or modulation format of the incoming signal, either in continuous-mode or in burst-mode. Previously, a multi-rate receiver for ON-OFF keying (OOK) signals was reported on [2]. More recently, a flexible multi-format receiver using data-aided digital signal processing (DSP) was reported [3]. However, in [3] the format of the optical modulation should be known *a priori* besides the data-aided DSP is modulation-format agnostic. In this work, we propose and numerically investigate a novel multi-rate coherent burst-mode receiver for flexible hybrid optical networks with PDM-QPSK payloads, capable to detect the line-rate on a packet-by-packet basis. The receiver uses a standard optical front-end and includes a line-rate detection algorithm that can be efficiently implemented. Line-rate estimator sizes of 64 and 128 samples yield packet error rates (PERs) of less than 10^{-2} and $2.5 \cdot 10^{-3}$, respectively, when operating at an optical signal to noise ratio (OSNR) corresponding to a payload bit error rate (pBER) of 10^{-3} for the lowest line-rate.

II. ALGORITHM DESCRIPTION AND BURST-MODE RECEIVER ARCHITECTURE

The optical packet header consists of the consecutive repetition of an 8-symbol header sequence, whose constellation is shown in Figure 1.A. The transmitted symbols in the header sequence are {1 0 3 0 4 0 2 0}, and both polarizations transmit the same header sequence. This resembles an OOK Ethernet 1010... preamble, with the difference that here there is phase modulation on each of the '1' symbols. The header sequence block is repeated several times to generate a packet header with a desired number of symbols, hence the total number of symbols in the header is a multiple of 8.

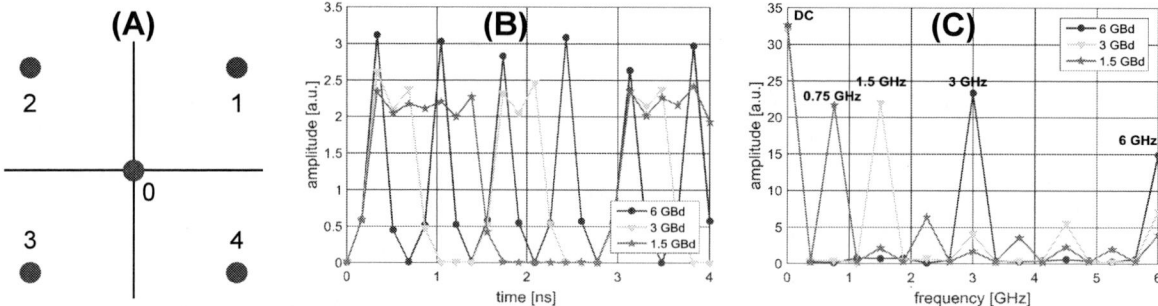

Fig. 1: (A) Constellation diagram of the optical packet 8-symbol header sequence. Numbers are the encoded constellation symbols. (B) Temporal waveform of signal S[n] for line-rates of 6, 3, and 1.5 GBd. (C) Absolute value of the DFT of S[n] for line-rates of 6, 3, and 1.5 GBd.

When an incoming optical packet is detected, the line-rate detection algorithm first computes the signal $S[n]$, defined as,

$$S[n] = |\tilde{X}[n]|^2 + |\tilde{Y}[n]|^2$$

where $\tilde{X}[n]$ and $\tilde{Y}[n]$ are the complex envelope components of the $\tilde{X}$ and $\tilde{Y}$ polarizations electrical field. Let A be the highest line-rate in GBd, then define B = A/2, and C = A/4, etc. To demonstrate the line-rate detection algorithm, and without loss of generality, we chose line-rates of A = 6, B = 3, and C = 1.5 GBd so all the signals can be generated with a digital to analog (DAC) converter with sampling frequency of 12 GS/s. Fig. 1.B shows the sampled signal $S[n]$ in the time domain for optical packets with line-rates of A (circles), B (triangles), and C GBd (stars). Due to the fact that $S[n]$ discards the phase information of $\tilde{X}[n]$ and $\tilde{Y}[n]$, the shape of $S[n]$ is invariant if there is frequency offset (difference

392

between the transmitter and receiver lasers) and/or state of polarization (SOP) rotation. For each line-rate, $S[n]$ exhibits a strong frequency component located at half the baud rate. This is shown in Fig. 1.C, where the modulus of the 32-point discrete Fourier transform (DFT), for the positive frequencies when line-rates A (circles), B (triangles) and C (starts) are input, are plotted. Let's define F_A, F_B and F_C as the absolute value of the DFT for frequencies A/2, A/4 and A/8 GHz. The line-rate detection algorithm then uses a bank of three comparators to compute the digital values $F_A>F_B$, $F_A>F_C$ and $F_B>F_C$. These digital values are input to a combinational logic consisting of inverters and AND gates that produce a digital output corresponding to the detected line-rate. Since the line-rate detector only needs 3 output coefficients of the DFT, it is not required to compute the whole DFT coefficients. Efficient implementations like the Goertzel algorithm are available [4]. Note that the DFT at a particular frequency is equivalent to computing the matched filter for a sinusoidal signal at that frequency [2]. In addition, due to the fact that the DFT is the discrete sampling of the continuous discrete-time Fourier transform (DTFT) [4], this algorithm will perform best when the half-baud rate scanned frequencies of the DFT correspond to the exact frequency samples of the DTFT. This condition is always fulfilled if the sampling frequency is $f_s = 2 \cdot A$, the line-rates are as B = A/2, C = A/4, etc., and the DFT size is 2^N.

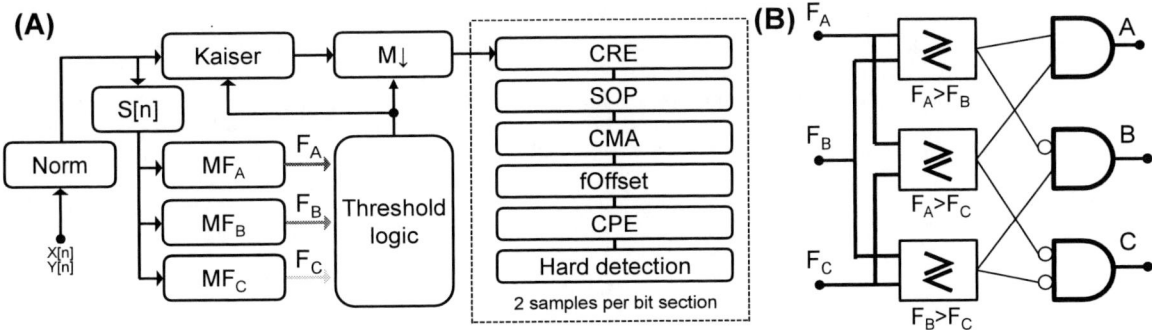

Fig. 2: (A) Coherent burst-mode receiver DSP architecture. (B) Detail of the line-rate threshold logic.

Figure 2.A shows the multi-rate coherent burst-mode receiver DSP architecture. Firstly, the incoming signal is normalized. Then, the signal $S[n]$ is computed and fed into a bank of matched filters. The output of these matched filters, denoted as F_A, F_B, and F_C, are input to a threshold logic which consist of three comparators, three inverters and three AND gates. For an input line-rate {A, B, C}, only the corresponding output will have a logical '1' and the rest will have a logical '0'. The output of the threshold logic is used to select the precomputed taps of a Kaiser filter, used as anti-aliasing filter and to remove out-of-band noise. After the Kaiser filter, line-rates B and C are decimated with ratios of 1:4 and 1:8, respectively. For line-rate A, both the Kaiser filter and the decimator are bypassed.

After the decimator, the DSP receiver always works at a constant 2 samples-per-bit, which implies that the DSP clock frequency, and hence the energy consumption, is reduced by the same rate as the decimation. First, there is a clock recovery estimation stage (CRE). Then, the SOP is estimated and this estimation used to initialize the taps of a constant modulus equalizer (CMA) which eliminates the CMA singularity [5]. Then, frequency offset and carrier phase are estimated and compensated both using the Viterbi and Viterbi algorithm. Finally, the signal is hard-detected and BER computed by error counting.

III. SIMULATION SETUP AND METHODOLOGY

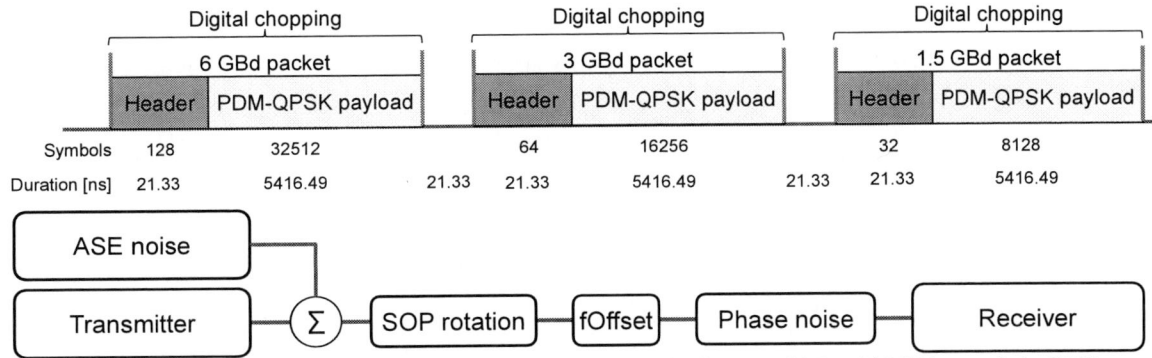

Fig. 3: (A) Diagram of the simulated optical traces. Each trace contains 3 packets, having line-rates of 6, 3, and 1.5 GBd, respectively. (B) Diagram of the main simulator stages.

Figure 3.A shows a diagram of the simulated optical traces, each having 3 optical packets at line-rates of 6, 3, and 1.5 GBd. The guard band between packets was 21.33 ns. The header duration also was 21.33 ns, which corresponds to 128, 64 and 32 symbols at 6, 3, and 1.5 GBd line-rates, respectively. The custom-made simulator was implemented in

MATLAB with some modules like the CMA equalizer written in C++. Figure 3.B shows a diagram of the main simulation stages. First, an ideal PDM-QPSK signal was generated. The ideal signal was then filtered with a 4th order Bessel shaping filter. Then, additive white Gaussian noise (AWGN) noise was added to set the desired OSNR. SOP rotation was simulated by applying a 2x2 rotation matrix, where the rotation angles θ and φ were random and uniformly distributed between [-90, 90] and [-180, 180] degrees, respectively [5]. Frequency offset was random and uniformly distributed over the range [-100, 100] MHz. Finally, 200 KHz phase noise was simulated, which is equivalent to have ECL lasers at the transmitter and receiver. In this work we analyzed the performance of the line-rate detector only. Hence, to simulate perfect packet detection, the optical data signal was then digitally chopped and the 6, 3 and 1.5 GBd packets were processed sequentially using the multi-rate coherent receiver. For every OSNR value, 400 data traces were generated to compute PER statistics.

IV. RESULTS AND DISCUSSION

With the simulator described in the previous section, the OSNR was swept ranging from -2 to 10 dB in steps of 1 dB, measured with a standard bandwidth of 0.1 nm. The size of the line-rate matched filters was swept in the range {32, 64, 128} samples. Figure 4.A shows the measured pBER for line-rates of 6 GBd (triangles), 3 GBd (circles) and 1.5 GBd (squares), averaged for all packet whose line-rate was correctly detected. At a pBER of 10^{-3}, the OSNR is 7.76, 4.73, and 1.73 dB, for line-rates of 6, 3, and 1.5 GBd, respectively, and the implementation penalties with respect to the theoretical curves are 1.18, 1.15 and 1.15 dB. As expected, the OSNR tolerance of the receiver is higher for lower line-rates, due to out-of-band noise being removed digitally by the Kaiser filter before the decimation process.

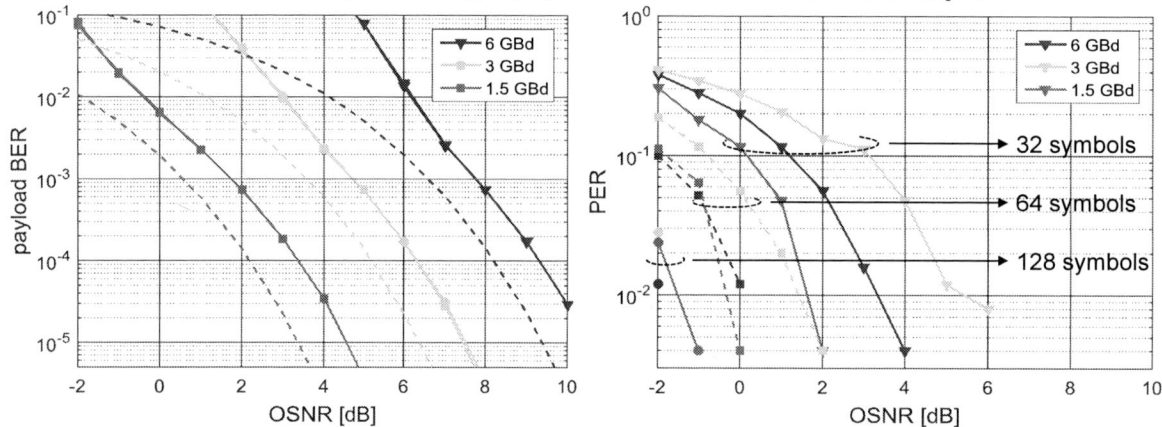

Fig. 4: (A) Payload BER for line-rates of 6, 3, and 1.5 GBd (solid lines) and theory curves (dashed lines). (B) Packet error rate for estimator sizes of 32, 64 and 128 samples, for line rates of 6, 3 and 1.5 GBd.

Finally, Figure 4.B shows the packet error rate (PER) for DFT/matched filter sizes of 32, 64 and 128 symbols, as a function of the OSNR. As expected, for a given amount of noise a longer estimator yields a lower PER, at a cost of increased complexity. For an estimator size of 64 samples, at a OSNR of 1.73 dB (10^{-3} BER for 1.5 GBd) the PER is lower than 10^{-2}. For an estimator size of 128 samples, only errors were recorded for OSNR values of -2 and -1 dB and at a OSNR of 1.73 dB PER is lower than 2.5·10^{-3}.

V. SUMMARY

In this work, we have proposed a novel multi-rate coherent burst-mode receiver for future flexible optical packet networks. The receiver line-rate detection is based on the matched filtering/FFT of a squared signal, which can be efficiently implemented and it is SOP rotation and frequency offset insensitive. Using a custom-made simulator, we assessed the performance of the algorithm in terms of PER and pBER. For 1.73 dB OSNR, which yields 10^{-3} pBER for the lowest line-rate, estimators of size 64 and 128 samples produce PERs lower than 10^{-2} and 2.5·10^{-3}, respectively.

REFERENCES

[1] J. Perelló et al., "*All-Optical Packet/Circuit Switching-based Data Center Network for Enhanced Scalability, Latency and Throughput,*" IEEE Network Magazine, vol. 27, pp. 14-22, 2013.

[2] J. M. Delgado Mendinueta et al., "*Digital dual-rate burst-mode receiver for 10G and 1G coexistence in optical access networks,*" OPEX, vol. 19, n. 15, pp. 14060-14066, July 2011.

[3] R. Elschner et al., "*Experimental demonstration of a format-flexible single-carrier coherent receiver using data-aided digital signal processing,*" OPEX, vol. 20, pp. 28786-28791, 2012.

[4] A. V. Oppenheim et al., "*Discrete-Time Signal Processing,*" Prentice-Hall, 2nd edition, 1999.

[5] J. M. Delgado Mendinueta at al., "*Fast Equalizer Kernel Initialization for Coherent PDM-QPSK Burst-mode Receivers Based on Stokes Estimator,*" in Proc. OSA Advanced Photonics/SPPCom, Jun. 2013.

ThE1-3

An InP Monolithically Integrated Multiwavelength Transmitter with Direct Modulation

N. Andriolli[1,2], P. Velha[1,2], P. Tommasino[3], M. Chiesa[1], G. B. Preve[2], A. Trifiletti[3,4], M. Romagnoli[2], and G. Contestabile[1,2]

[1]Scuola Superiore Sant'Anna, TeCIP Institute, Via Moruzzi 1, I-56124, Pisa, Italy
[2]CNIT, Photonic Networks National Lab, Via Moruzzi 1, I-56124, Pisa, Italy
[3]"La Sapienza" University of Rome, Via Eudossiana 18, I-00184 Rome, Italy
[4]Evoelectronics s.r.l., Via dei Castelli Romani 12a, I-00071 Pomezia, Italy
Corresponding author email: contesta@sssup.it

Abstract: *We report preliminary results on a multiwavelength (8 channels) transmitter integrated on a generic InP integration platform. The circuit is made by 8 tunable directly modulated DFB lasers, 8 monitor photodetectors and MMI couplers.*
Keywords: *InP monolithic integration, Photonic integrated circuits (PICs), Distributed feedback lasers (DFBs).*

I. INTRODUCTION

Multiwavelength photonic transmitters at high speed are key elements for various present and future network and microwave applications including access networks, mobile backhaul, antenna remoting and others [1-3]. The use of directly modulated lasers (DMLs) enables the implementation of simplified transmitter architectures where the monolithic integration on the same chip can lead to very compact photonic integrated devices (PICs).

In this contribution a novel photonic integrated transmitter with 8 channels based on broadband directly modulated distributed feedback (DFB) lasers is reported. The PIC fabricated using InP generic integration technologies [4] includes 8 thermally tunable DFBs, 8 monitor photodetectors (PDs), multi mode interference (MMI) couplers arranged in various various architectures (a cascade of 2×1, 2×1 and 4×1 and 8×1 MMIs) and a spot size converter (SSC) at the output. The output facet is anti-reflection (AR) coated. Ad-hoc broadband printed circuit boards (PCBs) including surface mounted resistors for laser load matching, SMA ports for laser biasing and modulation, and a GPIO connector for laser wavelength tuning have been designed and fabricated for PIC operation. A preliminary characterization of the transmitter including wavelength tunability and frequency response of the DFBs is reported.

II. INTEGRATED TRANSMITTER DESCRIPTION AND CHARACTERIZATION

A picture of one of the integrated transmitters is reported in Fig.1. The footprint is 6x4 mm^2. 8 directly modulated DFB lasers are butt coupled on both sides to passive waveguides having around 1 dB/cm insertion loss, one output is connected to an integrated monitor PD the other one to the coupling MMIs. We have realized different samples with various MMI architectures for the laser coupling: a cascade of 2×1 MMIs; four 2×1 MMIs followed by a 4×1 MMI, and a single 8×1 MMI. Fig. 1 reports the PIC version made by the cascade of 2×1 MMIs. After waveguide coupling, the output waveguide is connected to a SSC.

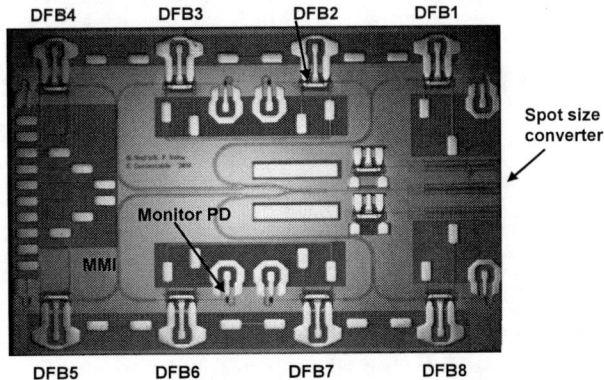

Fig. 1. Picture of a fabricated 8 channels transmitter. Footprint: (6x4) mm^2.

The SSC enlarges the waveguide mode to a size that gives maximum overlap with a single mode fiber. This is done adiabatically for minimum distortion and loss. A number of additional surface electrical contacts are also present in order to ease the wire bonding and the chip mountin. The DMLs are complex-coupled distributed feedback laser, which incorporates a twin-waveguide formed by active InGaAsP multi-quantum wells and a quaternary bulk layer, and is 200 µm long [5]. The emission wavelength can be obtained between 1530 and 1570 nm. An integrated heater allows an output wavelength tuning of around 4 nm.

The PICs have been mounted on ad-hoc designed high frequency electrical PCBs with RF lines (bandwidth ~ 20 GHz) and an interface connector for laser wavelength tuning. Pictures showing details of the PIC bonding and mounting on the PCB and the whole PCB set on a heat-sink and a Peltier cooler are reported in Fig. 2.

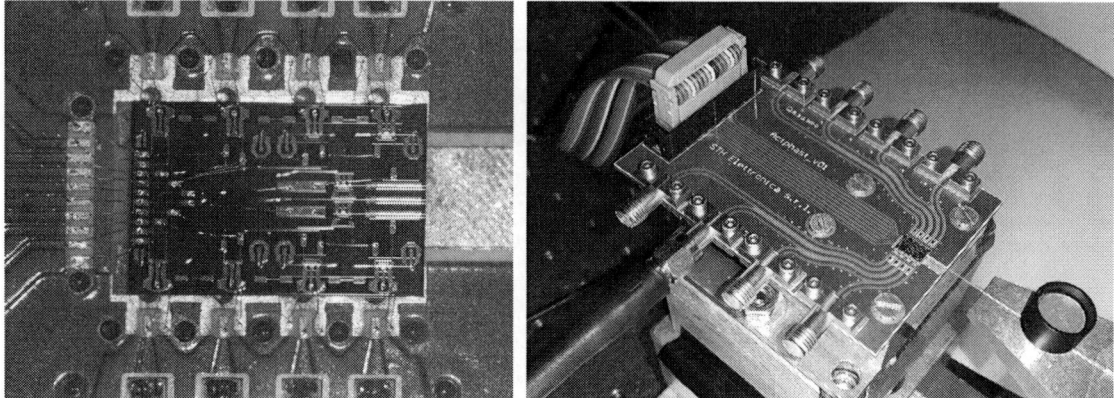

Fig. 2. Pictures of the PIC has mounted on the PCB (left part) and of the whole PIC-PCB assembly during characterization (right part).

Laser bias and modulating signals are supplied through the SMA connectors using bias-tees. The measured laser tunability through the integrated heaters is reported in the left part of Fig. 3 as a function of the supplied electrical powers. A tunability of 3.8 nm has been measured following a linear trend as function of the supplied power. The 8 DFBs have been designed having a 100 GHz (0.8 nm) spacing ranging from 1541.4 to 1547 nm. At 20 °C, the resulting emission wavelengths are as reported in Fig. 3, so that thermal tuning is required to set the wavelengths on a 100 GHz grid as shown in the lower part of the figure. The threshold current is 8 mA. The nominal output power from each DFB is 4 dBm while the measured output power from the chip is -10 dBm for each laser. This gives 14 dB total insertion loss for the cascade of MMIs and the fiber coupling loss.

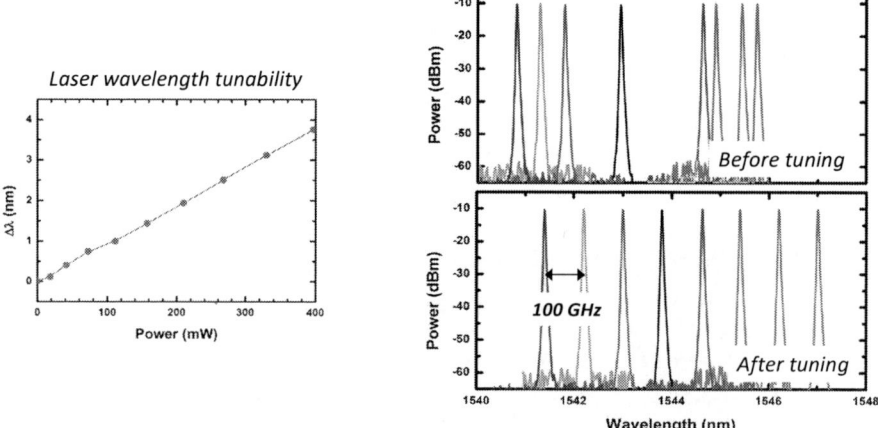

Fig. 3. Tunability feature of the DFBs (left part). Output DFB emission wavelengths before and after thermal tuning (right part).

The frequency response of the DMLs has been measured through a 50 GHz Network Analyzer using an external 40 GHz pin photodiode. The results, at varying the laser bias current from 50 to 150 mA, are reported in Fig. 4. We found a 3 dB bandwidth increasing from 12 to 14 GHz at increasing the bias current and recognized the typical frequency relaxation-oscillation peaks of DMLs. All the lasers have very similar frequency response. The measured DFB bandwidth is lower than expected from fabrication specifications. This is most probably due to a non-optimal ohmic p-contact formation on the wafer during manufacturing, and an improved version of the chip is currently under fabrication. Back-to-back transmission and BER measurements are ongoing at the time of the paper preparation and will reported at the conference presentation.

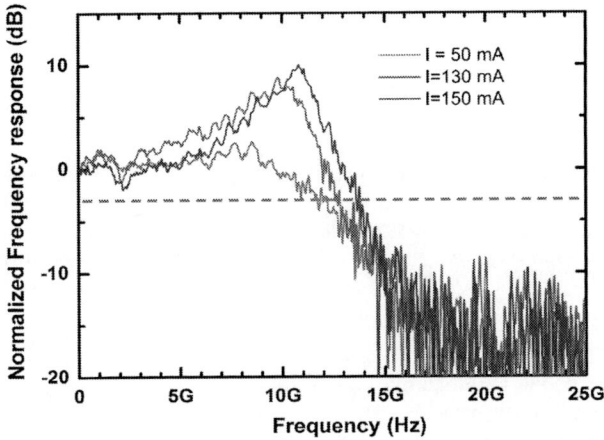

Fig. 4. DFB modulation response at varying the bias current.

III. CONCLUSIONS

We have reported a preliminary characterization of a novel 8-channels-transmitter monolithically integrated exploiting an InP generic integration platform. The circuit, made by 8 tunable DMLs, 8 monitor PDs and MMIs for signal multiplexing has 8 emitting wavelengths in C-band, ranging from 1541.4 to 1547 nm with 100 GHz spacing. Maximum modulation response of the DMLs is 14 GHz limited by a fabrication non-optimal manufacturing of the ohmic contacts. Back-to-back digital transmission with BER measurements, currently ongoing, will be reported in the conference presentation. Also the characterization of an improved version of the PIC, currently under fabrication, will be reported at the conference.

ACKNOWLEDGMENT

The authors would like to acknowledge Mr. M. Baier, Dr. F. Soares and Dr. N. Grote of the Fraunhofer Institute for Telecommunication, Heinrich Hertz Institut, Berlin for PIC fabrication.

This work has been partially supported by ACTPHAST (**A**ccess **C**en**T**er for **PH**otonics innov**A**tion **S**olutions and **T**echnology Support) the unique "one-stop-shop" for supporting photonics innovation by European companies, which is financially supported by the European Commission under the FP7 framework (Grant Agreement No. 619205).

References

[1] J. G. Andrews et al., "What Will 5G Be?, " *IEEE J. Sel. Areas Comm.*, vol. 32, pp. 1065 - 1082 (2014)

[2] V. J. Urick et al. "Long-Haul Analog Photonics," *J. Lightw. Technol.*, vol. 29, pp. 1182 - 1205 (2011)

[3] A. J. Seeds and K. J. Williams, "Microwave Photonics," *J. Lightw. Technol.*, vol. 24, pp. 4628 - 4641 (2006)

[4] M.K. Smit et al. "An introduction to InP-based generic integration technology," *Semiconductor Science and Technology*, vol. 29, p. 083001-1/41 (2014)

[5] M. Baier et al. "50 Gbit/s PAM-4 Transmission using a Directly Modulated Laser made on Generic InP Integration Platform," *International Conference on Indium Phosphide & Related Materials (IPRM) 2015*, Santa Barbara, CA (USA) 2015

ThE1-4

OECC/PS2016

Self-Clocking Synchronized Optical Demultiplexing Using Four-Wave Mixing in a Quantum-Dot SOA

Liang Yang, Tomoya Yatsu, and Motoharu Matsuura

Department of Communication Engineering and Informatics, The University of Electro-Communications
1-5-1 Chofugaoka, Chofu, Tokyo 18208585, Japan
ryou.you@uec.ac.jp

Abstract: *We demonstrated the optical demultiplexing using self-clocking synchronization. The demultiplexing was based on four-wave mixing in a quantum-dot semiconductor optical amplifier. We successfully achieved error-free operations and low power penalties for all extracted channels.*
Keywords: *Optical demultiplexing, optical time-division multiplexing (OTDM), quantum-dot semiconductor optical amplifiers (QD-SOA), clock recovery, self-clocking.*

I. INTRODUCTION

Optical demultiplexing is one of the key technologies in optical time-division multiplexing (OTDM) systems to realize ultrahigh-speed transmission in optical networks. In such systems, optical demultiplexing is required to overcome the limitation of electronic signal processing. Semiconductor optical amplifiers (SOAs) are useful not only for signal amplification but also optical signal processing such as wavelength conversion and optical demultiplexing [1]. In particular, four-wave mixing (FWM) is useful for various kinds of advanced modulation formats, because it enables us to preserve phase information in the process of the optical signal processing. The use of SOAs also have many advantages such as smaller footprint and lower switching energy than other nonlinear elements such as optical fibers. Recently, quantum-dot SOAs (QD-SOAs) had attracted much attention for their higher gain, faster gain recovery time, and broader bandwidth [2]. These characteristics are also useful for optical signal processing. Indeed, 320-to-40 Gb/s demultiplexing of on/off keying (OOK) signal using FWM and cross-gain modulation in a QD-SOA has been reported [3], [4]. However, no optical demultiplexing of phase-shift-keying signals has been demonstrated so far.

Clock recovery for OTDM systems is an indispensable technique to obtain a synchronized clock with the transmitted data at the receiver. In conventional OTDM systems, the required optical clock is extracted from the transmitted data signal. However, if we use ultra-high speed OTDM signals, it is impossible to extract the optical clock, because conventional clock recovery is excused by the electrical signal processing, which is much slower than the operating speed of the OTDM signals. Another approach is to simultaneously transmit an optical clock with a different wavelength from the data signal in the same transmission line. However, to obtain high transmission performance in a long-haul OTDM transmission, the chromatic dispersion and transmission loss have to be compensated in a wide transmission bandwidth, which includes the optical clock and data signal.

In this paper, we have demonstrated 40-to-10-Gbit/s optical demultiplexing of differential phase-shift-keying (DPSK) signals using self-clocking synchronization scheme. The self-clocking scheme is to insert the optical clock into the spectrum of the data signal and reduce the transmission bandwidth [5], [6]. The demultiplexing of this work is based on FWM in a QD-SOA. We have successfully achieved error-free operations. The obtained power penalties of less than 0.2 dB to the 10-Gbit/s base back-to-back signal have shown that the presented scheme has high demultiplexing performances.

II. EXPERIMENTAL SETUP

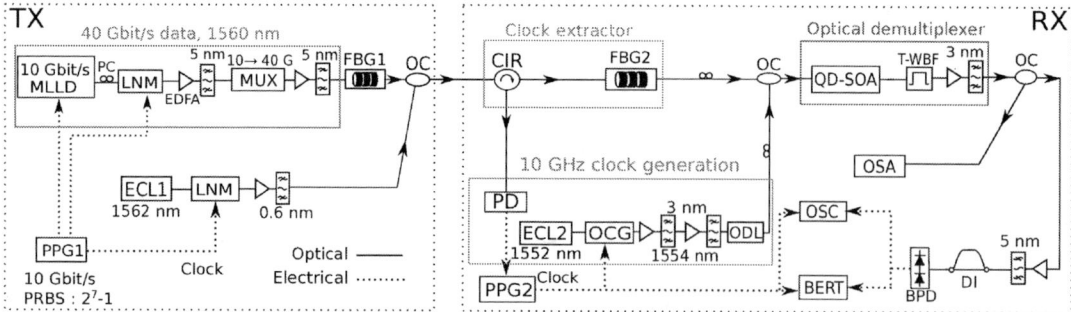

Fig. 1.   Experimental set up for optical demultiplexing of 40 Gbit/s DPSK signal using a QD-SOA and self-clocking synchronization.

398

The experimental setup for optical demultiplexing of 40 Gbit/s DPSK signal using a QD-SOA and self-clocking synchronization is shown in Fig.1. In the transmitter (TX), a 10 GHz clock signal at a wavelength of 1560 nm was generated from a mode-locked laser-diode (MLLD). A LiNbO3 modulator (LNM) and pulse pattern generator (PPG) were used to generate a 10 Gbit/s DPSK data signal with a 2^7-1 pseudorandom bit sequence (PRBS) data pattern. The data signal was amplified by an erbium-doped fiber amplifier (EDFA), and the amplified spontaneous emission (ASE) noise was removed by the following band pass filter (BPF) with a 3-dB bandwidth of 5 nm. The data signal was multiplexed to 40 Gbit/s by an OTDM multiplexer (MUX). After amplification, the 40 Gbit/s data signal passed through a fiber Bragg grating (FBG1) to filter-slice the spectrum component with a 3-dB bandwidth of 0.3 nm at around 1562 nm. The spectrum slicing was for inserting a 10 GHz optical clock at around 1562 nm. An external-cavity laser-diode (ECL1) and a LNM were used to generate a 10 GHz clock signal at a wavelength of 1562 nm. After amplification, the 10 GHz clock was combined with the 40 Gbit/s data signal by an optical coupler (OC). At the clock extractor, the FBG (FBG2) was used to pass through the data signal and reflect the clock. Thus, by inserting a circulator (CIR) at the input of the FBG2, the clock was extracted from the data signal spectrum component. The FBG2 had a 3-dB bandwidth of 0.3 nm and a center reflection wavelength of 1562 nm. In the receiver, the extracted clock signal was converted to the electrical clock by a photo-diode (PD). The electrical clock was injected into the PPG (PPG2) to generate an electrical clock synchronized with the optical clock for optical demultiplexing generated by an optical comb generator (OCG) and a bit-error-rate (BER) tester and an oscilloscope (OSC). In this way, the optical demultiplexing system was synchronized with the transmitted data signal. To generate a control signal for optical demultiplexing, a 10 GHz clock was generated by the ECL (ECL2) with a wavelength of 1520 nm and the OCG synchronized with the transmitted data signal. To obtain high quality and stable pulse train, the control signal was filtered by cascaded BPFs with a center wavelength of 1554 nm. An optical delay line (ODL) was used to adjust the timing between the 40 Gbit/s data and control signals for optical demultiplexing. After adjusting the states of polarization of the data and control signals by polarization controllers (PCs), these signals were coupled and injected into a QD-SOA. The QD-SOA was a sample device based on InAs Stranski-Krastanov (SK) growth. The gain recovery time was less than 10 ps. The driving current and temperature of the QD-SOA was set to 1722 mA and 25°C, respectively. The powers of the data and control signals injected into the QD-SOA were 1.3 dBm and 3.5 dBm, respectively. The pulse width of the data and control signals were 5.09 ps and 3.44 ps, respectively. The optical demultiplexer consisted of the QD-SOA and a tunable-wavelength bandwidth filter (T-WBF). The T-WBF is a rectangular filter shape, simply used to filter out the demultiplexed signal without any special filtering effect. After optical demultiplexing and amplification, one of the selected 10 Gbit/s signal was injected into the DPSK receiver, which consisted of a delayed interferometer (DI) and balanced photo-diode (BPD). The BER characteristics of the demodulated electrical signals were measured by a BER tester (BERT). And the spectrum of the signal was monitored by an optical spectrum analyzer (OSA).

III. EXPERIMENTS

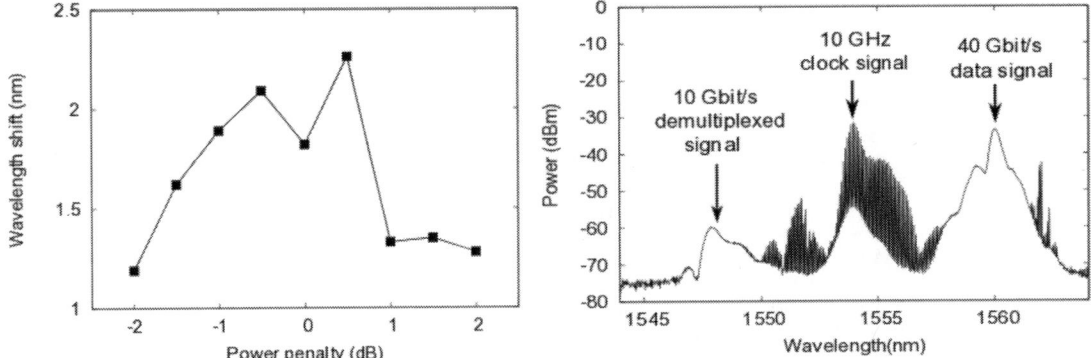

Fig. 2.   BER of wavelength shift of the clock signal from the center wavelength of the 10 Gbit/s data signal.

Fig. 3.   Signal spectra at output of the QD-SOA.

To determine the optimum inserted position of the optical clock into the spectrum of the data signal, we investigate the signal quality of the 10 Gbit/s data signal in terms of BER measurement when the optical clock with an arbitrary wavelength was inserted. Fig. 2 shows the power penalty at the BER=10^{-9} to the original data signal of the data signal as a function of the wavelength shift of the clock signal from the center wavelength of the data signal. As the absolute wavelength shift was increased, the power penalty tended to become smaller, because the signal spectrum carving and clock insertion at a position far from the center wavelength did not give large influence on the signal quality of the data signal. On the other hand, much larger wavelength shift will cause the degradation of the FWM efficiency of the optical demultiplexing. For these reasons, in the following experiment, the wavelength of the clock signal was set to 1562 nm (+1.5 nm wavelength shift).

Fig.3 shows the signal spectra at the output of the QD-SOA. The demultiplexed (conjugated) signals was generated at the symmetric position to the 40 Gbit/s data signal. The power ratio between the conjugated and 40 Gbit/s data signal was approximately −26.3 dB.

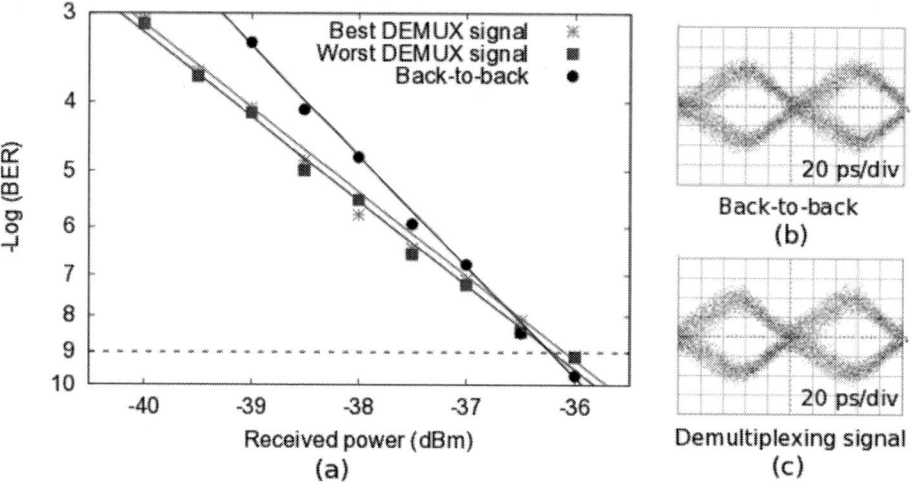

Fig. 4. (a) BER characteristics of BtoB and demultiplexed signals in all the 10 Gbit/s tributaries. The eye-patterns of the BtoB signal (b) and one of the demultiplexed signals (c).

Fig. 4 (a) shows the BER characteristics of the back-to-back (BtoB) and demultiplexed signals. In all 10 Gbit/s tributaries, error-free (BER < 10^{-9}) operations were successfully achieved. The average power penalty to 10 Gbit/s base BtoB signal of the demultiplexed signals was less than 0.2 dB. Figs.4 (b) and (c) show the eye-patterns of the BtoB and demultiplexed signals. Each eye-patterns had clear eye-open. These results show that the presented optical demultiplexer had high demultiplexing performance.

IV. CONCLUSIONS

We have successfully demonstrated the error-free operation of self-clocking synchronized optical demultiplexing of 40 Gbit/s DPSK signal using FWM in a QD-SOA. We also have shown the high demultiplexing performance with the low power penalties of all demultiplexed signals.

REFERENCES

[1] E. Tangdiongga et al., "All-optical demultiplexing of 640 to 40 Gbits/s using filtered chirp of a semiconductor optical amplifier," Opt. Lett., vol. 32, no. 7, pp. 835-837, 2007.
[2] T. Akiyama et al., "Quantum-Dot Semiconductor Optical Amplifiers," Proc. IEEE, vol. 95, no. 9, pp. 1757-1766, 2007.
[3] M. Matsuura et al., "320-to-40-Gb/s optical demultiplexing using four-wave mixing in a quantum-dot SOA," IEEE Photon Technol. Lett., vol. 24, no. 2, pp.101-103, 2012.
[4] M. Matsuura et al., "Error-free 320-to-40-Gbit/s optical demultiplexing based on blue-shift filtering in a quantum-dot semiconductor optical amplifer," vol. 38, no. 2, pp. 238-240, 2013.
[5] S. Zhang et al., "Fast-synchronization and low-timing-jitter self- clocking concept for 160 Gbit/s optical time-division multiplexing transmissions," Opt. Lett., vol. 35, no. 1, pp. 37-39, 2010.
[6] T. Yatsu, et al., "100-km transmission of 40 Gbit/s RZ-DPSK signal using self-clocking concept and optical phase conjugation in a QD-SOA," in Proc. OECC 2015, PWe.38, 2015.

ThE1-5

OECC/PS2016

Demonstration of Bi- and Multi-Stability in a High Order Ring Resonator

Li Jin[1], Alessia Pasquazi[2], Luigi Di Lauro[2], Marco Peccianti[2],
Edwin Y. B. Pun[1], David J. Moss[3], Roberto Morandotti[4], Brent E. Little[5] and Sai T. Chu[1]
[1]City University of Hong Kong, Kowloon Tong, Hong Kong, China, SAR
[2]Department of Physics and Astronomy, University of Sussex, Brighton, UK
[3]Centre of Micro-Photonics, Swinburne University of Technology, Melbourne, Australia
[4]INRS - EMT, Varennes, Québec, Canada
[5]State Key Lab of Transient Optics and Photonics, Xi'an Institute of Optics and Precision Mechanics, CAS, China
Email: saitchu@cityu.edu.hk

Abstract: *We experimentally demonstrated the bistability and multi-stability in a 5th order cascaded CMOS compatible ring resonator. The shape evolution of the filter under different states is also investigated.*
Keywords: *Nonlinear optics; Bi- and multi-stability; All-optical Switching; Integrated optics.*

I. INTRODUCTION

The development of integrated optical components for all-optical switching via bi- and multi-stability is of great importance in the advancement of future ultrafast optical networks. In these applications, ultrafast switches are required to have low operation threshold, wide band response and high on/off contrast. Owing to the low linear and nonlinear loss and maintaining a relatively high nonlinear figure of merit, CMOS compatible high index contrast doped silica glass and silicon nitride based integrated platforms have demonstrated excellent performances for the development of nonlinear optical devices [1]. This is due to their extremely low linear and nonlinear losses resulting in a relatively high nonlinear figure of merit, and they are generally considered viable device platforms for all-optical switching applications. For optical bi-stability, numerical studies of bi- and multi-stable behaviors in single, double and triple microrings have been thoroughly investigated [2-6]. Experimentally, demonstration of thermo-optical bi-stability and self-pulsation in silicon microring resonators have been reported [7]. Furthermore, multi-stability allows the generation of multiple temporal soltions [8] and it is predicted that multi-stability in super cavity can be used to generate super cavity solitons [9].

In this work, we report the evidence of bi- and multi-stability in a high order microring resonator filter based on a high-index contrast doped silica glass. Bi-stable and multi-stable response in a 5th order ring resonator filter could be observed by changing the input wavelength of an optical pump, for a power threshold of about 10mW. To further investigate the effect of the input power on the detuning of the ring resonator elements, we measured the change of the filter spectral shape at various input power levels by scanning the wavelength of a weak probe and measuring its transmission.

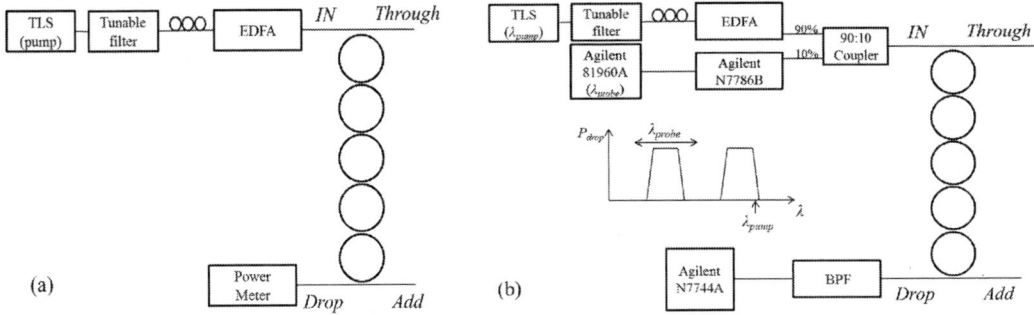

Figure 1. Schematic experimental setups (a) for observing the bi- and multi-stability and (b) for the measurement of the filter shape.

II. RESULTS AND DISCUSSIONS

A. Experiment setup

The high-order ring resonator filter consists of a cascade of five microring resonators with free spectral range (FSR) of 575GHz, with a Full Width Half Maximum (FWHM) bandwidth of 4.46GHz. The waveguide core composes of a low-loss, high-index (n=1.7) doped silica glass, buried within a SiO_2 cladding, and is fiber pigtailed to a PM fiber with a typical coupling loss of 1.5dB/facet. The waveguide cross section is 1.45μm x 1.45μm while the ring radius is 50μm. The advantages of this platform reside in its negligible linear (< 6 dB/m) and nonlinear losses as well as in a nonlinear parameter as high as 220W^{-1}km^{-1} [1]. The packaged device has a temperature control circuit which can control the chip temperature with a resolution of 0.01°C, corresponds to a wavelength resolution of 0.2pm. Figure 1 shows the schematics used in the experiment where Figure 1 a) is used to measure the bi-stable behavior and Figure 1 b) is used to measure the filter spectral shape dependence on input power. In the latter case we used a weak probe coupled in a passband resonance one FSR away from the pump, and measured its transmission for different wavelengths.

B. Bi- and multi-stability measurements

The setup used to measure the bi- and multi-stability behaviors in the high-order filter is shown in Figure 1 a) where an input laser pump is amplified and then coupled into the input port of the 5th order ring resonator filter with the drop port connected directly to an OSA, which also serves as a power detector. The response of the 5th order filter at the drop port is shown in Figure 2 a). The spectral response is obtained by scanning the input laser wavelength at a low power and recording the drop power using the same setup. Bi- and multi-stable response of the device is obtained by recording the transmitted power at the drop port at fixed wavelength. At each input signal wavelength λ_{pump}, the power is gradually increased and then decreased while tracking the drop port power. The power detected at the drop port as a function of pump power is shown in Figures 2 b), c) and d), at λ_{pump}=1556.058nm, 1556.060nm and 1556.062nm respectively. These wavelengths are chosen because they are at the edge of the filter passband, where bi-stability is most likely to occur [2]. Various degrees of bi- and multi-stability were observed at all the selected wavelengths, with larger openings between the two states as the wavelength moves further to the passband edge, due to the different amount of Kerr and thermal detuning of the resonances of the ring resonator elements.

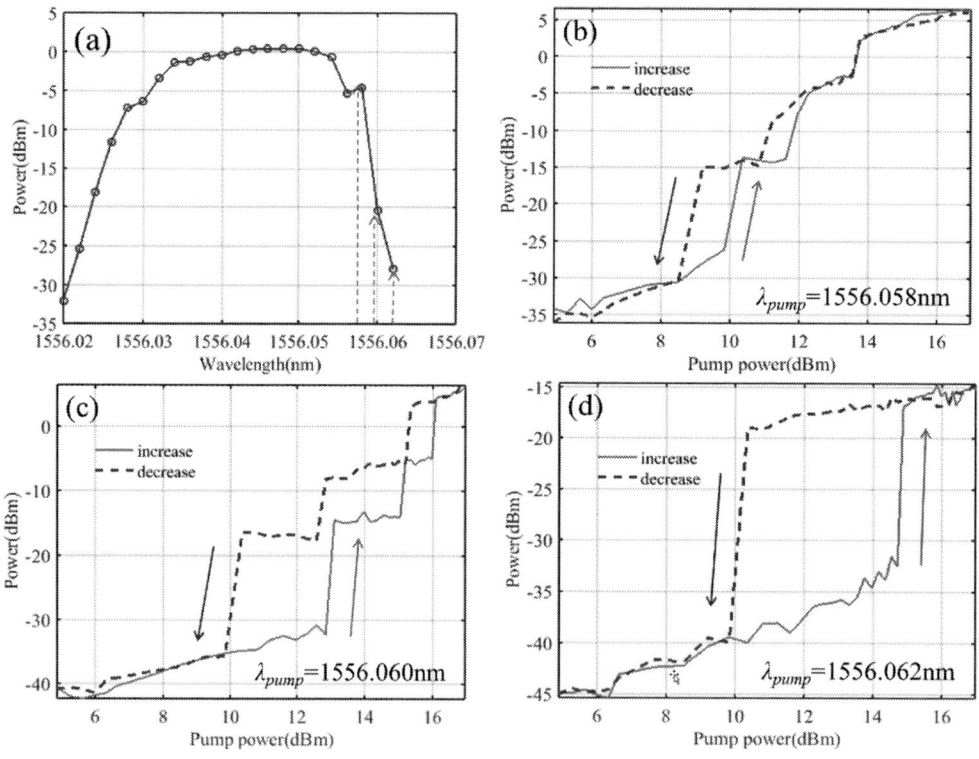

Figure 2. (a) Filter shape, the dash arrows indicate the pump wavelength for obtaining (b) , (c) multi-stability, and (d) bistability.

C. Evolution of filter shape

To further investigate the intrinsic behavior of the filter at the transitions, the dependence of the filter spectral shape on the pump power was measured using the setup shown in Figure 1 b). An additional probing signal λ_{probe} is coupled into the input along with the pump signal λ_{probe} which is scanned across the filter passband at one FSR away from the pump passband in order to limit the influence of the probe on the detuning. A bandpass filter was used to filter out the pump signal between the filter and the detector. Figure 3 shows filter shape slightly before and after the two transitions in Figure 2 d) for both the increased and decreased paths. It is possible to observe that such transitions are also connected to a dramatic change of the filter spectral response, due to the induced Kerr and thermal detuning.

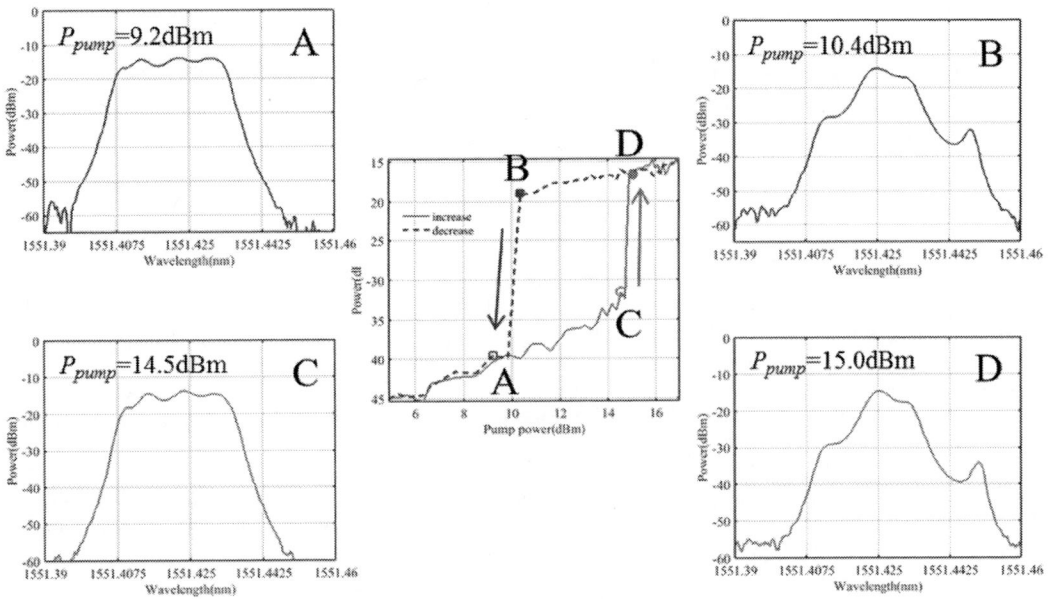

Figure 3. Measured filter response at pump power of (A) 9.2dBm and (B) 10.4dBm in the decrease process, (C) 14.5dBm and (D) 15.0dBm in the increase process, with λ_{pump}=1556.062nm.

III. CONCLUSIONS

In summary, bi- and multi-stability in a 5th order cascaded CMOS compatible ring resonator is observed on a pump signal with wavelength close to the passband edge. The spectral dependence of the filter on the bi-stability is investigated by measuring the transmission of a weak probe signal coupled one FSR away from the pump. It is observed that the filter shape drastically changed following the transitions in the measured pump transmission. A thorough analysis of the amount of Kerr and thermal induced detuning to the individual ring element will be presented at the conference.

REFERENCES

[1] D. J. Moss, R. Morandotti, A. L. Gaeta, R. Morandotti, and M. Lipson, "New CMOS-compatible platforms based on silicon nitride and Hydex for nonlinear optics," Nat. Photon., vol. 7, pp.597-607, 2013.

[2] Shaowu Chen, Libin Zhang, Yonghao Fei, and Tongtong Cao, "Bistability and self-pulsation phenomena in silicon microring resonators based on nonlinear optical effects," Optics Express, vol. 20, pp. 7454-7468, 2012.

[3] Libin Zhang, Yonghao Fei, Tongtong Cao, Yanmei Cao, Qingyang Xu, and Shaowu Chen, "Multibistability and self-pulsation in nonlinear high-Q silicon microring resonators considering thermo-optical effect," Phys. Rev. A, vol. 87, pp. 053810, 2013.

[4] F. Ramiro-Manzano, N. Prtljaga, L. Pavesi, G. Pucker, and M. Ghulinyan, "Thermo-optical bistability with Si nanocrystals in a whispering gallery mode resonator,"Opt. Lett. Vol. 38, pp. 3562-365, 2013.

[5] Yasa Ekşioğlu, Jiří Petráček, "Dynamical analysis of double-ring resonator with noninstantaneous Kerr response and effect of loss," Opt. Quant. Electron., vol. 47, pp. 3323-3335, 2015.

[6] Andrea Armaroli, Patrice Feron, and Yannick Dumeige, "Stable integrated hyper-parametric oscillator based on coupled optical microcavities,"Opt. Lett. Vol. 40, pp. 5622-5625, 2015.

[7] Libin Zhang, Yonghao Fei, Yanmei Cao, Xun Lei, and Shaowu Chen, "Experimental observations of thermo-optical bistability and self-pulsation in silicon microring resonators," J. Opt. Soc. Am. B, vol. 31, pp. 201-206, 2014.

[8] T. Herr, V. Brasch, J. D. Jost, C. Y. Wang, N. M. Kondratiev, M. L. Gorodetsky and T. J. Kippenberg, "Temporal solitons in optical microresonators," Nat. Photon., vol. 8, pp. 145-152, 2014.

[9] Tobias Hansson, and Stefan Wabnitz, "Frequency comb generation beyond the Lugiato–Lefever equation: multi-stability and super cavity solitons," J. Opt. Soc. Am. B, Vol. 32, pp. 1259-1266, 2015.

ThE2-2 (Invited)   OECC/PS2016

Optical Packet and Circuit Integrated Ring Network

Hideaki Furukawa

National Institute of Information and Communications Technology
4-2-1 Nukui-Kitamachi, Koganei-shi, Tokyo 184-8795, Japan
furukawa@nict.go.jp

Abstract: This paper shows optical packet and circuit integrated (OPCI) ring networks, which dynamically provide adequate optical paths or optical packets to satisfy QoS requests. We also describe our developed OPCI node and its operation.
Keywords: optical path, optical circuit switching, optical packet switching, wavelength resource allocation, QoS

I. INTRODUCTION

Enormous amounts of network resources and electrical power would be required to provide near-future network services such as data transmissions from a trillion sensors, entertainment, e-health, and ultrahigh-definition video delivery. Therefore, optical communication technologies are expected because they can increase the network performance and to decrease the power consumption of communication equipment. Recently, the transmission capacity per optical fiber will continue to grow with the suppression of power consumption through the introduction of advanced technologies such as multi-level modulation schemes and multiplexing techniques [1]. To efficiently utilize the large transmission capacity, optical networking technologies are important. Optical switching techniques are expected to improve the switching capacity without the increase of power consumption compared with a current electronic switching technique. Recently, more flexible and energy efficient optical networks simultaneously adopting not only optical circuit switching (OCS) but also optical packet switching (OPS) for datacenter or metro/core networks has been proposed by some groups [2]-[4].

We have been developing an optical packet and circuit integrated (OPCI) network, which realizes dynamic optical path, high-density packet multiplexing, and flexible wavelength resource allocation [4]. In the OPCI networks, a best-effort service and a QoS (Quality of service) -guaranteed service are provided by employing OPS and OCS respectively, and users can select these services. Different wavelength resources are assigned for OPS and OCS links, and the amount of their wavelength resources are dynamically changed in accordance with the service usage conditions. As shown in Fig.1, by configuring multiple logical networks based on optical packets or optical paths, various types of services can be provided efficiently.

Before now, we have developed a 2 × 2 OPCI node and constructed a single-ring OPCI network testbed with two 2 × 2 OPCI nodes [5]. We have confirmed basic operations such as add/drop and through operations for 14-wavelength 10 Gbps optical paths and 100 Gbps optical packets in the single-ring OPCI network testbed. Moreover, we have constructed a multi-ring OPCI network testbed consists of 2 × 2 OPCI nodes and one 3 × 3 OPCI node. We have also confirmed the operation of optical buffers which avoid packet collisions in more complicated networks such as multi-ring or mesh networks [6]. For the control plane, signaling and routing protocols for optical path setup and dynamic resource allocation between OPS and OCS links have been developed [7]. In this paper, we describe the overall architecture of OPCI node, and show the operation on the OPCI ring network testbed.

II. OPTICAL PACKET AND CIRCUIT INTEGRATED NETWORK

Figure 2 shows a block diagram of OPCI networks including OPCI nodes. Two wavelength selective switches (WSS) are used for combining and dividing OPS and OCS wavebands ('P' and 'C' in this figure). In principle, the WSS can flexibly move the boundary of wavelength resources between OPS and OCS links based on a resource control scheme.

In addition, WSSs also work as add/drop multiplexers for OCS links. Based on signaling and routing protocol of which information is exchanged among nodes, optical paths are set or released and the WSSs direct optical signals on an optical path to the correct output port. Optical switches and optical buffers are optical components for packet switching and contention resolution. An electronic header processor and a buffer controller control the optical system. Optical packet (OP) transponders encapsulate each IP packet into an optical packet and vice versa. Optical path transponders execute OTN (Optical Transport Network) wrapping. Network management system manages all OPCI nodes.

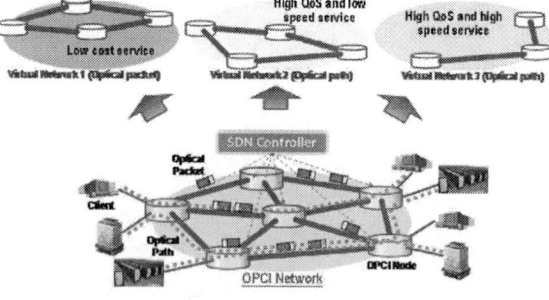

Fig. 1: Concept of OPCI network.

404

OECC/PS2016

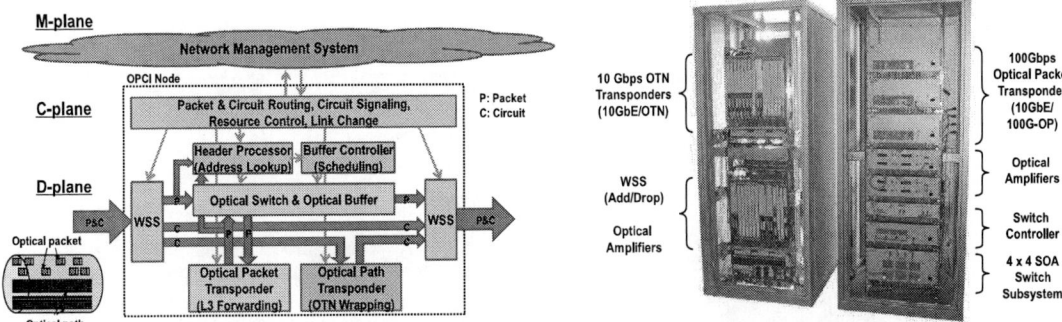

Fig.2. Block diagram of OPCInet including OPCI nodes. Fig.3. Snapshot of basic OPCI node.

Figure 3 shows a snapshot of 2 × 2 OPCI node [5], which mainly consists of six components: 10G-OTN transponders, a 100G-OP transponder, two WSSs, a 4 x 4 semiconductor optical amplifier (SOA) switch sub-system, a switch controller (header processor) and optical amplifiers. Dynamic routing and optical buffer have not been implemented (see Chapter III for optical buffer implementation). In OCS links, a 10G-OTN transponder encapsulates 10GBASE-LR Ethernet frames from the client side into OTU2e format. Because optical paths are established by control packets in advance, there is no need to read the IP destination address of incoming 10GbE frames.

In OPS links, a 100G-OP transponder encapsulates incoming IP packets from the client 10GBASE-LR Ethernet link into 100 Gbps "colored" optical packets. The length of Ethernet frames ranges from 64 byte to 9,604 byte. An optical packet consists of 10 optical lanes with different wavelength and each speed is 10.3125 Gbps. Each optical lane of optical packet has preamble for recovering clock and data and detecting start of optical packet. The optical packet length is variable corresponding to the Ethernet frame length. An 8-octet route header including a 16-bit optical packet (OP) address, packet length, and error correction code is attached to the first optical lane. The OPS system can be extended to 4 × 4 input/output ports and four optical buffers, each of which is attached to each output port, by using a number of SOA switches as shown in Fig. 4(a). One has 8 FDLs and others have 4 due to our resource constraint. We explain the operation of an optical buffer attached with output port 1 (OUT #1) consists of two 4 × 4 SOA switch subsystems and 8 FDLs with different lengths, and acts as both switching and buffering functions. The buffer size is 8 packets. We sent optical packets from input ports 1, 2, 3, and 4 (IN #1, #2, #3, #4) to OUT #1 at the same time. It means that packet collisions may occur. The packet traffic condition about average traffic and packet length of input optical packets was described in Fig.4(b). The length of each FDL was increased by 20 m, which corresponding time was about 100 ns. The optical buffer attached at OUT #1 delayed some packets to avoid packet collisions. Figure 4(a) also shows the temporal waveform of packet sequences at IN #1, #2, #3, #4, and OUT #1, which were measured at different timing, in condition of Case 3. Figure 4(b) also shows the total average traffic of input ports and the packet loss rates. We observed that the packet loss rate was dramatically improved by using 8-FDL optical buffer compared with 4-FDL one.

III. OPERATION ON OPCI NETWORK TESTBED

Figure 5 shows the configuration of the latest 2 x 2 OPCI node testbed for ring networks. In this OPCI node, two 100 Gbps optical transport network (100G-OTN) transponders were added. We demonstrated multi-format optical switching. We used 40 wavelength channels (named as λ1–λ40) ranging from 1531.90 to 1563.05 nm with 100 GHz grid spacing. The wavelengths were allocated to 10 Gbps OOK optical paths from node1 (λ11 – λ17) and node2 (λ18, λ19, λ32 – λ36), 100 Gbps DP-QPSK optical paths from node1 (λ37, λ38) and node2 (λ39, λ40), and 100 Gbps OOK optical packets (λ21 – λ30). Firstly, we established fourteen 10 Gbps OOK optical paths with a loop configuration. Network

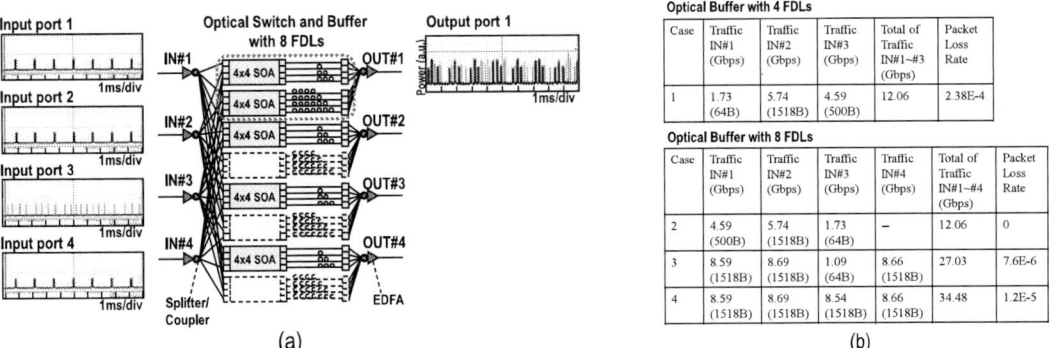

Optical Buffer with 4 FDLs

Case	Traffic IN#1 (Gbps)	Traffic IN#2 (Gbps)	Traffic IN#3 (Gbps)	Total of Traffic IN#1~#3 (Gbps)	Packet Loss Rate
1	1.73 (64B)	5.74 (1518B)	4.59 (500B)	12.06	2.38E-4

Optical Buffer with 8 FDLs

Case	Traffic IN#1 (Gbps)	Traffic IN#2 (Gbps)	Traffic IN#3 (Gbps)	Traffic IN#4 (Gbps)	Total of Traffic IN#1~#4 (Gbps)	Packet Loss Rate
2	4.59 (500B)	5.74 (1518B)	1.73 (64B)	–	12.06	0
3	8.59 (1518B)	8.69 (1518B)	1.09 (64B)	8.66 (1518B)	27.03	7.6E-6
4	8.59 (1518B)	8.69 (1518B)	8.54 (1518B)	8.66 (1518B)	34.48	1.2E-5

(a)   (b)

Fig. 4(a) Configuration diagram of 4× 4 OPS node with 8-FDL optical buffer, and optical packet sequences at input ports 1~4 and output port 1 in buffering operation. (b) Average traffic at each input port, the total of the average traffic of input ports and the packet loss rates in various conditions.

405

OECC/PS2016

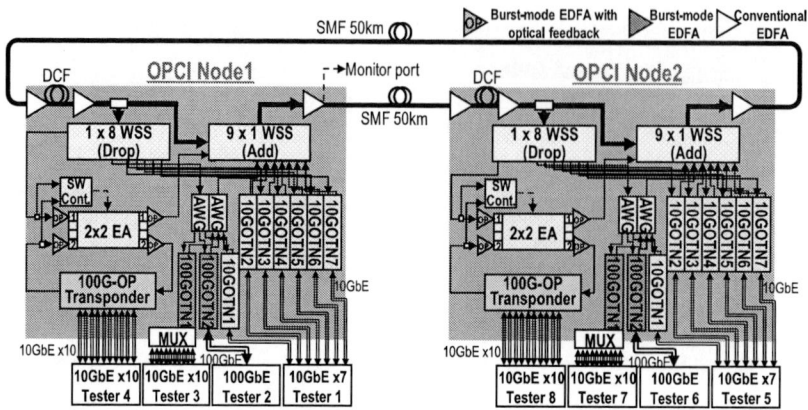

Fig. 5. Configuration of OPCI node and demonstration system of OPCI ring network.

testers 1 and 5 transmitted/received 1,518 byte 10GbE frames to/from each 10G OTN transponder in nodes 1 and 2. Also, we established four 100 Gbps DP-QPSK optical paths with a loop configuration. Testers 2 and 6 transmitted/received 64–1,518 bytes 10GbE frames to/from a 100G OTN transponder with 100GbE client-interface in nodes 1 and 2. Testers 3 and 7 transmitted/received 1,518 byte 10GbE frames to/from a 100G OTN transponder through ten 10GbE client-interfaces in nodes 1 and 2, respectively. 100 Gbps OOK optical packets were launched to input ports 2 of the electro-absorption (EA) based optical switch in node 1, which is used instead of SOA switch. These OOK packets encapsulated 1,518 byte 10GbE frames coming from ten 10GbE client-interfaces from tester 4. A switch controller read the route header on each optical packet to control the optical switch. Optical packets were switched to output port 1 in accordance with the routing table. These optical packets and paths were combined by a 9x1 WSS and transmitted over 50 km of single-mode fiber (SMF) to node 2. Figure 6 shows the spectral waveform of all optical signals in output port of node 1. In node 2, an 8 x 1 WSS split the optical packets and optical paths. In the EA switch of node 2, optical packets are switched into output port 1 and 2, respectively. A 100G OP transponder received the packets and forwarded the decapsulated the 10GbE frames into network tester 8. The frame error rate (FER) for 10GbE frames transmitted by 10 Gbps OOK and 100 Gbps DP-QPSK optical paths, and 100 Gbps OOK optical packets launched from node 1 were much less than 10^{-4}, which is regarded as high quality [8].

IV. CONCLUSIONS

This paper showed our optical packet and circuit integrated network employing OPS and OCS to satisfy various QoS requests. The overall architecture of OPCI node and the operation of the OPCI network testbed was reported.

REFERENCES

[1] B. J. Puttnam, R. S. Luís, W. Klaus, J. Sakaguchi, J.-M. Delgado Mendinueta, Y. Awaji, N. Wada, Y. Tamura, T. Hayashi, M. Hirano and J. Marciante, "2.15 Pb/s Transmission Using a 22 Core Homogeneous Single-Mode Multi-Core Fiber and Wideband Optical Comb," in Proc. 41st European Conference on Optical Communication (ECOC 2015), no. PDP3.1 (2015).

[2] J. Perelló, S. Spadaro, S. Ricciardi, and D. Careglio, " All-opti-cal packet/circuit switching-based data center network for en-hanced scalability, latency, and through, " IEEE Network, vol. 27, no. 6, pp. 14 – 22, Nov., (2013).

[3] Ryo Takahashi, Toru Segawa, Salah Ibrahim, Tatsushi Nakahara, Hiroshi Ishikawa, Atsushi Hiramatsu, Yue-Cai Huang, and Ken-ichi Kitayama, "Torus Data Center Network with Smart Flow Control Enabled by Hybrid Optoelectronic Routers," IEEE/OSA Journal of Optical Communications and Networking. vol. 7, no. 12, pp. B141-B152 (2015).

[4] H. Harai, H. Furukawa, K. Fujikawa, T. Miyazawa, N. Wada, "Optical Packet and Circuit Integrated Networks and Software Defined Networking Extension," IEEE/OSA Journal of Lightwave Technology, vol. 32, no.16, pp. 2751-2759 (2014).

[5] H. Furukawa, H. Harai, T. Miyazawa, S. Shinada, W. Kawasaki, and N. Wada, "Development of optical packet and circuit integrated ring network testbed," Optics Express, vol. 19, no. 26, pp. B242-B250 (2011).

[6] H. Furukawa, S. Shinada, T. Miyazawa, H. Harai, W. Kawasaki, T. Saito, K. Matsunaga, T. Toyozumi, and N. Wada, "A Multi-Ring Optical Packet and Circuit Integrated Network with Optical Buffering," Optics Express, vol. 20, no. 27, pp. 28764–28771, Dec (2012).

[7] T. Miyazawa, H. Furukawa, K. Fujikawa, N. Wada, and H. Harai, "Development of an Autonomous Distributed Control System for Optical Packet and Circuit Integrated Networks," IEEE/OSA J. Opt. Commun. Netw., vol.4, no.1, pp.25-37 (2012).

[8] ITU-T Recommendation Y.1541.

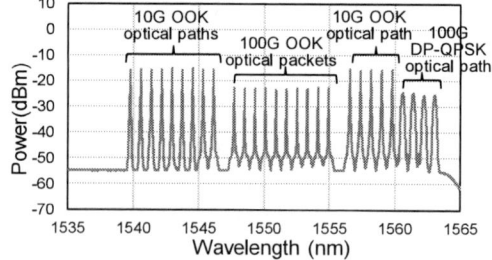

Fig.6. Spectrum of optical packets and paths in output of node 1.

4-Channel Silicon Photonic Mode Unscrambler

Andrea Melloni[1], Andrea Annoni[1], Emanuele Guglielmi[1], Marco Carminati[1], Giorgio Ferrari[1], Nicola Peserico[1], Stefano Grillanda[1], Marc Sorel[2] and Francesco Morichetti[1]

[1]Dipartimento di Elettronica, Informazione e Bioingegneria, Politecnico di Milano, via Ponzio 34/5, 20133 Milano, Italy
[2]School of Engineering, University of Glasgow, Glasgow, G12 8LT, UK
Author e-mail address: andrea.melloni@polimi.it

Abstract: We demonstrate all-optical unscrambling of four 10Gb/s mixed modes on a silicon photonic chip with less than -20 dB crosstalk and with a power penalty between 2 and 1dB at BER level of 10^{-8}.
Keywords: Mode Division Multiplexing, Silicon Photonics, Photonic Integrated Circuits

I. INTRODUCTION

Mode-division multiplexing (MDM) is considered a promising approach to boost the capacity of optical fiber transmission by launching several channels multiplexed over the spatial modes of few modes fiber [1]. This new paradigm requires however, optical components capable to manipulate signals encoded on different orthogonal modes and mixed after propagation though a multimode link. Multiple-input multiple output (MIMO) photonic integrated architectures realizing reconfigurable mode (de)multiplexing and unscrambling have been recently proposed [2] and realized [3]. In these schemes the possibility to monitor individual channels multiplexed in the MDM signal is an essential feature to enable simple and robust tuning strategies. However, channel monitoring in MDM systems must be non invasive in order to preserve mode orthogonality [2]. In this work we demonstrate all-optical unscrambling of four-mixed modes on a silicon photonic (SiP) chip. We exploit transparent photodetectors, realized though the non-invasive ContacLess Integrated Photonic Probe (CLIPPs) technology [4], to monitor the evolution of the modes intensity along a MIMO demultiplexer, without impairing mode orthogonality. Demultiplexing of 10 Gbit/s channels with less than -20 dB residual crosstalk is demonstrated.

II. DESIGN AND FABRICATION

The schematic of the realized 4-channel all-optical MIMO demultiplexer is depicted in Figure 1(a), the layout has been designed according to the architecture concept proposed in Ref. 2. Four input data channels (Ch. A – Ch. D) are mixed by means of multimode link realized with a multimode waveguide section and coupled into four single-mode optical waveguides. At this point, each waveguide carries a portion of the light intensity of each of the four input channels. Measurements performed on an identical mode mixer realized onto the same chip reveal that at this wavelength, at least 20% of the light intensity of each channel is distributed at each single-mode input waveguide of the demultiplexer.

The layout of the MIMO demultiplexer consists of an arrangement of 6 integrated Mach-Zehnder Interferometers (Si, i = 1, 2, ..., 6), that can be sequentially tuned in order to reconstruct and arbitrarily extract each channel at each of the four output ports (OUT1–OUT4). To reconstruct Ch. x (x = A, B, C, D) at port OUT1, stages S1, S2 and S3 need to be sequentially biased in order to minimize Ch. x power at the output lower branch of each stage. Transparent detectors are

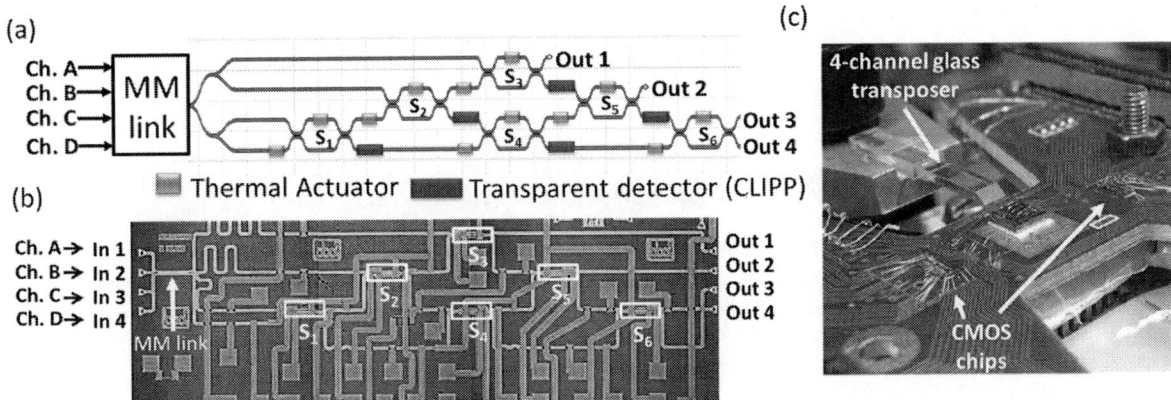

Figure 1: (a) Schematic and (b) top-view microphotograph of the 4-channel MIMO demultiplexer realized on a SiP platform. A multimode waveguide is used to induce mode mixing of the input channels (A,B,C,D). (c) Photograph of the SiP chip wire-bonded to a readout CMOS electronic ASIC and integrated onto the same printed circuit board. A glass based transposer is used to couple the four channels into the SiP circuit.

essential to monitor the switching state of each stage without introducing optical loss that would impair mode orthogonality.

The 4-channel MIMO demultiplexer was realized on a 220-nm SiP platform through LETIePIXfab multi-project-wafer run. A top-view photograph of the Si chip is shown in Fig. 1(b). Each switching stage has integrated two thermal actuators to control the relative phase of the optical field at the input ports and the relative phase between the inner arms of each MZI; the total footprint of the 4-Channel demultiplexer is 3.7 mm x 1.5 mm. With the CLIPP detector integrated at one output port of each MZI, we are able to monitor individually and in real time the switching state of each MZI. The four channels are coupled to the SiP through a glass based transposer [5] allowing to inject simultaneously the signals coming from four singlemode-fibers to four input grating couplers. To enable simultaneous read out of the integrated CLIPPs and to provide the driving voltage for the thermal actuators [6], the SiP chip is mounted on PCB and bridged to a CMOS electronic ASIC; underneath the PCB a thermoelectric cooler is user to keep stable the chip overall temperature. Figure 1 (c) shows the silicon photonic chip mounted on the PCB and wire bonded to the two CMOS electronic ASIC, and the glass transposer used to inject the four channels.

III. EXPERIMENTAL RESULTS

Figure 2 show the crosstalk level, for an input channel Ch. x, when the demultiplexer is configured to reconstruct Ch. x at the output port 1. To measure the crosstalk we sequentially tuned the switching stages S1-S3 by minimizing the light intensity of the Ch. x at the lower output branch of stages S1-S3 by using CLIPPs 1-3, respectively. The crosstalk over wavelength was obtained by measuring, with of a tunable laser source and an optical spectrum analyzer, the contribute to the output power of Out 1 for each of the four input channel. As visible in Fig. 2, the performances obtained vary between the different configuration of the demultiplexer; bandwidths with crosstalk lower than -20 dB and spanning over several nanometers were obtained for each of the four configurations.

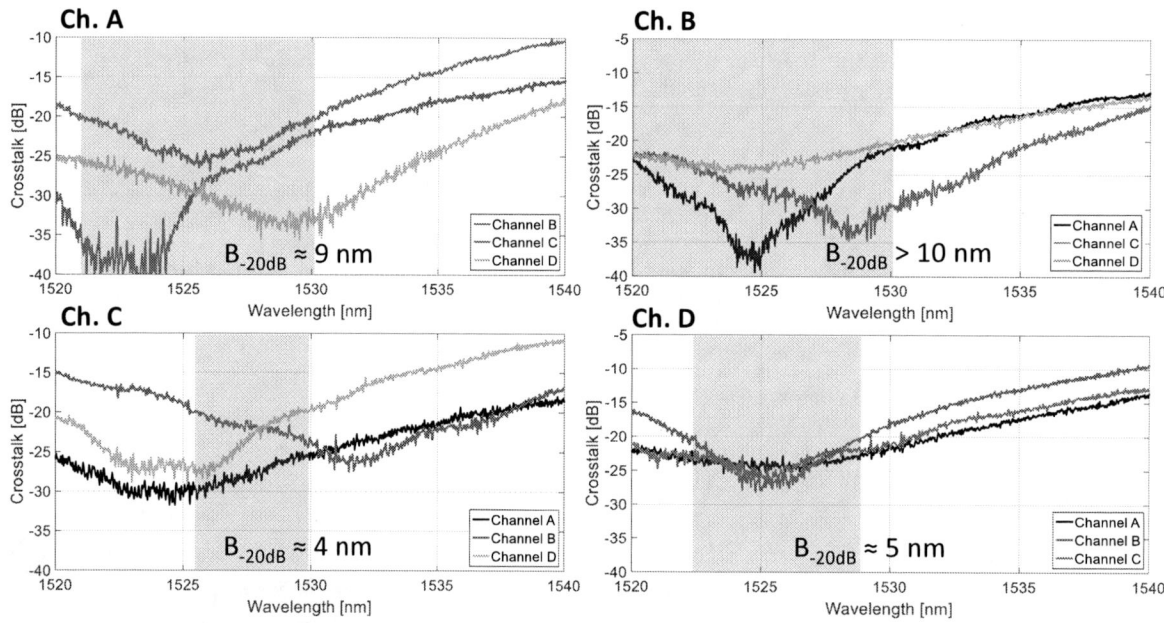

Figure 2: Crosstalk spectra for the four extracted channels (Ch. A - Ch. D) at port OUT1; the label at the top left corner of each graph identifies which channel is reconstructed at the output port 1; the curves show the crosstalk as a function of wavelength for the remaining "interfering" channels. Bandwidths with a crosstalk of -20dB are highlighted for each of the possible configuration.

Figure 3 shows the bit-error-rate (BER) of the demultiplexed channels in the cases considered in Fig. 2. Four 10 Gbit/s OOK NRZ channels sharing the same carrier wavelength of 1528 nm are achieved by splitting and decorrelating through a few-km-long fiber coils the data stream generated from a LiNb Mach-Zehnder modulator. The modulator is driven with a pattern $2^{31}-1$ bit long generated by a PRBS; the power levels of the four data channels were equalized by means of variable optical attenuators and polarization controlled independently.

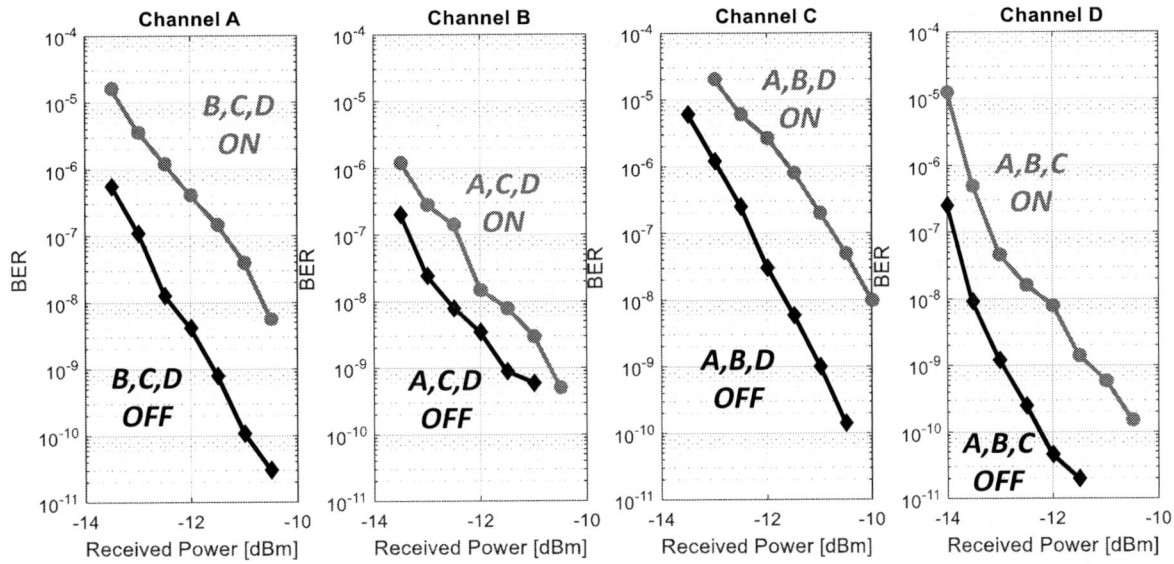

Figure 3: BER curves for the four extracted channels (Ch. A - Ch. D) at port OUT1. For each of the four channels the red curves with the circular markers is the measures conducted with the three "interfering" channels off while the black curves with the diamond marker is the reference BER when only the channel itself is injected into the demultiplexer.

To evaluate the performances of the device, we measured the power penalty to the bit error rate (BER) from the demultiplexing process, the measurements were performed for each of the four input channels. The penalty was obtained by measuring the reference BER when only the extract channel was present in the demultiplexer, and by comparing it with the BER obtained when all the three "interfering" channels were activated. As for the crosstalk spectra, the power penalty varies depending on which channel is extracted by the demultiplexer; for a BER level of 10^{-8}, the power penalty vary from 2 dB for Ch. A to 1 dB for Ch. B.

IV. CONCLUSIONS

We demonstrated a 4-channel SiP photonic MIMO demultiplexer performing all-optical unscrambling of four mixed modes. On-chip light monitoring through transparent CLIPP detectors was exploited to achieve accurate and robust sequential tuning of the demultiplexer without affecting mode orthogonality, thus enabling demultiplexing of 10 Gbit/s channels with less than -20 dB crosstalk over bandwidths spanning from 4 nm to more than 10 nm. The power penalty of the SiP demultiplexer varies from 2 dB to 1 dB depending on which channels is extracted.

V. REFERENCES

[1] P. J. Winzer, "Making spatial multiplexing a reality," Nat. Photon. 8, 345-348 (2014).
[2] D. A. B. Miller, "Self-configuring universal linear optical component," Photon. Res. 1, 1-15 (2013).
[3] N. K. Fontaine, "Space division multiplexing and all-optical MIMO demultiplexing using a photonics integrated circuit," Proceed. Optical Fiber Communication Conference (OFC 2012), Los Angeles (CA), 4–8 March 2012, paper PDP5B.1
[4] F. Morichetti, et al., "Non-invasive on-chip light observation by contactless waveguide conductivity monitoring," IEEE J. Sel. Topics Quantum Electron. 20, 292-301 (2014).
[5] For details on the glass transposer technology, refer to the website:
     http://www.plcconnections.com/documents/ICE%20R020815.pdf
[6] S. Grillanda et al. "Non-invasive monitoring and control in silicon photonics using CMOS integrated electronics," Optica 1, 129-136 (2014).

Mode Division Multiplexing Switch for On-Chip Optical Interconnects

Xinru Wu[1*], Ke Xu[2], Daoxin Dai[3] and Hon Ki Tsang[1]

[1]Department of Electronic Engineering, The Chinese University of Hong Kong, Shatin, N. T., Hong Kong, S. A. R., China
[2]Department of Electronic and Information Engineering, Shenzhen Graduate School, Harbin Institute of Technology, Xili, Shenzhen, China
[3]Centre for Optical and Electromagnetic Research, State Key Laboratory for Modern Optical Instrumentation, Zhejiang Provincial Key Laboratory for Sensing Technologies, Zhejiang University, Zijingang Campus, Hangzhou 310058, China
*E-mail: xrwu@ee.cuhk.edu.hk

Abstract: *We demonstrate a thermo-optic switch for on-chip mode division multiplexing. The crosstalk was less than 17dB for the nine possible switch states over a 40 nm wavelength band.*
Keywords: *Mode division multiplexing, Switch, Optical interconnects*

I. INTRODUCTION

Mode division multiplexing (MDM) is a potential approach to increase transmission capacity of on-chip optical networks by utilizing the different waveguide modes in a multimode waveguide to carry different channels of data. Unlike wavelength division multiplexing (WDM), MDM has the advantage of needing only a single wavelength source [1], thus reducing its cost. Due to the high bandwidth requirement for on-chip communications, WDM have been used in network on chips, with switching performed by cascaded active silicon microresonator array [2], single ring resonator [3] and active plasmonics [4]. However, direct switching of MDM signals is still challenging. Recently, an integrated switch for hybrid MDM-WDM operating in the single-mode regime was reported using microring resonator array for the optical switching [5].

Here we demonstrate an on-chip broadband switch in a multichannel MDM system utilizing cascaded thermal-optic Mach-Zehnder interferometer (MZI) optical switches with heaters for switch control. The proposed MDM switch is implemented by the integration of a three-channel mode multiplexer realized by cascading asymmetric directional couplers (ADCs) [6] to generate TE_0, TE_1 and TE_2 modes; a three-channel mode demultiplexer is used to achieve the conversion from a high order mode in the bus waveguide to the fundamental mode in a single mode waveguide to allow the independent processing of the modes using single-mode elements; a 3 x 3 single-mode optical switch is used to route the three modes independently to any of the three output ports. Fig. 1 shows the schematic of the implemented switching fabric. Less than 10 dB insertion loss and lower than -17 dB mode crosstalk over 40 nm wavelength band are measured for the nine data channels separately.

II. DEVICE STRUCTURE AND THERMO-OPTIC HEATER

The integrated MDM switch was fabricated on 220 nm silicon-on-insulator (SOI) wafer with 2 μm buried oxide using the multi-project wafer (MPW) foundry fabrication under the ISIPP25G technology at IMEC. The waveguides are patterned using 193nm DUV lithography with the dimensions of 450nm wide for single mode waveguides, 1.006μm

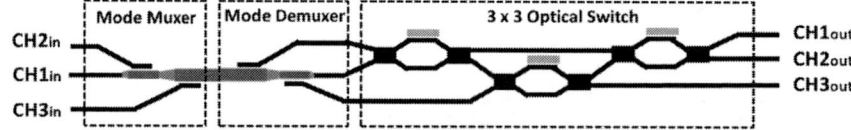

Fig. 1. Schematic of the integrated MDM switching fabric.

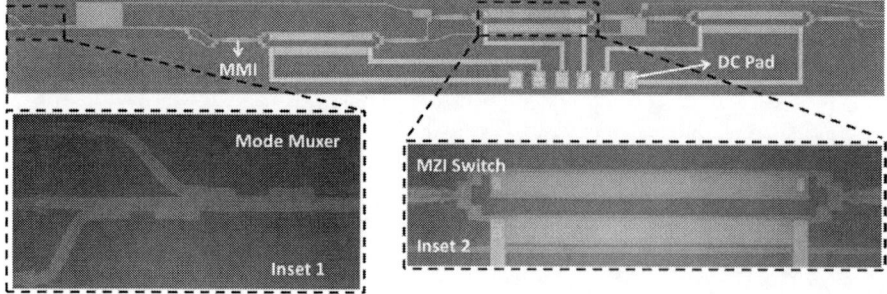

Fig. 2. Microscope images of the fabricated devices. Inset 1 is the mode multiplexer and inset 2 is the MZI switch cell.

and 1.556μm wide with 200nm gap for TE_1 and TE_2 (de)multiplexers respectively. Focusing grating couplers are used to do the light coupling between optical fiber and silicon chip.

The 2 x 2 switch cell depicted in Fig. 2 inset 2 consists of a symmetric MZI with two thermal-optic phase shifters and two multimode interferometers (MMI). As the path difference of the MZI is designed to be zero, the MZI switch can be operated for a wide range of wavelength within the grating coupler bandwidth. The thermo-optic phase shifter in the MZI relied on a p doped series resistance along the longitudinal direction of the waveguide as an electrical heater. Both arms of the MZI switch were doped to balance the doping induced optical loss, but only one arm had an electrical current applied. The electrical heating from the p doped strip will increase the silicon waveguide refractive index. Fig. 2 shows the microscope images of the on-chip multimode switch. The total area of this MDM switch is less than 1.5 mm^2, which could be further reduced by using compact switch configuration [7] or compact switch design [8] reported previously.

III. EXPERIMENTAL RESULTS

Mode conversion efficiencies of mode (de)multiplexers are characterized by measuring transmission responses at all three outputs when injecting light from inputs for TE_0 mode, TE_1 mode and TE_2 mode respectively. The optical transmission spectra as depicted in Fig. 3(a)-(c) have the grating couplers response normalized out. Over a 40 nm wavelength band around 1550nm, the measured insertion loss and mode crosstalk for TE_0 mode are ~1 dB and -25 dB, for TE_1 mode are ~2.5 dB and -25 dB and for TE_2 mode are ~3 dB and -23.5 dB.

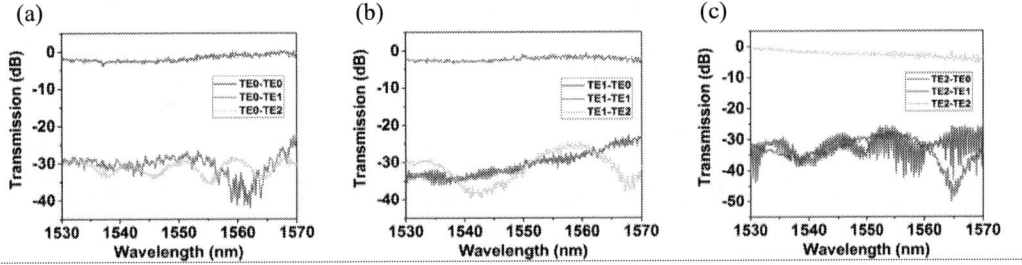

Fig. 3. Measured optical transmissions and crosstalk at three modes output ports when injecting light in the three modes input ports separately.

The resistance of the thermo-optic phase shifter in one arm is ~1.3 kΩ. Continuous-wave light was introduced to the input of a test structure which had unbalanced arm and the transmission at the bar and cross ports are measured respectively. The wavelength was fixed at 1550 nm and the switch states were controlled by the power applied on the phase shifter. Fig. 4 shows the signal and crosstalk at bar and cross output ports when a DC voltage applied on the phase shifter. The extinction ratio between on and off states is ~ 24 dB under 170 mW heating power. The loss for input and output gratings was about 8.5 dB at 1550nm, and the insertion loss of this switch is about 1.5 dB.

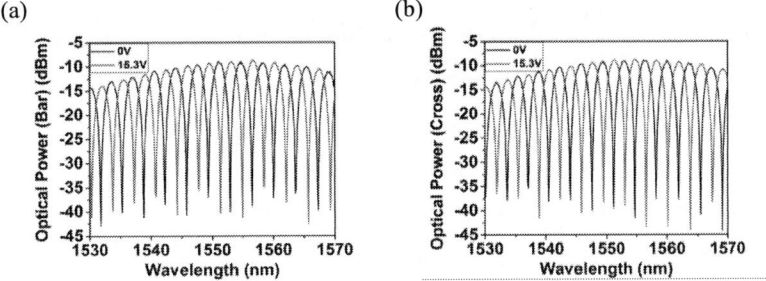

Fig. 4. Transmission and crosstalk as a function of wavelength at cross and bar output ports when injecting light in the input port of test structure.

We measured the intermodal crosstalk among the nine possible switch states from mode i to mode j (i, j =1,2 or 3) by tuning one input mode across the 40 nm wavelength band and measuring the power at three outputs respectively. Except for TE_2 mode route to the output of channel 3, all the other modes to each output ports will go through two switches, so at most two switches will be simultaneously applied to control the routing path. The measurement includes the crosstalk from other channels to the desired signal and the results are plotted in Fig. 4 (a)-(i) for TE_0, TE_1 and TE_2 data channels respectively. The measured insertion loss for the nine possible switch states range from 6 dB to 10 dB over the 40 nm wavelength band. The mode-dependent insertion loss mainly comes from the phase mismatch of the mode (de)multiplexers due to the fabrication errors. The crosstalk for each switch state channel is 17 dB to 23 dB over the whole wavelength band. Crosstalk values after switching are larger than that of the MDM system, which indicates that the 3 x 3 optical switch introduces additional crosstalk. This additional crosstalk may come from incomplete switching control as well as thermal crosstalk between neighboring switch units.

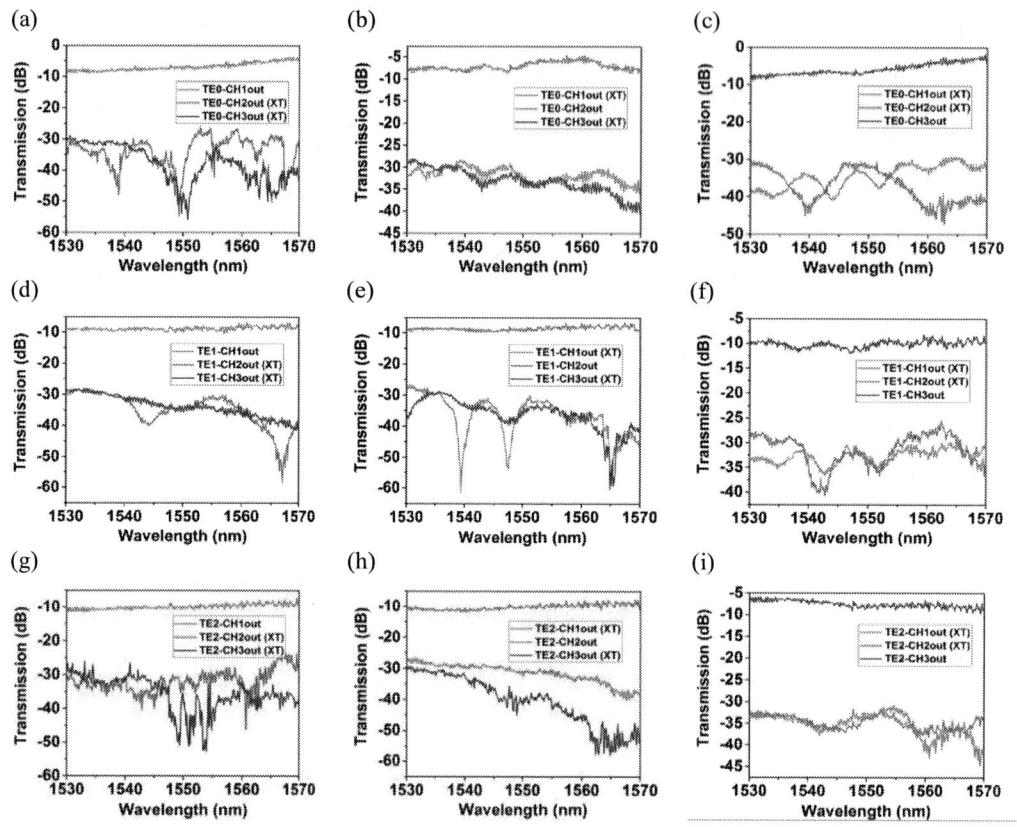

Fig. 5. Transmissions and crosstalk (XT) for (a)-(c) TE$_0$ mode; (d)-(f) TE$_1$ mode; (g)-(i) TE$_2$ mode switching to three outputs respectively.

IV. CONCLUSIONS

We have demonstrated a broadband on-chip MDM switch based on thermo-optic controlled MZI. The insertion loss varied between 6 dB to 10 dB and the crosstalk was better than 17 dB for all the switch states. In future the switch design may be further optimized by adding isolation trenches and substrate removal to suppress thermal leakage and thermal crosstalk. The switch can enable the use of reconfigurable MDM in on-chip optical interconnects.

ACKNOWLEDGMENT

We thank IMEC for device fabrication and funding support from RGC/NSFC Joint Research Grant N_CUHK404/14 and 61431166001.

REFERENCES

[1] S. Berdagué and P. Facq, "Mode division multiplexing in optical fibers," Applied optics, vol. 21, no. 11, pp. 1950-1955, 1982.

[2] A. W. Poon, X. Fang and X. Luo. "Cascaded active silicon microresonator array cross-connect circuits for WDM networks-on-chip", in proc. International Society for Optics and Photonics, vol. 6898, 2008.

[3] P. Pong, F. P. Stefan and M. Lipson, "All-optical compact silicon comb switch," Opt. Express, vol. 15, no. 15, pp. 9600-9605, 2007.

[4] S. Papaioannou, D. Kalavrouziotis, K. Vyrsokinos, J. C. Weeber, K. Hassan, L. Markey, T. Tekin, D. Apostolopoulos, H. Avramopoulos, and N. Pleros, "Active plasmonics in WDM traffic switching applications, ". Scientific reports, vol. 2, pp. 652, 2012.

[5] B. Stern, X. Zhu, C. P. Chen, L. D. Tzuang, J. Cardenas, K. Bergman and M. Lipson, "On-chip mode-division multiplexing switch," Optica, vol. 2, no. 6, pp. 530-535, 2015.

[6] D. Dai, "Silicon mode-(de) multiplexer for a hybrid multiplexing system to achieve ultrahigh capacity photonic networks-on-chip with a single-wavelength-carrier light," in proc. Communications and Photonics Conference (ACP), OSA Technical Digest, paper ATh3B.3, 2012.

[7] L. Chen and Y. Chen, "Compact, low-loss and low-power 8 x 8 broadband silicon optical switch", Opt. Exp., vol. 20, no. 17, pp. 18977-18985, 2012.

[8] Y. Vlasov, W. M. Green, and F. Xia, "High-throughput silicon nanophotonic wavelength-insensitive switch for on-chip optical networks.nature photonics", vol. 2, no. 4, pp. 242-246, 2008.

ThE3-1 (Invited)

OECC/PS2016

Plasmonic-Organic Hybrid (POH) Modulators

A. Melikyan[1], K. Koehnle[2], M. Lauermann[2], R. Palmer[3], S. Koeber[3], S. Muehlbrandt[3], P. C. Schindler[2],
D. L. Elder[4], S. Wolf[2], M. Sommer[3], L. R. Dalton[4], D. Van Thourhout[6], W. Freude[2], M. Kohl[3],
J. Leuthold[5], and C. Koos[2,3]

[1]*Now with* NOKIA – Bell Labs, 791 Holmdel Road, Holmdel, 07733 NJ, USA
[2]*Institute of Photonics and Quantum Electronics (IPQ), Karlsruhe Institute of Technology (KIT), Karlsruhe, Germany*
[3]*Institute of Microstructure Technology (IMT), Karlsruhe Institute of Technology (KIT), Karlsruhe, Germany*
[4]*University of Washington, Department of Chemistry, Seattle, USA*
[5]*Swiss Federal Institute of Technology (ETH), 8092 Zurich, Switzerland*
[6]*Photonics Research Group, Ghent University - imec, Gent, Belgium*

Abstract: An overview of high-speed plasmonic-organic hybrid (POH) modulators for BPSK and OOK signaling is presented. The optimum length of POH modulators resulting in maximum optical modulation amplitudes (OMA) are discussed.

Keywords: Optoelectronic devices, Plasmonics, Silicon photonics.

I. INTRODUCTION

Electro-optic modulators are key components in optical communications links [1]. They are required for electrical-to-optical signal conversions on the transmitters side of the optical link. To meet the future demands of large capacity optical links, the modulators should allow scaling in both number of parallel channels and the symbol rates [2]. Silicon photonics has been envisaged as the technology that fulfills both these requirements at reasonably low cost [3]. Most of the high-speed silicon photonic modulators use the plasma dispersion effect in silicon PN-junction and provide data rates up to 40 Gbit/s [1]. To generate sufficient modulation amplitudes, these devices are typically several millimeters long, which limits the integration density and imposes practical challenges associated with the group velocity matching and with the electrical/optical losses [4]. A more efficient phase modulation is reported for silicon-organic hybrid (SOH) modulators [5]. The high-speed operation of SOH modulators, however, is challenging because of the large RC-time constant of the devices [5]. To achieve modulation bandwidths in excess of 100 GHz, a vast amount of research has been long directed towards discovering novel and non-traditional solutions using metal optics – the so called plasmonics [6]. Plasmonic high-speed modulator designs, fundamentally relying either upon the Pockels [7] or the carrier dispersion [8-12] effects, have been subsequently reported.

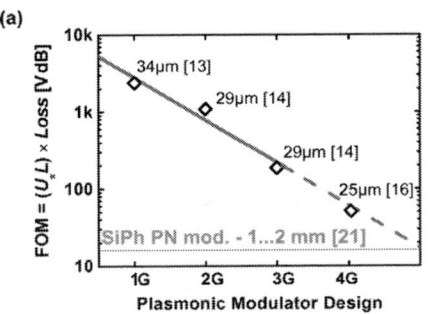
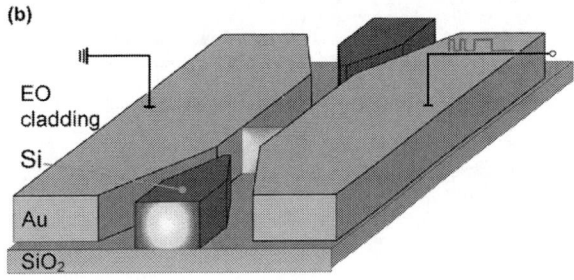

Fig. 1. Plasmonic-organic hybrid technology and an overview of the figure-of-merits (FOM) reported previously. (a) Performance improvement reported for POH modulators over the last several years. In the calculation of the FOMs, only the optical losses in the plasmonic section are only considered. (b) The schematic of the POH phase modulator comprising two metallic taper mode converters and metallic slot waveguide.

In 2014, a plasmonic-organic hybrid (POH) was demonstrated for the first time. It provides an electro-optic bandwidth of 64 GHz for the sub-50 μm long phase shifters [13, 14]. Moreover, recent developments in this area demonstrate the unequalled compactness of POH modulators, with footprints of only a few square micrometers [15]. The POH modulators have also the potential to outperform most of the conventional solutions in terms of the modulation bandwidths and the required drive voltages [13-20]. Fig. 1(a) summarizes the evolution of the POH modulators in terms of a figure of merit (FOM) defined as a product of $U_\pi \times L$ and optical loss [21, 24], where $U_\pi \times L$ is the voltage-length product that results in a phase modulation of π for a modulator of length L. Typical FOMs reported for state-of-the-art silicon modulators are given in a green solid line [21]. For given optical and electrical input powers, POH modulators might soon be able to provide the same optical modulation amplitudes at much higher speeds and with significantly smaller footprints. Provided that some of the technical challenges of POH modulators are solved in the future, the scaling in the symbol rates and in the numbers of parallel spatial and/or wavelength channels will become practical with POH modulators for a broader application range in optical communications. A step further is taken in Ref. [18, 19], by demonstrating the first POH modulator array for optical interconnect applications.

In this paper, we provide an overview of some of these latest advances in the field of plasmonic-organic hybrid (POH) integration [13-20]. The POH technology is first introduced followed by the discussion of the performance of POH modulators in direct detection and coherent systems [13,14]. First, we discuss error-free binary phase shift keying (BPSK) signaling with a single POH phase modulator for data rates up to 40 Gbit/s [14]. For low-cost direct detection applications, on-off keying (OOK) at the data rates of 40 Gbit/s is reported by incorporating two POH phase modulators in a silicon photonic Mach-Zehnder (MZ) interferometer [14]. As a fundamental limitation, the influence of the optical losses on the performance of a POH MZ modulator is discussed. The optimum POH phase shifter length resulting in a maximum optical modulation amplitude (OMA) is discussed.

II. PLASMONIC-ORGANIC HYBRID (POH) MODULATORS

The plasmonic section of a POH modulator is a metallic slot waveguide (MSW) comprising two metal electrodes separated by a sub-200 nm gap which is filled with an electro-optic organic cladding, see Fig. 1(b). Adiabatic metallic tapers convert the photonic mode to a plasmonic mode and vice versa [22]. Alternatively, an MSW can be constructed by stacking two metal electrodes separated by a horizontal slot filled with an electro-optic material. In this case, a directional coupler can be used for efficient photonic-to-plasmonic mode conversion [20]. In MSWs, both the optical and the RF fields are confined to the slot, resulting in a strong interaction between the two [6]. The modulating RF field changes the refractive index of the EO material inside the slot through the Pockels effect [23]. The refractive index change then produces a phase modulation of the optical signal propagating in the device.

A. Plasmonic-organic hybrid (POH) phase modulators

Fig. 2(a) shows the schematic of the POH phase modulator fabricated on silicon photonic circuit [13, 14]. Low-loss silicon waveguides are used for on-chip light routing. Light from a single mode fiber is launched to and collected from silicon waveguides with grating couplers (GC) having a coupling loss of 4…5 dB.

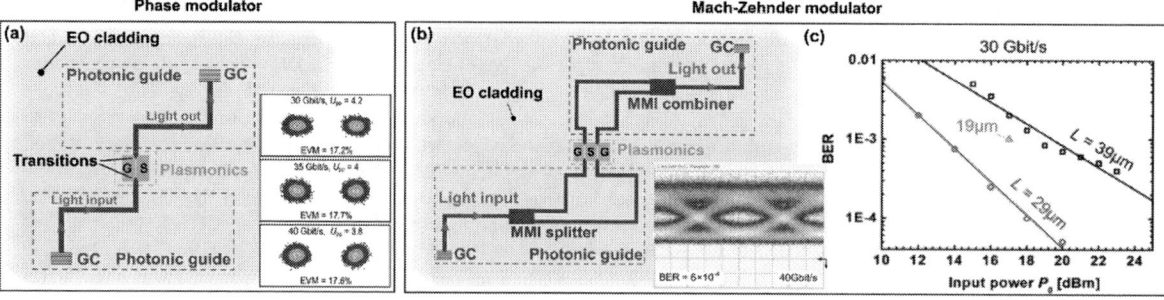

Fig. 2 Plasmonic organic hybrid (POH) phase and Mach-Zehnder modulators. (a) Schematic of the POH phase modulator. Grating couplers(GC) are used to couple light to the silicon waveguide (blue). The photonic-to-plasmonic conversion is performed by metallic tapers. Error-free BPSK constellation diagrams up to 40 Gbit/s generated with a coherent receiver are given as insets. (b) POH Mach-Zehnder (MZ) modulators. Two POH phase modulators with the length of $L = 29$ μm are incorporated in the arms of a silicon photonic MZ interferometer. The MZ interferometer comprises two 1×2 multimode interference couplers and nanowire waveguides. The footprint of the active part of the modulator is 21×29 μm². A 40 Gbit/s OOK NRZ eye diagram measured with a standard pre-amplified direct detection receiver is given as inset. (c) Performance dependence on the phase shifter length. For given optical and RF input powers, a POH phase shifter with the optimum length generates in a maximum optical modulation amplitude [14].

Binary phase shift keying (BPSK) modulation is reported with POH technology using a single phase modulator [6-7]. For a drive voltage swing of 3.8…4.2 V_{pp} and an input optical power of +10 dBm, error free BPSK reception with a bit error ratio of 10^{-10} are reported for 30 Gbit/s, 35 Gbit/s and 40 Gbit/s with absolutely no dependence on the data rate, see Fig. 2(a) [14].

B. Plasmonic-organic hybrid Mach-Zehnder modulators

The intensity modulation in a Mach-Zehnder (MZ) interferometric configuration has been reported by incorporating two plasmonic phase shifters inside a MZ interferometer, see Fig. 2(b) [14]. The OOK signals after the POH MZM have been received with a standard pre-amplified direct detection receiver. The 40 Gbit/s optical OOK eye diagram is given as an inset in Fig. 2(b) with the measured bit error ratio (BER) of 6×10^{-4}, which is below the threshold of the hard-decision forward error correction codes. We further study the influence of the length of the POH phase shifters on the optical signal quality [14]. A modulator with the optimum length realizes the best compromise between modulation depth and optical loss, and therefore leads to maximum optical modulation amplitudes, see Fig. 2(c).

III. CONCLUSIONS

We gave an overview of high-speed plasmonic-organic hybrid (POH) modulators for BPSK and OOK signaling for both direct detection and for coherent systems at data rates up to 40 Gbit/s. We also investigated the optimum length of the phase shifter sections inside the POH Mach-Zehnder modulator resulting in a maximum optical modulation amplitude and a minimum bit error ratio.

This work was supported by the European Research Council (ERC Starting Grant 'EnTeraPIC', number 280145), by the Alfried Krupp von Bohlen und Halbach Foundation, by the EU-FP7 projects BigPipes and Phoxtrot, by the Helmholtz International Research School for Teratronics (HIRST), by the Karlsruhe School of Optics & Photonics (KSOP), and by the Karlsruhe Nano-Micro Facility (KNMF).

REFERENCES

[1] G. T. Reed, G. Mashanovich, F. Y. Gardes, and D. J. Thomson, "Silicon optical modulators," Nature Photon., vol. 4, no. 8, pp. 518–526, Jul. 2010.

[2] P. J. Winzer, "Spatial Multiplexing in Fiber Optics: The 10X Scaling of Metro/Core Capacities," Bell Labs Tech. J., vol. 19, pp. 22–30, 2014.

[3] P. Dong, Y.-K. Chen, G.-H. Duan, and D. T. Neilson, "Silicon photonic devices and integrated circuits," J. Nanophoton., vol. 3, no. 4–5, pp. 215–228, 2014.

[4] X. Tu, K.-F. Chang, T.-Y. Liow, J. Song, X. Luo, L. Jia, Q. Fang, M. Yu, G.-Q. Lo, P. Dong, and Y.-K. Chen, "Silicon optical modulator with shield coplanar waveguide electrodes," Opt. Express, vol. 22, no. 19, pp. 23724–23731, 2014.

[5] J. Leuthold, C. Koos, W. Freude, L. Alloatti, R. Palmer, D. Korn, J. Pfeifle, M. Lauermann, R. Dinu, S. Wehrli, M. Jazbinsek, G. Peter, M. Waldow, T. Wahlbrink, J. Bolten, H. Kurz, M. Fournier, J. Fedeli, H. Yu, and W. Bogaerts, "Silicon-organic hybrid electro-optical devices," IEEE J. Sel. Top. Quantum Electron., vol. 19, no. 6, p. 3401413, 2013.

[6] D. K. Gramotnev and S. I. Bozhevolnyi, S. I. "Plasmonics beyond the diffraction limit," Nature Photon. 4, 83–91 (2010).

[7] W. Cai, J. S. White, and M. L. Brongersma, "Compact, high-speed and power-efficient electrooptic plasmonic modulators," Nano Lett., vol. 9, no. 12, pp. 4403–4411, 2009.

[8] J. A. Dionne, K. Diest, L. A. Sweatlock, and H. A. Atwater, "PlasMOStor: A metal-oxide-Si field effect plasmonic modulator," Nano Lett., vol. 9, no. 2, pp. 897–902, 2009.

[9] A. Melikyan, N. Lindenmann, S. Walheim, P. M. Leufke, S. Ulrich, J. Ye, P. Vincze, H. Hahn, T. Schimmel, C. Koos, W. Freude, and J. Leuthold, "Surface plasmon polariton absorption modulator," Opt. Express, vol. 19, no. 9, pp. 8855–8869, 2011.

[10] A. Melikyan, T. Vallaitis, N. Lindenmann, T. Schimmel, W. Freude, and J. Leuthold, "A surface plasmon polariton absorption modulator," in Conference on Lasers and Electro-Optics, 2010, p. JThE77.

[11] V. J. Sorger, N. D. Lanzillotti-Kimura, R.-M. Ma, and X. Zhang, "Ultra-compact silicon nanophotonic modulator with broadband response," J. Nanophoton., vol. 1, no. 1, pp. 17–22, 2012.

[12] V. E. Babicheva and A. V. Lavrinenko, "Plasmonic modulator optimized by patterning of active layer and tuning permittivity," Opt. Commun., vol. 285, no. 24, pp. 5500–5507, 2012.

[13] A. Melikyan, L. Alloatti, A. Muslija, D. Hillerkuss, P. C. Schindler, J. Li, R. Palmer, D. Korn, S. Muehlbrandt, D. Van Thourhout, B. Chen, R. Dinu, M. Sommer, C. Koos, M. Kohl, W. Freude, and J. Leuthold, "High-speed plasmonic phase modulators," Nature Photon., vol. 8, no. 3, pp. 229–233, 2014.

[14] A. Melikyan, K. Koehnle, M. Lauermann, R. Palmer, S. Koeber, S. Muehlbrandt, P. C. Schindler, D. L. Elder, S. Wolf, W. Heni, C. Haffner, Y. Fedoryshyn, D. Hillerkuss, M. Sommer, L. R. Dalton, D. Van Thourhout, W. Freude, M. Kohl, J. Leuthold, and C. Koos, "Plasmonic-organic hybrid (POH) modulators for OOK and BPSK signaling at 40 Gbit/s," Opt. Express, vol. 23, no. 8, pp. 9938-9946, 2015.

[15] C. Haffner, W. Heni, Y. Fedoryshyn, J. Niegemann, a. Melikyan, D. L. Elder, B. Baeuerle, Y. Salamin, a. Josten, U. Koch, C. Hoessbacher, F. Ducry, L. Juchli, a. Emboras, D. Hillerkuss, M. Kohl, L. R. Dalton, C. Hafner, and J. Leuthold, "All-plasmonic Mach–Zehnder modulator enabling optical high-speed communication at the microscale," Nature Photon., vol. 9, no. 8, pp. 525–528, 2015.

[16] W. Heni, A. Melikyan, C. Haffner, Y. Fedoryshyn, B. Baeuerle, A. Josten, J. Niegemann, D. Hillerkuss, M. Kohl, D. L. Elder, L. R. Dalton, C. Hafner, and J. Leuthold, "Plasmonic Mach-Zehnder modulator with >70 GHz electrical bandwidth demonstrating 90 Gbit/s 4-ASK." Proc. Opt. Fiber Commun. Conf. (OFC 2015), Los Angeles (USA), March 2015, paper Tu2A.2.

[17] W. Heni, C. Haffner, B. Baeuerle, Y. Fedoryshyn, A. Josten, D. Hillerkuss, J. Niegemann, A. Melikyan, M. Kohl, D. L. Elder, L. R. Dalton, C. Hafner, and J. Leuthold, "108 Gbit / s Plasmonic Mach – Zehnder Modulator with > 70-GHz Electrical Bandwidth," vol. 34, no. 2, pp. 393–400, 2016.

[18] C. Hoessbacher, W. Heni, A. Melikyan, Y. Fedoryshyn, C. Haffner, B. Baeuerle, and A. Josten, "Dense Plasmonic Mach-Zehnder Modulator Array for High-Speed Optical Interconnects." Proc. of Integrated Photonics Research, Silicon and Nanophotonics (IPR2015), Boston (USA), June 2015, paper IM2B.1.

[19] W. Heni, C. Hoessbacher, C. Haffner, Y. Fedoryshyn, B. Baeuerle, A. Josten, D. Hillerkuss, Y. Salamin, R. Bonjour, A. Melikyan, M. Kohl, D. L. Elder, L. R. Dalton, C. Hafner, and J. Leuthold, "High speed plasmonic modulator array enabling dense optical interconnect solutions," Opt. Express, vol. 23, no. 23, pp. 29746-29756, 2015.

[20] A. Melikyan, M. Kohl, M. Sommer, C. Koos, W. Freude, and J. Leuthold, "Photonic-to-plasmonic mode converter," Opt. Lett., vol. 39, no. 12, pp. 3488-3491, 2014.

[21] D. M. Gill, W. M. J. Green, S. Assefa, J. C. Rosenberg, T. Barwicz, S. M. Shank, H. Pan, and Y. A. Vlasov, "A figure of merit-based electro-optic Mach- Zehnder modulator link penalty estimate protocol," arXiv:1211.2419, 2012.

[22] D. F. P. Pile and D. K. Gramotnev, "Adiabatic and nonadiabatic nanofocusing of plasmons by tapered gap plasmon waveguides," Appl. Phys. Lett., vol. 89, no. 4, p. 041111, 2006.

[23] D. L. Elder, S. J. Benight, J. Song, B. H. Robinson, and L. R. Dalton, "Matrix-Assisted Poling of Monolithic Bridge-Disubstituted Organic NLO Chromophores," Chem. Mater., vol. 26, no. 2, pp. 872-874, 2014.

[24] C. Koos, J. Leuthold, W. Freude, M. Kohl, L. R. Dalton, W. Bogaerts, A. L. Giesecke, M. Lauermann, A. Melikyan, S. Koeber, S. Wolf, C. Weimann, S. Muehlbrandt, K. Koehnle, J. Pfeifle, W. Hartmann, Y. Kutuvantavida, S. Ummethala, R. Palmer, D. Korn, L. Alloatti, P. C. Schindler, D. L. Elder, T. Wahlbrink, J. Bolten, "Silicon-organic hybrid (SOH) and plasmonic-organic hybrid (POH) integration," J. Lightwave Technol., vol. 34, no. 2, pp. 256 – 268, 2015.

In-Plane Saturable Absorption of Graphene on a Silicon Slot Waveguide

Jiaqi Wang[1], Zhenzhou Cheng[2], Hon Ki Tsang[1], and Chester Shu[1]

[1]Department of Electronic Engineering, The Chinese University of Hong Kong, Shatin, N.T., Hong Kong
[2]Department of Chemistry, University of Tokyo, Hongo 7-3-1, Bunkyo-ku, Tokyo, Japan
jqwang@ee.cuhk.edu.hk

Abstract: *We studied the saturable absorption in 17-micrometer long graphene-on-silicon slot waveguide at 1550nm wavelength. Saturation behavior was observed beginning from 20 pJ input optical pulse energy. The transmission increased by 44% at 167 pJ input.*
Keywords: *(Silicon photonics, slot waveguide, graphene, nonlinear optics, integrated optics)*

I. INTRODUCTION

With the features of ultrahigh electron mobility, zero bandgap and tunable absorption [1], [2], graphene is promising for many photonic and optoelectronic applications, such as electro-absorption modulators [3], high-speed and broadband photodetectors [4-6], and mode-locked fiber lasers extending to the mid-infrared spectral region [7]. Graphene has been suggested as a material with giant $\chi^{(3)}$ nonlinearities. Strong third-order nonlinear response has been demonstrated in the graphene-on-silicon photonic crystal cavity [8]. Besides, the ultrafast carrier dynamics along with the linear electron/hole dispersion make graphene an ideal solution for ultrafast and broadband saturable absorbers. The band filling effect causes the absorption of graphene to decrease at high power levels. With chemical vapor deposition (CVD) grown graphene [9] or graphene-polymer composite [10] acting as a saturable absorber and adhered to the fiber end, ultrafast pulsed fiber lasers have been demonstrated.

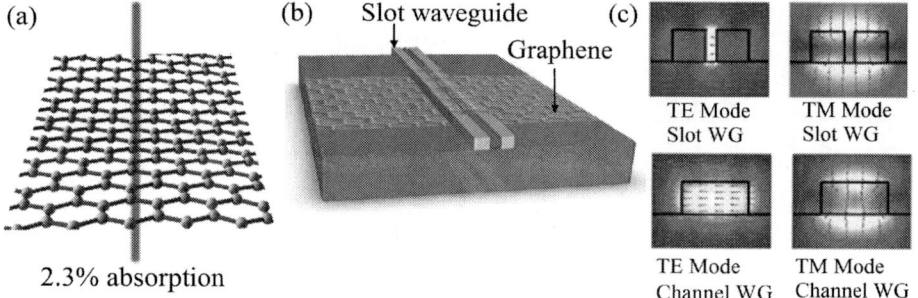

Fig. 1 (a) Graphene absorption for the normal incidence case. (b) Schematic illustration of the graphene-on-silicon slot waveguide. (c) Electrical field distribution of the TE/TM slot waveguides and channel waveguides with graphene integrated on top.

With the two-dimensional nature, graphene is suitable to be integrated on the photonic integrated circuits (PICs) [11]. The interaction of the evanescent field with graphene along the waveguide can dramatically increase the optical absorption towards ~100%, comparing to 2.3% absorption for the normal incidence case, as shown in Fig. 1 (a) and (b). In-plane nonlinear absorption of graphene-on-silicon channel waveguide structures has been previously studied [12-14]. Here we report the characterization of saturable absorption on silicon slot waveguides for possible application in mode-locked lasers on-chip [15-17] by using slot-waveguide integrated graphene saturable absorbers. Compared to conventional channel waveguides, the slot waveguide is able to guide and confine higher intensity of light in the nano-meter size low index slot due to the discontinuity of electrical field at the high-index contrast interface. The electrical field distributions of the fundamental silicon TE/TM slot waveguides and channel waveguides with graphene integrated on top are shown in Fig. 1 (c). The in-plane graphene-light interaction is stronger for the TE mode in graphene-on-slot waveguides, comparing to that in graphene-on-channel waveguides [18]. The graphene-on-silicon slot waveguide can further reduce the device footprint and eliminate the need on the growth of large-area high quality graphene. Moreover, the graphene-on-silicon slot waveguide is suitable for nonlinear applications, since there is no silicon present at the location of peak optical intensity, thus avoiding two-photon absorption (TPA) and TPA induced free carrier absorption loss from the silicon waveguide [19], [20].

We transferred CVD grown monolayer graphene onto the silicon vertical slot waveguide devices and tailored the length of graphene on top of the waveguide. The 17-μm long graphene integrated on top gives rise to ~15.6 dB absorption loss in the silicon slot waveguide. The nonlinear absorption behavior was studied by using a gain-switched short pulse laser at 1.55 μm wavelength. The transmission increased rapidly when the coupled pulse energy was larger than 20 pJ. When the pulse energy was above 167 pJ, the transmission was 44% larger than that at the low power regime.

II. FABRICATION AND EXPERIMENTAL RESULTS

The TE mode silicon slot waveguide consists of two silicon ridge waveguides separated by a 70-nm wide air slot. According to our previous study [18], the fundamental TE mode slot waveguide has the largest optical absorption loss among the TE/TM mode channel waveguides and TE/TM slot waveguides. A Y junction is used as the mode converter between the channel waveguide and the slot waveguide. The focusing subwavelength gratings [21], [22] are applied to couple the TE mode in and out of the waveguide. The silicon devices were fabricated on a commercial silicon-on-insulator (SOI) wafer which contains 250-nm silicon on a 3-μm buried oxide. The slot waveguides and subwavelength gratings were defined by electronic beam lithography (EBL) followed by deep reactive-ion etching. The commercial CVD grown graphene was wet transferred onto the silicon devices. The graphene on top of the waveguide was tailored to 17 μm by using EBL and O_2 plasma etching. The fabricated grating coupler and graphene-on-silicon slot waveguide are shown in the scanning electronic microscopy (SEM) images in Fig.2 (a) and (b), respectively. Through the contrast in Fig. 2 (b), the area with and without graphene can be clearly identified. Raman spectroscopy was used to verify the monolayer graphene. The Raman spectrum of the graphene integrated on the slot waveguide is measured using a 512 nm laser and the result is shown in Fig. 2 (c). The ratio of G peak (1588 cm^{-1}) to 2D peak (2690 cm^{-1}) is smaller than 0.5, indicating the monolayer graphene structure [23], [24]. The graphene introduces ~15.6 dB absorption loss in the low power measurement (-10 dBm input power).

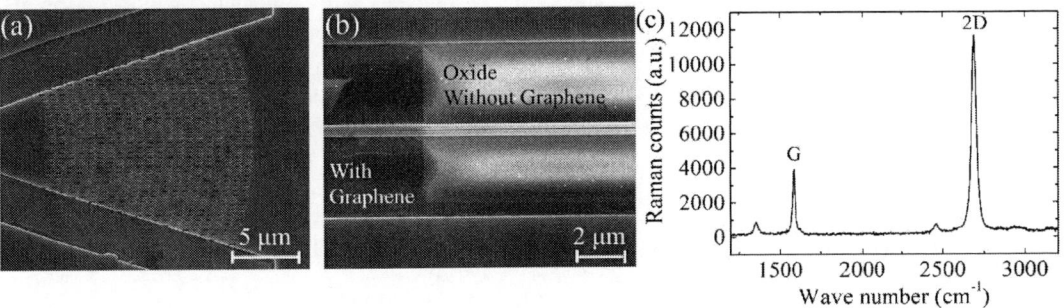

Fig. 2 (a) SEM image of the fabricated subwavelength grating coupler. (b) SEM image of the graphene-on-silicon slot waveguide. The area with and without graphene can be clearly identified through the contrast of the image. (c) Raman spectrum of the graphene integrated on top of the silicon slot waveguide.

The setup of the saturable absorption measurement is shown in Fig. 3 (a). The laser source was a gain-switched semiconductor laser with ~50 ps pulse width and 1 MHz repetition rate centered at 1555.65 nm. The laser was amplified and the output was subsequently filtered by an optical bandpass filter. A 5% tap was used to monitor the input power. The other 95% was coupled into the graphene-on-silicon slot waveguide through a TE mode subwavelength grating with a coupling efficiency of 30%. The transmission of the graphene-on-silicon slot waveguide as a function of pulse energy is shown in Fig. 3 (b). At low input pulse energy, the transmission is independent of the input energy. At high input level, the increased concentration of excited electrons results in band filling and begins to block some absorption. The transmission starts to increase rapidly at coupled pulse energy above 20 pJ (376mW peak power). When the coupled pulse energy is increased to 167 pJ (3.14 W peak power), the transmission was 44% larger than that in the low power regime.

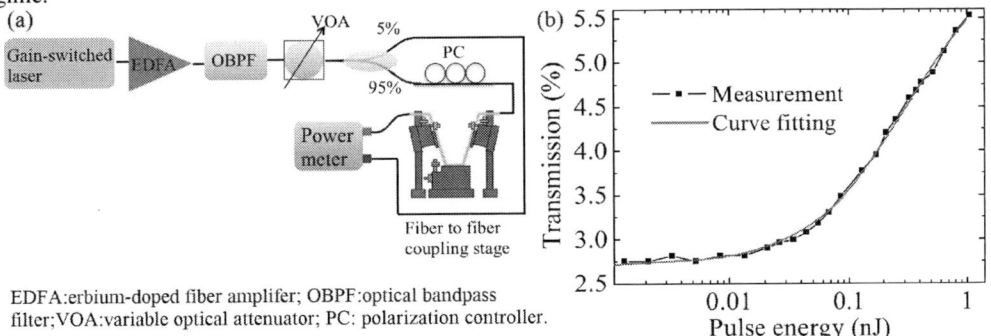

EDFA:erbium-doped fiber amplifer; OBPF:optical bandpass filter;VOA:variable optical attenuator; PC: polarization controller.

Fig. 3 (a) Schematic illustration of the saturable absorption measurement. (b) Transmission of the graphene-on-silicon slot waveguide at different input optical pulse energy.

The transmission of the graphene-on-silicon slot waveguide is described by $T=e^{-\alpha L}$, where α is the absorption coefficient and $L=17$ μm is the length of the graphene. The two-stage saturable absorption model is widely used to describe the nonlinear absorption in two-dimensional quantum wells. It can be adopted to approximate the nonlinear absorption in graphene [9]. The absorption coefficient $\alpha(E)$, where E is optical pulse energy, is described by

$$\alpha(E) = \frac{\alpha_S}{(1 + \frac{E}{E_S})} + \alpha_{NS} \qquad (1)$$

where α_S and α_{NS} are the saturable and non-saturable absorption components, E_S is the saturation pulse energy. The non-saturable absorption component is mainly caused by scattering loss in the silicon slot waveguide. The fitting curve is shown in Fig. 3 (b), and α_S, α_{NS} and E_S are obtained as 0.05 μm^{-1}, 0.019 μm^{-1}, and 213 pJ, respectively. The saturation energy is relatively low, comparing to the 3.9 nJ in the case of graphene-on-silicon channel waveguide [14]. The reason is that the optical interaction is significantly enhanced in the graphene-on-slot waveguide configuration and a smaller area of graphene is present on the slot waveguide.

III. CONCLUSION

A graphene-on-silicon slot waveguide structure is proposed and demonstrated to enhance the optical interaction between graphene and the evanescent field of the waveguide. The nonlinear absorption of the graphene-on-silicon slot waveguide is studied with a gain-switched short pulse laser. Saturable absorption is observed at relatively low input pulse energy of 20 pJ. The observation is explained by the enhanced optical interaction as well as the small area of graphene over the slot waveguide. The graphene-on-silicon slot waveguide can have potential applications in realizing ultrafast on-chip mode-locked lasers and nonlinear optical processing.

ACKNOWLEDGMENT

This work was supported by Hong Kong Research RGC GRF grants CUHK 416213 and 14206614.

REFERENCES

[1] M. Freitag, et al. "Photoconductivity of biased graphene." Nat. Photon., 7, 53 (2013).
[2] D. Kim, et al, "Work function engineering of graphene anode by bis (trifluoromethanesulfonyl) amide doping for efficient polymer light emitting diodes," Adv. Funct. Mater, 23, 5049-5055 (2013).
[3] M. Liu, et al, "A graphene-based broadband optical modulator," Nature, 474, 64-67 (2011).
[4] X. Gan, et al, "Chip-integrated ultrafast graphene photodetector with high responsivity," Nat. Photon., 7, 883-887 (2013).
[5] A. Pospischil, et al, "CMOS-compatible graphene photodetector covering all optical communication bands," Nat. Photon., 7, 892-896 (2013).
[6] X. Wang, et al, "High responsivity graphene/silicon heterostructure waveguide photodetectors", Nat. Photon.,7, 888-891 (2013).
[7] G. Zhu, et al, "Graphene Mode-Locked Fiber Laser at 2.8 µm," IEEE Phot. Tech. Lett.,28, 7-10 (2016).
[8] T. Gu, et al, "Regenerative oscillation and four-wave mixing in graphene optoelectronics," Nat. Photon., 6, 554-559 (2012).
[9] Q. Bao, et al, "Atomic layer graphene as a saturable absorber for ultrafast pulsed lasers," Adv. Funct. Mater, 19, 3077-3083 (2009).
[10] Z. Sun, et al, "Graphene mode-locked ultrafast laser," ACS Nano 4, 803-810 (2010).
[11] R. Kou, et al, "Influence of graphene on quality factor variation in a silicon ring resonator," Appl. Phys. Lett., 104, 091122 (2014).
[12] K. Alexander, et al, "Electrically controllable saturable absorption in hybrid graphene-silicon waveguides," In CLEO: Science and Innovations, STh4H-7, Optical Society of America (2015).
[13] Z. Cheng, et al, "In-plane optical absorption and free carrier absorption in graphene-on-silicon waveguides," IEEE J. Sel. Top. Quantum Electron., 20, 43-48 (2014).
[14] Z. Shi, et al, "In-plane saturable absorption of graphene on silicon waveguides," In Conference on Lasers and Electro-Optics/Pacific Rim, WA4_3, Optical Society of America (2013).
[15] Y. Urino, et al, "Demonstration of 12.5-Gbps optical interconnects integrated with lasers, optical splitters, optical modulators and photodetectors on a single silicon substrate," Opt. Express, 20, B256-B263 (2012).
[16] Y.D.Yang, et al, "Direct-modulated waveguide-coupled microspiral disk lasers with spatially selective injection for on-chip optical interconnects," Opt. Express, 22, 824-838 (2014).
[17] K. Ogawa, "High-speed silicon-based integrated optical modulators for optical-fiber telecommunications," In SPIE OPTO, 899010-899010, International Society for Optics and Photonics (2014).
[18] Z. Cheng, et al, "Graphene absorption enhancement using silicon slot waveguides," In Photonics Conference, 186-187, IEEE (2015).
[19] D. Marpaung, et al, "Nonlinear integrated microwave photonics," J. Lightwave Technol., 32, 3421-342720 (2014).
[20] M. Foster, et al, "Broad-band optical parametric gain on a silicon photonic chip," Nature, 441,960-963 (2006).
[21] L. Vivien, et al, "Light injection in SOI microwaveguides using high-efficiency grating couplers," J. Lightwave Technol., 10, 3810-381524 (2006).
[22] Z. Cheng, et al, "Focusing subwavelength grating coupler for mid-infrared suspended membrane waveguide," Opt. Lett., 37, 1217-1219 (2012).
[23] L. Xiao, et al, "Low-temperature Raman G-mode of plasmonic-graphene hybrid platform," In CLEO: Science and Innovations, JTu4A-9, Optical Society of America (2014).
[24] A. C. Ferrari, et al, "Raman spectrum of graphene and graphene layers," Phys. Rev. Lett., 97, 187401 (2006).

ThE3-3   OECC/PS2016

Graphene-induced On-demand Nanocavity Based on Si Photonic Crystal

Hisashi Chiba[1,2], Masaya Notomi[1,2,3]

[1]NTT Basic Research Laboratories, NTT Corporation, 3-1 Morinosato Wakamiya, Atsugi, Kanagawa 243-0198, Japan
[2]Department of Physics, Tokyo Institute of Technology, 2-12-1 Oookayama, Meguro, Tokyo, 152-8551, Japan
[3]Nanophotonics Center, NTT Corporation, 3-1 Morinosato Wakamiya, Atsugi, Kanagawa 243-0198, Japan
chiba.h.ag@m.titech.ac.jp

Abstract: We have numerically simulated graphene-loaded silicon photonic crystal waveguides. We found that it is possible to create a nanocavity mode induced by graphene, and Fermi energy modulation of graphene enables us to control cavity formation.
Keywords: (nanocavities, photonic crystals, graphene)

I. INTRODUCTION

Graphene's optical properties can be dramatically changed by controlling Fermi energy (E_F) resulting from Pauli blocking effects on interband transitions. Recently, this change was applied to optical modulators based on graphene-loaded Si-wire waveguides [1] and photonic crystal (PhC) nanocavities [2]. These devices are based on difference in the graphene absorption depending on Fermi energy.

On the other hand, even though graphene is a monoatomic film, the refractive index of graphene can also be changed by Fermi energy and this change was applied to shift the cavity resonance of PhCs [2]. It has been well established that that nanocavity modes can be created by local refractive-index tuning of PhC line defect waveguides as a result of local modulation of the mode gap (that is, PBGs in waveguide modes). When the mode gap is locally modulated by changing refractive index, strong 3D light confinement appears at the modulated area, which is the mechanism of the modulated mode-gap cavity formation. The required index modulation for cavity formation is extremely small ($\Delta n/n \sim 10^{-3}$) [3]. Thus, we expect that nanocavities can be formed not by difference in absorption, but by the local refractive index modulation in graphene.

In this study, we have numerically analyzed optical modes in graphene-loaded Si PhC waveguides, and found that nanocavities can be indeed created by local loading of graphene on PhC, and index modulation depending on Fermi energy. In addition, this nanocavity mode creation process can be controlled by modulating the Fermi energy of graphene, which may serve on-demand cavity formation. This study show new devices to control cavity formation by monoatomic film like graphene.

II. ON-DEMAND NANOCAVITY BASED ON GRAPHENE LOADED PHOTONIC CRYSTAL

A. The principle of light confinement with graphene loaded photonic crystal

First, we explain the principle of the modulated mode-gap nanocavity formation by graphene loading. Figure 1(a) shows a PhC waveguide without graphene, which does not have cavity modes. Figure 1(b) shows a graphene-loaded PhC waveguide. Graphene induces red shift in a Si PhC cavity, which means the effective refractive index increases. Hence, the mode gap in the graphene loaded shifts to a lower frequency as shown in Fig. 1 (b). Such local modulation of the mode gap is known to create a nanocavity mode [ref]. On the other hand, if the blue shift is possible, a nanocavity can also be created in a different configuration shown in Figure 1(c), where graphene is loaded at outer sections. The mode gap in the graphene-loaded area moves up to a higher frequency. The cavity is formed not in graphene loaded area, but in area without graphene. The blue shift modulation means the refractive index falls below air (n = 1.00), which is possible in at particular condition for graphene, as describe below.

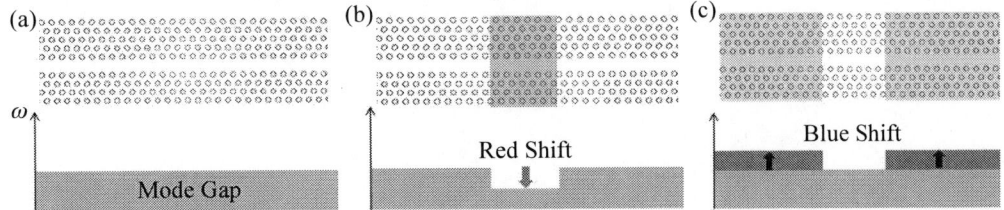

Fig. 1 (a) Si PhC waveguides without graphene. (b) Graphene loaded waveguide for red shift modulation. (c) Graphene loaded waveguide for blue shift modulation.

B. Nanocavity formation based on red shift modulation

Our model structure for cavity formation based on the red shift index modulation is shown in Fig. 2(a). It is a W1

419

line-defect waveguide in a 2D triangular air-hole silicon PhC slab. The lattice constant a, the hole radius r, the slab thickness t are 400nm, 100nm, 215nm, respectively. The width of the line defect is $0.98 \times \sqrt{3} \times a$. These parameters are for 1.5μm wavelength range. 2D graphene is placed on the top surface of Si photonic crystal shown as the purple area in Fig. 2(a). Graphene's length (parallel to the line) is $8 \times a$. Theoretical graphene's sheet conductivity

$$\tilde{\sigma}_{\mathrm{intra}} = i\frac{e^2 k_{\mathrm{B}} T}{\pi\hbar^2(\omega + i\tau^{-1})}\left[\frac{E_F}{k_{\mathrm{B}}T} + 2\ln\left(\exp\left(-\frac{E_F}{k_{\mathrm{B}}T}\right) + 1\right)\right] \qquad (1)$$

$$\tilde{\sigma}_{\mathrm{inter}} = i\frac{e^2}{4\pi\hbar}\ln\left[\frac{2|E_F| - \hbar(\omega + i\tau^{-1})}{2|E_F| + \hbar(\omega + i\tau^{-1})}\right] \qquad (2)$$

are used[4]. The relaxation time τ is 100 fs. In this section, we change graphene's Fermi energy from 0.20eV to 0.48eV, which induces a red shift when graphene is loaded on a Si PhC.

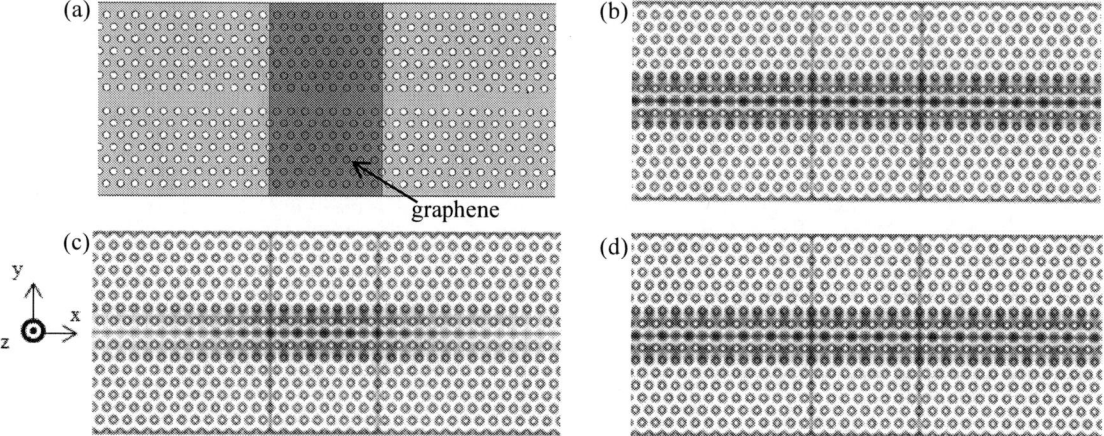

Fig. 2 (a)Geometry of Si PhC waveguides with graphene.(b)Optical field distribution (E_y, electric field parallel to the 2D plane) of Si PhC waveguide without graphene(λ=1538.90nm). (c)Graphene (E_F = 0.41eV, λ=1539.21nm) loaded waveguide. (d)Graphene (E_F = 0.48eV, λ=1538.91nm) loaded waveguide.

As a result of the finite element method electromagnetic simulation, we found that local refractive index modulation with monoatomic graphene (2D sheet) can create a nanocavity mode in our structures. Figure 2(b) (c) (d) show optical field distribution of a Si PhC waveguide with and without graphene with different E_F, and Fig. 3 shows corresponding Q and an effective mode volume V_{eff} of the cavity mode as a function of E_F. In the waveguide without graphene, the electric field is spread from end to end, thereby showing no cavity mode. In the waveguide with graphene (E_F=0.41eV), the electric field is concentrated around the graphene-loaded area. The calculated Q for E_F =0.41 eV is 9.2×10³, and the effective cavity mode volume is 0.332 μm³ (V_{eff} =3.7 (λ/n)³). We found that the cavity mode which has high Q and V_{eff} close to (λ/n)³ can be formed by local loading of graphene on PhC waveguides.

In addition, the nanocavity can be on-demand formed by Fermi energy modulation. When E_F rises from 0.41eV to 0.48eV, the cavity disappears. In this case, the waveguide is almost the same optical field distribution as a waveguide without graphene. Graphene's index becomes near 1.00 in E_F =0.48eV and no red shift by index modulation. We can control cavity formation by Fermi energy modulation in the 0.41 eV to 0.48eV range.

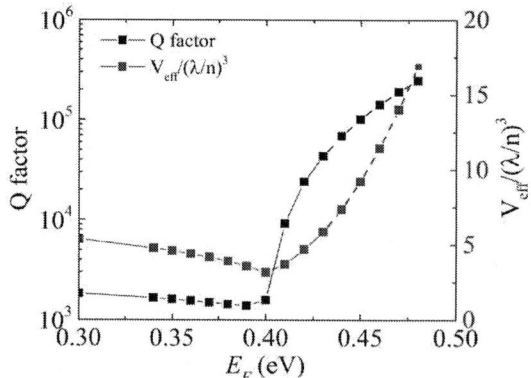

Fig. 3 Performance of nanocavities based on graphene loaded PhC. Q and V_{eff} as a function of E_F. Dash lines mean that nanocavities can't be identified for the limit of our calculation (E_F>0.43eV).

C.   Nanocavity formation based on blue shift modulation

The other structure is shown Fig. 4, which we expect cavity formation based on blue shift modulation. We use the same parameters as in the previous section, but graphene is placed on all top surfaces outside of the center area of the waveguide. The length (parallel to the line) of the center area is 8×a. In this section, we changed graphene's Fermi energy from 0.48 eV to 0.85eV. We confirmed that this range of Fermi energy induces a blue shift for graphene-loaded Si PhC. It was reported that Fermi energy of graphene can rise up to 0.85eV by polymer electorate gating [2].

As shown in Fig. 4(b), we confirmed that a nanocavity mode is created based on this blue shift index modulation. In addition, the obtain Q is even higher Q than that for the red shift case. The calculated Q for E_F =0.85 eV is 1.1×10^5 , which is ten times larger than the previous one, and the effective cavity mode volume is 0.305 μm^3 (V_{eff} =3.4$(\lambda/n)^3$) as shown in Fig. 4(c). The optical field as shown in Fig. 4(c) concentrates on the center area without graphene. We regard that the smaller overlap between the optical mode and graphene reduces the absorption and increase the Q factor. Thus the cavity by blue shift modulation obtains higher Q than the cavity by red shift modulation.

One of the interesting prospects for this structure is individual Fermi energy control of two graphene areas. We can selectively open and close left and right gates by individually changing the Fermi energy of two areas. Since high speed modulations within life time of the cavity can dynamically release trapped light in the cavity [5]. This structure can dynamically release the light to one desired direction because it can control one side only. However, the fabrication and operation processes for this structure would be more difficult than the previous structure based on red shift modulation.

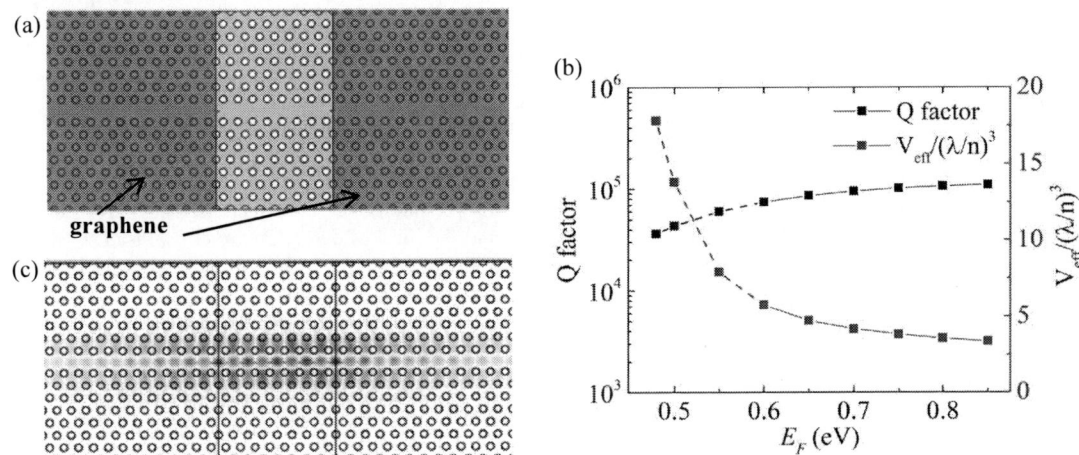

Fig. 4 (a)Geometry of Si PhC waveguides with graphene.(b)Optical field distribution (Ey, electric field parallel to the 2D plane) of Si PhC waveguide with graphene(E_F = 0.85eV, λ=1538.18nm). (c) Performance of nanocavities based on graphene loaded PhC. Q and V_{eff} as a function of E_F. Dash lines mean that nanocavities can't be identified for the limit of our calculation (E_F<0.60eV).

III. CONCLUSIONS

Even though graphene is a monoatomic film, nanocavities can be formed by local loading of graphene on PhC and Fermi energy modulation. We found that both of red and blue shift modulation can form nanocavities, and nanocavities by blue shift modulation obtain higher Q (1.1×10^5). Both of them can control cavity formation on demand by Fermi energy modulation of graphene. We hope that these devices apply to adiabatic frequency shifting and dynamic release of trapped light [5].

ACKNOWLEDGMENT

This work was supported by JSPS KAKENHI Grant Number 15H05735.

REFERENCES

[1] M. Liu, et al, "A graphene-based broadband optical modulator," *Nature*, vol. 474, no. 7349, pp. 64-67, 2011.

[2] X. Gan, et al, "High-contrast electrooptic modulation of a photonic crystal nanocavity by electrical gating of graphene," *Nano Letters*, vol. 13, no. 2, pp. 691-696, 2013.

[3] M. Notomi, and H. Taniyama, "On-demand ultrahigh-Q cavity formation and photon pinning via dynamic waveguide tuning," *Optics Express*, vol. 16, no. 23, pp. 18657-18666, 2008.

[4] G. W. Hanson, "Dyadic Green's functions and guided surface waves for a surface conductivity model of graphene," *Journal of Applied Physics*, vol. 103, no. 6, pp. 064302, 2008.

[5] T. Tanabe, M. Notomi, H. Taniyama, and E. Kuramochi, "Dynamic release of trapped light from an ultrahigh-Q nanocavity via adiabatic frequency tuning," *Physical Review Letters*, vol. 102, no. 4, pp. 043907, 2009.

Silicon slot waveguide ring resonator: from single resonance to envelope index sensing

Weiwei Zhang (*), Samuel Serna, Xavier Le Roux, Laurent Vivien, Eric Cassan

*IEF, CNRS, Univ Paris-Sud, Université Paris-Saclay, 91405, Orsay, France*Corresponding author: weiwei.zhang1@u-psud.fr*

[1] Institut d'Electronique Fondamentale, Université Paris-Sud, Université Paris-Saclay, CNRS UMR 8622, Bat. 220, 91405 Orsay Cedex, France,
[2] Laboratoire Charles Fabry, Institut d'Optique, CNRS, Université Paris-Sud, Université Paris-Saclay Avenue Augustin Fresnel, 91127 Palaiseau Cedex, France

Abstract: By tracking the spectrum envelope wavelength shift of silicon slot waveguide ring resonators, sensitivity of 1,300nm per Refraction Index Unit is reported when filling the slots by liquids with refraction index values close to 1.33.

Keywords: Silicon photonics, optical sensing, index sensing, integrated sensors

I. INTRODUCTION

Bio-molecule detection implemented with microfluidic channels assembled by Lab-On-Chip technologies is required for future healthcare diagnostics [1,2]. In this view, silicon photonics enabling large-scale fabrication of compact planar sensors by integrated CMOS processes is seen as a viable route for massive on chip bio-detection. Versatile structures, like single ring resonators (RR) [3–5], Vernier ring resonators [6], photonic crystals (PhC) [7–9], Mach-Zehnder interferometers (MZI) [10,11] and Bragg gratings [12] have been demonstrated as highly efficient sensors by the detection of refraction index changes from the surrounding fluidic channels [2]. To evaluate the performances of resonator-based integrated refractive index sensors, the sensitivity and the limit of detection are considered as two of the most important parameters. The sensitivity relates to how much the resonance shifts following a variation of the surrounding material refractive index, while the limit of detection assesses the minimum refractive index change that can be detected.

In this context, we report our recent results related to silicon slot waveguide ring resonators [13] (see Figs. 1 and 2). A series of microring resonators based on optimized silicon slot waveguides have been designed, fabricated, and characterized. Silicon slot waveguides were first fabricated in order to assess a practical and reliable propagation loss level. SOI wafers with a 220 nm thick Si film and 2 μm of buried silica were masked by a ZEP-520A resist. The patterns were written by a 80 kV e-beam lithography Nanobeam NB-4 system and then transferred by inductive coupling plasma reactive ion etching process using SF$_6$ gas.

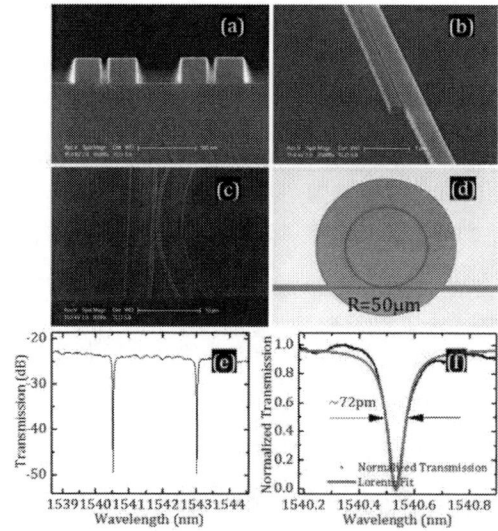

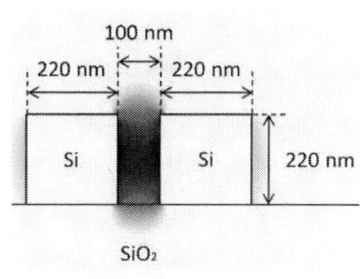

Fig. 1. Silicon slot waveguide (with typical dimensions): mode distribution at λ=1.55μm.

Fig. 2. (a) Cross-section of ring coupler with a coupling gap here of 310±10nm; **(b)** Mode converter used to couple light from strip waveguide to slot waveguide; **(c)** View of the straight bus-ring coupler region; **(d)** Optical overview of a fabricated slotted all pass ring resonator (APR); **(e-f)** Normalized transmission spectrum of slotted APR with coupler gap of 500nm with a Lorentzian fitting curve.

II. INDEX SENSING THROUGH THE RING RESONATOR SPECTRUM ENVELOPE SENSITIVITY

Relying on the experimental facilities described in section I., we prepared micro-ring resonators by specifically engineering their main critical part, i.e. the directional coupler. The tight control of the waveguide cross-section and of the slot waveguide/slot waveguide gap enabled by our technology is highlighted in Fig. 3, just hereafter. More specifically, we designed the directional couplers to operate as strongly wavelength dispersive components [14].

In the prepared conditions, the fabricated micro-ring resonators operate in the following conditions: the bus-waveguide/ring critical condition is made extremely sensitive to light wavelength. As a consequence, a weak perturbation of the slot waveguide easily brings a giant shift of the critical coupling condition.

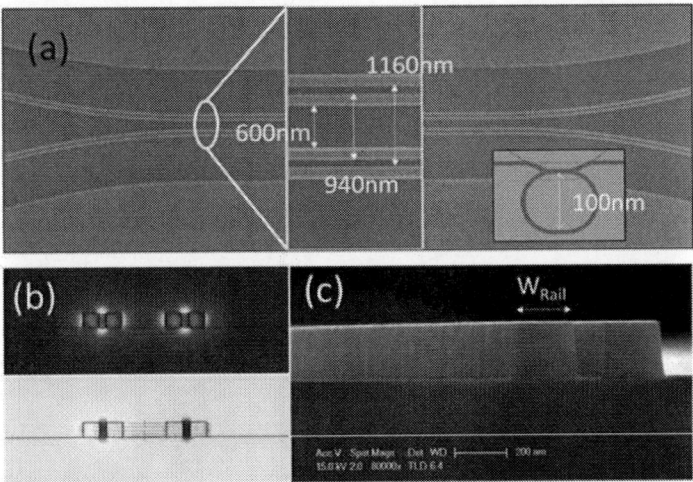

Fig. 3. Slot waveguide ring resonator: (a) SEM view of the directional coupler with typical dimensions. (b) Two supermodes of the direction coupler at the basis of the directional coupler design. (c) SEM view of the Si slot waveguide cross-section.

Our findings shown that the slot waveguide rail width (W_{Rail} in Fig. 3 (c)) is the primary geometrical parameter to obtain strongly dispersive directional couplers. Fig. 4 shows the direct consequence of the specifically achieved designs. We compare in Fig. 4 (a) the cases of slot waveguide ring resonators strictly identical excepted on one point: the rail widths of the two rings are 170nm and 235nm, respectively. As shown in Fig. 4, the ring transmission spectra strongly differ. As shown also, narrow rails lead to narrow V-shape ring resonator spectra.

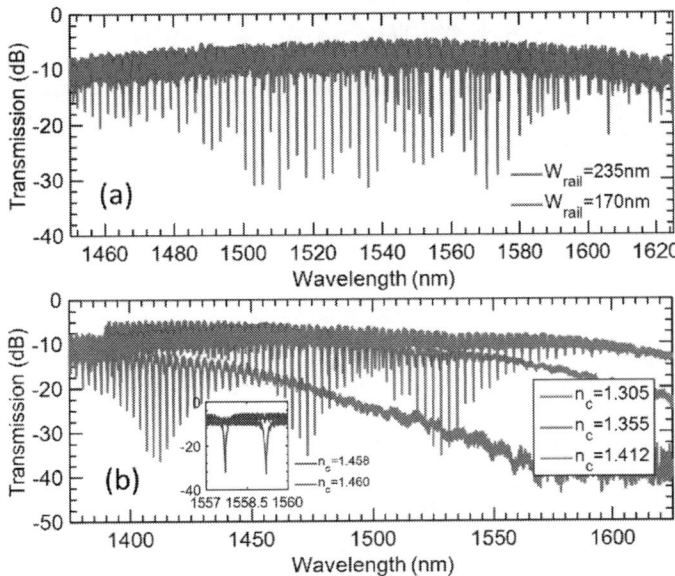

Fig. 4. (a) Transmission spectrum comparison between slot waveguide ring resonator with rail width 235nm and 170nm, respectively; the slot size of the two resonators is 110nm and the top cladding liquid RI is equal to 1.55 at λ=1.55μm. **(b)** Transmission spectra of micro-ring resonators with slot waveguide rail width of 170nm and for different top cladding liquids with refractive index values of 1.305, 1.355, 1.412, respectively. The inset shows the critical resonance (n_c) shifts when n_c is 1.458 and 1.460, respectively.

We then conducted a series of experiments by drop casting index liquids (from Cargille) on top of the silicon samples. A simple cleaning process allowed to clean them, enabling a new experiment with another liquid. We then conducted optical experiments by light injection into slot waveguides through butt-coupling from lenses fibers, while monitoring the micro-ring resonators transmission spectra. We focused our experimental investigations on narrow rail slot waveguide ring resonators. As shown in Fig. 4 (b) and Fig. 5, a giant shift of the ring resonator envelope spectrum is obtained when changing the liquid top cladding index value (n_c).

Our findings led to the following conclusions: i) The observed shift dependence on the cladding refractive index (n_c) is not linear for the whole n_c range from 1.3 to 1.5 (Fig. 5). The envelope sensitivity (S_{env}) decreases from 1300nm/RIU down to 400nm/RIU when n_c increases from 1.3 to 1.5, while the individual single-peak Q factor grows up to 6000 for n_c=1.46.In term of achievable optical sensor limit of detection (LOD), the proposed envelope-assisted wavelength peak approach is essentially efficient provided that two conditions can be fulfilled: i) A single resonance should be close to the critical coupling condition, ii) The top cladding index perturbation (δn_{clad}) is capable of inducing a shift of several microring resonator FSRs. It thus turns out that the device LOD is finally given by LOD=2FSR/S_{env}. Microring resonators based on cm-long spiral-shape slot waveguides can thus provide LOD values typically around 10^{-4}-10^{-5}. Another advantage of the proposed sensing configuration is to enable index sensing in a very wide index range from $n_{clad}{\approx}1.3$ to $n_{clad}{\approx}1.5$.

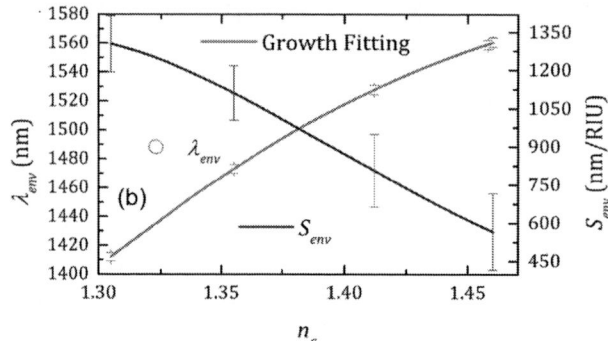

Fig. 5. Peak spectrum envelope resonance wavelength and envelope sensitivity as as function of the top cladding index (n_c).

III. CONCLUSIONS

We propose a new approach to on-chip index sensing using silicon-based micro-ring resonators. The proposed mechanism relies on the fast detuning of the bus-waveguide/ring critical coupling condition and mainly stems on a highly dispersive slot waveguide coupler design. This specific feature provides narrow resonance spectrum combs with few distinguished critical resonance peaks if compared with a standard design. The spectrum envelope peak wavelength can be easily figured out and treated as a sensing probe to monitor the cladding index changes. The concept is experimentally evidenced by characterizing a series of fabricated structures at around λ=1.55μm. Around the refractive index of water, we observed sensitivities S_{env} up to 1300nm/RIU. Additionally, the sensor Limit of Detection scales down as LOD =2 FSR/S_{env}, meaning that LOD typically lower than 10^{-4} can be obtained for spiral-engineered microring resonators in SOI photonics. This result brings improvement of Si slot-waveguide resonator sensors and opens interesting perspectives for on-chip optical sensing.

ACKNOWLEDGMENT

We would like to acknowledge the national ANR POSISLOT project, as well as the FP7 Cartoon European project.

REFERENCES

1. C. D. Chin, V. Linder, and S. K. Sia, Lab Chip 7, 41 (2007).
2. J. Wu and M. Gu, J. Biomed. Opt. 16, 080901 (2011).
3. K. B. Gylfason, C. F. Carlborg, A. Kaźmierczak, F. Dortu, H. Sohlström, L. Vivien, C. a Barrios, W. van der Wijngaart, and G. Stemme, Opt. Express 18, 3226 (2010).
4. T. Claes, J. G. Molera, K. De Vos, E. Schacht, R. Baets, and P. Bienstman, IEEE Photonics J. 1, 197 (2009).
5. X. Fan, I. M. White, H. Zhu, J. D. Suter, and H. Oveys, Lasers Appl. Sci. Eng. 64520M (2007).
6. T. Claes, W. Bogaerts, and P. Bienstman, Opt. Express 18, 22747 (2010).
7. C. Ca, S. F. Serna, W. Zhang, X. Le Roux, and E. Cassan, Opt. Lett. 39, 1 (2014).
8. J. Jágerská, H. Zhang, Z. Diao, N. Le Thomas, and R. Houdré, Opt. Lett. 35, 2523 (2010).
9. D. Yang, S. Kita, F. Liang, C. Wang, H. Tian, Y. Ji, M. Lončar, and Q. Quan, Appl. Phys. Lett. 105, 063118 (2014).
10. M. La Notte and V. M. N. N. Passaro, Sensors Actuators B Chem. 176, 994 (2013).
11. X. Tu, J. Song, T.-Y. Liow, M. K. Park, J. Q. Yiying, J. S. Kee, M. Yu, and G.-Q. Lo, Opt. Express 20, 2640 (2012).
12. X. Wang, J. Flueckiger, S. Schmidt, S. Grist, S. T. Fard, J. Kirk, M. Doerfler, K. C. Cheung, D. M. Ratner, and L. Chrostowski, J. Biophotonics 6, 821 (2013).
13. Weiwei Zhang, Samuel Serna, Xavier Le Roux, Carlos Alonso-Ramos, Laurent Vivien, and Eric Cassan, Opt. Lett. 40, 5566 (2015)
14. W. Zhang, S. Serna, X. Le Roux, L. Vivien, E. Cassan, *Optics Letters* Vol. 41, Issue 3, pp. 532-535 (2016).

Simultaneous Measurement of Strain and Temperature Using pi-Phase-shifted Fiber Bragg Grating on Polarization Maintaining Fiber

Jiageng Chen, Qingwen Liu*, Xinyu Fan, and Zuyuan He

State Key Laboratory of Advanced Optical Communication Systems and Networks, Shanghai Jiao Tong University,
800 Dongchuan Road, Shanghai 200240, China
*liuqingwen@sjtu.edu.cn

Abstract: A high resolution strain and temperature discriminative sensor has been developed, using a π-phase shifted FBG on polarization maintaining fiber. Resolutions of 0.104 με and 0.005 °C in strain and temperature are achieved.

Keywords: Optical fiber sensor, π-phase-shifted fiber Bragg grating, temperature and strain discrimination.

I. INTRODUCTION

Fiber Bragg grating (FBG) based optical fiber sensors have been widely used in strain and temperature measurement situations. For the static strain measurement in geophysics-related scenes such as crustal deformation observation, sensors must have excellent anti-thermal-interference ability because the measurement period can be as long as years. Besides, high sensing resolution is also required for the extremely small deformation.

For high resolution and long-term strain sensing, the sensors use fiber Fabry-Perot interferometers or π-phase-shifted FBGs which have narrow resonances as the sensing heads instead of common FBGs, and a reference grating is necessary to be introduced and be placed close to the sensing grating for temperature drift compensation [1]. However, this external compensation method is not strictly credible since it is based on the idealized assumption that the temperature on sensing grating and reference grating are always equal and synchronously changing. The intrinsic solution of the ambient temperature interference for strain sensing is the strain and temperature discrimination method [2], in which two optical parameters of the sensing head with different sensitivity to strain/temperature are measured, the result of strain and temperature can be calculated by a matrix equation without interference from each other. At present, many research works are about the issue of strain and temperature discrimination, however, the strain resolution can only reach a level of several με in most of the schemes.

In this paper, we present a high resolution strain and temperature discrimination scheme based on a π-phase-shifted FBG (π-PSFBG) on polarization maintaining fiber (PMF). Two variable frequencies of the grating's Bragg frequency and the mode birefringence introduced frequency difference are used to perform the discrimination. A fiber Fabry-Perot interferometer (FFPI) is employed to correct the laser frequency-sweeping nonlinearity, and a standard reference material gas cell for absolute frequency reference is introduced. With the proposed configuration, a strain resolution of 0.104 με is achieved, together with a temperature resolution of 0.005 °C. The long term stability of the sensor is also guaranteed.

II. PRINCIPLES AND SYSTEM CONFIGURATION

The π-phase-shifted FBG, which is a variant of classic FBG, has a narrow resonance at the center of the high reflection band [3]. With a π-PSFBG on panda-type PMF, we can measure both the Bragg frequency f_{Br} of the grating and the mode birefringence B (n_{slow}-n_{fast}) introduced frequency difference Δf_{Bi} simultaneously [4]. Since the f_{Br} and Δf_{Bi} approximately linearly response to the temperature T and strain ε, the classical solving matrix equation method in Ref. [2] can be used to perform the discrimination. In this method, the measurement error of both f_{Br} and Δf_{Bi} affects the sensing resolution.

The configuration of the sensor is shown in Fig. 1. A tunable narrow linewidth laser (Santec, TSL-710) is used as the frequency sweeping laser source. The sensor composes of a π-PSFBG on PMF for sensing, an HC¹³N gas cell for absolute frequency reference, and an FFPI with a free spectrum range (FSR) of approximate 10 MHz for frequency sweep nonlinearity correction. A 45-degree polarization-axis rotated splice is introduced to convert the linearly polarized lightwave on slow axis to both slow and fast axes in order to interrogate the π-PSFBG. The reflected lightwave of the grating on both slow and fast axes are separated by a polarization beam splitter (PBS) and converted to electrical signals by a pair of photo-detectors (PDs). The HC¹³N standard gas cell (Wavelength Reference, HCN-13-C-100) has absorption wavelength at 1549.7302 nm and 1550.5149 nm which are almost independent with temperature change and near the Bragg frequency of the π-PSFBG. Interrogation laser of the gas cell is modulated by a phase modulator (PM), together with a locked-in amplifier (LIA) to demodulate its transmission light signal, obtaining an "error signal" in the Pound-Drever-Hall (PDH) technique [5]. The output from FFPI appears as a sequence of pulses, which intrinsically corresponding to a strict equal frequency spaces. All the laser signals are sampled by the personal computer (PC) with data acquisition card, which performs the demodulation of strain and temperature on the sensing π-PSFBG.

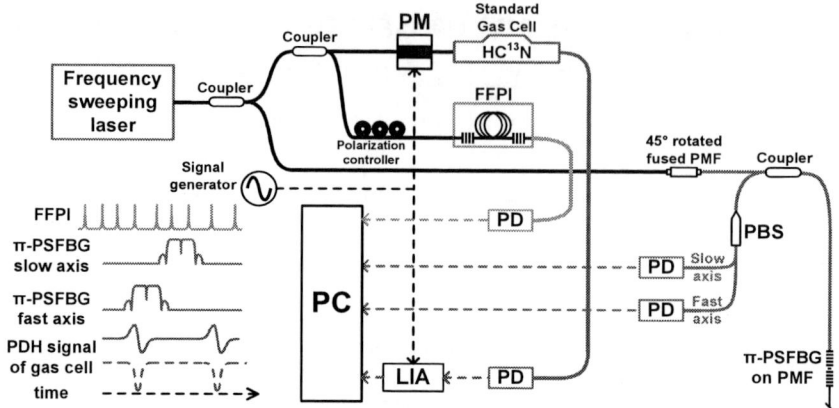

Fig. 1. Experimental setup of the sensing system. PM: phase modulator; FFPI: fiber Fabry-Perot interferometer; PBS: polarization beam splitter; LIA: locked-in amplifier; PD: photodetector; PC: personal computer.

Each frequency sweep of the laser will produce four curves from the four PDs as shown in Fig. 1, from which the Bragg frequency f_{Br} and the mode birefringence introduced frequency difference Δf_{Bi} of the π-PSFBG can be calculated. Although the frequency sweeping of the laser is nonlinear and with poor-repeatability, the peaks in the curve of FFPI have a constant frequency interval, which could be used for the resampling of the error signal curves of standard gas cell. By calculate the cross-zero points of the resampled gas cell error signal curve which precisely correspond to the absorption notch centers of the HC^{13}N gas, two absolute frequency points can be determined. The resonance center of π-PSFBG on both slow and fast axes, whose frequency equals to the grating's Bragg frequency of both axes, are obtained by curve fitting and then mapped to constant frequency interval data from FFPI by interpolation. Based on the two absolute frequency points and the constant frequency interval scale, Bragg frequency of the grating on slow axis f_{Br}^{slow} and fast axis f_{Br}^{fast} are obtained, and their frequency difference f_{Br}^{fast}-f_{Br}^{slow} is the birefringence introduced frequency difference Δf_{Bi}. To reduce the calculation error, we use the average value of f_{Br}^{slow} and f_{Br}^{fast} as the grating's Bragg frequency f_{Br}.

III. EXPERIMENTAL RESULTS

The strain and temperature sensitivity of the grating's Bragg frequency f_{Br} and the mode birefringence introduced frequency difference Δf_{Bi} are obtained by linear fitting of the measurement results in Fig. 2(a, b), which are $k_{Br}^{\varepsilon} = +112.69$ MHz/$\mu\varepsilon$, $k_{\Delta Bi}^{\varepsilon} = +0.79806$ MHz/$\mu\varepsilon$, $k_{Br}^{T} = +1236.9$ MHz/°C, and $k_{\Delta Bi}^{T} = -58.389$ MHz/°C, respectively. With the calculated Bragg frequency f_{Br} and birefringence frequency difference Δf_{Bi} during the measurement, the discrimination of strain ε and temperature T can be performed by solving the matrix equation in (1).

$$\begin{bmatrix} f_{Br} \\ \Delta f_{Bi} \end{bmatrix} = \begin{bmatrix} k_{Br}^{\varepsilon} & k_{Br}^{T} \\ k_{\Delta Bi}^{\varepsilon} & k_{\Delta Bi}^{T} \end{bmatrix} \cdot \begin{bmatrix} \varepsilon \\ T \end{bmatrix} \tag{1}$$

In the demonstrational experiment, the π-PSFBG is placed in a thermostat to eliminate the external vibration and to obtain a slowly varying temperature. Besides, in order to avoid the influence of the support bracket thermal expansion and contraction which could lead to an uncontrollable strain changes during a long-term measurement, the π-PSFBG is loosely placed, keeping a constant (zero) strain state on the grating.

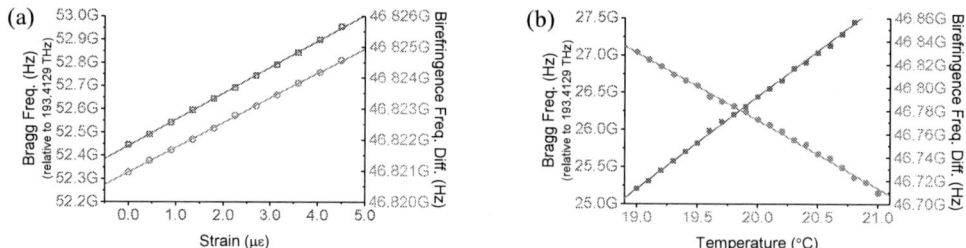

Fig. 2. Strain and temperature sensitivity of the grating's Bragg frequency and birefringence frequency difference.
(a) Strain sensitivity; (b) Temperature sensitivity.

During the measurement, an electronic thermal meter (Contec, PTI-4) with temperature resolution of 0.01 °C is used for comparison. The measurement period of the sensing system is about 1.5 s, mainly limited by the tuning speed of the laser. Fig. 3(a) shows the calculated data of Bragg frequency f_{Br} and birefringence frequency difference Δf_{Bi} in 10^4 seconds, a symmetric shape changes can be observed between the two curves since their temperature sensitivity are with opposite sign. The discriminated strain ε and temperature T are shown in Fig. 3(b). From this time domain result, a constant strain signal without temperature interference and a temperature signal consistent with the electronic thermometer readout are observed, the sensing resolutions are 0.104 με in strain and 0.005 °C in temperature, which are the standard deviation of measured strain ε and temperature T calculated from a selected relative flat segment in the result. The noise power density from 10^{-4} Hz to 0.3 Hz is also given in Fig. 3(c). For the frequency above 10^{-2} Hz, the noise power density reaches an up bound of 2×10^{-9} nε/Hz$^{1/2}$; in the 10^{-4} Hz to 10^{-2} Hz band, the noise power density behaves a typical $1/f$ distribution. The white noise in frequency band above 10^{-2} Hz is mainly composed of the fitting errors of the π-PSFBG resonance centers and the calculation errors of the FFPI transmission peaks. For the $1/f$ distribution noise in the lower frequency band, the measurement uncertainty of the standard gas cell absorption notches is the major cause.

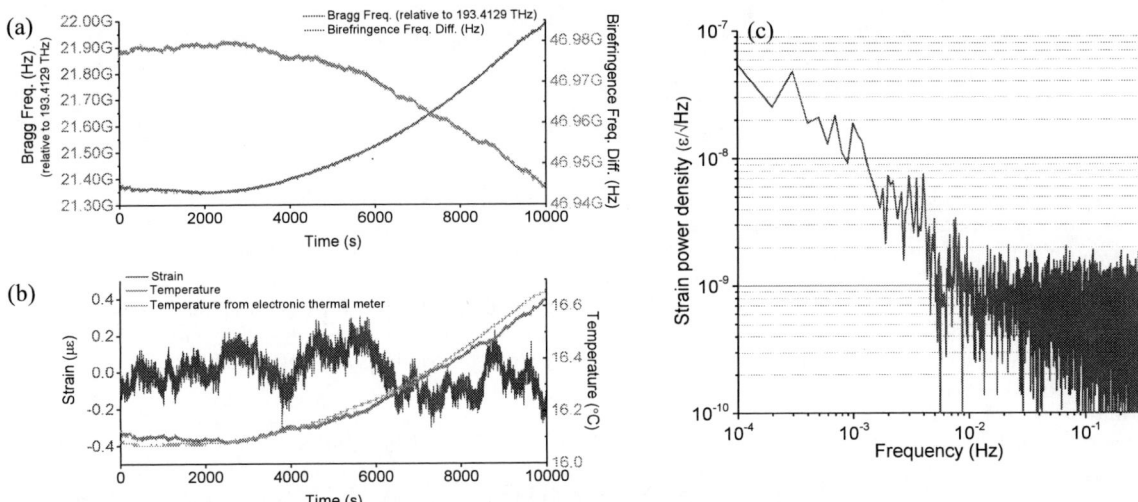

Fig. 3. Experimental results in the strain and temperature discrimination. (a) Measured Bragg frequency and birefringence frequency difference in an experiment; (b) Strain and temperature discrimination result; (c) Power density spectrum of the measured strain.

The measurement ranges of the sensor can achieve several mε in strain and more than 100 °C in temperature, since both the FFPI and standard gas cell have a working bandwidth larger than 10 nm. In the aspect of multiplexing ability, we can add sensing channels by serially connecting several π-PSFBGs with different working band as a wavelength division multiplexing scheme, or expanding the system with optical switches and additional gratings as a time division multiplexing scheme, or the combination of both.

IV. CONCLUSIONS

In this paper, a temperature and strain discriminative sensor with resolution of 0.104 με and 0.005 °C has been demonstrated, by using a π-PSFBG on PMF as the sensing head. The frequency sweep nonlinearity compensation scheme by a FFPI and the introduction of standard gas cell improve the sensing resolution and enhance the long-term stability. Due to these characteristics, the sensor shows great potential in applications requiring high resolution and good anti-thermal-interference ability such as crustal deformation and temperature observation in geophysical researches, especially in areas where geothermal effect is evident and under concern.

ACKNOWLEDGMENT

This work was supported by the National Natural Science Foundation of China under Grant 61327812, 61307106, 61411140038, and 61575001.

REFERENCES

[1] Q. Liu, T. Tokunaga, and Z. He, "Ultra-high-resolution large-dynamic-range optical fiber static strain sensor using Pound–Drever–Hall technique," Opt. Lett. **36**, 4044-4046 (2011).

[2] M. G. Xu, J.-L. Archambault, L. Reekie and J. P. Dakin, "Discrimination between strain and temperature effects using dual-wavelength fibre grating sensors," Electron. Lett. **30**(13), 1085-1087 (1994).

[3] T. Erdogan, "Fiber grating spectra," J. Lightwave Technol., **15**(8), 1277–1294 (1997).

[4] M. Sudo, M. Nakai, K. Himeno, S. Suzaki, A. Wada, and R. Yamauchi, "Simultaneous measurement of temperature and strain using PANDA fiber grating," in 12th International Conference on Optical Fiber Sensors, Post-conference Edition, 170-173 (1997).

[5] E. D. Black, "An introduction to Pound–Drever–Hall laser frequency stabilization," Am. J. Phys. **69**, 79–87 (2001).

TuA1-1

OECC/PS2016

Spectral Efficient Grouped Routing Network that Applies Dynamic Optical Path Grooming

Yuki Terada, Yojiro Mori, Hiroshi Hasegawa, Ken-ichi Sato

Nagoya University, Furo-cho, Chikusa-ku, Nagoya, 464-8603 Japan

y_terada@echo.nuee.nagoya-u.ac.jp, {mori, hasegawa, sato}@nuee.nagoya-u.ac.jp

Abstract: We present a novel dynamic optical path control algorithm for Grouped Routing that can greatly enhance fiber frequency utilization. Careful grooming is shown to effectively mitigate the routing penalty even with dynamic optical path control.

Keywords: WSS; Filter narrowing effect; Coarse granular routing; Dynamic path control; Grooming

I. INTRODUCTION

Broadband access is being rapidly adopted throughout the world and next-generation network services are emerging. One of the key requirements imposed on such networks is dynamic adaptability to the diverse traffic distribution changes. Dynamic network services including ultra-high definition TV distribution [1] and optical VPNs [2] will frequently set up and tear down optical paths, all of which are done in the electrical layer in present networks. High bit rate channels such as 100 Gbps or 400 Gbps will be used to support such services; for example, the 8k video bit rate is 144/72 Gbps. These channels can be transported using high-order modulation formats, however, they are sensitive to transmission impairment. Signal quality is degraded by not only long transmission distances but also the spectral narrowing effect caused by imperfect optical filtering at WSSs utilized in ROADMs [3]. To alleviate the filtering impairment, a sufficiently wide guard-band between adjacent channels is necessary, which degrades fiber frequency utilization. ROADMs will further penetrate deeper into the metro area and the hop count of optical paths will thus increase, which increases the number of WSSs traversed. In addition, high port count ROADMs are needed to connect the many fibers needed to support the traffic expansion. Such ROADMs will need the broadcast-and-select architecture which needs WSSs on both input and output fiber sides, which worsens the filtering effect. Furthermore, they need high port count WSSs, which have been studied [4,5], however, issues with their performance and the adjustment cost imposed by beam steering remain to be resolved. Instead, a large port count WSS can be created by concatenating WSSs, which expands the number of WSSs traversed at a node and hence further enhances filter narrowing.

In order to resolve the passband-narrowing problem created by traversing multiple WSSs and improve fiber frequency utilization simultaneously, we proposed a Grouped Routing [6,7] scheme that combines coarse granular routing and fine granular (channel by channel) add/drop. The fiber frequency bandwidth is divided into several sub-bands called Grouped Routing Entities (GREs). Routing at each ROADM node is done at the GRE level by setting the passband of each WSS to cover each GRE. An adequate guard band is inserted between GREs, while the optical channels are packed densely within each GRE. This simultaneously mitigates the passband narrowing effect and yields high spectral efficiency [8,9], which has been proved for a static network scenario [8] and for a traffic growth model that mirrors most of the present usages [9]. In this context, the coarse granular routing lowers the routing flexibility, but it can be effectively managed by developing new network control mechanisms. On the contrary, in the dynamic network scenario, where optical paths are frequently set up and torn down, the restricted routing flexibility imposed by GRE use can seriously degrade performance since the situation does not allow the use of sophisticated network control.

To support future dynamic services while attaining high frequency utilization, we propose a new network control scheme that both suppresses the filter narrowing effect and offers high routing flexibility. We extend Grouped Routing with a grooming operation that can substantially mitigate the routing restriction caused by the coarse granular routing of GREs. The optical grooming operation, however, incurs an additional filtering operation and hence careful path management criteria need to be established. We introduce here a new control algorithm that can effectively perform grooming operation taking account of the traffic condition. Numerical simulations prove that the enhanced spectral efficiency of Grouped Routing allows more optical paths to be accommodated than is possible in conventional networks.

II. GROUPED ROUTING NETWORKS THAT ADOPT GROOMING

In Grouped Routing networks, optical paths in a fiber are grouped as GREs. Every node performs routing operations at the GRE level. GREs are not divided/branched or joined on the way, so a Grouped Routing network is regarded as one where nodes are connected by virtual pipes (small bandwidth fibers) as shown in Fig.1. This virtual pipe is called "GRE pipe" and differs from the waveband paths of conventional optical path networks as GRE pipes do not offer path functions such as termination as defined by ITU-T and can form a loop configuration if necessary like a present fiber ring networks (Fig. 1). Furthermore, we need to distinguish a GRE pipe from a "super-channel". The latter is a "channel" and a sub-carrier channel cannot be dropped (/added) along the way from each super-channel source to destination node. It has been proven that freedom in path add/drop locations along GRE pipes can minimize the routing performance degradation caused by coarse granular routing [6,7].

Coarse granular routing can reduce the guard-band bandwidth [8,9], since significant guard-bands are necessary only between adjacent GREs (Fig.2). This broad path band filtering is possible with current LCOS-based WSSs with no changes. Within a GRE pipe, add/drop operations of a path cause signal impairment in the adjacent optical paths (Fig.3). However, by applying a constraint that bounds the frequency of such operations and an appropriate network design/control scheme that considers the constraint, optical paths can be densely packed within each GRE pipe with a minimum gap between channels sufficient for a small number of adjacent path add/drop operations. Figure 3 shows an example of the reduction in filtering impairment possible with path allocation (no filtering effect occurs with path allocations in the lower figure).

Once a GRE pipe is established, it is not released until all paths in the pipe are torn down. Moreover, a new path is accepted only if a GRE pipe that traverses the source and destination nodes of the path has sufficient spare capacity or a new GRE pipe can be established. Thus the inclusive relationship between GRE pipes and optical paths can be so restrictive so as to prevent effective dynamic path operation. Indeed, it was shown that Grouped Routing is much less effective in the dynamic path control scenario [7] than the static design scenario [6]. Therefore, it is necessary to develop a new architecture that can enhance optical path routing flexibility for dynamic path control. In this paper, we propose a new scheme that implements intermediate optical path grooming for GRE pipes. The "grooming" allows a path to transit multiple GRE pipes. The GRE pipes need to cross each other at the grooming node and to occupy the same frequency bandwidth (the same GRE index) on different fibers. Figure 4 shows an example of a grooming operation; one optical path in GRE Pipe1 is transferred to GRE Pipe2. The drawback of grooming is the additional filtering at WSSs needed for path granular switching.

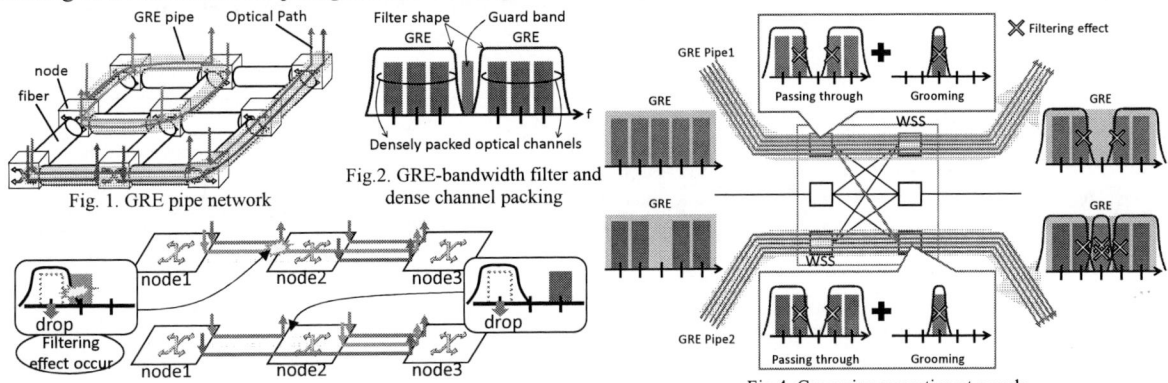

Fig. 1. GRE pipe network

Fig.2. GRE-bandwidth filter and dense channel packing

Fig.3. Filtering effect caused by adjacent channel drop

Fig.4. Grooming operation at a node

III. DYNAMIC PATH CONTROL OF GROUPED ROUTING NETWORK

Only the key points of the control algorithm are presented below due to the space limitation. For each optical path setup request, route candidates that connect the source and destination nodes are found (prior determination is possible). The following path-accommodation process is done on each route candidate. First try to accommodate the arriving optical path request in an existing GRE pipe. If no GRE pipe is available, then an attempt is made to extend existing GRE pipes to accommodate it. If this attempt fails, inter-GRE-pipe grooming is tried. If this fails then a new GRE pipe connecting the source and destination nodes of the path is established. If this fails, the set up request is blocked. In this process, if there are multiple route candidates that can successfully accommodate the path, the best route in terms of dispersing the traffic loads is selected.

IV. NUMERICAL EXPERIMENTS

The physical network topologies tested were the cost266 pan-European network (26 nodes, 51 links) [10] and the USnet network (24 nodes, 43 links) [11]. The available fiber bandwidth was assumed to be 4,400GHz (C-band). The path capacities are set at 100Gbps (28 Gbaud, polarization-multiplexed QPSK). Figure 6 shows the calculated OSNR penalties caused by the passband narrowing effect for the 100Gbps signals. Only the filtering effect is considered here to highlight the generic performance of the proposed algorithm. WSS filter passband shape was assumed to be a 3.5th-order super-Gaussian function. Minimum grid spacing was determined so as to keep the penalty under 1dB. Thus 37.5GHz channel spacing is used when the filtering number is 8 or less. When the number is more than 8, 50GHz channel spacing is used. Number of channels accommodated per fiber was calculated considering the channel spacing. Node architecture assumed was route-&-select (WSSs are located both input and output side) and hence paths traverse two WSSs at each node. In the conventional network where paths are individually routed, the number of filtering frequencies equals the number of WSSs traversed. The channel spacing must be 50GHz since the largest filtering number in the topologies tested always exceeded 8. The number of channels accommodated per fiber was 88 in the conventional network. On the other hand, in Grouped Routing, channel spacing can be set to 37.5GHz by limiting the worst value of filtering number to less than 8. Note that filtering effects can be triggered in Grouped Routing by adjacent path dropping (fig. 3) and grooming operation (fig. 4). Ten channels are bundled as a GRE. We assumed a 25GHz guard-band between each GRE. As a result, the Grouped Routing network has total fiber capacity of 110 channels. Spectrum usage of the conventional and proposed networks are depicted in Fig.5.

We evaluate routing performance under the dynamic traffic scenario. The traffic demand is uniformly and randomly distributed. Path-setup requests are generated in accordance with a Poisson process and the holding time of each connection follows a negative exponential distribution. The maximum detour from the minimum-hop path route is set to 1. Same fiber configurations (the configuration can accommodate the average number of optical paths of 10 between each node pair) are used for both networks (conventional and proposed), and various traffic demand intensities are examined.

Figure 7 plots overall blocking ratios as a function of the total number of average path requests. In the previous Grouped Routing method, shown by the grey lines, paths are allocated to the shortest GRE pipes first found and no grooming is used. Figure 7 shows that the proposed method accommodates most traffic when 6 filtering operations are allowed. When we set the target blocking ratio to 0.01, the number of paths can be increased by 57.2% in cost266 and 57.9% in USnet from the previous Grouped Routing network and by 15.2% and 16.3% from conventional routing method that does not use GREs.

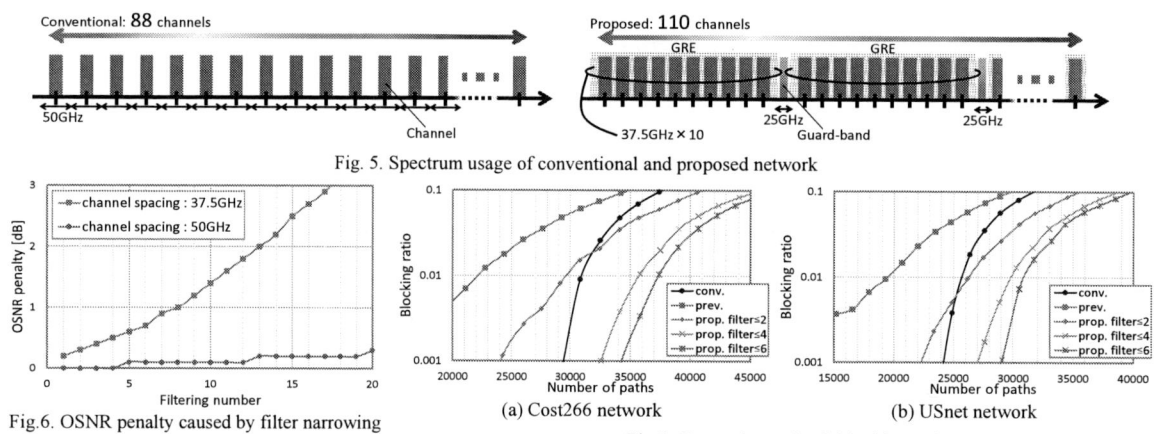

Fig. 5. Spectrum usage of conventional and proposed network

Fig.6. OSNR penalty caused by filter narrowing effect

(a) Cost266 network

(b) USnet network

Fig.7. Comparison of path blocking ratio

V. CONCLUSIONS

It has been shown that Grouped Routing networks are very efficient in static and semi-static (traffic growth model) traffic scenarios. In the future, however, dynamic operation of optical paths is envisaged, and the previous architecture provides very poor performance in this scenario. A novel network architecture based on Grouped Routing that incorporates optical grooming was proposed. It was demonstrated that optical grooming with a filtering number restriction can greatly mitigate the impairment stemming from the coarse granular routing of GREs. Numerical experiments showed that he proposed architecture can accommodate 15% more traffic than the conventional network. Please note that grouped routing and optical grooming operation do not require any WSS hardware changes. As a result, the Grouped Routing networks substantially improve frequency utilization for not only static and semi-static (traffic growth model) networks (around 30% [8,9]), but also for dynamic networks, which assures the seamless upgradability of Grouped Routing networks that can support the future dynamic traffic scenario and hence are future proof.

ACKNOWLEDGMENT

This work was partly supported by NICT and KAKENHI (26220905).

REFERENCES

[1] "Parameter values for ultra-high definition television systems for production and international programme exchange," ITU-R Recommendations BT. 2020-1, June 2014.
[2] K. Sato and H. Hasegawa, OSA J. Opt. Commun. Netw. vol. 1, issue 2, pp. A81-A93, June 2009.
[3] M. Filer and S. Tibuleac, Proc. of OFC, March 2014, paper Th11.2
[4] K. Suzuki et al., Proc. of OFC, March 2015, paper Tu3A.7.
[5] M. Iwama et al., Proc. of ECOC, Sept. 2015, paper Mo.4.2.2.
[6] Y. Taniguti et al., Proc. of OFC, March 2012, paper JW2A.2.
[7] Y. Taniguti et al., Proc. of OFC, March 2013, paper OM3A.5.
[8] Y. Terada et al., Proc. of (OFC, March 2014, paper W1C.6.
[9] Y. Terada et al., Proc. of ECOC, Sept. 2014, paper Mo.3.1.4.
[10] R. Inkretet al., "Advanced Infrastructure for Photonic Networks: Extended Final Report of COST Action 266," Faculty of Electrical Engineering and Computing, University of Zagreb, Croatia, 2003.
[11] S. F. Gieselman et al., Proc. of OFC, March 2005, paper OTuP3.

Field Trial of Virtual Transport Network Services Based on Hierarchical Control over Multi-Domain OTN Networks

Ruiquan Jing[1], Chengliang Zhang[1], Junjie Li[1], Yiran Ma[1], Qian Hu[1], Xiaoli Huo[1]
Yongli Zhao[2], Jiayu Wang[3], Baoquan Rao[4], Chen Qiu[5]

1. China Telecom Beijing Research Institute, Beijing, China   2. Beijing University of Posts and Telecommunications, Beijing, China
3.ZTE, Beijing, China   4.Huawei, Shenzhen,China   5.Fiberhome, Wuhan, China
Author e-mail address: jingrq@ctbri.com.cn

Abstract: BoD and VTS applications were developed based on the hierarchical SDON architecture. OpenFlow was extended to support Ethernet performance monitoring, protection and restoration in OTN networks. The applications have been verified in field trial network.
Keywords: SDN, VTNS, Multi-Domain OTN

I. Introduction

As new services emerge with the development of transport SDN technologies, new requirements and characteristics for transport networks have been identified. For example, BoD (Bandwidth on Demand) is a service that requires dynamic bandwidth provisioning to provide users with instantly created connectivity. Currently, Optical Internetworking Forum (OIF) is specifying Virtual Transport Network Services (VTNS) in [1].VTNS aims to become a main driver for the deployment of SDN in transport networks from a service perspective.

In [2], a novel NFV enabled virtualization mechanism supporting optical network + IT infrastructure slicing and on demand network control function virtualization was proposed. In [3], a hierarchical SDON (Software Defined Optical Networking) control mechanism is proposed for multi-domain and multi-vendor OTN networks. In this work, we have extended our experimental work to support more complex network functionalities such as Ethernet performance monitoring, protection and restoration etc. Besides that, we have developed two new applications which are BoD and VTS (Virtual Transport Service), as shown in figure 1. All the new functionalities and applications have been verified in a field trial network with commercial OTN equipments from three vendors.

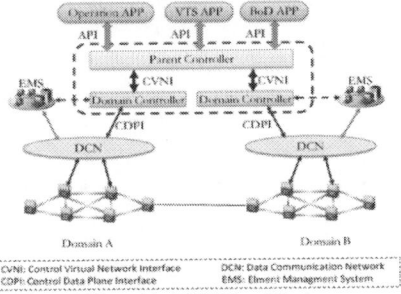

Figure 1 hierarchical SDON architecture

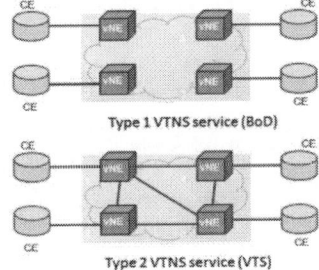

Figure 2 Virtual Transport Network Service Classification

II. Overview of Virtual Transport Network Services

VTNS is built on the virtualization of Layer 0~2 transport networks. The physical transport network resources are sliced and assigned with an abstract view to different users. The user request may include traffic matrix, SLA, topology, OAM, recovery etc. Users of VTNS may own the right to control and manage the assigned virtual network as if they were operating a physical transport network, more than just get managed connectivity. The virtual network can be reconfigured based on the request from the user, e.g. capacity expansion, topology change. The following two types of VTNS are developed and verified in our field trial network, as shown in figure 2:

Type 1: Dynamic connection service with limited customer control. Type 1 service defines dynamic connection service that is provided to customers based on their request, e.g. in this type of service, customers only have knowledge of the end points of the operator's network. Therefore, customers have very limited access to the operator's network. Type 1 VTNS service also known as BoD service.

Type 2: Dynamic virtual network topology with customer connection control. Type 2 service specifies a virtual network topology over which customers are allowed to have full control of connection setup/modification/deletion, e.g. include explicit routing on customer's demand. Initially, the virtual network topology could be pre-configured by operator, and dynamically configured by customer later. Operator stays in full control and responsibility over physical NE and how to physically realize the virtual network. Type 2 VTNS service is called VTS in this paper.

431

III. The generation of virtual network

Virtual network generation mechanism is the key enable technologies for providing VTS services. In our work, the virtual network could be pre-configured by operator, and dynamically configured by customer through VTS APP later. Customers can modify the virtual network topology according to their service requirements, such as adding/removing nodes and links.

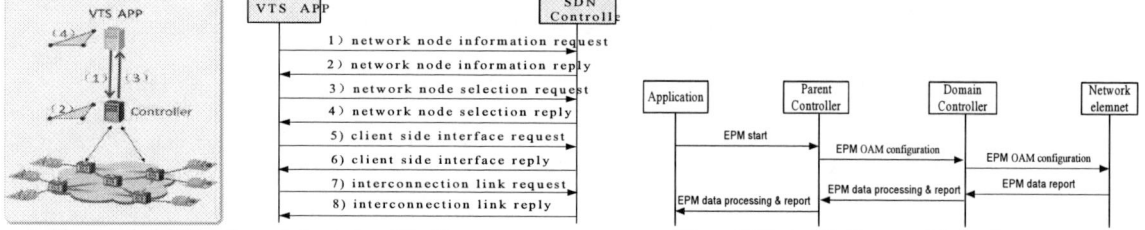

Figure 3 virtual network generation and modification processes   Figure 6 Ethernet Performance Monitoring processes

The basic process of virtual network generation is shown in left side of Figure 3. The main steps include: 1) VTS APP requests virtual network resource from the controller; 2) Controller makes resource allocation and reservation in routing database (RDB) for virtual network; 3) Controller feedback the resource allocation results to VTS APP; 4) VTS APP generates a virtual network topology based on the reply. The detailed message exchange process between VTS APP and the controller for virtual network generation and modification were shown in right side of Figure 3.

IV. Protection and restoration

Protection and restoration is an important feature of transport network. We have developed ODUk 1+1 protection and restoration based on Openflow [4] extensions. Although the proposed solutions were implemented and tested at CVNI between Parent Controller and Domain Controller, it can also be used at CDPI between Domain Controller and network elements. Two new ofp_flow_mod_flags, OFPFF_FLOW_TYPE and OFPFF_PROTECT_STATUS, were defined to distinguish the work and recovery path, and indicate the protection status.

For 1+1 protection, end-to-end unidirectional switching is used. After the failure is cleared, it will switch back to the work path. Parent controller is responsible for the work and protection paths computation and connection setup. The fault detection and protection switch is done within the transport plane. The work flow for 1+1 protection is shown in figure 4.

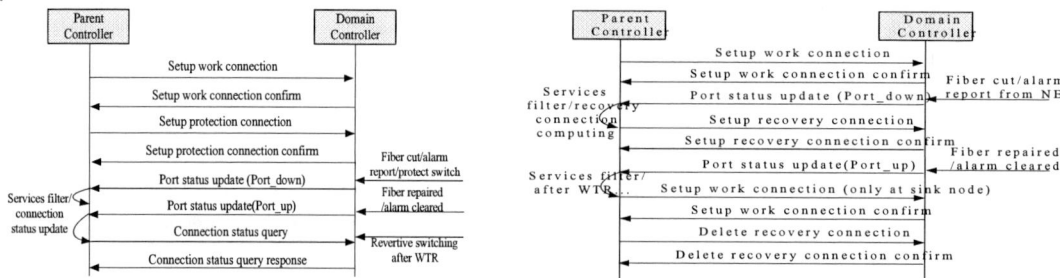

Figure 4 1+1 protection work flow   Figure 5 rerouting restoration work flow

For rerouting restoration, we use end-to-end bi-directional switching architecture. Parent controller is responsible for the work and recovery paths computation and connection setup. If the recovery is unsuccessful, the controller will retain the work connection and send a notification to application. The work flow for rerouting restoration is shown in figure 5.

V. Ethernet performance monitoring

Ethernet Performance Monitoring (EPM) is very useful for both network operator and customers. In our work, the monitored parameters include service throughput, delay and packet loss rate. The process of EPM was shown in figure 6. At CVNI, OFPT_SET_CONFIG message was extended to configure EPM related OAM parameters, e.g. MEP ID, MEG ID and CC interval; as well as the start and stop of EPM. The ofp_multipart_request & reply(OFPMP_QUEUE=5）messages were extend to request and report the EPM results. The reporting cycle (T) of EPM results is configurable, and the default value is 1 minute. The parent controller sends ofp_multipart_request message to source and sink node to get the measurement data for both directions.

The delay and packet loss rate are measured by using Ethernet OAM mechanisms as specified in ITU-T Y.1731. One-way service throughput for a point-to-point connection is calculated as follows:

$$Service\ throughput(Mb/s)=(tx\_bytes(t2)-tx\_bytes(t1)) \times 8bit/ (T \times 60s）/1000000$$

In which, tx_bytes (t2) and tx_bytes (t1) are the bytes sent at time T2 and T1 at one end of the connection.

VI. Field trial results

Figure 7 shows the field trial setup of BoD and VTS services over multi-domain SDON architecture. The trial is conducted in Fuzhou, Xiamen and Quanzhou, Fujian Province, China. The trial network is composed of 3 domains which were provided by 3 different vendors. The 10Gbit/s inter-domain links were provided by 40Gbit/s inter-city OTN network. The 10Gbit/s links were used for intra-domain links. Supported client interfaces include GE and 10GE. The Operation APP, VTS APP and BoD APP were connected to parent controller via RESTful API. To enable the controller to push topology update and performance monitor information to APPs, websocket protocol as defined in RFC6455 is running between APPs and parent controller.

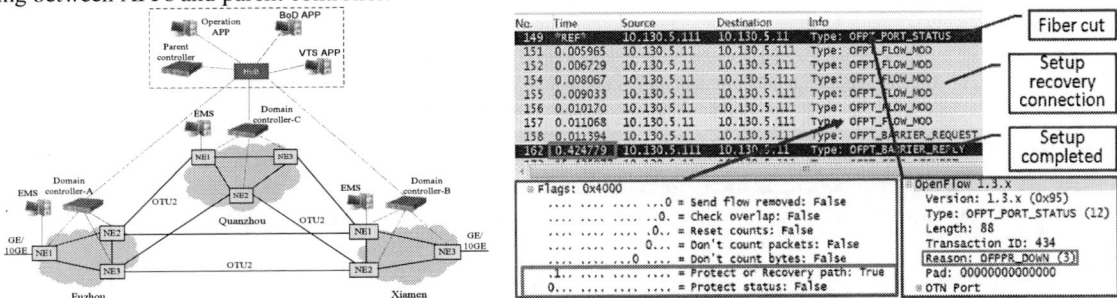

Figure 7 Field trial network setup

Figure 9 messages sequence for rerouting restoration

For the same physical network, Figure 8 shows the different level of network virtualization as seen in Operation APP, VTS APP and BoD APP. The network topology becomes more and more abstracted and virtual in order.

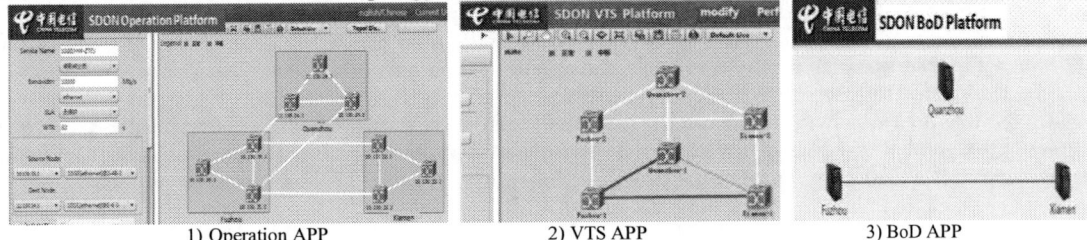

1) Operation APP   2) VTS APP   3) BoD APP

Figure 8 Different level of network virtualization

The proposed protection and restoration solutions were tested successfully for both single domain and multi-domain services. The 1+1 protection time is below 20ms, and the restoration time is about 500ms. The extended Openflow messages sequence for rerouting restoration were shown in Figure 9.

Figure 10 illustrates the traffic throughput monitoring result for a 10GE EPL service, and the extended Openflow messages for EPM OAM settings and results report. The bandwidth was adjusted based on oduflex's 1.25Gbit/s granularity. Transmitted and received traffic throughputs are reported by the controller on a time basis.

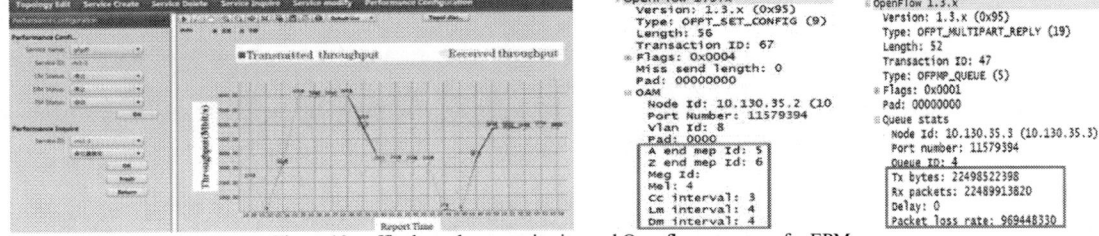

Figure 10 traffic throughput monitoring and Openflow messages for EPM

VII.Conclusions

The OpenFlow protocol was extended to support network functionalities such as Ethernet performance monitoring, protection and restoration in OTN networks. BoD and VTS applications were developed based on the hierarchical control over multiple domain OTN networks. All the new functionalities and applications have been tested and verified in a field trial network with commercial OTN equipments from three vendors. This work will be valuable for carriers providing innovative services in multi-vendor/ multi-domain environment with expected performance.

References

[1] OIF2014.269.05, "Programmable Virtual Transport Network Services Specification", July, 2015.
[2] R. Nejabati et al., "SDN and NFV Convergence a Technology Enabler for Abstracting and Virtualising Hardware and Control of Optical Networks (Invited)", OFC 2015.
[3] Ruiquan Jing et al., "Experimental Demonstration of Hierarchical Control over Multi-Domain OTN Networks Based on Extended Openflow Protocol", OFC 2015.
[4] ONF, "OpenFlow Switch Specification, Version 1.3.0 (Wire Protocol 0x04)", June 25, 2012.

TuA1-3

OECC/PS2016

A Static Traffic Grooming Algorithm for Elastic Optical Networks with Adaptive Modulation

Takafumi Tanaka, Tetsuro Inui, and Wataru Imajuku
NTT Network Innovation Laboratories, 1-1 Hikarinooka, Yokosuka, Kanagawa, Japan
tanaka.takafumi@lab.ntt.co.jp

Abstract: *We propose a novel static traffic grooming algorithm for elastic optical networks, which are aware of multiple modulation formats. Evaluation showed the algorithm significantly saves optical paths compared to other algorithms in various network conditions.*
Keywords: *Elastic optical network; Traffic grooming; Adaptive modulation*

I. INTRODUCTION

The elastic optical network (EON) is one of the promising architectures that can accommodate a wide variety of client traffic types efficiently. The EON raises several technical challenges worth addressing. Traffic grooming, the technique that efficiently shares multiple low-rate traffic flows among elastic optical paths by electrical switching, is one of these challenges. Thanks to adaptive modulation, which allows an elastic optical path in the EON to select multiple modulation formats such as BPSK, QPSK, and xQAM, it can have high flexibility in terms of bitrate and distance (optical reach). However, this flexibility increases the complexity of EON traffic grooming compared to traditional fixed-rate optical networks. There have been many studies on traffic grooming. For example, the dynamic traffic grooming algorithm that considers variable rate and distance of optical path [1]. Authors in [2] quantified the impact of adaptive modulation on EON traffic grooming, in which grooming-capable sites are arbitrary selected in advance and evaluated using Integer Linear Programming (ILP).

In this paper, we propose a heuristic EON traffic grooming algorithm that considers adaptive modulation. The algorithm yields a virtual topology that consists of optical path demands, and routes of traffic flows assuming the static traffic condition. To the best of our knowledge, this is the first paper to consider the heuristic approach for adaptive modulation in the static traffic condition. The following sections describe the proposed algorithm in detail and show evaluation results.

II. ADAPTIVE-MODULATION-AWARE TRAFFIC GROOMING ALGORITHM

A. Traffic Grooming in the Optical Networks with Adaptive Modulation

Determining the virtual topology, assigning route and spectrum (or wavelength) of virtual links to the physical topology (routing and spectrum assignment: RSA), and allocating traffic flows to virtual links in the virtual topology are the main processes in multi-layer network design, and each can be viewed as sub-problems of traffic grooming [4]. In the usual traditional network design, first the virtual topology is determined using traffic flows and then RSA is processed based on the topology. This approach of dealing with sub-problems separately is feasible only if the optical network has optical paths with fixed bitrate regardless of distance. On the other hand, we need to be aware of the parameters of optical paths in determining the virtual topologies if adaptive modulation is used. Usually the distance and bitrate of optical path have an inverse relationship. If we do not consider the relationship, more optical paths may be needed when high-rate traffic flows are allocated to small capacity long distance virtual links, while extra resources may be left unused if mid-rate traffic flows are allocated to large capacity short distance virtual links (Fig. 1). Therefore, the proposed algorithm considers the modulation format of optical paths in determining the virtual topology.

B. Heuristic Algorithm Description

An overview of the proposed grooming algorithm is illustrated in Fig. 2. The algorithm assumes that all nodes have grooming capability and no restrictions such as maximum number of transponders in a node are set. The strategy of the algorithm is simple; improve the accommodation rate of larger capacity virtual links as much as possible. If the bitrate of an optical path is fixed regardless of distance, offloading traffic to the optical layer and skipping electrical processing at the intermediate nodes would be useful techniques for saving cost and energy, but in the environment that this work assumes, grooming traffic flow into the large capacity virtual links is the more effective strategy.

The proposed grooming algorithm is as follows:
Step 1. Create initial virtual topology. Virtual links are allocated between every node pair if possible. The bitrate and distance parameters of each virtual link are set individually. The distance parameter denotes the distance of shortest

route along the physical topology between the end nodes of the virtual link, and the bitrate parameter is determined by the modulation format that can travel the distance. If the distance exceeds the maximum distance allowed by the optical path, no virtual link is created between the node pair. Next, virtual links which have the same modulation format are grouped in virtual layers. This work assumes that higher virtual layers corresponds to lower modulation levels, i.e. smaller capacity and longer reach, and lower virtual layers correspond to higher modulation levels.

Step 2. Allocate traffic flows to the virtual topology. Most traffic flows should be allocated with single hop on the virtual topology, and multiple virtual links with shortest hops are used for the traffic flows that can not be directly connected. If the bitrate of a traffic flow is larger than that of the virtual link that accommodates the traffic flow, multiple virtual links are employed.

Step 3. Pick a virtual link from the highest virtual layer and try to reroute its traffic flows to other virtual link(s) in the virtual topology. In the step there are two issues. The first is to avoid unnecessary traffic flow rerouting such as the case wherein a virtual link v_1 reroutes a traffic flow to virtual link v_2, and in turn virtual link v_2 reroutes the flow back to v_1. The second is to reroute them to virtual links in lower layers as much as possible. To satisfy these two requirements, in addition to the essential requirement that available bitrate of the virtual link(s) to be rerouted is equal to or larger than the bitrate of traffic flow, this work sets an additional requirement; total bitrate of assigned traffic flows in each virtual link to be rerouted is equal to or larger than that in the original virtual link. If there are no virtual links(s) that satisfy these requirements, rerouting for this traffic flow in the virtual link is terminated. This step is repeated until all traffic flows in the virtual layer are rerouted.

Step 4. Move on to the lower layer and repeat Step 3. The algorithm finishes when Step 3 is finished in the lowest layer.

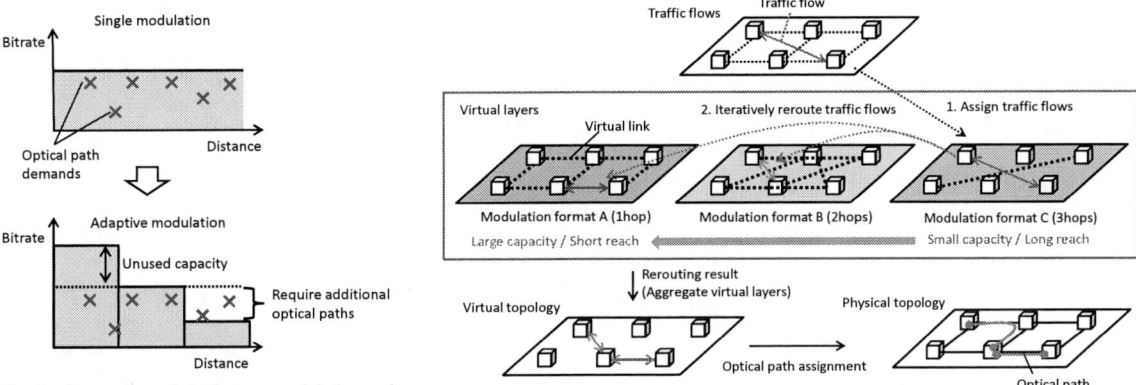

Fig. 1. Comparison of single type modulation and adaptive modulation.

Fig. 2. Overview of the proposed traffic grooming algorithm.

III. EVALUATION SETTINGS AND RESULTS

A. Network Models

In order to evaluate the performance of the proposed algorithm, we examined various topology and traffic patterns. This work compared four grooming approaches; 1) single-hop grooming approach, which terminates optical paths at every intermediate nodes traversed by traffic flows, 2) composition approach [4], which identifies the optimal topology by iteratively adding logical paths to minimum shortest tree topology, 3) proposed approach, and 4) ILP-based approach which is formulated based on [5]. Note that since grooming and RSA processes are clearly separated in the composition approach, we do not consider adaptive modulation in making virtual topology.

The relationship between the distance and bitrate of optical paths is shown in Table I. Transponders assumed to have single carrier and fixed symbol rate, and the modulation format can be adaptively changed to suit the virtual link.

This work considers three network topologies; JPN12 [6] which consists of 12 nodes and 17 links (average link distance is 437 km), 5x5 grid (link distance is 300 km), and JPN48 [6] which consists of 48 nodes and 82 links (average link distance is 154 km). Traffic flows are assigned to all node pairs, and these flows have uniform bitrate. We evaluated 10 Gb/s, 50 Gb/s, and 100 Gb/s traffic flow patterns for these topologies. The proposed approach can allow multiple traffic flows between each node pair, but the evaluation assumes single traffic flows, which can be viewed as the constraint that these multiple flows traverse the same route.

B. Evaluation Results

Figure 3 shows the required optical path numbers for each grooming approaches in the three topologies and three traffic types as described above. Note that the ILP result is plotted for only the JPN12 topology (Fig. 3(a)) because the approach is intractable with the larger scale topologies. In the JPN12 topology, we can observe that proposed approach requires fewer optical paths than the others in 50G and 100G traffic patterns, and that in these cases the result is close to the optimum as yielded by the ILP approach. In these traffic cases, the proposed approach requires 14-27% fewer optical paths. On the other hand, we can see that the proposed approach performed worse in the 10G case, which is because traffic flows of the higher virtual layer could not be rerouted efficiently. We have room for improvement in this regard. These findings are replicated by the 5x5 grid (Fig. 3(b)). In this topology, we can see that the single-hop approach set 16QAM for all virtual links, and that composition approach is worse than the single-hop approach because those virtual links were set without regard to the physical conditions. In contrast, the proposed approach appropriately selects different modulation formats, which yielded optical path savings by the cut-through of intermediate nodes if necessary. When we focus on the 50G and 100G traffic patterns, the proposed topology needed 15-45% and 34-55% fewer optical paths than the other two approaches for the 5x5 grid and JPN48 topology, respectively. Example of virtual topologies (JPN48, 50Gb/s) are shown in Fig. 4. Whereas the single-hop and composition approach makes multiple virtual links at the same link, the proposed approach makes virtual links between various varieties of nodes. We can conclude that the proposed approach can reduce the number of optical paths even as the topology and traffic grow.

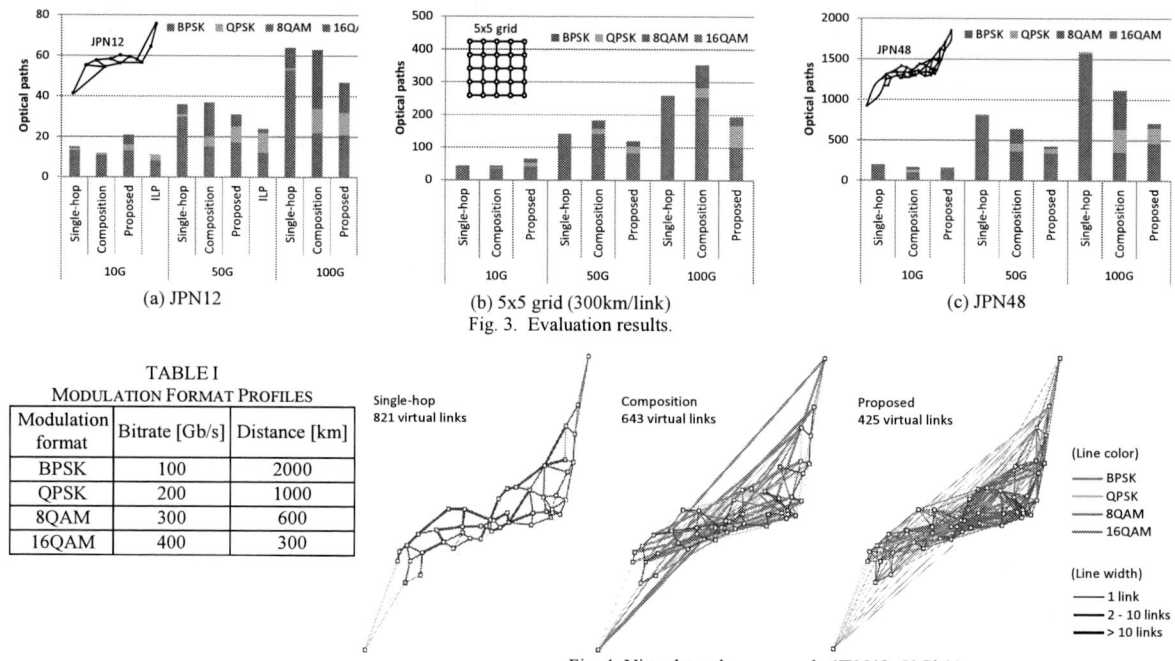

(a) JPN12 (b) 5x5 grid (300km/link) (c) JPN48
Fig. 3. Evaluation results.

TABLE I
MODULATION FORMAT PROFILES

Modulation format	Bitrate [Gb/s]	Distance [km]
BPSK	100	2000
QPSK	200	1000
8QAM	300	600
16QAM	400	300

Fig. 4. Virtual topology example (JPN48, 50Gb/s).

IV. CONCLUSIONS

We proposed a static traffic grooming algorithm for EON with adaptive modulation, which iteratively tries to reroute traffic flows in smaller capacity virtual links to larger capacity ones. Evaluation results showed that the proposed algorithm outperformed other heuristic algorithms in terms of optical path number in cases where the bitrate of traffic flows was an appreciable percentage of optical path capacity, and that the result was close to optimum in the small topologies.

ACKNOWLEDGMENT

This work was partly supported by the project "R&D of Elastic Optical Networking Technologies" of the National Institute of Information and Communication Technology (NICT), Japan.

REFERENCES

[1] X. Wang, M. Brandt-Pearce, and S. Subramaniam, "Grooming and RWA in translucent dynamic mixed-line-rate WDM networks with impairments," proc of Optical Fiber Communication (OFC) Conference, OTh1A.7, 2012.

[2] Y. Takita, K. Tajima, T. Hashiguchi, T. Katagiri, and T. Naito, "Impact of adaptive modulation on cost efficient traffic grooming in elastic optical networks," proc. of Optical Fiber Communication (OFC) Conference, Th1I.4, 2015.

[3] B. Mukherjee, "Optical WDM networks," Springer, 2006.

[4] T. Tanaka, A. Hirano, and M. Jinno, "Advantages of IP over elastic optical networks using multi-flow transponders from cost and equipment cost aspects," Optics Express, vol.22, no.1, 2014, pp. 62-70.

[5] P. Papanikolaou, K. Christodoulopoulos, and E. Varvarigos, "Multilayer flex-grid network planning," Proc. of 2015 Int. Conf. on Optical Network Design and Modeling (ONDM), 2015, pp. 151-156.

[6] Japan Photonic Network Model (JPNM), Available: http://www.ieice.org/cs/pn/jpn/jpnm.html

TuA1-4

OECC/PS2016

Hardware Scale Analysis and Prototype Development of Flexible Waveband Routing OXCs

Tomohiro Ishikawa[1], Masaki Niwa[1], Koh Ueda[1], Yojiro Mori[1], Hiroshi Hasegawa[1], Suresh Subramaniam[2], Ken-ichi Sato[1], Osamu Moriwaki[3]

[1] Nagoya University, Nagoya, Japan, [2] George Washington University, Washington DC, USA
[3] NTT Device Innovation Center, Atsugi, Japan
t_isikaw@echo.nuee.nagoya-u.ac.jp, {mori, hasegawa, sato}@nuee.nagoya-u.ac.jp, suresh@gwu.edu

Abstract: Hardware scale analysis shows that flexible waveband routing OXCs can substantially reduce the numbers of WSSs and EDFAs. Transmission experiments on developed prototype that adopts monolithic arrayed DC type matrix switches verify their technical feasibility.
Keywords: photonic network; elastic optical network; flexible waveband routing; optical cross-connect

I. INTRODUCTION

Recent developments in elastic optical path networks [1] are substantially enhancing the capacity of optical fibers, but the continuous explosive traffic growth exceeds the capacity enhancement. Increasing the number of optical fibers on each link and using them in parallel will cope with the traffic growth. Thus the scalability of optical cross-connects (OXCs) to efficiently accommodate more fibers is a crucial issue. Conventional OXCs need wavelength selective switches (WSSs) whose degrees are equal to or higher than OXC port count but cost-effective WSS degree expansion remains out of reach. Cascading existing WSSs (all of which have limited port counts) to realize higher degree WSSs is possible, however, this greatly increases the number of WSSs used.

As the number of paths will increase while the network topologies remain static, more paths will be routed together at each node. We proposed an OXC node architecture [2] that adopts a two-stage routing scheme; optical paths are bundled and routed as groups (flexible wavebands). Different from the conventional path hierarchy, any path group can form a flexible waveband regardless of center frequencies and bandwidths of paths. The path bundling operations are done by low port count WSSs as flexible demultiplexers. Delivery-and-coupling type matrix switches (DCSWs) with the same degree as the OXC are responsible for routing the flexible wavebands. Unfortunately, the high loss at optical couplers in large scale DCSWs limits the scalability of this original architecture. To further enhance scalability and hardware scale reduction, we proposed to split each large DCSW into arrayed small DCSWs [3]. A slight routing performance degradation relative to the original architecture is observed when the same parameter set as the original one is adopted.

In this paper, we mitigate the routing performance degradation by optimizing key parameter values; the number of flexible wavebands and the split switch size. The optimization involves not only the routing performance but also the transmission characteristics of the proposed OXC nodes. To this end, the total optical loss of a node is analyzed and the locations of EDFAs are specified. Next, the amount of hardware needed for given OXC degree is evaluated. Numerical experiments show that increasing the number of flexible wavebands will compensate the degradation caused by switch splitting and substantially reduce the number of EDFAs as well as WSSs. Finally, we develop a prototype using monolithically integrated arrayed small DCSWs on a PLC chip. Transmission experiments on the prototype demonstrate its excellent transmission characteristic.

II. FLEXIBLE WAVEBAND ROUTING OXCS THAT ADOPT ARRAYED SMALL SWITCHES

Figure 1 shows the original node architecture for flexible waveband routing [2]. This architecture consists of 1xB WSSs and B $N \times N$ DCSWs, where N is the degree of OXC port. Optical paths from an incoming fiber are first divided into a limited number ($\leq B$) of groups and these groups are then distributed to up to B output ports of the WSS. A flexible waveband can be created with any set of elastic optical paths and offers dynamic path addition and removal to/from that waveband. The scalability of the original flexible waveband node architecture depends on DCSW port scalability. The architecture has some redundancy in routing capability; different groups from an input port can be routed to the same output port. We note that when OXC node degree is large (e.g. $N \geq 32$), the optical power loss at 1xN couplers becomes critical. Figure 2 is an alternative architecture [3]; it splits the DCSWs into lower-order $m \times m$ DCSWs arrayed in parallel. To suppress the routing performance degradation caused by switch splitting, interconnections between WSSs and DCSWs are differentiated with regard to wavebands.

In a previous study [3], we fixed the number of flexible wavebands to B=4, which was shown to be sufficient for the original architecture [2]. Then different DCSW sizes, $m \times m$ ($m = 2,4,6,8$), were set. This resulted in a slight routing performance degradation, especially when small ($m = 2,4$) switches are used. Therefore, this paper optimizes (B, m), and clarifies that routing performance almost equivalent to that of the original architecture can be achieved. Then, we perform numerical experiments based on the design algorithm proposed in [3]. Figure 3 shows the relative fiber number variation normalized by that for conventional OXCs on the Pan-European network [4]. When the average number of paths between each node pair is 20, the largest OXC scale is over 60. The increment in B compensates the reduction in switch size. The shrinkage of DCSW size contributes to not only compact DCSW implementation but also the elimination of amplifiers, as will be shown in the next section.

437

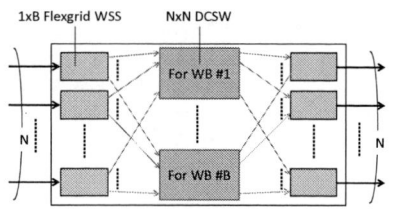

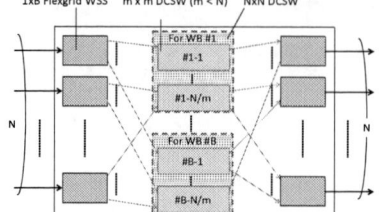

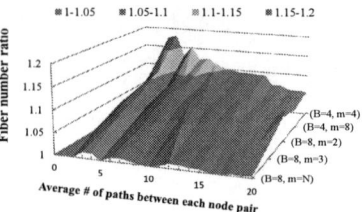

Fig. 1. Flexible waveband routing OXCs [2]　　Fig. 2. Flexible waveband routing OXCs with low-order DCSWs [3]　　Fig. 3. Fiber number ratio normalized by conventional OXCs on pan European network

III. HARDWARE SCALE ANALYSIS FOR FLEXIBLE WB ROUTING OXCs

The hardware scale of key components necessary in an NxN OXC is analyzed in this section. We assume that OXCs have pre/post amplifiers (EDFAs) at all input/output ports and omit consideration of the add/drop part for simplicity. All WSSs are assumed to be 1×9. Suppose that the WSS and the $m\times m$ DCSW have losses of 6.5dB and $\log_{10}(n+2)$ dB, respectively. Let the power budget be 20+dB. Whenever a higher degree WSS is necessary, multiple 1x9 WSSs are cascaded to fulfill the port count requirement. For conventional OXCs, a path can traverse up to three WSSs without using EDFAs, and thus, OXCs with cascaded WSSs at input and output sides need additional EDFAs (see Fig 4(a)). In this paper, EDFAs are located between all cascaded WSSs at the input side of the OXC. On the other hand, for flexible waveband routing OXCs, additional EDFAs are inserted just after WSSs at the input side (see Fig. 4(b)) if the DC type switch has high loss. In this case, 8×8 DCSWs need signal amplification whereas 3×3 DCSWs do not. Indeed, small DCSWs can be compactly integrated with the PLC technology [5, 6].

Based on the above, we analyze the hardware scale for conventional OXCs and proposed OXCs with (B, m) = (8, 3) and (B, m) = (4, 8). Figures 5 and 6 plot the number of WSSs and EDFAs per OXC with respect to OXC size, respectively. For conventional OXCs, 1x9 WSSs need to be cascaded to create a large scale OXC when N > 9. The necessary number of WSSs is given by $\lceil(N-1)/8\rceil$ x 2N. On the other hand, for the proposed flexible waveband routing OXCs, the necessary number of WSSs is always 2N. Thus the proposal holds the increases in WSSs and EDFAs to linear on N and smaller than the conventional one as shown in Fig. 5. Moreover, based on the above observation, the necessary number of EDFAs is shown in Fig. 6. For conventional OXCs, the number of EDFAs is given by $(\lceil(N-1)/8\rceil - 1)$ x N + 2N, where the second term stands for the number of essential pre/post amplifiers. On the other hand, for the proposed OXCs that need additional EDFAs to compensate loss at DCSWs, the number of EDFAs is given by NB + 2N. Thus the flexible waveband routing OXCs can reduce the number of EDFAs when the OXC size is large (e.g. N $\geq$ 42). If an additional EDFA is not necessary due to the use of small DCSWs, then the number of EDFAs will be 2N; only pre/post amplifiers are needed. The reduction is substantial, especially when (B, m) = (8, 3) (i.e. 3×3 DCSWs are used). Thus B=8 and 3×3 DCSW size are adopted for prototype development in this paper.

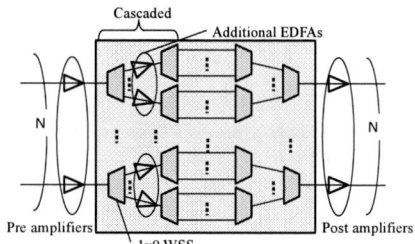

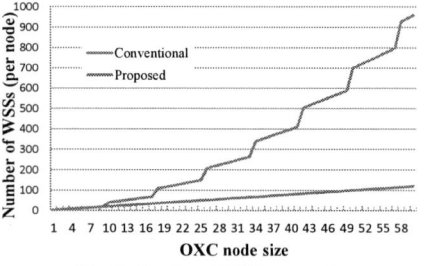

Fig. 4(a). Additional EDFAs for conventional OXCs

Fig. 5. Number of WSSs per node

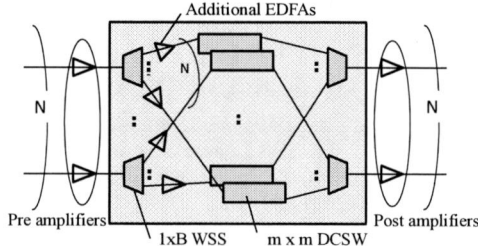

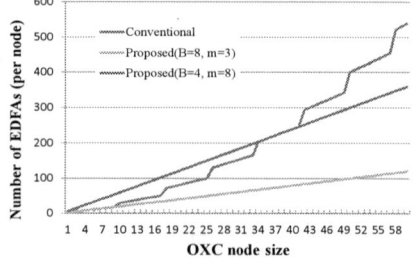

Fig. 4(b). Additional EDFAs for proposed OXCs

Fig. 6. Number of EDFAs per node

IV. PROTOTYPE DEVELOPMENT AND TRANSMISSION EXPERIMENTS

In order to confirm feasibility of our OXC architecture, we developed a flexible waveband routing OXC prototype with (B, m) = (8, 3). The experimental setup is depicted in Fig. 7. At the transmitter side, 80-wavelength 30-GBaud QPSK signals were formed with an IQ modulator (IQM) driven by a 2-channel pulse-pattern generator (PPG). Then, dual-polarization QPSK signals were emulated in a split-delay-combine manner and their optical signal-to-noise ratio (OSNR) was controlled with an amplified-spontaneous-emission (ASE) noise source. The signals thus obtained were incident on the developed node. The signals were then amplified with an erbium-doped fiber amplifier (EDFA), where its output power was set to 20 dBm. After that, the signals traversed a 1x9 WSS, 3x3 DCSW, and 9x1 WSS in this order. The 3x3 DCSW was developed with the planar-lightwave-circuit (PLC) technology and five 3×3 DCSWs were monolithically integrated on a chip with size of $13 \times 58 \, mm^2$ (Fig. 8). The losses of WSS and 3x3 DCSW were around 6.5 dB and 7.0 dB, respectively. In other words, the total loss of our node was around 20 dB. After the signals passed through N nodes, the target signal was dropped with a WSS and delivered to an offline digital coherent receiver. After the demodulation circuit, bit-error ratios (BERs) according to OSNR were counted.

Figure 9 shows the OSNR penalty measured at the forward-error-correction (FEC) threshold (BER=10^{-2}), where performance of the conventional 8x8 route-and-select OXC was used as a baseline. We observe that the OSNR penalty due to insertion of 3x3 DCSW is marginal, around 0.5 dB even after 4 hops. In contrast, our scheme can attain OXC scale of 60 as discussed in section 3.

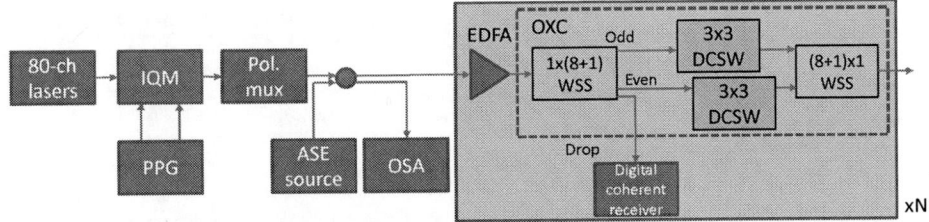

Fig. 7. Experimental setup

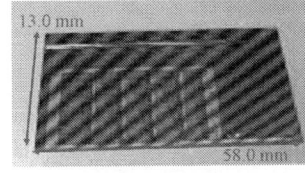

Fig. 8. 5-arrayed 3x3 DCSW PLC chip

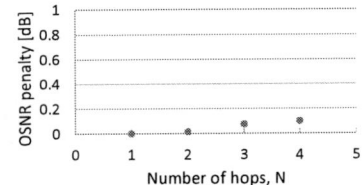

Fig. 9. OSNR penalty versus the number of nodes traversed (BER=10^{-2})

V. CONCLUSIONS

We showed that optimization of the number of wavebands and the switch scale enables compact DCSWs to maximally exploit flexible waveband routing, with only a marginal routing performance offset. Hardware scale analyses of our flexible waveband routing OXCs showed that the necessary numbers of WSSs and EDFAs can be substantially reduced, relative to conventional OXCs. We developed a prototype with using monolithically integrated arrayed small DCSWs created on a PLC chip. Transmission experiments on the prototype confirmed the technical feasibility of the OXCs.

ACKNOWLEDGMENT

This work was partly supported in part by NSF grant CNS-1406971 and NICT.

REFERENCES

[1] M. Jinno et al., "Spectrum-efficient and scalable elastic optical path network: architecture, benefits, and enabling technologies," IEEE Com. Magazine VoL. 47, issue. 11, pp. 66-73, Nov. 2009.

[2] H. Hasegawa et al., "Flexible Waveband Routing Optical Networks," ICC 2015.

[3] T. Ishikawa et al., "A compact OXC Node Architecture That Exploits Dynamic Path Bundling and Routing," submitted to ONDM 2016.

[4] R. Inkret et al., Advanced Infrastructure for Photonic Networks – Extended Final Report of COST266 Action. Zagreb, Croatia: Faculty of Electrical Engineering and Computing, University of Zagreb, 2003.

[5] T. Watanabe et al., "Silica-based PLC transponder aggregators for colorless, directionless, and contentionless ROADM," OFC 2012, OTh3D.1, Mar. 2012.

[6] S. Nakamura et al., "Compact and Low-Loss 8x8 Silicon Photonic Switch Module for Transponder Aggregators in CDC-ROADM Application," OFC 2015, paper M2B.6, Mar. 2015.

Scalable and Dynamic Optical Network Architecture[1]

Vincent W. S. Chan, *Fellow IEEE, OSA*
Joan and Irwin Jacobs Professor, EECS MIT
36-545 77 Mass Ave. Cambridge MA 02139
chan@mit.edu

Abstract: Future optical networks will integrate multi-layer functions, in a fast adapting infrastructure. When the data and control dynamics speed-up by 4 orders as projected, the entire network architecture must be redesigned to be scalable.

1. INTRODUCTION

New applications for cloud computing and data storage and the use of big data require an increase of network data rates of 10^{3-4} and beyond. The new architecture must meet bursty large flows (OFS, optical flow switching) of the future, [1] (Fig. 1). The architecture construct includes all layers of the network as indicated in Fig 2.

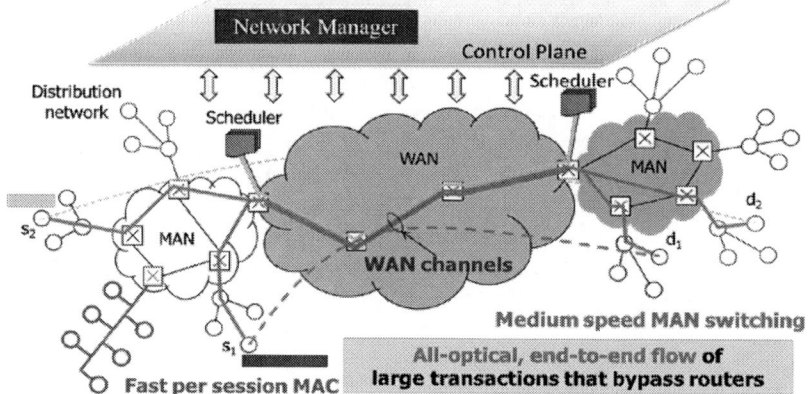

Figure 1. *OFS in context of WAN/MAN/LAN, [1], LAN/MAN version demonstrated in [2].*

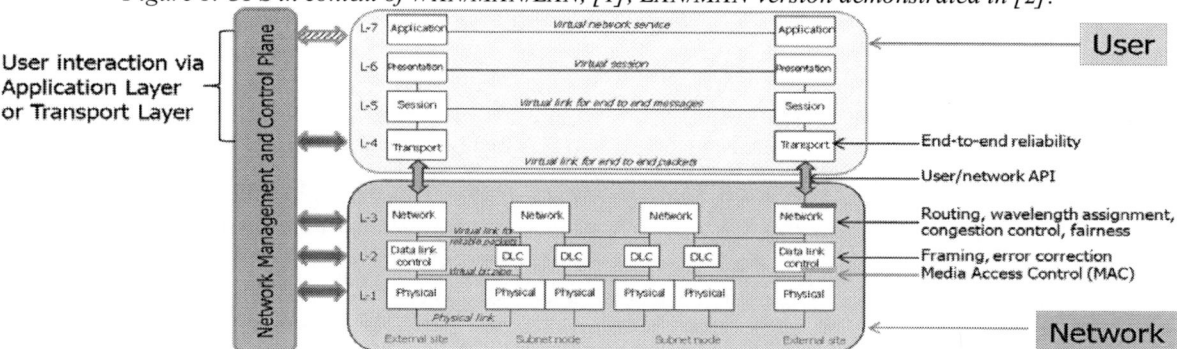

Figure 2. *Multi-layer network architecture and separation of layers and functions between user and network.*

2. ACCESS NETWORK AND MAN PHYSICAL ARCHITECTURE AND MAC.

The largest obstacle towards fast dynamic flow scheduling is the complexity of the network management and control system, (< 100 mS). Fig. 3 shows the control mechanism of OFS for switching elephant flows. In [1], we described the WAN architecture to have quasi-static switching using tunnels between MANs that last for minutes to hours and only changes for traffic trends and load balancing and never respond to per flow traffic inputs. In the LAN per flow media access is required and the candidate is a passive optical network architecture to lower cost by not

---

[1] Content previously presented in APC2015. Supported by NSF, OSD and the Claude Shannon Endowment Fund.

having any active components such that the MAC in the LAN is only scheduling time of transmission and assigning wavelengths, Fig.3. For the MAN the key issue is whether the MAN needs to be switched per flow, which needs a complex (as in fast) network management and control system and especially for switch setup and verification upon computing a schedule for the transaction or should the topology be quasi-static adjusting only to traffic trends. In [11], any time when the required delayed x traffic load > 2, it is cheaper to pick the no-routing architecture, Fig. 5.

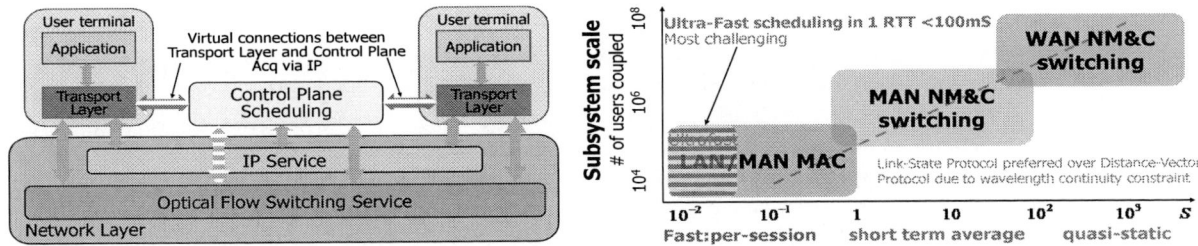

Figure 3. Off-band control of OFS for elephant flows and time scale and nature of control for LAN (fast per session), MAN (changes over short term, ~minutes) and WAN (quasi-static, ~ 10s of minutes – hours), [1].

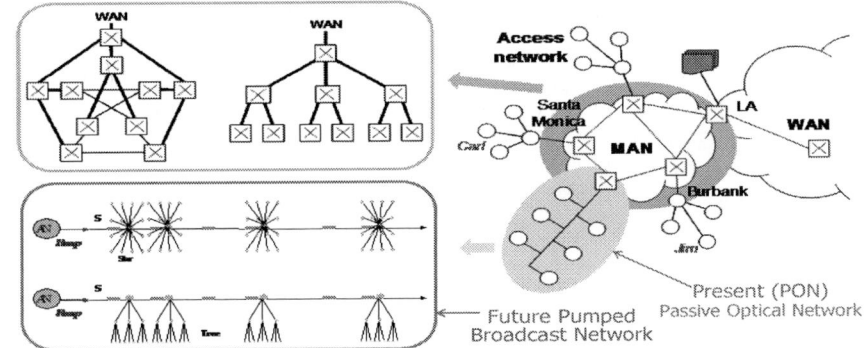

Figure 4. Remotely pumped amplified LAN and Generalised Moore Graph MAN architecture, [1].

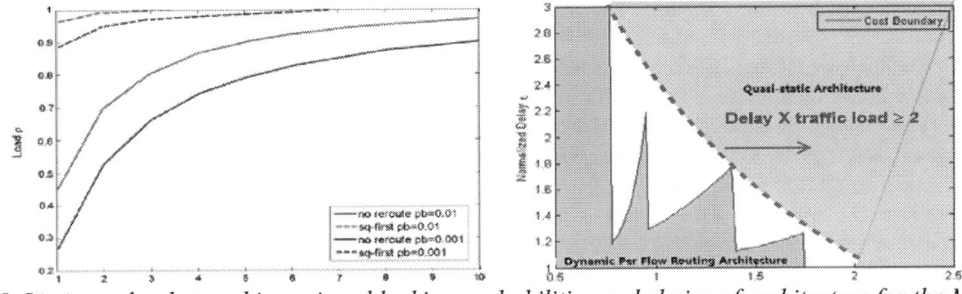

Figure 5. Limits on loads to achieve given blocking probabilities and choice of architecture for the MAN, [11].

3. CONTROL PLANE ARCHITECTURE: FUNCTIONS AND TIME SCALES, [13].

Fig.6 shows the control functions versus time scale in the management plane of this new agile optical network, including link state updates, traffic requests, scheduling, switching controls and session initiation messages, [10]. Two types of physical WAN architectures are shown: general mesh (NP complete $O(\exp[N^2])$ scheduling) and tunneled connections between MANs. The chosen architecture for the WAN is tunneled with centralized scheduling between the two MANs' schedulers.

4. TRANSPORT LAYER PROTOCOL

The tradition Internet Transport Layer protocol TCP will be very inefficient for large flows. Reference [9] describes an approach of encoding large blocks (~100Mb) and the ARQ is a reflow of the block instead of packet by packet retransmission. The efficiency is at least an order better than TCP.

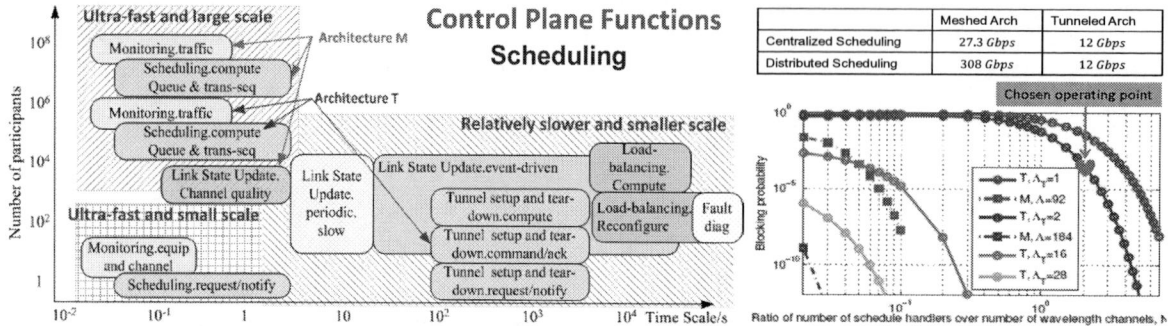

Figure 6. Control plane functions and their time scales, link state protocol traffic for centralized and distributed scheduling and blocking probability performance vs traffic loads between MAN pair, [10].

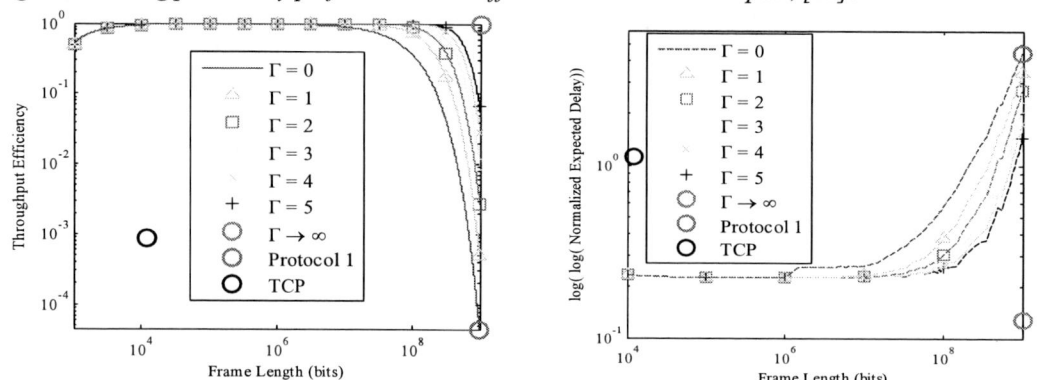

Fig. 7. Expected throughput and delay vs. frame length, Γ= number of burst erasure in frame.

5. CONCLUSIONS

The new agile optical network with growth in capacity of 3-4 orders needs a new architecture at all layers.

REFERENCES

[1] V.W.S. Chan, "Optical Flow Switching Networks," Proceedings of the IEEE , vol.100, no.5, May 2012.

[2] V.W.S. Chan, et.al., "A Precompetitive Consortium on Wideband All-Optical Networks," IEEE Journal of Lightwave Technology (Invited), Vol. 11, Issue 5, pp. 714 - 735, May 1993.

[3] V.W.S.Chan, et.al., "Architectures and Technologies for High-Speed Optical Data Networks," IEEE Journal of Lightwave Technology (Invited), Volume 16, Issue 12, pp. 2146 – 2168, Dec. 1998.

[4] M. Kuznetzov, N.M, Froberg, S.R. Henion, H.G. Rao, J. Korn, K.A. Rauschenbach, E.H. Modiano, V.W.S. Chan, "A Next-Generation Optical Regional Access Network," IEEE Com Magazine (Invited), Jan. 2000.

[5] G. Weichenberg, V. Chan, and M. Medard, ``Design and analysis of optical flow-switched networks," *Optical Communications and Networking, IEEE/OSA Journal of*, vol. 1, no. 3, pp. B81--B97, August 2009.

[6] Ganguly A. R., Guy Weichenberg, V Chan, . "Optical Flow Switching with Time Deadlines for High Performance Applications," *Global Telecommunications Conference, GLOBECOM 2009. IEEE*, Dec. 2009.

[7] Lei Zhang; V. Chan, "Fast Scheduling of Optical Flow Switching," *Global Telecommunications Conference (GLOBECOM 2010), 2010 IEEE* , vol., no., pp.1,6, 6-10 Dec. 2010

[8] H. Huang and V.W.S. Chan, "Transport Layer Protocol for Optical Flow-Switched Networks", IEEE ICC, Budapest, June 2013.

[9] H. Huang, "Transport Layer Protocol Design over Flow-Switched Networks", MIT Master's Thesis 2012.

[10] Zhang Lei, Vincent Chan ," Coupled Performance of Data and Control Planes for Optical Flow Switched Networks," IEEE ICC, June 2014 Sydney Australia.

[11] Anny Zhang, Zhang Lei, Vincent Chan, "Metropolitan and Access Network Architecture Design for Optical Flow Switching," IEEE Globecom 2014.

[12] Joseph Junio, Zhang Lei, Vincent Chan, "Physical Layer Characteristics and Design of Long Haul Fast Turn-on/off and Flow Switched All-Optical Networks," OFC San Francisco March 2014.

[13] V Chan, "Optical Network Architecture for "Elephant" Traffic," ICTON 2014.

3.5-Gbit/s QPSK Signal Transmission in Beam Forming of 60-GHz Integrated Photonic Array-Antenna

Kotoko Furuya*, Takayoshi Hirasawa, Masayuki Oishi, Shigeyuki Akiba, Jiro Hirokawa, and Makoto Ando

Dept. of Electrical and Electronic Eng., Tokyo Institute of Technology, 2-12-1 Ookayama, Meguro-ku, Tokyo 152-8552, Japan
*E-mail: furuya.k.ab@m.titech.ac.jp

Abstract: We demonstrate 3.5-Gbit/s QPSK signal transmission in radio-over-fiber system with beam forming of a 60 GHz-band integrated photonic array-antenna. The relationship between beam forming operation and signal quality is quantitatively confirmed.
Keywords: radio-over-fiber, millimeter-wave, array-antenna, beam forming, photodiode

I. INTRODUCTION

The high-speed mobile devices such as smart phones and data-intensive applications cause the rapid growth of mobile data traffic. In order to accommodate such explosive data traffic, the deployment of more wireless base stations (BSs) than ever is socially demanded. In this situation, however, the electromagnetic-wave interference among plural BSs will increase, which results in the degradation of signal quality. In addition, the installation of many BSs causes the increase of total cost, weight, and power consumption of radio access network. In order to resolve these problems, radio-over-fiber (RoF) transmission technology is a promising solution because it has capabilities of remote beam control and power feeding to antennas from the control site [1][2], which will contribute to significant simplification of the antenna sites. From these perspectives, we study a scheme for beam forming of array-antennas by RoF technology [3]. For further simplification of the current BS architecture, we proposed and fabricated an integrated photonic array-antenna (IPA) where the photodiodes and array-antenna elements were integrated into a single board, and experimentally demonstrated the beam forming operation for the fabricated IPA by using an RF tone [4]. In order to confirm the feasibility of the IPA-based beam forming in a practical use case, the data signal assuming the real mobile traffic should be used for the experimental demonstration.

In this paper, we demonstrate 3.5-Gbit/s QPSK signal transmission utilizing the fabricated IPA and experimentally investigate the change of signal quality in the beam forming operation by RoF.

II. TRANSMISSION EXPERIMENTS UTILIZING INTEGRATED PHOTONIC ARRAY-ANTENNA

A. Integrated Photonic Array-Antenna

Figure 1 shows a schematic overview of the IPA. The IPA consists of a PD-integrated 60 GHz-band patch antenna substrate (Fig. 1(a)) and an optical signal feeding jig (Fig. 1(b)). The antenna substrate consists of 60 GHz-band 4 × 2 patch antenna elements in the front side and uni-travelling carrier photodiode (UTC-PD) chips in the back side. The optical signal feeding jig consists of optical fibers, prisms and lenses. The input optical signals through fibers are condensed into the light receiving areas of each PD chip by the prisms and the lenses. The PDs convert the optical signals to RF signals, and then the RF signals are provided for each patch antenna element.

B. Experimental Setup for 3.5-Gbit/s Signal Transmission with Antenna Beam Forming by RoF

Figure 2 shows an experimental setup for 3.5-Gbit/s QPSK signal transmission with antenna beam forming of the fabricated IPA, where the frequency of RF carrier output from the IPA was set to be 60.48 GHz that corresponded to the

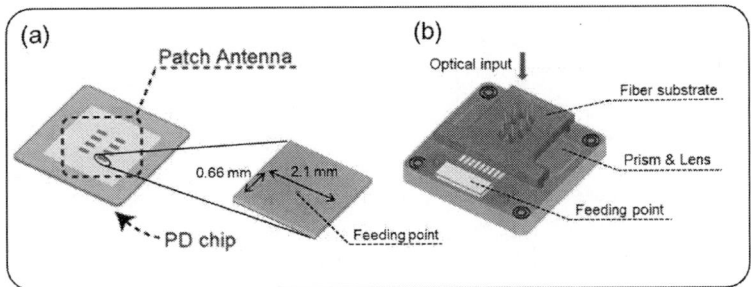

Fig. 1. Schematic overview of the integrated photonic array-antenna;
(a) PD-integrated antenna substrate, and (b) optical signal feeding jig.

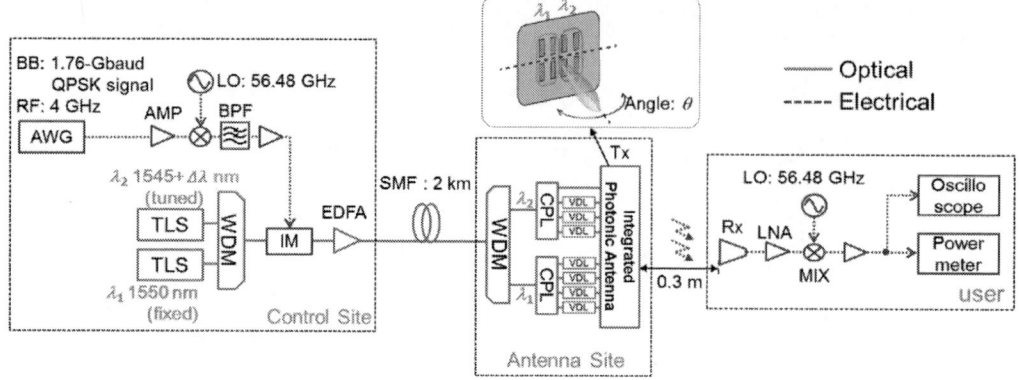

Fig. 2. Experimental setup.

central frequency of the channel 2 in the standardized 60-GHz band. In this experiment, two RoF signals having different optical wavelengths of λ_1 and λ_2 were used as the light sources of RoF signals. When the RF phases of the two RoF signals are initially synchronized, the array-antenna radiates a straight-forward beam toward the observation angle of 0 degree. In contrast, when the wavelength of λ_2 is slightly changed to $\lambda_2 + \Delta\lambda$, the relative RF phase between two RoF signals is changed by the effect of chromatic dispersion (CD) of the single-mode fiber (SMF), and therefore, the beam radiated from the array-antenna can be steered. Thus, the antenna beam can be remotely controlled by just changing the wavelengths of the RoF transmitter at the control site. The wavelength tuning range ($\Delta\lambda$) required for obtaining a relative RF phase shift of π is expressed as:

$$\Delta\lambda = \frac{1}{2LDf_{RF}} \qquad (1)$$

where L is a length of the SMF, D is a chromatic dispersion parameter of the SMF (typically, $D = 16 - 18$ ps/nm/km at 1550 nm), and f_{RF} is an RF carrier frequency.

A 1.76-Gbaud QPSK signal with the intermediate frequency of 4 GHz was generated by an arbitrary waveform generator (AWG), and then converted up to 60.48 GHz by being mixed with 56.48-GHz RF tone signal from a local oscillator (LO). After passing through a band-pass filter (BPF) of which the central frequency and the bandwidth were 60.48 and 1.76 GHz, the 60.48-GHz QPSK signal was input into an intensity modulator (IM). Two optical signals output from the tunable laser sources (TLSs), of which wavelengths were $\lambda_1 = 1550$ and $\lambda_2 = 1545$ nm, were multiplexed by a wavelength division multiplexing (WDM) coupler and amplitude-modulated in the IM by the 60.48-GHz RF signal. After being boosted by an erbium-doped fiber amplifier (EDFA), the two RoF signals were propagated over a 2 km-long SMF and demultiplexed to each wavelength by a WDM coupler in the antenna site. The two RoF signals were then divided into 4 signals by optical couplers (CPLs) and fed to each antenna element of the IPA. In order to compensate an RF phase shift due to the difference of two initial optical wavelengths, seven optical variable delay lines (VDLs) were inserted just before the IPA. The RoF signals were converted to the 60.48-GHz QPSK RF signal by the photodiodes of the IPA and the RF signal was radiated from the array-antenna of the IPA (Tx). Note that this experiment was conducted without supplying any bias voltages for all the PDs. After 0.3 m-long radio propagation, the 60.48-GHz RF signal was received by a receiver horn antenna (Rx), and then converted down to 4 GHz. The 4-GHz RF signal was injected into the oscilloscope and the power meter to evaluate quality of the QPSK signal.

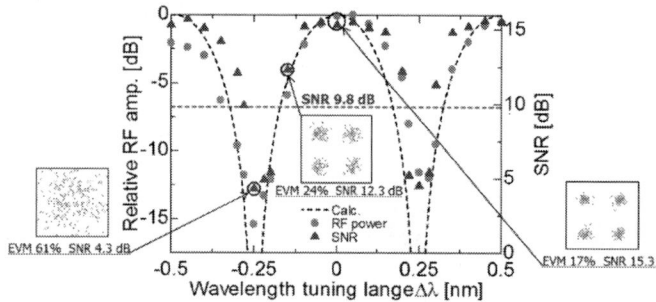

Fig. 3. Relationship between the received RF power and SNR when the IPA beam is controlled.

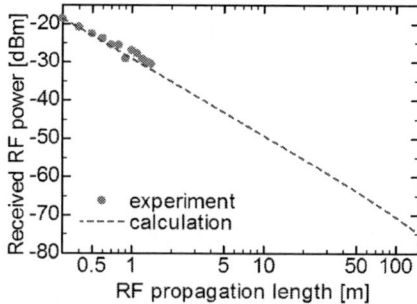

Fig. 4. Change of the received RF powers as a function of the RF propagation distance.

444

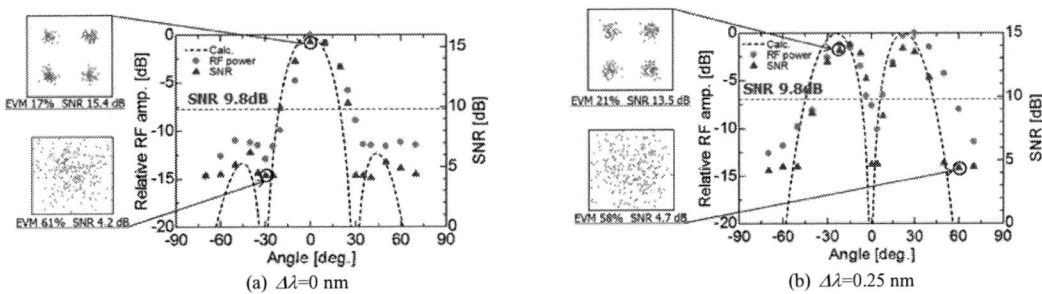

(a) $\Delta\lambda=0$ nm   (b) $\Delta\lambda=0.25$ nm

Fig. 5. RF radiation patterns with corresponding SNRs.

III. RESULTS AND DISCUSSION

Figure 3 shows the relationship between the received RF power and SNR of the QPSK signal at the observation angle θ of 0 degree. The constellations with the corresponding SNR, error vector magnitude (EVM) values are also shown in Fig. 3. The dotted line is the theoretical curve calculated from the array factor, which is described in Ref. [5]. In this measurement, the λ_2 was changed from 1544.5 to 1545.5 nm because the $\Delta\lambda$ for obtaining π RF phase shift was theoretically estimated to be 0.25 nm from Eq. (1). It is confirmed that the change of the measured RF powers coincided with the theoretical curve and the change of the SNR coincided with that of the received RF power. The constellations were also changed according to the wavelength tuning range of λ_2, and we obtained the SNR of more than 9.8 dB that corresponds to the bit error rate of 1.0×10^{-3} required for error-free transmission with a low-density parity-check (LDPC) code. Thus, the signal quality of the 3.5-Gbit/s QPSK signal in the IPA beam forming operation was experimentally confirmed. Figure 4 shows the change of the measured received RF powers for the 3.5-Gbit/s QPSK signal by changing the RF propagation distance between the IPA and the receiver horn antenna from 0.3 to 1.4 m. The dotted line in Fig. 4 is the calculated result of the received RF power as a function of RF propagation distance. In order to extend the propagation distance, the RF propagation loss must be compensated. In this experiment, however, the received RF power was restricted by the performances of the IM and PDs in the IPA. Hence, if we utilize UTC-PDs with higher saturation RF output power and an IM with higher modulation efficiency, the propagation distance will be able to be extended. The details of the quantitative analysis will be explained in the presentation.

Figures 5(a) and (b) show the RF radiation patterns at $\Delta\lambda = 0$ and 0.25 nm, respectively, where the change of relative RF power was plotted against the observation angle. The SNR with the constellations of the QPSK signals were also plotted as references. In the case of $\Delta\lambda = 0$ nm, the RF power at the observation angle of 0 degree was the maximum, while it changed to a null point in the case of $\Delta\lambda = 0.25$ nm. The dotted lines in Figs. 5(a) and (b) are calculated by High Frequency Simulation System (HFSS) that is a three-dimensional electromagnetic field simulator for antenna engineering. The measured RF powers and SNRs almost coincided with the dotted lines except for the measurement points at the low received RF powers, where the received RF signals were too weak to be clearly discriminated from noises.

IV. CONCLUSION

We experimentally demonstrated 3.5-Gbit/s QPSK signal transmission by RoF with beam forming of the 60 GHz-band IPA. Throughout the transmission experiments, we experimentally evaluated the relationship between the beam forming operation and the received signal quality of the QPSK signal. This study will be useful to design practical RoF systems with antenna beam forming functionality.

ACKNOWLEDGMENT

We thank Dr. K. Nishimura and Mr. T. Kobayashi at KDDI R&D Laboratories, Inc. and Mr. S. Maki at Tokyo Institute of Technology for their support in the preparation of measurement equipment. This research is partially supported by the Center of Innovation Program for Japan Science and Technology Agency, JST.

REFERENCES

[1] M. Y. Frankel and R. D. Esman, "True time-delay fiber-optic control of an ultrawideband array transmitter/receiver with multibeam capability," IEEE Trans. Microw. Theory and Techn., vol. 43, no. 9, pp. 2387-2394, Sept. 1995.

[2] M. Tadokoro et al., "Optically-controlled beam forming for 60 GHz-ROF system using dispersion of optical fiber and DFWM," Proc. 2007 IEEE/OSA Optical Fiber Communication conference (OFC), paper OWN2, Anaheim, CA, Mar. 2007.

[3] S. Akiba et al., "Effects of chromatic dispersion on RF power feeding to array antenna through fiber," Proc. the 2012 17th Opto-Electronics and Communications Conference (OECC), pp. 323-324, Busan, Korea, July 2012.

[4] T. Hirasawa et al., "Integrated photonic array-antennas in RoF system for MMW-RF antenna beam steering," Proc. the 2015 Int'l. Topical Meeting on Microwave Photonics (MWP), paper TuP-14, Paphos, Cyprus, Oct. 2015.

[5] H. Jasik and R. C. Johnson, Antenna Engineering Handbook, 3rd ed., McGraw-Hill, 1993.

TuA2-2

OECC/PS2016

Real-Time Transmission of 5-Gbit/s pi/4-shift DQPSK Signal over Millimeter-wave Radio-over-Fiber Links

Abdelmoula Bekkali[1], Takashi Kobayashi[1], Kosuke Nishimura[1], Nobuhiko Shibagaki[2], Kenichi Kashima[3] and Yosuke Sato[3]

[1] KDDI R&D Laboratories Inc., Saitama, Japan
[2] Hitachi Ltd. Tokyo, Japan
[3] Hitachi Kokusai Electric Inc., Tokyo, Japan
E-mail: ab-bekkali@kddilabs.jp

Abstract: *We demonstrate an error-free real-time transmission of π/4-DQPSK modulated Ethernet frames at 5-Gbit/s over a converged Radio-over-Fiber and millimeter-wave links. The obtained results show a stable and reliable transmission of Gigabit Ethernet data.*

Keywords: *RoF, Millimeter-Wave, Real Time Transmission, Gigabit Ethernet, EVM, Packet Loss Ratio.*

I. INTRODUCTION

Explosive growth of the demands for network capacity continues over a couple of decades mainly due to the expansion of mobile traffic. However, current networks are highly susceptible to outages from fiber cuts that can be caused by large scale disasters and thus it would have a significant impact on our social networks. Seamless integration of radio-over-fiber (RoF) and millimeter wave (mmWave) wireless links, referred to as mmWave-RoF links, is a potential candidate for highly resilient and agile communication systems, where high-capacity mm-wave can conveniently be used as fiber backup to bolster the network resiliency while satisfying the requirements of both high capacity and long transmission distance [1]–[3]. The mmWave-RoF systems are expected to have wide range of applications including broadband wireless access, quick disaster recovery and extension of the existing radio service coverage and capacity.

Recently, with the development of coherent optical communication technologies based on the high-performance digital signal processing (DSP), multilevel modulation format such as quadrature-phase-shift-keying (QPSK) and quadrature-amplitude-modulation (QAM), can be effectively generated and transferred by using RoF technologies [4]. Therefore, using a spectral efficient wideband mmWave-RoF systems, multi-Gigabit wireless transmission can be achieved [5]–[10]. In general, QAM and QPSK based systems require linear amplifiers, which are not power efficient, and thus makes them less attractive for mobile applications. For instance, the QPSK waveform does not have a constant envelope, where the 180-degree phase transitions will cause the signal's envelope to go through zero. As a result, employing non-linear amplifier, the QPSK signal will suffer from spectral spreading and interference into adjacent radio channels. To overcome this issue and have optimal usage of the power amplifiers, the π/4-shift differential quaternary phase-shift keying (DQPSK) modulation can be used.

In in this paper, and unlike the existing works which are based on offline DSP, we present a proof-of-concept of a real-time 10-Gbit Ethernet (10GbE) standard data connectivity over a converged RoF and mmWave links. We demonstrate an error-free real-time transmission of π/4-DQPSK modulated Ethernet frames at 5-Gbit/s over 20-km single mode fiber (SMF) and W-band (i.e. 96 GHz) links. In fact, the data rate is currently limited by the sampling rate of the low-cost analogue-to-digital converters (ADCs) and digital-to-analogue converters (DACs), where two parallel channels with a maximum rate of 1.25 GBd per channel can be obtained. The measured error vector magnitude (EVM) and packet loss ratio (PLR) are within the specifications after the transmission of Ethernet frames over 20-km RoF and 96-GHz W-band links. Moreover, the dependence of the signal performance on mmWave link path loss (i.e. attenuation) is experimentally investigated and evaluated. The obtained results demonstrate the potential use of mmWave-RoF links for a stable and reliable high bandwidth GbE data connectivity.

II. REAL TIME TRANSMISSION OVER MMWAVE-ROF SYSTEM EXPERIMENTAL SETUP

In order to investigate the feasibility of real-time 10Gbit Ethernet (10GbE) data connectivity over a converged RoF and mmWave links, an experiment setup was conducted at 5-Gb/s capacity as illustrated in Figure 1. Each π/4-shift DQPSK transmitter (Tx.) and receiver (Rx.) is equipped with a standard 10GbE interface connected to 10GbE traffic generator. Each Tx./Rx. contains a field-programmable gate array (FPGA)-based real-time transmission system, with 5Gbit/s

---

The research results have been achieved by "Agile Deployment Capability of Highly Resilient Optical and Radio Seamless Communication Systems," the commissioned research of National Institute of Information and Communications Technology (NICT).

transmission/reception modules composed of two parallel 1.25GBaud channels (i.e. Ch1, Ch2). They comprise a real-time DSP processor with embedded routines for synchronization, equalization, DC compensation, forward error correction (FEC) etc. The Ch1 in-phase/quadrature (I/Q) (resp. Ch2 I/Q) signal was up-converted to the intermediate frequency (IF) band at 23.5GHz (resp. 26.066GHz) by an IQ-mixer (IQM), and then both signals were coupled and amplified, where its spectrum is shown in inset (a). The IF signal was then used to drive a LiNbO Mach–Zehnder modulator (MZM) with 40-GHz bandwidth. The RF drive power and the optical modulation index (OMI) were adjusted to minimize the third order intermodulation product. The resultant modulated optical signal was then transmitted over 20km-long SMF with a dispersion compensating fiber (DCF) in order to suppress the interference between two sidebands, and then amplified by an erbium-doped fiber amplifier (EDFA) in conjunction with an optical band pass filters (OBPF) for suppressing amplified spontaneous emission (ASE) noise. Insets (b) illustrates the optical spectrum with 0.01-nm resolution bandwidth. The IF band signal, with an center frequency of 24.833GHz, was recovered using a photodiode (PD) as shown in insert (c), and then up-converted to 96GHz using double balanced mixer (DBM), by being mixed with a 71.217GHz tone output generated by 35.6085GHz local oscillator signal (i.e. LO3) and frequency doubler. Here, the 35.6085GHz LO signal is delivered through 20km SMF fiber using a wavelength division multiplexing (WDM), so that the IF/LO frequencies can be controlled remotely. The RF signal was transmitted over 96GHz W-band link with different attenuation level to emulate the link fading and distance separation. The received signal was amplified then down-converted to the IF band using a DBM by mixing the W-band signal with the output signal from frequency doubler of a tone generated and delivered from LO3. Inset (d) illustrates the received IF spectrum for different W-band link attenuation (i.e. 29dB, 33dB, 38dB and 42dB). Finally, the Ch1 and Ch2 IQ baseband signals were recovered using two IQMs by mixing the IF band signal with their corresponding local oscillator signal (i.e. LO4 and LO5). The baseband IQ signal for each channel was then real-time processed by the π/4-shift DQPSK receiver and their quality were evaluated.

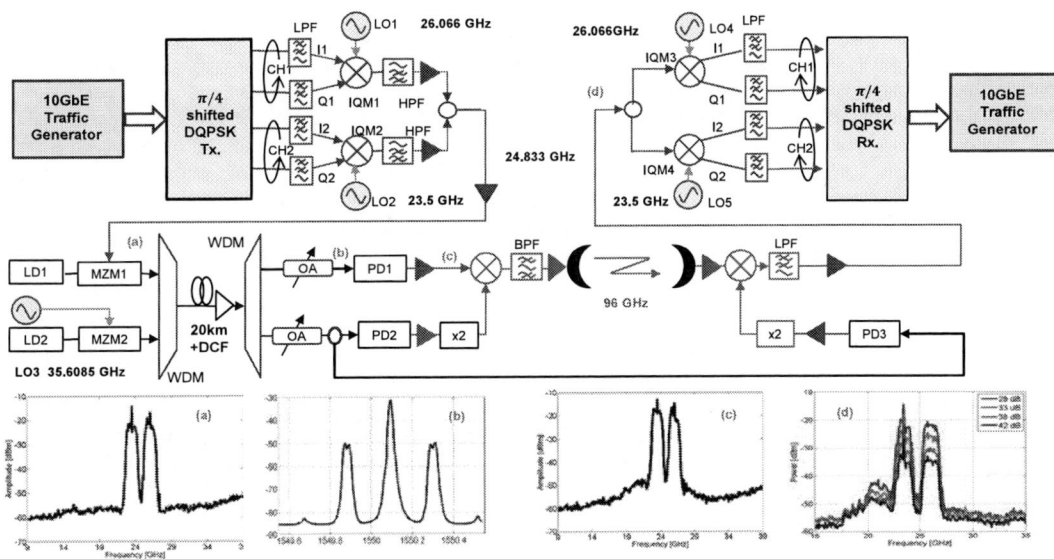

Fig.1: 5Gbit/s Ethernet frames transmission over mmWave-RoF system experimental setup

III. RESULTS AND DISCUSSION

In this experimental demonstration, the Ethernet frames transmission performance was evaluated using EVM before FEC and PLR after FEC. The PLR is defined as the number of lost packets divided by the total number of packets transmitted. Assuming Gaussian noise, the EVM threshold should be about 16% for π/4-shift DQPSK modulated signal, which corresponds to 3.8×10^{-3} FEC limit with 7% overhead. Whereas the PLR should be below 10^{-5} with an RS (255,239) FEC. The electrical back-to-back EVM was measured to be 1.5%.

Figure 2 illustrates the variation of the root-mean-square (rms) EVM with the received RF power per channel, for different values of the W-band link attenuation (i.e. 20dB, 28dB and 35dB). The attenuation values are used to assess the transmission loss and distance impact on the signal quality. As expected, when the Rx. power level is strong, the distortion components due to nonlinearities of the link as well as the receiver, will cause a strong degradation to the EVM as the signal level increases. At the midrange of Rx. signal levels, the link and receiver are behaving linearly, and the EVM has a tendency to reach an optimum level, where its value depends on the W-band link attenuation. For instance, after the RoF link transmission, the EVM of 4% can be achieved when the Rx. power lies between -15.5dBm and -12.5dBm. The optimum value reaches about 5%, 5% and 6.5% after transmission over W-band link with 20dB, 28dB and 35dB attenuation levels, respectively. Furthermore, it can be observed that the received power region that leads to the optimum EVM value for the RoF link defines the optimum value range for all W-band attenuation levels. For instance, the optimum EVM level can be achieved at Rx. power equal to -13dBm, -14dBm and -14.8dBm for 20dB, 28dB and 35dB attenuation levels, respectively. As the Rx. signal level decreases so that the noise is a major contribution, the EVM performance will exhibit a linear degradation with decreasing signal level. At lower signal levels, where noise is the dominant limitation, the EVM drops drastically as the receiver fails to decode the received data. An example of the received constellation map for different EVM values are shown in Figure 3.

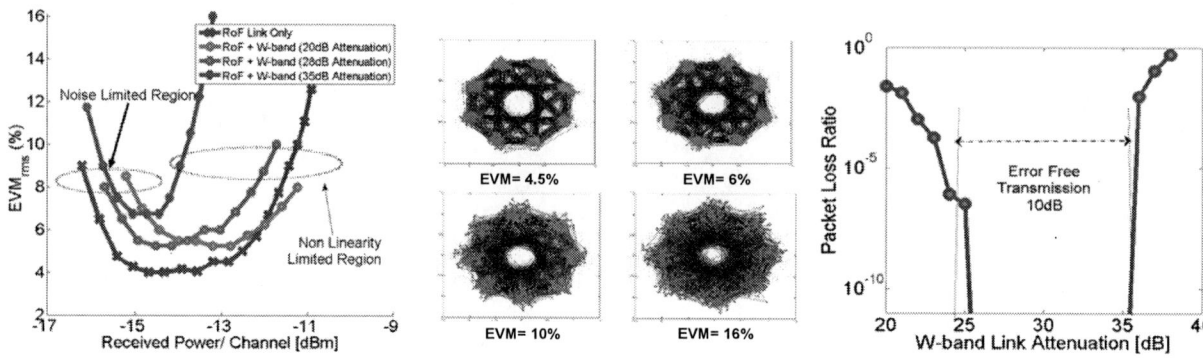

Fig.2: EVM vs Received RF Power for different W-band link attenuation.

Fig.3: Received constellations map and their corresponding EVM values.

Fig.4: Variation of Packet loss ratio with W-band link attenuation.

Figure 4 depicts the variation of the PLR with the W-band link attenuation that lies between 20dB to 40dB, when the Rx. optical power from the RoF link equal to -3dBm. The test Ethernet frames were recorded for 5 minutes using an Ethernet protocol analyzer, and selected to have a random content and size. From the figure, a zero packet loss at 5Gbit/s (net 4.633 Gbit/s) can be achieved over 10dB attenuation margin (i.e. 25dB~35dB). As expected, the packets loss are observed to increase especially for noise limited and non-linearity regions, induced by received power levels. However, the attenuation margin and signal quality can be drastically improved by using automatic gain controllers. It should be noted that, all transmitted frames have been received by the other port together with the flow control frames. Moreover no errors or losses have been recorded, such as frame undersize, oversized and errors in the cyclic redundancy check (CRC) field.

IV. CONCLUSION

In in this paper, we presented a proof-of-concept of a real-time 10GbE data connectivity over a converged RoF and mmWave links. We demonstrated an error-free real-time transmission of Ethernet frames at 5-Gbit/s over 20-km single mode fiber (SMF) and W-band (96 GHz) link. The measured EVM and packet loss ratio are within the specifications after the transmission of Ethernet frames over 20-km RoF and 96-GHz W-band links with different attenuation levels. The obtained results demonstrate the potential use of mmWave-RoF links for a stable and reliable high bandwidth GbE data connectivity.

REFERENCES

[1] J. Beas et al., "Millimeter-Wave Frequency Radio over Fiber Systems: A Survey," IEEE Com. Surveys & Tutorials, Dec. 2012.

[2] K. Kitayama, et al. "High-speed Optical and Millimeter-wave Wireless Link for Disaster Recovery ", in Proc. IEEE Globecom OWC workshop 2015, San Diego. USA.

[3] A. Bekkali, et al. "Seamless Convergence of Radio-over-Fiber and Millimeter- wave Links for highly resilient access Networks". IEEE WCNC 2016, Doha. Qatar.

[4] K. Kitayama, et al. "Digital coherent technology for optical fiber and radio-over-fiber transmission systems" IEEE/OSA J. Lightwave Technol., Vol.32, Issue.20, pp.3411-3420, 2014.

[5] A. Caballero, et al. "25 Gbit/s QPSK Hybrid Fiber-Wireless Transmission in the W-Band (75–110 GHz) With Remote Antenna Unit for In-Building Wireless Networks" IEEE Photonics Journal. 2012, Vol.4, No.3, p.691.

[6] A. Kanno, et al. "40 Gb/s W-band (75–110 GHz) 16-QAM radio-over-fiber signal generation and its wireless transmission," Opt. Exp. vol.19, no.26, pp. B56–B63, Dec. 2011.

[7] Y. Xu, et al. "Demonstration of 60 Gb/s W-Band Optical mm-wave Signal Full-Duplex Transmission Over Fiber-Wireless-Fiber Network" IEEE Com. Letters. 2014, Vol.18, No.12, p.2105.

[8] X. Pang, et. al. "100 Gbit/s hybrid optical fiber-wireless link in the W-band (75–110 GHz)", Opt. Exp. vol.19, no.25, pp. 24994–24949, Dec. 2011.

[9] S. Koenig et al. "100 Gbit/s Wireless Link with mm-Wave Photonics," in OFC/NFOEC 2013, paper PDP5B.4, 2013

[10] A. Bekkali, et al. "First Demonstration of Seamless Optical and Radio Transmission with Plural W-band Wireless Sections" in proc. of OFC, W2A.56, 2015.

TuA2-3

Mobile Fronthaul Optical Link for LTE-A System Using Directly-Modulated 1.5-µm VCSEL

Byung Gon Kim, S. H. Bae, Hoon Kim, and Y. C. Chung
School of Electrical Engineering, KAIST, 291 Daehak-ro, Yuseong-gu, Daejeon, Korea 34141
E-mail: ychung@kaist.ac.kr

Abstract: We demonstrate the transmission of LTE-A signals over 20 km of SSMF by using an uncooled 1.5-µm VCSEL. We achieve EVMs of 3.2% and 6.3% for twelve and twenty-four 20-MHz component carriers of LTE-A signals, respectively.

Keywords: mobile fronthaul, CRAN, radio-over-fiber, LTE-A, VCSEL, direct modulation;

I. Introduction

Recently, cloud radio access network (CRAN) has emerged as the most promising architectural alternative to the conventional distributed RAN for mobile communication systems. In this architecture, all the baseband units (BBUs) are placed together at a centralized location while the remote radio heads (RRHs) remain at their corresponding cell sites. Thus, by using the CRAN architecture, we can not only improve the resource utilization through the statistical multiplexing and BBU pooling, but also save the operation, maintenance, and upgrade costs of BBUs. The CRAN is also helpful to enhance the throughput of wireless systems and reduce the handover delays since it handles the digital signal processing (DSP) for the inter-cell interference and handover within the BBU pool [1]. However, the CRAN architecture poses a heavy burden to the fronthaul network connecting BBUs and RRHs. The fronthaul networks are typically implemented by using the Common Radio Public Interface (CPRI) or Open Base Station Architecture Initiative (OBSAI) protocol. The major technical challenges in such CPRI- or OBSAI-based fronthaul networks are the extremely high capacity (16~50 times higher data rate than the value required in the backhaul network) and the stringent latency requirements [1-3]. For example, the data rate of a CPRI-based fronthaul optical link can be as high as 29.4912 Gb/s, if a Long-Term Evolution-Advanced (LTE-A) system utilizes four component carriers (CCs) supporting an aggregated bandwidth of 80 MHz, 2×2 multiple-input multiple-output (MIMO) antennas, and 3 sectors. Also, the sub-frame processing delay in such a fronthaul network should be kept to be <1 ms, regardless of the propagation delay of the fiber [1]. Recently, a mobile fronthaul network based on the radio-over-fiber (RoF) technology has been proposed and demonstrated to relieve these data-rate and latency requirements [4-6]. In this network, the wireless signals destined for the same cell site are multiplexed by using the subcarrier multiplexing (SCM) technique and transported over an optical fiber without any format conversion at the BBU and RRH sites. Thus, by using this RoF-based fronthaul optical link, we can avoid the use of the large overhead required for the digitization of wireless signals and the extra latency incurred by the format conversions from the wireless signal to the CPRI-formatted digital signal and vice versa. The RoF technology can also simplify the fronthaul optical link as there is no need to utilize the analog-to-digital converters (ADCs) and digital-to-analog converters (DACs). This is an important benefit for the next-generation wireless communication systems since they are expected to utilize a large number of base stations due to the reduced cell size.

In this paper, we experimentally evaluate the possibility of utilizing low-cost uncooled 1.5-µm vertical-cavity surface emitting lasers (VCSELs) for the cost-effective implementation of the RoF-based mobile fronthaul optical link. We directly modulate such a VCSEL by using the SCM subcarrier frequencies of the LTE-A signals. The results show that we can achieve the error-vector magnitudes (EVMs) of 3.2% and 6.3% even after the transmission over 20-km long standard single-mode fiber (SSMF) for twelve and twenty-four 20-MHz CCs of LTE-A signals, respectively.

II. Experiment and Results

Fig. 1(a) shows the experimental setup. We calculated the waveforms of 20-MHz LTE signals compliant with the 3GPP specifications through off-line DSP, and ported them to an arbitrary waveform generator (AWG) operating at 12 Gsample/s. In the DSP, a random binary sequence was first mapped to the 64 quadrature amplitude modulation (64-QAM) signal, which is the highest modulation level specified in 3GPP Release 8 [7], and then converted into 1000 parallel data. For the generation of an orthogonal frequency-division multiplexing (OFDM)-based LTE signal, the parallelized 64-QAM signals together with 200 pilot signals were sent to the inverse fast Fourier transform (FFT) block. The size of the inverse FFT was 2048. Among the OFDM subcarriers, the unused 848 subcarriers were zero-padded. After the parallel-to-serial conversion, we added a 4.69-µs cyclic prefix to the signal. In order to transmit multiple CCs for LTE-A signals to a cell site, these CCs having different carrier frequencies were multiplexed by using the SCM technique. We investigated the performance of two different spectral configurations of the SCM subcarriers placed within multi-octave and one-octave frequency bands. In the case of multi-octave configuration, the SCM subcarriers were positioned from

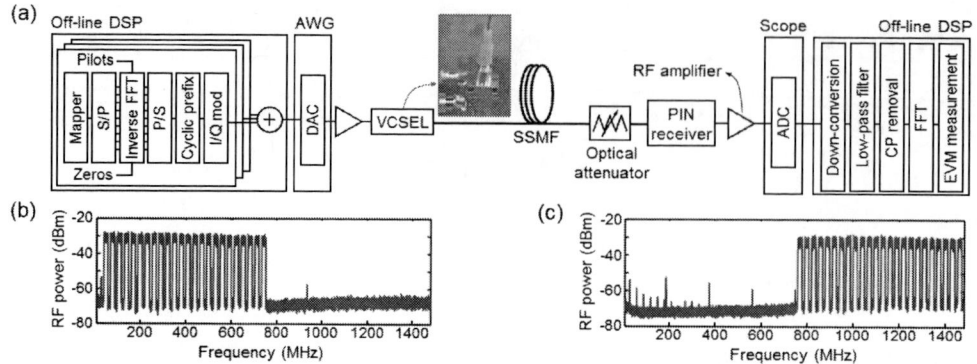

Fig. 1. (a) Experimental setup. (b) The RF spectrum of 24 SCM subcarriers arranged in a multi-octave frequency band. (c) The RF spectrum of 24 SCM subcarriers arranged in one-octave frequency band.

50 to $(20+30i)$ MHz, where i was the SCM subcarrier index. Fig. 1(b) shows the RF spectrum of 24 SCM subcarriers arranged in the multi-octave frequency band. On the other hand, the SCM subcarriers in one-octave frequency band were positioned from $(50+30N)$ to $\{20+30(N+i)\}$ MHz, where N was the total number of SCM subcarriers (i.e., the total number of CCs). Fig. 1(c) shows the measured RF spectrum of the 24 SCM subcarriers in an octave frequency band. In this spectral configuration, the even-order inter-modulation distortions (IMDs) would fall outside the signal band and thus did not affect the performance of the signals. Some spurious components observed below 600 MHz were generated by the AWG. However, these components did not affect the performance of the signals significantly. The subcarrier multiplexed LTE-A signals from the AWG (vertical resolution=8 bits) were amplified and directly fed to the uncooled, transistor-outline (TO)-can packaged VCSEL shown in the inset of Fig. 1(a). The output power and operating wavelength of this VCSEL were measured to be -5 dBm and 1546 nm, respectively, when it was biased at 6 mA. The 3-dB modulation bandwidth of this VCSEL was 7 GHz. The optical signal was sent over SSMF and detected by a PIN receiver. We set the link loss to be 10 dB, considering the fiber loss and the insertion losses of the coarse wavelength-division multiplexers (CWDMs) which would be located at the BBU and RRH sites for the duplex transmission and future capacity upgrade. Thus, the received signal power was fixed to be -15 dBm regardless of the transmission distance. The detected signal was digitized by using a digital sampling scope running at 20 Gsample/s and then processed off-line. After the down-conversion and low-pass filtering of the signals, we demodulated the signals by using FFT and measured their EVMs.

We first evaluated the linearity of the VCSEL. Fig. 2(a) shows the RF output power of the VCSEL link obtained by the two-tone analysis as a function of the RF input power. We fed two sinusoidal waves at 403 and 415 MHz to the VCSEL and measured the output powers of the signals as well as the 2nd- and 3rd-order IMDs. The results showed that the 2nd- and 3rd-order input intercept points were 10.3 and -4.8 dBm, respectively. The 2nd-order IMDs were measured to be larger than the 3rd-order IMDs when the signal-to-IMD-ratio was larger than 30 dB. However, these 2nd-order IMDs would not deteriorate the performance of the LTE-A signals when the SCM subcarriers were placed within an octave of bandwidth. Fig. 2(b) shows the EVM of the 24 CCs of LTE-A signals measured in the back-to-back condition as a function of the SCM subcarrier index, i. The RF input power of the LTE-A signals were -24 dBm. The EVM performance was degraded as the SCM subcarrier index increased and the multi-octave configuration slightly outperforms the one-octave configuration. This was mainly because high-frequency signals suffered more from the jitter of the ADC (the oscilloscope, in our case) than low-frequency signals.

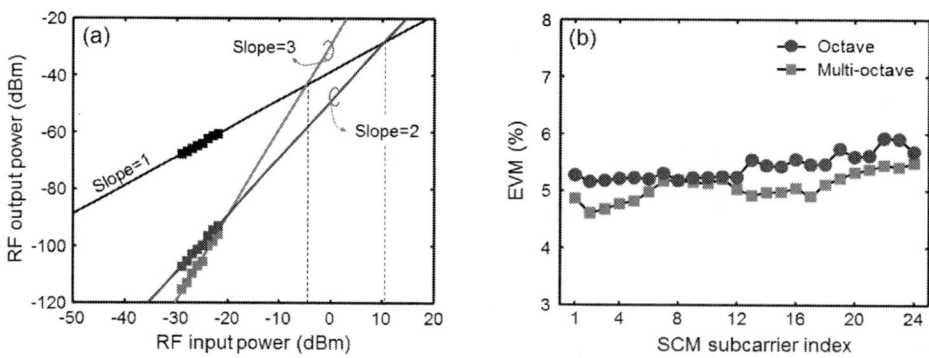

Fig. 2. (a) 2nd- and 3rd-order IMD characteristics of the VCSEL. (b) Measured EVM as a function of SCM subcarrier index.

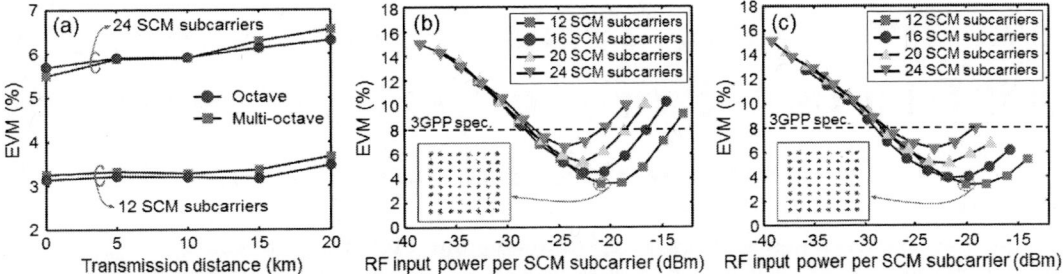

Fig. 3. (a) Measured EVM performance as a function of the transmission distance over SSMF. (b) EVM performance of the LTE-A signals after 20-km long SSMF transmission for the multi-octave configuration. (a) EVM performance of the LTE-A signals after 20-km long SSMF transmission for one-octave configuration.

Fig. 3(a) shows the EVM performance measured as a function of the transmission distance when the total RF input power applied to the VCSEL was -21 dBm. We plotted only the EVM performances of the highest-frequency CC since it exhibited the worst performance among the CCs. Despite the frequency chirp of the directly-modulated VCSEL, the 10- and 20-dB spectral widths of the optical signal generated by this VCSEL were measured to be 7.2 and 11.5 GHz, respectively. Thus, there was no significant degradation in the EVM performance caused by fiber's chromatic dispersion. Fig. 3(b) and (c) show the measured EVMs as a function of the RF input power per SCM subcarrier after the 20-km long SSMF transmission for the multi- and one-octave configurations, respectively. The results showed that, when the RF input power was low, the EVM performance was determined by the signal-to-noise ratio at the receiver. We achieved the best EVM performance when the RF input power was in the range of -24~-19 dBm (where the clipping and nonlinear distortions started to degrade the EVM performance of the LTE-A signals). For example, we achieved an EVM of 3.5% at the RF input power of -21 dBm when 12 SCM subcarriers were utilized in the multi-octave configuration. The constellations depicted in the insets of Fig. 3(b) and (c) also showed that the 64-QAM signals were recovered with high qualities. As the number of SCM subcarriers increased, the onset of the clipping and nonlinear distortions were observed at a lower RF input power per SCM subcarrier since the total RF power driving the VCSEL increased with the number of SCM subcarriers. Fig. 3(c) shows that the one-octave configuration is less susceptible to the clipping and nonlinear distortions. In particular, by comparing Fig. 3(c) with 3(b), it is evident that we can achieve slightly lower EVM values by using one-octave configuration when the input power is high. For example, when the RF input power was -18.5 dBm, the EVM was measured to be 7.8% for the 20 SCM subcarriers arranged in the multi-octave configuration, while it was only 5.8% for the signals placed in one-octave frequency band. Thus, although the CCs of the LTE-A signals should be placed at the higher SCM subcarrier frequencies in one-octave configuration, we could achieve a slightly better performance by using this configuration instead of the multi-octave configuration due to the reduced impacts of the even-order IMDs. Nevertheless, we could satisfy the 8% EVM requirement specified in 3GPP for 64-QAM signals and transmit 24 CCs for LTE-A signals over 20-km-long SSMF by using both configurations [8].

III. Conclusions

We have experimentally demonstrated the transmission of LTE-A signals generated by using a directly-modulated uncooled TO-can packaged 1.55-μm VCSEL. We achieved the EVM less than 6.3% after the 20-km long SSMF transmission for twenty-four 20-MHz component carriers of LTE-A signals, which was equivalent to 29.4912 Gb/s of data rate in CPRI-based mobile fronthaul optical links.

Acknowledgement

This work is supported by the IT R&D programs of MSIP (Ministry of Science, ICT and Future Planning), Korea [15ZI1300, Development of compact radio & dense digital base station technologies based on RoF technology].

References

[1] A. Checko et al., "Cloud RAN for mobile networks-a technology overview," IEEE Comm. Surv. Tut., vol. 17, pp. 405-426, 2015.
[2] K. Tanaka et al., "Next-Generation Optical Access Networks for C-RAN", OFC, paper Tu2E.1, 2015.
[3] S. H. Bae et al., "25-Gb/s TDM optical link using EMLs for mobile fronthaul network of LTE-A system," IEEE Photon. Technol. Lett., vol. 27, pp. 1825-1828, 2015.
[4] S. H. Cho et al., "Experimental demonstration of next generation cost-effective mobile fronthaul with IFoF technique," OFC, paper M2J.5, 2015.
[5] M. Zhu et al., "High-capacity mobile fronthaul supporting LTE-Advanced carrier aggregation 8×8 MIMO," OFC, paper M2J.3, 2015.
[6] F. Effenberger et al., "Power-efficient method for IM-DD optical transmission of multiple OFDM signals," Opt. Express, vol. 23, pp. 13571-13579, 2015.
[7] 3GPP TS 36.211 version 8.9.0 Release 8, 2010.
[8] 3GPP TS 36.104 version 11.9.0 Release 11, 2014.

TuA2-4

OECC/PS2016

Power-Over-Fiber Transmission Using 1.3-μm Dual-Channel Radio-Over-Fiber Signals in a Double-Clad Fiber

Akira Yoneyama, Yamato Minamoto, and Motoharu Matsuura

Department of Communication Engineering and Informatics, The University Electro-Communications
1-5-1 Chofugaoka, Chofu, Tokyo 182-8585, Japan.
akira.yoneyama@uec.ac.jp

Abstract: We successfully demonstrated 30-W power-over-fiber with 1.3-μm dual-channel radio-over-fiber signals using a 300-m double-clad fiber. The results showed that the signals had negligible error-vector magnitude penalties to the back-to-back signal.

Keywords: Power-over-fiber (PWoF), Radio-over-Fiber(RoF), Double-clad fibers (DCFs)

I. INTRODUCTION

Radio-over-fiber (RoF) is one of the attractive technologies for long distance transmissions of radio frequency (RF) signals with low loss and wide bandwidth. In wireless communication systems such as mobile communications, the demands on the RoF technologies will be rapidly increased in near future [1]. In RoF transmission systems, electrical power supply systems are required to drive each antenna base station (BS). The electrical power required for the BS is generally supplied from external batteries or power lines. However, as the number of BSs is increased, they increase the capital and operating expenditures.

To solve this problem, power-over-fiber (PWoF) for RoF systems using single-mode fibers (SMFs) [2] and multi-mode fibers (MMFs) [3] have been reported. These systems enable us to simultaneously transmit a RoF signal and feed light using a single optical fiber. However, it is difficult to transmit high-power feed light through SMFs because of its small core diameter. On the other hand, MMFs suffer from the limitation of the bandwidth due to the modal dispersion.

Recently, we have presented the PWoF system using a double-clad fiber (DCF). The core profile of DCFs consisted of a single-mode core and large inner-cladding. This scheme enables us to transmit RF signals with high bandwidth and high power feed light [4], [5]. Indeed, we have successfully demonstrated 60-W PWoF feed with an RoF signal using a DCF [6], [7]. However, the wavelength of the RoF signal (1550 nm) was far from the wavelength of the feed light (808 nm), and no wavelength spacing between RF signal and feed light dependence on the transmission performance were observed. Moreover, no multichannel operation with narrower wavelength spacing have been reported so far.

In this paper, we have demonstrated PWoF transmission using 1.3-μm dual-channel RoF signals in a 300-m DCF, for the first time. To show the feasibility of this scheme, we have evaluated the transmission performances of the RoF signals in terms of error-vector magnitude (EVM) measurement. As a result, we have successfully achieved the negligible power penalties with and without 30-W PWoF feed in the DCF in spite of the narrower wavelength spacing between the RoF signals and feed light than that of the previous works.

II. EXPERIMENTAL SETUP

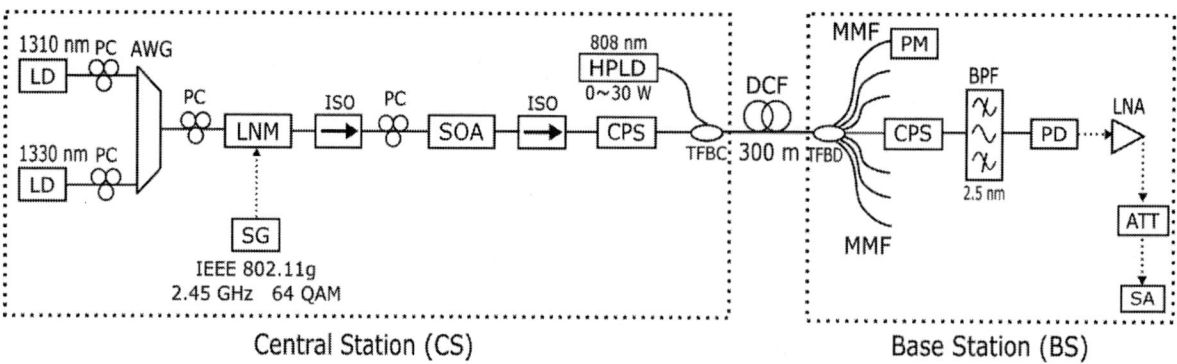

Fig:1. Experimental setup for PWoF transmission using 1.3-μm dual-channel RoF signals in a 300-m DCF. LD: Laser-diode, PC: Polarization controller, AWG: Arrayed-waveguide grating, LNM: LiNbO₃ intensity modulator, SG: Signal generator, ISO: Isolator, SOA: Semiconductor optical amplifier, CPS: Cladding power stripper, HPLD: High-power laser-diode, TFBC: Tapered fiber bundle combiner, DCF: Double-clad fiber, TFBD: Tapered fiber bundle divider, PM: Power meter, BPF: Band pass filter, PD: Photo-diode, LNA: Low noise amplifier, ATT: Attenuator, SA: Signal analyzer.

Fig. 1 shows the experimental setup for the PWoF transmission using 1.3-μm dual-channel RoF signals in a 300-m DCF. Two continuous-wave (CW) signals were generated from two laser-diodes (LDs) with wavelengths of 1310 nm and 1330 nm, respectively. These signals were combined using an array-waveguide grating (AWG) multiplexer. The combined signal was modulated using a $LiNbO_3$ intensity modulator (LNM) with an electrical data generated by a signal generator (SG). The data was based on IEEE 802.11g for conventional wireless local area network (WLAN) with orthogonal frequency division multiplexing (OFDM), 64 level quadrature amplitude modulation (64-QAM). The carrier frequency and the bit rate of the signal were 2.45 GHz and 54 Mbit/s, respectively. The isolators (ISOs) located at the input and output of the 1.3-μm semiconductor optical amplifier (SOA) were used for eliminating the laser oscillation. After passing through the first ISO, the data (RoF) signal was amplified by the SOA. A cladding power stripper (CPS) was used to remove residual reflected feed light in the inner clad. As feed light, a high-power laser-diode (HPLD) with a wavelength of 808 nm was used. The maximum output power was 30-W. The RoF signal and feed light were combined using a tapered fiber bundle combiner (TFBC) and transmitted into the 300 m DCF. After DCF transmission, they were divided to the RoF signal and feed light using a tapered fiber bundle divider (TFBD). The feed light power was almost equally divided to six multimode fiber (MMF) outputs with a core diameter of 105-μm, while the RoF signal was injected into the CPS and converted to the electrical signal using a photo-diode (PD). These configuration was almost same as our previous experiment. The details were explained in Ref. [6]. The delivered feed light power was measured using the power meters (PMs) at the outputs of the MMFs. The role of the band pass filter (BPF) at the output of the CPS was to select the RoF signal wavelength. To evaluate the transmission performance of one of the selected RoF signal, the EVM of the signal was measured by a signal analyzer (SA). A low noise amplifier (LNA) was used to compensate the low optical-to-electrical (O/E) conversion efficiency of the PD, while a variable electrical attenuator (ATT) was used to adjust the injected electrical power into the SA for the EVM measurement.

III. RESULTS AND DISCUSSION

Table 1 Delivered feed light power and power transmission efficiency as a function of output power of the HPLD.

Output power of HPLD (W)	Delivered feed light power (W)	Power transmission efficiency (%)
10.0	4.65	46.5
20.0	9.25	46.3
30.0	13.9	46.3

We measured the delivered feed light power and power transmission efficiency of the RoF link as a function of the output power of the HPLD as shown in Table 1. Here, the power transmission efficiency was defined as the power ration between the output power of the HPLD and delivered feed light power. The power transmission efficiency was almost constant in spite of the HPLD output power. When the HPLD output power was set to 30-W, the delivered feed light power of 13.9 W could be obtained. The main reason for the transmission loss was due to the insertion loss of the combiner, DCF link, and divider. In particular, the loss of the DCF link was closely related to the wavelength of the feed light. We think that it will be effective to decrease the DCF transmission loss if we use a feed light source with a wavelength of at around 1.55-μm.

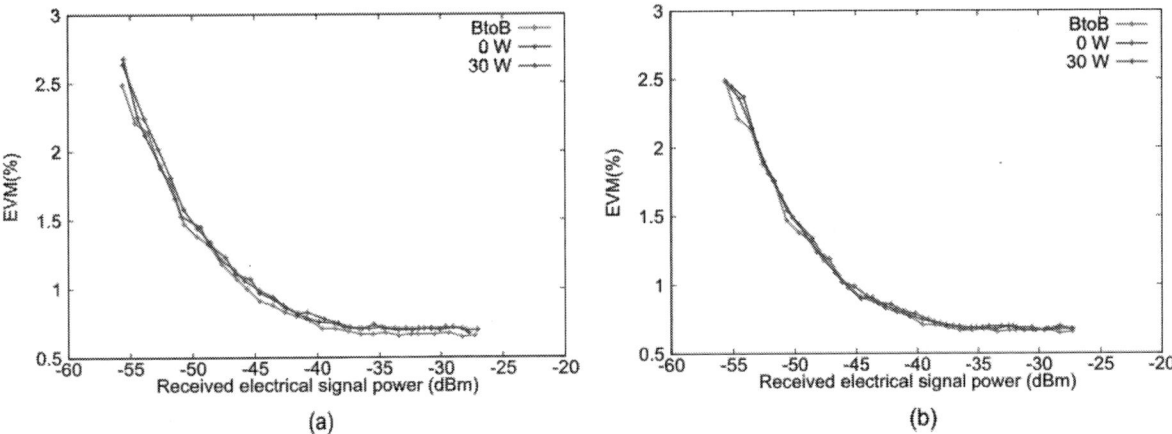

Fig.2 EVM characteristics of the back-to-back (BtoB) and transmitted signals at the wavelengths of (a) 1310 nm and (b) 1330 nm.

To compare the transmission performances with and without the high-power feed light, we measured the EVM characteristic of the back-to-back (BtoB) and dual-channel transmitted signals at the wavelengths of (a) 1310 nm and (b) 1330 nm as a function of the received electrical signal power. The results are shown in Fig. 2. As the received electrical signal power was increased, the EVM value became smaller. When the received electrical signal power was more than −35 dBm, the EVM value became almost constant at around 0.7%. Moreover, the EVM variations among the BtoB and transmitted signals were negligible for all the received electrical signal powers. This means that the high-power feed light does not give any influence on the transmission performance of the DCF link.

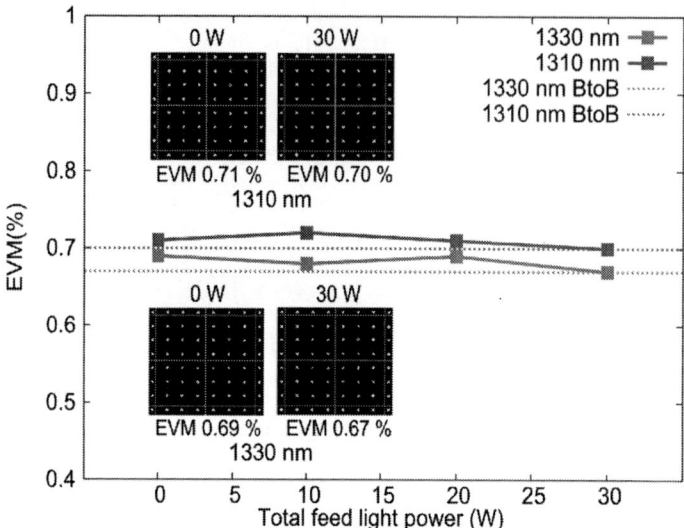

Fig.3. EVM characteristics of the dual-transmitted signals as a function of the total feed light power. Insets show the constellations of the transmitted signals at 1310 nm and 1330 nm. Dashed lines show the EVM values of the BtoB signals at 1310 nm and 1330 nm.

In order to evaluate more detailed transmission performances, we measured the EVM characteristics of the transmitted signals as a function of the total feed light power. Fig. 3 shows the obtained results when the received electrical signal power was set to −30 dBm. Even if the total feed light power was increased, the EVM values of the transmitted signals at the wavelengths of 1310 nm and 1330 nm were almost constant. In each signal wavelength, the EVM penalties to the BtoB signal (dashed lines) were less than 0.05%. This indicates that high transmission performances of dual-transmitted signals could be achieved under 30 W optical power feeding. The insets show the constellations of the 130 nm and 1330 nm transmitted signals without (0 W) and with the optical power feed of 30 W. In all cases, the constellations and its EVM values did not have significant difference. In this experiment, the wavelength spacing among the data signals and feed light was narrower than that of the previous work [6]. However, no degradation of the ducal-channel transmission performance was observed. These results show that the presented PWoF systems will be useful for multi-channel RoF transmission systems in a wide wavelength region.

IV. CONCLUSIONS

We demonstrated the PWoF transmission using 1.3-μm dual-channel RoF signals in the 300-m DCF. High transmission performances of the transmitted RoF signals could be successfully achieved under 30 W optical power feeding. The obtained results indicated that the PWoF systems using DCFs have high potentials for future optical and wireless integrated networks.

REFERENCES

[1] D. Wake, M. Webster, G. Wimpenny, K. Beacham, and L. Crawford, "Radio over fiber for mobile communications," *Proc. Microwave Photonics (MWP 2004)*, TA-1, 2004.
[2] T. Miki, K. Kawano, N. Nakajima, N. Kishi, M. Miyamoto, and T. Aoki, "Novel radio on fiber access eliminating electric power supply at base station," *Proc. Opto-Electronics and Communications Conference (OECC 2013)*, 16D3-4, 2004.
[3] D. Wake, A. Nkansah, N. J. Gomes, C. Lethien, C. Sion, and J.-P. Vilcot, "Optically powered remote units for radio-over-fiber systems," *J. Lightw. Technol.*, vol. 26, no. 15, pp. 2484-2491, 2008.
[4] J. Sato, M. Matsuura, "Radio-over-fiber transmission with optical power supply using a double-clad fiber," *Proc. Opto-Electronics and Communications Conference (OECC 2013)*, TuPO-8, 2013.
[5] M. Matsuura and J. Sato, "Bidirectional radio-over-fiber systems using double-clad fibers for optically powered remote antenna systems," *IEEE Photon. J.*, vol. 7, no. 1, pp. 1-9, 2015.
[6] M. Matsuura, H. Furugori, and J. Sato, "60 W power-over-fiber feed using double-clad fibers for radio-over-fiber systems with optically powered remote antenna units," *OSA Opt. Lett.*, vol. 40, no. 23, pp. 5598-5601, 2015.
[7] Y. Minamoto and M. Matsuura, "Optically controlled beam steering system with 60-W power-over-fiber feed for remote antenna units," *Proc. Optical fiber communication conference and exposition (OFC 2016)*, W3K.2, 2016.

Internet of Things with Optical Connectivity, Networking, and Beyond

Philip N. Ji and Ting Wang

NEC Laboratories America, 4 Independence Way, Princeton, NJ 08540, USA

pji@nec-labs.com

Abstract: Optical technologies play an important role in the Internet of Things networks, including providing high bandwidth connectivity, intelligent networking, and high sensitivity multi-phenomena sensing and imaging. Related optical technologies and IoT applications are discussed.
Keywords: Optical technologies, fiber optics, Internet of Things, networking, sensing

I. INTRODUCTION

The Internet of Things (IoT) is the network of interrelated physical objects (so called "things"), such as electronic devices, home appliances, vehicles, buildings, and so on, that can collect and exchange data with one another through embedded electronics, software, sensors, over the Internet. It extends Internet connectivity beyond traditional devices such as desktop and laptop computers, smartphones and tablets to a diverse range of devices and everyday things that utilize embedded technologies to communicate and interact with the external environment, all via the Internet.

Due to its many benefits, the scale of IoT is growing rapidly. A recent forecast projects that there will be 34 billion devices connected to the Internet by 2020, up from 10 billion in 2015, among which IoT devices will account for more than 70%, more than double of the traditional computing devices [1]. Some forecast even suggests that the connected device will reach 50 billion by 2020. It is projected that nearly $6 trillion will be spent on IoT solutions over the next five years, with businesses as the top adopter of IoT solutions to lower operating costs, increase productivity, and expand to new markets or develop new product offerings.

Optical technologies will be playing an important role in the growth of IoT, by providing connectivity, networking, sensing, and more. In this paper, the roles and applications of optical technologies in IoT networks are discussed.

II. ARCHITECTURE FOR THE INTERNET OF THINGS

There are three fundamental components in IoT, namely:
- Hardware technologies: Retrieving data from the physical world (such as key press action, temperature, location, motor speed, human behavior, etc.), and/or converting data instructions to physical responses (such as display, light switching on and off, robotics movement, vehicles operation, etc.). They can basically be viewed as sensors and actuators, in a general sense.
- Applications: Including various types of services, semantics, and software to utilize the data for automation at cities, buildings, homes, industrial facilities, etc.
- System infrastructure: Communication and networking infrastructure with protocols and technologies that enable data exchange among physical objects, and perform data management such as storage, processing, security management, etc.

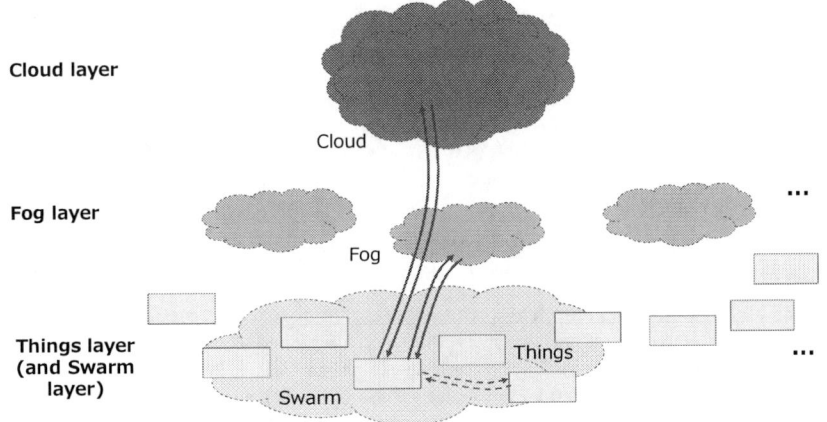

Fig. 1. Layers of the Internet of Things.

Fig. 1 shows the general structure of the IoT network. It mainly consists of three layers. The bottom layer is the "Things layer", which is made up of the large number of physical devices connected to the IoT. They generate the data from sensors and other input devices, and send the data to the "Fog layer", which is located locally and consists of many individual fog controllers. The information that requires real-time response is handled by the local fog controller, and the response is sent back directly to the relevant devices to be executed in real-time by the actuators and other output devices. Other data are sent to the top "Cloud layer", which interacts with more devices, handles large size data, and performs more complex processing for various applications. The security, mobility, and other network management issues are handled at the cloud layer. The instructions are then sent back to the respective devices to be executed.

Besides these three layers, a "Swarm layer" has also been proposed. In this layer, a large number of devices operate collectively as a decentralized, self-organized system. It can also be considered as part of the "Things layer".

III. OPTICAL TECHNOLOGIES IN THE INTERNET OF THINGS

Optical technologies, in particular fiber optic technologies, contribute to the IoT networks and applications in various aspects, ranging from data transportation, networking, and sensing and imaging.

A. Transmission and switching

IoT generates large amount of data. Even though some of the data are processed locally at the fog layer and do not require further analysis or storage, a large portion of data require processing in the cloud. Therefore transporting large amount of data between the devices and local fogs and the cloud is an important part of the IoT network. Besides high transmission bandwidth for large data volume, the data transmission in IoT has other requirements such as low latency, long distance, security, and flexibility. Fiber optic communication network provides the most suitable platform, due to its stable channel with high bandwidth, high transmission speed, low attenuation, multi-dimensional multiplexing capability. Optical circuit switching is by nature bit-date and protocol independent, and can support heterogeneous signal formats simultaneously. Flexibility can be further enhanced through utilizing CDC ROADMs, flexible grid WDM networks, and variable rate transponders, with the intelligent centralized control through transport SDN. Optical layer encryption adds physical level security to the existing security measures at the higher layers.

For the last segment of the network, i.e. between the devices and the rest of the network, wireless technologies are mostly used due to the mobility advantage. However, for devices that generate large amount of data, especially in industrial IoT applications, optical transmission is still a good solution. Fiber optic communication is also advantageous in locations where the RF interference is an issue or RF channels are unavailable. Current passive optical network and FTTx can be expanded to support more IoT data and applications. Free space optical communication (FSO) provides another alternative to connect the IoT devices to the network. This includes laser-based high-speed, long range FSO systems, and LED-based low-data-rate, short distance indoor visible light communication systems.

B. Data center networking

Cloud computing, a major part of IoT, is driving up the traffic volumes in the data center. It is forecasted that 83% of global data center traffic will come from cloud services and applications by 2019, with a total data center traffic volume of 10.4 zettabytes per year [2]. Optical technologies have been used in the data center networks (DCN) for high speed point-to-point links. The developments of CMOS technologies, photonic integrated circuit technologies, and DSP technologies, enable optical transceiver module hardware with higher data rate, longer transmission reach, and lower cost [3]. Spatial division multiplexing technologies will further increase the bandwidth capacity per fiber without increasing the fiber count [4].

In recent years, there have also been growing research interests in optical switching for DCN [5]. Optical circuit switching has good scalability, low power consumption, and is bit-rate and protocol independent. However the switching speed is slow and there is limited buffer capability. On the other hand, electrical switching provides faster switching speed and statistical multiplexing capability through finer packet-level switching granularity. It also can easily realize buffer and memory. However the bandwidth is not easily scalable, and the power consumption increases significantly with the data rate. Therefore hybrid optical-electrical switching presents the best solution to combine the advantages of both switching technologies simultaneously [6].

There is also an increasing need to achieve geo-distribution for the cloud service data centers, due to the requirements of low latency, redundancy, computation power, scalability, and location restriction. Inter-data center networks are almost entirely based on fiber optic technology.

C. Optical sensing and imaging

Besides transporting and routing data, optical technologies can be used to generate data in the IoT network, especially in terms of optical sensing and imaging. Optical sensors measure various physical phenomena, such as temperature, pressure, displacement, vibration, acceleration, electrical field, chemical, etc., by observing the optical property change in the light beam caused by the phenomena. The optical properties to be monitored include intensity, phase, wavelength/frequency, polarization, spectral distribution, etc. Optical sensing offers high sensitivity, low latency, and long sensing distance. It is also immune to electromagnetic interference, and can be implemented in harsh environment.

Based on how the medium that the light travels, optical sensing technologies can be divided into free space optical sensing and fiber optic sensing (including optical waveguide sensing). Fiber optic sensing does not require line-of-sight, and can perform distributed or quasi-distributed sensing over a long distance with remote monitoring capability. Distributed fiber optic sensing utilizes the backscatter of light pulses directed down a fiber optic cable. Because the backscatter occurs down the entire length of the cable, every single part of the optical fiber acts as a monitoring device. Common distributed fiber optic sensors include Rayleigh scattering-based vibration and acoustic sensors, Brillouin scattering-based temperature and strain sensors, and Raman scattering-based temperature sensors. There are also single point sensors that use sensing elements such as fiber Bragg grating to conduct sensing at a targeted location. Quasi-distributed sensors contain arrays of multiple sensing elements along the optical fiber.

Free space optical sensing and imaging does not require fiber installation, and is thus more flexible and non-intrusive. The sensing and imaging distance can be range from less than a millimeter (such as optical coherent tomography) to hundreds of kilometers (such as space lidar). The spectrum range in free space optical sensors is also broad. The optical source and receiver can be located at the same end (such as in most standoff detection) or different ends of the light path.

IV. APPLICATIONS OF INTERNET OF THINGS WITH OPTICAL TECHNOLOGIES

IoT is a ubiquitous network that connects huge amount of devices, aiming to cover every aspect of daily life and every business sector. Therefore it has wide range of applications. Here are a few examples where optical technologies play a key role.

- Utility network: The fiber optic broadband connections to homes and other buildings are used with wireless networks to connect individual smart meters to the utility company's network, allowing the utility operators to monitor the usage in real-time or near real-time, pinpoint outrages and reduce restoration time, automate billing validation, and better manage and balance energy load during different usage periods. The fiber optic support network also ensures always-on, gigabit per second speed Internet service required for supercomputers that constantly monitor the power grid throughout the city.

- Digital oil field (DOF): DOF is a direct example of industrial IoT. It consists of both the tools and the processes surrounding data and information management across the entire suite of oil/gas exploration and production activities. The combination of advanced fiber optic sensing technologies with integrated networking and big data analytic technologies allows more accurate underground resource exploration and smart well monitoring and management, realizing DOF with improved production and optimized facility performance. For example, distributed fiber sensors provide the operators with high resolution 3D or 4D vertical seismic profile images for accurate underground reservoir characterization.

- Automatic toll booth: Multi-height multi-species optical gas sensors and emission cameras equipped at toll booths screen the passing vehicles in real-time, and pick out the ones that fail emission regulation without slowing down the traffic. Aided by automatic plate recognition, the vehicles' information, including manufacturer, model and year can be linked with the emission screening results. Other information, such as vehicle occupancy can also been detected. The toll booths can also exchange information with the passing-by vehicle, such as extracting vehicle running status data, and sending test results and repair instruction.

V. CONCLUSIONS

IoT is growing rapidly, bringing automation and efficiency improvement to everyday life, business, and society. Optical technologies play an important role in the IoT network, including providing high bandwidth connectivity, intelligent networking, and high sensitivity multi-phenomena sensing and imaging. As optical technologies contribute to the development of IoT, IoT will also become a main driving force for the development of next generation optical technologies.

REFERENCES

[1] BI Intelligence, "The Internet of Things: Examining How the IoT Will Affect the World", market report (2015).
[2] Cisco, "Cisco Global Cloud Index: Forecast and Methodology, 2014-2019", White paper (2015).
[3] P. N. Ji, "Inter-datacenter optical networks", 7th International Symposium on VICTORIES Project and Workshop (2014).
[4] G. Milione, et al., "Real-time Bi-directional 10GbE Transmission using MIMO-less Space-division-multiplexing with Spatial Modes", OFC (2016), W1F.2.
[5] C. Kachris and T. Ioannis, "A survey on optical interconnects for data centers", IEEE Communications Surveys & Tutorials, (2012), pp. 1021-1036.
[6] P. N. Ji, "Hybrid optical-electrical data center networks", OSA Advanced Photonics Congress (2016), NeM3B.3.

TuA3-2

Standardization of Optical Access Technologies: Progress and Future Prospects

Jun-ichi Kani

NTT Access Network Service Systems Laboratories, NTT Corporation

Optical access network technologies have been supporting the massive deployment of Fiber To The Home (FTTH) services for the recent 15 years, and keep evolving to increase its capability and flexibility to cover various applications including the business services and the mobile backhaul/fronthaul.   Standardization of the optical access networks has been the key for the successful commercialization in term of promoting the cost reduction and the multi-vendor interoperability.

This tutorial firstly reviews progress of the optical access network standards with a focus on how key technologies have been successfully standardized.   Such key technologies include Time Division Multiple Access (TDMA) for the 1-Gigabit class and 10-Gigabit class Passive Optical Network (PON) systems, as well as Time and Wavelength Division Multiplexing (TWDM) for the recently standardized Next Generation PON 2 (NG-PON2); TWDM offers not only a higher capacity but also some advanced functions through the use of wavelength tunability.

This tutorial next discusses and tries to identity directions of the standardization activities in future.   The discussions will include the convergence of the optical access networks with the next-generation radio access networks as well as the softwarization of access-network functions for adapting to diverse requirements flexibly.

Biography

Junichi Kani

Jun-ichi Kani is Senior Research Engineer, Supervisor, and Distinguished Researcher in NTT Access Network Service Systems Laboratories, NTT Corporation. He received M.E. and Ph.D. in applied physics from Waseda University, Tokyo, in 1996 and 2005, respectively. He joined NTT Corporation in 1996. Since 2003, he has been with the NTT Access Network Service Systems Laboratories, where he has been engaged in R&D and standardization of optical communications systems for metro and access applications.

In 1996, he joined the NTT Optical Network Systems Laboratories, where he was engaged in research on optical multiplexing and transmission technologies.   Since 2003, he has been with the NTT Access Network Service Systems Laboratories, where he is engaged in the research and development of optical communication systems for metro and access applications.   He received the IEEE Communications Society Leonard G. Abraham Prize in the Field of Communications Systems in 2013, the Best Paper Award from the 18th Optoelectronics and Communications Conference (OECC) in 2013, and eight other technical/industrial awards.

Dr. Kani has been participating in ITU-T and the Full Service Access Network initiative (FSAN) since 2003.   He has been serving as Associate Rapporteur of ITU-T Q2/15 (optical systems for fibre access networks) since 2009 and Chair of FSAN Optical Access Network Working Group since 2015.   He is Chair of subcommittee N4 (optical access systems for fixed and mobile services) in IEEE/OSA Optical Fiber Communication Conference (OFC) 2017.

TuA4-1 (Invited)

OECC/PS2016

Tactile Internet Capable Passive Optical LAN for Healthcare

Elaine Wong, Maluge Pubuduni Imali Dias, and Lihua Ruan
Department of Electrical and Electronics Engineering, The University of Melbourne, Melbourne, Australia
ewon@unimelb.edu.au

Abstract: We present the underlying communications network that achieves rapid delivery of control/steering/sensor information in Tactile Internet capable healthcare facilities. Exploiting passive optical LANs with TWDM and predictive DBA, an end-to-end latency of 200µs is achieved.

Keywords: e-health; Tactile Internet; time and wavelength division multiplexing; passive optical local area network; predictive dynamic bandwidth allocation algorithm.

1. Introduction

Traditional copper-based healthcare local area networks (LANs) based on copper and multimode fiber, are inadequate in supporting the high bandwidth needs of modern healthcare facilities. Such facilities are rapidly embracing the use of WiFi connected mobile diagnostic equipment, advanced diagnostic image transfer and archiving, video conferencing, tele-medicine, IP camera surveillance for real-time patient monitoring and staff location, automation of building security and management services, digitized billing, clinical information, and patient records, in addition to Video-on-Demand and IPTV for patient and visitor entertainment. At the same time, traditional copper-based healthcare LANs cannot meet the stringent bandwidth, latency, and quality-of-service (QoS) requirements for critical care.

More importantly, healthcare is about to be revolutionized by the emergence of Tactile Internet, facilitating tele-diagnosis, tele-rehabilitation, and ultimately tele-surgery [1]-[2]. With Tactile Internet, patients/physicians are no longer confined to the same location since real-time human-machine interaction through visual, audio and tactile feedback can be realized anytime and anywhere. An example is tele-diagnostics whereby using a combination of sensors and actuators, a physician is able to diagnose a geographically-separated patient located in his/her healthcare facility through audio, visual, and tactile feedback. However, real-time visual-tactile feedback necessitates a system response time and hence end-to-end latency in the order of 1 ms [2]. As such, we must redesign the underlying communications infrastructure for ultra-low end-to-end latency transport. If one factors in (a) the use of control/steering servers located close to the tactile point of interaction (i.e. through tactile edge intelligence) and also (b) accounting for the additional delays arising from signal processing, protocol handling, switching and network delays, *future fiber-based communication infrastructures must be able to support end-to-end latencies of only 200 µs* [2].

In this paper, we discuss our on-going focus on designing low-latency optical networks to meet Tactile Internet capable healthcare needs. In particular, we discuss the use of a fiber optical based LAN, specifically the passive optical LAN to facilitate the convergence of bandwidth-intensive clinical-, patient-, and administrative- centric services, and the evolution of this network to constrain end-to-end latency through time and wavelength division multiplexing (TWDM) [3]. We also present our on-going work on predictive dynamic bandwidth allocation (DBA) algorithms designed to better accommodate dynamic service demands while maintaining ultra-low end-to-end latency [4].

2. Tactile Internet Capable Passive Optical LAN

Physical Topology: The optical network architecture of a Tactile Internet capable hospital campus providing inter- and intra-building connectivity, is shown in Figure 1. The passive optical LAN is built on single mode fiber (SMF) and connects the wide area network (WAN) to medical sensor networks based on the Smart Body Area Network (SmartBAN) specification. SmartBAN is specified for sensing patient biometric and context signals. More importantly, due to its inbuilt priority-based transmission [5]-[6], latency-sensitive sensor and actuator signals for Tactile Internet applications can be easily supported. The passive optical LAN comprises a Central Office (CO) with a control server, multiple passive splitters, and Optical Network Terminals (ONTs) that serve as a gateway to/from the SmartBANs. While traditional copper-based LANs comprise a layered architecture of distributed active elements [7], the passive optical LAN has a simpler physical and logical topology where ONTs are directly connected to a single CO.

In Fig. 1, the passive optical LAN carries multiple upstream and downstream wavelength channels. Conventional healthcare and sensor (non-Tactile Internet) traffic are transported on wavelengths channels that different to those of the Tactile Internet control/steering/sensor packets. Due to optical passive splitters in the network, each wavelength channel is also considered as a separate time-division-multiplexed (TDM) channel. Additionally, the number of utilized wavelengths can be dynamically varied to reduce network congestion and maintain ultra-low end-to-end latency. If upgrading from an already deployed passive optical LAN, active components/equipment are required to be implemented only at the CO and ONTs, thus keeping the distribution network purely passive. This, therefore, allows

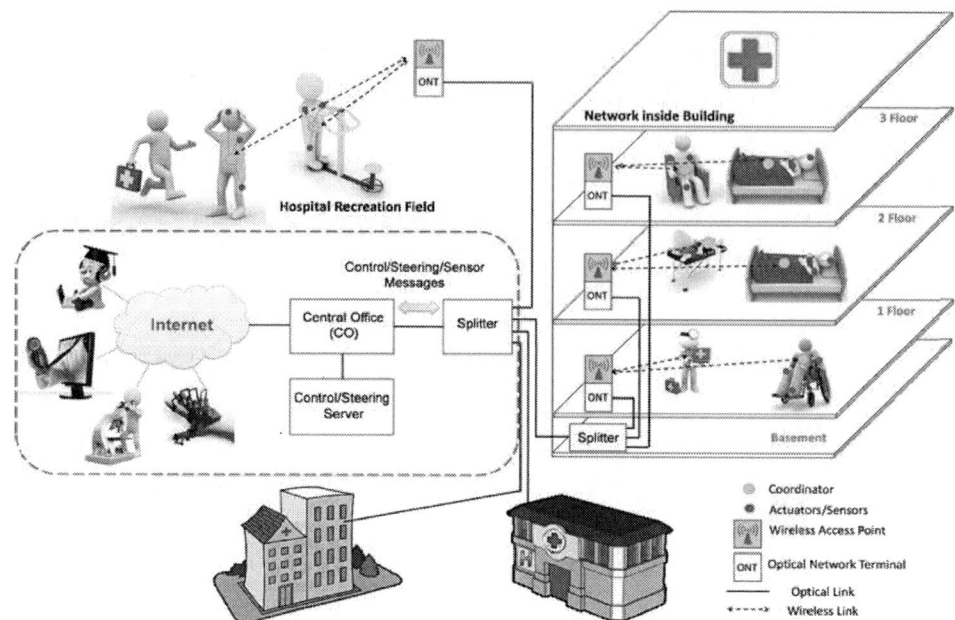

Fig 1. Tactile Internet capable passive optical LAN in a hospital campus.

the already deployed passive optical LAN to be reused and reconfigured for Tactile Internet with no need for "forklift" upgrades.

Drivers for passive optical LAN: The benefits for passive optical LAN include: (a) high-performance bandwidth over much greater distances (up to 20km vs 100m for copper or 550 m for MMF); (b) higher flexibility and scalability: (c) reduced capital costs from fewer equipment purchases, greater density and lower installation costs, (d) reduced overall lifecycle operating costs by eliminating climate control costs in the CO and power costs in the optical distribution network; (e) high environmental sustainability due to passive architecture; (f) simple administration with centralized management; and (g) high security and mobile diagnostic medical equipment operation with no interference from electromagnetic interference (EMI) and radio frequency interference (RFI). These significant technical and economic benefits are further cemented by the high network availability afforded by the passive optical LAN. Many healthcare functions are mission critical, necessitating the highest level of network availability and reliability. Compared to legacy copper with a three-9s availability (8 hours 45 minutes annual downtime), carrier-grade at five-9s availability (5 minutes 15 seconds annual downtime) or even at six-9s availability (30 seconds annual downtime) can be achieved using equipment redundancy and dual-homing, and at a fraction of the cost with passive optical LAN. Also, human error is significantly reduced through global management implemented centrally at the CO.

Edge intelligence: To meet the ultra-low end-to-end latency requirement of Tactile Internet, we consider the placement of a control server/engine at the CO, which also serves as a gateway to the WAN. The control/steering server is equipped with artificial intelligence to facilitate (a) predictive caching, (b) predictive bandwidth allocation for control/steering packets, and (c) predictive human actions [8]. This is similar in concept to the recently proposed Cloudlets in [9] where due to its proximity to the tactile edge, Cloudlets with artificial intelligence and caching can provide real-time execution of tactile applications.

Service Convergence: The most enticing feature of a passive optical LAN is its ability to simultaneously converge multiple clinical and patient-centric services on a single, efficient optical network infrastructure. Traditional copper-based LANs facilitate decentralized peer-to-peer traffic flows and multiple separate infrastructures for separate applications, e.g. nurse call, building automation, security, etc. have to be simultaneously supported. However, a passive optical LAN is more economically suited to future healthcare needs where traffic flow is centralized.

3. Tactile Internet capable Dynamic Wavelength Bandwidth Allocation (TI-DWBA)

To address this need to minimize the end-to-end latency of control/steering/sensor packets, we proposed a predictive DBA algorithm, termed Tactile Internet capable Dynamic Wavelength and Bandwidth Allocation algorithm (TI-DWBA), for the control and steering of Tactile Internet applications over the passive optical LAN [4]. As such, the end-to-end latency considered in this work is defined as the time taken between the reception of a sensor packet at the ONT, the propagation of this sensor packet through the passive optical LAN; processing of the sensor packet at the control/steering server and subsequent generation of a reaction through a control/steering packet; and the transmission of the control/steering packet back to the ONT through the passive optical LAN.

Our TI-DWBA is online in nature and its prediction of the uplink (sensor) and downlink (control/steering) traffic load is exploited to dynamically vary the number of utilized (active) wavelengths in the network such that network congestion is avoided and end-to-end latency is maintained to within a specified latency constrain, L_{cons}. Specifially, the estimation and traffic prediction techniques in TI-DWBA allows the uplink sensor packets to be allocated bandwidth in the same cycle as they arrive at the control server. In our performance evaluation of TI-DWBA, up to four wavelengths are used in each of the direction of transmission. This complies with TWDM-PON specifications, thus allowing standard equipment to be used for passive optical LAN deployment and avoiding additional procurement and maintenance cost of specialised equipment. Each wavelength operates at the maximum polling cycle time, $T_{max\,poll}$, which is defined as a function of L_{cons}, processing time, $T_{process}$, and round trip time, T_{rtt}, as follows:

$$T_{maxpoll} = 2 \times (L_{cons} - T_{process} - T_{rtt}) \qquad (1)$$

Further, in the performance evaluation of TI-DWBA, we consider a passive optical LAN with 10 ONTs, each with a different distance to the CO. As such, different ONTs have different T_{rtt}, and so for a given wavelength, $T_{max\,poll}$ is calculated based the ONT with the longest T_{rtt}. In any given polling cycle, an ONT can send multiple sensor packets to the control/steering server. At the end of this transmission, the ONT also sends a REPORT message, which contains a Bayesian estimated average inter-arrival time of sensor packets, to the control/steering server. The server then

Table 1: Network and Protocol Simulation Parameters

Test Networks	Configuration
Network 1	
Max/Min CO-ONT distance	100m/1 km
End-to-end latency constraint, L_{cons}	1ms
Network 2	
Max/Min CO-ONT distance	10 m/100 m
End-to-end latency constraint, L_{cons}	1ms
Network 3	
Max/Min CO-ONT distance	10m/100m
End-to-end latency constraint, L_{cons}	200 μs

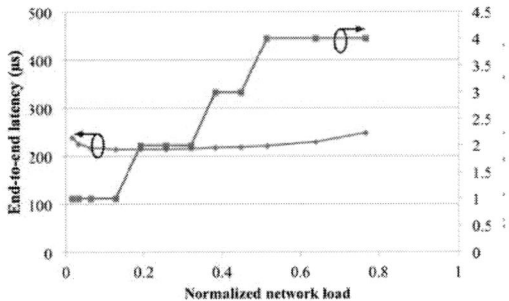

Fig 2. End-to-end latency and number of active wavelengths vs. normalized network load.

processes the sensor packets and sends the corresponding control/steering packets to the ONT for actuation. Further, using the estimated average inter-arrival time, the server also predicts the amount of sensor packets accumulated during $T_{max\,poll}$ and allocates bandwidth to each ONT as proposed in [10]. Finally, the control/steering server also determines the number of active wavelengths required in the network, their corresponding $T_{max\,poll}$, and the number of ONTs allocated to each active wavelength, based on this allocated bandwidth.

We tested three variations of the network covering intra- and inter-building deployments. Details of each network along with the corresponding end-to-end latency constraint are listed in Table 1. Fig. 2 plots the average end-to-end latency of Network 3 under TI-DWBA as a function of normalized network load. Also super-positioned on this plot is the number of active wavelengths required in each direction of transmission. Note that a normalized network load of 1 equals 40 Gbps (4 x 10 Gbps). When the normalized network load is approximately 0.6, all four wavelengths in each direction are fully loaded, and the end-to-end latency therefore gradually increases above L_{cons} (200 μs), respectively. Beyond 0.6 network load, additional wavelengths can be deployed to maintain end-to-end latency under L_{cons}.

4. Summary

In this paper, we discussed the urgent need to rethink the design of the underlying communications infrastructure to facilitate future healthcare needs and to support the emergence of Tactile Internet capable healthcare applications. To this end, we presented our combined solution of passive optical LAN, time and wavelength division multiplexing, edge intelligence, and predictive wavelength and bandwidth allocation to support both bandwidth intensive healthcare traffic and latency-sensitive control/steering traffic. In testing passive optical LAN scenarios representing intra- and inter-building connectivity, our results show that rapid delivery of control/steering/sensor information within the predetermined end-to-end latency constraint can be achieved.

5. References

[1] ITU-T, "The tactile internet," ITU-T technology watch report. https://www.itu.int/dms_pub/itut/oth/23/01/T23010000230001PDFE.pdf
[2] G. Fettweis, "The tactile internet: applications and challenges," IEEE Veh. Tech. Mag., 9(1), pp. 64-70, 2014.
[3] E. Wong, M. Mueller, and M.C. Amann, "Colorless operation of short cavity VCSELs in C-minus band for TWDM-PONs, *Optics Express*, vol. 21, 18, p. 20747-29761, 2013.
[4] M.P.I. Dias, L. Ruan, and E. Wong, "Predictive wavelength/bandwidth allocation for low-latency control and steering in Tactile Internet capable healthcare facility," submitted to ECOC 2016.
[5] L. Ruan, M.P.I. Dias, and E. Wong, "Time-optimized MAC for low-latency and energy-efficient SmartBAN", *Prof. INFOCOM*, p. 701, 2016.
[6] ETSI TC SMARTBAN, "Low complexity MAC for SmartBAN," TS DTS/SmartBAN-005 v1.1.1., April 2015.
[7] Bell Labs, "Passive optical LAN versus copper-based Ethernet," Nokia strategic white paper, 2015.
[8] X. Xu, B. Cizmeci, A. Al-Nuaimi, and E. Steinbach, "Point cloud-based model-mediated teleoperation with perception-based model updating," *IEEE Trans. Instrum. Meas.*, 63(11), p. 2558-2569, 2014.
[9] M. Satyanarayanan, Z. Chen, H. Kiryong, H. Wenlu, W. Richter, and P. Pillai, "Cloudlets: At the leading edge of mobile-cloud convergence," Proc. Conf. Mobile Cloud Comput. Appl. Serv., pp. 1–9, 2014.
[10] M. P. I. Dias, D.P. Van, L. Valcarenghi, and E. Wong, "An energy-efficient framework for wavelength and bandwidth allocation in TWDM PON", JOCN, vol. 7(6), pp. 496-504, 2015.

TuA4-2   OECC/PS2016

Analysis of Power Consumption in Mobile Backhaul Network with Densely Deployed Small Cells under Dynamic Traffic Behavior

Kyosuke Sone[1], Inwoong Kim[2], Xi Wang[2], Yasuhiko Aoki[1], Hiroyuki Seki[1], and Jens C. Rasmussen[1]

1) Fujitsu Laboratories, Ltd. 1-1, Kamikodanaka 4-chome, Nakahara-ku, Kawasaki, 211-8588, Japan
2) Fujitsu Laboratories of America, Inc., 2801 Telecom Parkway, Richardson, TX 75082, USA
sone.kyousuke@jp.fujitsu.com

Abstract: We present an analysis of energy efficiency by sleep mode control of small cell base station based on dynamic user distribution and traffic demand in mobile backhaul network design for dense small sell deployment.
Keywords: mobile backhaul, small cell deployment, network design, energy efficiency

I. INTRODUCTION

Mobile data traffic is growing at an unprecedented pace [1], with smart phones and tablet users driving the growth. In order to overcome radio spectrum limitation and user interference, mobile operators are now considering deployment of small cells which are overlaid in macro cell areas [2]. Deploying large numbers of small cells helps to offer more capacity and more spectrum reuse in a given area, but creates new challenge for the mobile backhaul (MBH) network. MBH must provide the connectivity between base stations (BSs) and central office (CO) with sufficient capacity and must carefully expand the network capacity to avoid significant cost increase. Therefore, efficient MBH network design tool [3] plays a key role in maintaining mobile operators' business competitiveness.

MBH network is designed to accommodate the peak traffic demand in a day, however, mobile traffic changes over time during the day and geographic location. So, it is not efficient from the energy consumption perspective. For the power saving, the small cell BSs are switched to sleep mode during low traffic hours, such as nighttime [4], but the MBH network design with high density small cell BSs under the dynamic traffic behavior haven't fully studied yet. In this work, we have studied the impact of the small cell BS power consumption with the sleep mode during nighttime in the MBH network designed on the basis of mobile user distribution and traffic demands in daytime. In addition, we have also estimated energy savings due to it.

II. MBH NETWORK DESIGN

We proposed a mobile x-haul network design method for small cell deployment to take into account such factors as the existing infrastructure, spectrum and license costs, availability of the equipment, and technology rules in Ref. [3]. The key idea is to construct a multi-technology auxiliary graph that incorporates all available network information, and then to find a minimum-cost technology path for each traffic demand from its cell site. Our mobile x-haul network design process is shown in Fig. 1. First, we build a network topology based on site locations, equipment and link technologies as shown in Fig. 1(a). Then we list all possible configurations between central office (CO) and cell sites supported by available technology choices as shown in Fig. 1(b). There may be multiple technology choices among sites and there may also be multiple physical routes. In order to find the best solution for each traffic demand, we construct a multi-technology auxiliary graph as shown in Fig. 1(c). The minimum-cost technology path can be found by finding the minimum cost path from a cell site to CO on the auxiliary graph. Prior to path search for each demand, links with insufficient capacity for the demand are deactivated. In addition, the links connected to the air interface termination point are also deactivated if the node cannot support the requested bandwidth. The available capacity of links and nodes is decreased after each demand

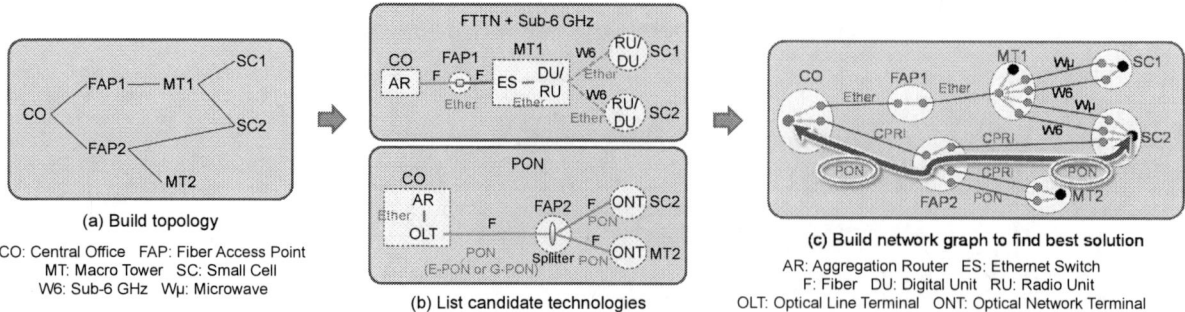

(a) Build topology

CO: Central Office   FAP: Fiber Access Point
MT: Macro Tower   SC: Small Cell
W6: Sub-6 GHz   Wμ: Microwave

(b) List candidate technologies

(c) Build network graph to find best solution

AR: Aggregation Router   ES: Ethernet Switch
F: Fiber   DU: Digital Unit   RU: Radio Unit
OLT: Optical Line Terminal   ONT: Optical Network Terminal

Fig. 1. Mobile x-haul network design process.

is assigned to a path. In this example shown in Fig. 1(c), the route shown as brown line provides the lowest cost solution for a demand between small cell BS 2 (SC2) and CO.

III. SIMULATION CONDITION OF THIS STUDY

Generally, small cell BSs are deployed to accommodate the peak traffic load during the day, so the MBH networks are also designed based on the peak traffic demand during the day. Thus, small cell BS resources are not utilized efficiently due to the imbalance of traffic load during the day. In this work, we have designed MBH network to accommodate the peak traffic demand in the daytime. After that, for lower nighttime traffic demand, we investigate how many small cell BSs can be switched to the sleep mode in the designed MBH, where the traffic demand of users in the area of sleep mode small cell BS is accommodated by the corresponding macro cell BS.

We assume a densely populated urban area and its nearby residential area, where 20 small cell BSs are overlaid in 7 macro cell areas. A total of 12 fiber access points (FAPs) and 189 links (64 fibers, 60 coppers, 65 microwaves) connecting various sites, such as CO, BSs (macro cell and small cell) and FAP, are assumed to be the candidates infrastructure to be installed in this assumption. Typically, BS utilization (traffic demand) changes during the day depending the area it serves. BSs located in office areas (urban area) will take the higher user density during working hours (daytime) as shown in Fig. 2(a). Conversely, BSs located in residential areas take the higher user density in the nighttime as shown in Fig. 2(b). In this scenario, the user density of the urban area is assumed 5 times of the residential area and the traffic demand per unit area of urban area is assumed 15-fold of the residential area in the daytime. On the other hand, the user density of the residential area is assumed 1.5 times of the urban area and the traffic demand per unit area is assumed approximately equal in both areas in the nighttime. Here, we refer to the daytime and nighttime population ratio of urban and residential areas in Japan [5].

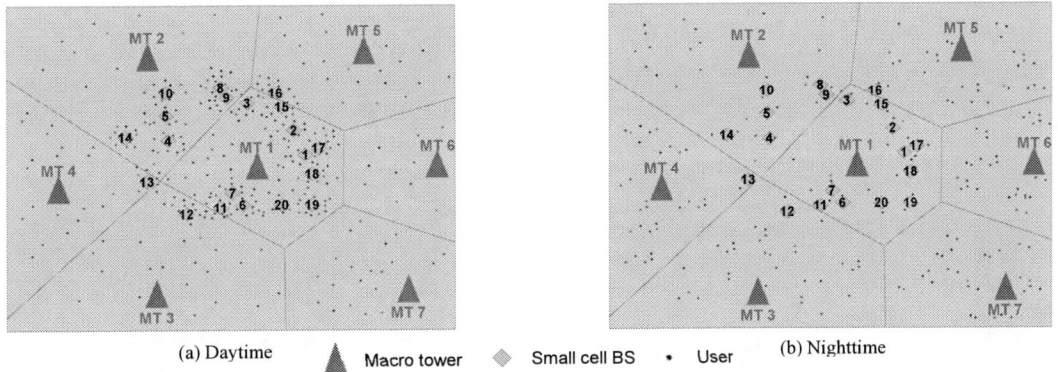

Fig. 2. User distribution in the assumption scenario at daytime (a) and nighttime (b).

IV. VERIFICATION RESULTS AND DISCUSSION

In this verification, we assumed the deployment of a hypothetical green field scenario. The number of users per small cell is fixed at 10 in the urban area while the number of users per macro cell is fixed at 20 in the residential area during the daytime. On the other hand, the number of users per small cell is fixed at 2 in the urban area while the number of users per macro cell is fixed at 30 in the residential area during the nighttime. In addition, the user traffic was verified in the three cases of 5 Mbps, 10 Mbps and 15 Mbps. The topological views of the MBH design results are shown in Fig. 3, in which the blue cube represents CO, green spheres represent macro cell BSs, blue spheres represent small cell BSs and

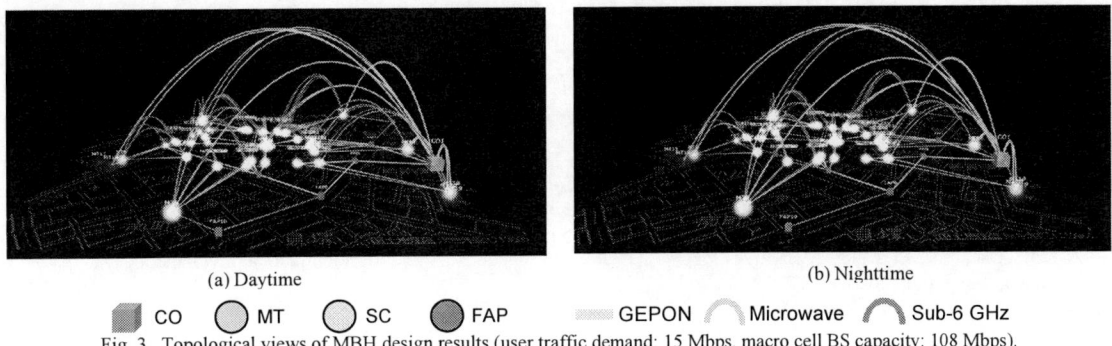

Fig. 3. Topological views of MBH design results (user traffic demand: 15 Mbps, macro cell BS capacity: 108 Mbps).

red points represent FAPs. Then, the straight lines (for wired link) and arcs (for wireless link) represent the paths decided by the proposed design method. Fig. 3(a) shows the design results in the case of the daytime and 15 Mbps traffic demand from each user. Various technologies are selected according to the most cost-efficient choices available at each site, such as GEPON, microwave and Sub-6 GHz. Then, based on the designed MBH for the daytime traffic demand, we investigated how many small cell BSs can be switched to a sleep mode for the nighttime traffic demand using our design tool. The limit of the capacity of macro cell BS or MBH determines whether the small cell BSs can be switched to sleep mode. We verify the three cases of the capacity of the macro cell BS per sector which are 27 Mbps, 54 Mbps and 108 Mbps. In each case, we switched the small cell BSs to sleep mode in turn until the accommodated traffic of macro cell BS reached the limit of its capacity or the MBH network cannot accommodate the traffic. Fig. 3(b) shows the design results at the nighttime in the case of 15 Mbps traffic demand from each user and 108 Mbps capacity at each macro cell BS. And, Fig. 4(a) shows the number of sleep mode small cell BSs when the traffic demand is gradually increasing from 5 Mbps to 15 Mbps in three cases of the capacity of macro cell BS. The results show the number of users that can be accommodated at the macro cell BS increases when the traffic demand of users is low and the number of small cell BSs that can be switched to the sleep mode also increases. Similarly, when the capacity of the macro cell BS is large enough, it can be seen that many small cell BSs can be switched to the sleep mode. From the results, we verified that the switching the small cell BSs brings how much reduction of the power consumption. Here, a list of the power consumption of the equipment used in the verification is shown in Fig. 4(b). In the list, each value of power consumption is referred to the literatures [6, 7]. In this verification, the power consumption of macro cell BSs is excluded from the total power consumption since it remains fixed irrespective of the supported traffic and we focus on a macro cell area covered by MT1, which is densely deployed small cell BSs. Fig. 4(c) shows the comparison of the total power consumption in a macro cell area covered by MT1 including the MBH network power consumption in the daytime and nighttime when the traffic demand of users and the capacity of the macro cell BS are changed. In this scenario, we obtained the power reduction of up to 62 % in the case of 5 Mbps user traffic and the power reduction of up to 29 % even in the case of 15 Mbps user traffic. In this verification, we demonstrated the importance of MBH design based on the mobile traffic behavior for dense small cell deployment from the energy saving perspective. In the future, since the number of small cell BSs per macro cell area will be increased and the requirement of MBH network capacity will become larger due to accommodating the small cell BSs traffic, so we believe the effect of the design considering the mobile user distribution and traffic demand becomes more crucial.

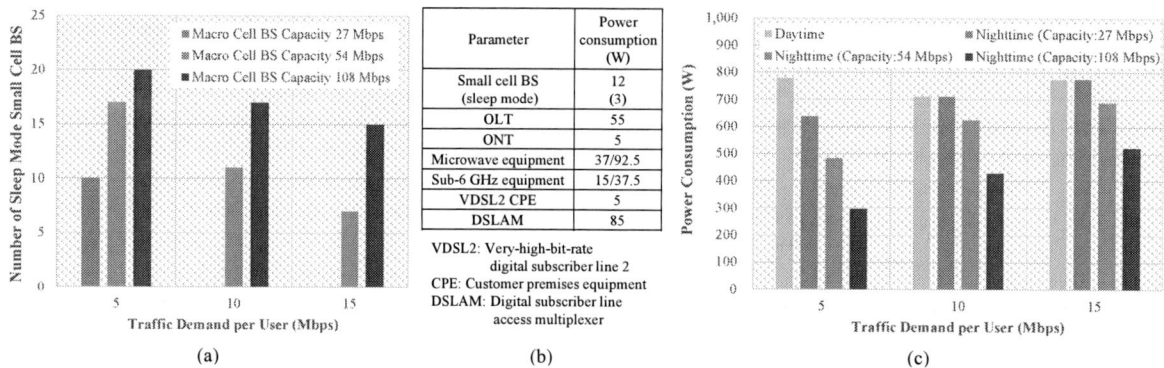

Fig. 4. (a) Number of sleep mode small cell BSs, (b) List of equipment power consumption,
(c) Total power consumption at macro cell area covered by MT1 (except macro cell BS power consumption).

V. CONCLUSIONS

We studied the impact of the small cell BS power consumption with the sleep mode during nighttime in the MBH network designed on the basis of mobile user distribution and traffic demands in daytime using our developed MBH design tool. In addition, we also verified energy savings due to the results and demonstrated the importance of MBH design according to the mobile user distribution and traffic demand for dense small cell deployment in the future.

REFERENCES

[1] Cisco, "Cisco Visual Networking Index: Global Mobile Data Traffic Forecast Update, 2015-2020," White Paper, 2016.
[2] NGMN, "Small Cell Backhaul Requirements," Next Generation Mobile Networks, 2012.
[3] K. Sone, et al., "New Mobile Backhaul Network Design Method for Dense Small Cell Deployment," OECC2015, JTuB.13, 2015.
[4] I. Ashraf, et al., "SLEEP Mode Techniques for Small Cell Deployments," IEEE Communications Magazine, pp. 72-79, 2011.
[5] http://www.stat.go.jp/data/kokusei/2010/final/pdf/01-11_4.pdf
[6] H. Claussen, et al., "Dynamic Idle Mode Procedures for Femtocells," Bell Labs Tech. J., vol.15, no.2, pp. 95-116, 2010.
[7] S. Tombaz, et al., "Is backhaul becoming a bottleneck for green wireless access networks?," ICC2014, pp. 4029–4035, 2014.

Optimal Design and Backhauling of Small-Cell Network: Implication of Energy Cost

Chathurika Ranaweera, Elaine Wong, Christina Lim, Chamil Jayasundara and Ampalavanapillai Nirmalathas
Department of Electrical and Electronics Engineering, The University of Melbourne, Melbourne, Australia
csr@unimelb.edu.au

Abstract: *We analyze the implication of energy cost on optimal design of small-cell and its fiber-backhaul networks. Our results provide insights into optimal deployment strategies for planning cost-effective and energy-efficient small-cell networks.*
Keywords: *Fiber, Backhaul, Small cell, Energy, Optimization*

I. INTRODUCTION

The continuous growth in mobile access and machine-centric applications will require high capacity, quality-of-service (QoS) guaranteed, ubiquitous, and continuous access to the Internet. New spectrum allocations along with technological advances are expected to use to satisfy these demands in mobile networks [1]. However, these solutions alone are insufficient to alleviate the capacity requirement entirely as mobile data traffic is predicted to increase eight fold over the next few years [2]. Densification of cells has been identified as one of the most efficient solutions to mitigate the capacity issue in mobile networks such as 5G networks [3]. Nevertheless, deployment of small cells has its own impediments as cost associated with backhauling and real estate can outweigh the benefits offered from small-cells networks. Therefore, small-cells and backhaul networks need to be planned simultaneously to achieve cost optimality.

Cost-optimal placement of small cells and its fiber-based backhaul network has been studied previously [4-6]. In particular, in [5], we proposed an optimization framework to cost optimally plan small-cell and point-to-point (PtP) fiber backhaul networks that leverage off resources associated with sparsely located fiber infrastructure. Whilst achieving a cost-optimal deployment, this framework is also capable of satisfying other network constraints such as coverage and capacity requirements. However, in that framework, we only considered the capital costs: that is we considered the total deployment cost which consists of the cost of fiber, equipment, and labor.

Recently, reducing energy consumption of network equipment has become an important research topic. The premise for its interest is the notion that energy cost is a significant expense for telecommunication network operators and also the ICT sector is expected to significantly contribute to the overall global energy consumption in the near future. One of the most effective ways to reduce the carbon footprint is to use renewable energy such as solar power. Therefore, in this paper we explore how the cost associated with small-cell and fiber backhaul networks can be further optimized by taking into account the energy cost and also the availability of renewable energy sources. In particular, we extend our optimization framework presented in [5], such that energy cost is also taken into account when planning cost-optimal small-cell network and its associated fiber backhaul. We demonstrate a practical example by using the proposed framework to plan a small-cell network for a suburban area of Victoria in Australia under diverse coverage requirements and cell radii. We also discuss the implications of energy cost on the optimal solution.

Fig.1 Small cell and fiber backhaul network deployment scenario

II. OPTIMAL PLACEMENT AND BACKHAULING OF SMALL-CELL NETWORK

A. Small Cell and Fiber Backhaul Network

Our objective is to achieve the required population coverage by cost-optimally deploying small cells and PtP fiber network to backhaul them. The small cell and PtP fiber backhaul deployment scenario that we consider in this study is shown in Fig. 1. Road intersections that have street light poles or traffic light poles are considered to be plausible locations for small-cell deployments. We consider that some of these locations are equipped with solar panels. Moreover, we use existing fiber access points (FAPs) to provide fiber backhaul for small cells. In particularly, a subset of local FAP locations are optimally chosen for connecting the backhauled fiber from small cells to the existing fiber network.

B. Energy Cost

Here, we consider the energy cost associated with small cells and its fiber-based backhaul networks. Since operational cost is calculated over a period of time that the network is expected to operate, we calculate the yearly energy cost and multiply this by the considered operational time to analyze its implication on the optimal solution. In our optimization framework, φ_{sc} and φ_f denote the total cost of energy consumed by a small cell and the equipment that resides at the local FAP, respectively. Moreover, φ_c denotes the cost of energy consumed by elements (e.g. equipment, building) at the FAP other than the communication equipment. This captures the energy costs associated with cooling and other overheads, and is chosen according to the metric of power usage of effectiveness (PUE). Furthermore, the locations that equipped with renewable energy sources are denoted by the binary parameter e_i where it is set to 1 if the ith location has renewable energy or 0 otherwise. Without loss of generality, we assume that the energy cost associated with a small cell is zero if it is deployed in a location with a renewable energy source.

C. Optimization Framework

Equation (1) shows the objective function of our proposed integer liner program (ILP) based optimization framework.

$$\min \; (\eta_s + \eta_{si}) \sum_{i \in V} x_i + \eta_t \sum_{i \in V} \sum_{j \in V} a_{i,j} d_{i,j} + \eta_f \sum_{i \in V} \sum_{j \in V} y_{i,j} d_{i,j} + \eta_e \sum_{i \in V} x_i + \varphi_{sc} \sum_{i \in V} x_i (1 - e_i) + \varphi_f \sum_{i \in V} x_i + \varphi_c \sum_{i \in V} r_i \sum_{j \in w} a_{i,j} \quad (1)$$

The objective is to minimize the total deployment cost as well as the costs arising from network energy consumption. The first four major cost components (summations) in (1) represents the deployment costs relating to new equipment, fiber, and labor [5]. The last three components correspond to the costs arising from the energy consumption of the network. In particular, $\varphi_{sc} \sum_{i \in V} x_i (1 - e_i)$ captures the total energy cost of small cells that are deployed in locations where renewable energy sources are unavailable, and $\varphi_f \sum_{i \in V} x_i$ captures the total energy costs of fiber backhaul connections at the FAP. In contrast, $\varphi_c \sum_{i \in V} r_i \sum_{j \in w} a_{i,j}$ captures the total energy cost associated with cooling of equipment at FAP. Our function takes this cost into account only if a particular FAP has fiber connectivity to one or more small cells.

III. RESULTS AND DISCUSSION

We use the commercially available CPLEX linear programing solver [7] to solve our proposed optimization framework for different deployment scenarios with diverse population coverage requirements and different small-cell radii. The costs of fiber, equipment, and labor used in our analysis are taken from different sources [5]. The energy and related costs are chosen according to industrial guidelines and manufacturer recommendations [8]. These energy cost components are calculated over a period of 10 years, unless stated otherwise. Furthermore, φ_c is calculated based on a PUE of 2 [9] and it is assumed that 30 % of potential small-cell locations are equipped with solar panels.

We compare the optimal solutions under three different scenarios: a) when renewable energy sources are available in some locations, b) when renewable energy sources are unavailable (i.e., where e_i is 0), and c) when energy cost is not considered (i.e., omitting the last three terms in (1)). Our observations point to the fact that when energy cost is considered in the optimization framework under the scenario (b), the total cost is increased in comparison to the scenario (c), though the optimal solution remain unchanged for both the scenarios. On the other hand, when the availability of renewable energy sources is taken into account under the scenario (a), the total cost is reduced and optimal placement of small cells are altered in comparison to the scenario (b). Figure 2(b) shows the total network cost of the scenario (b) under different coverage requirements and cell radii when energy cost is taken into account. The cost values shown in Fig. 2(b) are normalized with respect to the total cost of the 200m small-cell deployment scenario with 100% coverage such that the total cost of that scenario equals to 100. As shown, it is clear that the 200m small-cell deployment scenario has the largest cost for the entire range of the coverage requirements. Further, it was observed that the total network cost does not increase linearly with the percentage of population coverage. For an example, when the population coverage is increased to 100% from 50%, the total cost is increased by more than 100%. This is true for all considered cell radii.

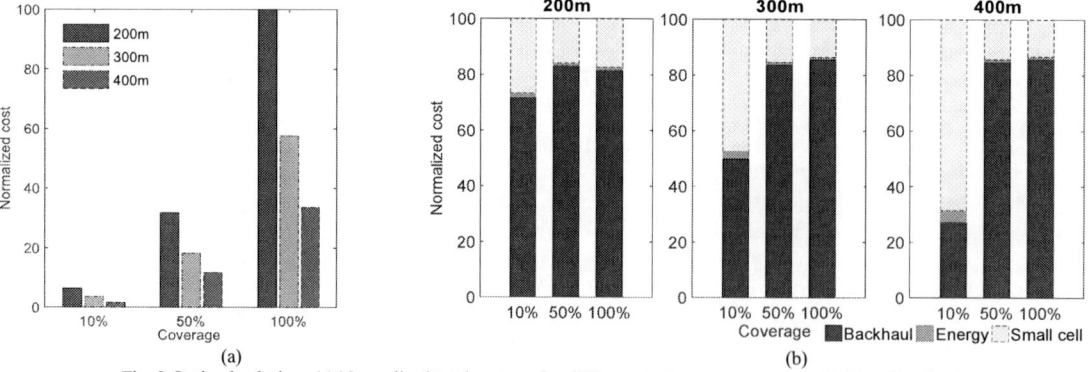

Fig. 2 Optimal solutions (a) Normalized total costs under different deployment scenarios; (b) Cost distributions;

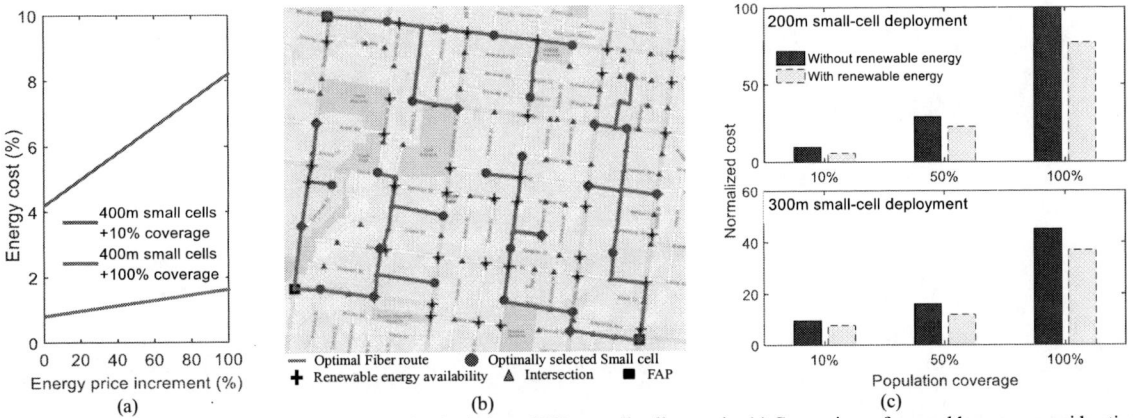

Fig. 3 (a)Variation of energy cost; (b) Map of the optimal solution of 200m small-cell scenario; (c) Comparison of renewable energy consideration;

In order to clearly understand the contribution of energy cost, we analyze the distribution of costs attributed to the backhaul, small cells, and energy of the optimal solutions. The results are shown in Fig. 2(b). Note that these cost values are normalized with respect to the total cost of each of the scenarios obtained in Fig. 2(a), such that the total cost equals to 100. As shown, in most cases, the largest cost contributor is the backhaul due to the deployment of new fiber routes. In comparison, the contribution of energy cost is only less than 5% in all cases considered. We then looked at the variation of fractional energy cost (i.e., energy cost / the total cost) as a function of energy price in order to understand how the price of energy can affect the achievable minimal cost. Figure 3(a) shows the fractional energy costs of 400m small-cell deployment scenario when the target coverage is 10% and 100%. As shown, when the energy price is doubled, the fractional energy cost increases to 8% and 2%, respectively. It is clear from these results that the contribution of energy cost to the achievable minimal total cost can vary considerably depending on the price of energy and deployment scenario.

Moreover, we also compare the optimal solutions and total costs of scenario (a) (with locations using renewable energy) and scenario (b) (all locations use non-renewable energy). It was observed that the optimal solution and the optimal cost vary depending on the availability of the renewable energy. The geographical map shown in Fig. 3(b) illustrates the optimal solution of an example deployment scenario where the small-cell radius is set to 200m and the coverage requirement is 100%, when the availability of renewable energy sources is taken into account. In particular, Fig. 3(b) shows the resulting optimal placement of fiber routes (red solid lines) and small-cell locations (maroon circles). The optimally selected FAPs are shown in black squares. Furthermore, Fig. 3(c) shows the comparison of total energy costs of scenario (a) and scenario (b). The cost values are normalized with respect to the total energy cost of the 200m small-cell deployment scenario with 100% coverage such that the energy cost of that scenario equals to 100. As can be seen, irrespective of the cell radii and achieved population coverage, a considerable amount of energy cost can be saved when locations that exploit renewable energy are considered in the planning stage. For an example, when the coverage is set to 100% for 200m deployment scenario, more than 20% of energy cost can be saved.

IV. SUMMARY

We analyzed the implication of energy cost on the optimal design of small-cell and fiber backhaul networks. Our results underscore the fact that taking the availability of renewable energy into account effects the achievable minimal cost and optimal placement of such a new network. The paper provided insight into a complete optimization framework that can be used to plan cost-optimal and energy-efficient small-cell and fiber backhaul networks.

REFERENCES

[1] A. Ghosh et al., "Heterogeneous cellular networks: From theory to practice," IEEE Com. Mag., vol. 50, no. 6, pp. 54–64, 2012.
[2] Cisco, "Cisco VNI: Global Mobile Data Traffic Forecast Update, 2015–2020", [Online], Available: http://www.cisco.com/.
[3] J. Hoydis, M. Kobayashi, and M. Debbah, "Green small-cell networks," IEEE Veh. Tech. Mag., vol. 6, no. 1, pp. 37–43, 2011.
[4] C. Ranaweera et al., "Design and optimization of fiber optic small-cell backhaul based on an existing fiber-to-the-node residential access network," IEEE Comm. Magazine, vol. 51, no. 9, pp. 62–69, Sep. 2013.
[5] C. Ranaweera et al., "Cost-Optimal Placement and Backhauling of Small-Cell Networks," Journal of Lightwave Technol. vol.33, pp. 3850-3857, 2015.
[6] F. Musumeci et al., "Optimal BBU placement for 5G C-RAN deployment over WDM aggregation networks," Journal of Lightwave Technology, vol..99, pp.1-1, 2015.
[7] IBM, "IBM ILOG CPLEX Optimizer," [Online], Available: http:// http://www.ibm.com/.
[8] "ICT code of conduct", [Online], Available: http://iet.jrc.ec.europa.eu/.
[9] W. Vereecken, et al., "Power consumption in telecommunication networks: overview and reduction strategies," IEEE Comm. Mag., vol.49, no.6, pp.62-69, 2011.

A Mobile Fronthaul System Architecture for Dynamic Provisioning and Protection

Qianmei Yang[1], Ning Deng[2], Xu Zhou[2] and Chun-Kit Chan[1]

[1]*The Chinese University of Hong Kong, Shatin, NT, Hong Kong*
[2]*Fixed Networks Research, Huawei Technologies Co., Ltd., Shenzhen, China*
yq013@ie.cuhk.edu.hk

Abstract: *We propose a mobile fronthaul architecture to handle the tidal effect for dynamic provisioning, and also protect the main baseband units, using injection locking of Fabry-Perot laser diodes. Downstream transmissions at 2.5 Gbit/s to 4 Gbit/s over 20-km have been experimentally demonstrated over the proposed architecture.*
Keywords: *fronthaul, injection locking, dynamic provisioning.*

I. INTRODUCTION

In next generation 5G mobile systems, cloud radio access network (C-RAN) is widely recognized as a practical approach to support mobile fronthaul. It centralizes the processing units, leaving only the radio access units at the remote cell sites [1]. It has been reported that there is a peak traffic load transition, recognized as tidal effect, between the office area and the residential area over different time periods in a day. With the centralization of the baseband units (BBUs), the C-RAN can achieve more efficient dynamic provisioning of the processing resources, based on the tidal effect, at both the office and the residential areas [1]. Therefore, the cost and power consumption can be reduced or optimized. Two approaches have been considered. One is to combine the channels of different remote radio heads (RRHs) into one at off-peak time for saving the processing resources [2] and the other one is to shut down some of the RRHs at off-peak time periods. The latter one is shown to be more effective in energy saving.

In this paper, we propose a mobile fronthaul architecture to simultaneously achieve dynamic provisioning, via shutting down the RRHs at off-peak time periods, as well as protect the BBUs in a cost-effective way, based on injection locking of Fabry-Perot laser diodes (FPLD). We have experimentally demonstrated 20-km downstream transmissions of six wavelength channels at 2.5-Gbit/s, 3-Gbit/s and 4-Gbit/s over the proposed architecture.

II. PROPOSED ARCHITECTURE

Fig. 1 shows our proposed mobile fronthaul architecture. The central office (CO) serves both the office and the residential areas, via two dedicated sets of static BBUs, sharing the same set of wavelengths, as the two areas do not overlap each other. One additional set of dynamic BBUs are designated to serve these areas according to the need of extra traffic demand. The wavelength multiplexers are the cyclic arrayed waveguide gratings (AWGs) such that each port supports two passband channels, one in blue band (1527.5-1542.5nm) and the other in red band (1547.5-1561nm). Wavelength channels in the blue band are allocated to the static BBUs and red-band wavelength channels are for the dynamic BBUs. At each dynamic BBU, a Fabry-Perot laser diode is employed as the optical source and it is injection-locked by an optical frequency comb source, or spectrally-sliced broadband light source. Its output port is connected to an optical red/blue (R/B) multiplexer for channel separation, via a 1×2 optical switch, in order to select a particular channel in either red band or blue band. All downstream channels are then optically combined and sent over the feeder fibers to reach different areas.

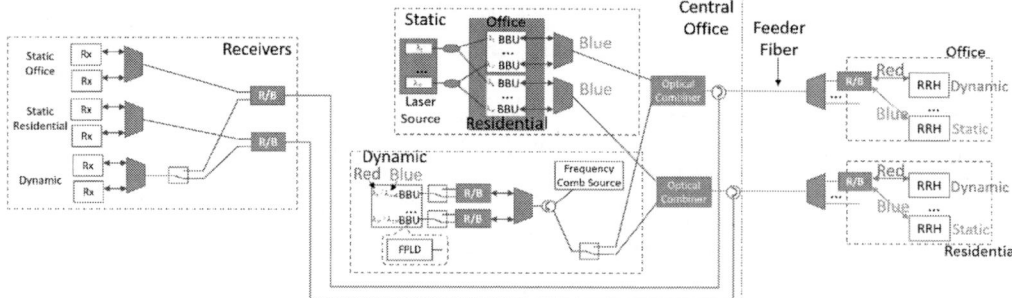

Fig. 1. The proposed mobile fronthaul architecture.

As illustrated in Fig. 2(a), for the downstream transmission, when the dynamic BBUs serve the RRHs for dynamic provisioning, they are connected to the red-band channels, so that the static BBUs and the dynamic BBUs occupy different sub-bands and wavelengths are multiplexed by an optical combiner.

Once a static BBU fails in the downstream transmission, the corresponding dynamic BBU is switched to the blue band to replace the failed static BBU and its red-band wavelength will be given up, as shown in Fig. 2(b). The failed static BBU is shut down and the alternative channel is multiplexed to the blue band, via the optical combiner. Thus, the RRHs in the static sub-network can still work normally, while the corresponding dynamic RRHs are disabled. The static BBUs are protected because the sub-network that the static BBUs are assumed to carry more important traffic than that the dynamic BBUs carry. This assumption is reasonable as the static BBUs serve a complete network supporting the whole area at off-peak time and the dynamic BBUs are supplementary to the static networks at peak time.

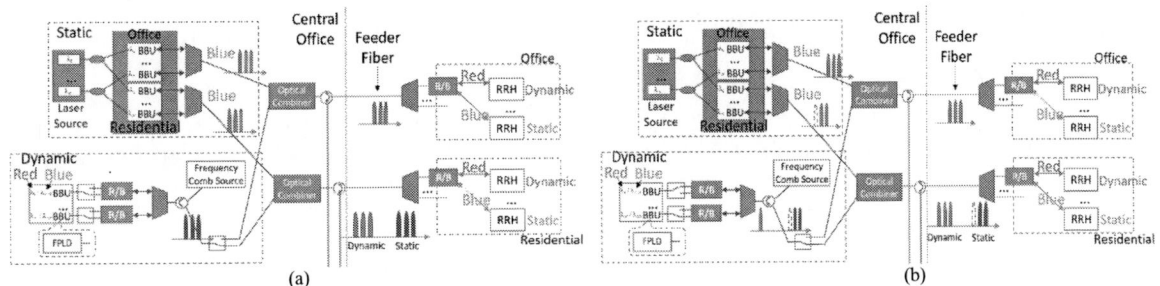

Fig. 2. The illustration of (a) the dynamic provisioning mode and (b) the protection mode.

III. EXPERIMENT AND RESULTS

At each dynamic BBU, a Fabry-Perot laser diode (FPLD) is employed as the optical source and it is injection-locked by an optical frequency comb source, or spectrally-sliced broadband light source. Here, we have analysed the performance of different channels of the FPLD in both red and blue bands, as a slave laser in optical injection locking. Fig. 3 shows the experimental setup. We employed two external cavity lasers (ECLs) as the master lasers to emulate two wavelengths in the optical comb source, as they were wavelength tunable. They provided different wavelengths in the red band and the blue band, separately, which could pass the same port of the cyclic AWG, simultaneously. The cyclic AWG had a channel spacing of 100 GHz, thus the free spectral range (FSR) at each port was 1.6 THz, as the dashed blue curve shown in Fig. 4(a). The red/blue multiplexer had the spectral range of 1547.6-1559nm and 1532-1542nm for the red band and the blue band, respectively, as dash-dot pink curve and the dot green curve, as shown in Fig. 4(a). We used a 2.5-Gbit/s FPLD as our slave laser and the FSR of the FPLD was around 146 GHz, shown as the solid red curve in Fig. 4(a). The FPLD was directly modulated by a pulse pattern generator (PPG) in non-return-to-zero (NRZ) format with a $2^{15}-1$ pseudorandom binary sequence (PRBS). The modulated signal passed through the optical switch, the red/blue multiplexer and the cyclic AWG and was then transmitted over 20-km single-mode fiber, before being detected by a PIN photodiode. The received electrical signal was amplified and filtered out by proper low-pass filter (LPF) and was analysed by a bit error rate tester (BERT).

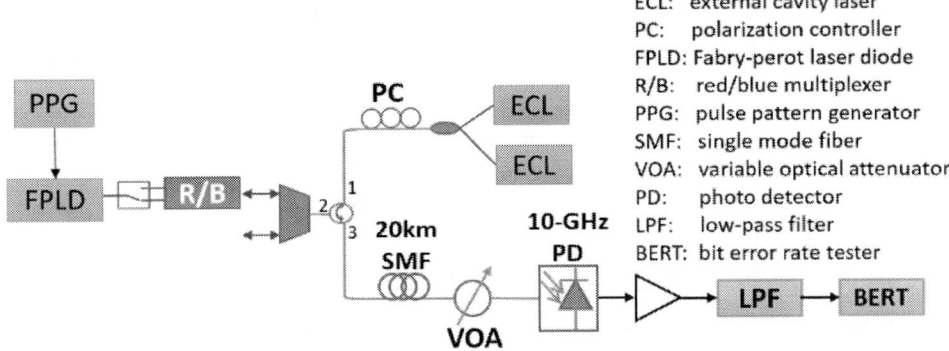

ECL: external cavity laser
PC: polarization controller
FPLD: Fabry-perot laser diode
R/B: red/blue multiplexer
PPG: pulse pattern generator
SMF: single mode fiber
VOA: variable optical attenuator
PD: photo detector
LPF: low-pass filter
BERT: bit error rate tester

Fig. 3. Experimental setup.

We have tested the performance of six different channels with wavelengths of 1534.9 nm, 1539.5 nm, 1541.9 nm, 1547.7 nm, 1552.5 nm and 1554.7 nm. They formed three pairs where each pair could pass through the same port of the cyclic AWG, while one was in the red band and the other was in the blue band. The pairs were chosen on purpose that the wavelength of 1541.9/1554.7 nm and 1534.9/1547.7 nm were at the edge of the red/blue multiplexer, while the last pair was randomly chosen within the range of the red and the blue bands. Three different transmission rates were studied which were 2.5 Gbit/s, 3 Gbit/s and 4 Gbit/s. The LPFs used at the receiver were with bandwidths of 2.4 GHz, 2.4 GHz and 3.4 GHz, respectively.

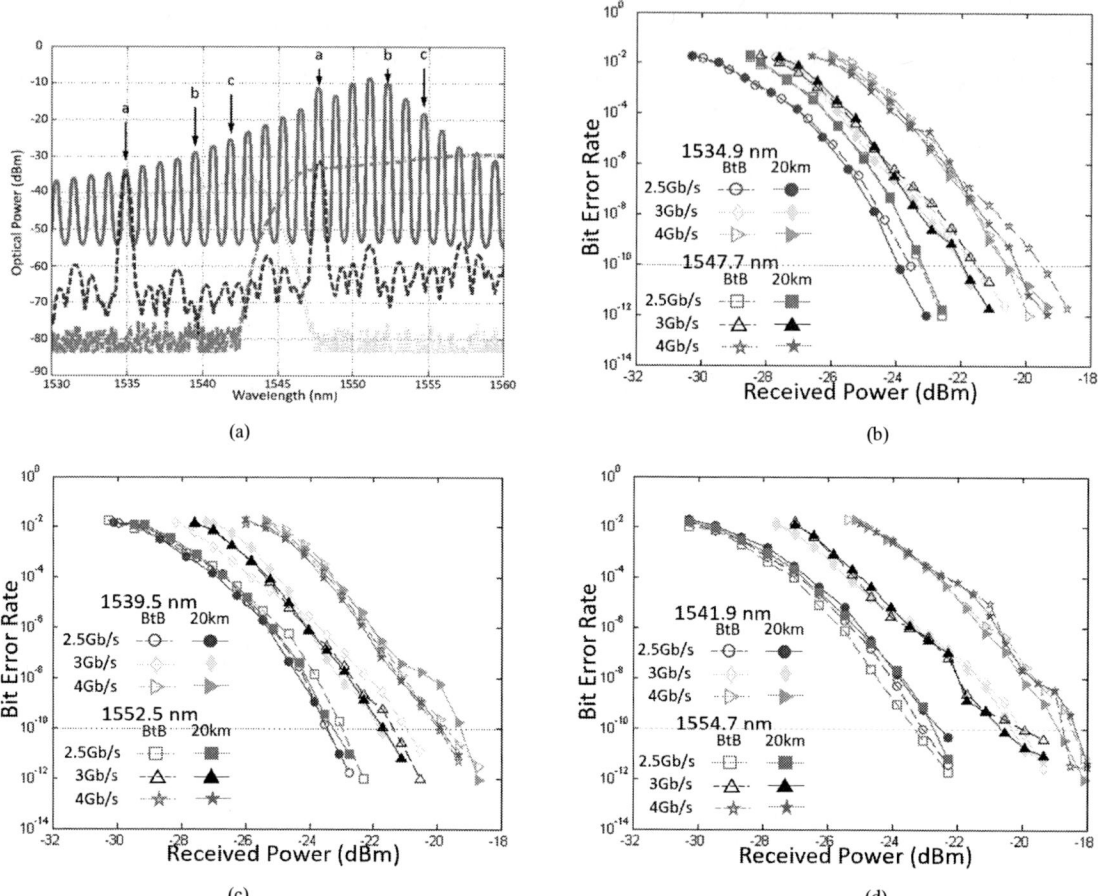

Fig. 1. (a) Spectra of free-run FPLD (solid red), red/blue multiplexer (dash-dot pink/dot green) and a port of cyclic AWG (dash blue); BER curves for (b) 1534.9nm, 1547.7nm (label a in (a)), (c) 1539.5nm, 1552.5nm (label b in (a)), (d) 1541.9nm and 1554.7nm (label c in (a)) channels.

Figs. 4(b)-(d) show the measured BER curves of the six channels. The six channels had very similar performance, in terms of the BER, even though the free-running FPLD had different output powers at these wavelengths. This might be attributed to the high-quality injection locking, as the side-mode suppression ratio of the FPLD was over 40 dB after injection locking. The 20-km transmission exhibited penalty free performance as compared to the back-to-back case, for the relatively low transmission rates. For the same channel, the power penalties between 2.5-Gbit/s and 3-Gbit/s transmission as well as between 3-Gbit/s and 4-Gbit/s, were both around 2 dB. Error free transmissions were achieved at all the three transmission rates, but could hardly be achieved at 5-Gbit/s, which was far beyond the FPLD modulation bandwidth. However, the transmission rate could be further improved by employing pre-emphasis or OFDM format.

IV. SUMMARY

We have proposed a mobile fronthaul architecture for dynamic provisioning in mobile networks and protecting the main BBUs against possible BBU failure. Injected locked FPLDs are employed as optical sources at the BBUs. With these optical sources, we have demonstrated 20-km transmission at the data rates of 2.5 Gbit/s, 3 Gbit/s and 4 Gbit/s, among six different channels, in both the red band and the blue band. The similar performances among the channels in the red and the blue bands have confirmed that our proposed architecture can provide high-quality and cost-effective downstream transmissions in both dynamic provisioning and protection modes in future mobile fronthaul systems.

REFERENCES

[1] China Mobile, "C-RAN: the road towards green RAN," White Paper, version 2, 2011.
[2] W. Du, H. Xin, H. He and W. Hu, "A resource sharing C-RAN architecture with wavelength selective switching and parallel uplink signal detection," Asia Communication and Photonic Conference (ACP 2015), Hong Kong, 2015.

TuA4-5

Effect of Chromatic Dispersion on Cable Re-routing Operation Support System with No Service Interruption

Kazutaka Noto[†], Masaaki Inoue[‡], Hiroshi Watanabe[†], Yusuke Koshikiya[†],

Keiji Okamoto[†] and Tetsuya Manabe[†]

† NTT Access Network Service Systems Laboratories, NTT, 1-7-1, Hanabatake, Tsukuba, Ibaraki 305-0805, Japan

‡ NTT EAST CORPORATION, NTT EAST, 3-19-2, Nishishinjuku, Shinjuku, Tokyo 163-8019, Japan

noto.kazutaka@lab.ntt.co.jp

Abstract: We report the effect of chromatic dispersion on a telecommunications system when duplicating signal transmissions in a cable re-routing operation support system with no service interruption.

Keywords: cable re-routing operation support system with no service interruption; chromatic dispersion; transmission line duplication.

I. INTRODUCTION

Optical network maintenance while retaining high communication quality has become an important social infrastructure with the spread of telemedicine, mobile phones and cloud computing. Therefore, the communication interruption caused by failures or maintenance must be minimized. When we change optical access line routes, customers are affected because service is interrupted when we disconnect the line.

A cable transfer splicing system for optical fiber cable switching has already been proposed [1]. However, this system interrupts the service while mechanically changing line connections. Therefore, we have proposed a cable re-routing operation support system with no service interruption that temporarily duplicates the current line with a detour line while we change the route [2]. The system has to calibrate the time delay between current and detour signals by using measurement results for test wavelengths. In the system, the optical path lengths and the wavelength characteristics of the group velocity of both the current and detour lines are required in order to calibrate the time delay. However, it is difficult to measure the wavelength characteristic of the group velocity without interrupting the service. Moreover, the group velocity cannot be uniquely determined, because the group velocity range is calculated from IEEE 802.3ah [3] and ITU-T G.652 [4]. Therefore, the time delay has been calibrated using the average group velocity value. However, time-delay calibration errors cause service interruption when the time-delay calibration is larger than the pulse width of the transmission signal. A system based on the above would have a limited application range and an extended work time. To cope with the above problems, we propose a method that does not require time-delay calibration. This method is designed to shorten the distance that signal lights of different wavelengths propagate through the same route. Therefore, this method makes it possible to expand the application region by reducing the calibration time delay.

II. THEORY

Fig. 1 shows the configuration of our cable re-routing operation support system with no service interruption. In this configuration, the measurement part for evaluating the propagation time difference between detour and current lines is divided into measurement parts for the upstream and downstream signals. The previously reported measurement part used a digital phase/frequency detector [5]. This configuration uses a variable electric delay line (DL). Small form factor pluggables were prepared as the optic-electric (O/E) or electric-optic (E/O) converters. For the downstream direction, the propagation time difference between the detour and current lines of the measurement part Δ and of optical network units (ONU) $\Delta + \Delta C$ are given by

$$\Delta = \frac{L2 - L5 + L7}{v_{olt}} - \frac{L8 + L7}{v_{test}} - \frac{L6}{v_{olt'}}, \tag{1}$$

$$\Delta + \Delta C = \frac{L2 + L3 - L5}{v_{olt}} - \frac{L3 + L6 + L8}{v_{olt'}}, \tag{2}$$

where, v_{olt} is the group velocity of the wavelength of an optical line terminal (OLT), v_{test} is the group velocity of the test wavelength, $v_{olt'}$ is the group velocity of the downstream wavelength of the cable re-routing operation support system with no service interruption, and $L1 \sim L8$ are the optical path lengths of each path. The time-delay calibration ΔC is given by using Eqs. (1) and (2),

$$\Delta C = L3 \left[\left(\frac{1}{v_{olt}} - \frac{1}{v_{olt'}} \right) + \frac{L7}{L3} \left(\frac{1}{v_{test}} - \frac{1}{v_{olt}} \right) + \frac{L8}{L3} \left(\frac{1}{v_{test}} - \frac{1}{v_{olt'}} \right) \right] \cong L3 \left(\frac{1}{v_{olt}} - \frac{1}{v_{olt'}} \right). \tag{3}$$

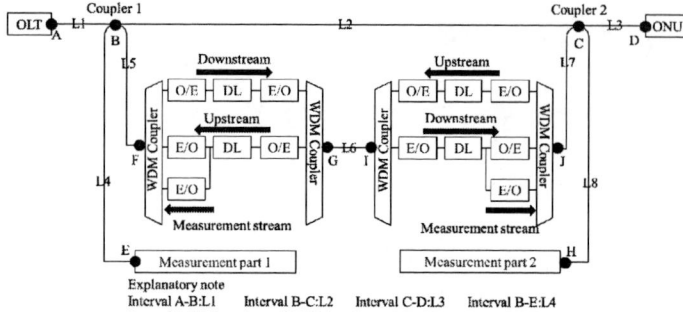

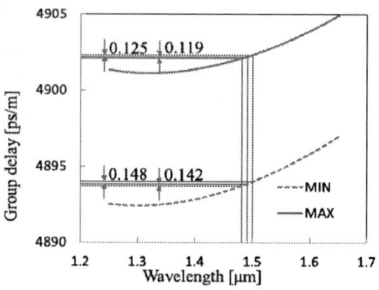

Fig. 1. Configuration of cable re-routing operation support system with no service interruption.

Fig. 2. Numerical analysis results when the maximum and minimum relative index differences are those specified in ITU-T G.652.

In Eq. (3), the first term is dominant because *L7* and *L8* are sufficiently smaller than *L3* in general. Moreover, any length of *L7* and *L8* optical fiber can be selected because the optical fiber can be retrofitted to the cable re-routing operation support system with no service interruption. The application range of the proposed system does not depend on the optical path length between the couplers because it is unaffected by optical path lengths *L2* and *L6*. However, Eq. (3) shows that the proposed system is affected by the optical path length *L3*. For example, when the optical path length L3 is 10 km, which is the maximum value specified in IEEE 802.3ah, the time-delay calibration *ΔC* is -1421 ps from 1484 ps by using Eq. (3) and Fig. 2, where, the OLT wavelength is 1480 nm from 1500 nm, the wavelength of the ONU is 1260 nm from 1360 nm, and the test wavelength is 1650 nm by using IEEE 802.3 ah. Fig. 2 shows numerical analysis results obtained with the maximum and minimum relative index differences specified in ITU-T G.652 [6]. When the signaling speed is faster, the time-delay calibration *ΔC* must be calculated. However, the group velocity cannot be measured without service interruption. Therefore, in this study, we confirmed experimentally the application range of our proposed system that does not use the time-delay calibration *ΔC*.

III. EXPERIMENT AND RESULTS

Fig. 3 shows the experimental setup for measuring the effect on the service of the optical path length between the couplers. For this experiment, the ONU and OLT conform to 1000BASE-BX10 of IEEE 802.3ah. The OLT wavelength is converted to 1471nm by using a 1.25 Gbps small form factor pluggable (SFP). A detour line was prepared between optical couplers 1 and 2 in which the signal wavelength was converted to 1491 nm which is substantially different from the original wavelength of 1471 nm. The transmitted optical power of the proposed system is adjusted so that the optical power received by the ONU is the same in both the current and detour lines. We evaluated the performance of the proposed system by measuring the frame loss of the transmitted and received signals using TestCenter [7]. We measured the bit error rate (BER) according to the optical path length between the couplers. Fig. 4 shows the dependence of the BER characteristics of the system on the optical path length between the couplers from 0 to 20 km. These results show that the power penalty of our proposed cable re-routing operation support system with no service interruption can be reduced to 1 dB or less, if the optical path length between the couplers is 20 km or less.

Fig. 5 shows the experimental setup we used for measuring the effect on the service of the optical path length between the coupler and ONU. We measured the BER dependence on the optical path length between the coupler and the ONU. Fig. 6 shows the dependence of the BER characteristics of the proposed system on the optical path length between the coupler and ONU from 0 to 4 km. These results show that the power penalty of our cable re-routing operation support system with no service interruption can be reduced to 1 dB or less, if the optical path length between

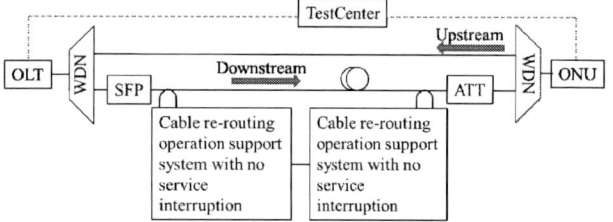

Fig. 3. Experimental setup for measuring the effect on the service of the optical path length between the couplers.

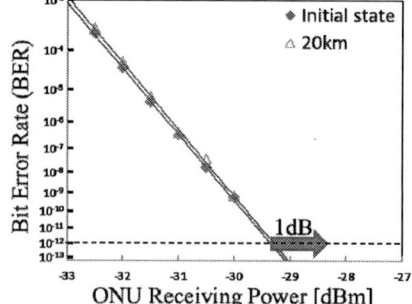

Fig. 4. Dependence of BER characteristics of the proposed system on the optical path length between the couplers from 0 to 20 km.

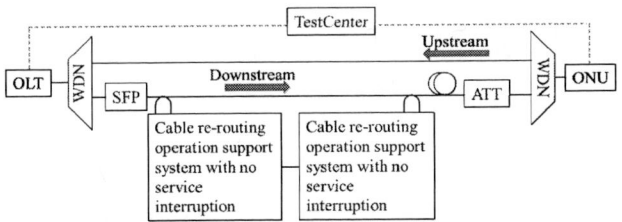

Fig. 5. Experimental setup for measuring the effect on the service of the optical path length between the coupler and ONU.

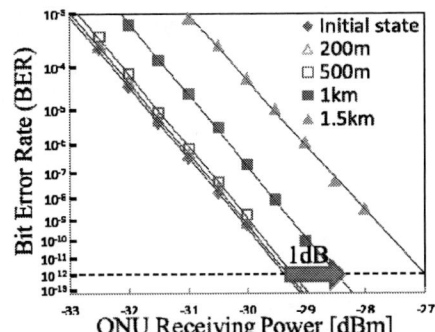

Fig. 6. Dependence of BER characteristics of the proposed system on the optical path length between the coupler and ONU from 0 to 4 km.

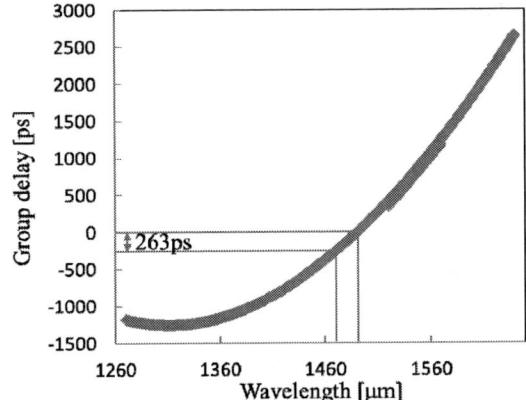

Fig. 7. Group velocity of 1 km of optical fiber.

the coupler and ONU is 1 km or less. Fig. 7 shows the group velocity of 1 km of optical fiber. The difference between the OLT signal arrival time at the ONU with the current line and the detour line is 263 ps according to Fig. 7. Therefore, these results show that the power penalty of our proposed system for changing optical access line routes without service interruption can be reduced to 1 dB or less, if the propagation time between the coupler and ONU is 263 ps or less. The proposed system without the time-delay calibration ΔC can be adapted to 1770 m in the wavelength range of IEEE 802.3ah in Fig. 2 because, 263 ps divided by 0.148 ps/m which is maximum group delay difference per unit length of IEEE 802.3ah in Fig. 2 is 1770 m.

IV. CONCLUSIONS

Our proposed cable re-routing operation support system with no service interruption is applicable to optical path lengths between couplers of up to 20 km. Moreover, the system is applicable to optical path lengths between couplers and ONUs of up to 1.8 km. This optical path length is sufficient for most buildings.

REFERENCES

[1] K. Tanaka, M. Zaima, M. Tachikura and M. Nakamura, "Downsized and enhanced optical fiber transfer splicing system", 51st The International Cable Connectivity Symposium (IWCS 2002), Florida (USA), Nov. 2002, pp.680-686.

[2] T. Manabe, K. Noto, M. Inoue, K. Katayama, N. Honda, and Y. Azuma, "System for Changing Optical Access Line Routes without Service Interruption using Photoelectric Converters and Electric Variable Delay Lines," 62nd The International Cable Connectivity Symposium (IWCS 2013), Providence (USA), Nov. 2013, pp.35-40.

[3] IEEE Std. 802.3ah (2004), "1000BASE-BX10", http://www.ieee802.org/21/doctree/2006_Meeting_Docs/2006-11_meeting_docs/802.3ah-2004.pdf, referenced in February 2016.

[4] ITU-T Std. G.652 (2009), https://www.itu.int/rec/dologin_pub.asp?lang=e&id=T-REC-G.652-200911-I!!PDF-E&type=items, referenced in February 2016.

[5] M. Inoue, T. Manabe, K. Noto, K. Katayama, N. Honda, and Y. Azuma, "High-resolution Delay Measurement between Duplicated Transmission Lines", 18th OptoElectronics and Communications Conference (OECC2013), Kyoto (Japan), Jun. 2013, pp. TuPT_7.

[6] K. Nakajima, "Optical Fiber Technologies for FTTH and Large Capacity Transmission Network", J. IEICE vol. 97, pp. 54-59, Jan. 2014.

[7] Spirent Communications, "Spirent TestCenter EDM-1003B", http://www.spirent.com/Networks-and-Applications/LTE_Testing/~/media/Datasheets/Broadband/PAB/SpirentTestCenter/STC_Series_1000-2000_GbE_Test_Modules_datasheet.ashx, referenced in February 2016.

TuB1-1 (Invited)

OECC/PS2016

Multidimensional Modulation Formats for Optical Communication

Magnus Karlsson

Photonics Laboratory, Dept. of Microtechnology and Nanoscience, Chalmers University of Technology, Sweden
magnus.karlsson@chalmers.se

Abstract: *We review the research on multidimensional modulation formats for coherent optical transmission systems. Examples of lattice-based formats in 4 and 8 dimensions are given, and a number of performance metrics are discussed.*
Keywords: *(coherent communication, modulation)*

I. INTRODUCTION

The emergence of coherent optical transmission systems in 2008 and onwards [1,2] enabled generation, transmission and detection of arbitrary signals over optical fibers. In particular, modulation in both quadratures and both polarization states became practical. Thus, coherent fiber transmission systems could be simply modeled as a four-dimensional signal space with additive white Gaussian noise from the inline amplifiers.

This also meant an increased freedom in designing modulation formats. Especially the limit of high signal to noise ratio (*SNR*) is interesting from a practical perspective. In such a limit the optimum formats (in the sense of requiring least *SNR* for a certain symbol error rate (*SER*)) becomes a simple geometric problem, namely that of packing M hard spheres in N-dimensional (N-d) space while minimizing their average (or alternatively maximum) distance to the origin. This contribution will discuss some of the resulting formats and review their performance in transmission experiments.

II. METRICS FOR COMPARISON OF MODULATION FORMATS

The first metric we need to compare various formats is the *spectral efficiency* β. For an M-ary constellation in N-d space, the spectral efficiency is given by $\beta=2 \log_2(M)/N$, in units of bits per symbol per dimension pair. For example 2-d quadrature shift keying, QPSK, has $M=4$, $N=2$ and $\beta=2$, which is the same as 4-d QPSK (e.g. polarization multiplexed, PM-QPSK) which has $M=16$, $N=4$ and the same $\beta=2$. The reason for this is that we define the spectral efficiency per channel (dimension pair) used, and more parallel channels will thus not affect the spectral efficiency.

The second metric we need is some measure of the required signal to noise ratio of the constellation to reach a certain symbol-error rate. This can be obtained in various ways; by either considering the average or the maximum power used in the constellation. The average symbol energy is given by the second moment of the constellation E_s, which is the average squared distance from the origin of the constellation. If the minimum distance between two constellation points is d_{min}, the *constellation figure of merit* is $CFM=d_{min}^2 N/(2E_s)$, [3]. It is a relevant measure when comparing constellations (or modulation formats) at the same total bandwidth, summed over all dimension pairs $N/2$. The other common metric is the *asymptotic power efficiency* $\gamma=d_{min}^2 \log_2(M)/(4 E_s)$, [4,5] which is the SNR gain over QPSK at the same data rate per dimension pair. It should be emphasized that these metrics are relevant in the high-SNR regime, where the minimum distance between two points in the constellation will limit the symbol error rate.

At a given cardinality M, dimension N, the dimensionless ratio d_{min}^2/E_s can be maximized by sphere packing simulations, i.e. keeping the sphere diameter (which is also the distance between two neighboring spheres) d_{min} constant and minimizing the squared average distance E_s. Such minimum constellations are referred to as *clusters* $C_{N,M}$. The clusters maximize γ and are referred to as the most power efficient constellations. They are usually unique (apart from rotations) and a discussion on the low-dimensional clusters are given in [6] and a list is available online [7]. The γ and *CFM* for the clusters in low dimensions are plotted vs. spectral efficiency in Fig 1. It can be shown that for every dimension N, there is at least one $M>2$ that maximizes γ, and for dimensions 2,4,8 they are 3, 8 and 16. It can be seen that the CFM, on the other hand, is monotonically decreasing with the number of constellation points.

Other metrics such a *mutual information* (MI) or *generalized mutual information* (GMI) are important to study ultimate limits for the modulation formats, such as the achievable data rate at a certain SNR, but they are more difficult to calculate, requiring often numerical integration. The MI depend only on the format and the channel, whereas the GMI accounts for also the bit-to symbol mapping, and is the relevant metric in a bitwise receiver (which is the case for almost all receivers in use today). At high SNR, MI and GMI approaches a data rate of $\log_2(M)$ bits per symbol per N dimensions. They can be interpreted as the limiting performance of the formats given an infinitely long code.

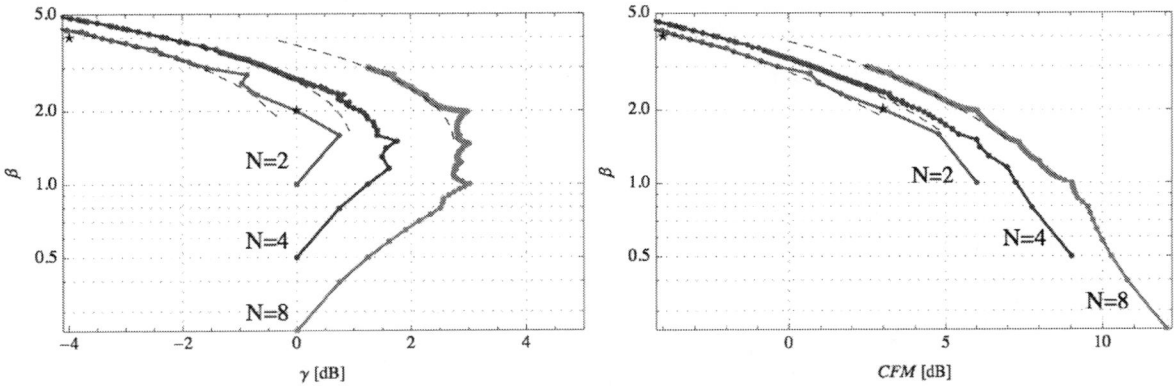

Fig. 1. Scatter plots of spectral efficiency β vs. asymptotic power efficiency γ (left) and *CFM* (right) for the clusters $C_{N,M}$. in dimensions N=2 (red), 4 (blue), and 8 (green). The black star denotes 2-d QPSK and 16-QAM. The dashed lines are the optimum lattice packings in respective dimension.

III. USEFUL MODULATION FORMATS

Clearly constellations where the number of points are a power of two are of most interest, as they can be used to transmit an integer number of bits. We restrict ourselves to regular structures, so called lattices, since they are often easier to generate and detect than irregular structures. The N-dimensional cubic lattice, Z_N, consists of all Nd vectors with integer coordinates, and it is important as it is particularly simple to generate and detect. The power efficiency for a spherical cut of M points from a lattice is $\gamma_{lat} = \log_2(M) (1+2/N) (\Delta/M)^{(2/N)}$, where Δ is the lattice density, i.e. the fraction of Nd space that is comprised by spheres in the lattice points. The corresponding $CFM_{lat} = \gamma_{lat} 2N/\log_2(M)$. This is plotted as dashed lines in Fig 1, using Δ-values of $\pi/(2\sqrt{3})$, $(\pi/4)^2$, $\pi^4/384$ for the respective A_2, D_4 and E_8 lattices which are the best lattice packings in 2,4 and 8 dimensions.

In 4d much of the research on power efficient constellations have been based on various cuts of M points from the D_4 lattice. The so-called set-partitioned (SP-)QAM formats can be obtained either as square cuts from the D_4 lattice, or as set partitions of the square 4d lattice Z_4 [8,9]. A third way of obtaining them is by applying a single parity check code on gray-coded square QAM constellations. Whereas the rectangular (square lattice) PAM formats have a relationship between the spectral efficiency and the power efficiency as $\gamma_{QAM} = 3\beta/(2(2^{\wedge}(\beta+1/2)-1))$ which holds for 4d and $\log_2 M=4,8,12...$, the SP-QAM formats have $\gamma_{SP-QAMa} = 3\beta/(2^{\wedge}(\beta+1/2)-1)$ for $\log_2(M)=3,7,11...$, and $\gamma_{SP-QAMb} = 3\beta/(2^{\wedge}(\beta+1/2)-1/2)$ for $\log_2(M)=5,9,13...$. Thus the SP-QAM formats fall in between the regular QAM formats in terms of spectral efficiency and have been suggested to be used in flexible networking where the data rate and reach can be adapted by changing the modulation format [10]. Experimental work on these 4d formats started with Sjödin's demonstration of PS-QPSK transmission [11], and implementations of other SP-QAM formats have been done by, e.g., Rios-Muller [12] for 32-SP-QAM and Eriksson [13] for 128-SP-QAM. Extensions to 8d can be done by using two subsequent time slots [14,15] or by transmitting correlated data on two parallel wavelengths [16]. The used format in [14-16] is the *biorthogonal* 8d modulation. Biorthogonal modulation means that all modulation level are of the form (±1,0,0,...), i.e. one coordinate is +1 or -1 and the rest are zero. The corresponding Nd polytope is called the *cross-polytope* and its bit-error rate and symbol error rate as a function of SNR can be derived analytically for any dimension [17]. As can be seen in Fig. 1, the conjectured most power efficient format in 8d is a 16-ary format, and just as for 4d, it is the biorthogonal modulation, and hence it use in [14-16]. The variant proposed by Shiner et al. [16] actually improved over [14-15] by rotating the format in the 8d. The idea was the force the relative polarization states for each symbol pair to be orthogonal, which also increased the tolerance to nonlinear distortions.

IV. FORMATS BASED ON REED-MULLER CODES

Another way of finding good modulation formats in higher dimension is to view them as codes [18]. This means that one starts from a simple Nd modulation format, e.g. BPSK in N parallel dimensions, and assigns levels as one does for N binary bits in an error-correcting code frame. Particular promising codes in that respect are Reed-Muller (RM) codes, which are described by two integers RM(r,u) $u \geq 1$ and $0 \leq r \leq$ u. The frame length n, the number of info bits k and the Hamming distance of the code words d_H of the RM(u,r) code is $(n,k,d_H)=(2^u, \sum_{i=0}^r \{u,i\}, 2^{(u-r)})$, where $\{u,i\}$ denotes the binomial coefficient. The resulting dimensionality is $N=n$, the number of constellation points is $M=2^k$, and the power efficiency will be $\gamma_{RM} = d_H k/n$, the spectral efficiency is $\beta=2k/n$, and this is plotted in Fig 2a for RM codes with lines connecting the same value of u ranging from 2...6. The corresponding dimensionality $N=2^u$ is marked in Fig 2a. Notably the known optima (in the sense of maximum gamma) for dimensions 4 and 8 are given by the RM(1,2) and the RM(1,3) codes. The codes RM(1,u) corresponds to the biorthogonal modulation discussed above. Clearly the RM codes seem to find the (known) most power efficient format in the lower

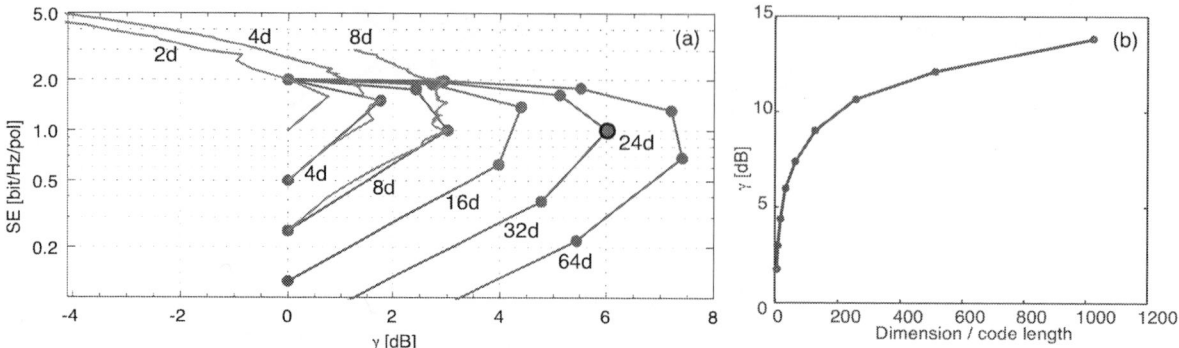

Fig. 2. (a): Performance of the Reed-Muller code RM(r,u)-based formats applied to Nd BPSK. The lines connect set with the same dimensionality 2^u, and r ranges from 0 (lower points, the repetition code) to u (upper points, uncoded). The black circle shows the 24d Golay code. The clusters from fig 1 is shown for reference in gray. (b): Plot of the maximum γ for each dimension 2^u (u ranging from 2 to 10) for the RM codes.

dimensions. A number of important classes of codes are actually special cases of the RM codes, for example the single-parity check (SPC) codes RM(u-1,u), and the extended Hamming codes RM(u-2,u). The trivial cases RM(u,u) and RM(0,u) are uncoded and the repetition codes, respectively. Another interesting code is the (24,12,8) extended 24d Golay code which is shown as the black circle in Fig 2(a) and discussed in [18,19]. Evidently it has the same performance as the RM(2,5) code, although the latter is 'less efficient' in that it requires a 32 dimensional space.

In fig 2(b) we show the highest power efficiencies for the RM-based modulation formats as a function of dimensionality. One may note that the asymptotic power efficiency increases slowly with dimensionality, but it is not clear what the upper limit is for so high dimensions. Obviously much better codes do exist, but to implement them as modulation formats is likely not as attractive for high dimensions. Rather, conventional encoders and decoders would be more practical for long codes, but there are actually benefits from using codes as modulation formats in this way. For example receiver signal processing algorithms converge better for more well-separated constellations, soft decision decoding (=maximum likelihood decoding) might be easier to implement, and the latency might be lower.

CONCLUSIONS

To conclude, we have reviewed the area of multidimensional modulation in fiber optical systems, and shown how modulation formats can be interpreted as codes and exemplified by using Reed-Muller codes in that application.

REFERENCES

[1] H. Sun et al., Real-time measurements of a 40Gb/s coherent system, Optics Express vol. 16, pp. 873–879, 2008.

[2] G. Charlet et al., Performance comparison of singly-polarised and polarisation-multiplexed coherent transmission at 10 Gbaud under linear impairments, Electronics Letters vol. 43, pp. 1109–1111, 2007.

[3] G. D. Forney, Jr. Multidimensional constellations—Part I: Introduction, figures of merit, and generalized cross constellations. IEEE J. Sel. Areas Commun. vol. 7, pp. 877– 892, 1989.

[4] S. Benedetto and E. Biglieri. *Principles of Digital Transmission: With Wireless Applications*. Kluwer Academic Publishers, 1999.

[5] E. Agrell and M. Karlsson. Power-efficient modulation formats in coherent transmission systems. Journal of Lightwave Technology vol. 27, pp. 5115–5126, 2009.

[6] M. Karlsson and E. Agrell. Power-efficient modulation schemes. In S. Kumar, editor, *Impact of Nonlinearities on Fiber Optic Communications*, chapter 5, pages 219–252. Springer, 2011.

[7] E. Agrell. Database of sphere packings. Online: http://codes.se/packings, 2014.

[8] L. Coelho and N. Hanik. Global optimization of fiber-optic communication systems using four-dimensional modulation formats. In *European Conference on Optical Communication*, page Mo.2.B.4. 2011.

[9] M. Karlsson and E. Agrell. Spectrally efficient four-dimensional modulation. In *Optical Fiber Communication Conference*, page OTu2C.1, 2012.

[10] J. K. Fischer et al., Bandwidth-variable transceivers based on four-dimensional modulation formats. Journal of Lightwave Technology vol. 32, pp.2886–2895, 2014.

[11] M. Sjödin et al., Comparison of polarization-switched QPSK and polarization-multiplexed QPSK at 30 Gbit/s. Optics Express vol. 19, pp. 7839–7846, 2011.

[12] R. Rios-Muller et al., Experimental comparison between hybrid-QPSK/8QAM and 4D-32SP- 16QAM formats at 31.2 GBaud using Nyquist pulse shaping. In *European Conference on Optical Communication*, page Th.2.D.2. 2013.

[13] T. A. Eriksson et al., Experimental demonstration of 128-SP-QAM in uncompensated long-haul transmission. In *Optical Fiber Communication Conference*, page OTu3B.2. 2013.

[14] T. A. Eriksson et al., Biorthogonal modulation in 8 dimensions experimentally implemented as 2PPM-PS-QPSK. In *Optical Fiber Communication Conference*, page W1A.5. 2014.

[15] A. D. Shiner et al., Demonstration of an 8-dimensional modulation format with reduced inter-channel nonlinearities in a polarization multiplexed coherent system. Optics Express vol. 22, pp. 20366–20374, 2014.

[16] T. A. Eriksson et al., Frequency and polarization switched QPSK. In *European Conference on Optical Communication*, page Th.2.D.4. 2013.

[17] E. Agrell and M. Karlsson. On the symbol error probability of regular polytopes. IEEE Transactions on Information Theory vol. 57, pp. 3411–3415, 2011.

[18] D. S. Millar et al., High-dimensional modulation for coherent optical communications systems. Optics Express vol. 22, pp. 8798–8812, 2014.

[19] D. S. Millar et al., Experimental demonstration of 24-dimensional extended Golay coded modulation with LDPC. In *Optical Fiber Communication Conference*, page M3A.5. Optical Society of America, 2014.

TuB1-2

Optimization of Networks Carrying Super-Channels with Different Modulation Formats for Maximum Spectral Efficiency and Reach

Olga Vassilieva[1], Tomohiro Yamauchi[2], Shoichiro Oda[2], Inwoong Kim[1], Motoyoshi Sekiya[2],
Takeshi Hoshida[2], Yasuhiko Aoki[2], Jens C. Rasmussen[2] and Tadashi Ikeuchi[1]

(1) Fujitsu Laboratories of America, Inc., 2801 Telecom Parkway, Richardson, Tx 75082, USA
(2) Fujitsu Laboratories Ltd., 1-1 Kamikodanaka 4-chome, Nakahara-ku, Kawasaki 211-8588, Japan
Olga.vassilieva@us.fujitsu.com

Abstract: *Optimization of optical network carrying mix of 1Tb/s 5sc-DP-16QAM and 400Gb/s 4sc-DP-QPSK superchannels reveals that subcarrier power pre-emphasis together with 12.5GHz guard band can maximize capacity-reach performance of each superchannel, while guard band alone can't.*
Keywords: *flexible grid, superchannel, power pre-emphasis, spectral efficiency, reach, nonlinearity, guard band*

I. INTRODUCTION

Optical networks are moving towards flexible and adaptive networks with variable modulation format, data rate and channel spacing to satisfy high capacity demands. The universal transceivers of such networks can provide adaptive modulation for improved utilization of capacity and reach [1]. And the flexible grid networks can deliver more efficient use of the optical bandwidth, whose gains strongly rely on the order of modulation format and spacing between channels [2]. To increase network capacity even further, the superchannels can be deployed, where multiple channels or so-called subcarriers (SC) are placed close to each other and transmitted through a network as single entity. However, nonlinear interactions between tightly spaced subcarriers can limit transmission reach [3, 4]. In our previous work we have confirmed numerically and experimentally that center subcarriers can experience larger OSNR penalties compared to edge subcarriers [5, 6]. We have proposed subcarrier power pre-emphasis (SPP) technique to equalize performance of all subcarriers and extend reach of center subcarriers. We have also shown that when multiple superchannels of the same modulation format were transmitted through the network, the performance of all subcarriers was affected by the neighboring channels and the guard band (GB) between them [7, 8]. We have experimentally confirmed that using optimum SPP together with small GB of 12.5 GHz, the maximum SE-reach performance of networks carrying uniform traffic with DP-16QAM and DP-QPSK superchannels could be improved by 62% and 45%, respectively [7].

However, as we mentioned above, in order to maximize network capacity, the superchannels with different modulation formats and data rates will propagate side by side throughout a network. Thus, these channels will experience different nonlinear (NL) interactions with each other. Therefore, it is very important to understand the range of NL interactions between superchannels of various configurations and, subsequently, optimize such networks for their mutual benefit. In this work, we extend experimentally confirmed physical layer optimization procedure of [7] to flexible grid network deploying mix of 1Tb/s 5sc-DP-16QAM and 400Gb/s 4sc-DP-QPSK superchannels. We demonstrate that due to different NL interactions between co-propagating superchannels, GB can effectively extend reach L of DP-16QAM superchannels, while it is not needed for DP-QPSK superchannels. We also reveal that SPP in combination with the small GB can maximize *SE-L* performance of each superchannel.

II. PHYSICAL LAYER NETWORK OPTIMIZATION WITH SUBCARRIER POWER PRE-EMPHASIS AND GUARD BAND

To determine the range of NL interactions between superchannels of various configurations, we consider an optical network with reconfigurable transceivers and reconfigurable optical add/drop multiplexers (ROADMs), as shown on Fig. 1(a). The universal transceiver is programmed to transmit DP-QPSK superchannels over longer distances and higher spectrally efficient DP-16QAM superchannels over shorter distances. When such superchannels co-propagate through the network, the performance of all subcarriers strongly depends on the type of the co-propagating channels and the GB between them. The channels with different modulation formats can impose different inter- and intra-superchannel nonlinearity. The larger GB between superchannels can effectively reduce inter-superchannel NL and impact of passband narrowing (PBN), while the smaller GB is highly desirable to keep a network at high capacity levels. The intra-superchannel NL, on the other hand, depends on the number of subcarriers [5] and imposes larger penalty on center subcarriers when all subcarriers are launched with equal power (see Fig. 1(b)). SPP technique, shown in Fig. 1(c), where edge subcarriers are launched at lower power levels compared to the center subcarriers, has been experimentally proven to reduce OSNR penalties and equalize performance of all subcarriers [5, 6]. Therefore, careful optimization of the GB and SPP can minimize both intra- and inter-superchannel NL, PBN and deliver the optimum performance in terms of the *SE* and reach L.

OECC/PS2016

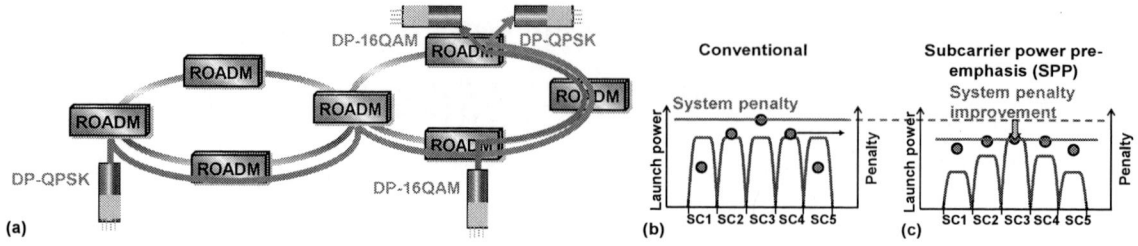

Fig. 1: (a) Schematic diagram of the network with variable modulation format, data rate and channel spacing; (b) Schematic diagram of the launch power per subcarrier and generated system penalty for conventional system and (c) when subcarrier power pre-emphasis is applied.

In this work, we transmit 400Gb/s 4sc-DP-QPSK and 1Tb/s 5sc-DP-16QAM superchannels, analyze their mutual interference as function of the GB and optimize the network for maximum SE and reach L performance. For this purpose, we consider 2 configurations, as shown in Fig. 2: (I) 400Gb/s DP-QPSK superchannel is surrounded by two 1Tb/s 5sc-DP-16QAM superchannels and (II) 1Tb/s 5sc-DP-16QAM superchannel is surrounded by two 400Gb/s DP-QPSK superchannels. The spectral occupancy of 5sc-DP-16QAM superchannel is set to 175 GHz and of 4sc-DP-QPSK superchannel is to 150 GHz, which is 140 GHz plus 5 GHz of unused GB on each side of the spectrum to fit into ITU-T DWDM frequency grid specification. The superchannels are spaced with variable GB ranging from 12.5 GHz to 87.5 GHz. Each subcarrier in both configurations is modulated at 32 GBaud, Nyquist pulse shaped with root raised cosine filter with 0.15 roll-off factor and spaced with 35 GHz. Then, the superchannels of both configurations were transmitted over 15 spans x 60 km (900 km) SMF fiber with uncompensated dispersion. The fiber input power of each subcarrier of Configuration-I without SPP was set to 0 dBm/sc and 4 dBm/sc for 5sc-DP-16QAM and 4sc-DP-QPSK superchannels, respectively. The higher fiber input power of DP-QPSK reflects its higher OSNR and NL tolerance compared to DP-16QAM. Thus, under this scenario, the transmission reach of 4sc-DP-QPSK superchannel can be assumed to be approximately 2.5 times longer than 5sc-DP-16QAM superchannels, which is 2250 km (2.5 x 900 km), at fiber input power 0 dBm/sc. This condition helps us to estimate NL impact of shorter traveling DP-16QAM superchannel on longer traveling DP-QPSK superchannels. On the other hand, the fiber input power of all subcarriers of Configuration-II was set to 0 dBm/sc, which was used to evaluate NL impact of DP-QPSK superchannels on DP-16QAM during their co-propagation over shorter reach (i.e. 900 km), as shown in Fig. 1(a). In addition, each superchannel was transmitted through 4 ROADM nodes with 8.5-th order Gaussian filter and 3-dB bandwidth (BW) as wide as superchannel BW plus GB to account PBN impact on edge subcarriers. At the end of the transmission line, the performance of each subcarrier was evaluated with coherent receiver by tuning local oscillator frequency to a specific subcarrier and DSP processing.

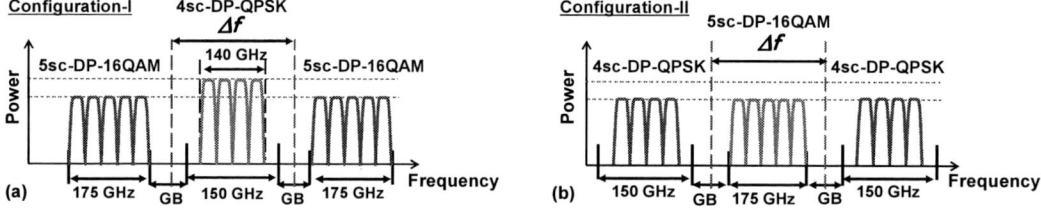

Fig. 2. Spectral allocation of superchannels without subcarrier power pre-emphasis.

Figures 3(a), 3(b) show OSNR penalties, received at BER = 10^{-3}, for two configurations shown in Fig. 2(a), 2(b). We can see that edge subcarriers (blue, solid lines) of 4sc-DP-QPSK superchannel of Configuration-I are less affected by PBN than 5sc-DP-16QAM of Configuration-II when GB=0 GHz. In addition, 5sc-DP-16QAM signals cause negligible inter-superchannel NL penalty on 4sc-DP-QPSK of Configuration-I, resulting in no OSNR penalty improvement due to larger GB. The NL impact of 4sc-DP-QPSK on 5sc-DP-16QAM superchannel of Configuration-II, on the other hand, can be reduced with larger GB. In addition, we clearly see the difference in performance between center and edge subcarriers in both configurations due to large intra-superchannel NL. SPP (shown as dotted lines), with optimum power per subcarrier shown in the insets on the plots can contribute to additional 0.8 to 0.9 dB OSNR improvement and equalize performance of all subcarriers. Thus, carefully optimized GB and SPP can extend reach of both configurations.

Next, the reach extension L in % with- and without SPP is demonstrated on Fig. 4(a). Once again we confirm that GB is not needed for reach extension in Configuration-I, while it can bring up to 25% reach extension with larger GB for Configuration-II. SPP is very effective for both configurations and can further extend reach by additional 21% and 15% for Configuration-I and Configuration-II, respectively. This plot can be used as a guideline for the choice of GB for a desired maximum system reach, assuming that the network is not capacity-hungry.

Next, we optimize capacity-hungry networks for maximum SE-L performance. To do so, we define SE as $SE = 2 \times M \times N \times SR/\Delta f$, where 2 is the number of polarizations, M is the number of bits per symbol in the constellation, N is the number of subcarriers in superchannel, SR is the symbol rate and Δf is the BW occupied by superchannel, including GB. The definitions of the GB and the BW Δf of each superchannel are shown on Fig. 2.

478

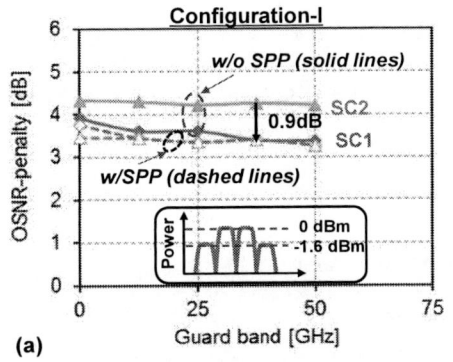 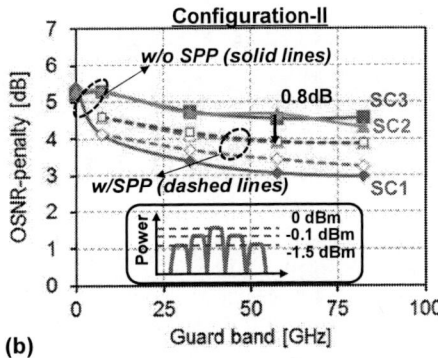

Fig. 3. OSNR penalty reduction with GB and SPP for: (a) 400GB/s 4sc-DP-QPSK superchannel is surrounded with 1Tb/s 5sc-DP-16QAM superchannels; (b) 1Tb/s 5sc-DP-16QAM superchannel is surrounded with 400GB/s 4sc-DP-QPSK superchannels.

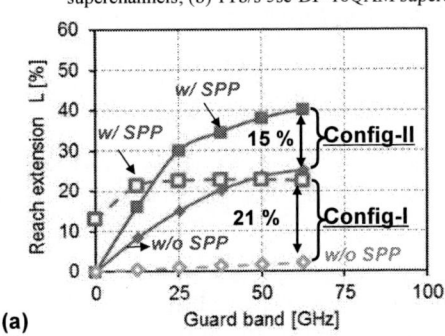 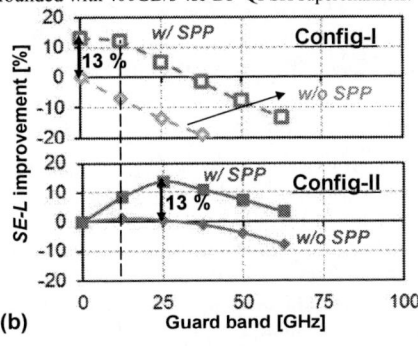

Fig. 4. (a) Rich extension and (b) SE x L improvement vs. GB for non-uniform traffic.

Figure 4(b) demonstrates that SPP is the only technique that can maximize *SE-L* product by 13% for Configuration-I, while SPP in combination with 25 GHz GB can improve maximum *SE-L* performance by 13% for Configuration-II. Applying GB alone is not enough to optimize the network of both configurations. However, considering the whole network, deploying mix of 1Tb/s 5sc-DP-16QAM and 400Gb/s DP-QPSK superchannels, the combination of the SPP and GB = 12.5 GHz can deliver the maximum *SE-L* performance of each superchannel (shown as black dotted line). This design optimization rule is necessary for mutual benefit of both configurations.

III. CONCLUSIONS

We have numerically investigated optimization of dynamic and flexible grid networks deploying mix of 1Tb/s 5sc-DP-16QAM and 400Gb/s 4sc-DP-QPSK superchannels. The combination of the SPP and GB was used to optimize the network for maximum reach *L* and/or *SE-L* product. We have shown that GB can extend reach of DP-16QAM superchannels by 25% due to smaller impact from co-propagating DP-QPSK superchannels, while it is not needed for DP-QPSK superchannels surrounded with DP-16QAM superchannels. SPP can extend reach even further for both superchannels. Moreover, we have also demonstrated that utilization of GB alone is not enough for reaching maximum *SE-L* product. In this case, the combination of SPP and small GB of 12.5 GHz (only one frequency slot) is necessary to maximize *SE-L* product for mutual benefit of both superchannels.

REFERENCES

[1] T. Hoshida et al., "Network innovations brought by digital coherent receivers", OFC/NFOEC 2010, paper NMB4.

[2] P. Wright et al., "Comparison of optical spectrum utilization between flexgrid and fixed grid on a real network topology", Proc. OFC, OTh3B.5, Los Angeles (2012).

[3] G. Bosco et al., "On the performance of Nyquist-WDM terabit superchannels based on PM-BPSK, PM-QPSK, PM-8QAM or PM-16QAM subcarriers", J. Lightw. Technol., vol. 29, No. 1, January 1, 2011.

[4] D. J. Ives et al., "Transmitter optimized optical networks", Proc. OFC, JW2A.64, Los Angeles (2013).

[5] O. Vassilieva et al., "Systematic analysis of intra-superchannel nonlinear crosstalk in flexible grid networks", Proc. ECOC, Mo.4.3.6, Cannes (2014).

[6] T. Yamauchi et al., "Experimental demonstration of nonlinear crosstalk mitigation in superchannels using subcarrier power pre-emphasis", Proc. OECC, Shanghai (2015).

[7] O. Vassilieva et al., "Flexible grid network optimization for maximum spectral efficiency and reach", Proc. ECOC, Tu.1.4.2, Valencia (2015).

[8] T. Yamauchi et al., "Nonlinear mitigation technique using subcarrier power pre-emphasis and digital nonlinear compensation in superchannel transmission", Proc. ECOC, P.5.02, Valencia (2015).

Optimum Capacity Utilization in Space-Division Multiplexed Transmission Systems with Few-Mode Fibers

Georg Rademacher and Klaus Petermann

Technische Universität Berlin, Fachbereich Hochfrequenztechnik, Einsteinufer 25, 10587 Berlin, Germany
georg.rademacher@tu-berlin.de

Abstract: For maximum capacity utilization, all spatial channels in space-division-multiplexed systems are desired to perform equally. We present a power-optimizing strategy for a three-mode fiber, enabling maximum performance, when including intra- and intermodal nonlinear transmission effects.

Keywords: Space-Division Multiplexing, Nonlinear Transmission Effects

I. INTRODUCTION

Space-Division Multiplexing (SDM) [1] using multi-mode [2] or multi-core fibers [3] is a very promising approach for a dramatic capacity increase in long-haul optical transmission systems. As in single-mode fiber systems, the maximum transmission capacity in multi-mode fibers is likely to be limited by Kerr-effect based nonlinear signal interaction, triggering an increased interest in studying such theoretically [4-7] and experimentally [8]. It has been shown that signals traveling in different fiber modes generally suffer from unequal levels of nonlinear signal distortion, due to different intra- and intermodal modal nonlinear coefficients that result from the unalike transverse field distributions of different modes. It has further been shown that the strength of intermodal signal interaction can be of similar magnitude as intramodal nonlinear interaction [4,5].

Theoretical studies have revealed [9,10] that the capacity of SDM transmission links significantly degrades, if some spatial channels experience a different loss than others. This effect is known as Mode-Dependent Loss (MDL). A similar situation appears when different spatial channels experience different levels of nonlinear signal distortion. Thus, this paper presents a method to balance all nonlinear signal distortions in SDM transmission systems with multi-mode fibers. In [11], a similar numerical analysis was carried out, however, low intermodal interaction was considered, leading to a negligible effect of the power optimization scheme.

II. ANALYTICAL MODEL FOR THE NONLINEAR INTERFERENCE IN MULTI-MODE FIBERS

In [4,5], we presented an analytical approach to evaluate the nonlinear signal distortion in SDM transmission links. The model assumes that all intra- and intermodal nonlinear distortions combine to one additional noise source of Gaussian shape. This can be included e.g. when calculating a nonlinear OSNR of mode p at the end of the transmission line as [4,5]:

$$OSNR_{NL}^{(p)} = \frac{P_{in}^{(p)}}{N_{SP}\left(G_{ASE} + G_{NLI}^{(p)}\right)12.5GHz} \tag{1}$$

$P_{in}^{(p)}$ is the input power per WDM/MDM mode, N_{SP} is the number of transmitted spans, G_{ASE} is the linear noise power spectral density, originating from inline fiber amplifiers, and 12.5GHz is the commonly used noise reference bandwidth. $G_{NLI}^{(p)}$ is the nonlinear noise power spectral density that is generally defined in [4] and can be simplified for this work as:

$$
\begin{aligned}
G_{NLI}^{(p)} = \left(G_{TX}^{(p)}\right)^3 \tilde{\gamma}_{pp}^2 \iint_{B_{opt}} \left|\eta(\Delta\beta^{(pp)}(\Delta f_1, \Delta f_2)\right|^2 d\Delta f_1 d\Delta f_2 \\
+ 2\sum_{q \neq p} \left(G_{TX}^{(p)}\right) \left(G_{TX}^{(q)}\right)^2 \left(\frac{\tilde{\gamma}_{pq}}{2}\right)^2 \iint_{B_{opt}} \left|\eta(\Delta\beta^{(pq)}(\Delta f_1, \Delta f_2)\right|^2 d\Delta f_1 d\Delta f_2
\end{aligned}
\tag{2}
$$

The first line in eq. (2) represents the impact of intramodal nonlinear interaction, the second line includes intermodal nonlinear integration. Thus, the sum needs to span over all modes that are excited in the multimode fiber. $G_{TX}^{(p)}$ is the power spectral density of the transmitted signal in the pth fiber mode. $\tilde{\gamma}_{pq}$ are the intra- and intermodal nonlinear coefficients that depend mainly on the fiber geometry. $\eta(\Delta\beta^{(pq)}(\Delta f_1, \Delta f_2)$ is the normalized four-wave mixing efficiency [4,6]. The transmission setup that is considered in this work is similar to that, reported in [4].

Nine pol-mux WDM channels of 28 GBaud, placed in Nyquist spacing with an overall optical bandwidth of approximately 250 GHz, are launched into each mode of a three mode parabolic-index fiber. We consider a scenario, where only the two polarizations of each fiber mode (mode 1: LP_{01}, mode 2: LP_{11a}, mode 3: LP_{11b}) experience strong linear mode coupling [5,6]. To incorporate this coupling scenario, the intra- and intermodal nonlinear coefficients are adapted as in [5,6]. Including these adaptations, the nonlinear coefficients that are used in the analysis are: $\widetilde{\gamma_{11}} = 1.38\ W^{-1}km^{-1}$, $\widetilde{\gamma_{22}} = \widetilde{\gamma_{33}} = 1.03\ W^{-1}km^{-1}$, $\widetilde{\gamma_{12}} = \widetilde{\gamma_{13}} = 1.03\ W^{-1}km^{-1}$ and $\widetilde{\gamma_{23}} = \widetilde{\gamma_{32}} = 0.51\ W^{-1}km^{-1}$.

We set the differential mode delay (DMD) to be zero. This leads to maximum intermodal nonlinear interactions, as they occur in most realistic transmission systems [4]. Chromatic dispersion is set equal in all modes at 15 ps/(nm km). All modes' attenuation is equal at 0.2 dB/km. We consider a setup with 10 links of 80 km each, where all amplifiers have a noise figure of 5 dB.

III. OPTIMIZATION TECHNIQUE

As it can be seen from eq. (2), signals in different modes generally experience a different level of intra- and intermodal nonlinear distortion, resulting in a different nonlinear OSNR. Thus, we want to optimize each modes input powers in a way that the nonlinear OSNR at the end of a transmission link has the same nonlinear OSNR and thus the same performance. The linear noise contribution, however, only depends on the gain of the amplifier. As the attenuation of all modes is considered to be equal, each mode requires the equal gain and thus accumulates an equal amount of noise power. As the nonlinear coefficients of the LP_{11a} and LP_{11b} are symmetric, it is reasonable to assume that the power in both modes should be set equal.

Figs. 1 (a) and (b) show the nonlinear OSNRs of the (a) first (LP_{01}) and (b) second/third ($LP_{11a/b}$) modes as a function of the modes' input powers. It can be seen in fig. 1 (a) that at a constant input power of the $LP_{11a/b}$ modes of e.g. -6 dBm, the OSNR of the LP_{01} mode increases with the input power (linear regime) until a maximum is reached at about -3 dBm, while further increase of the input power reduces the available nonlinear OSNR (nonlinear regime). This effect is very similar to the situation in single-mode fibers, however, with an increased power in the $LP_{11a/b}$ mode, the OSNR reduces as a result of stronger intermodal nonlinear interaction.

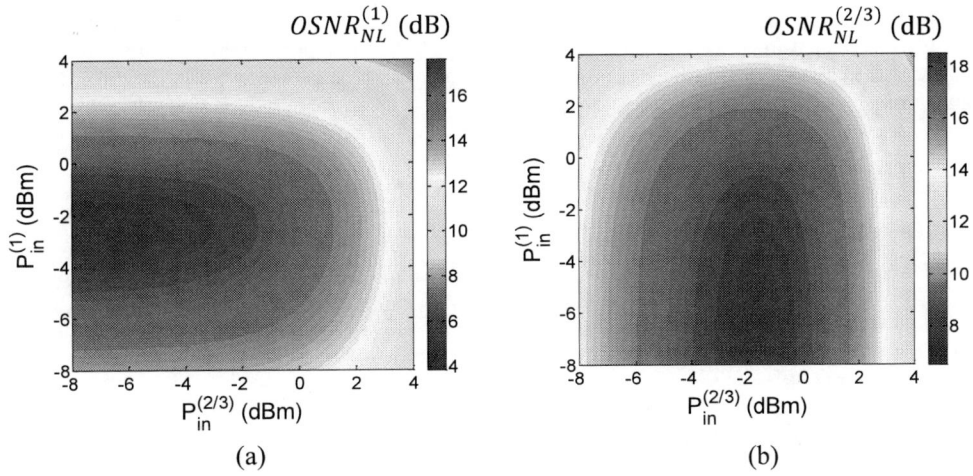

Figure 1: Nonlinear OSNR of the (a) first and (b) second/third mode as a function of the modes' input powers. (c) Difference of the nonlinear OSNR of the first and second/third mode.

From fig. 1 (a) and (b) , it can also be seen that it is not possible to reach maximum nonlinear OSNR in all modes, as e.g. the OSNR in the $LP_{11a/b}$ modes is at least 4 dB lower for the power combination that optimizes the nonlinear OSNR in the LP_{01} mode. Thus, it is necessary to find the two input powers, where the OSNR in all modes is equal and optimal. Fig. 2 (a) shows the difference of the two nonlinear OSNRs, defined as:

$$\Delta OSNR_{NL} = -\left| OSNR_{NL}^{(1)} - OSNR_{NL}^{(2/3)} \right| \tag{3}$$

In the linear regime, the nonlinear OSNRs are equal when all modes carry the same input power. At larger input powers, it can be observed that equal OSNR is reached at different input powers, e.g. when the LP_{01} mode carries -2 dBm and the $LP_{11a/b}$ modes carry -4 dBm. The launch power combinations for equal OSNRs lie on the blue marked trace in fig. 2 (a) with $\Delta OSNR = 0$. However, for optimal transmission configuration, it is necessary to find the power combination where the OSNRs are equal and maximal. Thus, fig. 2 (b) shows the OSNRs of all three modes as a funciton of the input power in the LP_{01} mode. Continous lines illustrate the OSNRs when chosing the power in the $LP_{11a/b}$ mode according to the blue line in fig. 2 (a). The dashed references graphs show the OSNRs at equal launch powers.

It can be seen that for equal launch powers, an OSNR difference of about 1 dB can be observed at the maximum OSNR of the LP_{01} mode compared to the LP_{11b} mode. However, when choosing and optimum launch power, all modes achieve an equal OSNR that lies in between the optima of both modes. The absolute optimum launch powers are found at -2.65 dBm for the LP_{01} and -3.85 dBm in the $LP_{11a/b}$ modes, leading to an equal nonlinear OSNR of 17.6 dB.

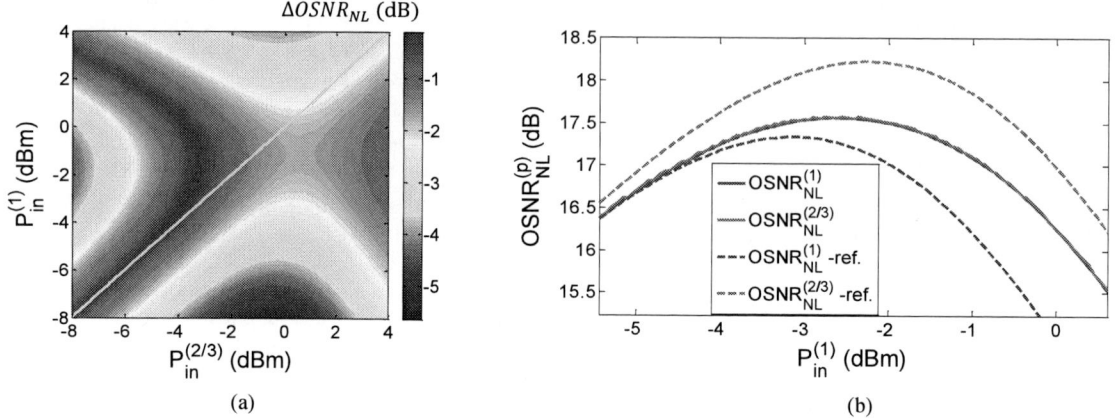

(a)

(b)

Figure 2: (a) Difference of the OSNRs in the LP_{01} mode (mode p = 1) and $LP_{11a/b}$ mode (mode p = 2/3) as defined in eq. (3). The blue trace corresponds to equal OSNRs in all modes, the light green graph to equal launch powers in all modes. (b) Absolute values of all modes' OSNRs when varying the power in the LP_{01} mode, with the power in the $LP_{11a/b}$ mode according to the traces in (a).

IV. CONCLUSIONS

We have shown an analytical method for launch power optimization in order to maximize the capacity of space-division multiplexed transmission links. It was used to find the optimum launch power setting for an exemplary transmission system with a three-mode fiber. At optimum performance, the launch powers in the three modes differ by about 1.2 dBm, eliminating an OSNR difference of at least 1 dB that occurs at the best condition with equal launch powers in all modes.

ACKNOWLEDGMENT

This work was partially funded by the German Research Foundation (DFG)

REFERENCES

[1] D. J. Richardson et al. "Space-division multiplexing in optical fibres," *Nature Photonics* 7.5 (2013): 354-362. J.

[2] N.K. Fontaine et al. "30× 30 MIMO transmission over 15 spatial modes," *Optical Fiber Communication Conference*. Optical Society of America, paper Th5C.1, 2015.

[3] B. Puttnam et al. "2.15 Pb/s transmission using a 22 core homogeneous single-mode multi-core fiber and wideband optical comb," in *European Conference and Exhibition on Optical Communication*, paper PDP 3.1, 2015

[4] G. Rademacher et al. "Nonlinear Gaussian Noise Model for Multi-Mode Fibers with Space-Division Multiplexing," accepted for publication in *Journal of Lightwave Technology*

[5] G. Rademacher et al. "Analytical Description of Cross-Modal Nonlinear Interaction in Mode-Multiplexed Multi-Mode Fibers," *IEEE Photonics Technolgy Letters*, vol. 24 pp. 1929-1932, 2012

[6] S. Mumtaz et al. "Nonlinear Propagation in Multimode and Multicore Fibers: Generalization of the Manakov Equations," *Journal of Lightwave Technology*, vol. 31 pp. 398--406, 2013

[7] C. Antonelli et al. "Modeling of nonlinear propagation in Space-Division Multiplexed fiber-optic transmission," *Journal of Lightwave Technology*, vol. 34, pp. 36-54, 2016.

[8] R.J. Essiambre et al. "Experimental Investigation of Inter-Modal Four-Wave Mixing in Multimode Fibers," *IEEE Photonics Technolgy Letters*, vol. 25, pp. 539-542, 2013.

[9] S. Warm et al. "Splice loss requirements in multi-mode fiber mode-division-multiplex transmission links," *Opt. Express*, vol. 21, pp. 519–532, 2013.

[10] P. J. Winzer et al. "MIMO capacities and outage probabilities in spatially multiplexed optical transport systems," *Opt. Express*, vol. 19, pp. 16680–16696, 2011.

[11] D. Rafique et al. "Impact of power allocation strategies in long-haul few-mode fiber transmission systems," *Opt. Express*, vol. 21, no. 9, p. 10801, Apr. 2013.

TuB1-4 (Invited)

OECC/PS2016

Comparison of Four Dimensionally Coded Versus Hybrid Modulation

Han Sun

Coherent Technologies, Infinera Canada, 555 Legget Drive Suite 222, Ottawa, Ontario, K2K 2X3, Canada
hsun@infinera.com

Abstract: A number of known four dimensional modulation formats at 3, 5, 7 bits per symbol spectral efficiency are compared against a hybrid mixture of conventional modulations having the same net spectral efficiency.

Keywords: (4D modulation, hybrid modulation, 3bits per symbol, 5bits per symbol, 7bits per symbol)

I. INTRODUCTION

Digital coherent optical transmission systems employing Polarization Multiplexed Quadrature Phase Shift Keying (PM-QPSK) modulation format has been studied extensively [1-3]. In recently years, four-dimensionally optimized modulation formats have been introduced and gained significant attention [4-6]. The fiber channel contains four degrees of freedom, namely the in-phase and quadrature component of TE and TM polarizations. Traditional modulation formats such as PM-BPSK, PM-QPSK, PM-8QAM and PM-16QAM are optimized using the in-phase and quadrature components of the carrier. By using four-dimensions (4D), Karlsson & Agrell [4,5] have shown that one can achieve larger Euclidean distance between modulated symbols. At high signal to noise ratio, better performance or higher spectral efficiency can be achieved. In the presence of a strong soft decision FEC (Forward-Error-Correction), these newly introduced modulation formats also compare well against the traditional formats. Fig. 1 shows the raw BER against SNR for a variety of modulations with spectral efficiencies ranging from 2 to 8 bits/dual-pol symbol. The spectral efficiency and modulation formats are shown in the legend. Included in here is the Polarization-Switched-QPSK (PS-QPSK) with 3 bits/dual-pol symbol [4,5], 32-SPQAM (Set-Partitioned QAM) with 5 bits/dual-pol symbol [6], and 128-SPQAM with 7 bits/dual-pol symbol [6]. Near the soft FEC threshold (at above 10^{-2} BER), PS-QPSK provides a SNR sensitivity half way between PM-BPSK and PM-QPSK. Similarly, the 32-SPQAM and 128-SPQAM each provides a good trade-off between spectral efficiency and SNR sensitivity. We denote these 4D formats as 4D-3bit, 4D-5bit, and 4D-7bit.

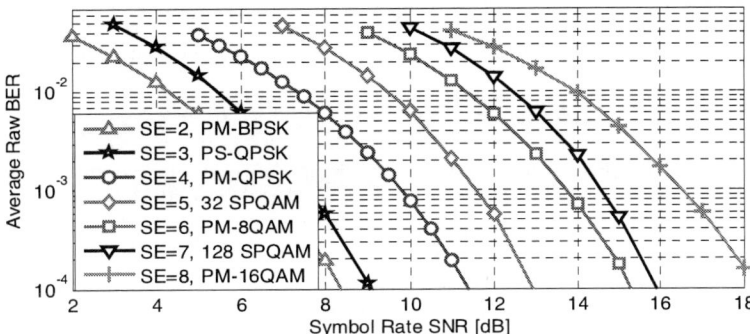

Fig. 1. Theoretical Coherent Performance for Various Modulations.

An even more flexible way is by time-multiplexing a mixture of PM-QPSK and PM-8QAM symbols [7-9]. With proper frame synchronization, fractional spectral efficiency between 4 bits/dual-pol symbol and 6 bits/dual-pol symbol can be achieved. At 5 bits/dual-pol symbol, half the transmitted symbols are PM-QPSK and other half is PM-8QAM. We refer to this modulation as Hybrid-5bit. Similarly, an even mixture of 16QAM and 8QAM symbols gives a net spectral efficiency of 7bits/dual-pol symbol. We refer to this modulation as Hybrid-7bit. Finally, Hybrid-3bit denotes an even mixture of QPSK and BPSK symbols.

The performance comparison of Hybrid versus 4D at 5bit spectral efficiency had been published in both simulations [10] and experiments [11]. In the following sections, we analyze these additional Hybrid modulations against their 4D counterparts in terms of their theoretical performance using AWGN channel model, and then their nonlinear performance in un-compensated SMF fiber. We focus on 3bit and 7bit modulations. While the results for 5bit modulations are repeated from before, all three results are plotted together for completeness.

II. PERFORMANCE IN ADDITIVE GAUSSIAN NOISE CHANNEL

The 4D formats are designed assuming AWGN channel only and not compatible with differential coding. Therefore we focus our comparison on coherent detection, without differential coding. Similar to the method illustrated by Zhou [9] & Zhuge [8], we first define a frame in which first 2 symbols are allocated as the pilot symbols, and they are known apriori in the receiver. The 2 pilot symbols are followed by 130 symbols worth of data. This frame format is also used in our earlier analysis [10]. For Hybrid-3bit modulation, it contains even symbols of QPSK and odd symbols of BPSK in adjacent time slots. Similarly for Hybrid-5bit and Hybrid-7bit, the two conventional modulations are even-odd interleaved. As shown in our earlier analysis [10], this provides the best nonlinear tolerance. For the 4D formats, the 130 symbols contain the 4-D optimized symbols. In the receiver, the 2 pilot symbols can be used to synchronize to this frame, and proper carrier recovery and decoding can be performed. Here we assume channel equalization had been achieved. A straight-forward application of CMA algorithm does not work, and channel acquisition may need additional pilot symbols for channel estimation. These symbols may be found in the overhead section of the OTN frame. In the steady state, a decision-directed LMS algorithm can be used to track polarization transients.

For the Hybrid modulations, the power of the symbol that carries higher spectral efficiency (higher order) needs to be optimized against the symbol with lower spectral efficiency (lower order). As an example, for the Hybrid-3bit, one can choose BPSK symbols from the QPSK constellation set, hence the transmitted symbols will generate binary levels on I and Q signal lanes. This can be attractive since one can use NRZ modulator optimized for binary signals. However, Fig. 2 below shows that the optimum is 3dB more power for the QPSK symbols and that may improve the average Q by 0.55dB near the soft FEC threshold. Also if 8QAM modulation is chosen as 8 symbols out of the 16QAM constellation set, then all the DSP can operate assuming 16QAM signaling format and that may simplify design, but at the cost of Q performance. In Fig. 2, for all three Hybrid formats, the higher order modulation needs to increase in power by 2 or 3dB compared to the simpler modulation. The Q improvement is too significant to ignore. Hence in the rest of this correspondence, we assume the power of the higher order modulation is increased by 3.0dB, 3.3dB, and 2.0dB for Hybrid -3bit, -5bit and -7bit respectively.

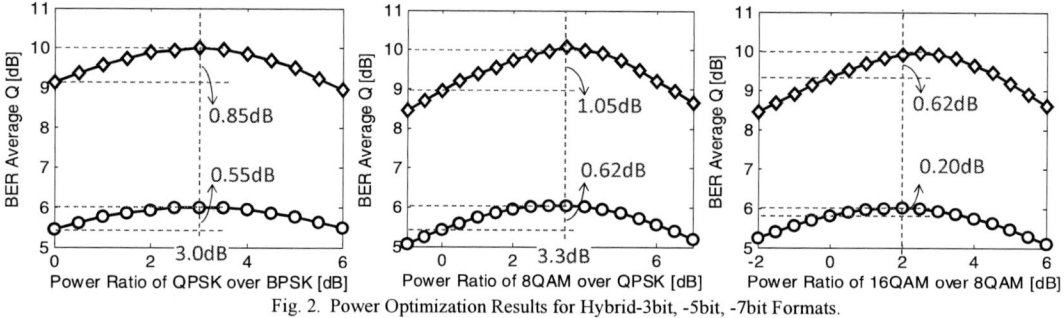

Fig. 2. Power Optimization Results for Hybrid-3bit, -5bit, -7bit Formats.

Fig. 3 below shows the AWGN channel performance comparison of 4D distance optimized formats versus power-ratio optimized Hybrid modulation. At high SNR, the 4D formats have better BER due to larger Euclidean distance. Near BER of 1×10^{-3}, 4D out-performs Hybrid by 1.0dB, 0.6dB, 0.75dB for 3bit, 5bit and 7bit respectively. Near BER of 2×10^{-2}, 4D out-performs Hybrid by 0.44dB, 0.15dB, 0.05dB for 3bit, 5bit and 7bit respectively. We note that at lower spectral efficiencies, the benefit of the 4D modulation is larger.

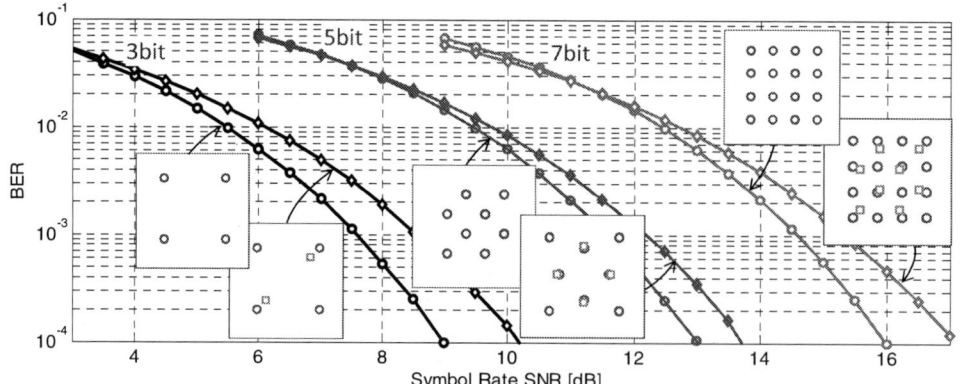

Fig. 3. 4D versus Hybrid Modulation in AWGN at 3bit, 5bit, and 7bit Spectral Efficiencies.

III. Simulated Performance over Uncompensated SMF Fiber

The modulated signals are each propagated over 60, 30, 15 spans (80km per span) of uncompensated SMF fiber for 3, 5, 7bit modulations respectively. 19 WDM channels are simulated in the middle of the C band at 32G symbol rate using channel spacing of 35GHz. Signals are near Nyquist shaped and multiplexed using raised cosine filtering with α factor of 1/16. In the receiver, white ASE noise is added to the signal at 14, 17, 20dB of OSNR normalized to 0dBm of per wave launch power. The middle 5 channels are decoded and processed to produce an average BER. The system Q value in dB is plotted against per wave launch power as shown in Fig. 4 below. The optimum launch power found is the same between 4D and Hybrid formats. At the optimum launch power, 4D formats out-perform Hybrids in Q value by 1.1, 0.4, 0.4dB for 3bit, 5bit and 7bit modulations respectively. Upon examination, these Q differences are very much similar to the differences observed in the AWGN channel model plotted in Fig. 3.

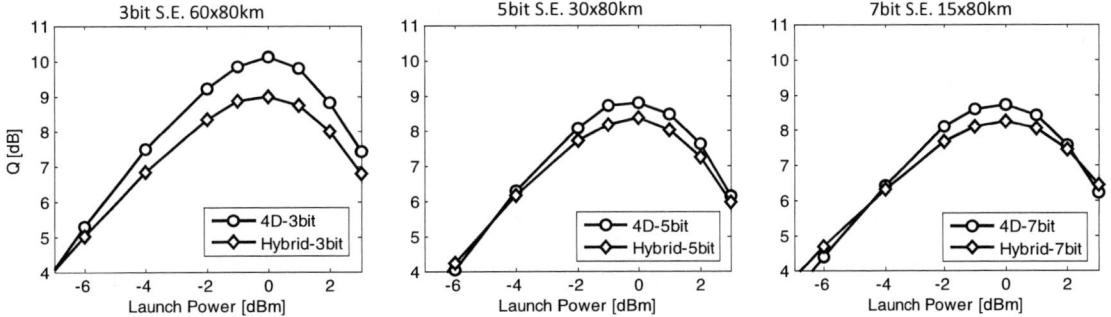

Fig. 4. Comparing 4D versus Hybrid Modulation over 60x80km, 30x80km and 15x80km of SMF for 3bit, 5bit and 7bit Spectral Efficiencies.

IV. Conclusions

In both linear and nonlinear regimes, four-dimensionally optimized formats are found to be better than the time-domain hybrid mixture of conventional modulations. The benefit of 4D modulation is larger at lower spectral efficiency such as at 3bit per symbol. Polarization-Switched QPSK outperforms hybrid mixture of QPSK and BPSK; 32-SPQAM and 128-SPQAM also outperform their hybrid counterparts.

References

[1] S. Savory, et. al., "Digital Coherent Optical Receivers: Algorithms and Subsystems", IEEE Sel. Topics in Quantum Electron., Vol. 16, Issue 5, page 1164-1179, Sept-Oct. 2010.

[2] H. Sun, et. al., "Real-time measurements of a 40Gb/s coherent system", Opt. Express Letters 16, pg. 873-879, 2008.

[3] J. Rahn, et. al., "Real-Time PMD Tolerance Measurements of a PIC-Based 500Gb/s Coherent Optical Modem", J. Lightwave Tech. Vol. 30, No. 17, September 2012.

[4] M. Karlsson, et. al., "Which is the most power-efficient modulation format in optical links?", Optics Express, Vol. 17, Issue 13, 2009.

[5] E. Agrell, et. al., "Power-efficient modulation formats in coherent transmission systems", J. Lightwave Tech., Vol. 27, No. 22, 2009.

[6] L. D. Coelho, et. al., "Global Optimization of Fiber-Optic Communication Systems using Four-Dimensional Modulation Formats", Proc. ECOC, Mo.2.B.4, 2011.

[7] W. Peng, et. al., "Hybrid QAM Transmission Techniques for Single-Carrier Ultra-Dense WDM Systems", Proc. OECC, pg. 824-825, 2011.

[8] Q. Zhuge, et. al., "Time domain hybrid QAM based rate-adaptive optical transmissions using high speed DACs", Proc. OFC OTh4E.6, 2013.

[9] X. Zhou, et. al., "4000km Transmission of 50GHz spaced, 10x494.85-Gb/s Hybrid 32-64QAM using Cascaded Equalization and Training-Assisted Phase Recovery", Proc. OFC, PDP5C.6, 2012.

[10] H. Sun, et al., "Comparison of Two Modulation Formats at Spectral Efficiency of 5 Bits/Dual-Pol Symbol", Proc. ECOC, 2013

[11] R. Rios-Muller, et. al., "Experimental Comparison between Hybrid-QPSK/8QAM and 4D-32SP-16QAM Formats at 31.2GBaud using Nyquist Pulse Shaping", Proc. ECOC, Th.2.D.2, 2013

TuB2-1 (Invited)

Joint Linear and Non-linear Adaptive Pre-Distortion of High Baud Rate Transmitters for High-order Modulation Formats

Bernhard Spinnler[1], Ginni Khanna[2], Stefano Calabrò[1], Erik De Man[1], Uwe Feiste[1], Thomas Bex[1], Heinrich von Kirchbauer[1]

(1) *Coriant R&D GmbH, Munich, Germany*, (2) *Lehrstuhl für Nachrichtentechnik, Technische Universität München*
Bernhard.Spinnler@coriant.com

Abstract: We review a digital pre-distortion technique to mitigate linear and nonlinear impairments from analog components. We assess the technique's effectiveness and accuracy for high baud rate signals and high-order modulation formats.
Keywords: Coherent communications, high-order modulation, digital pre-distortion

1. Introduction

To boost the throughput of an optical transponder it is tempting to increase the symbol rate and the order of the modulation scheme since, at first, these changes do not necessitate new analog and optical components and require only modest additional complexity in the digital signal processing. However, beyond a certain point, raising the symbol rate incurs severe linear distortions due to narrow-band filtering. Further, since high-order modulation schemes have higher peak-to-average power ratio (PAPR), the modulator needs to be driven deeply in the nonlinear region to get reasonable output power. This exacerbates nonlinear distortions precisely for high-order modulation schemes, which are more susceptible. In addition, sharp spectral shaping and high-order modulation formats strengthen the requirements to timing accuracy, equal signal delays, and path lengths among signal tributaries. Therefore, techniques are needed for simultaneous mitigation of nonlinear and linear distortions including tributary skew.

We propose to use state-of-the-art components and mitigate the analog components' lowpass characteristic, unequal signal delays, and nonlinear distortions using an adaptively adjusted Volterra digital pre-distortion (DPD) model [1,2]. This approach proves to be very effective and has been used to obtain several remarkable experimental results, from 200Gb/s QPSK at 57 GBaud [2] to 1 Tb/s 64QAM super channels using 4 sub channels at 45 GBaud [3] and 400 Gb/s single wavelength 128QAM at 42 GBaud [4]. In this paper we review the principle of the proposed algorithm, assess its estimation accuracy for high baud rate signals and high-order modulation formats and show exemplary results.

2. Adaptive Digital Pre-Distortion

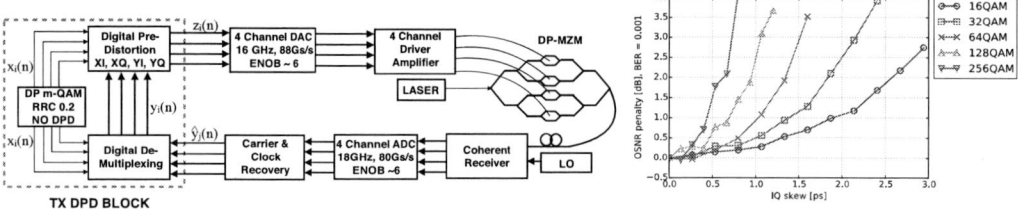

Fig. 1: a) Experimental setup implementing the adaptive digital pre-distortion. b) Skew tolerance of high order modulation formats.

The DPD algorithm is described in detail in [2]. The principle and implementation of the algorithm is shown in Fig. 1a. The technique relies on two entities: the DPD processor and an auxiliary coherent receiver providing the former with a local feedback signal tapped after the optical modulator. The analog part of the transmitter, including DAC, DA and DP-MZM, is modeled as a truncated Volterra series [7]. The i-th tributary (i = XI, XQ, YI, YQ) is represented as

$$z_i(n) = \sum_{p=1}^{P} \sum_{m=-(M-1)/2}^{(M+1)/2} h_{i,m,p} x_i^p(n-m),\qquad (1)$$

486

where $x_i(n)$ and $z_i(n)$ are the input and output samples and $h_{i,m,p}$ is the coefficient of the Volterra term of order p at memory tap m. The adaptation of the DPD coefficients proceeds as described in [1] according to the indirect learning architecture (ILA). The DPD coefficients are then used to pre-distort each tributary individually.

3. Results and Discussion

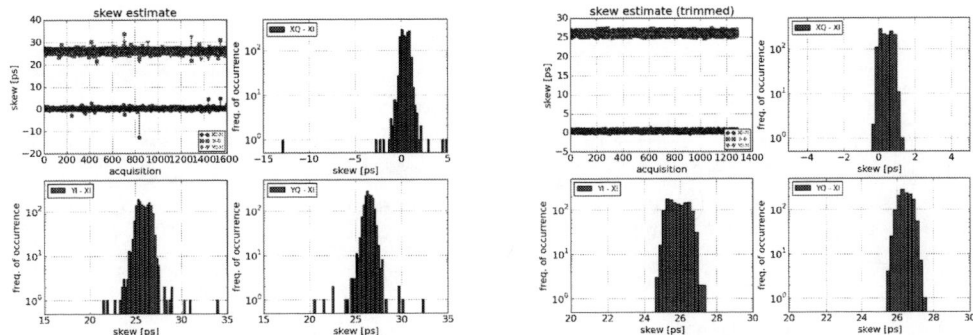

Fig. 2: Differential skew estimates over time and distributions a) without, b) with trimming of 20% outliers.

To assess the accuracy of the estimated pre-distortion coefficients we carried out 1600 independent measurements identifying the system's amplitude and phase response. We transmit standard 34 GBaud, root-raised cosine shaped (with roll-off 0.2) dual polarization 16QAM signals and use commercial DACs and ADCs running at 68 GHz and 56 GHz, resp., with memories of 16384 samples each. We assess the accuracy of the estimated DPD parameters on the basis of the identified differential delay between the signal tributaries. On one side this is a critical parameter for high-order modulation formats, on the other side it provides a compact figure of merit. Fig. 2a top left shows the estimated differential skew values $\tau_{XQ\text{-}XI}$, $\tau_{YI\text{-}XI}$, and $\tau_{YQ\text{-}XI}$ for all 1600 measurements with lane XI serving as reference. The remaining 3 subplots show the distribution of the differential skews. While the differential skew between XI and XQ is close to zero, our setup exhibits a substantial skew of around 26 ps between XI and both Y tributaries.

It is evident from Fig. 2a that there are several outliers with estimation errors in the range of 5 ps and beyond. These wrong estimates arise when the polarization and phase rotation matrix of the optical channel is accidentally close to the identity matrix, because in this case, the algorithm cannot distinguish transmitter from receiver skew. These events are, however, relatively rare and, after taking a couple of estimates, can be eliminated by sorting out results which deviate strongly from the majority. Fig. 2b shows the same plots after eliminating the 20% most deviating estimates. The estimation accuracy has improved substantially from elimination of outliers, and its standard deviation has dropped from 0.6 ... 1.0 ps to 0.3 ... 0.7 ps. Both in Fig. 2a and Fig. 2b each estimate is obtained from a single sequence of 16384 samples at 56 GHz.

Comparing the distributions in Fig. 2b to the skew tolerance of high order modulation schemes in Fig. 1b reveals that the obtained accuracy may be sufficient for 16QAM, which tolerates 2 ps IQ-skew at a 1 dB penalty, but not for 128QAM, which allows only 0.7 ps IQ-skew for the same penalty. To reach the necessary accuracy we combine the acquired sequences into 80 super-sequences of 20×16384 samples each and, after again discarding 20% of the most deviating estimates, use averaging to improve accuracy. The resulting cumulative mean skew estimates $\tau_{XQ\text{-}XI}$ and $\tau_{YQ\text{-}YI}$ and the corresponding standard deviations for the 80 super-sequences are shown in Fig. 3a. While the standard deviation of $\tau_{XQ\text{-}XI}$ is already 0.3 ps for a single estimate and reaches 0.1 ps after combining 16 sequences, the standard deviation of $\tau_{YQ\text{-}YI}$ reaches only 0.22 ps after combining 16 sequences. We attribute this disparity between the two polarizations to different amounts of noise injected by components such as driver amplifiers or modulator bias control.

Additionally, we validated our technique using 45.25 GBaud 64QAM and 42.0 GBaud 128QAM high baud rate signals. The DAC used for this measurement operates at a sampling rate of 88 GSample/s, has a 3 dB bandwidth of 16 GHz and an effective number of bits (ENOB) of ~6 [8]. Root raised cosine pulse shaping with roll off 0.2 is employed. At the receiver the optical signals are converted into the electrical domain using an integrated coherent receiver. Four ADCs with 18 GHz bandwidth, operating at 80 GSample/s, capture shots of $5 \cdot 10^5$ samples per tributary, which are further processed offline. Fig. 3b shows the back-to-back performance of the 45.25 GBaud 64QAM and 42.0 GBaud 128QAM signals both with and without DPD. We varied the number of linear as well as nonlinear taps of the DPD model and observed that in this scenario 33 linear and 7 higher order taps were sufficient to model the transmitter. DPD improves the 42.0 GBaud 128QAM performance by 2.5 dB at the FEC

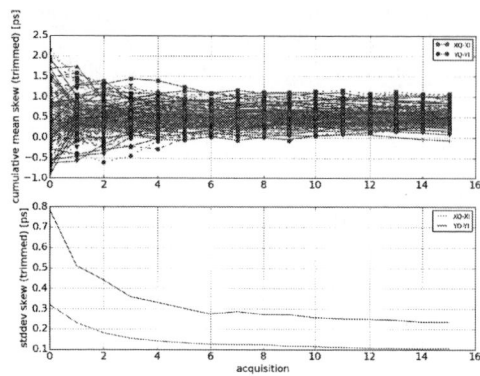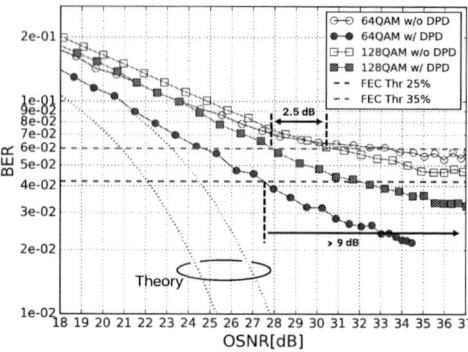

Fig. 3: a) Cumulative average differential skew $\tau_{XQ\text{-}XI}$ and $\tau_{YQ\text{-}YI}$ and corresponding standard deviation. b) Back-to-back performance of 45.25 GBaud 64QAM and 42.0 GBaud 128QAM

threshold. The improvement for the 45.25 GBaud 64QAM signal is even larger but cannot be given exactly since without DPD 45.25 GBaud 64QAM does not reach the FEC threshold at all. The poor performance without DPD is due to the extremely large baud rate of the signal compared to the small bandwidth of the components, which proves the necessity of DPD for high baud rate signals.

4. Conclusion

We described a fully adaptive digital pre-distortion technique and demonstrated its accuracy with respect to skew estimation between tributaries. Furthermore, we showed that this technique is indispensable for the generation of high baud rate signals offering gains in the range of several decibels for ~40 GBaud signals and making the transmission of higher baud rate signals with current state-of-the-art components even possible at all.

This work has been supported by the German Federal Ministry of Education and Research (BMBF) under support code 01BP12300A; EUREKA-Project SASER. We would like to thank Socionext for the DA and AD converters.

References

1. G. Khanna, S. Calabrò, B. Spinnler, E. de Man, and N. Hanik, "Joint Adaptive Pre-Compensation of Transmitter I/Q Skew and Frequency Response for High Order Modulation Formats and High Baud Rates," in *Opt. Fiber Commun. Conf.*, no. M2G.4, 2015.

2. G. Khanna, B. Spinnler, S. Calabrò, E. De Man, and N. Hanik, "A Robust Adaptive Pre-Distortion Method for Optical Communication Transmitters," *Photonics Technol. Lett. IEEE*, vol. 28, no. 7, pp. 752–755, 2016.

3. T. Rahman, D. Rafique, B. Spinnler, E. Pincemin, C. L. Bouëtté, J. Jauffrit, S. Calabrò, E. de Man, S. Bordais, U. Feiste, J. Slovak, A. Napoli, G. Khanna, N. Hanik, C. André, C. Okonkwo, M. Kuschnerov, A. M. J. Koonen, C. Dourthe, B. Raguénès, B. Sommerkorn-krombholz, M. Bohn, and H. D. Waardt, "Record Field Demonstration of C-band Multi-Terabit 16QAM, 32QAM and 64QAM over 762km of SSMF," in *Optoelectron. Commun. Conf.*, no. PDP1C.1, 2015.

4. G. Khanna, B. Spinnler, S. Calabrò, E. De Man, U. Feiste, T. Drenski, and N. Hanik, "400G Single Carrier Transmission in 50 GHz Grid Enabled by Adaptive Digital Pre-Distortion," in *Opt. Fiber Commun. Conf.*, 2016.

5. D. Rafique, A. Napoli, S. Calabro, and B. Spinnler, "Digital Preemphasis in Optical Communication Systems: On the DAC Requirements for Terabit Transmission Applications," *IEEE J. Light. Technol.*, vol. 32, no. 19, pp. 3247–3256, 2014.

6. P. W. Berenguer, T. Rahman, A. Napoli, M. Nölle, and C. Schubert, "Nonlinear Digital Pre-Distortion of Transmitter Components," in *Eur. Conf. Opt. Commun.*, 2015.

7. M. Schetzen, *The Volterra and Wiener theories of nonlinear systems*, ser. A Wiley - Interscience publication. Wiley, 1980.

8. "Fujistu Data Sheet." [Online]. Available: www.fujitsu.com/downloads/MICRO/fme/documentation/c60.pdf

OECC/PS2016

Field demonstration of modulation format adaptation based on pilot-aided OSNR estimation using 400Gbps/ch real-time DSP

Kazushige Yonenaga[1], Kengo Horikoshi[1], Seiji Okamoto[1], Mitsuteru Yoshida[1], Yutaka Miyamoto[1], Masahito Tomizawa[1], Takeshi Okamoto[2], Hidemi Noguchi[2], Jun-ichi Abe[2], Junichiro Matsui[2], Hisao Nakashima[3], Yuichi Akiyama[3], Takeshi Hoshida[3], Hiroshi Onaka[3], Kenya Sugihara[4], Soichiro Kametani[4], Kazuo Kubo[4], and Takashi Sugihara[4]

[1]NTT Network Innovation Laboratories, NTT Corporation, 1-1 Hikarinooka, Yokosuka, Kanagawa 239-0847, Japan
[2]NEC Corporation, 1753 Shimonumabe, Nakahara-ku, Kawasaki, Kanagawa, 211-8666 Japan
[3]Fujitsu Limited, 1-1, Kamikodanaka 4-chome, Nakahara-ku, Kawasaki, Kanagawa, 211-8588 Japan
[4]Mitsubishi Electric Corporation, 5-1-1 Ofuna, Kamakura, Kanagawa, 247-8501 Japan
yonenaga.kazushige@lab.ntt.co.jp

Abstract: We present a modulation format adaptation experiment on a real-time DSP and field-installed fiber using pilot-aided OSNR estimation. With modulation adaptation, 56-channel 400Gbps-2SC-PDM-16QAM and 200Gbps-2SC-PDM-QPSK signals are successfully transmitted over 216km and 3246km SSMF, respectively.

Keywords: adaptive modulation, field trial, real-time DSP, 400Gbps

I. INTRODUCTION

Driven by the rapid growth of broadband application services such as video data streaming and cloud computing in data centers, Internet traffic is continuously increasing. To accommodate the increasing traffic, digital coherent transmission technologies are indispensable. In 2011, we developed 100Gbps/ch digital coherent technologies and implemented them in digital signal processing circuits (DSPs)[1]. Many 100Gbps/ch digital coherent transmission systems are now being deployed in commercial networks.

As the next target, 400Gbps/ch transmission systems are being intensively studied[2-5]. In addition to raising the channel bit rate, flexible bandwidth utilization using modulation adaptation is also attracting much attention[6,7]. So far, we have proposed and demonstrated a pilot-aided optical signal-to-noise ratio (OSNR) estimation scheme for modulation format adaptation by offline signal processing[8,9].

In this paper, a modulation format adaptation experiment based on pilot-aided OSNR estimation is presented; we use a real-time DSP and field-installed fiber. Adapting the modulation format to suit estimated OSNR, 56-channel 400Gbps 2-Subcarrier Polarization-Division-Multiplexed 16-Quadrature-Amplitude-Modulation (2SC-PDM-16QAM) and 200Gbps 2SC PDM Quadrature-Phase-Shift-Keying (2SC-PDM-QPSK) signals are successfully transmitted over 216km field-installed standard single-mode fiber (SSMF) and 3246km SSMF (216km field fiber + 3030km laboratory fiber). Long-term stability in Q-factor is also confirmed for both modulation formats.

II. EXPERIMENTAL SETUP

Figure 1 shows a schematic of the experimental setup for modulation format adaptation. We constructed a network model consisting of three photonic nodes (Node#1-3) and two fiber transmission lines (216km field-installed SSMF, 3030km laboratory SSMF). Two real-time DSP based optical transponders were prepared for generating super-channel signals, e.g. 400Gbps-2SC-PDM-16QAM and 200Gbps-2SC-PDM-QPSK signals. At the transmitter, 112 wavelengths were prepared as background channels. Even and odd optical carriers with 75GHz spacing were independently modulated by separate IQ-modulators and polarization-multiplexed signals were generated by a self-delayed polarization multiplexer (Pol.-MUX). The even and odd wavelength division multiplexing (WDM) signals were optically multiplexed by a wavelength selective switch (WSS) #1; 37.5GHz-spaced background WDM signal was also generated. The wavelengths ranged from 1529.114nm to 1562.283nm. In WSS#1, two wavelengths of the background WDM signal were replaced by two wavelengths from two real-time DSP-based transponders. The resulting WDM signal including two real-time wavelengths was fed into the first transmission line (216km field-installed SSMF). The transmitted signal was dropped in Node#2 or Node#3 after transmission. At the receiver, one of the signals dropped in Node#2 or Node#3 was selectively received. The two wavelengths forming the super-channel were simultaneously received by two real-time DSP-based optical transponders.

The field SSMF used in this experiment is a part of the JGN-X testbed fiber cable installed between NICT Koganei Headquarters and TOKAI Chofu Repeater Station. The cable length is 18km including 10km aerial section. 12 fiber cores were used for the 216km transmission line, with 120km aerial section. A gain equalizer (EQ) was inserted mid span of the 216km field fiber. The transmission fiber in the laboratory consisted of 30 spans of 101km SSMF on a bobbin. Gain equalizers were inserted at every 4th span (404km).

The real-time DSP included the function of pilot-aided OSNR estimation[9]. The frame structure of the transmitted signal is shown in the upper/left inset of Fig. 1. It consists of pilot sequence (PS) and data. As the PS is an alternate sequence of an arbitrary complex number, S, and its opposite one, -S, its spectrum has two peaks as shown in the figure. This makes it possible to measure in-band OSNR, which stands for the power ratio of optical signal and noise in the same band. After PS timing detection, the noise power can be calculated from the frequency elements around DC in the PS spectrum. The signal power can be obtained from the data sequence spectrum at the timing without PS.

Figure 2 shows the optical WDM spectrum after 216km field-installed fiber transmission. Modulation format is PDM-16QAM. The 400Gbps super-channel consists of two adjacent wavelengths (λ_{2n-1}, λ_{2n}, n is an integer from 1 to 56). All 112 wavelengths are spaced by 37.5GHz and the number of super-channel is 56.

The demonstration scenario is as follows. First, we measured the estimated OSNR after 216km transmission at Node#2. The estimated OSNR is confirmed to be higher than the OSNR threshold (set to 20dB in this experiment). So we selected the modulation format of DP-16QAM for real-time and background channels. Then, upon changing the real-time channel (Ch.n: $\lambda_{2n-1}+\lambda_{2n}$) of the two real-time optical transponders from n=1 to n=56, we measured the bit error rates (BERs) before forward error correction (FEC) decoding (pre-FEC BER) and after FEC decoding (post-FEC BER). The two wavelengths of the super-channel were simultaneously received by two real-time DSP-based transponders and BERs were measured. Second, WSS in Node#2 was switched from drop to through mode. We measured the estimated OSNR after 3246km transmission at Node#3. The estimated OSNR is confirmed to be lower than the OSNR threshold (20dB). Thus we selected the modulation format of PDM-QPSK for real-time and background channels. We then measured pre-FEC and post-FEC BERs for all channels.

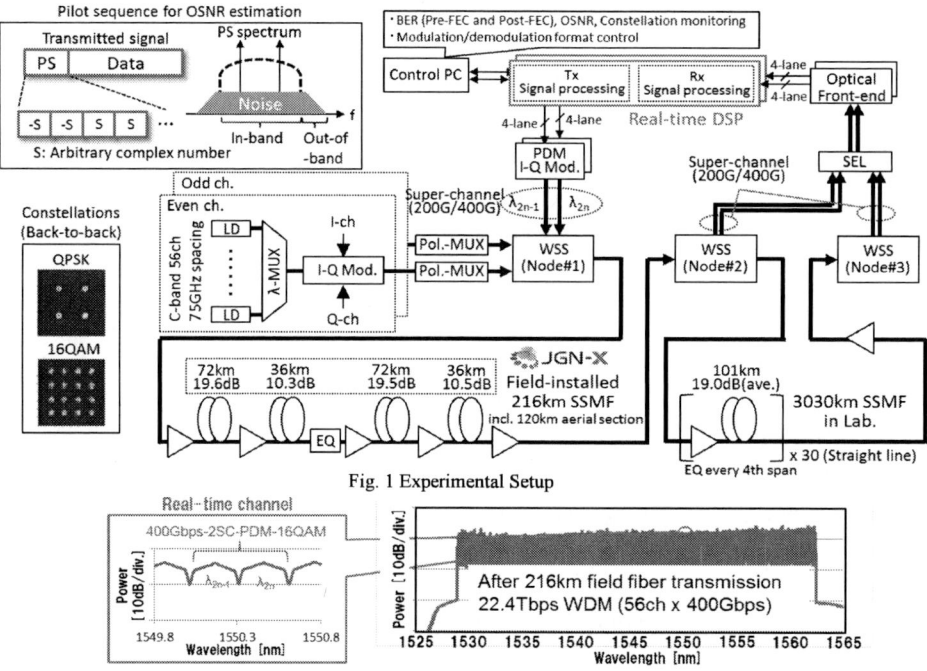

Fig. 1 Experimental Setup

Fig. 2 Optical WDM spectrum

III. RESULTS AND DISCUSSIONS

Figure 3 shows measured pre-FEC Q-factors and estimated OSNR for all 56 channels. Figure 3(a) shows data after 216km transmission (PDM-16QAM format). Blue diamonds indicate pre-FEC Q-factors and red squares indicate estimated OSNRs. Estimated OSNRs are higher than the OSNR threshold, 20dB, for all channels. This means PDM-16QAM can be transmitted without errors. Measured pre-FEC Q-factors were around 6dB for all channels and no error was observed after FEC decoding for all 400Gbps 2SC-PDM-16QAM channels. These results are consistent with the FEC performance where Q-limit is 5.0dB at post-FEC BER of 10^{-15}[10]. Figure 3(b) shows data after 3246km transmission (DP-QPSK format). Estimated OSNRs are lower than the OSNR threshold, 20dB, for all channels. This means PDM-16QAM cannot be transmitted without error. Thus the PDM-QPSK format was selected in this condition. Measured pre-FEC Q-factors were higher than 6dB for all channels and no error was observed after FEC decoding for all 200Gbps 2SC-PDM-QPSK channels. As shown in Fig. 3(a) and (b), the curves for estimated OSNR and pre-FEC Q-factor are similar. This means that the estimated OSNR is a suitable parameter for selecting the modulation format.

Figure 4 shows the results of the long-term stability measurement of pre-FEC Q-factors. Figure 4(a) indicates data for 400Gbps 2SC-PDM-16QAM signal over 216km field-installed SSMF including 120km aerial section. Pre-FEC and post-FEC Q-factors were measured for 64 hours. Figure 4(b) indicates data for 200Gbps 2SC-PDM-QPSK signal over 3246km SSMF (216km field-installed SSMF + 3030km laboratory SSMF). Pre-FEC and post-FEC Q-factors were measured for 111 hours. For both cases, no significant degradation in pre-FEC Q-factor and no error after FEC decoding were observed after transmission over 216km and 3246km SSMF including 120km aerial cable section.

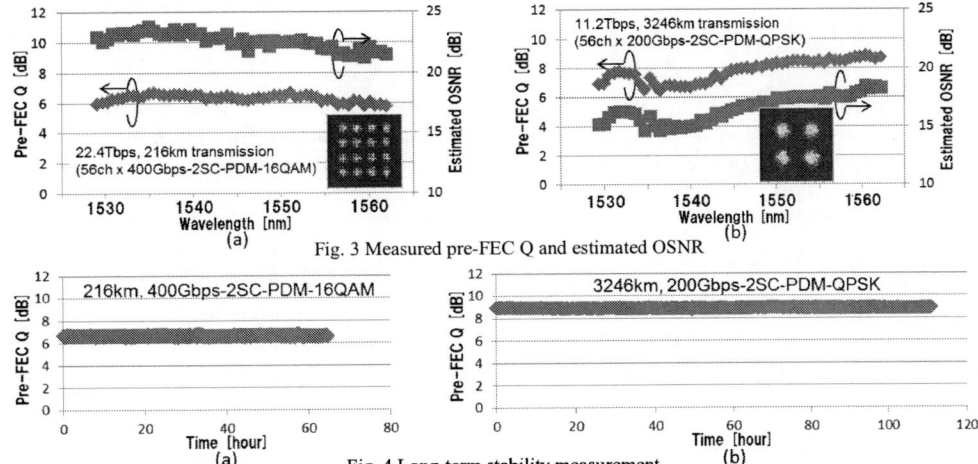

Fig. 3 Measured pre-FEC Q and estimated OSNR

Fig. 4 Long-term stability measurement

IV. CONCLUSION

A modulation format adaptation experiment based on pilot-aided OSNR estimation was conducted using real-time DSPs and field-installed fiber. Adapting the modulation format according to suit the estimated OSNR allowed 56-channel 400Gbps-2SC-PDM-16QAM and 200Gbps-2SC-PDM-QPSK signals to be successfully transmitted over 216km field-installed SSMF and 3246km SSMF (216km field-installed SSMF + 3030km laboratory SSMF), respectively. Total capacities were 22.4Tbps for 216km and 11.2Tbps for 3246km. Long-term stability measurements showed no significant degradation in pre-FEC Q-factor and no error after FEC decoding after transmission over 216km and 3246km SSMF including 120km aerial cable section.

ACKNOWLEDGMENT

This work is partly supported by the R&D project on "Research and Development of Ultra-high-speed and Low-power-consumption Optical Network Technologies" of the Ministry of Internal Affairs and Communications (MIC) of Japan and the R&D project on "Research and Development of Optical Transparent Transmission Technology (Lambda Reach)" of the National Institute of Information and Communications Technology (NICT). We also sincerely thank all parties concerned in the projects.

REFERENCES

[1] E. Yamazaki et al., "Fast optical channel recovery in field demonstration of 100-Gbit/s Ethernet over OTN using real-time DSP," Optics Express, vol. 19, no. 14, 2011.

[2] NTT press release, http://www.ntt.co.jp/news2014/1409e/140904a.html.

[3] H. Maeda et al., "Field trial of simultaneous 100-Gbps and 400-Gbps transmission using advanced digital coherent technologies," Proc. of OFC/NFOEC 2016, W1K.4.

[4] B. Lavigne et al., "400Gb/s trials on commercial systems using real-time bit-rate-adaptive transponders for next generation networks," Proc. of OFC/NFOEC 2015, W3.E.1.

[5] A. Pagano et al., "400Gb/s real-time trial using rate-adaptive transponders for next generation flexible-grid networks," Proc. of OFC/NFOEC 2014, Tu2B.4.

[6] D. J. Geisler et al., "First testbed demonstration of a flexible bandwidth network with a real-time adaptive control plane," Proc. ECOC 2011, Th.13.K.2.

[7] H. Y. Choi et al., "BER-adaptive flexible-format transmitter for elastic optical networks," Optics Express, vol. 20, no. 17, 2012.

[8] S. Okamoto et al., "Digital in-band OSNR estimation for polarization-multiplexed optical transmission," Proc. of OECC 2013, TuR2-4.

[9] S. Okamoto et al., "Field experiment of OSNR-aware adaptive optical transmission with pilot-aided bidirectional feedback channel," Proc. of OFC/NFOEC 2016, Th2A.2.

[10] K. Sugihara et al., "A spatially-coupled type LDPC code with an NGC of 12 dB for optical transmission beyond 100 Gb/s," Proc. of OFC/NFOEC 2013, OM2B.4.

QPSK Assisted Carrier Phase Recovery for High Order QAM

Xiaofei Su[1], Liang Dou[1], Zhenning Tao[1], Takeshi Hoshida[2,3], Jens C. Rasmussen[3]

[1] Fujitsu R&D Center, 3F, Space 8, Pacific Century Place, Gong Ti Bei Lu, Chaoyang District, Beijing, 100027, China,
[2] Fujitsu Limited, [3]Fujitsu Laboratories Ltd., 1-1 Kamikodanaka 4-chome, Nakahara-ku, Kawasaki, 211-8588, Japan.
suxiaofei@cn.fujitsu.com

Abstract: *With the assistance of inserted QPSK symbols, carrier phase recovery for high order QAM signal can be compatible with conventional Viterbi-Viterbi algorithm. For 64QAM, equivalent performance can be achieved but with extremely low hardware complexity.*
Keywords: *coherent optical system, carrier phase recovery, quadrature amplitude modulation*

I. INTRODUCTION

Higher order modulation formats beyond QPSK enable significant improvement in spectrum efficiency and thus transmission capacity. However, such higher order modulation formats suffer from limited tolerance to the noise and distortions generated both in the transmission link and the transceiver, resulting in a limited reach. Combination of intra-channel multiplexing technology either in time and/or frequency domain and flexible modulation formats can accommodate various requirements with different distance and throughput. One of the key enablers for such flexible transceiver is a universal carrier phase recovery (CPR) algorithm for arbitrary QAMs signal reception. Compared with the conventional Viterbi-Viterbi (V-V) CPR method that is widely used in DP-QPSK system [1], higher order QAM tends to require more sophisticated methods due to the higher constellation point density. The blind phase search (BPS) algorithm [2] can be applied to arbitrary high order QAM signals with good tolerance to laser phase noise, but it is known to suffer from large computational complexity even after introducing two-stage configuration [3] to reduce the test phase number. The algorithm based on QPSK partitioning was proposed in [4], which was derived from V-V algorithm, offers rather lower complexity, but its operation principle imposes inherent constellation dependency.

In this paper, we propose a feed-forward, low-complexity and flexible QPSK assisted CPR scheme, which is transparent to the higher order payload modulation formats. The linewidth tolerance and computational complexity of our proposed algorithm are investigated and compared with the other CPR algorithms.

II. PRINCIPLE OF THE PROPOSED QPSK ASSISTED CPR

The key idea of the proposed QPSK assisted CPR is shown as Fig. 1. k QPSK symbols are inserted periodically into the payload sequence in an N-symbol block at transmitter side, where the insertion ratio is defined as k/N. The detailed block diagram of the post processing procedure is shown in Fig. 2. Notice that we assume adaptive equalization and frequency offset compensation are already finished and 1Sa/Sym signal is fed into this algorithm. For each data block, QPSK symbols are first extracted and the conventional V-V algorithm is used to calculate the estimated phase noise θ_{est} on those time slots. Then the phase noise of the total sequence, θ'_{est}, is achieved through interpolation. Finally, CPR is realized by subtracting those estimated phase from the received symbols. Since the phase estimation is based on the inserted QPSK symbols, the proposed method is independent of the modulation format of the payload symbols.

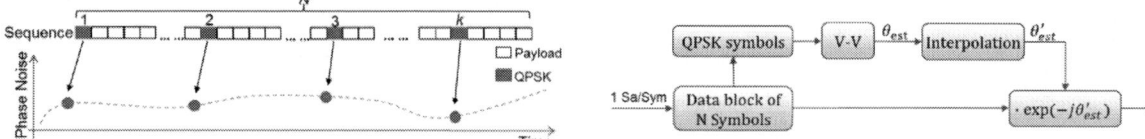

Fig. 1. Principle diagram of QPSK assisted CPR.   Fig. 2. DSP implementation of QPSK assisted CPR scheme.

Being different from the conventional blind CPR algorithms such as BPS and QPSK partition, the proposed method sacrifices the fraction of available channel throughput by altering the payload symbols with the QPSK symbols that conveys lower number of bits. To discuss this trade-off, capacity reduction ratio is defined by the QPSK insertion ratio (k/N) and bit number included in payload symbols (*Bits*), as equations (1) shows.

$$Capacity\ Reduction\ Ratio = 1 - \frac{k \cdot 2 + (N-k) \cdot \text{Bits}}{N \cdot \text{Bits}} = \frac{k}{N}\left(1 - \frac{2}{\text{Bits}}\right) \quad (1)$$

Fig. 3 describes the capacity reduction ratio under different modulation formats and QPSK insertion ratios. For example, if the payload is 64QAM and one QPSK symbol is inserted every 16 symbols, the capacity reduction ratio is approximately equal to 4.17%.

Although it is not a basic requirement for the proposed CPR, the QPSK symbols can be generated directly by choosing four of the QAM constellation points for compatibility in data generation circuit in transmitter when the QAM constellation has appropriate symmetry. For example there are 9 different choices for QPSK selection out of the regular 64QAM constellation (Fig. 4), on which we discuss in the following.

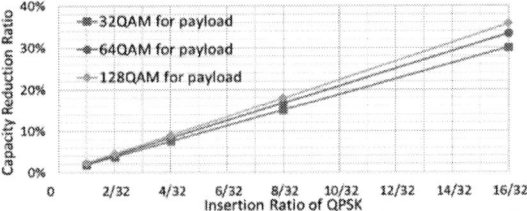

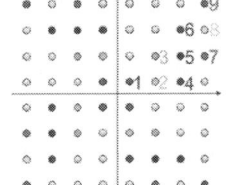

Fig. 3. Capacity reduction ratio vs. QPSK insertion ratio with different payload.

Fig. 4. QPSK power level choices out of 64QAM constellation.

III. PERFORMANCE ANALYSIS AND DISCUSSIONS

In order to investigate the performance of the QPSK assisted CPR, we construct a dual polarization 42GBaud 64QAM simulation platform. Both the transmitter and receiver lasers are with the same linewidth, where the phase noise is modeled as a Wiener process. White Gaussian noise is added in front of the receiver to emulate optical noise. Impact of various parameters, i.e. QPSK insertion ratio, QPSK amplitude, V-V block size, and laser linewidth is examined. To guarantee sufficient evaluation accuracy of the Q value below 10dB, more than 10^5 symbols are used for differential decoding and bit error ratio (BER) counting.

A. System Capacity Reduction Ratio

At first, we investigate the performance of the QPSK assisted CPR with different capacity reduction ratios as shown in Fig. 5. Here the linewidth of each laser and OSNR are 100kHz and 25dB, and the both QPSK amplitude and block size of V-V algorithm are optimized for each data point. Also shown by horizontal dashed lines are two references: one is for the QPSK partitioning algorithm in [4] with 120 symbols as block size, and the other one is for the two-stage BPS [3] with 8/8 test phase number and 90/25 block size for stage-1/stage-2. For the proposed method, the Q value increases along with the system capacity reduction ratio, and such phenomenon is brought by the reduced spectral efficiency. When the capacity reduction ratio is 2% that corresponds to 1 QPSK symbol in every 32-symbol block, the performance is almost the same as the QPSK partitioning method. If the ratio is doubled to 4%, the Q value is comparable with the BPS result. In the following, we discuss the performance of the QPSK assisted CPR with 4% capacity reduction ratio.

B. QPSK Amplitude

Another important parameter to be optimized is the QPSK amplitude. Fig. 6 shows the simulated Q value as a function of the QPSK amplitude for different laser linewidths at the fixed OSNR level of 25dB. Even without the laser phase noise, i.e. linewidth is set to 0kHz, there exists an optimal QPSK amplitude to achieve the best Q value. When QPSK amplitude level is increased, the phase noise estimated by CPR becomes more accurate but the power allocated to 64QAM symbols is lower, resulting a limited OSNR on payload symbols. So for fixed linewidth, the optimal QPSK level depends on the tradeoff between the phase estimation capability and reduction of payload OSNR level. When the laser linewidth is larger, the phase noise estimation capability based on QPSK symbol matters more and thus the optimal QPSK amplitude tends to be larger. As shown in Fig. 6, the optimal QPSK amplitude of different linewidth does not remain the same, and its value is around the 4th level (as defined in Fig. 4) for 100kHz laser linewidth, whereas it becomes the 6th level for 300kHz. It should be noted however that the difference in the optimal amplitude of QPSK is not significant (less than 0.2dB) for the case, even if the QPSK amplitude is fixed to the 4th level (or the 6th).

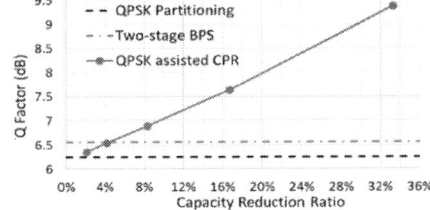

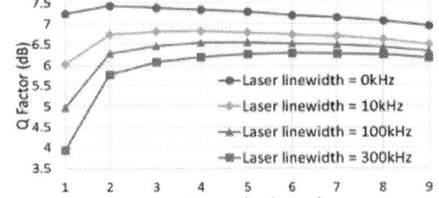

Fig. 5. Q performance vs. capacity reduction ratio with 100kHz laser linewidth at 25dB OSNR.

Fig. 6. Influence of QPSK amplitude level for 4% capacity reduction ratio 64QAM with different laser linewidth at 25dB OSNR.

C. Viterbi-Viterbi Block Size and QPSK Amplitude

The optimal block size (length of the sliding window) of V-V algorithm is determined by the tradeoff between Gaussian noise suppression and the phase noise tracking capability and thus is another critical aspect of the proposed CPR algorithm. The contour plot of the combined effect of block size and QPSK amplitude on Q factor is shown in Fig. 7. The grid of the contour line is spaced with 0.2dB. It can be concluded that the block size and QPSK amplitude settings are not extremely sensitive, and they can be selected from a large range of values without serious performance degradation exceeding 0.2dB.

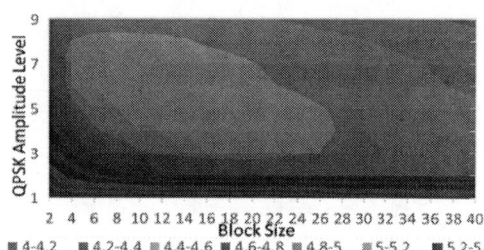

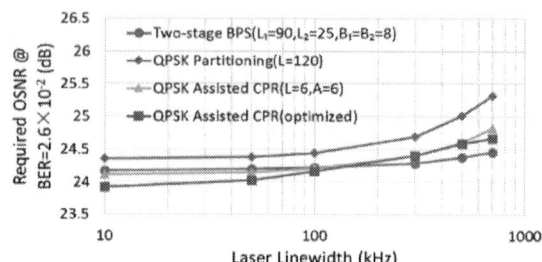

Fig. 7. Influence of QPSK amplitude and V-V block size to Q factor (unit: dB) for 4% capacity reduction ratio 64QAM with 100 kHz laser linewidth at OSNR=25dB.

Fig. 8. Required OSNR sensitivity vs. laser linewidth for 42GBaud/s 64QAM system using different CPR algorithms, 4% capacity reduction ratio for QPSK assisted CPR.

D. Performance Comparisons

Finally, the performance of different CPR algorithms is analyzed by calculating the required OSNR at the BER of 2.6×10^{-2}, which corresponds to 20% FEC overhead in [5]. In Fig. 8, L is the block size of the corresponding algorithms, and A is the QPSK amplitude level for QPSK assisted CPR. L_1/L_2 and B_1/B_2 are the block sizes and numbers of test phases for the first and second stage of BPS. The proposed method is evaluated in two versions, one is with fixed block size and QPSK amplitude, and the other is with them optimized. From Fig. 8, it can be seen that the two-stage BPS is the most ASE noise tolerant when laser linewidth is larger than 150kHz. The OSNR sensitivity of QPSK assisted CPR, however, is about at least 0.3dB better than that of QPSK partitioning regardless that it is with or without optimization.

E. Computational Complexity

The computational complexities of CPR algorithms analyzed in this paper are compared in Tab. 1. L_3 is the block size of QPSK partitioning algorithm, L_4 is the symbol number sent to the V-V operation ($L_4=L_3 \times 3/16$ due to the fact that 3 points are selected from the 64QAM constellation in each quarter), and L_5 is the block size of QPSK assisted CPR. Most hardware components of the function blocks in QPSK partitioning have been studied in [6]. Here, QPSK assisted CPR is similar to QPSK partitioning except that there is no partitioning operation required.

TABLE I COMPLEXITY COMPARISON OF VARIOUS CPR METHODS

Methods	Multipliers	Adders	Comparison	Decision	Look-up Table
Two-stage BPS	$6L_1B_1+6L_2B_2+4$ (5524)	$6L_1B_1-B_1+6L_2B_2-B_2+2$ (5506)	B_1+B_2-1 (15)	$L_1B_1+L_2B_2$ (920)	0
QPSK Partitioning	$2L_3+4L_4+5$ (355)	L_3+4L_4+1 (211)	$4L_3+1$ (481)	0	1
QPSK Assisted CPR	$4L_5+5$ (45)	$4L_5+1$ (41)	1	0	1

For clarity, the calculation of complexities for each symbol, as shown in Tab. 1, is based on optimum implementations for each algorithms of 42GBaud 64QAM with linewidth considerations ($L_1=90$, $L_2=25$, $B_1=B_2=8$, $L_3=120$, $L_4=22.5$, $L_5=10$). The complexity of the proposed QPSK assisted CPR with 4% capacity reduction ratio is about 1/6 of that for QPSK partitioning algorithm that is already much less complex than the two-stage BPS.

IV. CONCLUSIONS

We have proposed a universal CPR algorithm for high order QAM with the assistance of the inserted QPSK symbols. Several key parameters are examined by a 64QAM simulation platform. At the same Q performance, the complexity of the proposed CPR with 4% capacity reduction ratio is reduced by a factor of 6 compared with QPSK partitioning algorithm.

ACKNOWLEDGEMENT

This work was partly supported by the "Tera-bit optical network technologies towards big data era" of the Ministry of Internal Affairs and Communications of Japan.

REFERENCES

[1] R. Noé, "Phase noise tolerant synchronous QPSK/BPSK baseband-type intradyne receiver concept with feed-forward carrier recovery," J. Lightw. Technol., vol. 23, no. 2, pp. 802-808, February 2005.

[2] T. Pfau et al., "Hardware-efficient coherent digital receiver concept with feedforward carrier recovery for M-QAM constellations," J. Lightw. Technol., vol. 27, no. 8, pp. 989-999, April 2009.

[3] J. Li et al., "Laser-linewidth-tolerant feed-forward carrier phase estimator with reduced complexity for QAM," J. Lightw. Technol., vol. 29, no. 16, pp. 2358-2364, January 2011.

[4] I. Fatadin et al., "Laser linewidth tolerance for 16QAM coherent optical systems using QPSK partitioning," Photon. Technol. Lett., vol. 22, no. 9, pp. 631-633, May 2010.

[5] K. Sugihara et al., "A Spatially-coupled Type LDPC Code with an NCG of 12 dB for Optical Transmission beyond 100 Gb/s," OFC/NFOEC, OM2B.4, September 2013.

[6] Y. Gao et al., "Low-complexity and phase noise tolerant carrier phase estimation for dual-polarization 16-QAM systems," Optics Express, vol. 19, no. 22, pp. 21717-29, October 2011.

Block Carrier-Phase Recovery with Recursive Noise Adaptive Kalman Filtering for 16-QAM Signals

Zheng Bofang and Chester Shu
Department of Electronic Engineering and Center for Advanced Research in Photonics,
The Chinese University of Hong Kong, Shatin, N.T., Hong Kong   Email: bfzheng@ee.cuhk.edu.hk

Abstract: *A robust online block carrier-phase recovery scheme for 16-QAM signals is demonstrated via adaptive Kalman Filter. Less than 0.2dB Q-factor variation is achieved when the initial measurement noise variances deviate over 8 orders of magnitude.*
Keywords: *Adaptive Kalman filter; fiber optical communication; Coherent communications*

I. INTRODUCTION

Quadrature amplitude modulation in combination with optical coherent detection by digital signal processing attracts much attention thanks to the high spectral efficiency. To cope with the demand of the ever increasing bandwidth of the metro networks, 16-ary quadrature amplitude modulation (16-QAM) is a cost-efficient solution [1]. Robust blind digital carrier phase recovery (CPR) techniques are essential for the reliable reception of coherent 16-QAM signals. Optimal phase tracking can be achieved using the Kalman filter under the white frequency noise [2]. Unlike decision directed phase locked loop with fixed gain, the Kalman filter can be regarded as a feedback loop with variable gain. An extended Kalman filter has been experimentally demonstrated to outperform feed forward blind phase search (BPS) under low optical signal-to-noise ratio (OSNR) [3].

On the other hand, the mismatch between clock frequency of available digital signal processor (DSP) and the typical symbol rates requires hardware implementable schemes. The symbol-by-symbol algorithm is impractical, since the feedback delay due to parallel processing deteriorates the tolerance of the laser linewidth [4]. Block CPR based on the Kalman filter is a promising technique, which gives uncompromised performance with lower complexity compared to the improved BPS approach [5]. However, the Kalman filter requires exact knowledge of the noise statistics, which seems to be a plausible assumption in the reconfigurable and future dynamic optical networks. The nontrivial initialization of noise covariance gives rise to unstable system performance, which hinders the practical implementation. Noise Adaptive Kalman filter (AKF) can potentially solve this issue and improve the robustness of the algorithm.

In this paper, we incorporate the online measurement noise variance estimation with the block CPR for 16-QAM signals. The sufficient statistics of the measurement noise variances are estimated with a two-point iteration. We observe a noticeable low penalty (<0.2dB) in the operation of 16 Gbaud Nyquist 16-QAM signals even with over 8 orders of magnitude variation of the initial measurement noise covariance. The robust system performance positions the block AKF CPR as a measurement noise covariance-aware method for coherent optical links.

II. PRINCIPLE OF OPERATION

Generally, white frequency noise, which contributes to the Lorentzian lineshapes, is accompanied by time varying carrier frequency offset (CFO). In current digital coherent transmission systems, the symbol rate is around a few tens of gigahertz and the laser linewidth is typically hundreds of kilohertz. Moreover, the unavoidable frequency drift between the lasers in the transmitter and receiver side is below the microsecond scale. Consequently the evolvement of carrier phase can be linearly approximated within a certain number of symbol periods. The carrier phase of symbols in one block can be retrieved simultaneously by linear interpolation with the block state information $x_k = [\theta_k, \omega_k]^T$, where θ_k and ω_k are the estimated phase of the central symbol and the averaged frequency offset of the k-th block. In practice, the block state information has to be extracted from noise contaminated samples in a data-aided or decision aided style. Such noise eventually contributes to measurement noise in the framework of Kalman filter. Here we adopt variational Bayesian methods to joint estimation block state and measurement noise covariance [6], which enables linear Kalman filter structure without the need to know the covariance of measurement noise.

The state model of the linear Kalman filter is the same as that in [5]. For the k-th block with size M, the state-space model for x_k can be written as $x_k = A x_{k-1} + B q_k$. The transition-state-matrix $A = [1, M; 0,1]$ describes the carrier phase evolvement from block to block. The process noise q_k is assumed to be stationary and its covariance $Q = [0,0; 0, \sigma_s^2]$ is known. The measurement noise comes to play a role when the optimal estimation of x_k is based on noise samples. The measurement noise covariance $R_k = [\sigma_{k,\theta}^2, 0; 0, \sigma_{k,\omega}^2]$ is unknown and could be slowly time varying.

Fig.1 shows the algorithm diagram and the detailed steps. We assume that $\sigma_{k,\theta}^2$ and $\sigma_{k,\omega}^2$ in R_k can be approximated by two independent Gamma distributions and so $\sigma_{k,\theta}^2 \sim \Gamma(\beta_{k,1}, \alpha_{k,1})$ and $\sigma_{k,\omega}^2 \sim \Gamma(\beta_{k,2}, \alpha_{k,2})$. The covariance R_k can then be estimated jointly with block state x_k and error covariance P_k without leverage on the causality of the algorithm. Note that the receiver samples should undergo clock data recovery, equalization, and down-sample to one sample per symbol before the CPR process of the incoming M samples, $r_k^1, \cdots, r_k^M$ in the k-th block. Here, we use a superscript like x_k^- for a prior estimation and a superscript like $\hat{x}_k$ for a posterior estimation. At the prediction stage, the input samples are tentatively rotated and decoded using the calculated phase from the prior estimation x_k^-. The residual frequency and phase offset can be obtained consequently and used as innovation. Unlike x_k^- and P_k^-, the measurement covariance uncertainty is spread by a factor of $\rho^{-1}, \rho \in (0,1]$ while the expectation value remains the same as the last posterior. At the update stage, N-point iterations are used for the convergence, where the posterior of x_k, P_k and R_k are jointly estimated. As the measurement covariance can be estimated within each block, the initialization of R_k will not affect the algorithm performance.

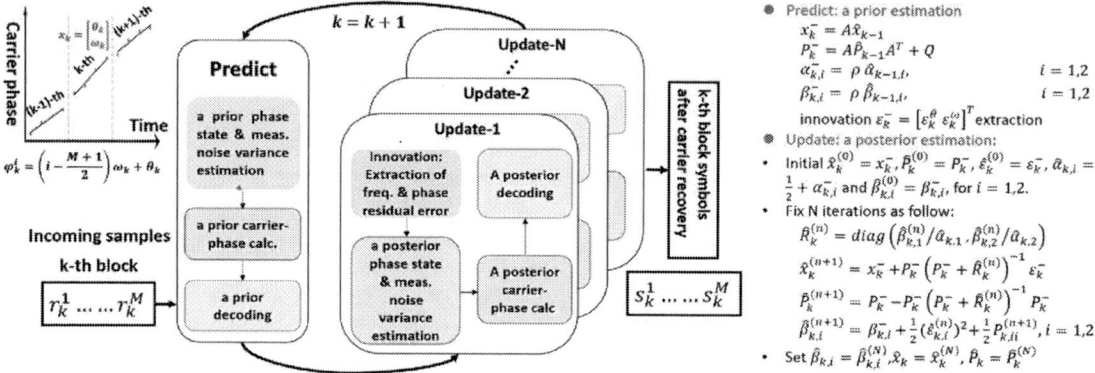

Fig. 1. Schematic diagram and algorithm structure for the proposed adaptive block CPR scheme

To test the proposed scheme for 16-QAM signals, the initial frequency offset is coarsely set by method [7] with 1k samples. The spread factor ρ equals 0.9. The error covariance $\hat{P}_0$ equals 0. Both $\alpha_{0,1}$ and $\alpha_{0,2}$ are set as 1. The initial measurement covariance can be setting by $\beta_{0,1}$ and $\beta_{0,2}$.

III. EXPERIMENT RESULTS

A 16-QAM system in the back-to-back condition is used to test the scheme. An external cavity laser (ECL) (<100 kHz linewidth) is modulated using an IQ modulator. The arbitrary waveform generator employed has a sample rate of 64 GS/s and a 20 GHz nominal bandwidth. The electrical data signals from the two channels are generated individually with a symbol length of 2^{12}. Nyquist pulse shaping with a 0.1 roll-off factor is adopted in the software. After the adjustment of OSNR, the signal was detected by a coherent receiver at a power level of -20dBm. The local oscillator is also an ECL with a 15dBm output power and less than 100 kHz linewidth. The output of the coherent receiver is sampled at 40 GS/s by the real time oscilloscope. Offline processing is used for data demodulation. The offline DSP consists of: I/Q imbalance compensation, clock recovery, polarization demultiplexing, channel equalization, proposed CPR and BER calculation. Note that although a single polarization is used, we still adopt the polarization diversity scheme in case of polarization drift. Besides, time-domain block equalization [8] is adopted and the block size is fixed as 16.

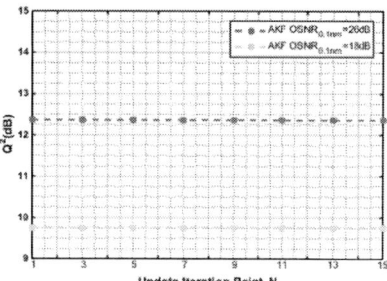

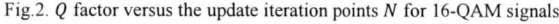

Fig.2. Q factor versus the update iteration points N for 16-QAM signals

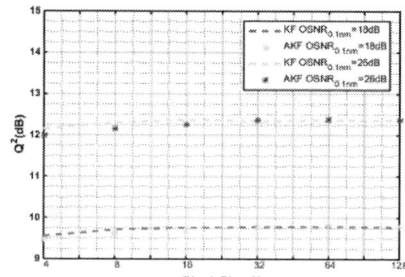

Fig.3. Q factor versus block size M for 16-QAM signals

The update iteration points determine the extra overhead for measurement noise covariance estimation. Fig. 2 plots the measured Q factor against the update iteration point for 16-QAM signals with different OSNR. The results show that only two-point iteration can get the same performance as 15-point iterations. So we adopt two-point iteration in the following tests to verify the effectiveness. Fig. 3 shows the measured Q factor against the block size. The linear Kalman filter scheme [5] is shown as a reference. The Q factor drops slightly as the block size is below 16 in both schemes. For the proposed AKF, the maximum Q factor variations are 0.30dB and 0.39dB when the OSNR is set at 18dB and 26dB, respectively. For the KF scheme, the corresponding variations are 0.22dB and 0.23dB. Under the condition of identical block sizes, the maximum Q-penalty for AKF against KF scheme is 0.17dB at an OSNR of 26dB. The penalty decreases further to 0.09dB at an OSNR of 18dB. The small penalty may originate from a higher adaptive noise in the steady state. A larger spread factor ρ will lead to more consistent results, since the AKF degenerates into KF when $\rho =1$, but the adaptiveness of the measurement noise will degrade. Fig. 4 shows the measured BER against the OSNR at different block sizes. The successful operation with a block size of 128 shows that AKF CPR inherits the capability for large block signal process from KF CPR.

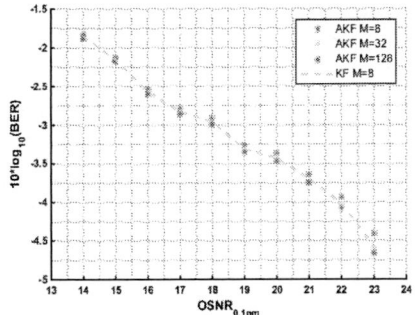

Fig.4. BER with different block size for 16-QAM signals

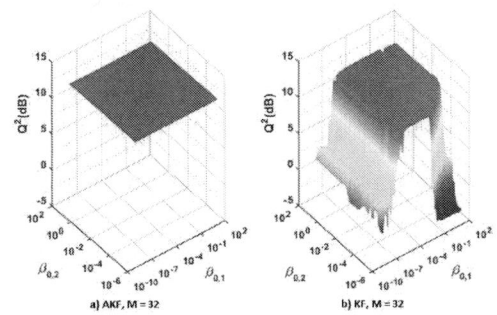

Fig.5. Robustness test with different initial measurement noise parameters using a block size M=32: a) AKF, b) KF

To investigate the robustness of our proposal algorithm, different settings of initial measurement noise parameters $\beta_{0,1}, \beta_{0,2}$ are tested with a block size of 32. In our analysis, $\beta_{0,1}$ sweeps from 10^{-9} to 10^1 and $\beta_{0,2}$ sweeps from 10^{-6} to 10^1. Fig.5 shows the Q factor variation for both schemes. Unlike the KF CPR, which failed with poorly initial measurement noise parameters, the proposed AKF CPR can successfully work over 8 orders of magnitude measurement covariance variations. The measured Q factor only varies by 0.185dB.

IV. CONCLUSIONS

We have proposed a robust block CPR scheme for 16-QAM signals based on adaptive Kalman filter. We experimentally demonstrate that the scheme can work for 16-GBaud 16-QAM signals. The adaptive nature offers an 8 orders of magnitude tolerance for the initial measurement noise covariance.

ACKNOWLEDGMENT

This work was supported by Hong Kong Research Grants Council through GRF grants CUHK 416213 and 14206614.

REFERENCES

[1] M. Iglesias Olmedo, X. Pang, M. Piels, R. Schatz, G. Jacobsen, S. Popov, I. Tafur Monroy, and D. Zibar, "Carrier Recovery Techniques for Semiconductor Laser Frequency Noise for 28 Gbd DP-16QAM," Optical Fiber Communications Conference and Exhibition, Los Angeles, Mar. 2015, Th2A.10.

[2] L. Barletta , M. Magarini and A. Spalvieri, "A new lower bound below the information rate of Wiener phase noise channel based on Kalman carrier recovery," Optics Express, vol. 20, no. 23, pp. 25471-25477, 2012.

[3] D. Zibar, L. Carvalho, M. Piels, A. Doberstein, J. Diniz, B. Nebendahl, C. Franciscangelis, J. M. Estaran Tolosa, H. Haisch, N. G. Gonzalez, J. F. R. De Oliveira and I. Tafur Monroy, "Bayesian Filtering for Phase Noise Characterization and Carrier Synchronization of up to 192 Gb/s PDM 64-QAM," European Conference on Optical Communications, Cannes, Sept. 2014, Tu.1.3.1.

[4] T. Pfau , S. Hoffmann and R. Noe, "Hardware-efficient coherent digital receiver concept with feedforward carrier recovery for M-QAM constellations", Journal of Lightwave Technology, vol. 27, no. 8, pp. 989-999, 2009.

[5] T. Inoue and S. Namiki, "Carrier recovery for M-QAM signals based on a block estimation process with Kalman filter", Optics Express, vol. 22, no. 13, pp. 15376, 2014.

[6] S. Sarkka and A. Nummenmaa, "Recursive noise adaptive Kalman filtering by variational Bayesian approximations", IEEE Trans. Autom. Control, vol. 54, no. 3, pp. 596-600, 2009.

[7] M. Selmi, Y. Jaouen, and P. Ciblat, "Accurate digital frequency offset estimator for coherent PolMux QAM transmission systems," European Conference on Optical Communications, Vienna, Sept. 2009, P3.08.

[8] O. Zia-Chahabi, R. L. Bidan, M. Morvan, and C. Laot, "Efficient frequency-domain implementation of Block-LMS/CMA fractionally spaced equalization for coherent optical communications," IEEE Photon. Technol. Lett., vol. 23, no. 22, pp. 1697–1699, Nov. 15, 2011.

Highly ASE Tolerant Pass-band Shape Monitor for Cascaded ROADMs

Guoxiu Huang[1], Shoichiro Oda[2], Tomohiro Yamauchi[2], Setsuo Yoshida[2], Goji Nakagawa[2], Yasuhiko Aoki[2], Zhenning Tao[3] and Jens C. Rasmussen[1]

1) Fujitsu Laboratories Ltd., 2) Fujitsu Limited, 1-1 Kamikodanaka 4-chome, Nakahara-ku, Kawasaki 211-8588, Japan
3) Fujitsu R&D Center, 355 Unit 3F, Gate 6, Space 8, Pacific Century Place, No.2A Gong Ti Bei Lu, Chaoyang District, Beijing 100027, P.R. China
E-mail: huang.guoxiu@jp.fujitsu.com

Abstract: Highly accurate optical pass-band shape monitor was proposed by employing FM-CW probe light immune to accumulated ASE noise. We experimentally show that this method can achieve high accuracy until 5dB optical probe to noise ratio.

Keywords: Pass-band shape monitor, Reconfigurable optical add/drop multiplexing, amplified spontaneous emission noise

I. INTRODUCTION

Flexibility to support mesh topologies, dynamic capacity allocation, and automated network control including light path setup are key elements in the design of next-generation networks [1, 2]. With the development of wavelength-selective switches (WSS), multi-degree reconfigurable optical add/drop multiplexing (MD-ROADM) node architectures that allow mesh configurations and wavelength routing have commonly deployed. Pass-band shape (PBS) of WSS is becoming an important consideration since filtering penalties for beyond 100Gb/s superchannel signals will become an issue with large cascaded ROADM nodes [3, 4]. In such a case fine tuning of central frequency and subcarrier spacing are indispensable for long-haul transmission. Since piece-to-piece variations of the PBS makes the subcarrier spacing optimization difficulty, practical PBS monitor will be required.

The PBS monitor employing frequency modulated (FM)-CW probe light has been proposed in [5]. In this proposal, the PBS absolute value and PBS slope can be achieved by detecting the transmitted FM-CW probe light power and power variance. The measurement speed can be enhanced by analyzing both PBS absolute value and slope value. However, the FM-CW probe light should be kept low power enough to suppress fiber nonlinearities to other living channels in real systems. In such a low optical probe to noise ratio (OPNR) systems, the accuracy of the previous method will be degraded drastically since the ASE noise generated from cascaded EDFAs mixing with probe light will affect the accuracy of detected power of probe light.

In this paper, an advanced analyzing method is proposed which can achieve higher accuracy even at low probe light power to solve above issue. In new proposal, processing switch is added to make sure that only appropriate detected transmitted power can be processed. The experimental demonstration has been done and the experimental results confirm that the proposed method can keep sufficiently high accuracy even in case of 5dB OPNR.

II. PRINCIPLE OF THE PROPOSED METHOD

The schematic diagram of the proposed optical PBS monitoring method is shown in Fig.1 (a). FM-CW probe light transmits through cascaded ROADMs. In PBS monitor at receiver side, the transmitted power P and power variance ΔP, which are proportional to PBS and its slope are detected. P_n and S_n is defined as the detected power of transmitted FM-CW light and PBS slope at n^{th} frequency f_n as shown in Fig.1 (b). In the previous method [5], spline cubic polynomial interpolation was employed to achieve PBS between each adjacent detected frequency point ($f_n \sim f_{n+1}$), where the polynomial coefficient was counted out with transmitted power and PBS slope at adjacent detected frequency point as (P_n, S_n), (P_{n+1}, S_{n+1}). The drawback of previous method is low tolerance to inaccuracy of detected power P_n due to the contribution of accumulated ASE noise.

The advanced data processing method proposed in this paper is interval integration of the detected slope value S instead of polynomial interpolation. The key point of the proposed method is that the detected PBS slope S is immune to ASE noise effect as shown in Fig.1 (a). Although the detected power is changed due to the contribution of ASE noise, the slope without ASE S_0 is equal to the one of with ASE S since the ASE noise power is considered to constant in the interest frequency range Δf. Another feature designed to improve the accuracy is that in order to avoid the error accumulation of the integration in the central part of PBS, i.e. high OPNR range detected power is simply adopted to the monitored PBS, while in the edge part, thus low OPNR range, the monitored PBS switched from the detected power to the one calculated by the integration. The detailed processing procedure is explained as follows (See Fig.1 (b)). (1) The transmitted power P_0 is measured at the central frequency of the PBS as an initial value and is assigned to $P_{process\_0}$. Here, $P_{process\_n}$ denotes the output PBS power processed by the proposed method. (2) At the next frequency point (f_1) both of the transmitted power and the slope are measured and we obtain the transmitted power P_1 and P'_1 calculated by the integration as $P'_1=$

$P'_{process\_0} + \Sigma S(f)\Delta f$ ($f_0 < f < f_1$). (3) If the absolute power difference between P_1 and P'_1 is within a threshold value θ, $P'_{process\_1}$ is determined to P_1. (4) Procedures (2) and (3) continues from the frequency point $n=2$. When the absolute power difference exceeds the threshold value θ, which means the ASE noise induced power error cannot be ignored, $P_{process\_n+1}$ is set to P'_{n+1}. (5) The above processing continues until the end of frequency range, f_{cut}. This switching functionality can make sure that only appropriate detected power can be processed at the start frequency for the integration.

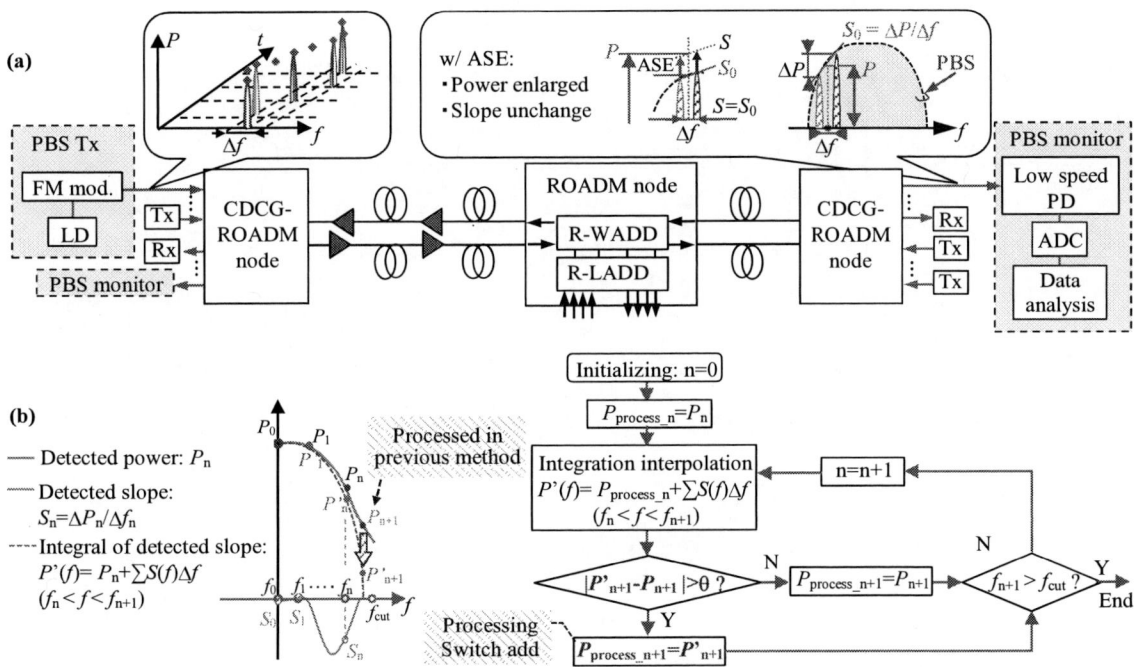

Fig.1 (a) Schematic diagram of proposed concept. (b) Detailed explanation of previous and proposed method and the flowchart of proposed data analysis. R-WADD: Reconfigurable wavelength add/drop devices, R-LADD: Reconfigurable local add/drop devices, CDCG-ROADM: Colorless directionless contentionless gridless – Reconfigurable optical add/drop multiplexer.

III. EXPERIMENTAL SETUP AND RESULTS

In order to confirm the accuracy of this advanced analyzing method, the experiment has been done. The experimental setup is shown in Fig.2 The integrable tunable laser assembly (ITLA) with FM function was used for FM-CW probe light generation. The FM frequency and FM index were set to be 30kHz and 82MHz, respectively. The ASE source followed by variable optical attenuator 2 (VOA2) was coupled with the probe light to achieve target OPNR. The OPNR is defined as the power difference between probe light and ASE noise within 0.1nm resolution in this experiment. The VOA1 followed with ITLA was used to adjust OPNR in the receiver side. As an in-service PBS monitor, 83 living channels of 32 Gbaud DP-QPSK were multiplexed with probe light by WSS. In the transmission part, 4-stage WSSs and EDFAs were included, VOA3 and VOA4 were used to adjust the input power of each EDFA and WSS. The EDFA gain was set to be 26dB and the input power was 0dBm. In the receiver side PBS monitor, an OBPF with bandwidth of 25GHz was employed to select the FM-CW probe signal. The structure of the PBS monitor Rx is as same as proposed one shown in [5]. The sampling speed of digital storage oscilloscope (DSO) was 1MSa/s. The slope value was obtained in the spectral domain by Fourier transform of DSO data. The optical spectrum in the input and output port of the last WSS for dropping detect channel are shown in Fig.2, when OPNR was set to be 15dB. The power difference between probe light and neighboring channel was 5dB.

Fig.3 (a) shows the analyzed results of PBS when OPNR is 5dB. The threshold value θ show in the flowchart was set to be 0.05dB and the detection resolution of frequency point was 2GHz. The normalized value of detected slope is also shown as blue line in Fig.3 (a). The slope curve was obtained by polynomial fitting of the detected slope value. The black solid line is a reference PBS of cascaded WSSs measured by commercial measuring instruments. The green dashed line was the results obtained by proposed method. We can observe that the dashed line agrees well with the reference black line until 20dB down point of PBS. The solid red line represents the results with previous method. We can find that the monitored accuracy of PBS was degraded since the transmitted probe power was affected by ASE noise in 5dB OPNR situation. The normalized mean square error (NMSE) versus OPNR is shown in Fig.3 (b) as comparison between proposed method and previous method which are shown by blue and orange curve, respectively. The definition of NMSE is as same as defined in [5]. From Fig.3 (b), we can observe that the NMSE of the previous method increased by 9dB with decreased

OPNR from 15dB to 5dB due to ASE noise, while the NMSE of the proposed method slightly improved by 1.5dB. The possible reason for the improvement of NMSE at lower OPNR thus lower power of the probe light may be attributed to the reduction of slope error induced by cross-gain modulation (XGM) effect in EDFAs. As a consequence 10.5dB improvement of PBS accuracy was achieved by the new proposal, when the OPNR was set to 5dB, which is sufficiently low to avoid nonlinear interference to in-service channels.

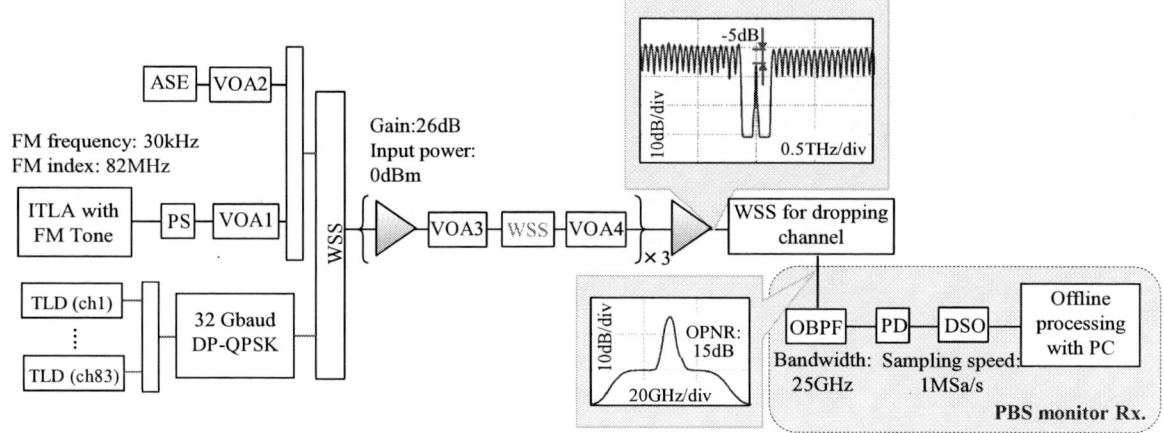

Fig.2. Experimental setup. ITLA: Integrable Tunable Laser Assembly. FM: Frequency Modulation. PS: Polarization Scrambler.
VOA: Variable Optical Attenuator. WSS: Wavelength Selective Switch. OPNR: Optical Probe to Noise ratio. OBPF: Optical Band Pass Filter.
PD: Photodiode. DSO: Digital Storage Oscilloscope. PC: Personal Computer

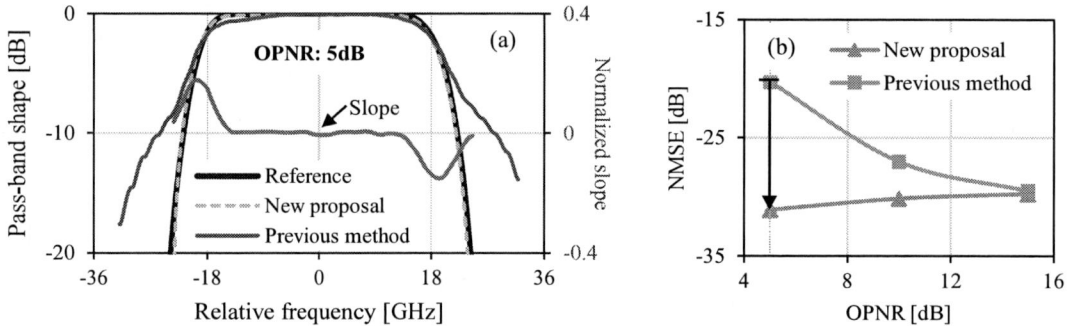

Fig.3. Experimental results. (a) The analyzed results of PBS and detected slope value.
(b) Normalized mean square error (NMSE) versus OPNR

IV. CONCLUSIONS

In this paper an advanced analyzing method for FM-CW probe light employed PBS monitor has been proposed. The proposed method can keep high accuracy even in low OPNR value. Comparing with the previous method, more than 10dB improvement of PBS accuracy was achieved when the OPNR was set to 5dB, which is sufficiently low to avoid nonlinear interference to in-service channels.

V. ACKNOWLEDGEMENTS

This work was partly supported by the National Institute of Information and Communications Technology (NICT), Japan.

REFERENCES

[1] K. Sato et al., ACP 2014, Workshop 7: Optical Datacentre Networks, Session 2.
[2] E. B. Basch et al., IEEE J. Sel. Topics in Quantum Elect., Vol. 12, No. 4, pp. 615-626, 2006.
[3] S. Gringeri et al., IEEE Communications Magazine, pp 40-50, July 2010.
[4] P. Jenneve et al, ACP 2015, AS3E.2.
[5] G. Huang et al., ACP 2015, paper AM1E.4.

Digital Nonlinear Distortion Compensation

Zhenning Tao[1], Liang Dou[1], Ying Zhao[1], Bo Liu[1], Lei Li[1],
Tomofumi Oyama[2], Takeshi Hoshida[3], Jens C. Rasmussen[2]

1. 3F, Gate 6, Space 8, Pacific Century Place, No.2A Gong Ti Bei Lu, Chaoyang District, Beijing, 100027, China,
2. Fujitsu Laboratories Ltd., 3. Fujitsu Limited, 1-1 Kamikodanaka 4-chome, Nakahara-ku, Kawasaki, 211-8588, Japan.
taozn@cn.fujitsu.com

Abstract: *Nonlinear distortion is considered as the ultimate limitation of optical transmission. Various digital nonlinear distortion compensation methods, including perturbation based method for long reach application and Volterra based method for short reach application are reviewed.*
Keywords: *Nonlinear compensation, Optical communication, Perturbation, Volterra*

I. INTRODUCTION

Digital signal processing is widely used in both long reach coherent detected transmission and short reach direct detected transmission. After the linear equalizer compensates the linear distortion, the nonlinear distortion is considered as the ultimate limitation of optical fiber communication. In long reach application, the major nonlinear distortion is the Kerr effect that the fiber refractive index changes with the optical intensity. Various compensation methods, such as back propagation [1], Volterra compensation [2], and perturbation based compensation [3-7] have been proposed. In short reach application, the major nonlinear distortions include the device nonlinearity and the square detection of photo diode. Various compensation methods have been proposed in [8-11].

In this paper, we review the perturbation based nonlinear compensation for coherent detection including the fundamental theory, the basic algorithm and advanced algorithm to reduce the complexity and to improve the performance. Then, we review the Volterra based compensation for direct detected short reach application.

II. PERTURBATION BASED NONLINEAR DISTORTION COMPENSATION FOR LONG REACH APPLICATION

A. Perturbation Based Nonlinear Model for Optical Fiber Transmission

The basic idea of nonlinear distortion compensation is to calculate the time-varying nonlinear noise waveform and to cancel it thereafter. In the long reach application, the major nonlinear distortion is the Kerr effect, and it can be described by the perturbation based nonlinear model [3, 12]. Figure 1 shows the perturbation based nonlinear model of arbitrary fiber transmission, where the nonlinear noise is modeled as an additive perturbation term $[\Delta u_x, \Delta u_y]^{\mathrm{T}}$.

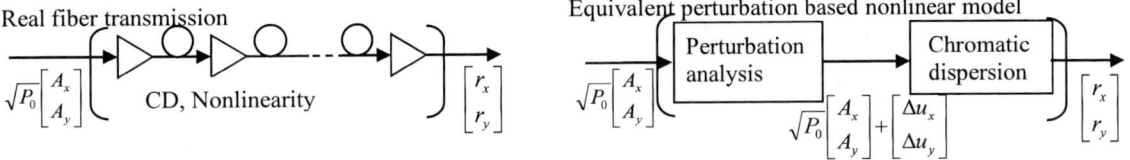

Fig. 1. The equivalent perturbation based nonlinear model for an arbitrary fiber transmission

With the assumptions of large chromatic dispersion and Gaussian input pulse, the perturbation is:

$$\Delta u_{0,x} = \sum_{m,n} P_0^{3/2}\left(A_{n,x}A_{m+n,x}^*A_{m,x}C_{m,n} + A_{n,y}A_{m+n,y}^*A_{m,x}C_{m,n}\right) \qquad \Delta u_{0,y} = \sum_{m,n} P_0^{3/2}\left(A_{n,y}A_{m+n,y}^*A_{m,y}C_{m,n} + A_{n,x}A_{m+n,x}^*A_{m,y}C_{m,n}\right) \quad (1)$$

where P_0 is the input power, $A_{m/n,x/y}$ is the normalized pulse complex amplitude. The perturbation coefficient $C_{m,n}$ is:

$$C_{m,n} = j\frac{8}{9}\frac{\gamma\tau^2}{\sqrt{3}|\beta_2|}\frac{L_{eff}}{L}E_1\left(-j\frac{mnT^2}{\beta_2 L}\right) \qquad m \neq 0, n \neq 0 \qquad (2\text{-}a)$$

$$C_{m,n} = j\frac{8}{9}\frac{\gamma\tau^2}{\sqrt{3}|\beta_2|}\frac{L_{eff}}{L}\frac{1}{2}E_1\left(\frac{(n-m)^2T^2\tau^2}{3|\beta_2|^2 L^2}\right) \qquad m \text{ or } n = 0 \qquad (2\text{-}b)$$

$$C_{0,0} = j\frac{8}{9}\frac{\gamma\tau^2}{\sqrt{3}|\beta_2|}\frac{L_{eff}}{L}\int_0^L dz\frac{1}{\sqrt{\tau^4/(3\beta_2^2) + z^2}} \qquad (2\text{-}c)$$

where β_2, γ, τ, T, L_{eff}, L are respectively the group velocity dispersion, the nonlinear coefficient, the pulse width, the inverse of pulse rate, the effective length, and the transmission distance, m/n is the pulse indexes, and $E_1(\,\cdot\,)$ is the exponential integral function. One essential fact is that the perturbation coefficient $C_{m,n}$ only depends on the parameters of the fiber link, so it could be calculated offline and stored in a look up table.

B. Basic Perturbation Based Nonlinear Compensation

One significant advantage of perturbation based compensation is the complexity reduction compared with conventional one stage per span digital back propagation because one perturbation based compensation stage could compensate multi-span fiber transmission or even the whole transmission link.

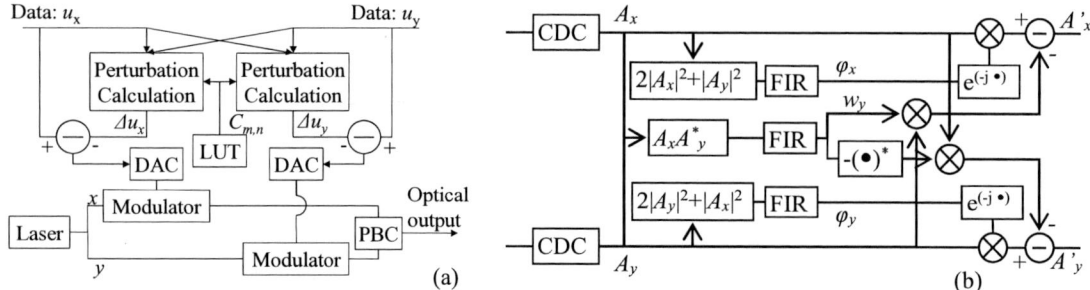

Fig. 2 (a) Block diagram of perturbation based pre-distortion, and (b) perturbation back propagation; LUT: look up table; DAC: digital-to-analog convertor; PBC: polarization beam combiner; CDC: chromatic dispersion compensation; FIR: finite impulse response.

Fig. 2 (a) shows the block diagram of perturbation based pre-distortion where the nonlinear noise waveform of the whole fiber link is calculated by perturbation analysis and is cancelled in transmitter [4]. This method is very good for QPSK modulation because the three pulse multiplication degenerates to logical operation and the complexity is significantly reduced. Similar method could also be implemented in receiver side where the pre-decided pulse symbol $A_{m/n,x/y}$ is used for perturbation calculation [13]. This method achieves same performance as one stage per span back propagation by just one nonlinear compensation stage. The major issue of this method is the high number of perturbation terms in (1) and corresponding complexity thereafter.

Fig. 2 (b) shows the block diagram of one compensation stage in perturbation back propagation [7]. In perturbation back propagation, the perturbation analysis is carried out for multi-span transmission so the compensation stages could be much less than the real span number. To reduce the complexity, only the intra-channel cross phase modulation term, i.e. the term $C_{m,n}$ with $mn=0$, is included. Because of the physical nature of Kerr effect, the additive perturbation term is converted to multiplicative phase modulation term by the approximation of $1+j\varphi \approx e^{j\varphi}$. It is interesting that three compensation stage could compensate the nonlinear distortion of 25 span transmission although lots of intra-channel four wave mixing terms are ignored [7]. Actually, conventional back propagation could be considered as a degenerated perturbation back propagation where only the $C_{0,0}$ term is included. Another degenerated perturbation back propagation is correlated back propagation, where the nonlinear polarization crosstalk term $A_{n,y}A^*_{m+n,y}A_{m,x}$ is ignored [5,6].

C. Further Effort to Reduce the Complexity and to Improve the Performance

One important method to reduce complexity is to combine the perturbation terms based on the symmetry of perturbation coefficient and the quantization of coefficient [14-16]. According to (2), the intra-channel four wave fixing term has the characteristic that all the coefficient $C_{m,n}$ are exactly the same if the product $m \times n = k$ are equal. Then, the perturbation calculation could be re-written as:

$$\Delta u_{0,x} = P_0^{3/2} \sum_k C_k \sum_{m \times n = k} \left(A_{n,x} A^*_{m+n,x} A_{m,x} + A_{n,y} A^*_{m+n,y} A_{m,x} \right); \quad \Delta u_{0,y} = P_0^{3/2} \sum_k C_k \sum_{m \times n = k} \left(A_{n,y} A^*_{m+n,y} A_{m,y} + A_{n,x} A^*_{m+n,x} A_{m,y} \right) \quad (3)$$

In the case of QPSK modulation, the inner summation of (3) could be converted to a simple logic circuit, then the complexity is very low. Taking 32 Gbaud DP-QPSK transmission over 60 km×25 span SMF link as an example, the required perturbation terms is 9756 if the truncated threshold is -34 dB. After combination, the number is reduced to 592. Such combination does not cause any performance degradation because equation (3) is exactly equal. The complexity could be reduced further if the perturbation coefficient is quantized. Fig. 3 (a) shows an example where the perturbation terms is reduced from 592 to 25 after quantization,. Fig. 3 (b) shows the system performance comparison. Although significant quantization occurs in Fig. 3 (a), the nonlinear compensation performance only reduces 0.1 dB.

The complexity could also be reduced by splitting single carrier high baud rate modulation into multicarrier low baud rate modulation where the compensation is carried out for each sub-carrier [17]. Owning to the nature of exponential integration function $E_1(\ )$ in (2), the coefficient reduces with m, n much faster if longer symbol duration T is used. Then, the required perturbation terms turns much less if the baud rate is reduced. Considering an example of 100-km × 30 spans SMF link, the numbers of the $C_{m,n}$ are 5365 and 433 for 32 Gbaud and 8 Gbaud signal respectively. In single carrier modulation, the nonlinear compensation is 5365 terms per symbol duration, i.e. 1/32Gbaud. In the parallel 4×8 Gbaud modulation, the nonlinear compensation for each sub-carrier is 433 terms per 1/8Gbaud, and the total compensation terms is 433×4 per 1/8Gbaud, i.e. 433 per 1/32Gbaud. Thus the overall complexity is reduced by 12.4=5365/433 times. As a penalty, some performance degradation occurs because only the nonlinear distortion within each sub-carrier is compensated, whereas the inter sub-carrier nonlinear distortion still occurs. Fortunately, such penalty is just 0.1 dB for the 4×8 Gbaud modulation [17].

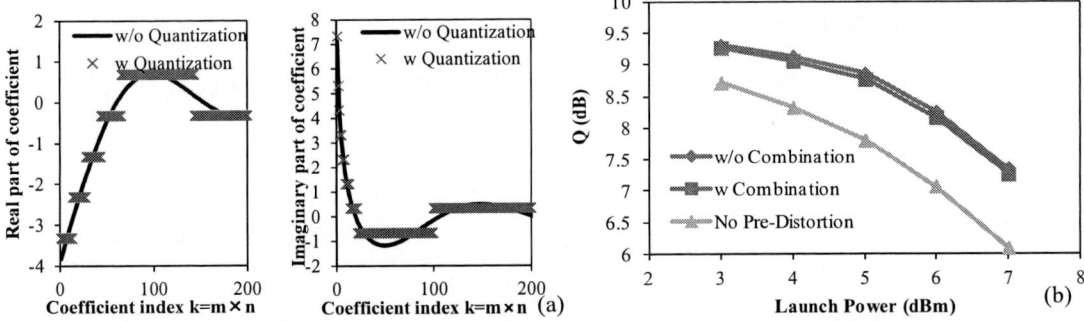

Fig. 3 (a) Quantization of perturbation coefficient (b) Nonlinear perturbation pre-distortion performance comparison transmission system: 32 Gbaud DP-QPSK, 60 km×25 span SMF, OSNR=14dB; terms before combination: 9756, terms after combination: 25

To improve the performance, an efficient way is to calculate the nonlinear noise more accurate. For example, the nonlinear model accuracy is increased from 80% to 95% by power weight nonlinear model and Nyquist nonlinear model [18-20]. In dispersion managed link, additional 0.5dB Q improvement could be achieved by power weighted nonlinear model [19]. It's worthy to notice that the new model only increases the complexity of perturbation coefficient calculation that could be carried out offline and the complexity of real-time processing does not change.

III. NONLINEAR DISTORTION COMPENSATION IN SHORT REACH APPLICATION

Nonlinear distortion even more severely occurs in low cost short reach application. Similar to that of long reach application, the key is how to calculate nonlinear noise waveform. Volterra model is widely used in nonlinear compensation [8-11]. Fig. 4 shows an example of 100Gb/s discrete multi-tone transmission in the directed detection short reach application [10]. Nonlinear compensation increases the capacity 25%. One essential way to reduce the complexity is to shorten the memory length of nonlinear compensation. Instead of a single 2nd order Volterra compensation stage with 15 tap memory length, the whole compensation is split into a 15 tap linear FIR filter and a 7 tap 2nd order Volterra compensation. As a result, the multiplier operation is reduced from 255 to 78.

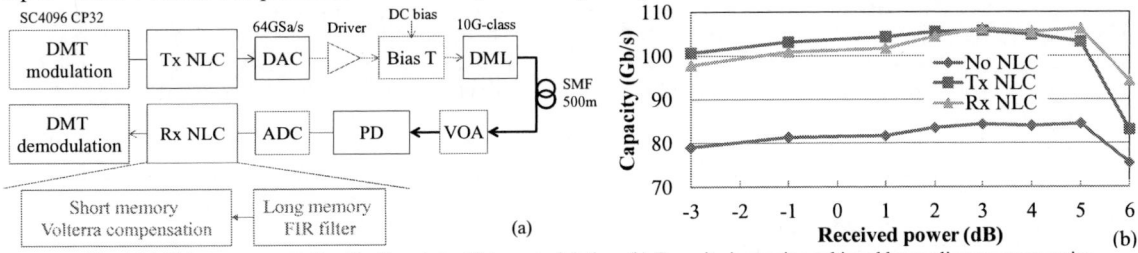

Fig. 4 (a) Volterra compensation for discrete multi-tone modulation, (b) Capacity increasing achieved by nonlinear compensation

IV. CONCLUSIONS

Nonlinear distortion turns to be the ultimate limitation of optical communication. Based on proper nonlinear distortion model, various digital nonlinear distortion compensations have been developed for both long reach coherent detection and short reach direct detection. Besides the basic compensation, the complexity is reduced by many methods.

ACKNOWLEDGEMENT

This work was partly supported by the National Institute of Information and Communications Technology (NICT).

REFERENCES

[1] X. Li, et al., OE, Vol. 16, No. 2, pp. 880–888, 2008.
[2] Y. Gao, et al., ECOC 2009, Paper 9.4.7.
[3] Z. Tao, et al., JLT, Vol. 29, No. 17, pp. 2570-2576, 2011
[4] L. Dou, et al., OFC2011, Paper OThF5
[5] L. Li, et al., OFC2011, Paper OWW3
[6] L. B. Du et al., OE, Vol. 18, No. 16, pp. 17075-17088, 2010
[7] W. Yan, et al., ECOC2011, Paper Tu3A2
[8] C.Xia et. al., JLT, Vol. 25, No. 4, pp. 996-1001, (2007).
[9] H. Chen et al., OFC2015, Paper Th1H2
[10] B. Liu et al., ECOC2013, Paper We.1.F.3
[11] R. Okabe et al., ECOC2015, Paper P.5.18
[12] A. Meccozi et al., PTL, Vol. 12, No. 4, pp. 392-394, 2000
[13] T. Oyama et al., OFC2014, Paper Tu3A3
[14] Q. Zhuge et al., OFC2014, Paper Th4D.7
[15] Y. Gao et al., ECOC2013, PDP 3.E.5
[16] Z. Tao, et al., OECC2013, Paper WR4-2
[17] T. Oyama, et al., OFC2015, Paper Th3D7
[18] Z. Tao, et al., JLT, Vol. 33, No. 10, pp. 2111-2119, 2015
[19] Y. Zhao, et al., ECOC2013, Paper P.4.15
[20] Y. Zhao, et al., ECOC2014, Paper P.5.8

TuB3-2

OECC/PS2016

Impact of Link Symmetry on Nonlinear Noise Mitigation using Spectral Inversion in Superchannel Transmission

Inwoong Kim[1], Olga Vassilieva[1], Paparao Palacharla[1], Motoyoshi Sekiya[2], Tadashi Ikeuchi[1]

(1) Fujitsu Laboratories of America, Inc., 2801 Telecom Parkway, Richardson, TX 75082, USA
(2) Fujitsu Laboratories Ltd., 4-1-1 Kamikodanaka, Nakahara-ku, Kawasaki, Kanagawa 211-8588, Japan
inwoong.kim@us.fujitsu.com

Abstract: *Spectral inversion can effectively mitigate intra-subcarrier nonlinear noise for uniform and non-uniform links, while inter-subcarrier nonlinear noise mitigation is more sensitive to the link symmetry. Spectral inversion mitigates significant nonlinear noise even with off-midspan placement.*
Keywords: *Optical communications, phase conjugation, nonlinear optical fiber*

I. INTRODUCTION

Nonlinear noise mitigation (NLNM) is a key technology to extend optical reach of superchannel transmission with higher order modulation format for high spectral efficiency. Digital back propagation (DBP) has been proposed for NLNM in electronic domain, while midspan spectral inversion (MSSI) for nonlinear noise mitigation is an alternative optical domain solution to DBP which needs higher power consumption than spectral inversion (SI) [1]. Furthermore, MSSI could be more economical by mitigating nonlinear noise of multi-subcarrier superchannels with a single device [2]. In experiment, NLNM has been demonstrated using MSSI implemented by optical parametric processing with low OSNR penalty [3]. On the other hand, the requirement of link symmetry with respect to SI for nonlinear noise and dispersion compensation has been an obstacle for using SI in optical mesh networks [4, 5]. However, with the advent of coherent receiver, hybrid solution combining SI and DBP has been proposed for more flexible placement of SI [5] in uniform transmission links, where DBP was used for compensation of residual nonlinear noise and dispersion caused by off-midspan placement of SI. We proposed a more aggressive approach by taking advantage of DSP for compensation of residual dispersion only and FEC correction without DBP in coherent receiver [6]. This way we were able to place SI more flexibly even in non-uniform transmission links [6]. Recently, NLNM in non-uniform transmission links with off-midspan SI was reported experimentally [7]. However, the detailed impact of link symmetry on intra- and inter-subcarrier nonlinear noise mitigation by SI in superchannel transmission has not been studied yet.

In this paper, we provide detailed analysis of the impact of link symmetry on NLNM in superchannels using SI. We show that significant NLNM of superchannel transmission can be achieved even with off-midspan SI in both uniform and non-uniform transmission links. We also demonstrate that mitigation of inter-subcarrier nonlinear noise is more sensitive to link symmetry with respect to SI than mitigation of intra-subcarrier nonlinear noise.

II. SIMULATION MODEL

Figure 1(a) shows the simulation model, where we transmit a five-subcarrier superchannel consisting of 32 GB DP-16QAM subcarriers with about 35 GHz subcarrier spacing (Fig. 1(b)). Each subcarrier is Nyquist pulse shaped with root raised cosine filter. We also transmit a single subcarrier channel for comparison with the superchannel transmission (Fig. 1(c)). The transmission line consists of total 20 spans of SMF fiber without dispersion compensation. Dispersion, nonlinear coefficient and attenuation of the fiber are 16 ps/nm/km, 1.3 /W/km and 0.2 dB/km, respectively. Two SI configurations are studied: conventional SI and pre-dispersed SI (PSI). In PSI configuration, the dispersion from the section of only the span prior to the SI excluding effective length, L_{es} (see Fig. 2(a)), is compensated [5]. Most of the nonlinear phase shift occurs in the effective length in each span (shown as yellow regions in power map or solid circles in dispersion map in Fig. 2). The nonlinear phase noise accumulation and compensation is better paired by having similar power levels and accumulated dispersions with PSI (red and blue circles in the blue oval) than with SI (red and green circles in the purple oval).

Two link configurations are studied for both SI options: (i) uniform transmission link, which consists of 20 spans of 60 km fiber and (ii) non-uniform transmission link, where 10 spans of 60 km fiber are followed by 10 spans of 75 km fiber. The PSI and SI are placed after M = 6, 7, …, 13 or14 spans for each transmission configuration. We set the fiber

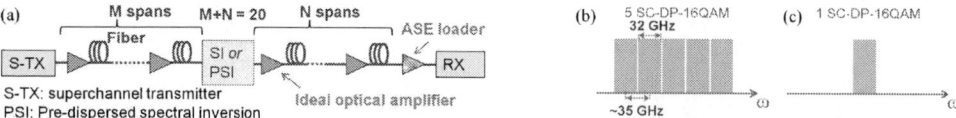

Fig. 1. (a) Schematic diagram of simulation model; (b) and (c) are spectra of 5-subcarrier superchannel and a single subcarrier channel, respectively.

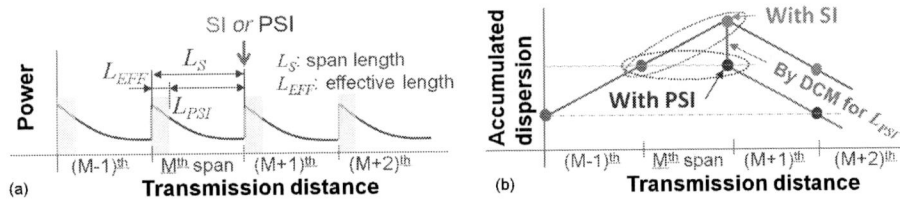

Fig. 2. (a) Schematic diagram of power map; (b) and the corresponding accumulated dispersion map (DCM is for compensation of dispersion from the transmission length of L_{PSI} in M^{th} span).

input power to 1 dBm per subcarrier channel. We assume an ideal SI with no OSNR penalty and no center-wavelength shift of optical signal by SI. This is because negligible OSNR penalty by SI and wavelength-shift-free SI have been reported in [3,8]. Lumped ASE noise is added at the receiver side with fixed received OSNR of 22 dB since it has been known that the Gordon–Mollenauer noise is not a major impairment for MSSI [9]. In this simulation study, the center or edge subcarrier is detected with coherent receiver, which also compensates the residual chromatic dispersion. To improve simulation accuracy, the BER is averaged over 200 runs of 2^{15} symbol transmission and converted to Q-factor.

III. NONLINEAR NOISE MITIGATION IN A SUBCARRIER CHANNEL AND A SUPERCHANNEL BY SI AND PSI

We start with the analysis of intra-subcarrier NLNM with SI and PSI by transmitting a single subcarrier channel. Figures 3(a) and 3(b) show Q-penalty due to nonlinear effect versus SI placement in uniform and non-uniform links, respectively. As expected, the majority of intra-subcarrier nonlinear noise is mitigated by midspan SI and PSI (after 10 spans) in the uniform transmission link. However, in non-uniform transmission link the best NLNM is achieved by off-midspan placement of SI and PSI, which is after 11 spans for this particular configuration, due to non-symmetric link configurations. This clearly demonstrates that in mesh optical networks with arbitrary link configurations the best possible NLNM can be realized at some specific location in the network, depending on the link configurations. In addition, we observe that both SI and PSI provide comparable NLNM. We also show that the Q-penalty increases as SI and PSI is placed farther away from the best placement. Nevertheless, more than 1.2 dB Q-factor improvements is still achieved even when SI and PSI are placed offset from midspan by up to 4 spans in both uniform and non-uniform links.

Next, we investigate mitigation of intra- and inter- subcarrier nonlinear noise using SI and PSI by transmitting a five-subcarrier DP-16QAM superchannel. Nonlinear Q-penalty of center subcarrier channel as a function of SI placement in the uniform and the non-uniform transmission links are shown in Fig. 3(c) and 3(d), respectively. Here, we observe the following key points: (1) Similar to single carrier case, the best performance in uniform transmission links is achieved when PSI and SI are located at midspan (after 10 spans) and after 11 spans for non-uniform transmission links. Also, NLNM of more than 1 dB and 2 dB is achieved when SI and PSI are placed offset from midspan by 4 spans in both link configurations, respectively. This clearly demonstrates that SI and PSI can achieve significant NLNM of superchannel transmission in uniform and non-uniform transmission links even with offset placement from midspan. (2) PSI can

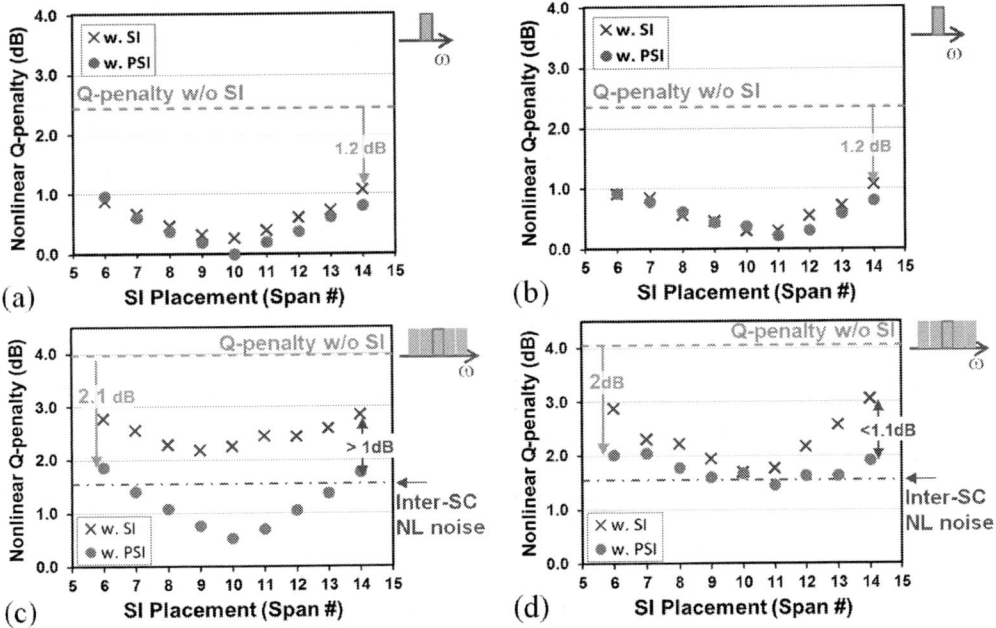

Fig. 3. Nonlinear Q-penalty vs. SI placement; (a) and (c): uniform links with 20 x 60 km spans, (b) and (d): non-uniform links with 10 x 60 km span + 10 x 75 km spans, (a) and (b): single subcarrier channel, (c) and (d): center channel of a 5-subcarrier superchannel.

(a) 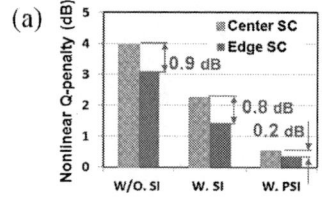 (b)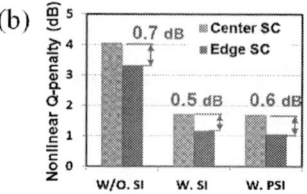

Fig. 4. Nonlinear Q-penalty of the center and the edge subcarrier channels for a superchannel transmission without SI, with MSSI and with midspan PSI; (a) uniform links with 20 x 60 km spans, (b) non-uniform links with 10 x 60 km span + 10 x 75 km spans.

compensate NLNM in uniform transmission link by more than 1dB better compared to SI (Fig. 3(c)). This difference mainly comes from the dependency of inter-subcarrier NLNM on link symmetry with respect to spectral inversion. To demonstrate this effect we calculate Q-penalty due to inter-subcarrier nonlinear noise (dash-dot line in Fig. 3(c)) by subtracting intra-subcarrier Q-penalty (dashed line in Fig. 3(a)), obtained in a single subcarrier transmission, from Q-penalty (dashed line in Fig. 3(c)) in a superchannel transmission. We can clearly see that the Q-penalty with SI is larger than the inter-subcarrier nonlinear Q-penalty (dash-dot line), while the Q-penalty with PSI is significantly smaller (especially for 6 spans < M <14 spans). This is because PSI provides better link symmetry which contributes to better inter-subcarrier NLNM than SI. In the non-uniform transmission links (Fig. 3(d)), however, the difference in Q-penalties between with SI and with PSI is minimal. In this case, PSI fails to provide better link symmetry and, as a result, the benefit of PSI over SI is significantly reduced. Furthermore, please note that slightly better compensation with SI in non-uniform links is observed (cross marks in Fig. 3(d) and 3(c)) compared to uniform links, which is the result of constructive or destructive interference of nonlinear noise from each link along the transmission line and which is described in more details in [10]. Thus, NLNM in non-uniform links may perform better than in uniform links. Based on the above observations, we can infer that intra-subcarrier nonlinear noise can be mitigated by SI or PSI effectively for both uniform and non-uniform links. However, the mitigation of inter-subcarrier nonlinear noise is more sensitive to the link symmetry with respect to spectral inversion placement and, thus, PSI is less effective in non-uniform links.

Finally, we investigate the dependency of NLNM mitigation using SI and PSI on subcarrier location, i.e. edge and center subcarriers, as shown in Fig. 4. We can clearly see that center subcarriers suffer from larger penalties compared to edge subcarriers. Similar behavior was reported in [11] without SI and was attributed to larger inter-subcarrier nonlinear impact on center subcarriers due to larger number of the most nearest interfering subcarriers. In this work we observe that in uniform transmission links (Fig. 4(a)), the Q-penalties of both center and edge subcarriers are mitigated by SI but the imbalance of 0.8 dB still remains and it is comparable to that without SI (0.9 dB) because there is no significant compensation of inter-subcarrier nonlinear noise with SI. However, the imbalance is dramatically reduced with PSI to as low as 0.2 dB because PSI compensates for inter-subcarrier nonlinear noise more effectively in addition to intra-subcarrier nonlinear noise compensation. In non-uniform transmission link configuration (Fig. 4(b)), the PSI advantage is reduced due to lack of link symmetry with respect to spectral inversion. In this case we observe similar Q-penalty imbalance (~0.5 dB) between center and edge subcarriers using both SI and PSI. Therefore, SI alone might be good enough for NLNM in optical mesh networks which mostly consist of non-uniform links.

IV. CONCLUSIONS

The impact of link symmetry on NLNM by SI and PSI was investigated for superchannel transmission over uniform and non-uniform transmission links. The simulation results demonstrated that the intra-subcarrier nonlinear noise can be mitigated efficiently by SI or PSI for any type of link configurations, while inter-subcarrier nonlinear noise can be mitigated better in symmetric link configurations with respect to placement of spectral inversion. Remarkably, significant NLNM (≥ 1dB with SI or ≥ 2dB with PSI) of superchannel transmission over 20 spans of uniform and non-uniform links was achieved even when SI or PSI was placed offset from midspan by 4 spans. This flexibility in placement of SI or PSI will enable practical deployments in mesh networks and extending optical reach for super-channel transmission.

REFERENCES

[1] D. Rafique et. al., "FEC overhead and fiber nonlinearity mitigation ...," Proc. OFC, W2A.32, San Francisco, 2014.
[2] L. B. Du et. al., "Fiber nonlinearity compensation for OFDM ...," Opt. Express vol. 20, pp. 19921-19927, 2012.
[3] S. L. Jansen et. al., "Optical phase conjugation for ultra long-haul ...," J. Lightw. Technol., vol. 24, pp. 54-63, 2012.
[4] S. Watanabe and et. al., "Compensation of pulse shape ...," IEEE Photon. Technol. Lett., vol. 5, pp. 1241-1243, 2011.
[5] D. Rafique and Ellis, D. Andrew, "Various nonlinearity ...," IEEE Photon. Technol. Lett., vol. 23, pp. 1838-1840, 2011.
[6] I. Kim et. al., "The impact of spectral inversion ...," Proc. IEEE Photonics Conference, MG3.1, San Diego, 2014.
[7] S. Yoshima et. al., "Nonlinearity mitigation through optical phase ...," Proc. ECOC, We2.6.3, Valencia, 2015.
[8] K. Mori et. al., "Wavelength-shift-free spectral inversion with ...," Opt Lett., vol. 21, pp. 110-112, 1996.
[9] P. Minzioni et. al., "Study of the Gordon–Mollenauer effect ...," IEEE Photon. J., vol. 2, pp. 284-291, 2010.
[10] I. Kim et. al., "Analytical Model for Nonlinear Noise Mitigation using Spectral Inversion for Superchannel Transmission", submitted to Optics Express
[11] O. Vassilieva et al., "Systematic analysis of intra-superchannel nonlinear crosstalk ...," Proc. ECOC, Mo.4.3.6, Cannes, 2014.

Achievable Rates Comparison for Phase-Conjugated Twin-Waves and PM-QPSK

Tobias A. Eriksson[1], Abel Lorences-Riesgo[1], Pontus Johannisson[1], Tobias Fehenberger[2],
Peter A. Andrekson[1], Magnus Karlsson[1]

[1]Chalmers University of Technology, 41296 Gothenburg, Sweden
[2]Technische Universität München (TUM), 80333 Munich, Germany
tobias.eriksson@chalmers.se

Abstract: Phase-conjugated twin-waves (PCTW)-QPSK is experimentally compared to PM-QPSK in terms of achievable information rate for bit-wise decoders. For typical long-haul transmission distances, PM-QPSK with soft-decision FEC achieves significantly higher spectral efficiency than PCTW-QPSK.

Keywords: Coherent communication, nonlinear mitigation, achievable information rate, phase conjugated twin waves.

I. INTRODUCTION

Coherent optical communication systems, as designed today, are limited in terms of transmission reach by Kerr nonlinearities. Different techniques for mitigation of the deterministic nonlinear distortion have been suggested such as digital back-propagation [1], mid-span spectral inversion [2], coherent superposition in phase sensitive amplification techniques [3], the nonlinear Fourier transform [4], and exploiting four-dimensional channel distributions in the receiver [5]. Another method, which is the topic of this paper, is the phase-conjugated twin-waves (PCTW) transmission scheme [6]. This method transmits a phase-conjugated copy of a constellation on, for instance, orthogonal polarizations [6] or different wavelengths [7]. The nonlinear distortion can then be compensated by adding the signal and conjugate using all-optical techniques [3] or in the digital signal processing (DSP) after detection [7]. When the signal and conjugated copy are transmitted in orthogonal polarizations, this method is equivalent to transmitting real-valued signals, such as PM-BPSK, where the nonlinear interference is squeezed during transmission [6]. Many of the experimental investigations of PCTW optimize the channel for the PCTW scheme, by for instance using in-line dispersion compensation [3] or fibers with low chromatic dispersion [6] [7], to enhance the nonlinearities, and compare it to a conventional modulation scheme such as PM-QPSK, over the same link. However, these types of channels are then typically far from optimized for a conventional modulation scheme.

In this paper we compare a single-channel 28 Gbaud PCTW-QPSK to a 28 Gbaud PM-QPSK system in a realistic upgrade scenario, i.e. no inline dispersion compensation is used and the dispersion map is optimized separately in the two cases by changing the pre- and post-dispersion compensation. We assume soft-decision decoding and compare the two systems in terms of achievable information rate (AIR) using generalized mutual information (GMI) [8]. It is found that for typical long-haul distances (< 20,000 km), PM-QPSK and a strong code achieves a significantly higher spectral efficiency than PCTW-QPSK. However, for extremely long transmission links, where conventional receiver algorithms tend to fail for PM-QPSK, the twin-waves approach could be an interesting alternative.

II. PHASE-CONJUGATED TWIN-WAVES AND ACHIEVABLE INFORMATION RATES

The constellation for conventional PM-QPSK can be written as $\left\{ [s_x, s_y]^T \right\}$, where $s_x = \{(\pm 1 \pm i)/2\}$ and $s_y = \{(\pm 1 \pm i)/2\}$. For PCTW-QPSK, the signal in the y-polarization is a conjugated copy of the signal in the x-polarization giving the symbols $\left\{ [s_x, s_y = s_x^*]^T \right\}$, where s_x is still a QPSK constellation. By applying a polarization rotation, using one of the Jones matrices

$$J_1 = \frac{1}{\sqrt{2}} \begin{bmatrix} 1 & 1 \\ -i & i \end{bmatrix}, \qquad J_2 = \frac{1}{2} \begin{bmatrix} 1-i & 1+i \\ 1+i & 1-i \end{bmatrix},$$

to the PCTW-QPSK symbol alphabet, the different representations shown in Fig. 1 can be obtained. Applying J_1, the PM-BPSK representation given by $\{[\pm 1, \pm 1]^T / \sqrt{2}\}$ is found [6], which is the representation used in our experimental realization. If J_2 is applied, PCTW-QPSK can be interpreted as polarization-switched (PS)-BPSK with the constellation $\{[\pm 1, 0]^T, [0, \pm 1]^T\}$. This representation is used in the DSP algorithms implemented for PCTW-QPSK in this paper.

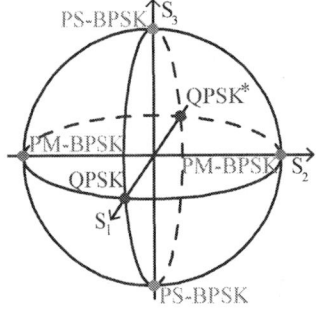

Fig. 1 - Poincaré-sphere representation of different interpretations of PCTW-QPSK.

This work was partly founded by the Swedish Research Council (VR) and the European Research Council under Grant ERC-2011-AdG-291618 PSOPA.

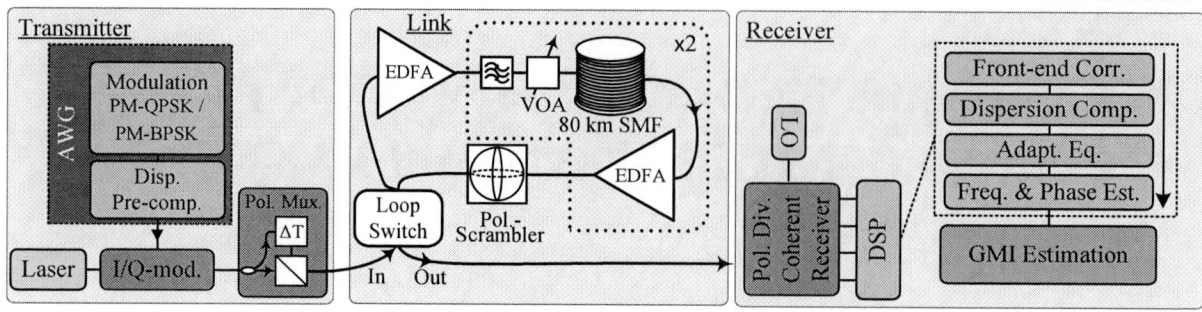

Fig. 2 - Experimental setup.

With todays' optical communication systems relying on sophisticated soft-decision forward-error correction (FEC) schemes, a relevant system figure-of-merit is the AIR estimated with either mutual information (MI) or GMI, depending on the FEC scheme that is intended to be used. It is important to note that for a known channel transition distribution, the PCTW-QPSK can never obtain a higher MI than PM-QPSK. The reason for this is that PCTW-QPSK is a subset of the PM-QPSK symbol alphabet. However, for the nonlinear fiber optical system the true channel transition distribution is not known. Typically, the concept of mismatched decoding is applied [9], which means that the received samples are decoded as if they were transmitted over a known auxiliary channel. This auxiliary channel is often assumed to be an additive white Gaussian noise channel with the same variance in all dimensions, which is the assumption in this paper and also in many realistic decoders.

The GMI and the log-likelihood ratios (LLRs) calculated for N received symbols are

$$\text{GMI} \approx m - \frac{1}{N}\sum_{k=1}^{m}\sum_{i=1}^{N}\log_2\big(1 + \exp\big((-1)^{b_{k,i}}LLR_{k,i}\big)\big), \qquad LLR_{k,i} = \log\frac{\sum_{s\in X_{k,1}}\exp\big(-\frac{1}{N_0}\|y_i-s\|^2\big)}{\sum_{s\in X_{k,0}}\exp\big(-\frac{1}{N_0}\|y_i-s\|^2\big)},$$

where $b_{k,i}$ is the transmitted bit sequence, $\mathbf{y}$ is the received symbol and s denotes a symbol from the constellation X with cardinality 2^m. The index i denotes the ith transmitted symbol and the index k the bit position [8]. Note that the GMI depends on the bit-to-symbol mapping and that we use Gray-mapping of the QPSK constellation. The GMI is normalized to two-dimensional (2D) symbols, i.e. PM-QPSK has a maximum GMI of 2 bit/2D-symbol and PCTW-QPSK a maximum GMI of 1 bit/2D-symbol. The GMI gives a good estimate of the post-FEC bit error rate (BER) for systems relying on bit-wise decoding, which is the case for many state-of-the-art FEC solutions for optical communication [8].

III. SYSTEM AND EXPERIMENTAL DESCRIPTION

The efficiency of the nonlinear mitigation for a PCTW system is determined by the design parameters of the transmission link such as power map, dispersion map, amplifier spacing, fiber type, etc. In this paper we investigate the suitability of PCTW in links that are typical for coherent PM-QPSK today, i.e., without inline dispersion compensation or Raman amplification. The available dispersion compensation is limited to pre-dispersion compensation on the transmitter side and post-dispersion compensation in the receiver, both done electronically. In this paper we investigate single channel transmission, but it should be noted that for WDM transmission the gains seen by PCTW are typically smaller, unless the phase conjugation can be applied over the full WDM spectra which is difficult [10].

The experimental setup is shown in Fig. 2. The transmitter is based on an arbitrary waveform generator (AWG) for electrical signal generation. Since PCTW-QPSK and PM-BPSK are equivalent for single-channel transmission, as explained in Section II, we implement the PCTW-QPSK scheme using PM-BPSK. We generate either 28 Gbaud PM-QPSK or 28 Gbaud PCTW-QPSK. When applicable, pre-dispersion compensation is applied in the AWG. Polarization multiplexing is emulated using a split and de-correlate stage. The signals are propagated over a recirculating loop consisting of two spans of 80 km standard single-mode fiber (SMF). Two erbium-doped fiber amplifiers (EDFAs) compensate the loss of the spans and a loop-synchronized polarization scrambler is used to avoid any nonrealistic accumulation of polarization effects. A third EDFA is used to compensate for the loss of the polarization scrambler and the loop switching components. The signal is detected using a coherent receiver and sampled by a four-channel oscilloscope with 50 GS/s sampling rate. In the digital domain, optical front-end correction is applied before dispersion compensation in the frequency domain is carried out. For PM-QPSK, a conventional DSP structure is used in which adaptive equalization and polarization demultiplexing is carried out by the constant modulus algorithm (CMA) and phase tracking is done with the Viterbi-Viterbi algorithm. The only DSP difference for PCTW-QPSK is that a modified CMA is used to enable polarization demultiplexing of the signals [11]. After the DSP, the AIR is estimated using GMI [8].

IV. RESULTS AND DISCUSSION

The attainable transmission distances for PCTW-QPSK at 20% FEC overhead, i.e. at an AIR of 1/1.2 = 0.83 bit/2D-symbol, as a function of launch power with different dispersion maps are shown in Fig. 3(a). For 0% and 100% pre-dispersion compensation, the optimal launch power is −1 dBm. The longest reach is obtained by a symmetrical dispersion map, i.e. 50% pre-dispersion compensation, which also increases the optimal launch power to 0 dBm. The transmission reach for 32% and 64% pre-dispersion compensation is shorter than that of 50% but longer than that obtained by 0% or 100%. For PM-QPSK there is no significant difference with regard to the dispersion maps, which can be seen in Fig. 3(b) where the plots for all the different dispersion maps are indistinguishable. For PM-QPSK (not plotted) the transmission

OECC/PS2016

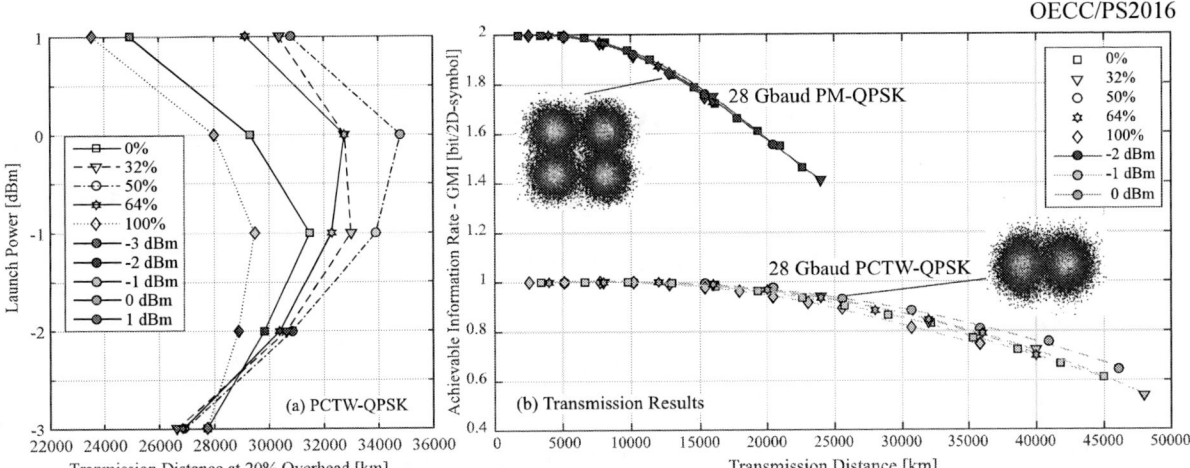

Fig. 3 – (a) Transmission distance for 28 Gbaud PCTW-QPSK at 20% FEC overhead for different launch powers and different amounts of dispersion pre-compensation. (b) Achievable information rate as a function of transmission distance using optimal launch power for each dispersion map for 28 Gbaud PM-QPSK and 28 Gbaud PCTW-QPSK. Note that for PM-QPSK, the differences between the curves for different dispersion maps are indistinguishable.

reach is approximately the same for −2 dBm and −1 dBm for all dispersion maps. This means that for PCTW-QPSK with a symmetric dispersion map, the optimal launch power is increased by roughly 1 to 2 dB over PM-QPSK.

In Fig. 3(b), the AIR, calculated using GMI, is plotted as a function of transmission distance for the optimal launch power for each scenario. Note that for PM-QPSK, the results for −1 dBm are very similar to those of −2 dBm but only the latter is plotted for clarity. PM-QPSK can be transmitted over slightly more than 20,000 km before the implemented DSP becomes unreliable. Note that we have removed measurement points where the DSP did not converge or phase-slips occurred in the phase tracking. As seen, for all transmission distances that the DSP for PM-QPSK can handle, PM-QPSK has a significantly higher AIR compared to PCTW-QPSK. Therefore, for these transmission distances, there is a clear loss in spectral efficiency by using PCTW-QPSK instead of decreasing the rate of the code that is used in combination with PM-QPSK. It is only for distances larger than 20,000 km, PCTW-QPSK is an interesting alternative since it can relax the requirements on the DSP. Note however, that if more noise-tolerant DSP algorithms are used and cycle slip mitigation is implemented, it should be possible to increase the distance over which PM-QPSK can be transmitted. For distances larger than 20,000 km, other modulation formats with increased sensitivity over PM-QPSK but with higher spectral efficiency than PCTW-QPSK, such as PS-QPSK [12], could also be an interesting alternative.

V. CONCLUSIONS

Although the phase-conjugated twin-wave approach has recently attracted a lot of attention for its capability to increase the distance significantly when compared at a fixed pre-FEC BER, we show that for typical long-haul transmission distances (<20,000 km), 28 Gbaud PM-QPSK outperforms 28 Gbaud PCTW-QPSK in terms of the achievable information rate. This means that for systems using soft-decision coding schemes with high overheads, PM-QPSK is a more spectrally-efficient solution. However, PCTW-QPSK could find niche applications such as low-complexity systems using hard-decision FEC schemes or systems where extreme transmission distances are required.

REFERENCES

[1]   E. Ip and J. M. Kahn, "Compensation of dispersion and nonlinear impairments using digital backpropagation," J. Lightw. Technol, vol. 26, no. 20, pp. 3416-3425, 2008.

[2]   S. Watanabe *et al.*, "Compensation of pulse shape distortion due to chromatic dispersion and Kerr effect by optical phase conjugation," IEEE Photon. Technol. Lett., vol. 5, no. 10, pp. 1241-1243, 1993.

[3]   S. L. I. Olsson *et al.*, "Phase-sensitive amplified transmission links for improved sensitivity and nonlinearity tolerance," J. Lightw. Technol, vol. 33, no. 3, pp. 710-721, 2015.

[4]   M. I. Yousefi and F. R. Kschischang, "Information transmission using the nonlinear Fourier transform, part III: Spectrum modulation," IEEE Trans. Inf. Theory, vol. 60, no. 7, pp. 4346-4369, 2014.

[5]   T. A. Eriksson *et al.*, "Impact of 4D channel distribution on the achievable rates in coherent optical communication experiments," J. Lightw. Technol, vol. PP, no. 99, 2016.

[6]   X. Liu *et al.*, "Phase-conjugated twin waves for communication beyond the Kerr nonlinearity limit," Nature Photonics, vol. 7, no. 7, pp. 560-568, 2013.

[7]   Y. Tian *et al.*, "Demonstration of digital phase-sensitive boosting to extend signal reach for long-haul WDM systems using optical phase-conjugated copy," Opt. Exp., vol. 21, pp. 5099-5016, 2013.

[8]   A. Alvarado *et al.*, "Replacing the soft-decision FEC limit paradigm in the design of optical communication systems," J. Lightw. Technol, vol. 33, no. 20, pp. 4338–4352, 2015.

[9]   D. Arnold *et al.*, "Simulation-based computation of information rates for channels with memory," IEEE Trans. Inf. Theory Theory, vol. 52, pp. 3498-3508, 2006.

[10] X. Liu *et al.*, "Fiber-nonlinearity-tolerant superchannel transmission via nonlinear noise squeezing and generalized phase-conjugated twin waves," J. Lightw. Technol, vol. 32, no. 4, pp. 766-775, 2014.

[11] P. Johannisson *et al.*, "Modified constant modulus algorithm for polarization-switched QPSK," Opt. Exp., vol. 19, no. 8, pp. 7734-7741, 2011.

[12] E. Agrell and M. Karlsson, "Power-efficient modulation formats in coherent transmission systems," J. Lightw. Technol, vol. 27, no. 22, pp. 5115-5126, 2009.

KNN-based Detector for Coherent Optical Systems in Presence of Nonlinear Phase Noise

Danshi Wang, Min Zhang, Meixia Fu, Zhongle Cai, Ze Li, Yue Cui, and Bin Luo

State Key Laboratory of Information Photonics and Optical Communications, Beijing University of Posts and Telecommunications, Beijing 100876, China. E-mail: mzhang@bupt.edu.cn

Abstract: *A machine learning-based detector, namely KNN, is proposed to mitigate the NLPN in a 16QAM coherent optical transmission system. The maximum transmission distance and nonlinear tolerance are improved by 240 km and 2.0 dBm.*

Keywords: *Machine Learning; Fiber optics communications; Nonlinear Phase Noise.*

I. INTRODUCTION

Fiber transmission systems based on coherent detection suffer from the noise-induced performance degradation. Amplified spontaneous emission (ASE) noise from inline amplifiers is a major source of noise and is referred to as linear noise. Nonlinear phase noise (NLPN) is induced by the interaction between the signal and ASE noise via the fiber Kerr nonlinearity, known as self-phase modulation [1]. Various approaches have been proposed to combat NLPN based on both optical field [2] and electronic field [3].

Recently, techniques from machine learning have been applied well to nonlinear dynamical systems, especially to the nonlinearity dominant fiber channel [4-6]. In [4], we have demonstrated the feasibility of support vector machine (SVM) for nonlinearity mitigation. However, the SVM is only a binary classifier, and thus many SVMs are necessary for the high-order modulation formats. In [5], the artificial neural network (ANN) requires much longer training time. In [6], the expectation maximum (EM) depends on the parameters of transmission link, which is not suitable for a dynamic optical network link. Hence, it is valuable to investigate a multiclass algorithm that is independent of the link information and without the training process.

The k-nearest neighbors (KNN) is one of the top ten machine learning algorithms and have been widely used to solve classification and clustering problems [7]. It is a straightforward and effective algorithm, especially in the case of large data sets and low dimensions. However, to the best of our knowledge, the application of KNN to nonlinear impairments mitigation in optical communication has never been reported before.

In this paper, we propose a KNN-based detector to combat NLPN in a 16QAM coherent optical system. Without any prior information and training process, KNN learns and captures the link properties from just a small set of labeled data. A nonlinear decision boundary is created to avoid errors caused by amplitude and phase noise. Compared with conventional method, KNN achieves longer transmission distance and higher nonlinear tolerance.

II. THEORY OF k-NEAREST NEIGHBORS

When KNN is used as a classifier, it creates a decision boundary to classify different kinds of patterns [7]. Unlike SVM, KNN is a multi-class classifier, thus it can classify multiple kinds of data simultaneously. The operation of the KNN algorithm when the patterns are points in a two-dimensional space is illustrated in Fig. 1(a). A majority voting strategy is adopted by KNN via using k of the nearest neighbors. There are two classes of labeled data, represented by "red triangles" and "blue circles," respectively. A query point x_q needs to be classified by the KNN algorithm. Then we need to calculate the Euclidean distances between the query point and each labeled data. Note that if $k=3$, the 3-nearest neighbors algorithm classifies x_q as a "blue circle" example, whereas if $k=7$, the 7-nearest neighbors algorithm classifies it as a "red triangle" example. Different values of k may lead to different results. Therefore, in order to obtain a high accuracy, it is important to select proper k according to the corresponding case. Different from other machine learning methods in [4-6], which have to learn the parameters of classifier through the training process, KNN can classify the test data directly based on the labeled data. In order to better understand the principle of KNN, the shape of the decision

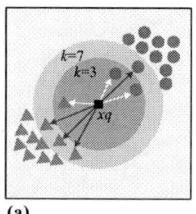 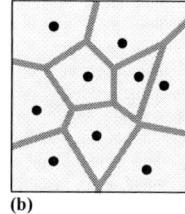 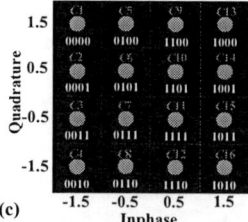 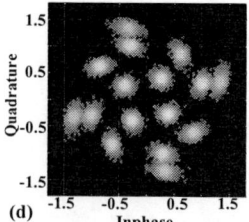

Fig. 1. (a) The principle of KNN for two classes of data; (b) the decision boundary induced by the 1-nearest neighbor; (c) gray-coded 16QAM constellations with the corresponding labels; (d) an example of 16QAM signal (at 0 dBm launch power over 1600 km) suffering from NLPN.

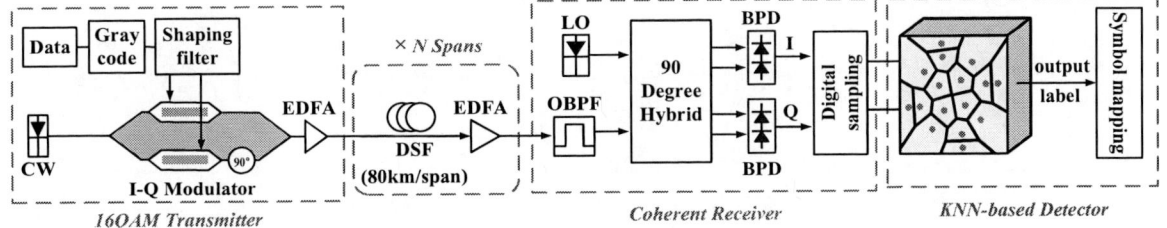

Fig. 2. Numerical simulation of 16QAM coherent transmission systems based on KNN-detector.

boundary induced by the 1-nearest neighbor over the entire pattern space is shown in Fig. 1(b).

In addition, an improved algorithm, named as distance-weight KNN (DW-KNN), is to weigh the contribution of each of the k neighbors according to their distance to the query point x_q, giving greater weight to closer neighbors. We can weigh the vote of each neighbor according to the inverse square of its distance from x_q [7].

As analyzed above, the KNN algorithm works as a classifier. If we regard the different constellations of 16QAM as different classes of data in a two-dimensional (2D) space, then KNN can be used to create the decision boundary for 16 classes of symbols, as shown in Fig. 1(c). Here, we select the in-phase and quadrature components of the received signal as the input feature vector. Some symbols with the corresponding labels ($C1,C2,..,C16$) are employed as labeled data to create the decision boundary. Next, the test data, which suffers from a degradation process similar to training data, can be classified by the KNN-created decision boundary. An example of 16QAM signal suffering from the NLPN is presented in Fig. 1(d). If without an effective decision processor, we can hardly identify the original signal properly. With the help of KNN, the nonlinear decision can be created successfully.

III. NUMERICAL SIMULATION AND RESULTS

The numerical simulation is set up to demonstrate the feasibility of KNN-based detector for 16QAM coherent optical transmission system, as shown in Fig. 2. The transmitter laser is a continuous wave (CW) at the fixed wavelength of 1550 nm. The electrical multilevel pulses shaping filter, whose raising time equals to 1/8 of the symbol duration, converts the in-phase and quadrature components into electrical signals with different voltage values. Through the IQ modulator, the 16QAM signal at 25 Gbaud (i.e. 100 Gbps) is modulated by the pseudo-random binary sequence (PRBS) with a length of 2^{16}-1.

In the fiber transmission subsystem, we follow the model in [3], in which we focus on the impacts of fiber NLPN and thus neglect the chromatic dispersion. Then the generated optical signal is sent into the transmission link consisting of $N×80$ km dispersion-shifted fiber (DSF) spans. We adopt split-step Fourier method to simulate the fiber link, in which the maximum nonlinear phase rotation per step is less than 0.003 rad. The fiber nonlinear coefficient is $\gamma=1.3$ $W^{-1}km^{-1}$ and the loss efficient is $\alpha=0.2$dB/km, corresponding to a 16 dB loss of each span. The EDFA with the noise figure of 6 dB is placed at the end of each span to compensate the fiber loss.

After the fiber transmission, a Gaussian optical filter with 28 GHz bandwidth is firstly used to reduce the ASE noise. Then the signal is mixed with the local oscillator (LO) via the optical 90° hybrid. The linewidths and phase noises of both CW and OL are assumed to be zero. The in-phase and quadrature components of electrical signals are down sampled to one sample per symbol and then processed by the KNN-based detector. The 1000 symbols are selected as the labeled data, and the rest of symbols are the test data. Finally, the output labels are mapped to the corresponding symbols and the bit-error rate (BER) is counted to evaluate the system performance.

First, we need to determine the value of k. As analyzed above, different values of k may lead to different classification accuracies. As a testing reference, the 16QAM signal with 0 dBm launch power over 1600 km transmission is detected by the KNN and DW-KNN, respectively. The values of k range from 1 to 30 and the corresponding BERs are calculated. From Fig. 3(a), we can see that DW-KNN outperforms KNN and that the BERs of DW-KNN have smaller fluctuations. For both KNN and DW-KNN, the small and large values of k deteriorate the decision performance. This is mainly because the model at small values is too rough, which causes an underfitting problem, whereas at large values it is too complex, which causes an overfitting problem. Therefore, we select the

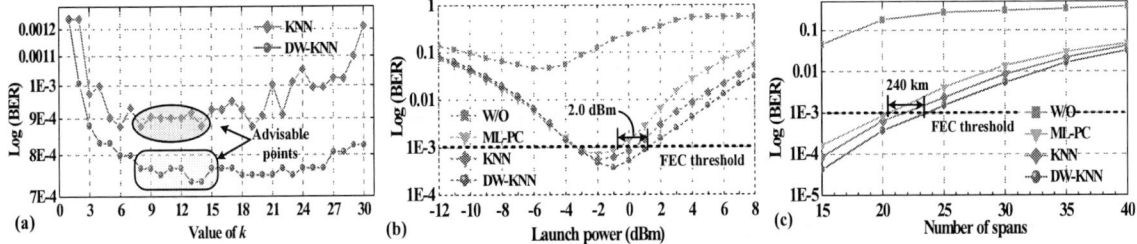

Fig. 3. BER as a function of (a) k values, (b) optical launch power, and (c) transmission distance.

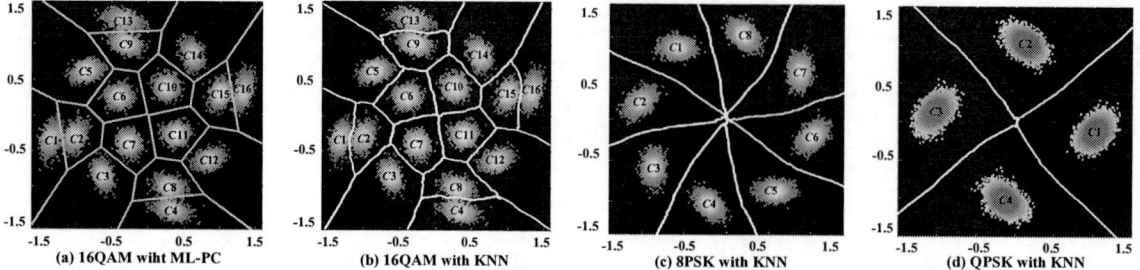

Fig. 4.(a) Decision boundaries created by (a) ML-PC for 16QAM; (b) KNN for 16QAM; (c) KNN for 8PSK; (d) KNN for QSPK (all the four signals are at the launch power of 0 dBm over 1600 km; 16QAM signals operating at 100 Gbps, 8PSK at 120 Gbps, and QPSK at 80 Gbps.)

moderate values from 8 to15 as the preferred points for 16QAM detection, as shown in Fig. 3(a).

Next, we set the value of k to be 11. In order to demonstrate the feasibility of the KNN, the popular maximum likelihood post-compensation (ML-PC) algorithm is selected as a comparison [3]. Then the BER performances of the received signals as functions of the launch power are measured by ML-PC, KNN, and DW-KNN, respectively. As a reference, the direct decision without any mitigation is also tested, as shown in Fig .3(b). The launch powers of 16QAM signal over 1600 km range from -12 dBm to 8 dBm. From Fig. 3(b), it is seen that without any nonlinear mitigation the BER performance is very poor. For the other three mitigation methods, as the launch power increases, OSNR of the received signal grows higher, which indicates a gradually decreasing ASE noise and contributes to a better BER performance. As the launch power exceeds the optimal value, the received signal is impaired by the remarkable nonlinear effect, which deteriorates the BER performance again. Here, the launch power dynamic range (LPDR), which denotes the power difference between the two points at a BER of 1×10^{-3}, is employed to evaluate the effectiveness of the other three algorithms. The numerical results show that DW-KNN achieves better performance than KNN; compared with ML-PC, LPDR is increased by about 2.0 dBm by DW-KNN. In addition, the BER as a function of transmission distance is also studied, as shown in Fig. 3(c). Here, we investigate the maximum transmission distance (MTD) where BER is equal to 1×10^{-3} for the signal at a given launch power of -1dBm. The BER increases gradually with the growth of the transmission distance from 15 to 40 spans. The MTD can be increased by up to 240 km by adopting the DW-KNN.

Compared with the ML-PC, the KNN methods achieve the larger launch power and longer transmission distance. This is because that the ML-PC only considers the deterministic channel effects, but ignoring the stochastic characteristics of nonlinearity. The probability density function (PDF) of the NLPN can be approximated analytically by a Gaussian distribution only under the given conditions. If in the practical system, the NLPN distribution may be more complicated. The decision boundary created by the ML-PC only performs the suboptimal results, as shown in Fig. 4(a). In contrast, the KNN based on the received symbols can not only suit any transmission link through its learning capacity, but also generate a better nonlinear decision boundary to classify the data more precisely, especially for the ones between $C1$ and $C2$, $C4$ and $C8$, $C9$ and $C13$, $C15$ and $C16$, as show in Fig. 4(b). Therefore, the KNN yields a larger improvement in nonlinear tolerance than the ML-PC does.

IV. CONCLUSIONS

An effective detector based on the KNN method was proposed to mitigate NLPN in the context of a coherent optical transmission system. Compared with the conventional ML-PC algorithm, LPDR was increased by 2.0 dBm and MTD improved by 240 km. The improved nonlinear tolerance demonstrates the feasibility of KNN. In addition, KNN can also be applied in other formats, such as 8PSK and QPSK etc., as shown in Fig. 4(c)-(d). Meanwhile, we believe that KNN have the great potential to solve other problems in optical communication.

ACKNOWLEDGMENT

This work is supported from the NSFC Project No.61372119, the Doctoral Scientific Fund Project of the Ministry of Education of China No. 20120005110010, and the BUPT Excellent Ph.D. Students Foundation No.CX2015306.

REFERENCES

[1] J. P. Gordon *et al.*, "Phase noise in photonic communications syetems using linear ...," *Opt. Lett.*, **15** (23), 1351-1353 (1990).
[2] A. E. Willner *et al.*, "All-optical signal processing," *J. Lightw. Technol.*, **32** (4), 660-680-755 (2014).
[3] A. Serdar Tan *et al.*, "An ML-based detector for optical communication in the presence of nonlinear phase noise," *IEEE ICC'11*,1-5 (2011).
[4] D. Wang *et al.*, "Nonlinear decision boundary created by a machine learning-based...," *Proc. ECOC'15*, P.3.16 (2015).
[5] M. A. Jarajreh *et al.*, "Artificial neural network nonlinear equalizer for...," *Photon. Technol. Lett.*, **27** (4), 387-390 (2015).
[6] D. Zibar *et al.*, "Nonlinear impairment compensation using expectation maximization ...," *Opt. Express.*, **20** (26), B181-B196 (2012).
[7] C. M. Bishop, Pattern recognition and machine learning. Springer (2006).

TuB3-5

OECC/PS2016

OSNR Monitoring by Deep Neural Networks Trained with Asynchronously Sampled Data

Takahito Tanimura[1,2], Takeshi Hoshida[1], Jens C. Rasmussen[1], Makoto Suzuki[2], and Hiroyuki Morikawa[2]

[1] Fujitsu Laboratories Ltd., 4-1-1 Kamikodanaka, Nakahara-ku, Kawasaki 211-8588, Japan

[2] Research Center for Advanced Science and Technology, The University of Tokyo, 4-6-1 Komaba, Meguro-ku, Tokyo 153-8904, Japan

tanimura.taka@jp.fujitsu.com

Abstract: We demonstrate a use of deep neural networks (DNN) for OSNR monitoring with minimum prior knowledge. By using 5-layers DNN trained with 400,000 samples, the DNN successfully estimates OSNR in a 16-GBd DP-QPSK system.

Keywords: machine learning, optical performance monitoring, coherent detection

I. Introduction

For enabling programmable and autonomous optical networks, optical physical layer monitoring is considered as an essential part of the network systems. The networks should handle various types of optical signals modulated in different signal format and baud rate to accommodate diverse transport requirements [1]. Thus the optical physical layer monitor should enable to extract physical information from incoming optical signals with minimum prior knowledge.

Aiming at a generic monitoring without a prior knowledge, the use of an artificial neural network (ANN) has been investigated [2-4]. Although all of existing ANN-based monitors could estimate physical condition of input signals, e.g. OSNR, chromatic dispersion and differential group delay, they require careful engineering to design a feature extractor that transforms the raw data into a feature vector that is a suitable internal representation. In [2], for example, a set of parameters including Q-factor, eye-closure, RMS-jitter, and crossing amplitude of eye-diagram was used for ANN input. The parameter set was chosen by a skilled engineer who holds expertise of modulated optical signal. In other words, the conventional ANN-based monitors used in [2-4] were limited in their scalability to more general set of signals with different symbol rates and modulation formats.

Recently, a deep neural network (DNN), which is an ANN with multiple hidden layers of units between the input and output layers, has attracted attention for various applications: such as visual object recognition and speech recognition [5]. The extra layers of the DNNs would enable automatic extraction of features of input, giving the potential of modeling complex data with a fewer units than similarly performing shallow networks [5].

In this paper, we propose and experimentally demonstrate the use of the DNNs for an OSNR monitor without specific feature engineering for raw data, as shown in Fig. 1(a). The feature vectors are learned in the DNNs from asynchronously sampled raw data by coherent receiver. As the number of training dataset is increased, the DNN starts to obtain a function of feature extractor of incoming signals. Varying the number of layer of the DNN, we experimentally evaluate a required number of layers in this specific case. By using 5-layer DNN trained with 400,000 dataset, OSNR of 16 GBd DP-QPSK signal is successfully estimated with the range of 7.5 to 31 dB.

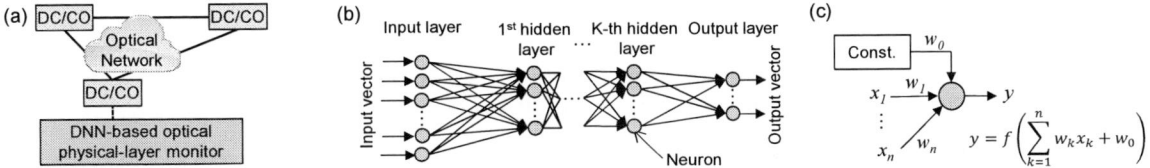

Fig. 1. Schematic of (a) optical networks with DNN-based monitors, (b) feed-forward deep neural networks (DNN), (c) illustrated neuron in DNN. DC: data center, CO: central office.

II. Concept of Deep Neural Network

Figure 1(b) shows a schematic of a feed-forward, full-connected multi-layer perceptron DNN used in this study. The DNNs are neuroscience-inspired information processing models that are composed of multiple processing layers to learn representations of data with multiple levels of abstraction. The first and last layers of the DNNs are an input and output layers, respectively. The other layers are called hidden layers. Each layer of the DNN contains multiple neurons that are connected to other neurons in neighboring layers by adaptive weights w_k. Each neuron has activation function $f(u)$ and calculates output of neuron from multiple inputs as shown in Fig. 1(c).

The connection weights of DNN are trained by input-output dataset, i.e. supervised learning. For example, input is measured raw values related with optical signals, and output is OSNR. The DNN learns a relationship between input

513

and output that is characteristic of the system under consideration. After calculations of output vectors, the output vectors are compared to the desired output vectors and errors are calculated. Error derivatives are then also calculated and summed for each weight. The error derivatives are used to update the connection weights for the neurons, and training continues until the errors reach enough low values. After training, the connection weights of DNN are fixed and the DNN can be tested by other set of data.

III. EXPERIMENTS

Figure 2(a) shows the experimental setup. At the transmitter, an external cavity laser (~25 kHz linewidth) was used as a light source for the channel at 193.3 THz. An integrated InP DP-IQ-modulator was driven by the drive signals generated by a four channel digital-to-analogue converters (DAC) with a sampling rate of 64 GSa/s, a physical resolution of 8 bits. The DAC generated Nyquist-filtered (roll-off factor = 0.01) 16 GBd DP-QPSK signal with pilot CW tones for carrier recovery (a detail of DSP is shown in [1]). The modulated signal was sent to an erbium-doped fiber amplifier (EDFA) and additional ASE noise was loaded to vary the received OSNR from 7.5 to 31 dB. The received OSNR was measured by optical spectrum analyzer (OSA). Chromatic dispersion was not imposed in this experiment, left for future work.

At the receiver, the local oscillator (linewidth ~25 kHz) was superimposed with the signal in a polarization-diversity optical 90° hybrid. The outputs of the hybrid were connected to four balanced photo-detectors. The resulting signals were digitized by four analog-to-digital converters (ADC) with a sample rate of 40 GS/s and a bandwidth of 16 GHz. The digital samples were processed with offline manner in desktop computer equipped with GPGPU.

The four-tributary dataset sampled by the ADCs was fed into the DNN, as shown in Fig. 2(b). Each tributary corresponding to HI/HQ/VI/VQ was 512 time-concatenated data sampled with 40 GS/s. Corresponding polarization states of both training and test dataset were randomly distributed on ten different polarization states. The DNN used in this study consists of 2,048 neurons in input layers and one neuron in output layers to predict OSNR. The activation function of output neuron was linear function. The number of hidden layer was set to 1, 3, and 5. Each hidden layer holds 500 neurons that have a activation function as rectified linear unit (ReLU), which is simply the half-wave rectifier $f(z) = max(z, 0)$, allowing training of deep supervised network without unsupervised pre-training [6]. The DNNs were trained by use of a Theano/Pylearn2 [7] software package. Batch gradient descent (BGD) and a conjugate-gradient technique were used for training. 4,000, 40,000, and 400,000 training dataset were used for training of the DNN, and another 10,000 test dataset was used for test.

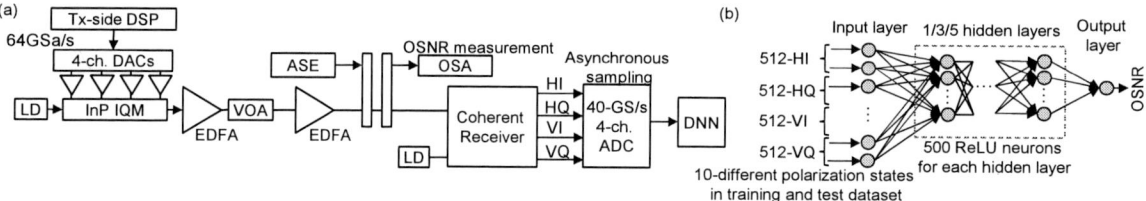

. Fig. 2. (a) Experimental setup, (b) details of the DNN used in this study. InP IQM: Indium phosphide IQ modulator, LD: laser diode, VOA: variable optical attenuator, ASE: amplified spontaneous emission noise source, ReLU: rectified linear unit.

IV. RESULTS AND DISCUSSION

Figure 3 shows the estimated OSNR by 3, 5, and 7 layers DNNs as a function of measured OSNR. Black circles in Fig. 3 show averaged value of estimated OSNR over test dataset. Each DNN was trained with 4,000, 40,000, and 400,000 dataset, respectively. Obviously, the 3 layers ANNs (or shallow neural networks) were failed to learn a relationship between input and output as shown in Fig. 3(a)-(c). Even with increased number of layer, the situation was not changed when the amount of training data was very limited such as 4,000 dataset as shown in Fig. 3 (a), (d), and (g). However, with increasing the number of both training dataset and layer, the situation was started to change as shown in Fig. 3 (e), (f), (h) and (i). The DNN might form themselves to a feature extractor by use of enormous number of training dataset such as 400,000 data. To the end, the DNN automatically learned a relationship between input raw data and OSNR as shown in Fig. 3(f) and (i).

For detailed discussion, we evaluated averaged errors between estimated and measured OSNR. Figure 4(a) shows averaged errors over test dataset or training dataset as a function of number of DNN layer with a fixed training dataset of 400,000. As the number of layer was increased, the both errors are reduced and saturated around 5 layers. According to Fig. 4(a), the 5 layer DNN seems likely to have enough deep layers in this specific case. Figure 4 (b) shows the averaged errors on both training and test dataset as a function of number of training dataset with 5 layers DNN. Since increasing of number of training dataset was beneficial for reducing the errors on test dataset, the over fitting of model may have not occurred yet in this case. Although more training dataset would provide an opportunity to additionally reduce the errors, the measured averaged error of 1.6 dB in this work (at the ranging from 7.5 to 31 dB OSNR) would be comparable to an averaged error of 1.23 dB (OSNR 12 - 32 dB) by the existing ANN-based monitor for RZ-DPSK signals with feature engineering [2].

OECC/PS2016

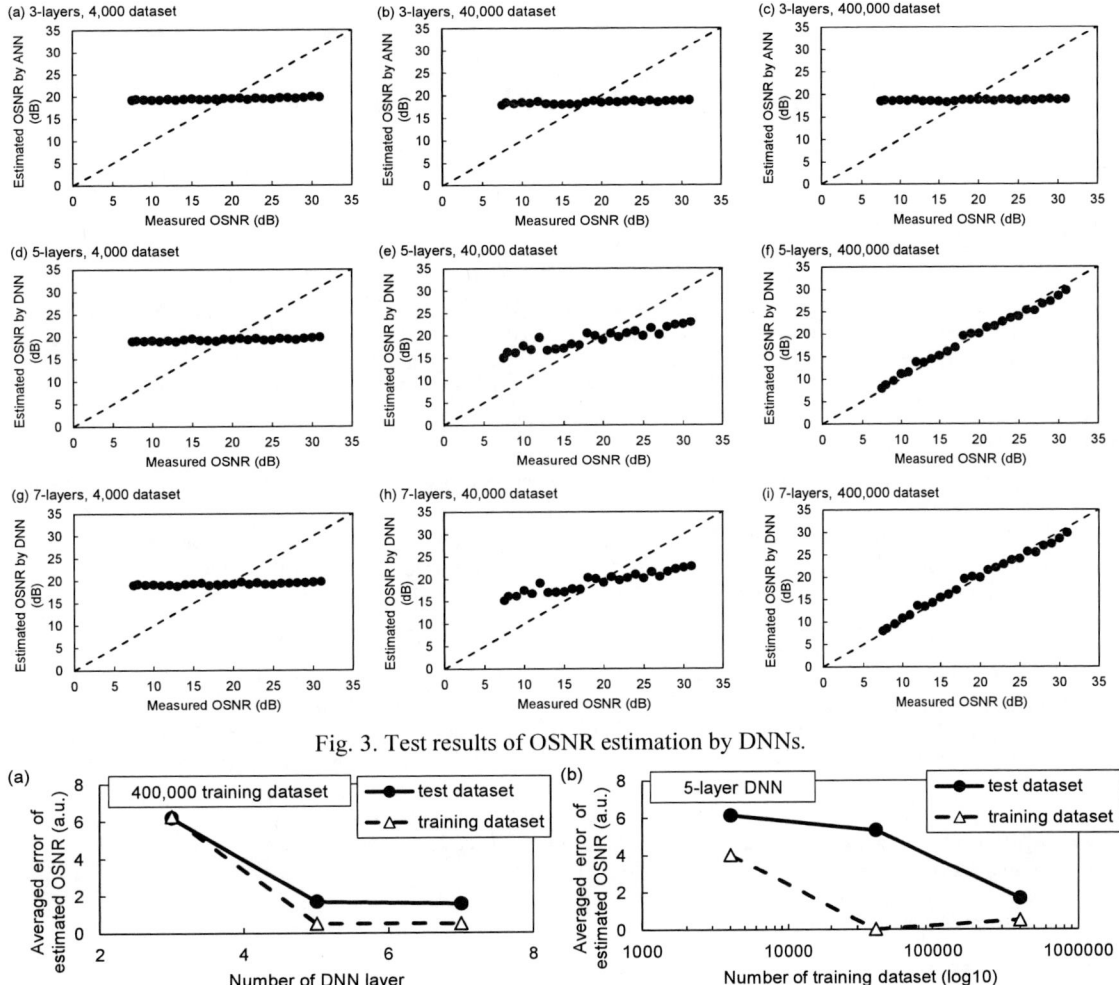

Fig. 3. Test results of OSNR estimation by DNNs.

Fig. 4. Averaged error of estimated OSNR as a function of (a) number of DNN layer, and (b) number of training dataset.

V. CONCLUSIONS

We experimentally demonstrated a use of deep neural networks (DNN) trained with raw data asynchronously sampled by coherent receiver in a 16 GBd DP-QPSK system. By using 5 layers DNN trained with 400,000 dataset, the DNN obtained a function of OSNR estimation without feature engineering depending on domain expertise. The DNN-based OSNR monitors were worked well against ten different polarization states of polarization division multiplexed input signals.

REFERENCES

[1] T. Tanimura, L. Dou, X. Su, T. Hoshida, Y. Aoki, Z. Tao, J. C. Rasmussen, M. Suzuki, and H. Morikawa, "Latency and bandwidth programmable transceivers with power arbitration among multi-tenanted signals," OFC 2016, W4A.6, 2016.
[2] X. Wu, J. A. Jargon, R. A. Skoog, L. Paraschis, and A. E. Willner, "Applications of artificial neural networks in optical performance monitoring, " IEEE J. Lightw. Technol. **27**, no. 16, 3580-3589, 2009.
[3] F. N. Khan, T. S. R. Shen, Y. Zhou, A. P. T. Lau, and C. Lu, "Optical performance monitoring using artificial neural networks trained with empirical moments of asynchronously sampled signal amplitudes," IEEE Photon. Technol. Lett. **24**, no. 12, 982-984, 2012.
[4] J. A. Jargon, X. Wu, H. Y. Choi, Y. C. Chung, and A. E. Willner, "Optical performance monitoring of QPSK data channels by use of neural networks trained with parameters derived from asynchronous constellation diagrams," OSA Optics Express **18**, no. 5, 4931-4938, 2010.
[5] Y. LeCun, Y. Bengio, and G. Hinton, "Deep learning," nature **521**, 436-444, 2015.
[6] X. Glorot, A. Bordes, and Y. Bengui, "Deep sparse rectifier neural networks," in Proc. of 14th International Conference on Artificial Intelligence and Statistics, 315-323, 2011.
[7] http://deeplearning.net/software/pylearn2/

Trellis Coded Optical Modulation Using QAM Constellations

Emmanuel Le Taillandier de Gabory, Tatsuya Nakamura, Hidemi Noguchi, Wakako Maeda, Sadao Fujita, Jun'ichi Abe and Kiyoshi Fukuchi

Green Platform Research Laboratories, NEC Corporation, 1753, Shimonumabe, Nakahara-ku, Kawasaki, 211-8666, Japan.
e-degabory@cb.jp.nec.com

Abstract: We review TCM with 16QAM constellations. Multiple state TCM widens design options and provides two additional settings for flexible transponders. 8-state TCM 12QAM increases transmission distance by 880km at 3b/s/Hz. TCM benefits substantially from BP.
Keywords: optical communications, long-haul transmission, digital signal processing, coded modulation

I. INTRODUCTION

Coded optical modulation using 4-Dimension (4D) constellations has been investigated for the improvement in sensitivity of transmitted signals, which it could provide in comparison to established Polarization Multiplexed (PM) formats using 2D constellations [1-2]. In this perspective, modulation formats with constellations optimized in 4D space to maximize the Euclidian distance between 4D symbols have been proposed [2]. Besides, Set Partitioning (SP) over 4D QAM constellations [3-4] has been investigated to be used in complement to PM-QAM formats. In this manner, SP offers additional operation settings for flexible systems, finely adapting signals and maximizing the capacity through spectral efficiency (SE) for specific transmission conditions.

Trellis Coded Modulation (TCM) has been used over SP technique [5], providing coding gain. For optical communications, it was first analyzed numerically with 8-state configuration and 4D QAM constellations [6]. In addition to changing the QAM constellation, we introduced multiple state optical TCM [7], where the number of states was also tuned for fine adaptation and higher gain, compared to SP. We studied its benefits for adaptive transponder.

In this paper, we review results of multiple state TCM, including experimentally evaluated gains in system sensitivity and application to short reach [7]. We also review reported long haul transmission results of TCM signals [8]. Notably TCM applied to 12QAM enables to increase the transmission distance compared to 8QAM at constant SE. Moreover nonlinear compensation substantially improves the transmission performance of TCM signals compared to PM signals.

II. TRELLIS CODED OPTICAL MODULATION

TCM was first proposed by G. Ungerboeck [5]. Here, we apply TCM to QAM constellations, were we map N+M bits on QAM constellation. N bits are used select one of 2^N cosets with SP technique and M bits are mapped on the 4D SP constellations. The TCM encoder generates redundancy bits used with remaining bits to select among 2^N states. TCM decoding can be performed with a Viterbi decoder.

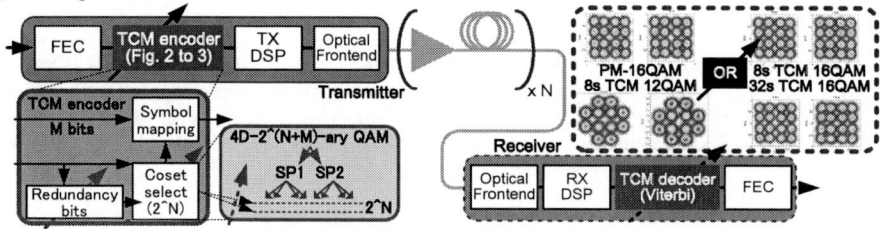

Fig. 1: Principle of multiple state TCM system

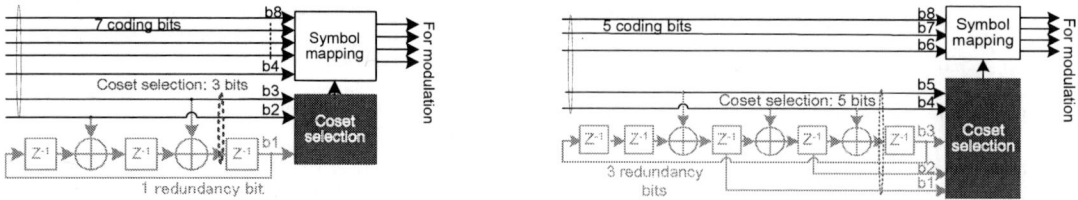

Fig. 2: 8-state TCM encoder for 16QAM constellation   Fig. 3: 32-state TCM encoder for 16QAM constellation

We studied multiple state TCM [7] for flexible transponders. It enables through configurable encoder and decoder of Fig. 1 to change the number of redundancy bits and cosets. More states enable higher sensitivity through higher gain at the expense of reduced SE, while a moderate number of states enables higher capacity through higher SE at with lower gain. Fig. 2 and Fig. 3 summarize reported encoders for 8-state and 32-state TCM over 16QAM constellations [7]. Further encoder efficiency has also been reported [9]. Moreover, we have reported use of TCM over 12QAM constellation [4], in order to obtain a TCM format with the same SE as a PM format, PM-8QAM. This enabled with high gain [8], compared to non-canonical SP, which could not clearly surpass PM-8QAM [3].

III. BASIC PERFORMANCE OF MULTIPLE STATE TCM SIGNALS

A. Improvement of Sensitivity of Flexible Transponders

We first studied the basic performance of multiple state TCM in back to back for flexible transponders [7] operating at 32Gbaud. Fig. 4 summarizes required OSNR to obtain a Q factor of 8.5dB, assuming system margin over a 6.4dB for SD-FEC limit, for 8-state and 32-state TCM over 16QAM and QPSK constellations with the corresponding PM formats.

Multiple state TCM enabled 3 additional operations for flexible transponders based on QPSK and 16QAM constellations. Moreover, 8-state TCM format enabled to improve required OSNR by 1.1dB for 16QAM and QPSK compared to the PM tradeoff. 32-state TCM underperforms the PM tradeoff due to difficulties to demodulate the 16QAM constellation in low OSNR region but it still offers one step of fine adaptation for flexible transponders.

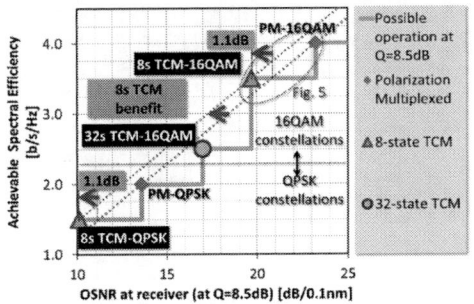

Fig. 4: SE versus OSNR at Q=8.5dB for 32Gbaud signals

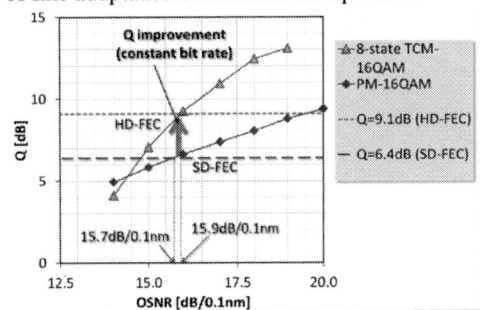

Fig. 5: Measured sensitivities for 100G signals

B. TCM for Short Reach Transceivers

Short reach transceivers based on 16QAM format are a candidate for high SE 100G low cost solutions. Comparing the two steps of higher SE of Fig. 4, 8-state TCM 16QAM, which would include a 8/7 overdriving to compensate for the TCM coding rate with 7% overhead for HD-FEC, and PM-16QAM with 20.5% overhead for SD-FEC would have very close bit rate for identical payload.

Fig. 5 summarizes results of evaluation in back to back of 100G 16QAM based signals, one with SD-FEC and the other with TCM and HD-FEC. The required OSNR for HD-FEC limit with 8-state TCM 16QAM was only 0.2dB higher than the OSNR required for SD-FEC limit with PM-16QAM. This shows that for close performance, TCM enables new design options for lower power: indeed, on-chip SD-FEC, which demands much resource, can be substituted with TCM decoder and be used with lower power off-chip legacy HD-FEC.

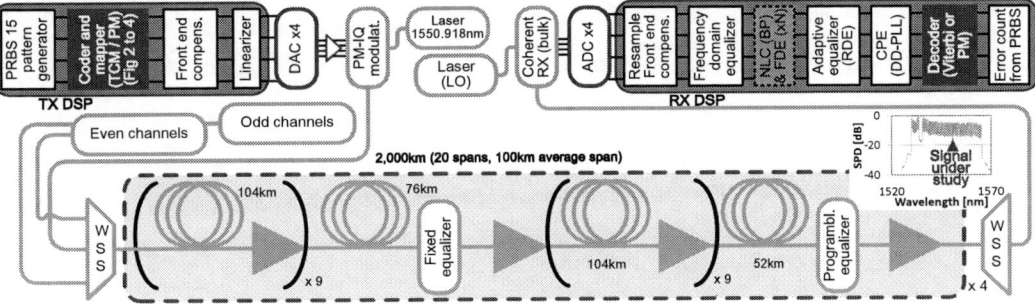

Fig. 6: Experimental setup for WDM transmission of TCM and PM signals

IV. TRANSMISSION PERFORMANCE OF MULTIPLE STATE TCM SIGNALS

Long-haul WDM transmission of 32Gbaud TCM signals was performed with the experimental setup of Fig. 6 [8]. 84 channels spaced by 50GHz were transmitted over a 8,000km-long straight line composed of 100km spans of pure silica core fiber, with EDFA amplification. All WDM channels had the same format and one channel was evaluated.

Considering formats with spectral efficiency of 3b/s/Hz from the point of view of a sweet spot for high capacity long-haul transmission [4], we compared 8-state TCM 12QAM to PM-8QAM. Results are reported on Fig. 7, without using compensation of nonlinearity. 8-state TCM 12QAM offered improvement of Q value over 6.2dB. Again, considering a Q value of 8.5dB, which assumed system margin over a 6.4dB SD-FEC limit, 8-state TCM 12QAM increased the transmission distance by 880km, compared to PM-8QAM.

Furthermore, investigating TCM for flexible transponders, we reported transmission distances estimated from interpolation between measured distances for the different TCM and PM formats, as shown on Fig. 8, for a Q value of 8.5dB. It should be noted that PM-QPSK was limited by the fiber length, namely 8,000km. First, we consider the cases without compensation of nonlinearities. TCM enabled to effectively increase the capacity at intermediate distances between PM-16QAM, PM-8QAM and PM-8QAM formats. Operation with a flexible transponder enabled to select the modulation format with the highest SE for a specific transmission distance. Therefore, TCM enables to increase capacity on intermediate distances, where it enables higher SE than PM formats only.

Finally, using Back-Propagation (BP) to compensate impairments due to nonlinear effects, further improvement was measured for TCM signals, as plot on Fig. 8. Considering PM signals, the transmission distances PM-16QAM and PM-8QAM were 1,280km and 3,650km with respective improvements of 7% and 11% with BP. Considering TCM signals, the transmission distances for 8-state TCM 16QAM, 8-state TCM 12QAM and 32-state TCM 16QAM were respectively 2,190km, 4,760km and 5,430km with respective improvements of 26%, 14% and 18%. This shows that TCM signals benefit more than PM signal from compensation of nonlinear impairments.

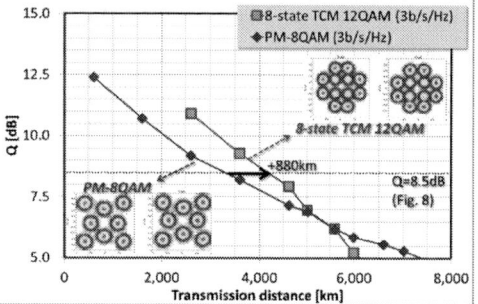

Fig. 7: WDM transmission results for 8-state TCM 12QAM and PM-8QAM

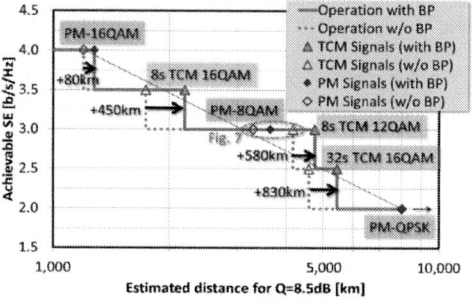

Fig. 8: SE versus estimated transmission distance

V. CONCLUSIONS

We have reviewed results of demonstrations of TCM signals using 16QAM and 12QAM constellations. We have shown that TCM can offer wider options of design for short reach 100G transceivers by enabling the possibility to use HD-FEC with comparable performance.

Furthermore, we have shown that multiple state TCM brings two operations settings for flexible transponders with 8-state TCM 16QAM and 32-state TCM 16QAM; it also enables to replace the step of PM-8QAM with 8-state TCM 12QAM improving the transmission distance by 880km. Finally, we have shown that BP enabled to increase transmission distance by 14% to 26% for TCM signals depending of SE, showing more benefit than PM signals, for which the improvement was below 11%.

ACKNOWLEDGMENT

A Part of the experimental results has been achieved using test equipment of the National Institute of Information and Communications Technology (NICT), Japan.

REFERENCES

[1] M. Karlsson et al., "Which is the most power-efficient modulation format in optical links?", Optics Express, Vol. 17, pp. 10814-10819 (2009)

[2] J. Karout et al., "Experimental demonstration of an optimized 16-ary four-dimensional modulation format using optical OFDM", Proc. of Optical Fiber Communications Conference and Exhibition (OFC 2013), paper OW3B.4.

[3] J. Renaudier et al., "Comparison of set-partitioned two-polarization 16QAM Formats with PDM-QPSK and PDM-8QAM for optical transmission systems with error-correction coding", Proc of European Conference on Optical Communication (ECOC 2012), paper We.1.C.5.

[4] T. Nakamura et al., "Long haul transmission of four-dimensional 64SP-12QAM signal based on 16QAM constellation for longer distance at same spectral efficiency as PM-8QAM", Proc. of European Conference on Optical Communication (ECOC 2015), paper Th2.2.2.

[5] G. Ungerboeck, "Channel coding with multilevel/phase signals", IEEE Transactions on Information Theory, Vol. IT-28, pp. 55-67 (1982).

[6] S. Ishimura et al., "8-state trellis-coded optical modulation with 4-dimensional QAM constellations", Proc. of OptoElectronics and Communication Conference (OECC 2014), paper TH12B-2.

[7] E. Le Taillandier de Gabory et al., "Experimental demonstration of the improvement of system sensitivity using multiple state trellis coded optical modulation with QPSK and 16QAM constellations", Proc. of Optical Fiber Communications Conference and Exhibition (OFC 2015). 2015, paper W3K.3.

[8] E. Le Taillandier de Gabory et al., "Demonstration of the improvement of transmission distance using multiple state trellis coded optical modulation", Proc. of Asia-Pacific Conference on Communications (APCC 2015), 2015, pp. 89-93.

[9] T. Nakamura et al., "Hardware-efficient Multi-redundancy Superposed Trellis Coded Modulation on 16QAM Constellation", Proc. of OptoElectronics and Communication Conference (OECC 2015), paper PWe.46.

TuB4-2

Differential Coding in Polarizations with Half-Symbol-Period Timing Offset for Coherent PM-QPSK to Improve Phase-Noise Tolerance

Guo-Wei Lu[1][2], Takahide Sakamoto[1], Yukiyoshi Kamio[1]

[1] *National Institute of Information and Communications Technology (NICT), 4-2-1 Nukui-Kitamachi, Koganei, Tokyo 184-8795 Japan*
[2] *Institute of Innovative Science and Technology, Tokai University, 4-1-1 Kitakaname, Hiratsuka, Kanagawa 259-1292 Japan*
gordon.guoweilu@gmail.com

Abstract: *We propose a phase-noise-tolerant coherent PM-QPSK system with differential coding and half-symbol-period timing-offset in polarizations. Owing to the improved phase noise tolerance, error-free operation of 10Gbaud PM-QPSK is experimentally demonstrated even using ~30MHz-linewidth DFB laser.*
Keywords: *Coherent communications; Phase modulation; Phase noise, Differential coding*

I. INTRODUCTION

With the development of high-speed analog-to-digital convertor (ADC) and digital signal processing (DSP) technologies, high-order multi-level signaling with digital coherent detection has been widely deployed in optical communications to realize ultra-high-speed and ultra-spectrally-efficient optical communication systems [1-2]. However, with the increase of the modulation level, the system becomes more sensitive to the phase noise originating from the finite laser linewidth of deployed laser diodes, referred to as linear phase noise, or the introduced nonlinear phase noise such as cross-phase modulation in the transmission link. The sensitivity to linear phase noise hinders the use of cost-effective distributed feedback laser (DFB), which usually has linewidth of several megahertz, as laser sources for signal or local oscillators in digital coherent transmission systems. To cope with the phase noise, several approached have been proposed, for example, self-homodyne detection by polarization-multiplexing pilot carrier [3], or differential delay detection implemented in either "optical" or "digital" manner [4-5]. In these approaches, un-modulated pilot carrier or previously-detected symbol severs as phase reference in "relative" phase detection from common laser source, showing improved phase-noise tolerance.

In this paper, we propose and experimentally demonstrate a phase-noise tolerant polarization-differential coding scheme for polarization-multiplexed quadrature phase-shift keying (PM-QPSK) with half-symbol-period timing-offset in polarizations, referred to as polarization-differentially-coded offset-PM-QPSK (PD-offset-PM-QPSK) here. Since the differential coding is realized in polarization domain, signals in each polarization work as phase reference mutually in detection, resulting in the improved tolerance against phase noise. Besides, in contrast to self-homodyne detection with polarization-multiplexed pilot carrier, with both polarizations modulated, the spectrum efficiency remains the same as that of the conventional PM-QPSK. The experimental results show that the proposed differential coding in polarizations enables the error-free operation of 10Gaud PM-QPSK system even using 30MHz-linewidth DFB as laser source. In contrast, error-floor is observed in the conventional PM-QPSK with DFB.

II. OPERATION PRINCIPLE

Figure 1 illustrates the operation principle of several different PM-QPSK schemes. In the conventional PM-QPSK shown in Fig. 1(a), data is modulated in orthogonal polarizations independently. No coding is applied in polarization domain. In the self-homodyne QPSK system shown in Fig. 1(b), one of polarization is left unmodulated as pilot carrier for phase reference in detection with improved phase-noise tolerance, but sacrificed spectrum efficiency. Alternatively, to improve the phase noise tolerance of PM-QPSK or PM- quadrature amplitude modulation (PM-QAM), delay-detection has been proposed by *N. Kikuchi*. The concept of "delay detection" is depicted in Fig. 1(c). With individually differential pre-coding for each polarization tributary at transmitter side, "delay detection" is performed at the receiver side to recover the original information using either "optical" or "digital" approach in each polarization. With the previous symbol as phase reference, it effectively improves the phase noise tolerance. In our proposed PD-offset-PM-QPSK scheme shown in Fig. 1(d), differential coding is performed in polarization domain. Since each polarization tributary serves as phase reference mutually, the tolerance against phase noise could be improved. In order to remain the same spectrum efficiency as that of conventional PM-QPSK, a half-symbol-period timing-offset is introduced between two polarizations. Since the differential coding is performed between two polarizations in the same symbol period, the proposed scheme is more tolerant to the fast phase noise compared with the "delay-detection" PM-QPSK. Moreover, as the differential de-coding could be performed after the polarization separation, it has the same requirement in hardware, such as bandwidth and ADC speed, as the conventional PM-QPSK at the receiver side.

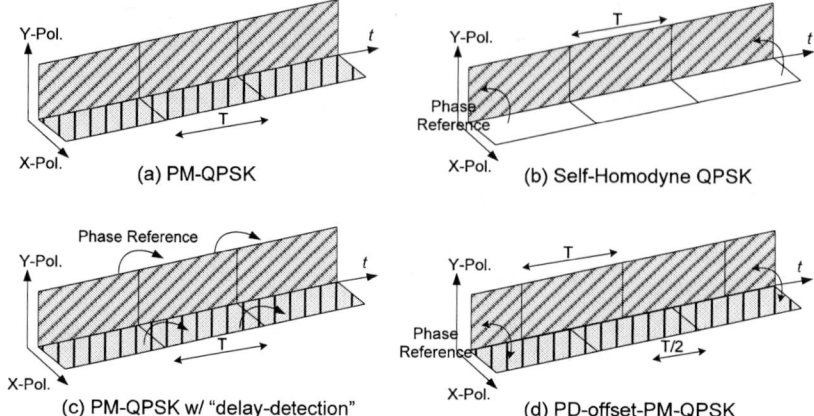

Fig. 1: Operation principle of (a) conventional PM-QPSK; (b) self-homodyne QPSK; (c) PM-QPSK w/ "delay-detection"; (d) PD-offset-PM-QPSK.

III. EXPERIMENT AND RESULTS

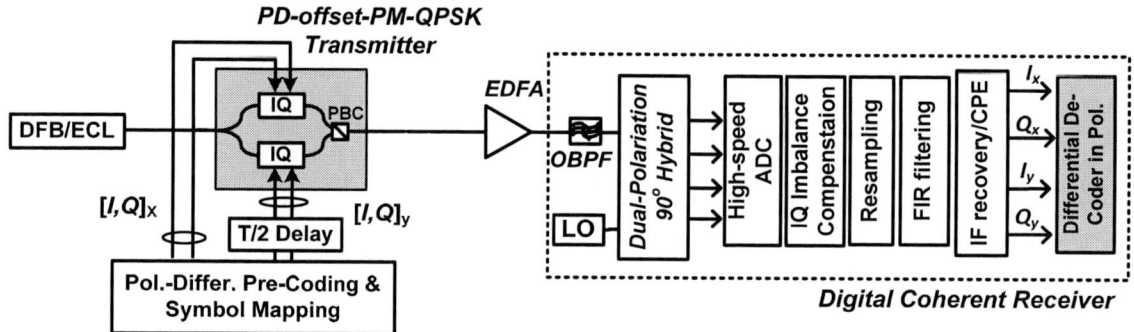

Fig. 2: Experimental setup.

To verify the feasibility of the proposed phase-tolerant PM-QPSK scheme, an experiment at 10Gbaud is carried out. Fig.2 depicts the experimental setup. At the transmitter side, an integrated dual-polarization in-phase/quadrature (DP-IQ) modulator is used for data modulation. Four independent data streams with length of 2^{15}-1 at 10Gbaud are generated to drive the modulator. In order to synthesize PD-offset-PM-QPSK, differential coding is performed in polarization domain with a 50-ps offset between polarizations. At the receiver side, the coherent receiver front-end consists of a local oscillator (LO) and a polarization-diversity 90-degree hybrid followed by four balanced photo-detectors (PDs) and ADCs. The LO is generated from a 100-kHz-linewidth external cavity laser (ECL). After detection via balanced PDs, data is digitized at 50 GSamples/s by using a digital storage oscilloscope with a 12.5-GHz analog bandwidth (Tektronix DPO71254). The captured data is then off-line processed for reconstructing constellation and calculating bit-error rate (BER). The deployed DSP flow is illustrated in Fig. 2. After polarization separation, differential de-coding is performed between two de-multiplexed polarizations, where the phase information in each polarization works as phase reference mutually. Hard decision is final performed for BER calculation and constellation reconstruction. In the experiment, a 30-MHz-linewidth DFB and a 100-kHz-linewidth ECL are used as laser sources individually for comparison. Besides, conventional PM-QPSK signals with DFB and ECL are also synthesized to provide baseline for performance evaluation.

Fig. 3 (a) and (b) show the measured constellations of conventional PM-QPSK at ~16-dB optical signal optical signal-to-noise ratio (OSNR) with ECL and DFB as laser sources, respectively. A clear constellation is observed for PM-QPSK with ECL. However, with a DFB as laser source, the presence of phase noise causes spreading of symbols around a unit circle, implying that the phase noise from DFB severely deteriorates the PM-QPSK signal. In contrast, the measured constellations with the proposed PD-offset-PM-QPSK are shown in Fig.3 (c)~(d) with DFB and ECL as laser sources. Different from the constellations in conventional PM-QPSKs, similar error vector magnitude (EVMs) are obtained for the PD-offset-PM-QPSK with DFB and ECL at OSNR of ~16dB. The improved phase-noise tolerance with differential-coding in polarization makes it is possible to obtain clear constellation even with DFB as laser source. However, the phase noise is enhanced compared to PM-QPSK with ECL laser under the same OSNR condition (Fig. 3(a)). This can also be confirmed by the measure BER curves as a function of OSNR (at 0.1nm), shown in Fig. 3. In the conventional PM-QPSK, error-free operation is obtained with ECL as laser source, whereas error-floor at BER of around 2×10^{-4} is observed with DFB laser due to the introduced phase noise from laser source. With the proposed polarization differential-coding, error free operations are observed for both ECL and DFB lasers with similar BER performance. However, as a cost of deployment of differential coding, around 2.3-dB OSNR penalty is obtained compared with the conventional PM-QPSK with ECL laser at BER of 10^{-3}. Further improvement is expected by applying the multi-symbol phase estimation to cope with the extra phase-noise caused by differential coding [4-5]. The error-free operation of the PD-offset-PM-QPSK with DFB laser verifies the feasibility of the proposed phase-noise-tolerant PM-QPSK scheme.

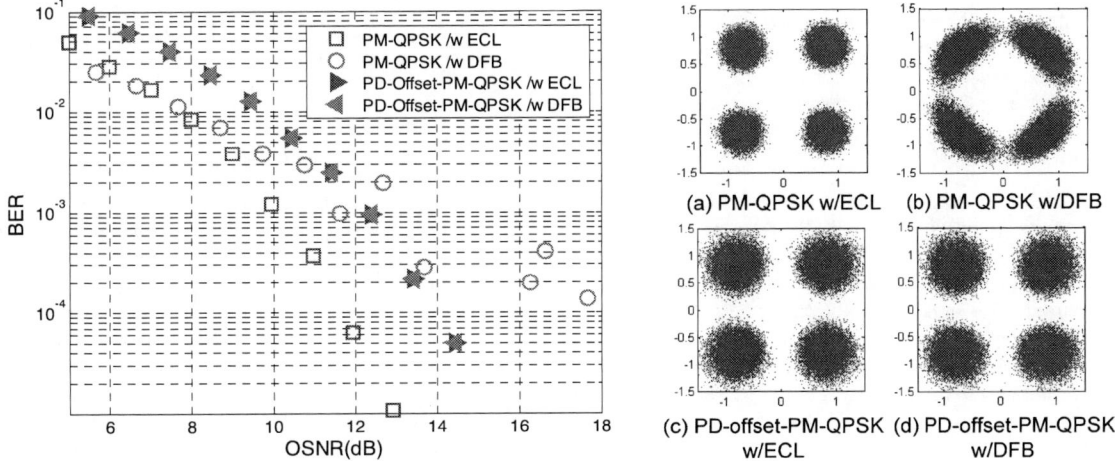

Fig. 3: Left: BER vs. OSNR curves of PM-QPSK and PD-offset-PM-QPSK. Right: constellations of the conventional PM-QPSK with (a) ECL and (b) DFB as laser source, and the proposed PD-offset-PM-QPSK with (c) ECL and (d) DFB as laser source.

IV. CONCLUSIONS

We have proposed and experimentally demonstrated a phase-noise tolerant PM-QPSK system with differential-coding and half-symbol-period timing-offset in polarizations. The improved phase-noise tolerance allows the use of ~30-MHz DFB as laser source in 10Gbaud PM-QPSK signals with error-free operation. It could be further extended to PM-QAM system with higher-order multi-level modulations.

ACKNOWLEDGMENT

The work was supported in part by Grant-in-Aid for Scientific Research (C) (15K06033) from the Ministry of Education, Science, Sports and Culture (MEXT), Japan.

REFERENCES

[1] K. Kikuchi, "Coherent detection of phase-shift keying signals using digital carrier-phase estimation," Proc. OFC, OTuI4 (2006).

[2] P.J. Winzer, et al., "Spectrally efficient long-haul optical networking using 112-Gb/s polarization-multiplexed 16QAM, " J. Lightwave Technol., Vol. 28, no. 4, (2012).

[3] M. Nakamura, et al., "30 Gbit/s 64-QAM transmission over 60 km SSMF using phase-noise cancelling technique and ISI suppression based on electronics digital processing," Electron. Letters, Vol. 45, no. 25 (2009).

[4] N. Kikuchi, et al., "Phase-noise tolerant coherent polarization-multiplexed 16QAM transmission with digital delay-detection," Proc. ECOC, Tu.3.A.5, London (2011).

[5] N. Kikuchi, et al., "Polarization-multiplexed 32QAM and 16PSK signaling with coherent receiver using digital-delay detection," Proc. OFC, OM2A.1, (2012).

TuB4-3

OECC/PS2016

Crossover Block Modulation with Complementary Codes Superposition

Tsuyoshi Yoshida[1], Keisuke Kojima[2], Toshiaki Koike-Akino[2], David Millar[2], Keisuke Dohi[1],
Keisuke Matsuda[1], Kieran Parsons[2], Kazuo Kubo[1], Kenichi Uto[1], and Takashi Sugihara[1]

[1] Information Technology R&D Center, Mitsubishi Electric Corporation, 5-1-1 Ofuna, Kamakura, Kanagawa, 247-8501 Japan
[2] Mitsubishi Electric Research Laboratories (MERL), 201 Broadway, Cambridge, MA 02139, USA
Yoshida.Tsuyoshi@ah.MitsubishiElectric.co.jp

Abstract: We propose novel 3-bit/symbol modulation schemes which cross-connects two sets of coded signals over time and polarization. Simulations show 0.8 dB higher tolerance against interplay of fiber nonlinearity and PDL compared with conventional PS-QPSK.
Keywords: Coherent Communication, Fiber Nonlinearity, Modulation.

I. INTRODUCTION

Coherent communication has been widely deployed with the assistance of digital signal processing. Dual-polarized quadrature phase-shift keying (DP-QPSK) is a standard solution for both long-haul and metro 100 Gb/s optical transport [1]. To optimize the spectral efficiency and the reach under each of these transmission conditions, both higher and lower order modulation have been investigated. Besides regular 2^N quadrature amplitude modulation with polarization multiplexing for $2N$ bit/symbol, intermediate solutions are considered since $2N$ bit/symbol may not have enough granularity to maximize the system capacity in all cases. Polarization-switched quadrature phase-shift keying (PS-QPSK) [2] is an early solution having an intermediate spectral efficiency of 3 bit/symbol, which is classified as part of a 4-dimensional modulation family. PS-QPSK shows good linear and reasonable nonlinear performance and efficiently fills the gap in spectral efficiency between DP-QPSK and DP binary phase-shift keying (DP-BPSK) [3]. Single parity-check (SPC) coded DP-QPSK is equivalent to PS-QPSK. The coding-based solution is hardware-efficient because it does not require changes to the optical transmitters of a standard DP-QPSK configuration.

In long-haul transmission, fiber nonlinearity and polarization dependency are critical factors limiting the reach. To reduce the degradation, state of polarization (SOP) management has been intensively investigated for non-coherent schemes [4]. Biasing the SOP increases degradations such as polarization dependent loss (PDL), polarization hole burning, and cross polarization modulation. We also have to be aware of this for polarization multiplexed coherent schemes [5]. PS-QPSK takes two SOPs with orthogonal linear polarizations and the SOP bias will depend on the bit sequence. Forward error correction (FEC) with bit-interleaving helps to mitigate the effects of PDL, but it is not enough. Diversity transmission such as phase-conjugate modulation and superposition demodulation [6–9] can help to reduce such degradation quite effectively. On the other hand, these signals have themselves so far provided either no or lower coding gain. The best mixture of diversity transmission would be with a concept like phase-conjugation modulation providing coding gain by increasing the Euclidean and/or Hamming distance by employing, e.g., high-dimensional modulation [10]. If we can realize not only SOP diversity and superposition but natural SOP management, additional performance improvement will be expected on transmission lines with strong nonlinearity and polarization dependent degradation. Although a solution for 2 bit/symbol has already been proposed [11], there has, however, so far been no reported solution for 3 bit/symbol transmission to the best of our knowledge.

In this paper, we propose three novel 3 bit/symbol modulation schemes, one of which realizes SOP management by crossover superposition of two sets of SPC coded signals over two time slots and two polarizations. Through numerical simulation, we observed 0.8 dB improvement in the presence of interplay of fiber nonlinearity and PDL.

II. PRINCIPLES OF THE PROPOSED MODULATION SCHEME

In this paper we examine the three types of modulation scheme shown in Figure 1. X1 and X2 are the electric fields of the first and second time slots of the X-polarization, while Y1 and Y2 are those of the Y-polarization. For efficient implementation, we use 3-bit to 4-bit conversion with SPC. Although the coding is 3 bits in 4-D, the modulation acts like 6 bits in 8-D. Using a block of 6-bit data, b_0, b_1, ..., b_5 are mapped onto two time slots and two polarizations. Table I shows the coding rules for the proposed scheme. To realize a spectral efficiency of 3 bit/symbol, we use not only an SPC code for converting 3 bits into 4 bits as shown in Table I (A), but also the complementary version of the SPC code shown in Table I (B). This complementary version is equivalent to partial phase conjugation [9]. Therefore, phase conjugate coding and coded modulation are strongly related to each other.

This work was partly supported by "The research and development project for the Tera-bit optical network technologies towards big data era" of the Ministry of Internal Affairs and Communications, Japan.

A. Cross Polarization- and Time-Block Coding (X-PTBC)

Figure 1 (a) shows the first case: cross polarization- and time-block coding (X-PTBC). Two sets of 3-bit 4-D block-coded signals are mapped onto two time slots and two polarizations. Electric fields A and B in Table I (A) are mapped onto X1 and Y2, and C and D in Table I (B) are mapped onto Y1 and X2. The SOPs of the two time slots are always different. One time slot is linearly polarized (at S2 on the Poincaré sphere; 0 or 180 degree optical phase difference between the X and Y polarizations) and the other time slot is circularly polarized (S3 on the Poincaré sphere; +/-90 degree optical phase difference between X and Y polarizations).

B. Parallel Time-Block Coding (P-TBC)

Figure 1 (b) shows the second case: parallel time-block coding (P-TBC). Two-set of 3-bit 4-D block-coded signals are mapped on two time slots, the first set on the X-polarization and the second set on the Y-polarization. Electric fields A and B in Table I (A) are mapped onto X1 and X2, and C and D in Table I (B) are mapped onto Y1 and Y2. One time slot is linearly polarized (at S2) and the other time slot is circularly polarized (at S3), just as with X-PTBC.

C. Parallel Polarization-Block Coding (P-PBC)

Figure 1 (c) is the third case: parallel polarization-block coding (P-PBC). Two sets of 3-bit 4-D block-coded signals are mapped onto two polarizations. Electric fields A and B in Table I (A) are mapped onto X1 and Y1, and C and D in Table I (B) are mapped onto X2 and Y2. The first time slot is linearly polarized (at S2 on the Poincaré sphere) and the second time slot is circularly polarized (at S3). The SOP randomization helps to reduce inter-channel polarization dependent degradation. However, there is no gain from superimposing over two time slots having different SOP.

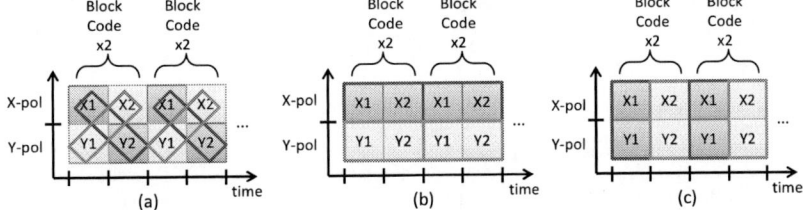

Fig. 1. The three types of modulation scheme considered; (a) cross polarization- and time-block coding (X-PTBC), (b) parallel time-block coding (P-TBC), (c) parallel polarization-block coding (P-PBC).

TABLE I
CODING RULES; (A) SINGLE PARITY CHECK AND (B) COMPLEMENTARY SINGLE PARITY CHECK

(A)

b_0	0	0	0	0	1	1	1	1
b_1	0	0	1	1	0	0	1	1
b_2	0	1	0	1	0	1	0	1
$A\ (E_1)$	(+1,+1)	(+1,+1)	(+1,-1)	(+1,-1)	(-1,+1)	(-1,+1)	(-1,-1)	(-1,-1)
$B\ (E_1\ or\ -E_1)$	(+1,+1)	(-1,-1)	(-1,+1)	(-1,+1)	(+1,-1)	(-1,+1)	(+1,+1)	(+1,+1)

(B)

b_3	0	0	0	0	1	1	1	1
b_4	0	0	1	1	0	0	1	1
b_5	0	1	0	1	0	1	0	1
$C\ (E_2)$	(+1,+1)	(+1,+1)	(+1,-1)	(+1,-1)	(-1,+1)	(-1,+1)	(-1,-1)	(-1,-1)
$D\ (E_2^*\ or\ -E_2^*)$	(+1,-1)	(-1,+1)	(+1,+1)	(-1,-1)	(+1,+1)	(-1,-1)	(+1,-1)	(-1,+1)

III. SIMULATION

To verify the transmission performance, we have conducted a set of numerical simulations whose parameters are listed below. The modulation schemes tested were X-PTBC, P-TBC, and P-PBC, and for comparison, PS-QPSK and DP-QPSK were also examined. The gross bit rate, including FEC parity, pilot symbols and frame overhead was 140 Gb/s. The baud rate was set to 46.67 Gbaud except for DP-QPSK (which used 35 Gbaud to ensure equal bit rate), and their signals were root-raised cosine (RRC) filtered with a roll-off factor of 0.1. Nine-channel wavelength division multiplexed with a channel spacing of 53.67 GHz (= 46.67 GHz x 1.15) were created. As we were focused on highly nonlinear conditions, the transmission line consisted 25 spans of non-zero dispersion-shifted fiber (NZDSF) having a loss of 0.2 dB/km, a span length of 80 km , a local chromatic dispersion (CD) of 3.9 ps/nm/km, and a nonlinear index of 1.6 W^{-1}km^{-1}. CD compensation was applied at a rate of 50% of residual dispersion at the transmitter and 90% in-line per span. Erbium-doped fiber amplifiers were used for span loss compensation with a noise figure of 5 dB. All amplified spontaneous emission noise was loaded at the receiver side to adjust optical signal-to-noise ratio (OSNR). The performance was evaluated using the span loss budget [12], calculated from Required OSNR(ROSNR), needed to achieve the target normalized generalized mutual information [13] of 0.85 assuming use of a soft-decision FEC [14].

Figure 2 shows the simulation result; Fig. 2 (a) is the span loss budget, and Fig. 2 (b) is the gain compared to PS-

QPSK. Under linear conditions where the launched power was less than -4 dBm/ch, there was no difference between the performances of the various 3 bit/symbol modulation schemes. On the other hand, under highly nonlinear conditions where the launched power was higher than -3 dBm/ch, all the proposed formats outperformed the conventional PS-QPSK. The maximum span loss budget was improved by 0.3 dB. Although SOP-managed X-PTBC shows the best performance of the three proposed types, the difference is not particularly significant.

In order to clarify the benefits of the SOP management of the proposed schemes, we simulated their tolerance to a combination of fiber nonlinearity and PDL. After adding 2 dB of PDL at the transmitter side on axes S1, S2, and S3 of the Poincaré sphere, the nonlinearity degradation was simulated in the cases of -10 dBm (linear condition) and -3 dBm (nonlinear condition) launched power. Table II shows the ROSNRs (in dB with 0.1 nm noise bandwidth) for each modulation format, launched power, and PDL condition. Here we can see that X-PTBC and P-TBC perform better than the others. When we consider the worst PDL conditions, the benefit of X-PTBC reaches nearly 0.3 dB and 0.8 dB under linear and nonlinear conditions, respectively. While all three proposed schemes have SOP diversity (traversing the S2 and S3 axes in two time slots), only X-PTBC and P-TBC superpose the different SOP signals at the receiver. Having both transmitter-side diversity and receiver-side superposition is key to reducing PDL and nonlinearity penalties.

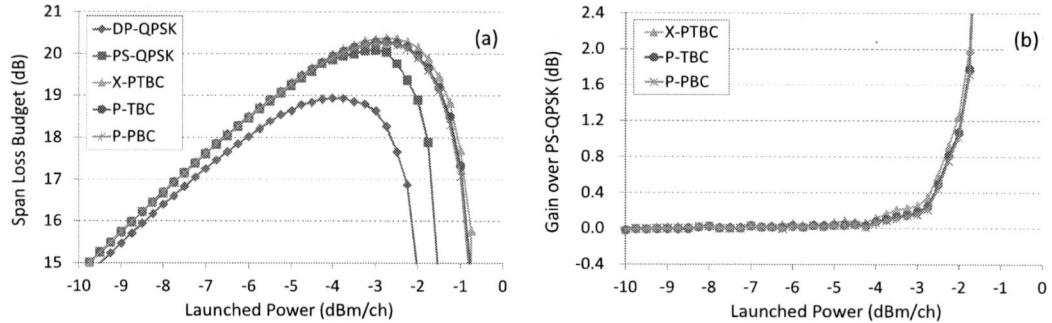

Fig. 2. Simulated transmission performance; (a) comparison of span loss budget, and (b) gain over PS-QPSK.

TABLE II
ROSNR WITH FIBER NONLINEARITY AND PDL

Launched Power		-10 dBm					-3 dBm				
PDL		0 dB	2 dB S1	2 dB S2	2 dB S3	2dB Worst	0 dB	2 dB S1	2 dB S2	2 dB S3	2dB Worst
Modn. Format	X-PTBC	9.49	9.69	9.64	9.64	9.68	10.91	11.08	11.29	11.31	11.31
	P-TBC	9.50	9.68	9.65	9.64	9.68	10.98	11.20	11.34	11.36	11.36
	P-PBC	9.50	9.84	9.66	9.64	9.84	11.00	11.34	11.35	11.41	11.41
	PS-QPSK	9.47	9.84	9.68	9.68	9.84	11.16	11.43	12.13	11.32	12.13
	DP-QPSK	9.73	10.00	9.91	9.91	10.00	12.62	12.92	13.64	13.72	13.72

IV. CONCLUSIONS

Novel 3 bit/symbol modulation schemes have been proposed. Due to SOP diversity at the transmitter side and superposing the signals across time slots and polarizations, the proposed schemes can reduce the degradation due to fiber nonlinearity and PDL. Our numerical simulations have shown that the performance of the proposed methods compared with PS-QPSK is the same under linear conditions and 0.3 dB better under highly nonlinear conditions comparing maximum span loss budget. In the presence of 2 dB of PDL with nonlinearity, the performance gain is increased to 0.8 dB.

REFERENCES

[1] K. Roberts, et al., *J. Lightw. Technol.*, vol. 27, no. 16, pp. 3546–3559 (2009).
[2] E. Agrell, et al., *J. Lightw. Technol.*, vol. 27, no. 22, pp. 5115–5126 (2009).
[3] M. Salsi, et al., *Bell Labs Tech. J.*, vol. 14, no. 4, pp. 131–148 (2010).
[4] F. Bruère, et al., *Photonics Technol. Lett.*, vol. 6, no. 9, pp. 1153–1155 (1994).
[5] K. Matsuda, et al., in *Proc. SPPcom*, paper SpS4D.1 (2015).
[6] X. Liu, et al., *J. Lightw. Technol.*, vol. 32, no. 4, pp. 766–775 (2014).
[7] T. Yoshida, et al., in *Proc. OFC*, paper M3C.6 (2014).
[8] H. Eliasson, et al., *Opt. Expr.*, vol. 23, no. 3, pp. 2392–2402 (2015).
[9] Y. Yu, et al. in *Proc. ECOC*, paper We.2.6.5 (2015).
[10] D. Millar, et al., *Opt. Expr.*, vol. 22, no. 7, pp. 8798–8812 (2014).
[11] A. Shiner, et al., *Opt. Expr.*, vol. 22, no. 17, pp. 20366–20374 (2014).
[12] P. Poggiolini, et al., *Opt. Expr.*, vol. 18, no. 11, pp. 11360–11371 (2010).
[13] A. Alvarado, et al., *J. Lightw. Technol.*, vol. 33, no. 10, pp. 1993–2003 (2015).
[14] K. Sugihara, et al., in *Proc. OFC*, paper OM2B.4 (2013).

Turbo Equalization for Duobinary-shaped Signals in Super-Nyquist WDM Systems

Shuai YUAN and Koji IGARASHI

Graduate School of Engineering, Osaka University, 2-1 Yamadaoka, Suita, Osaka, 565-0871 Japan
E-mail: ensui@procyon.comm.eng.osaka-u.ac.jp

Abstract: For Super-Nyquist WDM systems, performance of forward error correction (FEC) is remarkably degraded. In order to suppress the degradation, we propose turbo equalization, which improves performance of turbo code used as FEC by 3 dB.
Keywords: Wavelength division multiplexing, Pulse shaping, Forward error correction

I. INTRODUCTION

Pulse shaping of optical signals is essential for improving spectral efficiency in dense wavelength-division multiplexed (WDM) optical fiber transmission systems [1, 2]. Rectangular-shaped Nyquist shaping with bandwidth of the symbol rate is most suitable to suppress not only WDM crosstalk but also the inter-symbol interference (ISI) even when WDM spacing is equal to the symbol rate, called Nyquist WDM condition.

For further dense multiplexing beyond the Nyquist WDM condition, which is called Super-Nyquist WDM technique, duobinary-pulse shaping with an equalizer based on maximum likelihood sequence estimation (MLSE) has been proposed [3-7]. The duobinary shaping reduces the signal bandwidth of smaller than half the symbol rate, suppressing linear crosstalk from adjacent WDM channels. Although ISI is inevitable after duobinary shaping, it can be compensated for by MLSE equalization in the receiver. The Super-Nyquist WDM technique has been introduced to ultra-long-haul transmission experiments with capacity-distance product over 1 Exabit/s·km [6] and the ultra-high-capacity transmission with capacity of 2.05 Pbit/s [7]. In the Super-Nyquist WDM transmission experiments reported so far, the bit-error rates (BERs) of all channels of smaller than the BER threshold of the forward error correction (FEC) were achieved. Although FEC following MLSE equalizer is required, the actual FEC performance for the Super-Nyquist WDM transmission has not been investigated. Since duobinary shaping corresponds to addition of the present symbol and the delayed symbol in the time domain, the equalized symbols after the MLSE equalizer are interfered with the delayed symbols. In this case, the MLSE equalizer brings into excess continuous errors originated from noise after the transmission, and the continuous errors would remarkably degrade the FEC performance [8].

In this paper, we firstly show that the FEC performance in Super-Nyquist WDM systems is significantly degraded because of excess continuous errors after MLSE equalization for duobinary shaping. Next, we propose turbo equalization in order to suppress degradation of FEC performance in Super-Nyquist WDM systems. In the turbo equalization, soft information, namely logarithm of likelihood ratio (LLR), obtained from the decoder for FEC is fed back to equalization for duobinary shaping. The LLR feedback configuration can compensate for the continuous errors after the equalizer for duobinary shaping, suppressing degradation of FEC performance We numerically investigate performance of turbo equalization for turbo code used as high-performance FEC in Super-Nyquist WDM systems. Our simulation results suggest that the turbo equalization is effective to suppress performance degradation of turbo code for duobinary-shaped signals in Super-Nyquist WDM systems.

II. ERROR CHARACTERISTIC AFTER MLSE EQUALIZATION OF DUOBINARY-SHAPED SIGNALS

In Super-Nyquist WDM systems, duobinary shaping is usually introduced to suppress the WDM crosstalk. Figure 1(a) shows the discrete-time model of duobinary shaping, and the calculated power spectrum of the duobinary shaped signal is shown in Fig. 1(b). The signal bandwidth (full-width half maximum) after duobinary shaping is half the baudrate B. Although the narrow bandwidth characteristic is suitable for suppression of WDM crosstalk even with WDM spacing of smaller than the baudrate in the Super-Nyquist WDM systems, ISI due to duobinary shaping is unavoidable. Figure 1(c) shows the constellation of duobinary-shaped quadrature-phase-shift keying (QPSK) signals. The duobinary-shaped QPSK signals have nine types of symbols duo to ISI caused by duobinary shaping. For ISI compensation, MLSE is introduced in the receiver.

Here, we consider the error characteristic after the MLSE equalizer in Super-Nyquist WDM systems. Closed circles in Fig. 2 indicate calculated error distributions of single-channel duobinary-shaped QPSK signals after the MLSE equalizer when BERs are 10^{-2}, 10^{-3}, and 10^{-4}. The horizontal axis indicates the number of the continuous errors. Open circles indicate the results of the Nyquist shaping case without MLSE, and red lines present theoretical curves determined from additive white Gaussian noises (AWGN) model. Compared with the Nyquist shaping case without MLSE, we can see excess continuous errors of the duobinary shaping case with MLSE. Such continuous errors could degrade the FEC performance. Although the interleave technique is usually introduced to FEC, it is not effective in this case because the continuous errors after MLSE equalization are so long.

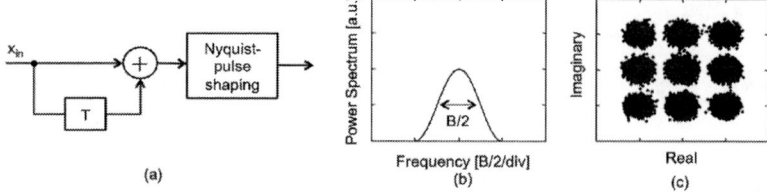

(a)

(b)

(c)

Fig. 1: (a) Discrete-time model of duobinary shaping, (b) power spectrum after duobinary shaping, and (c) constellation of the duobinary-shaped QPSK signal.

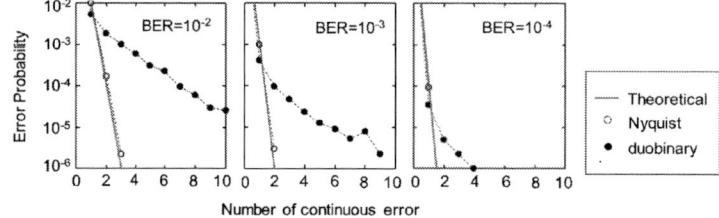

Fig. 2: The error probability of single-channel QPSK signals with duobinary shaping and Nyquist shaping.

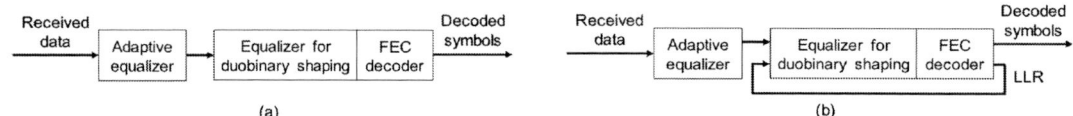

(a)

(b)

Fig 3: The configuration of (a) the conventional receiver and (b) the proposed receiver with turbo equalization.

III. TURBO EQUALIZATION

Here, we consider a receiver with equalization for duobinary shaping and decoding for FEC. The configuration of the conventional receiver is shown in Fig. 3(a). Adaptive equalization is performed for polarization multiplexing, signal equalization, and clock and phase recovery. After that, cascaded configuration with an equalizer for duobinary shaping followed by an FEC decoder is used. In this case, performance of the FEC decoder strongly depends on the error characteristics after equalization of duobinary shaping, and it would be degraded because of excess continuous errors after the equalizer of duobinary shaping. In order to suppress degradation of FEC performance, we propose turbo equalization in the receiver. The turbo equalization technique has been investigated in wireless communication systems [9, 10]. The configuration of the proposed scheme based on turbo equalization is shown in Fig. 3(b). After equalization for duobinary shaping, the equalized symbols are decoded for FEC. Here, note that soft-output information of LLR obtained from the FEC decoder is used as prior probabilities for the equalizer for duobinary shaping. Based on the feedback configuration between the equalizer for duobinary shaping and the decoder for FEC, the excess continuous errors after the equalizer for duobinary shaping can be compensated for, bringing out performance of the followed FEC.

IV. SIMULATION AND RESULTS

We numerically investigated performance of turbo equalization for turbo code used as FEC in Super-Nyquist WDM systems. The simulation model is shown in Fig 4(a). In the transmitter, a pseudo random bit sequences (PRBS) with a period of 2^{15}-1 is encoded to turbo code. The encoder of turbo code is shown in Fig. 4(b). It is based on parallel concatenated convolutional encoders with a random interleaver Π_1. After the encoded bit sequence is randomly-interleaved (Π_2), it is mapped to QPSK symbols. The square-root duobinary shaping is performed in the frequency domain, I and Q components of the processed samples are sent to digital-to-analog converters (DACs). Using an optical IQ modulator (IQM) driven by electrical duobinary-shaped signals generated from DACs, duobinary-shaped QPSK signals are obtained. In this simulation, linear modulation is assumed. Frequency-multiplexing three channels with frequency spacing Δf of smaller than the signal baudrate B, we obtain three-channel Super-Nyquist-WDM QPSK signals with duobinary shaping. In this simulation, Δf is fixed to 5/6 $\times$ B [5], as shown in Fig. 4(c).

After the WDM signals are contaminated by AWGN, the center channel is received based on homodyne detection. In this simulation, we assume that phase noise of coherent detection is neglected. After the received signals are equalized by square-root-duobinary shaped, maximum a posteriori probability (MAP) equalizer with Bahl-Cocke-Jelinek-Raviv (BCJR) algorithm is performed for I or Q component independently, and then the equalized samples are decoded for turbo code. Note that LLR obtained from the decoder for turbo code is used as prior probabilities for MAP equalizer. After that, errors of decoded samples are counted. For comparison, we calculated BERs of Super-Nyquist WDM QPSK signals without turbo equalization. In this case, we use MLSE for equalization of duobinary shaping, and then the soft output from MLSE are input to the decoder for turbo code.

OECC/PS2016

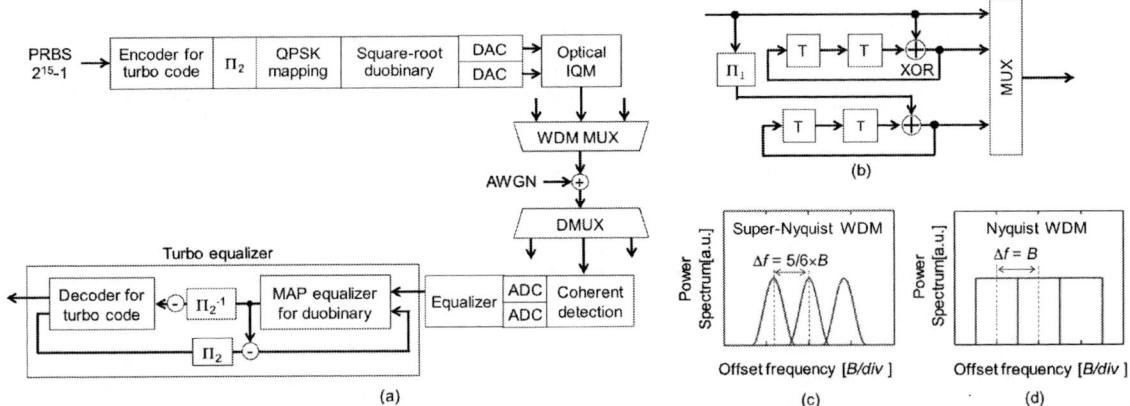

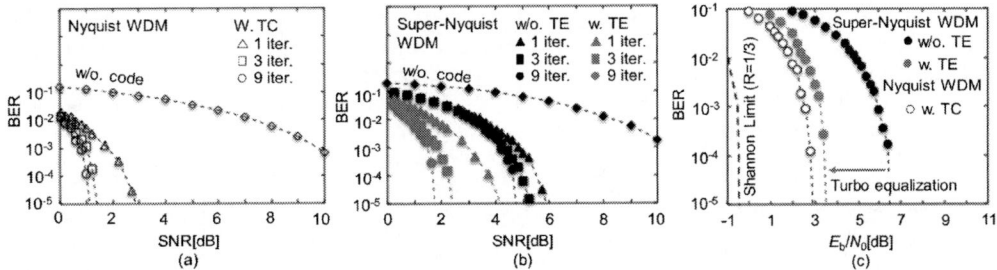

Fig. 4: (a) Simulation model of three-channel Super-Nyquist WDM system in back-to-back configuration. (b) Encoder of Turbo code. Optical spectra of (c) Super-Nyquist WDM signals based on duobinary shaping, and (d) Nyquist WDM signals based on Nyquist shaping. Π_1 and Π_2: interleaver, Π_2^{-1}: deinterleaver.

Fig. 5: Calculated BERs of (a) Nyquist WDM QPSK signals and (b) Super-Nyquist WDM QPSK signals. (c) Calculated BERs of Super-Nyquist WDM signals and Nyquist WDM signals with FEC as a function of energy per bit over spectral noise density, E_b/N_0.

The calculated BERs of Nyquist WDM QPSK signals are shown in Fig. 5(a). Open diamonds are the results without any FEC. Open triangles, open squares, and open circles indicate the results with FEC after 1 iteration, 3 iterations and 9 iterations, respectively. Figure 5(b) shows the calculated BERs of Super-Nyquist WDM QPSK signals. The results without FEC are plotted by closed diamond. Closed triangles, closed squares, and closed circles indicate the results using FEC without turbo equalization. Compared with Nyquist WDM signals, the required SNR for BER<10^{-5} is degraded by about 4 dB in the Super-Nyquist WDM case. This is because the continuous errors after the equalizer for duobinary shaping remarkably degrade FEC performance. With turbo equalization, the calculated BERs are plotted by red triangles, red squares, and red circles in Fig 5(b). We find that BER performance is drastically improved by the use of turbo equalization.

Figure 5(c) shows these calculated BERs with FEC as a function of the ratio of energy per bit to spectral noise density, E_b/N_0. The results of Nyquist WDM signals are plotted with open circles. Closed circles and red circles are the calculated BERs of Super-Nyquist WDM signals without and with turbo equalization, respectively. Using turbo equalization for Super-Nyquist WDM signals, the BER characteristic is improved by 3 dB. The results suggest that turbo equalization is effective to suppress the degradation of the FEC performance for Super-Nyquist WDM signals.

V. CONCLUSIONS

We investigated FEC performance in Super-Nyquist WDM systems. We showed errors trend to propagate after MLSE equalization for duobinary shaping, and such continuous errors remarkable degrade the FEC performance. In order to suppress degradation of FEC performance, we proposed turbo equalization for Super-Nyquist WDM systems with high-performance FEC. Using turbo equalization, FEC performance for Super-Nyquist WDM signals with turbo code is improved by 3 dB.

REFERENCES

[1] G. Bosco et al., IEEE Photon. Technol. Lett., **22**, 1129, 2010.
[2] M. Mazurczyk, IEEE J. Lightw. Technol., **32**, 2915, 2014.
[3] J. Li et al., IEEE J. Lightw. Technol., **30**, 1664, 2012
[4] J. Zhang et al., OFC2014, **Th5B.3**, 2014.
[5] K. Igarashi et al., IEEE J. Lightw. Technol., **34**, 1724, 2015.
[6] K. Igarashi et al., Opt. Express, **22**, 1220, 2014.
[7] D. Soma et al., ECOC2015, **PDP3.2**, 2015.
[8] Z. Jia et al., Opt. Express, **22**, 6047, 2014.
[9] R. Koetter et al., IEEE Signal Processing Magazine, **21**, 67, 2004.
[10] T. V. Souvignier et al., IEEE Trans. on Comm., **48**, 1297, 2000.

TuB4-5

Enhanced Blind Modulation Formats Recognition using Connected Component Analysis with Quadruple Rotation

Tianwai Bo, Jin Tang, Calvin Chun-Kit Chan
Department of Information Engineering, The Chinese University of Hong Kong
tianwai@ie.cuhk.edu.hk

Abstract: *An enhanced algorithm is proposed to improve the recognition accuracy of connected component analysis based modulation formats recognition method. It alleviates the requirements for both the optical signal-to-noise ratio and number of points simultaneously.*
Keywords: *Modulation format recognition; Software-defined network*

I. INTRODUCTION

Recently, management of optical networks is progressing more flexible and software-defined. Fully programmable bandwidth variable transponders are adaptive to the data rate and the modulation format, based on the transmission length and channel state information so as to increase the spectral efficiency and assure the quality of service. In particular, it is highly desirable for the receivers to have the cognitive ability of the signal's modulation format, hence proper digital signal processing (DSP) algorithm can be applied to achieve the optimum performance for the received optical signal. Another application that requires modulation format recognition (MFR) is the coherent burst mode transmission system, in which the coherent receiver is required to respond to the fast channel switching.

Several MFR schemes have been proposed, recently. It can be classified into Jones space method [1] and Stokes space methods based on the domain, where the MFR is performed. In general, Stokes space is intrinsically insensitive to laser phase noise and frequency. In [2, 3], K-means and other machine learning methods have been used to identify the modulation formats. However, the computation-extensive iterative operations hinder their practical implementation. In [4], we have proposed a novel blind MFR method based on image processing technique, in which connected component analysis (CCA) has been used to count the number of constellation points in the converted binary graph from the received symbols in Stokes space. It provides a one-step recognition of modulation formats other than iterative approaches. In this paper, we propose an enhanced blind MFR method based on our previous work. We show that the requirements for both optical signal-to-noise ratio (OSNR) and number of points can be significantly reduced, with only slight additional simple logic operations, making the MFR algorithm more robust for practical use.

II. PRINCIPLES

At the coherent receiver, it is common to perform analog-to-digital conversion, chromatic dispersion compensation and timing recovery, before handling the modulation format. After these pre-processing, in our proposed blind modulation recognition scheme, the dual polarization signal is first converted to the Stokes space, by [3]

$$[s_1, s_2, s_3] = \left[|X|^2 - |Y|^2, 2\,\mathrm{Re}\{X \cdot Y^*\}, 2\,\mathrm{Im}\{X \cdot Y^*\} \right] \qquad (1)$$

where X and Y represent the two orthogonal polarizations, while $\mathrm{Re}\{\cdot\}$ and $\mathrm{Im}\{\cdot\}$ stand for the real and the imaginary parts of a complex number, respectively. It can be easily seen that the absolute phase information has been removed in the transformation, i.e., the laser frequency offset and phase noise have no effect in the Stokes space. Another information that can be concluded from (1) is that phase-shift-keying (PSK) signals are distributed in the s_2-s_3 plane only as $s_1=0$, while the distribution of quadrature-amplitude-modulation (QAM) signals like 8QAM and 16QAM in Stokes space are more complicated and distributed in the planes that are parallel to s_0-plane, as depicted in Fig. 1.

Then, similar processes are performed as those in our previous work. First, the random-walked polarization should be tracked, which needs only several hundreds of points as demonstrated in [5]. Secondly, the points after normalization satisfying the constraints, as listed in Table I, are projected onto the s_2-s_3 plane. Then a density filter based on Voronoi density estimation is applied to remove the points with normalized density smaller than the threshold Th. After these DSP procedures, we convert the survived points to a binary graph, with a pre-defined resolution N, after which an averaging filter is used to smooth the image. Finally, the connected component analysis is employed to count the number of constellation points. It traverses each pixel with value "1" in the graph and checks whether it is connected to any of the surrounding 8 pixels. After labelling all the subsets, the number of subsets is obtained, obviously. The modulation format recognition is achieved based on the number of subsets. The detailed algorithm description could be found in [4].

528

OECC/PS2016

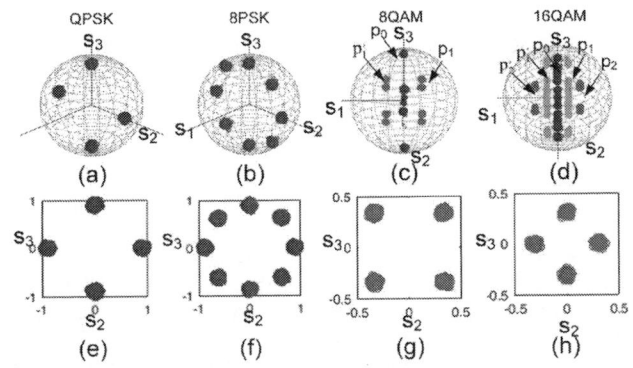

TABLE I
CONSTRAINS OF POINTS FOR PROJECTION
OF DIFFERENT MODULATION FORMATS IN
STOKES SPACE

Modulation Formats	Constrains of Points for Projection
QPSK	$|s_1|<0.1$
8PSK	$|s_1|<0.1$
8-QAM	$0.3<|s_1|<0.4$
16-QAM	$|s_1|>0.4$

Fig. 1 Signal representations in Stokes space for (a) QPSK, (b) 8PSK, (c) 8QAM. (d) 16QAM (e)-(h) are the projections on the plane of s_2-s_3 plane.

Now, we investigate how to alleviate the requirement of OSNR for the CCA-based MFR method. In principle, the parameter, which is the most relevant to OSNR sensitivity, is the threshold *Th* in the density filter. It is intuitive that increasing *Th* results in increasing the OSNR sensitivity, as it removes more data samples which are corrupted by the noise. Fig 2(b) shows the converted binary graph from a set of PM-8PSK signal (SNR =12 dB) with the threshold equal to 0.65 in the density filter. As the noise level is high, all the subsets in the converted binary graph mix together and CCA could not figure out the correct number of subsets. However, simply increasing *Th* leads to less survived points after the density filtering, which probably makes the subsequent CCA algorithm fail. In Fig.2 (c), the threshold in the density filter is set to be 0.8. It can be seen that even though the subsets can be distinguished now, there are two subsets missing, such that the output of CCA is still not correct. Fortunately, we can take advantage of the symmetry property of the patterns of these modulation formats in the binary graph, as shown in Fig. 1, and use simple logic AND between the original binary graph and the rotated graph to make up for the missing subsets. It is worth noting that the logic AND is employed in the binary graph, which is quite computation efficient in the DSP circuit. The additional logic operations are nearly negligible compared with the processing of complex numbers in other parts of the algorithm. Here ,we name the modified CCA-based algorithm as CCA-QR, in which QR is the short form for quadruple rotation, as in

$$I = I \& rot_{\pi/2}(I) \& rot_{\pi}(I) \& rot_{-\pi/2}(I) \qquad (2)$$

where I is the 0-1 matrix of the binary graph, & is the AND operation, and rot_θ is the function that rotates the matrix by an angle of θ. Fig. 2(d) shows the binary image after the QR process of Fig. 2(c). The output of CCA is correct now for recognition.

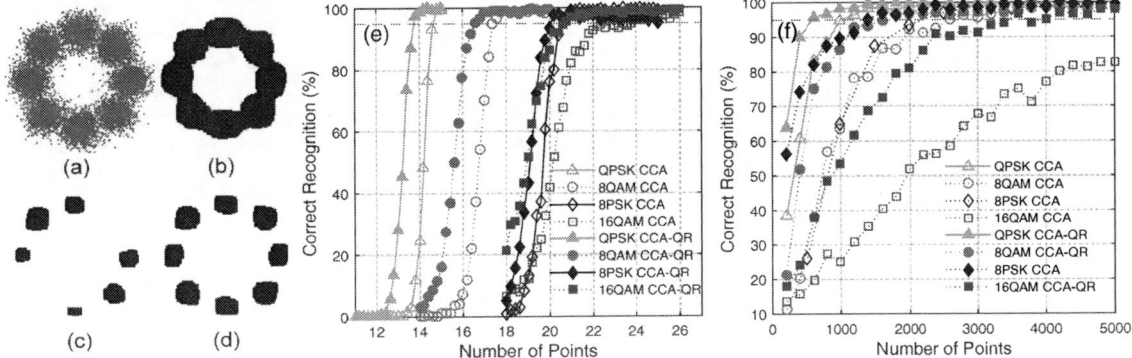

Fig. 2 (a) Constellation of 8PSK signal, SNR = 12; (b) Converted binary image, *Th* = 0.65; (c) Converted binary image, *Th* = 0.8; (d) Converted binary image, *Th* = 0.8, with quadruple rotation; (e) & (f) Simulation results of the correct recognition rate under different OSNR values and different number of points.

III. NUMERICAL SIMULATION

We have performed numerical simulations for the proposed MFR scheme. 32-Gbaud PM-QPSK, PM-8QAM, PM-8PSK and PM-16QAM signal were generated and passed through a channel with additive Gaussian white noise. The converted binary image had a size of 100*100 pixels (N = 100). The threshold for the Voronoi filtering was set to be 0.65 and 0.8, respectively, corresponding to our previous CCA-based method and the newly proposed CCA-QR method. First, the OSNR value was varied from 10 dB to 26 dB, with a step of 0.2 dB, each having 500 independent implementations. The correct recognition rate was then calculated. Fig. 2(e) and (f) show the correct recognition rate under different OSNR values, and different number of points involved in the MFR. The OSNR, in the latter case, was set to be 15 dB, 18 dB, 22 dB, and 24 dB for QPSK, 8QAM, 8PSK and 16QAM, respectively. The increase of OSNR sensitivity was clearly seen in Fig. 2(e), i.e., ~1 dB for QPSK, 8PSK and 8QAM, and ~2 dB for 16QAM. Meanwhile,

529

for each tested modulation formats, the number of required points to successfully recognize the modulation formats was reduced significantly, as depicted in Fig. 2(f). The case of 16QAM showed the most significant improvement, which only about half of the points (5000) are needed to reach a successful rate of 95%, compared with the previous scheme (~10000).

IV. EXPERIMENTAL RESULTS

The experimental setup and configurations were the same as those in [4]. 32-Gbaud polarization-multiplexed QPSK and 16QAM signals were transmitted in a general coherent optical communication testbed. Fig. 3(a) shows the experimental setup and the major DSP processes. CCA-based and CCA-QR-based MFR methods were employed to recognize the modulation formats of the received signal, with threshold Th = 0.65 and 0.8, respectively. Fig 3 (b) shows the back-to-back bit error rate (BER) performance. The orange and yellow color blocks represented the successful recognition range of the CCA and CCA-QR based MFR method. The OSNR sensitivity was increased by 1.2 dB and 4.3 dB for QPSK and 16QAM, respectively. Fig. 3(c)-(n) show the converted binary graphs of the CCA and CCA-QR based method. The performance improvement of CCA-QR was significant, as shown in Fig. 3(c) & (i), and (f) & (l).

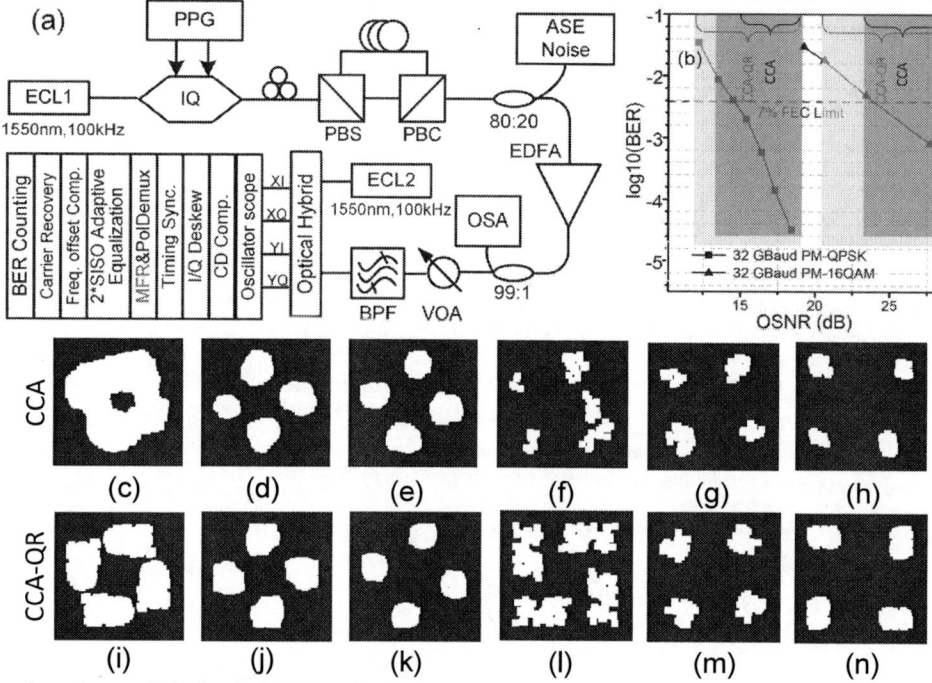

Fig. 3 (a) Experimental setup; (b) back-to-back BER result; (c)-(e) converted binary graph of QPSK signals using CCA, OSNR = 12.2 dB, 13.5 dB, and 14.5 dB, respectively; (f)-(h) converted binary graph of 16QAM signals using CCA, OSNR = 20.6 dB, 23.4 dB and 27.7 dB, respectively; (i)-(n): the corresponding CCA-QR results.

V. SUMMARY

We have proposed a modified modulation format recognition method based on image processing techniques. It shows that, via both simulation and experiment, about 1~3 dB OSNR sensitivity is improved, for coherent polarization multiplexed PSK and QAM signals, with requiring slight additional simple logic operations.

REFERENCES

[1] J. Liu, Z. Dong, K. Zhong, A. P. T. Lau, C. Lu, and Y. Lu, "Modulation format identification based on received signal power distributions for digital coherent receivers," in *Optical Fiber Communications Conference (OFC)*, San Francisco, California, USA, 2014.

[2] N. G. Gonzalez, D. Zibar, and I. T. Monroy, "Cognitive digital receiver for burst mode phase modulated radio over fiber links," in *Proc. ECOC*, 2010.

[3] R. Borkowski, D. Zibar, A. Caballero, V. Arlunno, and I. T. Monroy, "Stokes space-based optical modulation format recognition for digital coherent receivers," *IEEE Photonics Technology Letters*, vol. 25, pp. 2129-2132, Nov 2013.

[4] T. Bo, J. Tang, and C.-K. Chan, "Blind modulation format recognition for software-defined optical networks using image processing techniques " in *Optical Fiber Communications Conference (OFC)*, Anaheim, California, USA, 2016

[5] B. Szafraniec, B. Nebendahl, and T. Marshall, "Polarization demultiplexing in Stokes space," *Optics Express*, vol. 18, pp. 17928-17939, 2010

Broadband amplification characteristics on the cascade Bi-doped and Er-doped optical fiber amplifiers

Mikoto Takahashi[1], Daiki Higuchi[1], Mizuki Ohara[1], Yusuke Fujii[2], Naoto Yoshimoto[1] and Soichi Kobayashi[1]

[1] Chitose Institute of Science and Technology758-65 Bibi, Chitose, Hokkaido,Japan
[2] Photonic Science Technology, Inc.776-16 Kitashinano, Chitose, Hokkaido, Japan
e-mail : s-koba@photon.chitose.ac.jp

Abstract : Bismuth-doped fibers fabricated by the VAD method showed the wide fluorescent spectra in 950 nm-1480 nm by the 808 nm LD pumping. Fluorescence properties by 808nm light pumping in the cascaded BDF-EDF system showed the broadband spectra in 950 nm -1650 nm. In the BDF-EDF system, when the pumping light is launched into the BDF the total gain of 2.34dB was obtained and when the pumping light is launched into the EDF the total gain of 2.67 dB was obtained.

Keywords : Bi-doped fiber , Er-doped fiber, cascade BDF-EDF system, VAD method,

I. INTRODUCTION

Optical fiber amplifiers based on rare-earth-doped fibers play an important role in the optical transmission systems because they can compensate the insertion loss of the optical fibers in transmission lines and the loss of optical components used for optical nodes. The research on expanding transmission band was started immediately after developing C-band Er-doped fiber amplifier (EDFA) and Pr^{3+} doped fiber amplifiers [1]. 1450 nm or 1650 nm band with Tm^{3+} doped fiber amplifiers have been developed [2]. 1490 nm band optical signals are used in communication networks connecting telephone companies to subscribers. A 1310 nm band optical signal, however, is used for communication from subscribers to the telephone companies in Japan. For conventional subscriber lines, an average distance between subscribers and the telephone companies has typically been less than 10 km. The recent spread of optical communication networks into rural areas, however, has required optical fibers more than 20 km or more. Furthermore, the increase of subscriber networks is expected from the request of the high-speed delivery of video in urban areas. 1310 nm optical amplifiers are attractive for the use of future upstream long-distance access lines in telecommunication networks [3]. Bismuth-doped silica glass was proposed as an optical amplification and a laser oscillation at 1310 nm [4]. Bismuth-doped fiber lasers and amplifiers fabricated by the modified chemical vapor deposition (MCVD) were discussed by Dianov group [5, 6]. In this report the amplification in 1310 nm band based on bismuth-doped silica fiber (BDF) made by the vapor-phase axial deposition (VAD) method is presented and the cascade amplification with the BDF-EDF system for long distance optical access lines are proposed and gain characteristics of the systems are discussed [7, 8].

II. BISMUTH-DOPED & ERBIUM-DOPED FIBERS

The BDF is fabricated by the VAD method. The BDF consists of Si, Ge, Al, and Bi oxides. The core diameter is either 4 or 10 µm and the cladding diameter is 125 µm. The concentration of Bi in the core glass was measured to be 0.07~1 mol % by energy dispersive x-ray spectroscopy (EDX). The relative refractive index difference between the core and the cladding of the preform was measured as 0.5~0.95 % using an optical fiber preform analyzer. The fluorescence spectra of BDF pumped by an 808 nm laser diode (LD) with a pumping input power between 18 dBm and 32 dBm are shown in Fig. 1. The spectra showed broad band gain between 950 and 1500 nm without the dip because of OH ion absorption at 1390 nm. EDF is a commercially available rare-earth doped fiber (THORLABS, ER110-4/125). Mode field diameter of the fiber is 6.5 µm and suitable for both 980 nm and 1480 nm pumping. Both BDF and EDF are good spliceability to standard single mode fibers.

III. FLUORESCENCE SPECTRA OF THE CASCADE BDF-EDF SYSTEM

The cascaded BDF-EDF consists of the BDF which has core diameter of 10 μm and the fiber length of 30 cm and the EDF which has core diameter of 4 μm and fiber length of 3.5 m. Both fibers have the same clad diameters of 125μm and are fusion spliced in between. Fig .2 shows the fluorescence spectra of the cascade BDF-EDF. A pumpingwith 808 nm LD was performed with the measurement set-up as shown in Fig.3(a) without signal LD. It confirms that the gain spectra can be widely enlarged from the band in 950 nm-1480 nm to the 950 nm -1650 nm. Pumping power of 808nm LD are shown as inset right above in Fig2.

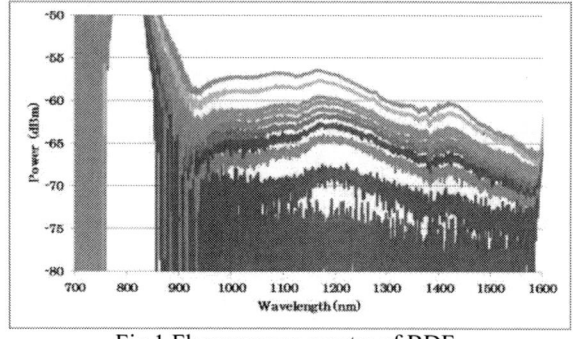
Fig.1 Fluorescence spectra of BDF.

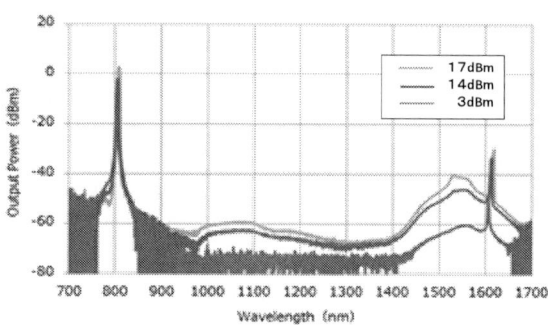
Fig.2 Fluorescence spectra of cascade BDF-EDF.

IV. GAIN CHARACTERISTICS ON THE CASCADE BDF-EDF SYSTEMS

Fig.3 (a) and (b) show the experimental setups for gain measurements of the cascade BDF-EDF system. Signal LDs are 1310nm and 1550nm LDs in CW operation. In Fig.3 (a) the signal is launched into the BDF and goes through the EDF. The pumping LD output is launched into the singlemode fiber and has the lasing wavelength of 808 nm. Amplified 1310nm signal is measured with the optical spectrum analyzer (OSA) as shown in Fig.4. The amplified gain was 2.34 dB measured with pumping power of 17 dBm. In Fig.3 (b) the signal is launched into the EDF and goes through the BDF. Amplified 1490 nm signal is measured with OSA as shown in Fig.5. The amplified gain was 2.67 dB measured with pumping power of 17 dBm. In fig.4 and 5 observed peaks at 1616 nm don't show the real signals but the ghost signals of 808 nm due to the OSA mechanism. Red peak values show the amplified signal outputs with 808 nm LD pumping and blue peaks show the input signals without pumping at 1310nm and 1490 nm in Fig.4 and Fig.5, respectively.

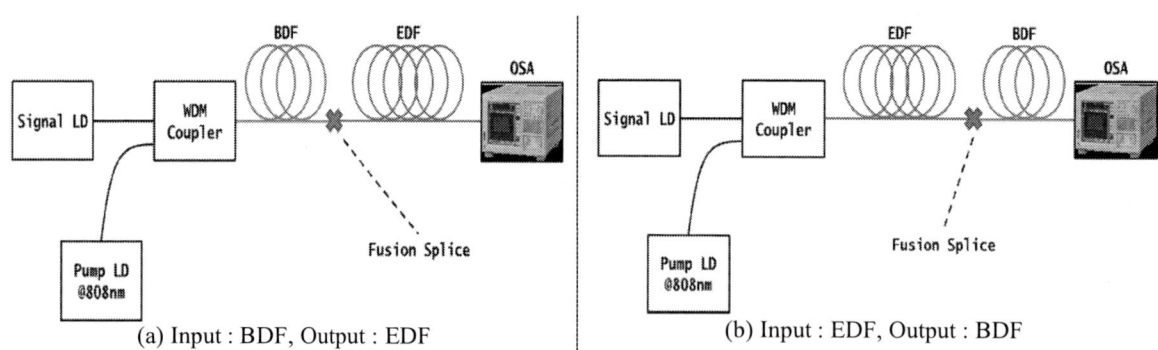

(a) Input : BDF, Output : EDF

(b) Input : EDF, Output : BDF

Fig3 Measurement setup for cascade BDF-EDF system

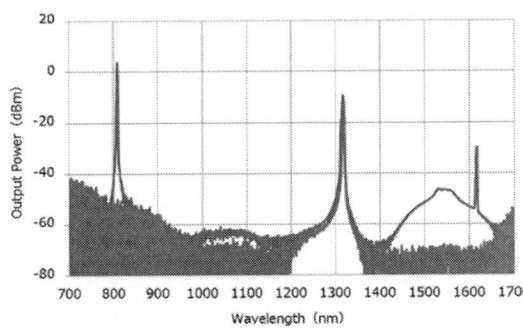

Fig.4 1310 nm signal amplification
Input : BDF, Output : EDF
Red: amplified signal, Blue: input signal

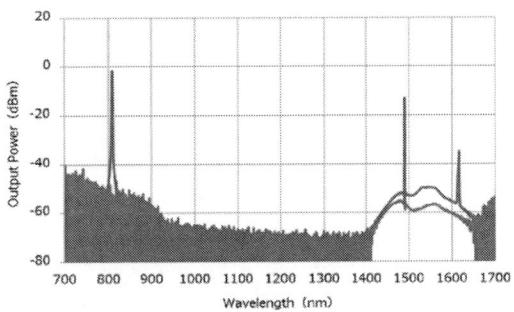

Fig.5 1490 nm signal amplification
Input : EDF, Output : BDF
Red: amplified signal, Blue: input signal

V. CONCLUSION

Bismuth-doped fibers fabricated by the VAD method showed the wide spread fluorescent spectra in 950 nm-1480 nm without OH absorption by the 808 nm LD pumping. Fluorescence properties by 808nm light pumping into the cascaded BDF-EDF system showed the more broadband spectra in 950 nm -1650 nm compared to the BDF spectra. In the BDF-EDF system, when the pumping light is launched into the BDF the total gain of 2.34dB was obtained. On the other hand when the pumping light is launched into the EDF the total gain of 2.67 dB was obtained. From the result it is shown that the BDF-EDF system has a potential as bi-directional optical amplifiers applied to the in-line amplifiers in rural areas and to the booster amplifiers in the urban areas.

ACKNOWLEDVGMENT

This research work was funded by the National Institute of Information and Communications Technology in Japan.

REFERENCES

[1] Ohishi, T. Kanamori, T. Kitagawa, S. Takahashi, E. Snitzer, and G. H. Sigel, Jr., "Pr3+-doped fluoride fiber amplifier operating at 1.31µm;" Opt. Lett., vol. 16, no. 22, 1747–1749, (1991).

[2] T. komukai, Y. Yamamoto, T. Sugawa and Y. Miyajima, "Upconversion pumped thulium-doped fluoride fiber amplifier and laser operating at 1.47 µm," IEEE J. Quantum Electron., Vol. 31, No. 11, pp. 1880-1889 (1995).

[3] K-I. Suzuki, Y. Fukada, N. Yoshimoto, K. Kumozaki and M. Tsubokawa "Automatic Level Controlled Burst-Mode Optical Fiber Amplifier for 10 Gbit/s PON Application," Optical Fiber Communication Conference and Exposition (OFC2010), OTuH1 (2010).

[4] Yasushi Fijimoto and Masahiro Nakatsuka, "Infrared Luminescence from Bismuth-Doped Silica Glass," Jpn. Appl. Phys. Vol.40, L279-L281(2001).

[5] V.V.Dvoyrin, V.M.Mashinsky, L.I.Bulatov, I.A.Bufetov, A.V.Shubin, A.Melkumov,.F.Kustov and E.M.Dianov, "Bismuth-doped-glass optical fibers-a new active medium for lasers and amplification", OPTICS LETTERS / Vol.31, No.20 / October 15, (2006).

[6] E. M. Dianov, V. V. Dvoryn, V. M. Mashinsky, A. A. Umnikov, M. V. Yashkov, and A. N. Guryanov, "CW bismuth fiber laser," Quantum Electron. Vol.35, 1082-1084, (2005).

[7] M. Takahashi, T. Fujii, Y. Saito, Y. Fujii, and S. Kobayashi, "Optical Amplification at 1.3 µm with Bi Doped Fiber Fabricated by VAD Method," CLEO-OECC Conference, Kyoto, Japan, TuPS-9 (2013).

[8] M. Takahashi, T. Fujii, Y. Saito, Y. Fujii, and S. Kobayashi, "1.3µm optical amplification with double-clad Bi doped silica fiber," SPIE Photonics West 2015, San Francisco, USA, 9344-88 (2015).

TuC3-2

OECC/PS2016

All-optical Dynamic Gain Control of Remotely Pumped Erbium-doped Fiber Amplifier

Kokoro Kitamura, Kenta Udagawa, and Hiroji Masuda

Interdisciplinary Graduate School of Science and Engineering. Shimane University, 1060 Nishi-Kawatsu, Matsue, Shimane, Japan
kitamura@ecs.shimane-u.ac.jp

Abstract: *We propose a novel all-optical automatic-gain-control scheme for a remotely-pumped erbium-doped fiber amplifier. Dynamic gain control has been experimentally achieved with a significant reduction of gain excursions from 2.1 to ~0.2 dB using the scheme.*
Keywords: *Automatic gain control, Remotely pumped erbium-doped fiber amplifier*

I. INTRODUCTION

Distributed optical amplifiers, which are the distributed Raman amplifier (DRA) and the remotely pumped erbium-doped fiber amplifier (RP-EDFA), are able to significantly enhance the transmission distances/capacities in ultra-long haul transmission systems (repeatered or unrepeatered systems) compared with conventional lumped EDFAs [1-4]. This is because the distributed optical amplifiers are able to achieve higher optical signal-to-noise ratios (OSNRs) of more than ~5 dB over the lumped EDFAs. It has been shown that a transmission system using an RP-EDFA achieves more than ~4 times higher pumping efficiencies than the one using a DRA at the designed OSNR improvements of ~5 to 6 dB [5-7]. The high pumping efficiencies are crucial for conducting multi-core fiber transmission experiments using remotely pumped multi-core EDFAs [5-8]. Moreover, the automatic gain control (AGC) function is indispensable in the wavelength routing/switching photonic networks of the future [9, 10]. We proposed an all-optical AGC (AO-AGC) scheme for the RP-EDFA, and clarified the AGC characteristics in a static condition, where the total input power for wavelength-division multiplexing (WDM) signal lights was slowly changed manually [11]. In this paper, we report an experimental study on the dynamic characteristics of the proposed AO-AGC scheme. An RP-EDFA module with a novel configuration is proposed and placed between the two spans of transmission fibers of the RP-EDFA transmission system. Using the RP-EDFA module, a dynamic AO-AGC operation with gain excursions of less than ~0.2 dB has been successfully achieved for the first time, to our knowledge.

II. EXPERIMENTAL CONFIGURATION

Figure 1 shows the experimental configuration of the RP-EDFA with the AO-AGC scheme. Single core fibers were used in the experiment for simplicity. The transmission line consisted of a variable optical attenuator (VOA-1), two transmission fibers (Fiber-1 and -2), and an RP-EDFA module placed between Fiber-1 and -2. The lengths of Fiber-1 and -2 were both 30 km. VOA-1 was employed instead of a transmission fiber for experimental convenience, and the transmission span loss (L_{span}) was modified by the attenuation of VOA-1. The RP-EDFA module consisted of a fiber loop with an RP-EDF, wavelength selective couplers (WSCs), an optical isolator, and another variable optical attenuator (VOA-2). Each WSC coupled/divided the signal, pump, and laser lights; and the insertion losses for the three lights were ~0.8 dB. The low loss transmission band of the WSCs for the laser light was ~1559 nm, and the transmission bands for the signal and pump lights ranged from ~1450 to ~1558 nm. The wavelength of the laser light in the fiber ring was 1559 nm, which had a long wavelength edge in the C-band. The gain of the RP-EDF was determined using VOA-2 to obtain the flat gain spectrum across the C-band. The net flat gain of the RP-EDFA module was ~14 dB.

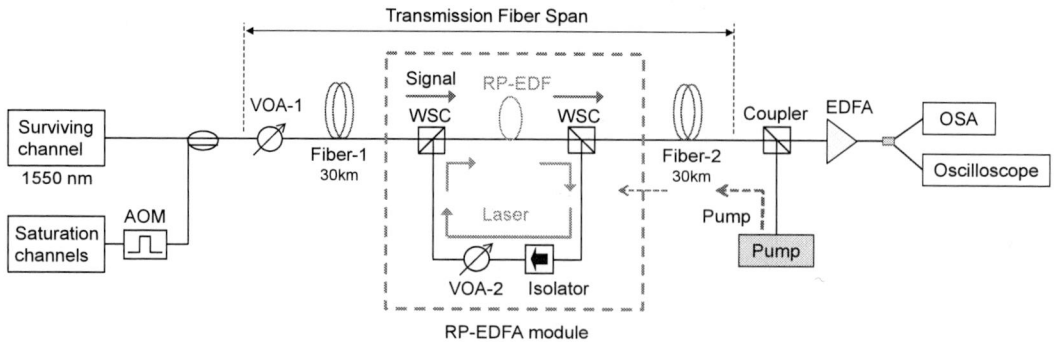

Fig. 1. Experimental configuration of the RP-EDFA system.

Acknowledgement: Part of this research use results from research commissioned by the National Institute of Information and Communications Technology (NICT) entitled "Research on innovative infrastructure of optical communications."

We assumed a 40 channel WDM system, in which a signal power (P_{sin}) of 1 mW/ch was launched into the transmission line. The WDM signal light consisted of a surviving channel (wavelength = 1550.0 nm) and the other channels, which were simulated by 4 saturation channels (wavelengths = 1532.7, 1539.0, 1546.9, and 1552.5 nm). The saturation channels were modulated by an acousto-optic modulator (AOM) to evaluate the dynamic AO-AGC characteristics. The pump light was launched into Fiber-2 using a coupler. The wavelength of the pump light was 1490 nm. The pump light power of the system with the AGC scheme nearly equaled that of the system without the AGC scheme at full WDM loading. The power penalty to employ the AGC scheme was the only loss in the WSC for the pump light (~0.8 dB). When the number of WDM channels decreased, the laser light power increased automatically so that the EDF gain was kept constant.

III. EXPERIMENTAL RESULTS

First, we evaluated the characteristics of the OSNR of the RP-EDFA system ($OSNR_s$). Figures 2 (a) and (b) show the measured OSNR spectra of the RP-EDFA systems at an L_{span} of 25.4 and 28.2 dB, respectively. In both cases, we measured the OSNRs of the systems with and without the RP-EDFA scheme ($OSNR_w$ and $OSNR_{wo}$) and calculated the OSNR improvements ($\Delta OSNR$): $\Delta OSNR = OSNR_w - OSNR_{wo}$. Here, the measuring conditions were P_{sin} = 0 dBm and a noise bandwidth = 0.1 nm. $OSNR_w$ ($OSNR_{wo}$) ranged from 30.9 to 32.7 dB (from 26.2 to 27.6 dB) at an L_{span} of 25.4 dB, and $\Delta OSNR$ ranged from 4.7 to 5.1 dB. On the other hand, $OSNR_w$ ($OSNR_{wo}$) ranged from 28.1 to 29.8 dB (from 23.3 to 24.6 dB) at an L_{span} of 28.2 dB, and $\Delta OSNR$ ranged from 4.7 to 5.3 dB. Therefore, OSNR improvements of ~5 dB were achieved by employing the RP-EDFA scheme in the C-band in both cases.

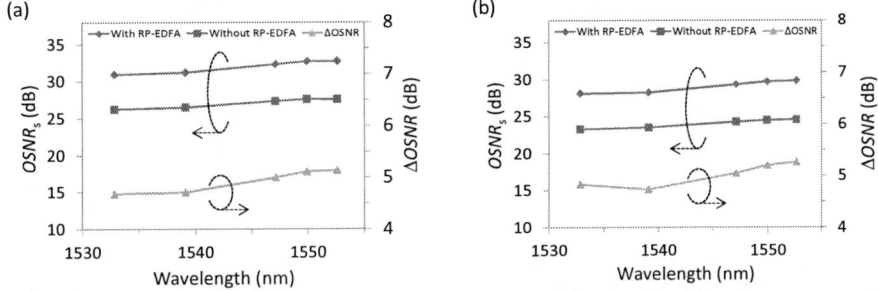

Fig. 2. OSNR characteristics of the RP-EDFA systems in the cases of (a) L_{span} = 25.4 dB and (b) L_{span} = 28.2 dB.

Next, the static gain characteristics of the AGC scheme were investigated. We measured the RP-EDFA module gains (G_s) with no use of Fiber-1 and -2 for simplicity in a static condition. The wavelength of the pump light was ~1475 nm in this measurement. Figure 3 shows the dependences of the gain difference (ΔG_s) on the total signal power (P_{sin}) and the pump light power launched into the RP-EDF (P_p). ΔG_s was defined as the difference from the gain at P_{sin} = 0.039 mW (G_{s0}): $\Delta G_s = G_s - G_{s0}$. Figures 3 (a) and (b) show the results at the signal wavelengths of 1550.0 and 1531.0 nm, respectively. In both cases, the absolute values of ΔG_s ($|\Delta G_s|$) increased with P_{sin}, and decreased with the increase of P_p. $|\Delta G_s|$ were ~2 times smaller at 1550.0 nm than at 1531.0 nm. This behavior can be explained by the spectral hole burning effect, which shows significant influence near the gain peak wavelength ~1532 nm in the C-band EDFA [11].

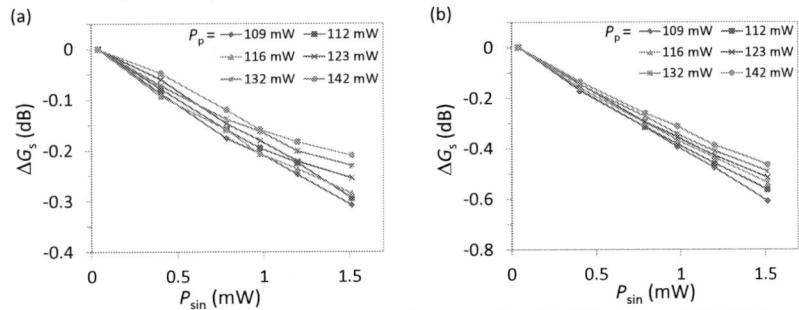

Fig. 3. Gain difference characteristics at the signal wavelengths of (a) 1550.0 nm and (b) 1531.0 nm.

Finally, we evaluated the dynamic characteristics of the AGC scheme. Considering the spectral hole burning effect shown in Fig. 3, the wavelength of the surviving channel was set at 1550.0 nm, which was a typical wavelength in the C-band. Figures 4 (a) and (b) show the transient gain excursion characteristics of the surviving channel at an L_{span} of 25.4 and 28.2 dB, respectively. L_{span} of 25.4 and 28.2 dB correspond to transmission fiber lengths of 134 and 149 km, respectively, if we assume the fiber loss coefficient at 1550.0 nm is 0.19 dB/km. Figures 4 (a) and (b) show the gain differences ($\Delta G_d(t)$) of the RP-EDFA as a function of time t with and without AGC scheme. The saturation channels were dropped and added at the time of 1.3 and 3.8 ms, respectively. $\Delta G_d(t)$ was defined as the gain at t ($G_d(t)$) minus the gain in the steady state in the case of full WDM loading (G_{d0}): $\Delta G_d(t) = G_d(t) - G_{d0}$. The power of the surviving channel launched into the RP-EDFA module was -17.7 dBm for both L_{span} for simplicity. The pump light powers launched into Fiber-2 were 170 and 315 mW at L_{span} of 25.4 and 28.2 dB, respectively. Raman gains in Fiber-2 were 2.3 and 4.3 dB at L_{span} of 25.4 and 28.2 dB, respectively.

$\Delta G_d(t)$ increased (decreased) in the case of dropping (adding) channels without the AGC scheme due to the gain saturation effect. The maximum absolute value of $\Delta G_d(t)$ (ΔG_{max}) was 2.1 dB at an L_{span} of 25.4 dB. On the other hand, the gain difference was significantly suppressed with the AGC scheme so that ΔG_{max} was ~0.22 dB. The steady state gain difference corresponds to the gain difference in the static characteristics of the PR-EDFA module shown in Fig. 3(a). The same tendency is seen in Fig. 4(b) at an L_{span} of 28.2 dB. ΔG_{max} with and without the AGC scheme were 1.7 and ~0.14 dB, respectively. The gain excursions were reduced by ~10 and ~12 times in dB at an L_{span} of 25.4 and 28.2 dB, respectively. Taking into account measurement accuracies, the reduction of the gain excursion was factor of more than ~10 in dB. It is considered that the gain difference decreases with the increases of L_{span} due to the decreased total signal power launched into the RP-EDFA module.

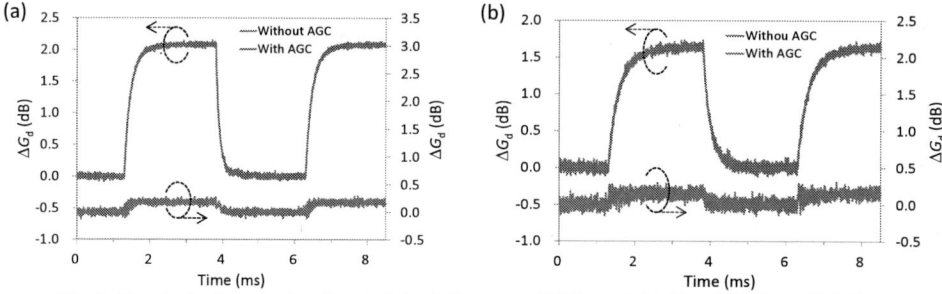

Fig. 4. Transient gain excursion characteristics in the cases of (a) L_{span} = 25.4 dB and (b) L_{span} = 28.2 dB.

IV. CONCLUSIONS

We have proposed a novel AO-AGC scheme for a RP-EDFA. Dynamic characteristics of the scheme for the RP-EDFA transmission system have been experimentally clarified. Employing the RP-EDFA, ONSR improvements of ~5 dB have been achieved across the C-band. Using the RP-EDFA module, the dynamic AO-AGC operation with gain excursions less than ~0.2 dB has been successfully achieved. The residual gain difference is considered to originate with the spectral hole burning effect. The gain excursion was reduced by ~10 times in dB (from 2.1 to ~0.2 dB) using the AO-AGC scheme.

REFERENCES

[1] H. Masuda, M. Tomizawa, Y. Miyamoto, and K. Hagimoto, "High-performance distributed Raman amplification systems with limited pump power," IEICE Trans. Commun., Vol. E89-B, No. 3, pp. 715-723, 2006.

[2] H. Takara, H. Ono, Y. Abe, H. Masuda, K. Takenaga, S. Matsuo, H. Kubota, K. Shibahara, T. Kobayashi, and Y. Miaymoto, "1000-km 7-core fiber transmission of 10 x 96-Gb/s PDM-16QAM using Raman amplification with 6.5 W per fiber," Optics Express, Vol. 20, No. 9, pp. 10100-10105, 2012.

[3] H. Kidorf, M. Nissov, and D. Foursa, "Ultra-long haul submarine and terrestrial applications," Chap. 17, in M. N. Islam ed., *Raman Amplifiers for Telecommunications 2*, Springer, 2004.

[4] H. Masuda, H. Kawakami, S. Kuwahara, A. Hirano, K. Sato, and Y. Miyamoto, "1.28-Tbit/s (32 x 43 Gbit/s) field trial over 528-km (6 x 88 km) DSF using L-band remotely-pumped EDF/distributed Raman hybrid inline-amplifiers," Elecron. Lett., Vol. 39, No. 23, pp. 1668-1670, 2003.

[5] H. Masuda, H. Nagaoka, and K. Tayama, "Optical SNR Characteristics of a Space-division-multiplexed Optically-amplified Transmission System using Remotely Pumped Multi-core EDFAs," OptoElectron. Communications Conf., Paper 6C2–4, 2011.

[6] Y. Yamauchi, H. Masuda, T. Nobukawa, and H. Nagaoka, "A Highly-Efficient Remotely-Pumped Multi-core EDFA Transmission System with a Novel Hybrid Wavelength-/Space-division Multiplexing Scheme," OptoElectron. Communications Conf., Paper 5C1–4, 2012.

[7] H. Masuda, H. Ono, H. Takara, Y. Miyamoto, K. Ichii, K. Takenaga, S. Matsuo, K. Kitamura, Y. Abe, and M. Yamada, "Remotely pumped multicore erbium-doped fiber amplifier system with high pumping efficiency," SUM 2013, IEEE Photonics Society Summer Topical Meeting on SDM for Optical Comm., WC3.3, 2013.

[8] H. Takara, T. Mizuno, H. Kawakami, Y. Miyamoto, H. Masuda, K. Kitamura, H. Ono, S. Asakawa, Y. Amma, K. Hirakawa, S. Matsuo, K. Tsujikawa, and M. Yamada, "120.7-Tb/s (7 SDM/180 WDM/95.8 Gb/s) MCF-ROPA Unrepeatered Transmission of PDM-32QAM Channels over 204 km," IEEE J. Lightwave Technol., Vol. 33, No. 7, pp. 1473-1478, 2015.

[9] K. Kitamura, H. Masuda, K. Tayama, and K. Ohnishi, "Novel all-optical feedforward automatic gain control scheme for multicore erbium-doped fiber amplifiers," OECC/ACOFT 2014, TU5C-2, pp. 310-311, 2014.

[10] K. Kitamura, H. Udagawa, and H. Masuda, "Dynamic characteristics of all-optical feedforward fast automatic gain control scheme for multicore erbium-doped fiber amplifiers," 20th Microoptics Conference (MOC '15), H65, pp. 226-227, 2015.

[11] K. Kitamura, H. Tanaka, K. Tayama, and H. Masuda, "All-optical gain control scheme for a remotely pumped multicore erbium-doped fiber amplifier," 18th Microoptics Conference (MOC '13), H21, 2013.

OECC/PS2016

Phase Adjustment by Wavelength Tuning of Parametric Pump on Raman Assisted Phase Sensitive Amplifier

Y. Cao[1], F. Alishahi[1], Y. Akasaka[2], M. Ziyadi[1], A. Mohajerin-Ariaei[1], A. Almaiman[1], T. Ikeuchi[2], S. Takasaka[3], R. Sugizaki[3], A. E. Willner[1]

1) University of Southern California, 3740 McClintock Ave, Los Angeles, CA 90089, USA
2) Fujitsu Laboratories of America, 2801 Telecom Parkway, Richardson, TX 75082, USA
3) Furukawa Electric Co., Ltd., 6 Yawatakaigandori, Ichihara, Chiba, 290-8555 Japan
yinwenca@usc.edu

Abstract: A Raman-assisted PSA scheme is proposed with the phase matching condition achieved by PSA pump wavelength tuning. More than 10dB signal gain and 2dB sensitivity improvement are experimentally demonstrated by a typical 20G-baud QPSK system.
Keywords: (290.5910) Scattering, stimulated Raman; (060.4370) Nonlinear optics, fibers.

I. INTRODUCTION

Phase sensitive amplifier (PSA) has been investigated as a key device for the next generation optical communication systems [1-3]. For the realization of phase sensitive amplification, it is required that phase relation between signal, idler and pump would be properly selected and kept stable [4]. Phase lock loop (PLL) was actively studied in the early stage of PSA research and recently new scheme using wavelength selected switch was demonstrated [5, 6]. Those techniques are effectively phase locked and they have own advantages and disadvantages at the same time. For the future practical usage of PSA, there are several things we may be able to improve. Among them, Raman assisted scheme has demonstrated its stablity without optical path separation [6]. For the further improvement of PSA performance, reduction of components inside PSA is desired. By using wavelength selective switch (WSS) in the scheme and tuning the power and phase difference between pump/signal/idler simultaneously, PSA performance would be optimized. However, if the gain profile of Raman amplifier could be used for power alignment, its positive adjustment against WSS's attenuation-based power adjustment would be effectively useful. Moreover, if phase adjustment could be realized without path separation or using WSS, it is prefered because of less attenuation inside PSA. In this study, we demonstrated positive power adjustment with Raman amplifier and phase adjustment with PSA pump power wavelength tuning for better PSA performance. More than 10dB signal net gain and 2dB sensitivity improvement are experimentally verified with a typical 20G-baud QPSK system.

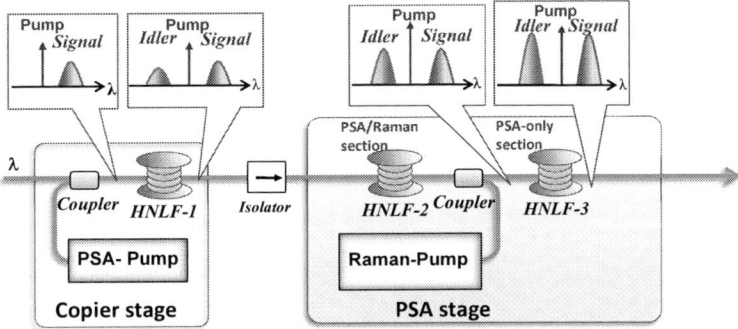

Fig. 1. Conceptual diagram of Raman-assisted PSA with phase matching by tuning the wavelength of PSA pump. Copier stage generates a conjugate signal copier (idler); in the PSA stage, the signal and idler are amplified by distributed interaction from Raman scattering and PSA to produce similar power level while the final part is the pure PSA section to have strengthen PSA gain. HNLF: highly nonlinear fiber.

The concept of the proposed Raman assisted PSA is shown in Fig. 1. First, a conjugated copy of the input signal is generated by launching a PSA pump and the signal in a highly nonlinear fiber (HNLF). In order to suppress the correlated noise generated by optical parametric amplification (OPA), the pump power is controlled to have moderate conversion efficiency (~-10dB). It should be noted that the wavelength of the PSA pump should be carefully selected for two reasons: 1) to compensate 10dB power difference between the signal and idler before the final PSA-only section, which is achieved by adjusting the PSA-pump wavelength (therefore the wavelength of the idler is also changed) to introduce more gain on the idler than signal in the following PSA/Raman section; 2) to ensure the in-phase constructive addition in PSA by changing the relative phase between signal and idler through keeping appropriate wavelength

537

difference. Next, the back-propagated Raman pump amplifies signal, PSA-pump as well as the idler. Because of the natural wavelength adjustment of the idler in the copier stage, the idler enjoys more gain and the power equalization of signal and the idler would give better PSA performance in the final HNLF of PSA-only amplification section in Fig. 1.

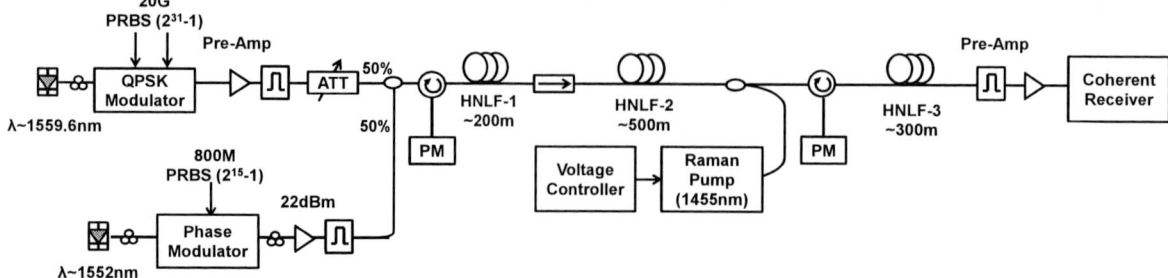

Fig. 2. Experimental setup: PSA pump is phase modulated by 800M-baud pseudorandom binary sequence (PRBS) to suppress stimulated Brillouin scattering (SBS). The attenuator (ATT) is for input signal power adjustment. By switching PSA pump on/off, the signal experience either Raman-only amplification or Raman-assisted PSA. Power meter (PM) is used for monitoring SBS power.

Figure 2 shows the experimental setup. At the transmitter side, a laser with the wavelength of 1559.6 nm is modulated with a 20G-baud QPSK data. Attenuation is added on the signal before sending to HNLF-1 together with a phase modulated pump at the wavelength of 1552nm. The three HNLFs have similar parameters with nonlinear coefficient of 21.4 $W^{-1}km^{-1}$, zero dispersion wavelength (ZDW) of 1551.5nm and a dispersion slope of 0.043ps/km/nm^2. By tuning PSA-pump power to be 22dBm, after 200m HNLF-1, a signal copy at 1544.4 nm is generated with -10dB conversion efficiency, which is shown in Fig. 3(a). Because of the low power level of the pump, there is only 0.7dB OPA gain for the signal. In the following PSA/Raman section, a Raman pump at 1455 nm propagates in the 500m HNLF2 with the opposite direction of parametric waves. The signal, pump and idler are amplified by Raman amplification. It is noted that since the power of all the components are increased, Raman gain and PSA might occur simultaneously. After this stage, the ultimate power imbalance between the signal and idler is decreased to be 1.4dB, which is shown in Fig. 3(b). After that, all the three components (signal, pump and idler) are sent to the 300m HNLF-3 for further PSA and the spectrum at the output is shown in Fig. 3(c). The ripple in the spectrum is the indication of PSA for the correlated part of the noise. The peak appears if the phase matching condition is met while the valley is the sign of the total phase mismatch. Peak and valley spectral positions are mainly determined by the wavelength of the PSA-pump. We have tuned that wavelength such that the signal and idler be located at the phase matching position (peak). In the end, the signal is filtered out and pre-amplified before sending to the coherent receiver for BER measurement.

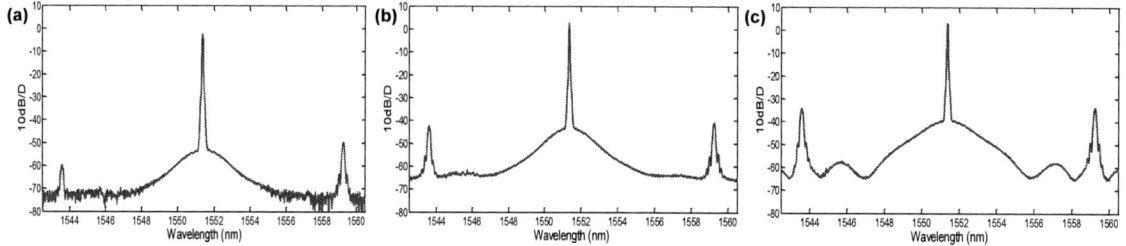

Fig. 3. Optical spectra after (a) copier section; (b) Raman/PSA section; (c) PSA-only section.

II. RESULTS AND DISCUSSIONS

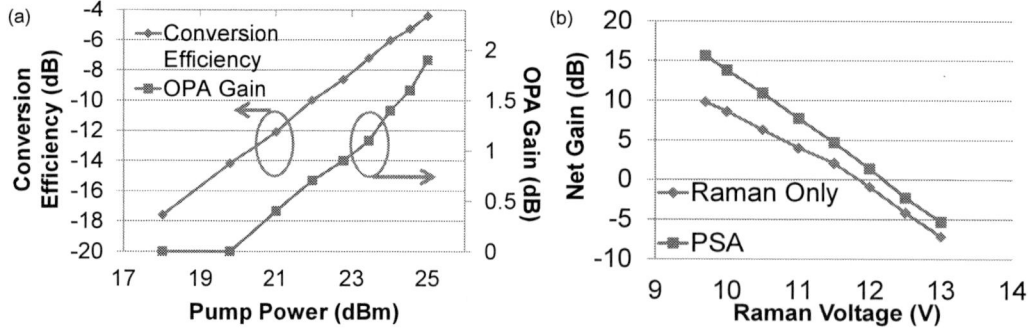

Fig. 4. (a) Conversion efficiency and OPA gain corresponding to the pump power at the copier stage; (b) power difference between input signal and output signal (net gain) along with different levels of Raman pump.

We first investigate the conversion efficiency for the idler in the copier stage, which is shown in Fig. 4(a). It can be seen that as the pump power increases from 18dBm to 25dBm, the conversion efficiency varies by about 14dB. Although low conversion efficiency gives no OPA gain as in the red line in Fig. 4(a), the low power idler would be more vulnerable to the amplified noise from the following stages; high conversion efficiency, on the other hand, would induce more OPA, which generates correlated noise and ultimately decreases the efficiency of PSA. From Fig. 4(a), it shows about 22dBm pump power would give -10dB conversion efficiency with 0.7dB OPA gain, which might be an acceptable tradeoff.

Then, we investigate the net gain performance of the system and compare with the scenario of Raman-only amplification. The net gain is defined as the power difference between the input signal and output signal. Since the overall link loss is fixed (about 9dB), if the gain is less than 9dB, the net gain would be negative. Fig. 4(b) shows the net gain along with the Raman pump level. It is worth mentioning that Raman pump increases along with the deceasing voltage of the controller shown in Fig. 2. If the PSA pump is turned off, the signal is only amplified by Raman scattering, which is the blue curve in Fig. 4(b). It can be seen that when the voltage is 9.7V, 10dB maximum gain is achieved. When PSA pump is turned on, PSA would add extra gain under the same level of Raman pump, which is denoted as the red curve in Fig. 4(b). For the voltage of 9.7V, more than 15dB net gain is observed. It is also noticeable that for the smaller Raman voltages (higher Raman pump), the gain difference becomes larger, which might be attributed to higher PSA efficiency due to the better power banlance between the signal and idler after PSA/Raman section.

Finally, BER is measured by changing the input signal power and we compare the performance between PSA and Raman-only scenarios. The signal power into the pre-amplifier right before the coherent receiver is kept moderate high (-28dBm) and constant to diminish the noise contribution from the pre-amplifier at the receiver side. In this case, the noise performance is almost entirely determined by PSA or Raman-only amplification. Two BER curves are shown in Fig. 5 and it can be seen PSA has more than 2dB sensitivity improvement compared with Raman-only amplification.

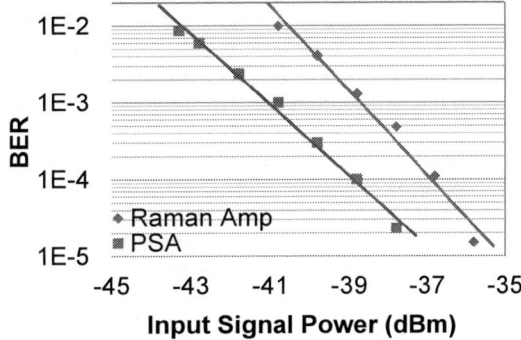

Fig. 5. BER performance comparison between PSA (red curve) and Raman-only amplification (blue curve).

III. CONCLUSIONS

A scheme of Raman-assisted PSA is proposed. The phase matching condition is achieved by tuning the wavelength of the PSA pump. More than 2dB sensitivity improvement compared with Raman-only amplification is experimentally demonstrated by a 20G-baud QPSK system.

REFERENCES

[1] Z. Tong, C. Lundstrom, P. A. Andrekson, C. J. McKinstrie, M. Karlsson, D. J. Blessing, E. Tipsuwannakul, B. J. Puttnam, H. Toda, and L. Gruner-Nielsen, "Towards ultrasensitive optical links enabled by low-noise phase-sensitive amplifiers," Nature Photonics, vol. 5, pp. 430-436, 2011.
[2] T. Umeki, O. Tadanaga, M. Asobe, Y. Miyamoto, and H. Takenouchi, "First demonstration of higher-order QAM signal amplification in PPLN-based phase sensitive amplifier," ECOC2013, PD1.C.5 (2013).
[3] J-Y. Yang, Y. Akasaka, M. Sekiya, A. Willner, A. Ariaei, M. Ziyadi, Y. Cao, "Investigation of Channel-based independent phase shifts for maximizing phase-sensitive amplification on WDM channels," OECC2015, PWe.39 (2015).
[4] Z. Tong, C. Lundstrom, P. A. Andrekson, M. Karlsson, and A. Bogris, "Ultralow noise, broadband phase-sensitive optical amplifiers, and their applications," IEEE J. Sel. Topics Quantum Electron., vol. 18, no. 2, pp1016-1032, March 2012.
[5] J. Touch, M. Ziyadi, A. Abouzaid, M. Chitgarha, S. Khaleghi, A. Ariaei, Y. Akasaka, J-Y. Yang, M. Sekiya, "Passive digital algorithmic stabilization of optical phase," CLEO2014, JTh2A.13 (2014).
[6] Y. Akasaka, J-Y. Yang, M. Sekiya, Y. Cao, A. Almaiman, M. Ziyadi, A. Ariaei, P. Liao, T. Ikeuchi, A. Willner, S. Takasaka, R. Sugizaki, "Experimental demonstration of Raman-assisted phase sensitive amplifier with negligible gain/power fluctuation," ECOC2015, Tu.1.1.4 (2015) .

Polarization Pulling in Fiber Optical Parametric Amplifiers

S. H. Wang[1*] and P. K. A. Wai[2]

[1] Department of Microelectronics, Fuzhou University, Qi Shan Campus, Fuzhou, 350108, China.
[2] Photonics Research Center, Department of Electronic and Information Engineering, The Hong Kong Polytechnic University,
Hung Hom, Hong Kong
*Phone: +(86) 591-87860838-820, Fax: +(86) 591-87382515, Email: shwang@fzu.edu.cn

Abstract: We found that polarization pulling to the idler is stronger than that to the signal in fiber optical parametric amplifiers. High conversion efficiency and strong polarization pulling can therefore be achieved simultaneously using long fibers.

Keywords: Nonlinear optics; Parametric oscillators and amplifiers; Polarization-selective devices.

I. INTRODUCTION

When a fiber optical parametric amplifier (FOPA) amplifies the input optical signal, it can simultaneously generate a phase conjugate idler [1]. This property promises potential applications of FOPAs in optical communication systems, such as all optical signal processing [1] and phase sensitive amplification [2]. However, in FOPAs both of the parametric amplification and the generation of the idler are highly polarization dependent, which is a challenge to commercial applications of FOPAs [1, 3-4].

Recently, nonlinear polarization pulling in FOPAs had been investigated to control all-optically the state of polarizations (SOPs) of polarization-scrambled signals such that the output SOPs of the signal and the idler become predictable [4-7]. Both the theoretical and measurement results indicate that the SOPs of the generated idler can match that of the parametric pump. This is because the idler can only be generated when the phase-matching condition is satisfied or nearly satisfied [1, 4]. Thus the polarization sensitive parametric process only generates co-polarized idler leading to strong polarization pulling.

However, the degree of polarization (DOP) of the output idler is determined by the parametric gain and the polarization mode dispersion (PMD) of the fiber [3]. Besides, in co-propagating FOPA the output SOPs of the pump cannot be predicted because of the randomly varying fiber birefringence, which makes the SOPs of the idler unpredictable. In this paper, we use a phase matching model recently proposed by us to investigate the effect of parametric gain on the polarization pulling to the idlers in degenerated FOPAs.

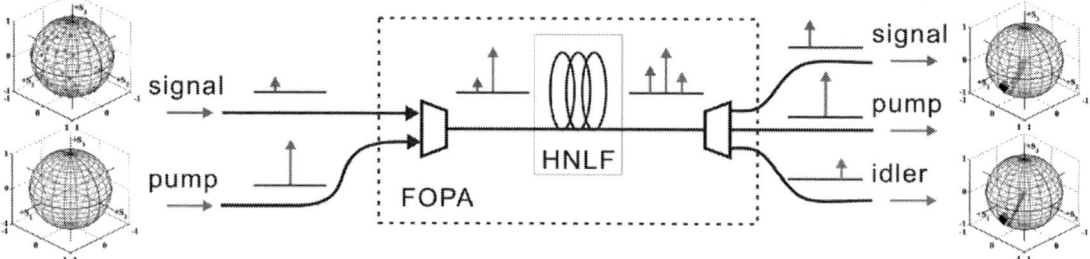

Fig. 1. Polarizing pulling in FOPAs. The of input polarization of the parametric pump (green) is kept at (1, 0, 0) for each fiber realization of the HNLF. The input SOPs of 100 signals (red) are random and uniformly distributed on the Poincaré sphere. The output SOPs of the 100 signals (red) and the corresponding idler (blue) are polarization pulled by the co-propagating parametric pump in a FOPA.

II. THEORETICAL MODEL AND NUMERICAL METHODS

In FOPAs, high conversion efficiency and broad conversion bandwidth can be achieved only when the phase matching condition is satisfied or nearly satisfied [1]. For polarized fiber optical amplification in randomly birefringent fibers, the phase matching condition can be broken by physical effects such as chromatic dispersion, randomly varying fiber birefringence, and nonlinear effects [4]. To investigate polarization pulling to the idlers in FOPAs, we used the model proposed in [4], which describes the evolution of the phase matching in the FOPAs and can well predict the general behaviors of polarized FOPAs in randomly birefringent fibers in the small signal region.

Figure 1 illustrates the effect of polarization pulling in FOPAs. As shown in Fig. 1, we fixed the input SOP of the pump to be (1, 0, 0), i.e. the parametric pump is linearly polarized and aligned with the birefringence axis of fiber at the input end. The FOPA in Fig. 1 is made up of one piece of highly nonlinear fiber (HNLF). In the simulation, 100 fiber realizations are generated for this fiber by using a random modulus model [4]. The generated 100 realizations have the same fiber parameters except for the profiles of randomly varying birefringence. In this paper, we investigated FOPAs consisting of two pieces of fibers, HNLF1 and HNLF2, as in [4]. The parameters of HNLF1 and HNLF2 are the same as that listed in Table 1 of [4]. As shown in Fig. 1, 100 signals (red) are simulated for each fiber realization. The input SOPs of the 100 signals are random and uniformly distributed on the Poincaré sphere. We adopt the numerical method in [4] to define a "principal" SOP referring to the state with optimum phase matching in FOPAs. The DOP of the signal (red) and idler (blue) are defined as $DOP_G = \sqrt{\sum_{i=1}^{3} \langle S_i(z) \rangle_{SOP}^{2}} \Big/ \langle S_0(z) \rangle_{SOP}$ and $DOP_{CE} = \sqrt{\sum_{i=1}^{3} \langle D_i(z) \rangle_{SOP}^{2}} \Big/ \langle D_0(z) \rangle_{SOP}$ respectively, where $\langle \cdot \rangle_{SOP}$ is the average taken over 100 signals the input SOPs. The fourth order Runge-Kutta method is used in the numerical integration with a maximum step size of 2 mm. Finally, we obtain $\langle DOP_G \rangle_{fiber}$ and $\langle DOP_{CE} \rangle_{fiber}$, the mean DOPs of the signal and the idler, by averaging over an ensemble of 100 fiber realizations [4].

III. POLARIZATION PULLING TO THE SIGNALS AND THE IDLERS

In co-propagating fiber Raman amplifiers (FRAs), FOPAs, and counter-propagating FRAs [4-5], the mean DOP_G of the output signal and average on-off gain are related by an interpolation function which can be written as

$$DOP_G = 1 - \exp\left[-G_{\text{on-off}}(dB)/\Gamma \right] \qquad (1)$$

Our previous work shown that Γ equals to 6.2 for co-propagating FOPAs, which is larger than that of co-propagating FRAs ($\Gamma = 4.3$) and smaller than that counter-propagating FRAs ($\Gamma \approx 10.2$ when $D_{PMD} = 0.08$ ps·km$^{-1/2}$ [5]).

Figure 2(a) shows the mean DOP_G of the output signal as a function of the average on-off gain over the gain band of the FOPA (HNLF1 and HNLF2) in open symbols. As shown in Fig. 2(a), the numerical results of eight FOPAs using different fibers and pump powers agree well with the interpolation function of Eq. (1) when $\Gamma = 6.2$. Here, we define the parametric gain factor σ as the product of the nonlinear coefficient γ of the fiber, the input parametric pump power P_{in}, and the fiber length L, i.e. $\sigma = \gamma P_{in} L$. In the co-polarized model, the peak parametric gain can be approximately written as a function of σ, $G_{max} = 1 + \sinh^2(\sigma)$ [1]. Table 1 lists the parameters of the fibers and the pumps used in the simulation. As shown in Table 1, for each fiber we change the value of the parametric gain factor σ only through changing the input pump power. In the simulation, 9/8 times the measured γ of fiber is used as in [4].

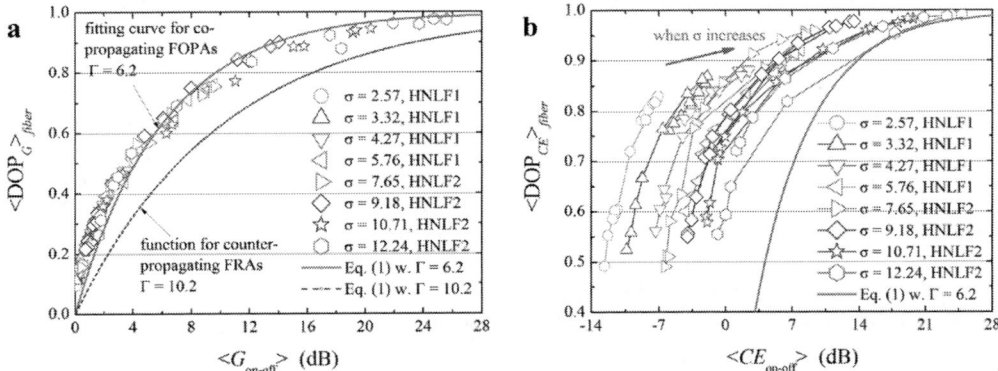

Fig. 2. (a) The mean DOP_G of the output signal as a function of the average on-off gain over the gain band of the FOPAs using HNLF1 and HNLF2 (open symbols). The interpolation functions of Eq. (1) for the FOPAs (red curves) and counter-propagating FRAs (blue dashed curves) are also shown, respectively. (b) The corresponding mean DOP_{CE} of the output idler as a function of the average conversion efficiency over the gain band of FOPAs.

TABLE I
THE PARAMETERS OF THE FIBERS AND PUMPS USED IN THE SIMULATION [4]

HNLF1 ($D_{PMD} = 0.07$ ps·km$^{-1/2}$)				HNLF2 ($D_{PMD} = 0.08$ ps·km$^{-1/2}$)			
$\sigma = \gamma P_{in} L$	γ (W^{-1}·km^{-1})	P_{in} (W)	L (km)	$\sigma = \gamma P_{in} L$	γ (W^{-1}·km^{-1})	P_{in} (W)	L (km)
5.76	9/8 × 11.2	0.933	0.49	12.24	9/8 × 13.6	0.8	1.0
4.27	9/8 × 11.2	0.691	0.49	10.71	9/8 × 13.6	0.7	1.0
3.32	9/8 × 11.2	0.537	0.49	9.18	9/8 × 13.6	0.6	1.0
2.57	9/8 × 11.2	0.417	0.49	7.65	9/8 × 13.6	0.5	1.0

Figure 1(b) shows the corresponding mean DOP_{CE} of the output idler as a function of the average conversion efficiency. The multiple data points in each curve map to the output conversion efficiencies over the bandwidth of FOPAs in each setup. In Fig. 2(b), all of the data points are on the left hand side of the curve described by Eq. (1) for Γ = 6.2. It indicates in general the polarization pulling to the idler is much stronger than that to the signal in FOPAs. This is because of the phase-matching condition; the polarization sensitive parametric process can only generate co-polarized idlers. Thus FOPAs shows strong polarization pulling to the output SOPs of the idler which are determined by the output SOPs of the parametric pump. But, in co-propagating setups, both of the output SOPs of the parametric pump and the idler cannot be predicted because of the randomly varying birefringence.

As shown in Fig. 2(b), when the parametric gain factor σ is small, the data points locate at the left side of the figure, which indicates when the parametric gain is small, the idler can still have high output DOP_{CE} with smaller than -7 dB mean conversion efficiency. When the parametric gain factor σ increases, the data points migrate toward the right side of the figure. In this case, although higher conversion efficiency is needed to obtain the same output DOP_{CE} than that with smaller σ, a larger output DOP_{CE} can be achieved at the parametric gain peak. If we keep increasing the parametric gain factor σ, the data points will approach Eq. (1) with $\Gamma = 6.2$. It indicates that when the parametric gain increases, the polarization pulling to the idler will be close to that of the signal.

We notice that the data points in Fig. 2(b) can be described by the interpolation function

$$DOP_{CE} = 1 - \exp\left\{-\left[CE_{\text{on-off}}\,(\text{dB}) + \Delta\right]/\Gamma\right\} \qquad (2)$$

where the coefficient Δ is a constant. Fig. 3(a) shows the mean DOP_{CE} as a function of the average conversion efficiency over the parametric bandwidth when σ equals to 4.27 in the FOPAs using HNLF1 and σ equals to 9.18 in the FOPAs using HNLF2, respectively. The corresponding curves of the interpolation functions of Eq. (2) with $\Delta = 9$ and 12 are also shown, where the parameter Γ is fixed to 6.2. As shown in Fig. 3(a), the numerical results agree well with the interpolation functions of Eq. (2).

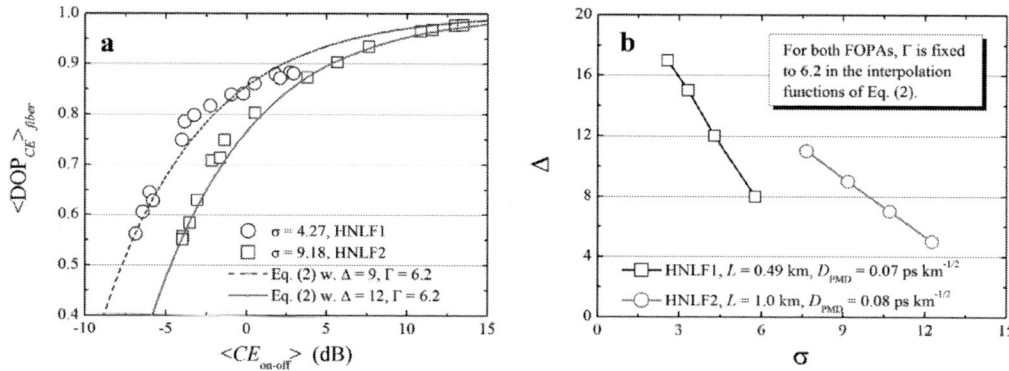

Fig. 3. (a) The mean DOP_{CE} as a function of the average conversion efficiency over the gain band of the FOPAs are shown in open symbols when σ equals to 4.27 and 9.18, respectively. The corresponding interpolation functions of Eq. (2) with different Δ are also shown. (b) The coefficient Δ as a function of the parametric gain factor σ in FOPAs using different fibers. Here, the parameter Γ is 6.2 in all of the interpolation functions.

Figure 2(b) shows the coefficient Δ as a function of the parametric gain factor σ in FOPAs using different fibers. Here the parameter Γ is 6.2. As shown in Fig. 3(a), the coefficient Δ decreases along with the parametric gain factor σ, which indicates the polarization pulling to the idler become weaker when the parametric gain increases. In Fig. 3(b), we also notice that the decrease rate of the coefficient Δ versus the parametric gain factor σ is lower in long fiber than that in short fiber, which indicates that high conversion efficiency as well as strong polarization pulling can be simultaneously achieved in FOPAs using long fibers.

In Fig. 3(b), we also notice that in FOPAs using short fiber such as the copier in the phase sensitive amplifiers [2], strong polarization pulling to the idler can be achieved even when the parametric gain is small.

IV. CONCLUSIONS

In conclusion, we use a phase matching model recently proposed by us to investigate the effect of the parametric gain on the polarization pulling to the idlers in degenerated FOPAs. Simulation results indicate that polarization pulling to the idler is stronger than that to the signal in FOPAs. We also found that high conversion efficiency and strong polarization pulling effect can be simultaneously archived in FOPAs using long fibers. In FOPAs using short fiber, strong polarization pulling to the idler can also be achieved even when the parametric gain is small.

ACKNOWLEDGMENT

This work was supported by National Natural Science Foundation of China, under Project 61205049.

REFERENCES

[1] M. E. Marhic, Fiber optical parametric amplifiers, oscillators and related devices, (Cambridge University, 2007).
[2] S. L. I. Olsson, B. Corcoran, Carl Lundström, T. A. Eriksson, M. Karlsson, and P. A. Andrekson, "Phase-sensitive amplified transmission links for improved sensitivity and nonlinearity tolerance," IEEE J. of Lightwave Technol. 33(3), 710-721, (2015).
[3] B. Stiller, P. Morin, D. M. Nguyen, J. Fatome, S. Pitois, E. Lantz, H. Maillotte, C. R. Menyuk, and T. Sylvestre, "Demonstration of polarization pulling using a fiber-optic parametric amplifier," Opt. Express 20(24), 27248-27253, (2012).
[4] S. H. Wang, Xinchuan Xu, and P. K. A. Wai, "Polarized fiber optical parametric amplification in randomly birefringent fibers," Opt. Express, 23(25), 32747-32758, (2015).
[5] F. Chiarello, L. Ursini, L. Palmieri, and M. Santagiustina, "Polarization attraction in counter propagating fiber Raman amplifiers," IEEE Photon. Technol. Lett. 23(20), 1457-1459, (2011).

TuC3-5 (Invited)

Nonlinear Impairments Mitigation aided by Low-noise Optical Frequency Combs

Ping Piu Kuo

Qualcomm Institute, University of California, San Diego, USA

Abstract

We present the role of mutual channel frequency stability in digital back-propagation-based nonlinear impairment mitigation methods. Elimination of mutual frequency uncertainty by using frequency-locked carriers has enabled unprecedented capacity extension for the first time.

TuC4-1

OECC/PS2016

1.6 ~ 1.8 μm band hybrid broadband light source consisting of cascaded SLD and TDFA

Kazuya OTA[1,2], XIAOEN Du[1], Sho TUJITA[1], Kousuke SENDA[1], Fumiki HANAFUJI[1], Jun ONO[1,3],
Kazuaki MISE[4], Yoshiharu SHIMOSE[4], Hiroshi MORI[4], Osanori KOYAMA[1], Hirotaka ONO[5],
Tatsuro ENDO[6] and Makoto YAMADA[1]

[1]Dept. of Elec. and Info. Systems, Osaka Prefecture University, 1-1 Gakuen-cho, Nakaku, Sakai, Osaka 599-8531, Japan
[2]Trimatiz Ltd., 801, La Pacifique B, 4 - 7 - 12 Minami Yawata, Ichikawa, Chiba 272-0023, Japan
[3]Anritsu Devices Co., Ltd., 5-1-1 Onna, Atsugi-shi, Kanagawa 243-8555 Japan
[4]Anritsu Co., Ltd., 5-1-1 Onna, Atsugi-shi, Kanagawa 243-8555 Japan
[5]NTT Device Technology Laboratories, NTT Corporation, 3-1 Morinosato-Wakamiya, Atsugi, Kanagawa 243-0198, Japan
[6]Dept. of Appl. Chemistry, Osaka Prefecture University, 1-1 Gakuen-cho, Nakaku, Sakai, Osaka 599-8531, Japan
e-mail: myamada@ eis.osakafu-u.ac.jp

Abstract: *We successfully improved the output characteristics of a hybrid broadband light source consisting of a cascaded super luminescent diode (SLD) and a Tm^{3+}-doped fiber amplifier (TDFA), which we propose as a candidate 1.7 ~ 1.8 μm band light source. The achieved bandwidth where the intensity exceeded -40 dBm/0.1 nm was 325 nm (from 1577 to 1902 nm), and the total output power was 16 dBm with a low ripple and a peak-to-peak value of less than 0.14 dB.*
Keywords: *Super luminescent diode, Tm^{3+}-doped fiber amplifier, 1.7 μm broadband light source*

I. INTRODUCTION

Infrared light sources operating at around 1.7 ~1.8 μm have been developed [1-8] for applications in a wide variety of fields including medical surgery, industrial machining, scientific experiments, and optical sensing [5, 9-11]. Moreover, they may be used for future optical transmission systems beyond 1.6 μm and in the development of their components if we employ hollow core photonic bandgap fiber to overcome the capacity limit of conventional systems [12, 13].

We have successfully developed the first ASE 1.7~1.8 μm band light source with a broadband spectrum and tunable fiber ring laser by using a Tm^{3+}-Tb^{3+}-doped fiber as the active fiber. Recently, we have also proposed a hybrid broadband light source consisting of a cascaded super luminescent diode (SLD) and a Tm^{3+}-doped fiber amplifier (TDFA) unit. An SLD is excellent as a compact and cost effective broadband light source. However, the highest operating wavelength of commercially available SLDs is limited to ~1.7 μm (there is no 1.8 μm band SLD), and ripples in the spectrum, which introduce noise into a measurement, increase as the SLD output power increases. The use of a hybrid configuration overcomes the issues with the SLD because the TDFA unit not only amplifies the SLD output light thus increasing the output power, but also adds ASE light to improve the broadband performance on the output spectrum. As a result, we achieved a wide bandwidth of 247 nm (from 1611 to 1858 nm) with low ripple characteristics.

In this paper, we report a way of improving the output characteristics of the hybrid broadband light source by tuning the peak wavelength of the SLD and the amplification characteristics of the TDFA unit. We realized a 1.6~1.8 μm band hybrid broadband light source by using a 1.65 μm SLD and a TDFA unit with a 3 m long Tm^{3+}-doped fiber, and broadened the spectrum characteristic of the optical components.

II. CONFIGURATION OF HYBRID BROADBAND LIGHT SOURCE AND EXPERIMENTAL ITEMS

A. Configuration of Hybrid Broadband Light Source

Figure 1 shows the hybrid broadband light source that we constructed using an SLD and a TDFA unit. The output light from the SLD is inputted into the TDFA unit, which consists of a pair of 1.2/1.8 μm band WDM couplers and optical isolators, a Tm^{3+}-doped fluoride fiber, and a 1.21 μm pump LD. The fiber was backward pumped with a maximum launched pump power of 232 mW. The relative refractive index difference, Tm^{3+} concentration and cut-off wavelength of the Tm^{3+}-doped fluoride fiber were 1.6%, 6000

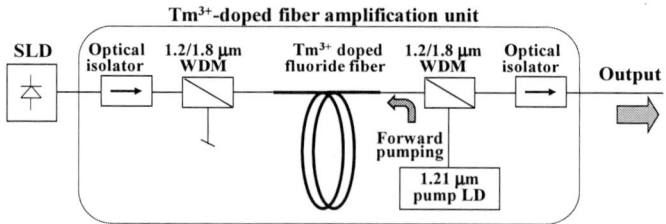

Fig. 1. Hybrid broadband light source employing an SLD and a TDFA unit.

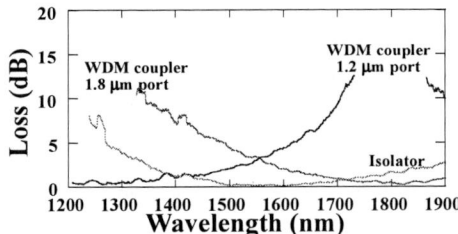

Fig. 2. Loss spectrum of WDM coupler and optical isolator.

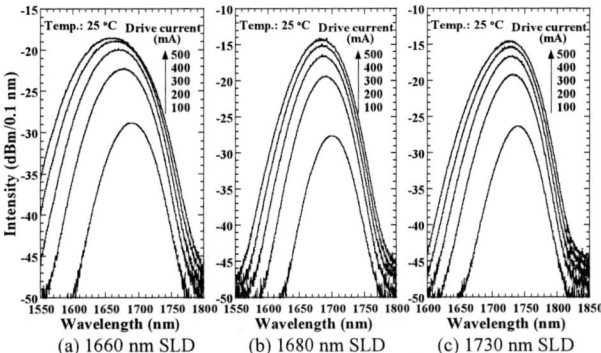

(a) 1660 nm SLD   (b) 1680 nm SLD   (c) 1730 nm SLD

Fig. 3. Output spectrum of each SLD for various drive currents

ppm and 1.0 μm, respectively. The loss spectrum of the WDM coupler and optical isolator is shown in Fig. 2. The insertion loss of the WDM coupler and the isolator in the 1600 to 1900 nm wavelength region, were <2 and <2.5 dB, respectively.

B. Experimental Items

To study ways of improving the output characteristics of the hybrid broadband light source, we used three TDFA units constructed with different lengths of Tm^{3+}-doped fiber, namely 1, 2 and 3 m, and three SLDs with different peak spectral wavelengths of 1660, 1690 and 1730 nm, as shown in Fig. 3 (a), (b) and (c), respectively. In the experiment, the intensities at the peak wavelengths were set at -20 dBm/0.1 nm for each SLD, and the ripple in the spectrum was <0.15 for the 1660 nm SLD, and <0.2 dB peak to peak (pp) for the 1690 and 1730 nm SLDs. Under the above condition, the output spectrum from the hybrid broadband light source, which consisted of a combination of one of the three SLDs and one of the three TDFAs with different lengths, were measured with an optical spectrum analyzer that had a resolution bandwidth of 0.1 nm.

III. EXPERIMENTAL RESULTS AND DISCUSSION

Figure 4 shows the output spectra of the hybrid light source, which consisted of various lengths of Tm^{3+}-doped fiber, with 1660, 1690 and 1730 nm SLDs. The backward pump power was 100 mW. The output characteristics improved at the longer wavelength but not at the short wavelength, with an increase in the TDF length. We found that a broad output spectrum can be achieved with a 1660 nm SLD.

The output spectra of the hybrid light source, which consisted of a 1660 nm SLD and a 2 m long Tm^{3+}-doped fiber, as a function of backward pump power are shown in Fig. 5. The output peak value and bandwidth of the output spectrum increased with increases in pump power, The bandwidth for which the output intensity exceeded -40 dBm/0.1 nm was 325 nm (from 1577 to 1902 nm) at a pump power of 200 mW, The ripple in the spectrum and the total output were <0.14 dB pp, and 16.8 dBm, respectively.

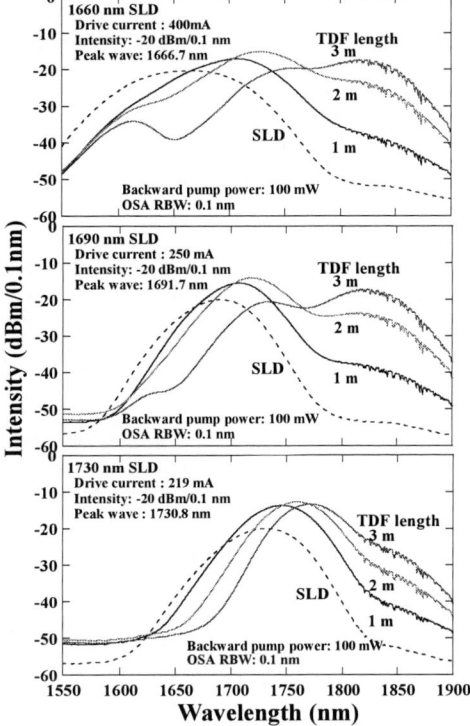

Fig. 4. Output spectrum of hybrid light sources consisting of various lengths of Tm^{3+}-doped fiber, with 1660, 1690 and 1730 nm SLDs.

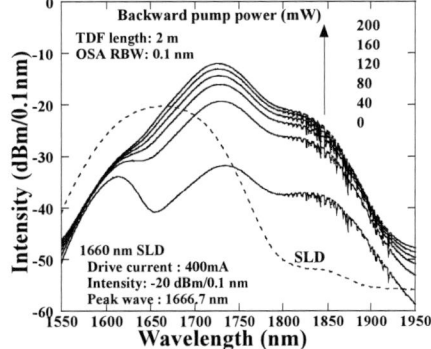

Fig. 5. Output spectrum of hybrid light sources consisting of 2 m long Tm^{3+}-doped fiber and 1660 nm SLD for various backward pump powers.

IV. CONCLUSIONS

The output characteristics of a hybrid light source consisting of a cascaded SLD and a Tm^{3+}-doped fiber amplifier were improved by tuning the peak wavelength of the SLD and the amplification characteristics of the TDFA unit. The achieved bandwidth for which the intensity exceeded -40 dBm/0.1 nm was 325 nm (from 1577 to 1902 nm) with a low ripple of less than 0.14 dB pp and a high total output power of 16.8 dBm. Our hybrid light source operating in the 1.6 ~ 1.8 μm wavelength region is attractive as a broadband light source for scientific experiments such as those involving near-infrared spectroscopy, and for the development of future optical transmission systems operating beyond 1.6 μm.

REFERENCES

[1] D.C. Hanna, R.M. Percival 1, R.G. Smart, and A.C. Tropper, "Efficient and tunable operation of a Tm-doped fibre laser," Optics Communications, Vol. 75, No. 3-4, pp. 283–286 (1990).

[2] W. A. Clarkson, N. P. Barnes, P. W. Turner, J. Nilsson, and D. C. Hanna, "High-power cladding-pumped Tm-doped silica fibre laser with wavelength tuning from 1860 to 2090 nm," Optics Lett., Vol. 27, No. 22, pp. 1989–1991 (2002).

[3] D. Y. Shen, J. K. Sahu, and W. A. Clarkson, "High-power widely tunable Tm:fibre lasers pumped by an Er,Yb co-doped fibre laser at 1.6 μm," Opt. Exp., Vol. 14, No. 13, pp. 6084-6090 (2006).

[4] Y. Tang, C. Huang, S. Wang, H. Li, and J. Xu, "High-power narrow-bandwidth thulium fiber laser with an all-fiber cavity," Opt. Exp., Vol. 20, No. 16, pp. 17539–17544 (2012).

[5] A.B. Rulkov, A.A. Ferin, J.C. Travers, S.V. Popov, and J.R. Taylor, "Broadband, low intensity noise CW source for OCT at 1800 nm," Optics Communications, Vol. 281, No. 1, pp. 154–156 (2008).

[6] M. Yamada, S. Aozasa, and H. Ono, "Broadband ASE light source for the 1800 nm wavelength region," Electron. Lett., Vol. 48, No. 23, pp. 1489-1490 (2012).

[7] M. Yamada, K. Senda, T. Tanaka, Y. Maeda, S. Aozasa, H. Ono, K. Ota, O. Koyama, and J. Ono, "Tm^{3+}-Tb^{3+}-doped tunable fibre ring laser for the 1700 nm wavelength region," Electron. Lett., Vol. 49, No. 20, pp. 1287-1288 (2013).

[8] M. Yamada, J. Ono, K. Mise, Y. Shimose, H. Mori, A. Yamada, K. Ota, K. Senda, Y. Maeda, O. Koyama, and H. Ono, "1.8 μm broadband light source using a super luminescent diode," Electron. Lett., Vol. 50, No. 20, pp. 1468-1470 (2014).

[9] M. Ebrahimzadeh, "Mid-infrared coherent sources and applications (NATO Science for Peace and Security Series)", Springer, (2008)

[10] Y. Maeda, M. Yamada, T. Endo, K. Ohta, T. Tanaka, M. Ono, K. Senda, J. Ono, and O. Koyama, "1700 nm ASE light source and its application to mid-infrared spectroscopy," Proc. OECC2014, June. 2014, paper TU6F.

[11] J. Ono, T. Endo, K. Ohta, H. Ono, Y. Maeda, K. Senda, O. Koyama, and M. Yamada, "Broadband light source and its application to near infrared spectroscopy," Sensors and Materials Vol. 27, No. 5, pp. 413-423 (2015)

[12] A.D. Ellis, "Current capacity limits and activities within the EU project MODE-GAP to overcome them," Proc. IEEE Summer Topical Meeting SDM Opt. Syst. Netw., Jul. 2012, pp. 169–170, paper MC1.1.

[13] D.J. Richardson, "Fiber amplifiers for SDM systems'," Proc. OFC/NFOEC2013, Mar. 2013, paper OTu3G.1.

TuC4-2   OECC/PS2016

1.8 μm High-Order Microring Resonator Mode-locked Laser Using a Carbon Nanotube

K.S. Tsang[1,4], Jie Wang[2], Li Jin[3], Victor Ho[3,4], Jack Cheung[4], Yanny Tsang[4], Alessia Pasquazi[5], Ray Man[4], Sai T. Chu[3], A. Ping Zhang[2], Hwa-yaw Tam[2], and P. K. A. Wai[1]

[1]Photonics Research Center and Department of Electronic and Information Engineering, The Hong Kong Polytechnic University, Hung Hom, Hong Kong
[2]Department of Electrical Engineering, The Hong Kong Polytechnic University, Kowloon, Hong Kong SAR, China
[3]Department of Physics and Materials Sciences, City University of Hong Kong, Kowloon Tong, Hong Kong
[4]AMONICS Ltd. 14/F, Lee King Industrial Building, 12 Ng Fong, San Po Kong, Hong Kong,
[5]Department of Physics and Astronomy, University of Sussex, Falmer, Brighton, BN1 9QH, United Kingdom

Abstract: We demonstrated a 1.81 μm mode-locked laser using a carbon nanotube and an integrated 11-th order microring resonator. A mode-locked pulse train centered at 1.81 μm with repetition rate of 12.6 MHz is achieved.
Keywords: 1.8 μm, Carbon Nanotube, Microring Resonator

I. INTRODUCTION

Pulsed lasers operating in the 2 μm region are of particular interest for applications in sensing, time-resolved spectroscopy, medical application, material processing, and LiDAR (Light Detection And Ranging) to detect wind velocity and track storms [1]. These lasers find applications in the environmental field because of their unique attributes of being eye-safe and the strong absorption by atmospheric gases and water vapor in this wavelength range.

Tm-doped fiber lasers cover the 2 μm wavelength region from 1.8 to 2.1 μm, depending on whether an optical filter is implemented. 2 μm mode-locked lasers using carbon nanotubes (CNTs) [2], SESAMs [3], and nonlinear loop mirrors (NOLMs) [4, 5] have been demonstrated. Previously, we reported stable mode-locked pulses in the 1550 nm range based on filter-driven four-wave-mixing (FD-FWM) by an 11-th order microring resonator [6]. We also demonstrated a 1.81 μm mode-locked figure-eight laser using the same microring resonator [4]. In this work, the same high-order microring resonator together with single-walled carbon nanotubes (SWCNTs) are used to generate optical pulse train in the 1.8 μm range. The SWCNTs serve as a passive mode-locker based on saturable absorption thus providing a simple configuration [7, 8] for ultrashort pulse generation.

II. TECHNICAL WORK PREPARATION

A. Background Theory

Since at 2 μm the microring resonator filter is in normal dispersion, it cannot provide sufficient FWM to generate stable mode-locked pulses in the 2 μm range [4]. Thus we proposed to introduce a saturable absorber into the laser cavity. SWCNTs are used as a mode-locker since they cover a wide range operation wavelength. In addition, it can be used in transmission mode making implementation of the ring configuration straightforward.

The wavelength selection is achieved by a wide band filter and the ring resonator filter. The tunable filter has a Gaussian shape and a wide band of 5 nm with 10 nm tuning range. The ring resonator acts as a narrow band filter, the bandwidth is ~0.6 nm. Figs. 2(a) and 2(b) show the transfer spectrum of the ring resonator. High FSR (~637 GHz) and high extinction ratio (>30 dB) can be achieved. The center wavelength can be tuned by tuning the wide band tunable filter and the ring resonator. The wavelength of the ring resonator can be changed by changing the temperature. The output laser pulse width ranges from several tens to hundreds of picoseconds.

B. Laser Configuration

From Fig. 1(a), the isolator ensures single direction operation of the ring laser. The SWCNTs are fabricated by inserting a small piece of home fabricated SWCNT/PVA composite film between two FC connector surfaces of two 1 m long Nufern SM1950 pigtails. The dispersion of the fiber is ~32 ps/nm/km. The gain is provided by a combination of a 1560 nm pump laser and a ~2.5 m Tm-doped fiber (Nuferm SM-TSF-9/125) using backward pumping. A tunable filter (Koshin Kogaku TFM-1800-S-FA) is used to select the specific filter passband chosen in the experiment. The tunable range of the filter is 10 nm from 1805 to 1815 nm. Thus, the signal will pass through the nonlinear resonator filter. Fig. 1(b) shows the geometry of the integrated resonator and an SEM image of the device cross section. From Fig. 1(b), there are two buses and four ports in the ring. Further details of the microring resonator can be found in [8]. The input port and drop port are connected in the ring cavity. The output is measured at the through port of the ring resonator. A polarization controller by Nufern SM1950 fiber is used to control the polarization. The total length of the cavity is ~15.8 meters, which gives a repetition rate of 12.6 MHz.

547

OECC/PS2016

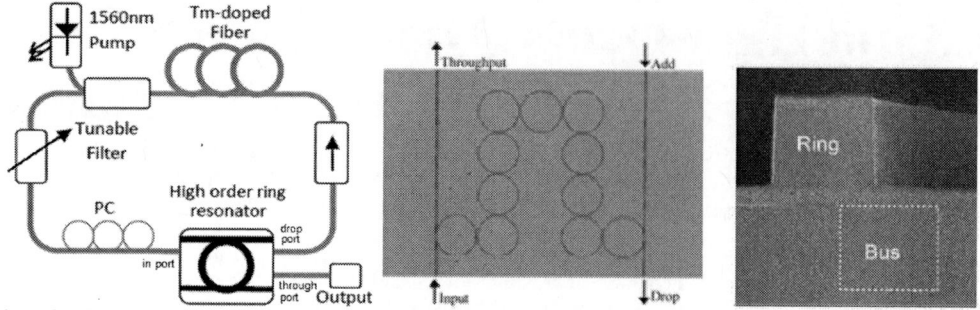

Fig. 1. (a) Schematic of the proposed mode-locked fiber laser. (b) Microscope image of the fabricated 11-th order microring resonator. (c) SEM image of the cross section of the microring before deposition of the upper SiO₂ cladding.

C. Experimental Results

Figs. 2(a) and 2(b) show the linear response of the 11-th order ring resonator at 1.81 µm. From Figs. 2(a) and 2(b), the bandwidth is approximately 45.5 GHz and FSR is 637 GHz. Fig. 2(c) shows the transfer function of the tunable filter. Fig. 3(a) shows the output OSA spectrum from the drop port. Fig. 3(b) shows the timing response of single pulse operation, and 3(c) shows the corresponding RF response. The pump power is ~1 W, and the output power is about 1.3 mW.

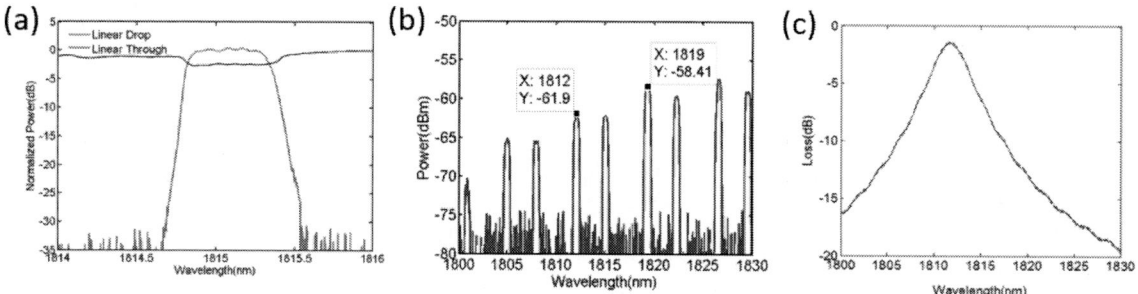

Fig. 2. (a) Linear response of the 11-th order ring resonator at 1.8 µm for both the drop and through ports, (b) The drop port at wider spectral span. (c) Linear response of tunable filter.

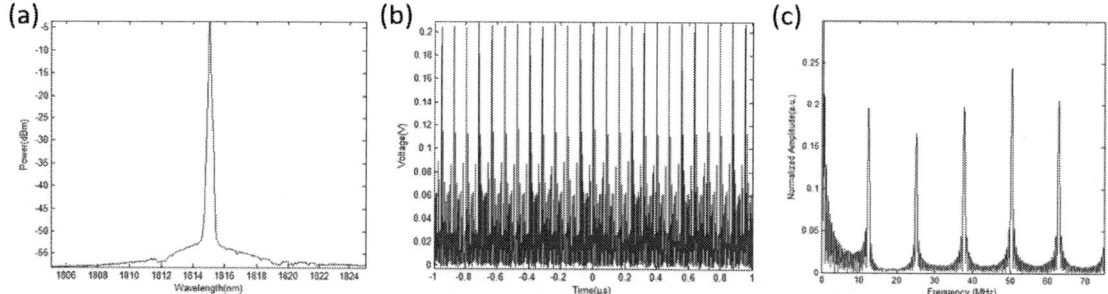

Fig. 3. (a) The optical spectrum measured at the through port. (b) The timing response measured at the through port. (c) The corresponding simulated RF response.

D. Modelling

We measured the transfer functions of the tunable filter and the microring resonator. We assume that the microring resonator functions as a narrow band filter [9]. We solved the coupled nonlinear Schrödinger (NLS) equations to determine the the mode-locked laser dynamics [10-12]. Only one polarization is taken into account for simplicity. Then,

$$\frac{\partial A(\xi,T)}{\partial \xi} + \frac{i}{2}\left(\beta^{(2)} + ig\frac{1}{\Omega_g^2}\right)\frac{\partial A(\xi,T)}{\partial T^2} = i\gamma|A(\xi,T)|^2 A(\xi,T) + \frac{g}{2}A(\xi,T)$$

(1)

where $A(\xi, T)$ is the electric field envelope, ξ is the propagation coordinate, T is pulse duration, $\beta^{(2)}$ is the group velocity dispersion, γ is the nonlinear coefficient, and Ω_g is the gain bandwidth. The gain g is given by

$$g = \frac{g_0}{(1 + P_{ave}/P_{sat})}$$

(2)

where g_0 is the small signal gain, P_{sat} is the saturation power, and P_{ave} is the average power. Here, we assume the effect of the SWCNTs works as a saturable absorber (SA), which is modeled by the intensity-dependent function T(I),

$$T(I) = 1 - \left[\frac{\alpha_0}{(1 + I/I_{sat})} + \alpha_{ns}\right]$$

(3)

where I is the pulse intensity, I_{sat} is the saturation intensity of the SA, α_n is the insertion loss, and α_0 is the modulation depth. The net cavity dispersion $\beta^{(2)}_{net}$ is

$$\beta^{(2)}_{net} = \sum_i L_i \times \beta^{(2)}_i$$

(4)

where L_i is the length and i-th fiber section. The gain bandwidth of the active fiber Ωg is 155 THz. Fig. 4 shows the simulated results with pulse width ~30-40 ps, which agrees with experimental results.

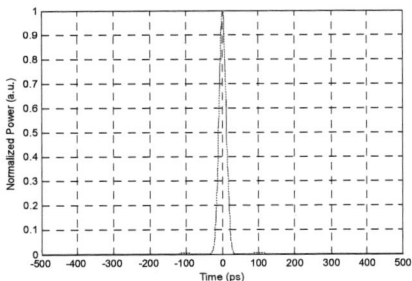

Fig. 4. The simulated results.

III. CONCLUSIONS

In summary, a mode-locked laser at 1.81 μm was achieved by using single-walled carbon nanotubes combined with an 11-th order microring resonator. Mode-locked pulses with pulse width less than 100 ps at a repetition rate of 12.6 MHz are demonstrated. In addition, simulated results show that the pulse width is ~30-40 ps which agreed with the experimental measurements.

REFERENCES

[1] Hui, Hu, et al. "Q-switched thulium-doped domestic silica fiber laser." Chinese Physics Letters 28.4 (2011): 044206.
[2] Harun, S. W., et al. "Self-starting harmonic mode-locked thulium-doped fiber laser with carbon nanotubes saturable absorber." Chinese Physics Letters 30.9 (2013): 094204.
[3] Gumenyuk, Regina, et al. "Dissipative Dispersion-Managed Soliton 2 um Tm-Ho Fiber Laser." The European Conference on Lasers and Electro-Optics. Optical Society of America, 2011.
[4] K.S. Tsang, et al. "Passive Mode-Locking at 1.8 μm using a High-Order Microring Resonator in a Figure Eight Fiber Laser." Asia Communications and Photonics Conference. Optical Society of America, 2015.
[5] Rudy, Charles W., et al. "Amplified 2-μm thulium-doped all-fiber mode-locked figure-eight laser." Journal of Lightwave Technology 31.11 (2013): 1809-1812.
[6] Jin, Li, et al. "Burst-mode operation of a 650GHz mode locked laser based on a high order microring resonator." OptoElectronics and Communications Conference (OECC) 2014.
[7] Mou, Chengbo, et al. "Passively harmonic mode locked erbium doped fiber soliton laser with carbon nanotubes based saturable absorber." Optical Materials Express 2.6 (2012): 884-890.
[8] Wang, Jie, et al. "Widely tunable mode-locked fiber laser using carbon nanotube and LPG W-shaped filter." Optics letters 40.18 (2015): 4329-4332.
[9] Geuzebroek, Douwe H., and Alfred Driessen. "Ring-resonator-based wavelength filters." Wavelength filters in fibre optics. Springer Berlin Heidelberg, 2006. 341-379.
[10] Liu, H. H., and K. K. Chow. "Enhanced stability of dispersion-managed mode-locked fiber lasers with near-zero net cavity dispersion by high-contrast saturable absorbers." Optics letters 39.1 (2014): 150-153.
[11] Fedotov, Y. S., et al. "High average power mode-locked figure-eight Yb fibre master oscillator." Optics express 22.25 (2014): 31379-31386.
[12] Li, Huihui, et al. "Pulse-shaping mechanisms in passively mode-locked thulium-doped fiber lasers." Optics express 23.5 (2015): 6292-6303.

TuC4-3

OECC/PS2016

Widely Tunable, Single-Longitudinal Mode Brillouin/Erbium-Doped Fiber Laser

Huan Wu, Bofang Zheng, and Chester Shu

Department of Electronic Engineering and Center for Advanced Research in Photonics,
The Chinese University of Hong Kong, Shatin, N.T., Hong Kong
Email: hwu@ee.cuhk.edu.hk

Abstract: *Using SBS in an 8-m highly nonlinear fiber and an unpumped EDF loop, a single-longitudinal mode fiber laser is demonstrated. Wide tuning over ~30 nm is achieved with a linewidth reduction ratio larger than 100.*

Keywords: *fiber laser, single-longitudinal mode, narrow linewidth, stimulated Brillouin scattering (SBS)*

I. INTRODUCTION

Wavelength tunable single-longitudinal mode fiber lasers have important applications in coherent communications, high-resolution spectroscopy, and optical sensing [1]. Stimulated Brillouin scattering (SBS) fiber laser is known for its narrow linewidth [2]. However, a critically coupled resonator together with a cavity that matches the frequency of the pump are indispensable due to the limited Brillouin gain and its polarization dependence [3]. A novel hybrid Brillouin/Erbium-doped fiber laser was reported in 1996 [4]. The need for a critically coupled resonator was overcome by using an EDF amplifier to compensate for the resonator losses. However, a long single-mode fiber is still required to initiate SBS due to the small Brillouin gain coefficient. The longitudinal mode of a Brillouin/Erbium-doped fiber laser with such a long cavity is unstable, as mode hopping occurs readily owing to environmental perturbations. In order to achieve a stable output and to eliminate mode hopping, the cavity should be substantially shortened.

In this paper, we demonstrate a single-longitudinal mode Brillouin/Erbium-doped fiber laser by incorporating SBS in an 8-meter long highly nonlinear fiber (HNLF) and spatial hole burning in a 1.2-meter long unpumped EDF loop. The HNLF has a relatively low SBS threshold, allowing the use of just a short segment to provide the necessary gain and to construct a compact cavity. Owing to the narrow gain bandwidth and linewidth narrowing effect of SBS [5], mode selection and linewidth reduction can be achieved. Also, due to the tunability of SBS and large bandwidth of the EDFA, the laser can be precisely tuned over a broad wavelength range. To improve the spectral purity, an unpumped EDF loop is used in the cavity to guarantee single-longitudinal mode operation [6].

II. EXPERIMENTAL SETUP

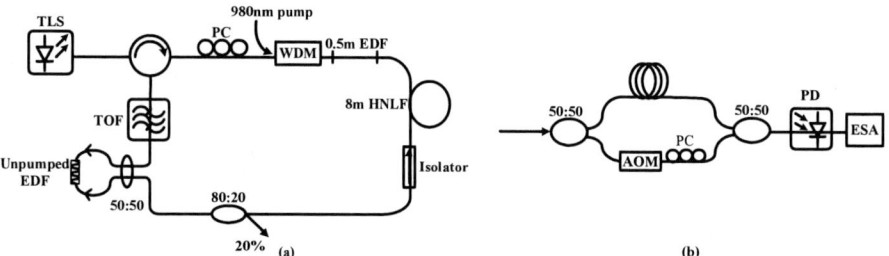

Fig. 1. (a) Experimental setup of the Brillouin/Erbium-doped fiber laser. TLS: tunable laser source; PC: polarization controller; EDF: erbium-doped fiber; HNLF: highly nonlinear fiber; TOF: tunable optical filter; (b) delayed self-heterodyne interferometer for linewidth measurement. AOM: acousto-optic modulator; PD: photodiode; ESA: electrical spectrum analyzer.

The experimental setup of the Brillouin/Erbium-doped fiber laser is illustrated in Fig. 1(a). The SBS pump is a tunable semiconductor laser. It is injected into the cavity through an optical circulator and is amplified by a homemade EDFA consisting of a 980 nm pump laser, a 1550/980 nm WDM, and 0.5 m long EDF. The EDFA amplifies both the SBS pump and the backward-scattered Stokes wave. The saturation power of the EDFA is 14 dBm. A polarization controller (PC) is used to control the polarization state of the SBS pump. After being amplified by the EDFA, the pump is directed into an 8-m long HNLF to initiate the SBS process. An isolator is added to the cavity to ensure the laser operates in the counter-clockwise direction. An optical filter with a 0.3 nm bandwidth is used in the cavity to filter out the amplified spontaneous emission noise. The total cavity length is ~18m. The Brillouin gain in the HNLF is described by [7]

$$G = \exp(\frac{g_B P_{pump} L_{eff}}{A_{eff}} - \alpha L) \tag{1}$$

where $g_B = 21 A_{eff}/P_{th} K L_{eff}$ is the Brillouin gain coefficient, P_{pump} is the SBS pump power, α is the fiber loss coefficient, $K = 0.5$ for both SMF and standard HNLF, L_{eff} and L are the effective length and physical length of the SBS

550

gain medium, and A_{eff} is the fiber effective area. The parameters of the HNLF are as follows: the effective area A_{eff}=11.6μm², loss coefficient α=0.9dB/km, and $P_{th} \times L_{eff} = 18W \cdot m$. Assuming the SBS pump power in the HNLF is equal to the saturation output power of the EDFA, i.e. 14 dBm, the Brillouin gain will be 1.584 according to equation (1). If we replace the HNLF by a SMF of the same length with g_B=4×10⁻¹¹m/W, α=0.2dB/km, and A_{eff}=90μm², the Brillouin gain will become 1.092 and is much smaller than that of the HNLF. In the following experiment, we employ an unpumped EDF loop with an optimized length of 1.2m to achieve single-longitudinal mode operation. The output of the Brillouin/Erbium-doped fiber laser is coupled out through an 80:20 optical coupler.

III. EXPERIMENTAL RESULTS AND DISCUSSION

The output optical spectra of the Brillouin/Erbium-doped fiber laser is depicted in Fig. 2. In Fig. 2 (a), the SBS pump laser is located at 1560.928 nm while the output wavelength is observed at 1561.008 nm, verifying that the cavity operates as a Brillouin/Erbium-doped fiber laser. Although Rayleigh scattered wave and anti-Stokes wave will also be generated in the HNLF and amplified by the EDFA, there is an extra loss at the splice between the HNLF and the SMF. The waves cannot be amplified by SBS and are not sustainable in the cavity. By tuning the SBS pump wavelength and the optical filter, the output can be precisely tuned over ~30 nm from 1532 to 1562 nm, as depicted in Fig. 2 (b).

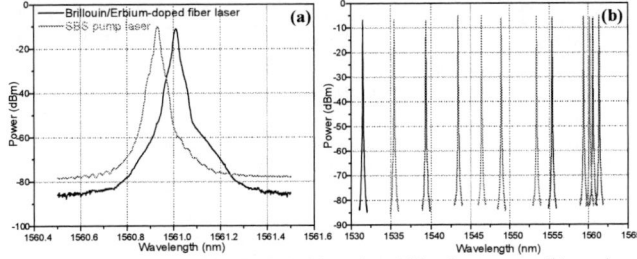

Fig. 2 Measured optical spectra. (a) SBS pump laser and Brillouin/Erbium-doped fiber laser output (b) precise tuning of the fiber laser output over ~ 30 nm wavelength range in the C-band.

The effect of mode-selection in the fiber laser is studied with delayed self-heterodyne interferometric detection shown in Fig. 1(b). The frequency shift in the AOM is 55 MHz and the delay fiber is about 10 km long. First, when the cavity lases without the SBS pump and the unpumped EDF loop, it acts as an EDF laser with an 11 MHz mode spacing. The measurement result is shown in two different frequency spans in Figs. 3(a) and 3(e). Next, when a SBS pump with -1 dBm power is injected into the cavity, most of the laser modes are strongly suppressed as seen from Figs. 3(b) and 3(f) over different frequency spans. The SBS pump depletes the EDFA gain. Consequently, the modes outside the SBS gain bandwidth cannot derive sufficient gain to lase while the modes within the SBS gain bandwidth are amplified by both the Brillouin gain and the EDFA. We also observe that the selected frequency mode is strengthened due to the SBS gain. Although the side modes near the selected frequency are not completely eliminated, they are suppressed by >47 dB. Typically, the 3-dB SBS gain bandwidth in a fiber is ~20 MHz. Due to the frequency drift of the SBS pump laser, the actual bandwidth will be larger. This is the reason why we can observe side modes beyond the 20MHz range. Therefore, to achieve single-longitudinal mode operation, an additional mode selection element is needed. Here, we use a 1.2-m unpumped EDF loop in the cavity for the purpose. To study the role of the unpumped EDF loop, the SBS pump laser is first removed and the laser output is shown in Figs. 3(c) and 3(g). Over a 50 MHz span depicted in Fig. 3(g), the side modes are almost completely suppressed. However, there are still several modes beyond the 50 MHz range as displayed in Fig. 3(c). Finally, when both the SBS pump and the unpumped EDF loop are included, a single-longitudinal mode fiber laser is achieved. The result is displayed over two different frequency spans in Fig. 3(d) and 3(h).

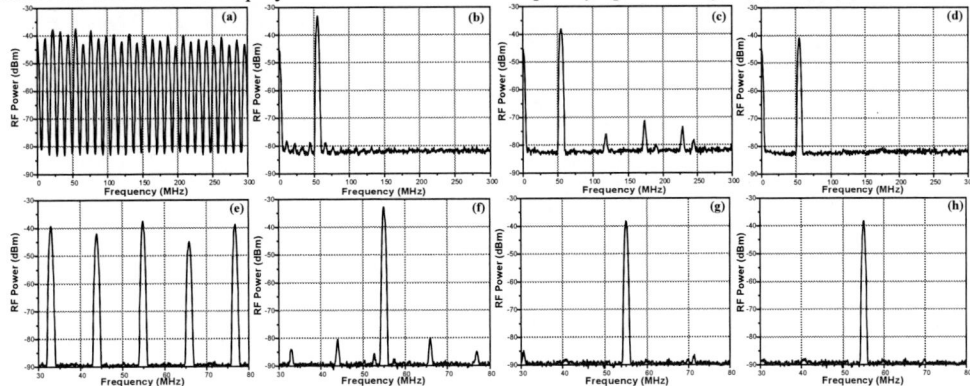

Fig. 3 Measured RF spectra over a 300 MHz span (top) and 50 MHz span (bottom). (a) (e) without SBS pump and unpumped EDF; (b) (f) with SBS pump and without unpumped EDF; (c) (g) without SBS pump and with unpumped EDF; (d) (h) with both SBS pump and unpumped EDF.

The linewidth of the single-longitudinal mode output is characterized. A strong linewidth reduction is observed in the Brillouin/Erbium-doped fiber laser as shown in Fig. 4 (a). Since the broadening effect of the 1/f noise is most pronounced near the center of the spectral lineshape [8], a more accurate linewidth measurement can be obtained at 20 dB down from the spectral peak. By Lorentzian fitting, the measured 20-dB linewidth of SBS pump is ~877.14 kHz and that of the Brillouin/Erbium-doped fiber laser is only ~7.85 kHz. A zoom-in view of the spectrum is shown in Fig. 4(b). Some ripples are observed because the fiber delay is shorter than the laser coherence time and the relative phase between the two interfering branches becomes important.

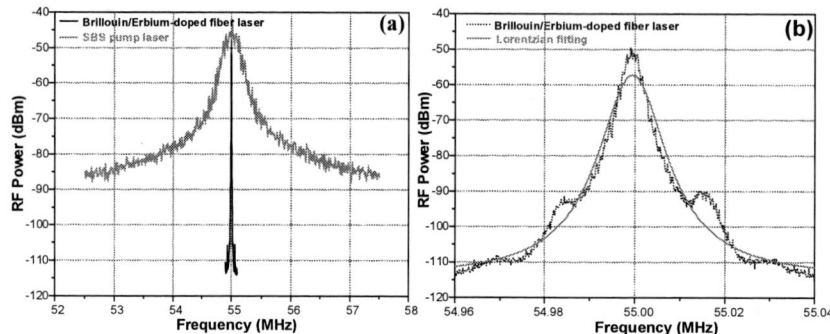

Fig. 4 Delayed self-heterodyne spectra of (a) SBS pump laser and Brillouin/Erbium-doped fiber laser output (b) zoom-in view of Brillouin/Erbium-doped fiber laser output and Lorentzian fitting.

IV. CONCLUSION

We have demonstrated a widely tunable, single-longitudinal mode Brillouin/Erbium-doped fiber laser based on an 8-m HNLF and an unpumped EDF loop. The laser can be precisely tuned over ~30 nm in the C-band by adjusting the SBS pump laser wavelength together with an optical filter. The linewidth narrowing effect in the HNLF is experimentally verified and the linewidth reduction ratio is larger than 100.

ACKNOWLEDGMENT

This work is supported by Hong Kong General Research Funds (CUHK 416213 and 14206614).

REFERENCES

[1] Park, Namkyoo, et al. "All fiber, low threshold, widely tunable single-frequency, erbium-doped fiber ring laser with a tandem fiber Fabry-Perot filter." Applied physics letters, 1991, pp. 2369-2371.

[2] Hill, K. O., B. S. Kawasaki, and D. C. Johnson. "CW Brillouin laser." *Applied Physics Letters*, 1976, pp. 608-609.

[3] Geng, Jihong, et al. "Highly stable low-noise Brillouin fiber laser with ultra-narrow spectral linewidth." *Photonics Technology Letter*, 2006, pp. 1813-1815.

[4] Cowle, Gregory J., and Dmitrii Yu Stepanov. "Hybrid Brillouin/erbium fiber laser." *Optics letters*, 1996, pp. 1250-1252.

[5] Debut, Alexis, Stéphane Randoux, and Jaouad Zemmouri. "Linewidth narrowing in Brillouin lasers: Theoretical analysis." *Physical Review A*, 2000.

[6] Cheng, Y., et al. "Stable single-frequency traveling-wave fiber loop laser with integral saturable-absorber-based tracking narrowband filter." *Optics letters*, 1995, pp. 875-877.

[7] Agrawal, Govind P. *Nonlinear fiber optics*. Academic press, 2007.

[8] Mercer, Linden B. "1/f frequency noise effects on self-heterodyne linewidth measurements." *Journal of Lightwave Technology*, 1991, pp. 485-493.

Continuously Tunable Microwave Photonic Filter Based on a Wavelength-spacing-tunable Multiwavelength laser

Seungmin Lee, Young Bo Shim, and Young-Geun Han
Department of Physics, Hanyang University
17 Haengdang-dong, Seongdong-gu, Seoul 133-791, Korea
yghan@hanyang.ac.kr

Abstract: We demonstrate a continuously tunable microwave photonic filter based on a wavelength-spacing-tunable multiwavelength laser. The free spectral range of the proposed microwave photonic filter was flexibly controlled in a range from 1.84 to 0.33 GHz.

Keywords: Multiwavelength fiber laser, Microwave filter, tunable filter

I. INTRODUCTION

There has been much research interest in microwave photonic filters because of their many advantages, such as high bandwidth, low loss, and immunity to electromagnetic interference (EMI) [1]. Microwave photonic filters have great potential in the application to Radio-over-Fiber (RoF) systems because they can provide by using microwave photonic filters the limitation of the traditional filters, e.g., the electric bottleneck problem and other sources of degradation. In general, microwave photonic filters consist of a multiwavelength laser and a dispersive medium. The wavelength spacing of the multiwavelength laser or the dispersion of the dispersive medium should be adjusted in order to control the free spectral range (FSR) of the microwave photonic filter. For the realization of multiwavelength laser with tunable wavelength spacing, tunable laser array was exploited [2]. The FSR of the microwave photonic filter was tuned discretely by switching on and off the laser source. In order to obtain multiwavelength in the single laser with the tunability of the wavelength spacing, tunable high birefringent Sagnac loop was used as a comb filter [3]. By using optical switches in the Sagnac loop mirror, the wavelength spacing was changed, and consequently the FSR of the microwave photonic filter was changed as well. However, since this scheme has only two wavelength spacings, the FSR of the microwave photonic filter could not be tuned continuously. Recently, multiwavelength laser with continuously tunable wavelength spacing by using a Mach-Zehnder interferometer (MZI) with an optical variable delay line for the microwave photonic filter was reported [4]. However, the continuous tunability of the microwave photonic filter was not shown. In this manuscript, we demonstrate a novel continuously tunable microwave photonic filter based on a polarization differential delay line (PDDL). The continuously tunable multiwavelength laser was realized by using the Saganc loop mirror with the PDDL. By controlling the PDDL, the wavelength spacing of the multiwavelength laser was continuously controlled while the stable lasing output was maintained. As the wavelength spacing was adjusted from 1.2 to 6.8 nm with the extinction ratio of ~ 40 nm, the FSR of the microwave photonic filter was continuously controlled from 1.84 to 0.33 GHz.

II. EXPERIMENTS AND DISCUSSION

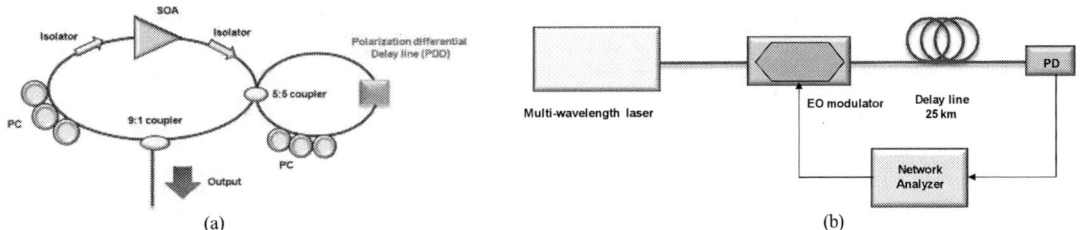

Fig. 1 (a) Experimental setup for the multiwavelength laser based on the PDDL, and (b) the microwave photonic filter.

Figure 1 (a) shows the continuously tunable multiwavelength laser based on the PDDL. The multiwavelength laser consists of a semiconductor optical amplifier (SOA), isolators for unidirectional lasing, a Sagnac loop mirror with the PDDL, a 9:1 coupler, and a polarization controller (PC). The time delay difference between two orthogonal polarization modes of the propagating light in the PDDL was precisely controlled and in turn the wavelength spacing of the Sagnac was tuned continuously. With the proposed multiwavelength laser, the microwave photonic filter was realized as shown in Fig. 1 (b). An electro-optic (EO) modulator driven by a signal generator is exploited to modulate the output of the multiwavelength laser. The modulated optical signal is transmitted through the 25-km single mode fiber as a dispersive medium. The optical power filtered by the microwave photonic filter can be expressed as [4],

$$P(f_{RF}) \propto \sum_{m=1}^{N} P_m \cos^2[\pi(m-1)f_{RF}D\Delta\lambda]$$

where f_{RF} is the modulation frequency, N is the number of the lasing wavelength. P_m is the output power of the mth lasing wavelength. D is the dispersion of the dispersive medium. $\Delta\lambda$ is the wavelength spacing.

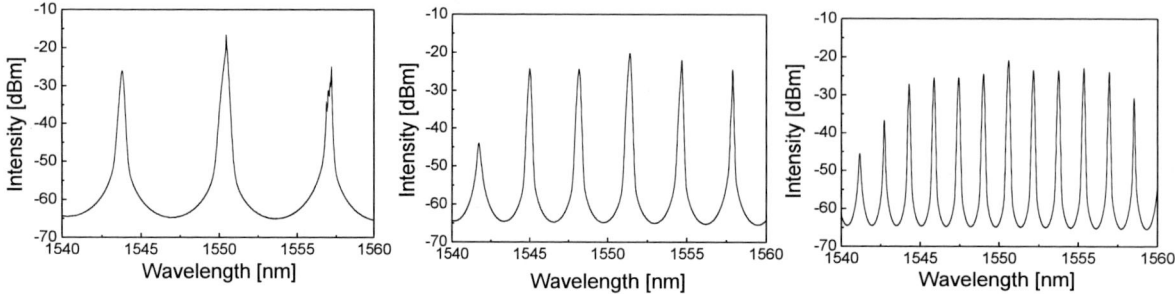

Fig. 2 (a) Output spectra of the multiwavelength fiber laser the with wavelength spacing tunability

Figure 2 shows the output spectrum of the multiwavelength laser based on the PDDL with the wavelength spacing tunability. An extinction ratio of the multiwavelength laser was measured to be ~ 40 dB. The frequency responses of the microwave photonic filters with respect to the wavelength spacing of 1.6, 3.2, 6.4 nm are shown in Fig 3. (a). As the wavelength spacing of the multiwavelength laser is increased, the FSR of the microwave photonic filter is decreased. The attenuation of the filter response at longer frequency is attributed to the degradation in the sensitivity of the photo-detector. Figure 3 (b) shows the theoretical and experimental results for variation of the FSR with respect to the wavelength spacing of the multiwavelength laser. As the wavelength spacing is adjusted from 1.2 to 6.8 nm with the extinction ratio of ~ 40 nm, the FSR of the microwave photonic filter was tuned from 1.84 to 0.33 GHz. The theoretical results are in good agreement with the experimental ones.

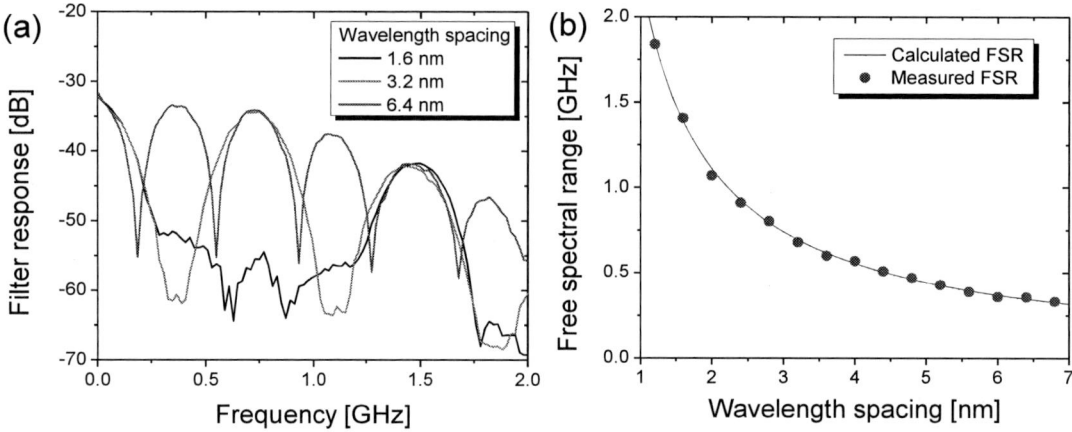

Fig. 3 (a) Frequency response of the microwave photonic filter and (b) the measured and calculated FSR as a function of the wavelength spacing.

III. CONCLUSIONS

To summarize, we presented a continuously tunable microwave photonic filter by a multiwavelength laser using a PDDL. The wavelength spacing of the multiwavelength laser was continuously controlled by the PDDL. The multiwavelength laser has a extinction ratio of ~ 40 dB and the continuous tunability of the wavelength spacing in a range from 1.2 to 6.8 nm. With the multiwavelength laser, the FSR of the microwave photonic filter was flexibly controlled in a range from 1.84 to 0.33 GHz. We believe that the proposed tunable microwave photonic filter is promising device in optical communication system, RoF, etc.

REFERENCES

[1] J. Capmany, B. Ortega, and D. Pastor, "A tutorial on microwave photonic filters," IEEE J. Lightwave Technol., 24, 201-29 (2006).

[2] J. Capmany, D. Pastor, and B. Ortega, "New and flexible fiber-optic delay-line filters using chirped Bragg gratings and laser arrays," IEEE T. Microw. Theory, 47, 1321-1326 (1999).

[3] L. R. Chen, and V. Pagé, "Tunable photonic microwave filter using semiconductor fibre laser," Electron. Lett., 41, 1183 - 1184 (2005).

[4] H. Ou, H. Fu, D. Chen, and S. He, "A tunable and reconfigurable microwave photonic filter based on a Raman fiber laser," Opt. Commun., 278, 48-51 (2007).

[5] X. Fang, and R. O. Claus, "Polarization-independent all-fiber wavelength-division multiplexer based on a Sagnac interferometer," Opt. Lett., 20, 2146-2148 (1995).

Real-Time Interrogation Technique Using Fourier Domain Mode-Locked Fiber Laser

Ik Su Jo, Seungmin Lee, Sanggwon Song, Kwang Wook Yoo, and Young-Geun Han

Department of Physics, Hanyang University
17 Haengdang-dong, Seongdong-gu, Seoul 133-791, Korea
yghan@hanyang.ac.kr

Abstract: *A real-time interrogation technique using a Fourier domain mode-locked fiber laser is investigated. We could successfully monitor the variation of the output powers and the detection time intervals of sensing signals at the photo-detector.*

Keywords: *Real-time monitoring, sensing signal interrogation, simultaneous measurement*

I. INTRODUCTION

Structural health monitoring system has been one of the essential aspects in the nuclear industry, as structures in nuclear power plants are subject to radiation damage or degradation which can cause a variety of changes in the physical properties of metals, concrete, and other materials [1–3]. The nuclear power sector is continuously looking into safety improvement and operations, in order to reduce the risks of accidents. Because there are many physical parameters to be determined in a nuclear power plant, such as temperature, pressure, strain, vibration, and radiation doses, multi-parameter monitoring techniques have been intensively investigated [4–6].

Fiber-optic sensors based on fiber Bragg gratings (FBGs) have been widely applied in multi-parameter sensing because an FBG reflects a specific wavelength which shifts sensitively, depending on various physical parameters that need to be determined in a nuclear power plant. It also has many advantages over other electrical and optical sensors, such as electromagnetic interference (EMI) resistance, possibility to carry multiplexed signals (time, wavelength multiplexing), possibility to monitor sites far away from the controller, and it is also suitable for distributed configuration [7–11]. However, FBGs-based multi-parameter monitoring system requires the numerous data process because it performs extensive data collection in the condition of the nuclear facilities and other parameters as well. Therefore, it is necessary to develop a high speed signal interrogation system.

Various types of FBG interrogation systems based on a passive optical filter [12–14], radio-frequency modulation [15], optical spectroscopy [16–18], and a wavelength-swept laser [19–21] have been introduced. Especially, wavelength-swept lasers have been prevalently employed due to their unique advantages such as broad wavelength range, high signal-to-noise ratio, high repetition speed and better resolution. However, standard approaches for conventional wavelength-swept lasers are usually inherently limited in the maximum achievable sweep speed due to build-up time limitations.

In this paper, we propose a high speed interrogators based on Fourier-domain mode-locked (FDML) swept laser for simultaneous measurement of radiation dose and strain. The applied strain changes the center wavelengths of multiple FBGs and the detection time intervals of the output signals at the photo-detector simultaneously. Since the operating wavelength of the proposed FDML fiber laser is continuously swept as a function of time, it is possible to simultaneously measure the variation multipoint strain in real time.

II. EXPERIMENTS AND DISCUSSION

Figure 1 shows the experimental setup and the photograph for the proposed FDML swept laser. The FDML swept laser is composed of a semiconductor optical amplifier (SOA, COVEGA SOA-2771) as a broadband gain medium, a 2.5 km dispersion compensating fiber (DCF) and a 5.2 km single mode fiber (SMF) for controlling the chromatic dispersion and the swept rate, a polarization controller (PC), a fiber Fabry-Perot tunable filter (FFP-TF, LambdaQuest, Inc) driven by a function generator, two isolators for a unidirectional rotating, and 9:1 coupler. For FDML operation, light from the SOA is launched into the FFP-TF driven with a sinusoidal waveform from the function generator. By synchronizing the driving frequency to the optical round-trip time of a cavity with a length of 7.7 km, The FDML can be readily realized. Because two optical wavelength sweeps are generated for each period of the sinusoidal drive waveform (bidirectional sweeping), the resulting effective sweep rate was 25.4 kHz and 10% of the intra-cavity power is extracted by 9:1 coupler.

Figure 2 shows the transmission spectra of the FDML laser for different bandwidth tuning ranges of the FFP-TF. As the tuning range of the FFP-TF increased from 60 to 75 nm at the same driving frequency, the time duration for which the filter transmits at a certain wavelength is shortened, and the transmission loss is increased. Therefore, the output bandwidth was widened, but the signal-to-noise ratio (SNR) was reduced from 52.2 to 40.8 dB. The optimized bandwidth was estimated to be ~70 nm. The driving frequency of the FFP-TF was set to 25.4 kHz with respect to a total length of 7.7 km in the cavity. The full width at half-maximum and average minimum extinction ratio were measured to be 70 nm and 45 dB at 1547 nm, respectively. The amplitudes of the temporal intensity profile are repeatedly generated for each period to satisfy the role of the FBG sensor interrogation. Because of bidirectional frequency sweeping, two pulses for forward and backward sweeping were generated in the pulse period of 39.3 μs, which is directly related to the driving frequency of 25.4 kHz. Consequently, the dispersion-compensated FDML fiber laser with a high sweep speed of 25.4 kHz and broad bandwidth of 70 nm can be expected to have less uncertainty in the center wavelength, which is suitable for accurate multiple sensor interrogation.

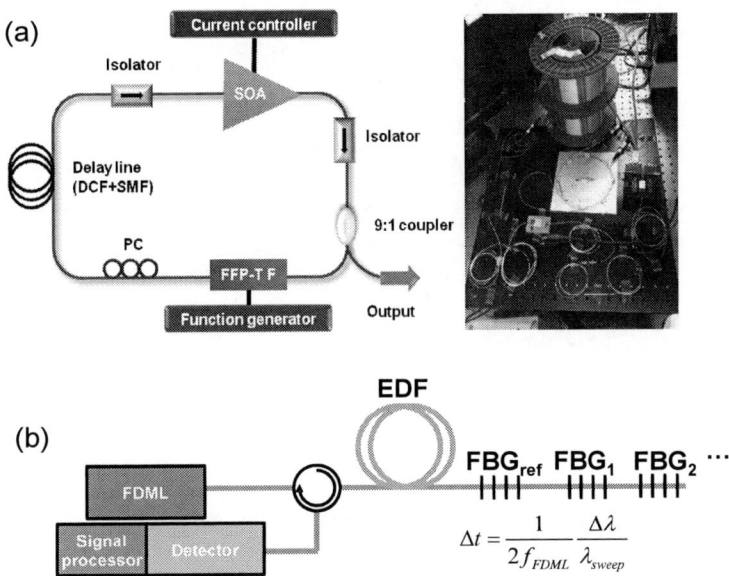

Figure 1. Experimental scheme for the proposed FDML swept laser.

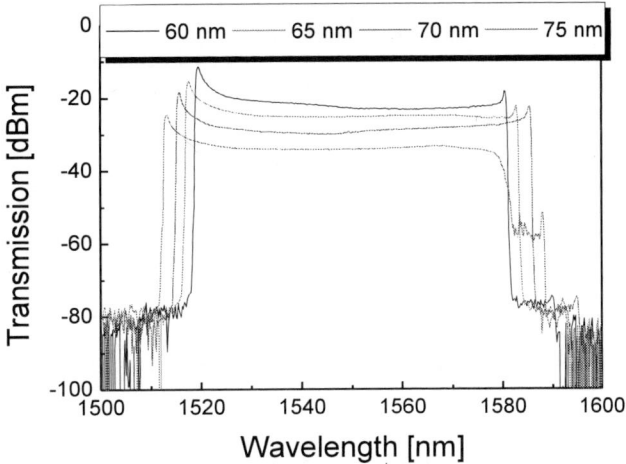

Fig. 2. Output spectra of the proposed FDML swept laser for different bandwidth tuning ranges of the FFP-TF.

Since the operating wavelength of the FDML is continuously swept as a function of time, the wavelength spacing ($\Delta\lambda$) between the reference FBG and the sensing probes like four FBGs effectively determines the detection time intervals (Δt) of the output signal at the photo-detector, which can be described as

$$\Delta t = \frac{1}{2f_{FDML}} \frac{\Delta\lambda}{\lambda_{sweep}},$$

where f_{FDML} and λ_{sweep} are the sweeping rate and the wavelength swept range of the FDML swept laser, respectively. The variation in external strain can be successfully monitored by measuring the variation of the detection time intervals of the sensing signals as a function of strain as seen in Fig. 1(b).

The strain applied to the FBGs can be measured by reading the variation of the time delay because the change in the wavelength spacing ($\Delta\lambda = \lambda_N - \lambda_{ref}$, N=1, 2, ... for FBG1, FBG2, ... , respectively) can be directly converted to the time delay ($\Delta\tau = \tau_N - \tau_1$),. Plots of the variation of the time delay ($\Delta\tau = \tau_N - \tau_1$) as strain is applied to FBG1 and FBG2 are shown in Fig. 3. When strain was applied up to 2000 $\mu\varepsilon$, the strain sensitivities of the FBGs to the variation of the time delay had the same value of -0.2 ps/mε. Consequently, the proposed multi-monitoring sensing technique makes possible to monitor the multi-point strain in real-time by measuring the normalized optical powers and the detection time intervals of the sensing signals.

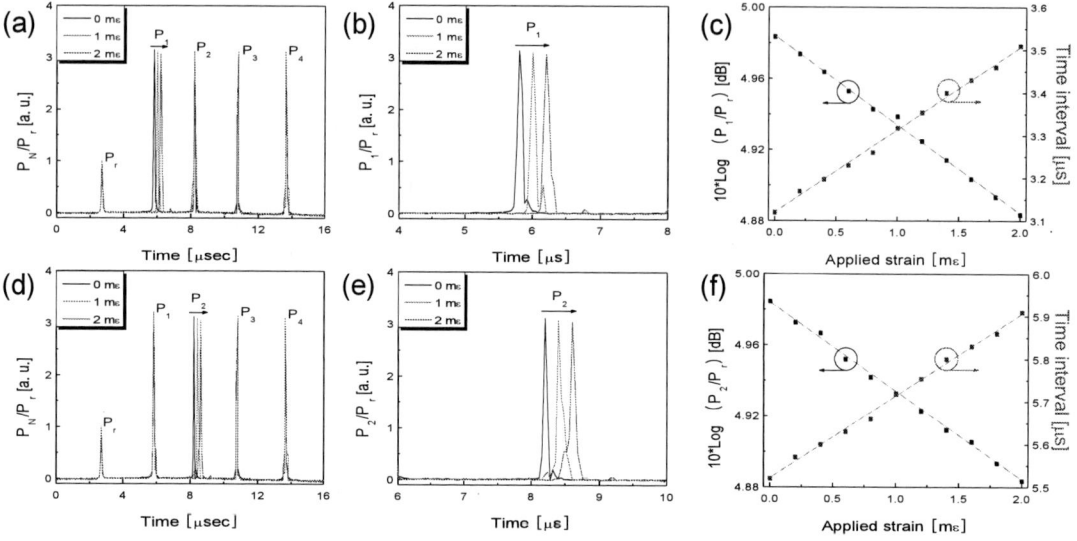

Fig 3. Normalized output signals at the photo-detector when strain is applied to the FBG1 (from (a) to (c)) and the FBG2 (from (d) to (f)) as a function of strain.

III. CONCLUSIONS

We experimentally demonstrated a multi-monitoring sensing interrogation technique based on a FDML swept laser. By using the continuously swept FDML fiber laser in time-domain, we could successfully monitor the variation of the output powers and the detection time intervals of the sensing probe signals at the photo-detector. The experiment results showed the significant improvements of the sensing measurement range, measurement time, and sensor multiplexing capabilities by employing the continuously swept FDML fiber laser which provides immunity to optical fiber losses and optical source power fluctuations.

REFERENCES

[1] Hatley, D. D., Watkings Jr, K. S., Chai, J., and Kim, W., "On-line intelligent self-diagnostic monitoring system for next generation nuclear power plants," Pacific Northwest National Laboratory, 2003.

[2] Jouan, B., Rudolph, J., and Bergholz, S. "Structural Health Monitoring Solutions for Power Plants," In EWSHM-7th European Workshop on Structural Health Monitoring, 2014.

[3] Glisic, B., and Inaudi, D., "Fibre optic methods for structural health monitoring," John Wiley and Sons, 2008.

[4] P. Ferdinand, S. Magne, O. Roy, V. D. Marty, S. Rougeault, and M. Bugaud, "Optical Fiber Sensors for the Nuclear Environment," In Optical Sensors and Microsystems, pp. 205–226, 2002.

[5] R. Lin, Z. Wang, and Y. Sun, "Wireless sensor networks solutions for real time monitoring of nuclear power plant," Intelligent Control and Automation, WCICA 2004 Fifth World Congress on., vol. 4, 2004.

[6] K. Nabeshima, T. Suzudo, K. Suzuki, and E. TÜRKCAN, "Real-time nuclear power plant monitoring with neural network," Journal of nuclear science and technology, vol. 35, pp. 93–100, 1998.

[7] Frazão, O., Ferreira, L. A., Araújo, F. M., and Santos, J. L., "Applications of fiber optic grating technology to multi-parameter measurement," Fiber and integrated optics, vol. 24, pp. 227–244, 2005.

[8] Rao, Y. J., "In-fibre Bragg grating sensors. Measurement science and technology," vol. 8, pp. 355, 1997.

[9] Chong, S. Y., Lee, J. R., Yun, C. Y., and Sohn, H., "Design of copper/carbon-coated fiber Bragg grating acoustic sensor net for integrated health monitoring of nuclear power plant," Nuclear Engineering and design, vol. 241, pp. 1889–1898, 2011.

[10] Mawatari, T., and Nelson, D., "A multi-parameter Bragg grating fiber optic sensor and triaxial strain measurement." Smart Materials and Structures, vol. 17, pp. 035033, 2008.

[11] Li, H. N., Li, D. S., and Song, G. B., "Recent applications of fiber optic sensors to health monitoring in civil engineering," Engineering structures, vol. 26, pp. 1647–1657, 2004.

TuD1-1 (Invited)

OECC/PS2016

Quantum Dot Lasers for Silicon Photonics

Yasuhiko Arakawa

Institute for Nano Quantum Information Electronics, Institute of Industrial Science
The University of Tokyo
e-mail address: email: arakawa@iis.u-tokyo.ac.jp

Abstract: *We discuss recent progresses in silicon photonics integrating quantum dot lasers. High temperature stability and high feedback-noise tolerance of the quantum dot lasers are advantageous features for application to silicon photonics. A silicon optical interposer with the bandwidth-density of 15Tbps/cm² at 125 C was demonstrated using flip-chip bonding method. Quantum dot lasers on silicon by wafer-bonding technique are also discussed.*

Keywords: *quantum dot lasers, silicon photonics, wafer bonding*

In the history of semiconductor lasers, the quantum-size effect is one of important physics which gave a great impact on enhancement of lasing characteristics. In 1982, the concept of quantum dot lasers was proposed with theoretical prediction of temperature insensitivity of threshold current by Arakawa and Sakaki [1]. In 2004, the temperature insensitivity was successfully demonstrated in a self-assembled InAs/GaAs quantum dot laser by Fujitsu and the Univ. of Tokyo [2]. This achievement was led to launch of a venture company named QD Laser Inc.. The quantum dot lasers have a variety of superior performance to conventional lasers such as high temperature operation, low-power consumption, and low-cost productivity. Silicon photonics is a promising application for the quantum dot lasers.

Evolution of computing systems is indispensable for realizing both cloud computing and autonomous distributed (edge) computing systems in the future cyber physical society. The CPU-CPU/CPU-memory inter-chip bandwidths in personal computers and servers are currently doubled every two years and are estimated to reach the 10 tera-scale bit rate around the end of the 2010s. In particular, the bandwidth for inter-chip interconnects in the future is required to be much higher than that for intra-chip ones. However, there are no solutions with electrical interconnects. Since photons can provide with wide bandwidth, low latency, low power consumption, and low mutual interference, optical interconnect based on silicon photonics is expected to overcome the bandwidth bottleneck and other problems and to open new directions for research into novel photonics function [3].

In this presentation, we discuss our recent progresses in quantum dot lasers for silicon photonics application. A silicon optical interposer integrating quantum dot laser arrays achieved the bandwidth-density of 15Tbps/cm² up to 125 C with a line speed of 20Gbps/channel (Fig.1) [4,5]. The quantum dot laser arrays was mounted by flip-chip bonding method. High temperature stability and high feedback-noise tolerance due to small α–parameter are advantageous features for application of the quantum dot lasers to isolator-free silicon optical interposers. Quantum dot lasers integrated on silicon substrates by wafer-bonding technique are also discussed [6].

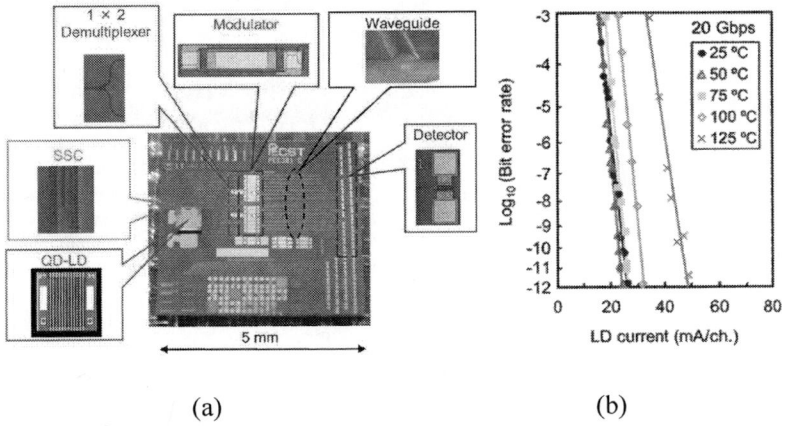

(a)   (b)

Fig.1: (a) Silicon optical integrated system with quantum dot laser array (b) 20 Gbps data transmission from 25 to 125 °C. The results demonstrate capability of the system with bandwidth density of 15Gbps/cm² without any temperature adjustment up to 125℃.

ACKNOWLEDGMENT

The author would like to thank Y. Urino, N. Hatori, K. Mizutani, T. Usuki, J. Fujikata, K.Yamada, T. Horikawa, T. Nakamura, S. Iwamoto, and K. Tanabe for their contribution. This research was partly supported by the New Energy and Industrial Technology Development Organization (NEDO) and FIRST program.

REFERENCES

[1] Y. Arakawa and H. Sakaki, Appl. Phys. Lett. 40, 939 (1982)

[2] Y. Arakawa et al., IEEE Communications Magazine, IEEE 51, 71 (2013)

[3] K. Otsubo, N. Hatori, M. Ishida, S. Okumura, T.Akiyama, Y. Nakata, H. Ebe, M. Sugawara and Y.Arakawa Jpn. J. of Appl. Phys. Vol. 43, L1124 (2004)

[4] N. Hatori, T. Shimizu, M. Okano, M. Ishizaka, T. Yamamoto, Y. Urino, M. Mori, T. Nakamura, and Y. Arakawa, IEEE J. Lightwave Tech., 32, 1329, (2014).

[5] Y. Urino, N. Hatori, K. Mizutani, T. Usuki, J. Fujikata, K.Yamada, T. Horikawa, T. Nakamura, and Y. Arakawa, IEEE Light wave Tech.30, 1223 (2015)

[6] K. Tanabe, T. Rae, K. Watanabe, and Y. Arakawa, Appl. Phys. Express 6, 2013-08-25 (2013)

TuD1-2

OECC/PS2016

Low temperature lasing characteristics of GaInAsP double-hetero laser integrated on InP/Si substrate using direct wafer bonding

Tetsuo Nishiyama, Keiichi Matsumoto, Junya Kishikawa, Toshiki Sukigara, Yuya Onuki, Naoki Kamada, Tomokazu Kanke, and Kazuhiko Shimomura

Dept. Engineering and Applied Sciences, Sophia University
7-1 Kioi-cho, Chiyoda-ku, Tokyo 102-8554, Japan
kshimom@sophia.ac.jp

Abstract: We have successfully obtained 2-inch mirror like surface InP/Si wafer by using direct bonding. GaInAsP-InP double-hetero structure laser was grown on InP/Si substrate by MOVPE. Low temperature lasing was attained and show their characteristics.

Keywords: Si photonics, InP, laser, MOVPE, direct wafer bonding

I. INTRODUCTION

Recently, there has been a great deal of attention for the integration of InP-based optical devices on Si as a light source for the realization of optical interconnection [1]. The integration techniques so far can be categorized into the monolithic integration and the hybrid integration. Monolithic integration demonstrating direct hetero-epitaxial growth of InP-based materials on Si would be the most desirable approach, but has been difficult to achieve due to the large lattice mismatch that causes the high density dislocations and resided stress. On the other hand, hybrid integration using variety of material bonding techniques between InP-based device chip and Si substrate is not subject to lattice matching limitations associated with epitaxial growth. So integration of InP-based light source on Si has been recently done by hybrid integration. However, as precise alignment has been required between InP-based device chip and Si, realizing a high density integration of the devices in an industrial scale may have challenges in its running cost, yield, and automation. Therefore, we have so far proposed monolithic integration of InP-based optical devices on wafer bonded InP/Si substrate using low pressure MOVPE [2-5]. This approach would enable to conquer the high density dislocations and anti-phased-disorder which have been inevitable in conventional monolithic approach and realize the growth of InP-based materials on Si with high crystalline quality. In this work, we report the epitaxial growth of GaInAsP-InP double-hetero laser structure on the InP/Si substrate. Successful lasing emission is obtained from the devices at cryogenic condition. This may be very promising for the integration of InP-based laser on Si as a light source for optical interconnection using our proposal.

II. EXPERIMENTS AND RESULTS

The preparation of InP/Si substrate using direct wafer bonding technique is explained as the following. GaInAs / n-InP / GaInAs were grown on (100)InP substrate by using low-pressure MOVPE. InP substrate was chemically etched using the hydrochloric acid and obtained the thin GaInAs/InP/GaInAs film. Si substrate was cleaned by HF and $NH_4OH:H_2O_2:H_2O$ solution and removed a native oxide and organic contaminants of the surface. Then GaInAs/InP/GaInAs film and Si substrate was dipped into the $H_2SO_4:H_2O_2:H_2O$ solution, and InP and Si surface terminated by OH group were contacted and these film and substrate were softly bonded by the van der Waals forces and the hydrogen bond. Finally, the InP/Si substrate were heated at 400-degree C with weight. Fig2 shows the photograph of fabricated InP/Si substrate of 2-inch wafer. We have successfully obtained the mirror like surface and was good in comparison with InP substrate.

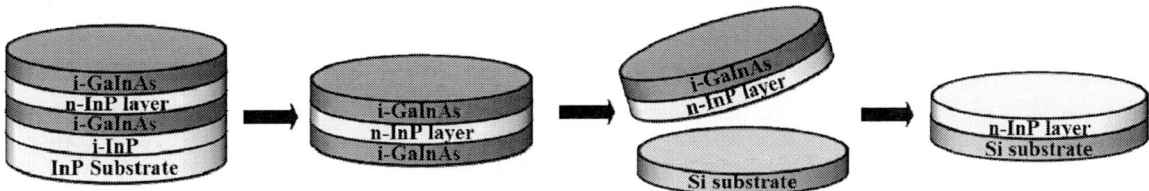

Fig.1 Schematic diagram of InP/Si fabrication process.

561

OECC/PS2016

Fig.2 A photograph of a thin InP epi-layer on a Si substrate demonstrating.

By using this InP/Si substrate, we have grown the GaInAsP-InP double-hetero structure using low pressure MOVPE as shown in Fig.3. The precursors were all organometallic compounds of trietyl-gallium (TEG), trimetyl-indium (TMI), tertiarybutyl-phosphine (TBP), tertiarybutyl-arsine (TBA), and dopants were dietyl-zinc (DEZn) and ditertiarybutyl-silane (DTBSi). The growth temperature, and pressure were 630 degrees C, 60 Torr, respectively. Fig.4 shows the PL spectrum of this structure grown on InP/Si substrate and InP substrate. The PL intensity on InP/Si substrate was a little lower than on InP substrate, however there were no significant difference of PL peak wavelength between two substrates. In the fabrication of laser, Au/Zn and Au/Al electrodes were evaporated to the p-GaInAs, n-Si, respectively. The laser cavity was made by cleavage of both facets, and lasing mode was Fabry-Perot mode.

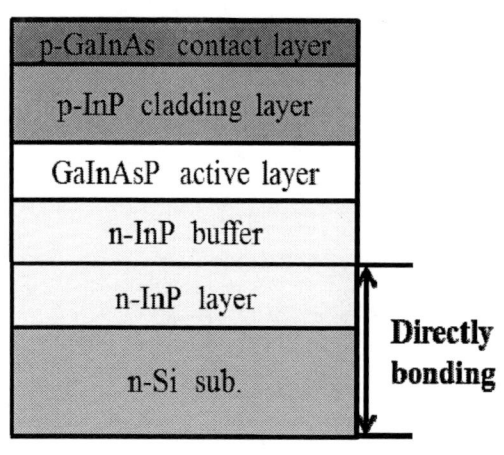

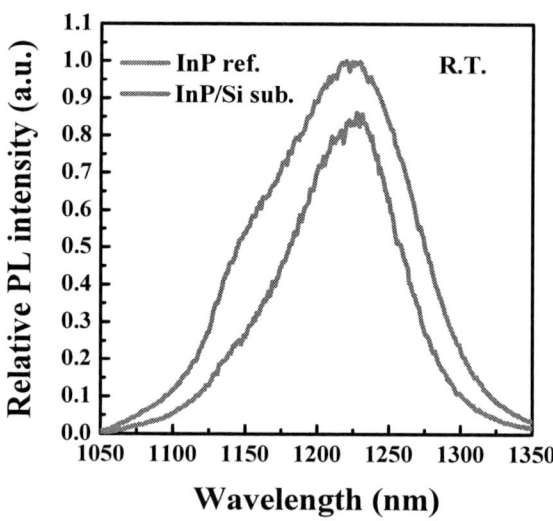

Fig.3 Layer structure of fabricated DH laser.

Fig.4 Relative PL characteristics.

Fig.5 shows the lasing spectrum of GaInAsP-InP double-hetero laser grown on InP/Si substrate. The laser chip mounted on the fixture was dipped into the liquid–nitrogen and measurements were performed under the room temperature atmosphere. The electrical current was pulsed current where the pulse width and duty ratio were 0.1μs, 0.1%, respectively. The edge emitted light from the cleaved facet coupled to the multi-mode fiber was observed by the optical spectrum analyzer. The dimensions of the device were $230*80\mu m^2$. By increasing the current density to the device, we have successfully obtained the lasing characteristics, and the lasing wavelength was 1175nm. Fig.6 shows the relative power and current density characteristics of this device, and relative power was obtained from the integrated power of spectrum in Fig.5. From this figure, the threshold current density was estimated about $4.5kA/cm^2$.

562

During MOVPE growth, we have grown the DH laser structure on the InP substrate at the same time. The threshold current density of the laser grown on InP substrate was 4.0kA/cm^2, hence the lasing characteristics on the InP/Si substrate was almost comparable with on the InP substrate. These results show our proposal technique to grow the InP-based layers on InP/Si substrate is effective for the integration of InP-based devices on the silicon substrate.

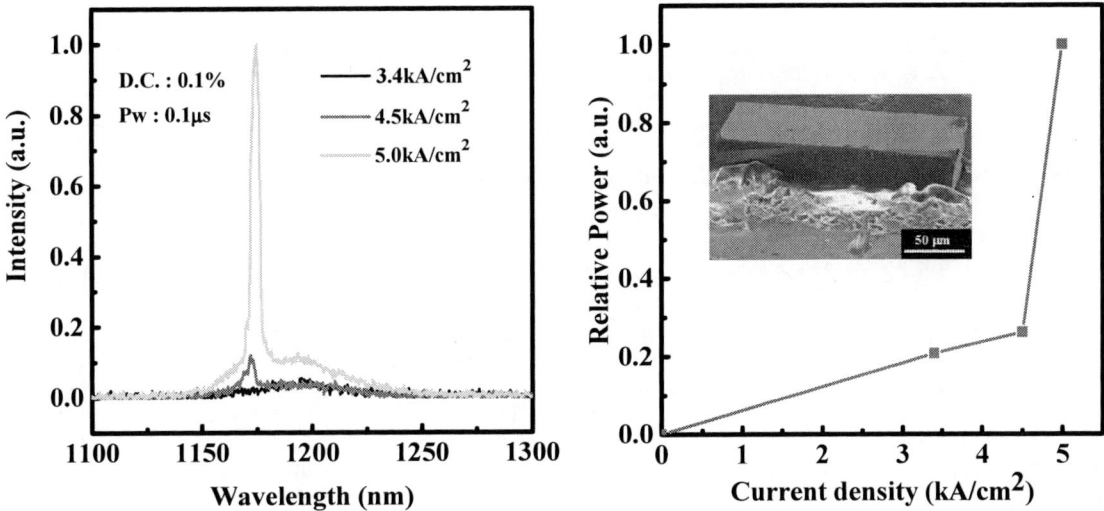

Fig.5 Lasing spectrum.   Fig.6 Light-current characteristics.

III. CONCLUSIONS

We have demonstrated the novel integration method of InP-based laser on Si substrate. Mirror like surface InP/Si substrate with 1μm thickness InP layer was successfully obtained using direct wafer bonding technique. GaInAsP/InP DH laser structure was grown on InP/Si substrate by low pressure MOVPE. By comparing the same structure on InP substrate, PL intensity was almost the same. Low temperature lasing was obtained in the edge emitted FP laser under pulsed current condition. The threshold current density of the laser on the InP/Si substrate and on the InP substrate was 4.5kA/cm^2, 4.0kA/cm^2, respectively. Lasing characteristics on the InP/Si substrate were almost comparable to the InP substrate. Our proposed integration method is very attractive for the integration of III-V devices on Si platform.

IV. REFERENCES

[1] D.A.B.Miller, "Rationale and Challenges for Optical Interconnects to Electronic Chips" Proc. IEEE, vol. 88, pp. 728-749, June 2000.
[2] K. Matsumoto, T. Makino, K. Kimura, and K. Shimomura, "Growth of GaInAs/InP MQW using MOVPE on directly-bonded InP/Si substrate", J. Crystal Growth, vol.370, pp.133-135, May 2013.
[3] K. Matsumoto, T. Makino, K. Kimura, K. Shimomura, "Extremely improved InP template and GaInAsP system growth on directly- bonded InP/SiO2-Si and InP/glass substrate", Phys. Staus Solidi C, vol.10, no.5, pp. 782-785, May 2013.
[4] K. Matsumoto, R. Kobie, and K. Shimomura, "Thermal treatment for preventing void formation on directly-bonded InP/Si interface", Jpn. J. Appl. Phys., vol.53, no.11, 116502, Oct. 2014.
[5] K. Matsumoto, Y. Kanaya, J. Kishikawa, Y. Yamamoto, T. Sukigara, T. Nishiyama, and K. Shimomura, "Epitaxial Grown GaInAsP-InP laser on wafer bonded InP/Si substrate", 42nd International Symposium on Compound Semiconductors (ISCS 2015), Santa Barbara, CA, USA, O6.6, July 1, 2015.

TuD1-3   OECC/PS2016

First Demonstration of Mode Selective Light Source by Using Active Multimode Interferometer Laser Diode

Bingzhou Hong*[1], Takuya Kitano[1], Akio Tajima[2], Haisong Jiang[1] and Kiichi Hamamoto[1]

[1]I-EggS, Kyushu Univ., 6-1, Kasuga-Koen, Kasuga, Fukuoka 816-8580, Japan
*2ES14054M@s.kyushu-u.ac.jp
[2]Green Platform Research Laboratories, NEC Corp., Kanagawa, Japan

Abstract: Mode selective light source is demonstrated by using active-multimode interferometer laser diode for the first time. Individual lasing of fundamental mode and first mode is successfully confirmed.

I. INTRODUCTION

Researchers have explored to multiplexing in time, wavelength, polarization and phase to increase data carrying capacity of SMF. Besides these, mode division multiplexing (MDM) is another approach to enhance transmission capacity of optical fibers [1-3]. Mode converter has been researched for MDM system, however, there is less research about mode-selective light source that emits in desired mode selectively on a light source itself. If such that mode selective light source is possible to be realized, the potential of MDM is expected to be extended toward short distance network including fiber to the home (FTTH) and data center due to its compactness and the possibility of low cost.

In this paper, we propose and demonstrate mode selective light source by using active-multimode interferometer laser diode (active-MMI LD) for the first time. By setting separated transverse mode path in one laser diode, the gain for different mode can be controlled individually. As a result, by adjusting current into mode-selector section, individual lasing of fundamental mode and first mode is successfully confirmed.

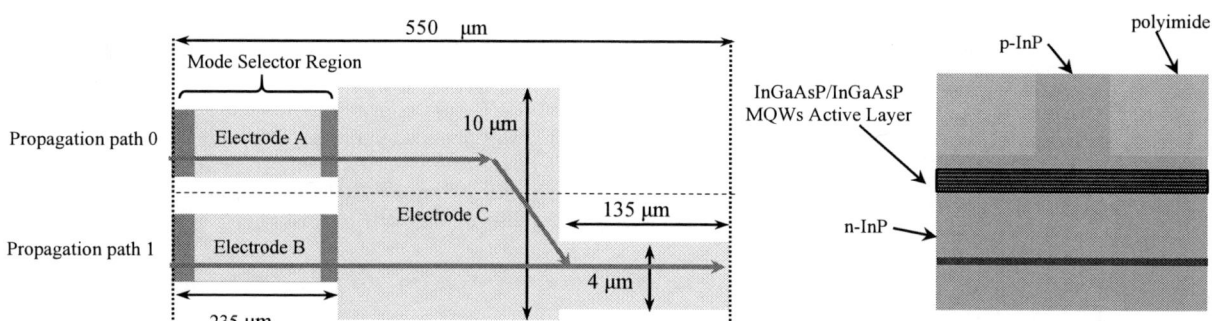

Fig. 1 (a) Schematic view and (b) layer structure of active MMI

II. DEVICE CONCEPT

Figure 1 (a) shows the schematic configuration of the implemented active-MMI LD. This structure was designed to support different mode paths inside one laser diode. The total length of the laser cavity is 550 μm. Width of MMI section was set as 10 μm. The length of MMI section was 180 μm. One edge of MMI was connected with one straight waveguide of 4 μm width and another edge of MMI was connected with two 5 μm off-centered straight waveguide of the same width. The device layer structure is illustrated in Fig. 1 (b). A ridge waveguide structure was used in our active-MMI LD fabrication. For the practically implemented device, 7 layers InGaAsP/ InGaAsP multiple quantum well (MQW) active layer has been used in order to obtain a high gain [4-7]. The side wall of the ridge waveguide was buried by polyimide material.

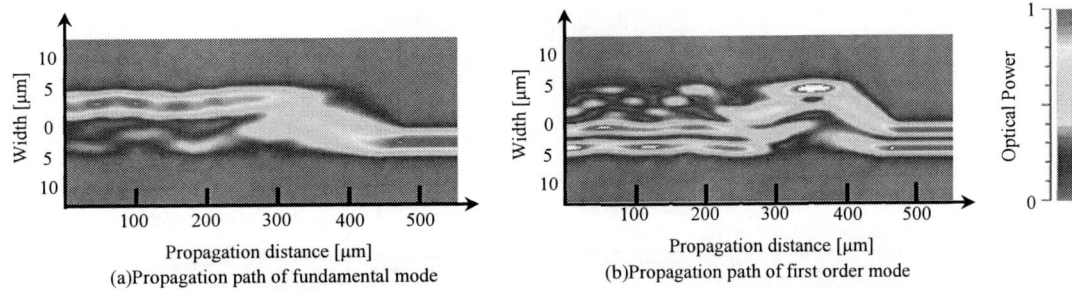

(a)Propagation path of fundamental mode   (b)Propagation path of first order mode

Fig. 2 BPM simulation results of individual mode for (a)fundamental mode and (b)first order mode

By using this configuration, two propagation paths are permitted within one laser structure. We verify our scheme of utilizing this waveguide configuration to obtain individual propagation path for each mode by using beam propagation method (BPM) [8]. Figure. 2 shows the simulation results of two different identical mode propagation paths. As can be seen in Fig. 2, fundamental mode and first order mode have the individual propagation paths. Propagation path 0 and propagation path 1 in Fig. 1 represent propagation path for fundamental mode and first order mode respectively. The two lateral modes share common propagation area in waveguide, which is indicated as section C in Fig. 1. The two straight waveguides (section A and B) at mode selector region in Fig. 1 (a) permit only fundamental mode and first order mode. In order to adjust injection current to each part, we have constructed separated electrode pad for different sections. By separating the electrode, injected current into each part was controlled, which brings the possibility of control the gain of different lateral mode.

III. RESULTS AND DISCUSSION

A. Operation principle

In order to obtain a selective lateral output mode, we have considered to control the injected current of each part. The common propagation path of fundamental mode and first order mode were covered by electrode pad C (in Fig. 1). Current injection through electrode C is to provide enough gain for the both modes. Current is injected only into electrode A in addition to pad C in case of fundamental mode oscillation. As propagation path for fundamental mode mainly cover these two sections, when the current is well adjusted, photons in fundamental mode propagation path have sufficient gain for the oscillation. First order mode does not get gain as there is no current injection in section B. In this case, section B acts as an absorption section. Similarly, in case first order mode is desired, current is injected into electrode B in addition to pad C. Section A absorbs photons and this current injection scheme prevent fundamental mode oscillation.

By selecting the current injected into each part properly, an only fundamental/first order mode lasing condition can be obtained.

TABLE I

CURRENT INJECTION PLAN AND CORRESPONDING NEAR FIELD PATTERN

Current injection Plan	Current injected into electrode pad C [mA]	Current injected into electrode pad B [mA]	Current injected into Electrode pad A [mA]	Near filed pattern in Fig. 3
1	Total 100			a
2	43	0	43	b
3	100	26	0	c

B. Experimental results and discussion

In this work, we have chosen the current injection plans shown in Table. I.

The threshold current of active-MMI LD is 55 mA, however in current injection plan 1, 100 mA current was injected into all 3 sections. In this situation, total current injected into the active MMI LD was 100 mA. We have monitored the near field pattern (NFP) form the back facet (left facet in Fig. 1 (a)). Observed near field pattern is shown in Fig. 3 (a). As can be seen in the figure, two laterally different modes are observed simultaneously. The right spot in NFP corresponds to fundamental mode while left spot corresponds to first order mode.

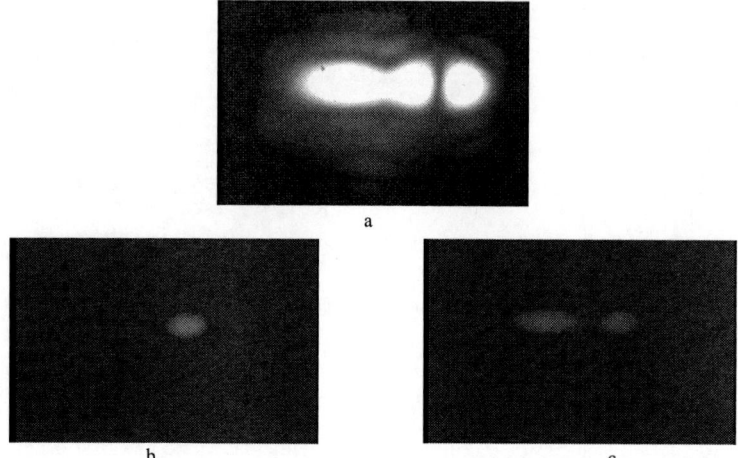

Fig. 3 Near field pattern for different current injection
(a) Plan 1 (b) Plan 2 (c) Plan 3

In current injection plan 2, 43 mA current was injected into section A in addition to C through electrode. Note that in this situation, the total current injected into the laser was 86 mA. No current was injected into section B. NFP of this plan is illustrated as Fig. 3 (b). As can be seen in the figure, the right side spot, which is the lasing spot of fundamental mode, was obtained. Meanwhile, there was no first order mode lasing spot observed. Figure 3 (c) was obtained in condition of current injection plan 3. 100 mA current was injected into the laser diode through electrode in section C. Current injected into electrode pad B and A was 26 mA and 0 mA respectively. NFP of first order mode observed in Fig. 3 (c) while lasing of fundamental mode was suppressed. In plan 2, when current of section A get increased to more than 43 mA, another lateral mode also starts lasing, which brings a NFP similar to Fig .3 (a). Same phenomenon happens if we increase current of section B in plan 3, therefore, these mode sectors may get saturation in photon absorption. For stable mode-selection, we need further oscillation blocker in addition to these mode-selector sections.

IV. CONCLUSIONS

We demonstrated a mode selective active MMI LD for the first time. Separated mode selecting electrode pad (namely, mode selector section) was designed and fabricated. As a result, individually lasing of fundamental mode as well as first order mode was obtained successfully.

REFERENCES

[1] S. G. L. Saval, "Photonic Lanterns for Mode Division Multiplexing," MOC, D3, pp. 46-47, October 2015.
[2] F. Ferreira, D. Borne, H. Silva and P Monteiro, "Crosstalk Optimization of Phase Masks for Mode Multiplexing in Few Mode Fibers", OFC, JW2A.37, March 2012.
[3] N. K. Fontaine, R. R. Ryf, S. G. L. Saval, J. B. Hawthorn, "Evaluation of Photonic Lanterns for Lossless Mode-Multiplexing," ECOC, Th.2.D.6, September 2012.
[4] K. Hamamoto, J. D. Merlier, M. Ohya, K. Shiba, K Naniwae, S. Sudo, T. Sasaki, "First demonstration of novel active multi-mode interferometer (MMI) LDs integrated with 1st order-mode permitted waveguides", IEICE Electron. Express, vol. 9, no. 12, pp. 1488-1453, July 2005.
[5] H. Jiang, H. A. Bastawrous, T. Hagio, S. Matsuo, and K. Hamamoto, "Low Hysteresis Threshold Current (39 mA) Active Multimode-Interferometer (MMI) Bistable Laser Diodes Using Lateral-Modes Bistability," IEEE Journal of Selected Topics in quantum electronics, vol. 17, no. 5, September 2011.
[6] H. Jiang, Y. Chaen, T. Hagio, K. Tsuruda, M. Jizodo, S. Matuso, J. Xu, C. Peucheret and K. Hamamoto, "All-optical flip-flop operation based on asymmetric active-multimode interferometer bi-stable laser diodes," OPEX, vol. 19, no. 26, October 2011.
[7] M. N. Uddin, T. Kizu, Y. Hinokuma, K. Tanabe, A. Tajima, K. Kato, K. Hamamoto, "Split pump region in 1.55 μm InGaAsP/InGaAsP asymmetric active multi-mode interferometer laser diode for improved modulation bandwidth," IEICE Transactions on Electronics, vol. E97-C, no. 7, July 2014.
[8] J. V. Roey, J. Donk, and P. E. Lagasse, "Beam-propagation method: analysis and assessment," JOSA, vol. 71, pp803-810, 1981.

TuD1-4 (Invited)

OECC/PS2016

Quantum-Cascade Lasers: Line-Narrowing and Suppression of Flicker-Noise

Masamichi Yamanishi*[1], Toru Hirohata[1], Saverio Bartalini[2], and Paolo De Natale[2]

[1]Central Research Laboratories, Hamamatsu Photonics KK, 5000 Hirakuchi, Hamakita-ku, Hamamatsu 434-8601 Japan
[2]Instituto Nazionale di Ottica (INO)-CNR, Largo Fermi 6, 50125 Firenze FI, Italy
*Author e-mail address: masamiya@crl.hpk.co.jp

Abstract: Our results on linewidths of quantum-cascade lasers are reviewed, clarifying physics underlying astonishingly narrow intrinsic linewidths ~100 Hz and showing possible narrowing of extrinsic linewidths down to ~30 kHz by suppression of electrical flicker-noise.
Keywords: quantum-cascade laser; linewidth; flicker-noise

I. INTRODUCTION

Quantum-cascade lasers (QCLs) have been attracting much attention as compact sources of mid-infrared or THz radiation for a variety of applications ranging from high-precision molecular gas spectroscopy and infrared astronomy to high-resolution coherent imaging and telecommunications. In addressing such application requirements, high frequency-stability of the sources is mandatorily demanded. The spontaneous emission-induced intrinsic linewidths (LWs) of QCLs are found to be unusually narrow down to ~100 Hz [1]–[3], compared with those of interband bipolar lasers. However, the free running LWs (150 kHz–10 MHz, reported so far for mid-infrared cases [2][4][5]) of existing QCLs are governed by flicker frequency-noise that originates from strong electrical flicker-noise in the devices via temperature fluctuations [4][6]. In the presentation, our experimental and theoretical results on both of the intrinsic and extrinsic LWs of mid-infrared QCLs running free of any type of feedback effect are exhibited. We place a stress on two interesting issues: role of nonradiative relaxation in intrinsic line-narrowing and controllability of flicker-noise. Namely, the astonishingly narrow intrinsic LWs, ~100 Hz are identified to originate from stable single-mode operation at very high photon flux levels of which stability is maintained by the unavoidable presence of ultra-fast nonradiative relaxation (~1 ps) for electrons at upper laser levels. Appropriate design of doping profile in injectors of a QCL with ridge-waveguide is demonstrated experimentally and theoretically to strongly weaken electrical flicker-noise and, as a consequence, is expected to narrow substantially the free running LW down to ~30 kHz.

II. INTRINSIC LINEWIDTH INDUCED BY SPONTANEOUS EMISSION

The spectral properties of a laser can be described either in terms of its emission line shape and associated LW or in terms of the power-spectral density (PSD) of the frequency noise. Both approaches are complementary, but the knowledge of the frequency-noise PSD provides us with much more information on the laser noise. Figure 1 shows a representative example of the frequency-noise PSD measured with a free running single-mode 4.3 μm-distributed feedback (DFB)-InGaAs/InAlAs-QCL driven by a DC current of I_0=219 mA (I_0/I_{th}=1.54) at a heat sink temperature of ~90 K [1]. The PSD obeys the $1/f_N$ power-law below 100 kHz, and then follows the $1/f_N^2$ in the medium frequency range between 100 kHz and 10 MHz. Finally, the PSD flattens asymptotically to a white-noise level of N_w=163 Hz²/Hz, which means the intrinsic LW (FWHM) of πN_w=510 Hz.

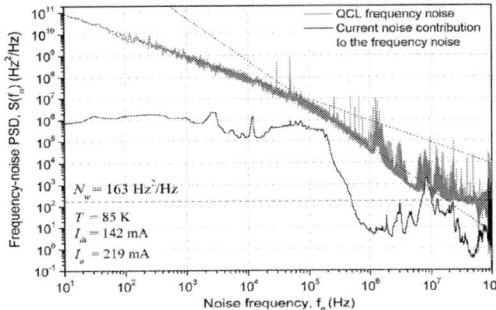

Fig. 1. Observed frequency-noise PSD (grey trace) of the mid-infrared (~4.3 μm) QCL at ~90 K acquired between 10 Hz and 100 MHz, compared with the expected contribution arising from the driving-current noise (black trace) [1].

The experimental data of intrinsic LW obtained with 4.3 μm-DFB-QCLs at ~90 K [1] and room temperature [2] both with (assumed) negligibly small α-parameters [7] are, as shown in Fig. 2, fitted by the theoretical curves estimated by using the modified Schawlow-Townes formula given in Ref. [8]:

567

$$\delta v = \frac{(1+\alpha_{\mathrm{c}}^{2})n_{sp0}\gamma}{4\pi}\left[\frac{\beta_{\mathrm{eff}}(1+\tau_{21}/\tau_{31})}{I_0/I_{\mathrm{th}}-1}+\frac{(\tau_{21}\beta/\tau_r)}{\eta}\right].$$

(1)

In the equation, γ is the photon decay rate $\sim 10^{11}$ 1/s in the QCL cavity, η the upper level pump efficiency, τ_{21} the depopulation time ~ 0.1 ps for the lower level ($\ll$the upper-level nonradiative relaxation time $\tau_t=[1/\tau_{32}+1/\tau_{31}]^{-1}\sim 1$ ps), and $n_{spo}=[1-(\tau_{21}/\tau_t)(1/\eta-\tau_t/\tau_{31})]^{-1}\sim 1.1$ the spontaneous emission factor. The effective coupling coefficient $\beta_{\mathrm{eff}}=\tau_t\beta/\tau_r\sim 10^{-9}\text{--}10^{-8}$ is much smaller than the coupling coefficient of spontaneous emission $\beta\sim 10^{-5}$ because of the upper-level relaxation time $\tau_t\sim 1$ps$\ll$the spontaneous emission lifetime $\tau_r\sim 10$ ns in mid-infrared QCLs.

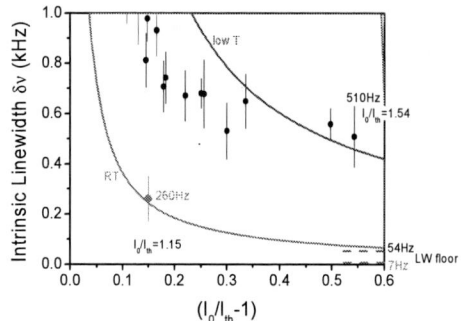

Fig. 2. Experimental data set for the intrinsic LWs of the mid-infrared (~ 4.3 μm) DFB QCLs at ~ 90 K and ~ 290 K versus the normalized pump current ($I_0/I_{\mathrm{th}}-1$) [1][2]. The two solid curves indicate theoretical ones estimated by using Eq. (1). The horizontal dashed-lines indicate the estimated LW floors. The numerical values for physical parameters indicated in Refs. [1][2] are used in the LW estimations. The LW enhancement factor is assumed to be negligibly small, $\alpha_{\mathrm{c}}^{2}\sim 0$.

The line narrowing of intrinsic LW, *still staying in the Schawlow-Townes limit*, in mid-infrared DFB-QCLs down to a few hundred hertz arises not only from their small α-parameters, $\alpha_{\mathrm{c}}^{2}\ll 1$ [7] but more profoundly from the *stable* single-mode operation at high photon flux levels, maintained by the unavoidable presence of very fast nonradiative relaxation process [8]. In fact, the high stimulated emission rate, for instance, $\sim 5\times 10^{18}$ 1/s (internal power ~ 230 mW) in the RT-device of a module number $M=25$ driven by $I_0/I_{\mathrm{th}}-1=0.15$, is still much lower than the overall relaxation rate for upper level electrons, $MN_3/\tau_t\sim 5\times 10^{19}$ 1/s, guaranteeing a weak spatial hole-burning in upper-level electron population, namely stable single-mode operation. (Note that mode-competition in multi-mode operation induces strong noise.)

III. ELECTRICAL FLICKER-NOISE FROM IMPURITY STATES

Substantial suppression of the flicker frequency-noise appearing in the low frequency range, <100 kHz as exhibited in Fig. 1is definitely required to take advantage of such narrow intrinsic LWs. In other words, suppression of electrical flicker-noise is demanded since the flicker frequency-noise originates from the electrical flicker-noise in the devices via temperature fluctuations [4][6]. In order to inspect technically-unavoidable flicker-noise, we have focused on an ad hoc model [9] in which impurity (donor) states are regarded as trapping states for noise generation in injector super-lattices of a QCL since a certain level of impurity doping ($\sim 10^{11}$ 1/cm^2 in a mid-infrared QCL and $\sim 10^{10}$ 1/cm^2 in a THz QCL) in the injectors is indispensable for smooth current transport in the cascade structure, namely for its stable lasing operation. Hence, impurity states might be regarded as an inherent (or ultimate) candidate for noise sources even in an ideal device, being free of influence of any other type of imperfection. In the model, so-called dipole-fluctuation model, fluctuating charge dipole consisting of an excited electron and ionized impurity state is assumed to induce fluctuating voltage. Moreover, an advanced noise-model named correlated dipole- and resistance-fluctuation (CDRF) model has been proposed recently, being still based on trapping/detrapping at impurity states of its injectors [10].

Figure 3 shows current-dependences of the measured voltage-noise power spectral densities (VNPSDs) at noise-frequencies of $f_N=100$ Hz and 1 kHz in a ~ 10 μm *single-mode* InGaAs/InAlAs DFB-QCL [10] with ridge-waveguide and with impurities (Si) of an areal density, $n_{\mathrm{inj}}\sim 2\times 10^{11}$ 1/cm^2 localized into the 3.6 nm-thick fourth well of its injector super-lattices. With recourse to the localized impurity-doping, the voltage-noises are well stabilized to be quite low above threshold, $\sim 4\times 10^{-16}$ V^2/Hz and $\sim 3\times 10^{-17}$ V^2/Hz at 100 Hz and 1 kHz, respectively. The inherent noise level, $\sim 3\times 10^{-17}$ V^2/Hz at 1 kHz are very much lower than previously reported ones, 1×10^{-15} V^2/Hz$-$ 6×10^{-13} V^2/Hz at 1 kHz [11] which primarily originate from additional imperfections such as surface states in ridge-waveguide QCLs or deep traps in Fe-doped InP layers of buried hetero-structure (BH) QCLs. The low noise levels at 1 kHz over the wide current range, 140–250 mA, obtained with our QCL of localized impurity-doping are well-interpreted in terms of the CDRF model. Furthermore, light-intensity-flicker-noise in the low noise-frequency range, 0.1–10 kHz of our QCLs with the designed doping profile is being confirmed to be quite low. Results on the relative intensity-noise in the low frequency range that gives us information on internal current-noise will be shown in the presentation. The weakened flicker-noise may result in substantial narrowing of free running LWs of QCLs down to ~30 kHz, acquired over an assumed measurement time of 10 ms, without assistances of any types of feedback schemes.

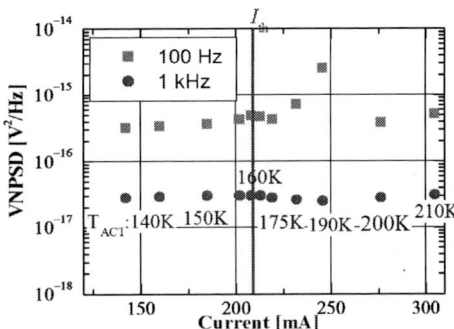

Fig. 3. Current-dependences of measured VNPSDs at 100 Hz and 1 kHz in a single-mode DFB-QCL with 14 μm ridge-waveguide and threshold current of 210 mA at an active region temperature of T_{act}=160 K [10].

IV. CONCLUSIONS

We have reviewed our results on LWs of QCLs, clarifying physics underlying astonishingly narrow intrinsic LWs~100 Hz and showing substantial reduction of electrical flicker-noise level in a ridge-waveguide QCL with localized doping profile, the reduction that may lead to substantial narrowing of extrinsic LWs down to ~30 kHz. Further reduction of flicker-noise level at a specific temperature range would become possible by adoption of delta (1nm) doping of impurities in QCL injectors.

ACKNOWLEDGMENT

The authors express their thanks to Syohei Hayashi, Kazuue Fujita, and Kazunori Tanaka, Hamamatsu Photonics KK, Japan for their collaboration in the flicker-noise project and Naota Akikusa, Hamamatsu Photonics KK, Japan and Simone Borri, INO-CNR, Italy for their collaborative efforts in the joint project on linewidth.

REFERENCES

[1] S. Bartalini, S. Borri, P. Cancio, A. Castrillo, I. Galli, G. Giusfredi, D. Mazzotti, L. Gianfrani, and P. De Natale, "Observing the intrinsic linewidth of a quantum-cascade laser: beyond the Schawlow-Townes limit," Phys. Rev. Lett., vol. 104, pp. 083904-1-083904-4, Feb. 2010.

[2] S. Bartalini, S. Borri, I. Galli, G. Giusfredi, D. Mazzotti, T. Edamura, N. Akikusa, M. Yamanishi, and P. De Natale, "Measuring frequency noise and intrinsic linewidth of a room-temperature DFB quantum cascade laser," Opt. Express, vol. 19, pp. 17996-18003, Sept. 2011.

[3] M. S. Vitiello, L. Consolino, S. Bartalini, A. Taschin, A. Tredicucci, M. Inguscio, and P. De Natale, "Quantum-limited frequency fluctuations in a terahertz laser," Nature Photon., vol. 6, pp. 525-528, Aug. 2012.

[4] L. Tombez, S. Schilt, J. Di Francesco, P. Thomann, and D. Hofstetter, "Temperature dependence of the frequency noise in a mid-IR DFB quantum cascade laser from cryogenic to room temperature," Opt. Express, vol. 20, pp. 6851-6859, March 2012.

[5] L. Tombez, S. Schilt, G. Di Domenico, S. Blaser, A. Muller, T. Gresch, B. Hinkov, M. Beck, J. Faist, and D. Hofstetter, "Physical origin of frequency noise and linewidth in mid-IR DFB quantum cascade lasers," presented at CLEO2013, June 9-14, 2013, San Jose, CA, USA.

[6] S. Borri, S. Bartalini, P. Cancio, I. Galli, G. Giusfredi, D. Mazzotti, M. Yamanishi, and P. De Natale, "Frequency-noise dynamics of mid-infrared quantum cascade lasers," IEEE J. Quantum Electron., vol. 47, pp. 984-988, July 2011.

[7] J. Faist, F. Capasso, C. Sirtori, D. L. Sivco, A. L. Hutchinson, S. G. Chu, and A. Y. Cho, "Continuous wave operation of quantum cascade lasers based on vertical transitions at λ=4.6 μm," Superlattices and Microstructures, vol. 19, pp. 337-345 April 1996.

[8] M. Yamanishi, T. Edamura, K. Fujita, N. Akikusa, and H. Kan, "Theory of the intrinsic linewidth of quantum-cascade lasers: hidden reason for the narrow linewidth and line-broadening by thermal photons," IEEE J. Quantum Electron., vol. 44, pp. 12-29 Jan. 2008.

[9] M. Yamanishi, T. Hirohata, S. Hayashi, K. Fujita, and K. Tanaka, "Electrical flicker-noise generated by filling and emptying of impurity states in injectors of quantum-cascade lasers," J. Appl. Phys., vol. 116, pp. 138106-1-138106-15, Nov. 2014.

[10] T. Hirohata, M. Yamanishi, S. Hayashi, K. Fujita, and K. Tanaka, "Flicker voltage-noise in quantum-cascade lasers below and above thresholds: experiments and correlated dipole- and resistance-fluctuation model," presented at ITQW2015, Sept. 6-11, 2015, Vienna, Austria.

[11] S. Schilt, L. Tombez, C. Tardy, A. Bismuto, S. Blaser, R. Maulini, R. Terazzi, M. Rochat, and T. Sudmeyer, "An experimental study of noise in mid-infrared quantum cascade lasers of different designs," Appl. Phys. B, published online: Feb. 10, 2015.

Improved Equalizing Characteristics in Pre-equalizing Electro-Optic Modulator with Polarization-Reversed Structures

Tomohiro OHNO, Hiroshi MURATA, and Yasuyuki OKAMURA

Graduate School of Engineering Science, Osaka University 1-3 Machikaneyama, Toyonaka, Osaka, 560-8531 Japan
E-mail : tomohirooono118@s.ee.es.osaka-u.ac.jp

Abstract: *We have proposed pre-equalizing electro-optic modulators using polarization-reversed structures, which can compensate for dispersion effect in long haul fiber communication systems. In this paper, new designs with improved equalizing characteristics are reported.*
Keywords: *Equalization, Electro-optic modulator, Optical fiber dispersion, Traveling-wave electrode, Polarization-reversed structures, Impulse response*

I. INTRODUCTION

Recently, the information and communication networks are drastically expanding in the world. With a rapid growth in internet users, terminals, machines, devices and sensors, communication systems with more and more capacity are required. Optical fiber networks can meet demands for such a huge capacity requirement. The potential bandwidth of a silica single mode fiber (SMF) is over 10 Pb/s and its propagation loss is about 0.2 dB/km, which is smaller than that of any other transmission line. However, dispersion in silica SMF causes optical signal distortion, which should be compensated for in high-speed data transfer.

One promising compensation method for the distortion caused by dispersion effect is based on digital coherent technology [1]. However, this might be affected by heating and power consumption problems; as a bit-rate becomes higher, higher electric power is required for signal processing and cooling.

We have proposed a pre-equalizing electro-optic (EO) modulator using polarization-reversed structures [2]. The proposed EO modulator can convert high-speed electrical signals to optical signals and can compensate for optical fiber dispersion effect at the same time. The impulse response of this EO modulator under a velocity mismatched condition corresponds to the polarization-reversed pattern itself. By utilizing this characteristic, it is possible to obtain the impulse responses for equalization easily. In this paper, new designs with improved equalizing characteristics are reported.

II. DEVICE STRUCTURE

Fig.1 shows the basic structure of the proposed EO modulator. It is composed of a Mach-Zehnder (MZ) waveguide and a coplanar traveling-wave electrode fabricated on a z-cut LiNbO$_3$ (LN) substrate. The traveling-wave electrode is fabricated along the MZ waveguide with a SiO$_2$ buffer layer. The polarization-reversed structures are fabricated in the substrate, where the polarization-reversed patterns are different between the two paths in the MZ waveguide for equalization. A $\pi/2$ optical phase shift is also added to the two paths in the MZ waveguide with a superposition of appropriate DC voltage to the modulation signal. By utilizing this scheme, pre-equalizing characteristic is obtainable in the EO modulator.

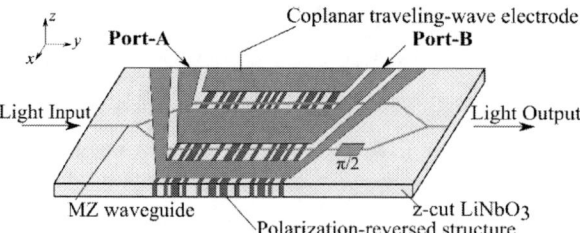

Fig. 1. The basic structure of the proposed EO modulator.
Modulation electric signal is supplied to Port-A or Port-B according to the fiber length to be compensated for.

III. BASIC THEORY

A. Impulse Response of Compensating for Optical Fiber Dispersion

Optical fiber dispersion characteristics can be expressed by use of a transfer function. Assuming a negligible optical loss in fiber, the transfer function for the fiber dispersion effect is expressed as following

$$H_{dis}(\omega) = \exp\left(-\frac{j}{2}\beta_2\omega^2 L_f\right), \tag{1}$$

where β_2 is the group velocity dispersion and L_f is the length of the fiber. Therefore, the transfer function for equalization can be expressed as $H_{cmp}(\omega)=1/H_{dis}(\omega)$. Impulse response of compensating for optical fiber dispersion effect is calculated by taking an inverse Fourier transformation of $H_{cmp}(\omega)$.

$$h_{cmp}(t) = \frac{1}{\sqrt{2\pi\beta_2 L_f}} \exp\left\{ j\left(-\frac{t^2}{2\beta_2 L_f} + \frac{\pi}{4} \right) \right\} \qquad (2)$$

In order to obtain equalizing characteristics in EO modulator, the real part of the impulse function, $\mathrm{Re}\{h_{cmp}(t)\}$, is set at the one path in the MZ waveguide and the imaginary part, $\mathrm{Im}\{h_{cmp}(t)\}$, is set at the other path in the MZ waveguide with a $\pi/2$ DC optical phase shift.

B. Impulse Response of EO Modulator with Polarization-Reversed Structures

In conventional high-speed EO modulators with a traveling-wave electrode, the velocity matching between the group velocity of the lightwave and the phase velocity of the modulation signal is standard. However, when we introduce polarization-reversed structures to a velocity mismatched EO modulator with a traveling-wave electrode, it is possible to obtain an impulse response for equalization. A simplified model of the EO modulator with a traveling-wave electrode and polarization-reversed structures is shown in Fig.2(a). For simplicity, it is assumed that the loss in the traveling-wave electrode is negligible.

Then, we consider the case when an impulse signal is applied to the electrode and the group velocity of the lightwave is larger than the phase velocity of the modulation signal. A lightwave propagating in a waveguide overtakes the impulse signal owing to the velocity difference. Therefore, the ligthwave is phase modulated when overtaking the impulse signal. If the lightwave overtakes the impulse signal in the polarization-reversed region, the sign of the modulation index becomes opposite. As the result, the impulse response of the EO modulator with the polarization-reversed structures is obtained as shown in Fig.2(b). We can see that the impulse response of the EO modulator corresponds to the polarization-reversed pattern itself.

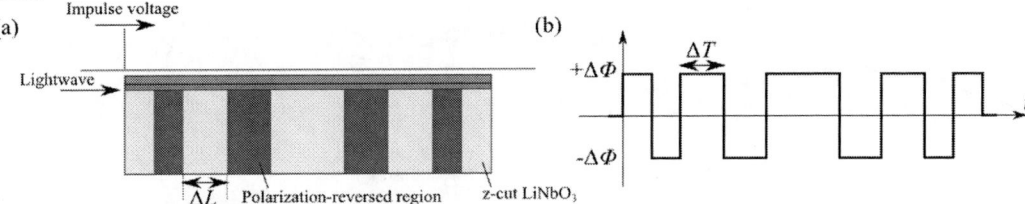

Fig.2. (a) Model of the EO modulator with a traveling-wave electrode and polarization-reversed structures and (b) its impulse response.

IV. IMPROVEMENT OF EQUALIZING CHARACTERISTICS

In conventional EO modulators with a traveling-wave electrode, electrical signals are supplied to Port-A for co-propagation with the lightwave. In this case, the time duration ΔT in the impulse response shown in Fig.2(b) is given by $\Delta T_A = (n_m - n_g)\Delta L/c$. Where ΔL is the interaction length in the modulator, c is the velocity of the lightwave in vacuum, n_g is the group index of the lightwave and n_m is the effective index of the modulation signal. By utilizing this relationship, the impulse response for equalization can be designed. However, the fiber length for the dispersion compensation is limited to about 20 km owing to the limitation on the velocity difference [4].

In order to extend the fiber length for the dispersion compensation, we propose a new scheme where electrical signals are supplied to Port-B for counter-propagation with the lightwave. By utilizing the new scheme, $\Delta T_B = (n_m + n_g)\Delta L/c$ and it is possible to drastically increase ΔT, since ΔT_B is proportional to the sum of the lightwave and the modulation signal velocities. Therefore, it is possible to greatly extend the fiber length for about 4 times.

Here, we also consider the modulation signal decay in the traveling-wave electrode. The electrode loss effect is not negligible when high-speed modulation signals are supplied to the electrode. In this case, the impulse responses with electrode loss effect are shown as solid lines in Fig.3. For comparison, the impulse responses without the electrode loss are also plotted. We can see that the electrode loss effect may cause degradation in equalizing characteristics. This electrode loss effect can be also compensated for by use of the delta-sigma transformation technique described in the next section.

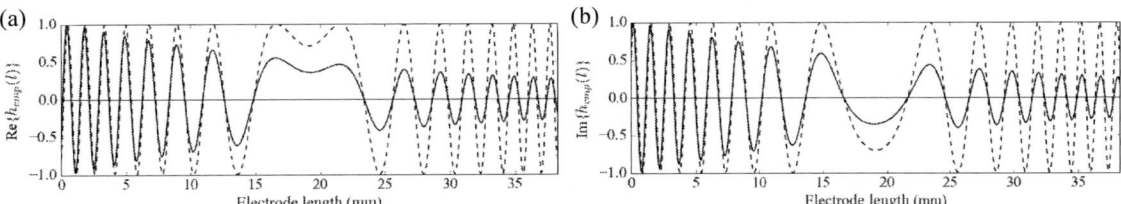

Fig. 3. Impulse responses with electrode loss effect (solid lines) and without electrode loss effect (dashed lines).
(a) Real part, (b) imaginary part ($L_f = 80$ km).

V. ANALYSIS AND DESIGN

We designed polarization-reversed patterns for improved equalizing characteristics using narrow polarization-reversed regions (~50 μm) and the delta-sigma transformation technique to obtain precise characteristics for equalization. For the design of the proposed EO modulator, n_m and n_g are calculated as n_m=3.98 by using HFSS ver.12 for Al coplanar traveling-wave electrodes and n_g =2.19 for a LN channel optical waveguide, respectively [3]. Then, impulse responses compensating for the degradation arising from the fiber dispersion and electrode loss effect were calculated as solid lines shown in Fig.4. The corresponding polarization-reversed patterns with a succession of narrow delta-regions for the precise compensations were decided by use of the delta-sigma transformation technique. Fig.4 shows also the designed polarization-reversed patterns.

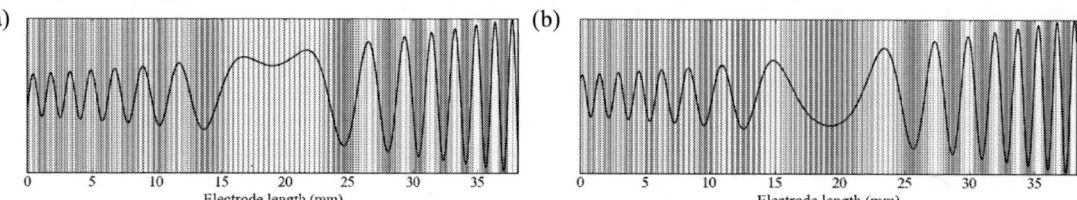

Fig.4. Impulse responses compensating for electrode loss effect (solid lines) and polarization-reversed patterns for equalization.
(a) Real part, (b) imaginary part.

The compensation for the signal distortion by using the proposed EO modulator was simulated. The light wavelength was set as 1.55 μm. Fig.5(a) and (b) show simulated eye diagrams when a 40 Gb/s NRZ ASK signal was supplied to the designed EO modulator. As can be seen from the compensated 40 Gb/s signal in Fig.5(b), the eye was fully restored, and error-free($<10^{-12}$) operation was obtained in Fig.5(c). These results are in good agreement with the designed characteristics.

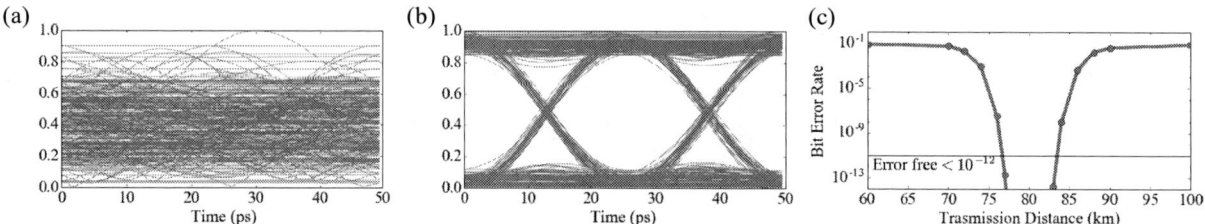

Fig.5. Simulated eye diagrams. (a) Back to back, (b) after 80 km transmission for a 40 Gb/s NRZ ASK signal through SMF.
(c) Variation of bit error rate depending on the fiber length.

VI. CONCLUSIONS

We have proposed new pre-equalizing EO modulators with polarization-reversed structures. The designs for an extension of a fiber length and for improvement of compensation characteristics with electrode loss effect are discussed. The analysis results show the compensation of a 40 Gb/s NRZ ASK signal for the propagation over a 80 km SMF is possible. Moreover, by using MEMS technology for the coplanar traveling-wave electrode, tuning of a fiber length for the compensation is possible. In addition, by utilizing photonic crystal waveguide in the proposed EO modulator, the further extension of fiber length up to 500 km is also expected [5]. Now, we are trying to fabricate the proposed EO modulator.

ACKNOWLEDGMENT

This research results was achieved in part by the Commissioned Research of National Institute of Information and Communications Technology (NICT), Japan. This was also partially supported by the Grants-in-Aid for Scientific Research from the Ministry of Education, Science, Sports and Culture, Japan under the grant no. 24360135.

REFERENCES

[1] S. Savory, "Digital signal processing for coherent systems," Optical Fiber Communication Conference, pp.OTh3C-7, Optical Society of America, 2012.

[2] H. Murata, and Y. Okamura, "High-speed signal processing utilizing polarization-reversed electro-optic devices," Journal of Lightwave Technology, vol.32, no.20, pp.3403-3410, 2014.

[3] D.E. Zelmon, D.L. Small, and D. Jundt, "Infrared corrected sellmeier coefficients for congruently grown lithium niobate and 5 mol.% magnesium oxide-doped lithium niobate," JOSA B, vol.14, no.12, pp.3319-3322, 1997.

[4] T. Mitsubo, H. Murata, and Y. Okamura, "Design of Pre-equalizing High-speed EO Modulator with Polarization-reversed Structure Using Delta-sigma Transformation," Microwave Photonics (MWP) and the 2014 9th Asia-Pacic Microwave Photonics Conference (APMP), 2014 International Topical Meeting on, pp.105-108IEEE, 2014.

[5] X. Wang, H. Tian, and Y. Ji, "Photonic crystal slow light Mach-zehnder interferometer modulator for optical interconnects," Journal of Optics, vol.12, no.6, p.065501, 2010.

TuD2-3

OECC/PS2016

Broadband Dual-Polarization Dual-Parallel Mach Zehnder Modulator based Photonic Microwave Phase Shifter

Tong Niu[1], Erwin H. W. Chan[2], Xudong Wang[1*], Xinhuan Feng[1], Bai-Ou Guan[1]

[1]Guangdong Provincial Key Laboratory of Optical Fiber Sensing and Communications, Institute of Photonics Technology, Jinan University, Guangzhou 510632, China

[2]School of Engineering and Information Technology, Charles Darwin University, Darwin NT 0909, Australia

Xudong Wang: txudong.wang@email.jnu.edu.cn

Abstract: A new photonic microwave phase shifter that is capable of realizing 360° phase shift with a flat amplitude and phase response performance over a wide frequency range of 4 to 40 GHz is experimentally demonstrated.

Keywords: *Microwave Photonics, Optical Signal Processing, Phase Shifters.*

I. INTRODUCTION

There has been a growing interest in applying photonic technologies in beamforming networks due to the advantages of wide bandwidth, low transmission loss for data remoting, electromagnetic interference immunity, excellent isolation and lightweight [1]. Many photonic microwave phase shifting structures have been reported [2-5]. However, very few of them [4], [5] have experimentally demonstrated a broadband (bandwidth > 30 GHz) phase shifting operation. These broadband phase shifters rely on the use of an optical filter to remove one sideband, which limits the phase shifter lower operating frequency due to the unwanted sideband at the frequencies close to the optical carrier frequency cannot be largely suppressed by the optical filter. This paper presents a new photonic microwave phase shifter based on a dual-polarization dual-parallel Mach Zehnder modulator (DP-DPMZM) to overcome this limitation. It also has the advantage of requiring only a single control voltage that has a linear relationship with the RF phase shift. Investigations of the tolerances to the design parameters in the phase shifter structure are also presented. Broadband 4 to 40 GHz phase shifting operation is demonstrated experimentally.

II. TOPOLOGY AND ANALYSIS

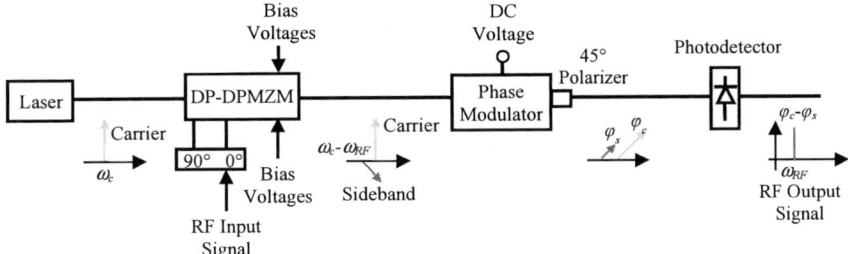

Fig. 1. Structure of the DP-DPMZM based photonic microwave phase shifter.

Figure 1 shows the structure of the DP-DPMZM based photonic microwave phase shifter. It consists of an integrated DP-DPMZM for RF signal modulation and an electro-optic phase modulator with a 45° polarizer to control the RF signal phase shift. The laser source provided a continuous wave light into an integrated DP-DPMZM that is formed by two DPMZMs connected in parallel, a 90° polarization rotator and a polarization beam combiner. The lower DPMZM inside the DP-DPMZM is driven by a pair of 90° phase difference RF signals from a 90° hybrid coupler. By designing the DP-DPMZM bias voltages, a single sideband RF modulated optical signal, in which the optical carrier and the single RF modulation sideband have an orthogonal polarization state, can be obtained at the modulator output as shown in Fig. 1. The DP-DPMZM output electric field can be expressed as

$$E_{out,DP-DPMZM} = \frac{\sqrt{2}E_{in}}{2}\sqrt{L}\left(\hat{y}e^{j\omega_c t} + \hat{x}J_1(\beta_{RF})e^{j((\omega_c - \omega_{RF})t - \pi/2)}\right) \qquad (1)$$

where E_{in} is the DP-DPMZM input electric field amplitude, L is the DP-DPMZM insertion loss, and $\hat{x}$ and $\hat{y}$ are the polarization states representing light travelling in the fast axis and slow axis respectively, ω_c and ω_{RF} are the angular frequency of the input light and input RF signal respectively, $J_m(x)$ is the Bessel function of mth order of first kind,

573

$\beta_{RF}=\pi V_{RF}/V_{\pi}$ is the modulation index, V_{RF} is the input RF signal amplitude, and V_{π} is the modulator switching voltage. The orthogonally polarized optical carrier and single RF modulation sideband pass through a phase modulator with a 45° linear polarizer at the output. The phase modulator is used as a polarization dependent optical phase shifter, which introduces different optical phases to the carrier (φ_c) and sideband (φ_s) by an input DC voltage. The polarizer converts the polarization state of the optical carrier and the RF modulation sideband to have the same polarization state with a 45° angle with respect to the slow axis. The carrier and sideband beat at the photodetector generating a photocurrent at the RF signal frequency and is given by

$$I_{RF}=\frac{1}{2}\Re L_{total}P_{in}J_1\left(\beta_{RF}\right)\sin\left[\omega_{RF}t+\left(\varphi_c-\varphi_s\right)\right] \qquad (2)$$

where $\Re$ is the photodiode responsivity, L_{total} is the product of the DP-DPMZM, phase modulator and polarizer insertion loss, and P_{in} is the optical power into the DP-DPMZM. Eq. (2) shows the phase of the output RF signal can be tuned by changing the phase difference between the optical carrier and the RF modulation sideband. It also shows the amplitude of the output RF signal is fixed during the RF phase shifting operation.

In practice, bias drift in the DP-DPMZM and imbalances between the two 90° hybrid coupler outputs affect the phase shifter performance. The bias drift problem can be overcome by using bias controllers. On the other hand, the 90° hybrid coupler output imbalances cannot be neglected. With the inclusion of the 90° hybrid coupler amplitude imbalance $A(\omega_{RF})$ and phase imbalance $\theta(\omega_{RF})$, the phase shifter output photocurrent at the RF signal frequency given in (2) can be rewritten as

$$I_{RF}=\frac{1}{4}\Re L_{total}P_{in}J_1\left(\beta_{RF}\right)\sqrt{M^2+N^2}\sin\left[\omega_{RF}t+\tan^{-1}\left(\frac{N}{M}\right)\right] \qquad (3)$$

where

$$M=-\sqrt{\left[-A\left(\omega_{RF}\right)\cos\theta\left(\omega_{RF}\right)\right]^2+\left[1-A\left(\omega_{RF}\right)\sin\theta\left(\omega_{RF}\right)\right]^2}\sin\left[\tan^{-1}\left(\frac{1-A\left(\omega_{RF}\right)\sin\theta\left(\omega_{RF}\right)}{-A\left(\omega_{RF}\right)\cos\theta\left(\omega_{RF}\right)}\right)-\varphi_c+\varphi_s\right]$$
$$-\sqrt{\left[A\left(\omega_{RF}\right)\cos\theta\left(\omega_{RF}\right)\right]^2+\left[-1-A\left(\omega_{RF}\right)\sin\theta\left(\omega_{RF}\right)\right]^2}\sin\left[-\tan^{-1}\left(\frac{-1-A\left(\omega_{RF}\right)\sin\theta\left(\omega_{RF}\right)}{A\left(\omega_{RF}\right)\cos\theta\left(\omega_{RF}\right)}\right)+\varphi_c-\varphi_s\right] \qquad (4)$$

$$N=\sqrt{\left[-A\left(\omega_{RF}\right)\cos\theta\left(\omega_{RF}\right)\right]^2+\left[1-A\left(\omega_{RF}\right)\sin\theta\left(\omega_{RF}\right)\right]^2}\cos\left[\tan^{-1}\left(\frac{1-A\left(\omega_{RF}\right)\sin\theta\left(\omega_{RF}\right)}{-A\left(\omega_{RF}\right)\cos\theta\left(\omega_{RF}\right)}\right)-\varphi_c+\varphi_s\right]$$
$$+\sqrt{\left[A\left(\omega_{RF}\right)\cos\theta\left(\omega_{RF}\right)\right]^2+\left[-1-A\left(\omega_{RF}\right)\sin\theta\left(\omega_{RF}\right)\right]^2}\cos\left[-\tan^{-1}\left(\frac{-1-A\left(\omega_{RF}\right)\sin\theta\left(\omega_{RF}\right)}{A\left(\omega_{RF}\right)\cos\theta\left(\omega_{RF}\right)}\right)+\varphi_c-\varphi_s\right] \qquad (5)$$

Note that in the ideal case, $A(\omega_{RF})=1$ and $\theta(\omega_{RF})=\pi/2$. Eqs. (3)-(5) show the imbalances between the two 90° hybrid coupler output ports cause the phase shifter to have a frequency dependent amplitude and phase response performance.

III. SIMULATION AND EXPERIMENTAL RESULTS

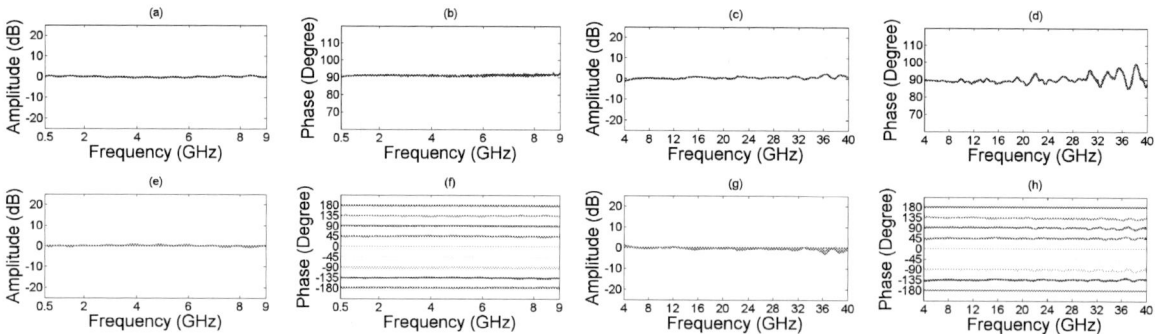

Fig. 2. Measured amplitude and phase response of a 0.5-9 GHz bandwidth ((a) and (b)) and 4-40 GHz bandwidth ((c) and (d)) 90° hybrid coupler. Simulated DP-DPMZM based photonic microwave phase shifter amplitude and phase response obtained using the measured 90° hybrid coupler frequency responses ((e), (f), (g) and (h)).

The imbalances between the 0° and 90° ports of two different bandwidth (0.5-9 GHz and 4-40 GHz) commercial 90° hybrid couplers were measured. Fig. 2(a)-(d) shows the amplitude and phase response of the two couplers measured at the 90° port after normalized with the corresponding coupler 0° port. The measurements reveal that the 0.5-9 GHz bandwidth coupler has ±0.7 dB and ±1.4° amplitude and phase imbalance respectively, and the amplitude and phase imbalance of the 4-40 GHz bandwidth coupler is ±1.9 dB and ±7.3° respectively. The DP-DPMZM based photonic microwave phase shifter amplitude and phase response were simulated using the measured 90° hybrid coupler frequency responses given in Fig. 2(a)-(d) together with (3)-(5). Fig. 2(e)-(h) show the -180° to 180° RF phase shifting operation, and the phase shifter has <1.3 dB amplitude variation and <2.4° phase variation when using the 0.5-9 GHz bandwidth 90° hybrid coupler to generate a pair of 90° phase difference RF signals into the DP-DPMZM, and the amplitude and phase variation become <3.8 dB and <7.3° when the 4-40 GHz bandwidth 90° hybrid coupler was used.

An experiment was set up as shown in Fig. 1 to verify the principle of the DP-DPMZM based photonic microwave phase shifter. A commercially available DP-DPMZM (Fujitsu FTM7977) was used to generate an orthogonally polarized optical carrier and RF modulation sideband. A Z-cut LiNbO$_3$ phase modulator (Covega Mach-10™ 053-10-S-A-A) was used to introduce different optical phases to the carrier and sideband. The output optical signal after a polarizer was detected by a 50 GHz bandwidth photodetector, and the output RF signal amplitude and phase response were measured using a network analyzer as shown in Fig. 3(a)-(d). It can be seen from the figure that the measured frequency response shapes are agreed with the simulation results given in Fig. 2(e)-(h). The experimental results demonstrate the new phase shifting structure has the ability to realize a broadband RF phase shifting operation. Fig. 3(a)-(d) show the phase shifter amplitude and phase variation is <2.4 dB and <4° over the 0.5 to 9 GHz frequency range when the 0.5-9 GHz bandwidth 90° hybrid coupler was used, and is <4.4 dB and <13° over the 4 to 40 GHz frequency range when the 4-40 GHz bandwidth 90° hybrid coupler was used. The difference between the measured and simulated results is mainly due to the RF cables and adapters that were used to connect between the 90° hybrid coupler and the DP-DPMZM had slightly different characteristics, e.g. the length, the insertion loss and the phase ripples. These deviations further increase the amplitude and phase imbalances between the two 90° phase difference RF signals into the DP-DPMZM, especially at high RF frequencies. The measured RF signal phase shift versus the DC voltage into the phase modulator is shown in Fig. 3(e). This demonstrates a linear RF phase shift to control voltage relationship.

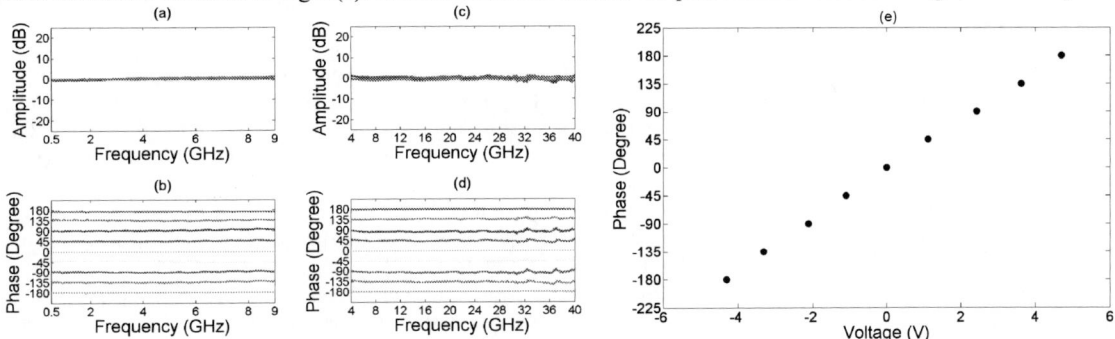

Fig. 3. Measured DP-DPMZM based photonic microwave phase shifter amplitude and phase response when using the 0.5-9 GHz bandwidth ((a) and (b)) and 4-40 GHz bandwidth ((c) and (d)) 90° hybrid coupler in the setup. (e) RF phase shift versus DC voltage into the phase modulator.

IV. CONCLUSIONS

A new photonic microwave signal processing structure that can realize a continuously tunable 0° to 360° RF phase shift over a wide frequency range has been presented. It has a simple structure, only requires a single control voltage that has a linear relationship with the RF phase shift, and is suitable for beamforming applications.

ACKNOWLEDGMENT

This work was supported by the National Natural Science Foundation of China (NSFC) (61475065, 61501205), the Guangdong Natural Science Foundation (2014A030310419, 2015A030313322) and the Fundamental Research Funds for the Central Universities (21615325).

REFERENCES

[1] R. A. Minasian, E. H. W. Chan, and X. Yi, "Microwave photonic signal processing," *Opt. Exp.*, vol. 21, no. 19, pp. 22918–22936, 2013.

[2] E. H. W. Chan, W. Zhang, and R. A. Minasian, "Photonic RF phase shifter based on optical carrier and RF modulation sidebands amplitude and phase control," *J. Lightw Technol.*, vol. 30, no. 23, pp. 3672–3678, 2012.

[3] X. Wang, E. H. W. Chan, and R. A. Minasian, "All-optical photonic microwave phase shifter based on an optical filter with a nonlinear phase response," *J. Lightw Technol.*, vol. 31, no. 20, pp. 3323–3330, 2013.

[4] S. Pan and Y. Zhang, "Tunable and wideband microwave photonic phase shifter based on a signal-sideband polarization modulator and a polarizer," *Opt. Lett.*, vol. 37, no. 21, pp. 4483–4485, 2012.

[5] W. Li, W. H. Sun, W. T. Wang, and N. H. Zhu, "Optically controlled microwave phase shifter based on nonlinear polarization rotation in a highly nonlinear fiber," *Opt. Lett.*, vol. 39, no. 11, pp. 3290–3293, 2014.

OECC/PS2016

Photonic radio frequency down-converter based on parallel electro-absorption modulators in Ku/Ku band for space applications

Jordane Thouras[1], Benoit Benazet[2], Herve Leblond[2] and Christelle Aupetit-Berthelemot[1]

[1]XLIM Laboratory, UMR CNRS 7252, University of Limoges, 16 rue Atlantis 87068 Limoges Cedex, France
[2]Thales Alenia Space, 26 avenue Jean François Champollion, 31100 TOULOUSE, France
jordan.thouras@ensil.unilim.fr ; christelle.aupetit-berthelemot@xlim.fr

Abstract: In this paper, we present a new photonic RF frequency down-convertor based on parallel electro-absorption modulators, and fiber Bragg grating filters, with reduced output spurious: only the converted IF signal and first three LO harmonics are present in the output spectrum. We give device performance in terms of conversion gain, noise factor, RF/IF and LO/IF isolations, carrier to third order intermodulation ratio (C/I_3) and spectral purity. We also proof that it can work properly in the entire Ku band without a significant variation of performance.

Keywords: satellite payload, down-convertor, photonic mixer, electro-absorption modulators

I. INTRODUCTION

Next-Generation Satellite telecommunication payloads need an improvement in terms of complexity and capability in order to extend functionalities and to present more flexibility. In particular, they require a large number of frequency converters (FC). FC is the key component of a satellite payload, as it allows the connection between the reception (RX) band and the transmission (TX) band by doing the frequency down-conversion. This separation is compulsory to prevent satellites from receiving their own emissions, which could disturb transmissions, and to respect IUT standards.

Nowadays microwave FC for space applications are based on nonlinear properties of diodes or transistors, which present quit low losses and noise factor (NF) but have also several drawbacks like parasite frequencies or spurious generation, and the need of a specific design for each frequency sub-band [1], [2].

In this context, we present a photonic RF frequency down-converter based on semiconductor optoelectronic technologies, and more precisely on parallel electro-absorption modulators (EAM), which is able to reduce significantly the number of parasite output frequencies and to work in the entire Ku band. Until now, photonic radio frequency down conversion has been explored by the use of Mach-Zehnder modulators [3], semi-conductor optical amplifiers [4], or with combination of different modulators and nonlinear devices, but never with only EAMs. Moreover, optical fibered sub-systems have shown a lot of interest as they present advantages including electromagnetic interference immunity, low transmission losses, low weight and size, and they can reproduce key functions of satellite payloads like local oscillator (LO) generation and distribution, frequency conversion, beam forming and signal routing [5].

In this paper, we explain how this module operates, give its performance in terms of conversion gain, noise factor, LO/IF and RF/FI isolations, carrier to third order intermodulation ratio or C/I3 ratio, and spectral purity. We also proof that performance does not change with the LO and RF input frequencies, staying in Ku band standards. Note that all these works have been realized with system simulation software VPIPhotonics®, in which we introduced realistic models issued from measurements on real devices.

II. ARCHITECTURE DESCRIPTION

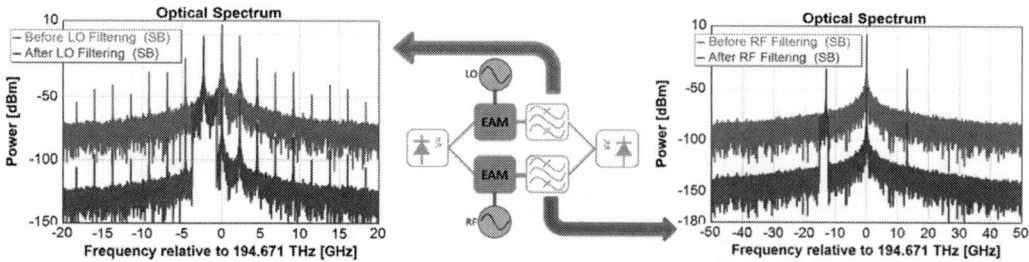

Fig. 1: Photonic RF frequency down-converter based on parallel electro-absorption modulators

We propose to use two EAMs in a parallel configuration, as represented in Fig. 1.

576

A DFB CW laser, with the parameters listed in TABLE I, emits the same optical carrier (P_{opt}=18 dBm) in the two modulators thanks to a power splitter. Each EAM receive an input power of 15 dBm: maximum allowed for such modulators.

The upper EAM modulates the light with a signal coming from a local oscillator, at 2.05 GHz (F_{LO}), and with an electrical power of 6.5 dBm. This value has been carefully optimized to get a good tradeoff between RF performance and spectral purity.

The lower modulator does the same with the RF signal, at 13 GHz (F_{RF}) with a RF power of -25 dBm, which is a common value for RF mixers dedicated to space applications. After each EAM we introduced a fiber Bragg grating to filter only one side band in each optical spectrum, initially composed of the optical carrier and RF or LO fundamentals and harmonics, as shown in Fig. 1. The two selected frequencies must be at the same side of the carrier. The bandpass filters have a total bandwidth of 3.0 GHz. We considered a frequency spacing between the FBG center and the optical carrier of 2.2 GHz for the FBG associated with the LO signal and of 13.8 GHz for the FBG associated with the RF signal. Note that stopband oscillations are not modeled in simulations. Finally, the signals are recombined with a standard optical coupler and electrically converted thanks to a photodetector with the characteristics listed in TABLE I. The quadratic respond of the photodiode will create, by beatings of the two optical sidebands, the IF (intermediate frequency) signal: $F_{IF} = F_{RF} - F_{LO}$.

TABLE I
Laser and photodiode parameters

Laser parameter	Value	Photodiode PIN parameter	Value
Wavelength	1540 nm	Responsivity @1540 nm	0.8 A/W
Linewidth	1 MHz	Electrical bandwidth	Wider than Ku band
RIN	-165 dB/Hz	Dark current	10 nA
Optical Power	18 dBm	Thermal noise	18 pA/Hz 1/2

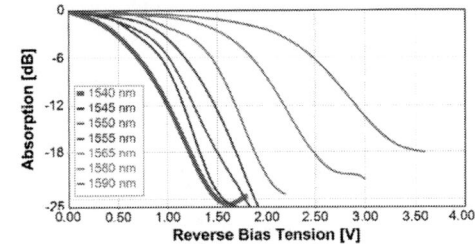

Fig. 2 : Tested absorption curves for EAMs

The EAMs have been modeled using the absorption curves presented in [6], which correspond to a unique modulator experimentally characterized for different wavelength. They have been fitted with 9^{nth} order polynomial functions and are represented in Fig. 2. The architecture study has been led with all of these curves and we chose the wavelength that gives the best conversion efficiency: 1540 nm. We considered the chirp parameter as a constant equals to 0.5: value to consider when polarized in linear zone, reported in [7]. We included total losses of 4 dB per modulator, which correspond to integrated devices.

III. RESULTS

In the following section, we present the performances (TABLE III) of all the previously depicted architecture according to the following well-known parameters (TABLE II).

TABLE II
Performance parameters considered

Parameter	Acronym	Definition
Conversion gain	G_c	Power ratio between the RF input signal and the IF output signal
Noise Factor	NF	Ratio between the input and the output SNR
RF/IF isolation	RF/IF	Power ratio between the RF and the IF signal at the output of the system
LO/IF isolation	LO/IF	Power ratio between the LO and the IF signal at the output of the system
Carrier to third order intermodulation ratio	C/I_3	Ratio between the IF power and a third order intermodulation product
Receive Optical Power	ROP	Optical power level received on the photodiode

TABLE III
Simulation results for FRF = 13 GHz and FLO = 2.3 GHz

OL tension bias	0.4 V
RF tension bias	0.8 V
Optical power on photodiode	-2,6 dBm
Conversion gain	-28 dB
C/I3 ratio	73 dBc
RF/IF Isolation	Inf dB
LO/IF Isolation	-21 dB
Noise factor	35 dB

The tension bias for EAM has been optimized to get the best performance and they are also reported. The output electrical spectrum is given by Fig. 3. The Tx Ku band is represented by a green band.

Finally, Fig. 4 to Fig. 7 show the evolution of NF and conversion gain of the architecture. Firstly, when RF frequency varies from 12.7 to 14.9 GHz (part of Ku Rx band) and LO frequency is fixed to 2.3 GHz, and secondly when LO frequency varies from 2 to 3.4 GHz and RF frequency is fixed to 13.8 GHz. Theses LO values allow the IF signal to stay in the TX Ku band (10.7 to 12.75 GHz).

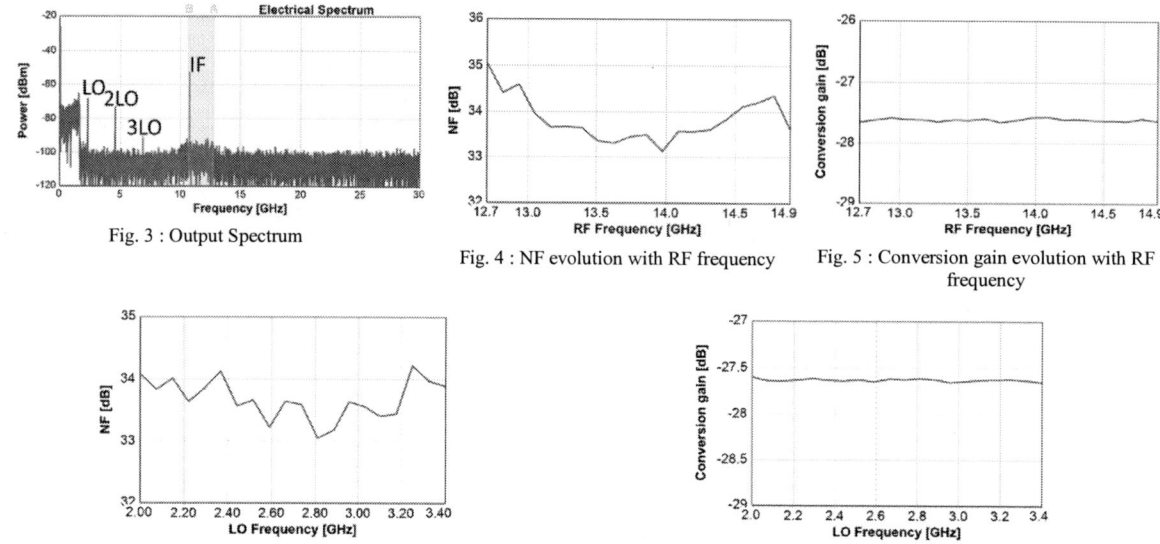

Fig. 3 : Output Spectrum

Fig. 4 : NF evolution with RF frequency

Fig. 5 : Conversion gain evolution with RF frequency

Fig. 6 : NF evolution with LO frequency

Fig. 7 : Conversion gain evolution with LO frequency

IV. CONCLUSIONS

In this paper, we presented a new photonic RF frequency down converter based on parallel EAMs. The output spectrum of such a structure contains only the IF signal and the first three LO harmonics, which are not problematic in Ku band, in contrary to the 4LO harmonic which appears in the useful output band. Here, there is no parasite frequency in the Tx Ku band. The conversion gain of the architecture is -28 dB, the NF is 35 dB, LO/IF isolation is -21 dB. RF/IF isolation can be considered as infinite, as the RF frequency doesn't appear in the output spectrum. As far as the C/I_3 ratio is concerned, it equals 73 dBc: the architecture shows a very good rejection of third order intermodulation products. The optical power before photo-detection is -2.6 dBm, so a standard PIN photodiode can be used for this application. Moreover, we shew that when LO and RF frequencies vary respectively from 2 to 3.4 GHz and from 12.7 to 14.9 GHz the variations on the conversion gain (flat) and NF (lower than 2 dB) are not significant. It means that this kind of module can work properly with any frequency sub-band of the Ku band: a standard design can be used for all sub-bands.

ACKNOWLEDGMENT

We want to thank Thales Alenia Space and the "Direction Générale de l'Armement", DGA, for their support during this study.

REFERENCES

[1] A. Suriani, et al., "Design of Microstrip Balanced Mixers for Spurious Outputs Suppression in Ku Band Satellite Repeaters," in Microwave Conference, 1997. 27th European , vol.2, no., pp.1076-1079, 8-12 Sept. 1997.

[2] Mordachev, et al., "Spurious and intermodulation response analysis of passive double-balanced mixers using the double-frequency scanning technique," in Electromagnetic Compatibility (EMC EUROPE), 2013 International Symposium on , vol., no., pp.737-742, 2-6 Sept. 2013.

[3] A. Mast et al.,"Extending frequency and bandwidth through the use of agile, high dynamic range photonic converters," Aerospace Conference, 2012 IEEE , March 2012.

[4] C. Bohémond et al., "Performances of a Photonic Microwave Mixer Based on Cross-Gain Modulation in a Semiconductor Optical Amplifier", JLT Vol. 29, issue 16, pp. 2402-2409, 2011.

[5] M. Sotom, et al., "Microwave Photonic Technologies for Flexible Satellite Telecom Payloads", ECOC 2009, Paper 10.6.3, 2009., in press.

[6] A. Konczykowska et al., "EAM DFF-driver optimization for 40 Gb/s transmitter," Microwave Symposium Digest, 2005 IEEE MTT-S International.

[7] French, ANR-VERSO project "MODULE" (nov. 2009-mai. 2013) convention N° 2009-VERS-006-05: « Source optique intégrée à MOdulation DUaLE pour réseaux locaux et métropolitains ».

TuD2-5

OECC/PS2016

Electro-Optic Modulator Using Millimeter-wave Gap-Embedded Patch Antenna with Stacked Structure

Hironori Aya[†], Yusuf Nur Wijayanto[††], Atsushi Kanno[††], Tetsuya Kawanishi[††,†††],
Hiroshi Murata[†], Yasuyuki Okamura[†]

[†] Graduate School of Engineering Science, Osaka University 1-3 Machikaneyama, Toyonaka, Osaka 560-8531 Japan
[††] NICT 4-2-1 Nukuikitamachi, Koganei, Tokyo 184-8795 Japan
[†††] Waseda University 3-4-1 Ohkubo, Shinjuku, Tokyo 169-8555 Japan
Author e-mail address: hironoriaya115@s.ee.es.osaka-u.ac.jp

Abstract: *We propose new electro-optic modulators using gap-embedded patch antennas with stacked structure. Utilizing stacked structure, the antenna gain, electric field for modulation, and interaction length increase, then modulation efficiency increases 3.65 dB at 60 GHz.*
Keywords: *Planar Patch Antenna, EO Modulation, Radio-Over-Fiber, LiNbO₃*

I. INTRODUCTION

Recently, traffic volume of wireless communication systems increases drastically. Therefore, millimeter-wave has been attracting a lot of attention because of its potential wide bandwidth of over several GHz. However, millimeter-wave is not suitable for long distance (>10 km) transmission since millimeter-wave has a large propagation loss in air. Radio-over-fiber (ROF) technology is a good solution to connect many cells of a small coverage area in wireless millimeter-wave communication systems. Moreover, ROF technology is useful for radar tracking, and sensing.

In millimeter-wave ROF systems, a converter from millimeter to lightwave is important. We have proposed several antenna-based optical modulation devices using planner antennas and electro-optic (EO) crystals [1], [2]. In this report, we discuss the improvement of modulation efficiency of patch-antenna-based conversion devices by utilizing a stacked structure.

II. STRUCTURE OF STACKED DEVICE

The structure of the proposed device and its cross sectional view are shown in Fig.1. This device has stacked patch antennas using multiple substrates combined with a thin EO crystal (LiNbO₃) and a low-dielectric constant material (SiO₂). In the lower level patch, a narrow (~5 μm) gap is introduced for optical modulation. When a wireless millimeter-wave signal with *x*-polarization is irradiated to the device from above, a strong millimeter-wave electric field (displacement current) is induced across the gap because of the continuity of the electric current. Therefore, by settling an optical waveguide close to the gap, the lightwave propagating in the optical waveguide is modulated by the millimeter-wave signal, and the signal conversion from the millimeter-wave to lightwave is obtained [1].

Utilizing the stacked structure as shown in Fig.1, the antenna gain and electric field across the gap are enhanced, since this structure can be regarded as a kind of Yagi-Uda antennas; the ground can be considered as a reflector and the patch antenna at the upper layer as a director. Then modulation efficiency can be improved [3], [4].

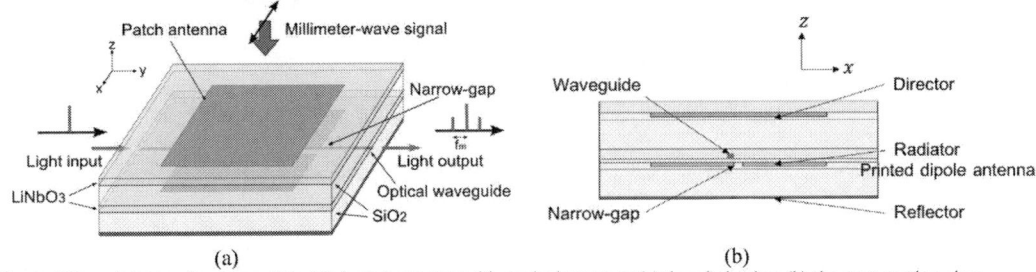

(a)   (b)

Fig. 1. EO modulator using a gap-embedded patch antenna with stacked structure (a) the whole view (b) the cross section view.

III. ANALYSIS

The proposed device was designed using 3-D electromagnetic field analysis software, HFSS. For the design operating in a 60 GHz band, we selected the multiple substrate structure composed of low-k (ε_{rL} = 4) SiO₂ of a thickness h_L = 250 μm and LiNbO₃ crystal ($\varepsilon_{rx} = \varepsilon_{ry} = 43$, $\varepsilon_{rz} = 28$) of a thickness of h_{EO} = 50 μm. We selected the two case for the

579

upper patch antenna: without a gap as shown in Fig.1 (the patch metal size LxW=770x770 μm), and with a 5 μm gap at the center (LxW=810x810 μm). The lower patch antenna is with a 5 μm gap at the center (LxW=770x770 μm, or LxW=810x810 μm depending on the upper).

Fig.2 (a), (b) shows the calculated result of the surface distribution of E_z/E_0 at 60 GHz for the structure of Fig.1. The field distributions of E_z/E_0 are mutually in out-of-phase between the upper layer and the lower layer. This is the same with the Yagi-Uda antennas. Fig.2 (c) shows the distribution of E_z/E_0 along ll' in Fig.2 (a). We can confirm the strong electric field is induced across the gap.

Utilizing the stacked structure, the magnitude of E_z/E_0 increased, and patch length become longer at the same operation frequency. It should be noted that we found an interesting characteristic in the stacked structure; both the LiNbO$_3$ layer and the SiO$_2$ layer are to be stacked for having the stronger electric field. If there is no LiNbO$_3$ layer on the top of the upper layer, the electric field at the waveguide is not so enhanced.

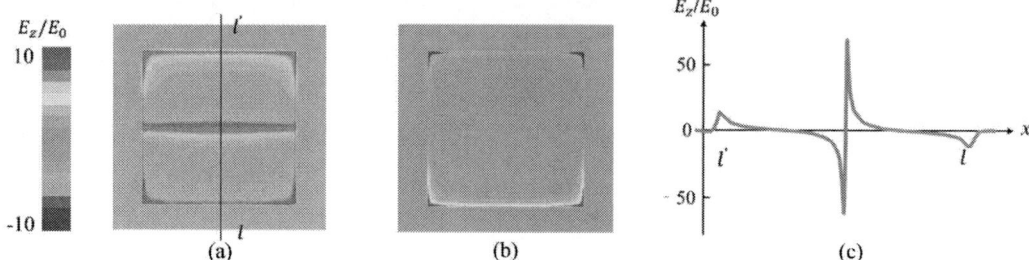

Fig. 2. Distribution of E_z/E_0 at LN layer at 60 GHz (a) at lower layer, (b) at upper layer, (c) at ll'.

Fig.3 represents the calculated radiation patterns of the stacked device and of the non-stacked device. Utilizing the stacked structure, directivity of the device becomes narrow. This characteristic is also the same as that of Yagi-Uda antennas.

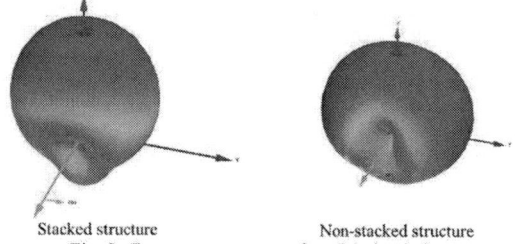

Stacked structure   Non-stacked structure

Fig. 3. Frequency response of modulation index

In the proposed device, the conversion efficiency from millimeter-wave to lightwave is proportional to the optical phase modulation index, $\Delta\varphi$. It can be expressed as

$$\Delta\varphi = \frac{\pi r_{33} n_e^3}{\lambda} \ \Gamma \int_0^L E_z \sin(n_g k_m y) dy \qquad (1)$$

where λ and n_g are the wavelength and the group index of lightwave propagating in the waveguide, respectively, r_{33} is the EO coefficient, n_e is the extraordinary refractive index of substrate, Γ is the overlapping factor between the induced millimeter-wave electric field and lightwave, and k_m is the wave number of the millimeter-wave in vacuum. Fig.4 shows the calculated frequency dependences of the modulation index; The blue one is for the 2 layer stacked structure embedded with a 5 μm gap for both patch metals. The green one is for the 2 layer stacked structure embedded with a 5 μm gap for only the lower metal, and the gray one is for the single layer structure (no-stacked) embedded with a 5 μm gap. The modulation index of the 2 layer stacked structure with a gap for both (blue) is increased by 3.65 dB, since the length of the patch becomes 1.28 times longer, and E_z/E_0 become 2.0 times higher compared with the conventional structure (gray).

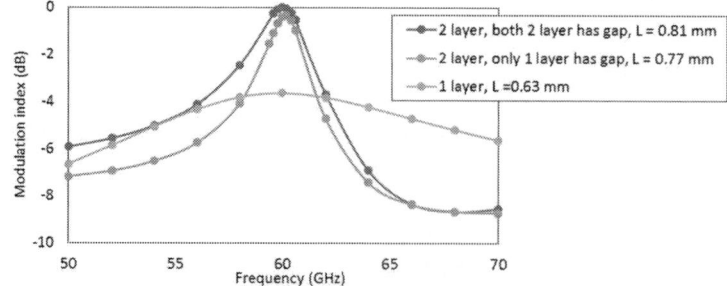

Fig. 4. Frequency response of modulation index

IV. CONTROL OF DIRECTIVITY

Adopting an array scheme to the proposed device, the directivity in the wireless millimeter-wave to lightwave conversion can be controlled. Fig.5 (a) shows the cross-sectional view of the device along the optical waveguide when a wireless millimeter-wave signal is irradiated to the device at an angle of θ. Then strong electric fields are induced across the gap in each patch antenna, and a lightwave propagating in the waveguide is modulated when passing through the patch. By considering the phase shift of the millimeter-wave electric field during the light propagation, the electric field as would be observed at the propagating light in the waveguide at h-th resonant electrode can be expressed by the following equation.

$$E_m^k(y, \theta, h) = E_z \cos[n_g k_m y - n_g k_m (h - 1) d_k \sin\theta + \Phi] \qquad (2)$$

where E_0 is amplitude of the electric field across the gap, n_0 is the refractive index of the millimeter-wave in the air, k_m is the wave number of the millimeter-wave in vacuum, n_g is the group index of the lightwave propagating in the waveguide, and Φ is an initial phase. Therefore, by setting the appropriate interval of the patch d_k, the modulation efficiency can be enhanced for the millimeter-wave signal of the irradiation angle, θ. As a result, the directivity in the conversion from millimeter-wave to lightwave signals can be designed. Fig 5 (b) shows the calculated examples of the directivities, where the total number of patch array was set 8 and maximum irradiation angles were set as -30, -15, 0, 15, and 30 degrees [2].

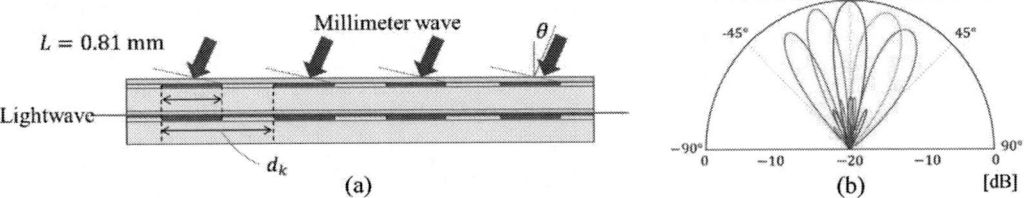

Fig. 5. (a) Cross sectional view of EO modulator, (b) Calculated directivity of modulation efficiency ($N = 8$)

V. CONCLUSION

In this paper, we proposed the electro-optic modulators using the stacked structure of the gap-embedded patch antennas. By stacking the patch antennas with the same size vertically, the patch length becomes longer, and the magnitude of E_z/E_0 becomes larger. These effects are contributed to the improvement of modulation efficiency, and 3.65 dB larger efficiency can be obtained compared with the simple non-stacked one. Moreover, adopting an array with the stacked patch antennas, we can obtain the controlled directivity. Now, we are trying to fabricate the proposed device. We are also analyzing more stacked structure devices.

ACKNOWLEDGMENT

The authors thank to Drs. Hidehisa Shiomi and Toshiyuki Inoue from Osaka University, Japan for their helpful advice and discussion on the analysis.

This research results was achieved in part by the research project of "Radio technologies for 5G using Advanced Photonic Infrastructure for Dense user environments" (RAPID), the Commissioned Research of National Institute of Information and Communications Technology (NICT), Japan.

REFERENCES

[1] Y. N. Wijayanto, H. Murata, Y. Okamura ''High-Speed Guided-Wave Electro-Optic Modulators Using an Array of Gap-Embedded Patch Antennas with Phase Reversal'', IEICE Technical Report MWP2011-72, 2012.

[2] N. Kohmu, H. Murata, Y. Okamura, "Electro-Optic Modulator Using LiNbO₃ Waveguide Suspended to Double-Antenna-Coupled Electrodes on Low-k Substrate for Millimeter-wave-Lightwave Signal Conversion", PA-6, APMP, Gwangju Apr.23, 2013

[3] O. Kramer, T. Djerafi, K. Wu" Very Small Footprint 60 GHz Stacked Yagi Antenna Array, "*IEEE Trans. Antennas Propag.* vol. 59, no.9, pp.3204-3210, 2011.

[4] O. Kramer, T. Djerafi, K. Wu "Vertically Multilayer- Stacked Yagi Antenna With Single and Dual Polarizations" *IEEE Trans. Antennas Propag.* vol.58, no.4, pp.1022-1030, 2010.

Robustness for Tracking Error in FMF based FSO Receiver

T. Ishikawa, K. Hosokawa, S. Takahashi, M. Arikawa, Y. Ono, T. Ito, K. Fukuchi

Green Platform Research Laboratories, NEC Corporation
1753 Shimonumabe, Nakahara-ku Kawasaki, 211-8666, Japan
t-ishikawa@dv.jp.nec.com

Abstract: We have numerically analyzed the relaxing effect for required tracking accuracy by the introduction of FMF based FSO receiver, and clarified 6-mode FMF provides a 2.2-times larger tolerance compared to the direct coupling to SMF.

Keywords: free space optics, spatial division multiplexing, few mode fiber, coupling efficiency

I. INTRODUCTION

The data traffic between midair and ground are rapidly increasing due to the technology advancements in observation and communication on airplane and satellite. Free space optics (FSO) is expected as a break-through technology for insufficient radio frequency resource [1]. In terms of the potential of data transmission speed, the use of optical components based on SMF (Single Mode Fiber) is preferable because of the availability of the technologies developed for the optical fiber network systems. FSO receiver with direct coupling to the SMF, however, requires severe alignment [2] for coupling the turbulence-induced laser beam into core whose diameter is about 10 micro meters. FSO receiver generally has expensive tracking system which consists of coarse and fine angular compensation functions with beam pointing mirrors. However, the tracking system is difficult to operate ideally especially when one of the communication terminals is placed on the fast-moving and unstable platforms such as airplane.

We propose FMF (Few Mode Fiber) based FSO receiver using SDM (Spatial Division Multiplexing) technologies explained in detail in the next section. The FSO receiver couples non-ideal laser beam into the FMF with proper efficiency, and SMF based high-speed optical components can be used owing to the low-loss FMF-SMF mode converter. This paper focuses on the relaxing effect for the required tracking accuracy by this proposed receiver, and the relationship between acceptable tracking error and number of available modes in FMF is analyzed with numerical simulation. The results show that proposed FSO receiver with 6 modes allows tracking error of 44.6 micro radians, which is equivalent to 2.2 times the tracking error for the direct coupling to SMF.

II. PROPOSED FMF BASED FSO RECEIVER

The proposed FSO receiver has two benefits. The first one is the better fiber coupling efficiency by using FMF which has multi propagation modes referred to as linear polarization (LP) modes. Second one is a capability of the high-speed operations with SMF based components. SDM technologies such as FMF-SMF mode-converter and digital signal processing (DSP) techniques will simultaneously enable these two benefits.

Fig.1 shows the configuration of the FSO system with proposed FSO receiver [3]. Transmitter outputs modulated laser beam to free space as an optical signal. The laser beam distorted by atmospheric turbulence is detected by receiver which is controlled with a tracking system, and coupled into FMF. The optical signal propagating in the FMF is separated according to LP modes, and converted respectively into SMF output in mode demultiplexer. These SMF outputs are combined and demodulated by DSP in appropriate method [4].

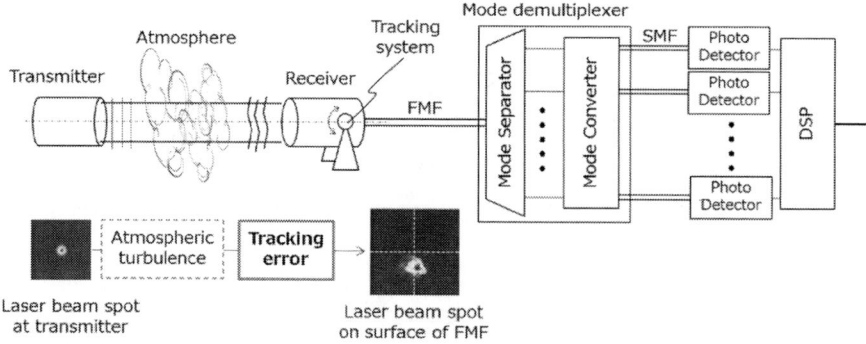

Fig. 1. Configuration of FSO communication system using SDM technologies.

III. MODELING AND SIMULATION OF COUPLING EFFICIENCY

In the application between midair and ground, aperture diameter above 50 mm of receiver optics causes significant distortion of the beam spot by atmospheric turbulence [5]. In this session, we evaluated the influence of the coupling efficiency caused simply by tracking error, and the evaluation model only considers the displacement of the beam spot and tilt of incident angle on the fiber by tracking error. Estimation of the coupling efficiencies of the distorted laser beam by atmosphere turbulence, excluded in this study, will be a future work.

A. Displacement of the laser beam spot due to tracking error

The FSO receiver has a tracking error while transmitter outputs the laser beam in the accurate direction. The laser beam propagates without disturbance and is focused onto the fiber with displacement and tilt due to incident angle misalignment as shown in Fig.2. Receiver optics can be assumed to be a paraxial lens which focuses the laser beam onto the fiber without aberration or absorption. The displacement is calculated with the product of the tracking error and focal length of the receiver optics. Since fiber core is very small, the laser beam spot can easily shift from the center with the slight tracking error.

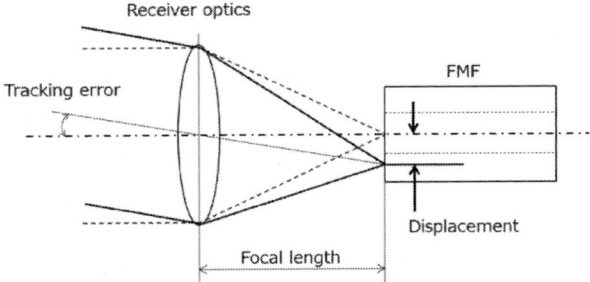

Fig. 2. Model of the laser beam spot displacement on fiber surface by tracking error.

B. Coupling efficiency into each LP modes in existence of tracking error

We analyzed the dependence of the fiber coupling efficiency on the tracking error. The coupling efficiency is defined as a normalized overlap integral between electrical field of the beam and mode field profile of the fiber. Calculation conditions for both SMF and FMF cases are as follows; wavelength: 1.55 um, receiver optics aperture diameter: 50 mm, receiver optics focal length: 250 mm, MFD of LP_{01} in FMF: 12 micro meters, number of LP modes calculated: 21 (maximum), tracking error: 0 to 150 micro radians. The mode demultiplexer has ideal characteristics with no intrinsic loss and no crosstalk between different modes. The laser beam is shifted and tilted depending on tracking error angle and parameters of the receiver optics, and then the coupling efficiency is calculated for individual LP modes.

Fig.3 shows the calculation results of the coupling efficiencies into each LP modes against tracking error. Some modes have two degenerate modes, for example LP_{11} has 'LP_{11a}' and 'LP_{11b}'. As shown in figure captions (6, 10, 21 modes), we counted the degenerate modes as independent two propagation modes, because the number to be counted with such manner is so essential that it determines the amount of resources like DSP. They are plotted, however, in these graphs with one line which is derived from the sum of coupling efficiencies into degenerate modes. This is because coupling efficiency into the LP mode against tracking error is determined by their sum into degenerate modes, and branching ratio of these two just depends on the direction of tracking error. Black lines in Fig. 3 (a) – (c) indicate the results of the coupling efficiencies into LP_{01}. As tracking error arises, coupling efficiency into LP_{01} decreases and one into LP_{11} increases. Similarly, the increase of tracking error decreases the coupling efficiency of lower modes and generates the higher ones in order. As a result, the tolerance to the tracking errors is expanded in accordance with numbers of available modes.

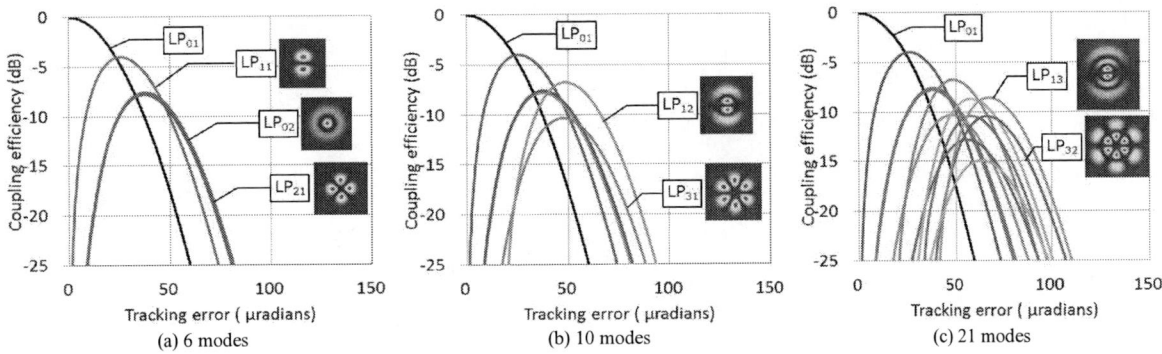

Fig. 3. Simulation results of coupling efficiency into each LP modes against tracking error.

IV. DEPENDENCE OF REQUIRED TRACKING ACCURACY ON NUMBER OF MODES

The coupling efficiency into LP_{01} is regarded as the one of SMF. Fig. 4(a) shows the coupling efficiency of FMFs against tracking error, in which the coupling efficiency of FMF is determined as sum of them into each LP modes. We defined a threshold at the angle causing the 3-dB degradation of the coupling efficiency; SMF based FSO receiver requires the tracking accuracy of 20.3 micro radians, and as shown in Fig. 4(b), FMF enables to expand the tolerance of the tracking error as number of modes increases. For example, FMF with 6 modes accepts 2.2 times larger tacking error (44.6 micro radians) and 3.6 times with 21 modes.

According to Fig.4(b), the tolerance is monotonically expanded as the number of modes increases, however, the receiver design relating to number of modes should be with considerations of costs and complexities coming from the photo-detectors and DSP. At this moment, the number of modes is limited by the availability of low-loss mode demultiplexer, and the recent technologies focus on up-to 15 modes [6], [7].

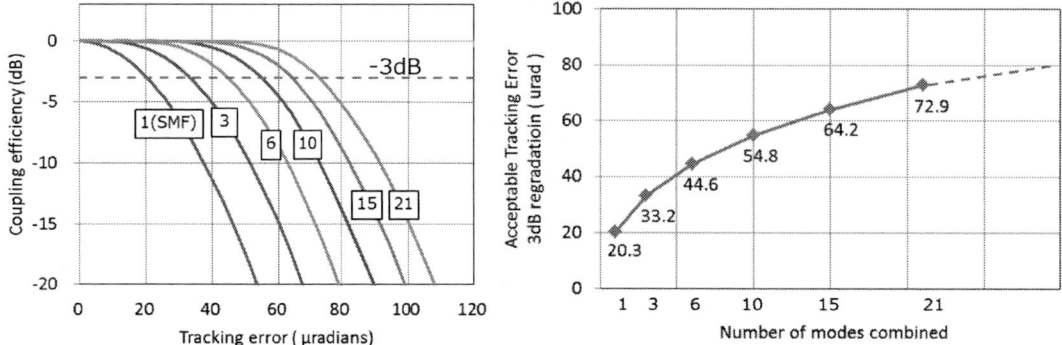

Fig. 4. (a): Dependence of coupling efficiency into FMF on number of modes against tracking error.
(b): Acceptable tracking error against number of modes for proposed receiver

The relaxing ratio in the tracking accuracy depends not on the aperture diameter but on the number of modes in FMF. The enhanced tolerance due to FMF has more important meanings especially for the receivers with larger aperture diameters. Note that the investigations with such receivers should include the influences of atmospheric turbulence.

V. CONCLUSIONS

We have numerically analyzed the relaxing effect for required tracking accuracy by the introduction of FMF based FSO receiver, and clarified the dependence of the relaxing effect on the number of modes. In condition that the aperture diameter is 50 mm, SMF based FSO receiver requires a tracking accuracy of 20.3 micro radians, while the proposed FSO receiver with 6 modes allows 2.2 times larger tracking error of 44.6 micro radians. The robustness for tracking error provided by the proposed receiver is expected to stabilize the data communication and reduce the costs.

REFERENCES

[1] D.O. Caplan, M.L. Stevens, B.S. Robinson "Free-space Laser Communications: Global Communications and Beyond," ECOC 2009, Paper 9.6.1.

[2] K. Yoshida, T. Tsujimura, "Seamless Transmission between Single-Mode Optical Fibers Using Free Space Optics System," SICE Journal of Control, Measurement, and System Integration, Vol.3, No.2, pp.94-100, March 2010.

[3] S. Takahashi et al., "Studies for Free-space Optical Receiver using SDM Technology (1) :Coupling Efficiency Improvement by a Few Mode Fiber," IEICE general conference 2015, 275(in Japanese)

[4] K. Hosokawa, et al., "Free-space Optical Receiver using SDM Technique to Overcome Atmospheric Turbulence," the Space Sciences and Technology Conference, 2015(in Japanese).

[5] Proceedings of the International Workshop on Ground-to-OICETS Laser Communications Experiments 2010 – GOLCE, Japanese Nationals Institute of Information and Communications Technology, 2010.

[6] G. Labroille et al., "Efficient and mode selective spatial mode multiplexer based on multi-plane light conversion," Optics Express, Vol. 22, Issue 13, pp.15599-15607, 2014.

[7] N. K. Fontaine et al., "30x30 MIMO Transmission over 15 Spatial Modes," OFC 2015, Th5C.1.

Impedance Matched GaN LD Package for Direct OFDM Communication at 14 Gbps

Yu-Fang Huang[1], Tsai-Chen Wu[1], Yu-Chieh Chi[1], Cheng-Ting Tsai[1], Wei Wang[2], Tien-Tsorng Shih[2], Hao-Chung Kuo[3], and Gong-Ru Lin[1,*]

[1]Graduate Institute of Photonics and Optoelectronics, and Department of Electrical Engineering,
National Taiwan University, No. 1, Roosevelt Rd, Sect. 4, Taipei 10617, Taiwan, ROC
[2]Department of Electrical Engineering, National Kaohsiung University of Applied Sciences,
No.415, Jiangong Rd., Sanmin Dist., Kaohsiung City 807, Taiwan, ROC
[3]Department of Photonics, National Chiao Tung University,
No.1001 Ta Hsueh Rd., Hsinchu 30050, Taiwan, ROC
*grlin@ntu.edu.tw

Abstract: A transmission line aided impedance matching package for 450-nm GaN laser diode is employed to increase the directly encoded 16-QAM OFDM transmission up to 14-Gbps over 7 m in free-space after modulation throughput intensity optimization.
Keywords: VLC, Impedance Matching, QAM-OFDM.

I. INTRODUCTION

Recently, the visible light-emitting diodes (LEDs) based visible light communication (VLC) system becomes popular because it enables both data transmission and indoor lighting with high brightness and low energy consumption. Limiting by the modulation bandwidth of LED at 10s-100s MHz, the LED based VLC system has a challenge for increasing its data transmission capacity. To meet this demand, Biagi *et al.* improved the data rate to 600 MHz by using pulse position modulation (PPM) for in a multiple-in multiple-out (MIMO) VLC system [1]. Later on, Zhan *et al.* demonstrated a 1.2-Gbps VLC system based on the LED MIMO orthogonal frequency-division multiplexing (OFDM) [2]. Tsonev *et al.* implemented a 3-Gbps wireless VLC system with direct current optical (DCO) 32-quadrature amplitude modulation (QAM) OFDM based on a single 50-μm gallium nitride (GaN) μLED [3]. Although the μLED biased at higher current density enables to reveal wider modulation bandwidth, such an operation may induce efficiency droop due to the electron overflow. To release the bandwidth limitation, the laser diode (LD) is raised to replace the LED for implementing the VLC system because of the LD modulation bandwidth of up to GHz. In addition, the visible LDs also have the advantages of high directivity and pumping efficiency. Hussein *et al.* demonstrated a 5-Gbps angle diversity receiver (ADR) LD VLC system using on-off keying (OOK) modulation format. The use of LD and ADR can effectively increase the transmission capacity and suppress the inter symbol interference (ISI) problem [4]. Recently, Chi *et al.* accomplished a 9-Gbps QAM OFDM VLC system based on a To-can packaged GaN blue LD over 5-m free space [5]. However, the impedance matching of such a To-can packaged GaN blue LD has not been implemented previously, which may limit the transmission capacity of proposed VLC system.

In this work, an impedance-matched 450-nm GaN blue LD is demonstrated for 14-Gbps/7-m pre-leveled QAM OFDM optical wireless communication. The frequency response of the 450-nm GaN blue LD with an impedance matching circuit board is revealed. By using the pre-leveling technique, the allowable OFDM data bandwidth for the impedance-matched 450-nm GaN blue LD is also discussed. The average signal-to-noise ratio (SNR), bit error rate (BER) and error vector magnitude (EVM) of 7-m free-space transmitted 16-QAM OFDM data are analyzed in detail.

II. EXPERIMENTAL SETUP

The homemade transmission line circuit board is shown in Fig. 1(a), which helps to optimize the modulation bandwidth of used GaN LD. In addition, the impedance-matched and temperature-controlled GaN LD package for further VLC application is shown in Fig. 1(b). The experiment setup of the TO-can packaged 450-nm GaN blue LD with impedance matching for 7-m/14-Gbps QAM OFDM communication is shown in Fig. 1(c). The used 16-QAM OFDM data with a bandwidth of 3.2 GHz ranging from 0.14 to 3.34 GHz represents 69 subcarriers. The blue LD with a threshold current of 35 mA and a P-I slope of 0.97 W/A was operated at a room temperature of 25°C. The 16-QAM OFDM data sent from a 24-GS/s arbitrary waveform generator (AWG, Tektronix 70001A) after passing through a 10-dB pre-amplifier (Picosecond, 5828A) and a bias tee (Mini-circuit, ZX85-12G-S+) merges the DC biased current to directly modulate the blue LD. Then, the output blue laser beam incidents on an aspheric objective lens to form a parallel light for 3.5-m free-space communication, and a mirror was used to extend the distance of the optical wireless communication to 7 m. For data receiving, another aspheric objective lens refocuses the parallel blue light into a p-i-n PD (Thorlabs, FDS025) driven by a bias tee (Picosecond, 55530B), and the received data was amplified by a 21-dB amplifier (Mini-circuits, ZX60-V63+). Finally, the data was decoded by a 100-GS/s digital serial analyzer (DSA, Tektronix, 71604C) for further analyzing the data performances.

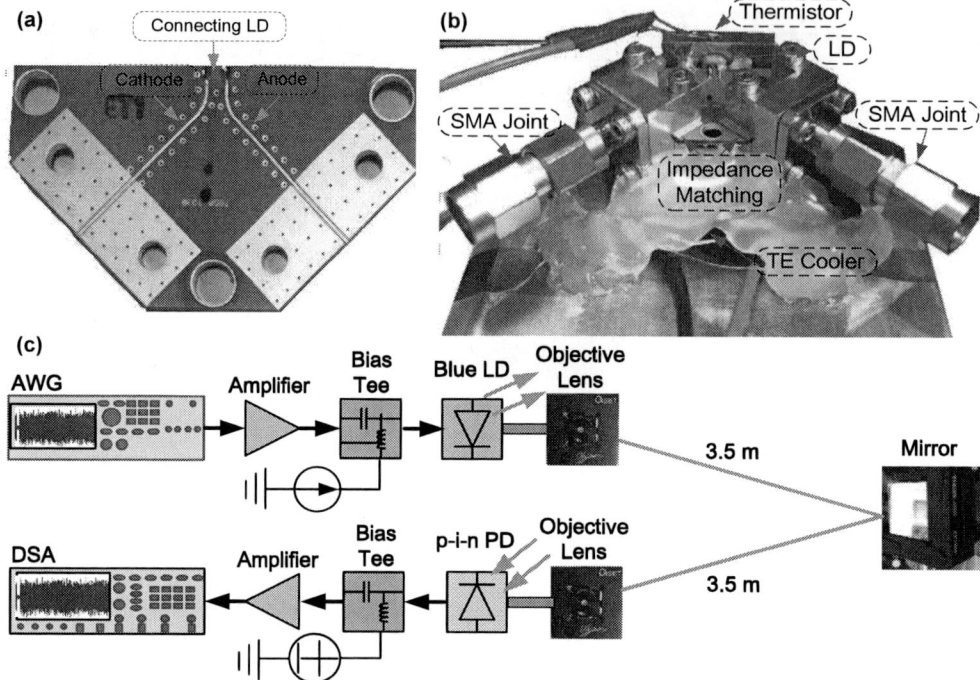

Fig. 1. (a) The transmission line circuit board designed for impedance matching with GaN LD; (b) the impedance matched and temperature controlled GaN LD package; (c) the impedance matched GaN blue LD with 16-QAM OFDM data over 7-m.

III. RESULTS AND DISCUSSIONS

Firstly, the frequency response of the impedance-matching 450-nm GaN blue LD at a biased current of 80 mA and a free-space distance of 0.5 m is shown in Fig. 2(a). Note that the -6 dB modulation bandwidth of 1.3 GHz is observed. For OFDM data encoding, the constellation plot of the 1.5-GHz 16-QAM OFDM data carried by the 80-mA biased impedance-matching GaN blue LD is shown in Fig. 2(b), and the related BER and EVM of 1.2×10^{-6} and 9.6% are observed, respectively. If the biased current is decreased to 70 mA, the BER and EVM would degrade to 1.7×10^{-6} and 10.1%, respectively, because of the data clipping effect. Besides, continuously increasing the current to 90 mA also degrades the transmitted BER to 3.3×10^{-6} and EVM to 10.1% because of the reduced OFDM data extinction ratio. In addition, the subcarriers response of the 80-mA biased impedance-matching GaN blue LD carried 1.5-GHz 16-QAM OFDM data is shown in Fig. 2(c), and an average SNR of 20.3 dB is obtained, which is significantly higher than the forward error correction (FEC) criterion of 15.2 dB.

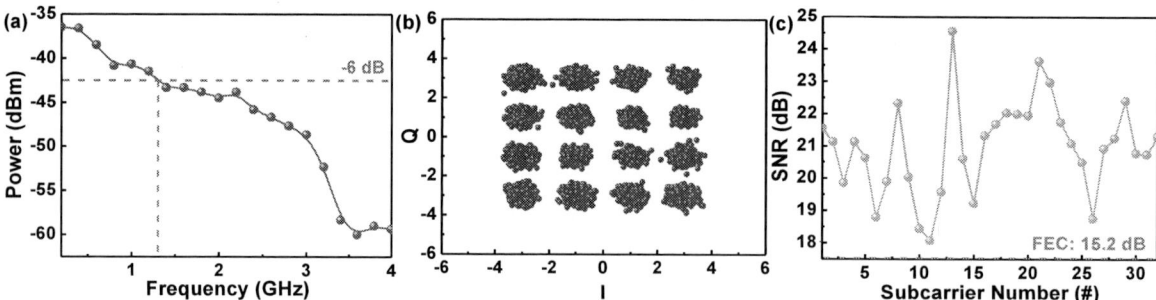

Fig. 2. (a) The frequency response of the impedance-matching 450-nm GaN blue LD biased at 80 mA; (b) the constellation plot and (c) the subcarrier SNRs of the 1.5-GHz 16-QAM OFDM data carried by the 80-mA biased impedance-matching GaN blue LD.

After optimizing the biased current for the impedance-matching 450-nm GaN blue LD, the free-space distance is extended to 7 m for practical application. To compensate the declined frequency response of used LD and the free-space transmission induced RF power fading, the pre-leveling technique which sacrificing the low-frequency subcarrier energy compensates the high-frequency one is employed. Without using the pre-leveling technique, a 16-QAM OFDM data bandwidth of 3.2 GHz at a raw data rate of 12.8 Gbps over a 7-m free-space can be achieved for the impedance-matching 450-nm GaN blue LD, which reveals FEC qualified SNR of 15.3, BER of 3.4×10^{-3} and EVM of 17.1%. To further extend the OFDM data bandwidth to 3.5 GHz, the transmitted BERs at different pre-leveling slopes are shown in Fig. 3(a). By increasing the pre-leveling slope from 0.2 to 0.38 dB/GHz, the BER could be improved from 6.8×10^{-3} to 3.5×10^{-3}. However, continuously increasing pre-leveling slope to 1 dB/GHz inversely degrades the BER to 9.2×10^{-3} because too much energy sacrificed at low frequency causes to worse the average SNR. After pre-leveling optimization, the allowable 16-QAM OFDM data bandwidth of 3.5 GHz is thus observed to reveal a data rate of up to 14 Gbps for the impedance-matching 450-nm GaN blue LD, and the related EVM of 17.2% and average SNR of 15.3 dB are also obtained, as shown in Figs. 3(a) and 3(b).

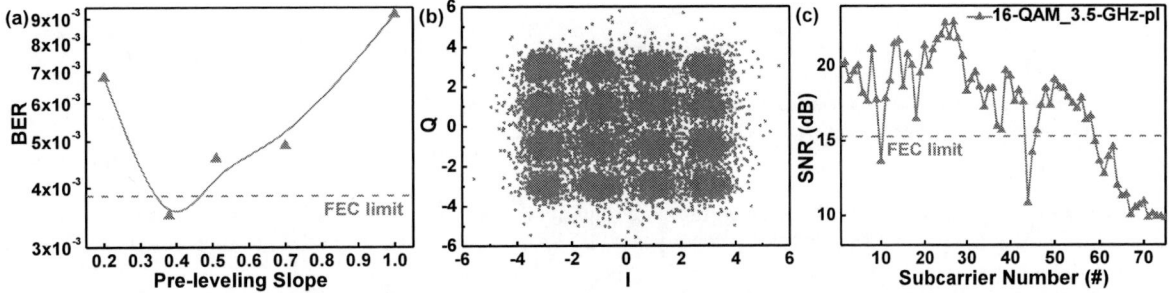

Fig. 3. (a) The BERs response of the impedance-matching 450-nm GaN blue LD carried 14-Gbps 16-QAM OFDM data at different pre-leveling slopes; (b) The constellation plot and (c) subcarrier SNRs of transmitted 14-Gbps 16-QAM OFDM data at a pre-leveling slope of 0.4 dB/GHz.

IV. CONCLUSION

With a GaN blue LD at wavelength of 450 nm, the directly encoded 16-QAM OFDM transmission up to 14-Gbps over 7 m in free-space is successfully implemented, which enables the higher data rate of communication than ever by adding a transmission line based impedance matching package for the To-can packaged 450-nm GaN blue LD. Without using the pre-leveling or bit-loading technique for the pre-emphasis of QAM-OFDM data stream in frequency domain, the 16-QAM OFDM data with a bandwidth covering 3.2 GHz and corresponding raw data bit rate at 12.8 Gbps can easily be achieved to reveal the FEC qualified EVM of 17.1%, SNR of 15.3 dB, and BER of 3.4×10^{-3}. After the optimization on modulation throughput intensity, the transmission capacity can be further increased to 14 Gbps, providing an EVM of 17.2%, an average SNR of 15.3 dB, and a BER of 3.5×10^{-3}, which is still decodable under the FEC criterion after passing through 7 meters in free space.

REFERENCES

[1] M. Biagi, and A. M. Vegni, "Enabling High Data Rate VLC via MIMO-LEDs PPM," Globecom Workshops (GC Wkshps), 2013 IEEE, Atlanta (GA), Dec. 2013, pp. 1058-1063.

[2] Z. Zhan, M. Zhang, D. Han, P. Luo, X. Tang, Z. Ghassemlooy, L. Lang, "1.2 Gbps Non-Imaging MIMO-OFDM Scheme based VLC over Indoor Lighting LED Arrangments," Opto-Electronics and Communications Conference (OECC), 2015, Shanghai (China), June 2015, pp. 1-3.

[3] D. Tsonev, H. Chun, S. Rajbhandari, J. J. D. McKendry, S. Videv, E. Gu, M. Haji, S.T. Watson, A. E. Kelly, G. Faulkner, M. D. Dawson, H. Haas, and D. O'Brien, "A 3-Gb/s Single-LED OFDM-Based Wireless VLC Link Using a Gallium Nitride μLED," IEEE Photonics Technology Letters, vol. 26, Apr. 2014, pp.637-640.

[4] A. T. Hussein and J. M. H. Elmirghani, "High-Speed Indoor Visible Light Communication System Employing Laser Diodes and Angle Diversity Receivers," Proc. of 17th Int. Conf. on Transparent Optical Networks (ICTON 2015), Budapest, Jul. 2015, pp. 1-6.

[5] Y. C. Chi, D. H. Hsieh, C. T. Tsai, H. Y. Chen, H. C. Kuo, and G.-R. Lin, "450-nm GaN laser diode enables high-speed visible light communication with 9-Gbps QAM-OFDM," Opt. Express, vol. 33, May 2015, pp. 13051-13059.

OCT Precoding for OFDM-based Indoor Visible Light Communications

Yang Hong and Lian-Kuan Chen

Department of Information Engineering, The Chinese University of Hong Kong, Shatin, N. T., Hong Kong
yanghong@ie.cuhk.edu.hk, lkchen@ie.cuhk.edu.hk

Abstract: An orthogonal circulant matrix transform (OCT) precoding scheme is proposed for indoor visible light communications. We show the proposed scheme has much reduced complexity but exhibits comparable performance with adaptive bit and power loading.

Keywords: Visible light communications; OFDM; OCT precoding

I. INTRODUCTION

Visible light communications (VLC), emerging as an alternative solution for the RF communications, has gained interests from both academia and industry in recent years [1-2]. One of the main challenges for high-speed VLC is the limited LED's modulation bandwidth, typically several megahertz [2]. Blue filtering or RGB-type LED is commonly used in recent works to improve the 3-dB system bandwidth. The electrical circuit based compensation is another way to extend the system bandwidth, achieving a relatively flat spectrum or signal to noise ratio (SNR) over a larger signal bandwidth. Various design of multi-resonant electrical circuits are reported for pre-distortion [1] or for post equalization [2] to improve the performance of VLC systems. However, besides the requirement of pre-knowledge for channel state information (CSI), the designed circuits cannot achieve adaptive adjustment to the changes of system setup or channel properties without incurred high complexity. Alternatively, the SNR equalization can also be achieved by the discrete Fourier transform (DFT) precoding, which is also known as single carrier frequency division multiple access (SC-FDMA) [3]. The DFT precoding is channel independent and only requires one linear transformation at transmitter and receiver, respectively. However, the conventional DFT precoding is more sensitive to inter-symbol interference (ISI). Therefore, a large cyclic prefix (CP) length is needed to combat the ISI, leading to lower spectral efficiency.

The discrete multitoned modulation (DMT) with bit and power loading [4-5] is generally known an effective solution to boost system capacity, since the bits and power are allocated adaptively according to the pre-known channel state information (CSI). Despite of its optimal performance, the adaptive bit and power loading scheme requires the support of multiple modulation levels at both transmitter and receiver. The allocation results of bits and power also need to be transmitted to receiver accurately. All these issues increase the complexity and cost for the implementation of adaptive bit and power loading scheme in practical VLC systems. We have recently proposed a channel-independent OCT precoding scheme for VLC systems, with proof-of-concept point-to-point experiments demonstrated in [6]. The results show that with the same spectral efficiency, the performance of OCT precoding outperforms that of conventional DFT precoding. To apply OCT precoding to the emerging indoor VLC applications, it is essential to investigate the feasibility of the proposed OCT precoding in indoor environments.

In this paper, we focus on the performance investigation of OCT precoding with the consideration of multi-reflection effect, as well as location influence of the different indoor scenarios. We show that by utilizing the proposed OCT scheme, relatively flat SNR with ~1-dB fluctuation for different subcarriers can be obtained, resulting in significant bit error rate (BER) performance improvement of the system. Numerical simulation results show that the proposed OCT precoding have comparable BER performance with the adaptive bit and power loading, but exhibits significantly reduced implementation complexity, thus is more preferable for practical VLC applications.

II. SYSTEM MODEL

Fig. 1 shows schematic diagram of a VLC system with the proposed OCT precoding. After serial to parallel (S/P) conversion and mapping, OCT precoding is performed, i.e., the mapped signal $[X_1, X_2, ..., X_N]$ is multiplied by an orthogonal circulant matrix given by

$$\mathbf{F} = \tfrac{1}{\sqrt{N}} \times \left[c_1, c_2, ..., c_N; c_N, c_1, ..., c_{N-1}; ...; c_2, c_3, ..., c_1 \right], \tag{1}$$

where each entry c_i $(1 \le i \le N)$ of $\mathbf{F}$ is the corresponding element of Zadoff-Chu (ZC) sequence with a length of N. We utilize the ZC sequence in this work because of its ideal periodic auto-correlation property, i.e., the periodic auto-correlation is zero for all time shifts other than zero. Therefore, the constructed circulant matrix $\mathbf{F}$ is orthogonal and $\mathbf{F}^*\mathbf{F} = \mathbf{I}$, where $(\cdot)^*$ denotes Hermitian transpose. In order to obtain real-valued OFDM signal, after the OCT precoding, subcarrier assignment is needed to constrain the input of the inverse fast Fourier transform (IFFT) operation to have Hermitian symmetry. Then, parallel to serial (P/S) conversion, CP insertion, pilot insertion and normalization are performed. Subsequently, the signal is fed into digital to analog converter (DAC). After DAC, the output is combined

This work was supported in part by RGC, Hong Kong, under GRF 14204015.

with DC-bias to drive four LEDs. The properties of LED transmitter are the same as that in [4].

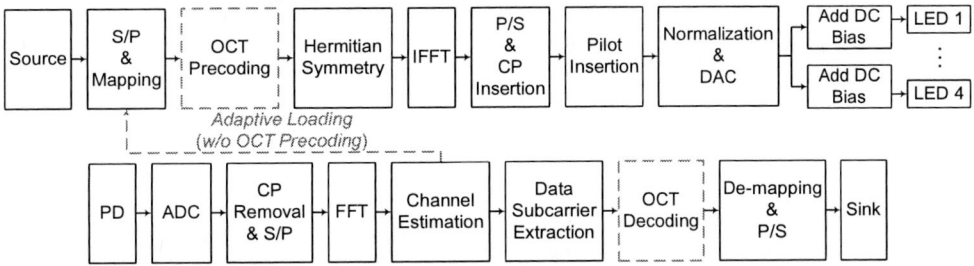

Fig. 1. Block diagram of indoor VLC system with (i) OCT precoding or (ii) adaptive loading.

Indoor VLC channel property from the i-th LED transmitter to the receiver can be characterized by the corresponding impulse response, which is described by

$$h_i(t) = h_i^{(0)}(t) + \sum_{k=1}^{\infty} h_i^{(k)}(t),$$ (2)

where the two terms on the right side of Eq. (2) represent the line-of-sight (LOS) contribution and the non-LOS contribution, respectively. Monte Carlo method used in [4] is utilized to model the indoor multi-reflection VLC channel, and the simulation parameters used are given in Table I.

TABLE I SIMULATION PARAMETERS

Parameter	Value	Parameter	Value
LED locations (in meter)	(3.5,3.5,3); (1.5,3.5,3); (3.5,1.5,3); (1.5,1.5,3)	Max. reflection order	5
Receiver location (Scenario A - center) (in meter)	(2.5,2.5,0.85)	Receiver field of view	60°
Receiver location (Scenario B - corner) (in meter)	(0.5,0.5,0.85)	LED power	3 W
Reflection coef. (walls, floor, ceiling)	0.83; 0.63; 0.4	Number of Rays	200,000

Assume that the output signal of LED is denoted by $x(t)$. The received signal $y(t)$ is

$$y(t) = \gamma \cdot \sum_{i=1}^{4} h_i(t) \otimes x(t) + n(t),$$ (3)

where γ is the photodiode responsivity and $n(t)$ denotes the noise with zero mean, with $n(t)$'s variance containing shot noise and thermal noise contributions. After data subcarrier extraction, OCT decoding is performed, i.e., the received signal is multiplied by the inverse matrix of $\boldsymbol{F}$, to recover the original complex signal. The error vector magnitude (EVM) of the recovered complex signal is then used to estimate each subcarrier's SNR.

III. SIMULATION RESULTS

The indoor configuration is set as follows. Four LEDs are mounted on the ceiling and the distance from ceiling to the receiving plane is 0.85 m. Two scenarios, as given in Table I, are considered in the performance investigation of the proposed scheme. Note that DCO-OFDM is utilized in the following simulations, with an IFFT/FFT size of 256 and a CP length of 8.

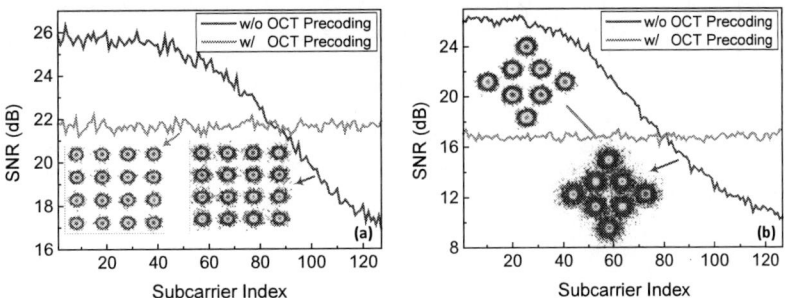

Fig. 2. SNR comparison of the VLC system with and without OCT precoding: (a) 16QAM-OFDM for scenario A, (b) 8QAM-OFDM for scenario B.

Fig. 2 shows the SNR performance comparison of the indoor VLC system with and without the proposed OCT precoding scheme in the two scenarios. The bandwidth of the DCO-OFDM signal is 60 MHz. Note that 16-QAM is used for all data subcarriers of DCO-OFDM symbols in scenario A, whereas 8-QAM is used for the data subcarriers of DCO-OFDM symbols in scenario B. The BER performance of scenario A and scenario B without the proposed OCT precoding scheme is 5.12×10^{-5} and 2.31×10^{-3}, respectively. As shown in Fig. 2, SNR can be equalized to around the median value with ~ 1dB fluctuation for different data subcarriers by using the OCT precoding scheme; and a BER of less than 10^{-5} can be achieved for both scenarios. Besides, it is clear that the performance improvement is related to SNR conditions of the system, therefore, it is desirable to investigate the proposed OCT precoding under the situation that data subcarriers are modulated with different order QAM signals. Fig. 3 shows the corresponding BER comparison.

589

The BER of scenario A is superior to that of scenario B because of larger received signal intensity and less multi-path influence. Generally, significant improvement in BER performance can be obtained for both scenarios when the signal bandwidth is lower than 100 MHz. The performance of the proposed scheme may be reduced for high-order QAM modulation format at larger bandwidth, where the BER is significantly above the 3.8×10^{-3} FEC limit.

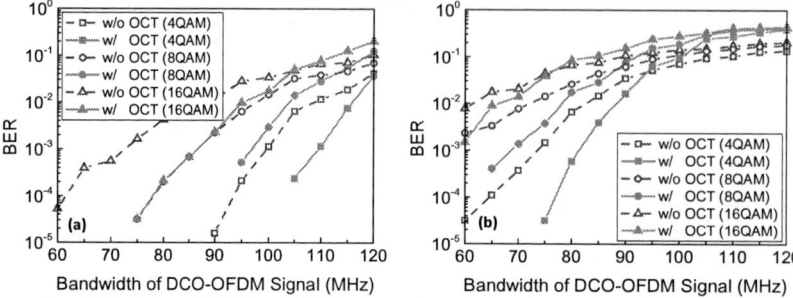

Fig. 3. BER comparison of the indoor OFDM-based VLC system with and without OCT precoding: (a) Scenario A, (b) Scenario B.

In order to further investigate the performance of the proposed scheme, we compare the BER of adaptive bit and power loading scheme with the results in Fig. 2. The main idea of bit and power loading scheme is to adaptively allocate the bits and power to different subcarriers according to CSI, which is estimated before data transmission with the help of training symbols. The estimated SNR of each subcarrier at the receiver along with the signal constellation diagrams are shown in Fig. 4. Basically, the estimated SNR from EVM over different subcarriers are positive correlated with the number of allocated bit. In both scenarios, a BER of less than 10^{-5}, similar to that of the proposed OCT precoding scheme, can be achieved. However, considering the complicated implementation of the adaptive bit and power loading scheme, the proposed OCT precoding, which offers significantly reduced complexity, is more robust for practical VLC applications.

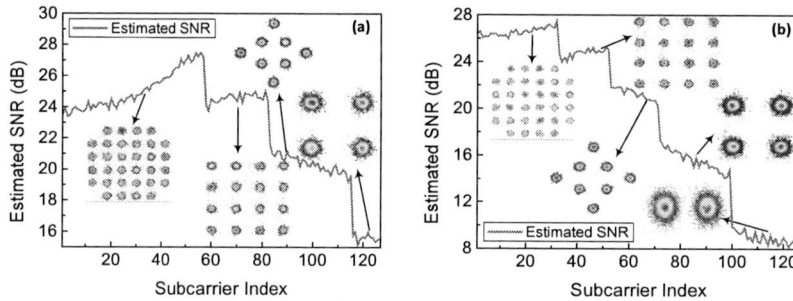

Fig. 4. SNR of the data subcarriers and corresponding signal constellations at the receiver: (a) Scenario A, (b) Scenario B.

IV. CONCLUSIONS

We propose and investigate an OCT precoding scheme to combat the bandwidth limitation issue in OFDM-based indoor VLC systems. Significant BER improvement compared to the conventional OFDM transmission, as well as a relatively flat SNR condition with ~1-dB fluctuation, can be achieved by using the proposed scheme. The robustness of the OCT precoding is verified by the BER performance improvements in both center and off-center of indoor coverage. In addition, simulation results show the proposed scheme's BER performance is comparable to that of the adaptive bit and power loading scheme. However, considering the implementation complexity, the proposed scheme is more suitable for practical VLC systems. It is worth noting that besides VLC applications, the proposed scheme is applicable to other bandwidth-limited OFDM transmission systems.

REFERENCES

[1] X. Huang, Z. Wang, J. Shi, and N. Chi, "1.6 Gbit/s phosphorescent white LED based VLC transmission using a cascaded pre-equalization circuit and a differential outputs PIN receiver," Opt. Express, vol. 23, no. 17, pp. 22034–22042, 2015.

[2] H. Le Minh, D. O'Brien, G. Faulkner, L. B. Zeng, K. Lee, D. Jung, Y. Oh, and E. T. Won, "100-Mb/s NRZ visible light communications using a postequalized white LED," IEEE Photon. Technol. Lett., vol. 21, no. 15, pp. 1063–1065, 2009.

[3] D. D. Falconer, "Linear Precoding of OFDMA Signals to Minimize Their Instantaneous Power Variance," *IEEE Trans. Commun.*, vol. 59, no. 4, pp. 1154-1162, 2011.

[4] Y. Hong, T. Wu and L.K. Chen, "On the Performance of Adaptive MIMO-OFDM Indoor Visible Light Communications," IEEE Photon. Technol. Lett., vol. 28, no. 8, pp. 907-910, 2016.

[5] D. Bykhovsky and S. Shlomi, "An experimental comparison of different bit-and-power-allocation algorithms for DCO-OFDM," J. Lightw. Technol., vol. 60, no. 4, pp. 1559–1564, 2014.

[6] Y. Hong, X. Guan, L.K. Chen and J. Zhao, "Experimental Demonstration of an OCT-based Precoding Scheme for Visible Light Communications,", in Proc. of OFC, Anaheim, CA, USA, Paper M3A.6, 2016.

Nonlinear Compensation of 850 nm VCSEL with Discrete Multi-Tone Modulation Employing a Volterra-Wiener Filter

Ta-Ching Tzu[1], Chia-Chien Wei[2*], Jun-Jie Liu[1], Chun-Yen Chang[1], Kai-Lun Chi[3], Xin-Nan Chen[3], Jin-Wei Shi[3], and Jyehong Chen[1]

1. Department of Photonics, National Chiao Tung University, Hsinchu 300, Taiwan
2. Department of Photonics, National Sun Yat-sen University, Kaohsiung 804, Taiwan
3. Department of Electrical Engineering, National Central University, Taoyuan 320, Taiwan
ccwei@mail.nsysu.edu.tw

- *Abstract: A DMT transmission with 32% and 21.2% data rate improvement by nonlinear Volterra-Wiener compensation respectively for 68.94Gb/s OBTB and 58.39Gb/s over 105m is demonstrated by a 850nm VCSEL with short ($\lambda/2$) cavity and mode controlled.*
  Keywords: Semiconductor laser; Optical interconnect; Optical communication

I. INTRODUCTION

Recently, due to the rapid growth of the global internet traffic and the quickly ramping demands for the internet of things (IoT), modern data centers play a vital role and keep realizing lots of applications such as high performance computing, cloud systems and electronic commerce. Consequently, desiring to provide such high data capacity, future data centers require interconnects that can transmit high data rate within links of few kilometers. Accordingly, optical interconnect (OIs) becomes a promising candidate for short-range communication. Especially, a combination of vertical-cavity surface-emitting lasers (VCSELs), owing to its low cost, high energy efficiency and high modulation bandwidth [1], and multi-mode fibers (MMFs) is regarded as a popular solution for short-range communication. Lately, the state of the art of 850 nm VCSELs for 64-Gb/s non-return-to-zero (NRZ) transmission over 57 m MMF [2] and 70-Gb/s 4-level pulse-amplitude modulation (4-PAM) transmission over MMF [3] has been reported. In addition, discrete-multi tone (DMT) format gradually becomes another potential candidate for higher bit-rate distant product in the data communication.

In our previous work, a record high bit-rate distance product (107.6 Gb/s·km) at nearly 50-Gb/s DMT transmission over 2.2-km OM4 fibers [4] has been demonstrated by the use of single-mode VCSEL. Generally, in the optical DMT transmission system, there are many advantages such as high spectral efficiency, simple equalization, and tolerance to dispersion-induced-inter-symbol interferences (ISI); besides, it is easy to optimize DMT transmission over a frequency-selective fading channel by the means of bit-loading. On the other hand, besides the two major disadvantages of optical DMT system; namely frequency offset and phase noise sensitivity, the high peak-to-average power ratio (PAPR) [5], which is also one of the most detrimental drawbacks in optical DMT system, will cause nonlinear distortion due to the limited linear operational region of VCSEL.

In this work, by intentionally over-driving our 850nm VCSEL to high bias current and large peak-to-peak modulation voltage (V_{pp}) with DMT modulation, nonlinear distortion is significant. The time-domain Volterra-Wiener filter was applied in this work to overcome the distortion. After transmission and equalization by Volterra-Wiener filter, 32% (52.03 Gb/s to 68.94 Gb/s) and 21.2% (48.14 Gb/s to 58.39 Gb/s) data rate improvement can be achieved respectively for optical back-to-back (OBTB) and over 105-m OM4 transmission.

II. STATISTIC AND SMALL SIGNAL CHARACTERISTIC

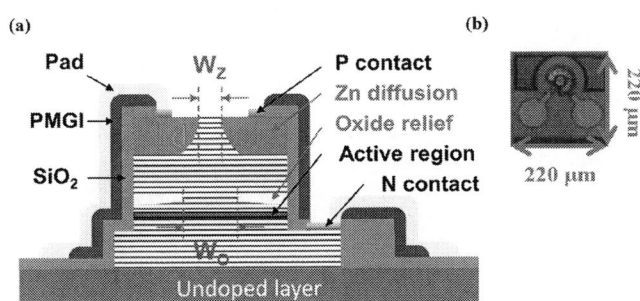

Fig. 1. (a) Conceptual cross-sectional view and (b) top view of VCSEL

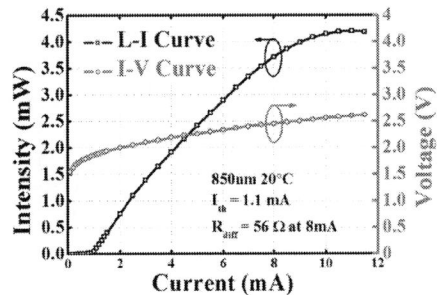

Fig. 2. LIV characteristic of VCSEL

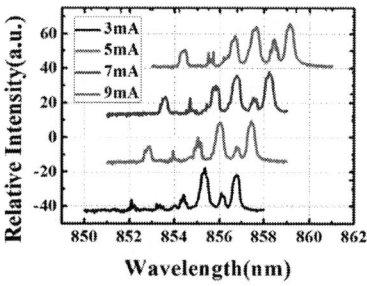

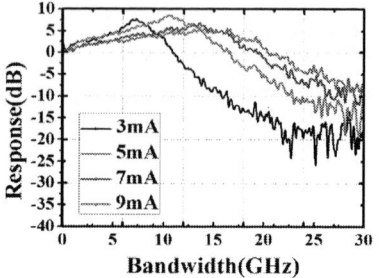

Fig. 3. Optical spectra at 25 °C   Fig. 4. E-O responses at 25 °C

Fig. 1 depicts top and conceptual cross-sectional view of the 850 nm VCSEL. The whole chip has a die size of 220 μm×220 μm and the probing length of 125 μm. The epi-layer structure includes three strained $In_{0.1}Ga_{0.9}As/Al_{0.3}Ga_{0.7}As$ quantum-wells (QWs) sandwiched between a 36-pair n-type and a 26-pair p-type $Al_{0.9}Ga_{0.1}As/Al_{0.12}Ga_{0.88}As$ Distributed-Bragg-Reflector (DBR) layers with an $Al_{0.98}Ga_{0.02}As$ layer. To improve the speed of the device shown in the Fig. 1(b), the aperture size (3 μm in diameter) in oxide-relief process was carefully controlled. Furthermore, in order to reduce the differential resistance and limit the transverse modes number of the device, proper size of the aperture (8 μm in diameter) and depth (1 μm) in Zn-diffusion process is chosen. Fig. 2 shows the L-I-V characteristics of the VCSEL at 25°C, and threshold current is 1.1 mA and resistance is 56 Ω at bias current of 8 mA. Fig. 3 plots the optical spectra of our VCSEL at bias currents of 3 to 9 mA at 25°C. Fig. 4 presents the E-O frequency responses at bias currents of 3 to 9 mA at 25°C, and the maximum 3-dB bandwidth is 24 GHz at bias current of 9 mA.

III. VOLTERRA-WIENER FILTER

In the 850 nm VCSEL DMT transmission system, the major cause of nonlinear distortion is due to the limited linear operational region of the 850 nm VCSEL. Aiming to lessen the nonlinear distortion while increasing the modulation amplitude, the well-known time-domain Volterrra-Wiener filter is used in this work. After balancing the complexity and efficiency of the equalization algorithm, a 3rd-order Volterra series is employed [6,7], and the nth sample $y(n)$ is

$$y(n) = \sum_{l_1=0}^{N-1} w_{01}(l_l)x(n-l_1) + \sum_{l_1=0}^{N-1}\sum_{l_2=0}^{N-1} w_{02}(l_1,l_2)\prod_{i=1}^{2} x(n-l_i) + \sum_{l_1=0}^{N-1}\sum_{l_2=0}^{N-1}\sum_{l_3=0}^{N-1} w_{03}(l_1,l_2,l_3)\prod_{i=1}^{3} x(n-l_i) \quad (1)$$

where $x(n-l_i)$ is the $(n-l_i)^{th}$ sample of received signal, $w_{0k}(l_q)$ is the weighting factor of the kth order, and N is the tap number. The 1st term in Eq. (1) indicates a linear filter, and the others are nonlinear filters. The weighting factors are determined by the Wiener solution [6,7], of which the required expectation values are replaced by sample means of training symbols.

IV. EXPERIMENTAL SETUP AND RESULTS

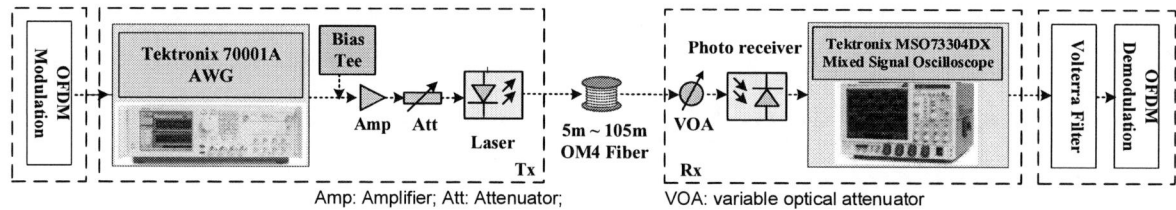

Fig. 5. Experimental setup

Fig. 5 illustrates the experimental setup for the 850 nm VCSEL. The baseband electrical DMT signal is generated using offline Matlab® with FFT size of 512, CP of 3.03%, and the subcarrier number of 138. The electrical DMT signal is then carried out using an arbitrary waveform generator (Tektronix® AWG70001A) at sampling rate of 50 GSample/s corresponding to the signal bandwidth of 13.5 GHz, and the digital-to-analog (D/A) conversion resolution is 10 bit. An electrical amplifier (Picosecond 5867) and an attenuator are used to adjust the optical modulation index (OMI) and the V_{pp}, and the optimized condition will be discussed later. The DMT signal is fed to our VCSEL through a 26.5-GHz bias-T (Agilent 11612A) via a GS probe, and the output light is butt-coupled into a lensed fiber with a core diameter of 62.5 μm, the coupling efficiency is approximately 60 %. After the OM4 fiber transmission of 5 m to 105 m, the optical signal is detected directly by a 9.5-GHz photoreceiver (PICOMETRIX PT-12B). The DMT signal is retrieved and digitized by a real-time oscilloscope (Tektronix® MSO 73304DX) with a sample rate of 50 GSample/s and a 3-dB bandwidth of 33

GHz. After Volterra filtering and demodulation by offline DSP, the signal-to-noise ratio (SNR) is estimated and bit error rates (BERs) are measured based on a bit-by-bit comparison between transmitted and received data.

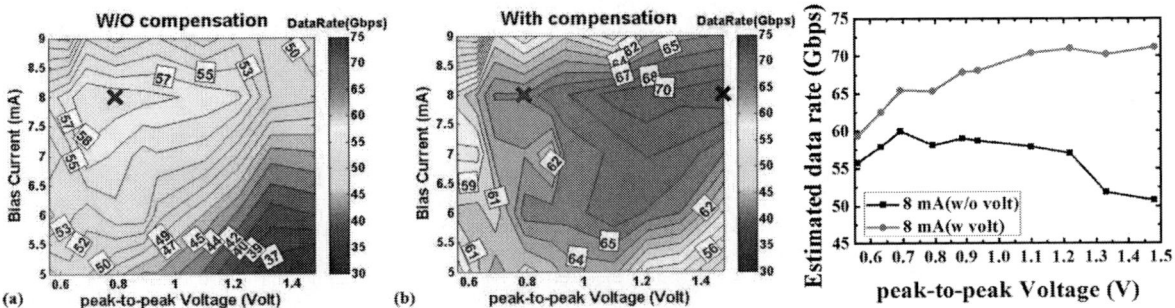

Fig. 6. Estimated data rate contour (a) without compensation (b) with compensation   Fig. 7. Estimated data rate versus V_{pp}

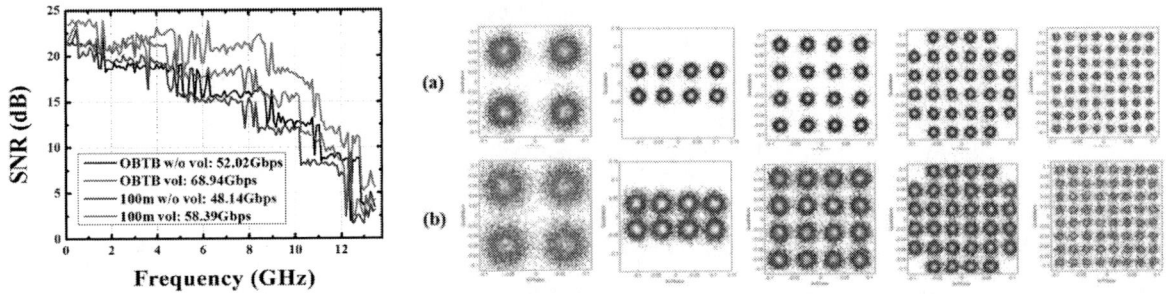

Fig. 8. SNR OBTB&100m w&w/o compensation   Fig. 9. (a)(b) OBTB constellation after bit-loading with and without compensation respectively

In order to have the full understanding of the influence of nonlinear distortion on the transmission performance, V_{pp} is controlled in the range of 0.55 V to 1.48 V, and we evaluated the signal performance by maximum achievable data rate under the FEC limit (i.e., BER of 3.8×10^{-3}). The maximum data rate is estimated using the bit-loading scheme in this work. Fig. 6(a) is the estimated data rate by adjusting V_{pp} and bias current before compensation. At bias current of 8 mA and V_{pp} of 0.79 V, the optimal result (blue cross) with the 59.96-Gb/s estimated data rate can be achieved. Fig. 6(b) shows the estimated data rate by adjusting V_{pp} and bias current with Volterra compensation. The estimated data rate of optimal result (black cross), which is at bias current of 8 mA and V_{pp} of 1.48 V, increases to 71.19-Gb/s. Fig. 7 plots the estimated data rate with and without Volterra equalization when bias is set at 8 mA. It is clear that the Volterra equalization can effectively reduce the non-linear distortion, such that the system can be operated at larger modulation voltage to achieve higher data rate. After nonlinear compensation, we achieve a 32% (OBTB: 52.02 Gb/s to 68.94 Gb/s) and 21.2% (105 m: 48.14 Gb/s to 58.39 Gb/s) improvement in data rate. Fig. (8) shows the SNR, after bit-loading algorithm, with and without nonlinear compensation for both OBTB and 105 m transmission. Fig. 9 (a), (b) are the corresponding constellations.

V. Conclusions

In this paper, Volterra-Wiener filter is used for nonlinear compensation in DMT transmission system with bit-loading algorithm by employing our novel 850 nm VCSEL with λ/2 cavity, oxide-relief and Zn-diffusion techniques. After optimizing the transmission performance by adjusting peak-peak modulation voltage and bias current, 38% and 22% data rate improvement in OBTB and 105m transmission can be achieved respectively under the FEC limit.

References

[1] L. A. Coldren et al., "Diode Lasers and Photonic Integrated Circuits", Wiley, New York, NY, USA, 2012.
[2] D. M. Kuchta et al., OFC, Th3C.2, 2014.
[3] K. Szczerba et al., IEEE J. Lightw. Technol., vol. 33, no.7, pp. 1395-1401, 2015.
[4] I.-C. Lu et al., IEEE J. Quantum Electron., vol. 21, no. 6, Article 1701009, 2015.
[5] S. William et al., "OFDM for Optical communications", Academic Press, MA, USA, 2010.
[6] H. Y. Chen, et al., ECOC, P.3.21, 2015.
[7] W. Yan, et al., ECOC, Mo.1.B.2, 2012.

Pre-leveled 16-QAM OFDM Modulation of an 850-nm VCSEL for 56-Gbit/s Transmission

Cheng-Ting Tsai[1], Chun-Yen Pong[1], Yun-Chen Wu[1], Shan-Fong Leong[1], Yu-Chieh Chi[1], Chao-Hsin Wu[1,*],
Tien-Tsorng Shih[2], Jian Jang Huang[1], Hao-Chung Kuo[3], Wood-Hi Cheng[4], and Gong-Ru Lin[1,*]

[1]Graduate Institute of Photonics and Optoelectronics, and Department of Electrical Engineering,
National Taiwan University
No. 1, Roosevelt Rd, Sect. 4, Taipei 10617, Taiwan ROC
[2]Department of Electronic Engineering,
National Kaohsiung University of Applied Sciences
No. 415, Chien Kung Rd, Sanmin District, Kaohsiung 80778, Taiwan ROC
[3]Graduate Institute of Electro-Optical Engineering, and Department of Photonics,
National Chiao Tung University
No. 1001, University Rd, Hsinchu 30100, Taiwan ROC
[4]Graduate Institute of Optoelectronic Engineering, and Department of Electrical Engineering,
National Chung Hsing University
No. 250, Kuo Kuang Rd, Taichung 402, Taiwan ROC
Corresponding Author E-mail address: grlin@ntu.edu.tw, chaohsinwu@ntu.edu.tw

Abstract: An 850-nm VCSEL carried 16-QAM OFDM transmission at 56 Gbit/s is demonstrated by using the pre-leveling technology, which reveals an optimized signal-to-noise ratio, error vector magnitude and bit-error-rate of 16.4 dB, 15.1% and 1.2×10^{-3}, respectively.

Keywords: Vertical cavity surface emitting lasers; Fluctuations, relaxations, and noise; Fiber optics, infrared; Optical interconnects

I. INTRODUCTION

Nowadays, the information explosion seriously impacts the data traffic of network systems. A data center with cloud storage can play an important role in the release of the traffic load between versatile applications. For rapidly switching and computing the stored data between numerous users, the future data center requires to be constructed by an optical interconnect with a demanded data rate of up to 400 Gbit/s. To meet the requirement of the cost effect, an 850-nm vertical cavity surface emitting laser (VCSEL) based 16×25 or 8×50 Gbit/s transmitter array due to its low threshold current, high power conversion and efficient end-face coupling features is thus employed as compact light source to establish high-speed interconnect links for the data center [1]. By directly modulating the 850-nm VCSEL with the non-return-to-zero (NRZ) data format, a high-speed transmission link at 40 Gbit/s can be constructed for the data center [3]; however, such a NRZ data stream requires to occupy most of analog bandwidth to achieve the desired data rate. To effectively use the limited bandwidth of the VCSEL, the orthogonal frequency division multiplexing (OFDM) combined quadrature amplitude modulation (QAM) with high spectral efficiency can be employed as a potential data format to directly modulate the VCSEL.

In this work, the 16-QAM OFDM transmission at 56 Gbit/s is demonstrated by homemade and conventional VCSELs at wavelength of 850 nm. The power-to-current and frequency responses of the VCSELs are surveyed for obtaining their output power performance and 3-dB analog bandwidth. The suppression of the relative intensity noise (RIN) in these VCSELs is also examined by optimizing the bias current. The bit-error-rate (BER) performances of the homemade and conventional VCSELs carried 16-QAM OFDM data are also optimized at different bias current. After employing the pre-leveling technology, the improvement on the signal-to-noise ratio (SNR), the error vector magnitude (EVM) and the BER performances of the VCSELs carried QAM OFDM data are obtained.

II. EXPERIMENTAL SETUP

In experiments, homemade (VCSEL$_1$) and commercial (VCSEL$_2$) VCSELs with central wavelength of 850 nm were employed as the transmitting light source for carrying the 16-QAM OFDM data, which was driven by a homemade probe station. To stable the output wavelength, temperatures of VCSELs are controlled at 22°C with a water cooling system. Note that the commercial VCSEL is manufactured by Oclaro. The threshold current and aperture size of each VCSEL is 1.2 mA/9 μm and 0.8 mA/10 μm, respectively. The experimental setup of the VCSEL carried 16-QAM OFDM transmission is illustrated in Fig. 1. The 16-QAM OFDM data was produced by a homemade program, which occupies 14 GHz bandwidth with 144 subcarriers so as to provide a raw data rate of 56 Gbit/s. The electrical OFDM data was synthesized by an arbitrary waveform generator (AWG, Tektronix, 70001A) with a sampling rate of 50 GS/s, which was directly modulated the VCSEL with a peak-to-peak amplitude of 500 mV by using a high-speed bias-tee (Anritsu V255, 65-GHz). The output of the VCSEL with the carried OFDM data was collected and received by a conventional 50/125 μm multi-mode fiber (MMF, OM2) and a 25-Gbps GaAs PIN photodiode (PD, GCS, DO351), respectively. To improve the detection sensitivity at the receiving end, a post-amplifier pair (AMP, Keysight, 83006A;

New Focus, 1422) with a gain of 20 and 18 dB was added after the PD. The received OFDM data was re-sampled by a digital signal oscilloscope (DSO, Tektronix 71604C) with a sampling rate of 100 GS/s, which was examined by a homemade demodulation program to obtain its BER, SNR and EVM performances.

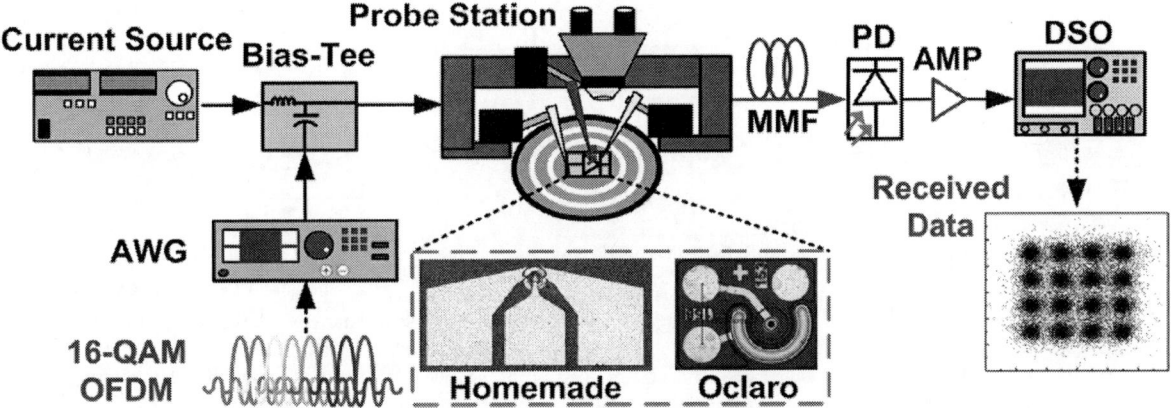

Fig. 1 The experimental setup of the VCSEL based 16-QAM OFDM transmission at 56 Gbit/s.

III. RESULTS AND DISCUSSION

Figure 2(a) shows the comparison of power-to-current responses between the $VCSEL_1$ (homemade) and $VCSEL_2$ (Oclaro). For the $VCSEL_1$, the maximum linear output power of 5 mW can be obtained at a bias current of 12 mA, which becomes saturated by further enlarging the bias current to 18 mA. In comparison, the maximum linear output power of the $VCSEL_2$ is reduced to about 3.5 mW at a bias current of 9 mA, which is easier saturated when increasing its bias current. The frequency responses of the $VCSEL_1$ and $VCSEL_2$ are illustrated in Fig. 2(b). At an optimized bias current of 13 mA, the 3-dB analog bandwidth of the $VCSEL_1$ and $VCSEL_2$ is 16.6 and 15.03 GHz, respectively. When observing the output intensity of these VCSELs, the relaxation oscillation induced resonance peak of the $VCSEL_1$ (1.5 dB) is larger than the $VCSEL_2$ (0.6 dB).

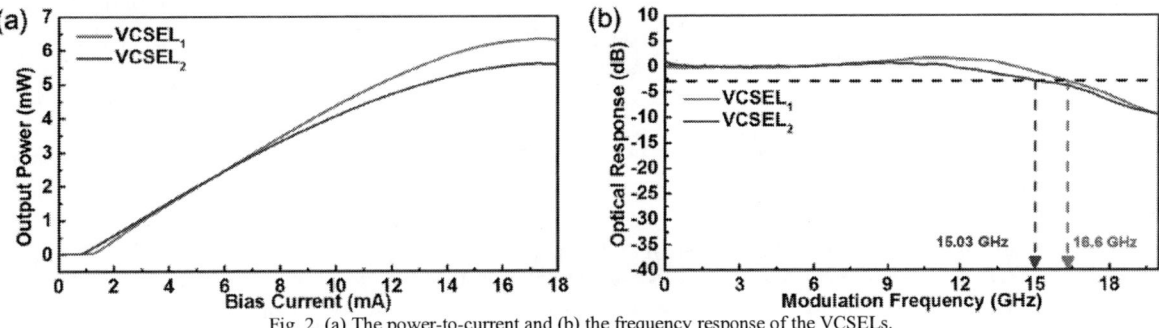

Fig. 2. (a) The power-to-current and (b) the frequency response of the VCSELs.

To verify the location of the relaxation oscillation, the RIN spectrum of these VCSELs is also examined, as shown in Fig. 3(a). By increasing the bias current from 3 to 8 mA, the relaxation oscillation related RIN peak of the $VCSEL_1$ is up-shifted and suppressed from 5 to 9 GHz and from -131.8 to -138 dBc/Hz. In compression, the RIN peak of the $VCSEL_2$ can be further up-shifted and suppressed from 5.2 to 9.5 GHz and from -138 to -140 dBc/Hz under same operation. Although the analog bandwidth of the $VCSEL_1$ is wider than the $VCSEL_2$, it also reveals a strong RIN peak at high frequencies, which will impact the SNR performance of its carried OFDM data. As shown in Fig. 3(b), to optimize the transmission performance during the direct modulation, the BER performances of the VCSEL carried 16-QAM OFDM data is surveyed at different bias currents. By increasing the bias current of the $VCSEL_1$ from 8 to 11 mA, the waveform clipping and the RIN induced intensity noise of the carried 16-QAM OFDM data can be effectively mitigated so as to improve its related BER performance from 5.5×10^{-3} to 1.9×10^{-3}, which can meet the FEC criterion of 3.8×10^{-3}. In comparison, the lowest BER of the $VCSEL_2$ carried 16-QAM OFDM data can be optimized from 1.6×10^{-4} to 6.4×10^{-4} by enlarging the bias current from 8 to 9 mA. The lower BER of the $VCSEL_2$ carried OFDM data is resulted from its RIN level that can be greatly suppressed with the increasing bias current. On the other hand, when further enlarging the bias current of the $VCSEL_2$ from 9 to 11 mA, the BER performance of the carried OFDM data is inversely degraded from 6.4×10^{-4} to 8.5×10^{-4}. That is, when the bias current of the $VCSEL_2$ is larger than 9 mA, the saturation on the output power of the $VCSEL_2$ is emerged, which can affect the SNR of the carried OFDM data with the induced nonlinear modulation distortion.

595

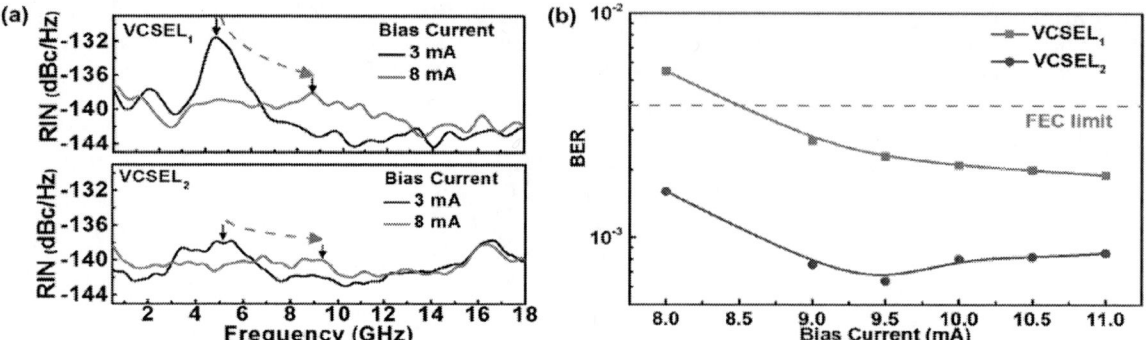

Fig. 3. (a) the RIN responses of the VCSELs; (b) The BER performances of the VCSELs carried 16-QAM OFDM data.

The transmission performance of the VCSEL carried 16-QAM OFDM data can be further improved by employing the pre-leveling technology. Figure 4(a) shows the SNR and constellation performances of the $VCSEL_1$ carried 16-QAM OFDM data. By increasing the pre-leveling slope from 0 to 0.4 dB/GHz, the SNR performance of the carried OFDM data at low frequencies is sacrificed for compensating the SNR performance at high frequencies, which improves the average SNR from 16 to 16.4 dB so as to enhance the corresponded EVM and BER from 15.9% to 15.1% and from 1.8×10^{-3} to 1.2×10^{-3}, respectively. On the other hand, as the RIN induced intensity noise of the $VCSEL_2$ is lower than $VCSEL_1$, it reveals the higher SNR trend of the carried OFDM data. By implementing the pre-leveling with slope of 0.2 dB/GHz, the average SNR the EVM and the BER of the $VCSEL_2$ carried OFDM data can be improved from 17.1 to 16.8 dB, from 13.9% to 13.6% and from 5.1×10^{-4} to 3.9×10^{-4}, respectively.

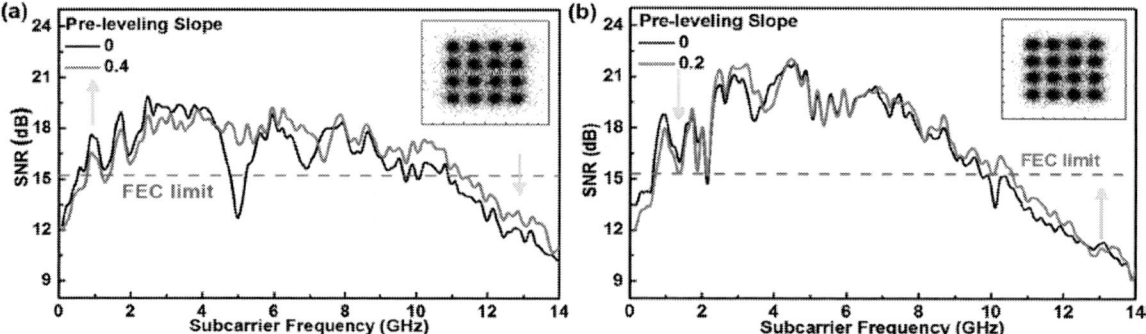

Fig. 4. The SNR and constellation performances of the (a) $VCSEL_1$ and (b) $VCSEL_2$ carried 16-QAM OFDM data without and with pre-leveling.

IV. CONCLUSIONS

This work demonstrates the 56-Gbit/s 16-QAM OFDM transmission by using homemade and conventional VCSELs at same wavelength of 850 nm. By surveying power-to-current responses, the conventional VCSEL is easily saturated than the homemade VCSEL at high bias condition. By examining the frequency responses of homemade and conventional VCSELs, their -3 dB analog bandwidth are 16.6 and 15.03 GHz, respectively, but the homemade VCSEL reveals stronger relaxation oscillation than the conventional VCSEL. After enlarging the bias current to 8 mA, the relaxation oscillation induced RIN peaks of homemade and conventional VCSELs can be up-shifted and suppressed to 9 GHz/138 dBc/Hz and 9.5 GHz/-140 dBc/Hz, respectively. The lowest BERs of homemade and conventional VCSELs carried 16-QAM OFDM data can be optimized to 1.9×10^{-3} and 8.5×10^{-4} by increasing the bias current to 11 and 9.5 mA, respectively. By employing the pre-leveling with slopes of 0.4 and 0.2 dB/GHz, the related EVM, SNR and BER performances of homemade and conventional VCSELs carried 16-QAM OFDM data can be further improved to 16.4/16.8 dB, 15.1%/13.6% and $1.2 \times 10^{-3}/3.9 \times 10^{-4}$, respectively. These results confirm that our homemade VCSEL reveals a competitive and qualitied transmission performance as compared the conventional one for 56-Gbit/s/channel data center application.

REFERENCES

[1] A. Larsson, "Advances in VCSELs for Communication and Sensing," IEEE J. Sel. Top. Quantum Electron., vol. 17, pp. 1552-1567, April 2011.

[2] P. Westbergh, R. Safaisini, E. Haglund, J. S. Gustavsson, A. Larsson, M. Geen, R. Lawrence, and A. Joel, "High-speed oxide confined 850 nm VCSELs operating error-free at 40 Gbit/s up to 85°C," IEEE Photon. Technol. Lett., vol. 28, pp. 768-771, March 2013.

[3] F. Tan, M.-K. Wu, M. Liu, M. Feng, and N. Holonyak, "850 nm Oxide-VCSEL With Low Relative Intensity Noise and 40 Gb/s Error Free Data Transmission," IEEE Photon. Technol. Lett., vol. 26, pp. 289-292, January 2014.

TuD4-1 (Invited)

Multimode fibers for telecommunications and imaging

Joel Carpenter[1], Benjamin J. Eggleton[2], Jochen Schröder[3]

[1] *School of Information Technology and Electrical Engineering, The University of Queensland, Brisbane, Queensland 4072, Australia*
[2] *Centre for Ultrahigh bandwidth Devices for Optical Systems (CUDOS), Institute of Photonics and Optical Science (IPOS), School of Physics, The University of Sydney, New South Wales 2006, Australia*
[3] *School of Electrical and Computer Engineering, RMIT University, Melbourne, Victoria 3000, Australia*
*j.carpenter@uq.edu.au

Abstract: *In this paper, we discuss the characterization of light propagation in multimode fiber and its application to telecommunications, imaging and fundamental physics.*

I. INTRODUCTION

As one of the basic fiber types, multimode optical fiber is used in just about any application which requires light to be routed from one location to another. These include telecommunications[1], [2] and imaging[3], [4] as well as fiber lasers[5] to name just a few. The theory of optical waveguide is well established and has been taught to undergraduates for decades. However it is only recently that advances in the area of mode division multiplexing[6]–[8] and biomedical imaging[3], [9] have given us the experimental techniques necessary to precisely measure the way light propagates through multimode optical fiber in a rigorous manner. As light travels through multimode optical fiber, it becomes scattered spatially and temporally as modes couple in complicated ways. This has traditionally been one of the main impediments to the use of multimode fiber. However the ability to accurately quantify this mode coupling allows applications such as MDM and multimode fiber imaging possible as well as the observation of new phenomena and experimental verification of established theories[3], [8].

II. TRANSFER MATRIX MEASUREMENT AND INVERSION

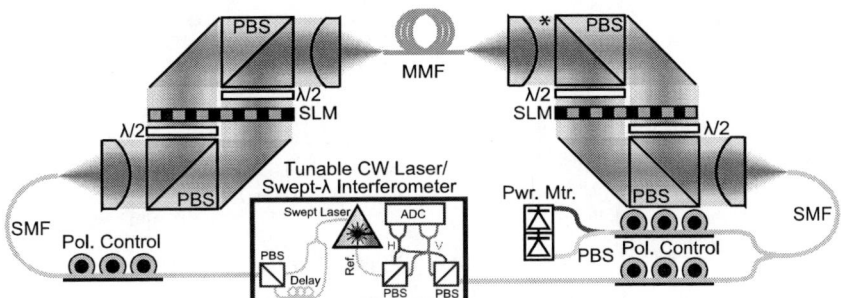

Fig. 1. Experimental setup used to measure the mode transfer function of a multimode fiber[8].

The experimental system of Figure 1 can be used to measure the entire optical transfer function of a multimode optical fiber. That is, the impulse response of every spatial/polarization mode in and out; a complete linear description of the fiber. From this, any linear property of interest can be calculated and the results of any linear experiment predicted. Figure 1 consist of two spatial light modulator (SLM) systems at either end of the fiber attached to a swept-wavelength interferometer (SWI). Together, the measure a complete spatiotemporal description of light propagation through the fiber, with the SLMs addressing the spatial components (modes) and the SWI addressing the temporal components (wavelength-dependent mode coupling). The system works by performing an impulse response measurement for combination of the N modes in and N modes out supported by the fiber. This constructs the wavelength-dependent optical transfer matrix.

Once the matrix is measured, it is possible to extract information about the fiber-under-test such as mode-dependent loss, but it is also possible to verify the validity of the measured matrix by inverting it and exciting the required superposition at the input required to generate a desired output. Examples of this are shown in Fig. 2. In

597

Fig. 2 (a), the illustrated amplitude, phase and polarization at the input, result in a $LP_{0,4}$ and $LP_{4,2}$ mode exiting from different ports of a polarizing beamsplitter. Demonstrating full spatial and polarization control. More sophisticated examples are shown in Fig 2(b)-(f).[4]

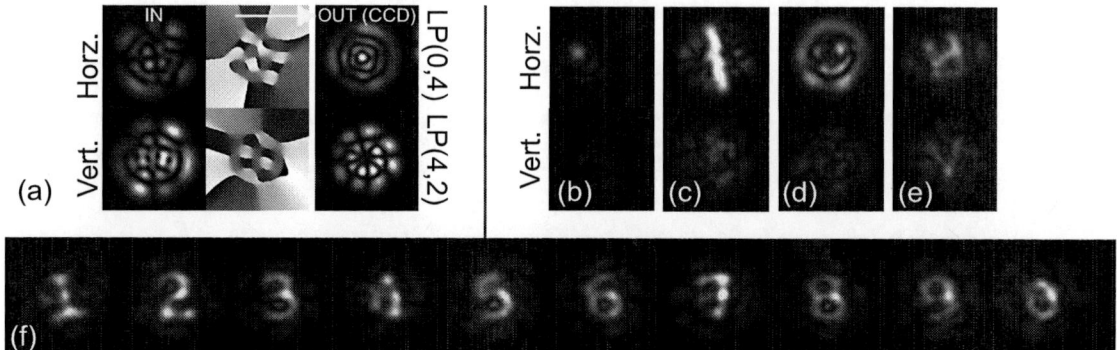

Fig. 2. (a) Input field required to generate orthogonally polarized $LP_{0,4}$ and $LP_{4,2}$ modes at the output. (b) Horizontally polarized spot (c) Horizontally polarized vertical line (d) Horizontally polarized 'smiley face', (e) Horizontally polarized 'H' and vertically polarized 'V' (f) Numbers 1,2,3,4,5,6,7,8,9,0.

III. PRINCIPAL MODES

From the measured optical transfer function it is also possible to find the principal modes[8], [10] of the fiber. Principal modes can be thought of as 'temporal eigenmodes' in that they enter and exit the fiber as the same pulse shape. They are the basis of modes for the fiber which have the least wavelength dependence. Such modes were theoretically predicted decades ago, but were only observed experimentally recently. Fig. 3 below illustrates an example comparison of a Laguerre-Gaussian mode (red) launched into the fiber and a principal mode (green). It can be seen that as the wavelength of the source is changed, the output modes observed on the camera have significantly less wavelength dependence for the principal modes when compared with the LG modes.

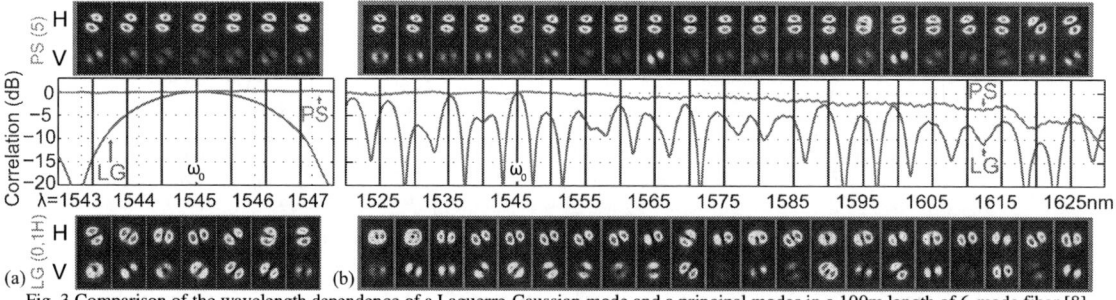

Fig. 3 Comparison of the wavelength dependence of a Laguerre-Gaussian mode and a principal modes in a 100m length of 6-mode fiber.[8]

REFERENCES

[1] R. Ryf, S. Randel, N. K. Fontaine, M. Montoliu, E. Burrows, S. Chandrasekhar, A. H. Gnauck, C. Xie, R.-J. Essiambre, and P. Winzer, "32-bit/s/Hz spectral efficiency WDM transmission over 177-km few-mode fiber," in *Optical Fiber Communication Conference*, 2013, p. PDP5A–1.

[2] J. Carpenter, B. C. Thomsen, and T. D. Wilkinson, "Degenerate mode-group division multiplexing," *J. Light. Technol.*, vol. 30, no. 24, pp. 3946–3952, 2012.

[3] M. Plöschner, T. Tyc, and T. Čižmár, "Seeing through chaos in multimode fibres," *Nat Phot.*, vol. 9, no. 8, pp. 529–535, Aug. 2015.

[4] J. Carpenter, B. J. Eggleton, and J. Schröder, "110X110 Optical Mode Transfer Matrix Inversion," *Opt. Express*, vol. 22, no. 1, p. 96, 2013.

[5] D. J. Richardson, J. Nilsson, and W. A. Clarkson, "High power fiber lasers: current status and future perspectives," *J. Opt. Soc. Am. B*, vol. 27, pp. B63–B92, 2010.

[6]   N. K. Fontaine, R. Ryf, M. A. Mestre, B. Guan, X. Palou, S. Randel, Y. Sun, L. Gruner-Nielsen, R. V Jensen, and R. Lingle, "Characterization of Space-Division Multiplexing Systems using a Swept-Wavelength Interferometer," in *Optical Fiber Communication Conference/National Fiber Optic Engineers Conference 2013*, 2013, p. OW1K.2.

[7]   J. Carpenter, B. J. Eggleton, and J. Schröder, "Reconfigurable spatially-diverse optical vector network analyser," *Opt. Express*, vol. 22, no. 3, pp. 2706–2713, 2014.

[8]   J. Carpenter, B. J. Eggleton, and J. Schröder, "Observation of Eisenbud–Wigner–Smith states as principal modes in multimode fibre," *Nat Phot.*, vol. 9, no. 11, pp. 751–757, Nov. 2015.

[9]   T. Cizmar, K. Dholakia, T. Čižmár, and K. Dholakia, "Exploiting multimode waveguides for pure fibre-based imaging," *Nat Commun*, vol. 3, p. 1027, Aug. 2012.

[10]   S. Fan and J. M. Kahn, "Principal modes in multimode waveguides," *Opt. Lett*, vol. 30, pp. 135–137, 2005.

Next-Generation Multimode Fibers

Pierre Sillard, Marianne Bigot-Astruc, Denis Molin, Adrian Amezcua-Correa

Prysmian Group, Parc des Industries Artois Flandres, 644 boulevard Est, Billy Berclau, 62092 Haisnes Cedex, France
pierre.sillard@prysmiangroup.com

Abstract: We show how multimode fibers, that have always been flexible transmission media, can be adapted to new wavelength-division-multiplexing techniques in data communications and mode-division-multiplexing techniques in telecommunications to keep up with capacity increase.
Keywords: multimode fiber; data communication; wavelength-division multiplexing; mode-division multiplexing.

I. INTRODUCTION

Wavelength Division Multiplexing (WDM) around 850nm has recently been proposed to reduce the number of parallel Multimode Fibers (MMFs) and keep up with the exponential growth of bandwidth in data centers [1-4]. Standard MMFs, however, have high modal bandwidth over a small wavelength range, which limits their WDM capabilities. MMFs with a broader operational window, or Wide-Band MMFs (WB-MMFs), are thus needed for WDM operation [1,5,6].

In telecommunications, Mode Division Multiplexing (MDM) has recently drawn much attention due to its potential to multiply capacity by the number of modes that one fiber can support. 1st Multiple-Input Multiple-Output (MIMO) MDM experiments were actually performed over MMFs [7], but the cross-talk between the high number of modes ultimately limited the capacity improvements. It is only recently, with the use of selective mode (de)multiplexing devices and advanced MIMO processing, that MMFs have been reconsidered for MIMO MDM [8,9].

In this paper, we will present next-generation MMFs adapted to these 2 techniques that allow to keep pace with capacity increase. The bandwidth and Bit Error Rate (BER) tests of a WB-MMF [1], that offers OM4 performance from 850 to 950nm and that allows for speeds of 100Gbps (4 WDM channels at 25Gbps) and above, will be detailed. And an MMF optimized to transmit up to 20 LP modes at 1550nm [10] will be shown to perform better than Few-Mode Fibers (FMFs) when the number of LP modes used in MIMO MDM is ≥ 9.

II. MULTIMODE FIBERS FOR WIDE-BAND DATA COMMUNICATIONS

Data communications based on MMFs and VCSEL technology are not only limited by modal dispersion but also by chromatic dispersion. The Effective Bandwidth (EB), that is a function of both Effective Modal Bandwidth (EMB) and Chromatic dispersion Bandwidth (CB), is thus a good predictor of system performance [1]. WB-MMFs should offer an EB from 850 to 950nm higher than that of standard OM4 MMFs at 850nm to ensure same system performance over this wavelength range. Using the worst-case VCSEL spectral width of 0.65nm and a non-favorable interaction between modal and chromatic dispersions, it is possible to define the minimal EB guaranteeing OM4 performance. The minimal EB was found to be 2000MHz.km [1]. Because CB increases from 2,800MHz.km at 850nm to 4,450MHz.km at 950nm, due to the reduction of the absolute value of the chromatic dispersion from ~100ps/nm/km to ~65ps/nm/km, the EMB requirements decrease from 4,700MHz.km to 2,800MHz.km, as shown in Fig.1(a).

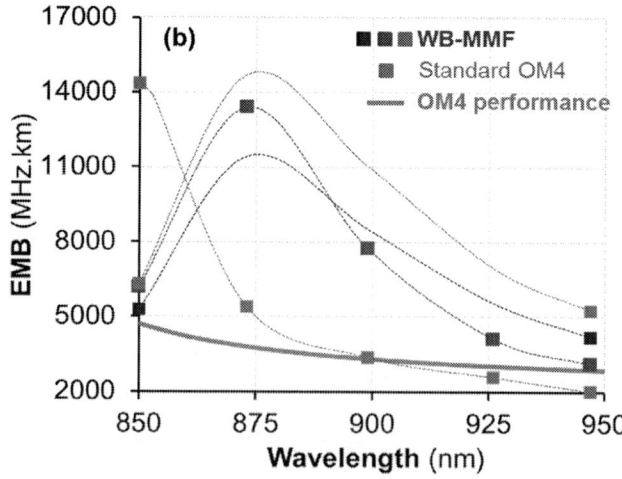

(a)

EB >2000MHz-km over 850-950nm		
Wavelength (nm)	EMB (MHz.km)	CB (0.65nm RMS) (MHz.km)
850	>4700	2800
875	>3700	3150
900	>3300	3500
925	>3000	3950
950	>2800	4450

Fig.1: EMB specifications for OM4 performance from 850 to 950nm (a); measurements of EMB for representative WB-MMFs as a function of wavelength (lines are guides for the eye) (b).

Despite the usual spectrally narrow bandwidth of MMFs around 850nm, it is possible to make WB-MMFs that meet these OM4 requirements from 850 to 950nm by optimizing the Alpha parameter of the graded-index core and by taking full advantage of the versatile plasma chemical vapor deposition process. Such WB-MMFs were fabricated and Differential Mode Group Delays (DMGDs) were measured at different wavelengths from 850 to 950nm using a tunable Titanium-Sapphire laser. The resulting EMBs are reported in Fig.1(b) for representative WB-MMFs and for a standard OM4 MMF for sake of comparison. The plots clearly show that WB-MMFs are optimized to have large EMB optimized at 875nm when standard OM4 MMFs have narrow EMB optimized at 850nm. As a result, WB-MMFs fulfill the EMB specifications of Fig.1(a), while standard OM4 MMFs fail at ~900nm.

In order to demonstrate the capability of such WB-MMFs to transmit 100Gbps (4 WDM channels at 25Gbps), when 4 standard OM4 MMFs are currently needed to reach this speed, we first performed BER tests at 28Gbps at 850 and 980nm over 100m. 980nm is outside the targeted operating window of 850-950nm, but BER sensitivity indicated an acceptable power penalty after 100m transmission [2]. Then, BER tests using a 40GbE duplex transceiver (2 WDM channels at 20Gbps at 850 and 900nm) showed that up to 300m error-free transmissions were possible over WB-MMFs, which is twice the specified reach of this transceiver [3]. Finally, 4 WDM channels at 25.8Gbps from 850 to 950nm with 30 nm spacing achieved error free transmission after 200m and BER $<10^{-9}$ after 300m of WB-MMF [4].

III. Multimode Fibers for Multiple-Input Multiple-Output Mode Division Multiplexing

50µm MMFs support 30 guided LP modes (55 spatial modes) that can be divided into 10 mode groups at 1550nm. But the 2 highest-order mode groups suffer from too high bend losses at 1550nm (>10dB/turn at 10mm bend radius) for proper use in MDM configuration. This leaves 8 mode groups for a total of 20 LP modes (36 spatial modes) that all exhibit low bend losses and that allow to multiply the capacity by a factor of 36 compared to standard single-mode operation. The effective index differences between the different mode groups are higher than 1.4×10^{-3}, which minimizes mode coupling between the last mode group that is used for MDM and the first that is not used. This is essential to guarantee efficient MIMO that is performed over the usable modes only. This way, 50µm MMFs can be used to transmit 2 to 20 LP modes. The DMGDs between the LP modes, however, have to be minimized at 1550nm for optimum MIMO MDM operation. This is done by tuning the Alpha parameter of the graded-index core and by slightly adjusting the position of the depressed-index region in the cladding. We have fabricated such a MIMO MMF [10]. The experimental index profile, measured on preform and set to the scale of the fiber is shown in Fig.2(a). Max|DMGD| of 50ps/km for 2 LP modes and of 160ps/km for 20 LP modes were obtained. The attenuation and the A_{eff} of the LP_{01} mode were measured at 0.22dB/km and 170µm², respectively, at 1550nm. The A_{eff} are very large, ranging up to ~700µm² for the LP_{33} mode of the 8th mode group, thus significantly limiting nonlinear effects.

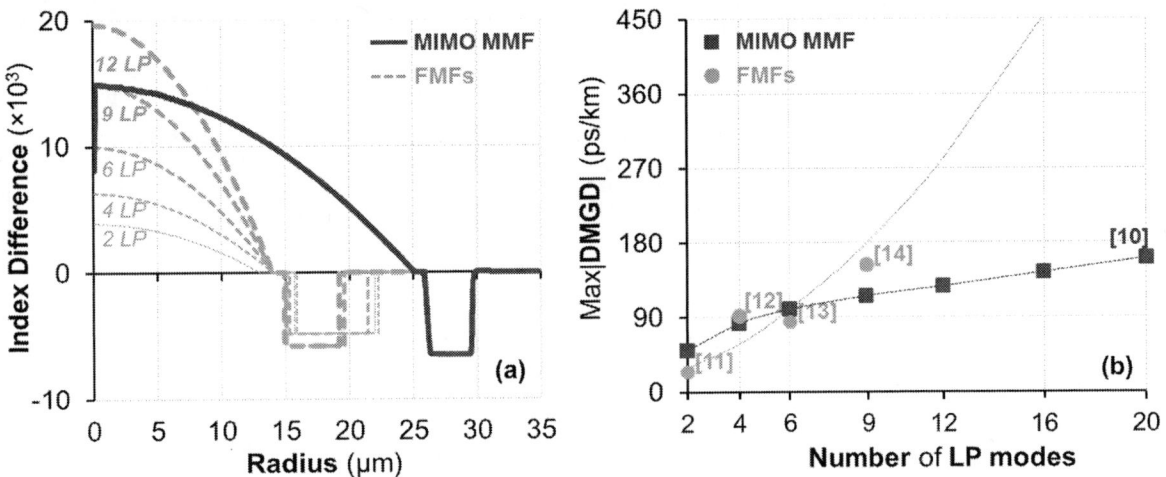

Fig.2: Index profiles: experiment (solid line for MIMO MMF) and calculations (dashed lines for FMFs) (a); max|DMGD| at 1550nm as a function of the number of LP modes: experiments (solid symbols for MIMO MMF and FMFs, lines are guides for the eye) (b).

We have compared these results to those obtained with low-DMGD FMFs [11-14]. Fig.2(b) shows how max|DMGD| increases with the number of LP modes. What is noticeable is that this increase is much steeper for the FMFs than for the MIMO MMF, resulting in higher values when the number of LP modes is ≥9. FMFs are designed so that all their LP modes have low bend losses which imposes small core radii [13], contrary to the MIMO MMF for which the 2 highest-order modes groups have high bend losses. In addition, when the number of LP modes is ≥9, their core indexes become equal or higher than that of the MIMO MMF (see Fig.2(a)). These 2 features make them more sensitive to process variability and eventually give higher DMGDs. Note that DMGD compensation (concatenation of fibers with LP modes with DMGDs with opposite signs) [11-14] can also be used to reduce these values. We have realized such DMGD-compensated MMF links and measured DMGDs $\leq$50ps/km, which is equivalent to what is expected from simulations for DMGD-compensated 9-LP-mode links (45ps/km) and better than for DMGD-compensated 12-LP-mode links (105ps/km) [15]. Attenuations of FMFs also increase as a function of LP modes, mainly because of the increase of Rayleigh scattering related to higher core indexes required to support more LP modes (see Fig.2(a)). The attenuation becomes higher than that of the MIMO MMF when the core index also becomes higher, i.e. when the number of LP modes is $\geq$12. What also matters is the Differential Mode Attenuation (DMA). It is comparable for both fiber types for 2 to 6 LP modes and then becomes higher for FMFs with 9 LP modes or more ($\geq$0.020dB/km compared to <0.015dB/km for the 9 LP modes of the MIMO MMF and 0.020dB/km for the 20 LP modes). DMA is a limiting factor because it has a direct impact on mode-dependent loss. Finally, the A_{eff} of FMFs decrease when the number of LP modes increases because higher core indexes impose tighter mode confinements. The A_{eff} of the most-confined LP_{01} mode decreases from ~190µm² for the 2-LP-mode fiber to ~90µm² for the 12-LP-mode fiber. This is still acceptable to limit nonlinearities if we compare to the ~80µm² of standard single-mode fibers but this is much smaller than the 170µm² of the LP_{01} mode of the MIMO MMF.

IV. CONCLUSION

Since the beginning of optical communications, MMFs have appeared as efficient and flexible transmission media that kept evolving to better fit capacity increase. This is all the more true today when capacity seems to reach its limits. We have shown how next-generation MMFs can be optimized to ensure OM4 performance from 850 to 950nm for WDM operation in data communication systems. Such WB-MMFs support speeds of 100Gbps (4 WDM channels at 25Gbps) and above over up to 300m, when at least 4 standard OM4 MMFs are currently needed for such speeds. 50µm MMFs can also be used in MIMO MDM communications if properly designed and optimized. Such MIMO MMFs can transmit up to 20 LP modes (capacity increase by a factor of 36) with selective excitation and detection and offer better performance than those of low-DMGD FMFs when the number of LP modes used in MDM is $\geq$9.

REFERENCES

[1] D. Molin *et al.*, "Wide-Band OM4 Multi-mode Fiber for Next-Generation 400Gbps Data Communications," ECOC'14, P.1.6
[2] D. Molin *et al.*, "850-950nm WideBand OM4 Multimode Fiber for Next-Generation WDM Systems," OFC'15, M.3.B1.
[3] R. Sambaraju *et al.*, "Extended Distance Transmission over Wideband MMF using Multi-wavelength VCSEL based Transcievers," Photonics in Switching, p. 336, 2015.
[4] I. Lyubomirsky *et al.*, "100G SWDM4 Transmission over 300m Wideband MMF," ECOC'15, P.5.4.
[5] M. Bigot *et al.*, "Extra-Wide-Band OM4 MMF for Future 1.6Tbps Data Communications," OFC'15, M.2.C4
[6] R. Shubochkin *et al.*, "New Generation Wideband Multimode Fiber for Shortwave Wavelength Division Multiplexing in Datacom Links," IWCS'15, p. 338.
[7] H.R. Stuart, "Dispersive multiplexing in multimode fiber," ECOC'00, ThV2.
[8] R. Ryf *et al.*, "Mode-multiplexed transmission over conventional graded-index multimode fibers," Opt. Exp., vol. 23, p. 235, 2015.
[9] J.J.A. van Weerdenburg *et al.*, "10 Spatial Mode Transmission over 40km 50µm Core Diameter Multimode Fiber," OFC'16, Th4C.3.
[10] P. Sillard *et al.*, "50µm Multimode Fibers for Mode Division Multiplexing," J. of Lightw. Technol., vol. 34, p.1672, 2016.
[11] L. Grüner-Nielsen *et al.*, "Splicing of Few Mode Fibers," ECOC'14, P.1.15.
[12] T. Mori *et al.*, "Few-Mode Fibers Supporting More Than Two LP Modes For Mode-Division-Multiplexed Transmission With MIMO DSP," J. of Lightw. Technol., vol. 32, p. 2468, 2014.
[13] P. Sillard *et al.*, "Few-Mode Fibers for Mode-Division-Multiplexed Systems," J. of Lightw. Technol., vol. 32, p. 2824, 2014.
[14] P. Sillard *et al.*, "Low-Differential-Mode-Group-Delay 9-LP-Mode Fiber," J. of Lightw. Technol., vol. 34, p. 425, 2016.
[15] P. Sillard, "Next-Generation Fibers for Space-Division-Multiplexed Transmissions," J. of Lightw. Technol., vol. 33, p. 1092, 2015.

AUTHOR INDEX

Abbott, John 603
Abe, Jun-Ichi 489, 516
Abedin, Kazi S. 326
Absil, P. 1104
Ackert, J. J. 622
Adachi, Fumiyuki 154
Adachi, Koichiro 1113
Agata, Akira 284
Aikawa, Kazuhiko 39, 45
Aikawa, Yohei 130
Akahane, K. 380
Akasaka, Y. 537
Akiba, Shigeyuki 443
Akimoto, Ryoichi 1173
Akiyama, Yuichi 489
Alferness, Rod C. 121
Alishahi, F. 537
Almaiman, A. 537
Alonso-Ramos, C. 622, 1149
Altabas, Jose A. 688
Amann, Markus-Christian 347
Amemiya, Tomohiro 642, 1119
Amezcua-Correa, Adrian 600
Amino, Kenta 832
Amma, Yoshimichi 39, 45
Ando, Makoto 443
Andrejew, Alexander 347
Andrekson, Peter A. 507
Andriolli, N. 395
Annoni, Andrea 88, 407
Aoki, Yasuhiko 254, 462, 477, 498
Aono, Yoshiaki 142
Arai, Shigehisa 642, 1119
Arakawa, Taro 383, 1146
Arakawa, Yasuhiko 148, 559
Arao, Hajime 627
Arikawa, M. 582
Aruga, Hiroshi 1131
Asaka, Kota 676
Asakura, Hideaki 712
Asakura, Keita 955
Asano, Takashi 1167
Asobe, Masaki 1152, 1158
Asuka, Shun 1065
Aupetit-Berthelemot, Christelle 576
Awaji, Y. 48, 655, 658
Aya, Hironori 579
Azuma, Yoshiyuki 949
Baba, Toshihiko 1164
Bae, S. H. 275, 449
Baeuerle, B. 30
Bai, Wei 874, 1042
Balakrishnan, S. 1104
Barré, Nicolas 181
Bartalini, Saverio 567

Bauwelinck, Johan 389
Beausoleil, R. G. 377, 1137, 1191
Bekkali, Abdelmoula 446
Benazet, Benoit 576
Benedikovic, D. 622, 1149
Ben-Ezra, S. 30
Berciano, M. 1149
Bergman, Keren 196
Bernier, Eric 1188
Betoule, C. 30
Bex, Thomas 486
Bigot-Astruc, Marianne 600
Blau, Miri 718
Blown, Patrick 124
Bo, Tianwai 528
Boehm, Gerhard 347
Bofang, Zheng 495
Bowers, John E. 121
Bradley, T. 225
Brenot, R. 931, 1116
Buchali, Fred 362
Cai, Yufeng 820
Cai, Zhongle 510
Calabretta, Nicola 208, 1203
Calabrò, Stefano 486
Cao, Bingyao 793
Cao, Xiaoyuan 236, 673, 1039
Cao, Y. 537
Cao, Zheng 661
Carminati, Marco 407
Carpenter, Joel 597
Carrier, Pilot 272
Casellas, Ramon 1036
Cassan, E. 422, 1149, 1170
Cavaliere, Fabio 724
Celo, Dritan 1188
Chan, C. C.-K. 528, 784
Chan, Cheng-Ta 940
Chan, Chun-Kit 468, 802
Chan, Erwin H. W. 573
Chan, Vincent W. S. 440
Chang, Chun-Yen 591
Chang, C-M. 1116
Chang, Gee-Kung 169
Chang, Hung-Ying 919
Chang, Kuo-Chun 1134
Chang, Wei-Kang 1074
Channa, Hin 1152
Cheben, P. 622
Chen, Bo-Rui 778
Chen, Chin-Hui 1137, 1191
Chen, Gang 236, 673
Chen, H. 187
Chen, Hongwei 103, 811
Chen, Huizhong 850

AUTHOR INDEX

Chen, Hung-Yu 781
Chen, Jiageng 425
Chen, Jianping 79
Chen, Jyehong 591, 859
Chen, Kaige 338
Chen, Lian-Kuan 588
Chen, Minghua 103, 811
Chen, Nan-Kuang 895, 910
Chen, Oscal T.-C. 940
Chen, Rongrong 754
Chen, Xi 817
Chen, Xin 603
Chen, Xin-Nan 591, 859
Chen, Xue 760
Chen, Y. 225
Chen, Y. K. 931, 1116
Chen, Yanxu 760
Chen, Yuan 1155
Chen, Zhangyuan 311, 715
Cheng, Gia-Ling 895
Cheng, Haiquan 850
Cheng, Lin 169
Cheng, Mengfan 260
Cheng, Wood-Hi 594, 895, 910
Cheng, Xiaoyang 1030
Cheng, Zhenzhou 416
Cheung, Jack 547
Chew, Suen Xin 703
Chi, Kai-Lun 591, 859
Chi, Yu-Chieh 585, 594, 976
Chiang, Jung-Sheng 919
Chiang, Kin Seng 630, 648
Chiba, Hisashi 419
Chida, Yasuyuki 51
Chiesa, M. 395
Chiu, Yi-Jen 1134
Cho, Seong-Ho 964
Cho, Sungmin 163
Choi, Woo-Young 964
Chou, Hsi-Hsir 618
Choudhary, Amol 139
Chow, C. W. 685
Chow, Chi-Wai 781
Chu, Ann-Kuo 67
Chu, Daping 97
Chu, Sai T. 401, 547, 736
Chuang, Chun-Yen 859
Chujo, Wataru 988
Chulok, Pacharapon 748
Chung, Hung-Ching 991
Chung, Kun-Lung 865
Chung, Y. 175
Chung, Y. C. 275, 449
Cincotti, G. 106, 109, 112, 609
Clarke, Ian 124

Coleman, Doug 603
Cong, Guangwei 1110
Contestabile, G. 127, 395
Coudyzer, G. 389
Cui, Yue 233, 510
Dahlem, Marcus S. 1009
Dai, Daoxin 410
Dai, Songyuan 296
Dai, Yitang 706
Dalton, L. R. 413
Damas, P. 1149
Daniel, Luca 1021
Dat, Pham Tien 763
De Heyn, P. 1104
De Man, Erik 486
De Natale, Paolo 567
De Valicourt, G. 931, 1116, 1194
De Waardt, Huug 1203
De Wolf, I. 1104
Deng, Lei 260, 814
Deng, Ning 308, 468
Denolle, Bertrand 181
Descos, Antoine 1191
Di Lauro, Luigi 401
Dias, M. P. I. 459
Ding, Huixia 874
Doerr, Chris 3
Dohi, Keisuke 522
Dong, P. 624, 1116
Dong, Shuai 1015
Dou, Liang 492, 501
Downie, John D. 142
Du, Jiangbing 1077
Du, Xiaoen 880
Duan, Xiaofeng 973, 982
Dumais, Patrick 1188
Dun, Han 278, 793
Edwards, S. 133
Effenberger, Frank J. 266
Eggleton, B. 139, 597
Eiselt, Michael H. 251
Elder, D. L. 413
Ellis, A. 30
Elschner, Robert 27
Endo, Tatsuro 544, 880
Eriksson, Tobias A. 507
Evans, P. 133
Fàbrega, Josep M. 302
Fan, Sujie 823
Fan, Xinyu 425, 1086, 1089, 1101
Fan, Yuting 706
Fang, Jian 323
Fang, Wenjing 982
Fehenberger, Tobias 507
Fei, Jiarui 973

AUTHOR INDEX

Feiste, Uwe 486
Feng, Da 745
Feng, Jijun 1173
Feng, Xinhuan 573
Feng, Xue 636
Feng, Zhenhua 814, 817
Fengkai, Bian 946
Ferran, J. F. 30
Ferrari, Giorgio 407
Fiorentino, Marco 377, 1137, 1191
Foggi, Tommaso 118
Fokoua, E. N. 225
Fontaine, N. K. 187
Fresi, Francesco 118, 724
Freude, W. 413
Fu, Meixia 510
Fu, Songnian 260, 814, 817, 877, 952, 1000
Fu, Teng-Wei 922
Fu, Xin 907, 1083
Fujii, Shohei 670
Fujii, Takuro 612
Fujii, Yusuke 531
Fujimura, Yuki 697
Fujisawa, Ryohei 1158
Fujisawa, Takeshi 51, 633
Fujita, Sadao 516
Fujiwara, Naoki 1122
Fukai, Chisato 317
Fukano, Hideki 916
Fukuchi, K. 516, 582
Fukui, Takayoshi 281
Fukushima, Akira 667
Fukutoku, Mitsunori 290, 1182
Fumagalli, A. 133
Furukawa, Hideaki 392, 404, 655
Furuya, Kotoko 443
Galili, Michael 136
Gao, Mingyi 889
Gao, Shuang 802
Gao, Tao 691
Garces, Ignacio 688
Geng, Dongyu 1188
Ghosh, Samir 1143
Goi, Kazuhiro 1107
Gonda, Tomohiro 42
Gong, Xiaoxue 263
Goodwill, Dominic J. 1188
Goto, Nobuo 115, 733, 925
Grani, Paolo 661
Grillanda, Stefano 88, 407
Gu, Wanyi 691, 808
Gu, Xiaodong 73, 76
Guan, Bai-Ou 573
Guan, Pengyu 136, 790
Guan, Xun 784

Gubenko, Alexey 1137
Guglielmi, Emanuele 407
Gui, C. 1116
Guo, Bingli 691, 808
Guo, Hongxiang 236, 673, 721
Guo, Lei 263, 904
Guo, Qiang 103
Gurusamy, Mohan 862
Halir, R. 622
Hamamoto, Kiichi 100, 564, 651, 1030
Hamaoka, Fukutaro 368
Han, Il-Ki 642
Han, Pin 922, 1074
Han, Sang-Kook 868
Han, Sangyoon 91
Han, Won-Taek 1080
Han, Young-Geun 553, 556
Hanafuji, Fumiki 544, 880
Hanawa, Masanori 296
Hanzawa, Nobutomo 54, 633
Harai, H. 133, 658, 709
Harako, Koudai 12
Haruki, Jun 615
Hasebe, Kazuhiko 341
Hasebe, Koichi 1122
Hasegawa, Hideaki 1128
Hasegawa, Hiroshi 428, 437, 1185
Hasegawa, Kiyotomo 1131
Hasegawa, M. 106, 112, 609
Hashimoto, Toshikazu 94, 368, 1182
Hatori, Nobuaki 148
Hattori, K. 106, 112, 239, 248, 609
Hayashi, Michiaki 1051
Hayashi, Neisei 341, 1068, 1095, 1098
Hayashi, Tetsuya 178
Hayashi, Yusuke 642
Hayashiguchi, S. 985
Hayes, J. R. 225
Hazama, Masaya 949
He, Jiale 260, 814
He, Jifang 1188
He, Zuyuan 425, 1077, 1086, 1089, 1101
Hicks, D. 133
Higuchi, Daiki 531
Hillerkuss, D. 30
Himbele, Luke 100, 1030
Himeno, A. 106, 112, 609
Hirai, Riu 281
Hirano, Akira 368
Hiraoka, M. 112, 742
Hirasawa, Takayoshi 443
Hiratani, Takuo 1119
Hirayama, Takahiro 709
Hirayama, Tomoki 1146
Hirohata, Toru 567

AUTHOR INDEX

Hirokawa, Jiro 443
Hirooka, Toshihiko 12, 33
Hirota, Yusuke 667, 670
Hisata, Yudai 826
Ho, Victor 547
Homemoto, Toru 245
Hong, Bingzhou 564
Hong, Seungjoo 163
Hong, Xiaobin 1092
Hong, Yang 588, 784
Hori, Kento 383
Horikoshi, Kengo 368, 489
Hoshi, Takuya 374
Hoshida, Takeshi 27, 477, 489, 492, 501, 513
Hoshino, Hiroki 943
Hosokawa, K. 582
Hosokawa, Tsukasa 1062
Hou, Weigang 904
Hou, Yinan 799
Hsieh, Chang-I 1033
Hsu, C. W. 685
Hsu, Chin-Wei 781
Hsu, Feng-Cheng 67
Hsu, Y. 685
Htein, Lin 928
Hu, Hao 136
Hu, Qian 431
Hu, Qikai 257, 841
Hu, Weisheng 739, 745
Hu, Weiwei 715
Hua, Bingchang 760
Huang, Guoxiu 498
Huang, Jian Jang 594
Huang, Liangkai 1057
Huang, Lingchen 802
Huang, Pi-Ling 895
Huang, Po-Chia 919
Huang, Shanguo 691, 808
Huang, Sheng-Lung 895
Huang, Tianye 952
Huang, Wei-Jhih 865
Huang, Yi-Chung 895
Huang, Yidong 636, 1015
Huang, Yongqing 973, 982
Huang, Yue-Cai 664
Huang, Yue-Kai 142, 145
Huang, Yu-Fang 585
Huang, Zhihong 377
Hung, Hung-Wen 679
Huo, Xiaoli 431
Hurley, Jason 142
Ibrahim, Salah 205, 664
Ichii, Kentaro 1062
Idler, Wilfried 362
Igarashi, Koji 190, 525

Igarashi, Shota 60
Iida, Mamoru 898
Iiyama, Koichi 967
Ikeda, Kazuhiro 82, 151
Ikeda, N. 985
Ikeuchi, T. 537
Ikeuchi, Tadashi 477, 504, 1048
Ikuma, Yuichiro 1182
Imajuku, Wataru 434
Imamura, Katsunori 42
Imansya, Ryan 1030
Imansyah, Ryan 100
Inaba, Takahiro 1012
Inada, Yoshihisa 142
Inafune, Koji 1027
Inoshita, Kensuke 733
Inoue, Daisuke 1119
Inoue, Gen 335
Inoue, Masaaki 471
Inoue, Shunya 70
Inoue, T. 18, 21, 139
Inoue, Yoshiaki 937
Inui, Tetsuro 434
Ip, Ezra 142, 145
Isaji, Y. 133
Ishihara, Hiroki 1107
Ishii, Hiroyuki 350, 1122
Ishii, K. 133, 1185
Ishii, Yuzo 1182
Ishikawa, T. 437, 582
Ishikura, Norihiro 1107
Ishizaka, Yuhei 633
Ito, Fumihiko 57
Ito, Kazuto 642
Ito, T. 582
Itoh, Mikitaka 1122, 1182
Iwamoto, Satoshi 148
Iwashita, Katsushi 970
Izawa, Tatsuo 211
Izquierdo, David 688
Izutsu, Masayuki 386, 1140
Jang, Bongyong 148
Jasion, G. T. 225
Jayasundara, Chamil 465
Jeong, Seok-Hwan 645
Jepsen, Peter U. 136
Ji, P. 145, 455, 1054
Ji, Yuefeng 242, 775
Jia, Shi 136
Jian, Pu 181
Jiang, Haisong 100, 564, 651, 1030
Jiang, Jia 1188
Jiang, Kai 877
Jiang, Wen 835, 847
Jiang, Xiaoshun 359, 1155

AUTHOR INDEX

Jiang, Xinyue 1083
Jin, Li 401, 547
Jin, Wei 648
Jin, Zhiqiang 359
Jing, Ruiquan 431
Jo, Ik Su 556
Johannisson, Pontus 507
Josten, A. 30
Ju, Seongmin 1080
Kai, Yutaka 302
Kakegawa, N. 133
Kako, Satoshi 148
Kam, Pooi-Yuen 257, 862
Kamada, Naoki 561, 958
Kametani, S. 133, 489
Kamikawa, Kosuke 766
Kamio, Yukiyoshi 519
Kanai, Shunsuke 682
Kanakubo, W. 85
Kanamori, Hiroo 213
Kanazawa, Shigeru 350, 1122
Kang, Joonhyun 642
Kani, Jun-Ichi 458
Kanke, Tomokazu 561
Kanno, A. 24, 380, 386, 579, 763, 766, 1140
Karanov, B. 112
Karinou, Fotini 287
Karlsson, Magnus 474, 507
Kasai, Keisuke 33, 305
Kashima, K. 380, 446
Katagiri, T. 133
Katayama, Masaru 239, 245, 248
Kato, Kazutoshi 615, 697
Kato, Tomoyuki 27
Katoh, Akira 1158
Katsuyama, T. 985
Kawabata, Yuto 1143
Kawaguchi, Naoki 383
Kawaguchi, Yu 365
Kawahara, Hdeaki 787
Kawahara, Hiroki 290
Kawai, Shingo 290
Kawanishi, T. 380, 386, 579, 763, 1140
Kawashima, Hitoshi 82, 151
Khanna, Ginni 486
Khilo, Anatol 1009
Khokhar, A. Z. 622
Khope, Akhilesh S. P. 121
Kihara, Mitsuru 317
Kikuchi, Kazuro 1060
Kikuchi, Nobuhiko 281
Kikuchi, Takahiro 751
Kim, B. G. 175
Kim, Bok Hyeon 1080
Kim, Byoung Yoon 323

Kim, Byung Gon 449
Kim, Gyu-Tae 901
Kim, Hoon 175, 257, 275, 293, 449
Kim, Inwoong 462, 477, 504, 1048
Kim, Jeongjun 1071
Kim, Minkyu 964
Kim, Yong Hyun 1071
Kishida, Tomoki 787
Kishikawa, Hiroki 115, 733, 925
Kishikawa, Junya 561, 958
Kishimoto, Tadashi 1027
Kitamura, Kokoro 534
Kitano, Takuya 564
Kitaoka, Ryotaro 988
Kitayama, Ken-Ichi 7, 127, 664
Klaus, W. 48
Klonidis, D. 30
Knights, A. P. 622
Kobayashi, Hirokazu 970
Kobayashi, Shuko 1045
Kobayashi, Soichi 531
Kobayashi, Takashi 284, 446
Kobayashi, Wataru 350, 1122
Koda, Katsutoshi 239, 245, 248
Kodama, Yutaro 386
Koeber, S. 413
Koehnle, K. 413
Koga, Masafumi 826
Kohl, M. 413
Koike-Akino, Toshiaki 522
Kojima, Keisuke 522
Kokubun, Yasuo 36, 320, 383
Komori, Yuki 356
Kondo, Tomoki 988
Kong, Deming 15, 823, 850
Kong, Qian 691
Konishi, T. 106, 112, 609, 742
Kono, Naoto 57
Konoike, Ryotaro 1167
Koos, C. 413
Koshikiya, Yusuke 471, 727
Koyama, Fumio 64, 70, 73, 76
Koyama, Osanori 544, 880, 898, 934
Koyama, Ryo 317
Kuang, Caixia 754
Kubo, Kazuo 489, 522
Kubo, Ryogo 751
Kubota, Hirokazu 60, 838, 1065
Kubota, Manabu 682
Kuchta, Daniel M. 63
Kuno, Yuki 642
Kuo, Hao-Chung 585, 594
Kuo, Ping Piu 543
Kuramochi, Eiichi 612
Kurashima, Toshio 317

AUTHOR INDEX

Kurata, Kazuhiko 199
Kuroda, Keiji.......................................892
Kurosu, Takayuki 889
Kusama, Akihiro 898
Kuwaki, Nobuo....................................57
Kuwatsuka, Haruhiko 1185
Kwon, Hong...1071
Kwong, Dim-Lee 1107
Labroille, Guillaume............................181
Lai, Chih-Hsien 1033
Lai, Yinchieh 883
Lauermann, M. 413
Lawin, Mirko.......................................251
Lazaro, Jose A. 688
Le Liepvre, A. 931
Le Roux, X. 422, 1149, 1170
Le Taillandier De Gabory, E....................516
Leblond, Herve 576
Lee, Cheng-Ling.........................922, 1074
Lee, D. 193
Lee, Heeyoung 1068, 1095
Lee, Jeong-Min 964
Lee, Joon-Woo....................................868
Lee, Kwanil 871, 901
Lee, Sang Bae 871, 901
Lee, San-Liang 679
Lee, Seungmin 553, 556
Leong, Shan-Fong 594, 976
Lepage, G. 1104
Leuthold, J. 30, 413
Li, Cheng 377
Li, Chung-Yi 778
Li, Chunsheng 886
Li, Hongpu 335
Li, Jianqiang.......................................706
Li, Junjie 431
Li, Kuan-Ting 919
Li, Lei 501
Li, Liwei 703
Li, Meifeng...1018
Li, Ming 1188
Li, Ming-Jun................................222, 603
Li, Ming-Jung 145
Li, Qing 124
Li, Shangyuan 269, 805, 946
Li, Wenzhe..691
Li, Xin 691, 808
Li, Yan 15, 823, 850
Li, Ze 233, 510
Li, Ziang 805
Liang, Di 377
Liang, Kevin 781
Liao, Hao 1083
Liaw, Shien-Kuei 910, 925
Lillieholm, Mads 790

Lim, Christina 465
Lin, Chun-Yu 778
Lin, Fang-Zeng 67
Lin, Gong-Ru 585, 594, 976
Lin, Hung-Hsien 778
Lin, Jun-Han 922
Lin, Kuan-Hao 910
Lin, Rui 814
Lin, Rujian 754
Lin, Wenqiao 1092
Linganna, Kadathala 1080
Liou, Jia-Hong 913
Liow, Tsung-Yang 1107
Lischke, Stefan....................................964
Little, Brent E. 401, 736
Littlejohns, C. J. 622
Liu, Aijun 242
Liu, Bo 501
Liu, Chun-Nien 895
Liu, D............260, 814, 817, 877, 907, 1000, 1018, 1083
Liu, Jun-Jie 591, 859
Liu, Kai 973, 982
Liu, Li 1083
Liu, Ling 772
Liu, Qingwen 425, 1086, 1089, 1101
Liu, Tangqing 850
Liu, Wanyuan 1188
Liu, Wen-Fung.....................................919
Liu, Yang..781
Liu, Yinping 1077
Liu, Youxin 973
Liu, Yu-Chang 679
Liu, Z. 225, 928
Livshits, Daniil 1137
Lo, Guo-Qiang 1107
Loo, R. 1104
Lorences-Riesgo, Abel..........................507
Lou, Yiming 811
Lu, Chao...928
Lu, Guo-Wei 519
Lu, Hai-Han 778
Lu, Liangjun 79
Lu, Luluzi 1000
Lu, Ping 907, 1083
Luís, R. S. 48
Luo, Bin 510
Luo, Chao 907
Luo, Jinhui 269, 805
Lv, Qiang..706
Ma, Jiyang..1155
Ma, Lin 1077
Ma, Yiran 431
Maeda, Hideki 829
Maeda, Wakako 516
Maegami, Yuriko..................................1110

AUTHOR INDEX

Maho, A. 931, 1116
Malacarne, Antonio 118
Man, Ray 547
Manabe, Tetsuya.........................471, 727
Mao, Mingzhi 793
Marcaud, G. 1149
Marco, T. 133
Marom, D. M....................30, 184, 718
Marpaung, David 139
Marris-Morini, D. 1149
Martínez, Ricardo 1036
Maruta, Akihiro 127, 853
Maruyama, Hiroaki 934
Maruyama, Ryo 57
Maruyama, Takeo................................967
Mashanovich, G. Z..............................622
Maštera, Radek 994
Masuda, Akira 290
Masuda, Hiroji 534
Masumoto, Kana 239
Mateo, Eduardo 142
Matsuda, Keisuke................................522
Matsuda, Toshiya.................239, 245, 248
Matsui, Junichiro 489
Matsui, Takahiro 341
Matsui, Takashi....................54, 60, 633
Matsumoto, A....................................380
Matsumoto, Jun 937
Matsumoto, Keiichi.......................561, 958
Matsumoto, Ryosuke 127
Matsumoto, S....................................133
Matsumoto, Yukihiro 341
Matsunaga, Akira 160
Matsuo, Shinji 612
Matsuo, Shoichiro 45, 54
Matsushita, Asuka 368
Matsutani, Akihiro....................70, 73, 76
Matsuura, Hiroyuki 82, 151, 1185
Matsuura, Motoharu 398, 452, 943
Matsuzaki, Hideaki 374
Mayoral, Arturo 1036
Mazurczyk, Matt.................................371
Mehrvar, Hamid 1188
Mei, Chao.......................................1161
Melati, Daniele 88, 1021
Melikyan, A. 413, 1116
Melloni, Andrea 88, 407, 1021
Meloni, Gianluca 118, 724
Mendinueta, J. M. D...........................392
Meyer, J. 133
Miao, Wang 208, 1203
Mikhrin, Sergey 1137
Mikhrin, Vladimir 1137
Milione, Giovanni...............................145
Millar, David 522

Minamoto, Yamato............................452
Minato, Naoki 1045
Mino, S. 106, 112, 609
Mise, Kazuaki 544
Mishra, Snigdharaj 142
Mitsuno, Hiroya...............................967
Mittal, V. 622
Miura, Seiya 36
Miyabe, M. 133
Miyamoto, Tomoyuki.....................353, 356
Miyamoto, Y. 45, 193, 299, 368, 606, 1182
Miyazawa, T..............................133, 655
Miyoshi, Yuji 60, 838, 1065
Mizuno, T. 193
Mizuno, Yosuke 341, 1068, 1095, 1098
Mizutori, Akira 826
Mochizuki, Keita 1131
Mohajerin-Ariaei, A. 537
Mohan, Gurusamy 293
Molin, Denis 600
Molina-Fernandez, I. 622
Monroy, Idelfonso Tafur 166
Morandotti, Roberto 401
Moreolo, Michela Svaluto 302
Mori, Hiroshi 544
Mori, Kazuya 115
Mori, Takayoshi 329
Mori, Yojiro.................82, 428, 437, 1185
Morichetti, Francesco 88, 407
Morikawa, Hiroyuki 513
Morioka, Toshio 45, 136, 790
Morita, Itsuro 190, 673, 1039
Morita, Kohei..................................997
Morito, Ken....................................645
Moriwaki, Osamu 437
Morizumi, Yuki 970
Morizur, Jean-François 181
Moss, David J..................................401
Muehlbrandt, S. 413
Muñoz, Raul 1036
Murai, Hitoshi 1027
Murakami, Toshinori 898
Murakawa, T. 106, 112, 609
Muramoto, Yoshifumi 350
Muranaka, Yusuke.............................1197
Murano, Akihiro 24
Murata, Hiroshi...............570, 579, 949
Murugan, G. S.................................622
Nabika, Kengo 787
Nada, Masahiro 374
Nadal, Laia 302
Nagashima, T........106, 112, 609, 742
Nagatani, Munehiko.......................368, 606
Nagatomi, Ken 670
Naka, Akira....................................829

AUTHOR INDEX

Nakadai, Masahiro 1167
Nakagawa, Goji 254, 498
Nakagawa, Masahiro 239, 248
Nakahama, Masanori 70, 73, 76
Nakahara, Kouji 1113
Nakajima, Hirochika 386, 1140
Nakajima, K. 54, 60, 227, 329, 633, 985
Nakajima, Mitsumasa 94, 1182
Nakamura, A. 133
Nakamura, Hitoshi 341
Nakamura, Kazuki 1152
Nakamura, Kentaro 341, 1068, 1095, 1098
Nakamura, Moriya 769, 832, 844
Nakamura, Ryoichiro 832
Nakamura, Seiki 1125
Nakamura, Takahiro 148, 199
Nakamura, Tatsuya 516
Nakanishi, Akira 1113
Nakanishi, Yasuhiko 350
Nakano, Yoshiaki 1143
Nakao, A. 985
Nakashima, Hisao 489
Nakatsuhara, K. 85
Nakayama, Yu 682
Nakazawa, Masataka 9, 12, 33, 305
Nakpeerayuth, Suvit 748
Namiki, S. 18, 21, 82, 139, 151, 889, 1179, 1185
Naoe, Kazuhiko 1113
Nedeljkovic, M. 622
Nguyen, Linh 703
Ni, Wenjun 907, 1083
Nicho, J. 133
Niihara, Takumi 934
Ninomiya, Norihiko 943
Nirmalathas, Ampalavanapillai 465
Nishi, Hidetaka 148
Nishide, Kenji 219
Nishihara, Masato 302
Nishimura, Kosuke 284, 446
Nishiyama, Nobuhiko 642, 1119
Nishiyama, Tetsuo 561, 958
Nishizaki, Itaru 341
Niu, Tong 573
Niwa, Masaki 437
Noda, Susumu 621, 1167
Nogami, Masamichi 1125, 1131
Noguchi, Hidemi 489, 516
Noguchi, Masataka 148
Nosaka, Hideyuki 368
Noto, Kazutaka 471, 727
Notomi, Masaya 419, 612, 1176
Nozaki, Kengo 612
Numata, Hidetoshi 639
Oba, Jinsei 766
Ochi, Hirotaka 856

Ochi, Yutaka 341
Oda, Shoichiro 254, 477, 498
Oe, Shota 100
Ogawa, Kensuke 1107
Ogawa, Yoh 1027
Ogino, Takehiro 955
Ogoshi, Haruki 216
Oguro, Takahiro 838
Ohara, Mizuki 531
Ohara, Y. 133, 787
Ohashi, Masaharu 60, 838, 1065
Ohno, Morifumi 1110
Ohno, Tomohiro 570
Ohshima, T. 133
Ohtsuki, Tomoaki 862
Oishi, Masayuki 443
Okabe, Ryo 302
Okada, Mao 925
Okamoto, Keiji 471
Okamoto, Satoru 133, 937
Okamoto, Seiji 489
Okamoto, Takeshi 489
Okamura, Yasuyuki 570, 579, 949
Okano, Makoto 1110
Okuda, Tadahiro 949
Okuno, M. 106, 112, 609
Onaka, Hiroshi 489
Ono, Hirotaka 544, 880, 1062
Ono, Jun 544, 880
Ono, Masaaki 612
Ono, Y. 582
Onuki, Yuya 561, 958
Ortega-Monux, A. 622
Osato, Kazunori 769
Oshiba, Saeko 787
Oshida, Shohei 353
Ota, Kazuya 544, 880, 934
Otaka, Akihiro 172, 682
Owaki, Shotaro 844
Oxenløwe, Leif K. 136, 790
Oyama, Tomofumi 501
Palacharla, Paparao 504, 1048
Palmer, R. 413
Pang, Jiangchuan 15, 823
Pantouvaki, M. 1104
Paredes, Bruna 1009
Parsons, Kieran 522
Pasquazi, Alessia 401, 547
Peccianti, Marco 401
Pelusi, Mark 21, 139
Penades, J. Soler 622
Peng, Chun-Yen 67
Peng, Gaozhu 145
Peng, Huanfa 715
Peng, Jiangde 1015

AUTHOR INDEX

Peng, Xiaofeng...715
Penkler, David..1191
Peserico, Nicola 407
Petermann, Klaus..480
Petrovich, M. N. 225
Pincemin, E...30
Poletti, F...225
Pong, Chun-Yen 594, 976
Poole, Simon B...124
Popescu, Ion...1039
Potì, Luca 118, 724
Powers, Dale 603
Prajzler, Václav...994
Preve, G. B...395
Proietti, Roberto..661
Pukhrambam, Puspa Devi.................................679
Pun, Edwin Y. B...401
Puttnam, B. J...48
Qi, Haifeng..1024
Qian, Chen..793
Qin, Yingchun 359
Qiu, Chen 431
Qiu, Feng 961
Qiu, Kun 841
Quack, Niels..91
Rademacher, Georg...480
Ramachandran, Siddharth................................1061
Ranaweera, Chathurika 465
Rao, Baoquan 431
Rasmussen, J. C. 254, 302, 462, 477, 492,
      498, 501, 513
Raz, Oded 208
Razo, M...133
Ren, Xiaomin...973
Richardson, D. J...225
Richter, Thomas...27
Rivas, J. M...30
Robertson, Brian..97
Røge, Kasper Meldgaard...................................790
Romagnoli, M...395
Ruan, Lihua...459
Rudnick, R...30
Rumley, Sébastien..196
Ryf, R..187
Ryu, Gukbeen..901
Saito, Hiroyuki...1045
Saito, Kohei 829
Saito, Kotaro 317
Saito, Tsunetoshi 314
Saitoh, Kunimasa 45, 51, 54, 633
Saitoh, Shota...39, 54
Sakaguchi, Takahiro...70, 76
Sakamoto, Junji..94
Sakamoto, Shinichi...1107
Sakamoto, Taiji.................................54, 60, 329, 633

Sakamoto, Takahide...519, 796
Sakano, Goki...615
Sakata, Ryosuke 651
Sakuma, Kazuki 615
Saleh, Adel A. M...121
Sampath, K. I. Amila 272
Sandoghchi, S. R. 225
Sang, Xinzhu 886, 1161
Sanjoh, Hiroaki 350
Sano, Akihide 368, 606
Sano, Tomomi 627
Saridis, G. M..48
Sasago, Hiroki 730
Sasaki, Keiichi 1179
Sasaki, Shinya 856
Sasaki, Yusuke 39, 45
Sato, Ken-Ichi.................82, 230, 428, 437, 1185
Sato, Manabu 272
Sato, T. 133
Sato, Yosuke..446
Satou, I...133
Sawamura, Taketsugu 1128
Schindler, P. C..413
Schmidt-Langhorst, Carsten...............................27
Schröder, Jochen 597
Schubert, Colja 27
Segawa, Toru 1197
Seki, Hiroyuki 462
Seki, Kazuki...1140
Sekine, Kawori...832
Sekiya, Motoyoshi..477, 504
Senda, Kousuke 544, 880
Seok, Tae Joon...91
Serna, Samuel 422, 1170
Seyedi, M. Ashkan..1137, 1191
Shakoor, Abdul 612
Shao, C. 133
Sharma, Prateeksha 700
Shen, Gangxiang...889
Shen, Mingya...1024
Shen, Shin-Wei 1134
Shi, Jin-Wei 591, 859
Shi, Lu 260
Shibagaki, Nobuhiko 446
Shibahara, K. 193
Shieh, Wern-Yarng..865
Shieh, William 323
Shih, Tien-Tsorng...585, 594
Shikama, Kota 1182
Shim, Young Bo 553
Shimada, Hirokazu..254
Shimada, Shumpei 341
Shimada, Tatsuya 172
Shimakawa, Osamu..627
Shimizu, S. 106, 109, 112, 609

AUTHOR INDEX

Shimizu, Tatsuya.............................172
Shimomura, K....................561, 955, 958
Shimose, Yoshiharu.........................544
Shinada, Satoshi.............................392
Shindo, Takahiko...........................1122
Shiraiwa, Masaki.......................655, 658
Shirao, Mizuki.............................1125
Shizuka, Makoto...........................1098
Shoji, Akihisa...............................694
Shu, Chester..................416, 495, 550
Shukla, Vishnu.................................1
Shum, P....................260, 814, 877
Sillard, Pierre...............................600
Simeonidou, D..........................48, 654
Slavik, R....................................225
Sobu, Yohei..................................645
Solis-Trapala, K..................18, 21, 139
Soma, Daiki..................................365
Sommer, M...................................413
Son, Yong-Hwan.............................868
Sone, Kyosuke.........................254, 462
Song, Jue....................................877
Song, Kwang Yong.....................901, 1071
Song, M......................................30
Song, Sanggwon.............................556
Song, Shijie.................................703
Song, Tianyu.................................862
Song, Yingxiong.......................754, 820
Sorel, Marc..................................407
Sotobayashi, Hideyuki.................24, 766
Spiga, Silvia................................347
Spinnler, Bernhard..........................486
Srinivasan, S. A...........................1104
Stabile, Ripalta...........................1200
Stankovic, S.................................622
Stojanovic, Nebojsa.........................287
Stolte, Ralf.................................124
Stone, Jeffery...............................145
Su, Xiaofei..................................492
Subramaniam, Suresh.........................437
Subramanian, Ramanathan.....................335
Suda, Satoshi...........................82, 151
Sugawara, Mitsuru...........................148
Sugihara, Kenya.............................489
Sugihara, Seitaro...........................670
Sugihara, Takashi.....................489, 522
Sugimoto, Y..................................985
Sugiyama, H..................................133
Sugiyama, Koki...............................712
Sugizaki, R.............................42, 537
Suhara, Masashi..............................353
Sukigara, Toshiki.....................561, 958
Sumida, Y....................................133
Sun, Chuanbowen..............................275
Sun, Han.....................................483

Sun, Nai-Hsiang..............................919
Sun, Weiqiang................................745
Sun, Yongmei...........................242, 775
Sung, J. Y...................................685
Sung, Yuan-Yuan..............................679
Suo, Jing...................................1015
Suzuki, Daiki.................................12
Suzuki, Junichi..............................642
Suzuki, Keijiro.........................82, 151
Suzuki, Keita...............................1146
Suzuki, Kenya...........................94, 1182
Suzuki, Makoto...............................513
Suzuki, Takanori............................1113
Sygletos, S...................................30
Szelag, Bertrand............................1191
Taga, Hidenori...............................332
Tajima, Akio.................................564
Takahara, Tomoo..............................302
Takahashi, Kazuto............................925
Takahashi, Mikoto............................531
Takahashi, Naoki............................1140
Takahashi, Ryo..............205, 664, 1197
Takahashi, S.................................582
Takano, Katsumi..............................272
Takano, Kohei................................955
Takasaka, S..................................537
Takashi, Goh................................1182
Takasuka, Syo................................898
Takata, Kenta...............................1176
Takeda, Koji.................................612
Takeda, M.....................................85
Takei, Aki..................................1113
Takemasa, Keizo..............................148
Takenaga, Katsuhiro...............39, 45, 54
Takenouchi, Hirokazu...............1152, 1158
Takeshima, Koki.......................365, 1039
Takizawa, Motoyuki...........................254
Tam, Hwa-Yaw...........................547, 928
Tamai, Hideaki..............................1045
Tamura, K. R...............................1113
Tan, H. N.......................18, 21, 139
Tan, Yuanlong................................874
Tan, Zhongwei..........................338, 1003
Tanabe, Katsuaki.............................148
Tanabe, Kazuhiro.............................651
Tanaka, Hiroki...............................341
Tanaka, Kazuki...............................284
Tanaka, S..............................985, 1113
Tanaka, Takafumi.............................434
Tanaka, Toshiki..............................302
Tanaka, Yoshiaki.............................757
Tanaka, Yoshinori...........................1167
Tanaka, Yosuke...............................341
Tanaka, Yu...................................645
Tanemura, Takuo.............................1143

AUTHOR INDEX

Tang, Jin 528
Tang, Ming 260, 814, 817
Tang, Xiaosheng............................979
Tanigawa, Yosuke..............................667
Tanimura, Takahito..................27, 513
Tanizawa, Ken..........................82, 151
Tao, Xiaofeng..................................157
Tao, Yemeng......................................1077
Tao, Yiming......................................1092
Tao, Zhenning 492, 498, 501
Taue, Shuji.......................................916
Ten, Sergey......................................344
Terada, Jun 172
Terada, Yuki 428
Thouenon, G.......................................30
Thouras, Jordane..............................576
Tobita, Yuki......................................54
Tode, Hideki 667, 670
Tokura, Akio......................................1158
Tomita, Kazuki...................................829
Tomizawa, Masahito.........................489
Tomkos, I..30
Tommasino, P.....................................395
Trifiletti, A.......................................395
Tsai, C.-Y..618
Tsai, Cheng-Ting............585, 594, 976
Tsai, Zong-Yu....................................865
Tsang, H. K. 410, 416, 685
Tsang, K. S..547
Tsang, Yanny....................................547
Tseng, Chung-Hao 1074
Tseng, Shuo-Yen 991
Tsuchida, Junichi.............................199
Tsuchizawa, Tai 148
Tsuda, Hiroyuki 712, 1012
Tsuda, Nobuaki................................916
Tsujikawa, Kyozo..............................633
Tsunashima, Satoshi 350
Tsuritani, T............. 133, 190, 236, 332, 365, 673, 1039
Tsutsumi, Takuya 682
Tu, Xiaoguang...................................1107
Tu, Xin 1188
Tucker, Rod..5
Tujita, Sho 544, 880
Tzu, Ta-Ching 591, 859
Udagawa, Kenta................................534
Ueda, Koh 437, 1185
Ueda, Yuta..350
Ueno, Fumiaki 949
Ueno, Yuto......................................1131
Uenohara, H. 106, 112, 130, 609, 730, 997
Uetsuka, Hisato................................1179
Umeki, Takeshi.................................1152
Umezawa, T.......................................380
Uto, Kenichi......................................522

V., Dinesh Kumar 700
Van Campenhout, J............................. 1104
Van Kerrebrouck, J. 389
Van Thourhout, D. 413, 1104
Vassilieva, Olga 477, 504
Velázquez, A. 931
Velha, P. 395
Verheyen, P. 1104
Vilalta, Ricard 1036
Vílchez, F. J. 302
Villafranca, A. 622
Vincent, François............................... 1191
Vivien, L. 422, 1149, 1170
Von Kirchbauer, Heinrich.................. 486
Wada, Kazuyuki 832
Wada, Masaki 329
Wada, N........ 48, 106, 109, 112, 392, 609, 655, 658, 748
Wagner, Christoph 251
Waho, Takao....................................... 955
Wai, P. K. A. 540, 547
Wakayama, Yuta 332
Wang, Bin 1086, 1089, 1101
Wang, Danshi................................ 233, 510
Wang, Guanzhong.............................. 359
Wang, Hongxiang 242, 775
Wang, Jiachen 871
Wang, Jiaqi 416
Wang, Jiayu 431
Wang, Jie.. 547
Wang, Jinghao 1018
Wang, Jun 982
Wang, Kuiru 1161
Wang, Min 754, 793
Wang, Ming 979
Wang, Peng 335
Wang, Qian 862
Wang, Ruoxu 814
Wang, S. H. 540, 736
Wang, Sheng-Min 883
Wang, Shuai..................1077, 1086, 1101
Wang, Shun 1083
Wang, Ti-Ho 865
Wang, Ting 145, 455
Wang, Wei.................................. 585, 1057
Wang, Xi 462, 1048
Wang, Xinyi 79
Wang, Xinyue 338, 1003
Wang, Xudong 573
Wang, Yixin 305
Wang, Yu 636
Wang, Yuanwu 1018
Wang, Yuxi 103
Wang, Ziyu 338, 1003
Wanguemert-Perez, G......................... 622
Watanabe, Hiroshi 471, 727

AUTHOR INDEX

Watanabe, Kengo..............................314
Watanabe, Shigeki 27
Watanabe, Takashi............................670
Watanabe, Tatsuhiko..........................36
Watanabe, Toshio 1185
Weerasekara, Gihan 853
Wei, Chia-Chien.................591, 859, 1134
Wei, L. Y...................................685
Wei, Liang-Yu...............................781
Wei, Pengjiang..............................736
Wei, Yuming................................1188
Weng, Tsui-Wei.............................1021
Wenjing, Fang..............................1006
Wheeler, N. V..............................225
Wijayanto, Yusuf Nur 579
Wilkinson, J. S.............................622
Wilkinson, Peter............................97
Willner, A. E...............................537
Wolf, S.....................................413
Wong, Elaine 459, 465
Wood, William A.............................142
Worasucheep, Duang-Rudee....................748
Wu, Chang-Jen...............................778
Wu, Chao-Hsin..........................594, 976
Wu, Huan....................................550
Wu, Jian.............. 15, 236, 673, 721, 823, 850, 1092
Wu, Jingjing................................263
Wu, Jui-Pin................................1134
Wu, Meng-Shan..............................1074
Wu, Ming C..................................91
Wu, Ming-Wei...............................862
Wu, Qiong...................................814
Wu, Tsai-Chen..............................585
Wu, Tsu-Shiu................................67
Wu, Weiliang..........................278, 820
Wu, X. R....................................685
Wu, Xiaojuan................................886
Wu, Xinru...................................410
Wu, Yun-Chen..........................594, 976
Wu, Yunfei..................................630
Wu, Zhichao.................................877
Xiao, Min..............................359, 1155
Xiao, Shilin..........................772, 799
Xiaoen, Du..................................544
Xiaomin, Ren..........................982, 1006
Xie, Changsong..............................287
Xie, Shizhong...............................811
Xin, Haiyun.................................772
Xin, Xin....................................775
Xu, Ke................................410, 685
Xu, Kun.....................................706
Xu, Sheng...................................757
Xu, Sugang............................658, 757
Xu, Yongchi.................................715
Xu, Zhaopeng...............................1003

Xue, Yuankai................................799
Yamada, Koji...............................1110
Yamada, Makoto 544, 880, 898, 934, 1062
Yamada, Shoko...............................24
Yamaguchi, Keita.......................94, 1182
Yamaguchi, Minoru..........................934
Yamaguchi, Shigeru........................1158
Yamaguchi, Yuya.......................386, 1140
Yamamoto, Fumihiko.....................54, 633
Yamamoto, N..................24, 380, 763, 766
Yamamoto, Shuto............................290
Yamamoto, Takashi..........................329
Yaman, Fatih...............................142
Yamanaka, N...........................133, 937
Yamanaka, Yusuke...........................697
Yamanishi, Masamichi.......................567
Yamasaki, Y................................742
Yamashita, Yoko............................633
Yamauchi, Tomohiro....................477, 498
Yamazaki, Hiroshi 350, 368, 606
Yan, Binbin................................886
Yan, Fulong................................208
Yan, Haozhe................................946
Yan, Shengyong............................1188
Yanan, Guo................................1006
Yang, Bingliang............................739
Yang, Chuanchuan..........................1003
Yang, Dung-Chin.............................67
Yang, Guangyao 1086, 1089, 1101
Yang, Haining...............................97
Yang, Hui 874, 1042
Yang, Liang................................398
Yang, Qianmei..............................468
Yang, Se-Hoon..............................868
Yang, Sigang 103, 811
Yang, Tingting.............................721
Yang, Wenjian..............................703
Yang, Yanfu...........................835, 847
Yang, Yu..................................1003
Yang, Yuan-Jie.............................922
Yang, Zhisheng............................1092
Yao, Yong.............................835, 847
Yatsu, Tomoya..............................398
Ye, Zilong................................1054
Yeh, C. H..................................685
Yeh, Chien-Hung............................781
Yeh, Tung-Yuan.............................922
Yi, Xiaoke.................................703
Yi, Xingwen................................841
Yin, Feifei................................706
Yin, Shan..................................808
Yin, Xin...................................389
Yin, Zhang-Qi..............................359
Ying, Wang................................1006
Yokokawa, S................................985

AUTHOR INDEX

Yokouchi, Noriyuki 1128
Yokoyama, C. 133
Yokoyama, Shiyoshi 961
Yonenaga, Kazushige 368, 489
Yoneyama, Akira 452
Yongqing, Huang 1006
Yoo, Kwang Wook 556
Yoo, S. J. B. 202, 661
Yoshida, Junji 1128
Yoshida, Masato 33, 305
Yoshida, Mitsuteru 489
Yoshida, S. 133, 254, 498
Yoshida, Tsuyoshi 522
Yoshida, Yuki 127, 664
Yoshikane, N. 133, 673, 1039
Yoshikuni, Yuzo 892
Yoshimoto, Naoto 531, 694
Yu, Ao 1042
Yu, Changyuan 257, 293, 841
Yu, Chin-Ping 913
Yu, Chongxiu 1161
Yu, Cunqian 904
Yu, Kunzhi 377
Yu, Miao 15, 823
Yu, Xianbin 136
Yu, Yi-Lin 925
Yu, Yinghong 772
Yu, Zhao 287
Yu, Zhenming 811
Yuan, Jinhui 1161
Yuan, Shuai 525
Yue, Lei 15, 823
Zafar, Humaira 1009
Zakharian, Aramais 142
Zang, Zhigang 979
Zeng, Heping 1173
Zervas, G. 48
Zhang, A. P. 547, 928
Zhang, Bingbing 760
Zhang, Cheng 715
Zhang, Chengliang 431
Zhang, Chunhui 1188
Zhang, Chunshu 1188
Zhang, Dongxu 236, 673, 721
Zhang, Fan 311
Zhang, Hanyi 269, 805, 946
Zhang, Jiangshan 907
Zhang, Jiawei 775
Zhang, Jie 808, 874, 1042, 1057
Zhang, Jing 841
Zhang, Junjie 278, 793, 820
Zhang, Kuo 739
Zhang, Lu 772, 799
Zhang, Min 233, 510, 808
Zhang, Minming 260, 877, 952, 1000, 1018

Zhang, Qianwu 278, 754, 793, 820
Zhang, Qiong 1048
Zhang, Qun 835, 847
Zhang, Shaojie 772
Zhang, Shaoliang 142
Zhang, Wei 1015
Zhang, Weiwei 422, 1170
Zhang, Wenjia 1077
Zhang, Yangan 973
Zhang, Yunchuan 1024
Zhang, Yunhao 772, 799
Zhang, Zhen 278, 820
Zhang, Zhiguo 760
Zhang, Zitian 739
Zhao, Fei 1188
Zhao, J. 30
Zhao, Mingming 359, 1155
Zhao, Mingxiao 1155
Zhao, Peng 636
Zhao, Shuoyi 79
Zhao, Weiqian 1024
Zhao, Ying 501
Zhao, Yongli 431, 808, 874, 1057
Zheng, Bofang 550
Zheng, Xiaoping 269, 805, 946
Zheng, Yue 706
Zhong, Kangping 835, 847
Zhou, Bin 928
Zhou, Bingkun 269, 805, 946
Zhou, Feiya 1000
Zhou, Huibin 817
Zhou, Jianwei 706
Zhou, Jingjing 293
Zhou, Linjie 79
Zhou, Xian 835, 847
Zhou, Xu 308, 468
Zhou, Yu 691
Zhu, Lixin 715
Zhu, Xiaoxu 874
Zhu, Yixiao 311
Zimmernman, Lars 964
Ziyadi, M. 537
Zong, Yue 904
Zou, Kaiheng 311

2016 21st OptoElectronics and Communications Conference (OECC 2016) held jointly with 2016 International Conference on Photonics in Switching (PS 2016)

Niigata, Japan
3-7 July 2016

Pages 603-1205

IEEE Catalog Number:	CFP1699A-POD
ISBN:	978-1-5090-2147-5

Copyright © 2016, The Institute of Electronics, Information and Communication Engineers
All Rights Reserved

***This publication is a representation of what appears in the IEEE Digital Libraries. Some format issues inherent in the e-media version may also appear in this print version.

IEEE Catalog Number: CFP1699A-POD
ISBN (Print-On-Demand): 978-1-5090-2147-5
ISBN (Online):   978-4-88552-305-2
ISSN: 2155-8507

Additional Copies of This Publication Are Available From:

Curran Associates, Inc
57 Morehouse Lane
Red Hook, NY 12571 USA
Phone: (845) 758-0400
Fax: (845) 758-2633
E-mail: curran@proceedings.com
Web: www.proceedings.com

TABLE OF CONTENTS

OPTICAL TRANSPORT TO OPEN TRANSPORT IN CARRIER NETWORKS 1
Vishnu Shukla

SILICON PHOTONIC INTEGRATED CIRCUITS FOR COHERENT COMMUNICATIONS 3
Chris Doerr

ENERGY LIMITATIONS IN DATA TRANSMISSION AND SWITCHING 5
Rod Tucker

OPTICAL PACKET SWITCHING: MYTH, FACT, AND PROMISE 7
Ken-ichi Kitayama

ULTRAHIGH SPEED AND HIGH SPECTRAL EFFICIENCY TRANSMISSION USING OPTICAL NYQUIST
PULSES 9
Masataka Nakazawa

ROLL-OFF FACTOR DEPENDENCE OF SYSTEM PERFORMANCE IN 1.28 TBIT/S/CH-525 KM NYQUIST
PULSE TRANSMISSION 12
Koudai Harako ; Daiki Suzuki ; Toshihiko Hirooka ; Masataka Nakazawa

A NOVEL DEMULTIPLEXING SCHEME FOR NYQUIST OTDM SIGNAL USING A SINGLE IQ
MODULATOR 15
Lei Yue ; Deming Kong ; Yan Li ; Jiangchuan Pang ; Miao Yu ; Jian Wu

1.76-TBIT/S SUPERCHANNEL GENERATION BASED ON INP IQ MODULATOR AND OPTICAL
FREQUENCY COMB 18
K. Solis-Trapala ; T. Inoue ; H. Nguyen Tan ; S. Namiki

C-BAND SPANNING FREQUENCY COMB BASED ON AN OUTPUT PHASE STABILIZED MODE-LOCKED
LASER 21
Mark Pelusi ; Hung Nguyen Tan ; Karen Solis-Trapala ; Takashi Inoue ; Shu Namiki

10-GBAUD QPSK SIGNAL SIMULTANEOUS TRANSMISSION IN THE T AND C BANDS FOR ULTRA-
BROADBAND PHOTONIC TRANSPORT SYSTEM 24
Akihiro Murano ; Shoko Yamada ; Atsushi Kanno ; Naokatsu Yamamoto ; Hideyuki Sotobayashi

SUBCARRIER POLARIZATION MANIPULATION USING ORTHOGONAL DUAL PHASE-MODULATION
IN FIBER 27
*Tomoyuki Kato ; Takahito Tanimura ; Takeshi Hoshida ; Thomas Richter ; Robert Elschner ; Carsten Schmidt-Langhorst ; Colja
Schubert ; Shigeki Watanabe*

CASCADED ALL-OPTICAL SUB-CHANNEL ADD/DROP MULTIPLEXING FROM A 1-TB/S SUPER-
CHANNEL HAVING 2-GHZ GUARD-BANDS 30
*M. Song ; E. Pincemin ; B. Baeuerle ; A. Josten ; D. Hillerkuss ; J. Leuthold ; R. Rudnick ; D. M. Marom ; S. Ben-Ezra ; J. F.
Ferran ; S. Sygletos ; A. Ellis ; J. Zhao ; G. Thouenon ; C. Betoule ; J. M. Rivas ; D. Klonidis ; I. Tomkos*

1024 QAM, 7-CORE FIBER/MULTI-CORE EDFA TRANSMISSION OVER 100 KM WITH AN
AGGREGATED SPECTRAL EFFICIENCY OF 109 BIT/S/HZ 33
Masato Yoshida ; Keisuke Kasai ; Toshihiko Hirooka ; Masataka Nakazawa

ACCURATE ANALYSIS OF CROSSTALK BETWEEN LP11 DEGENERATE MODES DUE TO OFFSET
CONNECTION USING EXACT EIGENMODES 36
Yasuo Kokubun ; Seiya Miura ; Tatsuhiko Watanabe

NEW METHOD FOR MEASURING INTER-CORE CROSSTALK IN MULTI-CORE FIBERS USING NEAR-
INFRARED CAMERA 39
Shota Saitoh ; Yoshimichi Amma ; Yusuke Sasaki ; Katsuhiro Takenaga ; Kazuhiko Aikawa

MULTICORE FIBER FOR BI-DIRECTIONAL TRANSMISSION 42
Tomohiro Gonda ; Katsunori Imamura ; Ryuichi Sugizaki

SINGLE-MODE MULTICORE FIBER FOR DENSE SPACE DIVISION MULTIPLEXING 45
*Yusuke Sasaki ; Yoshimichi Amma ; Katsuhiro Takenaga ; Shoichiro Matsuo ; Kazuhiko Aikawa ; Kunimasa Saitoh ; Toshio
Morioka ; Yutaka Miyamoto*

DYNAMIC SKEW MEASUREMENTS IN 7, 19 AND 22-CORE MULTI CORE FIBERS 48
G. M. Saridis ; B. J. Puttnam ; R. S. Luís ; W. Klaus ; Y. Awaji ; G. Zervas ; D. Simeonidou ; N. Wada

IMPULSE RESPONSE ANALYSIS OF AIR-HOLE ADDED COUPLED SIX-CORE FIBERS 51
Yasuyuki Chida ; Takeshi Fujisawa ; Kunimasa Saitoh

HIGH SPATIAL DENSITY FEW-MODE MULTI-CORE FIBER WITH LOW DIFFERENTIAL MODE DELAY
CHARACTERISTICS 54
*Taiji Sakamoto ; Takashi Matsui ; Kunimasa Saitoh ; Shota Saitoh ; Katsuhiro Takenaga ; Shoichiro Matsuo ; Yuki Tobita ;
Nobutomo Hanzawa ; Kazuhide Nakajima ; Fumihiko Yamamoto*

ULTRAFAST COMPLEX IMPULSE RESPONSE MEASUREMENT OF 2-MODE FIBERS BY USING LINEAR
OPTICAL SAMPLING 57
Naoto Kono ; Fumihiko Ito ; Ryo Maruyama ; Nobuo Kuwaki

PMD MEASUREMENTS OF TWO-MODE FIBERS USING MODE COUPLER BASED ON FIXED
ANALYZER TECHNIQUE 60
Shota Igarashi ; Masaharu Ohashi ; Yuji Miyoshi ; Hirokazu Kubota ; Taiji Sakamoto ; Takashi Matsui ; Kazuhide Nakajima

HIGH SPEED VCSEL-BASED LINKS FOR USE IN DATA CENTERS AND HPC 63
Daniel M. Kuchta

HIGH SPEED MODULATION OF TRANSVERSE COUPLED CAVITY VCSELS 64
Fumio Koyama

OSA MODULE WITH ×4 MINI MT AND 45° FIBER MIRRORS FOR 48 GB/S OPTICAL LINK 67
Feng-Cheng Hsu ; Tsu-Shiu Wu ; Fang-Zeng Lin ; Dung-Chin Yang ; Chun-Yen Peng ; Ann-Kuo Chu

FABRICATION OF HCG MEMS VCSELS FOR ATHERMAL OPERATIONS 70
Shunya Inoue ; Masanori Nakahama ; Akihiro Matsutani ; Takahiro Sakaguchi ; Fumio Koyama

VCSEL-INTEGRATED BRAGG REFLECTOR WAVEGUIDE AMPLIFIER WITH SINGLE-MODE OUTPUT POWER OVER 10 MW 73
Xiaodong Gu ; Masanori Nakahama ; Akihiro Matsutani ; Fumio Koyama

HIGH POWER NON-MECHANICAL BEAM SCANNER BASED ON VCSEL AMPLIFIER 76
Masanori Nakahama ; Xiaodong Gu ; Akihiro Matsutani ; Takahiro Sakaguchi ; Fumio Koyama

RECONFIGURABLE OPTICAL ROUTERS AND BUFFERS BUILT ON THE SILICON-ON-INSULATOR PLATFORM 79
Linjie Zhou ; Liangjun Lu ; Shuoyi Zhao ; Xinyi Wang ; Jianping Chen

POLARIZATION-INDEPENDENT C-BAND TUNABLE FILTER BASED ON CASCADED SI-WIRE ASYMMETRIC MACH-ZEHNDER INTERFEROMETER 82
Keijiro Suzuki ; Ken Tanizawa ; Satoshi Suda ; Hiroyuki Matsuura ; Kazuhiro Ikeda ; Yojiro Mori ; Ken-ichi Sato ; Shu Namiki ; Hitoshi Kawashima

FUNDAMENTAL OPERATION OF A PHASED ARRAY SWITCH USING FERROELECTRIC LIQUID CRYSTAL CLADDINGS 85
W. Kanakubo ; K. Nakatsuhara ; M. Takeda

CONTROL, CONFIGURATION AND STABILIZATION OF PHOTONIC INTEGRATED CIRCUITS 88
Andrea Melloni ; Andrea Annoni ; Stefano Grillanda ; Daniele Melati ; Francesco Morichetti

LARGE-SCALE SILICON PHOTONIC SWITCHES 91
Ming C. Wu ; Tae Joon Seok ; Sangyoon Han ; Niels Quack

WAVEFRONT TRANSMISSION FOR REMOTE OPTICAL BEAM FORMING 94
Mitsumasa Nakajima ; Junji Sakamoto ; Keita Yamaguchi ; Kenya Suzuki ; Toshikazu Hashimoto

STACKED WAVELENGTH SELECTIVE SWITCH DESIGN FOR LOW-COST CDC ROADMS 97
Haining Yang ; Brian Robertson ; Peter Wilkinson ; Daping Chu

FIRST DEMONSTRATION OF DYNAMIC MODE-SWITCHING BY USING OPTICAL MODE SWITCH 100
Ryan Imansyah ; Luke Himbele ; Shota Oe ; Haisong Jiang ; Kiichi Hamamoto

ULTRA-FAST SUPER-RESOLUTION IMAGING BY TIME-STRETCH STRUCTURED ILLUMINATION 103
Yuxi Wang ; Qiang Guo ; Hongwei Chen ; Minghua Chen ; Sigang Yang

TIME-FREQUENCY PHOTONIC SIGNAL PROCESSING 106
G. Cincotti ; T. Konishi ; T. Murakawa ; T. Nagashima ; S. Shimizu ; M. Hasegawa ; K. Hattori ; M. Okuno ; S. Mino ; A. Himeno ; N. Wada ; H. Uenohara

ALL-OPTICAL NYQUIST-WDM TO NYQUIST-OTDM CONVERSION FOR FLEXIBLE OPTICAL NETWORKS 109
Satoshi Shimizu ; Gabriella Cincotti ; Naoya Wada

OPTICAL SERIAL-TO-PARALLEL CONVERSION BASED ON FRACTIONAL OFDM SCHEME 112
M. Hiraoka ; T. Nagashima ; B. Karanov ; G. Cincotti ; S. Shimizu ; T. Murakawa ; M. Hasegawa ; K. Hattori ; M. Okuno ; S. Mino ; A. Himeno ; N. Wada ; H. Uenohara ; T. Konishi

MODULATION FORMAT CONVERSION FROM QPSK TO 16QAM USING DELAY LINE INTERFEROMETER AND SPECTRAL SHAPING FILTER 115
Kazuya Mori ; Hiroki Kishikawa ; Nobuo Goto

TIME-FREQUENCY PACKED VCSEL-BASED IM/DD TRANSMISSION FOR WDM ACCESS NETWORKS 118
Antonio Malacarne ; Francesco Fresi ; Gianluca Meloni ; Tommaso Foggi ; Luca Potì

ELASTIC WDM SWITCHING FOR SCALABLE DATA CENTER AND HPC INTERCONNECT NETWORKS 121
Adel A. M. Saleh ; Akhilesh S. P. Khope ; John E. Bowers ; Rod C. Alferness

PROGRAMMABLE OPTICAL PROCESSORS FOR SIGNAL PROCESSING AND MONITORING 124
Simon B. Poole ; Patrick Blown ; Qing Li ; Ralf Stolte ; Ian Clarke

PATTERN-INDEPENDENT WAVELENGTH CONVERSION OF PAM SIGNALS IN SOAS 127
Ryosuke Matsumoto ; Giampiero Contestabile ; Yuki Yoshida ; Akihiro Maruta ; Ken-ichi Kitayama

EXPERIMENTAL DEMONSTRATION OF ALL-OPTICAL FEC CODING SCHEME WITH CONVOLUTIONAL CODE 130
Yohei Aikawa ; Hiroyuki Uenohara

FIRST DEMONSTRATION OF GEOGRAPHICALLY UNCONSTRAINED CONTROL OF AN INDUSTRIAL ROBOT BY JOINTLY EMPLOYING SDN-BASED OPTICAL TRANSPORT NETWORKS AND EDGE COMPUTE 133
N. Yoshikane ; T. Sato ; Y. Isaji ; C. Shao ; T. Marco ; S. Okamoto ; T. Miyazawa ; T. Ohshima ; C. Yokoyama ; Y. Sumida ; H. Sugiyama ; M. Miyabe ; T. Katagiri ; N. Kakegawa ; S. Matsumoto ; Y. Ohara ; I. Satou ; A. Nakamura ; S. Yoshida ; K. Ishii ; S. Kametani ; J. Nicho ; J. Meyer ; S. Edwards ; P. Evans ; T. Tsuritani ; H. Harai ; M. Razo ; D. Hicks ; A. Fumagalli ; N. Yamanaka

THZ PHOTONICS-WIRELESS TRANSMISSION OF 160 GBIT/S BITRATE 136
Xianbin Yu ; Shi Jia ; Hao Hu ; Pengyu Guan ; Michael Galili ; Toshio Morioka ; Peter U. Jepsen ; Leif K. Oxenløwe

LOW NOISE, REGENERATION OF OPTICAL FREQUENCY COMB-LINES FOR 64QAM ENABLED BY SBS GAIN 139
Mark Pelusi ; Amol Choudhary ; Takashi Inoue ; David Marpaung ; Benjamin Eggleton ; Hung Nguyen Tan ; Karen Solis-Trapala ; Shu Namiki

20.7-TB/S REPEATER-LESS TRANSMISSION OVER 401.1-KM USING QSM FIBER AND XPM COMPENSATION VIA TRANSMITTER-SIDE DBP...........142
Yue-Kai Huang ; Ezra Ip ; Shaoliang Zhang ; Fatih Yaman ; John D. Downie ; William A. Wood ; Aramais Zakharian ; Jason Hurley ; Snigdharaj Mishra ; Yoshiaki Aono ; Eduardo Mateo ; Yoshihisa Inada

1.2-TB/S MIMO-LESS TRANSMISSION OVER 1 KM OF FOUR-CORE ELLIPTICAL-CORE FEW-MODE FIBER WITH 125-μM DIAMETER CLADDING...........145
Giovanni Milione ; Ezra Ip ; Yue-Kai Huang ; Philip Ji ; Ting Wang ; Ming-Jung Li ; Jeffery Stone ; Gaozhu Peng

DEMONSTRATION OF A HYBRID SILICON EVANESCENT QUANTUM DOT LASER...........148
Bongyong Jang ; Katsuaki Tanabe ; Satoshi Kako ; Satoshi Iwamoto ; Tai Tsuchizawa ; Hidetaka Nishi ; Nobuaki Hatori ; Masataka Noguchi ; Takahiro Nakamura ; Keizo Takemasa ; Mitsuru Sugawara ; Yasuhiko Arakawa

SILICON PHOTONIC 32 × 32 STRICTLY-NON-BLOCKING BLADE SWITCH AND ITS FULL PATH CHARACTERIZATION...........151
Ken Tanizawa ; Keijiro Suzuki ; Satoshi Suda ; Hiroyuki Matsuura ; Kazuhiro Ikeda ; Shu Namiki ; Hitoshi Kawashima

WIRELESS NETWORK EVOLUTION TOWARD 5G NETWORK...........154
Fumiyuki Adachi

5G WIRELESS NETWORK RESEARCH IN CHINA - NON-UNIFORM WIRELESS DENSE NETWORKS...........157
Xiaofeng Tao

ACTIVITIES OF THE FIFTH GENERATION MOBILE COMMUNICATIONS PROMOTION FORUM (5GMF) IN JAPAN...........160
Akira Matsunaga

HIGH CAPACITY MOBILE FRONTHAUL AND BACKHAUL NETWORK RESEARCH IN KOREA...........163
Seungjoo Hong ; Sungmin Cho

CONVERGENCE OF PHOTONICS AND ELECTRONICS FOR TERAHERTZ WIRELESS COMMUNICATIONS - THE ITN CELTA PROJECT...........166
Idelfonso Tafur Monroy

FIBER-WIRELESS FRONTHAUL: THE LAST FRONTIER...........169
Gee-Kung Chang ; Lin Cheng

OPTICAL ACCESS NETWORK TECHNOLOGY FOR 5G WIRELESS FRONT/BACKHAUL NETWORK...........172
Jun Terada ; Tatsuya Shimada ; Tatsuya Shimizu ; Akihiro Otaka

COST-EFFECTIVE OPTICAL TRANSMITTERS FOR NEXT-GENERATION MOBILE FRONTHAUL NETWORKS...........175
B. G. Kim ; Hoon Kim ; Y. Chung

ULTRA-DENSE SPACE-DIVISION-MULTIPLEXING OPTICAL FIBERS...........178
Tetsuya Hayashi

RECENT PROGRESS IN MODE MULTIPLEXING DEVICES USING MULTI-PLANE LIGHT CONVERSION...........181
Guillaume Labroille ; Pu Jian ; Nicolas Barré ; Bertrand Denolle ; Jean-François Morizur

SWITCHING SOLUTIONS FOR SPATIAL AND SPECTRAL MULTIPLEXED NETWORKS...........184
Dan M. Marom

MODE-MULTIPLEXED TRANSMISSION OVER MULTIMODE FIBERS...........187
R. Ryf ; H. Chen ; N. K. Fontaine

ULTRA-HIGH-CAPACITY TRANSMISSION OVER FEW-MODE MULTI-CORE FIBERS...........190
Koji Igarashi ; Takehiro Tsuritani ; Itsuro Morita

ADVANCED MIMO SIGNAL PROCESSING FOR DENSE SDM TRANSMISSION USING MULTI-CORE FEW-MODE FIBERS...........193
K. Shibahara ; T. Mizuno ; D. Lee ; Y. Miyamoto

OPTICAL SWITCHING PERFORMANCE METRICS FOR SCALABLE DATA CENTERS...........196
Keren Bergman ; Sébastien Rumley

ADVANCED OPTICAL INTERCONNECTION TECHNOLOGIES BASED ON SILICON PHOTONICS FOR FUTURE DCS...........199
Takahiro Nakamura ; Junichi Tsuchida ; Kazuhiko Kurata

THE ROLE OF PHOTONICS IN FUTURE EXASCALE DATA SYSTEMS...........202
S. J. Ben Yoo

BURST-MODE OPTICAL PACKET PROCESSING TECHNOLOGIES...........205
Salah Ibrahim ; Ryo Takahashi

HIGH PERFORMANCE DCN ARCHITECTURE BASED ON FLOW-CONTROLLED OPTICAL SWITCHING SYSTEM...........208
Nicola Calabretta ; Wang Miao ; Fulong Yan ; Oded Raz

EARLY DAYS OF VAD METHOD...........211
Tatsuo Izawa

FIBER AND FIBER BASED TECHNOLOGY AFTER VAD DEVELOPMENT - CONTRIBUTION TO TRANSOCEANIC SUBMARINE NETWORKS -...........213
Hiroo Kanamori

DEVELOPMENT OF OPTICAL FIBERS FOR LONG-HAUL TRANSMISSION...........216
Haruki Ogoshi

SPECIALTY FIBERS AND OPTICAL FIBER DEVICES FOLLOWING DEVELOPMENT OF VAPOR-PHASE AXIAL DEPOSITION...........219
Kenji Nishide

NEW OPTICAL FIBER DEVELOPMENT BASED ON OVD TECHNOLOGY...........222
Ming-Jun Li

HOLLOW CORE FIBRES FOR DATA TRANSMISSION 225
D. J. Richardson ; Y. Chen ; N. V. Wheeler ; J. R. Hayes ; T. Bradley ; S. R. Sandoghchi ; G. T. Jasion ; E. Numkam Fokoua ; Z. Liu ; R. Slavik ; M. N. Petrovich ; F. Poletti

OPTICAL FIBERS FOR FUTURE COMMUNICATION SYSTEMS 227
Kazuhide Nakajima

OPTICAL NETWORKING AND NODE TECHNOLOGIES FOR CREATING COST EFFECTIVE BANDWIDTH ABUNDANT NETWORKS 230
Ken-ichi Sato

RECONFIGURABLE WDM MULTICAST SUPPORTING CONTENT DELIVERY FOR CONTENT DELIVERY NETWORK BASED ON SOA AND TB-WSS 233
Ze Li ; Min Zhang ; Danshi Wang ; Yue Cui

METRO-EMBEDDED CLOUD PLATFORM WITH ALL-OPTICAL INTERCONNECTIONS FOR VIRTUAL DATACENTER PROVISIONING 236
Hongxiang Guo ; Gang Chen ; Dongxu Zhang ; Xiaoyuan Cao ; Jian Wu ; Takehiro Tsuritani

FLEXIBLE AND COST-EFFECTIVE OPTICAL METRO NETWORK WITH PHOTONIC-SUB-LAMBDA AGGREGATION CAPABILITY 239
Masahiro Nakagawa ; Kana Masumoto ; Kyota Hattori ; Toshiya Matsuda ; Masaru Katayama ; Katsutoshi Koda

A SCALABLE OPTICAL NETWORK ARCHITECTURE BASED ON WSS FOR INTRA DATA CENTER 242
Aijun Liu ; Yongmei Sun ; Hongxiang Wang ; Yuefeng Ji

FRAME LENGTH AVERAGING METHOD FOR MULTI-CARRIER AGGREGATION TRANSPORT 245
Toru Homemoto ; Toshiya Matsuda ; Masaru Katayama ; Katsutoshi Koda

METHOD TO SHARE A BURST-MODE RECEIVER BETWEEN CONTINUOUS- AND BURST-MODE TRANSMITTER FOR VM MIGRATION OF CLOUD EDGE 248
Kyota Hattori ; Masahiro Nakagawa ; Toshiya Matsuda ; Masaru Katayama ; Katsutoshi Koda

REMOTELY CONTROLLABLE WDM-PON TECHNOLOGY FOR WIRELESS FRONTHAUL/BACKHAUL APPLICATION 251
Michael H. Eiselt ; Christoph Wagner ; Mirko Lawin

DYNAMIC RECONFIGURATION OF COORDINATED MULTI-POINT TRANSMISSION IN HIGH DENSITY SMALL CELL NETWORK USING CENTRALIZED WAVELENGTH SWITCHING IN DWDM PON 254
Goji Nakagawa ; Kyosuke Sone ; Setsuo Yoshida ; Shoichiro Oda ; Motoyuki Takizawa ; Hirokazu Shimada ; Yasuhiko Aoki ; Jens C. Rasmussen

25-GB/S ASE-SEEDED WDM PON 257
Qikai Hu ; Changyuan Yu ; Pooi-Yuen Kam ; Hoon Kim

MITIGATION OF OPTICAL BEAT INTERFERENCE IN N-DIMENSIONAL CAP-PON UPLINK TRANSMISSION 260
Jiale He ; Lei Deng ; Lu Shi ; Mengfan Cheng ; Ming Tang ; Songnian Fu ; Minming Zhang ; Deming Liu ; Perry Ping Shum

PERFORMANCE INVESTIGATION FOR LDPC-CODED UPSTREAM TRANSMISSION SYSTEMS IN IM/DD OFDM-PONS 263
Xiaoxue Gong ; Lei Guo ; Jingjing Wu

MOBILE BACKHAUL AND FRONTHAUL SYSTEMS 266
Frank J. Effenberger

HIGHLY LINEAR W-BAND TRANSMITTER BASED ON OFC AND HETERODYNE UP-CONVERSION 269
Jinhui Luo ; Shangyuan Li ; Xiaoping Zheng ; Hanyi Zhang ; Bingkun Zhou

SELF-PHASE MODULATION BASED SIGNAL DISTORTIONS OF OPTICAL SSB-SC SIGNAL WITH PILOT CARRIER 272
Pilot Carrier ; K. I. Amila Sampath ; Katsumi Takano ; Manabu Sato

TRANSMISSION OF 28-GB/S DUOBINARY SIGNALS OVER 45-KM SSMF USING 1.55-μM DIRECTLY MODULATED LASER 275
Chuanbowen Sun ; S. H. Bae ; Hoon Kim ; Y. C. Chung

A NOVEL AMPLITUDE DECISION BASED SYMBOL SYNCHRONIZATION METHOD FOR REAL-TIME IMDD-OOFDM SYSTEMS 278
Weiliang Wu ; Junjie Zhang ; Zhen Zhang ; Han Dun ; Qianwu Zhang

STUDY ON OPTICAL MODULATION AMPLITUDE OF 4-LEVEL PULSE AMPLITUDE MODULATION WITH ELECTROABSORPTION MODULATOR 281
Riu Hirai ; Nobuhiko Kikuchi ; Takayoshi Fukui

NOVEL ELECTRICAL DISPERSION COMPENSATION TECHNIQUE FOR IMDD-BASED SYSTEMS 284
Kazuki Tanaka ; Takashi Kobayashi ; Akira Agata ; Kosuke Nishimura

TRANSMITTER DIGITAL PRE-DISTORTION TECHNIQUES FOR BAND-LIMITED OPTICAL NETWORKS 287
Fotini Karinou ; Zhao Yu ; Nebojsa Stojanovic ; Changsong Xie

IMPROVEMENT OF BANDWIDTH LIMITATION AND CHROMATIC- DISPERSION TOLERANCES USING 4-LEVEL/7-LEVEL CODING PAM FOR DIRECT-DETECTION SYSTEM 290
Akira Masuda ; Shuto Yamamoto ; Hiroki Kawahara ; Shingo Kawai ; Mitsunori Fukutoku

POWER BUDGET IMPROVEMENT OF 25-GB/S OPTICAL LINK BASED ON 1.5-μM 10G-CLASS VCSEL 293
Jingjing Zhou ; Changyuan Yu ; Gurusamy Mohan ; Hoon Kim

RZ-PAM4 TRANSMISSION WITH QUASI-FOURIER-TRANSFORM-LIMITED GAIN-SWITCHED PULSE SOURCE 296
Songyuan Dai ; Masanori Hanawa

OVER 400 GBIT/S DIGITAL COHERENT CHANNELS FOR OPTICAL TRANSPORT NETWORK 299
Yutaka Miyamoto

EXPERIMENTAL DEMONSTRATION OF A PROGRAMMABLE 400-GBPS DMT TRANSCEIVER WITH POLICY-BASED CONTROL 302

Yutaka Kai ; Ryo Okabe ; Masato Nishihara ; Toshiki Tanaka ; Tomoo Takahara ; Jens C. Rasmussen ; F. Javier Vilchez ; Laia Nadal ; Josep M. Fàbrega ; Michela Svaluto Moreolo

320 GBIT/S, 256 QAM LD-BASED COHERENT TRANSMISSION OVER 160 KM WITH AN INJECTION-LOCKED HOMODYNE DETECTION TECHNIQUE 305

Yixin Wang ; Keisuke Kasai ; Masato Yoshida ; Masataka Nakazawa

COHERENT INTERFERENCE REDUCTION IN SINGLE-FIBER BIDIRECTIONAL SYSTEM FOR 100 GB/S SHORT DISTANCE APPLICATIONS 308

Xu Zhou ; Ning Deng

200GBIT/S NYQUIST 16-QAM HALF-CYCLE SUBCARRIER MODULATION TRANSMISSION WITH DUAL-POLARIZATION DIRECT DETECTION 311

Kaiheng Zou ; Yixiao Zhu ; Fan Zhang ; Zhangyuan Chen

CONNECTIVITY TECHNIQUES OF MCF, FOR DEPLOYMENT TO PRACTICAL USE 314

Tsunetoshi Saito ; Kengo Watanabe

INVESTIGATION OF CONNECTOR DAMAGE CAUSED BY HIGH POWER TRANSMISSION IN OPTICAL CONNECTOR WITH DUST 317

Chisato Fukai ; Kotaro Saito ; Ryo Koyama ; Mitsuru Kihara ; Toshio Kurashima

INPUT/OUTPUT CHANNEL COUPLING DEVICES FOR SDM AND MDM 320

Yasuo Kokubun

SPINNING EFFECT IN FEW-MODE FIBER WITH DISTRIBUTED LONG PERIOD GRATINGS 323

Jian Fang ; Byoung Yoon Kim ; William Shieh

RECENT DEVELOPMENTS OF MULTICORE MULTIMODE FIBER AMPLIFIERS FOR SDM SYSTEMS 326

Kazi S. Abedin

4-LP MODE DISTRIBUTED RAMAN AMPLIFICATION TECHNIQUE WITH GRADED-INDEX MULTI-MODE FIBER TRANSMISSION LINE 329

Masaki Wada ; Taiji Sakamoto ; Takayoshi Mori ; Takashi Yamamoto ; Kazuhide Nakajima

MFD MEASUREMENT OF A SIX-MODE FIBER WITH LOW-COHERENCE DIGITAL HOLOGRAPHY 332

Yuta Wakayama ; Hidenori Taga ; Takehiro Tsuritani

FABRICATION OF HELICAL FIBER GRATING AND ITS APPLICATION TO FLAT-TOP BAND-REJECTION FILTER 335

Peng Wang ; Gen Inoue ; Ramanathan Subramanian ; Hongpu Li

TRI-SECTION SIDE-POLISHED POLARIZATION MAINTAINING FIBRE POLARIZER 338

Xinyue Wang ; Kaige Chen ; Zhongwei Tan ; Ziyu Wang

ULTRASONIC WELDING OF PLASTIC OPTICAL FIBERS ONTO COMPOSITE MATERIALS 341

Shumpei Shimada ; Hiroki Tanaka ; Kazuhiko Hasebe ; Neisei Hayashi ; Yutaka Ochi ; Takahiro Matsui ; Itaru Nishizaki ; Yukihiro Matsumoto ; Yosuke Tanaka ; Hitoshi Nakamura ; Yosuke Mizuno ; Kentaro Nakamura

ULTRA-LOW LOSS, ULTRA-LARGE AEFF OPTICAL FIBERS FOR UNDERSEA NETWORKS 344

Sergey Ten

HIGH-SPEED 1.55-µM VCSELS FOR DATACOM AND TELECOM APPLICATIONS 347

Silvia Spiga ; Alexander Andrejew ; Gerhard Boehm ; Markus-Christian Amann

EQUALIZER-FREE 2-KM SMF TRANSMISSION OF 106-GBIT/S 4-PAM SIGNAL USING OPTICAL TRANSMITTER/RECEIVER WITH 50 GHZ BANDWIDTH 350

Shigeru Kanazawa ; Satoshi Tsunashima ; Yasuhiko Nakanishi ; Yoshifumi Muramoto ; Hiroshi Yamazaki ; Yuta Ueda ; Wataru Kobayashi ; Hiroyuki Ishii ; Hiroaki Sanjoh

LOW THERMAL RESISTANCE VCSEL ARRAY ADOPTED TUNNEL JUNCTION DESTRUCTION USING PROTON IMPLANTATION 353

Shohei Oshida ; Masashi Suhara ; Tomoyuki Miyamoto

INFLUENCE OF WAVELENGTH DEVIATION ON PHASE LOCKED VCSEL ARRAY USING TALBOT EFFECT 356

Yuki Komori ; Tomoyuki Miyamoto

DEMONSTRATION OF AN ULTRA-LOW THRESHOLD PHONON LASER WITH COUPLED MICROTOROID CAVITIES IN VACUUM 359

Mingming Zhao ; Guanzhong Wang ; Zhiqiang Jin ; Yingchun Qin ; Zhang-qi Yin ; Xiaoshun Jiang ; Min Xiao

FLEX RATE TRANSMISSION AND CHALLENGES FOR LONG HAUL TRANSMISSION 362

Fred Buchali ; Wilfried Idler

95.2TB/S TRANSOCEANIC SEVEN-CORE FIBER TRANSMISSION USING OPTICAL PRE-FILTERED 200GBIT/S-BASED PDM-QPSK WDM SIGNALS 365

Yu Kawaguchi ; Koki Takeshima ; Daiki Soma ; Takehiro Tsuritani

96GBAUD NYQUIST-PDM-QPSK SIGNAL TRANSMISSION OVER 12,120KM USING DP-AM-DAC AND DECISION-FEEDBACK EQUALIZER 368

Kengo Horikoshi ; Fukutaro Hamaoka ; Asuka Matsushita ; Munehiko Nagatani ; Hiroshi Yamazaki ; Akihide Sano ; Toshikazu Hashimoto ; Hideyuki Nosaka ; Kazushige Yonenaga ; Akira Hirano ; Yutaka Miyamoto

ENABLING TECHNOLOGIES FOR HIGH SPECTRAL EFFICIENCY AND HIGH CAPACITY TRANSMISSION OVER TRANSOCEANIC DISTANCES 371

Matt Mazurczyk

HIGH-SPEED AVALANCHE PHOTODIODE FOR DATA CENTER NETWORKS 374

Masahiro Nada ; Takuya Hoshi ; Hideaki Matsuzaki

LOW VOLTAGE HIGH SPEED SI-GE AVALANCHE PHOTODIODES 377

Zhihong Huang ; Cheng Li ; Di Liang ; Kunzhi Yu ; Marco Fiorentino ; Raymond G. Beausoleil

11-GBPS 16-QAM OFDM RADIO OVER FIBER DEMONSTRATION USING 100 GHZ HIGH-EFFICIENCY PHOTORECEIVER BASED ON PHOTONIC POWER SUPPLY 380
T. Umezawa ; K. Kashima ; A. Kanno ; K. Akahane ; A. Matsumoto ; N. Yamamoto ; T. Kawanishi

DESIGN FOR HIGH SPEED OPERATION OF DOUBLE MICRORING RESONATOR-LOADED MACH-ZEHNDER 2×2 QUANTUM WELL OPTICAL SWITCH 383
Naoki Kawaguchi ; Kento Hori ; Taro Arakawa ; Yasuo Kokubun

MODE-DIVISION MULTIPLEXING LINBO₃ MODULATOR USING DIRECTIONAL COUPLER 386
Yutaro Kodama ; Yuya Yamaguchi ; Atsushi Kanno ; Tetsuya Kawanishi ; Masayuki Izutsu ; Hirochika Nakajima

MULTI-LEVEL HIGH SPEED BURST-MODE RECEIVERS 389
Xin Yin ; J. Van Kerrebrouck ; G. Coudyzer ; Johan Bauwelinck

MULTI-RATE COHERENT BURST-MODE PDM-QPSK OPTICAL RECEIVER FOR FLEXIBLE OPTICAL NETWORKS 392
José Manuel Delgado Mendinueta ; Hideaki Furukawa ; Satoshi Shinada ; Naoya Wada

AN INP MONOLITHICALLY INTEGRATED MULTIWAVELENGTH TRANSMITTER WITH DIRECT MODULATION 395
N. Andriolli ; P. Velha ; P. Tommasino ; M. Chiesa ; G. B. Preve ; A. Trifiletti ; M. Romagnoli ; G. Contestabile

SELF-CLOCKING SYNCHRONIZED OPTICAL DEMULTIPLEXING USING FOUR-WAVE MIXING IN A QUANTUM-DOT SOA 398
Liang Yang ; Tomoya Yatsu ; Motoharu Matsuura

DEMONSTRATION OF BI- AND MULTI-STABILITY IN A HIGH ORDER RING RESONATOR 401
Li Jin ; Alessia Pasquazi ; Luigi Di Lauro ; Marco Peccianti ; Edwin Y. B. Pun ; David J. Moss ; Roberto Morandotti ; Brent E. Little ; Sai T. Chu

OPTICAL PACKET AND CIRCUIT INTEGRATED RING NETWORK 404
Hideaki Furukawa

4-CHANNEL SILICON PHOTONIC MODE UNSCRAMBLER 407
Andrea Melloni ; Andrea Annoni ; Emanuele Guglielmi ; Marco Carminati ; Giorgio Ferrari ; Nicola Peserico ; Stefano Grillanda ; Marc Sorel ; Francesco Morichetti

MODE DIVISION MULTIPLEXING SWITCH FOR ON-CHIP OPTICAL INTERCONNECTS 410
Xinru Wu ; Ke Xu ; Daoxin Dai ; Hon Ki Tsang

PLASMONIC-ORGANIC HYBRID (POH) MODULATORS 413
A. Melikyan ; K. Koehnle ; M. Lauermann ; R. Palmer ; S. Koeber ; S. Muehlbrandt ; P. C. Schindler ; D. L. Elder ; S. Wolf ; M. Sommer ; L. R. Dalton ; D. Van Thourhout ; W. Freude ; M. Kohl ; J. Leuthold ; C. Koos

IN-PLANE SATURABLE ABSORPTION OF GRAPHENE ON A SILICON SLOT WAVEGUIDE 416
Jiaqi Wang ; Zhenzhou Cheng ; Hon Ki Tsang ; Chester Shu

GRAPHENE-INDUCED ON-DEMAND NANOCAVITY BASED ON SI PHOTONIC CRYSTAL 419
Hisashi Chiba ; Masaya Notomi

SILICON SLOT WAVEGUIDE RING RESONATOR: FROM SINGLE RESONANCE TO ENVELOPE INDEX SENSING 422
Weiwei Zhang ; Samuel Serna ; Xavier Le Roux ; Laurent Vivien ; Eric Cassan

SIMULTANEOUS MEASUREMENT OF STRAIN AND TEMPERATURE USING PI-PHASE-SHIFTED FIBER BRAGG GRATING ON POLARIZATION MAINTAINING FIBER 425
Jiageng Chen ; Qingwen Liu ; Xinyu Fan ; Zuyuan He

SPECTRAL EFFICIENT GROUPED ROUTING NETWORK THAT APPLIES DYNAMIC OPTICAL PATH GROOMING 428
Yuki Terada ; Yojiro Mori ; Hiroshi Hasegawa ; Ken-ichi Sato

FIELD TRIAL OF VIRTUAL TRANSPORT NETWORK SERVICES BASED ON HIERARCHICAL CONTROL OVER MULTI-DOMAIN OTN NETWORKS 431
Ruiquan Jing ; Chengliang Zhang ; Junjie Li ; Yiran Ma ; Qian Hu ; Xiaoli Huo ; Yongli Zhao ; Jiayu Wang ; Baoquan Rao ; Chen Qiu

A STATIC TRAFFIC GROOMING ALGORITHM FOR ELASTIC OPTICAL NETWORKS WITH ADAPTIVE MODULATION 434
Takafumi Tanaka ; Tetsuro Inui ; Wataru Imajuku

HARDWARE SCALE ANALYSIS AND PROTOTYPE DEVELOPMENT OF FLEXIBLE WAVEBAND ROUTING OXCS 437
Tomohiro Ishikawa ; Masaki Niwa ; Koh Ueda ; Yojiro Mori ; Hiroshi Hasegawa ; Suresh Subramaniam ; Ken-ichi Sato ; Osamu Moriwaki

SCALABLE AND DYNAMIC OPTICAL NETWORK ARCHITECTURE 440
Vincent W. S. Chan

3.5-GBIT/S QPSK SIGNAL TRANSMISSION IN BEAM FORMING OF 60-GHZ INTEGRATED PHOTONIC ARRAY-ANTENNA 443
Kotoko Furuya ; Takayoshi Hirasawa ; Masayuki Oishi ; Shigeyuki Akiba ; Jiro Hirokawa ; Makoto Ando

REAL-TIME TRANSMISSION OF 5-GBIT/S PI/4-SHIFT DQPSK SIGNAL OVER MILLIMETER-WAVE RADIO-OVER-FIBER LINKS 446
Abdelmoula Bekkali ; Takashi Kobayashi ; Kosuke Nishimura ; Nobuhiko Shibagaki ; Kenichi Kashima ; Yosuke Sato

MOBILE FRONTHAUL OPTICAL LINK FOR LTE-A SYSTEM USING DIRECTLY-MODULATED 1.5-μM VCSEL 449
Byung Gon Kim ; S. H. Bae ; Hoon Kim ; Y. C. Chung

POWER-OVER-FIBER TRANSMISSION USING 1.3-μM DUAL-CHANNEL RADIO-OVER-FIBER SIGNALS IN A DOUBLE-CLAD FIBER 452
Akira Yoneyama ; Yamato Minamoto ; Motoharu Matsuura

INTERNET OF THINGS WITH OPTICAL CONNECTIVITY, NETWORKING, AND BEYOND 455
Philip N. Ji ; Ting Wang

STANDARDIZATION OF OPTICAL ACCESS TECHNOLOGIES: PROGRESS AND FUTURE PROSPECTS 458
Jun-ichi Kani

TACTILE INTERNET CAPABLE PASSIVE OPTICAL LAN FOR HEALTHCARE 459
Elaine Wong ; Maluge Pubuduni Imali Dias ; Lihua Ruan

**ANALYSIS OF POWER CONSUMPTION IN MOBILE BACKHAUL NETWORK WITH DENSELY
DEPLOYED SMALL CELLS UNDER DYNAMIC TRAFFIC BEHAVIOR** 462
Kyosuke Sone ; Inwoong Kim ; Xi Wang ; Yasuhiko Aoki ; Hiroyuki Seki ; Jens C. Rasmussen

**OPTIMAL DESIGN AND BACKHAULING OF SMALL-CELL NETWORK: IMPLICATION OF ENERGY
COST** 465
Chathurika Ranaweera ; Elaine Wong ; Christina Lim ; Chamil Jayasundara ; Ampalavanapillai Nirmalathas

**A MOBILE FRONTHAUL SYSTEM ARCHITECTURE FOR DYNAMIC PROVISIONING AND
PROTECTION** 468
Qianmei Yang ; Ning Deng ; Xu Zhou ; Chun-Kit Chan

**EFFECT OF CHROMATIC DISPERSION ON CABLE RE-ROUTING OPERATION SUPPORT SYSTEM
WITH NO SERVICE INTERRUPTION** 471
Kazutaka Noto ; Masaaki Inoue ; Hiroshi Watanabe ; Yusuke Koshikiya ; Keiji Okamoto ; Tetsuya Manabe

MULTIDIMENSIONAL MODULATION FORMATS FOR OPTICAL COMMUNICATION 474
Magnus Karlsson

**OPTIMIZATION OF NETWORKS CARRYING SUPER-CHANNELS WITH DIFFERENT MODULATION
FORMATS FOR MAXIMUM SPECTRAL EFFICIENCY AND REACH** 477
*Olga Vassilieva ; Tomohiro Yamauchi ; Shoichiro Oda ; Inwoong Kim ; Motoyoshi Sekiya ; Takeshi Hoshida ; Yasuhiko Aoki ;
Jens C. Rasmussen ; Tadashi Ikeuchi*

**OPTIMUM CAPACITY UTILIZATION IN SPACE-DIVISION MULTIPLEXED TRANSMISSION SYSTEMS
WITH FEW-MODE FIBERS** 480
Georg Rademacher ; Klaus Petermann

COMPARISON OF FOUR DIMENSIONALLY CODED VERSUS HYBRID MODULATION 483
Han Sun

**JOINT LINEAR AND NON-LINEAR ADAPTIVE PRE-DISTORTION OF HIGH BAUD RATE
TRANSMITTERS FOR HIGH-ORDER MODULATION FORMATS** 486
Bernhard Spinnler ; Ginni Khanna ; Stefano Calabrò ; Erik De Man ; Uwe Feiste ; Thomas Bex ; Heinrich von Kirchbauer

**FIELD DEMONSTRATION OF MODULATION FORMAT ADAPTATION BASED ON PILOT-AIDED OSNR
ESTIMATION USING 400GBPS/CH REAL-TIME DSP** 489
*Kazushige Yonenaga ; Kengo Horikoshi ; Seiji Okamoto ; Mitsuteru Yoshida ; Yutaka Miyamoto ; Masahito Tomizawa ; Takeshi
Okamoto ; Hidemi Noguchi ; Jun-ichi Abe ; Junichiro Matsui ; Hisao Nakashima ; Yuichi Akiyama ; Takeshi Hoshida ; Hiroshi
Onaka ; Kenya Sugihara ; Soichiro Kametani ; Kazuo Kubo ; Takashi Sugihara*

QPSK ASSISTED CARRIER PHASE RECOVERY FOR HIGH ORDER QAM 492
Xiaofei Su ; Liang Dou ; Zhenning Tao ; Takeshi Hoshida ; Jens C. Rasmussen

**BLOCK CARRIER-PHASE RECOVERY WITH RECURSIVE NOISE ADAPTIVE KALMAN FILTERING
FOR 16-QAM SIGNALS** 495
Zheng Bofang ; Chester Shu

HIGHLY ASE TOLERANT PASS-BAND SHAPE MONITOR FOR CASCADED ROADMS 498
*Guoxiu Huang ; Shoichiro Oda ; Tomohiro Yamauchi ; Setsuo Yoshida ; Goji Nakagawa ; Yasuhiko Aoki ; Zhenning Tao ; Jens C.
Rasmussen*

DIGITAL NONLINEAR DISTORTION COMPENSATION 501
Zhenning Tao ; Liang Dou ; Ying Zhao ; Bo Liu ; Lei Li ; Tomofumi Oyama ; Takeshi Hoshida ; Jens C. Rasmussen

**IMPACT OF LINK SYMMETRY ON NONLINEAR NOISE MITIGATION USING SPECTRAL INVERSION
IN SUPERCHANNEL TRANSMISSION** 504
Inwoong Kim ; Olga Vassilieva ; Paparao Palacharla ; Motoyoshi Sekiya ; Tadashi Ikeuchi

ACHIEVABLE RATES COMPARISON FOR PHASE-CONJUGATED TWIN-WAVES AND PM-QPSK 507
Tobias A. Eriksson ; Abel Lorences-Riesgo ; Pontus Johannisson ; Tobias Fehenberger ; Peter A. Andrekson ; Magnus Karlsson

**KNN-BASED DETECTOR FOR COHERENT OPTICAL SYSTEMS IN PRESENCE OF NONLINEAR PHASE
NOISE** 510
Danshi Wang ; Min Zhang ; Meixia Fu ; Zhongle Cai ; Ze Li ; Yue Cui ; Bin Luo

**OSNR MONITORING BY DEEP NEURAL NETWORKS TRAINED WITH ASYNCHRONOUSLY SAMPLED
DATA** 513
Takahito Tanimura ; Takeshi Hoshida ; Jens C. Rasmussen ; Makoto Suzuki ; Hiroyuki Morikawa

TRELLIS CODED OPTICAL MODULATION USING QAM CONSTELLATIONS 516
*Emmanuel Le Taillandier de Gabory ; Tatsuya Nakamura ; Hidemi Noguchi ; Wakako Maeda ; Sadao Fujita ; Jun'ichi Abe ;
Kiyoshi Fukuchi*

**DIFFERENTIAL CODING IN POLARIZATIONS WITH HALF-SYMBOL-PERIOD TIMING OFFSET FOR
COHERENT PM-QPSK TO IMPROVE PHASE-NOISE TOLERANCE** 519
Guo-Wei Lu ; Takahide Sakamoto ; Yukiyoshi Kamio

CROSSOVER BLOCK MODULATION WITH COMPLEMENTARY CODES SUPERPOSITION 522
*Tsuyoshi Yoshida ; Keisuke Kojima ; Toshiaki Koike-Akino ; David Millar ; Keisuke Dohi ; Keisuke Matsuda ; Kieran Parsons ;
Kazuo Kubo ; Kenichi Uto ; Takashi Sugihara*

TURBO EQUALIZATION FOR DUOBINARY-SHAPED SIGNALS IN SUPER-NYQUIST WDM SYSTEMS 525
Shuai Yuan ; Koji Igarashi

ENHANCED BLIND MODULATION FORMATS RECOGNITION USING CONNECTED COMPONENT ANALYSIS WITH QUADRUPLE ROTATION 528
Tianwai Bo ; Jin Tang ; Calvin Chun-Kit Chan

BROADBAND AMPLIFICATION CHARACTERISTICS ON THE CASCADE BI-DOPED AND ER-DOPED OPTICAL FIBER AMPLIFIERS 531
Mikoto Takahashi ; Daiki Higuchi ; Mizuki Ohara ; Yusuke Fujii ; Naoto Yoshimoto ; Soichi Kobayashi

ALL-OPTICAL DYNAMIC GAIN CONTROL OF REMOTELY PUMPED ERBIUM-DOPED FIBER AMPLIFIER 534
Kokoro Kitamura ; Kenta Udagawa ; Hiroji Masuda

PHASE ADJUSTMENT BY WAVELENGTH TUNING OF PARAMETRIC PUMP ON RAMAN ASSISTED PHASE SENSITIVE AMPLIFIER 537
Y. Cao ; F. Alishahi ; Y. Akasaka ; M. Ziyadi ; A. Mohajerin-Ariaei ; A. Almaiman ; T. Ikeuchi ; S. Takasaka ; R. Sugizaki ; A. E. Willner

POLARIZATION PULLING IN FIBER OPTICAL PARAMETRIC AMPLIFIERS 540
S. H. Wang ; P. K. A. Wai

NONLINEAR IMPAIRMENTS MITIGATION AIDED BY LOW-NOISE OPTICAL FREQUENCY COMBS 543
Ping Piu Kuo

1.6 ~ 1.8 µM BAND HYBRID BROADBAND LIGHT SOURCE CONSISTING OF CASCADED SLD AND TDFA 544
Kazuya Ota ; Du Xiaoen ; Sho Tujita ; Kousuke Senda ; Fumiki Hanafuji ; Jun Ono ; Kazuaki Mise ; Yoshiharu Shimose ; Hiroshi Mori ; Osanori Koyama ; Hirotaka Ono ; Tatsuro Endo ; Makoto Yamada

1.8 µM HIGH-ORDER MICRORING RESONATOR MODE-LOCKED LASER USING A CARBON NANOTUBE 547
K. S. Tsang ; Jie Wang ; Li Jin ; Victor Ho ; Jack Cheung ; Yanny Tsang ; Alessia Pasquazi ; Ray Man ; Sai T. Chu ; A. Ping Zhang ; Hwa-yaw Tam ; P. K. A. Wai

WIDELY TUNABLE, SINGLE-LONGITUDINAL MODE BRILLOUIN/ERBIUM-DOPED FIBER LASER 550
Huan Wu ; Bofang Zheng ; Chester Shu

CONTINUOUSLY TUNABLE MICROWAVE PHOTONIC FILTER BASED ON A WAVELENGTH-SPACING-TUNABLE MULTIWAVELENGTH LASER 553
Seungmin Lee ; Young Bo Shim ; Young-Geun Han

REAL-TIME INTERROGATION TECHNIQUE USING FOURIER DOMAIN MODE-LOCKED FIBER LASER 556
Ik Su Jo ; Seungmin Lee ; Sanggwon Song ; Kwang Wook Yoo ; Young-Geun Han

QUANTUM DOT LASERS FOR SILICON PHOTONICS 559
Yasuhiko Arakawa

LOW TEMPERATURE LASING CHARACTERISTICS OF GAINASP DOUBLE-HETERO LASER INTEGRATED ON INP/SI SUBSTRATE USING DIRECT WAFER BONDING 561
Tetsuo Nishiyama ; Keiichi Matsumoto ; Junya Kishikawa ; Toshiki Sukigara ; Yuya Onuki ; Naoki Kamada ; Tomokazu Kanke ; Kazuhiko Shimomura

FIRST DEMONSTRATION OF MODE SELECTIVE LIGHT SOURCE BY USING ACTIVE MULTIMODE INTERFEROMETER LASER DIODE 564
Bingzhou Hong ; Takuya Kitano ; Akio Tajima ; Haisong Jiang ; Kiichi Hamamoto

QUANTUM-CASCADE LASERS: LINE-NARROWING AND SUPPRESSION OF FLICKER-NOISE 567
Masamichi Yamanishi ; Toru Hirohata ; Saverio Bartalini ; Paolo De Natale

IMPROVED EQUALIZING CHARACTERISTICS IN PRE-EQUALIZING ELECTRO-OPTIC MODULATOR WITH POLARIZATION-REVERSED STRUCTURES 570
Tomohiro Ohno ; Hiroshi Murata ; Yasuyuki Okamura

BROADBAND DUAL-POLARIZATION DUAL-PARALLEL MACH ZEHNDER MODULATOR BASED PHOTONIC MICROWAVE PHASE SHIFTER 573
Tong Niu ; Erwin H. W. Chan ; Xudong Wang ; Xinhuan Feng ; Bai-Ou Guan

PHOTONIC RADIO FREQUENCY DOWN-CONVERTER BASED ON PARALLEL ELECTRO-ABSORPTION MODULATORS IN KU/KU BAND FOR SPACE APPLICATIONS 576
Jordane Thouras ; Benoit Benazet ; Herve Leblond ; Christelle Aupetit-Berthelemot

ELECTRO-OPTIC MODULATOR USING MILLIMETER-WAVE GAP-EMBEDDED PATCH ANTENNA WITH STACKED STRUCTURE 579
Hironori Aya ; Yusuf Nur Wijayanto ; Atsushi Kanno ; Tetsuya Kawanishi ; Hiroshi Murata ; Yasuyuki Okamura

ROBUSTNESS FOR TRACKING ERROR IN FMF BASED FSO RECEIVER 582
T. Ishikawa ; K. Hosokawa ; S. Takahashi ; M. Arikawa ; Y. Ono ; T. Ito ; K. Fukuchi

IMPEDANCE MATCHED GAN LD PACKAGE FOR DIRECT OFDM COMMUNICATION AT 14 GBPS 585
Yu-Fang Huang ; Tsai-Chen Wu ; Yu-Chieh Chi ; Cheng-Ting Tsai ; Wei Wang ; Tien-Tsorng Shih ; Hao-Chung Kuo ; Gong-Ru Lin

OCT PRECODING FOR OFDM-BASED INDOOR VISIBLE LIGHT COMMUNICATIONS 588
Yang Hong ; Lian-Kuan Chen

NONLINEAR COMPENSATION OF 850 NM VCSEL WITH DISCRETE MULTI-TONE MODULATION EMPLOYING A VOLTERRA-WIENER FILTER 591
Ta-Ching Tzu ; Chia-Chien Wei ; Jun-Jie Liu ; Chun-Yen Chang ; Kai-Lun Chi ; Xin-Nan Chen ; Jin-Wei Shi ; Jyehong Chen

PRE-LEVELED 16-QAM OFDM MODULATION OF AN 850-NM VCSEL FOR 56-GBIT/S TRANSMISSION 594
Cheng-Ting Tsai ; Chun-Yen Pong ; Yun-Chen Wu ; Shan-Fong Leong ; Yu-Chieh Chi ; Chao-Hsin Wu ; Tien-Tsorng Shih ; Jian Jang Huang ; Hao-Chung Kuo ; Wood-Hi Cheng ; Gong-Ru Lin

MULTIMODE FIBERS FOR TELECOMMUNICATIONS AND IMAGING 597
Joel Carpenter ; Benjamin J. Eggleton ; Jochen Schröder

NEXT-GENERATION MULTIMODE FIBERS 600
Pierre Sillard ; Marianne Bigot-Astruc ; Denis Molin ; Adrian Amezcua-Correa

STATISTICAL TREATMENT OF IEEE SPREADSHEET MODEL FOR VCSEL-MULTIMODE FIBER TRANSMISSIONS 603
Xin Chen ; John Abbott ; Dale Powers ; Doug Coleman ; Ming-Jun Li

SINGLE-CARRIER 1-TB/S SIGNAL GENERATION USING HIGH-SPEED INP MUX-DACS AND AN INTEGRATED CSRZ-OTDM MODULATOR 606
Hiroshi Yamazaki ; Akihide Sano ; Munehiko Nagatani ; Yutaka Miyamoto

COST EFFECTIVE FRACTIONAL OFDM RECEIVER USING ARRAYED WAVEGUIDE GRATING 609
T. Nagashima ; G. Cincotti ; T. Murakawa ; S. Shimizu ; M. Hasegawa ; K. Hattori ; M. Okuno ; S. Mino ; A. Himeno ; N. Wada ; H. Uenohara ; T. Konishi

SUB-FF-CAPACITANCE PHOTONIC-CRYSTAL PHOTODETECTOR TOWARDS FJ/BIT ON-CHIP RECEIVER 612
Kengo Nozaki ; Shinji Matsuo ; Takuro Fujii ; Koji Takeda ; Masaaki Ono ; Abdul Shakoor ; Eiichi Kuramochi ; Masaya Notomi

4-CHANNEL SYNCHRONOUS THZ-WAVE GENERATORCOMPOSED OF ARRAYED UTC-PDS AND ANTENNAS 615
Goki Sakano ; Jun Haruki ; Kazuki Sakuma ; Kazutoshi Kato

DEMONSTRATION OF MICRO-PROJECTION ENABLED SHORT-RANGE COMMUNICATIONS FOR 5G 618
Hsi-Hsir Chou ; C. -Y. Tsai

THERMAL EMISSION CONTROL BY PHOTONIC CRYSTALS 621
Susumu Noda

MID IR APPLICATIONS OF SI PHOTONICS 622
G. Z. Mashanovich ; J. Soler Penades ; V. Mittal ; G. S. Murugan ; A. Z. Khokhar ; C. J. Littlejohns ; S. Stankovic ; A. Ortega-Monux ; G. Wanguemert-Perez ; R. Halir ; I. Molina-Fernandez ; C. Alonso-Ramos ; D. Benedikovic ; A. Villafranca ; P. Cheben ; J. J. Ackert ; A. P. Knights ; J. S. Wilkinson ; M. Nedeljkovic

SILICON PHOTONIC INTEGRATED CIRCUITS FOR HIGH-CAPACITY OPTICAL COMMUNICATIONS 624
Po Dong

LENS-INTEGRATED FAN-IN/FAN-OUT DEVICE FOR MULTI-CORE FIBER 627
Osamu Shimakawa ; Hajime Arao ; Tomomi Sano

FOUR-MODE-SELECTIVE PHOTONIC LANTERN BASED ON TWO-LAYER POLYMER WAVEGUIDE BRANCHES 630
Yunfei Wu ; Kin Seng Chiang

EXCITATION OF LP_{21B} AND LP_{02} MODES WITH PLC-BASED TAPERED WAVEGUIDE FOR MODE-DIVISION MULTIPLEXING 633
Yoko Yamashita ; Yuhei Ishizaka ; Nobutomo Hanzawa ; Takeshi Fujisawa ; Taiji Sakamoto ; Takashi Matsui ; Kyozo Tsujikawa ; Fumihiko Yamamoto ; Kazuhide Nakajima ; Kunimasa Saitoh

WIDELY SWITCHING THE ORBITAL ANGULAR MOMENTUM MODES WITH INTEGRATED COBWEB EMITTER 636
Yu Wang ; Peng Zhao ; Xue Feng ; Yidong Huang

COMPLIANT POLYMER INTERFACE FOR FIBER CONNECTION OF NANOPHOTONIC WAVEGUIDES 639
Hidetoshi Numata

HIGHLY EFFICIENT CIRCULAR HOLES ADDED A-SI:H GRATING COUPLER WITH METAL MIRROR FOR 3D OPTICAL INTERCONNECTS 642
JoonHyun Kang ; Yuki Kuno ; Kazuto Ito ; Yusuke Hayashi ; Junichi Suzuki ; Il-Ki Han ; Tomohiro Amemiya ; Nobuhiko Nishiyama ; Shigehisa Arai

300-MM ARF-IMMERSION LITHOGRAPHYTECHNOLOGY BASED SI-WIRE GRATING COUPLERS WITH HIGH COUPLING EFFICIENCY AND LOW CROSSTALK 645
Yohei Sobu ; Seok-Hwan Jeong ; Yu Tanaka ; Ken Morito

GRATING-BASED MODE CONVERTERS FABRICATED BY ONE-STEP PHOTOLITHOGRAPHY 648
Wei Jin ; Kin Seng Chiang

SIGNIFICANT LOSS REDUCTION OF 3.0 DB/CM ON CORE-TOP ETCHED WAVEGUIDE FOR VERTICAL MULTI-MODE INTERFERENCE 651
Ryosuke Sakata ; Kazuhiro Tanabe ; Haisong Jiang ; Kiichi Hamamoto

OPTICAL SYSTEM ARCHITECTURES AND CONTROL FOR DATA CENTER NETWORKS 654
Dimitra Simeonidou

EXPLICIT WAVELENGTH RESOURCE REALLOCATION TO ENSURE CRITICAL COMMUNICATION LINES IN EMERGENCY SITUATIONS 655
Masaki Shiraiwa ; Takaya Miyazawa ; Hideaki Furukawa ; Yoshinari Awaji ; Naoya Wada

REMOTE EXPERIMENTS WITH TEST-PLANE ON A DISTRIBUTED OPTICAL SWITCHED NETWORK TESTBED 658
Sugang Xu ; Masaki Shiraiwa ; Hiroaki Harai ; Yoshinari Awaji ; Naoya Wada

FLEXIBLE-BANDWIDTH OPTICAL INTERCONNECTS FOR DATACOM NETWORKS 661
Roberto Proietti ; Paolo Grani ; Zheng Cao ; S. J. Ben Yoo

LOAD BALANCING IN SWITCH-FABRIC TYPE OF TORUS OPS DATA CENTER NETWORKS WITH HYBRID OPTOELECTRONIC ROUTERS 664
Yue-Cai Huang ; Yuki Yoshida ; Salah Ibrahim ; Ryo Takahashi ; Ken-ichi Kitayama

SPECTRUM ASSIGNMENT AND UPDATE METHOD BASED ON ADAPTIVE SOFT RESERVATION IN ELASTIC OPTICAL NETWORKS 667
Hideki Tode ; Akira Fukushima ; Yosuke Tanigawa ; Yusuke Hirota

A STUDY ON SPECTRUM ASSIGNMENT AND RESOURCE SHARING METHOD IN ELASTIC OPTICAL PACKET AND CIRCUIT INTEGRATED NETWORKS 670
Ken Nagatomi ; Seitaro Sugihara ; Shohei Fujii ; Yusuke Hirota ; Hideki Tode ; Takashi Watanabe

OPTIMIZED MULTICAST SCHEDULING IN DATACENTER OPTICAL BURST RING NETWORK 673
Gang Chen ; Dongxu Zhang ; Hongxiang Guo ; Jian Wu ; Xiaoyuan Cao ; Noboru Yoshikane ; Takehiro Tsuritani ; Itsuro Morita

TUNABLE OPTICAL TECHNOLOGIES FOR NEXT GENERATION OPTICAL ACCESS SYSTEM 676
Kota Asaka

SYMMETRIC 4×25-GBIT/S TWDM-PON TRANSMISSION BY USING SPECTRUM RESHAPING 679
Yuan-Yuan Sung ; Yu-Chang Liu ; Puspa Devi Pukhrambam ; Hung-Wen Hung ; San-Liang Lee

AUTO-CONFIGURATION OPTICAL AMPLIFIER FOR FLEXIBLE ACCESS NETWORK DESIGN 682
Takuya Tsutsumi ; Yu Nakayama ; Shunsuke Kanai ; Manabu Kubota ; Akihiro Otaka

INTEGRATED SOI-BASED RECEIVER MODULE FOR TWDM-PON 685
Y. Hsu ; X. R. Wu ; L. Y. Wei ; K. Xu ; C. W. Hsu ; J. Y. Sung ; C. H. Yeh ; H. K. Tsang ; C. W. Chow

1GBPS FULL-DUPLEX 5GHZ FREQUENCY SLOTS UDWDM FLEXIBLE METRO/ACCESS NETWORKS BASED ON VCSEL-RSOA TRANSCEIVER 688
Jose A. Altabas ; David Izquierdo ; Jose A. Lazaro ; Ignacio Garces

SURVIVABLE MULTIPATH PROVISIONING WITH CONTENT CONNECTIVITY IN ELASTIC OPTICAL DATACENTER NETWORKS 691
Tao Gao ; Shanguo Huang ; Bingli Guo ; Xin Li ; Qian Kong ; Yu Zhou ; Wenzhe Li ; Wanyi Gu

DISTRIBUTED 4K-VIDEO CAMERA MONITORING SERVICE ON ETHERNET PON SYSTEM USING HYBRID BANDWIDTH ALLOCATION FOR SECURE COMMUNITY 694
Akihisa Shoji ; Naoto Yoshimoto

DEMONSTRATION OF CONTROLLING CHROMATIC DISPERSION AT SOA WITH MACH-ZEHNDER INTERFEROMETRIC MEASUREMENT SYSTEM 697
Yusuke Yamanaka ; Yuki Fujimura ; Kazutoshi Kato

HYBRID METAL INSULATOR METAL PLASMONIC WAVEGUIDE AND RING RESONATOR 700
Prateeksha Sharma ; Dinesh Kumar V.

TUNABLE SINGLE BANDPASS MICROWAVE PHOTONIC FILTER BASED ON PHASE COMPENSATED SILICON-ON-INSULATOR MICRORING RESONATOR 703
Wenjian Yang ; Xiaoke Yi ; Shijie Song ; Suen Xin Chew ; Liwei Li ; Linh Nguyen

FULL-BAND DIRECT-CONVERSION RECEIVER USING MICROWAVE PHOTONIC I/Q MIXER 706
Yue Zheng ; Jianqiang Li ; Qiang Lv ; Jianwei Zhou ; Yuting Fan ; Feifei Yin ; Yitang Dai ; Kun Xu

OUTPUT TIMING ADJUSTMENT MECHANISM OF OPTICAL AND ELECTRONIC COMBINED BUFFER FOR OPTICAL PACKET SWITCHING 709
Takahiro Hirayama ; Hiroaki Harai

DESIGN OF A 1×2 WAVELENGTH SELECTIVE SWITCH USING AN ARRAYED-WAVEGUIDE GRATING WITH FOLD-BACK PATHS ON A SILICON PLATFORM 712
Hideaki Asakura ; Koki Sugiyama ; Hiroyuki Tsuda

SUPPRESSION OF PHASE NOISE INDUCED BY OPTICAL INTERFERENCE IN OPTOELECTRONIC OSCILLATORS 715
Huanfa Peng ; Xiaofeng Peng ; Yongchi Xu ; Cheng Zhang ; Lixin Zhu ; Weiwei Hu ; Zhangyuan Chen

SDM NETWORKING SCHEMES COMPARED FOR COMPUTATION COMPLEXITY AND EFFICIENCY 718
Miri Blau ; Dan M. Marom

TOPOLOGY RECONSTRUCTION STRATEGY WITH THE OPTICAL SWITCHING BASED SMALL WORLD DATA CENTER NETWORK 721
Tingting Yang ; Dongxu Zhang ; Hongxiang Guo ; Jian Wu

TOLERANCE TO LASER FREQUENCY DEVIATION OF NYQUIST-WDM AND TIME-FREQUENCY-PACKING MODULATION FORMATS 724
Francesco Fresi ; Gianluca Meloni ; Fabio Cavaliere ; Luca Potì

BRANCHING RATIO OF OPTICAL COUPLER FOR CABLE RE-ROUTING OPERATION SUPPORT SYSTEM WITH NO SERVICE INTERRUPTION 727
Hiroshi Watanabe ; Kazutaka Noto ; Yusuke Koshikiya ; Tetsuya Manabe

INVESTIGATION OF SWITCH REDUCTION OF A MULTI-DIMENSIONAL OPTICAL NODE FOR SPATIAL DIVISION MULTIPLEXING NETWORKS 730
Hiroki Sasago ; Hiroyuki Uenohara

BIT-ERROR-RATE PERFORMANCE IN OPTICAL 16QAM RECOGNITION BY MAXIMUM OUTPUT WITH OPTICAL WAVEGUIDE CIRCUIT 733
Kensuke Inoshita ; Hiroki Kishikawa ; Nobuo Goto

ANALYSIS OF A SI-NANOCRYSTAL STRIP-LOADED WAVEGUIDE FOR NONLINEAR APPLICATIONS 736
Pengjiang Wei ; S. H. Wang ; Brent E. Little ; Sai Tak Chu

INTEGRATION OF MICRO DATA CENTER WITH OPTICAL LINE TERMINAL IN PASSIVE OPTICAL NETWORK 739
Bingliang Yang ; Zitian Zhang ; Kuo Zhang ; Weisheng Hu

AMPLIFIED SPONTANEOUS EMISSION NOISE INFLUENCE ANALYSIS ON OPTICAL QUANTIZATION USING SOLITON SELF-FREQUENCY SHIFT 742
Y. Yamasaki ; T. Nagashima ; M. Hiraoka ; T. Konishi

LARGE DATA TRANSFERS IN WDM NETWORKS WITH NODE STORAGE 745
Da Feng ; Weiqiang Sun ; Weisheng Hu

20 GB/S OPTICAL SWITCHED DQPSK TRANSMISSION OVER 50 KM SSMF AND 7 KM DCF 748
Pacharapon Chulok ; Suvit Nakpeerayuth ; Duang-rudee Worasucheep ; Naoya Wada

VARIABLE GAIN PI-BASED CYCLIC SLEEP CONTROL WITH ANTI-WINDUP TECHNIQUE FOR QOS-AWARE AND ENERGY-EFFICIENT ETHERNET PONS 751
Takahiro Kikuchi ; Ryogo Kubo

EXPERIMENTAL DEMONSTRATION OF A COST EFFECTIVE BIDIRECTIONAL COHERENT OFDM-PON 754
Rongrong Chen ; Caixia Kuang ; Yingxiong Song ; Min Wang ; Rujian Lin ; Qianwu Zhang

SUB-CARRIER SHARING IN OFDM-PON FOR 5G MOBILE NETWORKS SUPPORTING RADIO-OVER-FIBRE 757
Sheng Xu ; Sugang Xu ; Yoshiaki Tanaka

CHANNEL-REUSE IMDD-BASED 40 GB/S/λ 16-QAM NYQUIST-SCM DOWNSTREAM AND 20 GB/S/λ NYQUIST 4 PAM UPSTREAM WDM-PON 760
Zhiguo Zhang ; Bingbing Zhang ; Yanxu Chen ; Bingchang Hua ; Xue Chen

EFFICIENT MOBILE FRONTHAUL FOR SIMULTANEOUS TRANSMISSION OF 4G AND FUTURE MOBILE SIGNALS 763
Pham Tien Dat ; Atsushi Kanno ; Naokatsu Yamamoto ; Tetsuya Kawanishi

FREQUENCY-STABILIZED MULTI-TONE SIGNAL GENERATION BY OPTICAL FREQUENCY LOCKED-LOOP AND OPTICAL FREQUENCY COMB 766
Jinsei Oba ; Kosuke Kamikawa ; Atsushi Kanno ; Naokatsu Yamamoto ; Hideyuki Sotobayashi

THIRD-ORDER HARMONICS SUPPRESSION IN TWO-TONE SIGNAL GENERATION USING A DUAL-PARALLEL MACH-ZEHNDER MODULATOR 769
Kazunori Osato ; Moriya Nakamura

SELF-INTERFERENCE CANCELLATION FOR 2×2 MIMO IN-BAND FULL-DUPLEX RADIO-OVER-FIBER SYSTEMS 772
Yunhao Zhang ; Shilin Xiao ; Yinghong Yu ; Shaojie Zhang ; Lu Zhang ; Ling Liu ; Haiyun Xin

DYNAMIC VIRTUAL OPTICAL NETWORK MAPPING BASED ON SWITCHING CAPABILITY AND SPECTRUM FRAGMENTATION IN ELASTIC OPTICAL NETWORKS 775
Hongxiang Wang ; Xin Xin ; Jiawei Zhang ; Yongmei Sun ; Yuefeng Ji

FIBER-WIRELESS AND FIBER-IVLLC CONVERGENCES 778
Bo-Rui Chen ; Hung-Hsien Lin ; Chang-Jen Wu ; Chun-Yu Lin ; Chung-Yi Li ; Hai-Han Lu

VISIBLE LIGHT ENCRYPTION SYSTEM USING CAMERA IMAGE SENSOR 781
Chin-Wei Hsu ; Kevin Liang ; Hung-Yu Chen ; Liang-Yu Wei ; Chien-Hung Yeh ; Yang Liu ; Chi-Wai Chow

NON-ORTHOGONAL MULTIPLE ACCESS WITH MULTICARRIER PRECODING IN VISIBLE LIGHT COMMUNICATIONS 784
Xun Guan ; Yang Hong ; Calvin Chun-Kit Chan

6.6-GBIT/S PHASE MODULATED IMPULSE-RADIO GENERATION FROM RZ-OOK OPTICAL SIGNALS FOR OPTICAL-WIRELESS SYSTEMS 787
Yuri Ohara ; Hdeaki Kawahara ; Tomoki Kishida ; Kengo Nabika ; Saeko Oshiba

SIMULTANEOUS ALL-CHANNEL OTDM DEMULTIPLEXING BASED ON COMPLETE OPTICAL FOURIER TRANSFORMATION 790
Pengyu Guan ; Mads Lillieholm ; Kasper Meldgaard Røge ; Toshio Morioka ; Leif Katsuo Oxenløwe

EXPERIMENTAL DEMONSTRATION OF 7.09GB/S DML BASED REAL-TIME OPTICAL OFDM TRANSMISSION WITH SPECTRAL EFFICIENCY UP TO 6.93BITS/S/HZ OVER 50KM SSMF 793
Qianwu Zhang ; Han Dun ; Chen Qian ; Bingyao Cao ; Mingzhi Mao ; Junjie Zhang ; Min Wang

DISPERSION COMPENSATION SCHEME FOR TIME-FREQUENCY-DOMAIN MULTIPLEXED 4X/8X ULTRA-WIDEBAND MULTI-CARRIER SIGNALS 796
Takahide Sakamoto

REAL-TIME VISIBLE LIGHT COMMUNICATION SYSTEM BASED ON 2ASK-OFDM CODING 799
Yuankai Xue ; Yinan Hou ; Shilin Xiao ; Yunhao Zhang ; Lu Zhang

SPECTRAL OVERLAP OF TWO BANDWIDTH VARIABLE NYQUIST-WDM SIGNALS TO RESOLVE WAVELENGTH CONFLICT IN ELASTIC OPTICAL NETWORKS 802
Shuang Gao ; Lingchen Huang ; Chun-Kit Chan

HIGHLY LINEAR OPTICAL W-BAND RECEIVER FRONT-END BASED ON OPTICAL PROCESSING 805
Ziang Li ; Shangyuan Li ; Xiaoping Zheng ; Jinhui Luo ; Hanyi Zhang ; Bingkun Zhou

DISTANCE-ADAPTIVE ROUTING, MODULATION LEVEL AND SPECTRUM ALLOCATION (RMLSA) IN K-NODE (EDGE) CONTENT CONNECTED ELASTIC OPTICAL DATACENTER NETWORKS 808
Xin Li ; Shanguo Huang ; Shan Yin ; Bingli Guo ; Yongli Zhao ; Jie Zhang ; Min Zhang ; Wanyi Gu

EXPERIMENTAL DEMONSTRATION OF 10-GB/S DIRECT DETECTION OPTICAL OFDM TRANSMISSION WITH TRELLIS-CODED 8PSK SUBCARRIER MODULATION 811
Yiming Lou ; Zhenming Yu ; Minghua Chen ; Hongwei Chen ; Sigang Yang ; Shizhong Xie

POWER EFFICIENT OPTICAL OFDM TRANSMISSION WITH PHASE MODULATION AND DIRECT DETECTION 814
Zhenhua Feng ; Qiong Wu ; Ming Tang ; Rui Lin ; Ruoxu Wang ; Jiale He ; Songnian Fu ; Lei Deng ; Deming Liu ; Perry Ping Shum

A 3-D ADAPTIVE LOADING ALGORITHM FOR DIRECT DETECTION OPTICAL OFDM SYSTEM 817
Xi Chen ; Zhenhua Feng ; Ming Tang ; Huibin Zhou ; Songnian Fu ; Deming Liu

A ROBUST TIMING ESTIMATION METHOD FOR DDO-OFDM SYSTEMS 820
Zhen Zhang ; Yingxiong Song ; Junjie Zhang ; Yufeng Cai ; Weiliang Wu ; Qianwu Zhang

PERFORMANCE COMPARISON OF NYQUIST-4PPM-QPSK AND QPSK IN UNREPEATERED TRANSMISSION SYSTEMS 823
Jiangchuan Pang ; Yan Li ; Miao Yu ; Lei Yue ; Sujie Fan ; Deming Kong ; Jian Wu

SIGNAL LIGHT CARRIER AUTOMATIC PHASE-LOCK OPERATION TO OPTICAL FREQUENCY GRID COMB 826
Yudai Hisata ; Akira Mizutori ; Masafumi Koga

MAXIMUM RATIO COMBINING CHARACTERISTICS AFFECTED BY LASER PHASE NOISE FOR WAVELENGTH DIVERSITY DIGITAL COHERENT SYSTEM 829
Akira Naka ; Kohei Saito ; Kazuki Tomita ; Hideki Maeda

MULTI-LEVEL PRE-EQUALIZATION USING ANALOG FIR FILTERS BASED ON 28-NM FD-SOI FOR 20-GB/S 4-PAM MULTI-MODE FIBER TRANSMISSION 832
Ryoichiro Nakamura ; Kenta Amino ; Kawori Sekine ; Kazuyuki Wada ; Moriya Nakamura

BLIND POLARIZATION DE-MULTIPLEXING OF TIME DOMAIN HYBRID QAM SIGNALS BASED ON RADIUS-DIRECTED LINEAR KALMAN FILTER 835
Qun Zhang ; Wen Jiang ; Yanfu Yang ; Xian Zhou ; Kangping Zhong ; Yong Yao

RELATIONSHIP BETWEEN ROLL-OFF FACTOR AND TRANSMISSION DISTANCE IN NYQUIST OTDM SCHEME BASED ON CORRELATION DETECTION WITH EDFA REPEATERS 838
Takahiro Oguro ; Yuji Miyoshi ; Hirokazu Kubota ; Masaharu Ohashi

ELASTIC-BANDWIDTH ACCESS WITH SPECTRUM-SLICED INCOHERENT LIGHT SOURCE IN WDM-PON 841
Jing Zhang ; Qikai Hu ; Changyuan Yu ; Xingwen Yi ; Kun Qiu

EQUALIZATION OF OPTICAL NONLINEAR WAVEFORM DISTORTION USING NEURAL-NETWORK BASED DIGITAL SIGNAL PROCESSING 844
Shotaro Owaki ; Moriya Nakamura

PERFORMANCE ANALYSIS OF EKF-BASED POLARIZATION AND PHASE TRACKING FOR HIGH ORDER QAM SIGNALS 847
Wen Jiang ; Yanfu Yang ; Qun Zhang ; Kangping Zhong ; Xian Zhou ; Yong Yao

POST-FEC PERFORMANCE COMPARISON OF PILOTAIDED PHASE UNWRAP AND TURBO DIFFERENTIAL DECODING FOR CYCLE SLIP MITIGATION 850
Tangqing Liu ; Yan Li ; Huizhong Chen ; Haiquan Cheng ; Deming Kong ; Jian Wu

CHARACTERIZATION OF SOLITON FUSION PHENOMENON BASED ON SOLITONS' EIGENVALUE 853
Gihan Weerasekara ; Akihiro Maruta

FEASIBILITY STUDY OF 100G ETHERNET WITH CARRIERLESS AMPLITUDE AND PHASE MODULATION 856
Hirotaka Ochi ; Shinya Sasaki

BANDWIDTH ENHANCEMENT EQUALIZATION ENABLING A 40-GBPS PAM4 TRANSMISSION VIA A 9.5-GHZ PHOTORECEIVER 859
Jun-Jie Liu ; Chia-Chien Wei ; Ta-Ching Tzu ; Chun-Yen Chuang ; Kai-Lun Chi ; Xin-Nan Chen ; Jin-Wei Shi ; Jyehong Chen

INFLUENCE OF POINTING ERRORS ON ERROR PROBABILITY OF INTER-SATELLITE LASER COMMUNICATIONS 862
Qian Wang ; Tianyu Song ; Ming-Wei Wu ; Tomoaki Ohtsuki ; Mohan Gurusamy ; Pooi-Yuen Kam

TEST AND VERIFICATION OF VEHICLE POSITIONING BY INFRARED SIGNAL-DIRECTION DISCRIMINATION 865
Wern-Yarng Shieh ; Ti-Ho Wang ; Kun-Lung Chung ; Zong-Yu Tsai ; Wei-Jhih Huang

NOISE TOLERANT OPTICAL DETECTION IN CMOS IMAGE SENSOR BASED VISIBLE LIGHT COMMUNICATION 868
Joon-Woo Lee ; Se-Hoon Yang ; Yong-Hwan Son ; Sang-Kook Han

A CW, ALL-FIBER 2100 NM HO DOPED FIBER LASER PUMPED AT 1950 NM 871
Jiachen Wang ; Sang Bae Lee ; Kwanil Lee

SECURITY STRATEGY AGAINST MULTIPOINT EAVESDROPPING IN ELASTIC OPTICAL NETWORKS 874
Wei Bai ; Hui Yang ; Yongli Zhao ; Jie Zhang ; Yuanlong Tan ; Xiaoxu Zhu ; Huixia Ding

SWITCHABLE PASSIVELY MODE-LOCKED THULIUM-DOPED FIBER LASER BASED ON SINGLE-WALL CARBON NANOTUBES 877
Zhichao Wu ; Songnian Fu ; Minming Zhang ; Kai Jiang ; Jue Song ; Ping Shum ; Deming Liu

EVALUATION OF ALCOHOL CONCENTRATION IN SAKE USING 1.7 µM BAND TM^{3+}-TB^{3+}-DOPED TUNABLE FIBER RING LASER 880
Fumiki Hanafuji ; Kousuke Senda ; Xiaoen Du ; Sho Tujita ; Kazuya Ota ; Jun Ono ; Osanori Koyama ; Hirotaka Ono ; Tatsuro Endo ; Makoto Yamada

GENERATION OF 100 GHZ PULSE TRAIN FROM A PHASE MODULATED HYBRID MODE-LOCKED ER-FIBER LASER WITH AN INTRA-CAVITY ETALON 883
Sheng-Min Wang ; Yinchieh Lai

INFRARED LUMINESCENCE INVESTIGATION OF BISMUTH AND ERBIUM CO-DOPED FIBER 886
Chunsheng Li ; Binbin Yan ; Xiaojuan Wu ; Xinzhu Sang

ENERGY-EFFICIENT DUAL-PUMP DEGENERATED PHASE SENSITIVE AMPLIFIERS 889
Mingyi Gao ; Takayuki Kurosu ; Shu Namiki ; Gangxiang Shen

GAIN SATURATION AND RECOVERY OF AN ERBIUMDOPED FIBER AMPLIFIER MEASURED BY TEMPORALLY RESOLVED PROBE TECHNIQUES 892
Keiji Kuroda ; Yuzo Yoshikuni

HIGH GROSS GAIN OF SINGLE-MODE CR-DOPED FIBERS EMPLOYING ON-LINE GROWTH SYSTEM 895
Chun-Nien Liu ; Gia-Ling Cheng ; Nan-Kuang Chen ; Yi-Chung Huang ; Pi-Ling Huang ; Sheng-Lung Huang ; Wood-Hi Cheng

DOUBLE INSCRIBING METHOD WITH CO_2 LASER FOR SUITABLE SPECTRUM OF LPFG USED IN MULTIPOINT TEMPERATURE SENSOR 898
Akihiro Kusama ; Mamoru Iida ; Osanori Koyama ; Syo Takasuka ; Toshinori Murakami ; Makoto Yamada

SIMULTANEOUS INTERROGATION OF MULTIPLE CORRELATION PEAKS IN A BOCDA SYSTEM BY TIME-DOMAIN DATA PROCESSING 901
Gukbeen Ryu ; Kwang Yong Song ; Gyu-Tae Kim ; Sang Bae Lee ; Kwanil Lee

RISK-AWARE VIRTUAL NETWORK EMBEDDING IN OPTICAL DATA CENTER NETWORKS 904
Weigang Hou ; Lei Guo ; Cunqian Yu ; Yue Zong

SIMULTANEOUS MEASUREMENT OF CURVATURE AND TEMPERATURE BASED ON THIN CORE ULTRA-LONG PERIOD FIBER GRATING 907
Wenjun Ni ; Ping Lu ; Chao Luo ; Xin Fu ; Deming Liu ; Jiangshan Zhang

METAL-ASSISTED HIGH SENSITIVITY MICRO MACH-ZEHNDER FIBER TEMPERATURE SENSORS 910
Kuan-Hao Lin ; Nan-Kuang Chen ; Shien-Kuei Liaw ; Wood-Hi Cheng

THE LIQUID LENGTH EFFECTS ON THE MACH-ZEHNDER INTERFEROMETER INDUCED BY TWO LIQUID SECTIONS IN A PHOTONIC CRYSTAL FIBER 913
Jia-Hong Liou ; Chin-Ping Yu

HIGH-SENSITIVITY HUMIDITY SENSOR COMPOSED OF OPTICAL FIBER COATED WITH SOL-EL DERIVED POROUS SILICA 916
Nobuaki Tsuda ; Hideki Fukano ; Shuji Taue

A HIGH-SENSITIVITY TWO-DIMENSIONAL INCLINOMETER BASED ON TWO ETCHED-CHIRPED FIBER GRATINGS 919
Hung-Ying Chang ; Kuan-Ting Li ; Po-Chia Huang ; Jung-Sheng Chiang ; Nai-Hsiang Sun ; Wen-Fung Liu

A POLYMER-COATED HOLLOW CORE FIBER FABRY-PÉROT INTERFEROMETER FOR SENSING LIQUID LEVEL 922
Teng-Wei Fu ; Yuan-Jie Yang ; Jun-Han Lin ; Tung-Yuan Yeh ; Pin Han ; Cheng-Ling Lee

MULTI-CHANNEL LASING CHARACTERISTICS FOR LINEAR-CAVITY FIBER SENSOR SYSTEM USING SOA AND FIBER BRAGG GRATING ELEMENTS 925
Kazuto Takahashi ; Mao Okada ; Hiroki Kishikawa ; Nobuo Goto ; Yi-Lin Yu ; Shien-Kuei Liaw

NOVEL SOFT-CLADDING OPTICAL FIBER FOR DISTRIBUTED PRESSURE SENSING 928
Bin Zhou ; Lin Htein ; Zhengyong Liu ; A. Ping Zhang ; Chao Lu ; Hwa-yaw Tam

NOVEL BIDIRECTIONAL REFLECTIVE SEMICONDUCTOR OPTICAL AMPLIFIER 931
G. de Valicourt ; A. Maho ; A. Le liepvre ; R. Brenot ; A. Velázquez ; Y. K. Chen

SEMICONDUCTOR OPTICAL AMPLIFIER IN AWG-STAR NETWORK WITH WAVELENGTH PATH RELOCATION FUNCTION 934
Takumi Niihara ; Minoru Yamaguchi ; Osanori Koyama ; Hiroaki Maruyama ; Kazuya Ota ; Makoto Yamada

PROPOSAL OF AN ORCHESTRATOR-TO-ORCHESTRATOR INTERFACE USING THE CONTROL ORCHESTRATION PROTOCOL 937
Yoshiaki Inoue ; Jun Matsumoto ; Satoru Okamoto ; Naoaki Yamanaka

40GB/S OPTICAL RECEIVER USING HIGH-GAIN MULTI-LEVEL ACTIVE FEEDBACK WITH SERIAL INDUCTOR PEAKING 940
Cheng-Ta Chan ; Oscal T. -C. Chen

FREQUENCY CHIRP PROPERTIES WITH DATA PATTERN DEPENDENCE IN QUANTUM-DOT SOAS 943
Hiroki Hoshino ; Norihiko Ninomiya ; Motoharu Matsuura

OPTICAL SENSOR BASED ON MACH-ZEHNDER INTERFEROMETER USING ORBITAL ANGULAR MOMENTUM 946
Haozhe Yan ; Shangyuan Li ; Bian FengKai ; Xiaoping Zheng ; Hanyi Zhang ; Bingkun Zhou

PRECISE MEASUREMENT OF MICROWAVE EVANESCENT FIELDS ALONG FIBERGLASS-REINFORCED PLASTIC MORTAR PIPE USING ELECTRO-OPTIC SENSOR FOR NONDESTRUCTIVE INSPECTION 949
Yoshiyuki Azuma ; Fumiaki Ueno ; Hiroshi Murata ; Yasuyuki Okamura ; Tadahiro Okuda ; Masaya Hazama

PLASMON-INDUCED TRANSPARENCY BASED ON SIDECOUPLED STUB AND HEXAGONAL RESONATORS AND ITS SENSING CHARACTERISTICS 952
Tianye Huang ; Minming Zhang ; Songnian Fu

OPTICAL CHARACTERISTICS OF INP/GALNAS CORE-MULTISHELL NWS GROWN BY SELF-CATALYTIC VLS MODE 955
Kohei Takano ; Takehiro Ogino ; Keita Asakura ; Takao Waho ; Kuzuhiko Shimomura

S-K GROWTH OF INAS QUANTUM DOTS ON DIRECTLY-BONDED INP/SI SUBSTRATE USING MOVPE 958
Naoki Kamada ; Toshiki Sukigara ; Keiichi Matsumoto ; Junya Kishikawa ; Tetsuo Nishiyama ; Yuya Onuki ; Kazuhiko Shimomura

HYBRID ELECTRO-OPTIC POLYMER MODULATORS 961
Feng Qiu ; Shiyoshi Yokoyama

PHOTODETECTION FREQUENCY RESPONSE CHARACTERIZATION FOR HIGH-SPEED GE-PD ON SI WITH AN EQUIVALENT CIRCUIT 964
Jeong-Min Lee ; Minkyu Kim ; Stefan Lischke ; Lars Zimmernman ; Seong-Ho Cho ; Woo-Young Choi

SUB-µM ELECTRODE SPACING SOI-PIN PHOTODIODE FABRICATED BY CMOS COMPATIBLE PROCESS 967
Hiroya Mitsuno ; Takeo Maruyama ; Koichi Iiyama

ARBITRARY OUTPUT PORT SELECTION IN MULTI-MODE FIBER NETWORKS USING MODE DIVISION MULTIPLEXING 970
Yuki Morizumi ; Hirokazu Kobayashi ; Katsushi Iwashita

COMPARISON OF TWO PHOTODETECTOR LINEARITY CHARACTERIZING SYSTEMS 973
Youxin Liu ; Yongqing Huang ; JiaRui Fei ; Yangan Zhang ; Xiaomin Ren ; Kai Liu ; Xiaofeng Duan

OPTIMIZATION OF TEMPERATURE-DEPENDENT 850 NM VCSELS WITH DIFFERENT OXIDE-CONFINED APERTURE SIZES 976
Chun-Yen Pong ; Cheng-Ting Tsai ; Yun-Chen Wu ; Shan-Fong Leong ; Yu-Chieh Chi ; Gong-Ru Lin ; Chao-Hsin Wu

ENHANCEMENT OF LIGHT EXTRACTION EFFICIENCY OF INGAN LIGHT-EMITTING DIODES WITH MICROHOLES ARRAY AND NANO-ROUGHENED ZNO STRUCTURE 979
Ming Wang ; Zhigang Zang ; Xiaosheng Tang

CONCENTRIC CIRCULAR HIGH INDEX CONTRAST GRATINGS REFLECTOR WITH FOCUSING ABILITY 982
Wenjing Fang ; Yongqing Huang ; Xiaofeng Duan ; Kai Liu ; Ren Xiaomin ; Jun Wang

FABRICATION OF WAVEGUIDE-TYPE MIRRORS FOR RED-GREEN-BLUE LASER BEAM MULTIPLEXERS 985
S. Tanaka ; A. Nakao ; S. Yokokawa ; S. Hayashiguchi ; T. Katsuyama ; K. Nakajima ; N. Ikeda ; Y. Sugimoto

MULTIPLE-ACCESS AND TWO-WAY VISIBLE LIGHT COMMUNICATION WITH IMAGE SENSOR AND LED ARRAY 988
Tomoki Kondo ; Ryotaro Kitaoka ; Wataru Chujo

BROADBAND POLYMER 3-DB COUPLER USING SHORTCUT TO ADIABATICITY BASED OPTIMIZATION 991
Hung-Ching Chung ; Shuo-Yen Tseng

MULTIMODE THREE BRANCH POLYMER SPLITTER 994
Václav Prajzler ; Radek Maštera

2-STAGE CASCADED SILICON PHOTONIC PBS BASED ON MACH ZEHNDER DELAY INTERFEROMETERS 997
Kohei Morita ; Hiroyuki Uenohara

AN ULTRA-COMPACT AND LOW-LOSS WAVELENGTH DEMULTIPLEXER EMPLOYING THE PHOTONIC-CRYSTAL-LIKE METAMATERIAL STRUCTURE 1000
Feiya Zhou ; Luluzi Lu ; Minming Zhang ; Songnian Fu ; Deming Liu

ENERGY ANALYSIS FOR DYNAMIC CACHE STORAGE IN VIDEO ON DEMAND SERVICES 1003
Zhongwei Tan ; Chuanchuan Yang ; Yu Yang ; Zhaopeng Xu ; Xinyue Wang ; Ziyu Wang

POLARIZING BEAM SPLITTER WITH FOCUSING ABILITY BASED ON SUB-WAVELENGTH GRATINGS 1006
Wang Ying ; Huang Yongqing ; Guo Yanan ; Fang Wenjing ; Ren Xiaomin

SILICON PHOTONIC TE POLARIZER USING ADIABATIC WAVEGUIDE BENDS 1009
Bruna Paredes ; Humaira Zafar ; Marcus S. Dahlem ; Anatol Khilo

DESIGN OF A SI ARRAYED-WAVEGUIDE GRATING USING DISTRIBUTED BRAGG REFLECTORS 1012
Takahiro Inaba ; Hiroyuki Tsuda

FREQUENCY/ENERGY-TIME HYPER-ENTANGLED PHOTON PAIR GENERATION BASED ON A SILICON MICRO-RING RESONATOR 1015
Jing Suo ; Wei Zhang ; Shuai Dong ; Yidong Huang ; Jiangde Peng

ON THE CONTROL OF THE MICRORESONATOR OPTICAL FREQUENCY COMB IN TURING PATTERN REGIME VIA PARAMETRIC SEEDING 1018
Jinghao Wang ; Minming Zhang ; Meifeng Li ; Yuanwu Wang ; Deming Liu

GENERALIZED POLYNOMIAL CHAOS EXPANSION FOR PHOTONIC CIRCUITS OPTIMIZATION 1021
Daniele Melati ; Tsui-Wei Weng ; Luca Daniel ; Andrea Melloni

A GAUSSIAN BEAM WRITTEN SAMPLED FIBER GRATING FOR SUB-PS TIME DELAY LINES 1024
Weiqian Zhao ; Haifeng Qi ; Yunchuan Zhang ; Mingya Shen

PERIODICALLY POLED LINBO3 RIDGE WAVEGUIDE FOR HIGH-GAIN PHASE-SENSITIVE AMPLIFIER 1027
Tadashi Kishimoto ; Koji Inafune ; Yoh Ogawa ; Hitoshi Murai

PRELIMINARY RESEARCH OF MCF COUPLING TO OPTICAL MODE SWITCH CONFIGURATION 1030
Xiaoyang Cheng ; Luke Himbele ; Ryan Imansya ; Haisong Jiang ; Kiichi Hamamoto

NUMERICAL STUDY OF POWER COUPLING BETWEEN INDEX-ANTIGUIDED SLAB WAVEGUIDES 1033
Chang-I Hsieh ; Chih-Hsien Lai

SDN/NFV ORCHESTRATION OF MULTI-TECHNOLOGY AND MULTI-DOMAIN NETWORKS IN CLOUD/FOG ARCHITECTURES FOR 5G SERVICES 1036
Ricard Vilalta ; Arturo Mayoral ; Ramon Casellas ; Ricardo Martínez ; Raul Muñoz

SOFTWARE-DEFINED OPTICAL TRANSMISSION AND NETWORKING WITH FUNCTIONAL SERVICE DESIGN 1039
Xiaoyuan Cao ; Noboru Yoshikane ; Koki Takeshima ; Ion Popescu ; Takehiro Tsuritani ; Itsuro Morita

MULTI-STRATUM OPTIMIZATION WITH ROUTING RADIO AND SPECTRUM ALLOCATION FOR CLOUD RADIO OVER FIBER NETWORKS 1042
Hui Yang ; Wei Bai ; Ao Yu ; Jie Zhang

HIGHLY EFFICIENT ADAPTIVE BANDWIDTH ALLOCATION ALGORITHM FOR ELASTIC LAMBDA AGGREGATION NETWORK 1045
Hiroyuki Saito ; Naoki Minato ; Shuko Kobayashi ; Hideaki Tamai

RESILIENT VIRTUAL OPTICAL NETWORK PROVISIONING OVER SOFTWARE-DEFINED OPTICAL NETWORKS 1048
Xi Wang ; Inwoong Kim ; Qiong Zhang ; Paparao Palacharla ; Tadashi Ikeuchi

NFV AND SDN FOR NEXT TELECOM CLOUD AND CORE NETWORKING 1051
Michiaki Hayashi

MULTILAYER VIRTUAL INFRASTRUCTURE MAPPING IN IP OVER WDM NETWORKS 1054
Zilong Ye ; Philip N. Ji

WEIGHTED ATTACK-EVASION ROUTING AND WAVELENGTH ASSIGNMENT ALGORITHM AGAINST HIGH POWER JAMMING IN OPTICAL NETWORKS 1057
Liangkai Huang ; Yongli Zhao ; Wei Wang ; Jie Zhang

COHERENT OPTICAL COMMUNICATION TECHNOLOGY 1060
Kazuro Kikuchi

ON THE ORBITAL ANGULAR MOMENTUM (OAM) OF LIGHT IN FIBER.. 1061
Siddharth Ramachandran

GAIN CONTROL IN MULTI-CORE ERBIUM/YTTERBIUM-DOPED FIBER AMPLIFIER WITH HYBRID
PUMPING 1062
Makoto Yamada ; Hirotaka Ono ; Tsukasa Hosokawa ; Kentaro Ichii

EFFECTIVE AREA MEASUREMENT OF TWO-MODE FIBER USING BIDIRECTIONAL OTDR
TECHNIQUE... 1065
Masaharu Ohashi ; Shun Asuka ; Yuji Miyoshi ; Hirokazu Kubota

FIBER-OPTIC GUIDED-ACOUSTIC-WAVE BRILLOUIN SCATTERING OBSERVED WITH PUMP-PROBE
TECHNIQUE... 1068
Neisei Hayashi ; Heeyoung Lee ; Yosuke Mizuno ; Kentaro Nakamura

DISTRIBUTED HYDROSTATIC PRESSURE MEASUREMENT BASED ON BRILLOUIN DYNAMIC
GRATING IN POLARIZATION MAINTAINING FIBERS 1071
Yong Hyun Kim ; Hong Kwon ; Jeongjun Kim ; Kwang Yong Song

A MULTICAVITY FIBER FABRY-PÉROT INTERFEROMETER FOR SENSING MULTIPLE PARAMETERS 1074
Wei-Kang Chang ; Meng-Shan Wu ; Chung-Hao Tseng ; Pin Han ; Cheng-Ling Lee

HIGH SENSITIVITY FIBER OPTIC CURRENT SENSOR BASED ON RECIRCULATING FIBER LOOP 1077
Yemeng Tao ; Jiangbing Du ; Lin Ma ; Shuai ; Yinping Liu ; Wenjia Zhang ; Zuyuan He

FABRICATION AND CHARACTERIZATION OF LANTHANUM BOROALUMINOSILICATE GLASS FIBER
FOR MAGNETO-OPTICAL DEVICE APPLICATIONS.. 1080
Kadathala Linganna ; Seongmin Ju ; Bok Hyeon Kim ; Won-Taek Han

HIGH SENSITIVITY CURVATURE SENSOR BASED ON MODAL INTERFEROMETER FOR VIBRATION
DETECTION 1083
Li Liu ; Ping Lu ; Shun Wang ; Hao Liao ; Wenjun Ni ; Xin Fu ; Xinyue Jiang ; Deming Liu

ULTRA-HIGH-RESOLUTION OTDR BASED ON LINEAR OPTICAL SAMPLING 1086
Shuai Wang ; Xinyu Fan ; Bin Wang ; Guangyao Yang ; Qingwen Liu ; Zuyuan He

POLARIZATION FADING ELIMINATION IN PHASE-EXTRACTED OTDR FOR DISTRIBUTED FIBER-
OPTIC VIBRATION SENSING.. 1089
Guangyao Yang ; Xinyu Fan ; Bin Wang ; Qingwen Liu ; Zuyuan He

ORTHOGONALLY-POLARIZED PULSE PAIR BOTDA SENSOR WITH THREE-TONE PROBE...................... 1092
Yiming Tao ; Xiaobin Hong ; Zhisheng Yang ; Wenqiao Lin ; Jian Wu

PROOF OF CONCEPT FOR BRILLOUIN OPTICAL CORRELATION-DOMAIN REFLECTOMETRY
ASSISTED BY SPECTRAL SLOPE 1095
Heeyoung Lee ; Neisei Hayashi ; Yosuke Mizuno ; Kentaro Nakamura

SIMPLIFIED OPTICAL CORRELATION-DOMAIN REFLECTOMETRY WITHOUT REFERENCE PATH 1098
Makoto Shizuka ; Neisei Hayashi ; Yosuke Mizuno ; Kentaro Nakamura

PHASE NOISE MITIGATION FOR LONG-RANGE OFDR USING ULTRAFAST FREQUENCY SWEEP 1101
Bin Wang ; Xinyu Fan ; Shuai Wang ; Guangyao Yang ; Qingwen Liu ; Zuyuan He

HIGH-SPEED GERMANIUM-BASED WAVEGUIDE ELECTRO-ABSORPTION MODULATOR........................ 1104
P. De Heyn ; S. A. Srinivasan ; P. Verheyen ; R. Loo ; I. De Wolf ; S. Balakrishnan ; G. Lepage ; D. Van Thourhout ; M. Pantouvaki ; P. Absil ; J. Van Campenhout

LOW-VOLTAGE CARRIER-DEPLETION SILICON MACH-ZEHNDER MODULATOR AT HIGH
TEMPERATURES WITHOUT THERMO-ELECTRIC COOLING 1107
Norihiro Ishikura ; Kazuhiro Goi ; Hiroki Ishihara ; Shinichi Sakamoto ; Kensuke Ogawa ; Tsung-Yang Liow ; Xiaoguang Tu ; Guo-Qiang Lo ; Dim-Lee Kwong

HIGH-EFFICIENCY SILICON OPTICAL MODULATOR USING A SIN-STRIP LOADED WAVEGUIDE ON
THE PHOTONIC SOI PLATFORM 1110
Guangwei Cong ; Yuriko Maegami ; Morifumi Ohno ; Makoto Okano ; Koji Yamada

1.3-µM DFB LASER µ-PLATFORM; LIGHT SOURCE SUITABLE FOR SILICON PHOTONICS PLATFORM.................... 1113
Takanori Suzuki ; K. R. Tamura ; Koichiro Adachi ; Aki Takei ; Akira Nakanishi ; Kazuhiko Naoe ; Kouji Nakahara ; Shigehisa Tanaka

HYBRID SILICON-BASED TUNABLE LASER WITH INTEGRATED REFLECTIVITY-TUNABLE MIRROR.................... 1116
G. de Valicourt ; C. Gui ; A. Melikyan ; P. Dong ; C-M. Chang ; A. Maho ; R. Brenot ; Y. K. Chen

MEMBRANE DISTRIBUTED-REFLECTOR LASERS 1119
Shigehisa Arai ; Nobuhiko Nishiyama ; Tomohiro Amemiya ; Takuo Hiratani ; Daisuke Inoue

ULTRA LOW POWER CONSUMPTION OPERATION OF SOA ASSISTED EXTENDED REACH EADFB
LASER (AXEL) 1122
Wataru Kobayashi ; Naoki Fujiwara ; Takahiko Shindo ; Shigeru Kanazawa ; Koichi Hasebe ; Hiroyuki Ishii ; Mikitaka Itoh

UNCOOLED 25.78 GB/S TRANSMISSION OVER 10 KM USING A 1.3 µM DIRECTLY MODULATED DFB
LASER IN A TO-CAN PACKAGE 1125
Seiki Nakamura ; Mizuki Shirao ; Masamichi Nogami

STABILIZATION OF 14XX-NM HIGH POWER SEMICONDUCTOR LASER BY SINGLE FIBER BRAGG
GRATING CONFIGURATION 1128
Hideaki Hasegawa ; Taketsugu Sawamura ; Junji Yoshida ; Noriyuki Yokouchi

FAST WAVELENGTH SWITCHING OF DFB LD... 1131
Yuto Ueno ; Keita Mochizuki ; Kiyotomo Hasegawa ; Masamichi Nogami ; Hiroshi Aruga

OPTICAL CHIRP AND AMPLITUDE PROCESSING USING EAM INTEGRATION 1134
Yi-Jen Chiu ; Shin-Wei Shen ; Jui-Pin Wu ; Kuo-Chun Chang ; Chia-Chien Wei

CONCURRENT DWDM TRANSMISSION WITH RING MODULATORS DRIVEN BY A COMB LASER WITH 50GHZ CHANNEL SPACING..1137
M. Ashkan Seyedi ; Chin-Hui Chen ; Marco Fiorentino ; Daniil Livshits ; Alexey Gubenko ; Sergey Mikhrin ; Vladimir Mikhrin ; Raymond G. Beausoleil

MEASUREMENT OF VECTORIAL RESPONSE OF IQ MODULATOR USING OPTICAL INTERFERENCE......................1140
Yuya Yamaguchi ; Kazuki Seki ; Naoki Takahashi ; Atsushi Kanno ; Tetsuya Kawanishi ; Masayuki Izutsu ; Hirochika Nakajima

INTEGRATED STOKES VECTOR ANALYZER ON INP..1143
Samir Ghosh ; Yuto Kawabata ; Takuo Tanemura ; Yoshiaki Nakano

PROPOSAL OF COMPACT TE/TM POLARIZATION SWITCH BASED ON MICRORING RESONATOR...........1146
Keita Suzuki ; Tomoki Hirayama ; Taro Arakawa

ADVANCES IN SECOND ORDER NONLINEAR EFFECT IN SILICON..1149
P. Damas ; X. Le Roux ; M. Berciano ; G. Marcaud ; C. Alonso-Ramos ; D. Benedikovic ; E. Cassan ; D. Marris-Morini ; L. Vivien

MULTIPLE OPTICAL CARRIER GENERATION USING MULTIPLE QPM DEVICE................................1152
Kazuki Nakamura ; Hin Channa ; Masaki Asobe ; Takeshi Umeki ; Hirokazu Takenouchi

FABRICATION OF HIGH OPTICAL QUALITY FACTOR FREE-STANDING AS2S3 MICRODISK RESONATORS ON A SILICON CHIP..1155
Mingxiao Zhao ; Mingming Zhao ; Xiaoshun Jiang ; Yuan Chen ; Jiyang Ma ; Min Xiao

WAVELENGTH MODULATION SPECTROSCOPY OF FORMALDEHYDE USING 3μM DFG LASER............1158
Ryohei Fujisawa ; Masaki Asobe ; Akira Katoh ; Shigeru Yamaguchi ; Akio Tokura ; Hirokazu Takenouchi

CHIRP-FREE SPECTRAL COMPRESSION OF PARABOLIC PULSES IN SILICON NITRIDE CHANNEL WAVEGUIDES..1161
Chao Mei ; Jinhui Yuan ; Kuiru Wang ; Xinzhu Sang ; Chongxiu Yu

SLOW LIGHT DEVICES IN SILICON PHOTONICS..1164
Toshihiko Baba

DESIGN OF DOUBLE-SLOTTED HIGH-Q PHOTONIC CRYSTAL NANOCAVITY FILLED WITH ELECTRO-OPTIC POLYMER..1167
Masahiro Nakadai ; Ryotaro Konoike ; Yoshinori Tanaka ; Takashi Asano ; Susumu Noda

EXPERIMENTAL REPORT FOR DISPERSION ENGINEERING OF SILICON SLOT PHOTONIC CRYSTAL WAVEGUIDES..1170
Samuel Serna ; Weiwei Zhang ; Xavier Le Roux ; Laurent Vivien ; Eric Cassan

SILICON POLARIZING BEAM SPLITTER BASED ON ASYMMETRIC SLOT WAVEGUIDE............1173
Jijun Feng ; Ryoichi Akimoto ; Heping Zeng

PARITY-TIME SYMMETRIC COUPLED RESONATOR WAVEGUIDE WITH PHOTONIC CRYSTAL NANOCAVITIES..1176
Kenta Takata ; Masaya Notomi

NXN WAVELENGTH SELECTIVE SWITCHES..1179
Hisato Uetsuka ; Shu Namiki ; Keiichi Sasaki

WAVELENGTH SELECTIVE SWITCH FOR MULTI-CORE FIBER BASED SPACE DIVISION MULTIPLEXED NETWORK WITH CORE-BY-CORE SWITCHING CAPABILITY......................................1182
Kenya Suzuki ; Mitsumasa Nakajima ; Keita Yamaguchi ; Goh Takashi ; Yuichiro Ikuma ; Kota Shikama ; Yuzo Ishii ; Mikitaka Itoh ; Mitsunori Fukutoku ; Toshikazu Hashimoto ; Yutaka Miyamoto

DEMONSTRATION OF 1,440×1,440 FAST OPTICAL CIRCUIT SWITCH FOR DATACENTER NETWORKING..1185
Koh Ueda ; Yojiro Mori ; Hiroshi Hasegawa ; Hiroyuki Matsuura ; Kiyo Ishii ; Haruhiko Kuwatsuka ; Shu Namiki ; Toshio Watanabe ; Ken-ichi Sato

32×32 SILICON PHOTONIC SWITCH..1188
Dritan Celo ; Dominic J. Goodwill ; Jia Jiang ; Patrick Dumais ; Chunshu Zhang ; Fei Zhao ; Xin Tu ; Chunhui Zhang ; Shengyong Yan ; Jifang He ; Ming Li ; Wanyuan Liu ; Yuming Wei ; Dongyu Geng ; Hamid Mehrvar ; Eric Bernier

SILICON PHOTONICS OPTICAL SWITCH BASED ON RING RESONATOR..1191
Antoine Descos ; M. Ashkan Seyedi ; Chin-Hui Chen ; Marco Fiorentino ; François Vincent ; David Penkler ; Bertrand Szelag ; Raymond G. Beausoleil

OPTICAL SWITCHING FUNCTIONS USING INTEGRATED SILICON-BASED DEVICES................................1194
G. de Valicourt

INTEGRATED FAT-TREE OPTICAL SWITCH WITH CASCADED MZIS AND EAM-GATE ARRAY................1197
Yusuke Muranaka ; Toru Segawa ; Ryo Takahashi

INTEGRATED INP OPTICAL SWITCH MATRICES PERFORMANCE FOR PACKET DATA NETWORKS......................1200
Ripalta Stabile

1310NM HIGH-CAPACITY WAVEBAND SWITCH NODE FOR FLAT OPTICAL DATA CENTER NETWORKS..1203
Wang Miao ; Huug de Waardt ; Nicola Calabretta

Author Index

Statistical Treatment of IEEE Spreadsheet Model for VCSEL-multimode Fiber Transmissions

Xin Chen, John Abbott, Dale Powers, Doug Coleman and Ming-Jun Li
Science and Technology Division, Corning Incorporated, Corning, NY 14830
chenx2@corning.com

Abstract: A statistical treatment is applied to IEEE spreadsheet link model for VCSEL and multimode fiber transmission. The results reveal longer system reach capability compared to using 'worst' case parameters for the majority of cases.
Keywords: IEEE link model, VCSEL-multimode fiber transmission, statistical treatment, system reach

I. INTRODUCTION

In short distance communication, multimode fiber (MMF) has been the primary optical medium used with VCSEL-based transceivers for up to several hundreds of the meters depending on data rates. The system reach is dependent on a set of parameters related to the transmitter, receiver, MMF and connectivity attributes. These parameters have been captured by the standard in the IEEE spreadsheet link model [1-2] to determine the system reach expectation. The spreadsheet model includes the 10G model that has been used by IEEE 802.3ae 10 Gb/s Ethernet Task Force [1] and the 25G model used more recently by the IEEE 802.3bm task force that specifies VCSELs transmission over OM3 and OM4 fibers with 70 m and 100 m system reaches respectively [2].

We observe that in arriving at the system reach for a particular application, the IEEE standard provides the values for the transmitter, receiver, and fiber parameters that are worst case values for each of them. However, in practice, each attribute would be part of a distribution and only have the worst case value at the tail of the distribution. The parameters from different components would be independent and therefore unlikely to be simultaneously worst case by nature. Therefore, the modeled system reach value may be quite conservative. This observation motivates us to look at the system reach capability from statistical point of view by introducing distributions for some of the key parameters to see the effect on the overall system reach distribution when the statistical parameters are varied independently. In some other cases of optical communications, statistical analysis has been introduced in the problem of polarization mode dispersion and the study of multiple path interference [3-4] in order to make the system impairment assessment more realistic. Through this study, we would expect to get better understanding of the system reach capability when more practical distributions are considered.

In the next section, we will identify a few parameters and provide detailed analyses to generate their distributions. These parameters include the modal bandwidth of the MMF and connector loss. We then apply the statistical distributions of these parameters to the IEEE 10 Gb/s and 25 Gb/s spreadsheet models to generate a system reach distribution for different scenarios. The distribution is then compared with the results obtained using the default or 'worst' case parameters.

II. STATISTICAL DISTRIBUTIONS OF MMF MODAL BANDWIDTH AND CONNECTOR LOSSES

In the current study, we specifically used the version 3.1.16a of the 10G model [1] and May 3, 2013 version of the 100G Example model [2]. Each parameter can vary statistically, but some have larger variations and play more significant roles than others in affecting the system reach. Among all the input parameters, we focus on the modal bandwidth of MMF and the connector loss. Some attributes from the transmitter such as launch power and laser line width could also be taken into account.

We first take a close look at the modal bandwidth. For MMFs, the refractive index profile may vary from fiber to fiber, which results in different modal bandwidth values. In addition, the modal bandwidth of the system also strongly depends on the light launching condition into the fiber, adding further variation. The commonly used OM3 and OM4 fibers are defined by TIA and IEC standards [5, 6]. In the current work, we use the 5000 fiber sample set from the Monte Carlo study done by the TIA task force in developing the fiber requirements for 10 Gb/s Ethernet [6]. Each 'fiber' is defined by 19 mode group delays at 850 nm. The delay sets were generated in Ref.[7] to cover typical perturbations seen in measured differential mode delays (DMDs). The set is not meant to be an exact replica of any particular manufacturing distribution, but serves as a broad set that can be used for modeling. Each 'laser' is in effect defined by a mode power distribution (MPD) for the 19 mode groups. The variation in the fibers can be understood by relating the 850 nm bandwidth against the 1300 nm bandwidth as explained in Ref.[8]. Some fibers are more optimized for 850 nm, some are more optimized toward 1300 nm, while others have index perturbations that are large enough that they never have 'high' bandwidth at any wavelength.

For each of the 5000 fiber samples, the modal bandwidth values are available for the 10 TIA laser weights. The commonly used effective modal bandwidth (EMB) is the minimal modal bandwidth value (minEMBc) from the 10 laser weights multiplied by 1.13. For OM3 fiber, the EMB needs to be at least 2000 MHz-km, while for OM4 fiber, the EMB value is at least 4700 MHz-km. The values of modal bandwidth of 5000 fibers vary over a wide range, as noted above. We grouped them into OM3 and OM4 categories based on their EMB value at 850 nm and whether they meet 1300 nm OFL bandwidth requirement of 500 MHz-km. For the Monte Carlo study in the current work, OM3 fibers are chosen to be those with EMB ranging from 2000 MHz-km to 4700 MHz-km (labeled as 'OM3'). For OM4 fibers, we have chosen those fibers (labeled as 'OM4') with EMB ranging from 4700 MHz-km to 7000 MHz-km. We truncated the OM4 distribution to have a maximum EMB of 7000 MHz-km to focus on fibers just meeting the OM4 requirement. Among the OM3 and OM4 fibers, each laser weight would generates a hypothetical modal bandwidth appropriate to the laser launch condition. By assuming equal probabilities among the 10 laser weights, the laser launch conditions will yield a distribution of modal bandwidths, as shown in Fig. 1(a). We found that an individual laser will generally yield a modal bandwidth higher than the EMB value, and as shown in Fig. 1(b). Note that the bandwidth percentage above a certain value is the inverse of the cumulative density function (CDF). The majority of the modal bandwidth values are significantly higher than minimally allowed values.

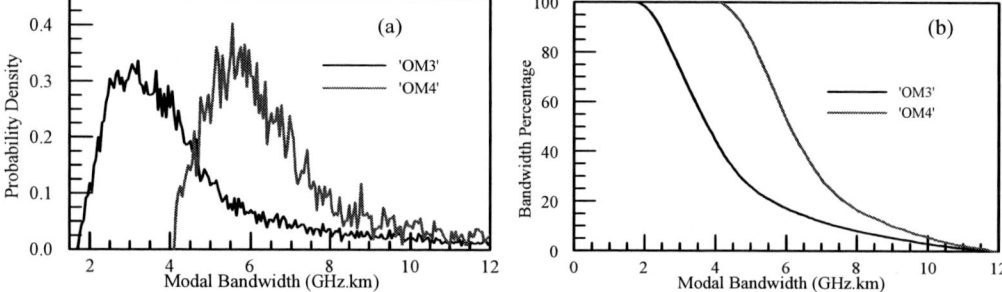

Figure 1(a). The probability density function of 'OM3' and 'OM4' fibers. (b) Bandwidth percentage of 'OM3' and 'OM4' fibers meeting and exceeding a given bandwidth value.

The connector loss also has statistical distributions, and more loss is incurred when higher counts of connector pairs are used. For structured cable used in data center, one trunk cable and two trunk cable configurations, which utilize two connector pairs and four connector pairs, respectively, account for the 90% of the cases [9], and the total connector loss increases with the number of connector pairs. Here we use the standard grade (Grade B) connector loss from the pending IEC MMF connector insertion loss guidance used in Ref. [9]. The statistical distribution of connector loss is shown in Fig. 2(a) and Fig. 2(b). Note that for six connector pairs, a part of the distribution exceeds 1.5 dB total connector loss as required by IEEE link model.

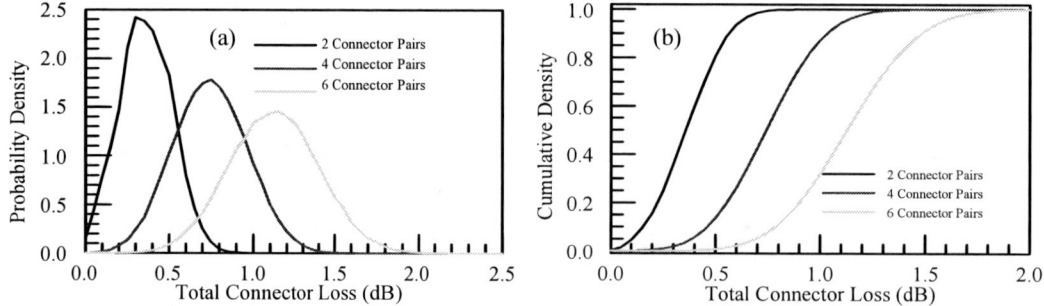

Figure 2. The probability density function (a) and (b) the cumulative density function (b) of connectors with 2, 4 and 6 connector pairs.

III. STATISTICAL DISTRIBUTIONS OF SYSTEM REACHES FOR 10G AND 25G TRANSMISSIONS

With the modal bandwidth distributions and connector loss distributions obtained in Section 2, we can now study how they contribute to the system reach distributions. Here we focus on the two IEEE link models for MMF transmissions at 10 Gb/s and 25 Gb/s respectively. In order to simulate the system reach performance for provided parameters in large quantity, we have coded the IEEE spreadsheets into two functions in C/C++ language to calculate ISI penalty and power margin. The system reach is calculated for system lengths meeting the requirements of having power margin equal or above 0 dB and an ISI penalty equal or below 3.6 dB.

Fig. 3(a) shows the results using 10G model for both 'OM3' and 'OM4' fibers. The modal bandwidth and connector loss values were generated based on the distribution in Fig. 2 with 100,000 independent instances. We used the default values for all other parameters. With the default values, the minimum system reach for OM3 and OM4 fibers are 330m and 500m respectively. However as shown in Fig. 3(a), in an overwhelming majority of the cases, the system reach is above 330 m. Specifically in more than 80% of the cases, the 'OM3' system reach is above 420 m. These results suggests the spreadsheet model is quite conservative. The situation is similar but is less pronounced for 'OM4' fibers, since they all have much higher modal bandwidth and the system performance is generally limited by chromatic dispersion. Also for 'OM4', the different connector losses associated with different number of pairs plays a role, and a reduced number of connector pairs can result in longer system reach since the system is more power budget limited in this regime. For 'OM3' fibers, the number of connector pairs plays a less important role in determining the system reach because the system is more ISI penalty limited. Fig. 3(b) shows the results for the 25G model. Note that the system reach is 70 m and 100 m for OM3 and OM4 as obtained from spreadsheet using default values. For 'OM3' fiber, in 80% of the cases, the system reach is above 96 m for two and four connector pairs and above 91 m for six connector pairs. For 'OM4', the system reach is above 122 m, 117 m and 106 m for two, four and six connector pairs, respectively. For the majority of the cases, the system reach significantly exceeds the worst case results, illustrating that the spreadsheet model is very conservative.

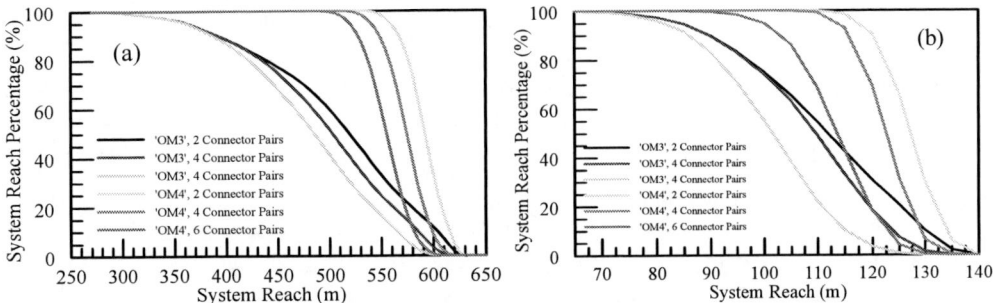

Figure 3 (a) the system reach of 'OM3' and 'OM4' fibers with 2, 4 and 6 connector pairs using 10 Gb/s IEEE link model; (b) the system reach of 'OM3' and 'OM4' fibers with 2, 4 and 6 connector pairs using 25 Gb/s IEEE link model.

Other parameters can also be treated by statistical distributions. For example, the optical power of an actual transmitter can be above the guidance by up to 1 dB or more. Commercial VCSELs for 25 Gb/s can also have laser linewidth as low as 0.45 nm, although 0.6 nm is specified. All these can be incorporated into the statistical analysis to reach more realistic system reach distribution using independent variations of the fiber, laser and connector parameters.

IV. CONCLUSIONS

We have conducted a statistical analysis for IEEE spreadsheet link models for VCSEL-MMF transmission at 10 Gb/s and 25 Gb/s. For modal bandwidth, the encircled flux variability of the laser is coupled with variations of MPD of the MMF, resulting in a wide distribution that is often better than 'worst' case values. Similar treatment was applied to the connector loss using IEC standard grade connector loss guidance. The system reach distribution was then calculated as the results of contributing parameters. While the spreadsheet results using worst case values can ensure a robust solution to be in place, we find that in 80% of the cases, the system reach can be much higher than worst case. This finding is particularly relevant for 'OM3' fibers, but less significant for 'OM4' fibers. We believe that the statistical study adopted in the current paper provides new insights into the IEEE link model and can help to understand better the system robustness.

REFERENCES

[1] http://grouper.ieee.org/groups/802/3/ae/public/index.html
[2] http://www.ieee802.org/3/bm/index.html
[3] Mou-Tion Lee et al, "Statistical PMD Specification – Evolution, Utilization and Control," OFC/NFOEC 2005, Paper NWE2
[4] Ming-Jun Li, et al, "Statistical Analysis of MPI in Bend-Insensitive Fibers", OFC-NFOEC 2009, paper OTuL1
[5] TIA-492AAAC-A, "Detailed Specification for 850nm Laser Optimized 50-mm Core Diameter/125-mm Cladding Diameter Class 1a Graded-Index Multimode Optical Fiber," (January, 2003).
[6] IEC 60793-2-10: 2011, "Optical fibres - Part 2-10: Product specifications - Sectional specification for category A1 multimode fibres"
[7] P. Pepeljugoski, et al., "Development of System Specification for Laser-Optimized 50-μm Multimode Fiber for Multigigabit Short-Wavelength LANS," Journal of Lightwave Technology, **21**, pp.1256-1275 (2003).
[8] J. S. Abbott et al., "Fibers for Short-Distance Applications," Chapter 7, Optical Fiber Telecommunications VIA. New York, Elsevier, 2013.
[9] Doug Coleman, "Statistical MMF Connector Insertion Loss," FC-PI-6/7 Ad Hoc Group, August 2015 meeting, 15-265V0.

TuE1-1 (Invited)

OECC/PS2016

Single-Carrier 1-Tb/s Signal Generation Using High-Speed InP MUX-DACs and an Integrated CSRZ-OTDM Modulator

Hiroshi Yamazaki[1,2], Akihide Sano[1], Munehiko Nagatani[1,2], and Yutaka Miyamoto[1]

[1]NTT Network Innovation Laboratories, NTT Corporation, 1-1 Hikari-no-oka, Yokosuka, Kanagawa, 239-0847 Japan
[2]NTT Device Technology Laboratories, NTT Corporation, 3-1 Morinosato-Wakamiya, Atsugi, Kanagawa, 243-0198 Japan
yamazaki.hiroshi@lab.ntt.co.jp

Abstract: We generated a 1-Tb/s single-carrier PDM-16QAM optical signal using InP MUX-DAC modules and an integrated CSRZ-OTDM modulator. The high symbol rate of 125 Gbaud was achieved by combining electronic and optical time-domain multiplexing.
Keywords: Single-carrier transmission, ETDM, OTDM, high-speed DAC, integrated optical modulator

I. INTRODUCTION

To keep up with the increasing demands for more optical transmission capacity, technologies for increasing per-channel line rates in optical transmission systems are being intensively studied [1]. In the last few years, technologies for achieving high-symbol-rate single-carrier multilevel modulation in digital coherent transmission systems have been gaining a lot of attention as simple and potentially cost-effective solutions for increasing line rates. Transmission with 72-Gbaud (864-Gb/s) polarization-division-multiplexed 64-level quadrature amplitude modulation (PDM-64QAM) was achieved by using high-speed SiGe digital-to-analog converters (DACs) [2]. Nyquist-shaped 124-Gbaud (1.24-Tb/s) PDM-32QAM transmission was also demonstrated by using a signal synthesis from four spectral slices in the optical domain [3].

Recently, we reported on a 125-Gbaud (1-Tb/s) PDM-16QAM signal using high-speed InP multiplexer-DAC (MUX-DAC) modules and an integrated modulator for optical time-domain multiplexing (OTDM) [4]. To the best of our knowledge, this was the first demonstration of single-carrier 1-Tb/s transmission with a single modulator (with external PDM emulation). In this paper, we review the technologies that enabled us to achieve the 1-Tb/s transmission.

II. MUX-DAC MODULE

Fig. 1(a). is a functional-block diagram of the MUX-DAC IC with 6-bit resolution [5]. The MUX-DAC operates as follows: First, the twelve input digital signals are synchronized with each other and reshaped by twelve D flip-flops (DFFs). Then, the outputs of the six pairs of DFFs are multiplexed by six 2:1 MUXs, respectively, which generates six digital signals with a doubled symbol rate. Finally, the outputs of the MUXs are converted to an analog signal with a resolution of 6 bits. Since the frequency of the clock signal input to the DFFs and MUXs is half the symbol rate of the output analog signal, the MUX-DAC is suitable for high-symbol-rate applications. For the DAC section, we employed the R-2R ladder configuration to achieve a high-speed operation with low power consumption. A microphotograph of the MUX-DAC IC is shown in Fig. 1(b). We fabricated the MUX-DAC IC with 0.5-µm-emitter InP heterojunction bipolar transistor (HBT) technology, which yields a cutoff frequency of 290 GHz and a maximum oscillation frequency of 320 GHz. The output analog 6-dB bandwidth of the MUX-DAC module, in which the IC is packaged with RF connectors, is larger than 40 GHz.

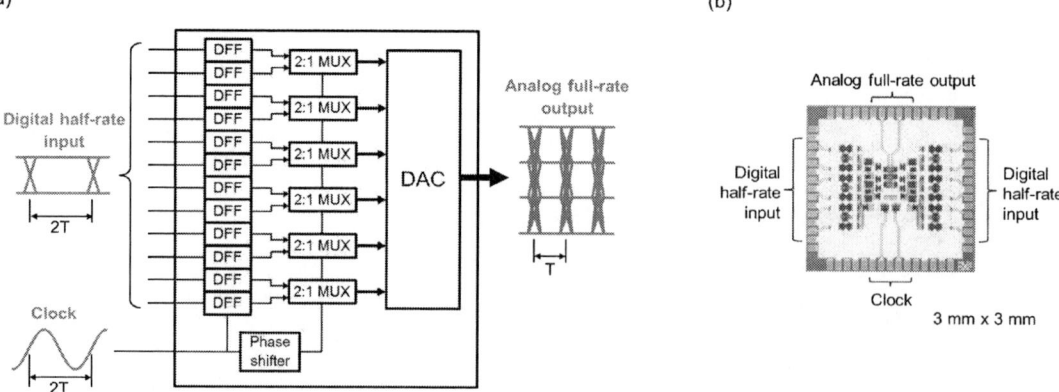

Fig. 1. (a) Block diagram and (b) microphotograph of the MUX-DAC IC.

III. CSRZ-OTDM MODULATOR

As shown in Fig. 2(a), the optical circuit of the CSRZ-OTDM modulator consists of a pulse generator, a quad-parallel IQ modulator, and a polarization-multiplexing (PM) circuit [4, 6]. The pulse generator is a dual-parallel Mach-Zehnder modulator (MZM) with a 2x2 output coupler. To obtain a final output symbol rate of B, the two MZMs are driven with sinusoidal clock signals with a frequency of B/4 and a relative delay of 1/B. The relative optical phase between the outputs from the two MZMs is set to zero. From the two output ports, A and B, we obtain two CSRZ pulse trains, respectively, which are orthogonal to each other as shown in Fig. 2(b). The CSRZ pulse train has a twin peak with a spacing of B/2 in the frequency domain. Since each IQ modulator is driven with a B/2-baud NRZ data signal, the final output of the CSRZ-OTDM modulator has the total spectral bandwidth of around 1.5B. As shown in Fig. 2(c), we fabricated the modulator with a hybrid configuration of silica planar lightwave circuits (PLCs) and a LiNbO$_3$ (LN) chip. The LN chip has ten push-pull pairs of straight phase modulators in an array, which corresponds to the ten MZMs (MZM1-10 from top to bottom): two for the pulse generator and eight for the four IQ modulators. All other passive components, such as couplers, static phase shifters, and the PM circuit, are fabricated in the PLCs. The pulse generator and the IQ modulators are connected via U-turn waveguides, and the input and output ports of the module are laid on the same side. The modulator's static insertion loss for both polarizations is 9.5 dB at 1550-nm wavelength. All the MZMs have electro-optic 3-dB bandwidths of ~23 GHz. The half-wave voltage, Vπ, of each MZM is ~3.5V.

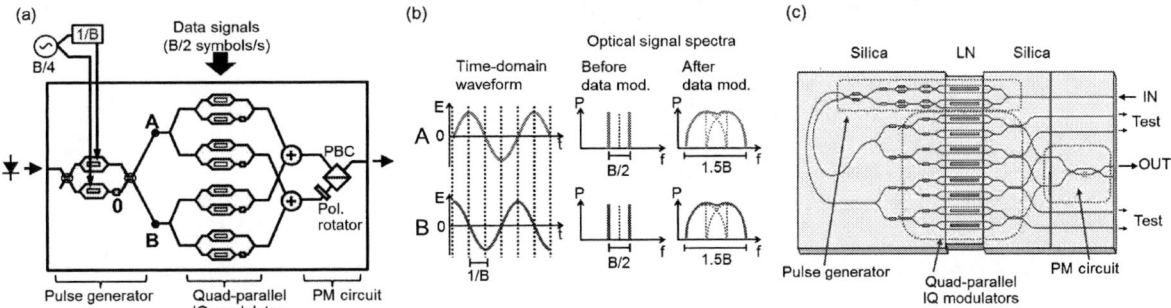

Fig. 2. (a) Optical-circuit diagram, (b) principle, and (c) configuration of the CSRZ-OTDM modulator.

IV. GENERATION AND TRANSMISSION OF 1-TB/S SIGNAL

Fig. 3 shows the experimental setup for the 1-Tb/s single-carrier transmission. Continuous-wave (CW) light from an external-cavity laser (ECL) with a wavelength of 1550 nm and a linewidth of ~30 kHz was modulated with the integrated CSRZ-OTDM modulator. The modulator was driven with two 31.25-GHz clocks and four 62.5-Gbaud four-level signals generated by two MUX-DACs, each of which was driven with four decorrelated 31.25-Gb/s 2^{15}-1 pseudo-random bit sequences (PRBSs) from a bit-pattern generator (BPG). Since only two MUX-DACs were available for this experiment, we used only two IQ modulators corresponding to one of the two polarization channels in the modulator. The output 125-Gbaud single-polarization 16QAM signal from the modulator was converted to a PDM signal by an emulation circuit. A programmable optical equalizer (OEQ) was optionally inserted between the modulator and PDM emulator. The PDM signal was transmitted over 80-km standard single-mode fiber (SSMF). At the receiver side, we used an offline digital coherent receiver consisting of another ECL as a local oscillator (LO), a dual-polarization optical hybrid (DPOH), four 70-GHz balanced photodiode modules (BPDs), and two synchronized digital storage oscilloscopes (DSOs), each with two input channels operating at 160 GSample/s with an analog bandwidth of 63 GHz. No optical time-domain demultiplexing was used. The stored digital data were analyzed by using offline processing similar to that described in [7].

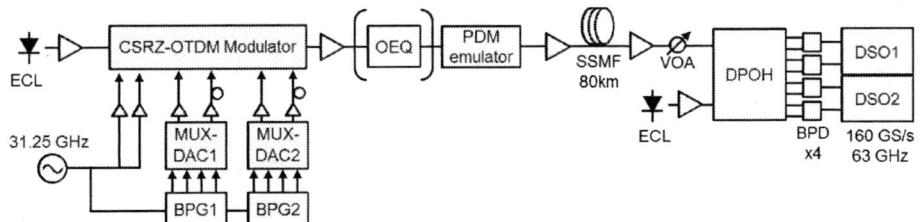

Fig. 3. Experimental setup for 1-Tb/s PDM-16QAM transmission.

Fig. 4(a) shows the bit-error ratio (BER) versus optical signal-to-noise ratio (OSNR) curves measured in a back-to-back configuration (the 80-km SSMF was bypassed and optical noise was loaded using an amplified-spontaneous-emission source). Constellations for the two polarizations with an OSNR of 37 dB are also shown. We successfully demodulated the 125-Gbaud 1-Tb/s PDM-16QAM signal without using any optical or electrical time-domain demultiplexing. The required OSNR is 28.4 dB without OEQ at the BER of 2.7×10^{-2}, which is the threshold of the low-density parity-check convolutional forward error correction (FEC) code using a layered decoding algorithm with 20% overhead [8]. Thus, the OSNR penalty with respect to the theoretical limit is about 6.4 dB. The use of OEQ reduced the penalty by about 1 dB. The result of the 80-km SSMF transmission experiment is shown in Fig. 4(b), where the received Q-factor is plotted against the optical power launched into the 80-km SSMF. This data was obtained without using the OEQ. The received power at the DPOH was kept constant (+4.2 dBm) by using a variable optical attenuator (VOA). The LO input power into the DPOH was 15.6 dBm. At the launched power of about +9 dBm, the received Q-factor is 6.5 dB, which corresponds to a margin of 1.8 dB from the threshold of the FEC mentioned above. The Q-factor drastically degrades when the launched power exceeds +14 dBm, probably as a result of nonlinear effects in the SSMF.

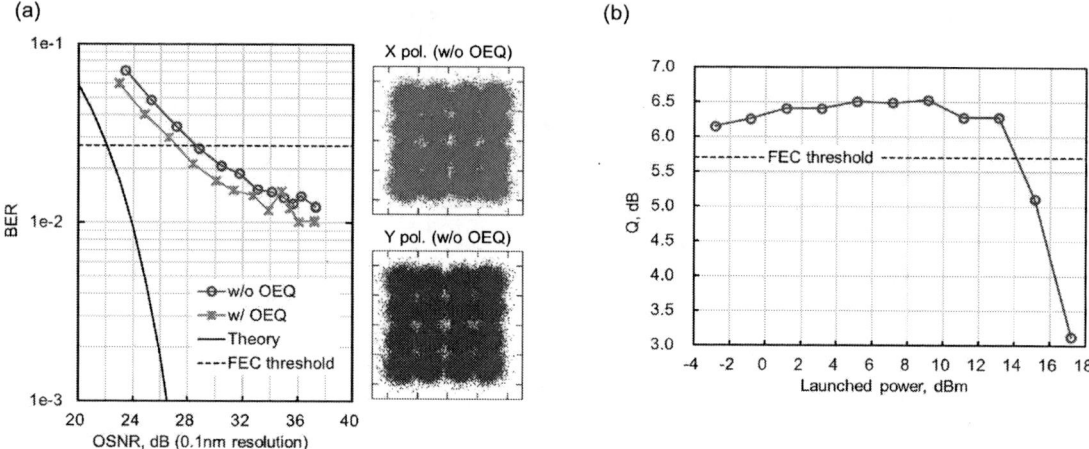

Fig. 4. (a) Back-to-back BER curves and constellations of the 1-Tb/s PDM-16QAM signal. (b) Received Q-factor versus optical power launched into the 80-km SSMF.

V. CONCLUSIONS

We demonstrated single-carrier 1-Tb/s PDM-16QAM optical transmission using high-speed InP MUX-DACs and an integrated CSRZ-OTDM modulator. The 125-Gbaud signal was generated from 31.25-Gbaud source data signals through the combination of electronic and optical TDM. This technology is promising for achieving 1-Tb/s-class single-carrier transmission with balanced complexities in the electronics and optics.

REFERENCES

[1] Y. Miyamoto, A. Sano, and T. Kobayashi, "The Challenge for the Next Generation OTN Based on 400Gbps and Beyond," in Proc. OFC/NFOEC 2012, Los Angeles, CA, USA, paper NTu2E.3.

[2] S. Randel, D. Pilori, S. Corteselli, G. Raybon, A. Adamiecki, A. Gnauck, S. Chandrasekhar, P. Winzer, L. Altenhain, A. Bielik, and R. Schmid, "All-electronic flexibly programmable 864-Gb/s single-carrier PDM-64-QAM," in Proc. OFC 2014, San Francisco, CA, paper Th5C.8.

[3] R. Rios-Müller, J. Renaudier, P. Brindel, H. Mardoyan, P. Jennevé, L. Schmalen, and G. Charlet, "1-Terabit/s Net Data-Rate Transceiver Based on Single-Carrier Nyquist-Shaped 124 GBaud PDM-32QAM," in Proc. OFC 2015, Los Angeles, CA, paper Th5.B.1.

[4] H. Yamazaki, A. Sano, M. Nagatani, and Y. Miyamoto, "Single-carrier 1-Tb/s PDM-16QAM transmission using high-speed InP MUX-DACs and an integrated OTDM modulator," Opt. Express vol. 23, no. 10, pp. 12866-12873 (2015)

[5] M. Nagatani, H. Wakita, H. Nosaka, K. Kurishima, M. Ida, A. Sano, and Y. Miyamoto, "75 GBd InP-HBT MUX-DAC module for high-symbol-rate optical transmission," Electron. Lett. 51(9), pp. 710-712..

[6] H. Yamazaki, T. Goh, T. Hashimoto, A. Sano, and Y. Miyamoto, "Generation of 448-Gbps OTDM-PDM-16QAM signal with an integrated modulator using orthogonal CSRZ pulses," in Proc. OFC 2015, Los Angeles, CA, paper Th2A.18.

[7] T. Kobayashi, A. Sano, H. Masuda, K. Ishihara, E. Yoshida, Y. Miyamoto, H. Yamazaki, and T. Yamada, " 160-Gb/s polarization-multiplexed 16-QAM long-haul transmission over 3,123 km using digital coherent receiver with digital PLL based frequency offset compensator," in Proc. OFC/NFOEC 2010, San Diego, CA, paper OTuD1.

[8] D. Chang, F. Yu, Z. Xiao, N. Stojanovic, F. N. Hauske, Y. Cai, C. Xie, L. Li, X. Xu, and Q. Xiong. "LDPC convolutional codes using layered decoding algorithm for high speed coherent optical transmission," in Proc.OFC/NFOEC 2012, Los Angeles, CA, paper OW1H.4.

Cost Effective Fractional OFDM Receiver using Arrayed Waveguide Grating

T. Nagashima[1], G. Cincotti[2], T. Murakawa[1], S. Shimizu[3], M. Hasegawa[1], K. Hattori[4],
M. Okuno[4], S. Mino[4], A. Himeno[4], N. Wada[3], H. Uenohara[5], T. Konishi[1]

[1] Osaka University, Graduate School of Engineering, Japan,
[2] University Roma Tre, via V. Volterra 62, I-00143 Rome, Italy
[3] NICT 4-2-1, Nukui-Kitamachi, Koganei, Tokyo 184-8795, Japan
[4] NTT Electronics Co. Ltd. 6700-2 to,Naka-shi,Ibaraki,311-0122, Japan
[5] Tokyo Institute of Technology 4259 Nagatsuta, Midoriku, Yokohama 226-8503, Japan
nagashima@photonics.mls.eng.osaka-u.ac.jp

Abstract: We experimentally demonstrate a cost effective fractional OFDM receiver using an arrayed waveguide grating. The eye diagrams of the demodulated 12×10 Gbit/s DBPSK fractional OFDM signals are clearly open.
Keywords: Arrayed waveguide grating, Fractional Fourier transform, Optical OFDM, Wavelength selective switch, Peak to average power ratio.

I. Introduction

High spectrum efficiency multiplexing techniques such as orthogonal frequency division multiplexing (OFDM) and Nyquist-optical time division multiplexing (N-OTDM) are key technologies for elastic optical network [1, 2]. High speed digital signal processing (DSP) and digital to analog conversion electrically generate optical signals with strong functions such as a chromatic dispersion compensation, a polarization mode dispersion compensation, and an insertion of cyclic prefix. To reduce the power consumption due to high bandwidth electric circuits, the implementation of IFFT/FFT operation or Nyquist-filtering by passive optical components is attracting much attention [3]. Although the modulation bandwidth and the functions are fixed, integrated waveguide devices could be mass-produced and reduce the system cost and complexity. Since the presently interested software defined network transmits various signals to various directions [4], the transmission conditions should be flexibly adjusted at all times. Therefore, wavelength selective switch (WSS) which could be implemented the reconfigurable filter functions in frequency domain should be used in optical line terminal in passive optical network and integrated waveguide devices should be used in user side optical network unit for the cost reduction.

OFDM and N-OTDM are complementary approaches for time or frequency multiplexing and have pros and cons. OFDM that aligns sinc shape spectra along frequency axis has an advantage of tolerances of residual dispersion and bandwidth mismatch in receiver side with cyclic prefix [5]. Since, however, multiple sinusoidal subcarriers are coherently superposed in time domain, peak to average power ratio (PAPR) of OFDM signal becomes high. The high PAPR signal may induce a lot of nonlinear effects in optical fiber and degrade the signal quality [6]. Although DSP can be used to compensate nonlinear effects, it is not cost effective for access network. N-OTDM can reduce PAPR because it aligns sinc shape waveforms along time axis. Although N-OTDM achieves the same spectral efficiency as OFDM, the receiver requires an ultra-short time gate. In accordance with the performance of receivers and the characteristic of the transmission fibers, the most feasible approach should be selectively used. For the ultimate efficiency of physical resource use, an intermediate fractional axis is useful in addition to time and frequency axes.

Fractional OFDM (FrOFDM) using fractional Fourier transform (FrFT) in place of FFT has an intermediate characteristic between OFDM and N-OTDM [7, 8]. Previously, we have demonstrated a fundamental process of transmit and receive of FrOFDM with FrFT operation in optical domain by WSSs [9, 10]. FrFT is also implemented by integrated devices adjusting phase shifters [11].

In this work, we demonstrate a cost effective FrOFDM receiver using an arrayed waveguide grating (AWG) in combination with a flexible WSS based transmitter. We observe the clear eye diagrams of demodulated 12×10 Gbit/s DBPSK FrOFDM signals with a four wave mixing based optical time gate.

II. Fractional Fourier Transform based OFDM

In a conventional OFDM scheme, complex data s_n with high-order modulation are transmitted in parallel using N subcarriers, within a symbol of duration T.

$$\phi_n^1(t) = s_n \cdot rect\left(\frac{t}{T}\right) \cdot e^{j2\pi\frac{n}{T}t} \quad n = 1,2,..,N. \tag{1}$$

where are the data symbols. On the other hand, N-OTDM symbol is a series of sinc pulses, delayed of Δt, that satisfy the orthogonal condition in time axis,

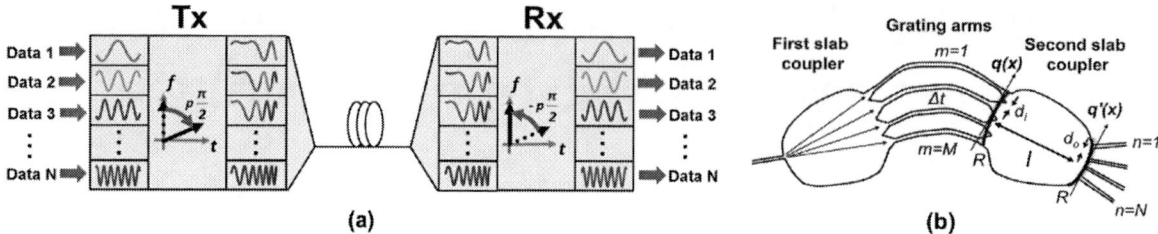

Fig. 1. (a) Schematic diagram of FrOFDM system, (b) The configuration of FrFT AWG.

$$\phi_n^0(t) = s_n \cdot \operatorname{sinc}\left(\frac{t}{\Delta t}\right) * \delta(t - n\Delta t) \qquad (2)$$

It is possible to generalize these multiplexing approaches, by introducing a new set of subcarriers, corresponding to the FrFT kernels, which are orthogonal over a symbol duration T.

$$\phi_n^p(t) = s_n \cdot \operatorname{rect}\left(\frac{t}{T}\right) \cdot e^{j\pi\left\{\left[n^2\sin^2\left(p\frac{\pi}{2}\right) + \frac{t^2}{T^2}\right]\cot\left(p\frac{\pi}{2}\right) - 2\frac{n}{T}t\right\}} \qquad (3)$$

where p is the fractional parameter ($p=1$ for conventional Fourier transform). FrOFDM symbol is rotated an angle p from OFDM symbol onto an intermediate axis between time and frequency. FrOFDM symbol could be demultiplexed by applying a complementary FrFT such as $-p$ at the receiver as shown in Fig. 1(a). FrFT could be implemented by WSS with transfer function of equation (3) or integrated waveguide devices with the suitable phase shift. The implementation of the FrFT by an AWG have already proposed by modifying the second slab coupler from a conventional FFT AWG as shown in Fig. 1(b). The relationship between the input field $q(x)$ and output field $q'(x)$ in the second slab coupler is described by Fresnel diffraction. The parameters of the FrFT AWG satisfy $l = \overline{d}\sin(p\pi/2)$, $R = \overline{d}\cot(p\pi/4)$, M N, $\Delta t = 1/FSR$, $d_i = \sqrt{\lambda\overline{d}}/N$, and $d_o = \sqrt{\lambda\overline{d}}\sin(p\pi/2)$, where $\overline{d}$ is a real-valued scale parameter with dimension of length, l is the distance between the two surfaces, R is their curvature radius, d_i is the input waveguide pitch, d_o is the output waveguide pitch, M is the number of grating arms, N is the number of output ports, and FSR is the free spectral range. The discretized output field is

$$q'\left(\frac{nd_o}{\sqrt{\lambda\overline{d}}}\right) = B_p \sum_{m=1}^{M} q\left(\frac{md_i}{\sqrt{\lambda\overline{d}}}\right) e^{j\pi\left\{\left[n^2\sin^2\left(p\frac{\pi}{2}\right) + \frac{m^2}{N^2}\right]\cot\left(p\frac{\pi}{2}\right) - \frac{2mn}{N}\right\}}, \qquad B_p = e^{j\frac{2\pi l}{\lambda}}\left|\sin\left(p\frac{\pi}{2}\right)\right|^{\frac{1}{2}} e^{-j\frac{\pi}{4}\operatorname{sign}\left[\sin\left(p\frac{\pi}{2}\right)\right]} \qquad (4)$$

The channel spacing is FSR/N similar as in a conventional OFDM. After the FrFT AWG, the output signals are sampled at appropriate time windows.

III. EXPERIMENT

Figure 2(a) shows the experimental setup for a combined WSS-AWG FrOFDM system. We took $p=-0.0529$ at the transmitter by WSS and $p=0.0529$ at the receiver by FrFT AWG. The fabricated FrFT AWG has these parameters, $N=12$ ports, $FSR=120$ GHz, $l=10.1$ mm, $R=2.92$ m, $d_i=30.0$ um and $d_o=29.8$ um. Figure 2(b) shows the transfer function at port 5, 6, and 7. Although the spectra of each channel of FrOFDM signal are broader than the conventional OFDM signal, the interval of each channel is equal to the conventional OFDM. Since the WSS (finisar4000S) has only 4 ports, Ch1-5-9, ch2-6-10, ch3-7-11, and ch4-8-12 transmitted the same data. A 10 GHz mode locked laser diode (MLLD) emitted 1.5 ps pulse at 1541 nm. The pulse train was modulated by a 10 Gbit/s pseudo random bit sequence (PRBS)

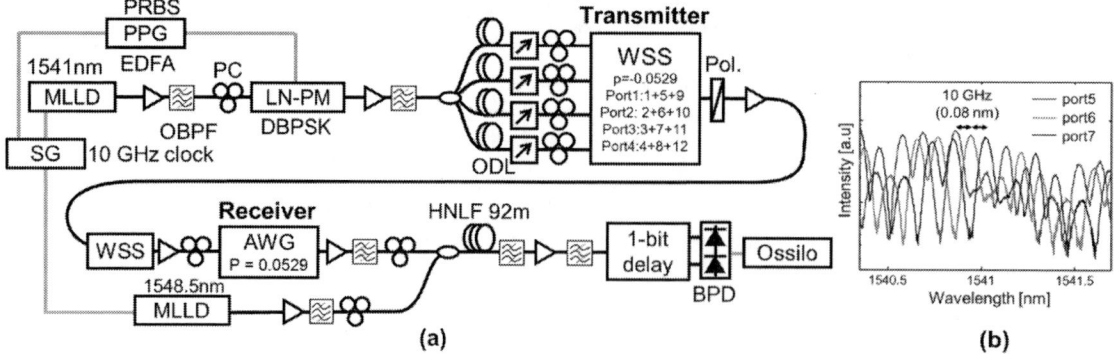

Fig. 2. (a) Experimental setup, (b) the transfer functions of FrFT AWG at port 5, 6, and 7.

OECC/PS2016

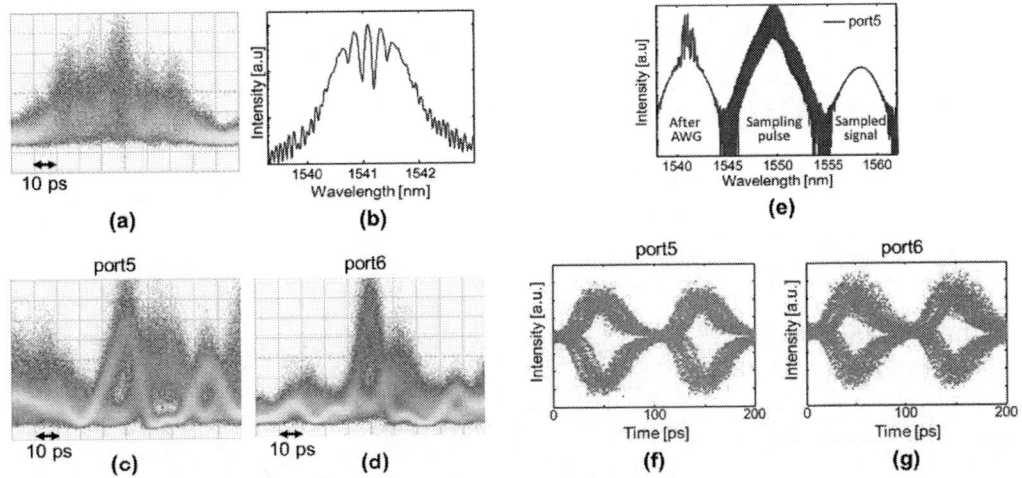

Fig. 3. (a) The waveform of the multiplexed signal, (b) the spectrum of the multiplexed signal, (c) the output signals of the FrFT AWG at port 5 and (d) port 6, (e) the result of four wave mixing, (f) the eye diagrams of demodulated signals at port 5 and (g) port 6.

with differential binary phase shift keying (DBPSK). The modulated signal was split into four signals, and different delays were applied by optical delay lines (ODLs), to synchronize them with pattern decorrelation. The four signals were fed to the input ports of 4×1 WSS at the transmitter, which generates a 120 Gbit/s FrOFDM signal. The polarizations of each port were aligned by a polarizer and polarization controllers (PCs). The multiplexed signal looks like a random shape as shown in Fig. 3(a), and Fig. 3(b) shows the spectrum. To reduce the influence of the wavelength dependency of the system, we applied the compensation filter using an additional WSS. Figure 3(c, d) show the output signals of the FrFT AWG at 5 port and 6 port. We can confirm the open eyes of the signals. A 1.5 ps four wave mixing based optical time gate by another MLLD, a highly nonlinear fiber (HNLF), and an optical bandpass filter (OBPF) sampled the output signals. Figure 3(e) shows the result of four wave mixing. The conversion efficiency was -24 dB. A 1 bit delay line and a balanced photodetector demodulated the sampled signal. Since the electrical oscilloscope had large timing jitter, the detected waveform was averaged. The eye diagram have a clear open at 5 port and 6 port as shown in Fig. 3(f, g).

IV. CONCLUSIONS

We have experimentally demonstrated the WSS based multiplexing and the AWG based demultiplexing for the 12×10 Gbit/s DBPSK fractional OFDM signal. This combination provides advantages for both the flexibility of the network architecture and the reduction of the system cost and complexity.

ACKNOWLEDGMENT

This work was supported by JSPS KAKENHI, Grant-in-Aid for JSPS Fellows, 261585. This work was supported by STARBOARD project of MIC Strategic Harmonized International R&D Promotion Programme (SHIP).

REFERENCES

[1] M. Jinno, et al., "Spectrum-efficient and scalable elastic optical path network: architecture, benefits, and enabling technologies," Communications Magazine IEEE., Vol. 47, no. 11, p. 66, 2009.

[2] M. Nakazawa, et al., "Ultra-speed "orthogonal" TDM transmission with an optical Nyquist pulse train," Opt. Exp., Vol. 20, no. 2, p. 1129, 2012.

[3] D. Hillerkuss, et al., "26 Tbit s-1 line-rate super-channel transmission utilizing all-optical fast Fourier transform processing," Nat. Photon., Vol. 5, p. 364, 2011.

[4] W. Shieh, et al., "OFDM for Optical Communications," Academic Press, 2009.

[5] J. Schroder, et al., "All-optical OFDM With Cyclic Prefix Insertion Using Flexible Wavelength Selective Switch Optical Processing," J. Lightwave Technol., Vol. 32, no. 4, p. 752, 2014.

[6] J. Pan, et al., "Nonlinear Electrical Compensation for the Coherent Optical OFDM System," J. Lightwave Technol., Vol. 29, no. 2, p. 215, 2011.

[7] G. Cincotti, "Optical OFDM based on the fractional Fourier transform," Proc. SPIE 8284,828409, 2012.

[8] T. Nagashima, et al., "PAPR Management of All-Optical OFDM Signal using Fractional Fourier Transform for Fibre Nonlinearity Mitigation," Proc. ECOC 2015, Valencia, 2015.

[9] T. Murakawa et al., "Fractional OFDM transmitter and receiver for time/frequency multiplexing in gridless, elastic networks," Proc. OFC 2015, Los Angeles, 2015.

[10] T. Nagashima, et al., "Cyclic Prefix Insertion for All-optical Fractional OFDM," Proc. PS 2015, Florence, 2015.

[11] G. Cincotti, "Enhanced Functionalities for AWGs," J. Lightwave Technol., Vol. 33, no. 5, p. 998, 2015.

TuE1-3

OECC/PS2016

Sub-fF-capacitance photonic-crystal photodetector towards fJ/bit on-chip receiver

Kengo Nozaki,[1,2,*] Shinji Matsuo,[1,3] Takuro Fujii,[1,3] Koji Takeda,[1,3] Masaaki Ono,[1,2] Abdul Shakoor,[2] Eiichi Kuramochi,[1,2] and Masaya Notomi[1,2]

[1]*Nanophotonics Center,* [2]*NTT Basic Research Laboratories,* [3]*NTT Device Technology Laboratories*
NTT Corporation, 3-1, Morinosato Wakamiya Atsugi, Kanagawa 243-0198, Japan
nozaki.kengo@lab.ntt.co.jp

Abstract: *A photonic-crystal photodetector having a <1-fF junction capacitance, 1-A/W responsivity, and 40-Gbit/s eye opening was demonstrated. Its resistor-loaded configuration unveiled a light-to-voltage conversion with a kV/W efficiency without amplifiers, promising for an fJ/bit-energy on-chip receiver.*
Keywords: *Photonic crystal, photodetector, photonic network on chip*

I. INTRODUCTION

Future microprocessors will need an unprecedented many-core architecture including a chip-scale optical communication which should be, especially for an on-chip-com network, fully integrated with laser sources, photoreceivers, and other functional nanophotonic devices with ultralow-power consumption. A photoreceiver, generally consisting of a photodetector (PD) and a trans-impedance amplifier (TIA) to generate sufficient voltage to drive the subsequent CMOS circuits, amounts to a sub-pJ/bit level energy cost and therefore will constitute a significant bottleneck when establishing such a dense on-chip photonic network [1]. One of the challenges with PDs is to realize an ultrasmall capacitance and thus allow the resistance-capacitance (RC) bandwidth to be kept at a high level even during connection to a high impedance receiver circuit. This would lead to a reduction of electrical amplification or even its elimination (referred as a receiver-less PD [1, 2]), promising an ultralow-energy on-chip photoreceiver.

Photonic crystals (PhCs) are promising as nano-PDs because of their strong light confinement in an ultrasmall dimension. We have already reported PhC-PDs embedded in an InGaAs absorption layer in an InP-based PhC waveguide, which we obtained using an ultracompact buried-heterostructure (BH) formation [3, 4]. This structure can confine both photons and carriers in an ultrasmall space that cannot be achieved by any other PDs, and hence promise a good applicability for nano-PDs with ultrasmall junction capacitance.

In this paper, we describe an InGaAs-embedded PhC-PD which an absorber length was reduced down to only 1.7 μm and a junction capacitance was less than 1 fF. Even with such ultrasmall size, it still exhibits a high responsivity of 1 A/W and a clear eye opening for a 40-Gbit/s signal. Furthermore, we fabricated a PhC-PD integrated with a several-kΩ load resistor to demonstrate an on-chip light-to-voltage conversion. We employed an electro-optic (EO) probing for the first time in testing the voltage generation in nano-PDs, and revealed a conversion efficiency as high as 4 kV/W. The expected bandwidth when removing the parasitic wiring elements would be more than 10 GHz. This suggests that the optical energy required for generating CMOS voltage level is less than 1 fJ/bit, which can be obtained without electrical amplification and therefore dominates the total energy consumption. These results reveal a successful way of realizing an ultrasmall/ultralow-energy photoreceiver that can be densely integrated on a chip.

II. ULTRASMALL PHOTONIC CRYSTAL PD

Figure 1(a) shows a schematic of our PhC-PD, consisting of an InP PhC waveguide, a BH for embedding the InGaAs absorber, and a lateral p-i-n junction [5]. The absorber was designed with a thickness of 150 nm, a width of 400 nm, and lengths of only 1.7 μm. A lateral p-i-n junction was formed by Zn diffusion and Si ion implantation for the p- and n-type doping, respectively. The PhC airholes were formed by EB lithography and Cl₂-based dry etching. After metallization, the InAlAs sacrificial layer beneath the PhC slab was etched to form an air-bridge structure. Figure 1(b) shows a cross-sectional SEM image, indicating a flat surface thanks to the successful butt-joint regrowth. The small dimensions of absorber and the p-i-n junction, the junction capacitance should be down to the fF level, as shown in Fig. 1(c). Parallel-plate capacitance (= $\varepsilon_0 \varepsilon_{InGaAsP} L_{abs} T_j / d_j$)

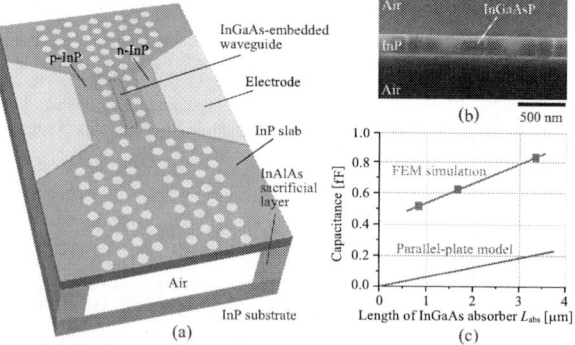

Fig. 1 PhC-PD structure. (a) Schematic of PhC-PD. (b) Cross-sectional view SEM images of fabricated device. (c) Calculated capacitance of PhC-PD. The blue curve is calculated from the parallel-plate model. The red plots are the results simulated by FEM with a 3-D model.

is less than 0.2 fF for L_{abs} < 3 μm. However, for an ultrasmall junction, the fringing field contribution of the junction also becomes significant, and hence it is important to include the fringe capacitance [6]. This contribution was simulated by the FEM with a full 3-D model excluding the large electrical pads, and the total capacitance becomes higher than that of the parallel plate model. The total capacitance of our PhC-PD is still < 1 fF, which is still much smaller than those of the Ge-waveguide PDs with 4–5 fF [7, 8].

We evaluated a photoresponse of the PhC-PD. The dark currents were approximately <100 pA and 15 nA for bias voltages of −2 and −10 V, respectively. We successfully estimated a large optical responsivity of 0.98 A/W even for a short absorber length of 1.7 μm. Specifically, our BH formation does not increase the non-radiative carrier recombination loss thanks to the successful butt-joint epitaxial growth. Figure 2 shows the operation dynamics of our PD, into which we injected an intensity-modulated optical signal with a peak power of 100 μW. The clear eye opening was observed for 40-Gbit/s non-return-to-zero (NRZ) signals generated with a $2^{31}−1$ pseudo-random bit sequence. As shown in the small-signal responses for different reverse bias voltages, the 3-dB bandwidth was 28.5 GHz when the bias voltage was −12 V. This bandwidth suggests the capability for a bit rate of around 50 Gbit/s for an NRZ signal, which agrees with the observed eye diagram.

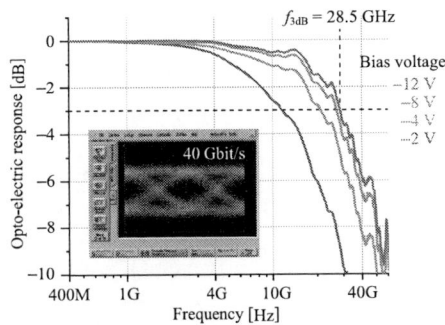

Fig. 2 Small signal responses for different reverse-bias voltages. Inset shows the eye diagram for 40 Gbit/s NRZ optical signals. The input wavelength was 1536.7 nm and the optical peak power was 100 μW.

III. A RESISTOR-LOADED PD FOR ON-CHIP LIGHT-TO-VOLTAGE CONVERSION

The ultrasmall capacitance of our PD enables us to connect it with a high load resistance without amplifiers to convert photocurrent to voltage while keeping a large RC bandwidth. However, there has never been a report evaluating the on-chip light-to-voltage conversion dynamics of resistor-loaded nano-PDs. The experimental difficulty is that a conventional measurement using an oscilloscope/network analyzer with an additional electrical pad would hinder correct device evaluation, because their impedances are generally lower than the device load, or 50 Ω in most cases. This makes it difficult to measure the voltage across the load. (Note that direct connection with a high-impedance CMOS gate would be available as a photoreceiver in on-chip-com application.) In our measurement, we employed an electro-optic (EO) probing technique [9]. When we prepared the sample for EO probing, our PhC-PD was connected to a load resistor R_{load} with a gold strip line with a length L_{strip}, as shown in Fig. 3(a).

The experimental setup for EO probing is shown in Fig. 3(b). Sinusoidal modulated light (50 MHz) was injected into the PhC-PD. Photocurrent flows into the load resistor, and generates a modulated electric field (proportional to the voltage) between the strip lines. An EO probe consisting of an optical fiber with an EO crystal (ZnTe) was brought towards the strip line. CW light (λ = 1.55 μm) was separately injected into the EO probe and sensed the modulated electric field via the EO crystal. By combining a polarization beam splitter and a balanced photoreceiver, the polarization change of CW light was detected as an EO probing voltage. Before the device measurement, the EO probing voltage for AC voltage applied to the strip line was acquired to obtain the correspondence between two voltages. Thereafter, we replaced the reference with a PhC-PD sample to evaluate the photo-generated voltage.

Figure 3(c) shows the light-to-voltage conversion efficiency η_{LV} for PDs with different R_{load}. The generated AC voltage V_{pp} clearly increased when R_{load} was larger. A maximum η_{LV} = 3.95 kV/W was achieved for R_{load} = 8.8 kΩ. These results show that an optical power of 50 μW can generate the V_{pp} = 200 mV that is required for a CMOS inverter [10]. On the other hand, the operation bandwidth should be limited by RC, although a gold strip line and a pad with a much larger capacitance than the PhC-PD were included in our sample because they were necessary for EO probing. Figure 3(d) summarizes the 3-dB bandwidth (blue plots for left vertical axis) as a function of $1/(R_{load} + R_{pd})$. These plots have a linear relation as they are mainly determined by $f_{RC} = [2\pi(R_{pd}+R_{load})C]^{-1}$, where C consists of both the PhC-PD capacitance and the parasitic capacitance caused by the strip line and pads. The dashed lines are the theoretical curves obtained by assuming C = 16 and 110 fF, which are dominated by parasites, and fit well with the experimental plots.

Another figure we evaluated was the product of η_{LV} and f_{RC}, which are in a trade-off relationship, because they are proportional to R_{load} and $(R_{pd}+R_{load})^{-1}$, respectively. This efficiency-bandwidth product (EBP) [V/W·Hz] (= [V/J]) can indicate the optical energy needed to generate the required voltage, regardless of the bit rate of the optical signal. The EBPs are denoted by green plots on the right vertical axis in Fig. 3(d). A shorter L_{strip} enhances the EBP because f_{RC} increases while η_{LV} remains constant (See Fig. 3(c)). The EBP values were in the 4–5 × 10^{11} and 2–3 × 10^{12} V/J ranges for L_{strip} = 2.5 and 0.2 mm, respectively. As a result, they can be translated to a required optical energies of 200 and 33 fJ/bit for L_{strip} = 2.5 and 0.2 mm, respectively, to obtain V_{pp} = 200 mV with an NRZ optical signal. We also theoretically discuss an ideal case where there is no parasitic capacitance. The bold dashed curves in Fig. 3(d) denote f_{RC} and EBP in an ideal situation calculated by assuming only a PD junction capacitance of C = 0.6 fF. This makes the bandwidth higher than 10 GHz, which should be practically acceptable. Subsequently, the expected EBP exceeds 10^{14} V/J, corresponding to a required optical energy of less than 1 fJ/bit. These performance levels significantly surpass the performance of a conventional PD-TIA circuit. Such a situation can be realized by removing the strip line and the pads used in the experiment, because they were just needed for the EO-probing measurement. Our experimental and theoretical results for an ultrasmall PhC-PD have revealed the feasibility of an amplifier-less photoreceiver on a chip with a practically acceptable size, efficiency, bandwidth, and power consumption.

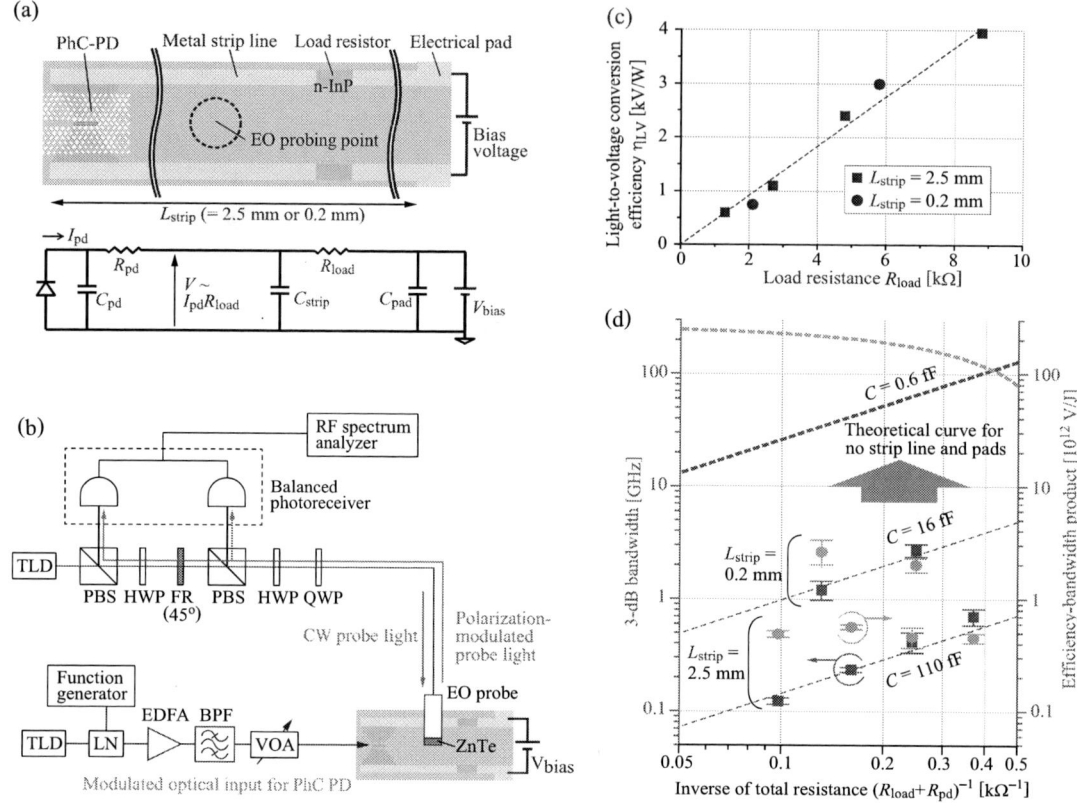

Fig. 3   Resistor-loaded PhC-PD and EO probing measurement. (a) Schematic of the sample (top) and corresponding equivalent circuit (bottom). (b) Experimental setup for EO probing measurement. (c) Light-to-voltage conversion efficiency for different load resistances R_{load}. Square and circle plots denote the results for L_{strip} values of 2.5 and 0.2 mm, respectively. (d) 3-dB bandwidth (square plots for left axis) and the efficiency-bandwidth product (circle plots for right axis). The plots show the experimental results, and the dashed curves show the calculated results considering both the PD junction capacitance and the parasitic capacitances. The bold dashed curves are calculated under the assumption of no parasitic capacitances.

ACKNOWLEDGMENT

This work was supported by CREST, Japan Science and Technology Agency.

REFERENCES

[1]   D. A. B. Miller, "Device requirements for optical interconnects to silicon chips," Proceedings of the IEEE, vol. 97, pp. 1166-1185, 2009.

[2]   C. Debaes, et al., "Receiver-less optical clock injection for clock distribution networks," IEEE Journal of Selected Topics in Quantum Electronics, vol. 9, pp. 400-409, Mar-Apr 2003.

[3]   S. Matsuo, et al., "High-speed ultracompact buried heterostructure photonic-crystal laser with 13 fJ of energy consumed per bit transmitted," Nature Photonics, vol. 4, pp. 648-654, 2010.

[4]   K. Nozaki, et al., "InGaAs nano-photodetectors based on photonic crystal waveguide including ultracompact buried heterostructure," Optics Express, vol. 21, pp. 19022-19028, Aug 12 2013.

[5]   K. Takeda, et al., "Few-fJ/bit data transmissions using directly modulated lambda-scale embedded active region photonic-crystal lasers," Nature Photonics, vol. 7, pp. 569-575, Jul 2013.

[6]   A. Shakoor, et al., "Compact 1D-silicon photonic crystal electro-optic modulator operating with ultra-low switching voltage and energy," Optics Express, vol. 22, pp. 28623-28634, Nov 17 2014.

[7]   R. Going, et al., "Germanium wrap-around photodetectors on Silicon photonics," Optics Express, vol. 23, pp. 11975-11984, May 4 2015.

[8]   L. Virot, et al., "Germanium avalanche receiver for low power interconnects," Nature Communications, vol. 5, Sep 2014.

[9]   T. Nagatsuma, "Measurement of High-Speed Devices and Integrated-Circuits Using Electrooptic Sampling Technique," Ieice Transactions on Electronics, vol. E76c, pp. 55-63, Jan 1993.

[10] S. Assefa, et al., "CMOS-Integrated Optical Receivers for On-Chip Interconnects," IEEE Journal of Selected Topics in Quantum Electronics, vol. 16, pp. 1376-1385, Sep-Oct 2010.

TuE1-4

4-channel synchronous THz-wave generator composed of arrayed UTC-PDs and antennas

Goki Sakano, Jun Haruki, Kazuki Sakuma, and Kazutoshi Kato
Graduate School of Information Science and Electrical Engineering, Kyushu University,
744 Motooka Nishi-ku, Fukuoka 819-0395, Japan
E-mail: sakano@optoele.ed.kyushu-u.ac.jp

Abstract: We configured THz wave generator consistiong of four-channel arrayed photomixers/antennas. We demonstrated that the powers are combined with the directional gain of 5.9 dB and proportional to the square of the number of photomixers.

Keywords: terahertz wave, uni-traveling carrier photodiode, photomixing, arrayed antenna.

I. INTRODUCTION

In recent years, data traffic is increasing dramatically because of diffusion of the mobile PCs, smart phones, tablet terminals, and so on. Along with this trend, the amount of data dealt with the network of wireless access is required to be larger. To cope with these demand, the capacity and speed of wireless transmission should be improved. High frequency carrier is one of the effective media to meet these demands. Especially, the use of the terahertz wave (THz wave), whose frequency is defined as that from 100 GHz to 10 THz, is expected to be one of the solutions to realize high data-rate wireless transmission. An effective approach to generate coherent THz wave is phomixing two different lightwaves by using a photomixer[1,2]. The uni-traveling carrier photodiode (UTC-PD) has been used as the photomixer [3]. The UTC-PD generate higher output power than the typical pin-PD, but its maximum output power is limited to about 100 μW at 300 GHz [4]. THz wave of 100 μW propagates only several meters because its power attenuates proportional to the square of distance. Thus, it is necessary to improve the power of THz wave to achieve the practical THz-wave transmission. We have proposed the system to combine THz waves which consists of arrayed photomixers and demonstrated that the powers are combined with a directional gain of 4.5 dB at three-arrayed photomixers[5].

In this paper, to obtain the THz wave with much higher power, we increase the number of photomixers and investigate the relationship between the directional gain and the number of photomixers. We demonstrate that the powers are combined with the directional gain of 5.9 dB at four arrayed photomixers and it is in good agreement with theoretical values.

II. EXPERIMENTAL CONFIGURATION

Our proposed system is configured to combine THz waves which are generated at the photomixers by photomixing two lightwaves and radiated from the antennas as shown in Fig. 1. For synchronizing the THz waves, their phases are tuned by the optical delay lines (ODLs). In the experiment, the single mode lightwave output from a laser diode (LD) was phase-modulated at the frequency of 25GHz with an optical modulator to generate the optical frequency comb (OFC) with 25 GHz interval. The two lightwaves (f_1 and f_2) whose frequency difference is 300 GHz were extracted from the OFC by the arrayed waveguide grating (AWG) filter. They were coupled with an optical coupler (OC) and amplitude-modulated at the frequency of 10 MHz. The lightwaves were amplified by an EDFA and split into four optical paths with the optical splitters (OSs). Thus, each optical path contains a pair of lightwaves (f_1 and f_2). These four pairs of lightwaves were introduced through micro lens array (MLA) into four channels of 500-μm-spaced UTC-PD array with the bowtie-antennas as shown in Fig. 2(a). The enlarged picture of the single UTC-PD/antenna is shown in Fig. 2(b). Each UTC-PD generated the THz wave with the frequency of 300 GHz and it was radiated to the air from the bowtie-antenna. The phases of the THz waves were tuned to be synchronized with each other by the ODLs which were located between the OSs and the MLA. The power of the combined THz wave was detected by the schottky-barrier diode (SBD) and measured at the 10 MHz component with the spectrum analyzer.

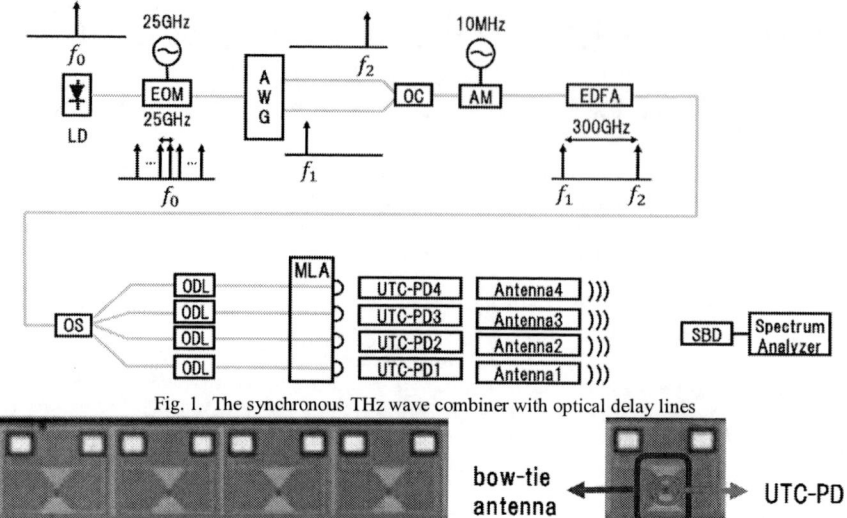

Fig. 1. The synchronous THz wave combiner with optical delay lines

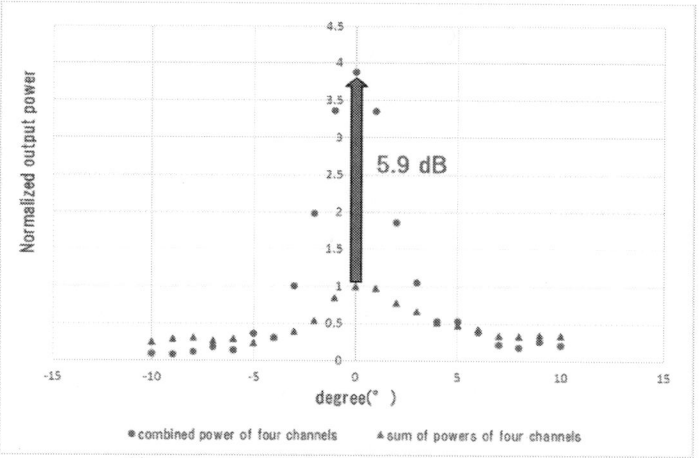

Fig. 2(a). UTC-PD/bowtie antenna array (b) Enlarged view of UTC-PD/bowtie antenna

III. EXPERIMENTAL RESULT

We measured the angular dependence of the radiated THz wave by revolving the SBD. In Fig. 3, the blue circles show the power of the THz wave combined from four channels and the red circles show the arithmetic sum of each power from each channel. These values are normalized with the peak power of the arithmetic sum of each power. The radiation profile of the combined THz wave has a single sharp peak and the peak power is larger by 5.9-dB than the arithmetic sum of each power from each channel. Furthermore, the peak width of the power of combined THz wave is narrower than that of the power of the arithmetic sum of four channels. These results mean that synchronously combining THz wave results not only in sum of the THz powers but also in directional gain which is caused by power concentration into the vertical direction.

Fig. 3. Angular dependence of the radiated THz waves

In general, the directional gain G_d is approximated by the equation (1)

$$G_d \approx D_f = \cfrac{1}{\cfrac{1}{N} + \cfrac{2}{N^2}\sum_{m=1}^{N-1}\cfrac{N-m}{mk_0d}\sin mk_0 d \cos m\alpha} \qquad (1)$$

where D_f is directivity depending on the antennas' layout, k_0 is the wavenumber of the electromagnetic wave, d is the spacing between each antenna, α is the excitation phase of each antenna and N is the number of the antennas (in this experiment, N equals to the number of the photomixers).

Assuming that the spacing between each antenna is 0.8λ or less (λ is the wavelength of electromagnetic) and α is zero (perpendicular radiation), G_d is approximated by equation (2).

$$G_d \approx 2N\frac{d}{\lambda} \qquad (2)$$

Here, in our experiment, the channel spacing is 500 μm and which equals to a half of wavelength at 300GHz. So, the directional gain equals to the number of channels N, and consequently, the relative peak power of the combined THz wave is expected to be N^2 such as equation (3).

$$relative\ peak\ power = (the\ number\ of\ channels) \cdot (directiond\ gain)$$

$$= N \cdot N = N^2 \qquad (3)$$

We also measured the combined power of two channels and three channels as well as that of four channels. In Fig. 4 the red circles show the relative peak powers of the combined THz waves, and the blue circles and the dotted line show the theoretical relative THz power ($=N^2$) as a function of the number of the channels. The measured powers are in good agreement with the theoretical values and increasing proportionally to the square of the number of channels.

These results indicate that our proposed configuration has a potential to achieve a THz-wave power for future practical wireless transmission by increasing the number of the arrayed photomixers.

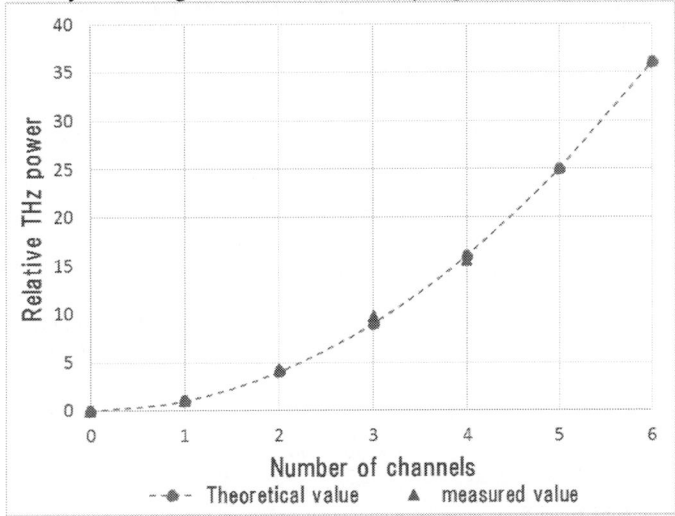

Fig. 4. The relative THz power as a function of the number of channels

IV. CONCLUSIONS

We proposed and demonstrated the THz-wave generator with tuning the phases of the THz waves so as to synchronously combined with each other. The experimental result showed that the combined 300-GHz wave had directional gain and it increased up to 5.9dB in the case of four arrayed photomixers. We also showed the relative peak combined power is increasing in proportion to the square of the number of photomixers which is good agreement with the arrayed antenna's theory.

ACKNOWLEDGMENT

A part of this work was supported by the Strategic Information and Communications R&D Promotion Programme (SCOPE) 2015, from the Ministry of Internal Affairs and Communications, Japan, the Collaborative Research Based on Industrial Demand/JST, and CREST/JST. The authors thank NTT Device Technology Laboratories for their experimental support.

REFERENCES

[1] A. Hirata, and M. Yaita, "Ultrafast Terahertz Wireless Communications Technologies", IEEE Trans. Terahertz Science and Technology, Vol.5, pp. 1128-1132 , 2015

[2] M. Inoue, M. Hodono, S. Horiguchi, K. Arakawa, M. Fujita, and T. Nagatsuma, Proc. Int. Symp. Electromagnetic Theory., pp. 211, 2013

[3] T. Ishibashi. Y. Muramoto, T. Yoshimatsu, and H.Ito, "Unitraveling-Carrier Photodiodes for Terahertz Applications" IEEE J. of sel. Topics in Quantum Electronics, Vol. 20, pp. 3804210, 2014

[4] A. J. Seeds, H. Dhams, M. J. Fice, and C. C. Renaud, J. Lightwave Technol, vol.33, pp. 579-587, 2015

[5] K. Sakuma, J. Haruki, G. Sakano, K. Kato, "Coherent THz wave combiner composed of arrayed-uni-traveling carrier photodiodes and planar lightwave circuit", SPIE Photonics West 2016, 9747-59, 2016

[6] W. L. Stutzman, G. A. Thiele, "Antenna Theory and Design", John & Winley sons, p292-294, 2013

Demonstration of Micro-projection enabled Short-Range Communications for 5G

Hsi-Hsir Chou[1,2,*] and C.-Y. Tsai[1]

[1]Department of Electronic and Computer Engineering, National Taiwan University of Science and Technology, Taipei, 106 Taiwan
[2]Department of Engineering, Cambridge University, Cambridge, CB3 0FA U.K
*E-mail address:hsi-hsir.chou@trinity.cantab.net

Abstract: *A micro-projection enabled short-range communication (SRC) system using red-, green- and blue-based light-emitting diodes (LEDs) which offers simultaneously micro-projection and high-speed data transmission, for personal communication device in 5G application is reported and experimentally demonstrated.*

Keywords: *Free-space optical communication; Light-emitting diodes; Spatial light modulators.*

INTRODUCTION

With the applications of portable communication devices (PCDs) i.e. smartphones for consumer use which have been extremely rapid in the last decade, it is widely expected that in 2017, there will be approximately 2 billion active tablets and 4~5 billion smartphones to access Internet service [1]. Although these PCDs are getting great, one of their limitations is the size of the screen, which of necessity has to be small. In order to solve this major problem, the idea of building a projector into a PCD so that images can be displayed on any nearby flat surface was raised by makers. However such PCDs proved were unfortunately too bulky for users to accept until Samsung firstly announced its new smartphone, Galaxy Beam [02], which is a new, small, thin and lightweight phone that has a projector built into it that allows users to project whatever is on the screen onto any nearby surface. As a precursor to the adoption of the next generation wireless access (5G) which is projected to be in place by 2020, digital services are being upgraded allowing for more data bandwidth transmission since the aim of 5G is to provide connectivity for any kind of device and any kind of applications that may benefit from being connected, including mobile connectivity for people and various objects in user's environment [03]. However this raises new challenges for the design of a new generation PCD to meet the aim of 5G since those currently available short-range communication (SRC) technologies on PCDs such as Near Field Communication (NFC) and Bluetooth which have a limited data transmission rate less than 5 Mb/s seems not competitive.

With the recent development of optical wireless communication technologies, a white light emitting diode (LED) based visible light communication (VLC) technology [4] offers simultaneously illumination and high-speed data transmission, making it a good candidate as an alternative short-range communication technology for 5G PCD application since LEDs have widely used on PCD either for camera or for micro-projector application. In order to increase the functionality and simultaneously reduce the physical size of a PCD, it is widely believed that VLC technology using the light sources of micro-projector system i.e. RGB-based LEDs will be integrated into the micro-projector module on PCD [5, 6] to provide a new communication approach. This will dramatically provide not only a new efficient but also a faster and a secure communication approach for PCD applications. In this paper, VLCs using RGB-based LEDs with an Liquid Crystal on Silicon (LCoS) based micro-projector system serving as a new SRC system for 5G PCD application is experimentally investigated and reported. Complex modulation schemes such as Multilevel Pulse Amplitude Modulation (M-PAM), M-ary Phase Shift Keying modulation (MPSK) and M-ary Quadrature Amplitude Modulation (M-QAM) which represented the amplitude, phase and quadrature amplitude were used to investigate the highest possible data transmission rate of a VLC within an LCoS-based micro-projector system.

EXPERIMENTAL SETUPS

In our experimental setup, a low cost commercially available component, Philip Luxeon Z series RGB-based LEDs as illustrated in Fig.1 was used as the light source. It has three LED chips which had the peak wavelengths of 650nm (Red), 550nm (Green) and 450nm (Blue) respectively. These chips were fixed on a 20 mm x 20 mm metal base printed circuit board (metal core PCB, MCPCB) serving as a RGB-based LEDs transmitter (TX) in our experimental architecture in order to minimize the physical size of TX and to provide a sufficient illumination efficiency for micro-projection application simultaneously. The measured 3-dB modulation bandwidth of the Red, Green and Blue LEDs within the LCoS-based micro-projector system are approximately 8 MHz, 16MHz and 12 MHz respectively. The signals for TX transmission were randomly generated and modulated offline by a Matlab program on a personal computer (PC) and uploaded to the memory of an arbitrary waveform generator (AWG, Keysight 33621A, 1GSa/s sampling rate). Since intensity modulations were utilized to encode data on the LEDs for SRC links, the emitted light of each LED chip on TX were then individually driven by modulated signals from AWG and separately biased by DC currents through low-frequency Bias Tees. A DC current of 350 mA which has been observed to have the best performance from our previous experiments was used in this research.

In order to experimentally demonstrate the feasibility of establishing SRC links while performing micro-projection simultaneously, a reflective LCoS (Liquid Crystal on Silicon) based micro projector system for micro-projection application which was designed and simulated in [6] was modified and applied to this experimental works. This modified micro-projector system to perform micro-projection was composed of a collimated lens, L1 (Thorlabs, 50 mm diameter), a polarization beam splitter (PBS, Thorlabs, 50:50), an LCoS device, and an image lens, L2 (Thorlabs, 50 mm diameter). The incident light from RGB-based LEDs was firstly collimated by a L1 and was then selected by a PBS. Although the LCoS device is a polarization sensitive device, in which only one polarization state of the incident light selected by the PBS can be used in our proof of concept research works, this disadvantage can be further compensated through a conventional polarization conversion system (PCS) in order to reduce the system light loss in the practical applications. The LCoS device used was JD9554 manufactured from Jasper Display Corp. (JDC), Taiwan. In our experiments, a computer generated hologram (CGH) composed of sub-holograms was uploaded to the LCoS device. For the proof of the concept research, one of the sub-holograms (800 x 600 pixel) corresponding to a super video graphics array (SVGA) resolution was designed to perform the micro projection function. The rest of the other sub-holograms on the LCoS device were simply used to perform the polarization modulation reflecting the light into the RX to establish SRC links.

At the receiver side (RX), incident lights from LCoS-based micro-projector system was firstly collimated by a focus lens, L3 (Thorlabs, 50 mm diameter) before focusing on a Si transimpedance amplified photodetectors (Thorlabs, PDA10A). The photodetector used have responsivities of 0.385 A/W, 0.25A/W and 0.175 A/W for Red, Green and Blue wavelength respectively and has an active area of 0.8 mm^2. The received signals were recorded by a real-time oscilloscope (Keysight DSOX 4104A, 5 GSa/s sampling rate) and analyzed directly through a vector signal analysis software (Keysight VSA89600). Although a transmission distance of 0.65m between TX and RX was initially used for our proof of concept research work, it could be further extended by using an array of identical LED emitters.

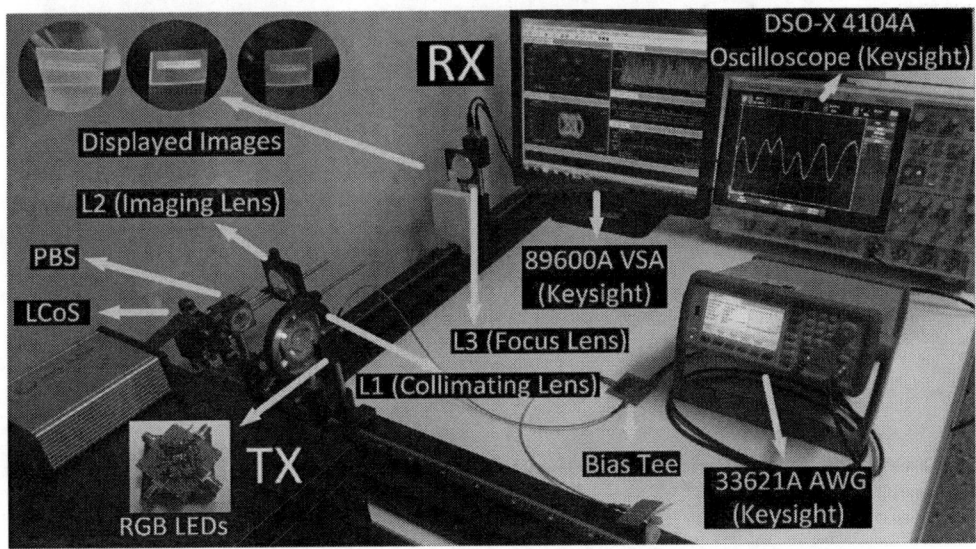

Fig. 1. Experimental demonstration of VLCs within an LCoS-based micro-projector system

EXPERIMENTAL RESULTS AND DISCUSSIONS

In our first experiment, SRCs were performed based on an LCoS-based micro-projector system and micro-projection was not performed simultaneously. The data transmission performance of M-PAM (i.e. 4-PAM), MPSK (i.e. QPSK) and M-QAM (i.e. 16-QAM) modulation schemes was investigated and the results are shown in Fig. 2-4 for Red, Green and Blue LEDs respectively. The transmission performance of each modulation scheme was evaluated through the estimated and calculated BERs from the measured eye diagrams and constellation diagrams [7]. The results shown that the highest possible aggregative data transmission rates of 4-PAM, QPSK and 16-QAM modulation schemes at a BER of 10^{-6} which are highly superior to the limitation of Forward Error Correction (FEC) standard (BER $\leq 3.8 \times 10^{-3}$) are 81 Mb/s, 420 Mb/s and 288 Mb/s respectively. A higher aggregative data transmission rate up to 560 Mb/s is possible by using QPSK modulation scheme, but the BER will be increased from 10^{-6} to 10^{-3}. In the second experiment, SRCs were performed in the same LCoS-based micro-projector architecture and micro-projection was also performed simultaneously. The measurement results shown that the highest possible aggregative data transmission rates of 4-PAM, QPSK and 16-QAM modulation schemes at a BER of 10^{-6} are 71 Mb/s, 344 Mb/s and 240 Mb/s respectively. A higher aggregative data transmission rate up to 480 Mb/s is possible by using QPSK modulation scheme, but the BER will be increased from

10^{-6} to 10^{-3}. Although there is a slightly performance degradation, compared with the pure SRCs without performing micro-projection simultaneously, our proposed works presented will potentially offer an alternative and faster communication approach for the application of a new PCD in 5G since to the best of our knowledge, this is the first time that a micro-projection enabled SRC system has experimentally implemented and demonstrated without any bandwidth improvement techniques and offline signal processing.

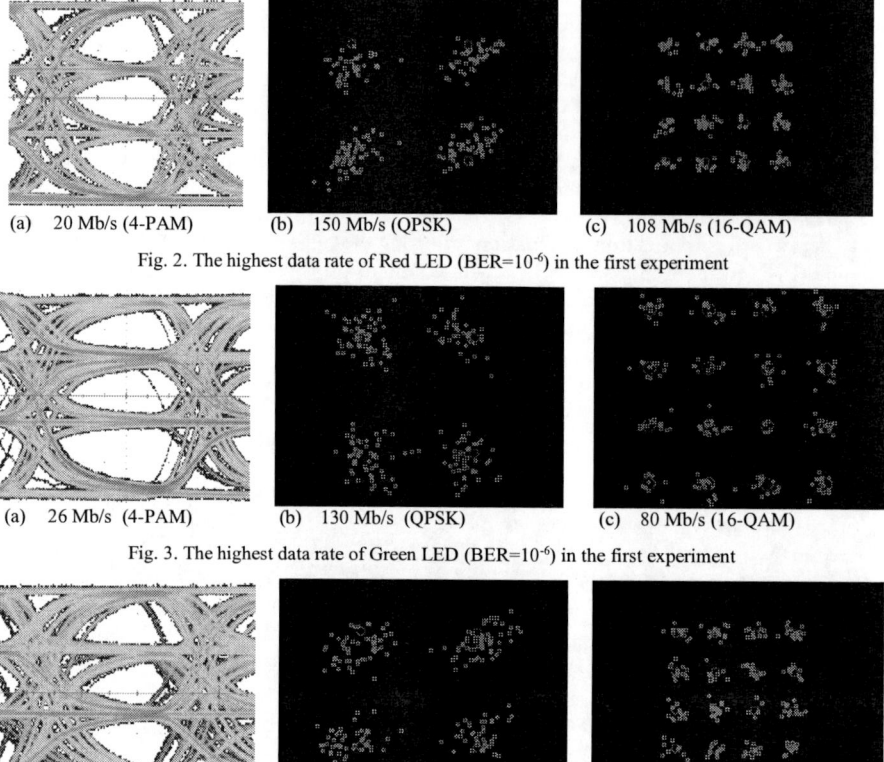

    (a)   20 Mb/s (4-PAM)   (b)   150 Mb/s (QPSK)   (c)   108 Mb/s (16-QAM)

Fig. 2. The highest data rate of Red LED (BER=10^{-6}) in the first experiment

    (a)   26 Mb/s (4-PAM)   (b)   130 Mb/s (QPSK)   (c)   80 Mb/s (16-QAM)

Fig. 3. The highest data rate of Green LED (BER=10^{-6}) in the first experiment

    (a)   35 Mb/s (4-PAM)   (b)   140 Mb/s (QPSK)   (c)   100 Mb/s (16-QAM)

Fig. 4. The highest data rate of Blue LED (BER=10^{-6}) in the first experiment

ACKNOWLEDGMENT

The research work presented in this paper was in part supported by Ministry of Science and Technology (MOST), Taiwan under grant numbers MOST 103-2221-E-011-037. The authors would also like to appreciate part of the financial supports from the Department of Electronic and Computer Engineering and Taiwan Building Technology Center, National Taiwan University of Science and Technology, Taiwan. Acknowledgements also go to the Department of Engineering, and Trinity College, Cambridge University.

REFERENCES

[1] Corcoran, P., "The Internet of Things: Why now, and what's next?," in Consumer Electronics Mag., IEEE , vol.5, no.1, pp.63-68, Jan. 2016

[2] Samsung Electronics. Samsung GALAXYbeam, http://www.samsung.com/global/microsite/galaxybeam/

[3] "5G: What is It,?" Ericsson white paper, available at http://www.ericsson.com/res/docs/2014/5g-what-is-it.pdf

[4] D. C. O'Brien, et al., "Visible light communications: Challenges and possibilities," Personal, Indoor and Mobile Radio Communications, 2008. PIMRC 2008. IEEE 19th International Symposium on, Cannes, 2008, pp. 1-5.

[5] A. Jovicic, Junyi Li and T. Richardson, "Visible light communication: opportunities, challenges and the path to market," in IEEE Communications Magazine, vol. 51, no. 12, pp. 26-32, December 2013.

[6] H. H. Chou, S. K. Liaw, M. J. Chien, C. J. Wu and C. Teng, "Experimental study of a portable FLCOS projector with visible light communication (VLC) technology," Next-Generation Electronics (ISNE), 2015 International Symposium on, Taipei, 2015, pp. 1-3.

[7] F. Xiong, "Digital modulation techniques," Artech House, Boston, London, 2000.

TuE2-1

Thermal Emission Control by Photonic Crystals

Susumu Noda

Department of Electronic Science and Engineering, Kyoto University, Japan

*corresponding author: snoda@kuee.kyoto-u.ac.jp

Converting from a broadband to a narrowband thermal emission spectrum with minimal loss of energy is important in the creation of efficient environmental sensors and biosensors as well as thermo-photovoltaic power generation systems. In the first part of my talk, I would like to discuss about such thermal emission control. It is shown that the emission peak intensity can be much greater than that of a blackbody sample and the emission bandwidth are narrowed significantly (Q>30~100) by controlling the electronic and photonic states, under the same input power and thermal management conditions. Another issue in thermal emission is that high-speed modulation is difficult to achieve because the intensity of thermal emission is usually determined by the temperature, and the frequency of temperature modulation is limited to 10–100 Hz even when the thermal mass of the object is small. In the second part of my talk, I would like to discuss about the dynamic control of thermal emission via the control of emissivity (absorptivity), at a speed much faster than is possible using the conventional temperature-modulation method.

Biography

Susumu Noda

Susumu Noda received B.S., M.S., and Ph.D. degrees from Kyoto University, Kyoto, Japan, in 1982, 1984, and 1991, respectively, all in electronics. In 2006, he has received an honorary degree from Gent University, Gent, Belgium. From 1984 to 1988, he was with the Mitsubishi Electric Corporation, and he joined Kyoto University in 1988. Currently he is a full Professor with the Department of Electronic Science and Engineering and a director of Photonics and Electronics Science and Engineering Center (PESEC), Kyoto University. His research interest covers physics and applications of photonic nanostructures based on photonic crystals. He is the recipient of various awards, including the IBM Science Award (2000), the Japan Society of Applied Physics Achievement Award on Quantum Electronics (2005), Optical Society of America Joseph Fraunhofer Award/Robert M. Burley Prize (2006), 1st the Japan Society of Applied Physics Fellow (2007), IEEE Fellow (2008), The Commendation for Science and Technology by the Minister of Education, Culture, Sports, Science and Technology (2009), IEEE Nanotechnology Pioneer Award (2009), The Reo-Esaki Award (2009), Medal with Purple Ribbon (2014), and the Japan Society of Applied Physics Outstanding Achievement Award (2015).

Mid IR Applications of Si Photonics

G. Z. Mashanovich[1], J. Soler Penades[1], V. Mittal[1], G. S. Murugan[1], A. Z. Khokhar[1], C. J. Littlejohns[1], S. Stankovic[1], A. Ortega-Monux[2], G. Wanguemert-Perez[2], R. Halir[2], I. Molina-Fernandez[2], C. Alonso-Ramos[3], D. Benedikovic[3], A. Villafranca[4], P. Cheben[5], J. J. Ackert[6], A. P. Knights[6], J. S. Wilkinson[1], M. Nedeljkovic[1]

[1] Optoelectronics Research Centre, University of Southampton, Southampton, SO17 1BJ, UK
[2] Departamento de Ingenier a de Comunicaciones , Universidad de Malaga, 29071 Malaga, Spain
[3] Inst. Elect. Fondamentale (IEF), Univ Paris Sud, CNRS UMR 8622, Université Paris-Saclay F-91405 Orsay, France
[4] Institute of Optics, Spanish National Research Council, Madrid 28006, Spain
[5] National Research Council Canada, Building M-50, Ottawa, K1A 0R6 Canada
[6] Department of Engineering Physics, McMaster University, 1280 Main St. West, Hamilton ON, L8S 4L7 Canada
g.mashanovich@soton.ac.uk

Abstract: *A review of passive and active photonic devices in three mid-IR silicon photonics material platforms suitable for communications and sensing, silicon on insulator (SOI), suspended silicon and germanium on silicon, is presented.*
Keywords: *mid-infrared, silicon, germanium, detector, sensor*

I. INTRODUCTION

Mid-infrared (MIR) silicon and germanium photonic devices and systems could be useful in a range of applications [1]. Two most important areas are sensing, particularly in the so-called fingerprint region (8-15μm), and telecom/ datacom applications, especially in the 2-3μm range [2]. The latter application area can utilize silicon on insulator (SOI), which is beneficial as advanced design, fabrication and testing techniques of SOI devices have been developed over the years and SOI is the most mature group IV technology available. We review some of our recent results on this platform here. The former application area does require new material platforms, because silicon dioxide is lossy beyond ~4μm and therefore SOI cannot be used in the fingerprint region. Several alternatives have been investigated such as silicon on sapphire (SOS), silicon on nitride (SON), or silicon on porous silicon. We report our recent results on suspended silicon and germanium on silicon (Ge-on-Si) platforms in this paper.

II. THREE MATERIAL PLATFORMS

A. Silicon on insulator

We have recently designed, fabricated and characterised a library of passive SOI devices at wavelengths as long as 3.8μm. In terms of waveguides, ~1dB/cm rib [2], strip [3] and slot [4] waveguides have been demonstrated in 400 and 500nm SOI. Also, low loss MMIs (0.1-0.2dB/MMI) have been fabricated and used in Mach-Zehnder interferometers (MZIs). Both standard and spiral waveguide designs were used for the demonstration of the first MIR thermal optical modulators in this platform [5]. In addition, single and cascaded ring/racetrack resonators have been reported [6]. Also, a 6-channel DEMUX based on angled MMIs (AMMI) has been reported with insertion loss of 2.5dB and cross talk better than -15dB [7]. Moreover, a MIR Fourier-transform spectrometer (Fig. 1) comprised of waveguides, couplers, bends, MMIs, and MZIs has also been realised [8]. Finally, by implanting boron into 220nm SOI rib waveguides and creating mid-bandgap states, a high speed detection (~28Gb/s) was achieved at ~2μm. The MIR silicon detector operated in avalanche mode with responsivity of 0.3A/W [9].

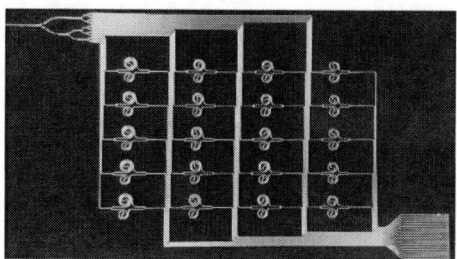

Fig. 1. MIR Fourier-transform spectrometer in SOI.

B. Suspended silicon

To fully exploit the transparency of Si, oxide can be removed and suspended waveguides created. This can be performed via two etch steps to form a rib waveguide and holes on the side for HF removal of oxide [10], or by fabricating photonic crystal waveguides [11]. In our novel approach, we are using subwavelength gratings as lateral claddings and also as access points for removal of the buried oxide layer, thus creating suspended silicon waveguides [12] (Fig. 2). Our approach involves only one dry etch step. Waveguides with propagation loss of <1dB/cm have been demonstrated at a wavelength of 3.8μm as well as low loss bends, MZIs, directional couplers and MMIs.

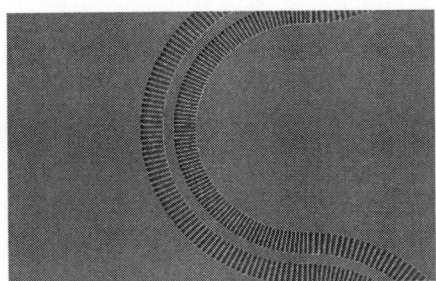

Fig. 2. Suspended Si waveguide with subwavelength grating claddings.

C. Germanium on silicon

Germanium is interesting photonic material that has larger transparency range, refractive index, optical modulation, non-linear effects and carrier mobility compared to silicon. In our work we have used Ge-on-Si wafers with 2 and 3μm thick Ge layers. The wafers were fabricated by the reduced pressure chemical vapor deposition (RPCVD). By using such wafers we have designed and characterised low loss waveguides (0.6 dB/cm at 3.8μm), MMIs (0.2dB/MMI), MZIs, AMMIs, and grating couplers [13]. In addition, we have demonstrated for the first time all optical modulation via free carrier absorption in Ge waveguides, two photon absorption (TPA) has been characterized in these waveguides showing the same trend with previously published results in bulk Ge, and picosecond TPA-based cross absorption modulation scheme was implemented [14]. Finally, for the first time an extensive theoretical investigation of free carrier plasma dispersion effects in Ge has been conducted [15] predicting that electro-absorption will likely dominate in Ge free carrier effect modulators.

III. CONCLUSIONS

The SOI platform is a good choice for shorter MIR wavelengths due to the well developed fabrication processes already in use for NIR applications. It is therefore the most suitable platform for high speed communications at the 2-3μm range. For wavelengths beyond 4μm, we have demonstrated a novel suspended Si platform that utilise subwavelength gratings as lateral claddings. After extensive design and fabrication stages, we have characterised sub dB/cm waveguides, and low loss passive devices at 3.8μm. Our future work will involve characterisation of these devices at longer wavelengths in order to explore the full transparency range of the platform and to investigate its suitability for sensing. Whilst silicon is dominating shorter wavelengths, germanium is becoming attractive for sensing applications in the fingerprint region due to its large transmission range (2-15μm). We have shown low loss Ge-on-Si waveguides and other passive devices at wavelengths ranging from 2 to 9μm. In order to use the entire MIR transmission range of Ge, novel Ge-based material platforms will need to be developed. The focus of the current research effort is on the realization of active devices and their integration with already demonstrated passive devices, in order to unlock the potential of MIR silicon photonics for a range of applications.

REFERENCES

[1]   R. Soref, Nat. Photon., vol. 4, pp. 495-497, 2010.
[2]   D. J. Richardson, Science, vol. 330, pp. 327-328, October 15, 2010.
[3]   G. Z. Mashanovich et al., J. Sel. Top. Quant. Electron., vol. 21, 8200112, 2015.
[4]   J. Soler Penades et al., IEEE Photon. Technol. Lett., vol. 27, pp. 1197-1199, 2015.
[5]   M. Nedeljkovic et al., IEEE Photon. Technol. Lett., vol. 26, pp. 1352-1355, 2014.
[6]   B. Troia et al., Opt. Exp., vol. 22, pp. 23990-24003, 2014.
[7]   Y. Hu et al., Opt. Lett., vol. 39, pp. 1406–1409, 2014.
[8]   M. Nedeljkovic et al., IEEE Photon. Technol. Lett., vol. 28, pp. 528-531, 2016.
[9]   J. J. Ackert et al., Nat. Photon., vol. 9, pp. 393-396, 2015.
[10] Z. Cheng et al., Photonics Journal, IEEE , vol.4, pp. 1510-1519, 2012.
[11] C. Reimer et al., Optics Express, vol. 20, pp. 29361-29368, 2012.
[12] J. Soler Penades et al., Opt. Lett., vol. 39, pp. 5661-5664, 2014.
[13] M. Nedeljkovic et al. IEEE Photon. Technol. Lett., vol. 27, pp. 1040-1043, 2015.
[14] L. Shen et al., Opt. Lett., vol. 40, pp. 2213-2216, 2015.
[15] M. Nedeljkovic et al., IEEE Photon. J., vol. 7, 2419217, 2015.

TuE3-1 (Invited)

Silicon Photonic Integrated Circuits for High-Capacity Optical Communications

Po Dong
Bell Labs, Nokia, 791 Holmdel Road,
Holmdel, NJ 07733, USA
E-mail address: po.dong@nokia.com

Abstract— We review silicon photonic integrated circuits for high-capacity optical communications. These circuits include highly integrated dual-polarization in-phase/quadrature modulators, single-chip polarization-diversity receivers, multi-channel discrete multi-tone modulation and detection circuits, and Stokes vector receivers. The high-degree integration of silicon photonics leads to low-cost and compact high-capacity optical modules.

Keywords—Optical communications, Silicon photonics, Optical interconnects, Photonic integrated circuits, Coherent transmission, Wavelength-division multiplexing, Stokes vector receiver.

I. INTRODUCTION

Silicon photonics exploits established CMOS fabrication infrastructure and is emerging as a disruptive optical technology for data communications, with a wide range applications such as intra-chip interconnects, short-reach communications in datacenters and supercomputers, optical access networks, and long-haul optical transmissions [1]. Its principal technical merits include high levels of optical integration and simple implementation of polarization diversity, making it an important platform to implement dual-polarization in-phase/quadrature (I/Q) modulators, coherent optical receivers, multiple-channel wavelength-division multiplexing (WDM) transmitters, WDM polarization diversified receivers, and Stokes vector (SV) receivers. We review demonstrated silicon photonic integrated circuits (PICs) for these applications.

II. SILICON PHOTONIC INTEGRATED CIRCUITS FOR COHERENT OPTICAL TRANSCEIVERS

Advanced modulation formats with coherent detection are key technologies for long-haul and metro optical networks. In 2012, quadrature phase shift keying (QPSK) modulation based on silicon microring modulators [2] and Mach-Zehnder modulators (MZMs) [3] were first reported. By further integrating two I/Q modulators and an on-chip polarization rotator (PR) and polarization beam combiner (PBC) in silicon, a monolithic single-chip dual-polarization (DP) coherent modulator to generate a 112-Gb/s DP-QPSK [4] and a 224-Gb/s 16-quadrature amplitude modulation (16-QAM) signal [5] were reported in 2012 and 2013, respectively [with a PIC picture shown in Fig. 1(a)]. Single-polarization (SP) and DP silicon I/Q modulators were further reported from other groups [6-10].

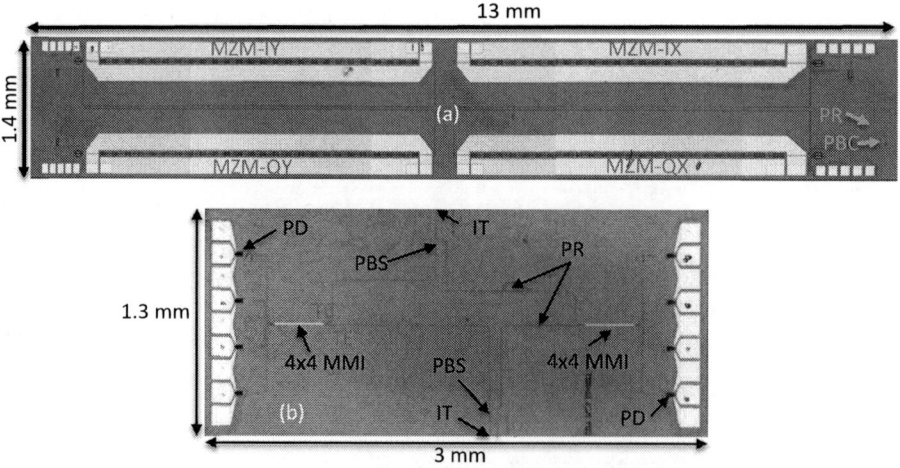

Fig. 1. Monolithic dual-polarization silicon-PIC I/Q modulator and coherent receiver. (a) Photograph of the first dual-pol I/Q modulator on silicon PIC in [4]. MZM: Mach-Zehnder modulator; PR: polarization rotator; PBC: polarization beam combiner. (b) Photograph of the PIC for coherent receiver in [5]. PD: photo detector; IT: inverse taper; PBS: polarization beam splitter; MMI: multimode interference coupler.

624

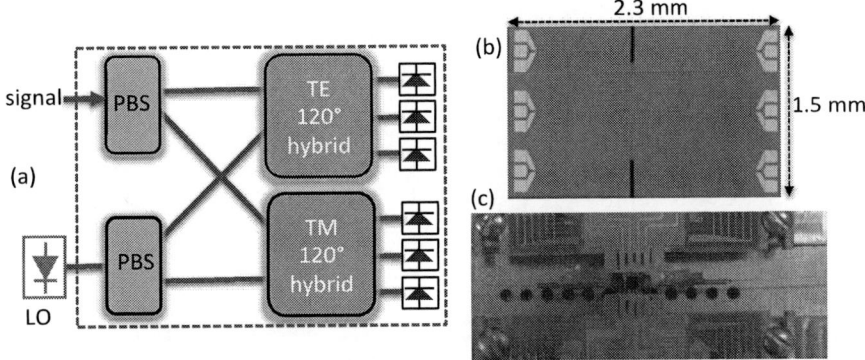

Fig. 2. Dual-polarization coherent receiver based on 120-degree hybrids reported in [12]. (a) Optical circuit diagram. (b) and (c) Photo of a fully fabricated PIC and the packaged PIC with fibers and circuit boards. PBS: polarization beam splitter, LO: local oscillator.

·For the coherent receiver, grating-based monolithic silicon PICs have been reported for DP coherent detection as early as in 2011 [11]. In 2013, a monolithic silicon-PIC DP coherent receiver based on on-chip PRs, on-chip polarization beam splitters (PBSs) and edge coupling was demonstrated in [5] [with a PIC picture shown in Fig. 1(b)]. In [12], a monolithic polarization diversity coherent receiver by employing 120-degree optical hybrids on a silicon PIC was reported, shown in Fig. 2. In [13], Doerr et al. demonstrated a monolithic silicon PIC that contains all the optical frontend for a 100-Gb/s coherent transceiver except the laser.

III. SILICON PHOTONIC INTEGRATED CIRCUITS FOR MULTI-CHANNEL DISCTETE MULTI-TONE MODULATION AND DETECTION

Intensity modulation with direct detection (IMDD) with complex modulation formats receives significant intention in recent years, pushing the channel rate to 100-200G by employing multi-level pulse amplitude modulation (PAM), discrete multi-tone (DMT) modulation or carrierless amplitude phase (CAP) modulation. The DD technology is particularly attractive for short-reach interconnects operating at 100 Gb/s and beyond for metro and data center applications, since these applications are very sensitive to cost.

In [14], silicon photonics four-channel DMT integrated circuits demonstrated net channel rates of 70 Gb/s and 100 Gb/s detected by integrated germanium receivers and commercial receivers, respectively. The transmitter PIC integrates four silicon Mach-Zehnder modulators (MZMs) and WDM multiplexers using thermally-tuned second-order microring filters, showns in Figs. 3(a) and (b). For the receier, an integrated polarization-diversity WDM receiver chip was demonstrated based on an inverse taper for edge coupling, an on-chip polarization beam splitter (PBS), wavelength-tunable second-order ring filters, and monolithic germanium photodiodes, shown in Figs. 3(c) and (d).

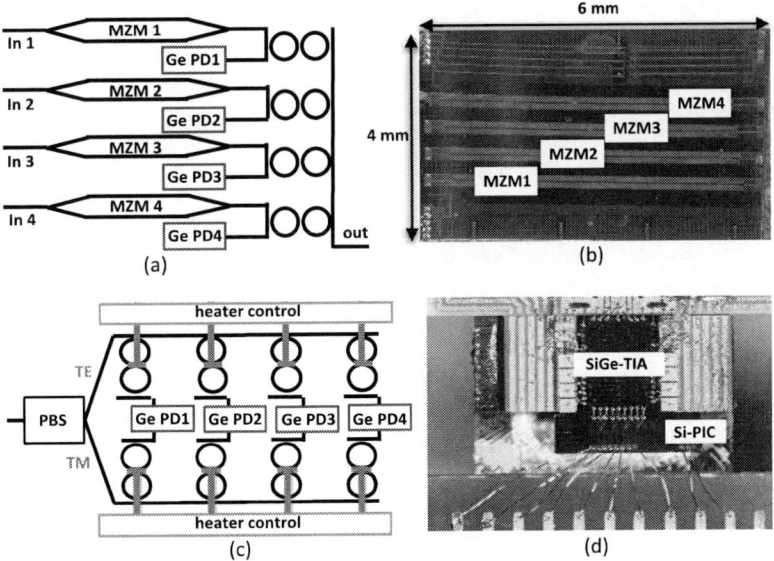

Fig. 3. 400G DMT modulation and detection using silicon PICs in [14]. (a) and (b) Optical circuit and picture of 4-channel silicon MZM transmitter chip. (d) and (d) Optical circuit and photo picture of a packaged 4-channel WDM receiver.

IV. SILICON PHOTONIC INTEGRATED CIRCUITS FOR STOKES VECTOR RECEIVER AND MODULATOR

Multi-dimensional IMDD can further increase of the channel rate without the use of coherent detection. Recently, there is increasing interest in utilizing Stokes vector (SV) receiver, which is a DD technique with the capability to digitally track the polarization changes in fibers and decode information in two to four dimensions. In a recent paper [15], a highly compact and monolithically integrated silicon photonics SV receiver was reported. Paired with a silicon I/Q modulator incorporating a power-tunable carrier in the orthogonal polarization, transmission at 128-Gb/s over 100-km fiber was achieved. Fig. 4 presents the optical circuits and the photographs of the silicon PICs.

REFERENCES

[1] P. Dong, Y.-K. Chen, G.-H. Duan, and D. T. Neilson, "Silicon photonic devices and integrated circuits," Nanophotonics 3, 215-228 (2014).

[2] P. Dong, C. Xie, L. Chen, N. K. Fontaine, and Y.-K. Chen, "Experimental demonstration of microring quadrature phase-shift keying modulators," Opt. Lett., vol. 37, pp. 1178-1180, 2012.

[3] P. Dong, L. Chen, C. Xie, L. L. Buhl, and Y.-K. Chen, "50-Gb/s silicon quadrature phase-shift keying modulator," Opt. Express, vol. 20, pp. 21181-21186, 2012.

[4] P. Dong, C. Xie, L. Chen, L. L. Buhl, and Y. Chen, "112-Gb/s Monolithic PDM-QPSK Modulator in Silicon," Opt. Express, vol. 20, pp.B624-B629, 2012.

[5] P. Dong, X. Liu, C. Sethumadhavan, L. L. Buhl, R. Aroca, Y. Baeyens, and Y. Chen, "224-Gb/s PDM-16-QAM Modulator and Receiver based on Silicon Photonic Integrated Circuits," in Optical Fiber Communication Conference/National Fiber Optic Engineers Conference (OFC 2013), paper PDP5C.6, 2013.

[6] K. Goi, H. Kusaka, A. Oka, Y. Terada, K. Ogawa, T. –Y. Liow, X. Tu; G.-Q. Lo, and D. –L. Kwong, "DQPSK/QPSK Modulation at 40-60 Gb/s using Low-Loss Nested Silicon Mach-Zehnder Modulator," in Optical Fiber Communication Conference/National Fiber Optic Engineers Conference (OFC 2013), paper OW4J.4, 2013.

[7] B. Milivojevic, C. Raabe, A. Shastri, M. Webster, P. Metz, S. Sunder, B. Chattin, S. Wiese, B. Dama, and K. Shastri, "112Gb/s DP-QPSK Transmission Over 2427km SSMF Using Small-Size Silicon Photonic IQ Modulator and Low-Power CMOS Driver," in Optical Fiber Communication Conference/National Fiber Optic Engineers Conference (OFC 2013), paper OTh1D.1, 2013.

[8] K. Goi, H. Kusaka, A. Oka, K. Ogawa, T. Liow, X. Tu, P. G. Lo, and D. L. Kwong, "128-Gb/s DP-QPSK using low-loss monolithic silicon IQ modulator integrated with partial-rib polarization rotator," in Optical Fiber Communication Conference (OFC 2014), paper W1I.2, 2014.

[9] D. Korn, R. Palmer, H. Yu, P. C. Schindler, L. Alloatti, M. Baier, R. Schmogrow, W. Bogaerts, S. K. Selvaraja, G. Lepage, M. Pantouvaki, J. M. D. Wouters, P. Verheyen, J. Van Campenhout, B. Chen, R. Baets, P. Absil, R. Dinu, C. Koos, W. Freude, and J. Leuthold, "Silicon-organic hybrid (SOH) IQ modulator using the linear electro-optic effect for transmitting 16QAM at 112 Gbit/s," Opt. Express, vol. 21, pp. 13219-13227, 2013.

[10] M. Lauermann, P. C. Schindler, S. Wolf, R. Palmer, S. Koeber, D. Korn, L. Alloatti, T. Wahlbrink, J. Bolten, M. Waldow, M. Koenigsmann, M. Kohler, D. Malsam, D. L. Elder, P. V. Johnston, N. Phillips-Sylvain, P. A. Sullivan, L. R. Dalton, J. Leuthold, W. Freude, and C. Koos, "40 GBd 16QAM modulation at 160 Gbit/s" in European Conference on Optical Communication (ECOC 2014), paper We.3.1.3, 2014.

[11] C. R. Doerr, L. L. Buhl, Y. Baeyens, R. Aroca, S. Chandrasekhar, X. Liu, L. Chen, and Y.-K. Chen, "Packaged Monolithic Silicon 112-Gb/s Coherent Receiver," , IEEE Photonics Technology Letters, vol. 23, no. 12, pp.762-764, June 2011.

[12] P. Dong, C. Xie, and L. L. Buhl, "Monolithic polarization diversity coherent receiver based on 120-degree optical hybrids on silicon," Opt. Express 22, 2119-2125, 2014.

[13] C. R. Doerr, L. Chen, D. Vermeulen, T. Nielsen, S. Azemati, S. Stulz, G. McBrien, X. Xu, B. Mikkelsen, M. Givehchi, C. Rasmussen, and S. Y. Park, "Single-Chip Silicon Photonics 100-Gb/s Coherent Transceiver," in Optical Fiber Communication Conference (OFC 2014), paper Th5C.1, 2014.

[14] P. Dong, J. Lee, Y. Chen, L. L. Buhl, S. Chandrasekhar, J. H. Sinsky, and K. Kim, "Four-Channel 100-Gb/s per Channel Discrete Multi-Tone Modulation Using Silicon Photonic Integrated Circuits," in Optical Fiber Communication Conference Post Deadline Papers, OSA Technical Digest (online) (Optical Society of America, 2015), paper Th5BP.

[15] P. Dong, X. Chen, K. Kim, S. Chandrasekhar, Y.-K. Chen, and J. H. Sinsky, "128-Gb/s 100-km transmission with direct detection using silicon photonic Stokes vector receiver and I/Q modulator", accepted by Optics Express, 2016.

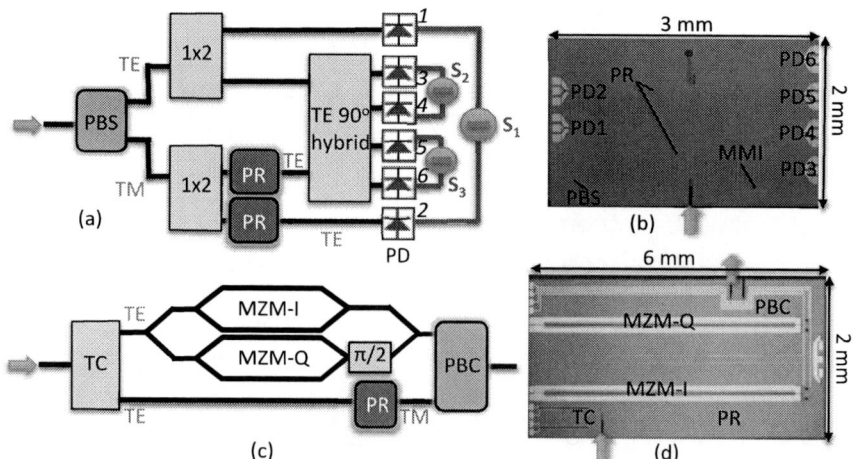

Fig. 4. Silicon photonic integrated circuits for Stokes vector receiver and modulator in [15]. (a) and (b) Optical circuit and photograph of SV receiver. PBS: polarization beam splitter, PR: polarization rotator, PD: photodetector, MMI: multimode interference coupler. (c) and (d) Optical circuit and photograph of a silicon I/Q modulator with a power-tunable carrier in the orthogonal polarization. TC: tunable coupler, PBC: polarization beam combiner.

TuE3-2

OECC/PS2016

Lens-integrated Fan-in/Fan-out Device for Multi-core Fiber

Osamu Shimakawa, Hajime Arao, Tomomi Sano

Sumitomo Electric Industries, Ltd., 1, Taya-cho, Sakae-ku, Yokohama, 244-8588 Japan
shimakawa-osamu@sei.co.jp

Abstract: *We design and fabricate a multi-core fiber fan-in/fan-out using lenses. The structure allows it to be assembled with only one active alignment process. IL less than 0.87dB and RL more than 47dB were achieved.*

Keywords: *Multi-core fiber, Fan-in/Fan-out, Grin lens, Micro-lens array*

I. INTRODUCTION

The data traffic in optical communication networks has been exponentially increasing ever since the deployment of the dense wavelength division multiplexing (DWDM) optical network in the 1990's. An optical input power per fiber is also increasing and is reaching its capacity limit. One of the candidates to overcome the crunch is a spatial division multiplexing (SDM) system using multi-core fibers (MCFs) [1,2]. A fan-in/fan-out (FIFO) devices that connects an MCF to single-core fibers (SCFs) is indispensable to construct the SDM system. Several FIFOs have been proposed up to now, utilizing such as 3D-waveguide [3], spatial coupling with lens [4], fused taper fibers [5] and fiber bundle [6]. As for spatial coupling with lens, we reported a compact optical coupling device for 7 single-mode core MCF and 7 single-mode SCFs [7,8]. This FIFO device consisted of several optical elements including a GRIN lens and a micro-lens array (MLA). Since these elements were directly bonded with each other, the FIFO device was compact and easy in assembling compared to conventional spatial coupling devices using lenses. However, the return loss of it was around 30 dB and was not high enough to put into practical use when SDM system is deployed in the future.

In this report, we change the optical design and the structure to improve the return loss, then, fabricate a FIFO with a new design and measure a return loss, an insertion loss and a crosstalk. The return loss more than 47 dB across all ports was achieved.

II. OPTICAL DESIGN AND STRUCTURE

A schematic of the FIFO is shown in Fig.1. It consists of an MCF in a ferrule, a GRIN-lens, a glass spacer, a MLA and a fiber array with seven single-core fibers (SCF-array). The MCF has seven single-mode cores whose core-to-core pitch is 45 μm as shown in Fig.2 (a) [9]. The SCF-array shown in Fig.2 (b) has hexagonally-arranged seven single-mode fibers with core-to-core pitch of 0.5 mm. The GRIN-lens has a focal length f of 1 mm and the fractional pitch of 0.25 so as to collimate the beams from the MCF. The MLA has hexagonally-arranged seven lenses and each of them has aspheric surface. All seven lenses have a focal length f of 1 mm as well as the GRIN-lens. The lens-to-lens pitch is approximately 0.455mm and several μm of each lens position is adjusted to compensate for the misalignment by eight degrees end face. It should be noted that the optical system should be confocal, because the MCF and the SCF have the almost same mode field diameter. Therefore, as shown in Fig.3 (For simplification, optical elements do not have angled end faces in this figure.) the outer port lens position should have an offset of 0.045 mm as with the MCF outer port. Then, the center to outer lens pitch was designed to be not 0.5 mm but 0.455 mm. The glass spacer is put between the GRIN-lens and the MLA. This spacer enables the two lenses to be relatively positioned without any Z-axis active alignment.

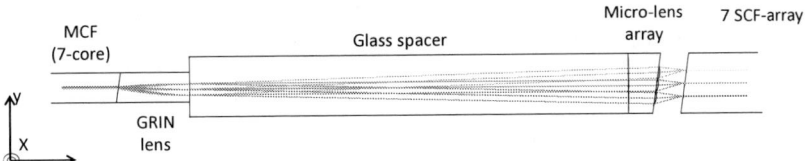

Fig.1 Schematic of the proposed FIFO

The major difference from the former design is end face angles of each optical element. The MCF and the GRIN-lens have different refractive indexes each other. Therefore, both angles of the ferrule with the MCF and the GRIN-lens are set to eight degrees to reduce back-reflection. There is an air gap of 1 mm between the MLA and the fiber array. Therefore, the end face angle of SCF-array is set to eight degrees. The MLA end face is also set eight degrees to avoid optical coupling efficiency decrease between the MLA and the SCF-array. On the other hand, end faces between the

627

GRIN-lens and the glass spacer, and between the glass spacer and the MLA do not have an angle. Since beams themselves go through the end faces with angle, they do not need to be angled.

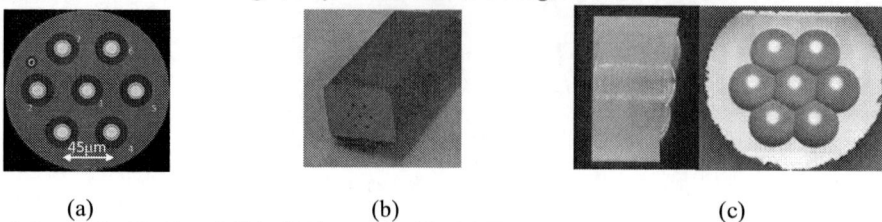

(a)   (b)   (c)

Fig.2 (a) Cross-sectional photograph of the 7-core MCF. (b) Photograph of the 7-SCF-array. (c) Photographs of the 7-MLA.

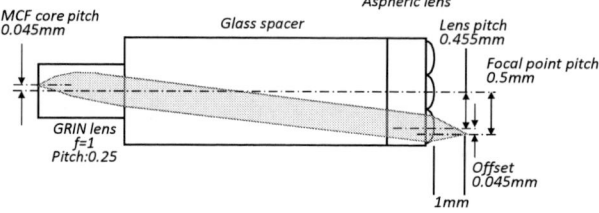

Fig.3 Schematic of an optical coupling system between the MCF and the SCF.
(For simplification, optical elements do not have angled end faces.)

Seven output beams from the MCF enter into the GRIN-lens and they are collimated. As the beams from the MCF outer cores enter the GRIN-lens with an offset of 0.045 mm, they propagate in different directions from center core beam. Seven beams are separated from each other after propagating through the glass spacer. The separated beams are respectively refocused by the MLA. As the MLA pitch is 0.045 mm shorter than propagated beam pitch on the MLA surfaces, outer port beams from the MLA become parallel to the center port beam and the refocused beam pitch become 0.5 mm as well as the SCF-array pitch.

III. FABRICATION

The proposed FIFO structure is designed to be assembled with a small number of active alignment process. The FIFO consists of three subassembly as shown in Fig.4. A subassembly 1 has the MCF inserted into a ceramic ferrule, the GRIN-lens and a glass tube. As the ceramic ferrule and the GRIN-lens have the same outside diameter, they are aligned with each other in X, Y axes by inserted into the glass tube whose inside diameter is slightly larger than the outside diameter of the GRIN-lens and the ceramic ferrule. A subassembly 2 has the glass spacer and the MLA. As they have the same outside diameter, they are also aligned and bonded with each other without any active alignment. The SCF-array consists of a metal ferrule and seven SCFs. The SCFs are precisely arranged by seven holes in the metal ferrule. These three subassemblies are aligned simultaneously with optical power monitoring and bonded. The fabricated FIFO is shown in Fig.5. The size is φ8 x 40 mm.

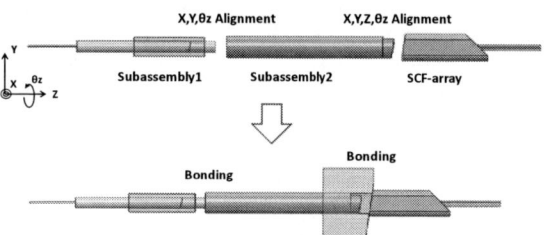
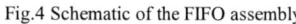
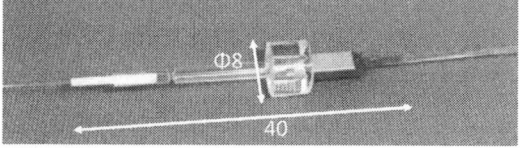

Fig.4 Schematic of the FIFO assembly   Fig.5 Photograph of the fabricated FIFO

IV. OPTICAL CHARACTERISTICS

The optical characteristics of the fabricated FIFO were measured at the wavelength of 1.55 μm. Refocused beam profiles by the MLA are shown in Fig.6. Left side shows port 1(Center port) and right side shows port 2 (Outer port). Dashed lines show the MCF output profiles and red and blue lines show X and Y cross-sectional profiles of refocused beams, respectively. The refocused profiles correspond well to MCF profiles. This result indicates the beam profiles at the end of the MCF are maintained at the focal point of the MLA. The measured insertion loss (IL) is shown in Fig.7 (a) and each port is less than 0.87 dB and average of seven ports is 0.71 dB. Each IL includes a reflection loss of 0.15 dB at the SCF-array end face because it has no anti-reflection (AR) coating. IL variation among seven ports is caused by a mismatch between the focal point arrangement by the MLA and fiber arrangement in SCF-array. Lower IL can be expected by improving the position accuracy of focal point arrangement by the MLA. The measured return loss (RL) is shown in Fig.7 (b). Each port has the RL of more than 47.8 dB and is more than 18 dB higher than the RL of the FIFO with the former design. The average RL of seven ports is 51.8 dB. This result shows the validity of the new design. The crosstalk (XT) measurement setup is shown in Fig.7 (c). A fiber bundle fan-in [6] was used to enter light into each core of the MCF. The measured result is shown in Fig.7 (d). The XT including the fiber bundle fan-in and the MCF is less than -56.9 dB.

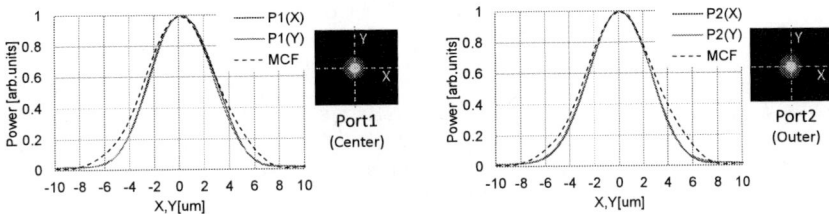

Fig.6 Refocused Beam profiles by the MLA. Left side shows port 1 (Center port) and right side shows port2 (Outer port).

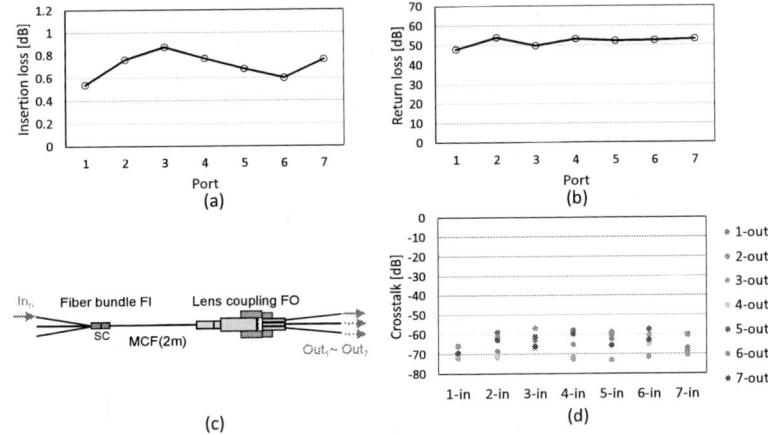

Fig.7 Optical characteristics of the FIFO. (a) Insertion loss. (b) Return loss. (c) Crosstalk measurement setup. (d) Crosstalk.

V. CONCLUSIONS

We fabricated a FIFO for seven-core MCF with a unique structure whose optical elements were directly bonded with each other. This structure allows the FIFO to be assembled with only one alignment process. The optical design was changed from former one to improve the RL. The measured RL was more than 47.8 dB and was more than 18 dB higher than former one. The measured IL was less than 0.87 dB. This IL is expected to be decreased by applying AR coating on the SCF-array.

ACKNOWLEDGMENT

This research is supported by the National Institute of Information and Communications Technology (NICT), Japan under "Research on Innovative Optical Communication Infrastructure Technologies" initiative.

REFERENCES

[1] T. Morioka, "New Generation Optical Infrastructure Technologies: "EXAT Initiative" Towards 2020 and Beyond," in Proc. OECC2009, FT4.
[2] R.J.Essiambre et al., "Capacity Limits of Fiber-Optic Communication Systems," Proc. OFC2009, OThL1.
[3] P. Mitchell et al., "57 Channel (19x3) Spatial Multiplexer Fabricated using Direct Laser Inscription," Proc. OFC2014, M3K.5.
[4] Y. Tottori et al., "Multi Functionality Demonstration for Multi Core Fiber Fan-in/Fan-out Devices using Free Space Optics," Proc. OFC2014, Th2A.44.
[5] H. Uemura et al., "Fused Taper Type Fan-in/Fan-out Device for 12 Core Multi-Core Fiber," in Proc. OECC/ACOFT2014, MO1E-4.
[6] O. Shimakawa et al., "Pluggable fan-out realizing physical-contact and low coupling loss for multi-core fiber," Proc. OFC2013, OM3I.2.
[7] H. Arao et al., "Compact Multi-core Fiber Fan-in/out Using GRIN Lens and Microlens Array," in Proc. OECC/ACOFT2014, MO1E-1.
[8] O.Shimakawa et al., "Compact Multi-core Fiber Fan-out with GRIN-lens and Micro-lens Array," Proc. OFC2013, OM3I.2.
[9] T .Hayashi et al., "Ultra-low-crosstalk multi-core fiber feasible to ultra-long-haul transmission," Proc. OFC2011, PDPC2.

TuE3-3

OECC/PS2016

Four-Mode-Selective Photonic Lantern Based on Two-Layer Polymer Waveguide Branches

Yunfei Wu and Kin Seng Chiang*

Department of Electronic Engineering, City University of Hong Kong,
83 Tat Chee Avenue, Kowloon, Hong Kong SAR, China
*eeksc@cityu.edu.hk

Abstract: We design and fabricate a four-core photonic lantern with two-layer asymmetric polymer waveguide branches to spatially (de)multiplex the LP_{01}, LP_{11a}, LP_{11b}, and LP_{21a} modes. The device operates over the C+L band with negligible polarization dependence.
Keywords: Mode multiplexer; photonic lantern; polymer waveguide.

I. INTRODUCTION

Photonic lanterns, which were originally proposed for interface between a multimode fiber and multiple single-mode fibers for astronomical instrumentation, have recently found applications as mode multiplexers in mode-division multiplexing (MDM) transmission [1]. Photonic lanterns can be formed with dissimilar single-mode fibers or 3D waveguides to select different mode groups of a few-mode fiber [2,3]. Mode-group selectivity can reduce the complexity of multi-input multi-output (MIMO) signal processing caused by differential mode delays. The same design permits multiplexing the individual spatial modes (including the degenerate ones) of a few-mode fiber, i.e., to achieve one-to-one mapping between the fundamental modes of the single-mode fibers and the various spatial modes of the few-mode fiber [4]. To further suppress couplings between degenerate spatial modes, all the single-mode cores of the photonic lantern can be made dissimilar [5]. High spatial-mode selectivity is required for low-cost direct detection in short-reach transmission with mode-preserving fibers, such as elliptical-core fibers, and other mode-selective applications, such as mode-dependent loss compensation.

Recently, we demonstrated a three-core photonic lantern based on two-layer polymer waveguides with high spatial-mode selectivity [6]. The polymer waveguide technology provides a simple process for the fabrication of three-dimensional structures, which is necessary for the construction of waveguide-based photonic lanterns, and offers the flexibility in the choice of waveguide material and the precision in the control of waveguide dimensions. In this paper, we propose and experimentally demonstrate a new design of polymer-waveguide photonic lantern, which is based on spatially splitting a few-mode core into four dissimilar single-mode cores adiabatically in two geometric layers. Our device is capable of (de)multiplexing the LP_{01}, LP_{11a}, LP_{11b}, and LP_{21a} modes in the C+L band with negligible polarization dependence.

II. DESIGN OF THE FOUR-MODE-SELECTIVE PHOTONIC LANTERN

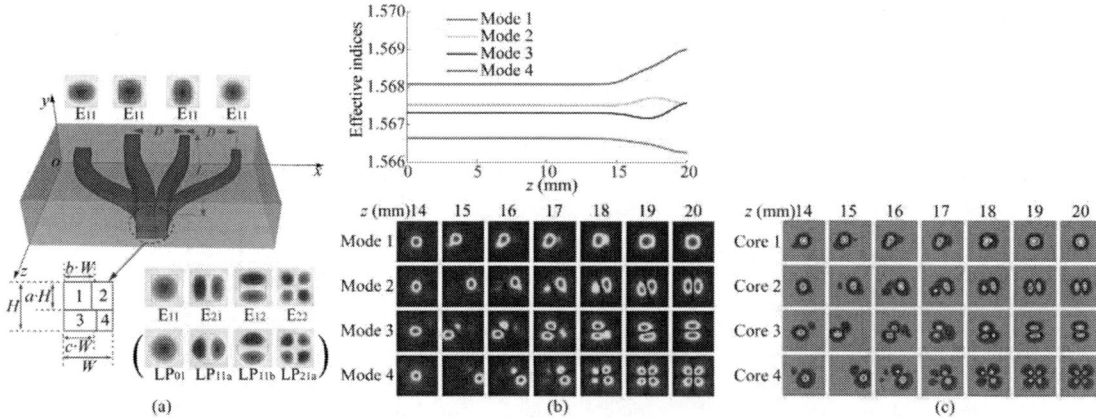

Fig. 1. (a) The structure of the proposed four-mode-selective photonic lantern. (b) The effective indices and the corresponding intensity distributions of the local modes along the device calculated with the finite-element method (COMSOL). (c) Intensity distributions of light propagating along the device calculated with the scalar beam-propagation method (RSOFT) when the E_{11} modes are launched into the four cores individually. The simulation is performed at the wavelength 1550 nm.

Our proposed four-mode-selective photonic lantern is shown schematically in Fig. 1(a). A rectangular few-mode core, which has width W and height H, is divided into upper and lower layers according to the ratio a. Each of the upper and

630

lower layers is further divided into two segments, according to the ratios b and c. The four segments of the few-mode core branch out into four separate single-mode cores, Core 1, 2, 3 and 4, with cosine S-bends of length L to a final core-to-core separation D. The few-mode core supports the E_{11}, E_{21}, E_{12}, and E_{22} modes, which correspond to the LP_{01}, LP_{11a}, LP_{11b}, and LP_{21a} modes of a few-mode fiber, respectively. The device operates on the principle of adiabatic mode transition, where the fundamental modes of the four single-mode cores evolve respectively into the individual spatial modes of the few-mode core. The early demonstrations of the mode transition effect employ only a single waveguide layer [7], which poses a serious restriction on the (de)multiplexing of certain high-order modes. Although a two-layer structure based on two cascaded Y junctions has been proposed to solve the problem [8], the resultant structure is complex and has been implemented for only three spatial modes [9]. Here, by allowing a single rectangular few-mode core to spatially branch out into two layers, we can achieve good mode matching with a circular few-mode fiber and (de)multiplex modes with symmetric and anti-symmetric field distributions in both the horizontal and vertical directions, which is difficult to achieve with conventional planar waveguide structures. As a design example, we take $W = 15$ μm, $H = 15$ μm, $a = 0.65$, $b = 0.60$, $c = 0.65$, $D = 62.5$ μm, and $L = 20$ mm. The refractive indices of the core and the cladding are 1.570 and 1.566. Fig. 1(b) shows the effective indices and the corresponding intensity distributions of the local modes calculated with the finite-element method (COMSOL) at the wavelength 1550 nm. The four local spatial modes are non-degenerate along the device, except at the few-mode end ($z = 20$ mm), where the E_{12} and E_{21} modes are degenerate. Keeping the local modes non-degenerate helps to suppress couplings among these modes. Fig. 1(c) shows light propagation along the device calculated with the scalar beam-propagation method (RSOFT) at 1550 nm, where the E_{11} modes are launched into the four single-mode cores individually. A comparison of Fig. 1(b) and Fig. 1(c) shows excellent agreement between the local-mode results and the beam-propagation results and thus confirms excellent adiabatic mode transition. Our simulation shows that the LP_{01} modes of Core 1, 2, 3, and 4 evolve into the E_{11}, E_{21}, E_{12}, and E_{22} modes of the few-mode core, respectively, with a maximum crosstalk of −20 dB among the four spatial modes at 1550 nm.

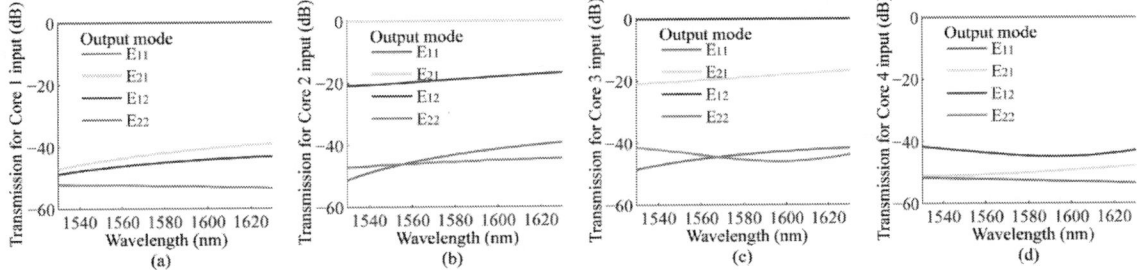

Fig. 2. Transmission spectra of the E_{11}, E_{21}, E_{12}, and E_{22} modes in the few-mode core, when the E_{11} modes of (a) Core 1, (b) Core 2, (c) Core 3, and (d) Core 4, are excited, respectively.

Fig. 2 shows the transmission spectra of the E_{11}, E_{21}, E_{12}, and E_{22} modes in the few-mode core, when the E_{11} modes of Core 1, 2, 3, and 4 are excited, respectively. As shown in Fig. 2, the E_{11}, E_{21}, E_{12}, and E_{22} modes of the few-mode core are selectively generated from the E_{11} modes launched into Core 1, 2, 3, and 4, respectively. Over the C+L band (1530 nm – 1630 nm), the selectivity among the three mode groups (i.e., E_{11}, E_{21}/E_{12}, and E_{22}) is larger than 39 dB, as shown in Fig. 2(a)-(d), and the selectivity between the degenerate E_{21} and E_{12} modes is larger than 17 dB, as shown in Fig. 2(b) and Fig. 2(c). We also confirm by semi-vector beam-propagation simulation that the performance of the device is polarization-insensitive.

III. FABRICATION AND MEASUREMENT

We fabricated the device by successively forming two layers of polymer waveguides with spin-coating and photolithography using two properly designed waveguide masks. The fabrication process was similar to that reported in [10]. The polymers used were EpoCore and EpoClad (Micro Resist Technology). The waveguide cores were formed with EpoCore, while the cladding was formed with a 3:2 mass-ratio mix of EpoCore and EpoClad. The refractive indices of the cores and the cladding, measured with a Metricon 2010 prism coupler for separately prepared thin-film samples on glass substrates, were 1.5712 (1.5704) and 1.5665 (1.5669) for the TE (TM) polarization at 1536 nm. Microscopic images of the few-mode and single-mode cores of a typical fabricated device are shown in Fig. 3. The parameters of the device were measured to be $W = 15.8$ μm, $H = 16.0$ μm, $a = 0.67$, $b = 0.60$, $c = 0.65$, $D = 62.5$ μm, and $L = 20$ mm. The total length of the device was 24 mm, which included 4 mm long waveguide leads at the two ends.

To demonstrate the operation of the device, we launched light into Core 1, 2, 3, and 4 individually with a C-band tunable laser and measured the corresponding near-field intensity patterns at the few-mode end of the device with an infrared camera. Our measurement results in Fig. 4 confirm that the E_{11} modes of Core 1, 2, 3, and 4 evolve, respectively, into the E_{11}, E_{21}, E_{12}, and E_{22} modes of the few-mode core. The few-mode patterns remain practically unchanged over the wavelength range for the measurement (limited by the light source available) and are similar for the x and y polarizations. The apparent 45-degree tilt of the orientations of the E_{21} and E_{12} modes may be caused by small imperfections in the long transition structure of the device. Nevertheless, this tilt does not affect the application of the device and, if necessary, may be corrected by using a rectangular few-mode core with a larger aspect ratio.

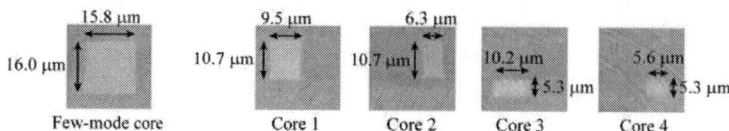

Fig. 3. Microscopic images of the few-mode core at one end and the four split cores at the other end in a typical fabricated device.

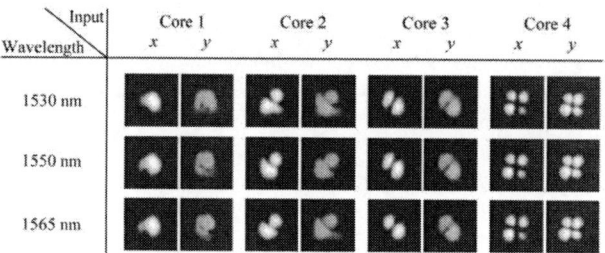

Fig. 4. Near-field images of the intensity patterns taken at the few-mode end of the device at different wavelengths, when the x- or y-polarized E_{11} modes were launched into Core 1, 2, 3, and 4, respectively.

IV. CONCLUSIONS

We have designed and fabricated a simple, compact four-core photonic lantern with two-layer asymmetric polymer waveguide branches to spatially (de)multiplex the LP_{01}, LP_{11a}, LP_{11b}, and LP_{21a} modes of a fiber. Such a device offers broadband operation with negligible polarization dependence and could be further developed for practical applications in high-capacity MDM communication systems.

ACKNOWLEDGMENT

The authors thank Jiangli Dong and Wei Jin for their technical assistance. This research was supported by a grant from the Research Grants Council, University Grants Committee, The Hong Kong Special Administrative Region, China, under Project CityU 112113.

REFERENCES

[1] T. A. Birks, I. Gris-Sánchez, S. Yerolatsitis, S. G. Leon-Saval, and R. R. Thomson, "The photonic lantern," Adv. Opt. Photon., vol. 7, pp. 107-167, 2015.

[2] B. Huang, N. K. Fontaine, R. Ryf, B. Guan, S. G. Leon-Saval, R. Shubochkin, Y. Sun, R. Lingle, and G. Li, "All-fiber mode-group-selective photonic lantern using graded-index multimode fibers," Opt. Express, vol. 23, pp. 224-234, 2015.

[3] B. Guan, B. Ercan, N. K. Fontaine, R. P. Scott, and S. J. B. Yoo, "Mode-group-selective photonic lantern based on integrated 3D devices fabricated by ultrafast laser inscription," Optical Fiber Communication Conference (OFC 2015), Mar. 2015, p. W2A.16.

[4] A. M. Velazquez-Benitez, J. C. Alvarado, G. Lopez-Galmiche, J. E. Antonio-Lopez, J. Hernández-Cordero, J. Sanchez-Mondragon, P. Sillard, C. M. Okonkwo, and R. Amezcua-Correa, "Six mode selective fiber optic spatial multiplexer," Opt. Lett., vol. 40, pp. 1663-1666, 2015.

[5] S. Yerolatsitis, I. Gris-Sánchez, and T. A. Birks, "Adiabatically-tapered fiber mode multiplexers," Opt. Express, vol. 22, pp. 608-617, 2014.

[6] Y. Wu and K. S. Chiang, "Broadband photonic lantern mode multiplexers based on multilayer polymer waveguides," OptoElectronics and Communications Conference (OECC 2015), Jun. 2015, pp. 1-3.

[7] J. B. Driscoll, R. R. Grote, B. Souhan, J. I. Dadap, M. Lu, and R. M. Osgood, "Asymmetric Y junctions in silicon waveguides for on-chip mode-division multiplexing," Opt. Lett., vol. 38, pp. 1854-1856, 2013.

[8] A. M. Bratkovsky, J. B. Khurgin, E. Ponizovskaya, W. V. Sorin, and, M. R. Tan, "Mode division multiplexed (MDM) waveguide link scheme with cascaded Y-junctions," Opt. Commun., vol. 309, pp. 85-89, 2013.

[9] T. Watanabe and Y. Kokubun, "Demonstration of mode-evolutional multiplexer for few-mode fibers using stacked polymer waveguide," IEEE Photonics J., vol. 7, pp. 1–11, 2015.

[10] J. Dong, K. S. Chiang, and W. Jin, "Mode multiplexer based on integrated horizontal and vertical polymer waveguide couplers," Opt. Lett., vol. 40, pp. 3125-3128, 2015.

TuE3-4

OECC/PS2016

Excitation of LP$_{21b}$ and LP$_{02}$ modes with PLC-based tapered waveguide for mode-division multiplexing

Yoko Yamashita[1], Yuhei Ishizaka[1], Nobutomo Hanzawa[2], Takeshi Fujisawa[1], Taiji Sakamoto[2], Takashi Matsui[2], Kyozo Tsujikawa[2], Fumihiko Yamamoto[2], Kazuhide Nakajima[2], Kunimasa Saitoh[1]

[1]Graduate School of Information Science and Technology, Hokkaido University, Sapporo 060-0814, Japan
[2]NTT Access Network Service Systems Laboratories, NTT Corporation, 1-7-1 Hanabatake, Tsukuba, Ibaraki 305-0805, Japan
yamashita@icp.ist.hokudai.ac.jp

Abstract: *Multi-stage mode conversion technique, using asymmetric directional couplers and tapered mode converters, for launching LP$_{21b}$ and LP$_{02}$ mode on PLC for six mode multi/demultiplexer is proposed. Successful excitation of these modes are demonstrated.*
Keywords: *Mode multi/demultiplexer, Mode-division multiplexing, Planar lightwave cirucit*

I. INTRODUCTION

In order to expand the transmission capacity per fiber, the mode-division multiplexing (MDM) transmission has attracted a lot of attention [1]. The mode multi/demultiplexer (MUX/DEMUX) is an important component for MDM transmission to excite the multiple modes. Recently, various types of mode MUX have been proposed, for example, free space optics [2], photonic lanterns [3], and planar lightwave circuit (PLC) [4]-[6]. The PLC-based mode MUX has advantages such as compactness, mass-productivity, and large coupling ratio. We have proposed the PLC-based four-mode (LP$_{01}$, LP$_{11a}$, LP$_{11b}$, and LP$_{21a}$) MUX [5] on one chip. These four modes can be excited by using asymmetric directional coupler (ADC) by matching the effective indices of LP$_{01}$ mode of input waveguide and desired higher order mode of bus waveguide. In order to multiplex six modes, the next higher order modes, LP$_{21b}$ and LP$_{02}$ modes, have to be excited. Recently, we fabricated six-mode MUX PLC chip for demonstrating the feasibility of six mode operation [6]. Here, we report detailed design of mode conversion technique for LP$_{21b}$ and LP$_{02}$ modes. It consists of multi-stage mode converter, consisting of ADCs and tapered waveguides. Launched LP$_{01}$ mode is converted to the desired mode via LP$_{11a}$, E$_{13}$ and E$_{31}$ modes.

II. LP$_{21B}$ AND LP$_{02}$ MODE MUX AND SUCCESSFUL EXCITATION

Figure 1 shows the schematic of our PLC-based LP$_{21b}$ and LP$_{02}$ mode MUX. We use two types of ADCs (Coupler 1

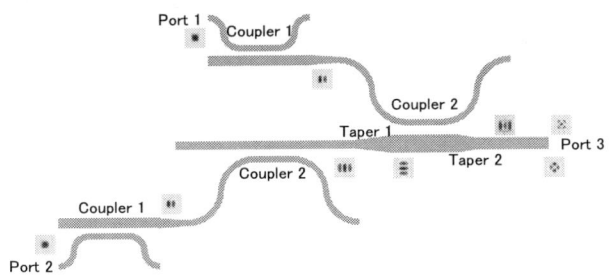

Fig. 1. The schimatic of our PLC-based LP$_{21b}$ and LP$_{02}$ mode MUX.

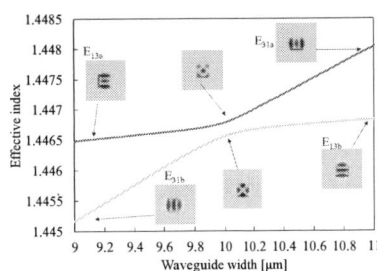

Fig. 2. The effective indices of LP$_{21b}$ and LP$_{02}$ mode as a function of waveguide width.

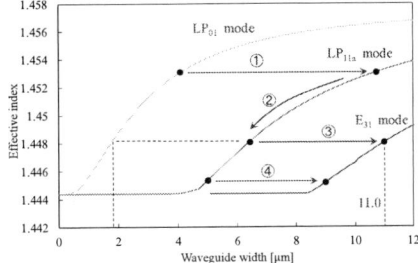

Fig. 3. The effective indices of LP$_{01}$, LP$_{11a}$, and E$_{31}$ mode as a function of waveguide width.

633

OECC/PS2016

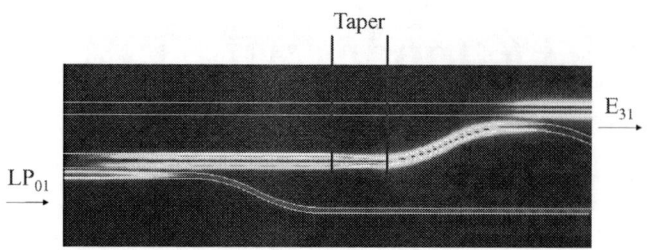

Fig. 4. The field distribution showing the mode coupling from LP_{01} to E_{31} mode.

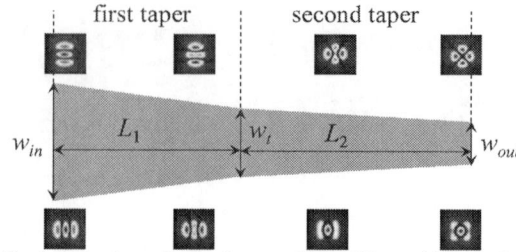

Fig. 5. A structure of tapered waveguide for LP_{21b} and LP_{02} modes excitation.

and Coupler 2), and two tapered waveguides (Taper 1 and Taper 2) for LP_{02} and LP_{21b} mode excitation. Figure 2 shows the effective indices of LP_{21b} and LP_{02} modes as a function of waveguide width. The wavelength is 1550 nm. Here, the refractive index difference is $\Delta = 1.0\ \%$ and the waveguide height is 10.0 µm. Although the LP_{21b}- and LP_{02}-like modes exist for square waveguide (width = 10 µm), their effective indices are very close and it is difficult to excite these modes separately by simple ADC. Therefore, we use the mode transition from E_{13} and E_{31} modes to LP_{21b} and LP_{02} modes by using tapered waveguide.

First, we describe how to excite E_{31} mode. Figure 3 shows the effective indices of LP_{01}, LP_{11a}, and E_{31} modes as a function of waveguide width. If we want to excite E_{31a} mode from LP_{01} mode directly by using single ADC, the width of input waveguide is around 2.0 µm, and it is difficult to fabricate such high-aspect ratio waveguide. Also, such waveguide has small fabrication tolerance. Therefore, we excite E_{31a} mode by two steps. First, LP_{01} mode is converted to LP_{11a} mode by using ADC (Coupler 1), where the input waveguide width is 4.08 µm and the bus waveguide width is 10.73 µm (Fig. 2 ①). Next, the width of bus waveguide where LP_{11a} mode exists is tapered down from 10.73 µm to 6.44 µm (Fig. 2 ②). Then, LP_{11a} mode is converted to E_{31a} mode similarly (Coupler 2), where the other waveguide width is 11.0 µm (Fig. 2 ③). By using this process, we can avoid to use too narrow waveguide. Figure 4 shows the field distribution showing the mode conversion process from LP_{01} to E_{31} modes. It can be seen that LP_{01} mode is coupled to LP_{11a} mode and LP_{11a} mode coupled to E_{31} mode. LP_{02} mode can be excited by tapering down the waveguide widths to 12.0 µm, as shown later.

For LP_{21b} mode excitation, the same technique can be used. To excite LP_{02} and LP_{21b} modes simultaneously, LP_{21b} mode should be converted from E_{13b} mode with the waveguide width of 11.0 µm. However, E_{13b} mode cannot be excited in the same plane by ADC due to the symmetry of the field distribution. Therefore, we excite E_{31b} mode by

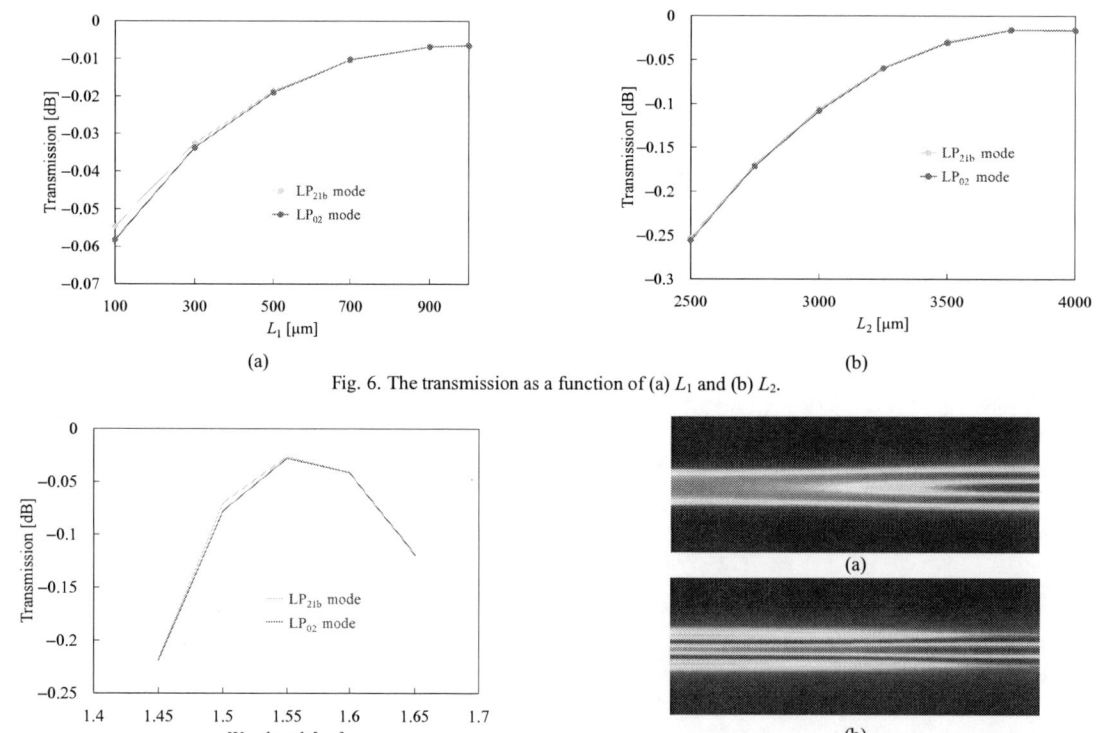

Fig. 6. The transmission as a function of (a) L_1 and (b) L_2.

Fig. 7 The transmission spectra of LP_{21b} and LP_{02} modes when E_{13b} and E_{31a} modes are launched.

Fig. 8. The field distribution showing the mode converting (a) from E_{13} to LP_{21b} mode and (b) from E_{31} to LP_{02} mode.

634

setting the width of the waveguide as 9.0 µm (Fig. 3 ④), and the widths is tapered up to 11.0 µm to convert to E_{13b} mode. Thus, we can excite E_{31a} and E_{13b} modes for the waveguide width of 11.0 µm at Coupler 2 in Fig. 1.

Next task is to convert these modes to LP_{02} and LP_{21b} modes by tapering down the waveguide width to 10.0 µm (Taper 2). Figure 5 shows a structure of tapered waveguide for LP_{21b} and LP_{02} mode excitation. Two-step taper structure is employed to make the size of taper compact. In the first region (the widths change from w_{in} to w_t, first taper), the mode coupling between E_{31a} and E_{13b} modes is weak due to large index difference and the length can be short. On the other hand, in the second region (from w_t to w_{out}, second taper) the index difference is so small that the length of taper has to be long sufficiently. w_{in}, w_t, , and w_{out} are 11.0, 10.2, and 10.0 µm. Figure 6 (a), and (b) show the transmission of first and second taper waveguides as a function of their lengths (L_1 and L_2). The green lines show the transmission of LP_{21b} mode when E_{13b} mode is launched and the red lines show the transmission of LP_{02} mode when E_{31a} mode is launched. From Fig. 6, it can be seen that 1000 µm is enough for L_1 and L_2 should be more than 4000 µm for adiabatic mode transition.

Figure 7 shows the transmission spectra of two-step taper waveguide for LP_{21b} and LP_{02} modes when E_{13b} and E_{31a} modes are launched. Here L_1 is 2000 µm, and L_2 is 4000 µm. This tapered waveguide has transmission more than −0.1 dB within C-band, and more than −0.25 dB within the wavelength range from 1450 nm to 1650 nm. The wavelength dependence of this mode converter is small. Figure 8 shows the field distribution showing the mode converting (a) from E_{13} to LP_{21b} modes and (b) from E_{31} to LP_{02} modes. Each mode is converted adiabatically, as shown in Fig. 8.

Table 1 shows the near field patterns (NFPs) measured for the wavelength of 1560 and 1570 nm of the fabricated PLC-based mode MUX. For LP_{21b} and LP_{02} mode excitation, LP_{01} mode is launched from port 1 and 2 in Fig. 1. A seen in the Table, LP_{21b}- and LP_{02}-like field distributions are successfully observed, showing the validity of our concept.

Table 1. NFPs of the fabricated PLC-based mode MUX

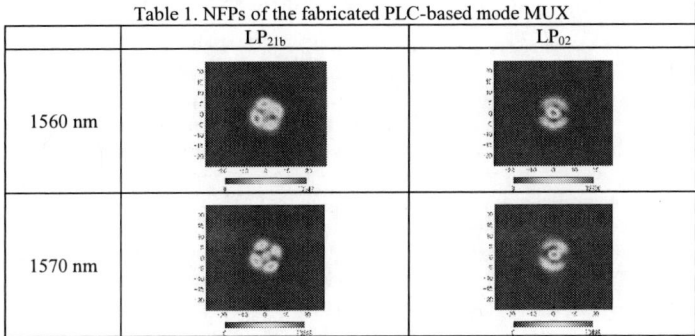

	LP_{21b}	LP_{02}
1560 nm		
1570 nm		

III. CONCLUSIONS

Multi-stage mode conversion technique for launching LP_{21b} and LP_{02} modes on PLC for six mode multi/demultiplexer is proposed. These two nearly degenerate modes can be excited separately by using tapered mode converter from E_{13} (E_{31}) to LP_{21b} (LP_{02}) modes. From the numerical result, a high transmission more than −0.1 dB is possible and successful excitation of these modes are demonstrated.

REFERENCES

[1] T. Morioka, D. T. U. Fotonik, R. Ryf, and P. Winzer, "Enhancing optical communications with brand new fibers," *IEEE Commun. Mag.*, vol. 50, no. 2, pp. s31–s42, Feb. 2012.

[2] E. Ip, N. Bai, Y.-K. Huang, E. Mateo, F. Yaman, M.-J. Li, S. Bickham, S. Ten, Y. Luo, G.-D. Peng, G. Li, T. Wang, J. Linares, C. Montero, and V. Moreno, "6x6 MIMO transmission over 50+25+10 km heterogeneous spans of few-mode fiber with inline erbium-doped fiber amplifier," *Optical Fiber Communication Conference*, OTu2C.4, 2012.

[3] J. D. Love and N. Riesen, "Mode-selective couplers for few-mode optical fiber networks," *Opt. Lett.*, vol. 37, no. 19, pp. 3990–3992, Oct. 2012.

[4] N. Hanzawa, K. Saitoh, T. Sakamoto, T. Matsui, K. Tsujikawa, M. Koshiba, and F. Yamamoto, "Two-mode PLC-based mode multi/demultiplexer for mode and wavelength division multiplexed transmission," *Opt. Express*, vol. 21, no. 22, pp. 25752–25760, Nov. 2013.

[5] N. Hanzawa, K. Saitoh, T. Sakamoto, T. Matsui, K. Tsujikawa, T. Uematsu, and F. Yamamoto, "PLC-based four-mode multi/demultiplexer with LP_{11} mode rotator on one chip," *J. Lightw. Technol.*, vol. 33, no. 6, pp. 1161-1165, Mar. 2015.

[6] N. Hanzawa, K. Saitoh, T. Sakamoto, T. Matsui, K. Tsujikawa, T, Fujisawa, Y. Ishizaka, and F. Yamamoto, "Demonstration of PLC-based six-mode multiplexer for mode division multiplexing transmission," *European Conference and Exhibition on Optical Communidation*, P2.5, 2015.

TuE3-5

OECC/PS2016

Widely Switching the Orbital Angular Momentum Modes with Integrated Cobweb Emitter

Yu Wang, Peng Zhao, Xue Feng, and Yidong Huang
Department of Electronic Engineering, Tsinghua University, Beijing, 100084
x-feng@tsinghua.edu.cn

Abstract: An integrated cobweb emitter with a wide switching range of orbital angular momentum modes (l = −4~4) is demonstrated. The unique independence of micro-ring cavity and scattering unit provides the flexibility to design and optimize.
Keywords: *Orbital angular momentum; Silicon photonics; Grating*

I. INTRODUCTION

Pioneered by Allen *et al* in 1992 [1], the orbital angular momentum (OAM) of light has been acknowledged to be independent with the spin angular momentum (SAM), constituting a complete orthogonal basis as a new degree of freedom. Due to the unique nature of infinite dimensionality, OAM has widely expanded the scopes of substantial optical applications such as classical and quantum information, imaging and etc [2].

The aforementioned implementations are mainly based on spatial optics, with manipulated OAM modes generated by spatial light modulator (SLM), Q-plate, or spiral phase plate (SPP) [2]. Recently, integrated OAM emitters have been actively investigated as a fruitful combination of silicon photonics and OAM, showing huge potentials of being compact and dynamic, compared with spatial optics [3]–[5]. When considering the potential applications of integrated OAM emitters, such as ultra-fast inter-chip communication [6] and high dimensional quantum state manipulation, the more OAM modes are involved, the higher dimensionality would be explored. Therefore, it is the number of available OAM modes that actually matters. However, as far as we know, the best experimental result ever reported is only of five OAM modes in total [4]. Thus integrated emitter with more switching OAM modes is crucial for real applications.

In this work, an integrated cobweb emitter with a potentially wide switching range of OAM modes is theoretically developed, numerically simulated, and experimentally verified. In our design, the micro-ring cavity and the scattering unit are spatially separated so that they could be designed and optimized independently. A larger micro-ring cavity is beneficial to achieve a wider switching range, so the radius of 200μm is adopted. Meanwhile, the scattering unit is designed to be a cobweb like structure so that lightwaves around 1550nm can be efficiently and vertically scattered. The dynamic switching of nine azimuthally polarized OAM modes (l = −4~4) with electrically controlled thermo-optic effect is demonstrated.

II. PRINCIPLE AND DESIGN

The integrated cobweb emitter, which is fabricated on a 220nm SOI substrate, is shown in Fig. 1. This device is based on a micro-ring cavity with a bus waveguide and $N = 16$ downloading waveguides. The fundamental mode of a micro-ring cavity is the whispering gallery mode (WGM), which has annular intensity and ascending/descending phase profiles. With properly placed gratings, an OAM mode can be scattered from the micro-ring cavity into free space. The topological charge is resulted from the phase shift between any two adjacent gratings, which is determined by the WGM order in micro-ring cavity. Consequently, varying the WGM order could switch the emitted OAM modes. The whole scattering unit of all 16 gratings looks like a cobweb so that our proposed structure is named as the cobweb emitter. As discussed in reference [5], the topological charge can be expressed as:

$$\Delta\varphi = \frac{2\pi M}{N} \bmod 2\pi - \pi \in [-\pi, \pi), \tag{1}$$

where M stands for the WGM order in the micro-ring cavity. Intuitively, there are 16 possible values for the topological charge when $N = 16$. However, when $\Delta\varphi$ equals to $-\pi$, the scattered lights would be an azimuthally standing wave, which is the so-called cogwheel beam and possesses total OAM of zero [7]. As a result, by excluding this particular situation, the topological charge could only have 15 values as:

$$l = \frac{16\Delta\varphi}{2\pi} = \frac{8}{\pi}\Delta\varphi \in \{0, \pm 1, \pm 2, \dots, \pm 7\}. \tag{2}$$

Compared with some previous work reported in Ref [4] and [8], in which the scattering unit is right on or adhered to the micro-ring cavity, the most significance of our design is to separate the scattering unit and the micro-ring cavity apart as such two components actually play relatively different roles for generating the desired OAM mode. The micro-

This work is supported by the National Basic Research Program of China (No. 2011CBA00608 and 2011CBA00303), the National Natural Science Foundation of China (No. 61307068 and 61321004), and the Opened Fund of the State Key Laboratory on Integrated Optoelectronics, China (No. IOSKL2013KF09).

ring cavity tells the extinct phase shifts between the gratings and results in different topological charges. The scattering unit, meanwhile, determines the beam characteristics, such as the state of polarization and directionality. Therefore, within our proposal, these two components could be designed and optimized independently.

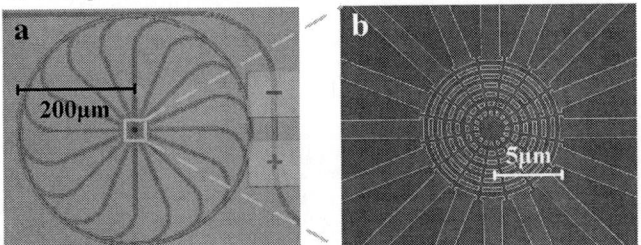

Fig. 1. (a) The micrograph and (b) the scanning electron micrograph of the integrated cobweb emitter from top view.

As shown in Eq. (1-2), the topological charge of OAM modes could be tuned by varying the WGM order in micro-ring cavity, which is given by

$$N = \frac{2\pi r}{\lambda} n_{eff},\qquad(3)$$

where r, λ, and n_{eff} stand for the radius of micro-ring cavity, the operating wavelength (in vacuum), and the effective refractive index of silicon, respectively. Both r and λ are relatively fixed in a fabricated OAM emitter, the preferred solution to realize switching is varying n_{eff} by electrically controlled thermo-optic effect, as implemented by the electrode shown in Fig. 1. The switching range of the topological charge (rounded down) is given by

$$\Delta l = \Delta N = \left\lfloor \frac{2\pi r n_{eff}}{\lambda} \eta \right\rfloor.\qquad(4)$$

Here, a parameter of the modulation ratio $\eta = \Delta n_{eff} / n_{eff}$ is introduced to evaluate the variation extent of n_{eff}. Generally, a larger modulation ratio is related to a higher temperature in thermo-optic effect. It is obvious that a larger micro-ring cavity is helpful to achieve a wider switching range. So in this work, the radius is designed to be as large as 200μm.

After the micro-ring cavity is settled, the scattering unit would be designed. A quasi-TE mode is assumed to be excited at the incident side of one downloading waveguide. Figure 2(a-b) illustrate the structural parameters and behaviors of one grating at the center. The period of gratings is designed as 630nm with duty cycle of 50%, which is quite suitable for vertically scattering lightwave around 1550nm. The electric field of the fundamental quasi-TE mode propagating along the ridge waveguide is transversal, resulting in the scattered OAM mode as azimuthally polarized. By duplicating the obtained electromagnetic field with proper phase shifts as outputs from all 16 different gratings, OAM modes with determined topological charge could be generated. Figure 2(c-d) illustrate intensity and phase (of the azimuthal component) profiles of an OAM mode with settled topological charge of −3.

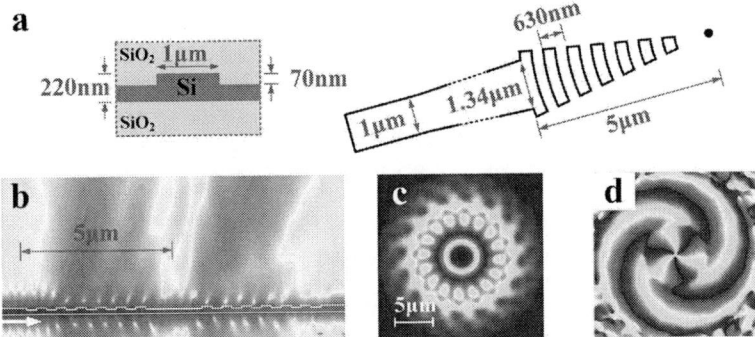

Fig. 2. (a) The structural parameters and (b) the lateral view of the intensity profile of scattered lightwave of one grating. The top view of (c) the intensity and (d) the phase profiles of $l = -3$ OAM mode from the cobweb emitter.

III. MEASUREMENTS

The generated OAM mode is azimuthally polarized and could be decomposed into two parts. The carried SAM and OAM per photon of left-handed circular polarized (LHCP) part would be $-\hbar$ and $(l+1)\hbar$, while those of the right-handed circular polarized (RHCP) part would be $\hbar$ and $(l-1)\hbar$. The interference pattern would be a spiral figure of $l+1$ (LHCP) or $l-1$ (RHCP) arms with a coherent LHCP (RHCP) beam as reference.

With the incident wavelength fixed at 1550.49nm, the dynamic characteristics of the device are investigated. By continuously increasing the voltage applied on the electrode, the local temperature of micro-ring cavity and the effective

This work is supported by the National Basic Research Program of China (No. 2011CBA00608 and 2011CBA00303), the National Natural Science Foundation of China (No. 61307068 and 61321004), and the Opened Fund of the State Key Laboratory on Integrated Optoelectronics, China (No. IOSKL2013KF09).

refractive index would be significantly varied. Both the experimental and simulation results are shown in Fig. 3. Under meticulously controlled driving voltage, different OAM modes could be generated. With a maximum modulation ratio of only 0.4% ($l = 4$, the last column), nine OAM modes is dynamically tunable. By subtracting the resistance of contacting electrode, the residual power on the heating electrode is referred as the effective power. We then conclude the linear dependence between the topological charge of OAM mode and the effective power. Linear fitting result shows that the effective switching power is about 20mW per mode.

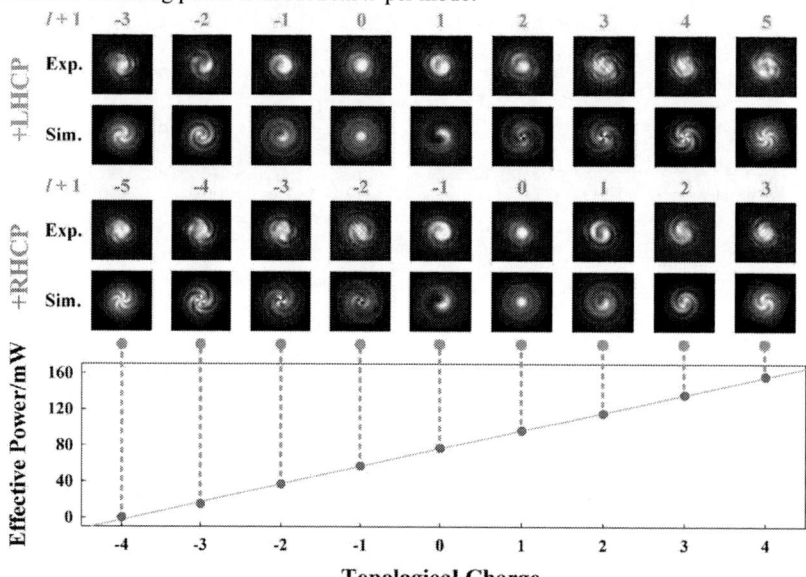

Fig. 3 Dynamic switching of nine OAM modes has been implemented with the effective switching power of 20mW per mode.

IV. CONCLUSIONS

In summary, an integrated cobweb emitter is demonstrated in this work. As the micro-ring cavity and the scattering unit could be optimized independently, a large micro-ring cavity for a wide switching rang and a scattering unit for azimuthal polarization are adopted. By further utilizing the above special characteristics of our design, we believe that OAM modes with polarization diversity, desired directionality of emission and a much wider switching range can be achieved, bringing far more potentials in real applications. This work would be published on a journal soon in detail [9].

ACKNOWLEDGMENT

We would like to thank Dr. Wei Zhang, Dr. Fang Liu, and Dr. Kaiyu Cui for their valuable discussions and comments.

REFERENCES

[1] L. Allen, M. W. Beijersbergen, R. J. C. Spreeuw, and J. P. Woerdman, "Orbital angular momentum of light and the transformation of Laguerre-Gaussian laser modes," *Phys. Rev. A*, vol. 45, no. 11, pp. 8185–8189, 1992.

[2] A. M. Yao and M. J. Padgett, "Orbital angular momentum: origins, behavior and applications," *Adv. Opt. Photonics*, vol. 3, no. 2, p. 161, Jun. 2011.

[3] D. Zhang, X. Feng, and Y. Huang, "Encoding and decoding of orbital angular momentum for wireless optical interconnects on chip," *Opt. Express*, vol. 20, no. 24, pp. 26986–26995, 2012.

[4] M. J. Strain, X. Cai, J. Wang, J. Zhu, D. B. Phillips, L. Chen, M. Lopez-Garcia, J. L. O'Brien, M. G. Thompson, M. Sorel, and S. Yu, "Fast electrical switching of orbital angular momentum modes using ultra-compact integrated vortex emitters," *Nat. Commun.*, vol. 5, p. 4856, Sep. 2014.

[5] Y. Wang, P. Zhao, X. Feng, K. Cui, and Y. Huang, "Integrated emitters for optical vortices with a cobweb structure," in *Opto-Electronics and Communications Conference (OECC), 2015*, 2015, pp. 1–3.

[6] W. Liu, G. Chen, Y. Wang, Y. Wang, X. Feng, Y. Xie, Y. Huang, and H. Yang, "Exploration of Electrical and Novel Optical Chip-to-Chip Interconnects," *IEEE Des. Test*, vol. 31, no. 5, pp. 28–35, Oct. 2014.

[7] Y. Wang, X. Feng, D. Zhang, P. Zhao, X. Li, K. Cui, F. Liu, and Y. Huang, "Generating optical superimposed vortex beam with tunable orbital angular momentum using integrated devices," *Sci. Rep.*, vol. 5, Jul. 2015.

[8] R. Li, X. Feng, D. Zhang, K. Cui, F. Liu, and Y. Huang, "Radially Polarized Orbital Angular Momentum Beam Emitter Based on Shallow-Ridge Silicon Microring Cavity," *IEEE Photonics J.*, vol. 6, no. 3, pp. 1–10, Jun. 2014.

[9] Y. Wang, P. Zhao, X. Feng, Y. Xu, K. Cui, F. Liu, W. Zhang and Y. Huang, "Integrated photonic emitter with a wide switching range of orbital angular momentum modes," will be published on *Sci. Rep.*

This work is supported by the National Basic Research Program of China (No. 2011CBA00608 and 2011CBA00303), the National Natural Science Foundation of China (No. 61307068 and 61321004), and the Opened Fund of the State Key Laboratory on Integrated Optoelectronics, China (No. IOSKL2013KF09).

TuE4-1 (Invited)

Compliant polymer interface for fiber connection of nanophotonic waveguides

Hidetoshi Numata

IBM Research – Tokyo, NANOBIC, 7-7, Shin-Kawasaki, Saiwai-ku, Kawasaki, Kanagawa Japan
hnumata@jp.ibm.com

Abstract— We developed a low loss polymer waveguide components for near future silicon photonics packaging. We improved the structure of the polymer waveguide connector and realized a low loss single mode polymer waveguide MT compatible connector.
Keywords—silicon photonics, polymer waveguide connector

I. INTRODUCTION

There is a large demand for higher speed data communications for the off-chip I/O bandwidth of central processor unit in personal computers and server systems because the on-chip performance of CPU has become improved rapidly. A wiring density of electric circuits on a printed circuit board is restricted by the electromagnetic interference during the wirings, and date rate is restricted by the dielectric signal loss and the signal reflection for electrical interconnect. On the other hand, the optical interconnect is not restricted that speed by the propagation medium and wiring density, so the rapid development is expected as the next generation data communications. Recent progresses in optical interconnects realize low power, high data rate, and high density. Especially, a polymer waveguide is very useful wiring for the optical interconnect. Since an optoelectronic circuit with high wiring density can be realized by a polymer waveguide, it is considered as the important means of the high data rate communications in a short distance between elements on PCB such as CPU, memory, and optical connector. The communications between the boards through optical connectors can be also considered.

Especially, in recent years, silicon nano-photonics technology has begun to realize large advances to the optical interconnect. But this technology requires a low loss optical coupling with external single mode optical fiber cables. In this point of view, a single mode polymer waveguide is useful mediation device between the silicon nano-photonics circuit and the external optical fiber cables. Because a core size and a channel pitch of a polymer waveguide can be flexibly and precisely fabricated within +/- one micrometer accuracy. The next problem is how to realize a precise positioning accuracy of each core and a low loss coupling of the single mode optical signals between the silicon nano-photonics circuit and the external optical fiber cables through a single mode polymer waveguide.

We report the useful self-alignment method of a single mode polymer waveguide connector which connects a single mode polymer waveguide and external single mode optical fiber cables with high positioning accuracy within +/- two micrometers.

II. POLYMER WAVEGUIDE CONNECTOR ASSEMBLY

Polymer waveguide is essential for high density optical interconnect, and a polymer waveguide connector is necessary for the connection between a polymer waveguide and another polymer waveguide or a glass fiber. The "PMT" connector (JPCA–PE03-01-07S–2006) is a well-known standard as a multimode polymer waveguide connector. But it is not easy to realize a precise assembly under a few micrometers positioning error. And it cannot be applied for a single mode polymer waveguide connector which needs a positioning error smaller than one micrometer.

On the other hand, we have already realized two unique single mode polymer waveguide connector structures. The one contains the polymer waveguide flex which has two grooves on both sides, and the ferrule which has two studs on both sides [1]. These two combinations can realize a precise single mode polymer waveguide connector. The other single mode polymer waveguide connector structure of us contains the polymer waveguide flex which has many grooves between each core and the ferrule which has many studs (Fig.1) [2]. The top surface of the under clad layer acts as a reference plane of a polymer waveguide core positioning in the height direction. In this polymer waveguide connector, the pitch of the grooves and that of the studs should be equal each other. The original positioning error of the groove on the polymer waveguide itself is smaller than one micrometer because of a precise photolithography fabrication process. But the polymer easily shrinks and expands because of the relatively large CTE (coefficient of thermal expansion) of its material specification. And the positioning error of the groove on the polymer waveguide against the ferrule largely varies. If the CTE of the polymer waveguide is about 50ppm / deg C, that of the ferrule is 1ppm / deg C, and the distance between the outermost two grooves (A) is 3mm (3 micrometers smaller than the distance between two outermost studs (B) at 25 deg C), then, these two distances (A and B) become equal each other at 45 deg C and each core comes to the precise position.

In order to solve the CTE difference between a polymer waveguide and a ferrule and to realize the precise positioning of cores inner a polymer waveguide connector, we executed the following assembly process of our second polymer waveguide connector structure: (1) heat the polymer waveguide, (2) put a ultraviolet cure type glue and exposure ultraviolet light just on the outermost two sets of the groove and the stud, (3) stop the heating and fix the polymer waveguide and the ferrule (Fig.2). At the third process, the polymer waveguide wants to shrink, but both sides of the waveguide are already fixed. And the polymer waveguide layer becomes flat because of the shrinking tension force. And each core comes to the precise position.

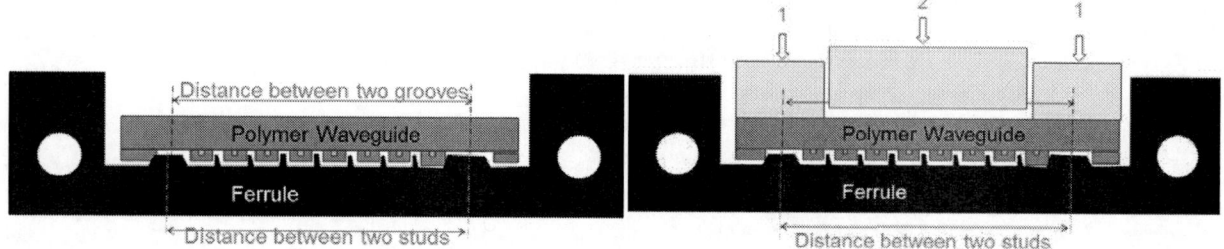

Fig.1: Our unique polymer waveguide connector structures. It contains the polymer waveguide flex which has many grooves between each core and the ferrule which has many studs. [2] Fig.2: Our assembly process of the polymer waveguide connector structure: (1) heat the polymer waveguide, (2) put a UV cure type glue and exposure ultraviolet light just on the outermost two sets of the groove and the stud, (3) stop the heating and fix the polymer waveguide and the ferrule.

III. IMPROVEMENT OF CONNECTOR ASSEMBLY (1)

There remains a severe problem for a polymer waveguide connector assembly. Many polymer waveguides have a back film on a single side or duals sides of these. In generally, glue is used between a polymer waveguide and a back film. In our polymer waveguide connector assembly process, a strong pressure is given from an upper side jig on a glass lid. The soft set of a polymer waveguide layer and a glue layer is sandwiched between a hard glass and a hard ferrule. Both the polymer waveguide layer and the glue layer become distorted and waved mutually. And the large positioning error occurs because of the large CTE difference between the hard back film and the soft polymer waveguide (Fig.3).

Fig.3: The soft set of a polymer waveguide layer and a glue layer is sandwiched between a hard glass and a hard ferrule. Both the polymer waveguide layer and the glue layer become distorted and waved mutually.

In order to solve this problem, we removed a back film from a polymer waveguide flex. We made four polymer waveguide MT compatible connector samples in order to compare positioning errors of cores which were influenced by the back film of the polymer waveguide flex. Figure 4 shows the surface of the polymer waveguide MT compatible connector sample 1 and 2. These two samples have a back film attached polymer waveguide flex. And Fig.5 shows the surface of the polymer waveguide MT compatible connector sample 3 and 4. These two samples have a back film less polymer waveguide flex. We measured the core positioning errors of these four polymer waveguide MT compatible connector samples. The measurement results of the core positions are shown in Figure 6. Figure 6 shows an absolute misalignment of the core positioning. In Fig.6, the back film less polymer waveguide flex samples show better core positioning alignment results (sample 3 and 4) than the back film attached polymer waveguide flex samples (sample 1 and 2). And we demonstrated the precise core positioning which was smaller than one micrometer for all channels for sample 3.

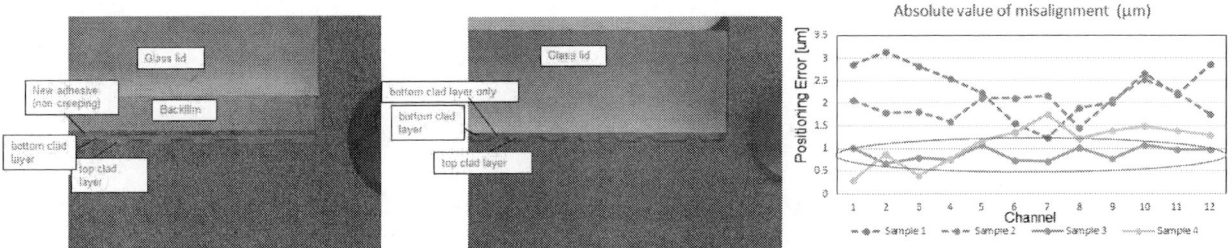

Fig.4: The surface of the polymer waveguide MT compatible connector sample 1 and 2. These two samples have a back film attached polymer waveguide flex.
Fig.5: The surface of the polymer waveguide MT compatible connector sample 3 and 4. These two samples have a back film less polymer waveguide flex.
Fig. 6: The measured absolute misalignment of the core positioning for the four polymer waveguide MT compatible connector samples

IV. IMPROVEMENT OF CONNECTOR ASSEMBLY (2)

In order to improve the positioning accuracy of the polymer waveguide connector, we did not only remove a back film from a polymer waveguide flex, but also increased a thickness of an under clad layer from 20 micrometers to 50micrometers. This new structure brings these advantages: (1) the stability of a precise core position is obtained while assembly process, (2) the easy manipulation of a polymer waveguide becomes realized while the assembly process, and (3) a physical strength of a polymer waveguide flex becomes obtained after a polymer waveguide connector assembly.

The designed under clad thickness is 50 micrometers without back film. At the polymer waveguide MT compatible connector surface, the measured polymer waveguide thickness was 73.431 micrometers and the under clad layer thickness was 49.996 micrometers. At the left side of the polymer waveguide MT compatible connector surface, the measured spacing between a groove of the polymer waveguide and the ridge of the MT ferrule in x direction was 32.792 micrometers at the left side of the measured ridge and 31.992 micrometers at the right side of the measured ridge in Fig.7. On the other hand, at the right side of the polymer waveguide MT compatible connector surface, the measured spacing between a groove of the polymer waveguide and the ridge of the MT ferrule in x direction was 31.992 micrometers at the left side of the measured ridge and 32.792 micrometers at the right side of the measured ridge in Fig.8. The assembly with precise positioning smaller than one micrometer was realized.

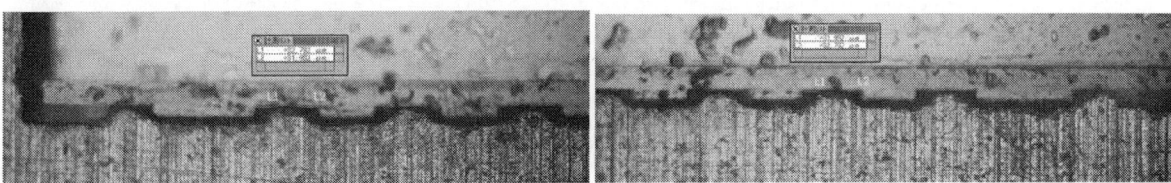

Fig.7: At the left side of the polymer waveguide MT compatible connector surface, the measured spacing between a groove of the polymer waveguide and the ridge of the MT ferrule in x direction was 32.792 micrometers at the left side of the measured ridge and 31.992 micrometers at the right side of the measured ridge.
Fig.8: At the right side of the polymer waveguide MT compatible connector surface, the measured spacing between a groove of the polymer waveguide and the ridge of the MT ferrule in x direction was 31.992 micrometers at the left side of the measured ridge and 32.792 micrometers at the right side of the measured ridge.

V. SUMMARY

We removed the back film from the polymer waveguide and increased the thickness of the under clad layer from 20 micrometers to 50micrometers. As the results, the uniformity of the under clad layer thickness and the alignment accuracy between the polymer waveguide and the ferrule became improved. The physical stability while the polymer waveguide MT compatible connector assembly process was obtained. The treatment of the polymer waveguide flex while assembly became much easier. The polymer waveguide MT compatible connector assembly with precise positioning which was less than one micrometer in x and y directions was realized, and the assembled polymer waveguide MT compatible connector became durable.

Part of this work has been supported by NEDO strategic energy-saving technology innovation program. The authors thank S.Takenobu of Asahi Glass Co. Ltd. for preparation of polymer waveguide flex samples and technical discussion.

REFERENCES

[1] T.Barwicz, Y.Taira, H.Numata et.al "Assembly of Mechanically Compliant Interfaces between Optical Fibers and Nanophotonic Chips" Proceedings of the IEEE Electronic Components and Technology Conference, May 27-30, 2014

[2] Y.Taira, H.Numata, T.Barwicz et.al. "Improved Connectorization of Compliant Polymer Waveguide Ribbon for Silicon Nanophotonics Chip Interfacing to Optical Fibers" Proceedings of the IEEE Electronic Components and Technology Conference, May 26-29, 2015

TuE4-2

OECC/PS2016

Highly efficient circular holes added a-Si:H grating coupler with metal mirror for 3D optical interconnects

JoonHyun Kang[1,2], Yuki Kuno[1], Kazuto Ito[1], Yusuke Hayashi[1], Junichi Suzuki[1], Il-Ki Han[2], Tomohiro Amemiya[3], Nobuhiko Nishiyama[1], and Shigehisa Arai[1,3]

1 Department of Electrical and Electronic Engineering, Tokyo Institute of Technology
2 Nanophotonics research center, Korea Institute of Science and Technology
3 Quantum Nanoelectronics Research Center, Tokyo Institute of Technology
Author e-mail address: n-nishi@pe.titech.ac.jp

Abstract: *A monolithically fabricated apodized a-Si:H grating coupler with metal mirror was demonstrated. The minimum groove length for the apodized grating was more than 100 nm and measured coupling efficiency was 70%.*
Keywords: *grating coupler, amorphous silicon waveguide, metal mirror*

I. INTRODUCTION

Grating couplers have been studied for the coupling between Si photonic waveguides and vertical single-mode fibers (SMFs) [1]. Through the grating coupler, spot size of the sub-μm waveguide can be converted to that of um-order fibers. Further, the grating coupler can be located anywhere on the surface of the chip since it can achieved vertical off-chip coupling without cleaving the chip. The coupling efficiency of the grating coupler is determined by following key factors; first, the beam profile of the radiated light should match with that of the SMFs. Second, the portion of the radiated light facing upward is needed to be maximized instead of downward radiation. There have been many novel structures for the efficient coupling including apodization of gratings [2,3], and DBR or metal mirrors under the grating coupler in order to maximize the upward radiated light [4,5]. Recently, Y. Ding et al. reported a grating coupler which introduced both the apodization of grating and the metal mirror but the peak coupling efficiency was limited by 81% and needed an additional wafer for flip chip bonding in order to introduce a metal mirror under the grating coupler [6].

We have proposed to use deposited a-Si:H multi-layered waveguides for optical interconnects and demonstrated highly efficient inter-layer grating couplers with metal mirrors [7, 8]. By applying the fabrication process used in Ref [7], monolithic integration of Si waveguides above the metal mirrors can be realized. In this paper, design procedures and experimental results of a-Si:H apodized grating coupler with a metal mirror will be presented.

II. TECHNICAL WORK PREPARATION

The grating coupler was designed for the coupling between SMFs (SMF-28, Corning Inc.) and a-Si:H waveguides. The width of the grating region was set to be 12 μm and then it was connected to 500-nm a-Si:H wire waveguide through 1-mm-long linearly tapered mode-converter. As mentioned earlier, the grating was apodized in order to shape the radiated beam profile by tuning the duty ratio and grating periods. Typically, the groove lengths of the first few grating periods are needed to be rather short which are tens of nm for the gradual increase in beam profile. However, those groove lengths are not preferable if we consider the fabrication tolerance or RIE lag effects. Figure 1(a) illustrates

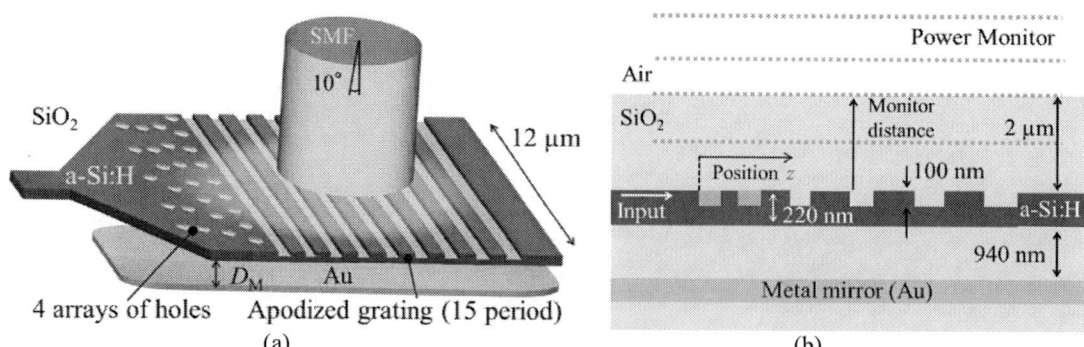

Fig. 1 (a) Schematics of grating coupler with holes. (b) Simulation model for grating design.

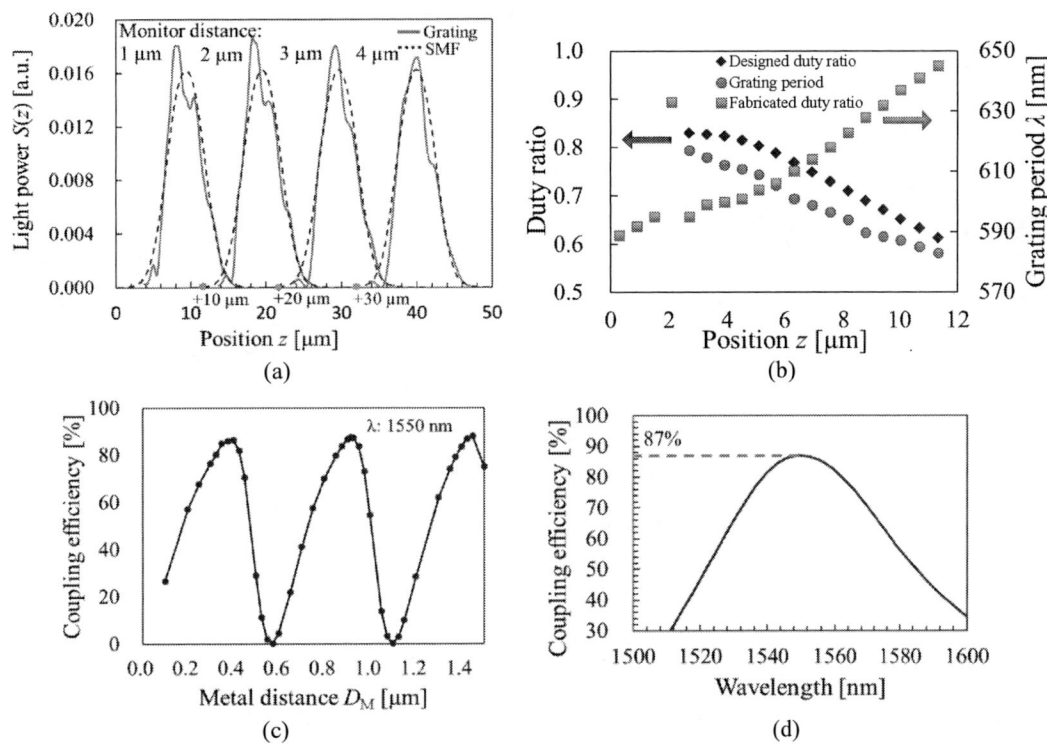

Fig. 2 (a) Radiated beam profile from the apodized grating and SMF. (b) Duty ratio and grating period of apodized grating coupler, blue diamonds: designed duty ratio, red circles: fabricated duty ratio, orange rectangles: grating period. (c) D_M and (d) wavelength dependence of the coupling efficiency.

the apodized grating coupler with the metal mirror. The grating consisted of 4 arrays of holes instead of the short grooves followed by 15 line-space gratings. By introducing 150 nm (diameter) holes, the minimum groove length of more than 100 nm was available. The grating depth of 100 nm was used in this work.

In order to design the apodized grating, we monitored the radiated beam profile while adjusting the duty ratio and period of the grating so it could match with mode profile of SMF. The simulation model used here is shown in Fig. 1(b). The monitors were located at every 1-μm distance in order to determine whether the radiation angle was constant in each position. Figure 2.(a) shows the radiated beam profile of the designed apodized grating coupler with metal mirror and mode profile of SMF with an angle of 10 degree. The both profiles were roughly correspond to each other at the monitor distance within 1~4 μm

The designed duty ratio and the grating period of apodized grating are shown in Fig. 2(b). For the holes, the apodization was applied by controlling the distance of neighboring holes in the same array which was ranged from 400 nm to 460 nm. Also, the grating periods were gradually increased according to the propagation direction to maintain the constant radiation angle. The grating period at the last hole-array was exceptionally long for the smooth connection between holes and line-space grating.

In addition to the apodization, the distance from the metal mirror to the grating (metal distance, DM) was also critical to the performance of the grating coupler. Therefore, we examined the effect of DM on the coupling efficiency, and the coupling efficiency was varied periodically as shown in Fig. 2(c). The periodic change was resulted from the phase matching condition between the reflected light by the metal mirror and the light passed through the grating [8]. Figure 2(d) shows the wavelength dependence of the coupling efficiency when the DM is 940 nm. The peak coupling efficiency was 87% at 1550 nm. The 1dB bandwidth was 37nm which can cover the C-band and 3dB bandwidth was 71 nm.

III. FABRICATION PROCESS AND COUPLING EFFICIENCY MEASUREMENT

The apodized grating coupler with the metal mirror was fabricated based on the design as described in section 2. The initial wafer was 2-inch Si wafer with 3-μm-thick layer of thermal SiO_2. The first step was evaporation of a 100-nm-thick Au film on the SiO_2 and the metal mirror and alignment mark for EB-lithography was formed by lift-off. The Au

OECC/PS2016

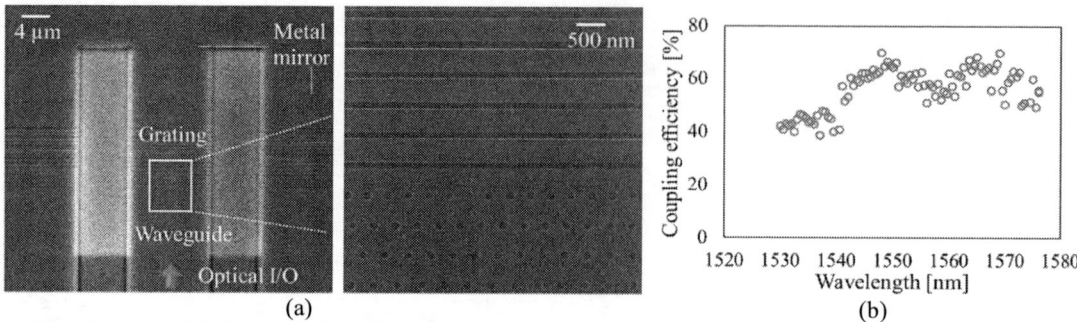

(a)   (b)

Fig. 3 (a) Top SEM image of apodized grating and its enlarged image. (b) Measured coupling efficiency of apodized grating coupler with metal mirror

patterns were buried by SiO_2 and then flattened by CMP process. The thickness of SiO_2 was controlled targeting 940 nm and then, 220-nm-thick a-Si:H layer was deposited on the SiO_2 at 300°C. The apodized grating with holes and waveguide patterns were formed by EB lithography and ICP-RIE system. Finally, 2 μm of SiO_2 overcladding layer was deposited and flattened again by CMP process. After the fabrication process, the thickness of DM was confirmed to be 915 nm. Figure 3 shows the top SEM image of the apodized grating. The apodized grating was placed above the metal mirror which can be seen through the waveguide patterns and the grating was located at the center of the metal mirror. The duty ratio of the fabricated apodized grating is plotted in Fig. 2.(b). There was 2~8% difference in the designed and fabricated values of duty ratio.

The measured coupling efficiency of the apodized grating coupler with the metal mirror is shown in Fig. 3(b). The coupling efficiency of the fabricated apodized grating coupler with the metal mirror was calculated by fiber to fiber transmission through a pair of grating couplers. The peak coupling efficiency was 70%, which was smaller than the designed coupling efficiency since the duty ratio of the fabricated grating was slightly different from the target value. Higher coupling efficiency is expected with correct groove-lengths control of apodized grating by adjusting the exposure amount in the EB lithography. Even though, we successfully demonstrated the monolithically integrated a-Si:H grating coupler with the metal mirror which is promising for the realization of 3D optical interconnect.

IV. CONCLUSIONS

In this work, a fiber-to-chip a-Si:H grating coupler was monolithically fabricated with a-Si:H multi-stacking method. Metal mirror was added under the a-Si:H grating coupler to improve the coupling efficiency. In order to optimize the mode profile of the radiated light from the grating to the SMF, apodization of grating was demonstrated by shallow etched grooves. 4 arrays of holes were introduced to the grating instead of any grooves shorter than 100 nm were needed. The peak coupling efficiency of 87% was calculated in a simulation and 70% was measured with fabricated grating coupler.

REFERENCES

[1] D. Vermeulen, S. Selvaraja, P. Verheyen, G. Lepage, W. Bogaerts, P. Absil, D. Van Thourhout, and G. Roelkens, "High efficiency fiber-to-chip grating couplers realized using an advanced CMOS-compatible Silicon-On-Insulator platform," Optics Express, vol. 92, pp. 18278-18283, Aug. 2008.

[2] R. Halir, P. Cheben, J. Schmid, R. Ma, D. Bedard, S. Janz, D. Xu, A. Densmore, J. Lapointe, and I. Molina-Fernandez, "Continuously apodized fiber-to-chip surface grating coupler with refractive index engineered subwavelength structure," Optics letters vol. 35, pp. 3243-3245, Oct. 2010.

[3] Y. Ding, H. Ou, and C. Peucheret, "Ultrahigh-efficiency apodized grating coupler using fully etched photonic crystals," Optics letters, vol. 38, pp. 2732-2734, Aug. 2013.

[4] C. Kopp, S. Bernabe, B. Bakir, J. Fedeli, R. Orobtchouk, F. Schrank, H. Porte, L. Zimmermann, and T. Tekin, "Silicon Photonic Circuits: On-CMOS Integration, Fiber Optical Coupling, and Packaging", J. Sel. Top. Quantum Electron. vol. 17, pp. 498-509, May/June 2010.

[5] W. Zaoui, M. Rosa, W. Vogel, M. Berroth, J. Butchke, and F. Letzkus, "Cost-effective CMOS-compatible grating couplers with backside metal mirror and 69% coupling efficiency, Optics express, vol. 20, pp. B238-B243, Dec. 2012.

[6] Y. Ding, C. Peucheret, and H. Ou, "Ultra-low coupling loss fully-etched apodized grating coupler with bonded metal mirror," in Proceeding of IEEE International Conference on Group Four Photonics, ThD2, 2014.

[7] J. Kang, Y. Atsumi, M. Oda, T. Amemiya, N. Nishiyama, and S. Arai, "Low-loss Amorphous Silicon Multilayer Waveguides Vertically Stacked on Silicon-on-Insulator Substrate", Jpn. J. Appl. Phys., vol. 50, p. 120208, Nov. (2012.

[8] J. Kang, Y. Atsumi, M. Oda, T. Amemiya, N. Nishiyama, and S. Arai, "Amorphous-Silicon Inter-Layer Grating Couplers with Metal Mirrors toward 3D Interconnect," J. Sel. Top. Quantum Electron. vol.20, p. 8202308, July/Aug. 2014.

TuE4-3

OECC/PS2016

300-mm ArF-immersion lithography technology based Si-wire grating couplers with high coupling efficiency and low crosstalk

Yohei Sobu, Seok-Hwan Jeong, Yu Tanaka, and Ken Morito

Photonics Electronics Technology Research Association (PETRA)

AIST West 7SCR, 16-1 Onogawa, Tsukuba, Ibaraki, Japan 305-8569

y-sobu@petra-jp.org

Abstract: *The feasibility of the grating couplers was characterized by using ArF-immersion lithography. The coupling loss of 1D and 2DGC was measured to be −2.2 dB and −4.1 dB with good spectral uniformity.*

Keywords: *grating couplers, optical coupling, polarization splitting, silicon photonics, optical waveguide*

I. INTRODUCTION

Recently, the demand for high-density interconnect technology has been increasing in high-end servers and high performance computing systems. Optical interconnect systems based on Si photonics are promising candidates for surpassing the processing bandwidth limit of electrical interconnects [1]. Silicon photonics enables realization of high-density optical circuits on a large scale by utilizing mature CMOS technology. In PETRA, we have been developing a Si-based WDM (wavelength division multiplexing) transceiver consisting of optical light sources [2], modulators, photodiodes, and multiplexer/demultiplexers [3].

Considering the optical coupling between a Si waveguide and a single-mode fiber (SMF), we developed two kinds of grating couplers (GCs) for WDM integrated transmitters and receivers: 1) the GC for the transmitter that is based on one dimensional corrugated grating (1D-GC) and 2) the GC for the receiver based on two dimensional square lattice grating (2D-GC). To date, several kinds of 1D and 2DGCs have been widely developed to minimize coupling loss from the viewpoint of directionality and mode matching between SMF and GC. As a way of improving directionality, it has been reported that the optimization of the thickness of the Si top layer and BOX is effective. In addition, a bottom mirror under the Si waveguide layer [4, 5] and Si overlay structure [6] has been reported. Meanwhile, as an alternative way of enhancing the degree of mode matching, the apodized (non-uniform) grating whose period and relative index change are non-uniformly modulated has been reported [7, 8]. However, although various reports of 1D and 2DGCs with low loss coupling have been demonstrated, there has been little research on systematic discussion of spectral uniformity on silicon-on-insulator (SOI) wafer and polarization crosstalk for the 2DGC.

In our previous work, we reported feasibility studies of the coupling loss and spectral uniformity of 1D/2DGC by using the 248-nm KrF lithography process on a 200-mm SOI wafer. In this paper, we report high-performance 1D and 2DGCs based on 300-mm wafer-scale ArF-immersion lithography technology. Low coupling loss for both the 1D and 2DGCs with good spectral uniformities across the SOI wafer were experimentally demonstrated. Additionally, we discuss polarization crosstalk for the 2DGCs and the dependence of their crosstalk on the shape of scattering elements.

II. DESIGN OF GRATING COUPLERS

The schematic drawings of 1DGC operating with a TE-mode are shown in Fig. 1(a) and (b). The grating structure is composed of a 220-nm-thick Si layer on a 2-μm-thick BOX layer and a SiO_2 top cladding layer. The depth of the shallow etching is 70 nm as shown in Fig. 1 (a). The SMF is tilted at a 10° angle with respect to the vertical axis to reduce the impacts on the back reflection of input signals. As mentioned in the previous section, to obtain a lower coupling loss, we need to obtain a Gaussian-like profile from 1DGC, which can be achieved by apodizing the coupling strength along the propagation direction. In this way, we designed the apodized grating by varying the grating filling factor (FF), which was defined as the ratio of the groove width to the grating period. It should be noted that the grating period was also optimized to maintain the radiation angle of 1DGC. Usually, the ideal apodization of GC requires the FF to be as low as possible to minimize the coupling strength, which inevitably requires a high-resolution lithography process. In this work, ArF-immersion lithography technology was utilized to obtain sufficient apodization effects.

The top view of 2DGC is illustrated in Fig. 2. 2DGC consists of square lattice scattering elements and two orthogonal output ports. When the input signals with an arbitrary polarization state are incident on the 2DGC, the signal is separated into two output ports with the same polarization in the Si-wire waveguides (i.e. TE-like mode), depending on its polarization state. The layer structure of 2DGC is the same as that of 1DGC (see Fig. 1(a)). The gray are as indicate the remaining silicon layer, and the white areas indicate where the silicon has been etched away by 70-nm-deep shallow etching, as shown in Fig. 2. In this case, the SMF is tilted in the same way as for the 1DGC: along the bisector direction between port 1 and port 2. It is important to note that the two output waveguides are mutually angled at 3.15 degrees on the inward side to avoid increasing the coupling loss caused by mismatch of the wavefronts of the scattering elements of 2DGC and the two output ports [9, 10]. In the optical receiver, lower coupling loss and lower polarization crosstalk are absolutely required for 2DGCs. Here, the optical polarization crosstalk is defined as the degree of optical leakage to another port for a certain linearly polarized input light in a SMF. To evaluate the optical polarization crosstalk, we fabricated two types of scattering elements called circular cylinders and rhomboid patterns as shown in Fig. 2.

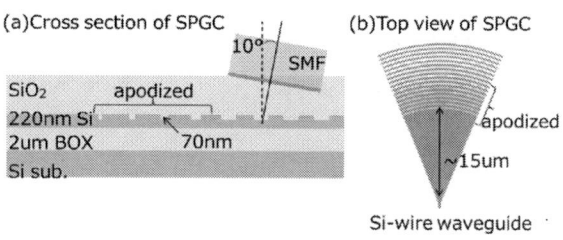

Fig. 1 (a) Cross-sectional view and (b) top view of 1DGC

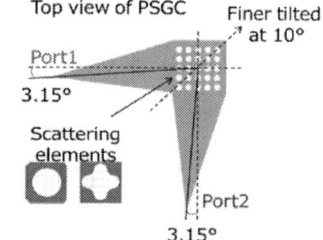

Fig. 2 Schematic drawing of 2DGC and two kinds of shapes of scattering elements

III. CHARACTERISTICS OF GRATING COUPLERS

First, the coupling loss of 1DGC was characterized. The input light was set to TE-mode, and the transmittance spectra of 1DGC were measured by a spectrum analyzer. The transmittance of 1DGC using apodized grating was shown in Fig. 3. The red and black lines indicate the coupling loss with apodized grating and uniform period grating, respectively. The coupling loss with apodized grating was measured to be −2.2 dB, which was 0.8 dB smaller than that with uniform period grating. This improvement was based on the mode matching effect by using apodized grating structures.

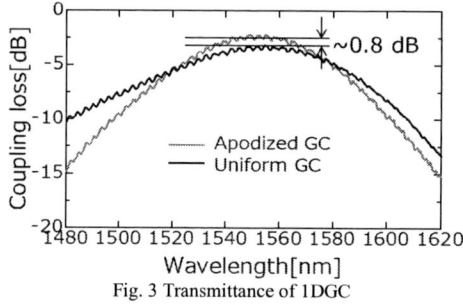

Fig. 3 Transmittance of 1DGC

Subsequently, the coupling loss and optical polarization crosstalk of the 2DGC were experimentally characterized. The measured transmittances for the scattering elements with the circular cylinders and the rhomboid patterns are shown in Fig. 4 (a) and (b). The red and blue lines correspond to the transmittance of output port 1 and port 2, respectively, as shown in Fig. 2. The coupling losses of 2DGCs with circular cylinders and rhomboid patterns were -4.2 dB and -4.1 dB, respectively. On the other hand, the optical polarization crosstalk was measured to be −11.2 dB and −13.8 dB, respectively. Compared to the characteristics of 2DGC with two types of scattering elements, the coupling loss and polarization crosstalk were improved in the case of 2DGC with rhomboid patterns.

In a similar way, we measured the 2DGC at five points of the wafer, as shown in Fig. 5 (a), to examine the uniformities of coupling loss and optical crosstalk. In Fig. 5 (b), the blue and red circles indicate coupling loss and optical crosstalk, respectively. The average and standard deviation of coupling loss and optical crosstalk were calculated using the results shown in Fig. 5 (b). The average and standard deviation of coupling loss and optical crosstalk were characterized to be −4.4 ± 0.48 dB and −10.98 ± 0.73 dB, respectively, in the case of using the circular cylinders. In the case of the rhomboid scattering elements, the coupling loss and optical crosstalk were characterized to be −4.23 ± 0.33 dB and −14.1 ± 0.8 dB. The uniformity of coupling loss and optical crosstalk in inter-dies were extremely stable due to ArF-immersion lithography on 300-mm SOI. From this result, the coupling loss and the optical crosstalk could be reduced to be 0.17 dB and 3.1 dB, respectively, by using rhomboid patterns. The reproducibility of this tendency was confirmed across the SOI wafer. Therefore, the characteristics of 2DGC were improved by changing the scattering elements.

Overall, through our feasibility study of GCs fabricated using ArF-immersion lithography process technology on a 300-mm SOI wafer, we experimentally demonstrated low loss coupling for 1D and 2DGCs and low polarization crosstalk for 2DGCs with high spectral uniformities. Additionally, we verified the dependence of the polarization crosstalk on the shapes of scattering elements.

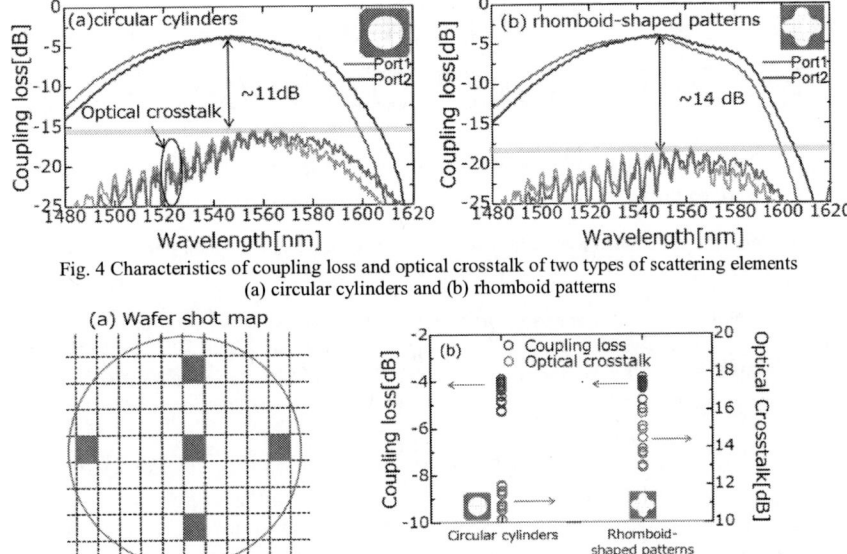

Fig. 4 Characteristics of coupling loss and optical crosstalk of two types of scattering elements
(a) circular cylinders and (b) rhomboid patterns

Fig. 5 (a) Measurement point of 2DGC (wafer shot map),
(b) Uniformities in inter-dies of coupling loss and optical crosstalk in two types of scattering elements at five points of wafer

IV. CONCLUSION

We examined the feasibility of 1D and 2DGC by ArF-immersion lithography process on 300-mm SOI wafer. The coupling loss for the 1DGC was measured to be −2.2 dB by adopting the apodized grating structures. In the 2DGC, the coupling loss and optical crosstalk of two types of scattering elements were characterized. The average and standard deviation of the coupling loss and the optical crosstalk were −4.23 ± 0.33 dB and −14.1 ± 0.8 dB, respectively, in the case of using rhomboid patterns. The spectral uniformity of the coupling loss and optical crosstalk in inter-dies were extremely stable due to ArF-immersion lithography on 300-mm SOI wafer. In the future, a lower GC coupling loss will be achieved to improve the directionality and optimize the grating shapes for the development of a more practical Si-based WDM transceiver.

ACKNOWLEDGMENT

The research is partly supported by New Energy and Industrial Technology Development Organization (NEDO).

REFERENCES

[1] G. Li, X. Zheng, J. Lexau, Y. Luo, H. Thacker, T. Pinguet, P. Dong, D. Feng, S. Liao, R. Shafiiha, M. Asghari, J. Yao, J. Shi, I. N. Shubin, D. Patil, F. Liu, K. Raj, R. Ho, J. E. Cunningham, and A. V. Krishnamoorthy, "Ultralow-power silicon photonic interconnect for high-performance computing systems," Proc. SPIE 7607,760703, 760703-15 2010

[2] S. Tanaka, T. Matsumoto, T. Kurahashi, M. Matsuda, A. Uetake, S. Sekiguchi, Y. Tanaka, and K. Morito, "Flip-chip-bonded, 8-wavelength AlGaInAs DFB laser array operable up to 70°C for silicon WDM interconnects," Proc. ECOC, Cannes, France, Tu. 1.1.4, 2014

[3] S-.H. Jeong, D. Shimura, Y. Tanaka, and K. Morito,"Novel Si-wire microring assisted multiple delayline based optical demultiplexers with the highest spectral flatness," Proc. ECOC, Cannes, France, We.1.4.3, 2014

[4] W. S. Zaoui, A. Kunze, W. Vogel, M. Berroth, J. Butschke, and F. Letzkus, "CMOS-compatible nonuniform grating coupler with 86% coupling efficiency," OSA Technical Digest, European Conference and Exhibition on Optical Communication, paper Mo.3.B.3, 2013

[5] L. Carroll, D. Gerace, I. Cristiani, and L. C. Andreani, "Optimizing polarization-diversity couplers for Si-photonics: reaching the -1 dB coupling efficiency threshold," Opt. Express, vol. 22, no. 12, pp. 14769-14781, June 2014

[6] D. Vermeulen, S. Selvaraja, P. Verheyen, G. Lepage, W. Bogaerts, P. Absil, D. V. Thourhout, and G. Roelkens, "High-efficiency fiber-to-chip grating couplers realized using an advanced CMOS-compatible Silicon-on-Insulator platform," Opt. Express, vol. 18, no. 17, pp. 18278-18283, August 2010

[7] A. Mekis, S. Gloeckner, G. Masini, A. Narasimha, T. Pinguet, S. Sahni, and P. D. Dobbelaere, "A grating-coupler-enabled CMOS photonics platform," IEEE J. Sel. Top. Quantum Electron, vol. 17, no. 3, pp. 597-608, May/June 2011

[8] X. Chen, C. Li, C. K. Y. Fung, S. M. G. Lo, and H. K. Tsang, "Apodized waveguide grating couplers for efficient coupling to optical fibers," IEEE Photonics Technol. Lett., 22(15), 1156-1158, August 2010

[9] F. V. Laere, W. Bogaerts, P.Dumon, G. Roelkens, D. V. Tourhout, and R. Baets, "Focusing polarization diversity grating couplers in silicon-on-insulator," J. Lightwave Technol., vol. 27, no. 5, pp. 612-618, March 2009

[10] Y. Sobu, S.-H. Jeong, S. Sekiguchi, Y. Tanaka, and K. Morito, "Si-wire grating couplers for integrated optical transceivers based on single-mode fiber connection," Proc. SPIE 9367, 936710, 936710-8, 2015

TuE4-4

OECC/PS2016

Grating-Based Mode Converters Fabricated by One-Step Photolithography

Wei Jin and Kin Seng Chiang*

Department of Electronic Engineering, City University of Hong Kong,
83 Tat Chee Ave., Kowloon, Hong Kong, China
City University Shenzhen Research Institute, 6 Yuexingyidao, Nanshan, Shenzhen, China
*eeksc@cityu.edu.hk

Abstract: We develop grating-based polymer-waveguide mode converters with a one-step photolithography process, which allows a grating to be formed anywhere on a waveguide. We demonstrate the flexibility of the process with LP_{01}-LP_{11b} and LP_{01}-LP_{21a} mode converters.
Keywords: *Grating; mode converter; polymer waveguide; waveguide fabrication.*

I. INTRODUCTION

Mode-division multiplexing (MDM), where different spatial modes of a few-mode fiber are allowed to carry different signal channels, is a promising technology to increase the transmission capacity of an optical fiber [1]. One of the key components in an MDM system is a mode converter, which performs conversion between two specific spatial modes. Mode converters have been realized with different materials and structures, such as bulk phase plates [2], long-period fiber gratings [3], and various waveguide devices, including Mach-Zehnder interferometers [4], graded-index structures [5], trench waveguide structures [6], and long-period waveguide gratings (LPWGs) [7]. An LPWG formed on the surface of a waveguide, in particular, can achieve LP_{01}-LP_{11b} mode conversion [7], which is difficult to achieve with planar waveguide structures [6]. The conventional method of making a surface grating needs two masks and hence involves two photolithography steps, one for the formation of the waveguide core and the other for the formation of the grating. In general, a conventional grating is formed across the whole core. Such a symmetric grating cannot be designed to convert modes with more complicated field distributions, such as the conversion between the LP_{01} and LP_{21a} modes. To remove this limitation, it is necessary to create more general grating structures, where corrugations can be introduced anywhere on the surface of the core. It is difficult, however, to precisely control the position of the grating with the conventional two-step photolithography process.

In this paper, we propose a one-step photolithography process for the fabrication of LPWGs, which requires only a single mask that contains both the core and the grating pattern. We take advantage of the different etching rates for the core (defined by a large window in the mask) and the grating (defined by much smaller apertures in the mask) to achieve simultaneous formation of the core and the grating. Using polymer as the waveguide material, we demonstrate the flexibility of this process with two mode converters: a LP_{01}-LP_{11b} mode converter, where the grating is formed along the central axis of the core surface, and a LP_{01}-LP_{21a} mode converter, where the grating is formed on one side of the core surface. The maximum conversion efficiencies obtained are ~98% (at 1550 nm) for the LP_{01}-LP_{11b} mode converter and ~95% (at 1570 nm) for the LP_{01}-LP_{21a} mode converter.

II. GRATING STRUCTURE AND FABRICATION PROCESS

We consider a waveguide that consists of a core embedded in a cladding. To form the waveguide core, a standard photolithography process can be applied, using a mask that defines the width of the core. To form a grating on the core, a second photolithography process can be applied, using another mask that defines the grating pattern. The corrugation depth of the grating is controlled by the etching process. This two-step photolithography process leads to a grating with corrugations across the entire core, as shown in Fig. 1(a). Such a symmetric grating can only be designed to achieve conversion between two horizontally symmetric modes, for example, between the LP_{01} and LP_{11b} modes [7]. To expand the mode-conversion functions, we propose a grating with corrugations formed locally on the core surface, e.g., on one side of the core, as shown in Fig. 1(b). In principle, such a grating can be designed to achieve conversion between any two spatial modes and thus offer more powerful mode-conversion functions. The conventional two-step photolithography process, however, does not provide the necessary precision to control the position of the corrugations.

To make the grating shown in Fig. 1(b), we propose a one-step photolithography process. The steps in the process for polymer waveguide fabrication are shown in Fig. 2. First, a low-index polymer is spin-coated on a Si substrate to form a lower cladding. A high-index polymer is next spin-coated on the lower cladding to form the core layer, which is then exposed to UV light through a specially designed chromium (Cr) mask that contains both the core and the grating patterns. The grating pattern consists of narrow periodic Cr strips deposited on the desired locations of the core area. The Cr strips, which have a width typically 5 or 6 times smaller than that of the core, cast periodic shadows on the core. As a result, the shadowed core area is under-exposed, while the remaining core area is fully exposed. In the development process, the etching rate for the shadowed core area becomes higher than that for the fully-exposed area,

648

but still much lower than that for the unexposed area. After development, the shadowed core area is only partially etched, which leaves periodic dents on the surface of the core. Finally, an upper cladding of low-index polymer is spin-coated onto the core to complete the process. A photo of a Cr mask used in the one-step photolithography process is shown in Fig. 2(b).

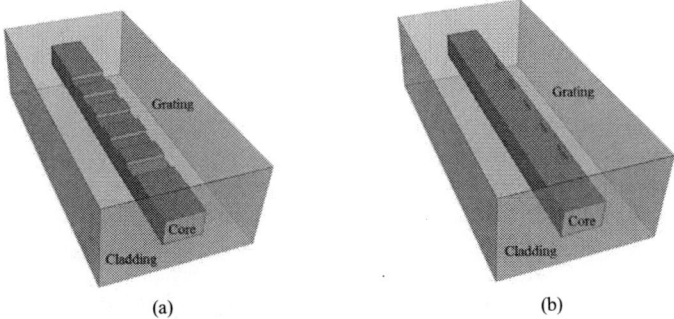

(a)   (b)

Fig. 1. Structures of (a) a grating formed by the conventional two-step photolithography process and (b) our proposed grating formed by the new one-step photolithography process.

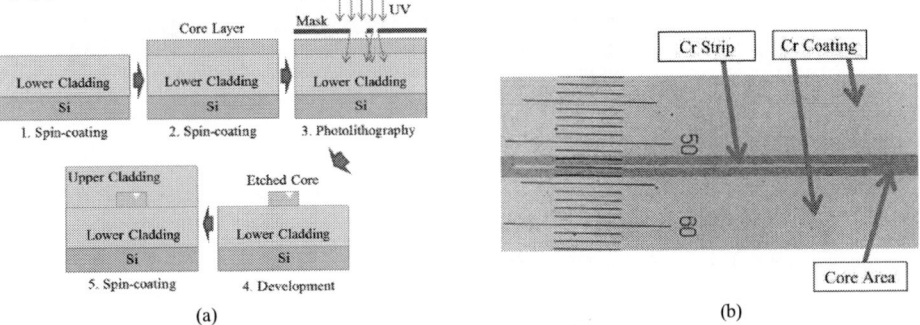

(a)   (b)

Fig. 2. (a) Steps in the one-step photolithography process. (b) Photo of a chromium (Cr) mask showing the core area covered by a narrow Cr strip.

We followed the process illustrated in Fig. 2(a) and fabricated two mode converters: a LP_{01}-LP_{11b} mode converter, where the grating is located along the central axis of the core surface, and a LP_{01}-LP_{21a} mode converter, where the grating is located on one side of the core surface. We designed the gratings by solving the modes of the waveguide and calculating the pitches from the phase-matching conditions [7]. We were able to control the corrugation depth of the grating by controlling the UV exposure time and the size of the grating strips on the mask. With the knowledge of the corrugation depth, we determined the optimum length of the grating. The polymer materials used were EpoCore and EpoClad (Micro Resist Technology) and their mix. The refractive indices of the core and the cladding, measured for thin-film samples with a prism coupler (Metricon 2010) at 1536 nm, were n_{co} = 1.571 and n_{cl} = 1.564, respectively, which were insensitive to the polarization state of light. For both mode converters, the thickness of the lower cladding was ~25 μm and the thickness of the core was about 11.6 μm. The width of the core, defined by the mask, was 12.5 μm. For the LP_{01}-LP_{11b} mode converter, the grating had a pitch of 630 μm and a length of 9.45 mm (15 periods). For the LP_{01}-LP_{21a} mode converter, the grating had a pitch of 370 μm and a length of 18.87 mm (51 periods). The total lengths of the fabricated devices were ~15 mm for the LP_{01}-LP_{11b} mode converter and ~25 mm for the LP_{01}-LP_{21a} mode converter. Microscopic images of the two mode converters taken before the upper cladding was applied are shown in Fig. 3(a) and 3(b), respectively. We can see a small dent at the center of the core in Fig. 3(a) and a small dent on one side of the core in Fig. 3(b). The surface profiles of the cores were measured with a step profiler (Ambios Technology, Model XP-2). The results are shown in Fig. 3(c). The maximum corrugation depths were ~0.4 μm for both converters.

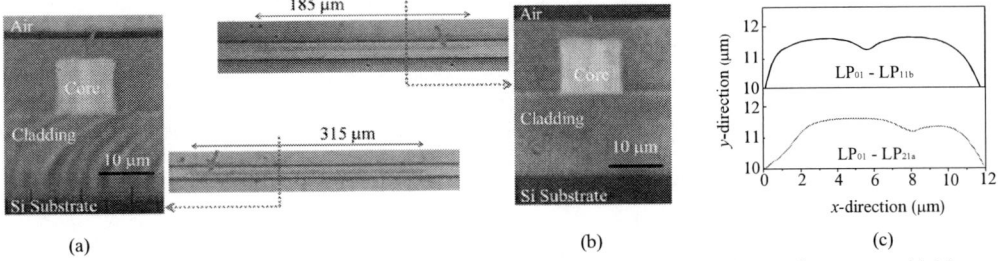

(a)   (b)   (c)

Fig. 3. Top views and cross-sectional views of (a) the LP_{01}-LP_{11b} mode converter and (b) the LP_{01}-LP_{21a} mode converter. (c) Measured surface profiles of the cores for the two mode converters.

III. MEASUREMENTS AND DISCUSSION

To characterize the mode converter, we launched only the LP_{01} mode into the devices with a lensed fiber and a broadband source (SuperK COMPACT, KOHERAS) and measured the transmission spectrum of the LP_{01} mode at the output end with another lensed fiber and an optical spectrum analyzer (AQ6370, Yokogawa). The polarization of light was controlled with a polarization controller and a polarizer placed at the input and output ends, respectively. The results measured for both the x- and y-polarizations are shown in Fig 4(a). The transmission spectra of the LP_{01}-LP_{11b} and LP_{01}-LP_{21a} mode converters show strong rejection bands at ~1545 and ~1575 nm, respectively. The maximum contrasts are about −18dB and −13dB, which correspond to maximum mode conversion efficiencies of 98% and 95%, respectively. As expected, the transmission characteristics of the gratings are insensitive to the polarization state of light. Fig. 4(b) and 4(c) show the near-field images taken at the output ends of the two mode converters, respectively, which fully confirm the functions of the two devices.

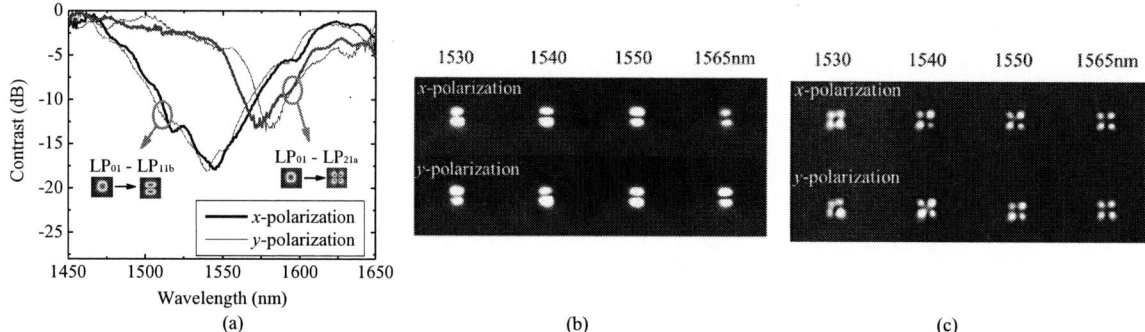

(a)   (b)   (c)

Fig. 4. (a) Transmission spectra of the LP_{01} mode for the two mode converters and output near-field images of (b) the LP_{01}-LP_{11b} mode converter and (c) the LP_{01}-LP_{21a} mode converter for the x and y polarizations.

To characterize the loss of the device, we compared the output powers from the mode converter and a reference waveguide of the same size and length but without any corrugations, using the same launching condition. Over the wavelength range 1530 − 1565 nm, the output power from the mode converter was smaller than that from the reference waveguide by ~0.5 dB for the LP_{01}-LP_{11b} mode converter and ~1 dB for the LP_{01}-LP_{21b} mode converter. The propagation loss of the reference waveguide, measured at 1550 nm by the cutback method, was ~2.0 dB/cm. Therefore, the losses for the LP_{01}-LP_{11b} and LP_{01}-LP_{21a} mode converters should be lower than ~3.5 dB and ~6 dB, respectively.

IV. CONCLUSIONS

We have proposed a one-step photolithography process for the fabrication of surface-corrugated gratings. This process allows a grating to be formed anywhere on the surface of a core and thus opens up many new possibilities for the design of grating-based mode converters for high-capacity MDM applications. We have demonstrated a LP_{01}-LP_{11b} and a LP_{01}-LP_{21b} mode converter fabricated with polymer material by this process. Both devices show excellent performance. This new process can greatly facilitate the development of grating-based waveguide devices.

This work was supported by the National Natural Science Foundation of China under Project 61377057.

REFERENCES

[1] G. F. Li, N. Bai, N. B. Zhao, and C. Xia, "Space-division multiplexing: the next frontier in optical communication," Adv. Opt. Photon., vol. 6, pp. 413-487, Dec. 2014.

[2] E. Ip, N. Bai, Y.-K. Huang, E. Mateo, F. Yaman, S. Bickham, H.-Y. Tam, C. Lu, M.-J. Li, S. Ten, A. P. T. Lau, V. Tse, G.-D. Peng, C. Montero, X. Prieto, and G. Li, "88 × 3 × 112-Gb/s WDM transmission over 50-km of three-mode fiber with inline multimode fiber amplifier," Proc. of 37th European Conference and Exhibition, Geneva (Switzerland), 2011, paper Th.13.C.2.

[3] J. Dong and K. S. Chiang, "Temperature-insensitive mode converters with CO2-laser written long-period fiber gratings," IEEE Photon. Technol. Lett., vol. 27, pp. 1006-1009, May 2015.

[4] Y. Huang, G. Xu, and S.T. Ho, "An ultracompact optical mode order converter," IEEE Photon. Technol. Lett., vol. 18, pp. 2281-2283, Nov. 2006.

[5] B. B. Oner, M. Turduev, I. H. Giden, and H. Kurt, "Efficient mode converter design using asymmetric graded index photonic structures," Opt. Lett., vol. 38, pp. 220-222, Jan. 2013.

[6] K. Saitoh, T. Uematsu, N. Hanzawa, Y. Ishizaka, K. Masumoto, T. Sakamoto, T. Matsui, K. Tsujikawa, and F. Yamamoto, "PLC-based LP_{11} mode rotator for mode-division multiplexing transmission," Opt. Exp., vol. 22, pp. 19117-19130, Aug. 2014.

[7] Y. Yang, K. Chen, W. Jin, and K. S. Chiang, "Widely wavelength-tunable mode converter based on polymer waveguide grating," IEEE Photon. Technol. Lett., vol., vol. 27, pp. 1985-1988, Sept. 2015.

TuE4-5

OECC/PS2016

Significant Loss Reduction of 3.0 dB/cm on Core-Top Etched Waveguide for Vertical Multi-Mode Interference

Ryosuke Sakata, Kazuhiro Tanabe, Haisong Jiang and Kiichi Hamamoto

Interdisciplinary Graduate School of Engineering Sciences, Kyushu University
6-1, Kasuga-koen, Kasuga, Fukuoka 816-8580, Japan
2ES14023Y@s.kyushu-u.ac.jp

Abstract: Core-top etched waveguide has been proposed to realize vertical multi-mode interference (MMI) toward higher order LP mode-converter. Significant loss reduction of 3.0 dB/cm on core-top etched waveguide is achieved in this work.

Keywords: Vertical Multi Mode Interference (MMI), Core-Top Etched Waveguide, Mode Converter

I. INTRODUCTION

Space division multiplexing (SDM) including multi-mode and multi-core, is an attractive way to overcome capacity-limit issue on optical fiber [1]. In order to implement especially multi-mode multiplexing transmission, one important issue is to realize mode-converter based on waveguide device due to its integration capability on a same wafer. Several attempts have been reported so far [2-4] to realize the mode-converter based on waveguide, and it has been relatively difficult to realize higher order LP-modes (more than LP_{21}) because of the necessity of scaling toward vertical direction (not only toward horizontal one). One possible way to scale up mode toward vertical direction is to utilize vertical multi-mode interference (MMI) [5-8], and we have proposed core-top etched waveguide to realize core height step along the propagation direction that leads to vertical MMI [7-8]. It was, however, reported that the propagation loss of the core-top etched waveguide was very huge because the core was dry-etched to realize precise height for vertical MMI phenomena and so called plasma damage was remained on the top [9]. To improve this high propagation loss issue, damaged layer removal is studied. With burying process using SiO_2, significant loss reduction of 3.0 dB/cm was achieved will be expected by extracting plasma damaged layer on core-top etched waveguide.

II. TWO STEP ETCHING WITH CORE-TOP ETCHING PROCESS

Figure 1 shows the fabrication process of core-top/non core-top etched waveguide. We use two-step etching process to realize core height difference in the waveguide. To realize precise height of the etched core, we use dry-etching for core-top etching as shown in Fig. 1.

First step) Lateral optical confinement is realized for the both high-core and low-core regions simultaneously (see Fig. 1 (a)). Waveguide width is defined with this process.

Second step) Vertical height difference is realized along propagation direction (see Fig. 1 (b)). High-core region is covered by photo-resist mask, and then core-top is etched only for low-core region.

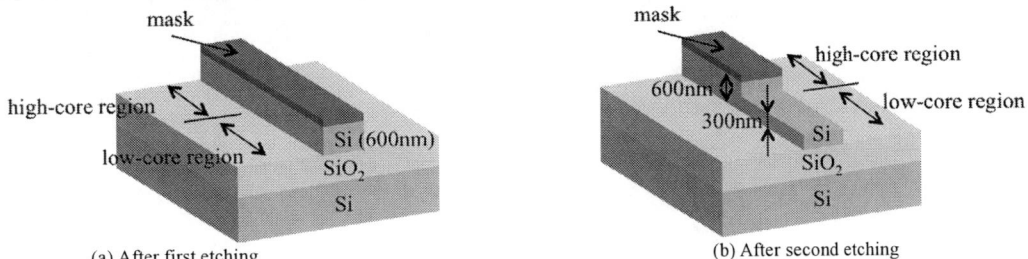

(a) After first etching   (b) After second etching

Fig. 1. Two step dry-etching process

For the low core region, the core-top is etched by using plasma process that may result in high propagation loss. To confirm the excess loss caused by the core-top etched process, we fabricated core top etched waveguide. The Si core was etched down from 0.6 mm to 0.3mm. For the dry-etching, we used inductively coupled plasma (ICP) reactive ion etching with using a mixture of SF_6 and C_3F_8. For comparison, we also fabricated non core-top etched waveguide with the same core height of 0.3μm. For the waveguide implementation, we have fabricated waveguides with different widths simultaneously to extract scattering loss from the evaluated propagation loss by using curve-fitting method.

Figure 2 shows the evaluated propagation loss as a function of waveguide width. We used Fabry-Perot resonance technique [10] for the evaluation. As shown in Fig. 2, the propagation loss of the both core-top etched waveguides

651

increases as the waveguide width decreases. This loss increase in terms of waveguide width corresponds to scattering loss at the waveguide sidewall in general [11]. In addition, the propagation loss of the core-top etched waveguide is higher than non core-top etched waveguide. Theoretical curve corresponding to 26nm side-wall roughness is fitted to the both waveguides, therefore, the loss difference between core-top etched and non core-top etched seems to be almost constant of 3.5 dB/cm apart from the waveguide width.

The excess loss of 3.5dB/cm is considered to be caused by so-called plasma damage, therefore, we have tried to extract the damaged layer to improve the propagation loss.

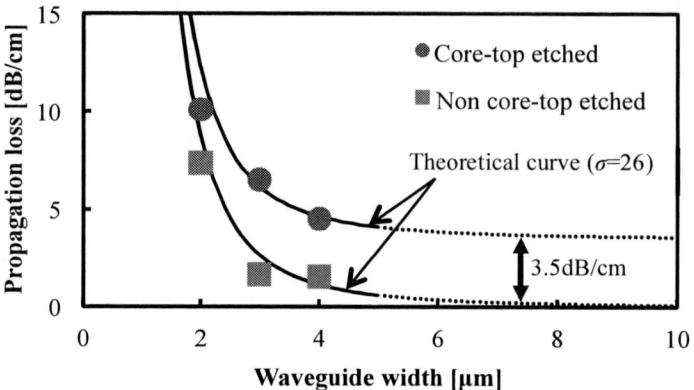

Fig. 2. Experimental results of propagation loss measurement.

III. LOSS REDUCTION ON CORE-TOP ETCHED WAVEGUIDE

It is known in general that the thickness plasma damaged layer, is several hundred angstrom from the etched surface (namely, core-top) [12]. In order to remove this damaged layer the surface of waveguide was oxidized by oxygen plasma and then wet etching was done slightly by using BHF. This is because the damaged layer is relatively easily oxidized and BHF is highly active for SiO_2. Another way, we buried the waveguide core with SiO_2 using sol-gel method [13] that does not require plasma process while burying the waveguide core. For comparison, we compared three different processed waveguides. The fabrication processes were summarized in Tab. 1.

TABLE I

Process	Process detail
a) Core-top etched	As dry etched core-top
b) Wet etch treatment	a) + Oxidize (20 sec) + BHF wet-etching (3sec)
c) Burying treatment	a) + using sol gel method

Figure 3 shows the evaluated propagation loss fabricated the above three different processes. Theoretical curves are fitted to the experimental results. The theoretical scattering loss curve of 26 nm side-wall roughness were fitted to the results of all the fabricated waveguides. The loss difference is considered to be the remaining plasma-damaged layer.

As shown in Fig. 3, wet-etching process (process b) is effective to reduce the damaged layer. Approximately 2.7 dB/cm loss reduction was confirmed by the process b. Moreover, stacking SiO_2 seemed to contribute to the loss reduction of approximately 3.0 dB/cm. Through this sol-gel method, SiO2 layer was stacked and then annealing process was carried. It seems this process (process *c*) is much effective to recover the damaged layer rather than process *b* because there is no plasma during burying treatment.

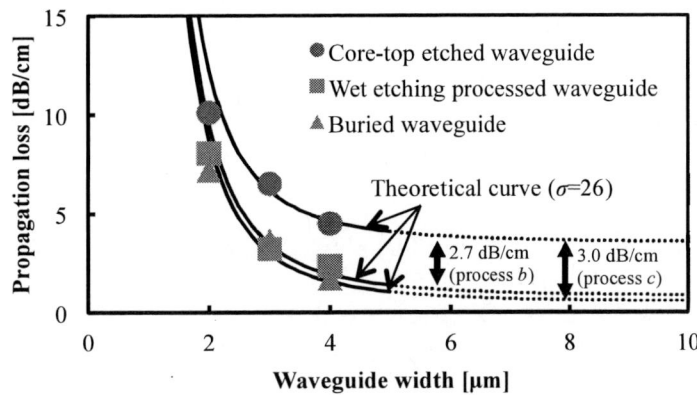

Fig. 3. Experimental results of propagation loss measurement of 3 kinds of waveguide.

IV. CONCLUSION

To realize vertical MMI toward higher order LP mode converter, we have researched core-top etched waveguide. To realize precise height control, plasma process was used to realize the core-top etched waveguide, however, the plasma damage layer induced high propagation loss of 3.5 dB/cm so far. To improve this high propagation loss issue, damaged layer removal was studied using SiO_2 sol-gel treatment, significant loss reduction of 3.0 dB/cm was achieved successfully.

ACKNOWLEDGMENT

This work has been supported by NICT.

REFERENCES

[1] T. Morioka, "New generation optical infrastructure technologies: "EXAT initiative" towards 2020 and beyond," Tech. Dig. OptoElectron. Commun. Conf., FT4, (2009)

[2] T. Uematsu, K. Saitoh, N. Hanzawa, T. Sakamoto, T. Matsui, K. Tsujikawa and M. Koshiba, "Low-loss and broadband PLC-type mode (de)multiplexer for mode-division multiplexing transmission," Tech. Dig. Opt. Fiber Commun. Conf., OTh1B.5 (2015)

[3] N. Hanzawa, K. Saitoh, T. Sakamoto, T. Matsui, K. Tsujikawa, M. Koshiba, and F. Yamamoto, "Mode multi/demultiplexing with parallel waveguide for mode division multiplexed transmission," Opt. Express, Vol. 22, No. 24, pp. 29321-29330 (2014)

[4] N. Hanzawa, K. Saitoh, T. Sakamoto, T. Matsui, K. Tsujikawa, T. Fujisawa, Y. Ishizaka and F. Yamamoto "Demonstration of PLC-based six-mode multiplexer for mode division multiplexing transmission," Proc. Europ. Conf. Opt. Commun., P.2.6 (2015)

[5] Ken Kirita and Fumio Koyama, "Vertical MMI coupler for 3D Photonic integrated circuits," IEEE LEOS Annual Meeting Conf. Proceedings, pp. 477-478 (2009)

[6] Chris J. Brooks, Andrew P. Knights, and Paul E. Jessop, "Vertically-integrated multimode interferometer coupler for 3D photonic circuits in SOI," Opt. Express, vol. 19, 4, pp. 2916-2921 (2011)

[7] Y. Chaen, R. Tanaka and K. Hamamoto, "Optical mode converter using multi-mode interference structure," Tech. Dig. Micoropt. Conf., H8, pp. 63-64 (2013)

[8] Y. Chaen, Z. Zhao, Y. Satou and K. Hamamoto,"Quasi-LP21 mode converter by using simple step-core structure," Tech. Dig. OptoElectron. Commun. Conf. / Photon. Switch., TuPL-14 (2013)

[9] R. Sakata, K. Tanabe, H. Jiang and K. Hamamoto, "Optical mode converter using multi-mode interference structure," Tech. Dig. Micoropt. Conf.,H52, pp. 63-64 (2015)

[10] R. G. Walker, "Simple and accurate loss measurement technique for semiconductor optical waveguide," Electron. Lett., vol. 21, No. 13, pp. 581-583 (1995)

[11] F. P. Payne and J. P. R. Lacy, "A theoretical analysis of scattering loss from planar optical waveguide," Opt. Quant. Electron., Vol 26, pp. 977-986 (1994)

[12] G. S. Oehrlein, R. M. Tromp, J. C. Tsang, Y. H. Lee, and E. J. Petrillo, "Near-Surface Damage and Contamination after CF_4/H_2 Reactive Ion Etching of Si," J. Electrochem. Soc., Vol. 132, Issue 6, pp. 1441-1447 (1985)

[13] S. Maekawa, K. Okude and T. Ohishi, "Synthesis of SiO2 thin films by sol-gel method using photoirradiation and molecular structure analysis," J. Sol-Gel Sci. Technol., Vol. 2, Issue 1, pp. 497-501, 1994

TuF1-1

Optical system architectures and control for data center networks

Dimitra Simeonidou
University of Bristol, UK

The global amount of digital information is growing at a staggering pace of 50% p.a. and will exceed 60 Zettabytes in 2020. While storing and processing of such massive data will offer new business opportunities, it will also require new Data Centre and Data Centre networking architectures to provide the necessary scalability, resource sharing and automation.

This tutorial will discuss requirements, challenges and solutions for next-generation Data Centres. Key results from our latest research as well as practical findings from deployments will be presented.

Specifically, the following topics will be addressed:
- What will be the role of Data Centres in future service & content provider networks?
- What network architectures & connectivity is required (intra-DC, inter-DC, user to DC)?
- Which role will open source software frameworks play?
- How can network resources be virtualized and where will they be allocated?
- How will next generation DCs look like?
- Which role will optics play inside future DCs (as well as inter-DC and user to DC)?
- What will be the granularity of interconnected resource, specifically with reference to needs and benefits of disaggregation?
- Use cases with referencing 5G, IoT and smart city deployments

Biography

Dimitra Simeonidou

Dimitra Simeonidou is a Full Professor of High Performance Networks a t the University of Bristol, Director of the University's Smart Internet Lab and the CTO of Bristol Is Open. Her researcher research focuses on the fields of high performance networks, data centre networking, Software Defined Networking and smart city infrastructures. Dimitra is in the editorial teams of leading Journals in the field and she chairs committees, conferences, standardisation groups and fora in the relevant bodies. She is the author and co-author of over 400 papers in peer reviewed journals and international conferences, book chapters, several standardisation documents and patents. She has been co-founder in two spin-out companies, the most recent being the University of Bristol spin-out Zeetta Networks (http://www.zeetta.com), which is delivering SDN software platforms for enterprise networks.

TuF1-2

Explicit Wavelength Resource Reallocation to Ensure Critical Communication Lines in Emergency Situations

Masaki Shiraiwa, Takaya Miyazawa, Hideaki Furukawa, Yoshinari Awaji, and Naoya Wada

National institute of information and communications technology (NICT)

4-2-1 Nukui-Kitamachi Koganei Tokyo 184-8795 Japan   E-mail: shiraiwa@nict.go.jp

Abstract: *To reliably ensure critical communication lines in emergency situations, we propose and experimentally demonstrate an explicit wavelength resource reconfiguration system on an optical packet and circuit integrated network, which achieves enough low frame error rates.*

Keywords: *Optical packet switching, Optical circuit switching, Wavelength resource reconfiguration*

I. INTRODUCTION

Optical communication networks will continue to be required to achieve higher-speed and larger-capacity in the future. Since efficient utilization of limited wavelength resources results in increase of communication capacity, wavelength resource control technologies such as elastic optical networks [1] have been widely investigaed for future optical networks. Moreover, some technologies to mitigate network congestion are necessary when the network traffic increases drastically in emergency situations such as a large-scale disaster, a large-scale accident and a big event. Under those situations, ensuring the critical communication lines for high-priority data transfers is crucial. For example, the government agencies might use such critical communication lines in order to obtain serious information and give instructions to a relevant emergency response office as soon as possible [2]. To ensure the wavelength resources to high-priority data transfers, wavelength resources preemption methods has been studied [3, 4]. However, it would be inefficient that the network preserves such wavelength resources for future unpredictable large-scale disasters and does not use the resources at all in normal cases.

In order to use the wavelength resources efficiently, we have been developing a distributed wavelength resource allocation [5] for an optical packet and circuit integrated (OPCI) network [6]. The OPCI network dynamically changes the amount of optical packet switching (OPS) and optical circuit switching (OCS) resources depending on the demands for high-quality services. OCS resources are often called optical paths. Concretely, there are three types of resources: dedicated OCS waveband, dedicated OPS waveband and shared waveband. Here, a waveband consists of multiple continuous wavelengths. The shared waveband is used for either OCS or OPS depending the status of optical path usage [5]. In the method in [5], the processing time increases in proportion to the number of the wavelength-channels because those wavelength-channels are established or removed one by one in turn. Thus, we have developed a system to decrease the wavelength resource reconfiguration time [7]. In addition to this method, we implement a control function driven by network service providers (NSPs) which can explicitly change the amount of OCS and OPS resources. The control function is considered to be also applicable in OCS networks.

In this paper, we propose a novel wavelength resource reconfiguration system to ensure critical communication lines in emergency situations such large-scale disasters. We implemented a centralized controller in addition to the existing distributed path control plane developed in [5] to execute explicit wavelength resource reallocation to OPS and OCS. Moreover, we experimentally demonstrate that the system can change the shared waveband from OPS resources to OCS resources forcibly in specified links and also achieve frame error rates (FERs) of less than 1×10^{-8} in data transmissions.

II. PROPOSED RECONFIGURATION SYSTEM

In Japan, when a large-scale disaster occurs, the general telephone communication is restricted to give priority to the disaster emergency telephone (including the public telephones) so that congestion in critical communication lines can be avoided. A priority signal is added to the disaster emergency call, which is connected preferentially in a network.

Our proposed system executes the distributed resource allocation developed in [5] in normal situations, which dynamically moves the boundary between OCS and OPS resources depending on demands for optical paths. When network congestion occurs due to some emergency situations, the shared wavebands in critical communication lines are used for critical data transfers associated with the situations. Since the quality of such critical communication needs to be guaranteed, the shared wavebands should be changed from OPS resources to OCS resources in relevant links if the network has been using the shared wavebands for OPS. The system ensures critical communication lines by the following processing in an emergency situation.

  0. In normal situations, dynamic resource allocation to OCS and OPS is executed on the basis of the distributed control function developed in [5].

OECC/PS2016

1. A NSP detects network congestion due to an emergency situation such as a large-scale disaster, a large-scase accident, and so on.

2. The NSP decides to execute explicit resource reallocation on the critical communication links between the source (e.g. Government agencies) and destination (e.g. Emergency response office). In this case, distributed resource allocation developed in [5] is not executed on those links.

3. In ciritial communication links, if the shared wavebands are being used for OPS, the NSP transfers all data from general users to the interface for dedicated OPS waveband in order to change the shared wavebands from OPS resources to OCS ones.

4. The NSP switches the shared waveband from the OPS to OCS resources by the explicit resource reallocation. The maximum number of OCS resources is fixed and set to x. (We set the value of x to 20 in this demonstration.)

5. The NSP ensures optical paths in response to the request of the critical communication lines.

6. The path establishment request from general users are rejected so that optical paths can be established preferentially for critical communication associated with the emergency situation. (General users might be allocated optical paths within the dedicated OCS waveband if available.)

7. When recovery from the emergency situation is done and network congestion does not occur any longer, the NSP makes the value of x variable by stoping the explicit resource reallocation in the critical communication links, and resumes the distributed resource allocation developed in [5] for these links.

8. Return to the process 0. (i.e. normal state)

Note that, in the above process 1, we require some management system to monitor network traffic especially on OPS links. Actually, we have developed such management system [8], which can be utilized in the proposed system. In this paper, we demonstrate the processes 2, 4 and 5 in the next session.

III. EXPERIMENTAL DEMONSTRATION

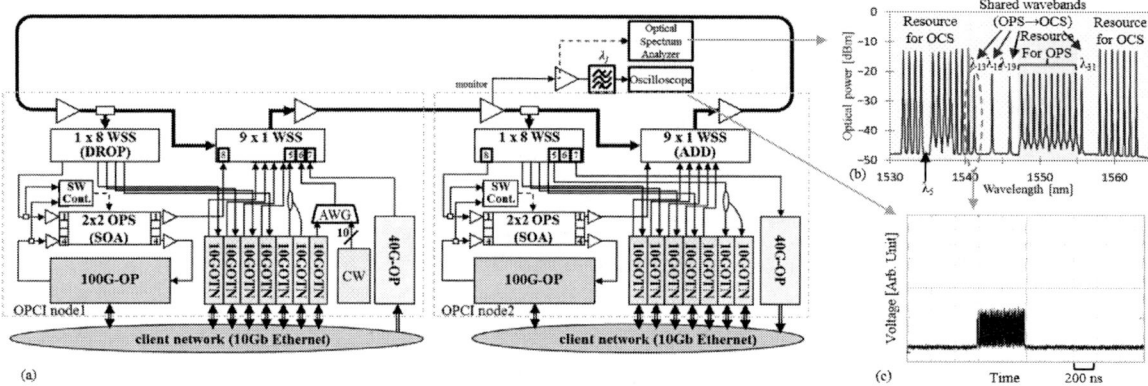

Fig. 1. (a) Experimental setup; (b) the initial state optical wavelength spectrum; (c) waveform of the λ_{13}.

An experimental setup is shown in Fig. 1 (a). The network utilized two OPCI nodes and 40 wavelengths (λ_1–λ_{40}) with 100 GHz spacing within a range from 1531.90 nm to 1563.05 nm. Each node has seven 10 Gbps optical path transponders (10GOTN), a 100 Gbps (10 Gbps x 10 wavelengths) optical packet transponder (100G-OP), a 40 Gbps (10 Gbps x 4 wavelengths) optical packet transponder (40G-OP), and WSSs for adding and dropping. Two of 10GOTN and a 40G-OP are transponders for the shared waveband. The OPCI node 1 has an arrayed continuous wave laser (CW) for dummy signals. These optical path or packet transponders accommodate 10 Gbit Ethernet frames coming from a client network (consisting of router testers) in optical paths or packets links, respectively. On the control-plane, we handle 40 wavelengths under the assumption that one waveband consists of 10 wavelengths. Meanwhile, on the data-plane, we handle three wavelength-groups of (λ_{13}, λ_{16}, λ_{19}, λ_{31}), (λ_{21}–λ_{30}), and (λ_1–λ_{10}, λ_{34}–$_{40}$) as the shared waveband, dedicated OPS waveband and dedicated OCS waveband, respectively. Both the discontinuity of wavelength number and the small number of wavelengths in the shared-WB (i.e. less than 10) are enough to show the effectiveness of our resource reconfiguration system. Figure 1 (b) shows the optical signal wavelength spectrum of the initial setting. In the default setting, the shared wavelength bandwidth is set to the OPS. Figure 1 (c) shows that the signal with λ_{13} has the waveform of optical packet signal.

Firstly, we demonstrated the system in the normal situation, in which only resource allocation developped in [5] is executed. If an optical path with λ_5 is established and the number of in-use paths in the dedicated OCS waveband exceeds the prescribed threshold, the shared waveband is changed from OPS resources to OCS resources [5]. However, if the number of in-use paths in the dedicated OCS waveband does not exceed the threshold, optical path with λ_{13} cannot be established because the shared waveband is used for OPS. Figure 2 (a) shows optical signal wavelength spectra in the case that the request of optical path establishment with λ_{13} was rejected because the optical path λ_5 had not

656

been established and the shared waveband was still used for OPS. In Fig. 2 (b), we can see that the control screen displayed "error" for the request of optical path establishment which means the request was rejected. Secondly, we demonstrated the proposed system in emergency situations, in which the NSP executes explicit resource reallocation in critical communication links. As in the case of normal situation, the shared waveband was initially used for OPS. The proposed system changes the shared waveband from OPS to OCS resources (i.e. removes the optical packet signals in the shared waveband) so that optical paths can be established within the shared waveband even if the number of in-use paths in the dedicated OCS waveband does not exceed the prescribed threshold (i.e. optical path with λ_5 has not been established). Figure 2 (c) shows optical signal wavelength spectra after switching the shared waveband from OPS resources to OCS resources. We confirmed that Fig. 2 (a) and Fig. 2 (c) are almost same characteristics. The result shows that the change does not affect the wavelength characteristic. Figure 2 (d) shows a console screen which displayed the result of explicit wavelength resource reallocation to change the shared waveband from OPS to OCS resources. We can confirmed that the resource reallocation was successfully executed (by checking "OK"). Figure 2 (e) shows the optical wavelength spectrum in the case that optical path signals with (λ_{13}, λ_{16}) were established within the shared waveband, and Figure 2 (f) shows the waveform of optical path signal with λ_{13}. We confirmed that the two optical paths were successfully established.

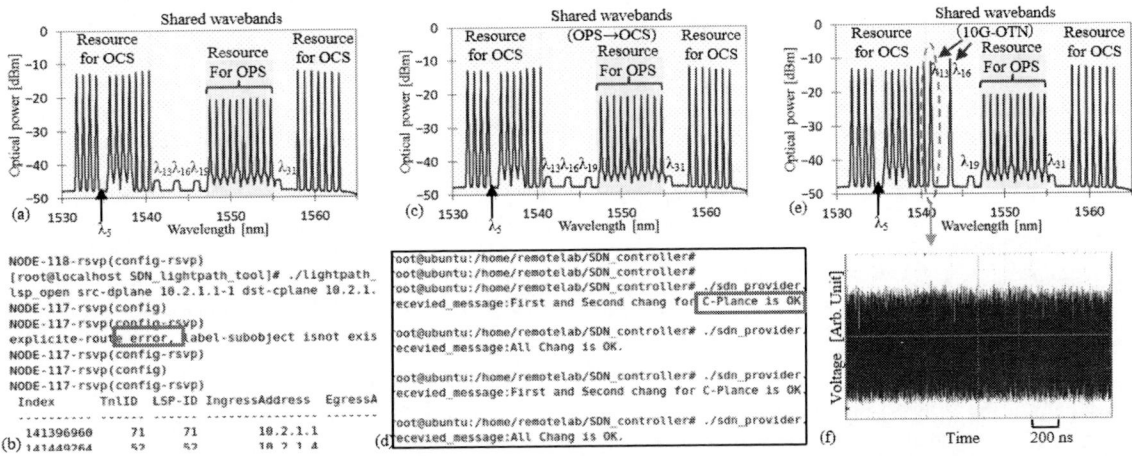

Fig. 2. Optical wavelength spectrums (a, c, e) , control screens (b, d) and continuous signal waveform (f).

Finally, we measured the FER of transmitted data. The optical signal consisting of one million frames of 1518 byte 10Gigabit Ethernet signal were sent from the OPCI node 1 to the OPCI node 2 without the forward error correction (FEC) technology by 40 Gbps optical packets or two optical paths before and after the reconfiguration. In both cases, the obtained FERs were less than 1×10^{-8}, which is regarded as high quality.

IV. CONCLUSIONS

We have developed a novel explicit resource reconfiguration system which can ensure critical communication lines in emergency situations while the wavelength resources are efficiently utilized in normal situations. We have demonstrated that the system can change the shared waveband from OPS resources to OCS resources forcibly to ensure lightpaths for high-priority data transfers, and also achieve frame error rates (FERs) of less than 1×10^{-8} in data transmissions. Our system can be applied not only to OPCI networks but also to OCS networks. In future works, we investigate how to handle data from general users; for example moving a lightpath from a waveband to another waveband.

ACKNOWLEDGMENT

The authors thank Wei Ping Ren, Ryo Mikami, and Takeshi Makino of NICT for their technical support.

REFERENCES

[1] M. Jinno, et al., IEEE Commun. Mag., 47, p. 66, (2009).
[2] Government of Japan, "Disaster management in Japan," (2015).
[3] T. Tachibana, et al., IEEE Trans. Commun., 6, p. 1439, (2004).
[4] A. Szymanski, et al., IEEE Commun. Mag., 45, p. 66, (2007).
[5] T. Miyazawa, et al., J. Opt. Commun. Netw., 4, p. 25, (2012).
[6] H. Harai, IEICE Trans. Commun., E95-B(3), p. 714, (2012).
[7] M. Shiraiwa, et al., Photonics in Switching (PS 2015), p. 136, (2015).
[8] T. Miyazawa, et al., IFIP/IEEE Integrated Network Management Symposium (IM 2015), p. 665, (2015).

TuF1-3

OECC/PS2016

Remote Experiments with Test-Plane on a Distributed Optical Switched Network Testbed

Sugang Xu, Masaki Shiraiwa, Hiroaki Harai, Yoshinari Awaji, Naoya Wada

NICT, 4-2-1, Nukui-Kitamachi Koganei, Tokyo 184-8795, Japan

{xsg, shiraiwa, harai, yossy, wada}@nict.go.jp

Abstract: We demonstrated a remote-lab environment for a distributed optical network testbed which enables the optical-layer through IP-layer experiments. The automated remote test and experiment capability is beneficial to both the testbeds and future networks.

Keywords: Optical circuit-switched; Optical packet-switched, Testbed, Test-Plane

I. INTRODUCTION

Large national scale multi-purpose optical network testbeds with different ranges such as NLR [1], GEANT [2], JGN-X [3] etc., have been created. The main purposes of such testbeds are the feasibility study of latest technologies. Recently established multi-layer testbeds such as ADRENALINE [4] etc., have the wide scope of the advanced researches from the optical and wireless transmission to the network control and management. Optical packet switching (OPS) is promising which offers network operators with the large capacity and great potential for energy saving [5-8]. We have been conducting the research and development of the optical packet and circuit integrated (OPCI) network [9], which offers both the optical packet switching and optical circuit switching (OCS) capabilities. OPCI-based multi-layer testbed has been established to facilitate the application studies and the future researches from the device level to the network level. However, as the testbed for the data-plane employs plenty of rare and latest advanced technologies, the researchers from different places normally need to conduct the experiments at the testbed site which are far from them. It is inconvenient for researchers to conduct the experiments, especially the future experiments that spanning multiple distributed testbeds. We established a remote-lab environment for the OPCI network testbed. It enables the remote experiments with the remotely-located testbed resources including not only the data-plane, control-plane, but also the measurement instruments from the optical layer to the IP layer. The automated remote test and experiment capability is not only helpful for testbeds, but also for the future networks, e.g., employing the network virtualization technology, which will need the automated service validation after network provisioning and configuration.

II. REQUIREMENTS IN THE REMOTE-LAB ENVIRONMENT

To ease the remote experiments, we first identify the key requirements in the remote-lab environment as follows:

(1) Remote control

Researchers can perform the remote control of both the OPS, OCS and IP systems from the element level to the network level. The control of these systems would be required to conduct in one of the following ways.

(i) Direct access of the control-plane of the testbed remotely with certain dedicated access connection (L2).

(ii) Indirect access of the control-plane through a server via either application programming interfaces (APIs) or graphical user interfaces (GUIs) e.g., the Web-based interface. This server offers secured connections (L3) to researchers outside of the testbed.

(2) Remote test and measurement

Researchers can perform the network test and measurement remotely by operating the test and measurement instruments. For example, the test is performed by generating and injecting the IP packet flows with the router testers (layer-3) to the testbed, and measuring the performance of the entire data-plane from the optical layer (layer-0) to the IP layer (layer-3) with corresponding measurement instruments. The operations of the test and measurement instruments would be required to perform in one of the following ways.

(i) Viewers of the test and the measurement instruments are executed in the researchers' sites which are outside of the testbed. The communication of these software systems and the instruments (within the testbed) can be directly achieved with certain dedicated access connection (L2).

(ii) The aforementioned communication can be achieved indirectly via an agent machine. This agent machine is implemented as the access point for the test and measurement purpose. This agent machine offers secured connections (L3) to researchers outside of the testbed, and relays the packets between the outside software systems (e.g., viewers) and the inside instruments.

(iii) Viewers of the test and the measurement instruments are executed in a server which is implemented within the testbed. Researchers can directly access to this server remotely. After logging in to this server remotely, researchers perform the operations of the test, measurement with certain dedicated access connection (L2).

(iv) The secured access to the aforementioned server can be achieved via an agent machine, which is similar to (ii).

(3) Programmable control-test-measurement experimental scenario

The experimental scenario consists of a collection of operations of the remote configuration, control, test and measurement. With remote-lab, the experimental scenario can be performed manually. Researchers can conduct these operations in experiments easily, e.g., via GUI. In addition to the manual operations, researchers can perform the automated control-test-measurement operations; and the experimental scenario is programmable. Note that the measurement will be performed from the optical layer to the IP layer, the operations of the multi-type multi-vendor measurement instruments should be taken into account.

(4)   Interconnection with other testbeds

In addition to the experiments which are conducted within one testbed, researchers would conduct the experiments with multiple distributed optical network testbeds in the future. The secured connections of the data-planes and the control-planes of these testbeds would be required to perform in one of the following ways.

(i)   All of the data-planes and the control-planes can be directly connected with the dedicated connections (L2).

(ii) Similar to that mentioned above, all of the data-planes and the control-planes can be connected indirectly via corresponding agent machines with secured connections.

III. IMPLEMENTATION OF THE REMOTE-LAB ENVIRONMENT AND INTRODUCTION OF A TEST-PLANE (T-PLANE)

Fig. 1 depicts the image of a remote-lab-enabled OPCI testbed. With the requirements addressed above, we implemented the testbed with three planes. (1) The data-plane consists of the OPCI nodes and the PC-based open virtual switches (OVS) [10] which are attached at the OPCI nodes' client side. For each OVS, two ports are employed to connect to the OPCI node; one is for OCS, another is for OPS utilization. (2) The control-plane (the management-plane related issues are outside the scope of this paper due to the space limitation) consists of the node controller of each OPCI node, and the OpenFlow controller (OFC). In addition, we propose the third plane (3) the test-and-measurement-plane (test-plane for short). It consists of the test/measurement instruments including optical spectrum analyzer (OSA), digital storage oscilloscope (DSO) and others for the optical layer measurement, and router tester (RT) for the IP flow generation and analysis.

As shown in Fig.1, to meet the aforementioned requirement (1), we implemented an OPCI control server (a SDN-controller) (requirement 1-ii), which also offers the APIs (e.g., via telnet) and the GUI (e.g., via Web application) to control the OPCI nodes (both of the OCS and OPS subsystems). The secured access to this server can be implemented through dedicated L2 connections or secured L3 connections (e.g., VPN, SSH Tunnels). In addition, we implemented an agent machine as the access point for the control-plane remote operation purposes. To the requirement (2), in addition to the introduction of the test-plane, we developed a remote-lab server. This server manages and controls the multi-type multi-vendor test-measurement instruments. In addition, we implemented an agent as the access point for the test-plane remote operation purposes. This is according to the requirement (2-iv). To the requirement (3) in the aforementioned remote-lab server, we have implemented a platform to enable the programmability of the experimental scenarios. All of the operations (sequential or parallel) in the test-measurement can be prepared and coded. To the requirement (4), we adopt the second approach (4-ii). We prepared three agent machines which are connected to the data-plane, control-plane and test-plane, respectively. Note that these agents can be applied for both remote-access and the testbed interconnection purposes. For the secured access, the access control and encrypted communication are adopted.

The test-plane is one of the key components in the remote-lab enabled network. For testbeds, this test-plane is helpful for the easy remote experiments. For future networks, we can take advantage of the network test capability of this test-plane to validate the network resource provisioning, network configurations, which may involve the complicated end-to-end test tasks across multiple layers. For this purpose, as shown in Fig.1, we connected the OPCI control server (a SDN controller) and the remote-lab server. After provisioning of lightpaths or routing configuration of OPS and OpenFlow networks, it is possible to call the test-measurement functions built in the test-plane from the OPCI control server directly, resulting in more powerful network control and management for future networks.

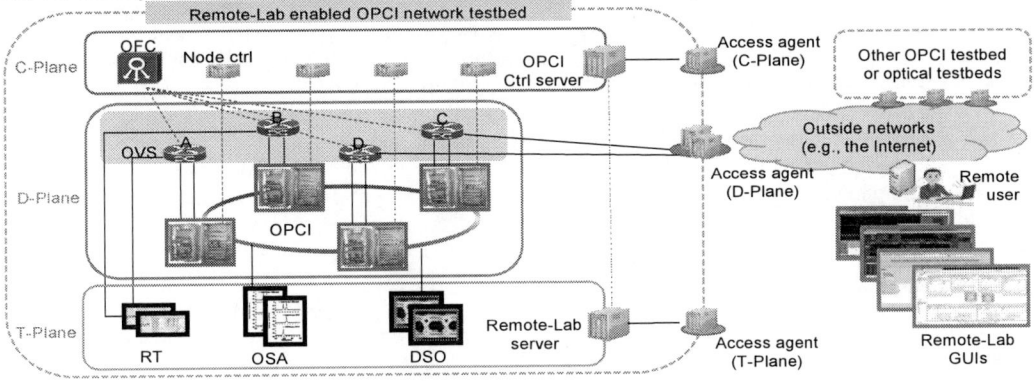

Fig. 1: Remote experiments with the remote-lab enabled OPCI network testbeds.

OECC/PS2016

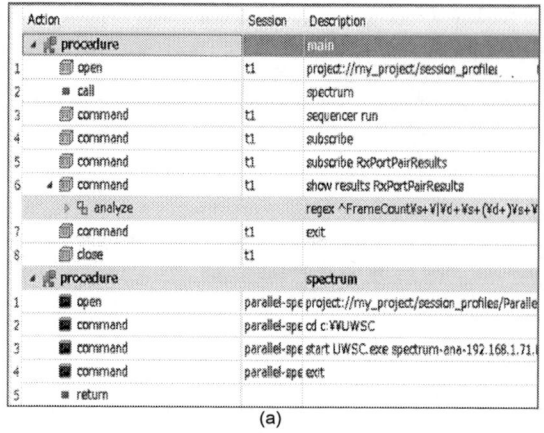
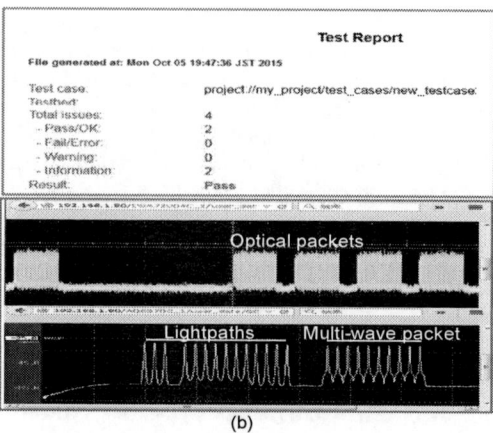

(a)   (b)

Fig. 2: An example of the remote experiments with remote-lab environment. (a) programmed test-measurement scenario including the previously stored IP packet flow traffic generation/analysis subroutines, and the waveform/spectrum measurement subroutines; (b) test results.

As multi-layer test-measurement information can be obtained for the network control, further applications of this test-plane are considered possible, and are treated as the future work.

IV. DEMONSTRATION OF THE REMOTE EXPERIMENTS WITH THE REMOTE-LAB ENVIRONMENT

In order to explain the capability of the remote experiment with the remote-lab environment, in this section we demonstrate a multi-layer network experimental scenario. As shown in Fig.1, the RT is connected to two OVS switches A and B. The OSA and DSO are attached at the output line ports of the OPCI node A and D, respectively, to measure the transmission in the optical layer. The RT generates one IP flow which is injected into the OVS-A and to be received from the OVS-B. Upon receiving test input packets, the OVS-A sends a request to the OFC for flow control. Upon receiving this request the OFC first triggers the lightpath provisioning function (which is built in the control-plane) to establish a bidirectional lightpath between the OPCI node A and B. After the establishment of this lightpath, the OFC calculates the routing and sends the flow configuration information to the OVS-A and OVS-B. Upon receiving the flow configuration information, the OVS-A and OVS-B execute the local flow configurations. With the lightpath creation and the OpenFlow control, the test IP packet flow is delivered to the RT. Fig.2 (a) shows the programmed remote experiment scenario which is implemented in the remote-lab server including the previously stored IP packet flow traffic generation/analysis subroutines, and the waveform/spectrum measurement subroutines. The waveform/spectrum measurement [the stored spectrum procedure shown in Fig.2 (a)] is firstly started as a forked parallel process (i.e., the step-2, call spectrum). Then the RT is started to generate/analyze the IP packet flow (i.e., the step-3, sequencer run). After a period of time for the packet transmission/receiving test, the IP packet generation at RT is stopped. Then the test results are analyzed at the RT (i.e., from the step-4 to the step-6). Finally, the experiment is terminated (i.e., the step-7) and closed (i.e., the step-8). The selected results of this experiment demonstration are shown in Fig.2 (b). The upper block shows the automatically generated report (trimmed) for this test. The lower block shows the simultaneously and automatically measured/captured waveform of the coexisting optical packet signals of other traffic, and the spectrum (including both the wavelengths employed by the optical packets and the lightpaths) which confirms the successfully established lightpath.

V. CONCLUSIONS

We established a remote-lab environment for the OPCI network testbed. This environment enables the remote experiments with the remotely-located testbed resources including not only the data-plane, control-plane, but also the measurement instruments from the optical layer to the IP layer. The test-plane is one of the key components in the remote-lab enabled network. For testbeds, this test-plane is helpful for the secured and easy remote experiments. For future network services, e.g., network virtualization services, we can take advantage of the automated network test capability with the test-plane to validate the network resource provisioning, network configurations, which may involve the complicated end-to-end test tasks across multiple layers. Further investigation is seen as the future work.

REFERENCES

[1] http://www.nlr.net/, [Online]
[2] http://www.geant.org,/ [Online]
[3] http://www.jgn.nict.go.jp/, [Online]
[4] http://networks.cttc.es/, [Online]
[5] A. Takada et al., IEEE J. Lightw. Technol., 20, 12, (2002).
[6] A. Carena et al., IEEE J. Sel. Areas Commun. 22, 8, (2004).

[7] S. J. Ben Yoo, IEEE J. Lightw. Technol., 24, 12, (2006).
[8] S. Yao et al., IEEE J. Lightw. Technol., 21, 3, (2003).
[9] H. Harai et al., IEEE J. Lightw. Technol., 32, 16, (2014).
[10] http://openvswitch.org/,[Online]

TuF2-1 (Invited)

OECC/PS2016

Flexible-bandwidth Optical Interconnects for Datacom Networks

Roberto Proietti, Paolo Grani, Zheng Cao, and S.J.Ben Yoo

Department of Electrical and Computer Engineering, University of California, Davis, CA, 95616, USA.
*rproietti@ucdavis.edu

Abstract: *This paper presents flexible bandwidth solutions for Datacom optical networks exploiting AWGR technology, DVFS and channel bonding techniques. Benchmarking simulations and experiments demonstrate up to 2× reduction in energy consumption and 1.77x throughput increase.*

Keywords: *Flexible Bandwidth; Data Centers; Arrayed Waveguide Grating Routers.*

I. INTRODUCTION

The exponential increase of data in today's data centers brings significant challenges in terms of power consumption. While optical interconnects enable a scalable bandwidth interconnection (independent of the communication distance) with low energy requirements below 1 pJ/bit, the power consumption due to communications alone still represents a significant portion of the overall power consumed in data centers today. A significant part of this communication power is actually wasted because the conventional communication systems cannot adapt the communication to the traffic patterns, which are bursty [1] with high peak-to-average ratios. As a result, lot of energy is consumed even when no meaningful bits are transmitted [2]. However, by employing flexible-bandwidth optical communication techniques, it is possible to significantly improve the energy efficiency of the communication systems. Within the rack and at the board level, architectures usually rely on shared memory and cache-coherency for the communication and synchronization between the working threads. Exploiting flexible bandwidth optical communications with this kind of traffic is challenging because of its burstiness and because of the very limited size of most of the packets (control packets). Between racks and clusters, where aggregation takes place, the link utilization and average packets or flow sizes can be higher, making it easier to implement flexible-bandwidth optical transmission and allocation schemes. Typical systems are designed to support either highest-peak-demands (energy and resource inefficient) or average-demands). In particular, several applications create hot-spots that can change dynamically over time, overloading the electronic switches at the higher hierarchy, resulting in limited system-wide performance. The next two sections will present two use cases for board-level (intra-rack) and inter-rack optical interconnection networks and demonstrate by simulation and experiments the benefits of using different flexible bandwidth assignment techniques [3, 4].

II. ALL-TO-ALL AWGR INTERCONNECTION WITH DYNAMIC VOLTAGE FREQUENCY SCALING

For board-level communication between multiple sockets, we propose to use all-to-all communication based on wavelength routing in Arrayed Waveguide Grating Routers (AWGRs), and Dynamic Voltage and Frequency Scaling (DVFS [5]) to dynamically adjust the transceivers bandwidth according to the link utilization. Figure 1 shows the proposed architecture.

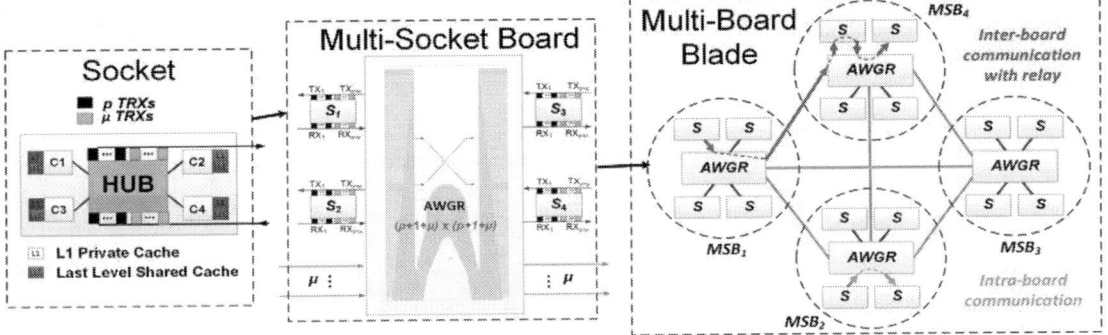

Figure 1. Hierarchical optical interconnected architecture for inter-socket communication within a board and between multiple boards. (Left): the Socket (S) topology with the Hub switch connecting the four computing cores with private and shared cache memory; (Center): the Multi-Socket Board (MSB) with four sockets based on passive AWGR all-to-all interconnection; (Right): a Multi-Board Blade (MBB) computing node with four MSBs. [3]

It is well known that the dynamic power of CMOS transistors scales as $\propto V_{dd}^2 * f$, where V_{dd} is the driving voltage and f is the clock speed. If V_{dd} can be lowered for circuits with low f, it is then possible to obtain significant improvement in energy efficiency by lowering the clock speed in combination with the driving voltage (nearly 2× improvements in power efficiency for 20% underclocking). We studied the benefits of using DVFS when applying this

This work was supported in part under DoD Agreement Number: W911NF-13-1-0090.

technique to the p and μ TRXs of each node (Hub switch) in the architecture of Figure 1. We modeled the overall system by using the GEM5 simulator with 64 cores distributed in 16 sockets. GEM5 boots the Linux 2.6.27 operating system and runs the PARSEC 2.1 benchmarks suite [6]. Figure 2 shows the benchmark results in terms of execution time (left) and Energy Delay Power (EDP, right). We evaluated the performance of the proposed AWGR-based architecture with DVFS using different transmission parallelism (4 and 8 bits). Specifically, we compared the DVFS approach with source-synchronous technique [7] against a conventional approach using optical links with Clock and Data Recovery (CDR). We also compared the results with a state-of-the-art Hyper Transport electronic system.

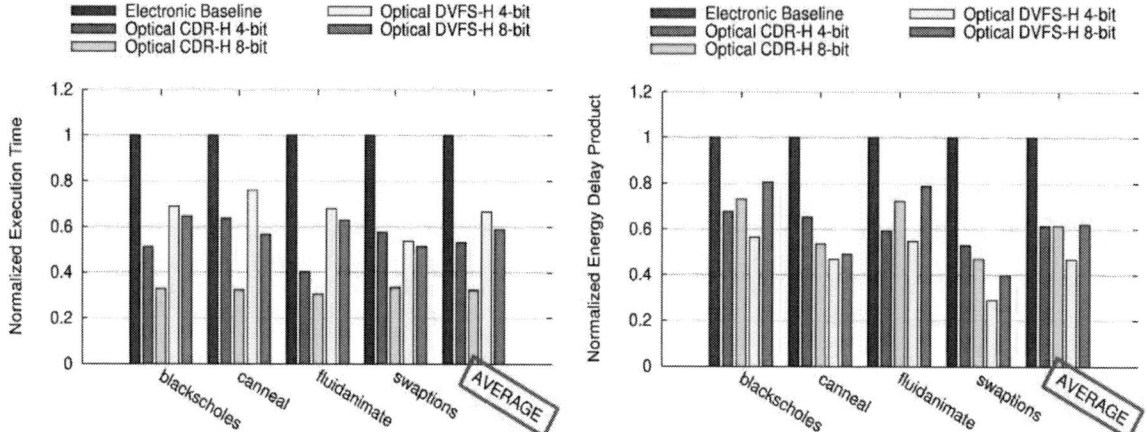

Figure 2. Execution time (left) and EDP (right) normalized to the electronic baseline in comparison with an optical hierarchical solution with CDR and DVFS and 4 and 8-bit parallelism (low values in the plots indicate superior results). [3]

Note that, the CDR system always transmits at the nominal speed (i.e., 10 Gb/s) while paying for the CDR circuitry consumption and power to send synchronization bits to keep the receivers locked. In terms of execution time, the performance of with the system with CDR is always the best. By increasing the communication parallelism from 4- to 8-bit helps to further improve the execution time, as shown in in the first two bars of Figure 2(left). When using DVFS, the system avoids the transmission of bits when the Hub buffers are empty. The system sets the transmitter frequency and voltage supply to a maximum and minimum values according to the traffic load in each Hub. DVFS introduces some latency [see Figure 2(left)] due to the burstiness of the considered benchmarking traffic. However especially when using only 4-bit parallelism, DVFS gives significant energy reduction compared to CDR system [see Figure 2(right)]. Note that, increasing the bit parallelism for the DVFS solution comports an average higher EDP value [fourth bars in Figure 2(right)] due to the higher number of transceivers as well as the higher execution time in comparison to the CDR case.

III. FLEXIBLE BANDWIDTH INTER-RACK ALL-TO-ALL OPTICAL NETWORK WITH CHANNEL BONDING

Figure 3(top) shows an example of rack-to-rack optical interconnect architecture exploiting wavelength routing in AWGR to perform "channel bonding" (see Figure 3(bottom)) to dynamically, rapidly, and flexibly assign additional bandwidth upon demand between hot spots. To support both high scalability and connectivity, μ AWGRs interconnect with each other in an all-to-all pattern and each AWGR connects with p clusters. The architecture scales to $p \times \mu$ clusters and the radix of AWGR is $p \times (p+\mu-2)+\mu-1$. Therefore, the full system can reach up to 103,680 servers using six 65-port AWGRs when $p = 6$, $\mu = 6$, 40 servers/racks, 72 racks/clusters. By using two intra-region transceivers (the grey TRXs in Figure 3), three clusters connected with the same AWGR can achieve contention-free all-to-all communication. Each cluster communicates with other AWGRs with two inter-region TRXs (the green ones). To achieve flexible-bandwidth reconfiguration between the hot spots, the TRXs make use of fast Tunable Lasers (TLs) which can achieve fast wavelength tuning in ~10 nanoseconds [8]. When hot spots arise, the control plane can the TRXs' wavelengths to increase the number of connections between the hot clusters. For instance, originally, the four clusters in Figure 3(bottom) connected with each other in an all-to-all fashion. When the bandwidth between C0 and C3 exceeds the peak bandwidth of a single link, TRX for C0→C2 with λ_1 (blue link) tunes to λ_2 (red link) for C0→C3. Eventually, the bandwidth between C0 and C3 is doubled by bonding the signals transmitted by the two red TRXs The TRXs that require tuning are selected based on the following rules: 1) all the Cluster are still reachable after tuning; 2) the tuning introduces limited additional forwarding to non-hot Clusters; 3) the TRXs to be tuned have light traffic load. We experimentally demonstrated the channel bonding on an 8-node demo using 8 FPGA boards forming two regions with four FPGAs per region. In particular, we demonstrated the following four scenarios: 40Gbps hot-spot traffic between FPGA 1 and 4 (intra-region) with and without flexible bandwidth adjustment by channel bonding; 40Gbps hot-spot traffic between FPGA 1 and 5 (inter-region) with and without flexible bandwidth adjustment.

This work was supported in part under DoD Agreement Number: W911NF-13-1-0090.

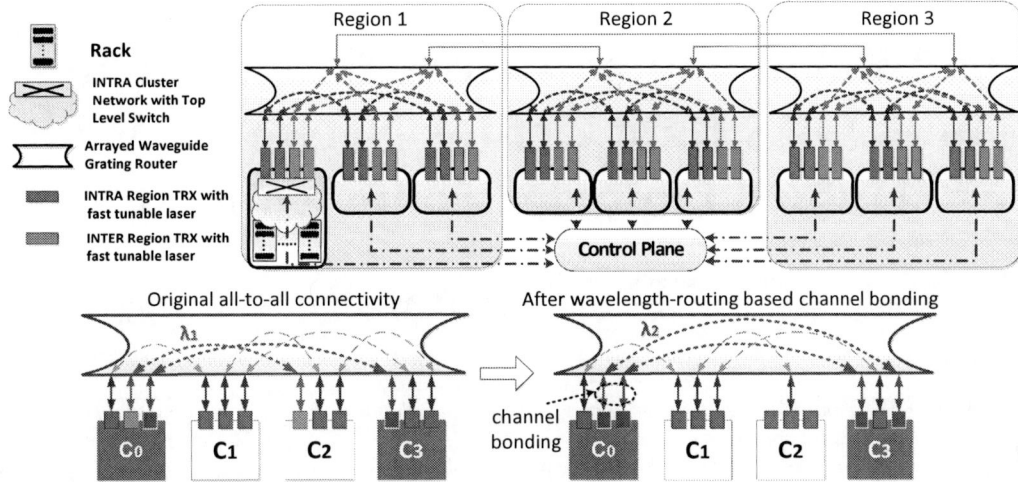

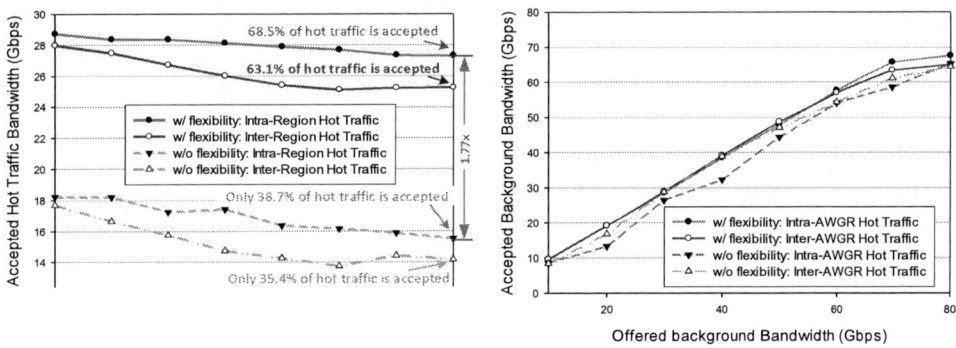

Figure 3. (Top) Proposed data center flat all-to-all optical interconnect architecture based on wavelength routing in AWGR; (bottom) The concept of wavelength-routing based channel bonding to achieve bandwidth flexibility [4].

Figure 4 shows the statistics (measured experimentally) for the four scenarios described above. Figure 4(left) shows that network with channel bonding achieves up to 1.77× improvement in accepted hot-spot traffic. Figure 4(right) shows how the links reconfiguration dedicated to certain clusters does not necessarily reduce but can actually increase the accepted background bandwidth. This can be explained considering that the congestion caused by the hot spot traffic is released even though the link reconfiguration causes additior

Figure 4. (Left) Experimentally measured accepted hot spot bandwidth. (Right) Experimentally measured accepted background traffic bandwidth [4].

IV. CONCLUSION

In this paper we demonstrated the benefits of applying flexible bandwidth techniques in optically-interconnected architectures for intra and inter-rack communication. Specifically, we performed benchmarking simulations and experimental measurements to prove that adapting the architecture capabilities to the traffic requirements is crucial to achieve good performances. This is fundamental to meet the requirements of next generation data center systems.

V. REFERENCES

[1] T. Benson, A. Anand, A. Akella, and M. Zhang, "Understanding data center traffic characteristics," *SIGCOMM Comput. Commun. Rev.,* vol. 40, pp. 92-99, 2010.

[2] C. Gray, D. Keezer, O. Liboiron-Ladouceur, and K. Bergman, "Multi-Gigahertz Source Synchronous Testing of an Optical Packet Switching Network," in *Mixed-Signals Test Workshop*, 2006.

[3] P. Grani, R. Proietti, and S. B. Yoo, "Hierarchical AWGR-based Computing Node Architecture: Performance under Realistic Benchmark Workload," presented at the Optical Interconnects 2016.

[4] Z. Cao, R. Proietti, M. Clements, and S. J. B. Yoo, "Experimental Demonstration of Flexible Bandwidth Optical Data Center Core Network With All-to-All Interconnectivity," *Journal of Lightwave Technology,* vol. 33, pp. 1578-1585, 2015.

[5] C. Xuning, W. Gu-Yeon, and P. Li-Shiuan, "Design of low-power short-distance opto-electronic transceiver front-ends with scalable supply voltages and frequencies," in *International Symposium on Low Power Electronics and Design*, 2008.

[6] C. Bienia, S. Kumar, J. P. Singh, and K. Li, "The PARSEC benchmark suite: Characterization and architectural implications," in *Proceedings of the international conference on Parallel architectures and compilation techniques*, 2008.

[7] C. Gray, D. Keezer, O. Liboiron-Ladouceur, and K. Bergman, "Multi-Gigahertz Source Synchronous Testing of an Optical Packet Switching Network," in *International Mixed-Signals Test Workshop*, 2006.

[8] R. Proietti, Y. Yin, R. Yu, C. J. Nitta, V. Akella, C. Mineo, *et al.*, "Scalable Optical Interconnect Architecture Using AWGR-Based TONAK LION Switch With Limited Number of Wavelengths," *J. Lightwave Technol.*, 2013.

This work was supported in part under DoD Agreement Number: W911NF-13-1-0090.

Load Balancing in Switch-Fabric Type of Torus OPS Data Center Networks With Hybrid Optoelectronic Routers

Yue-Cai Huang*, Yuki Yoshida*, Salah Ibrahim†,
Ryo Takahashi†, and Ken-ichi Kitayama*

*Osaka University, †NTT Device Technology Laboratories

e-mail address: *{hycsea; yuki; kitayama}@comm.eng.osaka-u.ac.jp }, †{ibrahim.salah; t.ryo\}@lab.ntt.co.jp}

Abstract: *Load balancing in a switch-fabric type of torus optical packet switching data center networks with hybrid optoelectronic routers shows the performance improvement such as packet dropping probability by decreasing the average number of hops.*

Keywords: *Data center networks, optical packet switching, load balancing*

I. INTRODUCTION

The rapid growth in demand of cloud-based services, has fueled the fast emergence of large scale data center networks (DCN), comprising of hundreds of thousands of servers. However, the current intra-DCN communication, based on electrical switches, is facing technical challenges in terms of power consumption and scalability [1].

In recent past, new DCN architecture using hybrid optical/electrical switching or all-optical switching have attracted much attention [2]—[9]. Optical packet switching (OPS) presents a potential solution for reducing power and size of the nodes while maintaining ideal characteristics of packet switched networks. However, the lack of practical optical buffer technology makes it difficult to seek all-optical solutions. The hybrid optoelectronic router (HOPR) with CMOS buffers has been proposed and demonstrated [10], and it has been anticipated that its energy efficiency could be reduced to almost one-fourth, compared to the current high-end routers.

In this paper, load balancing in a switch-fabric type of torus optical packet switching data center networks with hybrid optoelectronic routers shows the performance improvement such as packet dropping probability by decreasing the average number of hops.

II. ARCHITECTURE

A. HOPR

The OPS-based DCN are as in [6-9]. The router, illustrated in Fig. 1, is called a Hybrid OPtoelectronic Router (HOPR) [9], combining the optical switch with the optoelectronic shared buffer. Key notations in Fig. 1 and in the paper are listed in Table 1. HOPR supports 100 Gbit/s link rate. Arriving optical packets firstly go through the label processors which extract the label information for forwarding decisions. Then the optical switch is configured accordingly and forwards the packets to their scheduled output links. Fiber delay lines (FDLs) are equipped for contention resolution. The shared buffer serves as a bridge between the 40 Gbit/s electronic domain Top-of-Rack (ToR) switches and the 100 Gbit/s optical domain transponders. It buffers packets during injection/reception and during contention resolution. Each HOPR is connected to a number of ToR switches, where each ToR switch is connected to a rack of servers.

TABLE I
LIST OF NOTATIONS AND THEIR VALUES

Network	HOPR	Traffic
Dimension, $N=6$ Radix, $K=4$ # Total nodes, $K^N=4096$	# FDLs, $D=2$ # Buffer ports, $M=2,3$ Size of input queue, $Q=100$ # ToR switch groups, $A_g=4$ # ToR switches per group, $A_t = 10$	Additional TTL, $\alpha =20$ Packet length, $L=1500B$ Offered traffic, $R=0\sim300$ Gbit/s

B. Switch-Fabric Type of Torus Network

The network topology (interconnection between HOPRs) is torus. Denote by N and K the dimension and the radix, respectively. The total number of nodes is K^N. Each node connects $2N$ neighbors. The torus topology enables simple hence fast routing algorithms. Its high capability of deflection routing could reduce the use of energy-consuming buffers for contention resolution.

In our previous works [6-9], each ToR switch is connected to only one HOPR. We call this architecture as architecture-A, as shown in Fig. 2. In this paper, we consider architecture-B, where each ToR switch is connected to k HOPRs. As shown in Fig. B, each HOPR is connected to A_g (=k) groups of ToR switches, and each group is consists of A_t ToR switches. For ToR switches from the same group, they are connected to the same set of HOPRs and regard one HOPR as their master HOPR. Denote the address of the master HOPR as $(x_1, x_2, ..., x_N)$, then the set of routers are chosen as $[(x_1, x_2, ..., x_N) + (1,1,...,1)] \bmod K$. Figure 3 illustrates the case for K=3. ToR switch sends packets to its own master HOPR. If a master link is congested, packets are sent to other HOPRs with load balancing. ToR switch can receive packets from k HOPRs. The distance between this k destination HOPRs and the source HOPR can be different. Therefore, the destination HOPR can be chosen by minimizing the number of hops. In this way, the average number of hop counts can be reduced and therefore the performance can be improved.

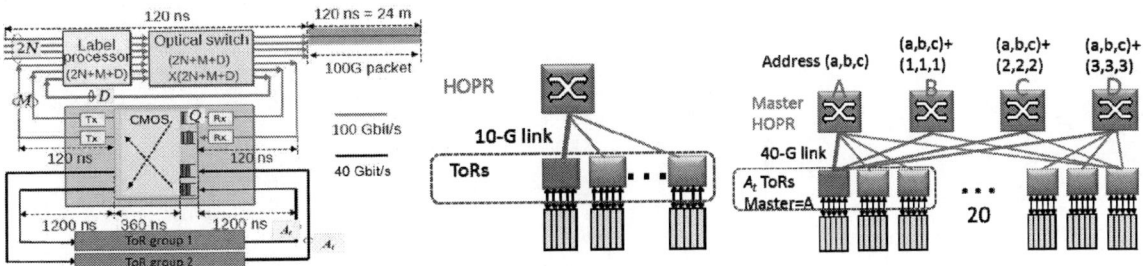

Fig. 1. Structure of HOPR. Fig. 2. Architecture-A Fig.3. Architecture-B

C. Routing and Contention Resolution

For the high speed OPS network, label processing should be down extremely fast. Therefore, complicated while time consuming routing algorithms are not appropriate. We choose simple routing and contention resolution methodologies detailed in [6]. The basic rules are as follows:

• Packet closer to its destination has higher priority.

• Each packet tries the following routing and contention resolution methods consecutively until being forwarded or dropped: shortest path (deflection) routing, FDL buffering, farther path deflection routing, electronic buffering.

• For (deflection) routing, apply consecutively: (1) dimension with longer distance has higher priority; (2) lower dimension has higher priority; (3) positive direction has higher priority than negative direction.

Besides, to avoid circulating packets infinitely and maintain the SINR, time-to-live (TTL), is set in the label of each packet before the transmission and is deducted by one each time the packet passes a label processor. When TTL =0, the packet will be dropped. We set the TTL as additional TTL (denoted by α) + minimum number of routers to pass.

III. PERFORMANCE EVALUATION

The DCN is simulated by a cycle-accurate (1 cycle = 1 time slot = 120 ns) simulator written in C++. The latencies inside different blocks of the HOPR are approximated as multiple of time slots, as shown in Fig. 1. We only consider ToR-to-ToR communication with Uniform traffic pattern and Bernoulli traffic injection processes. Performance metrics include the throughput which is defined as the traffic successfully received per HOPR per unit time, the packet dropping probability (PDP), and the end-to-end delay representing the average time for packets to reach the destination. For more details, please refer to [6-9]. For each scenario, 6 times of simulations are done and averaged.

Two sets of simulations are conducted. For Simulation-I, architecture A and B are compared. The simulation parameters are summarized in Table 1, except that the number of buffer ports, M, is two. The results are shown in Fig. 4. We can observe from Fig. 4 that for architecture B, the maximum throughput is increased; the PDP is decreased and especially, the average end-to-end delay is significantly reduced due to the reduction of hop counts.

Simulation II investigates the performance of architecture B with different number of buffer ports. Increasing buffer port counts allows more traffic to be injected to the network, while it requires larger optical switch. For the 6-dimensional torus networks with 4096 nodes, the average number of hop counts is 4.125 in case of no congestion, the maximum carried traffic in case of no traffic congestion is about 291 Gbit/s per router [9]. Therefore, M=3 is big enough and we choose M as 2 and 3 in the simulations. The results are shown in Fig. 5. We can observe that the throughput, PDP and average end-to-end delay are significantly improved for M=3. Combining Fig. 4 and Fig. 5, we can conclude a significant improvement in architecture B.

IV. CONCLUSION

We have proposed a new architecture for OPS data center networks, where routers are connected in a torus topology and ToR switches and routers are interconnected in a multi-by-multi manner. Simulation results demonstrated the performance improvement of this architecture compared with our previous architecture.

OECC/PS2016

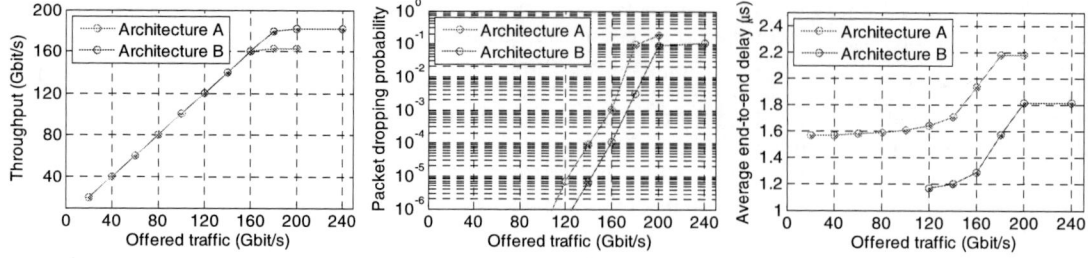

Fig. 4. Performance comparisons of architecture A and architecture B.

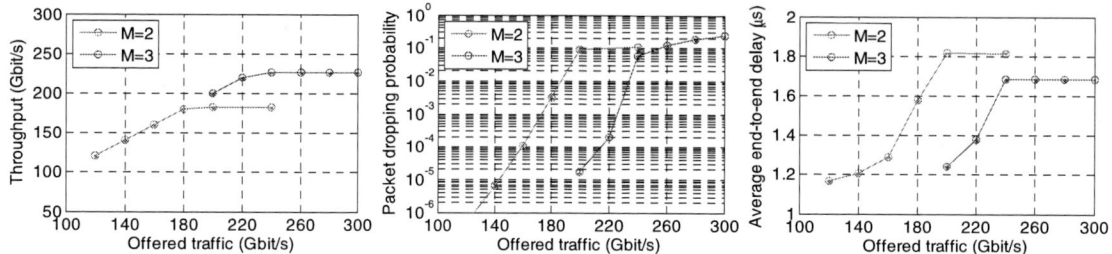

Fig. 5. Performance comparison of different input buffer port counts (M).

ACKNOWLEDGMENT

This work has been funded by the NICT R&D program, "Basic Technologies for High-Performance Optoelectronic Hybrid Packet Router" (2011~2016).

REFERENCES

[1] M. Al-Fares, A. Loukissas, and A. Vahdat, "A scalable, commodity data center network architecture," in Proc. ACM SIGCOMM, 2008.

[2] C. Kachris and I. Tomkos, "A survey on optical interconnects for data centers," IEEE Commun. Surveys Tuts., vol. 14, no. 4, pp. 1021–1036, 2012.

[3] X. Ye, Y. Yin, S. B. Yoo, P. Mejia, R. Proietti, and V. Akella, "DOS: A scalable optical switch for datacenters," in Proc. ACM/IEEE Symposium on Architectures for Networking and Communications Systems, 2010.

[4] G. Wang, D. G. Andersen, M. Kaminsky, K. Papagiannaki, T. Ng, M. Kozuch, and M. Ryan, "c-Through: Part-time optics in data centers," in Proc. ACM SIGCOMM, 2010.

[5] N. Farrington, G. Porter, S. Radhakrishnan, H. H. Bazzaz, V. Subramanya, Y. Fainman, G. Papen, and A. Vahdat, "Helios: a hybrid electrical/optical switch architecture for modular data centers," in Proc. ACM SIGCOMM, 2010

[6] Y.-C. Huang, Y. Yoshida, K. Kitayama, S. Ibrahim, R. Takahashi, and A. Hiramatsu, "Modeling and performance analysis of OPS data center network with flow management using express path," in Proc. Optical Network Design and Modeling (ONDM), 2014.

[7] K. Kitayama, Y.-C. Huang, Y. Yoshida, R. Takahashi, T. Segawa, and S. e. a. Ibrahim, "Torus-topology data center network based on optical packet/agile circuit switching with intelligent flow management," J. Lightw. Technol., vol. 33, no. 5, pp. 1063–1071, 2015.

[8] R. Takahashi, S. Ibrahim, T. Segawa, T. Nakahara, H. Ishikawa, Y. Suzaki, Y.-C. Huang, K. Kitayama, and A. Hiramatsu, "A torus datacenter network based on OPS/OCS/VOCS enabled by smart flow management," in Proc. Optical Fiber Communication Conference (OFC), 2015.

[9] Y.-C. Huang, Y. Yoshida, K. Kitayama, S. Ibrahim, R. Takahashi, and A. Hiramatsu, "OPS/agile-OCS data center network with flow management." J. Opt. Commun. Netw., vol. 7, no. 12, pp. 1109-1119, 2015.

[10] R. Takahashi, T. Nakahara, Y. Suzaki, T. Segawa, H. Ishikawa, and S. Ibrahim, "Recent progress on the hybrid optoelectronic router," in Proc. International Conference on Photonics in Switching (PS), 2012.

TuF2-3

Spectrum Assignment and Update Method based on Adaptive Soft Reservation in Elastic Optical Networks

Hideki Tode†, Akira Fukushima†, Yosuke Tanigawa†, and Yusuke Hirota‡

† Department of Computer Science and Intelligent Systems, Osaka Prefecture University, Japan 599-8531
‡ Department of Information Networking, Osaka University, Japan 565-0871

† {tode@, fukushima@com., tanigawa@}cs.osakafu-u.ac.jp, ‡ hirota.yusuke@ist.osaka-u.ac.jp

Abstract: *We propose a spectrum assignment method, called Soft Reservation, that improves the frequency utilization by introducing tentative reservation for forthcoming demand, and dynamic control to determine the reservation range by means of estimated traffic demand.*
Keywords: *Wavelength assignment, Soft reservation, Elastic optical network*

I. Introduction

Recently, studies on elastic optical networks(EON) [1] are making progress aggressively. In EON, frequency resources are densely multiplexed with selecting adequate modulation schemes according to optical reach, physical impairment condition, and requested bandwidth. As a result, it enhances transmission capability per fiber.

In EON, because frequency resources are used in a finer grained manner, so-called guard band is required to split adjacent optical paths on the frequency axis clearly. As a mean to utilize frequency resources more densely, if several paths are aggregated into one and handled as one broader bandwidth path, the guard band can be reduced. However, EON has two kinds of frequency continuity constraints and has different modulation schemes corresponding to required bandwidth and optical reach. Under dynamic network environment, especially, efficient path-aggregation considering these constraints would be difficult.

Under the above background, we have proposed a tentative reservation method of frequency resources, called Soft Reservation [2], which saves the required guard band, the number of required transponders, and the control overhead of path establishment by exploiting frequency resources that are tentatively reserved between the same source and destination (s-d) in advance for newly generated path set-up requests.

On the other hand, the previously proposed method [2] has the technical issue on flexibility; it simply makes reservation of static and fixed tentative resources, but not dynamic and adaptive ones for the forthcoming requests between the same s-d pair. In this paper, more flexible Soft Reservation method, as generalized extension of our previous framework, is proposed. Specifically, the proposed method sets up adequate number of tentatively reserved paths $N(s,d)$ between an s-d pair based on dynamic control using estimated traffic demand information. Also, more aggressive path aggregation can be attained with expanding the amount of tentative reservation for the s-d pair. These advanced control leads to the improvement of frequency utilization and high-quality network operation due to the reduction of request blocking rate.

II. Spectrum Assignment and Update Method based on Adaptive Soft Reservation

A. Outline

In the proposed Soft Reservation, frequency resources tentatively assigned for each s-d pair can be utilized as the soft-reserved frequency region prioritized for the s-d pair. Then, the blocking probability of path setup requests in case of using the prioritized region is presented in Eq.(1),

$$P_{bl}^{Soft}(s,d) = \frac{(\rho(s,d))^N/N!}{\sum_i (\rho(s,d))^i/i!}. \tag{1}$$

Let N be the number of paths possible to be set up by using the soft-reserved resources for the s-d node pair, and $\rho(s,d)$ be traffic demand of the s-d pair; namely, $\rho(s,d) = \lambda(s,d) \cdot h(s,d)$, where $\lambda(s,d)$ and $h(s,d)$ are the average arrival rate of new path set-up requests per s-d pair, and its average holding time, respectively. In the proposed method, based on Eq.(1) and the estimate of traffic demand derived every time interval T, the number of frequency slots satisfying $P_{bl}^{Soft}(s,d) < \varepsilon$ is set to the target value of soft-reserved frequency slots between s-d pair, $N_{all}(s,d)$. This is used for the proposed control. The specific procedure is described in Algorithm 1(Alg. 1), where ε in Alg. 1 is static input parameter, which tunes allowable request blocking probability.

B. Tentative Resource Reservation for a New Path

When a new path is set up, the number of soft-reserved frequency slots for not only the path but also future requests, $N(s,d)$, is determined by Algorithm 2(Alg. 2), so that target value $N_{all}(s,d)$ calculated by Alg. 1 can be accomplished in the entire network. In Alg. 2, first, a shortage Δ that equals $N_{all}(s,d) - N_{now}(s,d)$ is calculated, but when $\Delta \leq 0$, in other words, $N_{now}(s,d) > N_{all}(s,d)$, soft-reservation is not performed. Second, the number of soft-reserved frequency slots $N(s,d)$ is determined by the smaller value between Δ and the number of paths aggregatable into one transponder N_{MAX}. N_{MAX} is the fixed value corresponding to the path hop.

By means of this control, when the frequency resources assigned between a s-d pair lack, more soft-reserved slots are additively assigned. On the other hand, when the assigned resources are redundant, no additive extension of the soft-reservation performs, which avoids depriving the frequency resources from paths between the other s-d pairs.

C. Expansion of Tentative Resources Already Reserved for Existing Path

As for a path of a certain s-d node pair, joining both sides of frequency slots into the soft-reserved region adaptively gives further efficacy on frequency utilization. The soft-reserved area may be expanded when all the soft-reserved slots are filled, or they are preferentially deprived by paths of other s-d pairs because of their bandwidth starvation. The procedure of the expansion is described as follows.

Algorithm 1: Determination of $N_{all}(s,d)$

Input: $\varepsilon, \rho(s,d), Node$
Output: $N_{all}(s,d)$
```
/* s - source node, d - destination
   node                              */
```
while $s,d \in Node$ **do**

    initialization: $N_{all}(s,d) \leftarrow 1$;

    $P_{bl}^{Soft}(s,d) \leftarrow$
    $(\rho(s,d))^{N_{all}(s,d)}/N_{all}(s,d)!/(\sum_i(\rho(s,d))^i/i!)$;

```
       /* update N_all(s,d)          */
```
    while $\varepsilon \leq P_{bl}^{Soft}(s,d)$ **do**

        $P_{bl}^{Soft}(s,d) \leftarrow$
        $(\rho(s,d))^{N_{all}(s,d)}/N_{all}(s,d)!/(\sum_i(\rho(s,d))^i/i!)$;

        $N_{all}(s,d) \leftarrow N_{all}(s,d)+1$;

return $N_{all}(s,d)$

Algorithm 2: Determination of $N(s,d)$

Input: $N_{all}(s,d), N_{now}(s,d)$
Output: $N(s,d)$
```
/* s - source node, d - destination
   node                              */
```
$\Delta \leftarrow N_{all}(s,d) - N_{now}(s,d)$;
$N' \leftarrow Max(\Delta,0)$;
$N(s,d) \leftarrow Min(N',N_{MAX})$;
return $N(s,d)$

In advance, we make a path list composed of already set-up paths per each s-d pair, which is sorted in order of the larger vacant capacity of their bottleneck link under the condition that soft-reserved resources are dealt with in use.

Step1: Calculate additional soft-reserved slots for the s-d node pair by Alg. 2.

Step2: Select the expanded path candidate from the head of the path list and remove it from the list.

Step3: Check whether there are enough frequency resources in both sides of the selected path, on each link along the path, under condition that the soft-reserved resources are in use. This check is started from higher side of frequency axis, and if not found, shifted to lower side. If all links along the path have necessary frequency slots for expansion, then soft-reserved region is actually expanded. If there are not enough frequency slots but some slots available, then the region is expanded as much as possible. Here, the above expansion is performed within the upper limit of transponder capacity.

Step4: Repeat **Step2,3**. But if enough expansion of soft-reserved frequency region was already performed (i.e. successfully updated), or path candidates are starved (i.e. failed or partially updated), then the procedure for the s-d pair is terminated.

III. Performance Evaluation

We evaluated the performance of the proposed method by simulation experiment. In this evaluation, EWMA(Exponentially Weighted Moving Average) was applied to estimate traffic demand. We set up the following simulation parameters and conditions. Network topology is 4×4 grid. Traffic generation is homogeneous between any s-d node pairs. Holding time and inter-arrival time of requests follow the exponential distribution. Number of fibers per link $F = 1$. Frequency slots per fiber $W = 1000$. Frequency slots required as guard band are 2. Time interval for calculating EWMA $T = 1$[s]. Threshold in Alg. 1 $\varepsilon = 10^{-2}$. Routing follows fixed shortest routing by Dijkstra. For comparison, we adopt the two methods: Previous method [2](SoftReservation(N=3)), First-Fit not aggregating paths.

In addition, frequency slots each path requires are listed in Table 1, and capacity limit per transponder N_{MAX} is listed in Table 2. Also, traffic load is defined as the following equation.

$$\text{Traffic Load} = \frac{E \times Max_{link} \times Ave_{slot}}{W \times F \times T_a}, \tag{2}$$

where E is average holding time of optical path, W is the number of frequency slots per 1 fiber, F is the number of fibers per 1 link, T_a is average requesting time interval per s-d node pair, Ave_{slot} is average required frequency slots per s-d node pair, and Max_{link} is the number of optical paths traversing bottleneck link in case that one path per s-d node pair is set up according to the shortest routing.

Figures 1 and 2 show the request blocking rate with different traffic load, and the number of required transponders in case of Traffic Load = 1.0, respectively. From these graphs, request blocking rate is improved in comparison with the previous Soft-Reservation method and First-Fit. The transponders are also saved by our proposal. As a result, the proposed method enables to aggregate optical paths more efficiently than the compared methods, and to save frequency resources and network components.

Table 1. Required Frequency Slots

Hops	Modulation	Required Slots(GB=2)
1,2,3	16QAM	$1+2=3$
4,5	QPSK	$2+2=4$
6 or more	QPSK	$3+2=5$

Table 2. N_{MAX} of $400Gbps$ Transponder

Hops	Modulation	Added Slots
1,2,3	16QAM	5
4,5	QPSK	6
6 or more	QPSK	3

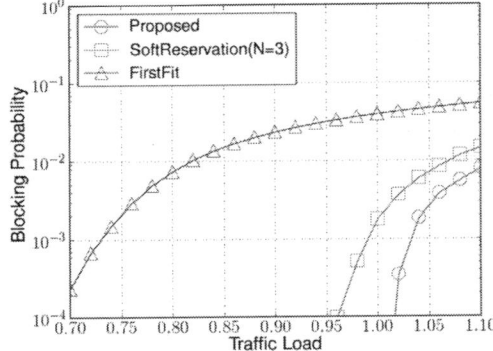

Fig. 1. Blocking Probability of Path Set-Up Requests

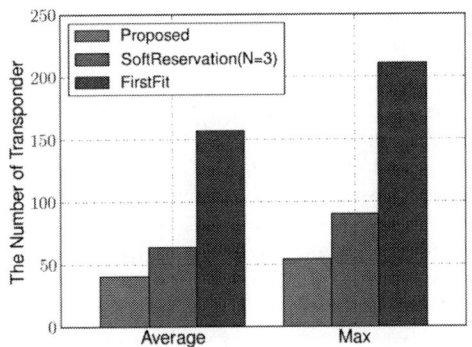

Fig. 2. The Number of Required Transponders

IV. Conclusion

In this paper, we proposed a frequency assignment and update method based on adaptive Soft-Reservation in Elastic Optical Networks, and described the specific algorithms. Through extensive computer simulation, we verified that the proposed method can reduce the request blocking events and the required number of transponders compared with static and tentative reservation.

Acknowledgment

This research is partly supported by the National Institute of Information and Communications Technology, Japan.

References

1. M. Jinno, H. Takara, B. Kozicki, Y. Tsukishima, Y. Sone, and S. Matsuoka, "Spectrum-efficient and Scalable Elastic Optical Path Network: Architecture, Benefits, and Enabling Technologies," *IEEE Comm. Mag.*, vol.47, no.11, pp.66–73, Nov. 2009.
2. N. Wakabayashi, Y. Hirota, H. Tode, and K. Murakami, "On-demand Path Provisioning with Tentative Spectrum Reservation in Elastic Optical Networks," *Proc. CLEO-PR&OECC/PS 2013*, TUT2-2, 2 pages, June-July 2013.

TuF2-4

A Study on Spectrum Assignment and Resource Sharing Method in Elastic Optical Packet and Circuit Integrated Networks

Ken Nagatomi*, Seitaro Sugihara*, Shohei Fujii*, Yusuke Hirota*, Hideki Tode[†] and Takashi Watanabe*
*Department of Information Networking, Osaka University, Japan
Email: {nagatomi.ken, sugihara.seitaro, hirota.yusuke, fujii.shohei, watanabe}@ist.osaka-u.ac.jp
[†]Department of Computer Science and Intelligent Systems, Osaka Prefecture University, Japan
Email: tode@cs.osakafu-u ac.jp

Abstract

We propose and evaluate a dynamic spectrum assignment method to reduce spectrum fragmentations and an adaptive resource sharing method to improve resource utilization efficiency for elastic optical packet and circuit integrated networks.

Keywords

Elastic optical networks; Optical packet and circuit integrated network; Spectrum allocation

I. INTRODUCTION

Many technologies have been intensively studied to cope with exponentially increasing network traffic. Elastic Optical Network (EON) is one of the promising technologies that can increase transmission capacity of optical networks [1], because it can flexibly exploit spectrum resources by selecting modulation format based on both requested bit rate and optical reach [2]. In dynamic EONs, repeated provisioning / leasing of connections having various sizes of bandwidths make non-contiguous small pieces within spectrum resources. These fragmentations waste the spectrum resources. Previous works on the spectrum fragmentations handle only the connection requests of optical circuit switching (OCS) which enables bandwidth-guaranteed data transmission. On the other hand, optical packet switching (OPS) which is suitable for best-effort services achieves efficient bandwidth utilization because of statistical multiplexing effect. More various services will coexist in future networks. Therefore, optical packet and circuit integrated (OPCI) networks that can deal with both OPS and OCS traffic have been studied [3].

In an OPCI network, the OPS plane and the OCS plane share spectrum resources with using a boundary on the frequency domain. However, a fixed boundary degrades spectrum utilization when the traffic volume of each plane changes. For more efficient resource usage, the boundary between two planes should be flexibly adjusted in response to user demand for best-effort and bandwidth-guaranteed services [4].

This paper proposes a novel spectrum assignment method to reduce spectrum fragmentations and two adaptive resource sharing methods to improve resource utilization efficiency. This method controls the boundary between the OPS and OCS planes based on the utilization of OPS and OCS. In addition, by limiting area of allocating spectrum resources depending on the number of required frequency slots, spectrum fragmentations are reduced. The proposed method is evaluated through computer simulations. We validate the improvement of link utilization and blocking probabilities by the proposed boundary control and resource arrangement.

II. RELATED WORKS

A. Elastic Optical Networks

The concept of EON is introduced as spectrum-sliced elastic optical path network (SLICE) in [1]. EON aims at fully utilizing the limited resources by flexible spectrum assignment for each connection in consideration of physical factors, such as optical reach. EON is expected to provide more efficient and highly-available optical network infrastructure for future networks [2]. Routing and Spectrum Assignment (RSA) problem is one of essential issues for realizing EONs. In RSA problem, there are two important constraints, which are called spectrum contiguity and continuity constraints for EONs. The former requires all the frequency slots assigned to a lightpath to be spectrally neighboring. The latter implies that same frequency slots are assigned on all the links on the path [5].

B. OPCI Networks

Figure 1 shows an image of OPCI network. The OPCI network can provide diversification of services. To diversify services, OPS plane enables best-effort data transfer, while OCS plane enables bandwidth-guaranteed data transmission. In the OPCI network, when spectrum resources used by the OPS and OCS are mixed, it is difficult to separate signals in core nodes. Therefore, spectrum resources for OPS and OCS should be separated with using a boundary on the frequency domain. Figure 2 (a) shows an outline of sharing resource with a fixed boundary. However, if the boundary is statically fixed, network utilization worsens when the traffic load of OPS and OCS changes. For efficient resource usage, the boundary between two planes should be flexibly moved in response to user demand for best-effort or bandwidth-guaranteed services [4]. By dynamic

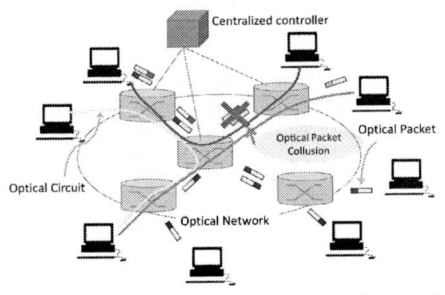

Fig. 1. An image of OPCI network.

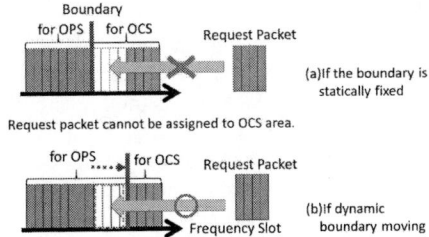

Fig. 2. Sharing resource and moving boundary.

boundary moving based on traffic situation and user requests, as shown in figure 2 (b), efficiency of resource utilization will be improved. Reference [4] demonstrates moving boundary technique without considering reduction of resource utilization by spectrum fragmentation in EONs.

III. The Proposed Method

We propose two novel dynamic resource sharing methods to improve resource utilization efficiency and an adaptive spectrum allocation method to reduce spectrum fragmentations for multi-fiber EONs. The proposed dynamic resources sharing and spectrum assignment method has two types; one is Dynamic Unified Boundary (DUB) method which unifies the boundary in the entire network, and the other is Dynamic Per Link Boundary (DPLB) method which controls the boundaries for each link.

A. Dynamic Unified Boundary (DUB)

In DUB method, boundary position is unified in the entire network. As an advantage of DUB, the control is simplified and spectrum contiguity and continuity are easily satisfied. Figure 3 (a) shows an example of a fiber configured by DUB. This method configures the dedicated areas based on the number of required frequency slots [6]. Aligning frequency slots reduces spectrum fragmentations for OPS. Spectrum alignment is executed based on B-PAC (Bottleneck-based Prioritized Area Configuration) proposed in Ref [6]. In each dedicated area in OPS, assigned spectrum slots are selected randomly. If First-Fit policy is adapted for OPS spectrum assignment, optical packets tend to collide with themselves in a congested link. Last-Fit policy with respect to the frequency domain is adapted for spectrum assignment in OCS plane. After a OCS connection is provisioned, or released the values V_{OPS} and V_{OCS} are calculated for boundary moving. Note that $V_{OPS} = \frac{U_{OPS}}{\log F_{OPS}}$ and $V_{OCS} = \frac{U_{OCS}}{\log F_{OCS}}$ where U_{OPS} and U_{OCS} mean resource utilization of each area of the bottleneck link and F_{OPS} and F_{OCS} represent the number of frequency slots of each area of the bottleneck link. When V_{OPS} is larger than $N_U \times V_{OCS}$, the boundary is moved to increase the frequency slot of OPS plane, and vice verse. N_{DUB} is a parameter of the proposed DUB method.

B. Dynamic Per Link Boundary (DPLB)

Although DUB unifies the boundary of all links in the network, DPLB controls boundaries for each link. As an advantage of DPLB, we can set boundaries depending on the traffic situation of each link. Figure 3 (b) shows an example of a fiber configured by DPLB. Because of the boundary control for each link, it is difficult to adapt B-PAC in [6]. DPLB selects frequency slots randomly among multiples of f ($f \times n(n = 0, 1, 2, ...)$), where f represents the number of request frequency slots. Frequency slots are aligned on demand to reduce spectrum fragmentations for OPS. Last-Fit policy with respect to the frequency domain is used in OCS spectrum assignment like DUB. After a OCS connection is provisioned, or released the values V_{OPS} and V_{OCS} are calculated for boundary moving. Note that $V_{OPS} = \frac{U_{OPS}}{\log F_{OPS}}$ and $V_{OCS} = \frac{U_{OCS}}{\log F_{OCS}}$ where U_{OPS} and U_{OCS} mean resource utilization of each area of all links used by the connection and F_{OPS} and F_{OCS} represent the number of frequency slots of each area of all links used by the connection. When V_{OPS} is larger than $N_U \times V_{OCS}$, the boundary is moved to increase the frequency slot of OPS plane in each link, and vice verse. N_{DPLB} is a parameter of the proposed DPLB method.

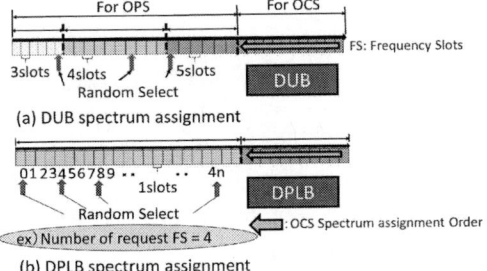

Fig. 3. Area configuration of the proposed methods.

TABLE I
NUMBER OF REQUIRED SLOTS

Number of hops	Modulation format	Number of required slots
1, 2	DP-16-QAM	3
3, 4, 5	DP-QPSK	4
6 or more	DP-QPSK	5

671

IV. PERFORMANCE EVALUATION

A. Simulation Model

We evaluate the performance of the proposed spectrum assignment methods through computer simulations. USA topology and JPN topology are adopted as the test networks. Figure 4 shows USA topology which has 28 nodes and 45 links. Figure 5 shows JPN-12 topology which has 12 nodes and 16 links. Each link is bidirectional and it has 8 fibers. Each fiber has 320 frequency slots (12.5 GHz within 4 THz). The load of traffic requests is defined as the ratio of the incoming demands to the bottleneck link capacity. The sum of the traffic load of OPS and OCS is set to 1.0 and the ratio of the traffic load for each plane is changed. The inter-arrival time of each connection follows exponential distribution. All connections require 100 Gbps and 3, 4, or 5 frequency slots are selected by adopting different modulation formats based on the hop counts. Table I shows the adopted modulation formats and required frequency slots for connections with each number of hops. Source and destination nodes are selected randomly, and the shortest path from source to destination is selected. The proposed spectrum assignment methods named DUB and DPLB are compared with two spectrum assignment methods. One is named *Fixed*. In *Fixed*, the boundary is statically fixed to the middle of the frequency domain. The other is named *RateStatic*, In *RateStatic*, the boundary is statically fixed to the position matching the ratio of OPS traffic load and OCS traffic load. The parameters in the proposed methods N_{DUB}, N_{DPLB} are 10, 3, respectively.

B. Simulation Results

Figures 4 and 5 show the blocking probability of the OPS and OCS. The blocking probability of OPS in DPLB is higher than that of *RateStatic* in USA topology and nearly same value in JPN topology. On the other hand, The blocking probability of OCS in DPLB is lower than that of *RateStatic* in both topologies.

Figures 6 and 7 show the link utilization that is the average of the ratio of the occupied frequency slots to all the slots in the entire network at some time points which are randomly selected. *Fixed* represents high link utilization when the traffic loads of OPS and OCS are equal. Because the boundary is statically fixed to a position according to the ratio of the traffic loads for each type of requests in *RateStatic*, the blocking probability for OPS and OCS are constant and the link utilization is high value. The link utilization of DPLB is higher than that of *RateStatic* in USA topology when the ratio of the OCS traffic load is high. DUB shows better performance in JPN topology than in USA topology. This is because the difference of the traffic situation between the bottleneck link and the other links gets bigger as the topology size gets larger. Since DPLB individually and adaptively moves boundaries for each links, the parformance is improved regardless of the topology size.

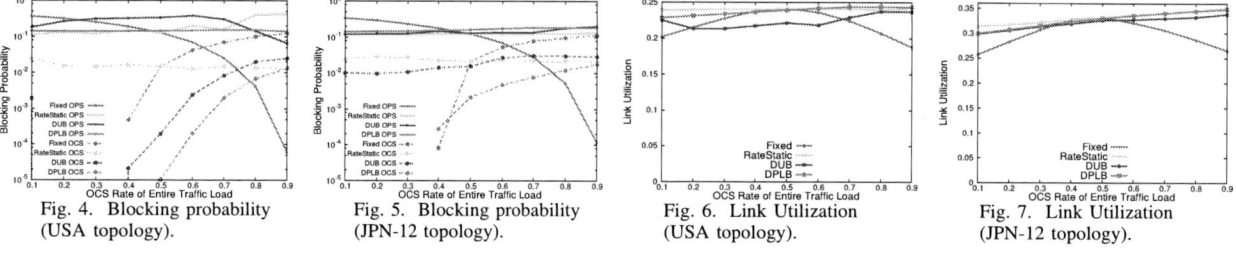

Fig. 4. Blocking probability (USA topology). Fig. 5. Blocking probability (JPN-12 topology). Fig. 6. Link Utilization (USA topology). Fig. 7. Link Utilization (JPN-12 topology).

V. CONCLUSION

In this paper, we investigated the challenge in the elastic optical networks and OPCI networks supporting OPS and OCS. We proposed two adaptive methods to reduce spectrum fragmentations and to improve the spectrum resource utilization. The computer simulations show that the proposed method, particulary DPLB, can achive better effectiveness in terms of blocking probability and link utilization by reducing spectrum fragmentations.

ACKNOWLEDGMENT

This research is partly supported by the National Institute of Information and Communications Technology, Japan.

REFERENCES

[1] M. Jinno, H. Takara, B. Kozicki, Y. Tsukishima, Y. Sone and S. Matsuoka, "Spectrum-efficient and scalable elastic optical path network: Architecture, benefits, and enabling technologies," *IEEE Commun. Mag.*, vol. 47, no. 11, pp. 66–73, Nov. 2009.

[2] M. Jinno, B. Kozicki, H. Takara, A. Watanabe, Y. Sone, T. Tanaka and A. Hirano, "Distance-adaptive spectrum resource allocation in spectrum-sliced elastic optical path network," *IEEE Commun. Mag.*, vol. 48, no. 8, pp. 138–145, Aug. 2010.

[3] H. Harai, H. Furukawa, K. Fujikawa, T. Miyazawa and N. Wada, "Optacal packet and circuit integrated networks and software defined networking extension." *Journal of Lightwave Technology*, vol. 32, no. 16, pp. 2751-2759, Aug. 2014.

[4] H. Furukawa, T. Miyazawa, N. Wada and H. Harai, "Moving the boundary between wavelength resources in optical packet and circuit integrated ring network." *Optics Express*, vol. 22, no. 1, pp. 4754, Sept. 2013.

[5] C. Wang, G. Shen and S. Bose, "Distance Adaptive Dynamic Routing and Spectrum Allocation in Elastic Optical Networks With Shared Backup Path Protection," *J. Lightwave Technology.*, vol. 33, no. 14, pp. 2955–2964, July. 2015.

[6] Y. Hirota, Y. Hatada, T. Watanabe, "Dynamic Spectrum Allocation Based on Connection Alignment for Elastic Optical Networks," *APSITT*, pp.34-36, Aug. 2015.

[7] T. Takagi, H. Hasegawa, K. Sato, Y. Sone, B. Kozicki, A. Hirano, and M. Jinno, "Dynamic Routing and Frequency Slot Assignment for Elastic Optical Path Networks that Adopt Distance Adaptive Modulation," OFC 2011, Mar. 2012.

Optimized Multicast Scheduling in Datacenter Optical Burst Ring Network

Gang Chen[1], Dongxu Zhang[1], Hongxiang Guo[1,*], Jian Wu[1],
Xiaoyuan Cao[2], Noboru Yoshikane[2], Takehiro Tsuritani[2], Itsuro Morita[2]

[1]Beijing University of Posts and Telecommunications, NO.10 Xitucheng Road, Haidian District, Beijing 100876, China
[2]KDDI R&D Laboratories Inc, 2-1-15 Ohara, Fujimino-shi, Saitama 356-8502, Japan
Email: {cgbupt, *hxguo}@bupt.edu.cn

Abstract: We propose an ILP-based model and a time-efficient heuristic algorithm for multicast scheduling in optical burst ring network. Simulation results verify the proposed methods could improve latency and throughput performance significantly compared to unicast transmission.
Keywords: optical burst switching, ring network, multicast, ILP.

I. INTRODUCTION

With the rapid development of cloud computing and big data services, datacenters (DCs) are widely deployed as fundamental IT infrastructures. In order to support fine-granularity, low energy consumption, low latency data transmission, and simplify the complexity of network control, the optical burst switching based on ring network has been regarded as a candidate solution for future datacenter network (DCN) [1,2]. Meanwhile, multicast transmission within DC is ubiquitous and increasingly important, in which data packets are sent from one source, replicated at the intermediate nodes and delivered to all destinations of the multicast group, such as virtual machines (VMs) cloning, data backup, and database synchronization among geographically distributed DCs [2]. Since optical burst rings inherently support multicast in the optical layer, it is necessary to consider the multicast scheduling in optical burst ring networks to support these multicast services. However, to the best of our knowledge, existing studies about optical multicast scheduling algorithms mostly focus on optical circuit switching network or passive optical network (PON)[3,4], but no one has investigated it in optical burst ring networks.

According to the aforementioned discussion, in this paper, we firstly introduce the general ring-based DCN architecture with optical burst sub-wavelength switching, and describe the control plane workflow for multicast bandwidth allocation. Then we propose an integer linear programming (ILP)-based multicast modeling and a first-fit (FF)-based heuristic algorithm. Finally, simulation results are presented to verify that the proposed scheme can significantly improve throughput and reduce latency compared to unicast transmission.

II. MULTICAST CAPABLE NETWORK ARCHITECTURE AND CONTROL PLANE WORKFLOW

The optical burst ring network-based DCN architecture with the ability of multicast is shown in the left part of Fig.1. Each node is composed of couplers and fast optical switches (FOSs). Assuming that node A sends multicast data to node B, C and D at a wavelength channel λ, a fraction of the optical signal power on λ will be split by the couplers at node B, C, and D, and dropped to the burst mode receivers. To realize collision-free data transceiving, all nodes need

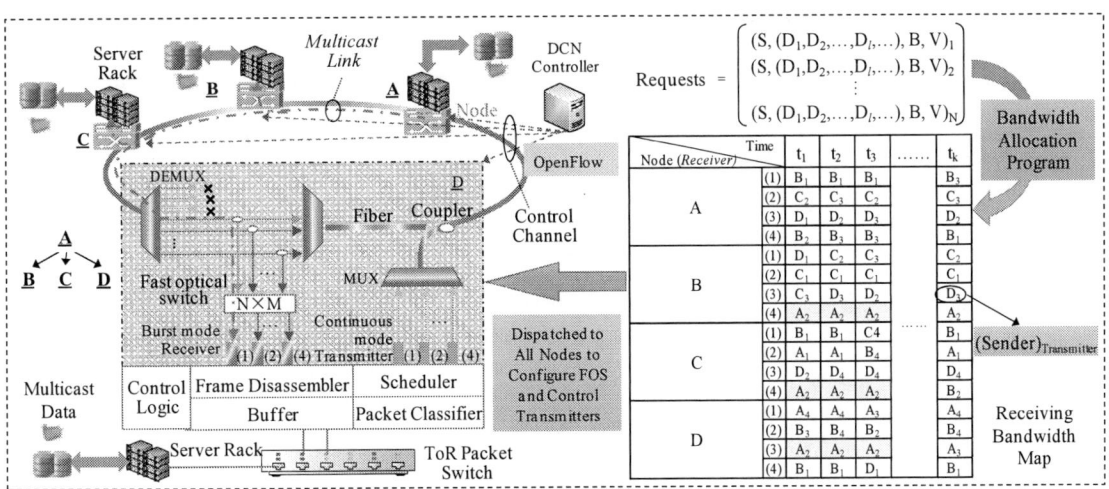

Fig.1 Optical ring-based DCN architecture for multicast communication and the corresponding control plane workflow

synchronization and adding/dropping optical bursts in the allocated time slots described by the *receiving bandwidth map*. As shown in Fig.1, the receiving bandwidth map ensures the data transmission without conflicts by allocating only one transmitter to a receiver in a specific timeslot. The colored entries in the form denotes that the 2nd transmitter of node A sends multicast data to the 4th receiver of node B, the 4th receiver of node C, and the 3rd receiver of node D.

To achieve more agile and efficient network scheduling, the control plane takes advantages of SDN technology such as OpenFlow [5]. Through a dedicated control channel, network controller periodically collects bandwidth requests information from all nodes, as shown in the right of Fig.1. A flow transmission request f_i can be described as $(S_i, (D_{i1}, D_{i2}, \dots, D_{iL}), B_i, V_i)$, where S and D_l denote source node and destination nodes ($L \geqslant 2$ implies multicast). B is the request bandwidth which can be described as the amount of time slots per bandwidth allocation period and V is the flow's data volume. Based on certain strategies, the controller generates a receiving bandwidth map and dispatches it to all nodes.

Designing an appropriate algorithm to compute bandwidth map is vital to achieve efficient and conflict-free multicast data transmission in optical burst ring network. With the objective of satisfying flows' bandwidth requests as much as possible, the allocation algorithm needs to tackle with some constraints such as one receiver can only receive data from one source node in each time slot. We further discuss our bandwidth allocation algorithm based on ILP and FF in section III.

III. OPTIMIZATION MODELING AND HEURISTIC ALGORITHM

We assume that requests will not be blocked but their bandwidth could be cut proportionally such that all the flows can be transmitted even if contention happens. To this end, we introduce a *cutting coefficient* (C) to ensure fairness across all requests, which is to say, all the requests would be cut by a same proportion (C), therefore maximizing C can be regarded as an objective of the bandwidth map computing problem. The ILP formulation is as follow.

Given:
- $\{F \mid f_i = (S_i, (D_{i1}, D_{i2}, \dots, D_{iL}), B_i, V_i)\}$: a set of flow transmission requests.

Variables:
- C, $0 \leqslant C \leqslant 1$: Cutting coefficient. The eventually allocated bandwidth for each flow is its expected bandwidth B_i multiply by C.
- $Node_{i\_m\_x\_n\_y\_k} \in \{0, 1\}$: If the yth transmitter in nth node and xth receiver in m node is allocated to request i in the kth time slot, $= 1$; otherwise, $=0$.
- $R_{i\_n\_y\_k} \in \{0, 1\}$: If the yth transmitter in nth node is allocated to request i in the kth time slot, $=1$; otherwise, $=0$.

Objective: Maximize C

Constraints:

$$\forall k, \forall m, \forall x, \sum_{i=1}^{I} \sum_{n=1}^{N} \sum_{y=1}^{Y} Node_{i\_m\_x\_n\_y\_k} \leq 1 \quad (1)$$

$$\forall k, \forall y, \forall n, \sum_{i=1}^{I} R_{i\_n\_y\_k} \leq 1 \quad (2)$$

$$\forall i, \forall k, \forall y, \forall l, \sum_{x=1}^{X} Node_{i\_(D_l)_i\_x\_S_i\_y\_k} = R_{i\_S_i\_y\_k} \quad (3)$$

$$\forall i, \sum_{y=1}^{Y} \sum_{k=1}^{K} R_{i\_S_i\_y\_k} \geq B_i \bullet C \quad (4)$$

The objective function maximizes the amount of transmitted data. Constraints (1) and (2) enforce a receiver to only receive data from one transmitter, and a transmitter to only respond to one request. Constraint (3) ensures that each destination node of the ith request has only one receiver to receive data from the yth transmitter of the source node. Constraint (4) ensures each request would be allocated with at least $B_i * C$ bandwidth.

Although the optimized allocation can be guaranteed, but solving the ILP model tends to be time-consuming, which is not very practical in real-time request-grant operation scenarios. So we also proposed a simple and time-efficient heuristic algorithm, in which flows requesting larger bandwidths are prioritized to be allocated with idle slots while slots are chosen in a first-fit manner. C is cutting coefficient as the same meaning within ILP formulation, and the heuristic algorithm is described as the following pseudocode.

Step 1: Sort all requests in the descending order according to their expected bandwidths.

Step 2: Set variable $i=1$, $C_0 = 0$, $C_1 = 1$, and an "ending condition" ε, which is a small positive number near 0.

Step 3: Adjust C to make it close to the optimal value:

    3.1: Let $C = C_i$, cut the requests' expected bandwidth values using C (for each flow f_i, let $B_i = B_i * C$), then try allocating time slots in a first-fit manner conforming to the constraints described in (1), (2) and (3), until no more idle slots can be used.

    3.2: If C_i is feasible (all the requests can be allocated), then check if $C_i - C_{i-1} < \varepsilon$ or $C_i \geq 1$. Yes → end algorithm; No → let $C_{i+1} = C_i + |C_i - C_{i-1}|/2$, then i++ and go to Step 3.1.
Else, C_i is not feasible, then Let $C_{i+1} = C_i - |C_i - C_{i-1}|/2$, i++ and then go to Step 3.1

Note that the basic principle of this algorithm is to try and find a feasible coefficient C and adjust it iteratively until the procedure converges to some point that C is near optimal. However, the first-fit slot allocation (Step 3.1) itself does not guarantee optimality, which trades some resource utilization efficiency for fast processing time.

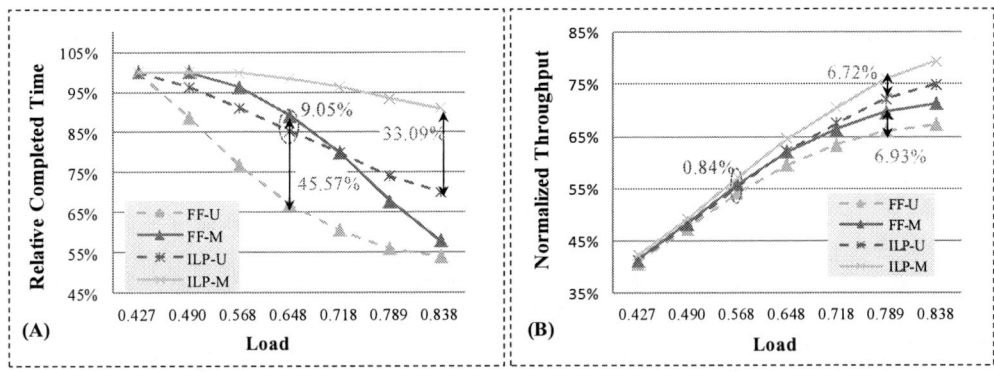

Fig.2 Simulation results of relative completed time (A) and normalized throughput (B)

IV. SIMULATION AND RESULTS DISCUSSION

Our ILP and First-Fit (FF) heuristics are implemented with a simulated scenario where the optical burst ring network is composed of 4 nodes as shown in Fig.1 and each node has 4 transceivers. The dynamic requests follow Poisson distribution, of which about 25% are unicast requests and the others are multicast requests. Each multicast request has 2 or 3 destination nodes that are selected randomly.

Given a certain load which is defined as the average total bandwidth requests divided by total link capacity, we generated 100 groups of requests each run. The ILP model described above is implemented with GUROBI [6] software. Two performance metrics are concerned: the *relative completed time* is the mean value of flows' expected transmission time divided by their actual transmission time, and the *normalized throughput* is the total allocated bandwidths (averaged over time) normalized to the total link capacity. The results shown in Fig.2 can be summarized as follow:

(1) With increasing load, the relative completed time and normalized throughput of both the unicast and multicast algorithms increase, but the performances of multicast are better than the unicast in any case.

(2) The relative completed time improvement of ILP multicast (ILP-M) can reach 33.09% compared to ILP unicast (ILP-U) under heavy load; For the multicast heuristics (FF-M), its improvement reaches 45.57% under moderate load (about 0.648) compared to unicast (FF-U).

(3) The normalized throughput improvement of ILP-M reached 6.72%, while for FF-M it can improve 6.93% under heavy load (about 0.8).

(4) Under light load, the performance of FF-M is better than ILP-U as the dotted circles show in Fig.2. When the load is 0.648, the relative completed time improvement of FF-M is about 9.05% compared to ILP-U and the normalized throughput improvement is about 0.84% under the load of 0.568. But under heavy load, ILP-U has a better performance than FF-M.

V. CONCLUSIONS

An ILP model and a heuristic algorithm were proposed for multicast resource scheduling in optical burst ring network. Via simulation, we compared the relative completed time and normalized throughput performance of our multicast algorithms with typical unicast ones and the results indicate that multicast algorithms performed better.

ACKNOWLEDGMENT

This work was partly supported by NSFC program (Grant No. 61331008, 61471054) and SRFDP program (No. 20130005110013).

REFERENCES

[1] Dongxu Zhang, Jian Wu, Hongxiang Guo, Rongqing Hui, "An optical switching architecture for intra data center interconnections with ultra-high scalability," Optical Interconnects Conference, paper PTu6, San Diego, USA, May 2014

[2] Gang Chen, et al., "First demonstration of holistically-organized metro-embedded cloud platform with all-optical interconnections for virtual datacenter provisioning", Post-Deadline Paper, Opto Electronics and Communications Conference (OECC), paper PDP2C.3, Shanghai, China, Jul.2015.

[3] Ze Li, Min Zhang, Danshi Wang, Yue Cui, "Simultaneous All-optical WDM Multicast and Unicast Scheme for WDM Optical Access Network Based on SOA and AWG," Optoelectronics Global Conference (OGC), pp1-4, Shenzhen, China, Aug.2015

[4] Kunitaka Ashizawa, et al., "Efficient Singlecast / Multicast Method For Active Optical Access Network Using PLZT High-speed Optical Switches," High Performance Switching and Routing (HPSR), pp14-19,Richardson, TX, Jun.2010

[5] SDN, https://www.opennetworking.org/index.php

[6] GUROBI, http://www.gurobi.com/

WA1-1 (Invited)

OECC/PS2016

Tunable Optical Technologies for Next Generation Optical Access System

Kota Asaka

NTT Access Network Service Systems Laboratories
asaka.kota@lab.ntt.co.jp

Abstract: Tunable optical components for next generation passive optical network stage2 (NG-PON2) and their uses are reviewed in terms of system requirements. Our developed transceiver provides high-speed wavelength switching of less than 200 ns.
Keywords: Tunable laser, Tunable photodetector, NG-PON2, TWDM-PON

I. INTRODUCTION

Although FTTH-based broadband services are found throughout the world, there is still a growing demand for a transmission capacity exceeding 10 Gb/s. In addition, network operators consider multiple-service accommodation realized by using existing optical access network systems to be very attractive in terms of CAPEX reduction. Responding to the above requirements, ITU-T completed specifications for the next generation passive optical network stage2 (NG-PON2) as the G.989 series in 2015 [1-5]. NG-PON2 offers a large bandwidth of 40 Gb/s by using time and wavelength division multiplexing (TWDM) – PON as shown in Fig.1. In addition, it can accommodate multiple services by using flexible wavelength allocation. As regards system cost, a colorless ONU is needed for NG-PON2 [3, 4]. Otherwise the inventory cost of fixed-wavelength ONUs and the operational cost generated by improper connection might be excessive. Therefore, colorless ONU transmitters and receivers must both offer tunability.

This paper shows our recent demonstration to review why NG-PON2 requires tunable ONUs and describes how tunability works in NG-PON2.

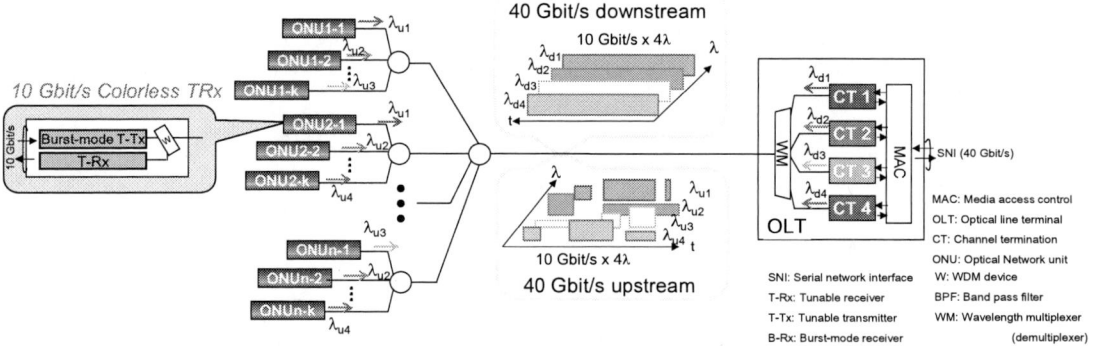

Fig. 1. Example of NG-PON2 system architecture.

II. USES OF TUNABLE COMPONENTS IN NG-PON2

A. Discovery process for newly connected ONUs

The wavelengths of ONUs must be tuned when they are first connected to a network. This procedure is specified as "discovery" in G.989.3 [5], and it requires the four downstream channels to be swept to obtain a discovery gate from an OLT because the newly connected ONU does not know which channel contains the gate. As soon as the ONU identifies the gate at a certain downstream channel, it starts to send upstream signals through the assigned upstream wavelength channel by the OLT after the discovery process. Therefore, the ONU also has to tune upstream wavelengths. A long (sub s) tuning time can be accepted in the ONUs since wavelength tuning is only required for initialization.

B. Dynamic wavelength allocation using in-service tuning

In addition to the discovery process, a wavelength tuning function in ONUs can make NG-PON2 a much more attractive system [6]. Fig. 2 shows the expected uses of in-service wavelength tuning; (a) power saving in OLTs (b) channel termination (CT) protection. As shown in Fig. 2(a), when there is less traffic, all the ONUs are connected to one (or a few) CT(s) and the other CTs can be forced to sleep. This results in power saving at the OLT. On the other hand, if the traffic becomes congested, the OLT turns on the inactive CT and assigns an ONU to re-tune to the corresponding wavelength. In Fig. 2(b), even if a CT has failed, the ONU can continue communication via other CTs by

676

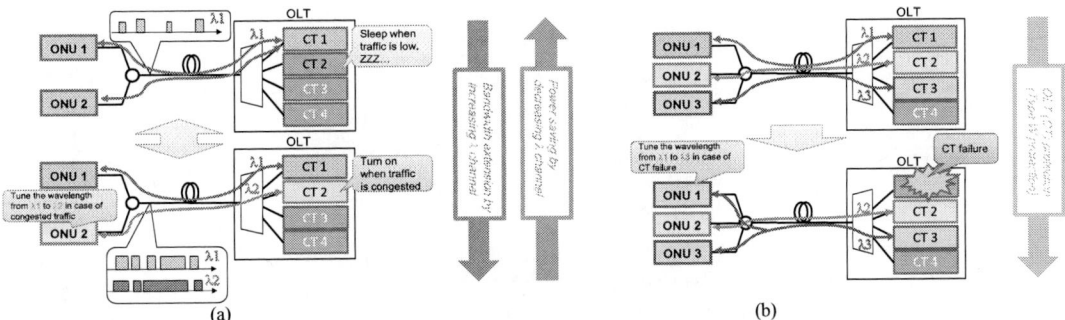

Fig. 2. Expected uses of in-service wavelength tuning. (a) Power saving (b) Protection.

changing its wavelength. This leads to high system reliability without any extra equipment, which will be attractive for business and mobile services. However, the protection function obviously needs tunable components with short (ns to ms) tuning times.

C. Requirements for tunable components

The use of wavelength tuning described above implies that an ONU should preferably be capable of a short tuning time of less than 10 ms. Moreover, ONUs in a TWDM-PON must have a 4-ch tunable transmitter that emits at a wavelength in the 1524-1544 nm range with a channel spacing of 50-200 GHz. The ONUs are also required to have a 4-ch tunable photodetector to receive downstream signals at 1596.34 nm (λ_{d1}), 1597.19 nm (λ_{d2}), 1598.04 nm (λ_{d3}), and 1598.89 nm (λ_{d4}) as specified in ITU-T G.989.2. Since a PON system requires a large loss budget of 29 dB or more, the minimum output power and the minimum received optical power for N1 class ONUs are specified as +4 and -26.5 dBm, respectively.

III. DEVELOPED PROTOTYPE TRANSCEIVER

Fig. 3 shows our newly developed transceiver (TRx) for ONUs [7]. The TRx consists of a 4-ch EML TOSA, a 4-ch APD/TIA ROSA, high-speed selector switches (SWs), and control circuits with an external wavelength band (C/L) filter. Tunability at a transmitter and a receiver is achieved by using a configuration of arrayed optoelectronics devices with a selector SW and a multiplexer/demultiplexer. Thanks to the configuration, we can expect stable high-speed wavelength tuning without mode hopping. The 4-ch EML TOSA mainly consists of four 10 Gbit/s EML chips and a power combiner. Each EML has a different emission wavelength with a 100-GHz channel spacing in the C-band. Moreover, to achieve a high output power, a booster SOA is integrated in each EML and the SOA is also used as an optical shutter to obtain burst-mode upstream signals. A100 GHz-spacing AWG and four 10 Gbit/s APD/ITA chips are packaged in the 4-ch APD/TIA ROSA. Thanks to the use of advanced integration technologies, the TOSA/ROSA package size is as small as that of the OSAs used for CFP2/4 TRx.

IV. EXPERIMENTAL RESULTS

The measured characteristics of the developed ONU transceiver are summarized in Fig. 4. As shown in Fig. 4(a), we confirmed the emission spectra at wavelengths of 1532.68 nm (λ_{u1}), 1533.47 nm (λ_{u2}), 1534.25 nm (λ_{u3}), and 1535.04 nm (λ_{u4}), which comply well with the example grids in ITU-T G.989.2. Fig. 4(b) shows a typical eye pattern of a transmission waveform obtained with our TRx. A clear eye opening and a high extinction ration of > 7 dB were achieved. Furthermore, high output powers of >+7.8 dBm were achieved for all wavelength channels thanks to the integrated booster SOA in each EML. Fig. 4(c) shows the back-to-back bit-error-rate (BER) characteristics of the receiver at the input of the C/L filter. The low-loss AWG resulted in a good received optical power of <-29 dBm for all channels (at a BER of 10^{-3}). Fig. 4(d) and 4(e) show high-speed wavelength switching waveforms of the transmitter and the receiver, respectively. The combination of a high-speed SW and 4-ch arrayed device technology resulted in very fast wavelength switching of less than 200 ns.

V. CONCLUSIONS

We have reviewed the uses of tunability in NG-PON2, and demonstrated our developed tunable optical components technologies. Our developed transceivers exhibited very fast wavelength switching of less than 200 ns in the transmitter and the receiver. Moreover, our prototype had characteristics that comply with the main specifications of G.989.2. Note that transceivers with a long tuning time could also be attractive if an operator requires wavelength tuning in an ONU solely for the discovery process.

OECC/PS2016

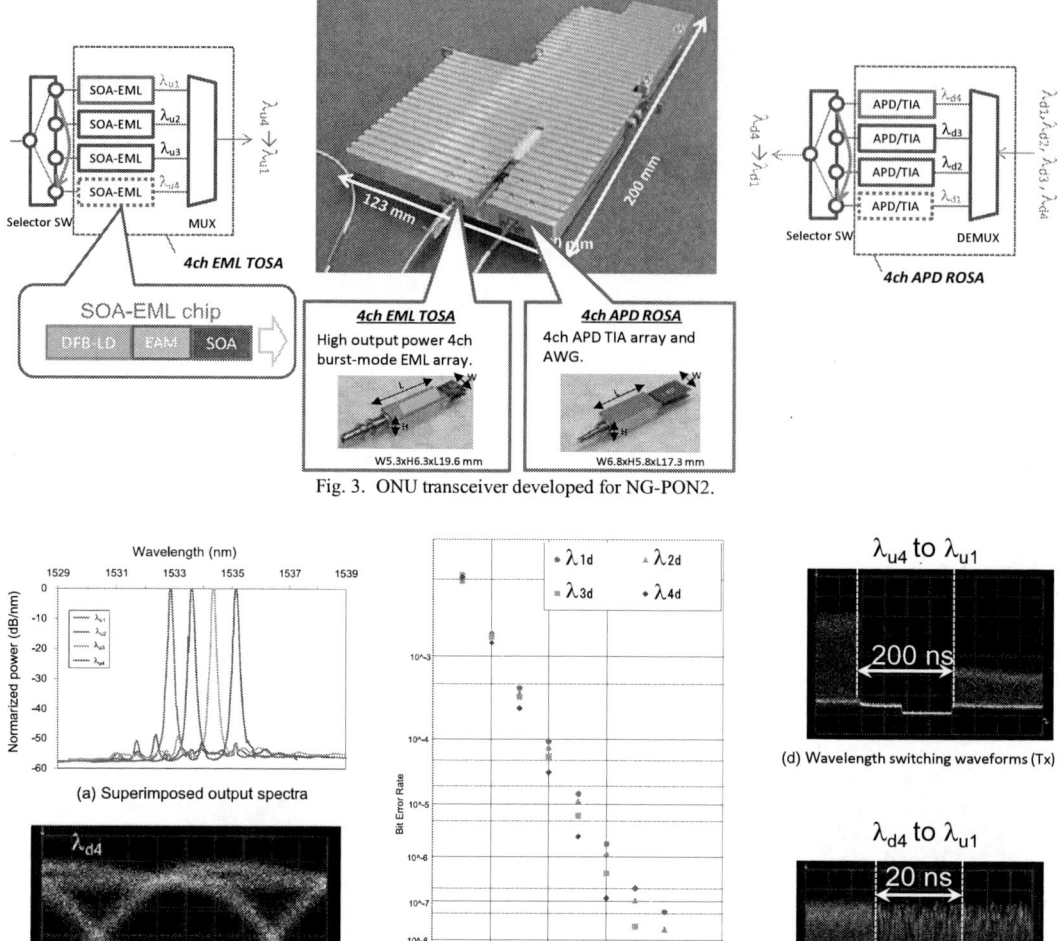

Fig. 3. ONU transceiver developed for NG-PON2.

Fig. 4. Measured characteristics of developed ONU transceiver.

ACKNOWLEDGMENTS

The author would like to thank Dr. Hirotaka Otaka, Dr. Shunji Kimura, Dr. Ken-Ichi Suzuki, Dr. Katsuhisa Taguchi, and Mr. Yuki Sakaue of NTT Laboratories for their continuous encouragement and discussions throughout this work. Part of this work was supported by the Ministry of Internal Affairs and Communications of Japan.

REFERENCES

[1] Y. Luo, F. Effenberger, X. Yan, G. Peng, Y. Qian, and Y. Ma, "Time and wavelength-division multiplexed passive optical network (TWDM-PON) for next-generation passive optical network stage 2 (NG-PON2)", J. Lightw. Technol., Vol. 31, no. 4, pp. 587-593, 2013.

[2] D. Nesset, "NG-PON2 technology and standards", J. Lightw. Technol., Vol. 33, no. 5, pp. 1136-1143, 2015.

[3] Recommendation ITU-T G.989.1 (Mar., 2013).

[4] Recommendation ITU-T G.989.2 (Dec., 2014).

[5] Recommendation ITU-T G.989.3 (Oct., 2015).

[6] K. Asaka, "Consideration of tunable components for next-generation passive optical network stage 2 (NG-PON2)", J. Lightw. Technol., Vol. 33, no. 5, pp. 1072-1076, 2015.

[7] K. Asaka, K. Taguchi, Y. Sakaue, K-I. Suzuki, S. Kimura, and A. Otaka, "High output power OLT/ONU transceivers for 40 Gbit/s symmetric-rate NG-PON2 systems", Proc. of the 41st European Conference on Optical Communication (ECOC 2015), Valencia (Spain), Mo.4.4.1, 2015.

678

WA1-2

OECC/PS2016

Symmetric 4x25-Gbit/s TWDM-PON Transmission by Using Spectrum Reshaping

Yuan-Yuan Sung[1], Yu-Chang Liu[1], Puspa Devi Pukhrambam[1], Hung-Wen Hung[2], and San-Liang Lee[1]

1.Department of Electronic Engineering, National Taiwan University of Science and Technology, No.43, Sec. 4, Keelung Rd., Taipei 106, Taiwan

2.Department of Electronic Engineering Tungnan University, NO.152, SEC. 3, BEISHEN RD., SHENKENG DIST., NEW TAIPEI CITY 222, TAIWAN (R.O.C.)

Author e-mail address:sllee@mail.ntust.edu.tw

Abstract: Spectral reshaping is used to both enhance the data transmission with insufficient laser bandwidth and provide wavelength-locking mechanism. 100-Gb/s over 20km SMF is demonstrated both experimentally and by simulation using four 10-Gb/s direct modulation lasers.

Keywords: Spectrum Reshaping, Fabry-Perot (FP) Etalon, Time- and Wavelength Division Multiplexed Passive Optical Network (TWDM-PON)

I. INTRODUCTION

Time and wavelength division multiplexed passive optical network (TWDM-PON) has been adopted for next-generation optical access by upgrading the total capacity from 10 Gb/s to 40 Gb/s with four wavelength channels [1]. To further boost the capacity, more wavelength channels and/or larger data rate per channel is needed. Directly modulated lasers (DMLs) are known to have serious dispersion penalty due to modulation induced frequency chirp, so external optical modulators and/or advanced modulation formats are often used in high-capacity links [2-3]. However, using external modulators might suffer from the reduced power budget and need optical amplification to meet the requirement of next-generation PONs. This will increase the system complexity and cost.

We demonstrate here the transmission of 25-G/s data using DMLs with limited bandwidth. The DMLs are originally used for transmitting 10-Gb/s data and have insufficient bandwidth for 25-Gb/s applications. An optical reshaping filter is used to enhance the transmission performance. This approach allows the transmission of high-speed data with low-cost transmitters. The transmission of 100-Gb/s (4x 25-G/s) TWDM-PON signals over 20km SMF is verified with VPI simulation and experiments. The same reshaping filter is also used for locking and stabilizing the four laser wavelengths.

II. SYSTEM ARCHITECTURE AND OPERATION PRINCIPLE

The proposed architecture is similar to one for standard 4-channel TWDM-PON system but with symmetric 100-Gb/s up-and down-stream capacity [1]. Fig. 1 shows the system architecture for the Optical Line Terminal (OLT) and Optical Network Unit (ONU). The OLT includes four transceivers with different wavelengths of 100-GHz channel spacing to carry four 25-Gb/s data, which will be simulated with pseudorandom binary sequence (PRBS) data in the experiments. Each ONU uses one 10-Gb/s DML laser to transmit 25-Gb/s upstream data. It is desired to realize the high data rate with low-cost solutions. We investigate here the feasibility of using a 10-Gb/s DML to transmit 25-Gb/s NRZ data. The laser has insufficient bandwidth and suffers from chirp-limited transmission distance. These two issues of using the DML is resolved with an optical spectral reshaping filter. Thus, a Fabry-Perot (FP) etalon as the reshaping filter is placed in the OLT to reshape the signal spectrum for both downstream and upstream transmission. This etalon can also be used to lock the wavelength of OLT and ONU DMLs. When the wavelength of DML shifts, the difference between transmission and reflection power of filter will be detected and the output wavelength of DML is adjusted by a temperature controller. A Tunable Optical Filter (TOF) is used at the ONU to select the assigned downlink channel.

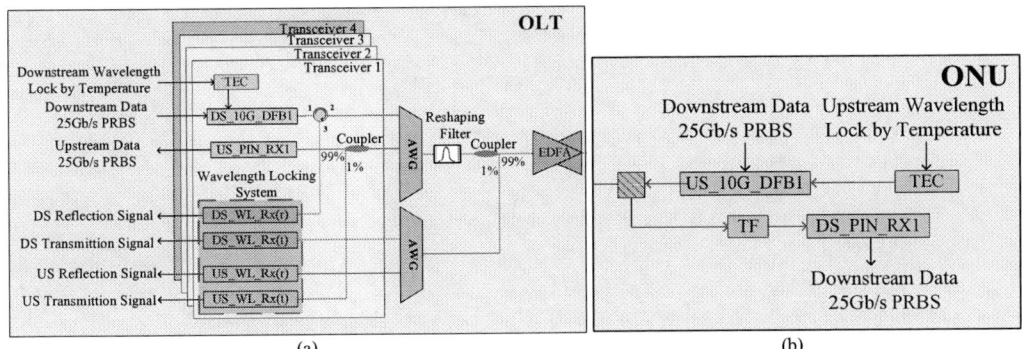

Figure 1 Architecture of (a) OLT, (b) ONU for the proposed 100Gb/s TWDN-PON.

The reshaping filter can provide three functions. Firstly, it allows modulation of the DML with smaller extinction ratio (ER) to reduce chirp and then enhance the ER with the slope of its filter response [4]. Secondly, it can induce an extra phase shift for neighboring bits to reduce the inter-symbol interference (ISI) by properly control the frequency chirp [5]. Thirdly, it can also improve the modulation bandwidth like an equalizer by using the frequency to amplitude conversion. Therefore, it is possible to over-modulate a 10-Gb/s DML to carry 25-Gb/s data.

III. SIMULATION AND EXPERIMENT DEMONSTRATION

The simulated and experimental setup for TWDM-PON system is similar to that shown in Fig.1. After the reshaping filter, the downstream signals are amplified by a bidirectional optical amplifier (OA) and sent via the feeder fiber. Four receivers are included in the OLT for monitoring the transmission and reflection power form the reshaping filter. During the simulation, the parameters for the DML and spectral filter are adjusted to match with the experimental device characteristics. The laser emits light at 193.5386THz of optical frequency with a linewidth enhancement factor of 2, linear gain coefficient of 3.3×10^{-20} m^2, transparent carrier density of 1.5×10^{18} cm^{-3}, and carrier lifetime of 2.1 ns. It has an active region with dimension of 2x0.13x250 μm^3. The etalon filter is designed to have a free spectral range (FSR) of 100 GHz and variable mirror transmission (MT). The devices used in the experiment include a reshaping filter (Micron Optics fiber Fabry-Perot tunable filter) with a FSR of 100 GHz and finesse of 8.33, and the DMLs made by NEL (NLK5C5EBKA). For simulation and experiments, the laser wavelength is aligned to the spectral response of the reshaping filter at which the slope is 0.53 and 0.58 dB/GHz, respectively. The output frequency of the DML is red-shifted from the peak frequency of the filter by 11 GHz. The laser is biased to have about 3.4-dBm optical power and 2.3-dB ER at the laser output. After reshaping the ER is increased to 5.8 dB.

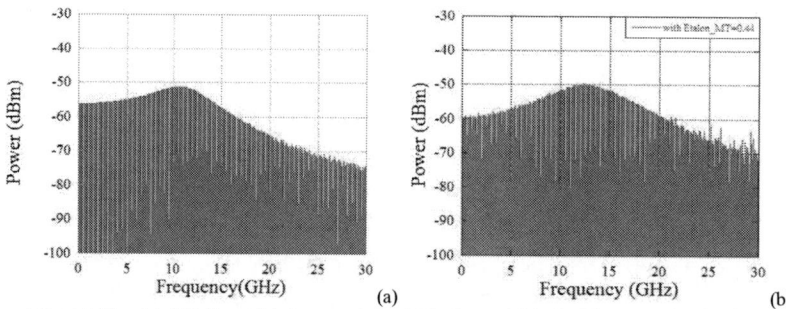

Figure 2 Simulated DML modulation spectrum (a) before and (b) after the spectral reshaping.

Fig. 2 compares the simulated modulation frequency response of the DML before and after the spectral reshaping with an etalon of 0.44 MT value. The 3-dB bandwidth is increased from 16 to 23 GHz by spectral reshaping through frequency to intensity conversion. The simulated and measured bit error rates (BER) for the downstream transmission over different fiber lengths are shown in Fig. 3(a) and (b), respectively. The lowest BER is found at around 20~30 km of SMF transmission due to the interaction between laser chirp fiber dispersion, and spectral reshaping. The SMF length with minimal BER depends on the laser/filter frequency shift and etalon parameter. When the link span is increased beyond 20-km, the ISI effect becomes serious and the BER degrades gradually. The simulated trend of BER variation matches well with the experimental one.

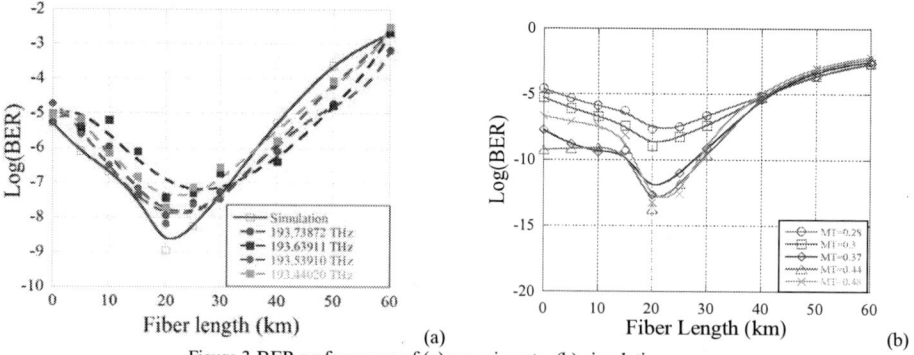

Figure 3 BER performance of (a) experiments, (b) simulation.

The BER performance of the downstream (DS) and upstream (US) transmission after 20 km SMF fiber is shown in Fig. 4(a). The DS BER is lower than the US one by two orders of magnitude. The receiver sensitivity is -13 dBm for 10^{-3} BER, which corresponds to 10^{-9} with forward error correcting (FEC) codes. Fig. 4(b) shows the BER variation with the wavelength drift of DMLs. With FEC, the tolerance on wavelength drift of the DS and US transmission is around 28 and 24 GHz, respectively. The eye diagrams for the DS and US transmission are shown in Fig.4(c) and (d). The BER is more sensitive to the wavelength shift to the higher frequency side due to the decrease in ER.

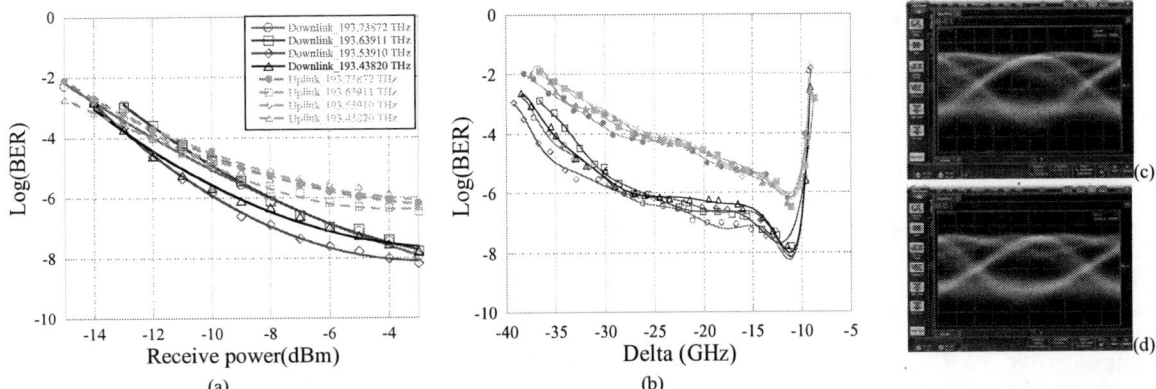

Figure 4 BER of 25 Gb/s over 20 km SMF for (a) DS and US, (b) the case with wavelength drift of the DML, and eye diagrams for (c) DS and (d) US.

The wavelength locking mechanism can also be implemented by using the same optical reshaping filter. The experimental setup is shown in Fig. 5(a). We can detect the variation of the transmitted and reflected optical power of the filter when laser wavelength drifts; and the results are shown as Fig 6(b) for the four wavelength channels. The ratio of the transmitted power over the reflected power is denoted as P in the plot. The best BER occurs at P = -6.3 dB. The laser wavelengths can be locked at the optimal positions based on the detected P ratios.

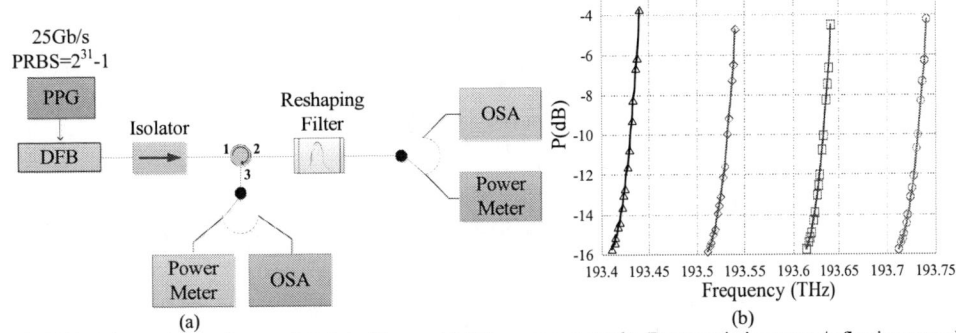

Figure 5 (a) Experimental setup for wavelength locking, and (b) Measurement results (P=transmission power/reflection power).

IV. CONCLUSIONS

We demonstrated a TWDM-PON system with symmetric 100-Gb/s (4 x 25Gb/s) transmission capacity over 20-km SMF by using DMLs with insufficient modulation bandwidth. Spectral reshaping by using a low-cost and compact FP etalon can enhance the high-frequency response like an equalizer and also mitigate the chirp-induced dispersion penalty. The reshaping filter can also be utilized to simultaneously lock the wavelength of DMLs by detecting the transmitted and reflected power from the reshaping filter. Furthermore, all the DMLs in the ONU and OLT can share the same FP etalon as the reshaping filter with our proposed scheme.

ACKNOWLEDGMENT

This work was supported in part by the National Science Council, Taiwan, under Grant 103-2221-E-011-058-

REFERENCES

[1] Y. Luo, *etal.*, "Time- and Wavelength-Division Multiplexed Passive Optical Network (TWDM-PON) for Next-Generation PON Stage 2 (NG-PON2)," Journal of Lightwave Technology, vol. 31, no. 4, pp. 587-593, 2013.

[2] J.-P. Wu, *etal.*, "40 Gb/s optical modulation using monolithically chain integration of semiconductor optical amplifiers (SOA) and electroabsorption modulators (EAM)," 22nd LEOS Annual Meeting Conference Proceedings, Belek-Antalya, Turkey, pp. 432-433, 2009.

[3] M. C. Cheng, *etal.*, "Direct QAM-OFDM Encoding of an L-band Master-to-Slave Injection-Locked WRC-FPLD Pair for 28 x 20 Gb/s DWDM-PON Transmission," Journal of Lightwave Technology, vol. 32, pp. 2981-2988, 2014.

[4] Z. Zhou; *etal.*, "Experimental Demonstration of Symmetric 100-Gb/s DML-Based TWDM-PON System", IEEE Photon. Technol. Lett., , . pp.470 -473, vol. 27, no. 5, 2015

[5] D. Mahgerefteh, *etal.*, "Chirp Managed Laser and Applications," IEEE J. Selected Topics in Quantum Electron., vol. 16, no. 5, 2010

WA1-3

OECC/PS2016

Auto-Configuration Optical Amplifier for Flexible Access Network Design

Takuya Tsutsumi, Yu Nakayama, Shunsuke Kanai, Manabu Kubota and Akihiro Otaka

NTT Access Service Systems Laboratories

Email: tsutsumi.takuya@lab.ntt.co.jp

Abstract: *We propose an auto-configuration optical amplifier (AC-OA) for passive optical networks; it can flexibly operate as either a relay or Central-Office-set OA by a controller synchronized with an external database.*

Keywords: *Passive optical network, Optical amplifier, Auto configuration, Field measurement*

INTRODUCTION

The passive optical network (PON) is widely applied to create cost-effective optical access networks that share a single optical line terminal (OLT) among many optical network units (ONUs). While Ethernet PON (EPON) or Gigabit PON (GPON) are being commercially deployed, 10 Gb/s EPON (10G-EPON), which was standardized in 2009 [1], is considered to be a strong candidate for the upcoming next generation PON as it can satisfy the explosive growth in Internet traffic. One of the main advantages of 10G-EPON is its cost-efficiency triggered by its interoperability with conventional 1 Gb/s-class PONs and large capacity ONUs. To further enhance its cost-efficiency, many institutes and organizations have been targeting the acceptable PON link budget by applying optical amplifiers (OAs) [2]. Increasing the link budget reduces operational expense (OPEX) or/and capital expense (CAPEX) by extending reachable distance or/and increasing the splitting ratio of the system.

TABLE I
COMPARISON OF OPTICAL AMPLIFIER ARCHITECTURES.

Item	Relay OA	CO-set OA
Setting location	Relay office	Central office
Extending reachable distance	OK	NG
Increasing splitting ratio	OK	OK
Eliminating buildings	NG	OK
Trunk distance (D_t)	> 0 km	= 0 km
Access distance (D_a)	> 0 km	> 0 km

OAs fall into two classes according to their architecture: relay OA and CO-set OA, see TABLE I and Fig. 1. Relay OAs, which are set in offices between central office (CO) and subscribers, are advantageous for extending the reachable distance and increasing the splitting ratio. However, they cannot reduce the number of COs since OAs have to be set at certain relay offices as shown in Fig. 1a. On the other hand, CO-set OAs, which are set in the CO, can increase the splitting ratio and reduce the number of offices. Unfortunately, they cannot extend the reachable distance as amplification levels are insufficient, see Fig. 1b.

The OLT-OA span is defined as the trunk distance, while the OA-ONU span is called the access distance. The CO-set OA architecture is essentially equivalent to the relay OA with D_t of 0. However, a fixed attenuator (ATT) is manually inserted for controlling the optical input power to suit the OA used. For instance, the CO-set OA needs large input ATT values for downstream transmission to avoid overloads since D_t is 0 (Output ATT is controlled by the auto-level-control (ALC) function). On the other hand, the input ATT of relay OA must be optimized so that the OA offers optimal operation while preventing overloads in the cases of small splitting ratios/short optical transmission lines. So far, relay and CO-set OAs are separately handled with different setting procedures. It complicates network management increase installation cost against an original purpose of OA for reducing the cost despite of the equivalence of architectures between relay and CO-set OA.

In this paper, we propose an auto-configuration OA (AC-OA) that simplifies network operation by operating as both relay OA and CO-set OA. AC-OA has an input VOA and a centralized calculator synchronized with an integrated database (DB) as shown in Fig. 1c. The calculator determines the input VOA value based on DB information and returns the value to OA controller. The mechanism realizes plug and play OA operation and reduces installation cost as it simplifies the control of the input VOA. The concept described in this paper can also be applied to future access networks such as the Next generation-PON stage 2 (NG-PON2) or video distribution systems.

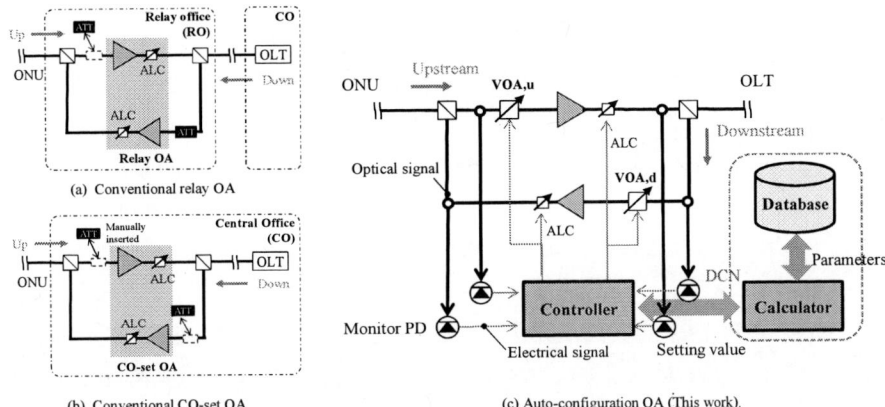

Fig.1. Architectures of optical amplifier (OA).

AUTO-CONFIGURATION OPTICAL AMPLIFIER

The AC-OA controller is connected to the calculator by a data communication network (DCN) for network management by the communication carrier as shown in Fig. 1c. First, the calculator extracts the required parameters from DB in order to decide the VOA value. The calculator sends the determined VOA value to the AC-OA controller. Finally, the controller sets VOA value in order to prevent optical overloading. DB parameters are shown in Fig. 2 and TABLE II. Actually, the parameters of optical fibers connected to the AC-OA are given by statistical analysis of measurement results in advance. The optical specifications or characteristics such as S, T, P are given in advance for each kind of ONU i. Actual DB also includes the affiliation information indicating how CO, OLT and ONU are connected each other.

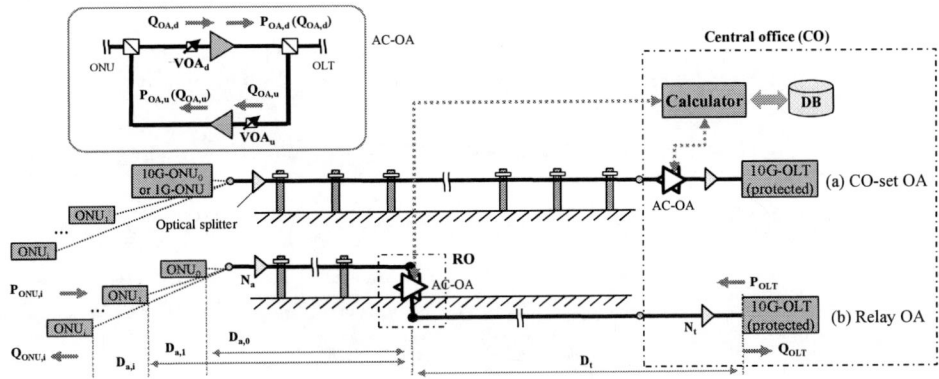

Fig.2. Explanation of infrastructure parameters.

TABLE II
COMPARISON OF OPTICAL AMPLIFIER ARCHITECTURES.

Parameter	Definition
D_t [km]	The distance of trunk span.
$D_{a,i}$ [km]	The distance of access span (Index i is identifier of ONUs).
N_t, N_a	The optical splitting ratio of trunk and access span.
α_d, α_u [dB/km]	Optical transmission loss per kilometer for downstream and upstream, respectively.
β_t, β_a [dB]	Miscellaneous optical loss by connectors and fusion points and so on.
$S_{OLT}, S_{ONU,i}, S_{OA,d}, S_{OA,d}$ [dBm]	Required minimum input power to OLT, ONU, OA for downstream and OA for upstream, respectively.
$T_{OLT}, T_{ONU,i}, T_{OA,d}, T_{OA,d}$ [dBm]	Damage threshold of OLT, ONU, OA for downstream and OA for upstream, respectively.
$P_{OLT}, P_{ONU,i}$ [dBm]	Optical output power of OLT and ONU$_i$, respectively.
$P_{OA,d}(Q), P_{OA,u}(Q)$ [dBm]	Optical output power of OA for input power Q.

The input powers of Q have to satisfy following condition in order to satisfy the minimum sensitivity and avoid overloading of the transceiver or OA in all ONU.

$$S_{OLT} < Q_{OLT} < T_{OLT} \quad (1) \qquad S_{ONU} < Q_{ONU} < T_{ONU} \quad (2)$$
$$S_{OA,d} < Q_{OA,d} < T_{OA,d} \quad (3) \qquad S_{OA,u} < Q_{OA,u} < T_{OA,u} \quad (4)$$

Input powers of Q are determined by the following formulae. VOA_d and VOA_u are determined so as to satisfy (1) - (4) for preventing overloading.

$$Q_{OA,d} = P_{OLT} - \{\alpha_d D_t + 10 \log_{10} N_t + \beta_t + VOA_d\} \tag{5}$$

$$Q_{ONU,i} = P_{OA,d}(Q_{OA,d}) - \{\alpha_d D_{a,i} + 10 \log_{10} N_a + \beta_{a,i}\} \tag{6}$$

$$Q_{OA,u,i} = P_{ONU,i} - \{\alpha_u D_{a,i} + 10 \log_{10} N_a + \beta_{a,i} + VOA_u\} \tag{7}$$

$$Q_{OLT,i} = P_{OA,u}(Q_{OA,u,i}) - \{\alpha_u D_t + 10 \log_{10} N_t + \beta_t\} \tag{8}$$

After setting VOA, OLT and ONU initiate communication via the Multi-point-control protocol (MPCP). These automated mechanisms allow AC-OA to operate as both relay and CO-set OA while avoiding overloading.

SIMULATION OF AUTO-CONFIGURATION OPTICAL AMPLIFIER

A. Condition

In this paper, we use the results (α and β) of an actual field measurement [2] to confirm the feasibility of AC-OA. TABLE III shows the contents of the infrastructure DB. In CO-set OA, D_t is assumed to be 0. In this table, though Relay 0 and CO-set 0 were not actually measured, their parameter values were assumed with the goal of preventing upstream overloading (Relay 0 and CO-set 0 are respectively calculated from Relay I and CO-set IV by increasing the upstream power of 3 dB). The simulation assumptions are described follows. The 10G-OLT implements N:1 protection [2] and conforms to the specifications of PR30. 10G-ONU and 1G-ONU conform to the specifications of PR30 and PX20, respectively. OA is a semiconductor-based OA (SOA), and implements the ALC function to control output VOA. The optical gain in the downstream signal is around 15 dB (10G) / 25 dB (1G) and in the upstream signal has optical gain of over 30 dB (10G/1G). Maximum and minimum levels of optical power to AC-OA are -10 and -30 dBm, respectively so optical signal quality is not degraded by non-linear distortion effect in the SOA. Optimized attenuation value, which satisfies formulae (1)-(8), is calculated by a genetic algorithm with solution size and mutation rate of 100 and 7.5 %, respectively.

B. Results

Figure 3 shows upstream and downstream input powers measured on the commercial infrastructure. With conventional OA, AC-OA is overloaded on downstream or/and upstream input power in most cases as shown in Fig. 3a. Figure 3b shows the result after optimization of input VOA. In all cases, input powers are lower than overload level (-10 dBm) and exceed the required level (-30 dBm). The optimization time is between 0.1 and 3.0 seconds on a general purpose PC (CPU: Core i5 2Cores 2.4GHz, RAM: 2GB). When input VOA does not have to be controlled, the calculation is finished instantly. Moreover, the reachable distance after setting input VOA is not decreased and corresponds to the distance described in TABLE III.

TABLE III
COMPARISON OF OPTICAL AMPLIFIER ARCHITECTURES.

Infra.	N_t	D_t [km]	N_a	D_a [km]
(Relay0)	2	20	8	0.5, 5, 10, 18.5
Relay I	2	20	16	0.5, 5, 10, 18.5
Relay I	1	20	64	0.5, 5, 10, 18.5
Relay I	1	20	32	0.5, 5, 10, 18.5
Relay I	2	20	64	0.5, 5, 10, 18.5
(CO-set 0)	2	0	8	0.5, 5, 10, 18.5
CO-set I	8	0	16	0.5, 5, 10, 18.5
CO-set II	4	0	16	0.5, 5, 10, 18.5
CO-set III	4	0	32	0.5, 5, 10, 18.5
CO-set IV	2	0	32	0.5, 5, 10, 18.5
CO-set V	4	0	8	20.5, 25, 30
CO-set VI	2	0	16	20.5, 25, 30

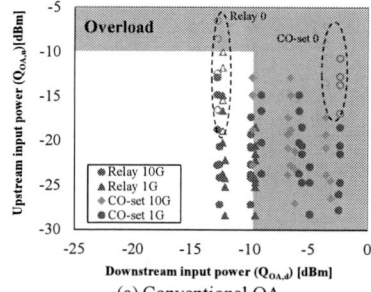

(a) Conventional OA.

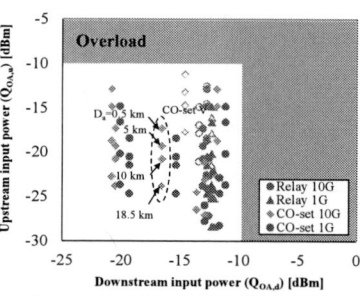

(b) AC-OA with calculated input VOA.

Fig.3. Upstream and downstream input power measured on commercial infrastructure.

CONCLUSIONS

For raising PON efficiency, we proposed AC-OA, which can flexibly operate as either relay OA or CO-set OA; it simplifies the network operation by merging two kinds of OA into one device. AC-OA has an input VOA and a controller that calculates the optimized value of input VOA according to the entries in an infrastructure DB. Also, we confirmed AC-OA operation by using the field measurement results from a commercial access network. The concept described in this paper can also be applied to future access networks such as Next-generation-PON stage 2 (NG-PON2), as well as video distribution systems.

REFERENCES

[1] IEEE Std 802.3-2012, "*10Gb/s Ethernet Passive Optical Network*," 2012.

[2] T. Tsutsumi, T. Sakamoto, Y. Sakai, T. Fujiwara, H. Ou, Y. Kimura, and K.-I. Suzuki, "Long-reach and high-splitting-ratio 10g-epon system with semiconductor optical amplifier and N:1 OSU protection," Journal of Lightwave Technology, vol. 33, no. 8, pp. 1660-1665, 2015.

WA1-4

OECC/PS2016

Integrated SOI-based Receiver Module for TWDM-PON

Y. Hsu[1], X. R. Wu[2], L. Y. Wei[1], K. Xu[3], C. W. Hsu[1], J. Y. Sung[1], C. H. Yeh[4], H. K. Tsang[2], and C. W. Chow[1]

[1]Department of Photonics and Institute of Electro-Optical Engineering, National Chiao Tung University, Hsinchu, 30010, Taiwan
[2]Department of Electronic Engineering, The Chinese University of Hong Kong, Hong Kong
[3]Department of Electronic and Information Engineering, Harbin Institute of Technology, Shenzhen, China
[4]Department of Photonics, Feng Chia University, Taichung 40724, Taiwan
Author e-mail address: matrtihewl0937@gmail.com

Abstract: We demonstrate a SOI-based integrated receiver (Rx) module for the TWDM-PON. The downstream can be de-multiplexed by a tunable de-multiplexer using cascaded silicon-micro-ring and each channel is received by germanium-on-silicon photodiode.
Keywords: Integrated Receiver Module; TWDM-PON; SOI

I. INTRODUCTION

As the bandwidth demand is increasing very rapidly, the bandwidth provided by conventional time-division-multiplexed passive optical networks (TDM-PON) is not enough. As a result, the second stage next-generation PON (NG-PON2) [1] has been proposed. Among these architectures, time-wavelength-division-multiplexed (TWDM) PON [2] is regarded as a promising architecture for the NG-PON2 since it can reuse the existing optical distribution network (ODN) of the PON. Besides, orthogonal frequency division multiplexed (OFDM) PON [3] has been proposed and demonstrated for the future access since OFDM has a high spectral efficient and a strong tolerance to fiber chromatic dispersion. Silicon photonics has become an attractive technology to enable the miniaturization of photonic integrated circuits down to micrometer length scale, with submicron cross-sections for individual waveguides. It can be potential low-cost and compatible with complementary metal-oxide-semiconductor (CMOS) processes. Hence, it is suitable for the cost-sensitive optical networking units (ONUs) for the future access networks [4].

In this work, we propose and demonstrate a semiconductor-on-insulator (SOI)-based integrated receiver (Rx) module for the TWDM-PON. The downstream signals can be first de-multiplexed by a tunable de-multiplexer (Demux) using a cascaded silicon micro-ring optical filters. Then each downstream channel can be received by the germanium on silicon (Ge-Si) photodiode (PD). The experimental data rate of the Rx module can be 4×16 Gbit/s = 64 Gbit/s by using OFDM signal; and 4×25 Gbit/s = 100 Gbit/s by using non-return-to-zero (NRZ) signal, satisfying the forward error correction (FEC) limit.

II. ARCHITECTURE AND INTEGRATED RX MODULE DESIGN

Figs. 1(a) show the architectures of the proposed TWDM-PON using the integrated SOI-based Rx module. To simplify the figure, only one optical line terminal (OLT) and one ONU are shown. Inside the OLT, 4 sets of transmitter (Tx) modules are multiplexed via a tunable multiplexer (MUX) and sent to the ONU. In our experiment, the downstream signal is in OFDM and NRZ formats produced by a Mach-Zehnder modulator (MZM). At the ONU, the downstream signals are de-multiplexed by a tunable de-mux using a cascaded silicon micro-ring optical filter. Then each downstream channel can be received by the Ge-Si PD for bit-error-rate (BER) analysis.

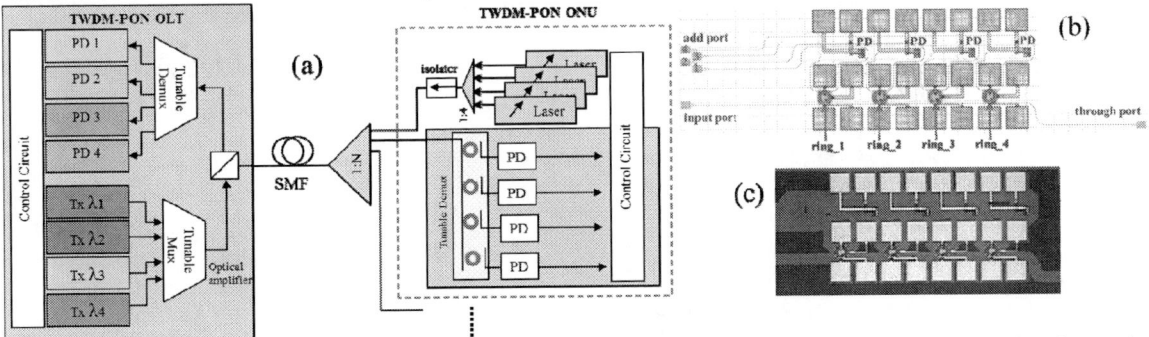

Fig. 1. (a) Architectures of the proposed TWDM-PON using the integrated SOI-based Rx module, (b) design layout and (c) top-view photograph of the Rx module.

685

The cascaded silicon micro-ring includes an integrated four silicon micro-rings and 4 Ge-Si PDs at the drop-ports. The design layout and the top-view photograph of the Rx module is show in Fig. 1(b) and 1(c) respectively. The device was fabricated on SOI substrate with 220 nm top silicon and 2 μm buried oxide. The grating coupler and the waveguide were defined by UV photolithography and followed by anisotropic dry etching. The radius of each silicon micro-ring is 20 μm, with free-spectral range (FSR) of 4.8 nm, Q-value of ~ 5500. Each micro-ring is lightly doped to form p-n junction, so that each can be thermally tunable with high speed. The thermal tuner is a p-type resistor near the coupler region in order not to sacrifice the useful modulation length. Grating coupler is used to couple the downstream signal to the Rx module. As shown in Fig. 1(b), the downstream signal is launched at the "input-port" at the lower left corner. Then the four wavelength channels will be de-multiplexed by the silicon micro-rings, and each channel is finally detected by the Ge-Si PD. The "add-port" shown in the upper left corner in Fig. 1(b) can be used to launch the upstream signals and sending back to the OLT. The "through-port" shown in the lower right corner in Fig. 1(b) is not used in this experiment.

III. EXPERIMENTAL RESULTS AND DISCUSSIONS

We first characterize of the cascaded silicon micro-ring using a C-band erbium-doped fiber amplifier (EDFA) and detected by an optical spectrum analyzer (OSA, Agilent 86142B) with resolution 0.06 nm. The characteristic of the drop port of a silicon micro-ring is shown by Fig. 2, by thermal tuning the p-n junction of the micro-ring with voltages from 0 V to 7.5 V, we can observe the wavelength is increased by ~ 2.3 nm. Hence the thermal tuning efficiency is ~ 0.31 nm/V. In Fig. 2(a), we can observe the FSR is 4.8 nm. Fig. 2(b) is the drop-ports of the four micro-rings in the cascaded silicon micro-ring module, by applying different thermal tuning voltages to the four micro-rings, their drop-port spectra will not overlap; hence the downstream signals can be successfully de-multiplexed. It is also worth to point out that the cascaded micro-ring is scalable, for example, two cascaded silicon micro-ring modules can be connected to de-multiplex 8 wavelength channels with arbitrary wavelengths.

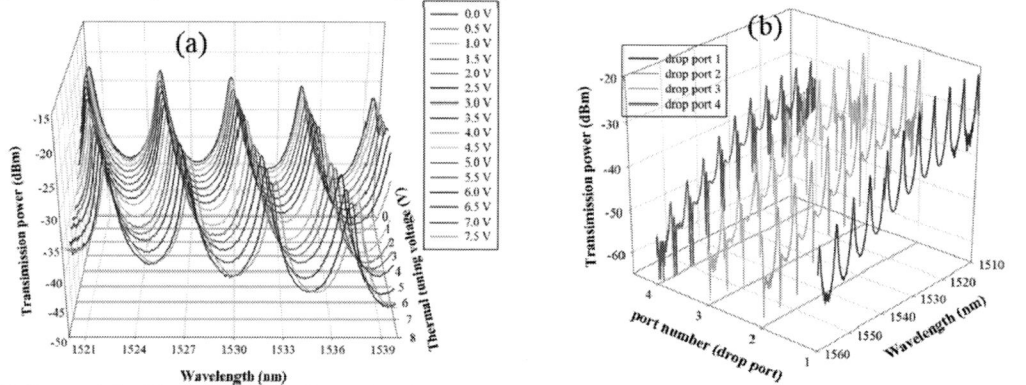

Fig. 2. (a) Characteristic of the drop-port of a silicon micro-ring. (b) The drop-port spectra of the 4 silicon micro-rings in the cascaded micro-ring module.

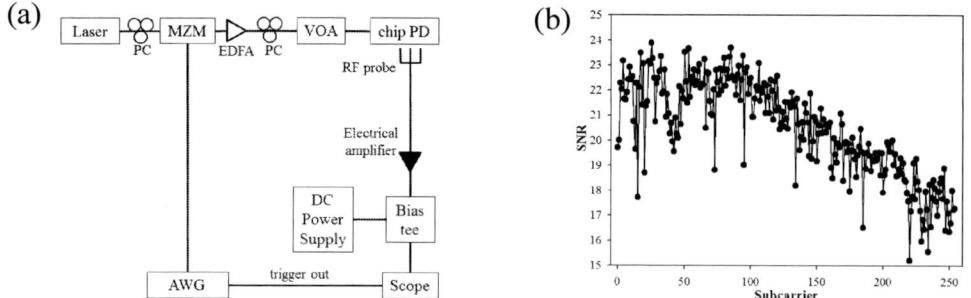

Fig. 3. (a) Experimental setup for the OFDM transmission system. (b) The SNR of the 255 OFDM subcarriers.

Then, we evaluate the performance of the Ge-Si PD. First, an optical OFDM signal is launched into the Ge-Si PD, and the experimental setup is shown in Fig. 3(a). As shown in Fig. 3(a), a continuous-wave optical signal produced by a distributed feedback (DFB) laser is launched into a MZM, which is driven by an electrical baseband OFDM signal generated by an arbitrary waveform generator (AWG) (Tektronix AWG 7082C). The AWG has a sampling rate of 8 GSample/s and analog bandwidth of 3.5 GHz. The optical OFDM signal is then launched into the Ge-Si PD via a variable optical attenuator (VOA) and grating coupler. After the optical-to-electrical (OE) conversion by the Ge-Si PD, a high speed radio-frequency (RF) probe is used to retrieve the detected OFDM signal and send to the real-time oscilloscope (RTO) (Tektronix DPO 7354C). The RTO has an analog bandwidth of 3.5 GHz and sample rate of 40

GSample/s. A bias tee is used to apply a negative direct-current (DC) base to the Ge-Si PD and retrieve the received electrical data signal. Fig. 3(b) shows the measured signal-to-noise (SNR) of the 255 OFDM subcarriers. Based on the received SNR, bit-loading technique can be used to further enhance the bit-rate. Hence, higher level of modulation, such as 64-quadrature amplitude modulation (QAM) can be applied for the higher SNR subcarrier; while 16-QAM can be applied for the lower SNR subcarrier. The bit-loading of different OFDM subcarriers are shown in Fig. 4(a). By using the bit-loading technique, the bit-rate of 16 Gbit/s can be achieved and satisfying the FEC limit (BER = 3.8×10^{-3}) in a single Ge-Si PD, as shown in the measured BER performance of the optical OFDM signal by the Ge-Si PD in Fig. 4(b). Hence the total supported data rate of the integrated Rx module is 4×16 Gbit/s (= 64 Gbit/s). The signal quality can be improved by minimizing the insertion loss of the grating coupler which will enhance the optical SNR.

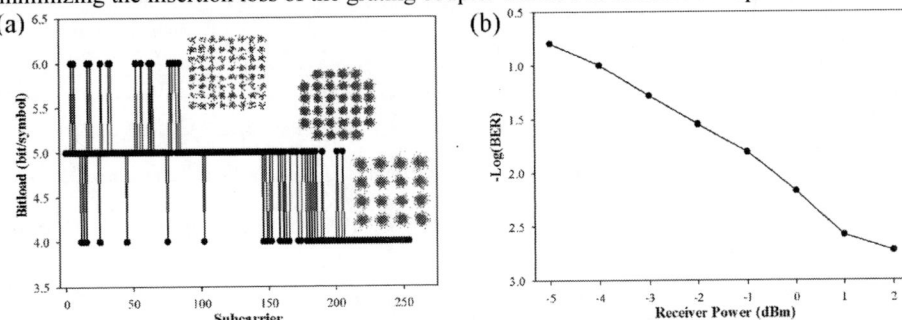

Fig. 4. (a) Bit-loading used in different OFDM subcarriers, with the corresponding constellation diagrams. (b) Measured BER performances of the optical OFDM by the Ge-Si PD.

As shown in Fig. 3(b), there is a drop of the SNR at higher frequencies. This is mainly due to the limited analog bandwidth of the AWG and RTO available in the laboratory, and not by the Ge-Si PD. Hence another experiment similar to Fig. 3(a) is performed. Instead of sending the optical OFDM signal, we launch optical NRZ signal to the Ge-Si PD at 15 Gbit/s and 25 Gbit/s respectively. FEC limit can be satisfied at both cases with open eye-diagrams as shown in Fig. 5(a) and 5(b) respectively. Hence the total supported data rate of the integrated Rx module is 4×25 Gbit/s (= 100 Gbit/s). As OFDM signal is more spectral-efficient than the NRZ signal, and 16 Gbit/s data only occupy 3.5 GHz analog bandwidth. Each proposed Ge-Si PD may support at least 50 Gbit/s OFDM data if higher bandwidth AWG and RTO are available in the laboratory; hence the total data rate of the Rx module could be about 4×50 Gbit/s = 200 Gbit/s.

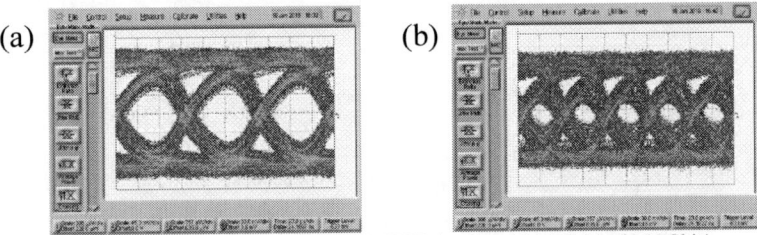

Fig. 5. Measured eye diagram of the NRZ signals at (a) 15 Gbit/s, (b) 25 Gbit/s

IV. CONCLUSIONS

We proposed and demonstrated a SOI-based integrated Rx module for the TWDM-PON. The downstream signals can be first de-multiplexed by a tunable de-mux using a cascaded silicon micro-ring optical filter and each downstream channel was received by a Ge-Si PD. The experimental data rate of the Rx module was 4×16 Gbit/s = 64 Gbit/s by using OFDM signal; and 4×25 Gbit/s = 100 Gbit/s by using NRZ signal.

ACKNOWLEDGMENT

This work was supported by the Taiwan Ministry of Education, Taiwan Ministry of Science and Technology under Grant MOST-104-2628-E-009-011-MY3, Taiwan Aim for the Top University Plan, and by Hong Kong ITF grant ITS/097/14.

REFERENCES

[1] R. W. Heron and E. Harstead, "FSAN NG-PON2 updates," in Proc. OFC 2013, paper OW4D.5.
[2] Y. Luo, X. Zhou, F. Effenberger, X. Yan, G. Peng, Y. Qian, and Y. Ma, "Time- and wavelength-division multiplexed passive optical network (TWDM-PON) for next-generation PON stage 2 (NG-PON2)," J. Lightw. Technol., vol. 31, pp. 587-593, 2013.
[3] N. Cvijetic, "OFDM in optical access networks," in Proc. OFC 2011, paper OMG3.
[4] A. Rickman, "The commercialization of silicon photonics," Nat. Photon., vol. 8, pp. 579–582, 2014.

1Gbps Full-Duplex 5GHz Frequency Slots uDWDM Flexible Metro/Access Networks Based on VCSEL-RSOA Transceiver

Jose A. Altabas[1,*], David Izquierdo[1,2], Jose A. Lazaro[3], and Ignacio Garces[1]

[1]Grupo de Tecnologías Fotónicas (GTF), Aragon Institute of Engineering Research (I3A), Universidad de Zaragoza, Mariano Esquillor ed. I+D+i, Zaragoza, 50018, Spain
[2]Centro Universitario de la Defensa (CUD), Academia General Militar, Carretera de Huesca s/n, Zaragoza, 50090.
[3]Grupo de Comunicaciones Ópticas (GCO), Universitat Politècnica de Catalunya, Jordi Girona 31, Barcelona, 08034, Spain
*jaltabas@unizar.es

Abstract: *1Gbps full-duplex 5GHz frequency slot is proposed for uDWDM Flexible Metro-Access Networks. The cost-effective ONU transceiver is based on VCSEL as LO for coherent reception and seed for phase-modulated RSOA, providing 40.5dB of power budget.*
Keywords: *Coherent communications; uDWDM; Phase modulation; Vertical Cavity Surface Emitting Lasers*

I. INTRODUCTION

New multimedia and cloud services, the deployment of Internet of Things (IoT) and the convergence between optical and wireless communications at the 5G paradigm [1] are pushing up the traffic demand at metro and access optical networks. These networks are converging and evolving to all-optical high-capacity flexible networks as the one shown in Fig. 1. Flexible ultra Dense Wavelength Division Multiplexing (uDWDM) metro-access networks using coherent techniques are being proposed as an efficient solution for this growing demand [2]. These networks require cost-effective devices, including coherent transceivers, for multiplexing several users inside a DWDM channel [3].

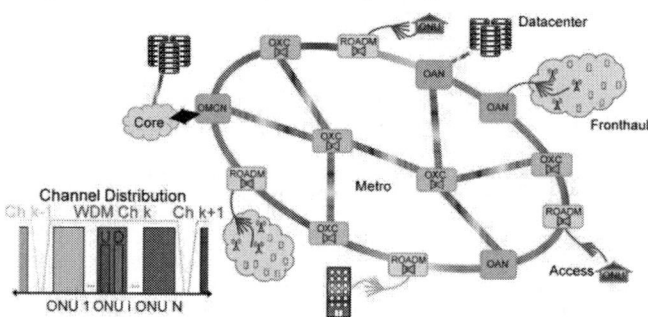

Fig. 1. Flexible 5G Metro-Access Network scenario. OMCN: Optical Metro-Core Node, OAN: Optical Aggregation Node, ROADM: Reconfigurable Optical Add-Drop Node, OXC: Optical Cross-connect, ONU: Optical Network Unit. Inlet: proposed flexible uDWDM full-duplex channel distribution.

In this paper, it is presented a full-duplex 1Gbps flexible uDWDM channel for its use in Passive Optical Networks (PON) implemented in scenarios as the one shown in Fig. 1. This channel is based on a novel cost-effective transceiver used at the Optical Network Unit (ONU) consisting on a Vertical-Cavity Surface-Emitting Laser (VCSEL) and a phase-modulated Reflective Semiconductor Optical Amplifier (RSOA) [3]. The VCSEL is used both as a seed source for the RSOA, and as Local Oscillator (LO) [4] for a single-detector heterodyne receiver. In addition, the downlink generated at the Optical Aggregation Node (OAN) uses an externally modulated laser with Nyquist shaped Differential Binary Phase Shift Keying (DPSK) for addressing the spectral efficiency required on the flexible uDWDM metro-access PON. We will show that the combination of cost-effective devices and spectrally efficient modulation formats allows a full-duplex 1Gbps communication in a 5GHz frequency slot, achieving a distribution of 10 user channels inside a 50GHz WDM ITU-T grid.

II. SETUP

The experimental setup is shown in Fig. 2. The ONU transceiver is based on a single low-cost free-running VCSEL. The transmitter uses half of the optical power generated by the VCSEL to seed a RSOA which is directly phase-modulated with a Non Return to Zero (NRZ) DPSK signal. The RSOA phase modulation is achieved by means of its own frequency chirp. The transceiver uses a 2:1 coupler and an isolator to inject the VCSEL seeding light into the RSOA and to couple the resulting modulated signal to the output fiber. The single-detector heterodyne coherent receiver at the ONU uses as LO part of the VCSEL output power (P_{LO}=-5.5dBm). This receiver can be easily upgraded to an independent polarization solution [5]. The input and output at the ONU transceiver are separated using a 50/50 coupler.

The OAN emitter is a 100kHz linewidth external cavity Tunable Laser Source (TLS) which is tuned 2GHz apart from the VCSEL central wavelength. The optical power from the TLS is split to use it as light source for the OAN transmitter and as LO for the coherent receiver (P_{LO} = 0dBm). The downlink is generated by a Mach Zehnder Modulator (MZM) biased at the null transmission point using a Nyquist-shaped DPSK modulation format. The receiver is based on the same single-detector heterodyne configuration used in the ONU. Uplink and downlink at the OAN transceiver port are separated using a circulator. The OAN transmitter can be cost-effectively implemented using a unique external modulator to generate the downlink of several users as shown in [6].

The optical power outputs at both transceivers (ONU and OAN) are -3dBm. The PON has been implemented using 50Km of Standard Single Mode Fiber (SSMF) and a 1:16 distribution splitter to share out the data to the ONUs.

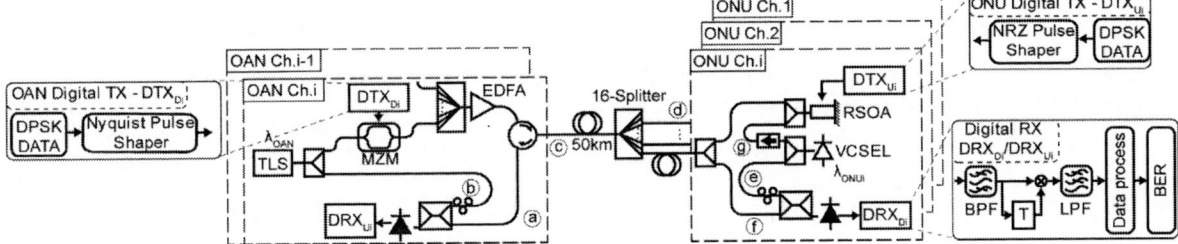

Fig. 2. Experimental setup for the evaluation of the optical link. P_{RX} at (a) and (e) points, P_{LO} at the (b) and (f) points, P_{TX} at (c) and (d) points, P_S at (g) point. (h) VCSEL spectrum

The Digital Transmitters (DTX$_{Di}$/DTX$_{Ui}$) are implemented using Mathworks Matlab™ and digitally generated with a 12 GS/s Arbitrary Waveform Generator. The data at both transmitters is encoded differentially to simplify the reception side and reduce the phase noise at the coherent receivers. The OAN DTX$_{Di}$ is based on a raised cosine pulse shaper with 12-symbols length and zero roll-off factor and the ONU DTX$_{Ui}$ encodes the signal with NRZ pulses. The received signals are digitalized at the Digital Receivers (DRX$_{Di}$/DRX$_{Ui}$) with a 40GS/s Real Time Oscilloscope with an electrical bandwidth of 2.5GHz. The digitalized received signals are bandpass filtered with a central frequency (f_c) of 2GHz and bandwidth (BW) of 1.5GHz at the ONU and f_c=2.35GHz and BW=2.3GHz at the OAN in order to reduce the noise [7]. Then, the filtered signals are demodulated multiplying them by the previous symbol and filtered with 1.25GHz cut-off frequency lowpass filters.

III. RESULTS

The sensitivity with bidirectional transmission, which is the minimum received power to ensure a BER of $2.2\cdot10^{-3}$ without FEC and 10^{-12} with a 7% overhead FEC [8], is -45dBm for the uplink and -43.5dBm for the downlink, as can be seen in Fig. 3. The resulting power budget for the uplink is 42dB and 40.5dB for the downlink. These power budgets may allow to reach distances higher than 100Km with a 1:16 distribution splitter, or higher than 50Km with splitting ratios greater than 1:128.

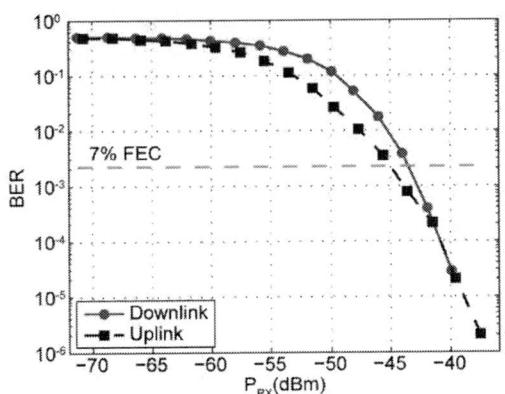
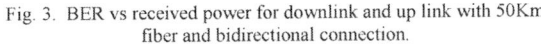

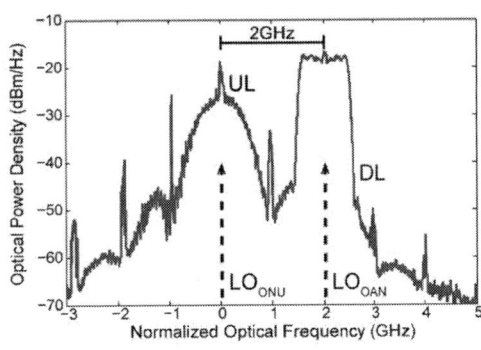

Fig. 3. BER vs received power for downlink and up link with 50Km fiber and bidirectional connection.

Fig. 4. Optical spectrum for both links of a single user channel and LO position of each link. The central frequency corresponds to the central emission wavelength of the VCSEL (1539.8nm).

The 1Gbps symmetrical channel optical spectrum is shown in Fig. 4 and it was obtained with a High Resolution Complex Optical Spectrum Analyzer (HRCOSA) from Aragon Photonics Labs which also was used for optimizing the RSOA phase modulation. On the center is the RSOA phased modulated uplink (UL) signal and on the right, the Nyquist shaped downlink (DL) signal. The separation between links (UL and DL) at the channel is 2GHz due to the wavelength reuse at the ONU.

The backscattering interference of an adjacent channel downlink with the main channel uplink at the OAN restricts the adjacent downlink central wavelength. As shown in Fig. 5, OAN BER curve, this central wavelength cannot be placed in the ranges from -1GHz to 1.5GHz and from 2.5GHz to 5GHz relative to the VCSEL wavelength. The adjacent channel downlink also interferences with the main channel downlink at the ONU and, as is shown in ONU BER curve at Fig. 5, it cannot be placed either in the range from -2.5GHz to -0.5GHz and from 1.5GHz to 2.5GHz. Thus, the adjacent channel downlink cannot be placed from -2.5GHz to 5GHz relative to the VCSEL wavelength.

The adjacent channel uplink allocation has not to interfere with the main channel uplink at the OAN, and, as is shown in Fig.6, it cannot be placed in the range of -2GHz to 4.5GHz. The backscattering of the adjacent channels uplinks will not interfere with the main channel downlink at the ONU because of the distribution splitter attenuation [7].

Thus, taking into account these measurements, the adjacent channel (Ch.i-1) uplink can be placed at -3GHz and the adjacent channel (Ch.i+1) downlink at 5GHz, when the main channel (Ch.i) downlink is placed at 0GHz and the uplink at 2GHz. These channels distribution results in a channel separation or frequency slot as small as 5GHz, allowing the allocation of 10 user channels inside a 50GHz WDM channel.

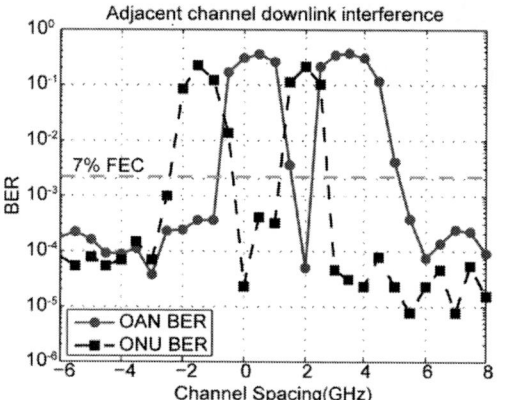

Fig. 5. BER degradation due to the adjacent channel downlink spectral position. Zero frequency corresponds to the VCSEL wavelength

Fig. 6. BER degradation due to the adjacent channel uplink spectral position. Zero frequency corresponds to the VCSEL wavelength

IV. CONCLUSIONS

The performance of cost-effective 1Gbps full-duplex links in uDWDM flexible 5G Metro-Access Networks is shown. The 1Gbps downlink is based on Nyquist-DPSK over a MZM. The 1Gbps uplink consists of NRZ-DPSK directly phase modulated RSOA seeded with a VCSEL. Both receivers are based on single photodiode heterodyne detection with a LO (VCSEL in ONU and TLS in OAN). The power budget is around 40.5dB for bidirectional uDWDM channels for the proposed cost-effective RSOA-VCSEL transceiver and compatible with a record minimum frequency slot of 5GHz. Thus, 10 users can be allocated inside a 50GHz ITU channels and the flexible 5G Metro-Access Network could reach distances greater than 100Km with 1:16 splitter or greater than 50Km with 1:128 splitter.

ACKNOWLEDGMENT

This work was supported in part by the DGA under grant T25, the Spanish MINECO projects muCORE (TEC2013-46917-C2-2-R) and SUNSET (TEC2014-59583-C2-1-R) co-funded by FEDER, Centro Universitario de la Defensa project CUD2013-05: SIRENA and FPU grant from MECD to the first author (FPU 13/00620).

REFERENCES

[1] M. Fiorani, et al., "Challenges for 5G transport networks", in *Proc. ANTS*, New Delhi. India, Dec. 2014, pp. 1-6.
[2] H. Rohde, et al., "Coherent Ultra Dense WDM technology for Next Generation Optical Metro and Access Networks", J. Lightw. Technol., vol. 32, no. 10, pp. 2041-2052, 2014
[3] J. A. Altabas, et al., "Cost-effective Transceiver based on a RSOA and a VCSEL for Flexible uDWDM Networks", Photonics Technol. Lett. Feb. 2016, accepted
[4] J. B. Jensen, et al., "VCSEL Based Coherent PONs", J. Lightw. Technol., vol. 32, no. 8, pp. 1423-1433, 2014
[5] B. Glance, "Polarization independent coherent optical receiver", *J. Lightw. Technol.*, vol. 5, no. 2, pp. 274-276, 1987.
[6] C. Kottke, et al., "Coherent UDWDM PON with joint subcarrier reception at OLT", *Opt. Express*, vol.22, no. 14, pp.16876-16888, 2014.
[7] J. A. Altabas, et al., "1Gbps full-duplex links for ultra-dense-WDM 6.25GHz frequency slots in optical metro-access networks" Opt. Express, vol. 24, no.1, 555-565, 2016
[8] ITU-T Recommendation, G.975.1 (2004).

WA2-1

OECC/PS2016

Survivable Multipath Provisioning with Content Connectivity in Elastic Optical Datacenter Networks

Tao Gao, Shanguo Huang, Bingli Guo, Xin Li, Qian Kong, Yu Zhou, Wenzhe Li, and Wanyi Gu

State Key Laboratory of Information Photonics and Optical Communications,
Beijing University of Posts and Telecommunications, Beijing 100876, China
shghuang@bupt.edu.cn; ghsshou@bupt.edu.cn

Abstract: We propose a survivable multipath routing and spectrum allocation with content connectivity scheme in OFDM-based elastic datacenter networks. This scheme achieves better performance in resource utilization compared with traditional protection schemes.

Keywords: Elastic optical datacenter networks; Content connectivity; Multipath provisioning; Routing and spectrum allocation

I. INTRODUCTION

The risk of link failure in datacenter networks is rising, which may cause enormous loss of data and revenue. Traditional protection methods may obviously deteriorate the performance of bandwidth utilization in networks especially in the case that traffic load is heavy. Hence, how to increase the bandwidth utilization and provide better service performance in networks is important. Technologies such as orthogonal frequency-division multiplexing (OFDM) [1] and distance adaptive modulation level [2] provide a new way to reduce spectrum consumption. Based on this, the authors in [3] proposed several algorithms to solve the Routing, Modulation level and Spectrum Allocation (RMLSA) problem efficiently. As opposed to traditional path protection that results in completing service restoration, partial protection can guarantee essential service is provisioned when any link fails [4]. In [5], the authors proposed a multipath provisioning (MP) scheme supporting full and partial protection with higher efficiency than traditional protection schemes in OFDM-based networks.

On the other hand, requests tend to be point-to-content, which can be served by any datacenter storing the same application service information, rather than traditional point-to-point requests served by establishing connection from source node to destination node [6]. The content connectivity can guarantee that the application service is provisioned when any link or datacenter breaks down [7], which improves the survivability greatly. On this basis, how to decrease the spectrum consumption in datacenter networks is worth being investigated further.

In this paper, we focus on the survivability of requests with content connectivity (CC) in elastic optical datacenter networks through multipath provisioning (MPC). For a request with a certain protection level, content connectivity is adopted to calculate multiple link-disjointed paths to different datacenters so that the content is still connected when a failure occurs on any path because traffic carried on the other paths is not affected. To minimize the utilized spectrum, distance adaptive modulation level is adopted in spectrum allocation. We employ binary shifted keying (BPSK), quadrature phase-shifted keying (QPSK), 8QAM and 16QAM for different paths with distinct distance respectively. An ILP formulation as well as an efficient heuristic algorithm is proposed to investigate this problem. Extensive simulations demonstrate that the proposed survivable MPC performs better than traditional protection methods in terms of spectrum utilization.

II. SURVIVABLE MULTIPATH PROVISIONING WITH CONTENT CONNECTIVITY

We represent a request as $r = <s, C, q>$, where s denotes the source node; C denotes the capacity requirement in Gbps and $q(0 \leq q \leq 1)$ denotes the flexible protection level requirement. When a link failure occurs, qC capacity must be available. To provision a connection request for content, we first compute $K \geq 2$ link-disjointed paths and then allocate spectrum with the goal to minimize the total utilized spectrum resources. If $K \geq 1/(1-q)$, C/K capacity is allocated on each path and $qC/(K-1)$ capacity is allocated otherwise. An example is given in Fig. 1, and the total required capacity for the request is 1.2C less than 2C in the traditional full protection (FP) method.

In spectrum allocation, distance adaptive modulation level is used to relieve the heavy traffic load around datacenters. While the modulation level of an OFDM signal becomes higher which means more bits per symbol, the transmission reach decreases. We employ the highest modulation level that the path can support within its transmission reach. The capacity of one spectrum slot is represented by C_{BPSK} in Gbps using the modulation format BPSK. Hence, the capacity of one spectrum slot for QPSK, 8QAM, 16QAM is 2, 3, 4 times of C_{BPSK} respectively. Finally, we use fist fit (FF) to assign spectrum for each computed path of the request according to the modulation level and the capacity need to be allocated.

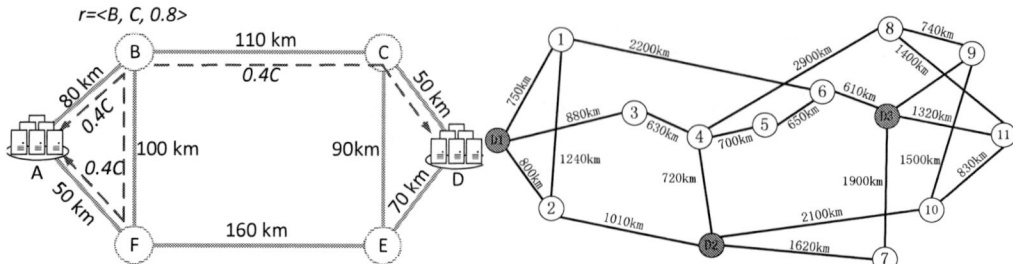

Fig. 1. Multipath provision with content connectivity   Fig. 2. 14-node NSFNET

A. Integer Linear Program (ILP)

In this part, an ILP formulation is proposed for the survivable MPC problem in elastic optical datacenter networks. In the model, we are given a datacenter network graph G (**L**, **N**, **D**) with $|\mathbf{L}|$ links, $|\mathbf{N}|$ user nodes and $|\mathbf{D}|$ datacenters and a set of requests **R**. For each request r_i, a set of candidate link- disjointed paths $\mathbf{P}_i$ ($|\mathbf{P}_i| = K$) is precomputed and for each path $p_{i,j}$, the highest modulation level supported is $m_{i,j}$. We assume that there are ϕ spectrum slots on each link and λ_G guard slots between traffic. The Boolean variables required are as follows: $x_{i,j,s}$ equals 1 if path $p_{i,j}$ uses spectrum slot s; $o_{i,j}$ equals 1 if the path $p_{i,j}$ is used; $l_{i,j,e}$ equals 1 if the path $p_{i,j}$ uses link e in **L**. The maximum index of occupied slots is represented by $MIOS$.

Objective:

$$\text{Minimize } MIOS \tag{1}$$

Subject to:

$$\sum_{j\in[1,K]}(\sum_{s\in[1,\phi]}x_{i,j,s}-2\lambda_G)o_{i,j}m_{i,j}C_{BPSK}\geq C_i \quad \forall r_i\in\mathbf{R} \tag{2}$$

$$\sum_{j\in[1,K]}^{j\neq k}(\sum_{s\in[1,\phi]}x_{i,j,s}-2\lambda_G)o_{i,j}m_{i,j}C_{BPSK}\geq q_iC_i \quad \forall r_i\in\mathbf{R},k\in[1,K] \tag{3}$$

$$o_{i,j}\geq \sum_{s\in[1,\phi]}x_{i,j,s}/\phi \quad \forall r_i\in\mathbf{R},j\in[1,K] \tag{4}$$

$$o_{i,j}\leq \sum_{s\in[1,\phi]}x_{i,j,s} \quad \forall r_i\in\mathbf{R},j\in[1,K] \tag{5}$$

$$(x_{i,j,s}-x_{i,j,s+1}-1)\bullet(-\phi)\geq \sum_{s'\in[s+2,\phi]}x_{i,j,s'} \tag{6}$$
$$\forall r_i\in\mathbf{R},j\in[1,K],s\in[1,\phi]$$

$$\sum_{r_i\in\mathbf{R}}\sum_{j\in[1,K]}x_{i,j,s}l_{i,j,e}\leq 1 \quad \forall e\in\mathbf{L},s\in[1,\phi] \tag{7}$$

$$MIOS\geq sx_{i,j,s}l_{i,j,e} \quad \forall r_i\in\mathbf{R},j\in[1,K],s\in[1,\phi],e\in\mathbf{L} \tag{8}$$

The objective (1) is to minimize the volume of spectrum utilization. Equation (2) and (3) ensure that the capacity and protection requirement of the request are satisfied, in the meantime, there are λ_G guard slots for the request at the start and end of the occupied slots. Equation (4) and (5) ensure that if any slot on a link is occupied, the path $p_{i,j}$ is selected. Equation (6) ensures that the spectrum slots allocated for a path are contiguous. Equation (7) ensures that any slot can be utilized once at most. Equation (8) denotes the maximum index of occupied spectrum slots in the networks.

B. Heuristic Algorithm

Since the ILP formulation above is for static network planning and cannot be applied for large amounts of traffic in the datacenter networks, we present an efficient heuristic algorithm to solve the MPC problem.

For better performance, the given requests are sorted in descending order so that the request with the largest capacity requirement is serviced first (LCF). Note that the number of paths used U may not be just equal to K with the goal to minimize the total utilized spectrum resources [5], U is determined as follows. First, the candidate paths are sorted in increasing order of path distance, and secondly, we calculate the gross occupied slots using first two, first three, $\cdots$, first K-1 paths and all candidate paths, that is $U=2,3,\cdots,K$-1,K. Finally, U is determined with the least occupied slots, after which we allocate contiguous spectrum slots using FF with the paths set selected and corresponding modulation level. The pseudo code is shown as follows.

Algorithm 1 MPC LCF Algorithm. Input: G(**L, N, D**), **R**, K
1.   **for** every request $r_i = <s_i, C_i, q_i>$ in **R** do
2.   Compute K paths to obtain candidate paths set $\mathbf{P}_i$ and determine the modulation level for each path
3.   Sort candidate paths in $\mathbf{P}_i$ in increasing order
4. $Total = \infty, U = 0, PathSum = 0, PathSet = \emptyset$
5.   **for** $U' = 2$ to K
6.   Calculate the capacity need to allocate on each path according to C_i, q_i, U'
7.   **for** each path $p_{i,j}$ in $\mathbf{P}_i$
8.   Calculate slots needed in $p_{i,j}$ according to $m_{i,j}$
9.   **if** slots on links along the path are available
10. $PathSum$++ , add the path to $PathSet_{U'}$
11.   **else**
12.   **break**
13.   **end if**
14.   **if** $PathSum == U'$
15.   **break**
16.   **end if**
17.   **end for**
18.   Sum up slots of each path in $PathSet_{U'}$ to get $Total'$
19.   **if** $Total' < Total$
20. $Total = Total', U = U', PathSet = PathSet_{U'}$
21.   **end if**
22.   **end for**
23.   Allocate spectrum slots using FF for paths in $PathSet$
24. **end for**

III. NUMERICAL RESULTS

We evaluate the results of our proposed ILP model and heuristic algorithm using the 14-node NSFNET topology (as shown in Fig. 2) in static traffic scenario. The requests are randomly generated offline with the capacity requirement at a range of 50-200 Gbps and protection level requirement between 0.4 and 1.0. The transmission reach for BPSK, QPSK, 8QAM and 16QAM is 10000 km, 5000 km, 2500 km and 1250 km respectively, and the data rate per slot of BPSK is 12.5Gbps. For every request scenario, we run the simulation ten times at different requests sets to get average values.

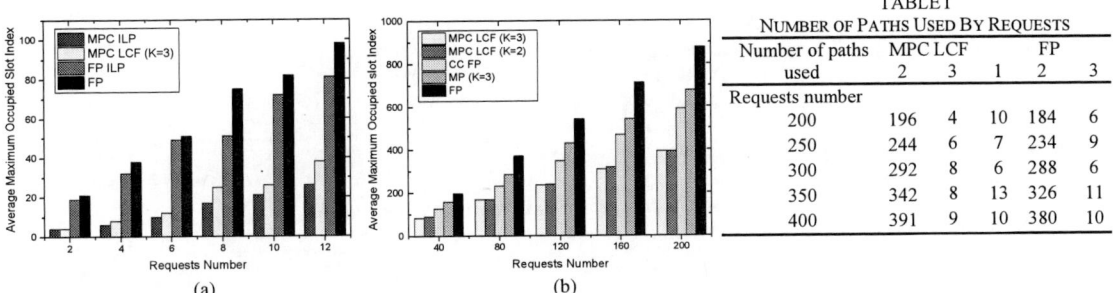

TABLE I
NUMBER OF PATHS USED BY REQUESTS

Number of paths used	MPC LCF		FP		
	2	3	1	2	3
Requests number					
200	196	4	10	184	6
250	244	6	7	234	9
300	292	8	6	288	6
350	342	8	13	326	11
400	391	9	10	380	10

Fig. 3. Average maximum occupied slot index of MPC, FP and MP.

Fig. 3 presents the results for our proposed MPC and other survivable provision schemes. Note that for both ILP model and heuristic algorithm shown in Fig. 3 (a), MPC decreases the spectrum slots consumption dramatically than traditional FP method. With the requests number increasing, the gaps between ILP model and heuristic algorithm become larger, while the result of ILP model is unavailable in reasonable time when traffic load is heavy. To check the performance in large amounts of requests, the illustrative results are shown in Fig. 3 (b). For all requests numbers, our proposed MPC LCP algorithm performs better than content connected requests provisioning with full protection (CC FP), MP scheme and FP scheme. The advantage becomes more remarkable with the increase of requests number. It is because that there are multiple shortest paths to select and modulation level is distance adaptive, which can obviously reduce the spectrum consumption. Moreover, the performance of MPC LCF is similar when the number of candidate paths is two and three since more paths mean more guard bandwidth and more occupied slots. Table I verifies our surmise that most requests use only two paths although there are more candidate paths. However, we observe that the FP scheme uses only one path in some conditions that is because in the 14-node NSFNET topology for point-to-point requests, there may be no paths link-disjointed with the shortest one such as from node 2 to datacenter D3. That means longer paths using more spectrum resources will be selected to guarantee the survivability of the request. As a contrast, our proposed scheme trades off the spectrum consumption and survivable guarantee very well.

IV. CONCLUSIONS

Survivability and spectrum resource utilization are both important issues in elastic optical datacenter networks because of the emerging high-rate applications. In this paper, we propose a survivable MPC scheme that reduces the spectrum resources consumption as well as guarantees the content connectivity. The simulation results for both ILP model and heuristic algorithm verify that our proposed scheme performs better in spectrum resource utilization comparing with traditional protection methods.

ACKNOWLEDGMENT

This work is supported in part by the NSFC under Grant 61331008, in part by the program for New Century Excellent Talents in University under Grant NCET-12-0793, in part by the Beijing Nova Program under Grant 2011065, and in part by the China Post Doctoral Science Foundation under Grant 2015M570979.

REFERENCES

[1] Q. Yang, et al., "Bit and Power Loading for Coherent Optical OFDM," IEEE Photon Technol. Lett., 20(15), 2008.
[2] H. Takara, et al., "Distance-Adaptive Super-Wavelength Routing in Elastic Optical Path Network (SLICE) with Optical OFDM," Proc. ECOC, Paper We.8.D.2, Sep. 2010
[3] K. Christodoulopoulos, et al., "Elastic Bandwidth Allocation in Flexible OFDM-based Optical Networks," J. Lightwave Technol., 29(9), 2011.
[4] A. Saleh and J. Simmons, "Evolution toward the nextgeneration core optical network," J. Lightwave Technol., 24(9), 2006.
[5] L. Ruan, et al., "Survivable multipath routing and spectrum allocation in OFDM-based flexible optical networks," 2013.
[6] X. Li, et al., " Design of K-Node (Edge) Content Connected Optical Data Center Networks," IEEE Communications Lett., vol.2,no. 3, 2016.
[7] S. Huang, et al., "Preconfigured polyhedron based protection against multi-link failure in optical mesh networks," Opt Express., vol. 22, no 3, pp. 2386-2402 2014.

WA2-10

OECC/PS2016

Distributed 4K-Video Camera Monitoring service on Ethernet PON System using Hybrid Bandwidth Allocation for secure community

Akihisa Shoji and Naoto Yoshimoto

Department of Opto-electronic system engineering, Chitose Institute of Science and Technology
758-65, Bibi, Chitose, Hokkaido, 066-8655, Japan
n-yoshi@photon.chitose.ac.jp

Abstract: *1G-Ethernet Passive Optical Network (PON) system based massively distributed 4K-video monitoring service is proposed. By using hybrid bandwidth allocation scheme, this service can be provide with low delay nevertheless co-existence with existing Internet services.*
Keywords: *Access Network Passive Optical Network, sensor network, bandwidth allocation, video monitoring, 4K*

I. INTRODUCTION

Recently, wired-wireless converged networks with sensors have been eagerly investigated to provide various cloud services using big data handling techniques. Video monitoring is a kind of attractive wireless sensors. High definition video technology such as 4K has been drastically developed and various 4K cameras are widely diffused. Since such high quality video cameras can be monitored human activities and traffic condition clearly, it will be expected to construct secure and safety community toward future IoT era. Although high definition video cameras constantly output large size of data (giga-bit class), quite efficient data compressed technique for high definition video data such as H.265/HEVC has been progressed [1]. Therefore, we can transmit data of 4K video camera compressed to moderate capacity (mega-bit class). However, to deploy optical fiber infrastructure to transmit 4K video data needs huge investment. In this paper, to overcome this issue, we propose Distributed 4K-Video camera monitoring on already deployed Ethernet PON (Passive Optical Network) system. This is some advantage. The first is cost-effectiveness. In Asia-Pacific region, 1G-Ethernet PON (1G-EPON) has massively been installed. We can easily and quickly utilize installed broadband infrastructure. The second is upgradability to 8K and uncompressed signal compared with wireless access such as Long Term Evolution (LTE), because PON system has authorized upgrade scenario toward 10G-Ethernet PON (10G-EPON) and Next-generation PON-2 (NG-PON2). The third is capability of service co-existence with current provided broadband services such as WiFi spot service. In broadband service, downstream traffic is generally large than upstream one. On the other side, in 4K-video monitoring service, upstream traffic is large than downstream one. Therefore, bandwidth can be efficiently used at not only downstream and but also upstream. Figure 1 shows an image of secure and ubiquitous community based on optical access network using widely distributed 4K video cameras. 4K video cameras are located in the office area and residential area as well as WiFi base stations (for outside Internet service), and are connected with optical fiber cable deployed for FTTH (Fiber to the Home).

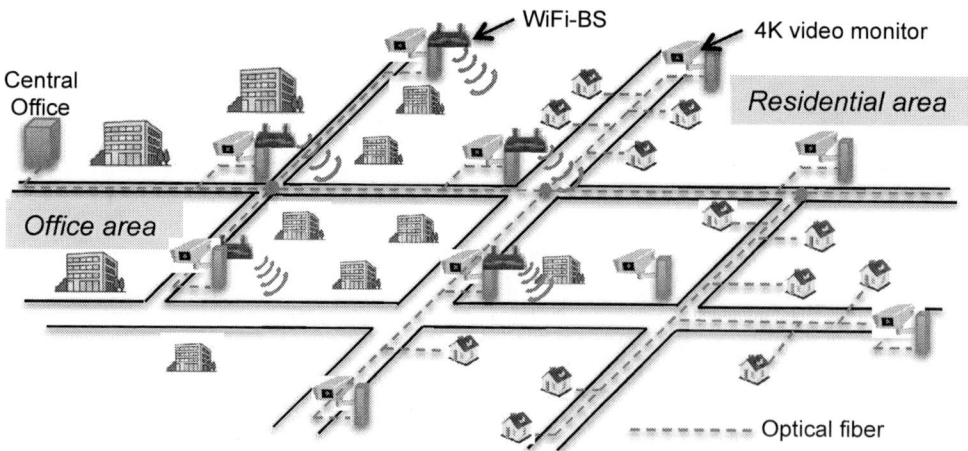

Fig.1 Secure and ubiquitous community based optical access network using widely distributed 4K video monitors and WiFi-BS

II. NETWORK CONFIGURATION

Figure 2 shows proposed 4K video monitoring and data transmission system based on 1G-EPON. Service area is up to 10 km as typical size of community and number of connected 4K video monitors is up to 32 as equal to typical splitting ratio of 1G-EPON systems. At the 4K-transmitter, compressed 4K video data is packetized and transmitted through gigabit Ethernet interface with variable frame size in accordance with traffic condition and buffering capacity. Since the data direction is mainly upstream, at the Optical Network unit (ONU), packetized 4K video data should be burst mode transmission on the Time-Division Multiplexing Access (TDMA). The burst cycle is set to 10 msec and the data load is controlled by changing the burst size flexibly as shown in the inserted right side block in Fig.2. In PON span, the upstream packets are controlled by the both fixed and dynamic bandwidth allocation technique.

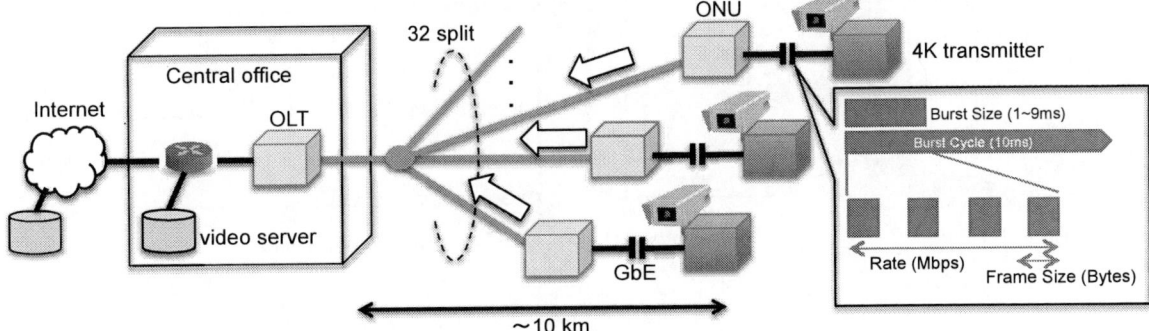

Fig.2 4K video monitoring and transmission system based on Gigabit Ethernet Passive Optical Network

III. SIMULATION

In order to confirm the feasibility of the proposed network configuration for 4K-video monitoring service, we evaluated the performance of 1G-EPON systems by using PON traffic simulation "SimOliver". The simulation parameters of 4K video monitoring service and Internet service are listed in Table 1. The queue set had two priorities; the data traffic of 4K-video monitoring service is assigned to "high priority" and that of Internet service is assigned "low priority". The bandwidth for high priority was reserved to 15 Mbps. The upstream data rate of 4K-video is set to 30 Mbps according to the typical transmission rate of compressed 4K video data by H.265. The flame size of the upstream data was fixed to 1518 Byte to easily control bandwidth allocation at the PON system for efficient transmit thorough the PON span. On the other side, the upstream and downstream data rate of Internet service was set to 1.5 Mbps and 30 Mbps, respectively, which values is according to the typical transmission rate of Internet service. The flame sizes of the both directions were varied from 64 to1518 Byte randomly.

Firstly, a case of dedicated 4K-video service (all 32 ONUs are connected to 4K-video) was simulated. Figure 3 shows the simulated result of mean packet delay as a function of network load. At this case, Fixed Bandwidth Allocation (FBA) mode is more suitable, because only 4K-video traffic continuously transmitted and the frame size was set to fixed value at the 4K-transmitter as mentioned above. As the network load increased, the mean packet delay gradually increased, and the network load exceeded to 0.8, the delay rapidly increases. As a result, the burst packet interval should be controlled as equal to the network load of less than 0.8.

Table 1 Simulation parameter of 4K monitoring and Internet service

Service	Priority	Bit rate	Frame size
4K monitoring	high	upstream 30 Mbps	upstream fixed 1518 Byte
Internet (WiFi spot)	low	upstream 1.5 Mbps downstream 20 Mbps	up/downstream random 64-1518 Byte

Secondly, a case of shared 4K-video service (32 ONUs are connected to 4K-video and the additional ONUs are connected to Internet service) was simulated for multiple service coexistence. Figure 4 shows the simulated result of mean packet delay as a function of connected Internet service users on the same PON branch. In this simulation, there are four types of bandwidth allocation scheme as listed in Fig. 4. The Priority Dynamic Bandwidth Allocation (P-DBA) means that 4K-video service is set priority and the total bandwidth is assigned by DBA mode. On the other hand, the Non Priority DBA (Non P-DBA) means that both services are fairly treated and the total bandwidth is assigned by DBA mode. P-FBA and Non P-FBA shows that the total bandwidth is assigned by FBA mode. As shown in Fig. 4, even priority based FBA, the value of the mean packet delay was high, because the traffic pattern of Internet service is assumed to be random access, not to be fixed rate access. By using DBA mode, as the number of Internet user increased, the mean packet delay was still low. Therefore, in case of both 4K-video and Internet service provisioning, DBA mode is more suitable for highly efficient traffic management. In such way, hybrid bandwidth allocation scheme by using both FBA and DBA according to various service provisioning and traffic condition is valuable for multiple service access networks in near future.

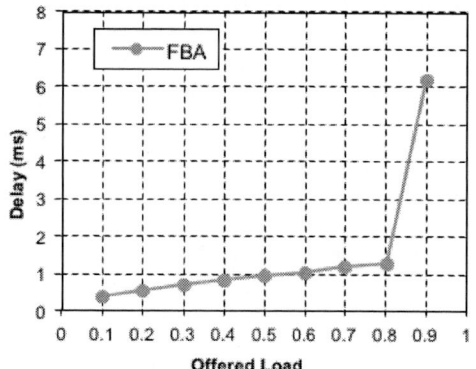

Fig.3 Simulated result of mean packet delay in case of dedicated 4K-video monitoring service

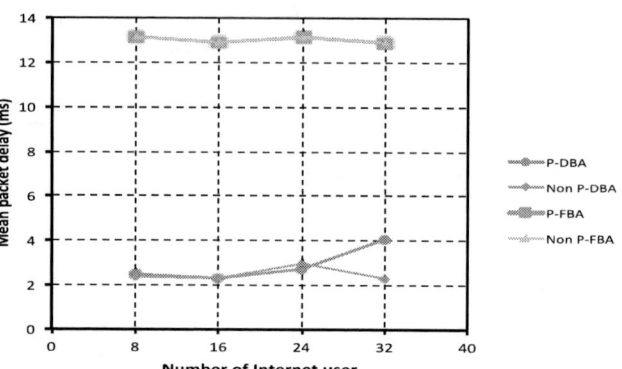

Fig.4 Simulated result of mean packet delay of 4K-video monitoring service in case of shared with Internet service as a function of number of Internet user

IV. EXPERIMENTAL RESULTS

Fig. 5 shows the experimental result of latency of 4K and Internet traffic. The traffic path of 4K and Internet were set to 16, respectively, and the traffic load was set to 0.7. The data was transmitted from the traffic generator (Spirent C1) on the commercialized 1G-EPON system with 32 splits ratio and 10 km transmission distance. The bandwidth was allocated by DBA mode in 1G-EPON OLT. As a result, the latency of both 4K and Internet service were kept low latency nevertheless traffic load increase as similar to the simulated results. Because suitable bandwidth allocation scheme, priority control, and traffic load control are valuable for multiple service provisioning with wide variety of traffic between high definition video through small data such as IoT sensing data.

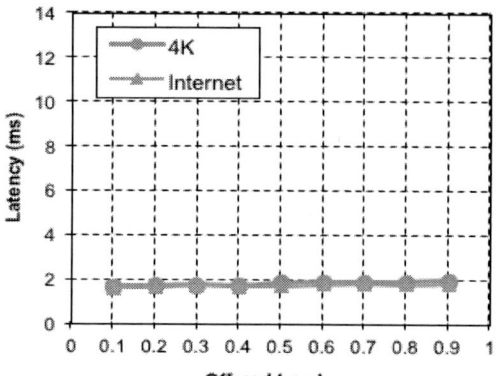

Fig.5 Experimental result of latency of 4K and Internet

V. CONCLUSIONS

We propose distributed 4K-Video Camera Monitoring service using currently existing 1G-Ethernet PON System to construct secure and safety community. The simulated and experimental result shows this service can be provided with low delay and has a good capability to co-exist current Internet services by using adequate priority control, traffic load control, and hybrid bandwidth allocation scheme.

ACKNOWLEDGMENT

The authors would like to thank Mr. Toshi Kusano of Oliver solutions Ltd and Mr. Toshiaki Mukoujima of Oki Electric Industry Co. Ltd for continuous technical support.

REFERENCES

[1] M. Ikeda, "H.265/HEVC encoder for UHDTV," 20th Asia and South Pacific Design Automation Conference (ASP-DAC2015), pp. 686-688, Chiba, Japan, January 2015.

[2] T. Yoshida et al., "Application Drivers and Trends for Future Broadband Access," OSA technical digest of Optical Fiber Communication conference (OFC2015), Th3B.1, Los Angels LA, March 2015.

[3] J. Xie et al., "A dynamic bandwidth allocation scheme for differentiated services in EPONs," IEEE Optical Communications, Vol. 2, No. 3, pp. 532-539, August 2004.

Demonstration of Controlling Chromatic Dispersion at SOA with Mach-Zehnder Interferometric Measurement System

Yusuke Yamanaka, Yuki Fujimura, and Kazutoshi Kato

Graduate school of Information Science and Electrical Engineering, Kyushu University,
744 Motooka Nishi-ku, Fukuoka 819-0395, Japan,
yamanaka@optoele.ed.kyushu-u.ac.jp

Abstract: We proposed the Mach-Zehnder interferometric system to measure the chromatic dispersion at an optical device. We demonstrated that the phase difference between two lightwaves can be controlled at the SOA with our system.

Keywords: semiconductor optical amplifier, chromatic dispersion, Mach-Zehnder Interferometer

I. INTRODUCTION

Terahertz (THz) wave, whose frequency is ranging from 100 GHz to 10 THz, is an effective media for ultra-high capacity wireless transmission and far infrared imaging. One of the successful approaches of generating the THz wave is photomixing of two different lightwaves by a phtomixer. Either for transmission and imaging, the THz wave is required to be with high power. For high power THz-wave generation, we proposed a power combiner of synchronous THz waves from arrayed photomixers [1]. In this system, we controlled the phase of one of two lightwaves so that the phase difference between two lightwaves is tuned to make a phase-controlled THz wave. We designed eight-channel phase shifters together with eight couplers integrated on a glass-based planar lightwave circuit and demonstrated synchronous power combination from the arrayed photomixers. For larger-scale power combination, the lightwave circuit is required to have less elements to maintain its size. Recently, we have invented a novel phase control system which enables the lightwave circuit to be simpler. It controls the phase difference between two lightwaves at a single waveguide device with utilizing its chromatic dispersion.

In this paper, at the semiconductor optical amplifier (SOA) as the waveguide device, we demonstrate the control of phase difference between 1-THz-spaced two lightwaves with using the Mach-Zehnder interferometric measurement system which we devise for measuring the chromatic dispersion of the waveguide device

II. CONCEPT OF PROPOSED INTERFEROMETRIC METHOD

The chromatic dispersion or the refractive index of the optical device changes with changing injection current or applied voltage, which, therefore, causes the change of the phase difference between two lightwaves. Fig. 1 shows the concept of our proposed measurement method. Two lightwaves with optical frequencies of f_1 and f_2 are introduced into the Mach-Zehnder Interferometer (MZI) consisting optical fibers. After passing the MZI two lightwaves are separated into two optical paths by an optical filter. In this configuration, two lightwaves share the same MZI. As for the lightwave of f_2, the phase or the effective path length L_a is changed with changing the refractive index of the optical device. If we control the phase shifter (PS) so that the intensity of the interfered lightwaves of f_2 to be constant, the difference between the phases or the optical path lengths L_a and L_b is kept at constant. On the other hand, as for the lightwave of f_1, the intensity of the interfered lightwave is not kept at constant because the chromatic dispersion of the device causes the phase change of f_1 (or the path length change for f_1) different from that of f_2. Thus, the change of the interfered power of the lightwave of f_1 indicates the change of the phase difference of two lightwaves.

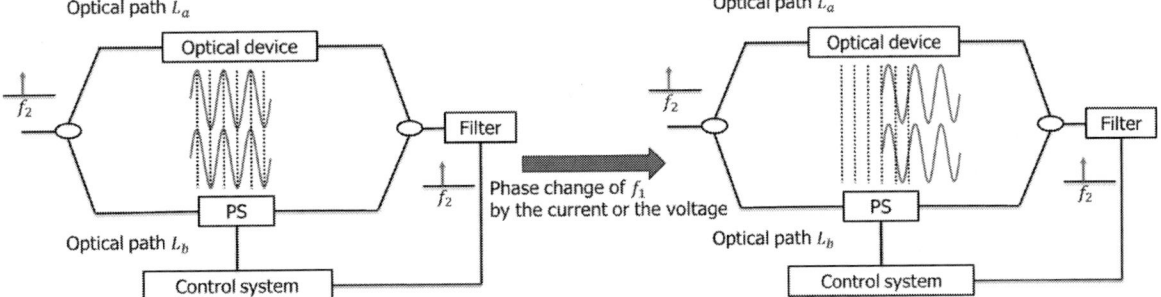

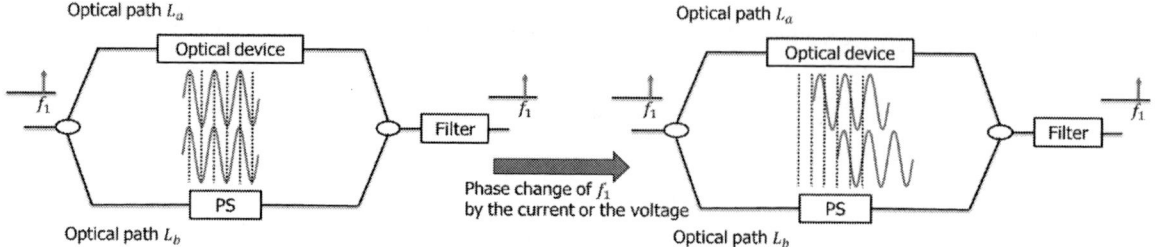

Fig. 1 Phase changes of f_2 (upper) and f_1 (lower)

III. EXPERIMENT

To make the change of the phase difference between two lightwaves, we utilized the chromatic dispersion at the SOA. The SOA was originally designed to amplify an optical power, and we expected that the SOA can change the phase difference between two lightwaves with changing its chromatic dispersion by injection current [2]. The measurement system based on the MZI is shown in Fig. 2. Two lightwaves (the optical frequencies f_1 and f_2 of 193.4THz and 192.4THz, respectively) emitted from narrow-linewidth lasers are coupled by an optical coupler and amplified by the Erbium-doped fiber amplifier (EDFA). These lightwaves are introduced into the MZI consisting of two fiber-based optical paths. One optical path contains the SOA and the variable optical attenuator (VOA) which keeps the optical power after the SOA at constant. The other optical path contains the PS to adjust the path lengths or the optical phase. After passing the MZI, two lightwaves are separated into two optical paths by the Fiber Bragg Grating (FBG) with the refraction frequency of f_2 and an optical circulator. We name the lightwave of the optical frequency of f_1 the measurement light and that of f_2 the controlling light. Here, both lightwaves are interfered after the MZI. The interfered power of the controlling light (f_2) is detected by the photodetector PD2 and the PID controller operates the PS to keep the power at constant [3]. On the other hand, the interfered power of the measurement light (f_1) is detected by the photodetector PD1. If the variation of the refractive index for the measurement light (f_1) at the SOA is exactly the same as that for the controlling light (f_2), the detected interfered power at PD1 also be kept at constant even with changing the injection current of the SOA. However, in fact, because of the chromatic dispersion of the SOA, the refractive index for the measurement light (f_1) is somewhat different from that of the controlling light (f_2) and the interfered power at PD1 varies with changing the injection current. In other words, by keeping the interfered power of the controlling light (f_2) at constant, the difference of the refractive index change can be measured as the change of the interfered power of the measurement light (f_1).

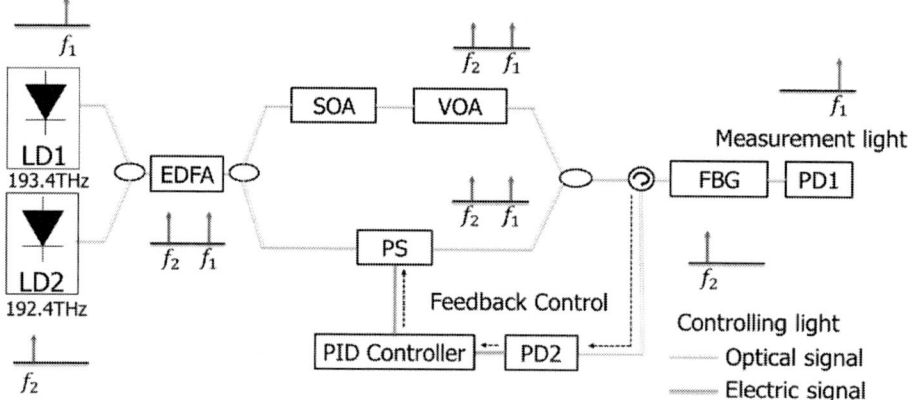

Fig. 2 Mach-Zehnder Interferometric system to measure the phase difference of two lightwaves

The blue circles in Fig. 3 show the measured interfered power of the measurement light (f_1) as a function of the injection current of the SOA (case 1). The interfered power is increasing with the injection current. This result implies the difference of the refractive index change between the measurement light (f_1) and the controlling light (f_2). However, this result also implies a possibility of the difference of the gain change between these two lightwaves. Thus, next, we slightly shift the path length difference of the MZI and measured the interfered power of the measurement light (f_1) (case 2). The red circles in Fig. 3 are the power as a function of the same injection current as that in the case of the blue circles. The interfered power also varies with the injection current but it is decreasing. From these results it can be concluded that the variation of the interfered power of the measurement light (f_1) is resulting from the difference between the refractive index changes of two lightwaves. These variations of interfered power represent some parts of the sinusoidal transmission characteristics of the MZI.

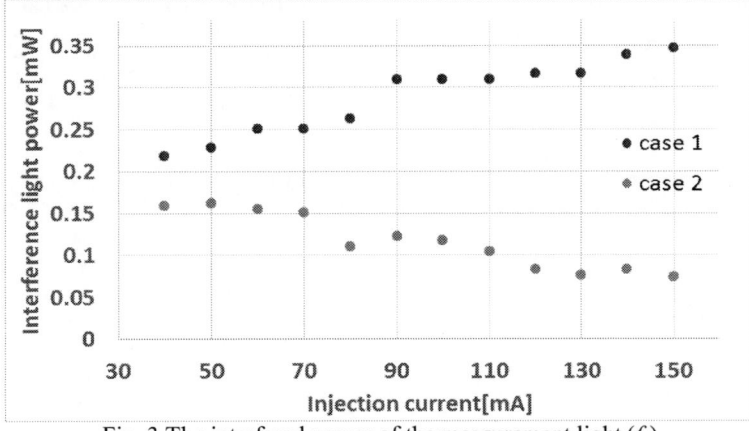

Fig. 3 The interfered power of the measurement light (f_1)

IV. CONCLUSIONS

We proposed the Mach-Zehnder Interferometric method for measuring the change of the chromatic dispersion at a waveguide device. Using this method we demonstrated that the chromatic dispersion of the SOA changes with increasing the injection current and, thus, the phase difference between 1-THz-spaced two lightwaves can be controlled at the SOA

ACKNOWLEDGMENT

A part of this work was supported by the Strategic Information and Communications R&D Promotion Programme (SCOPE) 2015, from the Ministry of Internal Affairs and Communications, Japan, the Collaborative Research Based on Industrial Demand/JST, and CREST/JST. The authors thank NTT Device Technology Laboratories for their experimental support.

REFERENCES

[1] J. Haruki, K. Sakuma, and K. Kato "Synchronous THZ wave combiner consisting of arrayed photomixiers" Microoptics Conference, Fukuoka, Japan, Oct. 25-28, 2015

[2] Michael J. Connelly, "Theoretical calculations of the carrier induced refractive index change in tensile-strained InGaAsP for use in 1550 nm semiconductor optical amplifiers," Applied Physics Letters 93, 181111, 2008

[3] Y. Fujimura, K. Sakuma, S. Takeuchi, K. Kato, S. Hisatake, and T. Nagatsuma "Compact and robust phase stabilization system for high-frequency carrier generation using an integrated lightwave circuit" Microoptics Conference, Fukuoka, Japan, Oct. 25-28, 2015

WA2-101

Hybrid Metal Insulator Metal Plasmonic Waveguide and Ring Resonator

Prateeksha Sharma, Dr. Dinesh Kumar V.
Indian Institute of Information Technology, Design and Manufacturing, Jabalpur
Prateekshasharma024@gmail.com

Abstract: Mode properties of hybrid metal insulator metal (HMIM) plasmonic waveguide has been analyzed and compared with hybrid metal insulator plasmonic waveguide. Transmission characteristics of HMIM ring resonator has also been investigated using CST microwave studio.
Keywords: Plasmonic, ring resonator, bends, waveguides, confinement.

I. INTRODUCTION

Surface plasmon polariton based waveguide components are receiving more attention in large scale integration due to its ability to confine light below the diffraction limit and low bending losses [1]. So many components based on plasmonic waveguide have already been investigated, like bends, connectors, mach zhender interferometer, ring resonators etc [2]. But still practical implementation of plasmonic waveguide components is not possible due to large ohmic losses. Solution to this problem is high dielectric contrast waveguides which is also called hybrid plasmonic waveguides, as components based on it offers strong confinement with fewer losses[3-5]. So far, various hybrid plasmonic waveguides and components based on it have been investigated [3-5]. It contains low dielectric insulator layer sandwiched between metal and high dielectric, which works on coupling mechanism of plasmonic and dielectric modes, hence light will be guided from low dielectric layer [3]. But still confinement of these types of waveguide will be poor which can be improved by multilayer structures [6]. In this study, multilayer hybrid metal insulator metal (HMIM) waveguide has been studied considering different structural aspects, then comparison of hybrid metal insulator(HMI) and HMIM has been done on the basis of propagation length and mode area, at last HMIM ring resonator has been investigated at different radius. These all numerical simulation has been done by frequency domain solver of CST microwave studio. Validation of simulation procedure has been done by re-implementing references [3-6] and it is found that results are in well agreement with the paper results; like in [6] we get 193nm cut off thickness for the reported planar waveguide which is much closer to 200nm, as given in the paper. Though fabrication of ring resonator is more challenging task, still there are many practices like; mirror plating technique, liquid phase coating, spliced fiber pressure filling, thermal process like; evaporation, drawing, blowing and many more, which might be supportive in fabrication process [7,8]. To the author's best knowledge, ring resonator based on HMIM has not been investigated so far, hence it is novel kind of study. Ring resonator is key component for nano scale electronic photonic integrated circuits, as it is used for applications like optical switching, electro-optical switching, wavelength conversion, and filtering [4].

II. THEORY AND OPTICAL PROPERTIES

HMIM (hybrid metal insulator) contains a set of insulator sandwiched between two metal layers, where set of insulators contains three dielectric layers; high dielectric in between two low dielectric layers. Energy will be guided through two layers of low dielectrics, due to the coupling behavior of plasmonic and hybrid modes [6]. This waveguide provides tighter confinement hence smaller mode area as compared to the three layered hybrid plasmonic waveguide. Comparison of both the waveguides i.e. HMIM and HMI has also been done using mode area and propagation length. Cross section of HMIM structure is shown in Fig. 1(a). Aluminium Gallium Arsenide (AlGaAs) and Poly tetra fluoro ethylene (PTFE) are used for high and low dielectric(spacer) layer with dielectric constants, 12 and 1.7 respectively [6]. Johnson's silver has been taken from CST template and its permittivity will be, ε_m= -128.924 + 2.703i at 1550nm. In mathematical form it can be expressed by Drude model as [9]:

$$\varepsilon_\infty = \varepsilon_m - \frac{\omega_p^2}{\omega^2 + j\omega\Gamma}$$

Where $\varepsilon_\infty = 1$; is the dielectric constant at infinite angular frequency, ω_p=1.39x10^{16} rad/s depicts the bulk plasma frequency; Γ=2.53x10^{13}s^{-1} represents the damping frequency of oscillation at 1550 nm. These all parameters are calculated using CST optical calculator. In case of HMIM, metal layer thickness has been represented as t_m, spacer layer as t_s, high dielectric thickness as t_h and air cladding has been used as shown in Fig. 1(a). Same notation has been used for standard three layered hybrid plasmonic waveguide. Simulation has been done using frequency domain solver, keeping perfect matched layer boundary conditions. These hybrid waveguides are the combination of plasmonics and dielectric waveguides and take the benefits of both, hence could scale down up to nanometers like plasmonic waveguides and provide fewer losses similar to dielectric waveguides [1]. 3-D electric field plot and E-field graph is shown in Fig. 1 (b) and (c) for t_m=100nm, t_s=20nm and t_h=200nm. There will be high peak of electric field in spacer layers due to the abrupt change of dielectric material and geometry as shown in Fig. 1(c). So, there will be not as much of contact of metal with energy, which outcomes, less losses and high propagation length [3-5]. Analysis of HMIM has been done at various structural aspects, as shown in Fig.1 (d) and (e). Real and imaginary part of effective index, mode area and propagation length have been calculated at t_s=5-200nm and t_h=100nm, 200nm, 300nm, keeping metal thickness fixed to 100nm. It is found that real part of effective index (Re(N_eff)=β/k_0, where β and k_0 are phase constant and free space wave number respectively[10]) will increase with increasing high dielectric thickness and decrease with spacer thickness and gets saturated after particular spacer thickness, shown in Fig. 1(d); it shows dispersion properties of waveguide. At very small spacer thickness and optimum high dielectric thickness, waveguide will function as plasmonic waveguide and modes will be SPP like mode but on the other hand, if high dielectric layer thickness will increase, it will start accommodating dielectric modes. Hybrid modes will generate at in-between thicknesses of spacer and high dielectric. Mode area can be numerically defined as; $A_{\mod e} = \dfrac{[IdA]^2}{I^2 dA}$ and it shows the confinement property; ideally this parameter should be small for large scale integration [11]. Mode area has been normalized by square of wavelength and it is found that it will always increase whether we increase high dielectric thickness or spacer thickness, as shown in Fig. 1(d). Imaginary part of effective index (Imag(N_eff)) depicts losses and numerically it can be defined as α/k_0, where α is attenuation constant [10]. It will always decrease on increasing the spacer or high dielectric thickness, since energy starts propagating through dielectric layer instead of metal, as shown in Fig. 1(e), but then it is harmful for mode area. Propagation length is inversely proportional to Imag(N_eff) and it can be defined as; 1/2α [6], It will follow opposite trend of Imag(N_eff), will increase on increasing dielectric thicknesses, as shown in Fig. 1(e). Maximum propagation length of 260μm has been observed at t_h=300nm, t_s=200nm. A comparison graph between propagation length and mode area has also been plotted for both the waveguides i.e. HMIM and HMI for parameters as; t_h=200nm, t_m=100nm and t_s=5-200nm and it is found that HMIM provides better figure of merits in comparison to HMI waveguide, since the ratio between propagation length and mode area should be higher for a good waveguide and as shown in the Fig. 1(f), this ratio is higher for HMIM compared to HMI.

It is found that HMIM provides better balance between propagation length and confinement in comparison to standard HMI waveguide; hence for designing of ring resonator, HMIM plasmonic waveguide has been used. Top view of multi layered planar HMIM ring resonator is shown in Fig. 2(a). Five layers are stacked on PTFE substrate and a ring waveguide is positioned in close proximity of a straight bus waveguide at the gap distance of g to permit the optical coupling between them. Parameters have been set as; inner radius, r=1μm and 2μm, high dielectric thickness t_h=200nm, metal layer thickness t_m=100nm, spacer thickness t_s=20nm, gap distance g=20nm, waveguide width w=200nm as shown in Fig. 2(a).

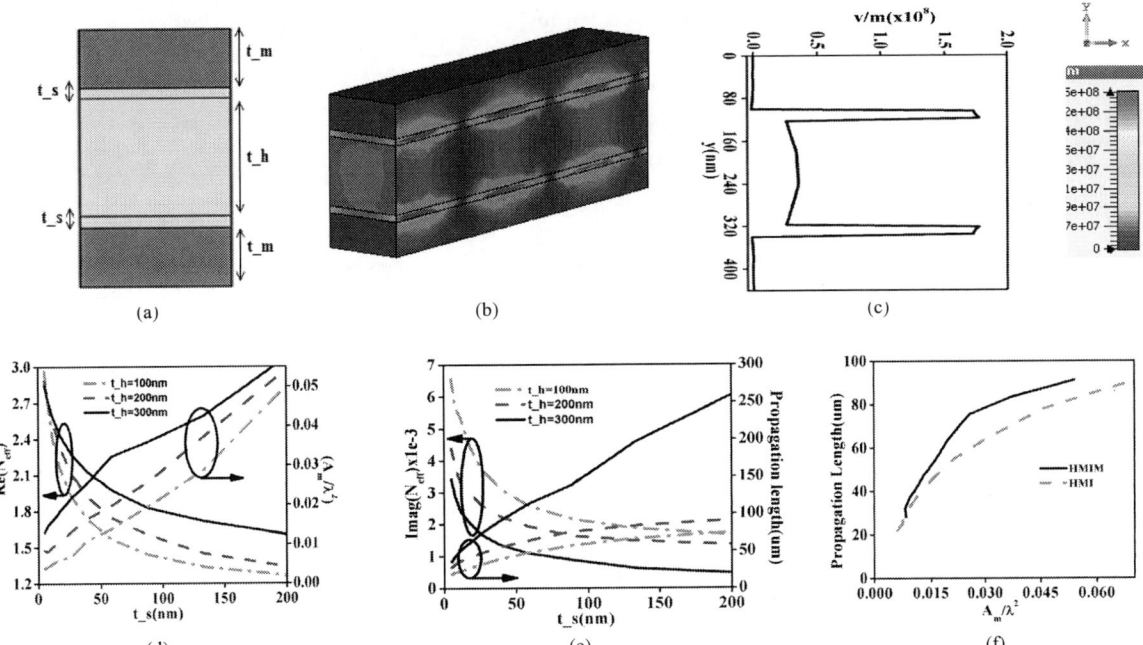

Fig.1 (a) Cross section and (b) 3-D electric field view (c) E-field plot (d) Real part of effective index and mode area (e) Imaginary part of effective index and propagation length of hybrid metal insulator metal (HMIM) waveguide (f) Comparison graph of HMIM and HMI plasmonic waveguides.

A part of optical power propagating in the bus waveguide will be coupled to ring and motivates a circulating mode of ring resonator. Filter properties of ring resonator comes into existence after the interference of propagating modes in the bus and circulating modes in the ring waveguides and effectiveness of coupling depends on ring radius and gap region between straight bus waveguide and ring [4]. Its filtering performance can be determined by extinction ratio and free spectral range. Extinction ratio is defined as the ratio between the minimum and maximum transmission outputs and it depends on the coupling between the ring and bus waveguide, bending losses around the ring and ohmic losses due to the metal [4]. Conversely, the free spectral range and ring resonator bandwidth depend on the working wavelength, effective mode index of ring waveguide and ring radius [4]. Its transmission is shown in Fig. 2(b) with respect to wavelength which shows periodicity and if radius is increased then number of resonating point will also increase. Extinction ratio till 25dB has been achieved for r=1μm, which is far better than previous literature, as in [4]. Free spectral range of 115nm and bandwidth of 23nm has been observed around 1.55μm resonance wavelength for r=1μm, as shown in Fig. 2(b). 3-D field plot at 1.55μm wavelength for r=1μm is shown in Fig.2 (c). At this wavelength, almost all energy couples to ring waveguide as shown in Fig. 2(c).

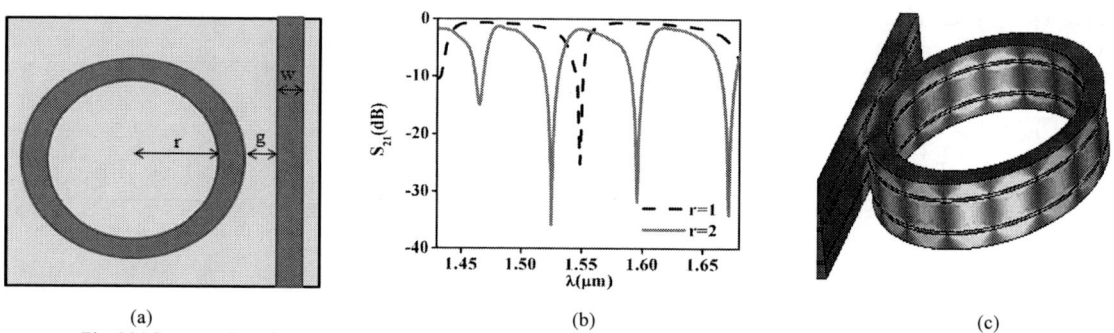

| (a) | (b) | (c) |

Fig. 2(a)Cross section of HMIM ring resonator (b) Transmission plot (c) 3-D electric field view for r=1μm at 1.55μm

III. CONCLUSION

We have described HMIM multilayered plasmonic waveguide and practiced it to project wavelength selective ring resonator. The propagation characteristics of HMIM waveguide have been investigated in terms of effective indices, propagation length and mode area. Comparison of HMIM and HMI has also been done and it is found that HMIM provide better balance between propagation length and mode area. It provides different propagation length and mode area, depending on the material and structural aspects and according to the required application it can be optimized. Ring resonator based on HMIM provides 115nm free spectral range, 23nm bandwidth and 25dB extinction ratio for r=1μm. This kind of study provides motivation for further investigation of other waveguide components for large scale integration.

REFERENCES

[1] Muhammad Z. Alam , J. Stewart Aitchison, and Mo Mojahedi," A marriage of convenience: Hybridization of surface plasmonand dielectric waveguide modes"Laser Photonics Rev. 8, No. 3, 394–408 (2014) / DOI 10.1002/lpor.201300168.

[2] Sergey I. Bozhevolny, Valentyn S. Volkov, Eloïse Devaux, Jean-Yves Laluet and Thomas W. Ebbesen,"Channel plasmon subwave length waveguide components including interferometers and ring resonators", *Nature* 440, 508-511 (23 March 2006) doi:10.1038/nature04594.

[3] Ruixi ZENG, Yuan ZHANG, Sailing HE, "Energy intensity analysis of modes in hybrid plasmonic waveguide," *Front. Opto electron.*5 (1): 68-72 (2012).

[4] Hong-Son Chu,Yuriy A. Akimov, Ping Bai, and Er-Ping Li,"Hybrid dielectric loaded plasmonic waveguide and wavelength selective components for efficiently controlling light at sub wavelength scale" Vol. 28, No. 12 , December 2011, J. Opt. Soc. Am. B.

[5] Hong-Son Chu, Er-Ping Li, Ping Bai, and Ravi Hegde, "Optical performance of single mode hybrid dielectric loaded plasmonic waveguide based components", Appl. Phys. Lett. 96, 221103, 2010.

[6] Mahmoud Talafi Noghani, Mohammad H., Vadjed Samiei, "Analysis and Optimum Design of Hybrid Plasmonic Slab Waveguides," *Plasmonics*, 8:1155-1168 (2013).

[7] Hart SD, Maskaly GR, Temelkuran B, Prideaux PH, Joannopoulos JD, Fink Y., "External reflection from omnidirectional dielectric mirror fibers," Science. 296, 5567, 510-3(2002).

[8] Ofer Shapira, Ken Kuriki, Nicholas D. Orf, Ayman F. Abouraddy, Gilles Benoit, Jean F. Viens *et al.*, "Surface-emitting fiber lasers," Optics Exp. 14, No. 9, 3929 (2002).

[9] P. B. Johnson and R. W. Christy, "Optical constants of the noble metals," *Phys. Rev. B* 6(12), 4370-4379(1972).

[10] Stefan A. Maier, *Plasmonics: Fundamentals and applications*, (Springer Science, 2007) Chap.7.

[11] https://www.rp-photonics.com/effective_mode_area.html.

WA2-102

OECC/PS2016

Tunable Single Bandpass Microwave Photonic Filter based on Phase Compensated Silicon-on-Insulator Microring Resonator

Wenjian Yang, Xiaoke Yi, Shijie Song, Suen Xin Chew, Liwei Li and Linh Nguyen

School of Electrical and Information Engineering, Institute of Photonics and Optical Science,
The University of Sydney, NSW, 2006, Australia
xiaoke.yi@sydney.edu.au

Abstract: *A tunable single bandpass microwave photonic filter exhibiting narrowband RF response with more than 20 dB sideband suppression ratio and an improved shape factor is proposed and experimentally demonstrated based on phase compensated silicon-on-insulator microring resonator.*
Keywords: *Microwave photonic filter, single bandpass filter, integrated optics, ring resonator.*

I. INTRODUCTION

Microwave photonic filters (MPFs) are designed to undertake the tasks of radio frequency (RF) filters, with their inherent advantages of wide bandwidth, high tunability and immunity to electromagnetic interference. The photonic processing of microwave signals using MPF has attracted extensive attention in past years with an important role in the generation, processing and distribution of RF signals [1, 2]. Most of the previously reported MPFs were based on discrete time signal processing techniques, but these approaches are fundamentally limited by the fact that discrete time signal processors intrinsically exhibit periodic passbands [1, 3, 4]. Various techniques have been presented to overcome the periodic nature of delay line filters and to achieve a single bandpass filter response including the use of fibre Bragg gratings [5], broadband source based RF filtering [6], and the use of stimulated Brillouin scattering [7]. Recent research trends have seen the evolution of conventional microwave signal processing to on-chip signal processors owing to the CMOS-compatible fabrication technology which promises a more cost effective, robust and compact solution for wide ranging applications. Although silicon-on-insulator (SOI) microring resonator has been used to implement MPFs in [8, 9], the RF filters show distorted passbands and limited selectivity with 3-dB bandwidths of several GHz, and sideband suppression of less than 10 dB in [8] and 12 dB in [9].

In this paper, we report a novel tunable microwave photonic single bandpass filter technique based on a phase compensated SOI microring resonator, in combination with a tunable laser source and a single photodetector. Experimental results show that the RF filter response exhibits a wideband frequency tuning range between 5 GHz to 40 GHz, while achieving high selectivity with a 3-dB bandwidth of just above 1 GHz, and a sideband suppression of over 20 dB, thus presenting the best performance of MPF based on phase-modulation to intensity-modulation in notch ring resonator.

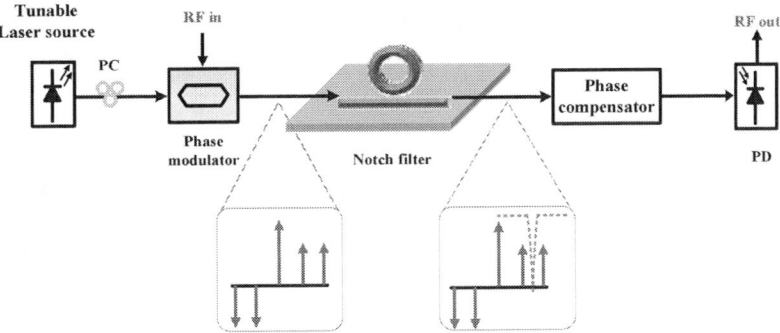

Fig. 1. Schematic diagram of the proposed filter.

II. OPERATING PRINCIPLE

The schematic diagram of the proposed single bandpass microwave photonic filter is shown in Fig. 1. The system consists of a tunable laser source, a phase modulator (PM), a microring resonator notch filter, a phase compensator and

OECC/PS2016

a photodetector (PD). The light from the tunable laser is launched into the PM through a polarization controller (PC). The optical field at the output of the PM under a small signal modulation condition can be described as

$$E_{PM}(t) = E_0 \left\{ J_0(m)e^{j\omega_0 t} + J_1(m)e^{j(\omega_0 - \omega_{RF})t} - J_1(m)e^{j(\omega_0 + \omega_{RF})t} \right\} \tag{1}$$

where E_0 is the optical field of the light source, ω_0 and ω_{RF} are the angular frequency of the optical carrier and RF signal, respectively, $J_n(\cdot)$ is the nth-order Bessel function of the first kind where n = 0, ±1, and m is the phase modulation index.

The PM produces an output signal with 180° out of phase between the upper and lower sidebands. Generally, if this signal is applied to the PD directly, there will be no RF signal detected at the output of the PD as the two sidebands beat out of phase with the carrier and cancels each other out. However, the presence of a notch filter breaks the out of phase symmetry of the modulated sidebands, where the modulated RF frequencies located at the notch filter position remains after photodetection, resulting in a single bandpass response. The value of the relative spacing between the notch filter position and laser wavelength determines the location of the filter in microwave domain. Therefore, the center frequency of the microwave filter can be tuned by changing the wavelength of the tunable light source while the filter shape is determined by the spectrum of the optical notch filter. Note that the optical phase of the notch filter will also affect the detected RF power as the phase variation of the sidebands can generate phase-modulation to intensity-modulation conversion. Therefore, a phase compensator is introduced before the PD to overcome the RF distortion induced by the unwanted phase variation.

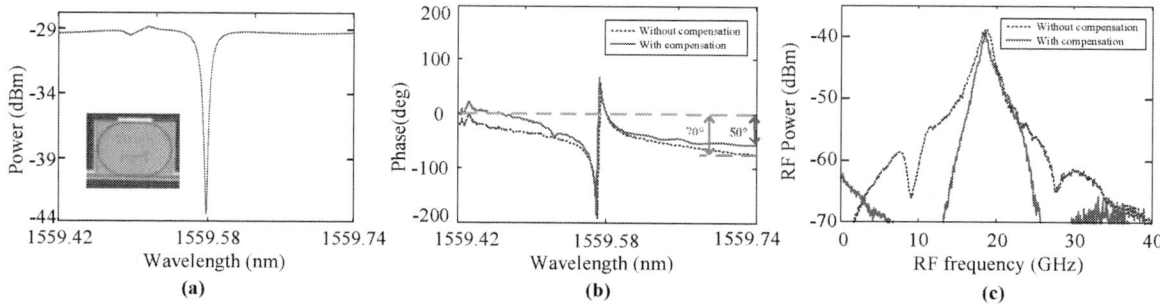

Fig. 2. (a) Measured optical notch filter response of the SOI microring resonator. Inset: microscope image of the fabricated filter. (b) Measured phase of the optical notch filter without and with phase compensation. (c) Measured microwave filter response without and with phase compensation.

III. EXPERIMENTAL RESULTS AND DISCUSSION

Proof of concept experiment was performed to verify the proposed filter. The experimental setup is shown in Fig. 1. A tunable laser source (Keysight 81960A) was used as the optical source. The notch filter employed in the system is based on a microring resonator fabricated on a SOI wafer via ePIXfab. The dimension of the silicon core waveguide is 220 nm by 450 nm for both the bus and racetrack waveguides. The gap between the waveguide and the ring is 520 nm. The optical spectrum of notch filter was measured as shown in Fig. 2(a), where the inset shows the microscope image of the fabricated SOI microring resonator notch filter measured by using an optical microscope (Olympus BX61). It can be seen that the notch is centered at 1559.58 nm with a 3-dB bandwidth of 0.33 GHz and a depth of 14.5 dB. The phase compensation function was performed by the Finisar WaveShaper in the experiment. Considering the bandwidth setting resolution of the WaveShaper is restricted to ±5 GHz, which is wider than the stopband region of the notch filter, and the impact of optical phase change on the RF filter response is significant at the bandpass region of the notch filter due to the high optical power of the remaining phase modulated sidebands. The phase compensation strategy is focused on reducing the phase variation at the bandpass region of the notch filter that affects the sideband suppression of the MPF. The center frequency of the phase compensator is aligned to that of the notch filter. Fig. 2(b) shows the measured optical phase responses. It can be seen that the remaining sidebands which fall within the bandpass region will experience an additional non-zero phase induced by the filter. In order to maintain 180° out of phase between the upper and lower sidebands, the phase compensator is designed to apply a phase offset to the filter so that the effective phase after compensation approaches 0° as illustrated in the green dash line in Fig. 2(b). A maximum of 20° phase variation reduction can be seen after introducing the phase compensator. The insertion loss of the filter is mainly contributed by the fibre-chip coupling loss from a pair of vertical grating couplers. In order to increase the received optical power at the PD, an erbium doped fibre amplifier (EDFA) with 20 dB optical gain was used in this experiment.

The measured MPF response is illustrated in Fig. 2(c). It can be noted that after the phase compensation, the filter response is significantly improved. The 3-dB bandwidth is just 1.05 GHz (solid line), which is 0.8 GHz narrower than the filter response with the uncompensated phase, as shown in the dotted line. At 20-dB bandwidth, by using the phase compensation, the filter bandwidth is reduced by over 7.5 GHz. The overall filter shape is much narrower as the phase

704

compensation is applied accordingly. This strongly indicates that the use of phase compensation significantly increases the filter performance. Due to the resolution limitation of the Waveshaper, the phase variation of the notch cannot be compensated completely, thus leading to the measured RF filter response is still broader than the 3-dB bandwidth of the optical notch filter.

The MPF response was measured using a 43.5 GHz vector network analyzer (Keysight VNA N5234A). The input sweeping RF signal was launched into the PM, and the receiving port captured the output signal after the PD. The measured RF filter responses are shown in Fig. 3(a). The single bandpass filter has an average 3-dB bandwidth of 1.59 GHz while the central frequency of the RF filter was tuned from 5 GHz to as high frequency as 40 GHz. Fig. 3(b) shows the shape factor of the MPF at 20-dB/3-dB bandwidth for the filter response with compensated and uncompensated phase response. It can be seen that there is large improvement in the filter response after the phase compensation which is notably evident as the shape factor is halved at a filter central frequency close to 40 GHz.

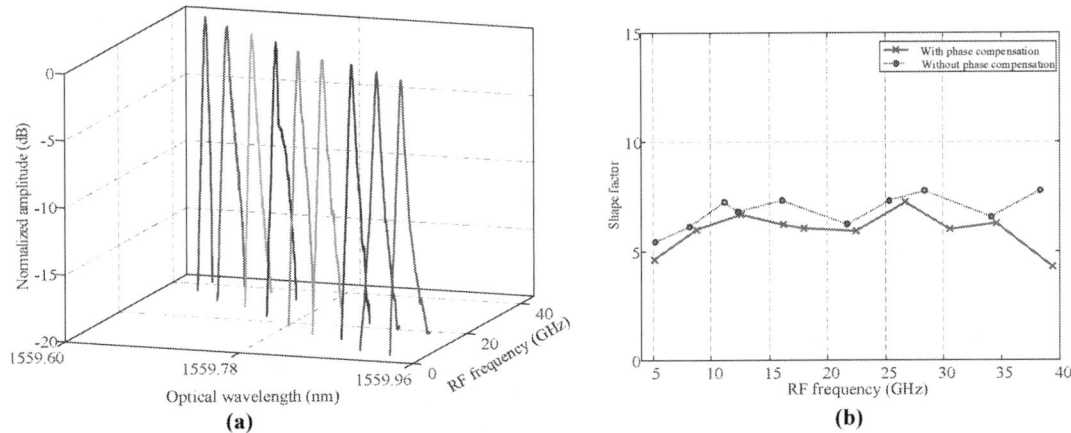

Fig. 3. (a) Tunable RF filter frequency response from 5 GHz to 40 GHz by altering the laser wavelength. (b) Filter shape factor at 20-dB/3-dB bandwidth

IV. CONCLUSION

A tunable single bandpass filter based on a phase compensated SOI microring resonator has been presented, and the filter RF distortions induced by phase variations of the optical sidebands has been significantly reduced. Experimental results have shown the capability to achieve narrow bandpass filter response with a 3-dB bandwidth of about 1 GHz, over 20 dB sideband suppression ratio and a wide tuning range from 5 GHz to 40 GHz while maintaining an improved shape factor.

Acknowledgement

This work was supported by the Australian Department of Defence.

References

[1] R.A. Minasian, E H Chan, and X. Yi, "Microwave photonic signal processing," invited paper, Optics Express, vol. 21, no. 19, pp. 22918-22936, 2013.

[2] R. W. Ridgway, C.L. Dohrman and J.A. Conway, "Microwave Photonics Programs at DARPA," J. Lightwave Technol., vol. 32, pp.3428 - 3439, 2014.

[3] X. Yi and R. A. Minasian, "Recent advances in single passband microwave photonic filtering," Invited Paper, Opto-Electronics and Communications Conference (OECC), Melbourne, Australia, July 2014.

[4] T. Chen, X. Yi, L. Li, and R. Minasian, "Single passband microwave photonic filter with wideband tunability and adjustable bandwidth," Optics letters, vol. 37, pp. 4699-4701, 2012.

[5] W. Li, M. Li, and J. P. Yao, "A Narrow-Passband and Frequency-Tunable Microwave Photonic Filter Based on Phase-Modulation to Intensity-Modulation Conversion Using a Phase-Shifted Fiber Bragg Grating, " IEEE Trans. Microw. Theory Tech., vol.60, pp. 1287–1296, 2012.

[6] L. Li, X. Yi, T.X.H. Huang and R.A. Minasian, "High-resolution single bandpass microwave photonic filter with shape-invariant tunability," IEEE Photonics Technology Letters, Vol. 26, No. 1, pp. 82-85, 2014

[7] S. Hu, L. Li, X. Yi and C. Yu, "Ultra-flat Widely Tuned Single Bandpass Filter Based on Stimulated Brillouin Scattering," IEEE Photonics Technology Letters, vol. 26, pp.1466-1469, 2014.

[8] J. Palací, G. E. Villanueva, J. V. Galán, J. Martí, and B. Vidal, "Single bandpass photonic microwave filter based on a notch ring resonator," IEEE Photonics Technology Letters, vol. 22, pp. 1276-1278, 2010.

[9] N. Ehteshami, W. Zhang, and J. Yao, "Optically tunable single passband microwave photonic filter based on phase-modulation to intensity-modulation conversion in a silicon-on-insulator microring resonator," in Microwave Photonics (MWP), 2015 International Topical Meeting on, 2015, pp. 1-4.

WA2-103

OECC/PS2016

Full-Band Direct-Conversion Receiver Using Microwave Photonic I/Q Mixer

Yue Zheng[1,2], Jianqiang Li[1,*], Qiang Lv[2,3], Jianwei Zhou[1], Yuting Fan[1], Feifei Yin[1], Yitang Dai[1] and Kun Xu[1]

[1] State Key Laboratory of Information Photonics and Optical Communications, Beijing University of Posts and Telecommunications, Beijing, 100876, China
[2] CETC Key Laboratory of Aerospace Information Applications, Hebei, Shijiazhuang, 050081, China
[3] The 54th Research Institute of China Electronics Technology Group Corporation, Hebei, Shijiazhuang, 050081, China
* jianqiangli@bupt.edu.cn

Abstract: *A full-band direct-conversion receiver using microwave photonic I/Q mixer is proposed and experimentally demonstrated in terms of RF frequency range, LO-RF port isolation, I/Q imbalance, conversion gain, noise figure, SFDR and EVM.*

Keywords: *direct-conversion receiver, microwave photonics, I/Q mixer, RF frontend*

I. INTRODUCTION

The growing use of radio spectrum requires electronic systems to operate at extended radio-frequency (RF) bands and signal bandwidth, putting forward great challenges to RF receiver design. Microwave photonic techniques are recently introduced to overcome these limitations due to several inherent advantages, such as potential full-band operation, large instantaneous bandwidth, high RF isolation, low-loss transmission and electromagnetic interference immunity [1]. As for frequency conversion that is a crucial function in an RF receiver, several microwave photonic schemes have been studied. A photonic method for wideband tunable RF frequency conversion was presented by Harris Corporation [2]. With different photodetection fashions, a reconfigurable photonic microwave mixer was proposed [3]. Further RF frontend design was done using an integrated ultra-high Q bandpass filter in [4]. In the above works, tunable narrow-band optical filters have to be used to reduce the spurs, raising the implementation complexity. In addition, the above works are mainly oriented to superheterodyne receivers. Compared to superheterodyne, the direct-conversion architecture with zero intermediate frequency can circumvent image problem, be more suitable to multi-standard software-defined radio, facilitate amplifier and filter design, and potentially decrease power consumption. More importantly, high port isolation property of microwave photonic mixers determines the reduction of self-mixing and local oscillator (LO) leakage which are considered as major drawbacks of direct-conversion receivers.

In this paper, we propose using a microwave photonic I/Q mixer to implement a full-band direct-conversion receiver. Optical I/Q modulators are employed instead of conventional Mach-Zehnder modulators (MZMs) to achieve single-sideband carrier-suppressed (SSB-CS) modulation, saving tunable narrow-band optical filters.

II. PRINCIPLE AND EXPERIMENTAL SETUP

The experimental setup of the proposed direct-conversion receiver based on microwave photonic I/Q mixer is shown in Fig. 1.

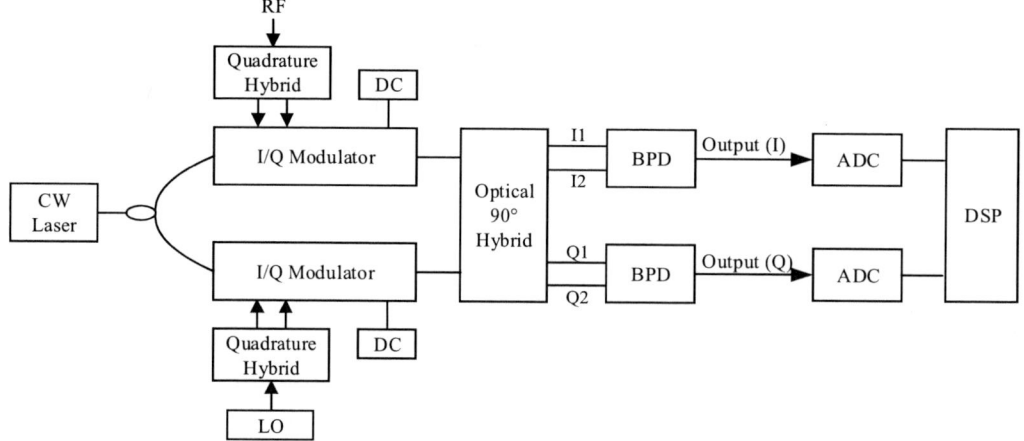

Fig. 1. Block diagram of the proposed full-band direct-conversion receiver based on microwave photonic I/Q mixer

---

This work was in part supported by NSFC Program (61431003), National 863 Program (2015AA016903), NSFC Program (61302086, 61401411), Innovation Foundation of China Electronics Technology Group Corporation (CETC), and Innovation Foundation of Key Laboratory of Aerospace Information Applications at CETC.

A continuous wave (CW) laser at C-band is first generated and then split into two paths. RF and LO signals are respectively applied onto the optical carriers on the two paths by two optical LiNbO$_3$ I/Q modulators both working at SSB-CS mode. Two microwave quadrature hybrids combined with proper bias of the two optical I/Q modulators enable SSB-CS modulation. The SSB-CS signals on the two paths are sent to a 90-degree optical hybrid generating two differential optical in-phase outputs (labeled as I1 and I2) and two differential optical quadrature outputs (labeled as Q1 and Q2). The I1 and I2 outputs are injected into a balanced photo-detector (BPD), producing the in-phase component (I) of the down-converted RF signal. Q1 and Q2 outputs were injected into the other BPD, producing the quadrature component (Q). Note that the DC-offset in I and Q baseband signals could be mitigated due to use of BPDs. Finally, the I and Q baseband signals are digitized for subsequent digital signal processing.

In the experiments, a vector signal generator (VSG) was used to generate a single-carrier quadrature phase-shift keying (QPSK) RF signal with 50 MHz bandwidth and 0.8 roll-off factor for test. A microwave source was used to provide the LO signal. The carrier frequency of both the VSG and microwave source can be tuned up to 20 GHz. The used two BPDs were both embedded with ~20 dB low-noise amplifiers to compensate for the conversion loss. The down-converted I and Q signals were captured by a two-channel data acquisition card with 200 MHz sampling rate and 14-bit resolution. Offline digital signal processing was done in Matlab [5]. After digital I/Q imbalance compensation based on the Gram–Schmidt algorithm, the sample streams were resampled to 2 samples/symbol. A blind adaptive equalizer with four butterfly 15-tap T/2-spaced FIR filters adapted by the classic constant modulus algorithm. Carrier recovery was done based on fourth power Viterbi-Viterbi algorithm. Finally, error vector magnitude (EVM) was calculated.

III. EXPERIMENTAL RESUTLS

A. Microwave Performance of the microwave photonic I/Q mixer

First, we experimentally evaluated the microwave photonic I/Q mixer as a four-port microwave component (i.e. Port RF, LO, I and Q). By the same test methods as for conventional microwave components, conversion gain, noise figure, spurious-free dynamic range (SFDR) in the carrier frequency range from 2 GHz to 20 GHz were measured. The LO-to-RF port isolation was measured to be larger than 100 dB over the entire frequency range, which avoid LO leakage and self-mixing. The rest of results are summarized in Fig. 2-4. Note that the carrier frequency under test is mainly limited by our available instruments. It can be seen that the microwave performance holds over the entire 20 GHz with insignificant fluctuation. This indicates the feasibility of full-band operation. The conversion loss is below 16 dB. The noise figure is below 42 dB. The SFDR keeps beyond 108 dB·Hz$^{2/3}$. Note that the performance can be further improved by increasing the laser power, using pre-LNA and low-Vpi optical modulators.

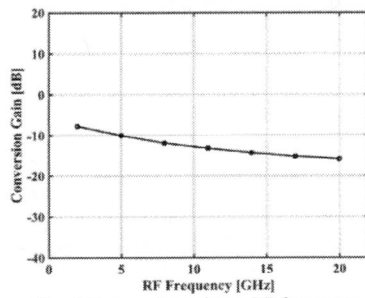 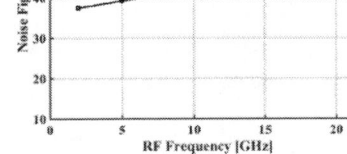 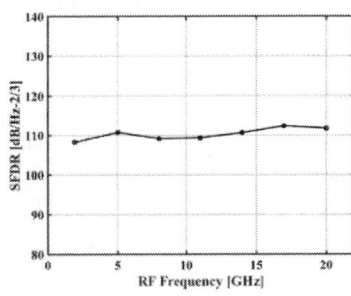

Fig. 2. Conversion gain vs. RF frequency   Fig. 3. Noise figure vs. RF frequency   Fig. 4. SFDR vs. RF frequency

In addition, we also measured the amplitude and phase imbalance of the I/Q mixer with the help of the digital signal processing. The typical amplitude imbalance between I and Q outputs is 0.3 dB and the phase imbalance is 2 degrees, which are lower than that of a commercial microwave I/Q mixer such as MLIQ-0218 made by Marki Microwave, Inc. This is because the I/Q imbalance mainly originates from the optical 90-degree hybrid whose specification is higher with respect to microwave frequencies. This might be another advantage of microwave photonic mixers for direct-conversion receivers.

B. EVM Performance

EVM performance was evaluated by directly converting 50 MHz single-carrier QPSK RF signals. Fig.5 and Fig.6 shows the typical spectrum and constellation of the converted QPSK signals with the help of offline digital signal processing. The EVM performance as a function of input RF power is shown in Fig.7. Given a 10% of EVM threshold, the EVM performance holds over 50 dB variation of input RF power, indicating a large dynamic range of the proposed direct-conversion receiver. Note that the input RF power ends at 30 dBm due to our limited experimental condition. As can be predicted from Fig. 7, he proposed direct-conversion receiver can tolerate higher than 30 dBm of input RF power. Fig. 8 shows the EVM performance as a function of RF carrier frequency. It is clear that the EVM performance also holds over the entire 20 GHz with less than 0.4% of EVM variation

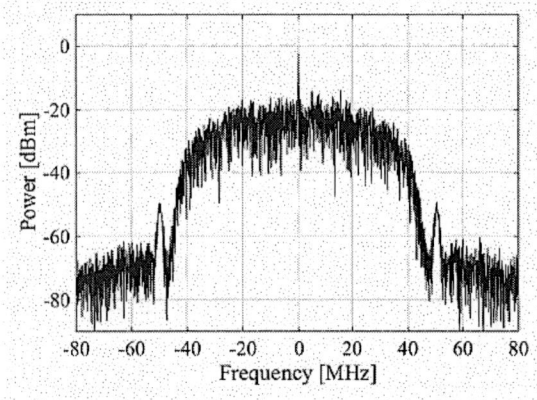

Fig. 5. Typical spectrum of converted QPSK signal

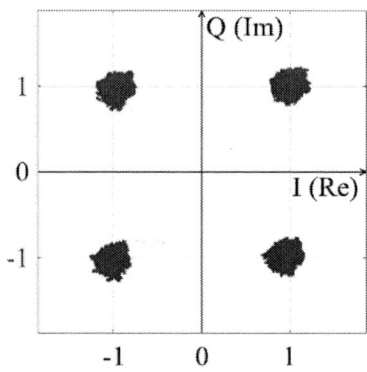

Fig. 6. Typical recovered QPSK constellation.

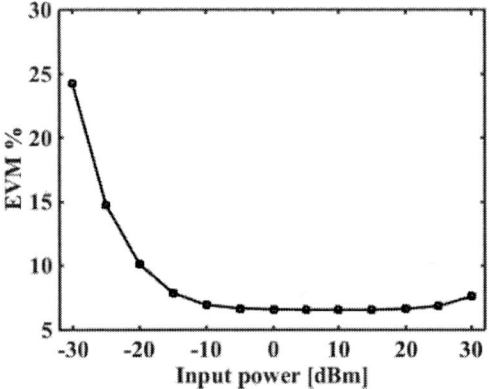

Fig. 7. EVM vs. RF input power.

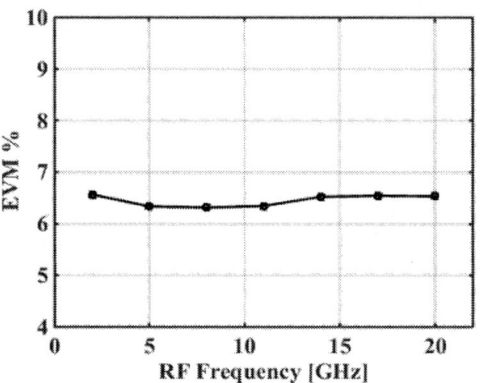

Fig. 8. EVM vs. RF frequency

IV. CONCLUSIONS

A full-band direct-conversion receiver using universal microwave photonic I/Q mixer is proposed. The full-band operation was testified in terms of RF carrier frequency range, conversion gain, noise figure, SFDR and EVM. The performance holds in the entire 20 GHz range under test. The performance can be further improved by increasing the laser power, using pre-LNA and low-Vpi optical modulators. Besides the well-known full-band operation, large instantaneous bandwidth, and electromagnetic interference immunity, the proposed microwave photonic I/Q mixer also shows mitigated LO leakage and I/Q imbalance which are recognized as major drawbacks of conventional direct-conversion receivers. Due to our limited experimental environment, the operating RF frequency was only demonstrated up to 20 GHz. However, the operating RF frequency range of the proposed mixer is mainly determined by the bandwidth of optical modulators which can potentially reach up to 40GHz or higher.

REFERENCES

[1] T. R. Clark and R. Waterhouse, "Photonics for RF Front Ends," IEEE Microwave Magazine, vol. 12, no. 3, pp. 87-95, May 2011.

[2] A. Mast, C. Middleton, S. Meredith and R. DeSalvo, "Extending frequency and bandwidth through the use of agile, high dynamic range photonic converters," presented at IEEE Aerospace Conference, Big Sky, MT, 2012.

[3] Z. Tang and S. Pan, "A reconfigurable photonic microwave mixer," presented at International Topical Meeting on Microwave Photonics (MWP) and the 2014 9th Asia-Pacific Microwave Photonics Conference (APMP), Sendai, 2014.

[4] H. Yu, M. Chen, Q. Guo, M. Hoekman, H. Chen, A. Leinse, R. G. Heideman, S. Yang and S. Xie, "A full-band RF photonic receiver based on the integrated ultra-high Q bandpass filter," presented at 2015 Optical Fiber Communications Conference and Exhibition (OFC), Los Angeles, CA, 2015.

[5] S. J. Savory, "Digital Coherent Optical Receivers: Algorithms and Subsystems," Journal of Lightwave Technology, vol. 16, no. 5, pp. 1164-1179, 2010.

Output Timing Adjustment Mechanism of Optical and Electronic Combined Buffer for Optical Packet Switching

Takahiro Hirayama and Hiroaki Harai

National Institute of Information and Communications Technology,
4-2-1 Nukui-kitamachi, Koganei-shi, Tokyo, 184-8795 Japan
E-mail: {hirayama, harai}@nict.go.jp

Abstract: Our optical and electronic combined buffer architecture copes with both extensibility and energy-efficiency in optical packet switching. In this paper, we propose an output timing adjustment mechanism for its electronic parts.

Keywords: Buffer architecture; FDL buffer; Optical packet switch; Switch Controller; Block Diagram

I. INTRODUCTION

Fiber delay line (FDL) buffering architecture is one of the feasible solutions for optical packet switch (OPS) systems. The FDL buffer provides various discrete-time delays to packets and avoids collisions among packets. The FDL buffer accomplishes high-speed packet forwarding and low power consumption without bit-rate independent Optical-Electronic-Optical (O/E/O) conversion. However, it is difficult to extend the size of the FDL buffer in case of a larger amount of traffic. As the number of FDLs increases, it requires a large chassis and a lot of highly-accurate optical devices such as optical splitters, couplers, and amplifiers. And also, increasing the number of FDLs does not improve throughput of TCP applications drastically against heavy traffic [1]. Due to the discrete-time nature of the FDL buffer, there is an extra space between adjacent outgoing packets. It leads to degradation of link utilization.

Then, we previously proposed optical and electronic combined buffer (OE buffer) architecture. The OE buffer has an FDL buffer and an electronic RAM buffer. An electronic RAM makes buffers more extendable and flexible. However, it requires bit-rate dependent O/E/O conversion and consumes high power under even low traffic. Therefore, the OE buffer ordinary accommodates packets with only the FDL buffer when traffic volume is low. The buffer, then, starts to use its electronic part when traffic volume increases and exceeds a certain threshold. By using two types of buffers adequately, this buffer architecture accomplishes high-speed and low-power packet storing under a low volume of traffic; besides, it improves throughput against a high volume of traffic. In Ref [1], we demonstrated that our new buffer architecture obtains 70% larger throughput than that of the existing FDL buffer against heavy traffic. We also shown that power consumption of the OE buffer architecture reduces power consumpton to 70% of an electronic RAM buffer under low volume of traffic.

When traffic increases, the OE buffer activates its electronic part and expands its capacity. When the electronic buffer is activated, packets are distributed to the FDL buffer or electronic buffer. That is, the OE buffer uses the FDL buffer and the electronic buffer simultaneously. The FDL buffer cannot perform a fine adjustment of output time due to the discrete-time nature. Hence, the electronic buffer controls the output time to avoid collisions of packets coming out from the two types of buffers. We already proposed the algorithm for electronic buffer controllers in Ref. [2].

In this paper, we discuss feasibility of the electronic buffer controller. Firstly, we show the block diagram of the processing unit for output timing adjustment of the electronic buffer controller. This processing unit consists of a few number of registers, comparators and adders. In addition, this unit also requires a minimum value detector to determine which FDL outputs a packet firstly. However, computation complexity of the minimum value detector does not exceed $O(\log_2 N)$. It does not influence its feasibility significantly.

This paper is organized as follows. In Sec. II, we explain the architecture of the OE buffer. And then, we describe their behavior. The architecture of the electronic buffer controller is described in details in Sec. III. Finally, we conclude this paper in Sec. IV.

II. OPTICAL AND ELECTRONIC COMBINED BUFFER ARCHITECTURE

Many studies on the performances of FDL buffers assume that the buffers have dozens or hundreds of FDLs. However, it is difficult to extend the size of FDL buffers to dozens or hundreds. To overcome this scalability issue, we have proposed optical and electronic (OE) buffer architecture, i.e. architecture where an optical buffer and an electronic buffer are combined [1]. The combination of two types of buffers is also discussed in such as Ref. [4].

An OE buffer consists of an FDL buffer and a supplementary electronic (Elc.) RAM buffer. Fig. 1 shows the architecture of our proposed OE buffer. N optical switches of size 1×2 transmit signals to the FDL buffer or the electronic buffer. The length of the N FDLs is a multiple of the unit length D [ns] (0, D, $2D$, ..., $(N-1)D$). Each input

OECC/PS2016

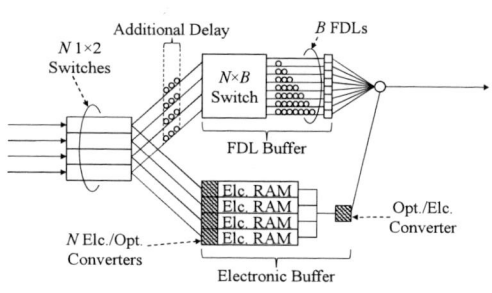

Fig. 1. $N\times1$ electronic and optical combined buffer architecture.

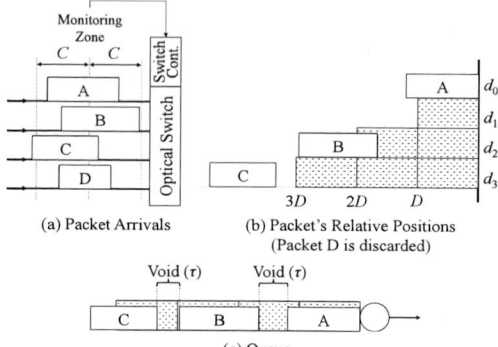

Fig.2. Behavior of FDL buffer in consideration of implementation.

port accesses different O/E converter to achieve high speed processing. Since an extension of the electronic buffer is easier than that of the FDL buffer, the OE buffer solves the scalability issue. The OE buffer can also establish both capacity enhancement and energy saving by controlling ON/OFF of power supply to the electronic buffer.

OE buffer only uses the FDL buffer when traffic volume is low. We first explain fundamental behavior of the FDL buffer. Fig. 2 illustrates an example of the packet behavior when packets arrive at a 4×1 dedicated FDL buffer. This figure shows a case where hardware implementation is considered [3]. The switch controller monitors packet arrivals at every C [ns] interval, i.e., the controller frequency is $1/C$ [GHz]. To detect all packets perfectly, C must be shorter than the minimum packet length. The controller receives packets' information, such as the arrival time and length of each packet. Then, the controller estimates the delays of packets sequentially at ports 1, 2, …, N to attain high-speed processing. Fig. 2(b) exhibits the assigned FDLs and the relative positions of packets when 4 packets arrive as shown in Fig. 2(a). In this situation, the packets A, B, and C are delayed in the FDL buffer and the packet D is discarded. Fig. 2(c) shows logical packet positions in the FDL buffer. Due to the discrete-time nature of the FDL buffer, there is a void space between a newly arrived packet and the previous packet. Note that the delay characteristics of the FDL buffers and the electronic buffer are different due to the differences in their architectural structure and processing capacities. The processing delay in the FDL buffer tends to be less than that of the electronic buffers because no O/E/O conversion is involved. Therefore, we introduce additional delay to all packets traversing the FDL buffer to avoid collision.

When traffic increases, the OE buffer activates its electronic part and expands its capacity. If packet would be stored in the electronic buffer only when the longer delay than the longest FDL length is required, the OE buffer cannot eliminate the large void spaces. Therefore, packets are distributed to the FDL buffer or electronic buffers according to some principles (Ex. Flow-based distribution [1] or random distribution). The electronic memory controller handles the order of packet arrivals at the electronic buffer and outputs them accordingly, i.e., in the FIFO order. The FDL buffer cannot perform a fine adjustment of output time due to the discrete-time nature. Hence, the electronic buffer controls the output time to avoid collisions of packets coming out of the two types of buffers as shown in Fig. 3.

Adding longer delay than the maximum packet length prevents conflict between packets outgoing from the two buffers, even if an unmonitored packet arrives at the FDL buffer during the electronic buffer is transmitting a packet. By assuming additional delay does not exist, packets may conflict if a next packet arrives at the FDL buffer and goes through the 0D-FDL. In Fig. 3, the packet C and the packet 2' conflict because that the electronic buffer starts to send a packet in the k-th monitoring cycle although the packet C arrivals in the $(k+1)$-th monitoring cycle. One of the other solutions for this conflict is that the FDL buffer controller also monitors the output port of the electronic buffer and adds delay to packets as necessary. However, use frequency of long FDLs increases as the electronic buffer stores more packets. It leads to worsening of capacity of the FDL buffer. In summary, the pseudo code of the electronic buffer controller is as shown in Fig. 4, where t_i, l_i, and D_i denote arrival time, length, and delay of the i-th outputted packet from the FDL buffer respectively. d_a denotes the additional delay of the FDL buffer.

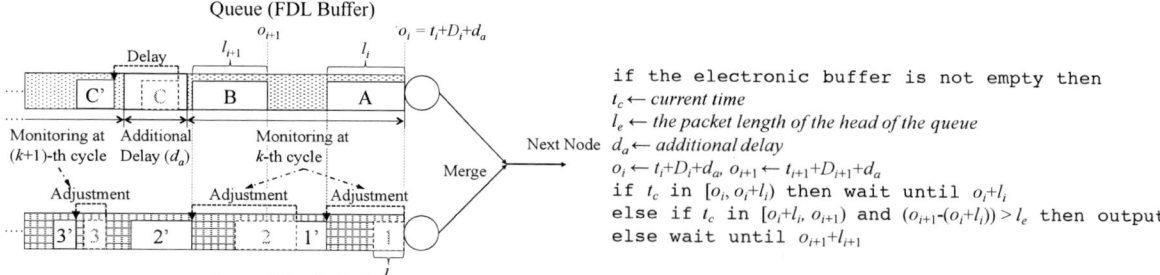

Fig. 3. Output time adjustment for the electronic buffer

```
if the electronic buffer is not empty then
    tc ← current time
    le ← the packet length of the head of the queue
    da ← additional delay
    oi ← ti+Di+da, oi+1 ← ti+1+Di+1+da
    if tc in [oi, oi+li) then wait until oi+li
    else if tc in [oi+li, oi+1) and (oi+1-(oi+li))>le then output
    else wait until oi+1+li+1
```

Fig. 4. Pseudo code of the electronic buffer controller

710

III. ARCHITECTURE AND BEHAVIOR OF THE ELECTRONIC BUFFER CONTROLLER

The electronic buffer controller monitors the FDL buffer and outputs packets via the E/O converter when enough space is detected between adjacent packets outgoing from the FDL buffer. As mentioned above, the FDL buffer controller has to collect packets' information. By using the information-collecting scheme for FDL buffer controller, the electronic buffer controller can monitor the information when the FDL buffer outputs packets. Fig. 5 shows the block diagram to execute the pseudo code of Fig. 4. This architecture consists of memory, registers, adders, and comparators. The memory stores packets' information of the FDL buffer and the electronic buffer. Information of the FDL buffer includes output time o_j and packet length l_j of the packet outputted from input port j. If no packet arrives at a certain port during a monitoring cycle, output time of the port is set to infinity and packet length is set to zero (Ex. the input port 2 in Fig. 5). Meanwhile, information of the electronic buffer are packet lengths sorted in the FIFO order. Three data registers store the information of output time and length of the head packet in two types of buffer. Two address registers keep the memory address for the next packet in the both buffers. Fig. 6 shows pipeline processing for output timing adjustment. Labels A1, A2, … F2 in Figs. 5 and 6 corresponds to the following processes;

A1, 2: Addressing and Loading the following packet from the FDL buffer.

B1, 2, 3: Addressing and Loading the head packet's information in the electronic buffer after address increment.

C1, 2: Determining whether collision will occur. If $(l_e + t_c - o_j)$ is larger than zero, collision occurs.

D: Adding o_j and l_j, i. e., the waiting time of the electronic buffer to avoid collision.

E: Output the waiting time. If collision occurs (condition of C2 is *true*) E outputs o_j+l_j, otherwise E outputs 0.

F1, F2: Determining the number of the port that outputs the following packet.

The Minimum Value Detector is used to determine the minimum value of o_j ($j = 1, 2, …,N$). As one of simple architecture (Fig. 7), the detector can be built by comparators in $\log_2 N$ steps. As shown in Fig. 6, F1 may become the bottleneck in the output timing adjustment. However, it does not degrade the feasibility unless the hundreds of input ports are equipped.

IV. CONCLUSION

Combining the two types of buffer can extend the buffer and save power at the same time. To achieve high speed processing, we have to consider the computation complexity and feasibility of the buffer controller. Our architecture of the buffer and its controller is feasible from the standpoint of them.

REFERENCES

[1] T. Hirayama, T. Miyazawa, H. Furukawa, and H. Harai, "Optical and Electronic Combined Buffer Architecture for Optical Packet Switches," Journal of Optical Communications and Networking, vol. 7, pp. 776-784, Aug. 2015.

[2] Takahiro Hirayama, Takaya Miyazawa, Hiroaki Harai, "Queueing Analysis of Optical and Electronic Combined Buffer for Optical Packet Switches" Optical Switching and Networking, vol. 18, p.p. 201-210, Nov. 2015.

[3] H. Harai and M. Murata, "High-Speed Buffer Management for 40 Gb/s-Based Photonic Packet Switches," IEEE/ACM Transactions on Networking, vol. 14, pp. 191-204, Feb. 2006.

[4] H. Uzawa, K. Terada, N. Ikeda, A. Miyazaki, M.Urano, and T. Shibata, "Energy-efficient Frame-buffer Architecture and It's Control Schemes for ONU Power Reduction," in Proceedings of IEEE GLOBECOM, Dec. 2011.

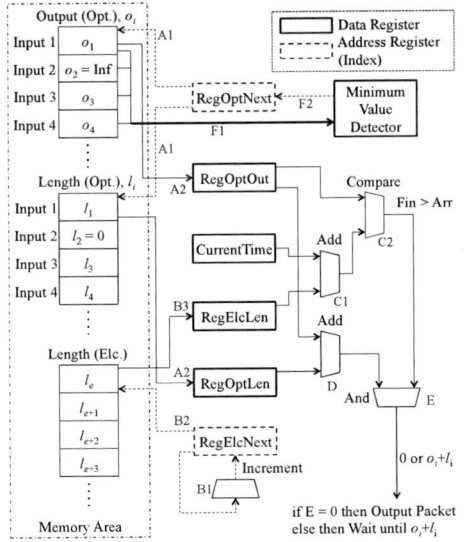

Fig. 5. Block diagram of output timing adjustment mechanism

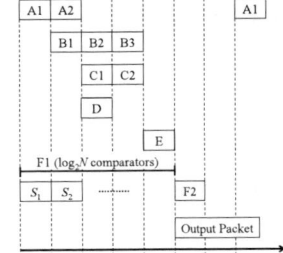

Fig. 6. Pipeline processing for timing adjustment

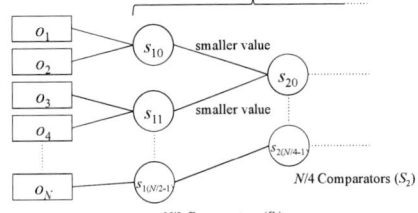

Fig. 7. An example of minimum value detector.

WA2-105

OECC/PS2016

Design of a 1x2 Wavelength Selective Switch Using an Arrayed-waveguide Grating with Fold-Back Paths on a Silicon Platform

Hideaki Asakura, Koki Sugiyama, and Hiroyuki Tsuda
Graduate School of Science and Technology, Keio University
3-14-1 Hiyoshi, Yokohama-shi, Kohoku-ku, Kanagawa, 223-8522 Japan
askr@tsud.elec.keio.ac.jp

Abstract: *A 1x2 wavelength selective switch was designed. It consists of an arrayed-waveguide grating with fold-back paths, three 1x4 interleavers, and 1x2 switches. The number of crossings is independent of the number of wavelength channels.*
Keywords: *Wavelength selective switch; Arrayed-waveguide grating; Fold-back; Interleaver; Silicon*

I. INTRODUCTION

Wavelength selective switches (WSS) are key components of the reconfigurable optical add/drop multiplexers (ROADM) in optical network systems. Free-space optics based WSSs are now commercially available, however are costly because of the complex optical assembly. In contrast, waveguide-type WSSs[1-2], especially on a silicon platform, have high density integration capability and are potentially low cost because they can be manufactured using high volume CMOS processing technology. To realize a WSS using a silicon (Si) arrayed-waveguide grating (AWG) [3], the center wavelength shift and crosstalk due to fabrication errors are to be minimized. The center wavelength mismatch between AWGs for demultiplexing and multiplexing leads to degradation of the performance.

Recently, we reported a 1x2 Si-WSS using an AWG with loop back paths[4]. The WSS had only one AWG to avoid the center wavelength mismatch. However, the extinction ratio of the fabricated WSS was significantly reduced due to the strong coherent interference between the signal and crosstalk of demultiplexing operation of the AWG. The fold-back configurations have been reported to reduce the coherent crosstalk [5-8]. The fold-back design requires only one AWG, and can also avoid the center wavelength mismatch. However, the reported configurations had a number of waveguide crossings, 3-dB couplers or optical circulators. A number of waveguide crossings generates non-negligible loss on a silicon platform, which increased with the number of wavelength channels. The 3-dB couplers generate intrinsic 3 dB loss, and integration of circulators is difficult.

In this paper, we propose a novel 1x2 WSS using an AWG with fold-back paths and interleavers. It has less waveguide crossings than that of the conventional design, and the number of crossings is independent of the number of wavelength channels.

II. OPERATING PRINCIPLE

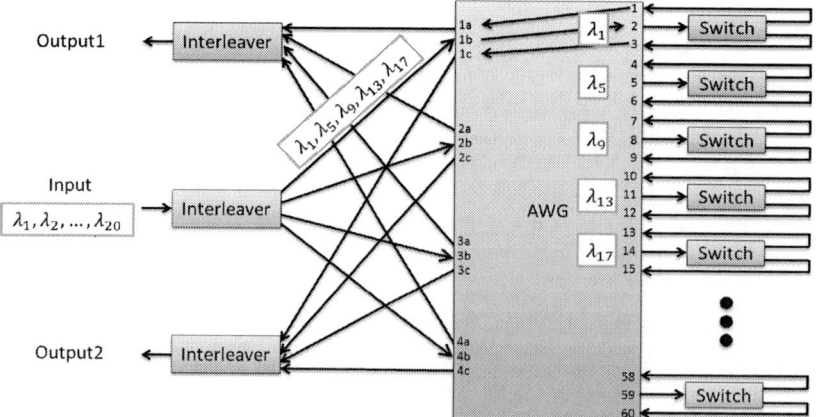

Fig. 1. Schematic diagram of the 1x2 WSS using a fold-back AWG.

The fold-back type 1x2 WSS, consisting of an AWG, three 1x4 interleavers, and 1x2 switch array, is shown in Fig. 1. The operating principle is as follows. The input wavelength division multiplexed (WDM) signal is decomposed into

712

four wavelength groups by a 1x4 interleaver. The channel spacing of the decomposed WDM signal is four times as wide as that of the input WDM signal. These four wavelength groups are coupled to the four input waveguides of an AWG, and are decomposed into individual signals by the AWG. The free spectral range (FSR) of the AWG is very wide, and therefore, the demultiplexing is done successfully. Each individual signal is switched by a corresponding 1x2 switch, and the switched signal is coupled to a waveguide of the AWG located on either side of the waveguide to which the demultiplexed signal was coupled. For example, a signal of λ_1 is coupled to either waveguide '1' or '3', as shown in Fig. 1. The switched signals through the AWG are multiplexed into wavelength groups, and each wavelength group is coupled to the waveguide located on either side of each input waveguide of the AWG. For example, switched signals of λ_1, λ_5, λ_9, λ_{13}, λ_{17} are coupled to either waveguide '1a' or '1c' depending on the each switching state. Finally, the interleaver combines the four groups into one.

III. DESIGN OF THE WSS WITH FOLD-BACK PATHS

Figure 2 shows the mask layout of 1x2, 200-GHz channel spacing, 20-λ WSS. The chip size is 5 mm x 10 mm. The designs of the AWG and the 1x4 interleaver are described in the following subsections.

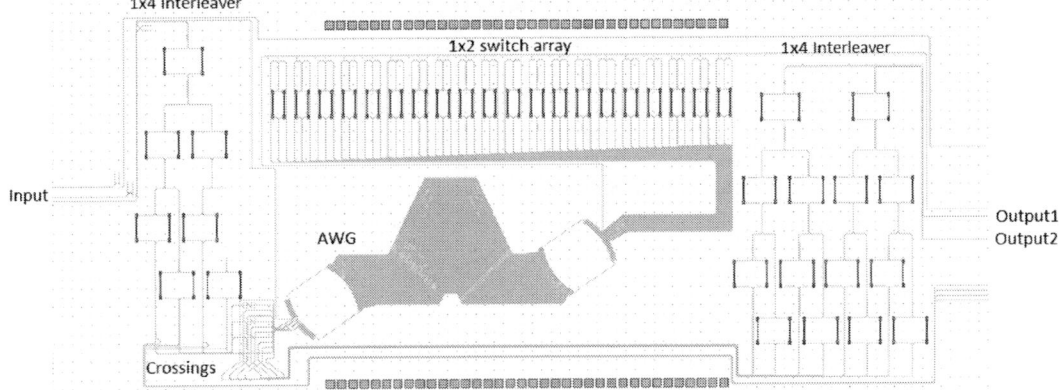

Fig. 2. Mask layout of the 1x2 fold-back WSS.

A. AWG

The channel spacing of the AWG is 800/3 GHz because the input signal to the AWG is 800-GHz-spacing and it demultiplexed to every three output waveguide, as shown in Fig. 1. The FSR of the AWG is required to be wider than 3 (the number of waveguides per wavelength) x 20 (the number of wavelength channels) x 800/3 GHz (the channel spacing). The FSR is set to be 23.1 THz in order to improve the loss uniformity of the AWG. The diffraction order, the path difference, and the slab length are 6, 3.39 μm, and 636.71 μm, respectively. The number of waveguides in the array is 270. As reported in [9], a demultiplexer consisting of an interleaver and AWGs had lower crosstalk. However, the larger FSR may degrade the crosstalk characteristics.

B. 1x4 Interleaver

The 1x4 interleaver consists of seven 1x2 interleavers and is a three-stage configuration as shown in Fig. 3. The FSR of the 1st stage interleaver is 400 GHz, and those of the 2nd and the 3rd stage interleavers are 800 GHz. Each interleaver is an asymmetric Mach-Zehnder interferometer consisting of a 1x2 multimode interference (MMI) coupler, a 2x2 MMI coupler, a path difference, ΔL, and heaters to compensate for the differences between each center wavelength. The 1st and 2nd stage interleavers demultiplex a 200-GHz-spacing signal to four 800-GHz-spacing signals, and the 3rd stage is utilized for reducing the crosstalk due to the non-steep passband. The ΔL of the 1st stage is 195.20 μm (ΔL_1), and those of the 2nd stage are 97.46 μm (ΔL_2) and 97.60 μm (ΔL_3), respectively. The ΔLs of the 3rd stage are equal to those of the corresponding 2nd stage. The core width of the path-difference-waveguides is 1.0 μm, which is the same as that of the ΔL of the AWG in order to reduce center wavelength mismatch. The crosstalk is -17 dB, and the 3-dB bandwidth is 168 GHz on calculation.

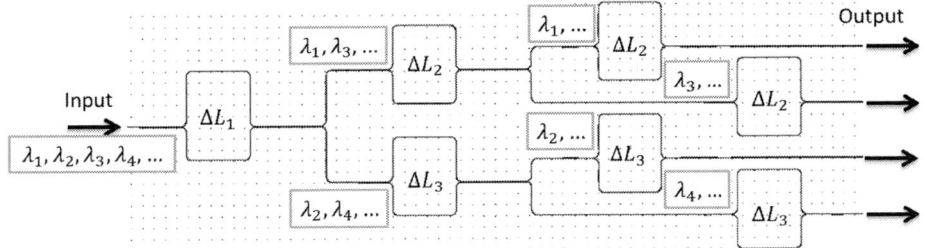

Fig. 3. Waveguide layout of the 1x4 interleaver.

IV. SIMULATION RESULTS OF THE WSS

Figure 4 shows the transmission spectra of the 1x2 WSS. The loss uniformity is about 4 dB, and the crosstalk is less than -30 dB when the extinction ratio of the 1x2 switches is -40 dB. All wavelength channels are switched successfully.

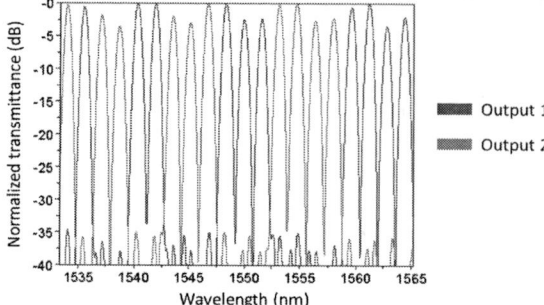

Fig. 4. Transmission spectra of the fold-back 1x2 WSS.

V. CONCLUSIONS

A 1x2 WSS using a fold-back AWG was designed. The channel spacing and the number of channels were 200 GHz and 20, respectively. The fold-back configuration enabled the WSS to be realized using one AWG, and the problem of center wavelength mismatch was solved. In this design, the number of waveguide crossings is independent of the number of wavelength channels without 3-dB couplers or circulators. This configuration will have lower coherent crosstalk than that of a loop-back configuration in a realistic device. The FSR of the AWG should be narrower to reduce the size of the AWG, which results in crosstalk reduction. A loss uniformity improved AWG [10] and an interleaver with large fabrication tolerance [11] are required in order to optimize the design of the fold-back path, with which the FSR can be reduced by one third.

ACKNOWLEDGMENT

A part of this work was supported by JSPS Core-to-Core Program.

REFERENCES

[1] C. R. Doerr, L. L. Buhl, L. Chen, and N. Dupuis, "Monolithic flexible-grid 1×2 wavelength-selective switch in silicon photonics," J. Lightwave Technol., vol. 30, no. 4, pp. 473-478, Feb. 2012.

[2] K. Miura, Y. Shoji, and T. Mizumoto, "Silicon waveguide wavelength-selective switch for on-chip WDM communications," Proc. of IEEE Photonics Conference (IPC 2012), Burlingame, CA (US), Sept. 2012, pp. 630-631.

[3] H. Okayama, D. Shimura, H. Takahashi, M. Seki, M. Toyama, T. Sano, K. Koshino, N. Yokoyama, M. Ohtsuka, A. Sugiyama, S. Ishitsuka, T. Tsuchizawa, H. Nishi, K. Yamada, H. Yaegashi, T. Horikawa, and H. Sasaki, "Si wire array waveguide grating with reduced phase error: effect of advanced lithography process," Proc. of OptoElectronics and Communications Conference held jointly with 2013 International Conference on Photonics in Switching (OECC/PS 2013), Kyoto (Japan), June-July 2013, WM2-1.

[4] H. Asakura, T. Yoshida, K. Suzuki, K. Tanizawa, M. Toyama, M. Ohtsuka, N. Yokoyama, K. Matsumaro, M. Seki, K. Koshino, K. Ikeda, S. Namiki, H. Kawashima, and H. Tsuda, "A 200-GHz spacing, 17-channel, 1x2 wavelength selective switch using a silicon arrayed-waveguide grating with loopback," Proc. of Photonics in Switching (PS 2015), Florence (Italy), Sept. 2015, pp. 22-26.

[5] O. Ishida, H. Takahashi, S. Suzuki, and Y. Inoue, "Multichannel frequency-selective switch employing an arrayed-waveguide grating multiplexer with fold-back optical paths," Photon. Technol. Lett., vol. 6, no. 10, pp. 1219-1221, Oct. 1994.

[6] J. K. Kim and Y. C. Chung, "Simple optical add-drop multiplexer using 2xN arrayed-waveguide grating," Proc. of Optical Fiber Communication Conference (OFC 2002), Anaheim, CA (US), March 2002, pp. 737-738.

[7] C. G. P. Herben, C. G. M. Vreeburg, D. H. P. Maat, X. J. M. Leijtens, Y. S. Oei, F. H. Groen, J. W. Pedersen, P. Demeester, and M. K. Smit, "A compact integrated InP-based single-phasar optical crossconnect," Photon. Technol. Lett., vol. 10, no. 5, pp. 678-680, May 1998.

[8] K. Lee, S. B. Lee, S. G. Mun and C. H. Lee, "Novel architecture for reconfigurable optical add-drop multiplexer using single arrayed waveguide grating," Proc. of the 6th International Conference on the Optical Internet and the 32nd Australian Conference on Optical Fibre Technology (COIN-ACOFT 2007), Melbourne, VIC (Australia), June 2007, pp. 1-3.

[9] S. Chen, X. Fu, J. Wang, Y. Shi, S. He, and D. Dai, "Compact dense wavelength-division (de)multiplexer utilizing a bidirectional arrayed-waveguide grating integrated with a Mach–Zehnder interferometer," J. Lightwave Technol., vol. 33, no. 11, pp. 2279-2285, June 2015.

[10] Z. Sheng, D. Dai, and S. He, "Improve channel uniformity of an Si-nanowire AWG demultiplexer by using dual-tapered auxiliary waveguides," J. Lightwave Technol., vol. 25, no. 10, pp. 3001-3007, Oct. 2007.

[11] H. Okayama, D. Shimura, H. Takahashi, M. Seki, M. Toyama, T. Sano, H. Yaegashi, T. Horikawa, and H. Sasaki, "Mach-Zehnder filter using multiple Si waveguide structure sections for width error tolerance," Electron. Lett., vol. 48, no. 14, pp. 869-870, July 2012.

WA2-106

OECC/PS2016

Suppression of Phase Noise Induced by Optical Interference in Optoelectronic Oscillators

Huanfa Peng, Xiaofeng Peng, Yongchi Xu, Cheng Zhang, Lixin Zhu, Weiwei Hu and Zhangyuan Chen*
State Key Laboratory of Advanced Optical Communication Systems and Networks, School of Electronics Engineering and Computer
Science, Peking University, Beijing 100871, P. R. China
*chenzhy@pku.edu.cn

Abstract: The influence of optical interference induced by reflection of photodiode on the phase noise
of optoelectronic oscillator is experimentally investigated. The phase noise at 10-kHz offset is
improved by 18-dB through optical interference suppression.
Keywords: phase noise; optical interference; optoelectronic oscillator

I. INTRODUCTION

Ultra-low phase noise microwave or millimeter-wave signal is desirable for many applications, such as radar, communication and metrology. In particular, the close-in (<1 MHz) phase noise of the microwave or millimeter-wave signal has great effects on the electronic systems. For instance, in a communication system, the close-in phase noise of the local oscillator (LO) would affect the adjacent-channel-interference (ACI) rejection and sensitivity of the receivers [1]. Beyond this, in a Doppler radar system, the close-in phase noise of the reference oscillator would degrade the performance of the probability of moving targets detection [2]. Optoelectronic oscillator (OEO) has been demonstrated ability to generate microwave and millimeter-wave signals with ultra-low close-in phase noise [3]. The most spectrally pure 10 GHz OEO has achieved a phase noise of -163 dBc/Hz at 6 kHz offset [4]. Conventionally, the phase noise sources of OEO are composed by two categories: the phase noise from RF components and optical components. Experimentally, different noise sources dominate the phase noise in different offset frequencies [5]. The flicker noise of photodiode (PD) and amplifier, Rayleigh scattering, and optical interference contribute to the close-in phase noise of OEO [6]. Due to the poor performance of optical return loss (ORL) of commercial PDs, the optical interferometric noise cannot be ignored in OEO.

In this paper, the influence of optical interference induced by PD on the phase noise of OEO is experimentally investigated. The optical interferometric noise arises from a pair of optical reflection interfaces at a PD and an optical connector. The phase noise of the laser is converted to the optical intensity noise by the optical interference [7]. Due to the intensity noise to RF phase noise conversion in the PD [9], the interferometric noise is converted to the close-in RF phase noise in the OEO [8], [9]. An optical interference suppression method is proposed by leaking the reflected light out of the fiber link by an optical circulator (OC). An 18 dB phase noise reduction at 10 kHz offset is experimentally obtained in an OEO.

II. OPTICAL INTERFERENCE INDUCED BY PHOTODIODE

A. Principle and Experimental Setups

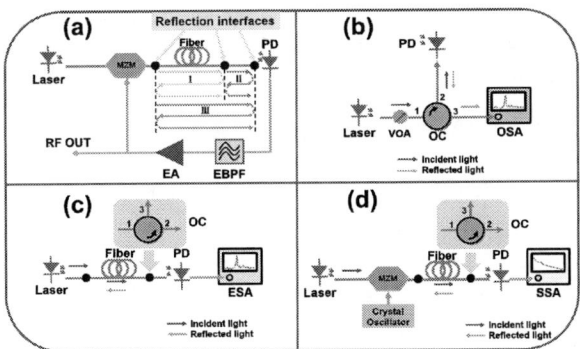

Fig. 1. Schematic diagrams of the optical interference and experimental setups for the investigation of the influence of optical interference on OEO. (a) Schematic diagram for the optical interference in OEO. (b) Setup for the measurement of optical reflection by PD. (c) Setup for the measurement of the influence of optical interference on the intensity noise of the transmitted light. (d) Setup for the measurement of the influence of optical interference on the phase noise of the output RF signal after an external modulated RF-photonic link. PD, photodiode; EBPF, electrical bandpass filter; EA, electrical amplifier; MZM, Mach-Zehnder intensity modulator; OC, optical circulator; VOA, variable optical attenuator.

715

Fig. 1(a) shows the schematic diagram of the optical interference in an OEO. Three reflection interfaces are formed by the two optical connectors of the fiber and the endface of PD. The reflection interfaces form three Fabry-Perot interferometers in the optical link. The double reflected optical field will be combined with the direct optical field which converts the laser's phase noise to the excess intensity noise. Typically, the ORL of the FC/APC type optical connector is about 60 dB, and about 30 dB for the commercial PDs. The optical interference formed by Fabry-Perot interferometers II and III have the most important influences. Due to the use of long fiber in the OEO, the round trip time of Fabry-Perot interferometer III is larger than that of the Fabry-Perot interferometer II. In order to investigate the optical interferometric noise in the OEO, the power of the reflected light by PD, the intensity noise of the transmitted light, and the phase noise of the photo-detected RF signal after the external modulated RF-photonic link are measured, as shown in Fig. 1(b), (c), and (d) respectively. In Fig. 1(b), the continuous wave (CW) light from a narrow-linewidth (<10 kHz) laser with fixed wavelength (1549.5 nm) and power (50 mW) is injected to an OC through a variable optical attenuator (VOA). The reflected light by the endface of PD is separated by the OC. An optical spectrum analyzer (OSA) is used to measure the optical power of the reflected light. In Fig. 1(c), an incident light from the laser is detected by the PD after going through a spool of fiber. An electrical spectrum analyzer (ESA) is utilized to monitor the intensity noise of the photo-detected electrical signal with different fiber length. For comparison, an OC is placed before the PD to reduce the power of the reflected light. In Fig. 1(d), the MZM is driven by a spectrally pure 125 MHz crystal oscillator output. By changing the length of the fiber used in the optical link, the single-sideband (SSB) phase noise of the photo-detected RF signals are measured. For comparison, the experimental setup is configured with and without the use of an OC before the PD. The optical connectors used in the experiments are all FC/APC types. The length of the pigtail of PD is fixed to approximately 1 meter.

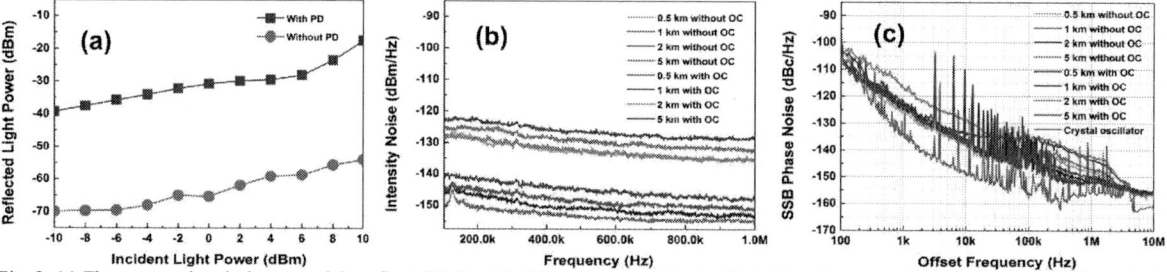

Fig. 2. (a) The measured optical power of the reflected light with different power of the incident light. For comparison, the setup is configured with and without the connection of PD, (b) Intensity noise of the transmitted light with different fiber length. For comparison, the setup is configured with and without optical circulator (OC), (c) The measured SSB phase noise of the external modulated RF-photonic link with different fiber length. For comparison, the setup is configured with and without OC.

B. Experimental Results

Fig. 2(a) shows the measured optical power of the reflected light. By tuning the VOA, the optical power of the incident light is changed from -10 dBm to 10 dBm with a step of 2 dB. The ORL of the PD is about 30 dB. In addition to the reflected light by PD, the reflected light caused by the optical connectors and OC is necessary to investigated. The power of the reflected light by replacing the PD with an optical isolator are also measured, as shown the red curve in Fig. 2(a). The power level of the reflected light by PD is about 30 dB larger than that induced by the optical connectors and OC. Fig. 2(b) shows the measured intensity noise of the transmitted light. The measurement range is from 100 kHz to 1 MHz. The single mode fiber (SMF) used in the optical link are 0.5, 1, 2, and 5 km, respectively. The measured intensity noise of the electrical signal for the setup configured without OC is approximately 20 dBm/Hz larger than that of the setup with OC. The OC greatly reduces the reflected light by the endface of PD. The intensity noise performance is improved with the suppression of the optical interference in the optical link. Fig. 2(c) shows the measured SSB phase noise of the output RF signal of the external modulated RF-photonic link. The setup is configured with different fiber length. The utilizing of 125 MHz crystal oscillator achieves a RF driving signal with an ultra-low phase noise, as shown the green curve in Fig. 2(c). The RF-photonic link degrades the phase noise performance of the RF driving signal due to introduction of the excess phase noise sources of the RF and optical components. The phase noise performance of the RF signal detected in the setup without OC has a greater degradation compared with that of the setup with OC, especially in the frequency offset range from 10 kHz to 1 MHz. It demonstrates that the optical interference has a great influence on the close-in phase noise of the modulated RF signal.

III. OPTICAL INTERFERENCE SUPPRESSION IN OEO

A. Experimental Setup

A single-loop OEO with a narrow-band (<2 MHz) electrical bandpass filter (EBPF) is designed to verify the effects of optical interferometric noise in OEO, as shown in Fig. 3(a). An optical interference suppression system formed by placing an OC between the fiber and PD of the OEO loop is implemented. The SSA has a high sensitivity phase noise measurement at low carrier frequency. An oscillation frequency of 100 MHz is chosen to obtain a precise phase noise measurement of the oscillation RF signal.

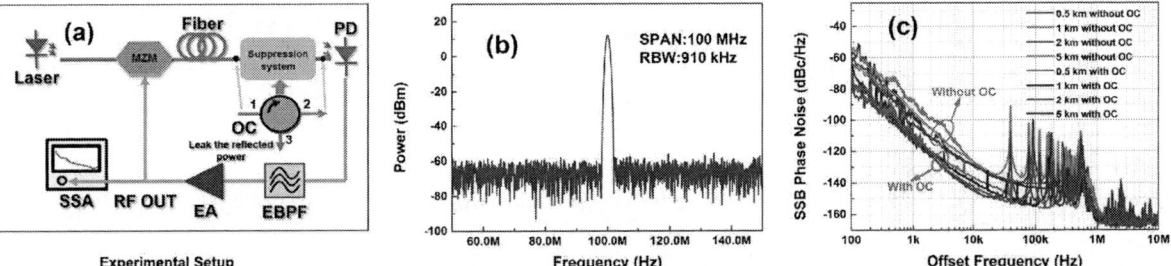

Fig. 3. (a) Schematic diagram of the optical interference suppression in the OEO, (b) The electrical spectrum of the generated RF signal. (c) The SSB phase noise of the OEO with different fiber length. For comparison, the OEO is configured with and without the utilizing of the optical interference suppression system.

B. Experimental Results

Fig. 3(b) shows the electrical spectrum of the generated 100 MHz RF signal. The frequency span (SPAN) and resolution bandwidth (RBW) of the ESA are 100 MHz and 910 kHz, respectively. The SSB phase noise of the generated RF signals are also measured with different fiber length. The close-in (<1 MHz) phase noise performance of the OEO are greatly improved with an optical interference suppression system, as shown in Fig. 3(c). For instance, with a configuration of 5 km fiber, the SSB phase noise of the OEO with OC is -144 dBc/Hz at 10 kHz offset, which is 18 dB lower than that of the OEO without OC. The low phase noise at offset frequency beyond 1 MHz is due to the utilizing of narrow-bandwidth EBPF in the OEO. The side-modes are caused by the long fibers used in the OEO loop. The optical interference induce by the PD strongly decreases the close-in SSB phase noise performance of the OEO. A high performance of ORL of PD is recommended to utilize in an OEO loop to achieve an ultra-low phase noise RF signal.

IV. CONCLUSIONS

In conclusion, the influence of the optical interferometric noise by the endface reflection of PD on the phase noise of OEO is experimentally investigated. An optical interference suppression scheme formed by an optical circulator (OC) is proposed. 18 dB reduction of the SSB phase noise at 10 kHz offset is experimentally achieved in the OEO.

ACKNOWLEDGMENT

This work was supported in part by the National Basic Research Program of China (973 Program 2012CB315606), the National Natural Science foundation of China (Grant No. 61401005and No. 61505002).

REFERENCES

[1] M. K. Nezami, "Evaluate the impact of phase noise on receiver performance," Microwaves & RF Magazine, pp. 1-11, June 1998.

[2] J. R. Vig, "Quartz Crystal Resonators and Oscillators," For Frequency Control and Timing Applications-A Tutorial, January 2004.

[3] X. S. Yao, and L. Maleki, "Optoelectronic microwave oscillator," J. Opt. Soc. Am. B. vol. 13, no. 8, pp. 1725-1735, Augest 1996.

[4] D. Eliyahu, D. Seidel, and L. Maleki, "RF Amplitude and Phase-Noise Reduction of an Optical Link and an Opto-Electronic Oscillator," IEEE Trans. Microw. Tech., vol. 56, no. 2, FEBRUARY 2008.

[5] W. Zhou, O. Okusaga, E. Levy, J. CAahill, A. Docherty, C. Menyuk, G. Carter, and M. Horowitz, "Potentials and Challenges for Optoelectronic Oscillator," Proceedings of SPIE, vol. 8255, 2012.

[6] O. Okusaga, J. P. Cahill, A. Docherty, C. R. Menyuk, W. Zhou, and G. M. Carter, "Suppression of Rayleigh-scattering-induced noise in OEOs," Opt. Exp., vol. 21, no. 19, September 2013.

[7] J. L. Gimlett, and N. K. Cheung, "Effects of Phase-to-Intensity Noise Conversion by Multiple Reflections on Gigabit-per-Second DFB Laser Transmission Systems," J. Lightw. Technol., vol. 7, no. 6, pp. 888-895, JUNE 1989.

[8] W. Shieh, and L. Maleki, "Phase Noise of Optical Interference in Photonic RF Systems," IEEE Photon. Technol. Lett., vol. 10, no. 11, pp.1617-1619, NOVEMBER 1998.

[9] K. Volyanskiy, Y. K. Chembo, L. Larger, and E. Rubiloa, "Contribution of Laser Frequency and Power Fluctuations to the Microwave Phase Noise of Optoelectronic Oscillators," J. Lightw. Technol., vol. 28, no. 18, pp. 2730-2735, SEPTEMBER 2010.

SDM networking schemes compared for computation complexity and efficiency

Miri Blau and Dan M. Marom
Applied Physics department Hebrew University of Jerusalem
danmarom@mail.huji.ac.il

Abstract: *We explore the SDM signal evolution through the network, in the strongly and weakly coupled regimes. Linear transmission and full network model are offered, and suggestions for improvement evaluated. Weakly and strongly coupled transmission is compared in terms of DSP complexity and efficiency.*

Keywords: *Space Division Multiplexing, optical communication*

I. INTRODUCTION

Over the last 40 years, network traffic has consistently grown at an exponential rate, and there is no indication this relentless trend will cease [1]. Global communication is mostly based on optical single mode fiber (SMF) as information carrier. At present, industry is hard-pressed to identify how future networks will continue to scale in capacity, energy consumption, and economic viability as present day technologies (Erbium Doped Fiber Amplifiers (EDFA), wavelength-division multiplexing (WDM) and complex modulation formats) are being stretched to their limits.

Space-domain multiplexing is suggested as a means to overcome the transmission capacity exhaust of SMF. The capacity increase (multiplier) offered by SDM channel varies with the SDM solution properties: number of SMF fibers, cores or spatial modes. Although these SDM solutions are different in their physical nature, each of them supports independent information channels. Even though modes and cores are prone to mixing throughout propagation, as long as the mixing operation is a linear transformation, the data is preserved, with the exception of loss caused by coupling to radiative modes. The problem of unraveling mixed information channels has been addressed successfully in wireless communications, polarization multiplexed transmission on SMF, and recently on SDM channels by digital signal processing (DSP), with the use of multiple input multiple output (MIMO) algorithm [2]. MIMO complexity is determined by the number of mixed channels, as well as the length of the temporal window required for the equalizer, i.e. MIMO computational load increases with the time delay between the fastest and slowest mode. Assuming a linear channel and frequency domain equalization, the computational complexity of the DSP processor [Kahn et al] results from the number of taps ($N_{taps} \propto \tau_{DMGD} f_s$ number of iterations at the filter) and the size of the transfer matrix to be inversed ($M \times M$):

$$Comput.Complex \propto M \cdot N_{taps} \cdot \log_2\left(N_{taps}\right) + M^2 N_{taps} \qquad (1)$$

The first term results from the M times N_{taps} IFTT calculations, while the second term denotes the complexity of the LMS equalizer. This expression summarizes the sources of computational complexity of the DSP processor. In order to decrease the complexity, either the number of mixed modes or the number of taps must be reduced. The properties of the received signal in terms of number of mixed channels (fig. 1) and DMGD (fig. 2) depends on SDM fiber properties. Although FMF and coupled MCF are prone to mode mixing throughout propagation, the nature of this mixing, whether it will be limited within mode group (results in weakly coupled SDM channels) or occurs across all guided mode

Fig. 1: Mode mixing schemes. (a) uncoupled fiber - no MIMO required (b) weakly coupled fiber - partial MIMO required (c) strong coupled fiber - full MIMO required

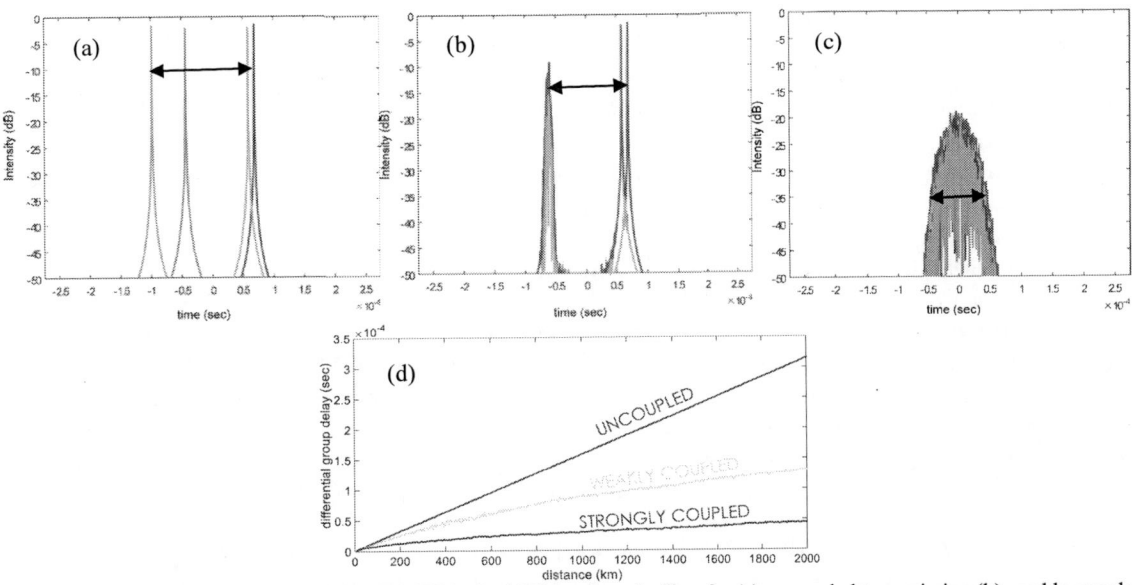

Fig. 2: Top: Differential mode group delay (DMGD) of a 1000 km 6 mode fiber for (a) uncoupled transmission (b) weakly coupled transmission (c) strongly coupled transmission. Bottom: DMGD vs. propagation distance (in km) for uncoupled, weakly coupled and strongly coupled transmission

strongly coupled SDM networks.

II. TRANSMISSION SIMULATIONS

As widely described in literature [3,4], the strong coupled transmission benefits from the shortest DMGD (fig. 2). Induced mode mixing within mode groups reduces DMGD in the group, combining the advantage of low number of taps (of the strong coupled transmission) with the advantage of smaller coupled matrix of the weakly coupled transmission. We induced a relatively strong mode mixing within the mode groups, in changing intervals over 1000 km transmission. Fig 3 shows the reduction of DMGD as a function of the occurrence rate of intentional mode mixing along the link. As mode moxing rate increases, the energy of each group is more concentrated and the noise between the different mode groups decreases. When aiming at partial MIMO, over the subgroups only, the DMGD denoting the required number of taps at the equalizer is within the mode groups only. The computational complexity becomes [5]:

$$CC \propto \tilde{M} \cdot N_{taps} \cdot \log_2\left(N_{taps}\right) + \tilde{M}^2 N_{taps}$$ while $N_{taps} \propto \tilde{\tau}_{DMD} f_s$. For a six mode fiber, $\tilde{M} = 3$, according to our simulations,

$\tilde{\tau}_{DMD}$ can be reduced to by ~20%. MCF-FMF hybrid fiber [6], may be a good SDM candidate, as it offers flexible group size, along with good separation between the uncoupled groups.

III. NETWORK SIMULATIONS

Optical communication network is constructed of links of optical fiber, as well as various optical components (mux/demux, optical switches, in-line amplifiers, etc.). In this section we offer a whole view of the different schemes of SDM transmission described earlier. The design of SDM optical component must agree with the transmission scheme, in order to preserve its unique features. For a weakly coupled transmission, the different devices must maintain the

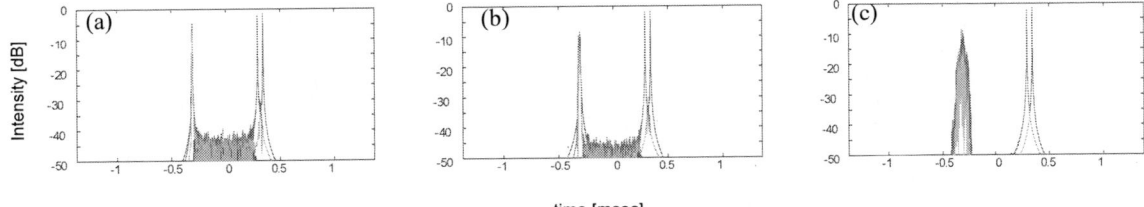

time [msec]
Figure 3: Induced mode mixing in a 1000 km. weakly coupled 6 modes fiber. right to left: increased occurrence rate of intentional mode mixing along the link. (a) every 10 km (b) every 5 km (c) every 2 km.

independent propagation of the coupled mode groups. In strongly coupled transmission, mode mixing may occur with no restrictions. The influence of uniform (as to optical frequency) devices, like splices, amplifiers, demultiplexers etc. is limited to mode mixing and mode dependent loss or gain (MDL or MDG). While MDL (or MDG) are performance degrading, mode mixing can be an advantage as we described in the previous sections. On the other hand, non-uniform devices (i.e. wavelength selective switches, WSS) introduce spectral trimming of the signal, leading us to the considerations bandpass width and flatness, which will also impact the efficiency of our network. In the last few years a few SDM-WSS were presented [7,8]. They can be roughly divided into two subgroups: in the first group, separately-switched SDM-WSS, the spatial modes are demultiplexed prior to switching, and the switching mechanism is very similar to that of the SMF one. In the second group, FM-WSS, the spatial modes are switched together, introducing mode mixing at the channel edges. While FM-WSS is more economical in means of optical components (no need of multiplexing and demultiplexing of the signal) the possible passband of the switch is limited due to crosstalk between the guided modes. However, the separately-switched SDM-WSS, suffers from possible errors of alignment between the different spatial channels, resulting in spectral shifts of the frequency response, which also limits the possible passband. The weakly guided network must use the first, independent switching method, to avoid mode mixing between mode groups. We compared the two switching options, also assessing the effect of induced mode mixing (in mode groups or across all modes) between switches, on the MDL and passband.

Induced mode mixing across all guided modes along transmission and before/after switching reduces the effect of the narrowed passband (fig. 4), improving the efficiency of the network. In the separately-switched SDM-WSS the impact of mode mixing was small, a statistical reduction of less than 0.5 dB MDL was obtained.

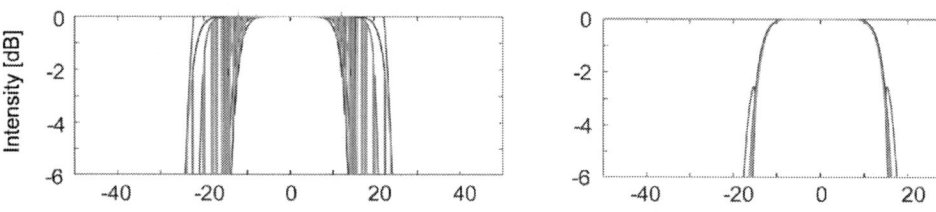

Figure 4: (a) Six mode FM-WSS passband. Right: 10 cascaded WSSs. Left: 10 cascaded WSSs with mode mixing after each switch.

IV. CONCLUSIONS

We presented a performance analysis of two SDM communication network transmission schemes. Strongly coupled and weakly coupled SDM networks were compared. Induced mode mixing was suggested to decrease DMGD and DSP computational complexity. Optical components for the two transmission schemes were suggested, and mode mixing effect on them was studied.

ACKNOWLEDGMENT

The authors gratefully acknowledge the funding by the European Community's Seventh Framework Program (FP7/2007-2013) under grant agreement n° 619732 (INSPACE).

REFERENCES

[1] R.-J. Essiambre, G. Kramer, P. J. Winzer, G. J. Foschini, and B. Goebel, "Capacity limits of optical fiber networks," J. Lightwave Technol. 28, pp. 662–701, 2010.

[2] D. Agrawal, V. Tarokh, and N. Seshadri. Space-time coded OFDM for high data-rate wireless communication over wideband channels. In Proc. IEEE Vehicular Technol. Conf. (VTC), Ottawa, ON, Canada, May 1998.

[3] K.P. Ho and J. M. Kahn. "Statistics of group delays in multimode fiber with strong mode coupling." *Lightwave Technology, Journal of* 29.21 (2011): 3119-3128

[4] C. Antonelli, A. Mecozzi, M. Shtaif, and P. J. Winzer, "Stokes-space analysis of modal dispersion in fibers with multiple mode transmission," Opt. Express 20, 11718-11733 (2012)

[5] N. P Diamantopoulos et al. "Low DSP complexity mid-haul mode-division multiplexing links utilizing wideband modal dispersion compensated two-mode fibers." Optics Communications 355 (2015): 411-418.

[6] D. Soma et al., "2.05 Peta-bit/s super-nyquist-WDM SDM transmission using 9.8-km 6-mode 19-core fiber in full C band," Optical Communication (ECOC), 2015 European Conference on, Valencia, 2015, pp. 1-3.

[7] Ryf, R., et al. "Wavelength-selective switch for few-mode fiber transmission." 39th European Conference and Exhibition on Optical Communication (ECOC 2013). 2013.

[8] Fontaine, Nicolas K., et al. "Heterogeneous space-division multiplexing and joint wavelength switching demonstration." Optical Fiber Communication Conference. Optical Society of America, 2015.

Topology Reconstruction Strategy with the Optical Switching based Small World Data Center Network

Tingting Yang, Dongxu Zhang, Hongxiang Guo*, Jian Wu
Beijing University of Posts and Telecommunications, 100876, Beijing, China
{ttyang, zhangdongxu, hxguo*, jianwu}@bupt.edu.cn

Abstract: *We proposed periodic and load-driven topology reconstruction strategies for the OpenScale architecture. Simulation results verified their effectiveness while load-driven reconstruction strategy performs better in reducing control plane's overhead and nodes' packet forwarding burden.*
Keywords: *topology reconstruction, strategy, OpenScale.*

I. INTRODUCTION

With the rapid growth of Internet business relying on cloud services, intra data center (DC) networks are facing more and more challenges. Traditional electrical switching networks have been struggling to meet the ever increasing demands, while the optical network technologies, with distinguished high bandwidth and low energy consumption merits, have spawned interests of researchers worldwide and been regarded as among the most fundamental technologies for the future network infrastructure deployments. Numerous researchers investigated optical switching based intra-DC network architectures which may bring high flexibility to adjust the logical topology without complex manual re-cabling. However, few discussed how this flexible reconfiguration ability can benefit network performance. We have proposed the OpenScale architecture [1][2], which uses fast sub-wavelength switching to support highly dynamic regional traffic and employs wavelength switching to provide high agility in topology reconfiguration. A traffic-optimized topology reconstruction algorithm was also proposed based on this architecture [3]. However, when and how to trigger the topology reconstruction is still not discussed. Topology reconstruction will cause large amounts of computation and rerouting, thus continuously adjusting topology under dynamic traffic might be unrealistic. It is necessary to consider a strategy that can reduce the reconfiguration frequency yet take the most of topology flexibility to improve system performance.

In light of this, based on the OpenScale architecture, we propose two topology reconstruction strategies, periodic reconstruction and network load-driven reconstruction. In the following parts of this paper, we firstly review the network architecture and explain the strategies in detail, and then we evaluate the network performance after applying the two strategies. Simulation results show that load-driven reconfiguration outperforms periodic reconstruction in both reconfiguration interval and load balancing.

II. NETWORK ARCHITECTURE AND TOPOLOGY RECONSTRUCTION STRATEGY

A. OpenScale Architecture

OpenScale is a novel intra-DC optical network architecture, whose design principles are basically inspired by the "small world topology". OpenScale network has the characteristics of high local clustering coefficient, short average path length and strong scalability. As shown in Fig. 1, OpenScale architecture interconnects network nodes into extendable hexagon-shaped rings using fibers. Each hexagon ring functions as an optical burst switching ring, which provides full mesh interconnections among six nodes. This means that, a node can reach multiple neighbors on the rings with only one-hop. Meanwhile, OpenScale nodes also support optical circuit switching, therefore long distance wavelength connections can be established between non-neighbor nodes. Fig. 1(a) illustrates the node structure, which contains three pairs of in and out ports connecting to three rings respectively. The optical burst add/drop module realizes dynamic bandwidth allocation between nodes without collisions through synchronized timeslot scheduling [4]. The wavelength switching fabric makes it possible to achieve direct yet reconfigurable lightpaths.

Here we assume that all OpenScale nodes use a consistent routing scheme (e.g., shortest path routing) on a given topology. Communications between non-neighbor nodes will contribute forwarding load at intermediate nodes. As network flows' bandwidths change time-to-time, there may exist highly loaded nodes (namely hotspots) which could lead to unacceptable latency or packet loss. However, the wavelength lightpaths among nodes can be viewed as adjustable "free degrees" of the network. Free degrees can be used to support dynamic establishment and removal of direct wavelength connection between any nodes as shown in Fig. 1(b). Such characteristics make it possible to adjust inter-rack logical topology dynamically according to varying traffic demands, mitigate the hotspot problem and improve the overall throughput.

Given that the intra-DC network typically has a large size (e.g., hundreds or even thousands of racks), it tends to be time-consuming to calculate an appropriate topology and then deploy it accordingly. Thus topology reconstruction is preferred to be performed strategically, seeking for a proper tradeoff between operation overhead and system performance.

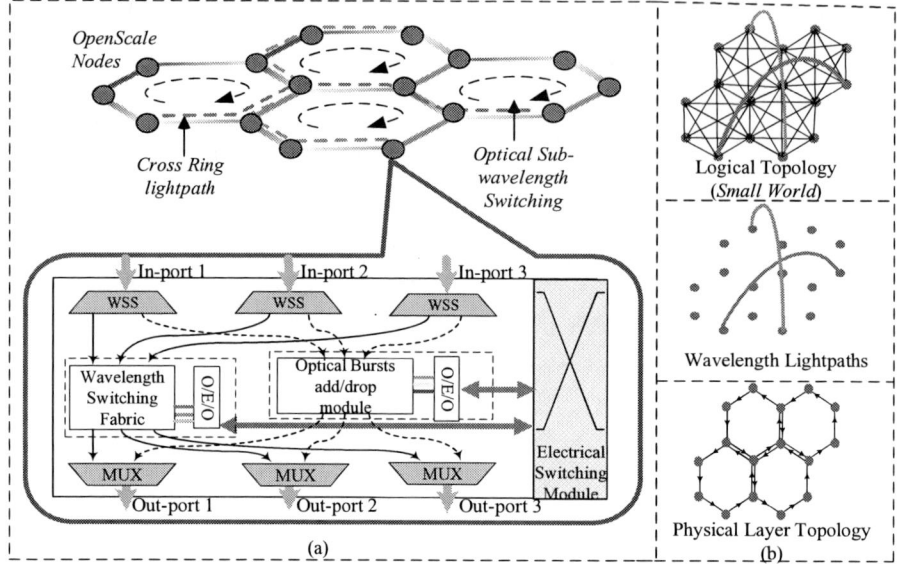

Fig. 1. OpenScale network architecture and the node structure.

B. Topology Reconstruction Strategy

We propose two strategies: periodic reconstruction and load-driven reconstruction.

The periodic reconstruction means that topology may change every a fixed interval. The centralized network controller collects traffic information periodically, then updates the topology using a specific algorithm. When the reconstruction period is short, the network has better adaptability to traffic fluctuations, but the control plane may suffer a heavier processing burden. On the contrary, a long reconfiguration period can mitigate control plane operation overhead, but may sacrifice the topology agility.

Since traffic fluctuates unpredictably, it may make more sense to trigger topology adjustments according to the traffic status dynamically. Consequently, we put forward the load-driven reconstruction strategy as follow:

First, the network controller continually monitors the forwarding load of each node. When the number of hotspots exceeds a certain level, namely network congestion starts to occur, then the topology reconstruction will be triggered. Apparently, there are two essential parameters deciding the triggering point of topology reconstruction. The first is *forwarding load threshold* which defines whether a node is a hotspot, the second is *hotspot count threshold* which indicates the maximum allowed number of highly loaded nodes in the network.

For simplicity, in this paper we set the *hotspot count threshold* as a fixed percentage (i.e., 20%) of the total number of nodes. Hence, the sensitivity to traffic fluctuation can simply be adjusted by tuning *forwarding load threshold*. Intuitively, setting a higher *forwarding load threshold* means that the network can tolerate heavier fluctuations thus the topology adjustment overheads can be mitigated at the expense of minor, network performance degradation.

As for the topology calculation algorithm, in this paper we use a *b-matching* based approach [5]. More specifically, the problem of updating wavelength connections in a meshed-hexagon topology can be mapped to a weighted b-matching model, in which the weights represent traffic demands among nodes, and *b* represents the number of "free degrees" (wavelength connections) originated from each node.

III. SIMULATION AND RESULTS

A. Simulation Setup

Studies have shown that network traffic with self-similarity can be emulated by aggregating multiple heavy-tailed ON/OFF flows, in which a parameter H (Hurst) can be used to indicate the degree of self-similarity [6]. Typically, higher H also implies more drastic traffic fluctuations. In this paper, we used 60 ON/OFF flows (whose ON & OFF intervals conform to Pareto distribution) to simulate the self-similar traffic between each node pair. The H parameter could be tuned from 0.6 to 0.9. In addition, we assumed a minimum traffic observation interval Δ, which means that the control plane could obtain traffic status from all nodes in no less than Δ. Therefore, as for the periodic topology reconstruction scheme, the operation periods (T) of 1~4 times of Δ were considered. In the load-driven strategy, the *forwarding load threshold* was set to 0.6 and 0.7 (normalized forwarding load). We adopt the definition of throughput in [7] to evaluate the network performance. For a 54 nodes OpenScale network with free degree d=1, we evaluated the throughput performance as well as normalized node forwarding load (the actual forwarding load normalized to the node's switching capacity) with different strategies.

B. Results Analysis

Fig. 2(a) shows the throughput analysis results, in which each data point is an average value of 15 runs. Solid lines represent the periodic reconstruction scheme with different period lengths, while the dashed lines represent results obtained with the load-driven topology reconstruction strategy.

For the periodic scheme, with the increase of H, the throughput decreases. This is because that higher H represents more intensively changing traffic, which makes it more difficult for the network to adapt to current traffic by changing its topology. When H is fixed, the longer the reconstruction period is set, the faster the throughput declines, which is easy to understand because longer topology adjustment periods lead to more mismatch between traffic and topology.

For the load-driven strategy, the trend of throughput changing with H is similar. However, if we concern the control plane overhead, the average reconstruction interval at threshold 0.6 and 0.7 are found be close to 3Δ and 5Δ respectively, but the throughput after applying the load-driven strategy with threshold 0.6 and 0.7 are significantly higher than their equivalent periodic topology reconstruction scenarios.

Meantime, in Fig. 2(b), we observed the probability distribution of node forwarding loads under both strategies (periodic reconstruction strategy with period = 3Δ and load-driven strategy with forwarding load threshold = 0.7, H parameter is set to 0.8). Results show that, with the load-driven topology reconstruction strategy, the amount of highly loaded nodes are significantly reduced compared with the periodic reconstruction strategy. Whenever there are an increasing number of highly loaded nodes, the load-driven topology reconstruction strategy would trigger topology adjustments, so that forwarding loads are sparsely spread across the whole network.

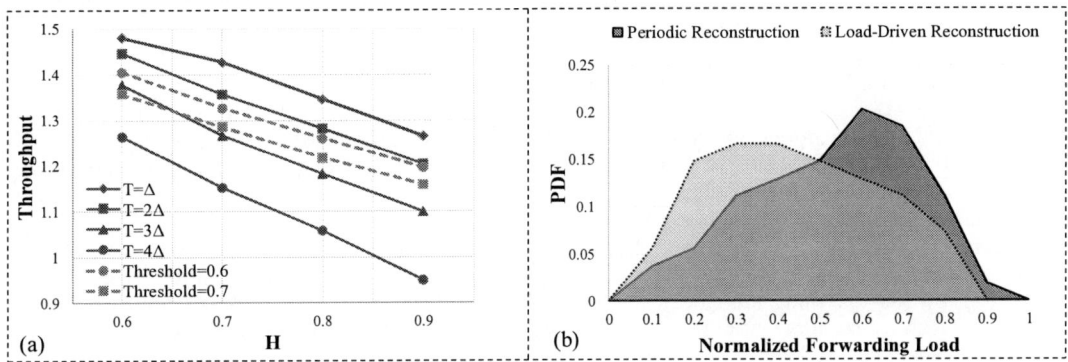

Fig. 2. (a) Throughput under two strategies with different H parameters. (b) Probability distribution of normalized forwarding load.

IV. CONCLUSIONS

To make the most of OpenScale network's topology flexibility and avoid adding too much system operation costs, we proposed the periodic and load-driven topology reconstruction strategies. Simulation results confirmed that the proposed schemes are all effective but the load-driven dynamic topology reconstruction scheme could outperform the periodic reconstruction strategy in terms of reducing control plane's overhead and improving network load balancing.

ACKNOWLEDGMENT

This work was partly supported by NSFC program (Grant No. 61331008, 61471054) and SRFDP program (No. 20130005110013).

REFERENCES

[1] D. Zhang, et al., "An Optical Switching Architecture for Intra Data Center Interconnections with Ultra-high Scalability," Proc. OI, pp. 45 (2014).

[2] D. Zhang, et al., "A Deterministic Small-World Topology based Optical Switching Network Architecture for Data Centers," Proc. ECOC, P.6.11 (2014).

[3] D. Zhang, et al., "Enabling Traffic Optimized Topology Reconstruction with the Optical Switching based Small World Data Center Network," Proc. ECOC, P.6.6.6 (2015).

[4] G. Porter, et al., "Integrating Microsecond Circuit Switching into the Data Center," Proc. SIGCOMM, pp.447-458 (2013).

[5] M.Müllerhannemann, et al., "Implementing weighted b matching algorithms: Insights from a computational study," J. Exp. Algor., vol. 5, p. 8 (2000).

[6] W. Willinger, et al., "Self-similar and Heavy Tails: Structural Modeling of Network Traffic," Preprint (1996).

[7] S. Jyothi, et al., "Measuring throughput of data center network topologies." Proc. ACM Conf. on Measurement and modeling of computer systems, pp. 597 (2014).

WA2-109

OECC/PS2016

Tolerance to Laser Frequency Deviation of Nyquist-WDM and Time-Frequency-Packing Modulation Formats

Francesco Fresi[1]*, Gianluca Meloni[1], Fabio Cavaliere[2] and Luca Potì, *Member, IEEE*[1]

1: CNIT, Via G. Moruzzi 1, 56124 Pisa, Italy
2: Ericsson Research Italy, Via G. Moruzzi. 1, 56124, Pisa, Italy
*francesco.fresi@cnit.it

Abstract: Nyquist-WDM and Time-Frequency-Packing tolerance to ICI arising from laser frequency deviation from nominal value is investigated through simulations. Performance comparison at 6.35bit/s/Hz Spectral Efficiency shows that TFP exhibits higher robustness to ICI and OSNR degradation.

Keywords: Nyquist-WDM, Time-Frequency Packing, Coherent communications, Optical Networks.

I. INTRODUCTION

Spectrally efficient coherent transmission over a flexible frequency grid is strongly increasing the capacity of deployed DWDM optical networks, that are rapidly evolving towards Elastic Optical Networks [1]. Nyquist WDM is nowadays imposing as the most straightforward technique to increases the Spectral Efficiency (SE) by placing Nyquist filtered WDM channels as close as possible reducing the required guard band size. In this context, a root raised cosine (RRC) filter with a sharp roll-off (i.e. ≤ 0.2) is used to confine signal bandwidth and guarantee the absence of Inter-Symbol Interference (ISI). For example, assuming ideal Nyquist-WDM transmission with channel spacing equal to the symbol rate, the actual bandwidth occupation of a 200 Gbit/s (240 Gbit/s gross with 20% FEC overhead, OH) dual polarization (DP) 16QAM channel could be, in principle, as low as 30 GHz. Another technique, called Time-Frequency Packing (TFP), has proved to provide comparable, if not higher, SE by utilizing lower order modulation formats [2]. This is obtained by transmitting pulses that significantly overlap in frequency or time or both, achieving SE and capacities beyond Nyquist by introducing a controlled amount of ISI and Inter-Channel Interference (ICI).

In optical networks, guard bands are often necessary also to limit the detrimental effect of filter edges in wavelength selective switches (WSSs). To mitigate this issue, WSS configuration can be optimized at each node, e.g. by grouping adjacent channels with common output port through a single wider filter. Similarly, large data-rate flows (e.g. 1Tb/s) are obtained as *superchannels* composed by multiple sub-carriers at lower rate and managed as a single flow from a source to a destination. An experimental comparison between 1Tb superchannels realized through Nyquist-WDM and TFP over the same test-bed has been reported in [3], confirming that TFP can outperform Nyquist-WDM in terms of SE and robustness to WSS filtering. Although sub-carriers are not individually subjected to multiple filtering at each network node, the use of guard band between adjacent sub-channels is also required as lasers emission frequency might deviate from nominal value, thus causing ICI worsening.

In this work, the tolerance of Nyquist-WDM and TFP to laser frequency deviation are compared, investigating the robustness to ICI of the two techniques and the requirements in terms of guard band. Realistic laser frequency deviation within +/- 1GHz range is assumed.

II. NYQUIST-WDM

Ideal Nyquist-WDM, having rectangular shaped channels in the frequency domain and spectral occupation equal to the symbol rate represents the minimum spectral occupancy that still preserves the constraints for orthogonal signaling. The sinc pulse shape in the time domain guarantees the absence of ISI at optimum sampling instant, while ICI is prevented even for channel spacing as low as the symbol rate. The orthogonality condition set a lower limit to time- and frequency-spacing (the Nyquist criterion), such that the achievable SE is limited by the number of levels of the underlying modulation format. In fact, higher SE requires higher-level modulation (e.g., 16QAM), with higher complexity and lower resilience to OSNR degradation and nonlinear effects. For DP-16QAM maximum achievable SE would be 8 bit/s/Hz. To increase robustness to OSNR degradation, the use of FEC is required. Here we consider the use of SD-FEC based on LDPC with 20% OH and uncorrected BER threshold of 2.4×10^{-2} as in [4]. As a consequence, the net spectral efficiency is reduced to 6.66 bit/s/Hz. Real implementations can only approximate the pure rectangular shape of ideal Nyquist-WDM. A good approximation is represented by a root-raised cosine profile with a very low roll-off (i.e. alpha = 0.05). This requires the use of high speed digital-to-analog converters (DACs) at the transmitter side, thus increasing complexity and cost of the transmitter. To mitigate the effect of non-ideal rectangular shape and to prevent transmission failures due to laser frequency deviations, guard bands may be adopted. As proposed in [5] we

724

consider DP-16QAM modulation, RRC shape with 0.05 roll-off and a guard band equal to 5% of the baud rate. This results in a further reduction of spectral efficiency to 6.35 bit/s/Hz.

III. TIME-FREQUENCY PACKING

The TFP approach, giving up the orthogonality condition, allows to overcome the Nyquist limit and achieve high SE with low-level modulations. By reducing pulse bandwidth for a fixed signaling rate below the Nyquist limit, some bandwidth is saved at the expenses of introducing ISI. Equivalently, signaling rate can be increased for a fixed bandwidth beyond the Nyquist limit. Similarly, channel spacing in the frequency domain can be chosen lower than Nyquist limit, and a certain ICI resulting from partial overlap among channels can be tolerated and successfully compensated by the decoding algorithm at receiver. TFP employs lower order modulation format (QPSK) and thus typically requires a higher symbol rate to maintain the same information rate of higher-level modulation formats (also depending on the applied coding). SE is then increased at expenses of introducing a certain amount of ISI and ICI, that have to be managed at receiver. TFP finds the optimum time- and frequency-spacing which provide the maximum achievable SE for the given input constellation and detector complexity.

In TFP, the choice of the channel spectral shape is relevant to find a good compromise between low spectral occupation and the introduced ISI and ICI. However, while Nyquist-WDM requires the use of advanced DACs to approximate the sharp edges and flat top of ideal rectangular channels, in TFP, as signal bandwidth is lower than the Nyquist limit, smoother spectral shape is preferable as it results in lower ISI (reduced oscillations of pulse response in the time domain) and ICI (reduced crosstalk in case of partial overlap between adjacent channels). Channel spectral shaping can be achieved either in the optical domain through optical filters after the IQ-MZM or in the electrical domain by means of electrical filters applied on the I and Q signals used to drive the IQ-MZM [6]. In this work we consider simple commercially available passive electrical low pass filters. Filters response has been modeled as Chebyshev Type I. In particular, cascade of 9th order and 2nd order has been considered. The first is used to strictly confine the channel thanks to its sharp edges and limit ICI; the latter softens the top of the spectrum thus reducing fluctuations in the time domain pulse response and limiting ISI. The use of passive devices for shaping the signals provides a big advantage in terms of power consumption, reliability and cost with respect of using a dedicated DAC.

IV. SYSTEM DESCRIPTION AND RESULTS

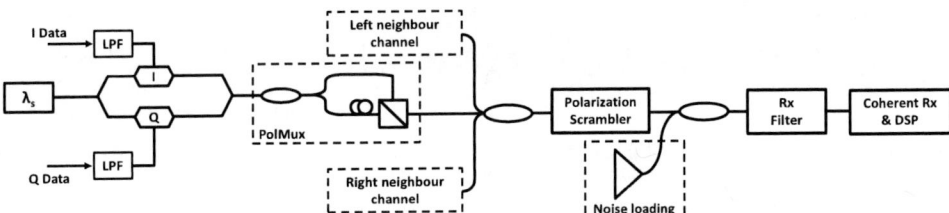

Fig. 1. System architecture considered for numerical simulations.

Fig.1 reports the considered scenario for simulations. Each WDM channel is obtained by in phase (I) and quadrature (Q) modulation of the optical carrier by means of a double nested IQ-MZM driven by two independent data streams. A low-pass filter (LPF) is used to apply the desired shaping to I and Q electrical signal, thus emulating the use of DAC for the Nyquist WDM and the Chebyshev filters for the TFP. Dual polarization is emulated through split-and-delay considering realistic polarization extinction ratio equal to 30 dB. Similarly, two neighbor channels are generated as decorrelated replicas and coupled together +/ 21 GHz apart from the central one. Polarization scrambling is applied and optical noise is added to assess system robustness to OSNR degradation. For the Nyquist-WDM case we consider DP-16QAM modulation at 20 GBaud. Shaping is obtained exploiting 10 GHz low-pass RRC filter with 0.05 roll-off. 1 GHz guard band (corresponding to 5% of the baud rate) is adopted, thus resulting in a channel spacing equal to 21 GHz. Assuming 20% OH for error protection, the net SE is then 6.35 bit/s/Hz (Fig.2a). For the TFP case, transmission parameters have been chosen so that net SE is same as Nyquist-WDM, adopting 39 Gbaud DP-QPSK with 12.5% OH LDPC concatenated with 4% OH HD FEC for eliminating LDPC floor located at BER = 10^{-7}. Channel spacing is maintained at 21 GHz, same as for the Nyquist WDM. Shaping is obtained through cascade of 6.5 GHz 2nd order and a 10.5 GHz 9th order Chebyshev filters (Fig.2b).

At the coherent receiver, the channel under investigation is extracted by means of a 21 GHz 4th order Gaussian filter. Analog-to-digital conversion is emulated considering a 50 Gsample/s, 20 GHz, 8-bit resolution ADC. 8×10^5 samples sequences are acquired for off-line processing. For the Nyquist-WDM case, sampling rate is then increased by interpolation up to 2 samples per symbol. The DSP chain includes an adaptive two dimensional feed forward equalizer (2D FFE) that converges to the matched filter, followed by asynchronous symbol-by-symbol detection strategy [7]. For the TFP case, as analog bandwidth is lower than 1/2T, symbol-time processing can be adopted without any performance degradation. The 2D FFE is followed by four parallel 8-state Bahl-Cocke-Jelinek-Raviv (BCJR) detectors working on the four signal quadratures that iteratively exchange information with 4 LDPC decoders according to the turbo principle, thus achieving maximum a posteriori (MAP) detection [8],[9].

To evaluate the impact of laser frequency deviations within +/- 1GHz range from nominal value, we consider three channels and focus on the performance of the central one. Two cases are considered; in the first (Fig.2c), the central channel (channel under investigation) remains stable at its nominal frequency, while both the neighbor channels progressively deviate from their nominal frequency moving closer to the channel under investigation. This results in a partial overlap of both the edges of the channel, reducing progressively channel spacing from 21 GHz to 20 GHz. For Nyquist-WDM, the adoption of the guard band (1 GHz, 5% of the baud rate) mitigates the effect of laser deviations. The resulting ICI is low and due to the use of a non-ideal rectangular shape. Performance has been evaluated in term of minimum OSNR required for pre-fec BER $\leq 2.4 \times 10^{-2}$, corresponding to SD-FEC threshold for 20% OH SD-FEC (Fig.2c, solid). For TFP, according to the TFP principle itself, a partial channel overlap is present even when channel lasers are at nominal frequency, but the resulting ICI is well managed by LDPC/BCJR turbo detection, guaranteeing error free operation for OSNR as low as 16 dB (1 dB better than Nyquist-WDM DP-16QAM at same SE). Impairments due to laser frequency deviations from nominal value are also limited, thanks to the spectral shape and the intrinsic robustness to ICI of the TFP technique (Fig.2c, dashed).

In the second case (Fig.2d), the central and left channels move towards and they experience an overlap that progressively increases up to 2 GHz. The right channel experiences the same detuning as the central, so their relative spacing is preserved at 21 GHz. For Nyquist-WDM, the guard band is not sufficient to prevent from strong ICI and the transmission suffers a huge penalty when the channels overlap (Fig.2d, solid). TFP, on the contrary, is much more tolerant to laser frequency deviations and the impact of channel overlap is limited also for the worst case of 2 GHz detuning, where the corresponding penalty is 1 dB (Fig.2d, dashed).

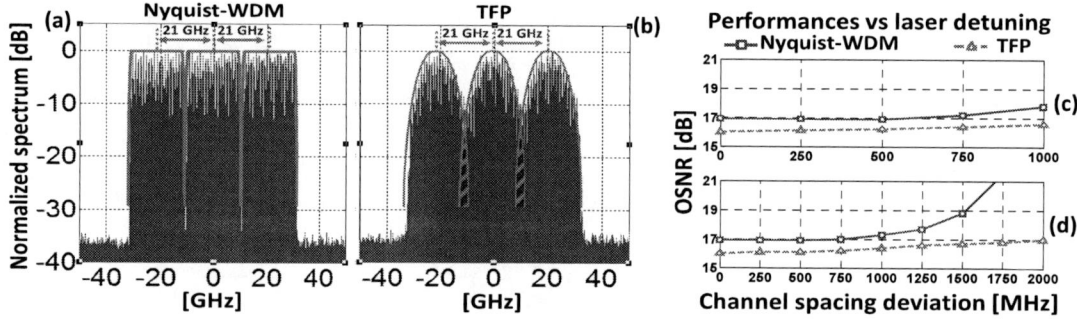

Fig. 2. Optical spectrum for the case of Nyquist-WDM (a) and TFP (b). Net spectral efficiency is 6.35 bit/s/Hz in both cases. Performance evaluation as a function of laser frequency deviation from nominal value demonstrates higher tolerance of TFP and lower OSNR requirements for error-free operations (c-d).

ACKNOWLEDGMENT

This work is partly supported by the European project "RAPIDO" (619806) and the H2020 project "ROAM" (645361).

REFERENCES

[1] N. Sambo et al., "Toward high-rate and flexible optical networks," Communications Magazine IEEE, vol.50, no.5, p. 66 (2012)

[2] M. Secondini et al., "Optical Time–Frequency Packing: Principles, Design, Implementation, and Experimental Demonstration," in Journal of Lightwave Technology, vol. 33, no. 17, pp. 3558-3570, Sept.1, 1 2015.

[3] G. Meloni et al., "Experimental Comparison of Transmission Performance for Nyquist WDM and Time–Frequency Packing," in Journal of Lightwave Technology, vol. 33, no. 24, pp. 5261-5268, Dec.15, 15 2015.

[4] T.J. Xia et al., "High Capacity Field Trials of 40.5 Tb/s for LH Distance of 1,822 km and 54.2 Tb/s for Regional Distance of 634 km", Proc. OFC, PDP5A.4, (2013).

[5] R. Cigliutti et al., "Ultra-Long-Haul Transmission of 16x112 Gb/s Spectrally-Engineered DAC-Generated Nyquist-WDM PM-16QAM Channels with 1.05x(Symbol-Rate) Frequency Spacing", Proc. OFC, OTh3A.3, (2012).

[6] F. Fresi et al., "Impact of Optical and Electrical Narrowband Spectral Shaping in Faster than Nyquist Tb Superchannel", Photon. Technol. Lett. vol.25, no.23, p.2301 (2013)

[7] G. Colavolpe et al., "Robust Multilevel Coherent Optical Systems With Linear Processing at the Receiver", J. Lightwave Technol., Vol. 27, no. 13, p.2357 (2009).

[8] L. Bahl et al, "Optimal decoding of linear codes for minimizing symbol error rate (Corresp.)", Information Theory, IEEE Transactions on, vol.20, no.2, p.284 (1974)

[9] J. Hagenauer, "The turbo principle: Tutorial introduction and state of the art", in Proc. International Symposium on Turbo Codes and Related Topics, p. 1-11 (1997)

WA2-11

OECC/PS2016

Branching Ratio of Optical Coupler for Cable Re-routing Operation Support System with No Service Interruption

Hiroshi Watanabe, Kazutaka Noto, Yusuke Koshikiya, Tetsuya Manabe

NTT Access Network Service Systems Laboratories, 1-7-1 Hanabatake, Tsukuba-city, Ibaraki, 305-0805, Japan
watanabe.h@lab.ntt.co.jp

Abstract: *We report the design and simulation of a suitable branching ratio for an optical coupler for a cable re-routing operation support system with no service interruption, and also report our experimental results.*

Keywords: *Optical coupler, branching ratio, cable re-routing operation support system*

I. INTRODUCTION

The optical access network is becoming increasingly important as a social infrastructure because it is used for several services, and the availability of communication needs to be improved. Therefore, we must reduce the time to interrupt communication during optical cable re-routing, and the quality of optical communication should be enhanced. We have been studying a cable re-routing operation support system that requires no service interruption. In other words, the system maintains telecommunication continuously when we perform cable re-routing work [1]. The communication stream provided by this support system passes through the current line and a detour line via two optical couplers. The system temporarily duplicates the current line with a detour line by using a special repeater with functions for measuring the difference between the current and detour optical lines and for adjusting the delay dynamically.

A characteristic of this cable re-routing support system is that communication signals travel between an Optical Line Terminal (OLT) and an Optical Network Unit (ONU), and between an OLT and a repeater via different branches of optical couplers. The optical losses of the optical couplers depend on the branching ratio. The optical coupler near the OLT was assumed to have been previously used in a Central Office (CO), and its branching ratio was fixed [2]. On the other hand, the optical coupler near the ONU must be newly installed. Therefore, it is necessary to study the most suitable branching ratio for the optical coupler of this system taking account of the acceptable maximum loss of the current and detour lines. However, this topic has not yet been sufficiently studied.

We report the design and simulation of a suitable optical coupler branching ratio for a cable re-routing operation support system with no service interruption, and also report experimental results showing the effectiveness of our simulation.

II. CONCEPT OF CABLE RE-ROUTING OPERATION SUPPORT SYSTEM WITH NO SERVICE INTERRUPTION

Fig. 1 shows the procedure of the cable re-routing operation support system with no service interruption. The procedure consists of four steps; Step 1: The streams of the media converter system between an OLT and an ONU are via two optical couplers. Step 2: Connect the detour line and adjust its length so that it is the same as that of the current line. Then, duplicate the signal transmissions of both the current and detour lines. Step 3: Disconnect the current line and replace it with the new line while the signal transmission is kept alive via the detour line. Then, adjust the detour line length to match that of the new line to make it possible to duplicate the signal transmission. Step 4: Stop the detour line signal transmission and remove the detour line and the repeater.

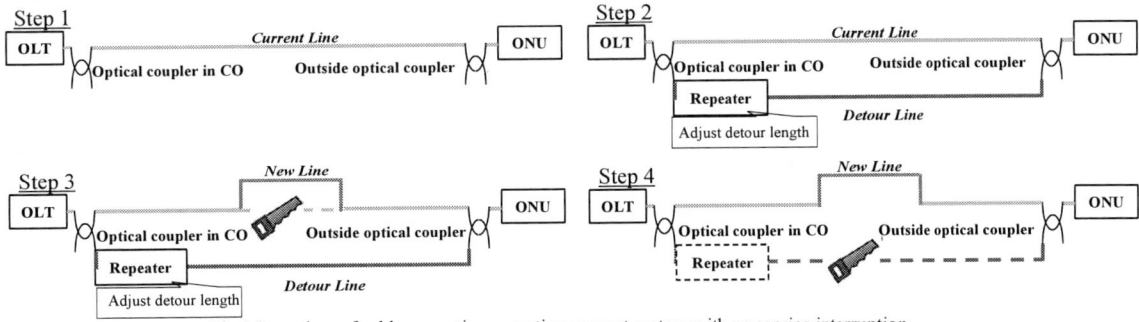

Fig.1 Procedure of cable re-routing operation support system with no service interruption

III. DESIGN OF SUITABLE BRANCHING RATIO FOR OPTICAL COUPLER

Fig. 2 shows the configuration of the system architecture for the cable re-routing operation support system with no service interruption. We propose a method for designing a suitable branching ratio for a coupler near an ONU to maximize the acceptable optical loss of the optical cable between two optical couplers. The current and detour lines are nearly the same length. The branching ratio of the outside optical coupler is assumed to be a certain value independent of the wavelength [3].

Step 1: Estimation of optical losses based on the branching ratio of the outside optical coupler

Optical losses $L_{coupler\_outside}$ (α) and $L_{coupler\_outside}$ (β), which constitute the branching ratio $\alpha : \beta$ of the outside optical coupler can be expressed as follows;

$$L_{coupler\_outside}(\alpha) = -10\log(\alpha) + L_{excess} \tag{1}$$

$$L_{coupler\_outside}(\beta) = -10\log(\beta) + L_{excess} \tag{2}$$

, where $\alpha+\beta$ equals one, and L_{excess} is the excess loss of the outside optical coupler.

Step 2: Estimation of the optical loss of the current and detour line with a power budget

The optical cable loss of the upstream current line $L_{upstream}$ is less than the maximum acceptable cable loss $L_{upstream\_max}$. Here we subtract the receive sensitivity of the OLT P_{rev_OLT} and the optical losses of the optical coupler in the CO $L_{coupler\_co}$ and that of the outside optical coupler for the current line $L_{coupler\_outside}(\alpha)$ from the launched power of the ONU P_{out_ONU}. As a result, the following equation can be derived.

$$0 < L_{upstream} < L_{upstream\_max} = P_{out\_ONU} - P_{rev\_OLT} - L_{coupler\_CO} - L_{coupler\_outside}(\alpha) \tag{3}$$

The upstream optical cable loss of the detour line $L'_{upstream}$ is also less than the maximum acceptable cable loss $L'_{upstream\_max}$. Here we subtract the receive sensitivity of the repeater P_{rev_rep} and the outside optical coupler for the detour line, $L_{coupler\_outside}(\beta)$ from the launched power of the ONU P_{out_ONU}. As a result, the following equation can be derived.

$$0 < L'_{upstream} < L'_{upstream\_max} = P_{out\_ONU} - P_{rev\_rep} - L_{coupler\_outside}(\beta) \tag{4}$$

The optical cable loss of the downstream current and detour lines $L_{downstream}$ and $L'_{downstream}$ are expressed as eq. (5) and (6) in the same way as in (3) and (4).

$$0 < L_{downstream} < L_{downstream\_max} = P_{out\_OLT} - P_{rev\_ONU} - L_{coupler\_CO} - L_{coupler\_outside}(\alpha) \tag{5}$$

$$0 < L'_{downstream} < L'_{downstream\_max} = P_{out\_rep} - P_{rev\_ONU} - L_{coupler\_outside}(\beta) \tag{6}$$

, where P_{out_OLT}, P_{rev_ONU} and P_{out_rep} are the launched power of the OLT, and the receive sensitivity of the ONU, and the launched power of the repeater, respectively.

Step 3: Comparison of the optical cable losses of the current and detour lines

Transmission is always possible on the current line at least in step 1 of Fig. 1. The maximum acceptable cable loss of the detour line both upstream and downstream must be more than that of the current line. Therefore, the following equation can be obtained.

$$L_{upstream\_max} < L'_{upstream\_max} \tag{7}$$

$$L_{downstream\_max} < L'_{downstream\_max} \tag{8}$$

Step 4: Estimation of the suitable ratio of the optical coupler by comparing upstream and downstream results

The most suitable branching ratio for the optical coupler can be decided as that where the acceptable optical cable loss is at its maximum value in the conditions in eqs. (3) to (8).

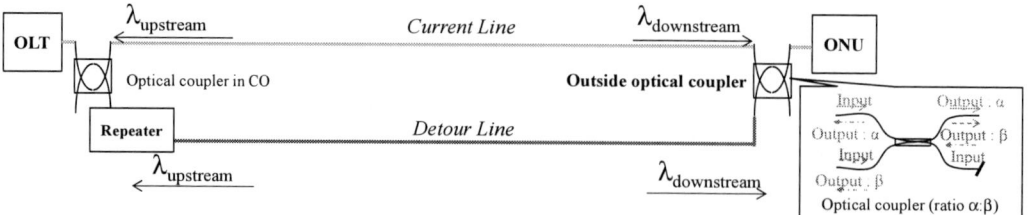

Fig.2 System architecture for the cable re-routing operation support system with no service interruption

IV. SIMULATION OF SUITABLE BRANCHING RATIO FOR OPTICAL COUPLER

Based on the design method discussed above, we simulated the most suitable branching ratio for the outside optical coupler. The characteristics of the OLT and ONU transmission comply with IEEE802.3ah [4], the data stream bitrates were 1.25Gbps, and the upstream and downstream wavelengths were 1.31, and 1.49 μm, respectively. Table 1 shows the parameters for the simulation, and each value was a typical value.

Fig.3 shows a simulation result from eq. (3) to eq. (6). In the upstream case, the suitable branching ratio is less than 0.80, according to eq. (7). In the downstream case, the suitable branching ratio is less than 0.65, according to eq. (8). Therefore, the most suitable branching ratio for the outside optical coupler is 0.65 because the acceptable optical cable loss between the optical couplers was at its maximum value.

V. EXPERIMENT TO CONFIRM A SUITABLE BRANCHING RATIO OF AN OPTICAL COUPLER

We conducted an experiment with the configuration shown in Fig. 2 to confirm the simulation result regarding the acceptable maximum loss of the optical cable. We measured the upstream and downstream bit error rates (BER) between an OLT and an ONU, or between an OLT and a repeater, using a commercially available optical coupler where the branching ratio of the outside optical coupler was 0.6. Fig. 4 shows the relationship between BER and optical cable loss for optical couplers, and the loss value at BER = 10^{-12} was estimated at 26.89 dB with a downstream detour line. Fig. 3 shows four experimentally obtained points whose loss values are at BER = 10^{-12}, and it also shows that the experimental values are close to the simulation values, namely they are from 0.2 to 0.5 dB more than the simulation values. Therefore, we confirmed the effectiveness of our simulation to determine a suitable branching ratio for an optical coupler.

TABLE 1 PARAMETERS FOR THE SIMULATION OF A SUITABLE BRANCHING RATIO FOR AN OPTICAL COUPLER

$P_{rev\ OLT}$	$P_{out\ OLT}$	$P_{rev\ ONU}$	$P_{out\ ONU}$	$P_{rev\ rep}$	$P_{out\ rep}$	$L_{coupler\_co}$	L_{excess_out}
-28.7 dBm	-2.4 dBm	-29.1 dBm	-3.1 dBm	-31.1 dBm	-1.4 dBm	1.2 dB	0.5 dB

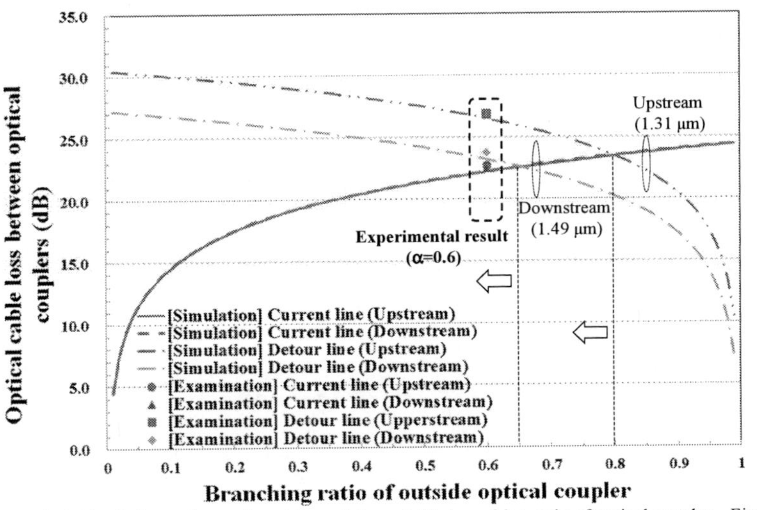

Fig.3 Simulation and experimental result for suitable branching ratio of optical coupler   Fig.4 BER result (detour line, downstream)

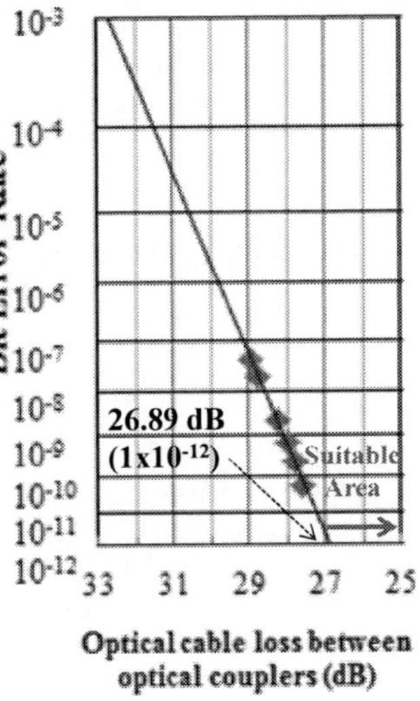

VI. CONCLUSIONS

We described our proposed method for designing a suitable branching ratio for an optical coupler for a cable re-routing operation support system with no service interruption. We simulated a suitable branching ratio, and showed that the most suitable value was 0.65. We confirmed our simulation was effective because the experimental values corresponded well with the simulation values.

REFERENCES

[1] T. Manabe, M. Inoue, H. Hirota, T. Kawano, T. Uematsu, K. Okamoto, T. Kiyokura, Y. Koshikiya, and K. Katayama, "Temporary optical coupler and dynamic delay adjustment technologies for optical cable re-routing operation support systems," proceedings of the 64th IWCS Conference, pp. 709-714, 2015.

[2] Y. Enomoto, H. Izumita, K. Mine, S. Uruno, and N. Tomita, "Design and performance of novel optical fiber distribution and management system with testing functions in central office," J. Lightw. Technol. Vol.29, No.12, pp.1818-1834, June 2011.

[3] Y. Inoue, M. Ishii, Y. Hida, M. Yanagisawa, and E. Enomoto, "PLC Components Used in FTTH Access," NTT Tech. Rev.,vol 3. No.7, pp.22-26, 2005.

[4] IEEE Std 802.3ah-2004.

Investigation of Switch Reduction of a Multi-Dimensional Optical Node for Spatial Division Multiplexing Networks

Hiroki SASAGO, Hiroyuki UENOHARA

Precision and Intelligence Laboratory, Tokyo Institute of Technology

uenohara.h.aa@m.titech.ac.jp

Abstract: *We propose multi-dimensional optical nodes with multicore fiber input and output. Compared with conventional WSS-based ROADM, reduction rate of switches of 23% and cumulative loss reduction from 96dB to 16dB are confirmed in simulation.*

Keywords: *Multi-dimensional ROADM, multi-core fiber, Wavelength Selective Switch*

1. INTRODUCTION

For meeting the demand of the increase of Internet traffic, high-speed and large capacity photonic network is necessary. In metropolitan area networks, reconfigurable optical add-drop multiplexers (ROADMs) that converges WDM data exchange and flexible path management are used as main subsystems. At present, single core-fibers are utilized in ROADM-based photonic networks. However multi-core fiber (MCF)-based networks would be introduced in the future to handle the increasing traffic [1], because it has already been intensively studied in the long-haul optical transmission systems. Under this situation, the size of the switch increases drastically because SDM as well as WDM data should be handled in the ROADMs [2] [3]. In addition, in order to configure multi-core fiber-based optical nodes that can be switched several types of granular size, optimization of the switch structure is required [4]. However, few reports have been investigated in detail, to the best of our knowledge.

In this report, we propose a novel multi-dimensional optical node to resolve the problem of switch scale, and we will present the effectiveness of the switch reduction in simulation.

2. NODE ARCHITECTURE

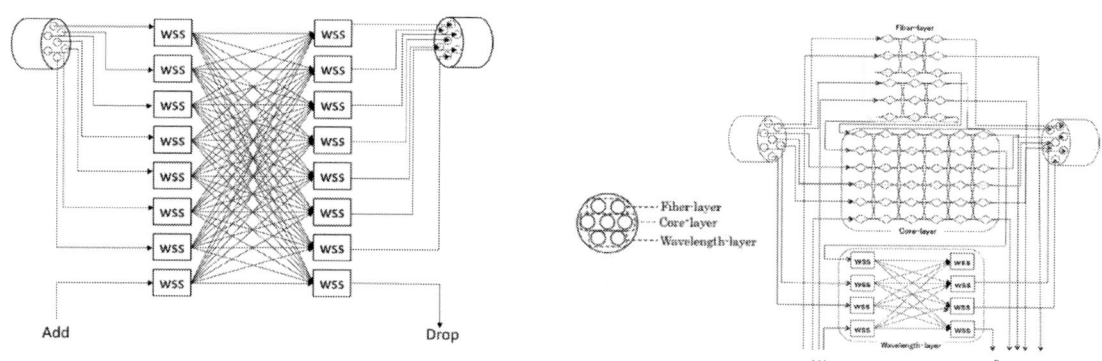

(a) WSS-ROADM with MCF   (b)Proposed Multi-dimensional ROADM node

Fig. 1. Architectures of (a) WSS-ROADM with MCF (b) Multi-dimensional optical node with MCF input/output

Figure 1 shows the architecture of (a) the conventional ROADM with MCF input and output, and (b) the proposed multi-dimensional optical node. In the proposed structure, it consists of 3-hierarchy, Fiber-layer, Core-layer, and Wavelength-layer that can switch different granularities to reduce the number of switches. It is divided into 3-layers. At Fiber –layer, it transmits WDM signals from specific core to specific core without WDM demultiplexing and interconnection between cores. At Core –layer, WDM signals can be cross-connected from specific core to arbitrary cores without WDM demultiplexing. At Wavelength –layer, arbitrary wavelength channels can be cross-connected to arbitrary outputs by WDM demultiplexing with WSSs.

The total number of switch components can be considered as follows: L, M, N, S and W represent the number of cores assigned to Fiber-layer, Core-layer, and Wavelength-layer, the number of cores in a MCF, and the number of multiplexed wavelengths, -respectively. Then the switch scale of conventional WSS-ROADM is expressed as $\{W(N + 1)\}^2$. On the other hand, the switch scale of the proposed structure is $(L + 2)^2 + (M + 2)^2 +$

$\{W(N + 2)\}^2$. "2" –in the equations means the number of add/drop port and drop port from higher layer taken into account.

3. SIMULATION METHOD

In this section, the number of cores in Core-layer and Wavelength-layer is optimized with ring network topology. We used simulation software OPNET (Riverbed). Fig.2 shows a model of a 2-layer ROADM node constructed on OPNET.

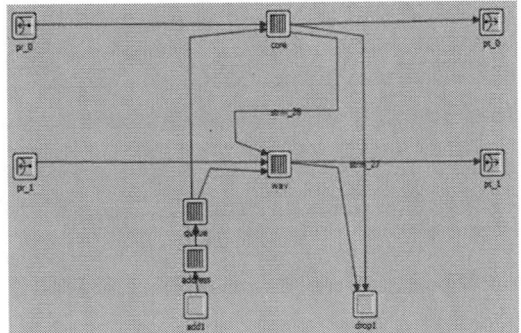

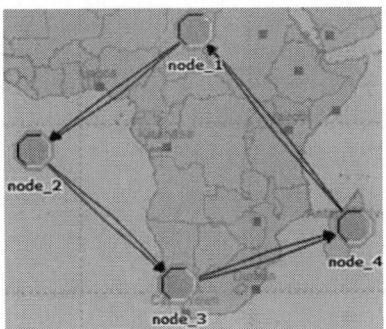

Fig.2 Model of a 2-layer ROADM node on OPNET™   Fig.3 Ring network on OPNET™ (N is 4)

Firstly, the signal is input from "add1" and the destination address is attached at "address". Secondly, the path computation is performed at "queue". The delay of path computation between Core-layers, or Core-layer and Wavelength-layer was set to be 100ms. In the case of path computation between Wavelength-layers, the delay of 200ms was assumed.

As for transmission between Core-layers, signals can be transmitted if core is not used or the destination of signal is the same as the signal that has already been transferred and there are available wavelength paths at the same time. On the other hand, transmission between Wavelength-layers, the signals can be sent even if the signal cannot be assigned in the Core-layer, and if there are available wavelength paths.

The arrival process of signal is assumed to be Poisson arrival and the transmission capacity (produce of bitrate of signal by number of wavelength) of 320Gbit/s~1.9Tbit/s were set. The number of multiplexed wavelength W was 8.

Optimized 2-layer ROADM node was compared with throughput and the amount of switches of a conventional ROADM node. To optimize the number of cores in Core-layer and Wavelength-layer, a simulation using a ring network was performed.

The number of nodes in ring network was set to be N. Fig.4 shows the example of a ring network when N is 4.

4. SIMULATION RESULTS

The simulation results of throughput as a function of traffic load are indicated in Fig.4 when N is 10(Fig.4 (a)) and N is 16(Fig4 (b)), respectively.

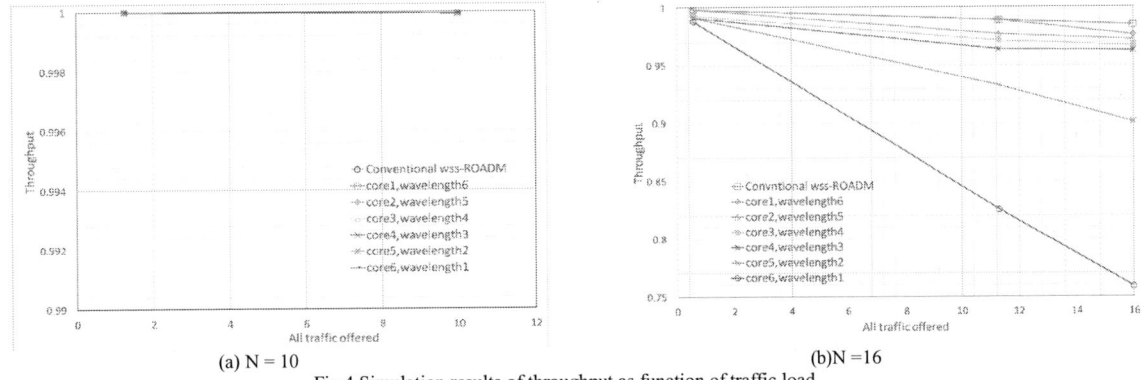

(a) N = 10   (b)N =16

Fig.4 Simulation results of throughput as function of traffic load

When the number of node N is 10, the throughput of all configuration is as same as that of the conventional WSS-ROADM. However, when the number of nodes N is 16, the throughput is close to that of WSS-ROADM in case of large number of cores assigned to wavelength-layer. This tendency is considered because signals cannot be transmitted unless the destination address is the same as the existent paths or no wavelength paths are assigned in Core-layers.

Fig.5 shows the dependence of throughput on the number of nodes with various number of Core-layer in the MCF.

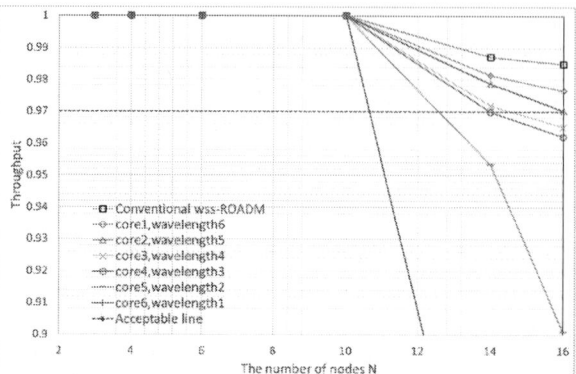

Fig.5 Simulation results of throughput (as a function of numbers of nodes in Ring network)

As can be seen in this figure, the throughput of all configuration is as same as that of the conventional WSS-ROADM when the number of nodes N is less than 10.

If the number of multiplexed wavelengths W is 8 and the number of cores S is 7, the switch scale of the WSS-ROADM is $\{8(7+1)\}^2 = 4096$. Meanwhile, the switch scale of the 2-layer ROADM node is $8 \times 8 + \{8(1+2)\}^2 = 640 (M = 6, N = 1)$.Therefore, the reduction rate of the switches is 84% at maximum.

When the number of nodes N is 16, on the other hand the optimized configuration for maintaining high throughput is the conditions of the numbers of Core-layer of 1, and Wavelength-layer of 6 or Core-layer of 2, and Wavelength-layer of 5.In this case the switch scale of WSS-ROADM is again $\{8(7+1)\}^2 = 4096$. In the latter case, the switch scale of 2-layer ROADM node is $(2+2)^2 + \{8(5+2)\}^2 = 640$.Therefore, the reduction rate of switches is 23%. When this simulation assumes a situation where traffic is concentrated, further reduction rate of switches could be expected.

Next, the cumulative loss of the nodes are considered. When the insertion loss of a WSS is 6dB and that of matrix switch is assumed to be 2dB, the cumulative loss of a WSS-ROADM is $8(6+6) = 96$dB.Minimum cumulative loss of a 2-layer ROADM node, on the other hand, is $2 \times 8 = 16$dB. This means that the proposed structure is also effective for suppressing the residual loss of the optical nodes.

5. CONCLUSIONS

In conclusion, we proposed a multi-dimensional ROADM node consisting of fiber-, core-, and wavelength-layer and we optimized the number of cores in each layer in simulation. By using the optimal configuration, the reduction rate of switches is 84% when the number of nodes is less than 10.When the number of nodes is 16, the reduction rate of switches is 23% and the cumulative loss can be reduced from 96dB to 16dB.

ACKNOWLEDGMENT

We would like to thank Prof. Emeritus K. Iga, Prof. Emeritus K. Kobayashi, Prof. F. Koyama and Assoc. Prof. T. Miyamoto for their encouragements and discussions.

REFERENCES

[1] L.E. nelson, M.D. Feuer, K. Abedin, X. Zhou, T.F. Taunay, J.M. Fini, B. Zhu, R. Isaac, R. Harel, G. Cohen, and D.M. Marom, "Spatial superchannel routing in a two-span ROADM system for space division multiplexing", J. Lightwave Technol., vol.32, No.4, pp.783-789(2014).

[2] S. Mitsui, H. Hasegawa, and K. Sato, "Hierarchical optical path cross-connect node architecture using WSS/WBSS", Photonics Switching 2008, S-04-1, Hokkaido, Japan August 4-7,2008.

[3] S. Kakehashi, H. Hasegawa, K. Sato, O. Moriwaki, and K. Kamei, "Optical cross-connect switch architecture for hierarchical optical path networks", IEICE Trans. Commun., vol.E91-B, No.10, October 2008, pp.3174-3184

[4] N. Amaya, and I. Henning. et al "Fully-elastic multi-granular network with space/frequency/time switching using multi-core fibers and programmable optical nodes", Opt. Express, 21(7),pp8865-pp8872, 2013

WA2-111

Bit-Error-Rate Performance in Optical 16QAM Recognition by Maximum Output with Optical Waveguide Circuit

Kensuke Inoshita, Hiroki Kishikawa, and Nobuo Goto

Dept. of Optical Science and Technology, Tokushima University, 2-1 Minamijosanjima-cho, Tokushima, 770-8506 Japan
{c501548002, kishikawa.hiroki, goto.nobuo}@tokushima-u.ac.jp

Abstract: We have proposed optical waveguide circuits for recognition of optical QAM codes by detecting the minimum or maximum output port. This report clarifies the BER performance for the device recognized by the maximum output port.
Keywords: optical QAM code, code recognition, optical waveguide circuit, BER, numerical simulation

I. INTRODUCTION

In photonic networks, optical label routing will be a practically promising solution because of its simplicity and flexibility. Optical label recognition is one of key functions in label routers. Various methods to recognize optical labels in on-off-keying (OOK) and multi-level phase-shift-keying (PSK) modulation formats have been investigated.[1,2] The authors have proposed waveguide circuits for recognizing optical quadrature amplitude modulation (QAM) codes. We reported two kinds of schemes depending on whether the detecting output port is exiting minimum or maximum output.[3,4]

In this report, we discuss noise tolerance in label recognition processing by detecting maximum-output port using numerical simulation, where the proposed integrated-optic circuit is modeled by a combination of optical discrete devices. We clarify the bit-error-rate (BER) performance.

II. RECOGNITION OF 16QAM CODES BY MAXIMUM OUTPUT

The optical electric field E_{16QAM} can be written as $E_{16QAM} = E_0 e^{j\pi/4} \left(2e^{jn\pi/2} + e^{jm\pi/2}\right)$, where n, m=0,1,2,3. We consider recognition of the 16QAM code by cascaded connection of two circuits for QPSK code recognition. Fig.1(a) shows a waveguide circuit for QPSK phase recognition circuit (QPRC) followed by phase adjustment waveguides [5]. This circuit consists of a 3-dB directional coupler, two Y-branches, and an asymmetric X-junction coupler. The output fields are related to the input fields by

$$\begin{pmatrix} E_{out}^{(1)} \\ E_{out}^{(2)} \\ E_{out}^{(3)} \\ E_{out}^{(4)} \end{pmatrix} = \frac{1}{2}\begin{pmatrix} 1 & -j \\ 1 & 1 \\ 1 & -1 \\ 1 & j \end{pmatrix}\begin{pmatrix} E_{in}^{(1)} \\ E_{in}^{(2)} \end{pmatrix}. \qquad (1)$$

Using this QPRC, a recognition circuit for 16QAM codes can be formed as shown in Fig.1(b). This circuit consists of an input signal part and a reference signal part. The incident signal and reference are assumed to be equal to $2\sqrt{2}E_{in}$ and $\sqrt{2}E_{ref}$, respectively. The input signal part consists of three Y-branches and an attenuator having amplitude attenuation

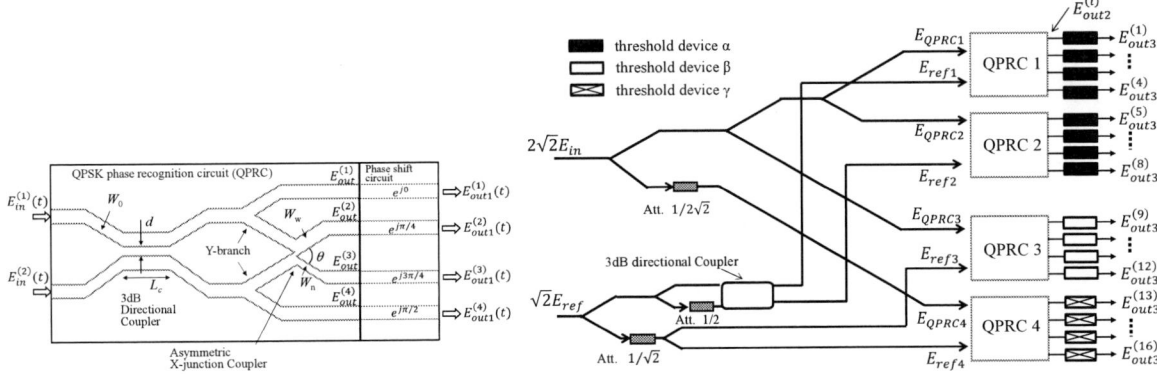

Fig.1 (a) Optical waveguide circuit (QPRC) for recognition of QPSK code and phase shift circuit. (b) The proposed 16QAM recognition circuit by the maximum output.

*This research was supported in part by JSPS KAKENHI (15H06443).

733

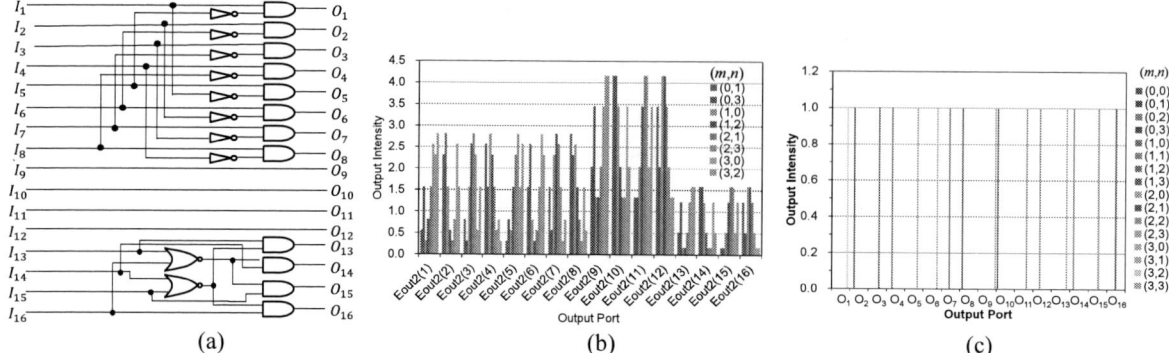

Fig.2 (a) Post-processing logic circuit, (b) output intensities for a part of codes before thresholder, and (c) outputs from the logic circuit.

coefficient of $1/(2\sqrt{2})$. The reference signal part consists of three Y-branches, attenuators having amplitude attenuation coefficient of $1/(\sqrt{2})$ and 1/2, and a 3-dB directional coupler. By feeding the output fields to the four QPRCs, sixteen outputs are obtained. Threshold devices α, β, and γ are connected at the output ports of QPRCs 1and 2, QPRC 3, and QPRC 4, respectively. The threshold devices α, β, and γ make the output zero for input signal intensity below $2.7|E_0|^2$, $6.0|E_0|^2$, and $0.5|E_0|^2$, respectively. Fig.2(a) shows the post-processing logic circuit. The output intensities before the thresholders and after the post-processing logic circuit are shown in (b) and (c), respectively. It is found that output signals for all of 16QAM codes exit at different output ports.

III. NOISE TOLERANCE IN RECOGNITION OF 16QAM CODES

Noise tolerance in the proposed code recognition was numerically simulated using OptiSystem (Optiwave Systems Inc.). An optical system setup is shown in Fig.3(a). The optical waveguide circuit of Fig.1(a) was modeled with discrete optical devices as shown in (b). A post-processing electric circuit and followed circuits of a combiner, multi-level thresholder, and a pulse amplitude demodulator can retrieve the incident 16QAM code. As optical 16QAM signal, RZ and NRZ pulse trains were considered. Fig.5(a) shows an example of simulated signals at various points in the recognition circuit. Incident 16QAM constellation is also shown. The simulated BER performance is shown in (b) as a function of the OSNR at the input of the recognition circuit, where the BER performance for the recognition circuit detected form the minimum-output port [6] is also shown for comparison. It is found that the OSNR required for BER of 10^{-3} is around 42dB for both RZ and NRZ signals, whereas the OSNR is 28dB and 32dB for RZ and NRZ signals in the minimum output based circuit. Therefore, From these results, it is concluded that the noise tolerance is better in the minimum output based circuit.

IV. CONCLUSION

The optical waveguide circuit for recognition of 16QAM codes was discussed from the viewpoint of noise tolerance. We will further investigate to improve the circuit having larger noise tolerance by improving the QPRC characteristics.

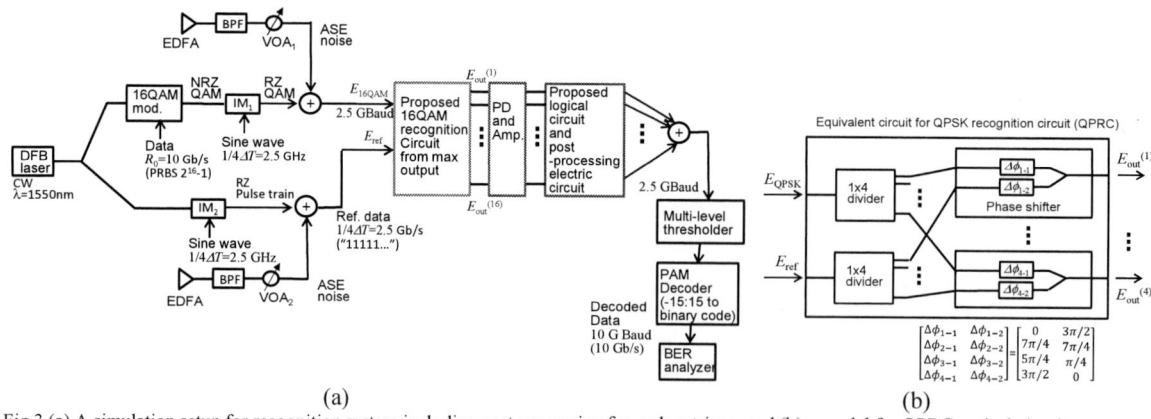

Fig.3 (a) A simulation setup for recognition system including post-processing for code retrieve, and (b) a model for QPRC optical circuit.

OECC/PS2016

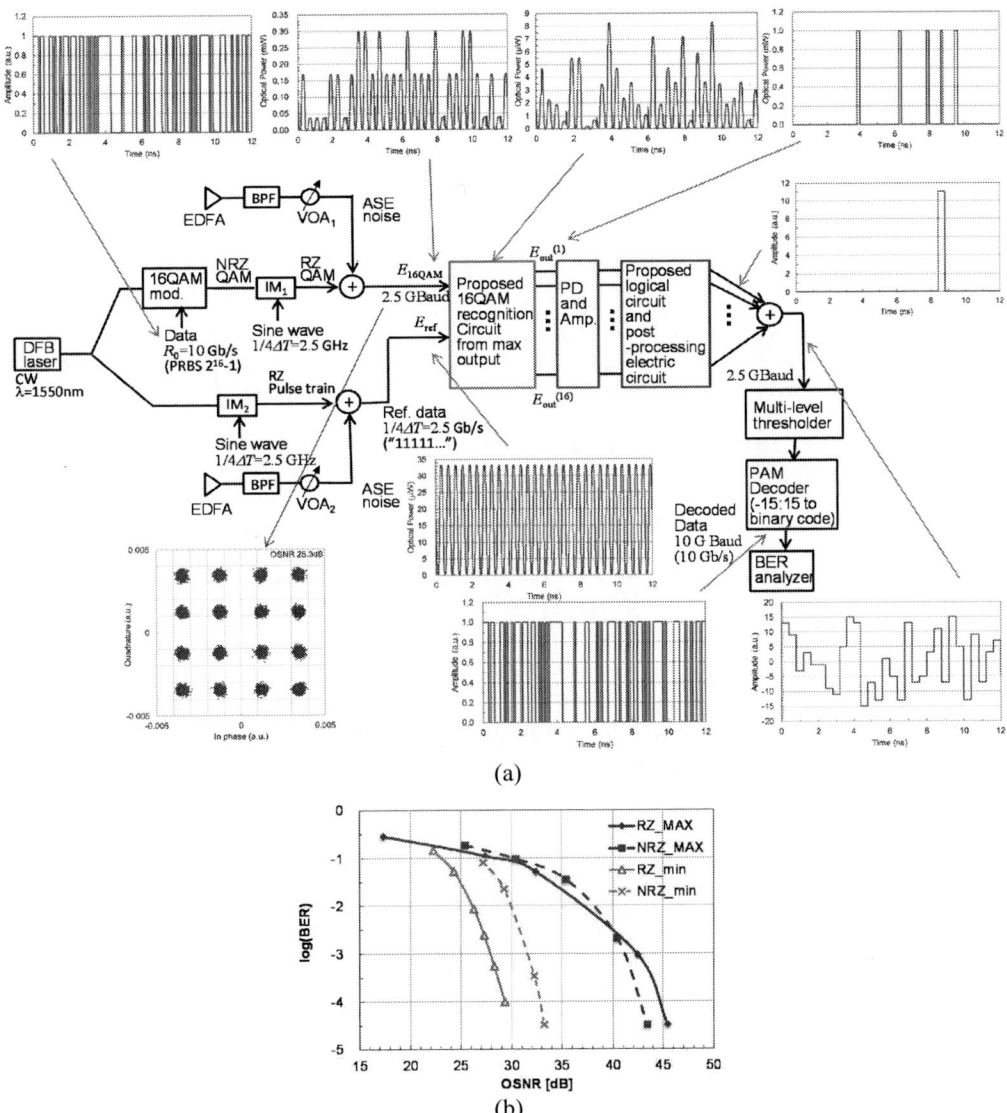

(a)

(b)

Fig. 4(a) Waveform and the constellation of the signals in the recognition circuit, and (b) BER performance for the maximum-output-based circuit in comparison with the minimum-output-based circuit.

REFERENCES

[1] H. Hiura, Y. Makimoto, N. Goto, & S. Yanagiya, "Wavelength-Insensitive Integrated-Optic Circuit Consisting of Asymmetric X-junction Couplers for Recognition of BPSK Labels."IEEE/OSA J. Lightwave Tech., vol.27, no.24, p.5543-5551, 2009.

[2] A. Ihara, H. Kishikawa, N. Goto, and S. Yanagiya, "Passive Waveguide Device Consisting of Cascaded Asymmetric X-junction Couplers for High-Contrast Recognition of Optical BPSK Labels," IEEE/OSA J. Lightwave Tech., vol.29, no.91, pp.1306-1313, 2011.

[3] K. Inoshita, H. Kishikawa, Y. Makimoto, N. Goto, and S. Yanagiya, "Proposal of Optical Waveguide Circuits for Recognition of Optical QAM Codes," IEEE/OSA J. Lightwave Tech., vol.31, pp.2271-2278, 2013.

[4] K. Inoshita, N. Goto, and S. Yanagiya,"Recognition of 16QAM Codes by Maximum Output with Optical Waveguide Circuits, Thresholders, and Post-Processing Logic Circuit," IEICE Trans. Electron, vol.E97-C, no.5, pp.448-454, 2014.

[5] Y. Makimoto, H. Hiura, N. Goto, and S. Yanagiya, "Waveguide-Type Optical Circuit for Recognition of Optical QPSK Coded Labels in Photonic Router," IEEE/OSA J. Lightwave Tech., vol.27, no.1, pp.60-67, 2009.

[6] A. Takahashi, K. Inoshita, Y. Hama, N. Goto, and S. Yanagiya, "Bit-Error-Rate Performance in Optical 16QAM Recognition with Integrated-Optic Circuit," ECIO-MOC 2014, Nice, P028, 2014.

WA2-112

Analysis of a Si-nanocrystal Strip-Loaded Waveguide for Nonlinear Applications

Pengjiang Wei[1], S. H. Wang[2], Brent E. Little[3] and Sai Tak Chu[1,*]

[1]Department of Physics and Materials Science, City University of Hong Kong, Hong Kong, China
[2]Department of Microelectronics, Fuzhou University, Qi Shan Campus, Fuzhou, 350108, China.
[3]State Key Lab of Transient Optics and Photonics, Xi'an Institute of Optics and Precision Mechanics, CAS, China
*E-mail address: saitchu@cityu.edu.hk

Abstract: We investigate the nonlinear properties of the doped silica glass integrated waveguide platform with a Si-nc strip-loaded structure. We show the nonlinear properties is drastically enhanced by adding a Si-nc layer in the waveguide core.

Keywords: High index contrast waveguide, nonlinear optics, integrated optics

INTRODUCTION

The high refractive index (n_0) and large Kerr nonlinearity (n_2) properties of silicon-on-insulator (SOI) waveguide have initiated the field of complementary metal-oxide-semiconductor (CMOS) compatible nonlinear integrated optical circuits. Nonlinear phenomena such as Raman gain and lasing [1, 2], time lensing [3], optical parametric amplification [4], in the SOI platform have been reported and a thorough review of the CMOS compatible platforms for nonlinear optics applications has been introduced in [5]. However, SOI nonlinear waveguide circuits suffer from two-photon absorption (TPA) which limits their potential commercial deployment in the telecommunications band [6]. To further increase the waveguide nonlinearity, one suggestion is to engineer the waveguide geometry by introducing a slot inside the SOI nanowire core region and fill the slot with other nonlinear materials such as silicon nanocrystals (Si-nc) or even SiO_2 [7-9]. It is to create an intense localized field at the slot area to enhance nonlinearity. However, challenges remain, especially in the need to reduce both the high coupling loss between these waveguides and optical fiber and the high propagation loss before they can become a viable solution. There are other CMOS compatible platforms that were introduced with similar nonlinear figure-of-merit as SOI, they are based on silicon nitride (Si_3N_4) and highly doped silica glass, respectively [10-12]. Although their intrinsic nonlinearities are far below materials such as Si and AlGaAs, they possess negligible TPA nonlinear loss and linear loss. The fact that highly doped silica glass devices which were originally developed for linear applications have such impressive nonlinear performance leads us to believe that there is a tremendous opportunity to improve their nonlinear properties with a combination of proper waveguide design, fabrication modification and material selection. In this work, we investigate enhanced nonlinear properties of highly doped silica glass based waveguides without sacrificing their low linear and nonlinear loss. The particular method is to embed a nonlinear material within the core of the highly doped silica glass waveguide to enhance the optical confinement without the large index discontinuity. A comparison of highly doped silica glass waveguide embedded with different materials is presented for the optimal material selection. Then with a geometric tuning on the proposed waveguide for instance, the nonlinear parameter has not only been enhanced as expected, but also flat dispersion can be realized.

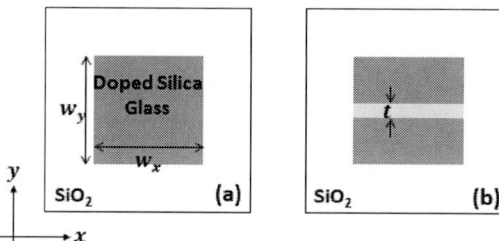

Fig. 1. Waveguide cross sections. (a) A highly doped silica glass channel ($w_x \times w_y$) waveguide with n_{core}=1.60 as the reference in the comparison. (b) A strip/slot ($w_x \times t$) waveguide within embedded material at the center of the core. The cover material is SiO_2.

I. COMPARISON OF THE STRIP MATERIAL IN THE STRIP-LOADED WAVEGUIDE PLATFORM

Fig. 1 shows the waveguide cross-section used in this analysis. A highly doped silica glass channel waveguide with dimension w_x by w_y with SiO_2 [5] cladding shown in Fig. 1(a) is used as the reference in our comparison. Fig. 1(b) is the schematic of the waveguide used in this study where the center strip/slot with thickness t of the core region is filled with different materials of interest, the structure is similar to the slot waveguide [7-9], except that the refractive index in the

736

strip/slot is not always lower than the core material. For evaluating nonlinear characteristics of optical waveguides, the nonlinear waveguide parameter γ is given as [13]:

$$\gamma = \frac{k_0}{Z_0^2} \frac{\int_{-\infty}^{\infty} \int_{-\infty}^{\infty} n_0^2 n_2 \left|\vec{E}\right|^4 dxdy}{\left|\int_{-\infty}^{\infty} \int_{-\infty}^{\infty} \mathrm{Re}\left[\vec{E} \times \vec{H}^*\right] \cdot \hat{z} dxdy\right|^2} \tag{1}$$

where Z_0 is the wave impedance in vacuum, k_0 is the free wave vector, Re[] means the real part, * represents the conjugate function and $\hat{z}$ is the unit vector along z-axes. γ indicates the accumulation of nonlinear phase per propagation length and power. In order to investigate waveguide properties, the finite element analysis (COMSOL Multiphysics) is used to calculate the effective mode index and both electric and magnetic fields, in which material information around $\lambda = 1.55\mu m$ is shown in Table I.

TABLE I

Material	n_0	$n_2(m^2/W)$	Ref
Highly doped silica glass	1.60	1.15×10^{-19}	[5]
SiO$_2$	1.45	0.25×10^{-19}	[5]
Si	3.45	5.0×10^{-18}	[14]
SiN	1.99	2.4×10^{-19}	[10]
Si-nc	1.95	4.8×10^{-17}	[15]

As a reference, the base structure shown in Fig. 1(a) with ($w_x = w_y = 2\mu m$, $t=0$) has a γ value around 0.13. The calculated γ of the TE mode at different strip/slot thickness t of the four selected embedded materials is shown in Fig. 2(a) for SiO$_2$, Si, SiN and Si-nc, and the inset is a magnified result inside the dashed ellipse. Fig. 2(b) shows their electric field distributions with the dominant component-E_x for a fixed thickness $t=0.07\mu m$. γ (the black square-line) decreases with increasing t, and is the result of weak localized field confinement at the interfaces between SiO$_2$ and highly doped silica glass layers, as t increases more power is trapped inside the low nonlinearity SiO$_2$ layer. In the comparison, waveguide with the Si strip shows the largest increase of γ, however its field profile shows that most of the power is confined inside the Si strip, thus it faces the same issues as the SOI waveguide. For the SiN and Si-nc cases, the enhancement of field confinement is more gradual, thus the mode profile exhibit a near circular mode profile, however the calculated γ of the Si-nc strip structure is close to two orders of magnitude larger than the SiN strip due to the very high intrinsic nonlinear property of Si-nc. Along with the fact the addition of the strip does drastically alter the field profile, we believe the placing of a very thin Si-nc strip in a conventional highly doped silica glass waveguide core will greatly enhance its nonlinearity without affecting other nice attributes of the highly doped silica glass platform.

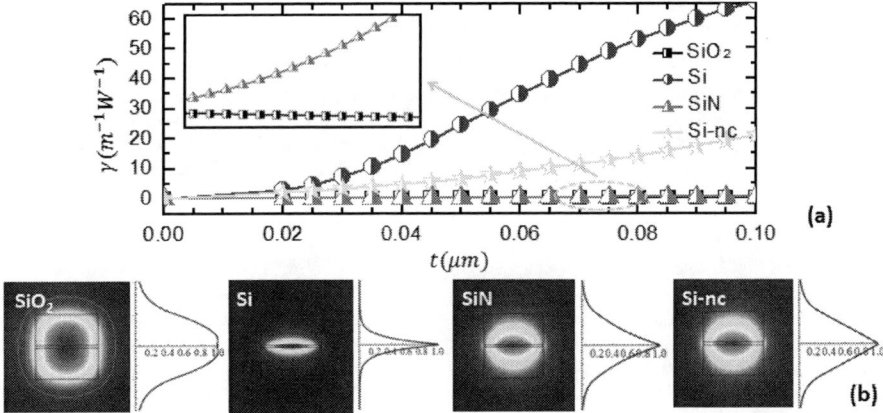

Fig. 2. (a) Calculated nonlinear waveguide parameter as a function of the strip/slot thickness t for different filling materials: SiO$_2$, Si, SiN, Si-nc. The inset is a magnified plot for lines by the dashed ellipse. (b) Electric modal fields with dominant component E_x for the fixed thickness $t=0.07\mu m$. $t=0$ means the waveguide (a) in Fig. 1 as the reference.

II. EXAMPLE: FLAT AND ZERO DISPERSION APPLICATION

We further investigate and engineer the dispersion properties of the proposed structure with the Si-nc strip at various strip sizes and positions suitable for nonlinear applications, such as flat and zero dispersion within the C-band. With a careful control in geometric parameters [16, 17] shown in Fig. 3(a), the strip inside the highly doped silica glass ($2 \times 4\mu m^2$) is designed with $0.55 \times 0.1\mu m^2$ and its embedded position δ (μm) can be defined by the distance between the core center and strip center. In order to demonstrate the validity of the modified structure, we simulate the group velocity dispersion $-(\lambda/c_0)(d^2 n_{eff}/d\lambda^2)$ as a function of strip position in Fig. 3(b). With the strip at the center (δ=0) of the core, the dispersion denoted by black square-line is low and flat. Nevertheless, while moving the strip along y-axes, the dispersion has become sharper, which is unwanted. Meanwhile, the nonlinear waveguide parameter is plotted in Fig. 3(c) and the result implies that with increasing δ less propagating power will be localized inside the strip, but its γ is still much larger than that of the reference. In this case, we obtain flat and low dispersion with the superior nonlinear performance, which can be further improved by adjusting the structural parameters. We will present more analysis result on the dispersion optimization at the conference.

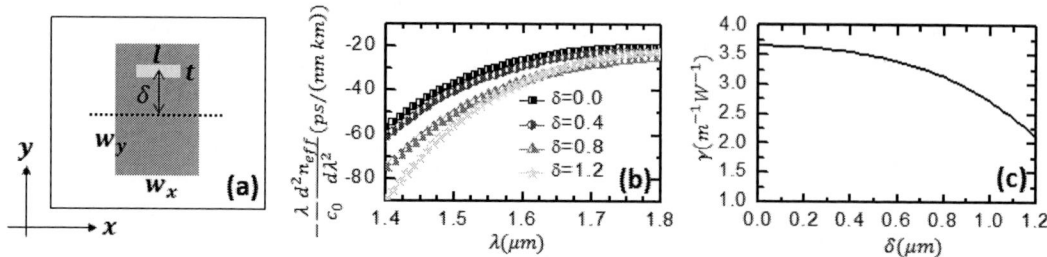

Fig. 3. (a) A Si-nc strip-loaded waveguide cross section. (b) Calculated group velocity dispersion of the waveguide of for different embedded strip position. (c) Nonlinear waveguide parameter vs the strip position δ.

III. CONCLUSION

Based on the proposed highly doped silica glass waveguide platform, we have demonstrated the embedded Si-nc strip is an optimal kind of material for improving nonlinear properties with altering the mode profile little. Further, by a careful control in geometric parameters, we can also obtain low and flat dispersion with the superior nonlinear performance. Our attempts provide a feasible method to improve nonlinear properties of highly doped silica glass waveguide platform, which is very useful for nonlinear optic applications

REFERENCES

[1] H. Rong, S. Xu, Ying-Hao Kuo, V. Sih, O. Cohen, O. Raday and M. Paniccia, "Low-threshold continuous-wave raman silicon laser," Nature Photon. vol. 1, pp. 232-237, April 2007.

[2] H. Rong, R. Jones, A. Liu, O. Cohen, D. Hak, A. Fang and M. Paniccia, "A continuous-wave ramman silicon laser," Nature vol. 433, pp. 725-728, February 2005.

[3] B. Jalali, D. R. Solli and S. Gupta, "Silicon photonics: Silicon's time lens," Nature Photon. vol. 3, pp. 8-10, January 2009.

[4] B. Kuyken, X. Liu, G. Roelkens, R. Baets, R. M. Osgood and W. M. J. Green, "50 dB parametric on-chip gain in silicon photonic wires," Opt. Lett. vol. 36, pp. 4401-4403, November 2011.

[5] D. J. Moss, R. Morandotti, A. L. Gaeta and M. Lipson, "New CMOS-compatible platforms based on silicon nitride and Hydex for nonlinear optics," Nature Photon. vol. 7, pp. 597-607, August 2013.

[6] T. K. Liang and H. K. Tsang, "Nonlinear absorption and raman scattering in silicon-on-insulator optical waveguides," IEEE J. Selected Topics in Quant. Electron. vol. 10, pp. 1149-1153, September 2004.

[7] J. Leuthold, W. Freude, J. M. Brosi, R. Baets, P. Dumon, I. Biaggio, M. L. Scimeca. F. Diederich, B. Frank and C. Koos, "Silicon organic hybrid technology-A platform for practical nonlinear optics," Proc. IEEE vol. 97, pp. 1304-1316, June 2009.

[8] Z. Kang, J. Yuan, X. Zhang, Q. Wu, X. Sang, G. Farrell, C. Yu, F. Li, H. Y. Tam and P. K. A. Wai, "CMOS-compatible 2-bit optical spectral quantization scheme using a silicon-nanocrystal-based horizontal slot waveguide," Sci. Rep. vol. 4, pp. 7177, November 2014.

[9] P. Muellner, M. Wellenzohn and R. Hainberger, "Nonlinearity of optimized silicon photonic slot waveguides," Opt. Express vol. 17, pp. 9282-9287, May 2009.

[10] K. Ikeda, R. E. Saperstein, N. Alic and Y. Fainman, "Thermal and Kerr nonlinear properties of plasma-deposited silicon nitride/silicon dioxide waveguides," Opt, Express vol. 16, pp. 12987-12994, August 2008.

[11] D. T. H. Tan, K. Ikeda, P. C. Sun and Y. Fainman, "Group velocity dispersion and self-phase modulation in silicon nitride waveguides," Appl. Phys. Lett. vol. 96, pp. 61101, February 2010.

[12] M. Ferrera, L. Razzari, D. Duchesne, R. Morandotti, Z. Yang, M. Liscidini, J. E. Sipe, S. Chu, B. E. Little and D. J. Moss, "Low-power continuous-wave nonlinear optics in doped silica glass integrated waveguide structures," Nature Photon. vol. 2, pp. 737-740, November 2008.

[13] T. Sato, S. Makino, Y. Ishizaka, T. Fujisawa and K. Saitoh, "A rigorous definition of nonlinear parameter γ and effective area Aeff for photonic crystal optical waveguides," J. Opt. Soc. Amer. B, Opt. Phys, vol. 32, pp. 1245-1251, June 2015.

[14] E. Dulkeith, Y. A. Vlasov, X. Chen, N. C. Panoiu and R. M. Osgood, "Self-phase-modulation in submicron silicon-on-insulator photonic wires," Opt. Express vol. 14, pp. 5524-5534, May 2006.

[15] L. Zhang, Y. Yue, Y. Xiao-Li, J. Wang, R. G. Beausoleil and A. E. Willner, "Flat and low dispersion in highly nonlinear slot waveguides," Opt. Express vol. 18, pp. 13187-13193, June 2010.

[16] L. Zhang, Y. Yue, R. G. Beausoleil and A. E. Willner, "Flattened dispersion in silicon slot waveguides," Opt. Express vol.18, pp. 20529-20534, September 2009.

[17] S. Mas, J. Caraquitena, J. V. Galan, P. Sanchis and J. Marti, "Tailoring the dispersion behavior of silicon nanophotonic slot waveguides," Opt. Express vol.18, pp. 20839-20844, September 2010.

WA2-113

OECC/PS2016

Integration of Micro Data Center with Optical Line Terminal in Passive Optical Network

Bingliang Yang, Zitian Zhang, Kuo Zhang, Weisheng Hu

State Key Laboratory of Advanced Optical Communication System and Networks, ShangHai Jiao Tong University,
Department of Electronic Engineering, Shanghai 200240, China
wshu@sjtu.edu.cn

Abstract: *We have combined Micro-DC and OLT by Leaf-Spine L2/L3 POD-based fabric, in order to solve the problem of bandwidth mismatch between PON and OLT uplink to core network.*
Keywords: *micro data center, OLT, PON, Leaf-Spine*

I. INTRODUCTION

With the rapid development of the Internet of Everything and the Industrial Internet, more and more smart devices need to access into network. The increase number of end device brings the increase of the data traffic. On the same time, there are more and more location based service being used, for example, the intelligent control of traffic lights. These services make a lot of local business data have to be uploaded to the cloud data center. Predictably, the data flow in the edge network will grow rapidly. Since the last mile of the network access is use the passive optical network (PON) technology to complete. In spite of the big bandwidth of PON, the constraints bandwidth of optical line terminal (OLT) uplink network (metro, core, etc) is still a problem. What's more, the I/O bottleneck between cloud data center and massive terminal equipment makes the transmission rate decreased. All these problems limit the bandwidth utilization of PON. So, when the metro or core network is congested, the data packet queue length at the optical network unit (ONU) may be very long and even overflow regardless of the bandwidth of PON. This will bring the increase of data delay and packet loss rate.

In order to solve this problem, we have proposed a scheme that integrates PON with micro data center (Micro-DC). The Micro-DC contains small servers and storage devices. When data go through the OLT, some of the local business data will be processed and stored at Micro-DC, and the remaining data will continue to be sent to upper layer. Thus, the amount of these data will be smaller than the beginning, and will relax the bandwidth demand of OLT uplink. The bandwidth utilization of PON will be increased with Micro-DC OLT. This scheme is also consistent with the concept of fog computing [1], which means deploy a lot of micro data centers in edge network. All these Micro Data Centers constitute the fog computing layer. These data centers deployed in the middle of ground and cloud computing layer, thus greatly reduce the pressure of the cloud data center. This scheme can also satisfy some other requirements of network, such as: a) real-time service requirement; b) diversity service requirement; c) the ability of collecting, processing and filtering the massive data which are generated by various smart devices [2]; e) Content Delivery Network, Peer to Peer and other network acceleration technology demand for edge servers [3]; etc.

The remaining of this paper is organized as follows. We will illustrate the rationality of OLT integrating with Micro-DC at first. Secondly, we will elaborate the specific internal architecture of Micro-DC OLT. Finally, the performance is verified by simulation, from two aspects of data delay and packet loss rate, we will prove that this scheme can reduce the average delay and loss rate of the packet and balance the peak and valley effect of traffic flow.

II. NETWORK MODEL AND PROBLEM SOLUTION

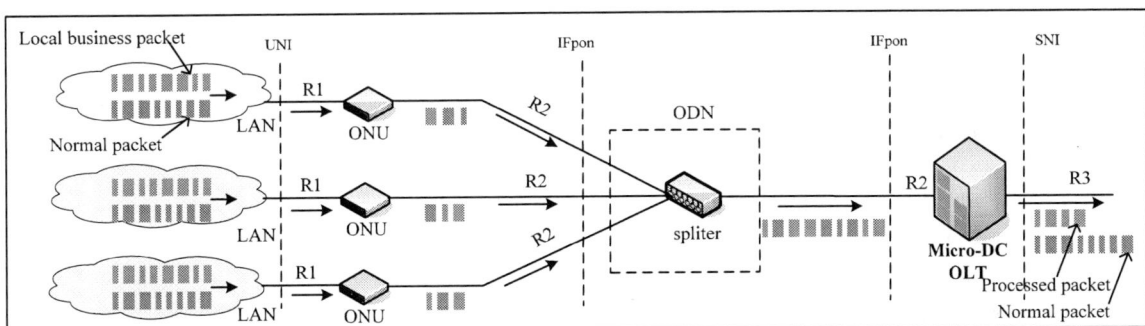

Fig. 1. System diagram with Micro-DC OLT

We have demonstrates the problem and the necessity of data center localization in part Ⅰ. Why should we choose to integrate micro data center with OLT? Except the bandwidth mismatch of the two sides of OLT, there are some other reasons as follow: a) OLT have completed the first data gathering from the terminal equipment of smart devices to the upper layer network, OLT is the first focal points. b) This point is most close to the user, and no other routers, the bandwidth between the user and the OLT can be guaranteed large enough, so that it can meet the needs of any real-time business. c) The central office that OLT deployed on can provide the condition what micro data center need, such as the floor space, power supply, etc. These conditions cannot be all met at other locations on the edge network.

The network model is shown in figure 1 above, including ONU, ODN and OLT, etc. Suppose the data rate of LAN access to ONU is R1, PON upstream data rate from ONU to OLT is R2, and the rate of line from the OLT to the upper layer aggregation switch is R3. Suppose there are only two types of packet need to be uploaded from end devices, they are use different color to represent as shown in figure 1, and distinguished by weather they will processed in Micro-DC OLT. When these packet going through the OLT equipment, the normal packet transported to upper layer directly, and the other packets need to going through the micro data center module. These packets be processed, and then, they will be continue transported to upper layer, but the amount of these packet is smaller than before.

The network model has being explained. Now, we will present the specific internal architecture of Micro-DC OLT. The internal architecture of it is shown in figure 2 below. When data is transmitted from ONU to Micro-DC OLT, they first arrived at the ODN interface, and then these data being demultiplexed and transported through transmission convergence to reach the packet switch. And then, with the help of leaf-spine fabric, some of these packets will switch to I/O device and being transported to upper layer directly, the other packets will be switch to servers and be locally processed, and some of the processed packets will be continue transferred to the upper layer.

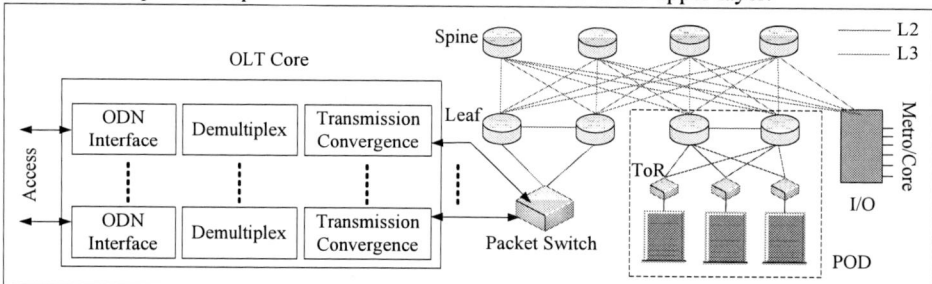

Fig. 2. Micro-DC OLT internal architecture

AT&T and Open Networking Lab have introduced a proposal of Central Office Re-architected as Datacenter (CORD) in 2015 [4]. In this proposal, the fabric of CORD is fulfilled by an open-source Leaf-Spine L3 Clos Fabric. But when we consider the real data flow in Central Office (CO), data packets enter from ODN interface and go out from I/O port. Data exchange between servers is not very frequently, it is no need for each server takes up a leaf switch. Besides, there is a large number of CO, it is necessary to integrate the servers and switch into a common infrastructure that can be easily deployed. As such, we have proposed a Leaf-Spine L2/L3 POD-based Fabric as shown in figure 2. This architecture support Layer 2 switch between server and leaf switch, and it also support Layer 3 routing between Leaf and Spine Switch. Servers and a couple of Leaf switch composed a POD which can modularize and manageable.

Integrate micro data center with OLT is more economical for the construction of data center [5], cause it is built by some white boxes or commodity devices. Besides, it is agility for new services provide since it can be rapid created and deployed on server [5].

Then, we will make some simulation to evaluate the performance of this scheme. Caused by the large bandwidth of PON, the data transmission rate in the edge network is very large and the delay is very small. Besides, when localized applications are deployed on the Micro-DC, some local data no longer need to be transmitted to the upper layer. In addition, many non real-time data can also take a part time storage in micro data center until the valley time of network traffic and be transmitted again. Hence, after adding the Micro-DC OLT, the data through it continue transmit to the upper layer in the peak time of traffic flow is bound to reduce. The following simulation is going to verify the effectiveness of this amount of data reduction for network performance. And we do simulation from two aspects of data delay and packet loss rate to evaluate.

III. SIMULATION AND PERFORMANCE EVALUATION

TABLE. 1. The parameters of simulation system

Parameter	Value	Parameter	Value
Number of ONU	16	Guard time for time slot	1 us
Stream rate of LAN to ONU	100 Mbps	Packet size	64-1518 bytes
Upstream rate of PON	10 Gbps	Packet number	1000000
Uplink rate of OLT	1 Gbps	Fiber transmission delay	5 ns/m
ONU cache size	1 M bytes	Distance from ONU to OLT	500m - 20km
Cycle time	2 ms		

Considering about the impact of the delay on Micro-DC OLT uplink data, it is related only to PON, so that the composition of the delay is from data arrive in ONU to the data leaves OLT. What's more, it is vital for the accuracy of the simulation results that the selection of the appropriate simulation model and parameters. The simulation model what we chose is EPON which is widely used currently, and data will be encapsulated in EPON frames. The network model what we use is shown in figure 1 above. Then, we choose a dynamic bandwidth allocation protocol from [6] called Interleaved Polling with Adaptive Cycle Time. The other parameters are consistent with the actual system as far as possible, consulting document [7] and we get a parameter table as shown in table 1 above. Besides, it is very important to get an appropriate traffic injected into the simulation system, and from document [8] we get such a traffic stream which can be characterized by self-similarity and long-range dependence. According to this set of parameters, we have made a simulation with the amount of normal packet is 80%, 90% and 100%, the result is shown below.

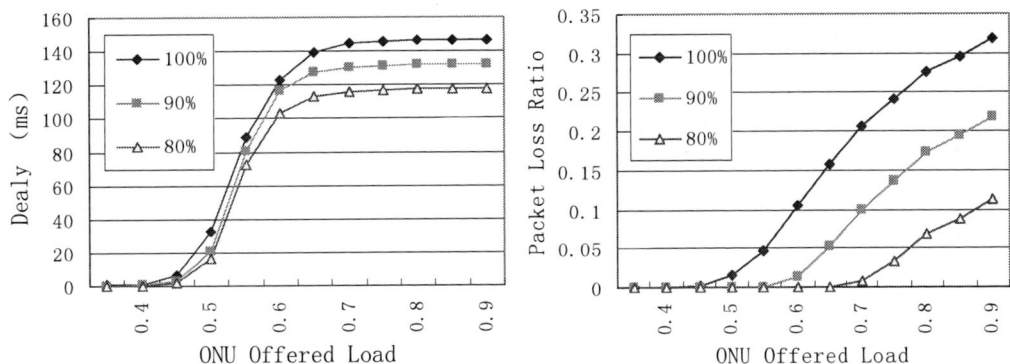

Fig. 3. Packet Delay and Loss Rate for packet need to upload in different proportions

The left side of the graph shows the packet delay. The delay is small at first, and with the increase of ONU load, it increases and becomes stable until the ONU buffer overflows. Correspondingly, there is a continuous increase of the packet loss rate as shown in right side. Besides, with the increase proportion of packet which need processed locally, the delay will be reduced accordingly as well as the packet loss rate.

IV. CONCLUSION

In this paper, we have proposed a data center localization scheme to solve the problem of bandwidth mismatch between PON and OLT uplink. We have given the detailed content of this scheme and designed a new device called Micro-DC OLT. At last, a simulation is conducted to verify the effectiveness of this scheme for network performance. The results show that deploying micro data center into edge network is necessary, and integrating micro data center with OLT is an economical and convenient scheme to complete the data center localization. This scheme can achieve the high speed data transmission on the edge network, reduce the packet loss rate and the data delay. What's more, plenty of localization services can be supported by Micro-DC OLT, this scheme also meet the concept of fog computing.

ACKNOWLEDGMENT

This work was supported by the National Science and Technology Major Project of the Ministry of Science and Technology of China (2015ZX03001021), and NSFC (61431009, 61371082).

References

[1] Bonomi F., Milito R., Zhu J. & Addepalli S., "Fog computing and its role in the internet of things." 1st ACM MCC Workshop on Mobile Cloud Computing, 13-16. 2012

[2] Stojmenovic, I., "Fog computing: A cloud to the ground support for smart things and machine-to-machine networks," in Telecommunication Networks and Applications Conference (ATNAC), 2014 Australasian , vol., no., pp.117-122, Nov. 2014.

[3] Jiang Zhu; Chan, D.S.; Prabhu, M.S.; Natarajan, P.; Hao Hu; Bonomi, F., "Improving Web Sites Performance Using Edge Servers in Fog Computing Architecture," in Service Oriented System Engineering (SOSE), 2013 IEEE 7th International Symposium on , vol., no., pp.320-323, 25-28 March 2013

[4] AT&T, ONOS. CORD white paper [EB/OL]. http://xosproject.org/wp-content/uploads/2015/04/Whitepaper-CORD.pdf.

[5] Larry, Peterson. Central Office Re-architected as a Datacenter (CORD) [EB/OL]. http://xos.wpengine.com/wp-content/uploads/2015/04/CORD-Arch-Alt.pdf.

[6] Kramer, G.; Mukherjee, B.; Pesavento, G., "IPACT a dynamic protocol for an Ethernet PON (EPON)," in *Communications Magazine, IEEE* , vol.40, no.2, pp.74-80, Feb 2002

[7] G. Kramer, Ethernet Passive Optical Networks, McGraw-Hill Professional, ISBN: 0071445625, Publication date: March 2005.pp.189-200

[8] W.Willinger, M.S.Taqqu, and A.Erramilli, "A bibliographical guide to self-similar traffic and performance modeling for modern high-speed networks," in Stochastic Networks, F.P.Kelly,S.Zachary, and I.Ziedins, eds., Oxford University Press, 1996, pp.339‐ 366.

Amplified Spontaneous Emission Noise Influence Analysis on Optical Quantization using Soliton Self-Frequency Shift

Y. Yamasaki[1], T. Nagashima[1], M. Hiraoka[1], and T. Konishi[1]

(1) Graduate school of Engineering, Osaka University, Suita, Osaka 565-0871, Japan

Author e-mail address: yamasaki@photonics.mls.eng.osaka-u.ac.jp

Abstract: We have carefully examined how amplified spontaneous emission noise affects optical quantization using soliton self-frequency shift for photonic analog-to-digital conversion in terms of integral nonlinearity and the influence of number of bit.

Keywords: ASE noise, optical signal processing, optical nonlinear effects, soliton self-frequency shift, photonic analog-to-digital conversion

I. INTRODUCTION

Recent tremendous growths of ultrawide-bandwidth applications demand high performance analog-to-digital convertors (ADCs) for high-speed and flexible signal processing [1-4]. Whereas current electrical ADCs have achieved excellent performances, there is a trade-off between sampling rate and resolution due to electrical jitter of the sampling aperture and ambiguity of the comparator [1-3]. Here, the above-mentioned trade-off sacrifices resolution in return for sampling rate improvement. Since the above timing jitter issue limit the bandwidth of electrical ADCs, photonic ADC has attracted much attention recently and very low jitter comes from optical sampling enables to drastically improve a jitter issue and realize a high speed ADC over 100 GS/s [5]. Various photonic approaches for high performance ADCs have been demonstrated with serial-to-parallel conversion techniques for bandwidth matching. To avoid the above issues due to serial-to-parallel conversion at an earlier stage, it is expected to develop optical approaches for quantization and coding with keeping a system as parallel-configuration-free as possible. We have also proposed a WDM-based approach for optical quantization using soliton self-frequency shift (SSFS) in a fiber [4-11] and the several coding schemes [5-10] that could follow the proposed optical quantization. To upgrade the resolution in the proposed optical quantization, the additionally-subsequent SPM-based spectral compression in a fiber is investigated and 6bit optical quantization and coding have already demonstrated [11]. On the other hand, effective generation of optical nonlinear effects like SSFS requires a signal with relatively high peak power and appropriate use of optical amplification is often necessary for a weak signal. However, amplified spontaneous emission (ASE) noise in an optical amplifier might induce degradation to optical nonlinear effects generation.

In this work, we carefully examine how ASE noise affects our proposed WDM-based optical quantization as a typical example of optical signal processing based on optical nonlinear effects. ASE noise power dependencies are examined in terms of integral nonlinearity (INL) and the influence of number of bit (NOB) for photonic ADC is discussed.

II. ASE NOISE POWER DEPENDENCY OF SOLITON SELF-FREQUENCY SHIFT

Figure 1 shows a schematic diagram for examination of ASE noise power dependency of soliton self-frequency shift (SSFS) in simulation. An optically sampled analog signal is fed to a high nonlinear fiber (HNLF) via an erbium-doped fiber amplifier (EDFA) for optical quantization using SSFS. Since SSFS in a HNLF can generate a different color signal in proportion to an input peak power, a peak power of each optically-sampled signal can be identified by an output signal color. Once each input peak power is identified by each color, each different colored signal can be easily separated as a level identification signal by an arrayed waveguide grating (AWG). In this scheme, ASE noise in an EDFA before SSFS generation might induce degradation to SSFS so as to reduce the resolution of optical quantization. In the proposed optical quantization, the resolution, that is the achievable number of bit; NOB, is described by following equation.

$$NOB = \log_2\left(\frac{\lambda_{shift} + \Delta\lambda_{FWHM}}{\Delta\lambda_{FWHM}}\right),\qquad(1)$$

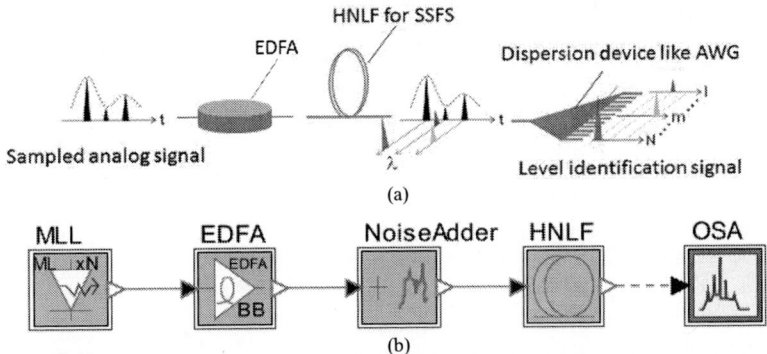

(a)

(b)

Fig. 1. (a) Conceptual diagram of system for investigation of ASE noise power dependency of SSFS. EDFA: Erbium-Doped Fiber Amplifier, HNLF: High Nonlinear Fiber, AWG: Array Waveguide Grating. (b) System used in simulation. The output pulse of Mode-Locked Laser (MLL) is amplified and ASE noise is added to the pulse. Wavelength shift is occurred by propagating HNLF and after that, Optical Spectrum Analyzer (OSA) acquires spectrum data.

where λ_{shift} and $\Delta\lambda_{FWHM}$ are the amount of the center wavelength shift and spectral width after SSFS. The compression of the spectral width $\Delta\lambda_{FWHM}$ enables the improvement of the resolution according to the spectral compression ratio. If the ASE noise affects the amount of λ_{shift}, quantization error like INL might be induced and it degrades NOB. To examine the influence of ASE noise, we calculate ASE noise power dependency of the amount of SSFS. We add a certain amount of ASE noise to an optically sampled analog signal and calculate fluctuation of λ_{shift}. We use a mode-locked laser; the repetition frequency and the center wavelength are 10 GHz and 1550 nm, respectively. Input sampled analog pulses were propagated in a 0.6km-HNLF (D=4.64 [ps/nm/km], S=0.022 [ps/nm²/km], γ=11 [/W/km]) for the generation of SSFS. Figure 2 shows a series of the simulation results of SSFS for 32-nm wavelength shift with changing the amount of ASE noise power. From Fig.2, the fluctuation of λ_{shift} gradually increases in proportional to the amount of ASE power while the influences are much weaker than those of the fundamental part around 1550 nm. It suggests that SSFS might have an effect of reduction of ASE noise influence in a signal after SSFS.

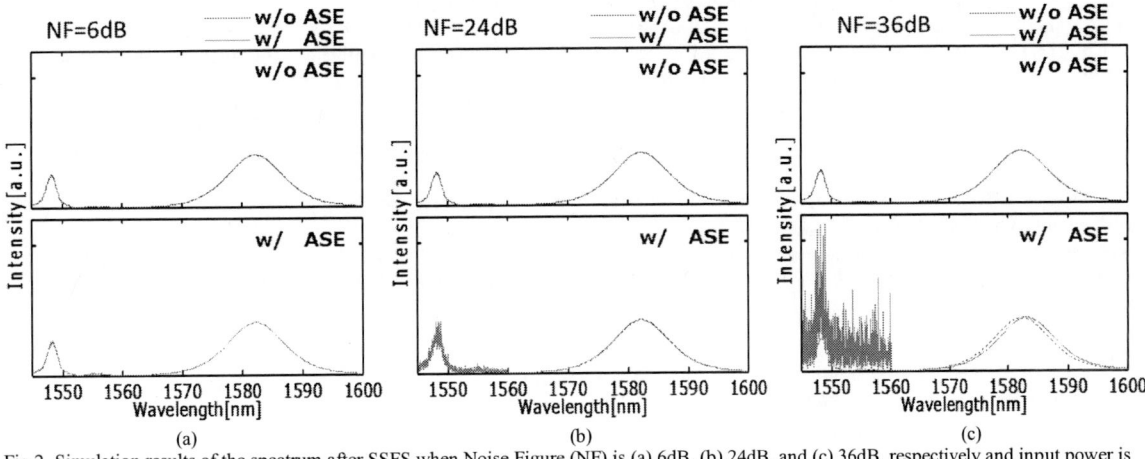

(a)   (b)   (c)

Fig.2. Simulation results of the spectrum after SSFS when Noise Figure (NF) is (a) 6dB, (b) 24dB, and (c) 36dB, respectively and input power is constant.

III. PERFORMANCE ANALYSIS IN TERMS OF INL AND ENOB FOR PHOTONIC ADC

To exmine how ASE noise affects our proposed WDM-based optical quantization, we achieve performance analysis in terms of INL from the above simulation results. The INL is derived from the obtained quantized function. In this analysis, we calculated the amount of wavelength shift as a function of input power at various NF (6 dB, 12 dB, 24 dB, and 36 dB), as shown in Fig. 3 (a). They are corresponding to quantized functions. In fact, since commercial available optical amplifiers can generally achieve ASE noise performance less than NF of 6 dB, these ASE noise figures are much worse than those of commercial available ones. Figure 3 (b) shows the calculated results of the INL for ASE noise figures, 6 dB, 12 dB, and 24 dB, respectively. If we assume that 6 bit optical quantization by the 32 nm-maximum λ_{shift}, the SSFS of 0.5 nm is equivalent to 1 level. From Fig. 3 (b), since the maximum fluctuation of λ_{shift}

due to 6 dB ASE noise is 0.01 nm, the maximum INL can be estimated to 0.3 least significant bit (LSB) at the worst level. Figure 3 (c) shows the calculated results of achievable NOB as a function of NF. From Fig. 3 (c), ASE noise influence of commercial available optical amplifiers is not serious in terms of NOB in a photonic ADC based on SSFS.

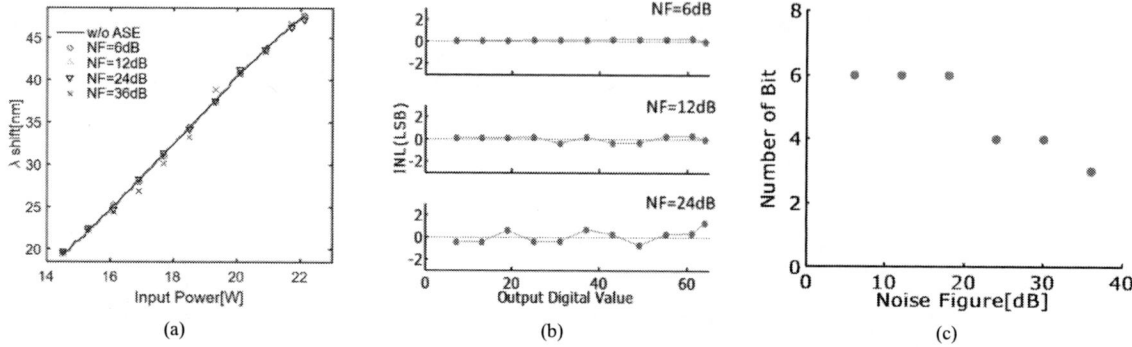

(a)   (b)   (c)

Fig.3. (a) Wavelength shift as a function of input power at various NF. (b) The relation between the output digital value and INL. (c) NOB determined from INL.

IV. CONCLUSIONS

We have carefully examined how ASE noise affects our proposed WDM-based optical quantization as a typical example of optical signal processing based on optical nonlinear effects. The obtained results suggest that optical nonlinear effects like SSFS might work as an ASE noise canceller and an optical amplifier could be used for optical signal processing based on optical nonlinear effects without caring ASE noise influence.

REFERENCES

[1] C. Laperle, "Advances in high-speed ADC, DAC, and DSP for Optical Transceivers", Proc. OFC/NFOEC'13, OTh1F (2013).
[2] B. L. Shoop,"Photonic Analog-to-Digital Conversion,"(Berlin, Springer-Verlag, 2001).
[3] G. C. Valley, "Photonic analog-to-digital converters," Opt. Exp., Vol.15, 1955-1982(2007).
[4] T. Konishi, K. Tanimura, K. Asano, Y. Oshita, and Y. Ichioka"All-optical analog-to-digital converter by use of self-frequency shifting in fiber and a pulse-shaping technique," J. Opt. Soc. Am. B, Vol.19, 2817-2823(2002).
[5] T. Nishitani, T. Konishi, and K. Itoh, "Optical coding scheme using optical interconnection for high sampling rate and high resolution photonic analog-to-digital conversion," Opt. Exp., Vol.15,15812-15817(2007).Y. Yorozu, M. Hirano, K. Oka, and Y. Tagawa, "Electron spectroscopy studies on magneto-optical media and plastic substrate interface," IEEE Transl. J. Magn. Japan, vol. 2, pp. 740-741, August 1987 [Digests 9th Annual Conf. Magnetics Japan, p. 301, 1982].
[6] T. Nishitani, T. Konishi, and K. Itoh, "All-optical M-ary ASK signal demultiplexer based on photonic analog-to-digital conversion," Opt. Exp., Vol.15, 17025-17031(2007).
[7] T. Nishitani, T. Konishi, and K. Itoh, "Resolution Improvement of All-Optical Analog-to-Digital Conversion Employing Self-frequency Shift and Self-Phase-Modulation-Induced Spectral Compression," IEEE J. Selected Topics in Quant. Electron., Vol.14, 724-732(2008).
[8] T. Nishitani, T. Konishi, and K. Itoh, "Demonstration of 4-bit Photonic Analog-to-digital Conversion Employing Self-frequency shift and SPM-induced Spectral Compression," ECOC 2007, Th2.4.3 (Berlin, 2007).
[9] T. Konishi, K. Takahashi, H. Matsui, T. Satoh, K. Itoh,"Five-bit parallel operation of optical quantization and coding for photonic analog-to-digital conversion," Opt. Express. 19, 16106-16114 (2011).
[10] T. Satoh, K. Takahashi, H. Matsui, K. Itoh and T. Konishi, "10-GS/s 5-bit Real-Time Optical Quantization for Photonic Analog-to-Digital Conversion," IEEE Photon. Technol. Lett., vol. 24, no. 10, 830-832, May 2012.
[11] K. Takahashi, H. Matsui T. Konishi and K. Itoh, "6 bit all-optical quantization using soliton self-frequency shift and multistage SPM-based spectral compression," 16th Opto-Electronics and Communications Conference (OECC 2011), pp. 814 - 815, Kaoshiung, Taiwan, July 8 (2011).

WA2-115

OECC/PS2016

Large Data Transfers in WDM Networks with Node Storage

Da Feng, Weiqiang Sun and Weisheng Hu
State Key Laboratory of Advanced Optical Communications Systems and Networks, SJTU, Shanghai, China
wshu@sjtu.edu.cn

Abstract: *We call a WDM network with node storage providing lightpaths for large data transfers the store-and-transfer WDM network and investigate resource provisioning in it under sliding scheduled traffic model, which can decrease blocking by 40%.*
Keywords: *Large Data Transfer, Advance Reservation, WDM Networks, Performance Analysis.*

I. INTRODUCTION

Large data transfer of Terabytes or Petabytes is required by applications, such as genetic sequencing, particle collider, data-center synchronization and content delivery. It takes hours and even days, but is limited by a given deadline. Wavelength-routed WDM networks with Advance Reservation (AR) are promising solutions for large data transfers [1]. Sliding scheduled traffic model (SSTM) is a variant of AR and has been investigated to provision lightpaths for large data transfers [2]. We call a WDM network with node storage providing lightpaths for large data transfers the Store-and-Transfer WDM Network (STWN).

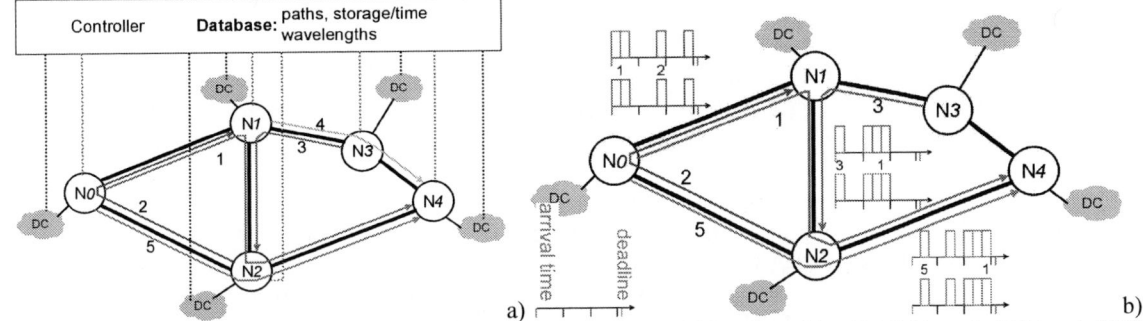

Fig. 1. Resource provisioning in a STWN. a) 5 large data transfer requests. b) Segmented lightpath provisioned by tDGA+ under TDM-mode SSTM. Centralized controller manages paths, wavelengths and storage for large data transfers between Data Centers (DCs). Both wavelengths and node storage can be used to resolve conflict. Two wavelengths are used and tDGA+ requires storage in embedded nodes N1, N2. Request 2, 3, 5 have resolved conflict with 1 time-slot per frame over each wavelength, but request 1 needs 2 time-slot per frame over each wavelength.

In Fig. 1 a), lightpaths for large data transfers between Data Centers (DCs) are provisioned with AR. Lightpaths from both branch and trunk can use the storage of STWN nodes to resolve conflict and those from trunk can store data in storage of destination STWN nodes. Every storage unit in a STWN node connects to a pair of incoming/outgoing wavelengths.

In this work, we extend the SSTM to use segmented lightpaths and to allocate fixed time-slots in continuous frames over multiple wavelengths for large data transfers. We propose two heuristic algorithms to evaluate the extension of SSTM. We use simulations to compare performance of the heuristic algorithms and also numerically analyze the influence of number of time-slots per frame on performance under TDM-mode SSTM.

II. PROBLEM DESCRIPTION AND DEFINITIONS

A. The System Model

Store-and-transfer denotes add/drop of wavelengths with wait and it uses sliding bandwidth allocation. In STWN, lightpaths optionally segmented with embedded storage nodes are established between source and destination storage nodes. The transfer along lightpath can use continuous time over a single wavelength or fixed time-slots in continuous frames over multiple wavelengths. Since source and embedded nodes read data from storage to transfer it over multiple wavelengths, the capacity per logical unit needs to be small (fits one request under SSTM and fits one time-slot under TDM-mode SSTM) and disk performance needs to be high. Since destination nodes write data to storage sequentially, the capacity per logical unit can be large and disk performance can be moderate. File system format should be optimized for high speed I/O and if input data rate is high, source and embedded storage nodes may discard data.

B. Problem Description

A WDM network is represented by a digraph $G(V,E,\Lambda)$, with $|V|$ nodes, $|E|$ links and Λ wavelengths. Under TDM-mode SSTM, bandwidth of wavelengths is organized into globally synchronized frames consisting of the same number of time-slots. *storage capacity = storage size * mean request size*. Data transfer requests are given as *{source, destination, arrival time, service time, deadline}*. We assume the Λ and storage capacity have been determined by dimensioning [3], thus blocking rate is low and there are extra resources reserved. We assume a percentage of wavelengths in all links are periodically preempted by high priority traffic, thus we investigate dynamic provisioning algorithms to decrease the blocking rate caused by traffic fluctuation and preemption of resources.

arrival time is the time when request is in storage, *deadline* is defined as maximum transfer time of request and *service time* is the transfer time of request over a single wavelength. Under SSTM, we use a single wavelength along one path to transfer one request and large requests can be partitioned [2]. Under TDM-mode SSTM, we allocate fixed time-slots in continuous frames over multiple wavelengths along one path to transfer one request. K-Shortest Paths (KSPs), *{paths}*, are used as fixed alternate paths by each source-destination pair (s-d pair) to transfer requests. For online analysis, each node s generates Poisson arrival of requests (λ) with exponentially distributed data size ($1/\mu$), ρ $=\lambda/\mu$, by randomly choosing s-d pairs from SD(X) with source = s and updating timing parameters. The λ, μ and ρ are assumed to be the same among all nodes. *deadline* is given as times of *mean service time* and *mean request size* is *mean service time* , because bandwidth of wavelength is constant.

C. Performance Metrics and System Parameters Definition

The performance metrics include *turnout*, *blocking* and *utilization*. *Turnout=total transferred data/complete time*, *blocking=total rejected data/total arrived data* and *utilization* is load of a wavelength. *Turnout* is bounded by *blocking* and *utilization* varies, because number of wavelengths per link changes periodically according to schedule of preemption. The system parameters include G, SD(X), Λ, M, T and Δ. G is the network, a digraph. SD(X) is the source-destination pairs set, generated by randomly choosing s-d pairs from V. M is number of time-slots per frame and T is duration of a frame. Δ is the duration of a time-slot.

III. PROVISIONING ALGORITHMS

A. General Idea

We investigate an extension of SSTM to provision resources for large data transfers. The extension is based on the fact that data transfers can be non-continuous [2] and blocking can be decreased by splitting transfer windows to decrease hops of lightpaths and by organizing bandwidth of wavelengths into frames to increase flexibility of resource allocations. First, we segment a lightpath with embedded storage nodes and split the transfer window into disjoint smaller ones and sequentially assign a smaller transfer window to each segment of lightpath. Second, we use TDM-mode transfer and organize bandwidth of wavelengths into frames. A time-slot in a frame is the smallest granularity of transfer time due to optical component reconfiguration time and signaling delay. We allocate multiple time-slots in continuous frames over multiple wavelengths along one path to transfer one large data request. The number of time-slots allocated is searched beginning with the minimum number required to meet deadline and is less than both the total available number over all wavelengths and the number required to transfer total data of the request. Finally, we combine the above two to use optionally segmented lightpaths and TDM-mode transfer along each segment of lightpath together. To evaluate the extensions of SSTM, we propose 2 heuristic algorithms: Deadline Guarantee Algorithm (DGA) and tDGA+. DGA implements the SSTM. tDGA+ implements the extension. We also propose parallel implementations of the TDM-mode first-fit resources allocation.

B. Algorithm Implementations

Fig.1 b) gives examples of tDGA+, which provisions segmented lightpath for TDM-mode transfer. The parallel implementations use work stealing to balance load over all available CPU. It searches for number of time-slots allocated to resolve conflict in parallel. The complexity of tDGA+ is $|E|*\Lambda^{max(|segments|)}*|V|*\lambda*deadline/(\mu*\Delta)*|KSP|$ and it is related with Δ because time-slots in frames are discretely allocated.

IV. SIMULATIONS AND DISCUSSIONS

Topology used is 24-nodes USNET, |KSP| is 15, Δ is 0.01, M is one of {5,10,15,20}, 800 time units are simulated (1 time unit is 1 hour) and max(|segments|) is 8. Both performance and computation complexity increase when number of segments increases. 30% of wavelengths are chosen and marked as used in [7,15) during every 24 hours period (preempted by high priority traffic).

A. Performance Comparison between DGA and tDGA+

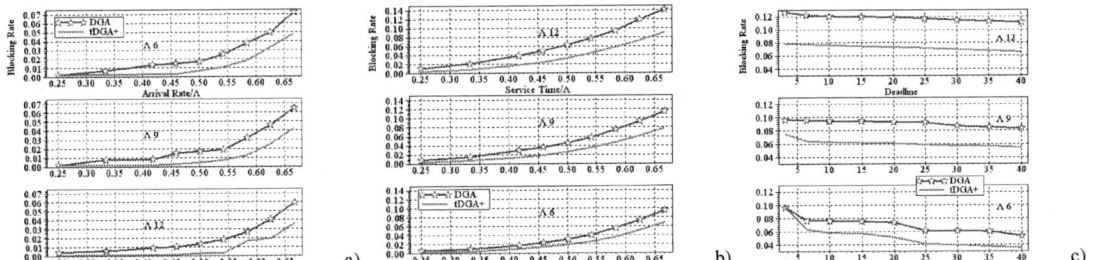

Fig. 2. Blocking vs. demand. a) Blocking vs. arrival rate. Service time is 1 and deadline is 20. b) Blocking vs. service time. Arrival rate is 1 and deadline is 20. c) Blocking vs. deadline. Arrival rate is 1 and service time is 0.625*Λ. Number of time-slots per frame is 5.

In comparison to DGA, tDGA+ decreases blocking by 40%. But it has higher complexity. Also, TDM-mode SSTM requires high performance storage device.

B. Influence of Number of Time-slots per Frame on Performance

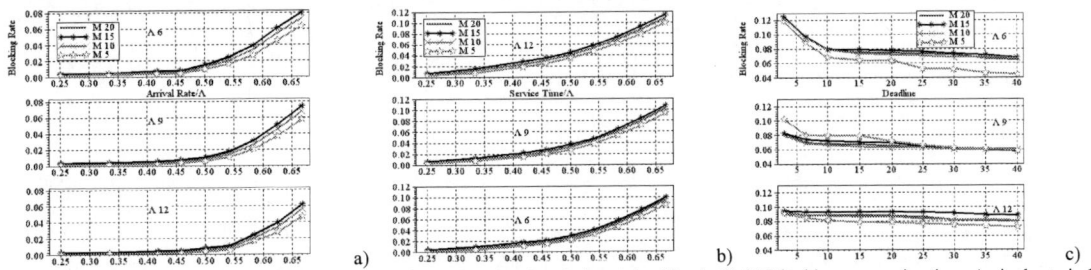

Fig. 3. Blocking vs. demand. a) Blocking vs. arrival rate. Service time is 1 and deadline is 20. b) Blocking vs. service time. Arrival rate is 1 and deadline is 20. c) Blocking vs. deadline. Arrival rate is 1 and service time is 0.625*Λ.

We set max(|segments|) to 1 to examine influence of M only. When M uses a wrong value, blocking can increase by 20% and M=20 isn't always the worst. Influence of M on performance is not monotonic. The reason is given as following. Service time is not integer multiples of allocated number of time-slots, so the last frame have wasted time-slots. With fixed Δ, to maintain service rate, large M requires more allocated time-slots per frame.

V. CONCLUSIONS

Under variable utilization, TDM-mode SSTM with segmented lightpaths can decrease blocking rate by 40%. Under high load, parallel algorithms with 8 CPU can increases speed by 6 times. Influence of frame size on performance is not monotonic.

REFERENCES

[1] N. Charbonneau and V. M. Vokkarane, "A survey of advance reservation routing and wavelength assignment in wavelength-routed wdm networks," in IEEE Comm. Surv. Tutor, vol. 14, no. 4, Fourth 2012.

[2] Y. Chen, A. Jaekel, and A. Bari, "Resource allocation strategies for a non-continuous sliding window traffic model in wdm networks," in BROADNETS, pp. 1–7, Sept 2009.

[3] D. Feng, W. Sun, P. Wu, X. Zhang, J. Wu, and W. Hu, "Dimensioning of store-and-transfer wdm network with limited storage," in NOMS, Apr 2016.

WA2-116

20 Gb/s Optical Switched DQPSK transmission over 50 km SSMF and 7 km DCF

Pacharapon Chulok, Suvit Nakpeerayuth, and Duang-rudee Worasucheep
Department of Electrical Engineering, Faculty of Engineering, Chulalongkorn University Bangkok, Thailand
pacharapon.pai@gmail.com, suvit.n@chula.ac.th, and duangrudee.w@chula.ac.th

Naoya Wada
Photonic Network Research Institute
National Institute of Information and Communications Technology (NICT), Tokyo, Japan
wada@nict.go.jp

Abstract: We achieve DQPSK transmissions over 105-km SSMF&12-km DCF, and 50-km SSMF&7-km DCF with optical switch at 1550-nm wavelength. Constellation, accumulated I-Q transition, eye diagram and BER are measured. Penalties versus residual CD distances are plotted.
Keywords: Differential Quadrature Phase Shift Keying, Chromatic Dispersion, Dispersion Compensation Fiber

I. INTRODUCTION

Majority of deployed backbone networks use binary Intensity Modulation, aka On-Off Keying (OOK), at transmitters and Direct Detection at receivers. Until after the year 2000 [1], the growing interests in alternative formats to improve spectral efficiency of WDM systems have led to an advent of multi-level phase modulations. For instance, Quadrature Phase Shift Keying (QPSK) format doubles spectral efficiency as compared to conventional Nonreturn-to-Zero (NRZ) OOK at the same symbol rate. However, as compared at the same bit rate, the QPSK uses half symbol rate while offers improved tolerance to Chromatic Dispersion (CD) and Polarization Mode Dispersion (PMD) [2-3].

To retrieve QPSK data, an expensive narrow-linewidth laser, aka a Local Oscillator (LO), is required at receiver to combine its stable-phase output with incoming signal using coherent detection. Instead, if a delayed replica of incoming signal is used in place of a LO's output, this detection is called self-coherent. And, at transmitter the QPSK data must be differential phase encoded, known as Differential QPSK (DQPSK). This format has gained more attention due to lower CAPEX by using simple devices, such as the Delay-Interferometer (DI) that converts phase information into amplitude variations for a balanced-photodetector to recover original data. Similar to QPSK, the DQPSK also offers tolerance to CD and PMD, as well as power fluctuation in optical packet switching [4].

The accumulated CD in a long Standard Single Mode Fiber (SSMF) transmission at 1550-nm wavelength will cause some irregularities to data constellation [5]. In the case of coherent QPSK, the high-speed digital signal processors are necessary to provide flexible CD compensation for any distance of SSMF. But, those processors operate at fixed bit rate and constantly consume large power, causing a higher OPEX. In contrast, Dispersion Compensation Fiber (DCF) works for all bit rates without power, requiring only an initial CAPEX. Once installed, DCF compensates a specific length of SSMF at a designated wavelength. Other neighboring WDM channels will be slightly under or over compensated, and have some power penalties due to their residual CDs.

In this paper, we transmit 2×10 Gb/s NRZ-DQPSK data over different fiber combinations: (1) only SSMF: 25, 40 and 50 km; (2) SSMF&DCF: 50-km SSMF&3-km DCF, 50-km SSMF&7-km DCF, 80-km SSMF&5-km DCF, and especially 105-km SSMF&12-km DCF; and lastly (3) SSMF&DCF with optical switch: bar & cross states over 50-km SSMF&7-km DCF. Data constellations, accumulated I-Q transitions and eye diagrams are measured by the Optical Modulation Analyzer (OMA). Finally, the Bit Error Rates (BER) are reported for different cases of perfect- and under-compensations. The corresponding power penalties evaluated at 10^{-9} BER versus their residual CD distances are plotted, as an indication for those WDM channels with imperfect CD compensations.

II. EXPERIMENTAL SETUP

In Fig.1, at transmitter, a tunable laser (+16 dBm at 1550-nm wavelength) is connected to DQPSK modulator via a Polarization Controller (PC). 2×10 Gb/s 2^{23} PRBS In-phase and Quadrature-phase (I and Q) data are generated by BER Tester, while Q signal from inverted data port passes through 11-bit delay to create uncorrelated I and Q signals. The DQPSK modulator, consisting of 2 parallel Mach-Zehnder modulators and $\pi/2$ phase shifter, has 8-dBm modulated output. Next, the DQPSK signal is split into 2 uncorrelated paths via a 3-dB coupler and 2-km SSMF difference. Then, the signal is switched by a 2×2 mechanical switch, and propagates through different length combinations of SSMF and DCF. To analyze the quality of transmitted DQPSK data, OMA is used to measure data constellation, accumulated I-Q transitions, eye diagrams, and important parameters such as Error Vector Magnitude (EVM).

At receiver, a Variable Optical Attenuator (VOA) is inserted for BER measurements by Error Detector (ED). We use 2 Erbium Doped Fiber Amplifiers: EDFA1 with 20-dB fixed gain and EDFA2 with adjustable gain to maintain 10-dBm output. A Tunable Optical Band Pass Filter (TOBPF) with 0.65-nm spectral width helps remove Amplified Spontaneous Emission noise. To demodulate DQPSK data, a 1-bit DI is either $+\pi/4$ or $-\pi/4$ phase tuned for I and Q data recovery. Finally, the balanced detector, consisting of 2 PIN photodetectors, converts DI's output to differential electrical signal. The eye diagram of recovered data is detected by Digital Communication Analyzer (DCA).

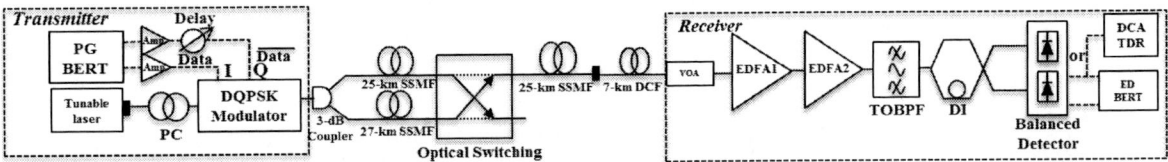

Fig.1 Block diagram of experimental setup.

III. EXPERIMENTAL RESULTS

According to the OMA results in Table I, as SSMF increases from 0 to 25, 50 and 80 km, their constellations spread outward like a cloud similar to simulated pattern in [5]. The accumulated I-Q transitions show a gradual counter-clockwise rotation while eye diagrams slowly close. But, when 5-km or 7-km DCF is inserted after 50-km SSMF (under- or perfect-compensation), their EVMs greatly reduce and constellations improve approaching the back-to-back case. Finally, when 2×2 switch (bar state) is included, some phase distortion are added and worsen its constellation.

TABLE I OMA MEASUREMENT RESULTS

Cases	Constellation	Accumulated I-Q Transition	Eye Diagram
1. Back-to-Back EVM =7.66% rms			
2. 25-km SSMF EVM = 16.85% rms			
3. 50-km SSMF EVM = 23.25 % rms			
4. 80-km SSMF+ 5-km DCF (Under-Compensation) EVM = 27.34% rms			
5. 50-km SSMF+ 7-km DCF (Perfect-Compensation) EVM = 16.46% rms			
6. 2×2 Switch (bar)+ 50-km SSMF+ 7-km DCF (Perfect-Compensation) EVM = 28.44% rms			

The accumulated CD is well-known to broaden optical pulses. The maximum distance L at data bit rate B is limited by Eq. (1) [1]. In this case, SSMF has D_{SMF} =16.5 ps/nm/km but DCF has negative D_{DCF} = −127 ps/nm/km. The spectral width σ of 10 Gb/s NRZ-DQPSK data is 0.0672 nm measured by an Optical Spectral Analyzer. Hence, the maximum distance of SSMF is calculated roughly 62.4 km. To reduce the accumulated CD according to Eq. (2), a matching length of DCF must be inserted before or after SSMF. For example, 6.69 km of DCF is needed for 50-km SSMF.

$$BL < \frac{0.7}{D\sigma} \tag{1}$$

$$D_{SSMF}L_{SMF} + D_{DCF}L_{DCF} = 0 \tag{2}$$

Fig.2 (a) shows the BER plot of different combinations of SSMF and DCF. As SSMF rises from 0 to 25 and 40 km, BER curve shifts rightward causing more penalties, labeled as "Uncompensated CD" in Fig.2 (b). But, when DCF is inserted, BER curve shifts leftward more or less depending on the matching length. There are 2 under-compensated cases: (1) "50-kmSSMF+3-kmDCF" (= 25-km residual CD distance) shifts its BER curve close to "over 25-km SSMF", and (2) "80-kmSSMF+5-kmDCF" (= 40-km residual CD distance) shifts its BER curve close to "over 40-km SSMF". Additionally, there are 2 nearly perfect-compensation cases ($\approx$ 0 residual CD distance): "50-kmSSMF+7-kmDCF" and "105-kmSSMF+12-kmDCF" shift their BER curves close to "DQPSK Back-to-Back". Finally, when the 2 × 2 switch is included as shown in Fig.1, their BER curves (bar and cross state) are close to "50-kmSSMF+7-kmDCF".

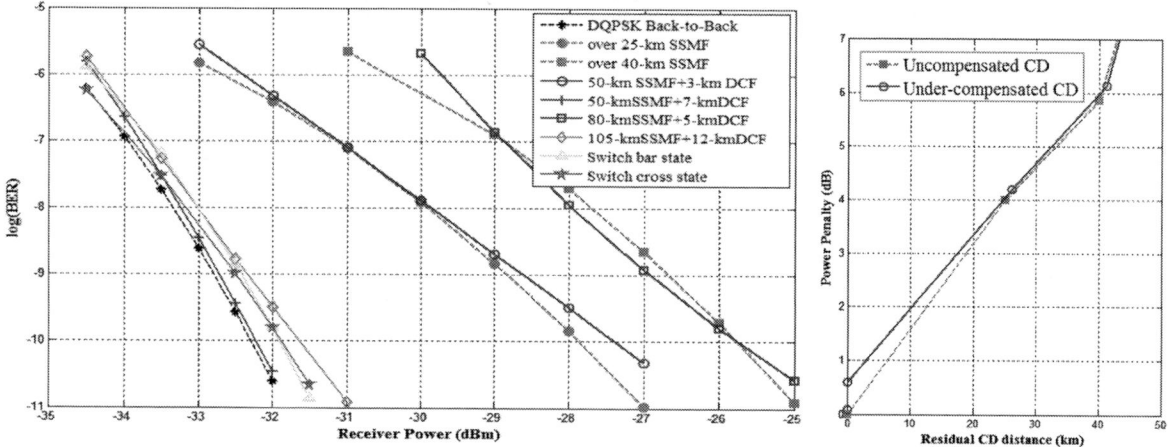

Fig.2 (a) BER of DQPSK transmissions over different combinations of SSMF and DCF

Fig.2 (b) Power penalties at 10^{-9} BER

IV. CONCLUSIONS

We successfully transmitted 2 × 10 Gb/s NRZ-DQPSK data over 105-km SSMF and 12-km DCF, plus the optical switched DQPSK data over 50-km SSMF and 7-km DCF at 1550-nm wavelength. The accumulated CD caused data constellation to spread outward and gradually rotate counter-clockwise. However, it was restored by adding DCF. When the DQPSK data was optically switched, some phase distortion occurred and worsen its constellation The BER plot of several under- and nearly perfect-compensation cases was reported. Moreover, the plot of power penalties versus residual CD distance was given as an indication for any WDM channel with imperfect CD compensation.

ACKNOWLEDGMENT

Special thanks to Mr. Virote Pirajnanchai and Dr.Wisit Lawtammajak, the Department of Electronic and Telecommunication Engineering, Rajamangala University of Technology Thanyaburi for lending the OMA Agilent N4392A. We would like to thanks Mr. Hiroyuki Sumimoto from NICT for his help and support in our experiment.

REFERENCES

[1] G. P. Agrawal, Fiber-Optic Communication Systems, 4th ed., Wiley, 2010.
[2] P. J. Winzer and R-J. Essiambre, "Advanced Optical Modulation Formats," Proceedings of The IEEE, Vol. 94, No. 5, pp. 952-985, May 2006.
[3] R. A. Griffin, R. I. Johnstone, R. G. Walker, J. Hall, S. D. Wadsworth, K. Berry, A. C. Carter, and M. J. Wale "10 Gb/s Optical Differential Quadrature Phase Shift Keying (DQPSK) Transmission using GaAs/AlGaAs Integration," Optical Fiber Communication Conference (OFC), postdeading papers FD6-1, March 2002.
[4] H. Furukawa, S. Shinada, and N. Wada "Tolerance of DQPSK Optical Packet for Power Fluctuation in Optical Packet Switching," 15th OptoElectronics and Communications Conference (OECC), Japan, pp. 406-407, July 2010.
[5] V. R. Arbab, X. Wu, L. C. Christen, J-Y. Yang, T. Dennis, P. Williams, and A. E. Willner "Analysis of Fiber Dispersion Effects on Phase Modulated Signals Using Constellation Diagram," Optical Fiber Communication Conference (OFC), JThA45, March 2009.

WA2-12

OECC/PS2016

Variable Gain PI-based Cyclic Sleep Control with Anti-Windup Technique for QoS-aware and Energy-Efficient Ethernet PONs

Takahiro Kikuchi, Ryogo Kubo

Department of Electronics and Electrical Engineering, Keio University
3-14-1, Hiyoshi, Kohoku-ku, Yokohama-shi, Kanagawa 223-8522, Japan
kikuchi.takahiro@kbl.elec.keio.ac.jp

Abstract: We propose a variable gain proportional-integral (PI)-based cyclic sleep controller with anti-windup technique for Ethernet passive optical network (EPON) systems and discuss the design strategy of the variable control gains.

Keywords: Optical Access, Passive Optical Network (PON), Power Saving, Cyclic Sleep, Quality of Service (QoS)

I. INTRODUCTION

Passive optical networks (PONs) have been deployed as fiber-to-the-home (FTTH) communication infrastructure. A PON consists of an optical line terminal (OLT), a passive remote node, and multiple optical network units (ONUs) as shown in Fig. 1. The OLT is located at a central office, and the ONUs are installed on customer premises. It was pointed out in [1] that the power consumption of time division multiplexing (TDM)-PONs accounted for 80% of the total number of public network systems. A next-generation time and wavelength division multiplexing (TWDM)-PON integrates multiple TDM-PONs on different wavelengths, and it will consume larger power than that of the TDM-PON. It is considered that the future PON systems require not only energy efficiency but also diversified quality of services (QoS) [2]. A service-aware cyclic sleep controller with variable sleep period for ONUs in TDM-PONs was proposed in [3]. Maneyama *et al.* [4] proposed a QoS-aware cyclic sleep controller on the basis of feedback control theory for Ethernet PONs (EPONs).

We have proposed a proportional-integral (PI)-based cyclic sleep controller with anti-windup (AW) technique in [5], i.e., PI/AW controller, to improve the QoS in terms of average downstream queuing delay in the OLT. However, the conventional fixed gain PI/AW controller generated some errors in QoS control. In this paper, we propose a variable gain PI/AW controller and discuss the design strategy of the variable control gains. Simulations confirm that the proposed variable gain PI/AW controller reduces the errors in QoS control compared to the conventional fixed gain PI/AW controller.

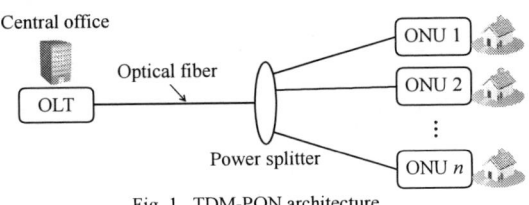

Fig. 1. TDM-PON architecture.

II. ONU POWER SAVING

This section presents the QoS-aware cyclic sleep control scheme, and proposes the variable gain PI/AW controller.

A. QoS-Aware Cyclic Sleep Control

An EPON system with the QoS-aware cyclic sleep controller of ONUs is illustrated in Fig. 2. The OLT has the downstream traffic monitor and cyclic sleep controller. The OLT notifies the ONU of the sleep period T_s, and makes the ONU enter the sleep state. The ONU enters the sleep state during the sleep period T_s. The information of the target queueing delay requested by executed applications, T_{qd}, is sent from the ONU to the OLT on ahead.

The downstream traffic monitor in the OLT measures the downstream queue length (QL) at time k, $q_{out}[k]$, and the average downstream frame arrival rate at time k, $\lambda[k]$. The cyclic sleep controller calculates an average QL at time k, $q_{ave}[k]$, as (1)

$$q_{ave}[k] = \alpha q_{out}[k] + (1-\alpha)q_{ave}[k-1], \quad (1)$$

where α is a smoothing factor. The cyclic sleep controller also calculates the target QL at time k, $q_{cmd}[k]$, as (2)

$$q_{cmd}[k] = T_{qd}\lambda[k]. \quad (2)$$

751

The conventional fixed gain PI/AW controller determines the sleep period at time k, $T_s[k]$, as (3)–(6)

$$e[k] = q_{cmd}[k] - q_{ave}[k],\qquad(3)$$

$$T_{s,tmp}[k] = \begin{cases} K_p e[k] + K_i \sum_{n=0}^{k} e[n] + T_s[k-1] & \text{if } x[k] = 0 \\ K_p e[k] + T_s[k-1] & \text{if } x[k] \neq 0 \end{cases},\qquad(4)$$

$$x[k] = T_{s,tmp}[k-1] - T_s[k-1],\qquad(5)$$

$$T_s[k] = \begin{cases} 0 & \text{if } T_{s,tmp}[k] < T_{s,\min} \\ T_{s,tmp}[k] & \text{if } T_{s,\min} \leq T_{s,tmp}[k] < T_{s,\max} \\ T_{s,\max} & \text{if } T_{s,\max} \leq T_{s,tmp}[k] \end{cases},\qquad(6)$$

where K_p, K_i, $T_{s,\min}$, and $T_{s,\max}$ denote the proportional gain, integral gain, minimum sleep period, and maximum sleep period, respectively.

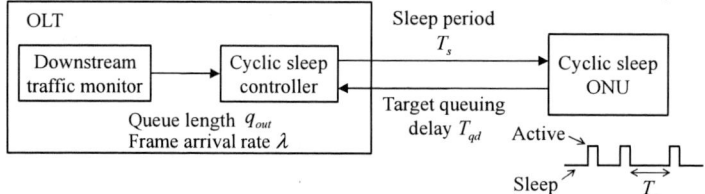

Fig. 2. QoS-aware cyclic sleep control scheme.

B. Variable Control Gains of the PI/AW Controller

The proposed variable gain PI/AW controller is illustrated in Fig. 3. In the proposed controller, the temporary sleep period at time k, $T_{s,tmp}[k]$, of (4) is modified as (7)

$$T_{s,tmp}[k] = \begin{cases} K_p(\lambda[k])e[k] + K_i(\lambda[k]) \sum_{n=0}^{k} e[n] & \text{if } x[k] = 0 \\ K_p(\lambda[k])e[k] & \text{if } x[k] \neq 0 \end{cases}.\qquad(7)$$

In the proposed variable gain PI/AW controller, optimum gains vary according to the downstream traffic rate $\lambda[k]$, and the criterial sleep period $T_s[k-1]$ is removed.

We discuss the design strategy of the variable control gains. First, we seek optimum values of K_p for 0.01, 0.025, 0.05, 0.075, 0.1, 0.25, 0.5, 0.75, 1, 2.5, 5, 7.5, 10, 25, 50, 75, 100, 250, 300, 400, 500, 600, 700, 800, 900, and 1000 Mbps. It is noted that the integral gains K_i is set to 0 when $\lambda[k] \leq 250$ Mbps and 2.0×10^{-9} when $\lambda[k] \geq 300$ Mbps. We define the optimum proportional gains $K_{p,op}$ as the values when the average downstream queuing delay stays within $T_{qd} \pm 0.2$ ms.

Since the obtained optimum proportional gains $K_{p,op}$ are discrete values at the measurement points, the interpolation of the optimum proportional gains is needed to determine K_p according to any downstream traffic rate. We discuss two interpolation methods, i.e., a floor/ceiling interpolation method and a linear interpolation method. In the floor/ceiling interpolation method, the controller adopts the optimum proportional gain at the nearest downstream traffic rate as K_p. In the linear interpolation method, the controller calculates K_p so as to linearly interpolate each discretized interval. The proposed variable gain PI/AW controller finally calculates integral gains K_i, as (8)

$$K_i = \begin{cases} 0 & \text{if } \lambda[k] < 275\,\text{Mbps} \\ 2.0 \times 10^{-9} & \text{if } \lambda[k] \geq 275\,\text{Mbps} \end{cases}.\qquad(8)$$

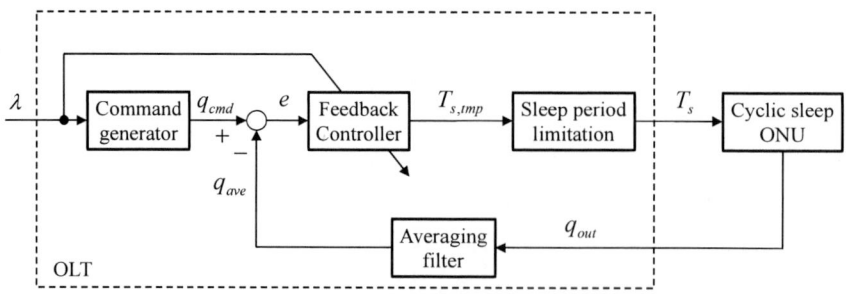

Fig. 3. Proposed variable gain PI/AW controller.

III. PERFORMANCE EVALUATION

The simulations compared the conventional fixed gain PI/AW controller and the proposed variable gain PI/AW controller. We assumed that the controllers were implemented in the 10 Gbps Ethernet PON (10G-EPON) OLT. The distance between the OLT and ONU was set to 20 km. The transition time from the sleep state to the active state was set to 10 ms [6]. The maximum sleep period $T_{s,max}$, minimum sleep period $T_{s,min}$, smoothing factor α, and target queuing delay T_{qd} were set to 50 ms, 10 ms, 0.2 and 20 ms, respectively. In the fixed gain PI/AW controller, the proportional gain K_p and integral gain K_i were set to 8.0×10^{-7} and 1.0×10^{-12}, respectively. In the simulations, we assumed that there was only downstream traffic whose arrival rate followed the Poisson distribution. The frame size was set to 1250 bytes. The simulation results are shown in Fig. 3. The fixed gain PI/AW controller, the proposed variable gain PI/AW controller using the floor/ceiling interpolation method, and the proposed variable gain PI/AW controller using the linear interpolation method are compared.

Fig. 3(a) shows the average downstream queuing delay. The fixed gain PI/AW controller maintained the average queuing delay at a constant level regardless of the amount of downstream traffic. However, the average queuing delay was accurate within 2.7 ms against the target queuing delay T_{qd}. The variable gain PI/AW controllers reduced the average queuing delay compared to the fixed gain PI/AW controller. In the low-rate range, the average queuing delay was in part smaller than 19 ms. In addition, the average queuing delay of the floor/ceiling interpolation method was smaller than 19.5 ms at 4, 40, and 400 Mbps, and the average queuing delay of the linear interpolation method was larger than 20.5 ms at 4 and 40 Mbps. The selection of a better interpolation method and measurement points needs further study. Fig. 3(b) shows the time occupancy of active periods of an ONU. The time occupancy of the floor/ceiling and linear interpolation methods were smaller in the low-rate range than the fixed gain PI/AW controller because removing feedback of the criterial sleep period $T_s[k\text{-}1]$ mitigated the impact of saturation.

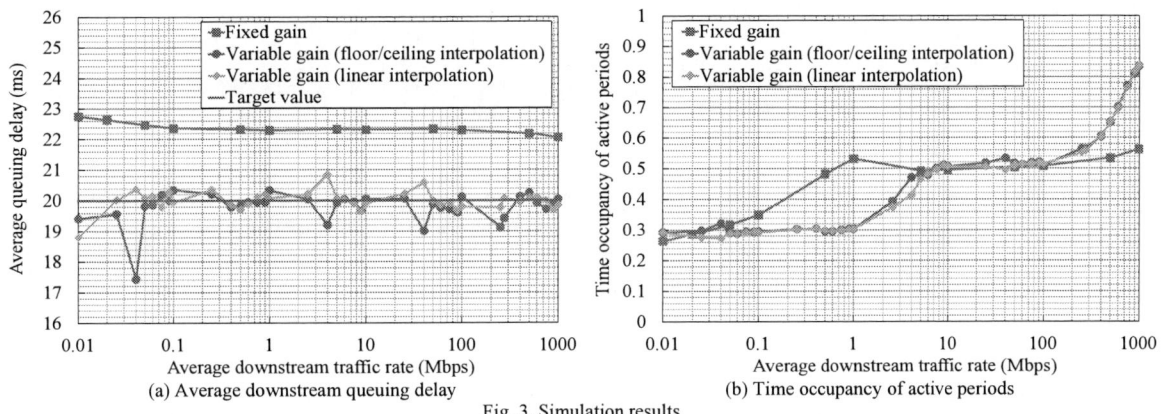

(a) Average downstream queuing delay
(b) Time occupancy of active periods
Fig. 3. Simulation results.

IV. CONCLUSIONS

We proposed the variable gain PI/AW controller for QoS-aware and energy-efficient EPON systems and discussed the design strategy of the variable control gains. The simulation results showed that the proposed controller reduced the errors in QoS control compared to the conventional fixed gain PI/AW controller.

ACKNOWLEDGMENT

This research was supported in part by the National Institute of Information and Communications Technology (NICT), Japan.

REFERENCES

[1] A. Otaka, "Power saving ad-hoc report," IEEE 802. 3av, 2008.
[2] IEEE Std 1904.1-2013, "Standard for Service Interoperability in Ethernet Passive Optical Networks (SIEPON)," June 2013.
[3] R. Kubo, J. Kani, Y. Fujimoto, N. Yoshimoto, and K. Kumozaki, "Adaptive power saving mechanism for 10 Gigabit class PON systems," IEICE Trans. Commun., vol. E93-B, no. 2, pp. 280–288, February 2010.
[4] Y. Maneyama and R. Kubo, "QoS-aware cyclic sleep control with proportional-derivative controllers for energy-efficient PON systems," J. Opt. Commun. Netw., vol. 6, no. 11, pp. 1048–1058, November 2014.
[5] T. Kikuchi and R. Kubo, "ONU power saving considering sleep period limitation in QoS-aware cyclic sleep control with PI controller," Proc. of 20th Microoptics Conf. (MOC 2015), Fukuoka (Japan), October 2015, pp. 258–259.
[6] H. Mukai, F. Tano, and J. Nakagawa, "Energy-Efficient 10G-EPON system," Proc. Optical Fiber Communication Conf. and Expo. and the Nat. Fiber Optic Engineers Conf. (OFC/NFOEC 2013), Anaheim (USA), March 2013, paper OW3G.1.

WA2-13

OECC/PS2016

Experimental Demonstration of a Cost Effective Bidirectional Coherent OFDM-PON

Rongrong Chen, Caixia Kuang, Yingxiong Song, Min Wang, Rujian Lin, and Qianwu Zhang*

The Key Laboratory of Specialty Fiber Optics and Optical Access Network, Shanghai University,
149 Yanchang Road, Shanghai 200072,China
qwzhang@shu.edu.cn

Abstract: An OFDM-PON system which employed cost effective bidirectional coherent detection with 100/64 Gb/s signal line rate and power budget of 34dB and 32dB for downstream and upstream based on optical comb is experimentally demonstrated.
Keywords: bidirectional, coherent, OFDM-PON, optical comb generator, RSOA

I. INTRODUCTION

The combination of coherent detection and OFDM technique in expensive long haul optical communications systems improves their performance on receiver sensitivity, spectral efficiency, and robustness against channel dispersion [1-2]. Coherent detection optical orthogonal frequency division multiplexing (CO-OFDM) system is also a promising multicarrier modulation solution for next generation passive optical network (NG-PON) systems for access field except that independent local oscillator (LO) laser in each ONU will increase the system cost in massive deployment. Moreover, CO-OFDM PON is sensitive to laser phase noise (PN) and frequency offset (FO) normally induced by instability or mismatch between independent laser sources in OLT and LO in each ONU. For PN compensation, in [3] Ly-Gagnon proposes an algorithm based on data aided (NA). A pilot subcarrier aided (PA) method is put forward by William Shieh in [4]. Then a perfect combination of NA and PA is presented by S Cao in [5]. S. Randel proposes a RF-pilot method based on FO estimation in [6]. Although many algorithms have been proposed to improve the performance, it increases the complexity of digital signal processing (DSP) and hardware.

In this paper, a novel cost effective CO-OFDM PON architecture which employed bidirectional coherent detection with 100/64 Gb/s signal line rate and power budget of 34dB and 32dB for downstream and upstream based on optical comb generation combined reflective semiconductor optical amplifier (RSOA) based colorless ONU are experimental demonstrated. Both upstream optical carrier and LO are delivered from a self-developed comb generator so that DSP algorithms can be simplified by canceling frequency offset estimation in both OLT and ONU side. IQ modulation with heterodyne coherent detection and intensity modulation homodyne detection are adopted in downlink and uplink, respectively.

II. EXPERIMENTAL SETUP

The proposed bidirectional CO-OFDM PON experimental setup is shown in Fig. 1. Self-developed Optical comb generator with 1MHz spectral line-width, 0.1nm channel space is employed as the light source in the OLT as shown in Fig. 1 point (a) [7]. Optical comb spectrum can be separated into 16 wavelengths by an Arrayed Waveguide Grating (AWG) whose insertion loss is 4.5dB. In the downstream transmitter, the downstream signal light λ_1 is followed by a pilot tone vector modulator (PTVM) [8]. In such structure, the signal is divided into two orthogonal polarization components by an integrated polarization beam splitter (PBS). One component can be modulated as signal portion and recombined with the other one by an integrated polarization beam combiner (PBC). The IQ modulator is driven by 2.5GHz bandwidth 16QAM-OFDM signal. The total power OFDM signal is set to <8dBm to avoid modulation distortion in IQ modulator. Wavelength λ_1 and λ_2 are combined into feeder fiber at power of 10dBm through the AWG. In the λ_1 path, a phase modulator (PM) is induced to alleviate the stimulated Brillouin scattering by increasing the SBS threshold to +12dBm [9].

A circulator is used to receive the upstream signal light. λ_1 and λ_2 is then used as the signal carrier and the LO light for coherent receiver in the downlink. At the ONU side, the LO light is amplified by an EDFA which is followed by an optical bandpass filter (OBF) used to erase the noise bandwidth. It is worth to mention that the EDFA is induced in order to investigate the LO dependent system performance. The downlink is operated at heterodyne mode with a central frequency of 12.5GHz plus upper sideband 2.5GHz OFDM signal as shown in Fig.1 point (b). It is found that SNR of OFDM signal is 20dB. Experimental scheme of RF down-conversion is also depicted in Fig. 1. In this scenario, 12.5GHz central carrier is respectively extracted from I and Q signals of coherent receiver by adopting a filter with 12.5GHz central frequency and a bandwidth of 12MHz. 12.5GHz central carrier is combined with I and Q signal by an electric mixer to achieve baseband OFDM signals as shown in Fig.1 point (c). 2.5GHz low-pass filter is used to remove the noise outside the signal bandwidth. The advantage of this down-conversion scheme is that OFDM upper sideband signal and 12.5GHz central carrier contain the same carrier phase noise, and phase noise can be canceled in the mixing process. Simultaneously wavelength λ_1 and λ_2 also have same optical phase noise because of coming from a same

754

optical source. The wavelength interval is constant with 0.1nm. Therefore subsequent DSP does not require frequency offset estimation algorithm which simplify the DSP procedure.

In the upstream, λ_2 is as well as a seed light upward through RSOA intensity modulation, amplification and reflection. The bandwidth of RSOA is about 1.6GHz at the bias current of 80mA and 0dBm input optical power. Therefore RSOA can be driven by 1.6GHz bandwidth 16QAM-OFDM signal as shown in Fig.1 point (d) for homodyne coherent detection. Given the superiority of the optical comb, 10 of 20 wavelengths can be used for downstream transmission, the rest for uplink transmission, which support transmission signal line rate of 100/64 Gb/s for downlink and uplink.

In the DSP procedure, driven signal of IQ are real and imaginary parts of the OFDM signal. In the downlink receiver, the two baseband signals are added in the form of complex. Because of uplink in intensity modulation, the receiver data processing is the square sum of two signals algorithm. Then the obtained baseband OFDM signal sampled by a digital oscilloscope is processed offline by MATLAB. After symbol synchronization, cyclic prefix removal, fast Fourier transform (FFT), phase estimation and QAM demodulation, The decoded output data is used to generate the BER performance estimation.

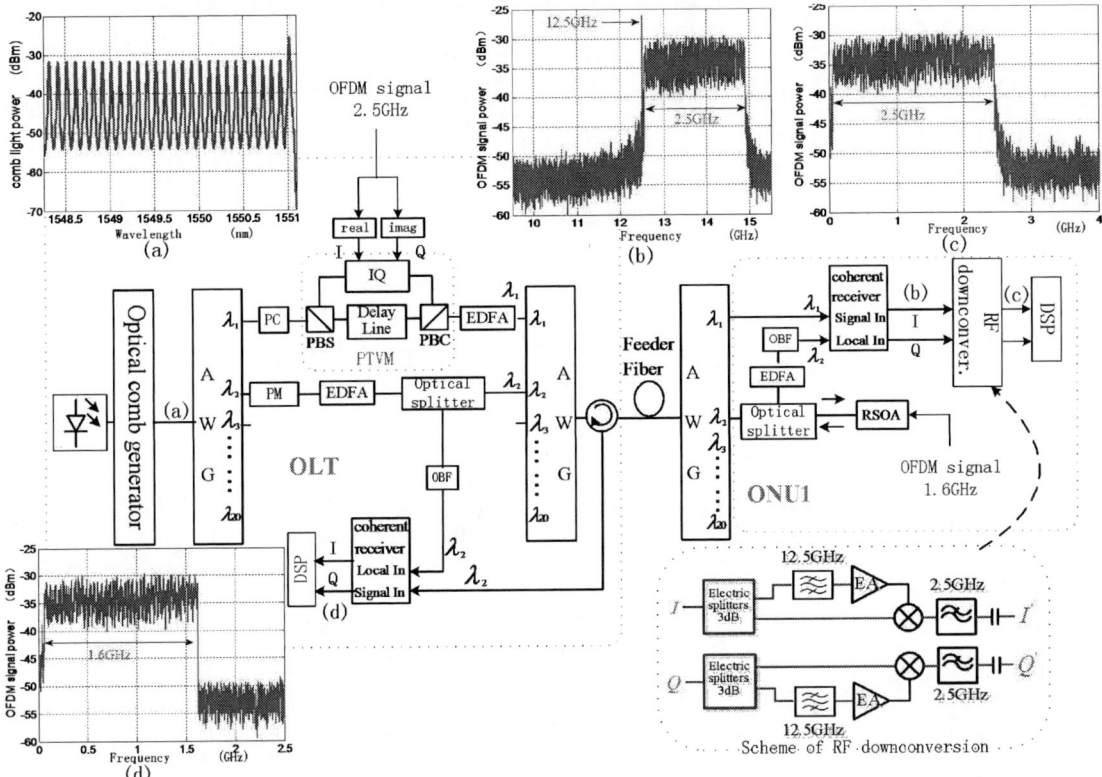

Fig. 1 The experimental schematic diagram of proposed bidirectional coherent OFDM-PON system

III. RESULTS AND DISCUSSIONS

In order to demonstrate the feasibility of the proposed system, we mainly make a study on the transmission capacity and the performance. Therefore it is necessary to measure BER performance. OFDM signals of coherent receiver are captured by a digital phosphor oscilloscope with sampling rate of 10GS/s. The data of 55 frames are collected for off-line processing. Fig. 2 shows the calculated BER of downstream and upstream as received signal power changes from -34dBm to -20dBm with +5dBm and 0dBm LO light power. Solid line displays BER of downstream. Dash line represents uplink BER curves. The overall performance of the downlink is better than the uplink, because of the superiority of IQ modulator and the effect of Rayleigh backscattering on the upstream.

With the 25km SSMF fibre span, receiver sensitivity of -27dBm can be achieved for downlink, and the uplink can reach -25dBm with the adopted FEC limit threshold of 3.8×10^{-3}. It can be found that when LO light power is 0dBm, receiver sensitivity of downstream and upstream respectively decreased to -24dBm and -23.3dBm as shown in Fig. 2. Fig. 3 indicates the relationship between bidirectional BER and local power range from -7dBm to +5dBm. In the signal power of -20dBm and -24dBm, the downstream receiver LO light power at FEC limit reaches -3dBm and +1dBm, and for the upstream it can reach -2dBm and +1dBm. By comparing Fig. 2 and Fig. 3, we may find that the signal light power has more influence on the BER performance than LO light power. The measured 16QAM constellations and

OECC/PS2016

corresponding BER values for downstream and upstream at different received optical signal power are shown in Fig. 4 which indicates that proposed scheme can support cost effective bidirectional coherent detection for OFDM-PON.

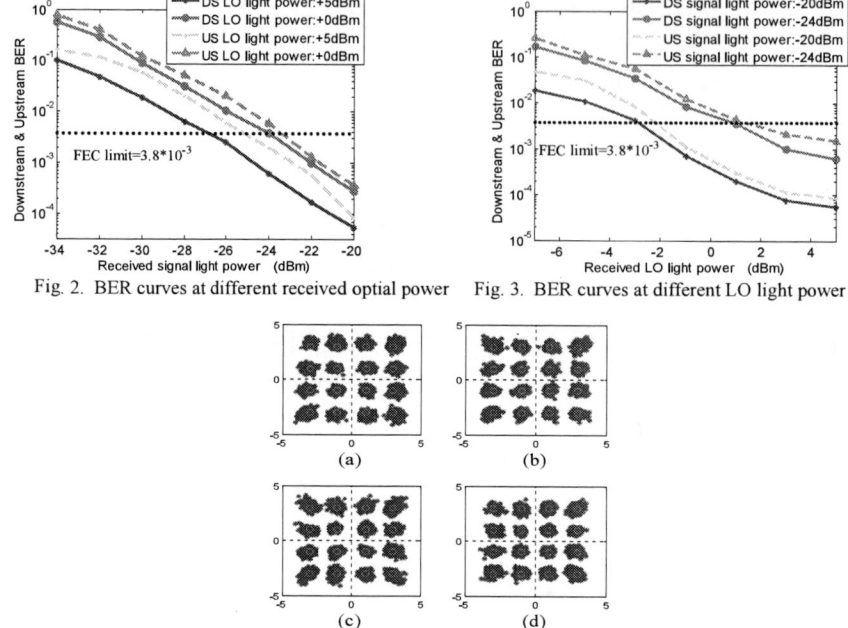

Fig. 2. BER curves at different received optial power   Fig. 3. BER curves at different LO light power

Fig. 4. Measured corresponding constellations with LO light power=+5dBm
(a)DS signal power=-24dBm, BER=4.39*10^{-4}; (b)DS signal power=-22dBm, BER=1.71*10^{-5};
(c)US signal power=-24dBm, BER=2.04*10^{-4}; (d)US signal power=-22dBm, BER=5.38*10^{-5};

IV. CONCLUSIONS

In this paper, a novel cost effective CO-OFDM PON architecture which employed bidirectional coherent detection with 100/64 Gb/s signal line rate and power budget of 34dB and 32dB for downstream and upstream based on optical comb generation combined reflective semiconductor optical amplifier (RSOA) based colorless ONU are experimental demonstrated. Both upstream optical carrier and LO are delivered from a self-developed comb generator so that DSP algorithms can be simplified by canceling frequency offset estimation in both OLT and ONU side.

ACKNOWLEDGMENT

This work was supported in part by Natural Science Foundation of China: (61132004, 61275073, 61420106011) and Shanghai Science and Technology Development Funds: (13JC1402600, 14511100100, 15511105400, 15530500600).

REFERENCES

[1] W. Shieh, "OFDM for adaptive ultra-high-speed optical networks," Optical Fiber Communication (OFC), pp. 1-51, 2010.
[2] X. Liu, S. Chandrasekhar, B. Zhu, P. J. Winzer, A. H. Gnauck, and D. W. Peckham, "Transmission of a 448-Gb/s reduced-guard-interval CO-OFDM signal with a 60-GHz optical bandwidth over 2000 km of ULAF and five 80-GHz-Grid ROADMs," Optical Fiber Communication (OFC), p. PDPC2, 2010.
[3] D. S. Ly-Gagnon, S. Tsukamoto, K. Katoh, and K. Kikuchi, "Coherent detection of optical quadrature phaseshift keying signals with carrier phase estimation," IEEE Journal of Lightwave Technology, vol. 24, pp. 12–21, 2006.
[4] X. Yi, W. Shieh, and Y. Tang, "Phase estimation for coherent optical OFDM," IEEE Photonics Technology Letters, vol. 19, pp. 919-921, Jun. 2007.
[5] S. Cao, P.Y. Kam, and C. Yu, "Decision-aided, pilot-aided, decision-feedback phase estimation for coherent optical OFDM systems," IEEE Photonics Technology Letters, vol. 24, pp. 2067-2069, 2012.
[6] S. Randel, S. Adhikari, and S. L. Jansen, "Analysis of RF-pilot-based phase noise compensation for coherent optical OFDM systems," IEEE Photonics Technology Letters, vol. 22, pp. 1288-1290, 2010.
[7] L. Zhang, Y.X. Song, S.H. Zou, Y.C. Li, J.J. Ye, and R.J. Lin, "Flat frequency comb generation based on mach-zehnder modulator and phase modulator," 12th IEEE international Conference On Communication Technology, 2010.
[8] R.S. Luís, B.J. Puttnam, J.M.D. Mendinueta, "Self-homodyne detection of polarization-multiplexed pilot tone signals using a polarization diversity coherent receiver," 39th European Conference and Exhibition on Optical Communication (ECOC), 2013.
[9] Y.N. Huang, M.Z. Mao, R.J. Lin, C.X. Kuang, R.R. Chen, Q.W. Zhang, Y.X. Song, Y.C. Li, J. Chen and M. Wang, "Experimental demonstration of 100/40 Gb/s OFDM-PON with bi-directional low-cost coherent detection," Asia Communications and Photonics Conference (ACP) 2015, AM1F.2.

WA2-14

OECC/PS2016

Sub-carrier Sharing in OFDM-PON for 5G Mobile Networks Supporting Radio-over-fibre

Sheng Xu[†], Sugang Xu[†], and Yoshiaki Tanaka[†,††]

†Global Information and Telecommunication Institute
††Department of Communications and Computer Engineering
Waseda University, Tokyo, Japan
xusheng@akane.waseda.jp, xusugang@m.ieice.org, ytanaka@waseda.jp

Abstract: *We propose a down-link delay-aware sub-carrier sharing method for future 5G mobile communications with OFDM-based passive optical network (PON) supporting radio over fibre (RoF). Simulation results verify the benefit of our proposal.*
Keywords: *wavelength; 5G; carrier aggregation; passive optical networks; OFDM; radio over fibre*

I. INTRODUCTION

In future 5G cellular networks, the single cell coverage will shrink so that the challenge on high efficient resource allocation is prominent to support high density user equipment (UE) in different micro/pico-cells for a high mobility, high access rate and low latency [1]. For example, numbers of UEs move frequently among a few of micro-cells, which results in idle wireless resource blocks (RB) [4] that emerge heavily over optical OFDM sub-carriers in RoF-OFDM-PON [2]. If these wasted RBs are fully re-utilized and sufficiently shared, it will greatly improve the performance of system. Hence, it is expected to seek a scheduling optimization of idle bandwidth resource to each micro-cell due to high UE mobility. To achieve these targets, we propose and observe a delay-aware sharing scheme for real-time services allowing that each sub-carrier of RoF-OFDM-PON [2] can be multi-cell shared by UEs accessed from different micro-cells. Namely, each UE is arranged to receive multiple data streams demodulated from different ONUs simultaneously. In this paper, we strictly employ a star architecture as shown in Fig.1. In this way, wireless OFDM radio frequency (RF) signals are carried over optical fibre, by means of optical OFDM (OOFDM) [3] transmission. Each optical network unit (ONU) is connected to a 5G micro-cell and shares bandwidth to neighboring cells for UEs.

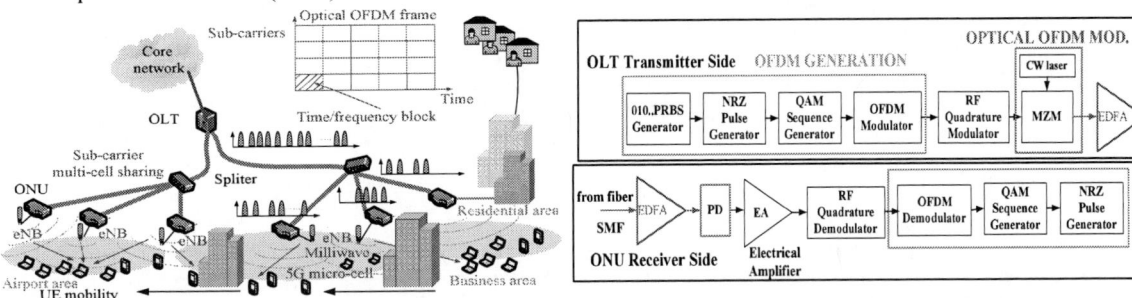

Fig.1. Network of RoF-OFDM-PON for 5G.   Fig.2. The downlink signal processing of RoF-OFDM-PON [2].

II. PHYSICAL LINK ARCHITECTURE IN RoF-OFDM-PON-BASED NETWORKS

Figure.2 illustrates an experimental link architecture for signal processing in RoF-OFDM-PON[2] to support our proposal and to be employed as a physical fundamental for one data stream in system. From the transmitter side, this implementation firstly modulates experimental data through an OFDM processing with performing of the PRBS, NRZ pulse and QAM sequence generation (e.g. 4-QAM) in advance [2]. After that, a RF-IQ mixer is used to deal with the OFDM signal to analog RF with a proper quadrature modulation. The output signal then experiences an optical OFDM (OOFDM) modulation with a 193.1 THz CW laser by LiNbO3 mach-zehnder modulator (MZM) [2], [3] and then are sent into fibre through EDFA to amplify the signal. On each receiver of ONU side, signals from fibre are received by photo-detector (PD) [2], [3] and are executed with a RF de-multiplexing and OFDM demodulation followed by QAM sequence generator and NRZ pulse generator in order to recovery the experimental data [2]. It is important to note that one set of optical OFDM sub-carrier on fibre could be modulated to accommodate different UEs belonging to two or more receivers/ONUs in cellular network, and the wireless data allocated to UEs belonging to any micro-cell could be transmitted by the broadcasting of multiple sharing streams from other ONUs with antennas in other micro-cells nearby (e.g. by a distributed massive MIMO [6]). In this paper, our work thus mainly consider these resource allocation problems, while detailed physical discussions on the control and configuration issues (e.g. protocol specification) for dynamic transmission from multiple ONUs for resource sharing are out of the scope of this paper.

III. HEURISTIC ALGORITHM FOR OPTICAL SUB-CARRIER MULTI-CELL SHARING

Considering the single UE k ($0 \leq k \leq K$) which is accommodated by multiple sharing paths from different ONUs, UE

757

OECC/PS2016

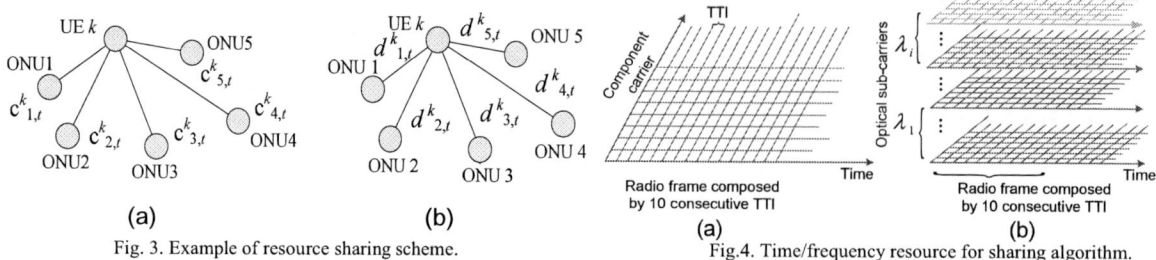

Fig. 3. Example of resource sharing scheme.

Fig.4. Time/frequency resource for sharing algorithm.

k can receive the data from each sharing ONU i ($0 \le i \le N$). For the data of UE k from ONU i, $d^k_{i,t}$ denotes the overall delay on sharing path through ONU i to UE k in time slot t. $c^k_{i,t}$ denotes the available sharing data capacity for UE k from ONU i in time slot t. For example, 5 ONUs share bandwidth resources to UE k in the illustration of a sharing tree in Fig.3(a). The rooted vertex represents UE k and any leaf vertex i represents ONU i, respectively. For each time slot t, the weight of each leaf is $c^k_{i,t}$. The weight of edge between the rooted vertex and any leaf vertex i is $d^k_{i,t}$, as shown in Fig.3(b). Therefore, the resource sharing problem could be briefly described mathematically as follows.

Given parameters:

-$\mathcal{G}$ ($\mathcal{V}$, $\mathcal{E}$) where $\mathcal{V}$: UE k and set of all ONUs, and $\mathcal{E}$: set of resource sharing paths through multiple ONUs to UE k

-Set of UE: $\mathcal{K} = \{k|k = 1, 2,..., K\}$ -Set of ONU: $\mathcal{N} = \{i|i = 1, 2,..., N\}$

-Matrix $C_{k,t} = [c^k_{1,t},..., c^k_{i,t},..., c^k_{N,t}]$, $\forall k$ in $\mathcal{K}$, $c^k_{i,t} > 0$ -Matrix $\mathcal{D}_{k,t} = [d^k_{1,t},..., d^k_{i,t},..., d^k_{N,t}]$, $\forall i$ in $\mathcal{N}$, $\forall k$ in $\mathcal{K}$

-$R_{k,t}$:Minimum data capacity requirement for UE k -$B_{k,t}$:Allocated data capacity to UE k

Objective: -Minimize the average delay of sharing data transmission by multiple ONUs to satisfy UE requirement.

The wireless spectrum resource can be illustrated by Fig.4 (a). In contrast to the single-layer radio frame which is illustrated in Fig.4 (a), by allocating more sub-carriers, multiple radio frames can be delivered on fibre to each ONU, forming the multi-layer radio frames which are shown in Fig.4 (b) for each ONU. The smallest radio resource unit is called resource block (RB). Each component carrier (CC) contains several RBs [4]. One UE can receive several CCs in a certain time slot simultaneously [4]. The concept of multi-layer frames will be adopted in resource sharing algorithm.

We propose a real-time sharing algorithm (RTSA) in Table I. Resource sharing for each UE, (e.g., RB of CC) is executed by assigning $\sigma^{(n,y,l)}_{k,l,p}$. Here, $\sigma^{(n,y,l)}_{k,l,p}$ is a binary variable to define whether or not the n-th RB of the y-th CC[4] on the l-th radio frame on p-th optical sub-carrier is assigned to the k-th UE in slot t for finding a proper $c^k_{i,t}$ subjected to the minimum delay $d^k_{i,t}$ from a set of ONUs. A sharing ONU combination could be found for UE k for a minimum delay under its data rate demand. For each p-th optical sub-carrier, it contains L layers of frames, as shown in Fig.4(b).

TABLE I
REAL-TIME SHARING ALGORITHM (RTSA)

Initialization: $G_k=\varnothing$ for all $k=1,2,...,K$; G_k: total RB set to UE k	**FUNCTION:** form G_k, for any k; $C_k(i)$: a RB set from ONU i to UE k
While ($B_{k,t} < R_{k,t}$) do	Initialize $G_k = \{C_k(1),C_k(2),C_k(i),....,C_k(N)\} = \varnothing$, set $B_{k,t} \leftarrow 0$.
1: Make a sharing graph $\mathcal{G}(\mathcal{V},\mathcal{E})$, $\mathcal{V}=\{1,...,N\}\cup\{k\}$,	**1:** Find ONU i for current link(i,k), where $i=$ argmin $d^k_{i,t}$, set $p \leftarrow 1$.
$\mathcal{E}=\{(i,k) \mid d^k_{i,t} \le D_k\}$; D_k is a maximum tolerable delay(MTD).	**2:** Set $l \leftarrow 1$; l is layer (radio frame) indicator.
2: Assign weights to $\mathcal{E}$ for links by matrix $\mathcal{D}_{k,t}$ and sort the edges in $\mathcal{E}$	**3:** Find the idle RB for any UE,where $\sigma^{(n,y,l)}_{k,l,p}=0$, $\forall k$.
according to its weight in graph $\mathcal{G}(\mathcal{V},\mathcal{E})$ in an ascending order.	**4:** Allocate a corresponding idle RB to UE k, put RB of $\sigma^{(n,y,l)}_{k,l,p}=0$ into
3: Find an edge (i,k) with minimum $d^k_{i,t}$ in the ascending order of $\mathcal{E}$ as	set $C_k(i)$,then for the UE k,$\sigma^{(n,y,l)}_{k,l,p} \leftarrow 1$. Increase capacity $B_{k,t}$, where,
a current link (i,k) for resource sharing.	$B_{k,t} = B_{k,t} + r^{(n,y,t)}_{\text{RB }k,l,p}$.
4: $B_{k,t} \leftarrow c^k_{i,t}$ by allocating $\sigma^{(n,y,l)}_{k,l,p}$ until $B_{k,t} \ge R_{k,t}$,otherwise, go to **STEP 3**	**5:** If $B_{k,t} \ge R_{k,t}$, **break**; $r^{(n,y,t)}_{\text{RB }k,l,p}$ is the capacity of RB.
5: Traverse all UEs in set $\mathcal{K}$ to allocate resource satisfying	**Else if** $\forall k$, $\forall \sigma^{(n,y,l)}_{k,l,p}=1$, *i.e.*, the set of RBs on current layer have
$i =$ argmin $d^k_{i,t}$, $B_{k,t} \leftarrow c^k_{i,t}$, find $c^k_{i,t}$ employing **FUNCTION:** form G_k	been occupied and can not be scheduled to UE k for sharing,
End while	and if $l \le L$ add l, go to **STEP 3**.
6: Traverse all ONUs in set $\mathcal{N}$ to allocate resources, repeat **STEP 1** to **5**.	**Else if** $p \le P_{\text{max-sub}}$, add p, go to **STEP 2**.
7: Update matrix $C_{k,t}$, and matrix $\mathcal{D}_{k,t}$.	**Else**, go to **STEP 1**.
End	**End IF**
Out put: $G_k=\{C_k(1),C_k(2),C_k(i),......C_k(N)\}$	**6:** Output $G_k = \{C_k(1),C_k(2),C_k(i),....C_k(N)\}$

IV. SIMULATION AND RESULTS

In simulations, a RoF-OFDM-PON with 240 cells as shown in Fig.1 is employed with different mobility ratios of UEs (a=number of mobile UEs/number of total UEs). Optical sub-carriers with per λ_i 100-Gb/s digital-equivalent data-rate are adopted. LTE-A-like wireless resources carried on sub-carriers are assigned to UEs corresponding to the scheduling model in the network simulator 3 (NS3) [5]. In addition, we define 8 different mobility ratios of UEs equaling to 0.02, 0.04, 0.06, 0.1, 0.2, 0.4, 0.6 and 0.8, respectively. The total UE number changes from 5000 to 10000 in simulations with the appending of optical OFDM sub-carriers. The detailed simulation settings are given by Table.II.

We simulate the average delay for 5 experimental UE groups (per 1000 UEs with totally 5000 UEs) in system for instance according to 5 different maximum tolerable delay (MTD) of UE k (D_k), which equals to 12ms, 14ms, 16ms, 18ms, and 20ms, respectively. UEs in each group run randomly among different cells in simulations by a setting of 3 different minimum UE data requirements (100Mbps, 150Mbps, and 200Mbps) in order to observe the change of delay time adopting our sharing proposal. Fig.5 (a), (b) and (c) show that the average delay time for all experimental UE groups become less and less by adding total optical sub-carrier number in system. The average delay herein is an average value of delay time ($d^{k}_{i,t}$) spent on all the sharing paths for all UEs in each experimental group, including the overall delay from the source (OLT) to destination (UE), e.g, scheduling delay on OLT side, packet delay on fibre, eNB processing delay, frame alignment delay, wireless data packet delay, HARQ re-transmission and UE processing delay.

In addition, we evaluate the average cell throughput in Fig.6 (a) under different cell numbers with total 10 sub-carriers by comparing RTSA with maximum throughput (MT) and proportional fair (PF) schemes [4]. Furthermore, in Fig.6 (b), throughput performance is observed in a single cell under different optical sub-carrier numbers. The results of Fig. 6 (a) and (b) reveal that RTSA with a higher mobility ratio of UEs has a higher cell throughput, and RTSA has better cell throughput compared to MT and PF respectively. The reason is that a higher UE mobility yields many more idle RBs on a single optical sub-carrier and the proposed RTSA re-utilizes these idle bandwidth resources sufficiently.

TABLE II
SIMULATION SETTINGS

Parameter	Value	Parameter	Value
LTE sub-carrier	15 kHz	Frame duration	10 ms
Resource block	180 kHz	TTI	1 ms
RB OFDM symbols	7	MCS	29
UE Received CC$_{max}$	4	Bandwidth of CC	5MHz
Single CC length	25 RBs	Testing MIMO per cell	4×4
BS TX power	30 dBm	Number of Cell	240
Noise spectral density	-174 dBm/Hz	Cell radius	500m
Pathloss (distance R),in dB	128.1+37.6lg$R$	SMF fibre distance	20km

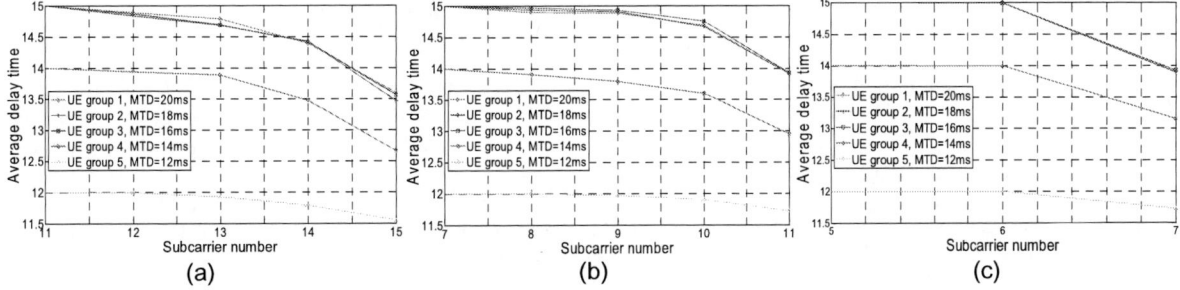

Fig.5. Delay (ms) performance comparison with RTSA, MT and PF for cell throughput. (a) UE minimum data requirements 200Mbps, (b) UE minimum data requirements150Mbps, (c) UE minimum data requirements 100Mbps.

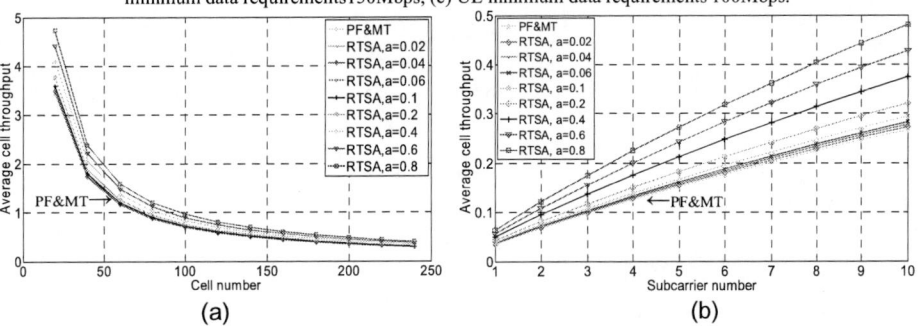

Fig. 6. Average throughput (per 10Gbps) performance comparison with RTSA, MT and PF. (a) Total cell average throughput under different cell numbers with total 10 sub-carrier, (b) Single cell average throughput under different sub-carrier numbers.

V. CONCLUSIONS

In this paper, an optical sub-carrier multi-cell sharing proposal considering minimum delay is given for future 5G cellular networks with OFDMA-based PON supporting radio-over-fibre. Simulation results prove our proposal is beneficial for cell throughput optimization and it also has an effective improvement on network latency.

REFERENCES

[1] S. Bi, R. Zhang, Z. Ding, and S. Cui, "Wireless communications in the era of big data," IEEE Communications Magazine, vol.53, no.10, pp.190-199, October 2015.

[2] F. Almasoudi, K. Alataw, and M. A. Matin, "Study of OFDM technique on RoF passive optical network," Optics and Photonics Journal, vol.3, pp.217-224, 2013.

[3] C.W.Chow, C.H.Yeh, C.H.Wang, C.L.Wu, S.Chi, and C.Lin, "Studies of OFDM signal for broadband optical access networks," IEEE Journal on Selected Areas in Commun., vol.28, no.6, pp.800-807, Aug. 2010.

[4] H.-S. Liao, P.-Y. Chen, and W.-T. Chen, "An efficient downlink radio resource allocation with carrier aggregation in LTE-Advanced networks," IEEE Trans. on Mobile Computing, vol.13, no.10, pp.2229-2239, Oct. 2014.

[5] NS-3 consortium, "NS-3 network simulator," http://www.nsnam. org/, NRIA and the University of Washington, accessed 2015.

[6] G.Xu, A.Liu, W.Jiang,H.Xiang,W. Luo,"Joint user scheduling and antenna selection in distributed massive MIMO systems with limited backhaul capacity," China Communications, vol.11, no.5, pp.17-30, May 2014.

WA2-15

OECC/PS2016

Channel-reuse IMDD-based 40 Gb/s/λ 16-QAM Nyquist-SCM Downstream and 20 Gb/s/λ Nyquist 4 PAM Upstream WDM-PON

Zhiguo Zhang, Bingbing Zhang, Yanxu Chen, Bingchang Hua, Xue Chen

State Key Laboratory of Information Photonics and Optical Communications (Beijing University of Posts and Telecommunications),
Beijing 100876, China
zhangzhiguo@bupt.edu.cn

Abstract: We experimentally demonstrate an optical intensity modulation transmitter and direct detection receiver-based channel-reuse WDM-PON. 40 Gb/s/λ downstream and 20 Gb/s/λ upstream transmissions over 45-km standard single-mode-fiber (SSMF) on a 100-GHz WDM grid are achieved.
Keywords: WDM-PON, Intensity modulation, Direct detection, Channel-reuse, Self wavelength management

I. INTRODUCTION

Wavelength-division-multiplexing passive optical networks (WDM-PONs) have been considered as a promising solution for next-generation optical access networks characterized by high security, easy maintenance, great flexibility, and broad bandwidth [1-2]. Driven by the ever-increasing user demands for broadband services to support high-quality Internet protocol television (IPTV), e-learning, interactive games, and peer-to-peer multimedia services, it is expected that the demand on the data transmission rate will continue to grow, and numerous access nodes will be deployed over the next decades. Technological developments for supporting longer reach and larger split are also expected [3]. As 40 Gb/s and 20 Gb/s per channel will become the typical channel rates for WDM-PONs in the near future [4-5], the main challenge associated with increasing the overall system capacity is to improve the corresponding spectral efficiency. Narrowing the inter-channel spacing and using WDM channel-reuse techniques are promising methods for increasing the overall system capacity of such WDM-PONs.

In this paper, we demonstrate a channel-reuse, 40 Gb/s/λ downstream and 20 Gb/s/λ upstream, 45-km-reach WDM-PON scheme on a 100 GHz WDM grid. Optical intensity modulation (IM) and direct detection (DD), employing 16 quadrature amplitude modulation (QAM) subcarrier modulation (SCM) for downstream and Nyquist 4 level pulse amplitude modulation (PAM) for upstream, are used in the system. We also propose a double branch fiber-based wavelength management method using Rayleigh backscattering (RB) noise to manage the wavelength of the tunable laser (TLD) or optical transmitter in colorless ONU, in which two branch fibers are used to improve the control accuracy in long-reach branch fiber configuration.

II. AUTOMATIC WAVELENGTH CONTROL METHOD

To determine the feasibility of the RB noise-based automatic wavelength control method, we experimentally measured the RB noise power reflected by the feeder fiber for various single or double branch fiber lengths. Fig. 1 shows the experimental setup. A 20 Gb/s electrical 4 PAM signal is generated in a Nyquist 4 PAM generator by combining two bit streams from a pulse pattern generator (PPG). The amplified electrical signal drove an electro absorption modulator (EAM) after a Bias-Tee with intensity modulation. The optical signal from the EAM was then sent to a 40 km-long standard single-mode fiber (SSMF).

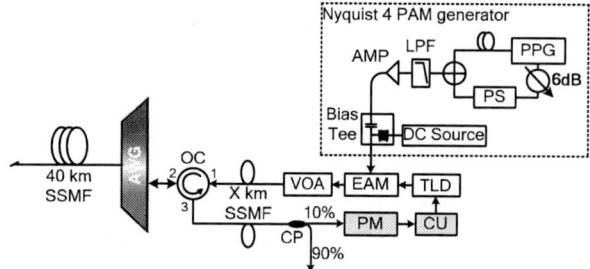

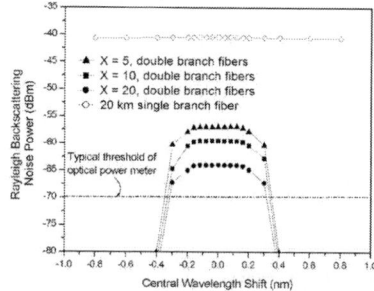

Fig. 1. Experimental setup for the wavelength control method.

Fig. 2. RB noise power VS. central WS.

In our experiment, the variable optical attenuator (VOA) output optical power was set at −2 dBm, based on the minimal actual power conditions in the proposed WDM-PON scheme. The port 2 of the three-port optical circulator

(OC) was connected to one of the array waveguide grating (AWG) WDM channels. The branch fiber lengths were 5 km (i.e., X = 5), 10 km (i.e., X = 10), and 20 km (i.e., X = 20), respectively. Fig. 2 shows the measured average RB noise power, versus the central wavelength shift (WS) between the TLD wavelength and the central wavelength of the WDM channel. The results show that the average RB noise power was ~−64 dBm, ~−59 dBm, and ~−57 dBm, when the double branch fiber lengths were set to 20 km, 10 km, and 5 km, respectively. In order to compare the difference of RB noise powers in double branch fibers setup and single branch fiber setup, we also measured the RB noise power with 20 km single branch fiber setup (i.e., X = 0 and the fiber length between the OC and AWG was set to 20 km). The result shows that the RB noise power with double branch fibers change obviously when the central WS changes; but it almost does not change with 20 km single branch fiber. Therefore, a double branch fibers setup is effective and necessary to manage the wavelength of the TLD.

III. CHANNEL-REUSE, 45-KM-REACH WDM-PON WITH 40 GB/S/Λ DOWNSTREAM AND 20 GB/S/Λ UPSTREAM

Fig. 3 shows the experimental setup of the channel-reuse WDM-PON system with the downstream transmission rate of 40 Gb/s/λ and upstream transmission rate of 20 Gb/s/λ. In the downlink direction of channel k, 40-Gb/s downlink 16-QAM Nyquist-SCM signal is generated by a 16-QAM Nyquist-SCM optical transmitter. In the receiver of the ONU, the converted photocurrent generated by a photodiode (i.e., PIN) is sampled and recorded by a digital oscilloscope with a sampling rate and resolution of 40 GSample/s and 8 bits, respectively. The related signal processing is implemented offline in a PC with a MATLAB-based DSP program. In the uplink direction of channel k, 20-Gb/s downlink Nyqusit 4 PAM signal is generated by a Nyquist 4 PAM optical transmitter. In the receiver of the OLT, the converted photocurrent generated by a photodiode (i.e., PIN) is also sampled and recorded by a digital oscilloscope with a sampling rate and resolution of 40 GSample/s and 8 bits, respectively. The related signal processing is also implemented offline in a PC by MATLAB-based DSP program. Moreover, a RB noise-based automatic wavelength control module consisting of an optical power meter (PM) and a control unit (CU) are added in the ONU to automatically control the wavelength of the TLD.

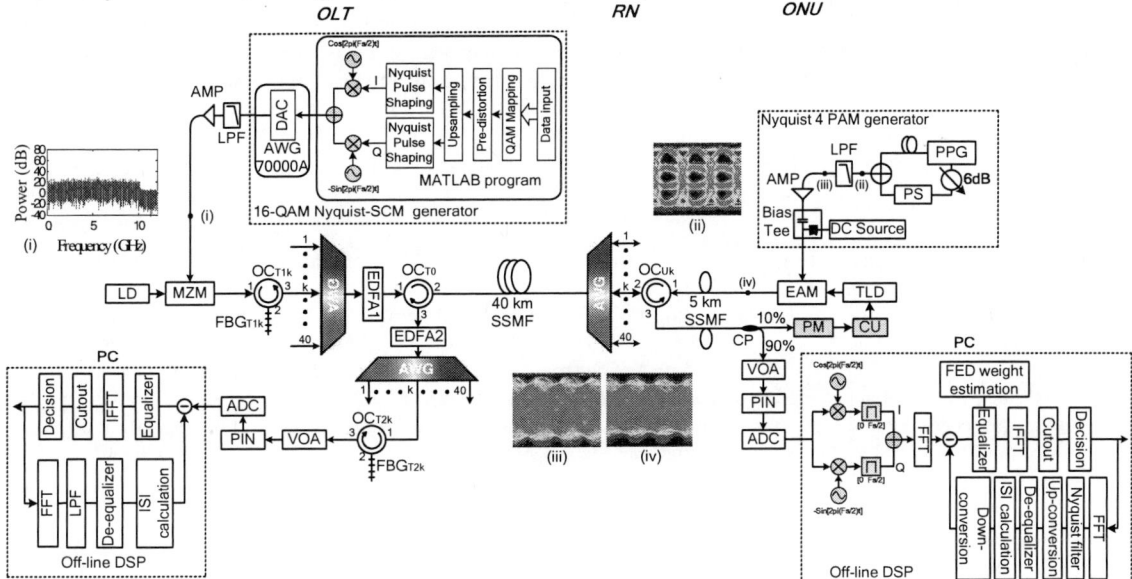

Fig. 3. Experimental setup of the channel-reuse WDM-PON.

The transmission performance for the downstream and upstream signals in the B-t-B configuration and after the 45-km-long transmission at channel 25 is shown in Figs. 4(a) and 4(b). In our measurement, the TLD operation wavelength was 1549.32 nm in the wavelength pass-band of channel 25, which was the same as the operation wavelength of the downlink 16-QAM Nyquist-SCM optical transmitter in channel 25. The output power of the power-tunable TLD was set at 13 dBm. As a result, the average output power of the EAM was ~5 dBm. The output power of the downlink Mach-Zehnder modulator (MZM) in channel 25 was ~0 dBm. The optical signal powers into downlink and uplink receivers can be adjusted by using two EDFAs (i.e., EDFA1, EDFA2). The black lines with triangles, circles, and squares in Fig. 4(a) show the measured downstream BERs in the B-t-B configuration, after the 45-km-long unidirectional transmission, and after the 45-km-long bidirectional simultaneous transmission, respectively. The lines in Fig. 4(b) show the upstream BERs for the same configurations. The lines with triangles and circles in Figs. 4(a) and 4(b) show that the downlink and uplink exhibit the transmission penalties of ~1.3 dB and ~1.5 dB, respectively, which originate from linear and nonlinear ISI. As shown in Fig. 4(b), compared with the 45-km-long unidirectional uplink transmission, a power penalty of ~5 dB (at the BER of 3.8×10^{-3}) exists for the 45-km-long bidirectional simultaneous transmission. As shown in Fig. 4(a), compared with the 45-km-long unidirectional downlink transmission, there is a power penalty of ~3.5 dB (at the BER of 3.8×10^{-3}) for the 45-km-long bidirectional simultaneous transmission.

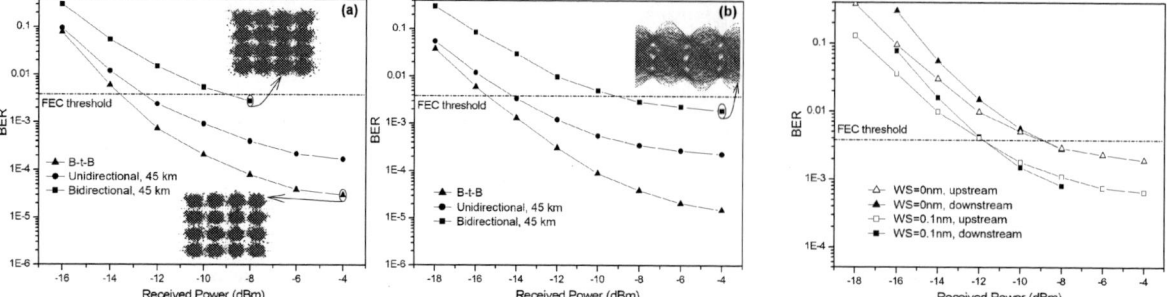

Fig. 4. The transmission performance   Fig. 5. The BERs versus WSs.

To analyze the impact of the Rayleigh backscattering noise, which could worsen the optical signal noise ratio (OSNR) and reduce the receiver sensitivity in the channel-reuse system, we measured the BER performance for the bidirectional transmission, for different central WS values in a WDM channel. The experimental setup was the same as that shown in Fig. 3. The BER performance for the different WS values is shown in Fig. 5. In this measurement, the average output power of the downlink MZM and the uplink EAM in channel 25 remained at ~0 dBm and ~5 dBm, respectively. As a result, the OSRBNR at port 1 of the OC_{T2k} in channel 25 remained at 20 dB, and the OSRBNR at port 3 of the OC_{Uk} in channel 25 also remained at 28 dB. The uplink and downlink receiver sensitivities both were −9 dBm at the BER of 3.8×10^{-3} when the central WS between the optical signals in both directions was 0 nm. Both the uplink and downlink receiver sensitivities improved when the WS increased. They reached −11.6 dBm and −11.8 dBm at the BER of 3.8×10^{-3}, respectively, when the central WS was 0.1 nm. The receiver sensitivities did not increase significantly when the central WS was above 0.1 nm. Therefore, the impact induced by the Rayleigh backscattering noise can be effectively reduced by mismatching the wavelengths of the upstream and downstream signals in the channel-reuse system: the greater the WS, the smaller the Rayleigh backscattering noise effect. For the channel-reuse WDM-PON system proposed in this paper, an intentional central WS can be set to reduce the impact of the Rayleigh backscattering noise.

IV. CONCLUSIONS

We proposed and investigated a long-reach, channel-reuse WMD-PON scheme, which exhibits the IMDD-based 40 Gb/s/λ downstream and 20 Gb/s/λ upstream transmissions. A double branch fibers-based automatic wavelength management method using Rayleigh backscattering noise was also demonstrated for managing the wavelength of the tunable laser in a colorless ONU. A channel-reuse, 45-km-reach, full-duplex, 40 Gb/s/λ downstream and 20 Gb/s/λ upstream transmissions on a 100 GHz WDM grid was experimentally demonstrated. The measurement results show that the impact of the Rayleigh backscattering noise decreases with increasing the central WS. Using a 3.8×10-3 uniform BER standard, the receiver sensitivities are improved by ~2.6 dB and ~2.8 dB in the uplink and the downlink directions, respectively, when the central WS is increased from 0 nm to 0.1 nm.

ACKNOWLEDGMENT

This study was supported by National Natural Science Foundation of China (No.61302079) and Fund of State Key Laboratory of Information Photonics and Optical Communications (Beijing University of Posts and Telecommunications), P. R. China.

REFERENCES

[1] Elaine Wong, "Next-Generation Broadband Access Networks and Technologies", IEEE JLT,30(4), 597-608, (2012).

[2] Dirk Breuer, Frank Geilhardt, Ralf Hülsermann, Mario Kind, Christoph Lange, Thomas Monath, and Erik Weis, "Opportunities for Next-Generation Optical Access", IEEE Comm. Magazine, 49(2), S16-S24, (2011).

[3] Philippe Chanclou, Anna Cui, Frank Geilhardt, Hirotaka Nakamura and Derek Nesset, "Network operator requirements for the next generation of optical access networks", IEEE Network, 26(2), 8-14, (2012).

[4] H.K. Shim, K.Y. Cho, U.H. Hong, Y.C. Chung, "Transmission of 40-Gb/s QPSK upstream signal in RSOA-based coherent WDM PON using offset PDM technique", Optics Express, 21(3), 3721-3725, (2013).

[5] Q. Guo, and A. V. Tran, "20-Gb/s single-feeder WDM-PON using partial-response maximum likelihood equalizer", IEEE Photonics Technology Letters, 23(23), 1802-1804, (2011)

Efficient Mobile Fronthaul for Simultaneous Transmission of 4G and Future Mobile Signals

Pham Tien Dat[1], Atsushi Kanno[1], Naokatsu Yamamoto[1], and Tetsuya Kawanishi[1,2]

[1]National Institute of Information and Communication Technology, Tokyo, Japan
[2]Waseda University, Tokyo, Japan

Abstract: We present an efficient mobile fronthaul system for co-transmission of radio signals in the microwave and millimeter-wave bands. We successfully transmit a WLAN and an LTE-A signal at 2.6 GHz and 33.6 GHz, respectively, over the system.
Keywords: Mobile fronthaul, radio over fiber, millimeter-wave, mobile small cells

I. INTRODUCTION

Recently, research on mobile fronthaul technologies has attracted much interest from both the academia and industry community. In current systems, oversampled digital baseband streams are transmitted from central stations (CSs) to remote sites via digital photonic links using interface protocols such as common public radio interface or open base station architecture initiative. However, future small cell-based wireless networks will pose many challenges to the digital photonic links, including high bit rate, strict jitter and synchronization requirements, and low latency. The simultaneous transmission of multiple wireless standards and radio access networks (RANs) over the same system presents another challenge. The use of high-frequency radio access networks in the millimeter-wave (MMW) bands will bring another challenge because very high speed analog-to-digital and digital-to-analog converters must be equipped at remote sites.

Radio-over-fiber (RoF) technology in which analog radio frequency signals are transmitted over a photonic fronthaul system to remote sites have recently been proposed [1, 2]. This system helps to simplify remote sites, reduce system cost and energy, support low-latency transmission, and enable co-existence of multiple radios. It should be noted that a co-transmission of multiple radios, including multiple bands, multiple services, and multiple operators, is very important to take advantage of the deployed infrastructure for fast and low-cost deployment of new services. In particular, the co-existence of the current 4G and future 5G RANs in the MMW bands would be of critical importance to support heterogeneous communications and management networks in the future. In this paper, we propose a simple and high-spectral efficiency system for the simultaneous transmission of a high-throughput WLAN signal and a high-speed mobile signal in the MMW band. The proposed system bases on an intermediate-frequency-over-fiber system (IFoF) [3] and remote delivery of high-quality local oscillator (LO) signals. We present a proof-of-concept demonstration on transmission of an IEEE 802.11ac and an LTE-A signal at 33.6 GHz over a 20-km fiber link. We confirm experimentally the successful transmission of high-speed signals over the system.

II. EXPERIMENTAL SETUP

The setup for the simultaneous transmission of a WLAN and an MMW mobile signal over a fiber link is shown in Fig. 1. In our system, instead of transmitting directly the MMW mobile signal via the fiber link, we transmit a much lower frequency signal at an intermediate frequency (IF) band to the remote radio head (RRH), and subsequently up-convert the signal to the desired MMW band using an electrical signal up-conversion. To reduce the system cost and remote site complexity, we also transmit an LO signal from the CS to the RRH for the signal up-conversion.

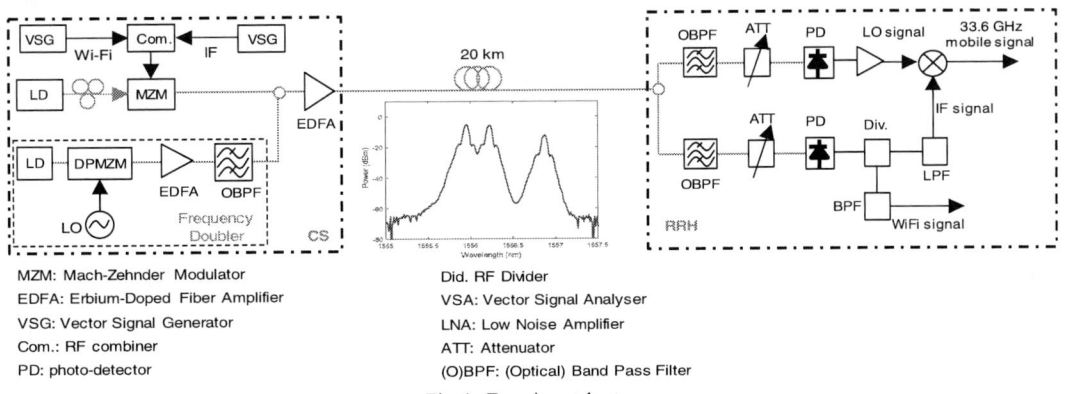

MZM: Mach-Zehnder Modulator
EDFA: Erbium-Doped Fiber Amplifier
VSG: Vector Signal Generator
Com.: RF combiner
PD: photo-detector

Did. RF Divider
VSA: Vector Signal Analyser
LNA: Low Noise Amplifier
ATT: Attenuator
(O)BPF: (Optical) Band Pass Filter

Fig. 1. Experimental setup.

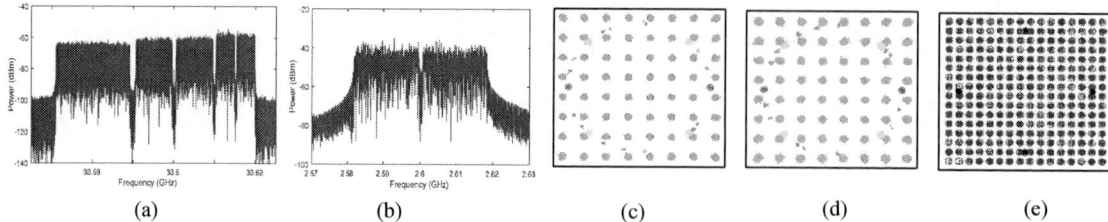

| (a) | (b) | (c) | (d) | (e) |

Fig. 2. Examples of 33.6-GHz LTE-A and WLAN signal after fiber transmission: (a) received LTE-A signal spectrum; (b) received WLAN signal spectrum; (c) 10-MHz CC LTE-A; (d) 20-MHz CC LTE-A; (e) 40-MHz 256-QAM WLAN signal.

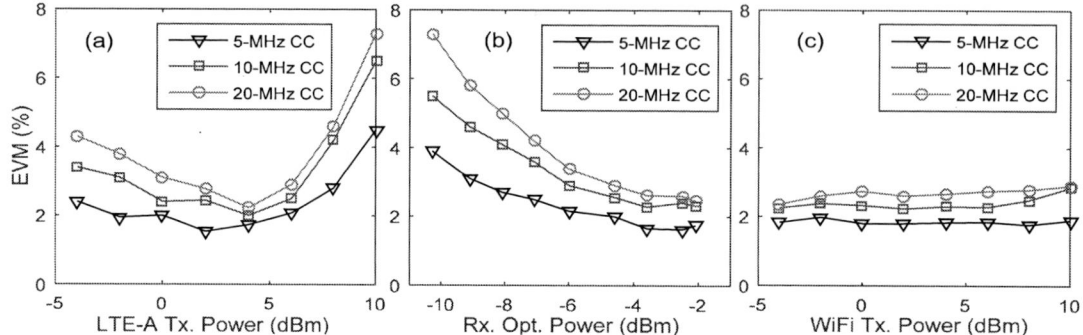

Fig. 3. Performance of LTE-A signal after fiber transmission and up-conversion to 33.6 GHz: (a) for different transmit powers; (b) for different received optical powers; (c) for different co-transmission WLAN signal powers.

At the CS, an IEEE 802.11ac signal at 2.6 GHz and a carrier aggregation LTE-A signal at 600 MHz are generated and combined before modulating a lightwave signal from a laser diode (LD). The modulated optical signal is then combined with an optical LO signal generated from an optical MMW signal generator using a high-precision optical modulation technology [4]. The combined signal is amplified and transmitted to the RRH via a 20-km single-mode fiber. The received optical signals are separated by a 3-dB optical coupler and optical band-pass filtered to recover the transmitted signals. After being converted to the electrical format by photo-detectors (PD), the electrical signals on the one branch is divided and electrically filtered to recover the transmitted WLAN and the IF LTE-A signals. On the other hand, the recovered LO signal from the optical LO signal is electrically amplified to a sufficiently high level before being mixed with the recovered IF LTE-A signal to form an MMW-band mobile signal. In our system, for a proof-of-concept demonstration, a 16.5-GHz electrical synthesized signal is transmitted from the CS to the RRH. At the RRH, because of the beating note between the two first-order sidebands at the PD, a 33-GHz LO signal is generated. After the signal up-conversion at the RRH, a mobile signal at 33.6 GHz can be created. The WLAN and the LTE-A signals at 33.6 GHz are then inputted to two vector signal analyzers (VSAs) and finally demodulated by a VSA software. Shown in the figure is the optical spectrum of the combined optical signals.

III. EXPERIMENTAL RESULTS

In this experiment, we transmit simultaneously a 40-MHz 256-QAM WLAN 802.11ac signal and a 50-MHz LTE-A signal over the system. To emulate the concept of using IFoF systems for transmission of multiple radio signals, such as multiple bands and multiple-input and multiple-output signals, we generate a five-carrier-component (CC) LTE-A signal aggregating two 5-MHz CCs, two 10-MHz CCs, and one 20-MHz CC. All the signals are standard-compliant ones and generated from a laptop using commercially available signal studios and downloaded to vector signal generators (VSGs). Figure 2 shows examples of received spectral and demodulated constellation maps of the WLAN and the LTE-A signal at 2.6 GHz and 33.6 GHz, respectively. We can receive the signals with very clear spectral and constellation maps. We then evaluate the transmission performance using root-mean-square (rms) EVM parameter. Figure 3 shows the LTE-A signal performance after transmission over the IFoF system and up-conversion to 33.6 GHz. In the actual system, this up-converted signal can be transmitted to end users via small-cell based RANs. The quality of this mobile signal at the RRH must follow the requirements for transmitter tests, such as less than 8% EVM for 64-QAM, 12.5% for 16-QAM, and 17.5% for QPSK signals [5]. The performance of the LTE-A signal carried by the 5-MHz, 10-MHz, and 20-MHz CCs for different transmit powers, different received optical powers, and different co-transmission WLAN transmit powers is shown in Figs. 3(a), (b), and (c), respectively. All the signals are successfully transmitted and received with EVM values much better than the requirements dictated in the standard. The increase in transmission power of the co-transmission WLAN signal has a relatively small effect on the LTE-A signal performance, confirming the possibility of the simultaneous transmission of multiple radio signals over the fronthaul system.

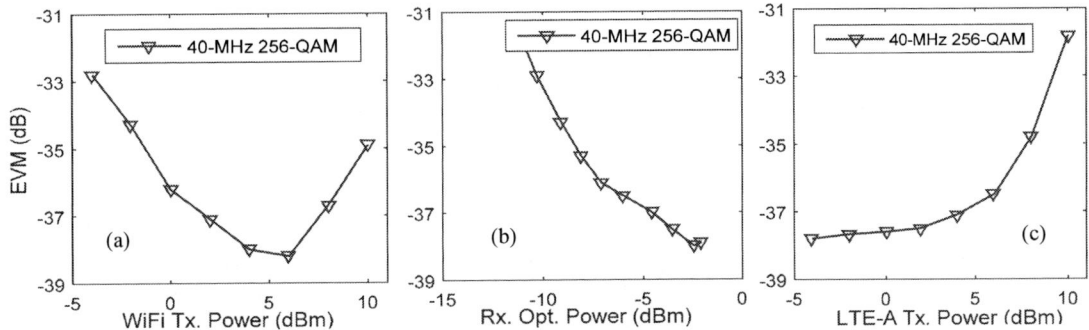

Fig. 4. Performance of WLAN 802.11ac signal after fiber transmission: (a) for different transmit powers; (b) for different received optical powers; (c) for different co-transmission IF LTE-A powers.

We also measure the performance of the WLAN signal and investigate the impact of the simultaneous transmission on the signal quality. Figures 4(a), (b) and (c) show the performance for the 40-MHz 256-QAM IEEE 802.11ac signal after transmission over the fiber link. The required EVM for a 256-QAM WLAN signal is -31 dB as dictated in the standard [6]. As shown in Figs. 4(a) and (b) for different transmit powers and different received optical powers, the transmission can easily satisfy the requirement with a sufficiently large power range. However, different to the LTE-A signal, the increase in transmission power of the co-transmission IF LTE-A signal has a relatively high effect on the WLAN signal performance. This should be caused by the distortion at the optical modulator because of the increase in the total electrical transmit power. Nevertheless, considering the optimum transmit powers for the IF LTE-A signal and the WLAN signal of approximately 4 dBm and 6 dBm, respectively, as shown in Fig. 3(a) and Fig. 4(a), we can confirm that there is a negligible effect of simultaneous transmission on the performance of both the LTE-A and WLAN signals.

IV. CONCLUSIONS

We have proposed a simple and efficient mobile fronthaul system for simultaneous transmission of radio access network signals in the conventional microwave and millimeter-wave bands. The system is based on an intermediate frequency over fiber transmission and remote signal up-conversion with a remote delivery of a local oscillator signal. We successfully demonstrate a co-transmission of a high-throughput IEEE 802.11ac signal and a high-speed mobile signal at 33.6 GHz over a 20-km fiber link. Satisfactory performance for both signals is experimentally confirmed with a negligible effect of simultaneous transmission on each signal performance. The proposed system can be a good solution for future heterogeneous communications using an overlap of macro cells for control plane and management signal transmission and small cells for high-speed user plane signal transmission.

ACKNOWLEDGMENT

This work was conducted as a part of the Research and development for expansion of radio wave resources, supported by the Ministry of Internal Affairs and Communications (MIC), Japan.

REFERENCES

[1] C. Liu et al., "A Novel Multi-Service Small-Cell Cloud Radio Access Network for Mobile Backhaul and Computing Based on Radio-over-Fiber Technologies," J. Lightwave Technol. 31(17), 2869-2875 (2013).

[2] P. T. Dat et al., "High-Capacity Wireless Backhaul Network Using Seamless Convergence of Radio-over-Fiber and 90-GHz Millimeter-Wave," J. Lightwave Technol. 32 (20), 3910-23 (2014).

[3] P. T. Dat et al., "High-Spectral Efficiency Millimeter-Wave-over-Fiber System for Future Mobile Fronthaul," Proc. of ECOC, 2015, P7.8.

[4] A. Kanno et al., "Coherent Radio-Over-Fiber and Millimeter-Wave Radio Seamless Transmission System for Resilient Access Networks," IEEE Photonics J, 4 (6), 2196-2204 (2012).

[5] 3GPP, Evolved universal terrestrial radio access (E-UTRA); physical channels and modulation, 3GPP TS 36.211 V10.4.0, Rel-10, 2011.

[6] IEEE Std 802.11™-2012, Part 11: Wireless LAN Medium Access Control (MAC) and Physical Layer (PHY) Specifications, March 2012.

WA2-17

OECC/PS2016

Frequency-Stabilized Multi-Tone Signal Generation By Optical Frequency Locked-Loop And Optical Frequency Comb

Jinsei OBA[†‡] Kosuke KAMIKAWA[†‡] Atsushi KANNO[‡] Naokatsu YAMAMOTO[‡]

and   Hideyuki SOTOBAYASHI[†]

[†]Aoyama Gakuin University 5-10-1 Fuchinobe, Chuo-ku, Sagamihara-shi, Kanagawa, 252-5258 Japan
[‡]National Institute of Information and Communications Technology, 4-2-1 Nukui-kitamachi, Koganei, Tokyo, 184-8795 Japan
[†]c5614062@aoyama.jp, sotobayashi@ee.aoyama.ac.jp   [‡]{kanno, naokatsu}@nict.go.jp

Abstract: We demonstrate frequency-stabilized millimeter-wave signal generation system using optical frequency locked-loop technique employed with an optical frequency comb. Observed frequency fluctuation system realized drastic reduction as compared with the fluctuation generated by two free-running lasers.

Keywords: Optical frequency locked loop, Optical frequency comb, Mach-Zehnder modulator, Double-sideband suppressed-carrier modulation

INTRODUCTION

Rapid growth of multimedia services on the Internet increases mobile traffics because smart devices including smart phones now become powerful "palm-top" computers, and therefore, a high-speed wireless communication system is highly demanded [1]. For realization of such high-speed wireless link, millimeter-wave and terahertz-wave communication technology have been actively developed owing to their broad bandwidth feature [2]–[4]. Especially, radio-over-fiber (RoF) technology in millimeter-wave bands has a candidate for high-speed wireless signal generation using high-speed optical devices adopted from an ultra-fast optical fiber communication technology. Basically, the wireless signal is regenerated by an optical heterodyning technique using the RoF signal; a beat note between two optical components forms the radio signal at the frequency of the difference of the two components. There are two methods for generation of the two optical components: one is based on an optical modulation technique with a laser, and the other is by two free-running lasers. In the former case, a frequency-locked signal, whose frequency stability corresponds to an electrical synthesizer connected to the modulator, can be easily generated [5]. This method is capable for a coherent detection scheme with advanced modulation formats such as quadrature phase shift keying (QPSK) and 16-ary quadrature amplitude modulation (QAM) with consistency to radio regulations. However, the obtained carrier frequency is limited by the bandwidth of the optical modulator, and thus, generation of the millimeter-wave and terahertz-wave signals at a frequency greater than 200 GHz is difficult by a straight-forward technique currently. In contrast, the latter technique can easily form a high frequency carrier signal with a high signal-to-noise ratio, however, the signal can only use a simple envelop-detection-based modulation scheme such as on-off keying (OOK) because the generated signal has small coherence [6]. The free-running laser has flexibility on installation: a great advantage to the two-tone generator based on the optical modulation technique. Synchronization of the lasers for high frequency stability is required for a higher frequency carrier signal generation in such submillimeter and terahertz waves for high-speed wireless transmission with advanced modulation formats.

In this study, we propose a frequency-stabilized millimeter-wave signal generation by the beat frequency generated by free-running lasers using an optical frequency locked loop (OFLL) technique configured with a feed-forward control [7]. A combination of optical frequency comb (OFC) and OFLL for higher frequency millimeter-wave signal generation is also discussed.

EXPERIMENTAL SETUP

Fig. 1(a) shows a block diagram for a frequency-stabilized millimeter-wave signal generation system. A fiber laser (FL) 1 and FL2 were utilized as two free-running lasers, whose frequency separation was set 6.3 GHz; a wavelength of the FL1 and FL2 was set at 1548.36 nm and 1548.41 nm, respectively. The linewidth of the FL1 and FL2 was 15 Hz. Fig. 1(b) shows a block diagram for an optical frequency comb (OFC) generation system. An optical signal from the FL2 was input into a dual-drive optical Mach-Zehnder intensity modulator (MZM). An OFC signal with several tens of sideband components was generated by the MZM operated at a radio frequency (RF) of 12.5 GHz. The RF signal from the synthesizer was split by a 180° hybrid coupler before an amplifier connected to the MZM. The amplitude and phase of the input RF signals were optimized by an attenuator (ATT) and a phase shifter. Finally, the a flat-top OFC signal was generated under optimized conditions [8]. Two optical components sliced by an arrayed waveguide grating with

OECC/PS2016

different output ports from the obtained OFC signal were utilized as an optical reference signal for launching to a photodiode (PD) 1 and as a millimeter-wave signal for a PD2. The optical two-tone signal that comprised of the signals from FL1 and the OFC component provided a beat note in the PD1 in a microwave band. Fig. 1(c) shows an electrical circuit for the OFLL control. The RF signal generated by the PD1 was down-converted by a heterodyne mixer, which was operated at a local oscillator frequency of 15 GHz with a high-pass filter (HPF). The down-converted signal was passed through an RF amplifier and a band-pass filter (BPF) with a bandwidth of 0.8 GHz and a center frequency of 8.7 GHz. An optical modulator under a double-sideband suppressed carrier (DSB-SC) operation was driven by the obtained RF signal at a frequency of approximately 8.7 GHz as a resultant error signal of the OFLL circuit under optimized conditions. It should be noted that the optical modulator is utilized as an optical frequency shifter. The modulator output was combined with an OFC component, and then, the combined signal was input to the PD2. A beat signal by the PD2 was measured by an electrical spectrum analyzer. It should be noted that optical and electrical delay times between each circuit is almost matched in the OFLL for feed-forward control.

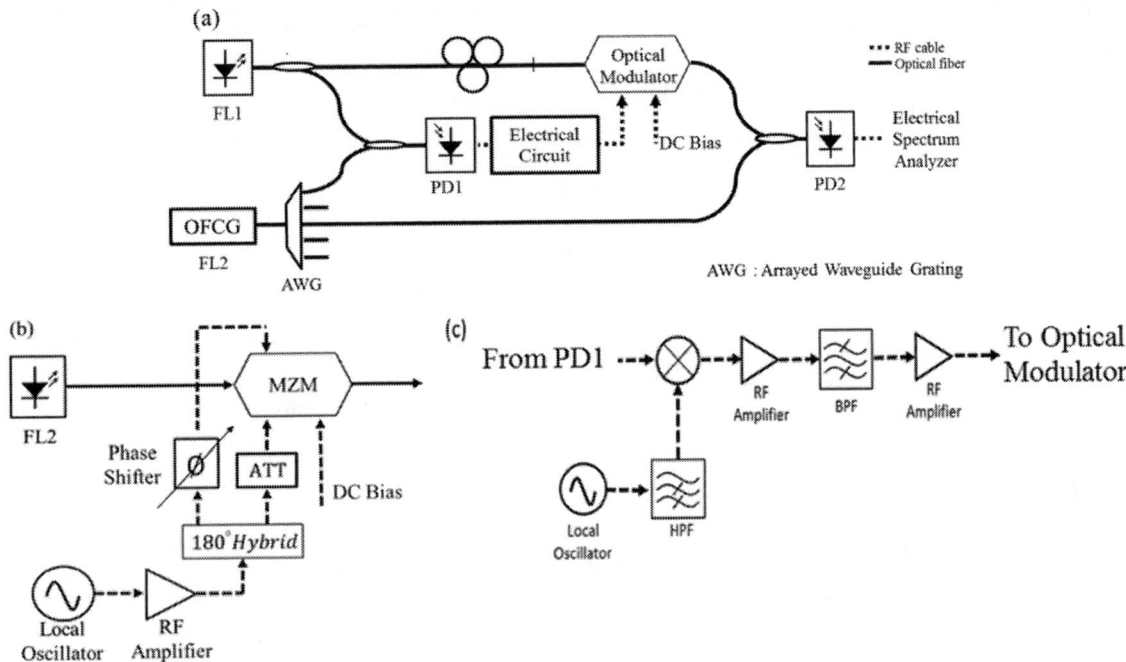

Fig. 1 (a) OFLL system (b) OFC signal generation system (c) Electrical Circuit for optical frequency control

DEMONSTRATION AND DISCUSSION

Figure 2 shows an optical spectrum of the OFC signal generated by the MZM. The obtained bandwidth is achieved 1.4 nm corresponding to an optical frequency of 175 GHz within 10-dB bandwidth, whose components have a frequency separation of 12.5 GHz. An optical signal picked from the OFC signal at a wavelength of 1548.41 nm is combined with the FL1 signal to input the OFLL circuit. A wavelength of the optical output from the MZM driven by the OFLL circuit is shifted by an error signal, and therefore, the frequency separation between the picked OFC signal and the wavelength-shifted FL1 would be locked. Figure 3 shows an optical spectrum observed at the input of the PD2. An upper-sideband component, which is utilized as the wavelength-shifted FL1 signal, and an OFC component were simultaneously observed. It should be noted that the residual components from the OFC signal exist because of a suppression ratio of the arrayed waveguide grating (AWG) of 30-dB. A frequency separation of two optical components is obtained 35 GHz. Figure 4 shows observed electrical spectrum of the frequency fluctuation generated by two free-running lasers (FL1 and FL2) with and without the OFLL circuit. The spectrum analyzer was operated under accumulation operation within 1 minute for evaluation of the frequency fluctuation. Obtained frequency fluctuation by the beat signal with and without the OFLL circuit was approximately 200 Hz and 10 MHz, respectively; approximately 10^5 times improvement on the fluctuation is demonstrated by the OFLL.

767

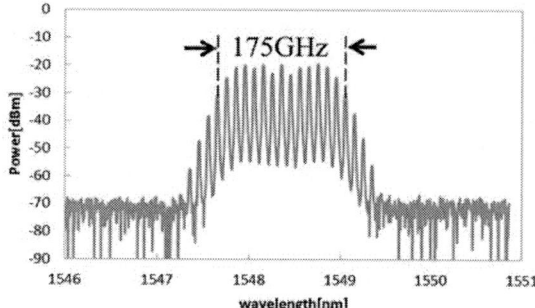

Fig. 2 Optical spectrum of OFCG output

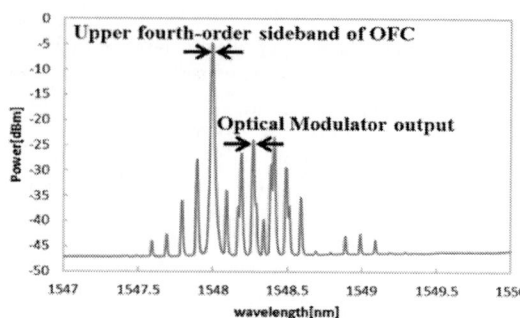

Fig. 3 Optical spectrum of the combined the Optical Modulator output and an OFC component

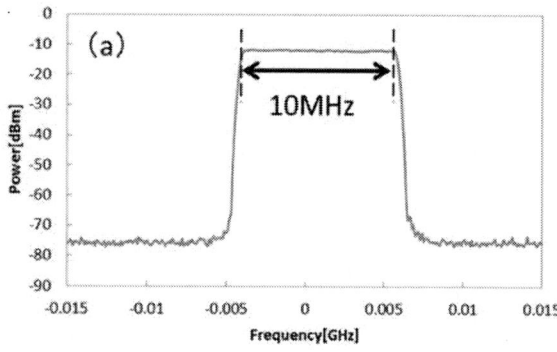

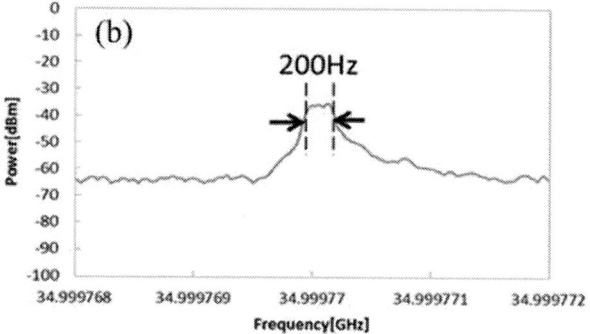

Fig. 4 Electrical spectrum of the beat signal frequency fluctuation generated by
(a) two free-running lasers and (b) the Optical Modulator output and the OFC component.

CONCLUSIONS

We successfully demonstrated frequency-stabilized millimeter-wave signal generation using the OFLL technique and OFC. Obtained millimeter-wave signal employed with the OFLL technique has frequency stability higher than that for two free-running lasers. The generation of multi-tone synchronization signals is useful not only for the generation of high RF signals but also to the generation of an arbitrary frequency signal.

REFERENCES

[1] H. J. Song, and T. Nagatsuma, "Present and Future of Terahertz Communications" IEEE Trans. Terahz. Sci. Technol., vol. 1, no. 1, pp.256-263, September 2011.

[2] X. Pang, A. Caballero, A. Dogadaev, V. Arlunno, R. Borkowski, J. S. Pedersen, L. Deng, F. Larinou, F. Roubeau, D. Zibar, X. Yu, and I. T. Monroy, "100 Gbitps hybrid optical fiber-wireless link in the W-band (75-110 GHz)," Optics Express, vol. 19, no. 25, pp.24944-24949, December 2011.

[3] A. Hirata, R. Yamaguchi, T. Kosugi, H. Takahashi, K. Murata, T. Nagatsuma, N. Kukutsu, Y. Kado, N. Iai, S. Okabe, S. Kimura, H. Ikegawa, H. Nishikawa, T. Nakayama, and T. Inada, "10-Gbit/s Wireless Link Using InP HEMT MMICs for Generating 120-GHz-Band Millimeter-Wave Signal," IEEE Trans. Microw. Theory Tech., vol. 57, no. 5, pp.1102-1109, May 2009.

[4] H. J. Song, K. Ajito, A. Wakatsuki, Y.Muramoto, N. Kukutsu, and Y. Kado, "Terahertz Wireless Communication Link at 300 GHz," Microwave Photonics, MWP 2010, October 2010.

[5] H. Kiuchi, T. Kawanishi, M. Yamada, T. Sakamoto, M. Tsuchiya, J. Amagai, and M. Izutsu, "High Extinction Ratio Mach-Zehnder Modulator Applied to a High Stable Optical Signal Generator," IEEE Trans. Microw. Theory Tech., vol. 55, no. 9, pp.1964-1972, September 2007.

[6] R. Yamanaka, R. Matsumoto, H. Sotobayashi, A. Kanno, and T. Kawanishi, "Highly frequency-stabilized millimeter-wave signal generation using optical phase-locked loop and flat optical frequency comb," Conference on Lasers and Electro-Optics Pacific Rim, CLEO-PR 2013, July 2013.

[7] J. Oba, A. kanno, N. Yamamoto, and H. Sotobayashi, "Frequency-Stabilized Millimeter-Wave Signal Generation by Optical Frequency Locked-Loop and Optical Frequency Comb Technologies" IEICE Tech. Rept., vol. 115, no. 435, MWP2015-93, pp. 205-209, January 2016 (In Japanese).

[8] I. Morohashi, T. Sakamoto, H. Sotobayashi, T. Kawanishi, and I. Hosako, "Broadband wavelength-tunable ultrashort pulse source using a Mach-Zehnder modulator and dispersion-flattened dispersion-decreasing fiber," Opt. Letters, vol. 24, no. 15, pp. 2297–2299, August 2009.

Third-Order Harmonics Suppression in Two-tone Signal Generation Using a Dual-Parallel Mach-Zehnder Modulator

Kazunori Osato and Moriya Nakamura

School of Science and Technology, Meiji Univ., 1-1-1 Higashi-Mita, Tama-ku, Kawasaki-shi, Kanagawa, 214-8571 Japan
ee21167@meiji.ac.jp

Abstract: *We experimentally investigated a novel two-tone signal generation method using a dual-parallel Mach-Zehnder modulator (DP-MZM). The proposed method can compensate nonlinear characteristics of the MZM. 8-dB improvement of third-order harmonics suppression was achieved.*

Keywords: *radio-over-fiber, two-tone signal generation, Mach-Zehnder modulator, non-linearity, third-order harmonics.*

I. INTRODUCTION

Broadband wireless communication networks require more capacity to accommodate increasing internet traffic due to the rapid spread of smart phones. In order to meet the demand of the bandwidth, Radio-over-Fiber (RoF) is regarded as one of the promising technology. In the RoF scheme, electronic signals are converted to optical signals, and transmitted along though optical fibers instead of coaxial cables. It enables lower system cost and longer-distance transmissions. In the scheme, generating two-tone lightwave signal is one of the key technologies. There are some ways to generate two-tone signal using Mach-Zehnder modulators (MZMs) [1-5]. Nonlinearity and asymmetry of splitting ratio in MZMs increase high-order harmonics of the generated two-tone signal and it causes waveform distortion. In this paper, we investigate two-tone signal generation method using a dual-parallel MZM (DP-MZM). By the proposed method, third-order harmonics of the generated two-tone signal can be suppressed based on simpler optical construction compared with conventional methods. We demonstrate the effectiveness of the proposed method by numerical simulation and experiment.

II. PRINCIPLE

Figure 1 schematically shows our proposed two-tone signal generation scheme using a DP-MZM [6]. Two-tone signal is basically generated by MZ1. Input lightwave with frequency f_c is modulated by radio frequency (RF) signal f_s by MZ1 which is biased at a bottom point. Here, MZ1 is driven by RF signal with the amplitude of $2V_\pi$, where V_π is half-wave voltage, to achieve maximum optical power. However, the generated two-tone signal by MZ1 includes high-order harmonics caused by saturation characteristics of MZM. Especially, third-order harmonics component seriously distorts the generated sinusoidal waveform. MZ2 is used to suppress this third-order harmonics. MZ2 is biased at bottom point, and driven by $3f_s$ which is generated by frequency tripler. The third-order harmonics is canceled by the output lightwave from MZ2, adjusting the driving amplitude and the phase of the output lightwave of MZ2.

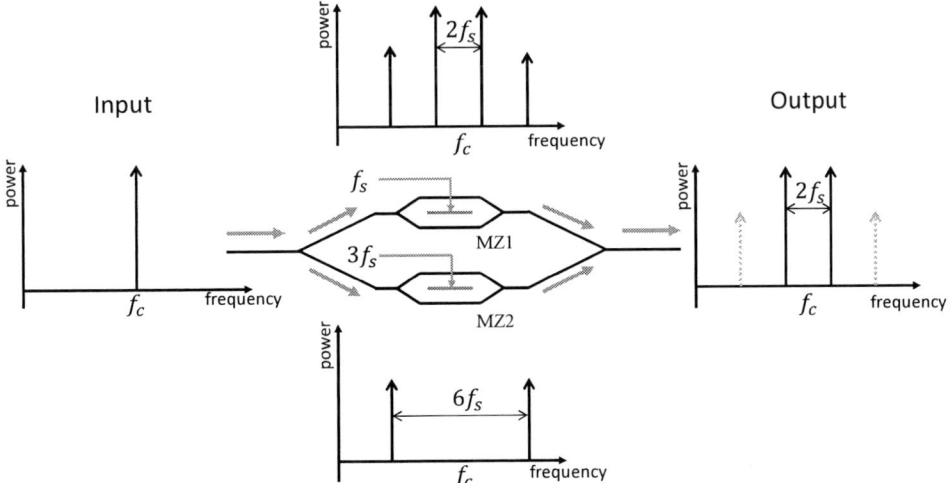

Fig. 1. Proposed scheme of third-order harmonics suppression in two-tone lightwave signal generation.

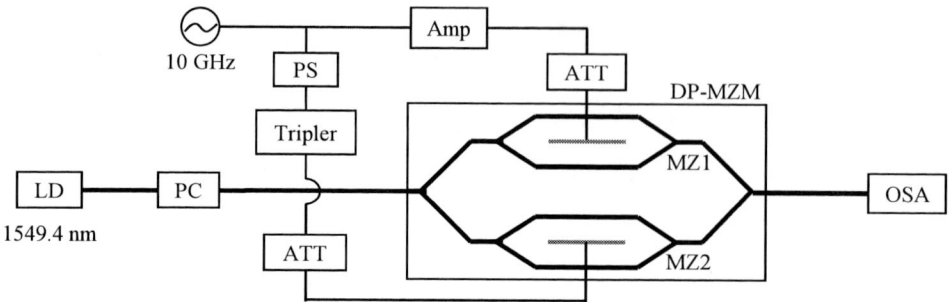

Fig. 2. System setup used in numerical simulation and experiment.

III. SYSTEM SETUP

Figure 2 schematically shows the system setup used in our numerical simulation and experiment. The lightwave of 1549.4-nm (193.62-THz) wavelength is polarization-controlled (PC) and modulated by DP-MZM. MZ1 is driven by RF signal of 10 GHz. At the same time, MZ2 is driven by 30-GHz RF signal generated by a frequency tripler. The phase of 30-GHz signal is controlled using RF phase shifter (PS). Attenuator (ATT) is used for adjusting amplitude of the driving RF signals. The driving RF amplitudes for the DP-MZM and bias points are adjusted, observing the output optical spectrum using optical spectrum analyzer (OSA) (Anritsu, MS9710C, 0.05-nm resolution). Extinction ratio of MZMs was assumed at 20 dB in the simulation.

IV. RESULTS AND DISCUSSION

First, we performed numerical simulation to investigate the ideal performance of our proposed scheme. Figures 3 show the result of numerical simulation. Figure 3(a) is the spectrum of output lightwave of MZ1 and Fig. 3(b) is the spectrum of output lightwave of DP-MZM. In Fig. 3(a), many high-order harmonics components are observed. Third-order harmonics is dominant in the components with the power level of about 20-dB smaller than required two-tone signal. However, the third-order harmonics is ideally canceled by our proposed scheme in the simulation as shown in Fig. 3(b). Figures 4 show the result of experiment. Figure 4(a) is the spectrum of output lightwave of MZ1. Power level of third-order harmonics was only about 32-dB smaller than required two-tone signal. This was caused by insufficient amplitude of 10-GHz signal of our RF amplifier. In Fig. 4(b), however, the power level of third-order harmonics is suppressed down to 40-dB smaller than two-tone signal, successfully achieving 8-dB improvement of the suppression.

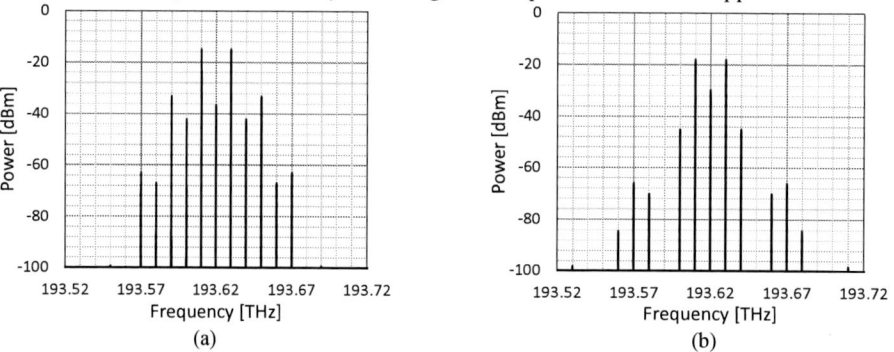

Fig. 3. (a) Spectrum of output lightwave of MZ1 and (b) spectrum of output lightwave of DP-MZM in numerical simulation.

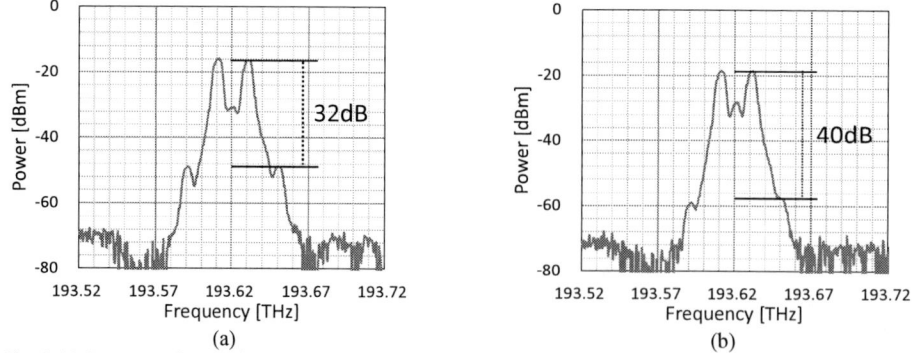

Fig. 4. (a) Spectrum of output lightwave of MZ1 and (b) spectrum of output lightwave of DP-MZM in experiment.

V. CONCLUSIONS

We proposed novel method to generate two-tone lightwave signal, in which the third-order harmonic components is suppressed using a DP-MZM. The performance was investigated by numerical simulation and experiment. In the simulation, we clearly showed that our method can ideally suppress third-order harmonics component of the generated two-tone signal. Also in the experiment, we successfully achieved 8-dB improvement of the suppression. The suppression ratio can be further improved with larger RF amplitude or using a DP-MZM with smaller V_π.

REFERENCES

[1] K. Seki, Y. Yamaguchi, A. Kanno, T. Kawanishi, M. Izutsu, and H. Nakajima, "Suppression of Third-order Harmonics in Two-tone Signals Using Cascaded Mach-Zehnder Modulators," MWP/APMP, TuEC-4, Oct. 2014.

[2] Y. Yamaguchi, S. Nakajima, A. Kanno, T. Kawanishi, M. Izutsu, and H. Nakajima, "Frequency-Quadruple Optical Two-Tone Signal Generation Using Integrated High Extinction-Ratio Mach-Zehnder Modulator," MWP/APMP, TuEC-5, Oct. 2014.

[3] A. Enokihara, T. Kawai, and T. Kawanishi, "Doubled Frequency Optical Two-tone Generation using Electro-optic Modulator and Suppression of Redundant Spectrum Components," Proc. 40th European Microwave Conf., pp. 125-128, Sep. 2010.

[4] W. Jiang, Q. Tan, W. Qin, D. Liang, X. Li, H. Ma, and Z. Zhu, "A Linearization Analog Photonic Link With High Third-Order Intermodulation Distortion Suppression Based on Dual-Parallel Mach–Zehnder Modulator," *IEEE Photon. J.*, vol. 7, no. 3, Jun. 2015, Art. ID. 7902208.

[5] J. Li, Y. Zhang, S. Yu, T. Jiang, Q. Xie, and W. Gu, "Intermodulation distortion elimination for analog photonics link based on integrated dual-parallel Mach-Zehnder modulator," CLEO, pp. 1-2, Jun. 2014.

[6] R. Imamura and M. Nakamura, "Two-tone lightwave generation using parallel MZ modulator," IEICE General Conference, C-3-72, Mar. 2015 (in Japanese).

Self-Interference Cancellation for 2×2 MIMO in-band full-duplex radio-over-fiber systems

Yunhao Zhang, Shilin Xiao*, Yinghong Yu, Shaojie Zhang, Lu Zhang, Ling Liu and Haiyun Xin

State Key Laboratory of Advanced Optical Communication System and Networks,
Department of Electronic Engineering, Shanghai Jiao Tong University, Shanghai 200240, China
*slxiao@sjtu.edu.cn

Abstract: *A self-interference cancellation (SIC) system based on dual-drive MZMs is proposed for 2×2 MIMO radio-over-fiber systems. Spectrum efficiency and system capacity are significantly promoted by employing SIC system for in-band full-duplex MIMO communications.*
Keywords: *self-interference cancellation, 2×2 MIMO, DD-MZM*

I. INTRODUCTION

Recently, the in-band full-duplex (IBFD) transmission have been widely investigated and considered as a promising air interface technique [1]. As for IBFD radio-over-fiber (RoF) systems, downlink (DL) and uplink (UL) radio frequency (RF) signals are simultaneously transmitted and received in the same frequency band between remote antenna units (RAUs) and user ends (UEs), which significantly doubles the spectrum efficiency of wireless system compared to frequency-division duplex (FDD) and time-division duplex (TDD) mode. Moreover, the combination of RoF and multiple-input-multiple-output (MIMO) has the potential to significantly enhance system capacity and reliability, thus representing an attractive trend for next generation mobile access networks [2-3].

However, in IBFD MIMO systems, each receive antenna gets to receive several high-power in-band self-interference (IBSI) signals from transmit antennas located nearby, and these IBSI cannot be removed by preselected band-pass filter. In this case, self-interference cancellation (SIC) systems are investigated to enable IBFD mode. Optical or optical/electrical mixed SIC schemes have been studied in order to overcome the limitation of cancellation bandwidth [4-6]. In our previous work [6], we have proposed a SIC system based on dual-drive Mach-Zehnder Modulator (DDMZM) for single-input-single-output (SISO) IBFD systems and experimentally demonstrated the viability with the available bandwidth range of 0-25GHz.

In this paper, we continue the IBFD research and propose a SIC system for 2×2 MIMO RoF systems. Based on DDMZMs, the SIC system cancels the self-interference from both transmit antenna 1 (Tx1) and transmit antenna 2 (Tx2), and successfully recovers the UL in-band signal from the UE. The results show good UL transmission performance with SIC for the 2×2 MIMO IBFD RoF system.

II. PRINCIPLE AND ARCHITECTURE

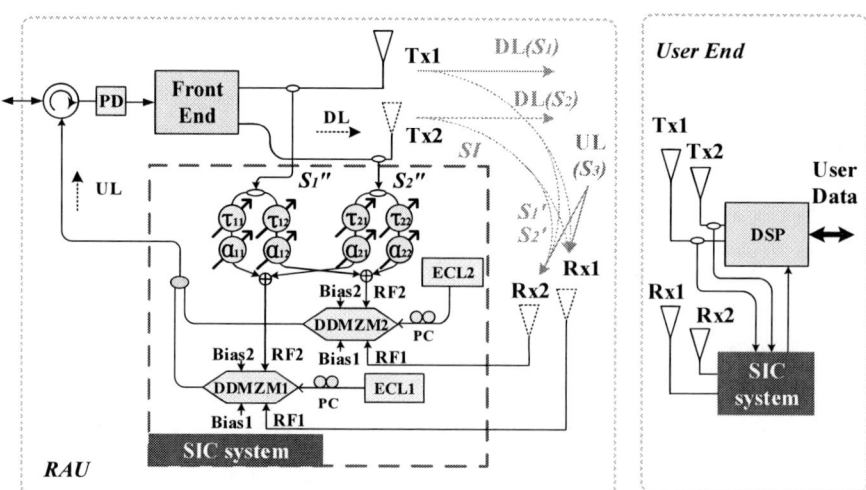

Fig. 1. Architecture of the IBFD system and proposed SIC system.

Figure 1 depicts the architecture of the proposed SIC system and 2×2 MIMO IBFD system. The SIC system has to remove self-interference signals from two adjacent transmit antennas. In RAU, the DL optical signal is demodulated in a photo-detector (PD) and the two DL radio frequency (RF) signals with the same RF carrier frequency are processed in front-end, represented as S_1 and S_2 from Tx1 and Tx2, respectively. S_3 represents the desired UL RF signal, which is the combination of two UL RF signals from two transmit antennas in UE. The processing of UL MIMO signals is conducted in central office (CO), which is not included in our research. The received signal from each receive antenna (Rx1 or Rx2) combines DL SI from Tx1 and Tx2, represented as S_1' and S_2' respectively, and the desired UL signal S_3.

The SIC system in RAU aims to remove S_1' and S_2' and to recover S_3. The two DL signals are tapped into SIC system before transmitting, represented as S_1'' and S_2''. In SIC system, S_1'' and S_2'' undergo splitting, delaying, attenuating, and coupling, then are entered into RF1 ports of the two DDMZMs, as shown in Fig. 1. In this paper, due to the same condition of the two receiving path, we only take Rx1 path as example to study. The condition in Rx2 path only has the difference of optical wavelength generated by ECL with that in Rx1 path. The signal transmitted into RF1 port of DDMZM1 is $S_1'+S_2'+S_3$, while into RF2 port is $\alpha_{11}S_1''(\tau_{11})+\alpha_{21}S_2''(\tau_{21})$. So the optical phase of bottom arm ϕ_1 and up arm ϕ_2 of DDMZM1 are shown in Eq. 1 and Eq. 2, respectively. Bias voltage of bottom arm is V_π and of up arm is 0 to keep the linear E/O modulation bias point.

$$\phi_1 = \frac{\pi}{V_\pi}V_1 = \frac{\pi}{V_\pi}(V_\pi + S_1' + S_2' + S_3). \qquad (1)$$

$$\phi_2 = \frac{\pi}{V_\pi}V_2 = \frac{\pi}{V_\pi}(\alpha_{11}S_1''(\tau_{11}) + \alpha_{21}S_2''(\tau_{21})). \qquad (2)$$

By precisely-tuned attenuated and delayed, S_1'' turns into $\alpha_{11}S_1''(\tau_{11})$, which is adjusted equal to S_1'. Likewise, S_2'' turns into $\alpha_{21}S_2''(\tau_{21})$, which is adjusted equal to S_2'. So that shown by Eq. 3, the S_1' and S_2' are subtracted and S_3 is remained.

$$E_{out} = E_{in}\cos\frac{\phi_1 - \phi_2}{2}e^{j\frac{\phi_1+\phi_2}{2}} = E_{in}\cos(\frac{V_\pi}{2}+\frac{\pi}{2V_\pi}S_3)e^{j\frac{\phi_1+\phi_2}{2}}. \qquad (3)$$

$$P_{out} = P_{in}\cos^2(\frac{V_\pi}{2}+\frac{\pi}{2V_\pi}S_3). \qquad (4)$$

E_{in} is input optical field and P_{in} is input optical power of DDMZM. The bias voltage $V_\pi/2$ is the linear modulation bias point of MZM, set for the best E/O modulation of S_3 in Eq. 4. After optical power detection in PD, the UL signal of interest S_3 is well recovered and treated in CO.

The SIC system is also employed in UE to recover in-band DL signal for realizing IBFD. Without modulation and transmission, recovered DL signal is photo-detected in SIC system and treated in digital signal processing (DSP) module in UE.

III. SIMULATION RESULTS AND DISCUSSION

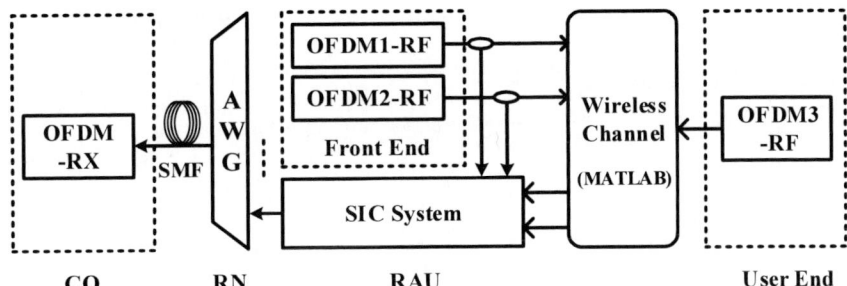

Fig. 2. Simulation architecture of IBFD RoF system

To prove the practicability of this SIC system for 2×2 MIMO IBFD systems, a simulation system is designed and real-time in-band signal transmission is performed over the band of interest, as depicted in Fig. 2. To meet the 3GPP standard [7], 64-QAM OFDM signals over 100-MHz bandwidth in 2.4-GHz wireless band are employed as S_1, S_2 and S_3, which are generated by OFDM1-RF, OFDM2-RF and OFDM3-RF modules, respectively. The wireless channel is simulated by MATLAB and the simulation is conducted by OptiSystem 13.0. For simplicity, only transmission delay, transmission attenuation and additive Gauss white noise (AWGN) are considered in wireless channel simulation. By properly adjusting all the delay and attenuation parameters towards wireless channel response, the self-interference is removed in SIC system. Recovered UL desired signal is transmitted through arrayed waveguide grating (AWG) in remote node (RN) and single mode fiber (SMF), then received in OFDM-RX module in CO. 64-QAM constellation diagrams are plotted and error vector magnitude (EVM) of them are calculated by off-line processing.

Firstly, the comparison of desired UL signal with and without cancellation is observed. The received constellation diagrams are shown in insert (i)-(iii) of Fig. 3. In the case of back-to-back optical transmission and about -3dBm received power, S_3 buried by self-interference cannot be received properly and the EVM is very large shown in insert (ii) of Fig. 3. After enabling the SIC system under the same condition, we can see from insert (iii) that clear 64-QAM constellation diagram is achieved and calculated EVM is only 1.923%, which shows the successful recovery and transmission of desired UL signal S_3. In real IBFD system, due to the variation of environment condition like channel change, the attenuation perameters in SIC system may not be precisely ajusted to the best cancellation condition quickly. So we investigate the ajustment budget of α_{11} and α_{21} shift towards the condition of α_{11} and α_{21} that insert (iii) of Fig. 3 indicates. Received EVM versus the shift of $\alpha11$ and $\alpha21$ is plotted in Fig.4. The specified requirement of EVM in the 3GPP standard for 64-QAM is 8% [7], so the simulation results show adjustment budget from -1 to +1.5 dB for both α_{11} and α_{21}. Therefore, this SIC system can support a spot of amplitude shift budget in b-t-b case so that it supports short-term change of system condition.

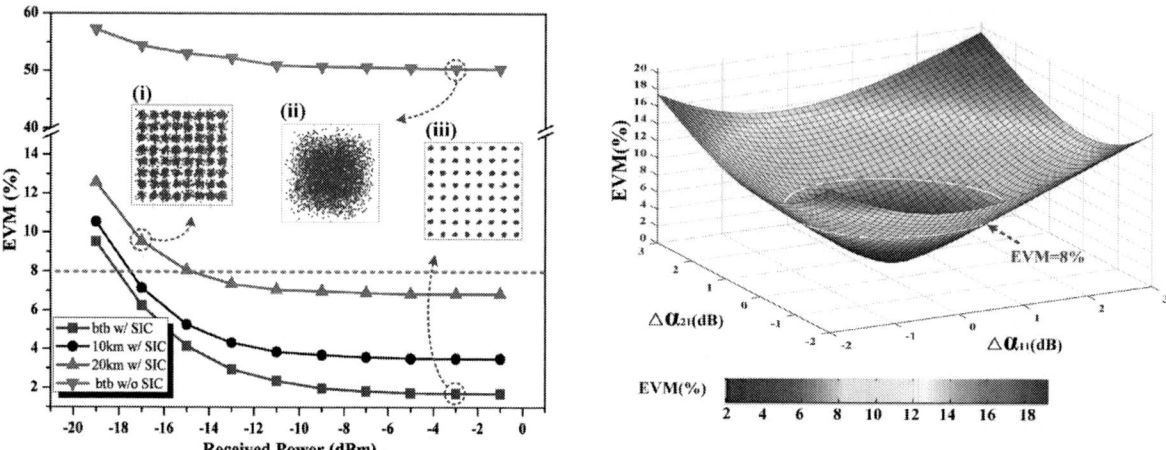

Fig. 3. EVM curves and constellation diagrams   Fig. 4. EVM versus the shift of $\alpha11$ and $\alpha21$

Then, the UL EVM performances with difference fiber lengths are evaluated. Figure 3 shows the EVM curves of b-t-b, 10-km and 20-km fiber reach, respectively. The requirements of EVM is 8%, marked as the red dot line in Fig. 3. Insert (i) of Fig. 3 shows the constellation diagram with 9.534%, larger than 8% in 20-km fiber transmission case. The simulation results show that the IBFD system supports 20-km SMF RoF transmission with received power of -15 dBm at least.

IV. CONCLUSIONS

In this paper, we propose the SIC system for 2×2 MIMO RoF systems based on DDMZMs. For the first time, SIC is introduced into MIMO systems in which transmission capacity is significantly enlarged than SISO systems. This SIC system can cancel the interference from two adjacent transmit antennas for IBFD communication, which significantly improves spectrum efficiency compared to TDD and FDD.

ACKNOWLEDGMENT

The work was jointly supported by the National Nature Science Fund of China (No.61271216, No. 61221001, No.61090393 and No.61433009), the National "973" Project of China (No. 2010CB328205, No.2010CB328204 and No. 2012CB315602) and the National "863" Hi-tech Project of China (No.2013AA013602 and No.2012AA011301).

REFERENCES

[1] S. Huberman, and T. Le-Ngoc. "MIMO Full-Duplex Precoding: A Joint Beamforming and Self-Interference Cancellation Structure." IEEE Transactions on Wireless Communications, vol. 14, No. 4, 2015, pp. 2205-2217.

[2] Q. Zhang, J. Yu, X. Li, M. Zhu, X. Xin, and G. K. Chang, "Photonic-aided pre-coding QAM signal transmission in multi-antenna radio over fiber system," Optics Communications, vol. 354, 2015, pp. 236-239.

[3] C. T. Lin, A. Ng'oma, W. Y. Lee, C. C. Wei, C. Y. Wang, T. H. Lu, J. Chen, W. J. Jiang, and C. H. Ho, "2× 2 MIMO radio-over-fiber system at 60 GHz employing frequency domain equalization," Optics Express, vol. 20, No. 1, 2012, pp. 562-567.

[4] M. P. Chang, M. Fok, A. Hofmaier, and P. R. Prucnal, "Optical analog self-interference cancellation using electro-absorption modulators," IEEE Microwave and Wireless Components Letters, vol. 23, No. 2, 2013, pp. 99-101.

[5] Q. Zhou, H. Feng, G. Scott, and M. P. Fok, "Wideband co-site interference cancellation based on hybrid electrical and optical techniques," Optics Letters, vol. 39, No. 22, 2014, pp. 6537-6540.

[6] Y. Zhang, S. Xiao, H. Feng, L. Zhang, Z. Zhou, and W. Hu. "Self-interference cancellation using dual-drive Mach-Zehnder modulator for in-band full-duplex radio-over-fiber system," Optics Express, vol. 23, No. 26, 2015, pp. 33205-33213.

[7] 3GPP, 3GPP TS 36.104 version 11.9.0 Release 11, 2014.

Dynamic Virtual Optical Network Mapping Based on Switching Capability and Spectrum Fragmentation in Elastic Optical Networks

Hongxiang Wang, Xin Xin, Jiawei Zhang, Yongmei Sun, Yuefeng Ji

State Key Laboratory of Information Photonics and Optical Communications, School of Information and Communication Engineering, Beijing University of Posts and Telecommunications. No.10, Xitucheng Road, Haidian District, Beijing, 100876, China.
wanghx@bupt.edu.cn

Abstract: *A dynamic virtual optical network mapping algorithm in elastic optical networks, which considers both switching capability and spectrum fragmentation, is proposed. Simulation results show that the proposed algorithm has better performance than the existing algorithms.*

Keywords: *virtual optical network mapping; elastic optical networks*

I. Introduction

With the booming of Internet-based applications, network virtualization is considered to be one of the key technologies in the future network. Its high utilization ratio of resources and characteristic of supporting network heterogeneity can bring efficient and reliable services to customers and eradicate the ossification of the current Internet [1]. Meanwhile, optical networks have abundant bandwidth resources, so network operators can rely on optical network technologies to satisfy the exponentially rising trend of bandwidth demands. Comparing with the traditional WDM optical networks, elastic optical networks (EONs) break the limit of fixed grid and have the characteristic of flexible spectrum [2]. Therefore, EONs are considered as the most potential physical infrastructures for network virtualization. As a key technology of optical network virtualization, virtual optical network mapping (VONM) allocates node resources and link resources in the physical optical network to each virtual optical network (VON) through node mapping and link mapping. In the process of VONM, how to achieve the maximum resource utilization of the physical optical network in order to reduce the blocking ratio of VON requests has become one of the major challenges.

VONM over EONs only starts to attract research interests recently, since it is more complicated. In [3], Zhao *et al.* proposed an ILP model and two heuristic algorithms over EONs. Considering the distance-adaptive modulation over EONs, Wang *et al.* investigated a distance-adaptive VONM in [4]. However, all the existing mapping algorithms only consider the constraint of node computing resource in the node mapping stage, and they avoid the constraint of peculiar attribute of optical nodes over EONs. In addition, the existing mapping algorithms over EONs regard the basic mapping conditions as the primary goal instead of considering the resource utilization of the underlying EONs. Therefore, there are lots of fragmentation of link resources over EONs, which causes the waste of the underlying network resources.

In this paper, we propose an efficient VONM algorithm over EONs in order to resolve the drawbacks of the existing algorithms. In the node mapping stage, we consider the node switching capability (NSC), so that we can ensure the effectiveness of switching information. In the link mapping stage, we focus on the fragmentation degree loss (FDL) of optical links in order to reduce the fragmentation of the underlying link resources. Thus, we can provide more contiguous vacant spectrum for more VON requests. Simulation results show that comparing with the existing algorithms, the proposed algorithm has better performance in terms of blocking ratio and revenue-to-cost ratio.

II. Problem Description

We model the substrate EON as a graph $G_s(N_s, L_s)$, where N_s is the set of substrate nodes, L_s is the set of substrate links. Each substrate node $n_s \in N_s$ has a node capacity (NC) of C_{n_s}. Each substrate link $l_s \in L_s$ has B subcarrier slots. We define a bit-mask b_{l_s} which contains B bits to describe the spectrum utilization of the link. When $b_{l_s}[j] = 1$, the jth slot on link l_s is occupied, otherwise, $b_{l_s}[j] = 0$. Similarly, $G_v(N_v, L_v)$ is modeled to represent a VON request, where N_v denotes the set of virtual nodes, L_v denotes the set of virtual links. The NC requirement of each virtual node $n_v \in N_v$ is defined as C_{n_v}. The bandwidth requirement of each virtual link $l_v \in L_v$, which is the number of contiguous slots, is BW_{l_v}.

In EONs, the NSC is measured by the number of ports and the number of vacant slots per port. The more the number of vacant slots on the optical links around the optical node is, the more easily the effective information can be switched and the stronger the NSC is. We denote the number of ports of optical node n as PN and the number of vacant slots on the link l which is around the node n as VSN. Then the NSC of node n can be defined as:

$$S_n = PN \times \sum\nolimits_{l \in adj\_link(n)} VSN^l \qquad (1)$$

where $\sum_{l \in adj\_link(n)} VSN^l$ denotes the sum of all the vacant slots on the links around node n. Fig.1(a) shows an example of the description of NSC.

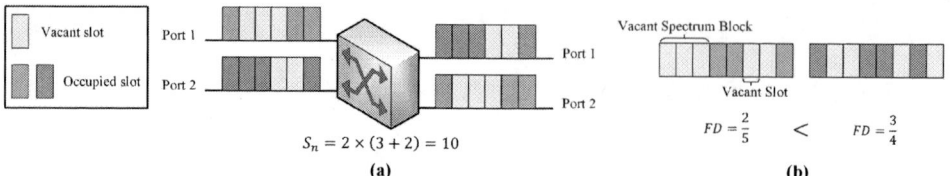

Fig. 1. (a) An example of the description of NSC. (b) Examples of the description of FD.

Due to the spectrum allocation constraints (such as the spectrum non-overlapping, continuity and contiguous constraints) over EONs, the existing mapping algorithms cause the fragmentation of the spectrum resource over underlying EONs. In order to keep the contiguous vacant spectrum blocks, we propose the concept of "Fragmentation Degree (FD)" according to the unique characteristic of spectrum over EONs. For a link, we denote the number of vacant spectrum blocks on the link as VSB and the total number of vacant slots on the link as VST. So FD can be defined as follows:

$$FD = VSB / VST \tag{2}$$

We can conclude that a larger value of FD means that the continuity of spectrum resource is worse. Fig.1(b) shows examples of the description of FD of link spectrum resource.

Then we propose the concept of FDL, which can be defined as follows:

$$FDL = FD_{After} - FD_{Before} \tag{3}$$

where FD_{Before} denotes the value of FD on the link before allocating spectrum, and FD_{After} denotes the value of FD on the link that assumes to allocate the available spectrum.

III. DYNAMIC VONM ALGORITHM

Comparing with most of the existing algorithms, which consider only NC in node mapping and use K-Shortest-Path (KSP) for routing and First Fit (FF) for spectrum allocation in link mapping, we propose an efficient VONM algorithm. We denote this proposed algorithm as "NSC-KSP-FDL" and it is described below.

Step 1: For each VON request G_v, we perform the node mapping first. We calculate the node rank $Rank(n_s)$ for each substrate node n_s in G_s and the node rank $Rank(n_v)$ for each virtual node n_v in G_v. The rank for substrate node n_s and the rank for virtual node n_v can be defined as:

$$Rank(n_s) = C_{n_s} \times S_{n_s} \tag{4}$$

$$Rank(n_v) = C_{n_v} \times PN_{n_v} \tag{5}$$

where C_{n_s} and S_{n_s} denote the NC and the NSC of substrate node n_s respectively. C_{n_v} is the NC requirement of virtual node n_v, and PN_{n_v} is the number of node port requirement of virtual node n_v.

Step 2: Sort the nodes in G_s and G_v according to their node rank in a decreasing order, and map the virtual nodes onto the substrate nodes from the top of the rank by following the rank order. If each virtual node in G_v satisfy the constraint that $C_{n_s} \geq C_{n_v}$, then the node mapping is successful. Otherwise, the VON request is blocked.

Step 3: Perform the link mapping after finishing the node mapping. For each virtual link l_v in G_v, we calculate K shortest paths using KSP algorithm in G_s.

Step 4: Find out all the contiguous vacant spectrum blocks in each path. If spectrum is satisfied, then we calculate the value of FDL_{sum} for each routing and spectrum allocation scheme among all the candidates. FDL_{sum} means the sum of FDL of each link along the path and can be defined as:

$$FDL_{sum} = \sum_{i=1}^{I} FDL_i \tag{6}$$

where I denotes the number of links along the path. Otherwise, the VON request is blocked.

Step 5: Choose the routing and spectrum scheme whose FDL_{sum} is the least among all the candidates.

IV. SIMULATION RESULTS

In the simulation, we use the typical NSF Network (14 nodes and 21 links) as the substrate EON. The NC of each substrate node is 200 units. The number of spectrum slots of each link is 200 and the bandwidth of each spectrum slot is 25GHz. We set the VON requests to follow a Poisson process whose arrival rate is λ and service rate is μ. The VON topologies, which the numbers of virtual nodes are 2, 3, 4, 5, 6 with the distribution probabilities 30%, 30%, 25%, 10%, 5% respectively, are generated randomly. The virtual nodes are randomly connected with probability 50%. The NC requirement of each virtual node is randomly distributed between 1 and 10 units, and the bandwidth requirement of each virtual link is randomly distributed between 1 and 10 spectrum slots and guard band between the traffics is 2 slots. In addition, the number of VON requests is 10000.

We consider the VON requests blocking ratio as the major evaluation objective. Besides, we consider the revenue-to-cost ratio as another evaluation objective by referring to IP network. In the process of node mapping, each virtual node can only map onto one substrate node and the node resource requirement of each virtual node is equal to the resource it is allocated by the substrate node. So we ignore the node revenue-to-cost ratio and regard the link revenue-to-cost ratio as the overall revenue-to-cost ratio. For a VON request, we define the revenue-to-cost ratio as follows:

$$RCR = \frac{\sum_{l_v \in L_v} BW(l_v)}{\sum_{l_v \in L_v} \sum_{l_s \in L_s} BW(f_{l_s}^{l_v}, l_v)} \tag{7}$$

where $BW(l_v)$ denotes the bandwidth requirement of virtual link l_v and $BW(f_{l_s}^{l_v}, l_v)$ denotes the bandwidth allocated to virtual link l_v from substrate link l_s. $f_{l_s}^{l_v} \in \{0,1\}$, $f_{l_s}^{l_v} = 1$ if substrate link l_s allocated bandwidth resource to virtual link l_v, otherwise, $f_{l_s}^{l_v} = 0$.

To evaluate the performance of the NSC-KSP-FDL VONM algorithm, we design two reference algorithms. The first reference algorithm is a typical VONM algorithm. In node mapping, this algorithm first sorts the nodes according to the NC in a decreasing order and then maps the virtual nodes onto the substrate nodes from top of the rank by following the rank order. And in link mapping, this algorithm uses the KSP-FF scheme. So we call it "NC-KSP-FF". The second reference algorithm uses the node mapping scheme proposed in this paper and the KSP-FF link mapping scheme, so we name it "NSC-KSP-FF".

Fig.2 shows the blocking ratio among different VONM algorithms in NSF Network. We observe that both NSC-KSP-FF and NSC-KSP-FDL reduce the blocking ratio comparing to NC-KSP-FF. The reason is that we consider not only NC but also NSC of optical nodes in node mapping stage. So we ensure the effectiveness of switching information in link mapping stage. Another observation is that NSC-KSP-FDL achieves better blocking performance than NSC-KSP-FF. This is because NSC-KSP-FDL doesn't use the KSP-FF link mapping scheme which NSC-KSP-FF uses. Instead, NSC-KSP-FDL considers all the K shortest paths and select the routing and spectrum allocation scheme which has the minimal FDL. Thus, it can reduce the fragmentation of the underlying link resources so that there are more high-capacity bandwidth serving for VON requests which require more bandwidth.

Fig.3 illustrates the revenue-to-cost ratio among different VONM algorithms in NSF Network. As shown, NSC-KSP-FF and NSC-KSP-FDL have better performance than NC-KSP-FF in terms of revenue-to-cost ratio. The reason is that NSC-KSP-FF and NSC-KSP-FDL choose the nodes with stronger NSC first in node mapping stage. So they have more opportunities to choose a shorter path among the K paths in link mapping stage. In addition, NSC-KSP-FF has better revenue-to-cost ratio performance than NSC-KSP-FDL. This is because NSC-KSP-FDL chooses the path with the minimal FDL instead of the shortest path among the K paths.

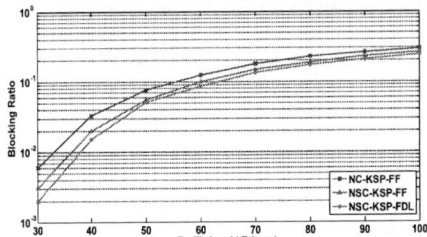

Fig. 2. Blocking ratio among different VONM algorithms.

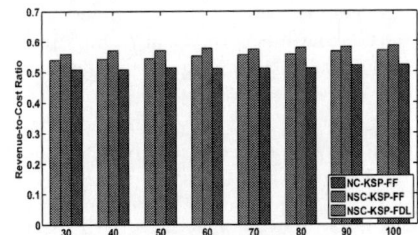

Fig. 3. Revenue-to-cost ratio among different VONM algorithms.

V. CONCLUSIONS

In this paper, we proposed a dynamic efficient VONM algorithm over EONs, which considered both NSC of optical nodes and FDL of optical links. The simulation results showed that the proposed VONM algorithm achieved good performance in terms of blocking ratio and revenue-to-cost ratio.

ACKNOWLEDGMENT

This work is supported by the National High Technology Research and Development program of China (863 Program) (No. 2013AA014501), NSFC (No.61501055), and NSFC project (No.61331008).

REFERENCES

[1] Anjing Wang, Iyer, et al., "Network Virtualization: Technologies, Perspectives, and Frontiers," in Lightwave Technology, Journal of , vol.31, no.4, pp.523-537, Feb.15, 2013.

[2] Guoying Zhang, De Leenheer, M., Morea, A., and Mukherjee, B., "A Survey on OFDM-Based Elastic Core Optical Networking," in Communications Surveys & Tutorials, IEEE , vol.15, no.1, pp.65-87, First Quarter 2013.

[3] Juzi Zhao, Subramaniam, S., and Brandt-Pearce, M., "Virtual topology mapping in elastic optical networks," in Communications (ICC), 2013 IEEE International Conference on , vol., no., pp.3904-3908, 9-13 June 2013.

[4] Xi Wang, Qiong Zhang, et al., "Flexible virtual network provisioning over distance-adaptive flex-grid optical networks," in Optical Fiber Communications Conference and Exhibition (OFC), 2014 , vol., no., pp.1-3, 9-13 March 2014.

Fiber-Wireless and Fiber-IVLLC Convergences

Bo-Rui Chen, Hung-Hsien Lin, Chang-Jen Wu, Chun-Yu Lin, Chung-Yi Li, and Hai-Han Lu*

Institute of Electro-Optical Engineering National Taipei University of Technology, Taipei, 106 Taiwan
*Email: *hhlu@ntut.edu.tw*

Abstract: Fiber–wireless and fiber–invisible laser light communication (IVLLC) convergences that adopt Mach-Zehnder modulator (MZM)-optoelectronic oscillator (OEO)-based broadband light source (BLS) for microwave (MW)/millimeter-wave (MMW)/baseband (BB) signal transmission is proposed and experimentally demonstrated.

Keywords: Broadband light source, Fiber–IVLLC convergence, Fiber–wireless convergence, MZM-OEO

I. INTRODUCTION

Through the large bandwidth of optical fiber and the flexibility of RF/optical wireless transmission, fiber–wireless and fiber–invisible laser light communication (IVLLC) convergences have progressed to meet multiple gigabit demands [1], [2]. Fiber–wireless and fiber–IVLLC convergences can utilize the advantages of both optical and wireless technologies; i.e., they can use the naturally enormous bandwidth of optical fiber and the unused bandwidth in microwave (MW) and millimeter-wave (MMW) bands. A bidirectional fiber–wireless and fiber–visible laser light communication (VLLC) transmission system that adopted an optoelectronic oscillator (OEO)-based broadband light source (BLS) and a reflective semiconductor optical amplifier (RSOA) to transmit downstream 10 Gbps/30 GHz MW, 10 Gbps/45 GHz MMW, and 10 Gbps/60 GHz MMW data signals, as well as upstream 5 Gbps data stream, was demonstrated in a previous study [3]. However, considerable improvement can still be achieved. Such bidirectional fiber–wireless and fiber–VLLC convergences are not flexible because of the same transmission rate for downstream MW and MMW data signals. Furthermore, given that 60-GHz MMW has a high atmospheric attenuation, fiber–IVLLC convergence is a promising substitute for fiber–wireless convergence in a 60-GHz MMW transmission. In addition, Mach–Zehnder modulator (MZM)–OEO-based BLS can be used as a substitute for OEO-based BLS to overcome the modulation bandwidth limitation caused by the distributed feedback (DFB) laser diode (LD) [4]. Moreover, for up-link transmission, fiber–VLLC convergence can be replaced with fiber–IVLLC convergence. System performance will be significantly improved because modal noise induced by the multimodal vertical-cavity surface-emitting laser (VCSEL) will not exist and the transmission rate limit imposed by the 5.2-GHz VCSEL will be overcome. Furthermore, given that both downstream and upstream signals are delivered by the same single-mode fiber (SMF), Rayleigh backscattering noise will severely restrict system performance. An intensity modulator can be used at the up-link transmission site to substitute for RSOA to reduce Rayleigh backscattering noise. In this paper, fiber–wireless and fiber–IVLLC convergences based on MZM–OEO-based BLS to transmit downstream 10 Gbps/30 GHz MW, 15 Gbps/50 GHz MMW, 20 Gbps/60 GHz MMW, and 25 Gbps/100 GHz MMW data signals, as well as upstream 25 Gbps baseband (BB) data stream, are proposed and experimentally demonstrated. To be the first one that employs a MZM-OEO-based BLS in such bidirectional fiber–wireless and fiber–IVLLC convergences, downstream light is optically promoted from a 15-GHz RF signal to 10 Gbps/30 GHz MW and 20 Gbps/60 GHz MMW data signals, whereas downstream light is also optically promoted from a 25-GHz RF signal to 15 Gbps/50 GHz and 25 Gbps/100 GHz MMW data signals. Furthermore, downstream light is reused and modulated using an intensity modulator with a 25-Gbps BB data stream for up-link transmission. The transmission rates of the downstream 30-GHz MW, 50-GHz MMW, 60-GHz MMW, 100-GHz MMW, as well as upstream BB signals are different between each other. It is flexible and practical for the real implementation of fiber–wireless and fiber–IVLLC convergences. Through a comprehensive investigation of such bidirectional fiber–wireless and fiber–IVLLC convergences, bit error rate (BER) performs efficiently in 40-km SMF and 10-m RF/25-m optical/100-m optical wireless transport scenarios.

II. EXPERIMENTAL SETUP

The configuration of the proposed fiber–wireless and fiber–IVLLC convergences based on MZM–OEO-based BLS is presented in Fig. 1. A BLS modulated with 15-GHz and 25-GHz RF signals is used to create multiple optical sidebands. For optical interleaver1 (OIL1), the zero optical sideband (central carrier) is utilized for the 25-Gbps BB up-link transmission, the −1 and +1 optical sidebands are utilized for the 30-GHz MW downlink transmission, and the −2 and +2 optical sidebands are utilized for the 60-GHz MMW downlink transmission. For OIL2, the −1 and +1 optical sidebands are utilized for the 50-GHz MMW downlink transmission, and the −2 and +2 optical sidebands are utilized for the 100-GHz MMW downlink transmission. All the optical signals are then combined using an optical combiner and amplified using an erbium-doped fiber amplifier (EDFA). For uplink transmission, the optical signal (central carrier) captured by an fiber Bragg grating (FBG2) with a central wavelength of 1540.16 nm is reused and modulated by an intensity modulator. The intensity modulator is modulated by a 25-Gbps pseudorandom binary sequence (PRBS) of $2^{15}-1$ generated by a PRBS generator.

OECC/PS2016

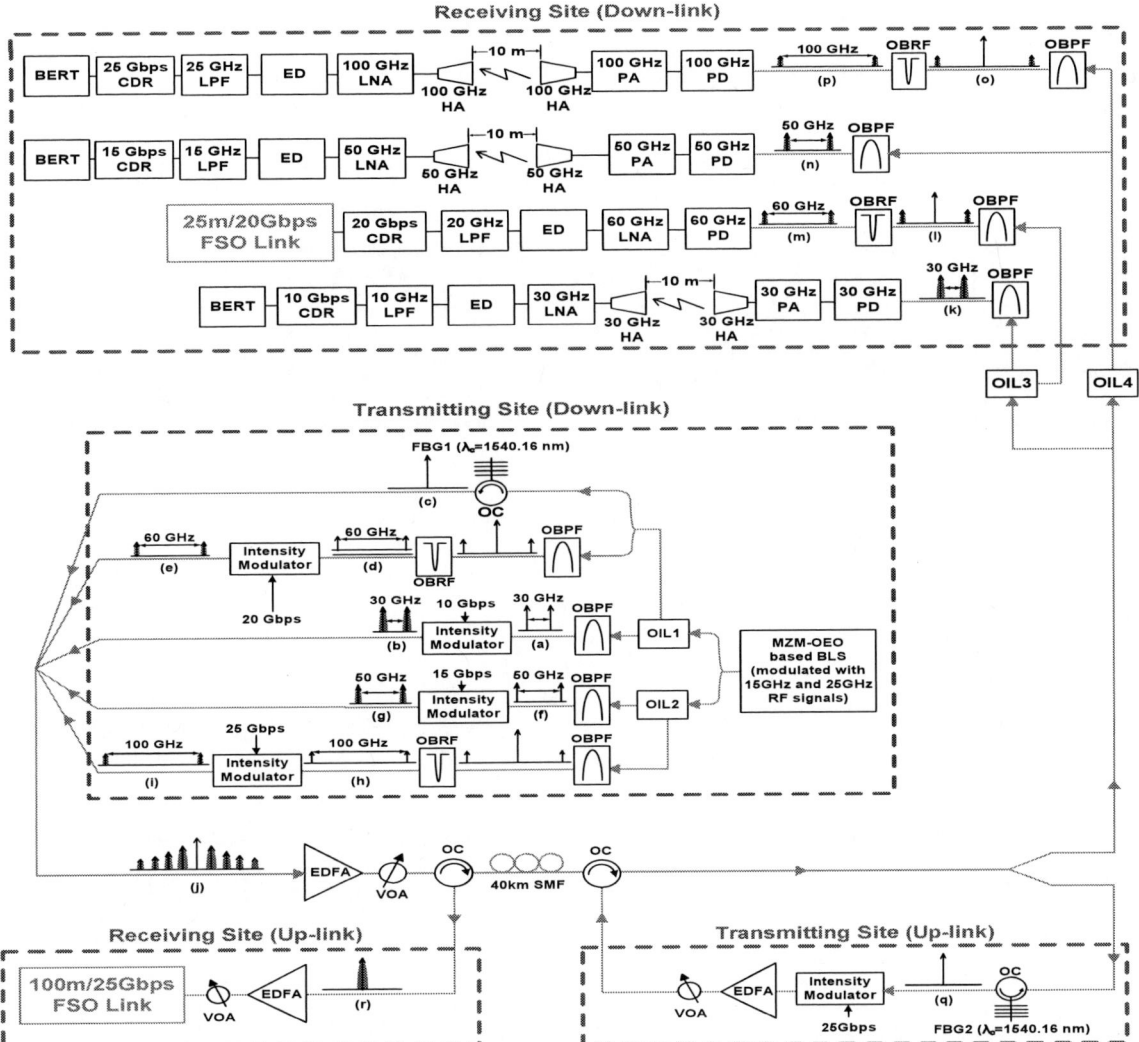

Fig. 1. The configuration of the proposed fiber-wireless and fiber-IVLLC convergences based on MZM-OEO-based BLS.

III. EXPERIMENTAL RESULTS AND DISCUSSION

As shown in Fig. 2, the BLS is composed of a MZM–OEO scheme. The MZM–OEO scheme is based on converting laser light to RF signals. The number of optical sidebands depends on the amplitude of the RF signals generated by the MZM–OEO scheme. Channel spacings are determined by the central frequencies of the RF BPFs used in the MZM–OEO scheme. The MZM is modulated by 15-GHz and 25-GHz RF signals, by which leading to the generation of multiple optical sidebands with channel spacings of 15-GHz and 25-GHz. The optical sidebands generated by the MZM–OEO are then launched into the OSNR enhancement scheme to ameliorate OSNR values. The optical spectra of the BLS before and after OSNR enhancement are shown in Fig. 3. When the OSNR enhancement scheme is applied, approximately 7 dB to 10 dB OSNR value enhancement is achieved for the optical sidebands. Fig. 4 shows the configuration of the proposed 25 m/20 Gbps FSO links that use a 1550-nm laser transmitter cascaded with an EDFA and a pair of fiber collimators. The FSO link has attracted considerable attention as a promising candidate for fiber–IVLLC convergence because of its numerous advantages over 60-GHz MMW fiber–wireless convergence. Fig. 5 illustrates the configuration of the proposed 100 m/25 Gbps FSO links which employ doublet lens scheme. This 100 m/25 Gbps FSO link provides the advantages of long-haul and high-speed optical wireless communications.

The measured BER curves of the 10 Gbps/30 GHz MW and 15 Gbps/50 GHz MMW data signals for back-to-back (BTB), over 40-km SMF as well as 10-m RF wireless transport scenarios are presented in Figs. 6(a) and 6(b), respectively. At a BER of 10^{-9}, power penalties of 4.4 dB and 4.8 dB, respectively, are observed. In addition, the measured BER curves of the 20 Gbps/60 GHz MMW data signal for the BTB, over 40-km SMF as well as 25-m free-space transport scenarios are presented in Fig. 6(c). At a BER of 10^{-9}, a power penalty of 5.3 dB is observed. Moreover, the measured BER curves of the 25 Gbps/100 GHz MMW data signal for the BTB and over 40-km SMF as well as 10-

779

m RF wireless transport scenarios are shown in Fig. 6(d). A large power penalty of 6.8 dB is observed, at a BER of 10^{-9}. For uplink transmission, the measured BER curves of the 25-Gbps BB data stream for BTB and over 40-km SMF as well as 100-m free-space transport scenarios are shown in Fig. 7. At a BER of 10^{-9}, a power penalty of 5 dB is observed.

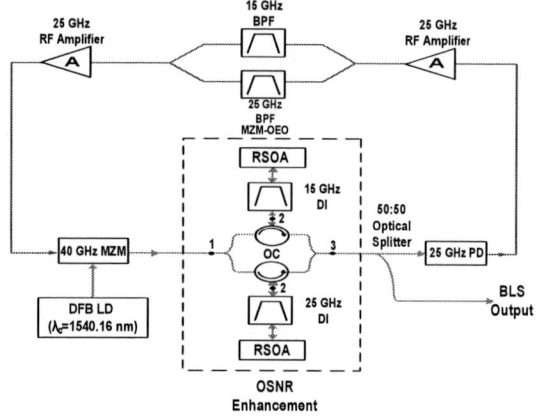

Fig. 2. The configuration of the proposed MZM-OEO-based BLS.

Fig. 3. The optical spectra of the BLS before and after the OSNR enhancement scheme.

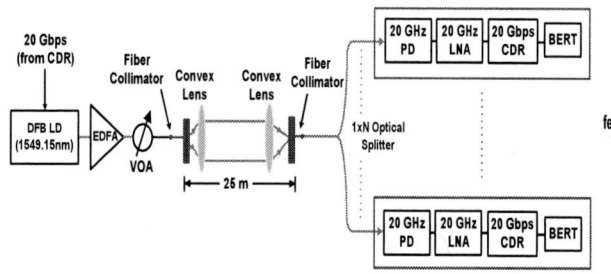

Fig. 4. The configuration of the proposed 25 m/20 Gbps FSO links.

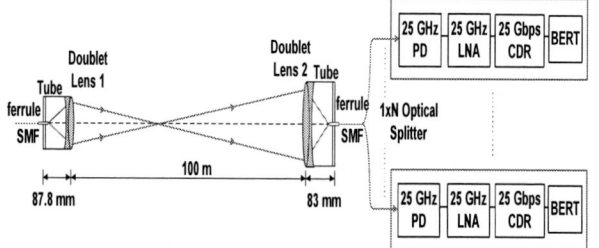

Fig. 5. The configuration of the proposed 100 m/25 Gbps FSO links.

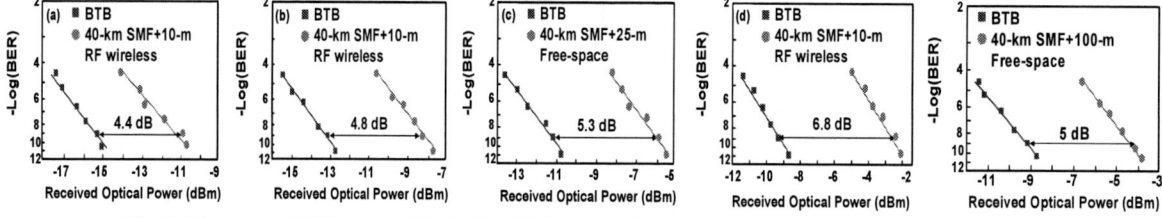

Fig. 6. The measured BER curves of (a) 10 Gbps/30 GHz MW, (b) 15 Gbps/50 GHz and (c) MMW, 20 Gbps/60 GHz MMW, (d) 25 Gbps/100 GHz MMW data signals.

Fig. 7. The measured BER curves of 25 Gbps BB data stream.

IV. CONCLUSIONS

Fiber–wireless and fiber–IVLLC convergences based on MZM–OEO-based BLS to deliver downstream 10 Gbps/30 GHz MW, 15 Gbps/50 GHz MMW, 20 Gbps/60 GHz MMW, and 25 Gbps/100 GHz MMW data signals, as well as upstream 25 Gbps BB data stream, is proposed and demonstrated. The results of the comprehensive examination indicated that BER performed efficiently over 40-km SMF and 10-m RF/25-m optical/100-m optical wireless transport. The proposed fiber–wireless and fiber–IVLLC convergences will be a highly attractive approach for integrating fiber backbone and RF/optical wireless feeder networks.

REFERENCES

[1] C. Y. Li, H. H. Lu, T. C. Lu, C. A. Chu, B. R. Chen, C. Y. Lin, and P. C. Peng, "A hybrid CATV/MMW/BB lightwave transmission system based on fiber-wired/fiber-wireless/fiber-VLLC integrations," Opt. Express vol. 23, no. 25, pp. 31807-31816, Dec 2015.

[2] T. Mochii, A. Shiva, H. H. Lu, C. J. Wu, T. C. Lu, C. A. Chu, and P. C. Peng, "A bidirectional wireless-over-fiber transport system," IEEE Photon. J. vol. 7, no. 6, pp. 7904409 (9 pages), Dec 2015.

[3] H. H. Lu, C. Y. Li, T. C. Lu, C. J. Wu, C. A. Chu, A. Shiva, and T. Mochii, "Bidirectional fiber-wireless and fiber-VLLC transmission system based on an OEO-based BLS and a RSOA," Opt. Lett. vol. 41, no. 3, pp. 476-479, Jan 2016.

[4] Q. Wang, L. Huo, Y. Xing, D. Wang, X. Chen, C. Lou, and B. Zhou, "Gaussian-like dual-wavelength prescaled clock recovery with simultaneous frequency-doubled clock recovery using an optoelectronic oscillator," Opt. Express vol. 22, no. 3, pp. 2798-2806, Feb 2014.

Visible Light Encryption System Using Camera Image Sensor

Chin-Wei Hsu[1*], Kevin Liang[1], Hung-Yu Chen[1], Liang-Yu Wei[1], Chien-Hung Yeh[2], Yang Liu[3], and Chi-Wai Chow[1]

[1]Department of Photonics and Institute of Electro-Optical Engineering, National Chiao Tung University, Hsinchu 30010, Taiwan
[2]Department of Photonics, Feng Chia University, Seatwen, Taichung 40724, Taiwan
[3]Philips Electronics Ltd., Hong Kong
*dicky0812@gmail.com

Abstract: We propose and experimentally demonstrate a light encryption scheme using camera image sensor. Besides, rolling-shutter effect of the CMOS camera in the mobile-phone can be used to enhance the data rate of transmission.

Keywords: *Visible light communication (VLC), Encryption, Rolling-shutter.*

I. INTRODUCTION

Visible light communication (VLC) [1] has been considered as one of the potential candidates for the next generation wireless networks because of its several advantages. Instead of using the congested traditional radio-frequency (RF) spectrum, it uses the extra electromagnetic spectrum (i.e. the visible light spectrum) for communications. Besides, since light is very directional, the communication zone of VLC can be confined in a small area. A high density and high capacity wireless communication can be achieved. Moreover, VLC is regarded as a relatively securer communication when compared with other wireless communications using RF, since VLC is directional and does not penetrate walls. Therefore, anyone outside the illuminance zone cannot receive the information. However, due to the visual nature, eavesdropping of the VLC signal is still inevitable when they are emitted by the light source.

In this work, a light encryption scheme is proposed and experimentally demonstrated. We use the device with both light emitting diode (LED) and camera image sensor receiver (Rx), such as a mobile phone, as a light encrypter. The original visible light signal sending from the ceiling lamp or desktop lamp can be first received by the proposed light encrypter, which can encrypt the information using a private key. Then, the encrypted signal can be emitted as visible light by this light encrypter. The light encrypter acts as an encryption gateway for signals in optical domain [2]. The Rx in this light encrypter can be any photo-detector such as photodiode, or a camera image sensor in the mobile phone. As mobile phone plays an important role in our daily lives, in this proof-of-concept demonstration, we use the mobile phone camera image sensor as the VLC light encrypter Rx.

Furthermore, rolling shutter effect of the CMOS camera [3] can be used to enhance the data rate higher than the frame rate of the camera. Then, by demodulating the rolling shutter pattern (bright and dark fringes received by the camera), the original data information can be obtained. Here, we also propose and demonstrate using Otsu thresholding scheme [4] to define the data logic in the rolling shutter pattern. Otsu method is widely used for segmenting a picture in image processing. Here, we show that Otsu scheme is also very effective to define the data logic in the rolling shutter pattern. Besides, we also apply the smoothing scheme to reduce the data pattern fluctuation. Finally, the bit-error-rate (BER) of the proposed scheme is estimated.

II. PROPOSED ARCHITECTURE

The proposed light encryption system is shown as Fig. 1. We use a desk lamp or other lighting equipment as a transmitter to send signal and utilize a mobile phone as a light encrypter to receive the signal. Here mobile phone can be used as the light encrypter since it has a camera and a white-light LED flash module. After the visible light signal is received by the light encrypter, the information can be encrypted using different encryption format. In this work, we demonstrate a XOR operation with a common private key for the data encryption. At the decryption side, the same key can be used to perform the XOR operation with the encrypted signal to retrieve the decrypted signal. In addition, the proposed scheme can support multiple devices with the correct key. In the symmetric-key scheme, besides the XOR operation, other schemes, such as shift-cipher (i.e. shifting the bits forward or backward in a fixed pattern) can be used.

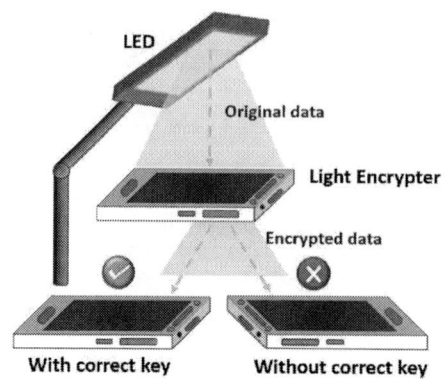

Fig. 1. The proposed lighting encryption scheme using mobile phone.

III. EXPERIMENT DEMONSTRATION

Fig. 2 shows the proof-of-concept experiment. A pseudo-random data is generated by a Matlab program and transmitted by an arbitrary waveform generator (AWG). The original signal from the lamp is then received by our proposed light encrypter. The original VLC signal is received by a CMOS camera image sensor with resolution of 640 x 480 pixels and frame rate of 28 frame/s. When the signal is received by a CMOS camera operated in video mode, rolling shutter effect can be observed. During the rolling shutter operation, the pixel row is activated without waiting for the scanning completion of the previous pixel row. This means there is an overlapping of exposure time of each pixel row; hence, the effective number of bits represent in each image frame is smaller than the vertical resolution of the image sensor. In this work, the net data rate (by removing the header and the successive packets) is 28 frame/s x 35 bit/frame = 0.98 kbit/s, which is limited by the resolution of the CMOS image sensor.

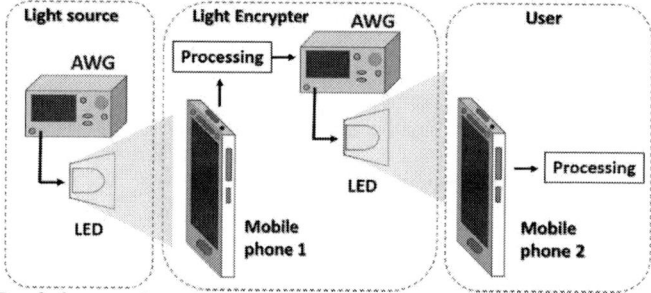

Fig. 2. Proof-of-concept demonstration of the light encryption and decryption. Inset: Rolling shutter patterns (bright and dark fringes) obtained by the CMOS image sensor.

A two minutes video is recorded for each BER measurement. In the demodulation of the rolling shutter pattern, the recorded movie file is converted into grayscale format, so that grayscale level of 255 represents total brightness and 0 represents total darkness as shown in the inset of Fig. 2. In order to enhance the extinction ratio (ER), we apply the smoothing scheme. The idea is to construct a second order polynomial fitting equation ("poly-smoothing 1" in Fig. 3(a)). For the grayscale values greater than the "poly-smoothing 1" equation are assigned equal to the values of that equation. Next, another polynomial fitting equation ("poly-smoothing 2" in Fig. 3(a)) is constructed, so that the grayscale values smaller than the "poly-smoothing 2" equation are assigned equal to zero. So that the ER fluctuation is significantly reduced, as shown in Fig. 3(b). Then, we use Otsu method to define the data logic. As mentioned in the introduction, Otsu method is very popular for segmenting a picture in image processing. Here, we show that it is also suitable for defining the data logic in the rolling shutter pattern.

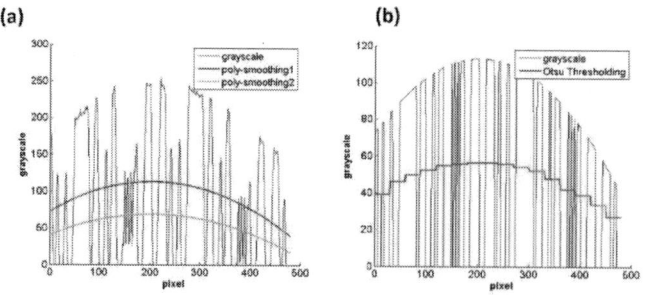

Fig. 3. Data pattern of received signal (a) before and (b) after applying the polynomial smoothing.

Then XOR operation will be performed between the received original signal and a private key to generate the encrypted data, which can then be emitted by a white-light LED built on the mobile phone. In this proof-of-concept demonstration, offline signal processing is used. Finally, the encrypted VLC signal is received by another mobile phone. Another XOR operation will be performed between the received encrypted signal and the private key; hence the decrypted signal can be retrieved.

IV. EXPERIMENT RESULTS AND DISCUSSION

Fig. 4 shows the comparison result of the original data and the encrypted data under different illuminance levels. We can observe that power penalty is negligible during the encrypting process. The results show that using the proposed smoothing scheme can significantly enhance the BER due to the reduction of ER fluctuation. A significant BER enhancement with up to 2 orders of magnitude can be observed at high illuminance cases. Furthermore, by using the proposed smoothing scheme, the standard BER of the forward error correction (FEC) limit can be achieved. In this proof-of-concept experiment, only a single white-light LED is used, and the distance between the LED and the image sensor at ~ 550 lux is 25 cm. If the light source is changed by a LED array or a brighter LED, the transmission distance can increase.

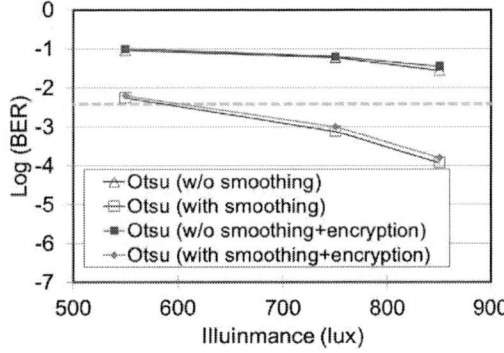

Fig. 4. Measured BER of the original data and the encrypted data under different illuminance levels with Otsu method.

V. CONCLUSIONS

In this paper, a novel light encryption scheme is proposed. The original visible light signal sending from the lamp can be first received by our proposed light encrypter. After encrypted, the encrypted data is emitted as visible light by this light encrypter. In this work, we also demonstrated using Otsu method as a thresholding scheme to define the data logic in the rolling shutter pattern. In the experiment, negligible power penalty was observed during the encrypting process. Besides, the proposed smoothing scheme can significantly enhance the BER with up to 2 orders of magnitude and achieve the standard of FEC due to the reduction of ER fluctuation.

ACKNOWLEDGMENT

We would like to thank for the support of Mr. Yen-Ting Chen and Institute for Information Industry (III),Taiwan.

REFERENCES

[1] C. W. Chow, C. H. Yeh, Y. Liu, and Y. F. Liu,"Digital signal processing for light emitting diode based visible light communication,"IEEE Photon. Soc. Newslett., vol. 26, pp. 9–13, 2012.

[2] Y. Liu, K. Liang, H. Y. Chen, L. Y. Wei, C. W. Hsu, C. W. Chow, and C. H. Yeh, "Light encryption scheme using light-emitting diode and camera image sensor," IEEE Photon. J., vol. 8, pp. 7801107, 2016.

[3] C. Danakis, M. Afgani, G. Povey, I. Underwood, and H. Haas, "Using a CMOS camera sensor for visible light communication," Proc. OWC'12, pp. 1244-1248.

[4] N. Otsu,"A threshold selection method from gray-level histograms," IEEE Trans. Syst., Man, Cybern., vol. SMC-9, pp. 62–66, 1979.

Non-Orthogonal Multiple Access with Multicarrier Precoding in Visible Light Communications

Xun Guan, Yang Hong, and Calvin Chun-Kit Chan
Department of Information Engineering, The Chinese University of Hong Kong, Hong Kong
guanxun@ie.cuhk.edu.hk

Abstract: *We propose a non-orthogonal multiple access method in visible light communications with multicarrier precoding and phase-amplitude predistortion. The proposed method eliminates the necessity of adaptive modulation while improving the system throughput.*
Keywords: *Multiple Access, NOMA, VLC, Precoding, Predistortion*

I. INTRODUCTION

Nowadays, visible light communication (VLC) is attracting wide attention among researchers as a potential key technology in the future communication systems, benefitting from its high volume, low-cost, reliability, and license-free [1]. Resembling other wireless communication technologies, appropriate multiple access strategies are indispensable for VLC to support multiple connections in one system. Recently in [2], we have proposed a novel scheme that can solve the dilemma between throughput and fairness, which is a main drawback of conventional multiple access technologies. In non-orthogonal multiple access (NOMA), the multiple access is granted according to different power levels of different users, particularly in every subcarrier of orthogonal frequency division multiplexing (OFDM) signal. We have also proposed a phase pre-distortion (PP) method to improve the system performance. However, the response across the whole OFDM spectrum is non-uniform. To accommodate different single-to-noise ratios (SNR) on different subcarriers, adaptive modulation formats are adopted on different subcarriers, thus largely complicates the NOMA process.

Recently, a multicarrier precoding (MP) method based on orthogonal circulant matrix transform (OCT) was proposed in [3] to flatten the SNR of all the subcarriers in an OFDM system. In this paper, we apply the MP method in NOMA to avoid the adoption of higher order modulation formats. A novel phase and amplitude predistortion (PAP) process is further proposed to enable the MP method under NOMA.

II. PRINCIPLES

The system model of NOMA is depicted in Fig. 1(a), which is the same to that in [2]. The photo diode (PD), acting as the NOMA receiver, lies near to LED_1 but far from LED_2. The signal PD received from LED_1 (denoted as S) is stronger than the weaker signal from LED_2 (denoted as W). S and W are combined into a composite signal, which is denoted as C. Both S and W are in DC-biased optical OFDM (DCO-OFDM), which means only half of the whole OFDM subcarriers are used.

In MP process proposed in [3], an orthogonal circulant matrix, Q, is multiplexed to H_S and H_W before modulation onto the subcarriers, where $Q = \left(1/\sqrt{N}\right) \times [c_1, c_2, \cdots c_N; c_N, c_1, \cdots c_{N-1}; \cdots; c_2, c_3, \cdots c_1]$, and each entry c_i corresponds to the element

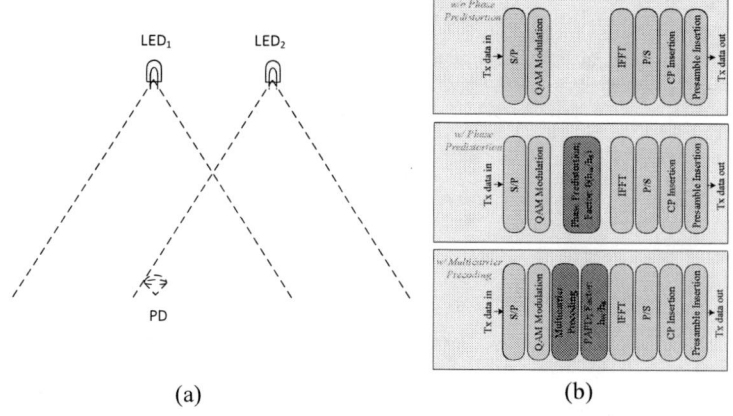

(a)　　　　　　　　　　(b)

Fig.1. (a) System Model (b) DSP of W in three cases.

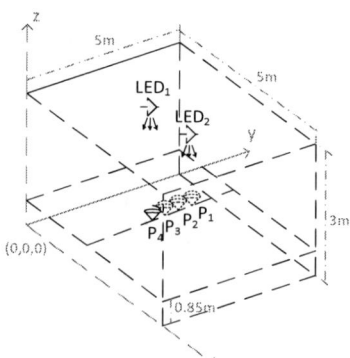

Fig.2. Simulation Setup

of the Zadoff-Chu (ZC) sequence in index of 1 and length of N. Unlike conventional NOMA, the introduction of MP interferes the decision of S in the first step of successive interference cancellation (SIC). As a solution, a PAP matrix, $H_W^{-1}H_S$, is multiplexed to W at the transmitter after the MP step. Fig. 1(b) shows the DSP process of the three cases: (1) NOMA without PP/MP, or conventional NOMA, (2) NOMA with PP, as in [2], (3) NOMA with MP and PAP.

III. NUMERICAL SIMULATIONS

(A) Simulation Setup

Fig. 2 shows the simulation setup of an indoor VLC system. Two LEDs are mounted on the ceiling for illumination and communication, while a PD is positioned on a plane with 0.85m in height to simulate the height of desk surface. Four PD positions are considered to simulate different receiving signal qualities. The simulation parameters are summarized in Table 1. The three DSP processes mentioned in the last section are studied in comparison. Two factors are controlled in the simulation: the location of PD, and the power of both two LEDs.

Table 1 Simulation Parameters

Parameter	Value
LED location	LED$_1$: (1.5,2.5,3); LED$_2$: (3.5,2.5,3)
PD location	P$_1$: (2,4,0.85); P$_2$: (2,3.5,0.85); P$_3$ (2,3,0.85); P$_4$: (2,2.5,0.85)
Detection area of PD	1 cm^2
Number of Rays	50000
Reflection coefficient (walls/floor/ceiling)	0.83/0.63/0.4
Max. reflection order	5
Receiver field of view	60°
System sample rate	200MSa/s
DCO-OFDM bandwidth	100MHz
FFT size	256
CP length	32

(B) Simulation Results

We first investigate the effect MP brought to the SNR performance of the whole spectrum of OFDM. Fig. 2(a) and (b) shows the SNR comparisons of all the OFDM subcarriers, with and without the MP. Solid lines represent the case without MP, while dashed lines are the multicarrier-precoded SNRs. It could be observed that the SNRs of all the subcarriers in one OFDM transmitter could be equalized by MP. On the other hand, PD at P$_4$ enjoys higher SNRs as well as larger SNR differences between S and W over the whole band, while the SNRs and SNR differences at P$_1$ are smaller. Fig. 2(c) and (d) shows the BER performance when PD is placed at different positions in the simulation setup. Although different PD positions lead to different SNRs as well as different power ratios between S and W, in most cases the adoption of MP decreases the BER of S and W compared to those without PP/MP and without MP. Specifically, we compare the case with PP to that with MP. When PD is located at P$_2$ and the LED power is 1.5W, the BER of S drops from 5×10^{-3} to 2.5×10^{-3} while the BER of W drops from 1×10^{-2} to 6×10^{-3}, indicating minor improvements. When the LED power increases to 2W, the BER of S changes from 2×10^{-3} to 2×10^{-4}, while the BER of W decreases from 4×10^{-3} to 4×10^{-4}. This conclusion is further verified by simulating the BER performance in regard to the LED power. Fig. 2(e)

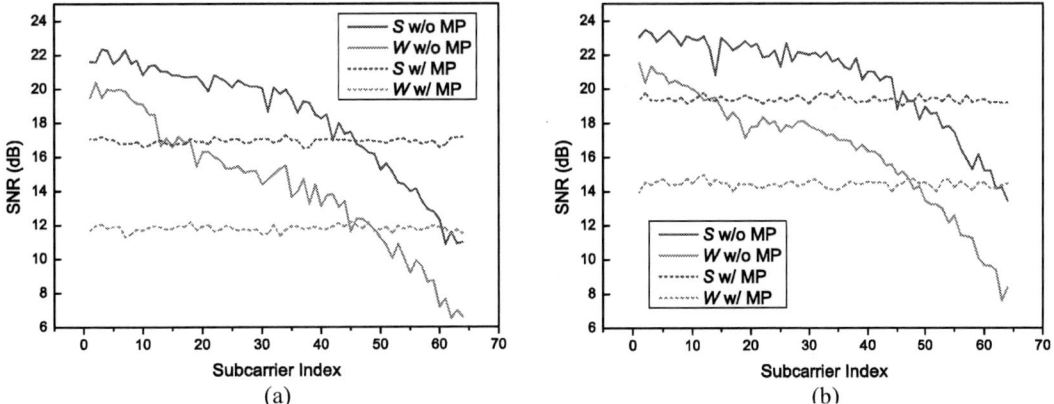

Fig.3. SNR of OFDM data subcarriers: (a) LED power=2W, PD at P$_2$, (b) LED power=2W, PD at P$_4$

and (f) shows the BER comparisons by fixing the PD at P_2 or P_4 and changing the power of LEDs. In both cases, the BER drops faster in MP case than PP case. It could be found that the improvement brought by MP is more significant under an appropriately better signal quality.

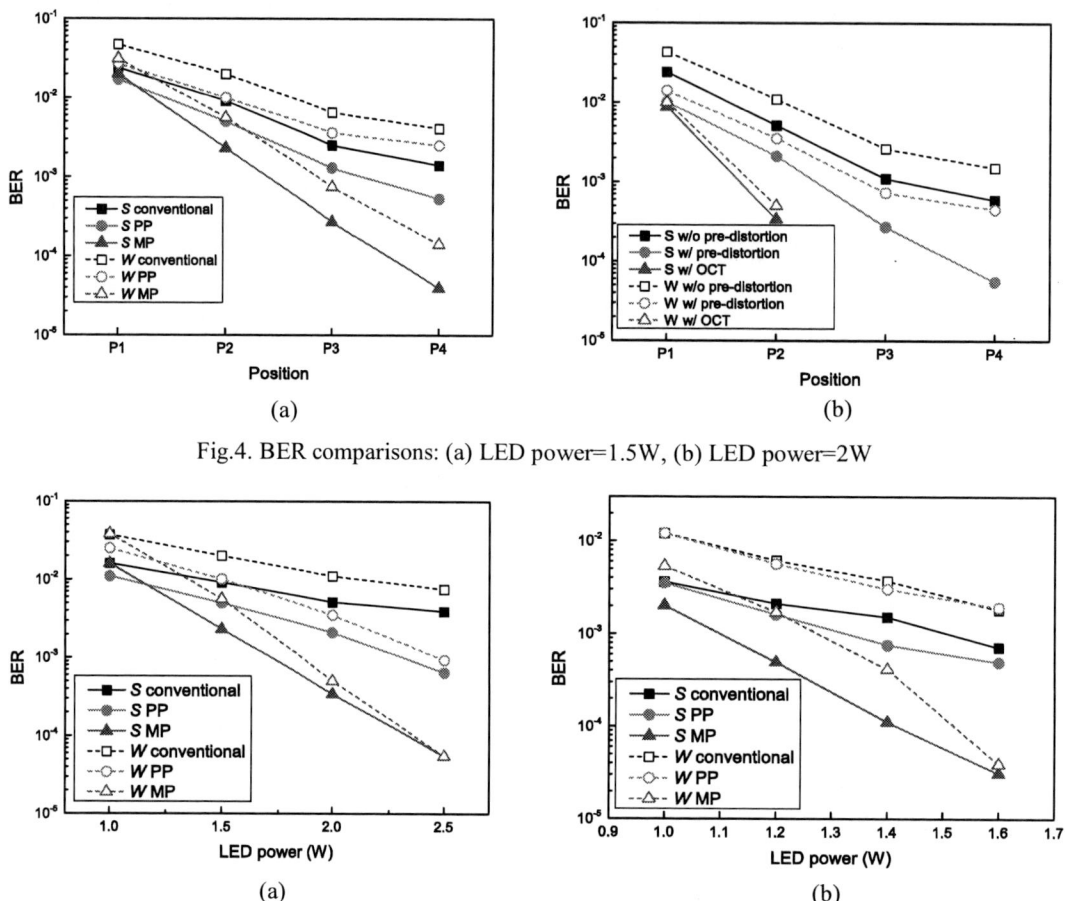

Fig.4. BER comparisons: (a) LED power=1.5W, (b) LED power=2W

Fig.5. BER comparisons: (a) PD at P_2, (b) PD at P_4

IV. SUMMARY

We have proposed a multicarrier precoding scheme for NOMA in VLC. The proposed scheme combats the low-pass nature of VLC systems by keeping low BER, high throughput without introducing adaptive modulation formats. Simulation studies have proved the effectiveness of the proposed scheme. This work was partially supported by a research grant from Hong Kong Research Grants Council (Project No. 14200614).

REFERENCES

[1] A. Jovicic et al., "Visible light communication: opportunities, challenges and the path to market," IEEE Communications Magazine, vol. 51, no. 12, pp. 26-32, 2013.
[2] X. Guan et al., "Phase pre-distortion for non-orthogonal multiple access in visible light communications," in Proc. OFC, Paper Th1H.4, 2016.
[3] Y. Hong et al., "Experimental demonstration of an OCT-based precoding scheme for visible light communications," in Proc. OFC, Paper M3A.6, 2016.

WA2-23

OECC/PS2016

6.6-Gbit/s Phase Modulated Impulse-Radio Generation from RZ-OOK Optical Signals for Optical-Wireless Systems

Yuri Ohara, Hdeaki Kawahara, Tomoki Kishida, Kengo Nabika, and Saeko Oshiba

A Department of Electronics, Graduate School of Science and Technology, Kyoto Institute of Technology
Goshokaidocho, Matsugaski, Sakyo-ku, Kyoto 606-8585, Japan
oshiba@kit.ac.jp

Abstract: *We investigated phase modulated impulse radio (IR) signal generated from RZ-OOK optical signals and its demodulation method by down-conversion with twice the phase shift. We examine to separate even bits and odd bits as orthogonal signals. We demonstrated the 6.6-Gbps IR signals with 3 m wireless radio transmission.*

Keywords: *Fiber optics communications, Radio frequency photonics.*

I. INTRODUCTION

Recently, more high-speed radio communication has been demanded owing to the increase in wireless data traffic according to the spread of mobile portable terminals such as smartphones. In addition, it is desirable that the cell size per access point be smaller because of the strict restrictions of the radio frequency. This requires a large number of access points to provide high-speed wireless network service. The combined optical and wireless link can be reduce the area of the wireless cells and maintains rapid radio communication by transmitting data to an access point from a central office.

We have proposed a combined fiber and wireless link system using impulse radio [1]. We converts short optical pulses into radio pulse signals called wavelets by band-limiting at several giga-hertz with a band-pass filter (BPF). Because a carrier wave is not required, the transceiver cost and power consumption can be reduced, as the configuration can be simpler. The BPF can be realized using waveguide antenna, or Radio over system, and the several types of EO antenna-coupled electrode modulator has been developed [2]. In a previous paper, we demonstrated the generation of a quasi-Nyquist IR signal waveform to avoid ISI for high-speed IR transmission by optimizing the optical RZ pulse width and filters [3]. As another method to increase transmission speed, we proposed phase modulation IR-UWB in wireless link and its demodulation method by down-conversion with twice phase shift [4],[5]. The phase modulation of IR signals is converted from pulse-position modulation of optical RZ-pulse in fiber link.

In this paper, we demonstrate the phase modulated IR-signal generated with an adjustment of the even and odd bits of the optical RZ-OOK signal as orthogonal IR signals. In our experiments with a 6.6-Gbit/s IR with UWB band of 7.25-10.25GHz, we realize a Q factor greater than six with error-free communication over a 3-m radio transmission distance.

II. IR-UWB SIGNALS GENERATION

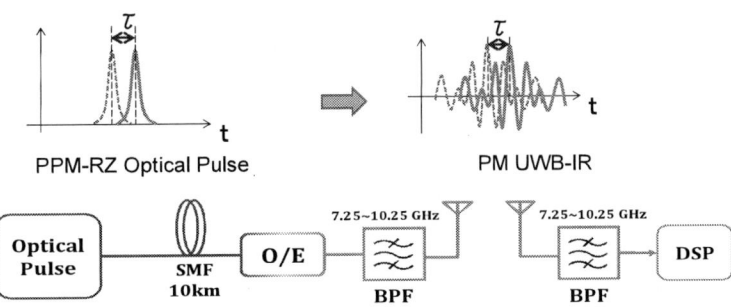

Fig. 1. Phase modulation IR signals generation process.

Fig.1 shows a phase modulation IR signals generation process. The IR signals are generated from OOK short RZ pulses by passing through a BPF (Impulse response $h(t)$, Center frequency f_B). The IR-UWB signal $x(t)$ with a time delay τ is expressed as in (1).

$$x(t) = h(t)\exp j2\pi(f_B t - f_B \tau) \qquad (1)$$

A IR signals have to set a time delay of odd pulse train to be phase is $2\pi f_B \tau = 0$, and a time delay of even pulse train to be phase is $2\pi f_B \tau = \pi/2$. So, we select $f_B = (2n+1)B/2$. B is bit rate, that the time delay between of even and odd bits of this signal was corresponding to $2\pi f_B \tau = \pi/2$.

III. EXPERIMENTS

Fig. 2 shows a diagram of the experiment. The RZ optical pulse train generated by 9.95328 GHz mode-locked laser is modulated in a pseudo-random signal of 3.31776 Gbit/s PRBS 23^{-1} by the EA modulator. The signal is divided by 3-dB optical coupler. One of divided signals transmits delay line, and amplified to same optical power level of another signal. Then, the divided signals are multiplexed as 6.63552 Gbit/s signal. The time delay between of even and odd bits of this signal was corresponding to $2\pi f_B \tau = \pi/2$. This signal was then converted into an RZ electrical pulse signal by an O/E converter PD. wavelets can be obtained by the RZ electrical pulse signal that has passed through the BPF. This wavelet was transmitted at distances of 0.5, 1.0, 1.5, 2.0, and 3.0 m with a directional antenna gain of 11 dBi. The power of the input signal of the antenna was −52.4 dBm/MHz, and the radiation power was within the limits of the spectral mask of UWB, even considering the gain of the antenna of −41.4 dBm/MHz. The signal passed through the BPF to remove noise after amplification by a broadband amplifier and observed by digital communication analyzer. The time waveform of the optical signal, wavelets and spectrum of wavelets were shown in fig.3. Then, signals were digitally processed by PC.

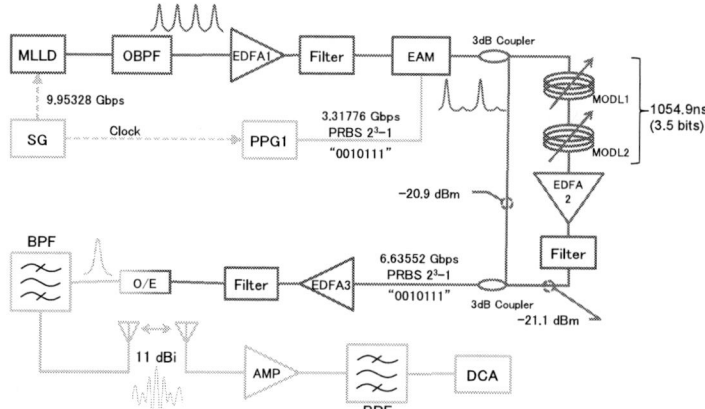

Fig. 2. Experiment configuration of IR signal generation and transmission.

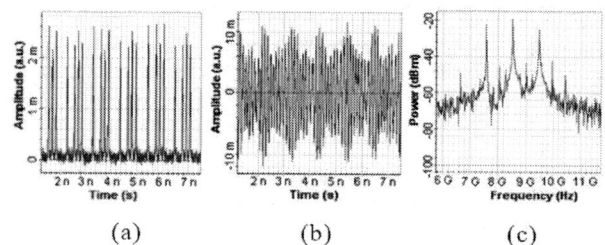

Fig. 3. (a) Time waveform of the optical signal, (b) wavelets and (c) spectrum of wavelets.

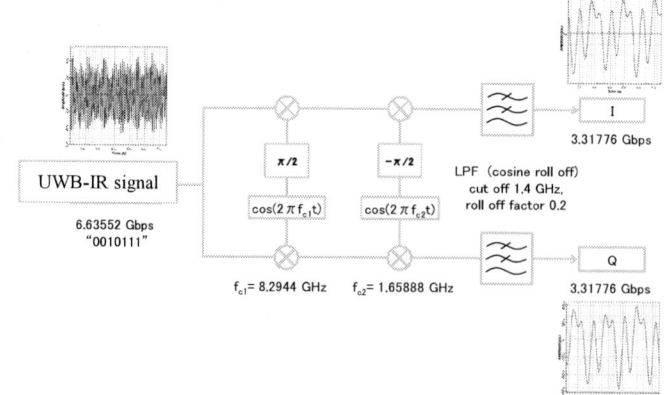

Fig. 4. Block diagram of the digital processing of down-conversion.

Fig.4 shows a block diagram of the digital processing of down-conversion in PC. The obtained in the experiment is multiplied by a sine wave with a frequency of 6.63552 GHz and input into a rectangular-type LPF (cut-off frequency: 2.9 GHz). For down-conversion with twice the phase shift in Fig. 4, one of the divided signals was multiplied by a frequency of 8.2944 GHz for down-conversion in the first stage. In the second stage, this signal described was multiplied by a frequency of 1.65888 GHz. Further, the other signal was multiplied by the phase differences of $\pi/2$ in the first stage and $-\pi/2$ in the second-stage. Then, these signals were input into the cosine roll-off LPF (cut-off frequency: 1.4 GHz). As a result, we obtained two signals at 3.31776 Gbit/s, which were separated into even- and odd-bit signals of 6.63552 Gbit/s. The time waveform of the signals are also shown in fig.4.

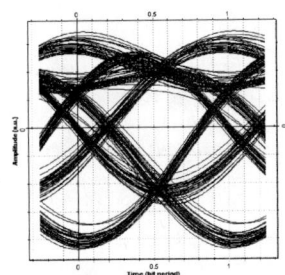

Fig. 5. Eye pattern after down-conversion.

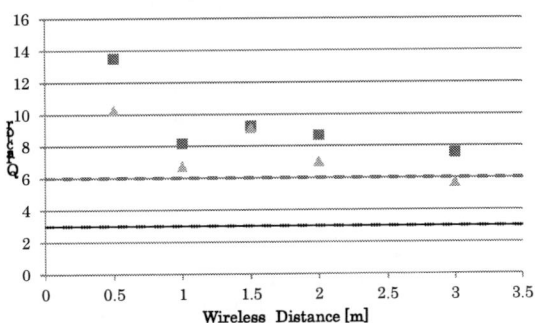

Fig. 6. Q factor vs wireless transmission distance.

Fig. 5 shows the eye pattern after down-conversion (wireless transmission experiment of 1.5 m). Good eye opening appears in the center.

Fig. 6 shows a graph of the Q factor and wireless transmission distance. Q factor greater than six for a wireless transmission distance of 3 m, this indicates that the communication can be error-free (equivalent BER : 10^{-9}). Therefore, these results demonstrate successful improvement of the communication quality by 6.6-Gbit/s IR signal.

IV. CONCLUSIONS

We have proposed a phase-modulation and demodulation method for IR signal generation from optical RZ- OOK signal. We considered the even and odd bits to be I and Q components and examined the ISI and high-speed communication with the configuration of a simple modulator. In our experiments with a 6.6-Gbit/s IR with UWB band of 7.25-10.25GHz, we realize a Q factor greater than 6 over a 3-m radio transmission distance.

REFERENCES

[1] S. Oshiba, Y. Kasai, H. Miura, and M. Akiyama, "RZ pulse-width dependence of impulse radio UWB over combined fiber and wireless link," in Proceedings of 2011 International Topical Meeting on & Microwave Photonics Conference, OCT.18-21, p.49 (2011).

[2] H. Murata, N. Suda, and Y. Okamura, "Electro-optic modulator using Patch antenna-coupled resonant electrodes and polarization-reversed structure for radio-on-fiber system" proceedings of CLEO-IQEC, CTuT5, (2009).

[3] Saeko Oshiba, Hiroshi Miura, Yuri Ohara, and Hitoshi Shimasaki, "3. 3 Gbps x 3TDM IR signal transmission for UWB over Combined Fiber and Wireless Link" in Technical Digests of CLEO-PR & OECC/PS 2013, TuPP-7, Kyoto, Jul. (2013).

[4] Yuri Ohara, Hiroshi Miura, Saeko Oshiba, "UWB-IR QPSK Using Time Delay of RZ Optical Pulses", IEICE TRANSACTIONS Vol.J97-B, No.2, pp. 78-85, Feb. (2014).

[5] Tomoki KISHIDA, Kengo NABIKA, and Saeko OSHIBA, "QAM UWB-Impulse Radio Using Optical Pulses Position Modulation for Optical Fiber-Wireless Links", Proceedings of APCC 2015, Kyoto, 14-PM1-C1, Oct, (2015).

WA2-24

Simultaneous All-channel OTDM Demultiplexing Based on Complete Optical Fourier Transformation

Pengyu Guan, Mads Lillieholm, Kasper Meldgaard Røge, Toshio Morioka, Leif Katsuo Oxenløwe
DTU Fotonik, Technical University of Denmark, Ørsteds Plads, 343, Kgs. Lyngby, 2800, Denmark
pengu@fotonik.dtu.dk

Abstract: *We demonstrate simultaneous OTDM demultiplexing of all 16-channels for 160-Gbit/s DPSK and 320-Gbit/s DQPSK signals based on complete OFT. Furthermore, numerical simulations show promising results for extending the proposed technique to spectrally efficient Nyquist-OTDM.*
Keywords: *Optical signal processing; Optical demultiplexing; Optical Fourier transformation.*

I. INTRODUCTION

Optical Time-Division Multiplexing (OTDM) is the time-interleaving of optical data signals generated from lower rate pulse trains at identical wavelengths. OTDM enables us to realize future ultrahigh-capacity optical networks with an increased bit rate per wavelength channel and decreased channel numbers. Thus, the number of components as well as the system complexity and power consumption, are reduced. Single-channel 10.2-Tb/s and 1.92-Tb/s OTDM transmissions have already been demonstrated with advanced modulation formats [1-2]. One of the key enabling technologies for such an ultrahigh-speed OTDM system is an optical demultiplexer, which can be realized by an optical temporal switch such as a nonlinear optical loop mirror [3], a Kerr switch [4] or a symmetric Mach–Zehnder device [5]. However, using these schemes to fully demultiplex an OTDM signal requires one optical temporal switch per channel. Therefore, the complexity of an OTDM receiver essentially scales with the number of OTDM channels. To overcome this problem, a new approach for OTDM demultiplexing based on a "partial" optical Fourier transformation (OFT) has been proposed [6]. Such a "partial" OFT allows several OTDM channels to be demultiplexed simultaneously. However, all-channel OTDM demultiplexing is still very challenging due to the inter-channel crosstalk produced by the "partial" OFT process [7].

Previously we have proposed a "complete" OFT scheme for all-channel serial-to-parallel conversion [8]. In this paper, we present the experimental demonstration of 160-Gbit/s DPSK and 320-Gbit/s DQPSK all-channel simultaneous OTDM demultiplexing using a complete OFT. Full system characterizations with bit error rate (BER) measurements are performed. For 160-Gbit/s OTDM-DPSK demultiplexing, error-free performance was achieved for all OTDM channels with an average power penalty of only 1.2 dB at BER = 10^{-9}. For 320-Gbit/s DQPSK signal, a BER performance below the FEC limit (2×10^{-3}) is achieved for all demultiplexed channels. In addition, detailed Monte Carlo BER simulations predict that the proposed scheme will also perform well for Nyquist-OTDM demultiplexing.

II. PRINCIPLE AND EXPERIMENTAL DEMONSTRATION

The OFT is based on the principle of a "time-lens", originating in the space-time duality of light. It can transfer the temporal profile of an optical signal into the frequency domain or vice-versa. The traditional OFT for OTDM demultiplexing consists of a dispersive medium with $D = \beta_2 L$ (β_2 is the second order dispersion and L is the length), followed by a quadratic phase-modulation stage ($\delta\phi = Kt^2/2$) with chirp rate K, satisfying the condition $K = 1/D$, and has been used in many demonstrations [6-7]. However, it is not suitable for all-channel simultaneous OTDM demultiplexing. Since the dispersive elements before the phase modulation stage will generally broaden the input waveform causing temporal clipping and hence spectral broadening and power loss of the edge channels after OFT, large inter-channel crosstalk and OSNR degradation occurs. To overcome this problem, we propose to use a new time-lens based complete OFT, which has already been used in demonstrations for WDM to Nyquist-OTDM conversion [9]

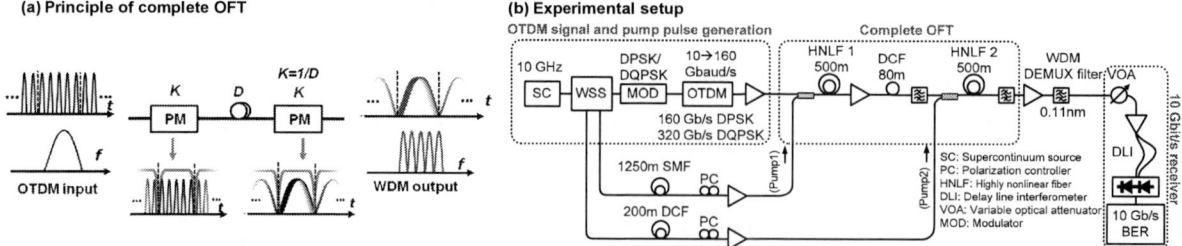

Fig. 1. (a) Schematic diagrams of a time-lens based complete OFT. (b) Experimental setup for 160 Gbit/s DPSK and 320 Gbit/s DQPSK all-channel OTDM demultiplexing

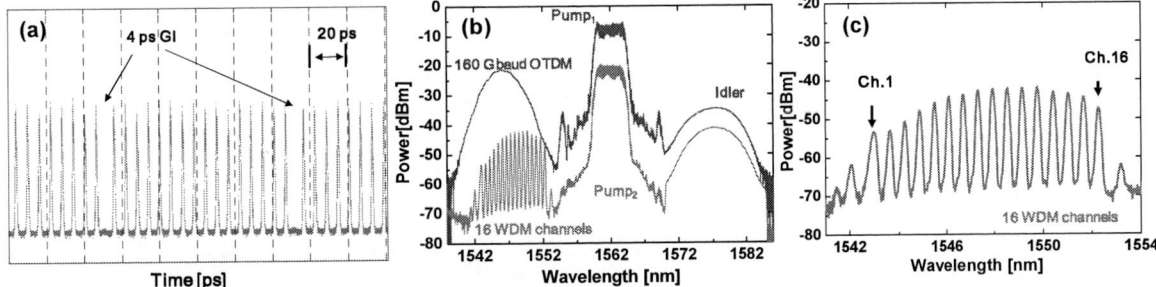

Fig. 2. Results of OTDM to WDM conversion. (a) 160 Gbaud OTDM signal waveform, (b) optical spectrum after the first (blue curve) and second (red curve) FWM process, (c) spectrum of the 16 obtained WDM channels.

and for OFDM to Nyquist-WDM conversion [10]. The schematic diagram of this OFT is shown in Fig. 1(a), where two quadratic phase-modulation stages with chirp rate K, separated by a medium with accumulated dispersion D, which satisfies the condition $K = 1/D$ (a K-D-K configuration). With this configuration, all input OTDM channels are pre-chirped by the first time-lens. The chirped signals propagating in the dispersive medium will not only experience waveform broadening, but also become aligned (focused) to the center of the second phase-modulation stages. This configuration thus confines the waveform to the time-lens apertures, making it suitable for a simultaneous, all-channel OTDM to WDM conversion (a "complete" OFT). The converted WDM signal can be simply demultiplexed by a passive WDM demultiplexer. The chirp rate K determines the scaling factor between the time- and frequency-domains according to $\Delta t = 2\pi\Delta f/K$.

The experimental setup is shown in Fig. 1(b). A supercontinuum signal is generated by a 10-GHz mode-locked laser and a 400-m dispersion-flattened highly non-linear fiber (DF-HNLF). The broadened spectrum is Gaussian-filtered at 1546 nm by a programmable wavelength selective switch (WSS). The obtained 1 ps pulses train is DPSK or DQPSK modulated using an IQ modulator. The 10-Gbaud modulated signal is then OTDM-multiplexed to 160 Gbaud, resulting in a 160-Gbit/s DPSK-OTDM signal or a 320-Gbit/s DQPSK-OTDM signal. Fig. 3(a) shows the obtained 160-Gbaud OTDM signal waveform with a tributary spacing of 6 ps. A 4 ps guard interval is inserted after every 16 tributaries to allow for the transition between consecutive quadratic phase-modulation windows. A complete OFT is used for the all-channel OTDM-to-WDM conversion. The quadratic phase modulation is based on a four-wave mixing (FWM) in a highly non-linear fiber (HNLF) using linearly chirped (quadratic phase) rectangular pump pulses at 10 GHz. Both pump pulses and the data signal are obtained from the same broadened spectrum using a WSS. To obtain linearly chirped pumps, the filtered pump1 and pump2 are subsequently propagated in a 1250 m SMF and 200 m dispersion compensating fiber (DCF), respectively. The chirp rate $K = 0.078$ ps^{-2} is set for conversion of the 6 ps temporal spacing to a 75-GHz (0.6 nm) frequency grid. The first FWM output is shown in Fig. 2(b). After extraction with a 15 nm optical bandpass filter (OBPF), the idler is dispersed in 80 m DCF corresponding to dispersion D, and subsequently combined with pump2 and coupled into HNLF$_2$ for the second FWM process. The resulting spectrum is also shown in Fig. 3(b). The generated idler is the 16-channel WDM signal converted from the 160-Gbaud OTDM signal. Fig. 3(c) shows a zoom-in of the generated idler, where 16 WDM channels with 75 GHz (0.6 nm) spacing can be observed. High extinction ratios of 25 dB and 17 dB, are observed for the center and left edge channels respectively. The skew between the channels is caused by the limited FWM bandwidth of the used HNLFs. This indicates insignificant spectral channel overlap, resulting in a crosstalk-less all-channel OTDM demultiplexing. In the receiver, a 14-GHz (0.11 nm) Gaussian filter is used to demultiplex each WDM channel. Finally, the BER of each demultiplexed DPSK or DQPSK channel is measured in a pre-amplified receiver including a delay-line interferometer (DLI) and a balanced photo-detector.

The BER measurements for all 16 demultiplexed 10-Gbit/s DPSK channels are shown in Fig. 3(a), where error free performance (BER $< 10^{-9}$) is achieved for all DPSK-OTDM channels with an average receiver sensitivity of -40.5 dBm. The average power penalty at BER $= 10^{-9}$ is only 1.2 dB compared to the 10 Gbit/s DPSK baseline. The BER measurements for 320-Gbit/s DQPSK-OTDM demultiplexing are shown are Fig. 3(b) for selected DQPSK channels. As

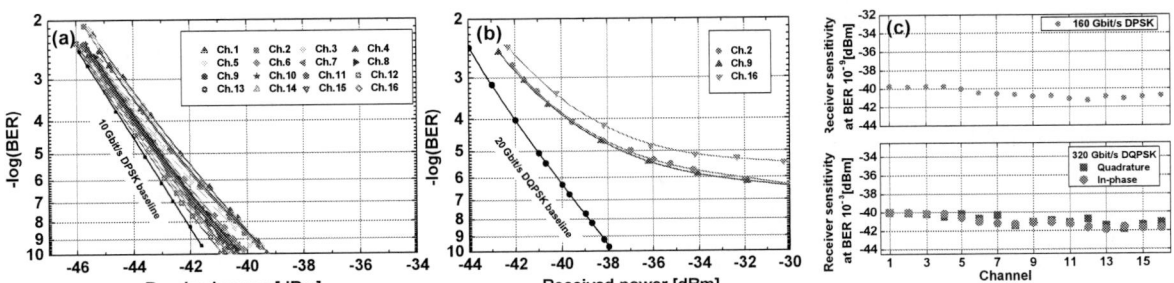

Fig. 3. Experimental results. (a) BER performance of all 16 demultiplexed 10-Gbit/s DPSK channels, (b) BER performance for selected demultiplexed DQPSK channels, (c) receiver sensitivities at BER = 10^{-9} of all demultiplexed DPSK channels, and receiver sensitivities at BER = 10^{-3} of all DQPSK channels.

791

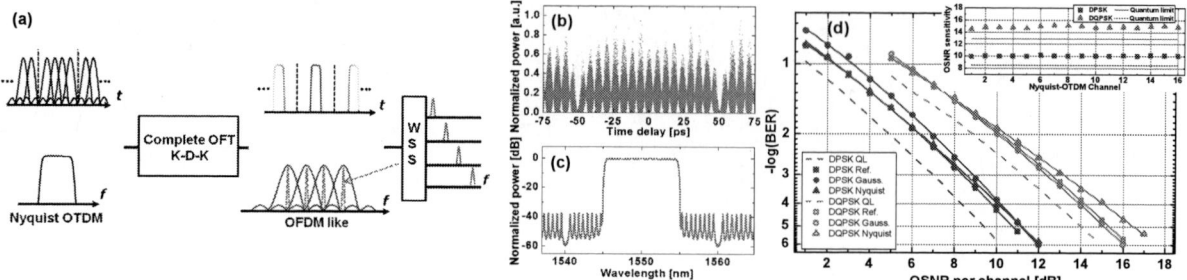

Fig. 4. Numerical simulations for all-channel Nyquist-OTDM demultiplexing, (a) schematic diagrams, (b) waveform of 160 Gbaud Nyquist-OTDM signal, (c) OFT output spectrum, (d) BER performance for 160 Gbit/s DPSK and 320 Gbit/s DQPSK signals. The inset shows the corresponding OSNR sensitivities at BER = 10^{-4} of all demultiplexed 10 Gbit/s DPSK and 20 Gbit/s DQPSK channels.

the DQPSK signals have a higher OSNR requirement and a lower phase noise tolerance than DPSK signals at the same baud rate, there are error floors around 10^{-5} and 10^{-6} due to the nonlinear phase noise and OSNR degradation introduced by the FWM processes. Fig. 3(c) shows the receiver sensitivities at BER = 10^{-9} of all demultiplexed DPSK-OTDM channels, as well as the receiver sensitivities at BER = 10^{-3} of all demultiplexed DQPSK-OTDM channels. The DQPSK demonstration confirms the fact that the proposed scheme can be applied to complex modulation formats.

III. NUMERICAL SIMULATIONS FOR NYQUIST-OTDM DEMULTIPLEXING

Nyquist-OTDM signals take advantage of orthogonality in the time domain, which enables high bit rates with high spectral efficiency [11]. We performed Monte Carlo simulations to investigate the proposed scheme for all-channel Nyquist-OTDM demultiplexing. Fig. 4(a) shows the schematic diagrams of the simulations. A 160-Gbaud Nyquist-OTDM signal with 5.9 ps (1/170 GHz) channel spacing is converted into 16 sinc-shaped multi-carriers at different frequencies simultaneously using a complete OFT. As the tributaries of the Nyquist-OTDM are all overlapping seamlessly, it is essential to insert a short guard-interval (5.9 ps) after every 16 tributaries for OFT operation. The waveform of the 160-Gbaud Nyquist-OTDM signal is shown in Fig. 4(b). The chirp rate, $K = 0.078$ ps^{-2}, is set for conversion of the 5.9 ps OTDM spacing to a 73 GHz (0.59 nm) frequency grid. Fig. 4(c) shows the OFT output spectrum. A bank of 24-GHz passive Gaussian filters with 73 GHz spacing is used to extract all converted channels at the crosstalk-less positions. The Nyquist-OTDM channels are independently data-modulated with random and uniformly distributed bits using either DPSK or DQPSK. The demultiplexed channels are detected using an ASE limited receiver consisting of a 10-GHz DLI and a BPD (no detector noise). To estimate the BER, iterations of 1024 symbols are repeated until at least 100 errors are observed. Each BER value is obtained by averaging over 5 simulations. The results are shown in Fig. 5 as the BER vs OSNR per subcarrier for both 160-Gbaud Nyquist-OTDM (sinc shaped) and normal OTDM (Gaussian shaped) signals. The inset shows the corresponding OSNR sensitivities at BER = 10^{-4} for all demultiplexed 10-Gbit/s DPSK and 20-Gbit/s DQPSK Nyquist-OTDM channels. As a reference, we calculate the BER performance in the absence of neighboring OTDM channels (denoted as Ref) after the OFT. The quantum limit for the reference (matched optical filter and no electrical filtering) is also shown (denoted as QL). The simulation results indicate that the proposed scheme enables all-channel Nyquist-OTDM demultiplexing with a BER < 10^{-5} for both DPSK and DQPSK signals. For DPSK, the OSNR penalty due to the crosstalk from neighboring Nyquist-OTDM channel is negligible relative to the reference and normal OTDM (down to BER~10^{-5}). For DQPSK, the OSNR penalty increases to ~1 dB at BER = 10^{-5}, due to the lower crosstalk tolerance for DQPSK signals.

IV. CONCLUSIONS

We have demonstrated 160-Gbit/s DPSK and 320-Gbit/s DQPSK simultaneous all-channel OTDM demultiplexing using a complete OFT. All-channel error-free performance was achieved for 160 Gbit/s DPSK signal. For a 320-Gbit/s DQPSK signal, a BER performance below the FEC limit (2×10^{-3}) is achieved for all demultiplexed channels. Furthermore, Monte Carlo simulations indicate that the proposed scheme can also be advantageously used for simultaneous all-channel Nyquist-OTDM demultiplexing.

ACKNOWLEDGMENT

OFS Denmark Aps, Danish Research Council: FTP project TOR (ref. no. 12-127224), FTP project LENS-COM (ref. DFF-5054-00184) and the DFF Sap. Aude Adv. Grant NANO-SPECs (DFF-4005-00558B).

REFERENCES

[1] T. Richter et al., in Proc. OFC 2011, PDPA9, (2011).
[2] D. O. Otuya et al., Opt. Express, 21(19), 22808 (2013).
[3] N. J. Doran et al., Opt. Lett., vol. 13, 56–58, (1988)
[4] T. Morioka et al., Electron. Lett., 23 (9), 453-454, (1987).
[5] T. Hirooka et al., IEEE PTL., 21(20), 1574-1576, (2009).
[6] H. C. H. Mulvad et al., Proc. OFC 2011, OThN2, (2011).

[7] K. G. Petrillo et al., Proc. CISS, 1-6, (2013).
[8] P. Guan et al., Proc. OFC 2016, W3D.2, (2016).
[9] P. Guan et al., Proc. ECOC, We.2.5.5, (2014).
[10] P. Guan et al., IEEE JLT, 34(2) 626-632 (2015).
[11] M. Nakazawa et al., Opt. Express, 20(2), 1129,(2012).

WA2-25

OECC/PS2016

Experimental Demonstration of 7.09Gb/s DML based Real-time Optical OFDM Transmission with Spectral Efficiency up to 6.93bits/s/Hz over 50km SSMF

Qianwu Zhang*, Han Dun, Chen Qian, Bingyao Cao, Mingzhi Mao, Junjie Zhang, and Min Wang

Key Laboratory of Specialty Fiber Optics and Optical Access Networks, Shanghai University, Shanghai 200072, China

qwzhang@shu.edu.cn

Abstract: 7.09Gb/s real-time optical OFDM transmissions with record high spectral efficiency up to 6.93bits/s/Hz are experimentally demonstrated in DML-based 50km SSMF IMDD systems in cooperated with DACs/ADCs whose sampling rates are as low as 2GS/s.

Keywords: spectral efficiency, optical OFDM, real-time

I. INTRODUCTION

With the rapid growth of the online data service (such as HDTV and VOD etc.), the bandwidth requirements have been rising at an alarming rate so that higher capacity is the priority in development of passive optical networks (PONs) for access networks. Optical orthogonal frequency division multiplexing (OOFDM) has been considered as a promising candidate for the next-generation PONs (NG-PONs), since it combined the advantages of wireless OFDM systems and optical communications which including high spectral efficiency, inherent resistance to chromatic dispersion, and provision of hybrid dynamic bandwidth allocation in both frequency and time domains. Intensity modulation and direct detection OFDM (IMDD-OFDM), for example directly modulated laser (DML) based OOFDM transceivers, are usually adopted in cost sensitive access scenario or data centers due to their simple structures and low DSP complexity [1]. Moreover, real-time realization of abovementioned OOFDM transceivers is critical for rigorously evaluating the practical implementation feasibility of OOFDM technique.

To further improve the signal line rate of real-time OFDM systems, two technical approaches can be applied: a) increase signal bandwidth, and b) improve the spectral efficiency. Wider signal bandwidth, however, requires higher sampling rate of DACs/ADCs, thus imposes significant difficulties on practical realization of OOFDM transceivers under mass deployment due to their unacceptable cost. On the other hand, high spectral efficiency approach is also one of the most promising strategies for satisfying the aforementioned demands of NG-PONs [2]. X. Q. Jin demonstrated 128-quaternatry amplitude modulation (QAM) encoded real-time end-to-end optical OFDM transmission with 5.25bits/s/Hz spectral efficiency over 25km SSMF [3]. Recently, Ming Chen developed a 1024-QAM OFDM transmitter with 2.5 GS/s DAC in short reach IMDD system by using expensive external cavity laser and MZM modulator, received signal is demodulated by a digital storage oscilloscope (DSO) with 20 GS/s sampling rate ADC [4].

In this paper, 7.09Gb/s real-time end-to-end OOFDM transmissions over 50km SSMF IMDD systems with adaptive modulation format up to 512-QAM without inline amplifier are experimentally demonstrated, by utilizing low-cost transceiver components including directly modulated DFB lasers and DACs/ADCs with sampling rates as low as 2GS/s.

II. EXPERIMENTAL SETUP

The real-time end-to-end OOFDM DML-based transmission system setup is shown in Fig. 1, transmitter and receiver is realized by using Virtex-6 FPGA from Xilinx, respectively. The transceiver design has a fully pipelined architecture so that data flow is continuous with very limited buffering of OOFDM symbols. In the transmitter, a 64 point IFFT is used to support 64 subcarriers, in which 32 subcarriers carry encoded user data. The highest 2 subcarriers and the 1st subcarrier are dropped due to combined system frequency roll-off and optical beat interference [5]. A parallel pseudo random data source feeds 29 active encoders, each of them encodes data by using a specific signal modulation format and produces a complex value in frequency domain. 64 real-valued samples are then generated by using Hermitian operators for all subcarriers. On-line adaptive bits and power loading is employed on all 29 data-carrying subcarriers with modulation formats taken from 16-QAM to 512-QAM according to the channel condition. After the IFFT, 12.5% CP and 304 points training sequence (TS) are added into OFDM time domain signals. Then the samples are clipped and quantized. The number of quantization bits is set to 10, which matches the resolution of the employed DAC. The clipping ratio also can be adjusted to optimize the transceiver performance. The internal system clock is 125MHz which offers a 125MHz symbol rate under parallel procedure, other detailed transceiver parameters are identical to those implemented in [6]. The 1GHz electrical OFDM signal generated by 2GS/s 10-bit DAC feeds a low pass filter followed by an electrical amplifier to directly drive a 3GHz bandwidth DFB laser at power of 2.1 Vpp. Without inline optical amplifier, 7.5dBm optical power is then injected into 25/50km SSMF.

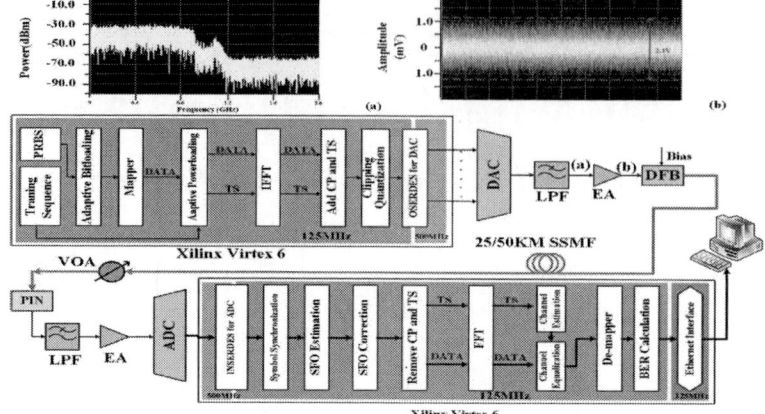

Fig.1 Experimental setup for the 512-QAM real-time OOFDM transmission system

At the receiver side, the optical signal passes through a variable optical attenuator (VOA) to adjust the received optical power (ROP) and is then directly detected by an 8 GHz PIN photodetector to convert the received optical signal into the electrical domain. The electrical OFDM signal passes to an anti-aliasing low pass filter followed by a RF amplifier for adjusting signal power that required by the 2GS/s 10-bit ADC. The receiver's RF power is manually adjusted according to the ROP to maintain an optimum peak-to-peak signal level for ADC and this can be achieved by an automatic gain control (AGC) circuit in practice. The digital signal emerging from the ADC is directed to the receiver's FPGA to proceed with the inverse DSP procedure compared to the transmitter side which including automatic symbol synchronization by using training sequence, sampling frequency offset (SFO) estimation and correction, 64 points FFT, channel estimation and equalization, decoding and error counting. Both transmitter and receiver are operated under asynchronous mode, the clock signals are derived from the on board oscillator of DAC/ADC. Self-developed clock synchronization algorithms are also applied [7].

Making use of the transceiver design and the experimental system setup illustrated in Fig. 1, adaptive bit and power loading are performed on the information-bearing subcarriers to maximize the obtainable signal line rate by effectively compensating for the system frequency response roll-off effect alone with optimizations of the transceiver and system operating conditions. The transceivers parameters and operating conditions are fixed during the system performance measurements.

III. RESULTS AND DISCUSSIONS

Figure 2 illustrates the obtained optimum subcarrier bits allocation profiles corresponding to a total raw signal bit rate of 7.09 Gb/s, of which 6.28 Gb/s can be employed to carry user data because of the 12.5% cyclic prefix. Considering the 906.25MHz effective signal bandwidth, it can be figured out that spectral efficiency is 6.93 bits/s/Hz. Fig. 2 also shows that lower signal modulation formats tend to be applied on the lower and higher frequency subcarriers, which is attributed by the combined effects of strong subcarrier intermixing upon direct detection in the receiver and the residual system frequency response roll-off [5]. The direct detection associated subcarrier intermixing effect is pronounced for subcarriers having low frequencies, and the residual system frequency response roll-off effect is pronounced for subcarriers having high frequencies. The SNR of each subcarrier are depicted in Fig. 2 as well, obviously developing trend of the bits loading profile envelop follows the corresponding system SNR.

For both optical back-to-back and standard single mode fibre (SSMF) IMDD system configurations, Fig. 3 presents the measured bit error rate (BER) performances as a function of ROP for the cases of employing 25km and 50km SSMF. Since EDFA is not involved so that the maximum received optical power can only reach -2.40dBm after 50km SSMF transmission when the launched optical power is fixed at 7.5dBm. BER of 2.7×10^{-3} can be achieved at the ROP of -6.45dBm for the 25km SSMF case and 3.2×10^{-3} for 50km case at the same ROP. Both of mentioned cases are lower than the adopted hard-decision forward-error-correction (HD-FEC) threshold of 3.8×10^{-3}.

In addition, by comparing the 25km case with the 50km, the difference in ROP at the adopted HD-FEC is a direct result of the extra chromatic dispersion caused by the longer fibre span. Due to the restrictions on receiving sensitivity of PIN detector and the gain limitation of the electrical amplifier (EA), BER performance falls sharply when the received optical power continues decreasing. Furthermore, Fig. 3 also shows that very similar BER evolutions for both 25km and 50km SSMF cases, indicating that our proposed scheme can support real-time transmissions of adaptively modulated OOFDM signals in a long-reach fibre link.

OECC/PS2016

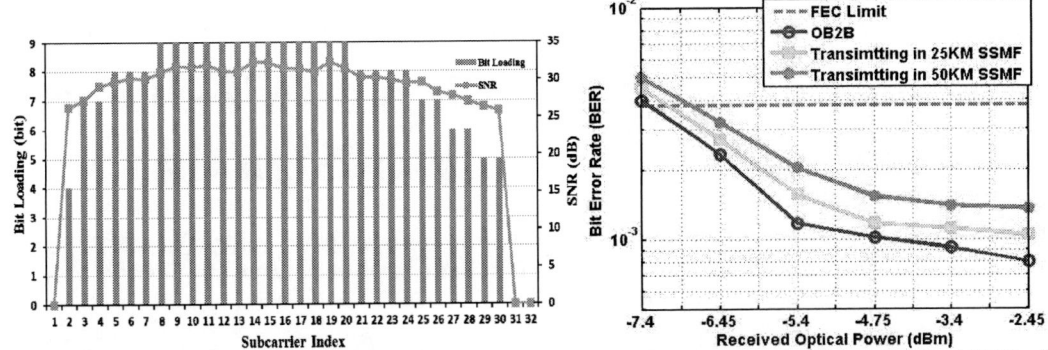

Fig. 2 Bit loading profile and corresponding SNR Fig. 3 BER performance versus ROP in both OB2B and 25/50km SSMF

Received signal constellations of representative subcarriers after channel equalization are plotted in Fig. 4 with modulation format from 16-QAM to 512-QAM. The constellations are measured at their minimum BERs after transmission over 50km SSMF. It also verifies that our current system configuration can achieve spectral efficiency up to 6.93bits/s/Hz.

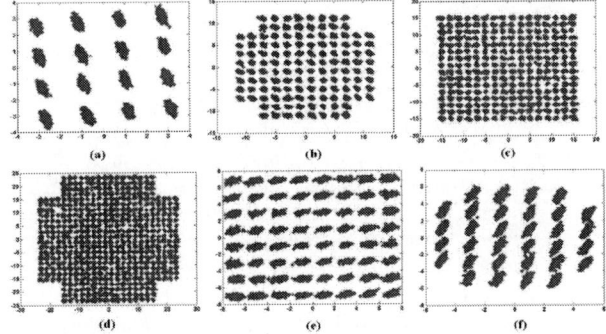

Fig.4 signal constellations for representative subcarrier (a) No. 2 (b) No. 3 (c) No. 5 (d) No. 15 (e) No. 27 (f) No. 30

IV. CONCLUSIONS

Real-time end-to-end OOFDM transmissions over 50km SSMF IMDD systems with adaptively modulated up to 512-QAM modulation format without oversampling or inline amplifier are experimentally demonstrated, by utilizing low-cost transceiver components including directly modulated DFB lasers and DACs/ADCs operating at sampling rates as low as 2GS/s.

ACKNOWLEDGMENT

This work was supported in part by Natural Science Foundation of China: (61132004, 61275073, 61420106011) and Shanghai Science and Technology Development Funds: (13JC1402600, 14511100100, 15511105400, 15530500600).

REFERENCES

[1] J.M. Tang, R.P. Giddings, X.Q. Jin, J.L. Wei, X. Zheng, E. Giacoumidis, E. Hugues-Salas, Y. Hong, C. Shu, J. Groenewald, and K. Muthusamy, "Real-time optical OFDM transceivers for PON applications," Optical Fiber Communication (OFC), p. OTuK3, 2011.

[2] N. Cvijetic, "OFDM for next-generation optical access networks," IEEE Journal of Lightwave Technology, vol. 30, pp. 384-398, 2012.

[3] X.Q. Jin, R.P. Giddings, E. Hugues-Salas and J.M. Tang, "Real-time demonstration of 128-QAM-encoded optical OFDM transmission with a 5.25 bit/s/Hz spectral efficiency in simple IMDD systems utilizing directly modulated DFB lasers," Optics Express, vol. 17, pp.20484-20493, 2009.

[4] M. Chen, J. He, J. Tang, X. Wu, and L. Chen, "Real-Time demonstration of 1024-QAM OFDM transmitter in short-reach IMDD systems," IEEE Photonics Technology Letters, vol. 27, pp. 824-827, 2015.

[5] Q.W. Zhang, E. Hugues-Salas, and Y. Ling, "17.125 Gb/s over 25km transmissions of real-time dual-band optical OFDM signals modulated by 1GHz RSOAs," Optical Fiber Communication (OFC), p. Th3G.6, 2014.

[6] Y.F. Cai, H. Dun, Y.C. Li, Z. Zhang, C. Qian, B.Y. Cao, and Q.W. Zhang, "An effective sampling frequency offset compensation method for OFDMA-PON," Asia Communications and Photonics Conference (ACP) 2015, AM1F.7.

[7] H. Dun, B.Y Cao, C. Qian, Y.C Li, Z. Zhang, and Q.W. Zhang, "A symbol synchronization method based on gold sequences for real-time upstream OFDMA-PON," Asia Communications and Photonics Conference (ACP) 2015, AM1F.3

WA2-26

OECC/PS2016

Dispersion Compensation Scheme for Time-Frequency-Domain Multiplexed 4x/8x Ultra-Wideband Multi-Carrier Signals

Takahide Sakamoto

National Institute of Information and Communications Technology, 4-2-1 Nukui-kitamachi, Koganei, Tokyo 184-8795, Japan
tsaka@nict.go.jp

Abstract: *We investigate an electrical dispersion compensation scheme for time-frequency-domain multiplexed ultra-wideband multi-carrier signals, deriving an equalizing matrix implemented in coherent matched detectors. Provided is numerical proof focusing on dispersion compensation for standard single-mode fiber transmission.*
Keywords: *Optical Multiplexing; Dispersion Compensation*

I. INTRODUCTION

Optical multiplexing techniques for multi-carrier transmission, such as superchannels, orthogonal frequency-division multiplexing (OFDM), enable high-bit-rate and high-spectral-efficiency optical fiber transmission [1–3]. Coherent matched detection is a useful approach for demultiplexing and detecting such ultra-wideband multi-carrier signals, where a target sub-channel is selectively matched detected among other sub-channels [4,5]. The detection technique does not rely on neither tight optical filters/optical FFT circuits for channel selection nor temporal sampling using ultra-short pulses, which greatly simplifies multi-carrier demultiplexing systems. The detection scheme can be applicable to any types of multi-carrier signals if we arrange multiple coherent matched detectors in parallel [5]; however, it fits well with orthogonal time-frequency domain multiplexing (OTFDM) systems because all sub-channels can be naturally demultiplexed through the detection process without additional signal processing [4].

A challenging point for such coherent matched detection is how to digitally implement a functionality of dispersion compensation in the electrical domain. Since all the frequency components at different wavelengths are simultaneously down-converted to the baseband region and they are mixed each other through the coherent matched detection, dispersion compensation technique conventionally used in digital signal processors (DSPs) cannot cancel the signal distortion caused by wavelength dispersion. In our previous paper, we investigated dispersion compensation technique in a special case, a case for dual-channel coherent matched detection, where we found a transfer function for dispersion compensation described in the form of 2x2 rotation matrix [6]. In this paper, we extend the analysis to the general case: a dispersion compensation technique for n-channel coherent-matched detection is clarified. After deriving nxn transfer matrix for dispersion compensation in n-channel coherent-matched detectors, some numerical proofs are provided through analysis on quad-/ octa-carrier OTFDM QPSK transmission over a standard single-mode fiber (SMF).

II. PRINCIPLES

Here, we derive an nxn transfer matrix required for electrical dispersion compensation in coherent matched detection systems.

Fig. 1 shows the basic construction of the n-parallel coherent matched detector, which consists of n sets of coherent matched detectors arranged in parallel. In the system, the coherent matched detectors are driven with n sets of local combs orthogonal each other; each detector independently matches to a particular waveform; the received signal is orthogonally down-converted and decomposed into n sets of slow-speed baseband signals through the n sets of orthogonal coherent matched detectors. The n sets of the baseband signals are digitized and input to n x n matrix section, consisting of three matrices cascaded. The first one is assigned for dispersion compensation, where the transfer matrix, **m**, derived below is applied. The second one has a role of channel-separation from the n-sets of the dispersion-compensated signals, where we can apply matrix tailored for different types of multi-carrier signals. The matrix can be set to unity if we receive OTFDM signals, because coherent matched detectors can exactly match to the received sub-channels even without this matrix in this case. Other impairments, such as residual dispersion, mismatched filtering, are mitigated with the third matrix by using multi-input-multi-output (MIMO) equalization techniques. Carrier phase drift is also canceled here or just after the matrix section using carrier phase estimation technique in the same way as single-carrier coherent receivers.

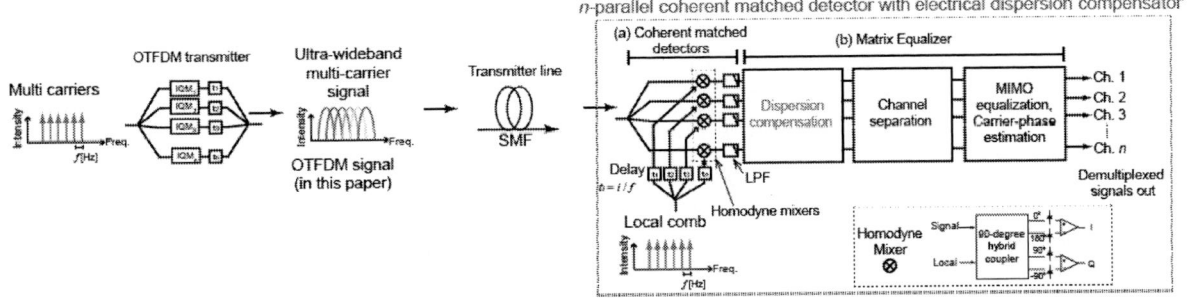

Fig. 1. *n*-parallel coherent matched detector with dispersion compensation functionality (in OTFDM transmission system)

For simplicity, in this analysis, we set the second and third matrices to unity, which means we assume reception of OTFDM signals, and no impairments other than wavelength dispersion are taken into account; however, the analysis below for dispersion compensation will not lose its generality.

An OTFDM signal transmitted over a dispersive fiber, $Y(\omega)$, is described in frequency domain as, $Y(\omega) = \sum_{l=-k}^{k} A(\omega) S_l(\omega) * C_l(\omega)$, where $S_l(\omega)$ is a baseband (IQ) data at the l-th channel; $C_l(\omega) \equiv \sum_{i=-k}^{k} \delta(\omega - \omega_i) e^{j\left(\theta_i + \frac{2il\pi}{n}\right)}$ describes multi carriers to carry the signal, where ω_i, θ_i are angular frequency and phase offset at the i-th carrier, respectively; n is the total number of the multi carriers, equivalent with number of sub-channels. In the equation, $A(\omega)$ stands for transfer function of the dispersive fiber defined as $A(\omega) \equiv \exp j\left(\frac{1}{2}\beta_2 L \omega^2\right)$, where β_2, L are group-velocity dispersion (GVD) and fiber length, respectively. The fiber transfer function is expanded around each carrier frequency, ω_i, like

$$A(\omega) = A(\Delta\omega) \exp j\left\{\beta_2 L\left(\omega_i \Delta\omega + \tfrac{1}{2}\omega_i^2\right)\right\} \equiv A(\Delta\omega) G_i(\Delta\omega). \tag{1}$$

Since coherent matched detected signal at the l-th sub-channel is described as $R_l(\omega) = Y(\omega) * C_l^*(\omega)$, the transfer function from the source signals to the coherent matched detected baseband signals yields,

$$\mathbf{T}(\omega) = A(\omega)
\begin{bmatrix}
\sum_k G_k & \sum_k G_k e^{j\frac{2k\pi}{n}} & \sum_k G_k e^{j\frac{4k\pi}{n}} & \cdots & \sum_k G_k e^{j\frac{2(n-1)\pi}{n}} \\
\sum_k G_k e^{-j\frac{2k\pi}{n}} & \sum_k G_k & \sum_k G_k e^{j\frac{2k\pi}{n}} & \cdots & \sum_k G_k e^{j\frac{2(n-2)\pi}{n}} \\
\vdots & & & \ddots & \vdots \\
\sum_k G_k e^{-j\frac{2(n-1)\pi}{n}} & \sum_k G_k e^{-j\frac{2(n-2)\pi}{n}} & \sum_k G_k e^{-j\frac{2(n-3)\pi}{n}} & \cdots & \sum_k G_k
\end{bmatrix}, \tag{2}$$

The transfer matrix for dispersion compensation applied in the coherent matched detectors should be the inverse of $\mathbf{T}$, which results in

$$\mathbf{M} = \mathbf{T}^{-1}(\omega) = \mathbf{T}^{\dagger}(\omega) = A^*(\omega)
\begin{bmatrix}
\sum_k G_k^* & \sum_k G_k^* e^{j\frac{2k\pi}{n}} & \sum_k G_k^* e^{j\frac{4k\pi}{n}} & \cdots & \sum_k G_k^* e^{j\frac{2(n-1)\pi}{n}} \\
\sum_k G_k^* e^{-j\frac{2k\pi}{n}} & \sum_k G_k^* & \sum_k G_k^* e^{j\frac{2k\pi}{n}} & \cdots & \sum_k G_k^* e^{j\frac{2(n-2)\pi}{n}} \\
\vdots & & & \ddots & \vdots \\
\sum_k G_k^* e^{-j\frac{2(n-1)\pi}{n}} & \sum_k G_k^* e^{-j\frac{2(n-2)\pi}{n}} & \sum_k G_k^* e^{-j\frac{2(n-3)\pi}{n}} & \cdots & \sum_k G_k^*
\end{bmatrix}. \tag{3}$$

III. NUMERICAL ANALYSIS

Here, through numerical analysis on OTFDM transmission over a SMF, we prove the dispersion compensation matrix digitally implemented in coherent matched detectors can restore distorted signal. Followings are the configuration and conditions. On the transmitter side, 4x 8x OTFDM QPSK signals are generated by using quad- or octa-parallel modulators. First, an optical comb is generated by using a Mach-Zehnder modulator based flat comb generator (MZ-FCG), where optical comb is generated from a continuous-wave (CW) laser source applying a large-amplitude driving signal on a dual-drive Mach-Zehnder modulator [7]. Frequency spacing of the generated comb is set at 10 GHz; the average induced phase shift and dc phase offset are assumed to be 4π and 0.25π, respectively. The generated comb is split in 4 or 8 and each of them is QPSK-modulated at a symbol rate of 10 Gbaud with an inphase-/quadrature- (IQ) modulator. The baseband signals for QPSK modulation in each arm is generated from independent 2^{15}-1 PRBS binary sources uncorrelated enough each other. The 4 or 8 sets of QPSK modulated comb lights are combined again with an optical combiner after giving delays of $25i$ (for 4x) or $12.5i$ (for 8x) [ps] to each light, where i is the channel index.

OECC/PS2016

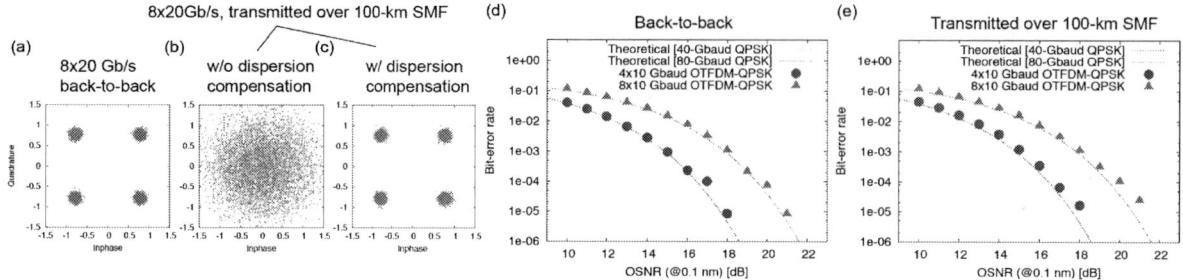

Fig. 2. (a)-(c): Constellations of 8x20-Gbps OTFDM-QPSK signals; (a) back-to-back, (b) 100-km SMF transmitted w/o dispersion compensation, (c) w/ dispersion compensation; (d) BER characteristics for back-to-back, (e) transmitted over 100-km SMF

The 4x/8x 10-Gbaud OTFDM-QPSK signal is transmitted over 100-km SMF with an input power of 1 mW. Followings are the parameters for the SMF. $\beta_2 = -20$ [ps²/km], $L = 100$ [km], $\alpha = 0.2$ [dB/km], $\gamma = 0.26$ [W⁻¹km⁻¹]. Split-step Fourier method is utilized for simulating SMF transmission.

On the receiver side, in the analysis, we assume to use 4- or 8-parallel coherent matched detector. In the coherent matched detector, optical local comb is generated with another MZ-FCG driven under the same condition mentioned above. All of the local combs for coherent matched detectors are generated from the common source and delays of $25i$ (for 4x) or $12.5i$ (for 8x) [ps] are given before input to each local port of the coherent matched detector. In each coherent matched detector, the received signal is homodyne mixed with the local comb by using a 90-degree hybrid coupler followed by balanced detectors for detecting I and Q components. The coherent matched detected signals (photo- detected signals) are digitized and led to the matrix equalizer section. In the section, the 4x4 or 8x8 equalizing matrix derived in Eq. 3 is applied first; then led to the third MIMO section skipping the second matrix for channel separation because OTFDM signal we receive does not require it. The MIMO section also does not contribute a lot because the model we analyze does not include impairments other than wavelength dispersion in a fiber; however, carrier phase is estimated at its output for canceling laser phase drift.

Figs. 2(a)-(c) are constellations calculated at an OSNR (at 0.1 nm) of 30 dB, *i.e.* 21 dB per channel. From the left side, the plots correspond to (a) back-to-back, (b) 100-km transmitted w/o dispersion compensation and (c) w/ compensation, respectively. Without the dispersion compensation, the signal was highly distorted. Applying the dispersion compensation, constellations were recovered as clear as those of back-to-back.

In Figs. 2(d)(e), bit-error rates (BERs) are calculated against OSNR (@0.1nm). Plots (d) and (e) in the figure indicate BER characteristics for back-to-back and 100-km SMF transmitted cases, respectively. In the analysis, OTFDM dispersion compensation is always applied after SMF transmission. In each plot, red triangles and blue dots stand for BERs for 8x and 4 x 10 Gbaud OTFDM-QPSK, respectively. Theoretically expected BER traces for QPSK at the same symbol rates as those of OTFDM-QPSK signals are also plotted in the same graph. It has been clearly seen that almost no OSNR penalty is appeared after the 100-km transmission and BERs stays close to the theoretical limit, in the analysis.

IV. CONCLUSIONS

We have investigated electrical dispersion compensation scheme in coherent matched detectors. Through numerical simulation on quad-/octa-carrier OTFDM signal transmission over 100-km SMF, it has been shown that dispersion is compensated well in a DSP using the analytically derived $n \times n$ matrix for equalization.

ACKNOWLEDGMENT

The author would like to thank to Dr. N. Yamamoto at NICT and Dr. Kawanishi, currently with Waseda Univ., for discussion and support. This work was partly supported by JSPS Grant-in-Aid for Scientific Research (B),15H04001.

REFERENCES

[1] A. Lowery, *et al, Optical Fiber Communication Conference (OFC2006)* PDP39 (2006).

[2] A. Sano *et al, European Conference on Optical Communication (ECOC 2007)*, PDP 1.7 (2007).

[3] S. Chandrasekhar *et al, European Conference on Optical Communication (ECOC 2009)*, PD2.6 (2009).

[4] T. Sakamoto *et al, European Conference on Optical Communication (ECOC2011)*, We.10.P1.77, (2011).

[5] T. Sakamoto *et al, OSA Optics Express*, 21, 16, 18602-18610 (2013).

[6] T. Sakamoto, *OSA Conference on Laser and Electro Optics (CLEO 2015)*, SW3M.3 (2015).

[7] T. Sakamoto *et al, Opt. Lett.* 32(11), 1515–1517 (2007).

Real-time visible light communication system based on 2ASK-OFDM coding

Yuankai Xue, Yinan Hou, Shilin Xiao*, Yunhao Zhang, Lu Zhang

State Key Laboratory of Advanced Optical Communication System and Networks,
Department of Electronic Engineering, Shanghai Jiao Tong University, Shanghai 200240, China
* Corresponding author: slxiao@sjtu.edu.cn

Abstract: *We demonstrate a real-time VLC system using 2ASK-OFDM coding for high-speed communication and communication in positioning simultaneously. The proposed 2ASK-OFDM modulation and demodulation are performed using FPGA and results verify the feasibility of the system.*

Keywords: *2ASK-OFDM, 2ASK, OFDM, real-time VLC system.*

I. INTRODUCTION

Due to the abundant frequency spectrum and no electro-magnetic interference etc. [1], visible light communication (VLC) has rapidly attracted significant attention as a promising candidate for optical wireless communication [2]. For VLC, high-speed communication [3] and high-accuracy positioning [4] are researched as two independent techniques and have made great achievements in both fields.

Since the demands for accessing multimedia services and positioning anywhere have grown rapidly, the tendency to combine high-speed communication and high-accuracy positioning in one system becomes inevitable. However, few efforts have been made to integrate these two techniques. For high-speed communication, high frequency orthogonal frequency division multiplexing (OFDM) signal etc. is employed, making the receiver end complicated. For high-accuracy positioning, low frequency signal with simple modulation is used as the beacon signal in most positioning schemes, which makes the receiver more portable. When combing two techniques, the two types of signals should be integrated and emitted by one LED lamp. Besides, for practical usage, the signals have to be demodulated separately in real time at the receiver ends. Although a coding method has been proposed for service integration in optical fiber [5], the method is designed for off-line system and hasn't been verified in VLC.

In this paper, we experimentally demonstrate a real-time VLC system based on 2ASK-OFDM coding. In our system, low frequency (600Kbit/s) 2ASK signal and high frequency (76Mbit/s) OFDM signal are integrated and modulated to the drive current of one LED. At receiver end, the signals are separately demodulated by two receivers for different applications. All signal processing is performed in Field Programmable Gate Array (FPGA). The feasibility of our system is verified by experiments and the results indicate that the resultant BERs of 2ASK signal and OFDM signal are 3×10^{-7} and 2.7×10^{-8} when the radial distance is 1.5m, much less than forward-error-correction (FEC) limit (3.8×10^{-3}). Therefore, our proposed system is promising to be used for high-speed communication and communication in high-accuracy positioning simultaneously.

II. PRINCIPLE AND IMPLEMENTATION OF 2ASK-OFDM CODING

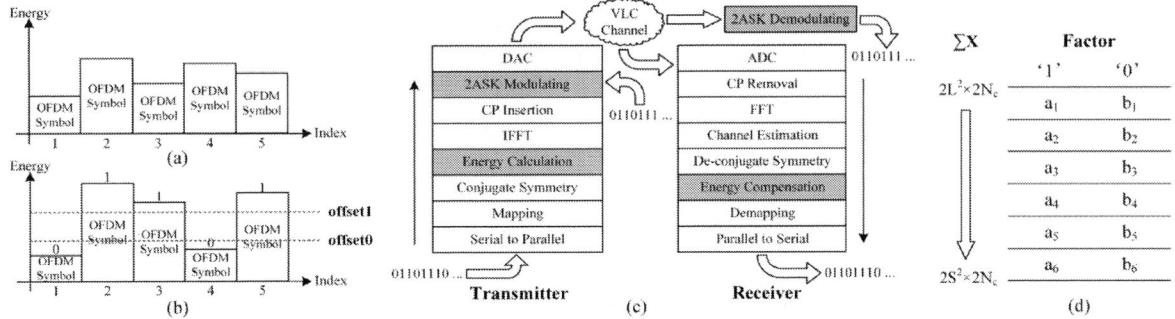

Fig. 1. (a) Energy of original OFDM symbols (b) Energy of OFDM symbols after 2ASK-OFDM modulating (c) Structure of the real-time VLC system (d) Scheme of choosing proper factors.

Fig. 1(a) and 1(b) illustrate the basic idea of 2ASK-OFDM modulating process. In Fig. 1(a), five OFDM symbols are assumed. To amplify or reduce the energy of OFDM symbols for denoting logic '1' and '0' as shown in Fig. 1(b), the data of each OFDM symbol are multiplied with a positive factor. Sometimes, the energy difference between '1' and '0' isn't large enough to ensure a correct decision. Therefore, two offsets (offset1 and offset0 in Fig. 1(b)) are employed to enlarge the difference, which will make the factors more proper. Based on the method above, a lower frequency 2ASK

799

signal can be modulated onto the OFDM signal without adding another modulating system and can be directly detected using envelope detection at the receiver end. Notice that we don't require the energy of '1's or '0's to be the same, which will avoid square root operation by not doing normalization. However, efforts will be made to reduce the energy difference between '1's or '0's in our scheme.

Although the principle of 2ASK-OFDM is quite simple, how to make it work in a real-time system remains a big problem. For FPGA, operations such as multiplication, division and square root are too complicated to realize. Therefore, only addition, subtraction and shifting are used in the system.

Figure 1(c) illustrates the structure of our proposed system. The blocks in white represent the conventional processes for OFDM, while the ones in gray are added for 2ASK-OFDM coding processes. For energy calculation, an approximation algorithm is proposed to avoid multiplication and division according to the following deduction. The energy of one OFDM symbol is defined in Eq. (1), where x[n] represents the data after inverse fast Fourier transform (IFFT). N and l are the length of IFFT and cyclic prefix (CP). E_{CP} denotes the energy of CP.

$$E_{OFDM\,Symbol} = \sum_{n=0}^{N-1}|x[n]|^2 + \sum_{n=N-l}^{N-1}|x[n]|^2 = \sum_{n=0}^{N-1}|x[n]|^2 + E_{CP} \quad (1)$$

Following Parseval's theorem of the discrete Fourier transformation (DFT) is defined in Eq. (2).

$$\sum_{n=0}^{N-1}|x[n]|^2 = \frac{1}{N}\sum_{k=0}^{N-1}|X[k]|^2 \quad\quad\quad (2)$$

where X[k] is the DFT of x[n] , both of length N. If the length of CP is short enough, we can use $\sum_{n=0}^{N-1}|x[n]|^2$ to estimate the symbol energy. Therefore, Eq. (1) can be simplified to be Eq. (3), where $\sum X$ is used to denote $\sum_{k=0}^{N-1}|X[k]|^2$ for simplicity.

$$E_{OFDM\,Symbol} \approx \frac{1}{N}\sum_{k=0}^{N-1}|X[k]|^2 = \frac{1}{N}\sum X \quad\quad\quad (3)$$

It is easy to calculate $\sum X$ because both real and imaginary components of X[k] are equal to mapping values and the square of mapping values can be calculated in advance. Since N is a constant for the system, we use $\sum X$ to perform the functionality of $E_{OFDM\,Symbol}$ for simplicity later. For this reason, only addition is needed for energy calculation.

Fig. 1(d) shows the scheme of choosing proper factors when modulating 2ASK signals. L and S represent the maximum and minimum of the absolute mapping values, respectively. N_c denotes the number of used subcarriers. As illustrated in Fig. 1(d), the value of $\sum X$ ranges from $2S^2 \times 2N_c$ to $2L^2 \times 2N_c$. Then we divide $[2S^2 \times 2N_c, 2L^2 \times 2N_c]$ into six equal parts and assign two factors a_n and b_n to each small interval (a_n is for '1' and b_n is for '0') to narrow the energy difference between '1's or '0's. The choice of a_n and b_n should follow two rules: (1) a_n (b_n) multiplied with the lower (upper) bound value of the small interval must larger (smaller) than offset1 (offset0); (2) a_n and b_n can be expressed as $x_1 + x_2/2 + x_3/4 + x_4/8$. The first rule ensures that the energy of each OFDM symbol located in the small interval will meet the requirements of two offsets. The second rule makes the 2ASK modulating process contain only addition and shifting.

To compensate the symbol energy, one additional pilot is added when doing conjugate symmetry. Then the receiver will know which factor has been multiplied to the current data according to the pilot value, and use the reciprocal of this factor to compensate the energy. To avoid using multiplication and division, we use $x_1 + x_2/2 + x_3/4 + x_4/8 + ...$ to approximate the reciprocal and the error should be smaller than 0.01 at the same time.

III. EXPERIMENT AND RESULT

The experimental setup is illustrated in Fig. 2(a). For digital parts, Xilinx Kintex-7 FPGA KC705 and 4DSP FMC150 are used as FPGA development board and DAC/ADC board respectively. In our system, we use digital method to do envelope detection for simplicity. For VLC link, a white LED (OSTAR R LE CW E3B) is deployed and a pre-equalization circuit is added to the LED transmitter to improve the frequency response of the optical channel. A commercially available photo detector (THORLABS PDA10A) is placed behind the blue filter and an optical convex lens. The 3dB bandwidth of VLC link can be achieved to 40MHz. When peak-to-peak value of the input signal for LED transmitter is fixed to 1V, the relationship between Signal-to-Noise-Ratio (SNR) and radial distance of the optical channel is shown in Fig. 2(b).

In our system, 16-QAM is used to do the mapping process. The size of FFT/IFFT is 128 and the length of CP is 12. 30 subcarriers are used for the data and one is used for the pilot. Two OFDM training sequences are taken every 50 OFDM symbols. The data rates of OFDM and 2ASK are about 76Mbit/s and 600Kbit/s. Absolute mapping values L and

S are set to 3 and 1 respectively. Then the value of the corresponding lines in Fig. 1(d) can be achieved as (from the top to the bottom): 1080, 920, 760, 600, 440, 280 and 120. And offset1 and offset0 are set to 920 and 280. The factors a1~a6 are 1, 1.125, 1.25, 1.5, 1.875, 2.875 and b1~b6 are 0.5, 0.5, 0.5, 0.625, 0.75, 1 by calculation.

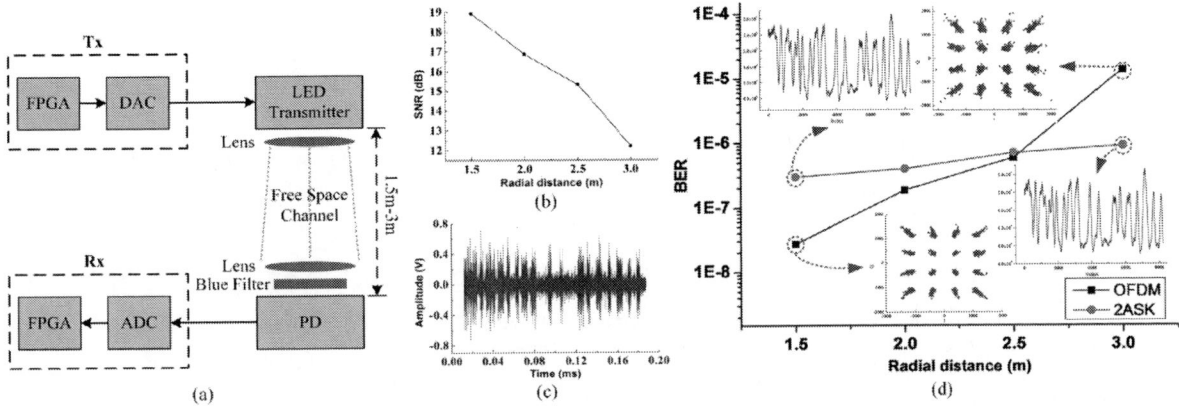

Fig. 2. (a) Experimental setup of the real-time VLC system (b) Measured SNR versus radial distance (c) Signal waveform of DAC's output (d) BER performances versus radial distance.

In our experiments, we use pseudorandom data as the original bit stream for both OFDM and 2ASK to test the system performance. The signal waveform of DAC's output is illustrated in Fig. 2(c). It can be seen the average magnitudes of OFDM symbols vary as expectation. The BER performances of the system at different radial distance are tested. As shown in Fig. 2(d), the resultant BERs of 2ASK signal and OFDM signal are 3×10^{-7} and 2.7×10^{-8} when the radial distance is 1.5m. And BERs of 9×10^{-7} and 1.3×10^{-5} are achieved at the distance of 3m. The accuracy loss of energy compensation has caused distortion of the constellation diagrams while the effect to demodulation is few. As the radial distance increases, the BER performances of both signals are getting worse. However, OFDM signal is much more sensitive to the distance than 2ASK signal. The BER values of both signals are still much smaller than FEC limit at the radial distance of 3m.

IV. CONCLUSIONS

In this work, we have experimentally demonstrated a real-time VLC system based on 2ASK-OFDM coding method. In our system, both low frequency 2ASK signal and high frequency OFDM signal are integrated into one 2ASK-OFDM signal and the two signals are demodulated from the integrated signal separately at the receiver ends. The modulation and de-modulation processes of the system are conducted using FPGA in real time. The resultant BERs of 2ASK signal and OFDM signal are 3×10^{-7} and 2.7×10^{-8} when the radial distance is 1.5m. The experimental results have verified that our proposed system is capable of combining 2ASK signal and OFDM signal to perform the communication simultaneously in high-speed communication and high-accuracy positioning of VLC.

ACKNOWLEDGMENT

The work was jointly supported by the National "863" Hi-tech Project of China (2013AA013602), the National Nature Science Fund of China (No. 61271216, No. 61221001, No. 61090393 and No. 60972032) and the National "973" Project of China (No.2010CB328205, No. 2010CB328204 and No. 2012CB315602).

REFERENCES

[1] P.P. Salian et al., "Visible light communication," 2013 Texas Instruments India Educators' Conference (TIIEC), pp. 379-383, April 2013.

[2] S.-M. Kim and S.-M. Kim, "Wireless visible light communication technology using optical beamforming," Optical Engineering, vol. 52, no. 10, pp. 106101, October 2013.

[3] L. Grobe et al., "High-speed visible light communication systems," IEEE Communications Magazine, vol. 51, no. 12, pp. 60-66, December 2013.

[4] S.-H. Yang, H.-S. Kim, Y.-H. Son, and S.-K. Han, "Three-dimensional visible light indoor localization using AOA and RSS with multiple optical receivers," Journal of Lightwave Technology, vol. 32, no. 14, pp. 2480-2485, July 2014.

[5] J.-Y. Sung, C.-W. Chow, C.-H. Yeh, and Y.-C. Wang, "Service integrated access network using highly spectral-efficient MASK-MQAM-OFDM coding," Optics Express, vol. 21, no. 5, pp. 6555-6560, March 2013.

WA2-28

OECC/PS2016

Spectral Overlap of Two Bandwidth Variable Nyquist-WDM Signals to Resolve Wavelength Conflict in Elastic Optical Networks

Shuang Gao[1]*, Lingchen Huang[2], Chun-Kit Chan[1]

(1) Department of Information Engineering, The Chinese University of Hong Kong, Shatin, Hong Kong SAR
(2) Center for Optical and Electromagnetic Research, Zhejiang University, Hangzhou, China
*Email address: gs012@ie.cuhk.edu.hk

Abstract: We investigate a simple spectral overlap method to resolve the wavelength conflict for two independent bandwidth variable Nyquist-WDM signals by relative power control. Individual signals can be separated and recovered by digital signal processing.
Keywords: *Optical communications; Networks, wavelength assignment*

I. INTRODUCTION

Nyquist signal based wavelength division multiplexing (Nyquist-WDM) is a promising solution to enable the future elastic optical networks (EON), since it occupies minimal bandwidth, which is close to the signal's baud rate, and requires relatively small inter-channel guard band. Hence, it achieves much higher spectral efficiency, compared with traditional modulation format without pulse shaping applied in optical system. In EON, the usable spectrum in fiber is quantized into a finite number of contiguous frequency slot units (FSUs) [1]. With the increasing demands for network capacity, wavelength contention may occur due to the limited spectrum resource for network routing.

To resolve the possible wavelength conflict issue, wavelength conversion technologies are proposed [2]. One conventional approach adopts the optical-electrical-optical (O-E-O) conversion, in which the optical signal needs de-multiplexing and regeneration onto other available FSUs with tunable laser. The nonlinear optical effect is another method for wavelength conversion, in which wavelength select switch (WSS) and pump lasers are required. The pump lasers demands precise polarization to achieve the optimal conversion efficiency, especially for polarization-multiplexed signals [3].

In this paper, a simple and cost-effective spectral overlap method of two independent optical Nyquist polarization-multiplexed quaternary phase-shift keying (Nyquist-PM-QPSK) channels is investigated, as a feasible solution to resolve wavelength conflict. The spectral overlap method is realized by combining two optical channels with relative power control as in [4]. Herein, we mainly analyze the impacts of time offset and relative polarization state on the overlapped system performance, via numerical simulations. Furthermore, since one optical channel can occupy more than more FSUs in EON, the system performance of the case of combining two optical channels with different bandwidths, is also characterized.

II. OPERATION PRINCIPLE

Fig.1 (a) illustrates an example of wavelength conflict in network routing. Two channels, denoted by S_d and S_w, transmit from their respective source nodes A and B and contend for the same output fiber link CD. At node C, their respective signal powers, P_d and P_w, are carefully adjusted such that the dominant channel S_d is set to have relatively higher power than the weak channel S_w, before being optically combined, via an optical coupler, to form the output overlapped composite signal S_o. Then, S_o is transmitted over the fiber link CD, before being power-split for further

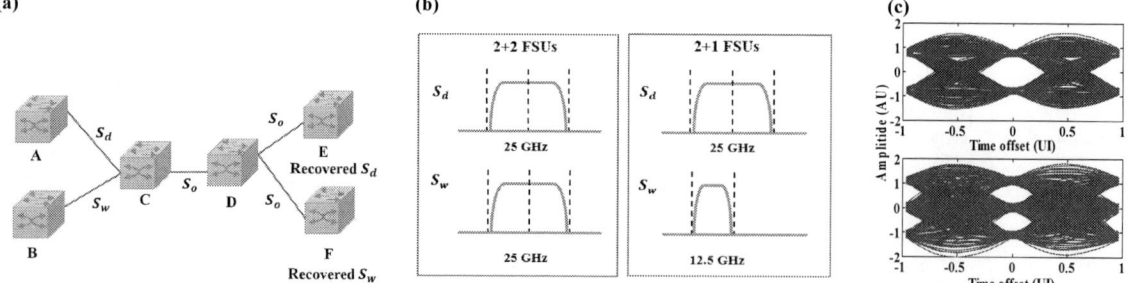

Fig. 1. (a) Network scenario for the conflicted two Nyquist signals; (b) Spectrum for S_d and S_w, for case 1 S_d and S_w occupy 2 FSUs, for case 2, S_d and S_w occupy 2 FSUs and 1 FSU, respectively; (c) eye diagrams for Nyquist-QPSK signal (up) and after 45-degree relative polarization rotation (down).

transmissions to their respective destination nodes E and F. As shown in Fig.1 (b), we consider two independent Nyquist-PM-QPSK signals with variable bandwidth. Both S_d and S_w occupy two FSUs as case "2+2 FSUs". In "2+1 FSUs" case, S_d and S_w occupy two FSUs and one FSU, respectively.

At node E, S_d is going to be recovered and demodulated, while the component of S_w contained inside S_o is treated as the interference. The recovery of S_d involves several steps. First, the acquired signal S_o is filtered by a digital fifth-order Bessel filter and is re-sampled to 2 samples per symbol, with respect to S_d. Timing phase recovery based on Gardner timing error detector (TED) is employed. Then, the signal is polarization de-multiplexed and equalized based on the constant modulus algorithm (CMA). The carrier is synchronized by Viterbi-Viterbi (VV) frequency estimation and blind phase search (BPS) method. Hence, S_d can be retrieved and forward error corrected. At node F, in order to recover S_w, the 2-samples-per-symbol version of the recovered S_d is reconstructed by Nyquist filtering, and is subtracted from the carrier synchronized signal S_o. At last, S_w can be recovered via the CMA, VV and BPS algorithms in sequence. For the "2+1 FSUs" case, additional frequency shift is applied before the second round of signal processing.

The two channels S_d and S_w suffer from strong interference after optical combination because of the spectral overlap. The signal-to-interference ratio (SIR) is defined as the optical signal power ratio of S_d to S_w ($SIR = P_d / P_w$). To ensure successful demodulation of S_d, the parameter SIR should be large enough. However, proper signal recovery for S_w requires small value of the SIR in S_o. Therefore, the SIR has to be optimally adjusted such that both tributary signals can be properly recovered, simultaneously.

III. NUMERICAL SIMULATION

In the numerical simulation, the FSU width was set to 12.5 GHz [1]. Two independent Nyquist-PM-QPSK signals, namely S_d and S_w, were generated, with the roll off factor of 0.2 and Nyquist filter length of 128 symbols. Then, the two channels were optically combined to form the composite signal S_o with relative power control. The OSNR of S_o was varied by adding white Gaussian amplified spontaneous emission (ASE) noise P_{ASE} (in 0.1-nm bandwidth), where $OSNR_o = (P_d + P_w)/P_{ASE}$, for performance characterization. At the receiver, S_o was acquired by coherent detection. The transmitter and local oscillator lasers were independent with 100-kHz linewidth.

As shown in Fig. 1(c), it could be noted that the relative time offset and polarization rotation between the two optical channels would influence the signal temporal characteristics, varying the interference amplitude between S_d and S_w. Thus, two set of conditions have been considered The first case, denoted as Case 1, had zero time offset and matched polarization state between S_d and S_w; while, in the second case, denoted as Case 2, time offset was half of the symbol period and the relative polarization rotation was 45-degree since the interference amplitude between S_d and S_w reached the maximum.

A. Spectral Overlap for 2+2 FSUs

For "2+2 FSUs", the baud rates of both S_d and S_w were 20 GBd with 0.2 roll off factor. Fig. 2(a) shows the bit-error-rate (BER) curves for S_o, S_d and S_w, as a function of $OSNR_o$. The BER of S_o is defined as the average BER of the recovered S_d and S_w, before applying forward error correction. It could be optimized by properly adjusting the SIR, at a given $OSNR_o$ value. As shown, under the optimal SIR, the required OSNR for S_o was 18.8 and 20.5 dB, at a BER of 3.8×10^{-3}, for both Case 1 and Case 2, respectively. Besides, the BER performance of the 20-GBd Nyquist PM-16QAM, as a function of OSNR was also depicted, for comparison. With respect to the case of PM-16QAM, the OSNR penalties for S_o were 0.3 dB and 2 dB, respectively. As shown in the insets of Fig. 2(a), the constellation diagrams for S_d and S_w in Case 1 were more clear than those in Case 2. The constellation of S_d in Case 1 was close to four "rings", since the two optical channels were timing and polarization aligned.

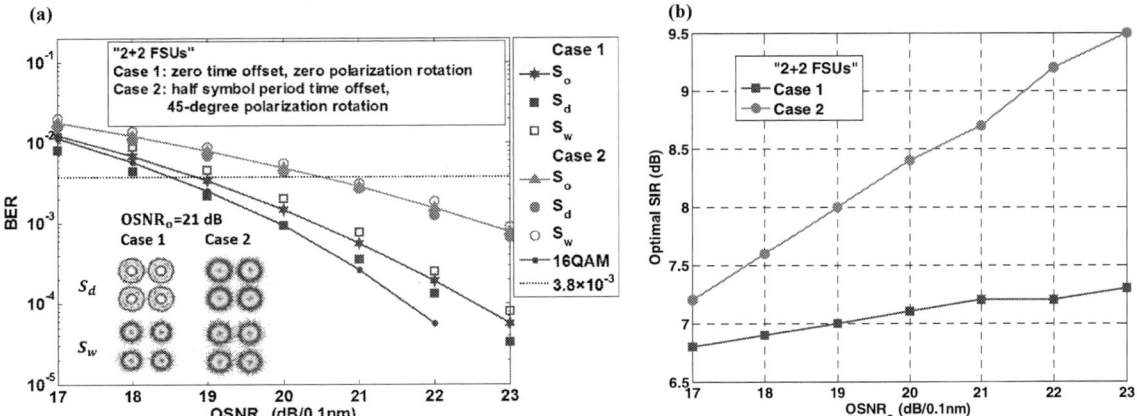

Fig. 2. Simulation results for "2+2 FSUs": (a) the calculated BERs for S_o, S_d and S_w as a function of $OSNR_o$ for Case 1 and Case 2, calculated BER for 20GBd Nyquist PM-16QAM case for reference, the constellations diagrams for two cases under the $OSNR_o$ 21 dB; (b) the optimal SIR as a function of $OSNR_o$ for Case 1 and Case 2.

As shown in Fig. 2(b), the optimal SIR for Case 1 was around 7 dB with less than 0.3 dB deviation. However, for Case 2, the optimal SIR increased significantly with respect to $OSNR_o$. From the results, it could be noticed that the two channels S_d and S_w could be separated and recovered from the overlapped composite signal S_o. Since the two channels had the same baud rate, we could see that the time offset and the relative polarization rotation would obviously influence the interference characteristics. With the increased interference amplitude in Case 2, a larger SIR was required at the same $OSNR_o$. Such optimal SIR in Case 2 was applied as the power control strategy, to assure successful recovery of S_d and S_w since the relative time offset and polarization state were unknown in most practical applications.

B. Spectral Overlap for 2+1 FSUs

For "2+1 FSUs", the baud rate of S_d was set to 20 GBd, while that of S_w was 10 GBd. Two set of conditions have been analyzed. That is, Case 1 referred to aligned polarization states for S_d and S_w; while Case 2 referred to 45-degree relative polarization rotation. The time offset between S_d and S_w had little impact on the BER performance since the two optical channels had different baud rates, and thus was not considered herein.

Fig. 3(a) shows the BER performance for S_o, S_d and S_w as a function of $OSNR_o$. Under the optimal SIR, the required OSNR for S_o was 18.8 and 19.3 dB at BER of 3.8×10^{-3}, for both Case 1 and Case 2, respectively. It could be observed that the effect of the relative polarization rotation was not significant. Since the two channels had different baud rates, misaligned timing resulted in high interference amplitude. The fact of having two optical channels, S_d and S_w centering at different wavelengths, is another reason leading to the relatively high OSNR requirement. As shown in Fig. 3(b), the optimal SIR increased with respect to $OSNR_o$ in both two cases. In Case 2, about 0.4 to 0.7 dB larger optimal SIR than that of Case 1 was required.

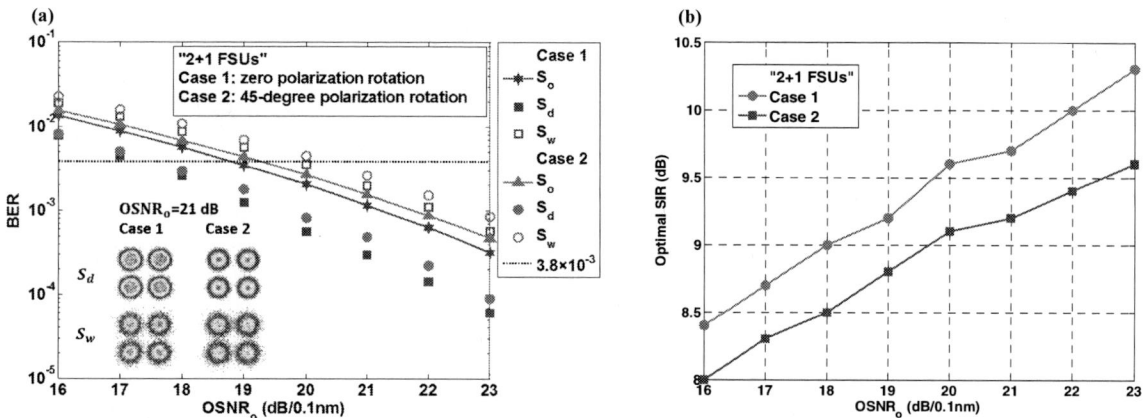

Fig. 3. Simulation results for "2+1 FSUs": (a) the calculated BERs for S_o, S_d and S_w as a function of $OSNR_o$ for Case 1 and Case 2, calculated BER for Nyquist PM-16QAM case for reference, the constellations diagrams for two cases under the $OSNR_o$ 21 dB; (b) the optimal SIR as a function of $OSNR_o$ for Case 1 and Case 2.

IV. SUMMARY

We investigate a cost-effective and simple complete spectral overlap method so as to resolve the possible wavelength contention issue during wavelength routing of two independent bandwidth variable Nyquist-PM-QPSK channels. The two channels can be recovered individually, via proper digital signal processing techniques. The performance of the overlapped system has been characterized, via numerical simulations, under the influence of time offset and relative polarization rotation between the two channels. The power control strategy for two optical channels has also been proposed. This work was partially supported by a research grant from Hong Kong Research Grants Council (Project No. 14200614).

REFERENCES

[1] O. Gerstel, M. Jinno, A. Lord and S. B. Yoo, "Elastic optical networking: a new dawn for the optical layer?," IEEE Commun. Mag., vol. 50, no. 2, pp. s12-s20, Feb. 2012.

[2] S. J. B. Yoo, "Wavelength conversion technologies for WDM network applications," J. Lightw. Technol., vol. 14, no.6, pp. 955-966, Jun. 1996.

[3] J. Lu, L. Chen, Z. Dong, Z. Cao and S. C. Wen, "Polarization insensitive wavelength conversion based on orthogonal pump four-wave mixing for polarization multiplexing signal in high-nonlinear fiber," J. Lightw. Technol., vol. 27, no.24, pp. 5767-5774, Dec. 2009.

[4] G. Meloni, T. Foggi, F. Paolucci, F. Fresi, F. Cugini, P. Castoldi, G. Colavolpe, and L. Potí, "First demonstration of optical signal overlap," Proc. NETWORKS, Jul. 2014, pp. 1-3, paper NM4D.4.

WA2-29

Highly Linear Optical W-band Receiver Front-End based on Optical Processing

Ziang Li, Shangyuan Li, Xiaoping Zheng,* Jinhui Luo, Hanyi Zhang, and Bingkun Zhou

Tsinghua National Laboratory for Information Science and Technology, Department of Electronic Engineering,
Tsinghua University, Beijing, 100084, China
*xpzheng@tsinghua.edu.cn

Abstract: We proposed a highly linear optical W-band receiver front-end based on optical processing. With phase-shift and attenuation on optical sidebands, an improved spur-free dynamic range of 108.6dB·Hz$^{2/3}$ is obtained in the two-tone test.
Keywords: W-band receiver, optical processing, down-conversion, spur-free dynamic range.

I. INTRODUCTION

The W-band (75-110GHz) millimeter-wave (MMW) which offers the atmosphere window around 94GHz and eliminates the inherent high propagation-loss of MMW, has remarkable advantages on MMW imaging [1], pulsed-radar [2] and high-capacity wireless backhaul network [3]. Unfortunately, difficult manufacturing processing and high loss of W-band devices severely reduce the sensitivity of the W-band receiver. To compensate such loss, multistage amplification is required in real applications, resulting in high cost, increased system complexity and poor expandability in existing systems. As some researchers noted [3, 4], optical W-band receiver font-end is a promising technique to overcome these issues by modulating microwave signal on optical carrier and processed them in optical domain.

The spur-free dynamic range (SFDR) is an important figure to estimate the performance of the W-band receiver, which is susceptible to nonlinear distortions, gain and noise. Numerous approaches have been approved to suppress the third order intermodulation distortion (IMD3) and improve SFDR in the last decade. Complex modulators, such as dual parallel MZM (DPMZM) [5] and dual-electrode MZM (DEMZM) [6], have been proved to eliminate the IMD3. Linearization proposals use electronic pre-distortion [7] to add distortions which can compensate intrinsic nonlinearity of the link. However, the fabrication and controlling tolerance of these linearization schemes are difficult to meet in W-band. In recent years, a few institutions concentrate on optical processing to suppress IMD3 [8, 9], with improved SFDR. Yet no experimental demonstration on W-band is proposed to the best of our knowledge.

In this paper, we propose a highly linear optical W-band receiver front-end with MMW down-conversion and direct optical processing. Optical spectrum contributors to IMD3 components is analyzed to investigate the nonlinear compensation strategy [8]. By imposing phase-shift and attenuation on optical sidebands (OSB), two parts of the beating products of IMD3 cancel each other. A 108.6 dB·Hz$^{2/3}$ SFDR is achieved experimentally, which is 16dB more than an un-compensate scheme.

II. OPERATIONAL PRINCIPLE

To measure the SFDR of our scheme, we generate a two-tone W-band MMW signal and set it to the antenna. An RF signal, donated as E_0, at the angular frequency of Ω_1 and Ω_2 is obtained after down conversion, and is used to drive the MZM. The output of the MZM can be expressed as:

$$E_{optical}(t) = E_0 e^{j\omega_0 t} \cos(\frac{\pi V_0}{2V_\pi}(\cos \Omega_1 t + \cos \Omega_2 t) + \varphi_b) \qquad (1)$$

where φ_b denotes the direct current bias of the MZM, V_0 is the amplitude of each RF tone, and V_π is the half-wave voltage of the MZM. With Bessel expansion, the sideband components of the output signal can be named ±1-OSB, ±2-OSB based on their order. We introduce the phase-shift φ_p to +2-OSB, $-\varphi_p$ to -2-OSB, and attenuation α_p of 0-OSB. The optical signal is sent to a square law photodetector (PD), of which the sensitivity is $\Re$. After optical processing, the IMD3 components $I_{2\Omega_{1,2}-\Omega_{2,1}}$ can be mathematically given as:

$$I_{2\Omega_{2,1}-\Omega_{1,2}} = \sin(2\varphi_b)\Re E_0^{\,2}$$
$$\cdot \{\sqrt{\alpha_p} \sum_{p=-\infty}^{+\infty} [J_{p+1}(m)J_{p-1}(m) - J_p(m)J_{p-2}(m)]J_p(m)J_{p-1}(m)\cos(2\Omega_{2,1}-\Omega_{1,2})$$
$$+ \sum_{p=-\infty}^{+\infty} [J_{p-1}(m)J_{p-3}(m) - J_{p+2}(m)J_p(m)]J_p(m)J_{p-1}(m)\cos(2\Omega_{2,1}-\Omega_{1,2}+\varphi_p)\} \qquad (2)$$

where $m=V_0/V_\pi$ is the modulation depth of the MZM. OSB higher than the second-order are ignored by taking small signal approximation.

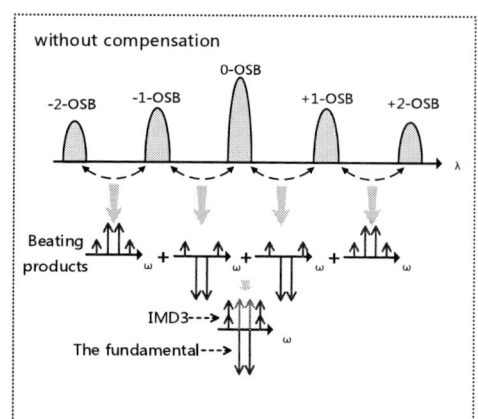 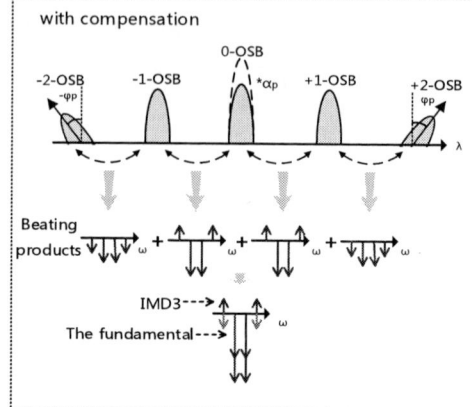

Fig. 1. Illustration of IMD3 suppression by optical processing.

By proper adjustment of the phase-shift φ_p and the OSB attenuation index α_p, the two kinds of IMD3 components $I2\Omega_{1,2}\text{-}\Omega_{2,1}$ cancel each other with invert phase and equal amplitude. It is obvious that the IMD3 can be completely suppressed theoretically, as illustrated in Fig. 1.

III. EXPERIMENT

Fig. 2 shows the experimental setup of the optical W-band front-end. The W-band MMW transmitter composes of a vector network analyzer (VNA, Rohde & Schwarz ZVA67), two W-band MMW converters (ZVA-Z110), a W-band coupler and a W-band MMW antenna. With the Z110 converters, the two-tone signal at the frequency of 94.98GHz and 95.02GHz are generated, coupled and emitted. The power level of the W-band two-tone signal has been calibrated to compensate the unbalanced transmission loss in the two different path.

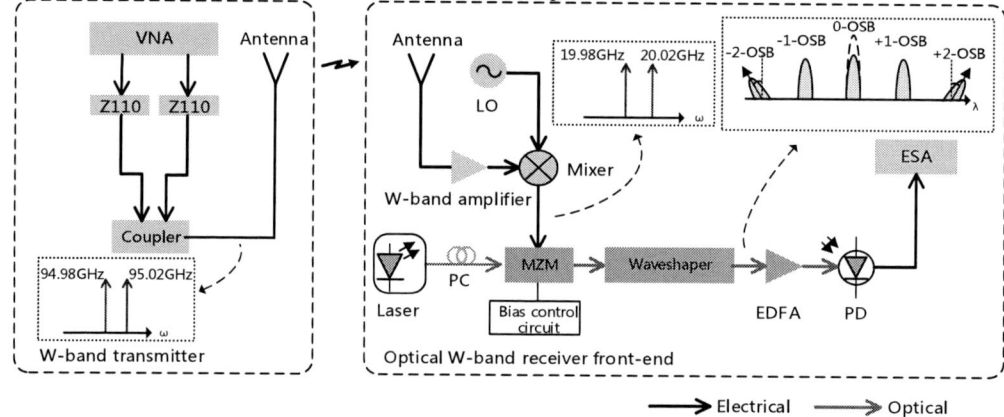

Fig. 2. Schematic diagram of the experimental setup.

In the optical W-band receiver front-end, the two-tone signal is firstly received and amplified at the Rx antenna, then introduced to a mixer, together with a sextupled local oscillating (LO) signal at 75.00GHz, subsequently down-converted to two-tone RF signal at 19.98GHz and 20.02GHz. The laser operating at a wavelength of 1550nm is applied as the optical carrier source, and the generated light beam is adjusted by a polarization controller before the following MZM (Eospace 40GHz MZM), which is driven by the two-tone RF signal mentioned above. The MZM works at quadrature mode with a bias control circuit. The modulated signal is then fed into the optical processor (Finisar Waveshaper 4000s) to perform phase-shift and attenuation, then imposed to a PD. The RF output is analyzed by an electrical signal analyzer (ESA, Agilent N9030A) to measure the SFDR and determine the optimal operating point.

The waveshaper is used as the optical processor, of which the frequency resolution is 18GHz, and the center wavelength is 1550nm. We can impose phase-shift and attenuation on OSB from 1527.4nm to 1567.5nm by manipulating the waveshaper through the controlling computer. Initially, the power of the W-band MMW signal is 5dBm and the phase-shift of ±2-OSB φ_p is adjusted to π. By properly sweeping the attenuation of 0-OSB α_p, from 0dB to −40dB, an optimal attenuation index $\alpha_p{}'$ that maximizes the fundamental to IMD3 ratio can been found. Then the sweeping of φ_p is also performed from 0 to π to get an optimal phase-shift to maximize the fundamental to IMD3 ratio. Alternate sweeps are repeated, until the optimal compensation working point $\alpha_p=-16.9dB$ and $\varphi_p=169°$ of the W-band front-end is obtained.

As shown in Fig.3, under the optimal compensation working point, the fundamental to IMD3 ratio has been improved from 33.9dB to 57.4dB when the input RF power is 10dBm. We have measured the performance for both compensated and uncompensated links by evenly varying the power level of VNA generators, as shown in Fig.4. The noise floor,

806

including shot noise and thermal noise, is -150.1dBm/Hz. After being compensated by optical processing, an SFDR of 108.6 dB·Hz$^{2/3}$ can be attained in this receiver front-end, which is 16dB more than the un-compensated.

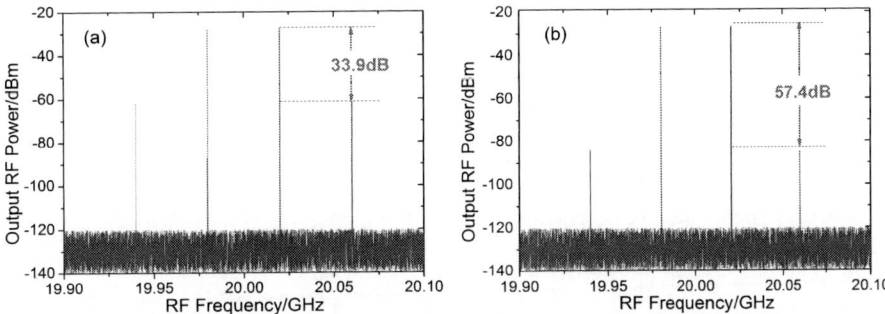

Fig. 3. Electrical spectra of the output RF signal (a) without compensation and (b) with compensation, under 10dBm MMW input power.

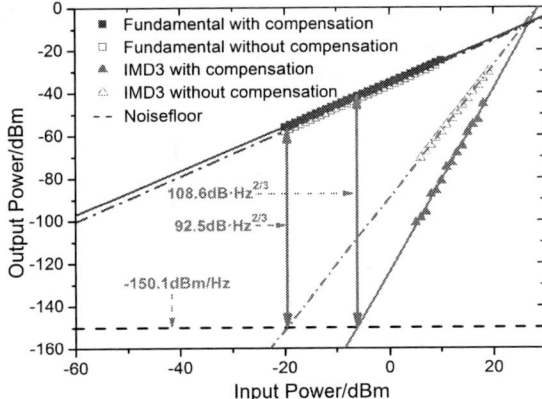

Fig. 4. Measured SFDR of the optical W-band receiver front-end. Both compensated and uncompensated links are tested.

IV. CONCLUSIONS

In summary, we analyze the optical spectrum contributors to IMD3 components in the optical RF receiver, and an approach to suppress the nonlinearity of MZM is proposed. By incorporating processing on OSB, two parts of IMD3 cancel each other. A highly linear optical W-band receiver front-end, of which the SFDR is 108.6dB·Hz$^{2/3}$, is achieved experimentally. The proposed scheme is promising for its prominent SFDR specification and simplicity compared with conventional W-band receiver.

ACKNOWLEDGMENT

This work was partly supported by National Basic Research Program of China under grant No. 2012CB315603-04, National Nature Science Foundation of China under grant No. 61321004, 61420106003, and 61435006.

REFERENCES

[1] R. Appleby, "Mechanically scanned real time passive millimeter wave imaging at 94 GHz," Proc. SPIE, vol. 5077, pp. 1–6, 2003.

[2] A. Arbabian et. al., "A 94 GHz mm-Wave-to-baseband pulsed-radar transceiver with applications in imaging and gesture recognition," IEEE J. Solid-State Circuits, 48(4), pp. 1055–1071, Apr. 2013.

[3] P. T. Dat et. al., "High-Capacity Wireless Backhaul Network Using Seamless Convergence of Radio-over-Fiber and 90-GHz Millimeter-Wave," Journal of Lightwave Technology, 2014.

[4] Sebastian Babiel, Atsushi Kanno, Tetsuya Kawanishi, "Radio-over-Fiber Photonic Wireless Bridge in the W-band," IEEE International Conference on Communications 2013, pp.838-842.

[5] S. Li, X. Zheng, H. Zhang, and B. Zhou, "Highly linear radio-over-fiber system incorporating a single-drive dual parallel Mach-Zehnder modulator," IEEE Photon. Technol. Lett. 22(24), 1775–1777, 2010.

[6] C. Lim, et. al., "Intermodulation Distortion Improvement for Fiber–Radio Applications Incorporating OSSB+C Modulation in an Optical Integrated-Access Environment," J. Lightwave Technol. 25(6), 1602–1612, 2007.

[7] V. Urick, M. Rogge, P. Knapp, L. Swingen, and F. Bucholtz, "Wide-band pre-distortion linearization for externally modulated long-haul analog fiber-optic links," IEEE Trans. Microw. Theory Tech. 54(4), 1458–1463, 2006.

[8] G. Zhang, S. Li, X. Zheng, H. Zhang, B. Zhou, and P. Xiang, "Dynamic range improvement strategy for Mach-Zehnder modulators in microwave/millimeter-wave ROF links," Opt. Express 20(15), 17214–17219, 2012.

[9] Y. Cui, Y. Dai, F. Yin, J. Dai, K. Xu, J. Li, and J. Lin, "Intermodulation distortion suppression for intensity-modulated analog fiber-optic link incorporating optical carrier band processing," Opt. Express 21(20), 23433–23440, 2013.

WA2-3

Distance-Adaptive Routing, Modulation Level and Spectrum Allocation (RMLSA) in K-Node (Edge) Content Connected Elastic Optical Datacenter Networks

Xin Li, Shanguo Huang, Shan Yin, Bingli Guo, Yongli Zhao, Jie Zhang, Min Zhang, Wanyi Gu

State Key Laboratory of Information Photonics and Optical Communication, Beijing University of Posts and Telecommunications, No 10, Xitucheng Road, Haidian District, Beijing, 100876, China

E-mail address: (xinli@bupt.edu.cn, shghuang@bupt.edu.cn)

Abstract: This paper addresses the RMLSA problem for end-to-content paths while satisfying k-node (edge) content connectivity requirement in elastic optical datacenter networks. Two distance-adaptive SWP-based RMLSA algorithms are proposed to improve the spectrum efficiency.

Keywords: RMLSA, Content Connectivity, Elastic Optical Networks, Datacenter Networks

I. INTRODUCTION

With the frequent occurrence of natural disaster and human-made intentional attacks on datacenter networks, a large number of data/services stored in or provided by datacenters are facing the risk of inaccessibility [1]. Traditional survivability techniques such as protection mechanism and restoration mechanism both rely on network connectivity to achieve traffic uninterrupted transmission. Note that network connectivity has a fixed value for a given topology, the survivability of optical datacenter network will reach bottleneck while merely relying on network connectivity. Nevertheless, the data/services can be replicated and maintained in multiple datacenters. The user can obtain the data/services from any reachable datacenter regardless of where it is located. Content connectivity which is defined as the reachability of content from any point of a network is proposed to improve the survivability of optical datacenter networks [2]. Content connectivity jointly takes advantage of network connectivity and content distributed backup. To quantitatively measure content connectivity, we propose the concept of k-node (edge) content connectivity which indicates the relationship between nodes (edges) cut set and content connectivity in [3]. Based on the k-node (edge) content connectivity, k-node (edge) content connected elastic optical datacenter networks (KC-EODN) is proposed to design disaster-resilient and spectrum-efficient datacenter networks. In KC-EODN, there exist k independent end-to-content paths between each source node and multiple datacenters where the content is hosted. We need to address the RMLSA problem for each end-to-content path. The RMLSA problem has obtained extensive research in elastic optical networks. In [4], the authors review and classify the most of existing routing and spectrum allocation (RSA) approaches including their pros and cons. In [5], heuristic algorithms based on spectrum window planes (SWP) are proposed for implementing distance and modulation format adaptive RMLSA to maximize spare capacity sharing. In this paper, we propose two distance-adaptive SWP-based algorithms to address the RMLSA for end-to-content paths in KC-EODN.

II. NETWORK MODEL

The proposed k-node (edge) content connectivity means that there does not exist a set of k-1 nodes (edges) whose removal disconnects the connectivity from any remaining node to content. Theoretical analysis shows that achieving k-node (edge) content connectivity is equivalent to searching k independent end-to-content paths between source node and multiple datacenters where the content is hosted [3]. The datacenter network which satisfies k-node (edge) content connectivity requirement is called k-node (edge) content connected datacenter network. Based on k-node (edge) content connectivity, we propose the disaster-resilient and spectrum-efficient KC-EODN.

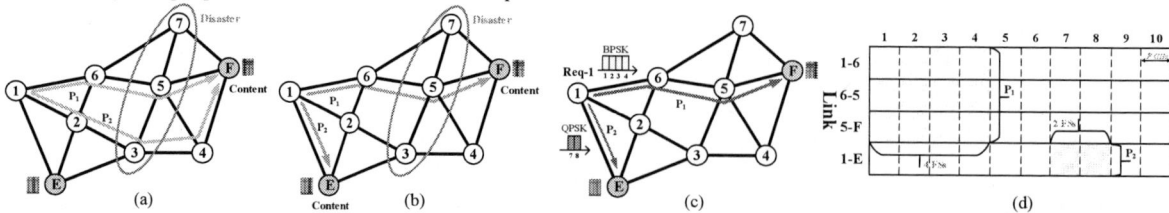

Fig. 1. (a) Network connectivity (b) Content distributed backup (c) RMLSA for end-to-content paths (d) FSs for end-to-content paths

The dual-node (edge) content connectivity can be realized merely by network connectivity as presented in Fig.1 (a) or jointly by content distributed backup and network connectivity as presented in Fig.1 (b). The comparison between Fig.1 (a) and Fig.1 (b) shows that the datacenter network cannot deal with natural disaster while merely relying on network connectivity. Content distributed backup provides a new method to improve the survivability of datacenter networks. We need to conduct RMLSA for each end-to-content path in KC-EODN. The RMLSA for each end-to-content path must satisfy spectrum contiguity and spectrum continuity requirements [5]. The spectrum contiguity constraint requires all the frequency slots (FSs) assigned to an end-to-content path to be spectrally neighboring. The spectrum continuity constraint requires all the fiber links traversed by an end-to-content path must use the same set of FSs. We assume that the transponder at each source node is tunable, so that any end-to-content path can use different sets of contiguous FSs. For request Req-1 in Fig.1 (c), two independent end-to-content paths are established. The paths P_1 and P_2 can use the different sets of contiguous FSs that path P_1 whose indices range from 1 to 4 adopts modulation level BPSK and path P_2 whose indices range from 7 to 8 adopts modulation level QPSK.

III. DISTANCE-ADAPTIVE HEURISTIC ALGORITHM

The concept of spectrum window (SW) represents a certain number of continuous FSs. For request (s,c) with n required FSs, the fiber link with FSs set $\{f_1, f_2, ..., f_N\}$ contains a total of N-n+1 SWs. The elastic optical datacenter network can be split into multiple spectrum window planes (SWPs), where in each plane a virtual link between a pair of nodes is connected if the SW in the corresponding fiber link is available [5]. In **EPSD**, $\lambda_{(s,c)}$ denotes the traffic demand for request (s,c), p_paths denotes the set of k shortest end-to-content path of the same request, p_paths_i denotes the i_{th} shortest end-to-content path, $indexs$ denotes the set of start indices of k end-to-content paths, $indexs\_i$ denotes the start index of path p_paths_i, $P_{(s,c)}$ denotes the shortest end-to-content path in current SWP, $P_{(s,d)}$ denotes the shortest path between source node s and datacenter d, ST represents the total of SWPs, B denotes the base capacity of a FS with single bit per symbol modulation (BPSK), $R=\{8QAM, QPSK, BPSK\}$ represents all optional modulation levels. In **EPMI**, p_paths denotes the set of k end-to-content paths with minimal indexed SWP of the same request, p_paths_i denotes the i_{th} end-to-content path with minimal indexed SWP.

EPSD: RMLSA for end-to-content path with shortest distance	**EPMI**: RMLSA for end-to-content paths with minimal index
Input: request (s,c)	**Input**: request (s,c)
Output: p_paths, $indexs$	**Output**: p_paths, $indexs$
1: set $p\_paths = \phi$, $indexs = \phi$	1: set $p\_paths = \phi$, $indexs = \phi$
2: **for** each modulation level $r \in R$ (from the highest to lowest level) **do**	2: **for** each modulation level $r \in R$ (from the highest to lowest level) **do**
3: **for** each $i=1: k$ **do**	3: **for** each $i=1: k$ **do**
4: set $p\_paths\_i = \phi$, $indexs\_i = \phi$	4: set $p\_paths\_i = \phi$, $indexs\_i = \phi$
5: $n = \lceil \lambda_{(s,c)} / (r * B) \rceil$	5: $n = \lceil \lambda_{(s,c)} / (r * B) \rceil$
6: initialize ST with N-n+1 SWPs	6: initialize ST with N-n+1 SWPs
7: remove all links which belong to p_paths	7: remove all links which belong to p_paths
8: set $P_{(s,c)}= Null$, $P_{(s,d)}= Null$	8: set $P_{(s,c)}= Null$, $P_{(s,d)}= Null$
9: **for** each SWP $\in ST$ (from the lowest to highest index) **do**	9: **for** each SWP $\in ST$ (from the lowest to highest index) **do**
10: remove links whose SWs are not available	10: remove links whose SWs are not available
11: **for** each datacenter d in set D **do**	11: **for** each datacenter d in set D **do**
12: find the shortest $s \to d$ route $P_{(s,d)}$ using Dijkstras's algorithm in current SWP	12: find the shortest $s \to d$ route $P_{(s,d)}$ using Dijkstras's algorithm in current SWP
13: **if** $(P_{(s,d)} \neq Null$&& $P_{(s,d)}$ is shorter than the transmission reach of current r)	13: **if** $(P_{(s,d)} \neq Null$&& $P_{(s,d)}$ is shorter than the transmission reach of current r)
14: **if** $(P_{(s,c)} == Null)$	14: **if** $(p\_path\_i == Null)$
15: $P_{(s,c)} = P_{(s,d)}$	15: $p\_path\_i = P_{(s,d)}$, $indexs\_i =$ startindex of current SWP
16: **else if** $(P_{(s,d)}$ is shorter than $P_{(s,c)})$	16: **else if** $(P_{(s,d)}$ is shorter than $p\_path\_i)$
17: $P_{(s,c)} = P_{(s,d)}$	17: $p\_path\_i = P_{(s,d)}$, $indexs\_i =$ startindex of current SWP
18: **end if**	18: **end if**
19: **end if**	19: **end if**
20: **end for**	20: **end for**
21: **if** $(P_{(s,c)} \neq Null)$	21: **if** $(p\_path\_i \neq Null)$
22: **if** $(p\_path\_i == Null)$	22: goto step 26
23: $p\_path\_i = P_{(s,c)}$, $indexs\_i =$ startindex of current SWP	23: **end if**
24: **else if** $(P_{(s,c)}$ is shorter than $p\_path\_i)$	
25: $p\_path\_i = P_{(s,c)}$, $indexs\_i =$ startindex of current SWP	
26: **end if**	
27: **end if**	

28: **end for**	24: **end for**
29: **if** ($p\_path\_i \neq Null$)	25: **end for**
30: **break**	26: $p\_paths = p\_paths \cup p\_paths\_i$, $indexs = indexs \cup indexs\_i$
31: **end if**	
32: **end for**	27: **end for**
33: $p\_paths = p\_paths \cup p\_paths\_i$, $indexs = indexs \cup indexs\_i$	28: **return** p_paths and $indexs$
34: **end for**	
35: **return** p_paths and $indexs$	

The time complexity of **EPSD** and **EPMI** is $O(|V|^2*|D|*(N\text{-}n+1)*|R|*k)$, where $|V|$ is the number of node in $G(V,E,D)$, $|D|$ is the number of datacenters, $N\text{-}n+1$ is the number of SWPs, $|R|$ is the number of all optional modulation levels. The two algorithms are guaranteed to run in polynomial time.

IV. NUMERICAL RESULTS

We assume that each fiber link in NSFNet consists of 80 FSs and each source node is equipped with tunable transponder. We assume that each content occupies 1G storage resource and we do not consider the datacenter's capacity limitation. The requests are generated at each source node according to a uniform distribution. The transmission rate of each request is selected from the set {40 Gbps, 100 Gbps, and 400 Gbps}.

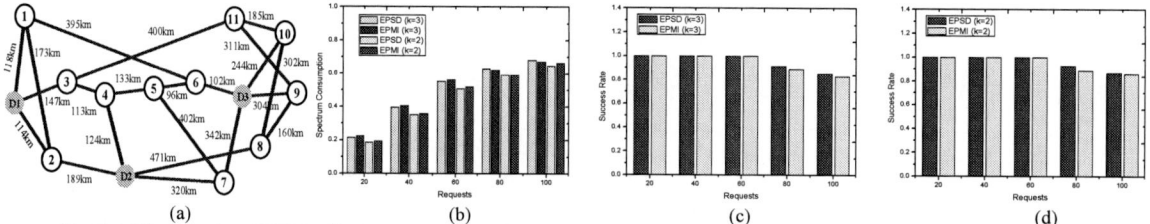

Fig. 2. (a) Test topology NSFNet (b) Spectrum consumption when k=3 and k=2 (c) Success rate when k=3 (d) Success rate when k=2

The NSFNet presented in Fig. 2 (a) has 11 network nodes, 3 datacenters, and the average node degree is 3.14. The transmission reach of modulation level {BPSK, QPSK, 8QAM} is 2000 km, 1000 km and 500 km respectively. Fig. 2 (b) shows the spectrum consumption of algorithm **EPSD** and algorithm **EPMI** when k=3 and k=2. The results show that algorithm **EPSD** has better spectrum efficiency than algorithm **EPMI**. Because algorithm **EPMI** has lower success rate in Fig.2 (c) when k=3, so the spectrum consumption of algorithm **EPSD** is higher than algorithm **EPMI** when the number of requests is 80 and 100. The comparison between Fig.2 (c) and Fig.2 (d) shows that algorithm **EPSD** has higher success rate and is more suitable for KC-EODN. In the simulation, we also verify the RMLSA algorithm which only adopts the lowest modulation level BPSK. The result shows that the proposed distance-adaptive **EPSD** algorithm and **EPMI** algorithm are far better than RMLSA algorithm which only adopts the lowest modulation level BPSK.

V. CONCLUSIONS

The proposed k-node (edge) content connectivity provides an effective metric for content connectivity. KC-EODN provides a new research direction for designing disaster-resilient and spectrum-efficient datacenter networks. The proposed distance-adaptive **EPSD** algorithm has better performance than **EPMI** algorithm in KC-EODN.

ACKNOWLEDGMENT

This work is supported in part by the NSFC (Nos. 61331008 and 61571058), the Hi-Tech Research and Development Program of China (863 Program) (No. 2012AA011302), the Program for New Century Excellent Talents in University (NCET-12-0793), and the Beijing Nova Program (No. 2011065).

REFERENCES

[1] M. Farhan Habib, Massimo Tornatore, Marc De Leenheer, Ferhat Dikbiyik, and Biswanath Mukherjee, "Design of Disaster-Resilient Optical Datacenter Networks," Journal of Lightwave Technology, Vol: 30, Issue: 16, pp:2563-2573, August 2012.

[2] M. Farhan Habib, Massimo Tornatore, Biswanath Mukherjee, "Fault-Tolerant Virtual Network Mapping to Provide Content Connectivity in Optical Networks," OFC, Anaheim, CA, USA, March 2013.

[3] Xin Li, Shanguo Huang, Shan Yin, etc, "Design of K-Node (Edge) Content Connected Optical Datacenter Networks," IEEE Communications Letters, DOI: 10.1109/LCOMM.2016.2517646, January 2016.

[4] Bijoy Chand Chatterjee, etc, "Routing and Spectrum Allocation in Elastic Optical Networks: A Tutorial," IEEE Communications Surveys & Tutorials, vol. 17, no. 3, pp: 1776–1800, August 2014.

[5] Chao Wang, Gangxiang Shen, etc, "Distance Adaptive Dynamic Routing and Spectrum Allocation in Elastic Optical Networks with Shared Backup Path Protection," Journal of Lightwave Technology, vol: 33, no: 14, pp: 2955-2964, July 2015.

Experimental Demonstration of 10-Gb/s Direct Detection Optical OFDM Transmission with Trellis-Coded 8PSK Subcarrier Modulation

Yiming Lou, Zhenming Yu, Minghua Chen*, Hongwei Chen, Sigang Yang, and Shizhong Xie

Tsinghua National Laboratory for Information Science and Technology (TNList)
Department of Electronic Engineering, Tsinghua University, Beijing, 100084, China
*chenmh@tsinghua.edu.cn

Abstract: Trellis-coded modulation direct detection optical orthogonal frequency division multiplexing transmission at 10-Gbit/s is experimentally demonstrated firstly. The required optical power for the 10-Gbit/s signal can be reduced by 1.5 dB at BER of 10^{-3}.

Keywords: Direct Detection, OFDM, Trellis-Coded

I. INTRODUCTION

Although the arrival of optical orthogonal frequency division multiplexing (OFDM) has been quite recent, it does inherit the major advantages of OFDM systems in Radio Freqency (RF) field, such as high spectral efficiency, high dispersion tolerance, and having Fast Fourier Transform (FFT) algorithm and so on. Direct detection optical orthogonal frequency division multiplexing (DDO-OFDM) has many more variants than the coherent counterpart. This mainly stems from the broader range of applications for direct detection OFDM due to its lower cost [1].

Trellis-coded modulation (TCM) which is first proposed by G. Ungerboeck in 1982 is a well-known technique in wireless communication to effectively increase the receiver sensitivity of high-level modulation formats [2]. TCM in which convolutional coding is applied through the expansion of signal constellation without requiring additional signal bandwidth is first applied in the field of optical communication in 2004 by H. Buelow et al [3], and then investigated by other groups. Q. Yang et al. experimentally demonstrated coherent optical orthogonal frequency division multiplexing (CO-OFDM) transmission at 1-Tb/s with Trellis-coded modulation and confirm the optical signal-to-noise ratio (OSNR) performance can be improved by 2.6 dB at BER of 10^{-3} [4]. X. Liu et al. demonstrated the transmission of a spectrally-efficient 44-Gb/s CO-OFDM signal whose subcarriers are trellis-coded with 32-QAM over a 99-km dispersion-managed standard single mode fiber (SSMF) link, and showed the performance improvements obtained through the coded-modulation and self-modulation compensation [5]. S. Ishimura and K. Kikuchi experimentally demonstrated that the 8-state Trellis-coded 4D-QPSK format can obtain 4 dB receiver-sensitivity improvement against DP-QPSK with a simple encoder structure. E. Le Taillandier de Gabory et al. evaluated experimentally 2, 8, 32Gbaud TCM-QPSK and TCM-16QAM, and found that TCM could offer finer configuration for flexible transponders with spectral efficiency increasing of 1.1 dB wider ranges [7].

In this paper, we first apply TCM technology to the DDO-OFDM system and experimentally verify the 4-State TCM-8PSK can obtain a considerable gain over the non-encoding QPSK in DDO-OFDM system. The gain is about 1.5 dB at the BER of 10^{-3} over a 60-km SSMF link.

II. EXPERIMENTAL SETUP

To evaluate the capability of our method in improving the BER performance of DDO-OFDM system, we conduct an experimental system as shown in Fig. 1. Offline transmitter digital signal processing (DSP), as shown in Fig. 1(a), is first performed to generate an OFDM signal with 512 subcarriers. In the offline DSP, data stream consisting of pseudo-random bit sequences (PRBS) of length 2^{15} is encoded by a rate-2/3 TCM encoder, as shown in Fig. 2(a), to generate 8PSK modulated subcarriers. In this step TCM encoder initially leaves the first one bit un-coded, but codes the last input bit into two coded bits through a 4-state convolutional encoder. The TCM encoder will then map the total 3 output bits onto trellis-coded 8PSK constellations. Thus, the QPSK modulation can be mapped into coded 8PSK modulation.

8PSK constellations must follow a set of set partitioning rules described in [8]. Fig. 2(b) shows the set partitioning process of TCM-8PSK where d_0 stands for the minimum Euclidian distance of the un-partitioned sct. Two-step set-partitioning process generates four partitioned sets where the minimum Euclidian distance is enlarged two times. Then we transform the 8PSK modulated subcarriers to time domain via inverse fast Fourier transform (IFFT) with size 2×2048 and later insert a cyclic prefix (CP) to the time domain signal to increase the dispersion tolerance. An intermediate frequency (IF) carrier is applied to turn the complex baseband OFDM signal to real after being converted into serial. Finally the signal will be stored in an arbitrary waveform generator (AWG) to generate an OFDM waveform.

A 10-Gbit/s DDO-OFDM signal is applied in this experiment as shown in Fig. 1(b). The transmission link is 60-km SSMF. At the receiver, a variable optical attenuator is used to control the received optical power as shown in Fig. 1(c).

Before the OFDM signals are directly photo-detected by a 50-GHz linear photodiode, we use an erbium-doped fiber amplifier (EDFA) to guarantee the response of the photodiode. The received RF OFDM signals are sampled by a real-time oscilloscope at 40 GS/s and processed off-line with MATLAB programs.

Offline receiver DSP unit is shown in insert (d) of Fig. 1. We first remove IF carrier in order to turn the signal into complex and then convert the serial signal taken from real-time sampling oscilloscope to parallel. We remove the CP then, after which we convert the time domain signal to frequency domain via fast Fourier transform with size 2×2048. After applying a rate-2/3 TCM decoder to decode and demodulate the frequency domain signal, we obtain the received sequence. Finally we can calculate the BER of the OFDM system via comparing the received sequence with transmitted sequence.

The TCM decoder applies Viterbi soft decision algorithm which can fully realize the advantages of TCM. Viterbi soft decision decoding algorithm fully utilizes the information of the received signal waveform using Euclidean distance as a metric, so that the decoder can be used to correct the code word, and improve the reliability of decoding. Soft decision decoding includes the following two steps: The first step is called "subset decoding". Un-coded bit results in the parallel state transitions in the TCM encoder. Subset decoding determines the signal closest to the received channel output by comparing the squared Euclidean distance between the received signal and the subset of the constellation points, and stores the signals together with their squared Euclidean distances from the channel output. In the second step, the signals chosen by subset decoding will be decoded using Viterbi algorithm which finds the signal path through the code trellis with the minimum sum of squared Euclidean distances from the sequence of noisy channel outputs received [8].

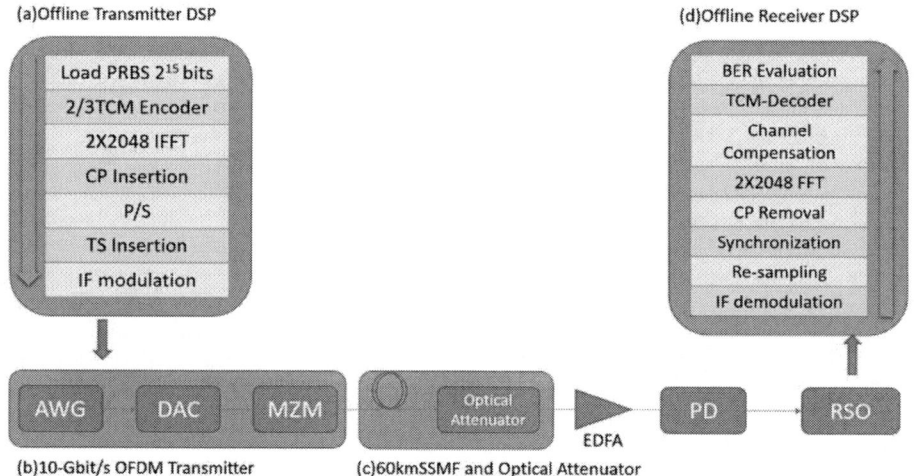

Fig. 1. Schematic of experimental setup of a 10-Gbit/s DDO-OFDM transmission system. Inserts: (a) Transmitter-side offline DSP modules including a rate-2/3 8PSK TCM encoder; (b) 10-Gbit/s DDO-OFDM transmitter; (c) Transmission channel; (d) Receiver-side offline DSP modules including a rate-2/3 TCM decoder. AWG: arbitrary waveform generator; DAC: digital to analog converter; MZM: Mach-Zehnder modulator; EDFA: erbium doped fiber amplifier; PD: photoelectric detector; RSO: real-time sampling oscilloscope;

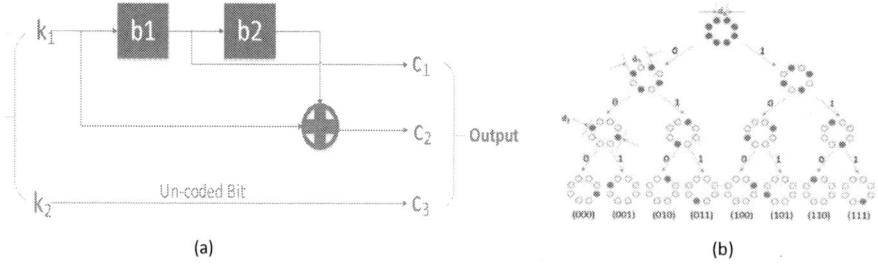

Fig. 2. (a) Rate-2/3 TCM encoder; (b) Set-partitioning process.

III. EXPERIMENTAL RESULT

Fig. 3 shows the BER performance of TCM-8PSK-DDO-OFDM system and QPSK-DDO-OFDM system. The insets show the typical constellation diagram for the un-coded and coded DDO-OFDM signal. The 8PSK constellation is much less clear, while can be recovered through TCM and obtain similar BER. The performance improvement for 4 state TCM is 1.5 dB against un-coded QPSK, and the BER performance can also be improved with the increase of the number of states.

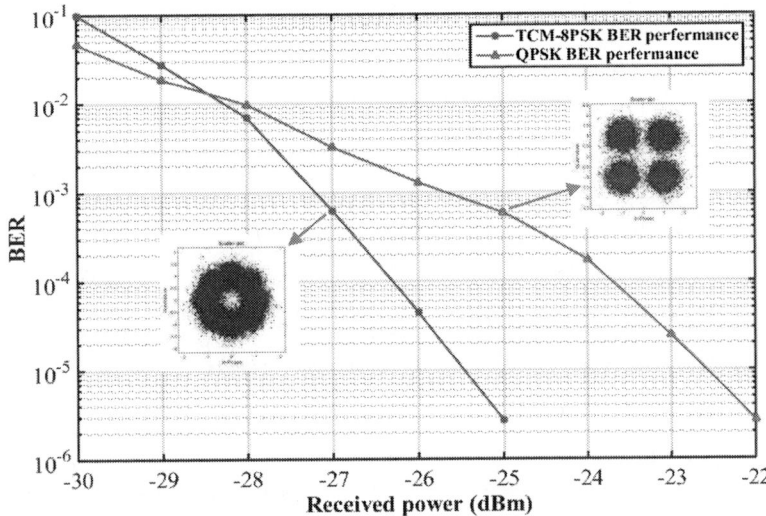

Fig. 3. OSNR sensitivity for 10-Gbit/s DDO-OFDM signal with/without TCM. Left inset: 8PSK constellation for TCM DDO-OFDM with received optical power of -27.0 dBm. Right inset: QPSK constellation for un-coded QPSK with received optical power of -25.0 dBm.

IV. CONCLUSION

We have conducted transmission of 10-Gbit/s DDO-OFDM with trellis-coded 8PSK subcarrier modulation over a 60-km SSMF link. The BER performance at different received optical power is measured in experiment. We confirm that trellis-coded 8PSK subcarrier modulation can offer about 1.5 dB sensitivity improvement at the BER of 10^{-3} compared with QPSK subcarrier modulation.

ACKNOWLEDGMENT

This work was supported by the NSFC under Contract, 61132004, 61090391, 61335002, 61322113, by the young top-notch talent program sponsored by Ministry of Organization, China, and by Tsinghua University Initiative Scientific Research Program.

REFERENCES

[1] W. Shieh, I. Djordjevic, "OFDM for Optical Communications". pp. 2, 272.
[2] G. Ungerboeck, "Channel Coding with Multilevel/Phase Signals". IEEE Trans. Inform. Theory, **IT-28**, no. 1, pp. 55-67 (1982).
[3] H. Buelow, G. Thielecke, F. Buchali, "Optical trellis coded modulation (oTCM)," OFC 2004, paper WM5.
[4] Q. Yang, Y. Ma, W. Shieh, "1-Tb/s single-channel coherent optical OFDM transmission with trellis-coded modulation," Electron. Lett. 45, 1045-1047 (2009).
[5] X. Liu, Q. Yang, S. Chandrasekhar, W. Shieh, "Transmission of 44-Gb/s Coherent Optical OFDM Signal with Trellis-Coded 32-QAM Subcarrier Modulation," OFC 2010, paper OMR3.
[6] S. Ishimura K. Kikuchi, "Experimental Demonstration of the 8-state Trellis-coded 4D-QPSK Optical Modulation Format," OSA 2015, paper W3K. 5.
[7] E. Le Taillandier de Gabory, T. Nakamura, H. Noguchi, W. Maeda, S. Fujita, J. I. Abe, K. Fukuchi, "Experimental Demonstration of the Improvement of System Sensitivity Using Multiple State Trellis Coded Optical Modulation with QPSK and 16QAM Constellations," OFC 2015, paper W3K. 3.
[8] G. Ungerboeck, "Trellis-coded modulation with redundant signal sets-Part I: Introduction." IEEE Commun. Mag. 25 5-22 (1987).

Power Efficient Optical OFDM Transmission with Phase Modulation and Direct Detection

Zhenhua Feng[1], Qiong Wu[1], Ming Tang[1*], Rui Lin[1, 2], Ruoxu Wang[1], Jiale He[1], Songnian Fu[1], Lei Deng[1], Deming Liu[1] and Perry Ping Shum[3]

(1) Next Generation Internet Access National Engineering Lab (NGIA), School of Optical and Electronic Information, Huazhong University of Science and Technology, Wuhan, 430074, China

(2) School of Information and Communication Technology, The Royal Institute of Technology, Electrum 229,164 40Kista, Sweden

(3) Photonics Centre of Excellence, School of EEE, Nanyang Technological University,637553, Singapore

*tangming@mail.hust.edu.cn

> **Abstract:** *Optical OFDM transmission with phase modulation and direct detection (PMDD) is verified by theoretical derivation and simulation. 26.12-Gb/s PMDD 16QAM-OFDM achieves comparable performance to single sideband IMDD 16QAM-OFDM with half of the optical modulation index.*
>
> **Keywords:** *Phase modulation (PM), direct detection (DD), intensity modulation (IM), OFDM.*

I. INTRODUCTION

Due to the emerging plentiful Internet services such as high-definition video downloads, social media and cloud computing, the bandwidth demand has driven the short-to-medium reach optical networks targeted at ever increasing data rates. In these cost-sensitive applications, intensity modulation and direction detection (IMDD) are usually preferred for its simplicity and cost-efficiency. Compared with intensity modulation (IM), phase modulation (PM) is more stable and simple without the need of DC bias which may drift with time when the operation environment changes [1]. Also, the insertion loss of a phase modulator is always lower than that of a commonly used Mach-Zehnder (MZ) intensity modulator [2]. Therefore, PM is a promising choice in short-reach optical communication systems if it can be directly detected. However, due to destructive interference of the conjugate sidebands, a PM signal is not available for direct detection (DD) unless PM to IM conversion is realized using either dispersive devices [3] or optical filter [4]. Inspired by the idea of PM to IM conversion in microwave photonics, phase modulation and direct detection (PMDD) can be penetrated into optical access networks to enable duplex transmission [5] and constant envelope OFDM systems to reduce PAPR and mitigate fiber nonlinearity [6].

In this paper, we apply PMDD to optical OFDM transmission systems and give full explanations to PMDD based OFDM using detailed theoretical derivation. To prove the feasibility, simulation is performed based on the platform of VPI 9.0 and Matlab. In our simulation, transmission performances between PMDD OFDM and single sideband (SSB) modulated IMDD OFDM are compared in terms of optimal optical modulation index (OMI), receiver sensitivity and dispersion tolerance.

II. OPERATION PRINCIPLE

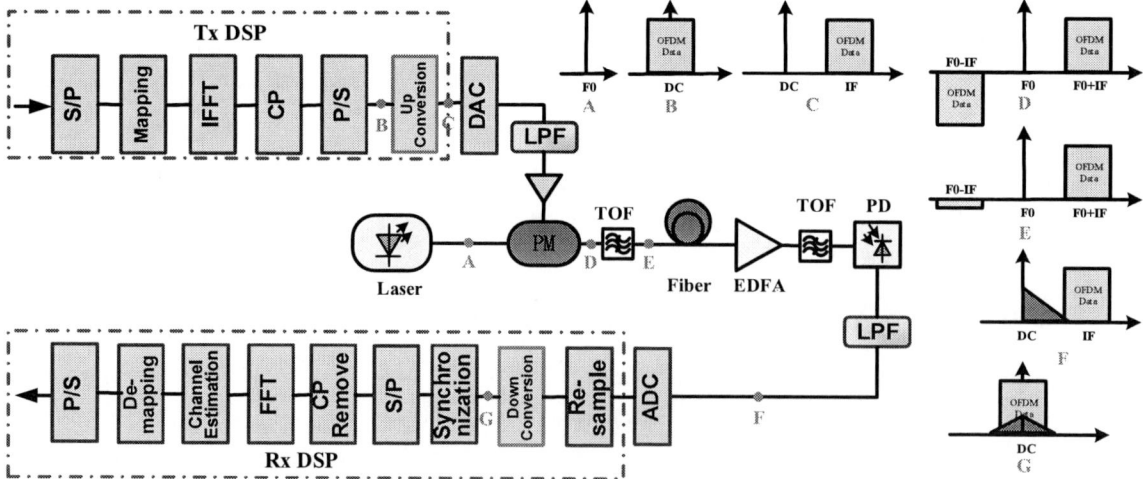

Fig. 1. Schematic of proposed optical OFDM transmission system with phase modulation and direct detection. The insets are the illustration of spectra for the corresponding signals.(PM: phase modulator, LPF: low-pass filter, CP: cyclic prefix, TOF: tunable optical filter, P/S: parallel to serial conversion, S/P: serial to parallel conversion, PD: photodiode, DAC: digital to analog conversion, ADC: analog to digital conversion)

As depicted in Fig. 1, the optical OFDM transmission system with PMDD is realized by suppressing one of the conjugated sidebands of the phase modulated optical OFDM signal using a tunable optical filter. In order to enable perfect sideband filtering, up-conversion in digital domain is used to keep a guard band (GB) between the optical carrier and the OFDM signal. The insets in Fig. 1 are the illustration of spectra for the corresponding signals of different locations. As can be seen, the generated baseband OFDM signal is firstly up-converted to an intermediate frequency (IF) and then fed to a phase modulator (PM) after digital to analog conversion, low-pass filtering and amplification. The filtered phase modulated OFDM signal is directly detected using a photodiode (PD) after fiber transmission and pre-amplification. In the receiver, the detected electrical signal is firstly digitalized by an analog to digital converter (ADC) and then down-converted in digital domain after resampling.

Theoretically, a complex QAM signal $s(t) = a + jb$ after up-conversion can be expressed as: $s_{up(t)} = a\cos(\omega_{IF}t) - b\sin(\omega_{IF}t)$, where ω_{IF} is the intermediate frequency. Then the optical field at the output of the phase modulator can be described by:

$$E_{PM}(t) = E_{in}exp\{jm[a\cos(\omega_{IF}t) - b\sin(\omega_{IF}t)]\} \qquad (1)$$

where $E_{in} = \sqrt{P_0}exp[j(\omega_0 t + \delta)]$ and $m = \frac{\pi}{V_\pi}$. P_0, ω_0 and δ are the optical power, angular frequency and phase noise from the laser, respectively. V_π is the half-wave voltage of the PM. Expanding equation (1) with Jacobi–Anger expansion we can get:

$$E_{PM}(t) = E_{in}\{\sum_{n=-\infty}^{\infty} j^n J_n(ma)\ exp[j(n\omega_{IF}t)]\} \cdot \{\sum_{n=-\infty}^{\infty} J_n(mb)\ exp[-j(n\omega_{IF}t)]\} \qquad (2)$$

where $J_n(x)$ is the nth order Bessel function of the first kind and $J_{-n}(x) = (-1)^n J_n(x)$. Under the assumption of small signal, we can neglect the higher order than 1st components and then it can be simplified as:

$$E_{out}(t) = E_{in}\{J_0(ma)J_0(mb) + 2j[J_0(mb)J_1(ma)\cos(\omega_{IF}t) - J_0(ma)J_1(mb)\sin(\omega_{IF}t)]\} \qquad (3)$$

If we keep the modulation index small enough, the Bessel function can be further approximated as: $J_0(x) \approx 1$, $J_1(x) \approx \frac{x}{2}$. So, we can obtain the expression of the signal for the output of the phase modulator.

$$E_{out}(t) = \sqrt{P_0}\left\{exp[j(\omega_0 t + \delta)] + \frac{1}{2}jms(t)exp[j((\omega_0 + \omega_{IF})t + \delta)] + \frac{1}{2}jms^*(t)exp[j((\omega_0 - \omega_{IF})t + \delta)]\right\} (4)$$

From equation (4), we note that phase modulated signal has similar spectrum as intensity modulated signal, except that the symmetric sideband pairs have complementary phases. Considering that a band-pass optical filter is used to select the upper sideband and the optical carrier while suppressing the other sideband, then the optical field after filtering is given by:

$$E_{out}(t) = \sqrt{P_0}\left\{exp[j(\omega_0 t + \delta)] + \frac{1}{2}jms(t)exp[j((\omega_0 + \omega_{IF})t + \delta)]\right\} \qquad (5)$$

After beating at the PD in the absence of fiber dispersion, the photocurrent can be described as:

$$I(t) = RP_0\left[1 + \frac{1}{4}m^2|s(t)|^2 - ms(t)\sin(\omega_{IF}t)\right] \qquad (6)$$

If the optical modulation index is small enough ($|s(t)| \ll 1$), the obtained electrical signal is proportional to the OFDM signal in the IF band, thus the information bearing in phase modulation can be recovered if it is down-converted to the baseband. We also notice that the laser phase noise item can be cancelled in the process of beating, so in principle the laser line-width has no influence on PMDD OFDM system.

III. SIMULATION RESULTS

In order to evaluate the transmission performance of the OFDM system with PMDD, we conduct simulation work on VPI Transmission Maker version 9.0 using the configuration shown in Fig. 1. For SSB IMDD OFDM transmission, the phase modulator is replaced by an MZ intensity modulator biased at quadrature point in order to achieve good linearity.

The transmitted signal is generated by MATLAB program originated from 2^{15}-1 pseudorandom binary sequence (PRBS), and then mapped into 16QAM modulation format. We adopt 128 points for the IFFT/FFT process, among which 100 effective subcarriers are activated to convey information. The first subcarrier is abandoned for eliminating noise near DC component. The length of cyclic prefix (CP) and frame are 13 and 139, respectively. The training sequence includes 11 OFDM symbols, in which one is used for frame synchronization and others are used for channel estimation. Compared with the traditional DDO-OFDM transmitter, the complex baseband OFDM signal is firstly resampled with an oversampling rate of 6 and then digitally up-converted to an intermediate frequency (IF) of 15GHz. After that, the up-converted OFDM signal is transformed to analog version via an 8-bit DAC, operated at a sampling rate of 60GS/s. A low-pass filter (LPF) with a 3-dB bandwidth of 20 GHz is followed to alleviate the image-band and emulate a bandwidth limited channel. Before fed to drive the phase modulator or MZ intensity modulator, the IF OFDM signal is linearly amplified. For both PM and IM cases, sideband filtering is performed by a 25GHz band-pass optical filter centered at 193.411 THz. The filtered OFDM signal with fixed optical power of -2 dBm is launched into the single mode fiber (SMF) while in optical back-to-back (OB2B) configuration the fiber is replaced by a variable optical attenuator (VOA) with equal loss. At the receiver side, the attenuated optical OFDM signal is firstly pre-amplified by an EDFA (with noise figure of 5 dB, the gain is set to 20 dB) and then filtered by a tunable optical filter (the same as the former one) to eliminate the out-of-band noise. Subsequently, the OFDM signal detected by a 20-GHz bandwidth photodiode (PD) (shot noise, thermal noise and current noise are considered) is oversampled by an 8-bit ADC. Demodulation and BER counting are implemented offline after resampling, down-conversion and frame synchronization. In our simulation, the center frequency of the laser is 193.4 THz with line-width of 1MHz, and the net data rate for 16QAM OFDM is 26.12-Gb/s.

We firstly study the influence of optical modulation index (OMI) on the Q-factor of the received signal in OB2B configuration for both PMDD and SSB IMDD 16QAM-OFDM. The OMI is defined as $OMI = \pi V_{rms}/V_\pi$, where V_π is the half-wave switching voltage of the phase modulator and MZ intensity modulator, while V_{rms} is the root-mean-square (RMS) amplitude of the electrical OFDM signal [7]. For both PM and IM situations, the received optical power (ROP) is fixed at -12 dBm during the OMI optimization. The simulation results are summarized in Fig. 2. We can conclude that the optimal OMI for IMDD OFDM is twice that of PMDD while the achieved Q-factor at the optimal OMI is almost the same, which means OFDM with PMDD is more power efficient than IMDD OFDM.

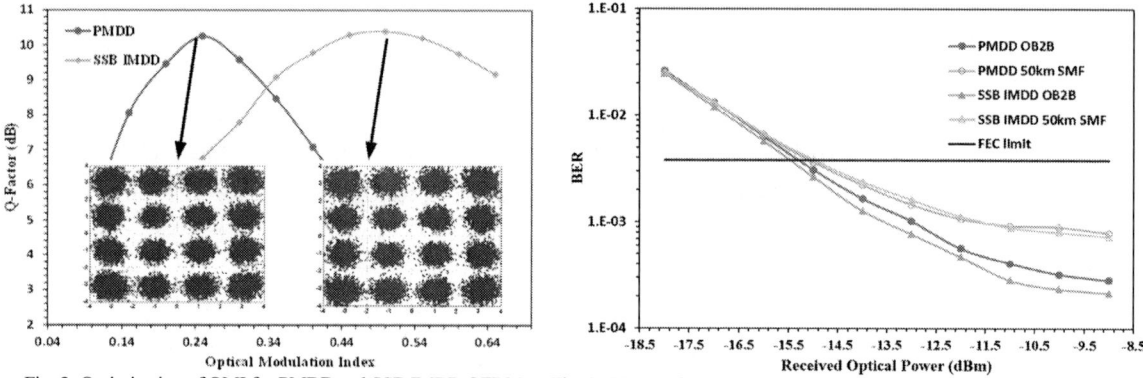

Fig. 2. Optimization of OMI for PMDD and SSB IMDD OFDM.   Fig. 3. BER performance Vs ROP for PMDD and SSB IMDD OFDM.

Subsequently, with the optimized OMI, we can evaluate the system transmission performance in both OB2B configuration and 50 km SMF link for PMDD and SSB IMDD OFDM. The BER performance at various ROPs is shown in Fig. 3. As can be seen, the receiver sensitivity of PMDD 16QAM-OFDM is nearly the same as that of SSB IMDD 16QAM-OFDM, except that a tiny distinction is observed in the large ROP region due to filtering penalty and smaller OMI used in PMDD OFDM. However, after the transmission over 50 km SMF, the BER of PMDD OFDM is slightly lower than that of SSB IMDD OFDM. This can be explained by the fact that for PMDD OFDM the residual sideband will weaken the detected OFDM signal because of the destructive interference in OB2B case, while in the presence of dispersion the π phase shift between the residual sideband and the filtered signal will break, thus totally destructive interference will not happen. We also notice that to reach the hard decision forward error correction coding (HD-FEC) limit of BER=3.8e^{-3}, the dispersion induced performance deterioration for these two OFDM systems is non-significant thanks to the adequate cyclic prefix (CP) we used and the sideband filtering. In the region of large ROP, there is a BER floor resulted from subcarrier to subcarrier beating noise (SSBN) and current noise in the receiver.

IV. CONCLUSIONS

In this paper, PMDD is applied to optical OFDM transmission system. Detailed theoretical derivation is given to explain the operation principle of PM to IM conversion. Transmission performances between PMDD 16QAM-OFDM and SSB IMDD 16QAM-OFDM are compared in terms of optimal OMI and receiver sensitivity in OB2B and 50 km SMF link. Simulation results show that 26.12-Gb/s PMDD 16QAM-OFDM can achieve almost the same receiver sensitivity and dispersion tolerance as that of SSB IMDD 16QAM-OFDM with only half of the OMI. This power efficient optical OFDM transmission technique based on PMDD is promising in high-speed short-reach optical networks providing with low cost and dispersion robustness.

REFERENCES

[1] Z. Feng, S. Fu, and M. Tang, "Investigation on agile bias control technique for arbitrary-point locking in lithium niobate Mach-Zehnder modulators," Acta Optica Sinica vol. 32, no. 12, 1206002, December 2012.

[2] J. Li, Y. Zhang, S. Yu and W. Gu, "Optical Sideband Processing Approach for Highly Linear Phase-Modulation/Direct-Detection Microwave Photonics Link," IEEE Photonics Journal, vol. 6, no. 5, pp. 1-10, October 2014.

[3] H. Chi, X. Zou, and J. Yao, "Analytical models for phase-modulation-Based microwave photonic systems with phase modulation to intensity modulation conversion using a dispersive device," J. Lightw. Technol., vol. 27, no. 5, pp. 511–521, March 2009.

[4] T. Chen, X. Yi, L. Li, and R. Minasian, "Single passband microwave photonic filter with wideband tunability and adjustable bandwidth," Optics Letters, vol. 37, pp. 4699-4701, November 2012.

[5] L. Cheng, M, Chen, H. Chen, S. Yang and S. Xie, "A Colorless ONU Scheme for WDM-OFDM-PON with Symmetric Bitrate and Low-cost Direct-detection Receivers," Proc. of CLEO 2015, California (America), paper JTh2A.69, May 2015.

[6] Z. Dong, Z. Cao, J. Lu, Y. Li, L. Chen, and S. Wen, "Transmission performance of optical OFDM signals with low peak-to average power ratio by a phase modulator," Opt. Commun., vol. 282, no. 21, pp. 4194–4197, 2009.

[7] W. R. Peng, X Wu, V. R. Arbab, K. M. Feng, B. Shamee, L. C. Christen, J. Y. Yang, A. E. Willner, and S. Chi, "Theoretical and experimental investigations of direct-Ddetected RF-tone-assisted optical OFDM systems," J. Lightw. Technol., vol. 27, no. 10, pp. 1332-1339, May. 2009.

WA2-32

A 3-D Adaptive Loading Algorithm for Direct Detection Optical OFDM System

Xi Chen, Zhenhua Feng, Ming Tang*, Huibin Zhou, Songnian Fu, Deming Liu

Next Generation Internet Access National Engineering Lab (NGIA), School of Optical and Electronic Information, Huazhong University of Science and Technology, Wuhan, 430074, China

*tangming@mail.hust.edu.cn

Abstract: We propose an adaptive loading algorithm to flexibly tailor each OFDM subcarrier in three dimensions including modulation formats, power and FEC codes. With look-up table operation introduced, the iterations are notably reduced without performance sacrifice.

Keywords: Adaptive Loading Algorithm, Adaptive Modulation and Coding, Optical OFDM

I. INTRODUCTION

Benefited from a lot of inherent advantages, for example, great resistance to dispersion, high spectral efficiency, and dynamic bandwidth allocation, optical orthogonal frequency division multiplexing (OOFDM) has attracted significant attention in last decade [1,4]. However, due to the extensive impairments existed in electrical components and optical links, the signal-to-noise ratio (SNR) of each subcarrier may vary significantly across the OFDM spectrum [2-4], degrading overall transmission performance. In this case, a widely adopted solution is to apply adaptive loading algorithms (ALAs) [3-5], which will sufficiently utilize channel capacity to enhance system performance. According to the loaded parameters, general ALAs assign specific modulation formats or power to each subcarrier with obtained channel state information (CSI). In order to achieve optimal loading results, massive iterations are introduced, which bring major increase of complexity.

In this paper, we introduce a new degree of freedom, FEC codes, to achieve the adaptive modulation and coding (AMC) or coded modulation (CM), which is rarely found in OOFDM system. To reduce the consequent increases of complexity, we adopt a look-up table (LUT) operation, which can notably reduce the iterations without performance sacrifice. Besides, in simulation results, we prove the feasibility of the proposed ALA and show the allocation results in bandwidth severely limited scenario.

II. PRINCIPLE OF THE PROPOSED ALA

The main principle is based on the fact that BER can be predicted from SNR under certain modulation formats and FEC codes in a certain channel [6]. Reversely, if the target BER is specified, the thresholds of SNR for special modulation formats and FEC codes can be obtained. Thus, with the SNR of each OFDM subcarrier estimated and compared with the SNR thresholds, the most suitable scheme can be selected, which can easily be done by LUT operation. Knowledge of previous iteration lookup results can be used to drastically reduce complexity [7] as some subcarriers converge rather quickly. Furthermore, by iteratively adjusting the power of each subcarrier with the total power constant, the expected SNR is normalized to the SNR thresholds, the least required power for the assigned schemes. To ensure the optimal allocation, we iteratively assign power to the subcarrier which achieves the maximum rate gain per unit power to upgrade the assigned scheme, which is similar to Hughes-Hartogs algorithm [5].

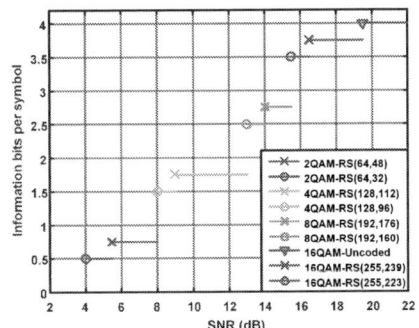

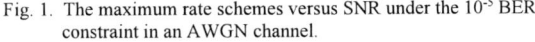

Fig. 1. The maximum rate schemes versus SNR under the 10^{-5} BER constraint in an AWGN channel.

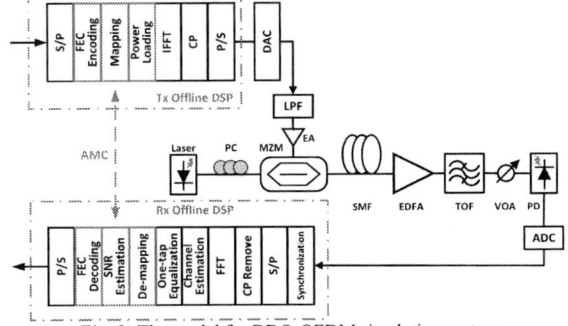

Fig. 2. The model for DDO-OFDM simulation system.

Noted that the ideal prediction of BER from SNR is generally under an AWGN channel, which is not completely consistent with practical optical channel. In order to enhance the robustness of the proposed ALA, we introduce the

SNR margin, which denotes the increment of the SNR thresholds. In this paper, we choose 4 different modulation levels of M-QAM ($M = 2,4,8,16$) as the modulation formats, and 2 extensively used FEC codes, RS (255,239) and RS (255,223), which are shortened to fit designed OFDM signal sequence when $M = 2,4,8$, thus, including uncoded schemes, there are total 12 available schemes. For the rate adaptive (RA) purpose, the schemes that achieve the maximum information bits per symbol (IBPS) at each SNR intervals are shown in Fig. 1, and will be put into the LUT, IBPS is defined as

$$IBPS = r_c \log_2(M) \tag{1}$$

where r_c and M denote the code rate of FEC codes and modulation orders, respectively.

III. SIMULATION RESULTS OF THE PROPOSED ALA

To examine the performance of the proposed ALA, we built a co-simulation system on the platform of VPI TransmissionMaker 9.1 and MATLAB. The numerical simulation model of DDO-OFDM transmission system is shown in Fig. 2.

The original serial data is generated from $2^{17} - 1$ pseudorandom binary sequence (PRBS), and then converted to parallel data. After that, each subcarrier is adaptively loaded with different modulation formats, power, and FEC codes, according to the allocation results from the proposed ALA. For the conventional schemes, all the subcarriers will be allocated with identical three parameters. To generate real-value OFDM signal, Hermitian symmetry must be ensured, thus for 128-points FFT, the max available number of subcarriers is 63 with the zero-padding DC subcarrier. 13 cyclic prefixes (CPs) are inserted to each OFDM symbol and 6 training symbols (TSs) are added to each OFDM frame with 36 symbols length. One of TSs is used for frame synchronization and others are used for channel estimation. Then, the digital signal is sampled by an 8-bit digital to analog converter (DAC) at 6.25 GSa/s sampling rate. Before converted to optical signal by the intensity modulation of Mach-Zehnder modulator (MZM) biased at quadrature point, the signal is passed by low-pass filter (LPF) and electrical amplifier (EA). The center frequency of laser is 193.4 THz, and the loss and the dispersion parameter for the SMF are 0.2 dB/km and 16 ps/km/nm, respectively. In the receiver, the OFDM signal is firstly pre-amplified by an EDFA with 20 dB gain and 5 dB noise figure, and then filtered by a tunable optical filter (TOF) with 30 GHz bandwidth to eliminate the out-of-band noise. Variable optical attenuator (VOA) is used to control received optical power. Lastly, the optical OFDM signal is directly detected by the photodetector and sampled by an 8-bit analog to digital converter (ADC) at 25 GSa/s sampling rate. 87702 points are stored to conduct a serial of digital signal processes (DSPs) to recover data and implement BER counting.

To prove the feasibility of our proposed AMC schemes, we investigate the performances of different modulation and coding schemes with various transmission distances and the allocation results in bandwidth severely limited scenario.

A. Performance Comparison with Fiber Length Variation

In order to verify the ALA resistance of chromatic dispersion induced fading, we vary the fiber lengths to measure the BER and effective data rate (EDR) variation while keeping the received optical power at -10 dBm.

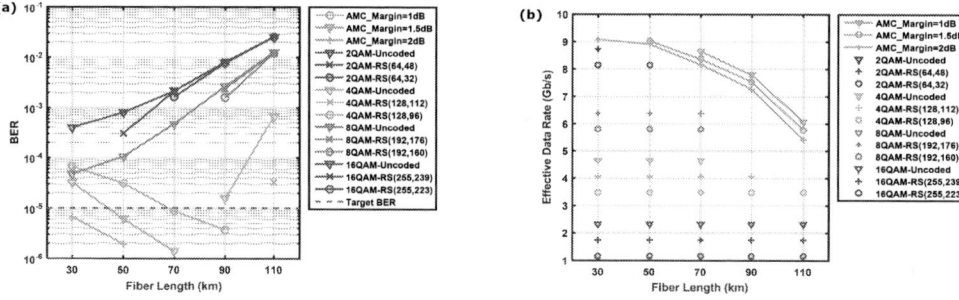

Fig. 3. Performance comparison between AMC schemes and conventional schemes: (a) BER versus fiber length, (b) EDR versus fiber length.

As depicted in Fig. 3(a), the BER of conventional schemes increases as fiber length increases. Due to the dispersion induced penalty, the 16QAM-OFDM signal cannot be transmitted as far as 70 km. Fortunately, with the help of the proposed ALA, the BER can be significantly decreased regardless of the SNR margin, especially in long fiber transmission scenario. We also note that the BER of AMC schemes exceeds the BER limit unless the margin reaches nearly 2 dB with fiber length of 30 km. Such phenomenon can be explained by the limitation of inherent signal–signal beat interference (SSBI) in direct detection of DSB-OFDM [8], which plays a dominant role at high SNR and makes the given SNR thresholds not precise.

The corresponding EDR is shown in Fig. 3(b), in which the points beyond the BER limit are not depicted. It is clear that the EDR drops significantly with the increase of fiber length for both conventional schemes and ALA. There is a trade-off between the SNR margin and the EDR. EDR slightly fades with enlarged SNR margin. However, the EDRs of AMC schemes with all measured margins are still higher than conventional schemes at all measured points. For the 50 km case, the EDR of AMC scheme with 1 dB margin achieves 7.78 Gb/s, which is 1.91 times the maximum EDR the conventional scheme can achieve under the BER constraint.

B. Allocation Results of the Proposed ALA in Bandwidth Severely-limited Scenario

In order to view the details, we evaluate the SNR variance and the 3-D allocation results of each subcarrier in a bandwidth severely limited channel with the LPF bandwidth set as 1.5625 GHz. During the simulation, the fiber length and received optical power are fixed at 50 km and -10 dBm, respectively. The SNR margin of AMC scheme is set to 1.5 dB.

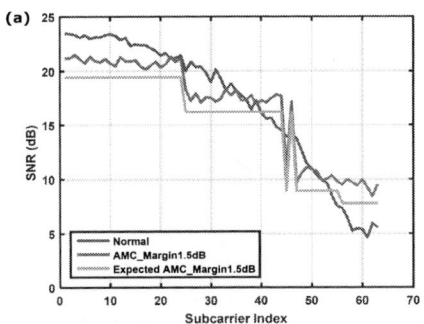 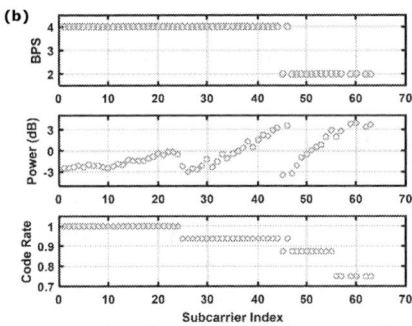

Fig. 4. (a) SNR variation with subcarrier index for conventional scheme and AMC scheme, the SNR of AMC scheme agrees well with the SNR expected from allocation results. (b) Allocation results by the proposed ALA when SNR margin is set to 1.5dB.

As shown in Fig.4(a), due to the system frequency response roll-off effect, the estimated SNR drops with subcarrier index. As a result of ALA, 16QAM (4 bits) and 4QAM (2 bits) are allocated to low-frequency and high-frequency subcarriers as Fig. 4(b), respectively. 2QAM (1 bit) and 8QAM (3 bits) do not appear in the final results since the upgrading process along with power loading just pass them. The 58th and the 61th subcarrier are abandoned for poor SNR. When BPS and FEC are constant, the power generally increases as subcarrier index increases, which continues until the change of either IBPS or FEC codes. It can also be observed from Fig. 4(a) that the subcarriers allocated with identical modulation formats and FEC codes roughly achieve the same SNR. Besides, the estimated SNR is about 1.5 dB larger than the expected SNR, which agrees well with the SNR margin we set.

IV. CONCLUSIONS

In this paper, we propose a 3-D ALA which can individually assign three parameters including modulation formats, power and FEC codes to each OFDM subcarrier. With the LUT operation introduced, the iterations of ALA are notably reduced. The simulation results show that the ALA has great potential to satisfy the various requirements of optical network services and applications by adaptively adjusting the SNR margin value.

REFERENCES

[1] W. Shieh, "OFDM for flexible high-speed optical networks," Journal of Lightwave Technology, vol. 29, pp. 1560-1577, May. 2011.

[2] Z. Feng, et al. "Performance-Enhanced Direct Detection Optical OFDM Transmission with CAZAC Equalization," IEEE Photonics Technology Letters, vol. 27, pp. 1507-1510, Jul. 2015.

[3] E. Giacoumidis, et al. "Statistical performance comparisons of optical OFDM adaptive loading algorithms in multimode fiber-based transmission systems," IEEE Photonics Journal, vol. 2, pp. 1051-1059, Dec. 2010.

[4] Q. Yang, W. Shieh, and Y. Ma. "Bit and power loading for coherent optical OFDM," IEEE Photonics Technology Letters, vol. 15. pp. 1305-1307, Aug. 2008.

[5] Hughes-Hartogs, "Ensemble modern structure for imperfect transmission media," U.S. Patent 4,679,227(July 1987), 4,731,816 (March 1988), and 4,883,706 (May 1989).

[6] K. Cho, and D. Yoon. "On the general BER expression of one-and two-dimensional amplitude modulations," Communications, IEEE Transactions on, vol. 50, pp. 1074-1080, Jul. 2002.

[7] Krongold, et al. "Section division operating point determination method for multicarrier communication systems," U.S. Patent No. 6,400,773. 4 Jun. 2002.

[8] W, Peng, et al. "Theoretical and experimental investigations of direct-detected RF-tone-assisted optical OFDM systems," Journal of Lightwave Technology, vol. 27, pp. 1332-1339, May. 2009.

WA2-33

OECC/PS2016

A Robust Timing Estimation Method for DDO-OFDM Systems

Zhen Zhang, Yingxiong Song*, Junjie Zhang, Yufeng Cai, Weiliang Wu, and Qianwu Zhang

Key Laboratory of Specialty Fiber Optics and Optical Access Networks, Shanghai University, Shanghai, 200072, China

E-mail: herosf@shu.edu.cn

Abstract: *A robust timing estimation method based on reverse correlation between the received signals is experimentally demonstrated which shows at least 4dB ROP sensitivity improvements compare to current method in asynchronous DDO-OFDM systems.*

Keywords: *Timing estimation, Direct-detection, Orthogonal frequency division multiplexing (OFDM)*

I. INTRODUCTION

Transmission capacity of the access networks are engaged form supporting end-users due to the rapidly emerging bandwidth-hungry multimedia services. Orthogonal frequency division multiplexing (OFDM)-based passive optical network (PON) has been considered as a promising candidate technology for satisfying the aforementioned demands because of its high spectral efficiency and powerful digital signal processing (DSP) technology enhanced flexibility [1]. Meanwhile, direct-detection optical (DDO) OFDM systems have attracted a lot of attention consider its low-complexity, stable performance and low-cost [2]. However, the DDO-OFDM system is very sensitive to symbol synchronization error which may be induced by imperfect timing estimation and often leads inter-symbol interference and bit-error-rate (BER) performance degradation [3]. Correlation of the two identical halves of training symbol is used to realize the timing estimation in Schmidl's method [4], however, the timing metric of this method has a plateau, which may cause a large variance in the timing estimate. Park *et al.* proposed a method based on the reverse autocorrelation of a repeated-conjugated-symmetric training symbol to avoid the ambiguity which occurs in Schmidl's method [5]. Park's method has an impulse-shaped timing metric, which enable it to achieve a more accurate timing offset estimation. In [6], they experimentally investigated and compared the performance of mentioned timing synchronization methods in DDO-OFDM standard single mode fiber (SSMF) transmission systems and present a modified method.

However, the timing metrics of these methods [5, 6] are not sharp enough at large sampling phases offset (SPO), which may leads to some uncertainty for detecting the start of OFDM symbols. In this paper, we propose a robust timing estimation method based on reverse correlation between the received signals which alleviate such uncertainty by smoothing signals. The experimental results show that the proposed method can improve at least 4dB on received optical power sensitivity than the method in [6] at the acquisition probability of 99% in DDO-OFDM transmission system with asynchronous sampling clock.

II. PROPOSED SYMBOL TIMING METHOD

Here, the proposed timing estimation method is based on a time-domain training symbol (TS). The TS is generated by modulating the sub-carriers with BPSK symbols. All the data (BPSK symbols and zeros) on the sub-carriers are constrained to have Hermitain symmetry. With the properties of IFFT, the result of IFFT will produce the real-valued time-domain sequence in the form of

$$P = [aA_{N/2-1}bB_{N/2-1}] \qquad (1)$$

Where a is a number and so it is b, N is the IFFT size. $A_{N/2-1}$ represents samples of length $N/2-1$, $B_{N/2-1}$ is symmetric with $A_{N/2-1}$. For example, if $A_{N/2-1} = [1,2,3]$, then $B_{N/2-1} = [3,2,1]$.

Similar with the method in [5,6], we defined timing metric as follows:

$$M_{\mathrm{Pro}} = \frac{|P(d)|^2}{R^2(d)} \qquad (2)$$

where

$$R(d) = \frac{1}{2} \sum_{n=1}^{N/2-1} \left[|r(d+n)|^2 + |r(d-n)|^2 \right] \qquad (3)$$

$$P(d) = \sum_{n=1}^{N/2-1} r(d+n)x(d-n) \qquad (4)$$

$$x(n) = r(n-1) + r(n) + r(n+1) \qquad (5)$$

where $r(n)$ is the discrete samples of the received OFDM signal, $x(n)$ is the smoothed signal of the received OFDM signal, $P(d)$ represents the reverse correlation between $r(n)$ and $x(n)$, $R(d)$ is the half-symbol energy in the N samples of the window, d is the time index corresponding to the center sample in a window of $N-1$ samples .

Thus the estimation of timing offset is

$$\hat{d} = \arg\max(M_{\mathrm{Pro}}(d)) - N/2 \qquad (6)$$

III. SIMULATION AND EXPERIMENTAL RESULTS

In the simulation, the size of IFFT/FFT is set at 128, and the length of CP is $N/8$. After serial-parallel (S/P) conversion, the 16-QAM modulation format is used for all sub-carriers carrying data. Fig. 1 is plotted to demonstrate the impacts of SPO on timing metric of method in [6] and proposed method. For simplicity, the effect of SPO only is considered in the numerical simulation. Firstly, a highly over-sampled OFDM signal with an over-sampling rate of 32 is generated. Then the OFDM signal with different SPOs can be obtained from it. The timing metric value under no noise at the correct point of method in [6] and proposed method versus SPOs are shown in Fig. 1. It is clear that the timing metric values of proposed method are stronger then the timing metric values of method in [6] when $-0.5T_S < \mathrm{SPO} \le 0.5T_S$. Especially when $\mathrm{SPO} = 0.5T_S$, the proposed method's timing metric curve has a very clear peak as seen in Fig. 2. However, the method in [6] do not have so sharp timing metric, thus the proposed method can achieve a more accurate timing offset estimation.

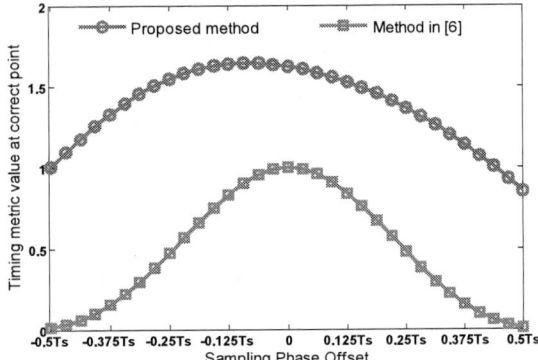

Fig. 1. The timing metric value at correct point versus sampling phase offset

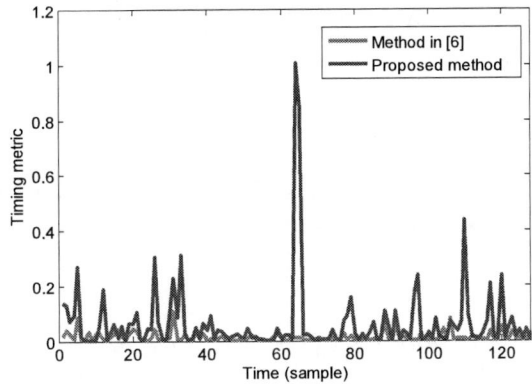

Fig. 2. Timing metric in numerical simulation (SPO=0.5Ts)

The experimental system setup is illustrated in Fig. 3. In this system, OFDM signal producing is operated on offline and coded by Matlab programming. The generated OFDM signal was downloaded to an Arbitrary Waveform Generator (AWG, Tektronix, 7122C) to generate electrical OFDM signal. The optical OFDM signal with an optical power of 7.5dBm modulated by DFB laser was transmitted over 25km SSMF.

In the receiver, after passing through a variable optical attenuator (VOA), the transmitted OOFDM signal is converted into the electrical baseband OFDM. After passing through an electrical low-pass filter with 3-dB bandwidth of 1GHz, the signal is sampled by a digital storage oscilloscope (DSO) and stored for off-line processing in Matlab. To avoid the difference of reference clock between AWG and DSO, we take the output 10MHz reference clock of ADC as the reference clock of AWG. In order to make the TS has different SPOs, we generate OFDM frames with only 2 OFDM symbols and set different sampling rates between AWG and DSO. Here, the sampling rates of DSO and AWG are set at 2GS/s and 2.000002GS/s, respectively.

OECC/PS2016

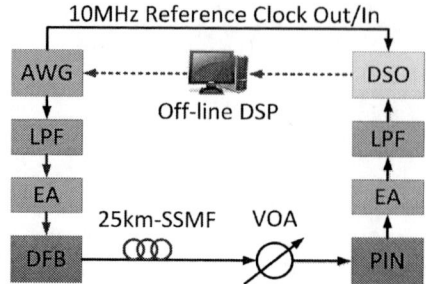

Fig. 3. Experimental setup of DDO-OFDM transmission system

The performance of the proposed method is evaluated by acquisition probability after 25km SSMF transmission, and is compared with the method in [6]. The acquisition probability curves versus the received optical signal power are shown in Fig. 4. It is clearly shown that the acquisition probability of the proposed method is higher than the other method though it will drop with the decreasing receiving power. From the picture, the acquisition probability of the proposed method is 99.37% when the received power is -24dbm, but the acquisition probability of the other method is only 99.04% when the received power is -20dbm, so the proposed method can improve more than 4dB optical power than the method in [6] at the acquisition probability of 99%.

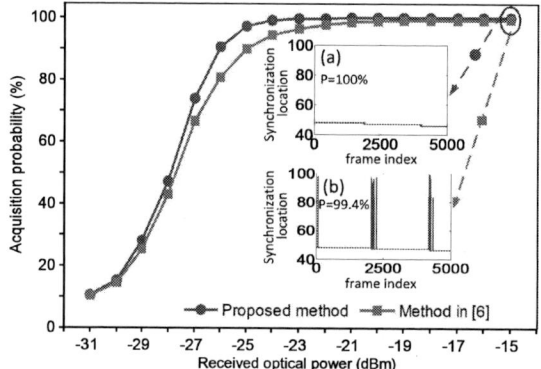

Fig. 4. Curves of acquisition probability of timing synchronization versus the received power for after transmission over 25-km SSMF

IV. CONCLUSIONS

A robust timing estimation method based on reverse correlation between the received signals is proposed. Theoretical and experimental investigations involved different SPOs show that the proposed method can improve at least 4dB on received optical power sensitivity than the method in [6] at the acquisition probability of 99% in DDO-OFDM transmission system with asynchronous sampling clock.

ACKNOWLEDGMENT

This work was supported in part by the Natural Science Foundation of China (Project No. 61132004, 61275073, 61420106011) and the Shanghai Science and Technology Development Funds (Project No.13JC1402600, 14511100100, 15511105400).

REFERENCES

[1] M. Chen, J. He, J. Tang, X. Wu, and L. Chen, "Experimental demonstration of real-time adaptively modulated DDO-OFDM systems with a high spectral efficiency up to 5.76 bit/s/Hz transmission over SMF links," Optics express, vol.22, pp.17691-17699, July 2014.

[2] H. Kimura, H. Nakamura, K. Asaka, S. Kimura, and N. Yoshimoto, "16QAM signal transmission experiment for dynamic SNR management on IM-DD OFDM-PON," 18th OptoElectronics and Communications Conference and Photonics in Switching (OECC/PS 2013), 2013, p.WP2_2

[3] C. R. N. Athaudage, "BER sensitivity of OFDM systems to time synchronization error," Proc. of 8th Int. Conf. on Communication Systems (ICCS 2002), Nov. 2002, pp.42 -46.

[4] T.M. Schmidl and D.C. Cox, "Robust frequency and timing synchronization for OFDM," IEEE Trans. Commun, vol. 45, pp.1613-1621, December 1997.

[5] B. Park, H. Cheon, C. Kang, and D. Hong, "A novel timing estimation method for OFDM systems," IEEE Commun. Lett, vol.7, pp.239-241, May 2003.

[6] X. Di, L. Chen, J. Xiao, M. Chen, J. He, J. Yu, and Y. Cheng, "A novel timing offset estimation method for direct-detection optical OFDM systems," Optical Fiber Technology, vol. 19, pp. 523-528, December 2013.

Performance Comparison of Nyquist-4PPM-QPSK and QPSK in Unrepeatered Transmission Systems

Jiangchuan Pang, Yan Li*, Miao Yu, Lei Yue, Sujie Fan, Deming Kong and Jian Wu

State Key Laboratory of Information Photonics and Optical Communications,
Beijing University of Posts and Telecommunications, 100876, Beijing, China
liyan1980@bupt.edu.cn

Abstract: Performance of 28 Gbit/s 4PPM-QPSK, Nyquist-4PPM-QPSK, QPSK and Nyquist-QPSK are investigated in unrepeatered transmission links. It is found that the maximum reach of Nyquist-4PPM-QPSK outperforms QPSK (@BER=1E-3) under the same spectrum efficiency.
Keywords: Nyquist-4PPM-QPSK, unrepeatered transmission, spectrum efficiency, power sensitivity.

I. INTRODUCTION

Unrepeatered transmission system is a cost-effective solution to transmit high capacity channels over moderate distances of several hundred kilometers without any in-line active elements such as submerged repeaters [1]. Spectrum efficiency and receiver sensitivity are two significant performance metrics for higher capacity and longer reach. And there is a continued quest to improve the receiver sensitivity without reducing spectrum efficiency in optical communication systems [2], particularly for unrepeatered transmission systems. Coherent detection of quadrature phase-shift keying (QPSK) is recognized as a promising modulation format to provide high sensitivity and spectrum efficiency [3]. Recently, combination polarization-division-multiplexed quadrature phase-shift keying (PDM-QPSK) with m-ary pulse position modulation (mPPM) is proposed to improve the receiver sensitivity, which demonstrated in unrepeatered transmission systems [4, 5] and free space optics communication system [6], however, the spectrum efficiency is lower than QPSK.

In [7], Nyquist-shaped mPPM-QPSK (N-mPPM-QPSK) is proposed to achieve higher power sensitivity compared with QPSK under the same spectrum efficiency. In this paper, the performance of N-mPPM-QPSK and QPSK in unrepeatered transmission systems is compared. Simulation results show that N-mPPM-QPSK outperforms QPSK in both unrepeatered single mode fiber (SMF) and ultra-large-area fiber (ULAF) links.

II. PRINCIPLE AND SYSTEM SETUP

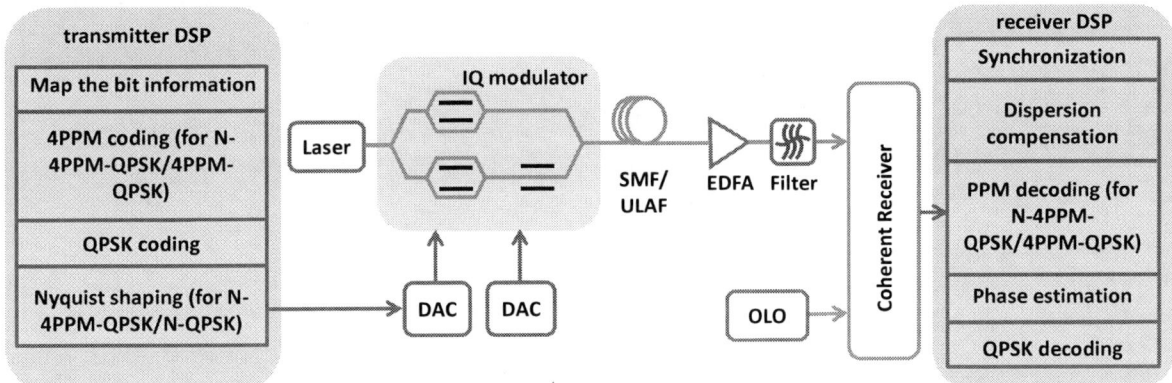

Fig. 1. Schematic of 28-Gb/s 4PPM-QPSK, Nyquist-4PPM-QPSK, QPSK, Nyquist-QPSK unrepeatered transmission systems. (DAC: digital-to-analog converter, EDFA: Erbium-doped fiber amplifier, OLO: optical local oscillator.)

Fig. 1 shows the schematic of the simulation setup for the generation and detection of 28 Gb/s N-4PPM-QPSK, 4PPM-QPSK, N-QPSK and QPSK signals. Each 4PPM-QPSK symbol contained 4 bits, of which the first 2 bits are encoded through 4PPM and the remaining 2 bits are encoded through QPSK. In the transmitter digital signal processor (DSP) diagram, for 4PPM-QPSK signal, the PRBS data with length of 2^{12}-1 is mapped into 4-bit symbols and then 4PPM-QPSK signal is generated after the 4PPM coding and QPSK coding. For QPSK signal, the PRBS data is mapped into 2 bits symbols to realizing the QPSK coding directly. For practical reasons, the N-4PPM-QPSK and N-QPSK signals are generated by shaping 4PPM-QPSK and QPSK signals utilizing the Nyquist filters which emulated by 4th order super Gaussian filters with bandwidths of 14 GHz and 7 GHz. Afterwards, the generated signals are oversampled to 20 samples per symbol. At the transmitter, the optical carrier is emitted from an external cavity laser (ECL) at 1550 nm with a linewidth of ~100 kHz and followed by an I/Q modulator. The optical spectrums after I/Q modulation are shown in Fig. 2, the bandwidths of 4PPM-QPSK, N-4PPM-QPSK, QPSK and N-QPSK are 56 GHz, 28 GHz, 28 GHz and 14 GHz, respectively, and the spectrum efficiency of them are 0.5 bit/s/Hz, 1 bit/s/Hz, 1 bit/s/Hz and 2 bit/s/Hz, respectively. It can be seen that the bandwidth of N-4PPM-QPSK is halved, compared with the 4PPM-QPSK, while equals to QPSK. Two different transmission links, unrepeated SMF link and unrepeated ULAF link, are used for comparison. Table I. lists the parameters of the SMF and ULAF.

TABLE I. PARAMETERS OF FIBERS

	SMF	ULAF
Attenuation @ 1550 nm	0.2 dB/km	0.176 dB/km
Dispersion @ 1550 nm	17 Ps/nm-km	21 Ps/nm-km
Effective Area	80 um²	156 um²
Nonlinear index	2.7e-20 m²/W	2.7e-20 m²/W

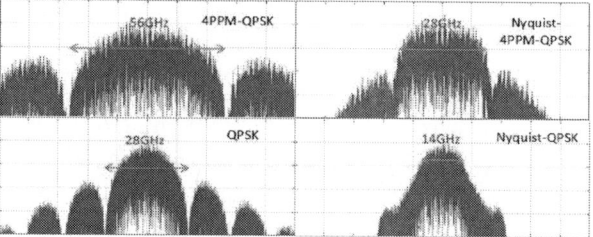

Fig. 2. Optical spectrums after I/Q modulation

At the receiver side, the received signal is firstly 40 dB amplified by an Erbium doped fiber amplifier (EDFA) with the noise figure of 4 dB and 50 GHz bandwidth before further filtered by a passive 0.4 nm optical bandpass filter (OBPF). After the coherent receiver, a digital 5th order Bessel low pass filter is applied, and the bandwidths of 4PPM-QPSK, N-4PPM-QPSK, QPSK, N-QPSK are optimized to 11.2 GHz, 11.2GHz, 8.4 GHz and 8.4 GHz respectively. The flow diagram of the receiver DSP is shown in Fig. 1. Firstly, timing synchronization and the digital dispersion compensation is realized. Then the processed signals are downsampled to 1 samples per symbol (or per slot). After that, a key step of 4-PPM decoding for 4PPM-QPSK and N-4PPM-QPSK signals is to find the time slot with the highest energy, out of the 4 slots of each N-4PPM-QPSK symbol. The position of the highest-energy slot is used to recover the first 2 bits associated with 4-PPM, the recovered optical field at this slot is used for phase estimation and recover the remaining 2 bits associated with QPSK. Common DSP algorithms are also used to recover the phase of the QPSK and N-QPSK signals. Finally, bit error rate (BER) values are evaluated by Monte Carlo Simulation.

III. RESULTS AND DISCUSSIONS

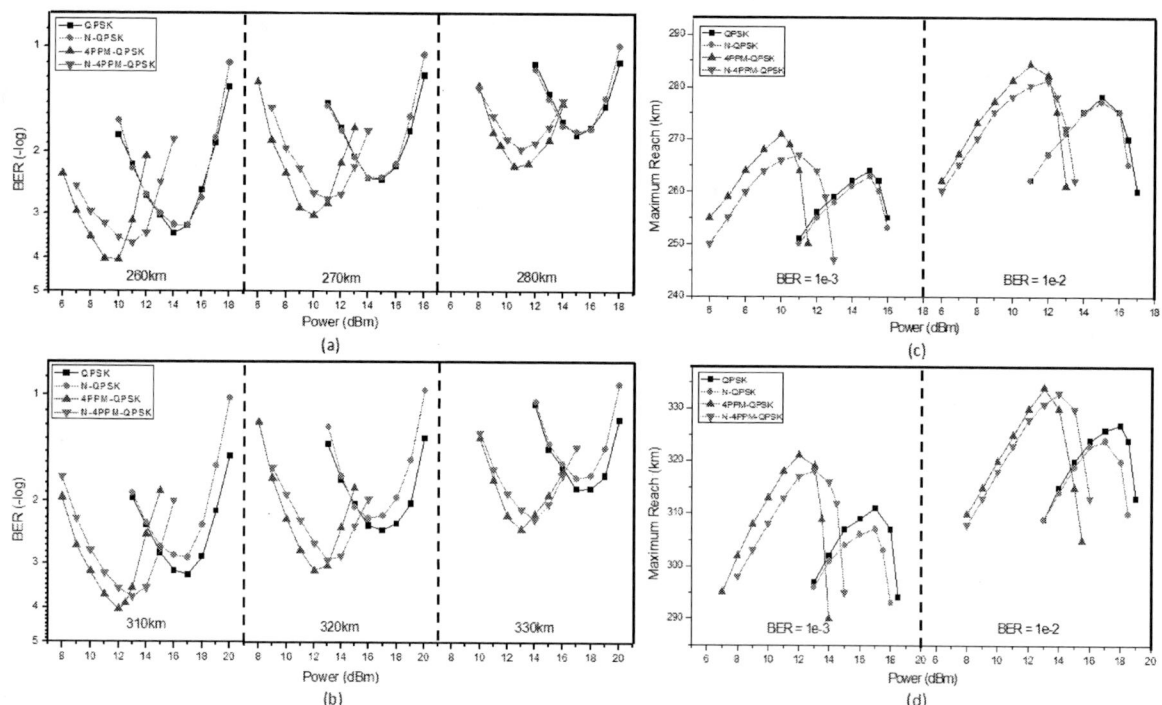

Fig. 3. (a) BER performance of four modulation formats after the unrepeated SMF transmission. (b) BER performance of four modulation formats after the unrepeated ULAF transmission. (c) Maximum reach of four modulation formats after the unrepeated SMF transmission. (d) Maximum reach of four modulation formats after the unrepeated ULAF transmission.

Fig. 3 (a) shows the BER performance of the 4PPM-QPSK, N-4PPM-QPSK, QPSK and N-QPSK formats as functions of launch power after 260 km, 270 km and 280 km unrepeated SMF transmission. Fig. 3 (b) depicts the BER results as functions of launch power after 310 km, 320 km and 330 km unrepeated ULAF transmission. Obviously, 4PPM-QPSK shows the best BER performance due to its widest bandwidth, and the N-QPSK with narrowest bandwidth shows the worst BER performance. N-4PPM-QPSK offers an advantage in BER performance at the optimal launch power compared with the QPSK, and the optimal launch power of the N-4PPM-QPSK is about 3~4 dB lower than that of the QPSK. This may be because of the bigger Peak-to-Average Power Ratio (PAPR) of the N-4PPM-QPSK leads to a worse performance in nonlinear tolerance compared with the QPSK. While the receiver sensitivity advantage of the N-4PPM-QPSK contributes to the better BER performance of N-4PPM-QPSK over QPSK, at the optimal launch power [7]. Therefore, N-4PPM-QPSK can not only improve the BER performance but also reduce the power consumption. In consideration of the ULAF has the smaller fiber nonlinearities and attenuation than the SMF, the improvement of the N-4PPM-QPSK is more significant.

As is shown in Fig. 3(c), at BER =1E-3, the N-4PPM-QPSK can reach 267 km when the launch power is 11 dBm in unrepeated SMF transmission, outperforms QPSK by 4 km, while the launch power of QPSK is as high as 15 dBm. The trend is the same for the BER of 1E-2. Fig. 3 (d) shows the maximum reach of the transmitted modulation formats in unrepeated ULAF transmission. As expected, the N-4PPM-QPSK outperforms QPSK by 8 km, the advantage is more obvious than that in unrepeated SMF transmission. Therefore, the N-4PPM-QPSK has advantages in both two unrepeated transmission systems compared with the QPSK.

IV. CONCLUSIONS

By numerical simulation, the performance of the 4PPM-QPSK, N-4PPM-QPSK, QPSK and N-QPSK formats are investigated under the same bit rate (28 Gb/s) and spectrum efficiency (1 bit/s/Hz) in both SMF and ULAF links. The maximum reach of N-4PPM-QPSK outperforms QPSK by 4 km and 8 km at FEC limit of 1E-3, for SMF and ULAF links respectively. N-4PPM-QPSK also takes 3~4 dB advantages in the optimal launch power over QPSK, which can reduce the power consumption of transmitters. It is anticipated that this power and spectrum efficient modulation format (N-4PPM-QPSK) may be attractive in future unrepeated transmission systems.

ACKNOWLEDGMENT

This work was partly supported by 863 program 2015AA015503, NSFC program 61475022, 61505011, 61331008, 973 program 2014CB340100, fund of state key laboratory of IPOC (BUPT), and the fundamental research funds for the central universities.

REFERENCES

[1] D. Chang, et al., "Unrepeated High-speed Transmission Systems," Opt. Fiber Commun. Conf., Paper W4E.3 (2015).

[2] D. O. Caplan, et al., "Free-space laser communications: Global communications and beyond," Eur. Conf. Opt. Commun., Tutorial, Paper 9.6.1 (2009).

[3] D. Chang, et al., "Ultra-long unrepeated transmission over 607 km at 100G and 632 km at 10G," Optics Express, 23(19): 25028-25033 (2015).

[4] X. Liu, et al., "Demonstration of record sensitivity in an optically pre-amplified receiver by combining PDM-QPSK and 16-PPM with pilot-assisted digital coherent detection," Opt. Fiber Commun. Conf., Paper PDB1 (2011).

[5] X. Liu, et al., "Demonstration of 2.7-PPB receiver sensitivity using PDM-QPSK with 4-PPM and unrepeated transmission over a single 370-km unamplified ultra-large-area fiber span," Eur. Conf. Opt. Commun., Paper Tu.3.B.4 (2011).

[6] W. Shi, et al., "Hybrid polarization-division-multiplexed quadrature phase-shift keying and multi-pulse pulse position modulation for free space optical communication," Optics Communications 334, 63-73 (2015).

[7] Miao Yu, et al., "Nyquist-mPPM-QPSK modulation for power and spectrum efficient optical communications," Opt. Fiber Commun. Conf., Paper W3D.6 (2016).

WA2-35

OECC/PS2016

Signal Light Carrier Automatic Phase-Lock Operation to Optical Frequency Grid Comb

Yudai Hisata, Akira Mizutori and Masafumi Koga

Oita University, 700, Dannoharu, Oita-city, Oita, 870-1192, Japan
Author e-mail address: {m.koga, mizutori}@oita-u.ac.jp

Abstract: This paper demonstrates signal light carrier automatic phase-lock operation by our newly developed microcomputer-controlled phase-lock loop circuit. It achieves 7.5msec automatic phase-locking time when the laser-diode frequency lies within the pull-in range of ±100MHz.
Keywords: Optical phase locked loop, optical carrier, automatic pull-in and phase- lock

I. INTRODUCTION

Due to the emergence of high-speed digital signal processors (DSP), interest in coherent communication has revived[1], and dramatic increases in spectral efficiency (SE) and transmission capacity have been reported in the last decade [2,3]. DSP-based schemes can calibrate local oscillator (LO) frequency offset and estimate the relative phase between carrier and LO lights. They can also precisely compensate the linear and nonlinear fiber characteristics as well as replicate the sharp cutoff characteristics of electrical filters that yield, approximately, the Nyquist minimum bandwidth [4]. However, the SE gain is limited to just a single channel and in wavelength-division multiplexing (WDM) transmission, the SE degrades because of the bandwidth unnecessarily reserved to counter frequency fluctuation in each optical signal [5], even though the sharp cutoff filter can realize such dense channel spacing that the optical carrier frequency spacing matches the bandwidth of the sharp cutoff filter. Furthermore, the amount of digital signal processing is becoming a significant burden.

We have also developed optical phase-lock loop technology to realize the optical synchronous network and demonstrated stable homodyne detection of 20Gbit/s QPSK signals [6]. If PLL technology becomes the enabling technology in not only long haul but also short distance optical transmission systems, we can enhance the light source frequency stability such that unnecessary bandwidth usage is suppressed, raise the symbol rate up to match the performance of pre-amplifiers and high-speed logic-ICs (not limited by analogue-to-digital conversion speed), and reduce the DSP requirements.

This paper demonstrates light carrier automatic pull-in and phase-lock operation to an optical frequency grid comb (OFGC). The microcomputer-controlled phase locked loop circuit (MC-PLL) is designed to achieve automatic pull-in of an optical signal carrier to an OFGC. The MC-PLL becomes active when the optical carrier frequency is set within ±100MHz from some frequency grid, and phase-locks the carrier to the OFGC in a few milliseconds. The loop filter used in the MC-PLL is the same integral, lag and lead loop filter (I&LL-LF) developed for 20Gbit/s QPSK signal homodyne detection described in reference [6]. The microcomputer makes the carrier frequency pull-in one of the spectral modes in OFGC by controlling the injection current of laser diode with a binary search algorithm (BSA) and phase-lock it with a phase error of 2.2°.

II. MICROCOMPUTER-CONTROLLED PHASE LOCK LOOP CIRCUIT

We have developed an MC-PLL circuit that realizes automatic pull-in and phase-lock operation. Fig.1 (a) shows the MC-PLL circuit while (b) shows a photograph. The MC-PLL consists of a microcomputer, I&LL loop filter, phase frequency detector (PFD), photo-diode (PD) followed with trans-impedance amplifier, and an LD that emits the signal carrier. The OFGC in Fig.1 (a) provides the ITU-T standardized grid frequencies of $f(n) = 193.1THz + 25GHz \times n$; where, n is integer from -15 to 15 [7]. The PD detects the beat frequency, f_{beat}, between the OFCR and LD lights and its phase error is sensed via the I&LL loop filter by comparing it to a 400MHz RF reference. By feeding back the phase error information to the injection current, I_{bias}, the LD oscillation frequency of the light carrier can be phase-locked to the OFGC. The red rectangle in Fig.1 (b) indicates the I&LL loop filter and the blue the microcomputer that runs the BSA. The BSA makes I_{bias} approach the value that corresponds to $f(n)$ by dividing the injection current variation, ΔI into two. Thus the LD carrier frequency pulls into $f(n)$ and becomes phase-locked to it.

The MC-PLL circuit becomes active when the LD frequency is set within the pull-in range of ±W. At this time, the PFD outputs positive or negative voltage ±v_0 according to the value of f_{beat}, as shown in Fig.1 (c). The BSA starts with the maximum variation of injection current ΔI_{max}, and the MC sets $I_{bias} + \Delta I_{max}$ if $f_{beat} > f(n)$ or $I_{bias} - \Delta I_{max}$ if $f_{beat} < f(n)$, as shown in Fig.1 (d). The PFD output is monitored and the MC sets ($I_{bias} + \Delta I_{max}$)-$\Delta I_{max}$/2 if $f_{beat} < f(n)$, and vice versa. This process is repeated to the count number set in advance. Upon discovering this state, MC activates the integral function of the I&LL loop filter.

OECC/PS2016

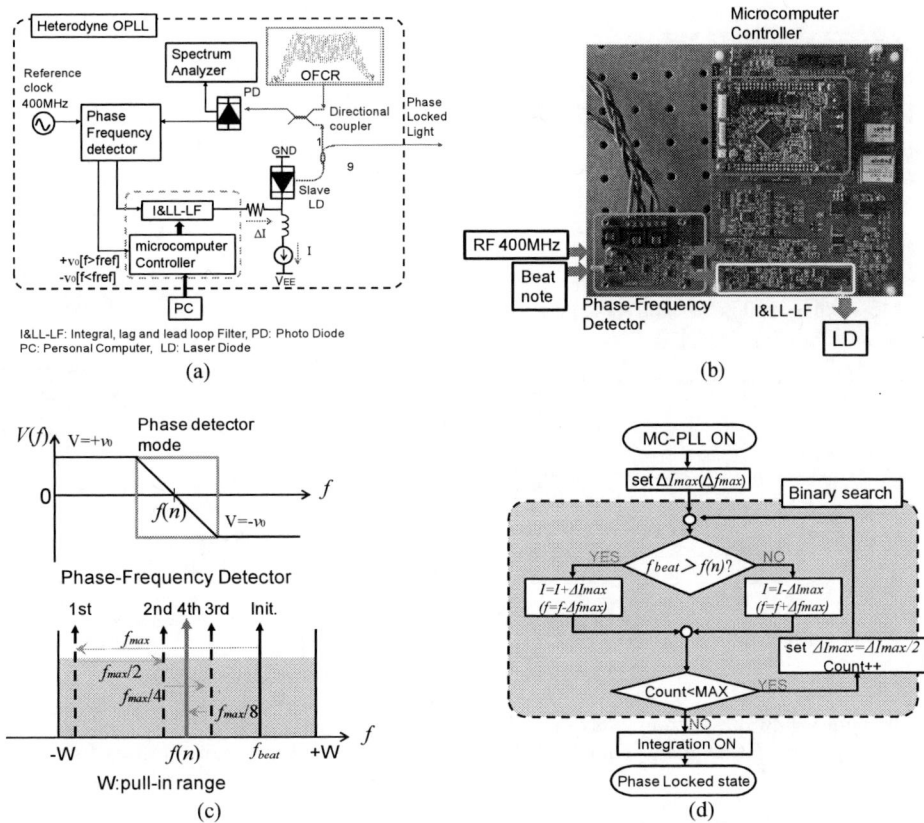

I&LL-LF: Integral, lag and lead loop Filter, PD: Photo Diode
PC: Personal Computer, LD: Laser Diode

(a)

(b)

(c)

(d)

Fig. 1 Microcomputer-controlled Phase-Lock Loop and its Binary Search Algorithm: (a) is MC-PLL circuit,(b) is photograph of prototyped MC-PLL circuit, (c) is PFD's characteristics and optical frequency behavior and (d) is pull-in binary search algorithm.

III. DEMONSTRATION OF AUTOMATIC PHASE-LOCKING OPERATION

We examined the automatic pull-in and phase-lock operation of our prototype circuit. The used LD was an external-cavity-structured LD (E-LD) that was identical to the one used in reference 6. Its FM response exhibited -20MHz/mA to -2.8MHz/mA between DC and 1MHz. The optical frequency pull-in operation was monitored by utilizing our optical frequency discrimination circuit that had been developed for 20Gbit/s QPSK homodyne detection, reported by reference 6, as shown in Fig.2 (a). The optical frequency discrimination circuit is comprised of 12.5GHz-spacing arrayed-waveguide grating (AWG), two pin-photodiodes followed by trans-impedance amplifiers and a differential amplifier. The one set of electrical amplifier forms a balanced amplifier that exhibits a common-mode rejection ratio (CMRR) of over 43dB for a measurement bandwidth between DC to 10MHz. The AWG's transmittance has a Gaussian profile of

$$f(x) = \exp\{-\frac{(x \pm x_0)^2}{2\sigma^2}\}$$

where, x_0=12.5/2=6.25GHz and σ=5.31GHz for the AWG used. The AWG crossover optical frequency was thermally controlled to $f(0)$=193.1THz, and the transmittance of two adjacent ports separated by 12.5GHz (3dB down) are shown in reference 6. The discrimination sensitivity at crossover frequency was 0.97MHz/mV when the AWG input optical power is 6.6dBm. Then the circuit output voltage when a frequency shifts by 100MHz was 103mV. Note that the common mode amplitude component, which was arisen from injection current variation, was cancelled out owing to the circuit CMRR and we could extract the true optical frequency behavior of which light was emitted from the LD.

Before processing the BSA, we set the LD optical frequency at about ±100MHz apart from $f(0)$ of OFCR by monitoring the beat frequency. After frequency setup, we triggered the process and confirmed automatic pull-in and phase lock operation. The result is shown in Fig.2.

When f_{beat} was set at -100MHz from reference clock frequency 400MHz, see the inset spectrum of Fig. 2 (b), the discriminator output showed the optical frequency automatically shifted from -100MHz to 0 in 7.5ms. When, on the other hand, set at +100MHz, Fig.2 (c), we could observe that the optical frequency shifted from 100MHz to 0 in 4ms. After transition, we confirmed the phase-lock state by observing the RF spectrum. The measured RF spectrum and SSB phase noise are shown in Fig.3 (a) and (b), respectively. We can see a smooth noise floor and 58dB beat power to noise level at around the beat spectrum. Note that the spur at 0.5MHz offset frequency observed in the beat spectrum is crosstalk from the E-LD's digital thermo controller. SSB phase noise has a standard deviation of only 2.1° and 2.2° at

827

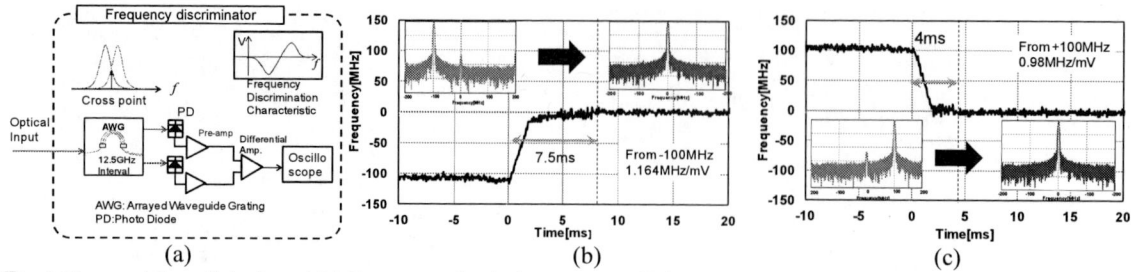

Fig. 2 Measured the pull-in times.(a) is Frequency discriminator setup , (b) is -100MHz, (c) is +100MHz apart from the 400MHz respectively.

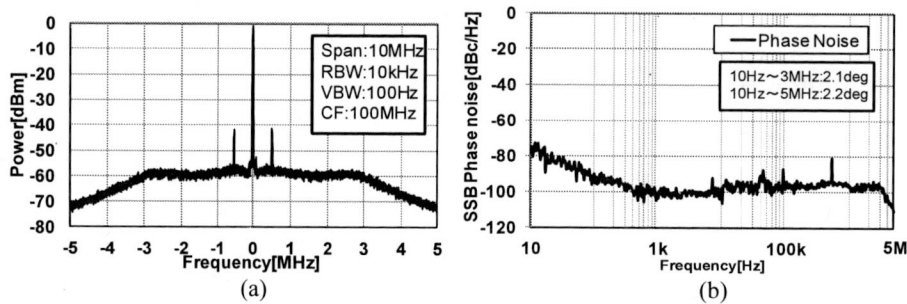

Fig. 3 Measured (a)RF spectrum and (b) is SSB phase noise

the measurement bandwidth of from 10Hz to 3MHz and 10 Hz to 5MHz, respectively (Agilent Technologies:N9010A), as shown by Fig. 3(b). These results confirm that our developed MC-PLL circuit achieves automatic pull-in and phase-lock operation.

IV. CONCLUSIONS

We successfully demonstrated signal light carrier automatic pull-in and phase-lock operation by our newly developed microcomputer-controlled phase-lock loop circuit. It achieves automatic phase-lock within 10ms when the laser-diode frequency lies within the pull-in range of ±100MHz. The achieved phase-lock state exhibited SSB phase noise of just 2.2° at 5MHz offset frequency.

ACKNOWLEDGMENT

The work was partly supported by National Institute of Information and Communication Technology (NICT) Japan and JSPS KAKENHI Grant Number 15H04009.

REFERENCES

[1] K. Kikuchi and S. Tsukamoto, "Evaluation of sensitivity of the digital coherent receiver," JLT, vol.26, no.13, pp.1817–1822, Jul. 2008.
[2] H. Takara, et al., "1.01-Pb/s (12 SDM/222 WDM/456 Gb/s) Crosstalk-managed Transmission with 91.4-b/s/Hz Aggregate Spectral Efficiency," ECOC2012, Th.3.c.1, Amsterdam, 2012.
[3] J. Sakaguchi, Y. Awaji, N. Wada, A. Kanno, T. Kawanishi, T. Hayashi,T. Taru, T. Kobayashi, and M. Watanabe, "109-Tb/s(7×97×172-Gb/s SDM/WDM/PDM)QPSK transmission through 16.8-km homogeneous multi-core fiber," OFC2011, PDPB6, 2011.
[4] A. Carena, V. Curry, G. Bosco, P. Poggiolini, F. Forghieri, "Modeling of the Impact of Nonlinear Propagation Effects in Uncompensated Optical Coherent Transmission Links " JLT, vol.30, NO.10,May 15, 2012
[5] OIF-TLMSA-01.0,Multi-Source Agreement for CW tunable Laser. May.2003.
[6] M. Koga, Y. Shigeta, F. Shirazawa, H. Ohta, A. Mizutori, "Costas Loop Homodyne Detection for 20-Gb/s QPSK Signal on the Optical Frequency Synchronous Network," JLT, Vol.33, No.23, Dec.1, 2015.
[7] ITU-T Recommendation G.694.1, "Spectral grids for WDM applications: DWDM frequency grid," Feb. 2012.

Maximum Ratio Combining Characteristics Affected by Laser Phase Noise for Wavelength Diversity Digital Coherent System

Akira Naka [1], Kohei Saito [2], Kazuki Tomita[1], and Hideki Maeda [2]

[1]Ibaraki University,   4-12-1, Naka-Narusawa-Cho, Hitachi-shi, Ibaraki, 316-8511, Japan

[2]NTT Network Service System Laboratories, NTT Corporation, 3-9-11, Midori-Cho, Musashino-Shi, Tokyo, 180-8585, Japan

akira.naka.dr@vc.ibaraki.ac.jp

Abstract: We numerically investigated MRC characteristics which deteriorate significantly by laser phase noise in adaptive MIMO equalization processing for wavelength diversity transmission, and verified that two-step-MIMO processing prevents deterioration.

Keywords: coherent communications, digital signal processing, MIMO, CPR

I. INTRODUCTION

Recent digital coherent optical technologies have contributed to realizing high-speed transmission systems of 100Gbps and beyond [1], and space division multiplex (SDM) transmission systems are expected to increase capacity further utilizing multi-core and multi-mode fibers where adaptive MIMO (multiple-input and multiple-output) equalization processing techniques are applied to de-multiplex signals [2][3][4]. In addition to higher capacity for point-to-point transmission, flexible optical path handling is another promising measure that aims to cope with increasing amounts of data traffic by deploying optical cross connection functions on flexible optical grids and creating a flexible optical transport network [5]. To achieve such a network, we have proposed and experimentally verified hitless operation of the optical path switching and spectrum defragmentation using the wavelength diversity systems, where MIMO processing is adopted to realize MRC (Maximum Ratio Combining) of two or more optical signals having the same data sequence which are separately transmitted on different optical grids in one fiber, or are transmitted along different routes [6] [7].

Experimental SDM systems as well as commercial 100Gbps systems deliberately avoid the effect of laser phase noise both by generating each polarization multiplexed signal from a single laser source at the transmitter and detecting the signals with several optical hybrid receivers sharing a single local oscillator followed by adaptive MIMO equalization process. The wavelength diversity system is, however, preferably composed of several independent laser sources to handle two or more optical wavelength channels with different colors for the transmitter and receiver in order to maintain scalability and avoid multiple optical path failure. Specifically, one phase-locked multi-carrier light source could generally be applied only for less than ten channels, and its failure could affect all related channels. Therefore, MIMO processing for multi-channels and laser phase noise effect in MIMO process should be more carefully investigated. In this paper, we numerically evaluate MRC characteristics realized by two configurations for adaptive MIMO equalization processing, namely, one-step MIMO and two-step MIMO. We show that the two-step MIMO offers stable MRC by eliminating the effect of phase noise, whereas one-step MIMO presents unstable Q-value, occasionally failing MRC subject to the laser phase noise condition.

II. NUMERICAL CALCULATION MODEL

Figure 1 schematically shows the overall configuration model for the numerical calculation. Two wavelength signals, CH1 and CH2, are respectively imposed on phase noise independent of each other to simulate two free-running laser

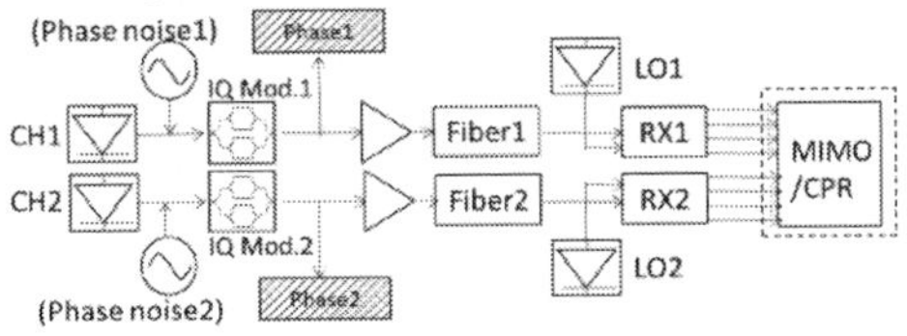

Fig. 1. Block diagram of calculation model

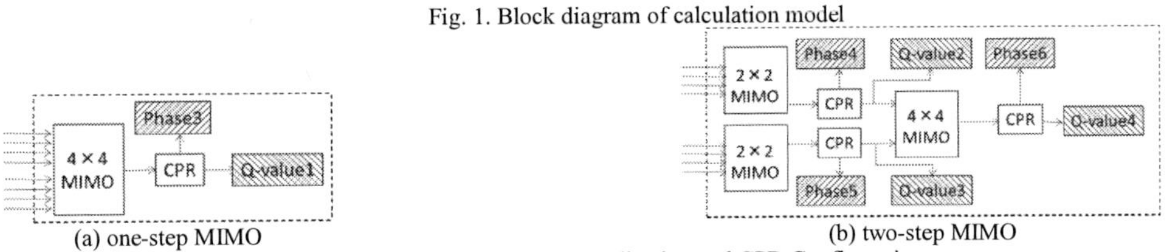

(a) one-step MIMO

(b) two-step MIMO

Fig. 2 Adaptive MIMO equalization and CPR Configurations

sources and generate two Dual -Polarization Quadrature Phase-Shift Keying (DP-QPSK) wavelength signals with the symbol rate of 32Gbaud. Each polarization channel of CH1 and CH2, whose data sequence is independently composed of a combination of PRBS patterns, is identical to the other. Amounts of phase noise on CH1 and CH2 are respectively monitored as Phase 1 and Phase 2 indicated in Fig. 1, which are estimated by the 4th–power phase estimation algorithm commonly adopted as a CPR process [8] [9]. We average 24 amounts of the estimated phase noise at each sampled symbol and accumulate them to monitor their continuous phase change.

These signals are respectively amplified by an optical amplifier with white optical noise as ASE (amplified spontaneous emission), the amount of which determines the Optical Signal Noise Ratio (OSNR) condition. Each of the signals is respectively transmitted on each fiber route, suffering different polarization rotations and differential delay between two polarization modes. We fix the delay of 1 bit and 2 bit for each route in this calculation. Chromatic dispersion is not considered here because the dispersion affects each polarization multiplexed signal in the same manner and can be removed by signal processing before MIMO. The transmitted signals are respectively detected by optical hybrid receivers followed by the MIMO and CPR process. Phase noise from the local oscillator, which would be simply added to each signal, is here ignored for simplicity of numerical analysis.

Figure 2 shows the two configurations of adaptive MIMO equalization and CPR for our analysis. Fig. 2 (a) shows one-step MIMO, in which one 4x4 MIMO is applied followed by one CPR process, whereas Fig. 2 (b) shows two-step MIMO, in which two 2x2 MIMO respectively detects each polarization multiplexed signal of CH1 or CH2 and removes its phase noise with the first CPR as a first step, and then 4x4 MIMO combines the two polarization multiplexed wavelength signals as a second step. For every MIMO process, we use CMA (Continuous Modulus Algorithm). Amounts of phase noise are monitored as Phase 3 through Phase 6 at the points indicated in Figs. 2 (a) and (b) as part of the CPR process with 4th–power algorithm as the same manner to estimate Phase 1 and Phase 2. Q-values are evaluated after 4x4 MIMO in both configurations as Q-value 1 and Q-value 4 to make MRC of CH1 and CH2, whereas Q-value 1 and Q-value 2 are measured after 2x2 MIMO and CPR to individually retrieve each CH1 and CH2 signal for references.

III. CALCULATION RESULTS

Figure 3 shows the amount of phase noise obtained as a function of time corresponding to the detected signal sequence. Continuous phase changes are observed with additive random noise in Fig. 3 (a) and (d) on the conditions of linewidths of 100 kHz and 1 MHz for CH1 and CH2, respectively. The amount of phase noise in Fig. 3 (d) is much larger than that in Fig. 3 (a) according to the linewidth conditions. The lines in between Phase 1 and 2 (yellow) in Figs.3 (a) and (d) respectively depict average amounts of Phase 1 and Phase 2 for references.

Figs. 3 (b) and (e) show the amounts of estimated phase noise on CH1 and CH2 indicated as Phase 3 through Phase 6. Each amount of Phase 4 and Phase 5 on the condition of 100 kHz in Fig.3 (b) respectively coincides well with each of Phase 1 and Phase 2 in Fig. 3 (a). Phase 3 and Phase 4 in Fig.3 (b) change more greatly than Phase 1 and Phase 2 to a small extent due to additional phase noise stemming from ASE on the OSNR condition of 16dB in this calculation. This result verifies that each 2x2 MIMO process respectively compensates the polarization difference delay and retrieves each initial optical signal well enough to adequately work CPR function on the way of two-step MIMO.

Amount of Phase 3 in Fig. 3 (b) also coincides well with the average amount of Phase 1 and Phase 2, which means one-step MIMO illustrated in Fig.2 (a) also correctly combines CH1 with CH2 on the condition of 100 kHz. To be noted, we numerically confirmed many times that the amount of Phase 3 coincides well with the average of Phase1 and

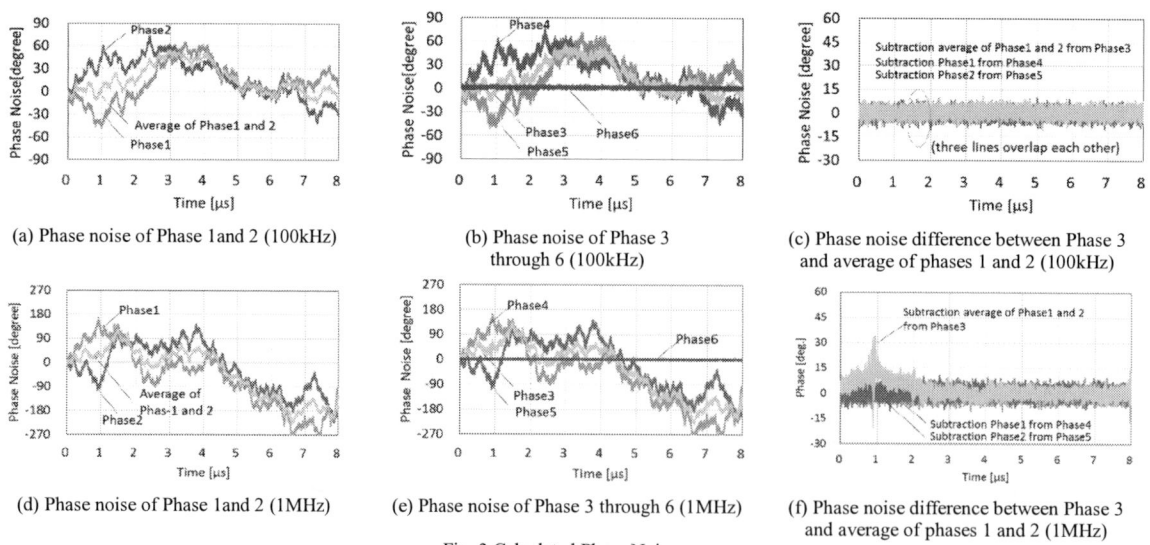

(a) Phase noise of Phase 1and 2 (100kHz)

(b) Phase noise of Phase 3 through 6 (100kHz)

(c) Phase noise difference between Phase 3 and average of phases 1 and 2 (100kHz)

(d) Phase noise of Phase 1and 2 (1MHz)

(e) Phase noise of Phase 3 through 6 (1MHz)

(f) Phase noise difference between Phase 3 and average of phases 1 and 2 (1MHz)

Fig. 3 Calculated Phase Noise

Phase 2 weighted in Q-value 2 and Q-value 3 at various combinations of OSNRs of CH1 and CH2 on this linewidth of 100 kHz.

In contrast, the amount of Phase 3 on the linewidth of 1MHz in Fig. 3 (e) occasionally differs from the average amount of Phase 1 and Phase 2. This difference is highlighted in Fig. 3 (f) as illustrated with the line in yellow that is obtained by subtracting the average of Phase 1 and Phase 2 from Phase 3. The value has large fluctuation especially at around 1μs, where Phase 1 and Phase 2 each has a large difference value from the other. This result suggests that MRC cannot be achieved in a one-step MIMO configuration when laser phase noise is 1MHz. On the other hand, the amount of Phase 6 has values around zero as lapse of time in Fig. 3 (e), which shows that phase noise is eliminated at the first CPR in two-step MIMO. Actually, the second step of the CPR in two-step MIMO can be omitted.

Figure 4 shows evaluated Q-values, namely Q-value 1, Q-value 2, and the average value of Q-value 2 and 3 as a function of linewidth on the condition of OSNR 16dB for both CH1 and CH2. Both obtained results of one-step MIMO and two-step MIMO improve nearly 3dB Q-value on wavelength conditions of less than 500 kHz. However, the Q-values deteriorates significantly in one-step MIMO configuration at linewidths larger than 500 kHz, whereas Q-values in two-step MIMO stably improves the signal quality by MRC.

Figure 5 shows the amount of Q-value improvement comparing Q-value 1 or Q-value 4 with average value of Q-value 2 and 3 on several OSNR conditions around 16 dB. The horizontal axis is the difference between Q-value 2 and Q-value 3. A theoretical curve, which is derived from averaged Q-values weighted in Q-value 2 and Q-value 3, almost coincides with all obtained results due to adequate MRC, except for the Q-value 1 in one-step MIMO on 1 MHz linewidth, under which condition adequate MRC is not correctly achieved.

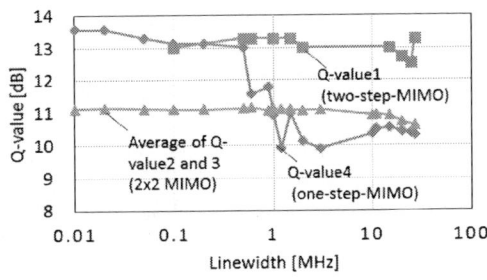

Fig. 4. Calculated Q-value as a function of linewidth

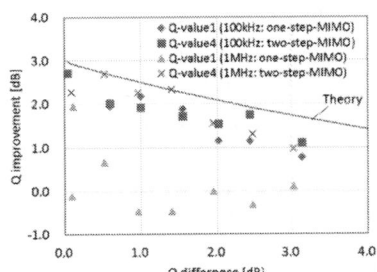

Fig. 5. Calculated Q-improvement by combining two signals

IV. CONCLUSIONS

We numerically investigated MRC characteristics affected by laser phase noise using two adaptive MIMO equalization processing configurations for wavelength diversity systems to realize a flexible optical transport network. Two-step MIMO is verified to offer stable MRC and improves Q-values of two channels regardless of amount of the phase noise, whereas one-step MIMO occasionally fails MRC and Q-value is significantly deteriorated by laser phase noise on the laser linewidth of more than 500 kHz. The obtained results would provide useful information on designing digital coherent systems for future application which uses adaptive MIMO process for multi-channels.

REFERENCES

[1] A. Sano et al., "Ultra-high capacity WDM transmission using spectrally-efficient PDM 16-QAM modulation and C- and extended L-band wideband optical amplification," J. Lightwave Technol. 29(4), pp.578–586 (2011).

[2] K. Shibahara et al., "Dense SDM (12-core × 3-mode) transmission over 527 km with 33.2-ns mode-dispersion employing low-complexity parallel MIMO frequency-domain equalization", Proc. OFC, Th5C.3 (2015).

[3] B. J. Puttnam et al., "2.15 Pb/s Transmission Using a 22 Core Homogeneous Single-Mode Multi-Core Fiber and Wideband Optical Comb", Proc. ECOC, PDP3.1 (2015)

[4] M. Koga et al., "Q-value improvement by electrical maximum ratio combining optical diversity transmission through multi-core fibre," Proc. OECC2014, WE9B-2, pp. 698-698 (2014).

[5] M. Jinno et al., "Demonstration of novel spectrum-efficient elastic optical path network with per-channel variable capacity of 40 Gb/s to over 400 Gb/s," Proc. ECOC, Th. 3. F. 6 (2008)

[6] S. Yamamoto et al., "Hitless spectrum defragmentation in flexible grid optical network using maximum ratio combining in wavelength diversity transmission", Proc. OECC2015, JThB.33 (2015).

[7] K. Saito et al., "Evaluation of maximum ratio combining in route diversity transmission and application to hitless optical path switching in field installed fibre", Proc. ECOC2015, P.4.1, (2015).

[8] Dany-Sebastien Ly-Gagnon et al., "Coherent Detection of Optical Quadrature Phase-Shift Keying Signals With Carrier Phase Estimation", J. Lightwave Technol. 24(1), pp.12-21 (2006).

[9] F. Hamaoka et al., "Transmission performance improvement in digital coherent system by suppressing cycle slip using statistical analysis algorithm", Elec. Lett., 49, (13), pp. 826-827 (2013).

Multi-level Pre-Equalization Using Analog FIR Filters Based on 28-nm FD-SOI for 20-Gb/s 4-PAM Multi-Mode Fiber Transmission

Ryoichiro Nakamura, Kenta Amino, Kawori Sekine, Kazuyuki Wada, and Moriya Nakamura
School of Science and Technology, Meiji Univ., 1-1-1 Higashi-Mita, Tama-ku, Kawasaki-shi, Kanagawa, 214-8571 Japan
ce51043@meiji.ac.jp

Abstract: We propose a novel multi-level pre-equalizer composed of multiple binary analog FIR filters, which achieves cost-effectiveness and low power consumption. The performance was investigated using 28-nm FD-SOI based CMOS circuits for 20-Gb/s 4-PAM MMF transmission.
Keywords: Equalization, pre-distortion, FIR filter, 4-PAM, Multi-mode fiber, 28-nm FD-SOI

I. INTRODUCTION

Higher speed and larger capacity data transmission in fiber optic communication link is required to accommodate rapidly increasing data traffic. 400 Gbit Ethernet is reported as the construction of high speed network link [1]. Multi-level modulation is one key-technology to increase the transmission capacity. In 400-Gbit Ethernet systems, four-level-pulse-amplitude-modulation (4-PAM) have been investigated as a cost-effective modulation scheme. One of limiting factor of transmission speed in fiber optic communication links is inter-symbol interference (ISI). ISI is caused by interference between adjacent symbols of transmitted signals. The factors of ISI in optical fiber transmission are, e.g., chromatic dispersion (CD), polarization mode dispersion (PMD), and differential modal delay (DMD). ISI caused by DMD is the dominant factor in case of multi-mode fibers (MMFs), which are used for short-haul low-cost systems. Various schemes are investigated to compensate DMD. One report employed optical domain compensation using mode filtering [2]. Electrical domain compensations using multipurpose digital-signal processors (DSPs) have also been studied, in which digital filters such as finite impulse response (FIR) filters are calculated by the DSPs [3,4]. On the other hand, analog FIR filter based on analog electronic circuit technology is also attractive scheme which realize cost-effectiveness and low power consumption [5-8]. We have investigated analog FIR filters using binary delay-line components based on complementary metal-oxide-semiconductor (CMOS) inverters, which can realize smaller integrated-circuit (IC) chip-size [9]. However, they can be used only for pre-distortion of binary signals for, e.g., binary phase-shift keying (BPSK) transmission, because they use binary delay-lines. In this paper, we propose a novel multi-level pre-equalizer using multiple binary analog FIR filters for multi-level modulation signals. The performance was investigated by numerical simulations of 28-nm fully depleted silicon on insulator (FD-SOI) based CMOS circuits for 20-Gb/s 4-PAM MMF transmission.

II. CONSTRUCTION OF MULTI-LEVEL PRE-EQUALIZER

A. Multi-level pre-equalizer using binary analog FIR filters

We designed multi-level pre-equalizer by using multiple binary analog FIR filters which consist of Gilbert Cells as multipliers and CMOS inverters as binary delay-lines. Figure 1 shows block diagram of the pre-equalizer for four-level signal composed of data1 and data2. The pre-equalizer consists of parallel binary FIR filters and 1/2-weighted adder. The two FIR filters have the same tap-coefficients h_k and tapped delay time T. Binary input signals data1 and data2 are pre-equalized by the two FIR filters independently, and added after 1/2-weighting. By this calculation, we can achieve the pre-equalization of the four-level signal, thanks to the linearity of the FIR filters.

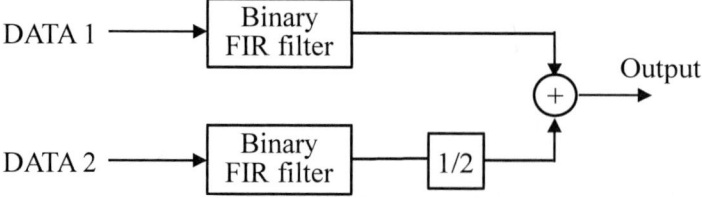

Fig. 1. Block diagram of four-level pre-equalizer using binary analog FIR filters.

B. Gilbert Cell as Multiplier

A FIR filter consists of multipliers and delay lines. Proposed multi-level pre-equalizer was composed of Gilbert Cells as multipliers and CMOS Inverters as delay lines. Figure 2 shows our designed Gilbert Cell. The output voltage of this circuit V_{out} can be expressed as

$$V_{out} = R_D(I_{D1} + I_{D3}) - R_D(I_{D2} + I_{D4}),\qquad(1)$$

where I_{D1}, I_{D2}, I_{D3}, and I_{D4} are currents flowing through MOSFET M_1, M_2, M_3, and M_4, respectively. The MOSFET pairs of M_1/M_2, M_3/M_4, and M_5/M_6 have the same W/L ratio. The currents can be adjusted by control voltages (V_{cont1} and V_{cont2}). The tap-coefficient of the multiplier can be proportionally adjusted by deferential voltage V_{cont1}-V_{con2} [9].

C. CMOS Inverters as Delay Line

Figure 3 shows a delay line which consists of CMOS inverters. Each CMOS inverter works as a binary switch. The rise-time t_r and the fall-time t_f of a CMOS inverter can be determined by on-state resistance of the MOSFETs and gate capacitance of the following inverter. The gate capacitance C_g and the on-state resistance R_{on} are described as

$$C_g = W \times L \times C_{ox},\qquad(2)$$

$$R_{on} = \frac{1}{\frac{W}{L}\mu C_{ox}(V_{gs}-V_{th})}.\qquad(3)$$

The tapped delay time can be determined by the rise-time and the fall-time of a CMOS inverter and the number of the cascaded inverters [9].

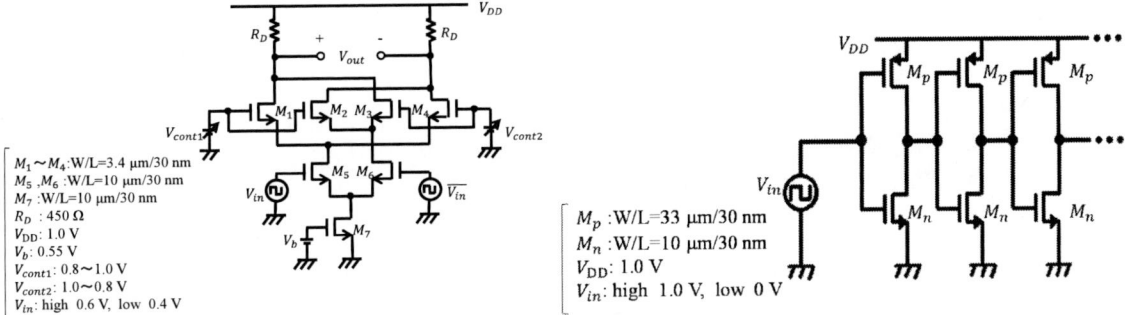

Fig. 2. Gilbert cell   Fig. 3. CMOS Inverters.

III. SYSTEM SETUP

Figure 4 schematically shows system setup of 20-Gb/s 4-PAM transmission over 500-m MMF for our numerical simulation. DFB laser of 1550-nm wavelength was directly modulated by the four-level signal which was pre-equalized as described in section II. *A.*, achieving pre-equalized optical 4-PAM signal. The pre-equalizer has 6 taps where tap-coefficients had been determined by least-mean-square (LMS) algorithm. The four-level signal was composed by 10-Gsymbol/s PRBS 2^9-1 binary data. The modulated optical signal was transmitted by a 500-m MMF with 50-μm parabolic-index core. A mode scrambler was used at the input of the MMF to evenly excite higher order modes. Received optical power was adjusted using an attenuator (ATT), and directly detected by a photodetector (PD). Eye diagram and Error Vector Magnitude (EVM) were measured by an oscilloscope.

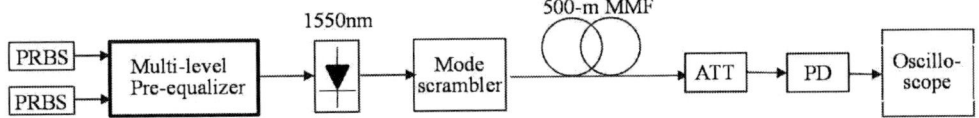

Fig. 4. System setup of 20-Gb/s 4-PAM transmission over 500-m MMF.

IV. RESULT AND DISCUSSION

Figures 5 show eye-diagram of the transmitted 4-PAM signals. Figure 5 (a) is eye-diagram in case that the pre-equalization was not employed. The waveform was completely distorted by DMD of the MMF. Figure 5 (b) shows eye-diagram in case that the pre-equalization was employed. Clear eye-opening could be achieved by using the multi-level pre-equalizer. Calculated EVM values versus optical received power are shown in Fig. 6. In the case without the pre-equalization, EVM did not decrease less than 40 %, even if we increased optical received power. In the case with the pre-equalizer, EVM of 9 % could be achieved. The results clearly shows the effectiveness of our proposed multi-level pre-equalizer.

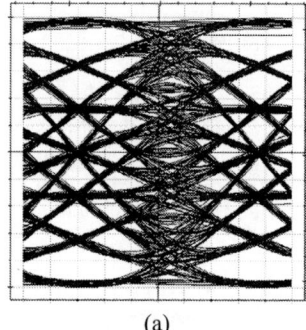
(a)

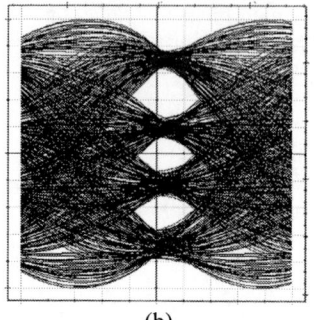
(b)

Fig. 5. Eye-diagram of transmitted 4-PAM signals. (a) Without equalization. (b) With equalization using 6-tap multi-level pre-equalizer.

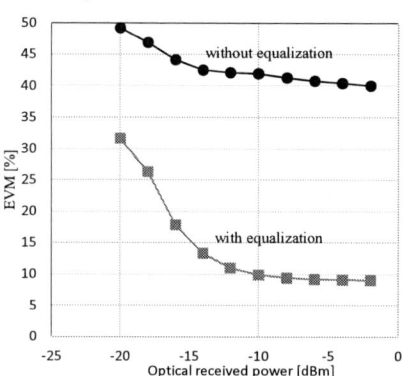

Fig. 6. EVM characteristics.

V. CONCLUSION

We proposed a novel multi-level pre-equalizer using multiple binary analog FIR filters. The analog FIR filters was composed of Gilbert Cells and CMOS inverters designed by 28-nm FD-SOI based CMOS circuit. The equalization technology should encourage the application of analog FIR filter to future cost-effective and low power-consuming high-speed optical data links.

REFERENCES

[1] R. Hirai and N. Kikuchi, "Current Status on Standardization of 400 Gigabit Ethernet," IEICE Technical Report, vol. 115, no. 407, pp. 17-22, Jan. 2016.

[2] K. M. Patel and S.E. Ralph, "Multimode Fiber Link Equalization by Mode Filtering via a Multisegment Photodetector," IEEE MTT-S Int. Microwave Symp. Dig, vol. 2, pp. 1343-1346, Jun. 2003.

[3] T. Mori, T. Sakamoto, T. Yamamoto, and S. Tomita, "Modal Dispersion Compensation by Using Digital Coherent Receiver with Adaptive Equalization in Multi-Mode Fiber Transmission," Optical Fiber Technology, vol. 19, Issue. 2, pp. 132-138, Mar. 2013.

[4] X. Zhao and F. S. Choa, "Demonstration of 10Gb/s Transmissions Over a 1.5-km-Long Multimode Fiber Using Equalization techniques," IEEE Photon. Technol. Lett., vol. 14, no. 8, pp. 1187-1189, Aug. 2002.

[5] C. Pelard, E. Gebara, A. J. Kim, M. Vrazel, E. J. Peddi, V. M. Hietala, S. Bajekal, S. E. Ralph, and J. Laskar, "Multilevel Signaling and Equalization over Multimode Fiber at 10 Gbit/s," in Proc. IEEE GaAs Integrated Circuits Symp., pp. 197-199, Nov. 2003.

[6] H. Wu, J. A. Tierno, P. Pepeljugoski, J. Schaub, S. Gowada, J. A. Kash, and A. Hajimiri, "Integrated Transversal Equalizers in High-Speed Fiber-Optic Systems," IEEE J. Solid-State Circuits, vol. 38, no. 12pp. 2131-2137, Dec. 2003.

[7] M. Maeng, Y. Hur, S. Chandramouli, F.Bien, H. Kim, C. Chun, E. Gebara, and J. Laskar, "Equalization and near-end crosstalk (NEXT) noise cancellation for 20 Gb/s 4-PAM backplane serial I/O interconnections," IEEE Trans. Microw. Theory Tech., vol. 53, no. 1, pp. 246-255, Jan. 2005.

[8] M. Maeng, F. Bien, Y. Hur, H. Kim, S. Chandramouli, E. Gebara, and J. Laskar : "0.18-μm CMOS Equalization Techniques for 10-Gb/s Fiber Optical Communication Links", IEEE Trans. Microw. Theory Tech., vol. 53, no. 11, Nov. 2005.

[9] K. Amino, S. Watanabe, K. Sekine, Y. Kitani, R. Nakamura, K. Wada, M. Nakamura, "An analog FIR filter Design Using Inverters and Gilbert Cells with 28-nm FD-SOI Process for Compensating Optical Transmission Distortion," to appear in IEEJ Technical Meeting on Electronic circuits, Mar. 2016. (in Japanese)

WA2-38

OECC/PS2016

Blind polarization de-multiplexing of time domain hybrid QAM signals based on radius-directed linear Kalman filter

Qun Zhang[1], Wen Jiang[1], Yanfu Yang[1*], Xian Zhou[2], Kangping Zhong[2], Yong Yao[1]

[1]Department of Electronic and Information Engineering, Shenzhen Graduate School, Harbin Institute of Technology, Shenzhen, Guangdong Province 518055, China
[2]Department of Electronic and Information Engineering, The Hong Kong Polytechnic University, Hung Hom, Kowloon, Hong Kong, China
*yangyanfu@hotmail.com

Abstract: *We propose a method of blind polarization de-multiplexing for TDHQ signals based on radius-directed linear Kalman filter. The influences of power ratio, format ratio, polarization rotation, residual chromatic dispersion, polarization mode dispersion are analyzed.*
Keywords: *TDHQ, Kalman filter, polarization de-multiplccexing.*

I. INTRODUCTION

Elastic optical networks aided with coherent technologies are a promising technology in dynamic transmission network, which can adaptively adjust link capacity or distance according to actual conditions of network and data rate cater and variable cloud computing requirements [1]. A transmitter using time domain hybrid quadrature amplitude modulation (QAM) (TDHQ) scheme can works well for different communication needs by configuring power ratio and frame length ratio [2]. Meanwhile, Efficient DSP algorithms are required for TDHQ signals on the receiver side. Several algorithms are adopted for polarization multiplexing in coherent receiver, including constant modulus algorithm (CMA), multi-modulus algorithm (MMA) and their modified form to de-multiplex and track polarization state. However, they have the disadvantages of low convergence/tracking speed and singularity problem [3]. Recently a simultaneous polarization and phase recovery algorithms based on extended Kalman filtering (EKF) for TDHQ signal is proposed, which has good convergence speed and no singularity problems whereas EKF is susceptible to frequency offset [4]. Yang proposed and experimentally demonstrated a blind radius-directed linear Kalman filter (RD-LKF) scheme to recover polarization state for quadratue phase shift keying (QPSK) or 16QAM. The ultrafast polarization state tracking ability are achieved with the immunity to carrier phase noise and frequency offset [5]. Here, we further applied RD-LKF scheme for polarization de-multiplexing for TDHQ signals.

In this paper, we proposed a blind polarization de-multiplexing scheme for TDHQ signals based on radius-directed linear Kalman filter (RD-LKF). The scheme is immune to carrier phase noise and frequency offset compared with the previous EKF method and can provide fast polarization de-multiplexing capability for time domain QPSK/16QAM hybrid signals. This feature can enable the implementation of flexible transceiver in future dynamic optical network. In the following, the principle of the RD-LKF based on polarization recovery of time-domain hybrid QPSK/16QAM signals is introduced. The proposed scheme is investigated through numerical simulation regarding the performance dependence on the hybrid format parameters and the system parameters, including power ratio (PR), format ratio (FR), polarization rotation, residual chromatic dispersion and polarization mode dispersion (PMD).

II. PRINCIPLE

Assuming PDL is not considered, the polarization multiplexing model [6] can be simplified as:

$$X(n) \cdot e^{j(\Delta\omega n)} \cdot e^{j\theta(n)} + \xi(n) = (J(n))^{-1} \cdot Z(n) = H(n) \cdot Z(n) \tag{1}$$

In which $X(n)$, $Z(n)$, $\Delta\omega$, $\theta(n)$, $\xi(n)$, $J(n)$ describe the polarization recovered signal, the received signal, carrier frequent offset, phase noise, additive white Gaussian noise and polarization rotation related Jones matrix, respectively. The proposed RDLKF scheme could mathematically expressed as the following equations:

$$H(n) = \begin{bmatrix} a(n) + jb(n) & c(n) + jd(n) \\ c(n) - jd(n) & a(n) - jb(n) \end{bmatrix} \tag{2}$$

$$S(n) = \begin{bmatrix} a(n) & b(n) & c(n) & d(n) \end{bmatrix}^{\mathrm{T}} \tag{3}$$

$$S(n) = S(n-1) + U(n) \tag{4}$$

$$Z_{out}(n) = H(n) \cdot Z(n) = M(n) S(n) + V(n) \tag{5}$$

OECC/PS2016

$$M(n) = \begin{bmatrix} Z_x(n) & jZ_x(n) & Z_y(n) & jZ_y(n) \\ Z_y(n) & -jZ_y(n) & -Z_x(n) & jZ_x(n) \end{bmatrix}$$

(6)

Eqs.(2-6) are the mathematical description of the Kalman filter, in which Eqs.(3-5) describe the state vector, the state equations and measurement equations, respectively. In the scheme single measurement method with one fixed target circle are employed in order to achieve a universal implementation in case of dynamic hybrid signals. The radius of the target circle are fixed at 1.0 and the received signal are normalized to have the power of 1.0. More details of Kalman filter could be found in [5]. It's worth noting that Q, R refers to process noise covariance and measurement noise covariance as well tuning parameters. In the following simulation, R is considered to be 0.1, and Q is settled as 1e-5.

III. SIMULATION SYSTEM

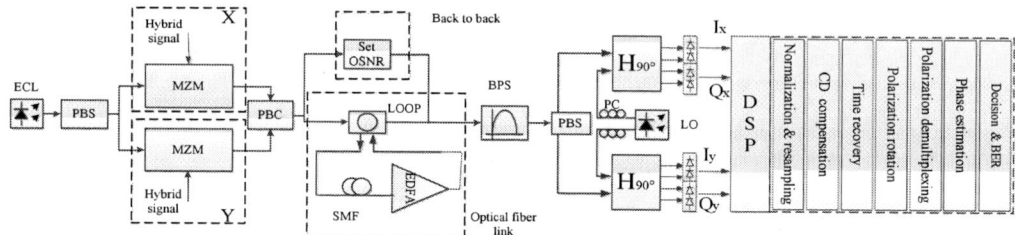

Fig. 1. Simulation scheme of 14 Gbaud QPSK/16QAM TDHQ coherent optical communication system.

Fig. 1 shows the simulation mode of a 14Gbaud QPSK/16QAM TDHQ coherent optical communication system. The IQ modulator is driven by electrical signals to generate QPSK/16QAM TDHQ signals at 14Gsymbol/s. Two modes of back to back and optical fiber link are considered. In the latter mode the polarization rotation, residual dispersion, polarization mode dispersion are taken into account. After polarization-diversity coherent detection, the data is processed offline by digital signal processing, including cubic spline interpolation based timing recovery, polarization de-multiplexing, and blind phase search (BPS) based carrier phase estimation.

IV. SIMULATION RESULTS

Firstly the influence of PR on polarization recovery performance for hybrid QPSK/16QAM signals are simulated. In the simulation the parameters of system and RDLKF are set as follows: the symbol rate is set as 14Gbaud, the laser line width is 100 kHz, RDLKF iterative update rate is equal to the system symbol rate (14GSymbol/s) and OSNR is 17 dB. The FR is fixed at 1:1 and the PR is set to three typical values of {1:5, 1:1, 9:5} to adjust relative position (inner circle, middle circle, outermost circle) of QPSK on 16QAM signal constellation diagrams. As shown in Fig.2, we use the product of the channel polarization related Jones matrix J and its inverse matrix H estimated by RDLKF ([A1 A2; A3 A4]) to plot convergence curves. The obtained curves illustrates the optimal choice would be PR at 1:5 in view of the most clear constellation and superior convergence results compared with other two alternatives.

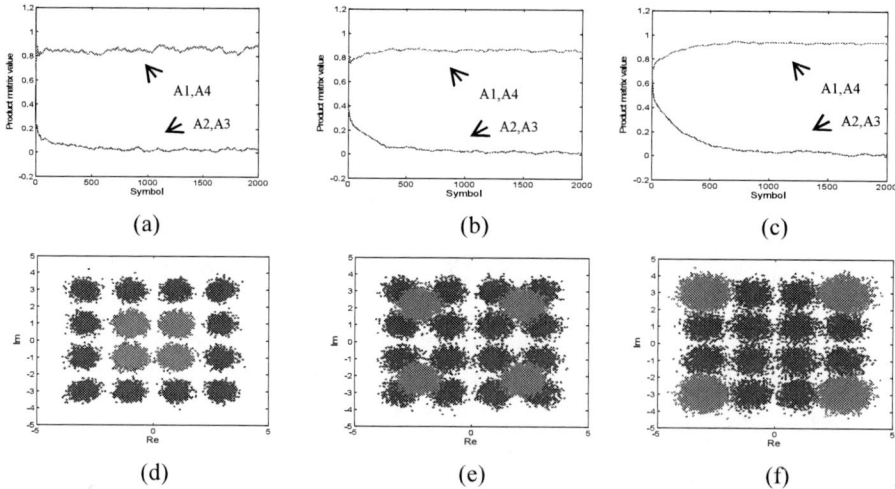

Fig. 2.Convergence performance of (a) PR 1:5 (b) PR 1:1 (c) PR 9:5; Recovery constellation of (d) PR 1:5 (b) PR 1:1 (c) PR 9:5 when FR=1:1.

In the following, the influence of different FR between QPSK and 16QAM on system BER performance are discussed under the optimal PR of 1:5. Fig.3 presents that the higher FR means better BER performance and less OSNR penalty caused by de-multiplexing process. When the target BER is set as 1e-3, the system at FR=1:1 corresponding to OSNR of 12.75 dB and roughly 0.05 dB OSNR penalty in comparison to FR=1:7 with OSNR of 17.6 dB and around 0.4dB penalty.

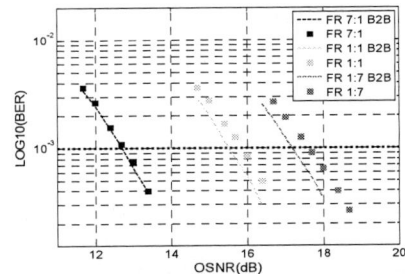

Fig. 3.BER vs. OSNR performance when PR 1:5.

Finally the OSNR penalty induced by system parameters are analyzed by considering the impact of polarization rotation, residual chromatic dispersion and differential group delay (DGD). We considered the optimal PR of 1:5 and the variable FR between QPSK and 16QAM of {7:1 1:1 1:7}. The target BER is 1e-3. In Fig. 4(a), with the polarization rotation increase, higher FR lead to less OSNR penalty that is to say, better polarization rotation tolerance. Similarly, those conclusions can be found in the curves of OSNR penalty vs. PMD and residual chromatic dispersion, as shown in Figure 4(b), 4(c). Those results can be understood by considering that higher FR means more QPSK are included in THDQ signals. The penalty induced by residual dispersion and polarization mode dispersion can be further mitigated by incorporating finite impulse response based channel equation in this scheme and the related work will be presented in the future.

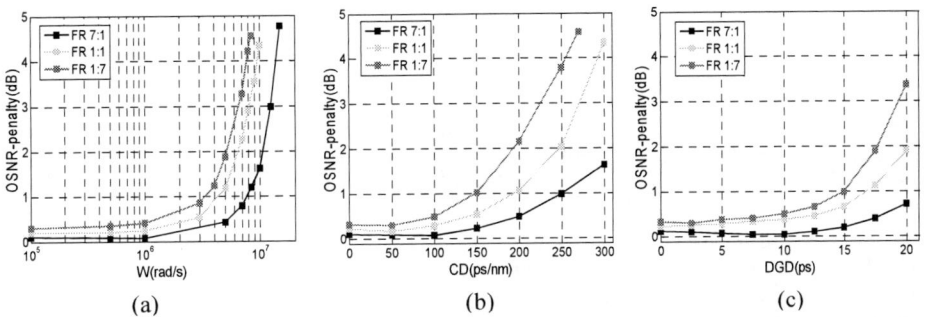

Fig. 4. The OSNR penalty as a function of (a) polarization rotation (b) residual chromatic dispersion (c) polarization mode dispersion when PR 1:5.

V. Conclusions

We proposed and demonstrated a RDLKF scheme to track polarization state for time domain hybrid QPSK/16QAM signals. The influence of PR between QPSK and 16QAM are analyzed and the confirmed optimal PR of 1:5. The results showed the higher FR values lead to better tolerance and superior BER performance against polarization rotation, residual chromatic dispersion and PMD as a result of larger format ratio of QPSK. The scheme can meet the need of dynamic communicate requirements by adjusting FR and PR value and has the application potentials for universal and blind polarization de-multiplexing for future dynamic optical network.

Acknowledgment

This work was supported by the National Natural Science Foundation of China (Contract No. 61205046, 61401020, 61575051 and 61505039) and Shenzhen Municipal Science and Technology Plan Project (JCYJ20150327155705357, KQCX2015032409501296, JSGG20150529153336124 and JCYJ20150529114045265).

References

[1]   A. P. T. Lau, Y. Gao, Q. Sui, et al.,"Advanced DSP Techniques Enabling High Spectral Efficiency and Flexible Transmissions: Toward elastic optical networks," IEEE Signal Process. Mag. 31(2), 82–92 (2014).

[2]   Y. Gao, et al.,"Blind and universal dsp for arbitrary modulation formats and time domain hybrid qamtransmissions,"In Optical FiberCommunication Conference (Optical Society of America, 2014, March), pp. Th3E-5.

[3]   K. Kikuchi, et al., "Performance analyses of polarization demultiplexing basedon constant-modulus algorithm in digital coherent optical receivers," Opt.Exp., vol. 19, no. 2, pp. 9868–9880, May. 2011.

[4]   W. Jiang, Q. Zhang, et al., "Blind and Simultaneous Polarization and Phase Recovery for Time Domain Hybrid QAM Signals Based on Extended Kalman Filtering," in Asia Communications and Photonics Conference, 2015, p. AS4F. 2.

[5]   Y. Yang, G. Cao, K. Zhong, X. Zhou, Y. Yao, A.P.T. Lau, C. Lu,"Fast polarization-state tracking scheme based on radius-directed linear Kalman filter,"Opt. Exp., 23 (2015) 19673-19680.

[6]   T. Marshall, et al., "Kalman filter carrier and polarization-state tracking," Optics letters 35.13, 2203-2205 (2010).

OECC/PS2016

Relationship between roll-off factor and transmission distance in Nyquist OTDM scheme based on correlation detection with EDFA repeaters

Takahiro Oguro, Yuji Miyoshi, Hirokazu Kubota and Masaharu Ohashi

Graduate School of Engineering, Osaka Prefecture University, 1-1 Gakuen-cho, Nakaku, Sakai, Osaka, 599-853 Japan

sv106003@edu.osakafu-u.ac.jp

Abstract: *We investigate the Nyquist OTDM scheme based on optical correlation detection in an EDFA repeater system. The maximum transmission distance and optimum transmission power are clarified as functions of roll-off factor and repeater spacing.*

Keywords: *OTDM; optical Nyquist pulse; optical correlation detection*

I. INTRODUCTION

A Nyquist optical-time-division-multiplexing (OTDM) scheme using an optical Nyquist pulse was proposed to realize low inter-symbol interference (ISI) and high spectral efficiency with an ultra-high-speed transmission [1,2]. In order to realize high optical-signal-to-noise ratio (OSNR) tolerance, a Nyquist OTDM scheme based on correlation-detection using optical root-Nyquist pulses was proposed [3, 4]. The correlation-detection scheme detect and de-multiplex OTDM signals without optical switches [5, 6]. An amplified spontaneous emission (ASE) noises of an erbium doped fiber amplifier (EDFA) degrades signal quality in a transmission line. The correlation detection scheme can suppress the signal quality degradation to the minimum limit. The scheme based on correlation detection achieves the high OSNR tolerance as a matched filter by digital signal processing [7]. In long haul transmission systems, many EDFA repeaters are used for compensating a transmission loss. The repeater spacing and transmission power affect signal quality due to Kerr effects and OSNR degradation in transmission lines. Additionally, a small roll-off factor of root-Nyquist pulses is required to increase the spectral efficiency. However, the effect of the roll-off factor on received signal quality in the EDFA repeater system is not clarified.

In this paper, we investigate the relationships of transmission properties among the roll-off factor, transmission power and repeater spacing. Then, we clarify the maximum transmission distance and optimum transmission power as functions of the roll-off factor and repeater spacing.

II. SIMULATION MODEL OF THE NYQUIST OTDM SCHEME WITH EDFA REPEATER

Fig. 1 shows a simulation model of the Nyquist OTDM scheme with EDFA repeaters. An optical modulator and signal with limited 3-dB bandwidth of MB_t by root-Nyquist filter (where M and B_t are the number of multiplexed signals and the baud rate of tributary signals, respectively) generated tributary signals with modulation format of binary phase shift keying (BPSK). The tributary signals were multiplexed by an OTDM multiplexer. The transition link consisted of a single mode fiber (SMF), a dispersion compensating fiber (DCF), and an EDFA. n is the number of loops in the EDFA based system. The optical correlation receiver using optical root-Nyquist pulses as the reference signal detected the transmitted signals. The output of the optical correlation receiver is expressed as follows:

$$u(t) = \int_{t-\frac{MT}{2}}^{t+\frac{MT}{2}} s(\tau)r^*(\tau)d\tau, \tag{1}$$

where s and r^* are the received signal and a phase conjugate of the reference signal, respectively. T is a time slot of the multiplexed signals. The detected signal had an average phase offset due to self-phase modulation (SPM), even if the phase noise and frequency offset between the received signal and the reference signal were ignored. The offset was compensated by a carrier phase estimation using the fourth power method [8].

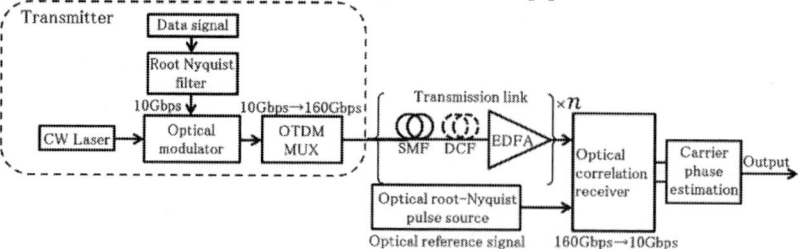

Fig. 1. Simulation model of the Nyquist OTDM scheme with an EDFA repeater.

838

OECC/PS2016

TABLE I. SIMULATION PARAMETERS

Signal parameters			SMF			
Wavelength[nm]		1550	Loss[dB/km]			0.2
Baud rate [Gbaud]		160	Chromatic dispersion[ps/(km·nm)]			17
OSNR before transmission[dB]		30	Nonlinear parameter[1/(W·km)]			2.0
Time slot[ps]		6.25	EDFA			
Power of reference signal [mW]		1	Noise figure[dB]			5.0
Multiplicity		16	Gain[dB]	(Loss[dB/km])*(Transmission distance[km])		

III. SIMULATION RESULTS

We investigated the relationships of the transmission properties among the roll-off factor α, transmission power P_{in} and repeater spacing L_{sp}. Table I shows the parameters used in the simulation. EDFA gain was determined so as to compensate the loss of the SMF. The transmission signal sequence used an M-sequence pseudo-random number sequence with a pattern length of 2^7-1 period. The loss and optical nonlinear effects in the DCF were ignored. The total chromatic dispersion of the SMF and the DCF were set at zero.

Fig. 2 shows the relationship between P_{in} and Q factor as a function of the loop number when the L_{sp} and α were 100 km and 0.5, respectively. The Q factor deteriorated due to the ASE noise when the transmission power P_{in} was low. Conversely, when the P_{in} was high, the Q factor deteriorated because of the nonlinear effect. The Q factor decreased as the number of loops increased.

Fig. 3 shows the relationship between the number of loops n and the transmission power P_{in} as a function of L_{sp}, when the Q factor was 15.6 dB (BER = 10^{-9}). α was set at 0.5. The hatching area referred to regions where the Q factor exceeded 15.6 dB. The allowable range of P_{in} decreased as the L_{sp} increased. There exists an optimum transmission power, which can realize the maximum number of loops.

Fig. 4 shows the maximum transmission distance and the optimum transmission power as a function of L_{sp} for the α of 0.5. As the L_{sp} became shorter, the optimum transmission power and maximum transmission distance became lower and longer, respectively. The transmission signal was susceptible to the Kerr effect when the L_{sp} was short. However, when the L_{sp} was long, the ASE noise had a large effect on the maximum transmission distance. This was because the signal power entering the EDFA weakened.

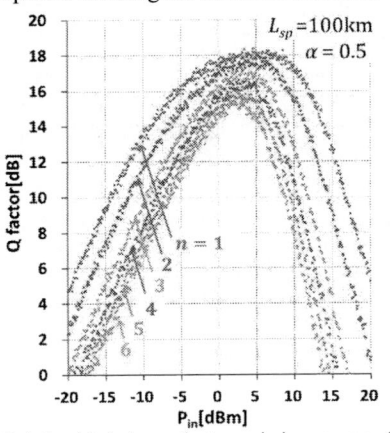

Fig. 2. Relationship between the transmission power and the Q factor.

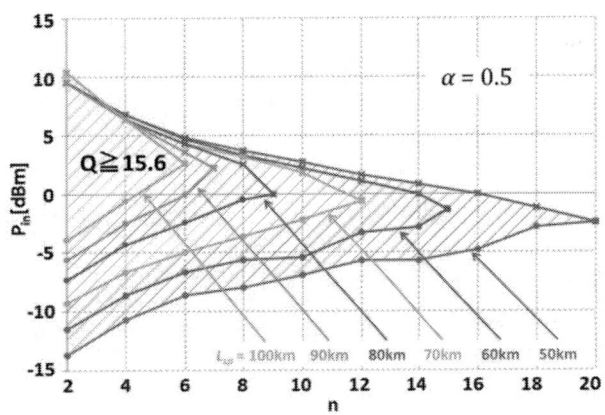

Fig. 3. Relationship between the number of loops and the transmission power at the Q factor of 15.6 dB as a function of repeater spacing.

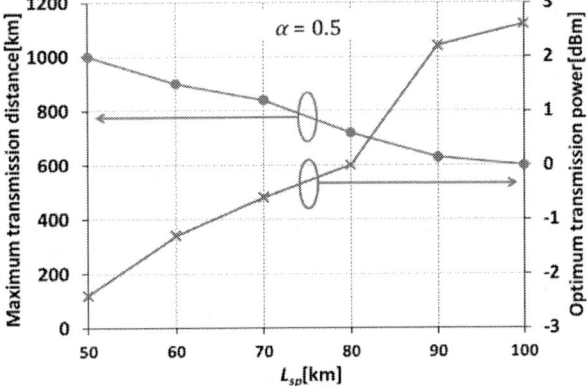

Fig. 4. Maximum transmission distance and optimum transmission power as a function of repeater spacing for the roll-off factor of 0.5.

Fig. 5(a) shows the relationship between the α and the maximum transmission distance. For a given L_{sp}, the maximum transmission distance was almost the same when the roll-off factor was greater than or equal to 0.1. The maximum transmission distance decreased when the α approached zero. Fig. 5(b) shows the relationship between α and the optimum transmission power. The optimum transmission power decreased with a decrease in L_{sp}. In order to increase the spectral efficiency without degrading the maximum transmission length, it is necessary for the α to be chosen as 0.1.

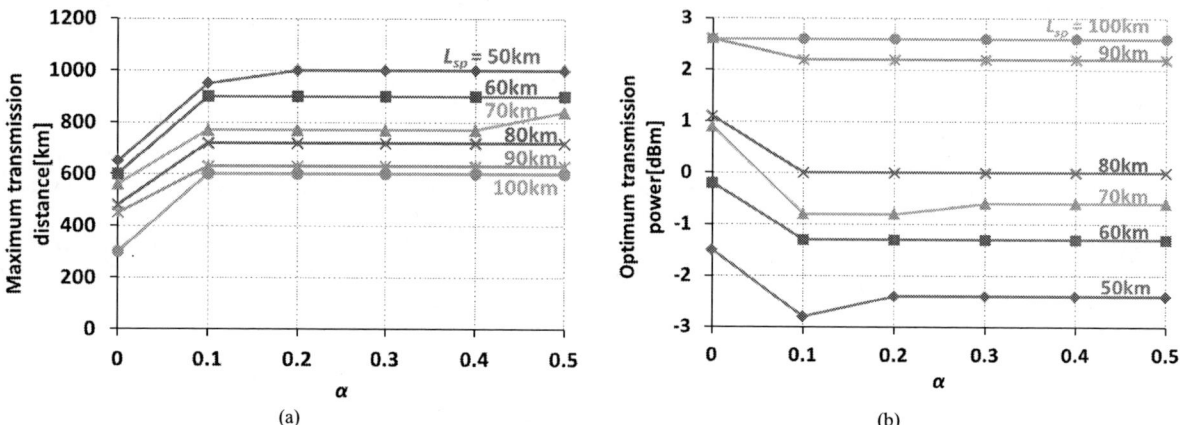

(a)   (b)

Fig. 5. Relationships (a) between roll-off factor and maximum transmission distance, and (b) between roll-off factor and optimum transmission power.

IV. CONCLUSIONS

The relationships of transmission properties among the roll-off factor, transmission power and repeater spacing was investigated with a signal distortion caused by Kerr effects and OSNR degradation by EDFA in the Nyquist OTDM scheme.

The maximum transmission distance and optimum transmission power dependence were specified by using the roll-off factor and repeater spacing in the Nyquist OTDM scheme. As the repeater spacing decreased, the maximum transmission distance and the optimum transmission power increased and decreased, respectively. When the roll-off factor was more than or equal to 0.1, the maximum transmission distance and optimum transmission power dependence of the roll-off factor were almost identical. Investigations also indicated that the spectral efficiency increased without degrading the maximum transmission distance when the roll-off factor was 0.1.

Additionally, the maximum transmission distances and optimum transmission powers for the repeater spacing of 50 km were obtained as 1000km and -2.4dBm, respectively. The maximum transmission distances and optimum transmission powers for the repeater spacing of 100 km were 600 km and 2.6 dBm, respectively.

ACKNOWLEDGMENT

This work was supported by JSPS KAKENHI Grant Number 15K18068.

REFERENCES

[1] M. Nakazawa, T. Hirooka, P. Ruan and P. Guan, "Ultrahigh-speed "orthogonal" TDM transmission with an optical Nyquist pulse train," OSA Opt. Express, vol. 20, pp. 1129-1140, 2012.

[2] T. Hirooka and M. Nakazawa, "Linear and nonlinear propagation of optical Nyquist pulses in fibers," OSA Opt. Express, vol. 20, pp. 19836-19849, 2012.

[3] Y. Miyoshi, H. Kubota, M. Ohashi, "Nyquist OTDM scheme using optical root-Nyquist pulse and optical correlation receiver," IEICE Electron. Express, vol. 11, pp. 20130943, 2014.

[4] Y. Miyoshi, H. Kubota, M. Ohashi, "Experimental demonstration of correlation detection with limited receiver bandwidth in Nyquist OTDM scheme," IEICE Communications Express, vol. 5, pp.33-38, 2016.

[5] F. Ito, "Demultiplexed detection of ultrafast optical signal using interferometric cross-correlation technique," J. Lightwave Technol., vol. 15, pp. 930–937, 1997.

[6] T. Richter, E. Palushani, C. Schmidt-Langhorst, R. Ludwig, L. Molle, M. Nölle, and C. Schubert, "Transmission of single-channel 16-QAM data signals at terabaud symbol rates," J. Lightwave Technol., vol. 30, pp. 504–511, 2012.

[7] J. G. Proakis and M. Salehi, "Digital Communications 5th ed.," Mc Graw Hill, New York, 2005.

[8] D. S. Ly-Gagnon, S. Tsukamoto, K. Katoh, K. Kikuchi, "Coherent detection of optical quadrature phase-shift keying signals with carrier phase estimation," J. Lightw. Technol., vol. 24, pp. 12-21, 2006.

WA2-4

OECC/PS2016

Elastic-bandwidth Access with Spectrum-Sliced Incoherent Light Source in WDM-PON

Jing Zhang[1,2], Qikai Hu[2], Changyuan Yu[2,3], Xingwen Yi[1], and Kun Qiu[1]

[1]Key Lab of Optical Fiber Sensing and Communications, Ministry of Education,
University of Electronic Science and Technology of China, Chengdu, 611731, China

[2]Dept. of Electrical & Computer Engineering, National University of Singapore, 4 Engineering Drive 3, Singapore 117576

[3] Dept. of Electronic and Information Engineering, The Hong Kong Polytechnic University, Hung Hom, Kowloon, Hong Kong
zhangjing1983@uestc.edu.cn

Abstract: We propose and experimental demonstrate an elastic-bandwidth access scheme using spectrum-sliced incoherent light (SSIL) source with adaptive OFDM modulation in WDM-PON. We optimize the frequency efficiency of SSIL source by adaptive algorithm.

Keywords: Fiber optics communications; passive optical network; spectrum-sliced incoherent light; orthogonal frequency division multiplexing

I. INTRODUCTION

With the rapid development of multimedia-based services, flexible bandwidth allocation, high capacity and low cost are required for the future optical access network systems [1-2]. For these reasons, wavelength-division-multiplexed (WDM) passive optical networks (PONs) has been introduced into NG-PON2. However, on one side, the widespread deployment of WDM is strictly limited by its expensive wavelength-specific light sources [3]. Currently, spectrum-sliced incoherent light (SSIL) sources have been considered as one of the most promising solutions due to its excellent cost effectiveness and simplicity to implement [3-7]. However, the application of SSIL source is mainly limited by the inherent excess intensity noise (EIN) of SSIL. Due to the EIN, SSIL with large optical linewidth is exploited to accommodate high-speed transmission [4]. Since large optical linewidth limits the system capacity [5] as well as the tolerance to chromatic dispersion (CD) [6], an ultra-narrow SSIL source has been proposed to enhance the dispersion tolerance. Nevertheless, this ultra-narrow SSIL results in relatively large EIN especially in low frequency region [2, 5]. Thus, a gain-saturated SOA [7] and optical offset filtering [8] techniques have been proposed to suppress the EIN of ultra-narrow SSIL. Even so, all previously demonstrated SSIL-based WDM PON system is still limited to 10-Gb/s/channel [3-6].

On the other side, inflexibility is another limitation for WDM-PON to adapt to future dynamic traffic flow. Therefore, orthogonal frequency division multiplexing (OFDM) is expected to provide dynamic bandwidth allocation due to its inherently high frequency efficiency and attractive flexible bandwidth assignment [2, 9]. Compared with single carrier modulation, we can simultaneously maximize the transmission capacity and ascertain the transmission performance by using adaptive algorithm according to different channel conditions. These processes are accomplished by software definition without any hardware modify.

In this paper, we propose and experimental demonstrate an elastic-bandwidth access scheme using ultra-narrow SSIL source with OFDM modulation in WDM-PON. With the aid of a gain-saturated SOA and optical offset filtering, we can improve the EIN of ultra-narrow SSIL. According to the EIN character, we can flexibly avoid the frequency bands with higher EIN and adaptively allocate the modulation format to optimize the frequency spectra efficiency and realize elastic bandwidth access by software.

II. EXPERIMENT AND RESULTS

The experimental setup is depicted in Fig. 1. An ASE light emitted from an erbium-doped fiber amplifier (EDFA) is first spectrally sliced by a fiber Fabry-Perot (FFP) filter, which has 0.005-nm (i.e. 700-MHz) bandwidth. Then one of the FFP mode at 1542.37 nm is selected by optical bandpass filter 1 (i.e., OBPF1). The EIN of selected SSILs is suppressed through a SOA, which has -15 dBm saturation input power and 1.5 dBm input power. Next, the EIN-

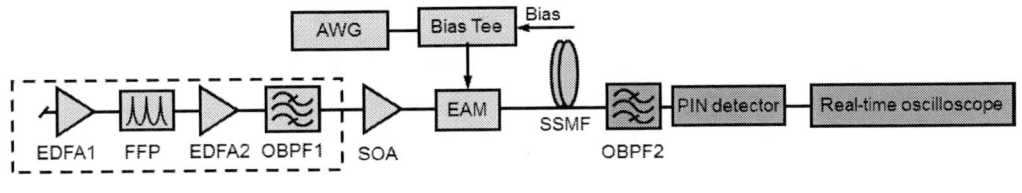

Spectrum-sliced light source

Fig. 1 Experimental setup.

841

smoothed incoherent light is fed to an electro-absorption modulator (EAM) for data modulation. This EAM is driven by an OFDM signal with Hermitian symmetry, which implies that the output of IFFT is real. The OFDM symbols are generated off-line and transferred to the time domain by an IFFT of size 256, followed by a 16-length cyclic prefix (CP) insertion. The electrical OFDM signal is generated by an arbitrary waveform generator operated at the sampling rate of 25 GS/s. After transmission over SSMF, the optical OFDM signal passes through OBPF2 and is detected by a PIN detector. This OBPF emulates an array waveguide grating at the remote node. To estimate BER performance, the detected electrical signal is sampled by a real-time sampling oscilloscope at 50 GS/s and processed offline.

To illustrate the efficacy of the ultra-narrow SSIL source, Fig. 2 (a) shows the relative intensity noise (RIN) spectra of SSIL source having 0.005-nm and 0.5-nm linewidth. No data modulation is applied to observe the RIN spectra. Due to the ultra-narrow linewidth, 0.005-nm SSIL source has considerable beating noise at the low frequency and much less beating noise located at high frequency [5]. For instance, the RIN of ultra-narrow SSIL is less than -130 dB/Hz at frequency >7.5 GHz, and increases up to -120 dB/Hz at frequency lower than 1GHz. Moreover, the RIN is still less than -120 dB/Hz beyond 5 GHz after 20-km transmission. On the other side, the RIN of wideband SSIL can be less than -120 dB/Hz only at frequency lower than 3GHz. After 20-km transmission, the RIN of 0.5-nm SSIL increases intensely since the EIN suppressed by the gain-saturated SOA is restored by optical dispersion [6]. Thus, the ultra-narrow SSIL outperforms wideband SSIL in terms of RIN and the tolerance to optical dispersion. This is also confirmed by the measured BER of QAM-4 as shown in Fig. 2(b). In the back-to-back operation, 0.005-nm SSIL has BER at 3.7×10^{-3} while the BER of 0.5-nm SSIL is only 7.9×10^{-2} and distorted to >0.1 after 6-km transmission.

To further suppress the EIN, we utilize the optical offset filtering. This optical offset filtering is realized by only detuning the center of the FFP passband from 1542.14 nm to 1542.37 nm without modifying other OBPFs. Thus, the offset filtering does not increase the system complexity. As indicated in Fig. 2 (a), offset filtering can reduce RIN by >5 dB at 4 GHz. This is because the gain-saturated SOA creates a negative correlation between the frequency chirp and intensity fluctuation of the SSIL. Hence, the optical offset filtering can help to convert frequency-modulation into amplitude modulation and produce destructive interference with the existing EIN [8]. The BER measurement in Fig. 2 (b) also demonstrates that offset filtering improves the BER from 3.7×10^{-3} to 1.9×10^{-3} in back-to-back condition.

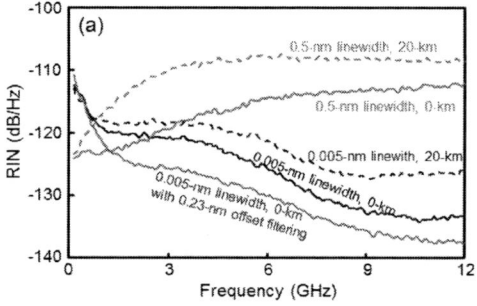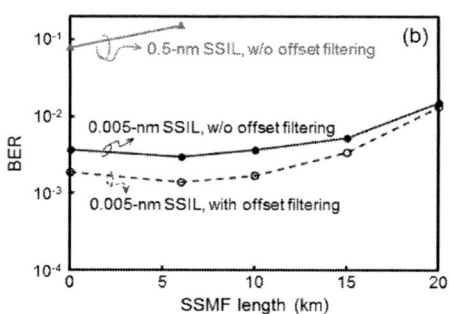

Fig. 2 (a) Measured RIN of SSIL sources having different linewidth. (b) Measured BER performance of 20.9 Gb/s QAM-4 signal as a function of SSMF length. The optical received power is kept to be -10 dBm.

According to Fig. 2 (a), the RIN distributions in our system are not uniform as traditional laser. Therefore, we use OFDM modulation with BPSK, QAM-4, QAM-8 and adaptive modulation mapping formats to study the transmission capacity of our system. Fig. 3(a) shows the BER performance at different transmission bit rate. When the frequency is smaller than 1 GHz, the RIN is larger than -120 dB/Hz, where the subcarriers limits the overall system performance. So, we fill with zero on these subcarriers with serious degradation in the low-frequency band. As shown in Fig.3 (a), the BER of BPSK (10.8 Gb/s) signal is less than 2.6×10^{-4} in back-to-back operation and increases to 1.9×10^{-3} when we change BPSK signal to QAM-4 (20.9 Gb/s) signal and nearly double the bandwidth efficiency. To further improve the bandwidth efficiency, we utilize adaptive modulation. We try two kinds of adaptive modulation formats to optimize the frequency spectra efficiency and transmission performance. One is the optimized transmission performance at 19.6 Gb/s (Adaptive1) and the other one has higher transmission bitrate at 24.1 Gb/s (Adaptive2). As demonstrated in Fig. 3(a), we can transmit 19.6-Gb/s signal with BER at 5.0×10^{-4}. Compared to QAM-4 modulation, this adaptive modulation improve the BER by >4.1 dB while the bit rate only reduces by 1.3 Gb/s. After 20 km SSMF transmission, BER performance is degraded considerably by CD. As explain in Fig. 2, CD restores the EIN suppressed by the gain-saturated SOA. Moreover, the optical OFDM are inherently double side band and frequency chirped result in CD-induced distortion which can modeled as subcarrier-to-subcarrier intermixing interference (SSII) [10]. The SSII degrades the signal-to-noise ratio (SNR) especially for subcarriers at high frequency bands. Thus, adaptive modulation applies lower order modulation on the subcarriers with larger EIN and SSII and hence, improve the system performance. As shown in Fig. 3 (a), after 20 km transmission, the BER of adaptive modulation at 24.1 Gb/s is < 0.013 which is better than the QAM-4 at 20.9 Gb/s.

Fig. 3 (b) shows the BER performance versus different transmission length. It clearly shows that BPSK has great performance except after 20 km transmission which may bring by SSII power increases at high frequency bands. The

BER of QAM-8 is always larger than $1 \times 10-2$. This can be explained as follows. During the direct detection, there is not only spontaneous-spontaneous beating between the different wavelength components inside SSIL (i.e. EIN) and subcarrier-to-subcarrier mixing interference, but also the beating between spontaneous components of SSIL and the subcarriers in other wavelength components within the SSIL. Hence, applying higher order modulation is more difficult in our system, even though, we optimize the adaptive modulation with the information of RIN and the channel condition. The BERs of OFDM signal at 19.6 Gb/s are $5.0 \times 10-4$ and $3.6 \times 10-3$ at back-to-back and after 20 km SSMF transmission, respectively. In Fig. 3(b), we also find that the BERs after 6-km and some even 10-km are better than that of back-to-back operation [6, 8]. We ascribe this to the SOA chirp induced by self-phase modulation during EIN smoothing [11]. This chirp may compress the spectrum of EIN-smoothed SSIL and ease the power fading or even provide gain for OFDM signal when the pulse passes through SSMF [10, 11].

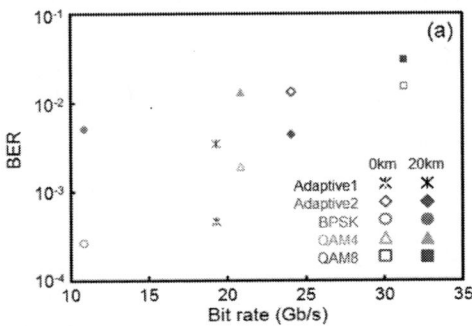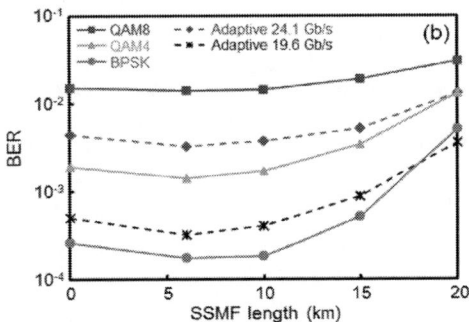

Fig. 3 (a) Measured BER performance as a function of bit rate. (b) Measured BER performance as a function of SSMF length for different modulation formats.

III. CONCLUSIONS

We have experimentally demonstrated an elastic-bandwidth access scheme using ultra-narrow spectrum-sliced incoherent light source with adaptive OFDM modulation in WDM-PON. With the aid of a gain-saturated SOA and optical offset filtering, we can transmit 19.6-Gb/s signal at BER of $5.0 \times 10-4$ and $3.6 \times 10-3$ at back-to-back and after 20-km SSMF transmission.

ACKNOWLEDGMENT

This work was supported in part by National High Technology Research and Development Program of China (863 Program) (2013AA013403), NSFC (No. 61405024 and No. 61420106011), China Scholarship Council and the Fundamental Research Funds for the Central Universities (No. ZYGX2014J004).

REFERENCES

[1] Wong E, "Next-generation broadband access networks and technologies," Journal of Lightwave Technology, vol.30, pp.597-608, 2012.

[2] J. von Hoyningen-Huene and C. Ruprecht, "OFDM for Optical Access," OFC2015, paper Th1H.1 (2015).

[3] F. Saliou, G. Simon, P. Chanclou, et al., "WDM PONs based on colorless technology," Optical Fiber Technology, vol.26, pp. 126-134, 2015.

[4] J. S. Lee, Y. C. Chung, D. J. DiGiovanni, "Spectrum-sliced fiber amplifier light source for multi-channel WDM applications," IEEE Photonics Technology Letters, vol. 5, pp. 1458-1461, 1993.

[5] Z. Al-Qazwini, H. Kim, "Ultra-narrow spectrum-sliced incoherent light source for 10-Gb/s WDM PON," Journal of Lightwave Technology, vol.30, 3157-3163, 2012.

[6] H. Kim, S. Kim, S. Hwang et al., "Impact of dispersion, PMD, and PDL on the performance of spectrum-sliced incoherent light sources using gain-saturated semiconductor optical amplifiers", Journal of Lightwave Technology, vol. 24, pp. 775-784, 2006.

[7] M. Munroe, J. Cooper, M G. Raymer, "Spectral broadening of stochastic light intensity-smoothed by a saturated semiconductor optical amplifier," IEEE Journal of Quantum Electronics, vol. 34, pp. 548-551, 1998.

[8] Q. Hu, H. Kim, "Performance improvement of ultra-narrow spectrum sliced incoherent light using offset filtering," IEEE Photonics Technology Letters, vol. 26, pp. 870-873, 2014.

[9] K. Ishii, Y. Yoshida, K. Onohara, et al., "Demonstration of No-guard-band Coherent IFDMA/OFDMA/SC-FDMA-PON Co-existence Uplink System using Real-time IFDMA Transmitter," OFC2015, paper Th1H.3, 2015.

[10] Chia-Chien Wei, "Analysis and iterative equalization of transient and adiabatic chirp effects in DML-based OFDM transmission systems," Opt. Express 20, 25774-25789, 2012.

[11] G. Agrawal, N. A. Olsson, "Self-phase modulation and spectral broadening of optical pulses in semiconductor laser amplifiers," IEEE Journal of Quantum Electronics, vol. 25, pp. 2297-2306, 1989.

WA2-40

OECC/PS2016

Equalization of Optical Nonlinear Waveform Distortion Using Neural-Network Based Digital Signal Processing

Shotaro Owaki and Moriya Nakamura

School of Science and Technology, Meiji Univ., 1-1-1 Higashi-Mita, Tama-ku, Kawasaki-shi, Kanagawa, 214-8571 Japan
ee21122@meiji.ac.jp

Abstract: *We studied novel nonlinear equalization scheme using digital signal processing based on a neural network (NN). The performance of a three-layer NN was investigated using 16QAM signals distorted by SPM, achieving BER less than 10^{-5}.*

Keywords: *Nonlinear equalization, Neural network, Digital signal processing, Self-phase modulation, 16QAM*

I. INTRODUCTION

Multi-level modulation schemes are an essential technology for handling the increasing amount of traffic on communication networks. The quadrature amplitude modulation (QAM) scheme has a large distance between symbols and therefore, compared with phase-shift keying (PSK), is more resistant to noise. On the other hand, the waveform of QAM signals are distorted by self-phase modulation (SPM), because the signal power varies according to the transmitted symbols, resulting in a large peak-to average power ratio (PAPR).

Conventionally, finite impulse response (FIR) filters have been used to compensate linear distortion caused by, e.g., chromatic dispersion [1], but they cannot be used to compensate for nonlinear distortion. Some methods have been proposed for compensating for nonlinear effects, including optical phase conjugation (OPC) [2], digital back propagation (DBP) [3], and the Volterra series transfer function (VSTF) [4]. However, these methods have the disadvantage that they involve an enormous amount of calculations.

Neural networks (NNs) have the merit that they can adaptively compensate for nonlinear distortion by supervised learning algorithm. A method of compensating for nonlinear effects in wireless communication has also been studied [5] [6]. In optical communication system, compensation methods using NNs have been conventionally proposed for nonlinearity of Intensity Modulation-Direct Detection (IMDD) [7]. In this paper, we investigated equalization performance of a three-layer NN to compensate nonlinear distortion caused by SPM in optical communication systems. Numerical simulation of 16QAM transmission shows that the NN can efficiently compensate the nonlinear distortion, and improves the performance of bit error rate (BER) and error vector magnitude (EVM).

II. NONLINEAR EQUALIZATION USING A NN

Figure 1 shows the construction of the NN which is used in the nonlinear equalization. Input layer of the NN has feedforward tapped delay lines and feedback tapped delay lines. Input signals of In-phase (I) and Quadrature (Q) components are fed into the feedforward lines. Neurons in hidden layer have sigmoidal output function. The neurons in input and output layers have linear function. Output signals from output layer are fed into the feedback lines.

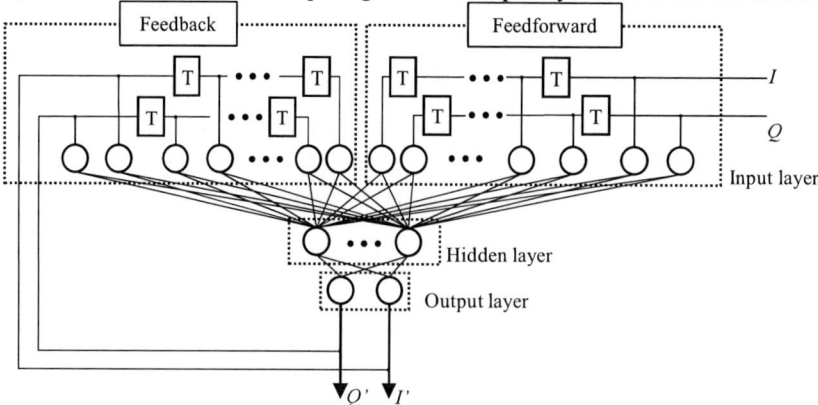

Fig. 1. Construction of NN.

844

Output of neurons y_i is described as

$$y = \sum_{k=1}^{n} f(x_k w_k + b), \qquad (1)$$

where x_k is input to neurons, w_k is the weight and b is the bias. The values of weight and bias are calculated by Back Propagation (BP) algorism, minimizing the error which is defined by the value of difference between output signals and supervised signals. The error e can be expressed as

$$e^2 = \sum_{k=1}^{n} (y_i - d_i), \qquad (2)$$

where d_i is supervised signals.

III. SYSTEM SETUP

Figure 2 shows 50-km 16QAM signal transmission system used in our simulations. 10-Gsymbol/s 16QAM optical signal was modulated by PRBS 2^{19}-1 data and transmitted by a standard single mode fiber (SSMF) and a dispersion compensation fiber (DCF) having a total length of 50 km, cancelling the chromatic dispersion. The input power to the optical fibers was 10 dBm. After the transmission, the optical signal was received by optical homodyne detection. Here, we assumed that the local oscillator (LO) was ideally synchronized to the optical signal. Then, the distorted signal after the transmission was equalized using an NN in a digital signal processor (DSP). The equalization performance was evaluated by BER and EVM. The NN used in the equalization include both the feedback unit and the feedforward unit having 6 neurons for I- and Q-signal components. The hidden layer has 16 neurons, and the output layer has 2 neurons. The NN was trained using the Levenberg–Marquardt algorithm that is a kind of BP [8].

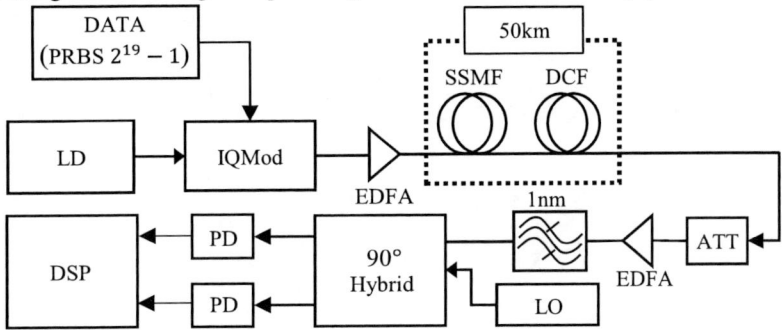

Fig. 2. System setup of 16QAM transmission.

IV. RESULT AND DISCUSSION

Figure 3(a) shows the constellation of the received 16QAM signal in back-to-back (BtB) configuration. Received optical power was -0.8 dBm. Figure 3(b) shows the constellation after the transmission .Due to the large input power, the outer symbols of 16QAM signals were rotated in the clockwise direction by SPM. Figure 3(c) shows the constellation after the equalization using a NN. The distorted symbols were successfully compensated. EVM was improved by about 25 %. Next, we investigated equalization performance in case that the received power is limited by attenuator (ATT). Figure 4(a) shows the constellation of the received 16QAM signal in BtB configuration when the received optical power was -32.7 dBm. Figure 4(b) shows the constellation of the attenuated signals after the transmission. Figure 4(c) shows constellation after equalization. EVM was improved by about 17 percent.

We calculated EVM and BER versus received optical power which was adjusted by the attenuator at the receiver side. Figure 5 shows EVM and BER characteristics calculated in the simulation. EVM less than 10 % was achieved when the received power was higher than about 32 dBm. BER of less than 10^{-5} was achieved by the equalization.

V. CONCLUSIONS

We investigated equalization performance of a three-layer NN to compensate nonlinear distortion in optical communication systems. Our numerical simulation of 16QAM transmission showed that the NN can efficiently compensate the nonlinear distortion caused by SPM, and improves the performance of BER and EVM.

OECC/PS2016

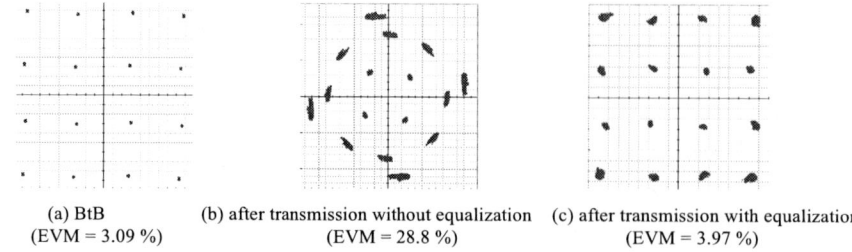

| (a) BtB | (b) after transmission without equalization | (c) after transmission with equalization |
| (EVM = 3.09 %) | (EVM = 28.8 %) | (EVM = 3.97 %) |

Fig. 3. Constellations when received power is -0.8 dBm.

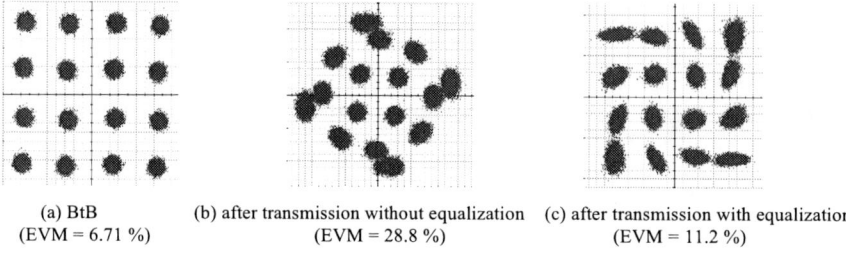

| (a) BtB | (b) after transmission without equalization | (c) after transmission with equalization |
| (EVM = 6.71 %) | (EVM = 28.8 %) | (EVM = 11.2 %) |

Fig. 4. Constellations when received power is -32.7 dBm.

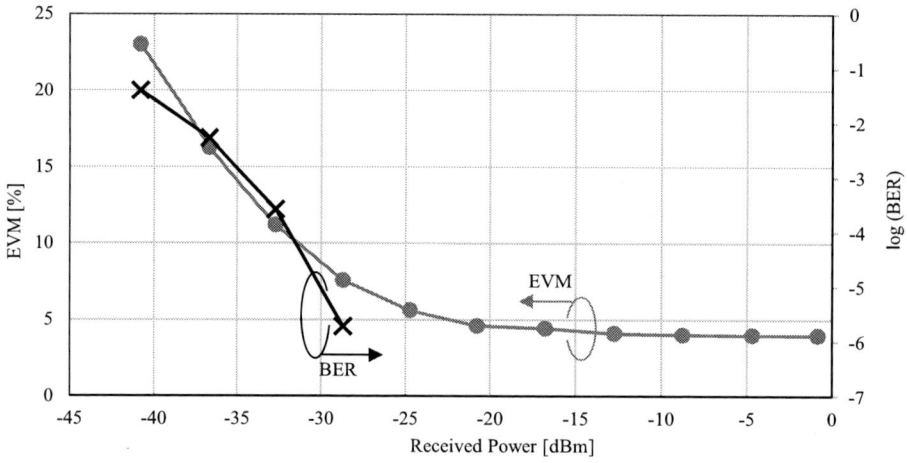

Fig. 5. EVM and BER performance improved by equalization using NN.

REFERENCES

[1] M. Nakamura and Y. Kamio, "30-Gbps (5-Gsymbol/s) 64-QAM Self-Homodyne Transmission over 60-km SSMF using Phase-Noise Cancelling Technique and ISI-Suppression based on Electronic Digital Processing," OFC/NFOEC, OWG4, Mar. 2009.

[2] X. Chen, X. Liu, S. Chandrasekhar, B. Zhu and R. W. Tkach, "Experimental Demonstration of Fiber Nonlinearity Mitigation Using Digital Phase Conjugation," OFC/NFOEC, OTh3C, Mar. 2012.

[3] E. Ip and J. M. Kahn, "Compensation of Dispersion and Nonlinear Impairments Using Digital Backpropagation," *J. Lightw. Technol.*, vol. 26, no. 20, pp. 3416-3425, Oct. 2008.

[4] L. Liu, L. Li, Y. Huang, K. Cui, Q. Xiong and F. N. Hauske, "Intrachannel Nonlinearity Compensation by Inverse Volterra Series Transfer Function," *J. Lightw. Technol.*, vol. 30, no. 3, pp. 310-316, Feb. 2012.

[5] R. Zayani, R. Bouallegue and D. Roviras, "Levenberg-Marquardt Learning Neural Network For Adaptive Pre-Distortion For Time-Varying HPA With Memory In OFDM Systems," EUSIPCO, Aug. 2008.

[6] P. Chang and B. Wang, "Adaptive Decision Feedback Equalizer for Digital Satellite Channels Using Neural Networks," *J. Select. Areas in Commun.*, vol. 13, no. 2, pp. 316-324, Feb. 1995.

[7] S. Warm, C.-A. Bunge, T. Wuth, and K. Petermann, "Electronic Dispersion Precompensation With a 10-Gb/s Directly Modulated Laser," Photon. Technol. Lett., vol. 21, no. 15, pp. 1090-1092, Aug. 2009

[8] D. W. Marquardt, "An Algorithm for Least-squares Estimation of Nonlinear Parameter," *J. Soc. Indust. Appl. Math.*, vol. 11, no. 2, pp. 431-441, Jun. 1963.

Performance analysis of EKF-based polarization and phase tracking for high order QAM signals

Wen Jiang[1], Yanfu Yang[1*], Qun Zhang[1], Kangping Zhong[2], Xian Zhou[2], Yong Yao[1]

[1]Department of Electronic and Information Engineering, Shenzhen Graduate School, Harbin Institute of Technology, Shenzhen, Guangdong Province 518055, China

[2]Department of Electronic and Information Engineering, The Hong Kong Polytechnic University, Hung Hom, Kowloon, Hong Kong, China

*yangyanfu@hotmail.com

Abstract: *The extended Kalman filter (EKF) employing double measurement are investigated for 16QAM and 64QAM signals with the filter parameter optimization strategy and the tracking tolerance against laser linewidth, frequency offset, and polarization rotation.*

Keywords: *High order QAM, coherent optical communication, Kalman filter, polarization tracking and phase recovery.*

I. INTRODUCTION

Recently Kalman filtering has attracted much attention for application in coherent optical communication systems. Due to its excellent estimation accuracy and tracking capability an extended Kalman filter (EKF) scheme has been firstly proposed for performance measurement and monitoring in coherent optical system[1]. Later the Kalman filtering has been applied for various purposes, including laser carrier recovery [2,3], polarization tracking [4-6], and simultaneous polarization/phase estimation [7,8]. Meanwhile, higher order quadrature amplitude modulation (QAM) has been employed for demonstrating high capacity optical communication systems. Regarding the accurate metrology of higher order QAM signals, fast tracking of polarization and phase against imperfect component parameters or time-varying channel is desired. The conventional blind multi-modulus algorithms (MMA) has the disadvantages of low convergence speed and thus not robust enough for tracking quick transmission channel change. Furthermore the tracking capability will be degraded along with the increased modulation order. The Kalman filter is a good candidate for high order QAM considering its inherent tracking and estimation accuracy. However, there are few analytical results on the dependence of EKF tracking performance on the modulation order, the filter and system operating parameters.

In this paper the EKF performance in both 16QAM and 64QAM case are investigated in detail by considering the influence of the filter tuning parameter under various system/channel conditions, including laser linewidth, frequency offset, and fiber channel polarization rotation. The configuration strategy are suggested regarding different parameter conditions and the dependence of EKF performance on modulation order are also studied. These results can provide meaningful guidelines to ensure optimized metrology of high order QAM signals under diverse operating conditions.

II. PRINCIPLE

The joint polarization and phase recovery for typical dual-polarization QAM optical communication system has been proposed and investigated for QPSK [1] and 16QAM [8]. Here we extend EKF further to 64QAM signals by employing double measurement method. The influence of the filter parameter on the tracking performance will be analyzed in both 16QAM and 64QAM signals at 14GSymbol/s and the comparison between 16QAM and 64QAM are presented for revealing the EKF performance dependence of modulation order. In both two formats, double measurement equations are adopted to ensure robust tracking and estimation performance. Specifically, one has four targeted symbols of QPSK constellation in order to coarse alignment and another is employed for approaching the symbols of 16QAM or 64QAM constellation. Noticeably, the targeted QPSK signals in the first measurement equation are configured to be identical to the four symbols located on the outermost circle of square 16QAM or 64QAM constellations. The same Kalman framework presented in [8] are used in two formats. The tuning parameters Q and R determines the estimation accuracy and tracking capability directly or indirectly by controlling the Kalman gain K and consequently adjusting the ratio of the prediction and the residual during the Kalman iteration. Q refers to the process noise covariance and R is the measurement noise covariance. In the paper the R are considered to have the constant value of $R_{16QAM} = 0.025$ and $R_{64QAM} = 0.00625$ for 16QAM and 64QAM, respectively. The left Q parameter are discussed by evaluating its influence on EKF performance under different system operation parameters in the following.

III. SIMULATION RESULTS

The following analysis was the evaluation of Q parameter on the filter performance referred to the laser linewidth, polarization rotation and frequency offset. The received optical signal-to-noise ratio (OSNR) is fixed at 18.6 dB and 24.8 dB for the cases of 16QAM and 64QAM, respectively. The operating condition of the fixed polarization rotation of 3Mrad/s and zero frequency offset are used. As shown in Fig. 1(a) and (b), the obtained curves illustrates the Q together with linewidth could make a significant impact on system performance and imply similar conclusion with respect to both standard formats: With the smaller Q, the BER performance is better in presence of lower linewidth. Nevertheless in the larger linewidth regime, a smaller Q means degraded tracking ability for phase noise or worse linewidth tolerance.

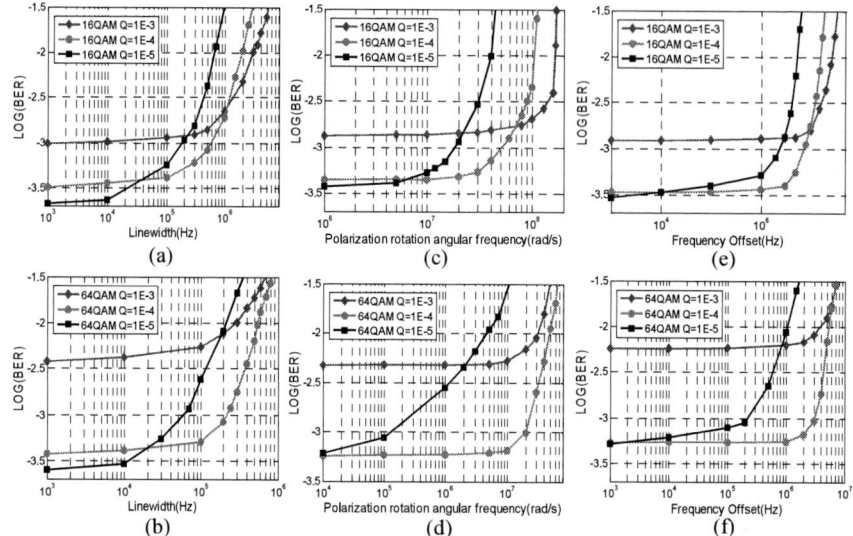

Fig. 1.The BER vs. Linewidth for the cases of (a) PDM-16QAM and (b) PDM-64QAM; The BER vs. polarization rotation angular frequency for (c) PDM-16QAM and (d) PDM-64QAM; The BER vs. frequency offset for (e) PDM-16QAM and (f) PDM-64QAM under different Q parameters.

Similarly, those conclusions can be found in the curves of BER vs. polarization rotation, as illustrated in Figure 1(c) and (d). The laser linewidth is fixed at 100 kHz. The result can be qualitatively explained based on the filter update equations. With the smaller Q, the prediction is dominant other than the residual during the state estimation update. This leads to the filter performance expectation: (1) with slow channel rotation, the stable and accurate estimation may be obtained; (2) However, the fast response to the residual caused by rotation can be achieved for rapidly-varying polarization. What's more, it is interesting to observe that the higher-order modulation is more sensitive to Q parameters, which can be attributed to smaller Euclidean distance in higher order modulation assuming the same average power for two formats. Regarding the format of PDM-64QAM, Q=1e-4 permits a far superior job on BER performance than that of Q=1e-3 situation in the range of polarization rotation angular frequency<40Mrad/s. This indicates that the filter tuning parameters should be set properly depending on the modulation order and the channel condition.

In the following, the influence of the frequency offset on system performance are discussed for different tuning parameters. The simulation has the configuration of the fixed polarization rotation of 3Mrad/s and the laser linewidth of 100 kHz. For two formats, there exists the same optimal filter tuning parameter of 1e-4. A roughly no penalty could be observed for PDM-64QAM at the frequency offset of 1.5MHz. Meanwhile, that tolerable frequency offset for PDM-16QAM could reach almost 2.5MHz.

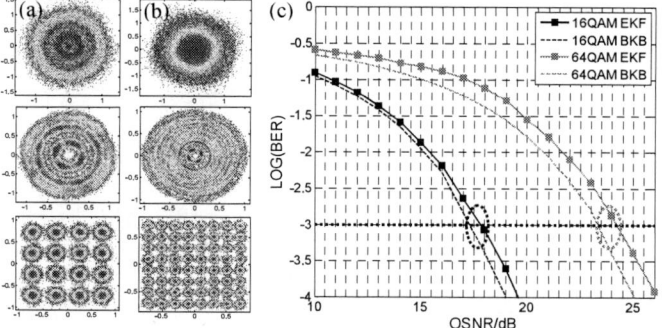

Fig. 2. The constellation diagrams of before demultiplexing, after demultiplexing, and carrier phase recovery for cases of (a) PDM-16QAM (b) PDM-64QAM; OSNR vs. BER performance analysis for (c) PDM-16QAM and PDM-64QAM.

With the above simulation results, the optimal choice would be Q 1e-4 compared with other two alternatives for achieving the expected balance between tracking speed and estimation accuracy. Thus, the following performance assessment would be done under the optimal Q of 1e-4. The demultiplexing results of PDM-16QAM, PDM-64QAM with the extended Kalman filter as presented in Fig. 2(a) and 2(b), respectively. In Fig. 2(c), the simulated Bit Error Rate (BER) for both constellation sizes in a Back-to-Back and extended Kalman filter system with the polarization rotation (5Mrad/s) and the laser linewidth (100KHz) are presented. The required OSNR sensitivities of $\{OSNR_{16QAM} = 17.9\text{dB}; OSNR_{64QAM} = 24.2\text{dB}\}$ at the target BER=1e-3 are obtained and the resultant penalty compared to BKB case is about 0.52dB and 0.88dB for 16QAM and 64QAM, respectively. Clear constellation diagrams and relatively lower OSNR penalty demonstrated the feasibility of the scheme for fast polarization tracking and phase estimation in both modulation formats with the same Kaman framework.

Finally the convergence performance of EKF are analyzed by considering 3Mrad/s polarization rotation and the linewidth of 100KHz under Q parameter of 1e-4. The product of the channel polarization related Jones matrix and the inverse one estimated by EKF is represented by [c1 c2; c3 c4]. The four elements [c1 c2; c3 c4] are plotted as a function of symbol period, as showed in Fig. 3. The fast convergence within 130 and 400 symbol periods under 3Mrad/s polarization rotation are successfully achieved for 16QAM and 64QAM, respectively.

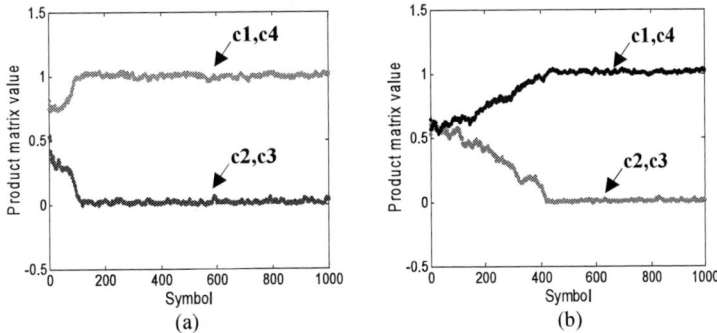

(a)   (b)

Fig.3.The elements of the product matrix in (a) PDM-16QAM, (b) PDM-64QAM

IV. CONCLUSIONS

In this paper, the EKF performance for 16QM and 64QAM signals are investigated by analyzing the influence of the filter parameters under the system operating condition including laser linewidth, frequency offset, and polarization rotation. The obtained results show that Q filter parameter should be configured at moderate value to give consideration to two sides: the estimation accuracy for low noise or slow channel variation and the tracking capability for quickly noise and channel fluctuation. The higher order modulation format is more sensitive to the filter parameters than the lower one. With the respective optimized parameters the tracking tolerance against laser linewidth, polarization rotation and frequency offset are presented for comparison between two formats. Considering dynamic network scenarios, adaptive parameter configuration relying on channel variation are desired and will be investigated in the future.

ACKNOWLEDGMENT

This work was supported by the National Natural Science Foundation of China (Contract No. 61205046, 61401020, 61575051 and 61505039) and Shenzhen Municipal Science and Technology Plan Project (JCYJ20150327155705357, KQCX2015032409501296, JSGG20150529153336124 and JCYJ20150529114045265).

REFERENCES

[1] T. Marshall, B. Szafraniec, and B. Nebendahl, "Kalman filter carrier and polarization-state tracking," Optics letters, vol. 35, pp. 2203-2205, 2010.

[2] T. Inoue and S. Namiki, "Carrier recovery for M-QAM signals based on a block estimation process with Kalman filter," Optics express, vol. 22, pp. 15376-15387, 2014.

[3] M. I. Olmedo, X. Pang, M. Piels, R. Schatz, G. Jacobsen, S. Popov, et al., "Carrier recovery techniques for semiconductor laser frequency noise for 28 GBd DP-16QAM," in Optical Fiber Communication Conference, 2015, p. Th2A. 10.

[4] N. J. Muga and A. N. Pinto, "Extended Kalman Filter vs. Geometrical Approach for Stokes Space-Based Polarization Demultiplexing," Journal of Lightwave Technology, vol. 33, pp. 4826-4833, 2015.

[5] W.-C. Ng, A. T. Nguyen, C. S. Park, and L. A. Rusch, "Enhancing Clock Tone via Polarization Pre-rotation: A Low-complexity, Extended Kalman Filter-based Approach," in Optical Fiber Communication Conference, 2015, p. Th2A. 19.

[6] Y. Yang, G. Cao, K. Zhong, X. Zhou, Y. Yao, A. P. T. Lau, et al., "Fast polarization-state tracking scheme based on radius-directed linear Kalman filter," Optics express, vol. 23, pp. 19673-19680, 2015.

[7] W. Jiang, Q. Zhang, G. Cao, K. Zhong, Y. Yao, and Y. Yang, "Blind and Simultaneous Polarization and Phase Recovery for Time Domain Hybrid QAM Signals Based on Extended Kalman Filtering," in Asia Communications and Photonics Conference, 2015, p. AS4F. 2.

[8] G. Cao, Y. Yang, K. P. Zhong, L. Cui, N. Rong, J. Gu, et al., "Assessment of Extended Kalman Filtering Based Simultaneous Polarization and Phase Tracking for PDM-16QAM," in Asia Communications and Photonics Conference, 2014, p. ATh3A. 134.

Post-FEC Performance Comparison of Pilot-aided Phase Unwrap and Turbo Differential Decoding for Cycle Slip Mitigation

Tangqing Liu, Yan Li *, Huizhong Chen, Haiquan Cheng, Deming Kong and Jian Wu

State Key Laboratory of Information Photonics & Optical Communication, BUPT, Beijing, 100876, China

* Corresponding author: liyan1980@bupt.edu.cn

Abstract: *Post-FEC BER-performance of pilot-aided-phase-unwrap(PAPU) and turbo-differential-decoding(TDD) have been investigated in different laser-linewidths and carrier-phase-estimation(CPE) filter-lengths. TDD outperforms PAPU by 0.3dB with sufficient filter-length, while PAPU has advantages of low or vanished error-floor with short filter-length.*

Keywords: *LDPC, Turbo differential decoding, Cycle Slip, FEC*

I. INTRODUCTION

Carrier phase estimation (CPE) is an important integral part of DSP-based receiver in coherent optical communications through which laser phase noise is compensated. Various methods have been proposed to eliminate the impact, among which VVPE is the most widely used blind CPE technique. The Viterbi &Viterbi phase estimation (VVPE) [1], the most widely used blind CPE algorithm, contains a phase unwrapping step which results in phase ambiguity. There will be a probability that phase rotation occurred when solving this ambiguity, that is cycle slip (CS) [2], and error bursts induced by CS cannot be corrected properly by the forward error correction (FEC) decoder [3]. Consequently, post-FEC BER performance should be taken into consideration when evaluating a CS elimination method.

Differential encoding (DE) is the most intuitive method to mitigate the impact of CSs. However, differential decoding introduces penalties to the pre-FEC BER, thus leading to a degraded post-FEC BER, which strongly depends on the FEC threshold and the decoding method. Turbo decoding of an outer low-density parity check (LDPC) code and an inner differential code, called turbo differential decoding (TDD), was proposed to reduce this DE-penalty [4]. But TDD exhibits an error floor for frequent phase slips which might cause total decoder failure [5]. Another promising way to avoid DE-penalty is pilot-aided phase unwrapping (PAPU) assisted CPE [6-9]. In our previous work [10] a pilot symbol-aided phase unwrapping (PAPU) scheme is proposed and demonstrated, in the combination with a concatenated SD-FEC coding scheme, post-FEC error-free performance is achieved. In this paper, we compared the post-FEC BER performance of TDD scheme with our proposed PAPU scheme in different equivalent linewidths and filter lengths of VVPE. A simulation based on 32 GBaud QPSK is carried out, and the results illustrate that TDD outperforms PAPU by 0.3 dB with sufficient filter length, however, TDD fails due to the serious error floor with short filter length, while PAPU can still work with low or vanished error floor. Considering the reduced complexity of PAPU, it might be a better candidate in contrast to TDD in coherent optical communication systems with high CS rate.

II. SIMULATION MODEL

We adopted the wiener phase noise channel as the model in both simulation schemes [11]. In this model, all the linear impairments are assumed to be fully compensated. After ideal timing recovery and synchronization, input symbols for CPE are composed of the transmitted symbols, accumulated noise phase and additive Gaussian white noise. The accumulated noise is made up of phase noise induced by laser equivalent linewidth. Since the length of filter has an enormous impact on the result of phase estimation in the process of VVPE, performance with different filter lengths is investigated.

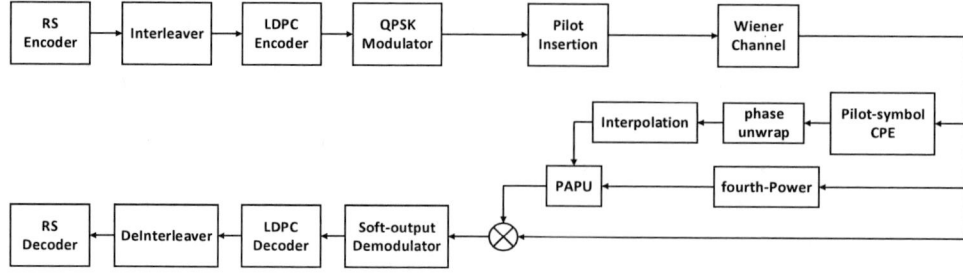

Fig. 1. PAPU transmission scheme.

Fig. 1 illustrates the block diagram of PAPU transmission link. The information data is encoded in sequence by outer encoder, RS (255,239), and inner encoder with a 1320×697 interleaver in between. The encoded data is modulated by QPSK. Then symbols are launched into the wiener channel after pilot insertion. In the receiver, carrier phase is recovered by PAPU VVPE. QPSK demodulation and decoding have been performed by soft-output demodulator and inner and outer decoder respectively. Inner encoder uses the most commonly-used soft-decision code, DVB-S2 standard low density parity check (LDPC), with 11.1% overhead [12] and its maximum iteration number is 50. Pilot symbols with known information of 0.78% [13] overhead are inserted periodically per 127 symbols at the transmitter side and the phase estimated by the pilot symbols is used as a reliable reference in the VVPE PU process at the receiver side. Total FEC overhead is 16.7%, lower than the 20% coding overhead upper limit for user which is suggested by the OIF's integrated photonics projects [14].

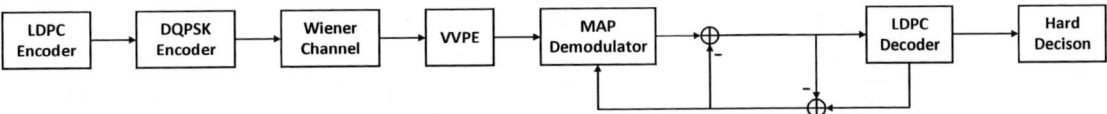

Fig. 2. TDD transmission scheme.

Fig. 2 shows the block diagram of the transmission link including a more detailed structure of the TDD engine [4]. The information data is encoded by DVB-S2 LDPC encoder with 16.7% overhead, modulated by DQPSK encoder. After the wiener channel transmission, carrier phase is recovered by VVPE. DQPSK demodulation and decoding are implemented by turbo decoder. Turbo decoder is constructed by an outer decoder, whose maximum iteration number is 10, and an inner MAP differential decoder. Maximum number of outer iteration, turbo iteration, is 5. Maximum total iteration number is 50, equals to the PAPU scheme. To achieve a BER of 10^{-7}, we validated post-FEC BER performance over 10^{-7} bits in both sceneries.

III. SIMULATION RESULTS AND DISCUSSIONS

Fig.3. shows the pre-FEC and post-FEC BER versus Eb/N0 curves of PAPU and TDD schemes under different equivalent linewidths and CPE filter lengths. Dash lines denote the pre-FEC BER and solid lines represent the post-FEC BER. CPE filter length defined as [2] is shown in the label.

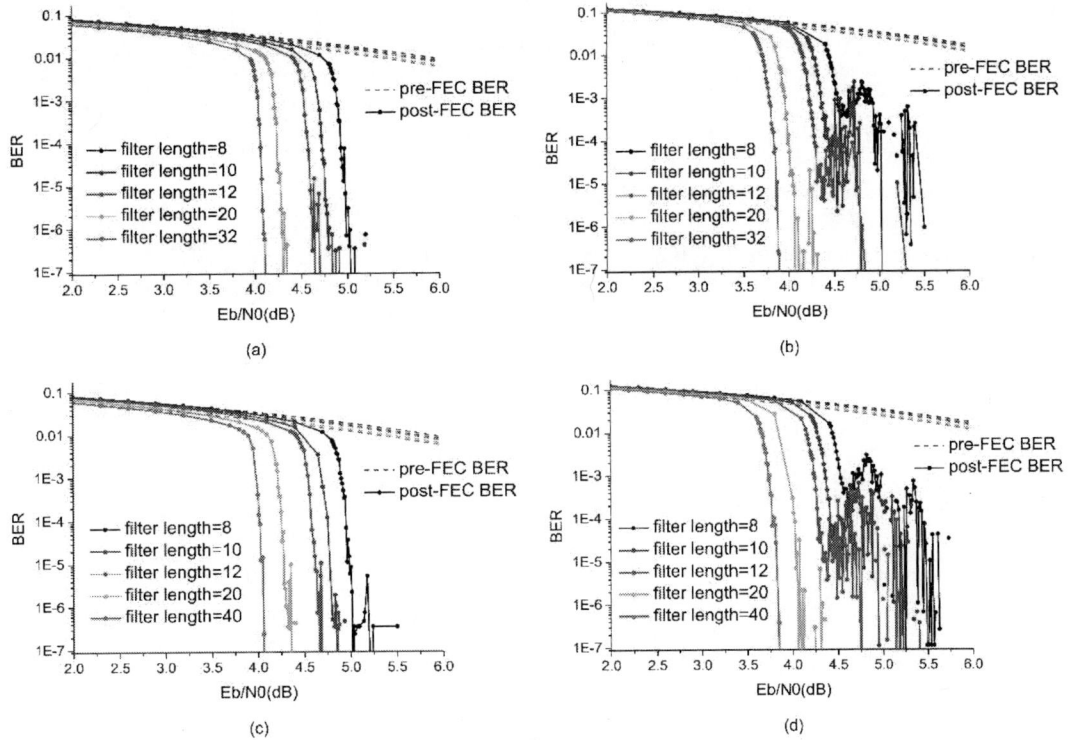

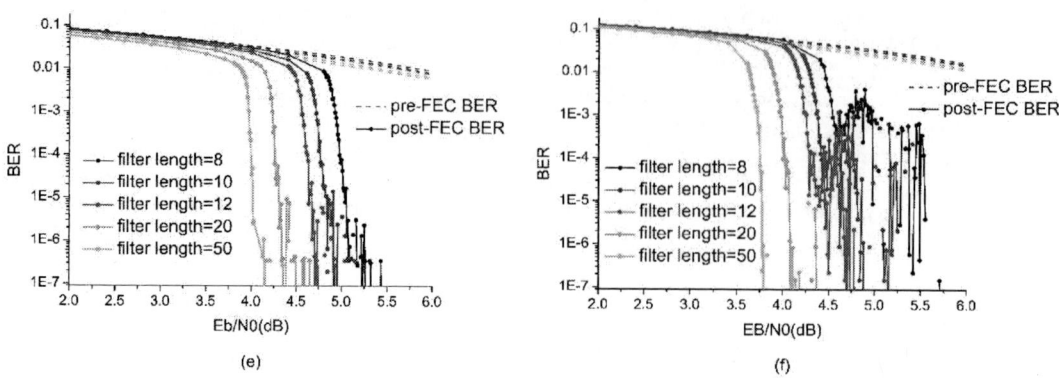

Fig. 3. Performance of PAPU scheme and TDD scheme in different linewidths.
(a) 200k PAPU; (b) 200k TDD; (c) 500k PAPU; (d) 500k TDD; (e) 1M PAPU; (f) 1M TDD.

It can be easily observed that slightly degeneration of pre-FEC BER may leads to rapidly deterioration of post-FEC BER due to cycle slip. As expected, TDD achieves about 0.3 dB Eb/N0 gain compared with PAPU under all the equivalent linewidths and filter lengths as shown in Fig.3. With filter length less than 12, both schemes suffer from early error floor at 10^{-6} to 10^{-3}. The error floor reduces with increasing filter length, and fades away at 10^{-7} with filter length increasing to 32, 40 and 50 for equivalent linewidth of 200 kHz, 500 kHz and 1 MHz, respectively. Nevertheless, PAPU outperforms TDD under shorter filter length benefits from its lower error floor. For instance, with equivalent linewidth of 200 kHz (a typical value of commercial available narrow line-width lasers) and filter length of 20, PAPU has no error floor, while TDD scheme has an error floor at 10^{-6}. When filter length is shortened to 8, error floor appears at 10^{-6} and 10^{-3} for PAPU and TDD. In short, TDD outperforms PAPU by 0.3 dB with sufficient filter length, however, TDD fails due to the serious error floor with short filter length, while PAPU is still able to work with low or vanished error floor. Another advantage of PAPU is that in contrast to TDD, PAPU has lower computing complexity. Therefore, PAPU outperforms TDD under short filter length. Allowing for reduced complexity of PAPU, it can be considered as a better candidate rather than TDD in coherent optical communication systems.

IV. CONCLUSIONS

We have investigated the post-FEC BER performance of PAPU and TDD schemes in various laser linewidths and different CPE filter lengths in 32 GBaud QPSK systems. Simulation results demonstrate that TDD outperforms PAPU by 0.3 dB with sufficient filter length, while PAPU takes the advantages of low or vanished error floor with short filter length. Allowing for reduced complexity of PAPU, it can be considered as a better candidate rather than TDD in coherent optical communication systems.

ACKNOWLEDGMENT

This work was partly supported by 863 program 2015AA015503, NSFC program 61475022, 61505011, 61331008, 973 project 2014CB340102, Fund of state key laboratory of IPOC (BUPT), and the fundamental research funds for the central universities.

REFERENCES

[1] A. Viterbi., IEEE Trans. Inf. Theory, 29, 543 (1983).
[2] A. Bisplinghoff et al., OFC, OTu3I. 1 (2013).
[3] E. Ibragimov et al., OFC, OWE2 (2010).
[4] F. Yu et al., ECOC, We.10.P1.70 (2011).
[5] A. Bisplinghoff, et al., ECOC, Mo.1.A5 (2012).
[6] H. Zhang et al., OFC, OMJ7 (2011).
[7] H. Cheng et al., Opt. Express, 21, 22166 (2013).
[8] M. Magarini et al., IEEE Photon. Technol. Lett., 24, 739 (2012).

[9] H. Cheng et al., Opt. Express 22, 20740 (2014).
[10] H. Chen et al., ICOCN,7203733 (2015).
[11] M. Magarini et al., Opt. Express, 19, 22455 (2011).
[12] DVB-S.2 Standard Specification, ETSI EN 302 307 V1.3.1 (2013-03).
[13] H. Cheng et al., OFC, Th4D.1 (2014).
[14] OIF-FEC-100G-01.0, 100G Forward Error Correction White Paper (May 2010).

Characterization of Soliton Fusion Phenomenon Based on Solitons' Eigenvalue

Gihan Weerasekara and Akihiro Maruta
Graduate School of Engineering, Osaka University,
2-1 Yamada-oka, Suita, Osaka, 565-0871 Japan
gihan@pn.comm.eng.osaka-u.ac.jp

Abstract: Soliton fusion phenomenon is recognized as a cause of optical rouge wave formation. The impact of initial pulse spacing and frequency separation between solitons for soliton fusion phenomenon has been analyzed based on solitons' eigenvalue.
Keywords: Nonlinear effect, Eigenvalue, Soliton fusion, Optical rogue wave

INTRODUCTION

Soliton fusion is a phenomenon that manifests itself in optical fibers due to the interaction between the co-propagating soliton pulses with small temporal and frequency separation in soliton transmission system. Pulse interactions are strongly influenced by the initial conditions [1]. Merge of solitons into one pulse during soliton collision is the main characteristic of soliton fusion. As a result of soliton fusion, large intensity robust light structures arise and propagate over significant distances.

Soliton fusion phenomenon itself can occur for solitons which have been discovered both theoretically and experimentally [1]. In recent years, research on rogue wave phenomena has been actively carried out in light waves. In fact soliton fusion has been observed in different fields, it has been reported that the one of the mechanisms to generate optical rogue waves is soliton fusion [2, 3] other than soliton collision process which we have already demonstrated [4]. Rogue wave structures are complex and interesting from the practical point of view of soliton fusion. Amplitude maintaining for a long propagation after merging of two solitons is the main characteristic of soliton fusion generated optical rogue wave. The objective of this work is to demonstrate an effective fusion-like processes resulting from interactions between solitons based on eigenvalue of the solitons to validate the soliton fusion generated optical rogue waves.

The process of soliton fusion is accompanied by the soliton parameters. Soliton fusion is mainly associated with the initial pulse spacing, frequency separation, phase difference, and amplitude difference between solitons. Since conditions for the effective fusion are very delicate, the fusion itself becomes a very rare one. The impact of pulse spacing and frequency separation are demonstrated for the generation of high-intensity pulses by utilizing eigenvalues of the solitons for the first time to the best of our knowledge. Nonlinear Schördinger equation (NLSE) has been used to model the pulse propagation.

SOLITON FUSION PROCESS AND THE EIGENVALUE ANALYSIS

When the two identical solitons launched into an optical fiber either attracted or repelled each other, depending on their relative phase. If the solitons have different center frequencies, similar effects have been predicted. In order to inspire a brief idea on soliton interaction, soliton collision process is schematically sketched in Figs. 1 and 2 along the propagation distance for different pulse spacing and frequency separation. The relative temporal and frequency separation of the two solitons determines their evolution. Attraction and repulsion can be seen after the collision according to pulse spacing and frequency separation of two fundamental solitons. Two soliton fission ejected solitons due to modulational instability will meet in interaction rather than quasi-elastic collision when the temporal and frequency separation between two solitons are small. Dissipation of the energy of solitons can be observed during the interaction process. Part of the energy is transferred to dispersive waves. As the result of multiple interactions both between solitons and dispersive waves, soliton fusion will be accelerated.

A numerical demonstration of soliton fusion can be obtained by solving the NLSE. The behavior of the complex envelope of a light wave propagating in a fiber in the presence of anomalous GVD and nonlinearity can be expressed as [5],

$$i\frac{\partial u}{\partial Z} + \frac{\partial^2 u}{\partial T^2} + |u|^2 u = 0,\qquad(1)$$

where Z, T, and $u(Z,T)$ represent the normalized quantities of propagation distance, time moving with the group velocity, and complex envelope of electric field, respectively. The eigenvalue equation associated to Eq. (1) can be represented as,

$$i\frac{\partial \psi_1}{\partial T} + u\psi_2 = \zeta\psi_1, \quad -i\frac{\partial \psi_2}{\partial T} - u^*\psi_1 = \zeta\psi_2, \tag{2}$$

where $\zeta = (\kappa + i\eta)/2$ is the complex eigenvalue and ψ_ℓ $(\ell = 1, 2)$ are the eigen functions [6]. As long as u is a solution of Eq. (1), eigenvalue ζ of Eq. (2) is invariant with Z. The real part and the imaginary part of soliton's eigenvalue are associated with nonzero group velocity and amplitude respectively. Eigenvalue ζ can be calculated by performing Fourier transformation of Eq. (2), discretization in frequency domain in order to compute the convolution integrals originated from the second term on the left-hand side of Eq. (2) [4], and solving the resultant matrix form eigen-value problem by a standard numerical procedure.

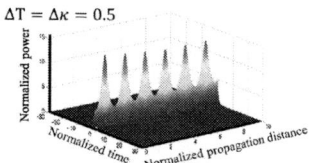

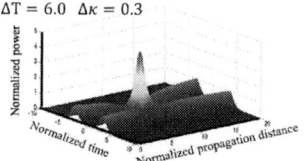

Fig. 1. Attraction after collision (Soliton interaction)

Fig. 2. Repulsion after soliton collision

NUMERICAL SIMULATION MODEL AND THE RESULTS

Eq. (1) was solved by using the split-step Fourier method to investigate the impact of temporal and frequency separation between two solitons. Computational window size and the sampling points were chosen as W=640 and N=2^{10} respectively. In order to suppress the effect of dispersive waves, traveling to the edges of the time window absorbing boundary conditions were used by applying slightly decreasing loss profile. Simultaneous launch of two solitons with different center frequencies were considered. Numerical simulations were performed by changing the pulse spacing ΔT and frequency separation $\Delta \kappa$ of the initial waveform given by,

$$u(0,T) = \text{sech}\,(T + \Delta T/2)\exp(i\,\Delta \kappa/2) + \text{sech}\,(T - \Delta T/2)\exp(-i\,\Delta \kappa/2) \tag{3}$$

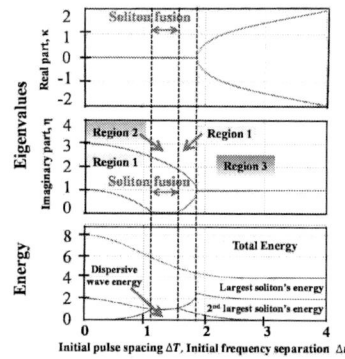

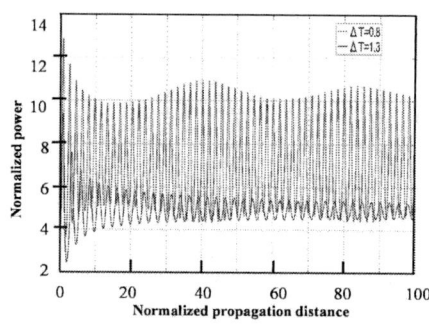

Fig. 3. Variation of eigenvalues and energy assignment

Fig. 4. Variation of peak power

Real and imaginary parts of discrete eigenvalues and energy assignment for $\Delta T = \Delta \kappa$ case are shown in Fig. 3. In order to describe the main properties of the soliton fusion phenomena, three particular regions were considered as shown in black dotted line. Eq. (3) has two discrete eigenvalues for extreme two cases; $(\kappa, \eta) = (0,1), (0,3)$ for $\Delta T = \Delta \kappa = 0$, and $(\pm \Delta \kappa/2, 1)$ for $|\Delta \kappa| \gg 1$, which are already known. Numerical simulations have been performed to know the behavior of eigenvalue in the unknown region. When the temporal and frequency separation becomes large, the amplitude of weaker soliton slightly decreases and tends to zero and collapse with the other soliton. Emergence of the dispersive wave can be associated to the dissipation of the second soliton. The high-amplitude soliton is stable. Generally, Eq. (3) has only one discrete eigenvalue for $1.13 \le \Delta T \le 1.56$. Complete fusion of two solitons can be seen in this region. Moreover, when the temporal and frequency separation becomes more larger, second soliton appears and amplitude becomes 1 and propagates independently. In conclusion, soliton fusion phenomena for $1.13 \le \Delta T \le 1.56$ and soliton interactions for $\Delta T < 1.13$ and $1.56 < \Delta T \le 1.86$ can be observed. When $\Delta T > 1.86$, solitons propagate with different group velocities. One approach for validate the simulation results is based upon the energy criteria. Collision of solitons is accompanied by strong irradiation of low-intensity dispersive wave. Our observation also corresponds to the energy conservation law and confirmed as shown in energy assignment in Fig. 3.

Variation of peak power along the propagation distance for two specific cases in region 1 and 2 are shown in Fig. 4. According to Fig. 4, peak power for $\Delta T = 1.3$ case tends to 5 which supports with the power calculated through amplitude $2.24^2 = 5.02$ along with the propagation distance. As for the $\Delta T = 0.8$ case, peak power oscillating with the propagation distance since they consists of two solitons.

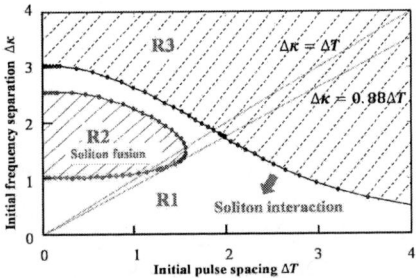

Fig. 5. Relationship between initial pulse spacing and
the initial frequency separation

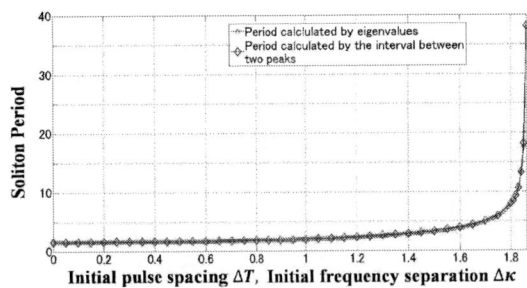

Fig. 6. Soliton period

In order to study soliton fusion phenomenon more precisely, relationship between the initial pulse spacing and the frequency separation was observed and results are shown in Fig.5. Region 2 starts from $\Delta\kappa \leq 0.88\Delta T$. Initial pulse spacing and frequency separation should be below the black solid line to emerge soliton interactions and should be in region 2 to emerge soliton fusion as shown in Fig.5. Optical solitons in region 2 can be related to generate optical rogue waves through soliton fusion process. This fact should be taken into account in the development of analytical models for the solitons' interactions governed by the NLSE. Based on the guiding center theory [7], the effect of fiber loss is negligible and it does not qualitatively change our conclusions for the propagation lengths which have been considered in our numerical simulation results.

Another approach for validate our results is based on the soliton period. Soliton period is given by [8],

$$Z_p = \frac{4\pi}{\eta_1^2 - \eta_2^2} \ , \tag{4}$$

where η_1 and η_2 are the amplitude of two solitons. Fig. 6 shows the soliton period calculated by eigenvalues and the interval between two peaks for above mentioned $\Delta T = \Delta\kappa$ case. Two values are well matched with each other and soliton period is increasing with the pulse spacing and $\Delta T \leq 1.86$ for the effective soliton fusion-like process.

CONCLUSIONS

The impact of pulse spacing and the frequency separation between solitons for the soliton fusion phenomena have been analyzed based on eigenvalue in this study for the first time to the best of our knowledge. Soliton interactions can be changed from attractive to repulsive after the collision according to pulse spacing and frequency separation. Due to the interactions, the solitons merge together, where the resultant amplitude is significantly greater than the amplitude of the other solitons. After the merging process, the dynamics of the resultant pulse is determined mainly by a high-amplitude soliton. Fusion of optical solitons can be considered as the new scenario to generate optical rogue waves.

ACKNOWLEDGMENT

This work was partially supported by λ-reach project conducted by NICT, Japan.

REFERENCES

[1] S. R. Friberg, "Soliton fusion and steering by the simultaneous launch of two different-color solitons," Opt. Lett., vol. 16, pp. 1484—1486, 1991.

[2] R. Driben and I. Babushkin, "Accelerated rogue waves generated by soliton fusion at the advanced stage of supercontinuum formation in photonic-crystal fibers," Opt. Lett., vol. 37, pp.5157—5159, 2012.

[3] M. Onorato, S. Residori, U. Bortolozzo, A. Montina, and F. T. Arecchi, "Rogue waves and their generating mechanisms in different physical contexts," Phys. Rep., vol. 528, pp.47—89, 2013.

[4] G. Weerasekara, A. Tokunaga, H. Terauchi, M. Eberhard, and A. Maruta, "Soliton's eigenvalue based analysis on the generation mechanism of rogue wave phenomenon in optical fibers exhibiting weak third order dispersion," Opt. Express, vol. 23, pp.143—153, 2015.

[5] G. P. Agrawal, Nonlinear fiber optics, 5th ed., Oxford, 2013.

[6] V. E. Zakharov and A. B. Shabat, Sov. Phys. JETP, "Exact theory of two-dimensional self-focusing and one-dimensional self-modulation of waves in nonlinear media," Sov. Phys. JETP, vol. 34, pp.118-134, 1972.

[7] A. Hasegawa and Y. Kodama, "Guiding-center soliton in optical fibers," Opt. Lett., vol. 15, pp.1443—1445, 1990.

[8] A. Hasegawa and Y. Kodama, Solitons in optical communications, Oxford, 1995.

WA2-44

OECC/PS2016

Feasibility Study of 100G Ethernet with Carrierless Amplitude and Phase Modulation

Hirotaka Ochi and Shinya Sasaki

Chitose Institute of Science and Technology 758-65 Bibi, Chitose, Hokkaido, 066-8655 Japan
e-mail: m2150050@photon.chitose.ac.jp

Abstract: *We present a system design of 100 Gb/s/λ, 10 km, 64-CAP system with numerical simulation including the DAC and ADC effects on the system performance to show the feasibility of 100 Gigabit Ethernet with the CAP modulation scheme. The results show that supposing the Tx output of 0 dBm, the system power margin is 7.8 dB.*

Keywords: *Optical fiber communication, Multi-level modulation, DA/AD converter, Digital signal processing*

INTRODUCTION

The amount of IP traffic in short reach networks is growing continuously. In order to support the IP traffic growth, 100 Gigabit Ethernet (100GE) standardization using wavelength division multiplexing (WDM) of 25 Gb/s×4λ has been completed [1]. However, this system is expensive due to the number of optical components. For this reason, higher-order modulation scheme which has higher spectral efficiency than NRZ is required (e.g. Quadrature Amplitude Modulation: QAM and Carrierless Amplitude and Phase modulation: CAP). In these modulation schemes, CAP is attractive for short reach applications, as it employs QAM by using digital finite impulse response (FIR) filter to enable low cost implementation [2]. Recently, 100 Gb/s CAP system has been proposed showing the power budget of 100 Gb/s, SMF: 500 m~5 km, 16/64-CAP with direct modulated laser (DML) by simulation [3] and experimental results of 102.4 Gb/s, SMF: 15 km, multiband CAP [4]. However these reports did not mentioned the details such as effects of DAC/ADC sampling rate and resolution on system performance and so on.

In this paper, we present a system design of 100 Gb/s/λ, 10 km, 64-CAP system with numerical simulation including the DAC/ADC effects on the system performance to show the feasibility of 100GE with the CAP modulation scheme.

SIMULATION

A. Simulation model

Figure 1 shows the simulation model we constructed. Detailed simulation parameters are listed in Table I. The model adopts the CAP modulation scheme in intensity modulation/direct detection (IM/DD) transmission system. On transmitter side, the bit sequence generated by the pulse generator is mapped onto a 64-CAP constellation using gray coding. The data sequence is up-sampled to 64 Sa/Symbol and divided into two paths. The two data sequences are sent to in-phase and quadrature shaping filters respectively to generate IQ components. In these filters, the square-root raised-cosine function with 15 % excess bandwidth is used to improve the spectral efficiency. The outputs of shaping filters are subtracted to produce single signal. Modulator non-liner compensation (NLC) is applied and this signal is down-sampled to the DAC/ADC sampling rate and then sent to DAC and pre-equalizer using least mean square (LMS) algorithm to compensate inter-symbol interference (ISI) caused by bandwidth limitation in transmitter and receiver. This bandwidth limitation in transmitter caused by the DAC, modulator driver and modulator is represented by LPF. After this process, 1.55 μm continuous wave (CW) light from semiconductor laser (LD) is modulated by the signal through a LN intensity modulator (LN-Mod.). A standard single mode fiber (SSMF) up to a length of 10 km is used as a transmission medium. Receiver consists of pin-PD and pre-amplifier in which shot noise and thermal noise are considered. The received signal is sent to LPF as bandwidth limitation in receiver caused by the ADC, pin-PD and pre-amplifier. In addition, the received signal is also sent to CDR to obtain clock signal. After LPF process, ADC and two matched filters is used to separate the IQ components. 64-CAP demapping is used to obtain the original bit sequence. Finally, we calculated BER and received power penalty at BER=10^{-6}.

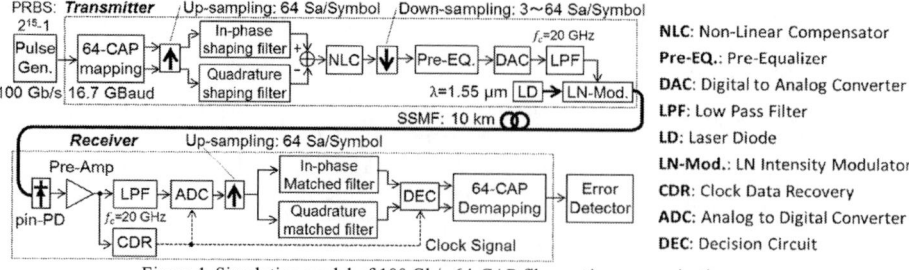

Figure 1. Simulation model of 100 Gb/s 64-CAP fiber optic communication system

NLC: Non-Linear Compensator
Pre-EQ.: Pre-Equalizer
DAC: Digital to Analog Converter
LPF: Low Pass Filter
LD: Laser Diode
LN-Mod.: LN Intensity Modulator
CDR: Clock Data Recovery
ADC: Analog to Digital Converter
DEC: Decision Circuit

TABLE I
Detailed simulation parameters

Device	Parameter	Values	Units
	Baud Rate	16.7	GBaud
	Transmission distance	10	km
LD	Wavelength	1.55	μm
	RIN (Relative Intensity Noise)	-155	dB/Hz
Shaping filter	Roll-off factor	0.15	
Matched filter	Number of taps	16∼1024	
DAC	Sampling rate (Down-sampling rate)	3∼64	Samples/Symbol
ADC	Resolution	3∼10	Bit
LPF (5th Bessel)	Cut-off frequency (@ -3 dB)	20	GHz
Optical Fiber	Chromatic Dispersion (@ λ=1.55 μm)	17	ps/nm/km
pin-PD	Responsivity	0.9	A/W
	Dark current	1	nA
Pre-Amp	Input equivalent noise current density	30	pA/√Hz
LN modulator	Peak to peak input voltage	1.0	V_π

B. CAP Modulation Scheme

CAP is one of the higher-order modulation schemes that have higher spectral efficiency than NRZ and is practical alternative to QAM using finite impulse response (FIR) in-phase and quadrature filters mostly implemented in the digital domain. This method makes CAP simpler than QAM achieving the same spectral efficiency and performance. The output of CAP transmitter $s(t)$ is given by the following equation,

$$s(t) = \sum_{k=-\infty}^{\infty} [a_k p(t - kT) - b_k \hat{p}(t - kT)] \tag{2.1}$$

where a_k and b_k are the input symbols and T is the symbol duration. In addition, $p(t)$ and $\hat{p}(t)$ are the passband in-phase and quadrature pulses respectively defined by,

$$p(t) = g(t)\cos(2\pi f_c t) \tag{2.2}$$
$$\hat{p}(t) = g(t)sin(2\pi f_c t) \tag{2.3}$$

where $g(t)$ is baseband pulse and f_c is carrier frequency. For the selection of $g(t)$, we used square-root raised-cosine function to improve the spectral efficiency.

SIMULATION RESULTS

A. Receiver Sensitivity

We calculate the receiver sensitivity of -14.1 dBm at B2B and BER=10^{-6} with NLC and pre-equalization using the DAC/ADC with the sampling rate of 64 Sa/Symbol and without quantization (i.e. infinite resolution). This receiver sensitivity is used as the reference value of the received power penalty

B. DAC/ADC Sampling Rate

Figure 2 shows the effects of the DAC/ADC sampling rate on the received power penalty at BER=10^{-6} for both B2B and 10 km transmission cases. The number of IQ filter taps in the simulation is 1024 tap to neglect the power penalty depending on the number of the IQ filter taps [5]. In addition, the DAC/ADC sampling rate is equal to the down sampling rate. This rate is denoted as samples per symbol (Sa/Symbol) and has integer value. When the DAC/ADC sampling rate is satisfied with Nyquist-Shannon sampling theorem (> 3 Sa/Symbol in this case), the power penalty depending on DAC/ADC sampling rate hardly changes. For instance, the power penalty at 3 Sa/Symbol is 0.2 dB (B2B) and 0.8 dB (10 km transmission). However, when the DAC/ADC sampling rate is 2 Sa/Symbol or less, BER does not reach 10^{-6} due to aliasing. The DAC and ADC having high speed sampling rate are not suitable for short reach fiber optic communication systems due to the cost of these devices. Therefore, we use DAC/ADC sampling rate of 3 Sa/Symbol on the following sections.

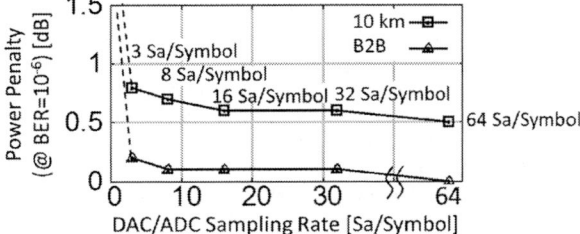

Fig. 2. The effects of DAC/ADC sampling rate on received power penalty
(@ 1024 IQ filter taps, w/o DAC/ADC quantization)

C. DAC/ADC Resolution

Figure 3 shows the effects of DAC/ADC resolution on the received power penalty at BER=10^{-6} for both B2B and 10 km transmission cases. The number of DAC/ADC sampling rate and IQ filter taps on the simulation is 3 Sa/Symbol and 1024 taps, respectively. These results indicate that in the case of 5 bit resolution, the power penalty and dispersion penalty rapidly increase. On the other hand, the DAC/ADC with 6 bit resolution and more does not give any significant power penalty. For instance, in the case of 6 bit resolution, the power penalties at B2B and 10 km transmission are 0.5 dB and 1.3 dB, respectively.

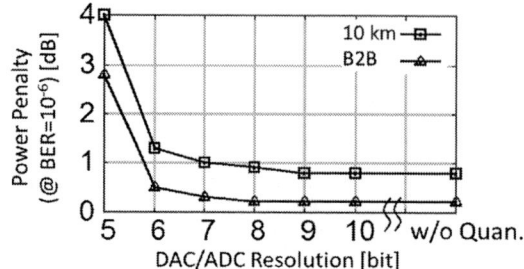

Fig. 3. The effects of DAC/ADC resolution on received power penalty
(@ 1024 IQ filter taps, the DAC/ADC sampling rate of 3 Sa/Symbol)

D. Power Budget

Figure 4 shows the power budget of simulated system after 10 km transmission at BER=10^{-6}. In this figure, reference receiver sensitivity of -14.1 dBm means the ideal one at B2B using NLC to compensate modulator non-linearity and pre-equalizer compensating ISI caused by bandwidth limitation. The total power penalties comprise contributions from the power penalties depending on DAC/ADC resolution of 6 bit (0.3 dB) and sampling rate of 3 Sa/Symbol (0.2 dB), connection loss of 2.0 dB (@ 0.5 dB/connector), dispersion penalty of 0.8 dB, and SSMF attenuation of 3.0 dB (@ 0.3 dB/km). Therefore, supposing the Tx output of 0 dBm, the system power margin (unallocated power penalty) is 7.8 dB.

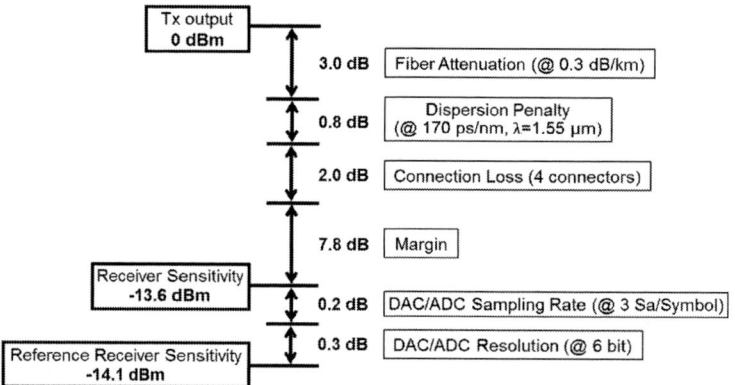

Fig. 4. The power budget of 100 Gb/s 64-CAP fiber optic communication systems with 10 km transmission (@ BER=10^{-6})

CONCLUSIONS

In this paper, we present a system design of 100 Gb/s/λ, 10 km, 64-CAP system with numerical simulation including the DAC/ADC effects on the system performance to show the feasibility of 100 Gigabit Ethernet. The simulation results show that supposing the Tx output of 0 dBm, the system power margin (unallocated power penalty) is 7.8 dB, which suggests the feasibility of 100 Gigabit Ethernet with the CAP modulation scheme.

REFERENCES

[1] IEEE, "IEEE Standard for Ethernet--Amendment 3: Physical Layer Specifications and Management Parameters for 40 Gb/s and 100 Gb/s Operation over Fiber Optic Cables," Mar. 2015
[2] Simon Haykin, "Communication Systems," Wiley, 4th edition, US, May 2000
[3] J. L. Wei *et al.*, "Study of 100 Gigabit Ethernet Using Carrierless Amplitude/Phase Modulation and Optical OFDM," *J. Lightwave Technol.*, Vol. 31, Issue 9, pp. 1367-1373, May 2013
[4] Miguel Iglesias Olmedo *et al.*, "Multiband Carrierless Amplitude Phase Modulation for High Capacity Optical Data Links," *J. Lightwave Technol.*, Vol. 32, Issue 4, pp. 798-804, Feb. 2014
[5] Hirotaka Ochi and Shinya Sasaki, "Short reach fiber optic communication systems with CAP modulation scheme," *IEICE Technical Report*, OCS2015-98, pp. 57-62, Jan. 2016

Bandwidth Enhancement Equalization Enabling a 40-Gbps PAM4 Transmission via a 9.5-GHz Photoreceiver

Jun-Jie Liu[1], Chia-Chien Wei[2*], Ta-Ching Tzu[1], Chun-Yen Chuang[1], Kai-Lun Chi[3], Xin-Nan Chen[3], Jin-Wei Shi[3], and Jyehong Chen[1]

1. Department of Photonics, National Chiao Tung University, Hsinchu 300, Taiwan
2. Department of Photonics, National Sun Yat-sen University, Kaohsiung 804, Taiwan
3. Department of Electrical Engineering, National Central University, Taoyuan 320, Taiwan
ccwei@mail.nsysu.edu.tw

Abstract: We demonstrate a 40-Gbps PAM4 optical system by employing a photoreceiver with bandwidth of only 9.5 GHz. Following 100 m multimode fiber, the BER of 1.8×10^{-4} and 1.2×10^{-4} after FFE and DFE equalizations are successfully achieved.
Keywords: PAM4, VCSEL, FFE, DFE, vertical-cavity surface-emitting lasers

I. INTRODUCTION

The continuous growth of global Internet traffic, fueled by on-demand Internet streaming video, remote office services, and electronic commerce, doubles almost every two years. Not only the number of Internet users is growing due to social and economic development, but also the number of devices that connected to Internet have been growing in even more astounding rate. It has estimated that the number of devices connected to Internet, i.e. Internet of things (IOT), will surpass 50 B by 2020 [1]. Because traffic pattern has changed from service provider to contend deliver network (CDN), modern data centers urgently require that interconnects can be operated at a high bit rate (of at least 25 Gbps before 2020 [2]), with better power efficiency and transmitted to longer distance [3]. Optical interconnects (OIs) have been widely studied and envisioned as a promising candidate for short-distance communication. In particular, vertical-cavity surface-emitting lasers (VCSELs) have unique advantages in terms of bandwidth, power efficiency, cost, manufacturability, and ease of packaging [4].

Four-level pulse amplitude modulation (PAM4) modulation format will soon be adopted as the standard for IEEE P802.3bs 400 GbE in 2 km and 10 km single-mode fiber (SMF) PMD (physical medium dependent) [5]. In the 100 m multi-mode fiber (MMF) PMD, however, still only 16×25 Gbps non-return-to-zero (NRZ) format is being considered. Aiming to further reduce the size, cost and power, 8×50 Gbps will soon be considered as a better alternative [5]. Two possible modulation formats are NRZ and PAM4. Both formats have their advantages and limitations. For example, NRZ has simpler electronic driving circuit that can use a limiting amplifier to reduce the nonlinearity stems from the temperature-dependent transfer curve, and it also has better receiving power sensitivity. Nonetheless, compared with PAM4 format, NRZ requires higher modulation bandwidth that will undoubtedly increase the cost of the system. On the other hand, with about only half bandwidth requirement of NRZ, PAM4 format does suffer about 4.7 dB higher receiving power and requires better linearity because four different levels are needed.

In this paper, we plan to provide the baseline information of the capability of current matured 10-Gbps OIs. We demonstrate, with the help of feed forward (FFE) and decision feedback equalizer (DFE), a 40-Gbps 100 m PAM4 multi-mode fiber transmission is successfully achieved.

II. CHARACTERISTIC OF HIGH-SPEED VCSELs

Figures 1(a) (a photo) and (b) respectively show the top and cross-section view of the 850-nm high speed VCSEL. Fig. 1(a) presents a 250×250 μm^2 chip composed of single high-speed VCSELs unit. Growth by LandMark®, the epi-layer structure is composed of three strained $In_{0.1}Ga_{0.9}As/Al_{0.3}Ga_{0.7}As$ quantum-wells (QWs) sandwiched between a 36-pair n-type layer and a 26-pair p-type $Al_{0.9}Ga_{0.1}As/Al_{0.12}Ga_{0.88}As$ Distributed-Bragg-Reflector (DBR) layer. On top of the MQWs is an $Al_{0.98}Ga_{0.02}As$ layer (50nm thick) for oxidation. As shown in Fig. 1(b), both Zn-diffusion and oxide-relief techniques applied to our VCSEL to improve maximum modulation speed and allowable transmission distance over MMF. Of Zn-diffusion (W_z), whose depth is ~1 μm, and oxide-relief (W_o), their apertures are ~8.0 and ~3.0 μm wide, respectively.

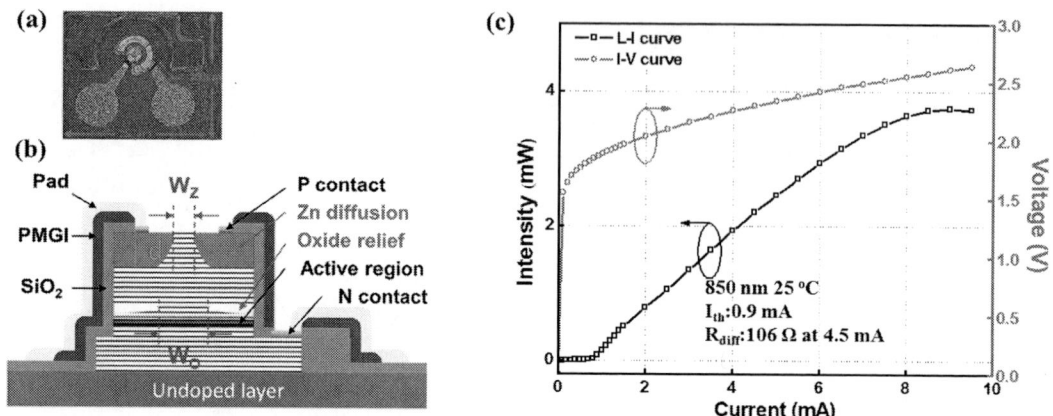

Fig. 1(a) Conceptual cross-sectional view, (b) top view of duo-mode VCSEL, (c) L-I and I-V curve of VCSEL.

A special coupling fiber tip, purchased from WT&T®, will collect as much output light as possible from VCSEL with 1.5-dB coupling loss. Fig. 1(c) indicates L-I-V characteristic of VCSEL, in which the threshold current is about 1.0 mA and the maximum power is ~3.75 mW at 25 °C, and resistance is 60 Ω at bias current of 9.5 mA. Fig. 2(a) plots the optical multi-mode spectrum at bias currents of 3 to 9 mA at 25 °C and Fig. 2(b) depicts the frequency response. The 3-dB bandwidth is around 25 GHz when the bias current is set at 9 mA.

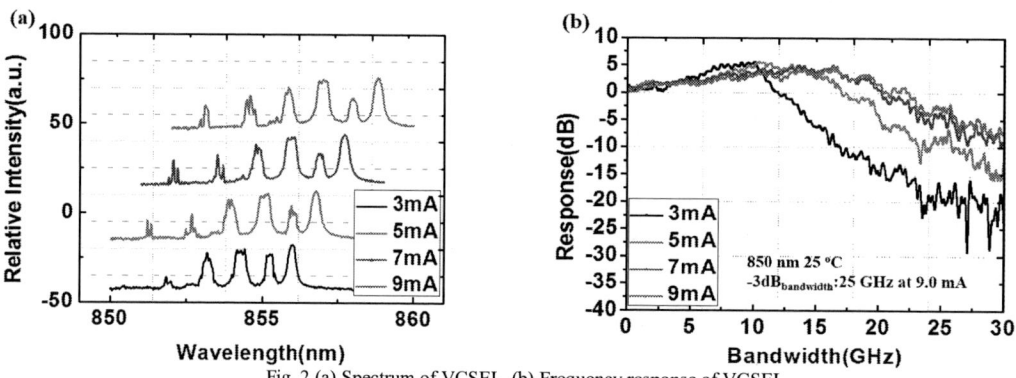

Fig. 2 (a) Spectrum of VCSEL, (b) Frequency response of VCSEL.

III. EXPERIMENTAL SETUP AND RESULTS

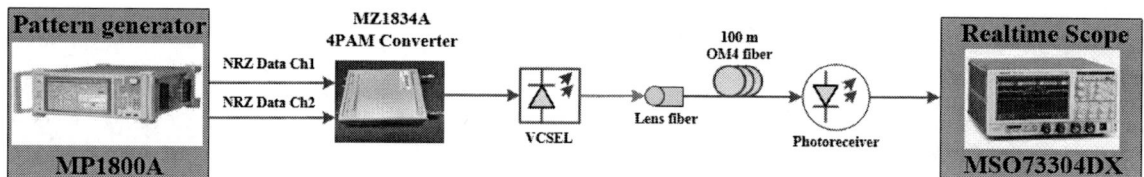

Fig. 4 Experimental transmission system scheme.

Fig. 4 shows the experimental transmission setup. The electrical signal is generated from Anritus® MP1800A Signal Quality Analyzer. Composed of independent pseudo-random-binary-sequence (PRBS) data signals of length $2^{15}-1$ at 20 Gbps, two NRZ signals are generated and sent into the MZ1834A 4PAM converter where the two NRZ signals will be integrated as one PAM4 signal. Then, the PAM4 signal is fed into the VCSEL through a 26.5-GHz bias-T (Agilent 11612A) and a GS probe. The output light is butt-coupled into a lensed fiber. Following 100 m of multi-mode fiber, the signal is directly detected by a 9.5-GHz photo-receiver (PICOMETRIX PT-12B). The PAM4 signal is retrieved and digitized by a real-time oscilloscope (Tektronix® DPO77002SX) with a sample rate of 100 GSample/s and a 3-dB bandwidth of 33 GHz. Offline Matlab® DSP programs is used to demodulate the signal, then signal-to-noise ratio (SNR), and the bit-error-ratios (BERs) can be calculated based on a bit-by-bit comparison between transmitted and received data.

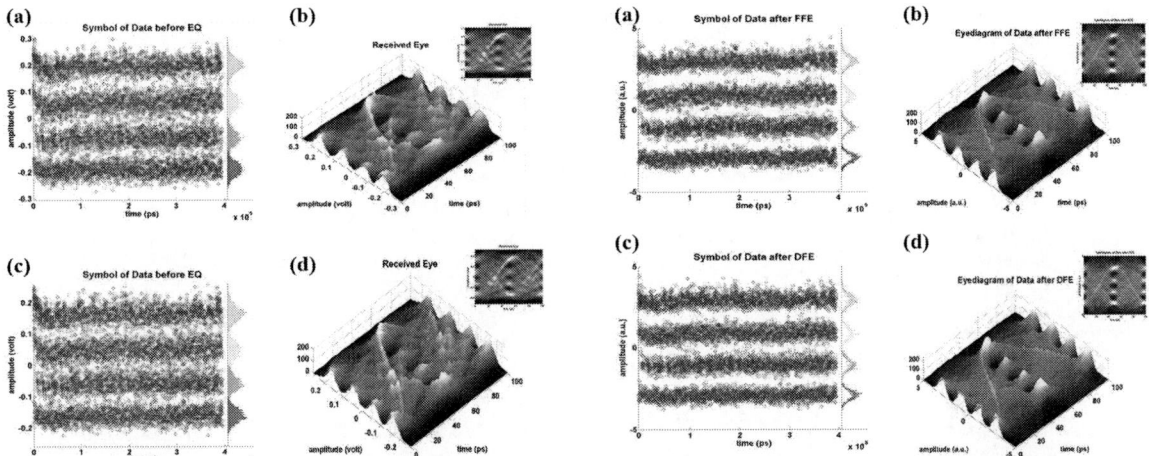

Fig. 5 (a) OBTB decision point amplitude distribution and histogram before equalization of OBTB, (b) 3D-eye diagram before equalization of OBTB, (c) decision point amplitude distribution and histogram before equalization through 100 m MMF, (d) 3D-eye diagram before equalization through 100 m MMF.

Fig. 6 (a) OBTB decision point amplitude distribution and histogram through 100 m MMF after FFE, (b) 3D-eye diagram through 100 m MMF after FFE, (c) decision point amplitude distribution and histogram through 100 m MMF after DFE, (d) 3D-eye diagram through 100 m MMF after DFE.

Fig. 5(a) plots OBTB-received-signal amplitude at the decision point and the corresponding histogram, while Fig. 5(b) shows the 3D-eye diagram and the calculated BER is 2.2×10^{-2}. Fig. 5(c)-(b) depicts amplitude at the decision point and the corresponding histogram and 3D-eye diagram after 100 m MMF, respectively. The BER is about 2.6×10^{-2} after 100 m MMF. Figs. 6(a)-(b) describe experimental results of better amplitude distribution, histogram and 3D-eye diagram after feed-forward equalizer (FFE), respectively. The BER is about 1.8×10^{-4}. The amplitude distribution and histogram after decision feedback equalizer (DFE) is shown in Fig. 6(c). Fig. 5(d) shows the 3D-eye diagram that gives an estimated 1.2×10^{-4} BER. In our experimental system, FFE taps are 13 and FFE+DFE taps are 10 as shown in Fig. 7 (a)-(b), respectively.

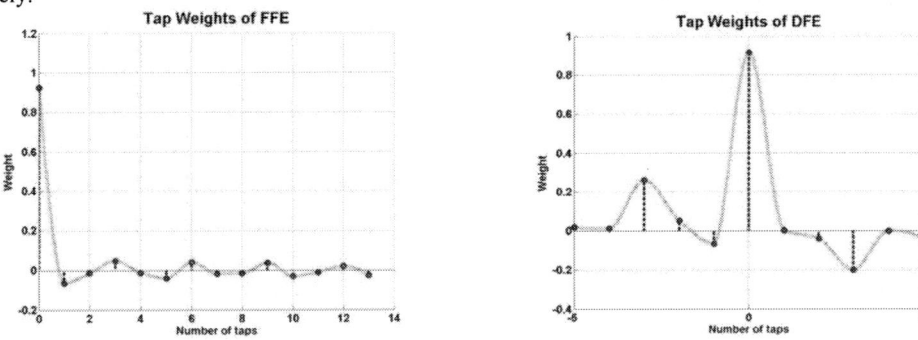

Fig. 7 (a) taps weights of FFE, (b) taps weights of DFE.

IV. CONCLUSIONS

In this paper, the lower bandwidth 9.5-GHz detector has presented a picture that the higher data rate 40-Gbps can be achieved, and that means insufficient bandwidth has vanished in our DSP demodulation system. We successfully demonstrate a 40-Gbps PAM4 signal transmission system with only 9.5-GHz bandwidth photo-receiver. BERs are respectively estimated 1.8×10^{-4} and 1.2×10^{-4} after DFE and FFE processing.

REFERENCES

[1] http://www.cisco.com/web/solutions/trends/iot/overview.html
[2] M. A. Taubenblatt, J. Lightw. Technol., vol. 30, no. 4, pp. 448-457, 2012.
[3] C. F. Lam et al., IEEE Communication Magazine, vol. 48, no. 7, pp. 32-39, 2010.
[4] L. A. Coldren et al., "Diode Lasers and Photonic Integrated Circuits," Wiley, New York, NY, USA, 2012.
[5] http://www.ieee802.org/3/bs/

Influence of Pointing Errors on Error Probability of Inter-Satellite Laser Communications

Qian Wang[1,2], Tianyu Song[1,2], Ming-Wei Wu[3], Tomoaki Ohtsuki[4], Mohan Gurusamy[1] and Pooi-Yuen Kam[1,2]

1. Dept. of Electrical and Computer Engineering, National University of Singapore, Engineering Drive 3, Singapore 117576,
2. Centre of Advanced Microelectronic Devices, NUS Suzhou Research Institute, Suzhou, China 215123
3. School of Information and Electronic Engineering, Zhejiang University of Science and Technology, China 310023
4. Dept. of Information and Computer Science, Keio University, 3-14-1, Hiyoshi, Kohokuku, Yokohama 223-8522, Japan

elesong@nus.edu.sg

Abstract: *The effect of pointing errors on the average bit error probability (ABEP) of inter-satellite laser communications is studied. The exact ABEP is derived using a closed-form channel gain expressed in terms of the Marcum Q-function.*
Keywords: *Pointing errors, inter-satellite laser communications, beam radius, Marcum Q-function, average bit error probability.*

I. INTRODUCTION

Long distance inter-satellite laser communication links are highly vulnerable due to the degrading effect of pointing errors [1-4]. The pointing errors are due to platform vibrations, which cause vibrations of the transmitter telescope and, therefore, misalignment between the transmitter and the receiver [1, 2]. Various statistical models have been proposed over the years to describe the pointing errors [1-3]. In these works, the effects of misalignment on the error performance have been investigated, and the optimization of system parameters is considered. The exact fraction of the power collected by the detector, i.e., the channel gain, is expressed as a double integral [3], and two approximations are assumed in the derivation of a simpler form in [3, Appendix]. Here, based on a geometric interpretation of the first-order Marcum Q-function introduced in [5], the exact channel gain is expressed in terms of the Marcum Q-function without any approximation. We analyze the effect of pointing errors on the average bit error probability (ABEP). The power penalties for different variances of pointing error angles are shown, and the explicit insights into how the beam radius affects the ABEP are discussed.

II. SYSTEM AND GAUSSIAN BEAM MODEL

For the kth symbol interval $((k-1)T_s, kT_s)$ where T_s is the symbol period, the received electrical signal $r(k)$ of an inter-satellite intensity modulation / direct detection system can be modeled as $r(k) = Am(k) + n(k)$ [6]. Since we only consider on-off keying (OOK) here, the transmitted data symbol $m(k)$ takes on any value from set {0, 1} with equal probability. The additive white Gaussian noise (AWGN) $n(k)$ has mean zero and variance $N_0/2$. We have $A = 2h_p\sqrt{T_s}RP_t$, where h_p denotes the channel gain due to pointing errors, i.e., the fraction of the power collected by the receiver, P_t is the transmit power, and R is the photodetector responsivity [6]. We have $T_s = 1/R_{\text{data}}$ where R_{data} is the system data rate, and $R = \eta e / h\upsilon = \lambda \eta e / hc$ where $\lambda = 1.55\mu$m is the optical carrier wavelength, $\eta = 1$ is the quantum efficiency, h is the Plank's constant, e is the elementary charge and c is the light speed in vacuum.

To study the channel gain h_p, we need to start from the Gaussian beam model, for which, the normalized spatial distribution of the transmitted intensity at a propagation distance z from the transmitter is given by [3, eq. (7)]

$$I_{\text{beam}}(\boldsymbol{\rho}; w_z) = \frac{2}{\pi w_z^2}\exp\left(-\frac{2\|\boldsymbol{\rho}\|^2}{w_z^2}\right), \qquad (1)$$

where $\boldsymbol{\rho}$ is the radial vector from the beam center, and w_z is the beam radius at which the intensity drops to e^{-2} at a link distance z [3]. The beam radius w_z is also referred to as the spot size, and achieves the minimum value w_0 at $z = 0$, known as the beam waist. The relation between w_z and w_0 is given by $w_z = w_0\sqrt{1 + \left(z\lambda/(\pi w_0^2)\right)^2}$ [3]. It should be noted that the Gaussian beam model fails if wave fronts are tilted by more than approximately 0.5 rad, which corresponds to $w_0 \le 2\lambda/\pi$ [7, P. 630]. This leads to $w_0 > 2\lambda/\pi$ as a constraint in finding the optimal beam waist. It is observed that the beam radius w_z increases almost linearly with z in the far field, i.e., where $z \gg \pi w_0^2/\lambda$, resulting in a cone-shaped beam. Thus, the divergence angle Φ of the laser beam in the far field is approximated by the ratio of the beam radius w_z to the distance z, which is $\Phi \approx w_z / z$. We can see that each of the three variables w_0, w_z and Φ can determine the other two. Thus, determining the optimum beam waist w_0 is equivalent to determining the optimum beam radius w_z, or the optimum divergence angle Φ which is discussed in [2].

III. THE EXACT CHANNEL GAIN AND ABEP

Here, we consider a circular optical detector $\mathbf{C}$ of radius a. We use the center of the Gaussian beam as the origin to build a coordinate system. Thus, with the pointing error vector denoted by $\mathbf{d}$, the exact fraction of the power collected by $\mathbf{C}$ can be expressed as

$$h_p(\mathbf{d}, w_z, a) = \iint_{\mathbf{C}} I_{\text{beam}}(\boldsymbol{\rho}; w_z)\, d\boldsymbol{\rho} \,. \qquad (2)$$

Let X and Y denote two independent and identically distributed Gaussian random variables with mean zero and variance $\sigma^2 = w_z^2 / 4$. If we let $\boldsymbol{\rho} = (X, Y)$, the joint probability distribution function (pdf) of X and Y is exactly the same as (1). Hence, the exact channel gain $h_p(\mathbf{d}, w_z, a)$ equals the probability of the point (X, Y) falling in the detector $\mathbf{C}$, i.e., $\Pr\left(\| (X, Y)\text{-}\mathbf{d} \| < a\right)$. Notation $\|.\|$ here denotes the Euclidean norm. Due to the circular symmetry of the beam cross-sectional shape and the detector area, $h_p(\mathbf{d}, w_z, a)$ depends only on the radial displacement $d = \| \mathbf{d} \|$. Hence, without loss of generality, we consider the case where the center of detector $\mathbf{C}$ is located at $(-d,\ 0)$ for simplicity, and $h_p(\mathbf{d}, w_z, a)$ is interpreted as the probability of event $\sqrt{(X+d)^2 + Y^2} < a$, i.e.,

$$h_p(\mathbf{d}, w_z, a) = \Pr\left(\| (X, Y)\text{-}\mathbf{d} \| < a\right) = \Pr\left(\sqrt{(X+d)^2 + Y^2} < a\right) = h_p(d, w_z, a),\ a > 0, d > 0. \quad (3)$$

According to the geometric interpretation of the first-order Marcum Q-function $Q^{Mar}(,)$ given in [5], $Q^{Mar}(,)$ can be used to represent the probability of event $\sqrt{(X+d)^2 + Y^2} > a$, i.e., $\Pr\left(\sqrt{(X+d)^2 + Y^2} > a\right) = Q^{Mar}(d/\sigma, a/\sigma)$.

Here, $Q^{Mar}(x, y)$ is defined as $Q^{Mar}(x, y) = \int_y^\infty r \exp\left(-r^2/2 + x^2/2\right) I_0(xr)\, dr$, where $I_0(.)$ is the modified Bessel function of the first kind of order zero. Thus, we have the channel gain $h_p(d, w_z, a)$ expressed as

$$h_p(d, w_z, a) = \Pr\left(\sqrt{(X+d)^2 + Y^2} < a\right) = 1 - \Pr\left(\sqrt{(X+d)^2 + Y^2} > a\right) = 1 - Q^{Mar}\left(\frac{2d}{w_z}, \frac{2a}{w_z}\right) \quad (4)$$

It should be emphasized that in deriving (4), no approximation is made. The calculations of $h_p(d, w_z, a)$ using (4) and (2) are exactly the same. Our result (4) is a simple, closed-form expression for the channel gain h_p. Since the Marcum Q-function $Q^{Mar}(,)$ is built in some commercial softwares, e.g., MATLAB®, the evaluation of (4) is much faster than that of the double integral given in (2), and thus (4) facilitates the numerical study later.

The radial displacement d between the detector center and the beam center is the product of z and $\tan\theta$, where θ is the pointing error angle. We thus have $d = z \cdot \tan\theta \approx z \cdot \theta$, since θ is very small. For OOK, the conditional BEP for a given value of θ (or h_p) is given as [6, eq.(17)]

$$P(e \mid \theta) = Q^G\left(h_p(\theta, w_z, z, a)\sqrt{\gamma}\right) = Q^G\left(h_p(\theta, w_z, z, a)\sqrt{2T_s(RP_t)^2 / N_0}\right). \quad (5)$$

Here, we have $\gamma = 2T_s(RP_t)^2 / N_0$ and $Q^G(x) = \left(1/\sqrt{2\pi}\right)\int_x^\infty \exp\left(-u^2/2\right) du$ is the Gaussian Q-function. Thus, in conjunction with (4) and (5), the exact ABEP $P(e)$ can be obtained by

$$P(e) = \int_0^\infty Q^G\left(h_p(\theta, w_z, z, a)\sqrt{\gamma}\right) p_\theta(\theta)\, d\theta \,. \qquad (6)$$

This result can only be evaluated numerically.

Pointing error influence on a laser communication link is discussed in [2-3], and it is assumed therein that the pointing error angles in azimuth and elevation are independent and identically Gaussian distributed, resulting in the pdf of the total pointing error angle θ, i.e., $p_\theta(\theta)$, to be a Rayleigh distribution with the scale parameter denoted by σ_θ.

A new idea proposed here is that by maximizing $h_p(\theta, w_z, z, a)$ in (4), we can find the optimum dynamic w_z to achieve the largest instantaneous optical power collected by the detector $\mathbf{C}$, when the pointing error angle θ is measured. This optimization problem can be formulated as $\{ \max\limits_{w_z}\ 1 - Q^{Mar}\left(\dfrac{2z\theta}{w_z}, \dfrac{2a}{w_z}\right),\ s.t.\ w_0 > 2\lambda/\pi \}$. We will simplify and discuss this in a future report.

OECC/PS2016

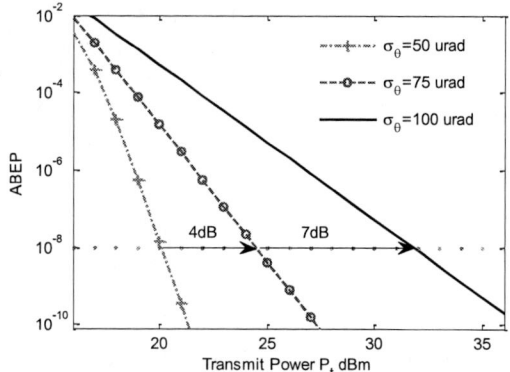
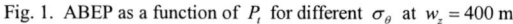

Fig. 1. ABEP as a function of P_t for different σ_θ at $w_z = 400$ m.

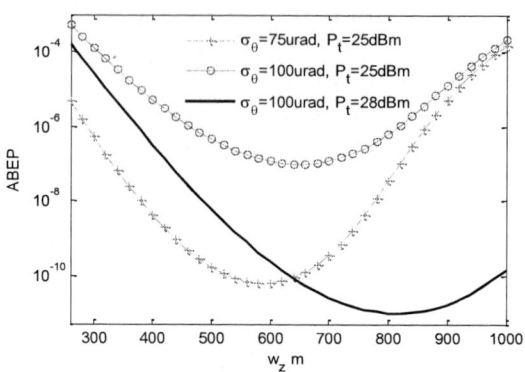

Fig. 2. ABEP as a function of w_z for different P_t and σ_θ.

IV. NUMERICAL RESULTS

The system parameters used for numerical illustrations in this paper are given as follows: $R_{\text{data}} = 1$ Gbps, $z = 1000$ km, $a = 0.125$ m, $N_0 = 2.2 \times 10^{-26}$ A^2/Hz (-174 dBm/Hz thermal noise passing through $179700 \, \Omega$ load resistor [8]), $\sigma_\theta \in [10\mu\text{rad}, 100\mu\text{rad}]$ (or the displacement standard jitter denoted as $\sigma_d (\approx z \cdot \sigma_\theta) \in [10\text{m}, 100\text{m}]$).

From Fig. 1, with the ABEP value of 10^{-8}, compared to the case of $\sigma_\theta = 50\mu\text{rad}$, the power penalties are 4dB and 11dB, respectively, for the cases of σ_θ=75 and 100 (μrad). Furthermore, we explicitly show the effect of the beam radius w_z on the ABEP. We find that the transmit power P_t and σ_θ jointly decide the optimum value of w_z, the adjustment of which can be done by adjusting the transmitter beam waist w_0. As shown in Fig. 2, the optimum values of w_z are 585m, 640m and 820m, respectively, for (σ_θ, P_t) combinations (75 μrad, 25dBm), (100 μrad, 25dBm) and (100 μrad, 28dBm). In some literatures, e.g., [2], there are wrong impressions that the ratio of w_z to σ_d, i.e., $w_z / \sigma_d = \Phi / \sigma_\theta$, can be optimized to a fixed value given a certain value of P_t. We hope to emphasize that this is not true from our observation. As given in our example, at $P_t = 25$dBm, the corresponding values of $\left(w_z / \sigma_d \right)$ are 7.8 and 6.4 for $\sigma_d = 75$m and $\sigma_d = 100$m, respectively, which are obviously non-identical.

V. CONCLUSIONS

A closed-form expression for the exact channel gain in the inter-satellite link with pointing errors is derived in terms of Marcum Q-function. Using this result, we have studied the effect of the beam radius on the system ABEP. We also observe that with a fixed transmit power, the ratio of the beam waist to the standard jitter cannot be optimized to a fixed value. Our result can be further used to optimize the dynamic beam radius to achieve the maximum instantaneous channel gain. This will be given in a future report.

ACKNOWLEDGMENT

The financial support by the Singapore MoE AcRF Tier 2 Grant MOE2013-T2-2-135, the National Natural Science Foundation of China (NSFC) with Grant nos. 61571316 and 61302112, and Qianjiang Talent Project (QJD1402023) is gratefully acknowledged.

REFERENCES

[1] H. Guo, B. Luo, Y. Ren, S. Zhao, and A. Dang, "Influence of beam wander on uplink of ground-to-satellite laser communication and optimization for transmitter beam radius," Optics letters 35, pp.1977–1979, 2010.

[2] Toyoshima, Morio, Takashi Jono, Keizo Nakagawa, and Akio Yamamoto. "Optimum divergence angle of a Gaussian beam wave in the presence of random jitter in free-space laser communication systems." JOSA A 19, no. 3, pp. 567-571, 2002.

[3] Harvard A. Farid, S. Hranilovic et al., "Outage capacity optimization for free-space optical links with pointing errors," J. Lightw. Technol., vol. 25, no. 7, pp. 1702–1710, 2007.

[4] H. G. Sandalidis, T. Tsiftsis, G. K. Karagiannidis, M. Uysal et al., "BER performance of FSO links over strong atmospheric turbulence channels with pointing errors," Commun. Lett., vol. 12, no. 1, pp. 44–46, 2008.

[5] P. Y. Kam and R. Li, "Computing and bounding the first-order Marcum Q-function: a geometric approach," IEEE Trans. Commun.,vol. 56, no. 7, pp. 1101–1110, 2008.

[6] T. Song and P.-Y. Kam, "Efficient symbol detection for the FSO IM/DD system with automatic and adaptive threshold adjustment: The multi-level PAM case," Proc. IEEE/CIC ICCC2015, Shenzhen, China, Nov. 2015 (arXiv:1505.02536).

[7] A.E. Siegman, Lasers, 1986, University Science Books.

[8] S. Dolinar, D. Divsalar, J. Hamkins, and F. Pollara, "Capacity of Pulse-Position Modulation (PPM) on Gaussian and Webb channels," Jet Propulsion Lab., Pasadena, CA, USA, TMO Progr. Rep. Aug. 2000.

WA2-47

Test and Verification of Vehicle Positioning by Infrared Signal-Direction Discrimination

Wern-Yarng Shieh, Ti-Ho Wang, Kun-Lung Chung, Zong-Yu Tsai, and Wei-Jhih Huang

Department of Electronic Engineering, St. John's University, New Taipei City, Taiwan
(shiehwy@mail.sju.edu.tw)

Abstract: With the aid of a simple geometric relation, vehicle positioning by infrared signal-direction discrimination is tested and verified with a receiver comprising four identical planar receiving modules with a simple symmetric structure to determine the coming direction of the signal. In our test, the position of the vehicle can be precisely located in the communication area of a typical infrared short-range vehicle-to-infrastructure communication system.

Keywords: amplitude comparison, angle of arrival, infrared communication, position location.

I. INTRODUCTION

Infrared communication is very suitable for short-range point-to-point and line-of-sight data transmission. Infrared communication system in the wavelength band 780-950 nm has the advantage of low cost and simple technique. It plays an important role in many intelligent-transportation-system (ITS) applications including intervehicle (V-to-V) communications and vehicle-to-infrastructure (V-to-I) communications. In some ITS applications it is necessary to track the trajectories of the vehicles, for example, in multilane-free-flow electronic-toll-collection (ETC) systems. For such V-to-I communications, the real-time position of the vehicle is very important. In this paper, we will propose a method to locate the position of the vehicle and track its trajectory for infrared short-range V-to-I communication systems. The technique is based on so-called *amplitude comparison*, i.e., comparing the signal strengths received by different parts of the receiver which was deliberately designed with a specific structure, to obtain the coming direction of the signal. Then the position of the vehicle can be located with a simple geometric relation while it travels through the communication area of the system.

II. INFRARED AMPLITUDE COMPARISON AND SIGNAL-DIRECTION DISCRIMINATION

In order to obtain the coming direction of the signal by means of comparing the signal strengths received by different parts of the receiver, we have to deliberately design a structure with several components such that for a signal incident from a definite direction relative to the receiver, each individual component has a different responsivitiy due to its specific geometric orientation in the receiver. In our previous work recently conducted, we have successfully constructed an infrared one-dimensional signal-direction discriminator for intervehicle communications. The direction of the target vehicle, i.e., the direction of the emitter carried by the target vehicle relative to the receiver, can be precisely determined up to a longitudinal range of 100 meters [1].

To determine the direction of the signal source relative to the receiver in two dimensions, i.e., to obtain the polar angle θ and azimuthal angle ϕ of the signal direction, where θ and ϕ are defined according to the conventional spherical polar coordinates, we have designed a receiver composed of four planar receiving modules divided into four quadrants with a symmetric structure. The detailed configuration of this signal-direction discriminator and the results of measurement in our laboratory have been reported in [2]. This receiver is able to determine the coming direction of infrared signal with high precision. Owing to the space restriction, here we just show its photograph in Fig. 1. As can be seen, each individual receiving module is composed of a 4×4 array of planar p-i-n photodiodes, and each module has a 45° backward tilt angle. In this paper, we utilize this receiver to locate the position of a vehicle while it travels through the communication area of a typical short-range V-to-I communication system.

III. VEHICLE POSITIONING FOR SHORT-RANGE VEHICLE-TO-INFRASTRUCTURE COMMUNICATION

Figure 2 shows a typical mounting configuration for short-range V-to-I communication systems, where α is the roadside unit (RSU) declining angle, β is the on-board unit (OBU) upward-inclining angle, and h is the vertical mounting height of the RSU relative to the horizontal plane of the OBU. An OBU (vehicle) travels through the communication area with a lateral distance x relative to the vertical plane (y-z plane) above the central line of the traffic lane. To facilitate the discussion, we set the coordinates of the RSU as $(0,0,h)$, and the OBU (vehicle) traveling in the y-direction on the x-y plane has the coordinates $(x,y,0)$, where y is the longitudinal distance of the OBU relative to the vertical plane (x-z plane) of the RSU. For the receiver of the RSU, (θ,ϕ) is the incident direction of the signal relative to its coordinate system $\hat{\imath}'$, $\hat{\jmath}'$, $\hat{k}'$. The influence of all the aforementioned mounting parameters α, β, h, and the position of the vehicle

OECC/PS2016

Fig. 1. Photograph of the signal-direction discriminator (receiver)

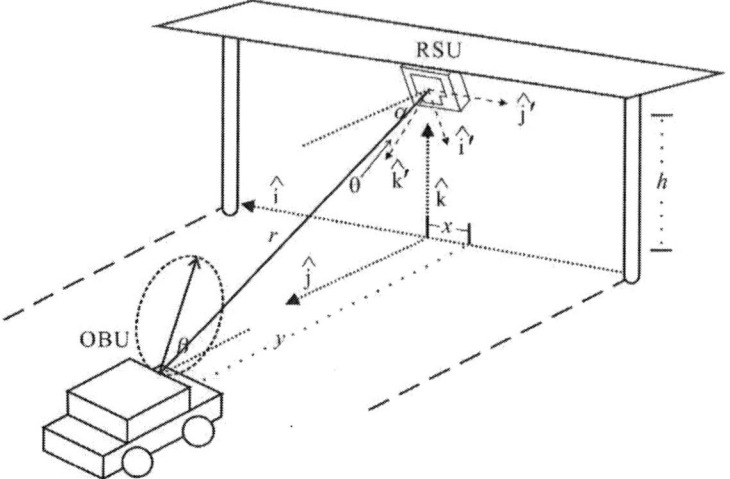

Fig. 2. Mounting configuration of a typical short-range V-to-I communication system

$(x, y, 0)$, on the performance of data transmission for short-range V-to-I communication systems has been thoroughly investigated, the results can be found in typical references such as [4]-[5]. We now discuss how to locate the position (x, y) of the vehicle from the measured signal direction (θ, ϕ). Referring to Fig. 2 it is easy to understand that with a simple geometric relation, from the information of the coming direction of the signal, the two-dimensional position of the signal source, i.e., the position of the vehicle on the ground, can be obtained without difficulty.

Figure 3 shows a photograph of our test field. During the measurement the vehicle, in our case a robot with wheels carrying the emitter, travels through the communication area of the system. The signal-direction discriminator was mounted with a vertical height $h = 4.93$ m and a down-inclined angle $\alpha = 45°$. For this mounting configuration, the relation between the position (x, y) of the signal source and the signal direction (θ, ϕ) relative to the receiver can be obtained straightforwardly, which reads as

$$\cos \theta = \frac{y \cos \alpha + h \sin \alpha}{(x^2 + y^2 + h^2)^{1/2}} \qquad (1)$$

$$\tan \phi = \frac{-x}{h \cos \alpha - y \sin \alpha} \qquad (2)$$

Hence, from the signal direction (θ, ϕ) detected by the receiver, together with the known mounting parameters h and α, this set of simultaneous nonlinear equations can be solved to obtain the position of the signal source (x, y) without much difficulty.

IV. Measured Results

We conducted both the straight-traveling conditions, i.e., the vehicle traverses straight through the communication area of the system with a definite lateral distance x relative to the central line of the traffic lane, and the lane-changing conditions, i.e., the vehicle changes the lateral distance x relative to the central line of the traffic lane when it traverses through the communication area of the system in order to change traffic lanes. In the measurement for lane-changing conditions, we let the vehicle travel along two slant straight trajectories from (-3 m, 20 m) to (2 m, 0 m), and from (2 m, 20 m) to (-3 m, 0 m), respectively, and call them LC_1 and LC_2. That is, during the 20-m longitudinal traverse, both trajectories have a total of 5-m lateral shift, leftward (LC_1) and rightward (LC_2), respectivly, where the value of 5 m is roughly the width of a typical traffic lane on highway. All the measurements are from the longitudinal distance $y = 20$ m to $y = 0$ m, and we took the data every 0.5 m. The positions of the vehicle are obtained by solving the simultaneous nonlinear equations Eqs. (1) and (2) from the coming direction of the signal which was detected by the receiver.

Figure 4 shows the measured trajectories of the vehicle when it traverses through the communication area of the system. In this figure the black dot-dashed line is the data for the vehicle traveling along the central line of the traffic lane, i.e., $x = 0$ m. For the lateral distances $x = -1$ m, -2 m, and -3 m, the results are sketched with blue solid, blue dashed, and blue dot-dashed lines, respectively. For the lateral distances $x = 1$ m and 2 m, the results are displayed with black solid and black dashed lines. For lane-changing conditions, the measured trajectories are depicted with red line for LC_1 and red dashed line for LC_2, respectively. Obviously, the measured trajectories, i.e., the positions, of the vehicle are in quite good agreement with the real values for both straight-traveling and lane-changing conditions as can be seen from the figure, despite the fact that the measured results for $y > 15$ m are not so accurate as that for $y < 15$ m. This is because

866

OECC/PS2016

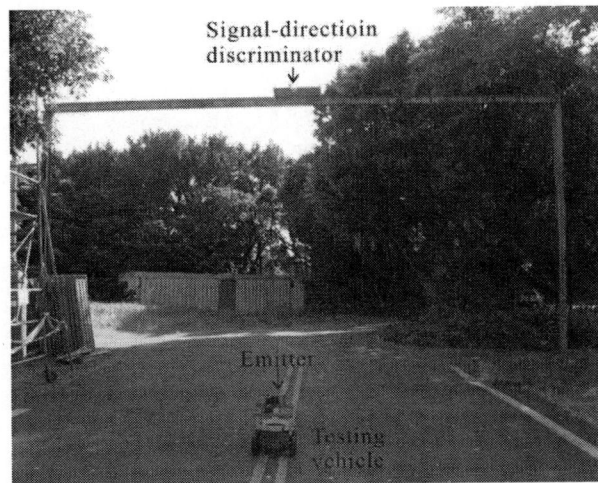

Fig. 3. Photograph of the test field

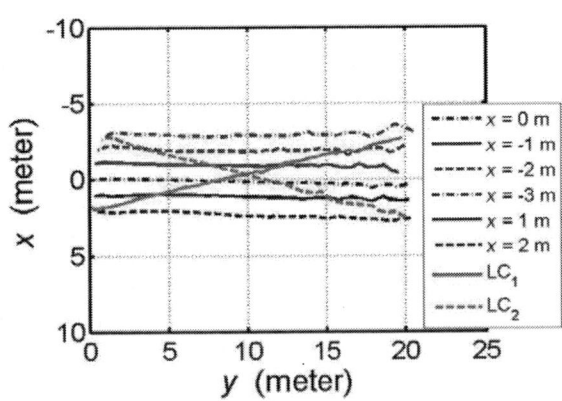

Fig. 4. Measured trajectories of the vehicle

the received signal strengths are very weak for signals coming from a remote region [4][5]. Under these circumstances it is very difficult to obtain the signal direction with high precision. However, we believe that our system meets the requirement of common short-range V-to-I communication systems such as ETC applications, for which the longitudinal communication range is generally shorter than 20 m.

V. CONCLUSIONS

With the knowledge of the coming direction of the signal originated from a vehicle and the aid of a simple geometric relation, the position of the vehicle can be located and the trajectory tracked without difficulty in infrared short-range vehicle-to-infrastructure communication systems. To determine the signal direction, we present a simple symmetric structure comprising four identical planar receiving modules to construct the receiver called the two-dimensional signal-direction discriminator. By comparing the signal strengths received by these individual receiving modules, the coming direction of the signal can be determined with high accuracy. From the detected signal direction, the positions, i.e., the trajectory, of the vehicle can be obtained by solving a set of two simultaneous nonlinear equations based on a simple geometric relation when the vehicle travels through the communication area of the system. Our signal-direction discriminator proposed in this paper can successfully locate the position of the vehicle in a communication area of 6-m width and 20-m length. Note that the width of 6 m is wider than a typical traffic lane, and we believe that the longitudinal length of 20 m will meet the general requirement of common short-range vehicle-to-infrastructure communication systems such as ETC applications.

ACKNOWLEDGMENT

This work was supported by the Ministry of Science and Technology, Republic of China (Taiwan), under Contract MOST 104-2221-E-129-008. The authors are also grateful to the National Center for High-Performance Computing for its support of facilities and software.

REFERENCES

[1] W.-Y. Shieh, C.-C. Hsu, H.-C. Chen, T.-H. Wang, and C.-C. Chen, "Construction of infrared signal-direction discriminator for intervehicle communication," IEEE Trans. Veh. Technol., vol. 64, pp. 2436-2447, June 2015.

[2] W.-Y. Shieh, T.-H. Wang, and C.-J. Wang, "Accuracy and Calibration of Infrared Signal-Direction Discriminator," in Proc. 2015 IEEE Int. Conf. on Consumer Electronics-Taiwan (ICCE-TW-2015), Taipei, Taiwan, June 6-8, 2015, pp. 348-349.

[3] W.-Y. Shieh, W.-H. Lee, S.-L. Tung, B.-S. Jenn, and C.-H. Liu, "Analysis of the Optimum Configuration of Roadside Units and Onboard Units in Dedicated Short-Range Communication Systems," IEEE Trans. Intell. Transport. Syst., vol. 7, pp. 565-571, Dec. 2006.

[4] W.-Y. Shieh, C.-C. Hsu, S.-L. Tung, P.-W. Lu, T-H. Wang, and S.-L. Chang, "Design of Infrared Electronic-Toll-Collection Systems with Extended Communication Areas and Performance of Data Transmission", IEEE Trans. Intell. Transport. Syst., vol. 12, pp. 25-35, March 2011.

[5] W.-Y. Shieh, C.-C. Hsu, H.-C. Chen, T.-H. Wang, and S.-L. Chang, "Design of Light-Emitting-Diode Array for Solving Problems of Irregular Radiation Pattern and Signal Attenuation for Infrared Electronic-Toll-Collection Systems", IET Intell. Transp. Syst., vol. 9, pp. 135-144, March 2015.

Noise Tolerant Optical Detection in CMOS Image Sensor Based Visible Light Communication

Joon-Woo Lee, Se-Hoon Yang, Yong-Hwan Son, Sang-Kook Han

Department of Electrical and Electronic Engineering, Yonsei University, Republic of Korea

junu0809@yonsei.ac.kr

Abstract: We propose optical detection with noise reduction process based on CMOS image sensor. By employing this process, a noise tolerant 3.8 kbps visible light transmission using CMOS image sensor was experimentally demonstrated.

Keywords: Noise in imaging system, Visible Light Communication, LED, CMOS image sensor

I. INTRODUCTION

Visible Light Communication (VLC) has a lot of advantage than Radio Frequency (RF) communication. At first, contrary to using RF bandwidth, a frequency license is not required to use bandwidth of visible light. Next, the VLC has excellent security. The transmission range is possible to recognize by our eyes, and there is little noise leaks to the outside. Also it does not have Electro-Magnetic Interference (EMI). So the VLC can be used near sensitive electronic devices. Finally, there is no hazard to the human body. The RF waves may cause harm to the human body due to the resonance of the water molecule. But, visible light don't have this demerit. For these reasons, there are lots of researches using LED as communication's transmitter with illumination.

Previous VLC studies have mainly used the Photo Diode (PD) to receive high rate data. However, the dongle has to be attached for communication in PD based VLC and this is critical disadvantage. In this paper, we use an image sensor embedded in smart phone to overcome this disadvantage. In communication by using CMOS image sensor, transmitted signal is captured with noise such as unwanted background image, AWGN noise, and etc. For these reasons, noise reduction process is important part of system performance in CMOS image sensor based VLC. Many researches have proposed to novel reception method by using rolling shutter effect, the modulation format for the prevention of flickering [1-3], measuring system performance by simulation about using rolling shutter method [1], using special image sensor [4] and so on. But noise reduction process was not importantly addressed in other researches.

In this paper, we propose noise tolerant detection in image sensor based VLC for improving capacity of data rate. By this method, we experimentally verify the feasibility of the data reception by the image sensor embedded in the smart phone. There are various noise in received signal. For noise reduction, a few processes are adopted in received image that contains signal with noise. We reduce shot noise or small image noise by using noise division process. Received signal has active threshold value because of LED shape and uneven brightness of background. For making stable threshold value, background ingredients have to be reduced. By low-pass filtering, polynomial curve fitting and normalize, we extract noise reduced signal and achieve performance improvement.

II. EXPERIMENT AND RESULTS

Fig. 1 shows experimental setup for using LED as transmitter and image-sensor as receiver. Arbitrary Waveform Generator (AWG) generates the random signal made from transmission code to measuring the system performance. And then this generated signal voltage flows to LED module that we made. LED module consists of three parts. Input signal is amplified at amplifier. And this signal is combined with DC voltage at a bias tee and then driven to a LED. A LED emits a signal with the light and an image sensor that is imbedded in a smart phone receives a signal.

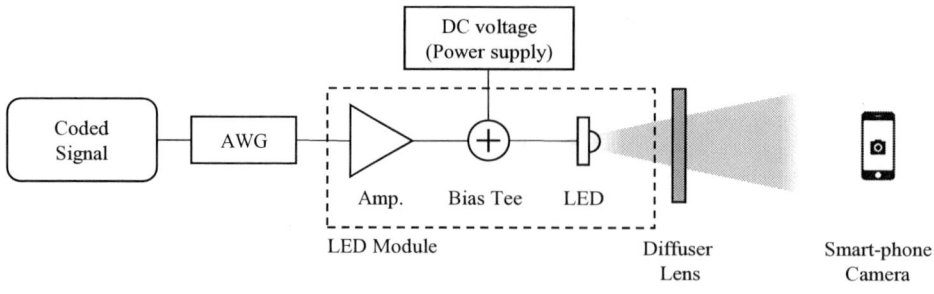

Fig. 1. Experimental setup

Camera setting is required in order to receive a signal from LED. We use LG-G3 smart phone and 'Sony IMX135w' is used as an image sensor in this smart phone. Many smart phones, including the LG-G3, don't support detailed camera setting in the basic camera application. So, android camera application, 'Open camera', is used in this experiment. Exposure, Focus, ISO, etc. are basically set to Auto. And 1920×1080 resolution is used. But auto setting about these values during video recording for data reception makes some troubles at data receiving. Due to the fluctuation of receiving power and the distortion of the image from auto setting, data detection is very hard. So these values must to be fixed at experiment.

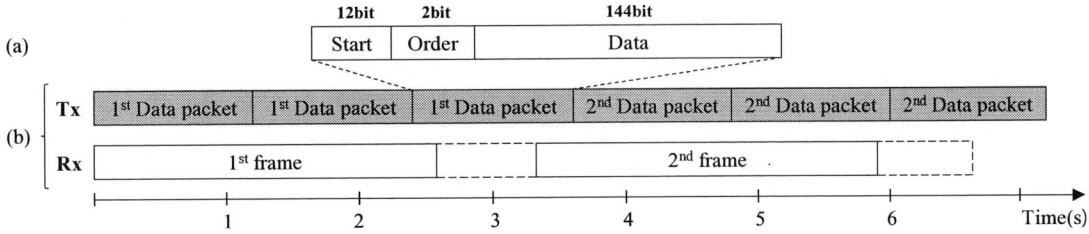

Fig. 2. (a) The structure of a data packet and (b) the length of the frame and data packet in time domain.

The frame rate of used smart phone in this experiment is 30fps. Fig. 2(b) shows the length of each frames isn't 0.033s. There is processing time between two frames, and the reception time of each frame is 0.0262s. For this reason, image sensor receives the discrete signal and we have to use proper transmission method. In this paper, we make data packet and repeat each data three times at transmitter. As shown in Fig. 2(a), data packet is consisted of three parts. First part is 'Start' part. This part contains the center level of level 0 and level 1. So the 'Start' part can be distinguished from other parts. Additionally, the data packet doesn't need to have end part because the end of the 'Data' part is in front of the next 'Start' part. The 'Order' part composed of two bits is repeated in order of precedence, 00→01→10→11→00. This part prevents to receive the same data packet and detects the missing of the data packets. The last part is 'Data' part. This part contains data of 144bits and each bit has a value of the 'Level 0' or 'Level 1'.

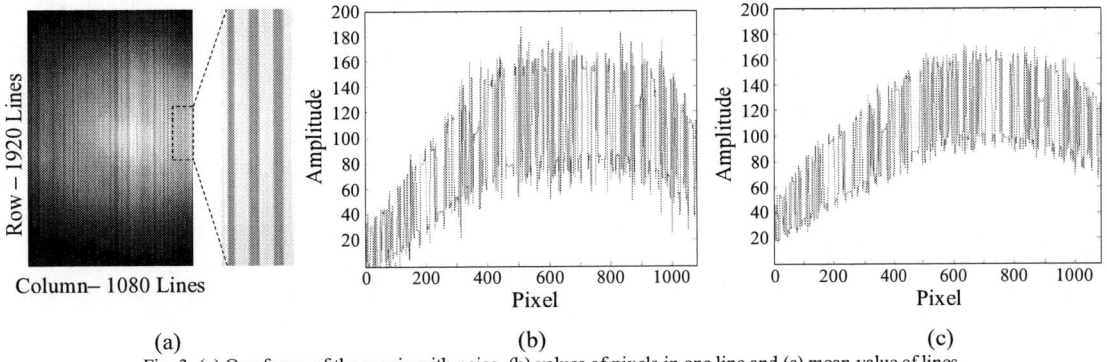

Fig. 3. (a) One frame of the movie with noise, (b) values of pixels in one line and (c) mean value of lines

At receiver part, the smart phone records the movie for receiving signal. Fig. 3(a) is one frame of the movie. In this picture, noise component is included such as background noise, LED shape, shot noise during light to electronic signal and so on. A signal can be shown as bright line in 1 and dark line in 0 because of rolling shutter effect. Fig. 3(b) is one line of a frame and values are uneven because of noise component. At Fig. 3(c), noise is divided by 1920 that is height value of a frames and reduced.

Fig. 4 represents the process of the background noise subtraction. We extract the background noise and then subtract it from the pixel values. The frequency components of the background noise are concentrated in low frequency than the signal. For these reason, the low-pass frequency filtering can subtract a large portion of signal values. And then the background noise are extracted from the polynomial curve fitting. The bold line of Fig. 4(b) is background noise that is finally extracted and we can get signal values by the background noise subtraction from pixel values. Extracted signal values are not flat as Fig. 4(c), so these values should be normalized. Finally Fig. 4(d) shows the signal values by the background noise subtraction.

Data can be detected from signal values. At first, a data packet can be extracted because a start part of a data packet has distinguishable level, center of 'Level 1' and 'Level 0' that is unused both the order and the data parts. After extracting one data packet, the width of one bit is calculated by using the width of a data packet and the number of bits in a data packet, and it makes bits of the 'Order' and 'Data' parts are detected.

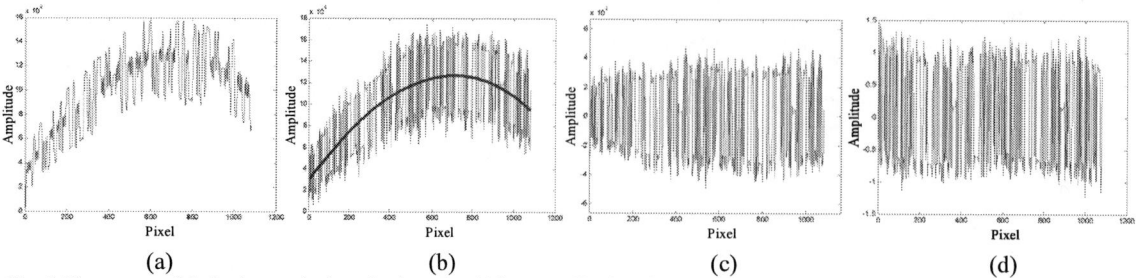

(a)   (b)   (c)   (d)

Fig. 4. The process of the background subtraction by using (a) low-pass filtering, (b) polynomial curve fitting, (c) subtraction and (d) normalize

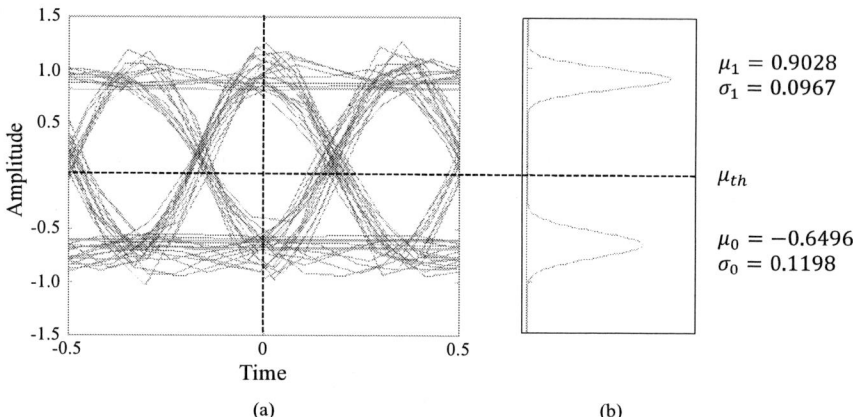

(a)   (b)

Fig. 5. The eye diagram of the received signal (a) and the corresponding distributions of 0s and 1s (b).

In this experiment, one chip is received from 3.3pixels in a frame and the image sensor received 3,797 bits per second. In experiment, we receive 21,600 bits and check all bits are all correct. Fig. 5 shows the eye-diagram of the received signal. In order to determine the Bit Error Rate (BER), the enough bits should be sent. But it takes so long time to send the enough bits to determine the BER. So we determine the Q-factor and BER by the calculation from eye-diagram. By the calculation using a mean and standard deviation of 'Level 0' and 'Level 1', we can determine that a Q-factor is 7.1735 and a BER is almost error free.

III. CONCLUSIONS

In VLC using an image sensor as a receiver, it is necessary to reduce noise because the signal intrinsically includes the background noise in the image. We experimentally tested the performance of a data transmission system with noise tolerant characteristics under various additional noise conditions. Our novel three-step noise-reduction process effectively reduced noise. The process achieved a transmission performance of 3.8 kbps error free transmission. Using the proposed method in CMOS image sensor based VLC, we not only improve the transmission performance, but also implement the VLC with ambient noise lights.

REFERENCES

[1] H. Aoyama, and M. Oshima, "Visible light communication using a conventional image sensor," Proc. of the Consumer Communications and Networking Conference (CCNC 2015), Las Vegas (USA), Jan. 2015, pp.103–108.

[2] R. D. Roberts, "Undersampled frequency shift ON-OFF keying (UFSOOK) for camera communications (CamCom)," Proc. of the Wireless and Optical Communication Conference (WOCC 2013), Chongging (China), pp.645–648.

[3] P. Ji, H. M. Tsai, C. Wang, and F. Liu, "Vehicular visible light communications with led taillight and rolling shutter camera," Proc. of the Vehicular Technology Conference (VTC 2014), Seoul (Korea), pp.1–6.

[4] I. Takai, S Ito, K. Yasutomi, K. Kagawa, M. Andoh, and S. Kawahito, "LED and CMOS image sensor based optical wireless communication system for automotive applications," Photonics Journal vol. 5, 2013, 6801418.

WA2-49

OECC/PS2016

A CW, All-Fiber 2100 nm Ho Doped Fiber Laser Pumped at 1950 nm

Jiachen Wang, Sang Bae Lee, and Kwanil Lee

Korea Institute of Science and Technology, Hwarang-ro 14-gil 5, Seongbuk-gu, Seoul 136-791, Republic of Korea

klee21@kist.re.kr

Abstract: A Ho doped fiber laser pumped at 1950 nm by Tm doped fiber laser is reported. With slope efficiency of 51.8%, the laser generates more than 7 W output at 2100 nm.

Keywords: fiber laser, Ho doped fiber, Tm doped fiber

I. INTRODUCTION

In recent years, 2 μm lasers drawn great attention in potential applications such as free space optical communication, remote sensing, plastic material processing, optical parameter oscillation, Lidar, and medical use [1-6]. Strong absorption by water and weak absorption by human tissues at this wavelength enables them to be ideal sources for biological and medical applications such as ophthalmic procedures, anthroscopy, and laser angioplasty. In addition, low atmospheric absorption and eye-safe property make it also desirable for material processing, remote sensing, and Lidar.

To date, fiber laser has been more intensely studied due to its compact configuration, low cost and excellent beam quality, compared to solid state bulk laser like Tm:YAP or Ho:YAG. Two major fiber gain medium for generating laser at 2 μm region are Tm doped fiber and Ho doped fiber. Among them, Tm doped fiber laser is more studied, but its wavelength is limited to a shorter range by its energy level structure, usually below 2050 nm. On the other hand, Ho doped fiber laser provides longer operation region which can go beyond 2100 nm and is important for many applications like optical parameter oscillation. There is still few research reports about Ho doped fiber laser, partly due to its rare pumping wavelength. The two major absorption peaks of Ho doped fiber are respectively at 1150 nm and 1950 nm, none of them locate in usual working range for commercial laser diode pump.

In this paper, we demonstrate an all-fiber configuration Ho doped fiber laser pumped at 1950 nm by a homemade Tm doped fiber laser. The laser is built with single mode fiber, and its operation is confined by a pair of FBG which stabilizes its output wavelength at 2100 nm. In experiments, the laser generates more than 7 W output, and both its spectrum performance and slope efficiency are considerable.

II. CONFIGURATION OF HO DOPED FIBER LASER

The setup of Ho doped fiber laser is shown in Fig. 1. The laser comprises two major parts, a homemade Tm doped fiber laser used as pump, and the main laser cavity built with Ho doped fiber and FBG reflectors.

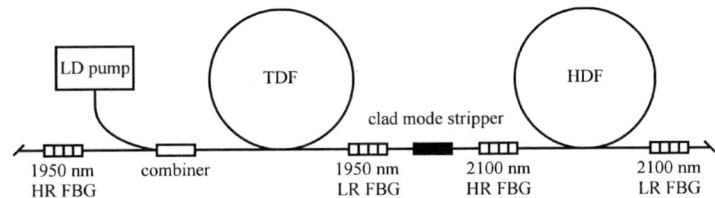

Fig. 1. Setup of Ho doped fiber laser. LD pump: 793 nm laser diode. TDF: Tm doped fiber. HDF: Ho doped fiber.

A. Tm Doped Fiber Laser as 1950 nm Pump

The homemade Tm doped fiber laser is built with all-fiber configuration. The gain fiber is 2.1 m double clad Tm doped fiber. A pair of FBG (HR: 99% reflectance, LR: 15% reflectance) is used to stabilize the output wavelength at 1950 nm. The FBG pair is inscribed on double clad passive fiber whose geometry size exactly matches Tm doped fiber. All these fiber are single mode at 2 micrometer range, with 10 μm diameter core and 130 μm diameter clad. A 793 nm laser diode is used as pump, which is delivered into Tm doped fiber by a combiner. Besides the pump input, the combiner has other pigtails as its signal input and output, both are matched with Tm doped fiber in size. When splicing the FBG pair with the gain fiber, HR FBG is spliced to combiner's signal input rather than directly spliced to Tm doped fiber. The Tm doped fiber is then spliced with combiner's output. The LR FBG is spliced to the other end of Tm doped fiber. Due to the limited length of our gain fiber (2.1 m), there is strong residual pump mixed in the laser's output. Since the residual pump mostly resides inside clad, an all-fiber clad mode stripper is used to eliminate it. The clad mode stripper is customized fabricated, and is exactly matched with FBG fiber.

This work has been partly supported by the R&D Program (Grant No: 10043479) for Industrial Core Technology funded by the Ministry of Trade, Industry and Energy (MOTIE), Republic of Korea.

B. Ho Doped Fiber Laser's Main Body

The Ho doped fiber laser adopts 2.5 m Ho doped fiber (SM-HDF-10/130 from Nufern) as its gain medium. The fiber has 10 μm diameter core and 130 μm diameter clad, which make it be single mode at 2 micrometer range. A pair of FBG (HR: 99% reflectance, LR: 15 reflectance) with reflection peak at 2100 nm is used as laser cavity's reflectors. These FBGs are inscribed on passive fiber whose geometry size matches Ho doped fiber. The Tm doped fiber laser's output is coupled into Ho doped fiber's core as its pump. The clad mode stripper inserted between two lasers' bodies effectively removes residual power inside fiber clad. The 2100 nm LR FBG and the 1950 nm HR FBG's un-spliced ends are angularly cleaved to suppress possible parasite lasing.

III. EXPERIMENT RESULTS

The output spectrum of Ho doped fiber laser is shown in Fig. 2. It indicates that the laser's output is strictly determined by the FBG pair. No parasite lasing is observed at any power level we have investigated. Furthermore, it is found that the 1950 nm pump generated by Tm doped fiber laser is mostly absorbed. At any power level there is a gap more than 23 dB between 2100 nm peak and 1950 nm peak, so the output power detected by power meter can be considered as pure signal power.

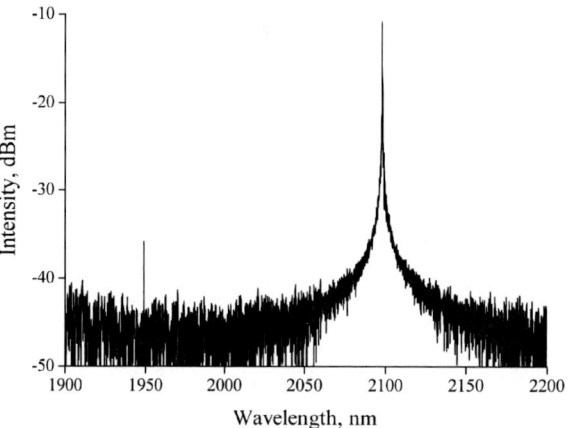

Fig. 2. Spectrum of Ho doped fiber laser's output.

The relationship between 2100 nm output of Ho doped fiber and its 1950 nm pump is demonstrated in Fig. 3. For the power of 1950 nm pump (which is the output of Tm doped fiber laser), it is measured after clad mode stripper, so any residual pump at 793 nm is filtered. We obtain a slope efficiency of 51.8%, and the optical-optical conversion rate is 51.2 % at the maximum power we have investigated (7.11 W).

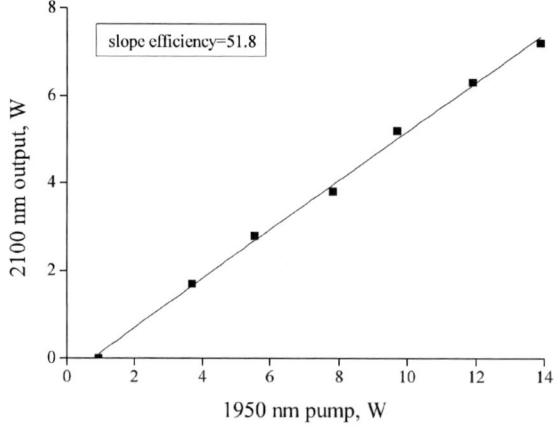

Fig. 3. Output of Ho doped fiber laser versus its 1950 nm pump.

Finally, we studied the total conversion efficiency between 2100 nm output and 793 nm pump. The result is shown in Fig. 4.

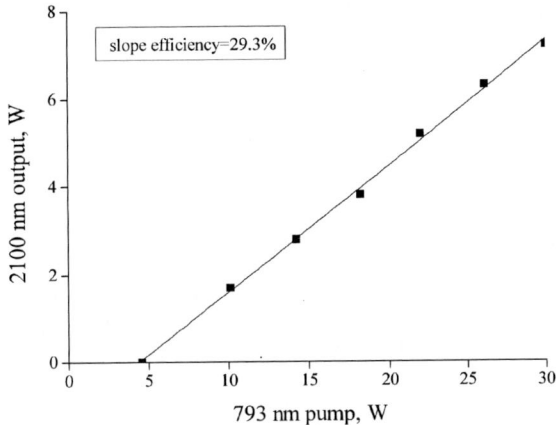

Fig. 4. Output of Ho doped fiber laser versus 793 nm pump.

It can be seen that the total system reach a slope efficiency of 29.3%, and the optical-optical conversion rate at maximum power is 23.7%.

IV. CONCLUSIONS

We reported an all fiber Ho doped fiber laser operating at 2100 nm. The laser is pumped at 1950 nm by a homemade Tm doped fiber laser. We obtained more than 7 W 2100 nm output from the Ho doped fiber laser with slope efficiency of 51.8%.

ACKNOWLEDGMENT

This work has been partly supported by the R&D Program (Grant No: 10043479) for Industrial Core Technology funded by the Ministry of Trade, Industry and Energy (MOTIE), Republic of Korea.

REFERENCES

[1] K. Scholle, S. Lamrini, P. P. Koopmann, and P. Fuhrberg, "2 µm laser sources and their possible applications," Frontiers in Guided Wave Optics and Optoelectronics, InTech: Croatia, February 2010, pp. 471-500.

[2] I. Mingareev, "Welding of polymers using a 2 µm thulium fiber laser," Opt. Laser Technol., vol. 44 (7), October 2012, pp. 2095-2099.

[3] Q. Wang, J. Geng, and S. Jiang, "2-µm fiber laser sources for sensing," Opt. Eng., vol. 53 (6), article id. 061609, June 2014

[4] G. Stoppler, C. Kieleck, M. Eichhorn, "High-pulse energy Q-switched Tm(3+):YAG laser for nonlinear frequency conversion to the Mid-IR," Proc. SPIE, vol. 7836, article id. 783609, 2010.

[5] M. Gebhardt, C. Gaida, P. Kadwani, A. Sincore, N. Gehlich, L. Shah, and M. Richardson, "Nanosecond Tm:fiber MOPA system for high peak power Mid-IR generation in a ZGP OPO," Advanced Solid-State Lasers Congress Technical Digest, 2013.

[6] R. L. Blackmon, P. B. Irby, and N. M. Fried, "Comparison of holmium:YAG and thulium fiber laser lithotripsy: ablation thresholds, ablation rates, and retropulsion effects," J. Biomed. Opt., vol. 16 (7), article id. 071403, July 2011.

Security Strategy against Multipoint Eavesdropping in Elastic Optical Networks

Wei Bai,[1] Hui Yang,[1] Yongli Zhao,[1] Jie Zhang,[1] Yuanlong Tan,[1] Xiaoxu Zhu,[1] Huixia Ding,[2]

[1] State Key Laboratory of Information Photonics and Optical Communication, Beijing University of Posts and Telecommunications,
Beijing, 100876, P. R. China
[2] China Electric Power Research Institute, Beijing 100192, P. R. China.
baiweidawn@foxmail.com

Abstract: *Against multipoint eavesdropping attack in Elastic Optical Networks, new objective criterion, security strategy and extended routing and spectrum allocation algorithm is proposed. The simulation results shows that better security guarantee and resource utilization are provided.*
Keywords: *Network Security, Elastic Optical Networks, Multipoint Eavesdropping Attack, Security Strategy.*

I. INTRODUCTION

Recently, plenty of malignant network attacks and information leakages burst and lead to serious harm to the network users. Network make an insecure impress to people frequently. Weakness and security issue of network have entered the spotlight. Elastic Optical Networks (EON) as a promising technology in data center optical interconnection has been widely studied for flexible and adaptive spectrum resource provisioning. Unfortunately, in which the research on the security is mentioned rarely, EON have no improvement in security and must surfer from the similar security threat to the previous optical networks.

The main network attacks in optical networks can be classified into two types according to the impact on the network behaviors. The first one will make the network running depart from the normal behaviors, such as Denial of Service (DoS). The other one with the intention of stealing information has little or no impact on the network behaviors, such as eavesdropping. Various detection methods against network attacks have been proposed based on the impact on the network action [1,2]. However, the eavesdropping would be detected in a long time or even not on account of the silent characteristic. Multipoint eavesdropping would lead to terrible information leakage through. Privacy and information security of users suffer from a serious threat.

In this paper, we present a new objective criterion for the security demands of a traffic request against multipoint eavesdropping. Based on this criterion, a security strategy using traffic flow slicing is proposed to prevent information leakage in EON. Furthermly, a security-aware routing and spectrum allocation (SA-RSA) algorithm is designed. With the analysis of simulation results, our strategy have an obvious effect to prevent information leakage in EON.

II. SECURITY STRATEGY AGAINST MULTIPOINT EAVESDROPPING

A. Network model:

The network model of EON can be presented as a graph G{**N, L, S**}. Where **N** is a set of nodes with bandwidth-variable optical cross-connects function, **L** denotes a set of links in the network and **S** is the set of spectrum sub-carriers on each fiber link. $TR_i(\mathbf{s}, \mathbf{d}, \omega)$ represents the ith traffic request in the network, **s** is the start node and **d** is the destination node, ω is the occupied resource of TR_i.

B. A new objective criterion: the minimum number of eavesdropping points (minNEP)

In the aspect of network security, to implement a more effective security strategy, more costs in the performance of network and economy is necessary in general. In the same way, the similar problem would exist in the aspect of network eavesdropping. To eavesdrop the complete information from a traffic flow, the more costs and risks eavesdropper would take. Therefore, a traffic flow is proved to have a higher security, which eavesdropper have to pay more cost for eavesdropping it. The costs of eavesdropping a traffic flow successfully can be used to evaluate the security of the flow.

According to the above-mentioned idea, we propose a new objective criterion: minNEP as the evaluation of security for a traffic flow. It can be interpreted as the minimum number of eavesdropping points (include only multipoint

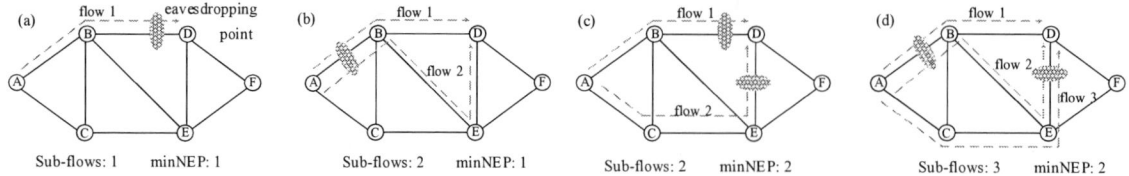

Fig. 1: cases of proposed security strategy against multipoint eavesdropping

eavesdropping in links and regardless of the eavesdropping in nodes) in the optical network to obtain the complete information of a traffic flow. The higher minNEP of a traffic flow with a higher cost for eavesdropping is more security.

Based on the criterion, when a traffic request with privacy security requirement arrivals in the network, the network can achieve the purpose of security through guaranteeing that the actual minNEP should not be less than the minNEP demand. So minNEP demands are added and the ith traffic request would be represented as $TR_i(\mathbf{s}, \mathbf{d}, \boldsymbol{\omega}, \mathbf{minNEP})$.

C. Proposed security strategy:

Obviously all the traffic flow with general transmission mode may lead that its minNEP would be 1. Because each traffic flow is assigned into one path, and only one eavesdropping point in the link of this path can obtain the complete information. To increase the minNEP, we propose a security strategy based on traffic flow slicing technology in EON.

Traffic flow slicing technology enables EON to slice the traffic flow into multiple independent partitions. Each sub-flow which carries a part of information can be transmitted independently. Only one eavesdropping point is not enough to obtain the complete information of a traffic flow, and its minNEP and security would be higher. In Fig.1, four examples to transmit a traffic flow are shown. Fig1.(a) shows the case in the general transmission mode. Fig1.(b)-(d) show the minNEP of a traffic flow in the cases of different sub-flow routes based on traffic flow slicing. According to these cases in Fig1, it is proved that the traffic flow can be sliced into several sub-flows and promote its minNEP. However, more sub-flows may not lead to a higher minNEP, just like Fig1.(b) and (d). Overlapped link of different sub-flows would make the traffic flow slicing ineffective for eavesdropping. Routing issue has a significant impact on the security guarantee.

On the other hand, the enhancement of security always leads to a degradation of performance in the networks. While the traffic flow is sliced into sub-flows, smaller granularity in the occupancy of spectrum resource is realized. The adjustable size of sub-flow can be sliced close to the size of spectrum fragmentation in the link. And the sub-flow would use the spectrum fragmentation resource that save spectrum resource [3]. A comparison between general transmission mode and proposed security strategy is shown in Fig.2.

In Fig.2 We assume that $L_{B,D}$ is eavesdropped and $TR_4(A, D, 10)$ is the last incoming traffic request with security requirement, rout $L_{A,B}$ and $L_{B,D}$ are employed as the transmission path according to the shortest path algorithm. The result of general transmission mode in Fig2.(a) is that the information in TR_4 is leaked out.

In Fig2.(b), using the proposed security strategy based on the traffic flow slicing technology, the incoming TR_4 is sliced into two flows. In the side of routing, the path consisting of $L_{A,B}$ and $L_{B,D}$ and the path consisting of $L_{A,C}$, $L_{C,E}$ and $L_{E,D}$ are employed to transfer the two traffic flows. In the side of spectrum allocation, the size of flow 2 in spectrum is sliced close to the size of spectrum fragmentation in $L_{A,C}$ and $L_{C,E}$. The size of flow 1 is decided as the remainder of TR_4.

The result of Fig2.(b) shows that the eavesdropping point in $L_{B,D}$ can not steal the complete information of TR_4 and the security of TR_4 is guaranteed effectively. In addition, the utilization of the spectrum resource in EON is enhanced and the problem of spectrum fragmentation is got valid settlement.

Algorithm: SA-RSA heuristic algorithm

Input: G(N, L, S), $TR_i(\mathbf{s}, \mathbf{d}, \boldsymbol{\omega}, \mathbf{minNEP})$,k;

Output: One or multiple spectrum path(s) for $TR_i(\mathbf{s}, \mathbf{d}, \boldsymbol{\omega}, \mathbf{minNEP})$;

R: the set of shortest paths;
asnR: the assigned route;
asnF: the assigned spectrum;
MinF: The minimum spectrum occupied by each sub-flow;
maxF(R_i): The maximum available spectrum resource in R_i;
num_asnR: the number of route in asnR;
total_sp: total available spectrum in asnR;

```
1:    begin
2:      If (minNEP=1)
3:        Break and Call for the general RSA;
4:      End if
5:      Find k shortest paths store in R(R₁,R₂, …,
           Rₖ);
6:      for (i=1; i<= minNEP; i++)
7:        for (j=1; j<= k; j++)
8:          if(maxF(Rⱼ)> MinF)
9:            Add Rⱼ into asnR and delete all the
               route having overlapped link with
               Rⱼ in R;
10:           Break;
11:         end if
12:         if(j==k||(num_asnR==i&&
             total_sp<ω))
13:           block the request and quit;
14:         end if
15:       end for
16:     end for
17:     Adjust slicing size of TRᵢ and take the
          spectrum fragment in asnR as the
          preferred subcarrier;
18:     Store assigned spectrum in asnF;
19:   End begin
```

Fig. 3: The pseudocode of SA-RSA heuristic algorithm

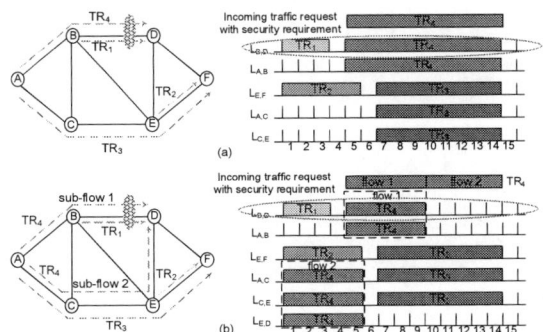

Fig. 2: A comparison between general transmission mode and proposed security strategy

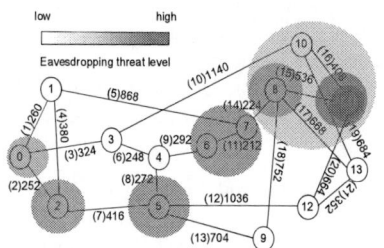
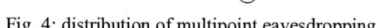

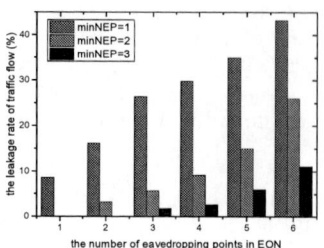

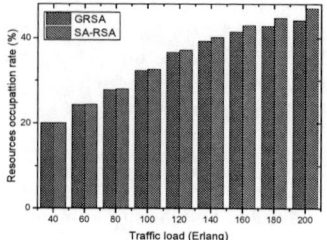

Fig. 4: distribution of multipoint eavesdropping   Fig. 5: the leakage rate of traffic flow   Fig. 6: resource occupation rate

The proposed security strategy can not only guarantee security but also improve the network performance. To achieve the above goals, a RSA algorithm is necessary.

D. Proposed SA-RSA:

To support this security strategy, a SA-RSA described in Fig.3 for EON is proposed. In the SA-RSA algorithm, spectrum continuity constraint and spectrum contiguity constraint two main constraints are employed. The minNEP of each traffic flow request should be guaranteed and the spectrum resource would be utilized effectively.

III. SIMULATION RESULTS AND DISCUSSION

As the differences of cities functions, the nodes in an optical network have different rates to be attacked by eavesdropper. As a result, we simulate our security strategy under EON on a 14-node NSFnet topology with a focus on eavesdropping attacks as shown in Fig.4. According to the cities functions in USA, we assume an eavesdropping attacks distribution proportion in the links of NSFnet in TABLE.1. In our simulation, there are 120 sub-carriers in each fiber and each sub-carriers has 12.5 GHz bandwidth.

TABLE. 1: Eavesdropping Attacks Distribution Proportion

Link	Pro (%)	Link	Pro (%)	Link	Pro (%)
1	5	8	5	15	7
2	6	9	5	16	6
3	5	10	3	17	5
4	5	11	6	18	5
5	5	12	5	19	3
6	1	13	5	20	5
7	6	14	6	21	1

The traffic requests arrive to the network following a Poisson process, each traffic request carry a minNEP demand. The proportion of requests are 85%, 10% and 5%, whose minNEP demands are 1, 2 and 3.

To verify the effect of the proposed strategy in security, we collect the leakage rate of traffic flow with different numbers of eavesdropping points show in Fig.5 when the traffic load is 100 Erlang. Through setting more eavesdropping points in the network, the leakage rate raise up. However, the traffic requests which has a higher minNEP demand is transmitted with a lower leakage rate and has a higher security. Fig.6 depicts the resource occupation rate of our proposed SA-RSA compared with the general RSA (GRSA) [4], we observe that the SA-RSA has a better resource occupation rate under the condition of high traffic load.

IV. CONCLUSIONS

To guarantee the security of optical network, we proposed an objective criterion and a security strategy against the multipoint eavesdropping in EON. To support the strategy, A SA-RSA algorithm is designed. The Simulation results show that our security strategy achieve to reduce the leakage rate of the traffic with a high security demand and use the source more efficiently.

ACKNOWLEDGMENT

This work was supported in part by NSFC project (61501049), Fund of State Key Laboratory of Information Photonics and Optical Communications (BUPT), P. R. China (IPOC2015ZT01), and funded by State Key Laboratory of Advanced Optical Communication Systems Networks, China, Research on Operation Quality Index System and Risk Assessment of Power Large Capacity Optical Transmission Networks.

REFERENCES

[1] Skorin-Kapov N, Chen J, Wosinska L, "A new approach to optical networks security: attack-aware routing and wavelength assignment," IEEE/ACM Transactions on Networking, 2010, vol. 18, no. 3, pp. 750-760.

[2] Lazzez A, "All-Optical Networks: Security Issues Analysis," Journal of Optical Communications and Networking, 2015, vol. 7, no. 3, pp. 136-145.

[3] Yang H, Zhang J, Zhao Y, et al. "Multi-flow virtual concatenation triggered by path cascading degree in Flexi-Grid optical networks," Optical Fiber Technology, 2013, vol. 19, no. 6, pp. 604-613.

[4] Wang Y, Zhang J, Zhao Y, et al. "Dynamic spectral defragmentation based on path connectivity in flexible bandwidth networks," European Conference and Exhibition on Optical Communication, Amsterdam (Netherlands), Sep 2012, pp. P5-10.

WA2-50

OECC/PS2016

Switchable passively mode-locked thulium-doped fiber laser based on single-wall carbon nanotubes

Zhichao Wu,[1] Songnian Fu,[1,*] Minming Zhang,[1] Kai Jiang,[2] Jue Song, [1]Ping Shum,[2] Deming Liu[1]

[1]National Engineering Laboratory of Next Generation Internet Access National Engineering Lab (NGIA), School of Optical and Electronic Information, Huazhong University of Science and Technology (HUST), Wuhan, China, 430074.
[2]Centre for Optical Fiber Technology (COFT), Nanyang Technological University (NTU), Singapore, 637553.
*E-mail: songnian@hust.edu.cn

Abstract: *A switchable passively mode-locked fiber laser is experimentally demonstrated by using single-wall carbon nanotubes as saturable absorber. The mode-locked fiber laser can be operated at the central wavelengths of 1947nm, 1945nm and 1943nm, individually.*
Keywords: *Ultrafast lasers; Fiber lasers; Nanomaterials.*

I. INTRODUCTION

$2\mu m$ light sources have been wide investigated in recent years due to their eye-safe property, as well as their diverse scientific applications in the fields including mid-infrared spectrum generation, remote sensing, free-space communication. In previous researches, mode-locked thulium-doped fiber lasers (TDFL) are usually based on nonlinear polarization rotation (NPR) [1], nonlinear amplifying loop mirror (NALM) [2], and semiconductor saturable absorber mirror (SESAM) [3]. As carbon allotropes draw much attention recently, Single-wall carbon nanotube (SWCNT) has been demonstrated to have ultrafast nonlinear optical responses [4,5], which makes it a popular alternative in ultrafast pulse generation. By varying SWCNT's diameter, many mode-locking operations have been realized over a wide wavelength range from $1\mu m$ to $2\mu m$ [6].

Compared with those lasers that emit at single center wavelength, multi-wavelength and wavelength-switchable mode-locked lasers can simultaneously generate pulse-trains at different central wavelengths. This kind of lasers can be used in wavelength division multiplexing (WDM) systems, optical signal processing, and precision spectroscopy. In previous studies, numerous researchers are focus on multi-wavelength or wavelength-switchable mode-locking in Ytterbium-doped and Erbium-doped fiber lasers [7,8]. As for $2\mu m$ region, switchable dual-wavelength continuous wave (CW) fiber has been realized by a cascaded filter structure [9]. However, such additional optical filter extends the length of cavity and increase the implementation cost. NALM and NPR techniques are appropriate alternatives to obtain mode-locking and filtering effect inside cavity to realize multi-wavelength or wavelength-switchable mode-locking [1,10]. Nevertheless, both NALM-based and NPR-based technologies generally need high pump power and careful adjustment, in order to realize stable mode-locking operations.

In this manuscript, we experimentally demonstrate a wavelength-switchable and dual-wavelength mode-locked TDFL by using SWCNTs as saturable absorber (SA). To the best of our knowledge, it is the first time that dual-wavelength mode-locking is realized in a SA-based TDFL. The mode-locked fiber laser can be operated at not only switchable single wavelength mode-locking among 1947nm, 1945nm and 1943nm, but also dual-wavelength mode-locking of 1947nm and 1945nm with a wavelength spacing of 2nm with long-term stability.

II. EXPERIMENTAL CONFIGURATION

The propose fiber laser is schematically showed in Fig.1. Driven by a 20 dBm light source, the SWCNTs are deposited on a standard FC/PC fiber end from dimethylformamide (DMF) solution. Compared with the techniques of CNT-film sandwiched between fiber connectors, the optically-driven deposition method possesses much higher optical damage threshold, which can experimentally bear high pump power launching as well as support high-energy pulse generation [11,12]. As for other optical components, a 3.5m commercial thulium-doped fiber (TDF, Nufern SM-TDF-10P/130-HE) is utilized as the gain medium, which is pumped via 1570/2000 WDM by an erbium-doped fiber amplifier (EDFA) with maximum output power of 5W. A CW of 1mW generated from a 1570nm laser is imported into the EDFA as a seed source. The polarization state of light is adjusted by a polarization controller (PC). The isolator is used to guarantee unidirectional propagation and suppress detrimental reflections. A 10:90 fiber optical coupler (OC) is utilized as the output port of the fiber laser. The total cavity length is around 15m. Finally, an optical spectrum analyzer (OSA, Yokogawa AQ6375) with a resolution of 0.05nm and a 350MHz oscilloscope (OSC, Agilent 54641A) together with a 12.5GHz photodetector (PD, EOT ET-5000F) are applied to monitor the optical spectrum and the temporal trace, respectively. The radio-frequency (RF) spectrum is measured by a 20Hz~26.5GHz signal source analyzer (R&S FSUP). Additionally, the pulse profile is measured by a commercial autocorrelator (FR-103XL).

OECC/PS2016

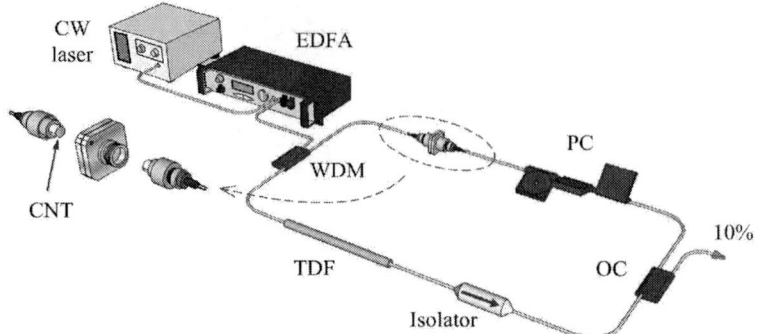

Fig. 1. Experimental configuration

III. RESULTS AND DISCUSSION

Firstly, CW laser emission at central wavelength of 1947nm is observed, when the output of EDFA is set above 200mW. Once the pump power is increased to 320mW, self-starting mode-locking at single central wavelength of 1947nm can be observed, as shown in Fig.2. Fig.2 (a) presents the typical mode-locked spectrum with a 3-dB bandwidth of 2.2nm. The Kelly sidebands indicate that the fiber laser is operated in traditional soliton region. The repetition period of the pulse-train is 73.5ns, corresponding to the cavity length of 15m, as shown in Fig.2 (b). The inset shows the pulse profile with the pulse-width of 2.8ps, corresponding to 1.8ps if a sech^2 pulse profile is assumed. Therefore, the time-bandwidth product of the pulses is ~0.32, indicating the pulse is almost transform-limited. The RF spectrum with a scanning range of 215MHz is illustrated in Fig.2 (c). There is no obvious fluctuation in the broad spectrum, which confirms the stable mode-locking without Q-switching modulation. The smooth noise floor also shows the low amplitude noise. The inset shows the fundamental frequency with a signal-to-noise ratio of about 60dB, when the resolution is set 1kHz.

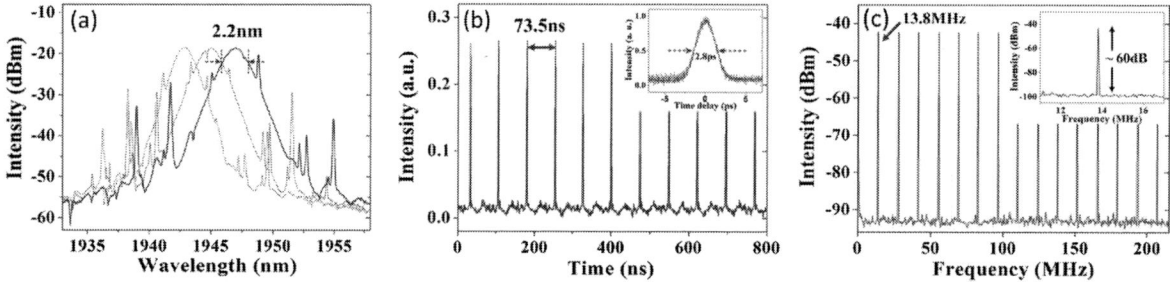

Fig. 2. Single wavelength mode-locking: (a) optical spectrum; (b) oscilloscope trace of the pulse-train and the autocorrelation trace (inset); (c) RF spectrum and fundamental frequency signal (inset)

Then, by adjusting PC, the central wavelength of mode-locking can be switched to 1945nm or 1943nm, as shown in Fig.2 (a) with the black and red dot line. These mode-locking operations present almost the same time-domain and frequency-domain characteristics as those in 1947nm. As the PC is further adjusted, a dual-wavelength mode-locking at both the central wavelength of 1945nm and 1947nm can be obtained. The two mode-locked wavelengths have almost the same peak power, as shown in Fig.3 (a) with the black solid line. We can conclude that optical spectra of dual-wavelength mode-locking is the superposition of independent mode-locking at two wavelengths. On this occasion, the pulse-trains move randomly on the oscilloscope screen, indicating the two solitons possess different group velocity within the cavity. Fig.3 (b) illustrates the snapshot of pulse-trains. In addition, the long-term stability of the dual-wavelength mode-locking is shown in Fig.3 (c), with 60-min intervals in six hours under laboratory condition. Consequently, the mode-locking operation is verified with good stability at room temperature.

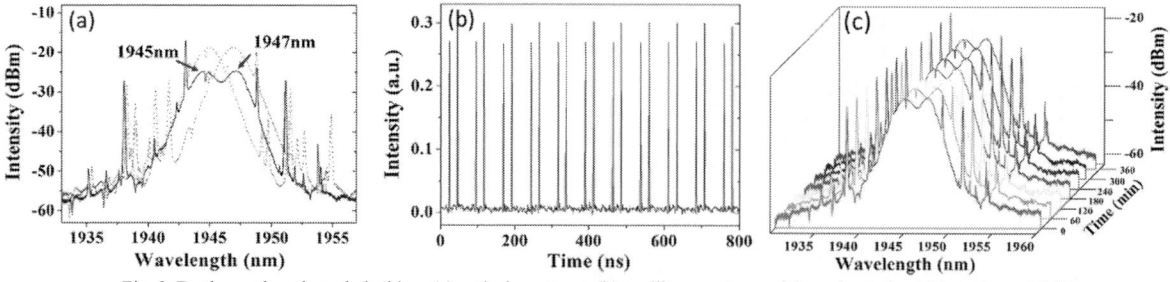

Fig. 3. Dual-wavelength mode-locking: (a) optical spectrum; (b) oscilloscope trace of the pulse-train; (c) long-term stability

For our fiber ring cavity, no polarization sensitive component is used. However, after experimental characterization, we identify that the WDM component cannot be treated as polarization-insensitive. With the help of a fiber pigtailed polarization beam splitter (PBS), we find the polarization dependent loss (PDL) of the used WDM is around 3dB. Therefore, there exists weak nonlinear polarization evolution (NPE) effect in the fiber ring cavity, which is helpful to generate a tunable comb filter within the cavity and pave the way for multi-wavelength emission. In particular, once the CWNT is removed from the fiber ring cavity, only CW emission can be observed. Therefore, we can conclude that the NPE effect is not sufficiently strong to realize mode-locking. Instead, it only contributes to the tunable comb filter generation. In our experiment, by finely rotating the PC to perturb the cavity birefringence, the loss of different wavelengths varies. With the saturable absorption effect of the SWNTs, when two wavelengths have equal intensities and are both within the effective gain bandwidth range, dual-wavelength mode-locking operation is successfully obtained. If the cavity loss for one mode-locked wavelength is quite large while the others are relatively low, switchable individual mode-locking operation among 1943 nm, 1945 nm and 1947 nm can be obtained, respectively.

According to the 2nm wavelength spacing, we can estimate the cavity birefringence to be 1.26×10^{-4}. Admittedly, due to the large channel spacing of intra-cavity birefringence-induced comb filter, only dual-wavelength mode-locking is achieved. If a comb filter with narrower channel is introduced into the cavity, it is possible to observe more mode-locked wavelengths, as long as all wavelengths are within the effective laser gain range. In addition, when the pump power is further increased to search for harmonic mode-locking, the dual-wavelength mode-locking is easily destroyed. Considering that the transmission spectrum formed by the birefringence comb filter is intensity-related, higher pump power will probably lead to the vanishing of the stable operation.

IV. CONCLUSIONS

A switchable mode-locked TDFL is experimentally demonstrated by using SWCNTs as SA. Besides the switchable single wavelength mode-locking among 1947nm, 1945nm and 1943nm, the fiber laser can be simultaneously mode-locked at both the central wavelengths of 1945nm and 1947nm with a wavelength spacing of 2nm. The two mode-locked wavelengths have the same peak intensity and remain stable in a long period time.

ACKNOWLEDGMENT

This work is supported by National Key Scientific Instrument and Equipment Development Project (2013YQ16048702); National Natural Science Foundation of China (NSFC) (61275069, 61331010).

REFERENCES

[1] Zhiyu Yan, Xiaohui Li, Yulong Tang, Perry Ping Shum, Xia Yu, Ying Zhang and Qi Jie Wang, "Tunable and swithable dual-wavelength Tm-doped mode-locked fiber laser by nonlinear polarization evolution," Opt. Express, vol. 23, pp. 4369-4376, Feb. 2015

[2] Yi Xu, Yu-li Song, Ge-guo Du, Pei-guang Yan, Chun-yu Guo, Guo-liang Zheng and Shuang-chen Ruan, "Soliton dynamic patterns of a passively mode-locked fiber laser operating in a 2μm region," Laser Phys. Lett., vol. 12, 045108, Mar. 2015

[3] Nan Yang, Yulong Tang and Jianqiu Xu, "High-energy harmonic mode-locked 2μm dissipative soliton fiber lasers," Laser Phys. Lett., vol. 12, 085102, Jul. 2015

[4] Amos Martinez, Kazuyuki Fuse, Bo Xu and Shinji Yamashita, "Optical deposition of graphene and carbon nanotubes in a fiber ferrule for passive mode-locked lasing," Opt. Express, vol. 18, pp. 23054-23061, Oct. 2010

[5] A. Y. Chamorovskiy, A. V. Marakulin, A. S. Kurkov and O. G. Okhotnikov, "Tunable Ho-doped soliton fiber laser mode-locked by carbon nanotube saturable absorber," Laser Phys. Lett., vol. 9, 602-606, June. 2012

[6] Xin Zhao, Zheng Zheng, Lei Liu, Ya Liu, Yaxing Jiang, Xin Yang and Jinsong Zhu, "Switchable, dual-wavelength passively mode-locked ultrafast fiber laser based on a single-wall carbon nanotube modelocker and intracavity loss tuning," Opt. Express, vol. 19, pp. 1168-1173, Jan. 2011

[7] X. Feng, H. Y. Tam and P. K. A. Wai, "Stable and uniform multiwavelength erbium-doped fiber laser using nonlinear polarization rotation," Opt. Express, vol. 14, pp. 8205-8210, Sep. 2006

[8] H. Zhang, D. Y. Tang, X. Wu and L. M. Zhao, "Multi-wavelength dissipative soliton operation of an erbium-doped fiber laser," Opt. Express, vol. 17, pp. 12692-12697, Jul. 2009

[9] S. Wang, P. Lu, S. Zhao, D. Liu, W. Yang and J. Zhang, "2-μm switchable dual wavelength fiber with cascaded filter structure based on dual-channel Mach-Zehnder interferometer and spatial mode beating effect," Appl. Phys. B, vol. 117, 563-569, Jun. 2014

[10] W. Peng, F. Yan, Q. Li, S. Liu, T. Feng and S. Tan, "A 1.97 μm multiwavelength thulium-doped silica fiber laser based on a nonlinear amplifier loop mirror," Laser Phys. Lett., vol. 10, 115102, Sep. 2013

[11] J. W. Nicholson, R. S. Windeler, and D. J. Digiovanni, "Optically driven deposition of single-walled carbon-nanotube saturable absorbers on optical fiber end-faces," Opt. Express, vol. 15, 9176-9183, Jul. 2007

[12] C. B. Mou, R. Arif, A. Rozhin, and S. Turitsyn, "Passively harmonic mode locked erbium doped fiber soliton laser with carbon nanotubes based saturable absorber," Opt. Express, vol. 2, 884-890, May. 2012

WA2-51

OECC/PS2016

Evaluation of alcohol concentration in sake using 1.7 μm band Tm³⁺-Tb³⁺-doped tunable fiber ring laser

Fumiki HANAFUJI[1], Kousuke SENDA[1], XIAOEN Du[1], Sho TUJITA[1], Kazuya OTA[1,2], Jun ONO[1,3], Osanori KOYAMA[1], Hirotaka ONO[4], Tatsuro ENDO[5], and Makoto YAMADA[1]

[1]Dept. of Elec. and Info. Systems, Osaka Prefecture University, 1-1 Gakuen-cho, Nakaku, Sakai, Osaka 599-8531, Japan
[2]Trimatiz Ltd., 801, La Pacifique B, 4 - 7 - 12 Minami Yawata, Ichikawa, Chiba 272-0023, Japan
[3]Anritsu Devices Co., Ltd., 5-1-1 Onna, Atsugi-shi, Kanagawa 243-8555 Japan
[4]NTT Device Technology Laboratories, NTT Corporation, 3-1 Morinosato-Wakamiya, Atsugi, Kanagawa 243-0198, Japan
[5]Dept. of Appl. Chemistry, Osaka Prefecture University, 1-1 Gakuen-cho, Nakaku, Sakai, Osaka 599-8531, Japan
e-mail: myamada@ eis.osakafu-u.ac.jp

Abstract: A 1.7 μm band Tm³⁺-Tb³⁺-doped tunable fiber ring laser with an output power of over - 3 dBm was realized. The lasing band was from 1654 to 1750 nm. We used the developed fiber laser to evaluate the ethanol concentration of sake, and confirmed its potential, because our entry-level evaluation determined the concentration to within 1.5% accuracy.
Keywords: Tm³⁺-Tb³⁺-doped fiber, Ring laser, Alcohol concentration evaluation

I. INTRODUCTION

The 1700 nm wavelength region is useful for a wide range of applications including medical, industrial machining, scientific experiments and optical sensing. Moreover, the region is the window wavelength of water and there are many absorption peaks based on molecular vibrations such as C-H, C-H₂, and S-H [1-4]. Recently, we have successfully developed 1.7 μm band broadband light sources (ASE light sources) by using Tm³⁺-Tb³⁺-doped fibre [5] and a hybrid broadband light source consisting of a cascaded super luminescent diode and a Tm³⁺-doped fiber amplifier [6]. The light sources have been applied to a near infrared (IR) spectroscopy system that allows us to measure low concentrations of organic solvents in water because the system can evaluate low volumetric concentrations (around 0.3%) of ethanol, methanol and dimethyl sulfoxide solution [2, 3]. We found that we could evaluate the concentrations of organic solvents in water by using a laser to measure the intensities of their intrinsic absorption peaks. We are now also developing fiber lasers [7] to realize a practical and cost effective system for evaluating organic solvents in water because the system can be constructed using an optical power meter rather than an optical spectrum analyzer, which is expensive.

In this paper, we realized a high output 1.7 μm band Tm³⁺-Tb³⁺- doped tunable fiber ring laser to evaluate the concentration of organic solvents in water, and applied it to an IR spectroscopy system to evaluate the ethanol concentration of sake as a practical example of concentration evaluation.

II. CONFIGURATION AND CHARACTERISTICS OF 1.7 μM BAND Tm³⁺-Tb³⁺- DOPED TUNABLE FIBER RING LASER

Figure 1 shows a Tm³⁺-Tb³⁺-doped tunable fiber laser that we constructed. We employed a ring laser configuration that consisted of two 1.2/1.8 μm band WDM couplers, two optical isolators, a tunable bandpass filter, a tap coupler, a Tm³⁺-Tb³⁺-doped fluoride fiber, a polarization beam splitter (PBS) and two 1.21 μm pump LDs. The tunable wavelength region and 3 dB bandpass bandwidth of the tunable filter were 1600 to 1800 nm and ~1.8 nm, respectively. The insertion loss of the WDM coupler, the bandpass filter and the isolator in the 1600 to 1800 nm wavelength region, were <1.4, ~3 and <1.7 dB, respectively. We used the Tm³⁺-Tb³⁺-doped fluoride fiber as a gain medium, [5-7]. The relative refractive index difference and cut-off wavelength of the Tm³⁺-Tb³⁺-doped fluoride were 2.5% and 1.0 μm, respectively. The gain was obtained as a result of the transition from the ³F₄ to the ³H₆ energy level of Tm³⁺ ions, which were doped in the core of the fiber, with 1.21 μm band pumping. Because the peak gain wavelength of the emission spectrum of the Tm³⁺-doped fiber was around 1.85 μm, the fiber cladding was doped with Tb³⁺ ions to achieve a gain spectrum peak around ~1.7 μm, and to suppress the gain of the 1700 to 2000 nm wavelength region by using the absorption of the Tb³⁺ ions from the ⁷F₆ to ⁷F₀ level [8]. We chose a Tm³⁺ concentration of 2000 ppm and a Tb³⁺ concentration of 1000 ppm. The maximum pump power launched into the fiber was 232 mW. The active fiber length was 5 m, and we estimated that the entire length of this long fiber could achieve sufficient excitation with the pump power we used.

Figure 2 (a) and (b), respectively, show the output power and lasing band of the ring laser as a function of launched pump power for various tap ratios. The excess loss of the tap coupler was less than 0.2 dB. The threshold pump power was 11 mW, except when the ratio was 80%. A high output of over -3 dBm was achieved with tap ratios of 50 and 80

%. The lasing band, which is between the maximum and minimum lasing wavelengths, decreased as the tap ratio increased, and was from 1654 to 1750 nm at a ratio of 50%, and 1677 to 1740 nm at a ratio of 80%.

III. EVALUATION OF ALCOHOL CONCENTRATION IN SAKE

The near IR spectroscopy system we constructed is shown in Fig. 3. To evaluate the ethanol concentration in sake, we used the deviation between two intrinsic absorption peak values at 1692.9 and 1729.9 nm [2, 3]. The specific absorption peak value is the deviation of the absorption from ultrapure water. We used two laser sources, one was our tunable fiber ring laser with a tap ratio of 50% and the other was a specially designed external cavity laser whose longest lasing wavelength was 1696 nm. The lasing wavelength was 1692.9 for the fiber laser, and 1729.9 nm for the ECL. Both laser sources had an output power of ~ 10 dBm. One output light was selected with a 1 x 2 optical switch, and observed with an optical power meter through an optical collimation device that incorporated a silica glass cell to evaluate water solutions containing organic solvent. The optical length and glass thickness of the glass cell were 5 and 1 mm, respectively. The beam diameter was 0.8 mm. The glass cell was located in a thermostatic chamber to maintain the temperature of the organic solvent and thus eliminate its temperature dependence. Figure 4 shows the deviation between two intrinsic absorption peak values at 1692.9 and 1729.9 nm as a function of ethanol concentration, which we used as the calibration curve to evaluate the ethanol concentration of sake. The curve was obtained by measuring the standard solution of the ethanol. The figure shows the absorption spectrum of the ethanol solution measured using the broadband light source as a reference.

Figure 5 shows photographs of the commercially available sake we evaluated. Table 1 shows the deviation between two absorption values at 1692.9 and 1729.9 nm obtained from measurements using our system, and concentrations estimated from the deviation values for each type of sake. This table also shows the ethanol concentration provided on the bottle. We could determine the ethanol concentration of sake to within 1.5% accuracy using our IR spectroscopy.

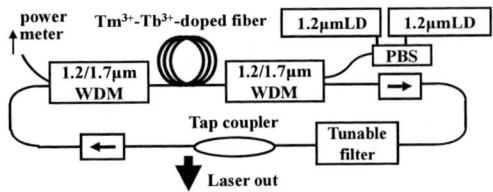

Fig. 1. Configuration of 1.7 μm band Tm^{3+}-Tb^{3+}- doped tunable fiber ring laser.

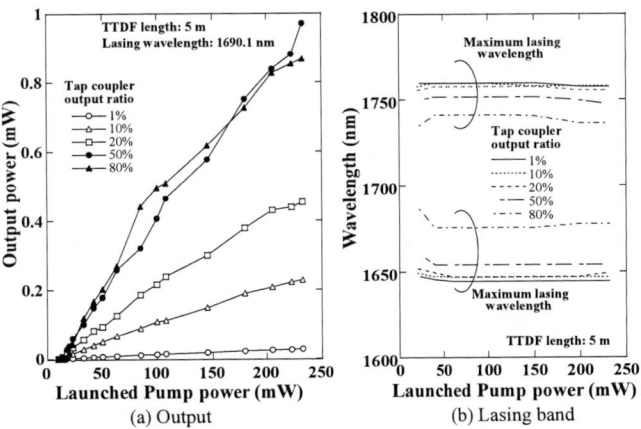

(a) Output   (b) Lasing band

Fig. 2. Output power and lasing band of ring laser.

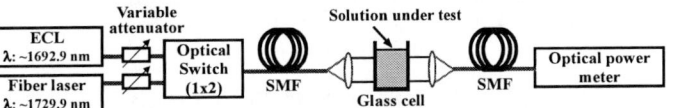

Fig. 3. Configuration of 1.7 μm band Tm^{3+}-Tb^{3+}- doped tunable fiber ring laser.

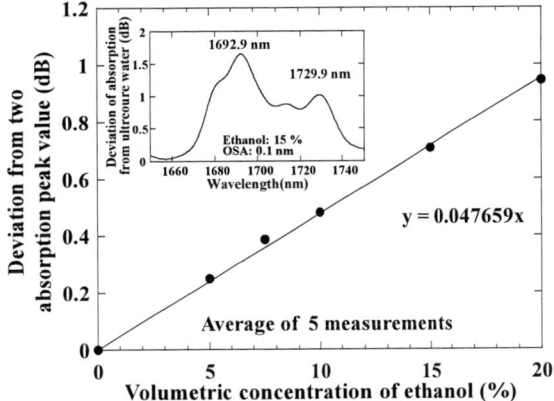

Fig. 4. Calibration curve of the ethanol concentration evaluation of sake.

Fig. 5. Photographs of commercially available sake we evaluated.

IV. CONCLUSIONS

We developed a high output 1.7 μm band Tm^{3+}-Tb^{3+}-doped tunable fiber ring laser by optimizing the tap ratio. The achieved output power was over -3 dBm, and the lasing band was from 1654 to 1750 nm. The developed fiber laser was used to evaluate the ethanol concentration of sake, and it was confirmed that the laser has the potential to be used for such a purpose, because our entry-level evaluation determined the concentration to within 1.5%. Our 1.7 μm band Tm^{3+}-Tb^{3+}- doped tunable fiber ring laser and an IR spectroscopy system employing it are attractive for use in scientific experiments such as those involving IR spectroscopy.

Table 1. Alcohol content evaluation results for sake.

Sample	Deviation between two absorption values at 1692.9 and 1729.9 nm (dB)	Concentration estimated from deviation of two absorption values (%)	Labeled concentration value (%)
A	0.72	15.11	14
B	0.68	14.27	13~14
C	0.73	15.32	13~14
D	0.69	14.48	13~14
E	0.77	16.16	14~15

REFERENCES

[1] M. Ebrahimzadeh, "Mid-Infrared labelled coherent sources and applications (NATO Science for Peace and Security Series)", Springer, (2008)

[2] Y. Maeda, M. Yamada, T. Endo, K. Ohta, T. Tanaka, M. Ono, K. Senda, J. Ono, and O. Koyama, "1700 nm ASE light source and its application to mid-infrared spectroscopy," Proc. OECC2014, June. 2014, paper TU6F.

[3] J. Ono, T. Endo, K. Ohta, H. Ono, Y. Maeda, K. Senda, O. Koyama, and M. Yamada, "Broadband light source and its application to near infrared spectroscopy," Sensors and Materials Vol. 27, No. 5, pp. 413-423 (2015)

[4] A.B. Rulkov, A.A. Ferin, J.C. Travers, S.V. Popov, and J.R. Taylor, "Broadband, low intensity noise CW source for OCT at 1800 nm," Optics Communications, Vol. 281, No. 1, pp. 154–156 (2008).

[5] M. Yamada, S. Aozasa, and H. Ono, "Broadband ASE light source for the 1800 nm wavelength region," Electron. Lett., Vol. 48, No. 23, pp. 1489-1490 (2012).

[6] M. Yamada, J. Ono, K. Mise, Y. Shimose, H. Mori, A. Yamada, K. Ota, K. Senda, Y. Maeda, O. Koyama, and H. Ono, "1.8 μm broadband light source using a super luminescent diode," Electron. Lett., Vol. 50, No. 20, pp. 1468-1470 (2014).

[7] M. Yamada, K. Senda, T. Tanaka, Y. Maeda, S. Aozasa, H. Ono, K. Ota, O. Koyama, and J. Ono, "Tm^{3+}-Tb^{3+}-doped tunable fibre ring laser for the 1700 nm wavelength region," Electron. Lett., Vol. 49, No. 20, pp. 1287-1288 (2013).

Generation of 100 GHz Pulse Train from a Phase Modulated Hybrid Mode-Locked Er-Fiber Laser with an Intra-Cavity Etalon

Sheng-Min Wang, and Yinchieh Lai

Department of Photonic and Institute of Electro-Optical Engineering, National Chiao Tung University, Hsinchu 300, Taiwan.
min.eo00g@nctu.edu.tw, yclai@mail.nctu.edu.tw

Abstract: *We demonstrate a 100 GHz hybrid mode-locked Er-doped fiber laser with the assistance of a high finesse etalon and phase modulation. Characteristics of the ultrahigh-repetition-rate laser under both passive and hybrid mode-locking operation are compared.*
Keywords: *Mode-locked fiber laser, Erbium-doped fiber lasers, Fabry-Perot interferometers.*

I. INTRODUCTION

Ultrahigh-repetition-rate laser sources have many applications in different fields, such as microwave signal generation [1], optical spectroscopy [2] and optical communication [3]. Many approaches have been proposed for the generation of high-repetition-rate pulse trains from mode-locked fiber lasers. The most common technique of creating hundred GHz pulses is to add a periodical spectral filter into the mode-locked laser for optical frequency selection, so that in the temporal domain high-repetition-rate pulse trains can be generated if the mode-locking of selected optical frequency components can be achieved [4]. In the literature, the birefringence interferometer composed of an in-line polarizer and a section of the PM fiber is able to provide a periodical spectral transmission window [5], which can then be utilized in a mode-locked fiber laser to help generating a high-repetition-rate pulse train [6]. Although the fiber birefringence filter can be integrated into the fiber laser cavity very well due to the all-fiber configuration, the wider line-width of the birefringence filter may result in the pulse bunch operation state instead of the continuous pulse train operation state. The silicon micro-ring resonator has also been utilized as a frequency selection device in mode-locked fiber lasers [7]. Typically the micro-ring resonator has a narrower line-width, thereby being able to prevent the laser from operating under a pulse bunch state, even though the larger coupling loss between the optical fiber and the micro-ring resonator may become an issue. Another appropriate type of periodic intra-cavity filters is an intra-cavity etalon. Its insertion loss can be lower and is not difficult to be aligned in the fiber laser cavity. Based on the intra-cavity etalon approach, a stable pulse train with the tens GHz repetition rate has been demonstrated in an active mode-locked fiber laser [8]. However, there are still less reported works on the characteristics of hybrid mode-locked fiber lasers assisted with a high finesse intra-cavity etalon.

In the present work, we demonstrate a 100 GHz hybrid mode-locked Er-doped fiber laser with the assistance of a high finesse 100 GHz FSR air-spaced etalon. The combinational effect of the high finesse etalon and nonlinear polarization rotation in the fiber cavity allows the laser to generate stable 100 GHz pulse trains under pure passive mode-locking. An active 10 GHz phase modulator is also added into the laser system for achieving the hybrid mode-locking operation. Even though the 10 GHz modulation frequency is 10 times less than the 100 GHz repetition rate, we experimentally find that for the studied laser configuration the applied active phase modulation still can influence the laser mode-locking performance greatly. Profound laser dynamics have been observed in the hybrid mode-locking regime and direct comparison with the results of pure passive mode-locking from the same laser has also been performed for deeper understanding.

II. *100 GHz HYBRID MODE-LOCKED ER-DOPED FIBER LASER*

A. Experimental Setup

The experimental setup of the 100 GHz hybrid mode-locked Er-doped fiber laser is shown in Fig. 1. One of the key components for ultrahigh-repetition-rate pulse train generation in this laser is an air-spaced Fabry-Perot etalon. The etalon provides the effect of spectral comb filtering to force the laser to operate at the repetition rate equal to the free spectral range of the installed etalon. The free spectral range of the etalon used in the experiment is 100 GHz, so that a 100 GHz continuous pulse train is expected to be observed from the laser output. The 2 m long Er-doped fiber is pumped with a bidirectional pumping configuration, in which the central pumping wavelength is at 976 nm. A section of 50 m highly nonlinearity fiber is added into the cavity and spliced right after the optical amplifier module for achieving high enough optical nonlinearity for 100 GHz passive mode-locking through the nonlinear polarization rotation effect implemented along with the polarization controllers and the in-line polarizer. The passive mode-locking mechanism is employed in order to lock the phases between the 100 GHz spacing frequency components selected by the etalon so as to generate a stable mode-locked pulse train. In addition, active phase modulation is also utilized in the experiment for achieving hybrid mode-locking. The LiNbO₃ phase modulator is driven by a synthesizer followed by a RF amplifier, which driving frequency is chosen to be about 10 GHz. The isolators ensure the propagation direction of the lights in the laser cavity and block the reflected lights from the etalon for avoiding unwanted effects. The central lasing wavelength is determined by a 3-nm bandpass tunable filter. The fundamental frequency of the cavity is measured to be about 3 MHz, by which the corresponding cavity length was calculated to be around 66.7 m. The typical laser output power extracted by the 20% port of the 20:80 tapered coupler is about 8 mW.

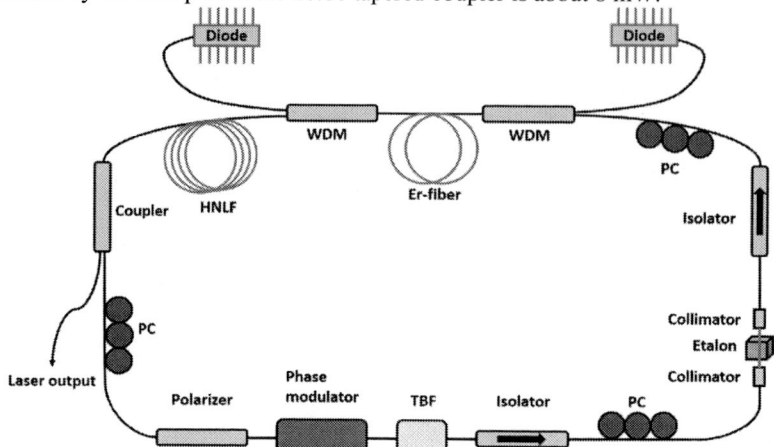

Fig. 1. 100 GHz hybrid mode-locked Er-doped fiber laser.

B. Experimental results

Since the effect of nonlinear polarization rotation is implemented in the laser, the laser is able to stably operate at the 100 GHz passive mode-locking state by carefully adjusting the polarization controllers under no active modulation. The optical spectrum of the passive mode-locked 100 GHz pulse train is shown in Fig. 2(a), in which the periodic frequency components spaced by 0.8 nm can be clearly observed. The finesse of the etalon is about 100 and thus the line-width of the etalon can be estimated to be about 1 GHz. An apparent consequence of the narrow filtering bandwidth is the high spectral contrast shown in the optical spectrum measurement. It should be noticed that the line-width of each peak shown in Fig. 2(a) should be limited by the resolution bandwidth of our optical spectrum analyzer. In Fig. 2(b) the corresponding autocorrelation trace of the 100 GHz passive mode-locked pulse train is plotted, and the pulse width from hyperbolic secant fitting is about 3.6 ps. The 10 ps spacing between each pulse indicates that the repetition rate of the continuous pulse train is 100 GHz. Moreover, the measurement of the optical frequency variation has been performed by detecting the beating signal with a frequency-stabilized laser reference source and its magnitude is apparently limited by the line-width of the installed etalon. We therefore believe that the narrow line-width filtering effect provided by the etalon is quite effective for reducing the optical frequency fluctuations of the studied mode-locked fiber laser.

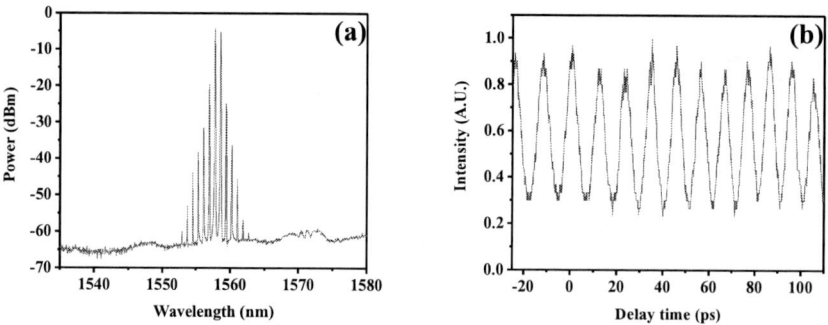

Fig. 2. (a) Optical spectrum and (b) autocorrelation trace of the 100 GHz pure passive mode-locked pulse train.

We then turn on the phase modulation when the laser has been operated under the passive 100 GHz mode-locking state. The driving RF frequency of the phase modulator is carefully adjusted near 10 GHz to match both the cavity harmonic frequency and the free spectrum range of the etalon for achieving stable hybrid mode-locking. Fig 3(a) shows the optical spectrum of the laser with 10 GHz phase modulation. The significant increase of the total longitudinal mode number compared with that of the passive mode-locking case is the first obvious effect. The spectral contrast is high and the line-width of each frequency component is also quite narrow. At this hybrid mode-locking case some 10 GHz-spacing frequency components may be generated close to each 100 GHz peak, but they cannot be clearly resolved in Fig 3(a) due to the resolution bandwidth and sampling point limitation of our optical spectrum analyzer under this span range. Because of the resulting broader optical bandwidth, the shorter pulse with the width of 1.2 ps is achieved as can be seen from the auto-correlation trace in Fig 3(b). The higher contrast and cleaner background level in Fig 3(b) indicate that the mode-locking quality of the hybrid case may be better than the pure passive mode-locking case. The envelope of the autocorrelation trace shows slight 10 GHz amplitude modulation, which should be caused by the applied 10 GHz phase modulation. However, since the variation magnitude of the envelope is not too large, the continuous pulse train is generated instead of the pulse bunch [6]. The sampling scope measurement also confirms that the laser is under a continuous mode-locked pulse train state. We believe that the continuous pulse train is obtained mainly because the narrow line-width of the intra-cavity etalon, which more severely limits the generation of the 10 GHz optical frequency components beside the 100 GHz etalon-selected frequency components. Moreover, we have found that the phase modulation strength is an important parameter here. The modulation depth of the pulse envelope and the bandwidth of the optical spectrum strongly depend on the phase modulation strength experimentally.

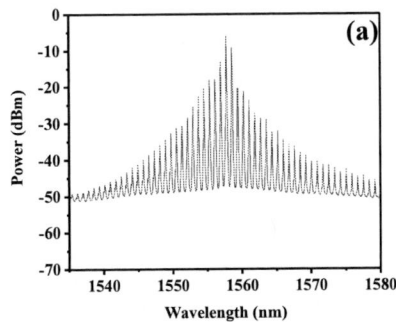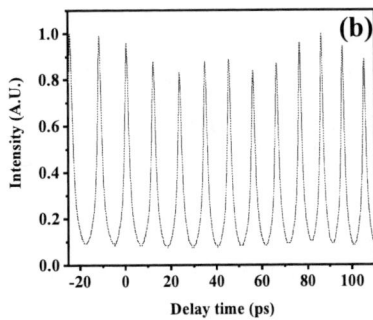

Fig. 3. (a) Optical spectrum and (b) autocorrelation trace of the 100 GHz hybrid mode-locked pulse train.

III. CONCLUSIONS

We have demonstrated a 100 GHz hybrid mode-locked Er-doped fiber laser with the assistance of a high-finesse Fabry-Perot etalon. The laser can stably generate a 100 GHz continuous pulse train under both the pure passive mode-locking regime and the hybrid mode-locking regime with 10 GHz active phase modulation. The lasing longitudinal modes are selected by the narrow line-width of the etalon and their phases are passively mode-locked through the effect of nonlinear polarization rotation. The narrow line-width filtering provided by the etalon is quite effective for reducing the optical frequency variation of the laser, which has been checked by detecting the beating signal with another frequency-stabilized reference source. When compared with the pure passive mode-locking case, the hybrid mode-locking case has exhibited a wider optical spectrum, shorter pulse width, and cleaner auto-correlation trace. These results indicate that for the studied laser configuration, the hybrid mode-locking approach should be able to improve the mode-locking quality when compared to the pure passive mode-locking. Profound laser dynamics have been observed for the studied hybrid mode-locked fiber laser, including the dependence of the pulse train envelope modulation on the phase modulation strength. More investigation will be carried out to clarify the observed new phenomena.

REFERENCES

[1] D. Zhichao, and Y. Jianping, "Photonic generation of microwave signal using a rational harmonic mode-locked fiber ring laser," IEEE Trans. Microwave Theory Tech., vol. 54, pp. 763-767, 2006.

[2] M. J. Thorpe, K. D. Moll, R. J. Jones, B. Safdi, and J. Ye, "Broadband cavity ringdown spectroscopy for sensitive and rapid molecular detection," Science, vol. 311, pp. 1595-1599, 2006.

[3] M. Nakazawa, T. Yamamoto, and K. R. Tamura, "1.28Tbit/s-70 km OTDM transmission using third- and fourth-order simultaneous dispersion compensation with a phase modulator," Electron. Lett., vol. 36, pp. 2027-2029, 2000.

[4] J. Schroder, T. D. Vo, and B. J. Eggleton, "Repetition-rate-selective, wavelength-tunable mode-locked laser at up to 640 GHz," Opt. Lett., vol. 34, pp. 3902-3904, 2009.

[5] K. Ozgoren, and F. O. Ilday, "All-fiber all-normal dispersion laser with a fiber-based Lyot filter," Opt. Lett., vol. 35, pp. 1296-1298, 2010.

[6] S.-M. Wang, S.-S. Jyu, and Y. Lai, "300 GHz bound pulse generation in a filter-assisted harmonic mode-locked Er-doped fiber laser with 20 GHz phase modulation," Proc. of Conf. on Lasers and Electro-Optics (CLEO: 2014), San Jose (the United Ststes of America), Jun. 2014, pp. STu1N.4.

[7] S. S. Jyu, L. G. Yang, C. Y. Wong, C. H. Yeh, C. W. Chow, H. K. Tsang, and Y. Lai, "250-GHz passive harmonic mode-locked Er-doped fiber laser by dissipative four-wave mixing with silicon-based micro-ring," IEEE Photonics J., vol. 5, pp. 7, 2013.

[8] M. Nakazawa, K. Kasai, and M. Yoshida, "C_2H_2 absolutely optical frequency-stabilized and 40 GHz repetition-rate-stabilized, regeneratively mode-locked picosecond erbium fiber laser at 1.53 μm," Opt. Lett., vol. 33, pp. 2641-2643, 2008.

WA2-54

Infrared Luminescence Investigation of Bismuth and Erbium Co-doped Fiber

Chunsheng Li,[1] Binbin Yan,[1] Xiaojuan Wu,[1] Xinzhu Sang,[1]

[1] State Key Laboratory of Information Photonics and Optical Communications, Beijing University of Posts and Telecommunications,

P.O. Box72 (BUPT), 100876, Beijing, China

E-mail: lichunsheng@bupt.edu.cn

Abstract: Broadband emission from 1400 to 1700 nm with a FWHM reaches 80 nm is demonstrated in Bi/Er^{3+} co-doped fiber and an abnormal temperature characteristic belongs to bismuth fluorescence center is observed.

Keywords: bismuth; optical fiber; near infrared fluorescence spectrum

I. INTRODUCTION

With the development of information superhighway and the promotion of "Internet of things", the development of materials for optical data transmission and amplification has been of continuous urgency. Recent years, a new type of active optical fiber, bismuth doped fiber, has been of great interest because of its broad band near-infrared luminescence, the Bi/Er^{3+} co-doped optical fibers also have been investigated in many papers [1, 2]. In this paper, we investigate a novel Bi/Er^{3+} co-doped Optical Fiber (BEDF), Broadband emission from 1400 to 1700 nm with a FWHM about 80 nm by 980 nm laser single pump and the luminescence intensity of Bi active center (BAC) increase with temperature (in our investigate temperature range) are observed.

II. EXPERIMENTAL

The Bi/Er^{3+} co-doped germanium- and aluminum- doped silica fiber sample is fabricated by modified chemical vapor deposition (MCVD) and the in-situ solution doping technique, the detected concentration: Si ~ 37.2, Al ~ 1.18, Er^{3+} ~ 0.2-0.3 mol % and Bi is about one order magnitude less than Er^{3+}, Ge is also Micro doping. The fiber has a core diameter about 8 μm and an outside diameter about 124 μm.

III. RESULTS AND DISCUSSION

Bi-relate active centers in BDFs with the simplest glass compositions have been made and characterized, enabling the identification of certain BACs, e.g. BAC-Si, BAC-Al, BAC-Ge and BAC-P, associated with Si, Al, Ge and P with their respective characteristic spectral properties [3].

A. the Emission Intensity of BEDF VS Pump Power

In this part, the ASE spectrums pumped by 830 and 980 nm laser diode VS pump power were investigated separately; a forward pumped device was used in the experiment, in this device, the pump source and OSA were connected on both ends of BEDF, respectively.

The results of a 30 cm long BEDF are shown in Fig.1. There are distinct fluorescence peaks exist at 1003, 1536 and 1555 nm as well as a shoulder peak at 1467 nm when pumped by 830 nm laser diode (see Fig.1 (a)). The emission band at 1536 and 1555 nm are linked to transition of $4I1^{1/2} \rightarrow 4I1^{5/2}$ which belongs to Er^{3+}, while the emission at 1003 nm band belongs to the intrinsic defect of glass network caused by BAC-Al, and the 1467 nm emission attributed to BAC-Si. the ASE spectrum pumped by 980 nm laser diode is characterized by emission peaks at 1467, 1539 and 1575 nm, the produced emissions around 1467 nm attributed to BAC-Si, however, the peaks at 1539 and 1575 nm which belong to Er^{3+} have an obviously red shift compare with 830 nm pump, especially for the 1575 nm which is far from the normal 1550 nm band peak but with the increase of pump power, the wavelength of the peak return back to the normal position (see Fig.1(b)). This phenomenon indicates that the ASE spectrum or self-absorption of BEDF is affected by pump power. With the increase of pump power of 830 or 980 nm lasers, the fluorescent intensity tends to saturation. The full width at half maximum (FWHM) of ASE spectrums are both nearly 80 nm whatever pumped by 830 or 980 nm diode laser which is much wider than the ordinary single Er^{3+} doped fiber in this band, and it is presented as an effective means to significantly enlarge the FWHM of Er^{3+} doped fiber devices in C band.

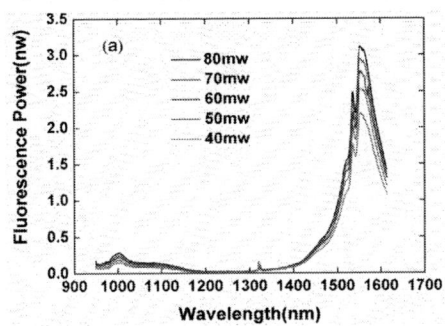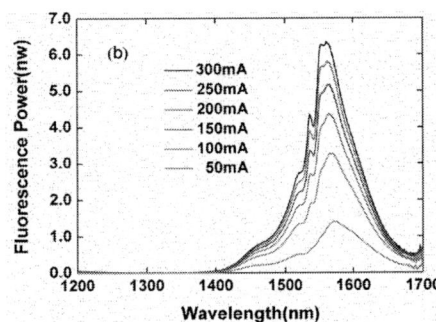

Fig.1. Forward ASE spectrum of BEDF by 830nm (a) and 980nm (b) pump VS pump power/current

B. the Emission Spectrum of BEDF for Different Length of Fiber

The variation of emission spectra is analyzed for different BEDF length at 80mW 830nm pumping as shown in Fig.2. To keep consistence of the splice loss, the backward pump device was chosen. Because of the nature insert loss of 830/1310nm WDM, the ASE spectrum is quite different from the forward pump spectrum, but this has no impact for indicating the rules. There are three obvious peaks, the band at 1120 and 1430nm related to BACs, and 1535nm belongs to Er^{3+}. In the experimental length range, the emissions intensity of BAC increase with the augment of fiber length and have not saturated yet, while the emission intensity of Er^{3+} doesn't shows obvious change, even acts a little decrease at 70cm length. So the optimum length should be found if want to use this kind of BEDF fiber sufficiently.

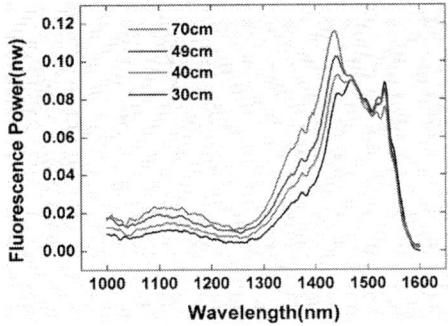

Fig.2. ASE spectrum （without compensate）varied with the length of BEDF when backward pumped by 830nm LD.

C. Temperature Effect on Fluorescence Intensity

The dependence of the fluorescence intensity on the BEDF temperature is measured for another sample of a 30 cm BEDF in the range of (34-64) ℃ with 980 nm pump. The temperature of all other elements remained constantly (27.7 ℃). The result is presented in Fig.3. Fig.3. (b) shows the partial enlarge drawing of 1550 nm band which belongs to Er^{3+} fluorescence center. It's obvious that the emission intensity decrease with the rise of temperature which is consistent with the normal theory. While the 1460 nm band of BAC-Si appears completely opposite phenomenon to Er^{3+} (in our investigate temperature range), as shown in Fig.3. (c), thus, two possible explanations are given: 1.There is

Energy transfer between Bi and Er^{3+} fluorescence center in BEDF. As the increase of temperature, the integrated power of the whole ASE spectrum nearly be constant, so the energy of 1550 nm band may be partially transfer to 1460nm band, after all, there have been a lot of reports on Bi/ Er^{3+} energy transfer [4,5]. 2. The unique properties of our investigated BEDF. There also other abnormal phenomenon of BAC been reported, such as the Bi emission peak blue-shift with increasing temperature while the normal circumstances is red shift [5].

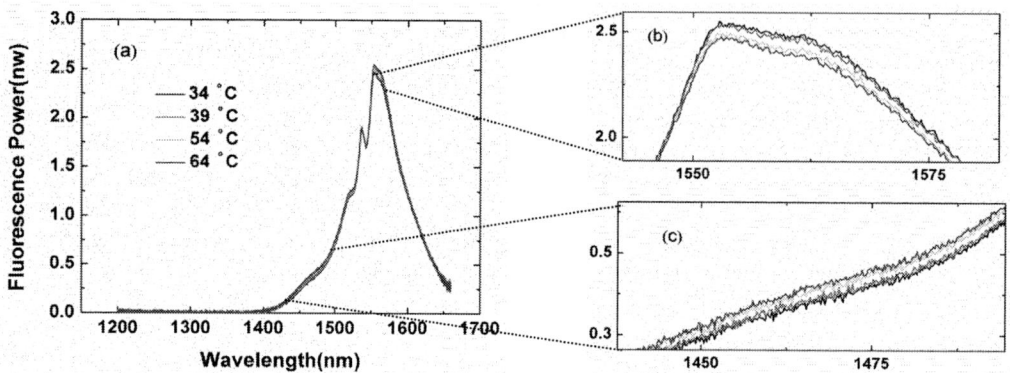

Fig.3. (a) ASE spectrum of BEDF upon 980nm excitation at different temperatures;
(b), (c) Partial enlarged detail at 1550 and 1460nm band, respectively

IV. CONCLUSIONS

The effect of pump power, fiber length and temperature on ASE spectrum of a kind of novel BEDF has been investigated in detail. Compare to the studies of others, the concentration of Er^{3+} ions is significantly higher than Bi element in our BEDF. The FWHM in 1550nm band of the ASE spectrum reaches nearly 80 nm, so it is presented as an effective means to significantly enlarge the band width of Er^{3+} doped fiber devices. An abnormal phenomenon that the fluorescence intensity increases with the rise of temperature is observed in 1460 nm band which belongs to BAC-Si. The studies of this phenomenon will help us further understand the mechanism of Bi fluorescence centers, and it's reasonable to believe that with the adjusting of the molar ratio between both species will provides an effective tool for further increasing emission bandwidth.

ACKNOWLEDGMENT

This work was supported by the National Natural Science Foundation of China (61405014).

REFERENCES

[1] K. E. Riumkin, M. A. Melkumov, I. A. Bufetov, A. V. Shubin, S. V. Firstov, V. F. Khopin, A. N. Guryanov, and E. M. Dianov, "Superfluorescent 1.44 μm bismuth-doped fiber source," Opt. Lett, vol. 37, pp.4817-4819, 2012.

[2] J. Zhang, Z. M. Sathi, Y. Luo, J. Canning, and G.-D. Peng, "Toward an ultra-broadband emission source based on the Bismuth and Erbium co-doped optical fiber and a single 830nm laser diode pump," Optics Express, vol. 21, pp. 7786-7792, 2013.

[3] A. Bufetov, M. A. Melkumov, S. V. Firstov, K. E. Riumkin, A. V. Shubin, V. F. Khopin, A. N. Guryanov, and E. M. Dianov, "Bi-Doped Optical Fibers and Fiber Lasers," IEEE JOURNAL OF SELECTED TOPICS IN QUANTUM ELECTRONICS, vol. 20, SEPTEMBER/OCTOBER 2014.

[4] T. M. Hau, X. Yu, D. Zhou, Z. Song, Z. Yang, R. Wang, and J. Qiu, "Super broadband near-infrared emission and energy transfer in Bi–Er co-doped lanthanum aluminosilicate glasses," Optical Materials, vol. 35, pp. 487–490 , 2013.

[5] Mingying Peng, Na Zhang, Lothar Wondraczek, Jianrong Qiu, Zhongmin Yang, and Qinyuan Zhang, "Ultrabroad NIR luminescence and energy transfer in Bi and Er/Bi co-doped germanate glasses," Opt. Express, vol. 19, pp.20799-20807, 2011.

WA2-55

OECC/PS2016

Energy-efficient Dual-pump Degenerated Phase Sensitive Amplifiers

Mingyi Gao[1*], Takayuki Kurosu[2], Shu Namiki[2] and Gangxiang Shen[1]

1 School of Electronic and Information Engineering, Soochow University,
No.1 Shizi Street, Suzhou, Jiangsu Province, 215006, China
2 Network Photonics Research Center, National Institute of Advanced Industrial Science and Technology,
Central 2, 1-1-1, Umezono, 305-8568, Tsukuba, Japan
*mygao@suda.edu.cn

Abstract: Dispersion effect in the dual-pump degenerated phase sensitive amplifiers has been investigated. More than 30-dB gain extinction ratio has been achieved at a nonlinear phase shift of only 0.73 rad by adjusting the fiber dispersion.

Keywords: Optical fiber communication, optical signal processing, phase sensitive optical amplifier

I. INTRODUCTION

Dual-pump degenerated phase-sensitive parametric amplifiers (PSAs) have been extensively investigated for the phase regeneration of differential phase shift keying (DPSK) signals and the phase de-multiplexing of quadrature phase shift keying (QPSK) signals utilizing the phase-sensitive characteristics of in-phase amplification and quadrature attenuation [1-6]. To implement effective phase regeneration, it is indispensable to enhance the gain extinction ratio (GER), defined as the ratio of phase sensitive amplification to de-amplification, to achieve a step-like phase response. It is intuitive that GER can be increased by increasing the nonlinear phase shift ($\gamma P_p L$), that is, the product of the pump power (P_p), the nonlinear coefficient (γ) and length (L) of the fiber. In comparison, several energy-efficient methods to enhance GER have been proposed based on sideband-assisted PSA, hybrid optical phase squeezer and polarization-assisted PSA, to achieve a high GER with a lower nonlinear phase shift [7-9].

The sideband-assisted dual-pump PSA is easy to implement, because the dual-pump PSAs with a low-dispersion-slope fiber always entail high-order sidebands and no extra complex components are required to generate sidebands. In the sideband-assisted dual-pump PSA, we observed a roll-over behavior of GER versus nonlinear phase shift curve and achieved a maximum GER of 30 dB at a nonlinear phase shift of around 0.9 rad [7]. Here we define the nonlinear phase shift at which the GER becomes maximum as "the optimum nonlinear phase shift". While a 30-dB GER at a nonlinear phase shift of 0.9 rad is remarkable, it is not well understood whether or not there is still a room to further reduce the optimum nonlinear phase shift, or to further increase the maximum GER in the sideband-assisted dual-pump PSA for its energy-efficient applications.

In this paper, we explore the effect of dispersion on the optimum nonlinear phase shift and the maximum GER. By varying the dispersion coefficients of the fiber, we find that the optimum nonlinear phase shift may only be 0.73 rad. Moreover, we plot the optimum nonlinear phase shift versus the fiber dispersion to investigate how to optimize the fiber dispersion in the dual-pump degenerated PSAs. Thus, we are able to design an energy-efficient dual-pump PSA for effective DPSK regeneration and QPSK de-multiplexing by optimizing the fiber dispersion.

II. THEORY AND SIMULATIONS

The 7-wave PSA numerical model is applied to investigate the signal behavior in the dual-pump degenerated PSA, where three input optical waves (two pumps and signal) generate two pump-pump FWM tones and two pump-signal FWM tones [7]. The numerical model includes all 13 non-degenerate and 9 degenerate FWM processes induced by the 7 waves. By integrating numerically the coupled nonlinear Schrödinger equations of these waves, the evolution of amplitudes and phases is obtained. The trajectories of the output signal vector in the complex plane with increasing nonlinear phase shift are plotted to determine the minimum optimum nonlinear phase shift.

In the simulations, we utilized a 600 m long highly nonlinear fiber (HNLF) with a zero-dispersion wavelength of 1542 nm, nonlinear coefficient of 10 $W^{-1}km^{-1}$ and dispersion slope of 0.026 $ps/nm^2/km$, respectively. Firstly, the signal wavelength is set to 1542 nm and the pump wavelength separation is 3.5 nm. Thus the reference HNLF dispersion is 0 ps/nm/km. The calculated output signal trajectories with increasing nonlinear phase shift in the complex plane for in-phase and quadrature components are shown in Fig. 1(a). The in-phase component increases slowly with the increment of nonlinear phase shift because the sidebands partly deplete the pump power. The quadrature component attenuates quickly and the output signal vector approaches closest to the origin at the nonlinear phase shift of approximately 0.89

889

rad. The corresponding gain and phase transfer function curves are shown in Fig. 1(b), the high GER more than 30 dB results in an almost ideal step-like phase transfer function.

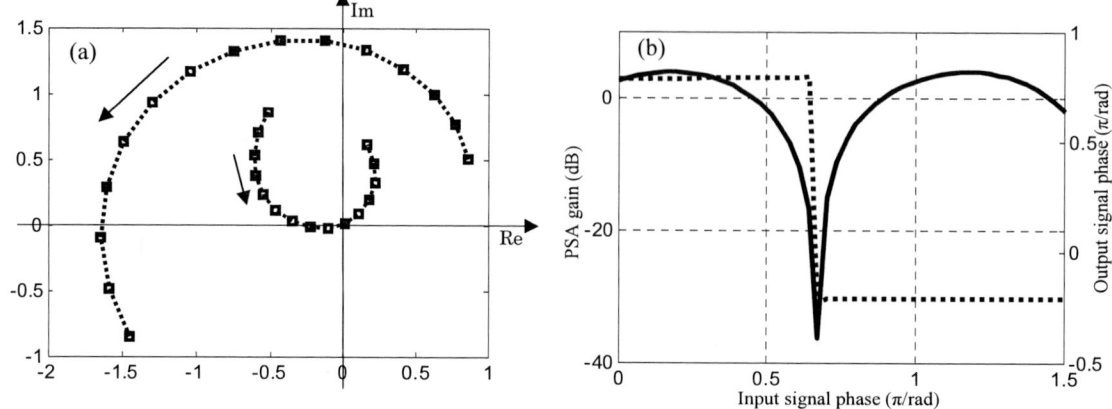

Fig.1. (a) The calculated output signal trajectories in the complex plane for increasing nonlinear phase shift with a step of 0.1 rad for in-phase case (outer) and quadrature case (inner) with dispersion of 0 ps/nm/km. (b) The corresponding PSA gain (dB) and output signal phase (π/rad) vs. input signal phase(π/rad)

In the sideband-assisted dual-pump PSAs, the GER enhancement stems from a large quadrature de-amplification. Therefore, in order to achieve a high GER, we should try to pass the trajectories of the output quadrature signal exactly through the origin. In the following simulations, we varied the HNLF dispersion and plotted the calculated output quadrature signal trajectories, as shown in Fig.2. For the HNLFs with different dispersions, the output quadrature signal trajectories have different curvatures, which results in various optimum nonlinear phase shifts. It is also important to note that the trajectories can always be made pass exactly through the origin by optimizing the input phase of the signal.

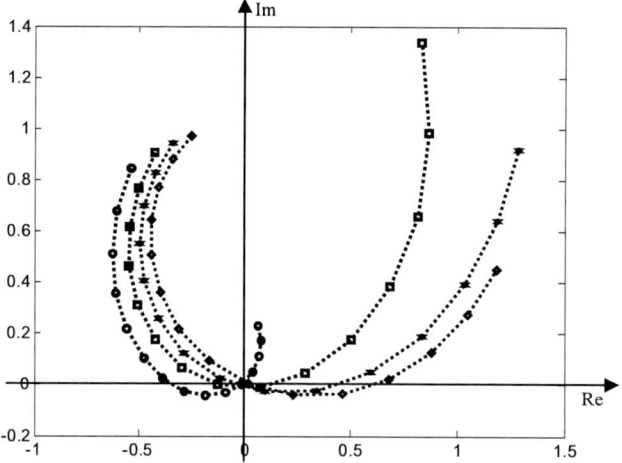

Fig.2. The calculated output quadrature signal trajectories in the complex plane for increasing nonlinear phase shift with a step of 0.1 rad with different dispersion of -0.208 ps/nm/km (circles), 0.482 ps/nm/km (squares), 0.832ps/nm/km (hexagram) and 1.04 ps/nm/km (diamonds)

In other words, an arbitrarily high GER is achievable by adjusting the input phase. However, the optimum nonlinear phase shift is determined by the system parameters. Figure 3 plots the optimum nonlinear phase shift versus the fiber second-order dispersion, and shows that when the second-order dispersion is close to -1 ps^2/km, the optimum nonlinear phase shift becomes minimum around 0.73 rad. Thus, we have shown that there is an optimum dispersion value to minimize the optimum nonlinear phase shift. These results can be exploited to design an energy-efficient dual-pump PSA for effective binary phase regeneration and quadrature phase de-multiplexing.

III. CONCLUSIONS

In the paper, we achieved an optimum nonlinear phase shift as low as 0.73 rad by adjusting the fiber dispersion in the sideband-assisted dual-pump degenerated PSAs. We have also addressed the design issues how to minimize the operating energy of PSA, that will be useful for realizing energy-efficient phase regenerators and de-multiplexers for the PSK signals.

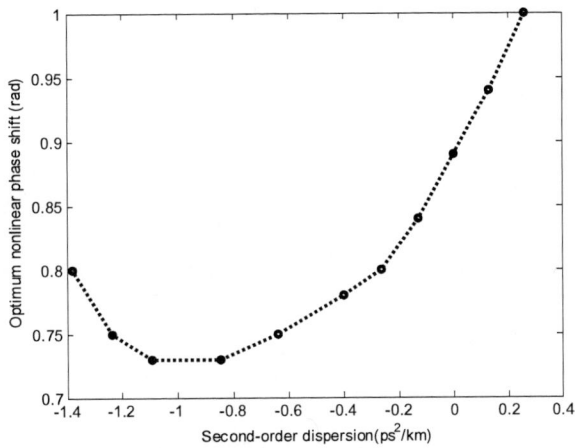

Fig.3. The calculated optimum nonlinear phase shift vs. second-order dispersion of the fiber

ACKNOWLEDGMENT

MG is grateful for the support from the National Natural Science Foundation of China (NSFC) (No. 61307082) and the Open Project (2013GZKF031305) from the State Key Laboratory of Advanced Optical Communication Systems and Networks, Shanghai JiaoTong University, China. TK and SN are grateful for the support from the Project for Developing Innovation Systems of MEXT, Japan.

REFERENCES

[1] K. Croussore and G. Li, "Phase and amplitude regeneration of differential phase-shift keyed signals using phase-sensitive amplification," IEEE J. Sel. Topics Quantum Electron., 14, 648–658 (2008).

[2] F. Parmigiani, R. Slavík, J. Kakande, C. Lundström, M. Sjödin, P.Andrekson, R. Weerasuriya, S. Sygletos, A. D. Ellis, L.Grüner-Nielsen, D. Jakobsen, S. Herstrøm, R. Phelan, J.O'Gorman, A.Bogris, D. Syvridis, S. Dasgupta, P. Petropoulos, and D. J. Richardson, "All-optical phase regeneration of 40Gbit/s DPSK signals in a black-box phase sensitive amplifier," in Optical Fiber Communication Conference, OSA Technical Digest (CD) (Optical Society of America, 2010), paper PDPC3.

[3] R. Slavík, F. Parmigiani, J. Kakande, C. Lundström, M. Sjödin, P. A. Andrekson, R. Weerasuriya, S. Sygletos, A.D. Ellis, L. Grüner-Nielsen, D. Jakobsen, S.Herstrøm, R. Phelan, J. O'Gorman, A. Bogris, D. Syvridis, S. Dasgupta, P. Petropoulos, and D. J. Richardson, "All-optical phase and amplitude regenerator for next-generation telecommunications systems, " Nat. Photon. 4, pp. 690 - 695, 2010.

[4] R. Slavík, A. Bogris, J. Kakande, F. Parmigiani, L. G.-Nielsen, R. Phelan, J. Vojtěch, P. Petropoulos, D. Syvridis, and D.J. Richardson, "Field-trial of an all optical PSK regenerator/multicaster in a 40 Gbit/s, 38 Channel DWDM transmission experiment," J. Lightwave Technol. 30, 512-520 (2012) .

[5] M. Gao, T. Kurosu, T. Inoue, S. Namiki., "Efficient phase regeneration of DPSK signal by sideband-assisted dual-pump phase-sensitive amplifier," Electron. Lett. 49, 140-141 (2013).

[6] M. Gao, T.Kurosu, K.Solis-Trapala, T.Inoue and S. Namiki, "Quadrature squeezing and IQ de-multiplexing of QPSK signals by sideband-assisted dual-pump phase sensitive amplifiers," IEICE Trans.Commun. vol. E98-B, No.11,2227-2237 (2015).

[7] M. Gao, T. Inoue, T. Kurosu, S. Namiki., "Evolution of the gain extinction ratio in dual-pump phase sensitive amplification," Opt. Lett. 37, 1439-1441 (2012).

[8] T. Kurosu, M. Gao, K. S. Trapala, and S. Namiki, "Phase regeneration of phase encoded signals by hybrid optical phase squeezer," Opt. Express 22, pp.12177-12188 (2014)

[9] F. Parmigiani, G. Hesketh, R. Slavík, P. Horak, P.Petropoulos, and D. J. Richardson, "Polarization-Assisted Phase-Sensitive Processor," J. Lightwave Technol. vol. 33, no. 6, pp.1166-1174, March 15, 2015.

WA2-56

Gain Saturation and Recovery of an Erbium-doped Fiber Amplifier Measured by Temporally Resolved Probe Techniques

Keiji Kuroda* and Yuzo Yoshikuni

Department of Physics, School of Science, Kitasato University, 1-15-1 Kitazato, Minamiku, Sagamihara, 252-0373, Japan
* e-mail address: kkuroda@kitasato-u.ac.jp

Abstract: We report on measurements of gain variations in an erbium-doped fiber amplifier. Transient gain saturation and recovery are measured using a self-probe technique and a pump-probe technique, respectively. Wavelength dependences of them are discussed.

Keywords: Erbium-doped fiber amplifier, Gain saturation, Gain recovery, Probe technique, Wavelength dependence

I. INTRODUCTION

Erbium-doped fiber amplifiers (EDFAs) have been widely used in optical networks. Substantial studies on their fundamental characteristics have been summarized [1]. For novel application of transient phenomena in EDFAs [2], precise evaluation of characteristic time constants of amplifier gain is required. In addition, to fully utilize their broadband nature, wavelength dependent characteristics of the time constants should be studied in detail.

In this presentation, we report on measurements of gain variations in an erbium-doped fiber amplifier. Transient gain saturation and recovery are measured using a self-probe technique and a pump-probe technique with a single DFB laser array, respectively. Wavelength dependences of the saturation and recovery times are discussed.

II. EXPERIMENTAL

A. EDFA

In the inset of Figure 1(a) shows an energy level diagram of erbium ions. Erbium ions are treated as a three-level system. By pumping to an excited level $^4I_{11/2}$, population inversion between level 1($^4I_{15/2}$) and level 2($^4I_{13/2}$) is formed. The two levels are broadened due to the Stark splitting. As a result, optical transitions over a wavelength range of 30 nm are possible at 1.5μm. Figure 1(a) shows an amplified spontaneous emission (ASE) spectrum of our 980 nm-pumped EDFA (Fiber Labs AMP-FL8013-CBDB-13). We can see that the spectrum covers the C-band from 1530 to 1560 nm.

B. DFB laser array

A tunable DFB (Distributed Feedback) laser array (NTT Electronics NLK1C7MBLG) was used as a light source [3] in experiments. This laser array consists of 12 channel LDs, a coupler, and a booster SOA (semiconductor optical amplifier). Each LD has a different oscillation wavelength that is determined by its grating period. The optical fields generated by the LDs are amplified by the SOA to compensate for the losses at the coupler. We can obtain single-frequency pulses by applying synchronous modulation voltages to both an LD and the SOA. In Figure 1(b), spectra of the laser array are shown. The emission wavelength ranges from 1528 to 1559 nm, covering the operation band of the EDFA as shown in Figure 1(a).

C. Temporally resolved probe techniques

In temporally resolved pump-probe measurements, rectangular pulse pairs are introduced as signal pulses, with the first being the pump pulse and the second being the probe pulse. Their wavelength is set to be resonant with the energy difference between level 1 and 2. Through stimulated emissions, the pulse pairs selectively change populations of the two levels. In usual pump-probe techniques, the probe pulse is weaker than the pump pulse to reduce the change in the population induced by the probe pulse itself. In our technique, the pulse width and peak intensity of the probe pulse are same as those of the pump pulse. To probe the transient gain without probe pulse-induced effects, we use only the pulse intensity at the probe pulse front I^0. This makes the probe photon number smaller than the pump photon number as in ordinary pump-probe methods. As a result, the amplification of the probe photon satisfies the small signal gain condition, in which the output intensity I^0_{out} is written by using the input intensity I^0_{in} as,

$$I^0_{out} = I^0_{in}\exp[\alpha L(\sigma_e N_2 - \sigma_a N_1)] \qquad (1)$$

where, α is a coefficient determined by the erbium concentration, interaction area, overlap factor. L is the fiber length. N_1 and N_2 are populations of level 1 and 2, respectively. σ_e and σ_a are the emission and absorption cross sections, respectively. From this expression, the pump-induced transient gain $\sigma_e N_2 - \sigma_a N_1$ can be calculated as,

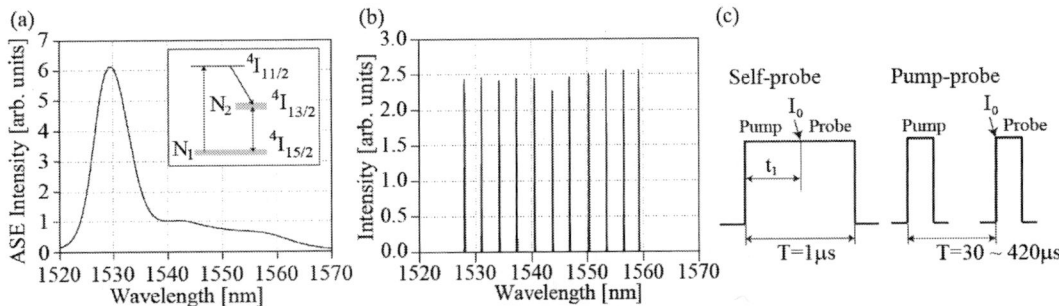

Figure 1(a) Amplified spontaneous emission spectrum of our EDFA. Inset: energy diagram of erbium ions. (b) Spectra of the DFB laser array. (c) Time chart of the self-probe and pump probe techniques.

$$\ln\left(\frac{I_{out}^0}{I_{in}^0}\right) \propto \sigma_e N_2 - \sigma_a N_1. \tag{2}$$

In the self-probe technique (see Figure 1(c)) to probe the gain saturation, a single pulse with a width of T is introduced to the EDFA. We assume a part of the pulse (from 0 to t_1) as the pump pulse, and the rest of the pulse (from t_1 to T) as the probe pulse [4-6]. The gain is calculated at t_1 to reduce the probe photon number. This allows us to calculate the transient gain saturation from t=0 to T in a single measurement. In the pump-probe technique (see Figure 1(c)) to probe the gain recovery, a pulse pair with an interval of T is introduced to the EDFA. The gain is calculated using the probe pulse front intensity [7, 8]. By changing the interval T, we can calculate the gain recovery after gain saturation induced by the pump pulse.

The pulse trains with various parameters used in the above techniques can be easily obtained by a laser diode directly modulated by a function generator. In addition, using the DFB laser array, we can perform the measurements over the operation wavelength range of the EDFA.

D. Experimental setup

Rectangular-shaped pulses from the laser modulated by a function generator were used as the signal pulses. These pulses were divided into two portions using a 99%/1% coupler. The 1% pulse was used to monitor the input pulse intensity. The 99% pulse was introduced into a 980 nm-pumped EDFA. The EDFA was 6.5 m long. The amplified pulse profiles were detected using a photodiode after attenuation and were stored using an oscilloscope. Details of the pulse conditions are given in Section III.

III. RESULTS AND DISCUSSIONS

A. Wavelength dependence of the gain saturation[5]

In the gain saturation measurements, a pulse width, repetition rate and intensity were 1μs, 1ms and 7mW, respectively. The 980 nm pump power for the EDFA was fixed at 50 mW. In figure 2(a), an output pulse shape at 1528 nm is shown. The instantaneous gain at the pulse front is approximately 25 dB. It can be seen that the shape is deformed from the rectangular input shape due to the gain saturation. In figure 2(b), a gray line indicates the gain $\ln(I_{out}/I_{in})$ calculated from the data in Figure 2(a). Black line is a fitting result using an exponential function [9]. Agreement between them is quite good. The gain saturation time was deduced through the fitting as 420 ns. Figure 2(c) shows a wavelength dependence of the gain saturation time by open circles. The saturation time increases to about 1500 ns as the wavelength increases. The black line is a fitting result using a model derived from rate equations [9] and McCumber relation [10]. The fitting line well reproduces the experimental results.

B. Wavelength dependence of the gain recovery[7]

In the gain recovery measurements, a pulse width, interval, repetition rate and intensity were 200 ns, 30 ~ 420 μs, 2ms and 8mW, respectively. The 980 nm pump power for the EDFA was fixed at 70 mW. The upper panel of Figure 3(a) shows pulse shapes of the pump (left) and probe pulses (right) for the interval of 30 μs at 1540 nm. The pulse shapes are deformed from the rectangular shape. In addition, the intensity of the probe pulse front is nearly same as that of the pump pulse end. This indicates that the gain does not recover during the interval. The lower panel of Figure 3(a) shows the pump pulse and the probe pulses for the interval of 30 ~ 420 μs. It can be seen that as the interval increases, the probe pulse intensity increases toward the pump pulse intensity. In figure 3(b), black circles indicate the gain $\ln(I_{out}/I_{in})$ calculated from the data in Figure 3(a). Black line is a fitting result using an exponential function [9]. Agreement between them is quite good. The gain recovery time was deduced through the fitting as 178 μs. Figure 2(c) shows a wavelength dependence of the gain recovery time by black circles. Although the recovery time decreases slightly for the longer wavelengths, it seems to be nearly constant. An average time was 176 μs. Independence of the recovery time on the wavelength is expected from the model given in [9].

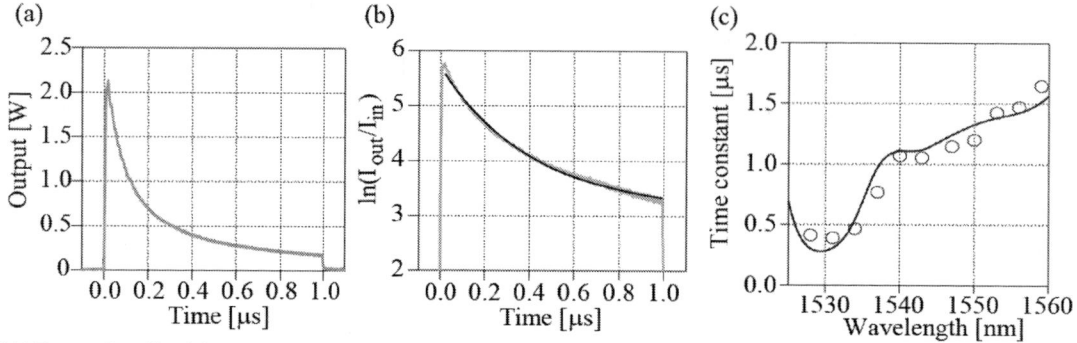

Figure 2(a) Temporal profile of the output pulse. (b) Gray line: Temporal gain $\ln(I_{out}/I_{in})$. Black line: fitting result. (c) Wavelength dependence of the gain saturation time.

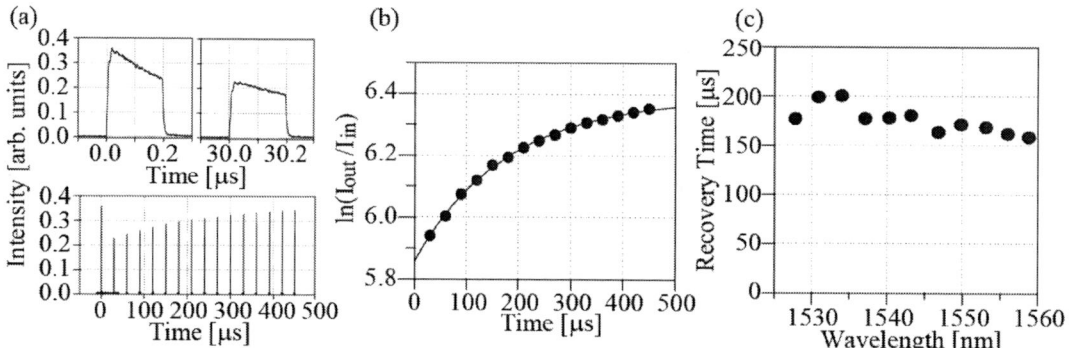

Figure 3(a) Upper panel: pulse shapes of the pump pulse (left) and the probe pulse (right) for the interval of 30 μs. Lower panel: the pump pulse and the probe pulses for the interval of 30 to 420 μs. (b) Black circles: temporal gain $\ln(I_{out}/I_{in})$. Black line: fitting result. (c) Wavelength dependence of the gain recovery time.

IV. CONCLUSIONS

We reported on measurements of the gain saturation and recovery of an Erbium-doped fiber amplifier. We applied the self-probe and the pump-probe technicues. The wavelendgth dependences of the gain saturation and recovery time were obtained using a tunable DFB laser array as a light source. It was found that the saturation time increased as the wavelength increased, on the other hand, the recovery time was nealy constant as the wavelength varied. The two dependences were explained by models, which we will discuss in detail at presentation.

REFERENCES

[1] E. Desurvire, Erbium-Doped Fiber Amplifiers, Wiley-Interscience, 1994.
[2] S. Stepanov, "Dynamic population gratings in rare-earth-doped optical fibers", J. Phys. D, vol. 41, pp. 224002-1-23, 2008.
[3] Y. Tohmori, H. Ishii, H. Oohashi, and Y. Yoshikuni, "Wavelength-Tunable Semiconductor Light Sources for WDM Applications", IEICE Trans. Electron. E85-C, pp.21-25, 2002.
[4] K. Kuroda, and Y. Yoshikuni, "Transient Population Inversion Induced and Probed by a Signal Pulse in a Continuous-Wave-Pumped Erbium-Doped Fiber Amplifier", Jpn. J. Appl. Phys., vol. 51, pp.120201-1-3, 2012.
[5] K. Kuroda, A. Suzuki, and Y. Yoshikuni, "Control and probe of population inversion using nanosecond pulse trains in an erbium-doped fiber amplifier", Opt. Fiber Technol. vol. 20, pp.483-486, 2014.
[6] K. Kuroda, K. Shibata, and Y. Yoshikuni, "Wavelength-dependent transition time of gain saturation in an erbium-doped fiber amplifier", Appl. Phys. B, vol. 120, pp.111-115, 2015.
[7] K. Kuroda, and Y. Yoshikuni, "Single wavelength pump-probe technique to measure population recovery in a continuously pumped fiber amplifier", Opt. Commun., vol. 300, pp.96-99, 2013.
[8] K. Kuroda, and Y. Yoshikuni, "Wavelength-resolved Measurement of Gain Recovery in an Erbium-doped Fiber Amplifier", Microw. Opt. Technol. Lett., vol. 58, pp.751-754, 2016.
[9] E. Desurvire, "Analysis of transient gain saturation and recovery in erbium-doped fiber amplifier", IEEE Photon. Technol. Lett., vol. pp.1, 196-199, 1989.
[10] W. J. Miniscalco, and R. S. Quimby, "General procedure for the analysis of Er3+ cross sections", Opt. Lett., vol. 16, pp. 258-260, 1991.

High Gross Gain of Single-Mode Cr-Doped Fibers Employing On-Line Growth System

Chun-Nien Liu[1], Gia-Ling Cheng[2], Nan-Kuang Chen[3], Yi-Chung Huang[4],
Pi-Ling Huang[1], Sheng-Lung Huang[2], and Wood-Hi Cheng[1],*

[1]Graduate Institute of Optoelectronic Engineering, National Chung Hsing University, Taichung, 402, Taiwan
[2]Graduate Institute of Photonics and Optoelectronics, National Taiwan University, Taipei, 10617, Taiwan
[3]Department of Electro-Optical Engineering, National United University, Miaoli, 36063, Taiwan
[4]CEO office, Brogent Technologies Inc., Kaohsiung, 806, Taiwan
*Corresponding author:whcheng@email.nchu.edu.tw

Abstract: *The fabrication and performance of single-mode Cr-doped crystalline core fiber (SMCDCCF) with high-index glass as a cladding employing on-line laser-heated pedestal growth (LHPG) technique are reported. A 3.9-dB gross gain with net gain of 1.9 dB for the SMCDCCF at wavelength of 1400-nm was obtained. The gross and net gains are the highest yet reported for SMCDCCF with high-index glass as a cladding layer. Further development on higher gain of the SMCDCCF may be functioned as a broadband fiber amplifier for use in the next-generation optical transmission systems.*

Keywords: *Fiber fabrication, Single-mode crystalline core fiber, Fiber laser, Gross gain.*

I. INTRODUCTION

Chromium-doped fibers (CDFs) have been shown the possible functions as broadband amplifiers for utilization over the entire low-loss transmission bandwidths from 1300-1600 nm [1]. In previous study using mode-matching technology to excite the predominantly fundamental mode of LP_{01} in a multimode fiber, a net gain of 1.2 dB of the chromium-doped fiber amplifiers (CDFA) was experimentally demonstrated [1]. However, at mode-matching condition in multimode fibers, the higher order modes are competing for the metastable state electronic energy, and it is difficult to obtain the predominant mode in fiber amplifiers. Further gain improvement of the CDFAs employing single-mode CDFs is necessary and crucial. The single-mode CDFs employing drawing-tower method in conjunction with rod-in-tube (RIT) technique and Cr:YAG as the core of the preform have been fabricated [2]. However, the fluorescent intensity of single-mode CDFs fabricated by drawing tower was reduced since the cerium oxide was highly volatile and Cr:YAG was highly reactive to SiO2, resulting in the inter-diffusion between core/cladding and lower gain [2]. Recently, a single-mode Cr-doped crystalline core fiber (SMCDCCF) cladded by high-index glass has been fabricated by the combination of the fiber drawing tower and laser-heated pedestal growth (LHPG) techniques [3]. The SMCDCCFs were measured with 2.7 dB gross gain and 0.7 dB net gain at the wavelength of 1.4 μm under a V-value of 2.4, which implies a single mode guiding could be realized in fiber [4]. However, a small core diameter (< 30 μm) with better uniformity and longer length of the SMCDCCFs were difficult to carry out through the LHPG process. In order to achieve a small core diameter and a longer length of the SMCDCCFs, it is essential to accurate aligning and positioning a small molten zone of solid-liquid shape to alleviate fluctuation of core diameter during the LHPG process.

In this study, we present a novel on-line growth system employing CCD and laser alignment to achieve small core and long length of the SMCDCCFs. This on-line growth system enables real time monitoring and controlling small molten zone of the solid-liquid shape to minimum fluctuations of core diameter during the LHPG process for achieving a small core diameter with better uniformity and longer length of the SMCDCCFs. The SMCDCCF exhibits a length of 10.6-cm, a core diameter of 25-μm, a gross gain of 3.9-dB, and a net gain of 1.9-dB at the wavelength of 1.4 μm. In comparison with the previous works without on-line growth technique on the SMCDCCFs which exhibited gross gain of 2.7-dB and net gain of 0.7-dB [4], this work has significantly improved the performance on gross and net gains. This indicates that the SMCDCCFs fabricated by an on-line growth system are more promising for developing higher gain used in fiber transmission system applications. This on-line growth system technique used to achieve higher gain may be one step further to being utilized as a new generation broadband fiber amplifier to cover the bandwidths in the entire low-loss window of silica fibers as well as for fiber communication applications.

II. FABRICATION OF SMCDCCFs BY ON-LINE GROWTH SYSTEM

A 3-D schematic of on-line growth system is shown in Fig. 1. A 100W and 10.6 μm polarized CO_2 laser with a polarizer-analyzer-attenuator for laser power adjustment was used as the heat source. The fluctuation of CO_2 laser power can be controlled < 1% by employing a feedback controlling device. The pulling and feeding mechanisms consist of computer-controlled linear stage driven by stepping motor with gearbox to reduce vibration. The CO_2 laser system has a collinear HeNe reference beam (visible) to indicate the travelling paths for the CO_2 laser beam (invisible). An optical focusing system within the growth chamber is designed to adjust the output of the CO_2 laser to an axially and azimuthally symmetrically focus on the molten zone. The molten zone was held in its place by surface tension during the growth process. This process helped to eliminate the crucible contamination problems associated with crystal fabrication. The x-axis and y-axis CCD can persistently observe the variation of molten zone. Furthermore, we positioned the source rod and single crystal in laser alignment at z-axis by green laser, as shown in Fig. 2. With the assistance of CCD and laser alignment, the physical locations of single crystal and source rod could be visually determined and real time adjusted a small molten zone of solid-liquid shape during LHPG process. Based on a relative low grow velocity ratio, the source rod can have more time to melt and control completely. By controlling the small molten zone of solid-liquid and the suitable growth ratio with CCD and laser alignment in the on-line LHPG process, we are able to achieve better uniformity and longer length for the SMCDCCFs with a small core diameter.

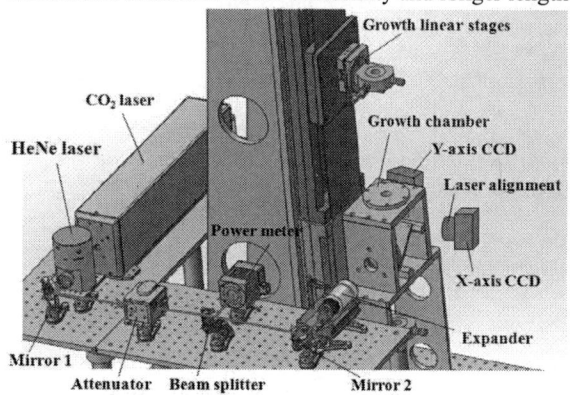

Fig. 1. 3D schematic of on-line LHPG system.

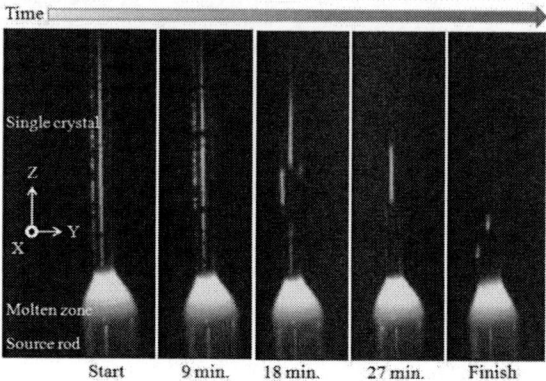

Fig. 2. The molten zone of solid-liquid shape during LHPG process.

The single-crystal-to-source-rod diameter ratio could be controlled by setting the pull/feed speed ratio [5]. At steady state, the volume of molten zone was constant due to conservation of mass. The volume of crystal added to the fiber must be equal to the volume of the source rod melted per unit time. Therefore, the ratio of fiber diameter to source rod diameter was given by $D_f / D_s = (V_s / V_f)^{1/2}$, where the D_f and D_s were the diameters of the single crystal and source rod, respectively. The V_f and V_s were the velocities of the single crystal and source rod, respectively. The rapid growth rates on the order of mm/min could be utilized to grow single crystals with the small size. In this work, the Cr^{4+}: YAG source rod diameter was 120 μm and expected single crystal was 25 μm. Figure 3(a) shows a photograph of a 25-μm-core-diameter single crystal of Cr^{4+}:YAG and the core diameter fluctuation is about ± 2 μm. A single crystal Cr^{4+}: YAG fiber with a length of 10.6-cm was successfully fabricated by this on-line LHPG process. Then, the single crystal Cr^{4+}: YAG inserted into a capillary of N-SF57 high-index glass with the inner and outer diameters of 70/260 μm fabricated by fiber drawing tower. A CO_2 laser beam was focused and shone around the capillary in order to heat it up to collapse the tightly-fitted glass capillary. The softening temperature of N-SF57 glass was 716°C, which was lower than the 1970°C melting temperature of Cr^{4+}:YAG single crystal. Therefore, the power of CO_2 laser was well controlled in order to get the high-index glass capillary softened and attached to the Cr^{4+}:YAG single crystal. Finally, a 10.6-cm of SMCDCCF has been successfully fabricated by on-line LHPG system. Figure 3(b) shows a polished end view of a SMCDCCF with 25 μm core (Cr^{4+}: YAG) and 260 μm cladding (N-SF57).

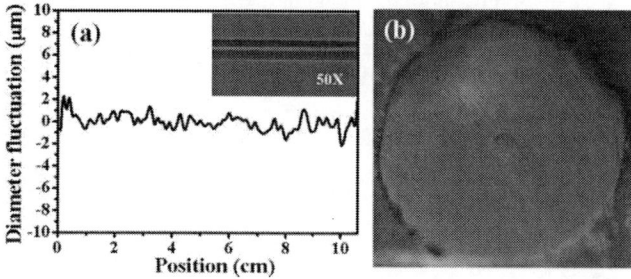

Fig. 3. (a) The core diameter variation of single crystal, and (b) the end-face of SMCDCCF.

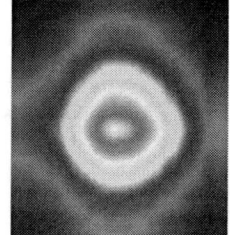

Fig. 4. A far-field pattern of SMCDCCF.

III. MEASUREMENTS AND RESULTS

The refractive indices of core and cladding of the SMCDCCF were measured 1.828 and 1.826, respectively. The index difference between core and clad cladding was 0.2%. The divergence angle of 2.29° was calculated from the far-field measurement, which varied with the distance between detector and the SMCDCCF. Numerical aperture of the 25-μm-core SMCDCCF was around 0.04, corresponding to normalized frequency V-value about 2.4 at 1.4 μm. Figure 4 shows a far-field pattern which indicates the single mode LP_{01} field pattern in the SMCDCCF by launching a wavelength of 1400 nm.

To characterize the gross gain of SMCDCCF with and without on-line growth system, we employed an Yb-doped fiber laser at 1064 nm as the pump source and a signal source from a 1400-nm tunable laser. The signal beam was injected into the SMCDCCF through a coupler. The signal beam was amplified by the pump source and then detected by an optical spectrum analyzer. The gross gain of fiber amplifiers is defined as $G = 10 \log [(P_{s+p} - P_p)/P_s]$, where the P_s, P_p, and P_{s+p} are the output power from the facet of SMCDCCF in the operation of injecting input signal light, optical pumping light, and combining input signal with pumping source, respectively. Figure 5 shows the gross gain and net gain of SMCDCCFs with and without on-line growth system. The highest gross gain of 3.9 dB and 2.8 dB with and without on-line growth system were obtained, respectively. The gross gain represents a component gain without considering the insertion loss of SMCDCCF. An insertion loss of 2 dB for the SMCDCCF was measured by a cutback method [4]. Therefore, by subtracting the insertion loss from the gross gain, we were able to obtain a net gain of 1.9 dB and 0.7 dB with and without on-line growth system for a 10.6-cm long SMCDCCF at the wavelength of 1400 nm.

Figure 6 shows the simulation of gross gain as a function of fiber length for SMCDCCFs at different pump powers. Figure 7 shows the simulation of gross gain as a function of pump power for SMCDCCFs with different lengths. In this simulation, the ground-state absorption and ESA cross section were set as 22×10^{-19} and 7×10^{-20} cm^2, respectively. The emission cross section at the center wavelength was 3.5×10^{-19} cm^2. The lifetime was 3.6 μs at the room temperature. Figures 6 and 7 indicates that a higher gross gain more than 12.5 dB may be achieved at pump power of 400 mW if a longer fiber length of 20 cm can be used.

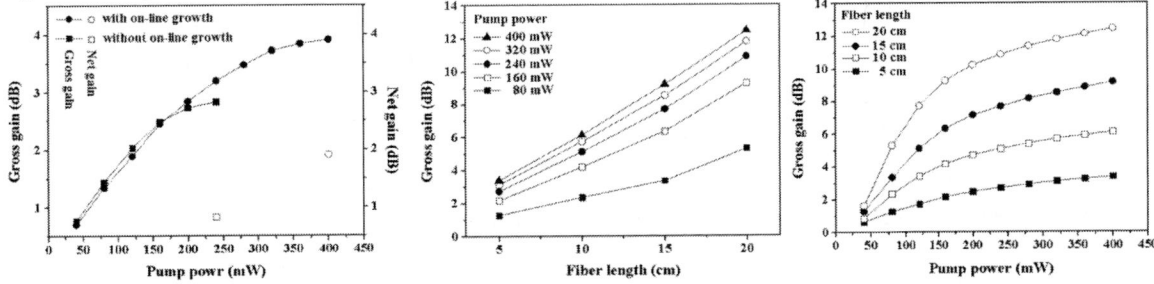

Fig. 5. Gross gain and net gain of SMCDCCF.   Fig. 6. Fiber length simulation of SMCDCCF.   Fig. 7. Pump power simulation of SMCDCCF.

IV. CONCLUSIONS

In summary, the SMCDCCF with cladded high-index glass of 25-μm in core diameter and 10.6-cm in fiber length has been successfully fabricated by on-line LHPG technique. The V-value of fiber was about 2.4, which could guide single mode (LP$_{01}$) in the fiber. The SMCDCCF showed a 3.9 dB gross gain and 1.9 dB net gain at the wavelength of 1.4 μm by pump power of 400 mW. In this study, the gross and net gains are the highest yet reported for SMCDCCF with high-index glass as a cladding layer. Further studies on higher gross and net gains improvement of the SMCDCCFs including a longer length and a lower loss in fibers are necessary to develop as broadband fiber amplifiers for use in fiber transmission system applications and are currently under investigation.

ACKNOWLEDGMENT

This work was partially supported by the Ministry of Science and Technology under the contract MOST 103-2221-E-110-003 and the MOE Program of the Aim for the Top University Plan.

REFERENCES

[1] S.M. Yeh, S.L. Huang, Y.J. Chiu, H. Taga, P.L. Huang, Y.C. Huang, Y.K. Lu, J. P. Wu, W.L. Wang, D.M. Kong, J.S. Wang, P. Yeh, and W.H. Cheng, "Broadband chromium-doped fiber amplifiers for next-generation optical communication systems," J. Lightw. Technol. Vol. 30, pp. 921-927, 2012.

[2] B. Y.C. Huang, J.S. Wang, Y.S. Lin, T.C. Lin, W.L. Wang, Y.K. Lu, S.M. Yeh, H.H. Kuo, S.L. Huang, and W.H. Cheng, "Development of Broadband Single-mode Cr-Doped Silica Fibers," IEEE Photon. Technol. Lett., Vol. 22, No. 12, pp. 914-916 , 2010.

[3] W.L. Wang, G.L. Cheng, Y.C. Huang, N.K. Chen, S.L. Huang, and W.H. Cheng, "Few-Mode Cr-Doped Fibers by Cladded High Index Glass for Broadband Fiber Amplifiers,". IEEE Photon. Technol. Lett., Vol. 26, No. 6, pp. 587-590, 2014.

[4] W.L. Wang, G.L. Cheng, Y.C. Huang, C.C. Wei, N.K. Chen, S.L. Huang, and W.H. Cheng, "Single-Mode Cr-Doped Crystalline Core Fibers for Broadband Fiber Amplifiers," IEEE Photon. Technol. Lett., Vol. 27, No. 2, pp. 205-208, 2015.

[5] R.S. Feigelson, "Pulling optical fibers," J. Crystal growth, Vol. 79, pp. 669-680, 1986.

[6] R. K. Nubling and James A. Harrington, "Optical properties of single-crystal sapphire fibers," Appl. Optics. Vol. 36, .No. 24, pp. 5934-5940, 1997.

WA2-58

OECC/PS2016

Double Inscribing Method with CO_2 Laser for Suitable Spectrum of LPFG Used in Multipoint Temperature Sensor

Akihiro Kusama, Mamoru Iida, Osanori Koyama, Syo Takasuka, Toshinori Murakami, Makoto Yamada
Department of Electrical and Information Systems, Graduate School of Engineering, Osaka Prefecture University
koyama@eis.osakafu-u.ac.jp, myamada@eis.osakafu-u.ac.jp

Abstract: A new method, the double inscribing method using a CO_2 laser, is proposed to make the loss spectrum of LPFGs more suitable when constructing a multipoint temperature sensor using several LPFGs. The method makes the main peak in the spectrum larger and suppresses associated sub peaks without greatly affecting the resonant wavelength of the LPFG in the range where the grating pitch of the LPFG is longer than 0.52 mm. Finally, the relationship between the grating pitch and a parameter of the method is clarified to obtain suitable spectrum for the multipoint sensor.
Keywords: Long Period Fiber Grating, CO_2 Laser, Temperature Sensor

I. INTRODUCTION

Optical fiber sensors are used in many advantages compared with electrical sensors such as remote sensing, non-necessity for power supply at the location where the sensors disposed, non-necessity for surge current provision, and non-electromagnetic interference. It is considered that they will be used in harsh environments that are not easily accessible to humans, for example in temperature sensing within oil wells. Long period fiber grating (LPFG) is a type of optical fiber sensor that can be fabricated by ultraviolet laser irradiation [1], mechanical micro-bending [2], electric discharge [3], and CO_2 laser irradiation [4]. However, as LPFG that is fabricated using a CO_2 laser has high stability in a high temperature environment [5,6] it is thus considered for use as a high temperature sensor.

We investigate a multipoint temperature sensor configured by cascading several LPFGs fabricated using a CO_2 laser [7]. The loss spectrum of this sensor has several optical loss peaks, and the temperature at each of the sensing locations can be measured by monitoring the amount of shift in the resonant wavelength (λ_R) at each loss peak within the spectrum of the LPFG. Before designing the multipoint temperature sensor, it is important to ascertain that the spectrum of each LPFG can be controlled, because the sensor specification depends strongly on the spectrum of the cascaded LPFGs (e.g., the number of sensing locations, the sensible temperature range at each location, and the measurement error). In order to obtain a suitable spectrum for the sensor, we have to make the main peak value (L_R) at λ_R in each spectrum much higher than the sub peak values, and have to suppress sub peak values, and also have to arrange λ_R accurately. The sub peak suppression is particularly important because measurement errors at other locations become larger when the sub peak values of a LPFG are high. However, it has not previously been possible to obtain low sub peaks using the conventional our fabrication method with the CO_2 laser in regions where the grating pitch Λ, of the LPFG is higher than 0.52 mm. In this paper, therefore, we propose the use of a double inscribing method with a CO_2 laser to obtain suitable spectrum of LPFG in the region of $\Lambda > 0.52$ mm. The proposed method enables enlargement of the main peak of each spectrum and suppresses the associated sub peaks without having too great an effect on λ_R. In this paper, we also clarify the relationship between the parameters of the proposed method and Λ. The relationship is considered to be useful in determining the desired spectrum of a multipoint temperature sensor.

II. MULTIPOINT TEMPERATURE SENSOR USING SEVERAL LPFGS

Figure 1 shows the LPFG fabrication system used in this study. LPFGs can be fabricated with common communications fiber (SMF-28) and a CO_2 laser using the conventional point-to-point technique [4]. In this conventional LPFG fabrication method, the fiber is fixed on a translation stage, and a weight (~13 g) is loaded on the other side of the fiber via another block to keep the fiber horizontal. After the refractive index of the fiber's core has been modulated by one-time irradiation, the stage is shifted to Λ along the fiber axis through the stage controller. The

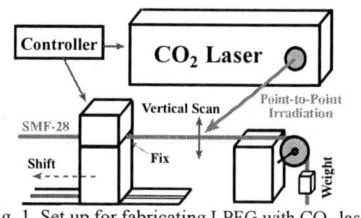

Fig. 1 Set up for fabricating LPFG with CO_2 laser.

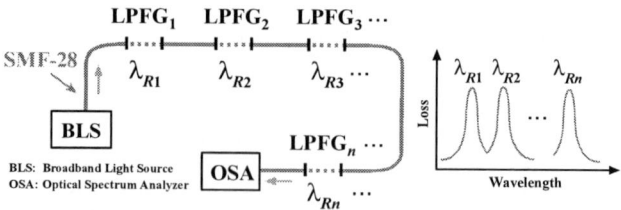

Fig. 2 Multipoint temperature sensor including several LPFGs.

fiber is then finally irradiated by the CO_2 laser again. The process is then repeated.

Figure 2 shows the multipoint temperature sensor configured by a cascade of several LPFGs that have been fabricated using the system shown in Fig. 1. Each LPFG has a main loss peak at λ_R and the value at each LPFG's λ_R is different from that of the other LPFGs. Sensors with different resonant wavelengths, λ_{R1}, λ_{R2}, ..., enable us to obtain different peaks in the loss spectrum at a reference temperature (room temperature), as shown in Fig. 2. We are thus able to measure the temperature at each of the locations where LPFGs are deployed, by monitoring the amount of shift occurring in comparison with each λ_{Rn} at the reference temperature.

III. DOUBLE INSCRIBING METHOD WITH CO_2 LASER

Some spectra of the LPFGs fabricated using the previously described conventional method with a CO_2 laser had a main loss peak of 7 to 8 dB (as shown in Fig. 3) when the Λ values were smaller than 0.52 mm. However, when Λ values were larger than 0.52 mm, the loss peaks were small and additional sub peaks appeared. To make an equal comparison of the characteristics, the LPFGs were fabricated in identical conditions, i.e., the irradiation power P of the CO_2 laser was equal to 3.6 W, and the grating number was set at 60 for each LPFG. Consequently, the LPFG length L values were equal at 28.8 mm, 29.52 mm, 30 mm, 30.6 mm, 31.8 mm, and 32.4 mm for LPFGs with $\Lambda = 0.48$ mm, 0.492 mm, 0.5 mm, 0.51 mm, 0.53 mm, and 0.54 mm, respectively. As shown in Fig. 3, the loss peaks of LPFGs with $\Lambda > 0.52$ mm were found to be unsuitable for cascading. In this research, we define a suitable loss peak L_R, as being at a sufficient height for monitoring and having associated low sub peaks, and where sub peak suppression has priority over the height of L_R. However, it was not possible to obtain suitable peaks in ranges where Λ was longer than 0.52 mm, as shown in Fig. 3. Therefore, we propose use of the double inscribing method (shown as Fig. 4) to obtain the suitable spectrum for an LPFG with $\Lambda > 0.52$ mm. In the proposed method, the stage is shifted slightly (from 0 mm to 0.14 mm) after the first irradiation, and the refractive index in the core of the fiber is then also modulated by the second irradiation. After the second irradiation, the stage is shifted to Λ. When this procedure is repeated, the refractive index modulated region is thus expanded. The width of the irradiated region is defined as d.

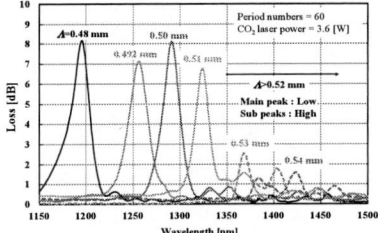

Fig. 3 Spectra of LPFG fabricated with conventional method.

Fig. 4 Double inscribing method with CO_2 Laser.

IV. SPECTRUM IMPROVEMENT BY DOUBLE INSCRIBING METHOD

CO_2 laser irradiation in the experiment was conducted with the following specifications: a scanning speed of 45 mm/s; power of 4.2 W; and a diameter at beam waist of 0.16 mm. The value of Λ was then changed from 0.52 mm to 0.60 mm, and d was also changed from 0.16 mm to 0.30 mm in each Λ, when the LPFGs were fabricated. The values of λ_R, L_R, and sub-peaks in the spectra of LPFG in each Λ and d were then measured.

As examples, results for the case of $\Lambda = 0.55$ mm are shown in Figs. 5 and 6 and those for $\Lambda = 0.59$ mm are shown in Figs. 7 and 8. Figure 5 shows that the values of λ_R underwent a slight change with respect to d, and the maximum loss value of L_R was 10.2 dB at $d = 0.23$ mm. Figure 6 then clarifies that it was possible to alter the proportion of the highest sub peak to the main loss peak (L_R) so that it was lower than 0.5 in the range of $d >= 0.22$ mm, and that it was possible to obtain the highest suppression of 0.35 at $d = 0.23$ mm. From these results, it was possible to obtain an optimal value of $d = 0.23$ mm in a case where $\Lambda = 0.55$ mm within a suitable spectrum of the multipoint temperature sensor. In the case of $\Lambda = 0.59$ mm, a main peak and two high sub peaks in the loss spectrum were obtained. Figure 7 shows that the maximum value of L_R was 9.7 dB at $d = 0.25$ mm. Figure 8 then shows that the proportion of the sub peak (1) to L_R was stable at around 0.2 in the range where $d >= 0.24$ mm, but that the sub peak (2) could not be measured in this range.

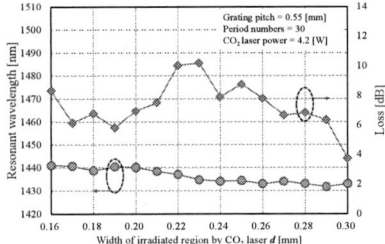

Fig. 5. Resonant wavelength and main loss peak ($\Lambda = 0.55$ mm).

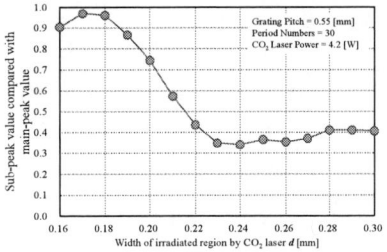

Fig. 6. Suppression of sub loss peak ($\Lambda = 0.55$ mm).

We also confirmed that two high sub peaks could be suppressed in the range of Λ by extending the refractive index modulated region. As shown in the results, the optimal value of d was 0.24 mm in the case where $\Lambda = 0.59$ mm, because suppression was the priority. Figure 9 shows the spectrum improvement when using the proposed method (red line) compared to the conventional method (open circles). The left and right sides of Fig. 9 show cases where $\Lambda = 0.55$ mm and $\Lambda = 0.59$ mm, respectively. We confirmed that use of the proposed method could make values of L_R larger, sub-peaks smaller, and enable the easier detection of λ_R in each spectrum. Finally, we clarified that optimum values of d were obtained in each of the grating periods ($\Lambda = 0.52$–0.60 mm). Figure 10 shows the best suppression obtained for sub peaks and high L_R. The relationship between d and Λ is useful when designing an optimum spectrum for a multipoint temperature sensor using several LPFGs.

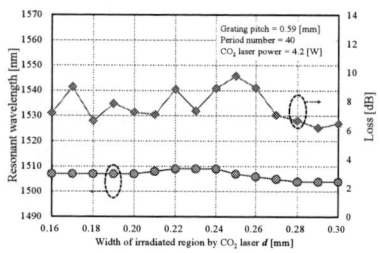

Fig. 7. Resonant wavelength and main loss peak ($\Lambda=0.59$ mm).

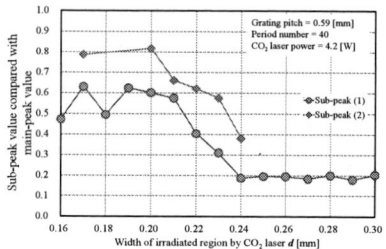

Fig. 8. Suppression of sub loss peaks ($\Lambda=0.59$ mm).

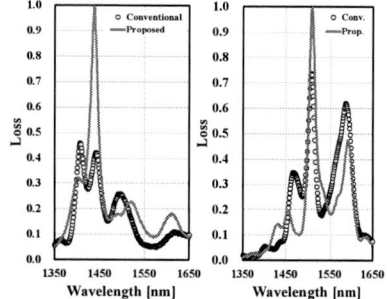

Fig. 9. Improvement of LPFG loss spectra using proposed method.

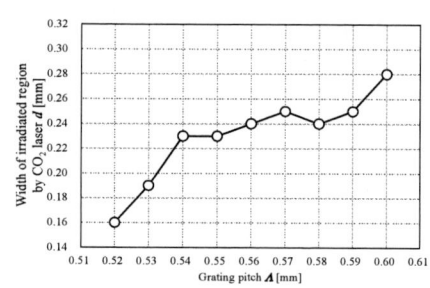

Fig. 10. Optimal width of irradiation region with CO_2 laser.

V. CONCLUSIONS

The double inscribing method with a CO_2 laser was proposed as a new fabricating method for a LPFG with a large grating pitch Λ (> 0.52 mm) that is intended for use in configuring a multipoint temperature sensor. The proposed method enabled enlargement of the main peak of the spectrum and suppression of the associated sub peaks, without greatly affecting the resonant wavelength λ_R. Finally, the optimum width d of the irradiated region was clarified using CO_2 laser for each grating pitch Λ. In future work, we will investigate the mechanism involved in spectrum improvement using the double inscribing method.

ACKNOWLEDGMENT

This work was supported by JSPS KAKENHI (25420336).

REFERENCES

[1] A.M. Vengsarkar, P.J. Lemaire, J.B. Judkins, V. Bhatia, T. Erdogan, and J.E. Sipe, "Long-period fiber gratings as band-rejection filters." Journal of Lightwave Technology, vol. 14, no. 1, pp. 58-65, 1996.

[2] C.H. Lin, Q. Li, A.A. Au, Y. Jiang, E. Wu, and H.P. Lee, "Strain-Induced Thermally Tuned Long-Period Fiber Gratings Fabricated on a Periodically Corrugated Substrate." Journal of Lightwave Technology, vol. 22, no. 7, pp. 1818-1827, 2004.

[3] G. Humbert, and A. Malk, "Characterizations at very high temperature of electric arc-induced long-period fiber gratings." Optics Communications, vol. 208, pp. 329-335, 2002.

[4] D.D. Davis, T.K. Gaylord, E.N. Glytsis, S.G. Kosinski, S.C. Mettler, and A.M. Vengsarkar, "Long-period fiber grating fabrication with focused CO_2 laser beams." Electronics Letters, vol. 34, no. 3, pp. 302-303, 1998.

[5] D.D. Davis, T.K. Gaylord, E.N. Glytsis, and S.C. Mettler, "Very-high-temperature stable CO2-laser-induced long-period fibre gratings," Electronics Letters, vol. 35, no. 9, pp. 740-742, 1999.

[6] M. Iida, O. Koyama, H. Sumiana, Y. Toyooka, and M. Yamada, "High-temperature attenuation peak behaviors of long-period fiber grating inscribed with CO_2 laser." Proc. of 19th OptoElectronics and Communications Conference (OECC), Melbourne (Australia), July 2014, pp. 484-489.

[7] O. Koyama, H. Sumiana, Y. Toyooka, and M. Yamada, "High temperature detection inside large-scale plants using distributed sensor fabricated by 10 LPG resonant wavelengths multiplexing." IEICE Electronics Express, vol. 10, no. 18, pp. 1-9, 2013.

Simultaneous Interrogation of Multiple Correlation Peaks in a BOCDA System by Time-domain Data Processing

Gukbeen Ryu[1,2], Kwang Yong Song[3], Gyu-Tae Kim[2], Sang Bae Lee[1] and Kwanil Lee[1]

[1]Nanophotonics Research Center, Korea Institute of Science and Technology (KIST), Seoul 02792, Republic of Korea
[2]School of Electrical Engineering, Korea University, Seoul 02841, Republic of Korea
[3]Dept. of Physics, Chung-Ang University, Seoul 06974, Republic of Korea
klee21@kist.re.kr

Abstract: *We introduce time-domain data processing to an ordinary Brillouin optical correlation domain analysis system for simultaneous interrogation of multiple correlation peaks. In experiment 32 correlation points are simultaneously interrogated in a 320 m optical fiber.*
Keywords: *Distributed measurement, stimulated Brillouin scattering, Optical fiber sensors*

I. INTRODUCTION

Brillouin scattering in an optical fiber has been effectively used for distributed sensing of various structures and materials thanks to the linear dependence of the Brillouin frequency shift on both local temperature and strain variation [1-2]. Among several types of Brillouin sensors, Brillouin optical correlation domain analysis (BOCDA) can easily provide high spatial resolution of an order of mm based on the synthesis of optical coherence function (SOCF) where sinusoidal frequency modulation is applied to continuous pump and probe waves to localize sensing position, called correlation peak (CP), within the fiber. This CW operation of the pump and probe waves also allows high sampling rate and random access of sensing position [3-4]. However, the modulation parameters (amplitude and frequency) of a light source need to be chosen in such a way that only a single CP lies within the sensing fiber, so the measurement range is limited to distance between adjacent CP's as a trade-off with the spatial resolution. For enlargement of the sensing range, various approaches have been proposed such as double modulation scheme [5], temporal gating scheme [6], and bidirectional measurement [7]. Recently, combined application of time- and correlation-domain analysis has been proposed where the pump and probe waves are modulated by a phase sequence and the pump wave is additionally modulated as a pulse to temporally separate signals from different CP's [8]. The combined system could effectively increase the sensing range and reduce the number of frequency scans for distributed measurement. However the demonstrated setup requires an arbitrary waveform generator with large bandwidth (~ 5 GHz) to apply the coded phase modulation for high resolution sensing, which could increase the cost and complexity of the sensor system.

In this work, we newly introduce the time-domain data processing to an ordinary BOCDA system, where MHz-order sinusoidal frequency modulation is applied to continuous probe wave and pulsed pump wave to produce multiple periodic CP's along the sensing fiber for simultaneous interrogation. In experiments, we successfully measure and analyze the Brillouin gain spectra from 32 CP's along a 320 m test fiber with a 3.5 cm spatial resolution. Our proposed scheme can substantially reduce the time required for scanning the whole fiber by simultaneous interrogation of multiple CP's of an ordinary BOCDA system without using high speed electronics.

II. PRINCIPLE AND EXPERIMENT

Figure 1 shows the proposed BOCDA system with simultaneous accessibility of multiple correlation peaks. In a similar way to the conventional BOCDA, the pump and probe waves are both frequency modulated, which enables the position selective excitation of the stimulated Brillouin scattering (SBS) at correlation peaks along the FUT. However, differences are introduced: Multiple correlation positions are located in the sensing fiber instead of one and pulsed pump wave is adopted instead of CW light. Thus, multiple correlation peaks are simultaneously interrogated by detecting the amplified CW probe light with time-resolved measurement as shown in Fig.1. It is notable that the width of pump pulse must be equal to the period of the sinusoidal modulation to analyze only single CP at each position. Finally, to achieve the distributed measurement, the multiple CP's are simultaneously position-scanned by simply sweeping the modulation frequency.

Figure 2 shows the experimental setup to demonstrate the proposed BOCDA scheme. A 1550 nm DFB-LD is used as a light source for pump and probe waves, where direct current modulation with a sinusoidal RF wave whose modulation frequency and modulation amplitude (Δf) were about 10 MHz and 2.8 GHz, respectively. With these parameters, the distance between CP's (d_m) and the spatial resolution (Δz) was calculated to be about 10.34 m and 3.5 cm, respectively from following well-known equations [3]:

$$d_m = \frac{v_g}{2f_m}, \qquad \Delta z = \frac{v_g \cdot \Delta v_B}{2\pi \cdot f_m \cdot \Delta f}$$

where v_g is the group velocity of light, f_m is the modulation frequency, Δf is the modulation amplitude, and Δv_B is the Brillouin gain bandwidth (~30 MHz).

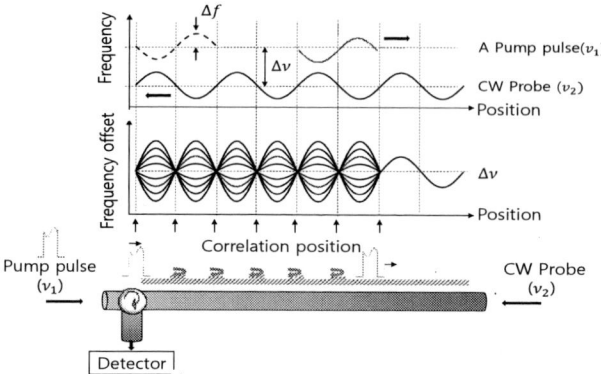

Fig. 1. Basic principle of Brillouin optical correlation domain analysis (BOCDA) using a pulse as a pump in time domain.

The output from the DFB-LD was divided into two beams by a 3 dB coupler. One of the output was injected into a single sideband modulator (SSBM) driven by a microwave signal generator and the lower sideband output was utilized as a probe wave. After passing through delay fiber (10km), it was amplified by an EDFA and propagated against pump wave in the sensing fiber. The other beam, after passing through a phase modulator for improvement of spatial resolution by differential measurement [9], was converted to the pulse light with an intensity modulator. The pulse light was amplified by an EDFA and was utilized as pump wave. A polarization scrambler was used in the system for suppressing the polarization induced fluctuation of the signal [10]. The pump waves was launched into the 330 m sensing fiber (Single mode fiber: Corning SMF-28) through a circulator in opposite direction to the probe wave and 32 CP's are generated in the sensing fiber. The probe wave was received by a 125 MHz photodetector and the average power of the single modulation period was recorded to calculate the Brillouin gain of each CP. It is notable that the lock-in detection of ordinary BOCDA systems is not adopted and two data sets with the pump phase modulation on and off are subtracted to obtain final signal through differential measurement.

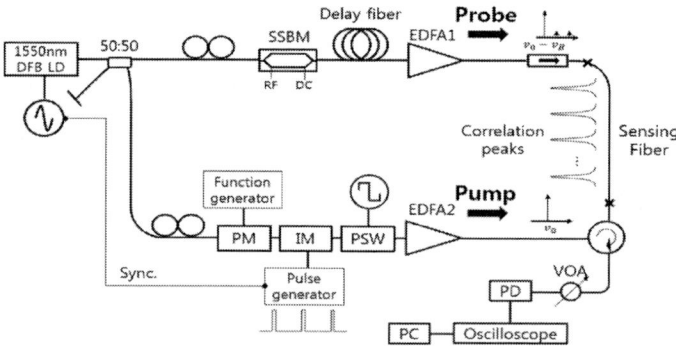

Fig. 2. Experiment setup of the BOCDA system with time-domain data processing: SSBM, sing-sideband modulator; EDFA, Er-doped fiber amplifier; PM, phase modulator; IM, Intensity modulator; PSW, polarization switch; VOA, variable optical attenuator; PD, photo detector.

Figure 3(a) shows an example of the simultaneously measured Brillouin gain spectra of all 32 CP's along the FUT with f_m=10 MHz fixed obtained by sweeping Δf from 10.8 GHz to 11 GHz, where one can see the signal amplitude is maintained through the sensing fiber. The BGS from single CP is depicted in Fig. 3(b) where one can clearly see the variation of Brillouin gain according to Δf. It should be noted that the Brillouin gain spectrum is the result of only 128 times of averaging at each Δf. Next, repeated measurement of the Brillouin gain signal distribution along the entire length of the FUT was conducted by changing f_m from 10 MHz to 10.02 MHz. For test of strain measurement, a 20 cm strain-applied section was located near the end of the FUT where 1.5 mε was applied, and the distributed measurement with at intervals of 10 cm is performed. The result is shown in Fig. 4, where the 3D map of the BGS and the distribution map of the Brillouin frequency are plotted in Fig. 4(a) and 4(b), respectively. One can clearly discern the strain applied section at 321 m, and the measurement error (σ) is estimated about 3 MHz, corresponding to strain resolution of 60 $\mu\varepsilon$.

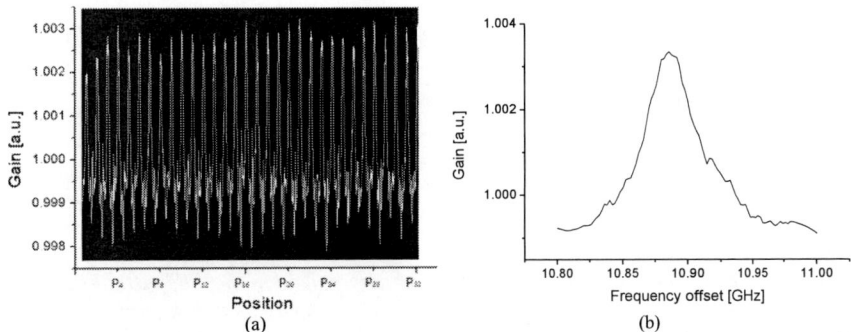

(a)   (b)

Fig. 3 (a) 32 Brillouin gain spectrum (BGS) measurement by distance side by side at once (with f_m=10 MHz fixed). (b) One of enlarged BGSs

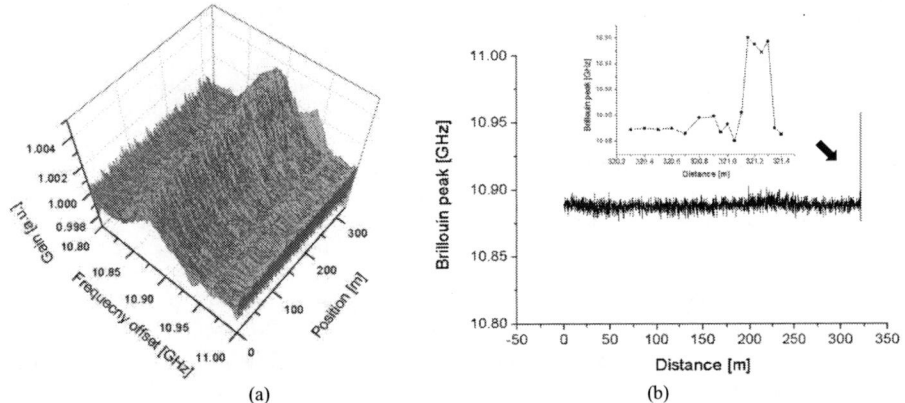

(a)   (b)

Fig. 4 (a) 3D plot of distribution map of Brillouin gain spectra with a 10 cm step along the sensing fiber obtained by simultaneously interrogating 32 CP's. The number of frequency scan is only 100 for monitoring of 3200 sensing points. (b) Distribution map of Brillouin frequency along the test fiber.

III. Conclusions

We proposed and experimentally demonstrated a BOCDA system with time-domain data processing for simultaneous interrogation of multiple correlation peaks. In the experiment, 32 correlation points were simultaneously interrogated with a 3.5 cm spatial resolution on a 320 m test fiber. We expect that the proposed BOCDA system has potential to reduce the measurement time in the fully distributed sensing without using high speed electronics.

References

[1] T. Horiguchi, T. Kurashima and M. Tateda, "Tensile strain dependence of Brillouin frequency shift in silica optical fibers", IEEE Photon. Technol. Lett., vol.1, pp. 107-109, 1989.

[2] T. Kurashima, T. Horiguchi, and M. Tateda, "Distributed temperature sensing using Stimulated Brillouin scattering in optial silica fibers", Opt. Lett., vol.15, pp. 1038-1040, 1990.

[3] K. Hotate and T. Hasegawa, "Measurement of Brillouin gain spectrum distribution along an optical fiber using a correlation based technique—proposal, experiment and simulation," IEICE Trans. Electron. vol.E83-C, 405-412, 2000.

[4] K. Y. Song and K. Hotate, "Distributed fiber strain sensor at 1 kHz sampling rate based on Brillouin optical correlation domain analysis," IEEE Photon. Technol. Lett., vol.19, pp. 1928-1930, 2007.

[5] Y. H. Kim, K. Lee, and K. Y. Song, "Brillouin optical correlation domain analysis with more than 1 million effective sensing points based on differential measurement," Opt. Express, vol. 23, pp. 33241-33248, 2015.

[6] K. Hotate and H. Arai, "Enlargement of measurement range of simplified BOCDA fiber-optic distributed strain sensing system using a temporal scheme," Proc. of 17th International Conference on Optical Fibre Sensors (OFS 2005), Bruges(Belgium), May. 2005, pp.184-187.

[7] J. H. Jeong, K. Lee, K. Y. Song, J.-M. Jeong, and S. B. Lee, "Bidirectional measurement for Brillouin optical correlation domain analysis," Opt. Express, vol. 20, pp. 11091-11096, 2012.

[8] D. Elooz, Y. Antman, N. Levanon and A. Zadok, "High-resolution long-reach distributed Brillouin sensing based on combined time-domain and correlation-domain analysis," Optics Express, vol. 22, pp. 6453-6463, 2014.

[9] J.H. Jeong, K. Lee, K. Y. song, J.-M. Jeong, and S. B. Lee, "Differential measurement scheme for Brillouin Optical Correlation Domain Analysis" Opt. Express, vol.20, pp. 27094-27101, 2012.

[10] K. Hotate, K. Abe, and K. Y. song, "Suppression of Signal Fluctuation in Brillouin Optical Correlation Domain Analysis System Using Polarization Diversity Scheme", IEEE Photon. Technol. Lett., vol.18, pp. 2653-2655, 2006.

WA2-6

Risk-aware Virtual Network Embedding in Optical Data Center Networks

Weigang Hou, Lei Guo*, Cunqian Yu, Yue Zong

School of computer science and engineering, Northeastern University, Shenyang, 110819, China
*guolei@cse.neu.edu.cn

Abstract: We developed the platform supporting real-time detection of risky virtual machines (VMs) in optical data center networks. To perform physical isolation between risky and ordinary VMs, we also proposed a risk-ware virtual network embedding heuristic.

Keywords: *optical data center network; virtual network embedding; risk detection and isolation*

I. INTRODUCTION

Data centers (DCs) are geographically distributed and the inter-DC traffic becomes bandwidth hungry. Since the optical interconnection (e.g., wavelength division multiplexing, WDM) has a high bandwidth provisioning, the optical data center network (ODCN) where DCs located at the edge of the optical backbone thus emerged. After virtualization, a service requirement owns a virtual network where the virtual machine (VM) node and the virtual link (VL) represent the requirement of computation and communication, respectively. Therefore, a service requirement is satisfied when its virtual network has successfully been embedded into the substrate ODCN. This virtual network embedding (VNE) operation on a service requirement includes two components: the VM mapping that puts a VM into an appropriate DC, and the VL mapping that finds one substrate path for a VL.

For VNE processes, with the increasing dependency of cloud computing, the ODCN security becomes very anxious. Designing a risk assessment based on historical data becomes the mainstream of recognizing VM threat and vulnerability. However, the existing models of risk assessment had a very limited adaption to a dynamic cloud computing environment, and the study of risk-aware VNE processes in ODCNs still remains untouched.

Therefore, with an assorted model of risk assessment, we first develop an ODCN platform through the establishment of different types of VMs so that potentially risky VMs will be recognized in a particular future time epoch based on historical data. We then design an off-line risk-ware VNE heuristic where potentially risky VMs are consolidated into specialized servers, in order to perform the physical isolation between potentially risky and ordinary VMs in future. As a result, the other servers holding ordinary VMs will be safe.

II. RISK ASSESSMENT MODEL

There are three elements in our risk assessment model: *asset* is the valuable resource in terms of the VM size, *threat* is the potential factor hazard to the ODCN security, and *vulnerability* is the resource preempted by threats. We consider every element is an integer lying between 1 and 5. For *threat*, the number of attacks on one type of VMs is measured within a period of time. We let F_j^i record the number of times the threat T_i appeared within the historical time period j, then the value of T_i is determined as follows: $T_i = 5$ (case: $F_j^i \geq 1, \forall j =$ 'one day'); $T_i = 4$ (case: $F_j^i \geq 1, \forall j =$ 'one week'); $T_i = 3$ (case: $F_j^i \geq 1, \forall j =$ 'one month'); $T_i = 2$ (case: $F_j^i \geq 1, \forall j =$ 'six months'); $T_i = 1$ (other cases). *Vulnerability* changes with the historical resource utilization of VMs. We let the average utilization ratio of the ith kind of VM resource (e.g., CPU) $Per_i = Ave_i / Max_i$. Here, Ave_i and Max_i are the average and maximal values of the ith kind of occupied resource in all VMs, respectively. The historical data of Ave_i and Max_i is obtained from PnP graph shown in Fig. 1(a), and we update the vulnerability V_i as follows: $V_i \leftarrow V_i + 1$ (case: $30\% < Per_i \leq 60\%$); $V_i \leftarrow V_i + 2$ (case: $60\% < Per_i \leq 80\%$); $V_i \leftarrow V_i + 3$ (case: $Per_i > 80\%$). If the updated value of V_i is bigger than 5, then $V_i = 5$. Next, the multiplication is utilized to compute risk values after the mapping relation $U(T) = \{V\}$ generates. Here, T is a threat and V is a set of vulnerabilities despitefully exploited by the threat T. As an example of the VM type with the asset A and mapping functions $U(T_1) = \{V_1\}$, $U(T_2) = \{V_2, V_3\}$ and $U(T_3) = \{V_4\}$, we first determine the set of threats $\{T_1, T_2, T_3\}$ and the set of vulnerabilities $\{V_1, V_2, V_3, V_4\}$ as mentioned above, and then this type of VMs will have four risk values: $R_1 = R(A, T_1, V_1) = \sqrt{T_1 \cdot V_1} \times \sqrt{A \cdot V_1}$, $R_2 = R(A, T_2, V_2) = \sqrt{T_2 \cdot V_2} \times \sqrt{A \cdot V_2}$, $R_3 = R(A, T_2, V_3) = \sqrt{T_2 \cdot V_3} \times \sqrt{A \cdot V_3}$, and $R_4 = R(A, T_3, V_4) = \sqrt{T_3 \cdot V_4} \times \sqrt{A \cdot V_4}$. Finally, potentially risky VMs with unacceptable risk values will be recognized in a particular future time epoch based on historical data.

III. RISK-AWARE VNE HEURISTIC (RVNE)

(Network model) In our study, the substrate ODCN infrastructure includes a set of optical cross-connect (OXC) nodes N and a set of bidirectional fiber links E within the optical backbone, as well as a set of DCs D at the edge of the optical backbone. Fiber links have the same number of wavelengths each with an initial bandwidth provisioning LC, and every DC has P networked servers each with an initial capacity SC. The entire system has T future time epochs, and accepts C types of different VMs during each future time epoch. Correspondingly, every service requirement can be represented by a 5-tuple: $<s,c,A_c,r_c,b_c>$ where s ($s \in N$) is a source OXC node, c is a type index, A_c is the asset proportional to the size r_c of type-c VMs, and b_c is the link bandwidth requirement lower than LC.

(Bound analysis) The potentially risky VMs with unacceptable risk values can be recognized in a particular future time epoch using our risk assessment model. Let M_c^t denote the number of type-c recognized risky VMs at a particular future time epoch t, then the total computing-resource requirement of recognized risky VMs is $\Re = \sum_T \sum_c \left(M_c^t \cdot r_c \right)$. Next, the minimal number of servers only holding risky VMs will be obtained by assuming those specified servers are replaced by one single server with aggregated computing resources so that there is no fragmentation, i.e., $\aleph_{min} = \Re / SC$. Finally, the maximal number of safe servers holding ordinary VMs is $\aleph_{max} = |D| \cdot P - \aleph_{min}$, which is the upper bound that will demonstrate the optimality of our RVNE whose pseudo code is shown in Table I.

(Our heuristic) The main step of executing RVNE is described as follows. At each future time epoch t: we first initialize the set δ of service requirements with C types of different VMs (line 3 of Table I). Here, $sr_{c,t}^i$ denotes the ith service requirement along with a type-c VM during the future time epoch t. Then, the service requirements along with risky VMs recognized by our risk assessment model will be deleted from δ and putted into the set ϖ of risky service requirements. This means that we separately make different VNE operations on ordinary and risky service requirements. According to lines 5-11, ordinary service requirements are first processed by $1^{\#}$ VNE operation mainly including expected DC selection, traffic grooming and server consolidation. More specifically, we first find a destination DC located at the edge of the optical backbone for an ordinary service requirement, and this expected DC dc should have one server p that has enough residual capacity rc_p of holding the corresponding VM. Then, a substrate path will be determined from the source OXC node to this expected DC using traffic grooming which can make several substrate paths share the same fiber link for the purpose of link bandwidth saving. Finally, the corresponding VM $vm_{c,t}^i$ ($vm_{c,t}^i \in sr_{c,t}^i$) will be consolidated into the first server that can hold this VM within the expected DC. Following, according to lines 12-17, risky service requirements are processed by $2^{\#}$ VNE operation which ensures that the server in the expected DC also should have a risky attribute or there is no VM launch in this server because one server will become dangerous once it has carried a risky VM, though using the similar process with $1^{\#}$ VNE operation. As a result, each dangerous server will only carry risky VMs as many as possible during the process of $2^{\#}$ VNE operation, and the other servers with only holding ordinary VMs during the process of $1^{\#}$ VNE operation will become safe.

As shown in Table 1, the time complexity of executing $1^{\#}$ VNE operation for each ordinary service requirement is $O(|D| \cdot P + |N| \cdot |D| + P)$, then the total time complexity of executing $1^{\#}$ VNE operation is $O[|\delta| \cdot (|D| \cdot P + |N| \cdot |D| + P)]$. Similarly, the time complexity of executing $2^{\#}$ VNE operation is $O[|\varpi| \cdot (|D| \cdot P + |N| \cdot |D| + P)]$.

TABLE I
PSEUDO CODE OF RVNE

Input: N, E, D, P, LC, SC, T, Settings of service requirements with C types of different VMs during each future time epoch, M (Number of each type of VMs during a particular future time epoch), W (Number of wavelengths per fiber link)
Output: $\aleph$ (Number of safe servers), θ (Total revenue of accepting services requirements)
1: Initialize $\aleph \leftarrow 0$, $\theta \leftarrow 0$, and the set of risky services requirements $\varpi \leftarrow Null$;
2: **For** future time epoch $t = 1$, $t \le T$, $t \leftarrow t+1$ **then**
3:   Initialize the set of service requirements with C types of different VMs $\delta \leftarrow \{sr_{c,t}^i | c \in [1,C], i \in [1,M]\}$;
4:   Update two sets: $\varpi \leftarrow \{sr_{c^*,t}^j | c^* \in [1,C], j \in [1,M_{c^*}^t]\}$ and $\delta \leftarrow \delta - \varpi$;
5:   **While** $\delta \ne Null$ **then**
6:     $sr_{c,t}^i \leftarrow \delta.top()$;
7:     Expected DC: $dc \leftarrow \{dc | dc \in D, p \in dc, rc_p \ge r_c\}$; /*Time complexity: $O(|D| \cdot P)$ */
8:     **VL mapping:** *traffic grooming*($s \rightarrow dc | s \in sr_{c,t}^i$); /*Time complexity: $O(|N| \cdot |D|)$ */
9:     **VM mapping:** *server consolidation*($vm_{c,t}^i \rightarrow p | vm_{c,t}^i \in sr_{c,t}^i$); /*Time complexity: $O(P)$ */
10:     $\delta.pop()$;
11:   **End While**
12:   **While** $\varpi \ne Null$ **then**
13:     $sr_{c^*,t}^j \leftarrow \varpi.top()$;
14:     Expected DC: $dc \leftarrow \{dc | dc \in D, p \in dc, rc_p \ge r_{c^*}, p\_property = risky$ or $dc \in D, p \in dc, rc_p = SC\}$;
15:     Execute lines 8 and 9;
16:     $\varpi.pop()$;
17:   **End While**
18: **End For**

IV. EXPERIMENT AND SIMULATION RESULTS

We first recognize potentially risky VMs with unacceptable risk values in a particular future time epoch through our risk assessment model embedded into the experimental ODCN platform where five types (i.e., $C=5$) of VMs (from guest0 to guest4) orderly have 3, 1, 2, 2, 1 CPUs, the memory settings of them are {2048, 1024, 1024, 2048, 2048}, and the assets of them are $A_1=5$, $A_2=1$, $A_3=2$, $A_4=4$, $A_5=3$, respectively. As shown in Figs. 1(b)-1(f), assume $T=24$, the 'guest0' type of VMs has two unacceptable risk values at 12th and 22th future time epochs, etc. Thus, risky VMs are recognized, and then our RVNE will be executed to perform the physical isolation between risky and ordinary VMs.

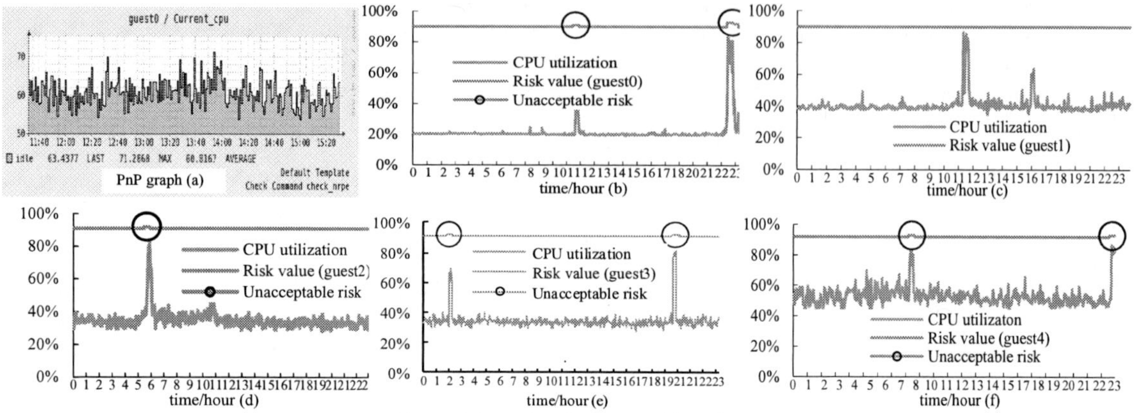

Fig. 1. PnP graph and experimental results.

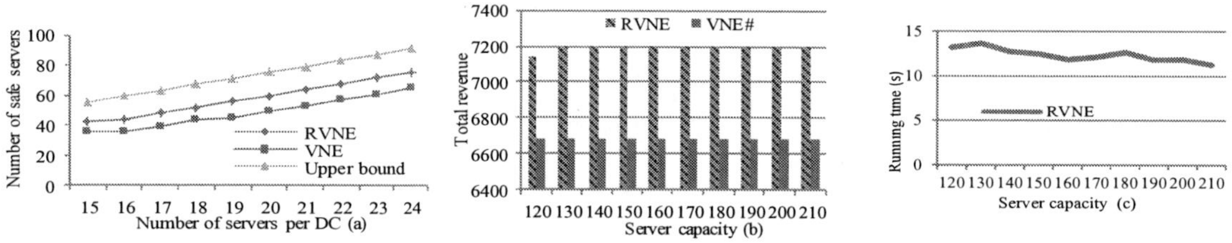

Fig.2. Simulation results.

To demonstrate the effectiveness of RVNE through simulations, we compare the number of safe servers between VNE benchmark [1] and RVNE, and compare the revenue between VNE benchmark with considering denial of risky service requirements (i.e., VNE#) and RVNE. The ODCN topology is the 14-node ($|N|$=14) and 21-edge ($|E|$=21) NSFNET with four DCs ($|D|$=4). We make a normalization processing for the size of five types of VMs: $r_1 = 3(\text{CPUs}) \cdot (2048/1024)(\text{Memory}) = 6$; $r_2 = 1 \cdot (1024/1024) = 1$; $r_3 = 2 \cdot (1024/1024) = 2$; $r_4 = 2 \cdot (2048/1024) = 4$; $r_5 = 1 \cdot (2048/1024) = 2$. The numbers of five types of service requirements per future time epoch follow {20, 20, 20, 20, 20}, i.e., M=20. We let LC=OC-96, W=12, and all service requirements have the same requirement of link bandwidth, i.e., OC-1. Note that, we serve the type of VMs with the largest asset (i.e., guest0) in prior since the revenue is one of our focuses. In Fig. 2(a), given SC=120, the number of safe servers obtained by RVNE and VNE benchmark is compared with the upper bound mentioned in section III, with the increasing P starting from $P_{\min} = \left\lceil T \cdot \sum_{c \in [1,C]} (M \cdot r_c) \right\rceil / (SC \cdot |D|) = 15$ (i.e., at least 60 servers in the infrastructure). The simulation results show that RVNE has the higher number of safe servers with the improvement ratio 20.5% over VNE benchmark. This is because dangerous servers only carry risky VMs as many as possible so that the other servers will become safe in RVNE. While in VNE benchmark, without the physical isolation between risky and ordinary VMs, many servers will become dangerous once they carry a risky VM. Moreover, the number of safe servers $\aleph$ obtained by RVNE is very close to the upper bound $\aleph_{\max}$ with the convergence ratio 79.3%, which also demonstrates the optimality of RVNE. In addition, the protection ratio of RVNE can arrive to $\aleph/(P \cdot |D|) \approx 74.1\%$. Next, in Fig. 2(b), given P=15, the revenue of RVNE and VNE# is compared with the increment of SC starting from $SC_{\min} = \left\lceil T \cdot \sum_{c \in [1,C]} (M \cdot r_c) \right\rceil / (P \cdot |D|) = 120$. The simulation results show that RVNE has the higher revenue with the improvement ratio 7.8% over VNE# because it lets dangerous servers to carry risky VMs without denial of services. Additionally, the revenue basically remains unchanged because there is no service requirements blocked due to scare resource provisioning. Finally, RVNE has an acceptable running time in the level of second as shown in Fig. 2(c).

V. CONCLUSIONS

In this paper, we established an ODCN platform to recognize risky VMs in a particular future time epoch. Then, a novel RVNE heuristic was proposed to perform the physical isolation between risky and ordinary VMs. The simulation results demonstrated that RVNE well guaranteed the ODCN security with the help of our risk assessment model.

REFERENCES

[1] J. Riera, A. Tzanakaki, A. Antonescu, et. al, "Virtual infrastructures as a service enabling converged optical networks and data centres,"Optical Switching and Networking, Vol. 14, No. 3, pp. 197–208, 2014.

Simultaneous Measurement of Curvature and Temperature Based on Thin Core Ultra-Long Period Fiber Grating

Wenjun Ni[1], Ping Lu[1,*], Chao Luo[1], Xin Fu[1], Deming Liu[1] and Jiangshan Zhang[2]

[1]National Engineering Laboratory for Next Generation Internet Access System, School of Optical and Electronic Information, Huazhong University of Science and Technology, Wuhan 430074, China;
[2]Department of Electronics and Information Engineering, Huazhong University of Science and Technology, Wuhan 430074, China
Email:pluriver@mail.hust.edu.cn

Abstract: We proposed and experimentally demonstrated a curvature sensor based on a thin core ultra-long period fiber grating, with high curvature sensitivity of $97.77dB/m^{-1}$. Simultaneous measurement of curvature and temperature without cross sensitivity are also demonstrated.

Keywords: fiber sensor; thin core long period fiber grating; Mach-Zehnder

I. INTRODUCTION

Temperature and curvature are of great importance in the fields of structure health monitoring and distributed sensing. Compared with conventional sensors, optical fiber sensors have shown dominant advantages such as compact structure, high sensitivity, low cost and immunity to electromagnetic interference. Most of fiber sensors such as long period fiber gratings (LPFGs)[1], inline Mach-Zehnder interferometer (MZI)[2], and few modes fiber (FMF)[3] etc have been widely applied to curvature measurement. To best of our knowledge, simultaneous measurement of temperature and curvature has attracted more attention than single parameter measurement[4]. Recently, different kinds of special fiber has been used for writing LPFG, such as photonic crystal fiber (PCF)[5], FMF[6] and TCF[7]. However, these LPFGs haven't been researched more in simultaneous measurement of dual-parameters. But to best of our knowledge no one has used TCF to inscribe ULPFG. In the latest report, most of special fiber mentioned above can also be used for dual-parameters sensing, but it is still necessary to be cascaded with other interference structure [8]. Thin core ultra-long period fiber grating (TC-ULPFG) can not only achieve dual-parameters simultaneous measurement, but also can avoid cross sensitivity.

In this paper, a novel fiber sensor based on a thin core ultra-long period fiber grating (TC-ULPFG) is proposed and experimentally demonstrated. Four resonant wavelengths are generated by different order of cladding modes and different diffraction order of TC-ULPFG. Therefore, the two measurands of curvature and temperature can be measured simultaneously due to each resonant wavelength corresponding to the different sensitivities. Moreover, cross sensitivity can be well overcome by the proposed fiber sensor.

II. CONFIGURATION OF SENSOR HEAD AND SPECTRUM ANALYSIS

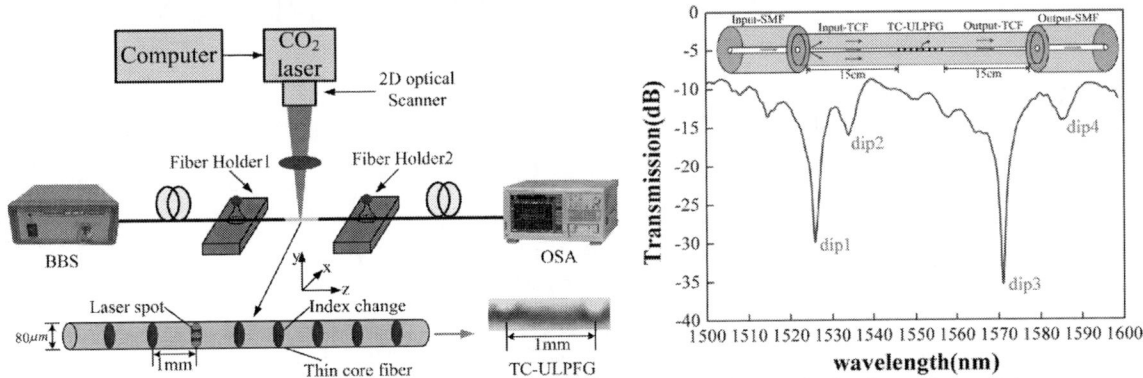

Fig. 1. (a) Fabrication process of the TC-ULPFG. (b) Schematic diagram of the sensor head and transmission spectrum

The TC-ULPFG is inscribed by the focused CO_2 laser pulse (SYNRAD, 48-series) on one side of TCF using point-by-point method, as shown in Fig. 1(a). The diameter of the TCF and grating pitch are $80\mu m$ and $1mm$, respectively. It can be seen from the inset that refractive index change is across whole the cross section of the fiber and has an unsymmetrical distribution. The transmission spectrum of this single optical device can be obtained real time on optical spectrum analyzer (OSA, Yokogawa AQ6370c). The schematic diagram of the sensor head with its transmission spectrum is shown in Fig. 1(b). The proposed structure is composed of five sections named Input-SMF(ISMF), Input-TCF(ITCF), TC-ULPFG, Output-TCF(OTCF) and Output-SMF(OSMF), which can be come true only by the fusion splicer(Fujikura, FSM-60S). The length of ITCF and OTCF are cut for 15cm. Cladding modes will be excited in ITCF at the first splicing point because of the mode field mismatch between ISMF and ITCF, and will suffer completely loss in ITCF because the length of ITCF is 15cm. Cladding modes will be excited again at the grating region due to the mode coupling between core and cladding modes in LPFG[9], and will disappear for the same reason. Therefore, an inline MZI can't be formed without cladding modes leading to the transmission spectrum with no modal interference pattern, the transmission spectrum in Fig. 1(b) is only generated by TC-ULPFG. Four dominated dips in Fig. 1(b) are named as dip1, dip2, dip3 and dip4 from left to right, respectively. Different dips are caused by the loss of different cladding modes which is excited by TC-ULPFG. The extinction ratio of dip3 is nearly 25dB, which can be well used for the wavelength and intensity demodulation. Dip1 has the same linear polarization mode and different diffraction order with dip3. Multi-parameters sensing can be realized simultaneously due to the different sensitivities of each dip. Curvature and temperature are measured in the following experiment.

III. EXPERIMENTAL RESULTS AND DISCUSSIONS

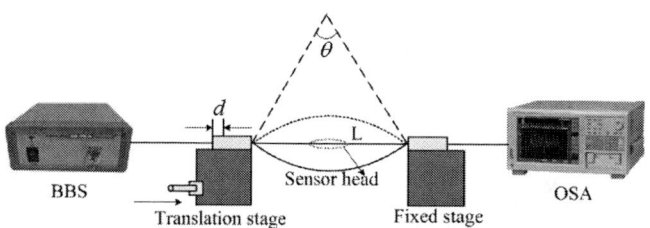

Fig. 2. The experimental setup of curvature measurement

The experimental setup of curvature measurement is displayed in Fig. 2. It contains broadband light source (BBS), OSA and two-dimensional (2D) translation stages. For the sensor head, the length of the TC-ULPFG is 3cm, which is kept straight line without external axial tension. The original interval of two stages is L=20cm. By adjusting the translation stage in Fig. 2, curvature measurement can be achieved. Fig. 3(a) illustrates the transmission spectrum with the curvature increasing from 0.424283m^{-1} to 0.547504m^{-1}. The sensor head with a length of 3cm is placed in the middle of the 2D translation stages. The extinction ratio of dip1 and dip3 decreases with the curvature increasing because part of the cladding modes leaks out. Thus, curvature can be detected by monitoring the intensity change of dip1 and dip3. When curvature increases from 0.424283m^{-1} to 0.547504m^{-1}, linear intensity variations of dip1 and dip3 with the sensitivities of 69.43dB/m^{-1} and 97.77dB/m^{-1} are achieved, as shown in Fig. 3(b). Different sensitivities of dip1 and dip3 are caused by the different order of cladding modes and different diffraction order. From Fig. 3(b), it can be seen that there are no wavelength shift and only intensity variation caused by dip1 and dip3.

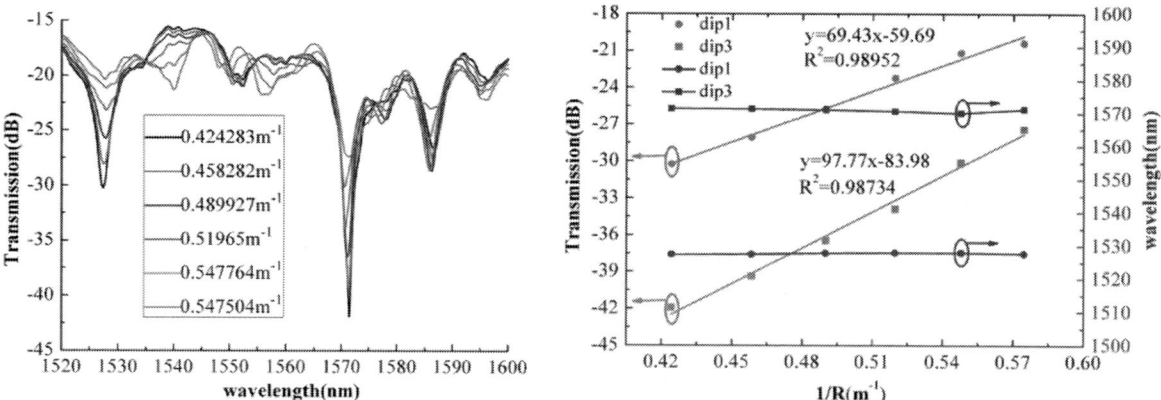

Fig. 3. (a) Transmission spectrum with curvature increasing from 0.424283m^{-1} to 0.547504m^{-1}. (b) Linear fit of dip1 and dip3 with sensitivities of 69.34 dB/m^{-1} and 97.77dB/m^{-1}, respectively.

Temperature sensing is realized by heating water solution from 25°C to 80°C with 5°C of the step interval. As shown in Fig. 4(a), the resonant wavelength of dip3 and dip4 undergoes a red shift with the temperature increasing from 25°C to 80°C. When temperature goes up, the resonant wavelength of dip3 and dip4 shifts to the long wavelength direction with sensitivities of 52.45pm/°C and 62.94pm/°C, respectively. From Fig. 4(b), it can be seen that the intensity of dip3 and dip4 are almost no variations, only several points have relatively big variation due to the light source shake. As a result, the measurand of temperature sensitivity can be obtained only by tracking resonant wavelength shift.

OECC/PS2016

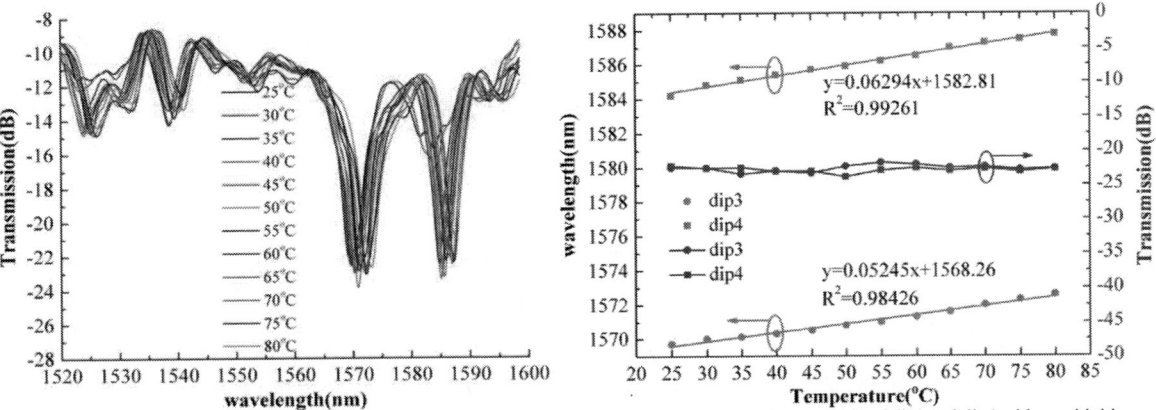

Fig. 4. (a) Transmission spectrum at the state of straight with temperature from 25°C to 80°C. (b) Linear fit of dip3 and dip4 with sensitivities of 52.45pm/°C and 62.94pm/°C, respectively.

From Fig. 3 and Fig. 4, by tracking the intensity variation and wavelength shift of the loss peak, simultaneous measurement of curvature and temperature can be achieved. Different loss peaks appear owing to the loss of different cladding modes excited by TC-ULPFG. Moreover, different cladding modes are corresponding to different thermal optic coefficient and elastic optical coefficient. Intensity and wavelength demodulation method is selected to measure curvature and temperature, respectively. Therefore, cross sensitivity can be overcome, then two different loss peaks are obtained for dual-parameters sensing, with the high curvature sensitivity of 97.77dB/m^{-1} by dip3 and temperature sensitivity of 62.94pm/°C by dip4.

IV. CONCLUSIONS

In conclusion, we have proposed and experimentally demonstrated a novel fiber sensor using a TC-ULPFG for simultaneous measurement of curvature and temperature. By tracking the power variation and wavelength shift of dip2 and dip3, curvature sensitivities of 97.77dB/m^{-1} and 69.43dB/m^{-1}, as well as temperature sensitivities of 52.45pm/°C and 62.94pm/°C are achieved. The proposed sensor shows many advantages including multi-parameters measurement, high sensitivities of curvature, easy fabrication, simple structure and low cost. Additionally, bending direction can be estimated at the same time because the inscribed method is point-by-point scanning on one side leading to the asymmetric distribution of refractive index in the TCF. Overall it has great potential for structural health monitoring and precision instrument processing.

ACKNOWLEDGMENT

This work is supported by a grant (No. 61275083, 61290315) from Natural Science Foundation of China and a grant (HUST: No.2014CG002) from the fundamental Research Funds for the Central Universities.

REFERENCES

[1] T. Yuan, X. Zhong, C. Guan, J. Fu, J. Yang, J. Shi, and L. Yuan, "Long period fiber grating in two-core hollow eccentric fiber," Optics Express, vol. 23, no. 26, pp. 33378-33385, 2015.

[2] Y. Gong, T. Zhao, Y. Rao, and Y. Wu, "All-fiber curvature sensor based on multimode interference," IEEE Photonics Technology Letters, vol. 23, no. 11, pp.679-681, 2011.

[3] J. Chen, P. Lu, D. Liu, J. Zhang, S. Wang, D. Chen, "Optical fiber curvature sensor based on few mode fiber," Optik-International Journal for Light and Electron Optics, vol. 125, no. 17, pp. 4776-4778, 2014.

[4] H. Gong, M. Xiong, Z. Qian, C. Zhao, X. Dong, "Simultaneous Measurement of Curvature and Temperature based on Mach-Zehnder Interferometer Comprising Core-offset and Spherical-shape Structures," IEEE Photonics Journal, vol. 8, no. 1, 2015

[5] F. Tian, Jiri Kanka, B. Zou, Kin Seng Chiang, and Henry Du "Long-period gratings inscribed in photonic crystal fiber by symmetric CO_2 laser irradiation," Optics Express, vol. 21, no. 11, pp. 13208-13218, 2013.

[6] B. Wang, W. Zhang, Z. Bai, L. Wang, L. Zhang, Q. Zhou, L. Chen, and T. Yan, "CO 2-laser-induced long period fiber gratings in few mode fibers," IEEE Photonics Technology Letters, vol. 27, no. 2, pp. 145-148, 2015.

[7] C. Fu, X. Zhong, C. Liao, Y. Wang, Y. Wang, J. Tang, S. Liu, and Q. Wang, "Thin-Core-Fiber-Based Long-Period Fiber Grating for High-Sensitivity Refractive Index Measurement," IEEE Photonics Journal, vol. 7, no. 6, pp. 1-8, 2015.

[8] Z. Wu, H. Zhang, P. P. Shum, X. Shao, T. Huang, Y. M. Seow, Y. Liu, H. Wei, Z. Wang, "Supermode Bragg grating combined Mach-Zehnder interferometer for temperature-strain discrimination," Optics Express, vol. 23, no. 26, pp. 33001-33007, 2015.

[9] E. M. Dianov, S. A. Vasiliev, A. S. Kurkov, O. I. Medvedkov, V. N. Protopopov, "IN-FTOER MACH-ZEHNDER INTERFEROMETER BASED ON A PAIR OF LONG-PERIOD GRATINGS,"in 22nd European Conference on Optical Communication(ECOC'96), pp. 65-68, 1996.

Metal-assisted high sensitivity micro Mach-Zehnder fiber temperature sensors

Kuan-Hao Lin[1], Nan-Kuang Chen[1,*], Shien-Kuei Liaw[2], and Wood-Hi Cheng[3]

[1]Department of Electro-Optical Engineering, National United University, Miaoli, 360, Taiwan
[2]Department of Electro-Optical Engineering, National Taiwan University of Science and Technology, Taipei, 106, Taiwan
[3]Graduate Institute of Optoelectronic Engineering, National Chung Hsing University, Taichung, 402, Taiwan
*Corresponding author: nankuang@gmail.com

Abstract: The bimetal is sensitive to ambient temperature variation and is attached against the miniaturized abrupt-tapered Mach-Zehnder interferometer for high sensitivity temperature sensing using an 8-mm-long highly Er/Yb codoped fiber. The two abrupt tapers are made using a micro hydrogen flame where the first abrupt taper coverts partial core mode into cladding modes which will return to meet the residual core mode at the second abrupt taper to achieve interferences. The maximal wavelength shift can be efficiently enhanced to be 12.78 nm in 4 °C temperature variations at around 1610 nm wavelength.

Index words: temperature sensors, Mach-Zehnder interferometer, abrupt tapers.

I. INTRODUCTION

In-line fiber temperature sensors are essential and fundamental components for sensing systems and have been extensively studied using various kinds of fiber components like fiber Bragg gratings, long period gratings, fiber couplers, short-pass filters, Fabry-Perot interferometers, Mach-Zehnder interferometers, Michelson interferometers, and so on [1-5]. Obviously, in contrast to the wavelength-coupling or wavelength-dispersive-guiding devices, the interferometric sensors can provide higher temperature sensitivity since the phase or index variations along the optical paths in sensors can be more accurately determined by optical interferences. However, the conventional interferometric sensors usually have the footprint of at least longer than a few centimeters especially for those having two physically separated optical arms like Mach-Zehnder or Michelson interferometers. Sometimes, a polarization controller could be needed for good interferences and high extinction ratios, which makes the footprint issues get even worse. This could be disadvantageous for satisfying the purposes like high integration or background noise reduction for some high precision measurement applications. Therefore, miniaturized fiber sensors without discrete components are important and is becoming the trend for optical internet of things.

In this work, the abrupt-tapered Mach-Zehnder interferometer (AT-MZI) is made on a short highly Er/Yb codoped fiber (EYDF) for an 8-mm-long fiber temperature sensors based on core-cladding modes interferences. For the AT-MZI, minimum device length can be down to less than 180 μm for picoliter microsensing [6]. Without using two physically separated fibers for sensing and reference arms, the background noise coming from the distributed temperature difference between two arms can be cancelled to improve the accuracy. The AT-MZI can be easily made by immersing the unjacked fiber into a micro hydrogen flame in which the heating temperature at the edge is higher than that at the inner portion. The bilateral ends of the heated fiber is stretched outwards to introduce two nonadiabatic tapers so that partial core mode can then be converted into cladding modes by the first taper and recombined again to generate interferences at the second taper which is about 4 mm away from the first one. The typical extinction ratio can be above 30 dB and the optical losses can be below 3 dB. In order to more rapidly and efficiently accumulate the optical path length difference (OPLD) within a short fiber length, the highly EYDF is employed to more clearly separate the core and cladding modes in the interferometer [6]. This AT-MZI can be served as a temperature sensor since the OPLD is dependent on the external temperature variations [7]. However, the temperature sensitivity is 0.013 nm/°C only due to the low thermo-optic and thermal expansion coefficients of the fiber silica glasses. In reference 7, it is interesting that the wavelength shift is much more sensitive to the bending angle and the sensitivity can reach 1.275 nm/°C. This is because a small bending angle variation can substantially change the cladding mode order number or the propagation length to obtain more OPLD. Accordingly, a thin and 1.5-cm-long bimetal strip with a thickness of 97 μm, made by two thin metals with very different thermal expansion coefficients, is used to change the bending angle of the AT-MZI when its two ends is individually fixed on the precisely controlled TE-cooler and the fiber abrupt tapers. The bimetal curls with ambient temperature variation to change the bending angle and OPLD as well. By doing so, The wavelength shift can reach to 12.78 nm in 4°C temperature variations at the wavelength around 1610 nm. Explicitly, the temperature sensitivity of this metal-assisted AT-MZI has been successfully improved to 3.195 nm/°C. This temperature sensor is compact, integrated, cost-effective, simple, and is promising for the high precision sensing systems.

II. FABRICATION OF TEMPERATURE SENSORS BASED ON AT-MZI

To make the metal-assisted temperature sensors, a bimetal strip with a thermal expansion coefficient of $15.5 \times 10^{-6}/°C$, a density of 8 g/cm³, and a dimension of 1.5 cm in length and 1.95 mm in width and 97 μm in thickness is chosen. The working temperature is from -70°C to 350°C. One of its end is fixed on the TE-cooler while the contact length is 5 mm. The working principle of the bimetal is shown in Fig. 1(a) where two metal films with different coefficient of thermal expansion (CTE) are tightly combined. When the ambient temperature is increasing, the metal 1 with higher CTE will stretch longer than that of the other metal 2. Thus, the bimetal curls towards the side of metal 2 and the bending angle is increasing with increasing temperature. Fig. 1(b) shows that the bending angle varies with heating temperature when the bimetal strip is fixed one the TE-cooler. The free end is then attached against the A, B, C points of the AT-MZI respectively to investigate the temperature sensitivity. A highly EYDF is used since the core size is smaller and the index difference between core and cladding is higher than the standard SMF-28 fiber [6]. The numerical aperture is 0.22. The erbium oxide and ytterbium oxide has a melting temperature of above 2600°C which is much higher than that of germanium oxide ~ 1500°C in SMF-28. Thus, the dopant is more difficult to be diffused for the EYDF when the fiber is heated by the hydrogen flame. The core and cladding boundary can be well-remained to be more clearly separated the propagating core mode and cladding modes. This is important to achieve a higher OPLD within a shorter fiber length. A sample of 8-mm-long AT-MZI was made by introducing two abrupt tapers by stretching the heated EYDF. The diameter of the two tapers is 49 and 53.47 μm and is denoted as point A and B, respectively, shown in Fig. 1(c). The free end of bimetal film is also placed at the center of untapered region, point C, to study the temperature sensitivity. The distance between two tapers is 4.11 mm determined by the size of micro hydrogen flame. It is important to keep the diameter difference to be less than 10 μm in order not to generate circular birefringence [8]. The broadband lightsource from superluminscent diodes covering 1250-1650 nm is launched into the AT-MZI and the corresponding spectral response is shown in Fig. 1(d). The best extinction ratio is 24.7 dB at around 1520 nm wavelength. The ripples at the oscillation curve, especially at the shorter wavelength, are due to the multiple interferences from the unwanted very high order cladding modes and can be further removed by modifying the shape of abrupt tapers. The two ends of AT-MZI are hanging by the clampers on the fiber tapering machine while the bimetal is placed right below the points A, B, C individually. A portion of bimetal with a contact length of 5 mm is fixed on the TE-cooler to control the heating temperature from 30°C - 34°C.

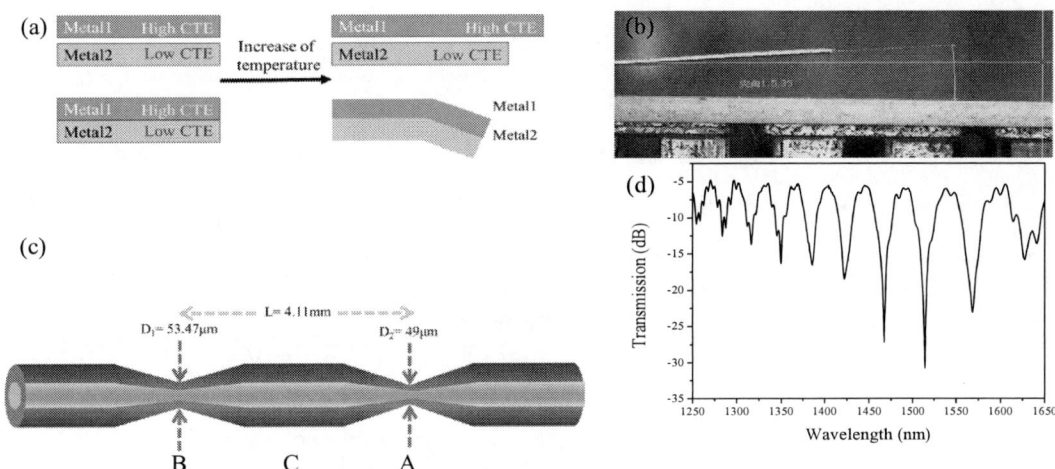

Fig. 1. (a) Working principle of the bimetal film. It curls towards the side of metal with a lower CTE when being heated. (b) The bending angle changes with temperature variations. (c) 8-mm-long AT-MZI is composed of two abrupt tapers on the highly Er/Yb codoped fiber. The free end of bimetal strip is attached against the A, B, C points respectively. (d) Spectral responses of the AT-MZI. The best extinction ratio is 24.7 dB.

III. MEASUREMENTS AND RESULTS

When the bimetal is heated, it curls upwards (upward bending) to change the position of points A, B, C to change the OPLD and blue-shift the resonant wavelengths as well. In many of our measurements, the maximal wavelength shift over 30°C - 34°C at points A and B are ranging from 3.31 to 3.89 nm for upward bending. The spectral responses are not shown here due to the length of pages. However, it is surprising that the interaction with C point can lead to the best performance of 12.78 nm, shown in Figs. 2(a)-2(b). The corresponding temperature sensitivity is thus improved to 3.195 nm/°C compared with the previous literatures in the references. The downward bending is also studied for points A, B, and C and similarly the temperature sensitivity is better at C point, shown in Figs. 2(c)-2(d). The maximal wavelength shift is 10.44 nm over 30°C - 34°C and the temperature sensitivity is 2.61 nm/°C. The temperature sensitivity at C point is much better than that at A and B points since the lateral displacement is larger under the same temperature variations. The diameter at points A and B (abrupt tapers) is around 50 μm whereas the diameter at point C (untapered region) is 125 μm. Accordingly, the bimetal end can give more displacement for point C for larger wavelength shift. This miniaturized metal-assisted AT-MZI temperature sensor uses the highly sensitive bimetal to change the bending of fibers for high sensitivity temperature sensing and could be promising for low-noise highly integrated sensing systems.

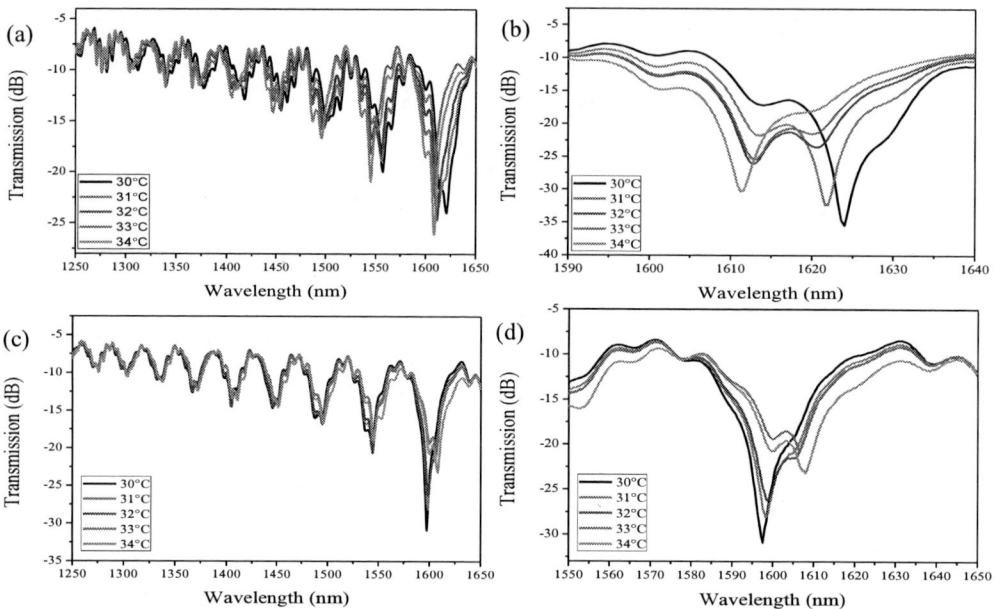

Fig. 2. Spectral responses of the metal-assisted AT-MZI temperature sensors over 30°C - 40°C at (a) point C and (b) its enlarged figure for upward bending and at (c) point C and (d) its enlarged figure for downward bending.

IV. CONCLUSIONS

In conclusion, the bimetal-assisted miniaturized abrupt-tapered Mach-Zehnder interferometer is made by intruding two abrupt tapers on an 8-mm-long highly Er/Yb codoped fiber. The temperature sensor is mounted on the bimetal to more efficiently improve the temperature to be above 3.195 nm/°C at around 1610 nm wavelength. This temperature is compact, simple, cost-effective, and highly sensitive to ambient temperature variations. It could be useful and promising for the highly integrated and low-noise sensing systems.

ACKNOWLEDGMENT

This work was partially supported by the Ministry of Science and Technology under the contract MOST 103-2221-E-239-002-MY3.

REFERENCES

[1] J. Yin, T. Liu, J. Jiang, K. Liu, S. Wang, S. Zou, and F. Wu, " Assembly-Free-Based Fiber-Optic Micro-Michelson Interferometer for High Temperature Sensing," IEEE Photon. Technol. Lett., Vol. 28, pp. 625-628 , 2016.

[2] A. Zhou et al., "Hybrid structured fiber-optic Fabry–Perot interferometer for simultaneous measurement of strain and temperature," Opt. Lett., Vol. 39, pp. 5267–5270, 2014.

[3] Y. Jung, S. Lee, B. H. Lee, and K. Oh, "Ultracompact in-line broadband Mach–Zehnder interferometer using a composite leaky hollow-optical- fiber waveguide," Opt. Lett., Vol. 33, pp. 2934–2936, 2008.

[4] S. J. Mihailov, " Fiber Bragg Grating Sensors for Harsh Environments," Sensors, Vol. 12, pp. 1898-1918, 2012.

[5] S. W. James and R. P. Tatam, " Optical fibre long-period grating sensors: characteristics and application," Meas. Sci. Technol. Vol. 14, pp. R49–R61, 2003.

[6] N. K. Chen, T. H. Yang, Z. Z. Feng, Y. N. Chen, and C. Lin, "Cellular-dimension picoliter-volume index microsensing using micro-abrupt-tapered fiber Mach-Zehnder interferometers," IEEE Photon. Technol. Lett. Vol. 24, pp. 842-844, 2012.

[7] N. K. Chen, Z. Z. Feng, J. J. Wang, S. K. Liaw, and H. C. Chui, "Interferometric interrogation of the inclination and displacement of tapered fiber Mach-Zehnder interferometers," IEEE Sens. J. Vol.13, pp.3437-3441, 2013.

[8] N. K. Chen, J. J. Wang, G. L. Cheng, W. H. Cheng, and P. P. Shum, "Fiber torsion sensor with directional discrimination based on twist-induced circular birefringence in unbalanced Mach-Zehnder interferometer," CLEO 2014 conference, San Jose, USA, Jun. 8-13, 2013. paper no.JW2A.35.

WA2-62

OECC/PS2016

The Liquid Length Effects on the Mach-Zehnder Interferometer Induced by Two Liquid Sections in a Photonic Crystal Fiber

Jia-Hong Liou and Chin-Ping Yu

Department of Photonics and Institute of Electro-optical Engineering,
National Sun Yat-sen University, Kaohsiung, Taiwan 80424, Taiwan
cpyu@faculty.nsysu.edu.tw

Abstract: *We investigate the liquid length effects on the interference properties of the MZI-based PCF sensor formed by filling two separated liquid sections. The temperature sensing sensitivities of variant liquid lengths are measured and discussed.*
Keywords: *Mach-Zehnder interferometer, liquid infiltration, photonic crystal fiber*

I. INTRODUCTION

Optical fiber sensors have attracted a lot of research attentions due to their compact size, low cost, simple configuration, high sensitivity, and immunity to electromagnetic interference [1-10]. In addition, optical fiber sensors can be used in distributed remote sensing and stand for harsh environment. As a result, in spite of physical and chemical sensing, optical fiber sensors can also be employed in structure strength monitoring and high-temperature sensing [3,4].

Several types of fiber sensors have been proposed based on evanescent wave sensing [1], fiber tapers [2], fiber gratings [3], Fabry-Pérot interferometer (FPIs) [4], and Mach-Zehnder interferometers (MZIs) [5-10]. Among them, the MZI-based fiber sensors demonstrate the impressive advantages of simple structures, easy fabrication, and high sensing sensitivities [5-11]. There are several methods to fabricate MZI-based fiber sensors, such as using taper regions [5], a pair of long-period fiber gratings [6] or core-offset splicing techniques [7] to simultaneously induce the core mode and cladding mode propagating along an optical fiber and recombine the two modes to form interference spectra. One can also utilize femtosecond laser micromaching [8] to form an air cavity on a single-mode fiber (SMF) or photonic crystal fiber (PCF) to realize a MZI fiber sensor. In addition, some researchers have demonstrated the MZI-based PCF sensors by using a fusion splicer to form two air-hole collapsed regions along a PCF [9]. Our group also shows that one can obtain a MZI-based PCF sensor by filling two separated liquid sections along a PCF [10]. In addition, the temperature sensing sensitivity of our proposed structure can be enhanced due to the infiltrated liquid [10].

In this paper, we will carefully investigate the effects of the liquid length on the MZI-based PCF sensor induced by two liquid sections. The interference properties of the MZI-based PCF sensors with different liquid lengths will be fabricated, measured, and discussed. In addition, the corresponding temperature sensing abilities for variant liquid lengths will also be presented.

II. INTERFERENCE MECHANISM AND FABRICATION

The structure diagram of our two-liquid-section-induced MZI PCF sensor is illustrated in Fig. 1(a). Input light launched from a broadband source (BBS) propagates along a SMF and is transmitted into a PCF with two separated liquid sections. Due to the liquid index is close to that of the silica, the mode field will be enlarged in the first liquid section. As light leaves the first liquid section, the core mode and cladding mode of the empty PCF will be simultaneously induced owing to the mode field mismatch. The two induced modes propagate along the empty PCF in the core region and cladding region, respectively, with different velocities. As they reach the second liquid section, the two modes will be combined together and transmitted to an OSA through another SMF, and we can obtain the interference spectrum resulted from the accumulated phase difference between the two modes.

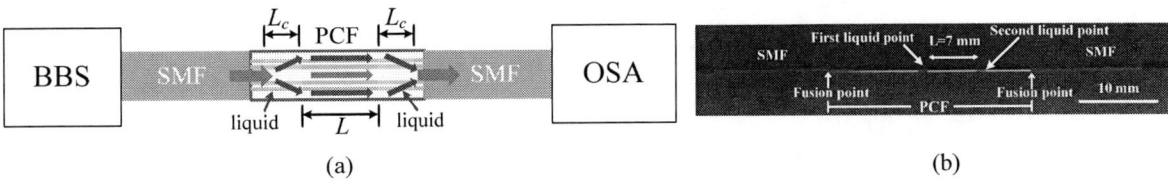

Fig. 1. (a) The structure diagram and (b) side view of the MZI-based PCF sensor induced by two separated liquid sections.

913

To form the MZI-based PCF sensor with two separated liquid sections, we first spliced a section of PCF with a SMF and placed the sample into a vacuum pumping chamber to perform the liquid infiltration. The end face of the PCF was first touched by the filling liquid to accomplish the infiltration of the first liquid section by capillary force. Please note that, to obtain different liquid lengths, the liquid infusion time should be carefully controlled. Then, we moved the filling liquid away from the PCF and slightly released the pressure of the chamber to push the first liquid section up to a desired distance. Following the same procedure, one can easily perform the infiltration of the second liquid section. Figure 1(b) shows the side view of the fabricated MZI-based PCF sensor with two separated liquid sections. The PCF we utilized is the commercial large mode area PCF (LMA-10) from *NKT Photonics* A/S. The filling liquid is from Cargille Laboratories with the liquid index n_q smaller but close to the silica index to efficiently enlarge the guided mode field. One can clearly observe two separated liquid sections along the PCF as indicated in Fig. 1(b). The liquid length L_c is about 1mm and the distance between the two liquid sections is $L = 7$mm.

III. RESULTS AND DISCUSSION

We first consider the sample with the liquid length $L_c = 1$mm and $L = 26$mm. The measured interference spectrum is demonstrated in Fig. 2(a). One can observe clear interference fringe between 1425nm and 1525 nm, which is attributed to the phase difference between the induced core mode and cladding mode. The corresponding free spectral range (FSR) is 33.0nm. We then fabricated another two samples with the same empty PCF length L but different liquid lengths, $L_c = 1.5$mm and $L_c = 2$m, by carefully controlling the filling time. The measured spectra of the two samples are shown in Figs. 2(b) and 2(c), respectively. Similar interference spectra can also be observed, which reveals that the lengthened liquid sections can also function as couplers to induce and combine the core mode and cladding mode of the empty PCF. The measured FSRs as $L_c = 1.5$mm and $L_c = 2$mm are 32.8nm and 32.7nm, respectively, which are very close to the FSR as $L_c = 1$mm. As a result, we can say that the liquid length has almost negligible effects on the interference characteristics of the MZI-based PCF sensor with two separated liquid sections.

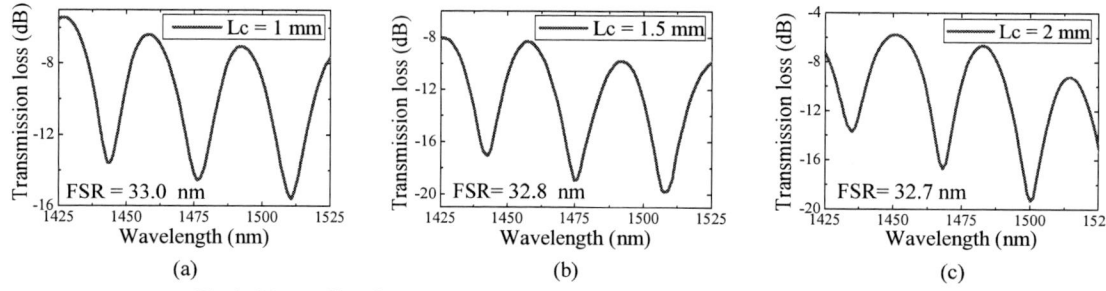

Fig. 2. Measured interference spectra as $L = 26$mm and $L_c = $ (a) 1mm, (b) 1.5mm and (c) 2mm.

To verify the effects of the liquid length on the temperature sensing properties, we consider two samples with $L_c = 1$mm and $L_c = 1.8$mm. The empty PCF length L is fixed at 26mm. During the raise of the operation temperature from 25°C to 35°C, the interference spectra of the two samples are recorded every 2°C and plotted in Figs. 3(a) and 3(b), respectively. It can be observed that the interference dips move toward shorter wavelength owing to the thermo-expansion coefficient of the liquid is larger than that of silica, which reduced the effective MZI length. In addition, owing to the negative thermo-optic coefficient of the filling liquid, the raised temperature decreases the liquid index. Therefore, the refractive index difference between the core and the liquid-filled cladding are enlarged, which reduces the guided mode field in the liquid section and the power of the induced cladding mode. As a result, the extinction ratio of the interference spectrum in Fig. 3(a) is decreased with the increased temperature, and the same phenomenon can also be observed in Fig. 3(b).

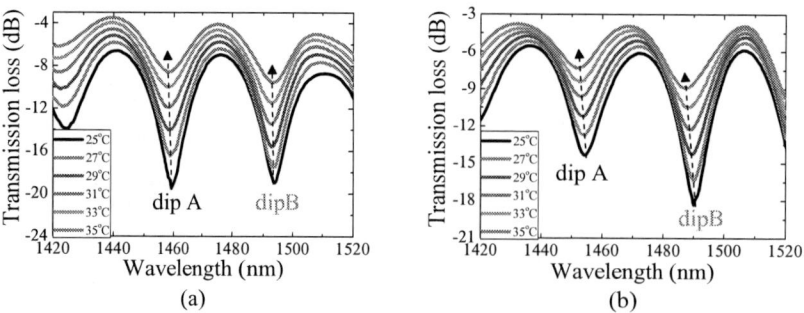

Fig. 3. The measured spectra with variant temperature as (a) $L_c = 1$mm and (b) $L_c = 1.8$mm.

However, if we record the positions of dip A in both Figs. 3(a) and 3(b) for variant temperature, the liquid length demonstrates very important influence. In Fig. 4(a), the black squares and red circles represent the dip positions as L_c = 1mm and L_c = 1.8mm, respectively. By applying linear fitting, the measured sensing sensibilities are 0.13nm/°C and 0.223nm/°C, respectively. It can be seen that the MZI-based PCF temperature sensor with longer liquid length possesses larger temperature sensing sensitivities. This is because longer liquid sections can make the MZI length shorter due to the larger thermal expansion, which raises the sensing sensitivity. We have also measured the results for dip B in Fig. 4(b). Similar results can be obtained, and the temperature sensing sensibilities for L_c = 1mm and L_c = 1.8mm are 0.08nm/°C and 0.281nm/°C, respectively. Again, longer liquid sections demonstrate more sensitive temperature sensing ability as shown in Fig. 4(b).

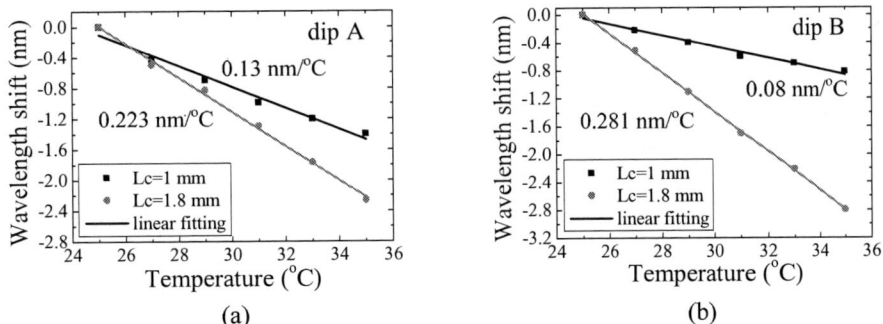

Fig. 4. The measured dip positions of (a) dip A and (b) dip B for L_c = 1mm and L_c = 1.8mm with variant temperature.

IV. CONCLUSIONS

We have fabricated MZI-based PCF sensor by infiltrating two separated liquid sections into a PCF. The measured results shows that the two liquid sections can indeed function as couplers to simultaneously induce and combine the core mode and cladding mode of the empty PCF to form clear interference spectra. We have also fabricated samples with different liquid lengths. The interference mechanism is kept the same for different liquid lengths. However, MZI-based PCF sensor with longer liquid length experience larger thermal expansion and possesses higher temperature sensing sensitivity.

ACKNOWLEDGMENT

This work was supported by the National Science Council of Taiwan under Grants No. NSC101-2221-E-110-074-MY3 and No. 104-2221-E-110 -066.

REFERENCES

[1] S. K. Khijwania, K. L. Srinivasan, and J. P. Singh, "An evanescent-wave optical fiber relative humidity sensor with enhanced sensitivity," Sens. Actuators B, vol. 104, pp. 217-222, 2005.

[2] L. Zhang, F. Gu, J. Lou, X. Yin, and L. Tong, "Fast detection of humidity with a subwavelength-diameter fiber taper coated with gelatin film," Opt. Express, vol. 16, pp. 13349-13353, 2008.

[3] J. M. Lopez-Higuera, L. R. Cobo, A. Q. Incera and A. Cobo , "Fiber optic sensors in structural health monitoring," J. Lightw. Technol. , vol. 29, pp. 587-608, 2011.

[4] Y. Zhang, L. Yuan, X. Lan, A. Kaur, J. Huang, and H. Xiao, "High-temperature fiber-optic Fabry-Perot interferometric pressure sensor fabricated by femtosecond laser," Opt. Lett. , vol. 38, pp. 4609-4612, 2013.

[5] Z. Tian, S. S.-H. Yam, J. Barnes, W. Bock, P. Greig, J. M. Fraser, H.-P. Loock, and R. D. Oleschuk, "Refractive index sensing with Mach-Zehnder interferometer based on concatenating two single-mode fiber tapers," IEEE Photon. Technol. Lett. , vol. 20, pp. 626-628, 2008.

[6] J. F. Ding, A.P. Zhang, L. Y. Shao, J. H. Yan, and S. L. He, "Fiber-taper seeded long-period grating pair as a highly sensitive refractive-index sensor," IEEE Photon. Technol. Lett. , vol. 17, pp. 1247-1249, 2005.

[7] D. W. Duan, Y. J. Rao, L. C. Xu, T. Zhu, D. Wu, and J. Yao, "In-fiber Mach–Zehnder interferometer formed by large lateral offset fusion splicing for gases refractive index measurement with high sensitivity," Sens. and Actuators B: Chemical, vol. 160, pp. 1198-1202, 2011.

[8] Y. Wang, M. Yang, D. N. Wang, S. Liu, and P. Lu, "Fiber in-line Mach-Zehnder interferometer fabricated by femtosecond laser micromachining for refractive index measurement with high sensitivity," J. Opt. Soc. Am. B, vol. 27, pp. 370-374, 2010.

[9] H. Y. Choi, M. J. Kim, and B. H. Lee, "All-fiber Mach-Zehnder type interferometers formed in photonic crystal fiber," Opt. Express, vol. 15, pp. 5711-5720, 2007.

[10] J. H. Liou and C. P. Yu, "All-fiber Mach-Zehnder interferometer based on two liquid infiltrations in a photonic crystal fiber," Opt. Express, vol. 23, 6946-6951, 2015.

OECC/PS2016

High-sensitivity humidity sensor composed of optical fiber coated with sol–gel derived porous silica

Nobuaki Tsuda, Hideki Fukano, and Shuji Taue
Graduate School of Natural Science and Technology,
Okayama University,
3-1-1 Tsushima-naka, Kita-ku, Okayama 700-8530, JAPAN
fukano@okayama-u.ac.jp

Abstract: We fabricated a humidity sensor by connecting single-mode–multimode–single-mode fibers and coating the sensing part with sol–gel derived porous silica, achieving a high sensitivity (0.748 nm/% at 0.4% relative humidity) and quick response.

Keywords: High sensitivity, humidity sensor, porous silica, Multimode interference

I. INTRODUCTION

Humidity sensors have been used in both industrial and environmental applications. In particular, low-humidity sensors have been employed in the fabrication process of products such as biomedical devices, electronic devices, and purified chemical gas [1, 2]. The majority of humidity sensors are capacitive and resistive [3][4]. However, since these operate using electrical signal in the sensing region, they are not suitable for use in inflammable and explosive environments. By contrast, optic fiber sensors only use a weak optical signal in the sensing region and have many distinctive characteristics such as high immunity to electromagnetic interference, strong tolerance to corrosive chemical substances, non-metallic conductivity, small size, and low weight. Therefore, they can be an ideal tool for sensing many physical properties. One of these properties is humidity [6, 7]. Z. Zhao et al. measured humidity by using a U-bend optical fiber, which was coated with a porous silica film doped with methylene blue using the sol–gel process. They estimated humidity by measuring the humidity-dependent absorbance spectra. They reported that both response and recovery can reach 90% of full scale within 10 s. A. A-Herrero et al. have side-polished a single-mode fiber to the vicinity of the core, and then attached a TiO_2 porous columnar nanostructure by electron beam evaporation. The refractive index (RI) of the porous TiO_2 column increased when it absorbed the water. Because the evanescent wave from the single-mode fiber was coupled to the TiO_2 nanostructure, transmission response exhibited sharp resonance whose wavelengths shifted with relative humidity (RH). The sensor sensitivity was 0.5 nm/% RH for low RH and 0.025 nm/% RH for 30–80% RH.

In this study, we fabricated a sensor structure consisting of an unclad multimode fiber (MMF) sandwiched between single-mode fibers (SMF), known as the SMS structure. The sensing part of the SMS structure was coated with a porous silica film using the sol–gel process. This sensor measures humidity by detecting changes in the RI of porous silica when absorbing moisture. The sensor has a simple structure with low cost and high sensitivity, and can operate in a wide humidity range. The sensor sensitivity is 0.748 nm/% RH even at 0.4% RH—the highest value reported so far—and 0.0278 nm/% RH at 85% RH. Response and recovery times, estimated from the intensity variation between 10 and 90% at humidity changes between 48.3% RH and 79.8% RH, are as short as 4.7 and 6.7 s, respectively.

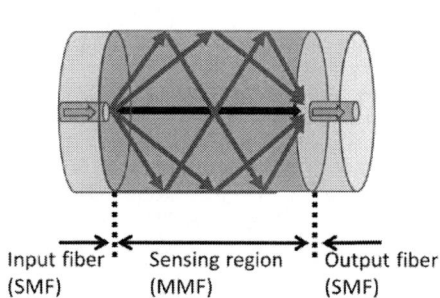

Fig. 1. Single-mode–multimode–single-mode fiber structure.

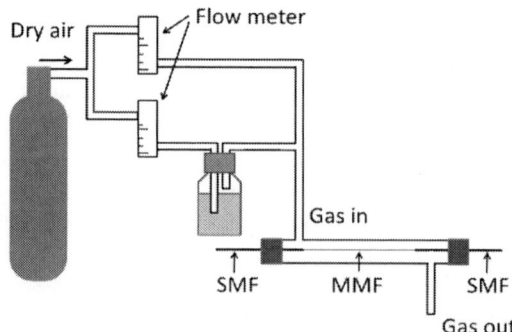

Fig. 2. Experimental setup.

916

II. OPERATION PRINCIPLE

In the SMS structure, when light enters from the input SMF to the MMF, it is diffracted and then transmitted as multimode light with total reflection at the outer surface of the MMF (see Fig. 1). The transmitting multimode lights interact, showing strong periodical interference. The interfered light is released from the output port of the SMF with periodic interference signal in the wavelength domain. There is an evanescent wave at the total reflection points. The penetration depth of the evanescent wave changes when the RI of the surrounding medium varies. Therefore, the phase of the light changes according to the Goos–Hanschen shift of evanescent waves, resulting in a wavelength shift of the interference signal [8]. As a result, this SMS structure becomes a high sensitivity RI sensor.

Many nanoscale pores in porous silica absorb moisture, then precipitating the water in the pores mainly by means of capillary condensation phenomena, inducing an increase in effective RI of porous silica. However, desorbing the moisture reduces its effective RI. By applying the porous silica at the outer surface of the MMF, the SMS structure RI sensor can work to measure humidity. Porous silica was fabricated by dip-coating the silica gel raw material produced by the sol–gel process using tetraethyl orthosilicate (TEOS), water, and ethanol, with hydrochloric acid as a catalyst [9].

III. EXPERIMENTAL SETUP

Figure 2 shows the experimental setup. Dry air from a cylinder is divided into two different ways: one becomes direct dry air whose flow is regulated with a meter; the other becomes saturated water vapor produced by flowing regulated dry air through a water bottle bubbler. Various RH of the tested water vapor are obtained by controlling two flow meters through adjusting the flow of dry air and saturated water vapor. The tested water vapor is sent to a 20 mL glass tube in which the sensor fiber is fixed by silicon plugs on both ends. An amplified spontaneous emission light source and an optical spectrum analyzer are used for evaluating the transmission spectra. Response time characteristics are evaluated by measuring the transmission intensity of the fixed-wavelength light emitted from the tunable light source using an optical power meter.

IV. SENSITIVITY TO HUMIDITY

Measurements at different RH of water vapor were performed to investigate the sensitivity of the sensor to humidity. At first, dry air was flowed for 3 h to provide a stable initial condition, and then the humidity was increased. Each measurement was carried out after maintaining a 10 min flow at different humidity levels. Figure 3 shows an example of the transmission spectrum at different RHs. Figure 4 shows the relationship between RH and the dip wavelength of the spectrum. The dip wavelength increases with RH, and its change in the low humidity region is more pronounced than that at high RHs. When RH is 0.4%, the sensitivity (which is given by the slope of the curve in Fig. 4) is as high as 0.748 nm/% RH. This is 50% higher than that reported in a previous study [7]. At high humidity regions, this sensor shows a constant sensitivity of 0.0278 nm/% RH.

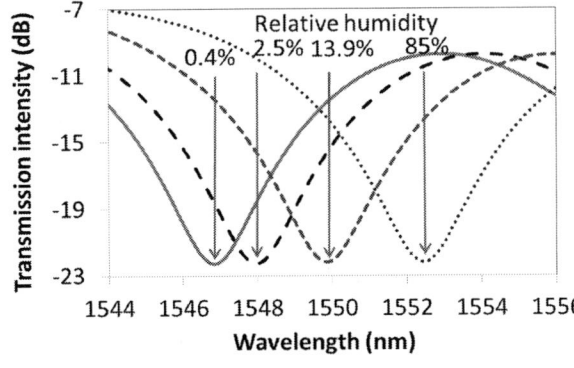
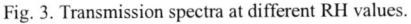

Fig. 3. Transmission spectra at different RH values.

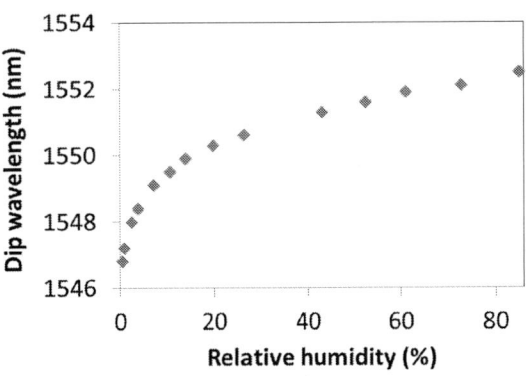

Fig. 4. Relative humidity dependence on dip wavelength.

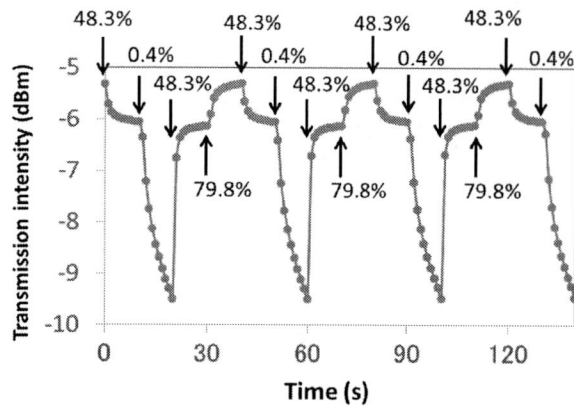

Fig. 5. Response at different relative humidity.

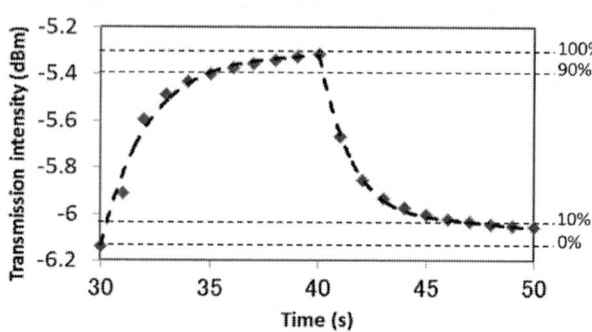

Fig. 6. Response and recovery characteristics between 48.3 and 79.8% RH.

V. RESPONSE CHARACTERISTICS

To investigate the response-time characteristics, several cycles with a period of 10 s were performed between three different RHs (0.4%, 48.3%, and 79.8%; see Fig. 5). In this measurement, we used light with a wavelength of 1545 nm, emitted from a tunable light source. According to Fig. 3, the transmission intensity at a wavelength of 1545 nm increases with RH. Figure 5 confirms very good reproducibility. When RH is increased from dry air to 48.3%, the transmission intensity becomes stable at 9 s. Figure 6 shows the detailed view of one cycle (48.3%→79.8%→48.3%). Also shown in this figure is a fitting curve (dotted line) expressed by an exponential function with a time constant. The response and recovery times, defined by the time interval between the intensity levels from 10 to 90%, are estimated to be 4.7 and 6.7 s, respectively. The response time is comparable to but recovery time is shorter than the values reported in previous work [6].

VI. CONCLUSIONS

We fabricated an optical fiber humidity sensor which is operable in a wide humidity range with very high sensitivity and good reproducibility. The sensitivity is as high as 0.748 nm/% even at 0.4% RH and 0.0278 nm/% RH at 85% RH. Response and recovery times at RH between 48.3 and 79.8% are as short as 4.7 and 6.7 s, respectively.

REFERENCES

[1] P. R. Story, D. W. Galipeau, and R. D. Mileham, "A study of low-cost sensors for measuring low relative humidity," Sens. Actuator B, vol. 24/25, pp. 681–685, 1995.

[2] P. G. Su, and Y. P. Chang, "Low-humidity sensor based on a quartz-crystal microbalance coated with poly pyrrole/Ag/TiO$_2$ nanoparticles composite thin films," Sens. Actuators B, vol. 129, pp. 915–920, 2008.

[3] R. J. Wu, Y. L. Sun, C .C. Lin, H. W. Chen, and M. Chavali, "Composite of TiO2 nanowires and Nafion as humidity sensor material," Sens. Actuators B, vol. 115, pp. 198–204, 2006.

[4] Z.M. Rittersma, "Recent achievements in miniatuarised humidity sensors–a review of transduction techniques," Sens. Actuators A, vol. 96, pp. 196–210, 2002.

[5] W. Cao, Y. Duan, "Optical fiber evanescent wave sensor for oxygen deficiency detection," Sens. Actuators B, vol. 119, pp. 363–369, 2006.

[6] Z. Zhao, and Y. Duan, "A low cost fiber-optic humidity sensor Based on silica sol–gel film," Sens. Actuators B, vol. 160, pp. 1340-1345, 2011.

[7] A. A. Herrero, H. Guerrero, and D. Levy, "High-Sensitivity Sensor of Low Relative Humidity Based on Overlay on Side-Polished Fibers," IEEE Sens. Journal, vol. 4, pp. 52-56, 2004.

[8] H. Fukano, Y. Kushida, and S. Taue, "Multimode-interference-structure optical-fiber temperature sensor with high sensitivity," IEICE Electronics Express, vol. 10, No.24, pp. 1-5, 2013.

[9] M. A. Fardada, E. M. Yeatman, E. J. Dawnay, M. Green, and F. Horowitz, "Effects of H$_2$O on structure of acid-catalysed SiO$_2$ sol-gel films," Journal of Non-Crystalline Solids, vol. 183, pp. 260–267, 1995.

A High-sensitivity Two-dimensional Inclinometer Based on Two Etched-chirped Fiber Gratings

Hung-Ying Chang[1], Kuan-Ting Li[3], Po-Chia Huang[2], Jung-Sheng Chiang[4], Nai-Hsiang Sun[4] and Wen-Fung Liu[2]

[1]Ph.D. Program of Electrical and Communications Engineering, Feng-Chia University, NO. 100, Wenhwa Rd., Taichung, Taiwan
[2]Department of Electrical Engineering, Feng-Chia University, NO. 100, Wenhwa Rd., Taichung, Taiwan
[3]Department of Automatic Control Engineering, Feng-Chia University, NO. 100, Wenhua Rd., Taichung, Taiwan
[4]Department of Electrical Engineering, I-Shou University, Kaohsiung, Taiwan
Author e-mail address: hungying.chang@gmail.com

Abstract: *we proposed a novel high-sensitivity dual-axis fiber inclinometer based on two etched chirped fiber Bragg gratings(CFBGs) with the CFBG1-sensitivity of 0.245 nm/degree and CFBG2-sensitivity of 0.215 nm/degree respectively.*
Keywords: *dual-axis; inclinometer; chirped fiber Bragg grating; etched*

I. INTRODUCTION

The sensor based on fiber gratings [1-4] becomes many potentially industrial applications in optical measurement fields due to its many advantages of small size, immunity to electromagnetic interference, anti-erosion and non-conducting, etc. A typical inclinometer [5, 6] is used for measuring angles of slope, elevation or depression of an object with respect to gravity with positive slopes and negative slopes by using different measuring units. For the traditional electronic inclinometer sensors, the temperature variation, vibration, and repeatability will affect their absolute accuracy. Therefore, for improving this phenomenon, the fiber type of inclinometer sensor based on two etched chirped fiber Bragg gratings [7-10] is proposed for real-time monitoring the landside and mudflow of mountains caused by the earthquake. They are also applied in platform leveling, boom angle indication, and in other applications requiring measurement of tilt. Thus, this dual-axis fiber inclinometer can provide a simple and high-sensitivity tilt-measurement technique.

II. BASIC PRINCIPLE

When a part of an etched chirped fiber Bragg grating is immersed into the liquid, the reflection spectrum of the grating immersed in the liquid will be shifted to the longer wavelength side, which is superimposed with the spectrum of the rest grating in air to create an overlapping wavelength peak. Therefore, the overlapped wavelength λ_p can be described as [11]:

$$\lambda_p = 2n_{eff}\Lambda_0 + Cz + 2\Delta n\Lambda_0 \qquad (1)$$

where n_{eff} is the grating effective index, Λ_0 is the grating initial period, C is the chirped value, z is the position of liquid level, and $\Delta n\Lambda_0$ is the variation of reflection wavelength.

For using an etched CFBG to measure the liquid level, Eq. (1) can simply be rewritten as:

$$\Delta\lambda_p(\Delta z) = C \times \Delta z \qquad (2)$$

where $\Delta\lambda_p$ is the wavelength shift of overlapped peak and Δz is the variation of liquid level.

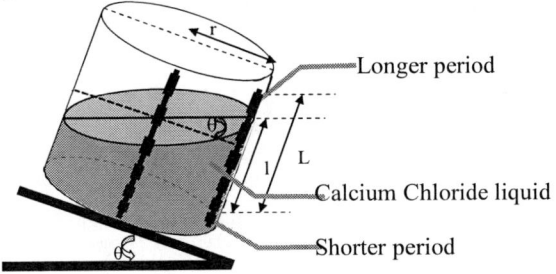

Fig. 1 The sensing head structure of fiber inclinometer

In this study, two etched CFBGs are installed in the wall of the cylindrical container to achieve a sensing head of fiber inclinometer as shown in Fig.1. The length of the two chirped fiber Bragg gratings is around 2 centimeters. The parameter of R (around 2.5 cm) is the diameter of the container and r (about 1.2 cm) is the distance from the fiber to the center. Then, twenty percent Calcium Chloride liquid (refractive index 1.41) is injected into the container with a half of the grating to be sunk in the liquid.

From Fig.1, we can see that two etched CFBGs (CFBG1 and CFBG2) are orthogonally installed in the wall of the cylindrical container. When the container is tilted in various angles and directions, the lengths of the two CFBGs immersed in the liquid will be changed. (φ is defined as the azimuth angle and θ is defined as the tilt angle.) For different liquid levels, the relationship between CFBG1 and CFBG2 can be described as:

$$\Delta z_{CFBG1} = r \tan \Delta \theta \cos \Delta \varphi \qquad (3)$$

$$\Delta z_{CFBG2} = r \tan \Delta \theta \cos(\Delta \varphi - 90) \qquad (4)$$

According to Eq. 3 and Eq. 4, the relationship between the power overlap wavelength shift ($\Delta\lambda$p), φ and θ can be respectively written as:

$$\Delta \theta = \tan^{-1}\left\{\frac{\Delta\lambda_{p-CFBG1}}{C_{CFBG1} r \cos\left[\cot^{-1}\left(\frac{\Delta\lambda_{CFBG1}}{\Delta\lambda_{CFBG2}} \times \frac{C_{CFBG1}}{C_{CFBG2}}\right)\right]}\right\} \qquad (5)$$

$$\Delta \varphi = \cot^{-1}\left(\frac{\Delta\lambda_{p-CFBG1}}{\Delta\lambda_{p-CFBG2}} \times \frac{C_{CFBG1}}{C_{CFBG2}}\right) \qquad (6)$$

III. EXPERIMENTAL SET-UP AND RESULTS

Fig.2 shows the experimental setup of dual-axis fiber inclinometer based on two etched CFBGs. By slanting platform and rotating the container simultaneously, both the tilted angle and the azimuth angle can be measured by the protractors. The range of tilted angle is from 0° to 35°, and the range of azimuth angle is from 0° to 360°. The two etched CFBGs are connected to the optical spectrum analyzer (OSA) and a wide-band amplified-spontaneous-emission (ASE) light source by means of the 3-dB coupler for observing the power overlap wavelength shift.

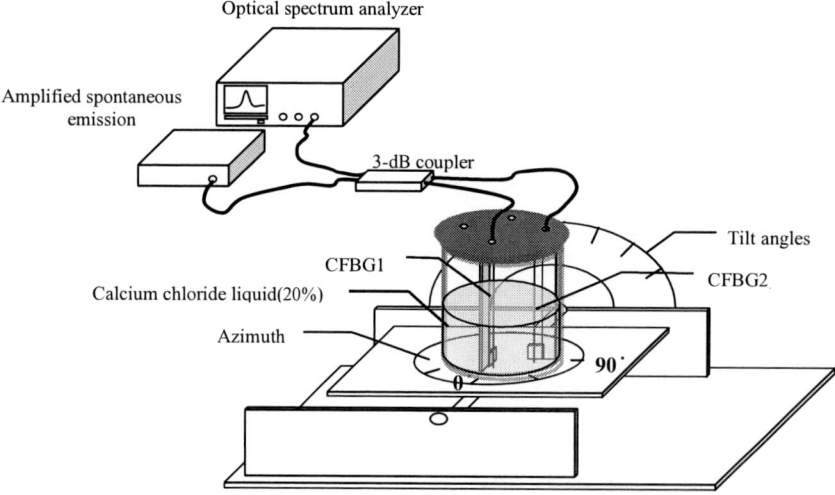

Fig. 2 The experimental setup of dual-axis inclinometer based on two etched CFBGs.

When the tilted angle and azimuth angle are changed, the lengths both of CFBG1 and CFBG2 immersed in the liquid are also varied. Thus, the tilted angle and azimuth angle can be obtained by means of measuring the power overlap wavelength shift of two CFBGs.

Fig.3 and Fig.4 show respectively the relationship curves for the wavelength shift versus the tilted angles in direction angles of the two CFBGs. By calculating the amount both of the red-shifted and blue-shift of the two CFBGs, two direction angles can be simultaneously obtained.

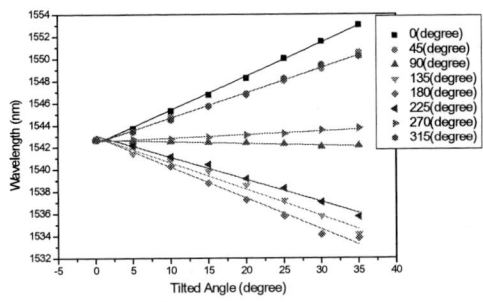

Fig. 3 The curves for the wavelength shift versus the tilted angle in different direction angles of CFBG 1.

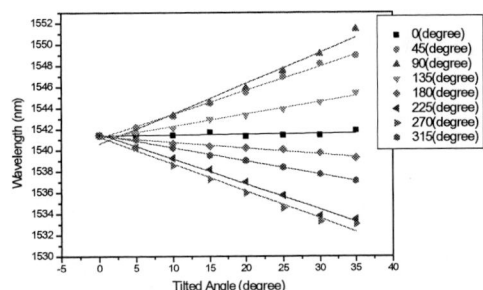

Fig. 4 The curves for the wavelength shift versus the tilted angle in different direction angles of CFBG 2.

IV. CONCLUSIONS

In this paper, a highly sensitive dual-axis inclinometer based on two etched chirped fiber Bragg gratings is experimentally demonstrated with the CFBG1-sensitivity of 0.24508 nm/degree and CFBG2-sensitivity of 0.21581 nm/degree respectively. This sensor can be applied in monitoring the gradient of mountain or ensuring the structure safety of the building. In the future, the material of the liquid can be replaced with an appropriate liquid to enhance its sensitivity.

ACKNOWLEDGMENT

The authors would like to specifically thank the Ministry of Science and Technology, Taiwan, for sponsoring this research under Contract No. MOST 103-2221-E-035-006-MY3 and MOST 104-2622-E-035-020-CC3

REFERENCES

[1] K.O. Hill, et al., "Photosensitivity in optical fiber waveguides: Application to reflection filter fabrication," Applied Physics Letters, Vol. 32, pp. 647–649, 1978.

[2] G. Meltz, et al., "Formation of Bragg gratings in optical fibers by transverse holographic method," Optics Letters, Vol. 14, Issue 15, pp. 823-825, 1989.

[3] A.D. Kersey, et al., "Fiber Grating Sensors", Journal of Lightwave Technology, Vol. 15, pp. 1442-11463, 1997.

[4] H.J. Sheng, et al., "High-sensitivity temperature-independent differential pressure sensor using fiber Bragg gratings", Optics Express, Vol. 16, pp. 16013-16018, 2008.

[5] H.J. Chen, et al., "Temperature-insensitive fiber Bragg grating tilt sensor," Applied Optics, Vol. 47, pp. 556-560, 2008.

[6] Y. Zhao, et al., "Experimental research on a novel fiber-optic cantilever-type inclinometer," Optics & Laser Technology, Vol. 37, pp. 555-559, 2005.

[7] S. Khaliq, et al., "Fiber-optic liquid-level sensor using a long-period grating," Optics Letters, Vol. 26, pp. 1224-1226, 2001.

[8] B. Yun, et al., "Highly Sensitive Liquid-Level Sensor Based on Etched Fiber Bragg Grating", IEEE Photonics Technology Letters, Vol. 19, pp. 1747-1749, 2007.

[9] K.R. Sohn and J.H. Shim, "Liquid-level monitoring sensor systems using fiber Bragg grating embedded in cantilever", Sensors and Actuators A:Physical, Vol. 152, pp. 248-251, 2009.

[10] T. Lu, et al., "Asymmetric Fabry-Perot fiber-optic pressure sensor for liquid-level measurement", Review of Scientific Instruments, Vol. 80, pp. 033104-033104-4, 2009.

[11] H. Y. Chang, et al., "An ultra-sensitive liquid-level indicator based on an etched chirped-fiber Bragg grating", IEEE P.T.L., Vol. 28, pp. 268-271, 2016.

WA2-65

OECC/PS2016

A Polymer-Coated Hollow Core Fiber Fabry-Pérot Interferometer for Sensing Liquid Level

[a]Teng-Wei Fu, [a]Yuan-Jie Yang, [a]Jun-Han Lin, [a]Tung-Yuan Yeh, [b]Pin Han, and [a]Cheng-Ling Lee.

[a] Department of Electro-Optical Engineering, National United University, Miaoli 360, Taiwan
[b]Institute of Precision Engineering, National Chung Hsing University, Taichung 402, Taiwan,
E-mail: cherry@nuu.edu.tw Phone: 886-37-381732 Fax: 886-37-351575

Abstract: This study presents a polymer-coated hollow-core-fiber-Fabry–Perot interferometer to measure liquid level. Wavelength shift of interference fringes of the proposed device exhibits linear response corresponding to the liquid level with a high sensitivity of −0.446nm/cm.

Keywords:Fiber-Fabry–Perot interferometer, Liquid level, Polymer, Hollow core fiber.

I. INTRODUCTION

Fiber-optic liquid level sensors (FOLLSs) with numerous smart and combination structures have been proposed in recent years [1-6]. Various FOLLSs based on the well known fiber Bragg gratings (FBGs) and long period fiber gratings (LPGs) have neen reported but expensive UV lasers were required in the fabrication [1-3, 5-6]. Therefore, other FOLLSs especially based on fiber Fabry-Pérot Interferometers (FFPIs) are attractive to much attention because of their highly sensitive, simple and convenvient properties. For example, Lu et al. have presented a fiber-optic based pressure sensor based on a fiber Fabry–Perot (FP) interferometer in which the liquid level was derived from the pressure-induced change in the FP cavity [4]. A dual-pressure-sensor system comprising a fiber Bragg grating (FBG) pressure sensor and a FP pressure sensor has been proposed [5-6]. The combination of the two devices can achieve the simultaneous measurement of unknown liquid level and specific gravity of the liquid. The liquid level sensitivities represented in the previous works are about 0.1125 nm/mm in the liquid of RI=1.333 [1], -0.141 nm/mm (Glycerin with RI=1.431, based on long period fiber grating (LPG)) [3], and 0.01491 nm/cm (in water, based on the FBG level sensor) [5], [6], respectively. For most liquid level fiber sensors listed above, they are useful but the fabrication processes are complicated. For the demands on the simplicity and costeffective for the fiber sensors, we had proposed some ultracompact and very simple structure FFPIs with an UV-cured NOA-polymer for sensitively measuring some parameters [7-8]. We believe that the proposed NOA-polymer material would also have high responses under the pressure and strain from the liquids. Therefore, in this study, we further develop a similar airgap-FFPI (AG-FFPI) device proposed in [7] with a very thin layer NOA65 polymer coating for sensing the pressure produced by the liquid level. Experimental results show that an approximately linear response over the measured range of water level with sensitivity of −0.446nm/cm can be achieved. Furtheromre, the proposed confiigriation also can be further applied to measure specific gravity of the liquids.

II. PRINCIPLE OF PREPARATION AND EXPERIMENTAL

Figure 1(a) shows experimental setup of the proposed liquid level sensor. Configuration and sensing scheme of the proposed AG-FFPI liquid level sensor is plotted in Fig.1(b). For enhancing the sensitivity, the thickness of the coating NOA65 monomer on the HCF endface is thin, as displayed in Fig. 1(c). After UV curing, the HCF tip with a solid layer of NOA65 polymer was around thickness of 12.81 μm, and HCF was closed to form an airgap was around 72.93 μm. When endface of the sensor is immersed in liquid, the resulting liquid pressure makes a change in the cavity length (*d*) of the AG-FFPI. The airgap of the interferometer is compressed as well as shifted the interference wavelength fringes. The wavelength shift ($\Delta\lambda$) at certain interference fringe resulting from a change in the applied pressure satisfy the following relation [5]:

$$\Delta\lambda = -\frac{3(1-v_m^2)a^4\rho g\lambda}{16E_m L^3 d}h \qquad (1)$$

922

Where the parameters of Eq.(1) are denoted as below, v_m: Poisson's ratio, a: radius of the sensor head, ρ: density of liquid, g: gravitational acceleration, λ: wavelength, h: liquid level, E_m: Young's modulus and L: thickness of NOA65.

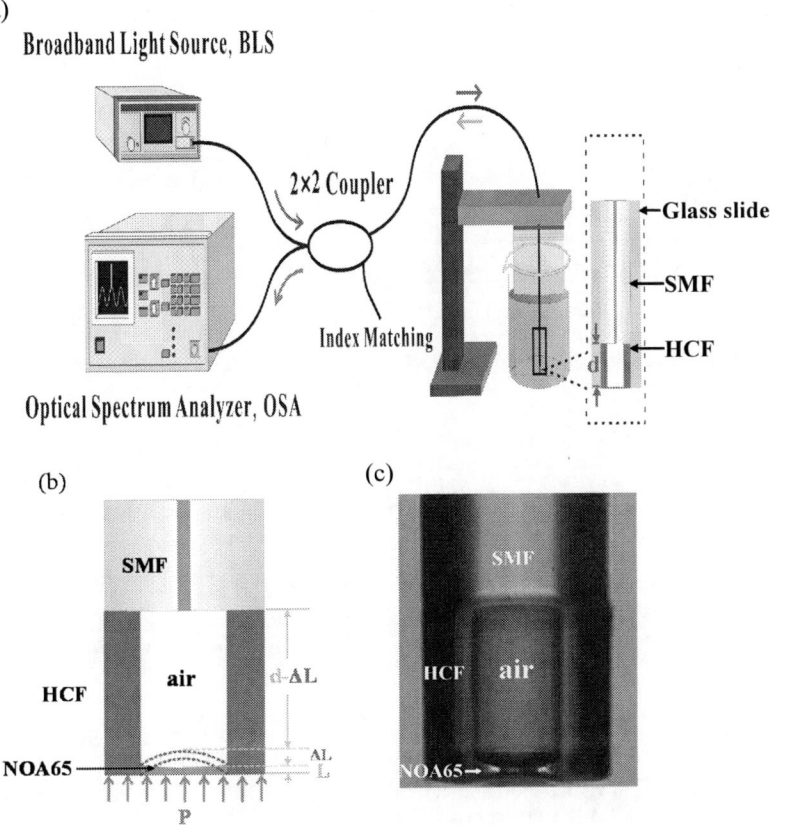

Fig.1(a) Experimental setup of proposed liquid level sensor, BLS: Broadband light source, OSA: Optical spectrum analyzer, inset shows the configuration of the sensor tip. (b) Diagram of the air gap variation ($-\Delta L$) in the proposed sensor by increasing pressure (P) from liquid level. (c) Micrograph of the device

III. EXPERIMENTAL RESULTS

In the experiment, the SMF was firstly spliced a very tiny section of HCF by using a special fusion method. Then, endface of HCF was coated a thin layer of monomer NOA65 by using a particular method. Based on the authors' experience, the UV exposure and aging of duration can be controlled to turn the NOA65 monomer into a solid material to stabilize the length of the airgap. In this work, the cavity length of the airgap (d) can be easily controlled by varying the length of HCF. The monomer, Norland Optical Adhesive 65 (NOA65), is a photo-polymerizable liquid that can be cured by ultraviolet (UV) light with a maximum absorption in the range 350-380nm. The solid edge of the polymer wall can fix the air-gap structure of the device and make the length of the airgap permanent. After the sensor has been fabricated, it was fully fixed with expoxy on the glass slide. Due to the property of NOA65 has inherently absorbing water vapor [8], thus the glass slide with the fiber sensor was entirely put into a protected plastic bag to ensure the whole sensor was separated from water. Experimental results demonstrate that highly sensitive response can be obtained as shown in Fig. 2. As shown in the Fig. 2(a), the interference minima shift to shorter wavelengths as the liquid level h increases. Figure 2(b) plots the sensitivity of the proposed liquid level sensor. Clearly, the experimental results for the variation of the cavity length with the pressure can be seen that the sensor has an approximately linear response over the considered pressure range with sensitivity of -0.446nm/cm. Herein, the negative sensitivity means that the polymer section is compressed by the liquid pressure that reduces the d, causing the wavelength to shift to the short wavelength side ($\Delta\lambda$ is negative).

OECC/PS2016

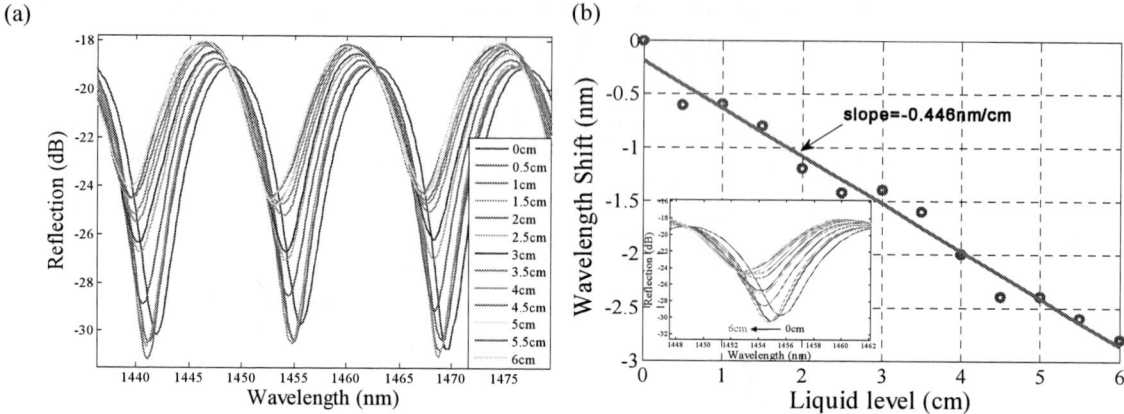

Fig.2 (a) Interference wavelength shifts of the proposed sensor in the water level range from 0 to 6cm, (b) Sensitivity of the proposed liquid level sensor based on the results of (a).

IV. CONCLUSIONS

The paper presents an optical fiber liquid-level sensor based on an a polymer-coated hollow-core-fiber-Fabry–Perot interferometer. The sensing element is based on the endface of HCF coated a thin layer of monomer NOA65, which is easily fabricated to accomplish an all-fiber liquid level sensor. The experimental results for the sensor has an approximately linear response over the considered pressure range with sensitivity of −0.446nm/cm. The developed device has very favorable characteristics and numerous advantages. We believe that it can be further applied to measure the specific gravity of other liquids for a wide range of applications.

V. ACKNOWLEDGMENT

The authors would like to thank the Ministry of Science and Technology of Taiwan, for financially supporting this research under Contract No. NSC 102- 2221-E-239-033-MY3.

REFERENCES

[1] K.R. Sohn, J.H. Shim, "Liquid-level monitoring sensor systems using fiber Bragg grating embedded in cantilever," Sens. actuators. A Phys., vol.152, no. 2, pp. 248–251, June 2009.

[2] Q. Jiang, D. Hu, M. Yang, "Simultaneous measurement of liquid level and surrounding refractive index using tilted fiber Bragg grating," Sens. actuators. A Phys., vol. 170, no. 1–2, pp 62–65, November 2011.

[3] H. Fu, X. Shu, A. P. Zhang, W. Liu, L. Zhang, S. He, and I. Bennion, "Implementation and characterization of liquid-level sensor base on a long-period fiber grating Mach-Zehnder interferometer," IEEE Sens. J., vol.152, no. 11, pp. 2878–2882, November 2011.

[4] T. Lü and S. Yang, "Extrinsic Fabry-Perot cavity optical fiber liquid-level sensor," Appl. Opt., vol. 46, no. 18, pp. 3682–3687, May 2007.

[5] C.W. Lai, Y.L. Lo, J.P. Yur, and C.H. Chuang, "Application of fiber Bragg grating level sensor and Fabry-Pérot pressure sensor to simultaneous measurement of liquid level and specific gravity," IEEE Sens. J., vol. 12 , no. 4, pp. 827–831, April 2012.

[6] C.W. Lai, Y.L. Lo, J.P. Yur, W.F. Liu, C.H. Chuang, "Application of Fabry-Pérot and fiber Bragg grating pressure sensors to simultaneous measurement of liquid level and specific gravity," Measurement, vol. 45, no. 3,pp. 469–473, April 2012.

[7] C.L. Lee, L.-H. Lee, H.-E. Hwang, and J.-M. Hsu, "Highly sensitive sir-gap fiber Fabry-Perot interferometers based on polymer-filled hollow core fibers," IEEE Photon. Tech. Lett., vol. 24, no. 2, pp. 149–151, January 2012.

[8] C.L. Lee, Y.W. You, J.H. Dai, J.M. Hsu, and J.S. Horng, "Hygroscopic polymer microcavity fiber Fizeau interferometer incorporating a fiber Bragg grating for simultaneously sensing humidity and temperature," Sens. actuators. B Chem., vol. 222, pp. 339-346, August 2016.

WA2-66

Multi-Channel Lasing Characteristics for Linear-Cavity Fiber Sensor System using SOA and Fiber Bragg Grating Elements

Kazuto Takahashi[1], Mao Okada[1], Hiroki Kishikawa[1], Nobuo Goto[1], Yi-Lin Yu[2], and Shien-Kuei Liaw[2]

[1]Dept. of Optical Science and Technology, Tokushima University, 2-1 Minamijosanjima-cho, Tokushima, 770-8506 Japan
[2]Dept. of Electronic Engineering, National Taiwan University of Science and Technology, Taipei 10617, Taiwan
{kishikawa.hiroki, goto.nobuo}@tokushima-u.ac.jp

Abstract: Multi-channel amplification with SOA is investigated for use in the proposed linear-cavity sensing system. The nonlinearity caused by gain saturation and FWM is analyzed. The lasing condition for multi-channel operation is also clarified.
Keywords: Optical nonlinearity, semiconductor optical amplifier, multiple wavelengths, fiber sensor.

I. INTRODUCTION

Optical fiber sensing has been extensively studied in various areas such as aging deterioration measurement of constructed building, seismic measurement, environmental measurement, etc [1-3]. We consider fiber lasing system including sensing element in the cavity, which has advantages such as higher resolution for wavelength-shift induced by sensing element, and higher signal-to-noise ratio (SNR) [4,5]. By employing a fiber Bragg grating (FBG) as the sensing element, reflecting center wavelength can be shifted due to environmental temperature or tension.

In fiber lasing systems, erbium-doped fiber amplifier (EDFA), Raman amplifier, and semiconductor optical amplifier (SOA) have been employed as the gain component. We consider multi-wavelength simultaneous lasing to sense multiple points. When the EDFA is employed in the system, the homogeneous broadening of erbium ions limits the number of lasing wavelengths. On the contrary, the SOA shows the inhomogeneous broadening properties, which makes it possible the lasing at multiple wavelengths.

In this report, we consider optical lasing at multiple wavelengths in linear-cavity sensing system consisting of an SOA, an arrayed waveguide grating (AWG), and FBGs. Multi-wavelength amplification in the SOA and multi-channel lasing are theoretically discussed.

II. LINEAR-CAVITY FIBER SENSOR USING SOA

The proposed fiber sensing system consists of multiple linear cavities lasing at different wavelengths as shown in Fig.1(a). The AWG plays a role as multi-/demultiplexer. The passband filtering response of the AWG is schematically illustrated in (b), where a flat-top filtering passband is assumed. The wavelength interval of the multi-channel lasing is $\Delta\lambda_{lasing}$. The AWG demultiplexes the optical signal propagating in the right-hand direction into N signals having different wavelength λ_i, $i=1, ...,N$. The signal at wavelength λ_i in port i is reflected by the FBG. The wavelength of each FBG is designed to be the center of the AWG passband of each channel. The wavelength multiplexed signals propagating in left-hand direction are amplified with the SOA. The left-end wavelength-independent loop mirror reflects the all channel signals, where a part of the signals are coupled out for detection. The lasing spectra are also shown in (b). Since the FBG reflection frequency depends on the environment, an environmental change results in wavelength-shift $\lambda_{shift,i}$ of the lasing wavelength. The maximum amount of $\Delta\lambda_{shift,i}$ has to be less than $\lambda_{AWG\_FT}/2$. When multi-wavelength signals are amplified with the SOA, gain saturation and optical nonlinearity become an important issue.

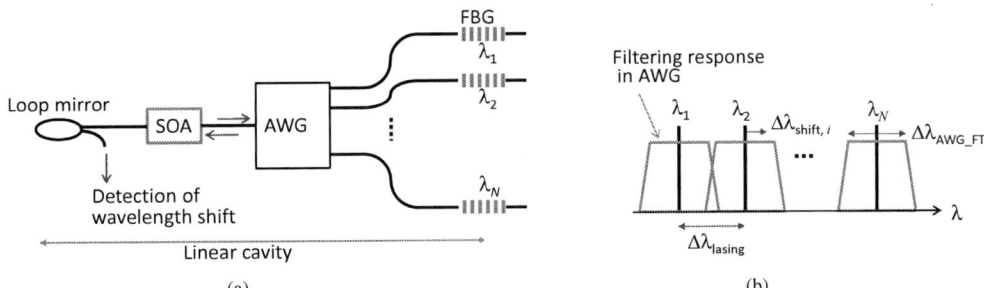

(a)   (b)
Fig.1 (a) Linear-cavity multi-channel fiber sensing system and (b) lasing spectrum with AWG filtering characteristics.

*This research was supported in part by JSPS KAKENHI (15H06443) and Collaborative Research Project between National Taiwan University of Science and Technology and Tokushima University.

III. ANALYSIS FOR NONLINEARITY IN MULTI-CHANNEL SOA AMPLIFICATION

Nonlinear phenomena for multi-wavelength signals are induced by gain saturation [6] and four wave mixing (FWM) [7,8]. The wavelength range available for amplification depends on the material gain. These factors cause restriction in the number of wavelength channels.

The mechanism of FWM in SOA is explained as follows: Interference of the multi-wavelength signals coupled in the SOA induces amplitude variation at the beat frequencies of the signals. Stimulated emission due to the amplitude varying signal induces carrier variation, which results in variation of refractive index as well as gain variation. These gain and refractive-index variations modulate the incident signals and generate sideband signals with the interval of the beat frequencies.

In this section, we analyze the multi-channel signal behavior through SOA based on the analysis reported by Connelly [9]. The analysis composed of the traveling-wave equations for signal fields and spontaneous emission, carrier-density rate equation, and material gain modeling. The numerical simulation was performed using OptiSystem (Optiwave Systems Inc.).

We consider amplification for two signals at different wavelengths. When the wavelength interval decreases, the gain for the longer wavelength signal increases whereas the gain for the shorter wavelength signal decreases due to FWM [10]. Fig.2(a) shows the simulated gain for the longer (fixed at λ=1570 nm) and shorter wavelength signals as a function of the wavelength interval. The input power is changed as a parameter. The induced FWM intensities are shown in (b). The generated FWM signals decrease with the wavelength interval. This gain deviation causes a limiting factor in the multi-channel sensing system.

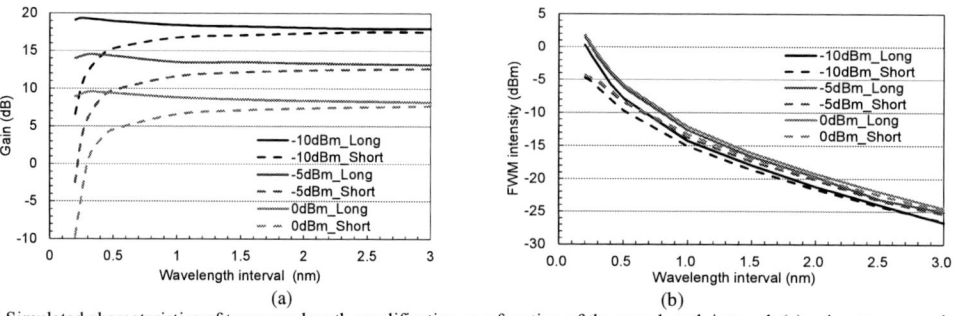

Fig.2 Simulated characteristics of two-wavelength amplification as a function of the wavelength interval; (a) gain at two wavelengths and (b) generated FWM intensities. Input power is changed as a parameter. The SOA current is 130mA.

Next, we investigate nonlinearity for amplification of multiple signals with equal interval. Fig.3(a) shows output signal intensities as a function of optical frequency for cases of 4, 8, and 16 channels. The multiple signals have equal frequency separation. It is found that FWM signals are generated. The gain at the wavelengths of the input signals is shown in (b). The gain decreases with the input power due to gain saturation. The maximum number of multi-channel amplification is restricted depending on the input optical power.

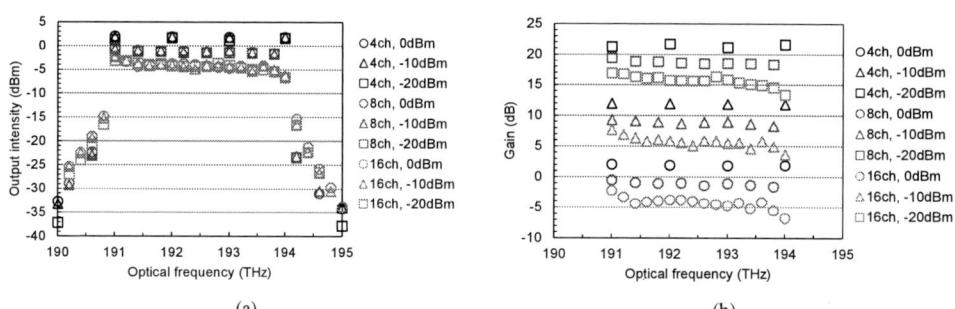

Fig.3 Simulated characteristics of multi-channel amplification as a function of optical frequency; (a) output intensities and (b) gain at input signal frequencies. Input power is changed as a parameter. The SOA current is 130mA.

IV. MULTI-CHANNEL LASING CONDITIONS

The proposed sensor system consists of a single SOA as the gain component for lasing. The number of channel at different wavelengths is limited due to gain saturation in SOA as discussed in previous section. The gain saturation, however, can be avoided by decreasing the incident optical power even if the channel number is large. In this section, we consider the optical power of each channel using a model shown in Fig.4(a).

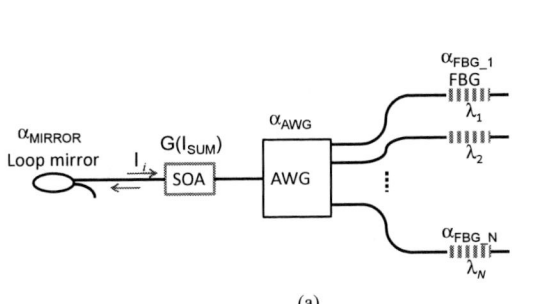 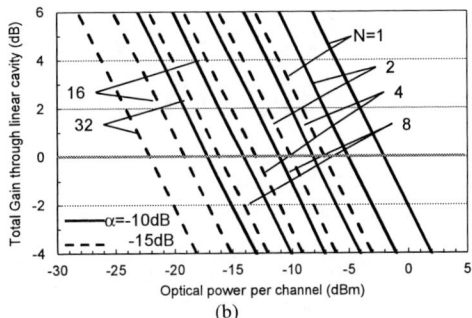

(a)   (b)

Fig.4 (a) A model to evaluate the lasing intensity. (b) Optical gain through the linear-cavity sensor as a function of optical intensity at the entrance of the SOA. The channel number of lasing lines is varied as a parameter.

Each signal propagating in the right-hand direction at wavelength λ_i has optical power of I_i, $i=1,...,N$ at the entrance of SOA. Since the each optical signal propagates in both directions, the total optical power I_{SUM} is given by

$$. I_{SUM} = 2 \sum_{i=1}^{N} I_i \qquad (1)$$

The gain in the SOA depends on I_{SUM} as denoted by $G(I_{SUM})$. The reflectance at the left mirror including the power splitting loss for detection is denoted by α_{MIRROR}. The transmittance of the AWG including the insertion loss is denoted by α_{AGW}. The reflectance at each FBGi is denoted by $\alpha_{FBG\,i}$. The total gain at λ_i through a cavity is written as

$$G_{cavity} = G^2(I_{SUM})\, \alpha_{AGW}^2\, \alpha_{FBG\,i}\, \alpha_{MIRROR}. \qquad (2)$$

The lasing condition in gain is given by $G_{cavity}=1$. We assume that $I_i = I_c$, $i=1,..., N$, and the total transmittance due to loss is $\alpha = \alpha_{AGW}^2\, \alpha_{FBG\,i}\, \alpha_{MIRROR}$. The total gain G_{cavity} is calculated as a function of I_c in Fig.4(b), where the channel number N is varied as a parameter. It is found that the lasing power per channel decreases with N. As an example, the lasing power at the entrance of the SOA is around -11dBm and -14dBm for $N=8$ with $\alpha = -$ 10dB and - 15dB, respectively.

V. CONCLUSIONS

Multi-channel amplification with SOA was investigated for use in the proposed linear-cavity sensing system. The nonlinearities caused by gain saturation and FWM were numerically discussed. The gain saturation limits the maximum number of channels. The FWM affects to the detection of wavelength shift due to sensing because the deviation of the lasing wavelength affects the lasing intensity when the neighborhood wavelength channels come close. The lasing condition for multi-channel operation was also discussed.

As a future work, we experimentally demonstrate multi-channel lasing and sensing.

REFERENCES

[1] S. Kim, J. Kwon, S. Kim, and B. Lee, "Multiplexed strain sensor using fiber grating-tuned fiber laser with a semiconductor optical amplifier," IEEE Photon. Technol. Lett., vol.13, no.4, pp.350-351, Apr. 2001.

[2] S. Diaz, D. Leandro, and M. Lopez-Amo, "Stable multiwavelength erbium fiber ring laser with optical feedback for remote sensing,'" IEEE/OSA J. Lightw. Technol., vol.33, no.12, pp.2439-2444, Jun. 2015.

[3] H. D. Lee, G. H. Kim, T. J. Eom, M. Y. Jeong, and C. -S. Kim, "Linearized wavelength interrogation system of fiber Bragg grating strain sensor based on wavelength-swept active mode locking fiber laser," IEEE/OSA J. Lightw. Technol., vol.33, no.12, pp.2617-2622, Jun. 2015.

[4] S. -K. Liaw, Y. -W. Lee, H. -W. Huang, and W. -F. Wu, "Multi-wavelength linear-cavity SOA-based laser array for long-haul sensing," IEEE Sensor J., vol.15, no.6, pp.3353-3358, Jun. 2015.

[5] S. Shin, S. -K. Liaw, and S. -W. Yang, "Post-impact fatigue damage monitoring using fiber Bragg grating sensors," Sensors, vol.14, no.3, pp.4144-4153, Mar. 2014.

[6] N. Pleros, C. Bintjas, M. Kalyvas, G. Theophilopoulos, K. Yiannopoulos, S. Sygletos, and H. Avramopoulos, "Multiwavelength and power equalized SOA laser sources," IEEE Photon. Technol. Lett., vol.14, no.5, pp.693-695, May 2002.

[7] R. J. Manning, A. D. Ellis, A. J. Poustie, and K. J. Blow, "Semiconductor laser amplifiers for ultrafast all-optical signal processing," J. Opt. Soc. Am. B, vol.14, no.11, pp.3204-3216, Nov. 1997.

[8] K. Inoue, T. Mukai, and T. Saitoh, "Nearly degenerate four-wave mixing in a traveling-wave semiconductor laser amplifier," Appl. Phys. Lett., vol.51, no.14, pp.1051-1053, Oct. 1987.

[9] M. J. Connelly, "Wideband semiconductor optical amplifier steady-state numerical model," IEEE J. Quant. Electron., vol.37, no.3, pp.439-447, Mar. 2001.

[10] G. P. Agrawal and I. M. Habbab, "Effect of four-wave mixing on multichannel amplification in semiconductor laser amplifiers," IEEE J. Quant. Electron., vol.26, no.3, pp.501-505, Mar. 1990.

Novel Soft-Cladding Optical Fiber for Distributed Pressure Sensing

Bin Zhou[1,2,*], Lin Htein[1], Zhengyong Liu[1], A. Ping Zhang[1], Chao Lu[3] and Hwa-yaw Tam[1,*]

[1] Photonics Research Centre, Department of Electrical Engineering, The Hong Kong Polytechnic University, Kowloon, Hong Kong SAR, China

[2] South China Academy of Advanced Optoelectronic, South China Normal University, Guangzhou 510006, P. R. China

[3] Photonics Research Centre, Department of Electronics and Information Engineering, The Hong Kong Polytechnic University, Kowloon, Hong Kong SAR, China

*Corresponding Author: zhoubin_mail@163.com, hwa-yaw.tam@polyu.edu.hk

Abstract: *A novel optical fiber with soft cladding is presented for surrounding pressure sensing application. The cladding is made of a kind of transparent silicone which can be compressed by and leads to extra loss. In the experiment the loss increment of a 0.5 meter long soft cladding fiber after applying high pressure up to 30 MPa is observed.*

Keywords: *optical fiber sensor, pressure measurement, silicone cladding*

I. INTRODUCTION

Optical fibers have many remarkable advantages in remote sensing applications. The neural like sensing, i.e. the distributed sensing is one of its most fascinating features, in which one single fiber can monitor a wide range of sensing area. The most common distributed optical fiber sensing technologies are based on nonlinear optical fiber scatterings, including Raman scattering effect and Brillouin scattering effect [1-4]. To stimulate these nonlinear scatterings, high power light source is required (especially for Raman scattering), and the readout modules are complicated. In this paper, a novel optical fiber with soft cladding is presented for distributed pressure sensing. The soft cladding is made of silicone rubber which can be substantially deformed by pressure, whereas the fiber core is the low-absorption silica for guiding the probe light. The thickness and the refraction index of the soft cladding changes significantly whenr the pressure is applied, which thus leads to an increment of the optical fiber transmission loss. To the best of our knowledge, it is the first time to use such a hybrid (polymer cladding / silica core) optical fiber for distributed pressure sensing.

II. OPERATING PRINCIPAL AND EXPERIMENT

In silica optical fiber, both core and cladding are made of rigid silica material. Therefore, the induced deformation is usually very small under pressure. In the present novel optical fiber with soft cladding, the cladding material is the transparent silicone and the core is still made of silica. Silicone is used to fabricate various optical devices because of its outstanding properties, such as optical transparency, stable, low price, flexibility and so on [5-6]. Since the Young's modulus and Poisson's ratio of the silicone is much smaller than the silica, a substantial change of the refractive index will be induced when the fiber is under pressure. Therefore, a change of the optical power distribution at the cross section of the fiber will be induced by the relative diameter and refractive index changes between the core and cladding. As the material losses of the silicone and silica are significantly different, such a deformation ultimately leads to an increment of the transmission loss.

Figure 1 shows the diagram of the pressure testing of the soft-cladding fiber. The fiber under test is immersed in high-pressure hydraulic oil which is generated by the oil pump. The pressure in the oil cylinder can be fast released by the valve. The left inset of Fig. 1 is the SEM photograph of the fiber, the core and cladding diameters are 75 um and 148 um, respectively. The right inset is the cross section of the fiber and the surrounding pressure. The broadband light source from SLED (super-luminescent LED) couples into the sensing fiber by using a fiber adapter. The optical transmission loss at different pressure was measured by a PD (photodetector), and the spectra were recorded by an optical spectrum analyzer (OSA).

Fig. 2 shows the measured loss of a 1.5-meter long soft-cladding fiber immersed in the oil for 132 hours. The pressure and temperature of the oil cylinder are 1 atm and 22 ℃, respectively. The loss quickly increases at the first few hours and then becomes nearly stable after about 10 hours (see the inset of Fig. 2). It means the molecular of the hydraulic oil permeates into the fiber becomes saturated and the fiber becomes very stable and suitable for sensing applications.

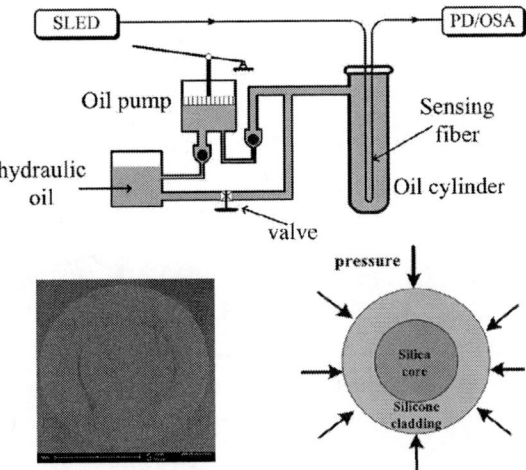

Fig. 1. The diagram of pressure testing for the soft-cladding fiber. The lower left image is the SEM photograph of the soft cladding fiber and the lower right figure shows the cross section structure of the fiber. SLED: super-luminescent LED, PD: photodetector, OSA: optical spectrum analyzer.

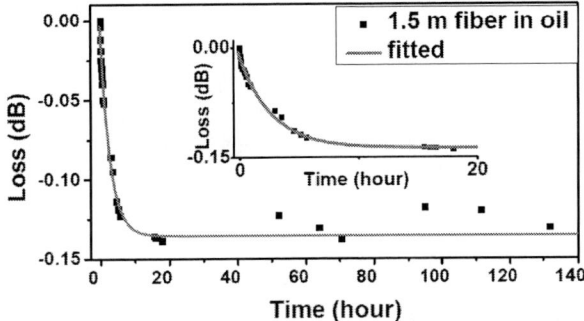

Fig. 2. The measured optical transmission loss of a 1.5-meter long soft-cladding fiber after immersed into the hydraulic oil. The inset shows the loss becomes stable after 10 hours.

Figure 3 shows the measured loss of a 0.5-meter sensing fiber with respect to the change of pressure up to 10 MPa. This fiber has not been pre-pressed and pre-immersed in hydraulic oil. The experimental results show a smooth exponential increment of the loss with respect to the raising pressure. More importantly, it was observed that the loss at zero pressure increases every time during the repeatability tests. It is mainly induced by the oil molecular permeation into the fiber and the Mullins effect, i.e. the softening effect of rubber [7]. Thus, the soft-cladding fiber has to be pre-pressed beyond its maximum range before use.

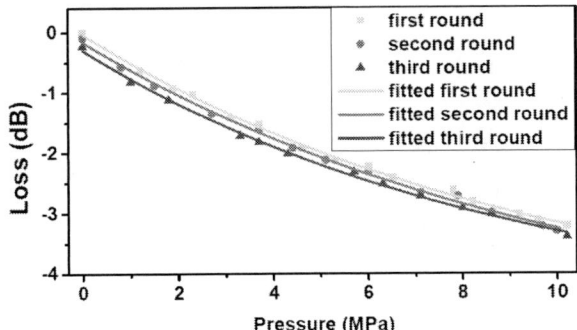

Fig. 3. The measured loss of a 0.5-meter long soft-cladding optical fiber (with no pretreatment) at different pressure.

In order to overcome the unrepeatable problem, the soft-cladding fiber was immersed into the hydraulic oil at 30 MPa for 15 minutes to allow the oil molecular saturated. The high pressure is to enhance the molecular permeability rate. After that, the fiber becomes stable. Fig. 4 shows the tested results after pretreating the fiber. One can see that the repeatabilities are very good both at large (up to 30 MPa, Fig. 4 (a)) and small (up to 5 MPa, Fig. 4(b)) pressure ranges.

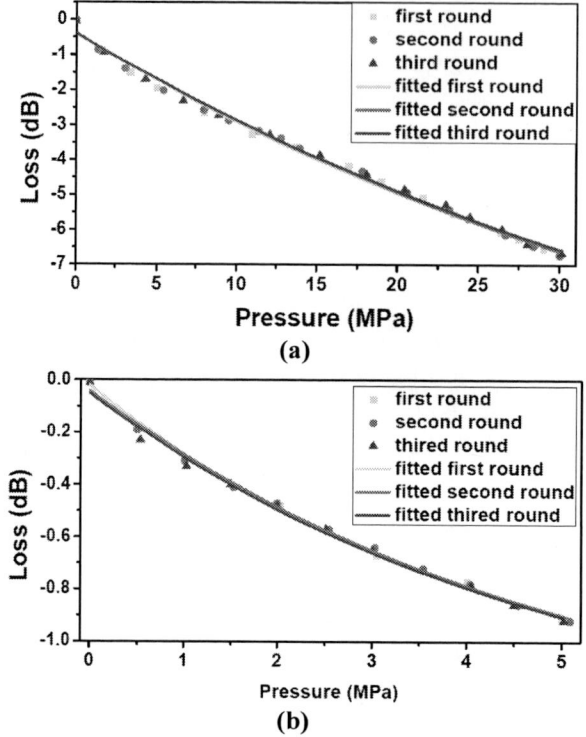

Fig. 4 The measured loss of a 0.5-meter long soft-cladding fiber under pressure up to 30 MPa (a) and 5 MPa (b) after pre-treated at 30 MPa for 15 minutes.

III. CONCLUSIONS

A novel silicone-cladding/ silica-core optical fiber has been demonstrated and proposed for distributed fiber sensing applications. Experimental results have revealed that the pretreatment in high-pressure hydraulic oil can improve the stability of the optical transmission loss of the soft-cladding fiber as well as its dependences on different pressures.

ACKNOWLEDGMENT

This work is supported by The Hong Kong Scholars Program (No. XJ2014027).

REFERENCES

[1] M. N. Alahbabi, Y. T. Cho, and T. P. Newson, 100 km distributed temperature sensor based on coherent detection of spontaneous Brillouin backscatter. Measurement Science and Technology, vol. 15, no. 8, pp. 1544, 2004

[2] A. Kobyakov, M. Sauer and D. Chowdhury, "Stimulated Brillouin scattering in optical fibers", Advances in optics and photonics, vol. 2, no. 1, pp. 1-59, 2010.

[3] J. P. Dakin, D. J. Pratt, G. W. Bibby, and J. N. Ross, "Distributed optical fibre Raman temperature sensor using a semiconductor light source and detector", Electronics letters, vol. 21, no. 13, pp. 569-570, 1985.

[4] Z. Amira, B. Mohamed, and E. Tahar, "Monitoring of Temperature in Distributed Optical Sensors: Raman and Brillouin Spectrum", Optik-International Journal for Light and Electron Optics, 2016.

[5] D. Fuard, T. Tzvetkova-Chevolleau, S. Decossas, P. Tracqui, and P. Schiavone, "Optimization of poly-di-methyl-siloxane (PDMS) substrates for studying cellular adhesion and motility". Microelectronic Engineering, vol. 85, no. 5, pp. 1289-1293, 2008.

[6] K. F. Lei, K. F. Lee, and M. Y. Lee, "Development of a flexible PDMS capacitive pressure sensor for plantar pressure measurement", Microelectronic Engineering, vol. 99, pp. 1-5, 2012.

[7] Leonard Mullins, "Softening of rubber by deformation." Rubber chemistry and technology, vol. 42, no. 1, pp. 339-362, 1969.

WA2-68   OECC/PS2016

Novel Bidirectional Reflective Semiconductor Optical amplifier

G. de Valicourt[1], A. Maho[2], A. Le liepvre[2], R. Brenot[2], A. Velázquez[1,3] and Y. K. Chen[1]

(1) Bell Labs, Nokia, 791 Holmdel Road, Holmdel, New Jersey 07733, USA
(2) III-V Lab-Common laboratory of Nokia Bell Labs France', 'Thales Research and Technology' and 'CEA Leti', 1, Avenue A. Fresnel, 91767 Palaiseau cedex, France
(3) Instituto de Investigaciones en Materiales, UNAM, A.P. 70-360, México D.F. 04510, México
devalicourt@bell-labs.com

Abstract: We propose a bidirectional reflective semiconductor optical amplifier as promising solution for on-chip amplification with silicon photonic integrated circuit. Small form factor device, wide optical bandwidth and high optical fiber-to-fiber gain are presented.

Keywords: Reflective Semiconductor Optical Amplifier, Silicon Photonic, Photonic Integrated Circuit

I. INTRODUCTION

Silicon photonic has been demonstrated to be a promising candidate in order to build complex photonic integrated circuit (PIC). Modulator, photodetector, attenuator, wavelength division (de)multiplexer, polarization combiner/splitter or rotator could be realized using such platform however significant losses occur as more and more advanced PIC are realized. Despite the above-mentioned functionalities, light emission/amplification is still an issue on the silicon photonic platform. Several solutions have been proposed in order to overcome the laser/amplifier challenge for silicon photonics. An external III-V gain medium coupled via an optical fiber or butt-coupling [1], bonding of III-V dies/wafers onto a processed Si wafer [2], using electrically pumped highly strained and heavily doped Ge materials [3] and the use of III-V on Si hetero-epitaxy [4] have been demonstrated as promising solutions. The first approach allows the independent optimization of the two materials, Silicon and III-V, however required the two chips to be accurately aligned. The emitted power for the III-V chip is distributed to the silicon chip via butt-coupling through a cleaved or etched facet. Reflective semiconductor optical amplifier (RSOA) is then flip-chipped onto the silicon chip where a hybrid laser is realized by sharing the cavity between the III-V and silicon material. High output power with low linewidth as well as high Wall-Plug efficiency (WPE) hybrid lasers has been demonstrated [5, 6]. High precision flip-chip bonding could be realized with accuracy less than one micron however multiple flip-chip alignments increase the final assembly cost (~80% of the final optoelectronic module cost). To overcome cost barriers in PIC module packaging, the obvious solution is to reduce the number of optical alignments.

In this paper, we propose a novel bidirectional-RSOA (B-RSOA) that allows having only one flip-chip step compared to two flip-chip steps with classic SOA configuration. Such device shares the same epitaxial growth as our standard RSOA device as well as same fabrication process allowing the integration of RSOA and B-RSOA together into a single chip for light generation and amplification respectively. The concept, design and characterization of the device are described. Chip optical gain up to 20 dB and 3dB optical bandwidth of 15 nm are demonstrated.

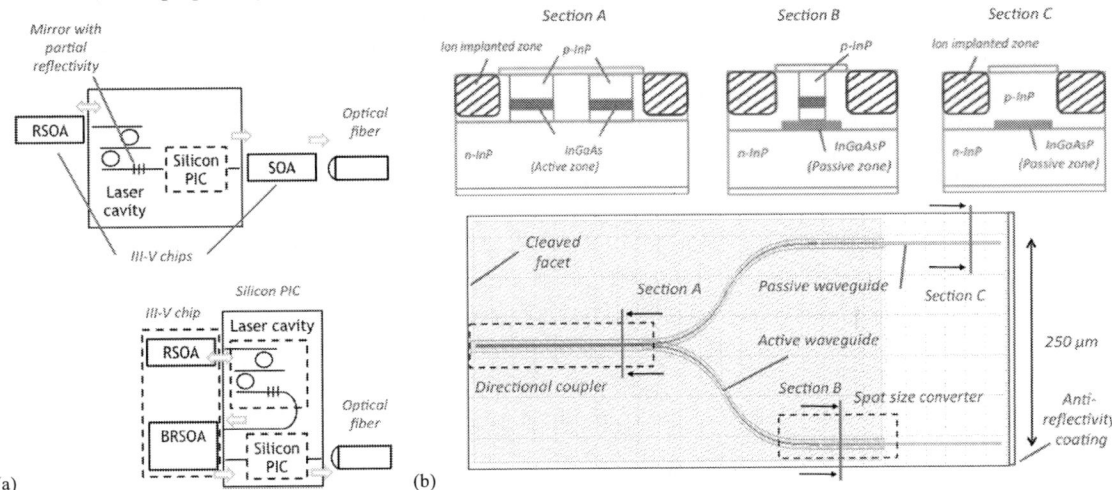

Fig. 1. Schematic of the hybridation (a) using RSOA and SOA devices or (b) RSOA and B-RSOA devices. (c) Laye-out of the B-RSOA. (Inset : cross section of the B-RSOA)

931

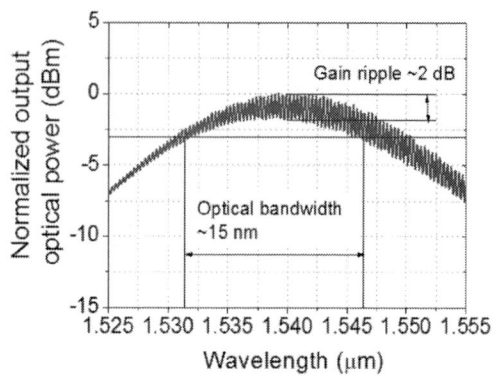

Fig. 3. Normalized ASE spectra of the B-RSOA at 50 mA bias current.

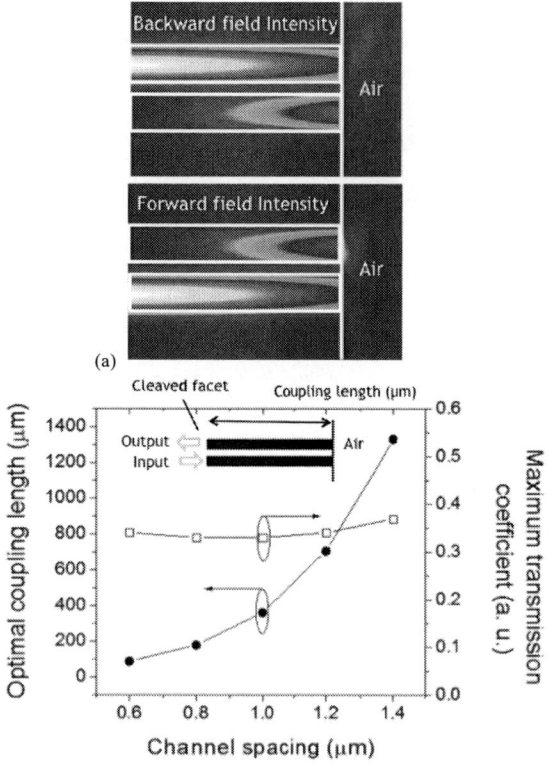

Fig. 2. (a) Forward and Backward field intensity. (b) Optimal coupling length and maximum transmission coefficient depending on the channel spacing

II. CONCEPT, DESIGN AND FABRICATION

Figure 1 (a) represents a SOA-based configuration of a hybrid PIC using butt-coupling. A RSOA is used with an external cavity on silicon in order to realize a hybrid tunable laser. A partial reflectivity mirror closes the laser cavity. A second PIC could be used to generate advanced modulation format generation, realize wavelength or polarization multiplexing or even more advanced all-optical signal processing. Then an extra SOA is used as booster before coupling into a single-mode (SMF) fiber. In such configuration, three alignments are required (RSOA/silicon, silicon/SOA, SOA/fiber) which increases the packaging costs accounting for 60% to 90% of the final optoelectronic module cost. Our concept based on our B-RSOA device is presented on the bottom part of figure. 1 (a). After closing the hybrid III-V/Si laser cavity with a partial-reflectivity mirror, the optical signal is re-directed on the same side of the silicon PIC and injected into the B-RSOA. In such configuration, only one alignment is required as the RSOA and B-RSOA could be fabricated into the same chip therefore reducing the number of coupling alignment between the silicon and III-V chips. The design of the B-RSOA is shown in figure 1 (b). We use a buried ridge structure as described in [7]. The active layer of the B-RSOA was grown using gas source molecular beam epitaxy (MBE), consisting of multi-quantum well (MQW) InGaAsP (Eg~0.78eV) surrounded by confinement layers of InGaAsP with a larger band gap (Eg~1.06eV). AR coating was deposited on the front facet. Ion implantation is realized around the active zone to ensure optimum electrical injection. A double core structure was chosen with active and passive waveguides integration by means of a taper. Such a spot size converter permits to optimize both the mode transfer and the output mode for the coupling between the B-RSOA and the silicon cavity [8]. The B-RSOA is designed such the input and output waveguides are separated by 250 um in order to be compatible with commercial fiber array waveguide for testing purpose. Reflective directional couplers are used to re-direct the optical signal from the input to the output waveguide. From the coupled mode theory, the optical mode bounces from one strip to another along its propagation in the coupling region. By choosing the right length and a 50:50 coupling factor, we can redirect the input from the first waveguide into the second waveguide at the output. To create contacts, the p-type pad is connected to a stripe on the top of the mesa via the gold-platinum layer, and the n-type pad is connected at the bottom of the chip. We simulated the transmission coefficient (from the input waveguide to the output waveguide in a reflective configuration) depending on the length of the active coupling region with different channel spacing based on a cleaved facet (air) therefore reaching a maximum transmission around 35%. Figure 2. (a) shows the forward and backward field intensity in the reflective directional coupler. We extrated the optimal coupling length and the maximum transmission coefficient as presented in figure 2 (b). As expected the coupling length strongly depends on the channel spacing and varies from 88 to 1330 µm for 0.6 and 1.4 µm channel spacing, respectively.

III. EXPERIMENTAL CHARACTERIZATION

We fabricate and evaluate the B-RSOA with a coupling length of 362 µm and a channel spacing of 1 µm according to previous simulations. In order to characterize the device on its own, we use a custom-made lensed fiber array with 250 µm pitch (using a commercial V-groove). Lensed fibers are used in order to match the output optical mode of the spot-size converter (3x4µm in the passive section). A peak gain at 1540 nm wavelength is observed along with ~15nm 3dB optical bandwidth (mainly limited by the reflective directional coupler) as represented in figure 3. The optical

OECC/PS2016

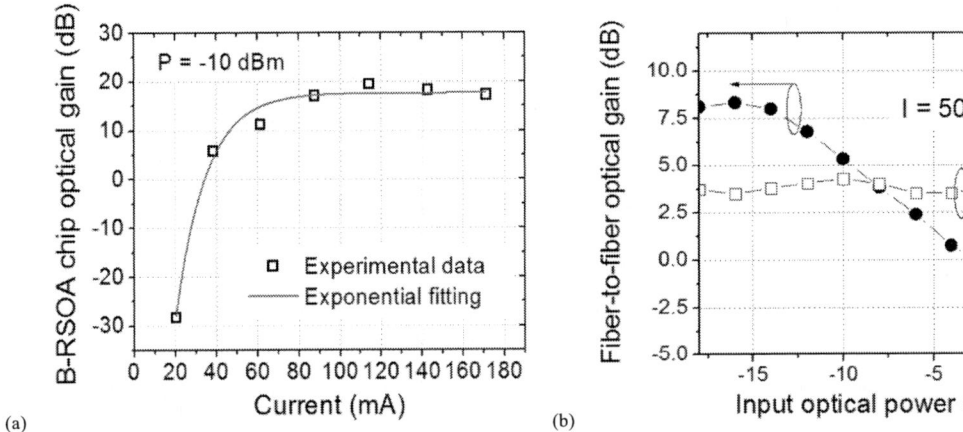

(a)   (b)

Fig. 4. (a) Chip optical gain depending on the input current (b) Fiber-to-fiber optical gain and noise factor depending on the input optical power.

bandwidth could be further increase by reducing the reflection and using reflective multimode interference (MMI) configuration for instance. The gain ripple around 2dB at a bias current of 50 mA is measured and can be reduced by tilting the output waveguide and still achieving effective coupling with the silicon chip has demonstrated in [8]. The B-RSOA chip optical gain depending on the injected electrical current is shown in figure 4 (a). The chip optical gain reaches a maximum value of 19.5 dB at 115 mA. We characterize the device at 50 mA in order to limit the gain ripple below 2 dB. Fiber-to-fiber optical gain values above 7.5 dB can be achieved in the small signal regime. Please, note that a cleaved facet is used at the back of the device allowing only 30 % of the signal to be reflected back to the structure. High reflection coating could be applied to provide 100% reflection however ultra-low reflectivity facet (tilted waveguide and AR coating) is required on the other end of the device in order to prevent lasing. The input saturation power is around – 9 dBm. Long active section induces carrier depletion due to guided amplified spontaneous emission (ASE) inducing reduced saturation power. Shorter devices are under investigation in order to reduce such effect. The internal noise factor of the device (i.e. considering coupling loss) was measured between 15.5 and 16.5 dB. Gain saturation takes place when the photon density in the active region overcomes the available carrier density. Therefore decreasing the optical confinement and designing more compact devices expect improving the saturation power and the noise factor.

IV. CONCLUSIONS

We propose novel bidirectional reflective semiconductor optical amplifier for on-chip amplification. The concept, design and fabrication are described in this paper. The device presents a 15 nm optical bandwidth and up to 19.5 dB chip optical gain. Further directions for improvements are identified and discussed as reducing the reflection in order to reduce the gain ripple on one end and increasing the reflection on the other end of device as well as designing lower optical confinement and compact devices for high saturation power B-RSOA.

ACKNOWLEDGMENT

We acknowledge support of J.-L. Gentner and M. Zirngibl at Bell Laboratories.

REFERENCES

[1] Y. Urino, et al., "First demonstration of high density optical interconnects integrated with lasers, optical modulators, and photodetectors on single silicon substrate", Opt. Express, vol. 19, No 26 2011.

[2] G.-H. Duan, et al., "New Advances on Heterogeneous Integration of III–V on Silicon", Journal of Lightwave Technology, vol. 33, No 05, 2015.

[3] J. F. Liu, et al., "A Ge-on-Si laser operating at room temperature", Optics Lett., vol. 35, pp. 679-681, March 2010.

[4] A. Y. Liu, et al., "High performance continuous wave 1.3 µm quantum dot lasers on silicon", Applied Physics Letters, Vol. 104 No 4, January 28, 2014.

[5] Kenji Sato et al,. "High Output Power and Narrow Linewidth Silicon Photonic Hybrid Ring-Filter External Cavity Wavelength Tunable Lasers", Proceeding of ECOC, PD2.3, Cannes, France, 2014.

[6] J.-H. Lee et al., "Demonstration of 12.2% wall plug efficiency in uncooled single mode external-cavity tunable Si/III-V hybrid laser", Opt. Express, vol. 23, No 9, 2015.

[7] G. de Valicourt et al., "High Gain (30 dB) and High Saturation Power (11dBm) RSOA Devices as Colourless ONU Sources in Long Reach Hybrid WDM/TDM -PON Architecture", Photon. Technol. Lett., Vol. 22, No. 3, p. 191 (2010).

[8] G. de Valicourt et al., "Dual Hybrid Silicon-Photonic Laser with Fast Wavelength Tuning", in Proc. OFC, M2C.1, Anaheim, USA, 2016.

WA2-69

OECC/PS2016

Semiconductor Optical Amplifier in AWG-STAR Network with Wavelength Path Relocation Function

Takumi Niihara[1], Minoru Yamaguchi[1], Osanori Koyama[1], Hiroaki Maruyama[1], Kazuya Ota[2], and Makoto Yamada[1]

[1]Osaka Prefecture University, Gakuen-cho 1-1, Naka-ku, Sakai, Osaka, 599-8531, Japan

[2]Trimatiz Ltd., 801, La Pacifique B, 4 - 7 - 12 Minami Yawata, Ichikawa, Chiba 272-0023, Japan

koyama@eis.osakafu-u.ac.jp, myamada@eis.osakafu-u.ac.jp

Abstract: *We constructed a semiconductor optical amplifier (SOA) unit with a signal gain greater than 20 dB, CWDM bandwidth amplification, low polarization dependence, and a low noise figure. To confirm the applicability of the SOA in an AWG-STAR network with our proposed wavelength path relocation function, we evaluated the power penalty of the SOA unit in an experimental AWG-STAR network. We found that amplification by the SOA unit was effective in making wavelength path relocation more flexible by compensating for accumulated optical losses due to optical devices used in the AWG-STAR network.*

Keywords: *semiconductor optical amplifier, arrayed waveguide grating, IP, Ethernet*

I. INTRODUCTION

Field trials using semiconductor optical amplifiers (SOAs) have been conducted to develop communication systems and infrastructures [1], and international standards are being drawn up [2]. In common use, SOAs can amplify wavelengths between 1.3 μm and 1.6 μm. To satisfy rapidly increasing traffic demands in recent years, various optical networks have been proposed, such as networks using arrayed waveguide gratings (AWGs), which can route light waves according to wavelength. We proposed and developed a novel AWG-STAR network with wavelength path relocation function [3-5]. It is necessary to compensate for optical losses due to wavelength path relocation to make reallocations more flexible. Thus, in this research, we constructed an SOA unit and applied it to our proposed AWG-STAR network. Moreover, we demonstrated some characteristics of the SOA when used for wavelength path relocation.

II. CONFIGURATION AND BASIC CHARACTERISTICS OF THE SOA UNIT

Figure 1 shows the configuration of our SOA unit. The SOA unit comprises an SOA module, optical isolators (insertion loss <0.8 dB), tap couplers for monitoring input and output signal powers (insertion loss <0.5 dB), and photodiodes. Figure 2 shows the amplification characteristics of the SOA unit. Figure 2(a) shows gain spectra and noise figures, Fig. 2(b)

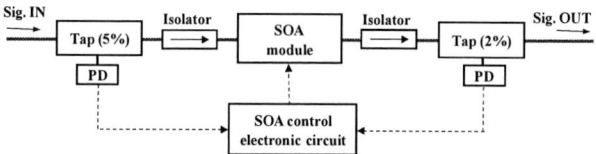

Fig. 1. Configuration of the semiconductor optical amplifier unit.

shows signal gain depending on driving current, and Fig. 2(c) shows saturation input signal power depending on signal gain. Here, saturation power was defined as the input signal power when signal gain reached 3 dB down from the maximum. As shown in Fig. 2, we found that the SOA can amplify the input signal more than 20 dB in CWDM bands, i.e., 1530 nm, 1550 nm, and 1570 nm. We also found that the SOA had a small polarization dependence (<1 dB) and noise figures less than 7 dB, and we confirmed that the saturation input signal power and signal gain had a relation of monotonic decrease. Generally, in the case of amplification using SOAs, it is believed that the decrease in the saturation input signal power with respect to the amplification deterioration is caused by noise figures and pattern effect [6].

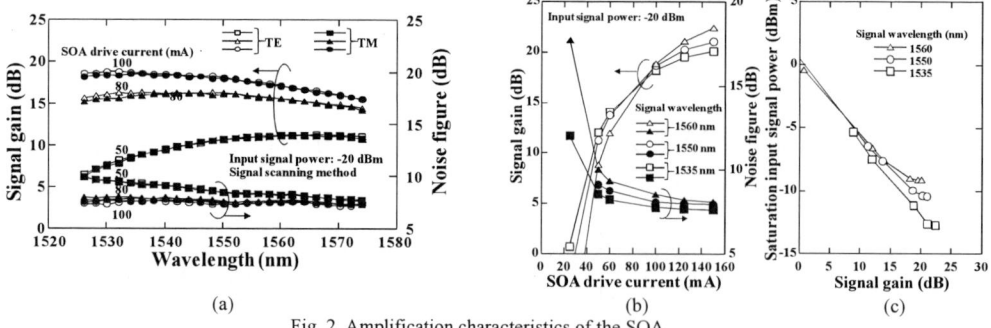

Fig. 2. Amplification characteristics of the SOA.

III. CHARACTERISTICS OF THE SOA UNIT IN AN IP NETWORK

We constructed an experimental setup based on an IP network as shown in Fig. 3 and then conducted IP packet transmission tests in order to clarify the characteristics of the SOA unit in this network. The signals from optical transceiver 1 (OTR1) to OTR2 carried IP packets, including echo requests, using the Internet Control Message Protocol (ICMP). The IP packets were generated using a personal computer 1 (PC1), then transmitted sequentially via optical attenuator 1 (ATT1), the SOA unit, a coarse wavelength division multiplexing (CWDM) filter, and ATT2. Three tap couplers were inserted to monitor the input power to the SOA unit, the signal gain by the SOA unit, and the received signal power at the OTR2, using optical power meters. After PC2 received the IP packets, these packets, including echo reply, were sent back to PC1, without passing through attenuation or amplification units. ATT1 and ATT2 were used to adjust the input power into the SOA unit and the received power at the OTR2, respectively. The optical filter in the CWDM had a transmission pass bandwidth of ~13 nm. To evaluate the power penalty of the SOA unit in the IP network, the IP packets containing echo requests and the replies were monitored with a LAN analyzer. The packet loss rate, which was defined as the ratio between the total requests and the replies, was measured. Figure 4 shows the relationship between the packet loss rate and the received signal power at OTR2 with, and without, the SOA unit, when the signal gain was 10 dB and the input power into the SOA unit was −20 dBm. The minimum receivable signal power was lowered by ~0.4 dB by introducing the SOA unit into the network. Figure 5 shows the power penalty of the SOA unit as a function of signal gain. We found that the SOA unit could be used in an optical IP network and that it has a relatively low power penalty when used below saturation input power.

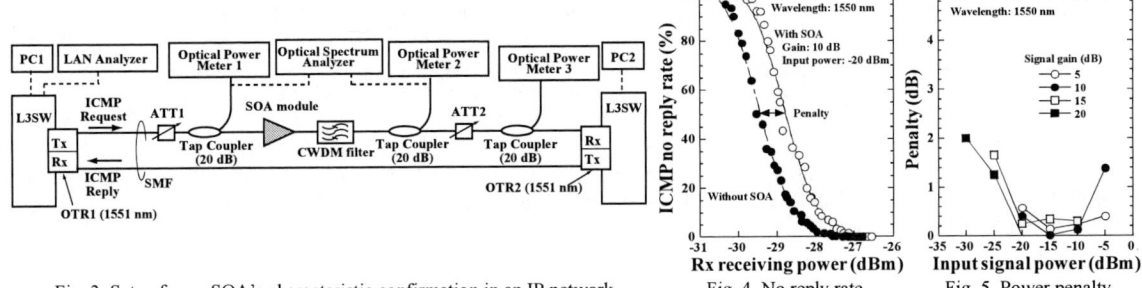

Fig. 3. Setup for an SOA's characteristic confirmation in an IP network. Fig. 4. No reply rate. Fig. 5. Power penalty.

IV. AWG-STAR NETWORK WITH WAVELENGTH PATH RELOCATION FUNCTION

Figure 6 shows a schematic of an AWG-STAR network with wavelength path relocation function incorporating the SOA unit. Communication nodes of which the number is N are physically connected to the AWG by a pair of single mode fibers. The AWG gives the network a full-mesh wavelength path topology logically. The gateway in each node consists of an SOA unit, an optical add-drop multiplexer (OADM), N optical switches (OSWs) corresponding to each wavelength, and a layer-3 switch (L3SW). The local network communicates with the other nodes through the gateway. The N wavelength signals are multiplexed into the fiber with the OADM, and reach each destination node via the AWG. Each OSW has one of two states, C_{PT} or C_{LB}. When the state is C_{PT}, the optical transceiver installed in the L3SW can send optical signals into the OADM. Conversely, when the state is C_{LB}, a wavelength signal sent from the OADM is looped back to the AWG through the OADM. By looping back, it is possible to allocate an appropriate transmission capacity in response to traffic demand changes, by relocating some wavelength paths among nodes where the traffic demand is relatively low [3]. However, when the number of loopbacks increases, it is impossible to receive IP packets exactly, because the optical losses accumulate. Therefore, we deployed the SOA unit in the AWG-STAR network to compensate for optical losses.

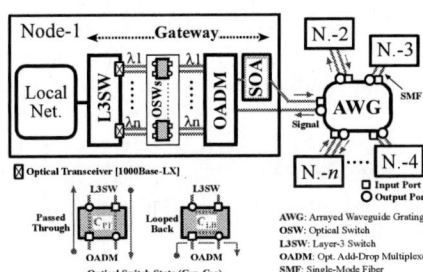

Fig. 6. AWG-STAR network with wavelength path relocation function including the SOA unit.

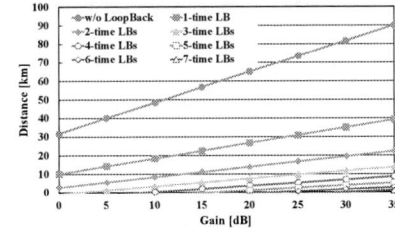

Fig. 7. Network scalability dependence on SOA unit signal gain.

V. NETWORK SCALABILITY WHEN USING AN SOA UNIT

Accumulated optical losses as described above restrict the scalability of the AWG-STAR network. We calculated the scalability when using the SOA unit. Figure 7 shows the required signal gain when multiple loopbacks are used, as a function of the distance between the AWG and each node. The following loss values that we obtained through measurements were used in the calculation: fiber, 0.3 dB/km; OSW, 0.6 dB; OADM, 1.07 dB; and AWG, 4.3 dB. It was assumed that the distances between the AWG and each node were equal. The number of nodes was chosen as 8 because we used an 8x8 AWG in our experimental network. Hence, the maximum loopbacks was 7. For example, we found that the optical loss compensation required was 15 dB when the number of loopbacks was 2 and the distance was 10 km.

VI. SUITABLE LOCATIONS FOR AMPLIFYING WITH AN SOA

It is necessary to consider suitable locations where the SOA unit should be placed, because the signal gain and the power penalty change, depending on the input signal power. The SOA unit's location also influences reception sensitivity. Figure 8 shows the power penalty of an SOA unit when the unit is placed in each node in turn, in the case where the signal from node 1 was transmitted to node 3 via nodes 5 and 7 (number of loopbacks = 2). We found that the power penalty was a minimum when the input signal power was between −20 dB and −10 dB. As a result, it is appropriate that the SOA unit should amplify the input signal in either node 5 or node 7.

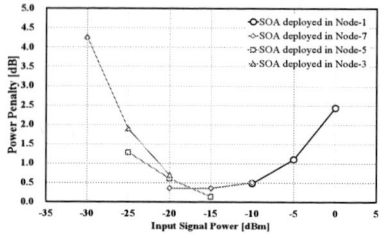

Fig. 8. Power penalty of an SOA in the AWG-STAR network.

VII. THROUGHPUT IMPROVEMENT BY WAVELENGTH PATH RELOCATION USING LOOPBACK

Figure 9 shows the throughput improvement between nodes 1 and 3 in our experiment. We used a direct wavelength path (no loopback) from nodes 1 to 3 and a looped back wavelength path from nodes 1 to 3 via nodes 7 and 5. Each wavelength path capacity was 1 Gbps. The distance between the AWG and the nodes was 10 km. The optical signals through the looped back wavelength path were not successfully received at node 3, due to optical loss accumulation when the SOA unit did not compensate for optical losses (SOA was deactivated). The SOA unit gave the signal 17 dB gain to compensate for the optical losses, taking into account the power penalty of the SOA unit. The SOA unit for the looped back wavelength path in this experiment was placed in node 5.

First, two IP packet streams, each at maximum output volume, were sent from nodes 1 to 3 with the SOA unit deactivated (gain 0 dB). Total throughput at the transmission side was 1974 Mbps. Total throughput at the receiving side was 987 Mbps, because the number of valid wavelength paths from nodes 1 to 3 was one (no loopback), which indicated that the loopback wavelength path was unusable due to loss accumulation and an IP packet loss rate of ~50 %. After 60 seconds, the SOA unit was activated, amplifying the looped back wavelength path by 17 dB gain. As a result, the total throughput at the receiving side increased to 1973 Mbps, because the looped back wavelength path was used successfully and each stream had a dedicated wavelength path. As seen in Fig. 9, we confirmed that the transmission capacity between nodes 1 and 3 increased by wavelength path relocation with the SOA unit. The results show that our SOA unit can be used for IP over a AWG-STAR network.

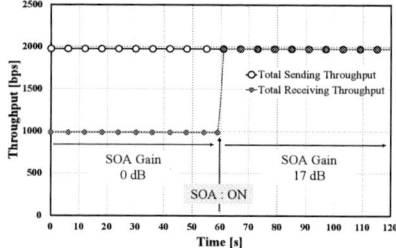

Fig. 9. Throughput improvement due to wavelength path reallocation using loopback with the SOA.

VIII. CONCLUSION

We constructed an SOA unit with a signal gain > 20 dB, CWDM bandwidth amplification (i.e., 1530 nm, 1550 nm, 1570 nm), polarization dependence < 1 dB, and a noise figure < 7 dB. We evaluated the power penalty of the SOA unit in an experimental IP over AWG-STAR network with wavelength path relocation function. It was found that the most suitable amplifying location of the SOA unit was in the middle nodes of a wavelength path. Furthermore, we confirmed that an unusable wavelength path (without amplification) became usable by utilizing the SOA. As a result, the SOA unit we constructed possesses a potential use in optical IP networks.

REFERENCES

[1] M. Fujiwara, T. Imai, K. Taguchi, K. Suzuki, H. Ishii, and N. Yoshimoto, "Field Trial of 79.5-dB Loss Budget, 100-km Reach 10G-EPON System Using ALC Burst-Mode SOAs and EDC." Proc. of OFC/NFOEC 2012, Los Angeles, CA (USA), March 2012, paper. PDP5D.8.

[2] IEC 86C/1144/CDV.

[3] M. Yamaguchi, R. Higashiyama, K. Toyonaga, O. Koyama, and M. Yamada, "AWG Star-network with Dynamically Reconfigurable Wavelength Paths via Node-Side Control." Proc. of OSA Advanced Photonics for Communications, San Diego, California (USA), July 2014, paper. JT3A.17.

[4] R. Higashiyama, M. Yamaguchi, K. Toyonaga, O. Koyama, and M. Yamada, "Dynamic enhancement of internode transmission capacity in IP over AWG-STAR network." Proc. of the 20th Asia-Pacific Conference on Communications (APCC), Pattaya (Thailand), October 2014, paper. F3A4_1017.

[5] M. Yamaguchi, O. Koyama, H. Maruyama, T. Niihara, and M. Yamada, "Matrix Representation for Wavelength Path Relocation in AWG-STAR Network with Loopback Function." Int. J. of ICIC, (Accepted), 2016.

[6] T. Akiyama, N. Hatori, Y. Nakata, H. Ebe, and M. Sugawara, "Pattern-effect-free semiconductor optical amplifier achieved using quantum dots." Electron. Lett., vol. 38, no. 19, pp. 1139-1140, 2002.

Proposal of an Orchestrator-to-Orchestrator Interface using the Control Orchestration Protocol

Yoshiaki Inoue, Jun Matsumoto, Satoru Okamoto, and Naoaki Yamanaka

Graduate School of Science and Technology, Keio University, 3-14-1 Hiyoshi, Kohoku-ku, Yokohama, Kanagawa, 223-8522 Japan

inoue@yamanaka.ics.keio.ac.jp

Abstract: In multi-carrier SDTN, Orchestrator-to-Orchestrator Interface is required since each carrier has its own orchestrator, whereas this interface is not defined. We propose a scheme to solve this issue and implement the proposed solution.

Keywords: SDTN, the Control Orchestration Protocol, Object-Defined Network OS, Orchestrator-to-Orchestrator

I. INTRODUCTION

The Software Defined Transport Network (SDTN) [1] which applies the Software Defined Networking (SDN) technology to the transport network control method has been emerging as a novel carrier network's architecture in order to manage heterogeneous networks such as different domains, layers, and vendors. SDTN consists of three hierarchical elements; (i), (ii), (iii). i) Physical networks, which are composed of various transport protocols and constructed multi-layer networks, ii) An SDN controller, which manages the physical network that forms a domain, iii) An orchestrator, which controls the whole network across multiple domains. The SDN controller has northbound and southbound interfaces (NBI and SBI), which communicate with the orchestrator and manage the network elements, respectively. The Object-Defined Network Operating System (ODENOS) [2, 3] has been developed as an orchestrator that can integrate and control heterogeneous networks by abstracting the physical network information.

NBI has a challenging issue in terms of standardization since each SDN controller uses own specific protocol. ODENOS supports OpenFlow, REST/JSON, and so on. The Control Orchestration Protocol (COP) [4] has been proposed as a common NBI/SBI protocol between SDN controllers and among orchestrator and SDN controllers. In case of forming SDTN with multiple carriers, each carrier has its own orchestrator. Therefore, a peer-to-peer Orchestrator-to-Orchestrator Interface (OOI) is required. This OOI is not defined yet.

In this paper, we propose a scheme that realizes an emulated OOI using COP. As an example, ODENOS communicates with SDN controllers and also ODENOSs using COP. The proposed scheme unifies all the orchestration functionalities into a single protocol paradigm.

II. ODENOS AND COP

A. Object-Defined Network Operating System (ODENOS)

ODENOS which is an orchestrator developed by the O3 project [5] has a driver component and four types of objects to integrate and control heterogeneous networks. The driver abstracts the actual network information (e.g. node, port, link, topology and flow) and converts them in common format. The four types of objects are Aggregator, Federator, Link Layerizer, and Slicer as shown in Fig. 1 [2]. Aggregator aggregates a whole network into a single logical node including path computation within a network. Federator glues multiple networks into a single network including path computation among networks. Link Layerizer converts lower layer flow into higher layer link. Slicer slices a network into multiple virtual networks having same topology and divides flow space. ODENOS can combine some of these objects for desired networks. As described above, ODENOS is based on the client/server communication model.

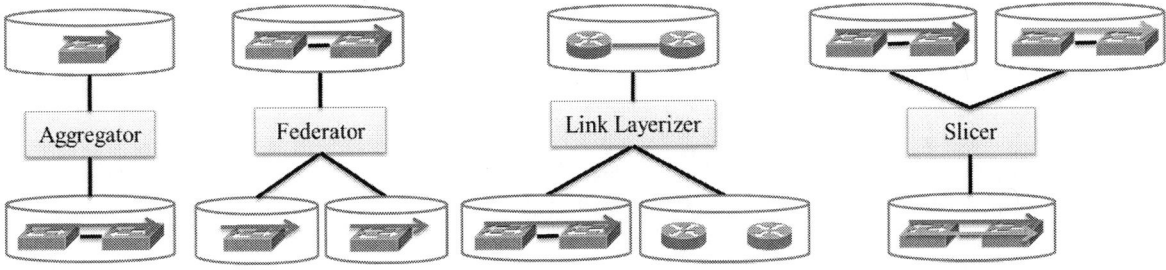

Fig. 1. The four types of objects: Aggregator, Federator, Link Layerizer, and Slicer.

B. Control Orchestration Protocol (COP)

COP which has been defined by STRAUSS project [6] abstracts a common set of control plane functions used by various SDN controllers and orchestrators, for the interworking of heterogeneous control plane paradigms such as OpenFlow and Generalized Multi-Protocol Label Switching (GMPLS)/Path Computation Element (PCE). COP has been defined using YANG model language and can be transported using RESTconf [4]. COP provides the necessary functions to bring the benefits of the programmable SDTN application programming interface (API). The latest SDTN API is designed towards COP objectives. OIF/ONF [7, 8] SDTN API is focused on standardization for orchestration of YANG/REST NBI that COP uses for SDN controllers. COP is also defined based on the client/server model.

III. PROPOSED ORCHESTRATOR -TO-ORCHESTRATOR INTERFACE

In this section, three OOI contributions are described A) Definition of the connection between two ODENOSs as an emulated peer-to-peer OOI, B) Implementation of OOI to realize ODENOSs communication and ODENOS and SDN controllers communication with COP, C) Conversion of the output information with NBI and SBI.

A. Emulated peer-to-peer OOI

Figure 2 shows a proposed SDTN architecture, an emulated peer-to-peer OOI and communication between ODENOSs and among ODENOS and SDN controllers with COP. The connection between ODENOSs using COP can be realized by emulating ODENOS as an SDN controller. Not only SDN controller but also ODENOS has NBI and SBI, which provide control information to applications and manages SDN controllers, respectively. One ODENOS's NBI and the other ODENOS's SBI connect to realize the emulated peer-to-peer OOI.

All layers over SDN controllers are defined as Control Plane in SDTN architecture, as Data Plane consists of three physical networks in Fig.2. While the conventional SDTN where OOI is not defined performs only end-to-end path within domains under one orchestrator, the proposed SDTN architecture enables operators to establish end-to-end paths by defining OOI and connecting between ODENOSs. Therefore, each carrier can have its own orchestrator in multi-orchestrator SDTN and it benefits the carrier in terms of confidentiality (described in subsection III-C). All orchestrators construct the global scale SDTN, exchanging via OOI the local scale SDTNs which are the abstracted network information. The orchestrators regenerate the link topology taking into account the overlaps since the only necessary information can be transmitted.

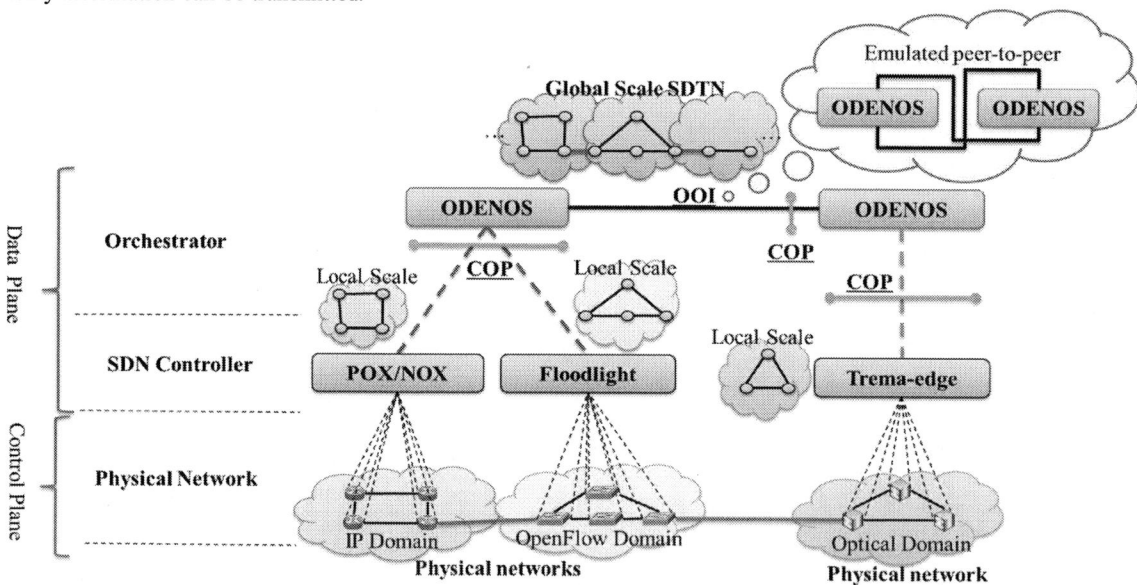

Fig. 2. SDTN architecture and communication between ODENOSs/ODENOS and SDN Controller with COP.

B. Implementation of COP to ODENOS

To communicate with ODENOS using NBI, the SDN controller uses Northbound API. We implemented modified API which outputs the information such as reachability and topology information in COP format instead of conventional Northbound API (e.g. NOX/POX API, Floodlight API and so on).

As for orchestrator, we also need to implement modified NBI. The ODENOS's NBI which exists in the lower layer needs also to be altered to COP so that the transported information to the higher layer's ODENOS is done conversion. Moreover, we have to develop ODENOS's modules because the conventional ODENOS does not have the ability to convert the input information into the output one with NBI and SBI.

C. Conversion of the output information with NBI and SBI

In multi-carrier SDTN, there is a need for conversion of the exchange information. The SBI has all information regarding the actual network. If one carrier does not want to transport a partial network topology to other carriers, ODENOS is able to eliminate it from the whole network information through NBI.

IV. CONCLUSIONS

The Control Orchestration Protocol (COP) based Orchestrator-to-Orchestrator Interface (OOI) is proposed. OOI is the important function to realize multi-carrier SDTN. The proposed scheme that emulates ODENOS as an SDN controller to achieve peer-to-peer OOI with COP helps carriers construct the global scale SDTN. The contributions of this paper are to define and implement an emulated peer-to-peer OOI, consider conversion of the output information with NBI and SBI and achieve end-to-end paths through OOI in multi-orchestrator SDTN. We verified the feasibility of the proposed scheme.

ACKNOWLEDGMENT

This work is partly supported by the "ACTION Project" funded by the National Institute of Information and Communications Technology (NICT) Japan.

REFERENCES

[1] National Institute of Information and Communications Technology (NICT) press release, "Successful interoperability among 100Gbit-class core, metro and access optical networks with Software Defined Transport Network technology," http://www.nict.go.jp/en/press/2014/05/27-1.html, May 2014.

[2] Y. Iizawa, M. Morimoto, T. Koide, Y. Ashida, and H. Shimonishi, "Network Abstraction and Control Models for Hierarchical SDN Controllers," Open Networking Summit (ONS) 2014, Santa Clara, CA, USA, Mar. 2014.

[3] ODENOS, http://www.o3project.org/en/index.html

[4] R. Vilalta, V. Lopez, A. Mayoral, N. Yoshikane, M. Ruffini, D. Siracusa, R. Martinez, T. Szyrkowiec, A. Autenrieth, S. Peng, R. Casellas, R. Nejabati, D. Simeonidou, X. Cao, T. Tsuritani, I. Morita, J. P. Fernandez-Palacios, and R. Munoz, "The Need for a Control Orchestration Protocol in Research Projects on Optical Networking," European Conference on Networks and Communications (EuCNC), pp.340-344, Paris, France, July 2015.

[5] The O3 project, http://www.o3project.org/en/index.html

[6] The STRAUSS project, http://www.ict-strauss.eu/en/

[7] The Open Networking Foundation, https://www.opennetworking.org/

[8] The Optical Interworking Forum white paper, "Framework for Transport SDN: Components and APIs," OIF-FD-Transport-SDN-01.0, May 2015.

WA2-70

OECC/PS2016

40Gb/s Optical Receiver Using High-Gain Multi-Level Active Feedback with Serial Inductor Peaking

Cheng-Ta Chan, and Oscal T.-C. Chen
Department of Electrical Engineering,
National Chung Cheng University,
Chiayi, 62102 Taiwan

Abstract: *In this work, a high-gain wide-bandwidth optical receiver consisting of a trans-impedance amplifier, a limiting amplifier, and an output buffer is developed. Especially in each gain stage of a limiting amplifier, the high-gain 4'th-order Multi-Level Active Feedback (MLAF) structure with serial inductor peaking is employed to effectively increase the bandwidth and the gain. The TSMC 90nm CMOS technology was used to implement the proposed optical receiver. With the use of inductor peaking applied in the 4'th-order MLAF to enlarge the bandwidth, the proposed optical receiver has a bandwidth of 35GHz, and a differential trans-impedance gain of 86dBΩ. Comparing to conventional optical receiver, the proposed optical receiver exhibits a wide bandwidth, a high gain and fairly good performance for applications of 40Gbps optical communications.*

Keywords: *Optical receiver, trans-impedance amplifier, limiting amplifier, active feedback structure, inductor peaking*

I. INTRODUCTION

To accomplish optical communications, the optical receiver including a trans-impedance amplifier and a limiting amplifier is commonly used to detect and transfer a photo current to a voltage signal, and then to enlarge a signal from several milli-volts to hundreds of milli-volts. In an optical receiver, the trans-impedance gain and voltage gain of a trans-impedance amplifier and a limiting amplifier are around 40dBΩ and 40dB, respectively, in order to achieve a wide bandwidth. With the increase of data rates, the advanced CMOS process technology with a high f_T is usually adopted to realize high-gain and wide-bandwidth optical receivers. However, the trans-conductance of MOSFET under a 100nm technology is likely less than that above a 100nm technology at the same width/length ratio [1]. Such phenomenon results in that the gain of an optical receiver may not be easily and straightforwardly raised up. For example, at the 0.18um CMOS technology, a common approach of an optical receiver is to employ inductor peaking to reach a data rate of 10Gbps. When the data rate is up to 40Gbps, the 45nm CMOS technology is likely preferred but its g_m is apparently decreased. It is little difficult to implement a high-gain and wide-bandwidth optical receiver under a nano-meter technology. Therefore, multiple bandwidth extension techniques need to be used together to design an optical receiver with a large gain and a wide bandwidth [2]-[10].

In this work, the high-gain MLAF structure with serial inductor peaking [7]-[9], is proposed and developed in an optical receiver which include a trans-impedance amplifier, a limiting amplifier and an output buffer. Particularly, the proposed limiting amplifier consists of two high-gain 4'th-order MLAF gain stages. The TSMC 90nm CMOS technology was used to implement the proposed optical receiver which can yield a frequency bandwidth of 35GHz, and the overall trans-impedance gain of 86dBΩ at 1.2V for 40Gbps communications. Notably, the optical receiver proposed herein demands a less number of inductors.

II. PROPOSED HIGH-GAIN WIDE-BANDWIDTH OPTICAL RECEIVER

The proposed optical receiver, which consists of a trans-impedance amplifier, a single-ended to differential circuit, a limiting amplifier and an output buffer, is shown in Fig. 1(a). The common-source amplifier with current source loading and feedback resister, R_f, connecting between input and output nodes is implemented in the proposed trans-impedance amplifier, where R_f is 200Ω to provide the trans-impedance gain of 200Ω (46dBΩ). The limiting amplifier is composed of two cascaded gain stages, where each gain stage has a gain of 20dB, and adopts the MLAF structure to extend the bandwidth. The N'th-order MLAF structure is shown in Fig. 1(b).

To realize the 40Gbps transmission rate in an optical receiver, Fig. 2(a) shows the curve of GBW_c multiplied by power dissipation as well as the curve of GBW_c where the total gain and bandwidth are 10 and 28GHz, respectively. Here, GBW_c denotes the gain-bandwidth product of an amplifier cell in MLAF, and $GBW_c - PD$ denotes the product of

GBW_c and power consumption, of which least value is preferred. At the 90nm CMOS technology, f_T approaches 120GHz. According to the concept of [10], N=4 is recommended in the proposed MLAF structure where $GBW_c - PD$ is the second best, and GBW_c is much lower than 120GHz. Figure 2(b) shows the circuit diagram of the 4'th-order MLAF structure in gain stages. To reach a high gain, two current sources, I_L, are connected with the loading resistors R_L of each differential amplifier of A_1, A_2, A_3, and A_4 in Fig. 2(b). The I_L shares the tailing current, and thus reduces the current through the loading resistor, R_L. Such arrangement can increase the voltage headroom of an output node, and then uses a large R_L to increase the gain. Additionally, serial inductors, L_s, connecting to the output nodes of A_1, A_2, A_3, and A_4 form serial peaking to further extend the bandwidth [7], [9].

Typically, the use of inductor peaking can enlarge the bandwidth efficiently [7]-[9], but a large value of inductor may cause the overshooting, and increase a group delay variation. In our design, the tailing current of A_1, A_2, A_3, and A_4 is 6mA where each amplifier has a differential gain of 5dB. With properly determining transistor sizes, R_L, and I_L, the ratios of input capacitance over output capacitance of A_1, A_2, A_3, and A_4 are around 3. According to the design concept in [7], adequate values of L_s used in A_1, A_2, A_3, and A_4 can be adopted to extend the bandwidth by around 2 times where a low group delay variation can be attained. Afterwards, the feedback circuits of A_{f_1}, A_{f2}, and A_{f3} are included to build a 4'th-order MLAF structure. Since there are series inductors in the MLAF structure, the frequency response may not likely remain at the Butterworth response. However, when gains of A_{f_1}, A_{f2}, and A_{f3} are fairly determined, the overall frequency response is close to the flat, and poles are still kept at the left-hand side of the S plane. Finally, the differential pair circuit is employed in the output buffer to drive 50Ω loading.

III. CIRCUIT SIMULATIONS OF PROPOSED OPTICAL RECEIVER

The proposed optical receiver was implemented using the TSMC 90nm CMOS technology. The current demanded by the trans-impedance amplifier is 6mA. The total current demanded by two gain stages is 60mA at 1.2V where the output buffer has a 12mA tailing current. Figure 3 shows the eye diagram of differential outputs with a swing of about $600mV_{pp}$, where the equivalent input current is $30uA_{pp}$. The total differential trans-impedance gain is about 86dBΩ. Figure 4(a) plots the frequency response of optical receiver. The bandwidth of the proposed optical receiver can achieve 35GHz. Additionally, the variation of the frequency response at the pass-band region is smaller than 3dB. Figure 4(b) displays the group delay of optical receiver where the group delay variation is about $\pm 10ps$ in the bandwidth of 35GHz. Table 1 compares the performance of the proposed amplifier and conventional TIAs used for optical receivers [4]-[6]. Of these amplifiers, the proposed LA has a wider bandwidth and a higher voltage gain as well as satisfying the required performance of 40Gbps optical communications. Particularly, the proposed optical receiver does not need a large number of inductors, and thus effectively minimizes the hardware cost.

IV. CONCLUSION

To effectively realize a 40Gbps optical receiver, the serial inductor peaking employed in the high-gain MLAF structure is proposed, and was implemented by using the TSMC 90nm CMOS technology. The −3dB cutoff frequency of the proposed optical receiver is 35GHz, and the total differential trans-impedance gain reaches 86dBΩ. The optical receiver proposed herein shows fairly good performance as compared to the conventional receivers at 40Gbps optical communications.

ACKNOWLEDGMENT

This work is partially supported by Ministry of Science and Technology, Taiwan, under the contract number of MOST 104-3011-E-194-002. Additionally the authors would like to thank the Chip Implementation Center, Hsinchu, Taiwan, providing the service of chip fabrication.

REFERENCES

[1] K. Lal Kishore, and V. S. V. Prabhakar, *VLSI Design*, New Delhi: I.K. International Publishing House, 2010.

[2] B. Razavi, *Design of Integrated Circuits for Optical Communication*, New York: McGraw-Hill, 2003.

[3] Eduard Sackinger, *Broadband Circuits for Optical Fiber Communication*, New Jersey: John Wiley & Sons, 2005.

[4] J.-D. Jin, and S. Hsu, "A 40-Gb/s trans-impedance amplifier in 0.18μm CMOS technology," *IEEE Journal of Solid-State Circuits*, vol. 43, no. 6, pp. 1449-1457, Mar. 2008.

[5] Chih-Fan Liao, and Shen-Iuan Liu, "40 Gb/s trans-impedance-AGC amplifier and CDR circuit for broadband data receivers in 90nm CMOS," *IEEE Journal of Solid-State Circuits*, vol. 43, no. 3, pp. 624-655, Mar. 2008.

[6] Joohwa Kim, and James F. Buckwalter, "A 40Gb/s optical tranceiver front-end in 45nm SOI CMOS, " *IEEE Journal of Solid-State Circuits*, vol. 47, no. 3, pp. 615-626, Mar. 2012.

[7] Joohwa Kim, and J. F. Buckwalter, "Bandwidth Enhancement with low group-delay variation for a 40-Gb/s transimpedance amplifier," *IEEE Trans. Circuits Syst. I*, vol. 57, iss. 8, pp.1964 -1972, Aug. 2010.

[8] B. Analui, and A. Hajimiri, "Bandwidth enhancement for trans-impedance amplifiers," *IEEE Journal of Solid-State Circuits*, vol. 39, iss. 8, pp. 1263-1270, Aug. 2004.

[9] Chia-Hsin Wu, Chih-Hun Lee, Wei-Sheng Chen, and Shen-Iuan Liu, "CMOS wideband amplifiers using multiple inductive-series peaking technique," *IEEE Journal of Solid-State Circuits*, vol. 40, iss. 2, pp. 548-552, Feb. 2005.

[10] O. T.-C. Chen, C.-T. Chan, and R.-P. Sheen, "Transimpedance limit exploration and inductor-less bandwidth extension for designing wideband amplifiers," *IEEE Transactions on VLSI Systems*, vol. 24, no. 1, pp. 348-352, Jan. 2016.

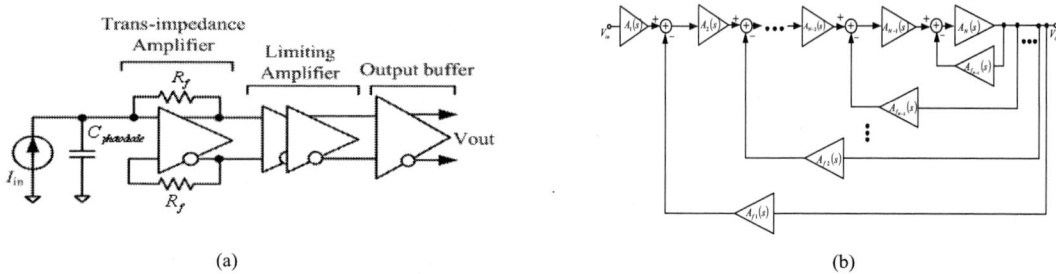

(a)   (b)

Fig. 1. Proposed optical receiver. (a) Block diagram. (b) N'th-order multi-level active feedback structure.

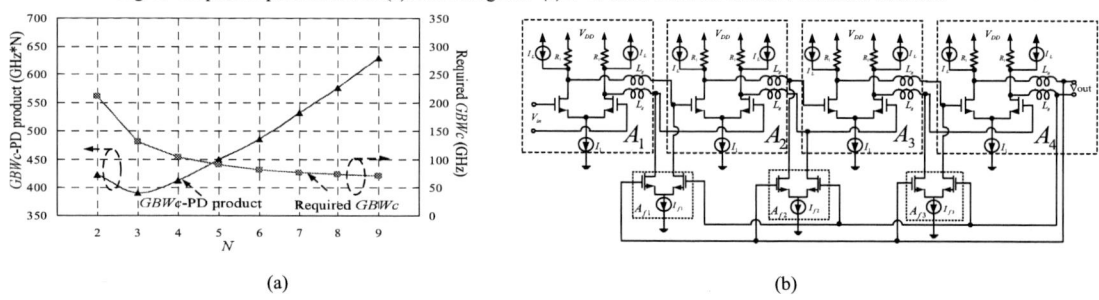

(a)   (b)

Fig 2. Proposed MLAF structure. (a) Curves of $GBW_C - PD$ and GBW_C versus N in MLAF under the total gain of 10 and a bandwidth of 28GHz. (b) Circuit diagram of 4'th-order MLAF with series inductor peaking.

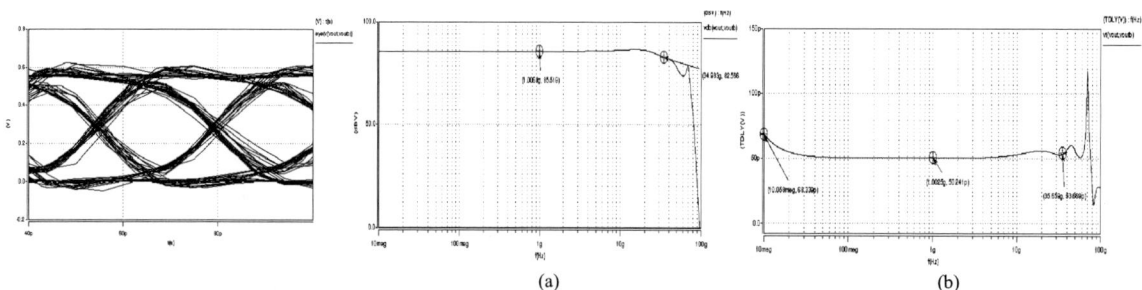

                           (a)   (b)

Fig. 3. Eye diagram of differential outputs from the proposed optical receiver at inputs of $2^{23} -1$ PRBS at 40Gbps.

Fig. 4. Proposed optical receiver. (a) Frequency response. (b) Group delay.

Table 1 Performance comparison of proposed and conventional 40Gbps optical receiver.

Specifications / Optical receivers	Technologies	Inductor peaking structure	Supply voltages	Bandwidth (GHz)	Z_T (dBΩ)	Group delay	Power dissipation
Proposed work	90nm CMOS	High-gain MLAF + series inductor peaking	1.2V	35	86 (differential)	±10 ps	93.6mW
Jin's work [4]	0.18μm CMOS	π-type inductor peaking	1.8V	30	51 (single-ended)	± 62.5 ps	60.1mW
Liao's work [5]	90nm CMOS	Reversed triple resonance network	1.2V	22	66 (differential)	N/A	75mW
Kim's work [6]	45nm SOI CMOS	Feedback +series inductor peaking	1V	30	55 (single-ended)	± 3.1 ps	9mW

OECC/PS2016

Frequency Chirp Properties With Data Pattern Dependence in Quantum-Dot SOAs

Hiroki Hoshino, Norihiko Ninomiya, and Motoharu Matsuura

Department of Communication Engineering and Informatics, The University of Electro-Communications
1-5-1 Chofugaoka, Chofu, Tokyo 182-8585, Japan.
hiroki.hoshino@uec.ac.jp

Abstract: We investigated the chirp properties using 10-Gbit/s signal with a fixed data pattern in quantum-dot semiconductor optical amplifiers. The results show that the properties depend on the data pattern affected by the gain recovery time.

Keywords: Quantum-dot semiconductor optical amplifiers (QD-SOAs), Frequency chirp, Gain recovery time

I. INTRODUCTION

Semiconductor optical amplifiers (SOAs) are attractive devices not only for optical amplification but also for optical signal processing such as optical switches and wavelength conversions [1, 2]. However, since the gain changes in SOAs give rise to refractive index change, the amplified signals by SOAs have frequency chirp at the leading and trailing edges of the signal pulse. As a result, the signals have unique chromatic dispersion in transmission line and degrade the transmission performance in a complex manner [1]. Therefore, it is important to investigate the dynamic frequency chirp properties in SOAs in detail.

Recently, quantum-dot SOAs (QD-SOAs) have attracted much attention because of their much higher gain, wider bandwidth, and faster gain recovery time, compare with bulk and quantum-well SOAs [3]. Also in QD-SOAs, since the gain changes give rise to refractive index change [4], it is important issue to investigate the frequency chirp properties induced in QD-SOAs and compare the properties with SOAs.

Until now, we have experimentally investigated the frequency chirp properties induced in SOAs and QD-SOAs using our proposed measurement method with a conventional bandpass filter (BPF) [5, 6]. It was thus found that QD-SOAs have stronger blue-chirp, constant gain, and chirp properties in a much wider wavelength region. On the other hand, although the previous experimental demonstrations showed that QD-SOAs have much faster gain recovery time than those of SOAs [7, 8], no frequency chirp properties considering the data pattern affected by the gain recovery time of the device has been reported so far, to the best of our knowledge.

In this paper, we investigate and compare the frequency chirp properties using 10-Gbit/s signals with various data patterns in a SOA and QD-SOA, taking into account the gain recovery times, for the first time. The results show that the frequency chirp properties depend on the device and the data patterns affected by the gain recovery times.

II. EXPERIMENTAL SETUP

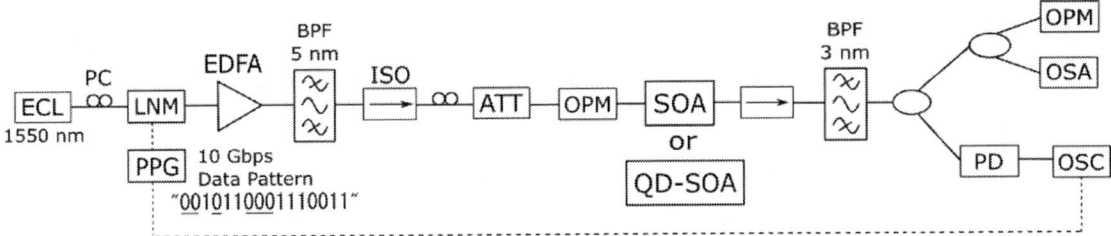

Fig. 1. Experimental setup for frequency chirp measurement using a SOA and QD-SOA. ECL: External cavity laser-diode, PC: Polarization controller, LNM: LiNbO3 intensity modulator, PPG: Pulse pattern generator, EDFA: Erbium-doped fiber amplifier, BPF: Bandpass filter, ISO: Isolator, ATT: Optical attenuator, OPM: Optical power meter, SOA: Semiconductor optical amplifier, QD-SOA: Quantum-dot SOA, OSA: Optical spectrum analyzer, PD: Photo-diode, OSC: Oscilloscope

The experimental setup for frequency chirp measurement using a SOA and QD-SOA is shown in Fig. 1. An external cavity laser-diode (ECL) and a LiNbO3 intensity modulator (LNM) were employed to generate a 10-Gbit/s non-return-to-zero on-off keying (NRZ-OOK) input data signal at a wavelength of 1550 nm. A fixed data pattern of the input data signal was generated by a pulse pattern generator (PPG) with a fixed 16-bit pattern "0010110001110011", which included single, double, and triple "0" and "1" data patterns. Thus, the data pattern enables us to evaluate the gain recovery time dependence on the frequency chirp of the SOA and QD-SOA. The input data signal was amplified by an erbium-doped fiber amplifier (EDFA), and the amplified spontaneous emission noise of the signal was removed by a BPF with a 3-dB bandwidth of 5.0 nm. To adjust the signal power injected into the SOA and QD-SOA, the power was controlled by an optical attenuator (ATT) and monitored by in-line optical power meter (OPM). The employed SOA was a commercially available SOA (INPHENIX IPSAD1503-5114), while the QD-SOA was a simple device based on Stranski Krastanov (SK) growth. After passing through an isolator (ISO) and a BPF with a 3-dB bandwidth of 3.0 nm at the output of the SOA or QD-SOA, the signal was injected into the chirp measurements scheme. The detailed setup and method for chirp measurement is explained in Ref. [5]. An OPM and optical spectrum analyzer (OSA) were used for adjusting the chirp measurement condition for the chirp property, while a photo-diode (PD) and oscilloscope (OSC) was used for monitoring the waveform and the frequency chirp of the signals. The frequency chirp properties were evaluated for various data patterns in the 16-bit fixed data pattern. "011(single zero)", "001(double zero)", "000111(triple zero)" in the data pattern were defined as "0" (Blue) or (Red), "00" (Blue) or (Red), and "000" (Blue) or (Red), respectively.

III. RESULTS AND DISCUSSION

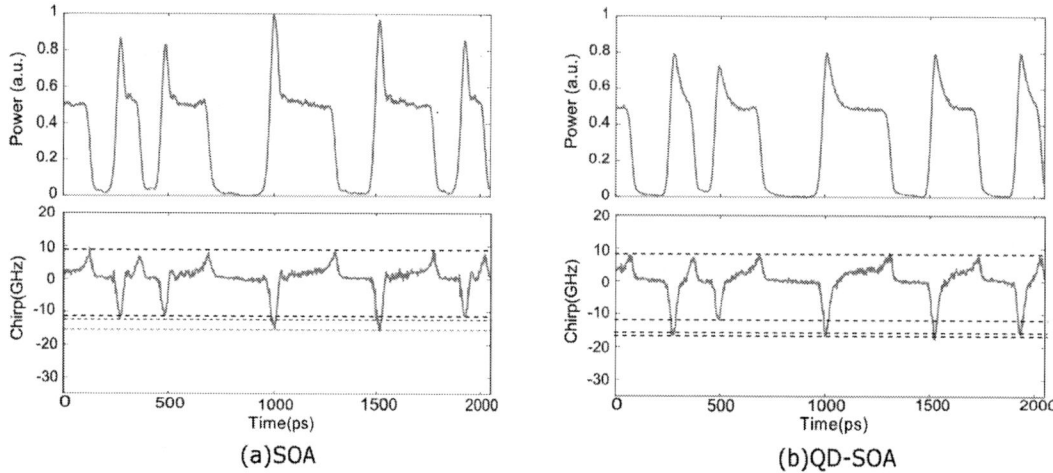

Fig. 2. Waveforms (upper) and time-resolved frequency chirp properties (bottom) of the signal amplified by (a) SOA and (b) QD-SOA. The forward currents of the SOA and QD-SOA were set to 350 mA and 1600 mA, respectively.

Fig. 2 shows the waveforms and time-resolved frequency chirp properties of the signal amplified by the SOA or QD-SOA. As shown in the dashed lines, the SOA and QD-SOA had the data pattern dependence on the red chirp peak. In particular, "00" and "000" data pattern had larger red chirp peak. This means that the amplitudes of the chirp peak depend the gain recovery time. Although it is well known that QD-SOAs have faster gain recovery time than the SOAs [3, 4], the QD-SOA also had the clear data pattern dependence in the case of the special data pattern, which is faster than the gain recovery time of the QD-SOA.

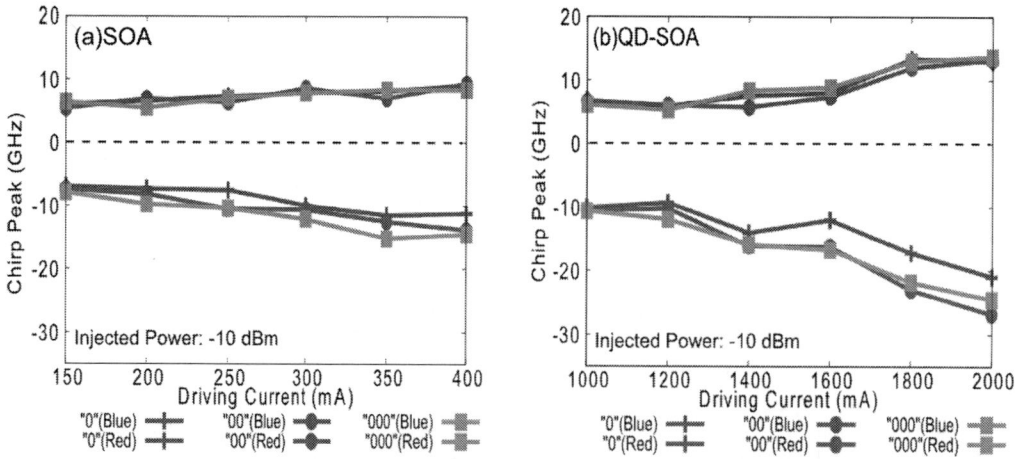

Fig. 3. Chirp peak characteristics for various data patterns of (a) SOA and (b) QD-SOA while changing the forward current.

Fig. 3 shows the chirp peak characteristics for various data patterns of (a) SOA and (b) QD-SOA while changing the forward current. The input signal power injected into the SOA and QD-SOA was set to −10 dBm. In the both cases, it was clearly seen that the amplitude of the blue chirp peak does not depend on the data pattern. Compared with the SOA, the blue chirp peak of the QD-SOA tended to be larger as the forward current was increased. The enhanced blue chirp was useful to further improve the gain recovery time with a blue-shifted BPF from the center wavelength of the pass-band. Actually, we have successfully achieved the error-free 320-Gb/s wavelength conversion based on cross-gain modulation using this technique [8]. Since the blue chirp peak does not depend on the data pattern related to the gain recovery time, it will be useful for various applications based on optical signal processing. On the other hand, in the case of the red chirp, the SOA had much smaller peak value than the QD-SOA, because the gain of the SOA was not completely recovered due to the slow gain recovery time even if the "000" data pattern with 300 ps time slot was used. On the other hand, the QD-SOA had large peak value in all data patterns, because the QD-SOA had much faster gain recovery time than the SOA. The obtained chirp characteristics were quite different from the previous work [6], because this work took into account the gain recovery time.

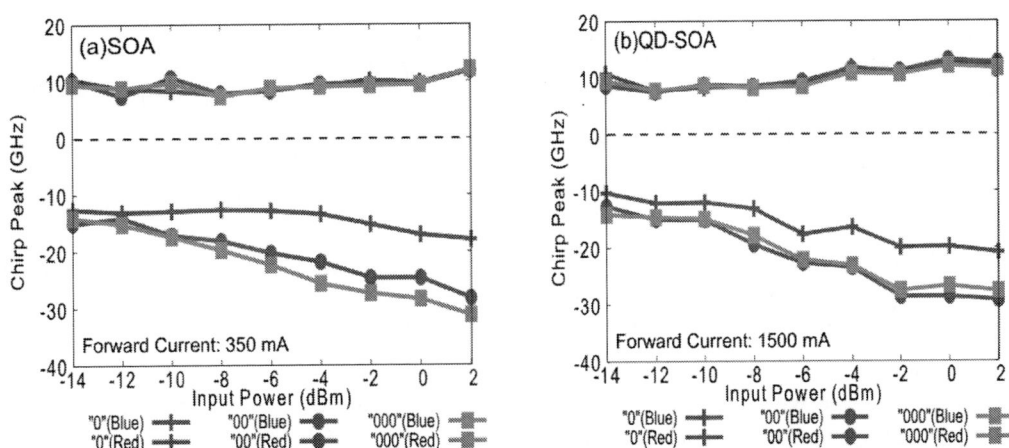

Fig. 4. Chirp peak characteristics for various data patterns of (a) SOA and (b) QD-SOA while changing the injected signal power.

Fig. 4 shows the chirp peak characteristics for various data patterns of (a) SOA and (b) QD-SOA while changing the injected signal power. The forward current of the SOA and the QD-SOA was set to 350 mA and 1500 mA, respectively. Compared with the previous result as shown in Fig. 3, it was not so clearly seen that the QD-SOA had larger red chirp peak and data pattern dependence affected by the gain recovery time than the SOA. The reasons for the similar peak value and data pattern dependence were due to the improved gain recovery time of the SOA injected by higher input power [9] and lower gain saturation power of the SOA. Nevertheless, the QD-SOA had lower data pattern dependence, owing to the original faster gain recovery time than the SOA.

IV. CONCLUSIONS

We investigated and compared the frequency chirp properties of a SOA and QD-SOA for various data patterns affected by the gain recovery time. The results showed that the blue chirp peak did not depend on the data pattern, while the red chirp peak strongly depended on the data pattern. These characteristics will be useful for various applications of optical signal processing using QD-SOAs.

REFERENCE

[1] T. Durhuus et al., "All-Optical Wavelength Conversion by Semiconductor Optical Amplifier," J. Lightwave Technol., vol. 14, no. 6, pp. 942-954, 1996.

[2] K. E. Stubkjaer, "Semiconductor optical amplifier-based all-optical gates for high-speed optical processing," IEEE J. Sel .Top. Quantum Electron., vol. 6, no. 6, pp. 1428-1435, 2000.

[3] T. Akiyama et al., "Quantum-Dot Semiconductor Optical Amplifiers," Proc. IEEE, vol. 95, no. 9, p. 1757-1766, 2007.

[4] Y. Ben-Ezra et al., "Theoretical Analysis of Gain-Recovery Time and Chirp in QD-SOA," IEEE Photon. Technol. Lett., vol. 17, pp. no. 9, pp. 1803-1805, 2005.

[5] M. Matsuura et al., "Time-Resolved Chirp Properties of SOAs Measured With an Optical Bandpass Filter," IEEE Photon. Technol. Lett., vol. 20, no. 23, pp. 2001-2003, 2008.

[6] M. Matsuura et al., "Experimental investigation of frequency chirp properties induced by signal amplification in quantum-dot semiconductor optical amplifiers," OSA Opt. Lett., vol. 40, no. 6, pp. 914-917, 2015.

[7] M. Matsuura et al., "320 Gbit/s wavelength conversion using four-wave mixing in quantum-dot semiconductor optical amplifiers," OSA Opt. Lett., vol. 36, no. 15, pp. 2910-2912, 2011.

[8] M. Matsuura et al., "Ultrahigh-speed and widely tunable wavelength conversion based on cross-gain modulation in a quantum-dot semiconductor optical amplifier," OSA Opt. Express, vol. 19, no. 26, pp. B551-B559, 2011.

[9] R. J. Manning et al., "Semiconductor laser amplifiers for ultrafast all-optical signal processing," J. Opt. Soc. Am. B, vol. 14, no. 11, pp. 3204-3216, 1997.

Optical Sensor based on Mach-Zehnder Interferometer using Orbital Angular Momentum

Haozhe Yan, Shangyuan Li, Bian FengKai, Xiaoping Zheng,*
Hanyi Zhang, and Bingkun Zhou

Tsinghua National Laboratory for Information Science and Technology, Department of Electronic Engineering,
Tsinghua University, Beijing 100084, China.
*email: xpzheng@tsinghua.edu.cn

Abstract: *A novel Mach-Zehnder interferometric optical sensor using orbital angular momentum (OAM) is proposed and experimentally demonstrated by high order OAM beam with topological charge up to 10.*

Keywords: *Optical Vortices; Fiber Optics Sensors; Mach-Zehnder Interferometer*

I. INTRODUCTION

In 1992, a significant breakthrough was made by Allen et al. [1] that an optical beam containing orbital angular momentum (OAM), so-called vortex beam has a helical phase front which can be expressed as $\exp(\pm il\theta)$ in polar coordinates, where l known as the topological charge can take arbitrary integer values. For any given l, the vortex beam has l intertwined helical phase fronts and OAM beams of different topological charge are distinguishable from one another because of their inherent orthogonality. Benefited from the unique properties, OAM has been widely used in optical communication [2], optical trapping [3], optical tweezers [4], and quantum optics [5].

As a helically phased beam, having l intertwined helical phase fronts, is made to interfere with a Gaussian beam, it produces a spiral intensity pattern with l spiral arms. The particular interference pattern, which is unlike the one of double Gaussian beams, makes OAM beam have different characteristics in optical sensing field. Especially after laser interferometers made great contributions to the detection of gravitational waves [6], the interferometric optical sensors have attracted lots of attention. It makes great sense to explore the potential applications of OAM in the field of optical sensor based on interferometer.

In this work, we propose, for the first time to the best of our knowledge, a structure of novel optical sensor based on Mach-Zehnder interferometer using OAM beam. Experiments are performed and the results show a Mach-Zehnder interferometric sensor using OAM beam have been realized, which can detect the direction and value of the changes in the optical path difference and then be used to monitor the variation of pressure, tension, and bend [7, 8].

II. EXPERIMENTAL AND DISCUSSION

The experimental apparatus are shown schematically in Fig. 1. A Gaussian (G) beam of 1550nm is splitted into two parts through a fiber splitter. The first part is reflected by a spatial light modulator (SLM) to generate an OAM beam. The second part goes through an optical fiber. A section of this fiber, which served as the measuring and sensing part in this optical sensor, is fixed on a rubber plate. The interference pattern (OAM-G interference pattern) is obtained by interference of the OAM beam with the Gaussian beam at a non-polarizing beam-splitter (BS1), and goes through the beam expander (BE). Then it is splitted by BS3 and BS4, and partly collected by two small aperture fiber collimators (CL3, CL4) connected to an optical power meter, respectively. Two Gaussian beams are interfered with each other at BS2. The OAM-G and G-G interference patterns are both captured by the CCD camera.

Figure 2(a) shows the relative position of expanded $l = -1$ OAM-G interference pattern with waist diameter of 2cm and two collimator with aperture of 2mm. To improve the measurement accuracy, the OAM-G interference pattern is expanded so that CL3 and CL4 only collect small part of the beam to gain a high extinction ratio. In order to detect the direction and value of the changes in optical path difference, a nonzero topological charge OAM beam should be used and the two collimators have to be located at a circle with a relative angle α meeting the condition [9]

$$\frac{-90^0}{|l|} < \alpha < \frac{90^0}{|l|}. \tag{1}$$

The optical path difference can be increased by tensing the rubber plate and decreased by compressing the plate, as shown in Fig. 1. The output optical powers from CL3 and CL4 are measured and sinusoidally fitted. For $l = -1$, when the optical path difference is reduced by 1550nm, the spiral arm of the OAM-G interference pattern rotates clockwise for 360° and the CL3 fitting curve maintains less than a quarter cycle ahead of CL4 fitting curve, shown in Fig. 2(c). The contrary is the case when the optical path difference is increased by 1550nm, shown in Fig. 2(d). Consequently, the value and direction of the variations of optical path difference are obtained at the same time. As illustrated in Fig. 2(b), the G-G interference pattern is shown in CCD simultaneously and serving as a comparison with the OAM-G interference pattern. The extinction ratio of the receiving system for $l = \pm 1$ is about 14.5dB, as shown in Fig. 2(c, d).

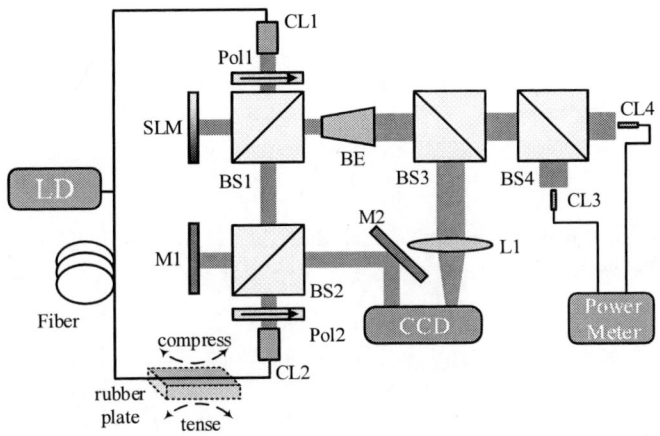

Fig. 1. Experimental setup of the optical sensor based on Mach-Zehnder interferometer using OAM. CL: fiber collimator; Pol: polarizer; BS: non-polarizing beam-splitter; SLM: spatial light modulator; BE: beam expander; M: mirror; L: lens.

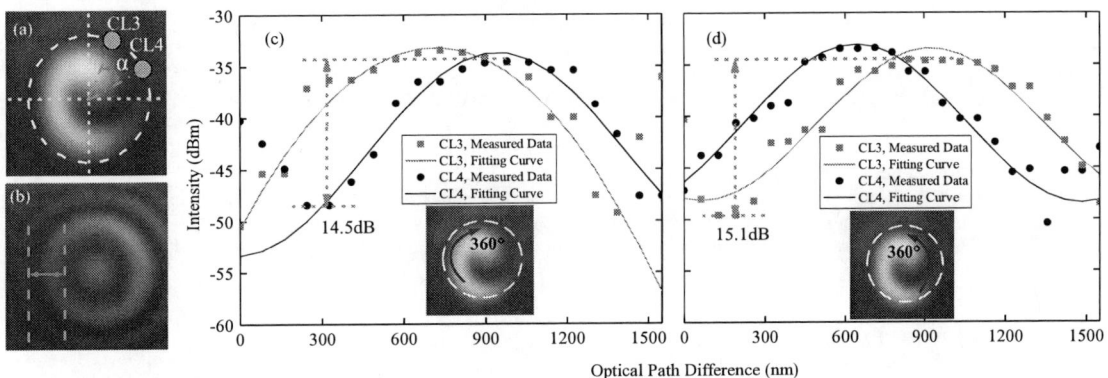

Fig. 2. Interferograms of (a) $l = -1$ OAM beam, (b) Gaussian beam with Gaussian beam. Measured data and fitting curves of CL3 (red) and CL4 (black) output intensity versus the changes of optical path difference for (c) $l = -1$ OAM as optical path difference reduce, (d) $l = -1$ OAM as optical path difference increase.

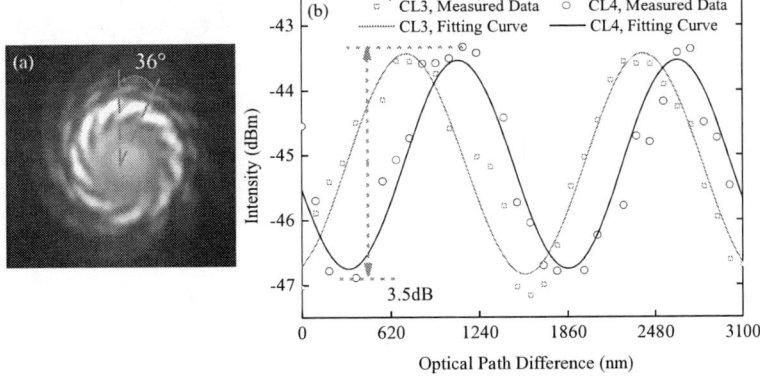

Fig. 3. (a) Interferograms of $l = 10$ OAM beam with Gaussian beam. (b) Measured data and fitting curves of CL3 (red) and CL4 (black) output intensity versus the changes of optical path difference for $l = 10$ OAM as optical path difference reduce.

In Fig. 3(a), the $l = 10$ OAM-G interference pattern with 10 spiral arms is observed, which will rotate 36° (one spiral arm) as the optical path difference changes by 1550nm. As shown in Fig. 3(b), CL3 (red sign) and CL4 (black sign) output data is measured during the interference pattern rotating clockwise for 72°, corresponding to optical path

947

difference increased by 2×1550nm. The extinction ratio of the receiving system for $l = \pm10$ is 3.5dB, which is 11dB less than the one for $l = \pm1$ because of the OAM topological charge increasing from 1 to 10.

III. CONCLUSION

A novel Mach-Zehnder interferometric optical sensor using OAM beam have been proposed and experimentally demonstrated. The simultaneous detection of the direction and value of the changes in the optical path difference have been realized experimentally. It has also been demonstrated that the extinction ratio of the sensor receiving system is inversely proportional to the topological charge of OAM beam used. This sensor can be used to monitor the variation of pressure, tension, and bend.

ACKNOWLEDGMENTS

This work was supported in part by 973 Program under the grant No 2014CB340003, National Nature Science Foundation of China (NSFC) under grant No. 61307081, 61321004, 61420106003.

REFERENCES

[1] L. Allen, M. W. Beijersbergen, R. J. C. Spreeuw, and J. P. Woerdman, "Orbital angular momentum of light and the transformation of Laguerre-Gaussian laser modes," Phys. Rev. A, vol. 45, pp. 8185-8189, 1992.

[2] Jian Wang, Jeng-Yuan Yang, Irfan M. Fazal, Nisar Ahmed, Yan Yan, Hao Huang, Yongxiong Ren, Yang Yue, Samuel Dolinar, Moshe Tur & Alan E. Willner, "Terabit free-space data transmission employing orbital angular momentum multiplexing," Nature Photon., vol. 6, pp. 488-496, 2012.

[3] H. He, M. Friese, N. R. Heckenberg, and H. Rubinsztein-Dunlop, "Direct observation of transfer of angular momentum to absorptive particles from a laser beam with a phase singularity," Phys. Rev. Lett., vol. 75, pp. 826–829, 1995.

[4] D. Grier, "A revolution in optical manipulation," Nature, vol. 424, pp. 810–816, 2003.

[5] B. Jack, A. M. Yao, J. Leach, J. Romero, S. Franke-Arnold, D. G. Ireland, S. M. Barnett, and M. J. Padgett, "Entanglement of arbitrary superpositions of modes within two-dimensional orbital angular momentum state spaces," Phys. Rev. A, vol. 81, pp. 043844, 2010.

[6] B. P. Abbott et al. "Observation of Gravitational Waves from a Binary Black Hole Merger," Phys. Rev. Lett., vol. 116, pp. 061102, 2016.

[7] Ming Deng, Chang-Ping Tang, Tao Zhu, Yun-Jiang Rao, "Highly sensitive bend sensor based on Mach–Zehnder interferometer using photonic crystal fiber," Opt. Commun., vol. 284, pp. 2849-2853, 2011.

[8] L. Jiang, J. Yang, S. Wang, B. Li, and M. Wang, "Fiber Mach–Zehnder interferometer based on microcavities for high-temperature sensing with high sensitivity," Opt. Lett., vol. 36, pp. 3753-3755, 2011.

[9] M Padgett, J Courtial, L Allen, "Light's orbital angular momentum," Phys. Today, vol. 5, pp. 35-40, 2004.

Precise Measurement of Microwave Evanescent Fields along Fiberglass-Reinforced Plastic Mortar Pipe Using Electro-Optic Sensor for Nondestructive Inspection

Yoshiyuki AZUMA   Fumiaki UENO   Hiroshi MURATA   Yasuyuki OKAMURA

Graduate School of Engineering Science, Osaka University   1-3 Machikaneyama-cho, Toyonaka, Osaka, 560-8531 Japan

E-mail address: yoshiyukiazuma113@s.ee.es.osaka-u.ac.jp

Tadahiro OKUDA   Masaya HAZAMA

Kurimoto LTD   1 Koyagi-cho, Higashi-Ohmi, Shiga, 527-0108 Japan

Abstract: We propose a new nondestructive inspection method for fiberglass-reinforced plastic mortar pipes using microwave guided-mode and photonic techniques. This method is based on precise measurement of microwave evanescent fields along the pipe-wall using electro-optic sensors.

Keywords: Electro-optic sensor, Microwave guided-mode, Nondestructive inspection, FRPM pipe

I. INTRODUCTION

Electro-optic (EO) sensor is a small invasive sensor for electro-magnetic fields in RF and microwave, since it can be composed of non-metal or minute metal components. Therefore, the EO sensor is used for application required small invasiveness such as near field measurement for antennas [1],[2].

Since fiberglass-reinforced plastic mortar (FRPM) has high mechanical strength and high chemical corrosion resistance despite being lightweight, FRPM is used in many application fields such as protecting tubes for power/communication cables, sewer pipes and agricultural water pipes. Especially, the total length of FRPM pipes used in agricultural water pipelines is approximately 50,000 km in Japan. Recently, nondestructive inspection for FRPM pipelines is required for regular testing and maintenance.

There are several candidates for inspection method for FRPM pipelines such as magnetic resonance imaging (MRI), X-ray and ultrasonic waves [3]. However, these methods are unsuitable for inspection of long and buried FRPM pipelines, since these methods require large and specific measurement machines. Therefore, there is no effective inspection method for long-range FRPM pipelines yet, as far as we know.

We have found that FRPM is a dielectric material with relatively small loss in the microwave frequency range of 1~10 GHz, and we have proposed a new inspection method for FRPM pipelines by using the microwave guided-mode propagating along pipe-walls of FRPM pipes and measurement of the change in microwave propagation characteristics caused by a deterioration, defects or cracks [4]. In this method, the invasiveness of the microwave measurement is very important, since a weak change in evanescent fields caused by cracks should be measured for inspection. Therefore, an EO sensor is suitable for the proposed method, since electro-magnetic fields can be measured with small distortion.

In this paper, the first experimental demonstration of the crack identification in a FRPM pipe using the EO sensor is reported.

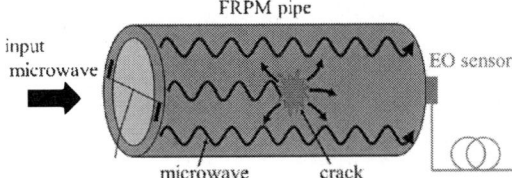

Fig.1 Schematic of the proposed inspection method.

II. BASIS OF INSPECTION METHOD

A schematic of the proposed inspection method is shown in Fig.1. In this method, an FRPM pipe is used as a kind of a microwave transmission line. Microwave signals are input from one end of the FRPM pipe, and propagated along the FRPM pipe-walls. Then the propagated microwave signals detected at the other end of the pipe. If there are some defects or cracks in the FRPM pipe, the microwave signals propagating along pipe-walls are distorted by defects or cracks, and then, the propagated microwave signal level and the distribution of the microwave fields are changed. Therefore, we can detect defects or cracks in the pipe as the change in microwave transmission characteristics and distributions of the microwave field.

We analyzed microwave transmission characteristics propagating along the FRPM pipe. An FRPM pipe is considered as a hollow cylindrical dielectric waveguide, as shown in Fig.2 (a), where a is an inner radius, b is an outer radius, ϵ_i ($i = 1, 2, 3$) is the permittivity and μ_i ($i = 1, 2, 3$) is the permeability in each region. From the Helmholtz equation, field distributions for transverse electric modes TE_{mn} can be described by use of the Bessel functions, and eigenvalue equations are derived from the continuity conditions of tangential components of electro-magnetic fields at the boundary.

By solving the eigenvalue equations, we derived the dispersion curves of the effective refractive indices and electric field distributions, as shown in Fig.2 (b), (c), where the regions 1 and 3 are air ($\epsilon_1 = \epsilon_3 = 1$), the region 2 is FRPM of permittivity $\epsilon_2 = 10.24$, an inner radius a is 125 mm and an outer radius b is 143 mm [4]. From these dispersion curves, we can see that there is only a single guided mode (TE_{00}) at 2.4 GHz while there are three guided modes (TE_{00}, TE_{01} and TE_{02}) at 6 GHz.

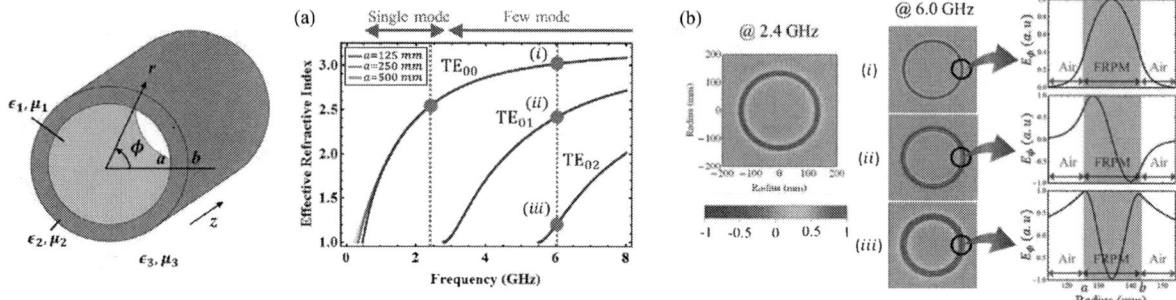

Fig.2 Structure of hollow cylindrical dielectric waveguide.

Fig.3 (a)Dispersion curve, (b)Electric field distributions in the cross section.

III. MEASUREMENT

A. The microwave transmission characteristics and field distributions

We measured the transmission characteristics of microwave propagating along the FRPM pipe-wall using the experimental setup as shown in Fig.4 (a). Microwave signals from a network analyzer were input to an FRPM pipe (length $l = 1000$ mm, inner radius $a = 125$ mm and pipe thickness $t = 18$ mm) from the end by use of a monopole antenna. The electric field emitted from the monopole antenna was set to be along ϕ direction in Fig.2 to excite TE-modes. The transmitted microwave signals along the FRPM pipe were detected using an EO sensor (SEIKOH GIKEN CS-1210) located at the other end of the FRPM pipe.

We also measured microwave electric field distributions by scanning the EO sensor along the direction of radius (the r direction in Fig.4 (b)) with a 1 mm-step at end of the pipe. We measured the microwave transmission and distributions both for a damaged FRPM pipe with some cracks and a normal FRPM pipe without cracks.

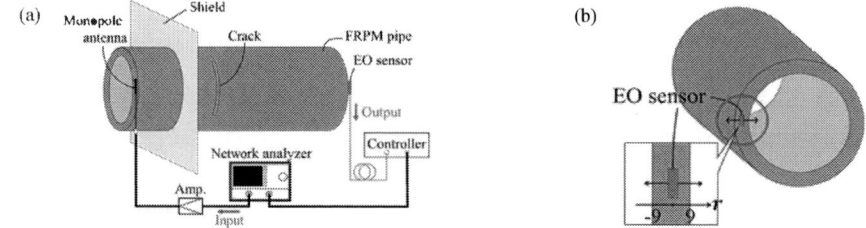

Fig.4 Experimental setup. (a)Whole view, (b)End view.

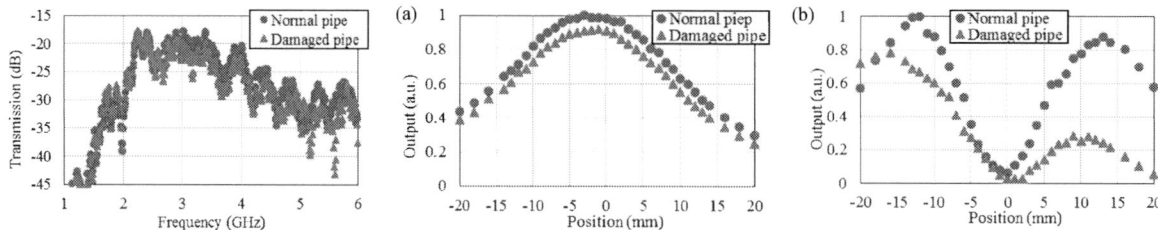

Fig.5 Microwave transmission characteristics

Fig.6 Electric field distributions at (a)2.4 GHz and (b)6 GHz.

Fig.5 shows measurement results of transmission characteristics of microwave propagating along the normal/damaged FRPM pipes when the EO sensor was set at the end of the pipe with the position of $r = 0$ as shown in Fig.4 (b). We can see a clear difference between the two results.

Fig.6 shows the measured electric field distributions at 2.4 GHz and 6 GHz by scanning the EO sensor at the end of the FRPM pipe, as shown in Fig.4 (b). In the results of the damaged pipe, the microwave signal levels were decreased. We believe that these changes were caused by defects or cracks in the FRPM pipe. Since there is only one microwave guided mode propagating along the FRPM pipe-wall at 2.4 GHz as shown in Fig.3 (a), the microwave amplitude was merely distorted by the defects or cracks. On the other hand, at 6 GHz there are three microwave guided modes propagating along the FRPM pipe-wall as shown in Fig.3 (a). Then the microwave guided modes can be not only distorted but also mutually coupled by defects or cracks. Therefore, the change in the field profile at 6 GHz was rather definite compared with that at 2.4 GHz.

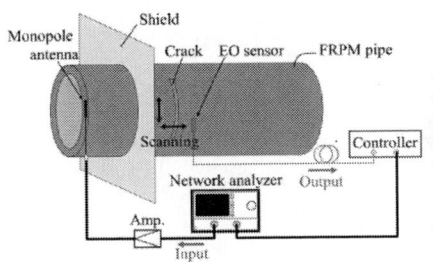

Fig.7 Experimental setup for measuring two-dimensional field distribution

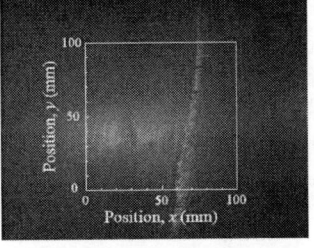

Fig.8 Surface of the FRPM pipe arround a crack

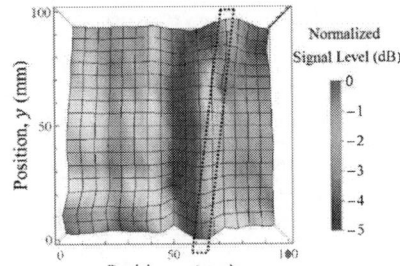

Fig.9 Two-dimensional electric field distribution on the surface of the damaged FRPM pipe.

B.   Two-dimensional field distribution around the crack on the surface of the FRPM pipe

We also measured the field distribution on the surface of the FRPM pipe by scanning the EO sensor two-dimensionally using the experimental setup as shown in Fig.7. Fig.8 shows photograph of the outer surface of the damaged FRPM pipe around the crack. The EO sensor scanned within the region of a square as shown in Fig.8 along the pipe axis direction (x direction) with 5 mm-step and along the circumferential (y direction) with 10 mm-step.

Fig.9 shows the microwave signal distribution on the outer surface of the damaged FRPM pipe around a crack, and the dotted lines indicate the crack position. In Fig.9, the microwave was propagated from left-hand side to right-hand side. We can see that the signal level is high on and near the crack and low just before the crack. These imply that the microwave was reflected and radiated by the crack. The reflected microwave was interfered with the following microwave coming from the inputted port, we believe. By scanning the EO sensor with small distortion on the surface of the pipe, we can identify the position, size and shape of cracks.

IV. CONCLUSIONS

We demonstrated the new nondestructive inspection method for the FRPM pipe. The microwave signals propagating along the pipe-wall were affected by cracks and surface condition of the pipe. The difference of microwave transmission and field distributions between the normal and the damaged FRPM pipes were clearly identified. Therefore, defects or cracks can be detected by measuring microwave propagation characteristics along the FRPM pipe. In addition, we can also experimentally identify the position of the crack by scanning the EO sensor with small invasiveness on surface of the FRPM pipe. Therefore, the FRPM pipelines can be inspected nondestructively by use of the microwave and photonic techniques.

ACKNOWLEDGMENT

The authors thanks to Drs. Hidehisa Shiomi and Toshiyuki Inoue from Osaka University, Japan for the helpful advices and discussion on the analysis and experiments. The authors thanks to Drs. Satoru Kurokawa and Masanobu Hirose from AIST, Japan for their valuable comments. The authors also thanks to Dr. Yoshikazu Toba from SEIKOH GIKEN for the provision of the EO sensor.

REFERENCES

[1] H. Togo, S. Mochizuki, and N. Kukutsu. "Optical fiber electric field sensor for antenna measurement," NTT Technical Review, vol.7, no.3, Mar. 2009.

[2] H. Togo, N. Shimizu, and T. Nagatsuma, "Tip-on-fiber Electro-optic Probe for Near-field Measurement," NTT Technical Review, vol.4, no.1, Jan. 2006.

[3] E. Marfisi, C. J. Burgoyne, L. D. Hall, and M. H. G. Amin, "Use of the MRI technique to study concrete and FRP reinforced concrete behavior," Research Leading to the Development of Design Guideline for the Use of FRP in Concrete Structure – 2nd ConFiberCrete Young Researcher Conference ,Corfu, Greece, June 2002.

[4] F. Ueno, H. Murata, T. Okuda, M. Hazama, and Y. Okamura, "New Nondestructive Measurement for Fiberglass-Reinforced Plastic Mortar Pipes Using Microwave and Photonic Techniques," MWP/APMP 2014, TuED-2, Oct. 2014.

WA2-74

OECC/PS2016

Plasmon-induced Transparency based on Side-coupled Stub and Hexagonal Resonators and Its Sensing Characteristics

Tianye Huang[1,2], Minming Zhang[2], and Songnian Fu[2,*]

[1]Faculty of Mechanical and Electronic Information, China University of Geosciences, Wuhan, China 430074.
[2]Next Generation Internet Access National Engineering Lab, School of optical and electronic information, and Wuhan National Laboratory for Optoelectronics, Huazhong University of Science and Technolgy, Wuhan 430074, China
Author e-mail address: songnian@mail.hust.edu.cn

Abstract: *A metal-insulator-metal (MIM) structure comprising stub and hexagonal resonator is proposed to realize plasmon-induced transparency (PIT) response. Benefit from high sensitivity and narrow transmission spectrum, sensing figure-of-merit as high as 178 RIU^{-1} can be achieved.*
Keywords: *Plasmonics, Plasmon-induced transparency, Integrated opitcs devices*

I. INTRODUCTION

Quantum interferences between different excitation pathways for a multilevel atoms system can dramatically modify the optical response and gives rise to a sharp transparency window within a broad absorption spectrum. This phenomenon is known as electromagnetically induced transparency (EIT) which possesses tremendous potential applications ranging from slow light, quantum correlation transferring, and nonlinear optics [1]. However, the demands of stable gas laser and rigorous environment temperature hamper the EIT implementation of integrated devices. To overcome this limitation, recently, it is found that the EIT effect can also occur in optical resonator system where coherent interference between two coupled resonators happens, instead of different atom levels [2]. Benefit from the strong confinement of light below diffraction limit offered by plasmonics [3], plasmon-induced transparency (PIT) devices possess several advantages of small footprint, room temperature operation, and tunability. On the other hand, plasmonic device is also regarded as potential candidate to realize nano-sensor for biological and chemical detection [4], because the optical response (i.e. resonant wavelength shift) of the plasmonic devices is highly sensitive to the variation of refractive index (RI) around the nanostructure. This enables the detection of target medium with different dielectric properties. It is no doubt that the narrow transparency window emerged from the stop-band in PIT devices can provide a potential solution for nano-sensor as well. For this purpose, several PIT-based nano-sensors have been reported with promising sensing performance [5, 6].

In this paper, a compact PIT device comprising metal-insulator-metal (MIM) bus waveguide, a stub resonator, and a hexagonal resonator is proposed. The system is then theoretically described by using temporal coupled mode theory (CMT) and subsequently verified by using finite-differential time-domain (FDTD) method. By selecting the hexagonal resonator as channel for the RI sensing, the narrow PIT window shows high wavelength sensitivity, indicating that the proposed device is able to monitor tiny RI change with high accuracy.

II. PIT SYSTEM DESIGN

As shown in Fig.1, the proposed two-dimensional PIT chip consists of a bus waveguide, a rectangular stub resonator, and a hexagonal resonator. The material filling in the bus waveguide and stub resonator are air with RI of 1, and the hexagonal resonator is acted as sensing channel with variable RI. The background material is silver whose permittivity is described by Drude model with $\varepsilon(\omega)=\omega_\infty-\omega_p^2/(\omega^2+j\omega\gamma)$, where $\omega_\infty=3.7$, $\omega_p=9.1$ eV, and $\gamma=0.018$ eV [5]. The widths of bus waveguide and stub resonator are defined as W and W_B, respectively. The lengths of the stub resonator and the side length of the hexagonal resonator are defined as L_B and L_D, respectively. Finally, g is the gap distance between two resonators. For the proposed chip, the resonant plasmonic mode in the stub resonator can be excited directly by the incident waves coming from bus waveguide. However, since the hexagonal resonator is far away from the bus waveguide, it can only be coupled to the stub resonator but cannot interact directly with the incident wave. Therefore, in this PIT chip, the stub resonator servers as the bright resonator, while the hexagonal resonator behaves as the dark resonator.

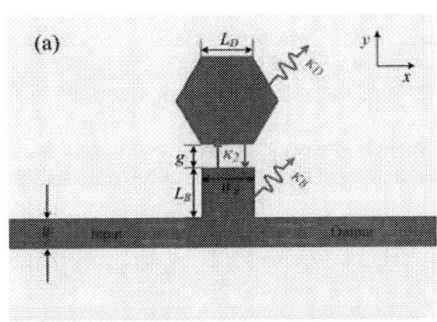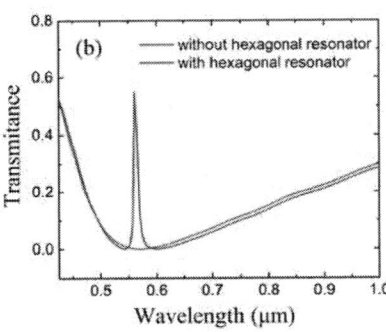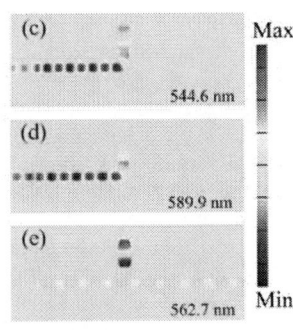

Fig. 1. (a) Schematic diagram of the proposed PIT nano-sensor. (b) System transmission spectra with or without hexagonal resonator, Hz distribution in the PIT chip at wavelength (c) 544. 6 nm, (d) 589.9 nm, and (e) 562.7 nm.

We consider the interaction between bright and dark resonators. According to the CMT, the transmittance of the system can be expressed as

$$T = \left| 1 - \kappa_1 \frac{j\delta_D + \kappa_D}{(j\delta_B + \kappa_B + \kappa_1)(j\delta_D + \kappa_D) + \kappa_2^2} \right|^2 \tag{1}$$

where κ_B and κ_D are the decay rates caused by the intrinsic loss of the bright/dark resonators, respectively, κ_1 denotes coupling between bus waveguide and bright resonator, and κ_2 denotes coupling between bright and dark resonators. $\delta_D = \omega - \omega_D$ and $\delta_B = \omega - \omega_B$ are the frequency detuning of the two resonators, respectively. Obviously, when dark resonator is absent ($\delta_D = 0$ and $\kappa_2 = 0$), the system demonstrates band-stop feature. However, the presence of dark resonator can significantly modify the transmission performance of the system. At frequency of $\omega \approx \omega_D \approx \omega_B$, Eq. (1) can be simplified to

$$T = \left| \frac{\kappa_B \kappa_D + \kappa_2^2}{\kappa_B \kappa_D + \kappa_1 \kappa_D + \kappa_2^2} \right|^2 \tag{2}$$

It is clearly that, in case the intrinsic decay rates of two resonators are weak enough, the transmittance is mainly determined by the coupling between two resonators. A pronounced transmission window can be generated under the condition of strong κ_2, leading to the so-called PIT-like effect. To verify the analysis, we add the hexagonal resonator with side length $L_D = W_B = 140$ nm and $L_B = 120$ nm above the stub resonator and set the gap distance $g = 30$ nm. By using FDTD method with mesh resolution of 2 nm, as shown in Fig. 2(b), a PIT-induced transparency window is emerged at the stop band of the bright resonator. Because of the transparency window, the broad stop band of the stub resonator is split into two bands. To further explain the properties of the PIT-like performance, the amplitude distribution of the magnetic field Hz at the wavelength of transmission window (562.7 nm) and two transmission dips (544.6 nm and 598.9 nm) are demonstrated in Fig. 2(c)-(e). It can be found that at two transmission dips, the incident plasmonic wave from bus waveguide excites the bright resonator directly, while the dark resonator is excited by coupling with the bright resonator. The field Hz distributions in the two resonators are different from each other, indicating resonant mode splitting in the system. However, at those two wavelengths, the fields Hz between waveguide channel and bright resonator are always out-of-phase, and therefore the plasmonic wave is prevented to pass through. On the other hand, at the transparency wavelength of Fig. 2(e), it is distinct that only the dark resonator is strong excited while its counterpart is suppressed. This is similar to the EIT effect in multilevel atoms system, cancelling of the bright resonator due to the destructive interference of two pathways. Particularly, in our chip the two pathways are bus waveguide → stub resonator and bus waveguide → stub resonator → hexagonal resonator → stub resonator. These two pathways interfere with each other destructively. As a result, a transparency window occurs.

III. SENSING PERFORMANCE

To investigate the sensing performance, the hexagonal resonator is operated as the sensing channel with different RI. Normally, there are two most important aspects to evaluate a sensor with wavelength interrogation method, namely, sensitivity and bandwidth of the resonance. High sensitivity ensures a drastic wavelength shift caused by change of the surrounding medium. While sharp resonant peak with narrow full width half maximum (FWHM) allows the detection instruments such as optical spectrum analyzer to effectively recognize even tiny wavelength shift so as to increase the accuracy. Here we define the wavelength sensitivity as $S = \Delta\lambda/\Delta n$, where $\Delta\lambda$ and Δn are the wavelength shift and RI variation, respectively. Moreover, we evaluate the sensor performance by using figure-of-merit (FOM) defined as FOM = S/FWHM. Fig. 2(a) shows the spectra under different ambient RI with $g = 40$ nm and $L_D = 140$ nm. The transparency window demonstrates red shift with larger RI. Fig. 2(b) demonstrates the resonant wavelength and FOM under different ambient RI. The FOM drops slightly with increased index. However, within the range under test, the sensitivity and FOM are always higher than 560 nm/RIU and 138 RIU^{-1}, respectively. Such high FOM is very suitable for sensing applications with high accuracy. The sensing performance near RI of 1 at different structure parameters are summarized in Fig. 2(c-d). When the gap distance varies from 10 nm to 50 nm, the sensitivity with value of 550 nm/RIU keeps nearly unchanged. However, the FOM can be enhanced with the growing gap distance, because of the FWHM shrinking, as shown in Fig. 2(c). It should be noted that narrower FWHM corresponds to lower transmission which may degrade the signal-noise-ratio (SNR) of the sensor, therefore, in practical application, there exists a trade-off between those two features. Particularly, with moderated gap distance of 40 nm, transmittance of ~34% together with a FOM of ~178 RIU^{-1} can be achieved, which shows great improvement comparing with previously reported PIT-based nano-sensors [5, 6] and plasmonic resonator-based nano-sensors [7-10]. The sensing performance dependences on side length L_D is also investigated, as shown in Fig. 2(d). It is found that the sensitivity increases with the size of the hexaagonal resonator but there is an optimal FOM at L_D=140 nm.

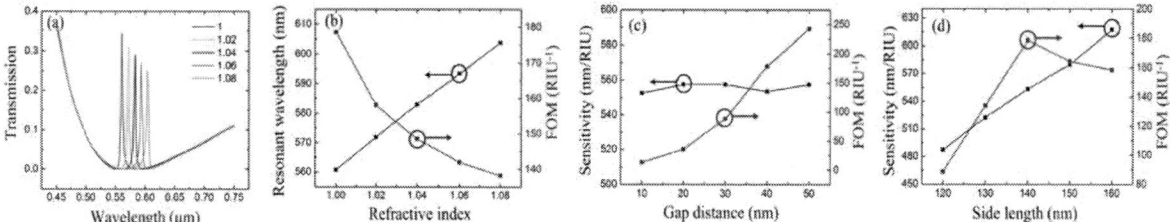

Fig. 2. (a) Transmission spectra with respect to index variation, (b) resonant wavelength and FOM at different ambient refractive indices, Sensitivity and FOM dependence on (c) gap distance and (d) side length.

IV. DISCUSSION AND CONCLUSIONS

In the proposed nano-sensor, it is desired that the stub resonator are well separated with the sensing channel (hexagonal resonator) so that the barrier region will not be easily crushed when filling with sensing medium. Therefore, the gap distance g should be large enough. Additionally, larger g also reduces the FWHM and subsequently increases the FOM. However, large gap distance reduces the coupling strength (κ_2) between two resonators and shrinks the transparency window, leading to degrade the SNR. Note that the coupling strength is determined by both gap distance and coupling length. In our design, because of the relatively long coupling between hexagonal resonator and stub resonator, the distance between them can be large while maintaining sufficient transmittance.

In conclusion, we have proposed and numerically investigated a PIT chip by employing MIM structure as bus waveguide, a stub resonator as bright resonator and a hexagonal resonator as dark resonator. The proposed device is further investigated as a RI nano-sensor. Benefit from the high wavelength sensitivity and narrow transmission window, FOM as high as 178 RIU^{-1} can be achieved.

ACKNOWLEDGMENT

This work was partially supported by the Fundamental Research Funds for the Central Universities, China University of Geosciences (Wuhan) and the 863 High Technology Plan (2015AA015502).

REFERENCES

[1] M. Fleischhauer, A. Imamoglu, and J. P. Marangos, "Electromagnetically induced transparency: Optics in coherent media," Rev. of Mod. Phys., vol. 77, pp. 633-673, Sep. 2005.

[2] S. Zhang, D. A. Genov, Y. Wang, M. Liu, and X. Zhang, "Plasmon-induced transparency in metamaterials," Phys. Rev. Lett., vol. 101, pp. 047401, Jul. 2008.

[3] M. I. Stockman, "Nanoplasmonics: past, present, and glimpse into future," Opt. Express, vol. 19, pp. 22029-22106, Oct. 2011.

[4] S. Lal, S. Link, N. J. Halas, "Nano-optics from sensing to waveguiding," Nat. Photon., vol. 1, pp. 641-648, Nov. 2007.

[5] C. Y. Chen, I. W. Un, N. H. Tai, and T. J. Yen, "Asymmetric coupling between subradiant and supperradiant plasmonic resonances and its enhanced sensing performance," Opt. Express, vol. 17, pp.15372-15380, Aug. 2009.

[6] B. X. Li, H. J. Li, L. L. Zeng, S. P. Zhan, Z. H. He, Z. Q. Chen, and H. Xu, "High-sensitivity sensing based on plasmon-induced transparency," IEEE Photon. J, vol. 7, pp. 4801207, Oct. 2015.

[7] Y. Y. Xie, Y. X. Huang, W. L. Zhao, W. Xu, and C. He, "A novel plasmonic sensor based on metal-insulator-metal waveguide with side-coupled hexagonal cavity," IEEE Photon. J., vol. 7, pp. 4800612, Apr. 2015.

[8] T. Wu, Y. Liu, Z. Yu, Y. Peng, C. Shu, and H. He, "The sensing characteristics of plasmonic waveguide with a single defect," Opt. Commun., vol. 323, pp. 44-48, Jul. 2014.

[9] T. Wu, Y. Liu, Z. Yu, Y. Peng, C. Sun, and H. Ye, "The sensing characteristics of plasmonic waveguide with a ring resonator," Opt. Express, vol. 22, pp. 7669-7677, Mar. 2014.

[10] X. Jin, X. Huang, J. Tao, X. Lin, and Q. Zhang, "A novel nanometeric plasmonic refractive index sensor," IEEE Trans. Nanotechnol., vol. 9, pp. 134-137, Mar. 2010.

Optical characteristics of InP/GaInAs core-multishell NWs grown by self-catalytic VLS mode

Kohei Takano, Takehiro Ogino, Keita Asakura, Takao Waho, and Kuzuhiko Shimomura

Dept. of Engineering and Applied Sciences, Sophia University,
7-1 Kioi-cho, Chiyoda-ku, Tokyo 102-8554, Japan.
kshimom@sophia.ac.jp

Abstract: We have successfully demonstrated the growth of InP/GaInAs core-multishell nanowires employed the self-catalytic VLS mode and VPE mode of MOVPE, and obtained the photoluminescence spectrum dependent on the thickness of GaInAs shell layer.
Keywords: InP, nanowire, core shell, MOVPE, self-catalytic VLS

I. INTRODUCTION

The potential of III-V compound semiconductor nanowires is immeasurable. Their electrical and optical characteristics are unique, and they are expected to be applied to lasers [1], and light-emitting-diodes (LEDs) [2]. In the many investigations, the vapor-liquid-solid (VLS) modes requiring metal particles such as gold (Au) as the catalyst and the selective area growth (SAG) are widely used. The SAG mode can grow nanowires with high controllability, but the SAG needs many steps of growth. The VLS modes are relatively simple, but using Au as the catalyst is generally undesirable because Au atoms are incorporated into the nanowire as impurities during growth and have a negative influence on the electronic and optical properties. In the self-catalytic VLS mode, group III particles are employed as the catalyst [3]. By heating an InP substrate prior to nanowire growth, the substrate surface is reconstructed and provides an In-rich surface. Tri-methyl-indium (TMI) is then supplied and indium droplets are formed on the substrate and act as the metal catalyst. In the self-catalytic VLS mode, there is no necessity of using Au as the catalyst for nanowire growth.

In this report, we adopted the self-catalytic VLS mode and show the growth of InP/GaInAs/InP core-multi shell nanowire on an InP(111)B substrate by low-pressure metal-organic-vapor-phase-epitaxy (MOVPE) and investigate the structure, photoluminescence properties dependent on the growth time of the GaInAs shell layer at the room temperature.

II. EXPERIMENT

The nanowire growth of InP-core, GaInAs-shell, and InP-shell were conducted in a 60 Torr low pressure MOVPE reactor using TMI and tri-ethyl-gallium (TEG) as the group III elements, and tertiary-butyl-arsine (TBA) and tertiary-butyl-phosphine (TBP) as the group V elements. Figure 1 shows the sequence of the growth of InP/$Ga_{0.47}In_{0.53}As$/InP core-multishell nanowires. First, the InP(111)B substrate was heated at 450 °C and annealed for 5 min under TBP at a flow rate of 353 μmol/min. The temperature was then decreased to 380 °C, and indium droplets were formed on the substrate by supplying TMI at a flow rate of 3.0 μmol/min for 5 min [4]. Then, when TMI and TBP were both supplied, the indium droplets were supersaturated, and InP nanowires were grown. The flow rates of TMI and TBP for nanowire growth were 15.0 and 400 μmol/min, respectively, for growth times of 60 min. After the growth of InP core nanowires, the temperature was increased to 560 °C for $Ga_{0.47}In_{0.53}As$/ InP multishell growth owing to change the growth mode from VLS mode to VPE mode. $Ga_{0.47}In_{0.53}As$ shells were grown at TEG, TMI, and TBA flow rates of 8.07, 8.86, and 176 μmol/min, respectively, and the growth times were changed between 10 s and 60 s. Then, InP shells were grown at TMI and TBP flow rates of 8.86 and 353 μmol/min, respectively where the growth time was 1 min.

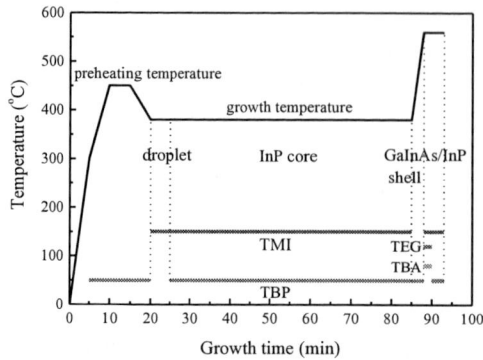

Figure 1 Growth sequence of InP/GaInAs/InP core-multishell nanowires.

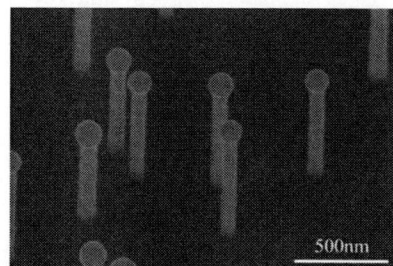

Figure 2 SEM image of InP nanowires grown on InP(111)B substrate.

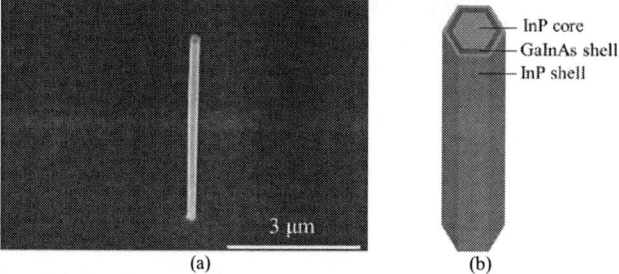

(a)   (b)

Figure 3 (a) SEM image of InP/GaInAs/InP nanowire grown on InP(111)B substrate.
(b) Structure of InP/GaInAs/InP core-multishell nanowire.

SEM image of InP core nanowires and InP/$Ga_{0.47}In_{0.53}$As/InP core-multishell nanowire are shown in Figure 2 and Figure 3 (a), and the schematic structure of a InP/$Ga_{0.47}In_{0.53}$As/InP core-multishell nanowire is shown in Figure 3 (b). The average diameter of core nanowires was about 150 nm, and the height, density of core-multishell nanowires were 10 μm, 10^7/cm^2, respectively.

We investigated the optical characteristics of InP/$Ga_{0.47}In_{0.53}$As/InP core-multishell nanowire dependent on the growth time of $Ga_{0.47}In_{0.53}$ shell layer using micro-photoluminescence (PL) analysis at room temperature at an excitation wavelength of 532 nm, and a spot diameter was as small as 25 μm by using an objective lens with 10× magnification.

III.RESULTS

We have grown the three InP/$Ga_{0.47}In_{0.53}$As/InP core-multishell nanowires where the growth time of the $Ga_{0.47}In_{0.53}$As shell layer was 10, 30, and 60s. Figure 4 shows SEM images of the cross section of a InP/$Ga_{0.47}In_{0.53}$As/InP nanowire for various growth times and chemically etched by H_2O_2 : H_2SO_4 : H_2O = 1 : 1 : 100. The $Ga_{0.47}In_{0.53}$As/InP multishell layer had a hexagonal crystal structure and the InP core had an unclear facet crystal structure. The unclear facet of the InP core was caused by the etching in the acid solution for the SEM observation.

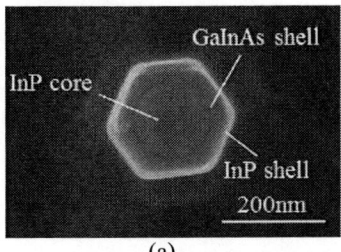 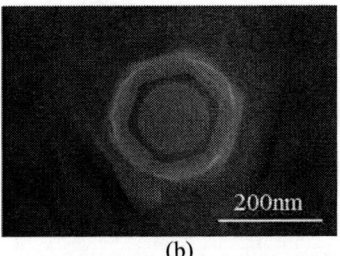 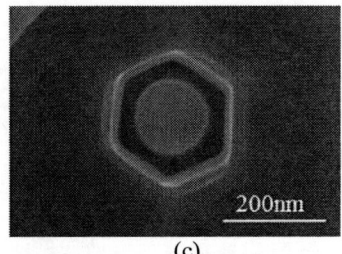

(a)   (b)   (c)

Figure 4 SEM images of cross-sectional slices of InP/$Ga_{0.47}In_{0.53}As$/InP nanowires prepared
at various $Ga_{0.47}In_{0.53}As$ shell layer growth times: (a) 10, (b) 30, and (c) 60 s.

From these SEM photograph, the measured thickness of the $Ga_{0.47}In_{0.53}As$ shell layer were 4.1, 9.5, and 23.3 nm where the $Ga_{0.47}In_{0.53}As$ shell layer growth time were 10, 30, and 60s, respectively. The PL spectrums of InP/$Ga_{0.47}In_{0.53}As$/InP core–multishell nanowires at various growth times of $Ga_{0.47}In_{0.53}As$ shell layer are shown in Figure 5. The PL peak wavelength was 1420, 1513, and 1624 nm where the $Ga_{0.47}In_{0.53}As$ shell layer growth times was 10, 30, and 60 s, respectively. The observed blue shift was attributed to the quantum confinement effect by the two dimensional quantum well made by InP-core, $Ga_{0.47}In_{0.53}As$ shell and InP outer shell along the side of the nanowires. These results show that we can control emission wavelength by changing $Ga_{0.47}In_{0.53}As$ growth time.

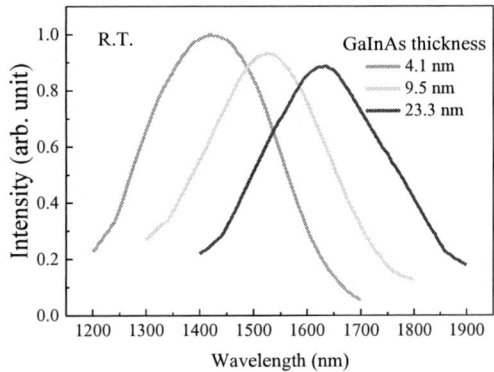

Figure 5 PL spectra of InP/$Ga_{0.47}In_{0.53}As$/InP nanowires prepared at the various growth time of $Ga_{0.47}In_{0.53}As$.

IV.CONCLUSION

We examined the PL characteristics dependent on the thickness of $Ga_{0.47}In_{0.53}As$ shell layer in the InP/$Ga_{0.47}In_{0.53}As$/InP core-multishell nanowires grown by self-catalytic VLS mode. The PL peak wavelength of InP/ $Ga_{0.47}In_{0.53}As$/InP core-multishell nanowire was 1420nm, 1513nm, and 1624 nm at the growth time of 10s, 30s, and 60 s, respectively. We can control the emission wavelength by changing the thickness of $Ga_{0.47}In_{0.53}As$ shell layer in the InP/ $Ga_{0.47}In_{0.53}As$/InP core-multishell nanowires.

ACKNOWLEDGEMENT

This work supported by the MEXT-Supported Program for the Strategic Research Foundation at Private Universities from the Ministry of Education, Culture, Sports, Science and Technology of Japan (MEXT).

REFERENCES

[1] X. Duan, Y. Huang, R. Agarwal, and C. M. Lieber, "Single-nanowire electrically driven lasers.", Nature 421, 241 (2003).
[2] W. Guo, M. Zhang, A. Banerjee, and P. Bhattacharya, "Catalyst-Free InGaN/GaN Nanowire Light Emitting Diodes Grown on (001) Silicon by Molecular Beam Epitaxy", Nano Lett. 10, 3355 (2010).
[3] G. Zhang, K. Tateno, H. Gotoh, and T. Sogawa, "Vertically Aligned InP Nanowires Grown via the Self-Assisted Vapor–Liquid–Solid Mode", Appl. Phys. Exp. 5 055201(2012).
[4] T. Ogino, M. Yamauchi, Y. Yamamoto, K. Shimomura, T. Waho, "Emission wavelength control of self-catalytic InP/GaInAs/InP core–multishell nanowire on InP substrate grown by metal organic vapor phase epitaxy", J. Cryst. Growth 414,161 (2015).

OECC/PS2016

S-K Growth of InAs quantum dots on directly-bonded InP/Si substrate using MOVPE

Naoki Kamada, Toshiki Sukigara, Keiichi Matsumoto, Junya Kishikawa, Tetsuo Nishiyama, Yuya Onuki
and Kazuhiko Shimomura
Department of Engineering and Applied Sciences, Sophia University,
7-1 Kioi, Chiyoda, 102-8554, Tokyo, JAPAN
kshimom@sophia.ac.jp

Abstract: *Stranski-Krastonogh QDs have been successfully grown on InP/Si substrate fabricated by wafer direct bonding. According to PL and AFM measurements, almost the same size and peak wavelength have been obtained with the InP substrate.*
Keywords: quantum dots, LD, InAs, InP/Si

I. INTRODUCTION

In order to achieve the high-speed and large-capacity lightwave transmission in the large-scale integrated circuits with low power consumption, we have studied InP-based optical devices on Si substrate. Epitaxial growth of III-V materials on Si substrate has been a technological challenge due to the material defects which stem from large lattice-mismatches of III-V materials and Si. For this purpose, many ideas have been proposed and hybrid integration using various bonding technique has been demonstrated. Employing the hybrid integration technique, however, the difference of height of each optical device was one of the bottlenecks to integrate several III-V based functional devices on Si substrate. To overcome these problems, we have developed InP/Si substrate employing conventional wafer direct bonding technique [1,2].

In this report, we show the growth procedure of Stranski-Krastonogh InAs/InP quantum dots (QDs) structure on the directly-bonded InP/Si substrate, and the results of AFM measurements and optical characteristics of QDs.

II. EXPERIMENTS AND RESULTS

The preparation of the InP/Si substrate and growth of InAs/InP QDs structure are explained as follows. The substrates used in this work were mirror-polished InP substrate and Si substrate. First, GaInAs / InP template layer / GaInAs was grown on InP substrate by low pressure MOVPE. After that, both as-grown GaInAs/InP and Si substrate were cleaned with ultra-sonic cleaning in order to avoid particles between substrates. Then the as-grown sample was dipped in $HCl:H_2O$ solution to etch-off the InP substrate. After that, both GaInAs / InP / GaInAs structure and Si substrate were cleaned with a $H_2SO_4:H_2O_2:H_2O$ solution to make their surfaces hydrophilic and obtain InP template, followed by de-ionized water rinse. Then surfaces were contacted in de-ionized water. After dried with N_2 gas flow in atmosphere, InP/Si substrate was realized. After that, the substrate was loaded into a furnace and heated in a N_2 atmosphere at 400 degree-C for 60 min. Thorough this process, InP/Si substrate was firmly fused.

By using InP/Si substrate, InAs QDs were grown by Stranski-Krastanov (S-K) grown mode using MOVPE growth technique and double cap procedure [3,4]. The growth temperature and pressure of QDs were 540 degree-C and 15 torr, and other layers were 630 degree-C and 100 Torr. The precursors used in this growth were tBA, tBP, TMI, and TEG.

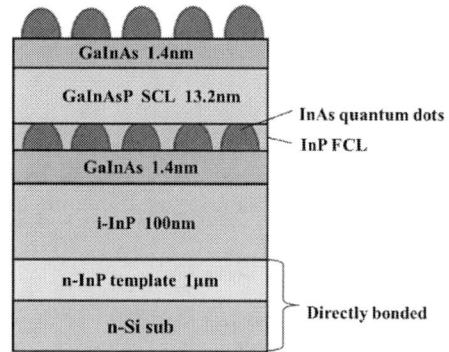

Fig. 1. Schematic layer structure of InAs QDs.

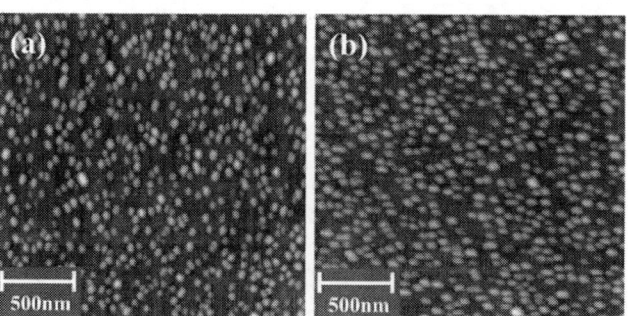

Fig. 2. AFM images of QDs on (a) InP and (b) InP/Si.

958

The fabricated layer structure is shown in Fig 1. In this experiment, the first cap layer (FCL) thicknesses were changed as 4nm, 6nm, 8nm, and 10nm, and this FCL thickness determine the height of QDs. Fig 2(a) and (b) show the atomic force microscopy (AFM) images of QDs grown on InP substrate and directly-bonded InP/Si substrate where the FCL thickness was 6nm. We have successfully grown QDs on InP/Si substrate. Fig.3 show the density, diameter, and height of QDs on InP and InP/Si substrate dependent on the FCL thickness. These figures are made from the AFM images of several points in one sample where the square symbol show the average value and error bar show the maximum and minimum values. Density and diameter of QDs were slightly reduced with the increase of FCL thickness, on the other hand the height was slightly increased. However, the differences between InP and InP/Si substrate were very small, and there were no relative merits between them. By using InP/Si substrate, additional lattice difference between InAs QDs and InP/Si substrate was occurred because of the thermal expansion difference between InP and Si. Table I shows the lattice constant (*a*) and thermal expansion coefficient (TEC) of InAs, InP and Si. Because of the thermal expansion coefficient of InP is larger than that of Si, compressive strain is received during QDs growth. By considering the thermal expansion coefficient, the lattice difference between InAs and InP were 3.14% on InP substrate and 3.24% on InP/Si substrate if we assumed the free standing temperature of InP/Si substrate was 200 degree-C [5]. Because of the lattice difference between two substrates was 0.1%, we have not observed the significant variations.

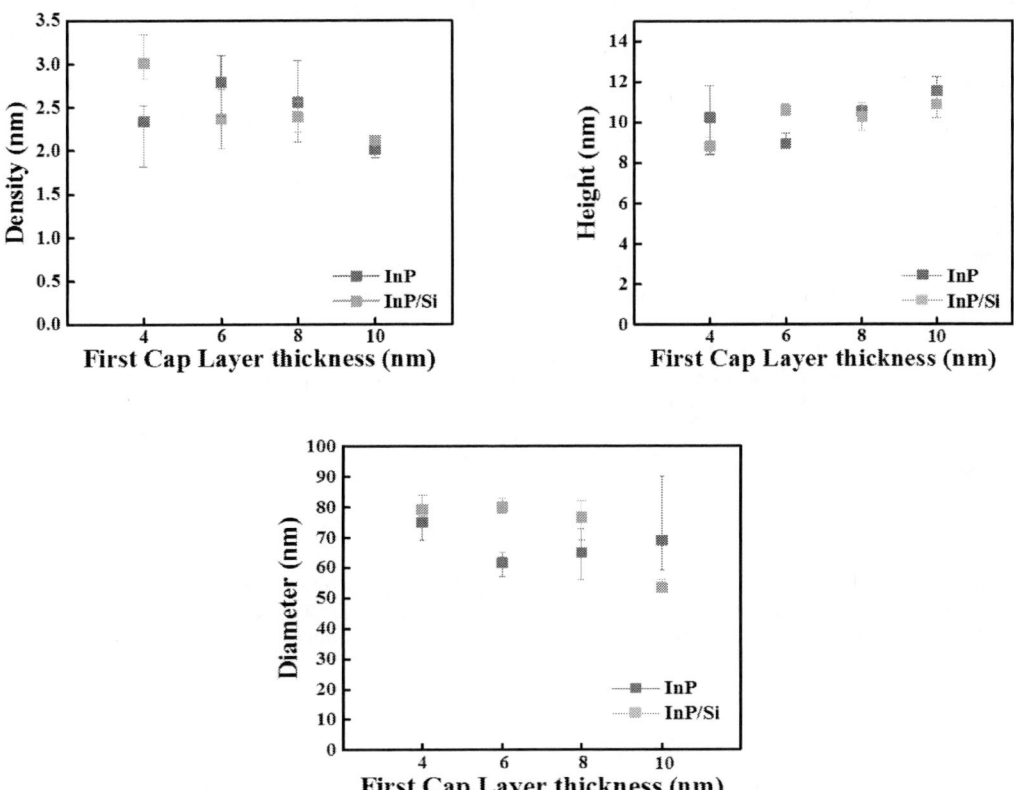

Fig. 3. Density, diameter and height of QDs dependent on FCL thickness.

TABLE I
Lattice constant (*a*) and thermal expansion coefficient (TEC) of InAs, InP and Si.

	a ($\times 10^{-10}$m)	TEC ($\times 10^{-6}$/K)
InAs	6.058	4.2
InP	5.869	4.6
Si	5.43	2.6

Fig.4 show the PL characteristics of the emission intensity and peak wavelength of QDs grown on InP substrate and InP/Si substrate dependent on FCL thickness. The emission intensity from 4nm and 6nm FCL thickness was larger on InP substrate and almost the same intensity from 8nm and 10nm FCL thickness. The error bar of intensity grown on

InP/Si substrate was relatively small compared to the InP substrate. When the FCL was 10nm, the intensity was remarkably reduced. The height of S-K grown InAs QDs were about 10nm, hence almost all the QDs were covered by InP with 10nm FCL thickness. This reduction was caused by the growth temperature difference between FCL of 540 degree-C and SCL of 630 degree-C, however further investigation is necessary. Emission wavelength was red shifted by increasing the FCL thickness because of the quantum size effect of QDs height. By comparing the InP and InP/Si substrate, the peak wavelength difference was small, and significant tendency was not observed.

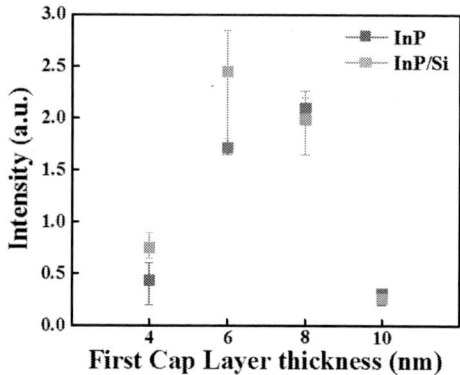 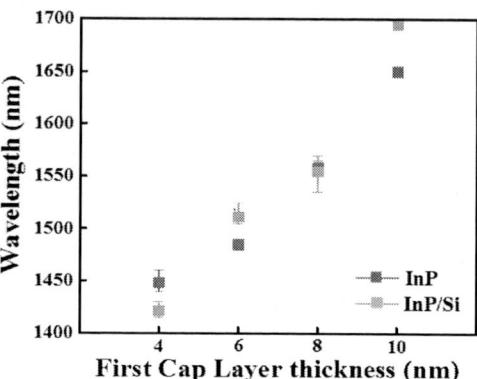

Fig. 4. PL intensity and peak wavelength of QDs dependent on FCL thickness.

III. CONCLUSIONS

We have shown the growth process of QDs grown on InP/Si substrate, and have successfully obtained S-K InAs QDs using double-cap procedure. The InP/Si substrate was obtained by direct wafer bonding technique. We have compared the density, diameter and height of QDs from AFM images, and also PL intensity and peak wavelength between InP/Si substrate and InP substrate. Because of the difference of thermal expansion of InP and Si, there were 0.1% lattice difference between two substrates at the QDs growth temperature, however, there was no significant difference between them. From these experimental results, we can grow QDs structure using double-cap procedure on InP/Si substrate with the same sequence on InP substrate.

REFERENCES

[1] K. Matsumoto, T. Makino, K. Kimura, and K. Shimomura, "Growth of GaInAs/InP MQW using MOVPE on directly-bonded InP/Si substrate", J. Crystal Growth, vol.370, pp.133-135, May 2013.
[2] K. Matsumoto, T. Makino, K. Kimura, K. Shimomura, "Extremely improved InP template and GaInAsP system growth on directly- bonded InP/SiO2-Si and InP/glass substrate", Phys. Staus Solidi C, vol.10, no.5, pp. 782-785, May 2013.
[3] C. Paranthoen, N. Bertru, O. Dehaese, A. Le Corre, S. Loualiche, and B. Lambert, "Height dispersion control of InAs/InP quantum dots emitting at 1.55 µm," Appl. Phys. Lett., vol. 78, no. 12, pp. 1751–1753, 2001.
[4] K. Matsumoto, X .Zhang, J. Kishikawa, and K. Shimomura, "Current-injected light emission of epitaxially grown InAs/InP quantum dots on directly bonded InP/Si substrate", Jpn. J. Appl. Phys., vol. 54, 030208, Jan. 2015.
[5] K. Matsumoto, R. Kobie, and K. Shimomura, "Thermal treatment for preventing void formation on directly-bonded InP/Si interface", Jpn. J. Appl. Phys., vol.53, no.11, 116502, Oct. 2014.

Hybrid Electro-optic Polymer Modulators

Feng Qiu*, and Shiyoshi Yokoyama

Institute for Materials Chemistry and Engineering, Kyushu University, 6-1 Kasuga-koen Kasuga-city, Fukuoka, 816-8580 Japan
Author e-mail address: drfqiu@cm.kyushu-u.ac.jp

Abstract: In this letter, we report a TiO_2/electro-optic polymer hybrid rib-waveguide with a low figure of merit of 3.3 V·cm, corresponding to 1.65 V·cm in a traditional push-pull Mach-Zehnder interferometer structure. This low figure of merit results from the 80% improved poling efficiency of our EO polymer in the hybrid structure. The waveguide also possesses a relatively low propagation loss of 3.0 dB/cm and a simple fabrication process.

Keywords: waveguide, modulator, polymer

I. INTRODUCTION

Waveguide modulators based on electro-optic (EO) polymer play a key role in telecom, datacom, and sensing areas. One of the most important characteristics of modulators is the figure of merit ($V\pi L$) required for operation, where $V\pi$ is half wave voltage and L is the active region length. $V\pi L$ is determined by the interelectrode gap, the integral between the applied electric field and the optical mode, and the EO coefficient (r_{33}) of the polymer.

In all polymeric waveguide modulators, the electrical resistivity of the commercially available polymeric cladding layers is usually on the order of 10^{11} - 10^{13} Ω ·cm. However, EO polymer core can be as low as 10^8 Ω ·cm. This makes the efficient poling of the EO polymer difficult. In addition, typical interelectrode distances in these devices range from $8 - 15$ μm, but the thickness of the EO core is $2 - 4$ μm. This means that large voltage in a modulator drops across the thick cladding layers. The above two drawbacks may result in the high $V\pi L$ for the all polymeric waveguide modulators.

In this work, we present a simple TiO_2/electro-optic hybrid rib waveguide modulator. The r_{33} of the EO polymer in the waveguide is 80% higher than that in a thin film. In addition, the low loss waveguide, including only a bottom cladding with the thickness of 1μm, is totally 3.8μm thick. These two factors induce the modulator with low $V\pi L$.

II. TECHNICAL WORK PREPARATION

The To demonstrate the poling efficiency, we first measured the r_{33} of an EO polymer thin film and a TiO_2/EO polymer double-layer film on ITO glass, as shown in the insets of Fig. 1 (a) and (b). The TiO_2 layer with a thickness of 250nm was obtained through the RF sputtering technique. The EO material used in this work was a chromophore-C60 (its structure shown in Fig. 1(c)) / PMMA guest-host polymer (33wt% C60 doped into PMMA). The 20wt% solution of this polymer in cyclopentanone was first filtrated through 0.2μm syringe filter, and then spin-coated onto ITO and TiO_2/ITO substrates, respectively. The polymer films were baked in a vacuum condition at 85°C for 48 hours, and the film thickness was measured to 2.4μm. The gold electrode was sputtered on top of the polymer films for contact poling and for the r_{33} measurement.

The r_{33} were measured by using the Teng-Man method at 1310nm, and were calibrated after taking into account the multiple-reflection effects. Fig. 1(a) and (b) exhibit the poling process of the two kinds of films. The poling temperatures for the two cases are both 96°C. In Fig. 1(a), the sample was heated up to poling temperature in ~1min. To avoid crack caused by the different heat expansion coefficients of EO polymer and TiO_2, it took ~1.6min to reach poling temperature in Fig. 1(b). After reaching the best poling voltage, the poling temperature was reduced to room temperature quickly (from ~1.2 to 1.4min in Fig. 1(a), and from 2.5 to 3.3 min in Fig. 1(b)). When reaching room temperature, the poling voltage was decreased to zero.

The poling fields are 78V/μm and 118 V/μm in Fig. 1 (a) and (b), respectively. However, the poling current in Fig. 1(b) is almost 3.5 times lower than that in Fig.1 (a), which means that the TiO_2 barrier effectively reduces leakage through current. Since the resistivity of the EO polymer (10^8 Ω ·cm) was much higher than that of the sputtered TiO_2 (~10^5 Ω ·cm), the poling voltage in Fig 1(b) could be applied nearly completely across the EO polymer layer. As a result, the r_{33} of the EO polymer in Fig. 1(b) can reach 125pm/V, which is ~80% higher than that in Fig.1 (a).

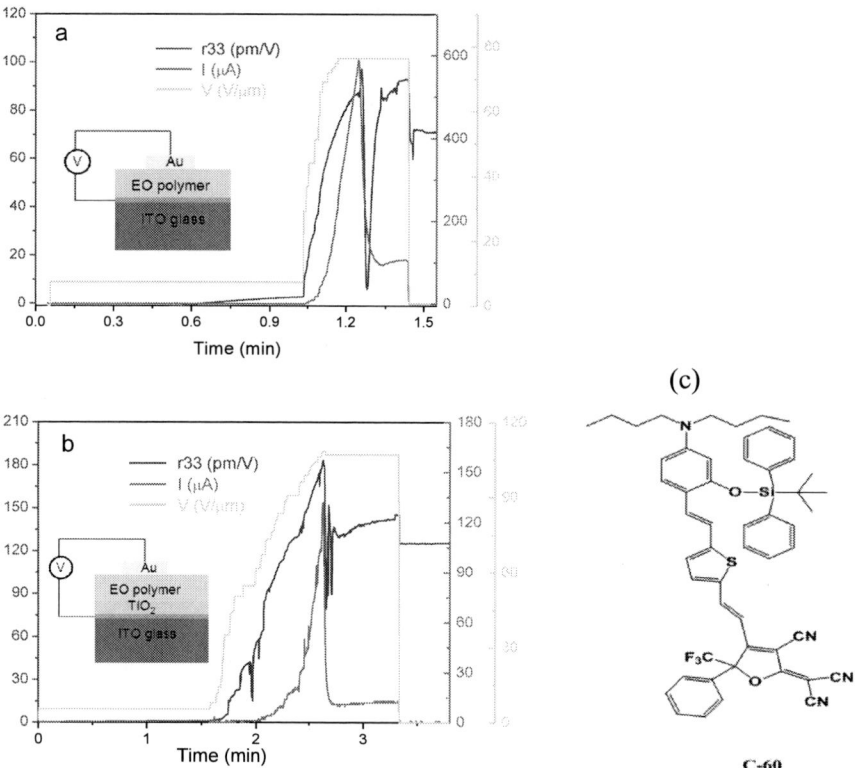

Fig.1 Poling process of (a) EO polymer thin film and (b) EO polymer/TiO2 double -layer film (insets show the structures); (c) Structure of chromophore-C60

Based on the TiO$_2$/EO polymer double-layer structure, a rib-waveguide was designed. An organic sol-gel SiO$_x$, consisting of ethylsilicate (ES) and triethoxythysilane with a mass ratio of 30 (ES) / 70%, was prepared for the waveguide bottom cladding. For the sol-gel SiO$_x$, acetic acid was used as a catalyst, and water and ethanol were the solvent. The resistivity of the sol-gel SiO$_x$ was measured almost the same as that of our EO polymer. The sol-gel SiO$_x$ with thickness of 1.1μm was coated on a 100nm Au lower electrode/silica (2μm)-on-silicon substrate. A 0.25μm TiO$_2$ layer was sputtered and then patterned by the photolithography and reactive ion etching. The width and height of the ridge section was 3.0 and 0.05μm, respectively. After spin-coating the EO polymer with thickness of 2.4μm, the upper Au electrode was sputtered [1-2].

The optical waveguide mode (TM) was simulated using the beam propagation method of Rsoft. The refractive indices of EO polymer, TiO$_2$ and SiO$_x$ are 1.61, 2.30 and 1.44 at the wavelength of 1.55μm, respectively. We estimated the waveguide coupling and propagation loss to be approximately 25dB and 3.0dB/cm respectively by an end-fire coupling system. The low material absorptions of TiO$_2$ (~1.0dB/cm), sol-gel (0.1-0.5dB/cm), and the EO polymer (~2.0dB/cm), and low optical absorption from the electrodes (~2.0dB/cm) should contribute to this relatively low propagation loss. To obtain the EO properties, the rib-waveguide phase modulator was observed as intensity modulation using a simple cross-polarization setup. Input light (1550nm) with a +45° linear polarization was coupled into the waveguide through a polarization maintaining fiber. Output light passing a -45° polarizer was collected by a detector. A driving voltage from a DC source was applied to contact the top and bottom electrodes. The Vπ of the modulator is ~15V. Considering the Au electrode length of 2.2mm, the VπL of our modulator is 3.3V·cm, which corresponds to the VπL of 1.65 V·cm in a push-pull MZI structure. The r$_{33}$ of the EO polymer was calculated by

$$r_{33} = \frac{\lambda d}{n^3 L \Gamma V_\pi} \qquad (1)$$

where λ is the work wavelength, d the interelectrode gap, n is the effective refractive index, L is the length of the electrodes, and Γ is the overlap integral between the applied electric field and the optical mode. According to equation (1), the r$_{33}$ of the EO polymer in the rib-waveguide structure is estimated to be ~120pm/V.

In summary, we demonstrated a hybrid TiO$_2$/EO polymer rib-waveguide modulator with VπL of 3.3 V·cm, corresponding to 1.65 V·cm in a push-pull MZI structure. In our simple waveguide structure, the poling efficiency was improved to 180%, resulting in this low VπL. The fabrication of the waveguide is simple and low-cost, which should be useful in industrial production.

REFERENCES

[1] F. Qiu, A. M. Spring, F. Yu, I. Aoki, A. Otomo, and S. Yokoyama, "Thin TiO$_2$ Core and Electro-Optic Polymer Cladding Wavaguide Modulators", Appl. Phys. Lett.102, 233504-1-3 (2013).

[2] F. Qiu, A. M. Spring, F. Yu, I. Aoki, A. Otomo, and S. Yokoyama, "Electro-Optic Polymer / Titanium Dioxide Hybrid Core Ring Resonator Modulators", Laser Photonics Rev. 7, 87 (2013).

WA2-78

OECC/PS2016

Photodetection Frequency Response Characterization for High-Speed Ge-PD on Si With an Equivalent Circuit

Jeong-Min Lee[1], Minkyu Kim[1], Stefan Lischke[2], Lars Zimmernman[2], Seong-Ho Cho[3], and Woo-Young Choi[1]

[1]Department of Electrical and Electronic Engineering, Yonsei University, Seoul, Korea
[2]IHP, Im Technologiepark 25, 15236 Frankfurt (Oder), Germany
[3]Mobile Health Care Group, Samsung Advanced Institute of Technology, Suwon-si, Gyeonggi-do, Korea
wchoi@yonsei.ac.kr

Abstract: *We characterize photodetection frequency response of a waveguide-type Ge-PD on Si having larger than 50-GHz photodetection bandwidth using an equivalent circuit model. Our model provides accurate frequency responses and allows clear identification of different contributions.*

Keywords: *Germanium photodetector, equivalent circuit model, Silicon photonics*

I. INTRODUCTION

Germanium photodetectors (Ge-PDs) realized on Si wafers are an essential component for Si photonic integrated circuits. Recently, Ge-PDs having photodetection bandwidth larger than 50-GHz have been reported [1] and high-performance monolithically integrated optical receiver circuits containing Ge-PDs have been demonstrated [2]. In order to achieve the optimal performance of integrated optical receivers, it is essential to have an accurate equivalent circuit model for Ge-PD that can be co-simulated with electronic circuits in the design stage. We have recently identified that the Ge-PD photodetection frequency response can be degraded with diffusion of photogenerated carriers and proposed an equivalent circuit model having two current sources, each of which respectively represents diffusion and drift of photogenerated carriers [3]. In this paper, we apply our modeling technique to the waveguide Ge-PD fabricated by IHP's photonic BiCMOS process, which has the unique capacity of integrating Si photonics devices with high-speed Si BiCMOS electronic circuits [4].

II. EQUIVALENT CIRCUIT MODEL

Fig. 1(a) shows the cross-section of the Ge-PD investigated in this paper. The intrinsic Ge layer is epitaxially grown on 220-nm thick, 750-nm wide Silicon-on-Insulator layer having 2-μm thick buried-oxide layer. The lateral PIN structure is realized with self-aligned implantation of P$^+$ and N$^+$ regions having peak concentrations of about 1×10^{18} cm^{-3} using 600-nm wide silicon nitride (SiN). The Ge-PD is 20-μm long. Details of the Ge-PD can be found in [1].

Fig. 1(b) shows the electron-hole pair generation rate due to absorption of 1.55-μm input light simulated with Lumerical 3-D FDTD, and Fig. 1(c) the electric-field distribution within our device biased at −1 V simulated with TCAD Sentaurus. As can be seen in the figures, a fair amount of electron-hole pairs are produced in the region where the electric field is not very strong and those carriers have to transport by slow diffusion. In order to accurately model the photodetection frequency response, consideration should be given to such this diffusion component as well as the drift process within the region having strong electric fields.

Fig. 2(a) shows the equivalent circuit model used in the present investigation. It has two current sources (I_1 and I_2) having different frequency responses for diffusion and drift of photogenerated carriers. Each current source has the single-pole frequency response with time constant τ_1 for I_1 and τ_2 for I_2, along with corresponding DC gain, A_1 and A_2,

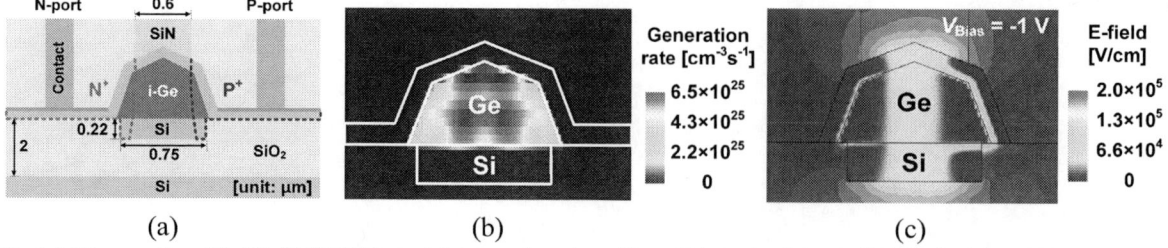

Fig. 1. (a) Cross-section of Ge-PD, (b) 3D-FDTD simulated generation-rate profile, and (c) simulated electric-field distribution at −1 V.

964

sum of which represents Ge-PD DC responsivity, normalized to one for simplicity in this paper. Z_{para} in the model represents passive electrical components due to interconnect, pad, parasitic resistances and capacitances. Specifically, R_{int} and L_{int} represent interconnect resistance and inductance, respectively, C_{pad} pad capacitance, C_{ox} oxide capacitance, R_{si} bottom silicon substrate resistance, and C_{c-c} capacitance between contacts. For modeling PIN junction, R_s represents series resistance, C_j depletion capacitance, and R_j depletion resistance.

S-parameters are measured for open and short test patterns on the same wafer with a vector network analyzer from 100 MHZ to 67 GHz, from which numerical values for R_{int}, L_{int}, C_{pad}, C_{ox}, and R_{si} are determined as 1.4 Ω, 56 pH, 16.7 fF, 30 fF, and 2 kΩ, respectively. Measured S-parameters of Ge-PD are used for extraction of R_s, C_j, R_j, and C_{c-c} values. The extracted values are listed in Table I.

To extract current source model parameters, the generation rate profile shown in Fig. 1(b) is imported into TCAD Sentaurus and two virtual generation rate profiles are created as shown in Fig. 3(a) and (b), one containing the generation rate only in the region where electric field is weak (< 2000 V/cm), representing the region where photogenerated carriers experience diffusion as shown in Fig. 3(a), and the other in the region where the electric field is strong (> 2000 V/cm), representing the region where photogenerated carriers experience drift in Fig. 3(b), respectively. Then we perform photodetection frequency response simulation for each case using TCAD Sentaurus and the results are fitted with single-pole frequency responses as can be seen Fig. 3(c). From these, we extract current source model parameters of τ_1 and A_1 for I_1, and τ_2 and A_2 for I_2 as listed in Table II. At −1-V bias voltage, about 9.2% of photogenerated carriers experience diffusion with the corresponding time constant of 15.9 ps.

III. PHOTODETECTION FREQUENCY RESPONSE CHARACTERIZATION

Fig. 4(a) shows the measured photodetection frequency response and the simulated result with our equivalent circuit. As can be seen, they agree well confirming the accuracy of our model. Using our equivalent circuit model, we can identify the contribution of each factor that influences the photodetection frequency responses. Fig. 4(b) shows the simulated results considering only τ_{RC} (without current sources in the equivalent circuit), τ_1 and τ_2 (without RC components), and τ_2 (without current source for diffusion and RC components). For these simulations, only the Ge-PD core is considered without Z_{para}. As can be seen in the figure, the photodetection bandwidth is limited by carrier transport and the diffusion of photogenerated carriers further degrades the photodetection bandwidth. It should be also noted that the bandwidth limitation due to parasitics is not very significant due to the optimized fabrication process providing very small parasitic resistances. This type of identification can be of great help for further device optimization.

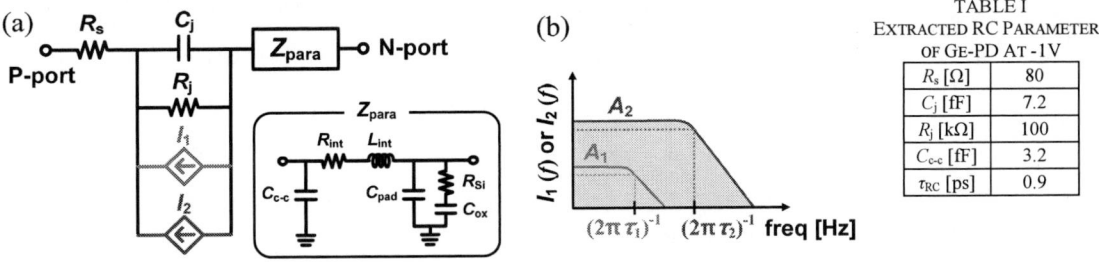

TABLE I
EXTRACTED RC PARAMETERS
OF GE-PD AT -1V

R_s [Ω]	80
C_j [fF]	7.2
R_j [kΩ]	100
C_{c-c} [fF]	3.2
τ_{RC} [ps]	0.9

Fig. 2. (a) A modified equivalent circuit model of Ge-PD and (b) frequency responses of photogenerated current source models.

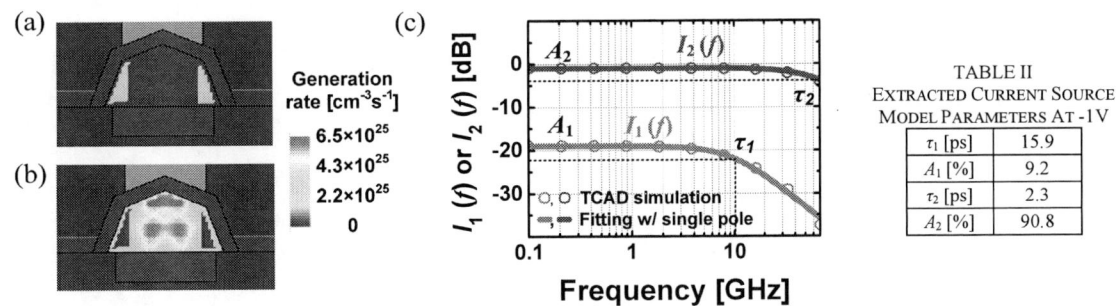

TABLE II
EXTRACTED CURRENT SOURCE
MODEL PARAMETERS AT -1V

τ_1 [ps]	15.9
A_1 [%]	9.2
τ_2 [ps]	2.3
A_2 [%]	90.8

Fig. 3. Virtual generation-rate profiles of photogenerated carrier (a) diffusion and (b) drift, and (c) simulated photodetection frequency responses of two current source models at −1 V.

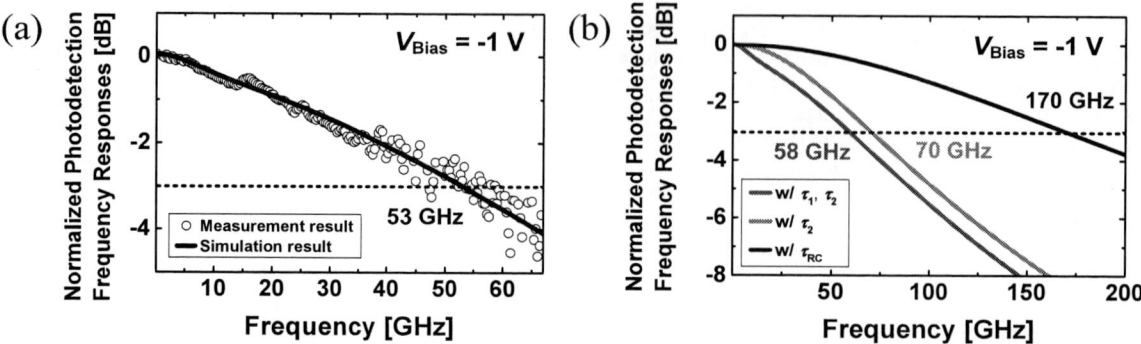

Fig. 4. (a) Measured and simulated photodetection frequency response and (b) simulated frequency responses with different time constant contributions.

IV. CONCLUSIONS

We present an equivalent circuit model for waveguide-type Ge-PD on Si having greater than 50-GHz photodetection bandwidth and show how to extract model parameters for Ge-PD. Using our equivalent circuit model, we can identify those factors that limit the photodetection frequency response. Our equivalent circuit can be of great help in designing high-performance monolithic integrated optical receivers.

ACKNOWLEDGMENT

Manuscript received. This work was supported by the National Research Foundation of Korea grant funded by the Korea government [2015R1A2A2A01007772].

REFERENCES

[1] S. Lischke et al., "High bandwidth, high responsivity waveguide-coupled germanium p-i-n photodiode," Opt. Exp., vol. 23, no. 21. pp. 27213-27220, Oct. 2015.

[2] A. Awny et al., "A 40 Gb/s monolithically integrated linear photonic receiver in a 0.25 μm BiCMOS SiGe:C technology," IEEE Microw. Compon. Lett., vol. 25, no.7, pp. 469-471, July 2015.

[3] J.-M. Lee and W.-Y. Choi, "An equivalent circuit model for germanium waveguide vertical photodetectors on Si," in Proc. MWP/APMP, 2014, pp. 139-141.

[4] L. Zimmermann et al., "BiCMOS silicon photonics platform," in Proc. Opt. Fiber Commun. Conf. (OFC), 2015, pp. 1-3.

WA2-79

Sub-µm Electrode Spacing SOI-PIN Photodiode Fabricated by CMOS Compatible Process

Hiroya MITSUNO, Takeo MARUYAMA and Koichi IIYAMA

School of Electrical and Computer Engineering, College of Science and Engineering, Kanazawa University

Author e-mail address: maruyama@ec.t.kanazawa-u.ac.jp

Abstract: SOI-PIN photodiodes were fabricated by CMOS compatible processes. The -3dB bandwidth of 13 GHz was obtained at electrode spacing of 0.6 µm, receiving area of 20x20 µm² and pad area of 30x30 µm².

Keywords: PIN-Photodiode, Silicon on insulator, CMOS compatible process

I. INTRODUCTION

The data transmission speed in an electronic system can be enhanced by utilizing an optical transmission technique. An optical interconnect technology has been studied thoroughly and silicon photonics have gained a lot of popularity in recent years [1], [2]. Silicon is generally used as an optical waveguide material in 1.3-1.55 µm wavelength region, while light sources and photodiodes are fabricated by Ge-on-Si or III-V-on-Si technology. On the other hands, it is an attractive approach to use silicon as a photodiode material in 0.8 µm wavelength region. High quality silicon film is suitable for electrical and optical performances of the devices such as a low dark current and high speed Si photodiodes [3]. It is important to eliminate the carrier generated in deep region of the substrate for high-speed operation. The spatially modulated type [4], a deep n-well type [5], and a silicon-on-insulator (SOI) type [6] are proposed and reported to solve the problem.

In this paper, narrow electrode spacing SOI-PIN photodiodes were fabricated by a CMOS compatible process for high speed operation. We measure the electrode spacing dependence of the frequency response at the wavelength of 830 nm.

II. DEVICE STRUCTURE AND MEASUREMENT SETUP

A. Device Structure of SOI-PIN Photodiodes

SOI-PIN photodiodes were fabricated through a foundry service of the institute of microelectronics (IME). Figure 1 shows a basic schematic cross-sectional view of SOI-PIN photodiodes fabricated on SOI substrate with a 210 nm thick Si absorption layer. The comb electrodes patterns for high speed measurement are formed. The Al electrode width is fixed on 1.0 µm and the electrode spacing is changing from 0.6 to 2.0 µm to investigate the high speed response. The optical receiving area and pad electrode size are 20×20 µm² and 60×60 µm², respectively (Fig. 2). The Si layer in the outside of the receiving area is etched completely to avoid the optical absorption.

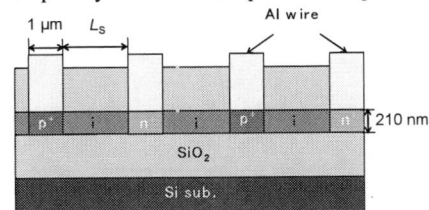

Fig. 1. Schematic cross-sectional view of SOI-PIN photodiodes

B. Measurement Setup

A laser light is normally illuminated on the SOI-PIN PDs via an optical fiber. The core diameter of the optical fiber is 9 µm. A semiconductor laser diode with a center wavelength of 830 nm was employed. When measuring the frequency responses, we used a 10 MHz - 40 GHz network analyzer, and the laser diode was intensity modulated by an electro-optic modulator with a bandwidth of more than 25 GHz. The frequency responses of the modulator and 40-GHz-RF cables were compensated by using a commercial GaAs PIN-PD with a bandwidth of 30 GHz.

OECC/PS2016

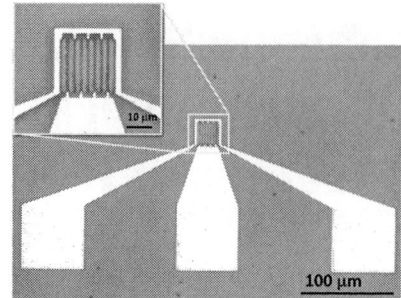

Fig. 2. Top view of the SOI-PIN photodiode fabricated by the CMOS compatible processes..

III. RESULTS AND DISCUSSION

Figure 3 shows a frequency response of the SOI-PIN photodiode with the electrode space of 0.6 μm at a reverse bias voltage of 10 V. The frequency response is normalized at the response of 100 MHz, and a -3 dB bandwidth of 13 GHz was obtained from low-pass filter (LPF) approximation.

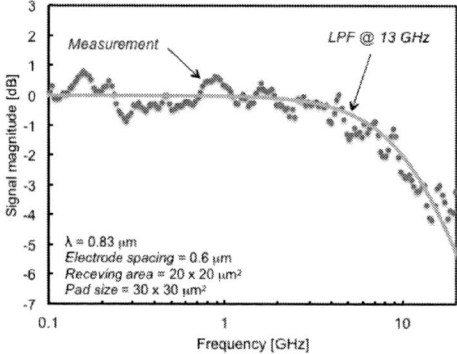

Fig. 3. The frequency response of the SOI-PIN photodiode at the reverse bias voltage of 10 V at the wavelength of 830 nm.

In order to investigate the finger electrode spacing dependence of the bandwidth with SOI-PIN photodiodes, the samples with electrode space of 0.6, 0.8, 1.0, 1.2 and 2.0 μm were measured at the reverse bias voltage of 10 V. The measurement value was shown in Fig. 4 as the points (red color).

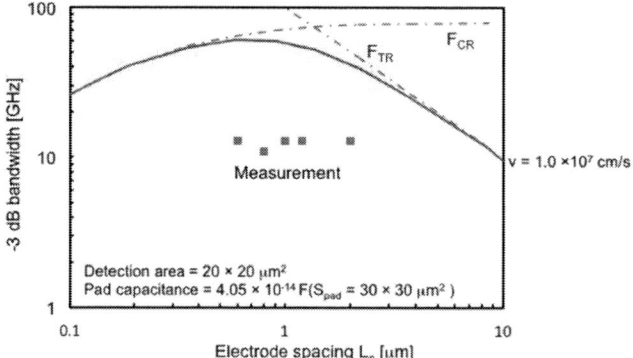

Fig. 4. Electrode spacing dependence of -3dB bandwidth of the SOI-PIN photodiode.

A estimated frequency response curve are also shown in Fig. 4. This trend is an important feature of SOI-PIN photodiodes. The response time of photodiode is primarily limited by two factors: (1) the transit time of photogenerated carriers to the electrode, shown as the green dashed line (velocity of hole was fitted to 1.0×10^7 cm/s as a saturation velocity); (2) a total capacitance of the depletion region capacitance of the semiconductor (as the same as a capacitance of the detector area calculated as around 10^{-14} F) and the electrode capacitance (a capacitance factor measured as about 0.045 fF/μm²), which was estimated and shown as the blue dashed line. An agreement between experimental data and estimated curve was not be obtained. This disagreement is due to exist another parasitic capacitance instead of the pad and the depletion layer. If this parasitic capacitance is removed, it is expected to realize over 40 GHz operation.

IV. CONCLUSIONS

We fabricated narrow electrode spacing SOI-PIN photodiodes by the CMOS compatible process for high speed operation. The -3 dB bandwidth of around 13 GHz was obtained in both of narrow and wide electrode spacing at reverse bias voltage of 10 V at 850 nm wavelength. The devices can be expected to realize the optoelectronic integrated circuits (OEIC) on Si-LSIs.

ACKNOWLEDGMENT

This research was partially supported by a Grant-in-Aid for Scientific Research (#15K06012) from the Ministry of Education, Culture, Sports, Science and Technology (MEXT).

REFERENCES

[1] M. A. Taubenblatt, "Optical Interconnects for High-Performance Computing," J. Lightwave Technol. Vol. 30, pp. 448-457, 30, Apr. 2012..

[2] D. A. B. Miller, "Device Requirements for Optical Interconnects to Silicon Chips," Proc. IEEE, Vol. 97, pp. 1166-1185, Jul. 2009.

[3] S. Y. Chou and Y. Liu, "32 GHz Metal-semiconductor-metal Photodetectors on Crystalline Silicon," Appl. Phys. Lett, Vo. 61 pp. 1760-1762, Jan. 1992.

[4] S. Radovanovic, A. J. Annema and B. Nauta, "A 3-Gb/s Optical Detector in Standard CMOS for 850-nm Optical Cmmunication," IEEE J. Solid-State Circuits Vol. 40, pp. 1706-1717, Aug. 2005.

[5] K. Iiyama, H. Takamatsu and T. Maruyama, "Hole-injection-type and Electron-injection-type Silicon Avalanche Photodiodes Fabricated by Standard 0.18 μm CMOS Process," IEEE Photon. Technol. Lett., Vol. 22, pp. 932-934, Jun. 2010.

[6] G. Li, K. Maekita, H. Mitsuno, T. Maruyama and K. Iiyama, "Over 10 GHz Lateral Silicon Photodetector Fabricated on Silicon-on-insulator Substrate by CMOS-compatible Process," Jpn. J. Appl. Phys., Vol. 54, pp. 04DG06 (4 pages) Mar. 2015.

WA2-8

OECC/PS2016

Arbitrary Output Port Selection in Multi-Mode Fiber Networks using Mode Division Multiplexing

Yuki Morizumi, Hirokazu Kobayashi and Katsushi Iwashita

Kochi University of Technology, 185 Miyanokuchi Tosayamada, Kami-Shi, Kochi 782-8502 Japan

e-mail:*morizumi.yuki@gmail.com*

Abstract: *We propose mode-forming networks which can transmit different signals to desired ports. Its feasibility is demonstrated by 2 channel different bit-rate signal transmission over 1 km multi-mode fiber to different ports by controlling input signals.*

Keywords: *Mode Division Multiplexing, MIMO, Subcarrier Multiplexing*

I. INTRODUCTION

In recent years, passive optical network (PON) technology is growing rapidly in order to reduce the overall cost and to spread the service area [1]. Wavelength division multiplexing (WDM) is effective method to enhance transmission capacity including extend reach. However, wavelength independence is preferable for future scalabilities and cost. There are some other PON architectures employing hybrid optical distribution networks, such as hybrid mode division multiplexing (MDM) by utilizing MDM transmission media [2, 3]. Recently, lossless remote controlled switch was proposed to transmit the desired data to the desired user by using time reversal array transmitter techniques as wavelength independent passive switching method for point-to-multipoint networks [4]. Time-reversal method requires extremely low propagation time differences within transmission paths because it utilizes optical carrier. We also proposed wavelength independent networks called mode-forming networks which utilized subcarrier multiplexing (SCM) techniques for remote switch networks [5]. The switching feasibilities of mode-forming networks were confirmed by switching the output port. By superimposing other signals, the desired data can be delivered to the desired output port using the proposed techniques.

In this paper, we demonstrate the possibilities of mode-forming networks. 2 channel subcarrier multiplexed optical signals are transmitted over multi-mode fiber and delivered different output ports using our proposed mode-forming network by getting the channel matrix.

II. PRINCIPLE OF MODE-FORMING NETWORKS

The basic concept of our proposed mode-forming networks is shown in Fig. 1. Networks consist of transmitters (*Tx n*) which can transmit superimposed different signals, multi-mode fiber (MMF) transmission networks, and receivers (*Rx n*). Mode division multiplexing is utilized to transmit several data with different modes. The transmitted signals can be combined and received by each receiver. If coherent system or SCM system is used, these interfered signals can be recovered by Multi-Input Multi-Output (MIMO) processing techniques. We propose to apply beam-forming techniques which are used for wireless communications. Subcarrier multiplexed signals are used in the proposed networks [6]. Transmitters send pre-distorted signals which are superimposed of different channel signals. Transmitted signals are coupled using a mode dependent coupler and divided by another mode dependent coupler. Several signals can arrive at the output port before *Rx n*, but the only desired signal can be received by *Rx n* while other signals are cancelled by controlling the phase and amplitude of SCM signals. We are not using MIMO processing for separating received signals, but utilize it to estimate the channel matrix of the MMF transmission system. The proposed system is using microwave subcarrier instead of optical carrier, then stable operation can be realized. The proposed method is independent to signal types such as data rate, modulation format and so on.

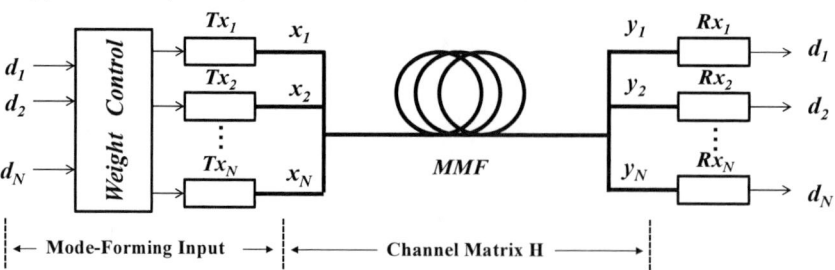

Fig.1. Configuration of the proposed Mode-Forming Networks.

970

To realize the proposed method, transmitter signals which are superimposed different channel's data should be clarified. The transmission characteristics of MMF network shown in Fig. 1 can be expressed as $y = Hx$ where H is the channel matrix which represents MMF characteristics, $y = [y_1 \quad y_2 \quad \cdots y_N]^T$ and $x = [x_1 \quad x_2 \quad \cdots \quad x_N]^T$ are vectores of received and transmitted signals, respectively, N is the number of transmitter or receivers, and $[\ ]^T$ denotes the transpose matrix.

If transmitter signals are superimposed by different channel data, the transmitter signal x is expressed as $x = wd$, where, w is the weight control matrix and d is the input data. Output y in Fig. 1 can be expressed as $y = Hx = Hwd$. If output y equals to d, $Hw(= E)$ should be identity matrix. As a result, the matrix w should be $w = H^{-1}$

This means that if the transmitter signals are produced by the linear combination of the channel data, the channel data can be divided into desired port at the output port without using MIMO processing.

The proposed method is clarified for 2 channel mode-forming network. The channel matrix and derived weight control matrix is expressed as

$$H = \begin{pmatrix} h_{11} & h_{12} \\ h_{21} & h_{22} \end{pmatrix} \quad , w = \frac{h_{22}}{\Delta} \begin{pmatrix} 1 & -h_{12}/h_{11}(h_{11}/h_{22}) \\ -h_{21}/h_{22} & h_{11}/h_{22} \end{pmatrix} \quad (1)$$

where $\Delta(= h_{11}h_{22} - h_{12}h_{21})$ is the determinant of the channel matrix.

The controlling values of the weight control matrix are derived by the absolute and phase of these values. In the proposed subcarrier multiplexing system, amplitudes and phases mean amplitudes and phases of the electric subcarrier signal.

III. EXPERIMENTAL SETUP

The experimental setup is shown Fig. 2. The electrical 100Mbps and 200Mbps BPSK (binary phase shift keying) subcarrier signal with 1GHz microwave are generated. Two signal amplitudes are controlled at the first stage of the weight control, each BPSK signal amplitude and phase are controlled at the second stage. Amplitude and phase controlled signals are applied to two Dual-drive Mach-Zehnder LiNbO$_3$ modulators, which are combined and converted into the optical amplitude modulated signal. The two lights with the almost same wavelength of 1550nm are used. The two optical signals are coupled by using a fiber-fused MMF coupler, the coupled signals are transmitted over graded index multi-mode optical fiber (GI-F) with the length of 1km. The output is divided using another fiber-fused type MMF coupler and directed to two photo detectors. The optical signals are converted into the electrical signals. The optical path lengths are the same from modulators and coupler, but the electrical path lengths from the pulse pattern generator (PPG) to modulator (MOD) is changed by applying 2 bits delayed coaxial cable for estimating channel matrix.

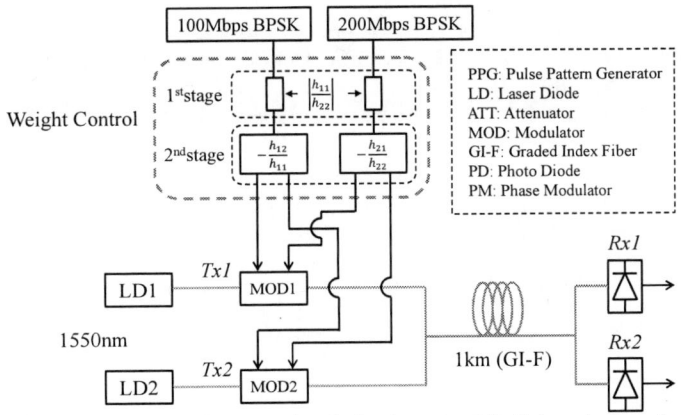

Fig.2. Experimental setup to confirm feasibilities of the proposed mode-forming network for 2 channel using subcarrier multiplexing scheme.
(Blue: SMF, Red: MMF)

The procedure for the simultaneous transmission is as follows. First, to obtain the channel matrix of the system, two bit shifted 100Mbps BPSK is transmitted by two modulators. The channel matrix is calculated by the MIMO processing by received signals [6]. Next, 100Mbps BPSK and 200Mbps BPSK signals are applied, the amplitude and phase are adjusted according to the values which are calculated from the channel matrix obtained by the previous experiment. The received signal are monitored by the $Rx\ n$ receivers and demodulated without MIMO processing.

IV. EXPERIMENTAL RESULTS

The channel matrix obtained from 2 channel MIMO transmission experiment with two bit shifted signals is shown in Table 1. The calculated values from the obtained channel matrix in Table 1 are shown in Table 2.

Table 1. Obtained channel matrix

Channel Matrix	Obtained value
h_{11}	-0.3+4.69i
h_{12}	3.25+7.27i
h_{21}	11.31-2.66i
h_{22}	4.68-2.13i

Table 2. Calculated values from Table 1

Channel Matrix	Calculated value	Amplitude	Phase	Experiment	
				amplitude	Phase
h_{11}/h_{22}	-0.43+0.81i	-1.9dB	-	-2dB	-
$-h_{21}/h_{22}$	-2.22+0.44i	8.15dB	-468.8ps	9dB	-440ps
$-h_{12}/h_{11}$	1.5-0.79i	-4.26dB	422.9ps	-3dB	400ps

Eye diagrams of controlled amplitude and phase in the transmitter derived from channel matrix $\boldsymbol{H}$ are shown in Fig. 3. The controlled amplitude and phase of 100Mbps BPSK or 200Mbps BPSK are adjusted according to Table 2. Optimum (experiment) values are a little different from calculated ones, because the fluctuation of the fiber mode condition. To avoid the fluctuation, feedback systems are required. These results indicated that it is possible to deliver the signals to the desired receiver by feedback the values of the channel matrix.

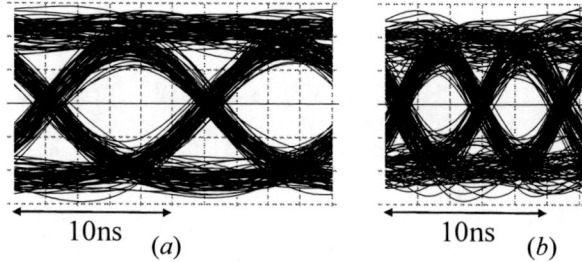

Fig. 3. Two receiver signal eye diagrams are shown, (a) 100Mbps BPSK from the output port Rx1, and (b) 200Mbps BPSK from the output port Rx2. These signals are simultaneously transmitted.

V. CONCLUSION

We have proposed mode-forming networks which can transmit desired data to the desired user without using WDM. We have demonstrated the feasibility of the proposed networks by transmitting 100M and 200M BPSK SCM signals to different ports at GI multi-mode fiber transmission system. Transmitted signals are successfully delivered by controlling the amplitude and phase of the transmitter signals.

This techniques can be applied to multiuser network by controlling each signal and superimpose to each transmitter.

REFERENCES

[1] L. G. Kazovsky, *et al.*, "Next Generation optical access network," *J. Lightw. Technol.*, vol. 25, no. 11, pp. 3428-3442, 2007
[2] T. Kodama, *et al.*, "Asynchronous MDM-OCDM-based 10G-PON over 40km-SMF and 2km-TMF Using Mode MUX/ DeMUX at remote Node and OLT," *Optical Fiber Communication Conference*, San Francisco, w2. A. 9, March 2014
[3] J. Li, *et al.*, "Hybrid Passive Optical Network Enabled by Mode-Division-Multiplexing," *Optical Communication and Networks*, pp. 1-3, July 2015
[4] Piels.M, *et al.*, "Focusing Over Optical Fiber Using Time Reversal" *IEEE Photonic. Technol. Lett.*, vol. 27, No6, pp.631-634, March 15, 2015
[5] Y. Morizumi, *et al.*, "Mode-Forming Remote Switch for Security Network," *Asia Communications and Photonics Conference*, ASu2A. 95, November, 2015
[6] Yang Zhang, *et al.*, "SCM-SS scheme for optical MIMO transmission using multimode fibers" *IEICE Communications Express*, vol.2, No6, pp.268-273, June 24, 2013

Comparison of Two Photodetector Linearity Characterizing Systems

Youxin Liu, Yongqing Huang*, JiaRui Fei, Yangan Zhang, Xiaomin Ren, Kai Liu, Xiaofeng Duan

State Key Laboratory of Information Photonics and Optical Communications

Beijing University of Posts and Telecommunications, Beijing, China

yqhuang@bupt.edu.cn

Abstract: *Two measurement techniques were investigated to characterize photodetector linearity. A model of the measurement system was developed to study the limitation of the two-tone method and the results correspond well to calculation results.*

Keywords: *OIP3 measurement systems; photodetector linearity*

I. INTRODUCTION

High saturation and high linearity photodiode (PD) is one of the key components in high performance analog optical links to handle high optical and electrical power [1]. Surface normal photodiodes with a high output third order intercept point (OIP3) of 52dBm at frequencies less than 1GHz have been reported [2]. Improvements in photodiode linearity create great challenges in linearity measurement. The third-order interception point (OIP3) measurement are important due to the fact that IMD3 appears very close to the fundamental signals and cannot be easily removed by filters. Many IP3 measurement systems have been employed including three-tone distortion measurement setup and two-tone setup [3-5]. The three-tone measurement is considered the most precise measurement system, but the measurement system is very complex. Two-tone setup measurement is simpler. However, the nonlinearities of the optical modulators and the optical source can affect the measurement results, because the second-order harmonics caused by the source or the bias drift from the quadrature point of the modulator will mix with another fundamental signal in the photodiode and contribute to the IMD3.

Previous works have studied the influence caused by the nonlinearity of the optical source [6]. In this work, we analyze the two-tone and three-tone measurement through the experiment and mathematical calculation. Additionally, a simulation model was developed to systematically investigate the influence of the bias shift from the quadrature point of the modulator on the two-tone setup. The simulation results correspond well to what we calculated and observed in the experiment.

II. EXPERIMENTS

The tested PD was a conventional uni-traveling carrier (UTC) photodiode. Fig.1 shows a block diagram of the three-tone measurement setup. The three lasers were operating at 1546nm, 1547nm, 1548nm, respectively. The RF signals were tuned at different frequencies. The three optical signals carrying RF modulation were combined and amplified by an Erbium Doped Fiber Amplifier (EDFA) followed by an optical attenuator. The fourth laser was used to maintain a constant average optical power illuminated to the photodiode. When the third laser source is turned off, this three-tone measurement setup turns into a two-tone measurement setup. The OIP3 measured by the two-tone setup is assumed 3dB larger than the three-tone setup [7]. OIP3 of the device was measured with both two-tone and three-tone measurement.

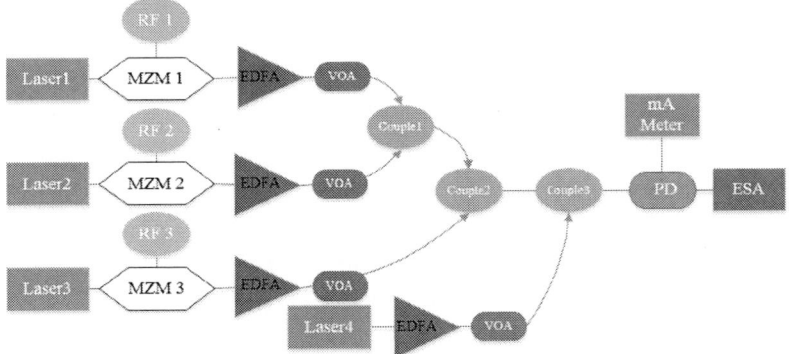

Fig.1. Block diagram of three-tone OIP3 measurement setup

Fig.2 and Fig.3 summarize the measured OIP3 results of different measurements as a function of reverse bias tested at 0.3GHz (three tone) and 0.4GHz (two tone) (f1=0.5 GHz, f2=0.6 GHz, f3=0.8 GHz) and 0.8GHz(three-tone) and

973

0.9GHz(two-tone) (f1=1 GHz, f2=1.1 GHz, f3=1.3 GHz). From the results, the OIP3 changes with the bias voltage due to the voltage swing effect and the voltage-dependent device capacitance [8]. Although the difference between the two-tone OIP3 and three-tone OIP3 is not quite the theoretical 3dB due to the bias shift from the quadrature point of the modulator, it clearly follows the theoretical trend. The influence of the bias shift of the modulator on the OIP3 will be systematically investigated through mathematical calculating and simulation.

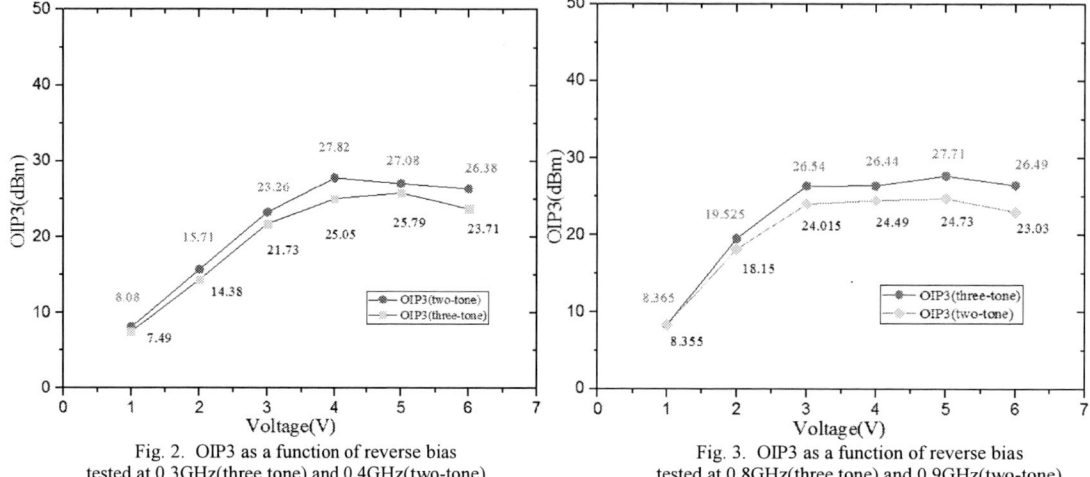

Fig. 2. OIP3 as a function of reverse bias tested at 0.3GHz(three tone) and 0.4GHz(two-tone)

Fig. 3. OIP3 as a function of reverse bias tested at 0.8GHz(three tone) and 0.9GHz(two-tone)

III. MATHEMATICAL RELATIONSHIPS

We assume the bias shift is $\lambda \times V_\pi/2$ in the two-tone measurement. When we assume $\lambda \ll 1$, the insertion loss for the modulator is t_x ($x=1, 2$), so the output optical power of the modulator is:

$$P_{out} = \frac{1}{2}P_{in}t_x[1+\cos(\frac{\pi}{V_\pi}(\frac{V_\pi}{2}+\lambda\frac{V_\pi}{2}+V_{RF}\sin\omega t))] \quad (1)$$

We use the Taylor's formula to simplify it and calculate the optical power after the combiner:

$$P_{out} = \sum_{x=1}^{2}\frac{1}{2}P_{in}t_xg_x\alpha[1-\phi_{RF}\cos(\omega t)+\frac{\pi}{8}\lambda(\phi_{RF})^2\cos(2\omega t)+\frac{1}{24}(\phi_{RF})^3\cos(3\omega t)] \quad (2)$$

Where g_x is the gain of the EDFA and α is the optical attenuation set through the optical attenuator, $\phi_{RF}=\pi/V_\pi\times V_{RF}$.
We assume the responsibility of the device is $\Re$ and the insertion loss of the combiner is t_c. So the output photocurrent of the fundamental RF signal is:

$$I_{fund} = \frac{1}{\sqrt{2}}\cdot\frac{1}{2}\cdot\Re\cdot t_c(P_{in}t_xg_x\alpha)\cdot\phi_{RF} \quad (3)$$

Now we calculate the photocurrent of intermodulation terms caused by PD and MZM separately, ρ and ν are the unknown constant representing the device nonlinearity due to the second and third order mixing.

$$i_{IMD3}^{PD} = \frac{1}{\sqrt{2}}\cdot\frac{1}{8}\cdot\nu\cdot\Re\cdot\alpha^3\cdot t_c(P_{in}t_xg_x)\cdot\phi^3_{RF} \quad (4)$$

$$i_{IMD3}^{mzm} = \frac{1}{\sqrt{2}}\cdot\frac{1}{16}\cdot\rho\cdot\Re\cdot\alpha^2\cdot\frac{\pi}{2}\cdot\lambda\cdot t_c(P_{in}t_xg_x)\cdot\phi^3_{RF} \quad (5)$$

$$p_{IMD3}^{mzm} = 20\log(\frac{5}{16}\cdot\rho\cdot\Re\cdot\alpha^2\cdot\frac{\pi}{2}\cdot t_c(P_{in}t_xg_x)\cdot\phi^3_{RF})+20\log(\lambda) \quad (6)$$

We can see that the IMD caused by PD changes by 6 with α, while IMD caused by MZM changes by 4, so as the photocurrent increases, the bias shift will have less influence on the measured OIP3. Moreover, from the equation (6), we can see the OIP3 versus bias shift λ is not symmetric. When λ is negative, the measured OIP3 will increase due to the decrease of the total i_{IMD} photocurrent, while a positive λ will decrease the measured OIP3.

IV. SIMULATED RESULTS AND DISCUSSION

In this section, the effect of the bias shift on the OIP3 was studied through simulation using the Optisystem simulation software with the help of Matlab. A nonlinear photodetector was defined with the nonlinearity function:

$$I_{RF} = a_1 P_{RF} + a_2 P_{RF}^2 + a_3 P_{RF}^3 + \cdots \quad (7)$$

The two-tone and three-tone OIP3 were found to be 23.38dBm and 20.38dBm. Fig.4 shows the simulated OIP3 for both the two-tone and three-tone case as the modulator bias shifts. On one hand, when λ is negative, the OIP3 increases as the bias shift tends to be larger, and as the λ arrive a certain point, the measured OIP3 tends to decrease due to the greatly loss of the fundamental power which can be concluded from the equation (5). On the other hand, when λ is positive, the OIP3 decreases with λ due to the decrease of fundamental power and increase of the IMD3. Additionally, Fig.5 shows that the IMD3 value changes by 20 with the λ (from 0.05 to 0.35) in log scale, which is in good agreement to what we have calculated.

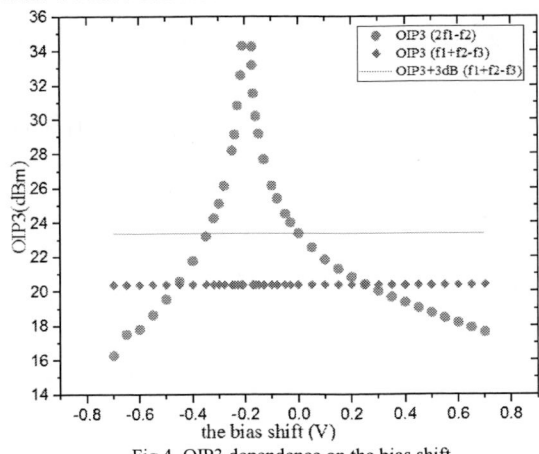

Fig.4. OIP3 dependence on the bias shift   Fig.5. IMD3$_{MZM}$(dBm) dependence on the bias shift (in log scale)

V. CONCLUSIONS

In conclusion, the influence of the bias shift from the quadrature point of the modulator on the OIP3 was systematically investigated through simulation and mathematical calculation. The simulation results show that the bias shift has a great impact on the two-tone measurement system and the three-tone OIP3 remains unchanged, which corresponds to the calculation results. In the simulation, when the bias shift is between 0V and -0.2V, OIP3 goes up as the bias shift becomes bigger. when the bias shift exceeds 0.2V, OIP3 drops off with the growth of negative bias shift. On the other hand, as the bias shift positively increases, the OIP3 falls off. Thus, as the linearity of photodiodes continues to increase, it is necessary to use a measurement technique such as the three-tone system to accurately characterize photodiode nonlinearities.

ACKNOWLEDGMENT

This work has been supported by the National Natural Science Foundation of China (Grant No.61574019, 4132069). the National Natural Science Foundation of China (Grant No.61274044) and the International Science Cooperation Program (Grant No.2011DFR11010).

REFERENCES

[1] J. S. Paslaski, P. C. Chen, J. S. Chen, C. M. Gee, and N. Bar-Chaim, "High-power microwave photodiode for improving performance of rf fiber optic links," in Proc. SPIE, vol. 2844, 1996, pp. 110–119.

[2] A. Beling, H. Pan, H. Chen and J. C. Campbell, "Linearity of modified uni-traveling carrier photodiodes," J. Lightw. Technol. 26, 2373-2378 (2008)

[3] M. N. Draa, J. Ren, D. C. Scott, W. S. Chang, and P. K. L. Yu, "Three laser two-tone setup for measurement of photodiode intercept points," Opt. Express 16(16), 12108–12113 (2008).

[4] H. Pan, A. Beling, and J. C. Campbell, "High-linearity uni-traveling-carrier photodiodes," IEEE Photon. Technol. Lett. 21(24), 1855–1857 (2009).

[5] A. Ramaswamy, J. Klamkin, "Three-tone characterization of high-linearity waveguide uni-traveling-carrier photodiodes," in 21st Annual Meeting of the IEEE Lasers and Electro-Optics Society, 2008. LEOS 2008 (IEEE LEOS, 2008), pp. 286–287.

[6] A. Ramaswamy, N. Nunoya, "Measurement of intermodulation distortion in high-linearity photodiodes," Optical Express 18(3):2317(2010)

[7] T. Ohno, H. Fukano, Y. Muramoto, T. Ishibashi, T. Yoshimatsu, and Y. Doi, "Measurement of intermodulation distortion in a unitravelingcarrier refracting-facet photodiode and a p-i-n refracting-facet photodiode," Photonics Technology Letters, IEEE, vol. 14, no. 3, pp. 375 –377, mar 2002

[8] A. Beling, H. Pan, C. Hao, and J. C. Campbell, "Measurement and modeling of a high-linearity modified unitraveling carrier photodiode," IEEE Photon. Technol. Lett. 20(14), 1219–1221 (2008).

WA2-81

Optimization of Temperature-Dependent 850 nm VCSELs with Different Oxide-Confined Aperture Sizes

Chun-Yen Pong, Cheng-Ting Tsai, Yun-Chen Wu, Shan-Fong Leong, Yu-Chieh Chi, Gong-Ru Lin
, and Chao-Hsin Wu[*]

Graduate Institute of Photonics and Optoelectronics, and Department of Electrical Engineering,
National Taiwan University
No. 1, Roosevelt Rd, Sect. 4, Taipei 10617, Taiwan ROC
Corresponding Author E-mail address: chaohsinwu@ntu.edu.tw

Abstract: *850 nm vertical-cavity surface-emitting lasers (VCSELs) with different aperture dimensions are fabricated and characterized at different temperatures. The characteristics of L-I-V, slope efficiency, temperature stability, small-signal measurement, and data transmission using OFDM format are presented.*
Keywords: *temperature-dependent, oxide-confined aperture, OFDM,*

I. Introduction

In recent years, the development of high performance computers, servers, and data centers are growing vigorously. In addition, the booming of "Big Data" and "Internet of Things" has led to the replacement of coaxial cable systems by optical fiber communication. Hence, there exists an urgent need on high-speed light sources such as edge-emitting diode lasers and VCSELs. Compared with edge-emitting diode lasers, oxide-confined VCSELs demonstrate lower threshold currents and robust fabrication [1-3]. However, the operating conditions and output characteristics of VCSELs depend heavily on the aperture sizes. The larger the diameters of oxide apertures, the stronger the output power is, resulting in the thermal rollover at a much larger current density. In addition to output power, at high current operation VCSELS with larger oxide aperture diameters can achieve larger modulation bandwidth (f_{3dB}) [4]. Actually, for the energy-efficient operation and commercial application, not only f_{3dB} needs to be considered, the power consumption and current density are of equal importance. Therefore, by investigating the effect of the aperture diameter on the temperature-dependent characteristics of 850 nm VCSELs with different transmission data format, we can balance it out to meet the requirement of different commercial applications.

In the work, we fabricate VCSELs with 5, 7, 9, and 11 μm aperture sizes and measure the DC characteristics such as L-I-V curve, slope efficiency, and temperature stability from 25 °C to 85 °C. Finally, the device with optimal aperture size is used to apply orthogonal frequency division multiplexing (OFDM) with highly spectral usage efficiency of encoding to investigate the effect of signal-to-noise (SNR).

II. Devices Fabrication And Measurement Experimental Setup

The VCSEL wafer is grown by metal-organic vapor phase epitaxy technology and prepared by a commercial vendor. Figure 1 shows the schematic cross section and optical microscope image of the VCSEL. The device is fabricated by first evaporating the p-type contact and followed by the mesa definition using inductively coupled plasma dry etching with different diameters (M_D) of 12, 14, 16, and 18 μm. The different oxide apertures (d_A) are then formed by wet oxidation on different DBR mesa size with careful oxidation time calibration and monitoring at the same time. After oxidation, the standard metallization of the n-contacts is then performed and annealed, followed by polyimide planarization, via-hole etching, and pad metallization to complete the fabrication of our devices.

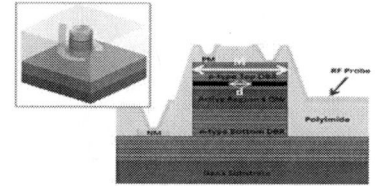 

Fig. 1 (a) schematic cross section of our VCSELs M_D: mesa diameter d_A: oxide apeature ;(b) Optical microscope images after pad metallization.

Figure 2 shows the experimental setup of directly-modulated VCSEL for 16-QAM OFDM transmissions, temperature-

---

[1] This work was supported by the funding of the Ministry of Science and Technology of Taiwan under Grant MOST 104-2218-E-005-004, MOST 102-2221-E-002-192-MY3, MOST 104-2622-E-002-024-CC3.

dependent characterization, and small-signal response measurement. A fiber (multi-mode fiber, MMF, OM4) connected to VCSELs with different aperture size is employed. During experiments, the homemade probe station is operated at specific temperature via a thermostat system. The OFDM data are exported by an arbitrary waveform generator (AWG, Tektronix, 70001A) and small signals are provided by vector network analyzers (PNA, Agilent, N5225A). Then, a high-speed bias-tee (Anritsu V255, 65-GHz) is used to combine the DC bias with the modulating data for directly encoding on the VCSEL. After MMF transmission, the OFDM optical data is received by a 25-Gbps GaAs PIN photodiode (PD, GCS, DO351). After passing through two post-amplifiers (AMP, Keysight, 83006A; New Focus, 1422) with power gain of 20 and 18 dB, respectively, the received data is re-sampled by a digital serial analyzer (DSA, Tektronix 71604C) with a sampling rate of 100 GS/s to obtain its signal-to-noise ratio (SNR), bit error rate (BER), and error vector magnitude (EVM) performances. In addition, the small signals are received by 25-GHz commercial photo detector (New Focus, 1414-50) and optical responses are directly measured by PNA.

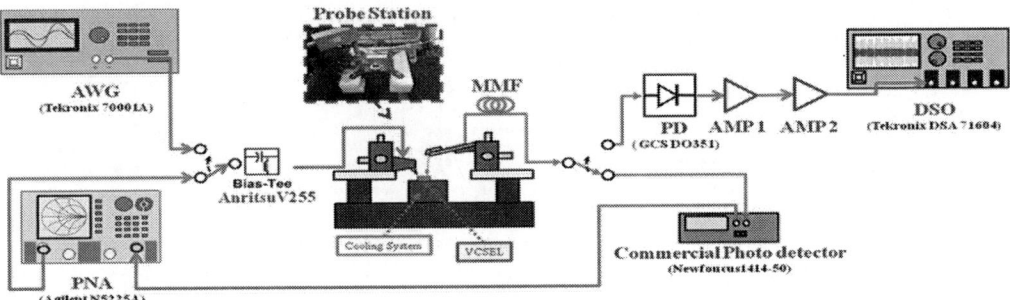

Fig. 2 The testing system of directly modulated VCSEL for 16-QAM OFDM and Small signal measurement

III. RESULTS AND DISCUSSION

In orde to characterize the optimal operation for commercial application, we study DC characteristics of different aperture diameters at varying temperatures. Figure 3 shows L-I-V curve in different conditions. We can find the maximum optical power and slop efficiency (SE_{max}) that change anlong with aperture diameter sizes. This is becaucse the maximum continuous-wave (CW) output power of VCSELs is limited by self-heating effects, and output power saturates and even decreases at high injection level. This is the commonly observed premature "thermal rollover." Hence, thermal factor allow us to estimate the thermal effect of temperature increase of ΔT:

$$\Delta T = R_{th} \times P_{diss} \quad (1)$$

where R_{th} is the theremal resistance and P_{diss} is the power dissipation. It represents that when theremal resistance is larger at same power consumption, the temperature variation increases together in the active region. Therefore, the self-heating effects will occur relative earlier, and the photon density will saturate, which reduces optical power at the same time. Fot the cylindrical oxide-confined VCSELs, the thermal resistance relationship can be represented as :

$$R_{th} = 1/2\xi d_A \; ; \; P_{diss} = I_b{}^2 R_{diss} \quad (2)$$

where ξ.is the equivalent material thermal conductivity material, and R_{diss} is equivalent resistance which follows with joule heating effect, assumes $R_{diss} \propto \frac{1}{d^{1\sim2}}$. Therefore, we can get a proportional relationship:

$$\Delta T \propto I_b{}^2 \times 1/d^{2\sim3} \quad (3)$$

It represents that smaller apeature will get smaller optical power and rollover current under same current injection. In addition, optical response bandwidth will simultaneously increase with the photon density. From the previous measurement and discussion, the larger apeature diameter has more advantages than smaller one. However larger apeature accompanies with much more transverse modes, less temperature stability, and high power consumption when VCSELs operating at high bias currents and high ambient temperatures. In order to address this issue, temperature-dependent results with different aperture sizes are performed to obtain the optical device for data transmission.

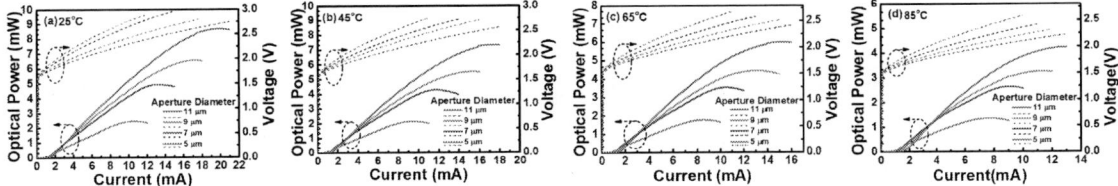

Fig. 3 VCSELs L-I-V curve in different temperature and aperture size (a) 25 °C (b) 45 °C (c) 65 °C (d) 85 °C

977

Figure 4 shows "Slope Efficiency, SE_{max}" and "Maximum Optical Power, P_{max}" characteristics at different temperatures. From differential quantum efficiencies we can understand that SE_{max} slope rely on injection efficiency and insertion loss. When temperature rises under current injection, carriers will escape from quantum well and narrow oxide aperture will cause the scattering loss to increase [5]. In the figure 4 we can see that SE_{max} declines form 66% to 33% when aperture diameter change from 11 μm to 5 μm at 25 °C. Furthermore, P_{max} decreases form 8.7 mW to 2.5 mW at the same time. Although the large-aperture device has wider operation range, it exhibits less temperature stability as shown in Fig. 4. If we consider the temperature differential rate together, the 7 μm aperture device has similar trend as 5 μm aperture device. VCSELs with lower temperature dependence can reduce the cost and complexity of compensation circuits and cooling systems for high temperature operation. Hence, we choose the aperture of 7 μm VCSEL to maintain high optical output, larger slope efficiency, and high temperature stability in the data transmission experiments.

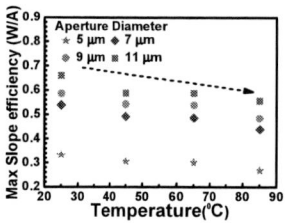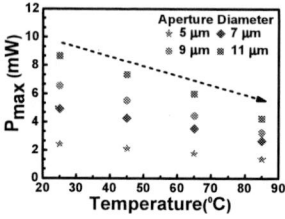

Fig. 4 (a) SE_{max} (b) P_{max} in different temperature and aperture diameter.

Figure 5 describes the small-signal response form 0.1-KHz to 20-GHz and the RF spectrum of 10-GHz 16-QAM OFDM data carried by the VCSEL at different bias currents. When the bias current drives from 5 to 11 mA, the f_{3dB} enhances from 12.5 to 16.8 GHz. The peak of relaxation frequency will be limited by non-linear gain saturation, so the optical response has large bandwidth and better flatness at larger bias condition. It is easily assumed that better signal quality can be achieved from small signal measurements in Fig. 5(a). However, SNR for OFDM shows different results in Fig. 5(b). When the bias current increases from 6 to 8 mA, the average SNR of OFDM data is improved from 14.6 to 15.4 dB and error vector magnitude is declined from 18.5 to 16.6 %. The bit-error-rate (BER) of 2.6×10^{-3} of OFDM data can pass the FEC criterion at 8 mA. This trend meets with previous small-signal experiment, however, the bias goes up to 9 mA and the average SNR drops to 15.2 dB. The extinction ration limits overall characteristics, and it just echoes our apertures balance design. For the highly spectral usage, not only bandwidths and relative intensity noises are considered criteria for low BER optical devices, but also the response flatness and high slope efficiency are necessarily included.

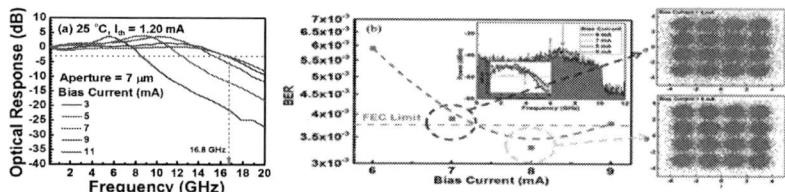

Fig. 5 (a) Optical Frequency Response of 7um aperture at 25 °C (b) 16-QAM OFDM summary results.

IV. CONCLUSIONS

The high-bandwidth and low-noise VCSEL is investigated in this report to meet new data format requirements that are different form conventional non-return-to-zero data stream. By operating VCSELs with different aperture sizes at different temperatures, we find the balanced characteristics between optical output power, SE_{max}, P_{max}, and temperature differential rate. The device with 7 μm aperture is applied in OFDM transmission systems to verify our predictions. The VCSEL shows SE_{max} of 53 %, P_{max} of 4.7 mW, stable temperature change rate dSE_{max} of -18.6 %, and dP_{max} of -46.2 %. Moreover, we demonstrate that the extinction ratio and bandwidth flatness are important in transmission systems.

REFERENCE

[1] J. M. Dallesasse, N. Holonyak, Jr., A. R. Sugg, T. A. Richard, and N. ElZein, Appl. Phys. Lett. 57, 2844 (1990).
[2] C. C. Hansing, H. Deng, D. L. Huffaker, and J. Sarathy, IEEE Photon. Technol. Lett. 6, 320 (1994).
[3] D. L. Huffaker, D. G. Deppe, K. Kumar, and T. J. Rogers, Appl. Phys. Lett. 65, 97 (1994).
[4] Moser, P., Lott, J. A., and Bimberg, D.,IEEE Journal of Selected Topics in Quantum Electronics, 19(4),1702212-1702212 (2013).
[5] Eric R. Hegblom., and Larry A. Coldren.,IEEE Journal of Selected Topics in Quantum Electronics, vol. 3, no. 2(1997)

WA2-82

OECC/PS2016

Enhancement of light extraction efficiency of InGaN light-emitting diodes with microholes array and nano-roughened ZnO structure

Ming Wang, Zhigang Zang[*], Xiaosheng Tang

Key Laboratory of Optoelectronic Technology & Systems (Ministry of Education), Chongqing University, Chongqing 400044, China
*Corresponding author: zangzg@cqu.edu.cn

Abstract: The light extraction efficiency of InGaN LEDs with microholes array and nano-roughened ZnO was increased, which resulted in the maximum output power enhancement of 58.4%. No degrading effect on the current–voltage (I–V) characteristics were observed.
Keywords: Light extraction efficiency; Microholes array; Nano-roughened ZnO; InGaN LEDs

I. INTRODUCTION

As an important optical device, light-emitting diodes (LEDs) play an important role and exhibit various wide applications in the daily life, such as traffic signals, full-color displays, and outside decoration, because of their promising advantages of high efficiency, high reliability, and long lifetime. In particular, GaN-based LEDs can be used to realize high-brightness white LEDs, which are considered as a potential candidate for the next generation new solid-state lighting platform. To support the next generation lighting development, high output power of GaN LEDs are extremely demanded. Until now, low light extraction efficiency is still the main issue that restricting the high performance of GaN LEDs, although high values of internal quantum efficiency with more than 70% have already been realized by optimizing the crystal quality and epitaxial layer structure. The low light extraction efficiency is mainly limited by the large difference in the refractive index between the GaN semiconductor (n_{GaN}= 2.5) and the surrounding air (n_{air} = 1.0), which leads to serious total internal reflection at the interface. With Snell's law ($n_{GaN}\sin(\theta c) = n_{air}$), only the photons that is within a critical angle of θc =23.5° can radiate into surrounding air, whereas most of the photons emitted from the active region are trapped by the total internal reflection between the GaN semiconductor and the surrounding air, and then are finally absorbed by the material and converted to heat, which conversely restricted the internal quantum efficiency and LEDs product life. Due to this reason, approximately only 6% of the internal light can be directly radiated into air from the top side of the GaN LEDs [1]. Therefore, it is necessary to enhance the light extraction efficiency in order to improve the performance of GaN LEDs. Recently, significant progress has been achieved in improving the light extraction efficiency of GaN LED, several approaches such as surface roughening [2,3], photonic crystals [4], flip-chip [5,6], and patterned sapphire substrate technology [7,8]. All these approaches effectively contribute to improving the light extraction efficiency. However, these approaches require complex fabrication process of dry etching, wet etching, or electrical beam lithography technology, which show some potential issues of high cost and non-uniformity.

Here in this paper, we report a novel technology of fabricating microholes array on the top surface of InGaN LEDs by using femtosecond laser, contributing to a significant enhancement of the light extraction efficiency. The femtosecond laser could directly fabricate controllable and high quality microholes array on LEDs without using complex etching and costly electrical beam lithography process, which is considered as a potential technology for realizing low cost and large scale production.

II. EXPERIMENTAL

The fabrication process of microholes array and nano-roughened ZnO on the top of InGaN LEDs chips is shown schematically from Fig. 1(b) to Fig1.(d). The total experimental process consists of three steps. The first step was high quality of ZnO coated on the p-side of InGaN LEDs chips by spin coating technique. The high quality of ZnO was fabricated by a double-phase method, and the detailed fabrication method could be found in our previous report. The thickness of ZnO film was 15nm. The second step was ZnO rough process by plasma bombardment. We moved InGaN LEDs chips into plasma-enhanced chemical vapor deposition (PECVD) system. Mixed argon (Ar) and hydrogen (H2) gas plasma (Ar: H_2=3:1) was bombarded on the ZnO films for 5 minutes. Thus, surface nano-roughened ZnO structures could be formed. The third step was direct writing of microholes array on InGaN LEDs chips by femtosecond laser. The InGaN LED chips were settled on a xyz- moving stage with the emission side face-up for the laser processing. The xyz-moving stage was used for controlling the position of InGaN LED chips. The stage movements were completely controlled by a high-precision positioning software system on a personal computer (PC) so that we could fabricate accurate microholes array on LED chips via the laser scanning back and forth. The femtosecond laser source used in the

979

experiments was a regeneratively amplified 800 nm Ti:sapphire laser with a repetition rate of 1kHz, a pulse duration of 120 fs and an output power of 0.3 W.

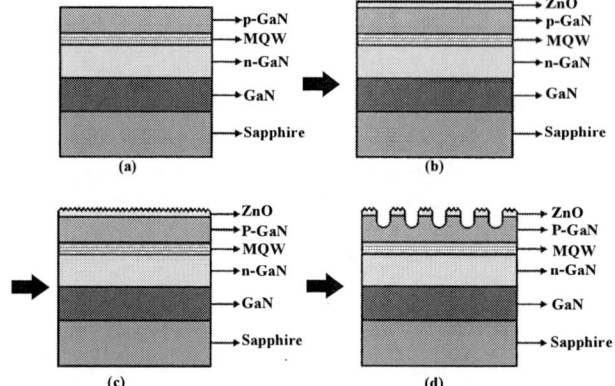

Fig.1. Process flow schematic of microholes array and nano-roughened ZnO on top of InGaN LEDs chip. (a) Schematic of InGaN LEDs structure, (b) schematic of InGaN LEDs coated with nano ZnO film, (c) nano roughened ZnO on InGaN LEDs, and (d) schematic of InGaN LEDs with microholes array and roughened ZnO.

III. RESULTS AND DISCUSSION

Figure 2 shows the SEM image of microstructure on InGaN LEDs chips with different magnification which was fabricated by adjusting the laser power to 3mW for a continuous process with three-pulsed laser irradiation at one point. The size and depth of the microholes could be controlled by the laser power. The periodic microholes with a diameter of 2μm and depth of 25nm were well produced. The surface surrounding the microholes is ZnO roughened area which was formed by the mixed Ar and H_2 gas plasma. As can be seen from Fig.2, all the arrays present a well-defined circular shape and are practically the same, which indicate high reproducibility.

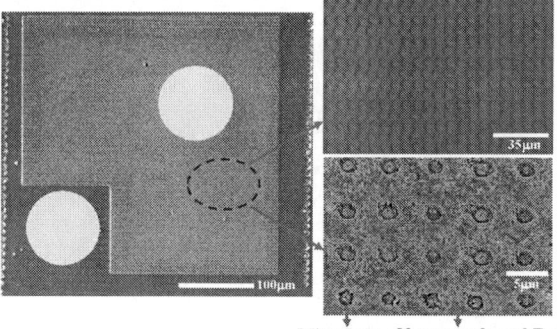

Microholes Nano roughened ZnO

Fig.2. Scanning electron microscopy images of micoholes array and nano-roughened ZnO on top of InGaN LEDs chip under different magnification.

Figure 3 shows the light output power as a function of injection current curves of the conventional InGaN LEDs and InGaN LEDs with microholes array and roughened ZnO for injection currents up to 220 mA. From the power–current curves, it was found that all the curves showed linear increases as a function of injection current and no intensity saturation was observed up to 220mA for all these LEDs. All the LED chips with microholes array and roughened ZnO have a significant power improvement compared to conventional InGaN LEDs. For the LED chips with microholes array and roughened ZnO, the maximum output power is 49.9 mW under injection current of 220 mA, while the maximum output power of conventional LEDs is 31.5 mW. This result corresponds to a significant output power improvement of 18.4 mW, i.e., 58.4%, compared to the maximum output power of the conventional LEDs. By combining the microholes array and roughened ZnO on the surface of LEDs, high output power is achieved due to the increase of the effective photon escape cone, which contributes to increasing the light extraction efficiency from the LEDs. Generally speaking, light extraction efficiency in the GaN-based LED is restricted mainly owing to the refractive index of GaN being higher than that of the surrounding air. According to Snell's law, these microholes array allows photons emitted from the active region to be escaped out of the LEDs with larger photon escape cone from 23.5° to 40.5°.

OECC/PS2016

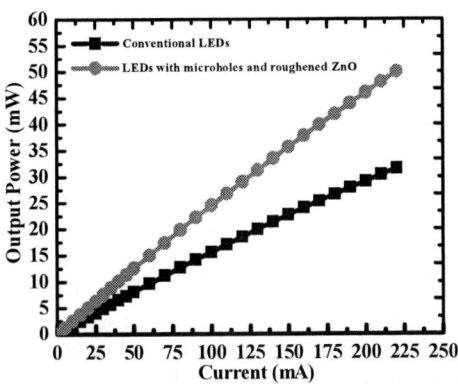

Fig.3. Light output power against injection current of InGaN LEDs.

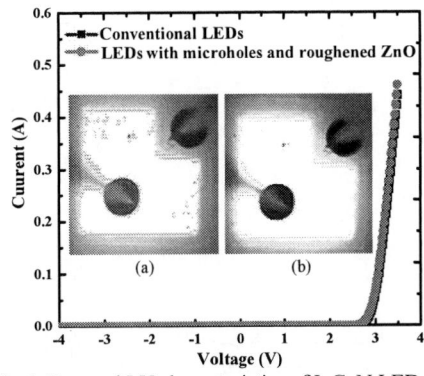

Fig.4. Forward I-V characteristics of InGaN LEDs.

Figure 4 shows the measured current–voltage (I–V) characteristics of all the LEDs. The turn on voltages of all the LEDs ware almost the same at the value of 2.9V. From the I –V curves, it was also found that all the LEDs show similar operation behavior, no significant difference of the I-V curves was found between the conventional LEDs and LEDs with microholes array and roughened ZnO. The inset shows the micrograph images of the conventional LEDs and LEDs with microholes array and roughened ZnO at a current injection level of 180 mA, respectively. The photomicrographs were taken with the same camera setting. As shown in the inset Fig4.(b), it is clearly observed that higher intensity light emission is coupled out in the top emitting LEDs with microholes array and roughened ZnO.

IV. CONCLUSION

In conclusion, the high quality of microholes array and roughened ZnO were successfully fabricated on LEDs. The microholes array and roughened ZnO could effectively improve the output power of LEDs. The maximum output power improvement of 58.4% was achieved for LEDs at an injection current of 220 mA, compared to the conventional LEDs. This microholes array fabrication method by using femtosecond laser shows flexible control, low cost, and time efficient process, which has great potential for realizing large scale production in industry.

ACKNOWLEDGMENT

This work is supported by National Natural Science Foundation of China (61404017), Chongqing Basic Science and Advanced Technology Fund (cstc2015jcyjA1055), the Fundamental Research Funds for the Central Universities, (106112015CDJZR125511, 0210005202057), initial funding of Hundred Young Talents Plan at Chongqing University (0210001104431).

REFERENCES

[1] A. David, T. Fujii, R. Sharma, K. McGroddy, S. Nakamura, S. P. Denbaars, E. L. Hu, C. Weisbuch, and H. Benisty, "Photonic-crystal GaN light-emitting diodes with tailored guided modes distribution," Appl. Phys. Lett. vol. 88, pp. 061124–061124, February 2006.

[2] T. Fujii, Y. Gao, R. Sharma, E. L. Hu, S. P. DenBaars, and S. Nakamura, "Increase in the extraction efficiency of GaN-based light-emitting diodes via surface roughening," Appl. Phys. Lett. vol. 84, pp. 855–857, December 2004.

[3] W.C. Peng and Y.C. S. Wu, "Improved luminance intensity of InGaN–GaN light-emitting diode by roughening both the p-GaN surface and the undoped-GaN surface," Appl. Phys. Lett. vol. 89, pp. 041116(1)-041116(3), 6 July 2006.

[4] T. A. Truong, L. M. Campos, E. Matioli, I. Meinel, C. J. Hawker, C. Weisbuch, and P. M. Petroff, "Light extraction from GaN-based light emitting diode structures with a noninvasive two-dimensional photonic crystal," Appl. Phys. Lett., vol. 94, no.2, pp. 023101(1)- 023101(3), January 2009.

[5] O. B. Shchekin, J. E. Epler, T. A. Trottier, T. Margalith, D. A. Steigerwald, M. O. Holcomb, P. S. Martin, and M.R. Krames, "High performance thin-film flip-chip InGaN-GaN light-emitting diodes," Appl. Phys. Lett. vol. 89, pp. 071109(1)-071109(3) August 2006.

[6] J. J. Wierer, D. A. Steigerwald, M. R. Krames, J. J.O'Shea, M. J. Ludowise, G. Christenson, Y.-C. Shen, C. Lowery, P. S. Martin, S. Subramanya, W. Goetz, N. F. Gardner, R. S. Kern, and S. A. Stockman, "High-power AlGaInN flip-chip light-emitting diodes," Appl. Phys. Lett. vol. 78, pp. 3379-3381, April 2001.

[7] Y. J. Lee, J. M. Hwang, T. C. Hsu, M. H. Hsieh, M. J. Jou, B. J. Lee, T. C. Lu, H. C. Kuo, and S. C. Wang, "Enhancing the output power of GaN-Based LEDs grown on wet-etched patterned sapphire substrates," IEEE Photon. Technol. Lett. vol. 18, pp. 1152–1154, MAY 2006.

[8] H. Chen, Y. Yao, C. Liao, C. Tu, C. Su, W. Chang, Y. Kiang, and C. Yang, "Light-emitting device with regularly patterned growth of an InGaN/GaN quantum-well nanorod light-emitting diode array," Opt. Lett. vol. 38, pp. 3370-3373, August 2013.

WA2-83

OECC/PS2016

Concentric Circular
High Index Contrast Gratings Reflector With
Focusing Ability

Wenjing Fang, Yongqing Huang*, Xiaofeng Duan, Kai Liu, Xiaomin, Ren, Jun Wang

Institute of Information Photonics and Optical Communications, Beijing University of Posts and Telecommunications (BUPT); State Key Laboratory of Information Photonics and Optical Communications, Beijing 100876, China
*Corresponding author: yqhuang@bupt.edu.cn;

Abstract: *A non-periodic concentric circular high index contrast grating (CC-HCG) focusing reflector is fabricated and demonstrated. We experimentally show that a spot of high concentration is appeared at 10.87mm at normal incidence for radially polarization.*

Keywords: *High index contrast gratings; Reflector; Focusing;*

I. INTRODUCTION

High index contrast grating (HCG) is a single layer of near subwavelength grating comprising high index bars fully surrounded by a low-index medium, and has attracted great attention recently due to its high reflectivity or transmissivity over a broad bandwidth and the ability to manipulate the phase front of reflected light or transmitted light. Since its simple and compact structure, the HCG used as reflector is an alternative to distributed Bragg reflectors (DBRs) especially for vertical-cavity surface-emitting laser (VCSEL) and reflection enhanced photodetectors [1-2]. Furthermore, it can bring novel characteristics including focusing ability and steering ability with non-periodic patterns, which is an important development in the application of HCGs. Recently, several kinds of non-periodic HCGs have been studied, aiming at focusing reflectors with high reflectivity. They include non- periodic strip patterns, non-periodic ring patterns and two-dimensional non-periodic patterns [3-6]. F. Lu et al. reported the design of the non-periodic strip HCG focusing reflectors and lens with low loss and high numerical aperture (NA) [3]. D. Fattal et al. showed the design and fabrication of flat, cylindrical and spherical HCG reflectors with a focal length of 50 nm at 1.55 μm wavelength [4]. For the HCG reflector with non-periodic ring patterns, a numerical study has shown that focusing reflector with high reflectivity is achievable and there has been no experimental study yet.

We investigated a focusing reflector composed of non-periodic concentric circular high index contrast gratings (CC-HCGs) on silicon-on-insulator (SOI) substrate. A CC-HCGs reflector with a focal length of 15 mm is fabricated and measured. Experimental results show that a spot of highest concentration is obtained at distance of 10.87 mm in front of grating, and the full width half maximum (FWHM) is decrease from the 400 μm to 260 μm, the intensity increases by a factor of 1.26 compared with incident beam.

II. DESIGN METHOD

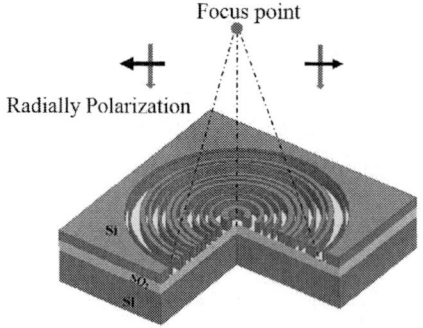

Fig. 1. Schematic representation of the CC-HCG reflector

Fig.1 shows a schematic representation of the non-periodic CC-HCG which consists of concentric circles of Si (n≈3.48) that are surrounded with air (n≈1) and SiO2 (n≈1.46) as the low-index cladding layers on the top and bottom. The SOI wafer consists of a 500 nm silicon layer and a 500 nm buried oxide layer. There are three physical parameters that control the reflectivity and phase: HCG bar width (s), air gap (a), and thickness (t). Changing the gating thickness is not feasible due to the multiple etching step it would imply, so only variations of period and width of grating bar are considered. When the bar width and air gap are locally changed, the properties of reflected beam, such as phase and reflectivity, will gradually adapt to these variation. The reflected light will be focused on a spot if the phase shift distribution of the reflected light meets Eq. (1).

$$\Phi(r) = \frac{2\pi}{\lambda}\left(\sqrt{r^2 + f^2} - f\right) + \Phi_0 \qquad (1)$$

where f is focal length, λ is the wavelength, and Φ_0 is the phase shift of the reflected light at the position of center.

Our design of the investigated CC-HCG reflector is for radially polarization at a wavelength of 1.55μm. Radial polarization for CC-HCGs, which electric field vector is perpendicular to the grating bar direction, is similar to TM polarization for the strip HCGs [5]. We follow Eq. (1) to calculate reflectivity and phase shift distribution of the periodic HCGs using rigorous coupled wave analysis (RCWA) simulation method for TM polarization [7]. The next step is to find out proper structural parameters which can meet the expected parabolic phase profile. It has been proved that the reflected property and phase shift of a concentric circular grating can be obtained from the corresponding periodic strip grating. Here, phase of the reflected light should cover a full 2π range of variation within the high reflectivity region. So by carefully selecting grating structure parameters, we can get a focusing reflector where the introducing phase profile adapts to Eq. (1).

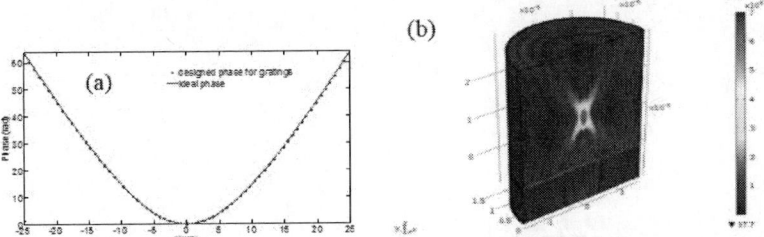

Fig. 2. (a) Phase distribution of a CC-HCG focusing reflector. The red line corresponds to the ideal phase distribution and the blue points correspond to discrete grating bars along r direction. (b) Normalized electric field intensity distribution for a CC-HCG focusing reflector

According to the design process, we take a CC-HCG structure with focal length of 5 μm for example. The COMSOL implementing finite-element method (FEM) numerical simulation is used to evaluate the focusing performance of the CC-HCG. As illustrated in Fig. 2(a), the red line is the ideal phase distribution along r direction for the CC-HCG focusing reflector calculated by Eq. (1), and the blue points are the each designed phase that correspond to each concentric circular grating unit. The simulation result with radially polarization at wavelength of 1.55 μm is plotted in Fig. 2(b), and it is obvious that the reflection beam can be focused on a spot of 4.9 μm, this small deviation is caused by discrete phase distribution.

III. FABRICATION

In order to facilitate the measurement of the reflection beam profile, a non-periodic CC-HCG structure with a diameter of 250 nm and a relatively long focal length of 15 mm was fabricated on SOI wafer. An EB resist (ZEP520) was spun on the surface of an SOI wafer. The grating patterns were defined by electron-beam lithography. Then, using the EB resist as a mask, the silicon grooves were formed by inductively coupled-plasma (ICP) etching using C_4F_8 and SF_6. Finally, the residual EB resist was removed with a 1:1 solution of H_2SO_4 and H_2O_2. The gratings were etched 500 nm into the top silicon layer. An optical microscope image of a fabricated CC-HCG reflector is shown in Fig. 3, together with scanning microscope images of the silicon grooves at various locations.

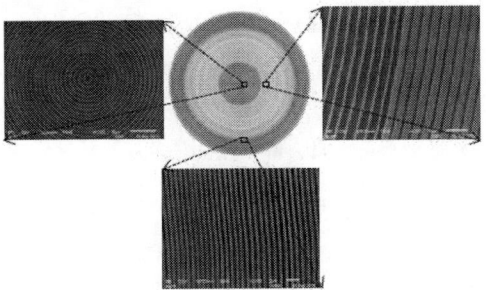

Fig. 3. An optical microscope picture and SEM images in various locations of the fabricated CC-HCG reflector.

IV. RESULTS AND DISCUSSION

Focusing properties of the CC-HCG reflector were experimentally studied, an Anritsu Tunics SCL tunable laser with a single-mode fiber pigtail was used as the light source. A fiber collimator, connected to the single-mode fiber, was used to output beam. The radially polarization converter was used to set polarization state of the input beam to radially polarization. The CC-HCG was illuminated by a transmitted beam though a cube beamsplitter at normal incidence. The reflection beam was separated from the incoming beam using the cube beamsplitter again. Different beam cross sections of the reflected beam were imaged using a CCD camera. The position of the imaged cross-sections was varied by moving the distance in the front of grating along z-direction.

Fig.4 illustrates the distribution of the focal spot at the different position along the z-axis. Here, the wafer was illuminated by a large beam matching the size of the CC-HCG at normal incidence as shown in Fig. 4(a). At the distance of 10.87 mm from the incoming end to the CC-HCG, the light is focused to a spot of highest concentration as seen in Fig. 4(b). As plotted in Fig.4 (c)-(d), when increasing the distance in front of the grating, the spot size increases. It can be seen that the CC-HCG focusing reflector reduces the beam spot size by nearly a factor of three – the FWHM of the incident light is decreased from 400 μm to 260 μm. The intensity increases by a factor of 1.26 compared to input beam intensity. We can calculate the focal length to be 10.87 mm for the fabricated CC-HCG reflector, smaller than the designed value, which is due to discrete phase distribution in design process and the effects in the electron-beam lithography step, as well as the surface roughness of the silicon grooves evident in Fig. 3.

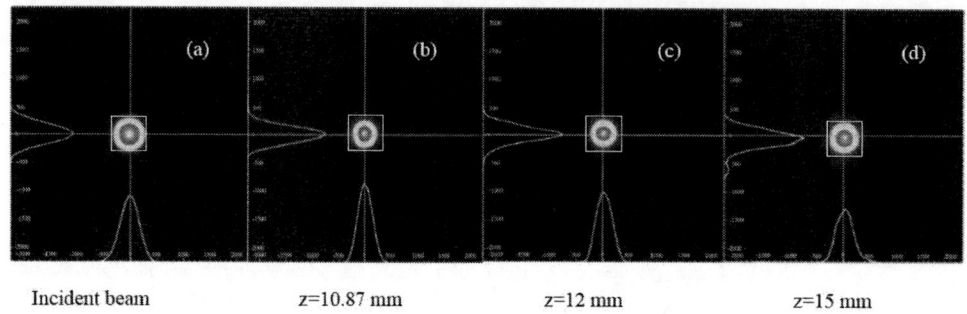

| Incident beam | z=10.87 mm | z=12 mm | z=15 mm |

Fig. 4. Intensity profiles at different positions measured from reflection beam.

V. CONCLUSIONS

We have demonstrated that non-periodic CC-HCG can realize beam focusing by manipulating the reflected light. Fabricated structure show the focal length of 10.87 mm at 1550 nm wavelength for radially polarization. The FWHM of the incident light is decreased from 400 μm to 260 μm. The intensity at the focal spot increases by a factor of 1.26 compared to incident beam. With the good focusing properties, such reflectors can be integrated seamlessly with photodetector, solar cells, microscopes, telescopes, and VCSELs to radically enhance their performance.

ACKNOWLEDGMENT

This work was supported by the National Natural Science Foundation of China (Nos. 61274044, 61574019 and 61020106007), the National Basic Research Program of China (No. 2010CB327600), Specialized Research Fund for the Doctoral Program of Higher Education (SRFDP) (No. 20130005130001), the Natural Science Foundation of Beijing (No. 4132069), Program of Key International Science and Technology Cooperation Projects (No. 2011RR000100), 111 Project of China (No. B07005), and the Program for Chang jiang Scholars and Innovative Research Team in University (No. IRT0609), MOE, China.

REFERENCES

[1] M.C.Y. Huang, Y. Zhou, and C.J. Chang-Hasnain, "A surface-emitting laser incorporating a high-index-contrast subwavelength grating," Nature Photonics, vol.1, pp. 119-122, April 2007.

[2] X. Duan, Y. Huang, X. Ren, Y. Shang, X. Fan, and F. Hu, "High-efficiency InGaAs/InP photodetector incorporating SOI-based concentric circular subwavelength gratings," IEEE Photon. Technol. Lett. vol. 24, pp. 863–865, May 2012.

[3] F. Lu, F. G. Sedgwick, V. Karagodsky, C. Chase, and C. J. Chang-Hasnain, "Planar high-numerical-aperture low-loss focusing reflectors and lenses using subwavelength high contrast gratings," Opt. Express, vol.18, pp.12606–12614, June 2010.

[4] D. Fattal, J. Li, Z. Peng, M. Fiorentino, and R. G. Beausoleil, "Flat dielectric grating reflectors with focusing abilities," Nat. Photonics, vol.4, pp. 466–470, May 2010.

[5] Xiaofeng Duan,* Guren Zhou, Yongqing Huang, Yufeng Shang, and Xiaomin Ren, "Theoretical analysis and design guideline for focusing subwavelength gratings," Opt. Express, vol.23, pp.2639–2646, Feb 2015.

[6] Changlian Ma, Yongqing Huang, Xiaofeng Duan, et al, "Polarizatioin –Insensitive Focusing Lens Using 2D Blocky High –Contrast Gratings", IEEE Photo. Technol. Lett., vol.27, pp. 697-700, April 2015.

[7] M. G. Moharam, E. B. Grann, and D. A. Pommet, "Formulation for stable and efficient implementation of the rigorous coupled-wave analysis of binary gratings" J. Opt. Soc. Am. Vol.12, pp.1068-1076, 1995.

Fabrication of Waveguide-type Mirrors for Red-green-blue Laser Beam Multiplexers

S. Tanaka[1], A. Nakao[1], S. Yokokawa[1], S. Hayashiguchi[1], T. Katsuyama[1,2],
K. Nakajima[3], N. Ikeda[3], and Y. Sugimoto[3]

[1]Graduate School of Engineering, University of Fukui, Bunkyo3-9-1, Fukui 910-8507, Japan

[2]Headquarters for Innovative Society-Academia Cooperation (UF-HISAC), University of Fukui,
Bunkyo3-9-1, Fukui 910-8507, Japan

[3]National Institute for Materials Science (NIMS), Sengen 1-2-1, Tsukuba, Ibaraki 305-0047, Japan
t-katsu@u-fukui.ac.jp (T. Katsuyama)

Abstract: Extremely-compact SiO_2–based waveguide-type mirrors composed of deeply-etched trenches are fabricated. The newly-developed deep and small trench enables us to fabricate extremely-compact integrated optical devices including red-green-blue multiplexers for laser projectors.

Keywords: Waveguide-type mirror, Etched trench, Laser beam multiplexer

I. INTRODUCTION

Recently, compact image projectors such as eyewear projectors have been extensively investigated [1]. So far, we have realized a new waveguide-type red-green-blue (RGB) laser beam multiplexer, which can be integrated with laser diodes (LDs) [2]. This integrated laser source is a promising candidate for the key element of the laser-beam-scanning-type image projectors. The laser source, however, should be compact, which requires a sophisticated layout of the LD positions. Thus, we have proposed a new configuration of the laser source (Fig. 1), which is advantageous to the heat dissipation and stray beam reduction [3]. The essential point of this configuration is that the laser beam of the green LD is directly inserted to the multiplexer central waveguide, while the blue and red LDs face each other and each laser beam enters the waveguide which is bent by 90 deg. from the multiplexer waveguides.

Here, we show a new waveguide-type mirror as such a bend waveguide. In the viewpoint of compactness and reproducibility, the mirror shown here is essentially improved compared to the waveguide-type mirror reported so far [4].

II. DESIGN OF THE WAVEGUIDE-TYPE MIRROR

The waveguide-type mirror studied here is composed of a deep trench. The light bending occurs by the internal total reflection between air and SiO_2 material composing the multiplexer waveguide. Figure 2 (a) shows the position of the waveguide-type mirror. The cross sectional view of the trench area is also shown in Fig. 2 (b). As shown, the mirror area is strictly restricted in the region between two waveguides for red (or blue) and green laser beams (distance is only 8 μm). Therefore, the trench should be as small as possible.

At first, we have simulated the reflection characteristics by using a finite-difference time-domain (FDTD) method. The waveguide in this case is made of SiO_2-GeO_2 glass. The cross section is 2 μm square and the refractive index difference is 0.5%. The important parameters are the mirror angle θ and the mirror position deviation d as shown in Fig. 3 (a). The transmittance from the mirror thus obtained is shown in Fig. 3 (b) as a function of θ and d. The maximum transmittance can be obtained when θ = 45deg. and d = 0.5 μm. Its maximum transmittance is over 70%. Furthermore, it can be estimated that the fabrication tolerances are within 0.5 deg. for θ and within 0.5 μm for d, which corresponds to the transmittance reduction of 5% only. The tolerance for the mirror perpendicularity (corresponding to the trench side wall perpendicularity) has the similar angle dependence as that for the mirror angle θ.

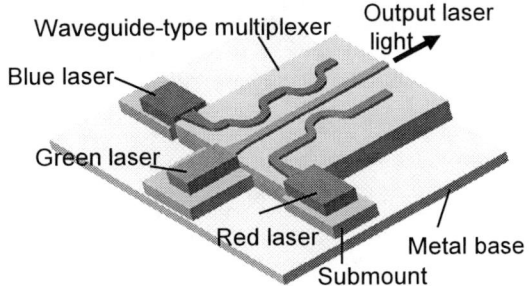

Fig. 1. Waveguide-type multiplexer integrated with RGB lasers.

OECC/PS2016

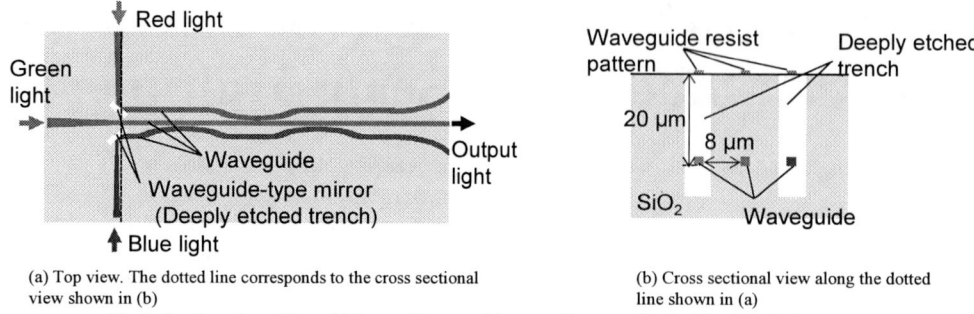

(a) Top view. The dotted line corresponds to the cross sectional view shown in (b)

(b) Cross sectional view along the dotted line shown in (a)

Fig. 2. Configuration of the multiplexer with waveguide-type mirrors consisted of deeply etched trenches.

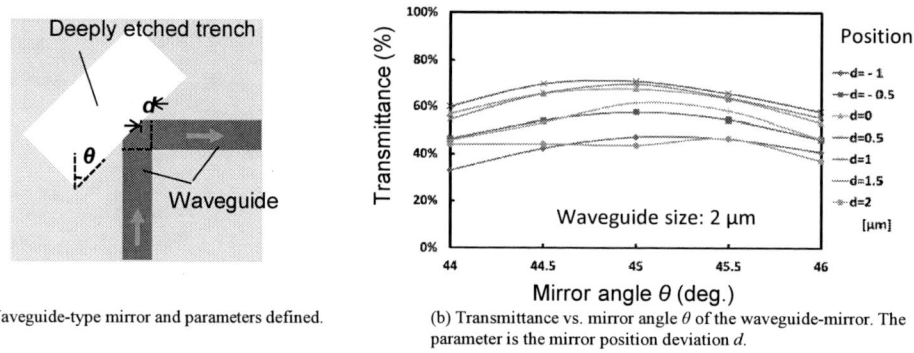

(a) Waveguide-type mirror and parameters defined.

(b) Transmittance vs. mirror angle θ of the waveguide-mirror. The parameter is the mirror position deviation d.

Fig. 3. Simulated results of the waveguide-type mirror composed of deeply etched trench.

III. FABRICATION CONDITION FOR THE WAVEGUIDE-TYPE MIRROR

We have tried to form the deep trench for the waveguide-type mirror by using a gallium (Ga) focused ion beam (FIB) etching technique, although the constituent SiO_2 material is a hard insulator. Usually, it is difficult to form a microstructure by using a FIB technique because such an insulator has a severe charge up problem. However, the compact formation of the deep trench inevitably requires a FIB technique. In this case, the important requirements for etching are (1) the large etching depth over 20 μm since the waveguide core is embedded below the thick cover clad SiO_2 layer (as shown in Fig. 2 (b)), (2) the small etching cross section because of the restricted mirror area, which means that the large aspect etching ratio is required, (3) the formation of the reflection wall which is just perpendicular to the waveguide (the angle tolerance is within 0.5 deg.), and (4) the precise position control of the trench area within 0.5μm.

In order to satisfy the above requirements, we have carefully studied the etching conditions. Furthermore, the re-deposition of the etched material elements was carefully eliminated since the aperture for the etching was so small and the aspect ratio was so large. Then, we have obtained the optimum conditions for the above requirements by introducing the following methods, i.e., (i) tilting the sample substrate to obtain the perpendicular side wall for light reflection and (ii) forming the waveguide resist pattern on the sample surface as the etching position marking. The pattern is just same as the embedded waveguide itself.

Figure 4 (a) shows the scanning electron microscope (SEM) image of the SiO_2 sample used for obtaining the optimum sample tilting angle. In this case, the half area of the etched trench was again etched to observe the side wall angle. As shown in the figure, the side wall of the trench with just 90 deg. perpendicular inclination was obtained when the tilting angle was 1.3 deg. In addition, it can be seen that the side wall surface was smooth, which expects the excellent reflection efficiency. On the other hand, Fig. 4 (b) shows the SEM image of the trench formation. As shown, the deeply etched trench was formed on the resist pattern showing the waveguide corner. In this case, the position accuracy was within 0.5 μm, as expected. As shown above, the requirements (3) and (4) for etching have been satisfied.

IV. FABRICATION RESULT OF THE WAVEGUIDE-TYPE MIRROR AND ITS CHARACTERISTICS

We have then fabricated the actual waveguide-type mirror, which is optimized by the simulation shown in Section II and the etching conditions shown in Section III. The accelerating voltage of the Ga ion was 30 kV. The fabricated waveguide-type mirror is shown in Fig. 5 (a). The trench was exactly formed on the waveguide corner. The trench depth was 27 μm and the trench area size was 7x5 μm. Those are corresponding to the aspect ratio of 5.4, which satisfies the requirements (1) and (2) for etching shown in the previous section. Figure 5 (b) shows the output laser

986

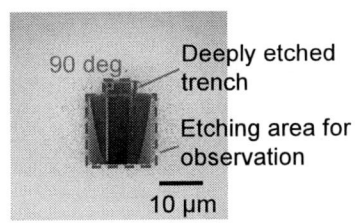

(a) Observation of the angle of the trench wall.

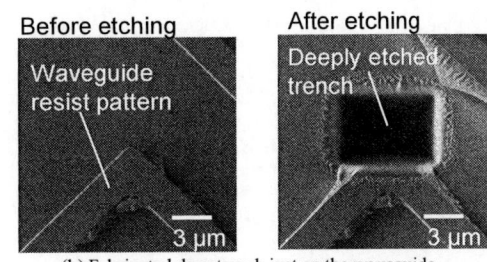

(b) Fabricated deep trench just on the waveguide corner (SEM image).

Fig. 4. Fabricated deep trench.

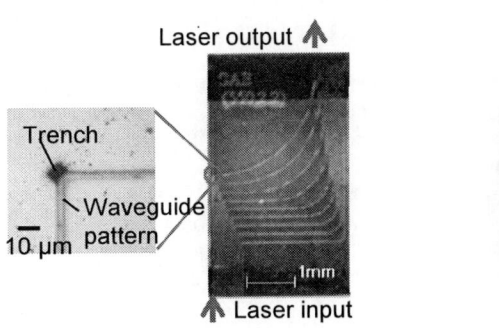

(a) Waveguide-type mirror fabricated for measurement. (Optical microscope image).

(b) Output laser beam pattern.

Fig. 5. Fabricated waveguide-type mirror and its characteristics.

beam pattern transmitted through the waveguide-type mirror thus fabricated. The light transmission efficiency in this case was approximately 55% of that for the waveguide without a mirror when the laser light with a 473 nm wavelength was used for the evaluation.

V. SUMMARY

We have successfully fabricated an extremely compact SiO_2–based waveguide-type mirror composed of a deeply etched trench. The trench depth was as large as 27 μm and the trench area size was as small as 7x5 μm. The side wall of the trench was almost perpendicular to the waveguide. Therefore, the waveguide-type mirror thus fabricated has an excellent compactness and a high reflection performance. Furthermore, the position control accuracy was within 0.5 μm, which means an excellent fabrication reproducibility. Therefore, the obtained performance is essentially improved even compared to the waveguide-type mirror reported so far. This waveguide-type mirror can be applied to the optical devices such as a compact RGB laser beam multiplexer.

ACKNOWLEDGMENT

This work was supported in part by SCOPE (152305002), Ministry of Internal Affairs and Communications, Japan.

REFERENCES

[1] B. Kress, "Optical Technologies for See through Wearable Displays: A Review", Frontiers in Optics 201, San Jose, USA, FTu2C.1, 2015.

[2] A. Nakao, R. Morimoto, Y. Kato, Y. Kakinoki, K. Ogawa, and T. Katsuyama, "Integrated waveguide-type red-green-blue beam combiners for compact projection-type displays", Opt. Commun. Vol. 330, pp. 45-48, 2014.

[3] A. Nakao, K. Tsujino, S. Yokokawa, S. Tanaka, S. Hayashiguchi, and T. Katsuyama, "Waveguide-type red-green-blue laser beam combiners integrated with semiconductor lasers", The 22nd International Display Workshops (IDW '15), Otsu Prince Hotel, Otsu, Japan, PRJp1-5L, 2015.

[4] J. W. Park, E. D. Sim, and Y. S. Beak, "Improvement of fabrication yield and loss uniformity of waveguide mirror", IEEE PHOTONICS TECHNOLOGY LETTERS, vol. 17, pp. 807-809, 2005.

WA2-85

OECC/PS2016

Multiple-access and two-way visible light communication with image sensor and LED array

Tomoki Kondo, Ryotaro Kitaoka, and Wataru Chujo

Meijo University, 1-501 Shiogamaguchi, Tempaku-ku, Nagoya 468-8502, Japan

Abstract: In order to realize multiple-access and two-way visible light communication with image sensor, isolation characteristics between LED and image sensor within the same transceiver are measured and sufficient isolations are obtained. In addition, BERs for multiple-access and two-way visible light communication with LED array and image sensor are experimentally evaluated for OOK and ASK and error-free operations were obtained.

Keywords: multiple-access, two-way, image sensor, visible light communication, LED array, OOK, ASK

I. INTRODUCTION

In LED visible light communication with 30-frames-per-second (fps) image sensor, synchronization algorithm applicable up to 25 symbols per second (sps) that is close to the frame rate of the image sensor was developed and verified experimentally [1]. Moreover, multiple-access visible light communication with image sensor and LED array were realized using space division multiplexing (SDM) [2]. In this presentation, multiple-access and two-way capability of visible light communication with image sensor and LED array are investigated. First, isolation characteristics between LED and image sensor within the same transceiver is measured while distance between the LED and image sensor are altered. Next, BERs for multiple-access and two-way visible light communication transceiver composed of image sensor and LED array are measured for OOK and ASK to verify multiple-access and two-way capability experimentally.

II. MULTIPLE-ACCESS AND TWO-WAY VISIBLE LIGHT COMMUNICATION FOR ON-OFF KEYING

A. Isolation between transmission and reception

Configuration of a transceiver used for isolation measurement between transmission and reception within the same transceiver is shown in Fig. 1. The transceiver consists of an LED and a CMOS image sensor for mobile object. Total luminous flux of transmitter LED is 100 lumens and half-power beam width of the LED is 16 deg. Frame rate of a CMOS image sensor is 30 fps and field of view of the image sensor is 68 deg. FPGA, Spartan-6, is used for signal processing. In order to measure the isolation characteristics, space between an LED and a CMOS image sensor in the same transceiver were set to 4, 5, and 6 centimeters and a distance from LED to CMOS image sensor position was altered within ± 5 centimeters. Access-point image captured at mobile-object image sensor is shown in Fig. 2. In both figures, LEDs of the access point are off. Fig. 2(a) and (b) show the access-point image when an LED of the mobile object is off and on, respectively. Since access-point image when the mobile-object LED is on is brighter than that when the LED is off, reflection of the mobile-object LED light at the access-point transmitter is affected. However, there are no differences in the background area in both figures, the mobile-object LED light does not affect to the image sensor directly. BERs for one-way visible light communication are measured while changing a space and distance between an LED and a CMOS image sensor within the same transceiver as shown in Fig. 3. Modulation is OOK, symbol rate is 25 sps, and communication distance is 1 meter. BERs of less than 10^{-5} are obtained regardless of the space and distance. Sufficient isolation between transmission and reception within the same transceiver was experimentally verified.

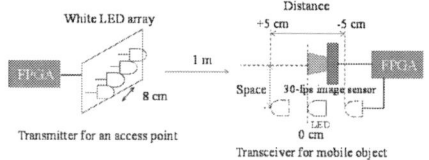

Fig. 1. Isolation measurement between transmission and reception within the same transceiver.

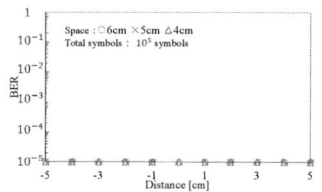

Fig. 3. BERs for OOK versus a space and distance between LED and image sensor.

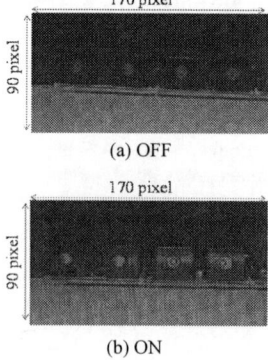

(a) OFF

(b) ON

Fig.2. Images of an access point captured at image sensor of mobile object. Space and distance between LED and image sensor is 4 and 2 cm, respectively and distance between access point and mobile object is 1 meter.

988

B. Multiple-access and two-way capability

Configuration of multiple-access and two-way visible light communication for OOK is shown in Fig. 4. Transceiver for an access point consists of 4-element LED array and a CMOS image sensor as shown in Fig. 5(a). Two transceivers for mobile objects consist of 1-element LED and a CMOS image sensor as shown in Fig. 5(b). In the access-point transceiver, space and distance between LEDs and image sensor is 8 and -4 centimeters, respectively. In the mobile-object transceivers, space and distance between LED and image sensor for is 4 and -2 centimeters, respectively. The access point receives LED signals from two mobile objects independently and synchronizes to each signal in parallel. Two LEDs in the access-point transceiver are used for one mobile object and the remaining two LEDs are used for the other mobile object. Synchronization algorithm applicable up to 25 symbols per second that is close to the frame rate of the image sensor, 30 frames per second, is used [1]. Two mobile-object LED images are acquired independently at the access-point transceiver as shown in Fig. 6(a). The access-point LEDs image is acquired at one mobile-object transceiver as shown in Fig. 6(b). Two LEDs in the access-point transceiver are acquired at one mobile object. BERs for OOK, multiple-access and two-way visible light communication were measured using the access-point transceiver and two mobile-object transceivers. Error free operation of less than 10^{-5} was achieved up to the symbol rate of 25 sps as shown in Fig. 7(a). In addition, BERs of less than 10^{-5} were achieved at the symbol rate of 25 sps for multiple-access and two-way communications up to a distance of 4 meters as shown in Fig. 7(b).

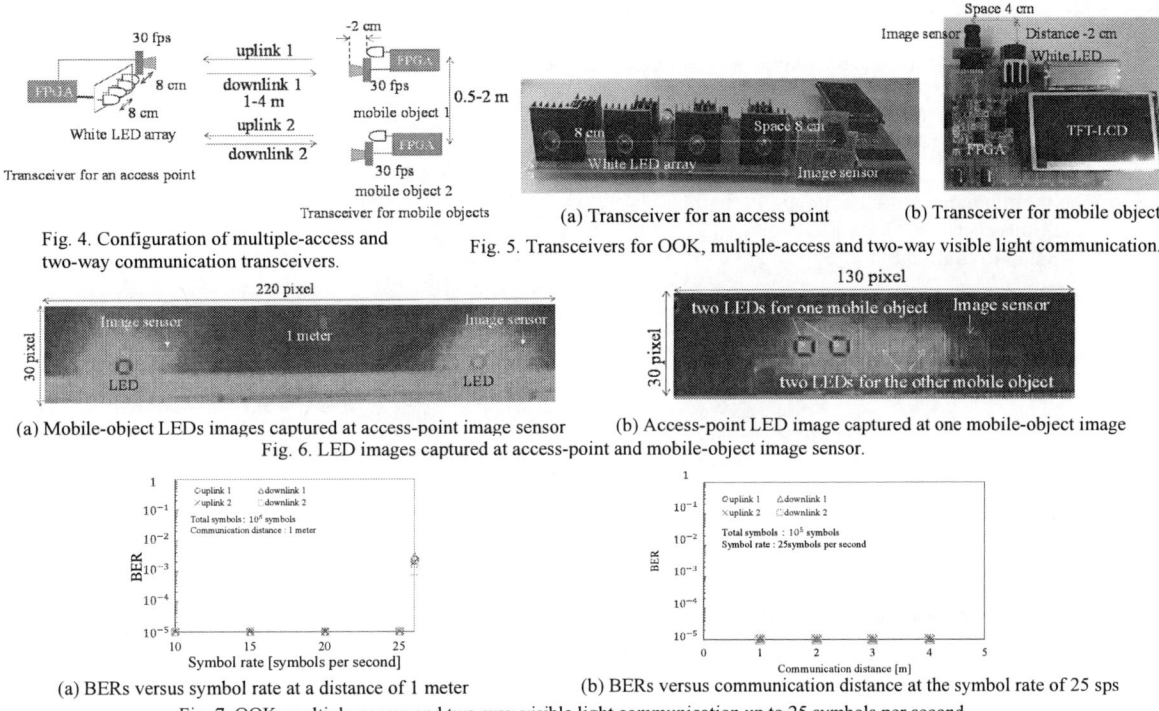

Fig. 4. Configuration of multiple-access and two-way communication transceivers.

Fig. 5. Transceivers for OOK, multiple-access and two-way visible light communication.

(a) Transceiver for an access point   (b) Transceiver for mobile object

(a) Mobile-object LEDs images captured at access-point image sensor   (b) Access-point LED image captured at one mobile-object image

Fig. 6. LED images captured at access-point and mobile-object image sensor.

(a) BERs versus symbol rate at a distance of 1 meter   (b) BERs versus communication distance at the symbol rate of 25 sps

Fig. 7. OOK, multiple-access and two-way visible light communication up to 25 symbols per second.

III. MULTIPLE-ACCESS AND TWO-WAY VISIBLE LIGHT COMMUNICATION FOR AMPLITUDE SHIFT KEYING

A. Isolation between transmission and reception

Isolation characteristics for Amplitude-Shift Keying (ASK) are also measured as shown in Fig. 8. LED for ASK is different from that for OOK. Total luminous flux of the LED is 100 lumens and half-power beam width of the LED is 22 deg. Access-point image captured at mobile-object image sensor is shown in Fig. 9. In both figures, LEDs of the access point are off. Fig. 9(a) and (b) show the access-point image when optical intensity ratio of the mobile-object LED is 50 and 100 %, respectively. Although LEDs of the access point are off, the LEDs look like on. Reflection of the mobile-object LED light at the access-point transmitter is clearly identified in both figures. However, since there are no differences in the background area, the mobile-object LED light does not affect to the image sensor directly in ASK as well as OOK. BERs for one-way visible light communication are measured while changing a space and distance between an LED and a CMOS image sensor within the same transceiver as shown in Fig. 10. Symbol rate is 25 symbols per second, communication distance is 1 meter, and optical intensity ratio of the transmitter LED is altered between 50 and 100 %. BERs of less than 10^{-4} are obtained regardless of the space and distance. Sufficient isolation between transmission and reception within the same transceiver was experimentally verified in ASK as well as OOK.

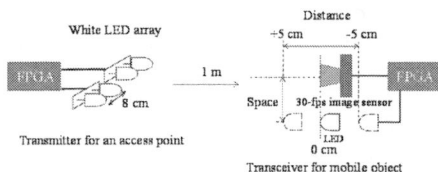

Fig. 8. Isolation measurement between transmission and reception within the same transceiver for ASK.

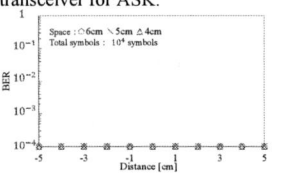

Fig. 10. BERs for ASK versus a space and distance between LED and image sensor within the same transceiver.

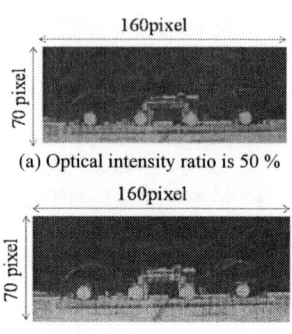

(a) Optical intensity ratio is 50 %

(b) Optical intensity ratio is 100 %

Fig. 9. Images of an access point captured at image sensor of mobile object. Space and distance between LED and image sensor of mobile object is 4 and 2 cm, respectively and distance between access point and mobile object is 1 meter.

B. Multiple-access and two-way capability

Configuration of multiple-access and two-way visible light communication for ASK is shown in Fig. 11. Access-point transceiver consists of 4-element LED array and a CMOS image sensor as shown in Fig. 12. Two mobile-object transceivers consist of 1-element LED and a CMOS image sensor. In the access-point transceiver, space and distance between LEDs and image sensor is 4 and 1 centimeters, respectively. In the mobile-object transceivers, space and distance between LED and image sensor for is 4 and 1 centimeters, respectively. Two LEDs of the access-point transceiver are used for one mobile object and the remaining two LEDs are used for the other mobile object as shown in Fig. 13. BERs for ASK, multiple-access and two-way visible light communication were measured. Error free operation of less than 10^{-4} was achieved up to the symbol rate of 25 sps for ASK as well as OOK as shown in Fig. 14.

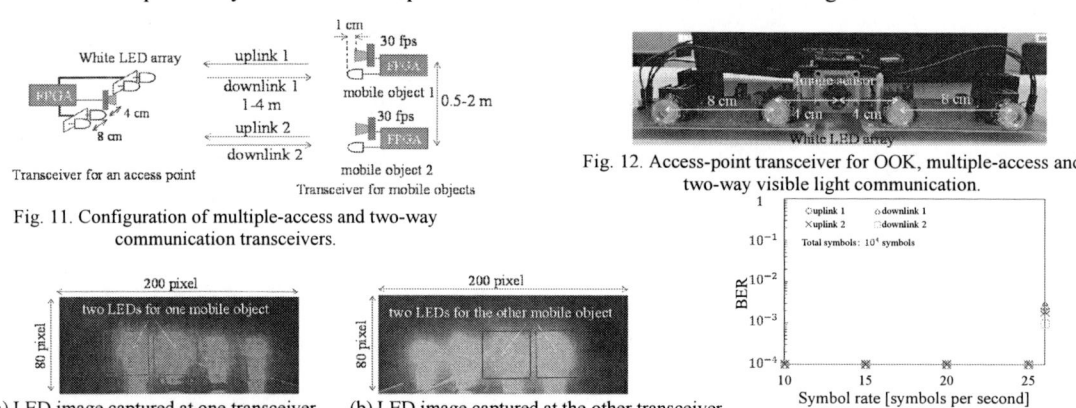

Fig. 11. Configuration of multiple-access and two-way communication transceivers.

Fig. 12. Access-point transceiver for OOK, multiple-access and two-way visible light communication.

(a) LED image captured at one transceiver   (b) LED image captured at the other transceiver

Fig. 13. Access-point LED images captured at two mobile-object transceivers.

Fig. 14. BERs for ASK versus symbol rate at a distance of 1 meter.

IV. CONCLUSIONS

In order to realize multiple-access and two-way visible light communication, isolation characteristics between LED and image sensor within the same transceiver were measured for OOK and ASK. BERs of less than 10^{-5} and 10^{-4} were achieved respectively and sufficient isolations were obtained regardless of space and distance between LED and image sensor. In addition, multiple-access and two-way visible light communication using an access-point transceiver and two mobile-object transceivers was experimentally verified for OOK and ASK. BERs of less than 10^{-5} and 10^{-4} were achieved for OOK and ASK, respectively up to the symbol rate of 25 symbols per second utilizing the multiple-access and two-way visible light communication transceivers with 30-fps image sensor.

REFERENCES

[1] W. Chujo, T. Kondo, and R. Kitaoka, "Improvement of Symbol Rate and Flicker-Free Performance of LED Visible Light Communication with Low-Frame-Rate CMOS Camera," International Conference and Exhibition on Visible Light Communication 2015, 1570199199CC, Oct. 2015.

[2] T. Kondo, R. Kitaoka, and W. Chujo, "Multiple-Access Capability of LED Visible Light Communication with Low-Frame-Rate CMOS Camera for Control and Data Transmission of Mobile Objects," 2015 IEEE/SICE International Symposium on System Integration, pp. 678-683, Dec. 2015.

WA2-86

Broadband Polymer 3-dB Coupler using Shortcut to Adiabaticity based Optimization

Hung-Ching Chung[1] and Shuo-Yen Tseng[1,2,*]

[1]Department of Photonics, National Cheng Kung University, Tainan, Taiwan
[2]Advanced Optoelectronic Technology Center, National Cheng Kung University, Tainan, Taiwan
*Author e-mail address: tsengsy@mail.ncku.edu.tw

Abstract: *We propose a scheme to optimize the adiabaticity of 3-dB couplers for broadband operation. Using shortcut to adiabaticity, a 3-dB coupler with a maximum imbalance of 0.2 dB from 1.25 to 1.6 μm is obtained.*

Keywords: *integrated optics, couplers, quantum physics*

I. INTRODUCTION

3-dB coupler is a key component in photonic circuits, and it also plays a significant role in modulators and switches. Because of its simplicity, 3-dB couplers are conventionally designed using directional couplers (DCs) in integrated optics [1]. However, owing to the sensitivity of DCs to wavelength and fabrication errors, alternative technologies such as multimode interference (MMI) couplers [2] and adiabatic couplers [3] have been developed. Although MMI couplers are robust against fabrication errors, they still have large wavelength dependency; and adiabatic couplers require long device lengths in general. To minimize length while maintaining good adiabaticity, the optimal design procedure of adiabatic coupled-waveguide devices has been proposed in different ways [3-5].

A family of techniques called shortcuts to adiabaticity (STA) has been proposed to speed up adiabatic passage [6]. This STA approach allows precise engineering of the system by describing the system evolution using the decoupled system state [7]. Many short and robust coupled waveguides have been designed using STA, and the sensitivity to wavelength and fabrication errors can be minimized [7-9]. These approaches, while being robust against particular errors by design, do not guarantee adiabaticity. Recently, a scheme to optimize system adiabaticity in coupled-waveguide devices using STA has been proposed [10]. In the new approach, optimal system adiabaticity is obtained by designing the system evolution to be as close to the adiabatic state as possible using the decoupled system states, thus guaranteeing both adiabaticity and 100 % efficiency. This is achieved by expanding the phase of the decoupled system state using Fourier series. This design approach gives simple analytic expressions for the desired adiabaticity that are applicable to a wide range of coupled-waveguide devices. In the paper, we apply this design approach to optimize the adiabaticity of a 3-dB coupler, resulting in a broadband device. The designed device is investigated by beam propagation simulations.

II. OPTIMIZING SYSTEM ADIABATICITY

We consider a weakly-coupled waveguide structure consisting of two waveguides, waveguide 1 and 2, placed in proximity with propagation constants β_1 and β_2. The separation between two waveguides and the widths of individual waveguides are allowed to vary along the propagation direction z. Input light is coupled into the device at $z=0$ and out at $z=L$. The guided-mode amplitudes $[A_1,A_2]^T$ changes with propagation distance and can be described by coupled-mode equation as

$$\frac{d}{dz}\begin{bmatrix} A_1 \\ A_2 \end{bmatrix} = -i\begin{bmatrix} -\Delta & \kappa \\ \kappa & \Delta \end{bmatrix}\begin{bmatrix} A_1 \\ A_2 \end{bmatrix} \qquad (1)$$

where κ is the coupling coefficient, and $\Delta=(\beta_1-\beta_2)/2$ describes the degree of mismatch between the waveguides. The solution of Eq. (1) can be described by [6,7]

$$\psi = \begin{bmatrix} \cos(\theta/2)e^{i\phi/2} \\ \sin(\theta/2)e^{-i\phi/2} \end{bmatrix} \qquad (2)$$

where $d\theta/dz=\kappa\sin\phi$ and $d\phi/dz=-\kappa\cot\theta\cos\phi-\Delta$. Different from the traditional adiabatic approaches where the evolution of κ and Δ are designed to satisfy the adiabatic criterion, the STA approach described here designs the system evolution using Eq. (2) to be as close to the adiabatic trajectory as possible. To describe 3-dB beam splitting at the device output using Eq. (2) in a device with length L, the initial and final states of the system are set as $\psi(0)=[1,0]^T$ and $\psi(L)=(1/\sqrt{2})[1,1]^T$ respectively. To satisfy the initial and final states, we choose a smooth function for θ as

---

This word is sponsored by the Ministry of Science and Technology (MOST) of Taiwan (104-2221-E-006-212).

$\theta(z)=(\pi/4)(1+\cos(\pi(L-z)/L))$. Applying l'Hopital's rule repeatedly, we can find an additional boundary condition for ϕ, $\phi(0)=\pi/2$. Full adiabaticity can be obtained by requiring $\phi(z)=0$ during the evolution. However, this would lead to an infinitely large κ. We instead set ϕ to a small value c that is directly related to a maximum obtainable coupling coefficient in the coupled-waveguide system under consideration.

The optimization scheme is now to design a smooth $\phi(z)$ function that satisfies $\phi(0)=\pi/2$, $\phi(L)=0$ and $\phi(z)=0$. Following [10], we expand $\phi(z)$ using Fourier series with sigma-approximation to eliminate the Gibbs phenomenon at the discontinuities

$$\phi(z) = \frac{\pi}{2} + \sum_{k=1}^{n} a_k \sin\left(k \frac{\pi z}{2L}\right) \ (n \text{ is odd}) \qquad (3)$$

In this work, we found that the third order expansion is sufficient to ensure good adiabaticity. Higher order expansions lead to κ's and Δ's with more oscillations, which result in sharp bends and quick width variations in the devices. So we use the third-order expansion with a_1=-1.6319 and a_3=-1.1813 corresponding to a maximum coupling coefficient of 13.79 (in unit of $1/L$) for the device design. The corresponding κ and Δ are then obtained using the relations under (2) and illustrated in Fig. 1.

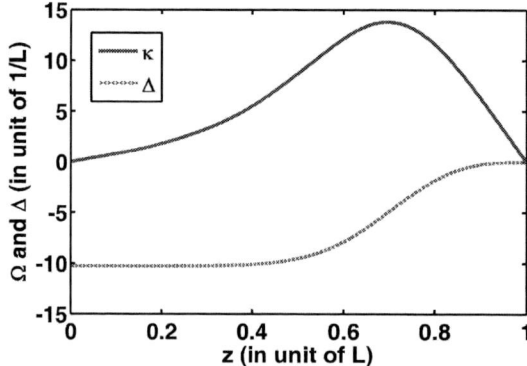

Fig. 1. The calculated coupling coefficient and mismatch for 3-dB coupler design (3rd order expansion using (3)).

III. NUMERICAL SIMULATIONS

The beam propagation method (BPM) code used in the simulations solves the scalar and paraxial wave equation using the finite difference scheme with the transparent boundary condition [1]. The polymer channel waveguide structure is the same as the one considered in [8], where 3 μm thick SiO$_2$ (n=1.46) on a Si (n=3.48) wafer is used for the bottom cladding layer, the core consists of a 2.4 μm layer of BCB (n=1.53), and the upper cladding is epoxy (n=1.50). The device length L is set at 1 mm and simulated at 1.55 μm input wavelength and the TE polarization. The geometry of the waveguide beam splitter is determined by adjusting the relative separation between the waveguides and the waveguide widths to satisfy the required coupling coefficient and mismatch in Fig. 1. The waveguide center-to-center spacing D and width difference δW are adjusted along the propagation direction to satisfy the designed set of κ and Δ functions. The resulting 3-dB coupler is shown in Fig. 2. In the same figure, we also show the BPM simulation results by exciting the bottom arm at z=0. Clearly, the input light is evenly split into the bar and cross arms at the output.

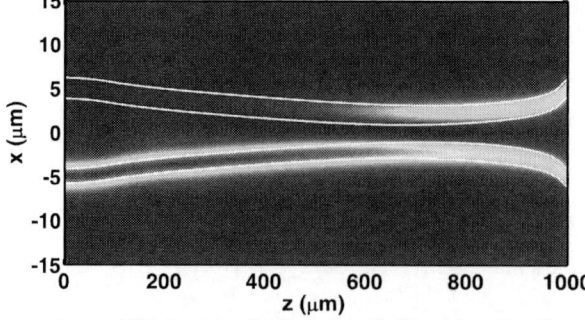

Fig. 1. Device geometry and BPM simulation for the designed 3-dB coupler using the κ and Δ in Fig. 1.

To investigate the device bandwidth, we change the input wavelength into the coupler from 1.2 to 1.7 μm in the BPM simulation. Material dispersion is not considered in the simulation. Figure 3 shows the simulated wavelength dependence of the transmission. It can be seen that for a wide span of over 350 nm from 1.25 μm to 1.6 μm, the imbalance of the two output arms is smaller than 0.2 dB. The 3-dB coupler designed by STA-based optimization exhibits the desired flatness as a result of our optimization. The device performance is much better than the DC beam splitter at the same device length of 1 mm. Also, we note that 3-dB splitting is achieved exactly at the designed wavelength of 1.55 μm due to the use of decoupled system states and boundary conditions. The same feature is not always obtainable using traditional adiabatic schemes.

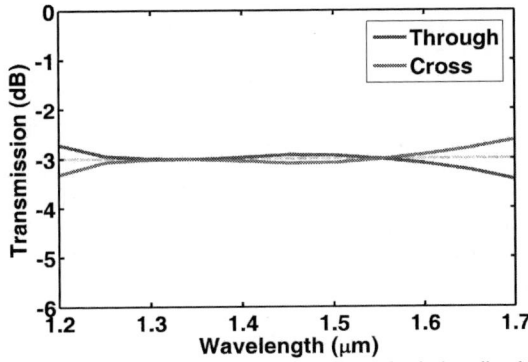

Fig. 3. BPM simulation results of the wavelength dependence for the broadband 3-dB coupler.

IV. CONCLUSIONS

We have successfully optimized the adiabaticity of 3-dB couplers using shortcuts to adiabaticity. Decoupled system states which are eigenvectors of Lewis-Riesenfeld invariants are used to precisely engineer the system evolution. The result is obtained by simply expanding the target phase function of the decoupled system state using Fourier series, without the need to optimize or nullify any specific functions. The result could also find applications in designing the time-dependent Rabi frequencies and detunings needed to control two-state quantum systems for coherent superposition of states [11].

ACKNOWLEDGMENT

We are grateful to J. G. Muga, X. Chen, S. Martínez-Garaot and R.-D. Wen for discussions.

REFERENCES

[1]   K. Okamoto, Fundamentals of optical waveguides, 2nd ed., Elsevier, 2006, Chap. 4.
[2]   L. B. Soldano and E. C. M. Pennings, "Optical multi-mode interference devices based on self-imaging: principles and applications," J. Lightw. Technol., vol. 13, pp. 615–627, 1995.
[3]   K. Solehmainen, M. Kapulainen, M. Harjanne, and T. Aalto, "Adiabatic and multimode interference couplers on silicon-on-insulator," IEEE Photon. Technol. Lett., vol. 18, pp. 2287-2289, 2006.
[4]   A. Syahriar, V. M. Schneider, and S. Al-Bader, "The design of mode evolution couplers," J. Lightw. Technol., vol. 16, pp. 1907-1914, 1998.
[5]   T. A. Ramadan, R. Scarmozzino, and R. M. Osgood, "Adiabatic couplers: design rules and optimization," J. Lightw. Technol., vol. 16, pp. 277-283, 1998.
[6]   E. Torrontegui, S. Ibáñez, S. Martínez-Garaot, M. Modugno, A. del Campo, D. Guéry-Odelin, A. Ruschhaupt, X. Chen, and J. G. Muga, "Shortcuts to adiabaticity," Adv. At. Mol. Opt. Phys., vol. 62, pp. 117–169, 2013.
[7]   S.-Y. Tseng, "Robust coupled-waveguide devices using shortcuts to adiabaticity," Opt. Lett., vol. 39, pp. 6600-6603, 2014.
[8]   S.-Y. Tseng, R.-D. Wen, Y.-F. Chiu, and X. Chen, "Short and robust directional couplers designed by shortcuts to adiabaticity," Opt. Express, vol. 22, pp. 18849-18859, 2014.
[9]   T.-H. Pan and S.-Y. Tseng, "Short and robust silicon mode (de)multiplexers using shortcuts to adiabaticity," Opt. Express, vol. 23, pp. 10405-10412, 2015.
[10] C.-P. Ho and S.-Y. Tseng, "Optimization of adiabaticity in coupled-waveguide devices using shortcuts to adiabaticity," Opt. Lett., vol. 40, pp. 4831-4834, 2015.
[11] M. Ndong, G. Djotyan, A Ruschhaupt, and S. Guérin, "Robust coherent superposition of states by single-shot shaped pulse," J. Phys. B: At. Mol. Opt. Phys., vol. 48, pp. 174007, 2015.

OECC/PS2016

Multimode Three Branch Polymer Splitter

Václav Prajzler, Radek Maštera

Dept. of Microelectronics, Faculty of Electrical Engineering, Czech Technical University in Prague,
Technická 2, 168 27 Prague, Czech Republic
Author e-mail address: xprajzlv@feld.cvut.cz

Abstract: *We report about three branch multimode polymer power splitters. New design with insertion of rectangle shaped spacing into the central part of the splitter ensures more evenly distribution of the output optical power.*
Keywords: *Optical splitter, Multimode waveguides, Polymer*

I. INTRODUCTION

Optical Y-branching waveguides are important key passive components in photonics integrated circuits used for power splitters and optical switches. Standard 1x2 Y-branch optical power splitter consists of a one single waveguide splitting symmetrically into two output waveguides [1]. Multiple Y-branch splitters are often cascading combined into $1x2^N$ optical power splitter where N is cascaded stages. Therefore these splitters have even number of output waveguides. Branching splitters with odd number with three of output waveguides with non-symmetrical optical power distribution have been also described [2]. These splitters have a uniform index distribution and therefore they have the most of the power concentrated in central branch. Up to now only few papers have been reported dealing with three-branch planar splitters with symmetrical power distribution [3-5]. Up to know all of the splitters with three output waveguides have been purposed for single mode waveguides and no multi-mode based structures with three output branch waveguides have been yet presented.

In this paper we are going to present our new approach to construct polymer large core power splitters with one input and three outputs plastic optical fiber (POF). The novelty of our approach is in inserting of rectangle shape spacing between the input and the central part of the splitter, which is to ensure more proportional distribution of the outgoing optical power.

II. 1x3Y BRANCH SPLITTERS WITH RECTANGLE SHAPE SPACING

Optical waveguide structure used for designing the optical multimode three branch polymer splitter is the step-index rectangular waveguide consisting of a waveguide core layer of refractive index (n_f), surrounded in all sides by materials having refractive index (n_c) lower than that of a core waveguide (see Fig. 1a).

Basic configurations of the proposed multimode three-branch optical splitter is shown in Fig. 1b while Fig. 1c shows our proposed splitter where the rectangle shape spacing is inserted between the input and the central output branches. Dimensions of the rectangular waveguide were set so that it could be connected to a standard plastic optical fiber (POF) with core diameter 980 µm and 20 µm thick cladding layer. For our 1x3Y branch splitter we chose acrylic-based polymers as a waveguide core material and poly(methyl methacrylate) (PMMA) as a substrate and cover layer. These materials were chosen due to suitable optical and mechanical properties (low optical absorption, sufficient temporal and temperature stability, well-controlled refractive indices and feasible fabrication processes) [6].

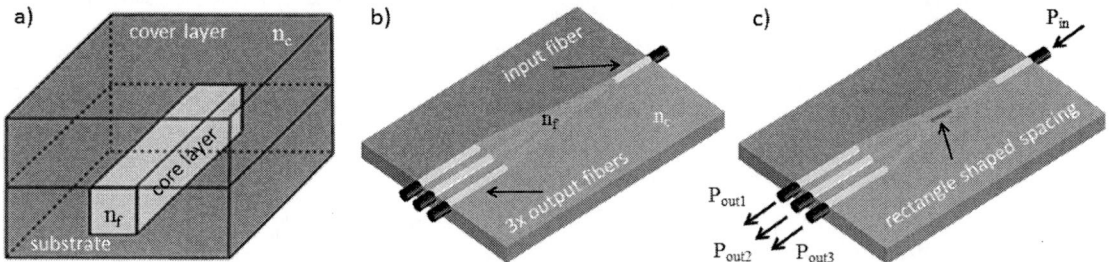

Fig. 1. a) Cross-sectional view of the proposed optical rectangular waveguide, b) geometrical structure of the basic configurations three-branch optical splitter, c) three-branch optical splitter with rectangle shaped spacing.

We started with designing of optical splitters with two output waveguides and prosecuted to three output splitters. Geometrical dimensions of the two-output splitters were calculated by analysis for a lossless Y-junction published by Beltrami (this design we published in [7]) and then we intended to extend the design to a three-output variety. To achieve it we had to widen the distance (gap) between two original output waveguides to insert the third central waveguide, as connecting three large core output fiber waveguides needs its space. For all our designed structures we used PMMA substrates and different core–waveguide layers (NOA - Norland Optical Adhesives) and the design was done by Beam Propagation Method (BPM) with help of BeamPROP program in the frame of the specialized simulation

994

package MOST, (*RSoft's, Multi-Variable Optimization and Scanning Tool*). For simulation we applied 2D dimensional channel with multimode source operating at 650 nm.

Results of the simulations of the basic multimode three-branch optical splitter for NOA73 core waveguide layer is shown in Fig. 2. The modeling was done for refractive indices of 1.558 (λ= 650 nm, core NOA waveguide layer) and 1.489 (λ= 650 nm, PMMA substrate and cover layer), respectively. As could be expected simulation shows that the power division is non-symmetrical the central waveguide branch of the NOA73 transmits more optical power (P_{out2}) than left (P_{out1}) and right (P_{out3}) output waveguides, which corresponds to their splitting ratio central branch 35.4% (P_{out2}) : left branch 31.6% (P_{out1}) : right branch 33.0% (P_{out3}) (see Fig 2a).

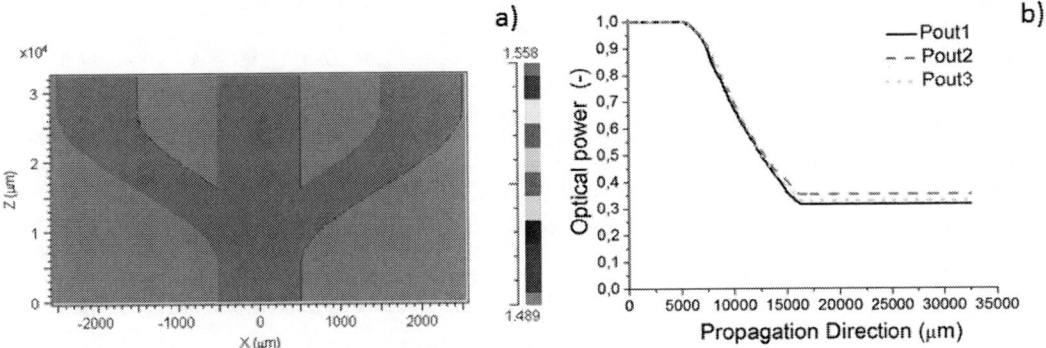

Fig. 2. Results of simulation of the basic configurations three-branch optical splitter (splitter without rectangle shaped spacing) for wavelength at 650 nm, a) computed index profile (top view), b) normalized optical signal propagation (P_{out1} – left, P_{out2} - central, P_{out3} right branch).

As was shown in Fig.2, normally in the three-branch optical splitter with a uniform refractive index distribution the most of the optical power will be predominantly driven via the central branch. To provide more proportional distribution of the power we proposed an optical splitter with inserted rectangle shaped spacing area between the input and the central output branch. This rectangle is made from same materials (PMMA) as the surrounding core materials. To achieve symmetrical distribution of the optical power between three output waveguides, the rectangle spacing needs to have optimal dimensions and it must be placed into an exactly defined position, which provides symmetrical distribution of electromagnetic field at the end of the tapered channel waveguide, where the signal is divided into left output branch, to the right output branch and to straight output branch. Optimization of location and dimensions of the rectangle is made in relations to distribution of electromagnetic field of the optical modes in the waveguiding structure.

The result of the simulations for the optimized dimension structure for NOA73 waveguides with rectangle shaped spacing area is shown in Fig. 3. Fig. 3a shows computed refractive index profile with the pertinent dimensions while the propagation of the signal is given in Fig. 3b. The modeling was done for a wavelength 650 nm and for the same value of refractive indices of the core NOA layer and PMMA substrate which were mention above. From Fig. 2b it is obvious that in this case the optical power division is proportional and this arrangement gave the distribution of the output power for waveguide to 33.1% (P_{out1}) : 33.4% (P_{out2}) : 33.4% (P_{out3}), i.e., left, central and right branches, respectively.

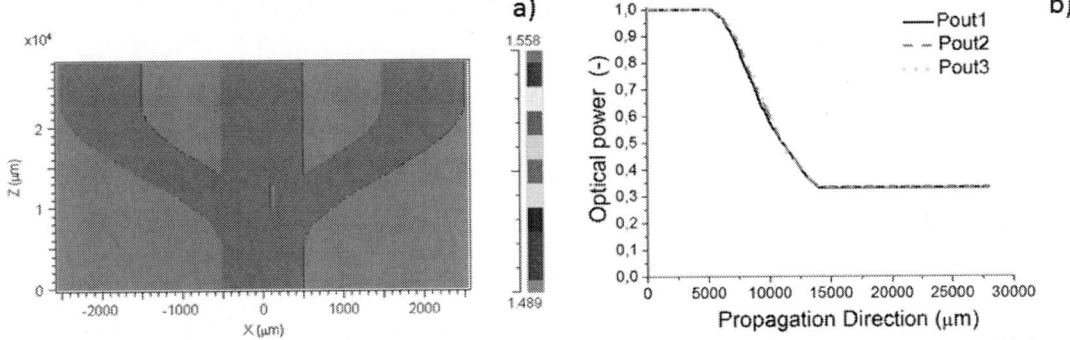

Fig. 3. Results of simulation of the three-branch optical splitter with rectangle shaped spacing for wavelength at 650 nm, a) computed index profile (top view), b) normalized optical signal propagation (P_{out1} – left, P_{out2} - central, P_{out3} right branch).

According to the simulation, the optimized structure with the inserted rectangle the output power from the output waveguides differing in less than 0.5%. For comparison, we also did modelling for a similar structure without inserted rectangle and, as expected, such structure was non-symmetrical with optical power distribution ratio 32% (P_{out1}) : 36% (P_{out2}) : 30% (P_{out3}) for right, central and left branches, respectively.

III. EXPERIMENTS AND RESULTS

Fabrication process of the designed optical splitters has been described in our papers [8]. The Y-groove for the waveguide core layer into PMMA substrate was made by using CNC (Computer Numerical Control) NONCO Kx3 milling machine. Then we inserted POF (PFU-UD1001-22V) waveguides, which were to serve as the input/output waveguides, into the groove. The faces of the POF fibers were polished prior to inserting into Y-groove. Next we filled up the taper region with NOA polymer and applied UV curing process. Finally, top cover PMMA is placed onto the structures.

The image of the splitter is shown in Fig. 4. Fig. 4a shows detail image of the splitter and Fig. 4b shows in a simple visual way how the final three-branch optical power splitter with the rectangle shaped spacing transmits the optical light at 635 nm (laser tester FLS-240).

We measured optical insertion losses for three wavelengths (green 532 nm Nd:YVO$_4$ laser; red light 650 nm laser Safibra OFLS-5 FP-650; 850 nm laser Safibra OFLS-5 DFB-850) using a method, which was described in [8]. The measurement proved that splitters had optical losses lower than 4 dB and the best sample has optical losses 2.8 dB at 532 nm, 2.1 dB at 650 nm and 2.4 dB at 850 nm. Comparing our splitters, which had inserted rectangles, with a reference structure without any rectangle, we found that the latter exhibited strongly non-proportional distribution of optical power and, as expected, the most of the energy was gathering in the central branch. The experimentally found non-proportionality of the splitter without inserted rectangle spacing was in reality much stronger than it followed from the simulation and the difference between the output power in central branch and that coming from the output left and right branches was around 21%. Our experimental splitters with rectangular shape spacing exhibited optical power division much more proportional and the output power in central branch differed from the side branches in less than 4%.

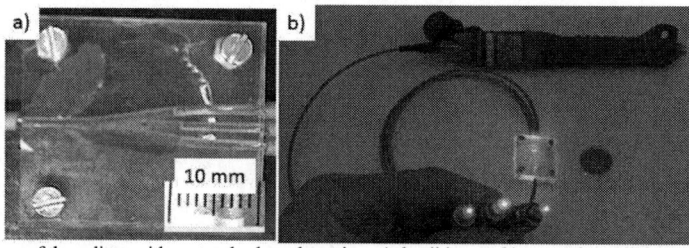

Fig. 4. Image of the splitter with rectangle shaped spacing, a) detail image, b) transmitting light at 635 nm.

IV. CONCLUSIONS

We designed, realized and measured properties of multimode polymer three-branch optical splitter with rectangle shaped spacing. The design was done by beam propagation method using RSoft software. The measurement of optical insertion loss proved that the three-branch optical splitter with rectangle shaped spacing had the lowest insertion loss 2.1 dB at 650 nm. Both, simulation and realization of the splitters showed the important role of insertion a rectangle into the central waveguiding region.

ACKNOWLEDGMENT

Our research is supported by the research program of Czech Technical University in Prague by project no. SGS14/195/OHK3/3T/13.

REFERENCES

[1] C. DeCusatis, "Handbook of Fiber Optic Data Communication: a practical guide to optical networking," Academic Press Elsevier, 2008.
[2] G.L. Yip, M.A. Sekerka-Bajbus, "Design of symmetric and asymmetric passive 3-branch power dividers by beam propagation method," El. Lett., vol. 24, pp. 1584-1586, 1988.
[3] S. Banba, H. Ogawa, "Novel symmetrical three-branch optical waveguide with equal power division," IEEE Microw. Guid. Lett., vol. 2, pp. 188-190, 1992.
[4] V. Prajzler, H. Tůma, J. Špirková, V. Jeřábek, "Design and modeling of symmetric three branch polymer planar optical power dividers," Radioengineering, vol. 22, pp. 233-239, 2013.
[5] T.J. Wang, C.F. Huang, W.S. Wang, "Wide-angle 1x3 optical power divider in LiNbO$_3$ for variable power splitting," IEEE Phot. Tech. Lett., vol. 15, pp. 2003.
[6] V. Prajzler, P.N. Kien, J. Špirková, "Design, Fabrication and Properties of the Multimode Polymer Planar 1 x 2 Y Optical Splitter," Radioengineering, vol. 21, pp. 1202-1207, 2012.
[7] D.R. Beltrami, J.D. Love, F. "Ladouceur, Multimode planar devices," Opt. Quant. El., vol. 31, pp. 307–326, 1999.
[8] V. Prajzler, R. Maštera, J. Špirková, "Large Core Three Branch Polymer Power Splitters," Radioengineering, vol. 24, pp. 885-891, December 2015.

2-stage Cascaded Silicon Photonic PBS based on Mach Zehnder Delay Interferometers

Kohei MORITA, Hiroyuki UENOHARA

Precision and Intelligence Laboratory, Tokyo Institute of Technology
4259 Nagatsuta, Midori-ku, Yokohama, Kanagawa 226-0026, Japan
morita.k.as@m.titech.ac.jp

Abstract: We demonstrate a 2-stage silicon photonic polarization beam splitter based on Mach Zehnder delay interferometer. Polarization extinction ratio greater than 20 dB over the full C-band for both polarizations was confirmed in simulation.
Keywords: Silicon photonics, Polarization Beam Splitter

I. INTRODUCTION

The emergence of smartphone and cloud computing is rapidly driving the growth of communication traffic at a growth rate of 35.8% per year [1]. Super High Vision, requiring an information capacity of up to 100Gbps per information stream has been standardized. Therefore, huge-capacity photonic networks are required urgently to realize SHV communication infrastructure. Expectations to optical network technologies also increases that can be used for low power with spectrally efficient and flexible network resources. Thus, digital coherent receivers and multi-degree CDC-ROADM [2], [3] are expected to be the key technologies to achieve such networks. In order to put them into practical use, much attention has been paid to Si photonics technology [4], [5] in recent years because of its potential for high density integration and low cost. In Si photonics technology, the use of silicon which has a high refractive index as optical waveguide material is able to achieve a large optical confinement. Therefore, it permits small waveguide bends, and compared with the optical integrated circuit using conventional silica-based waveguides, an optical circuit area can be reduced in size to 1/100. However, Si waveguides generally has a large polarization dependence, and then, the polarization beam splitter (PBS), which splits TE and TM modes, is one of the significant components in practical use for polarization control. In recent years, several kinds of PBSs in PICs have been investigated and demonstrated [6], [7]. But, most demonstrated devices have large excess losses, and narrow operating bandwidth. In this letter, we propose a 2-stage cascaded PBS based on Mach Zehnder delay interferometers(MZDIs) to solve these problems. Broadband as well as large isolation will be presented.

II. OPERATING PRINCIPLES

A. PBS based on Mach Zehnder delay interferometer

Figure 1 shows a PBS based on a MZDI formed by connecting two multi-mode-interference(MMI) with two arm waveguides. The power branch ratio of the output is able to be changed by varying the MMI length. By utilizing a birefringence between TE/TM mode in a MZDI, phase condition of the interferometer are designed to split into in-phase and out-of-phase in each polarization modes, respectively. The PBS can separated each modes by the following formula:

$$\Delta L = \frac{\lambda}{2(n_{TE} - n_{TM})} \cdot N \qquad (1)$$

where n_{TE} and n_{TM} represent the refractive indices of TE mode and TM mode, respectively, N is the odd number.

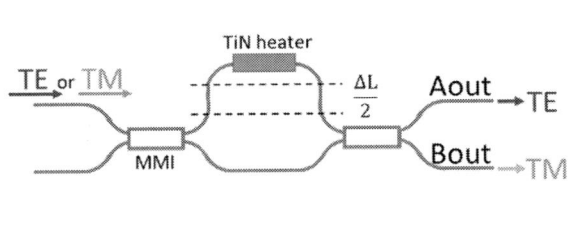

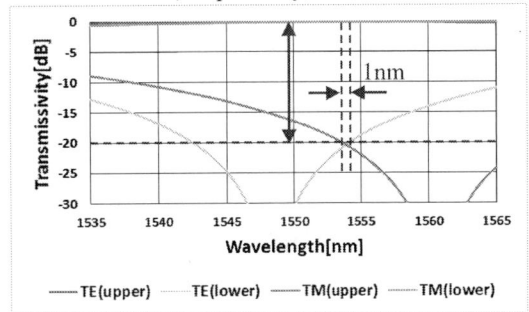

Fig.1 Structure of a PBS based on a MZDI.

Fig.2 Simulation results of transmissivity for TE/TM modes of a PBS shown in Fig.1.

The simulated transmissivity is indicated in Fig.2. In simulation, n_{TE} = 4.178, n_{TM} = 3.468, N = 3, ΔL = 3.144 μm were used, assuming waveguide width and thickness of 0.45 μm and 0.21 μm embedded between SiO$_2$ cladding layers. From this result, the separation of the TE and TM modes was confirmed. However, the device exhibits polarization isolation (or polarization extinction ratio, PER), which is defined as the ratio of the output power of one polarization mode to that of the other, at both two output ports of more than 20 dB over a small bandwidth of 1 nm within C-band and then, its isolation and bandwidth need to be improved further.

B. 2-stage cascaded PBS based on Mach Zehnder delay interferometers

To solve such problems, we propose a 2-stage cascaded structure as shown in Fig.3. Since transmissivity of the cascaded PBS is the product of the transfer function of two PBSs, wider broadband with larger PER is expected. In addition, the tolerance of the PER is significant in design when the power branching ratio of the MMI is deviated from ideal value. Thus, simulation of transmissivity with various power splitting ratio of the MMI was examined. The results are shown in Fig.4. It can be seen that PER of TE mode is not affected so much and only that of the TM mode is deteriorated as splitting ratio is deviated from bar : cross =50 : 50. The PER of TM mode becomes less than 20dB when MMI branching ratio is bar : cross = 30 : 70 and 40 : 60. The power splitting ratio of MMI is dependent on the MMI length and signal wavelength. In brief, it is necessary to determine the length of the MMI when the power splitting ratio of the MMI is bar : cross =50 : 50 for TM polarization.

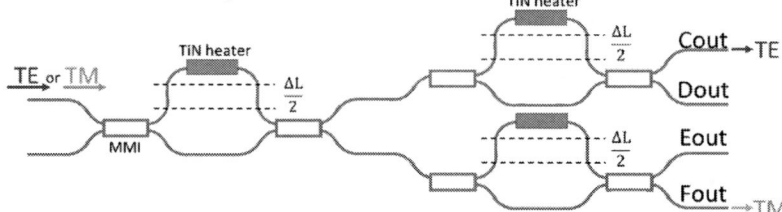

Fig.3 Schematic of a 2-stage cascaded PBS based on Mach Zehnder delay interferometers.

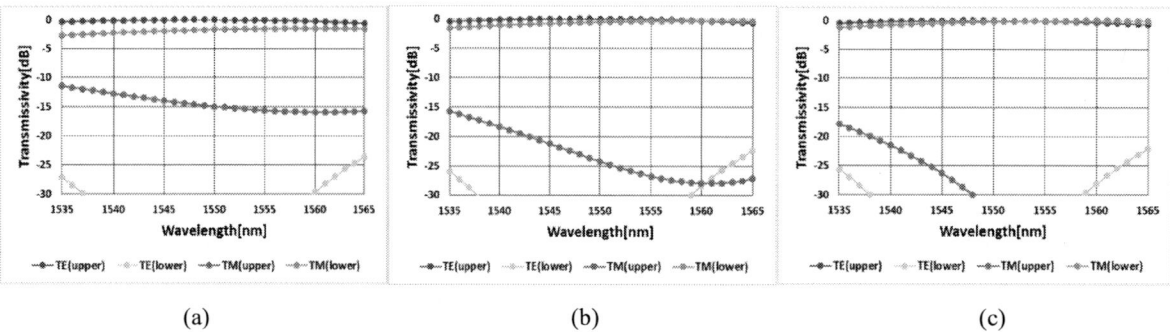

(a)   (b)   (c)

Fig.4 Simulation results of Fig.3 for power branch of the MMI [bar : cross = (a) 30 : 70 (b) 40 : 60 (c) 50 : 50]

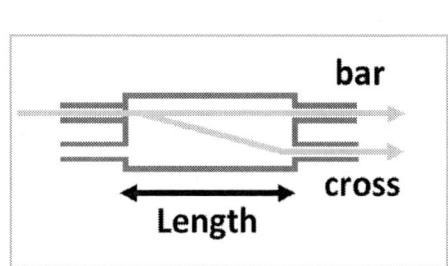

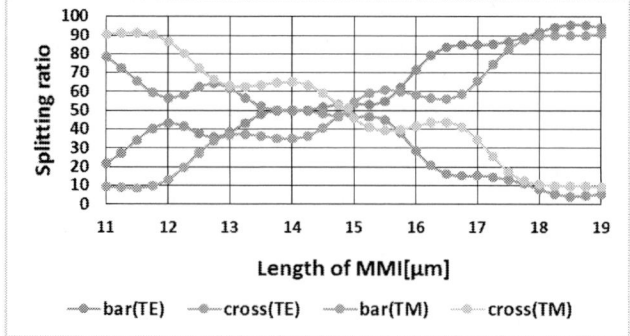

Fig.5 Schematic of a MMI splitter.   Fig.6 MMI splitter characterization for TE and TM polarization as a function of length at a wavelength of 1535nm.

From Fig.4(c), the extinction ratio of TM mode at a wavelength of 1535 nm is the worst in C-band. Thus, MMI length achieving power splitting ratio of MMI of bar : cross = 50 : 50 at a wavelength of 1535 nm is determined. Fig.6 shows MMI splitter characterization for TE and TM modes as a function of MMI length. From this result, MMI length is determined to 14.85 µm when power splitting ratio of MMI is bar : cross = 50 : 50 at 1535nm. Simulation results of such structure is shown in Fig.7. 20 dB extinction ratio for both TE and TM modes over a bandwidth of about 30 nm in entire C-band was achieved.

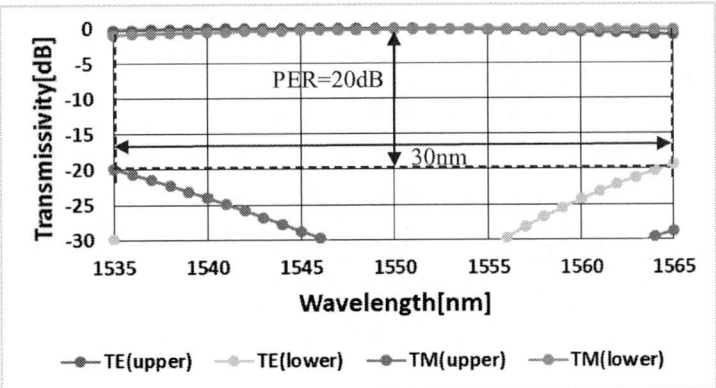

Fig.7 Simulation results of the 2-stage cascaded structure with MMI length of 14.85µm.

III. CONCLUSIONS

In conclusion, we have proposed a 2-stage cascaded PBS based on Mach Zehnder delay interferometers. The simulation results demonstrate that the proposed PBS has a large bandwidth, of more than 20 dB over a large bandwidth of 30 nm when splitting ratio of TM mode was set to 50:50 at 1535nm.

ACKNOWLEDGMENT

We would like to thank Prof. Emeritus K. Iga, Prof. Emeritus K. Kobayashi, Prof. F. Koyama and Assoc. Prof. T. Miyamoto for their encouragements and discussions.

REFERENCES

[1] Ministry of internal Affairs and Communications, "Grasp of the total traffic amount in the internet of our country," April 3, 2015.
[2] R. Jensen, A. Lord, and N. Parsons, "Colourless, directionless, contentionless ROADM architecture using low-loss optical matrix switches," Mo.2.D.2, ECOC, 2010.
[3] H. Takeshita, T. Hino, K. Ishii, J. Krumida, S. Namiki, "Prototype Highly Integtated 8×48 Transponder Aggregator Based on Si Photonics for Multi-Degree Colorless, Directionless, Contentionless Reconfigurable Optical Add/Drop Multiplexer," IEICE TRANS. ELECTRON., VOL.E96-C, NO.7 JULY 2013.
[4] H. Yamada, T. Chu, S. Ishida, and Y.Arakawa, "Si photonic wire waveguide devices," IEEE J. Sel. Top. Quantum Electron., vol.12, pp.1371-1379, 2006.
[5] S. Nakamura, S. Takahashi, I. Ogura, J. Ushida, K. Kurata, T. Hino, H. Takeshita, A. Tajima, M.-B. Yu, and G.-Q. Lo, "High extinction ratio optical switching independently of temperature with silicon photonic 1×8 switch," OTu2I.3, OFC/NFOEC, 2012.
[6] H. Qiu, Y. Su, P. Yu, T. Hu, J. Yang, and X. Jiang, "Compact polarization splitter based on silicon grating-assisted couples," Opt. Lett., vol. 40, No. 90, May 1, 2015.
[7] Dong Wook Kim, Moon Hyeok Lee, Yudeuk Kim, and Kyong Hon Kim, "Planar-type polarization beam splitter based on a bridged silicon waveguide coupler," Opt. Express, vol.23, No.2, January 26, 2015.

WA2-89

An ultra-compact and low-loss wavelength demultiplexer employing the photonic-crystal-like metamaterial structure

Feiya Zhou, Luluzi Lu, Minming Zhang[*], Songnian Fu and Deming Liu

School of Optical and Electrical Information, Huazhong University of Science and Technology, Wuhan, 430074, China
*mmz@mail.hust.edu.cn

Abstract: *A wavelength demultiplexer with the compact size of 2.6μm×3.8μm is proposed based on a novel photonic-crystal-like metamaterial structure. The device displays low loss (<-1.7dB), low crosstalk (<-13.6dB) and broad bandwidth (>59nm) with large fabrication tolerance.*
Keywords: *Metamaterials; Integrated optics devices; Wavelength demultiplexer.*

I. INTRODUCTION

Planar wavelength demultiplexers, such as Mach-Zehnder interferometer, ring resonator, array waveguide grating (AWG) and concave diffraction grating (CDG), are the basic building blocks in the optical wavelength division multiplexing (WDM) system. However, these devices currently suffer from the large size, high cost and difficulty of monolithic integration, more or less, prohibiting their wide applications [1].

Free-form metamaterial offers a feasible way of overcoming such limitations. By engineering the local refractive index at a deep sub-wavelength scale, high-performance mode-conversion functionality can be realized in a small area. Recently, a wavelength demultiplexer [2], as well as other integrated devices [3], has been demonstrated applying the metamaterial structure. Much attention has been attracted for their appealing properties including flexibility, compactness, CMOS-compatibility and so on. However, these metamaterial designs are confronted with challenges in the manufacture process. One is the unpatterned details. Sharp corners and complicated acute angles get lost inevitably as their dimensions are beyond the feature size of the lithography. The other is the dimension-relevant etching depth. As the defined shapes are of various fashions and distinct geometries, it is difficult to guarantee the full etching of the whole structure owing to the lag effect.

In this paper, we propose and characterize a wavelength demultiplexer employing the concept of metamaterial design. A novel photonic-crystal-like (PhC-like) structure, whose material composition is finely tuned by the discrete circle voids, is adopted to increase the accuracy of the lithography and the flexibility of the etching depth control. Benefiting from the optimization of the Direct-binary search (DBS) algorithm, this wavelength demultiplexer displays low loss, low crosstalk, broad operation bandwidth and fabrication robustness with the ultra compact footprint of only 2.6 μm×3.8 μm.

II. DESIGN

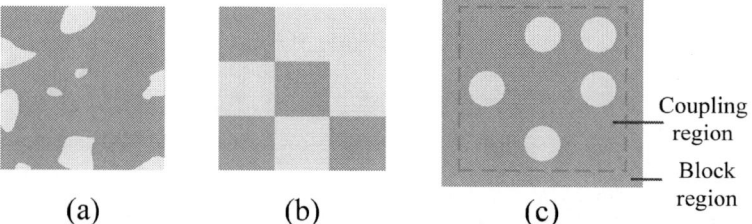

Fig. 1. Evolution of the etching shape from (a) the linear boundary cylinders, (b) the square pillars to (c) the circle voids.

We implement the demultiplexer on a SOI platform with the top silicon layer of 220 nm. Instead of the conventional metamaterial structures shown in Fig. 1 (a) and (b), the device is based on a novel PhC-like structure we have previously reported (see Fig. 1 (c)) [4], where the refractive index is engineered by altering the etching states of a series of circle voids. Since the definition of sharp corners is avoided, the fabrication accuracy is increased with more details reserved. Besides, the etching depth becomes controllable as the circles are unconnected and have fixed sizes, which also provides another freedom for the design.

To begin with, a 2.6 μm×3.8 μm region is separated into two regions: the coupling region and the block region. The coupling region is discretized into 20×30 pixels with the size of 120 nm×120 nm. A circle void with the radius r and the depth d is located in the center of the pixel for the engineering. Block waveguides, with the width of 100 nm, are

This work is partially supported by the National High Technology Research and Development Program of China (863 Program, Grand No. 2015AA015504), the Major Project of Science and Technology Innovation Program of Hubei Province of China (Grand No. 2014AAA006) and National Natural Science Foundation of China (Grand No. 61107051) .

attached around the coupling region to avoid the over-etch of the device edge in the fabrication. The input and output waveguides are 450 nm wide and two output ports are 1 μm apart.

Subsequently, we search for the optimal etching states of all the circle voids following the Direct-binary search (DBS) algorithm. A random 20×30 binary matrix is taken as the starting point of the design: 1 stands for no etching, and 0 stands for a circle void etched into silicon. During each step of the DBS algorithm, the circle state is toggled and a figure-of-merit (FOM) for the device is evaluated. Here, the FOM is defined as:

$$\text{FOM} = \alpha \cdot (T_{11} + T_{22}) + (1-\alpha) \cdot (1 - T_{21} + 1 - T_{12}) \qquad (1)$$

where T_{11} and T_{22} are the transmittances of two target wavelength at the desired output ports, T_{21} and T_{12} are the transmittances at the undesired ports, and α represents the weight of the transmission efficiency. If the FOM is improved, the circle state is retained. Otherwise, the circle state is toggled and the search continues. Each iteration of the DBS algorithm is comprised of the perturbation of all the pixels. The iteration stops when no device improvement is achieved or if the improvement is smaller than a previously specified threshold.

III. RESULTS AND DISCUSSION

The target wavelengths we expect to demultiplex are set to be 1.53 μm and 1.57 μm. The wavelength space of 40 nm is only 1/6 that of the demultiplexer in [2]. Assuming all the voids share the same radius of 40 nm, the DBS optimization is conducted at three etching depths: $d_1 = 70\,\text{nm}$, $d_2 = 140\,\text{nm}$ and $d_3 = 220\,\text{nm}$ with $\alpha = 0.5$. Determined by the lag effect, $d_1 = 70\,\text{nm}$ is exactly the etching depth of the void when the input and output waveguides are fully etched, in which case the device can be realized by single step etching [5]. All the simulations are conducted by 3D FDTD with a grid in the x-y-z directions of 20 nm - 20 nm - 20 nm and the material dispersion is also considered.

The transmission curves at various depths are depicted in Fig. 2. We notice that the device performances at $d_3 = 220\,\text{nm}$ are superior to those in the shallow etching cases. This comes from the fact that with the increase of the etching depth, the refractive index can be engineered in a larger range, which in turn improves the division effect. The optimized demultiplexer at $d_3 = 220\,\text{nm}$ exhibits low loss of -1.7 dB and -1.2 dB, low crosstalk of -15.6 dB and -13.6 dB and broad 3-dB bandwidth of 59 nm and 91 nm at the two target bands, respectively.

TABLE I
DEVICE PERFORMANCES AT DIFFERENT ETCHING DEPTHS

Depth (nm)	1.53 μm band			1.57 μm band		
	Loss (dB)	Crosstalk (dB)	3dB Bandwidth (nm)	Loss (dB)	Crosstalk (dB)	3dB Bandwidth (nm)
70	-3.4	-4.1	49	-2.6	-6.4	61
140	-1.4	-12	60	-1.4	-11.5	53
220	-1.7	-15.6	59	-1.2	-13.6	91

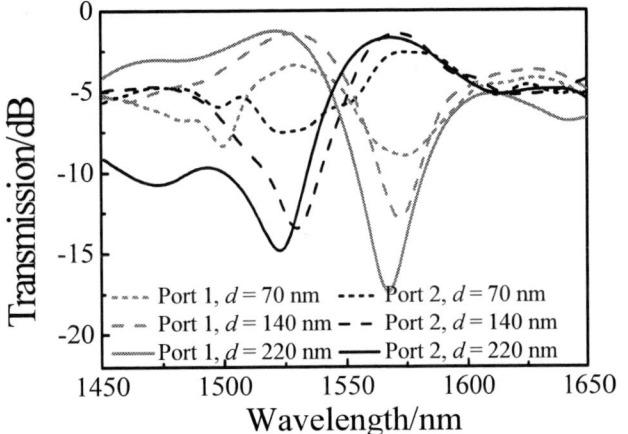

Fig. 2. Transmission of the output ports at different depths.

The circle distributions before and after the optimization at $d_3 = 220\,\text{nm}$ are illustrated in Fig. 3 (a) and (d), and their corresponding optical fields at the target wavelengths are shown in Fig. 3 (b), (c) and (e), (f) respectively. In the initial states, very little amount of light reaches the output ports, while after the DBS iterations, the rearranged circle voids efficiently guide the light to the desired ports along the relatively constrained paths. Note that the device is optimized to achieve a compromise between the insertion loss and the crosstalk. If the weight of the transmission efficiency in the FOM decreases, we would have different circle combinations and obtain lower crosstalk. In addition, the operation bandwidth at the 1.53 μm band can be further broadened if more wavelengths are involved to evaluate the FOM.

OECC/PS2016

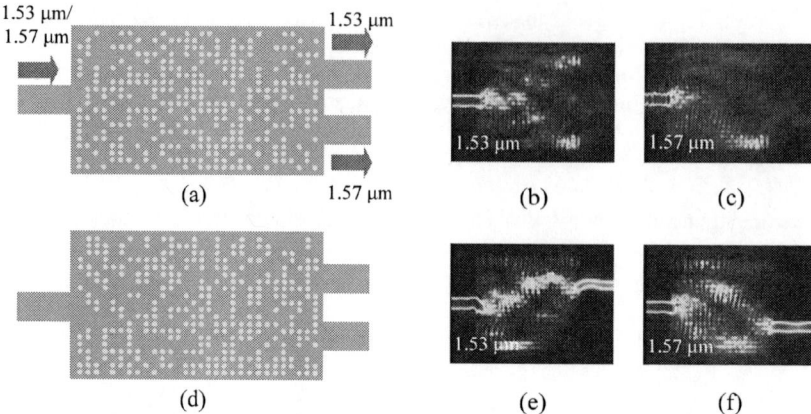

(a) (b) (c)

(d) (e) (f)

Fig. 3. (a) and (d) are the metametarial structures before and after the optimization. (b) and (c) are the optical fields of 1.53 μm and 1.57 μm before the optimization respectively. (e) and (f) are the optical fields after the optimization.

The fabrication tolerance is also analyzed here including the etching depth error and device thickness error. As indicated in Fig. 4 (a), the demultiplexer still functions efficiently when the device is etched 20 nm deeper, in which case the transmission remains almost constant around 1.53 μm and 1.57 μm bands. When the circles are 20 nm under-etch, the peak value slightly drops with the peak wavelength moves to the longer wavelength by 16 nm. The device thickness variation has a similar influence on the transmission curves. The increase of the thickness shifts the transmission curves towards longer wavelength, while the decrease of the thickness has an opposite effect. Up to 10 nm of the device thickness variation, the performance can be sustained before it falls below 60%. Therefore we can conclude that our design is robust against the fabrication imperfections.

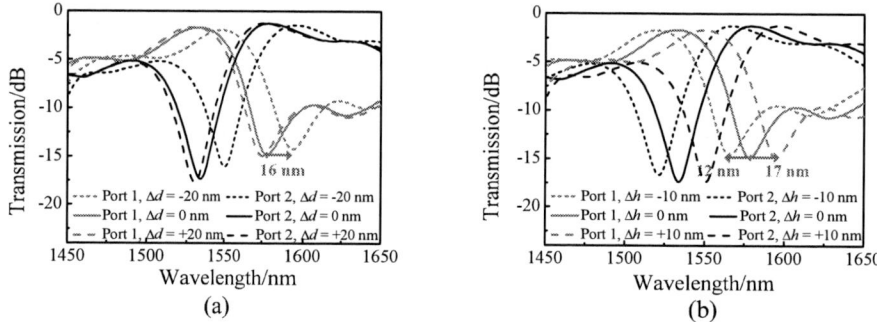

(a) (b)

Fig. 4 Transmission of the device as (a) the etching depth and (b) the device thickness changes.

IV. CONCLUSION

We propose and characterize a wavelength demultiplexer applying a novel PhC-like metamatirial structure. By altering the refractive index at the subwavelength scale, the device divides two wavelengths 40 nm apart with an ultra-compact footprint of 2.6 μm×3.8 μm. Low insertion loss (<-1.7 dB), low crosstalk (<-13.6 dB) and broad 3-dB bandwidth (>59 nm) are obtained after the DBS optimization. What's more, high pattern accuracy, flexible control of the etching depth and large fabrication tolerance are available thanks to the inherent property of the PhC-like structure. These characters make the metameterial-based demultiplexer a promising component for the next-generation chip-scale optical communications.

REFERENCES

[1] Pierre Pottier, Michael J. Strain, et al, "Integrated Microspectrometer with Elliptical Bragg Mirror Enhanced Diffraction Grating on Silicon on Insulator," ACS Photonics, vol. 1, pp. 430-436, April 2014.

[2] Alexander Y. Piggott, Jesse Lu, et al, "Inverse design and demonstration of a compact and broadband on-chip wavelength demultiplexer," Nature Photonics, vol. 9, pp. 374-377, June 2015.

[3] Bing Shen, Peng Wang, et al, "An integrated-nanophotonics polarization beamsplitter with 2.4 × 2.4 μm² footprint," Nature Photonics, vol. 9, pp. 378-382, June 2015.

[4] Luluzi Lu, Minming Zhang, Feiya Zhou, and Deming Liu, " An Ultra-compact Colorless 50:50 Coupler Based on PhC-like Metamaterial Structure," Optical Fiber Communication Conference 2016, Anaheim (USA), March 2016, pp. Tu3E.5.

[5] Jinghui Zou, Yu Yu, and Xinliang Zhang, "Single step etched two dimensional grating coupler based on the SOI platform," Optics Express, vol. 23, pp. 32490-32495, December 2015.

WA2-9

OECC/PS2016

Energy Analysis for Dynamic Cache Storage in Video on Demand Services

Zhongwei Tan[1], Chuanchuan Yang[1]*, Yu Yang[1], Zhaopeng Xu[1], Xinyue Wang[2], & Ziyu Wang[1]

[1]State Key Laboratory of Advanced Optical Communication Systems and Networks, Peking University, Beijing 100871, China
[2]College of Information and Electrical Engineering, China Agricultural University, Beijing 100083, China
E-mail: *yanchuanchuan@pku.edu.cn

Abstract: We evaluate the extra energy consumption for the adjustment of dynamic cache in VoD services, which is neglected by the former researches, and give guide to further optimization in the whole network.

Keywords: Network optimization; All-optical networks; Energy analysis; Video on demand services.

I. INTRODUCTION

In recent years, data traffic is growing in an exponential exploding trend, especially for online video streaming [1]. As new forms of access networks have been widely deployed, such as passive optical networks (PONs), large bandwidth makes it possible to the users enjoying innovative services through the Internet. Video services, including the internet protocol television (IPTV) with video on demand (VoD) service, are regarded as the most representative services in the network. To meet the demands of several services, content-delivery networks, or content-distribution networks (CDN), are being deployed over the telecom network infrastructures [2]. However, with the ever increasing energy consumption of communication technology, the networks must be designed to include energy-efficient measures.

Distributed storage systems are considered as one of the solutions that can save energy in transmission and satisfy users' requests at the same time. Content caching can help to mitigate the data growth by storing video content closer to the users in core, metro and access networks. Many strategies have been widely studied to optimize the transmission and storage costs in VoD systems. Jayasundara et al. first employ a method of hourly dynamically adjusting the cache storage to improve the energy efficiency in a VoD network [3]. Marco et al. propose a dynamic strategy in saving energy for VoD content delivery and replication in Integrated metro/access Networks [4]. However, most studies only consider the end users' requests and the storage in the most efficient way, which neglects the energy consumption caused by the dynamic adjustment of videos in different caches. In this paper, we analyze the energy of video adjustment among different caches by evaluating the transport energy in the process of adjustment, based on both weekday and weekend user behaviors. The result indicates that the extra energy occupies more than 15% in the whole network, which cannot be ignored in the calculation. As a consequence, an energy optimization direction is clear in the whole VoD distribution network according to our analysis, which builds the foundation for further energy saving.

II. SYSTEM AND ENERGY MODEL

Figure 1 illustrates a typical on-demand IPTV network model, which consists of Access Network, Metro Network, Core Network and Data Center. The passive optical network (PON) is employed in our Access Network part, in which optical network units (ONUs) and the optical line terminal (OLT) are connected with each other via optical fibers and passive splitters. The Metro network, providing connections to the Access Network and the Core network, includes edge routers and aggregation switches. The Core Network comprises several large core routers, which provide necessary routing functions.

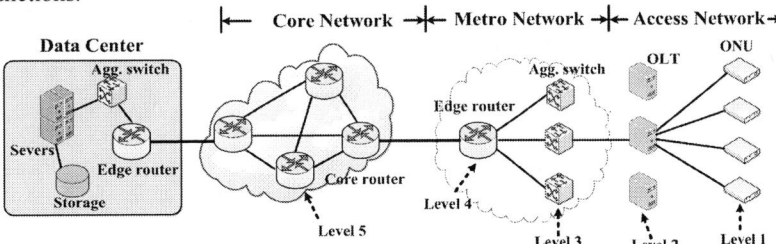

Fig. 1 VoD distribution storage network model.

The five level caches are set to store video content, as shown in Fig.1. They are in ONUs, OLTs, aggregation switches, edge routers, and core routers, respectively. When end users request video content from the caches, the transport energy is reduced. The energy consumption models in this paper refer to [3], which are as follows:

This work was supported by National Natural Science Foundation of China (No. 61275005), and State Key Laboratory of Advanced Optical Communication Systems and Networks, China.

$$P^{(1)} = \frac{P_{SD}'}{C_{SD}'} n_1 B , \qquad (1)$$

$$P^{(2)} = \frac{P_{SD}'}{C_{SD}'} n_2 B H_{CL} H_{RD} + \left[\frac{H_{CL} P_{OLT}}{C_{OLT}} + \frac{P_{ONU}}{C_{ONU}} \right] LD , \qquad (2)$$

$$P^{(3)} = \frac{P_{SD}}{C_{SD}} n_3 B H_{CL} H_{RD} + \left[\left(\frac{P_E}{C_E} + \frac{2P_{ES}}{C_{ES}} \right) \times H_{CL} H_{RD} H_{UU} + \frac{H_{CL} P_{OLT}}{C_{OLT}} + \frac{P_{ONU}}{C_{ONU}} \right] LD , \qquad (3)$$

$$P^{(4)} = \frac{P_{SD}}{C_{SD}} n_4 B H_{CL} H_{RD} + \left[\left(\frac{2P_E}{C_E} + \frac{2P_{ES}}{C_{ES}} \right) \times H_{CL} H_{RD} H_{UU} + \frac{H_{CL} P_{OLT}}{C_{OLT}} + \frac{P_{ONU}}{C_{ONU}} \right] LD , \qquad (4)$$

$$P^{(5)} = \frac{P_{SD}}{C_{SD}} n_5 B H_{CL} H_{RD} + \left[\left(\frac{2P_E}{C_E} + \frac{2P_{ES}}{C_{ES}} + \frac{hP_{WDM}}{C_{WDM}} + \frac{(h+1)P_C}{C_C} \right) \times H_{CL} H_{RD} H_{UU} + \frac{H_{CL} P_{OLT}}{C_{OLT}} + \frac{P_{ONU}}{C_{ONU}} \right] LD , \qquad (5)$$

where $P^{(i)}$ (i=1, 2, 3, 4, 5) is the total power consumption of a video served from Level i, the parameter B and D denote the average size of a single movie in bits and the average streaming rate in bits per second, respectively. L is the average number of simultaneous streams per second (i.e. the download rate). The parameters P_{SD}, P_{SD}', P_E, P_{ES}, P_{WDM}, P_C, P_{OLT}, and P_{ONU} are the power consumption of storage disk, hard disk driver (only in Level 1 and Level 2 caches), provider edge router, Ethernet Switch, WDM equipment, core router, OLT, and ONU, respectively. C_{SD} and C_{SD}' denote the capacities of the storage disk in bits, C_E, C_{ES}, C_{WDM}, Cc, C_{OLT}, and C_{ONU} denote the capacities of the corresponding equipment in bits per second. The factors H_{CL} and H_{RD} are employed to describe energy from the cooling and redundancy requirements, H_{UU} denotes the extra capacity deployed for the future growth by network providers, h is the average number of hops data traverse through in the core network before entering the metro network, and n_1, n_2, n_3, n_4, n_5 represent the numbers of content storage located at each level.

It is assumed that different users' behaviors are independent and the popularity has no relations with the video size. According to [3], the video popularity follows a Zipf-like distribution with parameter α approximately. For the kth most popular video, the request probability to is $P_k = \left(1/k^\alpha\right)/\sum_{i=1}^{N}\left(1/i^\alpha\right)$, and N is the number of the videos in the video library. The download rate of the kth most popular video can be described as $L_k = \lambda P_k T_k$, where λ is the number of requests to all videos and T_k is the time length of video k.

Particularly, this paper evaluates the extra energy consumption of storage adjustment among caches. We regard the cost for dynamic adjustment is mainly from the transport energy between different placement locations. From Eq. (1-5), the power consumption is divided into storage part and transmission part, thus we can calculate the energy caused by adjusting, using the latter. Moreover, download is the only way in this network, thus video contents can only transport from high level caches to low level caches. When a video is needed to be adjusted to a high level cache, it must be transported from the data center.

III. NUMERICAL ANALYSIS AND RESULTS

We analyze the storage distribution in caches and the adjusting costs in this section, based on weekday and weekend user behaviors. TABLE I shows the values of the parameters. In our simulation, the total number of users in the network is approach 10 million, thus n_1=9984000, n_2=312000, n_3=2000, n_4=100, and n_5=1. We set H_{CL}, H_{RD}, H_{UU} as 2, and h as 12. The Zipf skew parameter α is selected as 0.8. The total number of videos N in the library is 1000. The length of each video is set as a uniform distributed random value between 80min and 100min. The average streaming data rate D is taken as 10 Mbps.

TABLE I PARAMETERS SETTING [3]

Storage array	P_{SD}	4.9 kW	C_{SD}	240 TB	Core router	P_C	9.6 kW	C_C	640 Gb/s
Hard disk	P_{SD}'	6 W	C_{SD}'	1 TB	WDM	P_{WDM}	10.4 kW	C_{WDM}	7.04 Tb/s
Edge router	P_E	3.6 kW	C_E	128 Gb/s	OLT	P_{OLT}	48 W	C_{OLT}	2.5 Gb/s
Agg. switch	P_{ES}	2.4 kW	C_{ES}	360 Gb/s	ONU	P_{ONU}	5 W	C_{ONU}	75 Mb/s

Considering the different user behaviors in different days, we separate the weekday case and weekend case. The daily user download rate trends are chosen from [5], in which the file download rate in Wednesday is selected as the weekday case, while Sunday is the weekend case. To make download rate properly in this network, we just retain the trends of data stream and scale up the value. The maximum simultaneous streams per second in a weekday is set as 1.72 million, and the maximum in a weekend is set as 1.65 million (both at 23:00). In this paper, the simulation uses Poisson streams to describe the users' requests, which follow the daily user behavior in Fig. 2 and Fig.3 in average.

Figure 2 and 3 shows the percentage changes of videos stored in each level cache to achieve a dynamic storage, which is the most energy efficient strategy, regardless of the extra adjustment energy. The frequency for adjustment is once an hour. We can know that Level 1 and 5 caches store few videos for their large storage energy and transmission energy, respectively. It is clear that the most videos are stored and fluctuant in Level 3 and Level 4 caches along with the varying data stream. Overall, the percentage trends of videos in Level 3 cache are basically coincident with the simultaneous data streams at this level of data download rate in both cases.

The hourly energy consumption of the VoD services and the caches storage adjustment is presented in Fig. 4. In general, the VoD services cost more energy in weekend, especially in 1:00-5:00 and 9:00-21:00, and their energy consumption have similar trends to the simultaneous streams in both cases. We adjust the video storage in 5-level caches once an hour from 1:00 and the extra adjusting energy is only produced at the start of each hour. From Fig. 4, it is clear that the adjustment cost in some certain time cannot be neglected compared to the VoD service, such as 7:00, 13:00, and 17:00-23:00 for the weekday. Specifically, the energy consumption of VoD services in weekday and weekend are 4.22×10^{11} J and 4.78×10^{11} J, and the extra adjustment energy are 9.05×10^{10} J and 8.83×10^{10} J, respectively. This result indicates that the adjusting cost must be considered along with the end users' request when conducting the energy optimization for the whole network.

Considering the cost of cache storage adjustment, the optimization for the whole network becomes complex. In particular, the extra energy consumption in adjustment becomes obviously high for the content changes in Level 1 and Level 2 caches, often in the condition of large data stream. Thus, it's better to reduce dynamic storage in a busy network, which gives us guide to a further energy efficient strategy for the whole VoD distribution storage network.

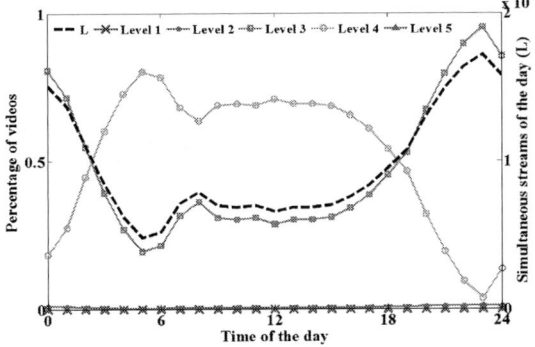

Fig.2 Percentage changes of videos stored in each level cache to make the network most energy efficient in a weekday.

Fig.3 Percentage changes of videos stored in each level cache to make the network most energy efficient in a weekend.

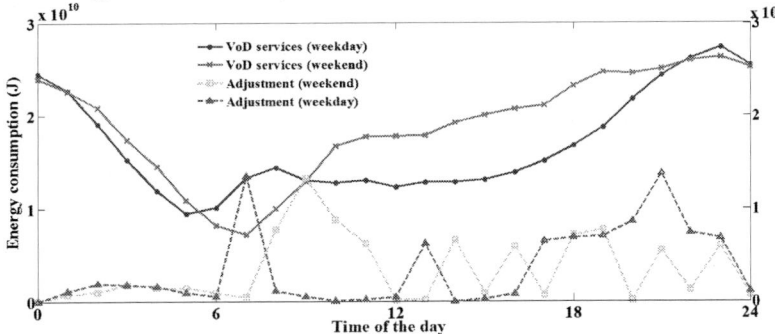

Fig.4 Energy consumption of VoD services and adjustment.

IV. CONCLUSIONS

In this paper, we evaluate the cost in dynamic cache adjustment in VoD services. The result shows that the adjustment costs occupy 17.7% and 15.6% in the whole energy consumption, under a typically parameter setting, for the weekday and weekend cases, respectively. It provides a direction for the entire network optimization.

REFERENCES

[1] Cisco, "Cisco Visual Networking Index: Global Mobile Data Traffic Forecast Update, 2014–2019," White Paper, Feb 2015.
[2] U. Mandal, P. Chowdhury, C. Lange, A. Gladisch, and B. Mukherjee, "Energy-efficient networking for content distribution over telecom network infrastructure," Opt. Switch. Netw., vol. 10, pp. 393-405, 2013.
[3] C. Jayasundara, A. Nirmalathas, E. Wong, and C. A. Chan, "Improving energy efficiency of video on demand services," J. Opt. Commun. Netw., vol. 3, pp. 870-880, 2011.
[4] M. Savi, G. Verticale, M. Tornatore, and A. Pattavina, "An Energy-Efficient Strategy for VoD Content Replication in Integrated Metro/Access Networks," IEEE Lat. Am. T., vol. 13, pp. 3627-3633, 2015.
[5] T. Bostoen, S. Mullender, and Y. Berbers, "LofoSwitch: An online policy for concerted server and disk power control in content distribution networks," Ad Hoc Netw., vol. 25, pp. 606-621, 2015.

WA2-90

Polarizing beam splitter with focusing ability based on sub-wavelength gratings

Wang Ying, Huang Yongqing*, Guo Yanan, Fang Wenjing, and Ren Xiaomin

State Key Laboratory of Information Photonics and Optical Communications, Beijing University of Posts and
Telecommunications, Beijing 100876, China
yqhuang@bupt.edu.cn

Abstract: *A polarizing splitter with focusing ability is proposed. The properties are numerically studied with Finite Element Method. Total transmittances are exceeding 78% at 1550 nm for both polarizations, and extinction ratio is about 10.3 dB.*

Keywords: *polarizing beam splitter, focus, sub-wavelength grating*

I. INTRODUCTION

The polarizing beam splitter (PBS), which can split an incident beam into two orthogonally polarized beams, is a key element in integrated photonics. It is widely used in optical switching [1], magneto-optic data storage [2], polarization-based imaging systems [3], and polarization-multiplexing [4]. The working principle of conventional PBSs is typically based on the use of the natural birefringence effect of crystals or the polarization selectivity of multilayer dielectric coatings [5]. Photonic crystal based PBS [6] is a plausible alternative, but the fabrication process is challenging. Sub-wavelength gratings (SWGs) function as efficient PBS devices have attracted more and more attention [7]. Their small sizes are advantageous to the miniaturization and integration of optical systems over conventional PBSs.

In this letter, we design a transmissive sub-wavelength grating PBS which simultaneously acts as a focusing element. The efficiencies and phase distributions for both polarized waves are precisely determined by using Rigorous Coupled-Wave Analysis (RCWA). The properties of the transmission and beam focusing are numerically studied with Finite Element Method (FEM). Simulation results show that the proposed PBS has a polarization extinction ratio of 10.3 dB. The entire transmittances are 93% for TM polarization and 78.3% for TE polarization, respectively.

II. THEORY AND STRUCTURE

The proposed PBS comprises two-layer 1-D SWGs (see Fig. 1). The upper layer is designed as a surface-etched periodic silica gratings structure, while the lower layer is non-periodic silicon gratings structure consisting of two different parts with focusing behaviors. The incident light with both polarizations can be separated by the upper gratings, and the two beams are respectively converged by the lower gratings of left and right sides.

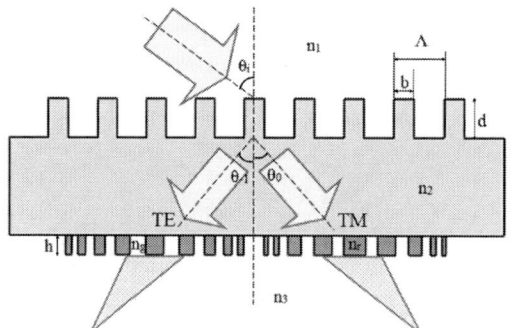

Fig. 1. Schematic of proposed two-layer gratings PBS device.

As is shown in Fig. 1, n_1, n_2, n_3, n_r and n_g are refractive indices of the air, silica layer, glass substrate, silicon grating bars and silicon grating grooves respectively. The major parameters of periodic SWG are period (Λ), bar width (b), and depth (d). The duty cycle f is defined as the ratio of bar width to period. A plane wave of wavelength λ is obliquely incident upon the upper periodic SWGs at Bragg angle of $\theta_i = \sin^{-1}(\lambda/(2\Lambda))$. The TE (electric field vector parallel to the grating grooves) and TM (electric field vector perpendicular to the grating grooves) polarized waves are diffracted in the -1st and 0th orders respectively. According to Equation (1), TE and TM beams have two distinct propagation directions due to the different diffraction angles.

$$n_2 \sin \theta_{d,m} = n_1 \sin \theta_i + m \frac{\lambda}{\Lambda}, \quad m = 0, \pm 1, \pm 2, \dots \tag{1}$$

where $\theta_{d,m}$ is the m-th diffraction angle.

While for the lower non-periodic structure, the focusing feature is realized by the phase front manipulation of transmitted beams [8]. Because of the large index contrast and sub-wavelength dimensions of the grating, the phase value as well as transmittance at a given point depends only on the local geometry. When a wave is incident on the non-periodic gratings, the transmitted light will develop a phase distribution along x axis. If the phase profile $\Phi(x)$ satisfies Equation (2), the beam can be focused.

$$\Phi(x) = \frac{2\pi}{\lambda}(\sqrt{x^2 + F^2} - F) + \Phi(x_0) \qquad (2)$$

where F is the focal length, and $\Phi(x_0)$ is the phase of the transmitted light at the center of grating with coordinate x_0. By designing the periods and duty cycles appropriately, the required phase conditions are fulfilled.

III. DESIGN PROCESS

In order to design the proposed device, both the periodic SWGs acting as a PBS and non-periodic SWGs as lenses will be considered. By using RCWA, diffraction properties of the upper periodic gratings can be calculated with the refractive indices of $n_1=1$ and $n_2=1.47$. The structural parameters are obtained with period of 900 nm, duty cycle of 0.45 and depth of 1.84 μm after optimization [9]. When a plane wave of wavelength λ=1550nm is incident at the angle of $\theta_i=58.91°$, the diffraction efficiencies of TE wave in the -1st order and TM wave in the 0th order are $\eta_{-1}^{TE}=93.12\%$ and $\eta_0^{TM}=99.61\%$, respectively.

The design process of the lower non-periodic SWGs is more complicated. To obtain the gratings which can map a particular phase profile on the transmitted beam, we should calculate the properties of the periodic gratings firstly. Fig. 2 shows the dependence of transmittance spectrum and phase on grating period (Λ) and duty cycle (f) with TM incidence using RCWA method. With the thickness (h) of Si ridges fixed at 620 nm and refractive indices of $n_r=3.48$ and $n_g=1$, the phase fully spans 2π range of variation within the high transmittance region (above 90%).

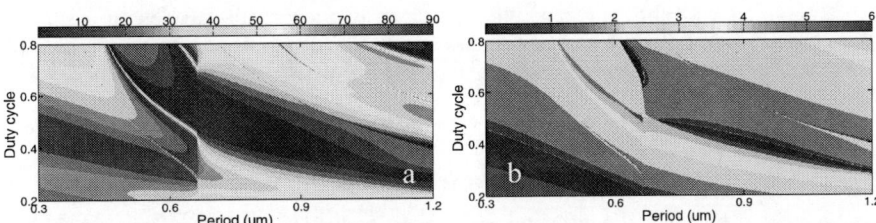

Fig. 2. (a) TM-transmittance spectrum and (b) phase shift at λ=1550nm for a periodic grating with h=620nm.

Referring to Fig. 2, we follow Equation (2) and design a 15 μm-width focusing element with focal length F=20μm. Fig. 3 (a) shows the grating structural parameters and the corresponding discrete phase distributions in detail. The method to design an SWG lens with TE polarized beam incidence is same as the case of TM. The detailed parameters are illustrated in Fig. 3 (b).

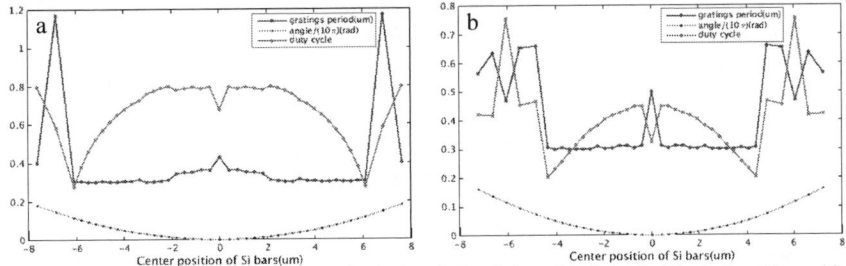

Fig. 3. The detailed grating periods (blue cycles), duty cycles (red cycles) and phase (black dots) of the designed lens with (a) TM, (b) TE polarization incidence.

IV. NUMERICAL SIMULATION AND DISCUSSIONS

The performance of the designed PBS is evaluated using the FEM, and responses of the device are separately calculated under TE and TM polarized plane waves incidence. The simulation results in Fig. 4 confirm a TE beam propagating to the left and a TM beam traveling to the right. The focal lengths from the bottom of Si bars are F=18.06μm for TM and F=16.83μm for TE. The full-width-half-maximum (FWHM) at focal plane is 1.75 μm for TM, while it's 1.91 μm for TE. The total transmittances are derived as 93% (for TM) and 78.3% (for TE) respectively. The polarization extinction ratios are 10.3 dB for TM beam and 13.8 dB for TE beam.

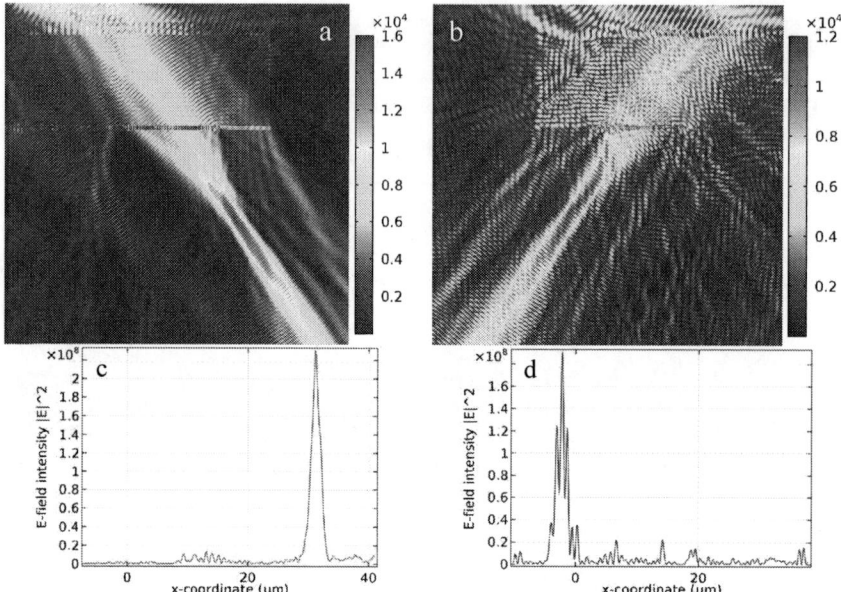

Fig. 4. Intensity distributions of E-field with (a), (c) TM incidence and (b), (d) TE incidence.

V. CONCLUSIONS

A polarizing beam splitter with focusing capability based on two-layer sub-wavelength gratings is proposed and demonstrated. We give a design approach in detail and obtain the focusing properties using FEM. Simulation results show that the total transmittances under TM and TE incidence are 93% and 78.3%, respectively. The theoretical polarization extinction ratios are 10.3 dB for TM and 13.8 dB for TE. This kind of PBS is easy to fabricate with standard photolithography, and the multifunctional concept may have a significant impact on integrated optics.

ACKNOWLEDGMENT

This work was supported by the National Natural Science Foundation of China (Nos. 61274044, 61574019 and 61020106007), the National Basic Research Program of China (No. 2010CB327600), Specialized Research Fund for the Doctoral Program of Higher Education (SRFDP) (No. 20130005130001), the Natural Science Foundation of Beijing (No. 4132069), Program of Key International Science and Technology Cooperation Projects (No. 2011RR000100), 111 Project of China (No. B07005), and the Program for Chang jiang Scholars and Innovative Research Team in University (No. IRT0609), MOE, China.

REFERENCES

[1] Q. W. Song, M. C. Lee and P. J. Talbot, "Polarization sensitivity of birefringent photorefractive holograms and its applications to binary switching," Appl. Opt. 31, 6240-6246 (1992).

[2] M. Ojima, A. Saito, T. Kaku, M. Ito, Y. Tsunoda, S. Takayama and Y. Sugita, "Compact magnetooptical disk for coded data storage," Appl. Opt. 25, 483-489 (1986).

[3] S. Eckhardt, C. Bruzzone, D. Aastuen and J. Ma, "3M PBS for high-performance LCOS optical engine," Proc. SPIE 5002, 106 (2003).

[4] H. Fukuda, K. Yamada, T. Tsuchizawa, T. Watanabe, H. Shinojima and S. Itabashi, "Silicon photonic circuit with polarization diversity," Opt. Express 16, 4872–4880 (2008).

[5] L. Li and J. A. Dobrowolski, "High-performance thin-film polarizing beam splitter operating at angles greater than the critical angle," Appl. Opt. 39, 2754-2771 (2000).

[6] V. Mocella, P. Dardano, L. Moretti and I. Rendina, "A polarizing beam splitter using negative refraction of photonic crystals," Opt. Express 13, 7699-7707 (2005).

[7] Y. Tang, D. Dai and S. He, "Proposal for a grating waveguide serving as both a polarization splitter and an efficient coupler for silicon-on-insulator nanophotonic circuits," IEEE Photon. Technol. Lett. 21, 242-244 (2009).

[8] F. Lu, F. G. Sedgwick, V. Karagodsky, C. Chase and C. J. Chang-Hasnain, "Planar high-numerical-aperture low-loss focusing reflectors and lenses using subwavelength high contrast gratings," Opt. Express 18, 12606-12614 (2010).

[9] B. Wang, "Numerical optimization of polarizing beam splitter gratings and modal explanation," Optica Applicata XL, NO. 1, 101-107 (2010)

WA2-91

Silicon Photonic TE Polarizer Using Adiabatic Waveguide Bends

Bruna Paredes, Humaira Zafar, Marcus S. Dahlem, and Anatol Khilo

Department of Electrical Engineering and Computer Science
Masdar Institute of Science and Technology
Abu Dhabi, United Arab Emirates

Abstract— A TE-pass silicon photonics polarizer based on adiabatic waveguide bends is proposed, with 30-55dB extinction and 0.3dB insertion loss predicted over 100nm-bandwidth. The fabricated proof-of-concept polarizer shows 20dB extinction and 1-2dB insertion loss over 50nm.

Keywords—polarizer, silicon photonics, bends, TE, TM

I. INTRODUCTION

Silicon photonic devices such as ring resonators are strongly polarization-dependent and are usually designed to work with TE polarization. Polarization diversity schemes can be used to achieve polarization-independent operation [1], [2], when the light is split into TE and TM components, the TM component is converted into TE, and both are processed by TE-optimized devices. In all these cases, the presence of undesired (TM) polarization leads to cross-talk and can degrade the system performance. An integrated on-chip polarizer can be used to remove the unwanted TM component of light, improving the system performance. For example, polarizers can be used as a part of a polarization splitter or rotator, in order to improve its extinction ratio [3].

There have been a number of polarizer designs proposed for silicon-on-insulator platform [4-13]. A polarizer can be implemented using a shallowly etched ridge waveguide [4]; the cross-section of the waveguide can be optimized to ensure high leakage loss of one of the modes [5-6]; tunnel into the silicon slab for one of the polarizations can be achieved [7]. The cross-sections of these structures are unconventional for silicon photonics and require specialized fabrication. A TM-pass polarizer can be realized using sidewall gratings without requiring specialized waveguide geometries [8]. Photonic crystals can work as polarizers [9], although with noticeable insertion loss. Plasmonics can be used to implement very short polarizers with very high extinction ratio [10-13], however, the insertion loss exceeds 1 dB. Plasmonic structures also require specialized fabrication (e.g. deposition of special metals), which can be

problematic for implementation in standard CMOS process. On-chip polarization splitters used in polarization diversity schemes can serve as polarizers too, if only one of their two outputs is used [1-3].

II. POLARIZER DESIGN

A. Concept

The goal of this work is to implement an integrated silicon photonic polarizer which does not require specialized fabrication and can be implemented in standard CMOS process. The polarizer needs to have high extinction ratio, i.e. high loss of the TM mode relative to the loss of the TE mode, and low insertion loss, i.e. low loss of the TE light.

The proposed polarizer is based on the fact that the bend loss for the TM mode can be much higher than for the TE mode, because the TE mode is much better confined in the waveguide for typical Si waveguide geometries, when the waveguide width is larger than its height. The bend radius can be optimized so that the loss of the TM mode is high, while the loss of the TE mode is low. A series of bends can be used in order to increase the extinction ratio.

B. Polarizer designs

Two versions of the layouts of the proposed polarizer are shown in Fig. 1. Note that in both of these layouts adiabatic bends are used, where the curvature of the bend linearly increases from zero to its maximum value, and then decreases to zero again. It is a key idea of the proposed polarizer that the adiabatic bends make it possible to achieve high TM loss while keeping the TE loss low. In case of non-adiabatic bends, the structures similar to the ones shown in Fig. 1 are implemented as series of arcs which have curvature with constant magnitude and alternating sign. For such bends, optical mode get scattered due to mode mismatch at the junction of arcs with different signs of the curvature, which leads to increased loss of the TE mode. Adiabatic bends eliminate this loss mechanism, leading to

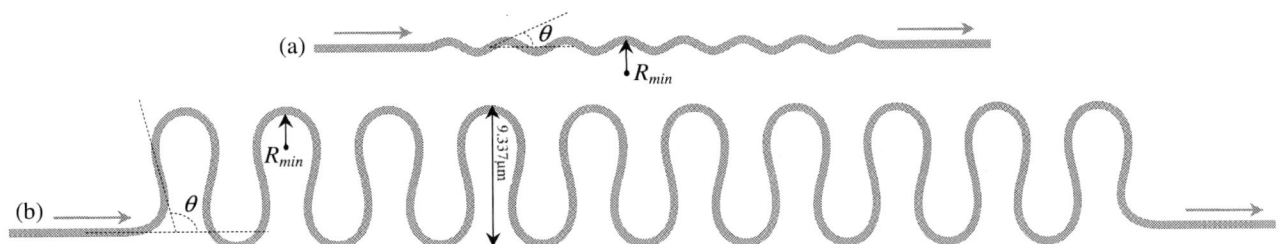

Fig. 1. The layouts of the proposed polarizer using adiabatic waveguide bends. (a) The layout of the proof-of-concept design, with the minimum bent radius $R_{min} = 1\mu m$ and angle $\theta = 30$ deg.; (b) the layout of the optimized polarizer, with $R_{min} = 1.5\mu m$ and angle $\theta = 105$ deg. The footprint of the optimized polarizer is approximately 68x10μm. The layouts are copied from the actual GDS files.

1009

much lower TE insertion loss for the same TM extinction ratio.

The two designs shown in Fig. 1 differ in the bending radius and the angles of the arcs. The polarizer shown in Fig. 1(a) has the minimum bend radius of adiabatic beds of R_{min}=1.0 μm, and the angle θ=30 degrees. This design was intended as a proof-of-principle demonstration and was not optimized for high performance.

The second design, shown in Fig. 1(b), has been optimized for high performance, and had the minimum bending radius R_{min}=1.5μm and the angle θ=105 degrees.

For both designs, the thickness of the Si layer was 220nm, and the core width was 500nm. The Si core was surrounded by silicon dioxide undercladding and overcladding, and the center wavelength was 1550 nm.

C. Simulation and Measurement Results

The proof-of-concept version of the polarizer, with the layout shown in Fig. 1(a), has been fabricated using 248nm optical lithography. The measurement results are presented in Fig. 2. The polarizer shows about 20 dB extinction ratio over 50 nm optical bandwidth, and about 15 dB extinction over 80 nm of bandwidth. The insertion loss is relatively high, varying form 1 to 2 dB over 100nm bandwidth. The high loss is attributed to the sub-optimal bend radius of R_{min}=1.0 μm, which was too low for the TE mode. While the measured extinction ratio is limited, it is expected that higher extinction ratio can be demonstrated for this polarizer with improved polarization control in the measurement setup.

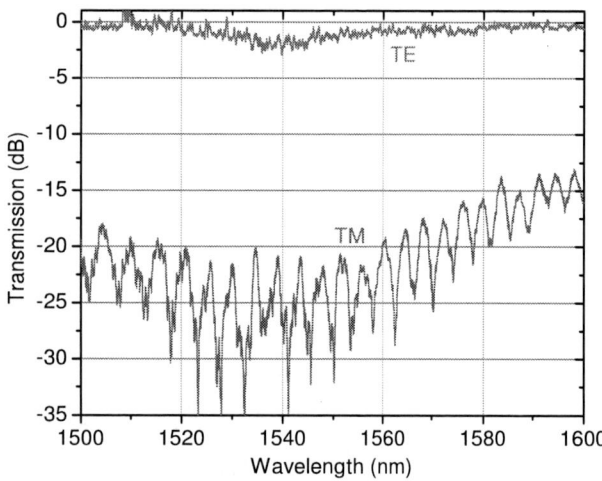

Fig. 2. The measured transmission of the TE and TM modes of the proof-of-concept polarizer design shown in Fig. 1(a), fabricated with 248nm optical lithography.

The polarizer design was further optimized resulting in the structure shown in Fig. 1(b). The performance of the polarizer simulated with 3D finite-difference time-domain method is presented in Figs. 3 and 4. It is clear that the design optimization made it possible to achieve very low insertion loss at the same time with high extinction ratio. According to Fig. 3, the TM mode extinction ratio varies within 30-55dB wange over 100nm of optical bandwidth. The insertion loss, with value too small to be discerned in

Fig. 3, is plotted in Fig. 4, showing about 0.3 dB loss over the entire 100nm band, with the worst-case loss of 0.36 dB.

The footprint of the optimized polarizer shown in Fig. 1(b) is approximately 68x10 microns. For applications requiring dense photonic integration this area might be too large, in which case the area of the structure can be reduced with some compromise in insertion loss and/or extinction ratio. If larger device area is available, higher extinction ratio can be achieved simply by increasing the number of bends.

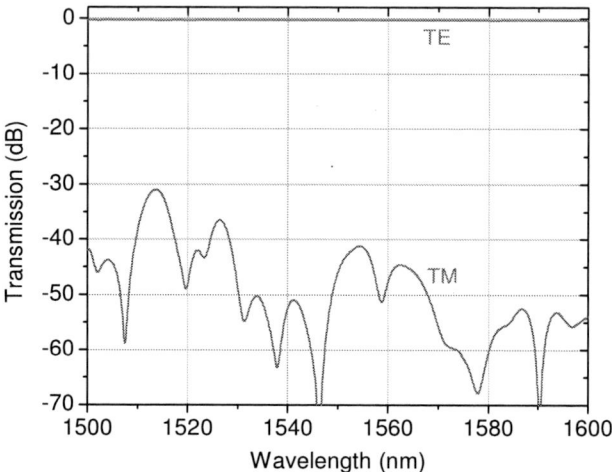

Fig. 3. The transmission of the TE and TM modes for the optimized polarizer design shown in Fig. 1(b), found using 3D FDTD simulations. The details of the TE insertion loss curve for this polarizer can be seen in Fig. 4.

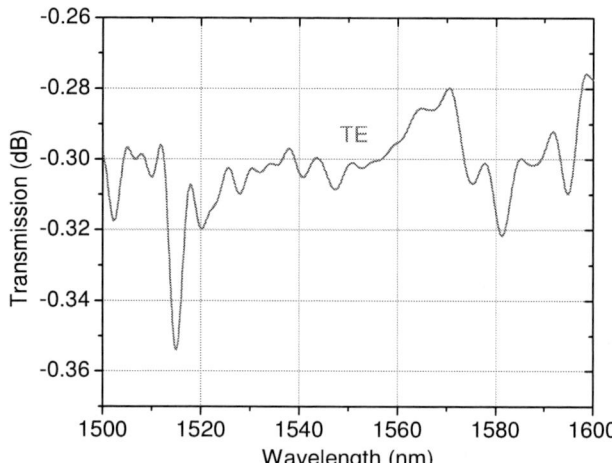

Fig. 4. The TE transmission of the optimized polarizer design shown in Fig. 1(b), found from 3D FDTD sumulations. The TE loss curve shown here is the same curve as in Fig. 3, only plotted in different scale.

III. CONCLUSIONS

This work presents a novel polarizer integrated on a silicon photonic chip. The design is simple, requires only one silicon layer, and does not require specialized fabrication techniques, which makes it suitable for integration on a CMOS chip. The polarizer is made of a series of waveguide bends, with different bend losses of the TE and TM modes leading to extinction ratio increasing commensurately to the number of bents. An essential idea of the proposed design is that the bends are adiabatic, which leads to significant reduction of the insertion loss. Measurements of the fabricated device prove feasibility of the proposed approach. The optimized polarizer design exhibits superior performance, with 30-55dB extinction ratio over 100nm of optical bandwidth, and the insertion loss as low as 0.3dB predicted with rigorous 3D FDTD simulations. Higher extinction ratios can be achieved by increasing the number of adiabatic bends.

IV. ACKNOWLEDGEMENTS

The authors are grateful to Jason S. Orcutt for the initial idea developed in this work. This work was partially funded by Mubadala Development Company (Abu Dhabi), Economic Development Board (Singapore) & GlobalFoundries, Singapore under the framework of 'Twinlab' project with A*STAR IME-Singapore.

REFERENCES

[1] T. Barwicz et al., "Polarization-transparent microphotonic devices in the strong confinement limit," Nature Photonics 1, pp. 57-60, 2007.

[2] H. Fukuda et al, "Silicon photonic circuit with polarization diversity," Opt. Express 16, 4872-4880, 2008.

[3] W. D. Sacher, T. Barwicz, B. J. F. Taylor, and J. K. S. Poon, "Polarization rotator-splitters in standard active silicon photonics platforms," Opt. Express 22, pp. 3777-3786, 2014.

[4] D. Dai, Z. Wang, N. Julian, and J. E. Bowers, "Compact broadband polarizer based on shallowly-etched silicon-on-insulator ridge optical waveguides," Opt. Express 18, pp. 27404-27415, 2010.

[5] S.I.H. Azzam et al., "Proposal of an Ultracompact CMOS-Compatible TE-/TM-Pass Polarizer Based on SoI Platform," IEEE Phot. Tech. Letters 26, pp.1633-1636, 2014.

[6] Q. Wang; S.-T. Ho, "Ultracompact TM-pass silicon nanophotonic waveguide polarizer and design," IEEE Photon. J. 2, pp. 49-56, 2010.

[7] S. Obayya and S. Azzam, "Ultra-compact resonant tunneling-based TE-pass and TM-pass polarizers for SOI platform," Optics Letters, in press.

[8] X. Guan, P. Chen, S. Chen, P. Xu, Y. Shi, and D. Dai, "Low-loss ultracompact transverse-magnetic-pass polarizer with a silicon subwavelength grating waveguide," Opt. Lett. 39, pp. 4514-4517, 2014.

[9] Y. Cui, Qi Wu, E. Schonbrun, M. Tinker, J.-B. Lee, Won Park, "Silicon-based 2-D slab photonic crystal TM polarizer at telecommunication wavelength," IEEE Photon. Tech. Lett. 20, p. 641, 2008.

[10] I. Avrutsky, "Integrated optical polarizer for silicon-on-insulator waveguides using evanescent wave coupling to gap plasmon–polaritons," IEEE J. Sel. Topics in Quant. Electron. 14, p. 1509, 2008.

[11] Guangyuan Li, Anshi Xu, "Analysis of the TE-Pass or TM-Pass Metal-Clad Polarizer With a Resonant Buffer Layer," J. of Lightwave Technology 26, pp. 1234-1241, 2008.

[12] M. Alam, J. Aitchison, and M. Mojahedi, "Compact hybrid TM-pass polarizer for silicon-on-insulator platform," Appl. Opt. 50, p.2294, 2011.

[13] X. Sun, M. Alam, S. Wagner, J. Aitchison, M. Mojahedi, "Experimental demonstration of a hybrid plasmonic transverse electric pass polarizer for a silicon-on-insulator platform," Opt. Lett. 37, p. 4814, 2012.

WA2-92

Design of a Si arrayed-waveguide grating using distributed Bragg reflectors

Takahiro Inaba, Hiroyuki Tsuda

School of Integrated Design Engineering, Graduate School of Science and Technology, Keio University, 3-4-1 Hiyoshi, Kohoku-ku, Yokohama, Kanagawa 223-8522, Japan
taka.ina@tsud.elec.keio.ac.jp

Abstract: *A silicon reflection type arrayed-waveguide grating using distributed Bragg reflectors is designed. 3 types of DBRs were optimized to have high reflectance, and those reflectivities were -1.81 dB, -0.22 dB, and -0.92 dB, respectively.*
Keywords: *Si arrayed-waveguide grating, distributed Bragg reflector*

I. INTRODUCTION

Arrayed-waveguide gratings (AWGs) have been used widely in wavelength division multiplexing (WDM) systems. The low crosstalk of less than -30 dB is usually required. The reflection type AWG (R-AWG) using Si waveguides was reported [1], [2], which has one slab region and reflectors at the edge of each waveguide in the arrayed waveguide. In comparison with a transmission type AWG (T-AWG), a R-AWG has smaller footprint, no bend in the arrayed waveguide, and shorter waveguide. Therefore, the phase error generated in the arrayed waveguide can be reduced, and the crosstalk performance can be improved. In this research, we proposed and optimized 3 types of distributed Bragg reflectors (DBRs) as a reflection facet and designed R-AWGs using them.

II. DESIGN OF DBR

The DBR structure is designed on the basis of Bragg condition as follows.

$$n_{core}(\Lambda - d) + n_{clad}d = \frac{k\lambda}{2} \quad (k = 1, 2, 3, ...) \qquad (1)$$

Where, n_{core} and n_{clad} are the effective refractive indices of the Si core and the SiO$_2$ clads, Λ is the period of the grating, d is the gap width, and λ denotes the central wavelength ($\lambda = 1.55$ µm). The schematic DBR is shown in Fig. 1, by arranging Si cores periodically to give Bragg reflection. Here, the refractive indices of SiO$_2$ and Si are 1.46 and 3.48, respectively.

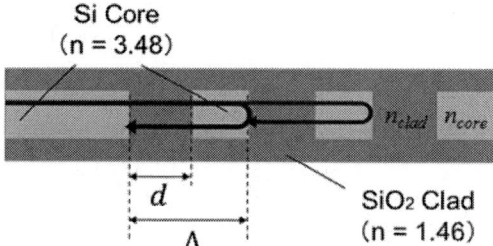

Fig. 1   Schematic structure of the DBR

Fig. 2(a), 2(b), and 2(c) show cross section drawings of the DBRs, the type-(a) DBR is the deeply etched Si wire waveguide, the type-(b) DBR is the shallowly etched Si rib waveguide, and the type-(c) DBR is the deeply etched Si rib waveguide. The width of the core, w, and the gap width, d were optimized to have high reflectance. A Si layer was 220-nm thick, and the height of Si-rib waveguide was 40 nm. In this case, w were set to be less than 0.8 µm for the type-(a) DBR, and less than 1.6 µm for the type-(b) and the type-(c) DBRs to cut off the higher-order modes of more than second-order. It is because the even-order mode is easily generated in a symmetric waveguide. The gap width d should be more than 0.3 µm for the deep etching and it should be more than 0.15 µm for the shallow etching due to the fabrication limitation. As shown in Fig. 3, the higher reflectance was obtained with smaller gap width. Therefore, d was set to be 0.3 µm in the type-(a) DBR and in the type-(c) DBR, and it was set to be 0.15 µm in the type-(b) DBR. The reflectances of the DBRs for the TE mode were calculated by finite difference time division (FDTD) method.

OECC/PS2016

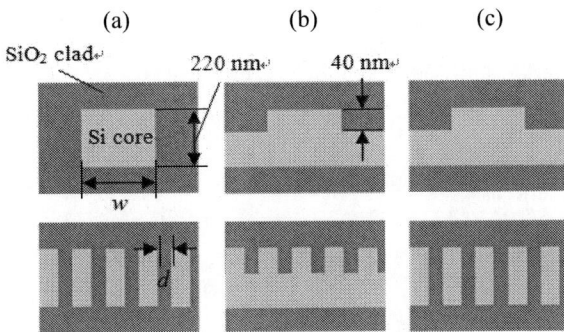

Fig. 2 Cross section drawings of (a) DBR by deeply etched Si wire waveguide, (b) DBR by shallowly etched Si rib waveguide, and (c) DBR by deeply etched Si rib waveguide

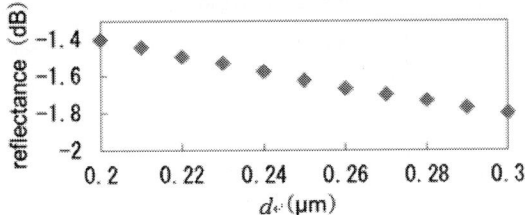

Fig. 3 Reflectance of the type-(a) DBR as a function of a gap width (w = 0.8 μm)

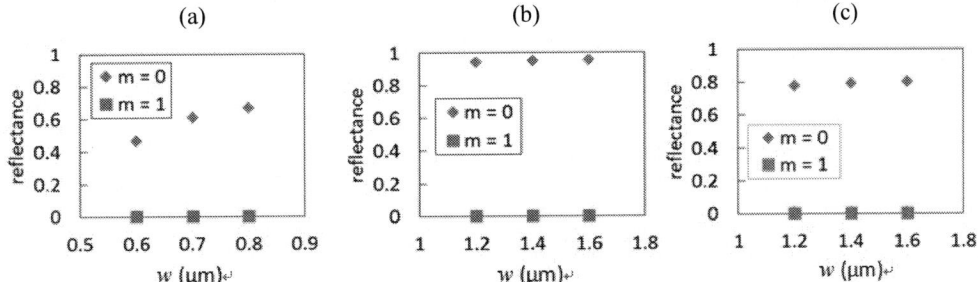

Fig. 4 Reflectance of DBRs when the fundamental mode light was input. as a function of a core width (m is the mode order)

Fig. 4(a), 4(b), and 4(c) show the reflectance of DBRs when the fundamental mode light was input. The higher reflectance obtained with wider core width for the fundamental mode. The highest reflectance for the type-(a) DBR was 67%, that for the type-(b) DBR was 96%, and that for the type-(c) DBR was 80%, respectively. In addition, any type of DBR didn't generate the reflection light of the first–order mode.

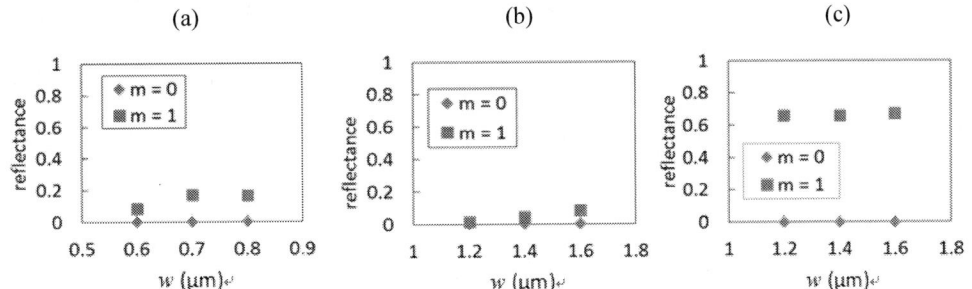

Fig. 5 Reflectance of DBRs when the first-order mode light was input, as a function of a core width

Similarly, the reflectance of DBRs when the light of the first-order mode was input, are shown in Fig. 5(a), 5(b), and 5(c). In the type-(a) DBR and the type-(b) DBR, the first-order mode reflection was suppressed to less than 17% and 8%. However, it was 67% in the type-(c) DBR.   Considering the high reflectance, w were chosen to be 0.8 μm in the type-(a) DBR, and 1.6 μm in the type-(b) DBR, and in the type-(c) DBR, respectively. Λ of those DBRs were 425 nm, 281 nm, and 419 nm, and the number of the periods to get higher reflectance enough were 6, 63, and 6, respectively.

1013

The reflectances as a function of wavelength for three types of DBRs are shown in Fig. 6. The type-(b) DBR has the high reflectance of about -0.22 dB at around the center wavelength; however the high-reflectance bandwidth is narrow. Type-(c) DBR needs much more periods than other two types. The type-(c) DBR has wider bandwidth with relatively high reflectance.

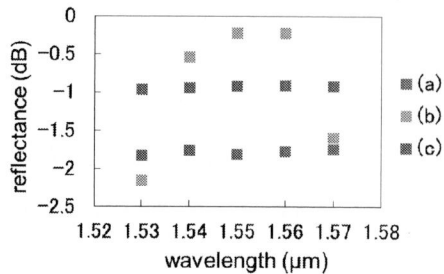

Fig. 6   Reflectances of the three types of DBRs as a function of wavelength

III. DESIGN OF REFLECTION TYPE AWG

A mask layout of the designed R-AWG is shown in Fig.7. The input waveguide was located at the outer side of the output waveguides to make the waveguide layout simple and compact. The AWG has 8 channels with 100 GHz channel spacing. The other parameters for the AWG are chosen as follows: the array length difference ΔL = 86.68 μm, the number of arrayed waveguides N = 24, the diffraction order m = 151, and the slab waveguide length f = 64.126 μm. The free spectral range was 920 GHz. The footprint of the AWG is about 1100 μm × 460 μm.

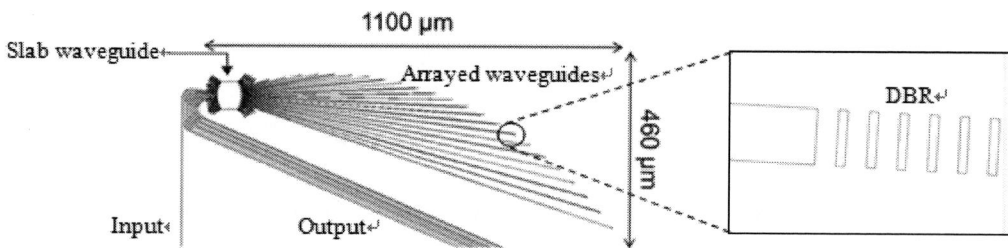

Fig. 7   Mask layout of the 8-channel, 100-GHz spacing R-AWG

IV. CONCLUSIONS

We have optimized the design of DBRs to have high reflectance and designed 8-channel, 100-GHz spacing R-AWGs using those DBRs. The reflectances of the three types of DBRs were -1.81 dB, -0.22 dB, and -0.92 dB, respectively, which were simulated by FDTD method. The total size of the AWG was 1100 μm × 460 μm.

REFERENCES

[1]   K. Okamoto, K. Ishida, Opt. Lett. , Vol. 38, No. 18, pp. 3530-3533, 2013.
[2]   D. Dai, X. Fu, Y. Shi, S. He, Opt. Lett. , Vol. 35, No. 15, pp. 2594-2596, 2010.

Frequency/energy-time Hyper-entangled Photon Pair Generation Based on a Silicon Micro-ring Resonator

Jing Suo, Wei Zhang*, Shuai Dong, Yidong Huang, and Jiangde Peng

Tsinghua National Laboratory for Information Science and Technology, Department of Electronic Engineering,
Tsinghua University, Beijing, 100084, P. R. China
*zwei@tsinghua.edu.cn

Abstract: We propose and demonstrate a scheme to realize frequency entanglement based on a silicon micro-ring resonator. The generated photon pairs also possess the property of energy-time entanglement. So hyper-entangled photon pairs are generated.

Keywords: quantum optics, nonlinear optics, integrated optics, four-wave mixing

I. INTRODUCTION

Correlated photon pairs can be generated in silicon micro-ring resonator through spontaneous four wave mixing (SFWM). In this process, two pump photons annihilate and a pair of photons is generated simultaneously. Thanks to the high nonlinear coefficient, narrow spontaneous Raman spectrum and resonance enhancement of silicon micro-ring resonator, correlated photon pairs of high quality can be realized[1]. In this work we propose and demonstrate a scheme, in which a silicon micro-ring resonator is placed in a Sagnac fiber loop, to realize frequency entangled photon pairs generation[2,3]. Since we use continuous wave lasers to generate pump lights, energy-time entanglement is an intrinsic property of the generated photon pairs. So we realized frequency/energy-time hyper-entangled photon pair generation based on a silicon micro-ring resonator[4].

II. THE SCHEME FOR THE HYPER-ENTANGLEMENT GENERATION

In our experiment, the silicon micro-ring resonator is placed in a Sagnac fiber loop constructed by a 1:1 fiber coupler, a manual alignment system and two polarization controllers(PC1 and PC2) to match the pump light into the quasi-TE mode of the silicon waveguide. Two pump lights coming from two tunable continuous wave lasers are injected into the fiber loop from two opposite direction, so the phase difference between the pump light in the clockwise and counterclockwise directions in the loop $\Delta\phi=0$. The photon pairs generated in the silicon micro-ring resonators in two opposite directions in the loop superpose at the 1:1 coupler, which leads to frequency entangled photon pairs coming out of the Sagnac fiber loop[5].

III. THE EXPERIMENT

The main part of the experiment setup is shown in Fig.1(a). Pump lights at resonance wavelengths 1551.5nm and 1556nm are used to generate photon pairs at resonance wavelengths 1542.58nm and 1565.08nm. Two circulators guide the generated photon pairs to the detection equipments. Before photon pairs are detected by single photon detectors (SPD1 and SPD2) and coincident counts are recorded by a time correlated single photon counter (TCSPC), a quantum state analyzer for frequency entanglement and energy-time entanglement is required to demonstrate the entanglement property. Quantum state analyzer for frequency entanglement is shown in Fig.1(b). A variable delay line, used to introduce a time difference, and PC3, used to maximize the interference at the 1:1 coupler, are inserted at two arms before the coupler. Tunable optical bandpass filters(TOBFs) are used to filter out signal and idler photons. When varying the time delay, the experiment of quantum spatial beating can be carried out. Quantum state analyzer for energy-time entanglement is shown in Fig.1(c). After being filtered out by TOBFs, signal and idler photons pass through two unbalance Mach-Zehnder interferometers (UMZI1 and UMZI2). The additional phases of the long arm of two UMZIs are α and β, respectively. The time difference between two arms of the UMZIs are larger than the coherence time of signal and idler photons and much shorter than the coherence time of the pump light, which can sort out the case that both photons pass through the long arms and both through short arms before reaching the SPDs. So there is a phase difference $\alpha+\beta$ between the two cases. When varying α while fixing β, the experiment of Franson-type interference can be carried out.

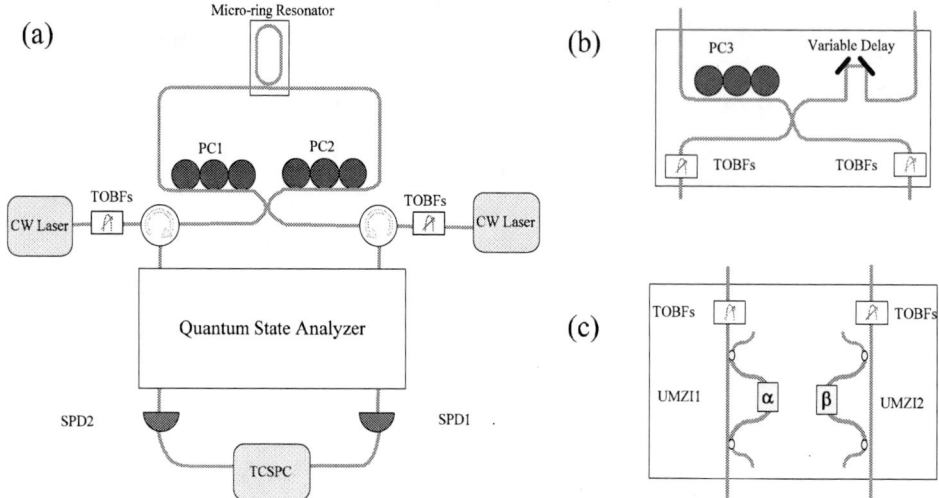

Fig. 1. The experiment setup. (a) The setup for the hyper-entanglement generation. (b) Quantum state analyzer for frequency entanglement. (c) Quantum state analyzer for energy-time entanglement.

The experiment result of frequency entanglement is shown in Fig.2(a). The fringe can be described by $y = A\left[1 - Ve^{-\omega_1 \tau} \sin(\omega_2 \tau + c)\right]$, where ω_1 and ω_2 are the spectral bandwidths of the resonances for generated photon pairs and the frequency difference between generated photon pairs. Since the fringe has a very broad envelope, we only measure it at three time delay periods. The raw visibility of the fringe is 66.7±3.4%.

The experiment result of energy-time entanglement is shown in Fig.2(b). We change α while β at 1.65rad and 2.86 rad, respectively. The raw visibilities of the two fringes are 78.8±2.6% and 85.8±3.5%, respectively, both higher than $1/\sqrt{2}$, the benchmark for the violation of Bell's inequality.

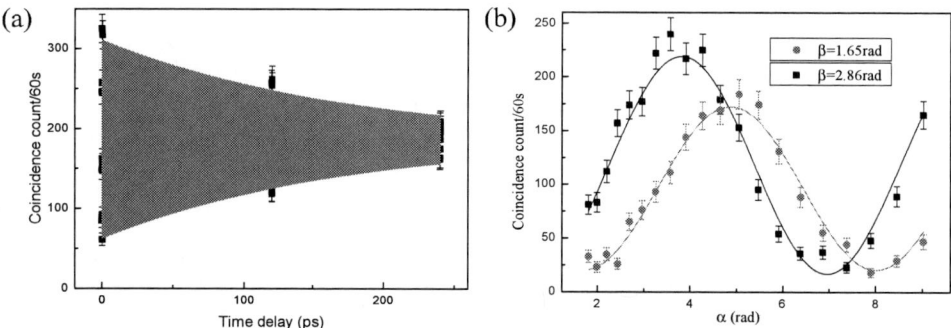

Fig. 2. The experiment result. (a) Result of quantum spatial beating (b) Result of Franson-type interferences under two non-orthogonal phases

IV. CONCLUSION

In this work, we realize the hyper-entanglement on frequency and energy-time based on a silicon micro-ring resonator placed in a fiber loop, showing that the silicon micro-ring resonators are ideal candidates for the complex photonic quantum state generation

ACKNOWLEDGMENT

This work was supported by 973 Programs of China under Contract No. 2013CB328700 and 2011CBA00303, the National Natural Science Foundation of China under Contract No. 61575102 and 61321004, Tsinghua University Initiative Scientific Research Program under Contract No. 20131089382.

REFERENCES

[1] Y. Guo, W. Zhang, N. Lv, Q. Zhou, Y. Huang and J. Peng, "The impact of nonlinear losses in the silicon micro-ring cavities on cw pumping correlated photon pair generation," Opt. Express 22(3), 2620-2631 (2014)

[2] Z. Y. Ou and L. Mandel, "Observation of spatial quantum beating with separated photodetectors," Physical Review Letters, 1988, Vol.61(1), pp.54-57

[3] X. Li, L. Yang, X. Ma, L. Cui, Z. Y. Ou and D. Yu, "All-fiber source of frequency-entangled photon pairs,"PHYSICAL REVIEW A 79, 033817(2009)

[4] S. Dong, L. Yu, W. Zhang, J. Wu, W. Zhang, L. You and Y. Huang, "Generation of hyper-entanglement in polarization/energy-time and discrete-frequency/energy-time in optical fibers," Scientific Reports 5, 9195 (2015)

[5] J. He, B. A. Bell, A. Casas-Bedoya, Y. Zhang, A. S. Clark, C. Xiong and B. J. Eggleton, "Ultracompact quantum splitter of degenerate photon pairs," Optica 2, 779-782 (2015)

WA2-94

On the Control of the Microresonator Optical Frequency Comb in Turing Pattern Regime via Parametric Seeding

Jinghao Wang, Minming Zhang*, Meifeng Li, Yuanwu Wang, Deming Liu
National Engineering Laboratory for Next Generation Internet Access System,
School of Optical and Electronic Information,
Huazhong University of Science and Technology, Wuhan 430074, China
mmz@hust.edu.cn and jinghaowong@hust.edu.cn

Abstract: We numerically investigate the influence of parametric seeding on the microresonator optical frequency comb in Turing pattern regime by using modified Lugiato-Lefever equation, and find it a powerful means of control of the comb state.
Keywords: optical frequency comb; microresonator; Turing pattern; parametric seeding; modulation

I. INTRODUCTION

Microresonator-based optical frequency comb (microcomb) has attracted wide interests from the scientific community since their first demonstration in an experiment a few years ago [1]. These frequency combs are generated by the four-wave mixing (FWM) process where new frequency components are created through the nonlinear interaction between the intracavity optical field and the host material of a monolithic microresonator [2]. Because of the simplicity of the generation mechanism, the compatibility of CMOS integration and the uniquely large tunable comb-line spacing from GHz to THz, microcomb has been proved to be a promising technique in applications such as precise frequency metrology, astronomical spectrograph calibration, ultrastable microwave generation, and so on [3].

The microcomb will generally experience a few chaotic stages before reaching a steady state, which can lead to nonequidistant combs with multiple radio-frequency beat notes that are not equal to single or multiple free-spectral ranges (FSR) [4]. It has been demonstrated that the optical frequency comb in Turing pattern regime has the most robust and coherent phase locking, which arises from the modulational instability [2], [5]. S. B. Papp et al. achieved a strictly equidistant microcomb spectrum through parametric seeding [3]. In this letter, we focus on the investigation of the influence of the frequency and intensity of the parametric seeding on the optical frequency comb in Turing pattern regime.

II. NUMERICAL ANALYSIS

To simulate the optical frequency comb formation with parametric seeding in microresonators, we modify the normalized Lugiato-Lefever equation [6] as follows:

$$\frac{\partial E}{\partial t} = \left[-1 - i\delta + i\sum_{k\geq 2}\frac{\beta_k}{k!}\left(i\frac{\partial}{\partial\theta}\right)^k \right] E + i|E|^2 E + f_0 + \sum_{n\geq 1} f_n \exp(j\Omega_n\theta + j\varphi_n) \tag{1}$$

where E is the normalized total intracavity field, t is the dimensionless slow time describing the evolution of the field over successive round trips, δ is normalized detuning between the pump and the closest resonant mode, β_k is the k-order dispersion coefficient, θ from $-\pi$ to π is the azimuthal angle along the circumference of the resonator, f_0 is the normalized external pump field and f_n is the normalized parametric seeding field with frequency Ω_n being an nonzero integer relative to the pump mode corresponding to Ω_n FSRs away from the pump mode and phase φ_n from 0 to 2π. Thus, there exist three additional free parameters f_n, Ω_n and φ_n in comparison to that of single pump case. For the sake of discussion, the simulation parameters are from a Silicon Nitride (Si3N4) microresonator with radius R = 100μm which has been used in both experimental and theoretical investigations [5-7]. Numerical simulations of (1) have been carried out by the split-step Fourier algorithm [6].

Firstly, for the sake of symmetry in the system, we inject a pair of symmetric seed signals which have the same amplitude and phase into the resonator and use $f = f_0 + f_1 \sin(\Omega\theta)$ to simulate the pump and seed signals. Fig. 1(a) shows the corresponding temporal and spectral evolutions. The normalized simulation parameters are: $\delta = 1$, $|f_0|^2 = 1.248$, $|f_1|^2 = 0.01248$, $\Omega = 2$, $\beta_2 = -0.04$ and the higher order dispersion items are neglected for simplification, with noisy initial condition. A Turing rolls Pattern are generated with 8 equal pulses corresponding to a Kerr comb with frequency space equal to 8 times the FSR before the seed signals are injected, as seen in Fig. 1(b). Subsequently, the seed signals are injected into the system at t = 440 ns and last 440ns before they are removed. An exciting change has been observed through this proceeding. Fig. 1(c) shows that new sub-combs have been generated with equidistant frequency space of 2 FSRs at ± 2, ± 4 modes, etc. and are superimposed on the primary combs. These new combs in the vicinity of the

This work is supported by the National High Technology Research and Development Program of China (863 Program, Grant No. 2015AA015504), the Major Project of Science and Technology Innovation Program of Hubei Province of China (Grant No. 2014AAA006), and National Natural Science Foundation of China (Grant No. 61107051).

primary comb lines get the largest gain because the parametric gain is centered around the primary combs. The temporal profile is heavily distorted but still has eight intensity peaks. Strikingly, the system returns to the initial state that coincides with the condition in Fig. 1(b) and the seeded sub-combs disappear after the seed signals are removed, as if the seeds have never been used, as seen in Fig. 1(d). It's notable that the same sub-combs can be generated when the same seed signals are injected into the system once again. What's more, we find that the space between the adjacent sub-combs is dependent on the frequency Ω of the seed signals. In the simulation above, 2 FSRs spaced sub-combs are generated that corresponds to Ω. Otherwise, one can get single FSR Kerr combs if seed signals with frequency $\Omega = 1$ are used. We attribute the phenomena above to the result of FWM between the primary combs and the parametric seeding.

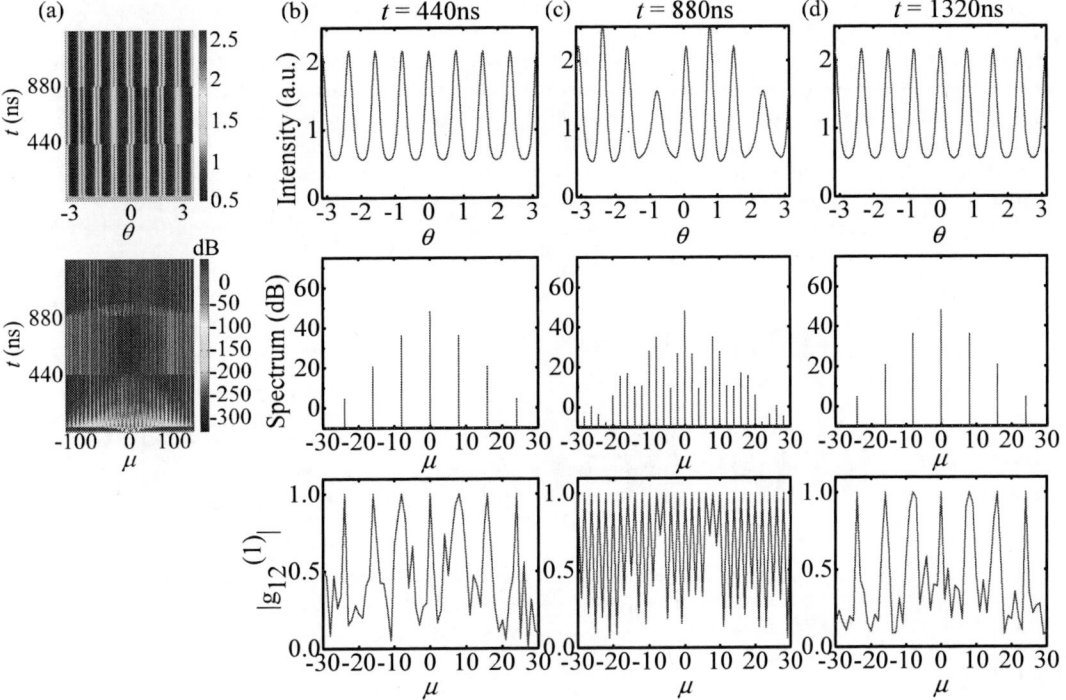

Fig. 1. (a) Temporal evolution of the waveform and spectrum with symmetric double-sideband parametric seeding. The seed signals are injected at $t = 440$ns and switched off at t = 880ns. The simulation parameters are: $\delta = 1$, $|f_0|^2 = 1.248$, $|f_1|^2 = 0.01248$, $\Omega = 2$, $\beta_2 = -0.04$ and the higher order dispersion items are neglected for simplification. (b-d) The corresponding temporal profile (top) and spectrum (center) at $t = 440$ns, 880ns and 1320ns, respectively. The line plots at the bottom show the degree of first-order coherence at a selected delay corresponding to 10 photon lifetimes, respectively.

To verify the stability of the simulated combs, we calculate the degree of first-order coherence using the formulas in [5]. The line plots at the bottom of Fig. 1(b)-(d) show the degree of coherence at a selected delay corresponding to 10 photon lifetimes. Before the seeds are injected, as seen in Fig. 1(b), the comb lines corresponding to the primary combs show high coherence, while other combs are incoherent. Once the sub-combs are generated by parametric seeding, their coherence rapidly increases to 1, i.e., full coherence, the same as that of the primary combs, as seen in Fig. 1(c). Subsequently, after the seed signals are removed from the system, the coherence of these sub-combs degrades rapidly and become incoherent as in Fig. 1(d). The simulation result manifests that symmetric double sideband seeds can generate fully coherent sub-combs and maintain them.

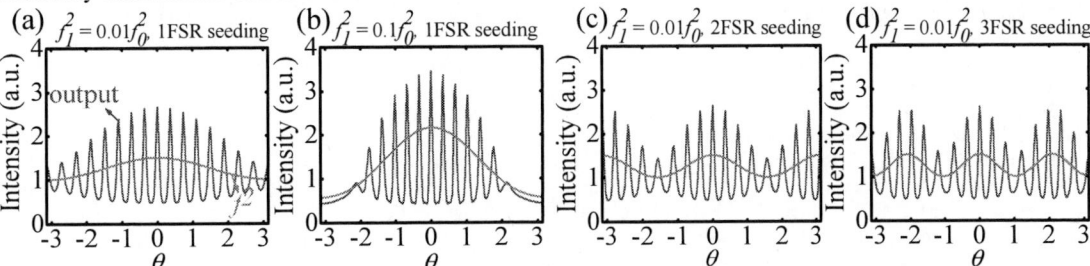

Fig. 2. Temporal profiles of the simulated frequency combs and pump corresponding to different frequency and intensity of the seed signals. The simulation parameters are the same as those in Fig. 1, except that $\beta_2 = -0.01$, which leads to a 16FSRs spaced Turing pattern.

Fig. 2 shows the effect of the seeded pump (in red) on the temporal envelope of the Kerr combs (in blue). The shape of the temporal envelope intensely depends on the intensity and frequency of the seed signals and the height of each pulse is proportional to the intensity of the seeded pump in the corresponding position in a way. But it should be noted that these pulses are still equidistant. Fig. 2(b) shows that the intensity of the equidistant pulses increases with increasing of the intensity of the seed signals, as compared with that in Fig. 2(a). When the frequency of the seed signals increases, the Turing pattern exhibits an additional periodicity, which is equal to the frequency of the seed signals, as shown in Fig. 2(c) and (d).

In order to reveal more general nonlinear dynamics behavior in the Turing pattern regime with parametric seeding, we simulate the single-sideband seeding situation subsequently. Now we use $f = f_0 + f_1 \exp(j\Omega\theta)$ to simulate the pump and seed signal. The density maps in Fig. 3(a) show the corresponding temporal and spectral evolutions for the same parameters as in Fig. 1. As expected, the single-sideband seed generates sub-combs similar to that in Fig. 1(b) as well. Different from the symmetric seeding situation, the intensity peaks in time domain are no longer fixed, but drift with time, somewhat analogous to the effect of higher-odd-order dispersion on Kerr combs [8]. The temporal profile will get back to the status where the seed signal is not injected, but with a time shift which depends on when we remove the seed signal. Fig. 3(b) shows the effect of the intensity and frequency of the seed signal on the temporal drift period. The higher the intensity or frequency of the seed signal is, the faster these temporal intensity peaks move. It's notable that the seed signal with reverse frequency but the same intensity induces an opposite drift behavior, that is, the intensity peaks move towards the different direction with the same velocity. This may explain why the symmetric double-sideband seed signals produce steady temporal dynamics as shown in Fig. 1.

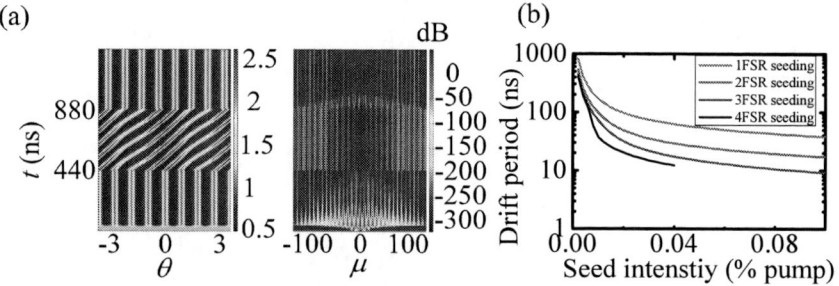

Fig. 3. (a) Temporal evolution of the waveform and spectrum with single-sideband parametric seeding. The seed signals are injected at t = 440ns and switched off at t = 880ns. The simulation parameters are the same as those in Fig. 1. (b) Temporal drift period versus the intensity and frequency of the seed signal.

III. CONCLUSIONS

We numerically investigate the influence of parametric seeding on the microresonator optical frequency comb in Turing pattern regime by using modified Lugiato-Lefever equation. Our result shows that parametric seeding can be a powerful means of control of the temporal profile and spectrum. Through double-sideband parametric seeding, coherent sub-combs can be generated with tunable comb interval dependent on the frequency of the seed signals. Meanwhile, the temporal envelope of the Kerr combs is modulated by the seeded pump, by which the height of the temporal intensity peaks can be controlled precisely. Otherwise, single-sideband parametric seeding can induce temporal drift of the intensity peaks, which will move faster with increasing of the intensity and frequency of the seed signal. By this way, the temporal position of Turing pattern can be tuned flexibly.

REFERENCES

[1] P. Del'Haye1, A. Schliesser, O. Arcizet, T. Wilken, R. Holzwarth & T. J. Kippenberg, "Optical frequency comb generation from a monolithic microresonator," Nature, vol. 450, pp. 1214–1217, Dec. 2007.

[2] Aurélien Coillet and Yanne Chembo, "On the robustness of phase locking in Kerr optical frequency combs," Opt. Lett., vol. 39, pp. 1529-1532, Mar. 2014.

[3] S. B. Papp, P. Del'Haye, and S. A. Diddams, "Parametric seeding of a microresonator optical frequency comb," Opt. Express, vol. 21, pp. 17615–17624, Jul. 2013.

[4] T. Herr, K. Hartinger, J. Riemensberger, C. Wang, E. Gavartin, R. Holzwarth, M. Gorodetsky, and T. Kippenberg, "Universal formation dynamics and noise of Kerr-frequency combs in microresonators," Nat. Photonics, vol. 6, pp. 480–487, Jul. 2012.

[5] M. Erkintalo and S. Coen, "Coherence properties of Kerr frequency combs," Opt. Lett., vol. 39, pp. 283–286, Jan. 2014.

[6] Coen S, Randle H G, Sylvestre T, et al., "Modeling of octave-spanning Kerr frequency combs using a generalized mean-field Lugiato–Lefever model," Opt. Express, vol. 38, pp. 37-39, Jan. 2013.

[7] Yoshitomo Okawachi, Kasturi Saha, Jacob S. Levy, Y. Henry Wen, Michal Lipson, and Alexander L. Gaeta, "Octave-spanning frequency comb generation in a silicon nitride chip," Opt. Lett., vol. 36, pp. 3398-3400, Sep. 2011.

[8] Shaofei Wang, Hairun Guo, Xuekun Bai, and Xianglong Zeng, "Broadband Kerr frequency combs and intracavity soliton dynamics influenced by high-order cavity dispersion," Opt. Lett., vol. 39, pp. 2880-2883, May. 2014.

Generalized polynomial chaos expansion for photonic circuits optimization

Daniele Melati[1], Tsui-Wei Weng[2], Luca Daniel[2] and Andrea Melloni[1]

[1] Dipartimento di Elettronica, Informazione e Bioingegneri, Politecnico di Milano, via Ponzio 34/5, Milano, Italy
[2] Research Laboratory of electronics, Massachusetts Institute of Technology (MIT), Cambridge, MA 02139, USA
daniele.melati@polimi.it

Abstract: A sparse combined generalized polynomial chaos model is proposed to characterize the impact of fabrication process variations in photonic circuits and perform design optimization. Simulations on a realistic example confirm the effectiveness of the technique.
Keywords: Photonic Circuits, Uncertainty Quantification, Polynomial Chaos Expansion, Stochastic Processes.

I. INTRODUCTION

Recent advances in photonic technologies are paving the way for market-ready applications, complex circuits, large production volumes and for the needs to reduce fabrication costs [1]. This evolution must face the requirements of yield and commercial quality standards, which make the management of the manufacturing process variations an urgent and unavoidable issue [2]. Hence, techniques for uncertainty quantification and for design optimization have become fundamental instruments for photonic circuit designers. Although Monte Carlo simulations are an approach commonly exploited to evaluate the impact of fabrication uncertainties on the functionality of the designed circuits [2], they suffer from slow convergence rate and they require long computation time when the complexity of the circuit grows beyond few building blocks and few uncorrelated random parameters. Stochastic spectral methods have recently been regarded as a promising alternative for statistical analysis. The key idea is to approximate the output quantity of interest with a set of polynomial basis functions, which is known as generalized polynomial chaos expansion. In addition, if the problem happens to have some special structure, such as sparsity, a sparse generalized polynomial chaos model can be constructed by solving L_1 minimization on generalized polynomial chaos coefficients [3]. The generalized polynomial chaos expansion is a surrogate that enables the fast post-computation of statistical information on the quantity of interest, such as probability density function (pdf) and moment information. The generated model is referred to as a combined generalized polynomial chaos model which expands the quantity of interest in terms of both process variation variables and design variables [4], where the cost function can be expressed in terms of design variables. This allows the model to be exploited also for circuit design optimization under process variations via gradient-based optimization algorithms. In particular, the cost function is a multivariate polynomial because generalized polynomial chaos bases are polynomials and a global polynomial optimization solver can be employed. In this work, we exploit the sparsity structure exhibited by many photonic circuits in order to construct a sparse combined generalized polynomial chaos model, analyze its related statistics and perform design optimization.

II. GENERALIZED POLYNOMIAL CHAOS FOR DESIGN ANALYSIS AND OPTIMIZATION

In order to define the analysis and optimization problem, let $u(\mathbf{x}, \xi)$ be the quantity of interest for a process subject to random variations. $\mathbf{x}$ is a constant vector of design or state variables while ξ is a random vector of process variation parameters. Generalized polynomial chaos expansion (generalized polynomial chaos) of $u(\mathbf{x}, \xi)$ is

$$u(\mathbf{x}, \xi) \approx \sum_{\alpha} C_{\alpha} \Psi_{\alpha}(\mathbf{x}, \xi) \qquad (1)$$

where $\Psi_{\alpha}(\xi)$ are multivariate orthogonal polynomials with respect to the joint pdf of ξ. $C_{\alpha}(\mathbf{x})$ are the corresponding coefficients with non-negative integer multi-index $\boldsymbol{\alpha}$. For some well-known distribution, such as uniform, gaussian, beta and gamma distribution, the corresponding bases are known to be Legendre, Hermite, Jacobi and Laguerre polynomial. The coefficient $C_{\alpha}(\mathbf{x})$ can be obtained either intrusively by stochastic Galerkin method or non-intrusively such as stochastic collocation method and regression method. The goal is to optimize the quantity of interest or a function of quantity of interest under given process variations and respecting design constraints. Since the quantity of interest is a random variable, it is reasonable to use its expectation or its associated functions as the objective in the optimization problem. In this case we minimize the expected mean-square-error with respect to the value originally designed u_0.

$$\text{minimize } E_{\xi}\left[\left(u(\mathbf{x}, \xi) - u_0\right)^2\right] \quad \text{subject to } a_i \leq x_i \leq b_i, i = 1, \ldots, m \qquad (2)$$

The mean-square-error can be expressed in terms of expectation $E_{\xi}[u(\mathbf{x}, \xi)]$ and variance of $V_{\xi}[u(\mathbf{x}, \xi)]$, which can be

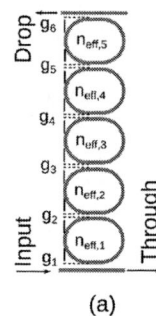
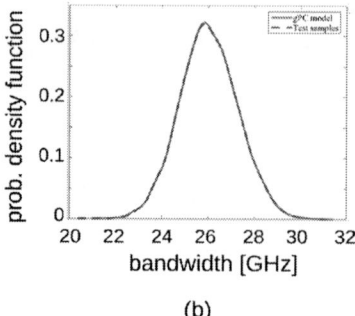

(a)   (b)

Fig. 1: (a) Sketch of the circuit considered in this work. (b) Comparison between the probability density function of the 3-dB bandwidth of the filter under process variations calculated with Monte Carlo samples and through the combined generalized polynomial chaos model.

directly derived from the generalized polynomial chaos expansion of $u(\mathbf{x},\xi)$. However, in this way $u(\mathbf{x},\xi)$ has to be re-expanded each time for each design point $\mathbf{x}$ inside the optimization loop, which could be very time consuming and only non-gradient based optimization algorithm such as evolutionary genetic algorithm employed. Alternatively, if $u(\mathbf{x},\xi)$ can be expanded both on $\mathbf{x}$ and ξ, then expectation and variance calculated through generalized polynomial chaos expansion will be functions of $\mathbf{x}$, and hence the cost function can be expressed as a function of design variables $\mathbf{x}$, where gradient-based optimizers can be employed. Since the constraints of x_i in Eq. (2) are boxed constraints, it is equal to express x_i as uniformly distributed variables in the interval $[a_i,b_i]$, and therefore $u(\mathbf{x},\xi)$ can be rewritten as

$$u(\mathbf{x},\xi) \approx \sum_{\alpha} C_{\alpha}\Psi_{\alpha}(\mathbf{x},\xi) = \sum_{\beta,\gamma} C_{\beta,\gamma}L_{\beta}(\mathbf{x})\Phi_{\gamma}(\xi) \qquad (3)$$

where L_{β} is a multi-variate Legendre polynomial with multi-index β, Φ is a multi-variate polynomial of ξ with multi-index γ and $\alpha = (\beta, \gamma)$. $u(\mathbf{x},\xi)$ is now referred to the combined generalized polynomial chaos model [4]. Under this form, the expectation and variance become a function of $\mathbf{x}$ and the optimization problem (2) becomes

$$\text{minimize}\quad V_{\xi}\big[(u(\mathbf{x},\xi))\big] + (E_{\xi}\big[(u(\mathbf{x},\xi))\big] - u_0)^2 \qquad \text{subject to}\quad a_i \le x_i \le b_i, i = 1,\dots,m \qquad (4)$$

Notice that now the objective functions are multivariate polynomials and a global polynomial optimization solver can be used to obtain global optimum with certificates [5,6].

III. OPTIMIZATION OF A FIFTH-ORDER COUPLED RING RESONATOR FILTER

The circuit considered here to demonstrate the potentiality of the proposed approach is the fifth-order ring coupled ring resonator filter shown in Fig.1 (a). The nominal design of the filter is obtained with a standard Butterworth synthesis technique [7]. With a bandwidth $BW_0 = 25.6$ GHz the resulting coupling coefficients for the six directional couplers are $K_1 = K_6 = 0.490$, $K_2 = K_5 = 0.042$, $K_3 = K_4 = 0.013$. All the five rings have the same length of 336.2 μm and the same the effective index $n_{eff,i} = n_{eff,0} = 2.23$. The nominal gap distance of the waveguides in the five coupling regions is $g_{0,i} = 0.3$μm. The free spectral range of the design is 400 GHz. In the described test case, random process variations are represented by a noise applied to the eleven relevant parameters of the filter, that is to the five effective indices and the six coupler gaps, being this a typical case also for real fabricated devices. Effective indices and gaps can hence be written as $n_{eff,i} = n_{eff,0} + \Delta n_{eff,i}$ and $g_i = g_{0,i} + \Delta g_i$. The vector of the random parameters is hence $\xi = [\Delta n_{eff,1},\dots, \Delta n_{eff,5}, \Delta g_1,\dots, \Delta g_6]$ with all the parameters considered in the example as zero-mean Gaussian distributed variables with standard deviation $\sigma_{neff,i} = 10^{-5}$ and $\sigma_{g,i} = 5$ nm, respectively. The design variables are $\mathbf{x} = [g_{0,1},\dots, g_{0,6}]$. Any other noise model can be used without affecting the proposed approach.

The considered quantity of interest $u(\mathbf{x},\xi)$ for the optimization problem (2) is the 3-dB bandwidth BW of the filter. In particular, the goal is to optimize the design value of the coupler gaps $g_{0,i}$ in order to minimize the fluctuations of the bandwidth under the described process variations, that is minimize the mean-square-error (MSE) of the bandwidth BW with respect to the designed value $u_0 = BW_0 = 25.6$ GHz. The design variables (gaps) must be comprised between 0.29 μm and 0.31 μm. Simulations were performed exploiting a commercial circuit simulator [8] and the combined generalized polynomial chaos model was prepared from a batch of 5000 Monte Carlo samples. Comparison between the probability density function (pdf) of the 3-dB bandwidth of the filter for the nominal design computed through Monte Carlo samples and surrogate model is shown in Fig.1 (b). As can be seen results are in perfect agreement, confirming that the generalized polynomial chaos model gives a good approximation of BW.

The generalized polynomial chaos model was hence used with a global optimizer to solve the problem of the minimization of the MSE [5,6]. The optimized solution resulted to be $\mathbf{x}_{opt} = [0.29, 0.31, 0.2947, 0.2931, 0.31, 0.29]$, which guaranteed a reduction of the MSE of about 18% when compared to the nominal design. A wider gaps range

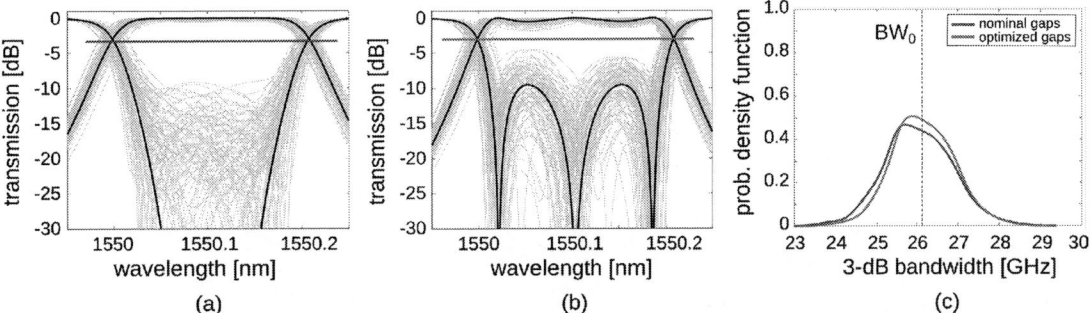

Fig. 2: Transfer function (without process variations) for the drop and through ports of the circuit in Fig.1(a) for the (a) nominal and (b) optimized design A (bold black lines). Gray lines represent 100 Monte Carlo simulations of the two transfer functions due to parameter uncertainty. Blue and red solid lines mark the 3-dB bandwidth in the two cases, respectively. (c) Probability density function of the 3-dB bandwidth of the filter for the nominal and optimized filter. $BW_0 = 25.6$GHz is the central frequency used for the nominal design.

would allow an optimized solution with a larger improvement of the MSE. Figure 2 (a) shows the transfer function of the filter at the drop and through ports for the nominal design while Fig.2 (b) refers to the optimized filter. Bold black lines represent the transfer function without any process fluctuations. As can be seen, the needs to reduce bandwidth fluctuations generated an optimized solution with predicted off-band isolation much smaller than the nominal case (about -10 dB instead of more than -50 dB). This is expected since no constrain was considered on this quantity during the optimization and the obtained solution has no guarantee on the performance of the off-band isolation. On the other hand it is interesting to notice how this difference largely reduces introducing parameter fluctuations in the simulation. In the two figures gray lines represent 100 Monte Carlo simulations of the two transfer functions (nominal and optimized) when each ring is affected by the described parameters fluctuations. Despite the small noise, in both cases the predicted fluctuations of the off-band isolation are quite broad. This effect largely exceeds the reduction due to the design and the minimum expected isolation is only 5 dB smaller for the optimized case compared to the nominal filter. Figure 2(c) shows the pdf of the 3-dB bandwidth in the two cases measured along the cuts marked with blue and red solid lines in Fig.2 (a,b), respectively. Dashed lines represents the design value BW0 = 25.6GHz. As can be seen the optimized filter has a bandwidth distribution less dispersed around the mean value, in agreement with the reduction of the MSE.

IV. CONCLUSIONS

In this work, we propose an approach to construct a sparse combined generalized polynomial chaos model that enables the efficient design optimization of photonic integrated circuits. We use it to analyze process variations problems in a coupled ring filter and to design the optimum gaps in the directional couplers of the filter in order to minimize expected fluctuations (MSE) of the 3dB Bandwidth.

ACKNOWLEDGMENT

The authors gratefully acknowledge the financial support of the Progetto Roberto Rocca Seed Funds.

REFERENCES

[1] Meint Smit, et al., "An introduction to InP-based generic integration technology," Semiconductor Science and Technology , 29(8):083001, 2014

[2] Daniele Melati, Eva Lovati, and Andrea Melloni, "Statistical Process Design Kits: analysis of fabrication tolerances in integrated photonic circuits," Integrated Photonics Research, Silicon and Nanophotonics, pages IT4A–5. Optical Society of America, 2015

[3] Emmanuel J Candès, Justin Romberg, and Terence Tao, "Robust uncertainty principles: Exact signal reconstruction from highly incomplete frequency information," Information Theory, IEEE Transactions on , 52(2):489–509, 2006.

[4] Brian M Adams, et al., "Dakota, a multilevel parallel object-oriented framework for design optimization, parameter estimation, uncertainty quantification, and sensitivity analysis," Technical report, Sandia National Laboratories (SNL-NM), Albuquerque, NM (UnitedStates), 2014

[5] Didier Henrion, Jean-Bernard Lasserre, and Johan Lofberg, "Gloptipoly 3: moments, optimization and semidenite programming," Optimization Methods & Software, 24(4-5):761-779, 2009

[6] Jean B Lasserre, "A semidenite programming approach to the generalized problem of moments," Mathematical Programming, 112(1):65-92, 2008.

[7] A. Melloni and M. Martinelli, "Synthesis of direct-coupled resonators bandpass filters for wdm systems," Lightwave Technology, Journal of, 20(2): 296–303, 2002

[8] Aspic, http://aspicdesign.com, http://www.phoenixbv.com

A Gaussian Beam Written Sampled Fiber Grating for Sub-ps Time Delay Lines

Weiqian Zhao[1], Haifeng Qi[2], Yunchuan Zhang[1], and Mingya Shen[1]

[1]Institute of Applied Photonic Technology, Yangzhou University.
180 Siwangting Rd., Yangzhou, China, 225002.
[2]Shandong Key Laboratory of Optical Fiber Sensing Technologies,
Laser Institute of Shandong Academy of Science, Shandong, China, 250014.
Email: myshen@yzu.edu.cn

Abstract: A sampled fiber grating made by Gaussian beam writing is reported. It has been designed for a signal processing function of WDM time delay lines with short delays. Experimental results are given for the grating spectra, and sub-ps time delays and linear profiles are achievable.
Keywords: Sampled fiber grating, signal processing.

I. INTRODUCTION

Signal processing using photonics provides functions and capability which conventional electronics cannot do [1]. In time domain signal processing, some applications require a large number of picoseconds (ps) or sub-ps time delays with linear time profiles. For example, when a phased array antenna system works for multiple beams and near the normal directions of the array at a frequency up to many GHz, it needs to compensate small time differences of the signal arriving at the antenna elements for high resolution detection [2-3]. Then, a time compensation profile with channel delays down to ps or smaller for a WDM-based signal processing is demanded. Some photonic methods [4-6] have been investigated to achieve short time delays with high performance on bandwidth, size, weight and electromagnetic compatibility guaranteed. The advantages of using sampled fiber gratings for delay lines include its intrinsic property of multiple wavelength channels, very small time delays, time profiles controllable by sampling function and grating parameters. This overcomes the disadvantages of using other photonic techniques such as superposed fiber gratings which have limited number of delay channels available from fabrication [7], or using a number of fiber gratings in parallel or series configuration or cutting fibers which cannot reach small enough fiber length differences for ps time delays [8]. This report presents an early result of a sampled fiber grating based on Gaussian beam writing in fabrication and measurement, and shows multiple sub-ps time delays with approximate linear profiles achievable.

II. SAMPLED FIBER GRATINGS FOR SUB-PS TIME DELAYS

Sampled fiber grating is formed by applying a sampling function defined refractive index changes to a uniform fiber grating [9]. Short time delays down to ps or smaller can be obtained from sampled fiber gratings. The time delay profiles can be linear or in other shapes determined by sampling function and grating parameters. Sampled fiber gratings have important applications, for example as time delay lines in high resolution signal processing in time domain, and ultra-high frequency signal processing by WDM-based transform from time domain to frequency domain.

A. Properties of Sampled Fiber Gratings for Short Time Delays

The refractive index change in the optical fiber of a sampled fiber grating can be expressed mathematically as:
$\delta n_e(z) = \delta n_{e0} S(z) Cos(4\pi n_e z / \lambda_0)$ where δn_{e0} is index change amplitude, $S(z)$ is sampling function, n_e is effective refractive index of the fiber, λ_0 is Bragg wavelength. The sampling function modifies the index distribution of a uniform grating, i.e. an index modulation. Using different sampling function, a sampled fiber grating has different index change distribution. According to coupled mode equations, mode coupling efficiencies are largely determined by sampling function as it controls index change amplitude. It can be known by Fourier transform, multiple reflectivity wavelength channels can be generated with the channel profile determined by the coupling efficiencies. The later is determined by the sampling function and other grating parameters.

When a grating is weak mode coupling in a single mode fiber of a certain length, according to the coupled mode equations, the reflectivity maximum of a fiber Bragg grating $R_{max}(\lambda) \propto Tanh^2[\kappa(\lambda)L]$, and minimum of time delay $\tau_{min}(\lambda) \propto [1/\kappa(\lambda)L]Tanh^2[\kappa(\lambda)L]$ (where L is grating length, and $\kappa(\lambda)$ is grating mode coupling function) [10]. In a sampled fiber grating, $\kappa(\lambda)$ is determined by both grating parameters e.g. index change and grating length, and sampling function. Investigating functions $R_{max}(\lambda)$ and $\tau_{min}(\lambda)$, it can be seen on the function curves that both R_{max} and τ_{min} change linearly with wavelength approximately in certain ranges of wavelength. Fig. 1 (a) and (b) show the approximate linear relationships of reflectivity and wavelength, and of time delay and wavelength for a 2cm long fiber Bragg grating. We also found such approximate linearity can be maintained well for grating reflectivity up to around 0.9. Therefore, if for a grating $R_{max}(\kappa)$ is linear, $\tau_{min}(\kappa)$ is also linear. This gives a prediction for a way to obtain many time delays with a linear profile of WDM channels from a fiber grating.

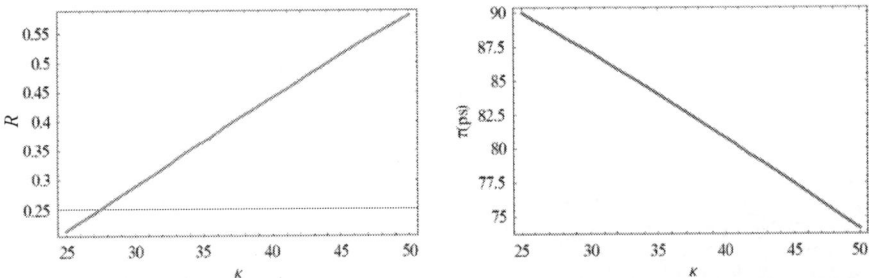

Fig. 1. Relationships of the grating for (a) reflectivity and coupling coefficient, and (b) time delay and coupling coefficient.

B. Gaussian Sampled Fiber Gratings

Based on the theory of Fourier transform, some functions have their Fourier transform in linear, approximate linear, or partly approximate linear functions. When such function is used as sampling function for sampled fiber gratings, a linear time delay distribution can be obtained from the grating. Some functions investigated for this purpose include Sinc squared, 2nd order rational, and Sinc Sinc-squared combined functions. However, in practice it is difficult to realize the WDM sampled fiber gratings by means of the functions which have feature of fast variations seen on the function curves. The Fourier transform functions of Gaussian function and Cosine 2nd rational combined function can offer approximate linear features in a certain variable range while the function does not vary fast with variable. In case of Gaussian function used in a sampled fiber grating, a section of linear time delay profile in the WDM channels in a Gaussian profile from the transform can be obtained. A Gaussian function sampled fiber grating can be made without the difficulty existing in realization of fast index variations in the fabrication of other gratings. Moreover, many lasers have the light beams in Gaussian distribution in intensity which offers simplification in the grating fabrication.

C. Fabrication and Measurement Results

In order to obtain many time delays with linear WDM channel profile, a sampled fiber grating based on Gaussian beam writing is investigated experimentally through the grating fabrication and measurement. First, a section of single mode fiber was processed by hydrogen loading in pressure of 12 Mpa for 2 weeks. A phase mask having a period of 1068.17nm was used for generating periodic effective index change along the fiber of 2cm length. A longitudinal force of 100g was applied on the fiber to stretch it so that the formed grating has Bragg wavelength around 1547.5nm. The grating was simply written by a laser beam with Gaussian intensity profile and wavelength 248nm. Light exposure for creating the grating was done by the laser shooting of 3×10^4 times. The laser beam was blocked using a slot of 0.3mm width for truncating the tails of the Gaussian beam so that the writing-grating areas were not disturbed by adjacent laser exposure operation. Thus, the profile of index changes in the fiber generated by Gaussian beam exposure is also in Gaussian, supposing that the index change is proportional to laser intensity. After 5 uniformly spaced fiber sections set as sampling areas were exposed, the Gaussian sampled fiber grating was fabricated. A Yokogawa AQ6370C optical spectrum analyzer was used to monitor and to measure the transmission and reflectivity spectra of the fiber grating created in the fabrication. Fig. 2 shows the fabricated fiber grating spectra with a Gaussian profile clearly displayed on

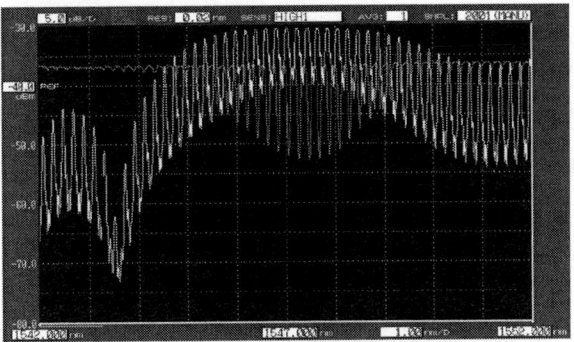

Fig. 2. A Gaussian sampled fiber grating measured for both transmission (purple) and reflectivity (yellow) spectra with a Gaussian profile seen on the transmission wavelength channels.

the transmission spectrum (purple colored). It can also be seen on the transmission spectrum that the wavelength channels near the center have strong reflectivity larger than 0.9. Since the light energy summation of transmission and reflection equals the energy input into the grating, the reflectivity spectrum (yellow colored) has a Gaussian profile, too.

Another Gaussian sampled fiber grating was made to investigate more linear-profile WDM channels to be obtained. Some fabrication parameters including the slot changed for 0.5mm in width, and the laser shooting was 1.5×10^4 times while kept others unchanged. A more linear reflectivity profile for more wavelength channels was obtained with the grating. This sampled fiber grating is weaker than the previous one which can be seen on the transmission spectrum (yellow colored). The measured results of the transmission and reflectivity (purple colored) spectra are shown in Fig. 3 (a), and the linear section of the reflectivity channels for clear observation is shown in Fig. 3 (b).

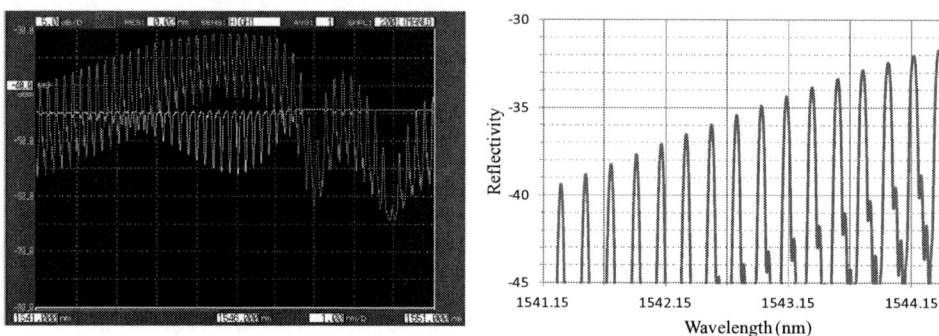

Fig. 3. (a) A Gaussian sampled fiber grating showing both transmission (yellow) and reflectivity (purple) spectra with a linear profile section on the reflectivity channels. (b) Clear observation for the linear profile of 16 wavelength channels on the reflectivity spectrum.

According to the analysis and conclusion in section "A" and Fig. 1, a time delay distribution with a linear profile on the wavelength channels can be obtained with this grating shown in Fig. 3. Previous work on the simulation of some sampled fiber gratings have shown 32 WDM channels with very linear time profiles and time delays of 0.25ps and 0.5ps generated by Sinc squared function sampled fiber gratings and 2nd order rational function sampled fiber gratings, respectively. Observing on Fig. 1 (a) and (b), with same coupling strength κ changes for example, from 34 to 48, when the reflectivity changes from 0.35 to 0.55, the corresponding time delay change range is about 9ps. Therefore, it is deduced that the time delays among the 16 wavelength channels in Fig. 3(b) are less than 1ps. This is to be confirmed by our next work, i.e. time delay spectrum measurement for the fabricated sampled fiber gratings.

III. Conclusions

A Gaussian sampled fiber grating for a time delay line has been investigated experimentally for achieving many short time delays with a linear profile in wavelength channels. The measured results on reflectivity spectrum show a linear channel profile, and a linear time delay profile obtainable among some wavelength channels. Simulation and analysis suggest that sub-ps time delays can be achieved from the sampled fiber gratings. The going-on work will be on high resolution time delay spectrum measurement, and investigating further the grating properties and their control in the fabrication.

References

[1] J. Capmany, J. Mora, I. Gasulla, J. Sancho, J. Lloret, and S. Sales, "Microwave Photonic Signal Processing", Journal of Lightwave Technology, vol. 31, no. 4, pp. 571 – 586, 2013.

[2] R. A. Minasian, and K. E. Alameh, "Optical-fiber grating-based beamforming network for microwave phased arrays," IEEE Transactions on Microwave Theory and Techniques, vol. 45, pp. 1513-1518, 1997.

[3] C. M. B. DePriest, A. Abeles, J. H. Delfyett, P. J. Jr., "10-GHz ultralow-noise optical sampling stream from a semiconductor diode ring laser," IEEE Photonics Technology Letters, vol. 13, pp. 1109-1111, 2001.

[4] R. T. Schermer, C. A. Villarruel, F. Bucholtz, and C. V. McLaughlin, "Reconfigurable liquid metal fiber-optic mirror for continuous, widely-tunable true-time-delay", OPTICS EXPRESS, vol. 21, no. 3, pp. 2741-2747, 2013.

[5] E. Udvary, T. Berceli, "New microwave / millimeter wave true time delay device utilizing photonic approaches", 2010 European Microwave Conference (EuMC), pp. 121-124, 2010.

[6] S. R. Davis, S. T. Johnson, S. D. Rommel, and M. H. Anderson, "Next Generation Photonic True Time Delay Devices as Enabled by a New Electro-Optic Architecture", Proc. of SPIE, vol. 8739, pp. 87390G1-15, 2013.

[7] M. Shen and R. A. Minasian, "Linearization Processing of a Novel Short Time-Delay WDM Superposed Fibre Bragg Grating", IEEE Photonics Technology Letters, vol. 14, no. 12, 2002, pp.1707-1709.

[8] A. Molony, L. Zhang, J. A. R. Williams, I. Bennion, C. Edge, and J. Fells, "Fiber Bragg-grating true time-delay systems: Discrete-grating array 3-b delay lines and chirped-grating 6-b delay lines," IEEE Transactions on Microwave Theory and Techniques, vol. 45, pp. 1527-1530, 1997.

[9] M. Ibsen, M. K. Durkin, M. J. Cole, R. I. Laming, "Sinc-sampled fiber Bragg gratings for identical multiple wavelength operation," IEEE Photonics Technology Letters, vol. 10, no. 6, pp. 842 - 844, 1998.

[10] E. Turan, "Fiber Grating Spectra", Journal of Lightwave Technology, vol. 15, no. 8, pp. 1277-1294, 1997.

Periodically Poled LiNbO₃ Ridge Waveguide for High-Gain Phase-Sensitive Amplifier

Tadashi Kishimoto[1,2], Koji Inafune[1], Yoh Ogawa[2], and Hitoshi Murai[1]

[1]Corporate Research and Development Center, Oki Electric Industry Co., Ltd., 1-16-8 Chuou, Warabi-shi, Saitama 335-8510, Japan.
[2]National Institute of Information and Communications Technology, 4-2-1 Nukui-Kitamachi, Koganei, Tokyo 184-8795, Japan.
kishimoto448@oki.com

Abstract: We have developed a periodically poled LiNbO₃ ridge waveguide device and experimentally demonstrated a phase-sensitive amplification with a high internal phase-sensitive parametric gain of 19.8 dB based on a cascaded SHG and DFG process.
Keywords: Periodically poled LiNbO₃, phase-sensitive amplifier, ridge waveguide

I. INTRODUCTION

Multilevel modulation formats such as a quadrature amplitude modulation (QAM) are very promising for future spectrally efficient transmission systems. Their transmission performances, however, might be more sensitive to the degradation of signal-to-noise ratio (SNR) and fiber-optic nonlinear effects. Against such difficulties, phase-sensitive amplifiers (PSA) are attracting much attention due to its low noise nature and phase regeneration capability [1]. PSA have been demonstrated based on a fiber-optic parametric amplifier (FOPA) by using a four-wave mixing (FWM) process [2], and OPA by a second-order nonlinear optical susceptibility $\chi^{(2)}$ effect in periodically poled LiNbO₃ (PPLN) waveguides [3]. Recently, the phase-sensitive signal processing for 16-QAM signal have been demonstrated by use of copier PSA [4], which enables both low noise amplification and phase-noise cancellation independently of modulation formats [5]. The PSA based on the PPLN waveguide device has several attractive features including the prospect of compact devices, ultrafast optical response, transparency to modulation formats, and immunity to the effect of stimulated Brillouin scattering of the pump light, which imposes performance limitations. The reported PPLN-based PSA scheme, however, required two separate PPLN waveguides for a second harmonic generation (SHG) and a deference frequency generation (DFG), respectively. In this report, we developed a PPLN waveguide device consisting of a ridge structure with an adhesion layer fabricated by mechanical lapping and polishing and diamond blade dicing. We experimentally demonstrated the phase-sensitive amplification and attained the internal phase-sensitive (PS) gain of 19.8 dB through a cascaded second harmonic generation and difference frequency generation (cSHG/DFG) process in the PPLN ridge waveguide. The cSHG/DFG scheme in the PPLN enables the parametric amplification such as PSA with single waveguide, and also has an advantage of the compatibility with optical fiber components in fiber-optic telecommunication, since the pump and signal lights are in the same wavelength band.

II. EXPERIMENTS AND DISCUSSION

The periodically inverted domain structure was fabricated by applying a high-voltage on a z-cut 5-mol% MgO-doped LiNbO₃ wafer through a liquid electrode of lithium chloride solution [6, 7]. We formed the 50-mm-long quasi-phase matched (QPM) grating of 18 μm period. After the fabrication of the periodically poled structure, the PPLN substrate was adhered to z-cut non-doped LiNbO₃ substrate with the adhesion layer as an under clad. The PPLN was reduced to 8 μm thickness by a lapping and polishing process. The ridge structure was formed by a precise diamond blade dicing. The ridge width is 8 μm as the waveguide core. Figure 1 represents the fabricated PPLN ridge waveguide device structure. Both end facets of the waveguide were anti-reflective coated for 1.5-μm-band wavelengths. We then fabricated the PPLN module with lens-coupled fiber pigtails. The average insertion loss of 1.2 dB is experimentally confirmed in the 1.5-μm-band range. The propagation loss is estimated to be below 0.1 dB/cm, which indicates smooth sidewalls of the ridge waveguide fabricated by the mechanical processing.

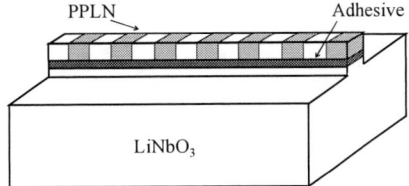

Fig. 1. PPLN ridge waveguide device structure.

OECC/PS2016

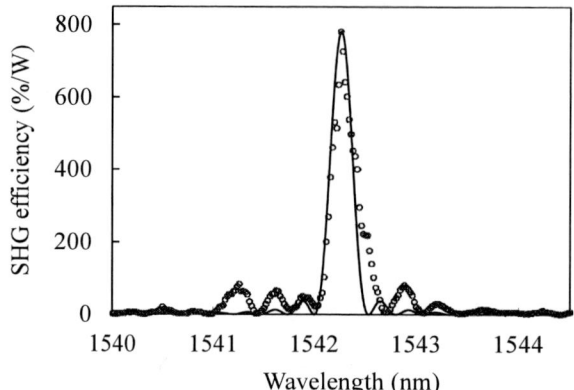

Fig. 2. The normalized SHG conversion efficiency as a function of the pump wavelength. Solid line is a calculated result.

Firstly, we investigated the SHG wavelength conversion efficiency of the PPLN waveguide. Figure 2 shows the dependence of the normalized SHG conversion efficiency on the pump wavelength. The QPM wavelength is 1542.2 nm and the normalized SHG conversion efficiency is 780%/W, which is estimated as the ratio of the output SHG power to the input pump power. The QPM bandwidth is 0.25 nm, which agrees well with a calculation as a function of $\text{sinc}^2(\Delta k L/2)$. Here, Δk is the wave vector mismatch and L is the interaction length. This consistency of the QPM bandwidth indicates that the uniform periodically poled structure was formed and fabricated waveguide was with a small fluctuation of the ridge width and thickness along the device length.

Next we demonstrated the PS operation and measured the PS gain. The experimental setup is illustrated in Fig. 3. 1.5-µm-band tunable continuous-wave lasers were used as signal and pump lights. The pump light was tuned to the QPM wavelength of 1542.2 nm. In the first stage, phase-correlated signal, pump, and idler waves were generated by using a FOPA through a non-degenerate FWM process in a highly nonlinear optical fiber (HNLF) as a phase-insensitive amplifier (PIA). The relative phase among signal, pump, and idler waves was controlled by their optical path lengths, respectively, after they were divided into different optical paths by a wavelength division multiplexing (WDM) coupler. Before these waves were fed into the PPLN module, the input pump power was amplified by an erbium-doped fiber amplifier (EDFA) up to 31.4 dBm. In the second stage, the phase-correlated signal, pump, and idler waves were injected into the PPLN module for the PSA through the cSHG/DFG interaction. Figure 4 shows the PSA output spectrum under the in-phase condition. The internal PS gain in the PPLN waveguide is 19.8 dB as compared with the output signal power with no injection of pump and idler waves used as a reference signal level. We also performed the PI operation by feeding pump and signal waves into the PPLN module. The internal PI gain of 14 dB is obtained at 31.4 dBm pump in comparison with the reference signal level. Therefore the additional gain of 5.8 dB is attained from PI to PS operation. The PS operation was successfully confirmed by using the highly efficient PPLN waveguide device. Taking into account the module insertion loss of 1.2 dB, we obtain the net PS gain of 18.6 dB estimated as the ratio of the output signal power to the input signal power of the PPLN module. To the best of our knowledge, this is the first demonstration of PPLN-based PSA with high gain enabling inline compensation of optical loss of 80-km-long optical fiber.

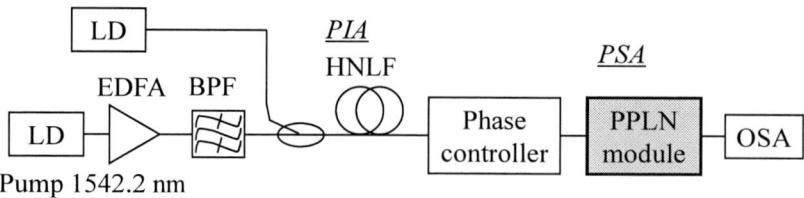

Fig. 3. Experimental setup for PSA. LD: Laser diode, OBF: Optical bandpass filter, OSA: optical spectrum analyzer.

OECC/PS2016

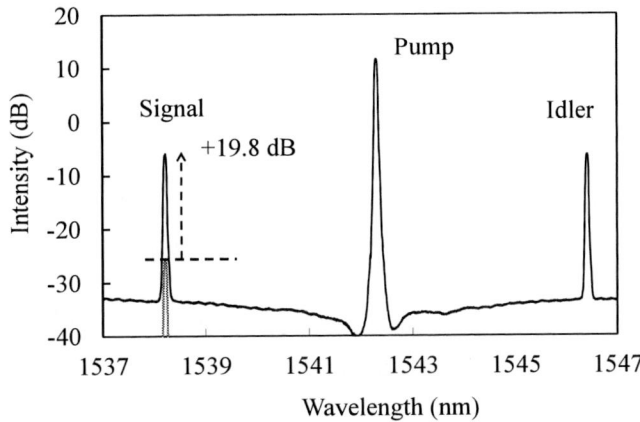

Fig. 4. In-phase PSA output spectrum of the PPLN module. The gray line represents the signal output spectrum without pump and idler waves

III. CONCLUSIONS

We have developed the PPLN ridge waveguide module and experimentally demonstrated the PSA based on the cSHG/DFG process. The internal PS gain of 19.8 dB with the additional gain of 5.8 dB from the PI operation is obtained in the PPLN ridge waveguide. The net PS gain is 18.6 dB, since the insertion loss of the PPLN module is 1.2 dB. The high efficiency of the optical nonlinear process and the low propagation loss are achieved by fabricating the ridge waveguide device with a 50-mm length. The investigation of the noise performance of the PSA and the further improvement of the PS gain of the PPLN waveguide for the practical use is future subjects. We believe that the developed PPLN waveguide module is useful for optical fiber transmission links.

REFERENCES

[1] J. A. Levenson, I. Abram, T. Rivera, and P. Grangier, "Reduction of quantum-noise in optical parametric amplification," J. Opt. Soc. Am. B, vol. 10, pp. 2233-2238, Nov. 1993.

[2] P. A. Andrekson, C. Lundström, and Z. Tong "Phase-Sensitive Fiber-Optic Parametric Amplifiers and Their Applications," Proc. of ECOC 2010, We.6.E.1.

[3] T. Umeki, O. Tadanaga, A. Takada, and M. Asobe, "Phase sensitive degenerate parametric amplification using directly-bonded PPLN ridge waveguides," Optics Express, vol. 19, pp. 6326-6332, Mar. 2011.

[4] T. Umeki, O. Tadanaga, M. Asobe, Y. Miyamoto, and H. Takenouchi, "First Demonstration of High-Order QAM Signal Amplification in PPLN-based Phase Sensitive Amplifier," Proc. of ECOC 2013, PD1.C.5.

[5] B. Corcoran, S. L.I. Olsson, C. Lundström, M. Karlsson, and P. A. Andrekson, "Mitigation of Nonlinear Impairments on QPSK Data in Phase-Sensitive Amplified Links," Proc. of ECOC 2013, We.3.A.1.

[6] T. Kishimoto and K. Nakamura, "Periodically Poled MgO-doped Stoichiometric LiNbO$_3$ Wavelength Converter With Ridge-Type Annealed Proton-Exchanged Waveguide," IEEE Photon. Technol. Lett., vol. 23, pp. 161-163, Feb. 2011.

[7] T. Kishimoto and K. Nakamura, "Loss-less Wavelength Converter with Ridge-Type Annealed Proton-Exchanged Periodically Poled MgO-Doped Congruent LiNbO$_3$ Waveguide," Jpn. J. Appl. Phys., vol. 51, pp. 012203, 2012.

WA2-98

Preliminary research of MCF Coupling to Optical Mode Switch Configuration

Xiaoyang Cheng, Luke Himbele, Ryan Imansya, Haisong Jiang and Kiichi Hamamoto
Interdisciplinary Graduate School of Engineering Sciences, Kyushu University
6-1, Kasugakouen, Kasuga, Fukuoka 816-8580, Japan
2ES14236Y@s.kyushu-u.ac.jp

Abstract: *Preliminary research of MCF coupling to optical mode switch is reported. Using single lens system like conventional SMF coupling, output mode coupling is confirmed between MCF and optical mode switch with coupling loss of 9dB.*
Keywords: optical mode switch, coupling loss, MCF

I. INTRODUCTION

Optical switches have been intensively studied for solving high power consumption issues [1, 2]. The optical router based on an optical switch can solve the extra energy exchanges problem between optical signals and electric signals [3]. The conventional type of optical space switch have had a matrix size problem, as it requires the extra S-bending region to integrate with fiber arrays [4, 5]. Based on this consideration, we have proposed the novel optical mode switch in which the optical mode is switched to replace the conventional port information. Instead of fiber arrays, optical mode switch only requires a single fiber connection which can be a few mode fiber or a multi core fiber (MCF) at both input and output ports [4, 5]. In this case, the coupling configuration between MCF and optical mode switch can work similarly like conventional SMF coupling. However, the coupling performance between optical mode switch and MCF has never previously been studied.

In this work, we report the research on the coupling performance between a 2 port optical mode switch and a MCF. A near field pattern with two separated light spots export from MCF is observed. The detection of such a near field pattern certificates that the optical mode switch is coupled to a "single" MCF. The estimated coupling loss evaluated by experimental result is 9dB.

II. OPTICAL MODE SWITCH AND MCF CONFIGURATION

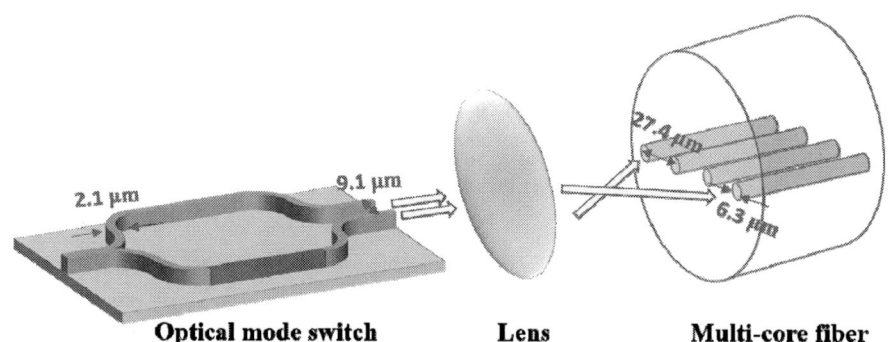

Optical mode switch　　　**Lens**　　　**Multi-core fiber**
Fig. 1. The coherent optical mode switch system.

As shown in Fig. 1, the optical mode switch between the 0th and 1st modes consists of Y-junction waveguide as a mode divider/combiner, with the π phase shifter region located on the split branch. At the phase shifter region, the refractive index is controlled by injecting current. The width of the input port is set to be twice that of the branch waveguide to allow 1st order mode. For the output mode coupling fiber, we utilize the laterally core-arranged multi-core fiber (4 cores) in this work. The layout of the cores is illustrated in Fig. 1. The mode field diameter of each core is 6.3 μm, while the gap between each two core is 27.4 μm. The whole fiber diameter is 125 μm. For the optical mode switch configuration especially at the device output, we newly design the output port waveguide as slightly separated arrangement as shown in Fig. 1, instead of using regular multi-mode waveguide. In order to couple the output mode from the slightly separated waveguide to the cores of MCF, the width of the slightly separated output waveguide is designed to be 2.1 μm, while the gap between the separated waveguide is 9.1 μm.

The optical lens is used to help convert the magnification of the output field toward the fiber facet. In order to get high coupling efficiency, the modal profiles (spot size) of each port, which is closely related to field distribution of the light output facet, have to be matched to the spot size of each core in the MCF with respect to phase and amplitude distributions. As a result, precise alignment of the system is required to reduce the avoidable coupling loss. Then the

1030

main problem turns to detecting the coupling performance and calculating the power coupling efficiency between the optical mode switch and the MCF through the lens.

Based on well researched similar coupling performance between arrayed waveguides and fiber arrays [6], the expression of coupling efficiency is given as follows,

$$\eta = \frac{4}{\left(\dfrac{m\omega_1}{\omega_2} + \dfrac{\omega_2}{m\omega_1}\right)^2}(\%) \qquad (1)$$

where ω_1 is the mode field diameter of the output port of the optical mode switch, ω_2 is the mode field diameter of each core in the MCF, m is the magnification of the system and η is the coupling efficiency. Here, we assume the whole system is perfectly aligned without any misalignment or mismatch of either the x or y optical axes. Based on Eq. 1, the expected coupling loss between optical mode switch and MCF is 2dB when ω_1 is 2.06 μm and ω_2 is 6.3 μm.

III. RESULTS AND DISCUSSION

As our target in this work is coupling light output from the optical mode switch to the "single" MCF, the near field patterns were observed for the both optical mode switch and MCF.

A. Power profile at the device facet

To verify the power profile of the device output, we utilized the hemispherical lensed-fiber with a 4 μm diameter to inject an optical power of 6dBm (λ=1.55 μm). As a result, a near field pattern with two bright spots is clearly detected. By using an IR camera, it was confirmed that each spot shared the same energy of -22dBm. The loss should be caused by coupling loss and propagation loss in the optical mode switch.

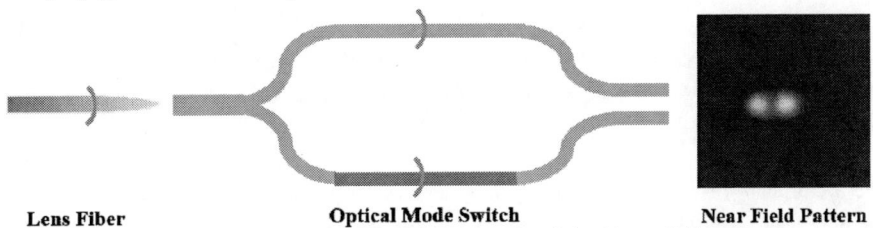

Lens Fiber   **Optical Mode Switch**   **Near Field Pattern**

Fig. 2. Schematic configuration of optical mode switch with near field pattern.

B. Power profile at the MCF facet

In order to choose a suitable lens, the coupling loss of the lens corresponding various numerical apertures (NA) is researched. In this case, the lensed fiber is directly coupled to the MCF by lens. It turns out that an NA of 0.8 shows the lowest coupling loss of 0.1dB among those three samples. In this case, we choose the lens with NA of 0.8 as the coupling lens between the optical mode switch and the MCF.

TABLE I
COUPLING LOSS CORRESPONDING TO NUMERICAL APERTURE

Numerical Aperture (NA)	Coupling Loss (dB)
0.20	2.2
0.50	1.0
0.80	0.1

In order to get a high coupling efficiency, the modal profile (spot size) of each output port, which is closely related to field profile of the light output facet, has to be matched to the modal profile of each core in MCF with respect to phase and amplitude distributions. The phase condition is met when the parallel wave front imaging of the laser beam waist is parallel to the wave front of the fiber beam waist. The amplitude condition requires a transformation of the elliptical laser beam waist into a circular fiber beam waist to match the sizes and amplitude distributions, respectively. To obtain these transformations, the lens should be normal to the device and MCF facet at the same time, while precise aligned in both the x and y optical axes.

Here, we assume d_0 is the distance between the optical switch device and the lens, while d_1 is the distance between the lens and the MCF. Considering the focal length (f) of the optical lens is 2.3, and the magnification we applied this time is 6, we can measure that d_0 is 2.68 mm and d_1 is 16.1 mm. In the actual experiment, the size limitation of the detecting equipment forces us to use the magnification of 6. In this case, considering the distance of the gap between two output ports is 9.1 μm, while the gap between each two cores is 27.4 μm in MCF, we can assume the light exit of the optical mode switch with 6 times magnification can match the two outside cores in MCF, but not the two inside ones. In the actual experiment, the near field pattern with two separated light spots is clearly observed. The power intensity of each spot of near field pattern here was -31 dBm, and the actual coupling loss was 9 dB. As for the coupling loss difference between theory and experiment, lens-magnification was one of the main reasons. This was caused by the possible distance between the device and the lens in our experimental set-up, and this problem must be overcome easily in the near future.

The detection of the final near field pattern certifies that output mode from the optical mode switch has been successfully coupled to "single" MCF. We believe that this result firstly shows the potential of higher scalability of the optical mode switch as the result proves that it is not necessary to couple to fiber-array like in the case of conventional optical matrix switch. As this time we could not choose proper magnification due to technical reason, we believe that proper coupling will be realized after optimization and technical improvements.

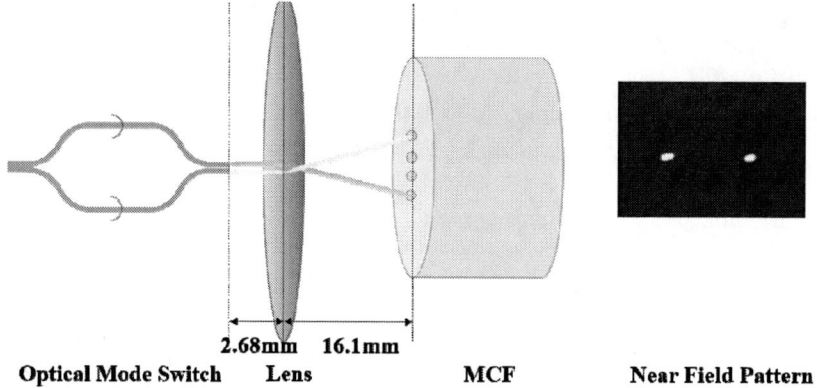

2.68mm   16.1mm

Optical Mode Switch   Lens   MCF   Near Field Pattern

Fig. 3. Schematic configuration of optical mode switch system with near field pattern.

IV. CONCLUSIONS

We report the preliminary research of coupling performance between an optical mode switch and a single MCF. The detection of the final near field pattern shows successful coupling of output mode to MCF. The result suggests higher scalability potential of the optical mode switch with single fiber coupling because it is not necessary to integrate S-bend regions toward fiber array, that prevent higher integration on conventional space switch.

ACKNOWLEDGMENT

This work has been supported by NICT, Japan. The authors thank to Dr. Shoichiro Matsuo, Fujikura Ltd., for providing the MCF. The authors also thank to Prof. Shiyoshi Yokoyama, Kyushu University, for his encouragement.

REFERENCES

[1] J. V. Campenhout, W. M. J. Green, S. Assefa and Y. A. Vlasov, "Low-power, 2x2 silicon electro-optic switch with 110-nm bandwidth for broadband reconfigurable optical networks," Opt. Express USA, vol. 17, No. 26, pp. 24020-24029, 2009.

[2] P. Dong, S. Liao, H. Liang, R. Shafiiha, D. Feng, G. Li, X. Zheng, A. Krishnamoorthy, M. Asghari, "High-speed and broadband electro-optic silicon switch with submilliwatt switching power," Tech. Dig. OFC. USA, OWZ4, ISBN: 978-1-55752-906-0, March 2011.

[3] D. J. Blumenthal, J. E. Bowers, L. Rau, H. F. Chou, S. Rangarajan, W. Wang and K. N. Poulsen, "Optical signal processing for optical packet switching networks," IEEE Opt. Comm. USA, vol. 41, No. 2, pp. S23-S29, 2003.

[4] K. Hamamoto, "Optical mode switch for high speed switching network," Tech. Dig. Japan, PS 2014, PTIB. 2, 2014.

[5] R. Imansyah, T. Tanaka, L. Himbele, H.S. Jiang and K. Hamamoto, "Mode Crosstalk Evaluation on Optical Mode Switch," IEEE International Broadband and Photonics Conference (2015), Bali, 23-25, April 2015.

[6] Y. Tahara, H. A. Bastawrous, H.S. Jiang, S. Matsuo and K. Hamamoto, "Fundamental research on dense integration of memory elements for optical random access memory (RAM) by using lens coupling with fiber-array," IEICE Tech. Report, Japan, 109, (401), pp. 101-106, Jan. 2010.

Numerical Study of Power Coupling between Index-Antiguided Slab Waveguides

Chang-I Hsieh and Chih-Hsien Lai

Department of Electronic Engineering, National Yunlin University of Science and Technology, Yunlin 64002, Taiwan
E-mail: chlai@yuntech.edu.tw

Abstract: *Power coupling between two index-antiguided slab waveguides is numerically investigated. The coupling length is found to vary periodically with the gap between the waveguide cores, owing to the Fabry-Perot-like behavior of the waveguide gap.*
Keywords: *Index-antiguided waveguides, directional couplers, Fabry-Perot*

I. INTRODUCTION

Development of high-power waveguide lasers usually faces some challenges such as nonlinear effect and optical damage. Increasing the waveguide core is a simple way to overcome the problems, but it will deteriorate the beam quality because the transverse mode number also increases accordingly. Hence, waveguides with large mode area (LMA) and single-mode operation are desired for high-power laser applications.

Index-antiguing (IAG) is one of the techniques for waveguide lasers to simultaneously achieve LMA and single-mode operation [1,2]. In the IAG waveguides, the waveguide modes confining in the low-index core are supported by the gain-guiding effect, in that the provided gain in the core compensates the leaky loss of the IAG structure. The IAG waveguide laser was first realized in fiber geometry and single-mode lasing with fiber core diameters up to 400 μm were demonstrated [3]. Recently, the IAG technique is applied to slab geometry and the IAG planar waveguide laser was reported [4].

Waveguide lasers may be placed closely in practical applications. If they are close enough, according to the waveguide theory, interaction among the waveguides would occur so that optical power would transfer among the waveguides [5]. Hence, to understand the coupling behavior of IAG waveguide lasers, it is important to study the power coupling of IAG waveguides (no gain in the core) in advance. Here, by considering two IAG slab waveguides placed in close proximity, we numerically investigate the power coupling occurring between the IAG waveguides.

II. STRUCTURE FOR NUMERICAL MODELING

Figure 1 plots two identical and parallel IAG slab waveguides which are close to each other. Index profile of the two-waveguide system is also shown. In this work, the following parameters are assumed: core index n_{co} = 1.5, cladding index n_{cl} = 1.6, core width w = 30 μm, and wavelength λ = 1.53 μm. We only consider the transverse electric (TE) polarization. For the IAG waveguide, since refractive index of the core is lower than that of the cladding, the waveguide has a complex propagation constant β and suffers a leakage loss. Modal index n_{eff} and attenuation constant α can be calculated from the real and imaginary parts of the complex propagation constant as $\mathrm{Re}(\beta)/k_0$ and $-2\mathrm{Im}(\beta)$, respectively, where k_0 is the free space wavenumber. By using a finite-difference mode solver [6], modal index and attenuation constant can be obtained. With the parameters assumed here, the values of the fundamental mode for a single IAG slab waveguide are n_{eff} = 1.49978 and α = 1.036 cm^{-1}, respectively.

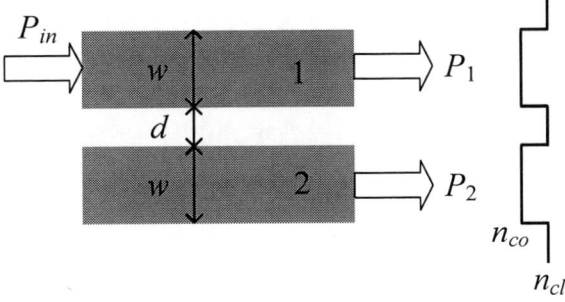

Fig. 1. Structure and index profile of two identical and parallel IAG slab waveguides.

In Fig. 1, the gap between the two waveguide cores is denoted as d. To simulate the power coupling, a Gaussian beam with unit power P_{in} = 1 is launched into the first waveguide. The beam width is assumed to be $2w_0$ = 22 μm. A finite-difference beam propagation method (BPM) [7] is utilized to simulate the power evolution along the two-waveguide system, and the powers in the first and second waveguides are denoted as P_1 and P_2, respectively.

III. Numerical Results and Discussions

Simulated power evolutions along the two-waveguide system (along the z direction) for gap $d = 1.5$ and 2.7 μm are shown in Figs. 2(a) and 2(b), respectively. In this work, it is assumed that there is no gain in the IAG wavegide core. Thus, the two-waveguide system suffers a leakage loss and the total power ($P_1 + P_2$) attenuates along the propagation. According to the coupled-mode theory [5], along the propagation distance, the power launched into the first waveguide will gradually transfer to the second one. For conventional index-guided (IG) waveguides where the core index is higher than the cladding index ($n_{co} > n_{cl}$), owing to the power conservation, the distance at which P_1 reaches minimum and the distance at which P_2 reaches maximum are the same, and the distance where total power transfer occurs is called the coupling length. However, for the IAG waveguides studied here, the behavior is different. For example, in Fig. 2(a), the distance at which P_1 becomes minimum is 16.0 mm, while the distance at which P_2 becomes maximum is 12.9 mm. They are not the same. Such a difference in distance is attributed to the leakage loss of the IAG waveguide [8].

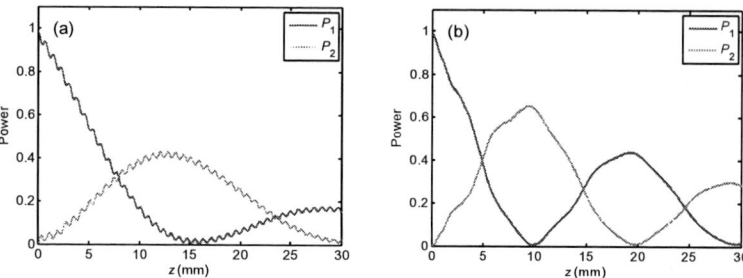

Fig. 2. Power evolution along the two IAG slab waveguides. (a) $d = 1.5$ μm. (b) $d = 2.7$ μm.

Let the coupling length (L_c) be the distance where P_1 becomes minimum (zero). From Fig. 2(a), $L_c = 16.0$ mm for waveguide gap $d = 1.5$ μm. However, when d increases to 2.7 μm, it is seen in Fig. 2(b) that L_c decreases to 9.8 mm. Again, this phenomenon is different from what usually observed in conventional IG waveguides, where the coupling length increases with the waveguide gap.

Figure 3 plots the coupling length L_c as a function of waveguide gap d. It is interesting to find that, instead of increasing with the waveguide gap as in IG waveguides, the coupling length in IAG waveguides (blue circle in Fig. 3) varies periodically with the waveguide gap, ranging from 3.9 to 61.0 mm.

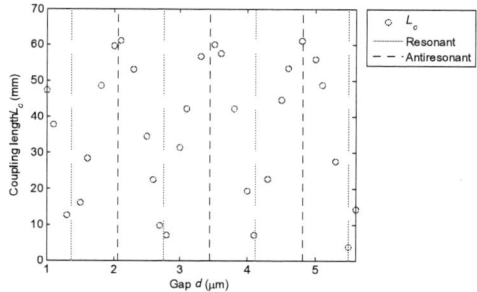

Fig. 3. Coupling length (blue circle) of the two-waveguide system as a function of waveguide gap. Red dashed lines and black dotted lines indicate the values where the waveguide gap meets the resonant and antiresonant conditions, respectively.

The periodic phenomenon of the coupling length can be explained in that the gap between the waveguide cores behaves like a Fabry-Perot etalon. For the Fabry-Perot etalon, its thickness plays an important role which determines the etalon being resonant or not. When the thickness of the gap meets the resonant condition, the transmission is 100%, i.e., the gap becomes nearly transparent. As a result, the optical power launched into the first waveguide core can penetrate the gap to the second waveguide core easily, leading to a shorter coupling length. On the other hand, when the thickness of the gap meets the antiresonant condition, the transmission becomes relatively small. That is to say, strong reflection occurs between the core-gap interface, so that it is difficult for the power to transfer from the first waveguide to the second one, thus the coupling length becomes longer. Since the resonant and antiresonant conditions appear periodically with the gap thickness, the coupling length as a function of the waveguide gap also exhibits a periodic feature.

Taking the gap as an etalon with the thickness being d, the following equation can be used to determine whether the gap thickness d meets the resonant or antiresonant conditions:

$$d = \frac{m\lambda}{4\sqrt{n_{cl}^2 - n_{co}^2}}, \qquad (1)$$

where m is a positive integer. When m is even, d meets the resonant condition; otherwise, the gap is antiresonant. According to Eq. (1), the d values for $m = 2, 4, 6,$ and 8 that make the gap resonant are 1.374, 2.748, 4.122, and 5.496 μm, respectively. These values are shown in Fig. 3 as red dashed lines. Clearly, in Fig. 3, the smallest coupling lengths are those near these d values. In other words, when the gap meets the resonant condition, power coupling between IAG waveguides is more likely to take place. On the other hand, when $m = 3, 5,$ and 7 are substituted into Eq. (1), the corresponding d values are 2.061, 3.435, and 4.809, respectively. These d values, shown as black dotted lines in Fig. (3), will make the gap antiresonant . As can be seen in Fig. 3, the coupling lengths around these antiresonant d values are the largest, showing that it is difficult for the power to couple between IAG waveguides under the antiresonant condition.

The behavior of the coupling length, i.e., it varies periodically with the waveguide gap, and has the shortest value when the gap is resonant, while the longest when the gap is antiresonant, can also be explained by the modal index. If the two-waveguide system is treated as a whole guiding structure, coupling length can be obtained from the modal indices as: $L_c = \lambda \; / \; |2(n_{eff,e} - n_{eff,o})|$, where $n_{eff,e}$ and $n_{eff,o}$ are modal indices of the even and odd system modes, respectively. Modal indices of the system modes as functions of waveguide gap are shown in Fig. 4. As observed from Fig. 4, when the gap meets the resonant condition ($d = 1.374$ and 2.748 μm), the difference between $n_{eff,e}$ and $n_{eff,o}$ is large, thus the coupling length is short. On the other hand, when the gap is antiresonant ($d = 2.061$ μm), the difference is small, which leads to a long coupling length.

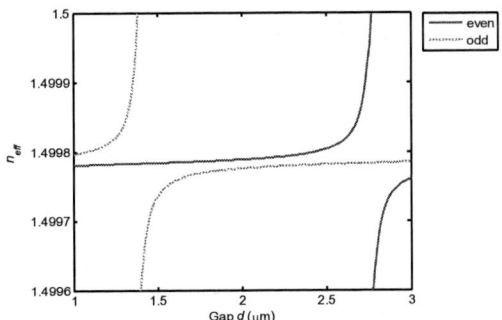

Fig. 4. Modal indices of the even and odd system modes as functions of waveguide gap.

IV. CONCLUSIONS

In this work, the power coupling behavior between two IAG slab waveguides is numerically investigated. Unlike what happens to conventional IG waveguides, the distance where one waveguide has the minimum power and the distance where another waveguide obtains the maximum power are different, because the IAG waveguides suffer leakage power losses. Moreover, owing to the Fabry-Perot-like behavior of the waveguide gap, it is found the coupling length varies periodically with the gap. This phenemenon reveals that power coupling between IAG waveguides takes place more easily when the gap meets the resonant condition, while with more difficulty when the gap is antiresonant. This work provides preliminary information useful for the design of IAG waveguide lasers. Results with a gain present in the low-index core of the IAG waveguides will be provided in the presentation.

ACKNOWLEDGMENT

This work was supported by the Ministry of Science and Technology of the Republic of China under grant MOST 102-2221-E-224-069-MY3.

REFERENCES

[1] A. E. Siegman, "Propagating modes in gain-guided optical fibers," J. Opt. Soc. Am. A, vol. 20, pp. 1617–1628, 2003.

[2] A. E. Siegman, "Gain-guided, index-antiguided fiber lasers," J. Opt. Soc. Am. B vol. 24, pp. 1677–1682, 2007.

[3] Y. Chen, T. McComb, V. Sudesh, M. Richardson, and M. Bass, "Very large-core single-mode, gain-guided, index-antiguided fiber lasers," Opt. Lett., vol. 32, pp. 2505–2507, 2007.

[4] Y. Liu, T.-H. Her, A. Dittli, and L. W. Casperson, "Continous-wave hybrid index-antiguided and thermal-guided planer waveguide laser," Appl. Phys. Lett., vol. 103, pp. 191103-1–191103-4, 2013.

[5] B. E. A. Saleh and M. C. Teich, Fundamentals of Photonics, 2nd, ed., Wiley, 2007.

[6] C.-P. Yu and H.-C. Chang, "Yee-mesh-based finite difference eigenmode solver with PML absorbing boundary conditions for optical waveguides and photonic crystal fibers," Opt. Express, vol. 12, pp. 6165–6177, 2004.

[7] J. Yamauchi, Propagating Beam Analysis of Optical Waveguides, Research Studies, 2004.

[8] V. R. Chinni, T. C. Huang, P.-K. A. Wai, C. R. Menyuk, and G. J. Simonis, "Crosstlak in a lossy directional coupler switch," J. Lightwave Technol., vol. 13, pp. 1530–1535, 1995.

WA3-1 (Invited)

OECC/PS2016

SDN/NFV Orchestration of Multi-technology and Multi-domain Networks in Cloud/Fog Architectures for 5G Services

Ricard Vilalta, Arturo Mayoral, Ramon Casellas, Ricardo Martínez, Raul Muñoz

Centre Tecnològic de Telecomunicacions de Catalunya (CTTC), Av. Carl Friedrich Gauss 7, Castelldefels, Spain

Author e-mail address: ricard.vilalta@cttc.es

Abstract: This paper presents an SDN/NFV architecture for delivery of future 5G services across multiple technological and administrative networks, including massive computing resources, which might be centralized or distributed, which are able to offer virtualized functions.

Keywords: Orchestration, Transport Networks, SDN, NFV, Cloud, Fog, IoT, 5G

I. INTRODUCTION

End-to-End (E2E) converged network and cloud/fog infrastructure becomes of the essence for the fifth generation of mobile networks technology (5G). This converged infrastructure, illustrated in Fig.1.a, is composed of: E2E heterogeneous network segments covering radio and fixed access, metro aggregation, core transport involving heterogeneous wireless and optical technologies; centralized cloud and distributed fog computing/storage/networking infrastructure; and large amounts of heterogeneous smart devices and terminals for traditional mobile broadband services and IoT services.

5G architecture needs to deliver high flexibility, low-latency, and high-capacity networks for supporting the estimated 1000x increase in mobile data traffic while providing sub-millisecond latency [1]. From the perspective of Management and Control Continuum, E2E connectivity services need to be provisioned between distributed cloud and fog infrastructures and end users. These requirements can only be fulfilled by efficiently integrating in using the Software Defined Networking (SDN) paradigm the heterogeneous access, metro aggregation packet networks and high-capacity optical core transport networks. For this integration, SDN Orchestration is proposed to coordinate [2], in a hierarchical/peer, logically centralized manner, the heterogeneous control plane technologies of the different network segments, which may remain separated as independent administrative domains.

From the centralized cloud and distributed fog infrastructure perspective, the demand of massive and decentralized computing/storage/network will dramatically be increased by new 5G services, which will require processing and storage capabilities (e.g., Big Data). In addition, the impending growth of Network Function Virtualization (NFV) and Mobile Edge Computing (MEC) also require the delivery of generalized Virtualized Functions (VFs) on top of cloud /fog infrastructure for the deployment of software functions (e.g., mobile Evolved Packet core (EPC), local cache, firewalls, databases, video analytics). Originally, cloud services have been implemented in core data centers (DCs) for high-computational or long-term processing.

This paper presents the necessary SDN/NFV architecture and how it has been deployed on top of ADRENALINE testbed. Special focus on hierarchical SDN orchestration and virtualized function orchestration at the edge of the network is given.

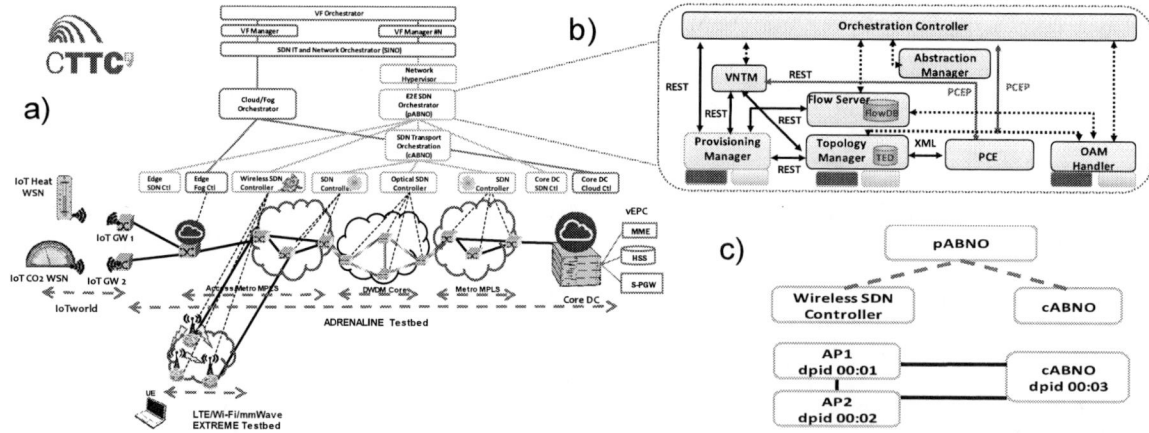

Fig. 1. a) Proposed SDN/NFV orchestration architecture, b) Detailed components of an SDN orchestrator (parent/child), c) Abtracted topological view as seen by E2E SDN orchestrator (pABNO).

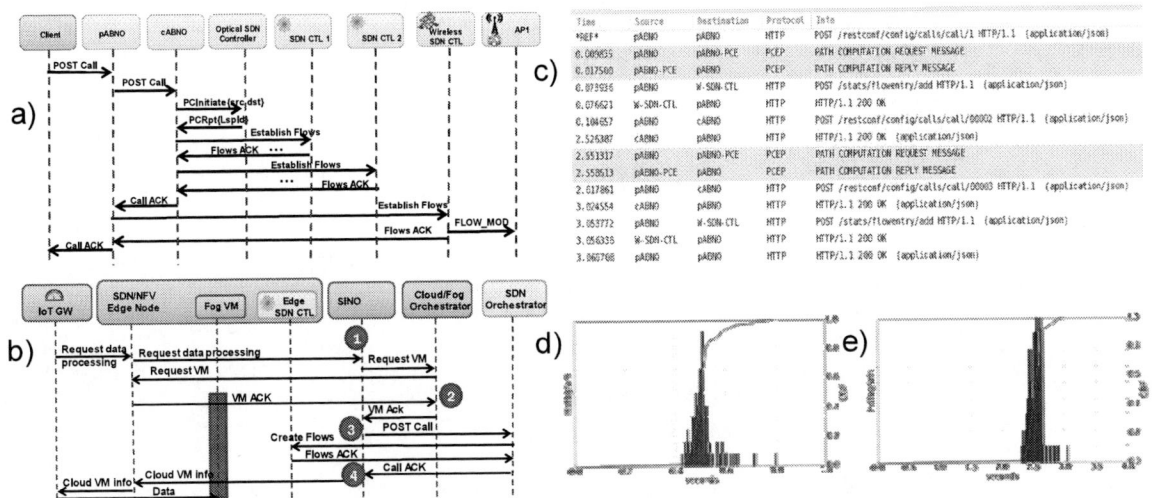

Fig. 2. a) Workflow for hierarchical SDN orchestration, b) Workflow for Virtualized Function Orchestration, c) Wireshark captures between pABNO-cABNO, d) Packet response time between IoT GW and edge VM, e) or core DC VM

II. SDN/NFV ORCHESTRATION AND THE ADRENALINE TESTBED

The SDN/NFV proposed architecture consists of four main building blocks: a Virtualized Functions Orchestrator (VF-O), a SDN IT and Network Orchestrator (SINO), a Cloud/Fog orchestrator, and an SDN Orchestrator (SDN-O) for control of heterogeneous networks. The VF-O is the responsible for the dynamic orchestration of generalized virtualized functions, which might include NFV, MEC and IoT services. The VF-O is responsible for allocating the necessary resources for the deployed services through the usage of the SINO. The SINO is the responsible for the joint orchestration of cloud/fog and network resources. The cloud/fog orchestrator enables the deployment of computing, storage and networking resources in distributed data centers and fog nodes. As networks are expected to span across several domains and technologies, a hierarchical SDN-O is introduced to manage this complexity.

The SDN-O follows a hierarchical approach where there is a parent SDN orchestrator (E2E SDN-O), child SDN-O, and SDN controllers for each domain. Each SDN controller might be in charge of the specifics of the underlying data plane technology. The SDN-O is based on the ABNO architecture proposed at IETF [3]. The main building blocks are shown in Fig.1.b. and they have been described previously in [4]. The Orchestration Controller handles all the processes (workflows) involved inside the NO to satisfy the provisioning of end-to-end connectivity. The Topology Server is the component responsible of gathering the network topology from each control domain and building the whole network topology which it is stored in the Traffic Engineering Databased (TED), which is used by the Path Computation Element (PCE) for calculating routes across the network. The Virtual Network Topology Manager (VNTM) is the responsible of the multilayer management. The Provisioning Manager implements the different provisioning interfaces to push the forwarding rules into the data plane. Flow server stores the connections established in the network into a FlowDB. Abstraction Manager is the responsible for providing network abstractions. Finally, OAM handler handler receives asynchronous notifications, such as topology updates, flow statistics, or failure alarms.

This proposed architecture has been deployed on top of the cloud computing platform and transport network of the ADRENALINE Testbed, located at CTTC premises in Castelldefels (Barcelona, Spain). The Cloud Computing platform of the ADRENALINE Testbed (Fig.1.a) includes OpenStack Havana release, which has been deployed into servers with 2 x Intel Xeon E5-2420 and 32GB RAM each. A fog server has been developed using Intel NUC NUC5i5RYH, with 16Gb RAM and 120 Gb SSD. Four OpenFlow switches have also been deployed. Each Data Center border switch has a 10 Gb/s XFP tunable transponder. Finally, the ADRENALINE transport network is controlled with an optical SDN controller, which interacts to a GMPLS-controlled optical network is composed of an all-optical WSON with 2 ROADMs and 2 OXCs providing reconfigurable end-to-end lightpaths.

III. HIERARCHICAL SDN ORCHESTRATION

In order to provide a feasible solution to handle the heterogeneity of different network domains, technologies, and vendors, we propose the introduction of hierarchical SDN Orchestration. It focuses on network control and abstraction of several control domains, whilst using standard protocols and modules. The need of hierarchical SDN orchestration has been previously presented in [5] to serve two basic purposes: the increase of security and scalability.

In this use case, the hierarchical SDN Orchestration is applied for the integration of wireless and optical transport networks, in order to provide E2E connectivity between the User Equipment (UE) and a cloud service deployed in the Core DC location. In the wireless segment, an SDN controller is in charge of the programming of the wireless network (access and backhaul). In the optical segment, implemented over the ADRENALINE Testbed, we consider an SDN-enabled MPLS-TP aggregation network, while the control of the core network relays on an optical SDN controller over a GMPLS distributed control plane.

A parent E2E SDN Orchestrator (pABNO) orchestrates several network segments: an SDN-enabled wireless segment and the MPLS/Metro and Core network segments orchestrated by a child ABNO (cABNO). The child SDN-O (cABNO) is responsible for abstracting the multi-domain transport segments, and it offers a simplified view to the pABNO, thus improving scalability and security. Fig.1.c shows the abtracted topological view as seen by pABNO.

The workflow is shown in Fig.2.a. It can be observed that an E2E connection is requested (POST Call) to the pABNO. The pABNO computes the involved network controllers (Wireless SDN/cABNO) and requests the underlying connection to them. Fig.2.c shows the wireshark captures of the exchanged messages at both the pABNO and cABNO. The bidirectional E2E service call request is received at the pABNO. The pABNO PCE is responsible for computing an E2E path. The pABNO VNTM decomposes the computed E2E path in order to request a call service to the cABNO. The cABNO is responsible for SDN orchestration towards the underlying network domains (SDN-CTL-1, AS-PCE, and SDN-CTL-2). After the cABNO provisions the requested call service, the pABNO requests the necessary flows to the wireless SDN controller. The setup delay for an E2E service call is around 3.06s.

IV. VIRTUALIZED FUNCTION ORCHESTRATION AT THE EDGE OF THE NETWORK

In this use case, we propose to extend the scope of NFV and introduce virtualized functions. As an example, we propose the deployment of a database services for the storage of Wireless Sensor Network (WSN) data that traverses the IoT gateways. We propose to analyze the necessary orchestration workflows for deploying the database services at the edge of the network (Fig.2.b).

The IoT gateway acts as the client which requests a database service to VF-O, which interacts with the SINO to request the creation of VMs to the Cloud/Fog Orchestrator (1). It is responsible for the creation of the instances. It will select the proper allocation of the computing and storage resources depending on the requested services. It is also responsible to attach the VMs to the virtual switch inside the host node (at the edge node or in a core DC). When the VMs creation is finished, the Cloud/Fog Orchestrator replies the VM's networking details to the integrated Cloud/Fog and network orchestrator (MAC address, IP address and physical computing node location) (2). The E2E SDN-O is responsible for provisioning of E2E network services (3). It will provide the E2E connectivity between the requested IoT gateway and the deployed VM (4). Finally, data from IoT gateway will flow to the processing resources, located in the proposed edge node.

We define packet response time as the necessary time it takes for a certain amount of data to travel from point A to point B and back (round trip). It can be observed that the average packet response time between the IoT GW and an edge VM is of 506 ms (2 switch hops) (Fig.2.d), while the average packet response time between the IoT GW and a VM located at the core DC is of 2552 ms (9 hops) (Fig.2.e). The allocation of processing resources at the edge results in better packet response time.

V. CONCLUSIONS

In this paper, we have presented an SDN/NFV orchestration architecture for the delivery of future 5G services. A Virtualized Function Orchestrator has been presented in order to generalize the necessity of orchestrating generic purpose applications and services (e.g., a database) through distributed cloud and fog resources, which are interconnected through heterogeneous network domains.

ACKNOWLEDGMENT

This work was partially funded by EU FP7 COMBO (317762), H2020 5g-Crosshaul (H2020-671598) and Spanish MINECO project DESTELLO (TEC2015-69256-R).

REFERENCES

[1] 5GPPP white paper, the 5G Infrastructure Public Private Partnership: the next generation of communication networks and services, March 2015.
[2] R. Muñoz, et al., The CTTC 5G end-to-end experimental platform integrating IoT, SDN, and distributed cloud , in Proceedings of Wireless World Research Forum Meeting 35 (WWRF), 14-16 October 2015, Copenhagen (Denmark).
[3] D. King, and A. Farrel, "A PCE-based Architecture for Application based Network Operations", IETF RFC 7491, 2015.
[4] R. Vilalta, et al., "Multi-Tenant Transport Networks with SDN/NFV", IEEE/OSA JLT, Vol. 34, No. 8, January 2016.
[5] R. Vilalta, et al., "Hierarchical SDN Orchestration of Wireless and Optical Networks with E2E Provisioning and Recovery for Future 5G Networks", in Proc. OFC 2016, Anaheim (CA), USA.
[6] R. Vilalta, et al., "End-to-End SDN Orchestration of IoT Services Using an SDN/NFV-enabled Edge Node", in Proc. OFC 2016, Anaheim (CA), USA.

OECC/PS2016

Software-defined Optical Transmission and Networking with Functional Service Design

Xiaoyuan Cao, Noboru Yoshikane, Koki Takeshima, Ion Popescu, Takehiro Tsuritani, Itsuro Morita

KDDI R&D Laboratories, Fujimino, Saitama, Japan

{xi-cao, yoshikane, ko-takeshima, io-popescu, tsuri, morita}@kddilabs.jp

Abstract: A service design concept is introduced into the optical transport network. Customized and orchestrated optical transmission is realized by coordinating and reconfiguring the sharable virtualized optical functions, while better flexibility is provided for optical networking.

Keywords: Optical transport network, Service design, SDN, Virtual network function

I. INTRODUCTION

The lack of flexibility has kept the optical transport network from fitting into the future software-defined network (SDN) [1] framework. Thus for a long time, the optical transport network has been playing a role of transmission and becomes a bottleneck for providing end-to-end elastic service. However, challenges may occur even for optical transmission as the future service becomes diversified and dynamic. Flexible and customized service provisioning would be difficult over the current dumb optical transmission and networking system. The introduction of SDN provides a solution for implementing a programmable control plane. In the meantime, the progress in electronic data processing enables software-defined optical (SDO) [2] transmission, which aims at shifting hardware functionality into the software layer. However, it is still challenging to find a pertinent control solution over the optical transport network and holistically coordinate the optical resource in a flexible and beneficial way.

We have proposed a functional "service design" concept in our previous work [3] in order to provide customized service and maximum network freedom, while reducing the network operation cost. As shown in Fig. 1(a), essentially, it decouples the hardware and network functionalities into virtual network functions (VNF) [4], which are then flexibly chosen and arranged in the SDN control plane for the traffic flows in order to provide different services. VNFs are deployed in an adaptive plug-in manner to be sharable and reusable among different controllers [1] or virtual networks (VN) [4]. Traffic flows are labeled with *flow_id* and the VNFs are authorized to the controller according to the *vnf_table*. *flow_id* could be defined with any unique and recognizable tags in the packet. For each service, it simply visits the chosen VNFs during the transmission, while the reconfiguration of service function chain (SFC) [5] conveniently creates the varied and customized service. In this paper, we extend the service design concept into the optical transport network and demonstrate the customized optical transmission and networking with multi-layer multi-functional orchestration. As shown in Fig. 1(b), hardware functionalities, transmission quality and networking control could be defined as new VNF modules and applied to the service in a flexible and reconfigurable way. The service-design based control plane allows the operator to optimize the optical system and virtual optical network (VON) [6] by orchestrating the VNFs, and also allows the client to request customized service with selectable VNFs. As a result, optical transmission with differentiated features and optical networking with better controllability will be achieved.

II. SERVICE-DESIGN BASED OPTICAL TRANSMISSION

Optical transmission has the direct impact on the provided service in terms of quality, latency and cost. With the development of the optical technologies, it is beneficial to add extra features into the transmitted signals with different modulation format/bit rate, flexible wavelength, quality and monitoring, etc., so that customized services could be provided for clients. Therefore, it is important to implement a programmable optical transmission system, which could be realized based on the proposed service design paradigm by virtualizing the various supporting optical techniques as VNFs into the SDN control plane and selectively reconfiguring the VNFs to provide varied optical transmissions for the

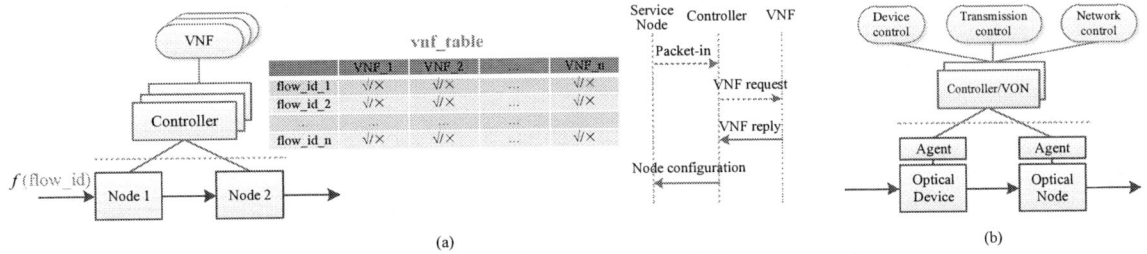

Fig. 1. (a) Service design; (b) service design in the optical transport network.

1039

clients.

A. Device Control

Each optical device usually has its own particular feature that could be decoupled as a corresponding VNF that is shared in the control plane. By adjusting the settings of VNFs in the control plane, we indirectly configure the device into different status through the controller and the device agent [3] which is a virtualized representative of the optical device. Such VNFs are implemented in a reconfigurable manner and sharable among all the authorized controllers. Therefore, although the deployment of an optical transmission system is static, its management becomes flexible with the introduction of service design paradigm as we are able to choose and configure the device functions dynamically in the software. Besides, as optical devices usually have distinguished and complicated functions which come at a fairly high cost, we could provide more pertinent and economic solutions for clients.

B. Orchestrated Transmission Control

A basic optical transmission system comprises of the transponders, the optical switches, the amplifiers and the transmission fiber. A complete transmission of the signal requires deliberative and cooperative setup of all the optical devices along the path. In a service-design based optical transmission system, the centralized SDN controller has the global knowledge of the overall optical resource and is able to decide which type of transmission is needed according to the user requirement. Then it coordinately chooses the path and the parameters for each device along the path. Such orchestrated transmission control guarantees the combinative result of the device configuration and thus the final transmission quality. Furthermore, in the case that the user requirement or the status of some device is changed, the controller is able to react accordingly and re-adjust the affected configuration.

C. Proof-of-concept Demonstration

As shown in Fig. 2(a), an experimental transmission system was set up with the transmitter/receiver pairs, a WSS (Finisar WaveShaper 4000s), two custom-designed EDFAs and the optical fiber with different transmission distances. Two physically-separated OFDM transmitters were employed with pre-configured QPSK and 16QAM modulation format respectively and an optical switch was controlled by the controller to select the optimum modulation format. The WSS was controlled by loading the WaveShaper profile, including the center wavelength, bandwidth and output port. The EDFA gain was controlled by adjusting the value of two pump currents using the proprietary commands. We were also able to read the EDFA status including the input optical power using the commands. We defined three VNFs in this transmission system: 1) *modulation*, which informs the controller of the modulation format for a labeled signal as well as the corresponding center wavelength and bandwidth; 2) *switching*, which maps the configuration of WSS according to the transmitted signal; 3) *amplification*, which calculates the value of the pump currents in order to provide enough gain based on the input optical power. By adjusting the setup of these self-defined functions, we were able to get different VNF chains and thus customized transmission results.

We firstly generated an OFDM signal (polarization-multiplexing) with bit rate of 17.0 Gbps modulated by QPSK format at wavelength of 192.75 THz with 5 GHz bandwidth, which was switched by the WSS to output 1, transmitted through a 40km SMF, amplified by the EDFA and transmitted to receiver_1, as shown in Fig. 2(b, c). Then the client requested a faster transmission with double bit rate (34.6 Gbps) and informed the controller. An Openflow switch (OFS) was used to trigger the reconfiguration process in the controller by sending a *packet-in* message [7] including a *flow_id*

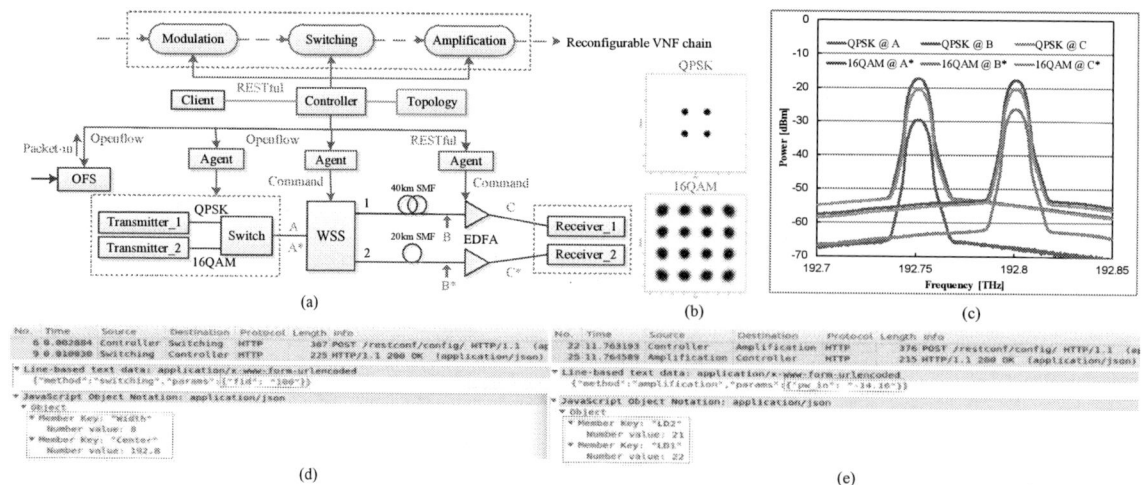

Fig. 2. Service-design based optical transmission demonstration: (a) optical transmission system setup; (b) transmitted QPSK and 16QAM signal constellations; (c) optical spectra of transmitted signals at different monitor spots (0.02nm resolution); (d) JSON messages between the controller and the switching function; (e) JSON messages between the controller and the amplification function.

of 100. Based on the available devices, the controller decided to transmit the signal in 16QAM format. However, since the signal with 16QAM has shorter transmission distance limit than that with QPSK, the controller recalculated the transmission path to the one through the 20km SMF based on the topology. Therefore, the traffic would go through the VNF chain with different settings consequently. In specific, the controller visited the modulation function via RESTful API [8] and was informed of the modulation format change. Then the controller visited the switching function and was informed to reconfigure a filter centering at 192.80 THz for the WSS (Fig. 2(d)), and the output 2 according to the adjusted path. The controller also requested the amplification function to calculate the value of the pump currents based on the input optical power in order to reach a proper output power for the receiver (Fig. 2(e)). The receiver was manually reconfigured to receive the signal. The new optical OFDM signal transmission with 16QAM (constellation diagram and the optical spectra at monitor spot A*, B*, C*) after the bit rate change was shown in Fig. 2(b, c).

III. SERVICE-DESIGN BASED OPTICAL NETWORKING

Due to the difficulties of control and lack of flexibility, optical transport network has become a bottleneck for the multi-domain end-to-end flexible service provisioning. With the extension of service design, we expect to realize a real optical "network" with a similar flexibility as the IP network. Similarly, if we decouple the networking functionalities based on the service design such as protection, multicast, constraint routing, etc., we are able to achieve a much better flexibility in the optical transport network [3]. An exemplary demonstration is shown in Fig. 3(a) where a "*protection*" function module was defined in an optical network comprising of four OXCs. The protection function was defined to calculate one or multiple backup paths for the service and inform the controller to setup the corresponding OXCs. The optical network was virtualized into two VONs using Flowvisor [9]: VON_1 (OXC_1~OXC_4) and VON_2 (OXC_2, OXC_4). As is shown in Fig. 3 (b), where both VON_1 and VON_2 was able to visit the same VNF, an important feature of service-design based optical networking is the ability to share the VNFs, which is beneficial as it reduces the complexity and the reconfiguration cost of VONs. Fig. 2(b) also shows the configuration of two paths for service protection, i.e. working path 1: OXC_1-OXC_3, and protection path 2: OXC_1-OXC_4-OXC_3.

IV. CONCLUSIONS

A service design concept is extended and introduced into the optical transport network. Customized transmissions were provided with differentiated features added into the transmitted signals via the configuration of VNFs. Orchestrated transmission control was realized and demonstrated by holistically adjusting the optical system in the service design control plane. Moreover, flexible optical networking was also demonstrated based on service design.

REFERENCES

[1] http://www.opennetworking.org/, [Online], "The Open networking foundation homepage for SDN and Openflow".
[2] W. Freude, et al., "Software-defined optical transmission," ICTON 2011, Stockholm, Sweden.
[3] X. Cao, et al., "Functional Service Design with SDN Orchestration across Heterogeneous Multi-domain Networks," OFC 2016, Anaheim, USA.
[4] N. Chowdhury, et al., "A survey of network virtualization," Computer Networks, 54 (5), 2010, pp. 862-876.
[5] P. Quinn, et al., "Service function chaining: creating a service plane via network service headers," Computer, 47 (11), 2014, pp. 38-44.
[6] A. Pages, et al, "Virtual network embedding in optical infrastructures," 14th ICTON, 2012.
[7] B. Heller, "Openflow switch specification, version 1.0.0," Dec 2009. [Online]. Available: http://www.openflow.org/documents/openflow-spec-v1.0.0.pdf.
[8] http://www.restapitutorial.com/, [Online], "The REST API tutorial".
[9] R. Sherwood et al., "Can the production network be the testbed?" USENIX OSDI, 2010.

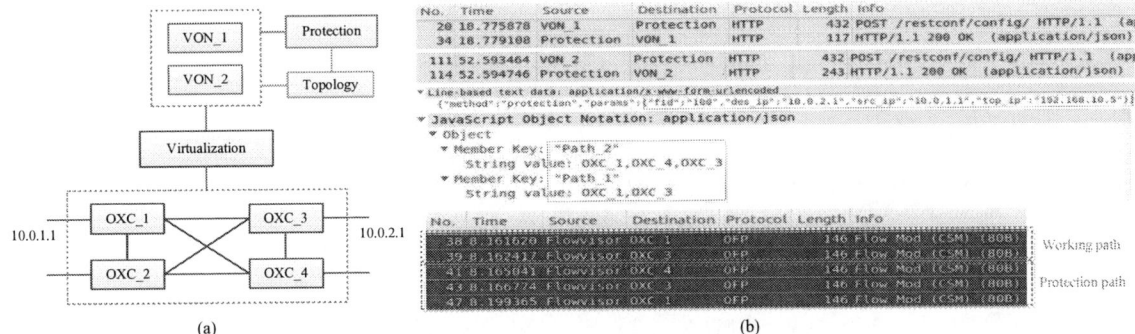

Fig. 3. Service-design based optical networking demonstration: (a) optical network setup; (b) JSON messages between the controllers and the sharable protection function and Openflow messages for OXC setup in VON_1 for path protection.

WA3-3

OECC/PS2016

Multi-Stratum Optimization with Routing Radio and Spectrum Allocation for Cloud Radio over Fiber Networks

Hui Yang, Wei Bai, Ao Yu, Jie Zhang

State Key Laboratory of IPOC, Beijing University of Posts and Telecommunications, No.10 Xitucheng Road, Beijing, 100876, China

Author e-mail address: yanghui@bupt.edu.cn

Abstract: We present a multi-stratum optimization architecture with routing, radio and spectrum allocation for cloud radio over fiber networks. The overall feasibility and efficiency of the proposed architecture are experimentally verified on our testbed.

Keywords: Software defined network, elastic optical network, radio over fiber, radio and spectrum allocation

I. INTRODUCTION

To adapt the rapid evolution of 5G mobile network, the cloud radio access network (C-RAN) is a paradigm introduced by operators which aggregates all base stations computational resources into a cloud baseband unit (BBU) pool, while the distributed radio frequency signals are collected by remote radio head (RRH) and transmitted to the cloud platform through optical transmission [1]. It aims to reduce capital and operating expenditure and enhance real-time cloud computing while offering better services [2]. Nowadays, the interaction between RRH and BBU or resource schedule among BBUs in cloud have become more frequent and complex with the development of system scale and user requirement. It can promote the networking demand among RRHs and BBUs and force to form elastic optical fiber switching and networking [3] due to the characteristics of high bandwidth, low cost and transparent transmission. In such network, the multiple stratum resources of radio, optical and processing unit have interweaved with each other. The traditional architecture cannot efficiently enough implement the resource optimization and scheduling for the high-level quality of service (QoS) guarantee. Recently, as a centralized software control architecture, software defined networking (SDN) can provide maximum flexibility and make a unified control over various resources for the joint optimization of functions and services with a global view [4, 5]. Therefore, it is very important to apply SDN technique to control and optimize the resource assignment in such environment.

In this paper, we propose a novel multi-stratum optimization (MSO) architecture for cloud radio over fiber networks (C-RoFN) with software defined networking. Based on this architecture, it abstracts C-RoFN infrastructure as various kinds of resource cloud. A routing, radio and spectrum allocation (RRSA) scheme is introduced in the proposed architecture. MSO can enhance the responsiveness to end-to-end user demands and globally optimize radio frequency, optical and BBU resources effectively to maximize radio coverage. The overall feasibility and efficiency of the proposed architecture are experimentally verified on SDN-enabled testbed [6].

II. MSO ARCHITECTURE FOR SOFTWARE DEFINED C-RoFN

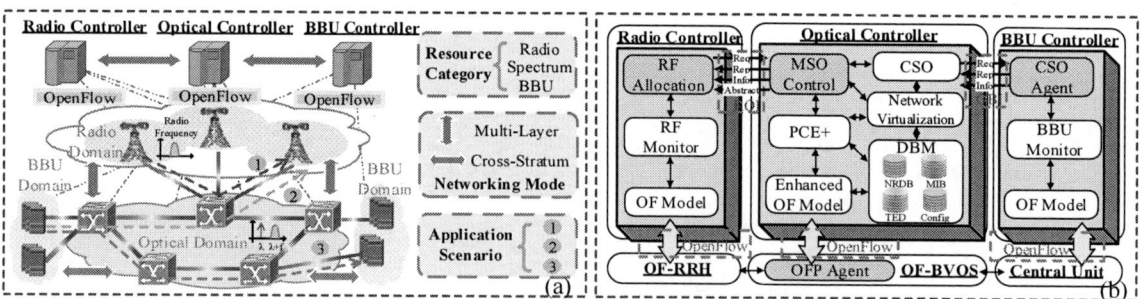

Fig. 1. (a) The architecture of MSO in C-RoFN, (b) The functional models of MSO.

MSO architecture. MSO architecture in software defined C-RoFN is illustrated in Fig. 1(a). The elastic optical network (EON) is used to interconnect the cloud BBUs, while the distributed RRHs are converged into EON, which allocates the customized spectrum with finer granularity for radio signals. C-RoFN consists of three stratums: radio, optical spectrum, and BBU resources. Networking mode for MSO extends in two directions. One is from the perspective of resource form. Optical and computing resources are interconnected cross EON and BBU stratums in *latitudinal direction*, which is established as "*cross-stratum*". The other is from the perspective of carrying capacity. The entity with small granularity is abstracted as high layer (e.g., radio), while large one is abstracted into low-layer network (e.g., EON). The networking of multiple layers is established along *longitudinal direction*, which is called "*multi-layer*". Three MSO applications are formed, i.e., interaction between RRHs, service from RRH to BBU and schedule among BBUs. Each stratum is software defined with OpenFlow protocol (OFP) and controlled by radio

1042

controller (RC), optical controller (OC) and BBU controller (BC) in a unified manner. To control MSO with OFP, OF-enabled RRH and bandwidth-variable optical switch with OFP agent software are required, which are referred to as OF-RRH and OF-BVOS [6]. The motivations for MSO architecture in C-RoFN are twofold. Firstly, MSO emphasizes the cooperation between RC and OC to overcome the obstacles from multi-layer networks and it realizes vertical integration. Secondly, in order to provide the end-to-end QoS, multi-stratum resources can be merged through controllers' interaction with horizontal merging, while achieving global CSO of optical and BBU resources.

Functional model for MSO. The RC, OC and BC are extended to support MSO and shown in Fig. 1(b). In OC, network virtualization is responsible for virtualizing the required optical network resources and interworks the information to perceive the EON, while RF monitor in RC manages virtual radio resource. MSO control decides which BBU is the destination for resource migration or accommodate for RRH, then provides this request to PCE in turn, including the request parameters (e.g., latency and bandwidth). The end-to-end path computation can be completed in PCE considering CSO of optical and BBU resources from BC, where the various strategies are alternative as a plug-in. Enhanced OF model and RF allocation perform continuous spectrum and radio frequency assignment for the computed path and provisions the path by using OFP. Data base management interacts with network virtualization and stores virtual network and BBU resources for MSO.

III. ROUTING, RADIO AND SPECTRUM ALLOCATION

Based on functional architecture described, we propose a routing, radio and spectrum allocation (RRSA) scheme in OC to choose the best destination according to multiple stratum resources using global evaluation factor, and perform service provisioning with continuous spectrum and radio frequency assignment for the destination.

Phase I. RRSA first selects the destination BBU based on processing status collected from BBU, radio and optical condition provided by RC and OC. To measure the choice rationality of service provisioning, we define α as the global evaluation factor which considers all multi-stratum parameters. Two processing parameters, CPU usage U_c^t and storage utilization U_m^t describe the current usage of BBU resource, while optical network parameters are comprised of the traffic engineering weight W_l of the current link and the hop H_p of the candidate path. Radio parameters contain the symbol rate B_r and radio frequency F_r of current radio signal. The overall BBU function is expressed as $f_{ac}(U_m^t, U_c^t, \phi) = \phi \times U_m^t + (1-\phi) \times U_c^t$, where ϕ is adjustable proportion between storage and CPU usage. In addition, optical network function is expressed as $f_{bc}(H_p, W_l) = \sum_{l=1}^{H_p} W_l$, while radio function is $f_{cc}(B_r, F_r) = B_r^2 / F_r$. $f_{a1}, f_{a2} \cdots f_{ak}$ are the BBU parameters among the K candidate BBU nodes, while $f_{b1}, f_{b2} \cdots f_{bk}$ and $f_{c1}, f_{c2} \cdots f_{ck}$ are the optical and radio parameters associated with the K candidate paths. So the global evaluation factor α meets the Eq. 1 as follows, where β and γ are the adjustable weights among the BBU, optical and radio parameters. According to BBU resource utilization, RRSA first chooses the best K candidate BBU nodes in BBU stratum for radio signals and spectrum path. In radio and optical stratums, the node with minimum α value based on global evaluation factor is selected from the K candidates.

$$\alpha = \frac{f_{ac}\left(U_m^t, U_c^t, \phi\right)}{\max\left\{f_{a1}, f_{a2} \cdots f_{ak}\right\}} \beta + \frac{f_{bc}\left(W_l, H_p\right)}{\max\left\{f_{b1}, f_{b2} \cdots f_{bk}\right\}} \gamma + \frac{f_{cc}\left(B_r, F_r\right)}{\max\left\{f_{c1}, f_{c2} \cdots f_{ck}\right\}}(1 - \beta - \gamma) \qquad (1)$$

Phase II. In radio and spectrum allocation phase, three dimensional resources are considered in this phase, which contain radio frequency, spectrum and link. We accommodate the service considering radio frequency assignment with available spectrum first, and then use other fit spectrum resource. Allocation process is illustrated shown in Fig. 2.

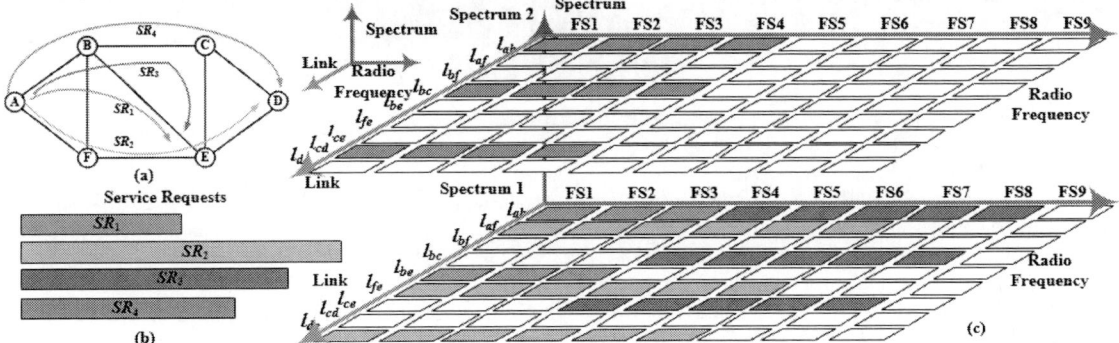

Fig. 2. (a) Simple network topology with six nodes. (b) The service requests. (c) Illustration of radio and spectrum allocation.

IV. EXPERIMENTAL DEMONSTRATION AND RESULTS DISCUSSION

To evaluate the overall feasibility and efficiency of the proposed architecture, we set up an elastic optical network with software defined C-RoFN based on our testbed, as shown in Fig. 3(a). In data plane, two analog RoF intensity modulators and detect modules are utilized, which driven by a microwave source working at 40GHz frequency to generate double sideband. Four OpenFlow-enabled elastic ROADM nodes are equipped with Finisar BV-WSSs in the EON. We use Open vSwitch (OVS) as software OFP agent according to the API to control the hardware and interact between controller and radio and optical nodes. In addition, OFP agents are used to emulate other nodes in data plane to support the MSO with OFP. The BBU pool and OFP agents are realized on an array of virtual machines created by VMware ESXi V5.1 running on IBM X3650 servers. The virtual operation system technology makes it easy to set up experiment topology for large scale extension. For OpenFlow-based MSO control plane, OC server is assigned to support the proposed architecture and deployed by means of three virtual machines for MSO control, network virtualization and PCE strategy as plug in, while the RC server is used as radio frequency resource monitor and assignment. BC server is deployed as CSO agent to monitor the computing resources from BBUs. Each controller server controls the corresponding resources, while the database servers are responsible for maintaining traffic engineering database (TED), connection status and the configuration of the database. We deploy the service information generator related with RC, which implements batch C-RoFN services for experiments.

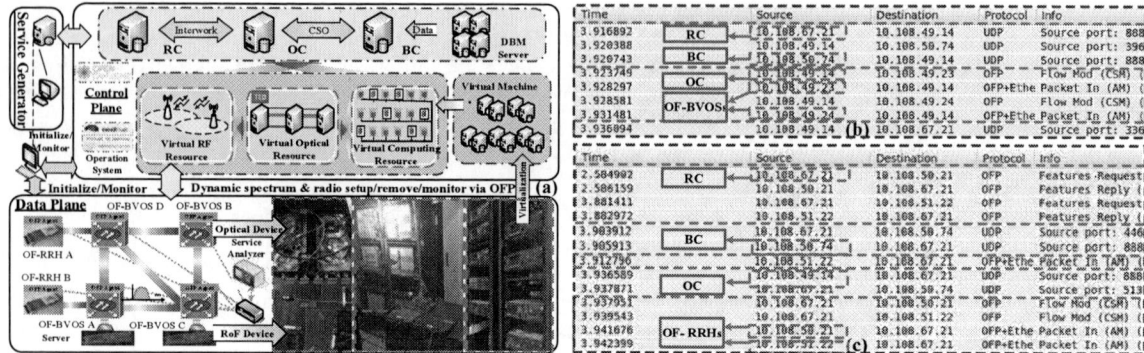

Fig. 3. (a) Experimental testbed for MSO, Wireshark capture of the message sequence for MSO in (b) OC and (c) RC.

Based on the testbed, we have designed and verified experimentally MSO for service in C-RoFN. Fig. 3(b)-(c) present the signaling procedure for MSO using OFP through a Wireshark capture deployed in OC and RC respectively. The spectrum of lightpath for analog C-RoFN is reflected on the filter profile, as shown in Fig. 4(a). The radio signals can be modulated on the spectral channel with MSO. We also evaluate the performances of MSO with RRSA scheme under heavy traffic load scenario and compare with the CSO scheme [6] using virtual machines. The requests are setup with bandwidth randomly distributed from 500MHz to 40GHz, while computing usage in BBU is selected randomly from 1% to 0.1% for each demand. Fig. 4(b)-(c) compare performances of two schemes in terms of resource occupation rate and path provisioning latency. As shown, RRSA outperforms CSO in the resource occupation rate significantly. The reason is that RRSA can globally optimize the radio, optical and BBU stratum resources to maximize radio coverage. Fig. 4(c) shows RRSA scheme reduces the provisioning latency compared to the other. That is because RRSA chooses the destination BBU before the service arrives, which leads to low computation and provisioning time being consumed.

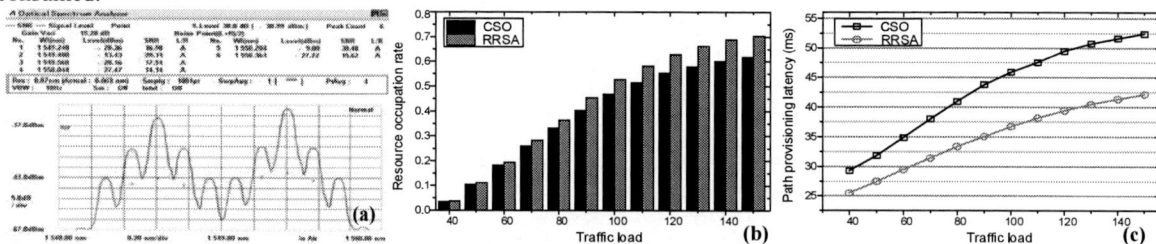

Fig. 4. (a) Filter output of spectrum, comparison on (b) resource occupation rate and (c) path provisioning latency in heavy traffic load scenario.

V. CONCLUSIONS

This paper presents a novel MSO architecture in software defined C-RoFN. Additionally, the RRSA scheme is introduced in the proposed architecture. Our experiments verify MSO with RRSA can enhance responsiveness and multiple stratum resources optimization of radio, optical spectrum and BBU processing resources effectively.

ACKNOWLEDGMENT This work has been supported by NSFC project (61501049), Fundamental Research Funds for the Central Universities (2015RC15), and Fund of State Key Laboratory of Information Photonics and Optical Communications (BUPT), P. R. China (IPOC2015ZT01).

REFERENCES

[1] X. Rao and V. Lau, "Distributed fronthaul compression and joint signal recovery in Cloud-RAN," IEEE Trans. Signal Process. 63, 1056-1065, 2015.

[2] J. Wu, et al., "Cloud radio access network (C-RAN): A primer," IEEE Network 29, 35-41, 2015.

[3] K. Tanaka and A. Agata, "Next Generation Optical Access Networks for C-RAN," Proc. of OFC, Tu2E.1, 2015.

[4] K. Kondepu et al., "An SDN-based integration of green TWDM-PONs and metro networks preserving end-to-end delay," Proc. of OFC, Th2A.62, 2015.

[5] A. Aguado, et al., "ABNO: a feasible SDN approach for multi-vendor IP and optical networks," Proc. of OFC, Th3I.5, 2014.

[6] H. Yang, et al., "CSO: Cross Stratum Optimization for Optical as a Service," IEEE Commun. Mag. 53, 130-139, 2015.

WA3-4

OECC/PS2016

Highly Efficient Adaptive Bandwidth Allocation Algorithm for Elastic Lambda Aggregation Network

Hiroyuki Saito, Naoki Minato, Shuko Kobayashi and Hideaki Tamai

Oki Electric Industry Co., Ltd., Corporate Research and Development Center, 1-16-8 Chuo, Warabi-shi, Saitama, 335-8510, Japan
saitou738@oki.com

Abstract: Adaptive bandwidth allocation algorithm is proposed for elastic lambda aggregation network. Simulation results show that the proposed algorithm is flexibly applicable to network conditions and can keep high spectrum efficiency in any traffic scenarios.

Keywords: WDM-OFDM-PON, elastic lambda aggregation network, spectrum efficiency

I. INTRODUCTION

Wavelength division multiplexed / Time division multiplexed (WDM/TDM)-PON was standardized by ITU-T/ FSAN as next-generation passive optical network stage 2 (NG-PON2) in 2015 and some demonstrations and field trials were recently reported [1, 2]. Meanwhile, in recent years, not only the bandwidth demand but also the variety of network services has explosively increased. Suppose a wavelength channel is allocated for each service, network operators would suffer from huge amount of operation and management costs and available wavelengths would run dry shortly. Therefore, highly efficient network is expected for next generation network.

Orthogonal frequency division multiple (OFDM) access based network systems [3-7] which have high frequency spectrum efficiency are studied for the next generation PON following on NG-PON2. We are developing WDM/OFDM-PON based Elastic Lambda Aggregation Network (EλAN) [8]. One of the advantages of EλAN is that the optical parameters, such as modulation format, a number of sub-carriers, symbol rate and wavelength, can be adaptively variable depending on demands of ONUs or network configuration. We proposed adaptive bandwidth allocation (ABA) algorithm for EλAN [8]. The point of the proposed algorithm is that if the total of required bandwidth of ONUs (BW_{alloc_total}) is greater than the available bandwidth (BW_{avail}), the algorithm decreases the bandwidth to be allocated to each ONU (BW_{alloc_i}) gradually until BW_{alloc_total} is less than BW_{avail}. However, the proposed algorithm has been studied only when BW_{alloc_total} is greater than BW_{avail}, in other words, network is crowded. To maximize the performance for EλAN, bandwidth allocation algorithm which is applicable for both overload and underload traffic scenario is desired.

In this paper, we propose an improved ABA algorithm. Simulation results shows that the proposed algorithm can achieve high spectrum efficiency steadily in any traffic scenarios.

II. PROPOSED ADAPTIVE BANDWIDTH ALLOCATION ALGORITHM

Fig.1 (a) shows a flow chart of proposed algorithm. The proposed algorithm consists of two phases. Phase 1 is to decide adaptive optical parameters and calculate BW_{alloc_i}. Phase 2 is to recalculate adaptive BW_{alloc_i} depending on network traffic. The optical parameters are decided adaptively by input parameters in steps 1-3 and calculate BW_{alloc_i} in step 4. Input parameters are defined as the distance of PON section, QoS and bandwidth demands of each ONU (ONU_{dem_i}). After that, BW_{alloc_total} is calculated. If BW_{alloc_total} is less than BW_{avail}, BW_{alloc_i} is increased gradually and recalculated until BW_{alloc_total} reaches BW_{avail} in steps 5-8. On the other hand, If BW_{alloc_total} is greater than BW_{avail}, BW_{alloc_i} is cut gradually and recalculated until BW_{alloc_total} is less than BW_{avail} in steps 9-12.

Following step 1 to 3 are executed for each ONU#i (i = 1 to N, where N is the maximum number of ONUs). In step 1, pairs of optical parameters which satisfy the constraint of the distance of PON section and QoS are selected by using a distribution diagram. The distribution diagram, of which example is shown in Fig.1 (b), shows the relationship between QoS and the distance of PON section as a function of the optical parameters. The diagram is simulation result using the network model of EλAN in Fig.2 (a) and prepared in advance. The model network is consists of WSS section and PON section. WSS section consists of two WSSs, a 20-km single mode fiber (SMF) and three optical amplifiers. The WSS and the SMF is assumed to power attenuator, of which insertion loss values are 7 dB and 4 dB, respectively. Gain of each optical amplifier is adjusted to compensate transmission losses. PON section consists of a power splitter and SMFs. The loss of power splitter is 28 dB. Tab.1 shows a list of the other parameters. A pair of optical parameters, a modulation format and the number of sub-carriers (SCs) of OFDM-band, is selected from QPSK/32SCs, QPSK/64SCs, QPSK/128SCs, 16QAM/32SCs, 16QAM/64SCs, 16QAM/128SCs, 64QAM/32SCs, 64QAM/64SCs and 64QAM/128SCs. In this simulation, QoS is defined as Symbol Error Rate (SER) and changes randomly from 10^{-4} to 10^{-9}. Also BW_{dem_i} is defined as a transmission bit rate and changes randomly from 1 Gb/s to 40 Gb/s. Transmission bit rate of QPSK/32SCs is 10 Gb/s. 256 ONUs are assumed to distribute uniformly between 0 km and 40km. 2 GHz guard band between neighboring OFDM-band is used. Pairs of optical parameters which satisfy BW_{dem_i} are selected in step 2. A pair of optical parameters is decided which satisfy both step 1 and step 2 in step 3. If multiple pairs satisfy this condition, then chose one which has the largest modulation format.

For example, when the value of the distance of PON section is 20km, QoS (SER) is 10^{-7}, and BW_{dem_i} is 40 Gb/s, In step 1, QPSK/32SCs, QPSK/64SCs, QPSK/128SCs, 16QAM/32SCs and 16QAM/64SCs are selected. In step 2, QPSK/128SCs, 16QAM/64SCs, 16QAM/128SCs, 64QAM/32SCs, 64QAM/64SCs and 64QAM/128SCs are selected. Therefore, adaptive optical parameter is 16QAM / 64SCs in step 3.

BW_{alloc_i} is calculated from modulation format and BW_{dem_i} in step 4. In step 5, fragmented sub-carriers are merged to reduce the number of OFDM-bands. BW_{alloc_total} is calculated in step 6. BW_{alloc_total} is compared with BW_{avail} in step 7. If BW_{alloc_total} is less than BW_{avail}, $BW_{alloc\_inc\_i}$ is calculated in step 8 and BW_{alloc_total} is recalculated. Calculation of step 5-8 is repeated until BW_{alloc_total} is greater than BW_{avail}. Calculated $BW_{alloc\_inc\_i}$ in step 8 is obtained as follows.

$$BW_{alloc\_inc\_i} = ceil[BW_{alloc\_initial\_i} \times (BW_{avail} \times (1 + (R \times n))) / BW_{alloc\_total\_initial}] \qquad (1)$$

where $BW_{alloc\_initial\_i}$ and $BW_{alloc\_total\_initial}$ are the initial values (i.e., $n=0$) of BW_{alloc_i} and BW_{alloc_total}, respectively. R is a constant of proportionality.

If BW_{alloc_total} is greater than BW_{avail}, $BW_{alloc\_dec\_i}$ is calculated in step 9 and BW_{alloc_total} is recalculated. Calculation of step 9-11 is repeated until BW_{alloc_total} is less than BW_{avail}. Step 5-6 and step 10-11 are same process. Calculated $BW_{alloc\_dec\_i}$ in step 9 is obtained as follows.

$$BW_{alloc\_dec\_i} = ceil[BW_{alloc\_inc\_i} \times BW_{avail} / (BW_{alloc\_total\_initial}(1 + (R \times n)))] \qquad (2)$$

If $BW_{alloc\_inc\_i}$ has never been calculated, $BW_{alloc\_inc\_i}$ is defined as BW_{alloc_i}.

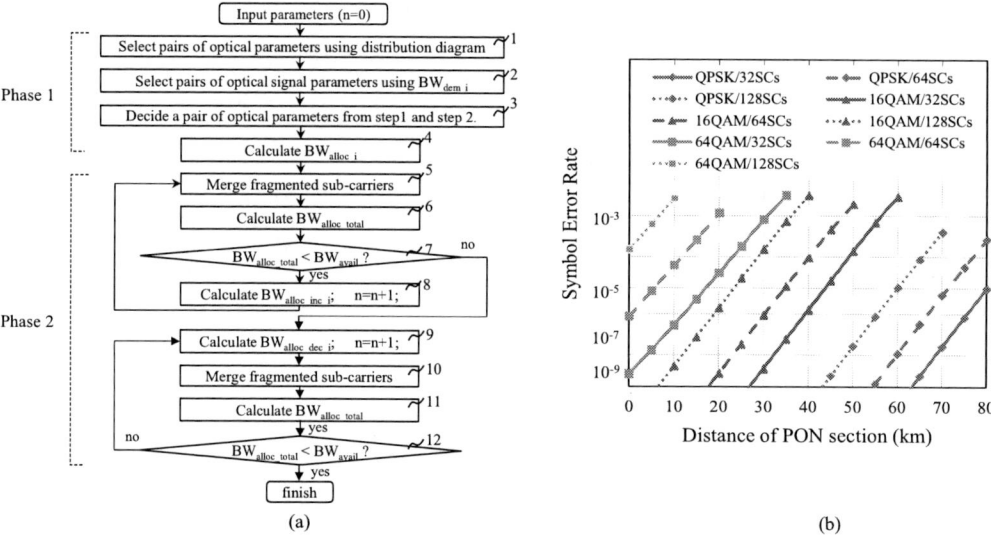

Fig.1. (a) Flow chart of ABA algorithm. (b) Distribution diagram.

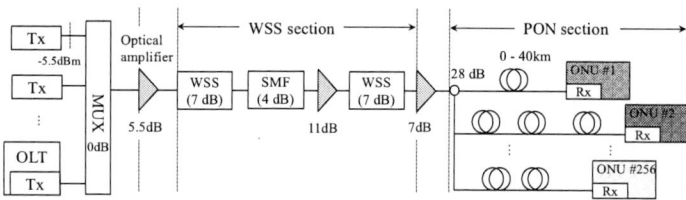

Fig.2. Network model of EλAN.

Tab.1. Parameter list.

Parameters	Values
Modulation format	QPSK / 16QAM / 64QAM
Number of sub-carriers (OFDM-band)	32 SCs / 64 SCs / 128 SCs
QoS (SER)	10^{-4} - 10^{-9}
BW_{dem_i} (Tx Bit Rate)	1-40 Gbps
Number of ONU	256
Guard band	2 GHz (0.016 nm)

III. SIMULATION RESULTS

Fig.3 (a) shows the calculation result of spectrum efficiency as a function of $BW_{alloc\_total\_initial}$ / BW_{avail}, which means how much overload or underload. The $BW_{alloc\_total\_initial}$ / BW_{avail} range is from 60 % to 170 %. Spectrum efficiency was obtained as follows.

$$spectrum\_efficiency = \frac{\sum_{i=1}^{256} BW_{alloc\_i}}{BW_{avail}} \qquad (3)$$

This result shows that the average of spectrum efficiency is 82 % and spectrum efficiency is steady independent of $BW_{alloc\_total\_initial}$ / BW_{avail} when R in Eq. (1) and Eq. (2) is 0.001. It indicates that proposed algorithm can keep high spectrum efficiency regardless of traffic scenario. Fig.3 (b) shows an example of the relationship between spectrum efficiency and loop count of proposed algorithm as a function of R in Eq. (2) when the value of $BW_{alloc\_total\_initial}$ / BW_{avail} is 162 %. This result shows that the smaller the value of R is, more the spectrum efficiency is high. On the other hand, loop count of the proposed algorithm, in other words the processing time, increases. It indicates that there is a trade-off between spectrum efficiency and processing time. Therefore, the value of n needs deciding adaptively depending on network situations. And so for example, in case of network is crowded, the value of n needs to be small. In case of response time is more important than spectrum efficiency, the value of n needs to be big. Thus, in these results, proposed algorithm is applicable flexibly to network conditions.

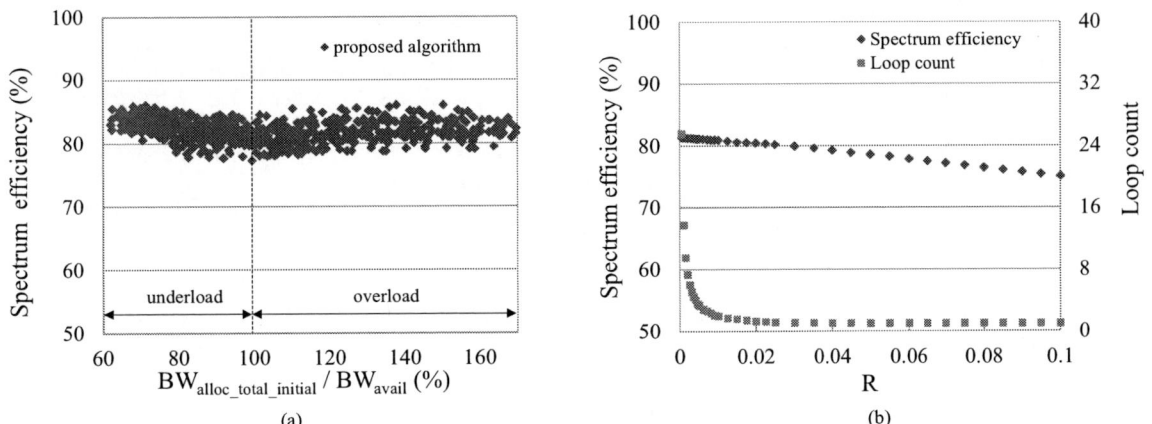

Fig.3. (a) Spectrum efficiency. (b) Relationship between spectrum efficiency and loop count.

IV. CONCLUSIONS

We proposed highly efficient adaptive bandwidth allocation algorithm for both overload and underload traffic scenario in WDM/OFDM-PON-based elastic lambda aggregation network. The point is that if when network is less crowded, the algorithm increases BW_{alloc_i} gradually until BW_{alloc_total} reaches BW_{avail}. On the other hand, if when network is crowded, the algorithm decreases BW_{alloc_i} gradually until BW_{alloc_total} is less than BW_{avail}. Simulation results showed that the proposed algorithm can keep high spectrum efficiency regardless of traffic scenario and there is a trade-off between spectrum efficiency and processing time. These results suggest that proposed algorithm is flexibly applicable to any traffic scenarios and network conditions.

ACKNOWLEDGMENT

A part of this work is supported by the National Institute of Information and Communications Technology (NICT) of Japan.

REFERENCES

[1] Y. Senoo, et al., "Demonstration of 512-ONU real-time dynamic-load-balancing with few wavelength reallocations for λ-tunable WDM/TDM-PON," in proc. OFC 2015, Tu.3E.1.

[2] K. Taguchi, et al., "Burst-Off-Level Power Reduction in λ-Tunable Transmitter Using a Reverse Bias Voltage Controlled Burst-Mode Booster SOA for 256-Split WDM/TDM-PON," in proc. OFC'2015, PDP, Tu3E.8.

[3] C. Ruprecht, et al., "37.5-km Urban Field Trial of OFDMA-PON using Colorless ONUs with Dynamic Bandwidth Allocation and TCM," in proc. OFC 2014, Th3G.5.

[4] X. Xiao, et al., "100-Gb/s Single-band Ream-time Coherent Optical DP-16QAM-OFDM Transmission and Reception," in proc. OFC'2014, PDP, Th5C.6.

[5] N. Cvijetic, et al., "What is Next for DSP-based Optical Access and OFDM-PON ?," in proc. ECOC 2014, We.1.6.2.

[6] K. Ishii, et al., "Demonstration of No-guard-band Coherent IFDMA/OFDMA/SC-FDMA-PON Co-existence Uplink System using Real-time IFDMA Transmitter," in proc. OFC'2015, Th1H.3.

[7] J. von Hoyningen-Huene, "OFDM for Optical Access," tutorial paper. OFC'2015, Th1H.1.

[8] H. Saito, et al., "Adaptive Bandwidth Allocation Algorithm for WDM/OFDM-PON-based Elastic Lambda Aggregation Network," in proc. OFC 2016, Th3C.2.

WA4-1 (Invited)

OECC/PS2016

Resilient Virtual Optical Network Provisioning over Software-Defined Optical Networks

Xi Wang, Inwoong Kim, Qiong Zhang, Paparao Palacharla, Tadashi Ikeuchi

Fujitsu Laboratories of America, Inc., Richardson, Texas USA   Email: xi.wang@us.fujitsu.com

Abstract: *Resiliency of the virtualization layer in software-defined optical networks is key for on-demand high-bandwidth services. We propose a resilient virtualization solution that can relax the reliability of physical optical nodes while providing highly-available services.*

Keywords: virtual optical network, software-defined optical network, node availability, optical whitebox

I. INTRODUCTION

Recently, an increasing number of geographically distributed datacenters are being interconnected by optical networks, forming wide-area Information and Communications Technology (ICT) infrastructures. Cloud applications such as video content distribution, social networking and online gaming rely heavily on such ICT infrastructures for improved service quality as well as user's quality of experience. Such demands in turn drive the need for optical networks with even higher capacity, flexibility and dynamic bandwidth reconfigurability on shorter time scale under cloud orchestration systems using Software-Defined Networking (SDN).

Software-Defined Optical Networks (SDONs) using colorless/directionless/flex-grid ROADMs and multi-modulation formats programmable transponders are expected to address these needs through their new capability of optical network virtualization. For multi-tenancy in inter-datacenter networks, optical network virtualization enables network operators to provision multiple coexisting and isolated Virtual Optical Networks (VONs) over the same physical infrastructure. The flexibility of VON empowers the network operators to optimize global resource usage while offering adequate level of agility and programmability tailored to individual services.

Meanwhile, resiliency continues to be a vital service attribute in optical networking. As such, VONs are expected to offer high level of service resiliency as well. Various VON resiliency solutions have been reported in literature [1-4], most of which assume 'reasonably reliable' physical optical networks, where failures usually occur one-at-a-time. However, this assumption may no longer hold true with the advent of optical whiteboxes [5]. Whiteboxes are similar to bare metal generic PCs, sold by multiple vendors, vastly different in quality and price. Although most vendors may strive to offer whiteboxes with excellent reliability, products from certain vendors or some low-end models may have less satisfactory (or even unpredictable) reliability. Accordingly, new challenge arises when the operators choose to build their optical networks with multi-vendor, mixed-reliability whiteboxes. More specifically, how can one provide resilient virtual network services on top of less reliable physical optical networks?

In this paper, we first study network failure events in whitebox optical networks with varying node availabilities. We identify the thresholds of node availabilities, beyond which simultaneous network failures become *common* events. We then present our new reliable virtual optical node solution that is capable of handling large number of simultaneous failures. Finally, we use reliable virtual optical nodes to provision virtualized Single Big Optical Switch (vSBOS) services, and evaluate service availabilities under different physical node availabilities and network topologies.

II. SIMULATION OF WHITEBOX OPTICAL NETWORK FAILURES

We use well-known DTnet (14 nodes, 19 links) and CORONET (75 nodes, 99 links) topologies to simulate network failure events caused by optical whitebox failures. For simplicity, in each simulation run, we assume that all nodes in the network are whiteboxes with the same level of Node Availability (NA). The NA is defined as NA = MTTF / (MTTF + MTTR), where MTTF stands for Mean Time To Failure and MTTR stands for Mean Time To Repair. For example, if each node operates continuously for an average of 99 days before failure occurs (MTTF = 99 days) and it takes an average of 1 day to repair/replace the node (MTTR = 1 day), then NA = 99 / (99 + 1) = 0.99. During simulation, each node's failure/recovery events occur independently. A node failure event follows Poisson arrival, and the time duration for a node recovery follows negative exponential distribution. Each simulation runs for 10,950 time units (corresponding to 30 years when 1 time unit = 1 day) with a certain NA, which is gradually lowered from 0.9999 (roughly 1 failure every 10,000 days = 27 years) to 0.95 (1 failure every 20 days) per simulation iteration. Note that in this study, we do not consider link failure (such as fiber cut, etc.) events independently. Instead, a link goes down when a node it connects to goes down, and comes back on when both nodes it connects to are back to working condition.

Fig. 1 and Fig. 2 show the histogram of the occurrence of simultaneous node failures in DTnet and CORONET, respectively. The histogram is obtained as follows. Every time the network encounters a new node failure/recovery event, the time interval between the current event and the previous event constitutes a time period. For each time period, we count the total number of nodes in the network that are currently in failure state. If during a simulation run there are 4 time periods where there is no node failure across the whole network, 3 time periods where there is one node failure, and 2 time periods where there are two simultaneous node failures, then the occurrences of 0, 1 and 2 simultaneous-node-failures are 4, 3 and 2, respectively. As shown in two figures, with NA=0.9999, both DTnet and CORONET have either failure-free operation, or encounter a small number of single-node failures *only*. Since optical network equipment today

1048

is usually engineered with such high reliability, it makes sense that existing resiliency solutions against single node/link failure would suffice. However, with NA=0.999, both networks start to experience 2-simultaneous-node-failures. With NA=0.99 and below, simultaneous node failures become *predominant* in both networks, and CORONET suffers more higher-count simultaneous-node-failures (such as five 9-simultaneous-node-failures with NA=0.97 and three 14-simultaneous-node-failures with NA=0.95) due to its larger node count. Node availability of 0.999 corresponds to around 3-year product lifecycle (similar to datacenter servers' replacement cycle), and simultaneous node failures become *common* in networks built with NA 0.999 and below. Consequently, it is imperative to have more robust resiliency solutions in place when operating potentially less-reliable whitebox optical networks.

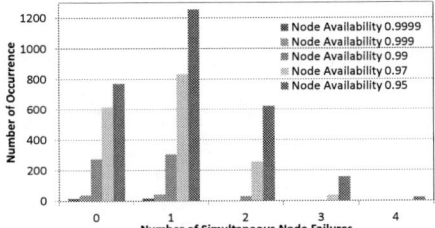

Fig. 1. Histogram of the occurrence of simultaneous node failures under different node availabilities in DTnet.

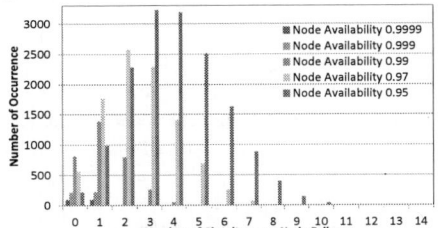

Fig. 2. Histogram of the occurrence of simultaneous node failures under different node availabilities in CORONET.

III. RESILIENT VIRTUAL OPTICAL NODES

Leveraging SDON's virtualization capability, our resiliency solution focuses on composing *resilient* virtual optical nodes from *less-reliable* physical optical nodes. A virtual optical node consists of virtual nodal degrees (vDegrees) and virtual nodal degree connections (vConnections) among these vDegrees. A vDegree is mapped to one or more physical nodes, and a vConnection linking a particular vDegree pair is mapped to a lightpath connecting the physical node pair mapped by this vDegree pair. Since both the vDegree-to-physical node mapping and the vConnection-to-lightpath mapping can be dynamically updated under network failures, the availability of the virtual node can be much improved compared to the underlying physical nodes. We then use such reliable virtual nodes to offer high-availability virtual optical network services. For example, a virtualized Single Big Optical Switch (vSBOS) service can be offered using one such virtual optical node, as illustrated in Fig. 3. In this example, a datacenter operator requests a vSBOS to interconnect four geographically distributed datacenters. All datacenters appear to be connected to a single 'optical' switching hub, with 100 Gbps connectivity for datacenter A-B & C-D and 200 Gbps connectivity for datacenter B-C & A-D. For this, the optical network operator provisions a 4-vDegree virtual optical node, with one 100 Gbps vConnection each for vDegree pairs A-B & C-D, two 200 Gbps vConnections each for vDegree pairs B-C & A-D. The vDegrees and vConnections are then mapped to physical nodes and lightpaths according to the locality information and resource availability. In Fig. 4, four such individual virtual optical nodes, each with different number of nodal degrees with full-mesh connectivity, are provisioned on top of the common physical network.

Although this paper focuses on isolated vSBOS provisioning using single virtual optical node at a time, one can also provision multiple virtual optical nodes connected by flexibly defined virtual links, thereby creating more sophisticated and versatile virtual optical network services.

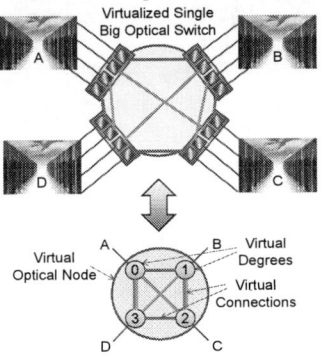

Fig. 3. A virtual optical node offering virtualized Single Big Optical Switch (vSBOS) service.

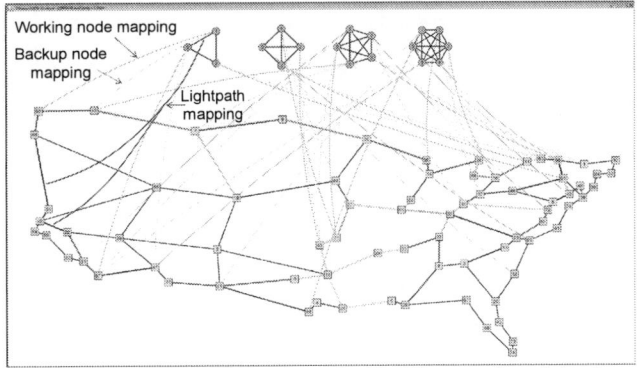

Fig. 4. Four virtual optical nodes, each with different number of virtual nodal degrees, are created on top of the physical optical network.

IV. RESILIENT VIRTUALIZED SINGLE BIG OPTICAL SWITCH SERVICE PROVISIONING

For each vSBOS service request, a dedicated virtual optical node is created. Each vDegree is mapped to its working and backup physical node(s) based on the connectivity from these physical nodes to the user site (e.g. dual homed to two nodes). Each vConnection is mapped to one or more lightpaths between the physical node pairs mapped by the vConnection's source and destination vDegrees. Once all vDegrees and vConnections of the virtual optical node are successfully mapped, the vSBOS service can be activated by the SDON controller.

A. Handling of vConnection failures (vConnection resiliency)

A physical node or link failure along a lightpath causes the lightpath failure, which results in its associated vConnection failure. The vConnection automatically remaps to its pre-calculated or dynamically re-calculated working lightpath. Successful remapping restores the vConnection back to working condition (here we assume that the switching time of vDegrees/vConnections is negligible compared to MTTR).

B. Handling of vDegree failures (vDegree resiliency)

For a physical node currently serving a vDegree, the physical node failure causes the vDegree failure, as well as failures of all vConnections associated with the vDegree. The vDegree automatically remaps to one of its backup physical nodes that is in working condition. A successful remapping restores the vDegree back to working condition. Upon vDegree recovery, the vConnections are also remapped and restored following the steps explained above.

V. PERFORMANCE EVALUATION

We evaluate vSBOS service availabilities under different virtual node configurations, physical node availabilities and network topologies. In the beginning of a simulation run, a small number of vSBOS service requests are offered to the network. A dedicated virtual optical node is created for each vSBOS service, and the number of its vDegrees ranges from 3 to 6 (as shown in Fig. 4). All virtual optical nodes require full mesh vConnections among vDegrees. Each vDegree is randomly mapped to a working physical node, as well as one backup physical node randomly chosen among the working node's neighbor nodes. Each vConnection is mapped to a lightpath routed on the shortest path (or shortest available path in case of handling vConnection failures). All vConnections are 400 Gbps and occupy 11 12.5 GHz spectrum slots per lightpath using first-fit slot assignment. Note that due to the small number of carried lightpaths in the network, there is no lightpath blocking due to spectrum slot exhaust. All vSBOS services continue for the whole duration of the simulation run that lasts 10,950 time units. The physical nodes in the network randomly fail/recover based on a given node availability as described in section II, which is varied from 0.95, 0.97, 0.99 to 0.999. Upon physical node failures, each affected virtual optical node attempts to restore vConnection and/or vDegree failures using the resiliency features described in section IV. We define a vSBOS service to be available only when all the virtual optical node's vConnections and vDegrees are available. At the end of a simulation run, the vSBOS's service availability is calculated by Total Up Time / (Total Up Time + Total Down Time). Furthermore, the result is averaged over 100 simulation runs with different random seeds (which result in different mappings).

Fig. 5 and Fig. 6 compare vSBOS service availabilities *with* vs. *without* virtual optical node resiliency in DTnet and CORONET. The resiliency features improve vSBOS service availabilities in both networks under all physical node availabilities, where higher physical node availability contributes to higher resiliency gain. The Service Availability (SA) improvement = (SA w/ Resiliency – SA w/o Resiliency) / (1 - SA w/ Resiliency). For example, (0.999-0.99) / (1-0.999) = 9, therefore 9 times improvement. Overall, resilient virtual optical nodes achieve 5-77 times higher service availability in DTnet and 5-739 times higher service availability in CORONET. The improvement is higher in CORONET due to its larger topological size, hence better chance of finding alternative paths under simultaneous node failures.

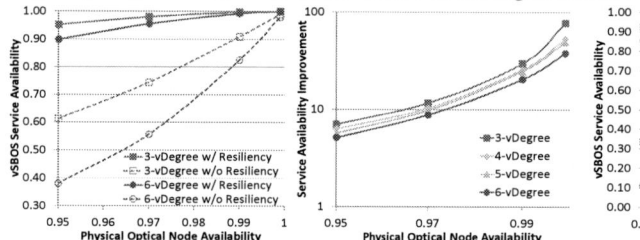

Fig. 5. vSBOS service availability and improvement *with* vs. *without* virtual optical node resiliency in DTnet.

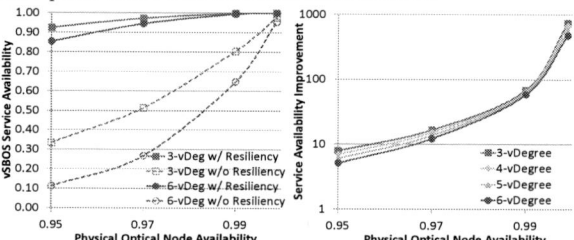

Fig. 6. vSBOS service availability and improvement *with* vs. *without* virtual optical node resiliency in CORONET.

VI. CONCLUSIONS

We present a resilient virtual optical networking solution that can maintain high level of service availability in optical networks with very low physical node availability (e.g. three 9s or less), where large number of simultaneous node failures is common. For example, with physical node availability 0.999 (corresponds to 1 failure per 3 years), the resilient virtual optical node realizes service availability above 0.9995 (1 outage per 5.5 years) in DTnet and above 0.9999 (1 outage per 27 years) in CORONET. These findings suggest that the proposed virtualization solution can relax the reliability requirement of optical whiteboxes, yet capable of offering high-availability virtualized optical mesh connectivity services.

REFERENCES

[1] W. Xie, J. P. Jue, Q. Zhang, X. Wang, Q. She, P. Palacharla, M. Sekiya, "Survivable Virtual Optical Network Mapping in Flexible-Grid Optical Networks," ICNC 2014, pp. 221-225, Feb. 3-6, Honolulu, Hawaii.

[2] X. Wang, M. R. Prasad, F. Yu, I. Ghosh, P. Palacharla, Q. Zhang, I. Kim, T. Ikeuchi, "Scalable Virtual Optical Network Mapping over Software-Defined Flexible Grid Optical Networks," ACP 2015, Paper ASu3H.2, Nov 19-23, Shanghai, China.

[3] J. Kong, S. Hong, J. P. Jue, I. Kim, X. Wang, Q. Zhang, H. C. Cankaya, W. Xie, T. Ikeuchi, "Availability-Guaranteed Virtual Optical Network Mapping with Selective Path Protection," OFC 2016, Mar 20-24, Paper W1B.4, Anaheim, California.

[4] A. Tzanakaki, M. P. Anastasopoulos, K. N. Georgakilas, "Dynamic Virtual Optical Networks Supporting Uncertain Traffic Demands," JOCN Vol. 5, Issue 10, pp. A76-A85 (2013).

[5] The Optical White Box Is Coming, https://www.sdxcentral.com/articles/editorial/the-optical-white-box-is-coming/2015/04/

WA4-2 (Invited)

OECC/PS2016

NFV and SDN for Next Telecom Cloud and Core Networking

Michiaki Hayashi

KDDI R&D Laboratories Inc.

mc-hayashi@kddilabs.jp

Abstract: Importance of NFV and SDN towards 5G is described with future scenarios and consequent issues. An automated, scalable common NFVI platform with a technique to manage packet processing performance considering data-intensive VNFs is introduced.

Keywords: NFV, SDN, 5G

I. INTRODUCTION

Telecom network infrastructure needs to be further compatible with cloud or computing architecture as a flexible, cost-effective service platform. Network function virtualization (NFV) and software-defined network (SDN) are essential enablers of porting telecom equipment into a cloud-based platform. Recently, technical issues from a network perspective toward 5G is being intensively discussed as well as radio access aspect, and the Telecommunication Technology Committee (TTC) of Japan has published white papers showing technical issues and key enablers especially from network technology aspect [1]. It implies the importance of network virtualization as a key technology. Indeed, a variety of proofs of concepts (PoC) [2] has shown the possibility of NFV and SDN, however, we still have two open questions: how to manage complex operational process, and to what extent we should virtualize network functions.

In this paper, the roles of NFV and SDN toward 5G network are introduced. Then, our technical contribution to achieving automated, scalable management of cloud-based NFV infrastructure (NFVI) is described to address the first question above. Since packet processing performance obtained from commodity computing platform significantly defines the extent of virtualization application, our effort to understand the performance is introduced to address the second question.

II. ROLE OF NFV AND SDN TO RESOLVE THE TECHNICAL ISSUES OF THE NEXT TELECOM INFRASTRUCTURE

To discuss the role of NFV and SDN toward the next telecom infrastructure, technical issues yielded from future scenarios expected in around 2023 are described with a summary shown in Table I [2]. From capacity, diversity, heterogeneity, and social infrastructure perspectives, four future scenarios are identified. The resulting technical issues from the future scenarios indicate that NFV and SDN are required to address these issues (as shown in bold face in Table I). Even achieving ultra-low latency may require edge-computing capabilities where NFV and SDN play an important role in the design and creation of the particular slice. To tackle increasing load, resilience, and complex operation, NFV and SDN are inevitable solutions as various past PoC claim.

TABLE I
SCENARIOS AND TECHNICAL ISSUES FOR 5G NETWORKS [2]
(**BOLD FACE**: NFV/SDN RELATED)

#	Future scenarios	Resulting technical issues
1	Expansion of mobile data	• Ultra large capacity in the U-plane • **Increasing load on C-plane** • Extreme power saving • **Complex operation and management**
2	Growing diversity of service types	• **Increasing load on C-plane** • **Achieving ultra low latency** • Limitations of transport layer protocols • **Complex operation and management**
3	Utilization of various types of RATs	• Inter-working among multiple RATs • Limitations of transport layer protocols • **Complex operation and management**
4	Increasing importance as social infrastructure	• Security • **Resilience against large-scale disasters, congestions, and failures** • **Dealing with various types of terminals, traffics, and operators**

1051

III. FRAMEWORK OF CLOSED-LOOP AUTOMATED MANAGEMENT

In a telecom cloud architecture, a common NFV platform should be applicable to various network functions from customer premises functions (e.g., customer premises equipment (CPE) and home gateway (HGW)) to core networking functions (e.g., evolved packet core (EPC) and routing). Network orchestration of such heterogeneous network functions is necessary, and we demonstrated orchestration from access to backbone network segments in order to make an end-to-end slice [3]. By integrating various network functions with the common NFV platform, barriers to achieving integrated management under unified rules, policies, and definitions are expected to be relaxed. Integrated management also enables operators to reduce maintenance costs by further introducing scheduled maintenance, which is achieved by sharing cloud resources among various services and also by introducing closed-loop automated management. Closed-loop management requires interworking between the monitoring system (e.g., operation support system (OSS)) and management and orchestration (MANO). This interwork needs clear definition of alarm/event transfer and life-cycle management between OSS and MANO, where the orchestrator plays an important role [4]. The interworking issue has been discussed in ETSI (e.g., IFA013), TMForum (e.g., ZOOM project), and 3GPP SA5 (e.g., TS28.525), and is nearly determined. We have contributed to these standard bodies through experiences including PoC. Last year, we successfully demonstrated the closed-loop management of the common NFV platform designed for virtualizing various mobile and fixed network functions provided by Nokia, Cisco, HP, and NEC [5]. In the demonstration, KDDI R&D Labs' orchestrator interworked with HP's OSS, as shown in Fig. 1 (a). To achieve scheduled maintenance, recovery process should be invoked in advance of detailed analysis of the problem itself. The modified operational process of eTOM (i.e., operational business process model defined by TMForum) needs to be changed as shown in Fig. 1 (b).

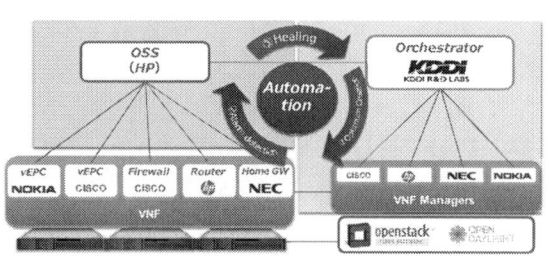

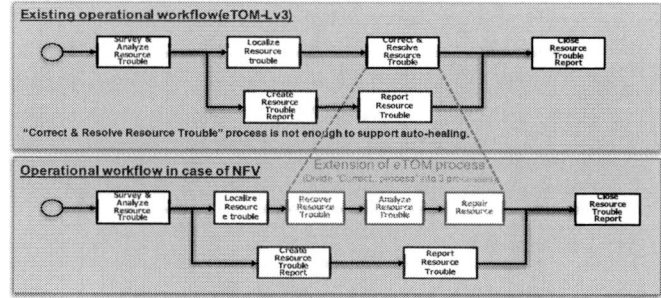

(a) Demonstration configuration of closed-loop automated management of the common NFV platform.

(b) Extension of the eTOM operation process (level 3) to achieve "recover first, resolve next" as an enabler of scheduled maintenance.

Fig. 1. Overview of the automated network management framework for the common NFV platform.

The number of management data of the virtualized environment tends to increase, and this sometimes causes delay in understanding an alarm, performance degradation, configuration change, and other anomaly situations. To tackle this issue, a distributed way of processing management data may help understand anomaly situations faster and finer. We have proposed a distributed management framework utilizing a computing resource of the NFV platform where management capability is embedded in a distributed manner. The distributed management framework is effective for achieving faster configuration change [6], attaining finer understanding of performance anomaly [7], and predicting fatal events [8]. Fig. 3 shows one implementation case of the distributed management agent (virtualized network management function (vNMF)) deployed as a virtualized network function (VNF) running on a virtual machine. vNMF collects statistical data related to VNF and physical host, and detects software anomaly such as memory leakage caused by VNF (Fig. 4) and traffic congestion caused by virtual switch (Fig. 5), which may sometimes fall into "silent" failures. These anomalies are detected before they result in a fatal situation by using the distributed machine learning of vNMF. Software-oriented anomalies (e.g., Fig. 2 (b) and (c)) were detected with a 94% true positive rate in around 30 minutes before actual failure.

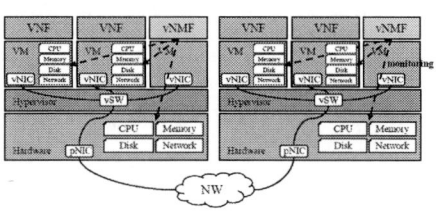

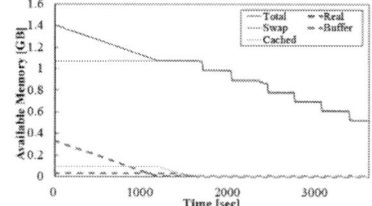

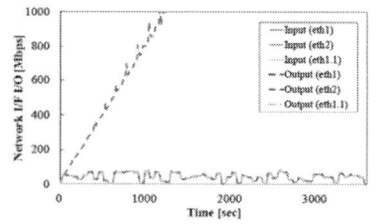

(a) Distributed management agents embedded into each host of NFVI as a virtual machine.

(b) Memory leakage anomaly caused by software bug of VNF.

(c) Traffic congestion caused by virtual switch.

Fig. 2. Framework of distributed network management and applied anomaly cases.

IV. PACKET PROCESSING PERFORMANCE FOR COMMON NFVI

Understanding packet processing performance is one of key factors in defining VNFs that can be accommodated to the common NFV platform as well as in assigning computing resources to VNF. One difficulty of the problem is the non-linearity of the performance against the number of assigned CPU cores. Fig. 3 (a) shows the non-linear trend obtained by the kernel-based packet processing. A short packet (e.g., 64 bytes) flow tends to limit throughput significantly because of the bound of CPU speed. On the other hand, a long packet (e.g., 1280 bytes) flow is instead bounded by NIC speed. Especially under a multi-core environment, cache contention among CPU cores must be considered in managing packet processing performance. As show in Fig. 3 (b), the last-level cache (LLC) is shared by the CPU cores, and thus cache contention occurs at the LLC through the processes of referencing instructions and coping data. Due to the cache contention, CPU spends more cycles because of longer latency to complete the same process. As indicated in Fig. 3 (c), contention at the LLC causes significant overhead by accessing the main memory, and this significantly affects the non-linearity of the performance. Based on the above understanding, we established a performance estimation model that shows a less-than-5% error to estimate the maximum throughput (i.e., packets per second) for 64-byte packet processing with four CPU cores [9]. This precise performance modeling helps us achieve stable provisioning of data-oriented VNF (e.g., packet processing at the core, edge, and gateway roles) on the common NFVI.

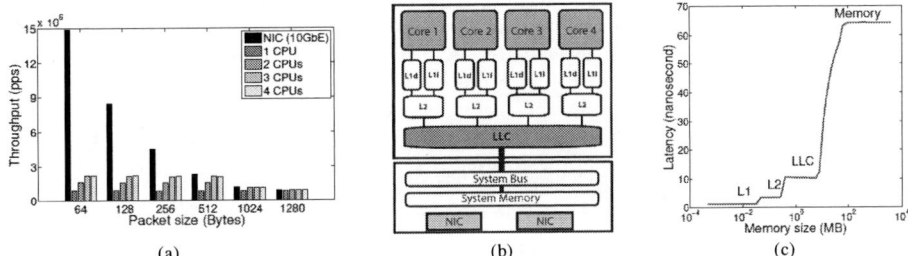

(a) (b) (c)

Fig. 3. Overview of packet processing behavior and mechanism of multi-core commodity servers.

V. CONCLUSIONS

Towards future networking including 5G, virtualization technologies should be introduced to the core platform of the telecom infrastructure. To further improve operational quality under complex operations, automation of the common NFVI and distributed management are promising techniques. In particular, understanding precise maximum throughput on the common NFVI is essential for appropriate design and determining extent of virtualization.

ACKNOWLEDGMENT

The author thanks the experts of the TTC ad hoc group on future mobile networking for their intensive discussions through making the white papers, and also thanks our laboratory members for the great achievements introduced in the paper.

REFERENCES

[1] TTC ad hoc group on future mobile networking, White paper ver. 1.0, Mar., 2015.
[2] NFV Proofs of Concepts, http://www.etsi.org/technologies-clusters/technologies/nfv/nfv-poc, ETSI, 2016.
[3] M. Hayashi et al., "Orchestration of heterogeneous virtualized resources for end-to-end service control," in Proceedings of NOMS 2008, 2008.
[4] K. Kuroki et al., "Framework of network service orchestrator for responsive service lifecycle management," in Proceedings of IM 2015, 2015.
[5] TMForum Catalyst, https://www.tmforum.org/about-tm-forum/awards-and-recognition/catalyst-team-awards/, 2015.
[6] M. Miyazawa et al., "RDStore: In-network resource datastore with distributed processing of resource graph," in Proceedings of IM 2013, 2013.
[7] M. Miyazawa et al., "In-network real-time performance monitoring with distributed event processing," in Proceedings of ManFI 2013, 2013.
[8] T. Niwa et. al., "Universal fault detection for NFV using SOM-based clustering," in Proceedings of APNOMS 2015, 2015.
[9] K. Suksomboon et al., "A dilated-CPU-consumption-based performance prediction for multi-core software routers," to be published in NetSoft 2016, 2016.

Multilayer Virtual Infrastructure Mapping in IP over WDM Networks

Zilong Ye[1] and Philip N. Ji[2]

[1]California State University, Los Angeles, Los Angeles, CA 90032
[2]NEC Laboratories America, Princeton, NJ 08540
zye5@calstatela.edu

Abstract: *We propose a novel algorithm to address multilayer virtual infrastructure mapping in IP over WDM networks. Through effectively aggregating traffic flows, the algorithm can reduce the blocking probability by 19% compared to the baseline algorithm.*
Keywords: *Virtual infrastructure mapping; IP over WDM;*

I. INTRODUCTION

Recent innovations in Software-Defined Networking (SDN) and Network Function Virtualization (NFV) offer major transformative changes and benefits for service providers as well as the entire IT and Telco eco-system. Given a number of existing virtualization tools, service providers can effectively virtualize all the elements in different network layers to provide flexible "on-demand" network services. In particular, the service providers can optimize their physical network resource utilization through a process called *virtual infrastructure* (VI) mapping, i.e., map virtual nodes (e.g., computing demands) to physical nodes (e.g., datacenters) that can provide the requested computing resources, and map virtual links (e.g., bandwidth demands) over physical links that can provide the requested bandwidth resources. Since the IP networks are supported by the underlying optical communication networks, a practical and fundamental issue is how to efficiently provision VIs onto multilayer physical networks (e.g., IP over WDM networks), which is referred to as the *Multilayer VI Mapping* (M-VIM) problem.

The M-VIM problem is more challenging than the traditional VI mapping problem [1-3] since we need to coordinate the operations across multiple network layers and make an optimal decision to efficiently utilize the physical resources. More specifically, in IP over WDM networks, multiple low-rate IP traffic flows from distinct VIs may be aggregated onto a lightpath to better utilize the wavelength resource. The M-VIM problem also differs from the traditional multilayer network design problem [4] and the traffic grooming problem [5] because a virtual topology (rather than a point-to-point link) needs to be mapped, and the source/destination of a traffic flow are not given in advance. Hence, the M-VIM problem is novel and the existing solutions [1-5] are not directly applicable.

Open challenges in provisioning M-VIM in IP over WDM networks are how to find suitable physical nodes to support virtual nodes, how to aggregate traffic flows of distinct virtual links onto lightpaths in the electrical/IP layer and how to determine the routing and wavelength assignment of lightpaths over physical links in the optical/WDM layer such that the blocking probability (the number of blocking requests divided by the total number of requests) is minimized. To the best of our knowledge, there is no existing works on M-VIM in IP over WDM networks.

In this paper, we propose an auxiliary-graph based heuristic algorithm, called *Grooming Assisted Mapping* (GAM), to efficiently solve the M-VIM problem. By jointly optimizing the VI mapping assignments in both IP layer and WDM layer networks, the GAM algorithm can significantly reduce the blocking probability compared to the baseline algorithm.

II. PROBLEM DESCRIPTION

In order to achieve multilayer network virtualization, a *multilayer switching node* (MSN) is used, which combines the benefits of both electrical switching (i.e., aggregating sub-wavelength traffic to fully utilize the wavelength capacity) and optical switching (i.e., bypassing the express traffic to reduce the OEO conversions). As shown in Fig. 1, the MSN consists of an IP router and a ROADM. The IP router receives (and sends) low-rate traffic flows and performs the functionality of multiplexing (and de-multiplexing) in the electrical domain. Each router port is connected to a transponder that enables a WDM transmission. The optical traffic flows can be routed to the destinations using ROADMs. Fig. 2 shows the architecture of a multilayer network, where the optical switching nodes (green circles) are physically interconnected by optical fiber links to form the WDM layer network, while the electrical switching nodes (orange circles) are logically connected by lightpaths to form the IP layer network.

The M-VIM problem can be defined as follows: We are given a physical network $G(N, L)$, where N is the set of physical nodes (PNs) and L is the set of physical links (PLs). For a given PN $n \in N$, it has a computing capacity of c_n and is equipped with a MSN with limited number of router ports. For a given PL $l \in L$, it has a wavelength capacity of w_l. We are also given a set of VI requests $R(V, E)$, where V is the set of virtual nodes (VNs) and E is the set of virtual links (VLs). For a given VN $v \in V$, it is associated with a computing demand of d_v and a geographic location constraint of m_v. For a given VL $e \in E$, it is associated with a bandwidth demand of b_e. For each VN, we need to find a PN that provides enough computing resources and lies in the VN's location range. For example, in Fig. 2, VN v_I is mapped to PN n_A. For each VL, we need to find a path that can provide the requested bandwidth resources. In particular, such a path can be a mixture of lightpaths and physical links. For example, in Fig. 2, VL e_{13} is mapped to path l_{ABDF}, where l_{ABD} is an existing lightpath and l_{DF} is a segment of physical link. Given the above inputs, the objective is to map VIs in the multilayer physical network such that the blocking probability is minimized. It is worth noting that the VL mapping process in M-VIM is more challenging than that in the traditional VI mapping since we need to jointly determine how to aggregate the bandwidth demands onto lightpaths, how to find an IP layer routing path over the existing lightpaths that have enough bandwidth resources, and how to decide the WDM layer routing and wavelength assignments to establish new lightpaths along a segment of physical links. In these processes, we need to interactively and cooperatively bridge the IP routing and the WDM routing (and wavelength assignments) in order to obtain an optimal assignment.

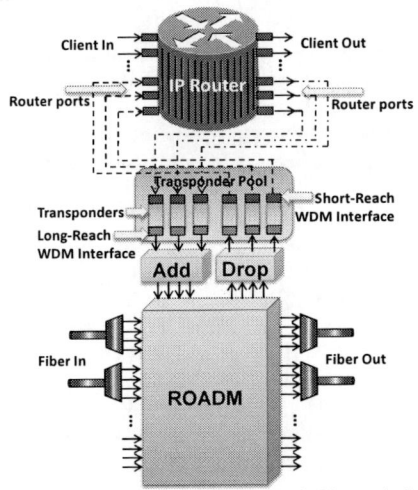

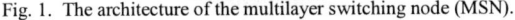

Fig. 1. The architecture of the multilayer switching node (MSN).

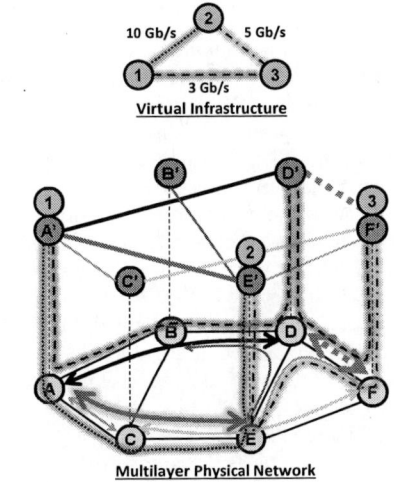

Fig. 2. Multilayer virtual infrastructure mapping (M-VIM).

III. THE GROOMING ASSISTED MAPPING ALGORITHM

We propose an auxiliary-graph based algorithm, called Grooming Assisted Mapping (GAM), to efficiently solve the M-VIM problem in IP over WDM networks. In GAM, the VL mapping is addressed first followed by the VN mapping. For a given VL, an auxiliary graph is constructed in the first step. As shown in Fig. 2, auxiliary links (ALs) are added between PNs that are connected by lightpaths that have enough remaining bandwidth resources to support the given VL's bandwidth demand. In the auxiliary graph, the weight of a PL is set to 1, while the weight of each AL is set to 0. In this way, the existing lightpaths are given a priority when shortest path algorithm is applied in the following steps, so that the existing router ports and the remaining bandwidth resources may have a higher chance to be reused. After the auxiliary graph is constructed, we enumerate all the possible combinations of source-destination pairs of PNs that have enough computing resources and are within the location constraints of the two end VNs of the given VL. For each combination, a shortest path is found between the source and destination. Then, the source-destination PNs and the corresponding shortest path that have the minimum weight will be selected to map the given VL. Note that the residual bandwidth resources will be updated on the ALs (if exist) and the first-fit approach will be applied for the wavelength assignment on the segments of PLs. Such a VL mapping strategy can result in a better utilization of the remaining bandwidth resources on the ALs. After the VL is mapped, the two end VNs are mapped accordingly by considering the computing load balancing policy. The detailed steps of the GAM algorithm are shown below.

The Grooming Assisted Mapping Algorithm (GAM)

Step 1: Select a VI that is not yet processed, but with the maximum cumulative bandwidth demands;

Step 2: Select a VL with the maximum bandwidth demand;

Step 3: Construct an auxiliary graph as follows. Establish an AL between a pair of PNs if a lightpath exists between them and it has enough remaining bandwidth resources to support the selected VL. The weight of each AL is set to 0 and the weight of each PL is set to 1;

Step 4: Enumerate all the possible source-destination pairs of PNs that are the candidate PNs for the two end VNs of the selected VL. Find the shortest path for each source-destination pair of the PNs in the auxiliary graph;

Step 5: Select the source-destination PNs and corresponding shortest path that have the minimum weight to map the selected VL;

Step 6: In the selected route, update the residual bandwidth resources of ALs (if exist). If there exists a segment of physical links, establish a lightpath along the PLs using first-fit approach for wavelength assignment (Note that the wavelength continuity constraint is also considered here);

Step 7: Map the two end VNs to the source and destination considering the computing load balancing;
Step 8: Repeat Step 1 to Step 7 until all VIs are provisioned;

For the comparison purpose, we also implement a baseline algorithm, called Shortest Path Based Mapping (SPBM), which also address the VL mapping first followed by the VN mapping. For a given VL, SPBM selects a source-destination pair of PNs and the corresponding shortest path on the original physical network. In the selected path, if there exist lightpaths which have enough remaining bandwidth resources, those lightpaths will be reused; otherwise, we will use first-fit approach to conduct the wavelength assignment to establish new lightpaths along the segments of PLs. The VN mapping process is similar to that of the GAM algorithm.

IV. NUMERICAL RESULTS

We evaluate the performance of the proposed algorithms in the 14-node NSF network. By default, each physical node is equipped with 1500 computing units and a MSN with 30 router ports. Each physical link is equipped with 20 wavelengths, each of which can provide a bandwidth of 40 *Gbps*. The VI requests arrive according to the incremental traffic pattern, and each VI consists of interconnected two to five virtual nodes. Each virtual node is associated with a computing demand uniformly distributed within 30 units, and each virtual link has a bandwidth demand within 1~20 *Gbps* with uniform distribution. For a given parameter setting, we randomly generate 100 test cases, and the statistics in this section are the average results.

Fig. 3(a) shows the cumulative distribution function (CDF) of the first blocking over the number of provisionable VI requests (the number of demands that can be accepted before first blocking occurs). We can see that GAM is more effective than SPBM since it can provision at least 9% more VI requests before first blocking occurs when the CDF of first blocking is set to 0.5. In Fig. 3(b), we compare the blocking probability as a function of the number of VI requests. It can be seen that the blocking probability of GAM is 19% smaller than that of SPBM. This is because GAM can jointly optimize the traffic aggregation and VI mapping assignments in both IP layer and WDM layer networks. To better understand that, we analyze how wavelength and router port resources will affect the blocking probability in Fig. 3(c) and (d), respectively. In Fig. 3(c), we can observe that when the wavelength resources are limited, GAM performs much better than SPBM. In Fig. 3(d), we can see that the relative performance improvement of GAM over SPBM becomes more obvious as the number of router ports increases. The reason behind Fig. 3(c) and (d) is because GAM prioritizes the opportunity to reuse the existing lightpaths, thus reducing the consumption of wavelengths and router ports for accepting more future requests.

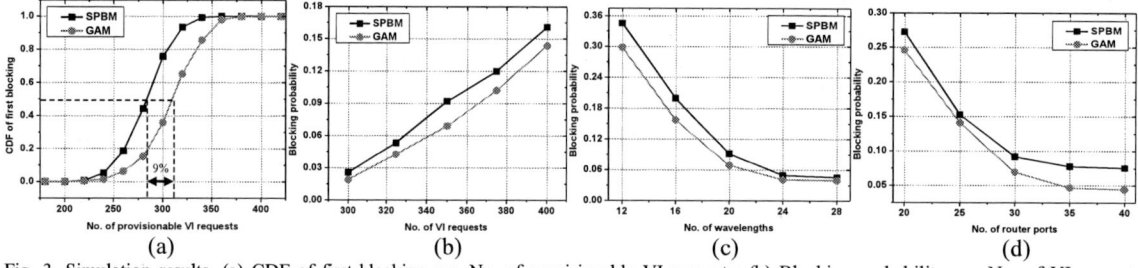

Fig. 3. Simulation results. (a) CDF of first blocking v.s. No. of provisionable VI requests; (b) Blocking probability v.s. No. of VI requests; (c) Blocking probability v.s. No. of wavelengths; (d) Blocking probability v.s. No. of router ports.

V. CONCLUSIONS

In this work, we have proposed an auxiliary-graph based algorithm to efficiently address the M-VIM problem in IP over WDM networks. The proposed algorithm can significantly reduce the VI request blocking probability, thus achieving an efficient multilayer network virtualization.

REFERENCES

[1] H. Yu *et al.*, "A cost efficient design of virtual infrastructures with joint node and link mapping," in *J. Netw. Syst. Mgmt*, 20(1): 97-115, 2012.

[2] Z. Ye *et al.*, "Virtual infrastructure embedding over software-defined flex-grid optical networks*"*, in *Proc. of GLOBECOM*, 2013.

[3] X. Wang *et al.*, "Flexible virtual network provisioning over distance-adaptive flex-grid optical networks," in *Proc. of OFC*, W3H.3, 2014.

[4] M. Nikolayev *et al.*, "An efficient model for the multilayer network planning of IP-over-WDM networks," in *Proc. of ECOC*, P.5.14, 2013.

[5] K. Zhu and B. Mukherjee, "Traffic grooming in an optical WDM mesh network," in *JSAC*, vol. 20, no. 1, pp. 122-133, 2002.

Weighted Attack-Evasion Routing and Wavelength Assignment Algorithm against High Power Jamming in Optical Networks

Liangkai Huang, Yongli Zhao, Wei Wang, Jie Zhang

State Key Laboratory of Information Photonics and Optical Communications,
Beijing University of Posts and Telecommunications, Beijing, 100876 China
lenkayhuang@sina.com, {yonglizhao, weiw, lgr24}@bupt.edu.cn

Abstract: An attack-evasion routing and wavelength assignment (AE-RWA) algorithm is first proposed in optical networks considering weighted attack probability. Numeric results show that AE-RWA can achieve better performance in terms of blocking probability and resource utilization.
Keywords: Routing and wavelength assignment, high-power jamming attack, weighted attack probability

I. INTRODUCTION

With the booming of variable network services, transparent optical networks (TONs) have become the fundamental infrastructure of the next generation backbone networks. Security issues and attack management in TONs are most important to network operators because of the high data rates involved and the vulnerabilities associated with transparency. Due to the inherent characteristics of optical fibers, amplifiers and some other optical components, transparent optical networks are vulnerable to physical layer attacks, one typical one of which is high-power jamming attack, including in-band jamming and out-band jamming. It will cause severe crosstalk beyond the legitimate signals.

In case of these types of attacks, some works have been done to reduce their damages. Aiming at the networks which have a large number of existing lightpaths, a heuristic algorithm is proposed to minimize the lightpath overlapping and improve the network reliability, and the optimal solution has a complexity that is polynomial to network size [1]. The idea of incorporating attack-awareness was first proposed for network planning problems to deal with out-of-band effects and using Tabu Search to limit the maximum disruption by minimizing Attack Radius [2]. The problem of jamming-aware shared path protection has been decomposed into routing and wavelength assignment sub-problem, and formulated each as an ILP, its basic principle is to ensure that the working and backup paths of each connection are not within reach of a same potential jamming signal [3]. In some Attack-Survivable (AS) routing and wavelength schemes, due to the fact that attacks can propagate through the network, a novel concept called Attack Group (AG) is defined. AG is applied to develop attack-aware dedicated path protection approaches which aim to establish AG-disjoint primary and backup paths in a cost-effective manner. Since providing full attack-protection may not be economically viable due to the large number of resources required, adding attack-awareness to standard survivability approaches is a resource efficient manner to reduce the number of attack-unprotected connections [4, 5]. However, all the above approaches attempt to avoid attack using backup paths, which need extra resources. This paper focuses on the problem of attack evasion without backup.

In this paper, we proposed an attack-evasion routing and wavelength assignment (AE-RWA) algorithm concentrating on the weighted attack probability (WAP) of optical link and optical node. For each service request, AE-RWA minimizes WAP of its lightpath to reduce the potential damage against high power jamming and updates network topology to make preparation for new service requests. Numeric results show that AE-RWA algorithm can efficiently reduce network blocking rate and improve network resource utilization. In a relatively low attack frequency situation, AE-RWA algorithm has almost identical protect ability as attack-survivable RWA algorithm.

II. DEFINITION OF WEIGHTED ATTACK PROBABILITY

High-power jamming can be realized by injecting a high-power signal on a legitimate wavelength. A legitimate signal carrying jamming signal is more destructive and can introduce potential attack probability on its lightpath. There are two main aspects of high-power jamming attacks in optical networks, in-band effect and out-band effect.

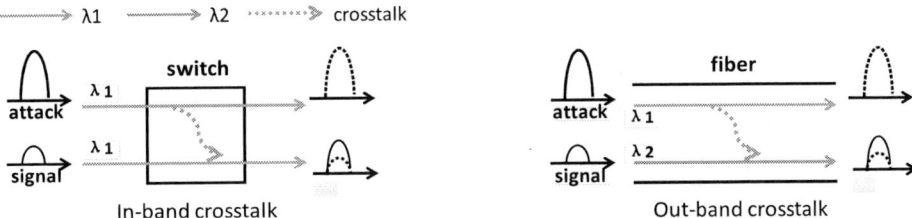

Fig.1 in-band effect in switch and out-band effect in fiber

As shown in Fig.1, when the attacking signals traverse by optical switches, user signals at the same wavelength are primarily affected via in-band crosstalk which happened in optical components. Signals on the different wavelengths from the attacker will be affected via out-band crosstalk which happened in optical fibers. In the algorithm model, the optical network is described as graph $G\{V, E\}$, V_p denotes node (switch) p and E_{qr} denotes link from node q to r. We define W_p^i as the in-band attack probability on V_p, and W_{qr}^o as the out-band attack probability on E_{qr}. When a new service request arrives, weighted attack probability (WAP) will be used in the routing process to evaluate the vulnerability of lightpaths under potential high-power jamming attacks. S denotes the queue which contains all service requests, t denotes a single service in S.

$$WAP(t) = \alpha \sum W_p^i + \beta \sum W_{qr}^o \ , V_p, E_{qr} \in G\{V,E\}, t \in S \qquad (1)$$

Equation (1) indicates that WAP for a service request t consists of in-band WAP (W^i) and out-band WAP (W^o), where V_p denotes all of the nodes and E_{qr} denotes all of the links on the lightpath from source to destination. Constant α and β can be changed in different situations depending on the relatively importance of in-band effect and out-band effect. The goal of AE-RWA is to find the lightpath with minimum WAP for each service and limit the propagation of attack. Under the assumption that legitimate signals may carry jamming attacks, the earlier arrived service should be routed via lightpath with smaller attack probability in order to reduce network WAP and destructiveness.

III. AE-RWA ALGORITHM

Attack-evasion problem consists of two stages, weighted attack probability calculation and attack-evasion routing. Taking Fig.2 for instant, there are three service requests: s1, s2 and s3 (time of arrival:s1<s2<s3), Fig.2 (a) shows the original routing map, $W_{BC}^o = \beta$, similarly $W_{AB}^o = \beta$, s1 and s2 with different λ shared common link E_{BC}, s1 and s3 with different λ shared common link E_{AB}, thus $WAP(s1) = 2\beta$, $WAP(s2) = \beta$. According to the principle to minimize WAP for each service, reroute s2 over links C-E-B (s3 to D-E-A) to reduce out-band attack potential. In this case as shown in Fig.2 (b), $W_E^i = \alpha$, s3 shared common node V_E with s2 and they occupy the same λ, $WAP(s2) = WAP(s3) = \alpha$, reroute s3 D-A to bypass V_E and make all $WAP = 0$ for each service, there is no high-power jamming attack propagation potential in the final result shown in Fig.2 (c).

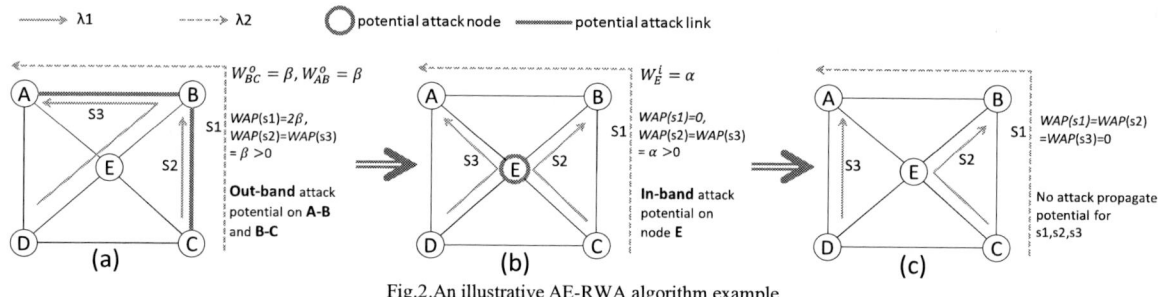

Fig.2.An illustrative AE-RWA algorithm example

Algorithm: Attack-evasion routing and wavelength assignment	
Input: network topology: G{V, E},service request queue: S, **output**: service result	
1 load G{V, E} and S;	10 **end for**;
2 **for** each t∈S **do**	11 **for** each t∈T
3 calculate WAP-valued K shortest lightpath P$_k$;	12 **if** t need to drop **then**
4 **for** P$_i$,i ∈ K	13 release w$_k$;
5 **if** wavelength WL$_k$ available **then**	14 update G{V, E} for Wi and Wo;
6 update G{V, E} for Wi and Wo, assign wl$_k$;	15 drop t;
7 put t to T for drop;	16 **end if**;
8 **else** t fail, drop t, **end if**;	17 **end for**.
9 **end for**;	

AE-RWA algorithm is simply described above. To coordinate service requests and the demand of avoiding high-power jamming. Firstly, we establish a request queue to store services in S, and import physical topology $G\{V, E\}$ for routing. Secondly, for request t, calculate WAP on all of the available lightpaths, using KSP-algorithm to find K shortest lightpaths considering WAP values. If there is a WAP-weighted shortest lightpath with available wavelength for t, choose it as the candidate lightpath, and update W^i and W^o of nodes and links on the lightpath of service t. If there is no available wavelength on all K lightpaths, service request t will be blocked. Thirdly, assign wavelength for t and update $G\{V, E\}$ for new coming services. Finally, when service t should be dropped, release the occupied wavelength, then update the W^i and W^o on its lightpath in $G\{V, E\}$ for new service requests. Algorithm cycles above process until request queue S becomes empty.

IV. NUMERIC ANALYSIS

We setup a simulation to evaluate the performance of AE-RWA on NSFNET (14 nodes, 21 links). Each fiber is configured with 100 wavelengths. Under each traffic load, we issue 2000 requests, the arrivals of which are modeled by a Poisson process (μ=0.4), the required wavelength of each request is randomly generated between 1 and 5. We set constant parameters as α=0 β=1 for basic attack-evasion scheme, and α=1 β=2 for the enhanced attack-evasion scheme with more attention on out-band effect. Attack frequency in Fig.5 denotes the ratio of attack signal quantity and all service requests signals quantity. Figures below show the result of AE-RWA compared to Attack Survivable RWA (AS-RWA) and non-optimization RWA (nop-RWA).

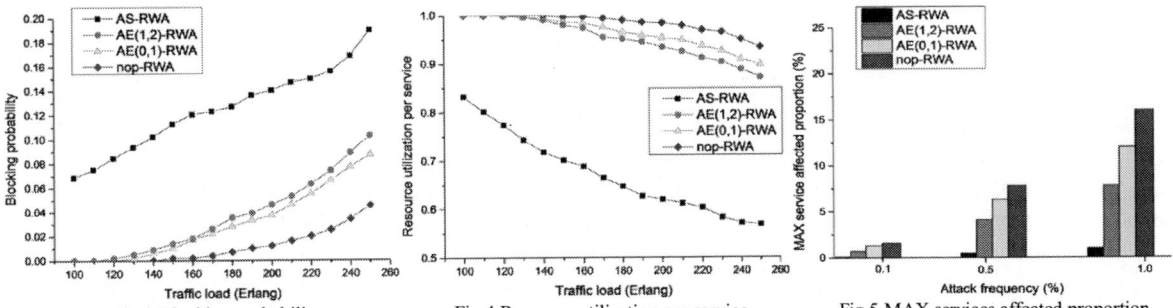

Fig.3.Blocking probability   Fig.4.Resource utilization per service   Fig.5.MAX services affected proportion

As shown in Fig.3, blocking probability of AE-RWA has a significant reduction compared with AS-RWA and is just a little higher than nop-RWA. This is because AS-RWA uses Attack Group disjoint (AG-disjoint) principle to ensure complete protection which limits links and nodes sharability, while other algorithm do not strictly limit network elements sharing in different service requests. In comparison with nop-RWA, AE-RWA add extra path weight in topology and may cause blocking on some small WAP value lightpaths because more requests are routed on it to avoid attack on other big WAP value lightpaths. Fig.4 presents the average resource utilization of each service. Because of sharing weighted links and nodes, AE-RWA makes full use of resources as nop-RWA, while the establishment of protection paths reduce resource utilization in AS-RWA. Fig.5 shows that AE-RWA with bigger α and β which indicates bigger WAP value on nodes and links, can provide more efficient protection for each service against high-power jamming attack propagation. In comparison to nop-RWA, AE-RWA (α=1, β=2) can avoid almost 50% of the attack damage. In a low attack frequency (less than 1%), AE-RWA with appropriate α and β can provide comparable protection with AS-RWA (whose affected proportion equals attack frequency).

V. CONCLUSION

In order to optimize conventional routing and wavelength assignment scheme against high-power jamming attack, we proposed attack-evasion routing and wavelength assignment (AE-RWA) algorithm in transparent optical networks, and evaluated them via simulations to verify the effectiveness of AE-RWA. Numeric result showed that AE-RWA can provide efficient protection against high-power jamming in low attack frequency, reduce blocking probability and improve network resource utilization.

ACKNOWLEDGMENT

This work has been supported in part by NSFC project (61271189, 61571058, 61302085), and the Fund of State Key Laboratory of Information Photonics and Optical Communications (BUPT).

REFERENCES

[1] Shengli Yuan, Daniel Stewart, "Protection of Optical Networks against Interchannel Eavesdropping and Jamming Attacks", 978-1-4799-3010-4/14, IEEE DOI 10.1109/CSCI.2014.14

[2] Nina Skorin-Kapov, Jiajia Chen, Lena Wosinska, "A New Approach to Optical Networks Security:Attack-Aware Routing and Wavelength Assignment", IEEE/ACM TRANSACTIONS ON NETWORKING, VOL. 18, NO. 3, JUNE 2010.

[3] Marija Furdek, Nina Skorin-Kapoy, Lena Wosinska, "Shared Path Protection Under the Risk of High-Power Jamming", 978-1-4799-3872-8/14, IEEE, NOC.2014.

[4] Marija Furdek, Nina Skorin-Kapoy, Lena Wosinska, "Attack-Aware Dedicated Path Protection in Optical Networks", JOURNAL OF LIGHTWAVE TECHNOLOGY, VOL. 34, NO. 4, FEBRUARY 15, 2016

[5] Marija Furdek, Nina Skorin-Kapoy, "Attack-Survivable Routing and Wavelength Assignment for high-power jamming", ONDM 2013, Brest, France.

WB1-1

Coherent optical communication technology

Kazuro Kikuchi
Department of Electronic Engineering and Information Systems
University of Tokyo

Coherent optical communications were studied extensively in the 1980s because of the high receiver sensitivity that could enhance the unrepeated transmission distance; however, since 1990, related R&D had been interrupted by the rapid advances in wavelength-division multiplexed (WDM) systems and erbium-doped fiber amplifiers (EDFAs).
In 2005, the demonstration of digital carrier-phase estimation in coherent receivers led to the renewal of widespread interest in coherent optical communications. The reason is that the digital coherent receiver enables us to employ a variety of spectrally efficient modulation formats, relying upon stable digital carrier-phase estimation.   In addition, since the phase information is preserved after detection, we can equalize linear transmission impairments such as group-velocity dispersion (GVD) and polarization-mode dispersion (PMD) of transmission fibers via digital signal processing (DSP). Recently, 100-Gbit/s transmission systems, which employ quadrature phase-shift-keying (QPSK) modulation, polarization-division multiplexing, and phase-diversity homodyne detection assisted with high-speed DSP, have been developed and introduced into commercial networks.
This talk reviews the history of coherent optical communications and describes the principle of coherent detection, including its quantum-noise characteristics. In addition, it discusses the role of DSP in mitigating linear transmission impairments, estimating the carrier phase, and tracking the state of polarization of the signal.

Biography

Kazuro Kikuchi

Kazuro Kikuchi was born in Miyagi, Japan, in 1952. He received his B.S. in Electrical Engineering and M.S. and Ph.D. in Electronic Engineering from the University of Tokyo. In 1979, he joined the Department of Electronic Engineering and is currently a Professor in the Department of Electrical Engineering and Information Systems. He also worked at Bell Communications Research from 1986 to 1987 as a consultant and presently serves on the board of directors of Alnair Labs Corporation.

Throughout his career, his research has focused on optical fiber communications including optical devices and systems. He is currently involved in coherent optical communication systems that realize multi-level modulation formats with digital signal processing.

Professor Kikuchi is a Fellow of the IEEE Photonics Society, a Fellow of OSA, and a Fellow of the Institute of Electronics, Information and Communication Engineers (IEICE). He is the recipient of awards, including Japan IBM Science Prize, Hattori Hokosho Prize, Ericsson Telecommunications Award, Japanese Prime Minister's Award for the promotion of academy-industry collaboration, C&C Prize, and John Tyndall Award.

WC1-1

On the Orbital Angular Momentum (OAM) of Light in Fiber

Siddharth Ramachandran
Boston University
sidr@bu.edu; http://people.bu.edu/sidr

In the last decade, some of the most extensively studied complex light beams are optical vortices, which possess phase or polarization singularities. These beams are interesting because they resemble the emission patterns of single molecule dipoles, or because they potentially represent an infinite set of eigenstates that can be constructed with light. Their use has been demonstrated in, or proposed for, several applications such as higher-dimensional quantum encryption, information capacity scaling, single-molecule spectroscopy and nano-scale imaging.

A recently developed fiber that has a ring-shaped core has enabled their stable generation and propagation in optical fibers for distances of up to kilometres. In doing so, we have learnt of several novel linear properties of these modes, including the role of conservation of angular momentum in providing mode stability. Aided by this stability, it is now possible to envisage fiber based versions of nanoscale microscopy systems that leverage such fibers' ability to offer spectro-spatial tailoring of light. Since fibers are well known for their ability to tailor nonlinearity and dispersion, studying and exploiting nonlinear phenomena with such beams present exciting prospects for future research.

We will discuss recent results and intriguing possibilities enabled by fiber propagation of beams that have long been considered interesting, but hitherto unstable in nature.

Biography

Siddharth Ramachandran

Dr. Siddharth Ramachandran obtained his Ph.D. in Electrical Engineering from the University of Illinois, Urbana-Champaign, in 1998. Thereafter, he joined Bell Laboratories as a Member of Technical Staff and subsequently continued with its spin-off, OFS Laboratories. After a decade in industry, Dr. Ramachandran moved back to academics in 2010, and is now a Professor in the Department of Electrical Engineering at Boston University.

Prof. Ramachandran's research focuses on the optical physics of guided waves. He has authored over 200 refereed journal and conference publications, more than 50 invited talks, plenary lectures and tutorials, 3 book-chapters, edited one book, and has been granted 38 patents. For his contributions in the field of fiber-optics, he was named a Distinguished Member of Technical Staff at OFS Labs in 2003, a fellow of the Optical Society (OSA) in 2010, and an IEEE Distinguished Lecturer for 2013-2015. He serves as a topical editor for Optica in addition to serving on numerous conference and advisory committees in the field of optics and applied physics, including as general chair for CLEO-2017.

Gain Control in Multi-Core Erbium/Ytterbium-Doped Fiber Amplifier with Hybrid Pumping

Makoto Yamada,[1] Hirotaka Ono,[2] Tsukasa Hosokawa,[3] and Kentaro Ichii[3]

[1]Graduate School of Engineering, Osaka Prefecture University, Sakai, Osaka 599-0814, Japan
[2]NTT Device Technology Laboratories, NTT Corporation, Atsugi, Kanagawa 243-0198, Japan
[3]Advanced Technology Laboratory, Fujikura Ltd., Sakura, Chiba 285-8550, Japan
E-mail address: myamada@eis.osakafu-u.ac.jp

Abstract: *A hybrid scheme employing cladding- and core-pumping in a multi-core erbium/ytterbium-doped fiber amplifier (MC-EYDFA) is investigated. Hybrid pumping achieves a small gain change against wavelength-division-multiplexing (WDM) signal power level and channel number changes.*
Keywords: multi-core fiber amplifier, cladding-pumping, core-pumping, gain control

I. INTRODUCTION

Multi-core fiber (MCF) transmission has attracted attention in recent years because it has the potential to overcome the capacity crunch and achieve an ultra-high exabit/s class capacity. For long-haul MCF transmission, multi-core erbium-doped fiber amplifiers (MC-EDFAs) have been realized [1]–[7], and long-haul transmissions employing MC-EDFAs have been demonstrated [8]–[13]. The MC-EDFAs described in Ref. [4]–[7], [13] use cladding pumping while the others use core pumping. The cladding-pumped MC-EDFAs have the benefit of a lower power consumption than a core-pumped MC-EDFA [7], [13] and the cladding-pumped multi-core erbium/ytterbium-doped fiber amplifier (MC-EYDFA) also has good pump efficiency [6]. However, a cladding-pumped MC-EDFA has the drawback that it cannot control the pump power for each core independently of the other cores because the pump laser diode (LD) is shared by multiple cores in a cladding-pumped MC-EDFA, Therefore, the cladding-pumped MC-EDFA has an issue as regards realizing the automatic gain control (AGC) that is indispensable for an MCF network with a wavelength-division-multiplexing (WDM) signal. In this paper, we propose hybrid pumping for an MC-EDFA that employs both cladding pumping and core pumping, and show that AGC is realized by hybrid pumping in an MC-EYDFA.

II. CONCEPT OF HYBRID PUMPING

The concept of hybrid pumping for an MC-EDFA is shown in Fig. 1. Figure 1 (a) and (b), respectively, illustrate hybrid pumping for a single-stage and a multi-stage MC-EDFA. In the single-stage hybrid-pumped MC-EDFA, a multi-core erbium-doped fiber (MC-EDF) has a double-cladding structure, and is pumped with cladding pumping from one side and with core pumping from the other side. The multi-stage hybrid-pumped MC-EDFA consists of cladding-pumped and core-pumped stages. In both the hybrid-pumped MC-EDFAs, the pump LD for cladding pumping is driven with a constant current, and the driving current of the LD for core pumping is adjusted to achieve gain control when the input signal power changes.

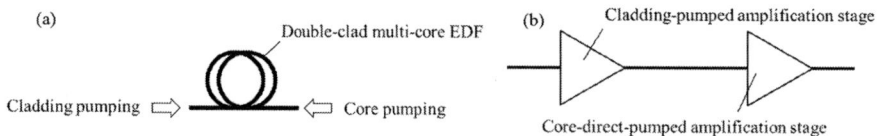

Fig. 1. Concept of hybrid pumping for MC-EDFA.

III. GAIN CONTROL BY USING HYBRID PUMPING

A. Hybrid pumping for single-stage MC-EYDF

Figure 2 shows the configuration of a single-stage MC-EYDFA. A 5 m-long double-clad erbium/ytterbium-doped fiber (DCMC-EYDF) was spliced to a free-space-coupling 980/1550-nm WDM module with MCF pigtails. The DCMC-EYDF was cladding-pumped with a 975-nm multi-mode LD for the forward direction through a WDM module spliced to the input end of the DCMC-EYDF. Fiber type fan-in (FI) and fan-out (FO) devices were spliced to both the input and output ends of the amplifier, and core #3 was used to characterize the gain control of the amplifier. A 1480/1550-nm WDM filter was connected to the FO port for core #3 so that the DCMC-EYDF was core-pumped with a single-mode 1480-nm LD in the backward direction. A five-channel WDM signal with wavelengths of 1543.7, 1548.5, 1553.3, 1556.2, and 1561.4 nm was used to characterize the gain controllability. The wavelength-averaged gain of the amplifier was 9.0 dB. We measured the gain controllability against the input power changes caused either by the input signal power per channel change or by the channel number change, i.e. channel add/drop. The pump power of the 1480-nm LD was adjusted against the changes while the pump power of the 975-nm LD was kept constant.

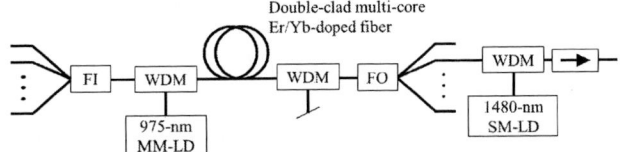

Fig. 2. Configuration of single-stage MC-EYDFA.

Figure 3 (a) and (b), respectively, show the gain and noise figure change for an input channel power change and a channel number change, respectively. The cladding-pumping power of the 975-nm LD was 1.27 W for all input signal conditions, and the core-pumping power of the 1480-nm LD was adjusted. Figure 3 (a) indicates that gain control was achieved with a gain change of less than 0.5 dB without degradation of the noise figure for an input signal power of less than −1 dBm. For input signal powers exceeding +4 dBm, the noise figure was degraded because of the low population inversion caused by the relatively large signal power input. For channel add/drop, we changed the input signal power by 15 dB (−6 ↔ −21 dBm) to emulate the 32 ↔ 1 channel add/drop. Figure 3 (b) suggests that gain control was achieved with gain and noise figure changes of less than 0.4 and 0.2 dB, respectively, for the 32 ↔ 1 channel add/drop.

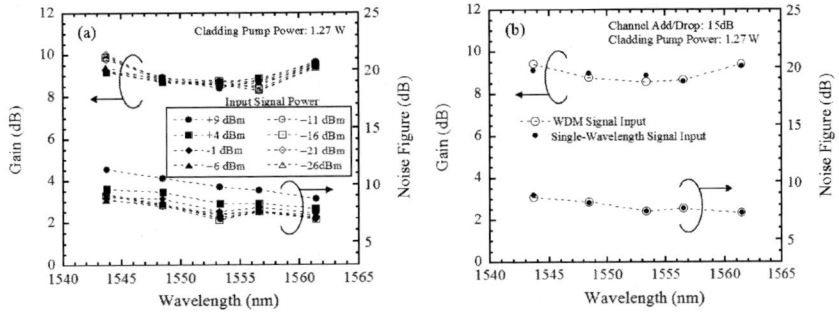

Fig. 3. Gain control characteristics of single-stage MC-EYDFA.

B. Hybrid pumping for multi-stage MC-EYDFA

Figure 4 shows the configuration of the multi-stage MC-EYDFA. The first stage is the cladding-pumped amplifier stage, which is the same as the single-stage MC-EYDFA mentioned in the previous section. The second stage consists of an 8.5-m long single-clad single-core erbium/ytterbium-doped fiber pumped in the forward direction with a single-mode 1480-nm LD, and it was connected to the output port of core #3 of the first stage. The total amplifier gain was 20 dB. We measured the gain controllability of the multi-stage MC-EYDFA in the same way as that of the single-stage MC-EYDFA.

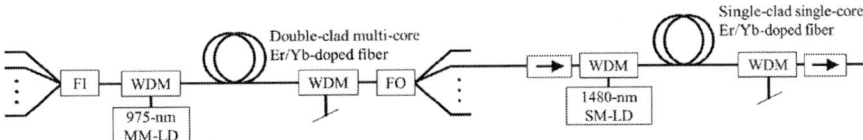

Fig. 4. Configuration of single-stage MC-EYDFA.

Figure 5 (a) and (b), respectively, show the gain and noise figure change for the input channel power change and channel number change, respectively. The cladding-pumping power of the 975-nm LD was 1.53 W for all the input signal conditions and the core-pumping power of the 1480-nm LD was adjusted. Figure 5 (a) indicates that gain control was achieved with a gain change of less than 0.6 dB without degradation of the noise figure for an input signal power of less than −6 dBm. For an input signal power larger than −1 dBm, the noise figure was degraded because of the low population inversion caused by the relatively large signal power input, which was as found for the single-stage MC-EYDFA. Figure 5 (b) suggests that the gain control was achieved for the 32 ↔ 1 channel add/drop with gain and noise figure changes of less than 0.7 and 0.6 dB, respectively.

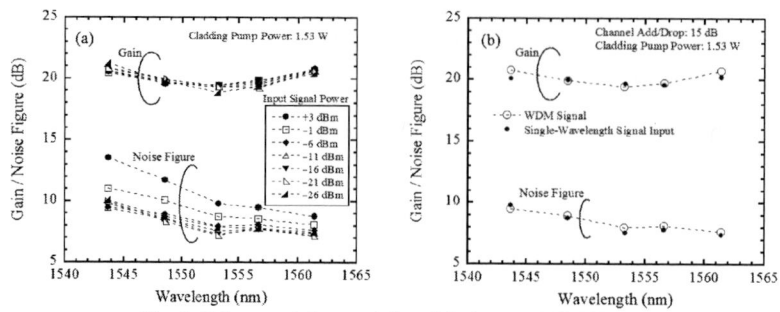

Fig. 5. Gain control characteristics of single-stage MC-EYDFA.

IV. CONCLUSION

We proposed a hybrid pumping technique for an MC-EYDFA that employed cladding pumping to achieve gain control for an input signal power change, and we confirmed experimentally the controllability of both a hybrid single-stage and a hybrid multi-stage MC-EYDFA. Hybrid pumping could be a way to achieve gain control in an MC-EDFA that employs cladding pumping.

ACKNOWLEDGMENT

This research is supported by the National Institute of Information and Communications Technology (NICT), Japan, as part of the "R&D of Innovative Optical Communication Infrastructure".

REFERENCES

[1] K. S. Abedin, T. F. Taunay, M. Fishteyn, M. F. Yan, B. Zhu, J. M. Fini, E. M. Monberg, F.V. Dimarcello, and P.W. Wisk, "Amplification and noise properties of an erbium doped multicore fiber amplifier," Opt. Express, vol. 19, no. 17, pp. 16715–16721, 2011.

[2] Y. Tsuchida, K. Maeda, K. Watanabe, T. Saito, S. Matsumoto, K. Aiso, Y. Mimura, and R. Sugizaki, "Simultaneous 7-core pumped amplification in multicore EDF through fibre based fan-in/out," Proc. ECOC2012, Amsterdam (The Netherlands), Paper, Tu.4.F.2.

[3] H. Ono, M. Yamada, K. Takenaga, S. Matsuo, Y. Abe, K. Shikama, and T. Takahashi, "Amplification method for crosstalk reduction in multi-core fibre amplifier," Electron. Lett., vol. 49, no. 2, pp. 138–140, 2013.

[4] K. S. Abedin, T. F. Taunay, M. Fishteyn, D. J. DiGiovanni, V. R. Supradeepa, J. M. Fini, M. F. Yan, B. Zhu, E. M. Monberg, and F.V. Dimarcello, "Cladding-pumped erbium-doped multicore fiber amplifier," Opt. Express, vol. 20, no. 18, pp. 20191–20200, 2012.

[5] Y. Mimura, Y. Tsuchida, K. Maeda, R. Miyabe, K. Aiso, H. Matsuura, R. Sugizaki, "Batch multicore amplification with cladding-pumped multicore EDF," Proc. ECOE2012, Amsterdam (The Netherlands), paper Tu.4.F.1.

[6] H. Ono, K. Takenaga, K. Ichii, S. Matsuo, T. Takahashi, H. Masuda, and M. Yamada, "12-core double-clad Er/Yb-doped fiber amplifier employing free-space coupling pump/signal combiner module," Proc. ECOC2013, London (UK), paper We.4.A.4.

[7] Y. Tsuchida, K. Maeda, K. Watanabe, T. Saito, S. Takasaka, M. Tadakuma, R. Sugizaki, and H. Ogoshi, "Cladding-pumped L-band multicore EDFA with reduced power consumption," Proc. IEEE SUM2014 Montreal (Canada), paper ME2.2, 2014.

[8] H. Takahashi, T. Tsuritani, E. L. T. de Gabory, T. Ito, W. R. Peng, K. Igarashi, K. Takeshima, K. Kawaguchi, I. Morita, Y. Tsuchida, Y. Mimura, K. Maeda, T. Saito, K. Watanabe, K. Imamura, R. Sugizaki, and M. Suzuki, "First demonstration of MC-EDFA-repeatered SDM transmission of 40 × 128-Gbit/s PDM-QPSK signals per core over 6,160-km 7-core MCF," Proc. ECOC2012, Amsterdam (The Netherlands), paper Th.3.C.3.

[9] A. Sano, H. Takara, T. Kobayashi, H. Kawakami, H. Kishikawa, T. Nakagawa, Y. Miyamoto, Y. Abe, H. Ono, K. Shikama, M. Nagatani, T. Mori, Y. Sasaki, I. Ishida, K. Takenaga, S. Matsuo, K. Saitoh, M. Koshiba, M. Yamada, H. Masuda, and T. Morioka, "409-Tb/s + 409-Tb/s crosstalk suppressed bidirectional MCF transmission over 450 km using propagation-direction interleaving," Opt. Express, vol. 21, no. 14, pp. 16777–16783, 2013.

[10] J. Sakaguchi, W. Klaus, B.J. Puttnam, J-M. D. Mendinueta, Y. Awaji, N. Wada, Y. Tsuchida, K. Maeda, M. Tadakuma, K. Imamura, R. Sugizaki, T. Kobayashi, Y. Tottori, M. Watanabe, and R. V. Jensen, "19-core MCF transmission system using EDFA with shared core pumping coupled in free-space optics," Proc. ECOC2013, London (UK), paper, Th.1.C.6.

[11] K. Igarashi, T. Tsuritani, I. Morita, Y. Tsuchida, K. Maeda, M. Tadakuma, T. Saito, K. Watanabe, K. Imamura, R. Sugizaki, and M. Suzuki, "1.03-Exabit/s·km super-Nyquist-WDM transmission over 7,326-km seven-core fiber," Proc. ECOC2013, London (UK), paper, PD3.E.3.

[12] T. Kobayashi , H. Takara, A. Sano, T. Mizuno, H. Kawakami, Y. Miyamoto , K. Hiraga, Y. Abe, H. Ono, M. Wada, Y. Sasaki, I. Ishida, K. Takenaga, S. Matsuo, K. Saitoh ,M. Yamada, H. Masuda, and T. Morioka, "2 × 344 Tb/s propagation-direction interleaved transmission over 1500-km MCF enhanced by multicarrier full electric-field digital back-propagation," Proc. ECOC2013, London (UK), paper, PD3.E.4.

[13] K. Takeshima, T. Tsuritani, Y. Tsuchida, K. Maeda, T. Saito, K. Watanabe, T. Sasa, K. Imamura, R. Sugizaki, K. Igarashi, I. Morita, and M. Suzuki, "51.1-Tbit/s MCF transmission over 2520 km using cladding-pumped seven-core EDFAs," J. Lightw. Technol., vol. 34, no. 2, pp. 761–767, 2016.

WC1-3

Effective Area Measurement of Two-Mode Fiber Using Bidirectional OTDR Technique

Masaharu Ohashi, Shun Asuka, Yuji Miyoshi, and Hirokazu Kubota
Osaka Prefecture University, 1-1, Gakuen-cho, Naka, Sakai, Osaka, 599-8531 Japan
ohashi@eis.osakafu-u.ac.jp

Abstract: We propose a simple technique for measuring the effective area of a two-mode fiber using bidirectional OTDR. We have successfully estimated the effective area distribution of the LP_{11} mode using our proposed technique.
Keywords: effective area, two-mode fiber, bidirectional OTDR technique

I. INTRODUCTION

The transmission capacity of a conventional fiber is approaching its fundamental limit because of fiber fuse and optical nonlinear effects. Recently, the research and development of novel optical fibers have been undertaken worldwide with the goal of overcoming the abovementioned limit [1], [2]. An innovative optical fiber is a few-mode fiber (FMF). In FMFs, the modal field broadening, cutoff wavelength, and optical loss of high-order modes are important parameters that should be considered when designing optical fiber transmission systems. However, there has been no report on the measurement techniques of the modal field broadening of high-order modes as well as its definition. In our previous study, we proposed an effective mode field diameter obtained from the effective area A_{eff} as a measure of the modal field broadening of high-order modes [3].

In this paper, we propose a simple technique for measuring the effective area of the LP_{11} mode using a mode coupler and an improved OTDR [4]. The measurement technique is based on the bidirectional OTDR using a double reference method [5]. To the best of our knowledge, this is the first study wherein the effective area distribution of LP_{11} and LP_{01} modes in a two-mode fiber is successfully estimated.

II. PRINCIPLE OF OPERATION OF PROPOSED TECHNIQUE

The experimental setup for measuring the effective area A_{eff} of a TMF is shown in Fig. 1. The transmission line consists of two reference TMFs, test fiber, and mode couplers. Mode couplers are inserted to launch each mode into the transmission line or to receive the backscattered power of each mode at the opposite end of the transmission line. In the measurements, two types of reference TMFs are employed to determine the absolute value of the effective area A_{eff} of the test TMF.

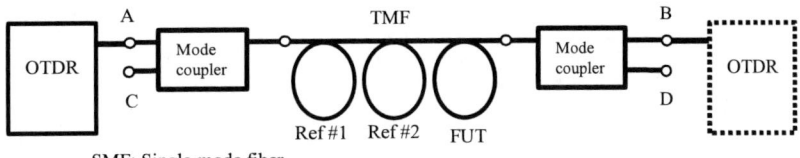

SMF: Single-mode fiber
TMF: Two mode fiber

Fig. 1. Experimental setup

The backscattered power P (λ, z) at a wavelength of λ from a given point z in the transmission line (as shown in Fig. 1) can be expressed as

$$P(\lambda,z) = \frac{P_0}{2}\alpha_s(z)B(\lambda,z)\exp(-2\alpha z)[1+\exp(-4hz)], \tag{1}$$

where P_0 is the input power, α_s (z) is the local scattering coefficient, B (λ, z) is the backscattered capture fraction, α is the attenuation coefficient, and h is the mode-coupling coefficient between LP_{01} and LP_{11} modes.

A reliable means of separating the effects of optical power and waveguide imperfections from OTDR signals has been reported in [6]. For OTDR signals of $S_1(\lambda,z) = 10\log P_1(\lambda,z)$ and $S_2(\lambda, L-z) = 10\log P_2(\lambda,L-z)$ launched via the mode coupler from opposite ends (subscripts 1 and 2) of a transmission line with a length of L, the imperfection contribution $I(\lambda,z)$ is expressed as

$$I(\lambda,z) = \frac{S_1(\lambda,z) + S_2(\lambda, L-z)}{2} \tag{2}$$

$$= 5\log P_1 P_2 + 10\log(\alpha_s(z)B(z)) - 2\alpha L(10\log e) + 5\log\left\{\frac{[1+\exp(-4hz)][1+\exp[-4h(L-z)]]}{4}\right\}.$$

The forth term on the right-hand side in Eq. (2) means the optical loss due to mode coupling between LP_{01} and LP_{11}

modes. Therefore, to obtain a pure imperfection contribution loss from Eq. (2), the mode coupling loss between LP_{01} and LP_{11} modes has to be measured. Here, the pure imperfection contribution loss $I(\lambda, z)$ can be redefined as

$$I(\lambda, z) = \frac{S_1(\lambda, z) + S_2(\lambda, L - z)}{2} - 5\log\left\{\frac{\left[1 + \exp(-4hz)\right]\left[1 + \exp\left[-4h(L - z)\right]\right]}{4}\right\} \tag{3}$$

$$= \frac{S_1(\lambda, z) + S_2(\lambda, L - z)}{2} - 5\log F(z),$$

where $F(z)$ is a function of h and length z.

The imperfection contribution $I_n(\lambda, z)$ normalized by the value at $z = z_0$ is

$$I_n(\lambda, z) = \left[\frac{S_1(\lambda, z) + S_2(\lambda, L - z)}{2} - 5\log F(z)\right] - \left[\frac{S_1(\lambda, z_0) + S_2(\lambda, L - z_0)}{2} - 5\log F(z_0)\right]. \tag{4}$$

The backscattered capture fraction $B(\lambda, z)$ for the LP_{01} mode is expressed using the refractive index of the core n and the mode field diameter $2w(\lambda, z)$ as [7]

$$B(\lambda, z) = \frac{3}{2}\left[\frac{\lambda}{2\pi n w(\lambda, z)}\right]^2. \tag{5}$$

Here, it was assumed that the backscattered capture fraction for all modes is in inverse proportion to the effective area as

$$B(\lambda, z) = \frac{K}{A_{eff}(\lambda, z)n^2(\lambda, z)}. \tag{6}$$

The normalized imperfection contribution $I_n(\lambda, z)$ can be rewritten as

$$I_n(\lambda, z) = 10\log\left[\frac{\alpha_s(z)}{\alpha_s(z_0)}\right] + 20\log\left[\frac{n(\lambda, z_0)}{n(\lambda, z)}\right] - 10\log\left[\frac{A_{eff}(\lambda, z)}{A_{eff}(\lambda, z_0)}\right]. \tag{7}$$

The variation in the local scattering coefficient $\alpha_s(z)$ is negligible compared to that in the effective area $A_{eff}(\lambda, z)$. Therefore, effective area $A_{eff}(\lambda, z)$ can be obtained utilizing the double reference technique [5], [8] as

$$A_{eff}(\lambda, z) = A_{eff}(\lambda, z_0)\left(\frac{A_{eff}(\lambda, z_1)}{A_{eff}(\lambda, z_0)}\right)^{\frac{I(z) - I(z_0)}{I(z_1) - I(z_0)}}, \tag{8}$$

where $A_{eff}(\lambda, z0)$ and $A_{eff}(\lambda, z1)$ represent the effective areas at the positions z_0 and z_1, respectively.

To obtain the pure imperfection contribution $I(\lambda, z)$, the mode coupling contribution $F(z)$ has to be estimated as shown in Eq. (3). When either pulse of the LP_{01} or LP_{11} modes is launched into the transmission line, $S_1(\lambda, z)$ and $S_3(\lambda, z)$ denote the backscattered powers of the input mode pulse and of the other mode pulse, respectively, in dB from a given position. Thus, the backscattered power ratio $\eta(\lambda, z)$ between $S_3(\lambda, z)$ and $S_1(\lambda, z)$ is expressed as

$$\eta(\lambda, z) = 10\log\left[\frac{S_3(\lambda, z)}{S_1(\lambda, z)}\right] = 10\log\left[\frac{1 - \exp(-4hz)}{1 + \exp(-4hz)}\right]. \tag{9}$$

By using Eq. (9), the mode coupling contribution $F(z)$ can be estimated. Therefore, the effective area A_{eff} can be measured by using Eqs. (3), (4), and (8).

III. Experimental Results

To confirm the effectiveness of the proposed technique, the effective area distributions of the LP_{01} and LP_{11} modes along a fiber link, composed of two reference TMFs and a test TMF were measured. Parameters of these fibers are summarized in Table I below. The effective area of the LP_{11} mode in Table I was calculated using the index profile.

Figure 1 shows the experimental setup for measuring the effective area of the LP_{11} mode. To launch either of the LP_{01} or LP_{11} modes into the TMF or receive the backscattered power of each mode, a phase plate based mode coupler was used. The improved OTDR [4] was used to detect the backscattered powers of the LP_{01} and LP_{11} modes.

Figure 2 shows the bidirectional backscattered powers at 1550 nm as a function of fiber length. The pulse width of the OTDR was 100 ns and the average time was 180 s. Taking the mode coupling contribution into account, Fig. 3 shows the relationship between the fiber length and the imperfection contribution $I(z)$ for the LP_{11} mode. It is seen from Fig. 3 that the imperfection loss slightly changes along the fiber length. The effective area A_{eff} can be estimated by Eq. (9). Figure 4 shows the effective area distribution A_{eff} of the LP_{11} mode at 1550 nm.

TABLE I. FIBER PARAMETERS OF REFERENCE FIBERS AND TEST FIBER

Fiber parameters	Reference #1	Reference #2	Test fiber
Effective area A_{eff} (μm^2) at 1550 nm			
For the LP_{01} mode	126.4	153.1	122
For the LP_{11} mode*	169.9	199.1	170.1
Fiber length (km)	3.0	3.2	0.9

* Calculated value

OECC/PS2016

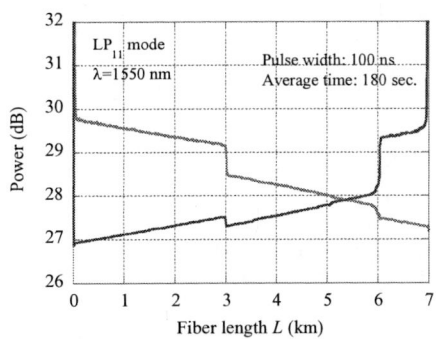

Fig. 2. Bidirectional backscattered powers of the LP$_{11}$ mode

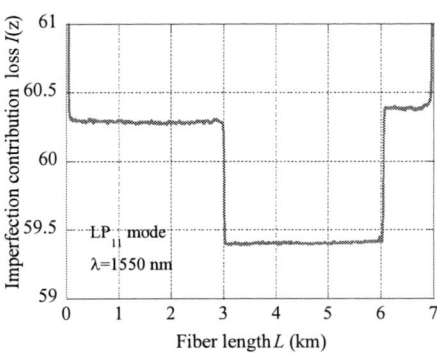

Fig. 3. Imperfection contribution loss for the LP$_{11}$ mode

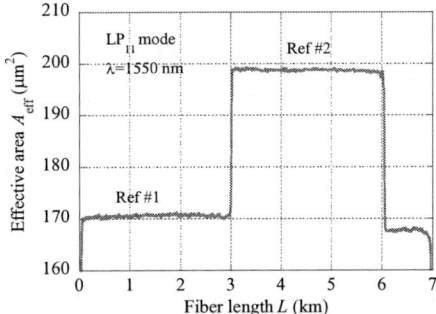

Fig. 4. Effective area distribution of the LP$_{11}$ mode

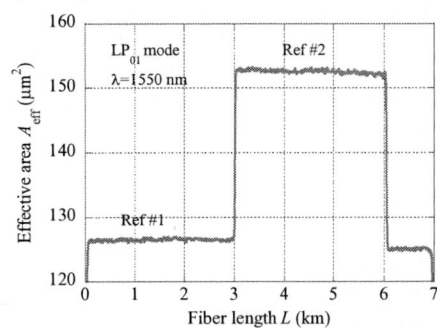

Fig. 5. Effective area distribution of the LP$_{01}$ mode

It was seen that the effective area of the LP$_{11}$ mode estimated by the proposed technique is in good agreement with the calculated values as shown in Table I. It was also seen that the effective area changes along the fiber length. Fig. 5 shows the effective area distribution of the LP$_{01}$ mode. This result is in good agreement with the values shown in Table I that were measured by the FFP method. As a result, the effective areas of the LP$_{01}$ and LP$_{11}$ modes in the TMF can be successfully estimated by this proposed technique.

IV. CONCLUSIONS

This paper proposed a simple technique for measuring the effective area of the LP$_{11}$ mode by using the phase plate based mode coupler and an improved OTDR. The effective area distribution along the fiber link, with a length of about 7 km, was successfully measured by the proposed technique. To the best of our knowledge, this is the first study wherein the effective area distribution of LP11 and LP01 modes in a two-mode fiber is successfully estimated. It was confirmed that this proposed technique can be useful for measuring the effective area of the fundamental and high order modes in a TMF.

ACKNOWLEDGMENT

The part of this research was results from research entitled "Research and development of innovative optical fiber for practical use" commissioned by the National Institute of Information and Communications Technology (NICT).

REFERENCES

[1] P. J. Winzer and G. J. Foschini, "MIMO capacities and outage probabilities in spatially multiplexed optical transport systems," Opt. Express, vol. 19, pp. 16680-96, 2011.

[2] M. Koshiba et al., "Heterogeneous multi-core fibers: proposal and design principle," IEICE Electron. Express, vol. 6, pp. 98-103, 2009.

[3] K. Ozaki et al., "Effective mode field diameter definition and splice loss estimation of LP$_{11}$ mode in few mode fibers," ACP2014, ATh3A.98, 2014.

[4] M. Ohashi et al., "Simple backscattered power technique for measuring crosstalk of multi-core fibers," in Proc. OECC2012, pp. 357-358, 2012.

[5] A. Rossaro et al., "Spatially resolved chromatic dispersion measurement by a bidirectional OTDR technique," IEEE J. Sel. Topics Quantum Electron., vol. 7, no. 3, pp. 475-483, 2001.

[6] M. S. O'Sullivan and J. Ferner, "Interpretation of SM fiber OTDR signature," in Proc. SPIE'86, vol. 661, p. 171, 1986.

[7] E. Brinkmeyer, "Analysis of backscattering method for single-mode fibers," J. Opt. Soc. Am., vol. 70, pp. 1010-1012, 1980.

[8] M. Ohashi et al., "Longitudinal fiber parameter measurements of multi-core fiber using OTDR," Opt. Express, vol. 22, pp. 30137 -30147, 2014.

WC3-1

OECC/PS2016

Fiber-optic guided-acoustic-wave Brillouin scattering observed with pump-probe technique

Neisei Hayashi[1], Heeyoung Lee[2], Yosuke Mizuno[2], and Kentaro Nakamura[2]

[1]Research Center for Advanced Science and Technology, The University of Tokyo, 4-6-1,
Komaba, Meguro-ku, Tokyo 153-8904, Japan; e-mail: hayashi@sonic.pi.titech.ac.jp
[2]Laboratory for Future Interdisciplinary Research of Science and Technology,
Tokyo Institute of Technology, 4259-R2-26, Nagatsuta-cho,
Midori-ku, Yokohama 226-8503, Japan

Abstract: We observe guided-acoustic-wave Brillouin scattering in optical fibers backward using a pump-probe technique and investigate its spectral dependence on pump/probe powers, proving the usefulness of this technique for distributed sensing with a high signal-to-noise ratio.

Keywords: Guided-acoustic-wave Brillouin scattering, optical fiber sensors, nonlinear optics, pump-probe technique

I. INTRODUCTION

Brillouin scattering in optical fibers, which is caused by the interaction between photons and acoustic phonons, is known to be categorized into two: (1) standard Brillouin scattering, in which incident light interacts with longitudinal acoustic waves, and (2) guided-acoustic-wave Brillouin scattering (GAWBS) [1] (also referred to as forward Brillouin scattering [2], cladding Brillouin scattering [3], and transverse stimulated Brillouin scattering [4]), in which incident light interacts with transverse acoustic waves. In standard Brillouin scattering, the scattered light propagates backward accompanying the Brillouin frequency shift (BFS) of the order of gigahertz (sometimes, several higher-order peaks appear), while in GAWBS, the scattered light propagates forward with multiple frequency shifts (which we named guided-acoustic-wave Brillouin frequency shifts (GAW-BFSs)) of lower than 1 GHz, corresponding to the resonance modes.

To date, standard longitudinal Brillouin scattering in optical fibers has been extensively exploited to develop various optical devices and systems, including not only lasers [5], slow light generators [6], optical comb generators [7], but distributed strain and temperature sensors [8-11]. However, no report has been provided on GAWBS-based distributed measurement, though the GAW-BFS exhibits strain and temperature dependence in the same way as the standard BFS [12], [13]. By using GAWBS, several physical parameters, such as fiber outer diameters [14] and acoustic impedances of the surrounding materials, can be additionally measured in a distributed manner. Thus, we aim at achieving GAWBS-based distributed sensing.

Some distributed sensing techniques have been developed for the systems based on standard Brillouin scattering, including time-domain [8]-[9] and correlation-domain techniques [10]-[11]. In order to directly apply these techniques to GAWBS-based systems, the scattering direction (backward or forward) turns out a serious problem. Namely, all the previously developed techniques exploit the optical path difference of the scattered light beams, and the scattered light needs to propagate backward. However, GAWBS basically generates forward-scattered light, which cannot be directly used to develop such distributed sensing systems. Therefore, the backward observation of the GAWBS light is extremely important.

One method of observing the GAWBS light backward is to exploit stimulated Brillouin scattering (SBS), which is a phenomenon that when the power of the incident pump light is higher than a certain threshold, the longitudinal Brillouin scattering becomes drastically strong. Tanaka *et al.* [15] and Dossou *et al.* [16] have reported the backward observation of the GAWBS light, which is made up of two components: the SBS-backscattered GAWBS light induced by high-power pump light, and the GAWBS light induced by SBS-backscattered high-power pump light. As the optical paths of the two components are the same, they need not be separated to develop GAWBS-based distributed sensors.

In practical applications of standard Brillouin sensors, a so-called pump-probe technique is often employed to improve a signal-to-noise ratio (SNR) of the measurement [17]. With this technique, SBS can be induced at much lower threshold power, leading to higher-power Brillouin-scattered signal under the same incident pump power. Although the pump-probe technique appears to be also effective in improving the SNR of the backward GAWBS measurement, there has been no detailed report.

In this paper, we employ the pump-probe technique to observe the GAWBS light backward, and investigate the GAWBS spectrum dependence on the pump and probe light powers, proving experimentally that the pump-probe technique is effective in improving the SNR. The SNR dependence on the probe power is non-monotonic, which indicates the existence of an optimal probe power in implementing actual GAWBS-based distributed sensors.

II. PRINCIPLE

GAWBS in an optical fiber, which is a forward scattering phenomenon, occurs by the interaction between incident light and the transverse modes of acoustic waves excited by thermally induced small vibration of glass molecules [1]. The refractive index of the fiber is perturbed by the acoustic modes, leading to the phase or polarization modulation in the scattering process. The GAW-BFSs are given by the resonance frequencies of the acoustic modes, which are mainly determined by the acoustic velocity (including longitudinal and transverse acoustic waves) and the fiber outer diameter. The transverse acoustic modes causing GAWBS are classified into two [1]: radial $R_{0,m}$ modes and mixed torsional-radial

$TR_{2,m}$ modes, where m is the order of the resonance. The $R_{0,m}$ modes have symmetric axes, while the $TR_{2,m}$ modes have symmetric planes along an optical fiber. The $R_{0,m}$ modes perturb the refractive index radially and maintain the polarization of the scattered light (referred to as polarized GAWBS). In contrast, the $TR_{2,m}$ modes perturb the birefringence and modulate the polarization (referred to as depolarized GAWBS). The GAW-BFS of the m-th acoustic mode $\nu_{GB,m}$ depends on the acoustic velocity in the fiber and the fiber outer diameter. The spectral powers of the $TR_{2,m}$ modes are reported to be higher than those of the $R_{0,m}$ modes [1].

The GAW-BFSs of the $TR_{2,m}$ modes are known to show strain and temperature dependence; for instance, when m is 5, the GAW-BFS linearly depends on strain and temperature with coefficients of ~1.9 MHz/% (at 1319 nm) [12] and ~11 kHz/K (at 1550 nm) [13], respectively. The linewidths of the GAWBS spectral peaks are also reported to be significantly dependent on the acoustic properties (acoustic impedance, etc) of the material attached to the optical fiber [12, 18]. For instance, the linewidth of the GAWBS peak ($TR_{2,5}$ mode) observed using a bare fiber is ~0.3 MHz, which increases to ~4 MHz when the fiber is attached to a polymer. This behavior may be applicable to the acoustic impedance measurement of the attached materials.

The backward observation of the GAWBS signal has been performed simply by injecting high-power pump light [3], [15], [16]. If the pump power becomes higher than the SBS threshold, we can observe part of the GAWBS signal in the backward direction. The backward GAWBS signal involves two components. One is the GAWBS light, which is first propagating forward induced by the high-power pump light and then backscattered by the SBS process. The other is the GAWBS light, which is induced by the SBS-backscattered high-power pump light. One useful technique to reduce the SBS threshold is the pump-probe technique [17], in which a probe light is additionally injected into the FUT from the other end. By setting the probe frequency to $\nu_0 - \nu_B$ (ν_0: pump frequency, ν_B: BFS of the FUT), the energy transfer from pump to probe is drastically accelerated.

III. EXPERIMENTAL SETUP

The experimental setup for observing the GAWBS light backward using the pump-probe technique is depicted in Fig. 1. A 10-km-long silica single-mode fiber (SMF) with a BFS of 10.86 GHz was used as an FUT. The laser output at 1550 nm, which had a power of 6.0 dBm and a linewidth of ~1 MHz, was divided into two using a 3-dB coupler. One light beam was used as pump light, which was polarization-controlled, amplified using an erbium-doped fiber amplifier (EDFA), and injected into the FUT. The pump light which transmitted through the FUT was guided to the air via an optical circulator in order not to be reflected and injected again into the FUT. The other light beam was used as probe light, which was frequency-downshifted by the BFS of the FUT (= 10.86 GHz) using a single-sideband modulator (SSBM; the carrier suppression ratio was >20 dB), amplified using another EDFA, and injected into the FUT from the other end. The GAWBS light propagating backward (i.e. in the opposite direction of the pump) was polarized and guided to a photo detector (PD) along with the transmitted

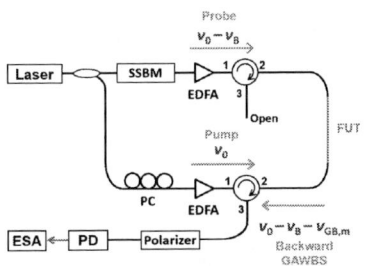

Fig. 1. Experimental setup for observing the GAWBS light backward using the pump-probe technique. EDFA: erbium-doped fiber amplifier, ESA: electrical spectrum analyzer, FUT: fiber under test, PC: polarization controller, PD: photo detector, SSBM: single-sideband modulator.

probe light. Using the PD, the beat signal between the GAWBS light and the transmitted probe light was converted into an electrical signal, which was observed using an electrical spectrum analyzer (ESA). The polarization state was optimized so that the observed GAWBS signal was maximal. The room temperature was 26°C.

IV. EXPERIMENTAL RESULTS

Figure 2 shows the GAWBS spectrum observed backward when the pump power was 12.0 dBm and the probe power was 0 dBm. Approximately 10 clear peaks appeared in the frequency range from 200 MHz to 800 MHz. Each GAW-BFS of these peaks well agreed with the theoretical value of each $TR_{2,m}$ mode [1].

Subsequently, we measured the pump power dependence of the GAWBS spectrum near the peak at ~320 MHz (corresponding to the $TR_{2,18}$ mode, which exhibits one of the clearest peaks). The probe power was 0 dBm. As the pump power increased, the spectral peak power increased, but the noise floor also increased (Fig. 3(a)). Therefore, here we define an SNR as the ratio between the spectral peak power and the spectral power at 340 MHz. Figure 3(b) shows the pump power dependence of the GAWBS spectrum, which was normalized so that the spectral power at 240 MHz was 0. The spectrum at the pump power of 3.6 dBm (regarded as the noise floor of the ESA) was subtracted from all the spectra. The SNR increased with increasing pump power. The SNR was then plotted as a function of the pump power, as shown in Fig. 3(c), where the result when the probe light was not injected was also presented. The Backward GAWBS started to be observed at the pump power of 3.6 dBm, which was

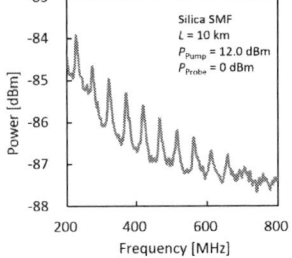

Fig. 2. Wide-range view of the observed GAWBS spectrum.

~10 dB lower than the value without the probe light. The SNR increased linearly with increasing pump power with a coefficient of 0.18 dB/dBm, which was almost twice of the value without the probe light. This result proves that the pump-probe technique is effective in improving the SNR of the backward GAWBS measurement.

Finally, we measured the probe power dependence of the GAWBS spectrum at ~320 MHz. The pump power was 12.0 dBm. As shown in Fig. 4(a), with increasing probe power, the spectral peak power as well as the noise floor increased. The probe power dependence of the GAWBS spectrum normalized using the spectral power at 340 MHz is then shown in Fig. 4(b). The spectrum at the probe power of −21.0 dBm was subtracted from all the spectra to suppress the influence

OECC/PS2016

of the noise of the ESA. Interestingly, the SNR obtained at –9.0 dBm probe power was higher than that at 0 dBm probe power. The SNR dependence on the probe power is shown in Fig. 4(c). The backward GAWBS signal started to be observed at the probe power of approximately –21 dBm. With increasing probe power up to –9.0 dBm, the SNR monotonically increased. As the probe power further increased above –9.0 dBm, the SNR decreased. This behavior originated from the fact that the increase in the noise floor (the influence of the noise caused by the ESA was negligible in this measurement) was more significant than the increase in the GAWBS signal. Note

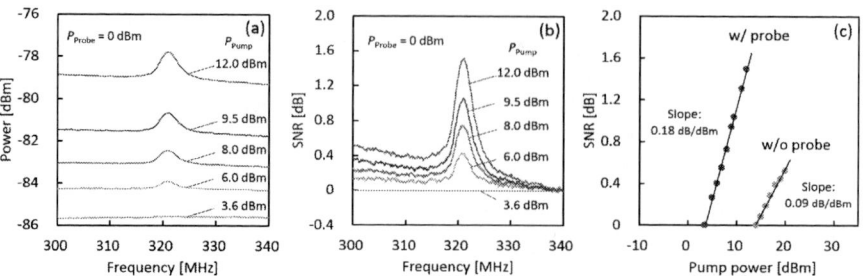

Fig. 3. (a) Pump power dependence of the GAWBS spectrum around 320 MHz. (b) Pump power dependence of the GAWBS spectrum normalized in order for the spectral power at 340 MHz to be 0. (c) Pump power dependence of the SNR of the backward GAWBS measurement with and without the probe light.

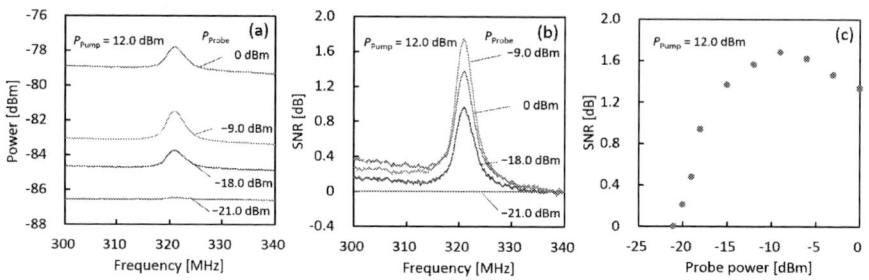

Fig. 4. (a) Probe power dependence of the GAWBS spectrum around 320 MHz. (b) Probe power dependence of the GAWBS spectrum normalized in order for the spectral power at 340 MHz to be 0. (c) Probe power dependence of the SNR of the backward GAWBS measurement when the pump power was 12.0 dBm.

that this noise floor is actually the foot of the beat signal between the probe light and the SBS light; the power of the SBS light is much higher than the GAWBS light. As it is infeasible in principle to suppress this noise floor while maintaining the peak power of the backward GAWBS light, this problem of the SNR deterioration is essential in this scheme. This result indicates that an optimal probe power exists in developing GAWBS-based sensors using the pump-probe technique.

V. CONCLUSIONS

The backward GAWBS observation was successfully performed for the first time using the pump-probe technique, and the GAWBS spectrum dependence on the pump and probe powers was investigated. The SNR was experimentally shown to drastically increase using this technique. A non-monotonic dependence of the SNR on the probe power (at a fixed pump power) suggested the existence of an optimal probe power in developing GAWBS-based sensors. These results will be a useful archive in demonstrating backward GAWBS-based distributed sensing of strain, temperature, fiber outer diameter, and acoustic impedance in the future.

ACKNOWLEDGMENTS

This work was supported by JSPS KAKENHI Grant Numbers 25709032, 26630180, and 25007652.

REFERENCES

[1] R. M. Shelby, M. D. Levenson, and P. W. Bayer, Phys. Rev. B, **31**, 5244 (1985).
[2] R. M. Shelby, M. D. Levenson, and P. W. Bayer, Phys. Rev. Lett., **54**, 939 (1985).
[3] I. Bongrand, A. Picozzi, and E. Picholle, IEEE Electron. Lett., **34**, 1769 (1998).
[4] I. Bongrand, E. Picholle, and C. Montes, Eur. Phys. J. D, **20**, 121 (2002).
[5] Y. Xu, D. Xiang, Z. Ou, P. Lu, and X. Bao, Opt. Lett., **40**, 1920 (2015).
[6] K. Y. Song, M. G. Herraez, and L. Thevenaz, Opt. Exp., **13**, 82 (2005).
[7] G. J. Cowle, D. Y. Stepanov, and Y. T. Chieng, J. Lightw. Technol., **15**, 1198 (1997).
[8] T. Horiguchi and M. Tateda, J. Lightw. Technol., **7**, 1170 (1989).
[9] T. Kurashima, T. Horiguchi, H. Izumita, S. Furukawa, and Y. Koyamada, IEICE Trans. Commun., **E76-B**, 382 (2014).
[10] K. Hotate and T. Hasegawa, IEICE Trans. Electron., **E83-C**, 405 (2000).
[11] Y. Mizuno, W. Zou, Z. He, and K. Hotate, Opt. Exp., **16**, 12148 (2008).
[12] Y. Tanaka and K. Ogusu, IEEE Photon. Technol. Lett., **11**, 865 (1999).
[13] Y. Tanaka and K. Ogusu, IEEE Photon. Technol. Lett., **10**, 1769 (1998).
[14] M. Ohashi, S. Naotaka, and S. Kazuyki, Electron. Lett., **28**, 900 (1992).
[15] Y. Tanaka, H. Yoshida, and T. Kurokawa, Meas. Sci. Technol., **15**, 1458 (2004).
[16] M. Dossou, D. Bacquet, A. Goffin, and P. Szriftgiser, "Observation of cladding Brillouin scattering (CBS) modes using backscattering technique," in Proc. Symp. IEEE/LEOS Benlux Chapter, Enschede, Netherlands, 2008, pp. 35–38.
[17] N. Shibata, Y. Azuma, T. Horiguchi, and M. Tateda, Opt. Lett., **13**, 595 (1988).
[18] A. J. Poustie, J. Opt. Soc. Am. B, **10**, 691 (1993).

Distributed hydrostatic pressure measurement based on Brillouin dynamic grating in polarization maintaining fibers

Yong Hyun Kim, Hong Kwon, Jeongjun Kim, and Kwang Yong Song*

Department of physics, Chung-Ang University, Seoul 06974, Korea

Abstract: Distributed measurement of hydrostatic pressure is demonstrated based on optical time domain analysis of Brillouin dynamic grating (BDG) in polarization maintaining fibers (PMF's). Spectral shift of the BDG is analyzed as a function of hydrostatic pressure from 14.5 psi (1 bar) to 884.7 psi (61 bar) using bow-tie PMF's and photonic crystal PMF (PM-PCF) with 1 m spatial resolution. The experimental results show the pressure sensitivities of +0.65 and +0.78 MHz/psi for bow-tie PMF's and -1.69 MHz/psi for the PM-PCF, respectively.

Keywords: Brillouin scattering, Fiber optics, Distributed sensing

I. INTRODUCTION

Real-time distributed sensing of hydrostatic pressure is necessary to detect the physical location of a blockage or obstruction along the pipelines or oil wells [1]. Although several schemes have been proposed for fiber-optic pressure sensors such as superstructure fiber grating, silica diaphragm, and PCF [2-4], most of them operates as a point sensor, and the method based on Brillouin optical time-domain analysis (BOTDA) is the only approach currently available for distributed measurement [1]. In this paper, we propose and experimentally demonstrate a novel distributed hydrostatic pressure sensor based on Brillouin dynamic grating in PMF's. The spectral shift of the BDG induced by local birefringence change is applied to measure the hydrostatic pressure by optical time-domain analysis with 1 m spatial resolution. Our test measurements on bow-tie fibers and PM photonic crystal fiber (PCF) show the pressure sensitivities of +0.65 and +0.78 MHz/psi for bow-tie PMF's with different cladding diameters, and -1.69 MHz/psi for the PM-PCF, respectively, which is about 30 to 70 times larger than that of the former BOTDA result.

II. PRINCIPLE AND EXPERIMENTS

The BDG represents the acoustic phonon playing the role of moving index grating in a birefringent medium which is generated by stimulated Brillouin scattering (SBS) of pump waves in one polarization and used to reflect probe wave in the orthogonal polarization at different optical frequency from the pump [5]. The frequency difference between the pump and the probe, called BDG frequency (ν_D), is determined by the local birefringence (Δn) of the medium through the phase matching condition of the SBS by the following equation:

$$\nu_D = \frac{\Delta n}{n_g} \cdot \nu_1,\qquad(1)$$

where ν_1, and n_g represents the optical frequency of the pump and the group index, respectively. The sensitivities of the BDG frequency shift to temperature and strain change are about 50 and 20 times larger than those of Brillouin frequency shift in conventional PMF's, respectively [6].

The experimental setup for the distributed pressure measurement by the BDG in PMF is shown in Fig. 1. A DFB-LD with center wavelength of 1550 nm was used as a light source, and two sidebands for the pump and probe waves were generated by electro-optic modulator (EOM) with the modulation frequency of f_{m1} before the amplification by Er-doped fiber amplifier (EDFA). To reduce the amplified spontaneous emission (ASE) noise, optical tunable filter (OTF) was used after the first EDFA. The pump and probe waves were divided by a fiber Bragg grating (FBG) before the pulse generation by another EOM, which is located at each arm of the pump and probe. The pump wave was divided by 50/50 coupler to generate the continuous wave (CW) pump2 and the 20 ns-pulsed pump1 by a single sideband modulator (SSBM) and an EOM, respectively. To maximize the reflection of the BDG, the frequency offset f_{m2} between the pump1 and pump2 was set to the Brillouin frequency (ν_B) of the test fiber. Both pump waves were amplified by EDFA's and launched to fiber under test (FUT) with x-polarization (slow axis) using polarization controller (PC) to generate the BDG. The probe pulse with 10 ns duration, which corresponds to 1 m spatial resolution, was launched to the FUT in the y-polarization (fast axis) after amplified by an EDFA. The signal of BDG reflection was filtered by another FBG and detected by a photo detector and data acquisition (DAQ) after fine tuning of the amplitude by variable optical attenuator (VOA). The spectrum of BDG was obtained while sweeping f_{m1} with a step of 4 MHz and 256 times of averaging. The configuration of the frequencies and polarizations of the pumps, probe, and BDG reflection is shown at the inset of Fig. 1. The time delay between the pump1 and probe pulses was set to 30 ns to maximize the signal of BDG reflection.

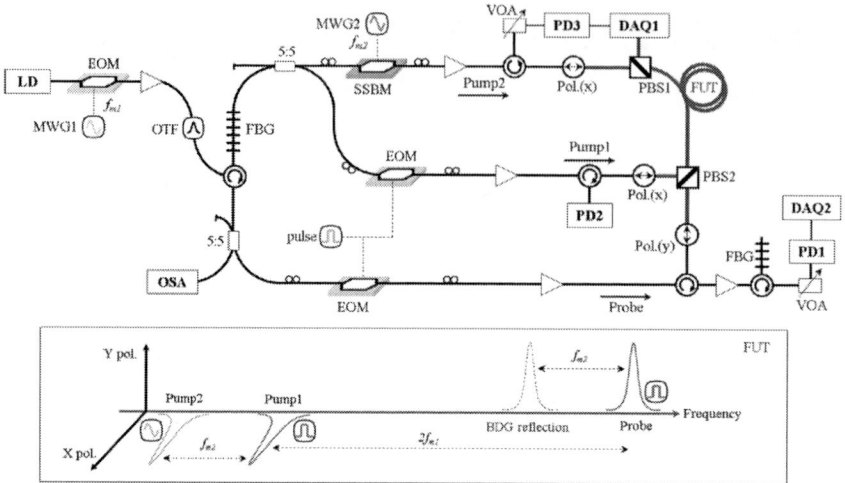

Fig. 1. Experimental setup for distributed measurement of pressure based on BDG: LD, laser diode; EOM, electro-optic modulator; MWG, microwave generator; OTF, optical tunable filter; FBG, fiber Bragg grating; OSA, optical spectrum analyzer; SSBM, single-sideband generator; PBS, polarization beam splitter; VOA, variable optical attenuator; DAQ, data acquisition.

The results of distributed pressure measurement using two types of Bow-tie PMF (HB1500 and HB1500G manufactured by Fibercore) are shown in Fig. 2. The detailed structure of FUT is described in Fig. 2(c). The ν_B and ν_D of the two PMF's at room temperature and atmospheric pressure are 10.534 and 59.8 GHz, 10.531 and 42.7 GHz, respectively. We applied cross correlation fitting to calculate the pressure dependent shift of the BDG spectra.

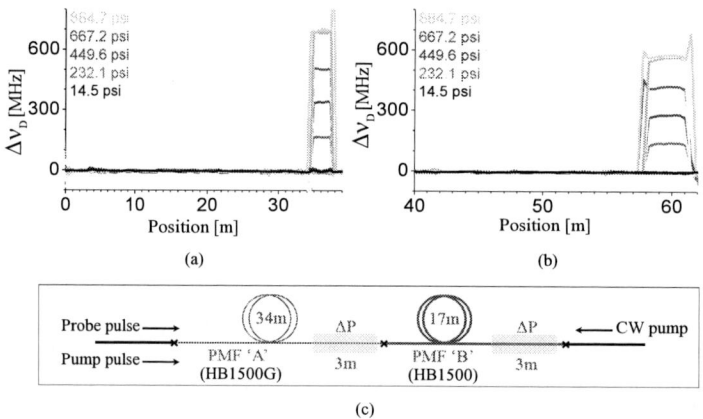

Fig. 2. Distributed BDG frequency variation as a function of pressure. (a) Distributed measurement data of type 'A' PMF. (b) Distributed measurement data of type 'B' PMF. (c) The structure of fiber under test.

The pressure coefficients of ν_D are +0.78 MHz/psi for PMF 'A' and +0.65 MHz/psi for PMF 'B', respectively, as depicted in Fig. 3. Applied pressure to the fiber is directly changed to the strain and the relation of both properties is well formulated in the early experiment as

$$\varepsilon = -\frac{P(1-2\mu)}{E} \quad , \tag{2}$$

where ε, P, μ, and E denotes strain, pressure, Poisson's ratio, and Young's modulus, respectively [7]. The ratio of pressure coefficients between PMF 'A' and 'B' is 0.827, which is very close to the ratio of strain coefficient between the two PMFs. The strain coefficient, in our former experiment, for PMF 'A' and 'B' was 1.28 and 1.06 MHz/με, respectively, so the ratio is 0.828 [8]. Temperature dependence of ν_D reported in our former experiment for PMFs 'A' and 'B' was about -60 and -43.6 MHz/°C, respectively. The measurement time spent for each pressure was about one minute, where the contribution of the temperature change to the BDG frequency was within that of the pressure error (~1.5 psi).

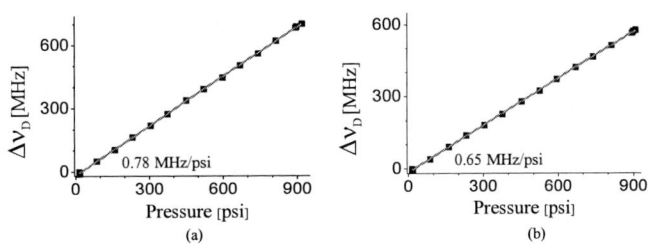

Fig. 3. Pressure dependence of the BDG frequency in (a) PMF 'A' and (b) PMF 'B'

In the BDG measurement of 3 m PM-PCF, the ν_D is decreased as the pressure is increased. Normalized BDG spectra at 884.7 psi (61 bar) and 14.5 psi (1 bar) are shown in Fig. 4(a). The slope of the ν_D variation according to pressure is measured to be -1.69 MHz/psi as depicted in Fig. 4(b). We think the negative slope might originate from the decreases of birefringence by the shrinkage of air hole under pressure. It is notable that the temperature compensation of ν_D is not necessary in the PM-PCF due to its temperature-insensitive birefringence. In our previous report, from 20°C and above, the change of ν_D is nearly negligible in the PM-PCF, and in current measurement the experiment was maintained at about 22°C. The deviation from the linear fitting result in Fig. 4(b) is within the effect of the pressure error (~1.5 psi).

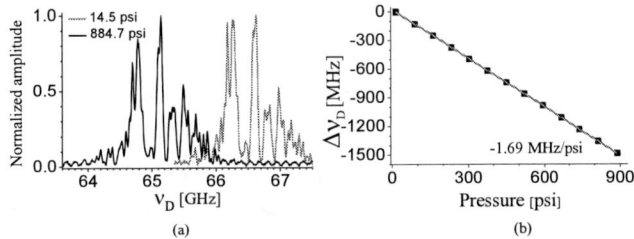

Fig. 4. (a) BDG spectrum at 14.5 psi and 884.7 psi in PM-PCF. (b) Pressure dependence of the BDG frequency of PM-PCF.

III. CONCLUSIONS

A novel distributed sensor of hydrostatic pressure is demonstrated on the basis of the BDG in two types of bowtie PMF and PM-PCF. Our test measurements show the pressure dependencies of +0.65 and +0.78 MHz/psi for bow-tie PMF's and -1.69 MHz/psi for the PM-PCF, respectively, which is about 30 to 70 times larger than that of the former BOTDA result. It is also confirmed that the ratio between the pressure coefficients of two bow-tie PMF's agrees well with the ratio of strain coefficients as theoretically expected. We believe the proposed scheme can be useful for applications which need high sensitivity monitoring of pressure distribution.

ACKNOWLEDGMENT

This work was supported by the National Research Foundation of Korea (NRF) grant funded by the Korea government (MSIP) (No. NRF-2015R1A2A2A01007078).

REFERENCES

[1] A. Mendez and E. Diatzikis, "Fiber optic distributed pressure sensor based on Brillouin scattering," in the 18th international Conference on Optical Fiber Sensors (OFS18), paper ThE46, Cancun (Mexico), 2006.

[2] C. M. Lin, Y. C. Liu, W. F. Liu, M. Y. Fu, H. J. Sheng, S. S. Bor, and C. L. Tien, " High-sensitivity simultaneous pressure and temperature sensor using a superstructure fiber grating," IEEE Sensors J., vol. 6, no. 3, pp.691-696, Jun. 2006.

[3] D. Donlagic and E. Cibula, "All-fiber high-sensitivity pressure sensor with SiO2 diaphragm," Opt. Lett., vol. 30, No. 16, pp. 2071-2073, Aug. 2005.

[4] W. J. Bock, J. Chen, T. Eftimov, and W. Urbanczyk, "A photonic crystal fiber sensor for pressure measurements," IEEE Trans. Instrument and Measurement, vol. 55, no. 4, pp. 1119-1123, Aug. 2006.

[5] K. Y. Song, W. Zou, Z. He, and K. Hotate, "All-optical dynamic grating generation based on Brillouin scattering in polarization maintaining fiber," Opt. Lett., vol. 33, no. 9, pp. 926-928, May 2008.

[6] K. Y. Song, "High sensitivity optical time-domain reflectometry based on Brillouin dynamic gratings in polarization maintaining fibers," Opt. Express, vol. 20, no. 25, pp. 27377-27383, Dec. 2012.

[7] G. B. Hocker, "Fiber-optic sensing of pressure and temperature," Appl. Opt., vol. 18, no. 9, pp. 1445-1448, May 1979.

[8] K. Y. Song, "Distributed fiber sensors based on Brillouin dynamic gratings," in IEEE SENSORS 2014, paper ID 1600, Valencia (Spain), 2014.

WC3-3   OECC/PS2016

A Multicavity Fiber Fabry–Pérot Interferometer for Sensing Multiple Parameters

[a]Wei-Kang Chang, [a]Meng-Shan Wu, [a]Chung-Hao Tseng, [b]Pin Han, [a]Cheng-Ling Lee*

[a] Department of Electro-Optical Engineering, National United University, Miaoli 360, Taiwan
[b]Institute of Precision Engineering, National Chung Hsing University, Taichung 402, Taiwan,
*e-mail: cherry@nuu.edu.tw

Abstract: We develop a microcavity-fiber-Fabry-Pérot-interferometer which can simultaneously measure external refractive-index (RI) and vibrations of surrounding. Experimental results show linear responses of RI measurement and good identified SNR over 30dB of vibration measurement can be obtained.

Keywords: Fiber-optic component; fiber-optic sensor; microcavity fiber Fabry-Pérot interferometer (MCFFPI); tapered-fiber plug; simultaneously sensor

I. INTRODUCTION

Miniature Fiber Fabry–Pérot interferometers (FFPIs) are extremely important in many sensing applications especially for exploring some physical and optical parameters in the micro-specimens. Numerous FFPIs with smart, simple and hybrid structures for many fields of applications on various parametric sensing have been proposed [1–6]. The FFPIs devices can be used to measure external refractive index (RI) [1–3], ambient Humidity/ temperature (T) [3-4], high frequency vibration [5-6]. The sensing characteristics of these tip-typed FFPI sensors are generally sensitive, probed design, ultracompact and convenient; such properties will be especially useful as a sensing probe for detecting in the harmful circumstances. The configuration of these FFPIs can achieve the well-known two modes interference mechanism and obtain low finesse interferometric characteristics. In this study, a novel, ultracompact and highly sensitive pendulum-typed microcavity fiber Fabry-Pérot interferometer (MCFFPI) based on a section of hollow core fiber (HCF) fused a tip-shaped tapered-fiber-plug with a flat cleaving endface to form a probe-type fiber sensor. The fabrication steps were immediately monitored by an optical microscope and they were shown in the Fig. 1(a)-1(e). First of all, in the step of (a): a single mode fiber (SMF) fused with a small section of HCF, then a tapered plug was inserted into endface of the HCF as shown in Fig. 1 (b-c). In the (d) step, we spliced the junction of the taper plug and the HCF by commercial fusion splicer for permanently fixed. Finally, cleave and remain a tiny section of taper plug in the end of HCF. The fused process was extremely required for the bonding strongly for the use as a vibration element. The micrograph of the MCFFPI is shown in Fig. 2 (a).

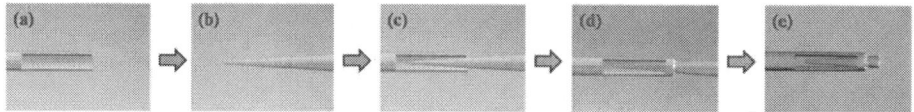

Fig. 1 Fabrication processes of the optical fiber sensor.

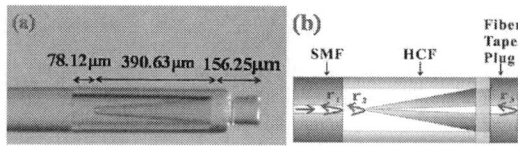

Fig. 2 (a) Micrograph of the MCFFPI device. (b) Configuration and operating principle of the device.

In Fig.2 (b), the light initially propagates from the SMF side. When the light arrives to the HCF, little part of the light is reflected then the other is transmitted and reflected at the tip of taper fiber plug. The reflected light r_3 from the endface of taper fiber plug is highly influenced by the surrounding liquids. The structure will generate interference of hybrid cavities with very low finesse with composition of multiple interferences. For the theoretical analysis, it is easy to speculate the multiple interferences of the proposed device by using the wave superposition. The analytical equations of the interference spectra $I(\lambda)$ are shown in the following equations, Eq.(1)~Eq.(3) for the surrounding of n_D (RI of liquids) > n_{SMF} (RI of fiber), $n_D = n_{SMF}$ and $n_D > n_{SMF}$ conditions, respectively.

$$I(\lambda) = r_1 + r_2 + r_3 - 2\sqrt{r_1 r_2} \cdot \cos(\phi_1) - 2\sqrt{r_2 r_3} \cdot \cos(\phi_2) + 2\sqrt{r_1 r_3} \cdot \cos(\phi_1 + \phi_2) \qquad \text{for } n_D < n_{SMF} \quad (1)$$

1074

$$I(\lambda) = r_1 + r_2 - 2\sqrt{r_1 r_2} \cdot \cos(\phi_1) \qquad\qquad \text{for } n_D = n_{SMF} \quad (2)$$

$$I(\lambda) = r_1 + r_2 + r_3 - 2\sqrt{r_1 r_2} \cdot \cos(\phi_1) + 2\sqrt{r_2 r_3} \cdot \cos(\phi_2) - 2\sqrt{r_1 r_3} \cdot \cos(\phi_1 + \phi_2) \qquad \text{for } n_D < n_{SMF} \quad (3)$$

Here, r_1, r_2 and r_3 denote Fresnel reflections from each interface of the structure. n_D and n_{SMF} are RIs of the liquids and fiber, respectively. The ϕ_1 and ϕ_2 represent the phase difference caused by the cavities of air gap (r_1 and r_2 cavity) and taper fiber plug (r_2 and r_3 cavity) respectively. Figure 3(a)-(c) show the three cases of (a) in air, in liquids of (b) n_D=1.456 and (c) n_D=1.6, respectively. The results demonstrate that multiple interferences by the hybrid cavities are achieved to further estimate the external refractive-index (RI) of the surrounding.

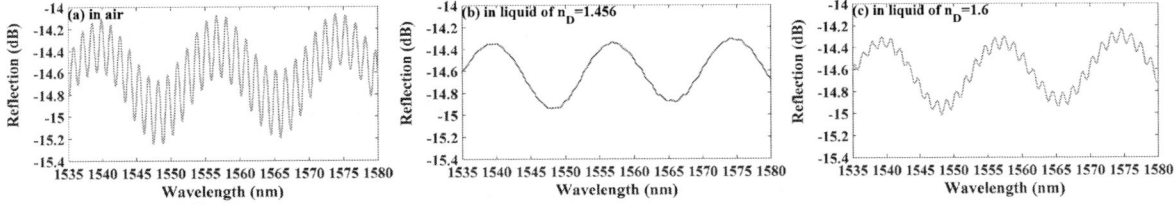

Fig. 3: Interference patterns of the proposed device (a) in air, in liquids of (b) n_D=1.456 and (c) n_D=1.6.

As the vibration measurement, operation principle of the proposed sensor is shown in Fig. 4. The tip of the tapered plug was the main part to oscillate with deviation from central of SMF for strongly reducing the light return to the core of SMF. Thus, optical light intensity received by the SMF presents a periodical variation with certain amplitude corresponding to the driving frequency and driving power/amplitude of the acoustic speaker.

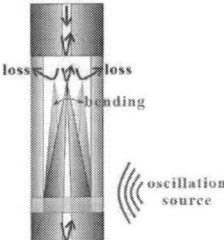

Fig. 4. Schematic diagram of proposed sensor under the vibration.

II. EXPERIMENTAL RESULTS AND DISCUSSION

The experimental setup for simultaneously sensing vibration frequency and refractive index (RI) is shown in Fig. 5. On the measurement of vibration sensing, the light from a tunable laser source with a wavelength of 1550nm propagates into the sensor, and the reflection power is detected by a power meter. The electronic signals are directly collected by an oscilloscope. The tapered fiber bridge is bent due to the vibration from the speaker to vary the output laser power. The dynamic vibration causes the power of the sensor synchronously oscillating with the current driving vibration frequency. Then the time domain power signals are converted to frequency spectra through the fast Fourier transform. In measure RI experiment, the room temperature is fixed and controlled by a TE cooler (resolution: ±0.01 °C). When a wideband light source (WLS) propagates to the device, the two reflective beams from tapered plug tip and terminal junction are combined in the SMF, producing interference patterns on the optical spectrum analyzer (OSA).

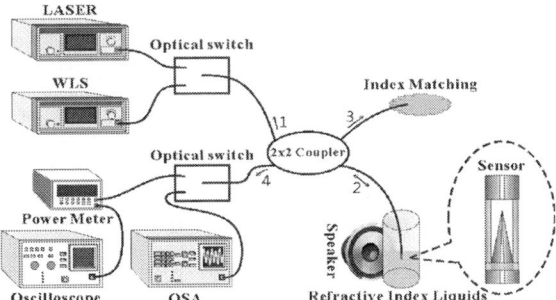

Fig. 5. Experimental setup of the proposed FFPI for making measurement. WLS: Wideband light source, OSA: Optical spectrum analyzer.

OECC/PS2016

Figures 6(a)-(b) respectively show the experimental results of the measured vibration in air and liquid with the same driving power of about 0.7W. The obtained frequency (f=5kHz) responses with the high order harmonics and their corresponding optical responses (output voltages) from the power meter are presented. The insets show time domain optical signals when under the corresponding vibration frequency, respectively. The results demonstrate that the sensor can detect the vibrations from the surrounding no matter in the air or in the liquid. When the fiber probe is immersed in the liquids, the contrast of the interference fringes is also getting changed.

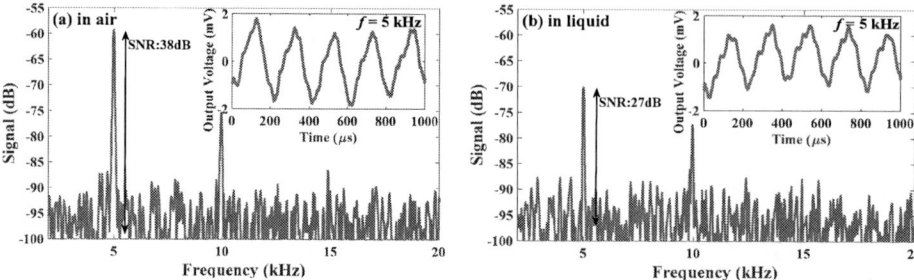

Fig. 6. The frequency-domain spectra of the proposed sensor oscillated in (a) air (b) liquid. The insets display the corresponding time domain spectra.

Figure 7(a) shows the extinction ratio of interference is getting lower also with higher loss of fringe peak when surrounding RI increases. The interference peaks are gradually vanished when RI of surrounding is approaching to that of the silica fiber ($\sim n_D$ = 1.456). This is because the Fresnel reflection (r_3) from the interface of fiber/liquid is extremely weak and is far different with r_2. Fig. 7(b) shows the relationship between the peak reflection and the external RI of the surrounding. The variations are measured with corresponding to the initial condition at n_D = 1(air). Experimental results also indicate a good linearity of response with the peak reflection with sensitivity of about −0.5576dB/RIU. The measured peak of spectral fringes is shown in the inset of Fig. 7(b).

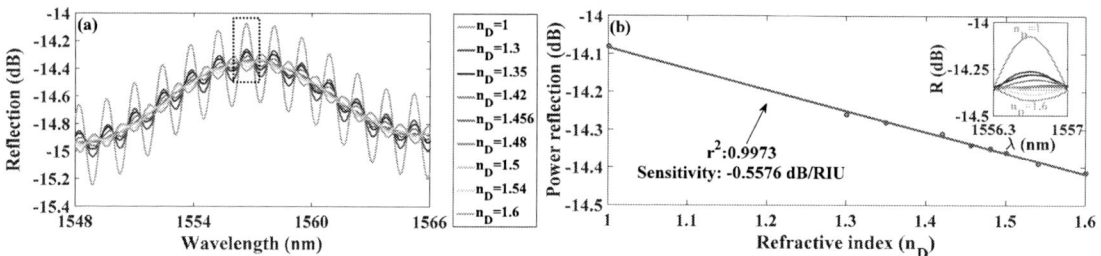

Fig. 7 (a) Interference spectra of experiment for different RI . (b) Experimental sensitivity of peak reflection for the sensor. Inset: the monitored fringe of the spectra.

III. CONCLUSIONS

This study has developed and demonstrated a Fabry–Pérot interferometer that can effectively utilized it to measure the external refractive index (RI) and vibration frequency. Measurements reveal that a RI sensitivity of −0.5576dB/RIU for the peak power variation measurement is achieved in the external RI sensing, and the sensor can detect the high frequency variation in any ambient liquids with a good identified SNR (about 30dB). We anticipate the proposed sensor will have a variety of applications in chemical and biological sensing fields.

REFERENCES

[1] C. L. Lee, J.M. Hsu, J.S. Horng, W.Y. Sung, and C.M. Li, "Microcavity fiber Fabry–Pérot interferometer with an embedded golden thin film," IEEE Photon. Technol. Lett., vol. 25, no. 9, pp. 833–836, May 2013.
[2] C. L. Lee, H.Y. Ho, J.H. Gu, T.Y. Yeh, and C.H. Tseng "Dual hollow core fiber-based Fabry–Perot interferometer for measuring the thermo-optic coefficients of liquids" Opt. Lett., vol. 40, no. 3, pp. 459–462, February 2015.
[3] Z. L. Ran, Y.J. Rao, J. Zhang, Z. Liu, and B. Xu, "A miniature fiber-optic refractive-index sensor based on laser-machined Fabry–Pérot interferometer tip," J. Lightw. Technol., vol. 27, no. 23, pp. 5426–5429, December 2009.
[4] C. L. Lee, Y.W. You, J.H. Dai, J.M. Hsu, and J.S. Horng, "Hygroscopic polymer microcavity fiber Fizeau interferometer incorporating a fiber Bragg grating for simultaneously sensing humidity and temperature," Sens. actuators. B Chem., vol. 222, pp. 339-346, August 2016.
[5] C. L. Lee, Y.N. Tsai, G.H. Chen, Y.J. Xiao, J.M. Hsu, and J.S. Horng, "Refined bridging of microfiber plugs in hollow core fiber for sensing acoustic vibrations," IEEE Photon. Technol. Lett., vol. 27, no. 22, pp. 2403-2406, November 2015.
[6] F. Guo, T. Fink, M. Han, L. Koester, J. Turner, and J. Huang, "High-sensitivity, high-frequency extrinsic Fabry–Pérot interferometric fiber-tip sensor based on a thin silver diaphragm," Opt. Lett., vol. 37, no. 9, pp. 1505–1507, May 2012.

WC3-4

High Sensitivity Fiber Optic Current Sensor based on Recirculating Fiber Loop

Yemeng Tao, Jiangbing Du*, Lin Ma*, Shuai Wang, Yinping Liu, Wenjia Zhang, and Zuyuan He

State Key Laboratory of Advanced Optical Communication Systems and Networks,
Shanghai Jiao Tong University, Shanghai 200240, China
*dujiangbing@sjtu.edu.cn, ma.lin@sjtu.edu.cn

Abstract: We propose high sensitivity fiber optic current sensor based on recirculating fiber loop architecture. We experimentally achieved a sensitivity of 11.5 degrees per ampere and a resolution of 10 mA using standard single-mode fiber.

Keywords: fiber optic current sensor, recirculating fiber loop, Faraday Effect, single-mode fiber.

I. INTRODUCTION

Fiber optic current sensor (FOCS), employing the magneto-optic effect (Faraday Effect) for current measurement, has been attracting a lot of attentions due to its distinctive advantages such as inherently insulation, light in weight, and immune to demagnetizing fields. There are two main obstacles which limit FOCS for practical applications: one is its susceptibility to many environment effects such as temperature variation and vibration [1]; the other is its limited sensitivity to the magnetic field which is determined by Verdet constant of silica material itself [2]. Many studies have been carried out to overcome these problems from aspects of improving fiber sensing media and system designs. There are some studies reported using circularly birefringent spun fibers to reduce its susceptibility to the environment [3], and using special fibers such as flint glass [4], Tb-doped glass fibers [2], and novel magnetic fluid with larger Verdet constants to improve sensitivity [5]. However, those specialty fibers are usually very expensive and very difficult to fabricate. On the other hand, sensing systems utilizing Sagnac or ring-down architectures have also been reported to improve the sensitivity by effectively increasing the interactive length between optical fiber and magnetic field [6, 7].

For ring-down architecture, pulsed light circulate inside the fiber of sensing head for certain times whose number is determined by the overall loss of each cycle. By increasing the time of circulation, the interactive length can be increased and a magnified Faraday Effect caused by the current can be observed. Conventional methods use one or two fiber couplers to build up the circulation loop. However, the use of fiber couplers inevitably affect the polarization stability of the system and limit the number of circulation cycles because some portion of the light power are coupled outside the loop for each cycle. Therefore, one has to get rid of the fiber coupler to overcome this issue and further improve the sensitivity.

In this work, we propose fiber optic current sensor with high sensitivity based on recirculating fiber loop architecture constructed by a 2×2 optic switch instead of fiber couplers. We use a fiber coil consists of 1 km-long SSMF which functions as both the sensing medium and the delay line. Instead of coupling out some portion of the light power after each cycle, the pulsed light can circulate inside the loop without any interruption until the desired circulation number has been reached. As a result, this architecture can effectively improve both the system stability, sensitivity, and reduce the complexity. We successfully achieved a sensitivity of 11.5 degrees per ampere and the minimum current intensity variation detected was 10 mA. The proposed method is promising for high sensitivity current monitoring applications.

II. PRINCIPLE

When linearly polarized light passes through a magnetic field parallel to the optical propagation orientation, the light will undertake the effect of polarization rotation. This effect is nonreciprocal, and only depends on whether the light is propagating along or against magnetic field. The rotation angle θ is determined by the length of optical path, magnetic field B and Verdet constant V of the material [7]. Their relationship can be expressed as:

$$\theta = V \int B dl \tag{1}$$

When polarization angle of linearly polarized light is changed, light intensity of fast axis and slow axis also changes. By detecting the intensity of polarized light out of polarization beam splitter (PBS), we can calculate the rotation angle. The intensities of two axes can be expressed as:

$$P_1 = P_0 \cos^2(\theta + \varphi) \tag{2}$$

$$P_2 = P_0 \sin^2(\theta + \varphi) \tag{3}$$

Where, P_0 is the detected power at PBS. P_1 and P_2 is the power of fast axis and slow axis, respectively. ϕ is the modulation angel generally set as 45 degrees to make P_1 and P_2 equal. By calculating the normalized intensity, θ can be obtained as:

$$\theta = -\frac{1}{2}\arcsin P_{BD} \qquad (4)$$

$$\text{Where,} \quad P_{BD} = \frac{P_1 - P_2}{P_0} = \cos(2(\theta + \varphi)) \qquad (5)$$

From equation (1), it can be observed that a larger θ can be obtained with larger l (longer fiber) or larger V. By employing a recirculating fiber loop architecture, pulsed light can circulate inside the fiber sensing head for several times and as a result, largely increase the equivalent length of sensing fiber. In our experiment, the system sensitivity is defined by degree per ampere and the resolution is the minimum current variation detected within the measurement uncertainty.

III. EXPERIMENTAL SETUP AND RESULTS

The experimental setup is shown in Fig. 1. We used a linearly polarized fiber laser (FL) operating at 1550 nm as the light source and a polarization controller (PC) to control its polarization state. The continuous-wave light from the FL was modulated by an acousto-optic modulator (AOM) to generate pulsed light with a pulse width of 2 μs. A 2×2 single-mode optic switch was used to build up the recirculating fiber loop architecture. When there is a bias applied to optic switch, port 1 is connected to port 4 and port 3 is connected to port 2. When there is no bias applied, port 1 is connected to port 2 and port 3 is connected to port 4. In this case, a fiber loop was formed and the pulsed light can circulate inside the loop for many times. We used a 1 km-long SSMF which satisfies the ITU-T recommendation G652.D as the sensing medium and the delay line as well. In order to minimize the influences of vibrations and noises, the fiber was wound and cured into a fiber coil with a diameter of 11 cm. The delay time of the fiber coil is about 5 μs with a total loss of 4 dB/cycle. When the pulse width from arbitrary waveform generator (AWG) is set as (N-1)×5 μs (N is the number of circulation), light circulates for N times inside the fiber loop. After reaching the desired circulation number, we can couple out the pulsed light from port 2 to the PBS, and analyze the signals detected by two photodiodes (PD).

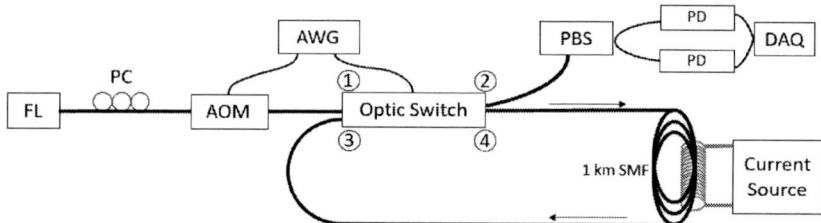

Fig. 1. The experimental setup. FL: fiber laser; PC: polarization controller; AOM: acousto-optic modulator; AWG: arbitrary waveform generator; PBS: polarization beam splitter; PD: photodiode; DAQ: data acquisition.

The measured light intensity as a function of current intensity is shown in Fig. 2 with an N=8. The power of the FL was set to 25 mW. Fig. 2. (a), (b) shows the measured light intensity dependence of two PDs on the current variation with an interval of 0.2 A. It can be observed that P_1 decreases, and P_2 increases linearly, respectively.

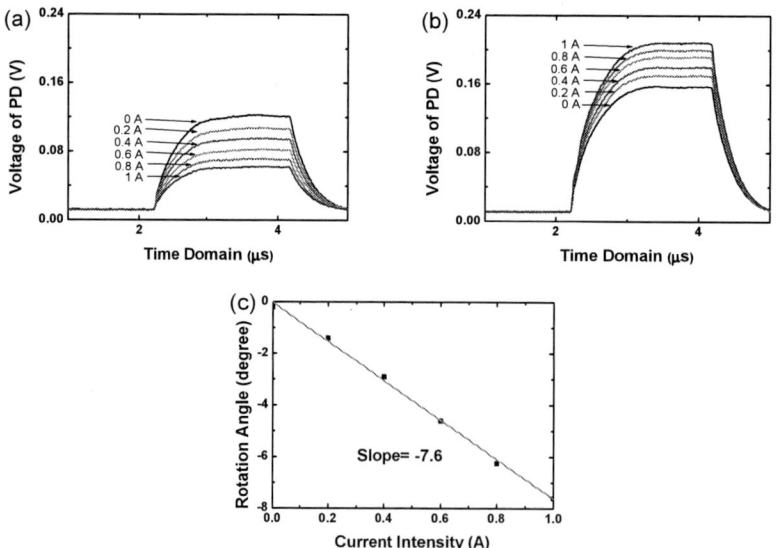

Fig. 2. The dependence of (a) P_1 and (b) P_2 with the change of current with N=8, (c) Relationship between current intensity and rotation angle

The calculated rotation angle as a function of current intensity is shown in Fig. 2. (c) according to equation (4) and (5). The rotation angle varies linearly with the increase of current intensity and the slope is -7.6, which means that the rotation angle varies 7.6 degrees per ampere. As a result, we can calculate the current intensity when the rotation angle and sensitivity are all known.

The current intensities and calculated rotation angles show good linearity when N was below 10. With the increase of N, however, it was necessary to increase the light power accordingly. When N exceeded 10, the system became unstable to obtain the accurate rotation angle because the calculated rotation angle badly fluctuated and its linearity became worse as shown in Fig. 3. One possible explanation for this phenomenon is that the light is actually elliptically polarized light, which makes the linearity deteriorate [8]. We optimized the parameters such as the power of FL and number of circulation. The highest sensitivity achieved was 11.5 degrees/A when the output power of FL was set to 35 mW with an N = 10, and the standard deviation of rotation angle was about 0.2 degree. We turned the current source on and off to test the variation of P_1 and P_2 when current intensity was set as 10 mA, angle variation was close to the standard deviation (0.2 degree). As a result, 10 mA can be defined as the system resolution of our experiment.

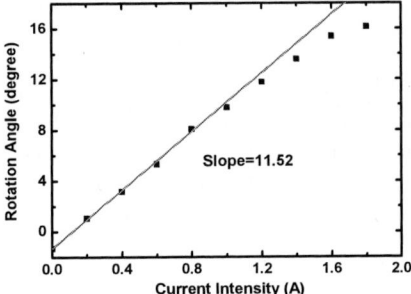

Fig. 3. Trend of angle variation when N=10

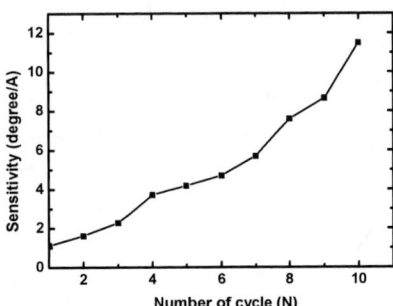

Fig. 4. Relationship between N and sensitivity

Figure 4 shows the relationship between N and system sensitivity. It can be observed that larger N leads to a higher sensitivity. However, due to the influences of initial polarization state of input light and environmental perturbations, the sensitivity does not increase linearly with the number of N. In our experiment the maximum sensitivity was achieved with an N=10 and the sensitivities may be further enhanced by system optimization such as improving the DAQ performance.

IV. CONCLUSION

In this paper, we analyzed the relationship between current intensity and Faraday rotation angle by using 1 km SMF and an optic switch to build recirculating fiber loop architecture. We achieved a sensitivity of 11.5 degrees/A and a minimum detectable current intensity variation of 10 mA.

ACKNOWLEDGMENT

This work was supported by the NSFC under Grant 61405113, 61307107, and Shanghai STCSM under Grant 15511103102 and 13ZR1456200.

REFERENCES

[1] K. Bohnert, G. Müller, L. Yang, and A. Frank, "Intrinsic Temperature Compensation of Interferometric and Polarimetric Fiber-Optic Current Sensors," CLEO, pp. JTu4A. 63, 2014.

[2] D. Huang, S. Srinivasan, and J. E. Bowers, "Compact Tb doped fiber optic current sensor with high sensitivity," Optics express, vol. 23, pp. 29993-29999, 2015

[3] N. Peng, Y. Huang, S. Wang, and L. Wang, "Fiber optic current sensor based on special spun highly birefringent fiber," Photonics Technology Letters, vol. 25, pp. 1668-1671, 2013.

[4] K. Kurosawa, S. Yoshida, K. Sakamoto, I. Masuda, and T. Yamashita, "An optical fiber-type current sensor utilizing the Faraday effect of flint glass fiber," in Proc. Int. Conf. Optical Fibre Sensors, vol. 2360, pp. 24 –27, 1994.

[5] W. Lin, H. Zhang, and B. Song, "Magnetic field sensor based on fiber taper coupler coated with magnetic fluid," International Conference on Optical Fibre Sensors, pp. 96347U-96347U-4, 2015.

[6] H. Zhang, Y. Dong, J. Leeson, and X. Bao, "High sensitivity optical fiber current sensor based on polarization diversity and a Faraday rotation mirror cavity," Applied optics, vol. 50, pp. 924-929, 2011.

[7] H. Zhang, Y. Qiu, H. Li, A. Huang, H. Chen, and G. Li, "High-current-sensitivity all-fiber current sensor based on fiber loop architecture," Optics express, vol. 20, pp. 18591-18599, 2012.

[8] R. Zhang, S. Yao, T. Liu, and L. Li, "The effect of linear birefringence on fiber optic current sensor based on Faraday mirror," International Society for Optics and Photonics, pp. 92741N-92741N-8, 2014.

Fabrication and characterization of lanthanum boroaluminosilicate glass fiber for magneto-optical device applications

Kadathala Linganna[1], Seongmin Ju[1], Bok Hyeon Kim[2], Won-Taek Han[1,*]

[1]School of Information and Communications, Gwangju Institute of Science and Technology, 123 Cheomdangwagi-ro, Buk-gu, Gwangju 500-712, South Korea
[2]Advanced Photonics Research Institute, GIST, 1 Oryong-Dong, Buk-Gu, Gwangju 500-712, South Korea
E-mail address: wthan@gist.ac.kr

Abstract: We fabricated and characterized a lanthanum boroaluminosilicate glass fiber for magneto-optical device applications and its Verdet constant was found to be -4.64 rad/(T·m), which is larger than those of the commercial and coreless multimode fibers.

Keywords: Lanthanum boroaluminosilicate glass fiber; Magneto-optical effect; Verdet constant.

I. INTRODUCTION

Optical isolator is an optical component, which allows the passage of light in only the forward direction but prevents it in the backward direction. Thus it blocks the back reflection of light into lasers or optical amplifiers (e.g. EDFA's) and reduces signal instability and noise. All-fiber optical isolators are highly desirable, particularly for high-power fiber laser system, where fiber termination and small free-space laser beams place restrictions on how much power can be transported through such components [1,2]. Conventional optical isolators are based on bulk optics, which require bulk optical devices such as polarizers, birefringent plates, launching lenses, and also need precision alignment and careful handling.

Magneto-optic glass fibers are attractive for the development of all-fiber Faraday rotation components such as optical isolators, Faraday mirrors, magnetic/current field sensors, etc. [3-6], because glass exhibits peculiar advantages such as the isotropy of its physical properties and the low cost of synthesis and preparation. Moreover, glass can be shaped easily into an optical fiber through a simple and highly productive manufacturing process. Aluminosilicate glasses containing La_2O_3 are attractive candidates for specific device applications due to their physical, thermal and mechanical properties such as high refractive index, high glass transition temperature, and large elastic moduli, on the other hand, low thermal expansion coefficient, small molar volume, and low electrical conductivity [7-10]. These glasses are significantly less explored relative to their counterparts based on traditional oxide modifiers such as alkali and alkaline-earth oxides.

In this study, we fabricated the lanthanum boroaluminosilicate glass fiber and studied its magneto-optical properties for magneto-optical device applications.

II. EXPERIMENTAL DETAILS

Fabrication of the lanthanum boroaluminosilicate glass rod and fiber

The lanthanum boroaluminosilicate glass rod with the composition of 54 SiO_2 · 20 Al_2O_3 · 12 B_2O_3 · 14 La_2O_3 was prepared by a conventional melt quenching technique. In this process, about 100 g of batch materials were weighed and crushed in an alumina mortar and the homogeneous mixture was transferred into an alumina crucible and heated in an electric furnace at 1600 °C for 8 hour. Then, the melt was air quenched by casting it onto a preheated cylindrical brass mold maintaining at 535 °C to avoid thermal shocks and cracks. Then, the sample was annealed at 600 °C for 2 hour in order to remove thermal stress inside the glass sample and subsequently, the sample was allowed to cool to room temperature. Finally, the glass rod was obtained with dimensions of 80 mm length and 10 mm diameter.

To draw a glass optical fiber, the fabricated glass rod was attached to a dummy silica glass rod and then the glass rod was drawn into a fiber with 125.1 µm core diameter at 1060 °C using the draw tower. During the drawing process, the fiber was coated with lower refractive index polymer (EFIRON UVF PC-375, n=1.3820) than that of the glass rod to induce total internal reflection for light transmission.

Absorption spectrum of the glass optical fiber was measured by a conventional cut back method using the white light source (AQ 4303B) and optical spectrum analyzer (AQ 6315B). The Faraday rotation angle was measured by using a He-Ne laser operating at 633 nm and a polarimeter (PA530: Thorlabs, USA) where the magnetic field strength up to 0.14 T was applied.

III. RESULTS AND DISCUSSION

A. Absorption spectrum of the lanthanum boroaluminosilicate glass fiber

Fig. 1 shows the absorption spectrum of the lanthanum boroaluminosilicate glass fiber. It is well known that La^{3+} has a completely empty 4f shell, and therefore, it has no absorption peaks in the visible region because ions without 4f electrons are reported to have no electronic energy levels that can induce excitation and luminescence processes. However, excited states can be arisen from a p-d transition leading to the electronic configuration $5s^2 5p^5 5d^1$ which brings about the singlet states of $^1P_1$, $^1D_2$ and $^1F_3$ and the triplet states of $^3P_1$, $^3D_2$ and $^3F_3$ [11]. Thus absorption band in the visible region at 415 nm was due to the La^{3+} ion [12] and the rest of the bands shown in the near infrared region, originate from the sharing of trace ion impurities within the raw materials.

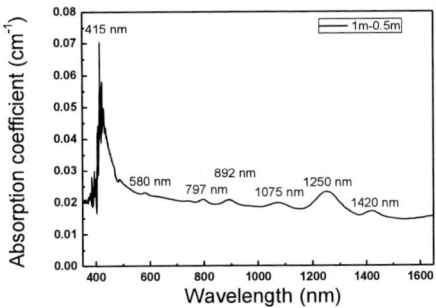

Fig. 1. Absorption spectrum of the lanthanum boroaluminosilicate glass fiber.

B. Determination of Verdet constant

The magneto-optical properties of any new material are dictated by its Verdet constant, and therefore, the rapid, easy, and reliable determination of such a parameter remains an important and challenging task. According to Faraday effect, the plane of a linearly polarized is rotated when it propagates through a magneto-optical material in the presence of magnetic field. Accordingly, the rotation angle (θ) is given by

$$\theta = VBL \qquad (1)$$

where V is the Verdet constant which depends on the wavelength of the light, B is the magnetic field in Tesla, and L is the length of the fiber in meter under the influence of magnetic field.

The experimental setup for the Faraday rotation measurement of the lanthanum boroaluminosilicate glass fiber is shown in Fig. 2. A solenoid (Walker scientific INC.) was used to provide the magnetic field for fiber samples. The solenoid's length, inside diameter and coil resistance were 0.7 m, 0.076 cm and 1.68 Ω, respectively. The magnetic field distribution with varying solenoid current was measured and reported in our earlier work [13].

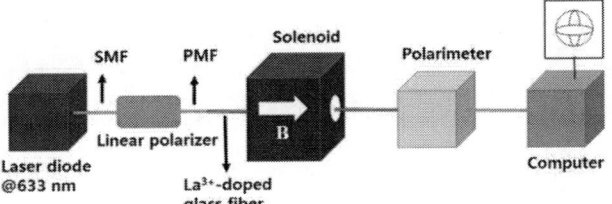

Fig. 2. Schematic diagram of the experimental setup for the Faraday rotation measurement of the lanthanum boroaluminosilicate glass fiber.

The lanthanum boroaluminosilicate glass optical fiber with length of 70 cm was centered and straight in the solenoid where a linearly polarized light at 633 nm was launched into the fiber kept under the variable magnetic field (Fig. 2). The magnetic field was varied by changing the current of the solenoid. The Faraday rotation angle was measured at room temperature (25 °C) by using a polarimeter (PA530:Thorlabs, USA) where the magnetic field strength up to 0.14 T was applied. In this work, the Faraday rotation angle of the fabricated lanthanum boroaluminosilicate glass fiber (core diameter 125/125 μm) was measured together with those of other silica based glass fibers of a commercial multimode fiber (105/125 μm) and a coreless multimode fiber as shown in Fig. 3. According to Eq. (1), the Faraday rotation angle, θ is linearly proportional to the applied magnetic field, B. Thus the Faraday rotation angles of three glass fibers were found to linearly increase with the applied magnetic field (Fig. 3). In real Faraday rotation component design, large magnetic field is always used to induce a larger rotation angle, which means a stronger signal to be easily detected from the background noise. However, high current through the solenoid will produce more heat which is not good for a precise measurement.

The Faraday effect appears both in diamagnetic and paramagnetic transparent media and is closely related to the magnetic behavior of the component ions. In a phenomenological approach, the optical rotation arises from the inequality of the refractive indices of right- and left-circularly polarized components in which the incoming linearly polarized light is decomposed. Hence, the Faraday effect is a magnetic field induced circular birefringence. The effect will be in fact an opposite rotation of the polarization vector in the two transparent media. Therefore, the Verdet constant is positive in diamagnetic media and negative in paramagnetic media. It has been reported that a significant Faraday rotation can be achieved if the diamagnetic matrix incorporates a high concentration of paramagnetic ions [3-6]. The Verdet constant for the present lanthanum boroaluminosilicate glass fiber, commercial multimode fiber, and coreless multimode fiber at 633 nm was found to be -4.64 rad/(T·m), 3.26 rad/(T·m), and 2.64 rad/(T·m), respectively. It is well known that high doping is not possible in conventional silica based fibers fabricated by modified chemical vapor deposition (MCVD) process. Thus the present lanthanum boroaluminosilicate glass fiber exhibited higher Verdet constant than those of the commercial and coreless multimode fibers due to larger concentration of paramagnetic La^{3+} ion.

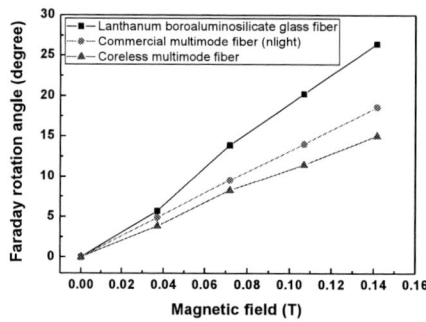

Fig. 3. Faraday rotation angle of the fibers with respect to the magnetic field.

IV. CONCLUSIONS

The magneto-optical properties of $54\ SiO_2 \cdot 20\ Al_2O_3 \cdot 12\ B_2O_3 \cdot 14\ La_2O_3$ glass fiber were investigated and compared to the commercial and coreless multimode silicate glass fibers. The Verdet constant of the lanthanum boroaluminosilicate glass fiber was found to be -4.64 rad/(T·m), which is larger than those of the commercial multimode fiber (3.26 rad/(T·m)) and the coreless multimode fiber (2.64 rad/(T·m)).

ACKNOWLEDGMENT

This work was supported by Basic Science Research Program through the National Research Foundation of Korea (NRF) funded by the Ministry of Education (No. 2013R1A1A2063250) and the Brain Korea-21 Plus Information Technology Project through a grant provided by the Gwangju Institute of Science and Technology, South Korea.

REFERENCES

[1] T. Sato, J. Sun, R. Kasahara, and S. Kawakami, "Lens-free in-line optical isolators," Opt. Lett., vol. 24, pp. 1337-1339, 1999.
[2] E. H. Turner, and R. H. Stolen, "Fiber Faraday circulator or isolator," Opt. Lett., vol. 6, pp. 322-323, 1981.
[3] L. Sun, S. Jiang, J.D. Zuegel, and J.R. Marciante, "All-fiber optical isolator based on Faraday rotation in highly terbium-doped fiber," Opt. Lett., vol. 35, pp. 706-708, 2010.
[4] L. Sun, S. Jiang, and J.R. Marciante, "All-fiber optical magnetic field sensor based on Faraday rotation in highly terbium doped fiber," Opt. Express, vol. 18, pp. 5407-5412, 2010.
[5] K. Shiraishi, S. Sugaya, and S. Kawakami, "Fiber Faraday rotator," Appl. Opt., vol. 23, pp. 1103-1106, 1984.
[6] J. Ballato, and E. Snitzer, "Fabrication of fibers with high rare-earth concentrations for Faraday isolator applications," Appl. Opt., vol. 34, pp. 6848-6854, 1995.
[7] M.J. Dejneka, B.Z. Hanson, S.G. Crigler, L.A. Zenteno, J. D. Minelly, D.C. Allan, W.J. Miller, and D. Kuksenkov, "La_2O_3-Al_2O_3-SiO_2 Glasses for High-Power, Yb^{3+}-doped, 980-nm Fiber Lasers," J. Am. Ceram. Soc., vol. 85, pp. 1100-1106, 2002.
[8] J.K.R. Weber, J.G. Abadie, T.S. Key, K. Hiera, and P.C. Nordine, "Synthesis and Optical properties of Rare-Earth-Alumino Oxide Glasses," J. Am. Ceram. Soc., vol. 85, pp. 1309-1311, 2002.
[9] S. Iftekhar, J. Grins, and M. Eden, "Composition-property relationships of the La_2O_3-Al_2O_3-SiO_2 glass system," J. Non-Cryst. Solids 356, pp. 1043-1048, 2010.
[10] S. Iftekhar, J. Grins, and M. Eden, "Structural characterization of lanthanum aluminosilicate glasses by ^{29}Si solid-state NMR," J. Non-Cryst. Solids, vol. 355, pp. 2165-2174, 2009.
[11] R.H. Abu-Eittah, S.A. Marie, and M.B. Salem, "The electronic absorption spectra of Lanthanum (III) Cerium (III) and Thorium (IV) ions in different solvents," Can. J. Anal. Sci. Spectrosc., vol. 49, pp. 248-257, 2004.
[12] M.A. Marzouk, "Optical characterization of some rare earth ions doped bismuth borate glasses and effect of gamma irradiation," J. Mol. Struct., vol. 1019, pp. 80-90, 2012.
[13] Y. Kim, S. Ju, S. Jeong, M.-J. Jang, J.-Y. Kim, N.-H. Lee, H.-K. Jung, W.-T. Han, "Influence of gamma-ray irradiation on Faraday effect of Cu-doped germane-silicate optical fiber," Nucl. Instrum. Methods Phys. Res. B, vol. 344, pp. 39-43, 2015.

WC3-6

OECC/PS2016

High Sensitivity Curvature Sensor Based on Modal Interferometer for Vibration Detection

Li Liu, Ping Lu*, Shun Wang, Hao Liao, Wenjun Ni, Xin Fu, Xinyue Jiang, Deming Liu

School of Optical and Electronic Information, National Engineering Laboratory for Next Generation Internet Access System
Huazhong University of Science and Technology, Wuhan, China
pluriver@mail.hust.edu.cn

Abstract: We proposed a novel in-line modal interferometer based on multimode-single mode-multimode fiber (MSM) structure. And the MSM structure with a very high curvature sensitivity of $141.6dB/m^{-1}$ is applied for vibration detection.

Keywords: Modal interferometer, curvature, vibration, fiber sensor

I. INTRODUCTION

The curvature and vibration detection are widely applied in the areas of composite structure, robotics, bridge and structural health monitoring. Fiber optic sensors have been studied extensively for curvature or vibration measurement due to their unique advantages over the traditional electrical or mechanical sensing techniques, such as high sensitivity, immunity to electromagnetic interference, remote detection and multiplexing capability [1]. Many fiber optic curvature or vibration sensors reported so far are based on fiber Bragg grating [2], long period grating [3] and tilted fiber Bragg grating [4]. However, the production process of these fiber gratings is complex and costly. Recently, the special optical fibers such as multi-mode Fiber (MMF) [5], photonic crystal fiber (PCF) [6], no core fiber [7] and thin core fiber [8] are used to fusion spliced with single mode fiber (SMF) to form modal interferometers for curvature or vibration measurement, yet having the deficiency of low sensitivity. In addition, some tapered SMF [9] or PCF [10] based in-fiber modal interferometers with high sensitivity have been proposed. But their un-robust structures limit their practical applications.

In this letter, we propose a novel MSM structure for curvature and vibration sensing based on fiber in-line modal interferometer. The MMF is employed as coupler waveguide with a half re-imaging length of 5.5m to excite the higher clad modes. And the maximum extinction ratio of MSM interference pattern can reach 40 dB by optimizing the length of middle SMF. The MSM sensor exhibits very high intensity sensitivity about $141.6dB/m^{-1}$ in the curvature sensing experiment and has a high signal-to-noise ratio (SNR) about 42 dB in the 1000Hz vibration detection experiment.

II. SENSING PRINCIPLE

The schematic diagram of the curvature sensor based on modal interferometer is shown in figure 1 in which a piece of SMF is spliced between two MMFs. When light transmits from MMF to SMF, part of light will leak from the core of MMF to the cladding layer of SMF due to the different core diameter between SMF and MMF. After they propagate to the second MMF, they re-couple with the core mode to construct in-line modal interferometers. The intensity of the transmitted light caused by the modal interference between the core mode and cladding mode can be expressed as:

$$I = E_{co}^2 + \sum_m E_{cl,m}^2 + 2E_{co}\sum E_{cl,m}\cos\frac{2\pi\Delta n^{eff,m}l}{\lambda}$$

$$\Delta n^{eff,m} = n_{co} - n_{cl,m}$$

(1)

where E_{co} and $E_{cl,m}$ are the magnitude of the electric field of the core mode and the excited m order cladding modes, respectively. l is the length of the middle SMF. $\Delta n^{eff,m}$ is the effective refractive index difference between the core mode and the excited m order cladding mode. n_{co} and $n_{cl,m}$ are the effective refractive index of the core mode and the excited m order cladding modes, respectively. λ represents the wavelength of light. When the MSM structure is bended, the inner part of SMF is compressed while the outer part is stretched. It will lead to the wavelength shift for the change of $\Delta n^{eff,m}$. And the coupling efficiency between different fiber core and cladding will also be modulated, which causes the contrast change of the modal interferences.

The curvature sensitivity of the proposed sensor can be improved by exciting the higher clad modes which are easier to leak out of the curving fiber. So the MMF (105/125, Corning) with a half re-imaging length of 5.5m is employed. The simulation results calculated by Rsoft demonstrate that the core mode will couple to the clad modes completely as shown in figure 1(a). That means the relatively high clad modes can be excited in the middle SMF for high sensitivity curving sensing. There is a 15μm separation distance between the SMF and MMF end faces when they are spliced by the fiber fusion splicer (FSM 60s, Fujikura). Hence, the refractive-index distribution of splice region will be a little different from that of simulation. So a part of core mode can still propagate into the core of middle SMF to form modal interferences. The length of middle SMF is optimized to be 1.5cm for getting a higher contrast ratio of MSM interferometer. Figure 1(b) shows the transmission spectrum of the proposed sensor and its corresponding spatial frequency spectrum. Two dominant cladding modes are excited to form the interferometer. And the interferometer has an interference pattern with a high extinction ratio about 40 dB, enabling it to be used as an intensity modulated curvature and vibration sensor with a very high sensitivity.

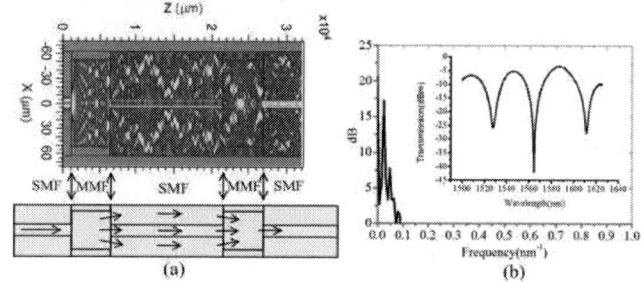

Fig. 1(a). Configuration of MZI and simulated light propagation within MSM Structure,
(b) transmission spectrum of the MSM Structure and its spatial frequency spectrum.

III. EXPERIMENT AND RESULTS

The experimental setup for curvature sensing is depicted in figure 2(a). Light emitted by a broadband source is employed to illuminate the interferometer and the output spectrum is acquired by an optical spectrum analyzer (YOKOGAWA AQ6370B). The two ends of MSM structure are fixed over two translational stages. Curvature change is induced by keeping one translational stage fixed and moving the other one stepwise. The fiber curvature for the setup can be expressed by [11]:

$$c = \sqrt{\frac{24d}{L^3}}$$

(2)

where d is the movement distance of translational stage and $L = 22.5$ cm is length of the fiber between two translational stages. The measured transmission spectrum variation of the proposed sensor under different d is shown in figure3 (a). It can be observed that with increase in curvature, the transmission intensity of dip 2 increases significantly for the change of coupling efficiency between different fiber core and cladding modes. The transmitted power of dip 2 as a function of the curvature is shown in figure 3(b). In curvature range 0.32 - 0.65 m^{-1}, the acquired data can be well fitted by a quadratic function. And when the curvature is in the range 0.32 - 0.46 m^{-1}, the curvature response of dip 2 can be seen as a linear response with a very high intensity sensitivity of 141.6dB/m^{-1}. This MSM structure can be used for vibration detecting with very high curvature sensitivity.

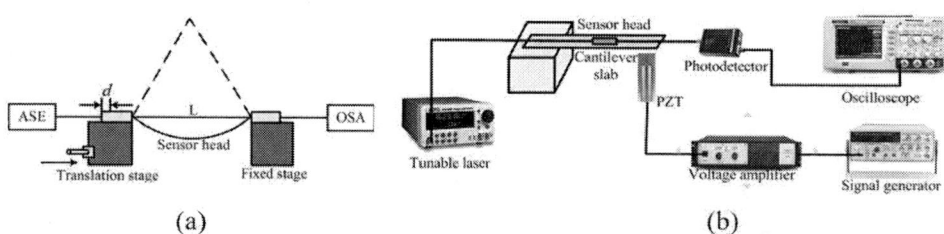

(a)   (b)

Fig. 2. Schematics of experimental Setup for curvature (a) and vibration (b) detection.

The experimental setup to utilize the MSM based vibration sensor is shown in figure 2(b). The sensor is fixed on a steel cantilever slab (length=10cm, width=2cm and thickness=1mm). When the cantilever is vibrated by standard piezoelectric transducer (PZT), it will cause a dynamic curvature variation on the sensor head. Thus it will result in periodical intensity decrease and increase of dip 2 with the vibration frequency. The tunable laser with working wavelength of 1564.5nm is employed to emit light into the MSM structure and the vibration signal of PZT is detected by direct measurement of the transmission power change by using a single photodetector (Newport 1623). The PZT is driven by the periodic sinusoidal signal with a peak to peak voltage of 100 mV. As shown in figure 4, the time domain waveform at 1000 Hz and the corresponding frequency spectrum after FFT transform are both recorded. A SNR of 42 dB with a noise floor about -58 dB can be observed in figure 4(b). Besides, the harmonics in figure 4(b) allow us to identify the detected vibration frequency more clearly. That means the external vibration interference can be interrogated by the proposed sensor well.

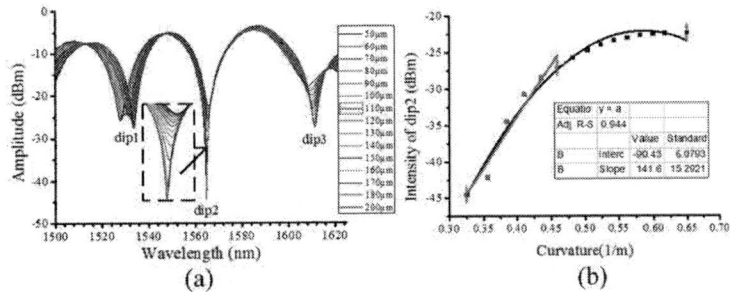

Fig. 3.(a) Transmission spectra variation under different curvature, (b) the intensity change of dip 2 with different curvature.

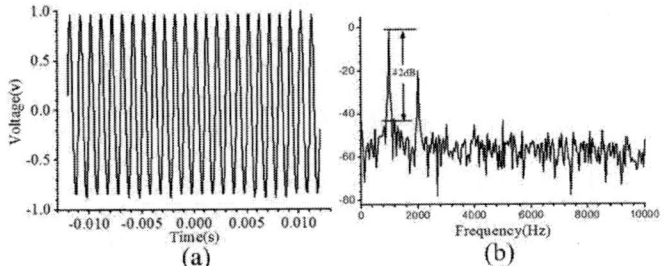

Fig. 4. The detected time waveform (a) and corresponding frequency spectrum (b) of the MSM sensor head at 1kHz.

IV. CONCLUSIONS

In summary, a novel fiber curvature and vibration sensors based on the MSM structure have been proposed and experimentally demonstrated. The MMF with the half re-imaging length is adopted to excite the higher clad modes. By optimizing length of middle SMF, the maximum extinction ratio of interference pattern can achieve to 40 dB. The experimental results demonstrate that the curvature sensitivity of proposed sensor can reach up to141.6dB/m^{-1} in the range of 0.32 m^{-1} to 0.46 m^{-1}. And this kind of compact fiber sensor exhibits a good property in the vibration detection test.

ACKNOWLEDGMENT

This work is supported by grants (Nos. 61290315, 61275083) from Natural Science Foundation of China and a grant (HUST: No. 2014CG002) from the Fundamental Research Funds for the Central Universities.

REFERENCES

[1] Kirkendall C K, Dandridge A. Overview of high performance fibre-optic sensing[J]. Journal of Physics D: Applied Physics, 2004, 37(18): R197.

[2] Kinet D, Mégret P, Goossen K W, et al. Fiber Bragg grating sensors toward structural health monitoring in composite materials: Challenges and solutions[J]. Sensors, 2014, 14(4): 7394-7419.

[3] Liu Y, Williams J A R, Bennion I. Optical bend sensor based on measurement of resonance mode splitting of long-period fiber grating[J]. Photonics Technology Letters, IEEE, 2000, 12(5): 531-533.

[4] Shao L Y, Laronche A, Smietana M, et al. Highly sensitive bend sensor with hybrid long-period and tilted fiber Bragg grating[J]. Optics Communications, 2010, 283(13): 2690-2694.

[5] Wu Q, Yang M, Yuan J, et al. The use of a bend singlemode–multimode–singlemode (SMS) fibre structure for vibration sensing[J]. Optics & Laser Technology, 2014, 63: 29-33.

[6] Deng M, Tang C P, Zhu T, et al. Highly sensitive bend sensor based on Mach–Zehnder interferometer using photonic crystal fiber[J]. Optics Communications, 2011, 284(12): 2849-2853.

[7] Ran Y, Xia L, Han Y, et al. Vibration fiber sensors based on SM-NC-SM fiber structure[J]. Photonics Journal, IEEE, 2015, 7(2): 1-7.

[8] Fu H, Zhao N, Shao M, et al. High-sensitivity Mach–Zehnder interferometric curvature fiber sensor based on thin-core fiber[J]. Sensors Journal, IEEE, 2015, 15(1): 520-525.

[9] Hernández-Serrano A I, Salceda-Delgado G, Moreno-Hernández D, et al. Robust optical fiber bending sensor to measure frequency of vibration[J]. Optics and Lasers in Engineering, 2013, 51(9): 1102-1105.

[10] Ni K, Li T, Hu L, et al. Temperature-independent curvature sensor based on tapered photonic crystal fiber interferometer[J]. Optics Communications, 2012, 285(24): 5148-5150.

[11] Wang Y, Richardson D, Brambilla G, et al. Intensity measurement bend sensors based on periodically tapered soft glass fibers[J]. Optics letters, 2011, 36(4): 558-560.

WC4-1   OECC/PS2016

Ultra-High-Resolution OTDR based on Linear Optical Sampling

Shuai Wang, Xinyu Fan*, Bin Wang, Guangyao Yang, Qingwen Liu and Zuyuan He

State Key Laboratory of Advanced Optical Communication Systems and Networks,
Shanghai Jiao Tong University, Shanghai 200240, China
*fan.xinyu@sjtu.edu.cn

Abstract: *We demonstrate an optical time domain reflectometry based on linear optical sampling (LOS). Taking advantage of its ultrahigh bandwidth and low timing jitter, we obtained a spatial resolution of 620 μm at 100 m.*

Keywords: *optical time domain reflectometry, linear optical sampling*

I. INTRODUCTION

Distributed fiber sensing technologies have been used to characterize optical links via backscattered signals, and optical time domain reflectometry (OTDR) is one of the most powerful tools for the diagnosis of optical fiber connectors, cracks and attenuation. For some high-end applications such as the failure prediction of airplane materials [1], it is of great importance to precisely locate the abnormal positions with a very strict requirement for high spatial resolution. However, the spatial resolution of commercial OTDR systems is limited to ~1 m because of the large noise and the much reduced Rayleigh scattering power while the optical pulse becomes narrow [2]. To improve the performance, photon-counting OTDR was proposed with an improvment to ~1 cm thanks to the ultra-high sensitivity of the photon counting technology [3-6]. But still, the spatial resolution is finally limited by the timing jitter of the photon counting detector (the state-of-the-art timing jitter is ~40 ps). To obtain a higher spatial resolution for OTDR system, it is necessary to have a detector with a higher response bandwidth, a higher detecting sensitivity and a low timing jitter.

Linear optical sampling technique, known for its high bandwidth and shot-noise limited sensitivity, has been used in many fields for monitoring the waveform in high speed transmission systems. The signal under detect interferes with the sampling signal emitted from the mode-locked laser (MLL) and is then detected by the ordinary photodetectors. Because the timing jitter of the MLL is ~10 fs, the sampling rate can be up to 100 TS/s and this makes it possible to measure ultrashort pulse emitted from pulsed lasers and its reflected lightwave from optical fibers, which has a capability of achieving a higher resolution in OTDR system.

In this work, we present an ultra-high-resolution OTDR system using ultrashort pulses of 6 ps width launched from a pulse generator using the detection based on linear optical sampling, and achieved a spatial resolution of 620 μm at 100m. The spatial resolution is broadened to 1.3 cm at 15 km due to the dispersion in the fiber, which we believe may be compensated at the successive computation stage.

II. PRINCIPLE AND EXPERIMENTAL SETUP

The experimental setup is shown in Fig. 1(a). For the ultra-high-resolution OTDR system, laser pulses with a central wavelength of 1550 nm from an optical pulse generator are launched into the FUT through a circulator, the repetition rate of the pulses is 500 MHz. The Rayleigh backscattered signal is coupled into the third port of the circulator, interfering with the pulses coming from the MLL in the 90 degree hybrid and then detected by a pair of 400 MHz

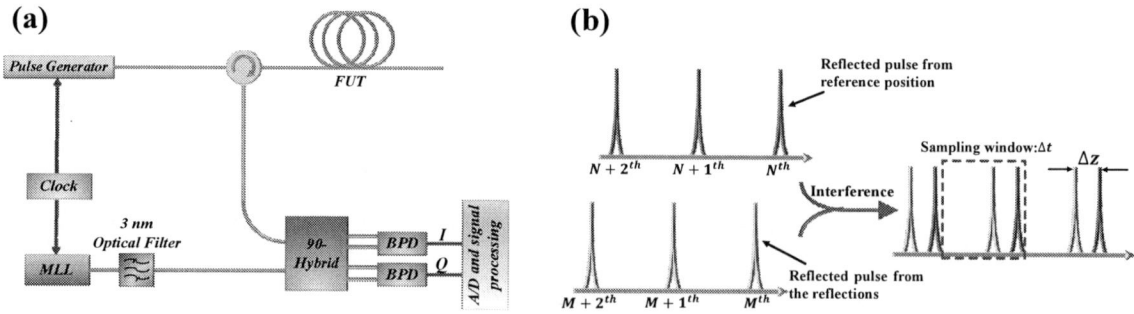

Fig. 1. (a) Experimental setup of ultra-high-resolution OTDR system. (b) Principle of locating the reflections.

FUT: fiber under test; BPD: balanced photodetector; MLL: mode-locked laser.

balanced photodetectors (BPDs) (Thorlabs PDB470C). The signals are sampled by an analog-to-digital card (ADC) and collected with a personal computer. An analog signal generator acts as a clock source to synchronize the pulse generator, the MLL and the ADC. The timing jitter of this system is less than 30 fs.

Considering the linear optical sampling process, the electric field of the pulse of sampling laser can be expressed as [7-9]:

$$E_s(t) = \sum_l a_s(t - lT_s)e^{j\omega_s t} \tag{1}$$

and that of the pulse generator is given by:

$$E_d(t) = \sum_l a_d(t - lT_d)e^{j\omega_d t} \tag{2}$$

where T_s and T_d are the period of pulse from sampling laser and pulse generator respectively. The output of the optical 90 degree hybrid and photodetector obeys the linear cross correlation relationship:

$$S(\tau) = \int E_s(t)E_d^*(t - \tau)dt \tag{3}$$

The sampling time interval Δt can be given by:

$$\Delta t = T_s - kT_d \tag{4}$$

where k is an integer.

Since the repetition rate of the optical pulse generator is about 500 MHz, the sampling window is T_s=2 ns, corresponding to a 20 cm measurement range. To locate reflections beyond this measurement range, it's necessary to set a reference position. As shown in Fig. 1(b), assuming at time t_0, the pulse reflected from reference reflector is the Nth pulse and the pulse reflected from the point to be located is the Mth pulse. The distance between the reference reflector and the located point can be expressed as:

$$\Delta L = \Delta z + cT_d|M - N|/n \tag{5}$$

If we change the repetition rate of the optical pulse generator, Δz changes and $|M-N|$ may be calculated, therefore the position of the reflection can be exactly located.

III. MEASUREMENT RESULTS

The ultrashort pulse was generated from an active MLL (Pritel UOC) which acts as a pulse generator with a pulse duration time of 6 ps, corresponding to a theoretical spatial resolution of 600 μm. By adjusting the repetition rates of the MLL and the pulse generator, the sampling rate is set to 10 TS/s, with a sampling interval of 100 fs, which is very narrow for us to clearly observe the pulses. To test the spatial resolution at different distances, the FUT is a combination of one 100-m, and four 5-km G.655 fibers fabricated by Yangze optical fiber and cables company (YOFC). As shown in Fig. 2, we found that the full width half maximum (FWHM) of the reflection peak at 100 m is 620 μm, which is slightly broadened from the original pulse coming from the laser. As the pulse propagates, when the FUT is 15 km, the reflection peak is broadened to 130 ps, which corresponding to a spatial resolution of 1.3 cm. The pulse broadening is thought to be caused by chromatic dispersion, which has an expression of

$$\tau(L) = \tau_0\sqrt{1 + (\frac{14.7DL^2}{\tau_0^2})^2} \tag{6}$$

Fig. 3 (a) shows the measured spatial resolution at different distance with a statistical illustration, and the result agreed well with Eq. 6 while choosing suitable dispersion coefficients D at different sections. Since the main limitation of spatial resolution at long distance is chromatic dispersion, we believe that a better resolution may be achieved if chromatic dispersion is properly compensated at the successive computation stage. Besides, since the timing jitter in our sampling system is much lower than the pulse-width of pulses from our optical pulse generator, if a pulse generator producing narrower pulses is used in the experiment, the spatial resolution is believed to be improved by another one

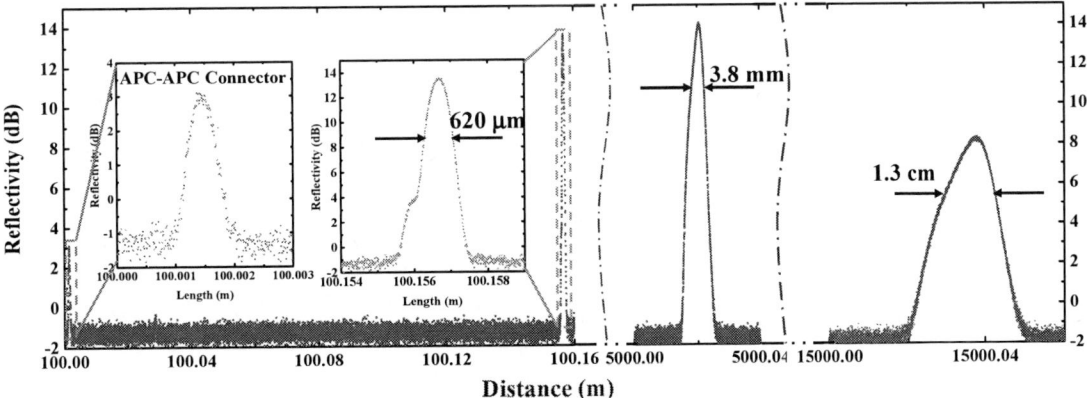

Fig. 2 Spatial resolution achieved at different distances.

OECC/PS2016

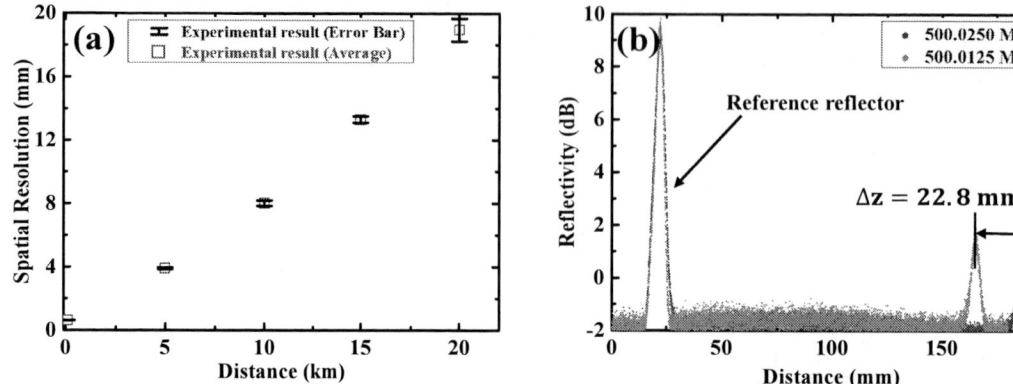

Fig. 3 (a) The spatial resolution versus detection distance. (b) Locating the refelctions by changing the repetition rate of the optical pulse generator

order in this system at short distance.

To locate the reflection, the experiment was done at different signal repetition rates, as shown in Fig. 3(b). The red line shows the two peaks from the reference position and the position to be located when the repetition rate of the optical pulse generator is $1/\tau_{d1}$=500.0125 MHz, and the distance between them in the same window is Δz_1=143.42 mm; the blue line shows the two when the repetition rate is $1/\tau_{d2}$=500.0250 MHz, and the distance between then in the same window is Δz_1=166.22 mm. According to Eq. 5, we calculated that the location of the peak is 912.021 m from the reference reflector. Note when the reflection is very close to the reference reflector, the repetition rate is required to be much changed for distinguishing the two peaks, and the repetition rate of the sampling laser should be changed as well to keep high sampling rate according to Eq. 4.

Another solution to locate the reflection is to select the pulses from the optical pulse generator to make a long sampling window. For example, if we decrease the repetition rate to 10 kHz, the corresponding detection distance would be increased to 10 km. However, the measurement time of a single window will increase from 80 μs to 4 s. Since the data in this experiment has been averaged for 100 times, the total time will be several minutes for this solution.

IV. CONCLUSIONS

An ultra-high-resolution OTDR system based on linear optical sampling is demonstrated with a spatial resolution of 620 μm at 100 m and 1.3 cm at 15 km. The deterioration is caused by the dispersion in the fiber, which we believe may be compensated at the successive computation stage.

ACKNOWLEDGMENT

This work was supported by the National Natural Science Foundation of China under Grant 61575001, 61275097, 61307106, 61307107, Shanghai STCSM Scientific and Technological Innovation Project under Grant 15511105401.

REFERENCES

[1] I. Kopacek et al., "Monitoring/Sensing Applications on AirPON," in SPIE2012, 8540 (2012).

[2] B. Josef. "Optical time domain reflectometer (OTDR) with improved dynamic range and linearity." U.S. Patent No. 5,621,518. 15 Apr. 1997.

[3] Q. Zhao et al. "Photon-counting optical time-domain reflectometry with superconducting nanowire single-photon detectors." 2013 IEEE 14th International Superconductive Electronics Conference (ISEC) 2013.

[4] M. Legré, R. Thew, H. Zbinden, and N. Gisin, . "High resolution optical time domain reflectometer based on 1.55 μm up-conversion photon-counting module." Optics express, Vol. 15, N0. 13, pp. 8237-8242, 2007

[5] LY. Herrera, G. Amaral, and J. Weid. "Ultra-High-Resolution Tunable PC-OTDR for PON Monitoring in Avionics." Optical Fiber Communications Conference and Exhibition (OFC), 2015.

[6] C. Schuck, W. Permice, X. Ma, and H. Tang."Optical time domain reflectometry with low noise waveguide-coupled superconducting nanowire single-photon detectors." Appl. Phys. Lett., vol. 102, no. 5, 2013, Art. ID. 191104

[7] C. Dorrer, D. C. Kilper, H. R. Stuart, G. Raybon, and M. G. Raymer, "Linear optical sampling," IEEE Photon. Technol. Lett., vol. 15, pp. 1746–1748, 2003.

[8] Okamoto, and F. Ito. "Ultrafast measurement of optical DPSK signals using 1-symbol delayed dual-channel linear optical sampling." IEEE Photon. Technol. Lett., vol. 20, no. 11, pp. 1041-1135, 2008.

[9] K. Okamoto, and F. Ito. "Dual-channel linear optical sampling for simultaneously monitoring ultrafast intensity and phase modulation." IEEE/OSA Journal of Lightwave Technol., vol. 27, no. 12, pp. 2169-2175, 2009.

OECC/PS2016

Polarization Fading Elimination in Phase-Extracted OTDR for Distributed Fiber-Optic Vibration Sensing

Guangyao Yang, Xinyu Fan*, Bin Wang, Qingwen Liu and Zuyuan He

State Key Laboratory of Advanced Optical Communication Systems and Networks,
Shanghai Jiao Tong University, Shanghai 200240, China
Author e-mail address: fan.xinyu@sjtu.edu.cn

Abstract: *We employed a polarization diversity system into phase-extracted coherent phase-sensitive OTDR. The results show that the amount of dead zones is significantly reduced after eliminating the influence of polarization fading with the polarization diversity system.*
Keywords: *phase-sensitive OTDR, phase extraction, fading, distributed fiber vibration sensing*

I. INTRODUCTION

Phase-sensitive optical time domain reflectometry (ϕ-OTDR) has been widely used as a distributed fiber vibration sensing technology [1] [2]. In a ϕ-OTDR system, highly coherent pulses from a laser source are injected into the fiber, resulting Rayleigh backscattered (RBS) traces in a jagged form [3], which is called Rayleigh fading phenomenon. When an external vibration is exerted on a fiber, the RBS trace changes, which is an effect that can be used for locating and measuring the vibration. Specially, for a coherent ϕ-OTDR, the polarization mismatch may also lead to a fading phenomenon, which is called polarization fading phenomenon, and it may deteriorate the vibration sensing result [4].

For the coherent ϕ-OTDR system, both the optical intensity and the optical phase of RBS lightwave can be extracted for vibration sensing. Here, we call the phase signal based ϕ-OTDR system as phase-extracted OTDR. The optical phase has a linear response to vibration magnitude, which allows the phase-extracted OTDR to have an excellent performance in vibration sensing compared to the intensity-based ϕ-OTDR system whose response is nonlinear [5]. However, due to the fading phenomenon, some points along the fiber have very poor signal-to-noise ratio (SNR), and are not suitable for vibration sensing in phase-extracted OTDR.

Both Rayleigh fading and polarization fading can affect the result of phase-extracted OTDR. Usually, the Rayleigh fading can be suppressed by a multi-phase pulses scheme [6] or taking measures to improve the SNR, but the polarization fading still deteriorates the results even the SNR is very high. In this paper, we employed the polarization diversity (PD) system into a phase-extracted OTDR to eliminate the influence of polarization fading. With the helps of PD system, we eliminated most of the dead zones, and successfully detected the vibration exerted at 10 km.

II. PRINCIPLE

In phase-extracted OTDR, the highly coherent lightwave from a narrow linewidth laser is divided into two beams by using an optical coupler: the probe lightwave for launching into the fiber under test (FUT), and the local reference lightwave for coherent detection. An acousto-optic modulator (AOM) is then used to modulate the probe lightwave into a series of pulses with a frequency shift of f_b. The RBS lightwave returned from FUT is mixed with the local lightwave in another 3dB coupler and the beat frequency is detected by using a balanced photodetector (BPD). The birefringence of FUT leads to a random polarization state of the RBS lightwave, and finally results in the polarization fading. Supposing that the polarization state of the local reference light remains unchanged while the polarization state of RBS lightwave change θ, the detected intensity output in a single measurement can be described as following:

$$I(t) \propto E_0 R(t) \cos\theta \cos\left(2\pi f_b t + \varphi(t)\right) , \tag{1}$$

where $R(t)$ and $\varphi(t)$ stand for the reflectivity and the phase change at FUT, respectively. Since the polarization mismatch angle θ has a range of $[0, \pi/2]$, when the polarization state between local reference lightwave and RBS lightwave is nearly mismatched, SNR comes to be less than 0 dB, and the error of phase extraction is unbearable.

Fig. 1(a) shows a schematic of the PD system. With a PD system, we obtain two signals, which can be described as following:

$$\begin{cases} I_A(t) \propto E_0 R(t) \cos\theta \cos\left(2\pi f_b t + \varphi(t)\right) \\ I_B(t) \propto E_0 R(t) \sin\theta \cos\left(2\pi f_b t + \varphi(t)\right) \end{cases} . \tag{2}$$

Here, $I_A(t)$ and $I_B(t)$ stands for the beat frequency signal from two orthogonal polarization states. Since the $\cos\theta$ and $\sin\theta$ can't fall to 0 at same time, at each point on fiber we can extract the phase signal from the polarization state with a better SNR, thus the error of phase extraction can be suppressed, as shown in Fig. 1(b). Without the PD system, the

1089

error of phase extraction increases rapidly with the polarization mismatch angle growing up, but with a PD system the phase extraction error keeps much lower with any polarization mismatch angles.

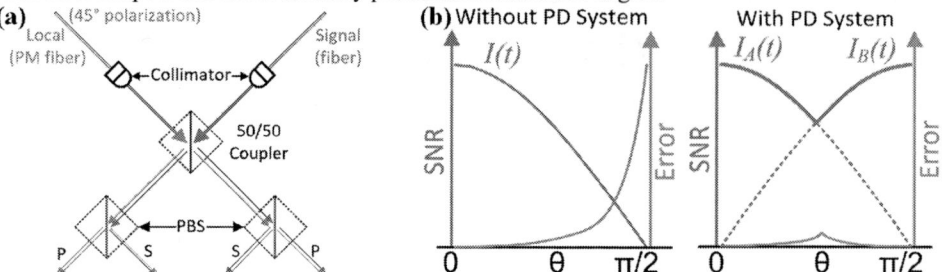

Fig. 1. (a) The schematic of the polarization diversity system. PBS: polarization beam splitter. (b) The relationship between SNR and the phase extraction error without and with PD system.

III. EXPERIMENTAL SETUP AND RESULTS

A. Experimental Setup

The experimental setup is shown in Fig. 2. All the polarization-maintaining fibers are highlighted by red thick line. A fiber laser is employed as a coherent light source with a linewidth of 1 kHz to meet the requirement of the long measurement range. The output power of the fiber laser is 16 dBm. The lightwave is then divided into two parts by a 3dB polarization maintaining optical coupler. The AOM produces pulsed lightwave with 100 ns duration and an 8 kHz repetition rate, and also introduces an 80 MHz frequency shift for heterodyne detection. The pulsed lightwave modulated by AOM is amplified by an EDFA, and then launched into the FUT through a circulator. The peak power of the amplified pulse is around 23 dBm, which is the threshold of nonlinear effects. A PZT is used for exerting vibration on the FUT. The local reference lightwave is launched into PD system through a polarization maintaining fiber to meet the requirement of polarization diversity. The RBS lightwave and local reference lightwave are mixed in the PD system, and the two outputs are detected by BPD respectively, for obtaining the beat frequency while blocking the DC component. The electric signals from BPD are sampled by an ADC (NI-5154) with a sampling rate of 1 GS/s and an accuracy of 8-bit. Finally, the phase signal is extracted from the sampled signal.

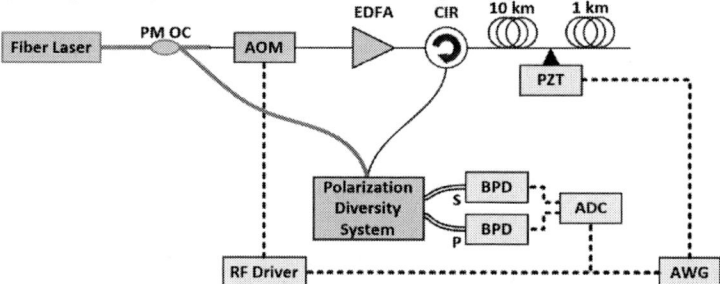

Fig. 2. The experimental setup. PM OC: polarization-maintaining optical coupler; EDFA: erbium-doped fiber amplifier; CIR: circulator; BPD: balanced photodetector; PZT: piezoelectric transducer; ADC: analog-to-digital convertor; AWG: arbitrary waveform generator.

B. Results

Some points on the fiber have very poor SNR due to the fading phenomenon. These points are called dead zones in this paper, since the external vibration is undetectable at these points. Fig. 3 shows one measurement result which is affected by fading noise. Fig. 3(a) shows that there are three dead zones on the distance from 2880 m to 2980 m, and Fig. 3(b) shows the corresponding extracted phase signal. It's very clear that at these dead zones, the error of extracted phase signal has overwhelmed the information of external vibrations.

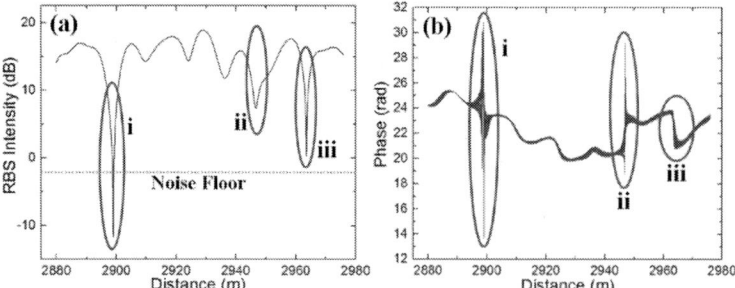

Fig. 3. The experimental results of (a) RBS intensity and (b) phase signal in one measurement.

Fig. 4 shows the results of polarization fading elimination with PD system. A strain of around 700 nε is exerted on a 1.5 m FUT by the PZT as the external vibration. To eliminate the influence of environmental disturbance, the differential phase is obtained by calculating the phase difference between two neighboring points with 10 m distance, which is equal to the spatial resolution. Fig. 4(a) and Fig. 4(b) shows that with only one polarization state, the fading phenomenon has a 16 dB range, but after the polarization diversity, the fading phenomenon has only 10 dB range, as shown in Fig. 4(c), which means that the polarization fading is eliminated while only Rayleigh fading exists. When only extracting phase signal from one polarization state, the dead zones greatly influence the vibration sensing results, and the vibration can hardly be detected, as shown in Fig. 4(d) and Fig. 4(e). After the PD system is used, the quantity of dead zones is greatly decreased, and the external vibration can be easily measured, as shown in Fig. 4(f).

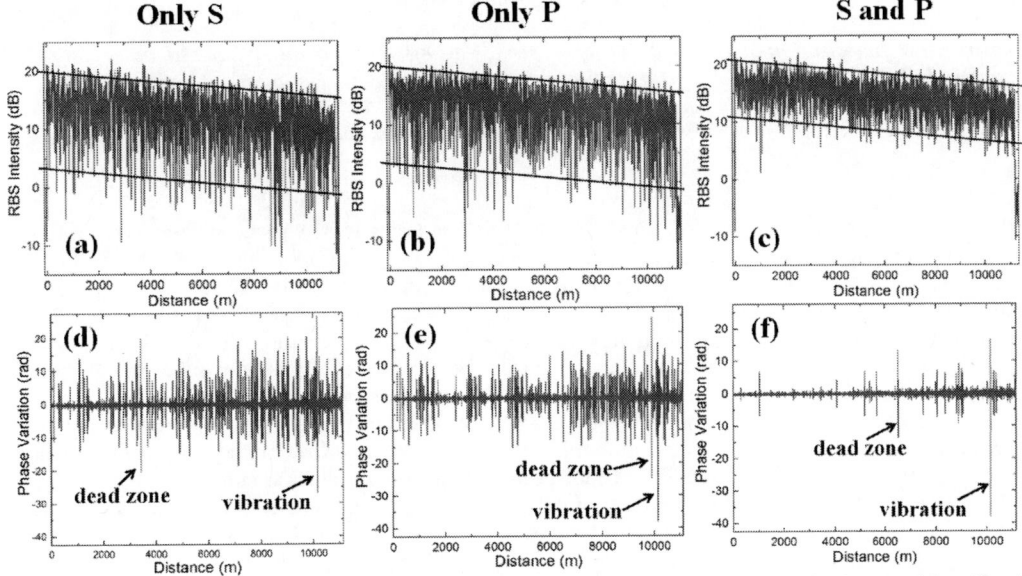

Fig. 4. The RBS intensity trace with (a) only S state, (b) only P state and (c) both S and P, as well as the phase signal extracted from (d) only S state, (e) only P state and (f) both S and P state.

Without a PD system, there are many dead zones even with a high SNR, since the range of polarization fading has no limitation. But after the PD system is employed, RBS intensity trace hardly falls to zero, which means that the dead zone can be eliminated by taking measures to enhance the SNR, since the range of Rayleigh fading can be controlled by using lasers with proper linewidths.

IV. CONCLUSION

In this paper, we analyzed the relationship between fading phenomenon and dead zones in phase-extracted OTDR, and employed a polarization diversity system in the system for eliminating the influence of polarization fading. With the helps of polarization diversity system, most of dead zones were eliminated, and the external vibration at a distance of 10 km was successfully detected.

ACKNOWLEDGMENT

This work was supported by the National Natural Science Foundation of China under Grant 61575001, 61275097, 61307106, 61307107, Shanghai STCSM Scientific and Technological Innovation Project under Grant 15511105401.

REFERENCES

[1] J. C. Juarez, E. W. Maier, K. N. Choi, and H. F. Taylor, "Distributed Fiber-Optic Intrusion Sensor System," Journal of Lightwave Technology, vol. 23, p. 2081, 2005.

[2] Y. Lu, T. Zhu, L. Chen and X. Bao, "Distributed Vibration Sensor Based on Coherent Detection of Phase-OTDR," Journal of Lightwave Technology, vol. 28, pp. 3243-3249, 2010.

[3] H. Izumita, S. Furukawa, Y. Koyamada, and I. Sankawa, "Fading noise reduction in coherent OTDR," IEEE Photonics Technology Letters, vol. 4, pp. 201-203, 1992.

[4] M. Ren, P. Lu, L. Chen, and X. Bao, "Theoretical and Experimental Analysis of φ-OTDR based on Polarization Diversity Detection," IEEE Photonics Technology Letters, vol. PP, pp. 1-1, 2015.

[5] G. Tu, X. Zhang, Y. Zhang, F. Zhu, L. Xia, and B. Nakarmi, "The Development of an φ-OTDR System for Quantitative Vibration Measurement," IEEE Photonics Technology Letters, vol. 27, pp. 1349-1352, 2015.

[6] Z. Pan, K. Liang, J. Zhou, Q. Ye, H. Cai, and R. Qu, "Interference-fading-free phase-demodulated OTDR system," in OFS2012 22nd International Conference on Optical Fiber Sensor, 2012, pp. 842129-842129-4.

WC4-3   OECC/PS2016

Orthogonally-polarized pulse pair BOTDA sensor with three-tone probe

Yiming Tao, Xiaobin Hong, Zhisheng Yang, Wenqiao Lin, Jian Wu

State Key Laboratory of Information Photonics & Optical Communications (Beijing University of Posts and Telecommunications),
P. O. Box 55, Beijing, 100876, China
ym.tao@foxmail.com

Abstract: A polarization-independent BOTDA sensor using orthogonally-polarized pulse pair is proposed. We experimentally demonstrate that the impact of the polarization noise can be minimized and only half of acquisition time is needed by using balance detection.
Keywords: distributed optic fiber sensor, balanced detection, polarization noise, Brillouin Scattering

I. INTRODUCTION

Distributed optical fiber sensors based on Stimulated Brillouin Scattering (SBS) has been demonstrated its excellent benefits especially for long sensing range and complicated electrically noisy environments. Since the Brillouin Optical Time Domain Analysis (BOTDA) came into being, distributed information can be obtained correctly by calculating Brillouin frequency shift (BFS) corresponding to the variation of strain or temperature along the sensing fiber. The enhancement of signal to noise ratio (SNR) is an important issue for BOTDA technology, especially when longer sensing range is applied. However, the state of polarization (SOP) of probe wave and pump pulse cannot keep be aligned all over the sensing fiber resulting from the random birefringence in single mode fiber (SMF) [1]. Because the intensity of SBS strongly depend on the SOPs when the probe wave is interacting with the pump pulse, only part of Brillouin gain can be detected during one acquisition time, which is called 'polarization fading'. And the other part can be obtained by orthogonally switching the SOP of probe beam but one more acquisition time is needed. In order to eliminate the polarization noise without requirement of polarization scrambling so that extra acquisition time is unnecessary, several polarization diversity solutions have been presented recently. J. Yang, C at el. [2] proposed a new method to solve this problem by using polarization beam splitters (PBSs) to split the pulsed pump wave into two beams with orthogonal polarization states and detect the two beams recombined by PBS. And then, J. Urricelqui, at el. [3] introduced a polarization diversity technique based on the use of two orthogonal pump pulses, which simultaneously interact with a phase-modulated probe wave. In 2013, A. Domínguez-López at el. [4] proved that BOTDA system can strongly benefit from the use of balanced detection among the two sidebands which can increase the SNR by 1/2. Moreover, A.Lopez-Gil at el. [5] proposed an another method that the time traces of orthogonally polarized by Faraday mirror Stokes and anti-Stokes CW probe pair separated by using a Dense Wavelength Division Multiplexer (DWDM) coupler are obtained by using the balanced detection. Later, I. Sovran at el. [6] presented a technique that polarization-independent and fast frequency-scanning method are applied by using a pair of CW probe obtained through a Differential Group Delay (DGD) module and counter-propagating pulsed pump wave, which SOPs are orthogonally-polarized. In order to obtain the Brillouin response with normal acquisition time (one acquisition process), while maintaining the advantages in original DPP-BOTDA methods (high spatial resolution and long sensing range), our groups [7] have proposed a novel DPP-BOTDA configuration based on 3-tone probe recently.

In this work, we propose and demonstrate a distributed Brillouin fiber sensor by using orthogonally-polarized pulse pair and detecting the probe wave by using balanced detection. Comparable results between the conventional BOTDA system using polarization scrambler and our method are presented to demonstrate the benefits of our proposal to eliminate the polarization noise with half acquisition time.

II. WORKING PRINCIPLE

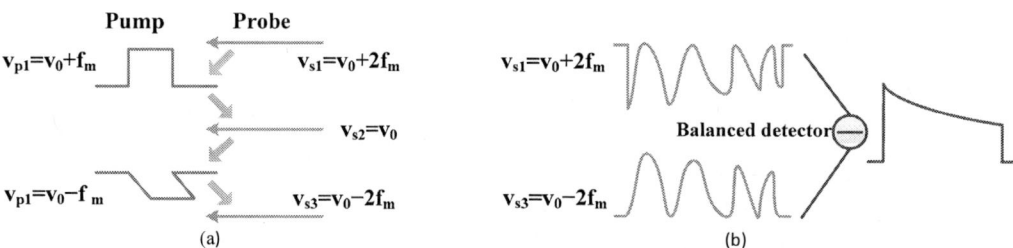

Fig.1. (a) The schematic model of the proposed technique. Red curve: pump wave; green and blue curve: probe wave. (b) Theoretical Brillouin response at peak gain for orthogonally-polarized pulse pair.

1092

In this paper, the upper tone and lower tone of 3-tone probe are directly detected to eliminate the effect of the state of polarization as shown in Fig. 1(a), instead of using polarization switch in traditional methods. Anti-Stokes and Stokes pump pulse with same power but orthogonal polarization in this paper are simultaneously launched into the fiber. They are spectrally separated by the modulation frequency f_m that is around the Brillouin frequency shift v_b. In the probe branch, the counter-propagating probe wave consists of three equalized spectral tones generated by intensity modulator biased at the quadrature transmission point. And the modulation frequency is synchronized with the one applied to the pump pulses, being simply multiplied by a factor 2 (i.e., $2f_m$). According to the well-known principle in SBS process that energy transfer from the high frequency wave to the lower one as the yellow arrows describe in Fig. 1(a), the upper tone at frequency v_{s1} and the lower tone at frequency v_{s3} experience the complementary Brillouin response resulting from the orthogonally-polarized pulse pair as mentioned in Fig. 1 (b). When the upper tone of probe beam experience the maximum response, the lower one experience the minimum response as theoretically described by the time traces in Fig. 1(b). Such two complementary time traces are obtained with twice of measurement time in the traditional method using polarization switch, however, they are obtained within one acquisition time in this proposal. And therefore, no extra acquisition time is needed in this paper. The final Brillouin response is detected by using a balance detector after filtering the probe wave consisted of the upper tone and lower tone.

III. EXPERIMENTAL SETUP AND RESULTS

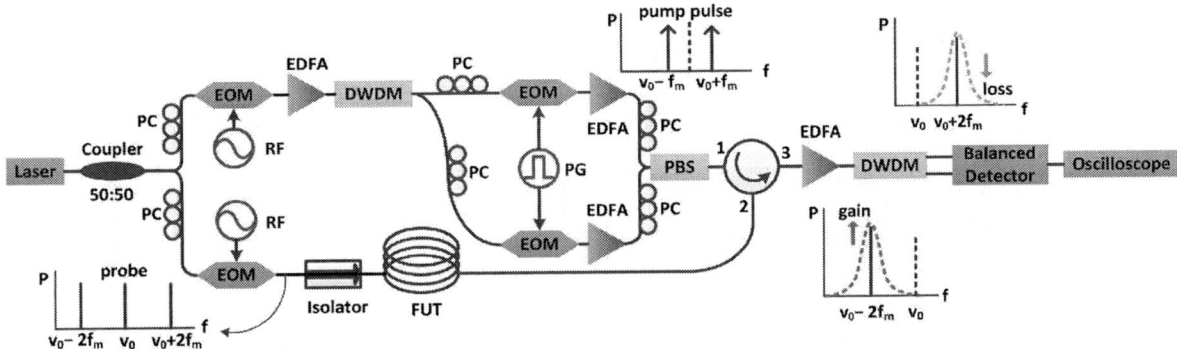

Fig. 2 Experimental setup of this proposal. EOM: electro-optic modulator, EDFA: erbium doped fiber amplifier, RF: radio-frequency generator, PC: polarization controller, PG: pulse generator, PBS: polarization beam splitter, FUT: fiber under test, DWDM: Dense Wavelength Division Multiplexer

Figure 2 describes the experimental setup to demonstrate the theoretical validation of this proposed method. The narrow linewidth tunable laser source with output power of 10 dBm at 1552.080 nm is split into two branches as pump pulse and probe beam respectively through a 50/50 coupler. On the upper branch, the CW beam is firstly modulated by an EOM to generate carrier-suppressed double-sideband (SC-DSB) with modulation frequency of f_m. And they are pulsed by two EOMs respectively with duration of 20 ns after being divided by a DWDM. And then, they are coupled by using a polarization beam splitter (PBS) to generate orthogonally-polarized pulse pair after being amplified to about 80 mW by two erbium doped fiber amplifiers (EDFAs). On the lower branch, the CW beam is modulated by another EOM with modulation frequency $2f_m$, which is synchronized with the one applied to the pump pulse. After through an isolator, the probe beam consisted of three equalized spectrum tones is injected into the sensing fiber interacting with the counter-propagated pump pulse pair with orthogonal polarization. Finally, the upper tone and the lower tone are divided by another DWDM, then, they are simultaneously captured by an oscilloscope with sample rate of 500 MSample/s using a balance detector

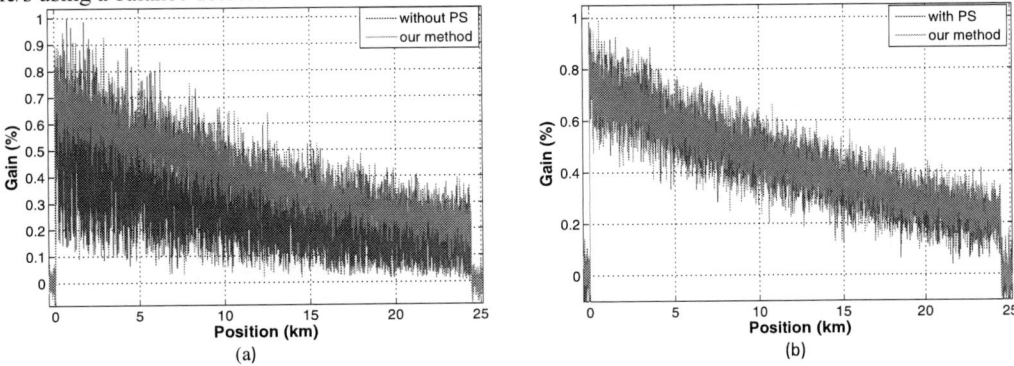

Fig.3 (a) Time traces at peak gain of our technique and basic system without polarization switch; (b) Time traces at peak gain of our technique and conventional system using polarization switch.

To demonstrate the validation of our proposal, a 24.444-km-long G.652D fiber is used as sensing fiber. The BFS of the fiber is 10.868 GHz at room temperature of 25°C. And the 2-m heated segment is loosely introduced at the end of the sensing fiber with temperature of 80°C. In order to avoid the spontaneous Brillouin scattering, the power of probe beam input to the fiber is limited down to -6 dBm per sideband.

The blue line in Fig. 3(a) is the result of conventional system with only one measurement time, in which typical 'polarization fading' exists. The red line in Fig. 3(a), is our result with the same measure conditions. The polarization noise is greatly reduced resulting from the subtraction between the two complementary Brillouin responses for the same measurement time in our method. For further demonstration of the validity of our polarization compensation technique, we also depict the time traces at peak gain both of our technique and traditional method using polarization scrambler. As described in Fig. 3 (b), similar Brillouin response can be observed in the two cases, indicating that both of them have almost the same polarization noise, but with the compromise of one more acquisition time in traditional technique using polarization scrambler or polarization switch.

Additionally, in order to check the ability of hot-spot detection in our technique, we plot in Fig. 4 (a) the measured BGS both at a few meters before the hot-spot and its precise location. As reported by Fig. 4 (a), the Lorentz profile can be easily distinguished between the two locations, which mean the distribution information can be accurately captured by detecting the probe beam consisted of upper tone and lower tone by using a balance detector in our proposal. Furthermore, Fig. 4 (b) shows the retrieved BFS profile around the hot-spot location where the 2-m hot-spot can be correctly demonstrated. And the 52 MHz frequency shift also matches our experimental condition well.

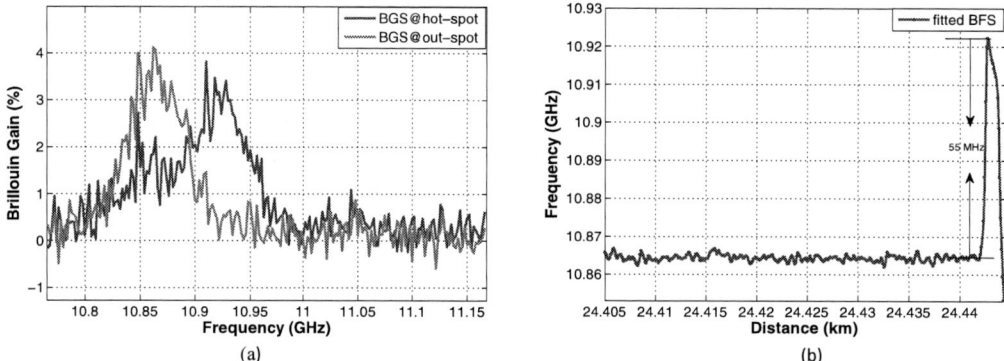

(a)   (b)

Fig .4 (a) Measured data and fitting curve at different locations; (b) Local BFS around the hot-spot location.

IV. CONCLUSIONS

In this work, we propose and demonstrate a new method based on balanced detection of orthogonally-polarized Stokes and anti-Stokes pump pulse pair to eliminate polarization noise without polarization scrambler or polarization switch. In this technique, the impact of the polarization noise is effectively eliminated in optical field without any extra processing. Moreover, half acquisition time for essential measurement precision is achieved under 24.444-km-long sensing fiber with spatial resolution of 2 m in this proposal.

ACKNOWLEDGMENT

This work was supported in part by 863 Program under Contract 2013AA014202, in part by Beijing Municipal Commission of Education. The authors thank Key Laboratory of Optical Fiber Sensing & Communications (Education Ministry of China) for their valuable research support.

REFERENCES

[1] M. O. Van Deventer and A. J. Boot, "Polarization Properties of Stimulated Brillouin Scattering in Single-Mode Fibers" Journal of Lightwave Technology, 12 (4), 585590 (1994).

[2] J. Yang, C. Yu, Z. Chen, J. Ng and X .Yang, "Suppression of polarization sensitivity in BOTDA fiber distributed sensing system," Proc. SPIE 7004 (700421), (2008)

[3] J. Urricelqui, F. López-Fernandino, M. Sagues, A. Loayssa, "Polarization diversity for Brillouin distributed fiber sensors based on a double orthogonal pump." in 23rd International Conference on Optical Fiber Sensors (2014).

[4] A. Domínguez-López, A. López-Gil, S.Martín-López, and M. González-Herráez, "Signal-to-noise ratio improvement in BOTDA using balanced detection." IEEE Photonic. Tech. Lett. Accepted (2013).

[5] A. Lopez-Gil, A. Dominguez-Lopez at el. "Simple Method for the Elimination of Polarization Noise in BOTDA Using Balanced Detection and Orthogonal Probe Sidebands." Journal of Lightwave Technology 33.12(2015):2605-2610.

[6] I. Sovran, et al. "An ultimately fast frequency-scanning Brillouin Optical Time Domain Analyzer." Optical Fiber Communications Conference and Exhibition (OFC), 2015 IEEE, 2015.

[7] Z. Yang, X. Hong, H. Guo, J. Wu, and J. Lin "Stokes and anti-Stokes differential pulse pair based distributed Brillouin fiber sensor with double-sideband probe wave. " Optics Express 22(3), 2881-2888 (2014).

WC4-4

Proof of Concept for Brillouin Optical Correlation-Domain Reflectometry Assisted by Spectral Slope

Heeyoung Lee[1], Neisei Hayashi[2], Yosuke Mizuno[1], and Kentaro Nakamura[1]

[1]Laboratory for Future Interdisciplinary Research of Science and Technology, Tokyo Institute of Technology,
4259-R2-26, Nagatsuta-cho, Midori-ku, Yokohama 226-8503, Japan
[2]Research Center for Advanced Science and Technology, The University of Tokyo,
4-6-1, Komaba, Meguro-ku, Tokyo 153-8904, Japan
hylee@sonic.pi.titech.ac.jp

Abstract: Exploiting the slope of the Brillouin gain spectrum, we develop a new configuration of Brillouin optical correlation-domain reflectometry, which can measure strain (or temperature) and optical loss distributions simultaneously with a high sampling rate.

Keywords: Brillouin scattering, optical fiber sensors, distributed measurement, correlation-domain techniques

I. INTRODUCTION

To monitor the conditions of civil infrastructures, a variety of fiber-optic distributed strain and temperature sensing techniques based on Brillouin scattering [1] have been extensively studied. They are categorized into two types: "analysis" systems [2]–[14], in which two light beams need to be injected into both ends of a fiber under test (FUT), and "reflectometry" systems [15]–[20], in which only one light beam is injected into one end of the FUT. The former include Brillouin optical time-, frequency-, and correlation-domain analysis (BOTDA [2]–[8], BOFDA [9],[10], and BOCDA [11]–[14]) systems, while the latter include Brillouin optical time- and correlation-domain reflectometry (BOTDR [15],[16] and BOCDR [17]–[20]) systems. In analysis systems, stimulated Brillouin scattering with a relatively high reflectivity can be exploited, leading to a high signal-to-noise ratio (SNR) of the measurement. Consequently, extremely high sensing performances, such as an ultimately fast measurement speed in BOTDA [8] and a nominal spatial resolution of as high as 1.6 mm in BOCDA [14], have been reported. However, such two-end-access systems reduce the degree of freedom in embedding the FUT into structures, and they do not function properly when a single breakage occurs in the FUT. From the viewpoint of the users' convenience, one-end-access reflectometry systems are preferable. Compared to BOTDR, where optical pulses are utilized to resolve the positions, continuous-wave-based BOCDR can achieve a relatively high SNR in principle, leading to a high spatial resolution and a sampling rate.

BOCDR operates based on the correlation control of propagating lightwaves. Even when optical loss occurs in the FUT, conventional BOCDR can provide the Brillouin frequency shift (BFS) distribution properly, which is derived from the Brillouin gain spectrum (BGS) distribution. This feature is an advantage from the aspect of stable strain and temperature measurement; however, in some practical applications, optical loss distribution should be simultaneously obtained.

In this work, to achieve simultaneous distributed measurement of strain (or temperature) and loss, we develop a new BOCDR configuration—which we named slope-assisted BOCDR (SA-BOCDR) in analogy with SA-BOTDA [5]–[7]—which operates with the assistance of the BGS slope. In this method, the whole BGS need not be observed to derive the BFS value, which leads to a higher sampling (or repetition) rate in principle.

II. PRINCIPLE

BOCDR is known as a distributed strain/temperature sensing technique with one-end accessibility, a high spatial resolution, a high sampling rate, and cost efficiency. To spatially resolve the sensing locations, we apply sinusoidal frequency modulation to the laser output to generate a so-called "correlation peak" in the FUT [11]. Using the correlation peak, the BFS at a specific position can be selectively detected. By sweeping the modulation frequency f_m, the correlation peak is scanned along the FUT, enabling a distributed BFS measurement. As sinusoidal frequency modulation periodically generates multiple correlation peaks along the FUT, their interval determines the measurement range d_m as [19]

$$d_m = \frac{c}{2nf_m},\tag{1}$$

where c is the velocity of light in vacuum and n is the refractive index of the fiber core. When f_m is lower than the Brillouin bandwidth Δv_B, the spatial resolution Δz is reported to be given by [19]

$$\Delta z = \frac{c\Delta v_B}{2\pi n f_m \Delta f},\tag{2}$$

where Δf is the modulation amplitude of the optical frequency.

OECC/PS2016

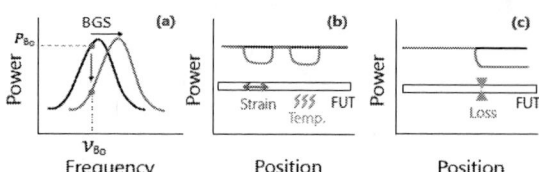

Fig. 1. (a–c) Schematic illustrations of operating principle.

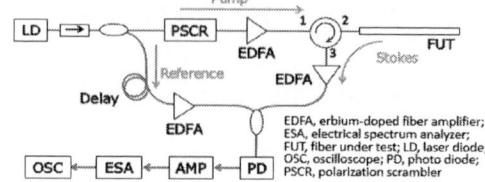

Fig. 2. Experimental setup of SA-BOCDR.

In conventional BOCDR systems, the BFS (i.e. strain or temperature) at one sensing point is derived after obtaining the whole BGS. In contrast, SA-BOCDR provides the strain/temperature information using the BGS slope. This scheme has been implemented for BOTDA systems [5]–[7]. As depicted in Fig. 1(a), the BFS is in one-to-one correspondence with the spectral power P_{B0} at a certain frequency v_{B0}, which is set at the high-frequency point in the linear region (lower-frequency side) of the BGS slope. Then, when the BFS slightly shifts to higher frequency according to strain and/or heat, P_{B0} decreases linearly; when the slight loss occurs in the FUT, P_{B0} also decreases linearly. Even if the BFS change is so large that v_{B0} gets out of the linear region, so long as P_{B0} is in one-to-one correspondence with the BFS, this system operates properly (with a reduced sensitivity) by simple nonlinear compensation. By analyzing a raw BGS with a BFS of 10.89 GHz (assuming an FUT used in the experiment below), we found that the optimal v_{B0} value should be 10.85 GHz to achieve a wide linear range (~50 MHz; corresponding to the strain of up to ~1035 µε and the temperature change of ~45 K), in which the theoretical strain and temperature dependence coefficients are 2.11×10^{-4} dB/µε and 4.27×10^{-3} dB/K, respectively. Figures 1(b) and (c) schematically show the changes in the P_{B0} distributions when strain (or temperature change) and loss are locally applied, respectively. Note that the strain and temperature effects cannot be separated in this method, but the loss effect can be discriminated from the two because the once decreased P_{B0} value does not return to the initial value. The P_{B0} change distributions (calculated by substituting the resultant P_{B0} distributions (red curves) from their initial distributions (black line)) are used as final measurement data.

III. EXPERIMENTAL SETUP

The FUT employed in the experiment was a 5.0-m-long silica single-mode fiber (SMF) with BFS of 10.89 GHz at 1.55 µm at room temperature. The experimental setup of SA-BOCDR is schematically shown in Fig. 2, which is basically the same as that of conventional BOCDR; the only essential difference lies in the final signal processing. The output light from a distributed-feedback laser diode (LD) at 1.55 µm was divided into two light beams, pump and reference. The pump light was amplified by an erbium-doped fiber amplifier (EDFA) and injected into the FUT. After passing through a 1-km-long delay line and an EDFA, the reference light was used for heterodyne detection with the Stokes light, which is the light Brillouin-scattered in the FUT. The heterodyned optical signal was converted into an electrical signal using a photo diode (PD) and was guided to an electrical spectrum analyzer (ESA). Using the narrowband-pass filtering function of the ESA, the P_{B0} change at a fixed frequency v_{B0} (= 10.85 GHz) was sequentially output to an oscilloscope (OSC).

A 0.1-m-long section around the distal open end of the FUT was bent to suppress the Fresnel reflection. The modulation frequency f_m and amplitude Δf were set to 7.975–8.055 MHz and 1.4 GHz, respectively, corresponding to the measurement range of 12.9 m and the theoretical spatial resolution of 88 mm according to Eqs. (1) and (2). The repetition rate was 100 Hz, and 16 times averaging was performed on the OSC. The room temperature was 26°C.

IV. EXPERIMENTAL RESULTS

First, we investigated the P_{B0} change dependence on strain. Strains of 0 to 850 µε were applied to a 0.2-m-long section (3.5–3.7 m away from the circulator) of the FUT. The measured P_{B0} change distributions along the FUT are shown in Fig. 3(a). With increasing strain, the P_{B0} change increased (P_{B0} itself decreased). The P_{B0} change dependence was almost linear in this range (Fig. 3(b)) and its coefficient was 1.95×10^{-4} dB/µε, which moderately agrees with the theoretical value (2.11×10^{-4} dB/µε). Note that, in Fig. 3(a), the SNR was so low that relatively small strains of < 300 µε were unable to be distinguished from

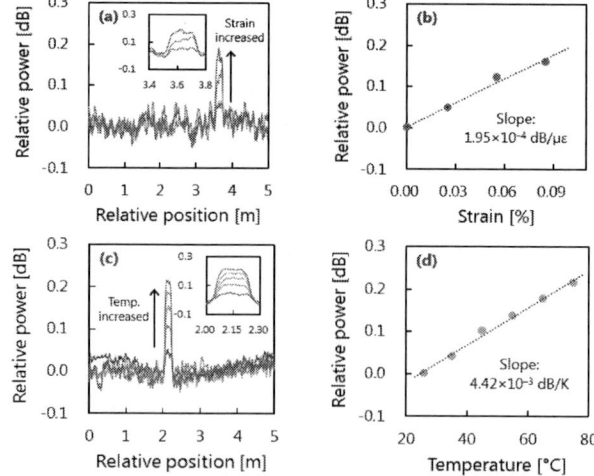

Fig. 3. (a) P_{B0} change distribution when strains were locally applied, and (b) P_{B0} change dependence on strain. (c) P_{B0} change distribution when temperature was locally changed, and (d) P_{B0} change dependence on temperature.

1096

the signal fluctuations in this measurement.

The P_{B0} change dependence on temperature was then measured. The result obtained when the temperature was locally changed to 75°C in a 0.2-m-long section (2.0–2.2 m) is shown in Fig. 3(c). With increasing temperature, the P_{B0} change increased. The P_{B0} change dependence was almost linear (Fig. 3(d)) with a coefficient of 4.42×10^{-3} dB/K, which agrees with the theoretical value (4.27×10^{-3} dB/K).

The loss dependence of the P_{B0} change distribution was also measured. Bending losses of 0 to 2.8 dB were applied at a midpoint of the FUT (2.5 m away from the circulator). As shown in Fig. 4(a), with increasing loss, the P_{B0} change increased on the distal side from the loss-applied point. The loss dependence of the P_{B0} change (averaged in the 2.5-m-long distal section (2.5–5.0 m)) was almost linear (Fig. 4(b)) with a coefficient of 0.191.

Finally, a proof-of-concept demonstration of

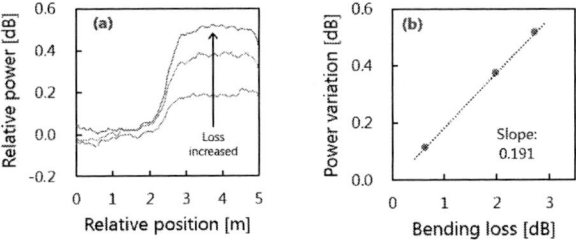

Fig. 4. (a) P_{B0} change distribution when losses were locally applied, and (b) P_{B0} change dependence on loss.

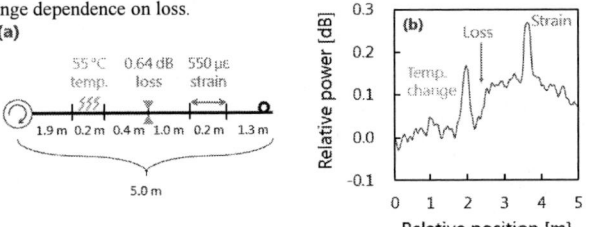

Fig. 5. (a) Structure of the FUT. (b) Measured P_{B0} change distribution.

SA-BOCDR was performed by simultaneous measurement of strain, temperature, and loss. The structure of the 5.0-m-long FUT is shown in Fig. 5(a); ambient temperature was changed to 55°C along the 1.9–2.1-m section, a 0.64-dB loss was applied at the midpoint, and a 550 με strain was applied to the 3.5–3.7-m section. Figure 5(b) shows the measured P_{B0} change distribution along the FUT. The P_{B0} changes corresponding to the temperature change, loss, and strain were observed at the expected sections. However, the SNR was not sufficiently high, which could be improved by optimal low-pass filtering and/or increase in the number of averaging.

V. CONCLUSIONS

We experimentally proved the concept of a new BOCDR configuration, named SA-BOCDR, which can perform the simultaneous distributed measurement of strain (or temperature) and optical loss by exploiting the slope of the BGS. After measuring the strain-, temperature-, and loss-dependence coefficients of the output signal (1.95×10^{-4} dB/με, 4.42×10^{-3} dB/K, and 0.191, respectively), we verified the basic operation of simultaneous measurement of the three parameters. The improvement of the low SNR is one of the most important future tasks. The limitation of the system performance, such as strain/temperature/loss dynamic ranges, measurement accuracy, and repetition rate, also needs to be clarified.

ACKNOWLEDGMENTS

This work was supported by JSPS KAKENHI Grant Numbers 25709032, 26630180, and 25007652.

REFERENCES

[1] G. P. Agrawal, *Nonlinear Fiber Optics* (Academic Press, Boston, 1995).
[2] T. Horiguchi and M. Tateda, J. Lightwave Technol. **7**, 1170 (1989).
[3] Z. Li, L. Yan, L. Shao, W. Pan, B. Luo, J. Liang, and H. He, IEEE Photon. J. **8**, 6800908 (2016).
[4] J. Urricelqui, A. Zornoza, M. Sagues, and A. Loayssa, Opt. Express **20**, 26942 (2012).
[5] Y. Peled, A. Motil, L. Yaron, and M. Tur, Opt. Express **19**, 19845 (2011).
[6] X. Tu, H. Luo, Q. Sun, X. Hu, and Z. Meng, J. Opt. **17**, 105503 (2015).
[7] A. Minardo, A. Coscetta, R. Bernini, and L. Zeni, J. Opt. **18**, 025606 (2016).
[8] I. Sovran, A. Motil, and M. Tur, IEEE Photon. Technol. Lett. **27**, 1426 (2015).
[9] D. Garus, K. Krebber, and F. Schliep, Opt. Lett. **21**, 1402 (1996).
[10] R. Bernini, A. Minardo, and L. Zeni, IEEE Photon. J. **4**, 48(2012).
[11] K. Hotate and T. Hasegawa, IEICE Trans. Electron. **E83-C**, 405 (2000).
[12] R. Cohen, Y. London, Y. Antman, and A. Zadok, Opt. Express **22**, 12070 (2014).
[13] C. Zhang, M. Kishi, and K. Hotate, Appl. Phys. Express **8**, 042501 (2015).
[14] K. Y. Song, Z. He, and K. Hotate, Opt. Lett. **31**, 2526 (2006).
[15] T. Kurashima, T. Horiguchi, H. Izumita, S. Furukawa, and Y. Koyamada, IEICE Trans. Commun. **E76-B**, 382 (1993).
[16] A. Masoudi, M. Belal, and T. P. Newson, Opt. Lett. **38**, 3312 (2013).
[17] Y. Mizuno, W. Zou, Z. He, and K. Hotate, Opt. Express **16**, 12148 (2008).
[18] N. Hayashi, Y. Mizuno, and K. Nakamura, J. Lightwave Technol. **32**, 3397 (2014).
[19] Y. Mizuno, W. Zou, Z. He, and K. Hotate, J. Lightwave Technol. **28**, 3300 (2010).
[20] N. Hayashi, Y. Mizuno, and K. Nakamura, Opt. Express **20**, 21101 (2012).

WC4-5

OECC/PS2016

Simplified Optical Correlation-Domain Reflectometry Without Reference Path

Makoto Shizuka[1], Neisei Hayashi[2], Yosuke Mizuno[1], and Kentaro Nakamura[1]

[1]Laboratory for Future Interdisciplinary Research of Science and Technology, Tokyo Institute of Technology,
4259-R2-26, Nagatsuta-cho, Midori-ku, Yokohama 226-8503, Japan
[2]Research Center for Advanced Science and Technology, The University of Tokyo,
4-6-1, Komaba, Meguro-ku, Tokyo 153-8904, Japan
mshizuka@sonic.pi.titech.ac.jp

Abstract: *We develop a simplified configuration of optical correlation-domain reflectometry without involving an explicit reference path. The Fresnel-reflected light generated at the distal open end of the sensing fiber is exploited as a reference light.*
Keywords: *Optical fiber sensors, reflectometry, Fresnel reflection, system simplification*

I. INTRODUCTION

Fiber-optic reflectometry is a fundamental technique for implementing multiplexed and distributed sensing systems [1]–[6], and its various configurations have been developed so far to detect many different kinds of physical parameters such as strain [2],[3], temperature [3],[4], pressure [5], humidity [6], etc. Among them, to detect bad connections (or splices) and other reflection points along fibers under test (FUTs) in a distributed manner, three types of fiber-optic reflectometry based on Fresnel reflection have been developed: optical time-domain reflectometry (OTDR) [7],[8], optical frequency-domain reflectometry (OFDR) [9],[10], and optical correlation (or coherence)-domain reflectometry (OCDR) [11]–[17]. It has been reported that OTDR commonly suffers from a relatively low spatial resolution and a low sampling rate, while OFDR generally suffers from phase fluctuations caused by environmental disturbance. Thus, we here focus on OCDR, which can mitigate these shortcomings.

OCDR operates by exploiting a synthesized optical coherence function (SOCF) [15], i.e. by controlling the correlation of propagating light beams through optical frequency modulation. The modulation methods can be categorized into two: sinusoidal modulation [12],[13] and stepwise modulation [14],[15] (including frequency-comb-based modulation [16],[17]). As the latter requires accurate frequency adjustment and/or frequency-comb generation, sinusoidal modulation is preferable for cost-efficient implementation [12],[13]. In a standard SOCF-OCDR system [12]–[17], an optical frequency shifter such as an acousto-optic modulator (AOM) is utilized so that the heterodyned Fresnel spectrum is shifted from DC by several tens of megahertz; otherwise the Fresnel reflection spectrum to be detected is overlapped by the low-frequency noise of the electrical devices. Then, in order to reduce the cost of the system, we have recently developed a new SOCF-OCDR configuration without using an AOM [18]. By exploiting the foot of the Fresnel reflection spectrum, a sufficiently high signal-to-noise ratio (SNR) was obtained. However, in both the standard and simplified SOCF-OCDR configurations mentioned above, two optical paths—the incident optical path including an FUT and the reference optical path for optical interference—were required. Removing the reference path will further simplify the system and boost the practical convenience.

By using a polymer optical fiber (POF) as an FUT, we have already demonstrated the SOCF-OCDR operation without involving an explicit reference path [19]. In this case, the Fresnel-reflected light generated at the boundary between the POF and a silica single-mode fiber (SMF; a pigtail of an optical circulator) was used as a reference light. However, when the FUT is composed not of a POF but of a silica SMF, relatively Fresnel-reflected light is not generated at the boundary of the two silica SMFs, which makes it difficult for us to demonstrate the similar operation.

In this work, we develop a simplified configuration of AOM-free SOCF-OCDR without involving an explicit reference path when a standard silica SMF is used as an FUT. The Fresnel-reflected light generated at the distal open end of the FUT is exploited as a reference light. In addition to the demonstration of the basic operation, we investigate the optimal incident power and the influence of the loss near the distal end on the measured results.

II. PRINCIPLE AND EXPERIMENTAL SETUP

Figure 1 depicts the experimental setup of the AOM-free SOCF-OCDR without involving a reference path, which is extremely simple compared to previous configurations [12]–[18]. The laser output at 1550 nm was amplified using an erbium-doped fiber amplifier (EDFA) and was injected into an FUT composed of sequentially connected multiple silica SMFs (detailed below in this section). The reflected light, which contained the optical beat signal between the light beams reflected at the SMF-to-SMF boundaries and the light beam Fresnel-reflected at the distal open end of the FUT (~4%), was guided to a photo detector (PD), and the beat signal converted into an electrical signal was input to an electrical spectrum analyzer (ESA). By using the ESA as an electrical narrow band-pass filter, the electrical spectral power at 2

OECC/PS2016

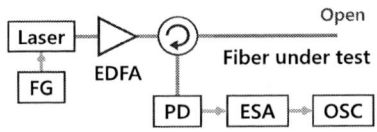

Fig. 1. Experimental setup of simplified OCDR without reference path. EDFA: erbium-doped fiber amplifier, FG: function generator, OSC: oscilloscope, PD: photo diode.

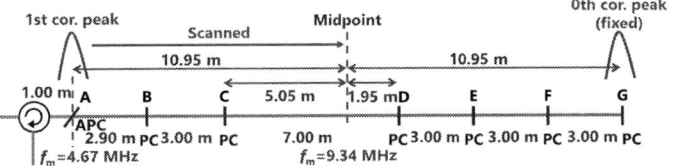

Fig. 2. Schematic structure of fiber under test.

MHz (at which a maximal SNR can be obtained [18]) was selectively transmitted to an oscilloscope (OSC). The resolution bandwidth and the video bandwidth of the ESA were set to 300 kHz and 1 kHz, respectively.

To perform distributed reflectivity (or reflection power) measurement, the output frequency of the laser was sinusoidally modulated by directly modulating the driving current, leading to the formation of a "correlation peak" in the FUT [15]. Using the correlation peak, the light reflected at a specific position can be selectively observed. By sweeping the modulation frequency, the correlation peak is scanned along the FUT, and thus the reflectivity distribution can be obtained. In a conventional SOCF-OCDR system involving a reference path, either with or without an AOM, sinusoidal frequency modulation generates multiple correlation peaks periodically. The measurement range D is then given by their interval as [20]

$$D = \frac{c}{2nf_m},\qquad(1)$$

where c is the optical velocity in vacuum, n is the refractive index of the fiber core, and f_m is the modulation frequency. According to detailed calculations, the spatial resolution Δz (which equals the 3-dB linewidth of the correlation peak) is theoretically given by [20]

$$\Delta z \cong \frac{0.76c}{\pi n \Delta f},\qquad(2)$$

where Δf is the modulation amplitude.

However, in a SOCF-OCDR system without involving a reference path, the measurement range D is limited to the proximal half of the FUT length rather than Eq. (1). This is because, in this configuration, the 0-th correlation peak, i.e. the zero-optical-path-difference point, is constantly located at the distal open end of the FUT, and the 1st correlation peak is utilized for distributed measurement. When the 1st peak reaches the midpoint of the FUT, the 2nd peak starts to enter the FUT at the optical circulator. Note that a similar configuration has been implemented in a Brillouin-based OCDR system [21]. Although the spatial resolution of this simplified Brillouin OCDR is a function of the sensing position, that of the SOCF-OCDR without a reference path is not dependent on the sensing position (constantly given by Eq. (2)).

The detailed structure of the FUT is shown in Fig. 2. A 1.0-m-long pigtail (silica SMF) of the circulator was connected to 2.9-m-, 3.0-m-, 7.0-m-, 3.0-m-, 3.0-m-, and 3.0-m-long silica SMFs sequentially using FC/PC or FC/APC connectors. The distal PC end of the FUT was kept open. The modulation frequency f_m was swept from 4.67 MHz to 9.34 MHz with a repetition rate of 33 Hz, which corresponds to the measurement range of $d = 0–11.0$ m (where d was defined as the length from Connector A toward the distal end). The modulation amplitude Δf was set to 0.75 GHz, corresponding to the spatial resolution of 66 mm according to Eq. (2).

III. EXPERIMENTAL RESULTS

A. Basic Operation and Optimization of Incident Power

First, the reflection power distributions along the FUT were measured with varying incident power P_{in}, as shown in Fig. 3. When P_{in} was higher than ~0 dBm, clear peaks corresponding to Connectors B and C were observed at $d = 2.9$ and 5.9 m, respectively, which verified the basic operation of this system. However, when P_{in} was lower than ~0 dBm, the peak corresponding to Connector C was buried by the noise floor. We then evaluated the P_{in} dependence of the SNR of each peak, which was defined as the ratio of the reflection power and the noise floor at a certain position. As shown in Fig. 4, as P_{in} increased, the SNR became maximal and then decreased for both the peaks. In this experimental condition, irrespective of the connectors, the optimal P_{in} value that gave the maximal SNR was approximately 8 dBm.

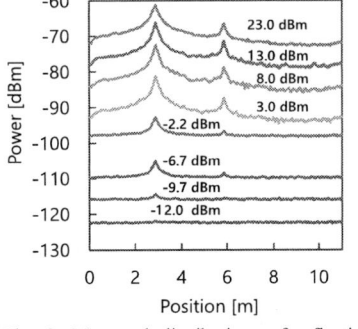

Fig. 3. Measured distributions of reflection power when the incident power was changed.

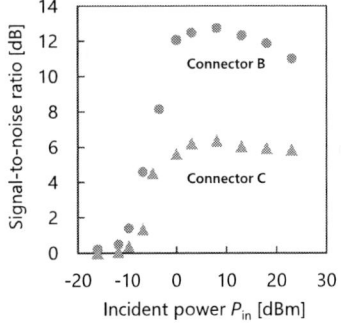

Fig. 4. SNRs measured with respect to the incident power.

B. Influence of Loss Near Fiber End

Subsequently, we evaluated the influence of the noise caused by the 0-th correlation peak. Note that, in a Brillouin-based OCDR system without involving a reference path, some amount of loss needs to be artificially applied near the distal open end to mitigate the influence of the 0-th correlation peak and thus to obtain a sufficiently high SNR [21]. A tunable bending loss was applied to the point 0.1 m away from the distal end. The P_{in} value was 8 dBm. The bending loss dependence of the measured reflection power distribution is shown in Fig. 5. As the bending loss

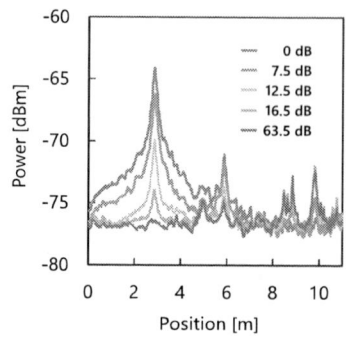

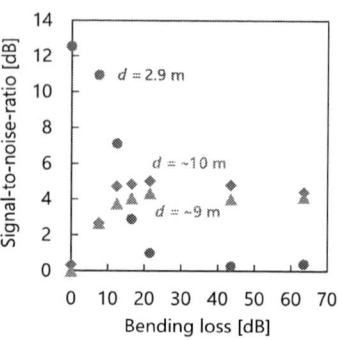

Fig. 5. Measured distributions of reflection power when the bending loss was changed.

Fig. 6. SNRs measured with respect to the bending loss.

increased, the power of the two peaks corresponding to Connector B and C decreased, while new peak started to grew at $d = {\sim}9$ and ${\sim}10$ m, where no connectors existed. These ghost peaks can be explained by the fact that, as the reflection power from the original 0-th correlation peak located at the distal end decreases, one of the SMF-to-SMF boundaries with relatively weak reflectivity starts to function as a new 0-th correlation peak; the multiple reflections among the connectors result in the appearance of the ghost peaks. Finally, the bending loss dependence of the SNR (this term is used also for the ghost peaks for expediency) at $d = 2.9, {\sim}9, {\sim}10$ m was measured (Fig. 6). As the bending loss increased, the SNR of the desired peak at $d = 2.9$ m decreased, while those of the ghost peaks increased. This result indicates that, unlike the Brillouin OCDR based on frequency information, we need not (or should not) apply an artificial loss near the open end in the Fresnel OCDR based on power information.

IV. CONCLUSIONS

A simplified configuration of SOCF-OCDR without involving an explicit reference path was developed. As a standard silica SMF was used as an FUT, the Fresnel-reflected light generated at the distal open end of the FUT was exploited as a reference light. Distributed reflection power measurement was demonstrated, and the optimal incident power was found to be approximately 8 dBm. We also showed that the loss near the distal end should not be applied, unlike the case of Brillouin-based OCDR. We believe that our further simplified AOM-free SOCF-OCDR system will be of great use in implementing cost-efficient distributed reflectivity sensors in future.

ACKNOWLEDGMENTS

This work was supported by JSPS KAKENHI Grant Numbers 25709032, 26630180, and 25007652, and by research grants from the Iwatani Naoji Foundation, the SCAT Foundation, and the Konica Minolta Science and Technology Foundation.

REFERENCES

[1] B. L. Danielson and C. D. Whittenberg, Appl. Opt. **26**, 2836 (1987).
[2] Y. Mizuno, W. Zou, Z. He, and K. Hotate, Opt. Express **16**, 12148 (2008).
[3] T. Kurashima, T. Horiguchi, H. Izumita, S. Furukawa, and Y. Koyamada, IEICE Trans. Commun. **E76-B**, 382 (1993).
[4] A. H. Hartog, A. P. Leach, and M. P. Gold, Electron. Lett. **21**, 1061 (1985).
[5] S. Binu, V. P. M. Pillai, and N. Chandrasekaran, J. Opt. **35**, 36 (2006).
[6] A. Kharaz and B. E. Jones, Sens. Actuators A: Phys. **47**, 491 (1995).
[7] M. K. Barnoski and S. M. Jensen, Appl. Opt. **15**, 2112 (1976).
[8] Q. Zhao, L. Xia, C. Wan, J. Hu, T. Jia, M. Gu, L. Zhang, L. Kang, J. Chen, X. Zhang, and P. Wu, Sci. Rep. **5**, 10441 (2015).
[9] W. Eickhoff and R. Ulrich, Appl. Phys. Lett. **39**, 693 (1981).
[10] B. Wang, X. Fan, S. Wang, G. Yang, Q. Liu, and Z. He, Opt. Commun. **365**, 220 (2016).
[11] R. C. Youngquist, S. Carr, and D. E. N. Davies, Opt. Lett. **12**, 158 (1987).
[12] K. Hotate, M. Enyama, S. Yamashita, and Y. Nasu, Meas. Sci. Technol. **15**, 148 (2004).
[13] Z. He, T. Tomizawa, and K. Hotate, IEICE Electron. Express **3**, 122 (2006).
[14] K. Hotate and O. Kamatani, Electron. Lett. **25**, 1503 (1989).
[15] K. Hotate, Meas. Sci. Technol. **13**, 1746 (2002).
[16] Z. He, H. Takahashi, and K. Hotate, Conference on Lasers and Electro-Optics 2010, CFH4.
[17] H. Takahashi, Z. He, and K. Hotate, 36th European Conference and Exhibition on Optical Communication 2010, Tu.3.F.4.
[18] M. Shizuka, S. Shimada, N. Hayashi, Y. Mizuno, and K. Nakamura, Appl. Phys. Express **9**, 032702 (2016).
[19] N. Hayashi, M. Shizuka, K. Minakawa, Y. Mizuno, and K. Nakamura, IEICE Electron. Express **12**, 20150824 (2015).
[20] K. Hotate and K. Kajiwara, Opt. Express **16**, 7881 (2008).
[21] N. Hayashi, Y. Mizuno, and K. Nakamura, IEEE Photon. J. **6**, 6802807 (2014).

WC4-6

OECC/PS2016

Phase Noise Mitigation for Long-range OFDR Using Ultrafast Frequency Sweep

Bin Wang, Xinyu Fan[*], Shuai Wang, Guangyao Yang, Qingwen Liu, and Zuyuan He

State Key Laboratory of Advanced Communication Systems and Networks,
Shanghai Jiao Tong University, Shanghai 200240, China
fan.xinyu@sjtu.edu.cn

Abstract: We demonstrate a novel technique to mitigate the influence of phase noise for long-range OFDR. A linear relationship between the spatial resolution and the frequency sweep time is verified before reaching the theoretical resolution.

Keywords: Phase noise, OFDR, ultrafast frequency sweep

I. INTRODUCTION

Optical reflectometry is a powerful technique for non-destructive diagnoses of optical devices and optical fiber networks. A well-known example is optical time domain reflectometry (OTDR), which is commonly used for measuring reflections in optical access networks and optical fiber links. However, the spatial resolution of OTDR is typically only several meters, which restricts its applications and refrains from meeting the requirements of high-resolution measurements.

Optical frequency domain reflectometry (OFDR) has been expected to achieve much higher spatial resolution and sensitivity as an alternative technique to conventional OTDR [1, 2]. However the measurement range is limited to the laser coherence length since the laser phase noise causes serious degradation of signal-to-noise ratio (SNR) as the measurement distance approaches the laser coherence length [3]. In order to employ the OFDR technique for applications that involve monitoring medium- or long-haul systems, phase-noise-compensated OFDR (PNC-OFDR) has been proposed. PNC-OFDR technique is capable of compensating the phase noise in OFDR, and realizing centimeter-level resolution over tens of kilometers [4]. However, the phase noise compensation process of PNC-OFDR is mainly realized by software and needs relatively tedious calculation to obtain the final results. In order to improve the processing speed, it is better to use optical configurations rather than software to extend the measurement range of OFDR. Recently a time-gated digital optical frequency domain reflectometry (TGD-OFDR) technique was proposed with an ultra-long measurement range of 110 km [5], and indicated that a higher frequency sweep rate is effective to suppress the phase noise of OFDR [6]. The spatial resolution of TGD-OFDR is limited to ~1.6 m because of the narrow frequency sweep span of acoustic optical modulator (AOM). In order to improve the spatial resolution of long-range OFDR, a wider frequency sweep span and a higher frequency sweep rate are required. Therefore, it is essential to solve the problems brought by ultrafast frequency sweep, such as a high bandwidth required for the detection and the subsequent electronics.

In this paper, we propose a novel technique to solve the above-mentioned problems in long-range OFDR when using ultrafast frequency sweep. A single-sideband (SSB) modulator rather than an AOM is employed to use a wide frequency sweep span and an ultrafast frequency sweep. Besides, we use time-gated receiving technique and I/Q demodulation technique to reduce the required detection bandwidth. By shortening the sweep time, we anticipate a better spatial resolution, and verified a linear relationship between the spatial resolution and the sweep time before reaching the theoretical spatial resolution. In our experiment, a 15 cm spatial resolution is realized at the distance of 60 km.

II. PRINCIPLE

OFDR is mainly composed of a tunable light source and an interferometer, and the measurement range is limited to the laser coherence length because of phase noise. In long-range OFDR, the spatial resolution and sweep rate follows a simple relationship as shown below

$$\frac{\Delta f}{\gamma \tau_{FUT}} = \frac{\Delta z}{L} \tag{1}$$

where Δf is the broadened frequency caused by phase noise, Δz is the spatial resolution, γ is the frequency sweep rate, τ_{FUT} is the delay time related to the reflection event, and L is the length of fiber under test (FUT). A higher frequency sweep rate is anticipated to suppress the phase noise of the laser source as well as the environmental perturbations. However, a higher frequency sweep rate generates a larger beat frequency for a long measurement range, which in most cases exceeds the sampling rate of available data acquisition devices. In other words, the allowable frequency sweep rate is inversely proportional to the measurement range due to the limitation of sampling rate of data acquisition.

1101

OECC/PS2016

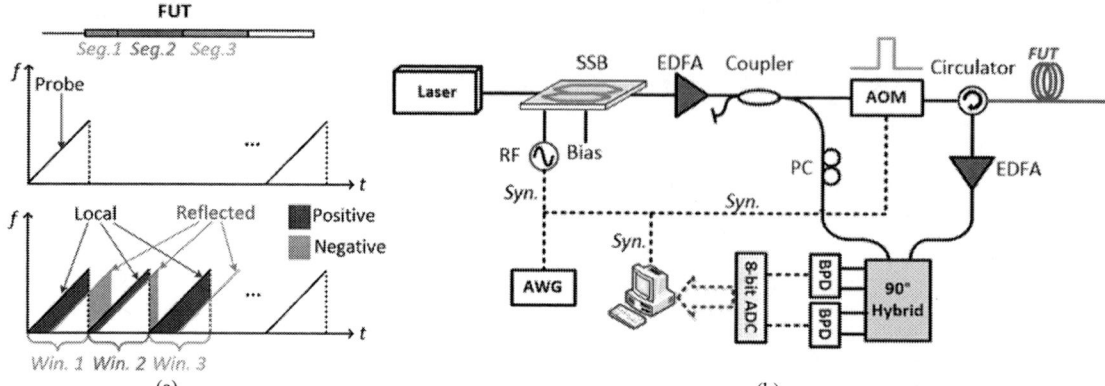

Fig. 1. (a) Schematic diagram and (b) experimental setup of the proposed long-range OFDR. SSB: single-sideband modulator; AWG: arbitrary waveform generator; EDFA: erbium-doped fiber amplifier; PC: polarization controller; AOM: acoustic optical modulator BPD: balanced photodetector; ADC: analog to digital converter.

In our proposed system, we adopt an ultrafast frequency sweep to mitigate the phase noise, and employ a time-gated receiving technique to reduce the required detection bandwidth. As shown in Fig. 1(a), the probe light is carved to be a pulse with the duration time equal to the frequency sweep time τ_p. The duration time of detection window is also set to be τ_p. In this way, we can obtain the backscattered signal from different segments of the FUT by controlling the delay time of the detection window. For a given duration time of τ_p, the length of segment1 is set to be $c\tau_p/4n$, and the length of other segments is the double of first segment, where c is the light speed in vacuum, and n is the refractive index of FUT. Therefore the required detection bandwidth is always low for these segments under detection. Besides, there is positive and negative beat frequency in every detection window except window 1. We use I/Q demodulation technique to separate the positive and negative beat frequency and further reduce the required detection bandwidth to its half.

III. EXPERIMENTAL SETUP AND RESULTS

The detailed experimental setup of the proposed system is shown in Fig. 1(b). A fiber laser (NKT, Adjustik E15) is employed as a light source and it is connected to an SSB modulator for external modulation. Ultrafast frequency sweep is realized by using the external modulation method with a linearly frequency swept RF signal generated by an arbitrary waveform generator (AWG). The frequency sweep span is 850 MHz, which corresponds to a spatial resolution of 12 cm theoretically, and we change the sweep time from 5 ms to 50 μs. Then the modulated lightwave is launched to an erbium-doped fiber amplifier (EDFA) to boost the power. The probe lightwave is carved to be a pulse by using an AOM, while the local lightwave is a continuous wave. The local lightwave and the backscattered lightwave are combined in an optical 90° hybrid and coherently detected by two balanced photodetectors (BPDs). The electronic signals are collected by an 8-bit analog to digital converter (ADC) and digitally processed using a computer.

The FUT length is ~60 km. In order to obtain the details of the reflection peak at the end of FUT, the delay time of the detection window is set to be 0.6 ms. We change the frequency sweep time from 5 ms to 50 μs to study the relationship of the frequency sweep rate and the influence of phase noise in long-range OFDR system.

Fig.2 shows reflectometric traces at different frequency sweep time. Fig. 2(a) shows that the spatial resolution is 13 m at the end of the FUT when the sweep time is set to 5 ms. In order to mitigate the phase noise, frequency sweep rate is increased, and the sweep time is set to be 300 μs, 200 μs, 100 μs, 50 μs in Fig. 2(b)-(e). The spatial resolution of reflection events at the end of FUT are improved as the sweep time decreases. A spatial resolution of 15 cm at the distance of 60 km was obtained when the frequency sweep time is set to be 50 μs. Fig. 2(f) shows that the relationship between the spatial resolution and the sweep time is almost linear before reaching the theoretical spatial resolution, which is consistent with the theoretical analysis. Therefore, we believe that a better spatial resolution may be obtained when an AWG with larger frequency span is employed in our system.

IV. CONCLUSIONS

We proposed and experimentally demonstrated a novel technique to mitigate the effect of phase noise in long-range OFDR by using an ultrafast frequency sweep. The preliminary experiment realized a 15 cm spatial resolution at the distance of 60 km. We verified a linear relationship between the spatial resolution and the sweep time before reaching the theoretical spatial resolution. This technique has great potential in applications where both long measurement range and high spatial resolution are required.

OECC/PS2016

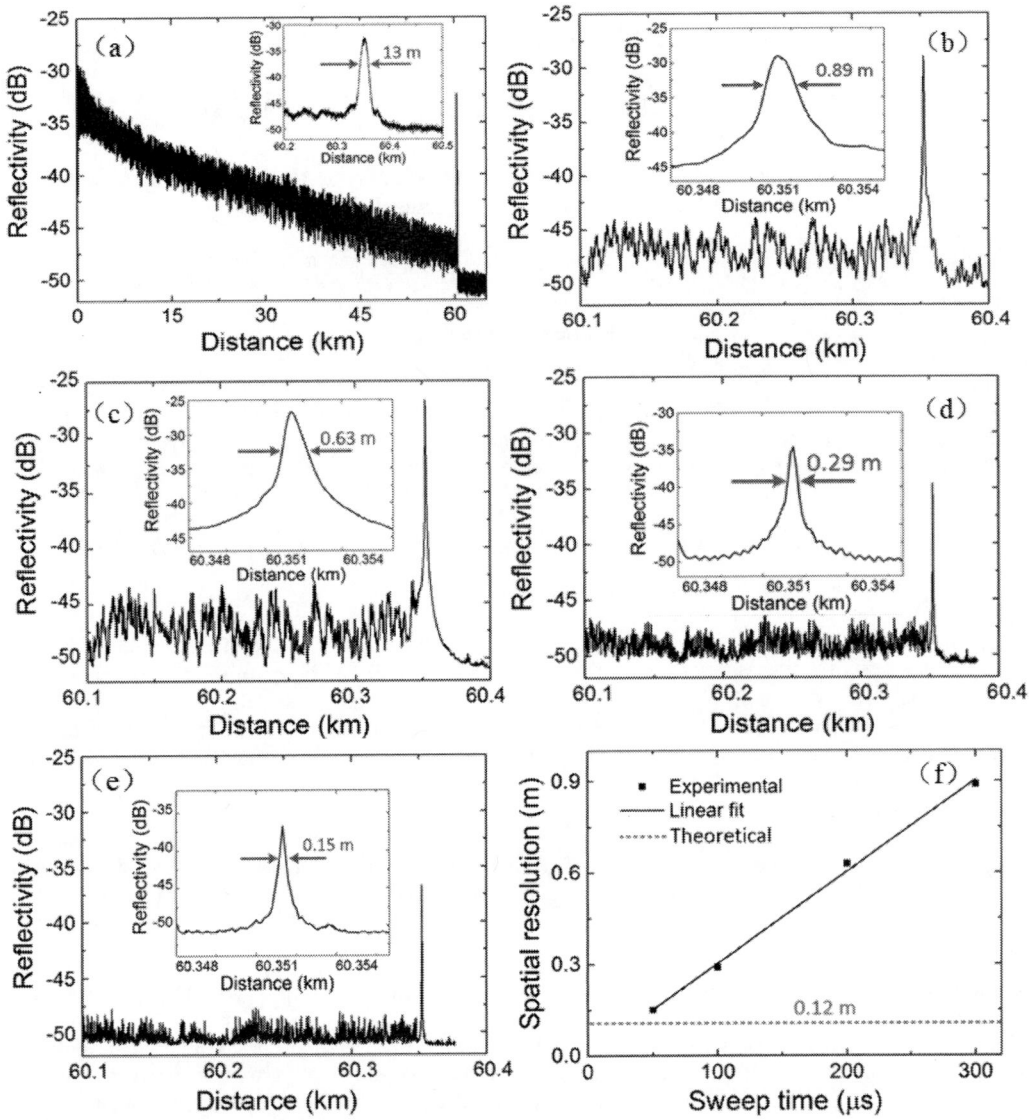

Fig. 2. Measured reflectometric traces at different frequency sweep time ((a):5 ms, (b): 300 μs, (c): 200 μs, (d): 100 μs, (e): 50 μs) and (f) the relationship between spatial resolution and sweep time.

ACKNOWLEDGMENT

This work was supported by the National Natural Science Foundation of China under Grant 61575001, 61275097, 61307106, 61307107, Shanghai STCSM Scientific and Technological Innovation Project under Grant 15511105401.

REFERENCES

[1] R. Passy, N. Gisin, J. P. von der Weid, and H. H. Gilgen, "Experimental and theoretical investigation of coherent OFDR with semiconductor laser sources," J. Lightwave Technol., vol. 12, pp. 1622–1630, Sep.1994.

[2] B. J. Soller, D. K. Gifford, M. S. Wolfe, and M. E. Froggatt, "High resolution optical frequency domain reflectometry for characterization of components and assemblies," Opt. Express, vol. 13, pp. 666–674, Jan. 2005.

[3] S. Venkatesh and W.V.Sorin, "Phase noise considerations in coherent optical FMCW reflectometry," J. Lightwave Technol., vol. 11, pp. 1694-1700, 1993.

[4] F. Ito and X. Fan, "Long-range coherent OFDR with light source phase noise compensation," J. Lightwave Technol., vol. 30, pp. 1015-1024, 2012.

[5] Q. Liu, X. Fan, and Z. He, "Time-gated digital optical frequency domain reflectometry with 1.6-m spatial resolution over entire 110-km range," Opt. Express, vol. 23, pp. 25988-25995, 2015.

[6] S. Wang, X. Fan, Q. Liu, and Z, He. "Distributed fiber-optic vibration sensing based on phase extraction from time-gated digital OFDR," Opt. Express, vol. 23, pp. 33301-33309, 2015.

High-Speed Germanium-Based Waveguide Electro-Absorption Modulator

P. De Heyn[1], S. A. Srinivasan[2], P. Verheyen[1], R. Loo[1], I. De Wolf[1,3], S. Balakrishnan[1], G. Lepage[1], D. Van Thourhout[2], M. Pantouvaki[1], P. Absil[1], and J. Van Campenhout[1]

[1]Imec, Kapeldreef 75, Leuven B-3001, Belgium
[2]Photonics Research Group, Dept. of Information Technology, Ghent University - Imec, St. Pietersnieuwstr. 41, 9000 Ghent, BE
[3]also at Dept. Materials Science (MTM), Faculty Engineering, KU Leuven, Belgium

Peter.DeHeyn@imec.be

Abstract: Germanium-based waveguide electro-absorption modulators are reported in C- and L-band wavelength operation at 56Gb/s (NRZ-OOK) with extinction ratio of >3dB at 2V peak-to-peak and insertion loss below 5dB. The device is implemented in a fully integrated Si photonics platform on 200mm silicon-on-insulator wafer with 220nm top Si thickness. Wafer-scale performance data confirms the manufacturability of the device. This demonstrates the great potential for realizing high-density and low-power silicon photonic transceivers for short range interconnect.
Keywords: Electro-absorption modulators, optoelectronics, optical interconnects, waveguide modulators

I. INTRODUCTION

Silicon Photonics is considered to be a key potential technology to enable the scaling of bandwidth and power density beyond the bottleneck of electrical interconnects [1]. As an important building block of the Silicon Photonics circuitry, modulators are required to have low power consumption, high modulation speed, small footprint, large optical bandwidth and robust thermal sensitivity [2]. In this paper, we review the waveguide-integrated Ge & GeSi electro-absorption modulator (EAM) operating in respectively L- and C-band showing wide open eye diagrams at 56Gb/s nonreturn-to-zero on-off keying (NRZ-OOK)[3,4]. The optimum operation point for Ge EAM devices is around 1610nm however using Ge with ~0.8% Si incorporation, the bandgap is sufficiently shifted for operation at 1550nm wavelength. The EAM devices are integrated in imec's iSiPP25G silicon photonics platform on 200mm silicon-on-insulator (SOI) wafers with 220nm top Si thickness. The Ge-based EAMs have higher optical bandwidth than Si based ring modulators and lower footprint and power consumption than the Mach–Zehnder based modulators [5].

II. DESIGN AND FABRICATION

The working principle of bulk EAMs is the Franz-Keldysh (FK) effect according to which the absorption coefficient near the band edge is increased due to band-tilting caused by an applied electrical field. To enable the C-band modulation around 1550nm wavelength at room temperature, ~0.8% of Si is incorporated into Ge to shift the bandgap from 0.785eV (1580nm) to 0.808eV (1535nm). The GeSi material was selectively grown in a recessed Si region with a gas mixture of GeH_4, diluted SiH_4 and H_2 using reduced pressure chemical vapor deposition (ASM EPSILON2000). High temperature anneal was applied to the wafer to reduce the threading dislocation density and afterwards, the over-grown GeSi was planarized by chemical mechanical polishing. The modulators were fabricated in imec's silicon photonics platform with other passive and active components.

The EAM is designed with poly-Si tapers to adiabatically butt couple the light to the active Ge or GeSi region with a width of 0.6μm and a length of 40μm, as depicted in Fig.1(a). A lateral p-i-n diode design is used to generate a high electrical field on GeSi waveguide, as shown in Fig. 1(b). Fig.1(c) shows the simulated electrical field distribution at -2V and the strength in the device center is up to 60kV/cm. The thickness of the epitaxial Ge is 400nm which is slightly reduced to 300nm for the epitaxial GeSi thickness to increase the electric field even further, as well as to lower the excess loss. In both cases, the light is clearly confined in the Ge/GeSi region with the fundamental mode shown in Fig.1(d). The absorption coefficient changes with the applied electrical field, clearly illustrating the FK effect near the band edge in 400nm thick Ge (Fig.1(e)) and 300nm thick GeSi (Fig.1(f)) EAM devices. The absorption spectra for different bias voltages cross around 1580nm and 1535nm, corresponding to the HH-Γ direct band gap for respectively epitaxial Ge and GeSi on Si. In both cases, a high quality of epitaxial grown material is obtained with dark current below 100nA at -2V. In Fig.1(g), a microscopic picture of the actual fabricated device is shown.

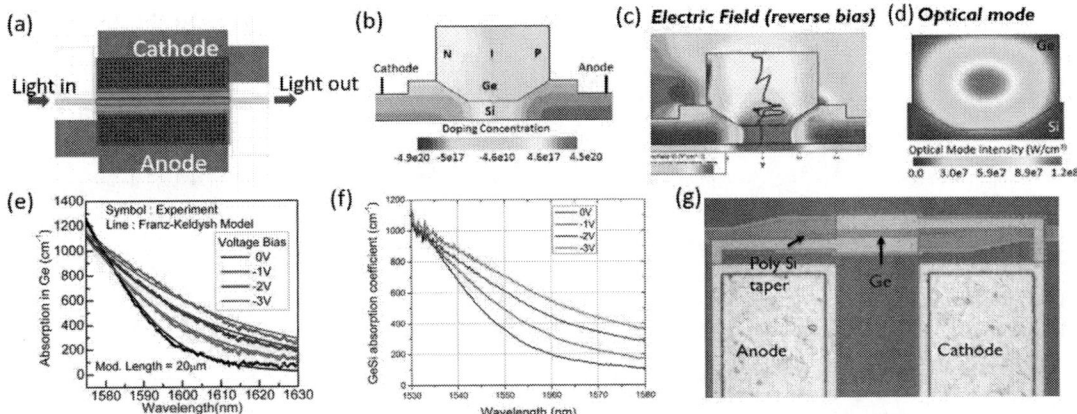

Figure 1: (a) schematic top overview (b) cross section p-i-n doping concentration (c) electric field at reverse bias (d) the main excited optical mode (e,f) measured and modelled FK effect in L- and C-band (g) microscopic picture of the fabricated device

III. STATIC MODULATOR PERFORMANCE

Fig. 2(a) shows the measured insertion loss (IL) and extinction ratio (ER) at different voltage swings measured on the GeSi EAM device in function of wavelength. The ER peaks around 1550nm. For a 2Vpp, the insertion loss is 4.8dB and the extinction ratio is 4.2dB at 1560nm. The extracted insertion loss includes the coupling loss between the Si waveguide and the GeSi modulator, the indirect band gap absorption, and FK absorption near the direct band gap due to the existing electrical field at 0V. The average ER is 4.2dB with a standard deviation of 0.3dB in 18 measured devices across a 200mm SOI wafer and the average IL is 4.4dB with a standard deviation of 0.6dB at 1560nm. Fig. 2(b) shows the extracted figure of merit (FOM) spectra at different voltage swings. The definition of FOM is ER divided by IL. The FOM peaks around 1560~1570nm. For a 2V swing, the FOM is 1.14 at 1560nm and therefore, the optimum operation wavelength for the EAM device is 1560nm. Another important device parameter is the Link Power Penalty (LPP) to quantify the operation bandwidth, defined in Ref. [4]. Fig. 2(c) shows the LPP spectra at different voltage swings. The minimum LPP value is 8.5dB and the 1dB optical bandwidth of 30nm at a 2Vpp.

The performance of the Ge EAM is very similar to the GeSi EAM with a 1dB LPP bandwidth at 2Vpp of >22nm and static ER and IL of respectively 4.6dB and 4.9dB.

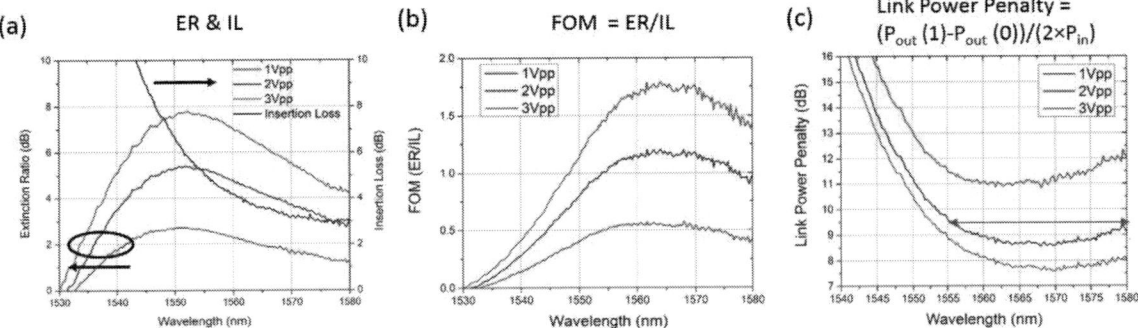

Figure 2: C-band GeSi static modulator performance in function of wavelength the GeSi EAM with in (a) the extinction ratio and insertion loss (b) the figure of merit and (c) the link power penalty. The Ge EAM has very similar performance in L-band [3].

IV. DYNAMIC MODULATOR PERFORMANCE

The 3dB bandwidth of the modulator was extracted from the electro-optic S21 measurement using a 50GHz lightwave component analyzer (LCA) with an input optical power of 3dBm. The bandwidth is greater than 50GHz for reverse bias above 1V as shown in Fig. 3(a). Since the FK effect is a sub-picosecond effect [5], the speed of the modulator is determined by the RC constant of the device. The junction capacitances is estimated to 13.8fF and 7.95fF at -1V and -2V, along with series resistances of 220Ω and 320Ω respectively. They are extracted with the help of the equivalent circuit model. These values suggest a RC limited bandwidth beyond 50GHz and further investigation is needed to explore the bandwidth limitation of these ultra-fast modulators.

(a)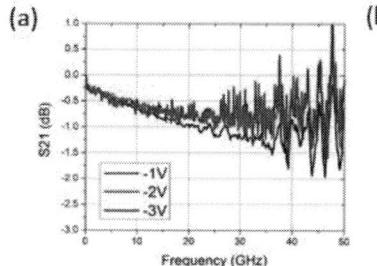
(b)
(c)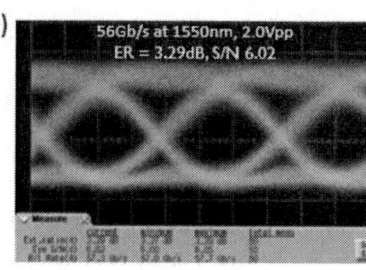

Figure 3: Dynamic modulator performance with (a) the EO - S21 measurements with varying bias at 1550nm for GeSi EAM and (b,c) 56Gb/s operation at 2Vpp and bias -1V for respectively the Ge and GeSi EAM

TABLE I: COMPARISON OF THIS WORK (*) WITH OTHER COMPETING DEVICE TECHNOLOGIES

Modulator type	Ref.	Footprint [μm^2]	Wavelength [nm]	Voltage swing [V]	Optical range [nm]	Static ER [dB]	IL [dB]	Power consumption static [mW] dynamic [fJ/bit]	3dB bandwidth GHz	Max. bit rate Gb/s
* GeSi FK	[4]	~40x10	1550	2.0	30	4.2	4.4	1.7 & 13.8	>50	50
GeSi FK	[5]	~50x10	1540	3.0	>40	5.9	4.8	11.3 & 147	38	28
* Ge Fk	[3]	~40x10	1615	2.0	>22.5	4.6	4.9	1.2 & 12.8	>50	56
Si MZM	[7]	~3000x500	1300	1.5	>80	3.4	7.1	20 & 450	30	50
Si ring	[8]	~10x10	1550	0.5	<0.1	6.4	1.2	<0.01 & 1	21	44
Si ring	[9]	~5x5	1566	2.5	<0.1	4.5	2.4	~ & 45	40	56
III-V on Si	[10]	>100x350	1300	2.2	>30	>10	4.8	6.2 & 484	74	50

V. CONCLUSIONS

We have demonstrated a 56Gb/s Ge and GeSi waveguide electro absorption modulator integrated in a 220nm SOI photonics platform for operation in respectively L- and C-band. The modulator has a 3-dB bandwidth modulation greater than 50GHz and a junction capacitance of 13.8fF at -1V. For a 2V swing, the static extinction ratio is 4.2±0.3dB, insertion loss is 4.4±0.6dB and link power penalty is 8.5dB. While operating at 56Gb/s data rate, we measure a >3.0dB dynamic extinction ratio at 2Vpp at 1610nm and 1550nm.

ACKNOWLEDGMENT

This work has been carried out as part of imec's industry-affiliation program on Optical I/O. The authors acknowledge imec's 200mm p-line for contributions to the device fabrication and imec's PDK team for mask preparation and tape-out. Device design was performed in Sentaurus TCAD, provided by Synopsys. Device layout was performed in IPKISS, provided by Luceda Photonics. The authors would like to thank Antoine Pacco and Veerle Simons for the help with GeSi material development.

REFERENCES

[1] D.A.B. Miller, "Device Requirements for Optical Interconnects to Silicon Chips", *Proc. IEEE*, vol. 97, no. 7, pp. 1166–1185, Jul. 2009.

[2] G.T. Reed, et al. "Silicon Optical Modulators", *Nat. Photon*, 4, 518 - 526 (2010).

[3] A. Srinivasan, et al. "56Gb/s Germanium Waveguide Electro-Absorption Modulator", *Lightwave Technology, Journal of*, vol.PP, no.99, pp.1-1.

[4] A. Srinivasan. et. al. "50Gb/s Germanium Waveguide Electro-Absorption Modulator", Tu3D.7, OFC 2016

[5] D. Feng, et. al. "High-speed GeSi electroabsorption modulator on the SOI waveguide platform," IEEE J. Sel. Topics Quantum Electron., vol. 19, no. 6, pp. 64–73, Nov./Dec. 2013.

[6] J.F. Lampin, et al. "Detection of picosecond electrical pulses using the intrinsic Franz–Keldysh effect", *APL* 78, 4103–4105 (2001).

[7] M. Streshinsky, et al. "Low power 50 Gb/s silicon traveling wave Mach-Zehnder modulator near 1300 nm", *Opt. Express* 21, 30350-30357, 2013.

[8] E. Timurdogan, et al. "An ultralow power athermal silicon modulator", *Nature Communications*, 5, 4008, 2014.

[9] M. Pantouvaki, et al. "56Gb/s Ring Modulator on a 300mm Silicon Photonics Platform", TH2.4.4, ECOC 2015

[10] Y. Tang, et al. "Over 67 GHz bandwidth hybrid silicon electroabsorption modulator with asymmetric segmented electrode for 1.3 μm transmission", *Opt. Express* 20, 11529-11535 (2012).

Low-Voltage Carrier-Depletion Silicon Mach-Zehnder Modulator at High Temperatures without Thermo-Electric Cooling

Norihiro Ishikura,[1] Kazuhiro Goi,[1] Hiroki Ishihara,[1] Shinichi Sakamoto,[1] Kensuke Ogawa,[1] Tsung-Yang Liow,[2] Xiaoguang Tu,[2] Guo-Qiang Lo[2] and Dim-Lee Kwong[2]

[1]Advanced Technology Laboratory, Fujikura Ltd., 1440 Mutsuzaki, Sakura, Chiba 285-8550, Japan
[2]Institute of Microelectronics, 11 Science Park Load, Singapore Science Park II, Singapore 117685, Singapore
Author e-mail address: norihiro.ishikura@jp.fujikura.com

Abstract: *High-extinction-ratio 10-Gb/s modulation in carrier-depletion silicon Mach-Zehnder modulator is demonstrated with RF amplitude as low as 3.6 V_{pp} at temperatures up to 130 °C without thermo-electric cooling. Algebraic representation of DC optical characteristics is presented.*
Keywords: *Optical modulator, Mach-Zehnder interferometer, Carrier depletion, Silicon waveguide*

I. INTRODUCTION

Silicon optical modulators play significant roles in small-footprint low-cost optical transceivers in datacenter, datacom and telecom networks. Carrier-depletion silicon Mach-Zehnder modulators (MZMs), which allow high-speed high-contrast optical modulation in broad spectral ranges, are suitable to small-footprint optical transceivers deployed in the high-capacity optical networks. Low-voltage operation of carrier-depletion silicon MZMs using electrical drivers of low power consumption is crucial for the realization of energy efficiency in the high-capacity optical networks. Low-voltage Si MZMs were reported using carrier-depletion PN-junction rib-waveguide phase shifters of 6-mm length with 3.1-V V_π and 1.86-V· cm $V_\pi L_\pi$ at 1550-nm wavelength, carrier-accumulation phase shifters of 0.4-mm length with 0.2-V · cm $V_\pi L_\pi$ at 1310-nm wavelength and carrier-injection phase shifters of 0.25-mm length with 0.29-V $V_\pi L_\pi$ with pre-emphasis drive at 1550-nm wavelength at room temperatures [1-3].

Optical modulation free from thermo-electric cooler (TEC) is essential to achieve further energy efficiency in the high-capacity optical networks. Carrier-plasma dispersion, which induces refractive-index modulation in silicon, has the significant advantage in TEC-free operation, because the mechanism is independent of spectral response near the band edge of silicon, thereby insensitive to thermal effects such as temperature dependence of the band-edge wavelength. However, only a limited number of reports have focused on high-temperature operation of Si MZMs. Operation of a carrier-depletion PN-junction rib-waveguide Si MZM with 1.5-V· cm $V_\pi L_\pi$ in 10- and 25-Gb/s nonreturn-to-zero on-off keying (NRZ-OOK) was studied at temperatures between 25 and 55 °C [4]. Operation of Si MZMs at much higher temperatures are required for the optical-network applications. Low-voltage TEC-free operation of carrier-depletion Si MZM having vertical PN-junction rib-waveguide phase shifters as short as 3 mm in length is demonstrated in this paper at temperatures between 26 °C and 150 °C in DC and high-speed characterization. In DC spectral characterization, V_π lower than 2.7 V is confirmed. This corresponds to 0.81-V· cm in $V_\pi L_\pi$. Extinction ratio (ER) higher than 10 dB is obtained in eye-diagrams of 10-Gb/s NRZ-OOK at temperatures up to 130 °C.

II. VERTICAL PN-JUNCTION PHASE SHIFTER

Vertical PN-junction silicon rib-waveguide phase shifter has been studied for optical modulators [5,6]. Efficient mode-field overlap with carrier-depletion region in vertical PN junction allow low V_π in optical modulation. Illustrated profile of vertical PN-junction silicon rib-waveguide phase shifter is presented with measured PN-junction and phase-shifter images in Fig. 1(a). Vertical PN-junction silicon rib-waveguide phase shifter was designed with cross-section of 500-nm width, 220-nm rib height and 95-nm slab height. A fabricated phase shifter was inspected in transmission electron microscopy (TEM), and has rib and slab heights close to those designed within 5-nm deviation as indicated in the TEM image. Microscopic distribution of p-type and n-type carriers in a fabricated vertical PN-junction phase shifter was visualized in the color-contrast image taken by scanning capacitance microscopy (SCM) [7]. Dopant concentration in the P and N areas in design is about 1×10^{18} cm³. Designed PN-junction profile was produced in the fabrication process as confirmed in the SCM image.

Dopant concentration and profile have been optimized for high-speed operation up to 32-Gbaud symbol rate in various modulation formats for intensity modulation and phase modulation. Traveling-wave vertical PN-junction Si MZMs having 3-mm vertical PN-junction phase shifters in each arm of single MZ interferometer has been characterized for on-off keying (OOK) operation at 10 Gb/s in this paper. Aluminum traveling-wave electrodes of coplanar waveguides were disposed on top of an asymmetric MZ interferometer (AMZI) waveguide of the Si MZM. For high-temperature operation, a Si MZM chip was attached on the top surface of a metal submount with a thermistor.

Temperature of the submount surface near the Si MZM chip was measured with the thermistor. A heater was attached on the back surface of the submount and submount temperature was increased up to 150 ˚C by injecting heater current. Temperature increase in the phase shifters was independently monitored as exponential increase in dark current through the reverse-biased PN junction in each of the phase shifters as described in Sect. V. Electro-optic bandwidth at 26 ˚C was 12 GHz and 14 GHz at DC reverse bias of 3 V and 5 V, respectively.

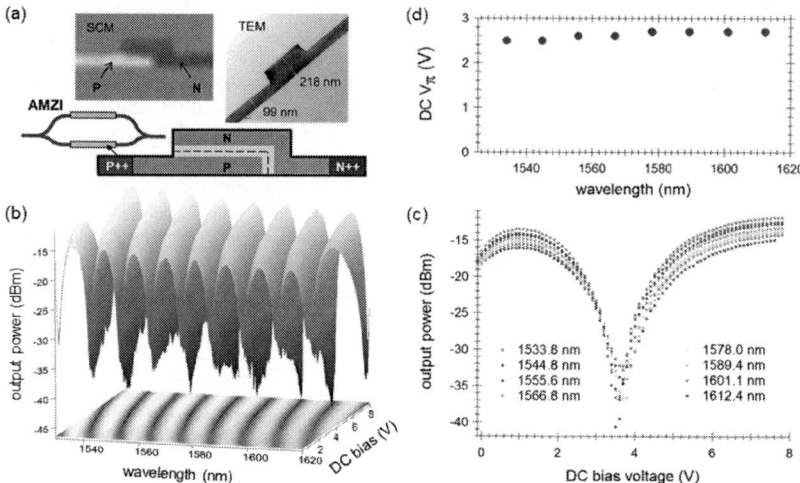

Fig. 1. Vertical PN-junction silicon rib-waveguide phase shifter and DC optical characteristics of Si MZM with the phase shifters.

III. FUNDAMENTAL SPECTRAL CHARACTERISTICS

Optical transmission spectra of the vertical PN-junction Si MZM at 26 ˚C under DC reverse bias from 0 V to 8 V are shown in Fig. 1(b). Negative DC voltage was applied to P_{++} contact with N_{++} grounded for carrier depletion. Free spectral range of the interferometric fringes yields very weak spectral dependence over C and L bands in the entire reverse-bias range. DC V_{π} has been measured in DC bias dependences of the output power in Fig. 1(c), which were obtained from slice cuts of the optical transmission spectra at respective wavelengths. As plotted in Fig. 1(d), DC V_{π} is 2.5-2.6 V in C band and 2.7 V in L band, respectively, thus DC V_{π} lower than 3 V achieved over C and L bands.

IV. COMPUTATIONAL ANALYSIS OF VERTICAL PN-JUNCTION PHASE SHIFTER

Computational analysis of the vertical PN-junction phase shifter has been performed on the basis of the simulation method described in the literature [8]. The results of the simulation are presented in Fig. 2(a). Electrons (e) and holes (h) are almost fully depleted in the central core at 3 V: voltage as low as 3 V is sufficient for driving the Si MZM. DC bias voltage dependences of computed phase shift and carrier-induced optical loss of the phase shifter are in good agreement with the measured DC optical characteristics.

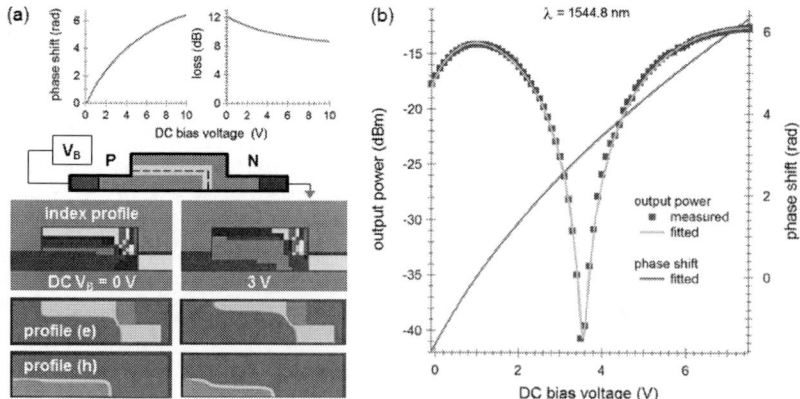

Fig. 2. (a) Computed carrier profiles and phase-shifter characteristics (b) curve fitting to the measured DC optical characteristics.

The computed phase shift and phase-shifter optical loss are represented by algebraic formulas. The computed voltage dependences of phase shift and phase-shifter optical loss are represented as a square-root function, $\sqrt{V_{\text{bias}}}$ and an inverse polynomial, $1/(V_{bias} - V_0)$, respectively. Numerical characteristics based on the algebraic formulas reproduces perfectly the measured DC bias dependence at a wavelength of 1544.8 nm as shown in Fig. 2(b), for instance, with phase-shift and power-transmittance offsets as fitting parameters. The algebraic representation will serve as a library module of process design kit for the vertical PN-junction phase shifter.

V. OPTICAL MODULATION AT HIGH TEMPERATURES WITHOUT THERMO-ELECTRIC COOLING

Almost constant DC V_π at 26-150 °C has been confirmed as presented in Fig. 3(a). TEC-free operation up to 150 °C has been demonstrated in 10-Gb/s OOK with ER higher than 10 dB using the apparatus depicted in Fig. 3(b) and eye diagrams are shown in Fig. 3(c). RF voltage applied to the phase shifter was as low as 3.3 V_{PP}. Increase in phase-shifter temperature was separately verified as exponential increase in dark current in the phase shifter more than a factor of 2×10^2 as the inserted graph in Fig. 3(a). The dark current reflects the PN-junction temperature since it is limited by the transport of intrinsic carriers in thermodynamic equilibrium in the reverse-biased PN junction [9]. Therefore, phase-shifter temperature was hold within 10-°C difference from the submount temperature.

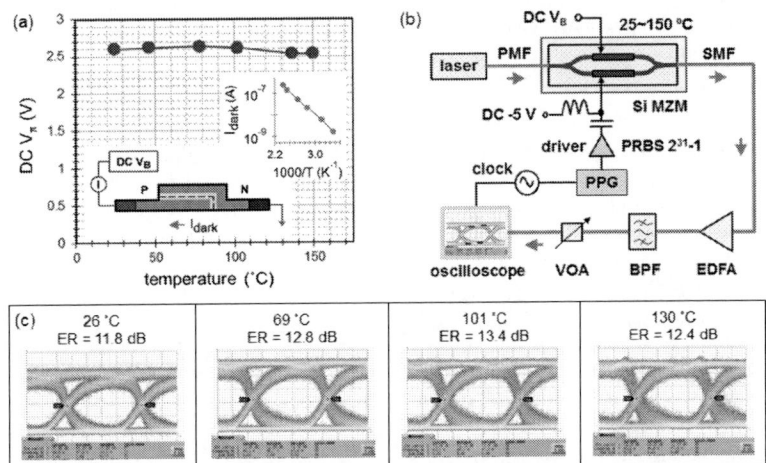

Fig. 3. DC V_π and 10-Gb/s eye-diagram characteristics at high temperatures.

VI. CONCLUSION

TEC-free low-voltage high-contrast modulation has been demonstrated in 10-Gb/s OOK using Si MZM having the vertical PN-junction phase shifters. The fundamental characteristics were well reproduced by an algebraic model, thereby the phase shifter will be useful as a potential library module for design and fabrication on silicon-photonic platform.

REFERENCES

[1] P. Dong, L. Chen, and Y.-K. Chen, "High-speed low-voltage single-drive push-pull silicon Mach-Zehnder modulators," Opt. Express, vol. 20, pp. 6163-6169, March 2012.

[2] M. Webster, C. Appel, P. Gothoskar, S. Sunder, B. Dama, and K. Shastri, "Silicon photonic modulator based on a MOS-capacitor and a CMOS driver," 2014 IEEE Compound Semiconductor Integrated Circuit Symposium, pp. 86-89, Oct. 2014.

[3] S. Akiyama, T. Baba, M. Imai, T. Akagawa, M. Takahashi, N. Hirayama, H. Takahashi, Y. Noguchi, H. Okayama, T. Horikawa, and T. Usuki, "12.5-Gb/s operation with 0.29-V·cm $V_\pi L$ using silicon Mach-Zehnder modulator based-on forward-biased pin diode," Opt. Express vol. 20, pp. 2911-2923, Jan. 2012.

[4] H.X. Yi, T.T. Li, J.L. Zhang, X.J. Wang, and Z. Zhou, "Temperature-independent broadband silicon modulator," Opt. Commun. vol. 340, pp. 107-109, 2015.

[5] G. T. Reed, G. Z. Mashanovich, F. Y. Gardes, M. Nedeljkovic, Y. Hu, D. J. Thomson, K. Li, P. R. Wilson, S.-W. Chen, and S. S. Hsu, "Recent breakthroughs in carrier depletion based silicon optical modulators," Nanophotonics vol. 3, pp. 229-245, Dec. 2013.

[6] J. C. Rosenberg, W. M. J. Green, S. Assefa, D. M. Gill, T. Barwicz, M. Yang, S. M. Shank, and Y. A. Vlasov, "A 25 Gbps silicon microring modulator based on an interleaved junction," Opt. Express vol. 20, pp. 26411-26423, Nov. 2012.

[7] K. Ogawa, "Silicon-based phase shifters for high figure of merit in optical modulation," Photonics West 2016, SPIE, San Francisco, USA, 9752-1, Feb, 2016.

[8] K. Ogawa, H. Ishihara, K. Goi, Y. Mashiko, S. T. Lim, M. J. Sun, S. Seah, C. E. Png, T.-Y. Liow, X. Tu, G.-Q. Lo, and D.-L. Kwong, "Fundamental characteristics and high-speed applications of carrier-depletion silicon Mach-Zehnder modulators," IEICE Electron. Express vol. 11, pp. 20142010, Dec. 2014.

[9] S. M. Sze, Physics of semiconductor devices, J. Wiley, 2nd ed., 1981, chaps. 1 & 2.

WD1-3

OECC/PS2016

High-efficiency Silicon Optical Modulator Using a SiN-strip Loaded Waveguide on The Photonic SOI Platform

Guangwei Cong*, Yuriko Maegami, Morifumi Ohno, Makoto Okano, Koji Yamada

Silicon Photonics Group, Electronics and Photonics Research Institute
National Inst. of Adv. Industrial Sci. and Tech. (AIST), Tsukuba, Ibaraki 305-8568, Japan
*E-mail address: gw-cong@aist.go.jp

Abstract: *We proposed a novel silicon optical modulator using the SiN-strip loaded waveguide instead of the shallow-etched rib waveguide. This modulator can show a ~2-times maximum efficiency enhancement than the conventional rib-waveguide based lateral PN modulator.*
Keywords: *silicon optical modulator, strip loaded waveguide, SOI*

I. INTRODUCTION

Then silicon optical modulator based on carrier depletion is one of the core devices in silicon photonics [1]. The low-voltage silicon modulator is the key in order to achieve the small footprint, low power driving, and high packaging density for the transmitter. Therefore, with the purpose to improve the performance of silicon modulators, great research efforts have been made to optimize the PN junction formation by varying the doping conditions [2,3]. However, from the waveguide point of view, the depletion type silicon modulators in these optimizations were almost all involved in the formation of shallow-etched rib structures in spite of different SOI (silicon on insulator) and slab thicknesses and lateral or vertical PN junctions [2-6]. The silicon modulators based on shallow-etched rib waveguides were seen a limited space for performance enhancement due to the following two technical difficulties in the fabrication. On one side, the etching depth that determines the rib height and the slab thickness cannot be precisely controlled even in the advanced CMOS foundry because of the middle etching stop in the SOI layer. Thus, the overlap between the optical mode field and PN junction always deviates from the design, resulting in the efficiency degradation. In addition, the concomitant waveguide parameter fluctuation from wafer to wafer greatly influences the property stability and uniformity. On the other, when the ion implantation under a certain amount of energy is done to the rib waveguide where a thickness difference between the slab and the core exists, the optimal doping profiles in the core and slab are difficult to be achieved simultaneously. In other words, both the optical mode field and the doping profiles are bound to one condition, the rib waveguide, which causes great difficulty in the device optimization. To overcome above difficulties, we propose a novel silicon modulator not based on a shallow-etched rib waveguide, but on a strip-loaded waveguide, without any etching to the SOI layer. In this proposed modulator, the mode field is controlled by the loaded strip and meanwhile the optimal doping profiles can be easily achieved because the SOI layer is kept flat as original so that the waveguide parameters and doping conditions can be separately optimized. The strip can use any compatible materials with suitable refractive indices, giving great flexibility to modulator fabrication. To form the strip by etching is also much easier due to the existing ending point between different materials.

In this study, we demonstrated a silicon optical modulator using a SiN-strip loaded waveguide on the photonic SOI platform and performed device simulation using a commercial opto-electronic simulation environment [7]. This modulator shows an about 2-times maximum efficiency enhancement compared to the lateral PN modulator based on the conventional shallow-etched rib waveguide. This strip-loaded type waveguide is beneficial for achieving the high-efficiency silicon optical modulator on the photonic SOI platform.

II. DEVICE STRUCTURE

The proposed modulator structure is shown in Fig. 1(a). The main feature is that for the phase shifter, no shallow-etched rib waveguide is formed in the SOI layer, but a SiN-strip is loaded on the top of SOI layer so that the optical confinement can be spontaneously induced in silicon below. In other words, the SOI layer in the phase shifter is kept as flat as it is, without experiencing any middle-stop etching. This SiN-strip loaded waveguide can be connected to the silicon channel waveguides using tapers that can be fabricated by only one-time full etching. For the fabrication of the phase shifter, ion implantation should be done prior to the SiN strip. Since the SOI layer is flat, uniform carrier distribution can be realized in top and bottom halves of the silicon layer to form a vertical PN diode, as shown in Fig. 1(a). In this study, a 220-nm SOI was used and a 10-nm-thick oxide was introduced between SiN and silicon, which is not shown in Fig. 1(a), and all calculation was done at the wavelength of 1.55 μm for the TE mode. As for the intrinsic waveguide characteristics without carrier, Fig. 1(b) and 1(c) show the optical mode field (SiN width = 1.5 μm, SiN height = 0.5 μm) and the SiN width dependent effective index of the fundamental mode, respectively. This waveguide remains the single mode for the SiN width from 0.8 to 2 μm. Two SiN heights of 0.5 and 0.8 μm show the same effective index, which indicates a very large SiN thickness allowance when the height is larger than 0.5 μm and thus to fabrication this modulator does not require the strict control on the SiN thickness.

In the device simulation, the SiN width and height was fixed to be 1.5 and 0.5 μm, respectively. The densities of N+ and P+ were 1.5×10^{18} cm^{-3}, and those of N++ and P++ were 1.0×10^{20} cm^{-3}. Highly-doped regions for contact electrodes were located 2-μm apart from the center of the waveguide. All doping region was assumed uniform. The length of phase shifter was set to 5 mm. For comparison, a conventional lateral PN modulator based on a shallow-etched rib waveguide (etching depth of silicon set to 110 nm) was also simulated under the same condition. Such lateral PN modulators were widely used as the phase shifters in various silicon modulators [8].

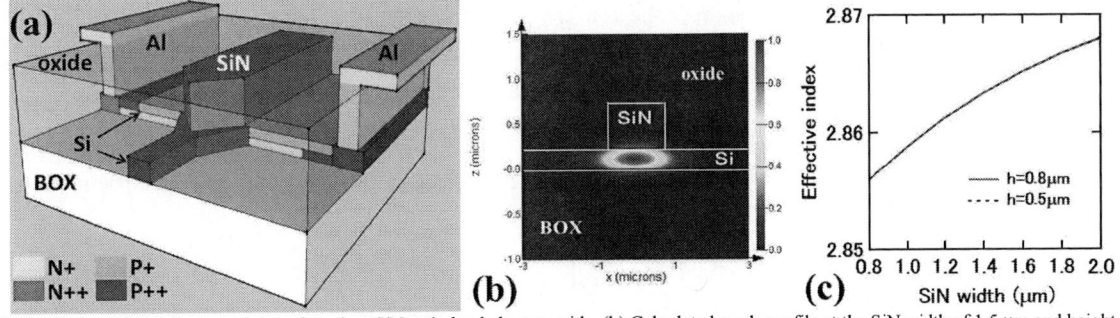

Fig. 1. (a) Schematic silicon modulator based on SiN-strip loaded waveguide. (b) Calculated mode profile at the SiN width of 1.5 μm and height of 0.5 μm. (c) SiN width dependence of effective index for the SiN height of 0.5 and 0.8 μm. Si: silicon. BOX: buried oxide. h: SiN heigth.

III. MODULATION EFFICIENCY

The modulation efficiency of carrier depletion based silicon modulators is usually described by the figure of merit, $V_\pi L$, where V_π and L denotes the voltage required to achieve a π phase shift and the length of phase shifter, respectively [5]. The phase shift is achieved by depleting the majority carrier in PN junction under a reverse bias and the efficiency is determined by the overlap between optical mode field and depletion region that can be tuned by the reverse bias. Smaller $V_\pi L$ means higher modulation efficiency of a phase shifter. Thus, the modulator with a smaller $V_\pi L$ will have a lower driving voltage and power. Given a length of phase shifter, the V_π can be figured out from the phase shift in relation to the applied reverse voltage. Fig. 2(a) compares the voltage dependent phase shift by taking -2.5 V as the DC bias between the conventional lateral PN modulator and the proposed one. Here, we varied the overlap distance between N+ and P+ regions (p-n overlap distance) from 1 to 3 μm. The V_π, the voltage difference between π/2 and -π/2, is about 1.28 for the proposed modulator with the p-n overlap distance of 3 μm; while the V_π is about 2.74 V for the conventional modulator. For the length of 5 mm, the corresponding $V_\pi L$ are 0.64 and 1.37 V·cm, indicating an about 2-times efficiency enhancement in the proposed modulator. The $V_\pi L$ increases to 0.95 V·cm when the p-n overlap distance is reduced to 1 μm due to the overlap reduction between optical mode and depletion region, which still denotes a ~1.4-times efficiency improvement. The DC bias dependent modulation efficiency is shown in Fig. 2(b). When the reverse bias increases, the $V_\pi L$ also increases for all modulators, however the efficiency enhancement at any overlap distances obtained in the proposed modulator remains effective within a wide DC bias range.

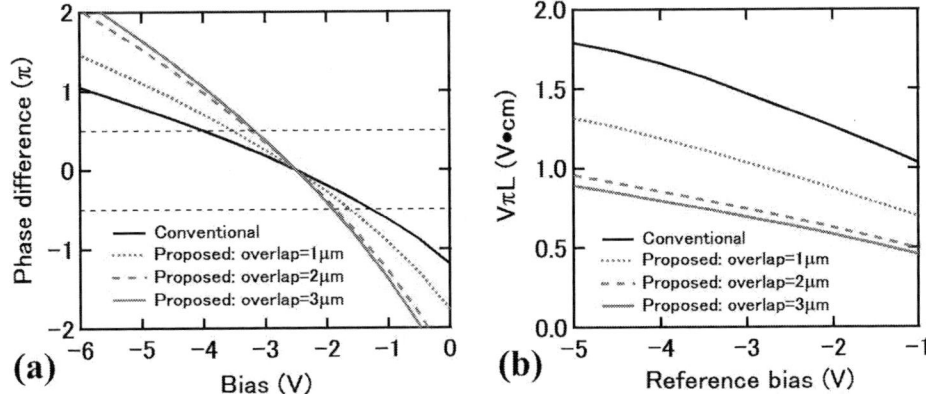

Fig. 2. (a) Voltage induced phase difference referred to -2.5 V. (b) Modulation efficiency ($V_\pi L$) at different bias voltages. Conventional: the conventional lateral PN modulator. Proposed: the proposed modulator in Fig. 1(a).

1111

IV. LOSS AND RESPONSE

The propagation loss in the phase shifter is also determined by the overlap between optical mode field and depletion region as for the same carrier density. The proposed modulator will have a lower loss since it has a larger optical-mode-depletion-region overlap which is also the origin of efficiency enhancement abovementioned. Fig. 3(a) shows the calculated propagation loss at different DC biases. At -2.5 V, the proposed modulator with the p-n overlap distance of 3 μm (maximum loss case) has a propagation loss of a ~31 dB/cm, which is about an 8 dB/cm less than the conventional one. For the conventional one, our calculated propagation loss (~39 dB/cm at 1.5×10^{18} cm^{-3}) well matches the reported result (12 dB/cm at 5×10^{17} cm^{-3}) in Ref. [5] if normalized to the same carrier density because the free carrier induced absorption is proportional to the carrier density [9]. Decreasing the p-n overlap distance will decrease the loss, which is much obvious on the low voltage side than on the high voltage side. In addition, with increasing the reverse bias, the propagation loss decreases with a larger speed in the proposed modulator. Therefore, the loss will be further decreased considering the swing driving involved in the high bias region.

The transient response of the phase was calculated for the p-n overlap distance of 1 μm using a step voltage pulse, as shown in Fig. 3(b). The voltage pulse has a width of 10 ps and the peak values of -1.8 and -3.3 V. The 10-90% response time of the phase is nearly 100 ps, which is possible for a >10 Gbps signal modulation. It is necessary to mention that the tradeoff among the efficiency, speed, and modulation ratio is unavoidable in silicon modulator [5,6] and for such a tradeoff, this proposed modulator scheme can give the maximum tunable range for further optimizing the doing density, operation DC bias, and waveguide parameters.

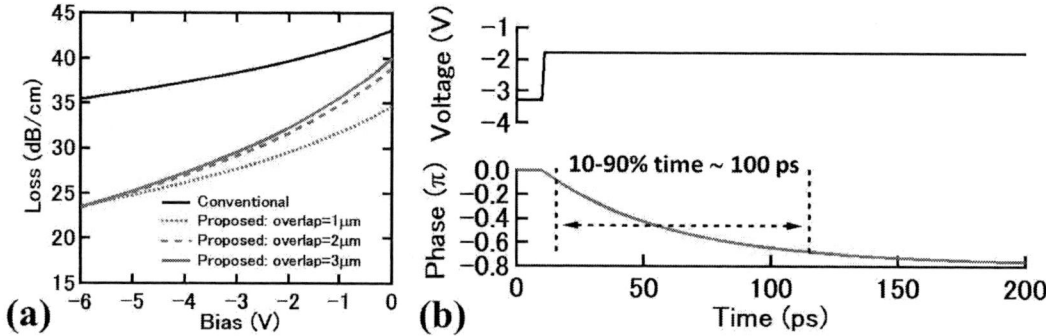

Fig. 3. (a) Voltage dependence of the propagation loss for both the conventional lateral PN modulator and the proposed modulator. (b) Transient response of the phase (bottom) to a step voltage pulse (top).

V. CONCLUSIONS

We proposed a novel silicon optical modulator using a SiN-strip loaded waveguide on the SOI layer instead of the shallow-etched rib waveguide. By adopting the vertical PN diode, this proposed modulator shows significant efficiency enhancement compared to the conventional lateral PN modulator based on the shallow-etched rib waveguide. At the same time, this novel modulator has lower propagation loss. This novel strip-loaded waveguide based modulator will be beneficial for achieving high-efficient silicon optical modulator on the photonic SOI platform.

REFERENCES

[1] G.T. Reed, G. Mashanovich, F.Y. Gardes, and D.J. Thomson, "Silicon optical modulators," Nature Photonics, vol. 4, pp. 518-526, July 2010.

[2] M.R. Watts, W.A. Zortman, D.C. Trotter, R.W. Yound, and A.L. Lentine, "Low-Voltage, Compact, Depletion-Mode, Silicon Mach-Zehnder Modulator," IEEE J. Select. Top. Quantum Electron., vol. 16, pp. 159-164, Jan/Feb. 2010.

[3] X. Xiao, H. Xu, X.Y Li, Z.Y. Li, T. Chu, Y.D. Yu, and J.Z. Yu, "High-speed, low-loss silicon Mach-Zehnder modulators with doping optimization," Opt. Express, vol. 21, pp. 4116-4126, Feb. 2013.

[4] G.T. Reed, G.Z. Mashanovich, F.Y. Gardes, M. Nedeljkovic, Y. Hu, D.J. Thomson, K. Li, P.R. Wilson, S.W. Chen, and S.S. Hsu, "Recent breakthroughs in carrier depletion based silicon optical modulators," Nanophotonics, vol. 3, pp. 229-245, Aug. 2014.

[5] P. Dong, L. Chen, and Y.K. Chen, "High-speed low-voltage single-drive push-pull silicon Mach-Zehnder modulators," Opt. Express, vol. 20, pp. 6163-6169, Mar. 2012.

[6] H. Yu, M. Pantouvaki, J.V. Campenhout, D. Korn, K. Komorowska, P. Dumon, Y. Li, P. Verheyen, P. Absil, L. Alloatti, D. Hillerkuss, J. Leuthold, R. Baets, and W. Bogaerts, "Performance tradeoff between lateral and interdigitated doping patterns for high speed carrier-depletion based silicon modulators, " Opt. Express, vol.20, pp.12926-12938, June 2012.

[7] Lumerical Solutions, Inc. http://www.lumerical.com/tcad-products/device/, and /mode/

[8] K. Goi, A. Oka, H. Kusaka, Y. Terada, K. Ogawa, T.Y. Liow, X. Tu, G.Q. Lo, and D.L. Kwong, "Low-loss high-speed silicon IQ modulator for QPSK/DQPSK in C and L bands," Opt. Express, vol. 22, pp. 10703-10709, May 2014.

[9] R.A. Soref and B.R. Bennett, "Electrooptical Effects in Silicon," IEEE J. Quantum Electron, vol. QE-23, pp.123-129, Jan. 1987.

1.3-μm DFB Laser μ-Platform; Light Source Suitable for Silicon Photonics Platform

Takanori Suzuki[1], K. R. Tamura[2], Koichiro Adachi[1], Aki Takei[1], Akira Nakanishi[2],
Kazuhiko Naoe[2], Kouji Nakahara[2], and Shigehisa Tanaka[2]

[1]Hitachi, Ltd.

1-280 Higashi-Koigakubo, Kokubunji, Tokyo 185-8601, Japan

[2]Oclaro Japan, Inc.

4-1-55 Oyama, Chuo-ku, Sagamihara, Kanagawa, 252-5250, Japan

takanori.suzuki.ym@hitachi.com

Abstract: We fabricated a light source based on a lens-integrated surface-emitting laser μ-platform for silicon photonics platforms. A low coupling loss to silicon photonics platform of 3.9dB was achieved by optimizing design of the laser μ-platform.

Keywords: Semiconductor lasers, Integrated optics devices

I. INTRODUCTION

Silicon photonics (SiP) has been widely studied, and various kinds of SiP devices (e.g., high-speed modulators, detectors, and passive components) are currently available from several silicon fabrication foundry services [1]. The development of a practical high-performance light source for SiP platforms, one of the most challenging parts of SiP, is still attracting a great deal of interest.

One approach is to use germanium-based light sources, which are compatible with CMOS processes [2][3]. The laser footprint is very small, so it is suitable for intra- or inter-chip applications. However, a large injection current is required when operating germanium-based lasers. Another approach involves heterogeneous or hybrid integration of a III-V laser on a silicon waveguide [4]-[6]. Hybrid integration of an InP-based distributed feedback (DFB) laser through a grating coupler (GC) on a SiP platform is a promising approach [7] in terms of reliability of the light source. This light source has been practically applied due to its well-established laser performance and high-volume manufacturing capability utilizing wafer-level packaging technology. The disadvantages of the hybrid-integration approach are, however, a large footprint and a large number of components (e.g., lens, mirror, and isolator).

We previously proposed the use of a DFB laser monolithically integrated with a mirror and a lens, which is called a lens-integrated surface-emitting laser (LISEL), as a light source for SiP [8]-[10]. Since the light of the LISEL is emitted from its surface, the light is directly coupled to the GC and its assembly is very simple. In this study, as a practical light source for SiP, the fundamental characteristics of a LISEL μ-platform which consists of a LISEL and a carrier only are shown. We optimized the design of the integrated lens of the LISEL and fabricated the LISEL μ-platform. As a result, we evaluated its optical coupling and lasing characteristics on SiP platform. A low coupling loss of 3.9 dB between the LISEL μ-platform and the SiP platform and low RIN less than -140 dB/Hz were experimentally achieved.

II. CONCEPTUAL STRUCTURE

Conceptual structure of a LISEL μ-platform for SiP platforms is shown in Fig. 1.

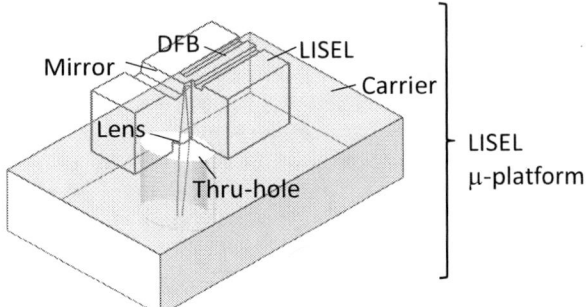

Fig. 1 Conceptual structure of LISEL μ-platform for silicon photonics platform

The LISEL μ-platform consists of a LISEL and a carrier. The LISEL is mounted in junction up configuration on the carrier. Since the carrier has a thru-hole, the integrated lens of the LISEL is aligned over the thru-hole so as to emit the light from the LISEL μ-platform through this hole. In order to obtain high heat dissipation efficiency, a material with high heat conductivity like AlN is utilized for the carrier. This LISEL μ-platform is finally aligned and mounted on the SiP platform by soldering or UV curable resin. The light emitted from the LISEL can be coupled directly to the GC fabricated in the SiP platform. Since the isolator and the external lens are not utilized in this configuration, there are big challenges to realize a LISEL which has both a high optical-feedback tolerance and high optical-coupling to the SiP platform.

III. OPTIMIZATION AND CHARACTERISTICS OF LISEL MICRO-PLATFORM

We fabricated a LISEL μ-platform and aligned it on a SiP platform. Figure 2 shows a photograph of the LISEL μ-platform on the SiP platform. In this SiP platform, optical integrated circuits consisting of modulators, GCs and 3-dB couplers are fabricated on a silicon-on-insulator (SOI) substrate. The light coupled to the GC is divided into four paths, which are connected to four modulators by three 3-dB couplers and is emitted outside from the four GCs which are connected to optical fibers.

In order to achieve a high coupling efficiency and a large alignment tolerance between the LISEL μ-platform and the SiP platform, the integrated lens of the LISEL and the carrier should be optimized. The mode profile of the beam emitted from the LISEL can be controlled by the radius of the curvature (ROC) and position of the integrated lens. In this study, two types of LISEL were fabricated: one LISEL emitted a divergent beam and the other emitted a focused beam. Figure 3 shows the coupling losses as a function of the distance between the LISEL and the GC. The incident angle to the GC was set to the diffraction angle of the GC. When the ROC was large (120 μm), the divergent beam was emitted from the LISEL and the coupling loss was high (about 8.5 dB) because the mode field of the beam emitted from the LISEL at the GC surface was larger than the size of the GC. In contrast, when the ROC was small (90 μm), the focused beam was emitted from the LISEL. In this case, a low coupling loss of about 3.9 dB was achieved because the mode field from the LISEL at the GC fitted the size of the GC. Although the beam waist position from the LISEL was about 250 μm which is shown in the calculation result, the minimum distance to the GC in this experiment was about 270 μm due to limitations of our experimental setup. This means that if the LISEL could be mounted closer to the GC, the coupling loss can be further reduced. From these experimental results, it is confirmed that the low coupling loss between the LISEL and the GC is obtained by the well-controlled ROC.

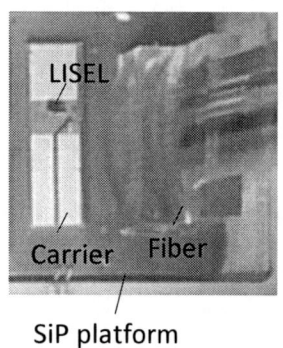

Fig. 2 LISEL μ-platform on SiP platform

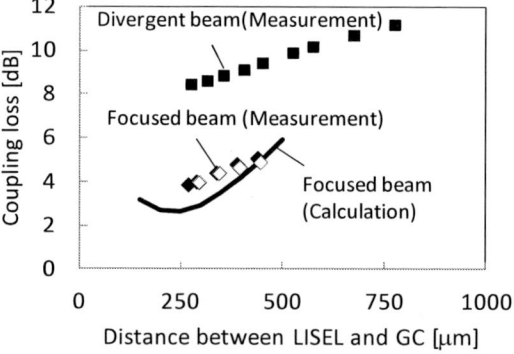

Fig. 3 Coupling loss between LISEL and SiP platform

The measured excess losses due to the misalignment are shown in Fig. 4(a) and (b). In this case, the distance between the LISEL and the GC was 270 μm. The X- and Z-directions are shown in Fig. 1. The optimized LISEL μ-platform has large alignment tolerance. If the alignment accuracy can be controlled to below ± 4 μm in both the X- and Z-directions, the excess coupling loss caused by misalignment can be suppressed less than 1 dB.

In order to estimate the stability of lasing characteristics of the LISEL μ-platform on the SiP platform, relative intensity noise (RIN) characteristics were measured. The light emitted from the LISEL was coupled into a RIN measurement system through the SiP optical integrated circuit which is mentioned in Fig. 2. Figure 5 shows the measured RIN characteristics at the room temperature. Low RINs less than -140 dB/Hz were confirmed even when the LISEL is coupled to the SiP platform.

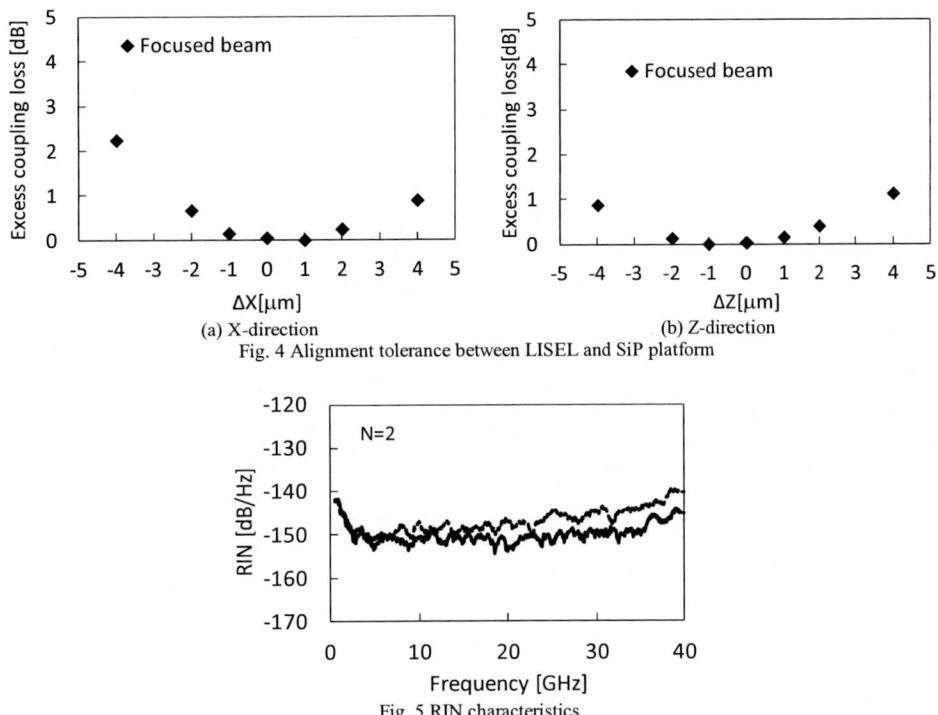

(a) X-direction   (b) Z-direction

Fig. 4 Alignment tolerance between LISEL and SiP platform

Fig. 5 RIN characteristics

IV. SUMMARY

A practical light source, namely, LISEL μ-platform is proposed for a SiP platform. The LISEL μ-platform consisting of only a laser and a carrier was fabricated and a low coupling loss of 3.9 dB between the LISEL μ-platform and the SiP platform was achieved.

ACKNOWLEDGMENT

The authors gratefully acknowledge STMicroelectronics for providing the silicon photonics technology.

REFERENCES

[1] F. Boeuf, et al., "Recent Progress in Silicon Photonics R&D and Manufacturing on 300mm Wafer Platform," *Proc. OFC2015*, W3A. 1, 2015.

[2] J. Michel, et al., "Electrically Pumped Ge-on-Si Laser," *Proc. OFC2012*, PDP5A.6, 2012.

[3] S. Saito, "Silicon and Germanium Quantum Well Light-Emitting Diode," *Proc. GFP2011*, ThA5, 2011.

[4] A. Fang, et al., "A Distributed Feedback Silicon Evanescent Laser," *Opt. Express*, vol. 16, No, 7, pp. 4413-4419, 2008.

[5] T. Shimizu, et al., "High Density Hybrid Integrated Light Source with a Laser Diode Array on a Silicon Optical Waveguide Platform for Inter-Chip Optical Interconnection," *Proc. GFP 2011*, ThB5, 2011.

[6] A. Narasimha, et al., "A 40-Gb/s QSFP Optoelectronic Transceiver in a 0.13 μm CMOS Silicon-on-Insulator Technology," *Proc. OFC2008*, OMK7. 2008.

[7] P. D. Dobbelaere, "External source approach for silicon photonics transceivers," *Proc. ECOC2014*, WS-1, 2014.

[8] T. Suzuki, et al., "A Light Source using 1.3-μm Lens-Integrated Surface-Emitting Laser for Silicon Platform," *IEEE Photon. Technol. Lett.*, vol. 26, No. 11, pp. 1089-1091, 2014.

[9] T. Suzuki, et al., "Capability of High Optical-Feedback Tolerance and Non-Hermetic-Packaging for Low-Cost Interconnections using Lens-Integrated Surface-Emitting Laser," *Proc. OFC2015*, M3B.4, 2015.

[10] T. Suzuki, et al., "Cost-Effective Optical Sub-Assembly Using Lens-Integrated Surface-Emitting Laser," *IEEE J. Lightwave Technol.*, vol. 34, No. 2, pp. 358-364, 2016.

WD1-5

OECC/PS2016

Hybrid silicon-based tunable laser with integrated reflectivity-tunable mirror

G. de Valicourt[1], C. Gui[1],[3], A. Melikyan[1], P. Dong[1], C-M. Chang[1], A. Maho[2], R. Brenot[2], and Y. K. Chen[1]

(1) Bell Labs, Alcatel-Lucent, 791 Holmdel Road, Holmdel, New Jersey 07733, USA
(2) III-V Lab-Common laboratory of Alcatel-Lucent Bell Labs France', 'Thales Research and Technology' and 'CEA Leti', 1, Avenue A. Fresnel, 91767 Palaiseau cedex, France
(3) Wuhan National Laboratory for Optoelectronics, Huazhong University of Science and Technology, Wuhan 430074, Hubei, China
devalicourt@bell-labs.com

Abstract: We propose a novel hybrid silicon-based laser with integrated variable reflectivity mirror. The single-ring laser is tunable over 10 nm with high SMSR. The adjustment of the mirror reflectivity allows controlling lasers characteristics.

Keywords: Hybrid III-V/Si laser, Reflective Semiconductor Optical Amplifier, Silicon Photonic, Photonic Integrated Circuit

I. INTRODUCTION

Silicon photonic has been viewed as a promising platform for next generation of photonic integrated circuit. Such platform has become available in multiple commercial foundries, which have lead to both high volume and low-cost manufacturing. However light emission and/or amplification is still not available and hybrid solutions combining III-V and silicon material are being investigated. There are several approaches for the laser source for silicon photonics as using an external III-V laser source coupled via an optical fiber or butt-coupling [1], bonding of III-V dies/wafers onto a processed Si wafer [2], using electrically pumped highly strained and heavily doped Ge materials [3] and the use of III-V on Si hetero-epitaxy [4]. All approaches use a shared cavity between the III-V and silicon material. In the first and most practical approach for now, each III-V source needs to be aligned with a silicon waveguide and butt-coupling through the facet is realized. This approach requires alignment between both chips however shows the best performance to date in order to realize tunable lasers as both chips could be optimized and fabricated independently [5]. A 100% reflectivity mirror and a partial reflectivity mirror close each sides of the cavity. Such last mirror could be realized using cleaved facet of the semiconductor optical amplifier (SOA) [6] or in the silicon photonic integrated circuit (PIC). Bragg grating [7], Sagnac loop mirror [8], ring reflector [9] or coupled resonator mirrors [10] have been proposed such as partial reflector. Controlling the facet reflectivity is important in order to optimize, the threshold

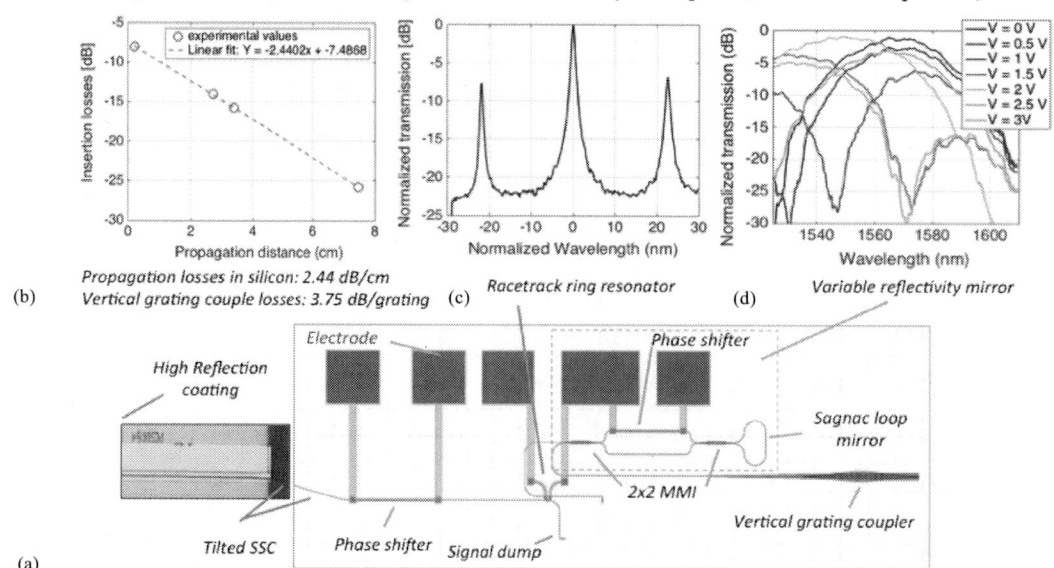

Fig. 1. (a) RSOA photography and Silicon-based external cavity schematic diagram including one tilted SSC, 1 phase shifters, 1 racetrack RR, 1 vertical grating coupler and 1 variable reflectivity mirror. (b) Propagation and coupling losses (from VGC). (c) Racetrack RR drop transmission. (d) Normalized transmission of an asymmetric MZ-based variable reflectivity mirror.

1116

OECC/PS2016

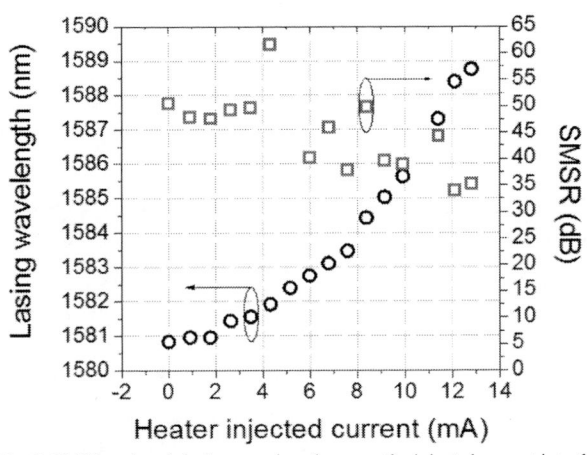

Fig. 2. SMSR and peak lasing wavelength versus the injected current into the ring filter

current as well as the wall plug efficicency [9] but also the linewidth [11] which are key metrics for hybrid laser.

In this paper, we propose and demonstrate a novel single-ring hybrid tunable laser with reflectivity-tunable integrated mirror. The concept and design of the laser is presented. The reflectivity is extracted from the asymmetric Mach-Zehnder (MZ) based mirror allowing full reflectivity tunability using a micro-heater-based phase shifter. First, the laser characteristics at R(V=0V) are investigated and 10 nm tuning range under single mode operation (Side mode suppression ratio [SMSR] higher than 30 dB) is measured. Then the impact of the integrated mirror reflectivity on the laser performances is then described. We demonstrate that the threshold current could be further reduced and the output power increases by choosing the optimum reflectivity condition. Such approach provides process-insensitive mirror with fully tunable reflectivity.

II. DESIGN AND CHARACTERIZATION OF HYBRID SILICON-BASED TUNABLE LASER

The schematic of our hybrid laser structure is presented in figure 1. (a). A reflective SOA (RSOA) is butt-coupled into a external silicon-based cavity. The RSOA structure used a double core structure as detailed in [12]. The external cavity is composed by one tilted inverted taper, one thermo-optic phaser shifter, one racetrack ring resonator (RR), a variable reflectivity mirror and a vertical grating coupler (VGC). The propagation and coupling losses from the VGC are measured to be 2.44 db/cm and 3.75 dB/VGC as measured in figure 1 (b). We use a single RR with large FSR of 20 nm as measured in figure 1. (c). This laser allows single mode operation as only one fabry-perot mode is expected to lase due to the superposition of the material gain function and the transmittance of such filter. The temperature dependency of refractive index allows controlling the peak wavelength of a resonant filter and wavelength tunability over 10 nm only using the ring resonator thermo-optical effect is obtained. In this design, we propose a novel variable reflective mirror (VRM) based on a reflective asymmetric MZ configuration. Two 2x2 MMIs are used to split/combine the two asymmetric arms of the MZ interferometer. The length difference between both arms is 2 μm and one thermo-optic phase shifter is placed in one of the arm. Asymmetric mirror design was chosen in order to guarantee single mode operation using a wavelength dependent reflectivity mirror. The transmission spectra of the VRM are measured in Figure 1. (d) for several bias voltage. A shift of the transmission spectra is observed at different voltage therefore controlling the reflectivity. First, the wavelength tunability is investigated without biasing the VRM. Figure 2 shows the 10 nm tuning range form 1580 to 1590 nm due to a RSOA gain peak centered at 1585 nm. Single mode operation is obtained over the full tuning range as confirmed by the SMSR measurement. The SMSR ranges from 30 to 50 dB. The impact of the reflectivity is then studied in order to optimize the lasing condition. No current is injected in the RR heater therefore the lasing laser is set to 1581 nm.

III. IMPACT OF THE MIRROR REFLECTIVITY

The reflectivity is extracted from the transmission measurements realized on a standalone variable mirror testing structure. The 3D reflectivity profile is represented in Figure 3. Due to the proposed reflective asymmetric MZ-based mirror, the reflectivity is wavelength dependent and could be adjusted by controlling the phase shifter inside the integrated mirror structure. Reflectivity could be continuously adjusted from 0 to 100%. The threshold current, wall-plug efficiency (WPE) and the output optical power (for one constant bias current) strongly depend on such parameter then we evaluate the impact of the reflectivity on such characteristics. At 1581 mA, the reflectivity variation only ranges from 50 to 100 %. Figure 4 (a) shows the light-intensity measurements for several bias voltages of the VRM (V). It can be seen that the L-I characteristics

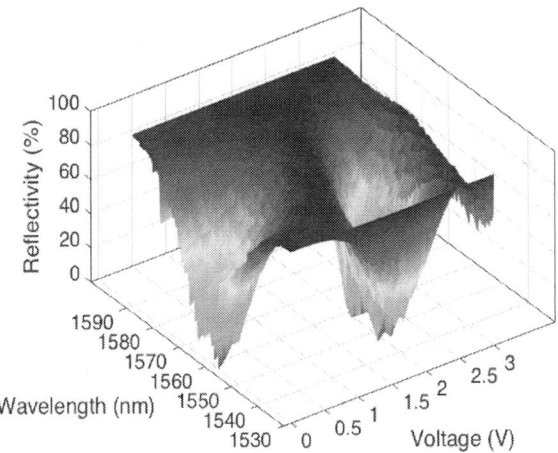

Fig. 3. 3D reflectivity mapping of Asymmetric MZ-based integrated mirror depending on the bias voltage of the phase shifter

1117

OECC/PS2016

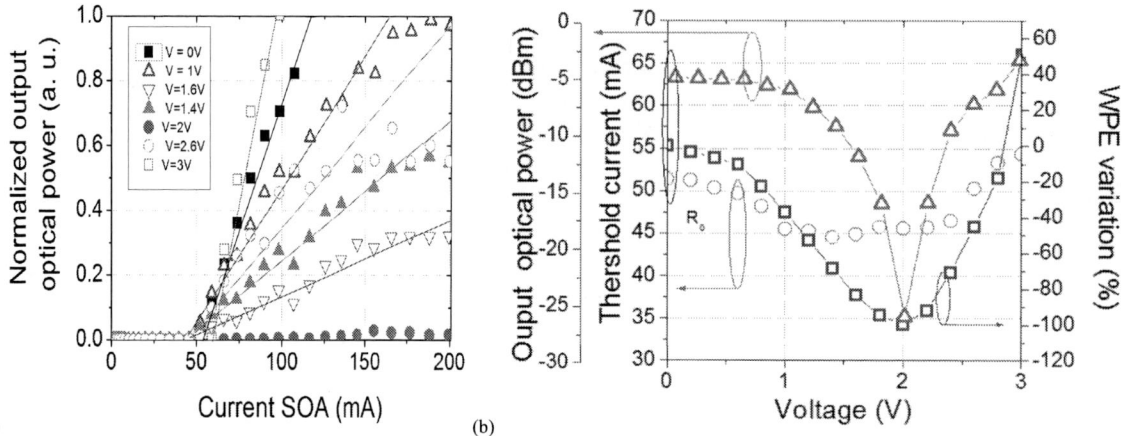

(a)   (b)

Fig. 4. (a) L-I curves for several mirror bias voltages. (b) Threshold current, output optical power and WPE variation depending on the reflectivity

are strongly impacted by the reflectivity of the partial mirror. In order to quantify such changes, we extracted the threshold current, WPE variation and output power at 100 mA RSOA bias current. The threshold current is shown in figure 4 (b). It varies from 44.5 mA to 55 mA therefore 10 mA variation is observed. Then the variation of the WPE is analyzed. From its original value, the WPE varies from − 100% up to more than +50 % for a range of V = 0 to 3 V. We believe that higher positive variation could be obtained using wavelengths centered around 1550 nm where the VGC transmission function does not limit the reflectivity range. Further work with different RSOA devices are under investigation. Finally, we analyze the output power depending on the reflectivity. The output power ranks from − 3 dBm down to − 26 dBm. Therefore, increasing the reflectivity above 50 %, reduces the threshold current by more than 10 mA however at the expense of the WPE as well as the output power.

IV. CONCLUSIONS.

We propose novel hybrid III-V/Silicon laser with integrated variable reflectivity mirror. Single-mode operation is obtained over more than 10 nm tuning range. We evaluated the reflectivity tuning range and shows that the reflectivity could be tuned over the full range of possible reflectivity value around 1550 nm. By adjusting the reflectivity, we could control the threshold current, WPE as well as the output power. Such novel integrated mirror could be applied to any type of hybrid laser providing a tunable mirror insensitive to the fabrication process.

ACKNOWLEDGMENT

We acknowledge G-H Duan and G. Levaufre for the technical discussions as well as the support of J.-L. Gentner and M. Zirngibl at Bell Laboratories.

REFERENCES

[1] Y. Urino, et al., "First demonstration of high density optical interconnects integrated with lasers, optical modulators, and photodetectors on single silicon substrate", Opt. Express, vol. 19, No 26, 2011.

[2] G.-H. Duan, et al., "New Advances on Heterogeneous Integration of III–V on Silicon", Journal of Lightwave Technology, vol. 33, No 05, 2015

[3] J. F. Liu, et al., "A Ge-on-Si laser operating at room temperature", Optics Lett., vol. 35, pp. 679-681, March 2010.

[4] A. Y. Liu, et al., "High performance continuous wave 1.3 μm quantum dot lasers on silicon", Applied Physics Letters, Vol. 104 No 4, January 28, 2014.

[5] Kenji Sato et al,. "High Output Power and Narrow Linewidth Silicon Photonic Hybrid Ring-Filter External Cavity Wavelength Tunable Lasers", in Proc. ECOC, PD2.3, Cannes, France, 2014.

[6] G. de Valicourt et al., "Dual Hybrid Silicon-Photonic Laser with Fast Wavelength Tuning", in Proc. OFC, M2C.1, Anaheim, USA, 2016.

[7] G. de Valicourt, et al., "Direct Modulation of Hybrid Integrated InP/Si Transmitters for short and long reach access network", Journal of Lightwave Technology, Vol. 33, No. 8, April 15, 2015.

[8] Y. Zhang et al., "Sagnac loop mirror and micro-ring based laser cavity for silicon-on-insulator", Opt. Express, vol. 22, No 15, 2014.

[9] J.-H. Lee et al., "Demonstration of 12.2% wall plug efficiency in uncooled single mode external-cavity tunable Si/III-V hybrid laser", Opt. Express, vol. 23, No 9, 2015.

[10] Tin Komljenovic et al., "Narrow linewidth tunable laser using coupled resonator mirrors", in Proc. OFC, W2A.52, 2015

[11] K. Aoyama et al., "Optical Negative Feedback for Linewidth Reduction of Semiconductor Lasers", Photon. Techn. Letters, Vol. 27, No 4, 2015

[12] G. de Valicourt et al., "High Gain (30 dB) and High Saturation Power (11dBm) RSOA Devices as Colourless ONU Sources in Long Reach Hybrid WDM/TDM -PON Architecture", Photon. Technol. Lett., Vol. 22, No. 3, p. 191, 2010.

Membrane Distributed-reflector Lasers

Shigehisa Arai[1,2], Nobuhiko Nishiyama[2], Tomohiro Amemiya[1], Takuo Hiratani[2], and Daisuke Inoue[2]

[1]Institute of Innovative Research (IIR), [2]Dept. of Electrical and Electronic Engineering, Tokyo Institute of Technology, Japan
e-mail: arai@pe.titech.ac.jp

Abstract: *1.55 μm wavelength GaInAsP/InP membrane distributed-reflector lasers for on-chip optical interconnects will be presented. Fairly low threshold current (250 μA) and asymmetric output characteristic were obtained with the DFB section length of 30 μm.*
Keywords: *Semiconductor laser; Distributed Bragg reflector; Distributed feedback; Membrane laser; GaInAsP/InP; Optical interconnect*

I. INTRODUCTION

We have proposed a distributed-reflector laser, DR laser for short, which consists of an active distributed-feedback (DFB) and a passive distributed-Bragg-reflector (DBR) sections so as to increase the output from one side of the cavity without interrupting superior single-mode characteristics [1]. Fundamental operation properties such as sub-mode suppression ratio (SMSR), an asymmetric output, and a narrow linewidth, have been reported for DR lasers with multiple-quantum-well (MQW) active region [2]-[4]. After demonstrating low threshold current density and low threshold current operations of DFB lasers by adopting wirelike active regions [5],[6], we have introduced the wirelike active regions into DR lasers [7] and achieved fairly low threshold current (0.9 mA) as well as relatively high external differential quantum efficiency (48%) from one side of the cavity [8]. A modulation current efficiency (MCEF) under a high-speed direct modulation, i.e. the slope of the resonance-like frequency f_r as a function of the square root of the bias current above the threshold $(I_b-I_{th})^{1/2}$, of this type of DR laser was reported to be 3.0 GHz/mA$^{1/2}$ [9]. Since DR lasers don't need high reflective facets for high efficiency operation, they are suitable for monolithic integration with other elements and the developments for high-speed optical fiber communications are undergoing [10],[11].

Even though DFB and DR lasers consisting of wirelike active regions showed low threshold current (< 1 mA) and moderately high differential quantum efficiency (~ 50% from one side of the cavity), they are suitable for low power-consumption operation at an output power of a few mW range. Then we proposed a new class of semiconductor lasers by utilizing a thin semiconductor membrane structure sandwiched by low refractive-index claddings so as to enhance an optical confinement into the active region and demonstrated low threshold membrane DFB laser [12]-[14].

On the other hand, ultra-low power consumption and high-speed operation of semiconductor lasers has been required for on-chip optical interconnects for next generation LSIs, however required pulse energy for the optical link was predicted to be 100 fJ/bit or the less [15] which corresponds to the power consumption of only 1 mW for the transmission speed [1]of 10 Gbit/s, and would be much lower in future [16]. For this purpose, high-speed (> 10 Gbit/s) optical modulators or current injection type lasers capable of ultra-low power consumption operation are indispensable. To meet such a requirement, we have been investigating injection-type membrane DFB and DR lasers bonded on another host InP or Si substrate. Here recent results obtained for these membrane lasers are reviewed.

II. STATIC OPERATION CHARACTERISTICS OF MEMBRANE DR LASER

Since the minimum receivable power of a typical PIN-photodiode (PD) used for long wavelength optical fiber communications is -13 dBm (approximately 50 μW) for 10 Gbit/s signals with a bit-error-rate (BER) of 10^{-9}, the required output power of the light source will be of the order of hundreds of μW, the optical pulse energy is in the order of tens of fJ/bit. When we consider the possibility of directly modulated semiconductor lasers for on-chip optical interconnects, a threshold current of much lower than 1 mA and the direct modulation speed higher than 10 GHz at a

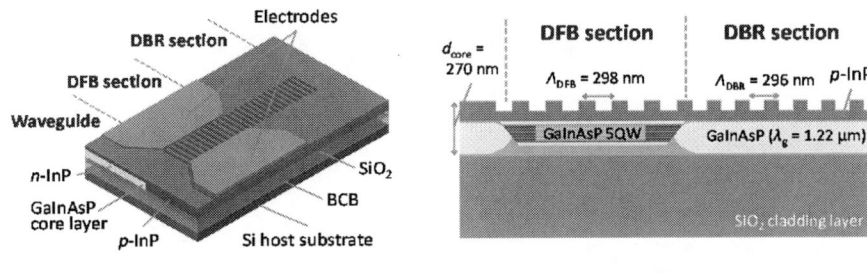

(a)   (b)

Fig. 1. (a) Schematic structure of a LCI-membrane-DR laser and (b) its cross-sectional structure [20].

ACKNOWLEDGMENT: This work was supported by the JSPS KAKENHI Grant numbers #15H02249, #25709026, #15J04654, #15J11776, and #16H06082.

bias current of 1 mA will be required, and the DFB laser with wirelike active regions can't meet this requirement. Therefore we need to reduce the operation current of semiconductor lasers such as an enhancement of the optical gain of the active region by well size-controlled low-dimensional quantum structures, or by a strong optical confinement into an extremely small laser cavity by photonic-crystal structures. We proposed to enhance it by using a thin semiconductor layer (membrane) sandwiched by low refractive-index materials such as BCB or SiO_2 so as to strongly confine the optical field into the active region of the semiconductor membrane structure [12]. The optical confinement factor of multiple-quantum-well structures is enhanced by reducing the core thickness d_{core} and it is approximately 3%/quantum-well which is around 3 times higher than that in conventional QW lasers [17]. This enhancement also increases an index-coupling coefficient κ_i of DFB and DBR structures, hence the threshold current of membrane DFB and DR lasers can be reduced to one order of magnitude lower than that of conventional ones.

Figure 1(a) shows a schematic of a lateral-current-injection (LCI)-membrane DR laser. It consists of an active DFB section and a passive DBR section. The passive waveguide was formed by the BJB regrowth process and the DFB and DBR structures were realized by a surface grating structure [18]. The DBR at one side of the DFB structure facilitates concentration of the light output at the opposite side of the DFB structure. Figure 1(b) shows the cross section along the cavity direction of the LCI-membrane DR laser. We adopted a period at the DFB section, Λ_{DFB}, of 298 nm and that at the DBR section, Λ_{DBR}, of 296 nm in order to match the lasing wavelength to the Bragg wavelength of the DBR section. A small reduction in the period at the DBR section of the membrane laser is needed because the membrane DFB laser with surface grating typically operates at the shorter wavelength side of the stop band because of its strong optical confinement to the low index region of the grating [19].

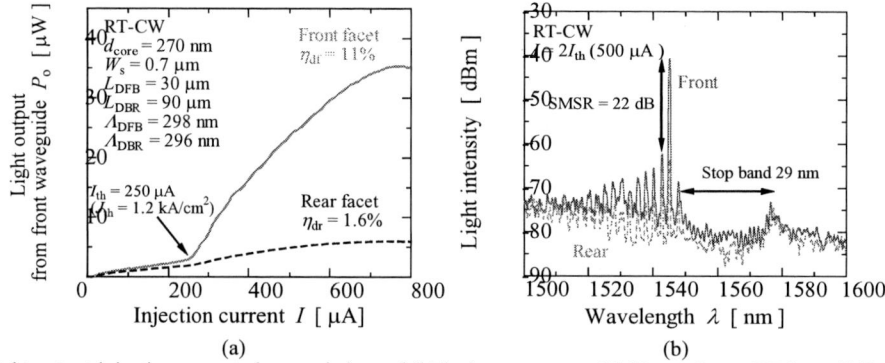

Fig. 2. (a) Light output-injection current characteristics and (b) lasing spectrum of LCI-membrane DR laser [20].

Figure 2(a) shows the light output characteristics of the LCI-membrane DR laser with d_{core}=270 nm, stripe width, W_s=0.7 μm, DFB section length, L_{DFB}=30 μm, and DBR section length, L_{DBR}=90 μm. The device was formed by cleavage and had an approximately 200 μm-long waveguide at the front side and a 100 μm-long waveguide at the rear side. A threshold current of I_{th}=250 μA, an external differential quantum efficiency at the front side of η_{df}=11% (indicated by the solid line) and that at the rear side of η_{dr}=1.6% (indicated by the dashed line) were obtained. An asymmetric output ratio of 6.7 can be attributed to high reflectivity DBR. The lasing spectrum of the device at a bias current of $2I_{th}$ is shown in Fig. 2(b), where a single mode operation at 1545 nm with a side-mode suppression-ratio (SMSR) of 22 dB, and a stopband width of 29 nm (κ_i was estimated to be 1300 cm⁻¹) were observed.

III. HIGH-SPEED DIRECT MODULATION CHARACTERISTICS OF DFB LASER

Since the enhancement of the optical confinement factor in the membrane structure is effective not only for a reduction of the threshold current but also for an increase of a modulation efficiency (slope of the 3dB bandwidth

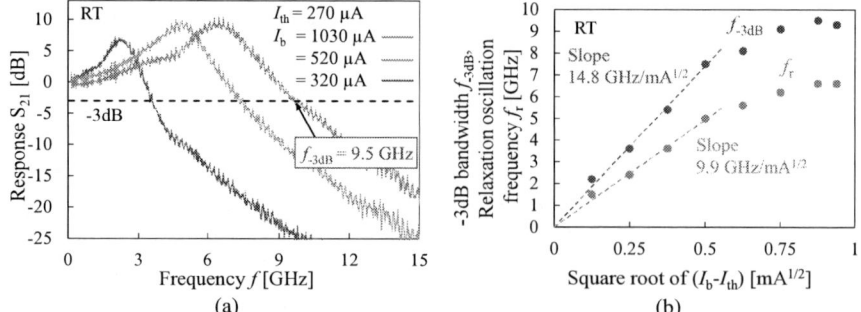

Fig. 3. (a) Small-signal frequency response of membrane DFB laser and its (b) relaxation oscillation frequency and -3 dB bandwidth frequency as a function of square root of the bias current obtained from small signal frequency response [21].

normalized by a square root of a bias current above the threshold), we measured direct modulation characteristics of a membrane DFB laser. The device consisting of d_{core}=270 nm, W_s=0.7 μm, L_{DFB}=80 μm, and 600 μm-long passive waveguide section showed I_{th}=270 μA and η_{df}=12%. As can be seen in Fi. 3(a), the maximum -3 dB bandwidth of 9.5 GHz was obtained at a bias current of 1.03 mA (3.8I_{th}). Figure 3(b) shows the relaxation oscillation frequency f_r and -3 dB bandwidth f_{-3dB} as a function of the square root of the bias current. The slope of relaxation oscillation frequency and -3 dB bandwidth are 9.9 GHz/mA$^{1/2}$ and 14.8 GHz/mA$^{1/2}$, respectively, which are approximately more than 2 times higher than those reported for conventional DFB lasers [21]. Much higher modulation efficiency was reported for a lambda-scale embedded active region photonic-crystal (LEAP) laser [22], which utilizes the membrane structure with 2D photonic-crystal structure.

IV. CONCLUSIONS

The present status of membrane-based DFB and DR lasers is reviewed. Even though there still remains problems to be solved for on-chip optical interconnects, the idea of strong optical confinement into a thin semiconductor membrane structure will lead to miniature photonic devices with ultra-low power-consumption operation in future.

REFERENCES

[1] K .Komori, S. Arai, Y. Suematsu, M. Aoki, and I. Arima, "Proposal of distributed reflector (DR) structure for high efficiency dynamic single mode (DSM) lasers," Trans. IEICE, Japan, vol. E71, no. 4, pp. 318-320, Apr. 1988.
[2] M. Aoki, K. Komori, Y. Miyamoto, S. Arai, and Y. Suematsu, "1.5 μm GaInAsP/InP distributed reflector (DR) lasers with high-low reflection grating structure," Electron. Lett., vol.25, no. 24, pp. 1650-1651, Dec. 1989.
[3] I. Arima, J. I. Shim, S. Arai, I. Morita, R. Somchai, Y. Suematsu, and K. Komori, "1.5 μm GaInAsP/InP distributed reflector (DR) lasers with SCH structure," Photon. Technol. Lett., vol. 2, no. 6, pp. 385-387, June 1990.
[4] J. I. Shim, K. Komori, S. Arai, I. Arima, Y. Suematsu, and R. Somchai, "Lasing characteristics of 1.5 μm GaInAsP-InP SCH-BIG-DR lasers," IEEE J. Quantum Electron., vol. 27, no. 6, pp. 1736-1745, June 1991.
[5] M. Nakamura, N. Nunoya, H. Yasumoto, M. Morshed, K. Fukuda, S. Tamura, and S. Arai, "Very low threshold current density operation of 1.5 μm DFB lasers with wire-like active regions," Electron. Lett., vol. 36, no. 7, pp. 639-640, Apr. 2000.
[6] N. Nunoya, M. Nakamura, M. Morshed, S. Tamura, and S. Arai, "High-performance 1.55 μm wavelength GaInAsP-InP distributed-feedback lasers with wirelike active regions," IEEE J. Sel. Top. Quantum Electron., vol. 7, no. 2, pp. 249-258, June 2001.
[7] K. Ohira, T. Murayama, S. Tamura, and S. Arai, "Low-threshold and high-efficiency operation of distributed reflector lasers with width-modulated wirelike active regions," IEEE J. Select. Topics Quantum Electron., vol. 11, no. 5, pp. 1162-1168, Sept./Oct. 2005.
[8] T. Shindo, S.H. Lee, D. Takahashi, N. Tajima, N. Nishiyama, and S. Arai, "Low-threshold and high-efficiency operation of distributed reflector laser with wirelike active regions," IEEE Photon. Technol. Lett., vol. 21, no. 19, pp. 1414-1416, Oct. 2009.
[9] S. H. Lee, D. Takahashi, T. Shindo, K. Shinno, T. Amemiya, N. Nishiyama, and S. Arai, "Low-power-consumption High-eye-margin 10 Gbit/s Operation by GaInAsP/InP Distributed Reflector Lasers with Wirelike Active Regions," IEEE Photon. Technol. Lett., vol. 23, no. 18, pp. 1349-1351, Sept. 2011.
[10] T. Ishikawa, T. Machida, H. Tanaka, Y. Oka, H. Shoji, T. Fujii, and S. Ogita, "A Novel High Output Power Full-band Wavelength Tunable Laser with Monolithically Integrated Single Stripe Structure," 33rd European Conf. on Opt. Commun., Berlin, Germany, PD2.4, Sept. 2007.
[11] M. Matsuda, A. Uetake, T. Shimoyama, S. Okumura, K. Takabayashi, M. Ekawa, and T. Yamamoto, "1.3-μm-Wavelength AlGaInAs Multiple-Quantum-Well Semi-Insulating Buried-Heterostructure Distributed-Reflector Laser Arrays on Semi-Insulating InP Substrate," IEEE J. Select. Topics in Quantum Electron., vol. 21, no. 6, p. 1502307 , Nov./Dec. 2015.
[12] T. Okamoto, N. Nunoya, Y. Onodera, S. Tamura, and S. Arai, "Low-Threshold Singlemode Operation of membrane BH-DFB lasers," Electron. Lett., vol. 38, no. 23, pp. 1444-1446, Nov. 2002.
[13] T. Okamoto, N. Nunoya, Y. Onodera, T. Yamazaki, S. Tamura, and S. Arai, "Optically pumped membrane BH-DFB lasers for low-threshold and single-mode operation," IEEE J. Select. Topics in Quantum Electron., vol. 9, no. 5, pp. 1361-1366, Sept./Oct. 2003.
[14] S. Sakamoto, H. Naitoh, H. Kawashima, M. Ohtake, Y. Nishimoto, S. Tamura, T. Maruyama, N. Nishiyama, and S. Arai, "Strongly index-coupled membrane BH-DFB lasers with surface corrugation grating," IEEE J. Select. Topics Quantum Electron., vol. 13, no. 5, pp. 1135-1141, Sept./Oct. 2007.
[15] D. A. B. Miller, "Rationale and challenges for optical interconnects to electronic chips," Proc. IEEE, vol. 88, no. 6, pp. 728-749, June 2000.
[16] D. A. B. Miller, "DeviceRequirementsforOptical Interconnects to Silicon Chips," Proc. IEEE, vol. 97, no. 7, pp. 1166-1185, July 2009.
[17] T. Okumura, T. Koguchi, H. Ito, N. Nishiyama, and S. Arai, "Injection-type GaInAsP/InP membrane buried heterostructure distributed feedback laser with wirelike active regions," Appl. Phys. Express, vol. 4, no. 4, p. 042101, Mar. 2011.
[18] T. Shindo, T. Okumura, H. Ito, T. Koguchi, D. Takahashi, Y. Atsumi, J. Kang, R. Osabe, T. Amemiya, N. Nishiyama, and S. Arai, "Lateral-current-injection distributed feedback laser with surface grating structure," IEEE J. Sel. Top. Quantum Electron., vol. 17, no. 5, pp. 1175−1182, Sept. 2011.
[19] D. Inoue, J. Lee, K. Doi, T. Hiratani, Y, Atsuji, T. Amemiya, N. Nishiyama, and S. Arai, "Room-temperature continuous-wave operation of GaInAsP/InP lateral-current-injection membrane laser bonded on Si substrate," Appl. Phys. Exp., vol. 7, no. 7, p. 072701, July 2014.
[20] T. Hiratani, D. Inoue, T. Tomiyasu, Y. Atsuji, K. Fukuda, T. Amemiya, N. Nishiyama, and S. Arai, "Room-temperature continuous-wave operation of membrane distributed-reflector laser," Appl. Phys. Express, vol. 8, no. 4, p. 112701, Oct. 2015.
[21] D. Inoue, T. Hiratani, T. Tomiyasu, Y. Atsuji, K. Fukuda, T. Amemiya, N. Nishiyama, and S. Arai, "High-modulation Efficiency Operation of GaInAsP/InP Membrane Distributed Feedback Laser on Si Substrate," Opt. Express, vol. 25, no. 22, pp. 29024−29031, Oct. 2015.
[22] K. Takeda, T. Sato, A. Shinya, K. Nozaki, W. Kobayashi, H. Taniyama, M. Notomi, K. Hasebe, T. Kakitsuka, and S. Matsuo, "Few-fJ/bit data transmissions using directly modulated lambda-scale embedded active region photonic-crystal lasers," Nat. Photonics, vol. 7 no. 7, pp. 569−575, May 2013.

WD3-2   OECC/PS2016

Ultra low power consumption operation of SOA assisted extended reach EADFB laser (AXEL)

Wataru Kobayashi, Naoki Fujiwara, Takahiko Shindo, Shigeru Kanazawa, Koichi Hasebe, Hiroyuki Ishii, and Mikitaka Itoh

NTT corporation, NTT Device Technology Laboratories, 3-1 Morinosato Wakamiya, Atsugi, kanagawa, Japan 243-0198
kobayashi.wataru@lab.ntt.co.jp

Abstract: We propose a novel structure to reduce the power consumption and extend the transmission distance of a 10-Gbit/s, 1.55-μm EADFB laser. The power consumption is reduced by 1/2 and the transmission distance is extended.

Keywords: electroabsorption modulator, semiconductor optical amplifier, DFB laser

I. INTRODUCTION

The increasing power consumption of datacenters due to the explosive increase in data traffic is becoming a social issue that is having a big impact on our total daily power consumption. To solve this problem, we must reduce the power consumption and increase the data transmission capacity of datacenters. This is motivating many developers and manufacturers to reduce optical transmitter power consumption and simultaneously boost the signaling rate.

An electroabsorption modulator integrated with a distributed feedback (EADFB) laser is used in many optical transmitters, because this light source has a clear eye waveform and a simple driving scheme. Two improvements are strongly required as regards the EADFB laser characteristics. One is to increase its signaling rate, and an increased frequency bandwidth and operation up to 100 Gbit/s have already been reported [1]. The other important requirement is to reduce its power consumption (P). There have been reports that the P of a thermo-electric cooler (TEC) has been reduced by increasing the setting temperature of the EADFB laser [2] or by eliminating the temperature control of the EADFB laser [3] by developing an EADFB laser with temperature tolerant characteristics. On the other hand, the P reduction of the EADFB laser itself has hardly been discussed. But to cope with the energy problem described above, an effort must be made to reduce the P of the EADFB laser itself because the P of the TEC cannot be reduced when an ambient operating temperature is required. Since the EADFB laser is most widely employed in the 10-Gbit/s market, we try to reduce the P of the EADFB laser at 10 Gbit/s in the 1.55-μm-wavelength window.

In this paper, we propose the integration of a semiconductor optical amplifier (SOA) with the EADFB laser while the LD and the SOA section are electrically shorted [4] as shown in figure 1 to reduce the P of the EADFB laser. We aim at automatically balancing the injection current between the LD and the SOA section according to the ratio of the LD length and SOA length and optimizing their driving current condition. The SOA also acts as an optical output power and chirp compensator with reduced P thanks to the managed current condition [5]. This approach can be used in two types of application. One is a very large power consumption reduction for the same transmission distance. The other is an extended transmission distance for the same power consumption.

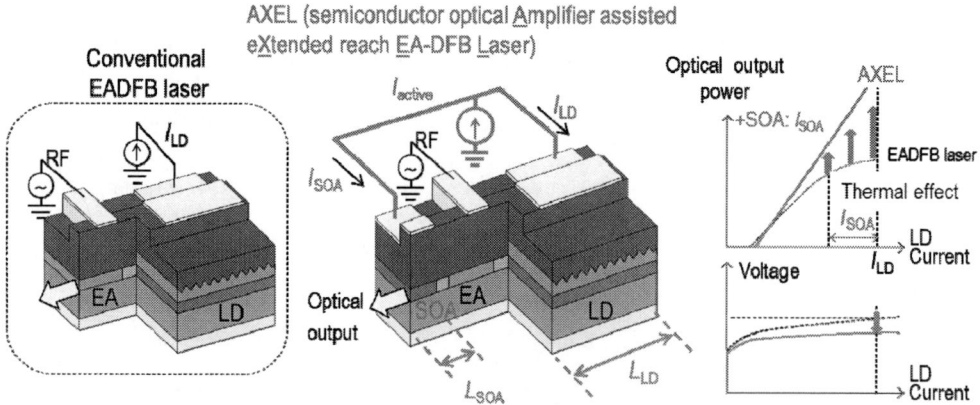

Fig. 1. Concept of AXEL.

II. CONCEPT AND DESIGN

Figure 1 shows the concept [4]. We call this device a SOA assisted extended reach EADFB laser (AXEL), because this device both amplifies an optical output power with the lower injection current of the LD (I_{LD}) and reduces the chirp parameter of the EA modulator with the integrated SOA. As indicated in the figure, a portion of the I_{LD} is automatically allocated to the SOA section, and this improves the current efficiency to optical output power ratio. Although the LD and the SOA are electrically shorted to avoid a termination increase compared with an EADFB laser, and we cannot independently set the I_{LD} and the injection current of the SOA (I_{SOA}), we can estimate these values. I_{LD} and I_{SOA} can be estimated by

$$I_{active} = I_{LD} + I_{SOA} \qquad (1)$$

$$I_{LD} = \frac{L_{LD}}{L_{LD} + L_{SOA}} \times I_{active} \qquad (2) \qquad I_{SOA} = \frac{L_{SOA}}{L_{LD} + L_{SOA}} \times I_{active} \qquad (3)$$

where I_{active} is the injection current to the total active layer, and L_{LD} and L_{SOA} are the LD and SOA lengths, respectively. In a previous study [5], we investigated the characteristics of a 50-μm-long SOA integrated with an EADFB laser while changing I_{LD} and I_{SOA}, independently. The 50-μm-long SOA can provide an advantage with the I_{SOA} was ranging from 10 to 25 mA. The general I_{LD} value of the EADFB laser is about 80 mA [6], and so we need to set I_{active} at a value of less than 80 mA. Based on the I_{SOA} and I_{active} values, we set L_{LD} and L_{SOA} at 300 and 50 μm, respectively.

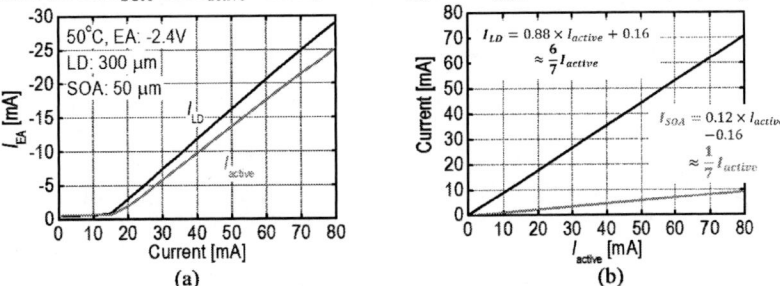

Fig. 2. (a) Measured I_{EA} versus Iactive and I_{LD}. (b) Estimated I_{LD} and I_{SOA} versus I_{active}.

We estimated the injected current of the LD and the SOA section when each section was electrically shorted. Figure 2 shows the measured photocurrent of the EA section (I_{EA}) for a bias voltage (V_{EA}) of -2.4 V. We measured the photocurrent of the EA modulator (I_{EA}) in two different ways, one is measured when the LD and the SOA are electrically shorted, and the other is measured when each section is not shorted and only using I_{LD}. We can measure the relationship between I_{active} and I_{LD} connected to I_{EA}, and estimate I_{SOA} by subtracting I_{LD} from I_{active}. Figure 2 (b) shows the estimated I_{active} value versus I_{LD} and I_{SOA}. As indicated in the figure, the current is actually divided almost according to the ratio of L_{LD} and L_{SOA}.

III. DEVICE PERFORMANCE

A. Low power consumption mode of AXEL

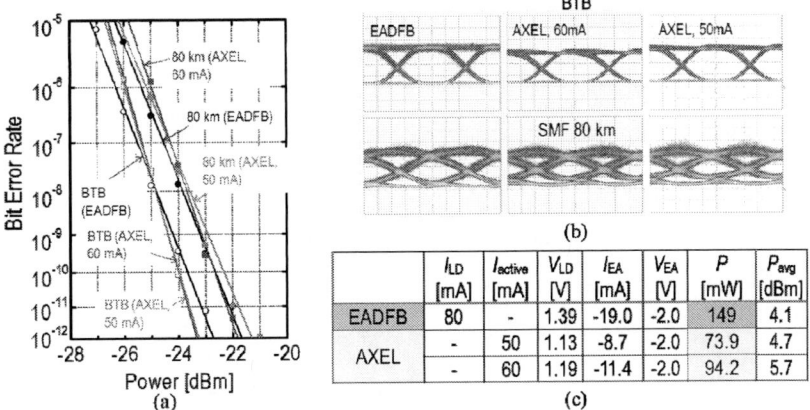

Fig. 3. (a) 10-Gbit/s BER characteristics at 50°C. (b) 10-Gbit/s eye diagrams of EADFB laser and AXEL at 50°C. (c) Typical characteristics of EADFB laser and AXEL at 50°C.

We evaluate whether or not the concept is valid for 10-Gbit/s operation, which constitutes the main use of the EADFB laser. Figure 3 (a) shows the bit error rate (BER) characteristics of the EADFB laser and the AXEL at 50°C. A 9.95-Gbit/s, non-return-zero (NRZ), pseudorandom binary sequence (PRBS) of 2^{31}-1 is used. The modulation voltage swing (V_{pp}) and V_{EA} are 2.0 and -2.0 V, respectively. The I_{active} of the AXEL is set at 50 and 60 mA. Figure 3 (b) shows the eye diagrams obtained for back-to-back (BTB) and 80-km transmissions over single mode fiber. Figure 3 (c) summarizes the typical characteristics of the EADFB and AXEL. The power consumption (P) is obtained by totaling the values of the LD, SOA and EA sections. The P of the EA section was calculated as $V_{EA} \times I_{EA}$, which does not include the P for V_{pp}. As shown in this figure, P was reduced to 1/2 that of the EADFB laser. In addition, the modulated output power (P_{avg}) was also increased.

B. Long transmission distance mode of AXEL

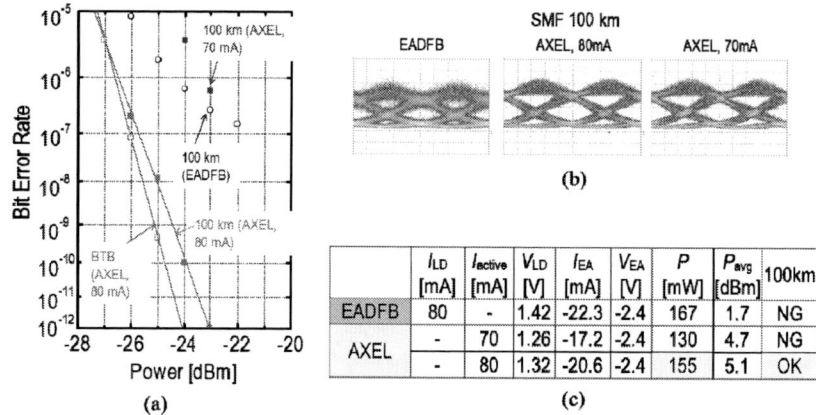

Fig. 4. (a) 10-Gbit/s BER characteristics at 50°C. (b) 10-Gbit/s eye diagrams of EADFB laser and AXEL at 50°C.
(c) Typical characteristics of EADFB laser and AXEL at 50°C.

We also measured the limit of the transmission distance of the AXEL. Figure 4 shows the BER characteristics, eye diagrams, and typical characteristics of the EADFB and AXEL. As shown in the figure, if P is compatible with that of the EADFB laser, the AXEL can extend the transmission distance compared with the EADFB laser.

IV. CONCLUSIONS

We demonstrated a novel approach to improving the characteristics of the EADFB laser. By integrating an SOA with the EADFB laser and with the LD and the SOA electrically shorted, the injection current of each section is balanced according to its length. By using the device, we reduced the power consumption by about 1/2 for an SMF transmission distance of 80 km. We also extended the transmission distance to 100 km when the power consumption was compatible. These results show that the AXEL will replace the EADFB laser.

ACKNOWLEDGMENT

We thank Prof. Hiroshi Yasaka of the Research Institute of Electrical Communication, Tohoku University for his valuable advice. We also thank Ms Marie-Aline Mattelin of Ghent University in Belgium, for her support in measuring the device characteristics and for helpful discussions.

REFERENCES

[1] S. Kanazawa, T. Fujisawa, K. Takahata, T. Ito, Y. Ueda, W. Kobayashi, H. Ishii, and H. Sanjoh, "Flip-Chip Interconnection Lumped-Electrode EADFB Laser for 100-Gb/s/λ Transmitter," IEEE Photonics Technology Letters, 27, 16, 1699-1701 (2015).

[2] K. Takada, S. Akiyama, M. Matsuda, S. Okumura, M. Ekawa and T. Yamamoto, "High-power 10-Gb/s semi-cooled operation of AlGaInAs electroabsorption modulator integrated λ/4-shifted DFB laser," in *Proc. ECOC2007*, We.8.1.6, Berlin, Germany (2007).

[3] S. Makino, K. Shinoda, T. Kitatani, T. Shiota, M. Aoki, N. Sasada, and K. Naoe, "Uncooled, electroabsorption modulator integrated DFB laser," in *Proc. OFC2008*, OthK6, San Diego, USA (2008).

[4] W. Kobayashi, Japan patent JP5823920B (16, October, 2015)

[5] W. Kobayashi, M. Arai, T. Fujisawa, T. Sato, T. Ito, K. Hasebe, S. Kanazawa, Y. Ueda, T. Yamanaka, and H. Sanjoh, "Novel approach for chirp and output power compensation applied to a 40-Gbit/s EADFB laser integrated with a short SOA," Optics Express, 23, 7, 9533-9542 (2015).

[6] H. Yamamoto, M. Hirai, O. Kagaya, K. Nogawa, K. Naoe, N. Sasada, and M. Okayasu, "Compact and low power consumption 1.55-μm electro-absorption modulator integrated DFB-LD TOSA for 10-Gbit/s 40-km transmission," in *Proc. OFC2009*, OThT5, San Diego, USA (2009).

WD3-3

OECC/PS2016

Uncooled 25.78 Gb/s Transmission over 10 km using a 1.3 μm Directly Modulated DFB Laser in a TO-CAN Package

Seiki Nakamura, Mizuki Shirao, and Masamichi Nogami

Information Technology R&D Center of Mitsubishi Electric Corporation, 5-1-1, Ofuna, Kamakura, Kanagawa 247-8501, Japan.
Nakamura.Seiki@eb.MitsubishiElectric.co.jp

Abstract: *We developed an uncooled TO-CAN packaged 1.3 μm directly modulated DFB laser for 25.78 Gb/s applications. Clear eye diagrams were obtained complying with the 100GBASE-LR4 mask from −5°C to 85°C over 10 km of SMF.*

Keywords: *transistor outline (TO)-CAN, direct modulation laser (DML), distributed feedback laser (DFB), uncooled operation, 100GBASE-LR4, CWDM4.*

I. INTRODUCTION

As discussed in the IEEE 802.3 Single Mode Fiber Study Group for 25 Gb/s Ethernet PMD(s), the need for high-speed Ethernet optical components for telecom, datacenter and mobile networks is increasing rapidly [1]. In particular, compact, cost-effective optical components are needed to increase the capacity and reduce the power consumption of such networks. An uncooled directly modulated laser (DML) in a compact TO-CAN package has the ability to reduce both power consumption and cost. Demonstrations of the operation of DML chips at more than 25 Gb/s over a wide temperature range have been described [2]–[5]. However, the uncooled operation of a 25 Gb/s DML in a TO-CAN package in the temperature range from −5°C to 85°C has not been reported [6]–[8].

In this paper, we discuss the impact of impedance mismatching of a flexible printed circuit (FPC) connection to a TO-CAN package at 25.78 Gb/s. Clear eye diagrams complying with the 100GBASE-LR4 mask [9] were experimentally demonstrated for temperatures ranging from −5°C to 85°C at 25.78 Gb/s. Also, error-free operation at a bit error ratio (BER) of 10^{-12} after transmission over 10 km of single-mode-fiber (SMF) was demonstrated under all temperature conditions. The 25.78 Gb/s DML in a TO-CAN package we have developed offers low cost and low power consumption in a small footprint for high-speed Ethernet applications.

II. CONSIDERATION OF RF DESIGN FOR 25 Ω DML

The 25.78 Gb/s TO-CAN DML and related components consist of a 1310 nm DML, a TO-header, an FPC and an evaluation board (EVB). Fig. 1 shows the schematic diagram of the simulation model. From the view point of the driver IC design, the transmission lines were given an impedance of 25 Ω. However, the transmission lines are terminated by the DML which has a serial resistance of about 10 Ω, and built-in matching resistors are avoided in the interest of low power consumption. The resulting impedance mismatch causes a large reflection that degrades the radio-frequency (RF) performance. In order to investigate the impact of the impedance mismatch, we assume a mismatched region with impedance Zx Ω in the middle of an ideal lossless transmission line with 25 Ω impedance.

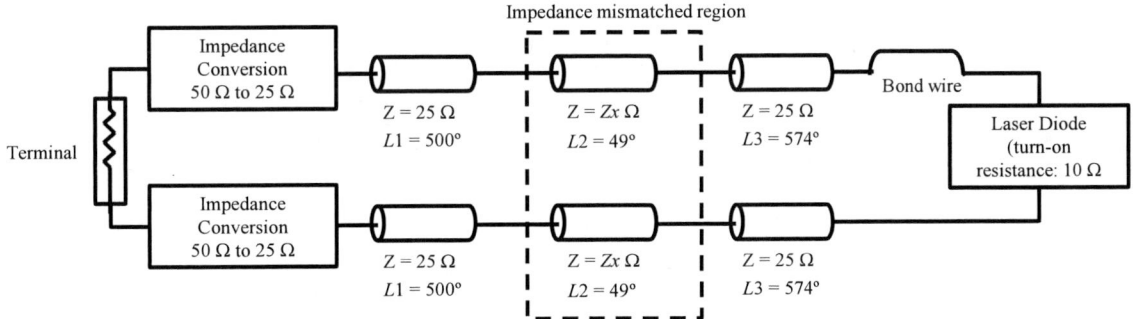

Fig. 1. Simulation model for the effect of impedance mismatch.

To evaluate the impact of the impedance mismatch of an EVB/FPC connection with a 1-mm long pad, electrical length $L2 = 49°$ at 25.78 GHz is used for the impedance mismatched region. In this case, $L3$ and $L1$ represent the FPC and TO header, and the EVB transmission lines, respectively. The calculated frequency responses (S21) and 25.78 Gb/s optical eye diagrams for $Zx = 15, 25$ and 35 Ω are shown in Fig. 2. Fig. 2(a) is for an impedance mismatch region 1 mm long, and (b) is for double that length (2 mm). The results indicate that the impedance mismatch affects the variation of

1125

S21, and that longer $L2$ results in worse performance. To avoid degradation, the impedance mismatch should be made as small as possible and/or the length of the mismatched section made as short as possible. We designed the impedance-matched EVB/FPC connection with 1-mm pad length shown in Fig. 3. With this design, we expected the negligible degradation shown in Fig. 2 (a).

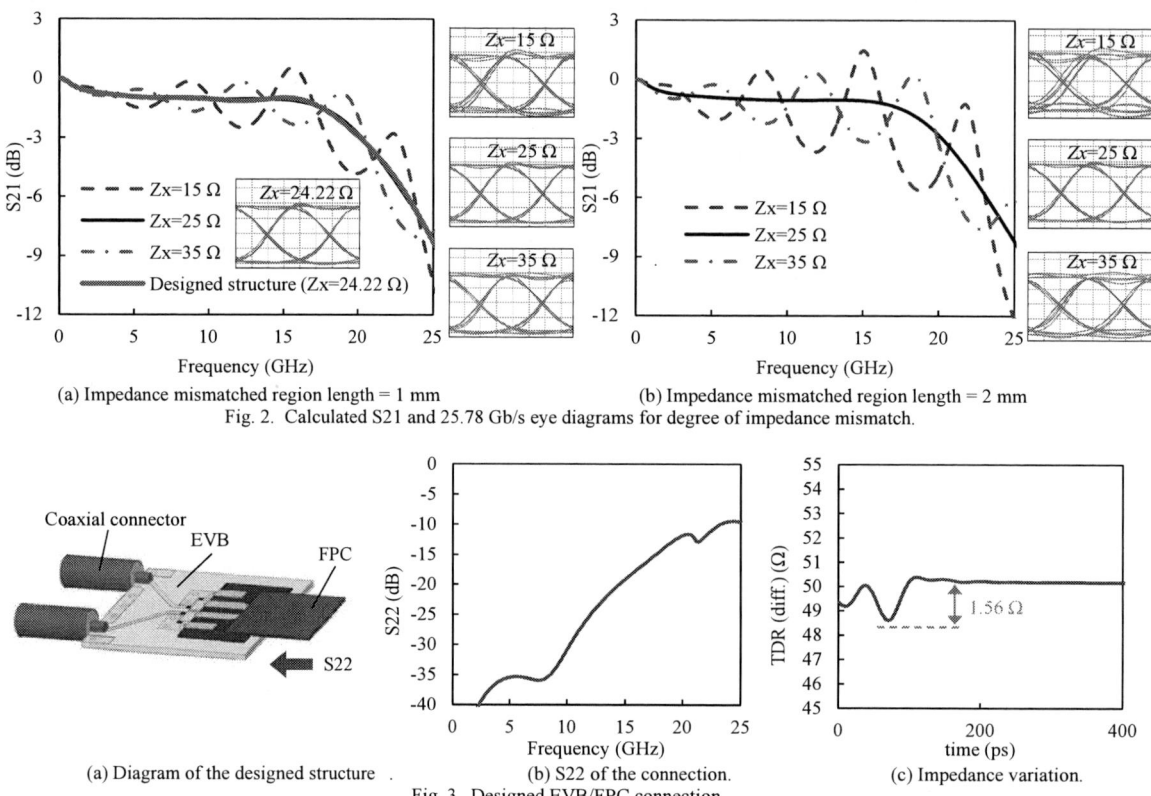

(a) Impedance mismatched region length = 1 mm

(b) Impedance mismatched region length = 2 mm

Fig. 2. Calculated S21 and 25.78 Gb/s eye diagrams for degree of impedance mismatch.

(a) Diagram of the designed structure .

(b) S22 of the connection.

(c) Impedance variation.

Fig. 3. Designed EVB/FPC connection.

III. EXPERIMENTAL SETUP AND RESULTS

The experimental setup for testing the 25.78 Gb/s eye diagrams and the BER after 10 km SMF transmission is shown in Fig. 4. We used a pulse pattern generator (Anritsu MU183020A) which generates a differential non-return-to-zero (NRZ) pseudorandom binary sequence (PRBS) 2^{31}-1 electrical pattern, an optical fiber amplifier (FIBERLABS AMP-FL8611-OB), a receiver (Mitsubishi FU-397SPP) with a responsivity of 0.6 A/W, and an oscilloscope (Agilent 86100C, 86116C-025). The extinction ratio was set to greater than 5.0 dB.

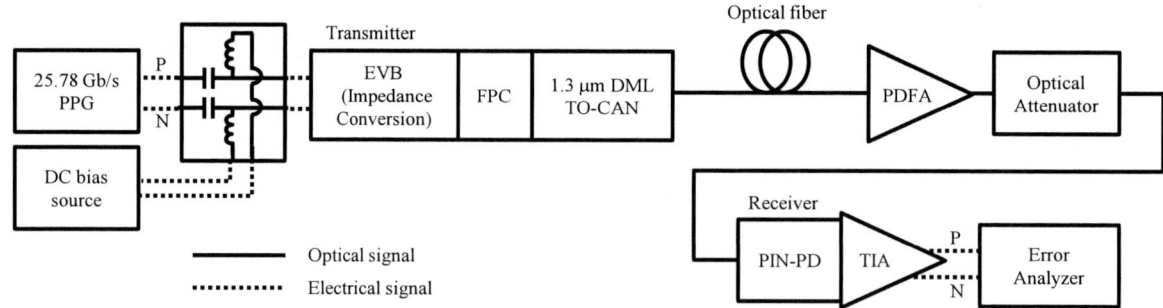

Fig. 4. Experimental setup for BER testing over 10 km fiber transmission.

Fig. 5 shows the 25.78 Gb/s eye diagrams following a 4th-order Bessel filter. The bias currents at −5°C, 25°C and 85°C were set to 30 mA, 35 mA and 90 mA, respectively. Clear eye diagrams with mask margin (MM) greater than 7% were obtained at all temperatures.

The BER test results for back-to-back connection (solid lines) and 10 km fiber transmission (dashed lines) are shown in Fig. 6. We estimated the minimum receiver sensitivity at BER = 10^{-12} after 10 km SMF transmission at average powers of −8.8 dBm at 25°C, −8.7 dBm at −5°C, and −8.3 dBm at 85°C. These results indicate that the dispersion penalty for 10 km SMF transmission is less than 0.2 dB and the power penalty over the temperature range is less than 0.7 dB.

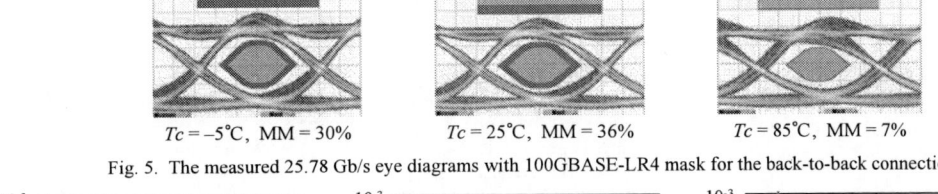

$Tc = -5°C$, MM = 30% $Tc = 25°C$, MM = 36% $Tc = 85°C$, MM = 7%

Fig. 5. The measured 25.78 Gb/s eye diagrams with 100GBASE-LR4 mask for the back-to-back connection.

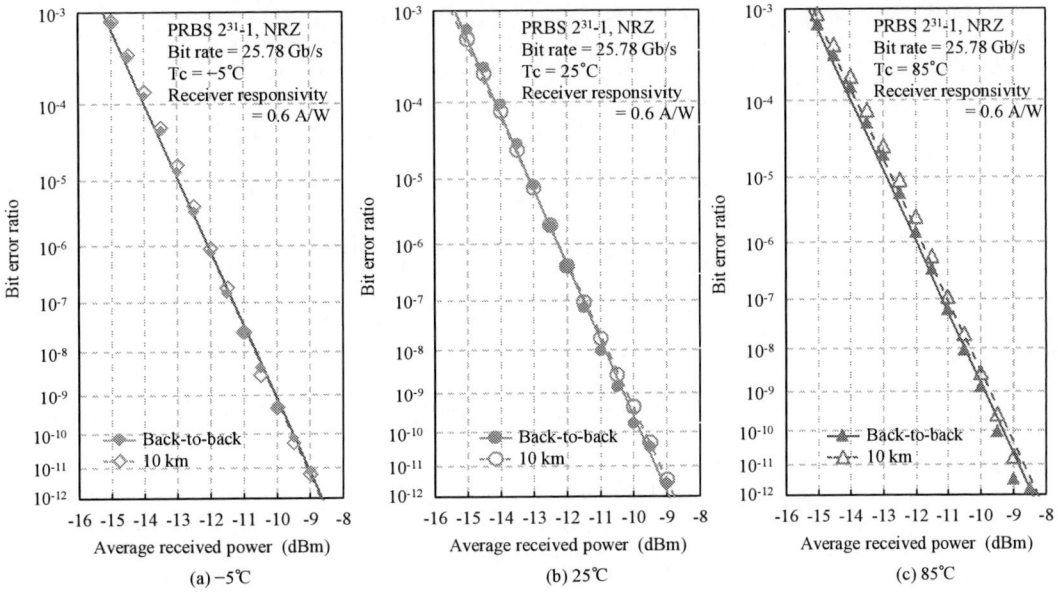

Fig. 6. BER test results for back-to-back connection (solid lines) and 10 km SMF transmission (dashed lines) at case temperatures (a) −5°C, (b) 25°C, and (c) 85°C.

IV. CONCLUSIONS

A 1.3 µm wavelength uncooled 25.78 Gb/s DML in a TO-CAN package was successfully developed. In experimental testing we obtained clear eye diagrams complying with the 100GBASE-LR4 mask at case temperatures ranging from −5 to 85°C. Transmission tests demonstrated a low power penalty of less than 0.7 dB over 10 km of SMF. This uncooled 25.78 Gb/s DML in a TO-CAN package is expected to enable cost-effective optical transceivers with low power consumption for high-speed Ethernet.

REFERENCES

[1] http://www.ieee802.org/3/25GSMF/

[2] G. Sakaino, et al., "25.8Gbps Direct Modulation AlGaInAs DFB Lasers with Ru-doped InP Buried Heterostructure for 70°C operation", OFC2012, Oth3F.3 (2012).

[3] N. Nakamura, et al., "25.8Gbps Direct Modulation AlGaInAs DFB lasers of low power consumption and wide temperature range operation for data center", OFC2015, W2A.53 (2015).

[4] K. Hasebe, et al., "Error-free 25-Gbit/s direct modulation of lateral-current-injection DFB laser", CLEO2014, STh1G.2 (2014).

[5] T. Fukamachi, et al., "Uncooled clear-eye-opening operation (25 to 95°C) of 25.8/28-Gbps 1.3-µm InGaAlAs-MQW directly modulated DFB lasers", OFC2014, Th3A.7 (2014).

[6] W. Kobayashi, et al., "40-Gbps direct modulation of 1.3-µm InGaAlAs DFB laser in compact TO-CAN package", OFC/NFOEC2011, OWD2 (2011).

[7] T.-T. Shih, et al., "A 25 Gbit/s Transmitter Optical Sub-Assembly Package Employing Cost-Effective TO-CAN Materials and Processes", J. Lightw. Technol., vol. 30, no. 6, pp. 834-840 (2012).

[8] T. Tadokoro, et al., "43 Gb/s 1.3 µm DFB Laser for 40 km Transmission", IEEE J. Lightw. Technol., vol. 30, no. 15, pp.2520 -2524 (2012).

[9] http://www.ieee802.org/3/ba/

OECC/PS2016

Stabilization of 14xx-nm High Power Semiconductor Laser By Single Fiber Bragg Grating Configuration

Hideaki Hasegawa, Taketsugu Sawamura, Junji Yoshida, and Noriyuki Yokouchi

Furukawa Electric Co., Ltd.

2-4-3, Okano, Nishi-ku, Yokohama, 220-0073, Japan

nyoko@yokoken.furukawa.co.jp

Abstract: *Output power stabilization of 14xx-nm high power semiconductor laser is achieved by using single fiber Bragg grating with relatively low reflectivity and large external cavity length. Well suppressed stimulated Brillouin scattering is confirmed for Raman amplifier applications.*

Keywords: *High Power Laser, 14xx-nm, Fiber Bragg Grating*

I. INTRODUCTION

Fiber Raman amplifiers are attractive for large capacity optical communication systems with advanced multi level formats due to their low noise characteristic as well as suppression of the nonlinearity of the transmission media. High power 14xx-nm semiconductor laser diodes (LDs) are used as pump sources for the amplifiers. Since the wavelength of the LD module is controlled by a fiber Bragg grating (FBG), the output power is essentially unstable due to the optical feedback [1, 2]. Although we already demonstrated that the output power can be stabilized by using double FBGs configuration [3], the single FBG configuration is more attractive from the view of the output power and fabrication cost.

There are 5 operation modes in a FBG stabilized LD module [4]. One is the mode locked by the FBG (FBG mode), another is the one where the LD chip oscillates solitary (Solitary LD mode). The low frequency fluctuation (LFF) mode exhibits switching operation between the FBG and Solitary LD modes with moderate speed (on the order of MHz). The coexistence mode is another switching phenomenon between the FBG and LFF modes with much slower repetition (~msec). The coherence collapse (CC) mode is recognized as the incoherent mixture of FBG and Solitary LD modes. The CC mode is considered to be the best mode to minimize output power fluctuation and have a broad lasing spectrum which is necessary to eliminate the stimulated Brillouin scattering (SBS) in the Raman amplifier systems.

In this study, we investigate the possibility of single FBG configuration to realize the stable operation for pumping of Raman amplifiers.

II. EXPERIMENTAL

We measure the FBG reflectivity and the external cavity length dependences of the lasing modes. The operating mode is assumed by measured power stability in time domain. In this experiment, we use a high speed photodetector (PD) with the response time of 10^{-9} sec. Throughout this experiment, the same LD module is used and the FBG reflectivity is changed by replacing the FBG. It is known that the operation mode moves from LFF to CC with increasing injection current [4]. In order to investigate the threshold of CC, we monitor the lasing mode at relatively lower current of 150mA corresponding to the output power of about 50mW. Figure 1 shows time resolved power fluctuations by changing the FBG reflectivity. The fluctuation seems to be changed from periodic to random with decreasing FBG reflectivity. This is consistent with the result of the previous works [4].

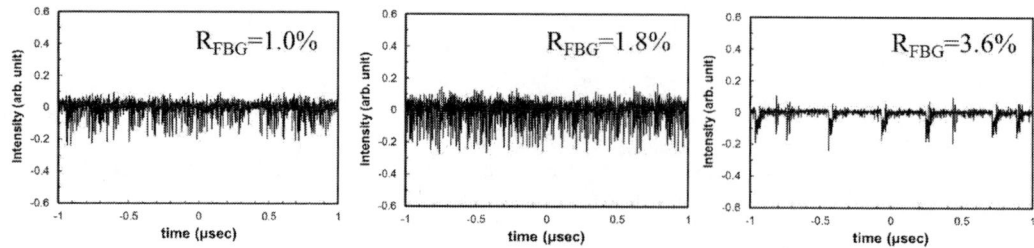

Fig. 1. Time resolved output power with different FBG reflectivity.

The same measurement is done on the module with different external cavity length, where the reflectivity of the FBG is fixed at 1.8%. The injection current is adjusted to be 110mA, where clear external cavity length dependence is observed since the operation mode is thought to be close to the boundary between LFF and CC modes. As shown in Fig. 2, the module having longer external cavity seems to be operating in CC mode.

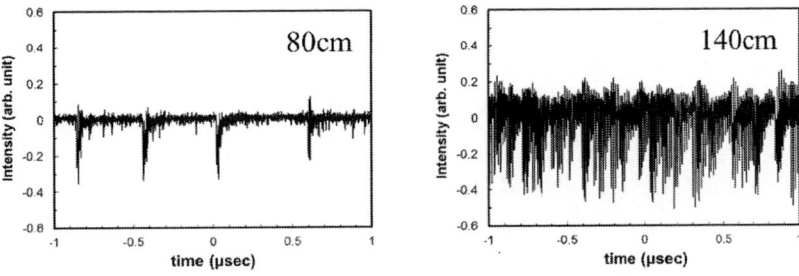

Fig. 2. Time resolved output power with different external cavity length.

We measure SBS to optimize the external cavity length. If the spectral linewidth of the LD module is below the SBS threshold, we will observe the amplified reflection due to SBS. On the other hand, if the LD module operates in CC mode, the linewidth should broad enough to suppress the SBS. Figure 3 shows experimental setup we use to measure a reflection power to confirm if SBS occurs in the transmission line. A LD module with FBG is connected with 100km True-Wave Fiber via 20dB coupler and output port is terminated under anti-reflection condition. The reflection light is monitored by high speed PD after the 20dB coupler. Reflection power fluctuation is defined as

$$\text{Reflection Power Fluctuation} = \frac{P_{\max} - P_{\min}}{P_{ave}} \qquad (1)$$

where subscripts of max, min and ave mean the maximum, the minimum and the average power, respectively. Figure 4 shows the injection current dependence of the reflection power fluctuation measured on the LD module with a FBG reflectivity of 1.8% and the external cavity length of 170cm. We don't see significant power increasing which is an evidence of suppression of the SBS even at the minimum current of 100mA, which corresponds to the output power of 20mW.

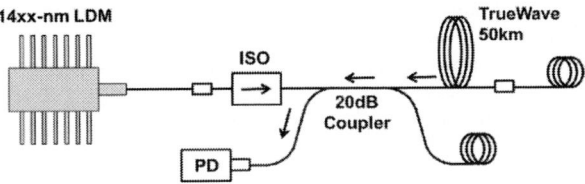

Fig. 3 The experimental setup for SBS evaluation.

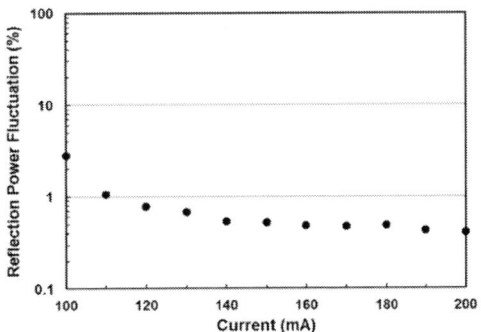

Fig. 4 The measured reflection power fluctuation.

Finally, a wavelength detuning $\delta\lambda$ dependence is evaluated by SBS experiment. The detuning is defined as the lasing wavelength λ_{FBG} relative to the gain peak λ_g as λ_g- λ_{FBG}, and is controlled by changing device temperature in this experiment. Figure 5 shows the measured reflection power with the device temperature of 25C, 30C and 35C, corresponding detuning is -14nm, -11nm and -8nm, respectively. As shown in the figure, the reflection power fluctuation of 35C condition is larger than 25 and 30C operation at 100mA and 160~180mA. The largest fluctuation as high as 1000% is observed at 100mA. This higher fluctuation is the evidence of SBS which may be caused by unstable operation under modes other than CC. From this experiment, the detuning should be below -8nm to avoid SBS. The lower limit of the detuning is determined by the wavelength lock-in characteristic. Much larger detuning results in the lasing at the gain peak, so called Fabry-Perot oscillation. In this experiment, material parameters such as gain, refractive indices, and so on, are changed as well as the detuning, so that the available detuning could be slightly different for the actual module design.

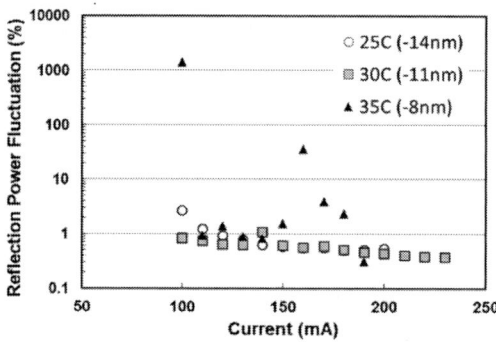

Fig. 5 The detuning dependence of the reflection power fluctuation.

III. CONCLUSIONS

We optimized the reflectivity and the external cavity length of the 14xx-nm LD module with single FBG configuration for fiber Raman amplifier applications. Single FBG with the appropriate reflectivity and the longer external cavity can eliminate SBS even at the lower output power of 20mW. This result will contribute to realization of improved LD module with higher output power and low cost.

REFERENCES

[1] R. Lang, and K. Kobayashi, "External optical feedback effect on semiconductor injection laser properties," IEEE J. of Quantum Electron., vol. QE-16, pp.347-355, March 1980.
[2] M.K. Davis, G. Ghislotti, S. Balsamo, D.A. Loeber, G.M. Smith, M.H. Hu, and H.K. Nguyen, "Grating stabilization design for high-power 980-nm semiconductor pump lasers," IEEE J. of Selected Topics in Quantum Electron. vol. 11, pp. 1197-1208, Sep. 2005.
[3] J. Yoshida, T. Sawamura, M. Miura, S. Irino, H. Itoh, M. Terada, T. Okada, H. Hasegawa, and N. Yokouchi, "Lasing stability of high power 14xx-nm pump lasers for Raman amplifiers," Optical Fiber Communication Conference 2016 (OFC2016), Anaheim, CA (USA), March, 2016, paper W2A.4.
[4] T. Heil, I. Fischer, and W. Elsaber, "Coexistence of low-frequency fluctuations and stable emission on a single high-gain mode in semiconductor lasers with external optical feedback," Phy. Rev. A, vol. 58, R2672, Oct., 1998.

Fast Wavelength Switching of DFB LD

Yuto Ueno, Keita Mochizuki, Kiyotomo Hasegawa, Masamichi Nogami, and Hiroshi Aruga

Information Technology R&D Center, Mitsubishi Electric Corporation, 5-1-1 Ofuna, Kamakura, Kanagawa, 247-8501 Japan

E-mail: Ueno.Yuto@ap.MitsubishiElectric.co.jp

Abstract: A fast wavelength switching method for DFB LDs is studied. The switching time is 86 ms. We also show that a wide tuning range can be obtained by improving the heat confinement in the LD.

Keywords: Tunable laser, Fast wavelength switching, DFB

I. INTRODUCTION

In wavelength division multiplexing (WDM) networks, tunable lasers are employed as the light sources [1]. When a fault occurs in a WDM network, optical path restoration with wavelength conversion is in general required to be completed within 100 ms.

For fast wavelength switching, a Distributed Bragg Reflector Laser Diode (DBR LD) or External Cavity Laser Diode (ECLD) is often employed. In these LDs, the wavelength tuning region is separate from the active region, and fast wavelength switching is realized by changing the refractive index of the wavelength tuning region by means of carrier density effects or localized heating [2][3]. However, these LDs tend to suffer from wavelength jumping due to mode-hopping. A Distributed Feedback Laser Diode (DFB LD) has higher stability and reliability, and is widely employed as a light source in optical communication systems [4]. In a DFB LD, the wavelength is usually switched by changing its temperature using a thermo-electric cooler (TEC). However, the switching speed is limited by the heat capacities of the LD submount and the TEC itself, and it can take several seconds to switch the wavelength. In order to achieve fast wavelength switching of a DFB LD, we propose a new switching method which combines manipulation of the LD current with thermal compensation [5]. In this previous work, we set a delay between the manipulation of the LD current and thermal compensation in order to realize optimum switching. Such a delay would be undesirable for an operating telecommunications system in which the time from command to switching is important.

In this paper, we evaluate the response of our switching method without any additional delay. We also study the tuning range and show that the tuning range can be significantly widened by improving the heat confinement in the LD.

II. CONCEPT BEHIND FAST WAVELENGTH SWITCHING

The conceptual structure of the proposed tunable light source is shown in Fig. 1. In our method, the wavelength is changed by manipulating the LD current. Changing the LD current changes the amount of heat generated in the active region. Since the region heated by this "on-chip heater" is very small, the tuning speed is high. The TEC and thermistor are used only to stabilize the temperature of the LD via an automatic temperature control (ATC) loop. The thermal compensation chip is used to "deceive" the thermistor at the time of switching to achieve fast wavelength switching. Fig. 2 shows the transient response of the LD temperature. When the compensation current is not changed at the time of switching as shown in (a), the change in the heat generated affects not only the LD temperature but also that of the thermistor. Since the ATC loop works to keep the thermistor's temperature constant, the TEC slowly alters both the LD temperature and the thermistor's temperature, which results in a slow wavelength drift. To suppress this drift, we adopted the thermal compensation shown in (b). If the effect of changing the LD current is thermally compensated so as not to alter the thermistor's temperature, the TEC can be prevented from reacting. As a result, the tendency to a slow wavelength drift after switching the wavelength is suppressed. By placing the compensation structure outside the chip, this method can be applied to a DFB LD without any need for special construction.

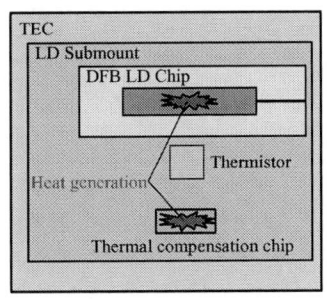

Fig.1. Structure of the light source

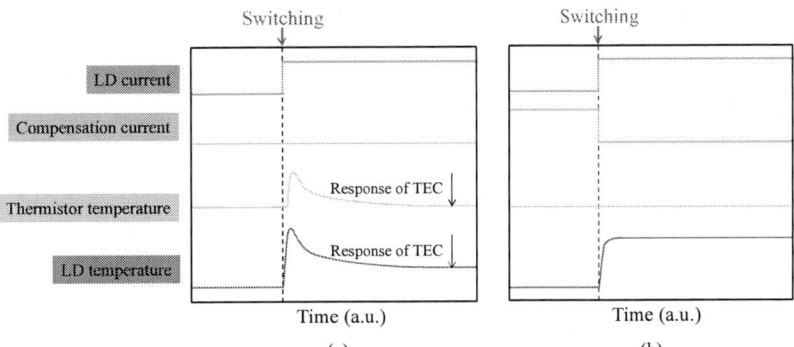

Fig.2. Transient response (a) w/o, and (b) with thermal compensation

III. SWITCHING TIME

Fig. 3 shows the structure of our prototype module. A DFB LD and a semiconductor optical amplifier (SOA) are integrated on the same chip. For thermal compensation, we employed a film resistor mounted on a ceramic substrate. In this module, the thermistor does not lie between the DFB LD and the thermal compensation chip due to constraints imposed by the components we had available.

The thermal resistance between the DFB LD and the thermistor, and that between the compensation chip and the thermistor depends on the layout of the module. Also, if multiple LDs are integrated on the same chip, the thermal resistance to the thermistor varies according to the LD selected. Therefore, it is important to set the compensation conditions in accordance with the situation. One guideline is that the thermal resistance is almost proportional to the distance if the element transferring the heat is the same. First, we established a temporary compensation condition based on the distances to the thermistor from the DFB LD and the compensating chip, and the change in the heat generated in the DFB LD. Then, based on that, we modified the conditions slightly to obtain the fastest switching. Fig. 4 shows the transient response of the frequency with the optimal compensation conditions obtained in that manner. The same data are used for both (a) and (b), but the frequency ranges covered by the graphs are different. The switching time was 86 ms with an allowable frequency tolerance of +/-0.1 GHz, which satisfies the general requirements for wavelength restoration. In the previous work [5], we switched the compensation current 50 ms prior to switching the LD current for optimum switching. However, for an operating telecommunications system, such a delay is undesirable because the time from command to switching is important. This time, we have been able to complete the switching operation in less than 100 ms without any additional delay.

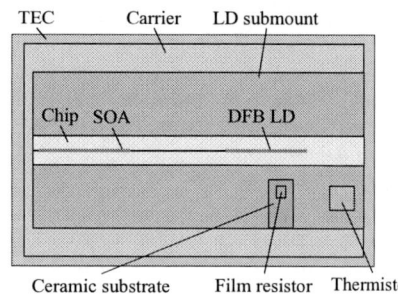

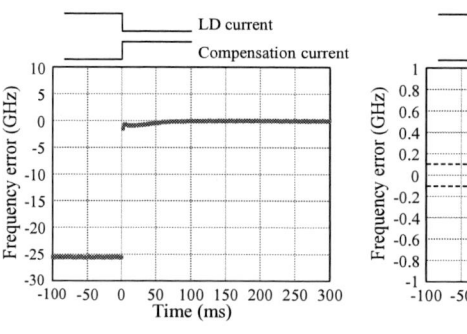

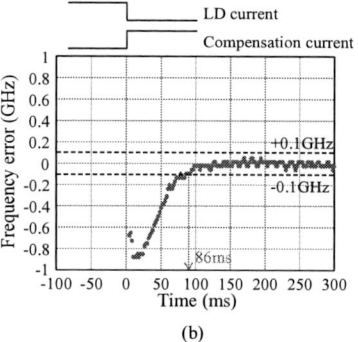

Fig.3. Structure of the prototype module   Fig.4. Transient response of frequency. Frequency range (a) -30 to +10 GHz, (b) -1 to +1 GHz

IV. TUNING RANGE

In order to evaluate the tuning range of our module, we measured the wavelength at various LD currents. Fig. 5 shows the results. Changing the LD current too much is impractical since the output power of the LD is insufficient at low LD currents, while the risk of damaging the LD increases at high LD currents. In this experiment, we adjusted the LD current within the range +/-80 mA from the nominal value. As for wavelength, we also plot the relative values. By changing the relative LD current ΔI_{LD} from -80 mA to +80 mA, the wavelength increased by 0.38 nm. For DFB LDs, it is usual to integrate multiple LDs with different wavelengths in order to obtain a wide tuning range. However, in order to fully cover the C band (1530 - 1565 nm) and the L band (1565 – 1625 nm) in our module, approximately 100 LDs for C band and more than 100 LDs for L band would have to be integrated in one chip, which is not realistic.

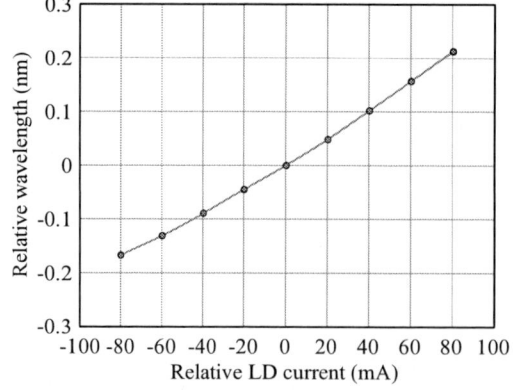

Fig. 5. Tuning range

For our method, the heat confinement inside the chip is an important factor in obtaining the wide tuning range. When the temperature is controlled by an on-chip heater as in our method, it is ideal that the heat generated in the heater does not flow into the LD submount and that it contributes only to the temperature rise of the LD. Therefore, from the view point of tuning range, a structure which can strongly confine the heat generated in the LD inside the chip is suitable for our method. We consider that the heat confinement is improved by employing a submount with a trench which covers the region under the LD, and conducted thermal analyses to see the effect of changing the structure of the submount.

Fig. 6 shows the cross sectional temperature distributions in our module obtained by thermal analysis for 4 different combinations of LD submount's structure (w/ or w/o a trench) and ΔI_{LD} (-80 mA or +80 mA). In this analysis, the LD current is converted into an amount of heat determined by the voltage-current characteristics of the LD. The TEC actually lies under the carrier although it is not shown in this figure. We set the temperature of the top side of the TEC to 15°C and the ambient temperature to 25°C. Note that the temperature ranges shown in (a), (b), (c), and (d) are different from one another since the temperatures around the LD are significantly different. When a submount without a trench is used, the LD temperature changes by 3.1°C when ΔI_{LD} changes from -80 mA to +80 mA (see (a) and (b)), which corresponds to 3.1 nm change in wavelength and accordingly is in good agreement with the experimental result shown in Fig. 5. On the other hand, when a submount with a trench is used, the temperature change as ΔI_{LD} changes from -80 mA to +80 mA is about 27°C (see (c) and (d)). This change in the LD temperature corresponds to a change in wavelength of about 2.7 nm. For example, by using a 16 DFB LD array [4], we obtain an approximately 43 nm tuning range, which we think is sufficient for practical use.

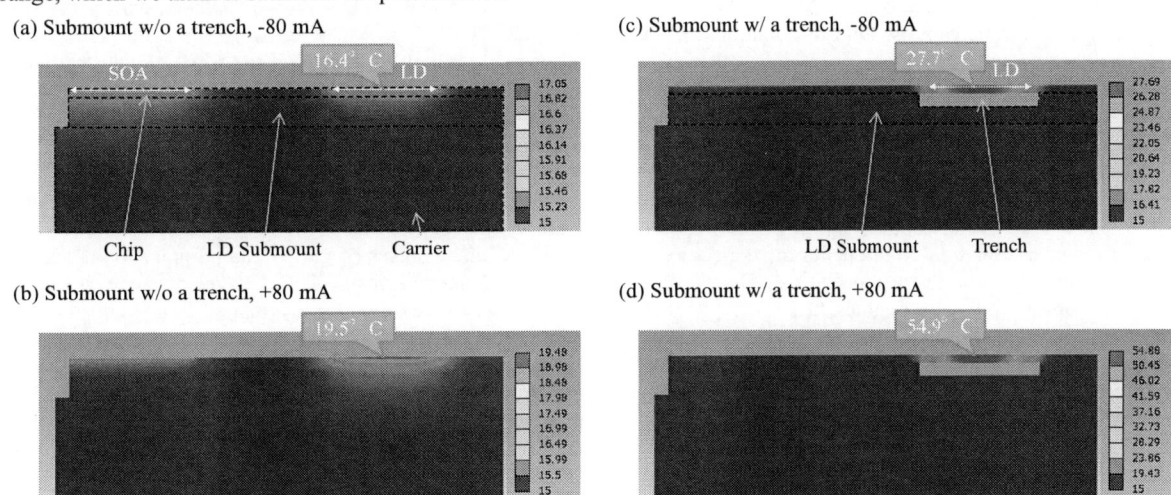

Fig. 6. Cross sectional temperature distribution (simulated), (a)–(d): LD submount's structure (w/ or w/o a trench) and relative LD current as indicated

V. CONCLUSIONS

A fast wavelength switching method for a DFB LD was studied. By optimizing the switching conditions, we obtained a switching time of 86 ms, which satisfies the overall requirement for wavelength restoration. We also showed that the tuning range can be significantly widened by improving the heat confinement in the LD. We believe this switching method will be useful for future WDM networks.

ACKNOWLEDGMENT

This research was in part supported by the "R&D on Optical Signal Transmission and Amplification with Frequency/Phase precisely controlled carrier" project of the National Institute of Information and Communication Technology (NICT) of Japan.

REFERENCES

[1] J. Buus et al., "Tunable lasers in optical networks," J. Lightwave Technol., vol. 24, no.1, pp. 5-11 (2006).

[2] H. Matuura et al., "6.25 GHz Flexible Grid Tuning of Fully Heater-tuned CSG-DR Lasers with Sub-millisecond Wavelength Switching," ECOC, no. Th1B4, London, UK (2013).

[3] K. Sato et al., " Demonstration of Silicon Photonic Hybrid Ring-Filter External Cavity Wavelength Tunable Lasers," ECOC, no. We.2.5.4, Valencia, Spain (2015).

[4] Y. Sasahata et al., "Tunable 16 DFB Laser Array with Unequally Spaced Passive Waveguides for Backside Wavelength Monitor," OFC, no. Th3A-2, San Francisco, USA (2013).

[5] Y. Ueno et al., "Fast Wavelength Switching with DFB Lasers Utilizing Thermal Compensation", OECC, no. JWeC.13, Shanghai, China (2015).

WD4-1 (Invited)

OECC/PS2016

Optical Chirp and Amplitude Processing Using EAM integration

Yi-Jen Chiu, Shin-Wei Shen, Jui-Pin Wu, Kuo-Chun Chang, and Chia-Chien Wei

Department of Photonics, National Sun Yat-Sen University, Kaoshiung 80424, Taiwan R.O.C
Tel: +886-7-525-2000 ext 4460, Fax: +886-7-525-4499
Email: yjchiu@faculty.nsysu.edu.tw

Abstract: *Optical modulation with pre-chirp control using electro-absorption modulator (EAM) and its integration have been reviewed. Device design, fabrication, and application of high-speed chirp-control intensity modulation are also discussed for OOK and OFDM modulation scheme.*
Keywords: *(EAM; modulation; chirp; high speed; integration)*

I. INTRODUCTION

Electroabsorption modulator (EAM) based on multiple quantum well (MQW) active region has been widely used in optical fiber communications. Due to quantum confines Stark effect (QCSE) in MQW and the waveguide structure, EAM has been found with high speed modulation, high extinction ratio, low driving power, and low chirp. In addition to such good properties, MQW based semiconductor layer intrinsically enables the monolithic integration with other semiconductor devices, for example, laser, semiconductor optical amplifier (SOA), leading to the key function in photonic integration. Therefore, in its application on optical transmission, EAM integration has arisen lots of attention for high-speed local area networks (LANs), such as 100Gb Ethernet (100GbE), and 25Gb/s based modulation [1-2].

However, during high-speed modulation for network application, the dispersion properties along long optical fiber will inevitably deteriorate optical signal waveform, suggesting the chirp becomes one of the major factors in developing transceiver or modulation components for the future. In single mode fiber, pulses of optical signal at 1550nm regime will experience serious distortion and also the power fading in long distance transmission. Longer distance or higher speed the optical pulse propagates, the more profound the chirp effect gives. High performance with long distance transmission has been found with setting negative chirp in transceiver [3], intriguing the interests of developing pre-chirp function of components. With respect to the modulation of EAM, the field-driven optical absorption in MQW will create intensity modulation, where the variation of optical absorption will also induce the optical index change (so called chirp). Through QCSE, exciton-enhancement optical absorption further generates near zero chirp or even negative chirp at high field regime. In the point view of optical network application, EAM enables more functions to be endorsed into optical signal processing. By integration EAM with other devices, such as SOA, low chirp performance of EAM can extend the transmission distance of on-off keying (OOK) data [4-5]. Likewise, it could be worth noting that such properties can be also applied to other famous modulation schemes in order to get higher spectral efficiency. Orthogonal Frequency Division Multiplexing (OFDM) modulation would be one of the best suitable applications. Through applying negative chirp, EAM has been further used for OFDM for high spectral efficiency [6]. By biasing EAM, chirp control can thus be realized for high capacity and long distance. Thereby, it is quite important to research and develop optical processing properties using EAM-related integration for future optical fiber communications.

In this paper, several design issues of EAM for high speed and low chirp application is described based on MQW active region. Some recent progress regarding the integration techniques of EAM will also be addressed for high-speed chirp-control intensity modulation, where the system application using OOK and OFDM schemes will also be included for demonstration template.

II. AMPLITUDE AND CHIRP MDOULATION BY EAM

By employing MQW as active region, the optical absorption can be greatly enhanced through QCSE and also the exciton transition from the attraction between electron and hole. Thus the electro-absorption can be utilized for efficient modulation, especially at the regime near the transition energy level of quantum well. Figure 1 shows the schematic QCSE of MQW. In order to further illustrate amplitude modulation and chirp variation, a 1550nm InGaAsP MQW with 200nm thickness is used as example. Figure 1 plots the simulated optical absorption spectrum for the first transition energy level of MQW. The significant peak of absorption spectrum occurs in exciton levels, thus greatly improving absorption change by applied electric field. By increasing bias (0, 0.5, and 1V), the peak of spectrum has the red shift, leaving the operation with the largest absorption change to be on the absorption edge of red site (one is near 1575nm from 0V to 0.5V, and the other is near 1590nm from 0.5V to 1V). Simultaneously, the field-driven MQW will induce the index change and thus the phase, where the index change is defined as relative to 0V case. To be noted that the chirp is defined as in equation (1) :

$$C = 2k_0 \cdot \frac{\Delta n_r}{\Delta \alpha} \qquad (1)$$

where $k_0=2\pi/\lambda$ is the propagation constant in vacuum, and Δn_r and $\Delta \alpha$ are the refractive index and absorption coefficient change with respect to the bias change. Thereby, the chirp parameter of EAM actually specifies the light phase variation as varying absorption. As can be seen in figure 1, the largest change in absorption could result in near zero chirps. Furthermore, with increasing bias, the more negative chirp will be formed. The large field-driven optical absorption with near zero chirp suggests that the good waveform quality of optical signal could be modulated though EAM.

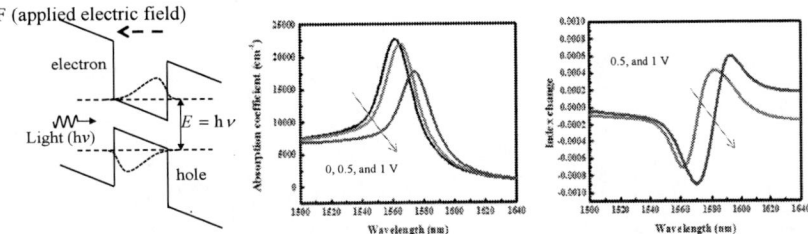

Figure 1. Left is the schematic diagram of optical absorption in MQW. Right is the calculated optical absorption coefficients of MQW for bias voltage of 0, 0.5, and 1V and the corresponding index change related to 0V.

III. APPLICATION OF OOK AND OFDM MODULATION BY EAM

As for aforementioned discussion, EAM can be treated as the chirp-control element, where swing the chirp from positive values to negative ones can be made just by increasing voltage. But, the operation point of negative chirp occurs at the high bias regime of EAM. In other words, high performance of waveform from negative chirp will be the price of high optical absorption for data transmission. In order to obtain high quality of optical signal transmission, signal processing need simultaneously consider both amplitude modulation and chirp control. Monolithic integration of SOA with EAM was proposed for optical signal transmission [5]. Since the integration with optical gain element can be used for compensating optical loss and reducing low coupling loss between devices, the output optical power can be enhanced with operating EAM at negative chirp side.

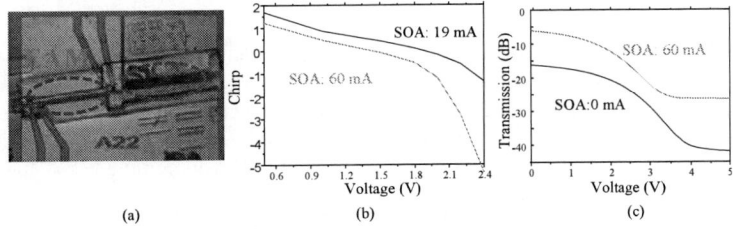

Figure 2, (a) the finished SOA-integrated EAM; (b) the measured chirp of EAM as SOA pump current is 19, and 60mA; (c) the measured optical transmission as SOA pump current is 0, and 60mA.

Figure 2(a) shows the finished SOA-integrated EAM. The same material structure was used for both EAM and SOA. 6 periods of InGaAsP-based MQW with transition wavelength of 1550 nm was grown as active region. To reduce the optical reflection on the facets, waveguide with 7^0 off alignment from cleaved facet is fabricated. Ion implantation is used to define the connection between EAM and SOA for electrical isolation. In order to get high-speed and low-loss waveguide for modulation, optical waveguide was defined by selectively undercutting etching active region [7]. Also, coplanar waveguide (CPW) was employed for loading and feeding microwave.

Figure 2(b) and 2(c) plot the measured chirp parameters and optical transmission of SOA-integrated EAM against with EAM voltage. In order to check the chirp effect from single EAM, the condition of SOA pumped with transparent current, 19mA, is used for comparison. High pump current of 60mA into SOA is used for amplifying signal. Optical wavelength centered 1582nm is used. By increasing the reversed bias of EAM, the chirp decreases from positive values to negative ones for all pump current levels. It can be seen that the zero chirp point occurs near at the operation point of maximum slope, suggesting that high modulation efficiency accompanying with low chirp can be realized through EAM for long distant transmission.

In order further confirming amplitude and phase modulation of EAM, OFDM as well as OOK modulation are used system demonstration. As shown in figure 3(a), OFDM modulation is applied for examining SOA-integrated EAM. 2V- and 1.5V- bias cases are compared, while the modulations show the same constellation of 16QAM in the back-to-back (BTB) conditions. However, as transmitting in 100km fiber, clear constellation plot is found at 2V bias. The higher loss operation (2V), yet, leads to the better signal-to-noise ratio than the lower loss one (1.5V) for transmitting 100 km fiber, indicating EAM chirp the major factor determining waveform quality. Above 20 Gb/s on 100km fiber transmission has been obtained in such SOA-integrated EAM. In addition, with such EAM-SOA serial integration, optical processing can also be beneficial from saturating both devices. By exciting elements with enough power, SOA saturation renders the overall chirp more negative. In addition, due to the mutually reversed carrier dynamic between EAM and SOA, i.e.

EAM is driven by field, and SOA is by current, the waveform pattern of optical signal can be maintained through controlling the bias points. As shown in figure 3(b), two 10Gb/s electrical pulse trains, one is 8 bit of '1'' and the other is 1 bit, are used to excite the devices. With 70mA current injection in SOA, the distorted optical pulse pattern can be recovered. OOK modulation is also employed for testing. As shown, 10 Gb/s eye diagrams without error floor transmission are measured for 3.5V and 4.5V biases. Through biasing to get negative chirp and low pattern dependence condition, the higher bias (4.5V) with high injection current (70mA) could realize successful longer transmission (36km), although the lower power condition occurs at 3.5V case.

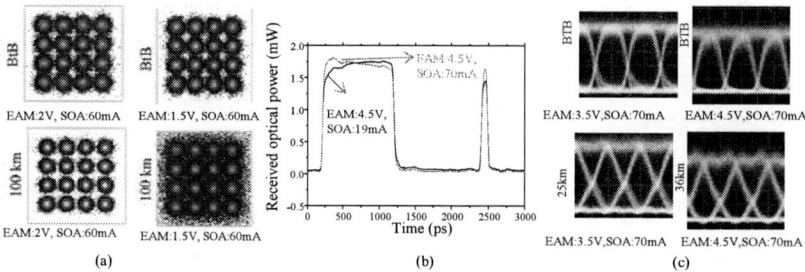

(a)   (b)   (c)

Figure 3, (a) the constellation diagrams using OFDM modulation on SOA-integrated EAM at 1.5V and 2V for back-to-back (BTB) and 100km transmission; (b) pulse evolution using OOK electrical pulse train; (c) the OOK eye diagram at 3.5V, 4.5V for different transmission distance.

There is another advantage of chirp control by using EAM integration. Owing to the compact size and high speed of EAM, serial connection of small EAMs can be easily obtained. In addition, with different biases on different devices, the combinational phase processing of signal can be set through the large chirp variation by sweeping chirp using voltage. Single side band (SSB) modulation for long distance transmission is one of important applications. The cascaded integration of two EAMs has been proposed for SSB [8]. The phase difference between double side band of modulated signal can be brought out by setting two large chirp differences. Thus, with adjusting bias, one of side band bandwidth could be removed due to the self-interference of signal. SSB modulation has been demonstrated through cascaded EAMS, leading to 13.5Gb/s and 200km of OFDM transmission.

IV. CONCLUSIONS

This work reviews the functions of EAM integrations for optical modulation. With employing MQW as active region, intensity as well as chirp modulation can be realized through the QCSE of EAM, enabling efficient optical processing. Because of the exciton property in MQW, strong field-driven optical absorption in EAM induces low chirp, enabling the long-distance transmission of optical fiber. Chirp-control intensity modulation based on OFDM and OOK modulations for long fiber transmission has been demonstrated by using SOA-integrated EAM and cascaded EAMs.

ACKNOWLEDGMENT

This work was supported by the financial supports from the National Science Council, Taiwan (NSC99-2221-E-110-097-MY3 and NSC101-2622-E-110-004-CC3). Also, the authors would like to thank the wafer growth from Land Mark Optoelectronic Corporation.

REFERENCES

[1] P. Torres-Ferrera, S.O. Vázquez, R. Gutiérrez-Castrejón,"4×100 Gb/s WDM DD-OFDM using EAM for next generation Ethernet transceivers over SMF," Optics Communications, vol.365, pp 86–92, 2016,.

[2] T. Fujisawa, K. Takahata, W. Kobayashi, T. Tadokoro,N. Fujiwara, S. Kanazawa and F. Kano,"A 1.3 μm, 50 Gbit/s electroabsorption modulator integrated with DFB laser for beyond 100G parallel LAN applications," Electronics Letters, vol.47, pp708-710, 2011.

[3] Wataru Kobayashi, Masakazu Arai, Member, Takayuki Yamanaka, Naoki Fujiwara, Takeshi Fujisawa, Takashi Tadokoro, Ken Tsuzuki, Yasuhiro Kondo, and Fumiyoshi Kano," Design and Fabrication of 10-/40-Gb/s, Uncooled Electroabsorption Modulator Integrated DFB Laser With Butt-Joint Structure," J. of Lightwave Technol. vol.28, pp164-171, 2010.

[4] M. N. Ngo, H. T. Nguyen, C. Gosset, D. Erasme, Q. Deniel, N. Genay, R. Guillamet, N. Lagay, J. Decobert, F. Poingt, and R. Brenot, "Electroabsorption modulated laser integrated with a semiconductor optical amplifier for 100-km 10.3 Gb/s dispersion-free-penalty-free transmission," J. Lightwave Technol. 31(2), 232–238 (2013).

[5] Jui-Pin Wu, Wei-Zun Ding, and Yi-Jen Chiu," Low-Pattern-Dependence Prechirp Optical Modulation by Using Saturation Behaviors of SOA-Integrated EAM," J. of Lightwave Technol. vol.31, pp3651-3657, 2013.

[6] K. C. Chang, S. W. Shen, M. C. Hsu, Y. J. Chiu, C. C. Wei, and C. K. Lee, "Negative chirped EAM-SOA for distance insensitive optical OFDM transmission in long reach OFDMA PONs," in Proceedings of OFC, Tu3H.4 (2014).

[7] Tsu-Hsiu Wu, Jui-Pin Wu, Yi-Jen Chiu," Low-Power-Driven and Low-Optical-Loss 40-Gb/s Electroabsorption Modulator Using Self-Aligned Two-Step Undercut-Etched Waveguide," IEEE Electron Device Letters, vol33, pp1021-1023, 2012.

[8] Hsuan-Lin Cheng,Wei-Hung Chen, Chia-ChienWei, and Yi-Jen Chiu, "Optical single-sideband OFDM transmission based on a two-segment EAM," Optics express, vol.23, pp982-990, 2015.

WD4-2   OECC/PS2016

Concurrent DWDM Transmission with Ring Modulators Driven By a Comb Laser with 50GHz Channel Spacing

M. Ashkan Seyedi[1,*], Chin-Hui Chen[1], Marco Fiorentino[1], Daniil Livshits[2], Alexey Gubenko[2], Sergey Mikhrin[2], Vladimir Mikhrin[2], and Raymond G. Beausoleil[1]

[1]Hewlett Packard Labs, Hewlett Packard Enterprise, Palo Alto, CA
[2]Innolume GmbH, Dortmund, Germany
*ashkan.seyedi@hpe.com

Abstract: A multi-channel ring-based transmitter is excited by a comb laser with 50GHz channel spacing and two channels are concurrently modulated at 10Gb/s. Bit error ratio tests show ~3dB optical power penalty for 50GHz relative to larger channel spacing.
Keywords: Optical Interconnects; Diode lasers; Modulators; Resonators

INTRODUCTION

In order to meet the data traffic demand in data centers, dense wavelength division multiplexing (DWDM) architectures that use silicon microring resonator modulators have been proposed[1], due to their small footprint, low insertion loss and inherent wavelength selectivity[2]. The bandwidth available via ring-based DWDM links is a complex trade-off between available optical power, receiver sensitivity, cavity quality factor Q, and free spectral range (FSR), which determine the channel count, modulation data rate and the associated crosstalk. Previously, a comb laser with 80GHz channel spacing was used to demonstrate concurrent modulation at 10Gb/s[3]. In order to progress towards the benchmark of Tb/s overall bandwidth, higher modulation data rate and small channel spacing designs are being investigated. Here we demonstrate concurrent transmission at 10Gb/s with 50GHz channel spacing comb laser and confirm the predicted power penalty as a step towards achieving the desired overall bandwidth.

EXPERIMENTAL

Comb lasers (CL) based on quantum dots have been used to successfully demonstrate error-free data transmission[4] at 10Gb/s and 12.5Gb/s with an 80GHz channel spacing. Denser channel spacing for these lasers can possibly enable narrower laser gain bandwidth, increasing the available power per optical tone[5]. Comb lasers are attractive options for the optical engine of a DWDM optical transmitter as they are able to concurrently excite multiple channels, which reduces the number of lasers needed. Furthermore, they are more resilient to temperature changes, have higher material gain and lower threshold current density compared to bulk lasers due to the quantum dot design. Another critical factor of the optical engine of a DWDM system is the relative intensity noise (RIN) which is a function of bias current, temperature, and can vary between the different tones of a comb laser. For the CL used in this experiment (Innolume GmbH), a RIN <-125 dB/Hz is measured at a bias of 165mA/25°C for the ~1310nm wavelength range. At this bias, output power of -1 to -2dBm is observed for each of the tones in the 1308-1315nm wavelength range. To confirm the high quality of this laser and its low RIN, the setup shown in fig. 1a is used to modulate five tones independently with a commercial Lithium Niobate modulator (MOD) by using a bandpass filter (BPF) to spectrally filter the laser. The BPF has a near-square shaped transmission spectrum with a 3dB bandwidth of 40GHz. The resulting spectrum from the drop port of the BPF (fig. 1b) shows the tone of interest that is being modulated at a lower power w.r.t to other tones and the output eye diagrams from the photodetector (PD) is shown in fig. 1c. Each of the eye diagrams show error-free transmission (bit error ratio (BER) of <1E10⁹[6]) and have a signal-to-noise ratio (SNR) of >9dB, confirming low-noise operation of the comb laser.

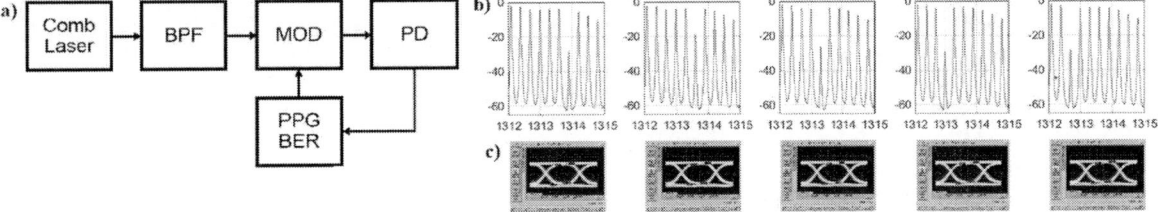

Figure 1 a) experimental setup used to modulate each optical tone. The b) spectrum at the drop port of the BPF is shown in and c) the resulting eye diagram as measured from the PD is shown.

1137

OECC/PS2016

A multi-channel transmitter based on ring resonators using a p-i-n carrier injection modulator was fabricated at CEA-Leti on a 250nm SOI wafer. The ring cavities have a 10 μm diameter and are formed with a 450nm wide waveguide and are coupled to a 350nm wide waveguide with a gap of 250nm and have a drop waveguide with a 350nm gap. The through-port transmission of this device is shown in fig. 2a, where resonances of the channels are denoted. The ring radii are designed to achieve 80GHz spacing, however to due to fabrication variation, the spacing is not exact. These ring resonators have a cavity quality factor, Q, of ~10,000 (FWHM of 0.13nm). The ring modulators also use thermal resistive tuners to precisely control the absolute resonance value. Ch. 3 initially has a detuning of 210GHz and is thermally tuned to the desired values with respect to channel 5 at a detuning of 200GHz, 150GHz, 100GHz, and 50GHz. The detuning is defined as the frequency of ch. 3 less the frequency of ch. 5, as shown by the dashed lines in fig. 2a. For reference, 50GHz corresponds to 0.3nm wavelength separation for the given wavelength range.

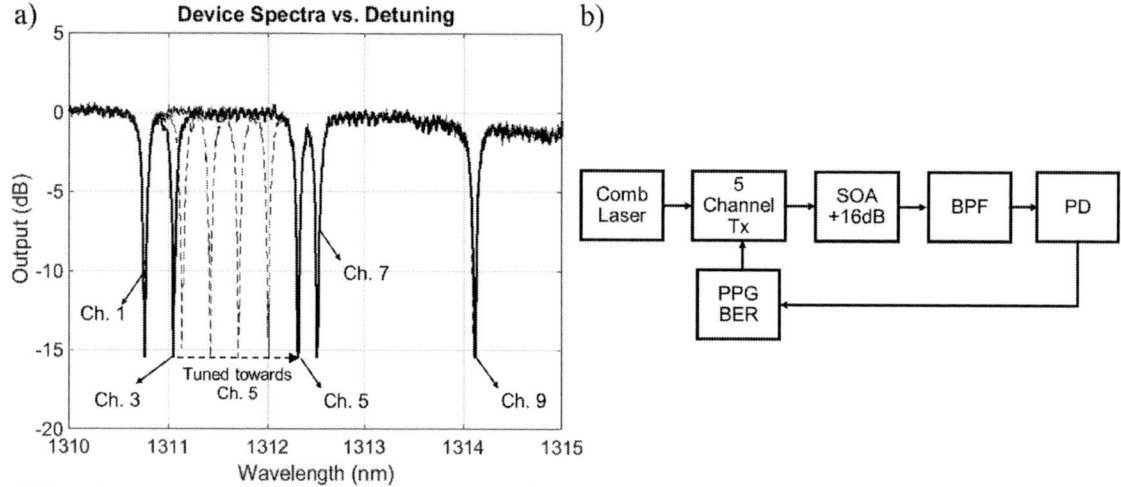

Figure 2 Through-port transmission of the multi-channel device shown in a) with the resonances of channels 1,3,5,7 and 9 marked in solid lines. The dashed lines show the resonance of ch. 3 as it's thermally tuned to various values w.r.t. to ch. 5. The b) optical setup is shown, including the SOA to offset coupling losses to the wafer.

The setup used to test the ring transmitter is shown in fig.2b. A semiconductor optical amplifier (SOA) is used to offset the high coupling loss to/from the wafer. Initially, channels 3 and 5 are optically excited by the CL and modulated independently to confirm single-channel operation and BER. A pre-emphasis signal is generated to electrically drive each of the rings using uncorrelated data sources with a 2^7-1 PRBS word length. Subsequently, the two channels are tuned to the desired value and modulated concurrently. The pre-emphasis signal, individual channel eye diagram, and the resulting eye diagram for each channel at the specified detuning value is shown in fig. 3. The output eye diagrams show wide jitter, which can be attributed to the rise and fall time variation of the pre-emphasis driving signal. Each eye diagram demonstrates SNR of >8dB, which can be expected from the previous experiments using the commercial modulator.

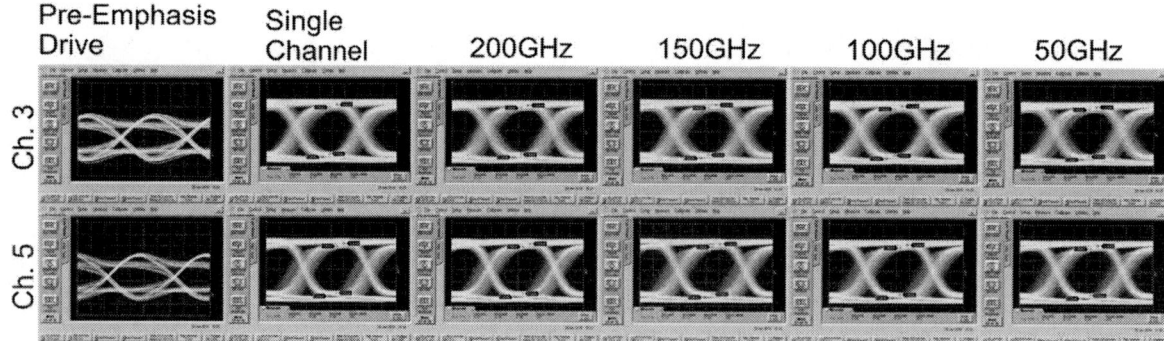

Figure 3 Pre-emphasized drive signal and optical eye diagrams for each channel are shown in the first two columns. Subsequent columns show the concurrent modulation eye diagrams for various detuning values.

In order to understand the cross-talk and associated optical power penalty (OPP) due to the trade-off between channel spacing and modulation data rate, the bit error ratio (BER) of each channel at varying detuning versus input power was measured. Previous experiments[4] using tunable lasers have quantified this cross-talk and predict no OPP for channel spacing of greater than 60GHz. At 50GHz, this same work predicts OPP between 1-3dB for 10Gb/s modulation data rate. Furthermore, we expect an asymmetry of this cross-talk due to the blue-shift of the ring resonator's Lorentzian

1138

transmission function due to the light-matter interaction with the electron plasma. Therefore, we can hypothesize that ch. 3 will see degradation of its BER due to the blue-shift from the modulation of ch. 5 and that no penalty will be observed on the BER of ch. 5. Fig. 4 shows the BER of both channels for various detuning values. We can see that ch. 5, fig.4a, shows no observable OPP and that ch. 3, fig.4b, shows an OPP of ~ 3dB for a detuning of 50GHz, as expected.

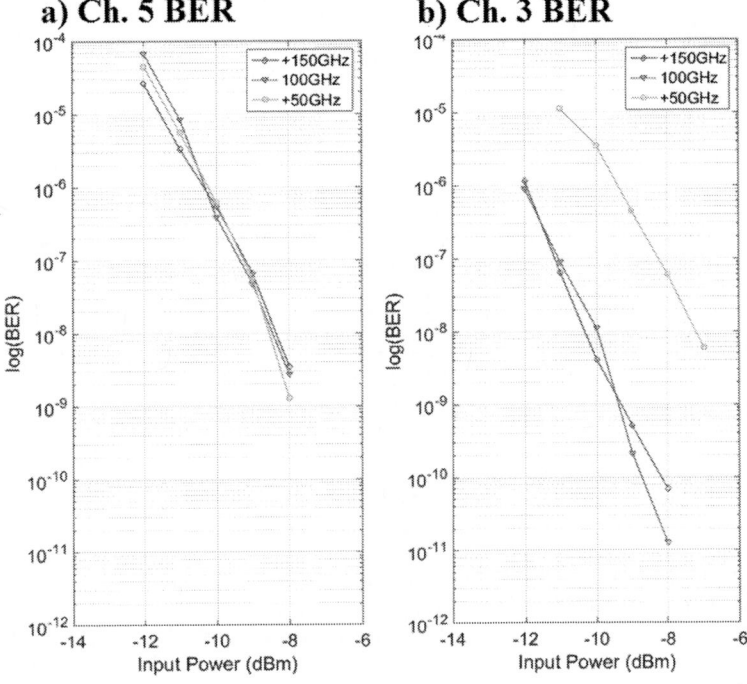

Figure 4 BER of both channels at various detuning values showing no observable OPP for a) ch. 5 and b) ~3dB OPP for ch. 3 at 50GHz channel separation.

CONCLUSIONS

We demonstrated the available bandwidth for state-of-the-art comb lasers that produce high optical power and narrow channel spacing. These lasers demonstrate low RIN and confirm the OPP associated with the cross-talk of concurrent modulation at 10Gb/s with a 50GHz channel spacing. As laser diode packaging improves we expect the extracted optical power to improve by 2-3dB, which will improve wall-plug efficiency and ameliorate strict requirements on receiver sensitivity.

REFERENCES

[1] A. Boletti, P. Boffi, P. Martelli, M. Ferrario, M. Martinelli, "Performance analysis of communication links based on VCSEL and silicon photonics technology for high-capacity data-intensive scenario," Opt. Express 23(2), 1806–1815 (2015).

[2] P. Dong, S. Liao, D. Feng, H. Liang, D. Zheng, R. Shafiiha, C.C. Kung, W. Qian, G. Li, X. Zheng, A. Krishnamoorthy, M. Asghari, "Low Vpp, ultralow-energy, compact, high-speed silicon electro-optic modulator," Opt. Express 17(25), 22484–22490 (2009).

[3] C.-H. Chen, M.A. Seyedi, M. Fiorentino, D. Livshits, A. Gubenko, S. Mikhrin, V. Mikhrin, R. Beausoleil, "A comb laser-driven DWDM silicon photonic transmitter based on microring modulators," Opt. Express 23(16), 21541–21548 (2015).

[4] M. A. Seyedi, C.-H. Chen, M. Fiorentino, R. Beausoleil, "Error-free DWDM transmission and crosstalk analysis for a silicon photonics transmitter," Opt. Express 23(26), 32968-32976 (2015).

[5] G. Wojcik, D. Yin, A. Kovsh, A. Gubenko, E. Alexey, I. Krestnikov, L. Igor, S. Mikhrin, D. Livshits, D. Fattal, A. David, M. Fiorentino, R. Beausoleil, "A single comb laser source for short reach WDM interconnects," Proc. SPIE 7230, 723012

[6] S. Haykin, An Introduction to Analog and Digital Communication (Wiley & Sons, 1989)

OECC/PS2016

Measurement of Vectorial Response of IQ Modulator Using Optical Interference

Yuya Yamaguchi[1], Kazuki Seki[1], Naoki Takahashi[1], Atsushi Kanno[2],
Tetsuya Kawanishi[1, 2], Masayuki Izutsu[1, 3], and Hirochika Nakajima[1]

[1]Waseda University, 3-4-1 Okubo, Shinjuku, Tokyo 169-8555, Japan
[2]National Institute of Information and Communications Technology, 4-2-1 Nukui-kitamachi,Koganei,Tokyo 184-8795, Japan
[3]Japan Society for the Promotion of Science (JSPS) San Francisco, 2001 Addison Street Suite 260 Berkeley, CA 94704, USA
yamaguchi@pic.phys.waseda.ac.jp

Abstract: We propose a new measurement method of modulator response in phasor domain for Mach-Zehnder modulators embedded in IQ modulator using simple equipments. The modulation curve with extinction ratio of 14.7 and 22 dB were measured.
Keywords: Mach-Zehnder modulator, modulation curve

I. INTRODUCTION

The multi-level quadrature amplitude modulation (QAM) and digital coherent detection have played an important role in high capacity optical fiber communication. Dual-parallel Mach-Zehnder modulators (MZMs) so-called IQ modulators are commonly used for generating the optical QAM signals from electrical multilevel signals, and a distortion of the generated optical signal relates on the specific parameters of the IQ modulator [1]. There are some methods to evaluate a performance as the intensity and phase changes of output lightwave corresponding to the applied voltage to the modulator by optical spectrum analysis , interference of BPSK signal, and coherent detection [2, 3]. But, especially, evaluating the performance of IQ modulator should be difficult because of the crosstalk between the two MZMs in both electrical and optical domain.

In this paper, we propose a new method to evaluate both amplitude and phase changes of the output from one MZM by calculation from the optical interference of the outputs from upper and lower MZM of IQ modulator. The proposed method requires only very simple equipment of a laser , an optical power meter and some DC sources.

II. PRINCIPLE OF PROPOSED MEASUREMENT METHOD

We consider the lightwave in the IQ modulator in phasor domain as shown in Fig. 1. When we measure the output characteristics of the upper MZM (In-phase component MZM), the lower MZM (Quadrature-phase component MZM) is used as the static reference lightwave for optical interference. The phase relation between the lightwaves from the upper and lower MZMs is controllable by applying electrical bias V_P to the phase shifter part of the IQ modulator. After we apply static DC bias V_I and V_Q to the upper and lower MZM, we can find the voltage V_{max} which the maximum output power P_{max} is obtained by applying to the phase shifter. Then, the phase of the lightwaves from the two MZMs should be same. On the other hand, we find also the voltage V_{min} which the minimum output power P_{min} is obtained by applying to the phase shifter, and then, the phase difference should be π. From the measured optical power P_{max} and P_{min}, we can calculate the optical amplitude E_I and E_Q in the upper and lower MZMs as follows.

$$\begin{cases} E_I + E_Q = \sqrt{P_{\max}} \\ E_I - E_Q = \sqrt{P_{\min}} \end{cases} \qquad (1)$$

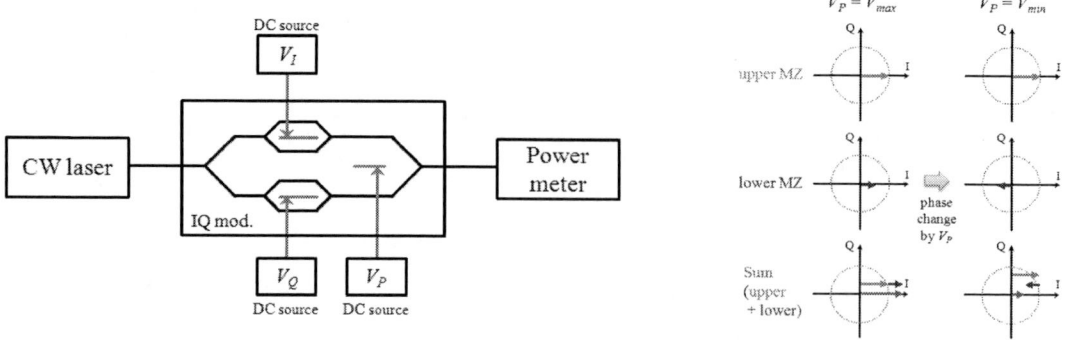

Fig. 1. Measurement set up for proposed method and the condition of the maximum and minimum optical power is obtained.

1140

We measure E_I with varying V_I, while E_Q would not depend on V_I. In the same way, we can measure the changes of E_Q depending on V_Q, by using the upper MZM as the reference light with static operation.

Next, we focus on the phase changes of the output from the MZM. In the ideal, the phase of the output from MZM is zero or π, which means minus sign in amplitude domain. However, MZM with finite extinction ratio (ER) shows the gradual phase change corresponding to the change of intensity [3]. Figure 2 shows a calculation results of optical power and phase of the output from MZM with ER of the infinite and 15 dB. We can find the gradual phase change in the 15 dB ER MZM as compared with the infinite ER MZM. As the consideration in the phasor domain, the trajectory of the output from the finite ER MZM should be a curve represented as dashed line, as shown in Fig.3. Figure 3 shows a difference of output lightwaves from the upper MZM in the case of applying 0V and 1V as V_I at the measurement shown above. By the change of applied voltage V_I, not only amplitude but also phase change θ is induced. This means the values of Vmax and V_{min} for the control of phase difference between the upper and lower MZMs to zero and π changes depending on V_I at the measurement, although the V_{max} and V_{min} should be constant with various V_I about the ideal modulator with infinite ER. The phase difference $\Delta\theta$ induced by changing the voltage V_I is calculated as

$$\Delta\theta \ [\mathrm{deg.}] = \frac{\Delta V_{min}}{V_{\pi P}} \times 180 \qquad (2)$$

where ΔV_{min} is a difference of the values of V_{min} before and after the change of V_I, and $V_{\pi P}$ is the half-wave voltage of the phase shifter of IQ modulator.

Thus, we can calculate the amplitude from the values of optical power P_{max} and P_{min} with various V_I, and also can calculate the phase from the values of ΔV_{min} focusing on the upper MZM. The lower MZM can be evaluated in the same way.

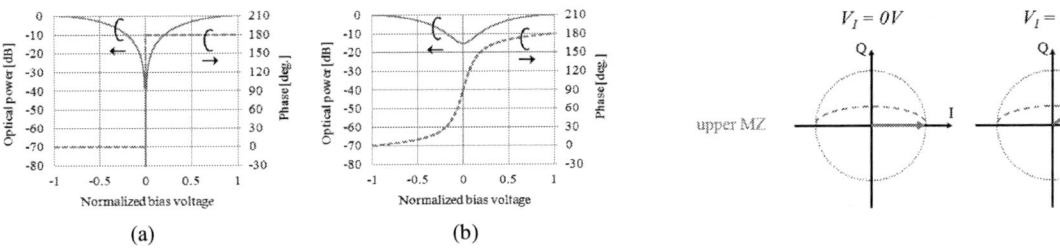

(a)   (b)

Fig. 2. Calculated output from MZM with ER of infinite (a) and 15dB (b).

Fig. 3. The difference of lightwave in upper MZM in phasor domain with change of V_I.

III. EXPERIMENTAL SETUP AND RESULTS

The experimental setup for the measurement is shown in Fig. 4. The electrical voltages input to each electrode of the modulator are supplied by some DC sources. The DC sources and optical power meter are connected to a programmable controller, and it memories the measured optical power with the values of DC sources according to the preset program. We used the extinction-ratio-tunable IQ modulator fabricated on x-cut LN as a device under test [5]. We evaluated the amplitude and phase of the output from upper MZM with different ER tuned by applying voltage to the active Y-branch is 0 and 20 V.

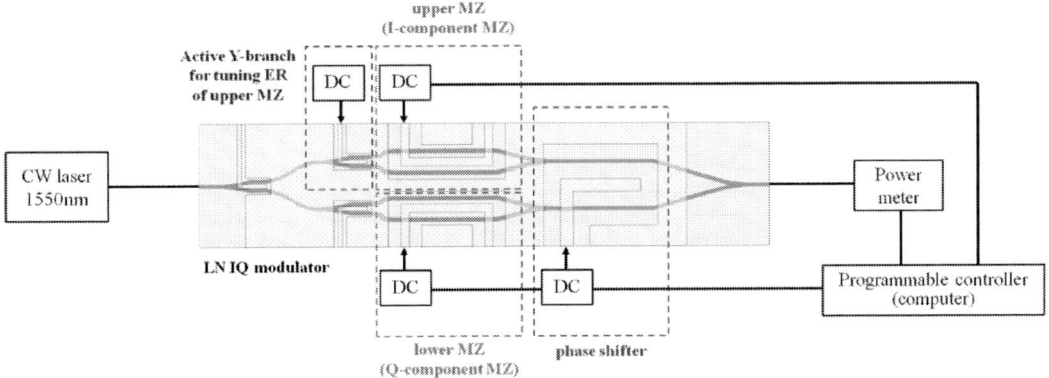

Fig. 4. Measurement setup with extinction-ratio-tunable IQ modulator.

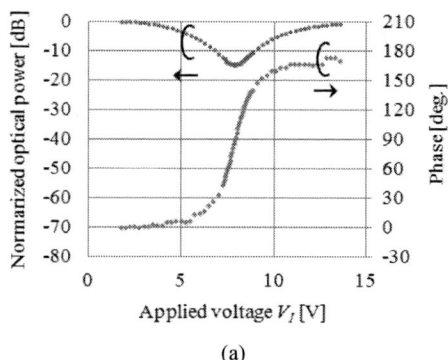 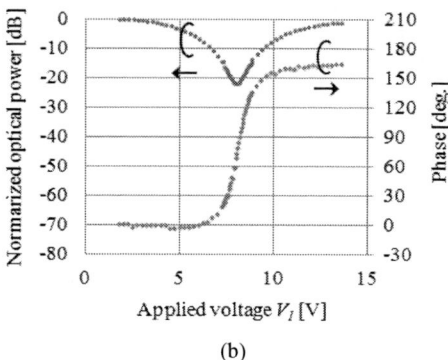

Fig. 5. Measured modulation curve as complex amplitude with ER of 14.7 (a) and 22 dB (b).

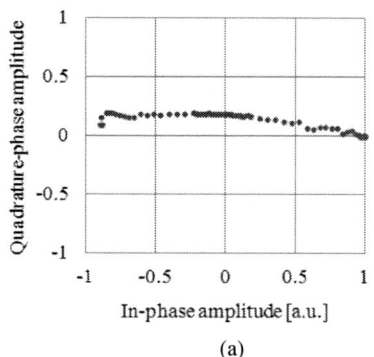 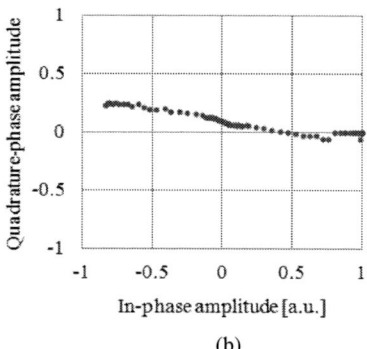

(a)   (b)

Fig. 6. Output characteristics in phasor domain with ER of 14.7 (a) and 22 dB (b).

The measured results are shown in Fig. 5. The half-wave voltage of upper MZM was 5.9 V, and ER was 14.7 and 22 dB with applying voltage of 0 and 20 V to the Y-branch, respectively. We can find the phase change of 22 dB ER is steeper than that of 14.7 dB ER in agreement with the theory. The trajectories of the output lightwaves represented in phasor domain are shown in Fig. 6. In this experiment, it is difficult to measure the complex amplitude of small power lightwave, but we successfully evaluated the behavior near the origin of phasor expansion.

IV. CONCLUSIONS

We proposed a measurement method of amplitude and phase of the output from an MZM in IQ modulator by calculation from the optical interference. The modulation curves in phasor domain under the condition of the ER of 14.7 and 22 dB were successfully measured with very simple equipments.

ACKNOWLEDGMENT

A part of this research is the result of research commissioned by National Institute of Information and Communications Technology (NICT) entitled "Agile Deployment Capability of Highly Resilient Optical and Radio Seamless Communication Systems."

REFERENCES

[1] T. Kawanishi, "Parallel Mach-Zehnder modulators for quadrature amplitude modulation," IEICE Electronics Express, vol. 8, pp. 1678-1688, October 2011.
[2] Y. Shi, L. Yan, and A. E. Willner, "High-Speed Electrooptic Modulator Characterization Using Optical Spectrum Analysis," J. Lightwave Technol., vol. 21, pp. 2358-2367, October 2003.
[3] T. Kawanishi, T. Sakamoto, A. Chiba, and M. Izutsu, "Study of precise optical modulation using Mach-Zehnder interferometers for advanced modulation formats," Proc. of European Conference on Optical Communication (ECOC), Berlin (Germany), Sep. 2007, 6.2.3.
[4] Y. Yamaguchi, A. Kanno, T. Kawanishi, M. Izutsu, and H. Nakajima, "Pure Single-Sideband Modulation Using High Extinction-Ratio Parallel Mach-Zehnder Modulator with Third-Order Harmonics Superposition Technique," Proc. of Conference on Lasers and Electro-Optics (CLEO), San Jose (USA), May. 2015, JTh2A.40.

Integrated Stokes Vector Analyzer on InP

Samir Ghosh, Yuto Kawabata, Takuo Tanemura, and Yoshiaki Nakano

Graduate School of Engineering, The University of Tokyo, 7-3-1 Hongo, Bunkyo-ku, Tokyo, 113-8656, Japan

ghosh.samir@hotaka.t.u-tokyo.ac.jp

Abstract: *Stokes-vector modulation formats are gaining increasing interest for next-generation high-speed optical communication networks. We propose and demonstrate potentially low-cost monolithically integrated InP/InGaAsP waveguide-based Stokes vector analyzer to retrieve the state-of-polarization without using coherent detection.*

Keywords: *(Polarization, Stokes parameter, Integrated optics)*

I. INTRODUCTION

Long-haul optical communication networks have gained a tremendous amount of capacity during the last decade due to evolutionary development in coherent communications [1]. All the fundamental attributes of light such as amplitude, phase, frequency and polarization have been utilized immensely for sending information from one place to another. At the same time, capacity upgradation per wavelength beyond 40 Gb/s or even 100 Gb/s is needed at short-reach or medium-haul communication networks spanning hundreds of kilometers to meet the demand for ever increasing traffic. Unlike long-haul communication networks, short-reach applications require a huge number of transceivers deployed across a wide geographical area. As a consequence, the cost of transceiver is a primary concern.

As a matter of fact, direct detection can lower the expenses as compared to coherent detection, making it ideal candidate for short-reach optical networks. While polarization-multiplexing techniques have been developed extensively to increase the data capacity per wavelength, there is an emerging interest in modulating the state-of-polarization (SOP) or the Stokes vector of light to send data efficiently and cost effectively with or without using coherent detection [2-4]. To this end, semiconductor waveguide-based Stokes-vector modulators have been reported [5, 6]. In order to retrieve the original information encoded in the Stokes vector of an optical signal at the receiver side, a low-cost integrated Stokes vector analyzer is highly desirable.

In this paper, we propose and design a potentially low-cost monolithically integrated InP/InGaAsP waveguide-based Stokes vector analyzer. It consists of half-ridge waveguide-based polarization converters (PCs) that can be integrated with InP photodetectors (PDs) to realize high-speed Stokes vector detection. Proof-of-concept device is fabricated to demonstrate that each Stokes vector component can be extracted without coherent detection.

II. DEVICE DESIGN AND WORKING PRINCIPLE

Stokes vector comprises of four parameters, namely S_0, S_1, S_2, and S_3, which are also known as Stokes parameters. Inside a single-mode waveguide, these parameters can be defined as: $S_0 = |A_{TE}|^2 + |A_{TM}|^2$, $S_1 = |A_{TE}|^2 - |A_{TM}|^2$, $S_2 = 2\text{Re}(A_{TE}^* A_{TM})$, and $S_3 = 2\text{Im}(A_{TE}^* A_{TM})$, where A_{TE} and A_{TM} are the complex amplitudes of the electric field for transverse electric (TE) and transverse magnetic (TM) modes, respectively. Note that S_1, S_2, and S_3 correspond to the relative fractions between x and y linearly polarized (or TE and) components, $+45^0$ and -45^0 linearly polarized components, and right/left-handed circularly polarized (RCP/LCP) components, respectively, in the free-space. A convenient way to visualize SOP of light is the Poincare sphere of radius S_0 as depicted in Fig. 1(a), satisfying $S_0 = \sqrt{S_1^2 + S_2^2 + S_3^2} = 1$, where S_1, S_2, S_3 are treated as Cartesian co-ordinates.

The layout of the proposed Stokes vector analyzer is shown in Fig. 1(b). It consists of a 1×4 multimode interference (MMI) splitter, standard ridge waveguide with a symmetrical cross-section [Fig. 2(a)], polarization converter (PC) with

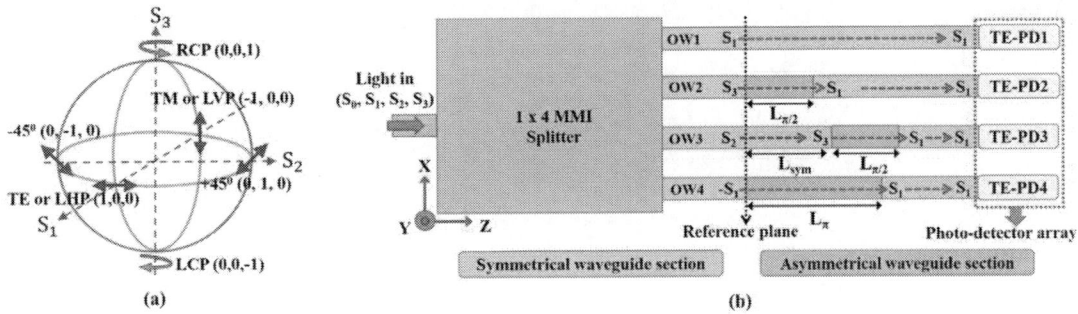

Fig. 1. (a) Poincare sphere (b) Schematic layout of the Stokes vector analyzer.

1143

asymmetric waveguide cross-section [Fig. 2(b)] [7], and polarization-dependent photo-detector (PD) that measures the optical intensity in the TE mode, i.e., the S_1 parameter of light [TE-PDs in Fig. 1(b)]. The length and the location of PC sections on the symmetrical waveguides are crucial and judiciously chosen to be different in each output waveguide (OW) of the MMI splitter as shown in Fig. 1(b). Due to the difference in effective refractive indices of the TE and TM

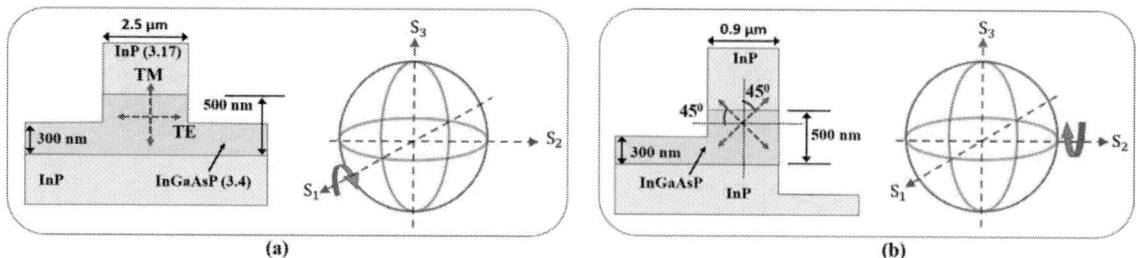

Fig. 2. Schematics of (a) symmetrical waveguide cross-section with corresponding SOP rotation about S_1-axis on Poincare sphere and (b) asymmetrical waveguide cross-section with corresponding SOP rotation about S_2-axis on Poincare sphere.

modes, SOP will rotate around S_1-axis on Poincare sphere along propagation inside the ridge symmetric waveguide [Fig. 2(a)]. In contrast, waveguide in the PC section has asymmetrical cross-section, in which the eigen modes can be tilted by an angle of 45° with respect to the InP substrate as shown in Fig. 2(b) [7]. In such a case, SOP rotates around the S_2- axis on Poincare sphere in the PC sections. Finally, the TE-mode polarization-dependent PDs may be realized by employing compressively strained multiple-quantum-well-based PDs and/or integrating waveguide-based polarizers [8-9].

The working principle of the device can be understood as follows. The MMI section splits the input power equally into all the four outputs from OW1 to OW4. Let us consider a reference plane in Fig. 1(b), where SOPs at all the four outputs of MMI are assumed to be identical in terms of S-parameters. The OW1 is just a straight waveguide with symmetrical cross-section, so that S_1 is unchanged. OW2 consists of a PC section followed by a symmetrical waveguide section until photodetector array. The length of the PC section is set to the quarter-beat length, $L_{\pi/2}$, which is around 70 µm for the waveguide profile shown in Fig. 2(b). In the OW2, therefore, the PC section transforms S_3 into S_1, which is preserved throughout rest of the symmetric waveguide. As the symmetrical waveguide section is located before the PC section in the case of OW3, the polarization transformation here is different from that in the OW2. Length of the symmetrical section, L_{sym} is designed in such a way that it gives $\pi/2$ rotation of SOP around S_1 axis. As a result, S_2 transforms into S_3 at this symmetrical section. For the structure shown in Fig. 2(a), L_{sym} is calculated to be 35 µm. The asymmetrical section afterwards rotates the SOP around S_2 axis and therefore the initial S_2 parameter is converted to S_1 at the final output. Finally, OW4 contains PC with a length equals to half-beat length L_π. As a result, it transforms $-S_1$ into S_1. While OW4 is redundant, it may be used to monitor the total intensity and to derive S_0. Hence, it is understood that all four S parameters of the incident light can be measured by monitoring the photo-current at TE-PDs.

III. FABRICATION AND MEASUREMENTS

For proof-of-concept demonstration, the passive section of the Stokes vector analyzer without the TE-PDs in Fig. 1(b) is fabricated on InP substrate with half-ridge InP/InGaAsP PCs. We employ self-aligned fabrication procedure to monolithically integrate PC sections with the MMI splitter and other symmetric waveguides, as presented previously [5, 7]. Designed waveguide dimensions for the symmetrical and asymmetrical are indicated in Fig. 2. Scanning electron microscopic (SEM) images of the respective waveguide cross-sections can be found in Fig. 3.

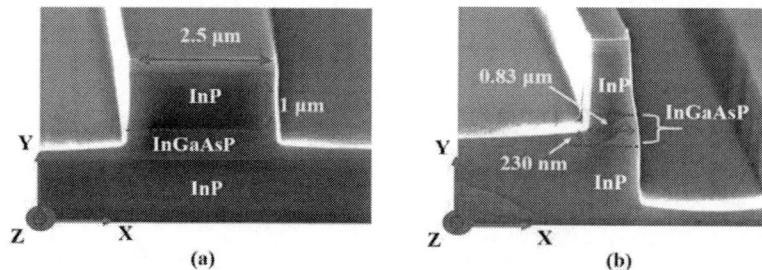

Fig. 3. SEM images of the (a) symmetrical ridge and (b) asymmetrical half-ridge PC waveguide cross-sections.

1144

A schematic of the optical setup in order to test the fabricated device is depicted in Fig. 4(a). A light with wavelength at 1550 nm from a tunable laser source (TLS) is sent through a polarizer, a half-wave plate (HWP), and a quarter-wave plate (QWP). Power as well as SOP of light from each output port OW1- OW4 are recorded by a commercial benchtop

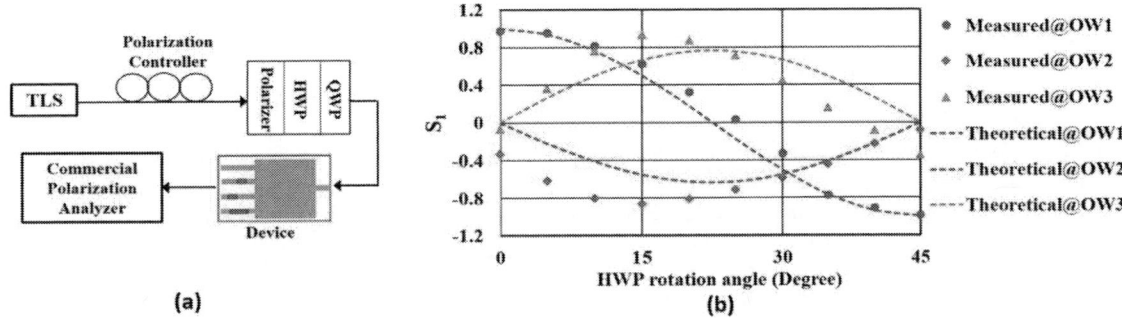

Fig. 4. (a) Schematic of measurement setup. (b) Measured (scattered points) and calculated (dotted lines) S_1 parameters of light at each output port as a function of HWP rotation angle.

polarization analyzer. At first, input SOP is aligned to the TE polarization by using a reference waveguide and adjusting the HWP and QWP. Then, input SOP is adjusted to an arbitrary state on the Poincare sphere by rotating the HWP and QWP. Any unitary transformation of SOP at the output fiber components is also evaluated before the measurement and calibrated from the obtained data.

Fig. 4(b) shows the measured and theoretically calculated values of S_1 parameter at each output port as we rotate the HWP relatively from 0° (TE) to 45° (TM). In calculating the theoretical curves, we derive one fitting parameter that describes unknown rotation around S_1-axis at the symmetrical waveguides located at the left hand side of the reference plane as shown in Fig. 1(b). Reasonable agreement is observed at the output of OW1, OW2 and OW3.

IV. CONCLUSION

A novel approach to unravel the SOP of light by a monolithically integrated waveguide-based device is proposed and demonstrated on InP. While the initial results prove the principle of operation of the proposed Stokes vector analyzer, the device design and fabrication should be improved to achieve required performance. Finally, monolithic integration of multiple-quantum-well-based photodetectors would allow on-chip high-speed detection of SOP, which should find attractive application in the future short-reach communication links.

ACKNOWLEDGMENT

This work was financially supported by the Grant-in-Aid of Japan Society for the Promotion of Science.

REFERENCES

[1] W. Shieh, Q. Yang, and Y. Ma, "107 Gb/s coherent optical OFDM transmission over 1000-km SSMF fiber using orthogonal band multiplexing," Opt. Exp., vol. 16, no. 9, pp. 6378-6386, 2008.

[2] M. Morsy-Osman, M. Chagnon, D. V. Plant, "Polarization division multiplexed intensity, inter polarization phase and inter polarization differential phase modulation with stokes space direct detection for 1λ×320 Gb/s 10 km transmission at 8 bits/symbol," European Conference on Optical Communication (ECOC), Valencia, Spain, Sept. 2015.

[3] D. Che, A. Li, X. Chen, Q. Hu, Y. Wang, and W. Shieh, "Stokes vector direct detection for short-reach optical communication," Opt. Lett., vol. 39, no. 11, pp. 3110-3113, 2014.

[4] K. Kikuchi and S. Kawakami, "Multi-level signaling in the Stokes space and its application to large-capacity optical communications," Opt. Exp., vol. 22, no. 7, pp. 7374-7387, 2014.

[5] Y. Kawabata, M. Zaitsu, T. Tanemura, and Y. Nakano, "Proposal and Experimental Demonstration of Monolithic InP/InGaAsP Polarization Modulator," European Conference on Optical Communication (ECOC), Cannes, France, pp. 21-25, Sept. 2014.

[6] M. A. Naeem, M. Haji, B. M. Holmes, D. C. Hutchings, J. H. Marsh, and A. E. Kelly, "Generation of High Speed Polarization Modulated Data Using a Monolithically Integrated Device," IEEE Journal of Selected Topics in Quantum Electronics, vol. 21, no.4, pp. 207-211, July-Aug. 2015.

[7] M. Zaitsu, T. Tanemura, A. Higo, and Y. Nakano, "Experimental demonstration of self-aligned InP/InGaAsP polarization converter for polarization multiplexed photonic integrated circuits," Opt. Exp., vol. 21, no. 6, pp. 6910-6918, 2013.

[8] J. T. Kim and C. G. Choi, "Graphene-based polymer waveguide polarizer," Opt. Exp., vol. 20, no. 4, pp. 3556-3562, 2012.

[9] A. Wieczorek, B. Roycroft, F. H. Peters, and B. Corbett, "TE/TM-mode pass polarizers and splitter based on an asymmetric twin waveguide and resonant coupling," Optical and Quantum Electronics, vol. 44, no. 3, pp. 175-181, 2012.

WD4-5

OECC/PS2016

Proposal of Compact TE/TM Polarization Switch Based on Microring Resonator

Keita Suzuki, Tomoki Hirayama, and Taro Arakawa

Graduate School of Engineering, Yokohama National University, 79-5, Tokiwadai, Hodogayaku, Yokohama, 240-8501, Japan
suzuki-keita-cx@ynu.jp, arakawa-taro-vj@ynu.ac.jp

Abstract: A compact TE/TM polarization switch using a quantum well microring resonator is proposed and its characteristics are theoretically discussed. Mode conversion can be controlled at low voltage with field-induced refractive index change in the microring waveguide.

Keywords: Microring Resonator, Polarization Switch, Quantum Well, Field-Induced Refractive Index Change

I. INTRODUCTION

For high capacity optical communication using polarization multiplexing, mode control devices such as polarization converters and splitters/combiners are essential, and various kinds of passive TE/TM mode converters and splitters/combiners have been proposed and developed [1-4]. In addition, a novel polarization modulator have also been proposed for efficient utilization of the state of polarization [5]. However, such an actively polarization controlling device tends to have a big footprint and complicated structure.

In this paper, we propose a compact TE/TM polarization switch based on a quantum well microring resonator (MRR) as an active polarization control device. Mode conversion can be controlled at low voltage with field-induced refractive index change in the MRR waveguide. Its polarization switching characteristics were calculated with a propagation matrix method incorporating the Jones calculus.

II. DEVICE STRUCTURE AND ANALYSIS METHOD

A. Device Structure

Fig. 1 shows the schematic top view of the proposed TE/TM polarization switch. It is composed of a multiple quantum well (MQW) MMR and two busline waveguides. The MRR and each busline waveguide is coupled through a polarization-independent directional coupler (DC). The structure of the DC is discussed in the next subsection.

A TE (TM) mode light incident at Input Port with a resonant wavelength of the MRR is normally transmitted through the MRR and outputted at Drop Port without any change in polarization. When the voltage is applied to the MRR and the refractive index of the MRR waveguide is changed by the quantum-confined Stark effect (QCSE) in the MQW, the TE (TM) mode light is converted into TM (TE) mode in the MRR and outputted. This is the polarization switching which is realized in the device. To obtain the phase shift required for switching polarization using the change in refractive index in the MQW, a several-mm long phase shifter is usually needed [5]. In our proposed device, however, the device size can be markedly reduced due to the phase-shift enhancement of the MRR [6,7].

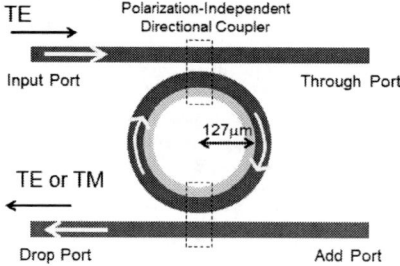

Fig. 1. Schematic top view of proposed TE/TM polarization switch.

Fig. 2(a) shows the schematic cross section of the waveguide in the MRR. It has a half-ridge structure for polarization rotation caused by birefringence in the MQW core [4]. The R parameter [4] of the half-ridge waveguide is designed to be 1.0. As the MQW core layer, a 300 nm-thick multiple InGaAs/InAlAs five-layer asymmetric coupled quantum well (FACQW) [8] is assumed. The radius of the MRR is 127 μm. The half of the round-trip length of the MRR is designed to be the same as the conversion length of the half-ridge waveguide from TE/TM to TM/TE modes. The calculated change in refractive index in the FACQW for TE and TM modes are shown in Fig. 2(b). When an electric field applied by reverse voltage is changed from -30 kV/cm to -34 kV/cm, only the refractive index for the TE mode changes by 0.0031. Here we define this index change is as Δn_{max}. Using this difference in the refractive index change in the core layer, polarization conversion length can be changed by changing the applied voltage.

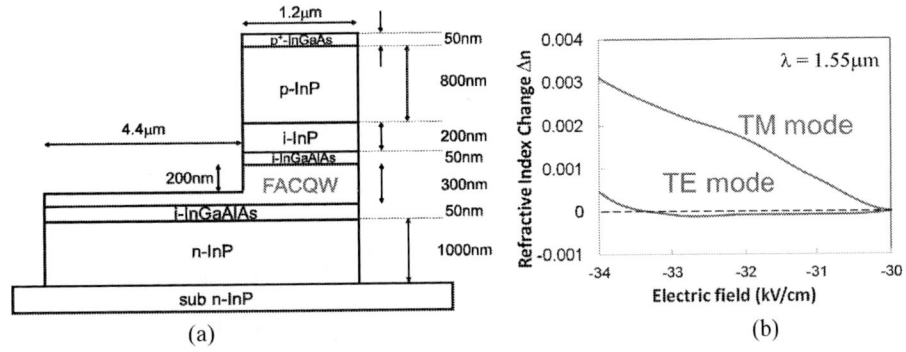

Fig. 2. (a) Schematic cross section of half-ridge waveguide in microring resonator.
(b) Calculated change in refractive index in five-layer asymmetric coupled quantum well (FACQW) for TE and TM modes.

B. Polarization-Independent Directional Coupler

To realize the proposed polarization switch, a polarization-independent DC between the MRR and a busline waveguide is essential. Because conventional DCs have great polarization dependence, we propose a new DC based on the combination of symmetric and asymmetric coupled waveguides [9] for polarization independence. Fig. 3 shows the schematic top view of the polarization-independent DC and the coupling efficiency for TE and TM mode lights as a function of the DC length. The DC is composed of 40 μm-long symmetric DCs connected to 19 μm-long asymmetric DC with 10 μm-long tapered DCs. When the total coupler length is between 79 and 100 μm, polarization-independent coupling is obtained, and the total coupling efficiency can be adjusted from 0.1 to 0.3 by changing the total length, as shown in Fig. 3. If the structure is optimized, the device length can be reduced further.

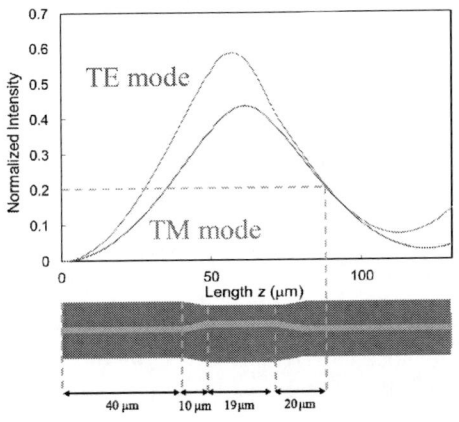

Fig. 3. Schematic top view of polarization-independent DC and calculated coupling efficiency for TE and TM mode lights as function of DC length.

Fig. 4. Schematic calculation model for the MRR with the busline waveguides.

III. Polarization Switching Characteristics

A. Calculation Method

To calculate the polarization switching characteristics of the proposed device, we developed a new propagation matrix method [10] incorporating the Jones calculus [11]. Fig. 4 shows the schematic calculation model for the MRR with the busline waveguides. The symbols E_I, E_T, E_D and E_A are the electric fields at the Input, Through, Drop, and Add ports, respectively. If x and y polarized components of E_i (i=I, T, D, A) are respectively denoted as E_{ix} and E_{iy}, the transfer matrix is given by

$$\left(E_{Dx} \quad E_{Dy} \quad E_{Ax} \quad E_{Ay}\right)^t = \mathbf{C}_b \mathbf{S}(\psi)\mathbf{R}\mathbf{S}(-\psi)\mathbf{C}_t\left(E_{Ix} \quad E_{Iy} \quad E_{Tx} \quad E_{Ty}\right)^t, \qquad (1)$$

where $\mathbf{C}_t$ and $\mathbf{C}_b$ are the transfer matrices for Input and Drop ports, respectively, and $\mathbf{R}$ is the transfer matrix for the MRR. The matrix $\mathbf{S}(\Psi)$ is the one for the polarization state rotated by Ψ, where Ψ represents the orientation of the principal axis regarding the x-axis.

B. Switching Characteristics

The polarization switching characteristics were calculated using the propagation matrix method described in the previous section. In the calculation, it is assumed for simplicity that the length of the polarization-independent DC is zero, that is, the MRR is a circular ring with a radius of 127 μm, and the coupling efficiency K is 0.2. The optical confinement factor and the filling factor of the FACQW are assumed to be 0.55 and 0.93, respectively. Propagation loss in the waveguides and the coupling loss at the DCs were neglected. It is also assumed that the inclination of the principal optic axis of the half-ridge waveguide does not change.

Fig. 5 shows the spectra of the transmitted light from Input from Drop Ports when the TE mode light is incident in Input Port. When the refractive index of the core layer is changed by $0.07\Delta n_{max}$ (Fig. 5(a)), the TM-mode light at the resonant wavelength of the MRR is mainly outputted. The TM/TE ratio is approximately 35 dB. Next, when the refractive index of the core layer is changed by $0.98\Delta n_{max}$ (Fig. 5(b)), the TM-mode light is mainly outputted. The TE/TM ratio is approximately 32 dB. The bandwidth of the resonant wavelength and the TE/TM ratio can be adjusted by K. The driving voltage is evaluated to be approximately 1.0 V. These results show that high-speed and low-voltage polarization switching can be realized with the proposed compact MRR polarization switch.

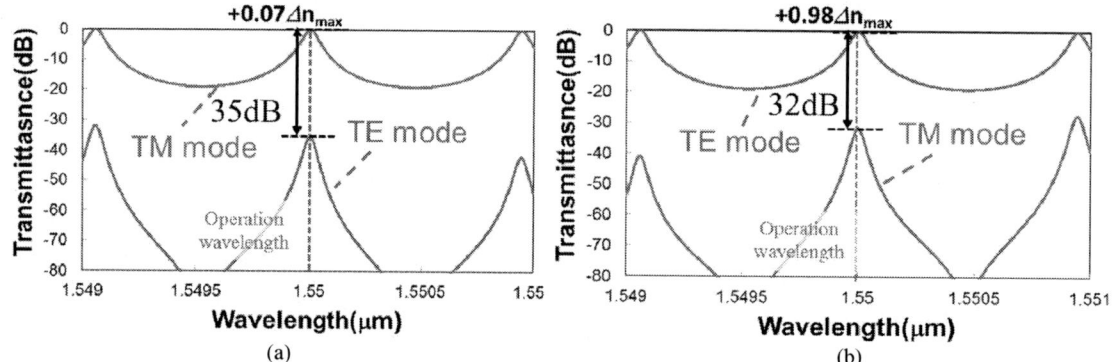

Fig. 5. Spectra of the transmitted light from Input from Drop Ports when TE mode light is incident on Input Port in case
(a) where refractive index of core layer of microring resonator is changed by $0.07\Delta n_{max}$ ($\Delta n_{max}=0.0031$).
(b) where refractive index of core layer of microring resonator is changed by $0.98\Delta n_{max}$.

IV. CONCLUSIONS

We proposed and theoretically demonstrated a compact polarization switch based on the MQW MRR as an active polarization control device. Mode conversion can be controlled at low voltage with field-induced refractive index change in the MRR waveguide. High-speed and low-voltage polarization switching can be realized with the proposed compact MRR polarization switch.

ACKNOWLEDGMENT

This work was partly supported by a Grant-in-Aid for Scientific Research B from the Ministry of Education, Culture, Sports, Science and Technology, Japan (No. 15H03577) and the Fujikura foundation, Japan.

REFERENCES

[1] J. Wang, B. Niu, Z.Sheng, A. Wu, W. Li, X. Wang, S. Zou, M. Qi, and F. Gan, "Novel ultra-broadband polarization splitter-rotator based on mode-evolution tapers and a mode-sorting asymmetric Y-junction," Opt. express, vol.22, no.11, pp.13565-13571, 2014.
[2] A. Xie, L. Zhou, J. Chen, and X. Li, "Efficient silicon polarization rotator based on mode-hybridization in a double-stair waveguide," Opt. express, vol.23, no.4, pp. 3960-3970, 2015.
[3] S. H. Kim, R. Takei, Y. Shoji, and T. Mizumoto, "Single-trench waveguide TE-TM mode converter," Opt. Express, vol.17, no.14, pp.11267-11273, 2009.
[4] M. Zaitsu, T. Tanemura, A. Higo, and Y. Nakano, "Experimental demonstration of self-aligned InP/InGaAsP polarization converter for polarization multiplexed photonic integrated circuits," Opt. Express, vol.21, no.6, pp. 6910-6918, 2013.
[5] Y. Kawabata, M. Zaitsu, T. Tanemura, and Y. Nakano, "Proposal and Experimental Demonstration of Monolithic InP/InGaAsP Polarization Modulator," European Conf. Opt. Com. (ECOC) 2014, Tu.4.4.4, 2014.
[6] J. E. Heebner and R. W. Boyd, "Enhanced all-optical switching by use of a nonlinear fiber ring resonator," Opt. Lett., no. 24, pp. 847–849, 1999.
[7] H. Kaneshige, R. Gautam, Y. Ueyama, R. Katouf, T. Arakawa, and Y. Kokubun, "Low-voltage quantum well microring-enhanced Mach-Zehnder modulator," Opt. Express, vol. 21, no. 14, pp.16888–16900, 2013.
[8] T. Arakawa, T. Hariki, Y. Amma, M. Fukuoka, M. Ushigome, K. Tada, "Low-Voltage Mach–Zehnder Modulator with InGaAs/InAlAs Five-Layer Asymmetric Coupled Quantum Well," Jpn. J. Appl. Phys., vol.51, 042203, 2012.
[9] Z. Lu, H. Yun, Y. Wang, Z. Chen, F. Zhang, N. A. F. Jaeger, and L. Chrostowski, "Broadband silicon photonic directional coupler using asymmetric-waveguide based phase control," Opt. Express, vol. 23, no.3, pp.3795-3806, 2015.
[10] T. Kato and Y. Kokubun, "Optimum coupling coefficients in second-order series-coupled ring resonator for nonblocking wavelength channel switch," J. Lightwave Technol., vol. 24, pp. 991–999, 2006.
[11] A. Yariv and P. Yeh, "Photonics," Oxford Univ. Press, 2005.

WE1-1 (Invited)

OECC/PS2016

Advances in second order nonlinear effect in Silicon

P. Damas, X. Le Roux, M Berciano, G. Marcaud, C. Alonso-Ramos,
D. Benedikovic, E. Cassan, D. Marris-Morini, L. Vivien

Centre for Nanoscience and Nanotechnology (C2N), University of Paris-Sud, University of Paris-Saclay, CNRS, Orsay, France
Laurent.vivien@u-psud.fr

Abstract: In this work, we present a theoretical model to determine the second order nonlinear coefficient under strain gradient in silicon. Furthermore, carrier effect due to applied electric field has also been taking into account to analyze the obtained experimental phase variation.

I. INTRODUCTION

Silicon-based photonics has generated a strong interest in recent years, mainly for optical communications and optical interconnects in CMOS circuits. The main motivations for silicon photonics are the reduction of photonic system costs and the increase of the number of functionalities on the same integrated chip by combining photonics and electronics, along with a strong reduction of power consumption[1]. However, one of the biggest constraints of silicon as an active photonic material is its vanishing second order optical susceptibility, the so called $\chi^{(2)}$, due to the centrosymmety of the silicon crystal. Without any second order nonlinear phenomena, fast and low power consumption optical modulation based on Pockels effect and wavelength conversions based on Second Harmonic Generation (SHG), other strategies, mainly based on either plasma dispersion effect in silicon[2] or electro-absorption in bulk Ge and Ge/SiGe quantum wells[3], have been developed for modulation. Even, impressive results have been obtained in terms of speed and efficiency; the power consumption is still too high for the next generation of photonic circuits. Indeed, the fact, not to have the possibility to exploit nonlinear effects in silicon for modulation is a very limiting factor when we expect silicon to be part of a solution to high performances and highly energy efficient devices. To overcome this limitation, strain (ε) has been used as a way to deform the crystal and destroy the centrosymmetry which inhibits $\chi^{(2)}$. In fact, over the last few years Pockels electro-optic modulation and SHG have been claimed and been demonstrated in devices where the silicon active region is strained by a stress overlayer, usually made of SiN[4-10]. This is the very motivation to the development of *strained silicon* devices for optical modulation: the prospect of a high speed, low loss, compact, low power consumption, with large optical bandwidth and silicon compatible modulator. The paper present the recent advances in the development of nonlinear is this exciting topic including discussions from fundamental origin of Pockels effect in silicon until its implementation in a real device.

II. THEORITICAL MODEL

We start our analysis to strained silicon by studying how strain affects the silicon crystal structure to understand how $\chi^{(2)}$ effects can be generated. To achieve that, we developed an original theoretical model based on the *Bond Orbital Model*, which is a quantum mechanical theory that describes the electrons in the bonds between the silicon. By using that approach together with symmetry arguments, we found that the spatial distribution of $\chi^{(2)}$ in the crystal can be given by

$$\chi^{(2)}_{ijk} = \Gamma_{ijk,lmn} \cdot \eta_{lmn}, \qquad (1)$$

where η_{lmn} is a *strain gradient* component defined by :

$$\eta_{ijk} \equiv \frac{\partial \varepsilon_{ij}}{\partial x_k}. \qquad (2)$$

This means that $\chi^{(2)}$ is defined by the contribution of all strain gradients, weighted by the Γ coefficients. The values of Γ can be uniquely determined by the Bond Orbital description, which defines Γ as a function of only two parameters α and β. In fact, being a tensor characteristic of the crystal, Γ depends on the crystal orientation and from our model, we could deduce the most relevant components of Γ in a [010] silicon wafer as a function of the angle φ ($\varphi = 0$ corresponds to [110] and $\varphi = \pi/4$ corresponds to [100] directions) as follows:

1149

$$\Gamma_{xxy,xxy}(\varphi) = \frac{d^6 K}{27\epsilon_0}\left[(5\beta - 3\alpha - (\alpha+\beta)\cos(4\varphi))\right]$$

$$\Gamma_{xxy,yyy}(\varphi) = -\frac{d^6 K}{27\epsilon_0}\left[2(\alpha - 3\beta)\right].$$

(3)

The only two unknowns in the previous set of equations are the constants α and β to be determined experimentally.

III. FABRICATION AND CHARATERIZATION

The silicon waveguide requires two main features: a source of strain and a source of electrostatic field. The former is achieved by placing a layer of straining material, under high internal stress (σ_0), on top of the waveguide. This highly stressed layer strains the waveguide underneath, creating the strain gradients required to generate $\chi^{(2)}$ in the waveguide. The stress layer we used was a SiN thin film, deposited by PECVD whose internal stress $\sigma_0 = 1.2$ GPa compressive stress. The electrostatic field is created by placing electrodes on top of the waveguide which are activated by the application of a voltage difference V_s. The cross-section of the final structure we designed is shown in Fig. 1

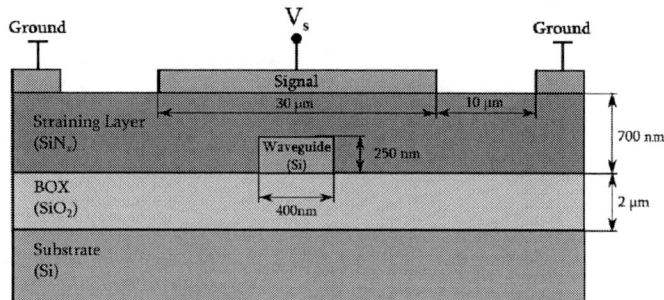

Fig. 1 : Schematic view of modulator based on Pockels effect including a stress layer and electrodes.

However, silicon is a semiconductor and its free-carrier concentration and distribution is dependent on the applied electric field (F)[11]. Therefore, there are two electro-optic effects that take place in the waveguide when an electric field is applied to the waveguide: the plasma- dispersion effect, an effect that corresponds to a variation of refractive index of silicon due to a variation of free-carrier concentration; and the Pockels effect, an effect that depends on the interaction of the electric field inside the waveguide and the strain gradients. These two effects, however, are deeply connected: the application of an electric field induces a change in free-carrier distribution, that in turn affects the electric field distribution inside the waveguide, which influences the Pockels effect. Furthermore, the SiN stress layer required to induce strain in the waveguide, is usually characterized by a considerable positive charge distribution (Q f) which also affects the free carrier concentration in the waveguide. Therefore, in order to study Pockels effect in strained silicon, we must include the study of the free-carriers inside the silicon waveguide together with the charging effects of the cladding and the strain effects. The final electro-optic effect is then a combination of the plasma dispersion effect ($\Delta n_{\text{eff-c}}$) and Pockels effect ($\Delta n_{\text{eff-P}}$), which can be reduced to:

$$\Delta n_{eff} = \Delta n_{eff_c} + \Delta n_{eff_P} =$$

$$= \Delta n_{eff_c} + \Gamma_{xxy,xxy}\left(\widehat{\eta_{xxy}^{xxy}} + \zeta\,\widehat{\eta_{yyy}^{xxy}}\right)$$

(4)

where $\quad \zeta = \Gamma_{xxy,xxy}/\Gamma_{xxy,yyy}$

Both Δn_{effc} and Δn_{effP} are voltage dependent and their values as a function of Vs are determined by simulating the structure using COMSOL multiphysics, which combines elasticity, optical and semiconductor effects in a single simulation. Furthermore, it is difficult to dissociate both effec

There are two main differences between both electro-optic effects: Pockels effect is a very fast effect limited only at the THz range, whereas plasma-dispersion effect can be limited at the GHz range. Furthermore, Pockels effect depends on the crystal orientation whereas carriers are not sensitive to crystallographic directions. To detect these two differences experimentally, we designed, fabricated and characterized Mach-Zehnder Interferometers compatible with RF electro-optical modulation and with the waveguides orientation. The fabrication of the devices was successfully performed in the Centre Technologique Universitaire (CTU) by using a wide range of techniques: e-beam lithography,

ICP etching, silicon thermal oxidation, PECVD, e-beam evaporation, among others. Obtained results will be discussed during the presentation and compared t the theoretical considerations.

IV. CONCLUSIONS

To sum up, we developed a new theoretical model to explain the origins of strain-induced $\chi^{(2)}$ in silicon which related the electro-optic effect with strain gradients. We used these concepts to simulate the electro-optic effects in an approach that includes semiconductor, strain and optical effects. Lastly, we designed, fabricated and measured some samples under several bias voltages. A discussion on the different effects involved in strained silicon as a function of the frequency and device geometry will be presented during the talk.

ACKNOWLEDGMENT

Authors would like to thank Frédéric Boeuf from STMicroelectronics (Crolles, France) for fruitful discussions. The authors also acknowledge STMicroelectronics for the financial support of the P. Damas' scholarship. This project has received funding from the European Research Council (ERC) under the European Union's Horizon 2020 research and innovation program (ERC POPSTAR Grant No. 647342)

REFERENCES

[1] Laurent Vivien and Lorenzo Pavesi, editors. *Handbook of Silicon Photonics*. CRC Press, 2013. ☐

[2] Marris-Morini Delphine, Virot Léopold, Baudot Charles, Fédéli Jean-Marc, Rasigade Gilles, Perez-Galacho Diego, Hartmann Jean-Michel, Olivier Ségolène, Brindel Patrick, Crozat Paul, Boeuf Frédéric, Vivien Laurent, 40 Gbit/s optical link in 300-mm silicon platform, Optics Express 22, 6674 (2014)

[3] Papichaya Chaisakul, Delphine Marris-Morini, Jacopo Frigerio, Daniel Chrastina, Mohamed-Said Rouifed, Stefano Cecchi, Paul Crozat, Giovanni Isella, and Laurent Vivien. Integrated germanium optical interconnects on silicon substrates. *Nat Photon*, 8(6):482–488, 2014

[4] Rune S Jacobsen, Karin N Andersen, Peter I Borel, Jacob Fage-Pedersen, Lars H Frandsen, Ole Hansen, Martin Kristensen, Andrei V Lavrinenko, Gaid Moulin, Haiyan Ou, Christophe Peucheret, Beáta Zsigri, and Anders Bjarklev. Strained silicon as a new electro-optic material. *Nature*, 441(7090):199–202, may 2006

[5] Bartos Chmielak, Michael Waldow, Christopher Matheisen, Christian Ripperda, Jens Bolten, Thorsten Wahlbrink, Michael Nagel, Florian Merget, and Heinrich Kurz. Pockels effect based fully integrated, strained silicon electro-optic modulator. *Optics express*, 19(18):17212–9, aug 2011

[6] M Cazzanelli, F Bianco, E Borga, G Pucker, M Ghulinyan, E Degoli, E Luppi, V Véniard, S Ossicini, D Modotto, S Wabnitz, R Pierobon, and L Pavesi. Second- harmonic generation in silicon waveguides strained by silicon nitride. *Nature materials*, 11(2):148–54, feb 2012

[7] Pedro Damas, Xavier Le Roux, David Le Bourdais, Eric Cassan, Delphine Marris- Morini, Nicolas Izard, Thomas Maroutian, Philippe Lecoeur, and Laurent Vivien. Wavelength dependence of Pockels effect in strained silicon waveguides. *Optics Express*, 22(18):22095, sep 2014 ☐

[8] Matthew W Puckett, Joseph S T Smalley, Maxim Abashin, Andrew Grieco, and Yesha- iahu Fainman. Tensor of the second-order nonlinear susceptibility in asymmetrically strained silicon waveguides: analysis and experimental validation. *Optics letters*, 39(6):1693–6, mar 2014 ☐

[9] Clemens Schriever, Federica Bianco, Massimo Cazzanelli, Mher Ghulinyan, Christian Eisenschmidt, Johannes de Boor, Alexander Schmid, Johannes Heitmann, Lorenzo Pavesi, and Jörg Schilling. Second-Order Optical Nonlinearity in Silicon Waveguides: Inhomogeneous Stress and Interfaces. *Advanced Optical Materials*, 3(1):129–136, 2015. ☐

[10] Bartos Chmielak, Christopher Matheisen, Christian Ripperda, Jens Bolten, Thorsten Wahlbrink, Michael Waldow, and Heinrich Kurz. Investigation of local strain distri- bution and linear electro-optic effect in strained silicon waveguides. *Optics Express*, 21(21):25324, oct 2013 ☐

[11] S Sharif Azadeh, F Merget, M P Nezhad, and J Witzens. On the measurement of the Pockels effect in strained silicon. *Optics letters*, 40(8):1877–1880, 2015 ☐

WE1-2

OECC/PS2016

Multiple optical carrier generation using multiple QPM device

Kazuki Nakamura[1], Hin Channa[1], Masaki Asobe [1], Takeshi Umeki[2], and Hirokazu Takenouchi[2]

[1] Tokai University, 4-1-1 Kitakaname, Hiratsuka, Kanagawa, 259-1292 Japan
[2] NTT Device Technology Laboratories, 3-1 Morinosato Wakamiya, Atsugi, Kanagawa, 243-0198 Japan
Author e-mail address: asobe@tokai-u.jp

Abstract: *Phase matching of multiple QPM LiNbO₃ waveguide was evaluated in the presence of high power second harmonic light. Based on the characterization, we demonstrate multiple carrier generation using multi-stage frequency mixing in multiple QPM device.*

Keywords: *Multiple optical carrier generation, Multi-stage frequency mixing, Quasi phase matching*

I. INTRODUCTION

The efficient optical frequency mixing in a periodically poled LiNbO₃ waveguide is widely used in many applications such as wavelength conversion and parametric amplification. By applying the phase modulation to periodically poled structure, multiple quasi-phase matching (QPM) can be obtained [1], thus the multi-stage frequency mixing becomes available. The multi-stage frequency in LiNbO₃ waveguide enables carrier phase recovery of multi-level phase modulated signal and multiple optical carrier generation in a compact device [2]. However, LiNbO₃ exhibit photorefractive effect in high output power condition. The variation of the phase matching condition can degrade the device performance. In this study, we characterized phase matching curve of multiple QPM device in the presence of high power second harmonic light. We also tried multiple carrier generation using multi-stage frequency mixing in the multiple QPM device.

II. PRINCIPLE OF MULTI-STAGE FREQUENCY MIXING

Figure 1 illustrates the principle of multi-stage frequency mixing. We utilize frequency mixing such as second harmonic generation (SHG), sum frequency generation (SFG), and difference frequency generation (DFG). To get high conversion efficiency, QPM condition should be satisfied. The multiple QPM device exhibits multiple QPM peaks as shown in Fig.1. Two lasers are input to the multiple QPM device. At least, the wavelength of a laser is matched to the QPM peak. Utilizing subsequent multiple frequency mixing through SHG/DFG processes, multiple idlers with uniform frequency spacing can be generated. Each idler preserves phase information of the inputs, so that each idler has coherent phase correlation. This feature will be useful for coherent WDM source. Carrier phase recovery from QPSK signal was also demonstrated using the same principle [2]. The phase matching condition is determined by the periodically poled structure and refractive index dispersion in LiNbO₃. The refractive index is changed by the photorefractive effect in the presence of high power SH light. Thus, conversion efficiency is affected by the SH output power condition. To evaluate the phase matching curve in the presence of high power SH light, we have conducted following experiment.

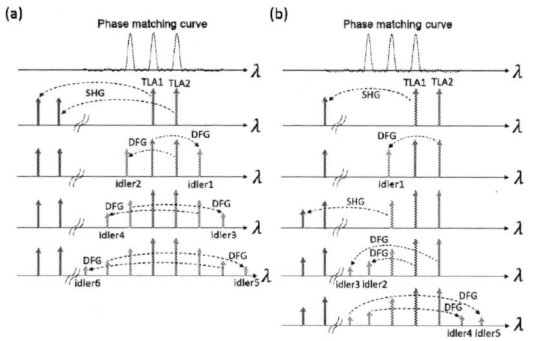

Fig.1. Process of multi-carrier generation
(a) Two inputs are matched to the QPM wavelength,
(b) One input is matched to QPM peak, and the other is detuned.

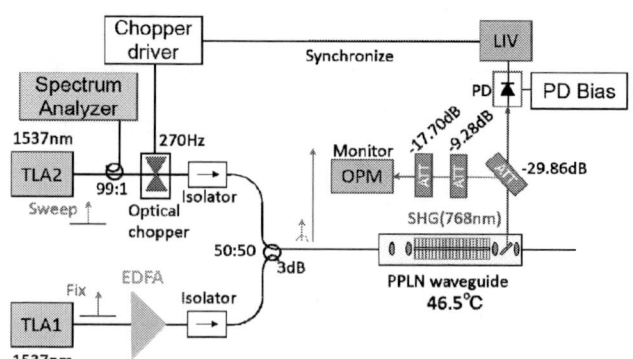

Fig.2. PM curve measurement set-up

1152

III. PM CURVE MEASUREMENT

Figure 2 shows the experimental setup. Two tunable lasers (TLA) were employed [3]. One TLA is modulated by optical chopper and the other is amplified by EDFA. They are combined with fiber coupler and injected to multiple QPM device. Wavelength of the amplified TLA was matched to a QPM peak, and wavelength of modulated TLA was scanned. The SH light generated from the multiple QPM device was detected by photodiode with lock-in voltmeter (LIV). By measurement of the intensity and phase of the SH light, we can distinguish the SH light generation through SHG/SFG and SH light depletion through DFG. For comparison, we also measured phase matching curve by scanning the wavelength of amplified TLA. The SH output was detected by an optical power meter (OPM) in that case.

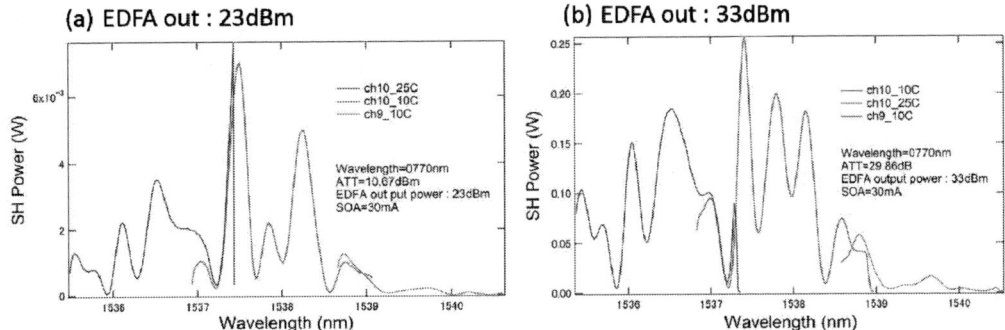

Fig.3. PM curve measured by scanning the amplified TLA.

Figures 3 show example of PM curves measured by scanning the wavelength of amplified TLA. The output power of the EDFA was set to 23dBm and 33dBm respectively. As the input power is increased, the PM curves have some distortion. It was not clear whether this is due to the photorefractive effect. In the case of high input power condition, the high power SH tends to convert to fundamental wavelength backwards. The backward conversion makes the PM curve complicated.

The phase measurement using modulated signal enables us to distinguish between forward and backward conversion. In the case of forward conversion, phase of the SH output follows the phase of the modulated input. In the case of backward conversion, the SH output exhibits reversed phase with respect to the input. We tried to characterize PM curve by measurement of the amplitude of the modulated SH output.

Figures 4 show PM curves measured by scanning the wavelength of modulated TLA with different power of amplified TLA. The positive value in Fig.4 corresponds to forward conversion due to SHG/SFG and negative value corresponds to a backward conversion due to DFG. As shown in Fig.4 significant backward conversion is obsereved as input power is increased. Even in the case of 30dBm output power of EDFA, shape of the PM curve is preserved. The multiple QPM device was fabricated by direct bonding method [4]. These results prove that direct bonded waveguide has high resistance to the photorefractive damage.

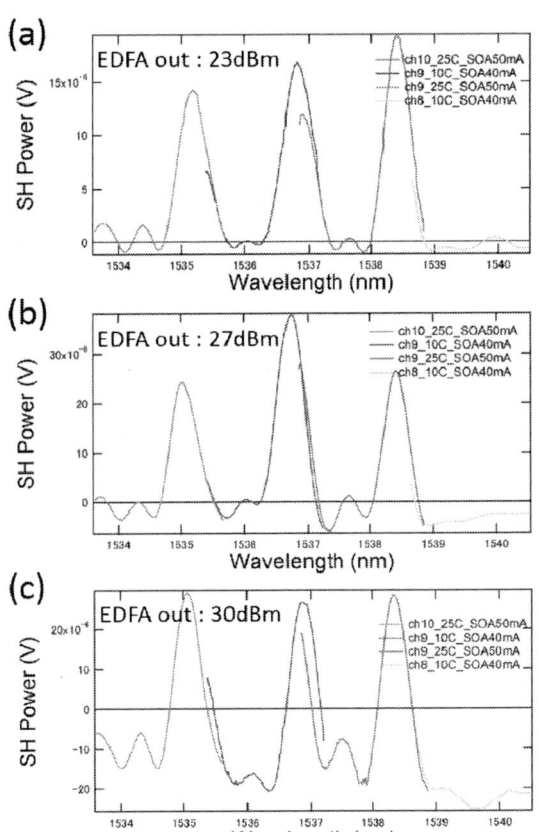

Fig.4. PM curves measured with lock-in voltmeter

IV. MULTIPLE OPTICAL CARRIER GENERATION

Based on the PM curve measurement, we have conducted multiple optical carrier generation. Figure 5 illustrates the experimental setup. Two TLA are combined with fiber coupler and they are amplified by EDFA simultaneously, and injected into the multiple QPM device. Figure 6 shows the optical spectra of the output with different wavelength assignment of the input. As the input power is increased, the number of the appreciable multiple carrier was increased. In Fig. 6 (a),(b),and (c) the wavelength of two TLAs are matched to the QPM peaks. In Fig. 6(d),(e),and (f) the wavelength of a TLA is matched to QPM peak, and the other TLA was detuned by 100 GHz. We obtained better efficiency in the former configuration. To the best of our knowledge, this is the first demonstration of optical comb generation using multiple QPM device.

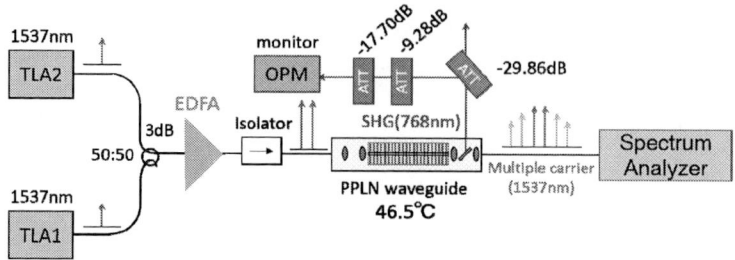

Fig.5. Experimental set-up for optical multiple carrier generation

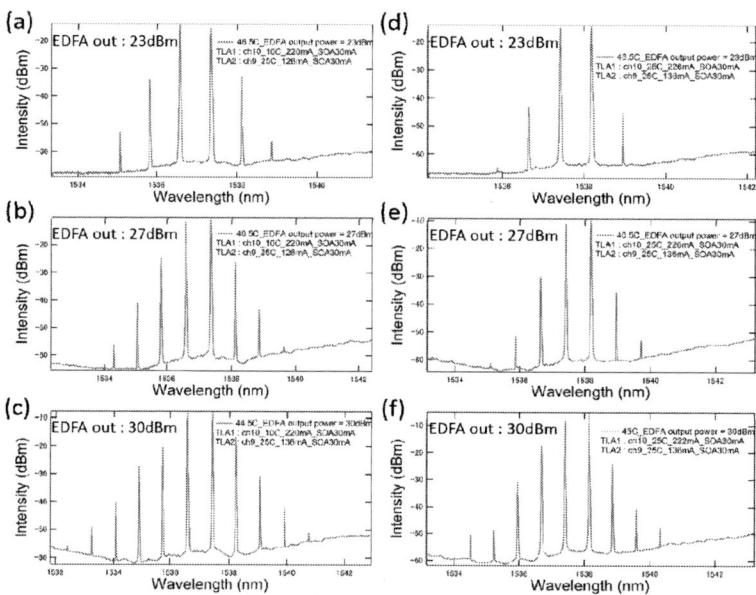

Fig.6. Optical spectra of multiple carrier with different input power

(a),(b),(c) : The wavelengths of two TLAs are matched to the QPM peaks

(d),(e),(f) : The wavelength of a TLA is matched to QPM peak, and the other TLA was detuned by 100 GHz

V. CONCLUSIONS

In conclusion, weak modulated signal enables PM curve measurement at the presence of high power input. The PM curve was preserved even with high power (~30dBm) input. Based on the measurement, we successfully demonstrated multiple carrier generation utilizing multi-stage frequency mixing.

REFERENCES

[1] M. Asobe, O. Tadanaga, H. Miyazawa, Y. Nishida, H. Suzuki, "Multiple quasi-phase-matched device using continuous phase modulation of χ(2) grating and its application to variable wavelength conversion" IEEE J. Quantum Electron. 41, 1540 – 1547 (2005)

[2] M.Asobe, T. Umeki, H. Takenouchi, and Y. Miyamoto "In-line phase-sensitive amplification of QPSK signal using multiple quasi-phase matched LiNbO3 waveguide" Optics Express, 22, 26642-26650 (2014)

[3] H. Ishii, K. Kasaya, H. Oohashi, Y. Shibata, H. Yasaka, and K. Okamoto, "Widely wavelength-tunable DFB laser array integrated with funnel combiner," IEEE J. Sel. Topics in Quantum Electron., 13, 1089-1094 (2007)

[4] T. Umeki, O. Tadanaga, and M. Asobe, "Highly efficient wavelength converter using direct-bonded PPZnLN ridge waveguide" IEEE J. Quantum Electron. 46, 1206-1213 (2010)

WE1-3   OECC/PS2016

Fabrication of High Optical Quality Factor Free-Standing As₂S₃ Microdisk Resonators on a Silicon Chip

Mingxiao Zhao,[1] Mingming Zhao,[1] Xiaoshun Jiang,[1,*] Yuan Chen,[1] Jiyang Ma,[1] and Min Xiao[1,2]

[1]National Laboratory of Solid State Microstructures and College of Engineering and Applied Sciences, Nanjing University, Nanjing 210093, China.
[2]Department of Physics, University of Arkansas, Fayetteville, Arkansas 72701, USA.
jxs@nju.edu.cn

Abstract: We demonstrate a chip-based free-standing As_2S_3 microdisk resonator with a Q-factor of 9.8×10^5. The critical coupling condition is achieved by efficiently coupling the modes of the microresonator with a tapered optical fiber.
Keywords: microcavity, As_2S_3

I. INTRODUCTION

Integrated silicon dioxide WGM resonators have been widely used for many applications in nonlinear optics, cavity QED, sensing and chip-based photonic devices due to their high quality factors, small mode volumes and compatibility with CMOS process [1]. However, silicon dioxide is opaque in the mid-infrared region (MIR). Recently, chalcogenide glasses including arsenic tri-sulfide (As_2S_3) have attracted much attention for the fabrication of integrated microcavities in the forms of microdisk, racetrack and micro-rings [3-7]. So far, the highest reported optical Q-factor of the chip-based As_2S_3 microcavities is smaller than 5×10^5 due to the fabrication limitations [3,7]. Here, for the first time we show a free-standing As_2S_3 microdisk resonator on a silicon chip with an intrinsic optical quality factor of $\sim 1 \times 10^6$.

II. FABRICATION PROCESS

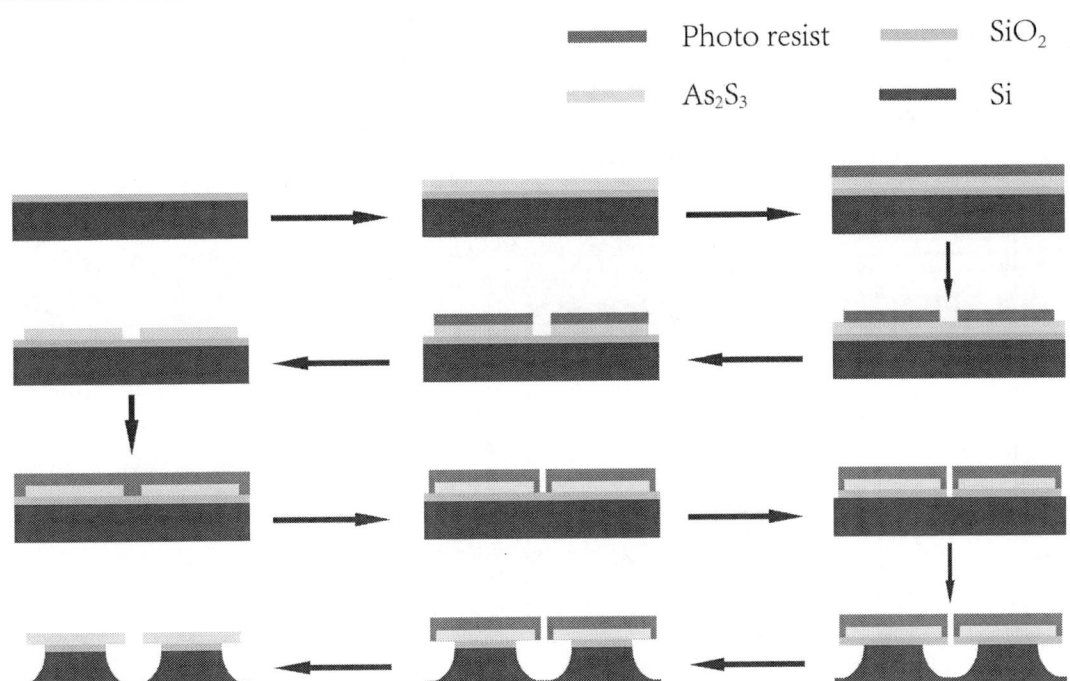

Fig. 1. Fabrication process of free-standing As_2S_3 microdisk resonators.

To fabricate the free-standing As_2S_3 microdisk resonator, we first prepare an As_2S_3 film on a thermal oxide film with a thickness around 1 μm using thermal evaporation deposition [6], as shown in figure 1. We then pattern the disk structure using standard optical lithography and develop the photoresist using MF319 developer. As the developer is NH_4^+ based, the uncovered As_2S_3 film is subsequently etched to leave the disk structure in the development process [9]. After removing the photoresist with acetone, we overlay the microdisks with new photoresist through another

1155

photolighography. Then we etch the silicon dioxide film into microdisk structures using the buffer oxide etcher (HF:NH$_4$F=1:5). After this step, we obtain a sandwich structure of an As$_2$S$_3$ microdisk covered by the photoresist and the silica microdisk without etching the As$_2$S$_3$ microdisk during the release of the silicon pillar. We undercut this sandwich structure using SF$_6$ chemical plasma. Last, we remove the silica layer outside of the silicon pillar using the buffer oxide etcher and the residual photoresist via photoresist remover (AZ remover 700). Figure 2 shows an optical microscopy image (Fig. 2(a)) and a scanning electron microscope image (Fig. 2(b)) of the fabricated freestanding As$_2$S$_3$ microdisk resonator.

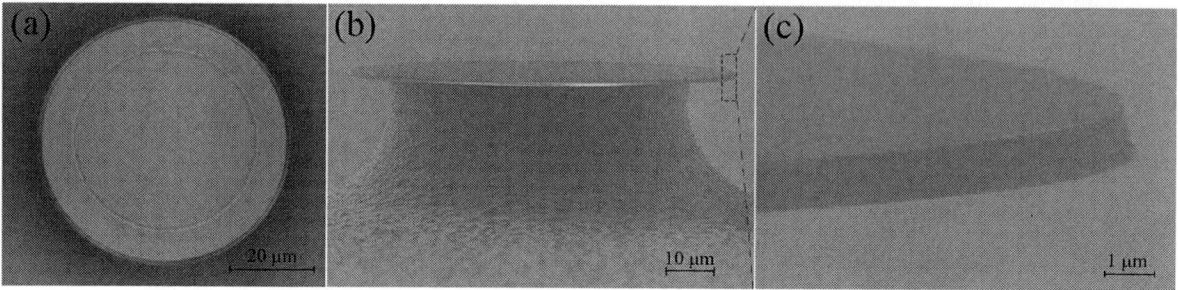

Fig. 2. (a) Top-view optical microscope image on an As$_2$S$_3$ microdisk. (b) Side-view SEM image of an As$_2$S$_3$ microdisk. (c) Zoomed in SEM image of the microdisk edge.

Figure 2 shows a free-standing As$_2$S$_3$ microdisk resonator, 60 μm in diameter and 900 nm in thickness. As illustrated in fig. 2(c), the fabricated As$_2$S$_3$ microdisk resonator has a relatively smooth surface and sidewall. This ensures low optical scattering and improves intrinsic optical quality factor [1].

III. CHARACTERIZATION

In order to characterize the optical quality factors of fabricated microdisk resonators, we measured the transmission spectra of the microdisk resonators at the telecommunication wavelength using the wavelength scanning method [10]. Light was coupled into a resonator through a tapered silica fiber with a diameter of ~1 μm. In the experiment, As$_2$S$_3$ microdisk resonators with diameters of 20, 30, 40, 50, 60 and 80 μm have been tested. The measured optical quality factors are in the range of 10^5 to 10^6. Figure 3(a) shows a typical optical transmission spectrum for an 50 μm-diameter microdisk with the thickness of 900 nm, indicating the intrinsic quality factor of the cavity mode is 9.8×10^5. In addition, optical critical coupling condition is achieved from a low order optical mode of another microdisk sample with an intrinsic optical quality factor of 2.5×10^5 (fig. 3(b)).

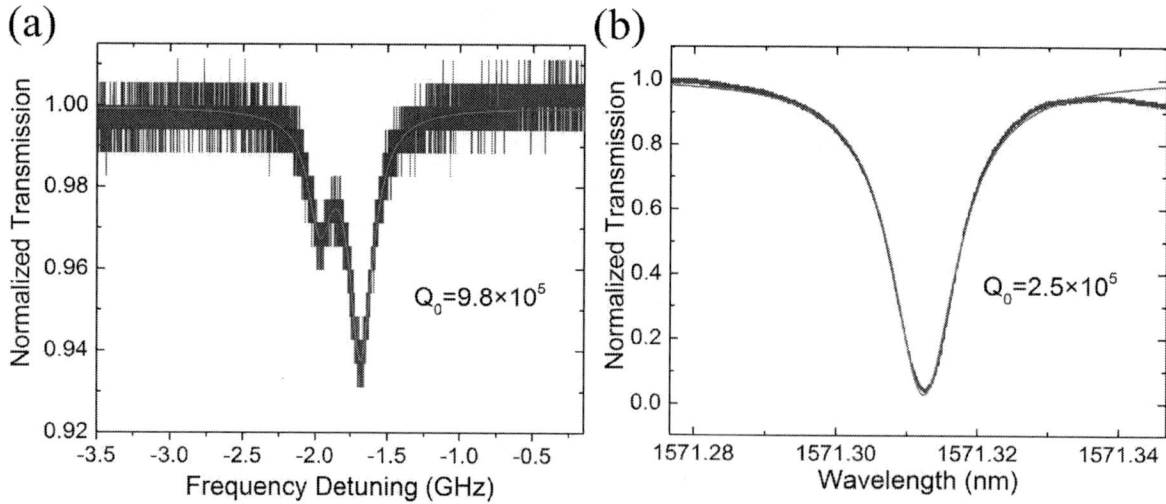

Fig. 3. (a) Normalized transmission spectrum of an As$_2$S$_3$ microdisk resonator with a diameter of 50 μm and thickness of 900 nm. (b) Normalized transmission spectrum of another As$_2$S$_3$ microdisk resonator with a diameter of 50 μm and thickness of 900 nm under the critical coupling condition.

IV. CONCLUSIONS

We have demonstrated a free-standing As$_2$S$_3$ microdisk resonator on a silicon chip using standard photolithography.

The resonators feature optical quality factors as high as 1×10^6 at the optical communication wavelength.

ACKNOWLEDGMENT

This work was supported by the National Basic Research Program of China (2012CB921804), the National Natural Science Foundation of China (nos. 61435007, 11574144 and 11321063) and the Natural Science Foundation of Jiangsu Province (BK20150015).

REFERENCES

[1] K. J. Vahala. "Optical microcavities," Nature, vol. 424, pp. 839-846, August 2003.

[2] B. J. Eggleton, B. Luther-Davies, and K. Richardson. "Chalcogenide photonics," Nat. Photonics, vol. 5, pp. 141-148, February 2011.

[3] L. Li, H. Lin, S. Qiao, Y. Zhou, S. Danto, K. Richardson, J. D. Musgraves, N. Lu, and J.Hu. "Integrated flexible chalcogenide glass photonic devices," Nat. Photonics, vol. 8, pp. 643-649, May 2014.

[4] H.Lin, L. Li, S. Danto, J. D. Musgraves, K. Rechardson, S. Kozacik, M. Murakowski, D. Prather, P. T. Lin, V. Singh, A. Agarwal, L. C. Kimerling, and J. Hu. "Demonstration of high-Q mid-infrared chalcogenide glass-on-silicon resonators," Opt. Lett., vol. 38, pp. 1470-1472, March 2013.

[5] J. Hu, M. Torregiani, F. Morichetti, N. Carlie, A. Agarwal, K. Richardson, L. C. Kimerling, and A. Melloni. "Resonant cavity-enhanced photosensitivity in As_2S_3 chalcogenide glass at 1550 nm telecommunication wavelength," Opt. Lett., vol. 35, pp. 874-876, March 2010.

[6] J. Hu, N. Carlie, L. Petit, A. Agarwal, K. Richardson, and L. Kimerling. "Demonstration of chalcogenide glass racetrack microresonators," Opt. Lett., vol. 33, pp. 761-763, March 2008.

[7] J. Hu, C. Nathan, N. Carlie, N. Feng, L. Petit, A. Agarwal, K. Rechardson, and L. Kimerling. "Planar waveguide-coupled, high-index-contrast, high-Q resonators in chalcogenide glass for sensing," Opt. Lett., vol. 33, pp. 2500-2502, September 2008.

[8] S. Song, D. Janesha, and C. B. Arnold. "Influence of annealing conditions on the optical and structural properties of spin-coated As_2S_3 chalcogenide glass thin films," Opt. Express, vol. 18, pp. 5472-5480, March 2010.

[9] A. B. Seddon. "Chalcogenide glasses: a review of their preparation, properties and applications," J. Non-Cryst. Solids, vol. 184, pp. 44-50, December 1995.

[10] J. C. Knight, G. Cheung, F. Jacques, and T. A. Birks. "Phase-matched excitation of whispering-gallery-mode resonances by a fiber taper," Opt. Lett., vol. 22, pp. 1129-1131, April 1997.

OECC/PS2016

Wavelength modulation spectroscopy of formaldehyde using 3μm DFG laser

Ryohei Fujisawa[1], Masaki Asobe[1], Akira Katoh[1], Shigeru Yamaguchi[1], Akio Tokura[2], Hirokazu Takenouchi[2]

[1] Tokai University, 4-1-1 Kitakaname, Hiratsuka, Kanagawa, 259-1292 Japan

[2] NTT Device Technology Laboratories, NTT Corporation, 3-1 Morinosato Wakamiya, Atsugi, Kanagawa, 243-0198 Japan

Author e-mail address: asobe@tokai-u.jp

Abstract: *Effect of modulation condition of 3μm DFG laser on intensity noise in wavelength modulation spectroscopy was studied. The reduction of the intensity noise enabled real time detection of formaldehyde without baseline subtraction.*

Keywords: *Nonlinear frequency conversion, gas spectroscopy, mid infrared laser*

I. INTRODUCTION

Recent advances in wavelength conversion using periodically poled $LiNdO_3$ (PPLN) waveguide facilitates the generation of a wide range of wavelengths using laser diodes (LDs) [1],[2]. A directly bonded $LiNbO_3$ waveguide allows us to generate mid-infrared light efficiently using difference frequency generation (DFG) [2]. In the 3-μm wavelength range, many hydrocarbon gases exhibit strong absorption due to the fundamental vibration mode of C-H bonds. Among hydrocarbon gases, formaldehyde is known as a causative agent of sick house syndrome. In this work, we studied the wavelength modulation spectroscopy of formaldehyde using a 3-μm DFG laser based on a PPLNwaveguide. The wavelength modulation of LD is converted to intensity noise due to the weak reflection at the facet of the PPLN waveguide. We found that intensity noise can be reduced by setting appropriate modulation condition. The reduction of noise allowed us to detect the formaldehyde concentrations in real time without baseline subtraction.

II. EXPERIMENT

Figures 1 shows the experimental setup for the gas detection system for the study of sick house syndrome. The 3-μm DFG laser consisted of 1.06-μm and 1.55-μm DFB LDs as well as a directly bonded PPLN waveguide. The output of the laser was divided by a half mirror, and transmitted through a White cell and reference cell which is filled with 100% formaldehyde. Transmitted lights were detected with InSb photodetectors (PDs). To obtain absorption spectra the wavelength of the DFG laser was scanned through current modulation of a LD. For wavelength modulation spectroscopy (WMS), a sinusoidal wave was applied to the driving current of a LD, and the second-harmonic component of the PD output was detected with lock-in amplifiers (LIAs) [3]. In this work, formaldehyde diluted by oxygen and air was introduced into the White cell. The gas sensor was designed to detect the concentration in sampled gas which is taken from a desiccator. This setup enables us to study the causality between the environment and a variety of vital signals of rats monitored by wireless sensors and a video camera.

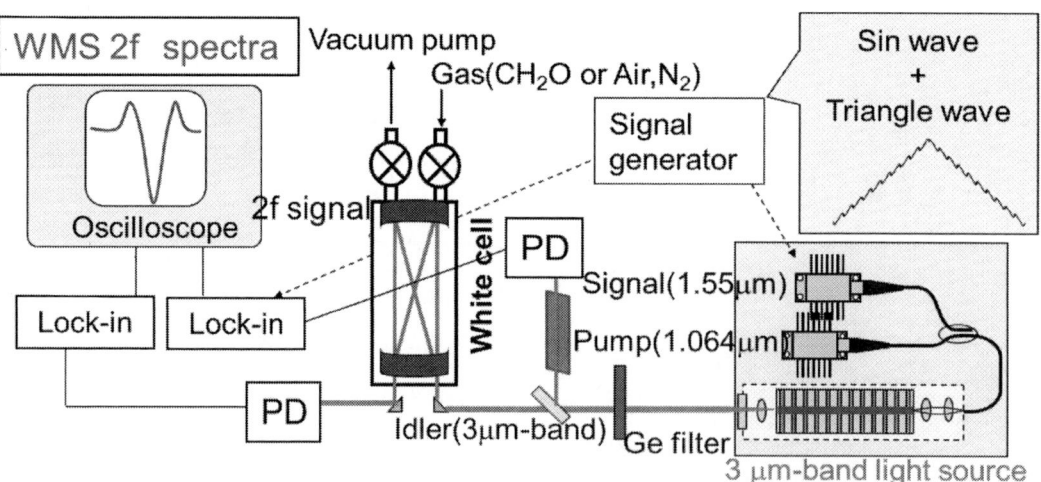

Fig. 1 Experimental setup

1158

III. EFFECT OF MODULATION CONDITIONS

Prior to measurement of diluted gas sample, we have studied the effect of modulation condition on the 2nd derivative waveform of absorption. Figures 2 show the waveforms with the reference cell and White cell filled with air. We observed periodic intensity noise in the absence of formaldehyde gas. As can be seen in Figs 2 (b) and (c), the amplitude of the noise depends strongly on the modulation depth. Figures 3 show the intensity of the 2nd derivative peak with the reference cell and amplitude of intensity noise with air as funcions of modulation depth. The 2nd derivative signal intensity is increased monotonously as modulation depth is increased. On the contrary, intensity noise exhibits minimum value with appropriate modulation depth.

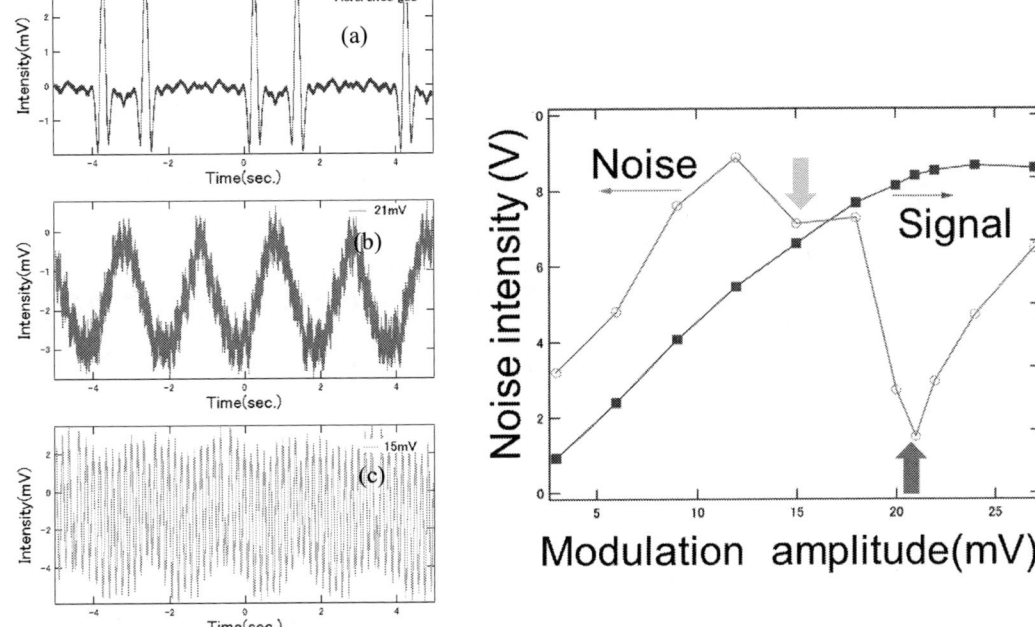

Fig. 2 WMS waveforms
(a) reference cell,
(b),(c) air with different modulation depth

Fig. 3 Peak and noise intensity vs. modulation depth

IV. ORIGIN OF INTENSITY NOISE

We attribute the intensity noise to FM-AM conversion caused by the reflection in PPLN waveguide. Although the PPLN waveguide had anti reflection coating on the facet, residual reflection causes Fabry-Perot resonance. The Fabry-Perot resonance causes periodic wavelength dependent trasmittance of the PPLN waveguide. The sinusoidal wavelength modulation of LD is converted to intensity modulation through the periodic trasmittance of the PPLN waveguide. The intensity modulation can be expressed as following equation.

$$v(t) \propto \cos\left(\beta \sin \omega t + \varphi\right) = \cos\varphi\cos\left(\beta \sin \omega t\right) - \sin\varphi\sin\left(\beta \sin \omega t\right) \quad (1)$$

Here, β is modulation depth., ω is modulation frequency and ϕ is center frequency relative to trasmittance peak period. Each term in eq.(1) can be expressed as as following equation.

$$\cos\left(\beta \sin \omega t\right) = J_0(\beta) + 2\sum_{n=1}^{\infty} J_{2n}(\beta)\cos(2n\omega t) \quad (2)$$

$$\sin(\beta \sin \omega t) = \sum_{n=0}^{\infty} J_{2n+1}(\beta)\sin[(2n+1)\omega t] \quad (3)$$

Here, J_m is m th order Bessel function. In our experiment, we observed second harmonic (n=1) of modulation frequency using LIA. This means that intensiy noise is proportional to the 2nd order Bessel function $J_2(\beta)$. Figure 4 shows the J_2 and it's absolute value as a function of modulation depth. From the comparison of Fig.3 and Fig. 4, we concluded that the intensity noise is caused by the FM-AM conversion and it could be minimized by a setting of appropriate modulation depth.

V. SPECTROSCOPY OF FORMALDEHYDE

Using the appropriate modulation condition, we tried to measure the absorption spectra of formaldehyde samples. Figures 5 show the expamles of the spectra with different concentration. Thanks to the reduced intensity noise we could observed absorption peaks without baseline subtraction even with a concentration of 0.47 ppm.

With the condition in Fig.5 (a), we could obtain strong transmission change even without sinusoidal modulation. From comparison with simulation using HITRAN data base we estimated the concentration to be 82.6ppm [4]. With the condition in Fig.5 (b), we estimated concentration of 0.47 ppm using a gas chromatgraphy.

The WMS peak intensity as a function of gas concentrations using these values exhibit good linearity, which assured that concentration measured in our experiment is in good agreement with the gas chromatography.

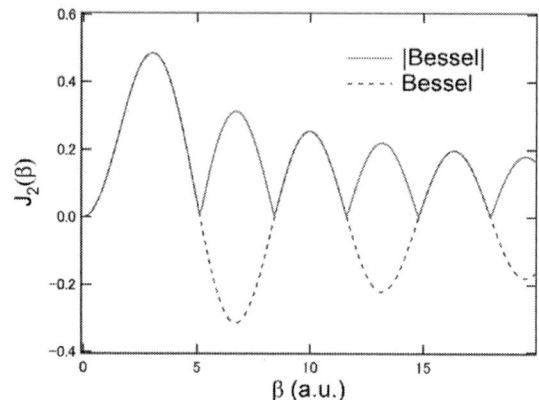

Fig. 4 Bessel function J_2 vs. modulation depth

To conduct automatic measurement we have prepared software which allows us to monitor the waveforms of reffrence gas and sample, as well as current concentration in real time. Thanks to the reduced intensity noise we could monitor the gas concentration in real time wiout baseline subtraction.

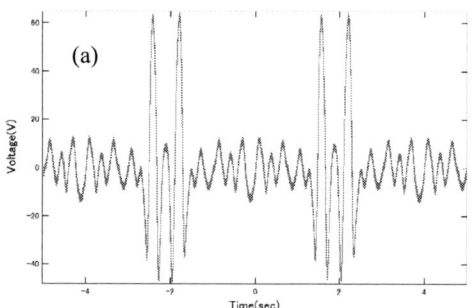
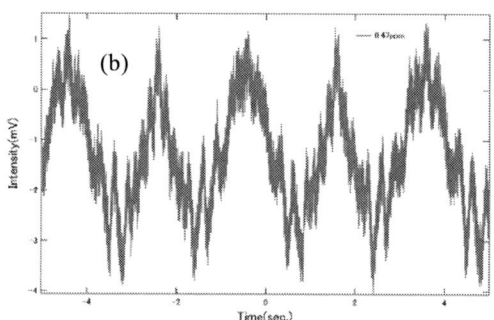

Fig. 5 Absorption spectra of formaldehyde, (a) 82.6 ppm, (b) 0.47 ppm

VI. CONCLUSIONS

We studied the wavelength modulation spectroscopy of formaldehyde using a 3-μm DFG laser based on a PPLNwaveguide. We found that intensity noise caused by reflaction at the facet of PPLN waveguide can be minimized by setting the appropriate modulation condition. The reduction of noise allowed us to detect the formaldehyde concentrations in real time without baseline subtraction. We expect that this technology is useful for the study of the causality between the environment and sick house syndrome.

REFERENCES

[1] T. Umeki, O. Tadanaga, and M. Asobe, "Highly efficient wavelength converter using direct-bonded PPZnLN ridge waveguide" J. Quantum Electron. 46, pp.1206-1213 (2010).

[2] O.Tadanaga, Y.Nishida, T.Yanagawa, K.Magari, T.Umeki, M.Asobe, and H.Suzuki : "Compact sub-mW mid-infrared DFG laser source using direct-bonded QPM-LN ridge waveguide and laser diodes,"Proc. SPIE, Vol.6455, 64550D, 2007.

[3] J. Reid and D. Labrie, "Second-harmonic detection with tunable diode lasers – Comparison of experiment and theory" Applied Phys. B 26, 203-210 (1981).

[4] The HITRAN Database http://www.cfa.harvard.edu/hitran/

WE1-5

Chirp-free Spectral Compression of Parabolic Pulses in Silicon nitride Channel Waveguides

Chao Mei, Jinhui Yuan, Kuiru Wang, Xinzhu Sang, Chongxiu Yu,

[1] State Key Laboratory of Information Photonics and Optical Communications, Being University of Post and Teleommunication,
No.10 Xitucheng Road, Haidian District, Beijing, China
[2] Photonics Research Centre, Department of Electronic and Information Engineering, Hong Kong Polytechnic University, Hung Hom,
Hong Kong SAR, China
[*] meichao@bupt.edu.cn

Abstract: *A practical scheme is presented to achieve spectral compression of highly chirped parabolic pulses in a 4-cm long Si_3N_4 channel waveguides. Chirp-free pulses with -15.5-dB pedestals and a compression ratio of 25.4 are obtained.*

Keywords: *optical pulse compression; optical waveguide; nonlinear optical devices*

I. INTRODUCTION

Spectral compression has found many applications in optical signal processing, such as the all-optical analog-to-digital converters based on power-to-wavelength conversion, soliton self-frequency shift effect [1] and coherent optical sources with high power density and narrow line width [2]. Spectral compression of parabolic pulses with a highly linearly negatively frequency chirp was numerically and experimentally investigated in a nonlinear photonic crystal fiber [3] and all-fiber configuration [4]. The advantage of pulses with parabolic profiles is to achieve chirp-free pulses with low pedestals comparing with hyperbolic secant and Gaussian profiles.

The spectral compression is important to be realized in the CMOS-compatible waveguides. The requirements of all-optical integration on chip should utilize small size waveguides. Recently, Si_3N_4 has attracted great interest as a promising candidate in nonlinear optics because it is compatible with the mature CMOS technology with the large three-order nonlinear refractive index [5]. Besides, comparing with Silicon, its nonlinear absorption at the telecommunication wavelength is negligible. Here, an efficient spectral compression of parabolic pulse is achieved in Si_3N_4 channel waveguide buried in the SiO_2. Besides, spectral compression of pulses with hyperbolic secant and Gaussian profile is also simulated numerically as a contrast. Results show that parabolic pulses can realize the lowest pedestals and highest energy utilization.

II. THEORETICAL MODELS

The nonlinear dynamic of optical pulse propagating in the waveguide is described by the generalized nonlinear Schrödinger equation (GNLSE), which is given as following [6]

$$\frac{\partial A}{\partial z}+\frac{\alpha_0}{2}A-\sum_{m\geq 2}\frac{i^{m+1}\beta_m(z)}{m!}\frac{\partial^m A}{\partial t^m}=i\gamma\left(|A|^2+\frac{i}{\omega_0}\frac{1}{A}\frac{\partial}{\partial t}\left(|A|^2 A\right)\right), \quad (1)$$

where A is the slowly varying envelope of the electric field, α_0 represents the linear loss and $\beta_m(z)$ is the m-th order dispersion coefficient. γ is the nonlinear coefficient describing the self-phase modulation (SPM) and self-steeping.

For parabolic pulse, the mathematical formula can be expressed as

$$\begin{cases} A(t)=\sqrt{P_0}\sqrt{1-\frac{2t^2}{T_F^2}}\exp\left(-i\frac{C}{2}t^2\right) & |t|\leq T_F/\sqrt{2} \\ A(t)=0 & otherwise, \end{cases} \quad (2)$$

where P_0 is the peak power of the parabolic pulse, T_F is the temporal full-width at half-maximum (FWHM) and C is the linear chirp coefficient. We assume that the initial temporal pulse width is T_0 ($T_0 = T_F/\sqrt{2}$).

The SPM introduces a temporal chirp for the case of a parabolic pulse to compensate the temporal negative chirp.

$$\gamma P_0 z+\frac{C}{2}T_0^2=C_s, \quad (3)$$

When the temporal negative is completely compensated, the pulse becomes chirp-free and the total chirp coefficient C_s is expected to be zero at a certain distance z_m.

$$z_m=\frac{|C|T_F^3}{3\sqrt{2}\gamma E_0}. \quad (4)$$

We can see that z_m is in inverse proportion of nonlinear coefficient γ and initial energy E_0. If a pulse with high energy propagates in a highly nonlinear medium, a chirp-free pulse is obtained in a short distance. It should note that in above analytical processes, the dispersion has been neglected, which requires the dispersion length (L_D) is much larger than the nonlinear length (L_{NL}).

III. SIMULATION RESULTS

A. Waveguides Design

The cross section of the waveguide used in our simulation is shown in Fig.1 (a). The width (W) and height (H) of the Si_3N_4 buried in the SiO_2 layer are 1000 nm and 750 nm respectively. The similar structure can be fabricated by using plasma enhanced chemical vapour deposition [7]. The inset shows the light is well confined in the Si_3N_4 waveguide at 1550 nm. The linear loss is 0.4 dB/cm. In Fig. 2 (a), curves of the effective refractive index, n_{eff} and group-velocity dispersion (GVD), β_2, versus wavelength are plotted. The second ZDP appears at 1525 nm, which makes the normal GVD at 1550 nm is relatively small (22.46 ps²/km). The effective mode area is calculated to be 0.78 μm² and the nonlinear coefficient γ is 1.3/W/m. These values agree well with the previous experimental results mentioned in [5].

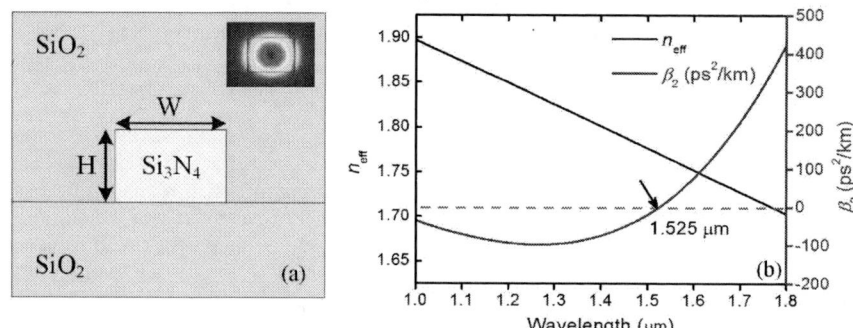

Fig. 1. (a) Cross-section sketches of the proposed Si_3N_4 waveguide which is buried in the SiO_2 layer. The inset is the mode field distribution of 1550-nm pump pulses. (b) n_{eff} and GVD versus wavelength are plotted with black solid and blue solid curves respectively.

B. Pulses Propagating in The Si_3N_4 Waveguide

To demonstrate the parabolic pulse evolutions in the Si_3N_4 waveguide clearly, the change of temporal and spectral pulses along the longitudinal direction are shown in Fig. 2 (a) and (b). The peak power and full width at half maximum (FWHM) of input pulse are 600 W and 5 ps respectively. For the parabolic pulse, the linear chirp coefficient, C is -5 THz/ps. The ratio of L_{NL} and L_D, is 2.27×10^{-6}. So the influence of dispersion can be neglected, which explains why the temporal profile remain unchanged in Fig. 2 (a). The spectral pulse reaches its minimum width at z = 4 cm when a high brightness appears in Fig. 2 (b). The result agrees with the theoretical prediction in (4) well.

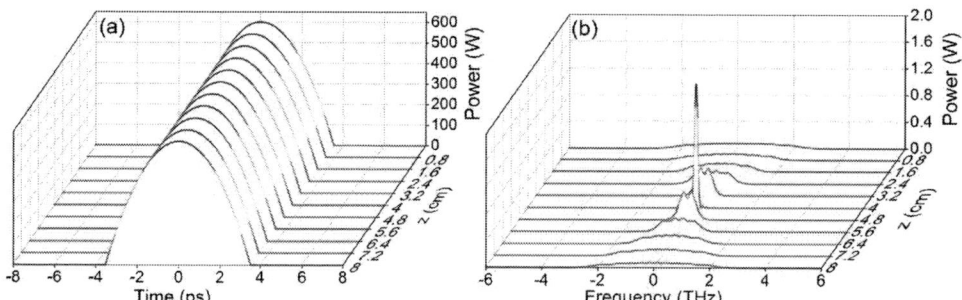

Fig. 2. Temporal (a) and spectral (b) pulse evolutions processes for the parabolic profile in the 8-cm long Si_3N_4 waveguide.

Nonlinear phases produced by the SPM are completely cancelled by the negatively linear chirp at z = 4 cm. In fact, for the parabolic pulse, the chirp is always zero because the total phase is a constant at a certain distance. However, the constant will changes as pulses propagate. When the constant is zero, the spectral compression ratio (SCR) is biggest. Fig. 3 (a) has confirmed the view. The SCR is greatly determined by the negatively linear chirp which can be introduced by the grating pairs. From Fig. 3 (b), the SCR is proportional to the chirp coefficient. However, from (4), the big SCR means long waveguides or high initial power, which are disadvantage factors for on-chip integrations.

The spectral compression of the Gaussian and hyperbolic secant pulses is also investigated for comparison. These pulses have the same initial energy of 4.8 nJ as parabolic pulses. We adjust the pulse width and initial chirp coefficient to ensure all kinds of pulses are with the same initial spectral width, 6.19 THz. Fig. 4 (a) shows the SCR of the three pulse profile versus propagating distance, z. Even the Gaussian and hyperbolic secant pulses reach their maximum SCR earlier, the SCR of the two kind of pulses is less than 15. Besides, the curves of Gaussian and hyperbolic secant pulses demonstrate irregular oscillation because their chirps are not completely cancelled by the initial linearly chirp.

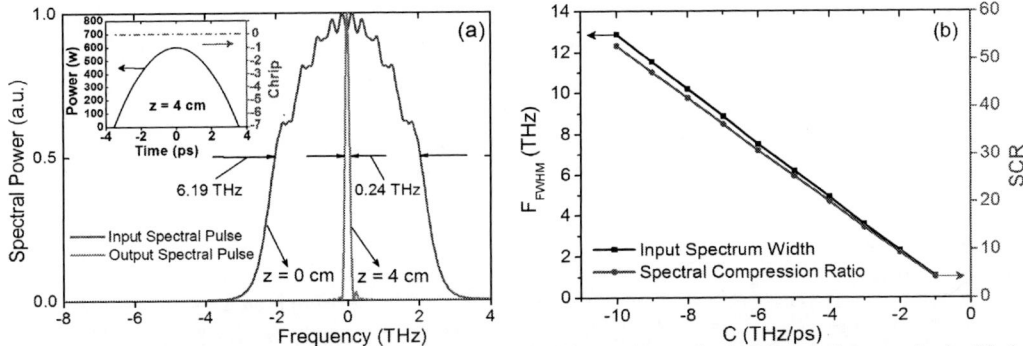

Fig. 3. (a) Comparisons of the input (blue solid curves) and output (red solid curves) spectral pulse. The inset is temporal pulse (black solid curves) and its chirp distribution (blue dashed dot line). (b) The input spectrum width (black solid curves) and SCR (blue solid curves) vary versus C.

The compressed pulses at the maximum SCR of the three profiles is shown is Fig. 4 (b). The ration between the peak power of the wings and main peak power is -5.1 dB, -7.1 dB and -15.5 dB for hyperbolic secant pulse, Gaussian pulse and parabolic pulse respectively. We can deduce that the parabolic pulse has the highest energy utilization.

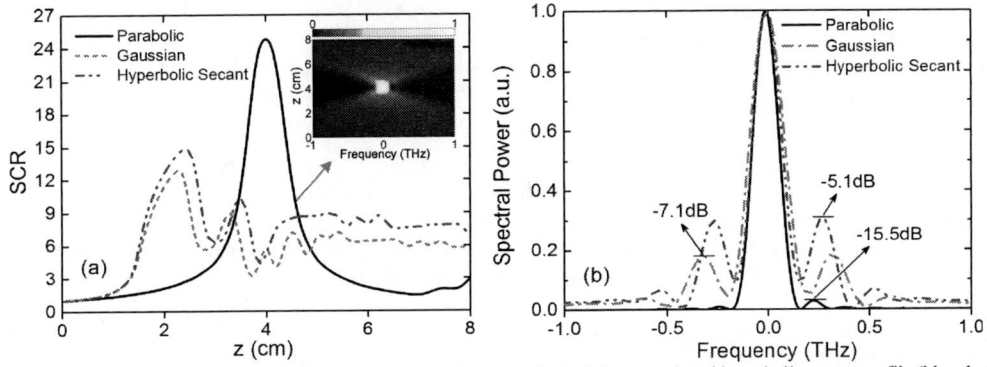

Fig. 4. (a) SCR of the parabolic profile (black solid curves), Gaussian profile (red dot curves) and hyperbolic secant profile (blue dashed dot dot curves). The inset is the depth map of parabolic pulses. (b) The output spectral pulse at the maximum SCR.

IV. CONCLUSIONS

In summary, a scheme to achieve efficient spectral compression of the parabolic pulses is presented in a 4-cm Si_3N_4 waveguide. The SCR is 25.4 and the peak power of pedestal is low to -15.5 dB. Comparing with the Gaussian pulse and hyperbolic secant pulse, the parabolic pulse exhibits bigger SCR and higher energy utilization.

ACKNOWLEDGMENT

This work is partly supported by the National Natural Science Foundation of China (61307109 and 61475023), Beijing Youth Top-notch Talent Support Program (2015000026833ZK08), the Natural Science Foundation of Beijing (4152037), the Hong Kong Scholars Program 2013 (PolyU G-YZ45), and the Research Grant Council of the Hong Kong Special Administrative Region China (PolyU5272/12E).

REFERENCES

[1] T. Nishitani, T. Konishi, and K. Itoh, "Demonstration of 4-bit photonic analog-to-digital conversion employing self-frequency shift and SPM-induced spectral compression," Proc. of 33th European Conference on Optical Communications (ECOC 2007), Berlin (Germany), Sept. 2007, vol.14, pp. 1-2.

[2] M. Rusua and O. G. Okhotnikov, "All-fiber picosecond laser source based on nonlinear spectral compression," Appl. Phy. Lett., vol. 89, pp. 091118-1-091118-3, Aug. 2006.

[3] C. F. Cheng, Y. Q. Wang, Y. W. Ou, and Q. H. Lv, "Transform-limited spectral compression by using negatively chirped parabolic pulse in all-normal dispersion photonics crystal fibers," J. Nonlinear Opt. Phys. Mater., vol. 21, pp. 1-11, Sep. 2012.

[4] J. Fatome, B. Kibler, E. R. Andresen, H. Rigneault, and C. Fiont, "All-fiber spectral compression of picosecond pulses at telecommunication wavelength enhanced by amplitude shaping," Appl. Opt., vol. 51, pp. 4547-4553, Jul. 2012.

[5] D. J. Moss, R. Morandotti, A. L. Gaeta and M. Lipson, "New CMOS-compatible platforms based on Silicon nitride and Hydex for nonlinear optics," Nature Photon., Vol. 7, pp. 597-607, Aug. 2013.

[6] G. P. Agrawal, Nonlinear Fiber Optics, 4rd ed., NewYork: Academic, 2009. pp. 11-60.

[7] K. Ikeda, R. E. Saperstein, N. Alic and Y. Fainman, "Thermal and Kerr nonlinear properties of plama-doposited silicon nitride/silicon dioxide waveguide. Opt. Express Vol. 16, pp. 12987-12994. Aug. 2008.

WE3-1 (Invited)

OECC/PS2016

Slow Light Devices in Silicon Photonics

Toshihiko Baba

Department of Electrical & Computer Engineering, Yokohama National University
79-5 Tokiwadai, Hodogayaku, Yokohama 240-8501, Japan
Author e-mail address: baba-toshihiko-zm@ynu.ac.jp

Abstract: *Photonic crystal slow light waveguides fabricated by CMOS process add values to Si photonics devices. 200-μm-long practical 25-Gbps MZ modulators, low-cost sub-bandgap Si photodiodes at wavelengths around 1.55 μm, and on-chip optical correlators are demonstrated.*
Keywords: *slow light, photonic crystal, silicon photonics, modulator, photodiode, optical correlator*

I. INTRODUCTION

In these ten years, silicon (Si) photonics has become a powerful platform of dense photonic integration [1]. However, the improvement of each device has almost been saturated due to the limited material properties of Si and unchanged waveguide structure. Slow light in photonic crystal waveguides adds values to this situation and contributes to the continuous progress of the devices [2,3].

This paper demonstrates three examples of slow light devices promising toward the practical use: 1) compact 25 Gbps Mach-Zehnder (MZ) modulator, 2) sub-bandgap Si photodiode (PD) at telecom wavelengths, and 3) on-chip optical correlator for picosecond pulse measurements. We discuss their merits and limitations expected.

II. SLOW LIGHT MODULATORS

In Si photonics, MZ and microring modulators have been developed, in which p-n-junction-loaded rib-waveguides are used as phase shifters. Microring modulators are very compact and less power consuming. However, their working spectrum is too narrow to use in optical interconnects without thermal control, which is a severe constraint toward the practical use. MZ modulators have a wide working spectrum suitable for the wide-temperature-range operation in optical interconnects, while the device is large in size (typically several millimeters) and comparably power consuming. Therefore, the photonic crystal waveguide which generates slow light is effective if it is used as a phase shifter [4]. Let us assume slow light to have a group index of n_g and a modal overlap factor with the p-n junction, ξ. The phase shift $\Delta\phi$ depends on a factor of $F_\phi \equiv (n_g\xi)/(n_{rib}\xi_{rib})$. This means that the change of the optical path length in the phase shifter, ΔnL, which is induced by the carrier plasma around the p-n junction can be reduced by the same factor. The ΔnL is proportional to the product of the junction capacitance C and applied voltage V. Therefore, V for the same $\Delta\phi$ can be reduced by a factor of $F_v \equiv (n_g\xi/C)/(n_{rib}\xi_{rib}/C_{rib})$ and the power consumption, a factor of $F_P \equiv (n_g^2\xi^2/C^3)/(n_{rib}^2\xi_{rib}^2/C_{rib}^3)$.

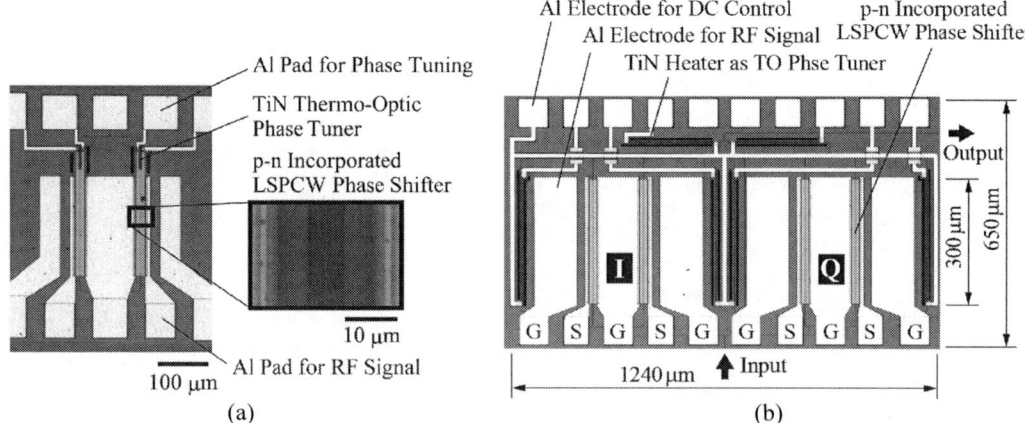

(a) (b)

Fig. 1. Si LSPCW MZ modulators with interleaved p-n junction. (a) OOK modulator. (b) QPSK modulator.

The on-off-keying (OOK) modulator in Fig. 1(a) was fabricated into SOI wafer (Si layer was 210 nm thick) using 180-nm standard CMOS process. We employed lattice-shifted photonic crystal waveguide (LSPCW) for slow light phase shifters, in which a single line defect waveguide was sandwiched by 210-nm-diameter holes arranged with a pitch of 400 nm, some rows of holes are shifted along the waveguide, and the interleaved p-n junction was formed at the waveguide center [5]. The processed Si and Al electrodes were finally covered with SiO_2 cladding. We have observed $n_g \approx 20$ with a working spectrum $\Delta\lambda \approx 20$ nm at telecom wavelengths for the third-row-shifted LSPCW. The slow-light

1164

mode large penetrates into the photonic crystal cladding, which decreases ξ. The interleaved p-n junction compensates for this, and consequently $\xi/\xi_{rib} \approx 1$. Thus F_ϕ takes a value of 5 for $n_g \approx 20$. In here, C is enhanced twice larger by the interleaved junction. Therefore, F_v and F_p are suppressed to 2.5 and 3.1, respectively. We have also observed $n_g \approx 40$ with $\Delta\lambda \approx 10$ nm and $n_g \approx 70$ with $\Delta\lambda \approx 5$ nm for the second-row-shifted LSPCW [6,7]. Since ξ and C are not so changed, F_ϕ, F_v and F_p will be 10, 5 and 13 for $n_g \approx 40$, and 18, 8.8 and 38, respectively.

One drawback is a large junction resistance R caused by the wider spacing between p^+- and n^+-doped regions for metal contact, which is set to avoid the absorption loss for the expanded slow light mode. The larger R attenuates the RF signals and hampers to elongate the device. Therefore, we fixed L to be 200 μm, for which $R \approx 100$ Ω (150 Ω in total with the internal resistance of the signal source) and $C \approx 60$ fF. It results in an RC time constant of 9 ps and a cutoff frequency f_{3dB} of 18 GHz. We actually measured a close value $f_{3dB} = 17$ GHz. It is sufficient for 25–32 Gbps operation but insufficient for ≥ 40 Gbps. If n_g is increased, f_{3dB} is further reduced by the phase mismatch between slow light and RF signals. For 25 Gbps, $n_g = 35$–40 will be an upper limit for standard RF electrodes.

In the $n_g \approx 20$ device, $V_\pi L$ was reduced to ~0.3 V·cm, reflecting the F_ϕ. For $L = 200$ μm, V_π becomes 16V and 8 V under single and push-pull drives. Roughly speaking, optical interconnects require an extinction ratio (ER) of 3 dB, which corresponds to $\Delta\phi = 0.25\pi$ and $V_{pp} = 2$ V under the push-pull drive. Figure 2 shows the obtained eye pattern for this voltage when the initial phase of the MZ circuit was shifted slightly. On this condition, a small excess modulation loss (ML) was added and the open eye was observed (Fig. 2(a)). The on-chip loss without modulation was typically 5 dB, and the total loss including the excess modulation loss ML was 6 dB. The passive loss includes a 2 dB coupling loss between input and output waveguides and LSPCW, which should be suppressed by optimizing the design.

The QPSK modulator was also fabricated using nested MZ circuits and the interleaved-junction-loaded LSPCW phase shifters (Fig. 1(b)) [8]. Here, L was elongated to 250–300 μm and V_{pp} was increased 3 V, which might be effectively higher than 5 V at the device due to the absence of electrical termination. The estimated $\Delta\phi$ was limited to 0.4π, but the clear constellation pattern was observed up to 28 Gbps with an EVM less than 30%, corresponding to the FEC error-free threshold. Fig. 2(b) shows the 32 Gbps constellation improved by EQ filter. L must be increased to 600 μm to obtain the complete QPSK operation with an ideal π phase shift, but it is also hampered by the RF transmission loss and phase mismatch.

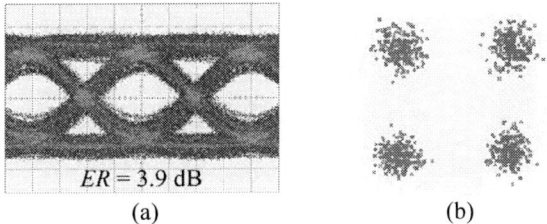

| (a) | (b) |

Fig. 2. Operation of LSPCW MZ modulators. (a) Eye pattern of 25 Gbps OOK modulation with $ML = 1$ dB. (b) Constellation pattern of 32 Gbps QPSK modulation (EVM = 19.6%) improved by EQ filter in a receiver.

III. SUB-BANDGAP PHOTODIODE

Si sub-bandgap photodiodes (PDs) have been studied for p-n-diode-loaded Si rib-waveguides, in which some exotic ions are doped to increase the defect states. The responsivity via the defect states is observed at telecom wavelengths when the reverse bias is higher than 10 V. Avalanche gain for a higher bias allows a high responsivity beyond 1 A/W although f_{3dB} is suppressed to several GHz and the dark current is usually higher than 1 μA. In addition, the exotic doping is not prepared in the menu of the standard CMOS process. We reduced these constraints by using slow light [9]. The device structure was similar to that for the above modulator. Only a difference was that the lattice was not shifted to investigate the responsivity for various n_g. The exotic doping was omitted and just the reverse bias was increased to -18 V to conduct the responsivity. The responsivity exceeded 0.1 A/W with the dark current still lower than 40 nA. The maximum value was 0.36 A/W for n_g higher than 50. Barely open eye was observed up to 30 Gbps.

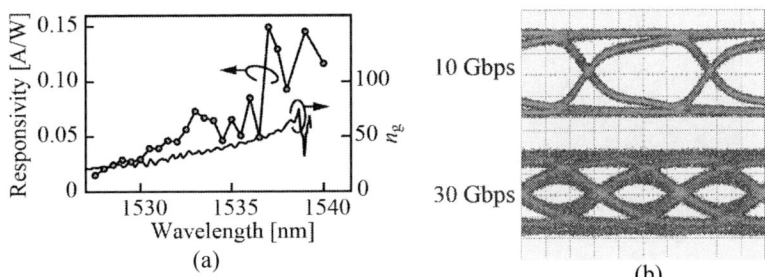

Fig. 3. Operation of LSPCW sub-bandgap PD. (a) Responsivity and n_g spectra. (b) Eye patterns.

IV. ON-CHIP OPTICAL CORRELATOR

Optical correlators are widely used for measuring the length and profile of picosecond-order short optical pulses. It is composed of a tunable delay line and nonlinear photodiode. Since a mirror which moves mechanically and changes the optical path length physically is used as the tunable delay line, the correlator usually becomes a box-sized equipment. Slow light in the LSPCW enables a tunable delay of up to several ten picoseconds by thermally tuning the photonic band of the LSPCW. Fig. 4(a) shows the top view of the device fabricated by CMOS process [10]. In here, two LSPCWs are integrated as a heater-driven tunable delay line and a two-photon-absorption (TPA) PD. The input waveguide is divided into two arms. One is used as a reference and the other is connected to the delay line. They are combined again and then inserted to the TPA PD. The TPA PD works when the pulse peak power is higher than 0.09 W (it works as a sub-bandgap photodiode for lower powers). The delay line could change the delay within 20 ps for a slow repetition of saw-tooth heating signal of 10 Hz. In the delay line, the phase of the incident pulse is changed simultaneously with the delay. Therefore, the correlation waveform exhibits an oscillation. The phase of the oscillation is changed by heating the reference branch, as shown by the dots of different colors in Fig. 4(b). The envelope function of these dots agree well with the theoretical curve for 7-ps pulse. Such coincidence was confirmed for 5–10 ps, while it was not for pulses shorter than 3 ps; the envelope function widely broadens due to the dispersion in the LSPCW used for the tunable delay line. The low dispersion tunable delay will be obtained by employing dispersion-compensated slow light produced by thermally chirping the delay peak in the LSPCW, which is usually lying between the dispersive wavelengths and low-dispersion wavelengths. The footprint of the device is as small as 1.0×0.3 mm^2. Therefore, the conventional box-sized equipment is drastically miniaturized and a palmtop pulse monitor will be available in future.

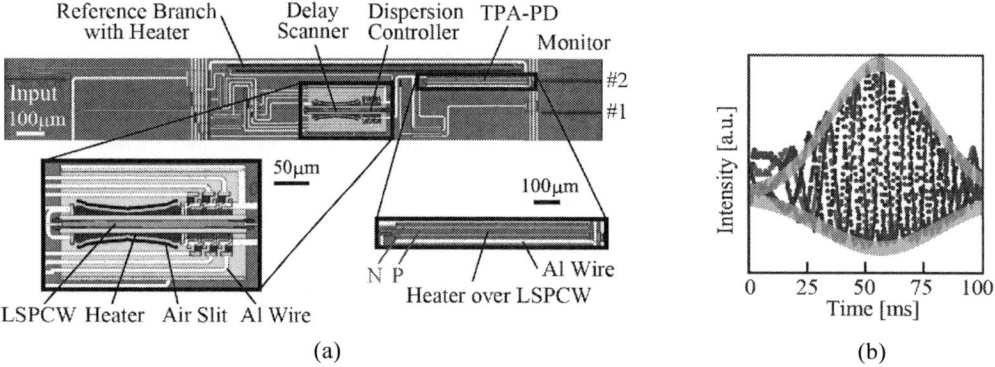

(a)　　　　　　　　　　　　　　　　　　　　　(b)

Fig. 4. On-chip optical correlator. (a) Fabricated device. (b) Theoretical correlation waveform (blue lines) and observed correlation intensities (dots) for three different phase conditions of reference branch.

ACKNOWLEDGMENT

This work was partly performed in New Energy and Industrial Technology Development Organization project.

REFERENCES

[1] L. Chrostowski and M. Hochberg, Silicon Photonics Design: From Devices to Systems, Cambridge University Press, 2015.
[2] T. Baba, "Slow light in photonic crystals," Nature Photonics, vol. 2, pp. 465-473, Aug. 2008.
[3] T. Baba, H. C. Nguyen, N. Ishikura, K. Suzuki, M. Shinkawa, R. Hayakawa and K. Kondo, "Si photonic crystal slow light devices," IEICE Electron. Express, vol. 10, pp. 1-15, Oct. 2013.
[4] T. Baba, H. C. Nguyen, N. Yazawa, Y. Terada, Hashimoto, T. Watanabe, "Slow-light Mach–Zehnder modulators based on Si photonic crystals," Sci. Technol. Adv. Mat., vol. 15, pp. 024602, Feb. 2014.
[5] Y. Terada, H. Ito, H. C. Nguyen and T. Baba, "Theoretical and experimental investigation of low-voltage and low-loss 25-Gbps Si photonic crystal slow light Mach-Zehnder modulators with interleaved p/n junction," Front. Phys., vol. 2, no. 61, pp. 1-9, Nov. 2014.
[6] T. Tamura, K. Kondo, Y. Terada, Y. Hinakura, N. Ishikura and T. Baba, "Silica-clad silicon photonic crystal waveguides for wideband dispersion-free slow light", J. Lightwave Technol., vol. 33, pp. 3034-3040, July 2015.
[7] T. Tamura, K. Kondo and T. Baba, "High group index silica-clad silicon photonic crystal slow light waveguides", Conf. Laser and Electro-Opt. Pacific Rim, Busan, no. 25I3-4, Aug. 2015.
[8] K. Hojo, Y. Terada, N. Yazawa, T. Watanabe and T. Baba, "Compact QPSK and PAM modulators with Si photonic crystal slow light phase shifters", IEEE Photon. Technol. Lett., vol. 28, 2016 (in press).
[9] Y. Terada, K. Miyasaka, H. Ito and T. Baba, "Slow-light effect in a silicon photonic crystal waveguide as a sub-bandgap photodiode", Opt. Lett., vol. 41, pp. 289-292, Jan. 2016.
[10] S. Kinugasa, N. Ishikura, H. Ito, N. Yazawa and T. Baba, "One-chip integration of optical correlator based on slow-light devices," Opt. Express, vol. 23, pp. 20767-20773, July 2015.

WE3-2   OECC/PS2016

Design of double-slotted high-Q photonic crystal nanocavity filled with electro-optic polymer

Masahiro Nakadai, Ryotaro Konoike, Yoshinori Tanaka, Takashi Asano, and Susumu Noda

Department of Electronic Science and Engineering Kyoto University

Kyoto Daigaku Katsura, Nishikyoku, Kyoto 615-8510, Japan

nakadai@qoe.kuee.kyoto-u.ac.jp, snoda@kuee.kyoto-u.ac.jp

Abstract: *We design a double-slotted Si-nanocavity embedded with electro-optic polymer (EOP) for realization of electronic control of resonant wavelength in high-Q nanocavities, which can achieve high-Q and strong concentration of light in EOP region.*

Keywords: *photonic crystal, nanocavity, waveguide, slot, nonlinearity,*

I. INTRODUCTION

High-quality (Q) factor photonic crystal (PC) nanocavities have attracted a lot of attention because PC nanocavities can confine light for a long time in tiny regions and enable strong interaction between light and matter. Recently, We have realized a two-dimensional (2D) Si-PC nanocavity with an ultra-high-Q of approximately 9million [1]. Such nanocavities are used for various applications such as coupled resonator optical waveguides, modulators, bistable-optical memories, etc. Furthermore, we have demonstrated advanced light manipulations, including dynamic control of Q factor [2], on-demand transfer of photons between distant nanocavities on a chip [3]. In such applications, dynamic control of refractive index is necessary to control light in time domain. So far, carrier plasma effect has been used to dynamically control the refractive index of the PC. Therefore, photon lifetime in the devices that utilize a dynamic control of refractive index are limited by the free-carrier absorption. It is desired to develop a method to control refractive index dynamically without the influence of free-carrier absorption in order to increase the performance of present devices and to develop on-chip quantum information processing devices.

In this abstract, we investigated Si PC nanocavities whose air-holes and slots filled with electro-optic polymer (EOP) show in Fig. 1 (a), where electro-optic coefficient of EOP is known to be very high (e.g. $r_{33} \sim 200\text{pm/V}$ [4]). By appropriately designing the structure to have ultra-high-Q factor and strong light confinement in EOP region, dynamic control of resonant wavelength of high-Q nanocavity is expected to be realized by using a nonlinearity such as the Pockels effect without an influence of free-carrier absorption.

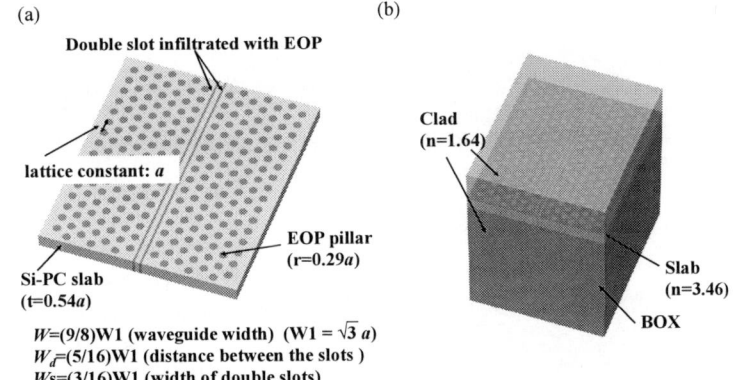

$W=(9/8)W1$ (waveguide width) (W1 = $\sqrt{3}\,a$)
$W_d=(5/16)W1$ (distance between the slots)
$W_S=(3/16)W1$ (width of double slots)

Fig. 1. Schematic diagram of double-slotted Si PC nanocavity. (a) PC slab structure, and (b) Overall view.

II. DESIGN CONCEPT

In fabricating a PC structure whose slots and holes are filled with EOP, it is considered necessary for the whole structure to be filled with EOP (Fig. 1 (b)). Since the difference of refractive index between Si (n=3.46) and EOP ($n \sim 1.6$) is smaller than that between Si and air, the light confinement in the EOP-clad nanocavity is smaller than that of air-clad nanocavity. In order to obtain a high-Q factor in EOP-clad nanocavity, we adopted a multi heterostructure (multi-step lattice constant modulation) design [5] based on a PC waveguide having a large group-velocity near the band edge [6]. Here, the latter is necessary to obtain a complete Gaussian-like envelope of the cavity light field [5], which can compensate low contrast of refractive index in EOP-clad structure.

1167

However, this requirement conflicts with the introduction of slot because the mode with a large group-velocity near the band edge is usually obtained with a light guiding by a lateral contrast of refractive index [6], and this lateral contrast is reduced when a slot filled with low index material is introduced. In order to avoid this conflict, we introduced two slots at the positions aside from the center of the PC waveguide as shown in Fig. 1 (a) because optical field of the guided mode is concentrated in the center of the waveguide and the influence of index reduction can be lowered by shifting the position of the slot aside.

III. CALCULATED RESULTS

A. Calculation of dispersion relation of the waveguide mode.

According to the design concept mentioned in II, we calculated dispersion relation in PC waveguide structures for various structural parameters by using 3-D finite difference time domain method. Figure 2 (a) shows the best result which were obtained for the structural parameters indicated in Fig. 1, where the waveguide width $W = 9/8$ W1 (W1=sqrt(3) a, and a is the lattice constant), the widths of the slots $W_s = 3/16$ W1, and the distance between the slots $W_d = 5/16$ W1. It is seen in the figure that the group-velocity (v_g) at wavevector $k=0.45$ [$2\pi/a$] is as large as $\sim 1.1 \times 10^{-1}$ c (c is the speed of light in vacuum). This large group velocity is comparable those obtained in non-slotted waveguide width $W = 0.62$ W1 which is used to obtain a nanocavity with a Q factor approaching 10^9 [5]. It is also confirmed that optical field of the mode is concentrated on double slots considerably (35% in energy) as shown in Fig.2 (b), which is important for the dynamic control of resonant wavelength.

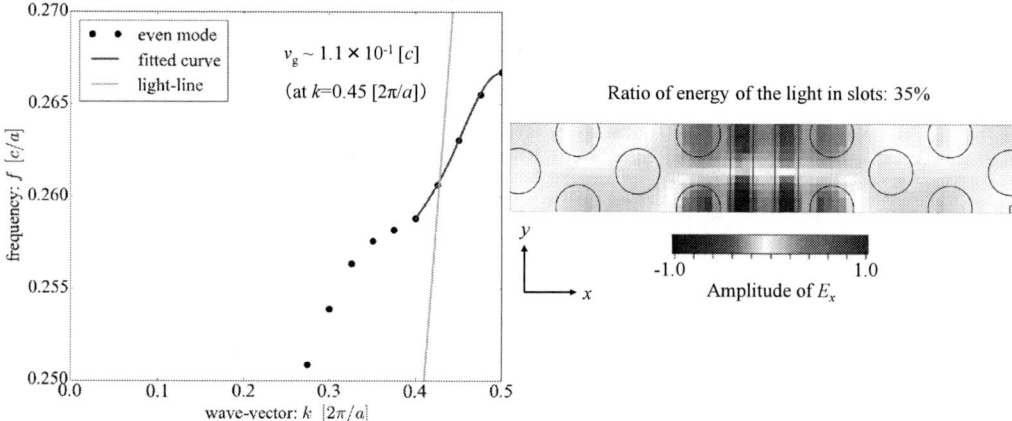

Fig. 2(a). Calculated band-diagram. (b).Calculated E_x distribution of the guiding mode.

B. Calculation of Electromagnetic field of the cavity mode and of Q factor.

We subsequently modulated the lattice constant of the waveguide, and calculated electromagnetic field of the cavity mode as shown in Fig. 3, where in the mode-gap region of the waveguide, propagation mode does not exist, but evanescent modes can be excited [5]. In this example the number of modulation step is ten and the lattice constants of PC_i (i=0, 1, 2, ..., 9) region is set to be Eq. 1.

$$a_i = a_0(1+0.025\frac{i^2}{9^2}) \qquad (1)$$

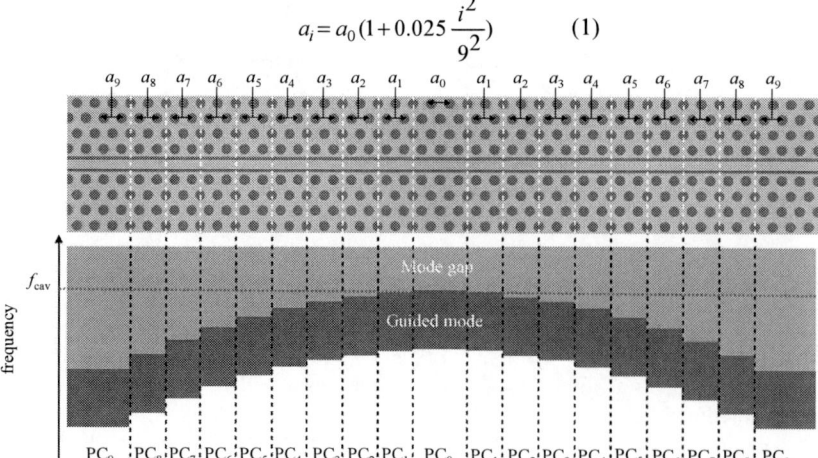

Fig. 3. Schematic picture of multi heterostructure nanocavity based on the waveguide mentioned in III.A and of its band diagram.

Figure 4 (a) shows E_y field distribution in real space. Owing to this multi heterostructure design with a high group-velocity waveguide, the components inside the light corn are very weak as can be seen in the E_y field distribution in Fourier space (Fig. 4 (b)). As a result, the Q factor of this cavity is as high as 5.9×10^6 regardless of the cladding with relatively high index EOP (n=1.64). The cross-sectional plot of the E_y field shown in Fig. 4 (c) indicates that optical field is concentrated in the slot region, where the energy concentration ratio is about 35%. When we change of the resonant wavelength of this nanocavity, static electric field is applied across the cavity, where the electrodes will be attached a few μm apart from the center of the cavity to avoid the absorption loss. Therefore, the control electric field is applied to not only EOP in the slots but also those in the holes. By taking into account this, the effective concentration ratio of the light energy in EOP is about 50%, and the expected change of the resonant wavelength is more than 0.1% for an applied field of 10 kV/cm with EOP of $r_{33} \sim 200$ pm/V [4]. This value of resonant wavelength change is enough for various applications.

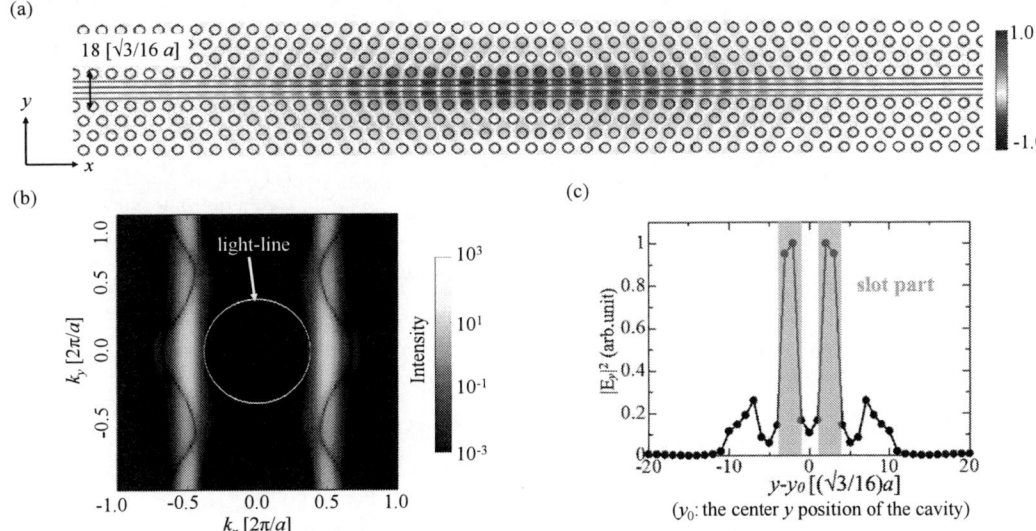

Fig. 4 (a) Calculated E_y distribution of the cavity mode. (b) 2-D Fourier transform spectra. (c) $|E_y|^2$ profile along the centerline of the cavity in the y direction.

IV. CONCLUSIONS

In summary, we have investigated a high-Q Si-PC nanocavity embedded with electro-optical polymer in order to control resonant wavelength of the cavity without using carrier plasma effect that accompanies free-carrier absorption. We have successfully obtained a multi heterostructure nanocavity based on a high group velocity double-slotted waveguide, which shows very high high-Q factor of $\sim5.9 \times 10^6$ even with polymer cladding. It has been shown that the effective concentration of light on EOP is about 50%, and a large resonant wavelength shift over 0.1% by Pockels effect is expected for an applied electric field of 10 kV/cm with polymers having electro-optic coefficient r_{33} of ~ 200 pm/V.

ACKNOWLEDGMENT

This paper is partially based on results obtained from a project supported by the New Energy and Industrial Technology Development Organization (NEDO).

REFERENCES

[1] H. Sekoguchi, Y. Takahashi, T. Asano, and S. Noda, "Photonic crystal nanocavity with a Q-factor of ~9million," Optics Express, vol. 22, pp. 916-924, 2014.

[2] Y. Tanaka, J. Upham, T. Nagashima, T. Sugiya, T. Asano, and S. Noda, "Dynamic control of the Q-factor in a photonic crystal nanocavity," Nature Materials, vol. 6, pp. 862-865, 2007.

[3] R. Konoike, Y. Sato, Y. Tanaka, T. Asano, and S. Noda, "Adiabatic transfer scheme of light between strongly coupled photonic crystal nanocavities," Physical Review B, vol. 87, pp. 1651381-1-6, 2013.

[4] Y. Enami, C. T Derose, D. Mathine, C. Loychik, C. Greenlee, R. A. Norwood, T. D. Kim, J. Luo, Y. Tian, A. K-Y. Jen, and N.Peyghambarian, "Hybrid polymer/sol-gel waveguide modulators with exceptionally large electro-optic coefficients," Nature Photonics, vol. 1, pp. 180-185, 2007.

[5] Y. Tanaka, T. ASano, and S. Noda, "Design of photonic crystal nanocavity with Q-factor of ~10^9," Journal of Lightwave Technology, vol. 26, pp. 1532-1539, 2008.

[6] B. S. Song, T. Aasano, and S. Noda, "Physical origin of the small modal volume of ultra-high-Q photonic double-heterostructure," New Journal of Physics, vol. 8, pp. 209-1-13, 2006.

OECC/PS2016

Experimental report for dispersion engineering of silicon slot photonic crystal waveguides

Samuel Serna[1,2], Weiwei Zhang[1], Xavier Le Roux[1], Laurent Vivien[1] and Eric Cassan[1]

[1]Institut d'Electronique Fondamentale, University Paris-Sud, CNRS UMR 8622, Université Paris Saclay, Bat. 220, 91405 Orsay Cedex, France ; [2]Laboratoire Charles Fabry, Institut d'Optique Graduate School, CNRS, Université Paris Saclay, 2 Avenue Augustin Fresnel, 91127 Palaiseau Cedex, France ; Author e-mail address: samuel.serna@u-psud.fr

Abstract: We present a rigorous numerical and the first experimental study of the dispersion properties of slot photonic crystal waveguides as a function of the refractive index cladding and the targeted group index.

Keywords: Silicon Photonics, Integrated Optics, Dispersion Engineering

I. INTRODUCTION

Hybrid silicon photonics rises as a promising platform to circumvent the drawbacks of silicon, such as its indirect bandgap, its absence of second order nonlinear susceptibility due to lattice symmetry or its high third order nonlinear losses in the telecom wavelength [1]. Several nonlinear experiments in the slow light regime require the dispersion compensation over a wide bandwidth. For instance, in W1 waveguides, the increase of the confined electric field due to spatial pulse compression has been used to improve the FWM efficiency or the photon generation of correlated photon pairs in the quantum regime. But the properties of silicon at the wavelengths around 1550 nm inhibit further enhancement of such phenomena. So the engineering of those hybrid waveguides opens a window to novel frontiers in the investigation of new materials and applications [2,3,4].

II. RESULTS AND DISCUSSION

In this section we will present the findings in three main stages: simulation, fabrication, and optical characterization.

A. *Simulations*

The basic geometry consists in a structure with a lattice constant of a=420 nm, and the hole radius r is 125 nm to ensure a wide TE bandgap [5,6]. In order to tune the mode and achieve changes in the dispersion curves, the first row of holes was shifted towards the slot by $0.20a$ and the second row shifted outwards by $0.35a$ (see Fig. 1). In this study, the radius of the second row of holes was fixed to 125 nm but the radius of the first row was shifted in order to quantify the experimental effects in small changes of only one parameter. The sweeping values are from 95 nm until 125 nm. We have guaranteed the single mode operation and some degree of freedom against possible fabrication imperfections. The dispersion engineered structures have been simulated by the plane wave expansion method using the MPB software [7] for a quasi-TE polarization. Eigenmodes have been calculated in a unit cell reproducing the geometrical parameters described in the fabrication. A mesh resolution $a/20 = 21$ nm and subpixel smoothing was used.

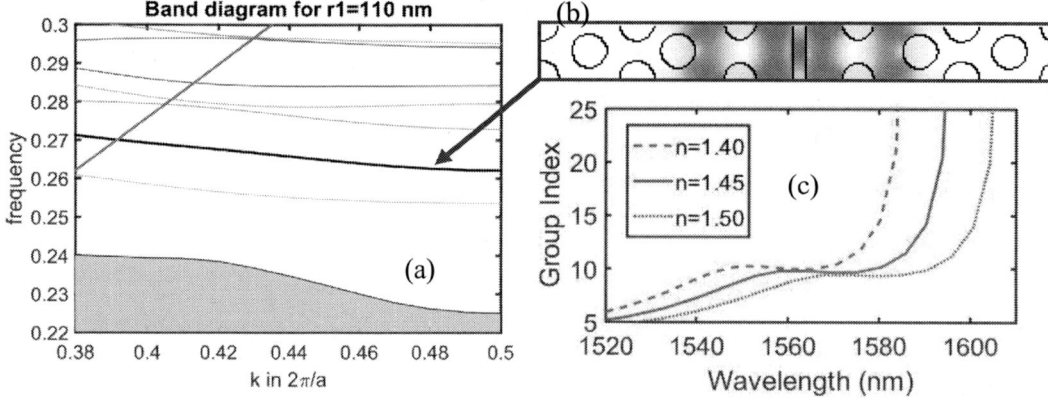

Fig. 1. Simulation results: (a) Band diagram of a dispersion engineered slot waveguide with the first row radius of 110nm and covered with a refractive index liquid of 1.45. The band of the used mode is highlighted in bold black. (b) Dielectric energy distribution of the structure at k=0.48. (c) Group index evolution as a function of the wavelength for different refractive index liquids.

1170

B. Fabrication

SOI wafers with a 260 nm thick Si film and 2 μm of buried silica were masked by a ZEP-520A resist. The patterns were written by a 80 kV e-beam lithography Nanobeam NB-4 system and then transferred by inductive coupling plasma reactive ion etching process using SF6 gas [8,9]. In Fig.2, we show two different waveguides, one under dispersion engineering where the shift of the first and second row of holes is evident as well as the decrease of the first hole radius. The second one is a non-engineered waveguide. A careful calibration of the different shapes has been performed in order to match the simulation parameters. The fabrication optimization will be described during the conference.

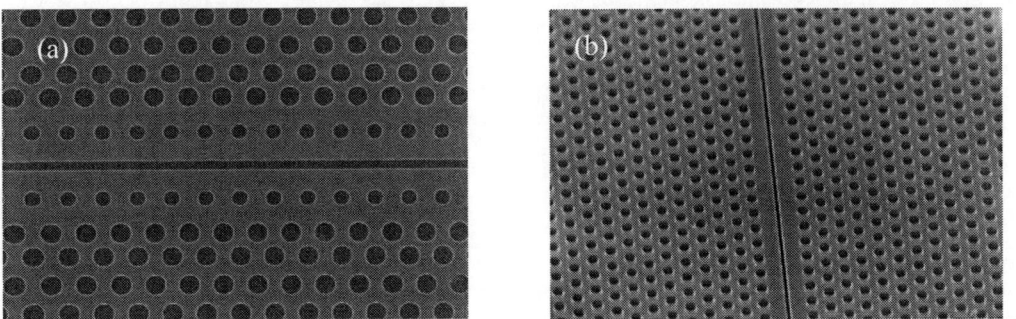

Fig. 2. SEM images: (a) Dispersion engineered waveguide. (b) Non dispersion engineered slot photonic crystal waveguide.

C. Experimental characterization

We have used several optical methods to characterize the samples. In this abstract, we only show the results related to the r_1=110nm radius structure. To obtain the results shown in fig 3 (a) an optical set-up consisting of a tunable external cavity laser filtered by a polarizer in order to excite only the TE mode and a monomode tapered lensed fiber was used to inject the light into the chip. At the output, the transmission was collected by a microscope objective and then sent to an all-band optical component tester CT400 to obtain the optical transfer function. In fig 3 (b) the general procedure used to change the filling refractive index is depicted.

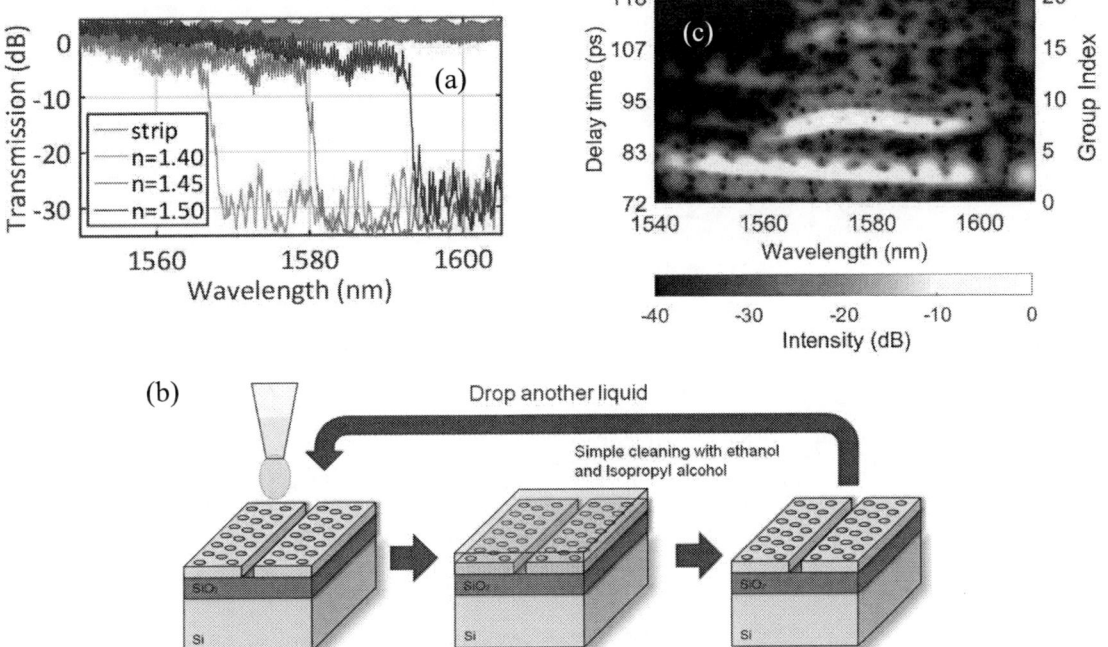

Fig. 3. Experimental results: (a) Bandgap shift as a function of the covering index for r_1=110nm for 50μm long slot photonic crystal waveguides. (b) Approach used to change the cladding optical material. (c) Reflectance map of the same geometric structure but with 700μm propagation length with a cover index of 1.45.

In a complementary fashion, we used an external interferometer experimental scheme inspired by the OCT-technique [10,11]. After the analysis of the interference fringes, the photon delay time of one arm with respect to the other was inspected. One example of these measurements is shown in fig. 3 (c). The set-up uses a tunable laser and two compensated photodiodes as detectors. As this technique is based on delays on the order of ps, the longer the structure, the easier the interpretation of the reflectance maps. Fig 3 (c) shows the reflectance map of a 700 μm propagation. Two clear transmission lines are present corresponding to the fundamental TE and TM modes, respectively. The latter is not affected in this range of wavelengths by the photonic crystal, so the transmission has an almost constant delay. On the other hand, the upper line presents a clear flat band at a group velocity of (c/8) for a wavelength range of around 30nm. The stop band can be clearly seen close to 1600 nm where the time of flight line inflects. Also, as expected in this region, the transmission decreases fast as the losses get proportional to the squared group index [12]. The sensitivity of the bandgap position was estimated with a 20 dB transmission decay basis and is in average of 250 nm/RIU, showing also the potential of this kind of structures for sensing proposes [9]. During the evaluation of SEM images, only statistical variations are observed around the calibrated values. Nevertheless, the optical characterization shows slight shifts in the edge wavelength for different structures with the same design geometry. Furthermore, the presence of the flat band is always observed with the same bandwidth. All those findings will be discussed during the conference.

III. CONCLUSIONS

We present the first experimental study of dispersion engineering in slot photonic crystals based in a careful compromise between realistic geometrical parameters, possibility to be fabricated by the use of electronic beam lithography, and waveguide lengths that allow the possibility of a resolution in a photon delay measurement with acceptable transmission losses. The proposed naturally highly dispersive structures are also analyzed under different cladding indices to study the effects in the bandgap and in the engineered dispersion properties. Important quantities, including the waveguide wavelength gap sensitivity and group index curves have been directly extracted by using different liquids and a simple methodology of recycling the waveguides without affecting their properties. The results show robust waveguides with the ability to host novel soft matter materials that are particularly interesting for quantum and nonlinear integrated photonics applications.

APPENDIX

The group index of the device under test is obtained by the deduction of the delay introduced by the access as

$$n_{gPC} = \frac{ct_T}{L_{PC}} - \frac{L_A n_{gA}}{L_{PC}}$$

where the subindices PC and A stands for Photonic Crystal and Accesses, respectively and t_T is the total delay measured with the spectrogram.

ACKNOWLEDGMENT

We would like to acknowledge the French Ministry of Higher Education, as well as the FP7 Cartoon European project.

REFERENCES

[1] G. P. Agrawal, Nonlinear fiber optics (Academic press, 2007).
[2] A. Di Falco, L. OFaolain, and T. Krauss, Applied Physics Letters 92, 083501 (2008).
[3] J.-M. Brosi, C. Koos, L. C. Andreani, M. Waldow, J. Leuthold, and W. Freude, Opt. Express 16, 4177 (2008).
[4] T. Baba, Nature Photonics 2, 465 (2008).
[5] J. Hou, H. Wu, D. Citrin, W. Mo, D. Gao, and Z. Zhou, Opt. Express 18, 10567 (2010).
[6] J. Wu, Y. Li, C. Peng, and Z. Wang, Optics Communications 283, 2815 (2010).
[7] S. Johnson and J. Joannopoulos, Opt. Express 8, 173 (2001).
[8] C. Caer, X. Le Roux, and E. Cassan, Optics letters 37, 3660 (2012).
[9] C. Caer, S. F. Serna-Otalvaro, W. Zhang, X. Le Roux, and E. Cassan, Optics letters 39, 5792 (2014).
[10] A. Parini, et al., J. Lightwave Technol. 26, 3794–3802 (2008).
[11] C. Caer, S. Combrie, X. Le Roux, E. Cassan, and A. De Rossi, Applied Physics Letters 105, 121111 (2014).
[12] S. Mazoyer, J.-P. Hugonin, and P. Lalanne, Physical Review Letters 103, 063903 (2009).

WE3-4

OECC/PS2016

Silicon Polarizing Beam Splitter based on Asymmetric Slot Waveguide

Jijun Feng[1*], Ryoichi Akimoto[2], and Heping Zeng[1,3]

[1]Shanghai Key Laboratory of Modern Optical System, Engineering Research Center of Optical Instrument and System (Ministry of Education), School of Optical-Electrical and Computer Engineering, University of Shanghai for Science and Technology, Shanghai 200093, China

[2]Electronics and Photonics Research Institute, National Institute of Advanced Industrial Science and Technology, Tsukuba 305-8568, Japan

[3]State Key Laboratory of Precision Spectroscopy, East China Normal University, Shanghai 200062, China.

*fjijun@usst.edu.cn

Abstract: Silicon polarizing beam splitter is designed and fabricated, without adopting strip-slot mode converters. An asymmetric slot waveguide structure can realize a TM light cross-coupling with a strip waveguide, while little TE light coupling can happen.

Keywords: Polarizing beam splitter, slot waveguide, silicon-on-insulator, integrated optics.

I. INTRODUCTION

Silicon-on-insulator (SOI) platform has been paid much attention due to its high refractive-index contrast enabling compact footprint with great potential for ultra-small optical circuits [1]. Its fabrication is compatible with the complementary-metal-oxide-semiconductor (CMOS) technology, facilitating cost-effective mass production. However, the high-index contrast in SOI platform causes the polarization-dependent loss and related wavelength dispersion. Thus polarization control and manipulation are crucial for the design and operation of SOI devices [2]. As one of the key polarization-handling devices, a polarizing beam splitter (PBS) has always attracted much attention, which can separate the orthogonally polarized transverse electric (TE) and transverse magnetic (TM) light into different pathway. Many methods have been demonstrated for on-chip polarization splitting to data, such as a multimode or two-mode interference coupler while a deviation from the optimum working length will cause an additional excess loss [3,4]. By using the birefringence of the arm waveguides, a Mach-Zehnder interferometer could be used to fulfill polarization splitting with a high extinction ratio but a relative large footprint [5]. As for the photonic crystal PBS [6], the design is complex and the fabrication is relatively difficult while the grating coupler PBS is not good for intra chip photonic integration [7]. Moreover, photonic crystal or grating structures usually have low coupling efficiency and can introduce a relatively large loss due to the scattering.

To enhance the waveguide birefringence, an SOI symmetric slot waveguide has already been used to realize a compact PBS by the coupling with a strip wire [8,9]. Since the birefringence properties in slot waveguides are dictated by the electric field boundary conditions and are thus almost independent of wavelength, high extinction ratios can be achieved over a broad wavelength range. And asymmetric slot devices would be less impacted by certain fabrication errors [10,11]. However, additional strip-slot waveguide mode converters are usually needed, which would induce additional insertion loss due to the limited fabrication accuracy. Thus, an improved asymmetric slot structure is proposed here without adopting any mode converter, and a wideband and efficient PBS is realized.

II. DEVICE DESIGN AND FABRICATION

The low refractive index slot area can concentrate optical power density through the electric field component discontinuity at the interface for an asymmetric slot waveguide, which is highly polarization-selective. Fig. 1 shows a schematic of the asymmetric slot based PBS device, with the right for the cross-sectional front view of the structure. A narrow satellite waveguide is placed asymmetrically between an ordinary DC, thus constitutes a slot waveguide and a strip waveguide to allow selectively coupling of only TM light. There is no additional strip-to-slot mode converter needed here. Little light power is stored in the narrow satellite waveguide, and it has low insertion loss for both polarizations.

A commercial available SOI wafer with a 220-nm-thick silicon core and 3-μm-thick buried oxide is adopted. The refractive indices are 1.46 and 3.476 for SiO_2 and Si, respectively. Narrow slot can confine more TE mode intensity, which is benefit for increasing the waveguide birefringence. However, considering the reactive ion etching lag effect where the etching depth depends much on the trench width for small feature size, the slot G_s is chosen to be 100 nm, with a trade-off to guarantee a roughly homogeneous silicon etching rate [12]. The narrow slot arm width W_{slot} is 140 nm with consideration of the ease of fabrication. The fundamental mode effective indices are calculated by *RSoft*, at a 1550-nm wavelength for a strip or slot waveguide with varying strip or wide slot arm width (W_{wg} or W_{wg_s}), as shown in Fig. 2(a). For a W_{wg_s} of 350 nm, the W_{wg} should be 398 nm to realize an effective index matching for the TM mode,

1173

both with the calculated effective index of about 1.707. While for the TE mode, large phase mismatch due to the effective index difference prevents power transfer between the strip and slot waveguide, with the effective index of 2.223 and 2.098, respectively. From the corresponding optical mode distribution of the TE and TM wave, it can be confirmed that the TM mode changes little after the introduction of a narrow slot waveguide while TE mode experiences strong filed enhancement in the slot area and there is little light power distributed in the narrow slot arm due to the asymmetric slot structure.

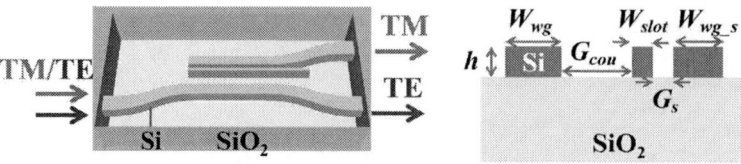

Fig. 1. Schematic illustration of the slot waveguide assisted PBS (left) with the right for the cross-sectional front view.

For the device length design, the TM even and odd DC supermode effective indices (n_{even} and n_{odd}) are calculated. The corresponding TM cross-coupling length ($\lambda/2/(n_{even}-n_{odd})$) can then be obtained. In a conventional directional coupler, a sizeable amount of coupling can take place in the output bends of the device [Fig. 1], apart from the central straight coupling region. To decrease the wave coupling influence in the bending area and compress the coupling of the TE light, a coupling gap G_{cou} of 450 nm is adopted. Moreover, a wide separation can improve the fabrication tolerance [13], though a long coupling length is usually needed. The calculated coupling length is about 21.7 μm based on the supermode method. A curvature with bending radius of 25 μm is placed at the end of the straight coupling region to separate the cross and bar port output.

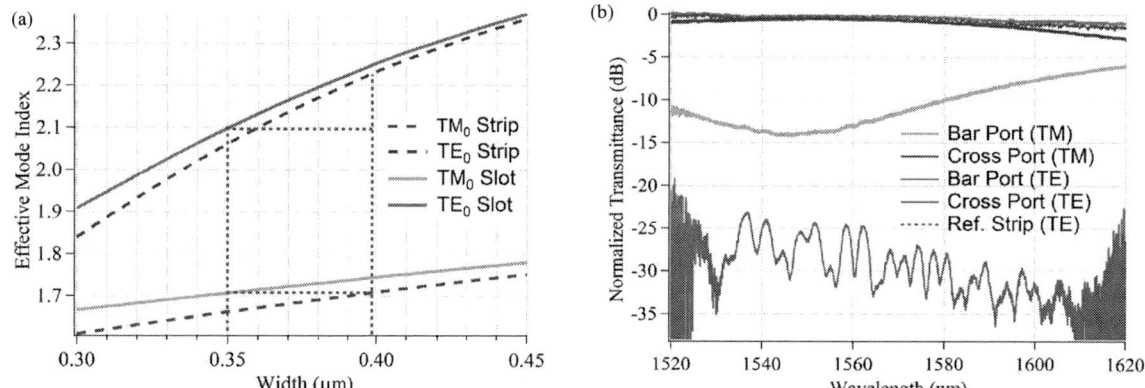

Fig. 2. (a) Effective index of a strip and slot SOI waveguide with varying strip (W_{wg}) or wide slot arm width (W_{wg_s}) for both polarizations. (b) Measured transmission spectra of the asymmetric slot PBS device for both polarizations.

Using three-dimensional finite-difference-time-domain (FDTD) simulation by *RSoft*, the DC performance is further confirmed. Optimum performance can be realized with a W_{wg} of 392 nm and a coupling length of 18.9 μm, with considering the contribution of the bending related coupling effect. The polarization extinction ratio (PER, defined as the transmittance difference between two polarizations at the output port) is about 34.43 dB and 33.66 dB at the bar and cross port, respectively. It can also be confirmed that for the optimum structure parameters, the device has almost no insertion loss.

A PBS device based on the above design is fabricated by using electron beam lithography and reactive ion etching as a proof-of-concept. After the sample cleaned by a wet chemical process, a 2-μm-thick top-cladding layer was deposited. Then the wafer was cut by dicing saw for measurement. Inverted taper coupler with a minimum waveguide width of 180 nm was used here for light coupling with single-mode fiber. We fabricated PBSs with varying coupling length.

III. DEVICE PERFORMANCE

Optimum polarization splitter can be realized with a coupling length of 18 μm, varying little from the designed parameters. Slight difference may be caused by the waveguide non-uniformity induced coupling gap variation and the non-perfect rectangular mesa profile. The transmission spectra of the TE and TM light for an optimum PBS were measured by an optical spectrum analyzer (OSA). The polarization state of the input-light is controlled by a polarizer with an extinction ratio of better than 50 dB. The TE/TM light was focused onto the input facet by an objective lens. After passing through the PBS, the output-light was collected by a lensed fiber and recorded by OSA. A reference 350-nm strip waveguide with the same configuration as that in the PBS was also fabricated at the same time.

As shown in Fig. 2(b), the measured spectra are normalized with considering the maximum TE output intensity of the reference strip waveguide. Compared with the reference spectrum, low insertion loss can be confirmed. The TE spectrum at the bar port is almost the same as that of the reference strip waveguide. For simplicity, the TM reference spectrum is not shown here, and the TM light cross-port output has a slightly low transmittance at the peak position compared with the reference line but with an excess coupling loss of only about 0.2 dB. The ripples in the spectra may originate from the Fabry-Perot interference between the chip facets. At a 1550-nm wavelength, a PER of more than 25 dB can be realized at the cross port, while it is only about 14-dB at the bar port. A 56-nm bandwidth (1520–1576 nm) can be fulfilled with the PER of more than 10 dB at the bar port, while more than 25-dB PER can be obtained almost in the whole $C+L$ band at the cross port.

The PER at the cross port has better performance than that at the bar port. This is due to the fabrication error induced waveguide width variations, which can translate into the effective index deviation and cause phase mismatch, and would further result in an incomplete power transfer for the TM light. To compensate for this fabrication error, devices can be fabricated with a small bending radius curve for the bar port right after the straight coupling region, which could help remove the residual TM polarized light as it experiences larger bending related loss [8]. Besides, the fabrication process can be further optimized with better waveguide width control. Therefore, there is still much room for the PBS performance improvement.

IV. CONCLUSIONS

Based on an asymmetric slot waveguide without using any additional strip-slot waveguide mode converter, a wideband and efficient SOI polarizing beam splitter is experimentally demonstrated. TM light can be coupled completely to the cross port, while TE light goes through the bar wire almost with no light coupling. The fabricated PBS device has a coupling length of only 18 μm, which shows polarization extinction ratios of 14 dB and 25 dB at the bar and cross port, respectively. Low insertion loss can also be realized for both polarizations. This device could be expected to have more application prospects such as constructing polarization-transparent or polarization-selective circuits in optical communications.

ACKNOWLEDGMENT

Jijun Feng acknowledges support by the Program for Professor of Special Appointment (Eastern Scholar) at Shanghai Institutions of Higher Learning.

REFERENCES

[1] V. R. Almeida, C. A. Barrios, R. R. Panepucci, and M. Lipson, "All-optical control of light on a silicon chip," Nature **431**, 1081–1084 (2004).

[2] T. Barwicz, M. R. Watts, M. A. Popović, P. T. Rakich, L. Socci, F. X. Kärtner, E. P. Ippen, and H. I. Smith, "Polarization-transparent microphotonic devices in the strong confinement limit," Nat. Photonics **1**, 57–60 (2007).

[3] X. Guan, H. Wu, Y. Shi, and D. Dai, "Extremely small polarization beam splitter based on a multimode interference coupler with a silicon hybrid plasmonic waveguide," Opt. Lett. **39**(2), 259–262 (2014).

[4] S.-H. Kim, G. Cong, H. Kawashima, T. Hasama, and H. Ishikawa, "Tilted MMI crossings based on silicon wire waveguide," Opt. Express **22**(3), 2545–2552 (2014).

[5] T. K. Liang and H. K. Tsang, "Integrated polarization beam splitter in high index contrast silicon-on-insulator waveguides," IEEE Photon. Technol. Lett. **17**(2), 393–395 (2005).

[6] X. Ao, L. Liu, L. Wosinski, and S. He, "Polarization beam splitter based on a two-dimensional photonic crystal of pillar type," Appl. Phys. Lett. **89**(17), 171115 (2006).

[7] D. Taillaert, H. Chong, P. I. Borel, L. H. Frandsen, R. M. De La Rue, R. Baets, "A compact two-dimensional grating coupler used as a polarization splitter," IEEE Photon. Technol. Lett. **15**(9), 1249–1251 (2003).

[8] S. Lin, J. Hu, and K. B. Crozier, "Ultracompact, broadband slot waveguide polarization splitter," Appl. Phys. Lett. **98**(15), 151101 (2011).

[9] D. Dai, Z. Wang, and J. E. Bowers, "Ultrashort broadband polarization beam splitter based on an asymmetrical directional coupler," Opt. Lett. **36**(13), 2590–2592 (2011).

[10] A. Spott, T. Baehr-Jones, R. Ding, Y. Liu, R. Bojko, T. O'Malley, A. Pomerene, C. Hill, W. Reinhardt, and M. Hochberg, "Photolithographically fabricated low-loss asymmetric silicon slot waveguides," Opt. Express **19**(11), 10950–10958 (2011).

[11] R. Ding, T. Baehr-Jones, W.-J. Kim, B. Boyko, R. Bojko, A. Spott, A. Pomerene, C. Hill, W. Reinhardt, and M. Hochberg, "Low-loss asymmetric strip-loaded slot waveguides in silicon-on-insulator," Appl. Phys. Lett. **98**(23), 233303 (2011).

[12] A. V. Velasco, M. L. Calvo, P. Cheben, A. Ortega-Moñux, J. H. Schmid, C. A. Ramos, Í. M. Fernandez, J. Lapointe, M. Vachon, S. Janz, and D.-X. Xu, "Ultracompact polarization converter with a dual subwavelength trench built in a silicon-on-insulator waveguide," Opt. Lett. **37**(3), 365–367 (2012).

[13] G. W. Cong, K. Suzuki, S. H. Kim, K. Tanizawa, S. Namiki, and H. Kawashima, "Demonstration of a 3-dB directional coupler with enhanced robustness to gap variations for silicon wire waveguides," Opt. Express **22**(2), 2051–2059 (2014).

WE3-5

OECC/PS2016

Parity-Time Symmetric Coupled Resonator Waveguide with Photonic Crystal Nanocavities

Kenta Takata[1,2] and Masaya Notomi[1,2]

[1]NTT Nanophotonics Center, NTT Corporation, 3-1 Morinosato-Wakamiya, Atsugi 243-0198, Kanagawa, Japan
[2]NTT Basic Research Laboratories, NTT Corporation, 3-1 Morinosato-Wakamiya, Atsugi 243-0198, Kanagawa, Japan
takata.kenta@lab.ntt.co.jp

Abstract: We propose a parity-time symmetric coupled resonator waveguide based on buried heterostructure photonic crystal nanocavities. We show theoretically the potential to greatly tune its group velocity using the parity-time phase transition for the first time.
Keywords: parity-time symmetry, coupled resonator optical waveguide, photonic crystal

I. INTRODUCTION

Exotic phenomena in optical devices with the parity-time (PT) symmetry [1] have been attracting attention. PT symmetric optical systems [2] are periodic structures that have a complex refractive index $n(\mathbf{r})$ with balanced gain and loss, where Re($n(\mathbf{r})$) is an even function and Im($n(\mathbf{r})$) is odd ($n(\mathbf{r}) = n^*(-\mathbf{r})$). In such systems, the detuning of the eigenmode frequency $\Delta\omega$ (or propagation constant $\Delta\beta$) undergoes an abrupt transition from real to imaginary values by the magnification of the gain and loss (called PT phase transition or PT symmetry breaking). Various phenomena around the PT phase transition point have been proposed and observed, for example, power oscillation [1, 3], double refraction [1], unidirectional reflectivity [4, 5], and the suppression and revival of lasing [6]. Here, while large PT-symmetric waveguide arrays have been reported in the time domain [3, 7], devices with spatial gain and loss structures remain limited in terms of their gain controllability and scalability.

In this paper, we propose a new type of PT-symmetric coupled resonator optical waveguide (CROW), based on buried heterostructure mode-gap (BHM) nanocavities in a photonic crystal (PhC) line defect, with potential advantages as regards scale and control. The coupled resonator optical waveguide [8] is a good platform for slowing the light. However, because it usually comprises passive optical cavities, it is difficult to tune its transport property widely with external signals. Although there have been many theoretical studies describing PT-symmetric CROWs, they have focused on their basic properties [9, 10]. In addition, the largest PT-symmetric CROW realized experimentally is a system of just two microtoroid resonators [11]. Here, we show the possibility of strongly controlling the group velocity in the device. We analytically study its spectral group velocity and group velocity dispersion using a tight-binding model. A numerical example indicates a more than tenfold change in the group velocity by switching between normal and PT-symmetric CROWs with external pumping control. We also perform a numerical simulation on a realistic structure composed of PhC BHM nanocavities with complex refractive indices, using the finite element method. The result demonstrates a clear PT phase transition in the system under gain and loss coefficients of the order of 100 cm⁻¹.

II. DEVICE STRUCTURE

Fig. 1 (a) shows a schematic of the PT-symmetric CROW proposed in this study. The system has a PhC slab, a line defect waveguide and pairs of BHM nanocavities with a pumping mechanism for each cavity. The photonic crystal structure is omitted in Fig. 1 (a) for simplicity. The blue and red boxes are BHM cavities with balanced loss and gain, respectively. The periodic loss and gain are achieved by the material absorption and the separate p-i-n electric pumping channels. Fig. 1 (b) shows the refractive index along with the PhC line defect. The buried heterostructures confine the light by a sharp index modulation and by the mode gap between them and the PhC line defect waveguide. We find that the index profile satisfies the condition for PT symmetry $n(x) = n^*(-x)$, when we take the reference of the x axis at the center of the device. This system enables the precise positioning of the nanocavities and hence accurate cavity coupling via sophisticated fabrication techniques. Also, tiny BHM cavities with independent pumping channels can potentially achieve a high Q, large coupling and controllability of the gain and loss simultaneously.

III. THEORETICAL ANALYSIS

A. Tight-Binding Model

Here, we study the transport property of the PT-symmetric CROW using the solution of the tight-binding model [9],

$$\Delta\omega = \pm\sqrt{4\kappa'^2 \cos^2 k - g'^2} \quad (1)$$

where $\Delta\omega$ is the eigenfrequency detuning from the single cavity resonance. g' is the magnitude of the periodic gain and loss rates. κ' is the coupling rate between the cavities. k is the normalized Block wavenumber. κ' and g' are described by,

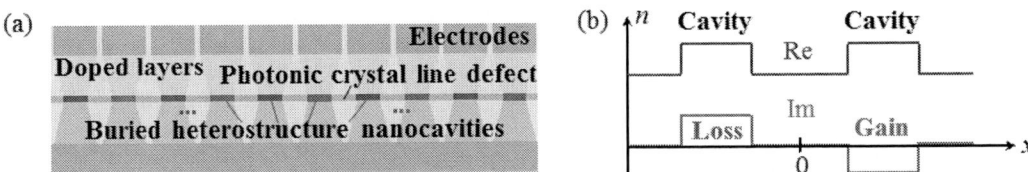

Fig. 1. (a) Schematic of the PT-symmetric CROW. (b) Spatial refractive index profile of the device, which holds PT symmetry.

$$\kappa' = \kappa \omega_0 / 2, \quad g' = gc /(2n_{BH}) \qquad (2)$$

where κ denotes the coupling coefficient, and ω_0 is the resonance frequency of a single BHM cavity. g is the spatial gain and loss coefficient, c the speed of light in a vacuum and n_{BH} the local refractive index in the cavity.

Fig. 2 (a) and (b) show the real and imaginary part of the detuning $\Delta\omega(k)$ (Eq. (1)) of the PT-symmetric CROW with 40 cavities for different spatial gain and loss coefficients g. Here, $g = 0$ (for the black points and curves) means a normal empty CROW. For a system with a finite number of cavities, possible eigenstates are limited to finite points in the k space. In the continuum limit, the empty CROW gives a folded cosine band with no imaginary component. On the other hand, real band curves for the PT-symmetric CROW have parabolic shapes. In addition, when the two branches coalesce at a single cavity resonance frequency ($\Delta\omega = 0$), $\Delta\omega$ switches from purely real to imaginary values (PT phase transition). The steep dispersion of the PT-symmetric CROW around this transition point, in sharp contrast to the empty CROW, is the consequence of non-Hermiticity, which leads to various interesting effects.

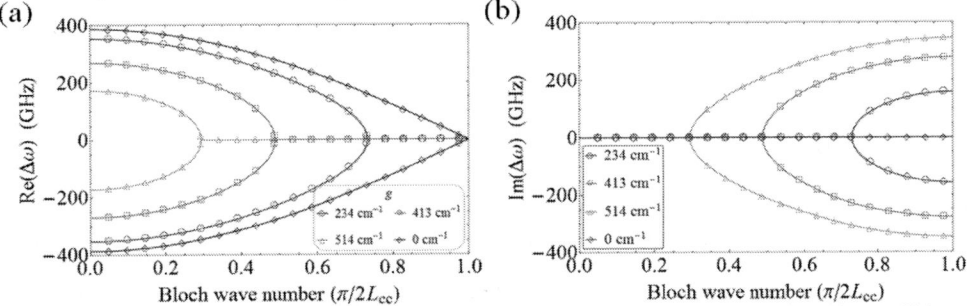

Fig. 2. (a) Real parts and (b) imaginary parts of the band curves for a PT-symmetric CROW with different gain and loss coefficients g (cm^{-1}). $\kappa = 2 \times 10^{-3}$, from [12]. $\omega_0 = c/1.55\mu m = 1.935 \times 10^{14}$ (Hz), $n_{BH} = 3.58$.

As an example application of the device, we present the potential controllability of its group velocity by gain and loss tuning. We have derived the spectral group velocity $v_g(\Delta\omega)$ and group velocity dispersion GVD($\Delta\omega$) analytically, and examined their parameter dependence. Fig. 3 (a) and (b) show numerical examples of $v_g(\Delta\omega)$ and GVD($\Delta\omega$). Due to the phase transition, v_g diverges at $\Delta\omega = 0$ for the PT-symmetric CROW. However, it converges at a finite value with the empty CROW ($g = 0$). This means that v_g can be significantly changed by controlling g around the PT phase transition. The eigenstate nearest to the transition point with $g = 413$ cm^{-1} achieves a v_g 13.1 times faster than the state of the empty CROW with a similar frequency. As seen in Fig. 3 (b), the GVD in the PT-symmetric CROW at $\Delta\omega = 0$ remains finite, in spite of the divergence of v_g. Nevertheless, the |GVD($\Delta\omega$)| of the PT-symmetric CROW is larger than that of the empty CROW. We have calculated the pulse broadening for the state closest to the transition point, and found that it is not significant for picosecond pulses propagating in the device with a length of about 100 μm.

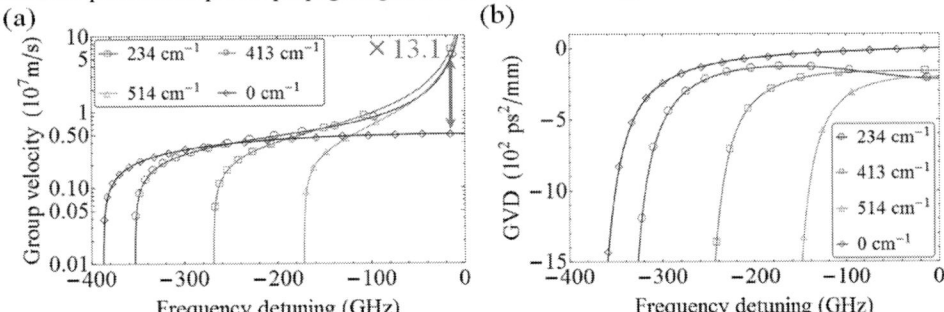

Fig. 3. (a) Group velocity $v_g(\Delta\omega)$ and (b) group velocity dispersion GVD($\Delta\omega$) of a PT-symmetric CROW for different g (cm^{-1}).

B. Device Simulation Based on Finite Element Method

Here, we show the result of the electromagnetic field simulation of a three dimensional PhC PT-symmetric CROW. The simulated structure is shown in Fig. 4 (a). Here, we consider an InP PhC slab containing a triangular lattice of air holes. The left and right InGaAsP BHM nanocavities in a line defect have loss (Im(n_{BH}) > 0) and gain (Im(n_{BH}) < 0), respectively. We place two symmetric boundary conditions to quarter the simulation cost and focus on the ground TE-like mode. The periodic boundary condition is introduced at the right and left ends of the structure. We performed an eigenfrequency analysis for different Bloch wavenumbers k. The PhC slab is sandwiched between air layers, and the whole structure is surrounded by perfectly matched layers, which are not shown in the figure.

Fig. 4 (b) illustrates the complex band structure of the simulated device with $\text{Im}(n_{BH}) = \pm 0.004$, corresponding to a reasonable gain coefficient of $g \sim 335\ \text{cm}^{-1}$. The black and red points give the wavelengths ($\lambda = c/\text{Re}(\omega)$) and loss rates ($\text{Im}(\omega)$), respectively. The PT phase transition is clearly seen around $k = 0.5\ \pi/L_{CC}$. The asymmetric shape of $\text{Re}(\omega)$ in the vertical axis and its finite slope after the phase transition are mainly because of the second nearest neighbor coupling ρ, which is non-negligible when the cavities are closely positioned. Here, we use the Rice-Mele model [10] including up to the second nearest neighbor coupling. The device parameters can be estimated from eigenfrequency data for $k = 0$ and $k = \pi/L_{CC}$ with a few different $\text{Im}(n_{BH})$ values. The resultant coupling coefficients are $\kappa = 4.7 \times 10^{-4}$ and $\rho = 4.1 \times 10^{-5}$. The theoretical curves (dashed and dotted lines in black and red) agree well with the simulation data. The small difference between them is possibly due to higher order couplings, non-reciprocity by the gain and loss and dissipation.

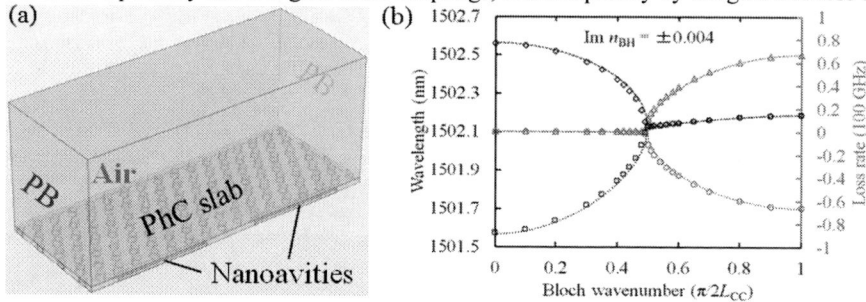

Fig. 4 (a) Device structure for the numerical simulation. PB: periodic boundary. (b) Simulated complex band structure (Black: λ vs. k, Red: $\text{Im}(\Delta\omega)$ vs. k). The dashed and dotted lines in black and red are theoretical curves obtained with the Rice-Mele model [10].

IV. CONCLUSIONS

In conclusion, we have proposed and theoretically studied a PT-symmetric CROW based on BHM photonic crystal nanocavities. Using a tight-binding model, we have obtained analytical expressions for the group velocity and GVD of the device and shown the potential for wide-range control of its group velocity with external pumping. We have also simulated a three-dimensional realistic structure including a PhC slab and BHM cavities with complex refractive indices, via the finite element method. The result indicates that a PT phase transition can be achieved with a reasonable gain and loss coefficient in the proposed device, and qualitatively supports the tight-binding analysis.

ACKNOWLEDGMENT

We thank Shota Kita for fruitful discussions about the simulation result. This work was supported by CREST, JST.

REFERENCES

[1] C. M. Bender and S. Boettcher, "Real spectra in non-Hermitian Hamiltonians having PT symmetry," Phys. Rev. Lett., vol. 80, pp. 5243–5246, June 1998.

[2] K. G. Makris, R. El-Ganainy and D. N. Christodoulides, "Beam dynamics in PT symmetric optical lattices," Phys. Rev. Lett., vol. 100, no. 103904, March 2008.

[3] A. Regensburger, C. Bersch, M.-A. Miri, G. Onishchukov, D. N. Christodoulides and U. Peschel, "Parity-time synthetic photonic lattices," Nature, vol. 488, pp. 167–171, August 2012.

[4] Z. Lin, H. Ramezani, T. Eighelkraut, T. Kottos, H. Cao and D. N. Christodoulides, "Unidirectional invisibility induced by PT-symmetric periodic structures," Phys. Rev. Lett., vol. 106, no. 213901, May 2011.

[5] L. Feng, Y.-L. Xu. W. S. Fegadolli, M.-H. Lu, J. E. B. Oliveira, V. R. Almeida, Y.-F. Chen and A. Scherer, "Experimental demonstration of a unidirectional reflectionless parity-time metamaterial at optical frequencies," Nature Mater., vol. 12, pp. 108–113, November 2013.

[6] B. Peng, S. K. Özdemir, S. Rotter, H. Yilmaz, M. Liertzer, C. M. Bender, F. Nori and L. Yang , "Loss-induced suppression and revival of lasing," Science, vol. 346, 328–332, October 2014.

[7] M. Wimmer, A. Regensburger, M.-A. Miri, C. Bersch, D. N. Christodoulides and U. Peschel, "Observation of optical solitons in PT-symmetric lattices," Nat. Commun., vol. 6, no. 7782, September 2014.

[8] A. Yariv, Y. Xu, R. K. Lee and A. Scherer, "Coupled-resonator optical waveguide: a proposal and analysis," Opt. Lett., vol. 24, pp. 711-713, June 1999.

[9] O. Vázquez-Candanedo, J. C. Hernández-Herrejón, F. M. Izrailev and D. N. Christodoulides, "Gain- or loss- induced localization in one-dimensional PT-symmetric tight-binding models," Phys. Rev. A, vol. 89, no. 013832, January 2014.

[10] S. Longhi, "Convective and absolute PT-symmetry breaking in tight-binding lattices," Phys. Rev. A, vol. 88, no. 052102, November 2013.

[11] B. Peng, S. K. Özdemir, F. Lei, F. Monifi, M. Gianfreda, G. L. Long, S. Fan, F. Nori, C. M. Bender, and L. Yang, "parity-time-symmetric whispering-gallery microcavities," Nature Phys., vol. 10, pp. 394–398, May 2014.

[12] M. Notomi, E. Kuramochi, and T. Tanabe, "Large-scale arrays of ultrahigh-Q coupled nanocavities," Nature Photon., vol. 2, pp. 741–747, November 2008.

WF1-1 (Invited)

OECC/PS2016

NxN Wavelength Selective Switches

Hisato Uetsuka, Shu Namiki and Keiichi Sasaki[*]

National Institute of Advanced Industrial Science and Technology (AIST), Central 2, 1-1-1 Umezono, Ibaraki 305-8568 Japan
[*]Kitanihon Electric Cable Corporation, LTD, 54-1, Shirasaka, Shibata-cho, Miyagi 989-1768 Japan
uetsuka.hisato@aist.go.jp

Abstract: *A 5x5 Wavelength Selective Switch (WSS) is proposed and demonstrated. The cross-connect optics has the orthogonal imaging systems operating in the switching plane and the dispersion one differently. The switching plane has the 2f-Fourier optics with Rayleigh length. On the other hand, the dispersion plane has the 4f- imaging optics. Two types of switching engine, as MEMS Mirrors and Liquid Crystal on Silicon (LCOS) are applied for the same cross-connect optics. The WSS with MEMS Mirrors has the 100GHz channel spacing compatible to the ITU-grid. On the other hand, the WSS with LCOS has the variable channel spacing.*
Keywords: *Wavelength Selective Switch, MEMS, Liquid Crystal on Silicon, ROADM*

I. INTRODUCTION

In order to respond the future exponential increase of power consumption due to the information traffic trend, we have developed dynamic optical path network which can realize the end-to-end path creation by the cut-through switching(1). In the cut-through switching, wavelength selective switch (WSS) is important. The future WSS should handle many optical channels in addition to the CDC (Color-less, Direction-less, Contention-less) and the flexible grid operations when considering next metro-traffic trend (2). The ROADM consists of the express traffic and the add/drop traffic. The main stream of their configuration is the route-and-select with 1xN WSSs and the multicast-and-select with 1xN optical splitters/1xN optical switches (3). On the other hand, the NxN WSS has been developed for other option. Here, we have developed the 5x5 WSS with different switching engines as MEMS mirror and Liquid Crystal on Silicone (LCoS).

II. 5x5 WSS STRUCTURE AND ITS OPERATION

The schematic structure of WSS module is shown in Fig.1. It consists of the input optics, the wavelength cross-connect (WXC) optics and the output optics. Fig.1(a) and Fig.1(b) show operations concerning the spectral plane (top view) and the switching plane (side view) respectively. In the spectral plane, the input various wavelength signals which come from GRIN fiber arrays are demultiplexed by grating and hit the switching engines. The WXC optics has 4-f (f is the focal length) optical system therefore the image on the each switching engine reappears on the opposite one. In the output optics, the various wavelength signals are multiplexed by the grating and output from the GRIN fiber array.

On the other hand, the WXC optics of the switching plane has 2-f Fourier optical system therefore the beam tilt change into the beam shift. The appropriate beam tilt by the 1st switching engine shifts the beam onto the target 2nd switching engine corresponding to the target output port. Tilt of the 2nd switching engine direct the beam toward the output port.

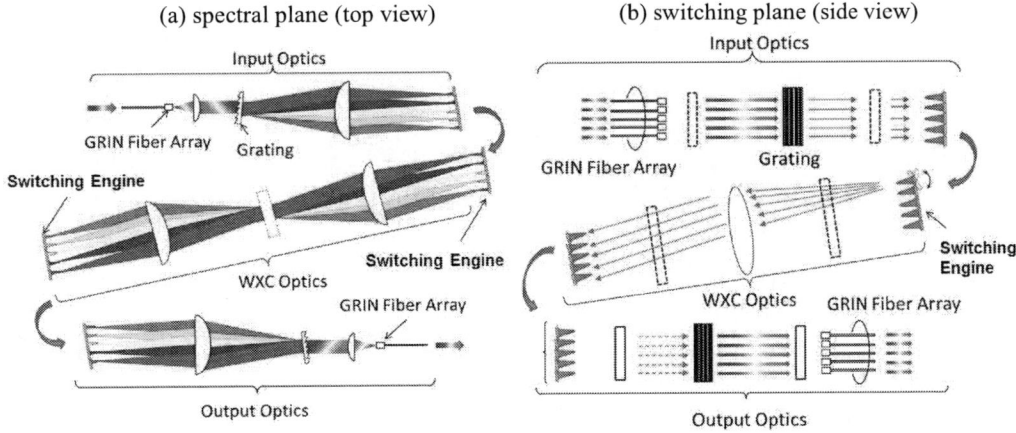

Fig. 1 5x5 WSS schematic structure

OECC/PS2016

Figure 2 shows the actual WSS module structure. It consists of optics layer and electrical layer. The electrical layer which includes both types of switching engines (MEMS and LCoS) are located underneath optical layer. They are connected optically bending the beam onto the switching engines by two prisms. All the lenses are spherical cylindrical lenses and some mirrors have Al-coating on a back facet in order to fold the beam, which can make the module compact. The spot size of the GRIN fiber array is ~40 micron, which is expanded asymmetrically in dispersion axis and switching axis by using cylindrical lenses. The spot size on the 1st and 2nd switching engine is ~100 micron (dispersion axis) x 460 micron (switching axis). The grating has the pitch of 300 lines/mm, which operates in diffraction order of three in our optics.

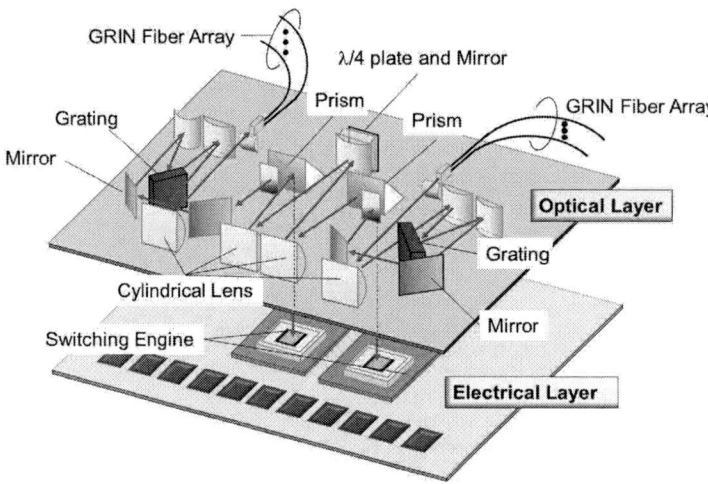

Fig. 2 Actual WSS module structure

III. MODULE CHARACTERISTICS

A. MEMS SWITCHING ENGINE

Figures 3 show the spectral response under interleaving operation in case of the MEMS switching engine. This module has normally-off operation, that is, when no voltage is applied, the module has high insertion loss in any ports combination. In case of switching from input port 3 to output port 3, the lowest loss is obtained when +0.1(V) and –0.1 (V) are applied to 1st MEMS mirror and 2nd MEMS mirror respectively. On the other hand, in case of switching from input port 3 to output port 4, the lowest loss is obtained when -0.5(V) and –0.1 (V) are applied to 1st MEMS mirror and 2nd MEMS mirror respectively. Among the channels, the optimum setting voltage is much the same. The insertion losses are around 8-10dB among the channels. The PDL is around 0.5dB over the pass band.

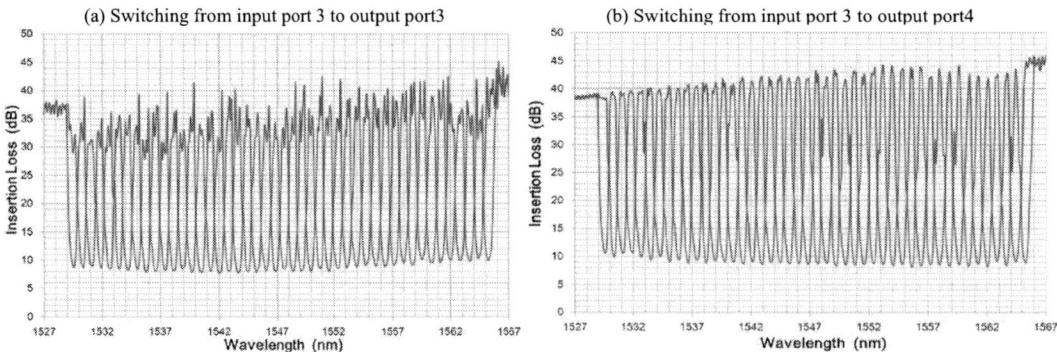

Fig. 3 Spectral response of the MEMS type WSS

1180

B. LCoS Switching Engine

Figures 4 show the spectral response in case of the LCoS switching engine. In this case, grid-less operation is possible. The LCOS has 1920x1080 pixels with 8 μm pixel pitch. The active area is 15.36 mm x 8. 64 mm. The channel spacing of 100GHz (a) and 50GHz (b) are demonstrated. Both of insertion loss are ~8 dB higher than MEMS type WSS, especially much worse in longer wavelength region, which is due to the two LCOS diffraction efficiency and alignment mismatch.

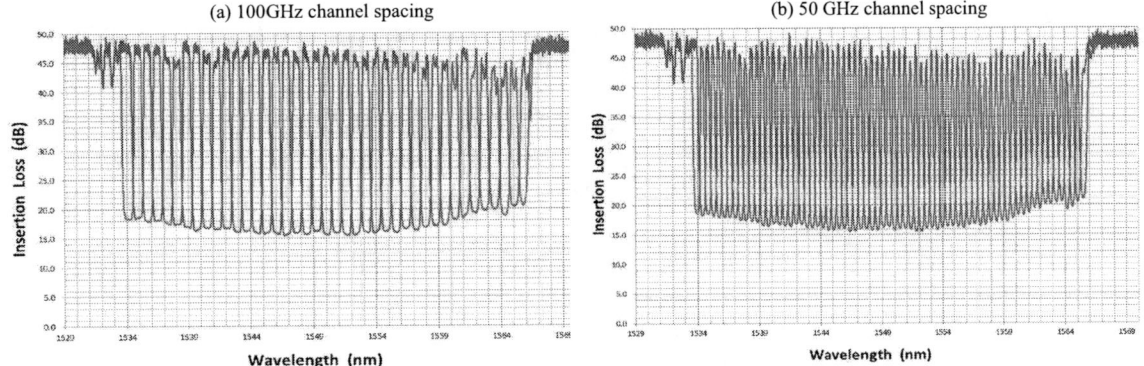

Fig. 4 Spectral response of the LCoS type WSS

IV. Conclusions

A 5x5 Wavelength Selective Switch (WSS) is proposed and demonstrated. Two types of switching engine, as MEMS Mirrors and Liquid Crystal on Silicon (LCoS) were applied for the same WSS optics. The WSS with MEMS Mirrors has 100GHz channel spacing compatible to the ITU-grid, low insertion loss (~8-10dB) and low PDL (~0.5dB). On the other hand, the WSS with LCoS has the variable channel spacing but the insertion loss was ~8 dB higher than MEMS type WSS which was mainly due to the two LCoS diffraction efficiency.

Acknowledgment

The authors acknowledge support from the Project for Developing Innovation Systems of the Ministry of Education, Culture, Sports, Science and Technology (MEXT), Japan.

References:

[1] J. Kurumida, K. Ishi, A. Takefusa, Y. Tanimura, S. Yamaguchi, H. Takeshita, A. Tajima, K. Fukuchi, H. Honma, W. Odashima, H. Onaka, K. Tanizawa, K. Suzuki, S. Suda, K. Ikeda, H. Kawashima, H. Uetsuka, H. Matsumura, H. Kuwatsuka, K. Sato, T. Kudoh, S. Namiki, "First demonstration of ultra-low-energy hierarchical multi-granular optical path network dynamically controlled through NSI-CS for video related applications," 40th European Conference on Optical Fiber Communications(ECOC 2014), paper PD.1.3, Cannes, Sep.2014

[2] R. Jensen, A. Lord, and N. Parsons, "Colourless, directionless, contentionless ROADM architecture using low-loss optical matrix switches," Proc. ECOC, Mo.2.D.2, 2010.

[3] P. Colbourne and B. Collings, "ROADM Switching Technologies," Proc. OFC2011, Paper OTuD1, 2011.

WF1-2

OECC/PS2016

Wavelength selective switch for multi-core fiber based space division multiplexed network with core-by-core switching capability

Kenya Suzuki[1,*], Mitsumasa Nakajima[1], Keita Yamaguchi[1], Goh Takashi[1], Yuichiro Ikuma[2], Kota Shikama[2], Yuzo Ishii[2], Mikitaka Itoh[1], Mitsunori Fukutoku[3], Toshikazu Hashimoto[3] and Yutaka Miyamoto[3]

[1]NTT Device Technology Laboratories, NTT Corporation. 3-1 Morinosato Wakamiya, Atsugi, Kanagawa, 243-0198 Japan.
[2]NTT Device Innovation Center, NTT Corporation. 3-1 Morinosato Wakamiya, Atsugi, Kanagawa, 243-0198 Japan.
[3]NTT Network Innovation Laboratories, NTT Corporation, 1-1 Hikari-no-oka, Yokosuka, Kanagawa, 239-0847 Japan
*s.kenya@lab.ntt.co.jp

Abstract: We demonstrate a 7-core-MCF 1x4 wavelength selective switch (WSS) for a space division multiplexed network with core-by-core switching capability by a WSS multiplexing technique using spatial and planar optical circuit platform with a laser-inscribed fan-in/fan-out.
Keywords: Wavelength selective switch, spatial division multiplexing, optical waveguide, free space optics

I. INTRODUCTION

The rapid increase in network traffic driven by, for example, video streaming services and the internet of things requires an extremely high transmission capacity for optical networks. The space division multiplexing (SDM) technique is expected to meet such demands where the capacity is multiplied by the spatial channel count [1, 2]. To achieve such an SDM network, we must consider a cost-effective integration technique for realizing a switching element as well as the SDM node architecture. There have already been several demonstrations of SDM optical switching systems [3-7]. Those switches support the "joint switching" of optical signals where the wavelength signals from an SDM fiber are routed in a certain direction at the same time. "Joint switching" is preferable for use with "coupled" type SDM fibers because they must transmit all the SDM channels in the ingress in a designated direction simultaneously where the MIMO processing unscrambles the information. The "joint switching" WSS is realized using the same simple optics as the WSS for single core/mode fiber.

In contrast, to make the optical network more flexible, it is preferable to switch the SDM and WDM channels independently in "uncoupled" SDM fibers, because it gives additional spatial channel granularity to the network [8]. A WSS for an "uncoupled" SDM network can be easily realized by combining an MCF fan-in/fan-out (FIFO) device and conventional 1xN WSSs for single mode fibers (SMFs). However, the configuration is expensive because the number of WSSs increases with the spatial channel count. In addition, many fiber connections are needed between WSSs. Recently, we have reported a WSS multiplexing technique based on a spatial and planar optical circuit (SPOC) platform that combines waveguide and free space optics [9, 10]. The technique appears suitable for use with MCF-WSSs to reduce the module count. The integration of an MCF FIFO device into the WSS module also makes it easy to handle an enormous number of optical fibers. The combination of WSS multiplexing and an integrated FIFO device will make the MCF-WSS reality. In this paper, we demonstrate a WSS for an "uncoupled" MCF network that can route the incoming SDM channels on a core-by-core basis. The device includes an array of 7 1x4 WSSs and an MCF FIFO device..

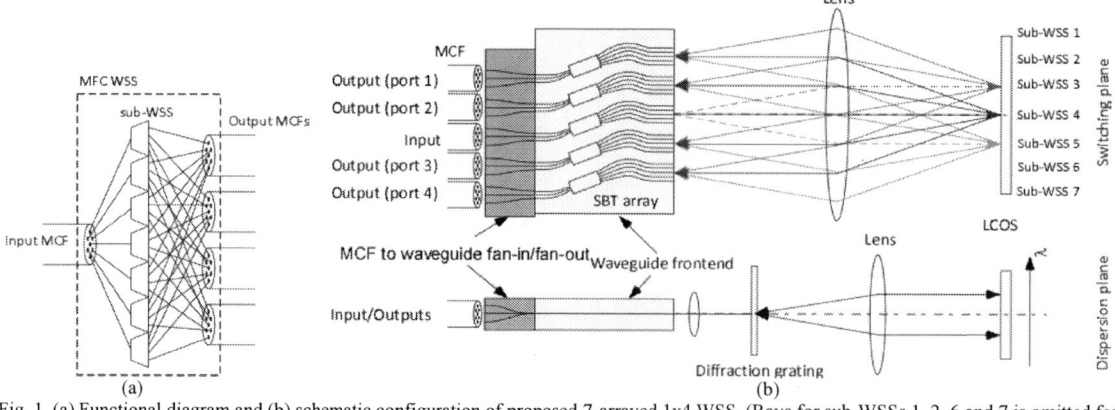

Fig. 1. (a) Functional diagram and (b) schematic configuration of proposed 7-arrayed 1x4 WSS. (Rays for sub-WSSs 1, 2, 6 and 7 is omitted for simplicity)

OECC/PS2016

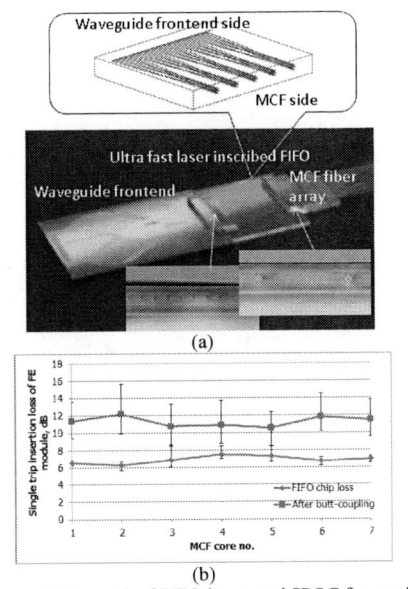

(a)

(b)

Fig. 2 (a) Photograph of FIFO integrated SPOC frontend sub assembly and (b) insertion loss variation over MCF core number.

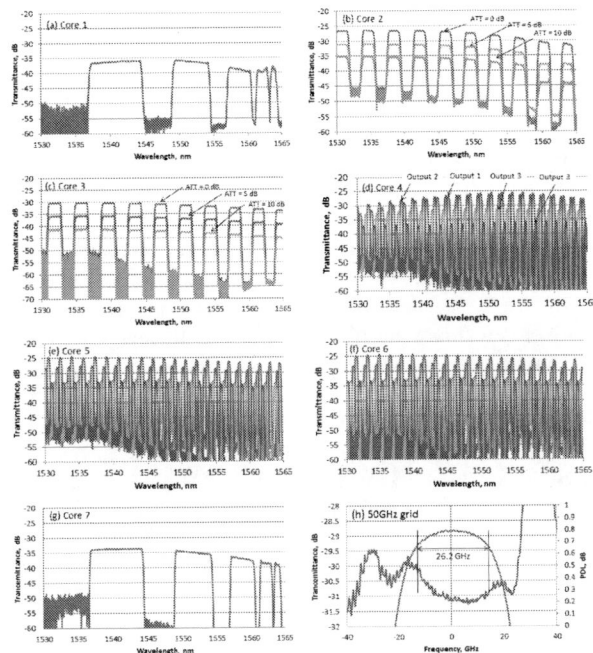

Fig. 3 (a-g) Transmission spectra of sub-WSSs 1 to 7 and (h) enlarged view of the spectrum.

II. DEVICE STRUCTURE AND OPERATING PRINCIPLE

Figure 1 (a) shows a functional diagram of a WSS for uncoupled MCF. It has sub-WSSs for each SDM channel. Each sub-WSS routes an incoming SDM channel in a designated output direction independently. Thus, SDM channel granularity is realized. The SPOC-based WSS multiplexing technique integrates multiple WSSs in a single module, which is suitable for the realization of such an MCF-WSS. Figure 1 (b) shows the schematic optics configuration of an MCF-WSS. In Fig. 1 (b), only the rays of sub-WSSs 3, 4, and 5 are depicted for simplicity. A spatial beam transformer (SBT) circuit element [11] is essential for WSS multiplexing, which has a similar circuit layout to an arrayed-waveguide grating (AWG) except that the path length difference of the arrayed waveguide is set at zero. It functions in the following way. The SDM channels carried through the input MCF are first coupled to the FIFO device. The FIFO device is a 3D waveguide, which converts the 2-dimensionally arranged MCF cores to 1-dimensionally arranged waveguide frontend cores. Then the SDM channels are fed into the waveguide frontend and therefore into the inputs of the SBT, which radiates the input signals in different directions in free space. In the free space region, SDM channel 3 (Core 3 of the MCF) propagates along the green dashed line, channel 4 (Core 4) propagates along the blue dashed line, and channel 5 (Core 5) propagates along the pink dashed line as shown in Fig. 1. Thus, the different SDM channels arrive at different positions in the LCOS. Finally, the LCOS reflects the signals back in different directions at the waveguide frontend so that they can couple with designated output ports. Thus, the WSS multiplexing function is realized. Spatial channel granularity is achieved by sharing the LCOS area between SDM channels in the switching plane while wavelength granularity is achieved by sharing the LCOS in the dispersion direction in the same way as a conventional WSS for SMF.

III. EXPERIMENT

We fabricated a waveguide frontend with 1.5-% index contrast silica-based planar lightwave circuit (PLC) technology and a FIFO by using the waveguide inscription technique with ultra-fast laser pulses in glass material [12-14]. 35 waveguide cores for 5-array MCF were written in a soda lime glass substrate with a Ti:sapphire laser whose wavelength, average power, repetition rate and duration were 800 nm, 950 mW, 250 kHz and 240 fs, respectively. We then butt coupled the waveguide frontend, the FIFO and the MCF fiber array. Therefore, the WSS was pigtailed with only 5 MCFs. Figure 2 (a) shows a photograph of the fabricated FIFO integrated frontend sub assembly. The chip sizes of the waveguide frontend and the FIFO were 28 x 42 mm and 26 x 20 mm, respectively. We measured the insertion loss (IL) of the FIFO itself and of the whole sub assembly. The ILs of the FIFO ranged from 5.6 to 8.6 dB over all 35 cores for 5 MCFs, however, they were degraded from 8.4 to 15.7 dB after the fabrication of the assembly. The IL increase was caused by positional error in the MCF fiber array, which requires axial alignment for all the MCFs. The insertion loss has to be improved by optimizing the waveguide writing condition and fine alignment of the MCF fiber array.

Figure 3 (a) to (g) show the transmission spectra of sub-WSSs 1 to 7, which correspond to the SDM channels or MCF core number. We set the LCOS patterns so that sub-WSSs 1 and 7 realized flexible grid operation with various

1183

passbands, the sub-WSSs 2 and 3 realized a variable optical attenuation function with attenuation levels of 5 and 10 dB, and the sub-WSSs 4, 5 and 6 routed the incoming WDM channels to different outputs, namely 1 to 4, with 50-GHz channel spacing, respectively. As shown in Fig. 3, basic WSS operation was realized for 7 sub-WSSs successfully. The insertion loss of the WSS ranged from 24.6 dB to 40 dB. It is broken down into the FIFO loss of 17 to 26 dB, the waveguide frontend loss of 1.5 to 2 dB and the loss in the free space optics of 6 to 12 dB. The intra and inter WSS crosstalks were typically 18.5 dB and 17.6 dB, respectively. Inter WSS crosstalk is an important parameter in an "uncoupled" MCF WSS. The crosstalk is assumed to be caused by the inter-waveguide crosstalk of the FIFO rather than of the waveguide front-end and the free space optics. Figure 3 (h) shows a typical spectrum of a WDM channel. The transmission bandwidth was 26.5 GHz with a 50-GHz channel spacing. The polarization dependent loss was 0.48 dB over the 26.5 GHz-passband.

We finally transmitted a 100-G DP-QPSK signal with 28-G baud rate through the device. The modulated signal was split into 7 with an optical splitter. Each signal component was given a time delay to eliminate any correlation with delay line fibers. They are fed into the input of the device, which routes all the SDM channels to the same output MCF. This routing condition emulates the worst transmission case, because the same wavelength crosstalk occurs severely. In addition, since the signal in the center core is most influenced by the signals in the outer cores in the FIFO device, we evaluated the bit error rate (BER) performance of the signals of the center core. The experimental setup and the BER result are shown in Fig. 4. As shown in Fig. 4 (b), no optical signal to noise ratio (OSNR) degradation was observed when only center core transmitted the signal while the OSNR penalty was 2.3 dB when all the SDM channels were transmitted in outer cores. This degradation is caused by the inter-sub-WSS crosstalk that occurs in the FIFO device. As mentioned above, the FIFO device and the coupling with MCF has a large insertion loss that leads to unwanted crosstalk to unintended cores. It is important to reduce the loss and crosstalk in the FIFO device for improvement of the performance.

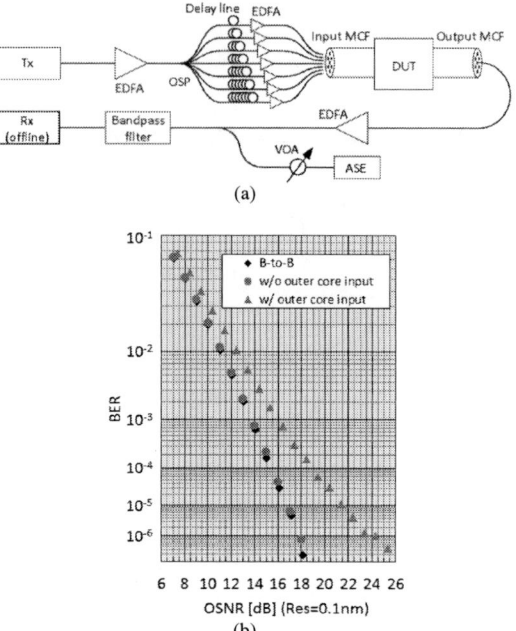

Fig. 4 (a) Experimental setup for transmission experiment and (b) bit error rate

IV. CONCLUSIONS

We demonstrated a 7-core-MCF 1x4 WSS for an SDM network with core-by-core switching capability. We utilized spatial and planar optic circuit technology, which provides the WSS multiplexing function with a spatial beam transformer circuit element. The WSS provides granularity not only in WDM channels but also in SDM channels. We also integrated a fan-in/fan-out device with the optical frontend so that the device has MCF pigtails. This overcomes the difficulty of fiber handling with a high port count MCF WSS. To the best of our knowledge, our WSS is the first demonstration of an "uncoupled" MCF network with high granularity.

REFERENCES

[1] T. Morioka, in Proc. OECC 2009, FT4, 2009, Vienna, Austria.
[2] H. Takara et al., in Proc. ECOC 2012, Th.3.C.1., 2012, Amsterdam, Holland.
[3] M. D. Feuer, et al., in Proc. OFC/NFOEC 2013, PDP5B.8, 2013, Anaheim.
[4] R. Ryf, et al., in Proc. ECOC 2013, PD1.C.4, 2013, London.
[5] J. Carpenter, et al., Optics Express, vol. 22, no. 3, pp. 2216-2221, 2014.
[6] N. K. Fontaine, et al., in Proc. OFC 2014, Th4A.7, 2014, San Francisco.
[7] D. M. Marom, et al., Opt. Express, vol. 23. no. 5, pp. 5723-5737, 2015.
[8] D. M. Marom, et al, IEEE Commun. Mag., vol. 53, no. 2, pp. 60-88, 2015.
[9] Y. Ikuma, et al., IEEE J. Lightwave Technol., vol. 34, no. 1, pp. 67-72, 2015.
[10] N. Nemoto, et al., in Proc. ECOC 2015, Tu.3.5.1, 2015, Valencia, Spain.
[11] K. Seno, et al., in Proc. OFC 2012, JTh2A.5. Anaheim.
[12] K.M. Davis, et al., Opt. Lett. 21, pp.1729, 1996.
[13] Y. Nasu, et al., Opt. Lett., vol.30, no.7, pp. 723-725, 2005.
[14] R. R. Thomson, et al., Opt. Express, vol. 15, no. 18, pp. 11691-11697, 2007.

WF1-3

OECC/PS2016

Demonstration of 1,440x1,440 Fast Optical Circuit Switch for Datacenter Networking

Koh Ueda[1], Yojiro Mori[1], Hiroshi Hasegawa[1], Hiroyuki Matsuura[2], Kiyo Ishii[2],
Haruhiko Kuwatsuka[2], Shu Namiki[2], Toshio Watanabe[3], and Ken-ichi Sato[2,1]

1: Nagoya University, Japan, 2: National Institute of Advanced Industrial Science and Technology (AIST), Japan,
3: NTT Device Innovation Center, Japan
k_ueda@nuee.nagoya-u.ac.jp

Abstract: We propose a fast and large-scale optical circuit-switch architecture for intra-datacenter applications. A 1,440x1,440 optical switch is designed and tested with 180-wavelength signals in the full C band; the worst switching time is 498 microseconds.
Keywords: Switching, circuit; Array waveguide devices

I. INTRODUCTION

With the recent rise of cloud-computing and big-data services, intra-datacenter traffic is growing at 22% a year [1]. In typical datacenters, many racks/pods including servers are interconnected via multi-tier electrical switches. The electrical switches necessitate power-consuming optical-to-electrical (OE) and electrical-to-optical (EO) conversion, and the power consumption increases with increase in the traffic. Most of the traffic volumes is occupied by elephant flows generated with virtual-machine migration, storage backup, and video-file transfer. To handle such traffic, optical circuit/flow switches that eliminate costly OE and EO conversion and can offload most of the elephant flows from electrical switches are being investigated [2-5].

For the cost-effective introduction of optical switches, two major requirements must be satisfied. First, the optical switch must have a high port count to enable flatter networks, which reduces total cost and control complexity. Second, the switching time of the optical switch must be so short that most traffic can be offloaded, e.g. 100 µs [6]. Among various optical-switch technologies, the micro-electro-mechanical system (MEMS) [2] and the semiconductor optical amplifier (SOA) [4] are being considered for intra-datacenter interconnections. However, the 3D-MEMS-based switch needs optical-path manipulations at the fabrication stage, and the number of adjustment operations increases with the square of the switch-port count. Moreover, the mechanical-driven switch is slow with millisecond-order switching. The SOA-based switch can realize nanosecond-order operation, however, its large power consumption severely limits the applicable scale, since the matrix type configuration needs increase in the number of switch elements to the square of the port count.

A wavelength-routing (WR) switch based on an NxN cyclic arrayed-waveguide grating (AWG) can create low-power switching systems [5], since the AWG itself is a passive device. However, the attainable port count of the WR switch is limited by the available AWG size. To overcome this limitation, we proposed an optical-switch architecture that utilizes a combination of MxM wavelength-independent delivery-and-coupling (DC) switches and WR switches based on NxN cyclic AWGs constructed of multi-stage AWGs [7]. This combination can drastically enlarge the switch scale, since the total port count becomes MNxMN. We fabricated a 270x270 optical-switch prototype using 3x3 DC switches and 90x90 WR switches and experimentally confirmed its effectiveness. To further enhance the switch scale needs enlargement of DC switch and/or WR switch. However, DC-switch scale increment inevitably increases the optical-coupler loss, and cyclic-AWG scale increment induces passband-frequency deviation which results in excessive filtering loss. Applying erbium-doped fiber amplifiers (EDFAs) can compensate such losses, however, the solution necessitates a substantial number of costly EDFAs since there are no effective EDFA insertion points where multiple wavelength signals are aggregated.

In this paper, we propose a large-scale optical-switch architecture realized by cost-effective wavelength-aggregated amplification and ultra-dense wavelength routing. The proposed architecture consists of a DC-switch part, an aggregation-amplification part, and a WR-switch part. Thanks to the wavelength aggregation, one EDFA can simultaneously amplify multiple wavelength signals and hence the per-port cost of the EDFA is significantly reduced, a key benefit of the proposed architecture. Furthermore, ultra-dense wavelength routing is realized by introducing two-stage wavelength routing that combines an interleaver with steep filtering and a pair of AWGs with relaxed filter shape but high port-counts. The result is a cost-effective high-port-count optical switch. We newly fabricated a pair of 1x90 AWGs on a monolithic planar-lightwave-circuit (PLC) chip, and developed tunable lasers [8] that conform to the ITU-T 25-GHz wavelength grid with the wavelength-tuning time of less than 436 µs. We constructed a part of a 1,440x1,440 optical switch by combining an 8x8 DC-switch part, a 180x1 wavelength-aggregated amplification part, and a 1x180 WR-switch part. Its good transmission characteristics and fast switching time are confirmed in transmission experiments. To the best of our knowledge, this is the first proof-of-concept demonstration of such a fast and large-scale optical switch.

II. Proposed Optical-Switch Architecture

Figure 1 shows the previously proposed MNxMN optical switch architecture [7] in which M NxN WR switches constructed by two-stage cyclic AWGs are parallelized by N MxM DC switches each of which consists of M 1xM Mach-Zehnder-interferometer (MZI) switches and M Mx1 optical couplers. To enlarge the switch scale, M and/or N must be enlarged; however, the intrinsic loss of the DC switch and/or excess filtering loss of the WR switch increase and hence available switch scale is rather limited [9]. Introducing EDFAs can resolve this problem, however, a costly EDFA is needed at each port (see Fig. 1) since there is no signal aggregation point. Furthermore, a narrow and flat-top passband AWG is difficult to create, which prevents the realization of dense wavelength routing.

Figure 2 is a schematic of the newly proposed high-port-count optical-switch architecture; it consists of N MxM DC switches, M Nx1 optical couplers, M EDFAs, M interleavers, and M pairs of 1xN/2 non-cyclic AWGs, where M is the degree of the DC switch, N represents the degree of the coupler and the number of wavelengths, and the product of M and N equals the overall switch scale. The operation is as follows. A wavelength on the ITU-T grid is selectively generated from a tunable laser. Note that we constrain ourselves to use tunable lasers whose frequencies conform to the ITU-T grid since we can utilize a widely available simple etalon-based frequency locker. The signal is then input to an MxM DC switch. The output from the DC switch is led to an Nx1 optical coupler. After the multiple (N) signals are aggregated, an EDFA post-compensates the loss of the DC switch and the optical coupler and pre-compensates for the losses of the following interleaver and AWG, simultaneously. With this configuration, the EDFA cost per port can be really small since each EDFA is shared by multiple wavelength signals, i.e. multiple ports. The signals are then de-interleaved by a 1x2 interleaver into odd-number channels and even-number channels. The frequency interval of each tributary is expanded from 25 GHz to 50 GHz. Finally, the signals of each tributary are further separated by a 1xN/2 non-cyclic AWG having a 50-GHz passband interval, where the passbands of the paired AWGs are interleaved with the 25-GHz offset. We can thus attain wavelength routing with fine granularity and hence the spectral efficiency of the wavelength-routing switch is enhanced. Thanks to the aggregation of wavelength signals and fine-granular wavelength routing, we can achieve a large-scale optical switch cost-effectively.

Table 1 summarizes the necessary number and the per-port cost contribution of each component for an MNxMN switch. Each value in parentheses corresponds to the case of M=8 and N=180. The costly EDFA and interleaver can be shared by N ports. Accordingly, the proposed architecture offers cost-effective optical switches that will suit cost-sensitive datacenter applications.

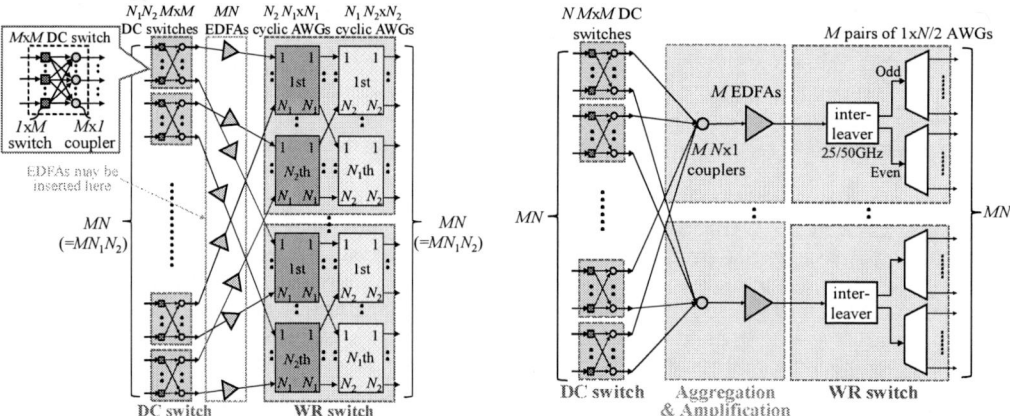

Fig. 1 Previously proposed optical-switch architecture [7].   Fig. 2 Newly proposed optical-switch architecture.

TABLE I

THE NUMBERS OF NECESSARY COMPONENTS FOR AN MNxMN OPTICAL SWITCH. THE NUMBERS IN PARENTHESES ARE THOSE EVALUATED HERE .

	MxM DC switch	Nx1 coupler	EDFA	Interleaver	Pair of 1xN/2 AWGs
Total number	N (180)	M (8)	M (8)	M (8)	M (8)
Per-port cost	1/M (1/8)	1/N (1/180)	1/N (1/180)	1/N (1/180)	1/N (1/180)

III. Experiments

To evaluate the transmission characteristics and switching time of the proposed switch architecture, we constructed a part of a 1,440x1,440 optical switch by using a fast-tunable laser [8], 8x8 DC switch, 180x1 coupler, EDFA, 1x2 interleaver, and pair of 1x90 AWGs. We measured the bit-error-ratio (BER) characteristics in both static and dynamic wavelength states. Figure 3 shows the experimental setup. The wavelength under test was generated by the fast-tunable laser. Regarding the wavelength switching time, we measured 32,220 (180x179) combinations and the average and worst values were 348 µs and 436 µs, respectively; and hence shutter time was set to 498 µs including ~60 µs margin. The intensity of laser output was modulated at 10 Gbps by an intensity modulator. The wavelength signal was then input to an 8x8 DC switch with input power Pin. The insertion loss of the DC switch was 11.6 dB including 1x8 coupler intrinsic loss (9 dB). The DC switch used an electro-optic effect switch and its switching time was around 200 ns, which is much faster than that of tunable lasers. As crosstalk sources, 180-wavelength signals on the 25-GHz ITU-T grid in the full C band were generated using commercially available continuous-wave (CW) sources and another intensity modulator, where an LCOS wavelength-selective switch (WSS) eliminated wavelength same as the target one and equalized the other-wavelength signal powers simultaneously. The target wavelength signal and the other wavelength signals in the full C band were then aggregated by a 256x1 coupler (in place of the 180x1 coupler) with 25.7-dB loss. After all signals were amplified by an EDFA, an interleaver with 1.5 dB loss de-interleaved the signals into odd channels and even channels and a pair of 1x90 AWGs routed the 180-wavelength signals according to their wavelengths. The pair of 1x90 AWGs were newly fabricated on a monolithic PLC chip (see Fig. 3 insertion) and its insertion loss was less than 4 dB. Finally, the number of bit errors was counted with a BER tester enabling burst-mode operation.

First, we measured the BER characteristics of 180-wavelength channels in the static wavelength state. We set input power Pin to 0 dBm; the input power was the minimum level that achieves BERs below the forward-error correction (FEC) limit (7% overhead was assumed). Figure 4 shows the BER characteristics measured as a function of the WR-switch port number. We confirm that BERs under the FEC threshold were obtained in all wavelength (180) channels when Pin was 0 dBm; the input power can easily be attained with commercially available transmitters.

Next, we measured BER transitions induced by switching operation, where each BER was calculated using a 104-length bit sequence. Input power Pin was set to 0 dBm. Figure 5 plots measured dynamic BER transitions, where laser wavelength was changed between edge and its adjacent channels (i.e. $\lambda 1$ and $\lambda 2$), edge and center channels (i.e. $\lambda 1$ and $\lambda 91$), and both edge channels (i.e. $\lambda 1$ and $\lambda 180$). During switching, BER becomes around 0.5 (i.e. log10(BER) ~ -0.3) since signal power was cut by a shutter to suppress crosstalk. In all cases, switching time was 498 µs and BERs under the FEC limit were confirmed.

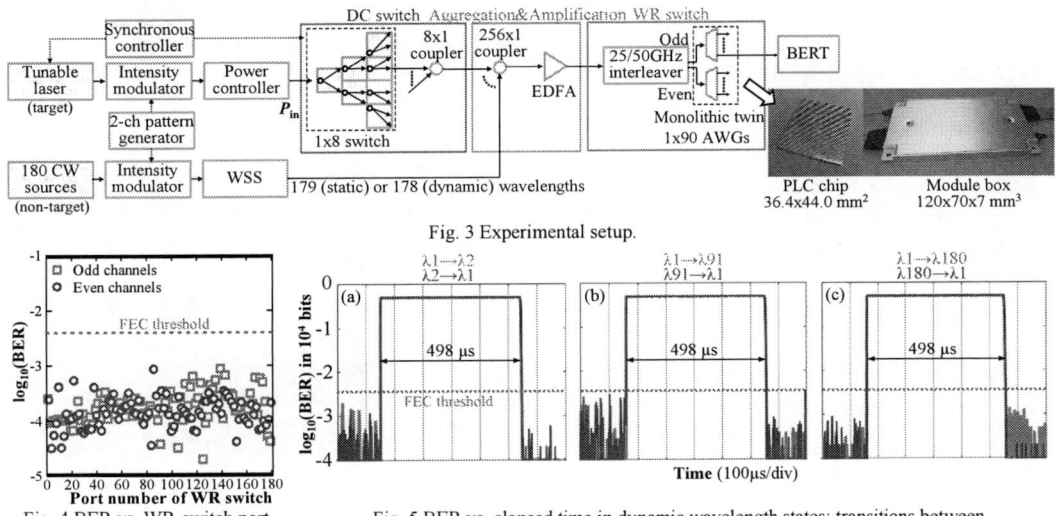

Fig. 3 Experimental setup.

Fig. 4 BER vs. WR-switch port number in static wavelength states.

Fig. 5 BER vs. elapsed time in dynamic wavelength states; transitions between (a) adjacent channels, (b) edge and center channels, and (c) both edge channels.

IV. CONCLUSIONS

We proposed a novel optical-switch architecture enabling high-port-counts for datacenter applications. We demonstrated a part of a 1,440x1,440 optical switch by combining an 8x8 DC-switch part, 180x1 aggregation-amplification part, and 1x180 WR-switch part. Overall transmission performance was evaluated both in static and dynamic wavelength states. Switching time of less than 498 µs was attained thanks to the use of fast-tunable laser. Our proposed switch offers high scalability in terms of hardware cost. The switching time is expected to be further reduced with continuing development.

ACKNOWLEDGMENT

This work was supported in part by NEDO and Project for Developing Innovation Systems of the MEXT in Japan.

REFERENCES

[1] Cisco System, Inc., White paper (2014).
[2] N. Farrington *et al.*, ACM SIGCOMM, pp. 339-350 (2010).
[3] C. Kachris *et al.*, IEEE Commun. **14**, pp. 1021-1036 (2012).
[4] A. Wonfor *et al.*, J. Opt. Commun. and Net. **3**, pp. A32–39 (2011).
[5] Y.-K. Yeo, Photonics in Switching, Fr-S37-I12 (2012).
[6] A. Vahdat, Proc. OFC, OTu1B.1 (2012).
[7] K. Sato *et al.*, IEEE Opt. Commun. **51**, pp. 46-52 (2013).
[8] H. Matsuura *et al.*, Proc. OECC/ACOFT, Tu5B1 (2014).
[9] K. Ueda *et al.*, Photonics in Switching, JT5C.3 (2014).

WF1-4

OECC/PS2016

32x32 Silicon Photonic Switch

Dritan Celo[1], Dominic J. Goodwill[1], Jia Jiang[1], Patrick Dumais[1], Chunshu Zhang[1], Fei Zhao[2], Xin Tu[2], Chunhui Zhang[2], Shengyong Yan[2], Jifang He[2], Ming Li[2], Wanyuan Liu[2], Yuming Wei[2], Dongyu Geng[2], Hamid Mehrvar[1], and Eric Bernier[1]

[1] Huawei Technologies Canada Co., Ltd., Ottawa, Ontario, Canada
[2] All-Optics Laboratory, Huawei Technologies Co. Ltd., Shenzhen, China
dominic.goodwill@huawei.com

Abstract: *Lightpaths are switched in a 32×32 silicon photonic switch, with 448 thermo-optic Mach-Zehnders and 864 monolithic monitor photodiodes. It is packaged in a CBGA with 68-fiber ribbon, and every cell has an off-chip control circuit.*

Keywords: *Photonic switching, silicon photonics, thermo-optic switch, Mach-Zehnder, integrated optics.*

I. INTRODUCTION

Photonic switches can reduce energy consumption and size, compared to electronic switches to support the continued growth in switching capacity requirements in telecom, datacenter, and high performance computing markets, and the growing pressure to reduce energy per bit. Silicon photonics (SiPh) can implement complex photonic switch circuits on a single substrate, to provide flexible lightpath connections that are compact and low cost. SiPh Mach-Zehnder thermo-optic matrices have previously been used to demonstrate lightpath circuit-switching in 32×32 [1] and 8×8 [2] switches. Other SiPh circuit-switches [3] and Mach-Zehnder switches in other material systems have also been reported [4]. The monolithic integration of switch control circuits with photonic components has been shown in a small circuit [5] and off-chip control circuits are also a viable option [1].

In this paper, we present experimental results of a packaged 32×32 SiPh circuit switch. Every switch cell of the matrix was connected to off-chip driving and monitoring circuits, and all of the optical I/O ports were coupled to a single 68-fiber array attached to the SiPh die. In comparison to [1], the most important advance in our work is monolithic monitor photodiodes within every cell. It also requires fewer cells at very large port counts.

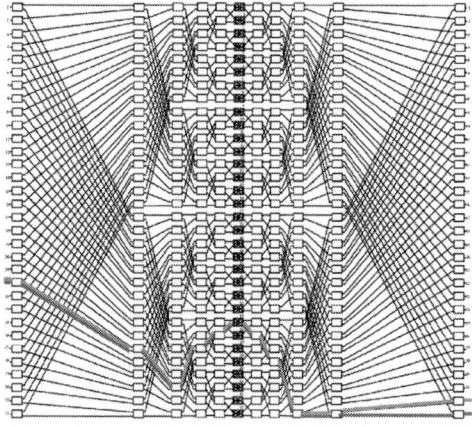

Fig. 1. a) Schematic of switch architecture. Each block in the middle column is a crosstalk-suppression stage comprising four 2×2 cells. Two lightpaths are highlighted in red and blue. b) Wire-bonded die on four-tier ceramic package, before fiber attach. Die size is 12.3 × 12.3 mm².

II. SWITCH IMPLEMENTATION

The switch matrix in the present work was implemented using 448 2×2 thermo-optic Mach-Zehnder switch cells on one die, with architecture as shown in Fig. 1a. This implements the architecture whose theory we presented in [6]. In the steady state, each cell carries 0 or 1 lightpaths, and never carries 2 lightpaths. As described in [7], strong suppression of crosstalk is critical for a multi-stage photonic switch circuit. When operated in a large matrix, the practical on/off ratio of a Mach-Zehnder thermo-optic switch cell is around 18 to 23 dB [8], and therefore first-order crosstalk creates a large impairment in the signal integrity. The architecture of Fig. 1a suppresses crosstalk, because its strongest crosstalk is second-order: for all 32-factorial connection maps, every aggressor light must pass the wrong way through at least 2 cells to attack any victim, using the routing algorithm described in reference [6].

The on/off drive current of each cell must be established by a training sequence. This has been previously accomplished using off-chip photodetectors after the optical outputs [9, 10]. However, off-chip detection requires additional equipment, and does not allow in-service adjustment to compensate for aging or dynamic thermal crosstalk. Most importantly, training by monitoring only the outputs is restricted to architectures for which there exists a training

1188

sequence [10] that can uniquely identify the lightpath from input to output at each step of the training sequence, which is not true for many architectures.

Therefore, we have incorporated on-chip monitor photodiodes with every switch cell, 864 in total (one column of cells lacked photodiodes), which are used to determine the on/off drive current of the individual switch cells during training, as proposed but not implemented in [11]. Cells are always trained to their off-state, as null detection is more precise than peak detection in a Mach-Zehnder transfer function.

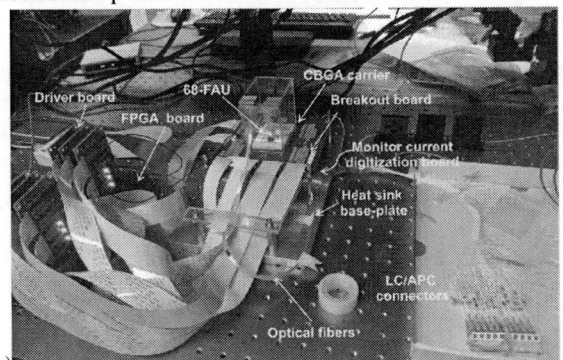
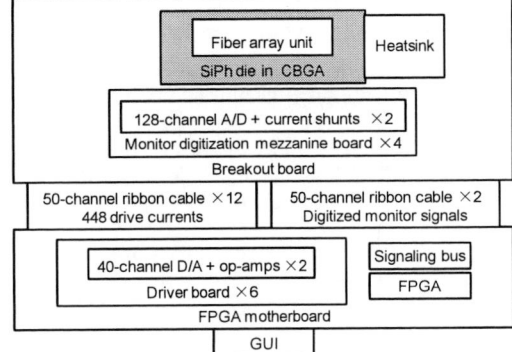

Fig. 2. Experimental setup. a) Packaged die with fiber attached on break-out board and control boards b) Schematic representation

III. DEVICE FABRICATION AND PACKAGING

A. Fabrication

The devices were fabricated on 200 mm SOI wafers, using 248 nm lithography. Details of the switch cell are provided in [12], and a block diagram of the switch cell is shown in Fig. 2b. The silicon waveguide layer thickness was 220 nm and the buried oxide layer was 2 um thick. Germanium monitor photodiodes were monolithically integrated in the SiPh die. Bare die testing showed a high yield of resistive heaters and photodiodes. Initial assemblies had high yield of bond wires, high yield of resistive heaters, but lower yield on photodiodes. Yield improvement is ongoing and will enable closed-loop control algorithms to be used on the system.

B. Optoelectronic packaging

Optoelectronic packaging and assembly must handle a very large number of optical and electrical I/Os. The SiPh die was attached on a custom cavity-up ceramic ball grid array (CBGA) carrier with 1560 bonding fingers and a CuW heat spreader for heat dissipation. Electrical connection is through 0.8 mil bond wires. Four rows of bonding pads on the SiPh die are connected to four tiers of bonding fingers along four edges of the package cavity, allowing for a large number of bonding pads on a compact die. The CBGA package with the die attached was mounted on a passive electrical breakout board using standard surface mount technology (SMT) with process peak temperature of 244°C at the die. The breakout board also includes off-the shelf passive components, electrical ribbon connectors, and mezzanine connectors, all assembled in the same SMT process.

Optical interconnect was achieved through surface grating couplers. A 68-channel fiber array unit was near-vertical permanently attached on the SiPh die surface using index-matching adhesive, as shown in Fig. 3c. Out of 68 channels, 64 are dedicated to switch optical I/Os and 4 to optical loops for alignment. All fibers were terminated with LC/APC connectors. Heat management was addressed using a conductive heatsink assembly which dissipates heat from the CuW heat spreader to the base plate. At 100% occupancy, the die with thermal undercut [12] is expected to dissipate < 1 W.

IV. RESULTS

The switch control and operation system, shown in Fig. 2, includes a fully optically and electrically assembled SiPh die in a CBGA carrier, attached to a passive breakout board, four photocurrent digitization mezzanine boards with 128-channel A/D chips and current shunts (not visible in the picture). Cable ribbons carried analog drive currents from the driver boards, and digitized monitoring signals to the FPGA motherboard. There are 6 driver (daughter) boards each with 40-channel D/A chips and op-amps, and a FPGA motherboard with an off-the shelf FPGA and signaling bus. All boards are custom designed with off-the-shelf components. The polarization of the C-band laser source was controlled to TE using an in-line polarization controller, and the polarizations of the output fibers were not controlled. The dark current was measured for all photodiodes, using the FPGA interface. The photodiode bias was uniformly set at -0.15 V, with a typical dark current < 0.5 μA per photodiode.

SiPh die with and without thermal undercut were fabricated and packaged. P_π of the switch cells with thermal undercut was measured to be less than 0.5 mW, with a rise and fall time of approximately 750 μs [12]. The optimal extinction ratio of a single 2×2 switch cell was 35.0 dB ± 0.5 dB over a large bandwidth.

A first lightpath, highlighted in red in Fig. 1a, was set up though the switch by stepping through each cell in the lightpath, in order. Each switch cell drive current was automatically scanned, while recording the photocurrent of the monitor photodiodes relevant to that switch cell. Then the switch cell drive current was set to the value that produced a null on the out-of-path photodiode (the photodiode on the nominally-OFF output of the cell), while the strong current observed on the in-path photodiode served to verify that the switch cell had been correctly actuated. A second lightpath, highlighted in blue in Fig. 1a, was set up using the same method, connecting the same input port to an alternate output. The input light was switched between the two outputs, by changing the drive current on the last common switch cell of th paths between the previously calibrated ON/OFF values. These lightpaths passed through twelve 2×2 switch cells and 35 and 34 crossings, respectively. The on-chip loss of these paths was measured relative to the transmission of reference loops. The on-chip loss transmission spectrum, shown on Fig. 3a, was measured to be between 23 and 28 dB over 1530 to 1565 nm wavelength range. This is consistent with our design estimate of the losses, which breaks down approximately as follows: 0.58 dB per switch cell, 2 dB for crossings, 1.5 dB for optical taps (leading to photodiodes), and a total routing loss ranging from 7 to 17 dB between the shortest and longest paths in the matrix.

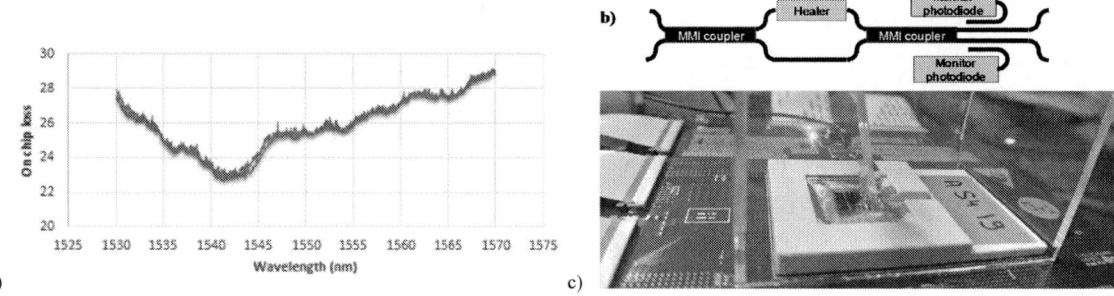

Fig. 3. a) On-chip transmission loss for both lightpaths highlighted in Fig. 1a. b) Schematic representation of each 2x2 Mach-Zehnder switch cell, illustrating the location of the taps and monitor photodiodes. c) Fiber array attached to packaged SiPh die.

V. CONCLUSION

We have demonstrated the operation of a silicon photonic 32×32 switch built from 2×2 Mach-Zehnder switch cells with on-chip monitor photodiodes. The switch control and packaging were implemented using commercially available technologies. In order to scale this switch matrix implementation to port counts larger than 32×32, or to satisfy system requirements with the present implementation, on-chip loss and coupling loss require improvement. Packaging requirements can be alleviated by using a sparse matrix of monitor photodiodes, reducing the amount of tapped power. Improvement of the on-chip loss will require an improvement of the baseline waveguide loss in addition to the simultaneous optimization of all passive components.

REFERENCES

[1] K. Tanizawa et al., "32x32 strictly non-blocking Si-wire optical switch on ultra-small die of 11x25mm^2," OFC 2015, M2B.5.

[2] L.Chen and Y. K. Chen, "Compact, low-loss and low-power 8x8 broadband silicon optical switch," Optics Express vol. 20, pp 18977-18985 (2012).

[3] S. Han et al., "Monolithic 50x50 MEMS silicon photonic switches with microsecond response time," OFC 2014, paper M2K,2.

[4] H. Kouketsu et al., "High-speed and compact non-blocking 8x8 InAlGaAs/InAlAs Mach-Zehnder-Type optical switch fabric, OFC 2014.

[5] B. G. Lee et al., "Monolithic silicon integration of scaled photonic switch fabrics, CMOS logic and device driver circuits," J. Lightwave Tech., vol. 32, pp743-751 (2014).

[6] Y. Qian et al., "Scalable photonic switch with crosstalk suppression for datacenters and optical networks," OECC 2015.

[7] P. Dumais, H. Mehrvar, D. J. Goodwill and E. Bernier, "Scaling up photonic switch fabrics," Group IV Photonics Conference (2015)

[8] R. Aguinaldo et al., "Energy-efficient, digitally-driven "fat pipe" silicon photonic circuit switch in the UCSD data-center network," CLEO 2014, paper STtu1J.2

[9] M.S. Hai et al., "Automated characterization of SiP MZI-based switches," IEEE Optical Interconnects Conf. 2015, pp. 94-95.

[10] S. Suda et al., "Fast and accurate calibration method for large-port-count Si-wire PILOSS optical switch," Asia Communications and Photonics Conference 2015, paper A54A.2.

[11] N. Dupuis et al., "Modeling and characterization of a nonblocking 4x4 Mach-Zehnder silicon photonic switch fabric," J. Lightwave Tech., vol. 33, pp. 4329-4337 (2015)

[12] D. Celo et al., "Thermo-optic silicon photonics with low power and extreme resilience to over-drive," IEEE Optical Interconnects conference 2016.

Silicon photonics optical switch based on ring resonator

Antoine Descos[1], M. Ashkan Seyedi[1], Chin-Hui Chen[1], Marco Fiorentino[1], François Vincent[2], David Penkler[2], [3]Bertrand Szelag and Raymond G. Beausoleil[1]

[1]Hewlett Packard Labs, Palo Alto, CA, USA, [2]Hewlett Packard Enterprise, Grenoble, France, [3]CEA-LETI, Grenoble, France
Author e-mail address: antoine.descos@hpe.com

Abstract: *Optical switches based on ring resonator cavities are fabricated in a silicon photonics foundry process and analyzed. Effective switching with P-i-N junction is shown for a 20Gbps signal with a bias voltage as low as 1 volt.*

Keywords: *Silicon Photonics, ring resonator, optical switch*

INTRODUCTION

With the increase of optical interconnects in datacenters and high-performance computers, the demand for a fully integrated optical switch technology is increasing. Several technology platforms have been developed with various disadvantages. Micro-electrical mechanical systems (MEMS) [1], with crosstalk as high as 50dB require high drive voltage (around 20V) and demonstrate low speed switching time (microsecond). Mach-Zehnder based switches [2] show high-speed operation with crosstalk of 25dB and a 5mW power consumption. Ring resonators have shown to be good candidates for high-speed modulation and high integration due to a small form factor. The carrier-injection microring [3] enables low optical loss inside the cavity as well as low drive voltage. The work presented in this paper shows the characterization of an optical switch based on a high quality factor (Q) ring resonator that is tuned by a p-i-n junction. Switching of a 20Gbps is obtained with a 1V drive voltage and less than 0.5dB power penalty is shown for a 40GHz channel spacing.

EXPERIMENTAL

The ring filter has a diameter of 14μm with a cavity quality factor (Q) of 46,000. It consists of a 450nm wide waveguide on a 300nm SOI wafer that is etched to a 50nm rib, fabricated at CEA-Leti. The add/drop waveguides of the ring cavity are 150nm away. The width of the add/drop waveguides are 450/400nm, respectively. The ring comprises a P-i-N junction with the inside of the ring is N-doped and the outside P-doped Carriers are injected through DC electrical probes. The microscope photograph of the device is shown in Fig.1b. The experimental setup is shown in Fig.1a where one tunable laser and modulator form the injected channel via grating couplers. The signal is either collected from the through port or the drop port, also via grating couplers. The fiber is connected to a variable optical attenuator (VOA) and photodetector (PD) to measure the bit error ratio (BER) versus input power. The following experiments quantify the BER at 20Gb/s for the ring filter using a PD which has a specification of BER $< 1 \times 10^{-12}$ at 25Gb/s with an optical input power noise floor of -10dBm.

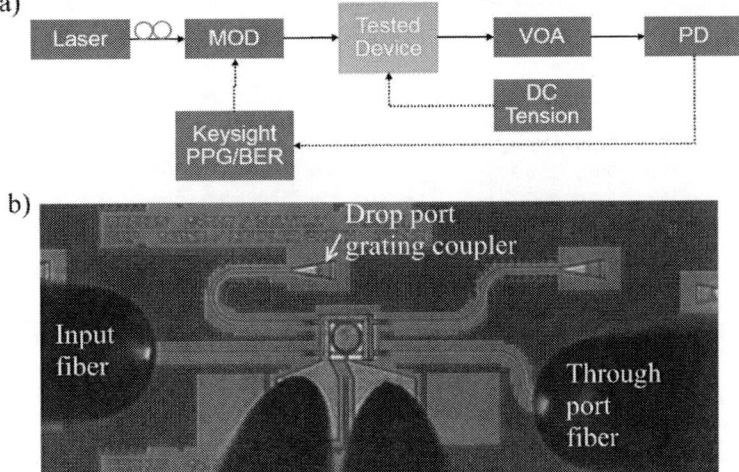

Figure 1, Figure showing and a) the experimental setup used for testing switch device, and b) microscope photography of the device.

SWITCHING OF A MODULATED SIGNAL

First, we characterized the ring response for different bias voltages in the P-i-N junction. The through port spectra are presented in the Fig.2a. At zero bias the ring has a Q factor of ~46,000 and a resonant wavelength of 1309.042nm. The resonant wavelength is blue-shifted as carriers are injected in the diode. This increases cavity losses and the Q factor is degraded. Fig.2b shows the Q factor and resonant wavelength shift for different bias points. With a bias of 1V, a shift of 1.31nm is achieved. The BER of the 20Gbps signal is measured at the PD. Measurements collected from the through port for a wavelength of 1310nm and collected from the drop port at the resonance are shown in Fig.2c. An optical power penalty of 2dB for a BER of 1×10^{-9} is measured for the drop port versus the through port. This is due to spectral filtering of the signal by the ring.

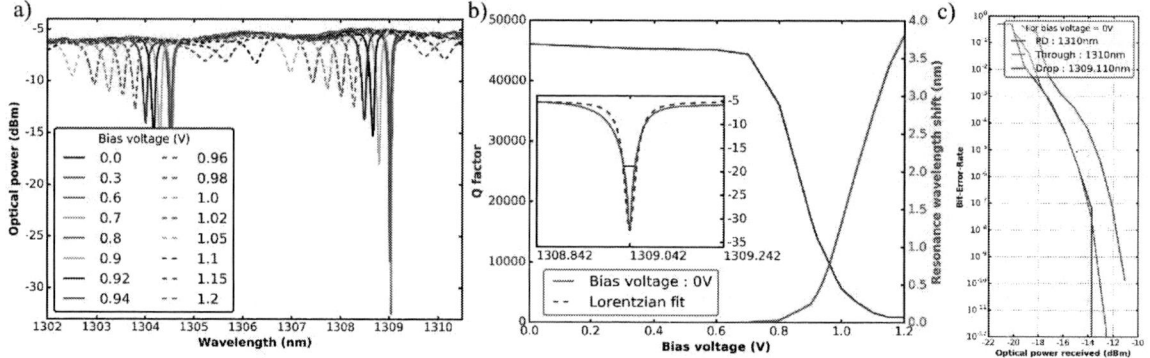

Figure 2, a) Spectrum of the through port for different bias voltage, b) Q factor and resonance wavelength shift for different bias voltage with inset of the lorentzian fit for a bias voltage of 0V, and c) BER measurement for the PD, through port outside resonance and drop port at resonance.

Finally, different BER measurements at 20Gbps where done from the drop port and through port for different bias values and wavelengths. Fig.3a shows the BER measurements at the resonant wavelength of 1309.042nm from the drop and through port for different biases. The signal is received from the through port with no optical power penalty for a bias of 1V.

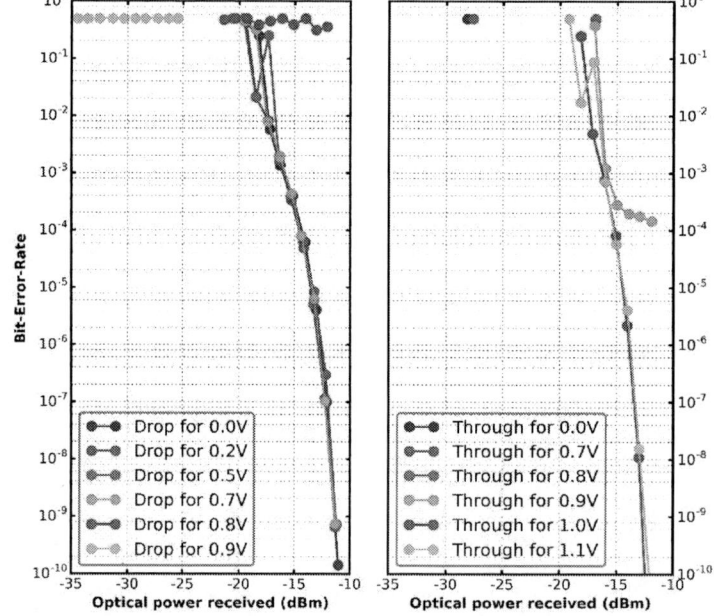

Figure 3, BER of the modulated signal for Drop and Through port @1309.042nm for different bias voltage.

CONCLUSIONS

This work presents the characterization of a high Q factor ring-based optical switch. Effective switching for a signal of 20Gbps is achieved with a drive voltage of 1V. The possibility of using an 80GHz spacing between two concurrent signals is demonstrated. These results show strong promise for the integration of these ring-based switches in optical data links that use 80GHz channel spacing with a 10Gbps modulation rate [5].

REFERENCES

[1] T. J. Seok, N. Quack, S. Han, and M. C. Wu, "50x50 digital silicon photonic switches with mems-actuated adiabatic couplers," in Optical Fiber Communication Conference. Optical Society of America, 2015, pp. M2B–4.

[2] N. Dupuis, "Technologies for fast, scalable silicon photonic switches," in Photonics in Switching (PS), 2015 International Conference on. IEEE, 2015, pp. 100–102.

[3] C.-H. Chen, C. Li, A. Shafik, M. Fiorentino, P. Chiang, S. Palermo, and R. Beausoleil, "A WDM silicon photonic transmitter based on carrier injection microring modulators," in Optical Interconnects Conference, 2014, pp. 121–122.

[4] B. Small, B. Lee, K. Bergman, Q. Xu, J. Shakya, and M. Lipson, "High data rate signal integrity in micron-scale silicon ring resonators," in Conference on Lasers and Electro-Optics. Optical Society of America, 2006, p. CTuCC4.

[5] M. A. Seyedi, C.-H. Chen, M. Fiorentino, and R. Beausoleil, "Error-free DWDM transmission and crosstalk analysis for a silicon photonics transmitter," Optics Express, vol. 23, no. 26, pp. 32 968–32 976, 2015.

WF3-1 (Invited)   OECC/PS2016

Optical Switching Functions Using Integrated Silicon-based Devices

G. de Valicourt

Bell Labs, Alcatel-Lucent, 791 Holmdel Road, Holmdel, New Jersey 07733, USA

devalicourt@bell-labs.com

Abstract: *We review the recent achievements done in optical slot switched technology using silicon-based photonic integrated circuit. Monolithic integrated slot-blocker and a dual-hybrid cavity laser for fast wavelength switching are reported.*

Keywords: *Reflective Semiconductor Optical Amplifier, Silicon Photonic, Photonic Integrated Circuit*

I. INTRODUCTION

The use of content delivery networks as well as the interconnections between large data centers that may be in separate geographical locations has contributed to increase the traffic in metropolitan-area network [1]. Such evolution is likely to create bursty and distributed traffic profiles such that flexible bandwidth-on-demand connections are required. To meet these requirements, elastic optical network (EON) [2] as well as optical packet switching (OPS) [3] have been proposed for the next-generation of ultrafast, energy- and resource-efficient data transport systems with fast reconfigurable connections. Optical packet switching refers to any technology that brings into optical communications the packet switching paradigm, which is well-known in electronics. The switching of data with sub-wavelength granularity enables statistical multiplexing, that is, the sharing of the transport medium between several data flows therefore providing highly efficient optical network. Furthermore opto-electronics conversions only occur at the source and destination nodes, translating to energy savings. Burst optical slot switching (BOSS) is a version of optical packet switching, where switched entities are slots of the same duration, as shown in Fig. 1. In the past few years Nokia Bell Labs has proposed several node architectures for high speed BOSS rings [4,5] using coherent signaling for high datarate transmission of 100 Gb/s per channel and above, targeted to metro or datacenter networks [6]. However, these network segments are more sensitive to cost and therefore relies on the development of low-cost photonic integrated circuits (PIC) such as a fast-tunable coherent transponder and a fast optical wavelength blocker to reuse the fiber capacity. Silicon photonics is a promising option to provide large-scale integration of photonic components with high-volume manufacturing compatibility. The BOSS proposed by Bell Labs rely on the three key building blocks:

• "Burst-mode" receivers, which are capable of receiving bursts of data (duration: a few microseconds) separated by short guard intervals (duration: well below 1 microsecond)

• Fast tunable lasers, which can change their emission wavelength on a per-slot basis, are used as local oscillators to make the coherent receivers fast wavelength-tunable. Fast tunable lasers are also used in the transmitters.

• Fast optical slot-blockers, implemented as a dynamic slot blocker that can physically erase slots after they are received so as to reuse the fiber capacity.

Recently, we demonstrated a several fast SOI-based optical wavelength blocker. A review of the different achievements

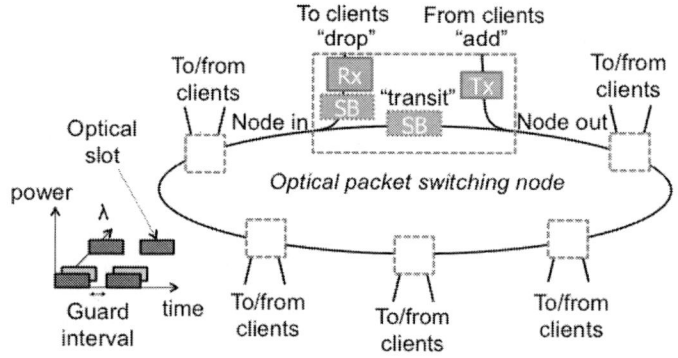

Fig. 1. Burst Optical Slot Switching (BOSS) ring network (SB : slot-blocker; the part in the dashed box may be integrated using silicon photonics).

has been presented in [7]. Fully integrated coherent receiver has been also demonstrated using silicon photonic technology [8] and one rising challenge is the development of fast-wavelength tunable laser as local oscillator compatible with the aforementioned material. Recent works focused on integrating a fast-wavelength-tunable hybrid silicon–based laser based on the use of two parallel widely tunable Vernier ring resonator (RR) laser cavities integrated into a single silicon chip. Such concept provides single mode operation and wavelength tunability

OECC/PS2016

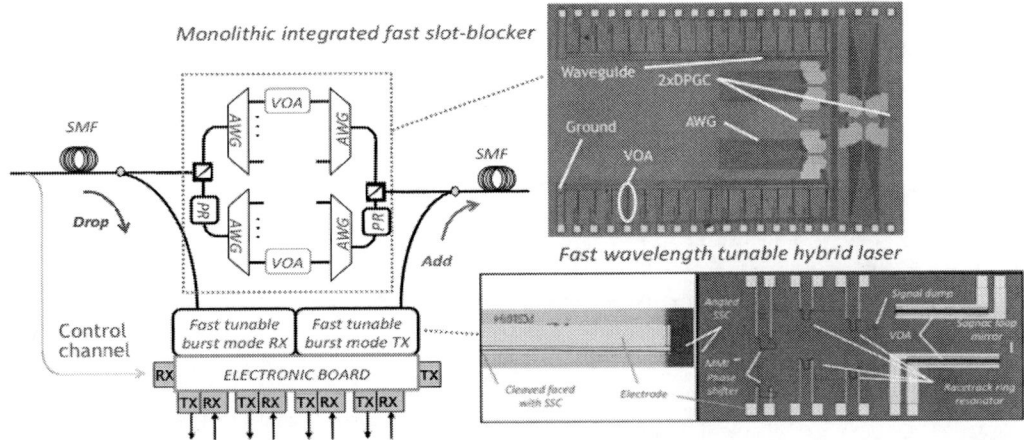

Fig. 2. Optical packet-switching node consisting on a fast slot-blocker, fast tunable burst mode receiver and transmitter. Inset: Integrated dual-Polarization slot-blocker and dual hybrid silicon-photonic laser photography.

over more than 35 nm with fast switching time in the tens of nanosecond range [9]. In this paper, we review the recent advances on the fast optical slot-blockers and fast-tunable hybrid lasers based on silicon-based PIC.

II. INTEGRATED SLOT-BLOCKER

A typical optical packet switching node is shown in Fig. 2 including the so-called slot-blocker. The wavelength-multiplexed signal is first demultiplexed, then each wavelength goes through an optical gate, which may block each channel independently based on control signals, and finally the wavelengths are multiplexed before exiting the slot blocker. We use vertical grating couplers (VGC) in order to couple the light into the PIC. AWGs are used to (de)multiplex the different wavelengths. Channel spacing, down to 100 GHz, have been demonstrated [10]. Slot blocking is performed by direct modulation of VOAs with switching time below 10 ns [11]. One optical gate per wavelength is required. For polarization diversity scheme, light is split/combined at the input/output into two orthogonal polarization components using dual-polarization vertical grating couplers (DPVGC). Each component is aligned with the TE mode of two identical PICs, side-by-side on the same chip. Polarization independent operation can be achieved by synchronously driving both PICs. A second AWG (per polarization) is used to re-multiplex the wavelengths before exiting the device through the output port. The SB was fabricated on an 8" silicon-on-insulator wafer with a device silicon layer of 220 nm. We then performed measurements to assess the performance of the slot blocker for add/drop operation. One tunable laser was used to launch a continuous-wave signal into the circuit. We used one QAM transmitter to form continuous flows of either 256 Gb/s or 320 Gb/s signals (PDM-16QAM or PDM-32QAM). During drop/add operation we measure the added channel after blocking the dropped channel, thus including in-band crosstalk from the blocked channel. The BER measurements are shown in Fig. 3. When dropping and adding a new channel, less that 1 dB OSNR penalty is observed for 256 Gb/s data-rates compared to back-to-back and less that 6 dB of OSNR penalty was measured for 320 Gb/s data-rates. In both cases, BER well below the soft-decision forward-error-correction (SD-FEC) limit (20% overhead) are achieved.

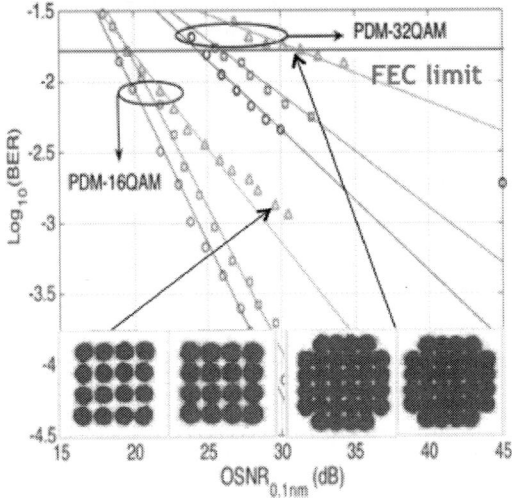

Fig. 3. Experimental results of BER with respect to the OSNR in Back-to-Back, through and add/drop configurations for 256 Gbit/s PDM-16QAM and 320 Gbit/s PDM- 32QAM using the polarization-diversity transmissive slot-blocker.

III. FAST WAVELENGTH TUNABLE HYBRID LASER

We propose a dual-selectable hybrid-cavities laser, which consist of a butt-coupled InP reflective semiconductor optical amplifier (RSOA) with a silicon photonic integrated circuit (PIC). One Vernier-based laser cavity is realized using two ring resonators (RR), one phase shifter, one VOA and closed with a 100 % reflection mirror, which is achieved using Sagnac loop mirror consisting of one 1x2 MMI and one waveguide loop. The difference of the ring FSRs creates a Vernier effect so this laser allows single-mode operation and wavelength tunability over more than 35 nm. Metal heaters are processed on the top of the ring resonators to allow thermal tuning of the ring resonator peak wavelengths. However this process is inherently slow taking tens of microseconds. One of the two cavities is selected by turning OFF the corresponding VOA while the

1195

OECC/PS2016

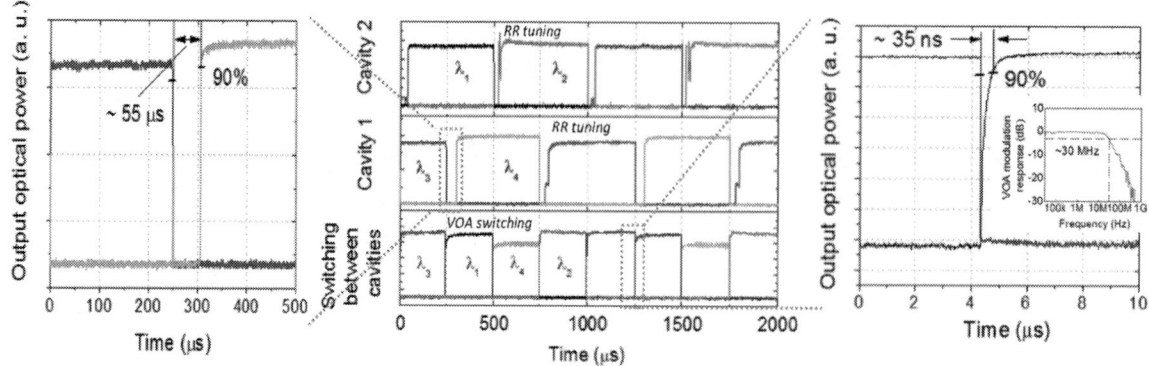

Fig. 4. Temporal response of the optical output for cavity 1 using RR tuning (top graph), for cavity 2 using RR tuning (middle graph) and for switching between both cavities using VOAs devices (bottom graph). Inset: rise and fall time for switching using VOAs.

other VOA is ON. The VOAs are p-i-n junctions based on carrier injection, when in the ON state, current in the VOA creates free carriers that absorb light. In contrast with the thermo-optic effect used to tune the ring, the carrier injection process is much faster as described above [11]. In order to demonstrate fast switching operation, a 4 channels arbitrary waveform generator is used to control one ring of each cavity and the two VOAs. To reduce the wavelength switching time, we employ VOA switching so that while the laser emits at a first wavelength, the second cavity RRs are pre-tuned to a second desired wavelength during a minimum slot duration. Then the emitting wavelength is switched to the second cavity by switching the VOAs. Fig. 4 shows the measured output power following a bandpass optical filter with 0.2 nm full width half maximum when switching a single cavity with thermo-optic tuning of one RR (for cavity 1 and 2) and between the two cavities by switching the VOAs. The switching time between two pre-set wavelengths using the VOAs is demonstrated to be less than 35 ns.

IV. CONCLUSIONS

We have demonstrated a 16-channel 100 GHz-spaced slot blocker using silicon photonic technology. High-speed VOAs allow nanosecond switching time as well as high extinction ratio. Add/drop operation of 256 Gbit/s PDM-16QAM and 320 Gbit/s PDM-32QAM signals is performed. We then demonstrated a novel fast wavelength-tunable hybrid laser using also silicon photonic technology. Dual-cavity configuration is proposed, where each cavity could be continuously tuned over more than 35 nm. Such technologies are viewed as promising blocks for BOSS networks in the metropolitan and datacenters network segments where cost, footprint, flexibility and network efficiency are true concerns.

ACKNOWLEDGMENT

The author acknowledges the teamwork with Y. Pointurier, M. A. Mestre, S. Bigo, J. E. Simsarian, S. Chandrasekhar, A. Shen, I. Ghorbel, G. H. Duan, A. Maho, R. Brenot, K. W. Kim, A. Melikyan, Po Dong, C-M. Chang and Young-Kai Chen at Bell Laboratories, J.-M. Fedeli at CEA-Leti, L. Bramerie, E. Borgne J.-C. Simon at Foton Laboratory and L. Vivien at Institut d'Electronique Fondamentale (IEF).

REFERENCES

[1] Cisco White Paper. San Diego, CA, USA. (2013, May 29). The Zettabyte Era-Trends and Analysis [Online]. Available: http://www.cisco.com

[2] Y. Yoshida et al., "First international SDN-based Network Orchestration of Variable-capacity OPS over Programmable Flexi-grid EON", Proc. OFC, Th5A.2, San Francisco, USA (2014).

[3] Y. Pointurier et al., "High data rate coherent optical slot switched networks: a practical and technological perspective", IEEE Communications Magazine, Vol. 53, No. 8, pp.124-129, Aug. 2015

[4] D. Chiaroni, et al., "Packet OADMs for the next generation of ring networks," Bell Labs Technical Journal, vol. 14, no. 4, pp. 265-283, Winter 2010.

[5] J. E. Simsarian, et al., "Fast-Tuning Coherent Burst-Mode Receiver for Metropolitan Networks," IEEE Photon. Technol. Lett., vol. 26, no. 8, pp. 813-816, April 2014.

[6] M. A. Mestre, et al., "Optical Slot Switching-Based Datacenters With Elastic Burst-Mode Coherent Transponders", in Proc. ECOC'14, Th.2.2.3,Cannes,France (2014).

[7] G. de Valicourt, et al., "Monolithic Integrated Slot-Blocker for High Datarate Coherent Optical slot switched networks", Journal of Lightwave Technology, Vol. 34, No. 8, pp. 1807-1814, 2016.

[8] P. Dong et al., "224-Gb/s PDM-16-QAM Modulator and Receiver based on Silicon Photonic Integrated Circuits," Proc. OFC, PDP5C.6, Anaheim (2013).

[9] G. de Valicourt, et al., "Dual Hybrid Silicon-Photonic Laser with Fast Wavelength Tuning", in Proc. OFC'16, M2C.1, Anaheim, USA (2016).

[10] G. de Valicourt, et al., "16-channel 100 GHz-spaced Integrated Polarization Diversity Silicon-based Slot-Blocker for High Data Rate Reconfigurable Networks", in Proc. OFC'16, Th2A.10, Anaheim, USA (2016).

[11] G. de Valicourt, et al. "Monolithic Integrated Silicon-based Slot-Blocker for Packet-Switched Networks", in Proc. ECOC 2014, We.3.5.5, Cannes, France (2014).

OECC/PS2016

Integrated Fat-Tree Optical Switch with Cascaded MZIs and EAM-Gate Array

Yusuke Muranaka, Toru Segawa, and Ryo Takahashi

NTT Device Technology Laboratories, NTT Corporation, Kanagawa, Japan
muranaka.yusuke@lab.ntt.co.jp

Abstract: *A 1×4 fat-tree optical switch with a high extinction ratio of more than 40 dB has been demonstrated which together with the low-intrinsic loss elucidate the advantage of the MZIs and EAM gates combination.*

Keywords: *optical switch; Mach-Zehnder interferometer (MZI); electro-absorption modulator (EAM); ·optical packet switching (OPS)*

I. INTRODUCTION

The growth of the Internet and Internet-related services has led to an increasing demand for network nodes that can support a large capacity with low power consumption. Recently, the capacity of optical transmission systems has been steadily increased by the introduction of various multiplexing technologies, such as digital coherent transmission enabled by digital signal processing (DSP), multicarrier transmission, and multi-mode/core fiber transmission. On the other hand, a major bottleneck remains at the network nodes due to throughput limitation, high power consumption, and high latency. Conventional electrical packet switching (EPS)-based networks are finding it increasingly difficult to cope with these conditions. Optical packet switching (OPS) technologies have been considered with a view to overcoming these problems [1].

Figure 1 shows a hybrid optoelectronic router (HOPR) that makes it possible to realize an OPS-based network. One of the key devices in an HOPR is an N×N optical switch (Fig. 2), which is used for forwarding incoming optical packets to the desired output ports [2]. The optical switch needs to be transparent with respect to high bit rates (100 ~ 400 Gbps), various packet formats (coherent/colored), wavelength and polarization, as well as providing a fast switching speed, power efficiency, high extinction ratios, low crosstalk, controllability and compactness. Various optical switches have already been demonstrated to meet these demands, including matrix switches such as cascaded 1×2 switches [3], phased-array switches [4], wavelength routing switches [5], and broadcast-and-select (B&S) switches [6].

We have previously reported an EAM-based B&S optical switch as a replacement for SOA deployment [7]. In the B&S switch, to enhance the optical coupling loss between the input/output fiber array and the switch's waveguide array, integrated spot size converters have been introduced. Moreover, to eliminate the electrical crosstalk in the device, a simple solution based on realizing a surface grounded electrode was demonstrated [8]. Although the B&S switch provides various advantages, its intrinsic large insertion loss caused by power splitting limits its port count scalability. In this paper, to reduce the branch loss of a large-port-count optical switch, we propose a Fat-Tree optical switch using a Mach-Zehnder interferometer (MZI) as to replace the 1×N MMI splitters. Our optical switch with integrated MZIs and EAM gates exhibits low loss and high extinction ratio.

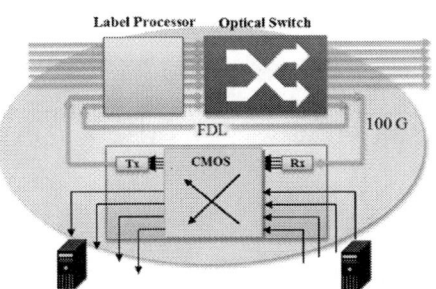

Fig. 1: Hybrid optoelectronic router architecture.

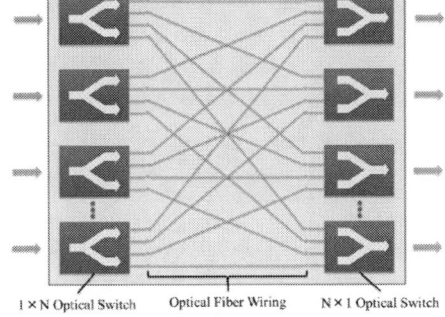

Fig. 2: N × N optical switch architecture.

II. DEVICE STRUCTURE

To realize an N×N optical switch with a conventional B&S structure, 1×N passive optical splitters and N×1 passive optical combiners were used to broadcast the optical signal. For a large port count, the intrinsic loss is high. Figure 3 compares the losses for the splitter and EAM gates combination and the cascaded MZIs and EAM gates combination. The loss difference increases as the port count increases since the loss of an MZI is much lower than 3 dB splitting loss.

1197

An MZI based on the carrier plasma (CP) effect has high-speed responsivity, and it is promising for use in selecting the light path. However, because of the insufficient extinction ratio of the MZI, we introduced a combination of cascaded MZIs and an EAM-gate array.

Figure 4 shows a photograph of a fabricated 1×4 optical switch. Three MZIs and four EAM gates are monolithically integrated on a single chip where the total chip size is 2.1 mm × 0.9 mm. The device has a p-i-n double hetero-junction structure with a 0.4-μm thick bulk InGaAsP (λgap=1.4 μm) layer as the junction core. The EAM section has a shallow-ridge waveguide structure, whereas the MZI section has a deep-ridge waveguide structure to allow a small bending radius. The bulk InGaAsP is used as the core layer for both waveguides. At the MZI section, the bulk InGaAsP acts as the phase control layer with a CP effect. In contrast, it also acts as the electro-absorption layer for the EAM section, which is based on the Franz-Keldysh (FK) effect.

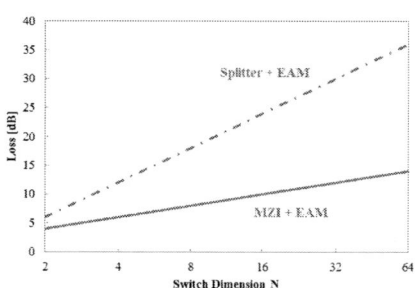

Fig. 3: Loss versus switch dimension .

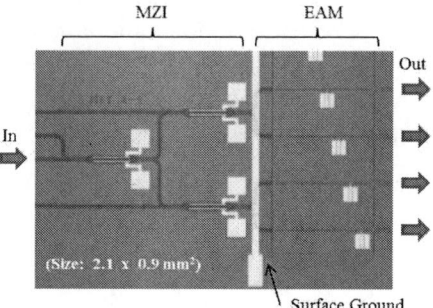

Fig. 4: Photograph of fabricated 1×4 optical switch.

III. MEASUREMENT RESULT

The MZI section consists of a 2×2 MMI splitter/coupler and two waveguide arms, where the electrode is 200 μm long. Figure 5 shows the transmission characteristics of the MZI for different values of injected current. By injecting current with a forward bias into one waveguide arm, the coupler is tuned and the output port is switched. At the switch operating point highlighted in Fig. 5, extinction ratios of 20 and 15 dB were obtained for the TE and TM modes, respectively. The MZIs control a light path with a small current, and exhibits low loss compared with an MMI splitter/coupler. Figure 6 shows the transmission characteristics of the EAM gate versus the applied reverse bias for wavelengths of 1540, 1550, and 1565 nm. A 20-dB extinction ratio was measured for a reverse bias of 3 volts, and by increasing the reverse bias to 4 volts an extinction ratio of 45 dB was achieved.

The integrated EAM and MZI sections share the same p-cladding layer. Therefore, when a voltage is applied to a given EAM, it affects the other EAMs and MZI section through the waveguide p-cladding layer, causing an undesirable change in the output power due to the unbalanced operation of MZIs and the electro-absorption caused by the FK effect in the passive waveguide. To suppress such electrical crosstalk, each EAM and MZI section should be electrically isolated, and one way to achieve this is to realize a high-impedance surface for passive waveguides by using Fe-doped InP instead of a low-impedance p-clad layer by incorporating an additional step in the regrowth process. However, the fabrication process becomes more complicated and the propagation loss of the optical waveguides is affected. Thus we have proposed a convenient approach that involves the simple addition of surface-ground electrodes that cover the MMIs of the MZI section [9]. By connecting the surface electrodes to the common ground, the isolation was enabled between the EAM section and the MZI section as well as among the EAMs.

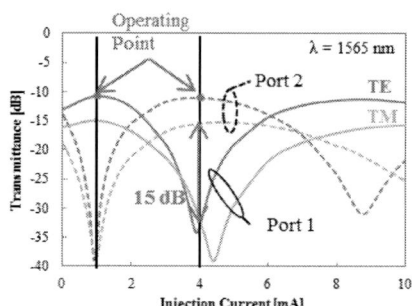

Fig. 5: MZI transmitted characteristics.

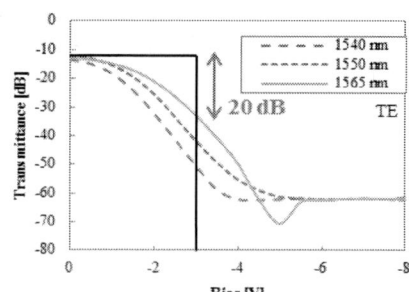

Fig. 6: EAM gate transmitted characteristics.

The 1×4 switching result of the fabricated switch is shown in Fig. 7, where a bright spot of light can be observed at the intended output port (see Fig. 7 (a)). The relationship between the EAM gate biasing and resulting loss is summarized in Fig. 7 (b). In this experiment 0.3 and 1.1 mA were injected into the MZIs at the first and second stages, respectively. The corresponding extinction ratio was measured to be 22 dB. Whereas by applying a bias of -3 V to the EAM gate, the extinction ratio was increased to 41 dB. Moreover, when a surface electrode was connected to a common ground, the optical intensity at the transmission port was further improved by 2 dB. Figure 8 plots the measured optical power for the 4 output ports of the switch when different currents injection supplied to the MZI. The two plotted curves for port II represent the measured output before and after using the corresponding EAM-gate array. In this set of experiments, the tandem operation of cascaded MZIs and EAM-gate array was demonstrated with a crosstalk of less than -40 dB. When an N×N optical switch is fabricated by using this configuration, the extinction ratio is doubled; therefore, the required swing voltage can actually be made smaller.

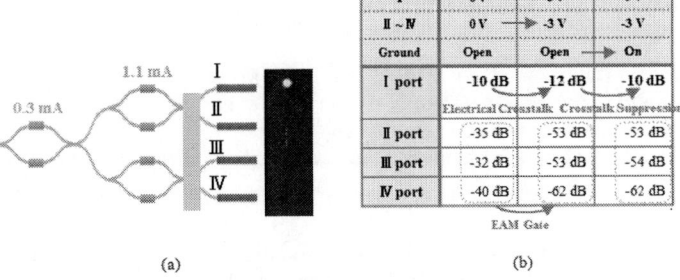

(a)   (b)

Fig. 7: (a) 1×4 optical switching operation. (b) Insertion loss dependence on EAM gate biasing and surface electrode grounding.

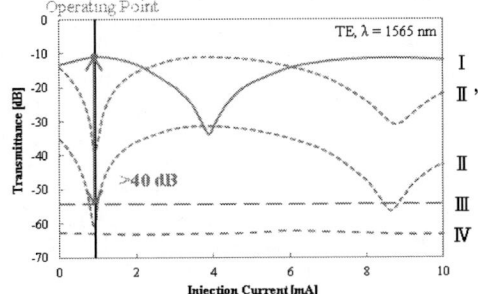

Fig. 8: 1×4 optical switch transmitted characteristics.

IV. CONCLUSIONS

We have described a low-loss and low-crosstalk integrated fat-tree optical switch based on cascaded MZIs and EAM-gate array, and confirmed the basic performance by fabricating a 1×4 optical switch. Low-loss switching was enabled by cascaded MZIs as a replacement for an MMI splitter. Moreover, the extinction ratio of the switch was improved by the EAM-gates. We have demonstrated a high extinction ratio of more than 40 dB, and a low-loss of 10 dB which was 5 dB lower than the conventional 1×4 B&S switch fabricated for the comparison. In addition, the electrical crosstalk was suppressed by a simple grounded surface electrode. For an N×N optical switch with a high extinction ratio, as it is composed of 1×N optical switch and N×1 optical switch, the required extinction ratio of the 1×N switch is only half the required value, and thus the required swing voltage can be made smaller. The low loss makes this integrated device promising as a large port-count optical switch.

ACKNOWLEDGMENT

This work was supported by the National Institute of Information and Communications Technology (NICT).

REFERENCES

[1] N. Calabretta, W. Miao, S. D. Lucente, J. Luo, and H. Dorren, "Scalable and low latency optical packet switching architecture for high performance data center networks," in Proc. Photonics in Switching (PS) 2014, PW4B.1.

[2] R. Takahashi, T. Segawa, S. Ibrahim, T. Nakahara, H. Ishikawa, A. Hiramatsu, Y. Huang, and K. Kitayama, "Torus datacenter network with smart flow control enabled by hybrid optoelectronic routers," J. Opt. Commun. Netw. 7(12), pp. B141-152, 2015.

[3] G. I. Papadimitriou, C. Papazoglou, and A. S. Pomportsis, "Optical switching," Wiley-Interscience, 2006.

[4] M-J. Kwack, T. Tanemura, A. Higo, and Y. Nakano, "Monolithic InP strictly non-blocking 8×8 switch for high-speed WDM optical interconnection," Proc. ECOC2012, Th.3.B.3.

[5] J. Gripp, M. Duelk, J. E. Simsarian, A. Bhardwaj, P. Bernasconi, O. Laznicka, and M. Zirngibl, "Optical switch fabrics for ultra-high-capacity IP routers," J. Lightwave Technol. Vol.21, No.11, p. 2839, Nov. 2003.

[6] K. Sone, S. Yoshida, Y. Kai, G. Nakagawa, G. Ishikawa, and S. Kinoshita, "High-speed 4×4 SOA switch subsystem for DWDM systems," Proc. 16th Opt-Electronics and Communication Conference (OECC), 776-777, 2011.

[7] T. Segawa, S. Ibrahim, T. Nakahara, Y. Muranaka, and R. Takahashi, "Low-power optical packet switching for 100-Gb/s burst optical packets with a label processor and 8×8 optical switch," J. Lightwave Technol., Vol.34, No.8, pp.1844-1850, 2016.

[8] Y. Muranaka, T. Segawa, Y.Ogiso, T. Fujii, and R. Takahashi, "Performance improved broadcast-and-select optical switch module based on EAM-gate array," IEEE Photonics Journal, to be published.

OECC/PS2016

Integrated InP Optical Switch Matrices Performance for Packet Data Networks

(Patty) Ripalta Stabile

COBRA Institute, Eindhoven University of Technology, Postbus 513, Eindhoven 5600MB, The Netherlands

Abstract: *The current status in InP integrated photonics for large-scale optical switch matrices is reviewed with regard to optical and electrical energy challenges. Projected performances are shown and strategies for improvements are discussed.*
Keywords: *Optical switching devices, photonic integrated circuits, large scale integration, optoelectronics*

I. INTRODUCTION

The incessant increase in capacity and connectivity in communication networks is not sustainable through packet routing and switching in the electronic domain [1]. This motivates research into hybrid solutions: photonic hardware is introduced for massive-bandwidth data transport and switching while routing is handled at a more modest GHz rates in the electronic domain. Planar integrated circuits offer an exciting opportunity to create single chip switching solutions for high capacity packet-compliant data routing [2], avoiding the high energy costs and delays associated with opto-electronic conversions and electronic de/serialization.

It is only recently that the technology has matured to the point that photonic circuits with sufficient numbers of components can be realised. InP integrated photonics offers high performance amplifiers, switches, modulators, detectors and de/multiplexers in the same wafer scale process. We have exploited SOA-based gate switches since they can enable passbands of several Terahertz, colourless and wavelength multiplexed routing [3]. Recently we have prototyped optical switching circuits using monolithic integration technology with up to several hundreds of integrated optical components per chip for high connectivity [4-5]. As the data throughput is not directly linked to the actuation energy, such broadband switches are expected to enable considerable energy savings with respect to electronic switching as the line-rate increases [6]. To date, the involved electrical input energy is approximately 16 W even for the all-active lossless 16×16 switch fabrics [7], but further electrical power reduction is desirable. New architectures and technologies must be conceived, which provide packet-scale switching functionality while being power efficient [8].

In this paper we discuss InP large-scale switch performances and describe critical aspects like noise figure and insertion losses, and we propose strategies for further performance improvement. We also explore complementary architecture based on resonant switching elements to offer a path to power efficient matrices.

II. INP SOA-BASED MATRICES: OPTICAL PERFORMANCES PROJECTION

We have exploited two different classes of SOA-base switching technologies. The first active-passive 16×16 three-stage SOA-based photonic space switch includes hundreds of components (Fig. 1): This can potentially route massive bandwidth signals independently of signal wavelength and in number of multiplexed wavelength channels [4]. Orders of magnitude increases in reconfigurability becomes possible when the use of wavelength division multiplexing is combined with on-chip wavelength selective routing: The 8×8 integrated cross-connect represents an elegant example of in plane circuit which provides a connectivity of up to 64 input and output channels [5].

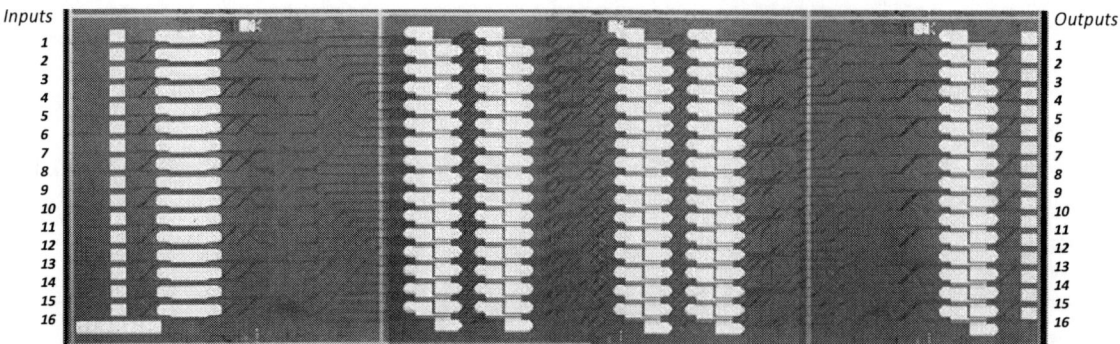

Figure 1. Composite image of the 16x16 space switching.

The use of low-loss passive waveguides in combination with high-contrast active gates leads to promising system level metrics in terms of optical signal to noise ratio of order of 28 dB/0.1nm in both classes of switches. However, while the optical loss figure in the cross-connect scheme is manageable because of the modest scale of integration, the space switch scheme presents far higher path losses, which derive from its moderate connectivity. The fiber-to-fiber signal losses increase as path length and complexity increase, because of the additional traversed components [4]. The inherent losses of about 30dB coming from the broadcast-enabled architecture are present in any given path, while the gain from the SOA gates can compensate much of the excess component losses. The excess losses can however be significantly reduced by using recent results obtained for single optimized components like mode-size adaption at the facets [9], low-parasitic-reflection splitters [10], low-loss waveguide crossing through a broadening in the intersecting waveguides [11], low waveguide losses of order 0.5 dB/cm minimizing p-doping level in the passive waveguides [12]. Such optimisations of component losses is shown to enable lossless circuit level operation for the 16×16 switch matrix and will also have a radical impact on optical signal to noise ratio, reducing it to about 49 dB/0.1nm. Fig. 2*a* and *b* show the projected performance for the same 16×16 switch matrix with optimised circuit elements.

The requirements for reduced energy use also impose the photonic components to be abstracted to enable tractable network level implementations. Multicast capability, packet-time-scale reconfigurability, power levelling irrespective of signal path, distributed signal quality monitoring and data rate transparency have been implemented at the chip level, to enable an abstracted digital control plane [13-15].

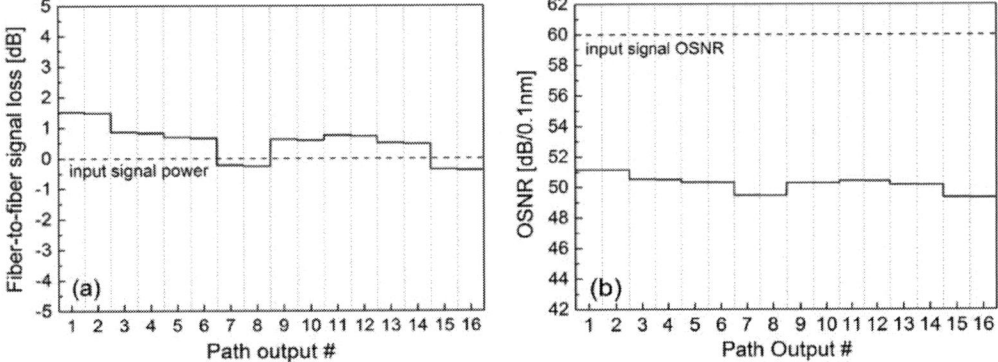

Figure 2: Projected performance for the broadcast and select 16x16 switching matrix as for (a) optical losses and (b) signal-to-noise ratio, for use of optimized components [3].

III. InP MRR-Based Matrices: Electrical Power Projection

Recently we have explored the use of photonic integrated circuits based on ring-resonators. By using InP integrated phase shifters for phase matching and switching activation, these switch matrices provides fast switching and low power consumption.

Prior work has focused on using first-order micro-ring resonators for this purpose [16-17], but higher-order resonators have been shown to decouple bandwidth and the signal extinction implying that both parameters can be optimised individually. As an added incentive, the enhanced bandwidth of such devices also increases their tolerance to variations in fabrication [18]. We have now designed and fabricated the first 4×4 InP switch matrix based on third-order resonators [8, 19]. The matrix is made of sixteen switching elements which are based on third-order ring resonators. The coupler regions are defined by close-spaced phase modulators which are electro-optically tuned. Fig. 3*a* shows a photo of the entire switch matrix.

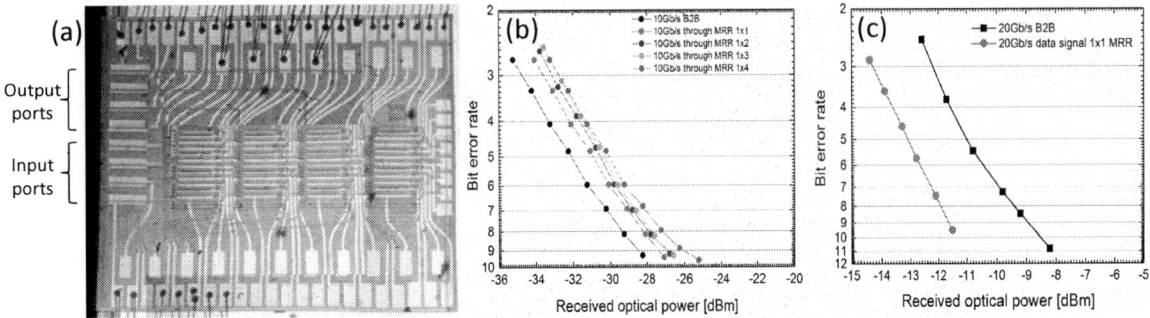

Fig. 3. (a) Composite optical microscope image of the full fabricated chip. (b) 10 Gb/s and (c) 20Gb/s BER curves.

The desired coupling coefficients are determined by varying the gap separating the directional couplers, while keeping fixed the coupling lengths. The total chip size is of about $4.5 \times 3.8\text{mm}^2$ and includes 500µm long SOAs placed after each input-output port to compensate for coupling losses and for power equalisation at the optical outputs. On-state on-chip losses for the single switching element are of order of 6dB. Techniques to further reduce the on-chip losses are currently being investigated: the use of high resolution lithography is being explored to enable higher density and lower on-chip losses. This is believed to enable further scaling up to the same order as for SOI ring-resonators based cross-point matrices [18]. Each switch element can switch by consuming a maximum power of only 150 μW. Data routing at 10Gbps and 20Gbps is also enabled (Fig. 3*b* and *c*) with a minimum power consumption of 30 fJ/b/s.

IV. Conclusions

Rapid recent developments in the complexity and functionality of monolithically integrated optical switch fabrics offers considerable promise for next generation networking components. Monolithic circuits with hundreds of components, tens of connections, and packet-reconfigurability are now demonstrated: Circuit connectivity has scaled to 16 ports. Further connectivity up to 64 channels is achievable with on-chip wavelength routing. However a carefully design of the amplifier chain together with loss compensation is needed. Active-passive integration and component optimizations will then lead to further electrical power reductions with lossless low-noise circuit level operation. However, further integration demands for new components which enable higher operation efficiencies. This is the case of the new cross-point architecture matrices based on higher-order ring resonators, which are voltage-driven and therefore potentially power efficient. Techniques to reduce on-chip losses are currently being explored: the use of DUV lithography is thought to enable higher density and lower losses, and therefore further scaling at low power consumption.

Acknowledgment

The work has been supported by the Dutch technology research agency STW. I thank prof. Kevin Williams for valuable discussions on large-scale integration approaches.

References

[1] "The zettabyte era: Trends and analysis", Cisco White Paper, 2015.
[2] C. Doerr, "Planar lightwave devices for WDM," in Optical Fiber Telecommunications IV, A Components, I.P. Kaminov, T. Li, ed., 402-477, 2002.
[3] R. Stabile, A. Albores-Mejia, A. Rohit and K.A. Williams, "Integrated optical switch matrices for packet data networks", Microsystems and Nano-Engineering, Nature Publishing Group, 2, 15042, (2016). Invited
[4] R. Stabile, A. Albores-Mejia, and K. A. Williams, "Monolithic active-passive 16×16 optoelectronic switch", Optics Letters, 37, 22, 4666-4668, 2012.
[5] R. Stabile, A. Rohit, K.A. Williams, "Monolithically integrated 8 × 8 space and wavelength selective cross-connect," Journal of Lightwave Technology, 32, 2, 201-207, 2013.
[6] A. Albores-Mejia, F. Gomez-Agis, H.J.S. Dorren, X.J.M. Leijtens, M.K. Smit, D.J. Robbins and K.A. Williams, "320Gb/s data routing in a monolithic multistage semiconductor optical amplifier switching circuits", ECOC, Invited paper We7E1, 2010.
[7] H. Wang, A. Wonfor, K.A. Williams, R.V. Penty and I.H. White, "Demonstration of a lossless monolithic 16x16 QW SOA switch", ECOC 2009, 1-2.
[8] R.Stabile, P. Dasmahapatra and K.A.Williams, "First 4×4 InP Switch Matrix Based on Third-Order Micro-Ring-Resonators", OFC 16, Th1C.3.
[9] L.N. Langley, D.J. Robbins, P.J. Williams, T.J. Reid, I. Moerman, X. Zhang, P. Van Daele, P. Demeester, "DFB laser with integrated wave-guided taper grown by shadow masked MOVPE", Electronics Letters, 32, 8, 738-739, 1996.
[10] E. Kleijn, D. Melati, A. Melloni, T. de Vries, M.K. Smit, and X.J.M. Leijtens, "Multimode interference couplers with reduced parasitic reflections", Photonics Technology Letters, 26, 4, 408-410, 2014.
[11] H. G. Bukkems, C. G. P. Herben, M. K. Smit, F. H. Groen, I. Moerman, "Minimization of the loss of intersecting waveguides in InP-based photonic integrated circuits", Photonics Technology Letters, 11, 1420-1422, 1999.
[12] D. D'Agostino, G. Carnicella, C. Ciminelli, H.P.M.M Ambrosius and M.K. Smit, "Low loss waveguides for standardized InP integration processes", Proceedings IEEE Photonics Society Benelux, 2014.
[13] R. Stabile, A. Rohit, K.A. Williams, "Dynamic multi-path WDM routing in a monolithically integrated 8 x 8 cross-connect", Optics Express, 22, 1, 435-442, 2014.
[14] Q. Cheng, R. Stabile, A. Rohit, A. Wonfor, R.V. Penty, I.H. White, K.A. Williams, "First demonstration of automated control and assessment of a dynamically reconfigured monolithic 8 × 8 wavelength-and-space switch", Journal of Optical Communications and Networking, 7, 3, A388-A395, 2015.
[15] R. Stabile and K.A. Williams, "Optical power meter co-integrated with a fast optical switch for on-chip OSNR monitoring", proceedings Photonics in Switching, 2015.
[16] R. Guzzon, E. Norberg, J. Parker, L. Johansson, and L. Coldren, "Integrated InP-InGaAsP tunable coupled ring optical bandpass filters with zero insertion loss," Opt. Express, 19, 7816-7826 (2011).
[17] R. Ji, L. Yang, L. Zhang, Y. Tian, J. Ding, H. Chen, Y. Lu, P. Zhou, and W. Zhu, "Five-port optical router for photonic networks-onchip", Opt. Express, vol. 19, no. 21, pp. 20 258-20 268, 2011.
[18] P. Dasmahapatra, R. Stabile, A. Rohit, K.A. Williams, "Optical cross-point matrix using broadband resonant switches", IEEE Journal of Selected Topics in Quantum Electronics, 20(4):5900410, 2014.
[19] R. Stabile, P. DasMahapatra, K.A. Williams, "Fast and energy efficient Micro-ring-resonator-based 4x4 InP Switch matrix", ECIO 2016, accepted.

WF3-4

OECC/PS2016

1310nm High-capacity Waveband Switch Node for Flat Optical Data Center Networks

Wang Miao, Huug de Waardt, and Nicola Calabretta

COBRA Research Institute, Eindhoven University of Technology, Eindhoven, the Netherlands
w.miao@tue.nl

Abstract: *Waveband switching capability of 1310nm SOA-based scalable and flow-controlled optical switching node is assessed. Experimental results show nanoseconds operation of 64-port switch node at 100 Gb/s WDM (4×25Gb/s) with >14dB dynamic range and <1.8dB penalty.*

Keywords: *Optical switch, data center network, wavelength division multiplexing*

I. INTRODUCTION

The increasing demand for faster device connectivity and diverse data-intensive applications has dramatically boosted the cloud-based IP traffic. As the major storage and computing resources, data centers (DCs) are experiencing the rapid speed upgrade of the servers from 1/10 Gigabit Ethernet (GbE) to 100GbE links in the coming years [1]. Considering the multi-Tb/s aggregation bandwidth at each top-of-the-rack (ToR) switch interconnecting 40+ servers, the DC network should be capable to handle Petabit/s inter-ToR communication traffic. This necessitates innovative optical interconnects and switching fabrics architecture. Compared with commonly used VCSEL-based multi-mode solution, single-mode (SM) optical interconnects at 1310nm are preferred when scaling the data rate (>25Gb/s) and the reach (>2Km) due to the superior bandwidth-distance product, future-proof scalability by exploiting wavelength division multiplexing (WDM), and close-to-zero dispersion [2]. A number of efforts exploring high capacity 1310 nm WDM interconnects have accelerated the migration beyond 1Tb/s bandwidth for DC interconnects applications [3].

With the pavement to Tb/s interconnecting solutions, more stringent challenges are exposed to the switching fabrics. In particular, to efficiently group several tens of Tb/s links, large port count electrical switch is limited by the switch ASIC I/O bandwidth (to roughly 8 Tb/s) due to the scaling issues of the ball grid array packages [4]. Larger-capacity can be realized in a hierarchical structure at the expense of high latency and power consumption caused by O/E/O converters. In addition, fueled by the demand for higher speed, it becomes beneficial for transmitters employing WDM signals to further scale the density, reach and data rates. Dedicated transceivers integrating parallel optics need to be further included as switch front-ends, which contribute unnecessarily to a larger latency and power consumption.

These scaling issues have justified the investigation of high capacity 1310nm optical switches. The transparency to data rate and format prevents the opaque O/E/O conversion, complex front-end interfaces and the I/O bandwidth bottleneck of the electrical switches. A 1310nm flow-controlled OPS architecture with sub-wavelength granularity and nanoseconds switching latency has been proposed [5]. The high bandwidth and scalable radix of the switch enable flat DC networking. However, the performance was only preliminarily investigated for single wavelength payload. Foreseeing the advancement in WDM interconnect technologies, 100GbE solutions containing WDM 4×25Gb/s lanes (e.g. QSFP28, and CFP4) will be brought into DCs aggressively to boost the capacity. The adaptability of the 1310nm optical switch to such traffic pattern in combination with the port-count scaling is therefore essential and should be properly addressed.

In this work, we perform an assessment of waveband switching capability of the 1310nm flow-controlled OPS architecture by employing 100 Gb/s WDM (4×25Gb/s) traffic. Results show error-free nanoseconds switching with < 1 dB penalty for all WDM channels. With the assistance of the pre-amplifier SOA, power loss due to the broadcasting and passive components has been fully compensated and the port-count scaling up to 64 ports with >14 dB dynamic range have been achieved, which proves the suitability for high-capacity and high-connectivity DC networks.

II. THE OPTICAL SWITCH OPERATION

The investigated 1310nm N×N fast OPS system for flat DC network is shown in Fig. 1. The traffic generated from the servers is grouped by the ToRs and transferred through the OPS switching node. At each ToR, the aggregation controller generates the binary label bits for the inter-ToR traffic (payload packet) which are coded into RF tones carried by the in-band optical label wavelength [5]. A copy of the transmitted packet is electrically stored in case of retransmission due to contention notified by the fast optical flow control between the OPS and the ToRs [5].

The OPS node consists of N identical modules each operating independently. It allows for a highly distributed control which makes the reconfiguration time of the switch port-count independent. This is especially important to ensure low latency as the switch scales to large radix. WDM packets are de-multiplexed and processed in parallel. The label is extracted by a narrow-band filter such as a fiber Bragg grating (FBG), and then processed by the label processor (LP). The recovered digital label bits are detected by the switch controller, according to which the SOA-based broadcast & select 1×N switch is configured. Payload is transparently forwarded to one of the OPS output ports by the 1×N

OECC/PS2016

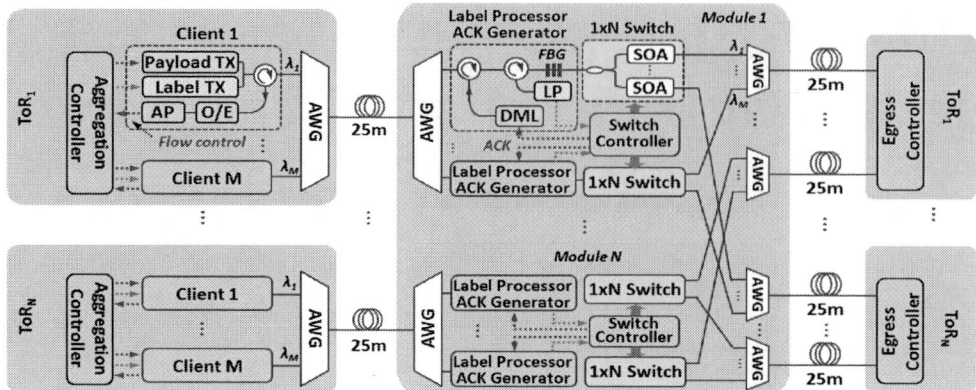

Fig. 1. Architecture of the 1310nm flow-controlled OPS switching node

switch. Here the O-band SOA has been used as the selecting gate. It can compensate the broadcasting loss and allows for waveband switching of the WDM traffic benefitted from the wideband operation. In case of contention, the packets with lower priority are blocked and an optical flow control signal (negative ACK) is sent back to the ToR side for packet retransmission. A low-speed directly-modulated laser (DML) at 1310 nm driven by the switch controller is used to generate the flow control signal. The nanoseconds switching time of the SOA in combination with the parallel processing of the label allows for port-count independent reconfiguration time at a few tens of nanoseconds scale.

III. EXPERIMENTAL ASSESSMENTS

Benefitted from the parallel modular architecture shown in Fig. 1, the performance assessment of the overall switching system can be investigated by considering one of the 1×N optical switch modules. Scaling the OPS is mostly limited by the splitting loss experienced by the payload during the 1×N broadcast and select stage. We experimentally assess the port-count scalability and performance of the optical switch for waveband WDM 4×25Gb/s NRZ-OOK traffic by using the experimental setup shown in Fig. 2. The broadcasting splitter is emulated by a variable optical attenuator (VOA) and another SOA has been added before the 1×N optical switch as pre-amplifier of the input signal. The use of the pre-amplifier SOA is twofold: it provides sufficient gain to compensate the losses when WDM channels are assessed. Secondly, it improves the optical dynamic range of the 1×N switch. The pre-amplifier SOA can be potentially photonic integrated with the optical switch to further lower the footprint and power consumption [6].

To generate the 100Gb/s multi-λ WDM waveband signal, two directly-modulated lasers (DMLs) and two electroabsorption-modulated lasers (EMLs) with 10GHz bandwidth are used. The wavelengths are allocated at 1302.20 nm, 1303.86 nm, 1307.74 nm and 1309.43 nm respectively. The four lasers are driven by the 25Gb/s PRBS 2^{11}-1 NRZ-OOK packetized payload from the bit pattern generator (BPG) and then de-correlated to generate the waveband 4×25Gb/s data. It is then fed into the emulated 1×N optical switch controlled by the FPGA switch controller after power adjusted by a VOA. The bias currents of the pre-amplifier SOA and the gate SOA have been varied to optimize the signal quality. At the output port, each wavelength channel is filtered out by an optical band-pass filter (OBPF) and then received by the 40Gb/s photo-receiver.

Figure 3(b) shows the optical spectra of the input waveband signal and the switched outputs with different bias current of the gate SOA (without pre-amplifier SOA). The total input power is 0 dBm and no obvious nonlinear effects have been observed. The BER curves and eye diagrams of the four wavelengths for a 1×4 optical switch configuration (6dB splitting loss) are presented in Fig. 3(a). Compared with B2B, error-free operation has been achieved for all wavelengths with less than 0.7 dB power penalty.

The pre-amplifier SOA is then added for the investigation of the optical power dynamic range as shown in Fig. 2. The splitting loss caused by the VOA is 9 dB, 12 dB, 15 dB, and 18 dB to emulate the scale of 1×8, 1×16, 1×32, and

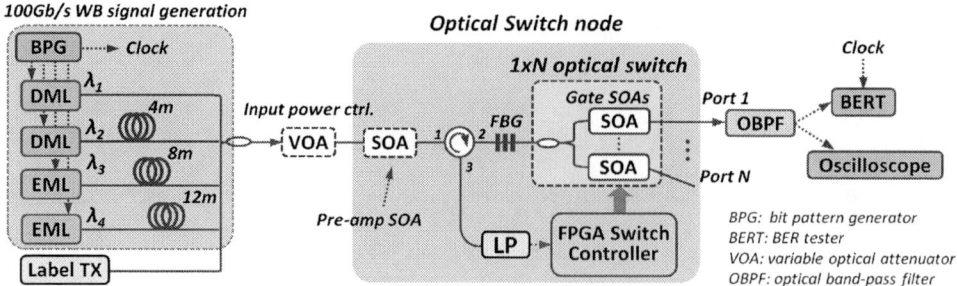

Fig. 2. Experimental set-up employing the emulated 1×N optical switch

1204

OECC/PS2016

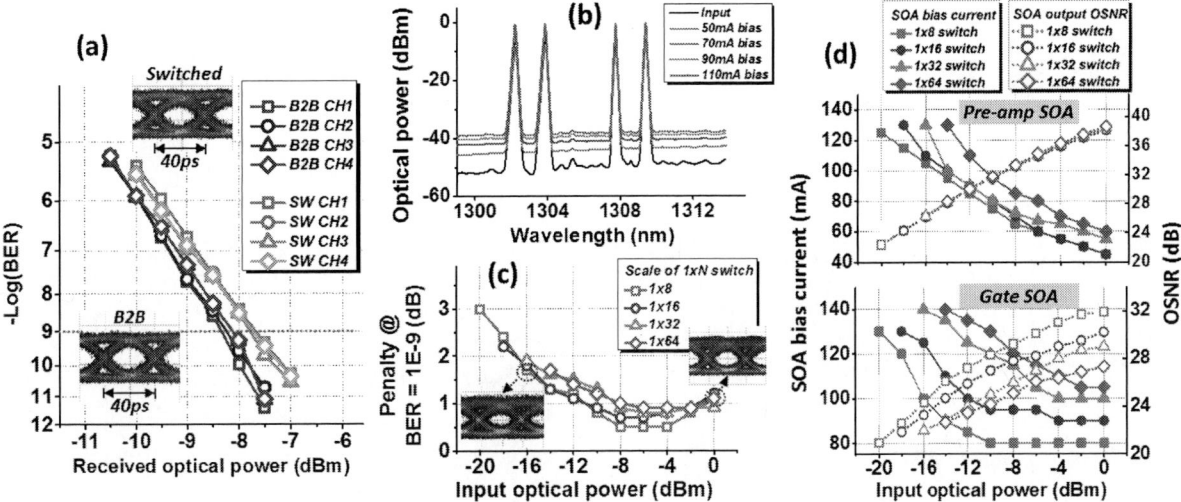

Fig. 3. (a) BER curves and eye diagrams; (b) optical spectra; (c) power penalty vs. input optical power; (d) SOA bias current and output OSNR

1×64 optical switch, respectively. The power penalty measured at BER = 1×10⁻⁹ when varying the input optical power between -20 dBm and 0 dBm is plotted in Fig. 3(c). The optimal input optical power is around -5 dBm where < 1dB power penalty is observed for all cases. Both lower and higher input power results in performance degradation mainly due to the noise and non-linear effect in the SOA, respectively. The loss caused by the broadcasting and passive components are compensated (even with amplification) by the two SOAs. 14 dB dynamic range is achieved with less than 1.8 dB power penalty for scale of 1×64. Error-free operation for lower input is limited by the inadequate power level at the switch output.

Figures (d) depicts the optimal bias current and the output OSNR for both SOAs. For pre-amplifier SOA, lower input power requires higher current to boost the signal and thus avoiding further OSNR degradation due to the low optical power at the gate SOA. With the same input optical power, higher amplification is needed at the pre-amplifier SOA for larger scale to pre-compensate the loss caused by broadcasting stage. Apart from forwarding/blocking the packets, the gate SOA also provides amplification especially for lower input power and larger scale (higher splitting loss). The OSNR of gate SOA output also confirms the deterioration of the signal caused by the broadcasting stage.

Considering the potential for photonic integration, adding the pre-amplifier SOA for each channel would be beneficial to improve the dynamic range of the switch. The bias current of the pre-amplifier SOA and gate SOA can be dynamically controlled to adapt to the input power. The assessments presented here serve as references for properly defining the SOA parameters. Better gain profile and noise figure could further increase the dynamic range and lower the bias current with limited power penalty.

IV. CONCLUSIONS

The waveband switching capability of the scalable 1310nm optical switching system for DC network has been assessed with 100Gb/s WDM (4×25Gb/s) traffic. Benefitted from the wide-band operation, loss compensation and nanoseconds switching time enabled by the SOA, error-free operation with less than 1 dB power penalty has been achieved. Scaling up to 64-port shows an input dynamic range of > 14 dB with limited 1.8 dB penalty which confirms the suitability of the optical switch for handling high-capacity multi-wavelength traffic in next-generation DC networks.

ACKNOWLEDGMENT

The authors would like to thank FP7 COSIGN project (NO. 619572) for supporting this work.

REFERENCES

[1] The 2015 Ethernet Roadmap, Ethernet Alliance white paper, 2015.
[2] R. Shubochkin et al, "Trends in Datacom optical links," in *Proc. 62nd Int. Cable Connectivity Symp.*, pp. 633-642, 2013.
[3] B. R. Koch et al, "Integrated silicon photonic laser sources for telecom and datacom," Paper PDP5C.8, OFC 2013.
[4] A. Ghiasi, "Large data centers interconnect bottlenecks," *Opt. Express*, vol. 23, no. 3, pp. 2085-2090, 2015.
[5] W. Miao et al, "1.3 µm SDN-enabled Optical Packet Switch Architecture for High Performance and Programmable Data Center Network," Paper Th2A.66, OFC 2015.
[6] N. Calabretta et al, "Monolithically Integrated WDM Cross-Connect Switch for Nanoseconds Wavelength, Space, and Time Switching," Paper Mo.3.2.2, ECOC 2015.

AUTHOR INDEX

Abbott, John 603
Abe, Jun-Ichi 489, 516
Abedin, Kazi S. 326
Absil, P. 1104
Ackert, J. J. 622
Adachi, Fumiyuki 154
Adachi, Koichiro 1113
Agata, Akira 284
Aikawa, Kazuhiko 39, 45
Aikawa, Yohei 130
Akahane, K. 380
Akasaka, Y. 537
Akiba, Shigeyuki 443
Akimoto, Ryoichi 1173
Akiyama, Yuichi 489
Alferness, Rod C. 121
Alishahi, F. 537
Almaiman, A. 537
Alonso-Ramos, C. 622, 1149
Altabas, Jose A. 688
Amann, Markus-Christian 347
Amemiya, Tomohiro 642, 1119
Amezcua-Correa, Adrian 600
Amino, Kenta 832
Amma, Yoshimichi 39, 45
Ando, Makoto 443
Andrejew, Alexander 347
Andrekson, Peter A. 507
Andriolli, N. 395
Annoni, Andrea 88, 407
Aoki, Yasuhiko 254, 462, 477, 498
Aono, Yoshiaki 142
Arai, Shigehisa 642, 1119
Arakawa, Taro 383, 1146
Arakawa, Yasuhiko 148, 559
Arao, Hajime 627
Arikawa, M. 582
Aruga, Hiroshi 1131
Asaka, Kota 676
Asakura, Hideaki 712
Asakura, Keita 955
Asano, Takashi 1167
Asobe, Masaki 1152, 1158
Asuka, Shun 1065
Aupetit-Berthelemot, Christelle 576
Awaji, Y. 48, 655, 658
Aya, Hironori 579
Azuma, Yoshiyuki 949
Baba, Toshihiko 1164
Bae, S. H. 275, 449
Baeuerle, B. 30
Bai, Wei 874, 1042
Balakrishnan, S. 1104
Barré, Nicolas 181
Bartalini, Saverio 567

Bauwelinck, Johan 389
Beausoleil, R. G. 377, 1137, 1191
Bekkali, Abdelmoula 446
Benazet, Benoit 576
Benedikovic, D. 622, 1149
Ben-Ezra, S. 30
Berciano, M. 1149
Bergman, Keren 196
Bernier, Eric 1188
Betoule, C. 30
Bex, Thomas 486
Bigot-Astruc, Marianne 600
Blau, Miri 718
Blown, Patrick 124
Bo, Tianwai 528
Boehm, Gerhard 347
Bofang, Zheng 495
Bowers, John E. 121
Bradley, T. 225
Brenot, R. 931, 1116
Buchali, Fred 362
Cai, Yufeng 820
Cai, Zhongle 510
Calabretta, Nicola 208, 1203
Calabrò, Stefano 486
Cao, Bingyao 793
Cao, Xiaoyuan 236, 673, 1039
Cao, Y. 537
Cao, Zheng 661
Carminati, Marco 407
Carpenter, Joel 597
Carrier, Pilot 272
Casellas, Ramon 1036
Cassan, E. 422, 1149, 1170
Cavaliere, Fabio 724
Celo, Dritan 1188
Chan, C. C.-K. 528, 784
Chan, Cheng-Ta 940
Chan, Chun-Kit 468, 802
Chan, Erwin H. W. 573
Chan, Vincent W. S. 440
Chang, Chun-Yen 591
Chang, C-M. 1116
Chang, Gee-Kung 169
Chang, Hung-Ying 919
Chang, Kuo-Chun 1134
Chang, Wei-Kang 1074
Channa, Hin 1152
Cheben, P. 622
Chen, Bo-Rui 778
Chen, Chin-Hui 1137, 1191
Chen, Gang 236, 673
Chen, H. 187
Chen, Hongwei 103, 811
Chen, Huizhong 850

AUTHOR INDEX

Chen, Hung-Yu 781
Chen, Jiageng 425
Chen, Jianping 79
Chen, Jyehong 591, 859
Chen, Kaige 338
Chen, Lian-Kuan 588
Chen, Minghua 103, 811
Chen, Nan-Kuang 895, 910
Chen, Oscal T.-C. 940
Chen, Rongrong 754
Chen, Xi 817
Chen, Xin 603
Chen, Xin-Nan 591, 859
Chen, Xue 760
Chen, Y. 225
Chen, Y. K. 931, 1116
Chen, Yanxu 760
Chen, Yuan 1155
Chen, Zhangyuan 311, 715
Cheng, Gia-Ling 895
Cheng, Haiquan 850
Cheng, Lin 169
Cheng, Mengfan 260
Cheng, Wood-Hi 594, 895, 910
Cheng, Xiaoyang 1030
Cheng, Zhenzhou 416
Cheung, Jack 547
Chew, Suen Xin 703
Chi, Kai-Lun 591, 859
Chi, Yu-Chieh 585, 594, 976
Chiang, Jung-Sheng 919
Chiang, Kin Seng 630, 648
Chiba, Hisashi 419
Chida, Yasuyuki 51
Chiesa, M. 395
Chiu, Yi-Jen 1134
Cho, Seong-Ho 964
Cho, Sungmin 163
Choi, Woo-Young 964
Chou, Hsi-Hsir 618
Choudhary, Amol 139
Chow, C. W. 685
Chow, Chi-Wai 781
Chu, Ann-Kuo 67
Chu, Daping 97
Chu, Sai T. 401, 547, 736
Chuang, Chun-Yen 859
Chujo, Wataru 988
Chulok, Pacharapon 748
Chung, Hung-Ching 991
Chung, Kun-Lung 865
Chung, Y. 175
Chung, Y. C. 275, 449
Cincotti, G. 106, 109, 112, 609
Clarke, Ian 124

Coleman, Doug 603
Cong, Guangwei 1110
Contestabile, G. 127, 395
Coudyzer, G. 389
Cui, Yue 233, 510
Dahlem, Marcus S. 1009
Dai, Daoxin 410
Dai, Songyuan 296
Dai, Yitang 706
Dalton, L. R. 413
Damas, P. 1149
Daniel, Luca 1021
Dat, Pham Tien 763
De Heyn, P. 1104
De Man, Erik 486
De Natale, Paolo 567
De Valicourt, G. 931, 1116, 1194
De Waardt, Huug 1203
De Wolf, I. 1104
Deng, Lei 260, 814
Deng, Ning 308, 468
Denolle, Bertrand 181
Descos, Antoine 1191
Di Lauro, Luigi 401
Dias, M. P. I. 459
Ding, Huixia 874
Doerr, Chris 3
Dohi, Keisuke 522
Dong, P. 624, 1116
Dong, Shuai 1015
Dou, Liang 492, 501
Downie, John D. 142
Du, Jiangbing 1077
Du, Xiaoen 880
Duan, Xiaofeng 973, 982
Dumais, Patrick 1188
Dun, Han 278, 793
Edwards, S. 133
Effenberger, Frank J. 266
Eggleton, B. 139, 597
Eiselt, Michael H. 251
Elder, D. L. 413
Ellis, A. 30
Elschner, Robert 27
Endo, Tatsuro 544, 880
Eriksson, Tobias A. 507
Evans, P. 133
Fàbrega, Josep M. 302
Fan, Sujie 823
Fan, Xinyu 425, 1086, 1089, 1101
Fan, Yuting 706
Fang, Jian 323
Fang, Wenjing 982
Fehenberger, Tobias 507
Fei, Jiarui 973

AUTHOR INDEX

Feiste, Uwe 486
Feng, Da 745
Feng, Jijun 1173
Feng, Xinhuan 573
Feng, Xue 636
Feng, Zhenhua 814, 817
Fengkai, Bian 946
Ferran, J. F. 30
Ferrari, Giorgio 407
Fiorentino, Marco 377, 1137, 1191
Foggi, Tommaso 118
Fokoua, E. N. 225
Fontaine, N. K. 187
Fresi, Francesco 118, 724
Freude, W. 413
Fu, Meixia 510
Fu, Songnian 260, 814, 817, 877, 952, 1000
Fu, Teng-Wei 922
Fu, Xin 907, 1083
Fujii, Shohei 670
Fujii, Takuro 612
Fujii, Yusuke 531
Fujimura, Yuki 697
Fujisawa, Ryohei 1158
Fujisawa, Takeshi 51, 633
Fujita, Sadao 516
Fujiwara, Naoki 1122
Fukai, Chisato 317
Fukano, Hideki 916
Fukuchi, K. 516, 582
Fukui, Takayoshi 281
Fukushima, Akira 667
Fukutoku, Mitsunori 290, 1182
Fumagalli, A. 133
Furukawa, Hideaki 392, 404, 655
Furuya, Kotoko 443
Galili, Michael 136
Gao, Mingyi 889
Gao, Shuang 802
Gao, Tao 691
Garces, Ignacio 688
Geng, Dongyu 1188
Ghosh, Samir 1143
Goi, Kazuhiro 1107
Gonda, Tomohiro 42
Gong, Xiaoxue 263
Goodwill, Dominic J 1188
Goto, Nobuo 115, 733, 925
Grani, Paolo 661
Grillanda, Stefano 88, 407
Gu, Wanyi 691, 808
Gu, Xiaodong 73, 76
Guan, Bai-Ou 573
Guan, Pengyu 136, 790
Guan, Xun 784

Gubenko, Alexey 1137
Guglielmi, Emanuele 407
Gui, C. 1116
Guo, Bingli 691, 808
Guo, Hongxiang 236, 673, 721
Guo, Lei 263, 904
Guo, Qiang 103
Gurusamy, Mohan 862
Halir, R. 622
Hamamoto, Kiichi 100, 564, 651, 1030
Hamaoka, Fukutaro 368
Han, Il-Ki 642
Han, Pin 922, 1074
Han, Sang-Kook 868
Han, Sangyoon 91
Han, Won-Taek 1080
Han, Young-Geun 553, 556
Hanafuji, Fumiki 544, 880
Hanawa, Masanori 296
Hanzawa, Nobutomo 54, 633
Harai, H. 133, 658, 709
Harako, Koudai 12
Haruki, Jun 615
Hasebe, Kazuhiko 341
Hasebe, Koichi 1122
Hasegawa, Hideaki 1128
Hasegawa, Hiroshi 428, 437, 1185
Hasegawa, Kiyotomo 1131
Hasegawa, M. 106, 112, 609
Hashimoto, Toshikazu 94, 368, 1182
Hatori, Nobuaki 148
Hattori, K. 106, 112, 239, 248, 609
Hayashi, Michiaki 1051
Hayashi, Neisei 341, 1068, 1095, 1098
Hayashi, Tetsuya 178
Hayashi, Yusuke 642
Hayashiguchi, S. 985
Hayes, J. R. 225
Hazama, Masaya 949
He, Jiale 260, 814
He, Jifang 1188
He, Zuyuan 425, 1077, 1086, 1089, 1101
Hicks, D. 133
Higuchi, Daiki 531
Hillerkuss, D. 30
Himbele, Luke 100, 1030
Himeno, A. 106, 112, 609
Hirai, Riu 281
Hirano, Akira 368
Hiraoka, M. 112, 742
Hirasawa, Takayoshi 443
Hiratani, Takuo 1119
Hirayama, Takahiro 709
Hirayama, Tomoki 1146
Hirohata, Toru 567

AUTHOR INDEX

Hirokawa, Jiro 443
Hirooka, Toshihiko 12, 33
Hirota, Yusuke 667, 670
Hisata, Yudai 826
Ho, Victor 547
Homemoto, Toru 245
Hong, Bingzhou 564
Hong, Seungjoo 163
Hong, Xiaobin 1092
Hong, Yang 588, 784
Hori, Kento 383
Horikoshi, Kengo 368, 489
Hoshi, Takuya 374
Hoshida, Takeshi 27, 477, 489, 492, 501, 513
Hoshino, Hiroki 943
Hosokawa, K. 582
Hosokawa, Tsukasa 1062
Hou, Weigang 904
Hou, Yinan 799
Hsieh, Chang-I 1033
Hsu, C. W. 685
Hsu, Chin-Wei 781
Hsu, Feng-Cheng 67
Hsu, Y. 685
Htein, Lin 928
Hu, Hao 136
Hu, Qian 431
Hu, Qikai 257, 841
Hu, Weisheng 739, 745
Hu, Weiwei 715
Hua, Bingchang 760
Huang, Guoxiu 498
Huang, Jian Jang 594
Huang, Liangkai 1057
Huang, Lingchen 802
Huang, Pi-Ling 895
Huang, Po-Chia 919
Huang, Shanguo 691, 808
Huang, Sheng-Lung 895
Huang, Tianye 952
Huang, Wei-Jhih 865
Huang, Yi-Chung 895
Huang, Yidong 636, 1015
Huang, Yongqing 973, 982
Huang, Yue-Cai 664
Huang, Yue-Kai 142, 145
Huang, Yu-Fang 585
Huang, Zhihong 377
Hung, Hung-Wen 679
Huo, Xiaoli 431
Hurley, Jason 142
Ibrahim, Salah 205, 664
Ichii, Kentaro 1062
Idler, Wilfried 362
Igarashi, Koji 190, 525

Igarashi, Shota 60
Iida, Mamoru 898
Iiyama, Koichi 967
Ikeda, Kazuhiro 82, 151
Ikeda, N. 985
Ikeuchi, T. 537
Ikeuchi, Tadashi 477, 504, 1048
Ikuma, Yuichiro 1182
Imajuku, Wataru 434
Imamura, Katsunori 42
Imansya, Ryan 1030
Imansyah, Ryan 100
Inaba, Takahiro 1012
Inada, Yoshihisa 142
Inafune, Koji 1027
Inoshita, Kensuke 733
Inoue, Daisuke 1119
Inoue, Gen 335
Inoue, Masaaki 471
Inoue, Shunya 70
Inoue, T. 18, 21, 139
Inoue, Yoshiaki 937
Inui, Tetsuro 434
Ip, Ezra 142, 145
Isaji, Y. 133
Ishihara, Hiroki 1107
Ishii, Hiroyuki 350, 1122
Ishii, K. 133, 1185
Ishii, Yuzo 1182
Ishikawa, T. 437, 582
Ishikura, Norihiro 1107
Ishizaka, Yuhei 633
Ito, Fumihiko 57
Ito, Kazuto 642
Ito, T. 582
Itoh, Mikitaka 1122, 1182
Iwamoto, Satoshi 148
Iwashita, Katsushi 970
Izawa, Tatsuo 211
Izquierdo, David 688
Izutsu, Masayuki 386, 1140
Jang, Bongyong 148
Jasion, G. T. 225
Jayasundara, Chamil 465
Jeong, Seok-Hwan 645
Jepsen, Peter U. 136
Ji, P. 145, 455, 1054
Ji, Yuefeng 242, 775
Jia, Shi 136
Jian, Pu 181
Jiang, Haisong 100, 564, 651, 1030
Jiang, Jia 1188
Jiang, Kai 877
Jiang, Wen 835, 847
Jiang, Xiaoshun 359, 1155

AUTHOR INDEX

Jiang, Xinyue 1083
Jin, Li 401, 547
Jin, Wei 648
Jin, Zhiqiang 359
Jing, Ruiquan 431
Jo, Ik Su 556
Johannisson, Pontus...507
Josten, A. 30
Ju, Seongmin 1080
Kai, Yutaka 302
Kakegawa, N. 133
Kako, Satoshi...148
Kam, Pooi-Yuen 257, 862
Kamada, Naoki..561, 958
Kametani, S. 133, 489
Kamikawa, Kosuke 766
Kamio, Yukiyoshi..519
Kanai, Shunsuke 682
Kanakubo, W. 85
Kanamori, Hiroo 213
Kanazawa, Shigeru 350, 1122
Kang, Joonhyun 642
Kani, Jun-Ichi 458
Kanke, Tomokazu 561
Kanno, A. 24, 380, 386, 579, 763, 766, 1140
Karanov, B. 112
Karinou, Fotini 287
Karlsson, Magnus 474, 507
Kasai, Keisuke 33, 305
Kashima, K. 380, 446
Katagiri, T. 133
Katayama, Masaru 239, 245, 248
Kato, Kazutoshi..615, 697
Kato, Tomoyuki 27
Katoh, Akira..1158
Katsuyama, T. 985
Kawabata, Yuto 1143
Kawaguchi, Naoki 383
Kawaguchi, Yu 365
Kawahara, Hdeaki 787
Kawahara, Hiroki 290
Kawai, Shingo 290
Kawanishi, T. 380, 386, 579, 763, 1140
Kawashima, Hitoshi.....................................82, 151
Khanna, Ginni 486
Khilo, Anatol 1009
Khokhar, A. Z. 622
Khope, Akhilesh S. P. 121
Kihara, Mitsuru 317
Kikuchi, Kazuro 1060
Kikuchi, Nobuhiko 281
Kikuchi, Takahiro 751
Kim, B. G. 175
Kim, Bok Hyeon 1080
Kim, Byoung Yoon..323

Kim, Byung Gon 449
Kim, Gyu-Tae 901
Kim, Hoon 175, 257, 275, 293, 449
Kim, Inwoong 462, 477, 504, 1048
Kim, Jeongjun 1071
Kim, Minkyu 964
Kim, Yong Hyun 1071
Kishida, Tomoki 787
Kishikawa, Hiroki 115, 733, 925
Kishikawa, Junya 561, 958
Kishimoto, Tadashi 1027
Kitamura, Kokoro 534
Kitano, Takuya 564
Kitaoka, Ryotaro 988
Kitayama, Ken-Ichi 7, 127, 664
Klaus, W. 48
Klonidis, D. 30
Knights, A. P. 622
Kobayashi, Hirokazu 970
Kobayashi, Shuko 1045
Kobayashi, Soichi 531
Kobayashi, Takashi 284, 446
Kobayashi, Wataru 350, 1122
Koda, Katsutoshi 239, 245, 248
Kodama, Yutaro 386
Koeber, S. 413
Koehnle, K. 413
Koga, Masafumi 826
Kohl, M. 413
Koike-Akino, Toshiaki 522
Kojima, Keisuke 522
Kokubun, Yasuo 36, 320, 383
Komori, Yuki 356
Kondo, Tomoki 988
Kong, Deming 15, 823, 850
Kong, Qian 691
Konishi, T. 106, 112, 609, 742
Kono, Naoto 57
Konoike, Ryotaro 1167
Koos, C. 413
Koshikiya, Yusuke 471, 727
Koyama, Fumio 64, 70, 73, 76
Koyama, Osanori................544, 880, 898, 934
Koyama, Ryo 317
Kuang, Caixia 754
Kubo, Kazuo 489, 522
Kubo, Ryogo 751
Kubota, Hirokazu........................ 60, 838, 1065
Kubota, Manabu 682
Kuchta, Daniel M. 63
Kuno, Yuki...642
Kuo, Hao-Chung 585, 594
Kuo, Ping Piu 543
Kuramochi, Eiichi 612
Kurashima, Toshio 317

AUTHOR INDEX

Kurata, Kazuhiko 199
Kuroda, Keiji...892
Kurosu, Takayuki 889
Kusama, Akihiro 898
Kuwaki, Nobuo......................................57
Kuwatsuka, Haruhiko 1185
Kwon, Hong...1071
Kwong, Dim-Lee 1107
Labroille, Guillaume............................181
Lai, Chih-Hsien 1033
Lai, Yinchieh 883
Lauermann, M. 413
Lawin, Mirko..251
Lazaro, Jose A. 688
Le Liepvre, A. 931
Le Roux, X. 422, 1149, 1170
Le Taillandier De Gabory, E.............516
Leblond, Herve 576
Lee, Cheng-Ling 922, 1074
Lee, D. 193
Lee, Heeyoung 1068, 1095
Lee, Jeong-Min 964
Lee, Joon-Woo 868
Lee, Kwanil 871, 901
Lee, Sang Bae 871, 901
Lee, San-Liang 679
Lee, Seungmin 553, 556
Leong, Shan-Fong 594, 976
Lepage, G. 1104
Leuthold, J.....................................30, 413
Li, Cheng 377
Li, Chung-Yi 778
Li, Chunsheng 886
Li, Hongpu..335
Li, Jianqiang...706
Li, Junjie 431
Li, Kuan-Ting...919
Li, Lei 501
Li, Liwei 703
Li, Meifeng...1018
Li, Ming 1188
Li, Ming-Jun..................................222, 603
Li, Ming-Jung 145
Li, Qing 124
Li, Shangyuan 269, 805, 946
Li, Wenzhe 691
Li, Xin 691, 808
Li, Yan 15, 823, 850
Li, Ze 233, 510
Li, Ziang 805
Liang, Di 377
Liang, Kevin 781
Liao, Hao 1083
Liaw, Shien-Kuei 910, 925
Lillieholm, Mads 790

Lim, Christina 465
Lin, Chun-Yu 778
Lin, Fang-Zeng 67
Lin, Gong-Ru 585, 594, 976
Lin, Hung-Hsien 778
Lin, Jun-Han 922
Lin, Kuan-Hao 910
Lin, Rui 814
Lin, Rujian 754
Lin, Wenqiao 1092
Linganna, Kadathala 1080
Liou, Jia-Hong 913
Liow, Tsung-Yang 1107
Lischke, Stefan......................................964
Little, Brent E. 401, 736
Littlejohns, C. J. 622
Liu, Aijun 242
Liu, Bo 501
Liu, Chun-Nien 895
Liu, D............260, 814, 817, 877, 907, 1000, 1018, 1083
Liu, Jun-Jie 591, 859
Liu, Kai 973, 982
Liu, Li 1083
Liu, Ling 772
Liu, Qingwen 425, 1086, 1089, 1101
Liu, Tangqing 850
Liu, Wanyuan 1188
Liu, Wen-Fung 919
Liu, Yang 781
Liu, Yinping 1077
Liu, Youxin 973
Liu, Yu-Chang 679
Liu, Z. 225, 928
Livshits, Daniil 1137
Lo, Guo-Qiang 1107
Loo, R. 1104
Lorences-Riesgo, Abel........................507
Lou, Yiming 811
Lu, Chao..928
Lu, Guo-Wei 519
Lu, Hai-Han 778
Lu, Liangjun 79
Lu, Luluzi 1000
Lu, Ping 907, 1083
Luís, R. S. 48
Luo, Bin 510
Luo, Chao 907
Luo, Jinhui 269, 805
Lv, Qiang..706
Ma, Jiyang..1155
Ma, Lin 1077
Ma, Yiran 431
Maeda, Hideki 829
Maeda, Wakako......................................516
Maegami, Yuriko...................................1110

AUTHOR INDEX

Maho, A. 931, 1116
Malacarne, Antonio 118
Man, Ray 547
Manabe, Tetsuya 471, 727
Mao, Mingzhi 793
Marcaud, G. 1149
Marco, T. 133
Marom, D. M. 30, 184, 718
Marpaung, David 139
Marris-Morini, D. 1149
Martínez, Ricardo 1036
Maruta, Akihiro 127, 853
Maruyama, Hiroaki 934
Maruyama, Ryo 57
Maruyama, Takeo 967
Mashanovich, G. Z. 622
Maštera, Radek 994
Masuda, Akira 290
Masuda, Hiroji 534
Masumoto, Kana 239
Mateo, Eduardo 142
Matsuda, Keisuke 522
Matsuda, Toshiya 239, 245, 248
Matsui, Junichiro 489
Matsui, Takahiro 341
Matsui, Takashi.........................54, 60, 633
Matsumoto, A. 380
Matsumoto, Jun 937
Matsumoto, Keiichi 561, 958
Matsumoto, Ryosuke 127
Matsumoto, S. 133
Matsumoto, Yukihiro 341
Matsunaga, Akira 160
Matsuo, Shinji 612
Matsuo, Shoichiro 45, 54
Matsushita, Asuka 368
Matsutani, Akihiro.......................70, 73, 76
Matsuura, Hiroyuki 82, 151, 1185
Matsuura, Motoharu 398, 452, 943
Matsuzaki, Hideaki 374
Mayoral, Arturo 1036
Mazurczyk, Matt 371
Mehrvar, Hamid 1188
Mei, Chao...1161
Melati, Daniele 88, 1021
Melikyan, A. 413, 1116
Melloni, Andrea 88, 407, 1021
Meloni, Gianluca 118, 724
Mendinueta, J. M. D. 392
Meyer, J. 133
Miao, Wang 208, 1203
Mikhrin, Sergey 1137
Mikhrin, Vladimir 1137
Milione, Giovanni...............................145
Millar, David 522

Minamoto, Yamato...............................452
Minato, Naoki 1045
Mino, S. 106, 112, 609
Mise, Kazuaki 544
Mishra, Snigdharaj 142
Mitsuno, Hiroya..................................967
Mittal, V. 622
Miura, Seiya 36
Miyabe, M. 133
Miyamoto, Tomoyuki...................353, 356
Miyamoto, Y. 45, 193, 299, 368, 489, 606, 1182
Miyazawa, T. 133, 655
Miyoshi, Yuji 60, 838, 1065
Mizuno, T. 193
Mizuno, Yosuke 341, 1068, 1095, 1098
Mizutori, Akira 826
Mochizuki, Keita 1131
Mohajerin-Ariaei, A. 537
Mohan, Gurusamy 293
Molin, Denis 600
Molina-Fernandez, I............................622
Monroy, Idelfonso Tafur 166
Morandotti, Roberto 401
Moreolo, Michela Svaluto......................302
Mori, Hiroshi 544
Mori, Kazuya 115
Mori, Takayoshi 329
Mori, Yojiro 82, 428, 437, 1185
Morichetti, Francesco 88, 407
Morikawa, Hiroyuki..............................513
Morioka, Toshio 45, 136, 790
Morita, Itsuro 190, 673, 1039
Morita, Kohei....................................997
Morito, Ken 645
Moriwaki, Osamu 437
Morizumi, Yuki 970
Morizur, Jean-François 181
Moss, David J....................................401
Muehlbrandt, S. 413
Muñoz, Raul 1036
Murai, Hitoshi 1027
Murakami, Toshinori............................898
Murakawa, T. 106, 112, 609
Muramoto, Yoshifumi...........................350
Muranaka, Yusuke...............................1197
Murano, Akihiro 24
Murata, Hiroshi.............570, 579, 949
Murugan, G. S. 622
Nabika, Kengo 787
Nada, Masahiro 374
Nadal, Laia 302
Nagashima, T. 106, 112, 609, 742
Nagatani, Munehiko 368, 606
Nagatomi, Ken 670
Naka, Akira 829

AUTHOR INDEX

Nakadai, Masahiro......................................1167
Nakagawa, Goji...................................254, 498
Nakagawa, Masahiro..........................239, 248
Nakahama, Masanori 70, 73, 76
Nakahara, Kouji..1113
Nakajima, Hirochika 386, 1140
Nakajima, K. 54, 60, 227, 329, 633, 985
Nakajima, Mitsumasa............................94, 1182
Nakamura, A...133
Nakamura, Hitoshi.....................................341
Nakamura, Kazuki....................................1152
Nakamura, Kentaro...........341, 1068, 1095, 1098
Nakamura, Moriya 769, 832, 844
Nakamura, Ryoichiro 832
Nakamura, Seiki.......................................1125
Nakamura, Takahiro 148, 199
Nakamura, Tatsuya...................................516
Nakanishi, Akira......................................1113
Nakanishi, Yasuhiko.................................350
Nakano, Yoshiaki.....................................1143
Nakao, A. 985
Nakashima, Hisao.....................................489
Nakatsuhara, K. 85
Nakayama, Yu...682
Nakazawa, Masataka....................9, 12, 33, 305
Nakpeerayuth, Suvit.................................748
Namiki, S. 18, 21, 82, 139, 151, 889, 1179, 1185
Naoe, Kazuhiko..1113
Nedeljkovic, M. 622
Nguyen, Linh..703
Ni, Wenjun..907, 1083
Nicho, J. 133
Niihara, Takumi..934
Ninomiya, Norihiko..................................943
Nirmalathas, Ampalavanapillai................465
Nishi, Hidetaka...148
Nishide, Kenji...219
Nishihara, Masato.....................................302
Nishimura, Kosuke 284, 446
Nishiyama, Nobuhiko 642, 1119
Nishiyama, Tetsuo.............................561, 958
Nishizaki, Itaru...341
Niu, Tong..573
Niwa, Masaki..437
Noda, Susumu....................................621, 1167
Nogami, Masamichi...........................1125, 1131
Noguchi, Hidemi................................489, 516
Noguchi, Masataka....................................148
Nosaka, Hideyuki......................................368
Noto, Kazutaka....................................471, 727
Notomi, Masaya 419, 612, 1176
Nozaki, Kengo...612
Numata, Hidetoshi.....................................639
Oba, Jinsei..766
Ochi, Hirotaka...856

Ochi, Yutaka...341
Oda, Shoichiro.............................254, 477, 498
Oe, Shota...100
Ogawa, Kensuke......................................1107
Ogawa, Yoh...1027
Ogino, Takehiro..955
Ogoshi, Haruki...216
Oguro, Takahiro..838
Ohara, Mizuki...531
Ohara, Y. 133, 787
Ohashi, Masaharu 60, 838, 1065
Ohno, Morifumi.......................................1110
Ohno, Tomohiro..570
Ohshima, T. 133
Ohtsuki, Tomoaki......................................862
Oishi, Masayuki..443
Okabe, Ryo 302
Okada, Mao...925
Okamoto, Keiji..471
Okamoto, Satoru 133, 937
Okamoto, Seiji...489
Okamoto, Takeshi 489
Okamura, Yasuyuki............570, 579, 949
Okano, Makoto..1110
Okuda, Tadahiro..949
Okuno, M. 106, 112, 609
Onaka, Hiroshi..489
Ono, Hirotaka 544, 880, 1062
Ono, Jun 544, 880
Ono, Masaaki...612
Ono, Y. 582
Onuki, Yuya 561, 958
Ortega-Monux, A. 622
Osato, Kazunori...769
Oshiba, Saeko..787
Oshida, Shohei..353
Ota, Kazuya 544, 880, 934
Otaka, Akihiro 172, 682
Owaki, Shotaro..844
Oxenløwe, Leif K. 136, 790
Oyama, Tomofumi.....................................501
Palacharla, Paparao 504, 1048
Palmer, R. 413
Pang, Jiangchuan 15, 823
Pantouvaki, M. 1104
Paredes, Bruna...1009
Parsons, Kieran...522
Pasquazi, Alessia 401, 547
Peccianti, Marco..401
Pelusi, Mark..21, 139
Penades, J. Soler.......................................622
Peng, Chun-Yen...67
Peng, Gaozhu...145
Peng, Huanfa...715
Peng, Jiangde...1015

AUTHOR INDEX

Peng, Xiaofeng...............715
Penkler, David...............1191
Peserico, Nicola...............407
Petermann, Klaus...............480
Petrovich, M. N...............225
Pincemin, E...............30
Poletti, F...............225
Pong, Chun-Yen...............594, 976
Poole, Simon B...............124
Popescu, Ion...............1039
Potì, Luca...............118, 724
Powers, Dale...............603
Prajzler, Václav...............994
Preve, G. B...............395
Proietti, Roberto...............661
Pukhrambam, Puspa Devi...............679
Pun, Edwin Y. B...............401
Puttnam, B. J...............48
Qi, Haifeng...............1024
Qian, Chen...............793
Qin, Yingchun...............359
Qiu, Chen...............431
Qiu, Feng...............961
Qiu, Kun...............841
Quack, Niels...............91
Rademacher, Georg...............480
Ramachandran, Siddharth...............1061
Ranaweera, Chathurika...............465
Rao, Baoquan...............431
Rasmussen, J. C...............254, 302, 462, 477, 492, 498, 501, 513
Raz, Oded...............208
Razo, M...............133
Ren, Xiaomin...............973
Richardson, D. J...............225
Richter, Thomas...............27
Rivas, J. M...............30
Robertson, Brian...............97
Røge, Kasper Meldgaard...............790
Romagnoli, M...............395
Ruan, Lihua...............459
Rudnick, R...............30
Rumley, Sébastien...............196
Ryf, R...............187
Ryu, Gukbeen...............901
Saito, Hiroyuki...............1045
Saito, Kohei...............829
Saito, Kotaro...............317
Saito, Tsunetoshi...............314
Saitoh, Kunimasa...............45, 51, 54, 633
Saitoh, Shota...............39, 54
Sakaguchi, Takahiro...............70, 76
Sakamoto, Junji...............94
Sakamoto, Shinichi...............1107
Sakamoto, Taiji...............54, 60, 329, 633

Sakamoto, Takahide...............519, 796
Sakano, Goki...............615
Sakata, Ryosuke...............651
Sakuma, Kazuki...............615
Saleh, Adel A. M...............121
Sampath, K. I. Amila...............272
Sandoghchi, S. R...............225
Sang, Xinzhu...............886, 1161
Sanjoh, Hiroaki...............350
Sano, Akihide...............368, 606
Sano, Tomomi...............627
Saridis, G. M...............48
Sasago, Hiroki...............730
Sasaki, Keiichi...............1179
Sasaki, Shinya...............856
Sasaki, Yusuke...............39, 45
Sato, Ken-Ichi...............82, 230, 428, 437, 1185
Sato, Manabu...............272
Sato, T...............133
Sato, Yosuke...............446
Satou, I...............133
Sawamura, Taketsugu...............1128
Schindler, P. C...............413
Schmidt-Langhorst, Carsten...............27
Schröder, Jochen...............597
Schubert, Colja...............27
Segawa, Toru...............1197
Seki, Hiroyuki...............462
Seki, Kazuki...............1140
Sekine, Kawori...............832
Sekiya, Motoyoshi...............477, 504
Senda, Kousuke...............544, 880
Seok, Tae Joon...............91
Serna, Samuel...............422, 1170
Seyedi, M. Ashkan...............1137, 1191
Shakoor, Abdul...............612
Shao, C...............133
Sharma, Prateeksha...............700
Shen, Gangxiang...............889
Shen, Mingya...............1024
Shen, Shin-Wei...............1134
Shi, Jin-Wei...............591, 859
Shi, Lu...............260
Shibagaki, Nobuhiko...............446
Shibahara, K...............193
Shieh, Wern-Yarng...............865
Shieh, William...............323
Shih, Tien-Tsorng...............585, 594
Shikama, Kota...............1182
Shim, Young Bo...............553
Shimada, Hirokazu...............254
Shimada, Shumpei...............341
Shimada, Tatsuya...............172
Shimakawa, Osamu...............627
Shimizu, S...............106, 109, 112, 609

AUTHOR INDEX

Shimizu, Tatsuya................................172
Shimomura, K. 561, 955, 958
Shimose, Yoshiharu 544
Shinada, Satoshi 392
Shindo, Takahiko 1122
Shiraiwa, Masaki 655, 658
Shirao, Mizuki 1125
Shizuka, Makoto 1098
Shoji, Akihisa 694
Shu, Chester 416, 495, 550
Shukla, Vishnu 1
Shum, P. 260, 814, 877
Sillard, Pierre 600
Simeonidou, D. 48, 654
Slavik, R. 225
Sobu, Yohei 645
Solis-Trapala, K. 18, 21, 139
Soma, Daiki 365
Sommer, M. 413
Son, Yong-Hwan 868
Sone, Kyosuke 254, 462
Song, Jue 877
Song, Kwang Yong 901, 1071
Song, M. 30
Song, Sanggwon 556
Song, Shijie 703
Song, Tianyu....................................862
Song, Yingxiong 754, 820
Sorel, Marc 407
Sotobayashi, Hideyuki 24, 766
Spiga, Silvia 347
Spinnler, Bernhard 486
Srinivasan, S. A. 1104
Stabile, Ripalta 1200
Stankovic, S. 622
Stojanovic, Nebojsa 287
Stolte, Ralf 124
Stone, Jeffery 145
Su, Xiaofei 492
Subramaniam, Suresh 437
Subramanian, Ramanathan 335
Suda, Satoshi 82, 151
Sugawara, Mitsuru 148
Sugihara, Kenya 489
Sugihara, Seitaro 670
Sugihara, Takashi 489, 522
Sugimoto, Y. 985
Sugiyama, H. 133
Sugiyama, Koki 712
Sugizaki, R. 42, 537
Suhara, Masashi 353
Sukigara, Toshiki 561, 958
Sumida, Y. 133
Sun, Chuanbowen 275
Sun, Han 483

Sun, Nai-Hsiang 919
Sun, Weiqiang 745
Sun, Yongmei 242, 775
Sung, J. Y. 685
Sung, Yuan-Yuan 679
Suo, Jing 1015
Suzuki, Daiki 12
Suzuki, Junichi 642
Suzuki, Keijiro 82, 151
Suzuki, Keita 1146
Suzuki, Kenya 94, 1182
Suzuki, Makoto 513
Suzuki, Takanori 1113
Sygletos, S. 30
Szelag, Bertrand 1191
Taga, Hidenori 332
Tajima, Akio 564
Takahara, Tomoo 302
Takahashi, Kazuto 925
Takahashi, Mikoto 531
Takahashi, Naoki 1140
Takahashi, Ryo 205, 664, 1197
Takahashi, S. 582
Takano, Katsumi 272
Takano, Kohei 955
Takasaka, S. 537
Takashi, Goh 1182
Takasuka, Syo 898
Takata, Kenta 1176
Takeda, Koji 612
Takeda, M. 85
Takei, Aki 1113
Takemasa, Keizo 148
Takenaga, Katsuhiro 39, 45, 54
Takenouchi, Hirokazu 1152, 1158
Takeshima, Koki 365, 1039
Takizawa, Motoyuki 254
Tam, Hwa-Yaw 547, 928
Tamai, Hideaki 1045
Tamura, K. R. 1113
Tan, H. N. 18, 21, 139
Tan, Yuanlong 874
Tan, Zhongwei 338, 1003
Tanabe, Katsuaki 148
Tanabe, Kazuhiro 651
Tanaka, Hiroki 341
Tanaka, Kazuki 284
Tanaka, S. 985, 1113
Tanaka, Takafumi 434
Tanaka, Toshiki 302
Tanaka, Yoshiaki 757
Tanaka, Yoshinori 1167
Tanaka, Yosuke 341
Tanaka, Yu 645
Tanemura, Takuo 1143

AUTHOR INDEX

Tang, Jin 528
Tang, Ming 260, 814, 817
Tang, Xiaosheng 979
Tanigawa, Yosuke 667
Tanimura, Takahito 27, 513
Tanizawa, Ken 82, 151
Tao, Xiaofeng 157
Tao, Yemeng...1077
Tao, Yiming 1092
Tao, Zhenning 492, 498, 501
Taue, Shuji 916
Ten, Sergey 344
Terada, Jun 172
Terada, Yuki 428
Thouenon, G. 30
Thouras, Jordane 576
Tobita, Yuki 54
Tode, Hideki 667, 670
Tokura, Akio 1158
Tomita, Kazuki 829
Tomizawa, Masahito 489
Tomkos, I. 30
Tommasino, P. 395
Trifiletti, A. 395
Tsai, C.-Y. 618
Tsai, Cheng-Ting.........................585, 594, 976
Tsai, Zong-Yu 865
Tsang, H. K. 410, 416, 685
Tsang, K. S. 547
Tsang, Yanny..547
Tseng, Chung-Hao 1074
Tseng, Shuo-Yen 991
Tsuchida, Junichi 199
Tsuchizawa, Tai 148
Tsuda, Hiroyuki 712, 1012
Tsuda, Nobuaki 916
Tsujikawa, Kyozo.......................................633
Tsunashima, Satoshi 350
Tsuritani, T. 133, 190, 236, 332, 365, 673, 1039
Tsutsumi, Takuya 682
Tu, Xiaoguang 1107
Tu, Xin 1188
Tucker, Rod 5
Tujita, Sho 544, 880
Tzu, Ta-Ching 591, 859
Udagawa, Kenta 534
Ueda, Koh 437, 1185
Ueda, Yuta 350
Ueno, Fumiaki 949
Ueno, Yuto 1131
Uenohara, H. 106, 112, 130, 609, 730, 997
Uetsuka, Hisato 1179
Umeki, Takeshi..1152
Umezawa, T. 380
Uto, Kenichi 522

V., Dinesh Kumar 700
Van Campenhout, J....................................1104
Van Kerrebrouck, J. 389
Van Thourhout, D. 413, 1104
Vassilieva, Olga 477, 504
Velázquez, A. 931
Velha, P. 395
Verheyen, P. 1104
Vilalta, Ricard 1036
Vílchez, F. J. 302
Villafranca, A...622
Vincent, François 1191
Vivien, L. 422, 1149, 1170
Von Kirchbauer, Heinrich 486
Wada, Kazuyuki 832
Wada, Masaki 329
Wada, N........48, 106, 109, 112, 392, 609, 655, 658, 748
Wagner, Christoph 251
Waho, Takao 955
Wai, P. K. A. 540, 547
Wakayama, Yuta 332
Wang, Bin 1086, 1089, 1101
Wang, Danshi..................................233, 510
Wang, Guanzhong 359
Wang, Hongxiang 242, 775
Wang, Jiachen 871
Wang, Jiaqi 416
Wang, Jiayu 431
Wang, Jie..547
Wang, Jinghao 1018
Wang, Jun 982
Wang, Kuiru 1161
Wang, Min 754, 793
Wang, Ming 979
Wang, Peng 335
Wang, Qian 862
Wang, Ruoxu 814
Wang, S. H. 540, 736
Wang, Sheng-Min 883
Wang, Shuai.....................1077, 1086, 1101
Wang, Shun 1083
Wang, Ti-Ho 865
Wang, Ting 145, 455
Wang, Wei.......................................585, 1057
Wang, Xi 462, 1048
Wang, Xinyi 79
Wang, Xinyue 338, 1003
Wang, Xudong 573
Wang, Yixin 305
Wang, Yu 636
Wang, Yuanwu 1018
Wang, Yuxi 103
Wang, Ziyu 338, 1003
Wanguemert-Perez, G..................................622
Watanabe, Hiroshi 471, 727

AUTHOR INDEX

Watanabe, Kengo 314
Watanabe, Shigeki 27
Watanabe, Takashi 670
Watanabe, Tatsuhiko...............................36
Watanabe, Toshio 1185
Weerasekara, Gihan 853
Wei, Chia-Chien 591, 859, 1134
Wei, L. Y. 685
Wei, Liang-Yu 781
Wei, Pengjiang 736
Wei, Yuming 1188
Weng, Tsui-Wei 1021
Wenjing, Fang 1006
Wheeler, N. V. 225
Wijayanto, Yusuf Nur 579
Wilkinson, J. S. 622
Wilkinson, Peter 97
Willner, A. E. 537
Wolf, S. 413
Wong, Elaine 459, 465
Wood, William A. 142
Worasucheep, Duang-Rudee 748
Wu, Chang-Jen 778
Wu, Chao-Hsin 594, 976
Wu, Huan 550
Wu, Jian 15, 236, 673, 721, 823, 850, 1092
Wu, Jingjing 263
Wu, Jui-Pin 1134
Wu, Meng-Shan 1074
Wu, Ming C. 91
Wu, Ming-Wei 862
Wu, Qiong 814
Wu, Tsai-Chen 585
Wu, Tsu-Shiu 67
Wu, Weiliang 278, 820
Wu, X. R. 685
Wu, Xiaojuan 886
Wu, Xinru 410
Wu, Yun-Chen 594, 976
Wu, Yunfei 630
Wu, Zhichao 877
Xiao, Min 359, 1155
Xiao, Shilin 772, 799
Xiaoen, Du 544
Xiaomin, Ren 982, 1006
Xie, Changsong 287
Xie, Shizhong 811
Xin, Haiyun 772
Xin, Xin 775
Xu, Ke 410, 685
Xu, Kun 706
Xu, Sheng 757
Xu, Sugang 658, 757
Xu, Yongchi 715
Xu, Zhaopeng 1003

Xue, Yuankai 799
Yamada, Koji 1110
Yamada, Makoto 544, 880, 898, 934, 1062
Yamada, Shoko 24
Yamaguchi, Keita 94, 1182
Yamaguchi, Minoru 934
Yamaguchi, Shigeru 1158
Yamaguchi, Yuya 386, 1140
Yamamoto, Fumihiko 54, 633
Yamamoto, N. 24, 380, 763, 766
Yamamoto, Shuto 290
Yamamoto, Takashi 329
Yaman, Fatih 142
Yamanaka, N. 133, 937
Yamanaka, Yusuke 697
Yamanishi, Masamichi...............................567
Yamasaki, Y. 742
Yamashita, Yoko 633
Yamauchi, Tomohiro 477, 498
Yamazaki, Hiroshi 350, 368, 606
Yan, Binbin 886
Yan, Fulong 208
Yan, Haozhe 946
Yan, Shengyong 1188
Yanan, Guo 1006
Yang, Bingliang 739
Yang, Chuanchuan 1003
Yang, Dung-Chin 67
Yang, Guangyao 1086, 1089, 1101
Yang, Haining 97
Yang, Hui 874, 1042
Yang, Liang 398
Yang, Qianmei 468
Yang, Se-Hoon 868
Yang, Sigang 103, 811
Yang, Tingting 721
Yang, Wenjian 703
Yang, Yanfu 835, 847
Yang, Yu 1003
Yang, Yuan-Jie 922
Yang, Zhisheng 1092
Yao, Yong 835, 847
Yatsu, Tomoya 398
Ye, Zilong 1054
Yeh, C. H. 685
Yeh, Chien-Hung 781
Yeh, Tung-Yuan 922
Yi, Xiaoke 703
Yi, Xingwen 841
Yin, Feifei 706
Yin, Shan 808
Yin, Xin 389
Yin, Zhang-Qi 359
Ying, Wang 1006
Yokokawa, S. 985

AUTHOR INDEX

Yokouchi, Noriyuki 1128
Yokoyama, C. 133
Yokoyama, Shiyoshi 961
Yonenaga, Kazushige 368, 489
Yoneyama, Akira 452
Yongqing, Huang 1006
Yoo, Kwang Wook 556
Yoo, S. J. B. 202, 661
Yoshida, Junji 1128
Yoshida, Masato 33, 305
Yoshida, Mitsuteru 489
Yoshida, S. 133, 254, 498
Yoshida, Tsuyoshi 522
Yoshida, Yuki 127, 664
Yoshikane, N. 133, 673, 1039
Yoshikuni, Yuzo 892
Yoshimoto, Naoto 531, 694
Yu, Ao 1042
Yu, Changyuan 257, 293, 841
Yu, Chin-Ping 913
Yu, Chongxiu 1161
Yu, Cunqian 904
Yu, Kunzhi 377
Yu, Miao 15, 823
Yu, Xianbin 136
Yu, Yi-Lin 925
Yu, Yinghong 772
Yu, Zhao 287
Yu, Zhenming 811
Yuan, Jinhui 1161
Yuan, Shuai 525
Yue, Lei 15, 823
Zafar, Humaira 1009
Zakharian, Aramais 142
Zang, Zhigang 979
Zeng, Heping 1173
Zervas, G. 48
Zhang, A. P. 547, 928
Zhang, Bingbing 760
Zhang, Cheng 715
Zhang, Chengliang 431
Zhang, Chunhui 1188
Zhang, Chunshu 1188
Zhang, Dongxu 236, 673, 721
Zhang, Fan 311
Zhang, Hanyi 269, 805, 946
Zhang, Jiangshan 907
Zhang, Jiawei 775
Zhang, Jie 808, 874, 1042, 1057
Zhang, Jing 841
Zhang, Junjie 278, 793, 820
Zhang, Kuo 739
Zhang, Lu 772, 799
Zhang, Min 233, 510, 808
Zhang, Minming 260, 877, 952, 1000, 1018

Zhang, Qianwu 278, 754, 793, 820
Zhang, Qiong 1048
Zhang, Qun 835, 847
Zhang, Shaojie 772
Zhang, Shaoliang 142
Zhang, Wei 1015
Zhang, Weiwei 422, 1170
Zhang, Wenjia 1077
Zhang, Yangan 973
Zhang, Yunchuan 1024
Zhang, Yunhao 772, 799
Zhang, Zhen 278, 820
Zhang, Zhiguo 760
Zhang, Zitian 739
Zhao, Fei 1188
Zhao, J. 30
Zhao, Mingming 359, 1155
Zhao, Mingxiao 1155
Zhao, Peng 636
Zhao, Shuoyi 79
Zhao, Weiqian 1024
Zhao, Ying 501
Zhao, Yongli 431, 808, 874, 1057
Zheng, Bofang 550
Zheng, Xiaoping 269, 805, 946
Zheng, Yue 706
Zhong, Kangping 835, 847
Zhou, Bin 928
Zhou, Bingkun 269, 805, 946
Zhou, Feiya 1000
Zhou, Huibin 817
Zhou, Jianwei 706
Zhou, Jingjing 293
Zhou, Linjie 79
Zhou, Xian 835, 847
Zhou, Xu 308, 468
Zhou, Yu 691
Zhu, Lixin 715
Zhu, Xiaoxu 874
Zhu, Yixiao 311
Zimmernman, Lars 964
Ziyadi, M. 537
Zong, Yue 904
Zou, Kaiheng 311